WATER QUALITY AND TREATMENT

WATER QUALITY AND TREATMENT

A Handbook of Community Water Supplies

American Water Works Association

Raymond D. Letterman *Technical Editor*

Fifth Edition

McGRAW-HILL, INC.

New York San Francisco Washington, D.C. Auckland Bogotá Caracas Lisbon London Madrid Mexico City Milan Montreal New Delhi San Juan Singapore Sydney Tokyo Toronto

McGraw-Hill

A Division of The ***McGraw-Hill*** *Companies*

1234567890 DOC/DOC 90432109

ISBN 0-07-001659-3

The sponsoring editor for this book was Larry Hager, the editing supervisor was Tom Laughman, and the production supervisor was Pamela Pelton. It was set in Times Roman by North Market Street Graphics.

Printed and bound by R. R. Donnelley & Sons Co.

CONTENTS

Chapter 7: Sedimentation and Flotation 7.1

Chapter 8: Granular Bed and Precoat Filtration 8.1

Chapter 9: Ion Exchange and Inorganic Adsorption 9.1

Chapter 13: Adsorption of Organic Compounds 13.1

Chapter 14: Disinfection 14.1

Chapter 15: Water Fluoridation 15.1

CONTRIBUTORS

Appiah Amirtharajah, Ph.D., P.E. *School of Civil Engineering, Georgia Institute of Technology, Atlanta, Georgia* (CHAP. 6)

Larry D. Benefield, Ph.D. *Department of Civil Engineering, Auburn Univeristy, Alabama* (CHAP. 10)

Paul S. Berger *U.S. Environmental Protection Agency, Washington, D.C.* (CHAP. 2)

Stephen W. Clark *U.S. Environmental Protection Agency, Washington, D.C.* (CHAP. 1)

John L. Cleasby, Ph.D., P.E. *Department of Civil and Construction Engineering, Iowa State University, Ames, Iowa* (CHAP. 8)

Dennis A. Clifford, Ph.D., P.E., DEE *Department of Civil and Environmental Engineering, University of Houston, Houston, Texas* (CHAP. 9)

Perry D. Cohn, Ph.D., M.P.H. *New Jersey Department of Health and Senior Services, Trenton, New Jersey* (CHAP. 2)

David A. Cornwell, Ph.D., P.E. *Environmental Engineering & Technology, Inc., Newport News, Virginia* (CHAP. 16)

Michael Cox, M.P.H. *U.S. Environmental Protection Agency, Washington, D.C.* (CHAP. 2)

John C. Crittenden, Ph.D., P.E., DEE *Department of Civil and Environmental Engineering, Michigan Technological University, Houghton, Michigan* (CHAP. 5)

James K. Edzwald, Ph.D., P.E. *Department of Civil and Environmental Engineering, University of Massachusetts, Amherst, Massachusetts* (CHAP. 7)

Edwin E. Geldreich, M.S. *Consulting Microbiologist, Cincinnati, Ohio* (CHAP. 18)

Ross Gregory *Water Research Centre, Swindon, Wiltshire, England* (CHAP. 7)

Charles N. Haas, Ph.D. *Drexel University, Philadelphia, Pennsylvania* (CHAP. 14)

David W. Hand, Ph.D. *Department of Civil and Environmental Engineering, Michigan Technological University, Houghton, Michigan* (CHAP. 5)

Alan Hess, P.E., DEE *Black & Veatch, Philadelphia, Pennsylvania* (CHAP. 3)

David R. Hokanson, M.S. *Department of Civil and Environmental Engineering, Michigan Technological University, Houghton, Michigan* (CHAP. 5)

Michael Horsley *Black & Veatch, Philadelphia, Pennsylvania* (CHAP. 3)

John A. Hroncich *United Water Management and Services Company, Harrington Park, New Jersey* (CHAP. 4)

Mark LeChevallier, Ph.D. *American Water Works Service Co., Voorhees, New Jersey* (CHAP. 18)

Raymond D. Letterman, Ph.D., P.E. *Department of Civil and Environmental Engineering, Syracuse University, Syracuse, New York* (CHAP. 6)

Gary S. Logsdon, D.Sc., P.E. *Black & Veatch, Cincinnati, Ohio* (CHAPS. 3, 8)

Joe M. Morgan, Ph.D. *Department of Civil Engineering, Auburn University, Alabama* (CHAP. 10)

Charles R. O'Melia, Ph.D., P.E. *Department of Geography and Environmental Engineering, The Johns Hopkins University, Baltimore, Maryland* (CHAP. 6)

Frederick W. Pontius, P.E. *American Water Works Association, Denver, Colorado* (CHAP. 1)

David A. Reckhow, Ph.D. *University of Massachusetts, Amherst, Massachusetts* (CHAP. 12)

Thomas G. Reeves, P.E. *Centers for Disease Control and Prevention, Atlanta, Georgia* (CHAP. 15)

Michael R. Schock *U.S. Environmental Protection Agency, Cincinnati, Ohio* (CHAP. 17)

Philip C. Singer, Ph.D. *University of North Carolina, Chapel Hill, North Carolina* (CHAP. 12)

Stuart A. Smith, C.G.W.P. *Smith-Comeskey Ground Water Science, Ada, Ohio* (CHAP. 4)

Vernon L. Snoeyink, Ph.D. *Department of Civil and Environmental Engineering, University of Illinois at Urbana-Champaign, Urbana, Illinois* (CHAP. 13)

R. Scott Summers, Ph.D. *Civil, Environmental, and Architectural Engineering, University of Colorado, Boulder, Colorado* (CHAP. 13)

J. S. Taylor, Ph.D., P.E. *Civil and Environmental Engineering Department, University of Central Florida, Orlando, Florida* (CHAP. 11)

Mark Wiesner, Ph.D. *Environmental Sciences and Engineering Department, Rice University, Houston, Texas* (CHAP. 11)

Thomas F. Zabel *Water Research Centre, Medmenham, Oxfordshire, England* (CHAP. 7)

PREFACE

The fifth edition of *Water Quality and Treatment* remains a complementary book to *Water Treatment Plant Design.* Each book covers its traditional area within the vast subject of water treatment theory, practice, and technology.

This new edition of *Water Quality and Treatment* retains the format of the fourth edition, with updating and expansion in all areas. In revising the book, the committee looked beyond its traditional readership of professionals in the field and college students to try to make this edition more useful to water treatment plant operating personnel. All chapters have been written especially for this edition or significantly rewritten, updated, and expanded by recognized experts in the subject areas. The book continues its tradition of providing authoritative, up-to-date information for evaluating and ensuring quality water supplies.

A committee from the Water Quality Division of the American Water Works Association guided preparation of this edition. Members included Clarence A. Blanck, American Water Works Service Co.; John E. Dyksen, United Water New Jersey; James S. Taylor, University of Central Florida; Stephen J. Randtke, University of Kansas; and Wendell R. Inhoffer, Delta Water Group.

The committee thanks all chapter authors for their patient and dedicated work on this edition. Additional thanks go to the past and present chairs of the Water Quality Division of the American Water Works Association. Patricia L. McGlothlin, City of Colorado Springs, and Gregg J. Kirmeyer, Economic & Engineering Services, also offered valuable support. The committee offers thanks to Clare Haas, AWWA staff secretary to the Water Quality Division, for the aid she provided in its work. Members of the AWWA publications staff also helped with this edition, including Mindy Burke, Kathleen Faller, and David Talley. Finally, the committee gratefully acknowledges the tireless efforts of this edition's technical editor, Raymond D. Letterman of Syracuse University, whose oversight and coordination expedited the development process and improved every chapter.

Clarence A. Blanck, chair
Revision Steering Committee
American Water Works Service Co. (retired)

CHAPTER 1

DRINKING WATER QUALITY STANDARDS, REGULATIONS, AND GOALS

Frederick W. Pontius, P.E.
American Water Works Association
Denver, Colorado

Stephen W. Clark
U.S. Environmental Protection Agency
Washington, D.C.

The principal law governing drinking water safety in the United States is the Safe Drinking Water Act (SDWA). Enacted initially in 1974 (SDWA, 1974), the SDWA authorizes the U.S. Environmental Protection Agency (USEPA) to establish comprehensive national drinking water regulations to ensure drinking water safety. The history and status of U.S. drinking water regulations and the SDWA are presented in this chapter. International standards for drinking water are also discussed briefly.

Drinking water regulations are issued by a regulatory agency under the authority of federal, state, or local law. Drinking water regulations established by USEPA typically require water utilities to meet specified water quality standards. Regulations also require that certain monitoring be conducted, that specified treatment be applied, and that the supplier submit reports to document that the regulations are being met.

To ensure that water quality regulations are not violated, a water supplier usually must produce water of a higher quality than the standard or regulation would demand. Hence, each water supplier must establish and meet its own water quality goals to ensure that applicable water quality regulations are met and that the highest-quality water possible is being delivered to the consumer within the financial resources available to the water supplier.

EARLY DEVELOPMENT OF DRINKING WATER STANDARDS

By the eighteenth century, removal of particles from water by filtration was established as an effective means of clarifying water. The general practice of making water clean was well recognized by that time, but the degree of clarity was not measurable (Borchardt and Walton, 1971). The first municipal water filtration plant started operations in 1832 in Paisley, Scotland (Baker, 1981). Aside from the frequent references of concern for the aesthetic properties of water, historical records indicate that standards for water quality were notably absent up to and including much of the nineteenth century.

With the realization that various epidemics (e.g., cholera and typhoid) had been caused and/or spread by water contamination, people learned that the quality of drinking water could not be accurately judged by the senses (i.e., appearance, taste, and smell). Appearance, taste, and smell alone are not an accurate means of judging the safety of drinking water. As a result, in 1852, a law was passed in London stating that all waters should be filtered (Borchardt and Walton, 1971). This was representative of new understanding resulting from an improved ability to observe and correlate facts. In 1855, epidemiologist Dr. John Snow was able to prove empirically that cholera was a waterborne disease. In the late 1880s, Pasteur demonstrated the particulate germ theory of disease, which was based upon the new science of bacteriology. Only after a century of generalized public health observations of deaths due to waterborne disease was this cause-and-effect relationship firmly established.

The growth of community water supply systems in the United States began in Philadelphia, Pennsylvania. In 1799, a small section was first served by wooden pipes and water was drawn from the Schuylkill River by steam pumps. By 1860, over 400 major water systems had been developed to serve the nation's major cities and towns.

Although municipal water supplies were growing in number during this early period of the nation's development, healthy and sanitary conditions did not begin to improve significantly until the turn of the century. By 1900, an increase in the number of water supply systems to over 3000 contributed to major outbreaks of disease because pumped and piped supplies, when contaminated, provide an efficient means for spreading pathogenic bacteria throughout a community.

EARLY HISTORY OF U.S. FEDERAL DRINKING WATER STANDARDS

Drinking water standards in the United States have developed and expanded over a 100-year period as knowledge of the health effects of contaminants increased and the treatment technology to control contaminants improved. Drinking water standards and regulations developed out of a growing recognition of the need to protect people from illness caused by contaminated drinking water.

Interstate Quarantine Act

In the United States, federal authority to establish drinking water regulations originated with the enactment by Congress in 1893 of the Interstate Quarantine Act (U.S.

Statutes, 1893). Under this act, the surgeon general of the U.S. Public Health Service (USPHS) was empowered

> ". . . to make and enforce such regulations as in his judgment are necessary to prevent the introduction, transmission, or spread of communicable disease from foreign countries into the states or possessions, or from one state or possession into any other state or possession."

This provision of the act resulted in promulgation of the interstate quarantine regulations in 1894. The first water-related regulation, adopted in 1912, prohibited the use of the common cup on carriers of interstate commerce, such as trains (McDermott, 1973).

U.S. Public Health Service Standards

The first formal and comprehensive review of drinking water concerns was launched in 1913. Reviewers quickly realized that "most sanitary drinking water cups" would be of no value if the water placed in them was unsafe. The first federal drinking water standards were adopted in 1914. The USPHS was then part of the U.S. Treasury Department and was charged with the task of administering a health care program for sailors in the Merchant Marine. The surgeon general recommended, and the U.S. Treasury Department adopted, standards that applied to water supplied to the public by interstate carriers. These standards were commonly referred to as the "Treasury Standards." They included a 100/cc (100 organisms/mL) limit for total bacterial plate count. Further, they stipulated that not more than one of five 10/cc portions of each sample examined could contain *B. coli* (now called *Escherichia coli*). Because the commission that drafted the standards had been unable to agree on specific physical and chemical requirements, the provisions of the 1914 standards were limited to the bacteriological quality of water (Borchardt and Walton, 1971).

The 1914 standards were legally binding only on water supplies used by interstate carriers, but many state and local governments adopted them as guidelines. Because local and state officials were responsible for inspecting and supervising community water systems, they inspected the carrier systems also. In 1915, a federal commitment was made to review the drinking water regulations on a regular basis.

By 1925, large cities applying either filtration, chlorination, or both encountered little difficulty complying with the 2 coliforms per 100 mL limit. The standards were revised to reflect the experience of systems with excellent records of safety against waterborne disease. The limit was changed to 1 coliform per 100 mL, and the principle of attainability was established. That is, to be meaningful, drinking water standards must consider the ability of existing technology to meet them. In addition to bacteriological standards, standards were established for physical and chemical (lead, copper, zinc, excessive soluble mineral substances) constituents (USPHS, 1925). The availability of adequate treatment methods and the risk of contracting disease from contaminated drinking water relative to other sources influenced development of the 1925 standards.

The USPHS standards were revised again in 1942 (USPHS, 1943), 1946 (USPHS, 1946), and 1962 (USPHS, 1962). The 1962 standards, covering 28 constituents, were the most comprehensive pre-SDWA federal drinking water standards at that time.

They set mandatory limits for health-related chemical and biological impurities and recommended limits for impurities affecting appearance, taste, and odor. All 50 states accepted these standards, with minor modifications, either as regulations or as guidelines (Oleckno, 1982). The regulations were legally binding at the federal level on only about 700 water systems that supplied common carriers in interstate commerce (fewer than 2 percent of the nation's water supply systems) (Train, 1974). As an enforcement tool, the 1962 standards were of limited use in ensuring clean drinking water for the vast majority of consumers.

In 1969, initial action was taken by the USPHS to review and revise the 1962 standards. The USPHS's Bureau of Water Hygiene undertook a comprehensive survey of water supplies in the United States, known as the Community Water Supply Study (CWSS) (USPHS, 1970a). Its objective was to determine whether the U.S. consumer's drinking water met the 1962 standards. A total of 969 public water systems were tested, most of which were community systems. At that time, this represented approximately 5 percent of the total national public water systems, serving a population of about 18.2 million people, or 12 percent of the total population served by public water systems.

The USPHS released the results of the CWSS in 1970 (USPHS 1970b). The study found that 41 percent of the systems surveyed did not meet the guidelines established in 1962. Many systems were deficient in aspects of source protection, disinfection, clarification, pressure in the distribution system, or combinations of these deficiencies. The study also showed that small water systems, especially those serving fewer than 500 people, had the most difficulty maintaining acceptable water quality. Although the water served to the majority of the U.S. population was safe, the survey indicated that several million people were being supplied water of an inadequate quality and that 360,000 people were being supplied potentially dangerous drinking water.

THE SAFE DRINKING WATER ACT

The results of the CWSS generated congressional interest in federal safe drinking water legislation. The first series of bills to give the federal government power to set enforceable standards for drinking water were introduced in 1970. Congressional hearings on legislative proposals concerning drinking water were held in 1971 and 1972 (Kyros, 1974).

In 1972, a report of an investigation of the quality of the Mississippi River in Louisiana was published. Sampling sites included finished water from the Carrollton water treatment plant in New Orleans. Organic compounds from the water were concentrated using granular activated carbon (GAC), extracted from the GAC using chloroform as a solvent, and then identified. Thirty-six organic compounds were isolated from the extracts collected from the finished water (USEPA, 1972). As a result of this report, new legislative proposals for a safe drinking water law were introduced and debated in Congress in 1973. In late 1973, the General Accounting Office (GAO) released a report investigating 446 community water systems in the states of Maryland, Massachusetts, Oregon, Vermont, Washington, and West Virginia (Symons, 1974). Only 60 systems were found to fully comply with the bacteriological and sampling requirements of the USPHS standards. Bacteriological and chemical monitoring programs of community water supplies were inadequate in five of the six states studied. Many water treatment plants needed to be expanded, replaced, or repaired.

Public awareness of organic compounds in drinking water increased in 1974 as a result of several events. A three-part series in *Consumer Reports* drew attention to organic contaminants in New Orleans drinking water (Harris and Brecher, 1974). Follow-up studies by the Environmental Defense Fund (EDF) (The States-Item, 1974; Page, Talbot, and Harris, 1974; Page, Harris, and Epstein, 1976) and by USEPA (USEPA, 1975a) identifying organic contaminants in New Orleans drinking water and their potential health consequences created further publicity. On December 5, 1974, CBS aired nationally in prime time a program with Dan Rather entitled *Caution, Drinking Water May Be Dangerous to Your Health.*

Also in 1974, researchers at USEPA and in the Netherlands discovered that a class of compounds, trihalomethanes (THMs), were formed as a by-product when free chlorine was added to water containing natural organic matter for disinfection (Bellar, Lichtenberg, and Kroner, 1974). Although unrelated, publicity surrounding the formation of THMs coincided with the finding of synthetic organic chemicals (SOCs) in the New Orleans water supply. On November 8, 1974, USEPA announced that a nationwide survey would be conducted to determine the extent of the THM problem in the United States (Symons et al., 1975). This survey was known as the National Organics Reconnaissance Survey, or NORS, and was completed in 1975 (discussed later). The health significance of THMs and SOCs in drinking water was uncertain, and questions still remain today regarding the health significance of low concentrations of organic chemicals and disinfection by-products. Chloroform, one of the more prevalent THMs found in these surveys, was banned by the U.S. Food and Drug Administration (USFDA) as an ingredient in any human drugs or cosmetic products effective July 29, 1976 (USFDA, 1976).

After more than four years of effort by Congress, federal legislation was enacted to develop a national program to protect the quality of the nation's public drinking water systems. The U.S. House of Representatives and the U.S. Senate passed a safe drinking water bill in November 1974 (Congressional Research Service, 1982). The SDWA was signed into law on December 16, 1974, as Public Law 93-523 (SDWA, 1974).

The 1974 SDWA established a cooperative program among local, state, and federal agencies. The act required the establishment of primary drinking water regulations designed to ensure safe drinking water for the consumer. These regulations were the first to apply to all public water systems in the United States, covering both chemical and microbial contaminants. Except for the coliform standard applicable to water used on interstate carriers (i.e., trains, ships, and airplanes), federal drinking water standards were not legally binding until the passage of the SDWA.

The SDWA mandated a major change in the surveillance of drinking water systems by establishing specific roles for the federal and state governments and for public water suppliers. The federal government, specifically USEPA, is authorized to set national drinking water regulations, conduct special studies and research, and oversee the implementation of the act. The state governments, through their health departments and environmental agencies, are expected to accept the major responsibility, called primary enforcement responsibility, or primacy, for the administration and enforcement of the regulations set by USEPA under the act. Public water suppliers have the day-to-day responsibility of meeting the regulations. To meet this goal, routine monitoring must be performed, with results reported to the regulatory agency. Violations must be reported to the public and corrected. Failure to perform any of these functions can result in enforcement actions and penalties.

The 1974 act specified the process by which USEPA was to adopt national drinking water regulations. Interim regulations [National Interim Primary Drinking

Water Regulations (NIPDWRs)] were to be adopted within six months of its enactment. Within about 2½ years (by March 1977), USEPA was to propose revised regulations (Revised National Drinking Water Regulations) based on a study of health effects of contaminants in drinking water conducted by the National Academy of Sciences (NAS).

Establishment of the revised regulations was to be a two-step process. First, the agency was to publish recommended maximum contaminant levels (RMCLs) for contaminants believed to have an adverse health effect based on the NAS study. The RMCLs were to be set at a level such that no known or anticipated health effect would occur. An adequate margin of safety was to be provided. These levels were to act only as health goals and were not intended to be federally enforceable.

Second, USEPA was to establish maximum contaminant levels (MCLs) as close to the RMCLs as the agency thought feasible. The agency was also authorized to establish a required treatment technique instead of an MCL if it was not economically or technologically feasible to determine the level of a contaminant. The MCLs and treatment techniques comprise the National Primary Drinking Water Regulations (NPDWRs) and are federally enforceable. The regulations were to be reviewed at least every three years.

The National Interim Primary Drinking Water Regulations

Interim regulations were adopted December 24, 1975 (USEPA, 1975b), based on the 1962 USPHS standards with little additional health effects support. The interim rules were amended several times before the first primary drinking water regulation was issued (see Table 1.1).

The findings of the NORS (mentioned previously) were published in November 1975 (Symons et al., 1975). The four THMs—chloroform, bromodichloro-

TABLE 1.1 History of the NIPDWRs

Regulation	Promulgation date	Effective date	Primary coverage
NIPDWRs (USEPA, 1975b)	December 24, 1975	June 24, 1977	Inorganic, organic, and microbiological contaminants and turbidity.
1st NIPDWR amendment (USEPA, 1976a)	July 9, 1976	June 24, 1977	Radionuclides.
2nd NIPDWR amendment (USEPA, 1979b)	November 29, 1979	Varied depending on system size	Total trihalomethanes.*
3rd NIPDWR amendment (USEPA, 1980)	August 27, 1980	February 27, 1982	Special monitoring requirements for corrosion and sodium.
4th NIPDWR amendment (USEPA, 1983a)	February 28, 1983	March 30, 1983	Identifies best general available means to comply with THM regulations.

* The sum of chloroform, bromoform, bromodichloromethane, and dibromochloromethane.

methane, dibromochloromethane, and bromoform—were found to be widespread in the chlorinated drinking waters of 80 cities studied. USEPA subsequently conducted the National Organics Monitoring Survey (NOMS) between 1976 and 1977 to determine the frequency of specific organic compounds in drinking water supplies (USEPA, 1978a). Included in the NOMS were 113 community water supplies representing different sources and treatment processes, each monitored three times during a 12-month period. The NOMS data showed that THMs were the most widespread organic contaminants in drinking water, occurring at the highest concentrations. From the NORS, NOMS, and other surveys, more than 700 specific organic chemicals had been identified in various drinking waters (Cotruvo and Wu, 1978).

On June 21, 1976, the EDF petitioned USEPA, claiming that the initial interim regulations set in 1975 did not sufficiently control organic compounds in drinking water. In response, USEPA issued an Advance Notice of Proposed Rulemaking (ANPRM) on July 14, 1976, requesting public input on how THMs and SOCs should be regulated (USEPA, 1976b).

On February 9, 1978, USEPA proposed a two-part regulation for the control of organic contaminants in drinking water (USEPA, 1978b). The first part concerned the control of THMs. The second part concerned control of source water SOCs and proposed the use of GAC adsorption by water utilities vulnerable to possible SOC contamination.

The next day, February 10, 1978, the U.S. Court of Appeals, District of Columbia Circuit, issued a ruling in the EDF case filed June 21, 1976 (U.S. Court of Appeals, 1978). The court upheld USEPA's discretion to not include comprehensive regulations for SOCs in the NIPDWRs; however, as a result of new data being collected by USEPA, the court told the agency to report back with a plan for amending the interim regulations to control organic contaminants. The court stated (U.S. Court of Appeals, 1978):

> In light of the clear language of the legislative history, the incomplete state of our knowledge regarding the health effects of certain contaminants and the imperfect nature of the available measurement and treatment techniques cannot serve as justification for delay in controlling contaminants that may be harmful.

The agency contended that the proposed rule published the day before satisfied the court's judgment.

Reaction to the proposed regulation on GAC adsorption treatment varied. Federal health agencies, environmental groups, and a few water utilities supported the proposed rule. Many state health agencies, consulting engineers, and most water utilities opposed it because of several technical concerns (Symons, 1984). USEPA responded to early opposition to the GAC proposal by publishing an additional notice on July 6, 1978, discussing health, technical, and operational issues, and presenting revised costs (USEPA, 1978c). Nevertheless, significant opposition continued based on several technical considerations (Pendygraft, Schegel, and Huston, 1979a,b,c). USEPA promulgated regulations for the control of THMs in drinking water on November 29, 1979 (USEPA, 1979b), but subsequently, on March 19, 1981, withdrew its proposal to control organic contaminants in vulnerable surface water supplies by GAC adsorption (USEPA, 1981). The NIPDWRs were also amended on August 27, 1980, to update analytical methods and impose special monitoring and reporting requirements (USEPA, 1980).

National Academy of Sciences Studies

As required by the 1974 SDWA, USEPA contracted with the NAS to have the National Research Council (NRC) assess human exposure via drinking water and the toxicology of contaminants in drinking water. The NRC Committee on Safe Drinking Water published their report, *Drinking Water and Health,* in 1977 (NAS, 1977). Five classes of contaminants were examined: microorganisms, particulate matter, inorganic solutes, organic solutes, and radionuclides. This report, the first in a series of nine, served as the basis for revised drinking water regulations. USEPA published the recommendations of the NAS study on July 11, 1977 (USEPA, 1977b).

The 1977 amendments to the SDWA called for revisions of the NAS study "reflecting new information which has become available since the most recent previous report [and which] shall be reported to the Congress each two years thereafter" (SDWA 1977). USEPA periodically funds the NAS to conduct independent assessments of drinking water contaminants; several studies have been completed or are in progress.

SAFE DRINKING WATER ACT AMENDMENTS, 1977 THROUGH 1986

The SDWA has been amended and/or reauthorized several times since initial passage (Table 1.2). In 1977, 1979, and 1980, Congress enacted amendments to the SDWA that reauthorized and revised certain provisions (Congressional Research Service, 1982).

USEPA's slowness in regulating contaminants from 1974 through the early 1980s and its failure to require GAC treatment for organic contaminants served as a focal point for discussion of possible revisions to the law. Reports in the early 1980s of drinking water contamination by organic contaminants and other chemicals (Westrick, Mello, and Thomas, 1984) and pathogens, such as *Giardia lamblia* (Craun, 1986), aroused congressional concern over the adequacy of the SDWA. The rate of progress made by USEPA to regulate contaminants was of particular concern. Both the House and Senate considered various legislative proposals beginning in 1982 that informed the SDWA debate and helped to shape the SDWA amendments enacted in 1986.

To strengthen the SDWA, especially the regulation-setting process and groundwater protection, most of the original 1974 SDWA was amended in 1986. Major provisions of the 1986 amendments included (Cook and Schnare, 1986; Dyksen, Hiltebrand, and Raczko, 1988; Gray and Koorse, 1988):

TABLE 1.2 SDWA and Amendments

Year	Public law	Date	Act
1974	P.L. 93-523	December 16, 1974	SDWA
1977	P.L. 95-190	November 16, 1977	SDWA amendments of 1977
1979	P.L. 96-63	September 6, 1979	SDWA amendments of 1979
1980	P.L. 96-502	December 5, 1980	SDWA amendments of 1980
1986	P.L. 99-339	June 16, 1986	SDWA amendments of 1986
1988	P.L. 100-572	October 31, 1988	Lead Contamination Control Act
1996	P.L. 104-182	August 6, 1996	SDWA amendments of 1996

Note: Codified generally as 42 U.S.C. 300f-300j-11.

- Mandatory standards for 83 contaminants by June 1989.
- Mandatory regulation of 25 contaminants every 3 years.
- National Interim Drinking Water Regulations were renamed National Primary Drinking Water Regulations (NPDWRs).
- Recommended maximum contaminant level (RMCL) goals were replaced by maximum contaminant level goals (MCLGs).
- Required designation of best available technology for each contaminant regulated.
- Specification of criteria for deciding when filtration of surface water supplies is required.
- Disinfection of all public water supplies with some exceptions for groundwater that meet, as yet, unspecified criteria.
- Monitoring for contaminants that are not regulated.
- A ban on lead solders, flux, and pipe in public water systems.
- New programs for wellhead protection and protection of sole source aquifers.
- Streamlined and more powerful enforcement provisions.

The 1986 amendments significantly increased the rate at which USEPA was to set drinking water standards. Table 1.3 summarizes the regulations promulgated each year in comparison with the number required by the 1986 amendments. Resource limitations and competing priorities within the agency prevented USEPA from fully meeting the mandates of the 1986 amendments.

1988 LEAD CONTAMINATION CONTROL ACT

On December 10, 1987, the House Subcommittee on Health and Environment held a hearing on lead contamination of drinking water. At that hearing, the USPHS warned that some drinking watercoolers may contain lead solder or lead-lined water tanks that release lead into the water they distribute. Data submitted to the subcommittee by manufacturers indicated that close to 1 million watercoolers containing lead were in use at that time (Congressional Research Service, 1993).

The Lead Contamination Control Act was enacted on October 31, 1988, as Public Law 100-572 (LCCA, 1988). This law amended the SDWA to, among other things, institute a program to eliminate lead-containing drinking watercoolers in schools. Part F, "Additional Requirements to Regulate the Safety of Drinking Water," was added to the SDWA. USEPA was required to provide guidance to states and localities to test for and remedy lead contamination in schools and day care centers. It also contains specific requirements for the testing, recall, repair, and/or replacement of watercoolers with lead-lined storage tanks or with parts containing lead. Civil and criminal penalties for the manufacture and sale of watercoolers containing lead are set.

1996 SAFE DRINKING WATER ACT AMENDMENTS

The 1986 SDWA amendments authorized congressional appropriations for implementation of the law through fiscal year 1991. Beginning in 1988, studies by environmental groups, the U.S. General Accounting Office (USGAO), USEPA, and the water industry groups drew attention to needed changes to the SDWA.

TABLE 1.3 Promulgation of U.S. Drinking Water Standards by Year

Year	Regulation	Incremental no. of contaminants regulated	Total no. of contaminants regulated*	Total no. required by 1986 SDWA amendments†	Reference
1975	NIPDWRs	18	18	—	USEPA, 1975b
1976	Interim radionuclides	4	22	—	USEPA, 1976a
1979	Interim TTHMs	1	23	—	USEPA, 1979b
1986	Revised fluoride	0	23	—	USEPA, 1986a
1987	Volatile organic chemicals	8	31	31	USEPA, 1987a
1988	—	0	31	62	—
1989	Surface Water Treatment Rule and Total Coliform Rule	4	35	96	USEPA, 1989b,c
1991	Phase II SOCs and IOCs and lead and copper	27	62	111	USEPA, 1991a,b,c,d
1992	Phase V SOCs and IOCs	22	84	111	USEPA, 1992c
1993	—	0	84	111	—
1994	—	0	84	136	—
1995	—	0	84	136	—
1996	Information Collection Rule (monitoring only)	0	84	—	USEPA, 1996b
1998	Consumer Confidence Reports	0	84	—	USEPA, 1998f
1998	Stage 1 D-DBP Rule	11	91	—	USEPA, 1998g
1998	Interim Enhanced Surface Water Treatment Rule	1	92	—	USEPA, 1998h

* NPDWRs for some contaminants have been stayed, remanded, or revised.
† Cumulative total at the time of promulgation.

Severe resource constraints made it increasingly difficult for many states to effectively carry out the monitoring, enforcement, and other mandatory activities to retain primacy. However, state funding needs represent only a fraction of the expenditures that public water systems must make to comply with SDWA requirements (USEPA, 1993a). Results of a survey released in 1993 by the Association of State Drinking Water Administrators (ASDWA) identified an immediate need of $2.738 billion for SDWA-related infrastructure projects (i.e., treatment, storage, and distribution) in 35 states (ASDWA, 1993).

In late March and early April of 1993, the largest waterborne disease outbreak in the United States in recent times occurred in Milwaukee, Wisconsin, drawing national attention to the importance of safe drinking water. More than 400,000 people in Milwaukee were reported to have developed symptomatic gastrointestinal infections as a consequence of exposure to drinking water contaminated with *cryptosporidium* (MacKenzie et al., 1994). More than 4000 people were hospitalized, and cryptosporidiosis contributed to between 54 and 100 deaths (Morris et al., 1996; Hoxie et al., 1997). *Cryptosporidium* was not regulated by USEPA at that time, and the Milwaukee outbreak became a rallying point for those seeking to make the SDWA stricter. The outbreak and its aftermath had a significant influence on USEPA and the Congress, who sought to ensure that another outbreak of this magnitude would not occur.

Discussion of SDWA reauthorization issues was intense throughout the 103rd and 104th Congresses. Environmental groups, water supplier representatives, USEPA, state regulatory agencies, governors, and elected officials contended with differences of opinion over how the SDWA should be changed. Unfunded federal mandates are of particular concern to state and local elected officials. These are laws passed by the U.S. Congress imposing requirements on state and local governments without providing adequate federal funds to implement those requirements. The cost of complying with environmental laws in general and the SDWA in particular, in the absence of federal, state, and local financial resources, caused many groups to pressure Congress for relief.

The House of Representatives and the Senate both passed SDWA reauthorization bills in the 104th Congress. A conference committee resolved the differences between the two bills, and the conference committee report was filed on August 1, 1996 (Conference Report 104-741). The House and Senate both approved the conference report on August 1, 1996, and the SDWA amendments of 1996 were signed into law as Public Law 104-182 on August 6, 1996 (SDWA, 1996).

The SDWA amendments of 1996 made substantial revisions to the SDWA, and 11 new sections were added (Pontius, 1996a,b). Provisions of the 1996 SDWA amendments include:

- Retention of the 1986 requirements regarding mandatory standards for 83 contaminants by June 1989
- Elimination of the 1986 requirement that 25 contaminants be regulated every three years
- Revision of the process for listing of contaminants for possible regulation
- Revision of the standard setting process to include consideration of cost, benefits, and competing health risks
- New programs for source water assessment, local source water petitions, and source water protection grants
- Mandatory regulation of filter backwash water recycle
- Specified schedules for regulation of radon and arsenic
- Revised requirements for unregulated contaminant monitoring and a national occurrence database
- Provisions creating a state revolving loan fund (SRLF) for drinking water
- New provisions regarding small system variances, treatment technology, and assistance centers
- Development of operator certification guidelines by USEPA.

In addition to revising the SDWA, the 1996 amendments contained significant provisions for infrastructure funding, drinking water research, and other items. A detailed review of the SDWA and the 1996 amendments is available (Pontius, 1997a,b,c).

DEVELOPMENT OF NATIONAL PRIMARY DRINKING WATER REGULATIONS

USEPA is given broad authority by Congress under the SDWA to publish MCLGs and NPDWRs for drinking water contaminants. The 1996 SDWA amendments require USEPA to publish an MCLG and promulgate an NPDWR for a contaminant that:

1. Has an adverse effect on the health of persons.
2. Is known to occur or there is a substantial likelihood that the contaminant will occur in public water systems with a frequency and at levels of public health concern.
3. In the sole judgment of USEPA, regulation of the contaminant presents a meaningful opportunity for health risk reduction for persons served by public water systems.

This authority is the primary driving force behind the establishment of new drinking water regulations. The adverse health effect of a contaminant need not be proven conclusively prior to regulation. This general authority to regulate contaminants in drinking water does not apply to contaminants for which an NPDWR was promulgated as of the date of enactment of the SDWA amendments of 1996 (August 6, 1996). In addition, no NPDWR may require addition of any substance for preventive health care purposes unrelated to drinking water contamination.

The 1996 SDWA amendments retained USEPA's authority to propose and promulgate the National Secondary Drinking Water Regulations (NSDWRs). The NSDWRs are based on aesthetic, as opposed to health, considerations and are not federally enforceable, but some states have adopted NSDWRs as enforceable standards. They may be established as the agency considers appropriate and may be amended and revised as needed.

Selection of Contaminants for Regulation

The 1986 SDWA amendments specifically required USEPA to set NPDWRs for 83 contaminants listed in the "Advanced Notice for Proposed Rulemakings," published in the *Federal Register* on March 4, 1982 (USEPA, 1982a), and on October 5, 1983 (USEPA, 1983b). Regulations were to be set for the 83 contaminants without regard to their occurrence in drinking water. USEPA was required to set regulations for these contaminants, although up to seven substitutes were allowed if regulation of the substitutes was more likely to be protective of public health. USEPA proposed (USEPA, 1987b) and adopted (USEPA, 1988) seven substitutes. Specific timelines were specified for regulation development, with at least 25 additional contaminants regulated every three years, selected from a Drinking Water Priority List (DWPL) (Pontius, 1997c).

The 1996 SDWA amendments revised the contaminant listing and regulatory process. However, the requirement to regulate the 83 contaminants imposed by the 1986 amendments was retained. Specific timelines specified in the amended SDWA for regulation development are summarized in Table 1.4.

The 1996 SDWA amendments require USEPA to publish a list of contaminants that may require regulation under the SDWA no later than February 6, 1998, and every five years thereafter. The agency is to consult with the scientific community, including the Science Advisory Board, when preparing the list, and provide notice and opportunity for public comment. An occurrence database, established by the 1996 SDWA amendments, is to be considered in determining whether the contaminants are known or anticipated to occur in public water systems. At the time of publication, listed contaminants are not to be subject to any proposed or promulgated NPDWR.

Unregulated contaminants considered for listing must include, but are not limited to, substances referred to in section 101(14) of the Comprehensive Environmental Response, Compensation, and Liability Act of 1980, and substances

TABLE 1.4 Water Quality Regulation Development Under the SDWA as Amended in 1996

Date	Action
	List of 83 contaminants
By June 19, 1987	USEPA must regulate at least 9 contaminants from the list of 83.
By June 19, 1988	USEPA must regulate at least 40 contaminants from the list of 83.
By June 19, 1999	USEPA must regulate the remainder of contaminants from the list of 83.
	Filtration for systems using surface water
By December 16, 1987	USEPA must propose and promulgate NPDWRs specifying when filtration is required for surface water systems.
	Disinfection
After August 6, 1999	USEPA must promulgate NPDWRs requiring disinfection for all public water systems, including surface water systems and, as necessary, groundwater systems. As part of the regulations, USEPA must promulgate criteria to determine whether disinfection is to be required for a groundwater system.
	Arsenic
By February 3, 1997	USEPA must develop a health effects study plan.
By January 1, 2000	USEPA must propose an NPDWR.
By January 1, 2001	USEPA must promulgate an NPDWR.
	Sulfate
By February 6, 1999	USEPA and CDC must jointly conduct a new dose-response study.
By August 6, 2001	USEPA must determine whether to regulate sulfate.
	Radon
By February 6, 1999	USEPA must publish a risk/cost study.
By August 6, 1999	USEPA must propose an MCLG and NPDWR.
By August 6, 2000	USEPA must publish an MCLG and promulgate an NPDWR.
	Disinfectants/disinfection by-products (D/DBPs)
According to schedule in Table III.13, 59 *Federal Register* 6361	USEPA must promulgate a Stage I and Stage II D/DBP rule.
	Enhanced Surface Water Treatment Rule (ESWTR)
According to schedule in Table III.13, 59 *Federal Register* 6361	USEPA must promulgate an interim and final ESWTR.
	New contaminants
By February 6, 1998, and every five years thereafter.	USEPA to list contaminants that may require regulation.
No later than August 6, 2001, and every five years thereafter.	USEPA to make a determination whether to regulate at least five contaminants.
No later than 24 months after decision to regulate.	USEPA to publish a proposed MCLG and NPDWR.

TABLE 1.4 Water Quality Regulation Development Under the SDWA as Amended in 1996 (*Continued*)

Date	Action
	New contaminants
Within 18 months of proposal. Deadline may be extended 9 months if necessary.	USEPA to publish final MCLG and promulgate NPDWR.
Three years after promulgation; up to two-year extension possible.	NPDWRs become effective.
	Recycling of filter backwash water
By August 6, 2000	USEPA must regulate backwash water recycle unless it is addressed in the ESWTR prior to this date.
	Urgent threats to public health
Anytime	USEPA may promulgate an interim NPDWR to address contaminants that present an urgent threat to public health.
Not later than five years after promulgation	USEPA is required to repromulgate or revise as appropriate an interim NPDWR as a final NPDWR.
	Review of regulations
Not less often than every six years	USEPA is required to review and revise as appropriate each NPDWR.

registered as pesticides under the Federal Insecticide, Fungicide, and Rodenticide Act. USEPA's decision whether to select an unregulated contaminant for listing is not subject to judicial review.

A draft of the Drinking Water Contaminant Candidate List (DWCCL) was published for public comment by USEPA in the *Federal Register* on October 6, 1997 (USEPA 1997). The final DWCCL was published on March 2, 1998, listing 50 chemical and 10 microbial contaminants/contaminant groups for possible regulation [see Table 1.5 (USEPA, 1998a)].

Determination to Regulate

No later than August 6, 2001, and every five years thereafter, USEPA is required to make determinations on whether to regulate at least five listed contaminants. Notice of the preliminary determination is to be given and an opportunity for public comment provided.

The agency's determination to regulate is to be based on the best available public health information, including the new occurrence database. To regulate a contaminant, USEPA must find that the contaminant may have an adverse effect on the health of persons, occurs or is likely to occur in public water systems with a frequency and at levels of public health concern, and that regulation of the contaminant presents a meaningful opportunity for health risk reduction for persons served by public water systems. USEPA may make a determination to regulate a contaminant that is not listed if the determination to regulate is made based on these criteria.

TABLE 1.5 Drinking Water Contaminant Candidate List*

	CASRN†	Priority‡
Chemical		
1,3-dichloropropene (telone or 1,3-D)	542-75-6	RD
1,1-dichloropropene	563-58-6	HR
1,2-diphenylhydrazine	122-66-7	AMR, O
1,1,2,2-tetrachloroethane	79-34-5	RD
1,1-dichloroethane	75-34-3	RD
1,2,4-trimethylbenzene	95-63-6	RD
1,3-dichloropropane	124-28-9	HR
2,4-dinitrophenol	51-28-5	AMR, O
2,4-dichlorophenol	120-83-2	AMR, O
2,4,6-trichlorophenol	88-06-2	AMR, O
2,2-dichloropropane	594-20-7	RD
2,4-dinitrotoluene	121-14-2	O
2,6-dinitrotoluene	606-20-2	O
2-methyl-phenol (*o*-cresol)	95-48-7	AMR, O
Acetochlor	34256-82-1	AMR, O
Alachlor ESA and other degradation products of acetanilide pesticides	—	AMR, O
Aldrin	309-00-2	RD
Aluminum	7429-90-5	HR, TR
Boron	7440-42-8	RD
Bromobenzene	108-86-1	RD
DCPA mono-acid degradate	887-54-7	HR, O
DCPA di-acid degradate	2136-79-0	HR, O
DDE	72-55-9	O
Diazinon	333-41-5	O
Dieldrin	60-57-1	RD
Disulfoton	298-04-4	O
Diuron	330-54-1	O
EPTC (*s*-ethyl-dipropylthiocarbonate)	759-94-4	O
Fonofos	944-22-9	AMR, O
Hexachlorobutadiene	87-68-3	RD
p-isopropyltoluene	99-87-6	RD
Linuron	330-55-2	O
Manganese	7439-96-5	RD
Methyl-*t*-butyl ether (MTBE)	1634-04-4	HR, TR, O
Methyl bromide (bromomethane)	74-83-9	HR
Metolachlor	51218-45-2	RD
Metribuzin	21087-64-9	RD
Molinate	2212-67-1	O
Naphthalene	91-20-3	RD
Nitrobenzene	98-95-3	O
Organotins	—	RD
Perchlorate	—	HR, TR, AMR, O
Prometon	1610-18-0	O
RDX (cyclo trimethylene trinitramine)	121-82-4	AMR, O
Sodium (guidance)	7440-23-5	HR
Sulfate	14808-79-8	RD
Terbacil	5902-51-2	O
Terbufos	13071-79-9	O
Triazines and degradation products (including, but not limited to cyanazine, and atrazine-desethyl)	—	RD
Vanadium	7440-62-2	RD

TABLE 1.5 Drinking Water Contaminant Candidate List* (*Continued*)

	CASRN†	Priority‡
Microbials		
Acanthamoeba (guidance expected for contact lens wearers)	—	RD
Adenoviruses	—	AMR, TR, O
Aeromonas hydrophila	—	HR, TR, O
Cyanobacteria (blue-green algae), other freshwater algae, and their toxins	—	HR, TR, AMR, O
Caliciviruses	—	HR, TR, AMR, O
Coxsackieviruses	—	TR, O
Echoviruses	—	TR, O
Heliobacter pylori	—	HR, TR, AMR, O
Microsporidia (Enterocytozoon and Septata)	—	HR, TR, AMR, O
Mycobacterium avium intracellulare (MAC)	—	HR, TR

* USEPA, 1998a.
† CASRN (Chemical Abstracts Service Registration Number).
‡ Priority: RD, regulatory determination; HR, health research; AMR, analytical methods research; O, occurrence; TR, treatment research.

A determination not to regulate a contaminant is considered final agency action subject to judicial review. Supporting documents for the determination are to be made available for public comment at the time the determination is published. The agency may publish a health advisory or take other appropriate actions for contaminants not regulated. Health advisories are not regulations and are not enforceable; they provide guidance to states regarding health effects and risks of a contaminant.

Regulatory Priorities and Urgent Threats

When selecting unregulated contaminants for consideration for regulation, contaminants that present the greatest public health concern are to be selected. USEPA is to consider public health factors when making this selection, and specifically the effect of the contaminants upon subgroups that comprise a meaningful portion of the general population. These include infants, children, pregnant women, the elderly, individuals with a history of serious illness, or other subpopulations that are identifiable as being at greater risk of adverse health effects due to exposure to contaminants in drinking water than the general population.

An interim NPDWR may be promulgated for a contaminant without making a regulatory determination, or completing the required risk reduction and cost analysis (discussed later), to address an urgent threat to public health. The agency must consult with and respond in writing to any comments provided by the Secretary of Health and Human Services, acting through the director of the Centers for Disease Control and Prevention (CDC) or the director of the National Institutes of Health. If an interim regulation is issued, the risk reduction and cost analysis must be published not later than three years after the date on which the regulation is promulgated. The regulation must be repromulgated, or revised if appropriate, not later than five years after that date.

Regulatory Deadlines and Procedures

For each contaminant that USEPA determines to regulate, an MCLG must be published and an NPDWR promulgated by rulemaking. An MCLG and an NPDWR must be proposed for a contaminant not later than 24 months after the determination to regulate. The proposed regulation may be published concurrently with the determination to regulate.

USEPA is required to publish an MCLG and promulgate an NPDWR within 18 months after proposal. The agency, by notice in the *Federal Register,* may extend the deadline for promulgation up to 9 months.

Regulations are to be developed in accordance with Section 553 of Title 5, United States Code, known as the Administrative Procedure Act. An opportunity for public comment prior to promulgation is required. The administrator is to consult with the secretary of Health and Human Services and the National Drinking Water Advisory Council (NDWAC) regarding proposed and final rules. Before proposing any MCLG or NPDWR, USEPA must request comments from the Science Advisory Board (SAB), which can respond any time before promulgation of the regulation. USEPA does not, however, have to postpone promulgation if no comments are received.

When establishing NPDWRs, USEPA is required under the SDWA, to the degree that an agency action is based on science, to use the best available, peer-reviewed science and supporting studies conducted in accordance with sound and objective scientific practices. Data must be collected by accepted methods or best available methods, if the reliability of the method and the nature of the decision justify the use of data. When establishing NPDWRs, information presented on health effects must be comprehensive, informative, and understandable.

Health Risks and Cost Analysis

Health risks and costs are to be considered when proposing MCLs. When proposing an NPDWR that includes an MCL, USEPA is required to publish, seek public comment on, and use an analysis of the health risk reduction and costs for the MCL being considered and each alternative MCL. This analysis is to evaluate the following:

- Health risk reduction benefits likely to occur
- Costs likely to occur
- Incremental costs and benefits associated with each alternative MCL considered
- Effects of the contaminant on the general population and on sensitive subgroups
- Increased health risks that may occur as the result of compliance, including risks associated with co-occurring contaminants
- Other relevant factors, including the quality and extent of the information, the uncertainties in the analysis of the above factors, and the degree and nature of the risk

When proposing an NPDWR that includes a treatment technique, USEPA must publish and seek public comment on an analysis of the health risk reduction benefits and costs likely to be experienced as a result of compliance with the treatment technique and alternative treatment techniques that are being considered. The aforementioned factors are to be taken into account, as appropriate.

Regulatory Basis of MCLGs

Maximum contaminant level goals (MCLGs) are nonenforceable, health-based goals. They are to "be set at a level at which no known or anticipated adverse effect on human health occurs and that allows for an adequate margin of safety," without regard to the cost of reaching these goals.

Risk Assessment Versus Risk Management. To establish an MCLG, USEPA conducts a risk assessment for the contaminant of concern, patterned after the recommendations of the National Research Council (NRC, 1983). The general components of this risk assessment, illustrated in Figure 1.1, are as follows:

- *Hazard identification.* Qualitative determination of the adverse effects that a constituent may cause
- *Dose-response assessment.* Quantitative determination of what effects a constituent may cause at different doses
- *Exposure assessment.* Determination of the route and amount of human exposure
- *Risk characterization.* Description of risk assessment results and underlying assumptions

The risk assessment process is used to establish an MCLG, whereas the risk management process is used to set the maximum contaminant level [MCL (or treatment technique requirement)]. Risk assessment and risk management overlap at risk characterization. Assumptions and decisions made in the risk assessment, such as which dose-response model to use, will affect the end result of the analysis. Hence, such assumptions must be made carefully.

Toxicology Reviews. An MCLG is established for a contaminant based on toxicology data on the contaminant's potential health effects associated with drinking water. Data evaluated include those obtained from human epidemiology or clinical studies and animal exposure studies. USEPA assesses all health factors for a particular contaminant, including absorption of the contaminant when ingested, pharmacokinetics (metabolic changes to a contaminant from ingestion through excretion), mutagenicity

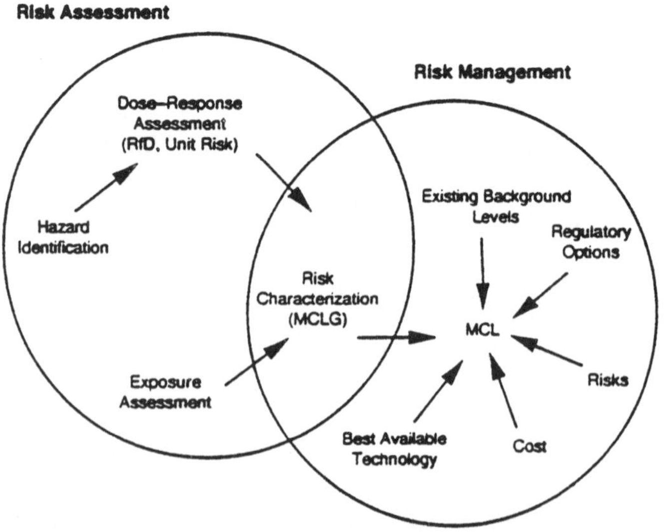

FIGURE 1.1 Differentiation between risk assessment and risk management in setting drinking water MCLGs and MCLs.

(capacity to cause or induce permanent changes in genetic materials), reproductive and developmental effects, and cancer-causing potential (carcinogenicity).

Because human epidemiology data usually cannot define cause-and-effect relationships and because human health effects data for many contaminants do not exist, a contaminant's potential risk to humans is typically estimated based on the response of laboratory animals to the contaminant. The underlying assumption is that effects observed in animals may occur in humans.

The use of animal exposure studies for assessing human health effects is the subject of ongoing controversy. In its assessment of irreversible health effects, USEPA has been guided by the following principles [recommended in 1977 by the NAS (NAS, 1977)], which are also the subject of continuing debate:

- Health effects in animals are applicable to humans, if properly qualified.
- Methods do not now exist to establish a threshold (exposure level required to cause measurable response) for long-term effects of toxic agents.
- Exposing animals to high doses of toxic agents is a necessary and valid method of discovering possible carcinogenic hazards in humans.
- Substances should be assessed in terms of human risk rather than as safe or unsafe.

In 1986, USEPA set guidelines (USEPA, 1986b) for classifying contaminants based on the weight of evidence of carcinogenicity (Table 1.6) (Cohrssen and Covello, 1989). These guidelines have been used by USEPA's internal cancer assessment group to classify contaminants of regulatory concern. For comparison, the classification systems of two other organizations that use weight-of-evidence criteria for classifying carcinogens are also shown in Table 1.6. On April 23, 1996, USEPA proposed new guidelines for carcinogen risk assessment (USEPA, 1996). The agency will be implementing these guidelines within the next few years.

A contaminant's carcinogenicity determines how USEPA establishes an MCLG for that contaminant. The USEPA Office of Science and Technology uses a three-category approach (Table 1.7) for setting MCLGs based on the weight of evidence of carcinogenicity via consumption of drinking water. In assigning a contaminant to one of the three categories, USEPA considers results of carcinogenicity evaluations by its own scientists as well as those by other scientific bodies, such as the International Agency for Research on Cancer and the National Academy of Sciences. Exposure, pharmacokinetics, and potency data are also considered. As a result, the USEPA cancer classification (Table 1.6) does not necessarily determine the regulatory category (Table 1.7). The agency is currently revisiting its three-category approach.

The three categories are as follows (USEPA, 1991a):

Category I. Contaminants for which sufficient evidence of carcinogenicity in humans and animals exists to warrant a carcinogenicity classification as "known probable human carcinogens via ingestion"

Category II. Contaminants for which limited evidence of carcinogenicity in animals exists and that are regulated as "possible human carcinogens via ingestion"

Category III. Substances for which insufficient or no evidence of carcinogenicity via ingestion exists

The approach to setting MCLGs differs for each category. For each contaminant to be regulated, USEPA summarizes results of its toxicology review in a health effects criteria document that identifies known and potential adverse health effects (called end points) and quantifies toxicology data used to calculate the MCLG.

TABLE 1.6 Weight-of-Evidence Criteria for Classifying Carcinogens*

Organization	Category	Criterion
USEPA	A	Human carcinogen (sufficient evidence of carcinogenicity from epidemiologic studies)
	B1	Probable human carcinogen (limited evidence of carcinogenicity to humans)
	B2	Probable human carcinogen (sufficient evidence from animal studies, and inadequate evidence or no data on carcinogenicity to humans)
	C	Possible human carcinogen (limited evidence from animal studies; no data for humans)
	D	Not classifiable because of inadequate evidence
	E	No evidence of carcinogenicity in at least two animal tests in different species or in both animal and epidemiologic studies
	1	Carcinogenic to humans (sufficient epidemiologic evidence)
	2A	Probably carcinogenic to humans (at least limited evidence of carcinogenicity to humans)
International Agency for Research on Cancer	2B	Probably carcinogenic to humans (no evidence of carcinogenicity to humans)
	3	Sufficient evidence of carcinogenicity in experimental animals
	a	Known to be carcinogenic (evidence from studies on humans)
National Toxicology Program	b	Reasonably anticipated to be a carcinogen (limited evidence of carcinogenicity in humans or sufficient evidence in experimental animals)

* USEPA, 1986b; Cohrssen and Covello, 1989.

Known or Probable Human Carcinogens. Language specifying how known or probable human carcinogens (category I) are to be regulated is not contained in the SDWA. Congressional guidance on setting MCLGs for carcinogens was provided in House Report 93-1185, which accompanied the 1974 SDWA (House of Representatives, 1974). USEPA must consider "the possible impact of synergistic effects, long-term and multistage exposures, and the existence of more susceptible groups in the population." The guidance also states that MCLGs must prevent occurrence of known or anticipated adverse effects and include an adequate margin of safety unless no safe threshold for a contaminant exists, in which case the MCLG should be zero.

With few exceptions, scientists have been unable to demonstrate a threshold level for carcinogens. This has led USEPA to adopt a policy that any level of exposure to

TABLE 1.7 USEPA Three-Category Approach for Establishing MCLGs*

Category	Evidence of carcinogenicity via ingestion	Setting MCLG
I	Strong	Set at zero
II	Limited or equivocal	Calculate based on RfD plus added safety margin or set within cancer risk range of 10^{-5} to 10^{-6}
III	Inadequate or none	Calculate RfD

* Barnes and Dourson, 1988.

carcinogens represents some level of risk. Such risk, however, could be insignificant (i.e., *de minimus*) at very low exposure levels depending on a given carcinogen's potency. Because the legislative history of the SDWA indicates that MCLGs should be set at zero if no *safe* threshold exists and because the courts have held that safe does not mean risk-free, USEPA has some discretion in setting MCLGs other than zero for probable carcinogens. In its proposals to regulate volatile organic contaminants (USEPA, 1985) and the Phase II group of organic and inorganic contaminants (USEPA, 1991a), USEPA outlined two alternative approaches for setting MCLGs for category I contaminants: Set the MCLG at zero (none) for all such contaminants or set it based on a *de minimus* risk level.

Because of the lack of evidence for a threshold for most probable carcinogens and because USEPA interprets MCLGs as aspirational goals, the agency chose to set MCLGs for category I contaminants at zero (USEPA, 1991a, 1987a). This expresses the ideal that drinking water should not contain carcinogens. In this context, zero is not a number (as in 0 mg/L) but is a concept (as in "none"). An MCLG of zero, however, does not necessarily imply that actual harm would occur to humans exposed to contaminant concentrations somewhat above zero. USEPA's revised cancer risk assessment guidelines allow a nonzero MCLG for carcinogens when scientifically justified (USEPA, 1998b).

Insufficient Evidence of Carcinogenicity. For noncarcinogenic (category III) contaminants, USEPA determines a "no effect level" [known as the reference dose (RfD)] for chronic or lifetime periods of exposure. The RfD represents the exposure level thought to be without significant risk to humans (including sensitive subgroups) when the contaminant is ingested daily over a lifetime (Barnes and Dourson, 1988).

Calculation of the RfD is based on the assumption that an organism can tolerate and detoxify some amount of toxic agent up to a certain dose (threshold). As the threshold is exceeded, the biologic response is a function of the dose applied and the duration of exposure. Available human and animal toxicology data for the contaminant are reviewed to identify the highest no-observed-adverse-effect level (NOAEL) or the lowest-observed-adverse-effect level (LOAEL).

The RfD, measured in milligrams per kilogram of body weight per day, is calculated as follows:

$$\text{RfD} = \frac{\text{NOAEL or LOAEL}}{\text{uncertainty factors}} \tag{1.1}$$

Uncertainty factors are used to account for differences in response to toxicity within the human population and between humans and animals (Table 1.8), illustrated in Figure 1.2. They compensate for such factors as intra- and interspecies variability, the small number of animals tested compared with the size of the exposed human population, sensitive human subpopulations, and possible synergistic effects. Using the RfD, a drinking water equivalent level (DWEL) is calculated. The DWEL represents a lifetime exposure at which adverse health effects are not anticipated to occur, assuming 100 percent exposure from drinking water:

$$\text{DWEL (mg/L)} = \frac{\text{RfD} \times \text{body weight (kg)}}{\text{drinking water volume (L/day)}} \tag{1.2}$$

For regulatory purposes, a body weight of 70 kg and a drinking water consumption rate of 2 L/day are assumed for adults. If the MCLG is based on effects in infants (e.g., nitrate), then a body weight of 10 kg and a consumption rate of 1 L/day are assumed.

TABLE 1.8 Uncertainty Factors Used in RfD Calculations*

Factor	Criterion
10	Valid data on acute or chronic human exposure are available and supported by data on acute or chronic toxicity in other species.
100	Data on acute or chronic toxicity are available for one or more species but not for humans.
1,000	Data on acute or chronic toxicity in all species are limited or incomplete, or data on acute or chronic toxicity identify an LOAEL (not an NOAEL) for one or more species, but data on humans are not available.
1–10	Other considerations (such as significance of the adverse health effect, pharmacokinetic factors, or quality of available data) may necessitate use of an additional uncertainty factor.

* Barnes and Dourson, 1988.

When an MCLG for a category III contaminant is determined, contributions from other sources of exposure, including air and food, are also taken into account. If sufficient quantitative data are available on each source's relative contribution to the total exposure, and if the exposure contribution from drinking water is between 20 and 80 percent, USEPA subtracts the exposure contributions from food and air from the DWEL to calculate the MCLG. If the exposure contribution from drinking water is 80 percent or greater, a value of 80 percent is used to protect individuals whose total exposure may be higher than that indicated by available data. If the drinking water exposure contribution is less than 20 percent, USEPA generally uses 20 percent. In cases in which drinking water contributes a relatively small portion of total exposure, reducing exposure from other sources is more prudent than controlling the small contribution from drinking water.

When sufficient data are not available on the contribution of each source to total exposure, a conservative value of 20 percent is used. Most regulated contaminants fall into this category. Once the relative source contribution from drinking water is determined, the MCLG is calculated as follows:

$$\text{MCLG} = \text{DWEL} \times (\text{percent drinking water contribution}) \tag{1.3}$$

Possible Human Carcinogens. Category II contaminants are not regulated as human carcinogens; however, they are treated more conservatively than category III contam-

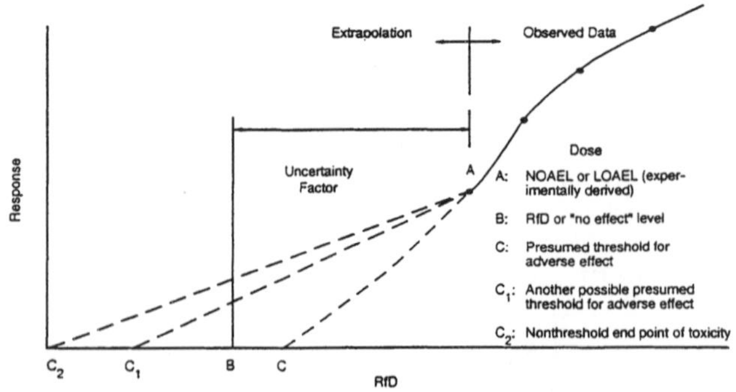

FIGURE 1.2 Example of RfD determination for noncarcinogenic effects.

inants. The method of establishing MCLGs for category II contaminants is more complex than for the other categories. Two options are the following (USEPA, 1991a):

1. The MCLG is calculated based on the RfD plus an additional uncertainty factor of 1 to 10 to account for the evidence of possible carcinogenicity.
2. The MCLG is calculated based on a lifetime risk from 10^{-5} (1 in 100,000 individuals theoretically would get cancer) to 10^{-6} (1 in 1 million individuals) using a conservative method that is unlikely to underestimate the actual risk.

The first option is used when sufficient chronic toxicity data are available. An MCLG is calculated based on an RfD and DWEL following the method used for category III contaminants, and an additional uncertainty factor is added. If available data do not justify use of the first option, the MCLG is based on risk calculation, provided sufficient data are available to perform the calculations.

Data from lifetime exposure studies in animals are manipulated using the linearized multistage dose-response model (one that assumes disease occurrence data are linear at low doses) to calculate estimates of upper-bound excess cancer risk (Figure 1.3). Because cancer mechanisms are not well understood, no evidence exists to suggest that the linearized multistage model (LMM) can predict cancer risk more accurately than any other extrapolation model. Figure 1.4 illustrates the differences between risk model extrapolations to low concentrations of vinyl chloride for four different models (Zavaleta, 1995). The LMM, however, was chosen for consistency. It incorporates a new-threshold carcinogenesis model and is generally more conservative than other models.

The LMM uses dose-response data from the most appropriate carcinogenicity study to calculate a human carcinogenic potency factor (q_1*) that is used to determine the concentration associated with theoretical upper-bound excess lifetime cancer risks of 10^{-4} (1 in 10,000 individuals theoretically would get cancer), 10^{-5}, and 10^{-6}.

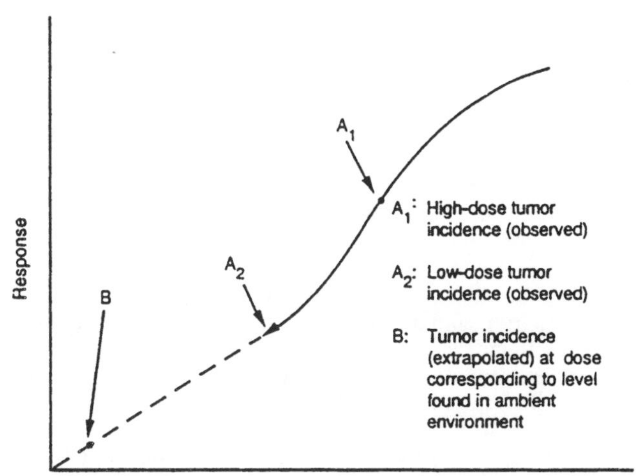

FIGURE 1.3 Examples of extrapolation using the linearized multistage dose-response model.

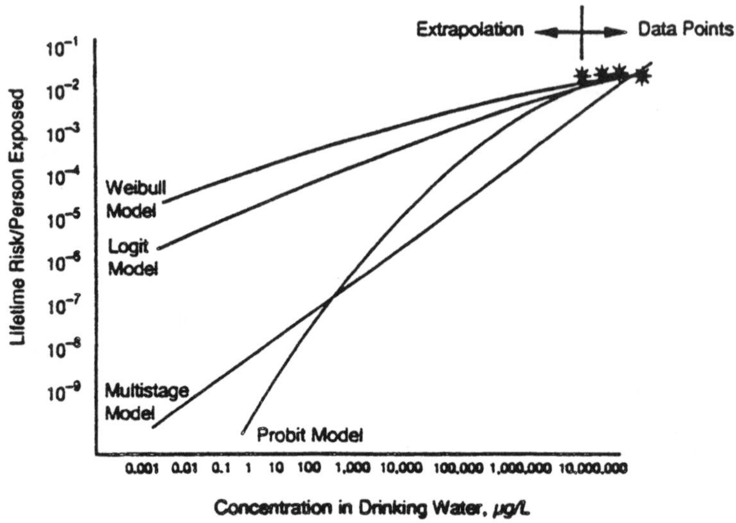

FIGURE 1.4 Comparison of risk model estimates for vinyl chloride.

$$\text{drinking water concentration} = \frac{(10^{-x})(70\text{ kg})}{(q_1^*)(2\text{ L/day})} \tag{1.4}$$

Where 10^{-x} = risk level ($x = 4, 5,$ or 6); 70 kg – assumed adult body weight
q_1^* = human carcinogenic potency factor as determined by the LMM in (micrograms per kilogram per day)$^{-1}$
2 L/d = assumed adult water consumption rate

Based on these calculations, the MCLG is set within the theoretical excess cancer risk range of 10^{-5} to 10^{-6}.

Microbial Contaminants. Historically, different approaches have been used to regulate chemical and microbial contaminants. Rather than conducting a formal risk assessment for microbials, regulators have established an MCLG of zero for pathogens. In general, analytical techniques for pathogens require a high degree of analytical skill, are time consuming, and are cost prohibitive for use in compliance monitoring. Indicator organisms, such as total or fecal coliforms, are used to show the possible presence of microbial contamination from human waste.

The use of indicator organisms serves well for indicating sewage contamination of surface waters. Unfortunately, viruses and protozoa can be present in source waters in which coliform organisms have been inactivated. Also, pathogens in source water may originate from sources other than human fecal material. *E. coli* appears to be best suited as an indicator for microbial contamination of surface waters, raw waters, recreational waters, and agricultural waters (USEPA, 1998d). USEPA is continuing to evaluate the use of indicators of both microbial contamination and treatment efficiency.

Epidemiology, the study of disease in populations, has historically been the major science used to study infectious disease transmission in drinking water. In recent years, however, the science of risk assessment for microbial contaminants has progressed sufficiently to become a useful tool in regulating microbial contaminants.

Microbial risk assessment methods were first used by USEPA in the development of the Surface Water Treatment Rule (USEPA, 1989b), which establishes a treatment requirement for 99.9 percent and 99.99 percent removal for *Giardia* and viruses, respectively. Regulators believe that this level of pathogen removal corresponds to an annual risk of no more than 1 infection per 10,000 people exposed over a year from drinking water (USEPA, 1989b). Use of risk assessment for microbial contaminants in drinking water requires data on exposure factors (i.e., transmission, environmental sources, survival, occurrence, treatment, and so forth) and health effects (i.e., outbreaks, endemic disease, immune status, methods for diagnosis, and so forth) (Haas et al., 1998). Because much of the needed data is lacking, indicator organisms in conjunction with treatment requirements (i.e., filtration and disinfection) continue to be used for regulating microbial contaminants.

Regulatory Basis of MCLs

The SDWA requires USEPA to set either an MCL or a treatment technique requirement for each contaminant that has an MCLG. Maximum contaminant levels (MCLs) are enforceable standards. They are to be set "as close to the MCLG as is feasible." *Feasible* is defined as feasible with the use of the best technology, treatment techniques, and other means that are available (taking cost into consideration). USEPA is required to examine treatment technologies, techniques, or other means for efficacy under field conditions and not solely under laboratory conditions when determining feasibility. Granular activated carbon (GAC) is designated as feasible for the control of synthetic organic chemicals (SOCs), and any technology, treatment technique, or other means designated as feasible for the control of SOCs must be at least as effective as GAC.

Maximum contaminant level goals for category I contaminants are set at zero, but because zero is neither measurable nor achievable using the best available technology (BAT), the MCLs for these contaminants cannot be zero. These MCLs, therefore, are usually set at either the practical quantitation level (the concentration detectable by certified laboratories under routine operating conditions) or at the level achievable using BAT; the resulting MCL typically falls within the relative risk range of 10^{-4} to 10^{-6}.

USEPA considers MCLs set within this range to be protective of public health and consistent with the recommendations of the World Health Organization. By setting the MCLG for category I contaminants at zero, USEPA has some flexibility to set the MCL within the relative risk range of 10^{-4} to 10^{-6}. If the MCLG was set at a specified risk level, then the MCL would always be driven toward that particular risk level.

Maximum contaminant level goals for category III contaminants typically fall above the practical quantitation level and the level achievable using BAT, and an MCL for one of these contaminants is most often identical to the contaminant's MCLG. An MCL for a category II contaminant is also often identical to the contaminant's MCLG depending on the method used to calculate the MCLG, the practical quantitation level, and the level achievable using BAT.

Treatment Techniques

USEPA may require the use of a treatment technique in lieu of establishing an MCL if it is determined that monitoring for the contaminant is not economically or technologically feasible. If an analytical method is not available for a contaminant, then in lieu of an MCL, a treatment technique will be established for that contaminant. Treatment techniques are typically set for contaminants that cannot be economically or feasibly measured, such as viruses and parasites.

Risk Balancing

USEPA may establish an MCL at a level other than the feasible level, if the technology, treatment techniques, and other means used to determine the feasible level would result in an increase in the health risk from drinking water by:

1. Increasing the concentration of other contaminants in drinking water
2. Interfering with the efficacy of drinking water treatment techniques or processes that are used to comply with other NPDWRs

If an MCL or a treatment technique for any contaminant or contaminants is set at other than the feasible level:

1. The MCL(s) or treatment techniques must minimize the overall risk of adverse health effects by balancing the risk from the contaminant and the risk from other contaminants, the concentrations of which may be affected by the use of a treatment technique or process that would be employed to attain the MCL or levels.
2. The combination of technology, treatment techniques, or other means required to meet the level or levels shall not be more stringent than is feasible.

Risk balancing and consideration of risk trade-offs are becoming more important in the development of drinking water regulations. For example, increase in waterborne microbial disease outbreaks may occur as MCLs for total trihalomethanes and other disinfection by-products are lowered. Controlling lead and copper by corrosion control through pH adjustment could increase formation of trihalomethanes. Treatment techniques to control a particular disinfection by-product could increase formation of other by-products.

Benefits Should Justify Costs

At the time an NPDWR is proposed, the agency must publish a determination as to whether the benefits of the MCL justify, or do not justify, the costs, based on a health risk reduction and cost analysis. If USEPA decides that the benefits of an MCL would not justify the costs of complying with the level, the agency may promulgate an MCL for the contaminant that maximizes health risk reduction benefits at a cost that is justified by the benefits. This authority cannot be used if the benefits of compliance with an NPDWR for the contaminant in question experienced by persons served by large public water systems and persons served by systems that are unlikely to receive a small system variance would justify the costs to the systems of complying with the regulation.

The authority may be used to establish regulations for the use of disinfection by systems relying on groundwater sources. However, the authority does not apply:

1. If the contaminant is found almost exclusively in small systems eligible for a small system variance

2. To establish an MCL in a Stage I or Stage II NPDWR for contaminants that are disinfectants or disinfection by-products
3. To establish an MCL or treatment technique requirement for the control of *Cryptosporidium.*

A determination by USEPA that the benefits of an MCL or treatment requirement justify or do not justify the costs of complying with the level is reviewable by the court only as part of a review of a final NPDWR that has been promulgated based on the determination. The determination will not be set aside by the court unless the court finds that the determination is arbitrary and capricious.

Best Available Technology

Each NPDWR including an MCL must list the technology, treatment technique, and other means feasible for purposes of meeting the MCL [referred to as best available technology (BAT)]. The amended SDWA does not, however, require the use of BAT. Systems may use any appropriate technology or other means that is acceptable to the state to comply with an MCL. When listing the technology, treatment techniques, and other means USEPA finds to be feasible for meeting an MCL or treatment technique, the agency must include "any technology, treatment technique, or other means that is affordable" for small systems. The quality of the source water to be treated must be considered when listing any technology, treatment technique, or other means. The agency is also required to list technologies for small systems.

Packaged or modular systems and point of entry (POE) or point of use (POU) may be included in the list (USEPA, 1998e). Point-of-entry and POU treatment units must be owned, controlled, and maintained by the public water system or by a person under contract with the public water system to ensure proper operation and maintenance and compliance with the MCL or treatment technique. Units must be equipped with mechanical warnings to ensure that customers are automatically notified of operational problems. Point-of-use treatment technology cannot be listed for a microbial contaminant or an indicator of microbial contamination.

If the American National Standards Institute (ANSI) has issued product standards applicable to a specific type of POE or POU treatment unit, individual units of that type must be independently certified against the ANSI standard before they can be accepted for compliance with an MCL or treatment technique requirement.

Monitoring and Analytical Methods

For each contaminant regulated, USEPA establishes monitoring requirements to ensure that water systems demonstrate compliance with MCLs or treatment technique requirements. Monitoring requirements typically specify the locations at which samples must be taken, frequency of sampling, sample preservation techniques and maximum sample holding times, and the constituents to be measured.

When a public water system supplies water to one or more other public water systems, the state may modify monitoring requirements to the extent that the interconnection of the systems justifies treating them as a single system for monitoring purposes. Modified monitoring must be conducted according to the schedule specified by the state, with concurrence by USEPA.

Analytical methods approved by USEPA must be used, and analyses must be performed by certified laboratories. Commercial and utility laboratories are certified by the state laboratory, which in turn is certified by USEPA. USEPA-approved analytical methods are specified in the *Federal Register* notice for each regulation. An updated list of USEPA-approved methods is published periodically by the agency.

Each analytical method has a limit of detection. Limits of detection are an important factor in the standard-setting process and can be defined in various ways. USEPA currently defines a method detection limit (MDL) as the minimum concentration that can be measured and reported with 99 percent confidence that the analyte concentration is greater than zero. Each USEPA-approved analytical method will have an MDL.

Method detection limits are not necessarily reproducible over time. This is true even in a given laboratory using the same analytical procedures, instrumentation, and sample matrix. To account for this variability, a practical quantitation level (PQL) is also specified for each USEPA-approved method. The PQL is the lowest concentration that can be quantified within specified limits of precision and accuracy during routine laboratory operations. It is intended to represent the lowest concentration measurable by good laboratories under practical and routine laboratory conditions, whereas the MDL represents the lowest detectable concentration under ideal conditions. USEPA sets PQLs based on interlaboratory studies using statistical methods. Practical quantitation levels usually range between 5 and 10 times the MDL (Cotruvo and Vogt, 1990). The methods used to set PQLs have certain limitations (Keith, 1992), and the agency is in the process of reconsidering its approach.

Reporting and Recordkeeping

Water systems must retain certain records and submit reports to the primacy agency following the requirements set by USEPA under SDWA Section 1445. Reporting may be required monthly, semiannually, or annually, depending on the contaminant regulated.

Water suppliers have certain deadlines by which specified information must be reported to the primacy agency. Except where a shorter period is specified for a particular contaminant or situation, 40 CFR 141.31 requires the supplier of water to report to the state the results of any required test measurement or analysis within the following:

- The first 10 days following the month in which the result is received
- The first 10 days following the end of the required monitoring period as stipulated by the state, whichever of these is shortest

Failure to comply with any NPDWR (including failure to comply with monitoring requirements) must be reported within 48 hours. Reporting of analytical results by water suppliers to the state is not required in cases in which a state laboratory performs the analysis and reports the results to the state office that would normally receive the supplier's notification.

Within 10 days of completion of each public notification, a representative copy of each type of notice distributed, published, posted, and/or made available to the persons served by the system and/or to the media must be submitted. The water supply system must submit to the state, within the time stated in the request, copies of any records required to be maintained under 40 CFR 141.33 or copies of any documents then in existence that the state or USEPA is entitled to inspect under the authority of SDWA section 1445 or the equivalent provisions of state law. All public water system owners or operators are generally required under 40 CFR 141.33 to retain on their premises or at a convenient location near the premises records of bacteriological analyses kept for not less than five years. Records of chemical analyses must be kept for not less than 10 years. Actual laboratory reports may be kept, or data may be transferred to tabular summaries, provided that the following information is included:

- Date, place, and time of sampling, and the name of the person who collected the sample
- Identification of the sample as to whether it was a routine distribution system sample, check sample, raw or process water sample, or other special-purpose sample
- Date of analysis
- Laboratory and person responsible for performing analysis
- Analytical technique/method used
- Results of the analysis

Records of action taken by the system to correct violations of primary drinking water regulations must be kept for not less than three years after the last action taken with respect to the particular violation involved.

Copies of any written reports, summaries, or communications relating to sanitary surveys of the system conducted by the system itself, by a private consultant, or by any local, state, or federal agency, must be kept for a period not less than 10 years after completion of the sanitary survey involved.

Records concerning a variance or exemption granted to the system must be kept for a period ending not less than five years following the expiration of such variance or exemption. Additional record retention requirements may be determined by a specific regulation.

Public Notification

Public notification requirements are intended to ensure that customers are informed of violations and the seriousness of any potential adverse health effects that may be involved. In general, public notification is required if any one of the following six conditions occurs (USEPA, 1987c, 1989a):

- Failure of the system to comply with an applicable MCL
- Failure to comply with a prescribed treatment technique
- Failure of the system to perform water quality monitoring as required by the regulations
- Failure to comply with testing procedures as prescribed by an NPDWR
- Issuance of a variance or an exemption
- Failure to comply with the requirements of any schedule that has been set under a variance or exemption

Public notice requirements apply to all drinking water regulations. The 1996 SDWA amendments revised the statutory requirements regarding public notification. USEPA is in the process of developing a rule to revise public notice requirements.

Effective Date and Review

A promulgated NPDWR and any amendment of an NPDWR take effect three years after the date of promulgation unless USEPA determines that an earlier date is practicable. However, USEPA, or a state (in the case of an individual system), may allow up to two additional years to comply with an MCL or treatment technique if USEPA or the state (in the case of an individual system) determines that additional time is necessary for capital improvements.

Each NPDWR will be reviewed and revised as appropriate not less often than every six years. Any revision of an NPDWR is to be promulgated in accordance with the SDWA as amended in 1996. Each revision must maintain or provide for greater protection of the health of persons.

Variances

Variances are allowed by the SDWA if a water system cannot come into compliance with an NPDWR within the required time. In essence, a variance is a time extension to comply with an NPDWR, in which interim action to install treatment or take other actions to lower the contaminant concentration is required. To receive a variance, the water system must agree to install the best technology, treatment technique, or other means, specified by USEPA, and show that alternative sources of water are not reasonably available. The MCL or treatment technique is not met, but the system must "do the best it can" to meet the NPDWR. The intent of a variance is to require the water utility to take a step toward full compliance, even though full compliance cannot be achieved. The 1996 SDWA amendments included special provisions for small system variances (USEPA, 1998c).

Exemptions

Public water systems can obtain exemptions if they show compelling factors for noncompliance with an NPDWR (USEPA, 1998c). The system must be in operation as of the effective date of the particular regulation, and an exemption must not cause an unreasonable risk to health.

A compliance schedule is issued at the same time the exemption is granted. Compliance is required as expeditiously as practicable, but not later than three years after the otherwise applicable compliance date.

Before receiving an exemption, a public water system must take all practicable steps to meet the standard and establish one of three conditions:

- The system cannot meet the standard without capital improvements that cannot be completed prior to the compliance date.
- Financial assistance is necessary and an agreement has been entered into to obtain such assistance, or SRLF or other assistance is reasonably likely to be available within the period of the exemption.
- An enforceable agreement has been entered into to become part of a regional public water system.

The 1996 SDWA amendments provide that systems serving not more than 3300 persons may renew an exemption if they are taking all practical steps and need financial assistance. Exemptions are limited to a total of six years. A water system cannot receive an exemption if it has received a variance.

CURRENT NATIONAL PRIMARY DRINKING WATER REGULATIONS

The 1996 SDWA amendments made sweeping changes to the SDWA, but most of the existing NPDWRs were not changed. The NPDWRs previously promulgated for arsenic (USEPA, 1975b), radionuclides (USEPA, 1976a), trihalomethanes (USEPA, 1979b), fluoride (USEPA, 1986a), volatile synthetic organic chemicals (USEPA, 1987a), surface water treatment (USEPA, 1989b), total coliforms (USEPA, 1989c), Phase II synthetic organic contaminants and inorganic contaminants (SOCs/IOCs) (USEPA, 1991a,d), lead and copper (USEPA, 1991c), and Phase V SOCs/IOCs (USEPA, 1992c) remain essentially unchanged. Any contaminant on the list of 83 not regulated as of August 6, 1996, and not addressed specifically in the 1996 amendments, must still be regulated. The 1996 amendments specifically addressed all such contaminants except for uranium, which must be regulated by November 2000 (U.S. District Court of Oregon, 1996).

Rulemaking Designations

Prior to enactment of the 1986 SDWA amendments, USEPA established a regulatory approach that placed contaminants being considered for regulation into groups, referred to as phases. After enactment of 1986 amendments, USEPA revised the grouping of contaminants to accommodate the additional contaminants mandated by Congress for regulation. In some cases, the original phase designation was retained as the common name for the regulatory package (e.g., Phase I). In many cases, however, a phase designation is not used (e.g., lead and copper). The use of phase designations is only for convenience of reference, is not based on statutory or regulatory requirements, and does not apply to rules set under the 1996 SDWA amendments.

Summary Listing of U.S. Drinking Water Standards

Current and proposed drinking water standards are summarized in Tables 1.9 to 1.11. The background and current status of anticipated USEPA drinking water regulations are reviewed annually in the *Journal of the American Water Works Association* (Pontius, 1990a, 1992, 1993a, 1995, 1996a, 1997b, 1998; Pontius and Roberson, 1994). In addition, information on current drinking water regulations can be found on the USEPA OGWDW web site (http://www.epa.gov/ogwdw).

USEPA Drinking Water Health Advisories

USEPA may publish health advisories (which are not regulations) or take other appropriate actions for contaminants not subject to any NPDWR. Health advisories are nonregulatory health guidance to assist states in dealing with incidences of drinking water contamination. They provide information on the concentrations of contaminants in drinking water at which adverse health effects would not be anticipated to occur. A margin of safety is included to protect sensitive members of the population.

Health advisories are developed from data describing carcinogenic and noncarcinogenic toxicity. For chemicals that are known or probable human carcinogens, nonzero, 1-day, 10-day, and longer-term advisories may be derived, with caveats. Pro-

TABLE 1.9 USEPA National Primary Drinking Water Contaminant Standards

Contaminant	MCLG (mg/L)	MCL (mg/L)	Potential health effects	Sources of drinking water contamination
			Fluoride Rule[a]	
Fluoride	4.0	4.0	Skeletal and dental fluorosis	Natural deposits; fertilizer, aluminum industries; drinking water additive
			Phase I Volatile Organics[b]	
Benzene	Zero	0.005	Cancer	Some foods; gas, drugs, pesticide, paint, plastic industries
Carbon Tetrachloride	Zero	0.005	Cancer	Solvents and their degradation products
p-dichlorobenzene	0.075	0.075	Cancer	Room and water deodorants and "mothballs"
1,2-dichloroethane	Zero	0.005	Cancer	Leaded gas, fumigants, paints
1,1-dichloroethylene	0.007	0.007	Cancer, liver and kidney effects	Plastics, dyes, perfumes, paints
Trichloroethylene	Zero	0.005	Cancer	Textiles, adhesives and metal degreasers
1,1,1-trichloroethane	0.2	0.2	Liver, nervous system effects	Adhesives, aerosols, textiles, paints, inks, metal degreasers
Vinyl chloride	Zero	0.002	Cancer	May leach from PVC pipe; formed by solvent breakdown
			Surface Water Treatment Rule[c] and Total Coliform Rule[d]	
Giardia lamblia	Zero	TT[e]	Gastroenteric disease	Human and animal fecal wastes
Legionella	Zero	TT	Legionnaire's disease	Natural waters; can grow in water heating systems
Heterotrophic plate count	N/A	TT	Indicates water quality, effectiveness of treatment	
Total coliform	Zero	<5%+	Indicates gastroenteric pathogens	Human and animal fecal waste
Escherichia coli	Zero	TT	Gastroenteric disease	Human and animal fecal waste
Fecal coliforms	Zero	TT	Indicates gastroenteric pathogens	Human and animal fecal waste
Turbidity	N/A	TT	Interferes with disinfection	Soil runoff
Viruses	Zero	TT	Gastroenteric disease	Human and animal fecal waste
			Phase II Rule Inorganics[f]	
Asbestos (>10 μm)	7 MFL[g]	7 MFL	Cancer	Natural deposits; asbestos cement in water systems
Barium	2	2	Circulatory system effects	Natural deposits; pigments, epoxy sealants, spent coal
Cadmium	0.005	0.005	Kidney effects	Galvanized pipe corrosion; natural deposits; batteries, paints
Chromium (total)	0.1	0.1	Liver, kidney, circulatory disorders	Natural deposits; mining, electroplating, pigments

Mercury (inorganic)	0.002	0.002	Kidney, nervous system disorders	Crop runoff; natural deposits; batteries, electrical switches
Nitrate	10	10	Methemoglobulinemia	Animal waste, fertilizer, natural deposits, septic tanks, sewage
Nitrite	1	1	Methemoglobulinemia	Same as nitrate; rapidly converted to nitrate
Nitrate + Nitrite	10	10		
Selenium	0.05	0.05	Liver damage	Natural deposits; mining, smelting, coal/oil combustion
			Phase II Rule Organics[h]	
Acrylamide	Zero	TT	Cancer, nervous system effects	Polymers used in sewage/waste water treatment
Alachlor	Zero	0.002	Cancer	Runoff from herbicide on corn, soybeans, other crops
Aldicarb	Delayed	Delayed	Nervous system effects	Insecticide on cotton, potatoes, other crops; widely restricted
Aldicarb sulfone	Delayed	Delayed	Nervous system effects	Biodegradation of aldicarb
Aldicarb sulfoxide	Delayed	Delayed	Nervous system effects	Biodegradation of aldicarb
Atrazine	Remanded	Remanded	Mammary gland tumors	Runoff from use as herbicide on corn and noncropland
Carbofuran	0.04	0.04	Nervous, reproductive system effects	Soil fumigant on corn and cotton; restricted in some areas
Chlordane	Zero	0.002	Cancer	Leaching from soil treatment for termites
Chlorobenzene	0.1	0.1	Nervous system and liver effects	Waste solvent from metal degreasing processes
2,4-D	0.07	0.07	Liver and kidney damage	Runoff from herbicide on wheat, corn, rangelands, lawns
o-Dichlorobenzene	0.6	0.6	Liver, kidney, blood cell damage	Paints, engine cleaning compounds, dyes, chemical wastes
cis-1,2-dichloroethylene	0.07	0.07	Liver, kidney, nervous, circulatory system effects	Waste industrial extraction solvents
trans-1,2-dichloroethylene	0.1	0.1	Liver, kidney, nervous, circulatory system effects	Waste industrial extraction solvents
Dibromochloropropane	Zero	0.0002	Cancer	Soil fumigant on soybeans, cotton, pineapple, orchards
1,2-dichloropropane	Zero	0.005	Liver, kidney effects; cancer	Soil fumigant; waste industrial solvents
Epichlorohydrin	Zero	TT	Cancer	Water treatment chemicals; waste epoxy resins, coatings
Ethylbenzene	0.7	0.7	Liver, kidney, nervous system effects	Gasoline; insecticides; chemical manufacturing wastes
Ethylene dibromide	Zero	0.00005	Cancer	Leaded gas additives; leaching of soil fumigant
Heptachlor	Zero	0.0004	Cancer	Leaching of insecticide for termites, very few crops
Heptachlor epoxide	Zero	0.0002	Cancer	Biodegradation of heptachlor
Lindane	0.0002	0.0002	Liver, kidney, nervous system, immune system, and circulatory system effects	Insecticide on cattle, lumber, gardens; restricted in 1983

TABLE 1.9 USEPA National Primary Drinking Water Contaminant Standards (*Continued*)

Contaminant	MCLG (mg/L)	MCL (mg/L)	Potential health effects	Sources of drinking water contamination
			Phase II Rule Organics[h]	
Methoxychlor	0.04	0.04	Growth, liver, kidney, and nervous system effects	Insecticide for fruits, vegetables, alfalfa, livestock, pets
Pentachlorophenol	Zero	0.001	Cancer; liver and kidney effects	Wood preservatives, herbicide, cooling tower wastes
PCBs	Zero	0.0005	Cancer	Coolant oils from electrical transformers; plasticizers
Styrene	0.1	0.1	Liver, nervous system	Plastics, rubber, resin, drug damage industries; leachate from city landfills
Tetrachloroethylene	Zero	0.005	Cancer	Improper disposal of dry cleaning and other solvents
Toluene	1	1	Liver, kidney, nervous system and circulatory system effects	Gasoline additive; manufacturing and solvent operations
Toxaphene	Zero	0.003	Cancer	Insecticide on cattle, cotton, soybeans; cancelled in 1982
2,4,5-TP	0.05	0.05	Liver and kidney damage	Herbicide on crops, right-of-way, golf courses; cancelled in 1983
Xyenes (total)	10	10	Liver, kidney, nervous system effects	By-product of gasoline refining; paints, inks, detergents
			Lead and Copper Rule[i]	
Lead	Zero	TT#[j]	Kidney, nervous system damage	Natural/industrial deposits; plumbing solder, brass alloy faucets
Copper	1.3	TT##[k]	Gastrointestinal irritation	Natural/industrial deposits; wood preservatives, plumbing
			Phase V Inorganics[l]	
Antimony	0.006	0.006	Cancer	Fire retardents, ceramics, electronics, fireworks, solder
Beryllium	0.004	0.004	Bone, lung damage	Electrical, aerospace, defense industries
Cyanide	0.2	0.2	Thyroid, nervous system	Electroplating, steel, damage plastics, mining, fertilizer
Nickel	Remanded	Remanded	Heart, liver damage	Metal alloys, electroplating, batteries, chemical production
Thallium	0.0005	0.002	Kidney, liver, brain, intestinal effects	Electronics, drugs, alloys, glass
			Phase V Organics[m]	
Adipate (di(2-ethylhexyl))	0.4	0.4	Decreased body weight	Synthetic rubber, food packaging, cosmetics
Dalapon	0.2	0.2	Liver, kidney effects	Herbicide on orchards, beans, coffee, lawns, roads, railways
Dichloromethane	Zero	0.005	Cancer	Paint stripper, metal degreaser, propellant, extractant

Dinoseb	0.007	0.007	Thyroid, reproductive organ damage	Runoff of herbicide from crop and noncrop applications
Diquat	0.02	0.02	Liver, kidney, eye effects	Runoff of herbicide on land and aquatic weeds
Dioxin	Zero	3×10^{-8}	Cancer	Chemical production by-product; impurity in herbicides
Endothall	0.1	0.1	Liver, kidney, gastrointestinal effects	Herbicide on crops, land/aquatic weeds; rapidly degraded
Endrin	0.002	0.002	Liver, kidney, heart damage	Pesticide on insects, rodents, birds; restricted since 1980
Glyphosate	0.7	0.7	Liver, kidney damage	Herbicide on grasses, weeds, brush
Hexachlorobenzene	Zero	0.001	Cancer	Pesticide production waste by-product
Hexachlorocyclo-pentadiene	0.05	0.05	Kidney, stomach damage	Pesticide production intermediate
Oxamyl (vydate)	0.2	0.2	Kidney damage	Insecticide on apples, potatoes, tomatoes
PAHs (benzo(a)-pyrene)	Zero	0.0002	Cancer	Coal tar coatings; burning organic matter; volcanoes, fossil fuels
Phathalate (di(2-ethylhexyl)	Zero	0.006	Cancer	PVC and other plastics
Picloram	0.5	0.5	Kidney, liver damage	Herbicide on broadleaf and woody plants
Simazine	0.004	0.004	Cancer	Herbicide on grass sod, some crops, aquatic algae
1,2,4-Trichlorobenzene	0.07	0.07	Liver, kidney damage	Herbicide production; dye carrier
1,1,2-Trichloroethane	0.003	0.005	Kidney, liver, nervous system damage	Solvent in rubber, other organic products; chemical production wastes
Interim (I) and proposed (P) standards for radionuclides[n] (USEPA 1976a, 1991e)				
Beta/photon emitters (I)	—	4 mrem/yr	Cancer	Natural and manmade deposits
Beta/photon emitters (P)	Zero	4 mrem/yr	Cancer	Natural and manmade deposits
Alpha emitters (I)	—	15 pCi/L	Cancer	Natural deposits
Alpha emitters (P)	Zero	15 pCi/L	Cancer	Natural deposits
Radium 226 +228 (I)	—	5 pCi/L	Bone cancer	Natural deposits
Radium 226 (P)	Zero	20 pCi/L	Bone cancer	Natural deposits
Radium 228 (P)	Zero	20 pCi/L	Bone cancer	Natural deposits
Uranium (P)	Zero	0.02	Cancer	Natural deposits
Disinfection by-products[o] (USEPA 1998g)				
Bromate	Zero	0.010	Cancer	Ozonation by-product
Bromodichloromethane	Zero	See TTHMs	Cancer, liver, kidney, and reproductive effects	Drinking water chlorination by-product
Bromoform	Zero	See TTHMs	Cancer, nervous system, liver and kidney effects	Drinking water chlorination by-product

TABLE 1.9 USEPA National Primary Drinking Water Contaminant Standards (*Continued*)

Contaminant	MCLG (mg/L)	MCL (mg/L)	Potential health effects	Sources of drinking water contamination
Disinfection by-products[o] (USEPA 1998g)				
Chlorite	0.8	1.0	Developmental neurotoxicity	Chlorine dioxide by-product
Chloroform	Zero	See TTHMs	Cancer, liver, kidney, reproductive effects	Drinking water chlorination by-product
Dibromochloromethane	0.06	See TTHMs	Nervous system, liver, kidney, reproductive effects	Drinking water chlorination by-product
Dichloroacetic acid	Zero	See HAA5	Cancer, reproductive, developmental effects	Drinking water chlorination by-product
Haloacetic acids (HAA5)[p]	Zero	0.060 (stage 1)	Cancer and other effects	Drinking water chlorination by-products
Trichloroacetic acid	0.3	See HAA5	Liver, kidney, spleen, developmental effects	Drinking water chlorination by-product
Total trihalomethanes (TTHMs)	Zero	0.080 (stage 1)	Cancer and other effects	Drinking water chlorination by-products
Interim Enhanced Surface Water Treatment Rule (USEPA 1998h)				
Cryptosporidium	Zero	TT	Gastroenteric disease	Human and animal fecal waste
Other interim (I) and proposed (P) standards[q]				
Sulfate (P)	500	500[r]	Diarrhea	Natural deposits
Arsenic (I)	—	0.05	Skin, nervous system toxicity, cancer	Natural deposits; smelters, glass, electronics wastes; orchards

[a] USEPA, 1986a, 1993b.
[b] USEPA, 1987a.
[c] USEPA, 1989b.
[d] USEPA, 1989c.
[e] TT, treatment technique requirement.
[f] USEPA, 1991a,d, 1992a.
[g] MFL, million fibers per liter.
[h] USEPA, 1991a,d, 1992.
[i] USEPA, 1991c, 1992b, 1994a.
[j] # (action level) = 0.015 mg/L.
[k] ## (action level) = 1.3 mg/L.
[l] USEPA, 1992c.
[m] USEPA, 1992c.
[n] USEPA, 1976a, 1991e.
[o] USEPA, 1998g.
[p] Sum of the concentrations of mono-, di-, and trichloroacetic acids and mono- and dibromoacetic acids. Alternatives allowing public water systems the flexibility to select compliance options appropriate to protect the population served were proposed.
[q] USEPA, 1994c, 1975b.
[r] For water systems analyzing at least 40 samples per month, no more than 5.0 percent of the monthly samples may be positive for total coliforms. For systems analyzing fewer than 40 samples per month, no more than one sample per month may be positive for total coliforms.

TABLE 1.10 USEPA National Primary Drinking Water Disinfectant Standards (USEPA 1998g)

Disinfectant	MRDLG* (mg/L)	MRDL† (mg/L)
Chlorine‡	4 (as Cl_2)	4 (as Cl_2)
Chloramines§	4 (as Cl_2)	4 (as Cl_2)
Chlorine Dioxide	0.3 (as ClO_2)	0.8 (as ClO_2)

* MRDLG, maximum residual disinfectant level goal.
† MRDL, maximum residual disinfectant level.
‡ Measured as free chlorine.
§ Measured as total chlorine.

jected excess lifetime cancer risks are provided to give an estimate of the concentrations of the contaminant that may pose a carcinogenic risk to humans. Over 200 drinking water health advisories have been prepared. Many of these documents are available for a fee from the National Technical Information Service, U.S. Dept. of Commerce, Washington, D.C., 800-487-4660.

TABLE 1.11 USEPA National Secondary Drinking Water Contaminant Standards

Contaminant*	Effect(s)	SMCL (mg/L)	Reference
Aluminum	Colored water	0.05–0.2	USEPA, 1991a
Chloride	Salty taste	250	USEPA, 1979a
Color	Visible tint	15 color units	USEPA, 1979a
Copper	Metallic taste; blue-green stain	1.0	USEPA, 1979a
Corrosivity	Metallic taste; corrosion; fixture staining	Noncorrosive	USEPA, 1979a
Fluoride	Tooth discoloration	2	USEPA, 1986a
Foaming agents	Frothy, cloudy; bitter taste; odor	0.5	USEPA, 1979a
Iron	Rusty color; sediment; metallic taste; reddish or orange staining	0.3	USEPA, 1979a
Manganese	Black-to-brown color; black staining; bitter, metallic taste	0.05	USEPA, 1979a
Odor†	"Rotten egg," musty, or chemical smell	3 TON	USEPA, 1979a
pH	Low pH: bitter metallic taste, corrosion; high pH: slippery feel, soda taste, deposits	6.5–8.5	USEPA, 1979a
Silver	Skin discoloration; greying of the white part of the eye	0.10	USEPA, 1991a
Sulfate	Salty taste	250	USEPA, 1979a
Total dissolved solids (TDS)	Hardness; deposits; colored water; staining; salty taste	500	USEPA, 1979a
Zinc	Metallic taste	5	USEPA, 1979a

* In the proposed Phase II rule, published on May 22, 1989, USEPA considered setting SMCLs for seven organic chemicals. They were not included in the final rule because of scientific concerns. The existing odor SMCL (3 threshold odor number) was retained. However, utilities should be aware that tastes and odors may be caused by the following organic chemicals at the levels indicated: *o*-dichlorobenzene, 0.01 mg/L; *p*-dichlorobenzene, 0.005 mg/L; ethylbenzene, 0.03 mg/L; pentachlorophenol, 0.03 mg/L; styrene, 0.01 mg/L; toluene, 0.04 mg/L; and xylene, 0.02 mg/L. These levels are below the MCLs for these contaminants, meaning that consumers may taste or smell them even though the MCLs are met.

† For more information on the identification and control of taste and odors, see AWWARF, 1987, 1995.

Direct and Indirect Drinking Water Additives

Public water systems use a broad array of chemical products to treat drinking water supplies and to maintain storage and distribution systems. Water suppliers may directly add chemicals, such as chlorine, alum, lime, and coagulant aids, in the process of treating water. These are known as *direct additives.* As a necessary function of maintaining a public water system, storage and distribution systems (including pipes, tanks, and other equipment) may be painted, coated, or treated with products that contain chemicals that may leach into or otherwise enter the water. These products are known as *indirect additives.*

The USEPA provided technical assistance to states and public water systems on the use of additives until the mid-1980s through the issuance of advisory opinions on the acceptability of many additive products. In 1985, USEPA entered into a cooperative agreement with NSF International to develop a voluntary, third-party, private-sector program for evaluating drinking water additives. Two voluntary standards were developed: ANSI/NSF standard 60 (covering direct additives) and ANSI/NSF standard 61 (covering indirect additives). USEPA terminated its additives advisory program, instead relying on manufacturers to seek certification of their products on a voluntary basis or as required by state regulatory agencies or water utilities. Many state regulatory agencies require water system products to be certified as meeting the appropriate NSF standard. Certification testing is provided by NSF International and Underwriter's Laboratories.

INTERNATIONAL DRINKING WATER STANDARDS

Drinking water standards have been set by a number of countries and international organizations. The number of standards published by these organizations and the frequency of their revision is increasing. Hence, only references for the current standards will be provided so that the most recent version can be obtained from the appropriate agency.

Canada

In Canada, provision of drinking water is primarily the responsibility of the provinces and municipalities. The federal Department of Health conducts research, provides advice, and in collaboration with the health and environment ministries of the provinces and territories, established guidelines for drinking water quality under the auspices of the Federal-Provincial Subcommittee of Drinking Water. A publication entitled *Guidelines for Canadian Drinking Water Quality* (Health and Welfare Canada, 1996) identifies substances that have been found in drinking water and are known or suspected to be harmful. For each substance, the *Guidelines* establish the maximum acceptable concentration (MAC) that can be permitted in water used for drinking. The MAC is similar to the USEPA MCL.

Mexico

In Mexico, the federal Secretariat of Health has the authority for setting drinking water standards. A national law analogous to the U.S. SDWA does not exist, but Mexico has set standards for a number of microbiological and chemical contaminants that they refer to as norms (Secretariat of Health, 1993). Compliance with the norms established by the federal government is mandatory. The norms are similar to the USEPA MCLs in that they are set at the federal level, and then the water purveyors are required to conduct monitoring in accordance with the norms and to meet the values set. The standards include sampling and analytical requirements as well as reporting requirements. Implementation is carried out by the Secretariat of Health, other government entities, and the National Water Commission.

World Health Organization

The World Health Organization (WHO) is a specialized agency of the United Nations with primary responsibility for international health matters and public health. In carrying out that responsibility, it assembles from time to time international experts in the field of drinking water to establish *Guidelines for Drinking-water Quality* (WHO, 1996). The primary aim of this publication is the protection of public health. These *Guidelines* are published primarily as a basis for the development of national drinking water standards, which, if properly implemented, will ensure the safety of drinking water supplies by eliminating known hazards. These *Guidelines* values are not mandatory limits. Each country must consider the *Guidelines* values in the context of local or national environmental, social, economic, and cultural conditions. They can then select which of the *Guidelines* are applicable to their situation and may choose to make adjustments to suit local conditions. The issues of monitoring, reporting, and enforcement are solely left to the discretion of the governmental entity using the WHO *Guidelines.* USEPA is an active participant in the development of the *Guidelines,* and the procedures used are in some ways similar to those used to develop the U.S. drinking water regulations.

European Union

The European Union (EU) is a voluntary economic alliance of member states. Through the European Commission, directives are created that must then be adopted and implemented by member states. Therefore, EU members must have enforceable standards that cannot be less stringent than the limit values set out in the directive. Of course, member states can set more stringent standards if they wish. In July 1980, the Commission adopted directive 80/778/EEC relating to the quality of water intended for human consumption (EEC, 1980). On January 4, 1995, the European Commission adopted a proposal to simplify, consolidate, and update the directive. The proposal reduces the number of parameters from 66 to 48 (including 13 new contaminants), obliges member states to fix values for additional health parameters as needed, adds more flexibility to redress failures, allows efficient monitoring by including a number of indicator parameters, and finally, adds a requirement for annual reports to the consumer (EC, 1996). This proposal was adopted by the Council of the European Union on December 19, 1997 (EU, 1997).

TRENDS FOR THE FUTURE

Legislative mandates, regulatory processes, and policies contribute to the development of national drinking water regulations and guidelines in the United States. Drinking water regulations specifically apply to public water suppliers, and are also used by regulatory agencies in other contexts in which protection of surface and groundwater sources, discharge controls, and cleanup requirements are established by federal or state actions.

USEPA's drinking water regulatory activity will increase in the next few years as the agency works to meet the various mandates and deadlines imposed by the 1996 SDWA amendments. The agency committed to meeting the deadlines and mandates imposed by the 1996 SDWA amendments. SDWA enforcement provisions and initiatives by USEPA underscore the importance of strategic planning for SDWA compliance (Pontius, 1999). As new regulations and programs are instituted, the responsibility for assuring the safety of drinking water at the tap will be shared by federal, state, and local authorities; the public water supplier; and consumers.

BIBLIOGRAPHY

ASDWA. "Over $2.7 Billion Needed for SDWA Infrastructure This Year." *ASDWA Update,* VIII:1 (February 1993).

AWWARF (AWWA Research Foundation) and Lyonnaise des Eaux. *Identification and Treatment of Tastes and Odors in Drinking Water,* J. Mallevialle and I. H. Suffet, editors. Denver, CO: AWWA Research Foundation and AWWA, 1987.

AWWARF (AWWA Research Foundation) and Lyonnaise des Eaux. *Advances in Taste-and-Odor Treatment and Control,* I. H. Suffet, J. Mallevialle, and E. Kawczynski, editors. Denver, CO: AWWA Research Foundation and AWWA, 1995.

Baker, M. N. *The Quest for Pure Water,* vol. I, 2nd ed. New York: McGraw-Hill and American Water Works Association, 1981.

Barnes, D. G., and M. Dourson. "Reference Dose (RfD): Description and Use in Health Risk Assessment." *Regulatory Toxicol. & Pharmacol.,* 8(4), December 1988: 471.

Bellar, T. A., J. J. Lichtenberg, and R. C. Kroner. "The Occurrence of Organohalides in Chlorinated Drinking Water." *Jour. AWWA,* 66(12), December 1974: 703.

Borchardt, J. A., and G. Walton. "Water Quality." In *Water Quality and Treatment,* 3rd ed., American Water Works Association, ed. New York: McGraw-Hill, 1971.

Cohrssen, J. J., and V.T. Covello. *Risk Analysis: A Guide to Principles and Methods for Analyzing Health and Environmental Risks.* Council on Environmental Quality. NTIS Order Number PB 89-137772, 1989.

Conference Report 104-741. "Conference Report on S. 1316, Safe Drinking Water Act Amendments of 1996." *Congressional Record* (*House*), (August 1, 1996): H9678–H9703.

Congressional Research Service. *A Legislative History of the Safe Drinking Water Act.* Washington, D.C.: U.S. Government Printing Office, 1982.

Congressional Research Service. *A Legislative History of the Safe Drinking Water Act Amendments 1983–1992.* Washington, D.C.: U.S. Government Printing Office, 1993.

Cook, M. B., and D. W. Schnare. "Amended SDWA Marks New Era in the Water Industry." *Jour. AWWA,* 78(8), August 1986: 66–69.

Cotruvo, J. A., and C. D. Vogt. "Rationale for Water Quality Standards and Goals. In *Water Quality and Treatment,* F.W. Pontius, ed. Denver, CO: AWWA, 1990.

Cotruvo, J. A., and C. Wu. "Controlling Organics: Why Now?" *Jour. AWWA,* 70(11), November 1978: 590.

Craun, G. F. *Waterborne Diseases in the United States,* Gunther F. Craun, ed. Boca Raton, FL: CRC Press, 1986.

Dyksen, J. E., D. J. Hiltebrand, and R. F. Raczko. "SDWA Amendments: Effects on the Water Industry." *Jour. AWWA,* 80(1), January 1988: 30–35.

EC (European Commission). *Briefing Note on the Proposal for a Council Directive on, Concerning the Quality of Water Intended for Human Consumption—the Revision of the Drinking Water Directive.* Brussels, Belgium: European Commission, 1996.

EEC. *Official Journal of the European Communities,* Official Directive No. L229/11-L229/23, 23, August 30, 1980.

EU (The Council of the European Union). *Council Directive 98/. . . /EC on the Quality of Water Intended for Human Consumption.* Brussels, Belgium: December 19, 1997. (Available from: European Commission, Mr. Papadopoulos, DG XI.D.1 Trmf 03/03, 174 Boulevard du Triomphe, B-1160, Bruxelles, Belgium.)

Gray, K. F., and S. J. Koorse. "Enforcement: USEPA Turns up the Heat." *Jour. AWWA,* 80(1), January 1988: 47–49.

Haas, C. N., J. B. Rose, and C. P. Gerber, editors. *Quantitative Microbial Risk Assessment.* New York: John Wiley and Sons, 1998.

Harris, R. H., and E. M. Brecher. "Is the Water Safe to Drink? Part I: The Problem. Part II: How to Make it Safe. Part III: What You Can Do." *Consumer Reports,* June, July, August 1974: 436, 538, 623.

Health and Welfare Canada. *Guidelines for Canadian Drinking Water Quality.* Ottawa, Ontario: Canada Communication Group—Publishing, 1996.

House of Representatives. House of Representatives Report No. 93-1185. 93rd Congress, Second Session. July 10, 1974.

Hoxie, N. J., et al. "Cryptosporidiosis-associated mortality following a massive waterborne outbreak in Milwaukee, Wisconsin." *Am. J. Public Health,* 87(12), December 1997: 2032–2035.

Keith, L. H. ACS/CEI Background and Perspectives on Redefining Detection and Quantitation. Presented at the 1992 AWWA WQTC, Toronto, Ontario. November 15–19, 1992.

Kyros, P. N. "Legislative History of the Safe Drinking Water Act." *Jour. AWWA,* 66(10), October 1974: 566.

LCCA. Lead Contamination Control Act of 1988. Public Law 100-572. October 31, 1988.

MacKenzie, W. R., et al. "A massive outbreak in Milwaukee of *Cryptosporidium* infection transmitted through public water supply." *N. Engl. J. Med.,* 331, 1994: 161–167.

McDermott, J. H. "Federal Drinking Water Standards—Past, Present, and Future." *Jour. Envir. Engrg. Div.—ASCE,* EE4(99), August 1973: 469.

Morris, R. D., et al. "Temporal Variation in Drinking Water Turbidity and Diagnosed Gastroenteritis in Milwaukee." *Am. J. Public Health,* 86(2), February 1996: 237–239.

National Academy of Sciences. Committee on Safe Drinking Water. *Drinking Water and Health.* Washington, D.C.: National Academy Press, 1977.

National Research Council. *Risk Assessment in the Federal Government: Managing the Process.* Washington, D.C.: National Academy Press, 1983.

Oleckno, W. A. "The National Interim Primary Drinking Water Regulations. Part I—Historical Development." *Jour. Envir. Health,* 44, 5 (May 1982).

Page, T., E. Talbot, and R. H. Harris. *The Implication of Cancer-Causing Substances in Mississippi River Water: A Report by the Environmental Defense Fund.* Washington, D.C.: EDF, 1974.

Page, T., R. H. Harris, and S. S. Epstein. "Drinking Water and Cancer Mortality in Louisiana." *Science,* 193, 55 (July 2, 1976).

Pendygraft, G. W., F. E. Schegel, and M.J. Huston. "The EPA-Proposed Granular Activated Carbon Treatment Requirement: Panacea or Pandora's Box?" *Jour. AWWA,* 71(2), February 1979a: 52.

Pendygraft, G. W., F. E. Schegel, and M.J. Huston. "Organics in Drinking Water: A Health Perspective." *Jour. AWWA,* 71(3), March 1979b: 118.

Pendygraft, G. W., F. E. Schegel, and M.J. Huston. "Maximum Contaminant Levels as an Alternative to the GAC Treatment Requirements." *Jour. AWWA,* 71(4), April 1979c: 174.

Pontius, F. W. "Complying With the New Drinking Water Quality Regulations." *Jour. AWWA,* 82(2), February 1990: 32–52.

Pontius, F. W. "A Current Look at the Federal Drinking Water Regulations." *Jour. AWWA,* 84(3), March 1992: 36–50.

Pontius, F. W. "Federal Drinking Water Regulation Update." *Jour. AWWA,* 82(2), February 1993: 42–51.

Pontius, F. W., and Roberson, J. A. "The Current Regulatory Agenda: An Update." *Jour. AWWA,* 86(2), February 1994: 54–63.

Pontius, F. W. "An Update of the Federal Drinking Water Regs." *Jour. AWWA,* 87(2), February 1995: 48.

Pontius, F. W. "An Update of the Federal Regs." *Jour. AWWA,* 88(3), March 1996a: 36.

Pontius, F. W. "Overview of the Safe Drinking Water Act Amendments of 1996. *Jour. AWWA,* 88(10), October 1996b: 22–27, 30–33.

Pontius, F. W. "Implementing the 1996 SDWA Amendments." *Jour. AWWA,* 89(3), March 1997a: 18.

Pontius, F. W. "Future Directions in Drinking Water Quality Regulations." *Jour. AWWA,* 89(3), March 1997b: 40.

Pontius, F. W. *SDWA Advisor on CD-ROM.* Denver, CO: AWWA, 1997c.

Pontius, F. W. "New Horizons in Federal Regulation." *Jour. AWWA,* 90(3), March 1998: 38–50.

Pontius, F. W. "Complying with Future Water Regulations." *Jour. AWWA*, 91(3), March 1999a: 46–58.

Pontius, F. W. "SDWA Enforcement Provisions, Initiatives, and Outlook." *Jour. AWWA*, 91(6), June 1999b: 14

SDWA. The Safe Drinking Water Act of 1974. Public Law 93-523. December 16, 1974.

SDWA. SDWA Amendments of 1977. Public Law 95-190. November 16, 1977.

SDWA. Safe Drinking Water Act Amendments of 1996. Public Law 104-182. August 6, 1996.

Secretariat of Health. *Official Mexican Norm NOM-012 et seq. -SSA1- 1993.* Mexico City, Mexico: Secretariat of Health, 1993.

Symons, G. E. "That GAO Report." *Jour. AWWA,* 66(5), May 1974: 275.

Symons, J. M. "A History of the Attempted Federal Regulation Requiring GAC Adsorption for Water Treatment." *Jour. AWWA,* 76(8), August 1984: 34.

Symons, J. M., T. A. Bellar, J. K. Carswell, J. DeMarco, K. L. Kropp, G. G. Robeck, D. R. Seeger, C. L. Slocum, B. L. Smith, and A. A. Stevens. "National Organics Reconnaissance Survey for Halogenated Organics." *Jour. AWWA,* 67(11), November 1975: 634.

The States-Item. "Cancer Victims Could Be Reduced—Deaths Tied to New Orleans Water." *The States-Item,* 98(129), November 7, 1974, New Orleans, LA: 1.

Train, R. S. "Facing the Real Cost of Clean Water." *Jour. AWWA,* 66(10), October 1974: 562.

U.S. Court of Appeals. *Environmental Defense Fund v. Costle,* No. 752224, 11 ERC 1214, U.S. Court of Appeals, D.C. Circuit, February 10, 1978.

U.S. District Court of Oregon. *Joseph L. Miller, Jr., et al., v. Carol Browner.* Stipulated Fed. R. Civ. P. 60(b)(6), Motion to dismiss and Stipulation Between Parties, November 18, 1996.

USEPA. Industrial Pollution of the Lower Mississippi River in Louisiana. Surveillance and Analysis Div., USEPA, Region VI, Dallas, Texas, 1972.

USEPA. New Orleans Area Water Supply Study. EPA Rept. EPA-906/9-75-003, December 9, 1975a.

USEPA. "National Interim Primary Drinking Water Regulations." *Fed. Reg.,* 40(248), December 24, 1975b: 59566–59588.

USEPA. "Promulgation of Regulations on Radionuclides." *Fed. Reg.,* 41(133), July 9, 1976a: 28402–28409.

USEPA. "Organic Chemical Contaminants; Control Options in Drinking Water." *Fed. Reg.,* 41(136), July 14, 1976b: 28991.

USEPA. "Recommendations of the National Academy of Sciences." *Fed. Reg.,* 42(132), July 11, 1977: 35764–35779.

USEPA. National Organics Monitoring Survey. Cincinnati, OH: USEPA Tech. Support Div., Office of Drinking Water, 1978a.

USEPA. "Control of Organic Chemicals in Drinking Water. Proposed Rule." *Fed. Reg.,* 43(28), February 9, 1978b: 5726.

USEPA. "Control of Organic Chemicals in drinking Water. Notice of Availability." *Fed. Reg.,* 43(130), July 6, 1978c: 29135.

USEPA. "National Secondary Drinking Water Regulations. Final Rule." *Fed. Reg.,* 44(140), July 19, 1979a: 42195–42202.

USEPA. "Control of Trihalomethanes in Drinking Water. Final Rule." *Fed. Reg.,* 44(231), November 29, 1979b: 68624.

USEPA. "Interim Primary Drinking Water Regulations; Amendments." *Fed. Reg.,* 45(168), August 27, 1980: 57332–57357.

USEPA. "Control of Organic Chemicals in Drinking Water. Notice of Withdrawal." *Fed. Reg.,* 46(53), March 19, 1981: 17567.

USEPA. "Advanced Notice of Proposed Rulemaking." *Fed. Reg.,* 47(43), March 4, 1982: 9350–9358.

USEPA. "National Interim Primary Drinking Water Regulations; Trihalomethanes. Final Rule." *Fed. Reg.,* 48(40), February 28, 1983a: 8406–8414.

USEPA. "Advanced Notice of Proposed Rulemaking." *Fed. Reg.,* 48(194), October 5, 1983b: 45502–45521.

USEPA. "Volatile Synthetic Organic Chemicals. Proposed Rule." *Fed. Reg.,* 50(219), November 13, 1985: 46902.

USEPA. "National Primary and Secondary Drinking Water Regulations; Fluoride. Final Rule." *Fed. Reg.,* 51(63), April 2, 1986a: 11396.

USEPA. "Guidelines for Carcinogen Risk Assessment." *Fed. Reg.,* 51, September 24, 1986b: 34006–34012.

USEPA. "Volatile Synthetic Organic Chemicals. Final Rule." *Fed. Reg.,* 52(130), July 8, 1987a: 23690.

USEPA. "Proposed Substitution of Contaminants and Proposed List of Additional Substances Which May Require Regulation Under the Safe Drinking Water Act." *Fed. Reg.,* 52(130), July 8, 1987b: 25720.

USEPA. "Public Notification. Final Rule." *Fed. Reg.,* 52(208), October 28, 1987c: 41534.

USEPA. "Substitution of Contaminants and Drinking Water Priority List of Additional Substances Which May Require Regulation Under the Safe Drinking Water Act." *Fed. Reg.,* 53(14), January 22, 1988: 1892.

USEPA. "Public Notification. Technical Amendments." *Fed. Reg.,* 54(72), April 17, 1989a: 15185.

USEPA. "Filtration and Disinfection; Turbidity, *Giardia Lamblia,* Viruses, *Legionella,* and Heterotrophic Bacteria. Final Rule." *Fed. Reg.,* 54(124), June 29, 1989b: 27486–27541.

USEPA. "Drinking Water; National Primary Drinking Water Regulations; Total Coliforms (including Fecal Coliforms and E. coli). Final Rule." *Fed. Reg.,* 54(124), June 29, 1989c: 27544–27568.

USEPA. "National Primary Drinking Water Regulations. Synthetic Organic Chemicals and Inorganic Chemicals; Monitoring for Unregulated Contaminants; National Primary Drinking Water Regulations Implementation; National Secondary Drinking Water Regulations. Final Rule." *Fed. Reg.,* 56(20), January 30, 1991a: 3526.

USEPA. "National Primary Drinking Water Regulations. Monitoring for Synthetic Organic Chemicals; MCLGs and MCLs for Aldicarb, Aldicarb Sulfoxide, Aldicarb Sulfone, Pentachlorophenol, and Barium. Proposed Rule." *Fed. Reg.,* 56(20), January 30, 1991b: 3600.

USEPA. "Lead and Copper. Final Rule." *Fed. Reg.,* 56, June 7, 1991c: 26460.

USEPA. "National Primary Drinking Water Regulations. Monitoring for Volatile Organic Chemicals; MCLGs and MCLs for Aldicarb, Aldicarb Sulfoxide, Aldicarb Sulfone, Pentachlorophenol, and Barium. Final Rule." *Fed. Reg.,* 56(126), July 1, 1991d: 30266.

USEPA. "Radionuclides. Proposed Rule." *Fed. Reg.,* 56(138), July 18, 1991e: 33050.

USEPA. "Aldicarb, Aldicarb Sulfoxide, and Aldicarb Sulfone. Notice of Postponement of Certain Provisions of the Final Rule." *Fed. Reg.,* 57(102), May 27, 1992a: 22178.

USEPA. "Lead and Copper. Final Rule Correction." *Fed. Reg.,* 57(125), June 29, 1992b: 28785.

USEPA. "National Primary Drinking Water Regulations. Synthetic Organic Chemicals and Inorganic Chemicals. Final Rule." *Fed. Reg.,* 57(138), July 17, 1992c: 31776.

USEPA. *Technical and Economic Capacity of States and Public Water Systems to Implement Drinking Water Regulations.* EPA 810-R-93-001, September 1993a.

USEPA. "Notice of Intent Not to Revise Fluoride Drinking Water Standard." *Fed. Reg.,* 58(248), December 29, 1993b: 68826–68827.

USEPA. "Lead and Copper. Final Rule Technical Corrections." *Fed. Reg.,* 59(125), June 30, 1994a: 33860–33864.

USEPA. "Disinfectants/Disinfection By-Products. Proposed Rule." *Fed. Reg.,* 59(145), July 29, 1994b: 38668.

USEPA. "Sulfate. Proposed Rule." *Fed. Reg.,* 59(243), December 20, 1994c: 65578.

USEPA. "Proposed Guidelines for Carcinogen Risk Assessment." *Fed. Reg.,* 61(79), April 23, 1996a: 17690–18011.

USEPA. "Monitoring Requirements for Public Drinking Water Supplies. Final Rule." *Fed. Reg.,* 61(94), May 14, 1996b: 24354–24388.

USEPA. "Announcement of the Draft Drinking Water Contaminant Candidate List (DWCCL)." *Fed. Reg.,* 62(193), October 6, 1997: 52193–52219.

USEPA. "Announcement of the Drinking Water Contaminant Candidate List; Notice." *Fed. Reg.,* 63(40), March 2, 1998a: 10273–10287.

USEPA. "Disinfectants and Disinfection Byproducts Notice of Data Availability; Proposed Rule." *Fed. Reg.,* 63(61), March 31, 1998b: 15673–15692.

USEPA. "Revision of Existing Variance and Exemption Regulations to Comply With Requirements of the Safe Drinking Water Act; Final Rule." *Fed. Reg.,* 63(157), August 14, 1998c: 43833–43851.

USEPA. Summary Report; EPA Workshop; Improved Indicator Methods of Pathogens Occurrence in Water, August 10–11, 1998, Arlington, Virginia. Washington, D.C.: USEPA Office of Water, 1998d.

USEPA. "Removal of the Prohibition on the Use of Point of Use Devices for Compliance with National Primary Drinking Water Regulations; Final Rule." *Fed. Reg.,* 63(112), June 11, 1998e: 31932–31934.

USEPA. "Consumer Confidence Reports; Final Rule." *Fed. Reg.,* 63(160), August 19, 1998f: 44512–44536.

USEPA. "Disinfectants and Disinfection By-products; Final Rule." *Fed. Reg.*, 63(241), Dec. 16, 1998g: 69390.

USEPA. "Interim Enhanced Surface Water Treatment Rule; Final Rule." *Fed. Reg.*, 63(241), Dec. 16, 1998h: 69478.

USFDA. "US Food and Drug Administration." *Fed. Reg.,* 41(126), June 1976.

USPHS. "Report of the Advisory Committee on Official Water Standards." *Public Health Rept.,* 40(693), April 10, 1925.

USPHS. "Public Health Service Drinking Water Standards and Manual of Recommended Water Sanitation Practice." *Public Health Rept.,* 58(69), January 15, 1943.

USPHS. "Public Health Service Drinking Water Standards." *Public Health Rept.,* 61(371), March 15, 1946.

USPHS. "Drinking Water Standards." *Fed. Reg.,* March 6, 1962: 2152–2155.

USPHS. *Community Water Supply Study: Analysis of National Survey Findings.* NTIS Pb214982. Springfield, VA: USPHS, 1970a.

USPHS. *Community Water Supply Study: Significance of National Findings.* NTIS PB215198/BE. Springfield, VA: USPHS, July 1970b.

U.S. Statues. "Interstate Quarantine Act of 1893," chap. 114. *U.S. Statutes at Large,* 27, February 15, 1893: 449.

Westrick, J. J., J. W. Mello, and R. F. Thomas. "The Groundwater Supply Survey." *Jour. AWWA,* 76(5), May 1984: 52.

WHO. *Guidelines for Drinking-Water Quality,* 2nd edition. Geneva, Switzerland: World Health Organization, 1996. [Available from: WHO Distribution and Sales, 1211 Geneva 27, Switzerland, or WHO Publication Center USA, 49 Sheridan Avenue, Albany, NY 12210.]

Zavaleta, J. O. Health Risk Assessment and MCLG Development. Informational Briefing Documents for Senate Staff. January 1995.

CHAPTER 2

HEALTH AND AESTHETIC ASPECTS OF WATER QUALITY[1]

Perry D. Cohn, Ph.D., M.P.H.
Research Scientist
Consumer and Environmental Health Services
New Jersey Department of Health and Senior Services
Trenton, NJ

Michael Cox, M.P.H.
Program Manager
Office of Ground Water and Drinking Water
U.S. Environmental Protection Agency
Seattle, WA

Paul S. Berger, Ph.D.
Senior Microbiologist
Office of Ground Water and Drinking Water
U.S. Environmental Protection Agency
Washington, D.C.

Health and aesthetics are the principal motivations for water treatment. In the late 1800s and early 1900s, acute waterborne diseases, such as cholera and typhoid fever, spurred development and proliferation of filtration and chlorination plants. Subsequent identification in water supplies of additional disease agents (such as *Legionella, Cryptosporidium,* and *Giardia*) and contaminants (such as cadmium and lead) resulted in more elaborate pretreatments to enhance filtration and disinfection. Additionally, specialized processes such as granular activated carbon (GAC) adsorption and ion exchange were occasionally applied to water treatment to control taste- and odor-causing compounds and to remove contaminants such as nitrates. In addition, water treatment can be used to protect and preserve the distribution system.

A variety of developments in the water quality field since the 1970s and an increasing understanding of health effects have created an upheaval in the water treatment field. With the identification in water of low levels of potentially harmful

[1] The views expressed in this paper are those of the authors and do not necessarily reflect the views or policies of the U.S. Environmental Protection Agency.

organic compounds, a coliform-free and low-turbidity water is no longer sufficient. New information regarding inorganic contaminants, such as lead, is forcing suppliers to tighten control of water quality within distribution systems. Increasing pressures on watersheds have resulted in a heavier incoming load of microorganisms to many treatment plants. Although a similarly intense reevaluation of the aesthetic aspects of water quality has not occurred, aesthetic quality is important. Problems, such as excessive minerals, fixture staining, and color, do affect consumer acceptance of the water supply. However, significant advances in the identification of taste- and odor-causing organisms and their metabolites have occurred within the last two decades.

This chapter summarizes the current state of knowledge on health and aesthetic aspects of water quality. Following an introductory discussion of waterborne disease outbreaks and basic concepts of toxicology, separate sections of the chapter are devoted to pathogenic organisms, indicator organisms, inorganic constituents, organic compounds, disinfectants and disinfection by-products, and radionuclides. Emphasis is placed on contaminants that occur more frequently and at levels that are of concern for human health. Because of major interest and pending changes regarding disinfection by-products as of this writing, more detail has been included for that section. Taste and odor, turbidity, color, mineralization, and hardness are discussed in a final section devoted to aesthetic quality.

Caution is necessary when applying the information presented in this chapter. The chapter is intended as an introductory overview of health effects and occurrence information. The cited references and additional sources must be studied carefully and public health officials consulted prior to making any decisions regarding specific contamination problems. Due to space considerations, citations have been limited to those most generally applicable, especially those from U.S. governmental organizations, such as the U.S. Environmental Protection Agency (USEPA) and the Centers for Disease Control and Prevention (CDC). Various state governments in the United States have more stringent standards based on different interpretations of studies or different referent studies. [General references for chemicals include the National Academy of Sciences series on drinking water, the Toxicological Profile series on individual chemicals produced by the Agency for Toxic Substances and Disease Registry (ATSDR), and the Integrated Risk Information System (IRIS) database of the USEPA.] For chemicals, emphasis has been given to possible health effects from long-term exposure rather than to the effects of acute poisoning, and, where the effects are markedly different, from oral exposure rather than inhalation. Other handbooks cover acute poisoning. If such poisoning has occurred, one should consult local poison control authorities.

WATERBORNE DISEASE OUTBREAKS

Drinking water quality has improved dramatically over the years because of better wastewater disposal practices, protection of ambient waters and groundwaters, and advances in the development, protection, and treatment of water supplies. However, these improvements are being threatened by the pressures of an increasing population and an aging infrastructure.

Despite many improvements, waterborne disease continues to occur at high levels. Between 1980 and 1996, 402 outbreaks were reported nationally with over 500,000 associated cases of waterborne disease (Figure 2.1, Table 2.1). However, the vast majority of waterborne disease cases undoubtedly are not reported. Few states have an active outbreak surveillance program, and disease outbreaks are often not recognized in a community or, if recognized, are not traced to the drinking water source. This situation is complicated by the fact that most people experiencing gastrointestinal illness (predominantly diarrhea) do not seek medical attention. For those who do, physicians generally cannot attribute gastrointestinal illness to any specific origin such as a drinking water source. An unknown, but probably significant, portion of the waterborne disease is endemic (i.e., not associated with an outbreak) and thus is even more difficult to recognize. Based on this information, the number of waterborne disease outbreaks and cases is probably much greater than that recorded or reported.

Yet, this difference between what is occurring and what is being reported may be decreasing. There appears to be greater public awareness of waterborne disease and a greater sensitivity to drinking water problems by the print and broadcast media. Moreover, newly recognized agents of waterborne disease are being identified and sought in the water. Improvements in sampling and analytical techniques have also improved the recognition of waterborne disease.

A number of microorganisms have been implicated in waterborne disease, including protozoa, viruses, and bacteria (Table 2.2). Waterborne disease is usually acute (i.e., rapid onset and generally lasting a short time in healthy people) and most often is characterized by gastrointestinal symptoms (e.g., diarrhea, fatigue, abdominal cramps). The time between exposure to a pathogen and the onset of illness may range from two days or less (e.g., Norwalk virus, *Salmonella,* and *Shigella*) to one or more weeks (e.g., Hepatitis A virus, *Giardia,* and *Cryptosporidium*). The severity and duration of illness is greater in those with weakened immune systems. These organisms may also produce the same gastrointestinal symptoms via transmission routes other than water (e.g., food and direct fecal-oral contact). The causative agent is not identified in about half of the waterborne disease outbreaks.

Most outbreaks are caused by the use of contaminated, untreated water, or are due to inadequacies in treatment; the majority tend to occur in small systems (Craun, 1986).

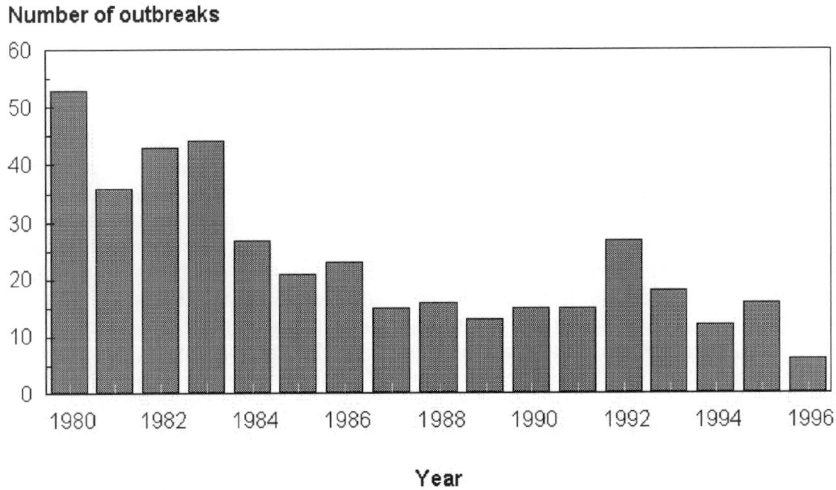

FIGURE 2.1 Waterborne disease outbreaks, 1980 to 1996.

TABLE 2.1 Waterborne Disease Outbreaks in the United States, 1980 to 1996*†

Illness	No. of outbreaks	Cases of illness
Gastroenteritis, undefined	183	55,562
Giardiasis	84	10,262
Chemical poisoning	46	3,097
Shigellosis	19	3,864
Gastroenteritis, Norwalk virus	15	9,437
Campylobacteriosis	15	2,480
Hepatitis A	13	412
Cryptosporidiosis	10	419,939‡
Salmonellosis	5	1,845
Gastroenteritis, *E. coli* 0157:H7	3	278
Yersiniosis	2	103
Cholera	2	28
Gastroenteritis, rotavirus	1	1,761
Typhoid fever	1	60
Gastroenteritis, Plesiomonas	1	60
Amoebiasis	1	4
Cyclosporiasis	1	21
TOTAL	**402**	**509,213**

* An outbreak of waterborne disease for microorganisms is defined as: (1) two or more persons experience a similar illness after consumption or use of water intended for drinking, and (2) epidemiologic evidence implicates the water as a source of illness. A single case of chemical poisoning constitutes an outbreak if a laboratory study indicates that the water has been contaminated by the chemical.

† Data are from CDC annual surveillance summaries for 1980 through 1985 and two-year summaries for 1986 through 1994, as corrected for several missing outbreaks by G.F. Craun (personal communications).

‡ Total includes 403,000 cases from a single outbreak.

PATHOGENIC ORGANISMS

Disease-causing organisms are called pathogens. Pathogens that have been implicated in waterborne disease include bacteria, viruses, protozoa, and algae. Table 2.2 lists known or suspected waterborne pathogens. Shown are the diseases they cause and where the organism is commonly found. The information given in this table is detailed in Sobsey and Olson (1983).

According to convention, every biological species (except viruses) bears a latinized name that consists of two words. The first word is the genus (e.g., *Giardia*), and the second word is the species (e.g., *lamblia*). The first letter of the genus name is capitalized, and both the genus and species are either italicized or underlined. After the full names of the genus and species names (e.g., *Escherichia coli*) are presented in a text, any further reference to the organism may be abbreviated (e.g., *E. coli*). Many of these organisms can be further differentiated on the basis of antigenic recognition by antibodies of the immune system, a process called serotyping.

Space limitations do not allow more than a sketch of the organisms listed in Table 2.2. Additional information on the waterborne pathogens listed may be found in the cited references.

Bacteria

Bacteria are single-celled microorganisms that possess no well-defined nucleus and reproduce by binary fission. Bacteria exhibit almost all possible variations in shape, from the simple sphere or rod to very elongated, branching threads. Some water bac-

TABLE 2.2 Potential Waterborne Disease-Causing Organisms

Organism	Major disease	Primary source
	Bacteria	
Salmonella typhi	Typhoid fever	Human feces
Salmonella paratyphi	Paratyphoid fever	Human feces
Other *Salmonella* sp.	Gastroenteritis (salmonellosis)	Human/animal feces
Shigella	Bacillary dysentery	Human feces
Vibrio cholerae	Cholera	Human feces, coastal water
Pathogenic *Escherichia coli*	Gastroenteritis	Human/animal feces
Yersinia enterocolitica	Gastroenteritis	Human/animal feces
Campylobacter jejuni	Gastroenteritis	Human/animal feces
Legionella pneumophila	Legionnaires' disease, Pontiac fever	Warm water
Mycobacterium avium intracellulare	Pulmonary disease	Human/animal feces, soil, water
Pseudomonas aeruginosa	Dermatitis	Natural waters
Aeromonas hydrophila	Gastroenteritis	Natural waters
Helicobacter pylori	Peptic ulcers	Saliva, human feces?
	Enteric viruses	
Poliovirus	Poliomyelitis	Human feces
Coxsackievirus	Upper respiratory disease	Human feces
Echovirus	Upper respiratory disease	Human feces
Rotavirus	Gastroenteritis	Human feces
Norwalk virus and other caliciviruses	Gastroenteritis	Human feces
Hepatitis A virus	Infectious hepatitis	Human feces
Hepatitis E virus	Hepatitis	Human feces
Astrovirus	Gastroenteritis	Human feces
Enteric adenoviruses	Gastroenteritis	Human feces
	Protozoa and other organisms	
Giardia lamblia	Giardiasis (gastroenteritis)	Human and animal feces
Cryptosporidium parvum	Cryptosporidiosis (gastroenteritis)	Human and animal feces
Entamoeba histolytica	Amoebic dysentery	Human feces
Cyclospora cayatanensis	Gastroenteritis	Human feces
Microspora	Gastroenteritis	Human feces
Acanthamoeba	Eye infection	Soil and water
Toxoplasma gondii	Flu-like symptoms	Cats
Naegleria fowleri	Primary amoebic meningoencephalitis	Soil and water
Blue-green algae	Gastroenteritis, liver damage, nervous system damage	Natural waters
Fungi	Respiratory allergies	Air, water?

teria are heterotrophic and use organic carbon sources for growth and energy, whereas some are autotrophic and use carbon dioxide or bicarbonate ion for growth and energy. Almost all the waterborne bacterial pathogens are heterotrophic, with the cyanobacteria (i.e., blue-green algae) as an exception. Autotrophic bacteria include some of the nitrifiers (e.g., *Nitrosomonas, Nitrobacter*), many iron bacteria, and many sulfur bacteria. Bacteria may also be aerobes (use oxygen), anaerobes (cannot use oxygen), or facultative aerobes (can grow in the presence or absence of oxygen). Only a few bacteria cause disease. Bacterial pathogens of current interest in drinking water are described below.

Salmonella. Over 2200 known serotypes of *Salmonella* exist, all of which are pathogenic to humans. Most cause gastrointestinal illness (e.g., diarrhea); however, a few can cause other types of disease, such as typhoid (*S. typhi*) and paratyphoid fevers (*S. paratyphi*). These latter two species infect only humans; the others are carried by both humans and animals. At any one time, 0.1 percent of the population will be excreting *Salmonella* (mostly as a result of infections caused by contaminated foods). Between 1980 and 1996, five outbreaks of waterborne salmonellosis occurred in the United States (including one outbreak of typhoid fever) with about 2000 associated cases.

Shigella. Four main species exist in this genus: *S. sonnei, S. flexneri, S. boydii,* and *S. dysenteriae.* They infect humans and primates, and cause bacillary dysentery. *S. sonnei* causes the bulk of waterborne infections, although all four subgroups have been isolated during different disease outbreaks. Waterborne shigellosis is most often the result of contamination from one identifiable source, such as an improperly disinfected well. The survival of *Shigella* in water and their response to water treatment is similar to that of the coliform bacteria. Therefore, systems that effectively control coliforms will protect against *Shigella.* Between 1980 and 1996, 19 outbreaks of waterborne shigellosis occurred with 3864 associated cases.

Yersinia enterocolitica. *Y. enterocolitica* can cause acute gastrointestinal illness, and is carried by humans, pigs, and a variety of other animals. Between 1980 and 1996, two outbreaks of waterborne yersiniosis were reported in the United States, with 103 associated cases. The organism is common in surface waters and has been occasionally isolated from groundwater and drinking water (Saari and Jansen, 1979; Schiemann, 1978). *Yersinia* can grow at temperatures as low as 39°F (4°C) and has been isolated in untreated surface waters more frequently during colder months. Chlorine is effective against this organism (Lund, 1996).

Campylobacter jejuni. *C. jejuni* can infect humans and a variety of animals. It is the most common bacterial cause of gastrointestinal illness requiring hospitalization and a major cause of foodborne illness (Fung, 1992). The natural habitat is the intestinal tract of warm-blooded animals, and *Campylobacter* is common in wastewater and surface waters (Koenraad et al., 1997). Between 1980 and 1996, 15 waterborne disease outbreaks with 2480 cases were attributed to *C. jejuni.* Laboratory tests indicate that conventional chlorination should control the organism (Blaser et al., 1986).

Legionella. Over 25 species of *Legionella* have been identified, and a substantial proportion can cause a type of pneumonia called Legionnaires' disease. *L. pneumophila* accounts for 90 percent of the cases of Legionnaires' disease reported to the CDC (Breiman 1993). The disease, which has a 15 percent mortality rate, is most probably the result of inhaling aerosols of water containing the organism. Little evidence currently exists to suggest that ingestion of water containing *Legionella* leads to infection (Bartlett, 1984). *Legionella* can also cause a milder, nonpneumonic illness called Pontiac fever.

L. pneumophila is a naturally occurring and widely distributed organism. In one study, it was isolated from all samples taken in a survey of 67 rivers and lakes in the United States. Higher isolations occurred in warmer waters (Fliermans et al., 1981). *Legionella* can also be found in groundwater (Lye et al., 1997), and in biofilms on water mains (Colbourne et al., 1988) and plumbing materials (Rogers et al., 1994). Multiplication is facilitated within *Acanthamoeba* and other aquatic protozoa

(States et al., 1990). Small numbers of *Legionella* can occur in the finished waters of systems employing full treatment. These organisms can colonize plumbing systems, especially warm ones, and aerosols from fixtures, such as showerheads, may cause the disease via inhalation (U.S. Environmental Protection Agency, 1989d). Aerosols from cooling towers containing *Legionella* have also been implicated as a route of infection. *Legionella* species associated with hot tubs and whirlpools have caused a number of cases of Pontiac fever (Moore et al., 1993). Direct person-to-person spread has not been documented (Yu et al., 1983). Because they are free-living in water, *Legionella* are not necessarily associated with fecal contamination. Ozone, chlorine dioxide, and ultraviolet light are effective in controlling *Legionella,* but data for chlorine are inconsistent (States et al., 1990).

***Pathogenic* Escherichia coli (E. coli).** Approximately 11 of the more than 140 existing serotypes of *E. coli* cause gastrointestinal disease in humans. One of these serotypes, *E. coli* O157:H7, is a prime cause of bloody diarrhea in infants. Between 1980 and 1996, three waterborne disease outbreaks with 278 cases and several deaths were caused by *E. coli* O157:H7. However, this organism has traditionally been more closely associated with cattle and sheep. Environmental processes and water disinfection are effective in controlling *E. coli;* its presence is an indication of recent fecal contamination from warm-blooded animals (Mitchell, 1972).

Vibrio cholerae. *V. cholerae* causes cholera, an acute intestinal disease with massive diarrhea, vomiting, dehydration, and other symptoms. Death may occur within a few hours unless medical treatment is given. *V. cholerae* has been associated with massive recent epidemics throughout the world, and most have been either waterborne or associated with the consumption of fish and shellfish taken from contaminated water (Craun et al., 1991). Some evidence suggests that only *V. cholerae* cells that have been infected with a virus (bacteriophage) can cause the disease (Williams, 1996).

Development of protected water supplies, control of sewage discharges, and water treatment have dramatically reduced widespread epidemics in the United States. However, one outbreak of cholera occurred in 1981, caused by wastewater contamination of an oil rig's potable water system, resulting in 17 cases of severe diarrhea (Centers for Disease Control, 1982). Another cholera outbreak, in 1994, may have been associated with bottled water taken from a contaminated well. *V. cholerae* is normally sensitive to chlorine, but may aggregate and assume a "rugose" form that is much more resistant to this disinfectant (Rice et al., 1993).

Helicobacter pylori. *H. pylori* is a bacterium that was, until recently, considered part of the genus *Campylobacter.* This organism has been closely associated with peptic ulcers, gastric carcinoma, and gastritis (Peterson, 1991; Nomura et al., 1991; Parsonnet et al., 1991; Cover and Blaser, 1995). Data about its distribution in the environment are scarce, but the organism has been found in sewage (Sutton et al., 1995) and linked to ambient water and drinking water by epidemiological tests, pure culture studies, and polymerase chain reaction studies (Klein et al., 1991; Shahamat et al., 1992; Shahamat et al., 1993; Hulten et al., 1996). The host range is thought to be narrow, possibly only humans and a few animals. The number of people in the United States that have antibodies against *H. pylori,* and thus have been exposed to the organism, is high, with the prevalence increasing with age (less than 30 years—about 10 percent; over 60 years—about 60 percent) (Peterson, 1991). About 50 percent of the world's population is infected with *H. pylori* (American Society of

Microbiology, 1994). However, most seropositive people have few or no symptoms throughout their lives (Cover and Blaser, 1995). The organism is easily inactivated by typical chlorine and monochloramine doses used in water treatment (Johnson et al., 1997).

Opportunistic Bacterial Pathogens. Opportunistic bacterial pathogens comprise a heterogeneous group of bacteria that seldom, if ever, causes disease in healthy people, but can often cause severe diseases in newborns, the elderly, AIDS patients, and other individuals with weakened immune systems. The opportunistic pathogens include strains of *Pseudomonas aeruginosa* and other *Pseudomonas* species, *Aeromonas hydrophila* and other *Aeromonas* species, *Mycobacterium avium intracellulare* (Mai or MAC), and species of *Flavobacterium, Klebsiella, Serratia, Proteus, Acinetobacter,* and others (Rusin et al., 1997; Horan et al., 1988; Latham and Schaffner, 1992; Kühn et al., 1997). These organisms are ubiquitous in the environment and are often common in finished waters and in biofilms on pipes. Although they have not been conclusively implicated in a reported waterborne disease outbreak, they have a significant role in hospital-acquired infections. *Pseudomonas aeruginosa* is commonly associated with dermatitis in hot tubs and pools (Kramer et al., 1996).

The Mai complex is common in the environment and can colonize water systems and plumbing systems (du Moulin and Stottmeier, 1986; du Moulin et al., 1988). It is known to cause pulmonary disease and other diseases, especially in individuals with weakened immune systems (e.g., AIDS patients). Drinking water has been epidemiologically linked to Mai infections in hospital patients (du Moulin and Stottmeier, 1986). Mai is relatively resistant to chlorine disinfection (Pelletier et al., 1988).

Viruses

Viruses are a large group of tiny infectious agents, ranging in size from 0.02 to 0.3 micrometers (μm). They are particles, not cells, like the other pathogens, and consist of a protein coat and a nucleic acid core. Viruses are characterized by a total dependence on living cells for reproduction and by no independent metabolism.

Viruses belonging to the group known as enteric viruses infect the gastrointestinal tract of humans and animals and are excreted in their feces. Most pathogenic waterborne viruses cause acute gastrointestinal disease. Over 100 types of enteric viruses are known, and many have been found in groundwater and surface water. Enteric virus strains that infect animals generally do not infect humans. Enteric viruses that have caused, or could potentially cause, waterborne disease in the United States are discussed next.

Hepatitis A. Although all enteric viruses are potentially transmitted by drinking water, evidence of this route of infection is strongest for hepatitis A (HAV). Hepatitis A causes infectious hepatitis, an illness characterized by inflammation and necrosis of the liver. Symptoms include fever, weakness, nausea, vomiting, diarrhea, and sometimes jaundice. Between 1980 and 1996, 13 waterborne disease outbreaks caused by HAV, with 412 associated cases, were reported (Table 2.1). Hepatitis A is effectively removed from water by coagulation, flocculation, and filtration (Rao et al., 1988). However, HAV is somewhat more difficult to inactivate by disinfection than some other enteric viruses (Peterson et al., 1983).

Norwalk Virus and Other Caliciviruses. The caliciviruses are a common cause of acute gastrointestinal illness in the United States. Between 1980 and 1996, 15 waterborne disease outbreaks with over 9000 associated cases caused by Norwalk virus and other caliciviruses were reported (Table 2.1). The illness is typically mild. The caliciviruses have generally been named after the location of the first outbreak (i.e., Norwalk agent, Snow Mountain agent, Hawaii agent, Montgomery County agent, and so on) (Gerba et al., 1985). Transmission is by the fecal-oral route. However, because adequate recovery and assay methods for the caliciviruses are not yet available, information about the occurrence of these viruses in water or their removal/inactivation during water treatment is lacking.

Rotaviruses. Rotaviruses cause acute gastroenteritis, primarily in children. Almost all children have been infected at least once by the age of five years (Parsonnet, 1992), and in developing countries, rotavirus infections are a major cause of infant mortality. During a two-year surveillance program, 21 percent of the stool samples submitted to virology laboratories in the United States were positive for rotaviruses (Ing et al., 1992). Rotaviruses are spread by fecal-oral transmission and have been found in municipal wastewater, lakes, rivers, groundwater, and tap water (Gerba et al., 1985; Gerba, 1996). However, only a single waterborne disease outbreak has been reported in the United States and only several have been documented outside the United States (Gerba et al., 1985). Filtration (with coagulation and flocculation) removes greater than 99 percent of the rotaviruses (Rao et al., 1988). Rotaviruses are readily inactivated by chlorine, chlorine dioxide, and ozone, but apparently not by monochloramine (Berman and Hoff, 1984; Chen and Vaughn, 1990; Vaughn et al., 1986, 1987).

Enteroviruses. The enteroviruses include polioviruses, coxsackieviruses, and echoviruses. Enteroviruses are readily found in wastewater and surface water, and sometimes in drinking water (Hurst, 1991). With one exception, no drinking water outbreaks implicating these viruses have been reported and, therefore, their significance as waterborne pathogens is uncertain. In 1952, a polio outbreak with 16 cases of paralytic disease was attributed to a drinking water source, but since then, no well-documented case of waterborne disease caused by poliovirus has been reported in the United States (Craun, 1986). Coxsackieviruses, and to a lesser extent echoviruses, produce a variety of illnesses in humans, including the common cold, heart disease, fever, aseptic meningitis, gastrointestinal problems, and many more, some of which are serious (Melnick, 1992).

Adenoviruses. There are 47 known types of adenoviruses, but only types 40 and 41 are important causes of gastrointestinal illness, especially in children. Other types of adenoviruses are responsible for upper respiratory illness, including the common cold. However, all types may be shed in the feces, and may be spread by the fecal-oral route. Although adenoviruses have been detected in wastewater, surface water, and drinking water, data on their occurrence in water are meager. No drinking water outbreaks implicating these viruses have been reported and, therefore, their significance as waterborne pathogens is uncertain. Adenoviruses are relatively resistant to disinfectants.

Hepatitis E Virus (HEV). Hepatitis E virus (HEV) has caused waterborne disease outbreaks and endemic disease over a wide geographic area, including Central and South America, Asia, Africa, Australia, and other parts of the world (Velazquez et al., 1990; Bowden et al., 1994; Ibarra et al., 1994; Mast and Krawczynski, 1996). There are no definitive reports that indicate that animals other than humans can be infected by HEV, but one study suggests that pigs can be infected (Dreesman and Reyes, 1992).

Hepatitis E virus is transmitted by the fecal-oral route (Dreesman and Reyes, 1992). It appears that a high percentage of the cases, probably a majority, are waterborne. To date, only one locally acquired case in the United States has been documented (Kwo et al., 1997), but the source of that case was not identified by epidemiological studies and six family members lacked antibodies (i.e., were seronegative) to the virus. In one study, 4.2 percent of 406 patients with hepatitis-like symptoms (all residents of the United States except for four Canadians) had antibodies (i.e., were seropositive) to HEV (Halling et al., personal communications). In another study, 21.3 percent of blood donors in Baltimore were seropositive (Thomas et al., 1997). Fecal shedding by infected humans may last more than one month (Nanda et al., 1995). Hepatitis E causes clinical symptoms similar to those caused by the hepatitis A virus, including abdominal pain, fever, and a prolonged lack of appetite. Infections are mild and self-limiting except for pregnant women, who have a fatality rate of up to 39 percent. No data from disinfection studies have been published.

Astroviruses. Astroviruses are small spherical viruses with a starlike appearance. They are found throughout the world and cause illness in 1- to 3-year-old children and in AIDS patients, but rarely in healthy adults (Kurtz and Lee, 1987; Grohmann et al., 1993). A few outbreaks have occurred in nursing homes (Gray et al., 1987). Symptoms are mild and typical of gastrointestinal illness, but the disease is more severe and persistent in the severely immunocompromised. Astroviruses are transmitted by the fecal-oral route. Several foodborne outbreaks have occurred (Oishi et al., 1994). They have been found in water and have been associated anecdotally with waterborne disease outbreaks (Cubitt, 1991; Pinto et al., 1996).

Protozoa

Protozoa are single-celled organisms that lack a cell wall. They are typically larger than bacteria and, unlike algae, cannot photosynthesize. Protozoa are common in fresh and marine water, and some can grow in soil and other locations (Brock and Madigan, 1991c). The few protozoa that are pathogenic to humans typically are found in water as resistant spores, cysts, or oocysts, forms that protect them from environmental stresses. The spores/cysts/oocysts are much more resistant to chlorine disinfection than are viruses and most bacteria. However, effective filtration and pretreatment can reduce their density by at least two logs (99 percent). Spores/cysts/oocysts of pathogenic protozoa are found typically in surface waters or groundwaters directly influenced by surface waters. The recognized human pathogens are described next.

Giardia lamblia. *G. lamblia* cysts are ovoid and are approximately 7.6 to 9.9 μm in width and 10.6 to 14 μm in length (LeChevallier et al., 1991). When ingested, *Giardia* can cause giardiasis, a gastrointestinal disease manifested by diarrhea, fatigue, and cramps. Symptoms may persist from a few days to months. Between 1980 and 1996, 84 outbreaks of waterborne giardiasis with 10,262 associated cases, were reported (Table 2.1). The infectious dose for giardiasis, based on human feeding studies, is 10 cysts or fewer (Rendtorff, 1979).

Water can be a major vehicle for transmission of giardiasis, although person-to-person contact and other routes are more important. In one large national study (262 samples), an average of 2.0 cysts/L were found in raw water. In this same study, when cysts were found in drinking water (4.6 percent of samples), the density averaged 2.6 cysts/100 L (LeChevallier and Norton, 1995). Humans and animals, particularly beavers and muskrats, are hosts for *Giardia*. *Giardia* cysts can remain infective in water for one to three months.

Giardia cysts are considerably more resistant to chlorine than are coliform bacteria. A properly operated treatment plant (either conventional or direct filtration) can achieve a 99.9 percent cyst removal when the finished water turbidity is between 0.1 and 0.2 nephelometric turbidity unit (NTU) (Nieminski and Ongerth, 1995). Diatomaceous earth filtration and slow sand filtration can also effectively remove cysts (Logsdon et al., 1984).

Entamoeba histolytica. *E. histolytica* cysts are about 10 to 15 μm in size (slightly larger than *Giardia* cysts). When ingested, *Entamoeba* can cause amoebic dysentery, with symptoms ranging from acute bloody diarrhea and fever to mild gastrointestinal illness. Occasionally, the organism can cause ulcers and then invade the bloodstream, causing more serious effects. However, most infected individuals do not have clinical symptoms. In contrast to the case for *Giardia* and *Cryptosporidium,* animals are not reservoirs for *E. histolytica,* so the potential for source water contamination is relatively low, especially if sewage treatment practices are adequate. According to the CDC, about 3000 cases of amoebiasis typically occur in the United States each year, and waterborne disease outbreaks caused by *E. histolytica* are infrequent. The last reported outbreak occurred in 1984 (four cases) as a result of a contaminated well (Centers for Disease Control, 1985).

Cryptosporidium parvum. *C. parvum* is one of several *Cryptosporidium* species found in many lakes and rivers, especially when the water is contaminated with sewage and animal wastes. In one study, *Cryptosporidium* oocysts were detected in 51.5 percent of 262 raw water samples (average density: 2.4 oocysts/L), and 13.4 percent of 262 filtered effluent samples (average density when oocysts were detected: 3.3 oocysts/100 L, with an average turbidity of 0.19 NTU) (LeChevallier and Norton, 1995). Some investigators have found higher levels (Ongerth and Stibbs, 1987; Rose et al., 1991). The data suggest that treated water systems with very low turbidity levels may still have viable, infective oocysts capable of causing an outbreak or endemic disease. The best currently available analytical methodology for waterborne *Cryptosporidium,* immunofluorescence, has deficiencies, including an inability to determine if oocysts are viable and infective to humans.

To date, *Cryptosporidium parvum* is the only *Cryptosporidium* species known to cause disease in humans. However, other animals are also known to harbor this species (Ernest et al., 1986). *C. parvum* is transmitted mostly by person-to-person contact, contaminated drinking water, and sex involving contact with feces. Contact with pets and farm animals, especially young ones, increases the potential for spread.

Cryptosporidium parvum has caused a series of waterborne disease outbreaks in the United States and elsewhere, including one outbreak where over 400,000 became ill and at least 50 people died (MacKenzie et al., 1994; Davis, 1996). Based on a human feeding study using healthy volunteers, it was calculated that 132 oocysts would infect 50 percent of the population. In this study, one person in five was infected by 30 oocysts (DuPont et al., 1995). Unlike the gastrointestinal illness caused by *Giardia* and most other waterborne pathogens, cryptosporidiosis is especially severe and chronic in those with severely weakened immune systems (e.g., those with AIDS) and may hasten death (Janoff and Reller, 1987).

A well-operated treatment facility that has optimized treatment should be able to remove well over 99 percent of the influent oocysts, which, at between 4 and 6 μm in diameter, are smaller than *Giardia* cysts and, thus, may be more difficult to remove. In one study, a small direct filtration pilot plant with multimedia filters was able to remove between 2.7 and 3.5 logs (Ongerth and Pecoraro, 1995). In another study,

Patania et al. (1995) observed that the median *Cryptosporidium* removal in four utilities that had optimized treatment for turbidity and particle removal was about 4.2 logs.

Cryptosporidium oocysts are more resistant to disinfection than are *Giardia* cysts, and normal disinfection practices, using chlorine or chloramines, may not kill oocysts. However, the oocysts are more sensitive to ozone and chlorine dioxide (Korich et al., 1990; Peeters et al., 1989). Some data suggest that a combination of disinfectants (e.g., ozone followed by monochloramine, or free chlorine followed by monochloramine) may be effective in killing the oocysts (Finch et al., 1995). More data are needed to assess adequately the effectiveness of a variety of treatment processes for removing and killing oocysts.

Naegleria fowleri. *N. fowleri* is a free-living amoeba, about 8 to 15 μm in size, found in soil, water, and decaying vegetation. Although it is common in many surface waters, it rarely causes disease. The disease, primary amoebic meningoencephalitis (PAM), is typically fatal, with death occurring within 72 hours after symptoms appear (Centers for Disease Control, 1992). According to the CDC (1992), in 1991 only four people contracted PAM in the United States. All disease incidents have been associated with swimming in natural or manmade, warm fresh waters; drinking water is not a suspected route of transmission. The route of infection is via the nasal passages leading to the brain. Chlorine may not kill the organism in the doses typically used (Chang, 1978).

Cyclospora. *Cyclospora* is a protozoan that has caused several known waterborne disease outbreaks in the world, including one in 1990 in the United States (Huang et al., 1995). This newly recognized pathogen (*C. cayetanensis*) was originally thought to be a blue-green alga. Human *Cyclospora* are round or ovoid and between 7 and 9 μm in diameter (Soave and Johnson, 1995), although some reports suggest the diameter may be as large as 10 μm (Ortega et al., 1993). They are larger than *Cryptosporidium* but morphologically similar (Knight, 1995). Disease symptoms include watery diarrhea, abdominal cramping, decreased appetite, and low-grade fever (Huang et al., 1995). In HIV-infected persons, the disease may be chronic and unremitting (Soave and Johnson, 1995). Their occurrence in natural waters and their animal host range are unknown. Foodborne outbreaks associated with contaminated berries have recently occurred. Chlorine may not be effective against *Cyclospora* (Rabold et al., 1994). Thus, effective filtration and watershed control may be needed to control this organism in drinking water.

Microspora. Microsporidia are a large group of protozoan parasites (phylum *Microspora*) that are common in the environment and live only inside cells (Cali, 1991). Microsporidia infect a large variety of animals, including insects, fish, amphibians, reptiles, birds, and mammals such as rodents, rabbits, pigs, sheep, dogs, cats, and primates (Canning et al., 1986; Cali, 1991). To date, five species of microsporidia have been reported to cause disease in humans, but only two are significant: *Enterocytozoon bieneusi* and *Encephalitozoon intestinalis,* both of which are common in people with AIDS (Goodgame, 1996). Microsporidiosis is considered an opportunistic infection, occurring chiefly in AIDS patients (Bryan, 1995), although it has been reported in otherwise healthy persons (Weber et al., 1994). Symptoms may include diarrhea (sometimes severe and chronic), and illness involving the respiratory tract, urogenital tract, eyes, kidney, liver, or muscles (Bryan, 1995; Goodgame, 1996; Cali, 1991).

Microsporidia that infect humans produce small (1 to 5 μm), very resistant spores (Waller, 1979; Cali, 1991) which are difficult to differentiate from bacteria and some yeasts. They are shed in bodily fluids, including urine and feces, and thus have a strong potential to enter water sources. However, no waterborne outbreak has yet been reported. Inhalation and person-to-person spread are apparently much less significant than the oral route of infection (Canning et al., 1986). Preliminary data suggest that microsporidium spores are more susceptible to chlorine than are *Cryptosporidium* oocysts (Rice et al., 1999).

Acanthamoeba. *Acanthamoeba* is a group of free-living amoeba (protozoa) that is common in soil and water, including drinking water (Sawyer, 1989; Gonzalez de la Cuesta et al., 1987). The organism has also been isolated from water taps (Rivera et al., 1979; Seal et al., 1992) and is able to protect pathogenic legionellae against disinfection by ingesting them (Barker et al., 1992). Some *Acanthamoeba* species are pathogenic, and are known to cause inflammation of the eye's cornea, especially in individuals who wear soft or disposable contact lenses (Seal et al., 1992), and chronic encephalitis in the immunocompromised population (Kilvington, 1990). To date, no case of disease from drinking water has been reported. However, *Acanthamoeba* cysts are relatively resistant to chlorine (see De Jonkheere and Van der Voorde, 1976). Apparently, effective water filtration is the primary means of control.

Toxoplasma. *Toxoplasma* causes a common infection of mammals and birds, but the complete life cycle only occurs in wild and domestic cats. The organism infects a high percentage of the human population (50 percent in some areas of the United States) but, although subclinical infections are prevalent, illness is rare (Fishback, 1992). Illness may be severe in fetuses and AIDS patients. Symptoms include fever, swelling of lymph glands in the neck, blindness and mental retardation in fetuses, and encephalitis in AIDS patients (Fishback, 1992). A waterborne outbreak of toxoplasmosis (*T. gondii*) occurred in 1979 among U.S. Army soldiers who were training in a Panamanian jungle. Clinical, serological, and epidemiologic studies revealed that consumption of water from jungle streams was the most likely source of infection (Benenson et al., 1982). An outbreak of toxoplasmosis in 1995 in British Columbia, which infected up to 3000 residents, has been linked (although not conclusively) to drinking water (American Water Works Association, 1995b). Iodine pills and chlorination of unfiltered surface waters are not effective against *Toxoplasma* (Benenson et al., 1982). Filtration should be effective in removing this organism, given that the oocyst is between 10 and 12 μm in diameter (Girdwood, 1995), about the same size as *Giardia*.

Algae

Unlike other waterborne pathogens, algae use photosynthesis as their primary mode of nutrition, and all produce chlorophyll. Algae do not typically pose a health concern. However, certain species may produce neurotoxins (substances that affect the nervous system), hepatotoxins (those that affect the liver), and other types of toxins that, if ingested at high enough concentrations, may be harmful. The most important neurotoxins from a health standpoint are those produced by three species of blue-green algae (also called cyanobacteria): *Anabaena flos-aquae, Microcystis aeruginosa,* and *Aphanizomenon flos-aquae.* High-toxin production generated during blue-green algal blooms has resulted in illness or death in mammals, birds, and fish that ingest the water (Collins, 1978). Toxins can cause toxemia and shock in immunosuppressed patients; however, little evidence exists that ambient levels found in most water supplies pose a health risk to the normal population. High concentrations of toxin associated with a bloom of *Schizothrix calciola* may have been responsible for an outbreak of gastroenteritis in 1975 (Lippy and Erb, 1976).

Fungi

Over 984 fungal species have been isolated from unchlorinated groundwater, systems using chlorinated surface water, and service mains (Highsmith and Crow, 1992). These include *Aspergillus, Penicillium, Alternaria,* and *Cladosporium* (Rosenzweig et al., 1986; Frankova and Horecka, 1995). In one enumeration study, using a membrane filtration procedure, the average year-long density in drinking water was 18 and 34 fungal colony-forming units per 100 mL in unchlorinated and chlorinated systems, respectively (Nagy and Olson, 1982). No waterborne disease outbreaks or cases have yet been documented, although a few fungi are pathogenic via the respiratory route, especially in immunocompromised persons. Inhalation of large numbers of spores can cause respiratory and other problems, including pneumonia, fever, and meningoencephalitis, but generally symptoms are mild (Bennett, 1994). In addition to directly causing infection, a number of the fungi produce toxins. Over 300 mycotoxins have been identified and are produced by some 350 fungal species (Pohland, 1993). Although a theoretical basis exists for waterborne disease, the primary concerns with fungi are the proliferation of fungi in water distribution systems and the resulting potential for taste and odor complaints (U.S. Environmental Protection Agency, 1992b). Fungal spores are relatively resistant to chlorine (Rosenzweig et al., 1983). Filtration (preceded by chemical coagulation) and disinfection can reduce, but not eliminate, them from raw water (Niemi et al., 1982).

INDICATORS AND INDICATOR ORGANISMS

It would be difficult to monitor routinely for pathogens in the water. Isolating and identifying each pathogen is beyond the capability of most water utility laboratories. In addition, the number of pathogens relative to other microorganisms in water can be very small, thus requiring a large sample volume. For these reasons, surrogate organisms are typically used as an indicator of water quality. An ideal indicator should meet all of the following general criteria:

- Should always be present when the pathogenic organism of concern is present, and absent in clean, uncontaminated water
- Should be present in fecal material in large numbers
- Should respond to natural environmental conditions and to treatment processes in a manner similar to the pathogens of interest
- Should be easily detected by simple, inexpensive laboratory tests in the shortest time with accurate results
- Should have a high indicator/pathogen ratio
- Should be stable and nonpathogenic
- Should be suitable for all types of drinking water

A number of microorganisms have been evaluated as indicators, but none are ideal. Some are sufficiently close to the ideal indicator for regulatory consideration. Each is briefly described in the following text and in Chapter 18. Information on the specific utility of other indicator organisms is available (Olivieri, 1983).

Total Coliforms

Total coliforms are a group of closely related bacteria (family *Enterobacteriaceae*) that have been used for many decades as the indicator of choice for drinking water. The group is defined as aerobic and facultatively anaerobic, gram-negative, non-spore-forming, rod-shaped bacteria that ferment the milk sugar lactose to produce acid and gas within 48 h at 35°C. Few bacteria other than coliforms can metabolize lactose; for this reason, lactose is used as the basis for identification (the hydrolysis of *o*-nitrophenyl-β-*d*-galactopyranoside, or ONPG, is also used for identification in some coliform tests). The total coliform group includes most species of the genera *Citrobacter, Enterobacter, Klebsiella,* and *Escherichia coli.* It also includes some species of *Serratia* and other genera. Although all coliform genera can be found in the gut of animals, most of these bacteria are widely distributed in the environment, including water, and wastewaters. A major exception is *E. coli,* which usually does not survive long outside the gut, except perhaps in the warm water associated with tropical climates.

Total coliforms are used to assess water treatment effectiveness and the integrity of the distribution system. They are also used as a screening test for recent fecal contamination. Treatment that provides coliform-free water should also reduce pathogens to minimal levels. A major shortcoming to using total coliforms as an indicator is that they are only marginally adequate for predicting the potential presence of pathogenic protozoan cysts/oocysts and some viruses, because total coliforms are less resistant to disinfection than these other organisms. Another shortcoming is that coliforms, under certain circumstances, may proliferate in the biofilms of water distribution systems, clouding their use as an indicator of external contamination. Coliforms are also often not of fecal origin. Despite these drawbacks, total coliforms remain a useful indicator of drinking water microbial quality, and the group is regulated under USEPA's Total Coliform Rule (USEPA, 1989e).

Fecal Coliforms and *E. coli*

Fecal coliforms are a subset of the total coliform group. *E. coli* is the major subset of the fecal coliform group. They are distinguished in the laboratory by their ability to grow at elevated temperatures (44.5°C) and by the ability of *E. coli* to produce the enzyme glucuronidase, which hydrolyzes 4-methyl-umbelliferyl-β-*D*-glucuronide (MUG). Both fecal coliforms and *E. coli* are better indicators for the presence of recent fecal contamination than are total coliforms, but they do not distinguish between human and animal contamination. Moreover, fecal coliform and *E. coli* densities are typically much lower than are those for total coliforms; thus, they are not used as an indicator for treatment effectiveness and posttreatment contamination. *E. coli* is a more specific indicator of fecal contamination than is the fecal coliform group. Under the Total Coliform Rule, all total coliform-positive samples must be tested for either fecal coliforms or *E. coli.* Fecal coliforms and *E. coli* are used by some states for assessing recreational water quality.

Heterotrophic Bacteria

Heterotrophic bacteria, as previously stated, are members of a large group of bacteria that use organic carbon for energy and growth. In the United States, laboratories often measure heterotrophic bacteria by the heterotrophic plate count (HPC) (American Public Health Association et al., 1995b). Because of its lack of specificity, the HPC has not been used to assess the likelihood of waterborne disease; a specific HPC level might contain many, few, or no pathogens. A sudden significant increase in the HPC may suggest a problem with treatment, including poor disinfection practice. The significance of the HPC lies in its indication of poor general biological quality of the drinking water. Under EPA regulations, systems that have no detectable disinfectant in the distribution system may claim, for the purposes of the regulation, that disinfectant is present if the HPC does not exceed 500 colonies/mL.

Clostridium perfringens

C. perfringens is a bacterium that is consistently associated with human fecal wastes (Bisson and Cabelli, 1980). The organism is anaerobic and forms spores (endospores) that are extremely resistant to environmental stresses and disinfection. *C. perfringens* is an agent of foodborne outbreaks, especially those associated with meats, but one has to ingest a high level of the organisms (1 million) to become ill (Fung, 1992), making it an unlikely agent of waterborne disease. The organism is considered a potential indicator of fecal contamination because it is consistently associated with the human gut and human fecal wastes at a high density (Cabelli, 1977; Bisson and Cabelli, 1980). Moreover, it is excreted in greater numbers than are fecal pathogens (Bitton, 1994). In addition, the survivability of *C. perfringens* endospores in water and their resistance to treatment compared with the pathogens is much greater than other indicators (Bonde, 1977). Analysis is simple and inexpensive, although anaerobic incubation is needed. The terms "sulfite-reducing *Clostridium*" or "anaerobic sporeformers" are often seen in the literature. These two groups include *Clostridium* species other than *C. perfringens,* as well as species of the genus *Desulfotomaculum,* but the primary organism is often *C. perfringens* (North Atlantic Treaty Organization, 1984).

One study (Payment and Franco, 1993) examined the densities of *Giardia* cysts, *Cryptosporidium* oocysts, culturable enteric viruses, *C. perfringens,* and coliphage in raw surface water and filtered water. The investigators reported that *C. perfringens* was significantly correlated with the densities of viruses, cysts, and oocysts in the raw water and with viruses and oocysts in the finished water. However, in another study, investigators reported that correlations between pathogens and *C. perfringens* were not strong (Water Research Centre, 1995). Thus, to date, insufficient data exist to support the use of *C. perfringens* as an indicator.

Coliphages

Coliphages are viruses that infect the bacterium *E. coli.* They are common in sewage and wastewater (Havelaar, 1993). Coliphages are often divided into two major categories: (1) somatic phages, which gain entry into *E. coli* cells through the cell wall, and (2) male-specific (or F-specific) phages, which gain entry only through short hair-like structures (pili) of those *E. coli* cells that have them (males). They are far easier to analyze than human or animal viruses, making them a promising indicator of fecal contamination. Two major issues regarding the use of coliphages as an indicator are (1) which of the many recognized coliphage strains to use, and (2) the most appropriate *E. coli* host(s) to use for optimum recovery.

Data on the relative resistance and removal of coliphage and human viruses are scarce and inconsistent. One study (Payment and Franco, 1993) reported that a reasonable correlation existed between enteroviruses and both somatic phages and male-specific coliphages in filtered water, but not in river water. Another study reported a correlation between viruses and male-specific phages in river and lake water, but not sewage (Havelaar et al., 1993). In spite of the large number of articles on coliphages, several major uncertainties exist, especially with regard to coliphage fate and transport in subsurface soil and water.

Bacteroides

The genus *Bacteroides* consists of a group of obligately anaerobic bacteria that grow only in the intestinal tract of humans and, to a lesser extent, in animals. The density of anaerobes in the intestines is about 10^{10} to 10^{11} cells per gram of intestinal contents, with *Bacteroides* accounting for the majority of microorganisms present. This density is far higher than that of *E. coli* (Brock and Madigan, 1991a). *Bacteroides* species do not survive well in aerobic fresh waters in temperate climates, and die off at a rate somewhat faster than that of *E. coli* (Fiksdal et al., 1985). The above information suggests that *Bacteroides* may be a more sensitive indicator of recent fecal contamination of fresh water than fecal coliforms or *E. coli.* Simple and inexpensive analytical methods are available, although anaerobic incubation is necessary. Viruses that infect the bacterium *Bacteroides fragilis* have shown promise as an indicator (Armon, 1993; Jofre et al., 1995).

Particle Counts

The concentration of particles in water, especially in selected particle size ranges, has been discussed as a possible indicator of water quality and treatment efficiency. Particle counts can aid in designing treatment processes, assessing operational problems, and determining treatment process efficiency (American Public Health Association, 1995a). Particle counts may be especially useful in estimating the removal efficiency of *Giardia* and *Cryptosporidium* in filtered water (LeChevallier and Norton, 1992). A variety of particle counters are available commercially, most of which rely on electronic, as opposed to manual, counting (American Public Health Association et al., 1995a). Although particle counting is more accurate, reliable, and reproducible than *Giardia* and *Cryptosporidium* counting, it has limitations. For example, the cost of the particle counter is high (several thousand dollars or higher) and there are a number of variables (e.g., run time, sensor, flow rate of instrument, different sensor optics). The problems associated with instrument variability can be partially overcome by measuring percent removal through treatment rather than absolute particle concentrations, but still a problem in interpreting measured values may exist. In addition, particle counts cannot distinguish between living and nonliving particles, and between particles from the raw water and those formed during treatment.

Turbidity

Turbidity is a nonspecific measure of the amount of particulate material in water (e.g., clay, silt, finely divided organic and inorganic matter, microorganisms) and is measured by detecting the amount of light scattered by particles in a sample, relative to the amount scattered by a reference suspension. Turbidity has been used for many decades as an indicator of drinking water quality and as an indicator of the efficiency of drinking water coagulation and filtration processes. Achieving adequate turbidity removal should at least partially remove pathogens in the source water, especially those pathogens which aggregate with particles. Turbidity is a relatively crude measurement, detecting a wide variety of particles from a wide assortment of sources; it provides no information about the nature of the particles. High turbidity levels can reduce the efficiency of disinfection by creating a disinfection demand. The particles may also provide absorption sites for toxic substances in the water, may protect pathogens (and coliforms) from disinfection by adsorbing or encasing them, and may interfere with the total coliform analysis (U.S. Environmental Protection Agency, 1995b; LeChevallier et al., 1981). Simple analytical methods for turbidity are available (American Public Health Association et al., 1995c; Letterman, 1994).

Aerobic Sporeformers

The aerobic sporeformers are a large group of bacteria, not always closely related, within the genus *Bacillus. Bacillus* species form spores (endospores) that allow them to survive in the environment, perhaps for thousands of years (Brock and Madigan, 1991b). These organisms are common and widespread in nature, primarily in the soil; thus, they are not closely associated with fecal contamination. A few species have been associated with food poisoning (Fung, 1992). Because simple, inexpensive analytical methods are available, and they outlive most, if not all, waterborne pathogens, *Bacillus* spores have been discussed as a possible indicator of treatment efficiency, especially for the removal of *Giardia, Cryptosporidium,* and other protozoa. One study showed that removal of *Bacillus* spores closely paralleled those of both total particle counts and counts in the 3.1- to 7-μm range, which is similar to the size range for *Giardia* and *Cryptosporidium* (Rice et al., 1996). Another study suggested that the removal of aerobic sporeformers during treatment is similar to that of *Cryptosporidium* and other small particles (Water Research Centre, 1995).

Microscopic Particulate Analysis

Microscopic particulate analysis (MPA) is a tool for examining groundwater samples to determine if they are under the direct influence of surface water and for evaluating the efficiency of filter treatment processes for systems using surface waters. The method consists of a microscopic examination of the water for the presence of plant debris, pollen, rotifers, crustaceans, amoebas, nematodes, insects/larvae, algae (including diatoms), coccidia (e.g., *Cryptosporidium*), and *Giardia* cysts. Microscopic particulate analysis guidance for groundwater defines five risk categories, based on the concentration of each of these bioindicators. The concentration of this material should be low in true groundwaters. The MPA is described in two documents, both by USEPA: the first for use in groundwaters (U.S. Environmental Protection Agency, 1992a), the second for use by surface water systems to measure the effectiveness of steps to optimize filter performance (U.S. Environmental Protection Agency, 1996b).

THE HEALTH EFFECTS OF CHEMICALS

Every chemical has an effect, some of them adverse, on living organisms exposed to it. Toxicology, the study of the adverse effects of chemicals on living organisms, provides a means of evaluating and understanding these effects. Epidemiology, the study of the distribution and determinants of diseases and injuries, can also provide evidence of chemical toxicity. Exposure to carcinogenic chemicals or materials like arsenic, benzene, vinyl chloride, asbestos, cigarette smoke, and radiation was first linked to the induction of cancer by epidemiologic studies. Of course, humans cannot be subjected to controlled tests with only one chemical at a time, which often makes the results of epidemiologic studies difficult to interpret. On the other hand, the use of epidemiologic data reduces the necessity of extrapolating from animal models. Although a complete review of toxicological and epidemiological principles is beyond the scope of this chapter, basic concepts needed to understand the information presented in this chapter follow.

Chemicals can cause clearly deleterious effects as well as changes (e.g., in enzyme levels) that, as of yet, are not considered adverse. Some adverse health effects in organisms are immediate (within 24 to 48 hours after exposure), but others are delayed (for example, 5 to 40 years or more for cancer in humans). Adverse effects may be reversible depending upon their nature, the severity of the effect, and the organ affected. Some effects may not appear until subsequent generations. Typically, more is known about the immediate effects of single or short-term, high doses than the delayed effects of long-term, low-dose exposure.

The response of a living organism to exposure to a chemical depends upon the chemical dose or exposure level. [Dose in laboratory experiments is typically measured as weight of chemical administered per body weight per day, such as milligrams per kilogram per day ($mg/kg \cdot day^{-1}$)]. The higher the dose, the more significant the effect. This is termed the dose-response relationship. Understanding this concept is important because simply knowing that a substance can have a particular toxicological property (e.g., carcinogenic) is not adequate alone to assess human health risk. The dose-response relationship should also be known, as well as information concerning human exposure, before a judgment can be made regarding the public health significance of exposure to that substance. In part the dose-response relationship depends on toxicokinetic parameters, which include absorption, distribution, metabolism, and excretion of a chemical. However, the traditional dose-response relationship may not adequately describe long-term or intermittent exposures, because the activity of metabolic enzymes enhancing or neutralizing toxicity can be increased or decreased in different human exposure scenarios. Furthermore, in adults the activity of key metabolic enzymes can vary as much as 200-fold among individuals, and because metabolism often requires two or more steps, combined variations of enzyme activities may result in much higher interindividual differences (Perera, 1996). In addition, cancer risk is also dependent on the activities of enzymes that repair damage to DNA, which can vary as much as several hundredfold among humans.

The dose-response relationship may be different in fetuses, children, and the elderly. Growing bodies may be more affected because key enzymes in certain tissues may be higher or lower and because children and pregnant women absorb many chemicals better. For example, it is estimated that lead and mercury salts are absorbed as much as 10 to 20 times more in infants and up to severalfold more in pregnant women (Centers for Disease Control, 1991; Agency for Toxic Substances and Disease Registry, 1994d). The ability to metabolize certain toxic chemicals can be diminished in the elderly.

A variety of adverse health effects are possible. In this chapter, the following general terms will be used to describe these effects.

Toxic. Causing a deleterious response in a biologic system, seriously disrupting function, or producing death. These effects may result from acute (short-term, high-dose), chronic (long-term, low-dose), or subchronic (intermediate-term and -dose) exposure. The biological system is either in vitro, such as liver cells in a culture dish, or in vivo (in laboratory animals or humans).

Carcinogenic (oncogenic). Causing or inducing uncontrolled growth of aberrant cells into malignant (or neoplastic) tumors. Some chemicals may initiate the series of changes in DNA necessary for carcinogenesis, while others may promote the growth of "initiated" cells. These chemicals are referred to as "initiators" and "promotors," respectively. Initiation is thought to involve mutations in the DNA of genetic segments (i.e., genes) controlling cell growth, whereas promotion may involve various mechanisms, such as altered DNA-repair enzyme function or increased levels of cellular growth activation enzymes that are products of "oncogenes." Mutations in certain "oncogenes" also play an important role. Certain benign tumors and foci of aberrant cells in organs of animals exposed to test chemicals may represent initiated cells whose growth could be increased by the presence of a promoter.

Genotoxic. Causing alterations or damage to the genetic material in living cells, such as deletions of portions of or entire chromosomes, but also including phenomena whose meaning is not fully understood, such as creation of micronuclei, sister chromatid exchanges, and unscheduled DNA synthesis (due to repair processes). Some changes may be lethal to the cell. Many carcinogens and mutagens are also genotoxic.

Mutagenic. Causing heritable alteration of the genetic material within living cells. A mutation is typically defined as the change in the genetic code of a specific gene that results in a change in a cellular or biochemical characteristic. Many carcinogens and genotoxic agents are also mutagens and vice versa.

Teratogenic. Causing nonhereditary congenital malformations (birth defects) in offspring.

In terms of chemical exposure, the focus of this chapter is on chronic (1 to 2 years) or subchronic (2 to 13 weeks) health effects in rodents, rather than the effects of acute poisoning. For noncarcinogenic endpoints, laboratory feeding or inhalation studies examine pathological changes in organs and blood, including organ weight, the ratio of organ-to-body weight, microscopic appearance, and enzyme activities. Assessments of toxics in drinking water focus on the oral exposure route, especially via food or water, but also dissolved in vegetable oil for organic chemicals with low water solubility. Among the organs most frequently affected in these studies are liver and kidney because of their role in metabolizing and removing toxic chemicals. For certain chemicals, neurotoxicity, immunotoxicity, teratogenicity, and reproductive toxicity (such as changes in fertility, birth weight, and survival) are also examined.

Studies of carcinogenic potential employ two procedures: the feeding study for the detection of carcinogens and tests for mutagenicity and genotoxicity. Many of these assays have been conducted by the National Cancer Institute (NCI) or National Toxicology Program (NTP). In a feeding study, rats, mice, or hamsters are fed the test chemical for two years. Tumors often found in laboratory studies include carcinomas and adenocarcinomas of liver, kidney, lung, mammary, and thyroid tissues, leukemias (of circulating white blood cells), and lymphomas (of white blood cells in lymph nodes). The typical benign tumor is an adenoma. Doses ideally include the maximally tolerated dose (MTD) and half MTD. The incidence of malignant tumors or total tumors (malignant and benign) in the experimental group is then statistically compared with a control group.

There is some disagreement on the human applicability of cancers, particularly liver cancers in certain mouse strains, following very high doses. However, humans have many years to develop cancers, compared with rodents, which have two-year average lifetimes. Thus, cancers must be induced quickly with high doses to grow sufficiently to be able to detect them microscopically. In addition, the economic practicalities of exposure group size require higher dosage to observe statistically significant increases in cancer. Although liver cancers are infrequent in humans, except with vinyl chloride exposure and certain types of chronic hepatitis, oncogene expression in mouse liver cancer and preneoplastic liver nodules (an apparent precursor of cancer in the liver) appears to be similar to expression in other mouse and human cancers. This suggests similarity in mechanism. One example is increased oncogene expression in liver tumors after exposure to di- and trichloroacetic acid (Nelson et al., 1990) and tri- and tetrachloroethylene (Anna et al., 1994).

The Ames test, also known as the *Salmonella* microsome assay, is one of many in vitro screening systems. Strains of the bacterium, *Salmonella typhimurium,* specially developed to have a histidine requirement, are exposed to the test chemical. Mutagenicity in that test system is measured by comparing reversions to histidine independence in the experimental group to those that would occur spontaneously in a control group. Usually, a chemical is tested with and without liver cell metabolic enzymes that might produce a mutagenic metabolic product (a technique called *activation*). There are many similar assays using other microorganisms (e.g., yeast and fungi) and endpoints. Similarly, fruit fly (*Drosophila*) and mammalian cells are also tested in vivo and in vitro to detect biochemical or morphologic changes in DNA structure or composition following chemical exposure. Such tests include mutagenicity, DNA strand breaks, chromosomal aberrations, and ability of a chemical to react with the DNA (adduct formation) and assays of other phenomena whose meaning is not well understood, but appear correlated with carcinogenicity (e.g., sister chromatid exchange and micronuclei). Some assays appear to be more sensitive to certain carcinogens. Because many known carcinogens test positive in these assays, they are used as screening tools.

Future carcinogenicity evaluations by USEPA will incorporate new developments, summarized in proposed guidelines (U.S. Environmental Protection Agency, 1996c).

Health effects information is compiled by the USEPA as a basis for drinking water regulations in the United States. The Agency uses a number of sources of data and peer review systems, including the National Academy of Sciences (NAS) and the International Agency for Research on Cancer (IARC), and its own external Science Advisory Board. Nevertheless, there are gaps in data that can make conclusions difficult. The assessment framework and associated nomenclature are described in Chapter 1. For inorganic and organic chemicals, respectively, Tables 2.3 and 2.4 present the regulatory maximum contaminant levels (MCLs), the maximum contaminant level goals (MCLGs), the various short-term and long-term health advisories for children and adults, the reference doses (RfDs) for noncarcinogenic end points, and drinking water equivalent levels (DWELs). The RfD is well described by Kimmel (1990). These terms are all described in more detail in Chapter 1.

In Tables 2.3 and 2.4, the health advisory sections also contain the carcinogenicity classification (see Chapter 1) and the concentrations of compounds corresponding to a 10^{-4} incremental lifetime cancer risk as estimated by the USEPA. Health advisories are periodically reviewed by USEPA as new data become available. Additional information on health effects of inorganics may be found in the cited references. Within this chapter, for consistency, USEPA values are used. In the assignment of cancer risks, the use of different models and assumptions can lead to significantly different figures. Based on the assumption of no minimum concentration (threshold) for a cancer response, USEPA has been using the upper bound of the 95 percent statistical confidence interval of the linearized low-dose slope of the cancer response for regulatory purposes. Although a discussion of the details is beyond the scope of this chapter, one consequence is that the estimated concentration for a lifetime risk of 1 in 10,000 (10^{-4}) is 10 times that for a risk of 1 in 100,000. As a matter of regulatory policy, the USEPA currently uses zero as its goal for carcinogens.

INORGANIC CONSTITUENTS

Inorganic constituents may be present in natural waters, in contaminated source waters, or, in some cases, may result from contact of water with piping or plumbing materials. Lead, copper, zinc, and asbestos are constituents that can derive from distribution and plumbing systems.

Inorganics found in drinking water represent a variety of health concerns. Some are known or suspected carcinogens, such as arsenic, lead, and cadmium. A number of inorganics are essential to human nutrition at low doses, yet demonstrate adverse health effects at higher doses. These include chromium, copper, manganese, molybdenum, nickel, selenium, zinc, and sodium, and are reviewed by the National Academy of Sciences Safe Drinking Water Committee (1980b) and National Research Council (1989). Two inorganics, sodium and barium, have been associated with high blood pressure. Numerous reports have also shown an inverse relationship between water hardness and hypertensive heart disease, but this is under continuing investigation.

Health aspects of inorganic constituents of interest in drinking water are summarized in this section. Drinking water regulations and health advisories are listed in Table 2.3. Inorganic disinfectants and disinfection by-products are discussed in a separate section of the chapter.

Aluminum

Aluminum occurs naturally in nearly all foods, the average dietary intake being about 20 mg/day. Aluminum salts are widely used in antiperspirants, soaps, cosmetics, and food additives (Reiber and Kukull, 1996). Aluminum is common in both raw and treated drinking waters, especially those treated with alum. It is estimated that drinking water typically represents only a small fraction of total aluminum intake (Reiber and Kukull, 1996).

Aluminum shows low acute toxicity, but administered to certain laboratory animals is a neurotoxicant (Ganrot, 1986). Chronic high-level exposure data are limited, but indicate that aluminum affects phosphorus absorption, resulting in weakness, bone pain, and anorexia. Carcinogenicity, mutagenicity, and teratogenicity tests have all been negative. Associations between aluminum and two neurological disorders, Alzheimer's disease and dementia associated with kidney dialysis, have been studied. Current evidence suggests that Alzheimer's disease is not related to aluminum intake from drinking water (Reiber and Kukull, 1996), but other sources of aluminum appeared to be associated with Alzheimer's disease (Graves et al., 1990). Dialysis dementia has been reasonably documented to be caused by aluminum (Ganrot, 1986; Shovlin et al., 1993). Most kidney dialysis machines now use specially prepared water.

Aluminum was included on the original list of 83 contaminants to be regulated under the 1986 SDWA amendments. USEPA removed aluminum from the list because it was concluded that no evidence existed at that time that aluminum ingested in drinking water poses a health threat (U.S. Environmental Protection Agency, 1988b). USEPA has a secondary maximum contaminant level (SMCL) of 50 to 200 μg/L to ensure removal of coagulated material before treated water enters the distribution system.

Arsenic

Arsenic concentrations in U.S. drinking waters are typically low. However, an estimated 5,000 community systems (out of 70,000) using groundwater and 370 systems (out of 6,000) using surface water were above 5 μg/L (Reid, 1994). These were primarily in the western states. Dissolution of arsenic-containing rocks and the smelting of nonferrous metal ores, especially copper, account for most of the arsenic in water supplies. Until the 1950s, arsenic was also a major agricultural insecticide.

Arsenic may be a trace dietary requirement and is present in many foods such as meat, fish, poultry, grain, and cereals. Market-basket surveys suggest that the daily adult intake of arsenic is about 50 μg, with about half coming from fish and shellfish (Pontius, Brown, and Chen, 1994). In fish, fruit, and vegetables, it is present in organic arsenical forms, which are less toxic than inorganic arsenic. However, arsenic is not currently considered essential (National Research Council, 1989). Extrapolating from animal studies, Uthus (1994) calculated a safe daily intake of between 12 and 40 μg.

In excessive amounts, arsenic causes acute gastrointestinal damage and cardiac damage. Chronic doses can cause Blackfoot disease, a peripheral vascular disorder affecting the skin, resulting in the discoloration, cracking, and ulceration. Changes in peripheral nerve conduction have also been observed.

Epidemiological studies in Chile, Argentina, Japan, and Taiwan have linked arsenic in drinking water with skin, bladder, and lung cancer (reviewed by Smith et al., 1992; Cantor, 1997). Some studies have also found increased kidney and liver cancer. Ingestion of arsenical medicines and other arsenic exposures have also been associated with several internal cancers, but several small studies of communities in the United States with high arsenic levels have failed to demonstrate any health effects (Pontius, Brown, and Chen, 1994). Micronuclei in bladder cells are increased among those chronically ingesting arsenic in drinking water (Moore et al., 1997). Inorganic arsenate and arsenite forms have been shown to be mutagenic or genotoxic in several bacterial and mammalian cell test systems and have shown teratogenic potential in several mammalian species, but cancers have not been induced in laboratory animals (Agency for Toxic Substances and Disease Registry, 1992a).

USEPA has classified arsenic as a human carcinogen, based primarily on skin cancer (U.S. Environmental Protection Agency, 1985). The ability of arsenic to cause internal cancers is still controversial. Under the NIPDWR regulations, an MCL of 50 μg/L had been set, but it is under review. Currently, USEPA's Risk Assessment Council estimates that an RfD (for noncarcinogenic skin problems) ranges from 0.1 to 0.8 μg/kg·day^{-1}, which translates into an MCLG of 0 to 23 μg/L (Pontius, Brown, and Chen, 1994). Based on a 1-in-10,000 risk of skin cancer, USEPA estimated that 2 μg/L might be an acceptable limit for arsenic in drinking water (Pontius, Brown, and Chen, 1994).

Asbestos

Asbestos is the name for a group of naturally occurring, hydrated silicate minerals with fibrous appearance. Included in this group are the minerals chrysotile, crocido-

TABLE 2.3 USEPA Drinking Water Regulations and Health Advisories for Inorganics

	Regulations				Health advisories								
					10-kg child			70-kg adult					
Chemical	Reg[a] status	MCLG[b] (mg/L)	MCL[b] (mg/L)	HA[a] status	1-day (mg/L)	10-day (mg/L)	Long-term (mg/L)	Long-term (mg/L)	RfD[b] (mg/kg·day^{-1})	DWEL[b] (mg/L)	Lifetime (mg/L)	10^{-4} cancer risk (mg/L)	USEPA cancer group
Aluminum	L	—	—	D	—	—	—	—	—	—	—	—	—
Ammonia	—	—	—	D	—	—	—	—	—	—	30	—	D
Antimony	F	0.006	0.006	F	0.01	0.01	0.01	0.015	0.0004	0.01	0.003	—	D
Arsenic	*[c]	—	0.05	D	—	—	—	—	—	—	—	0.002	A
Asbestos	F	7 MFL[d]	7 MFL	—	—	—	—	—	—	—	—	700 MFL	A
Barium	F	2	2	F	—	—	—	—	0.07	2	2	—	D
Beryllium	F	0.004	0.004	D	30	30	4	20	0.005	0.2	—	0.0008	B2
Boron	L	—	—	D	4	0.9	0.9	3	0.09	3	0.6	*[c]	D
Bromate	F	0	0.01	—	—	—	—	—	—	—	—	—	—
Cadmium	F	0.005	0.005	F	0.04	0.04	0.005	0.02	0.0005	0.02	0.005	—	D
Chloramine	F	4[e,f]	4[e,f]	D	1	1	1	1	0.1	3.3	4	—	D
Chlorate	L	—	—	D	—	—	—	—	—	—	—	—	—
Chlorine	F	4[e]	4[e]	D	—	—	—	—	0.1	—	—	—	D
Chlorine dioxide	F	0.8[e,g]	0.8[e]	D	—	—	—	—	0.01	0.35	0.3	—	D
Chlorite	F	0.8	1	D	—	—	—	—	0.003	0.1	0.08	—	D
Chromium (total)	F	0.1	0.1	F	1	1	0.2	0.8	0.005	0.2	0.1	—	D
Copper (at tap)	F	1.3	TT[g]	—	—	—	—	—	—	—	—	—	D
Cyanide	F	0.2	0.2	F	0.2	0.2	0.2	0.8	0.022	0.8	0.2	—	D
Fluoride (natural)	F	4	4	—	—	—	—	—	0.12	—	—	—	—
Lead (at tap)	F	0	TT[g]	—	—	—	—	—	—	—	—	—	B2
Manganese	L	—	—	D	—	—	—	—	0.14 in food	—	—	—	—
Mercury (inorganic)	F	0.002	0.002	F	—	—	—	0.002	0.0003	0.01	0.002	—	D

Molybdenum	L	—	—	D	0.02	0.02	0.01	0.05	0.005	0.2	0.04	—	D
Nickel	*[c]	0.1	0.1	F	1	1	0.5	1.7	0.02	0.6	0.1	—	D
Nitrate (as N)	F	10	10	F	—	10 *[c]	—	—	1.6	—	—	—	*[c]
Nitrite (as N)	F	1	1	F	—	1 *[c]	—	—	0.16 *[c]	—	—	—	*[c]
Nitrate + Nitrite (as N)	F	10	10	F	—	—	—	—	—	—	—	—	*[c]
Selenium	F	0.05	0.05	—	—	—	—	—	0.005	—	—	—	—
Silver	—	—	—	D	0.2	0.2	0.2	0.2	0.005	0.2	0.1	—	D
Sodium	—	—	—	D	—	—	—	—	—	20	—	—	—
Strontium	L	—	—	D	25	25	25	90	0.6	90	17	—	D
Sulfate	*[c]	500	500	D	—	—	—	—	—	—	—	—	—
Thallium	F	0.005	0.002	F	0.007	0.007	0.007	0.02	0.00007	0.0023	0.0005	—	—
Vanadium	T	—	—	D	—	—	—	—	—	—	—	—	D
White Phosphorous	—	—	—	F	—	—	—	—	0.00002	0.0005	0.0001	—	D
Zinc	L	—	—	D	6	—	3	10	0.3	10	2	—	D

[a] The codes for the regulatory and health advisory status are: F, final; D, draft; L, listed for regulation; P, proposed; T, tentative (not yet proposed).

[b] Abbreviations are as follows (see Chapter 1 for definitions): MCLG, maximum contaminant level goal; MCL, maximum contaminant level; RfD, reference dose; DWEL, drinking water equivalent level.

[c] Under review.

[d] Seven million asbestos fibers longer than 10 microns.

[e] Maximum residual disinfectant level goal (MRDLG) and maximum residual disinfectant level (MRDL) as discussed in text.

[f] Chloramine measured as chlorine.

[g] Treatment technique based on action level of 1.3 mg/L for copper and 0.015 mg/L for lead (see text).

Information based on:

U.S. Environmental Protection Agency. Drinking Water Regulations and Health Advisories. EPA 822-R-96-001. Washington, DC: USEPA, Office of Water, 1996.

U.S. Environmental Protection Agency. *Integrated Risk Information System, April 1998.* Cincinnati, OH: Office of Research and Development, Office of Health and Environmental Assessment, 1998.

U.S. Environmental Protection Agency. "National Primary Drinking Water Regulations: Disinfectants and Disinfection Byproducts; Final Rule." *Federal Register,* 63, 1998h: 69389–69476.

lite, anthophyllite, and some of the tremolite-actinolite and cummingtonite-grunerite series. All except chrysotile fibers are known as amphibole. Most commercially mined asbestos is chrysotile. Asbestos occurs in water exposed to natural deposits of these minerals, asbestos mining discharges, and asbestos-cement pipe (U.S. Environmental Protection Agency, 1985a).

The physical dimensions of asbestos fibers rather than the type are more important in health effects, with the shorter, thinner fibers more highly associated with cancers by inhalation. Human occupational and laboratory animal inhalation exposures are associated with the cancer, mesothelioma, found in lung, pleura, and peritoneum (U.S. Environmental Protection Agency, 1985a). An NTP study also observed gastrointestinal cancers in rats dietarily exposed to intermediate range fibers (65 percent of the fibers larger than 10 μm, 14 percent larger than 100 μm) for their lifetime (U.S. Environmental Protection Agency, 1985a). Epidemiologic studies of asbestos in drinking water have had inconsistent results, but there are suggestions of elevated risk for gastric, kidney, and pancreatic cancers (Cantor, 1997).

The USEPA based its 1/1,000,000 cancer risk estimate and the MCLG and MCL on 7×10^6 fibers/L > 10 μm observed in the NTP rat study (U.S. Environmental Protection Agency, 1991e).

Barium

Barium occurs naturally in trace amounts in most surface and groundwaters from their exposure to barium-containing minerals. Industrial release of barium occurs from oil and gas drilling muds, coal burning, and auto paints (U.S. Environmental Protection Agency, 1985a). It is also widely used in brick, tile, and ceramics manufacture. The insoluble and unabsorbed salt, barium sulfate, is used clinically as a radiopaque dye for X-ray diagnosis of the gastrointestinal tract.

Chronic exposure may contribute to hypertension. Rats ingesting 0.5 $mg/kg \cdot day^{-1}$ barium in drinking water (10 mg/L) for 16 months or 6 $mg/kg \cdot day^{-1}$ (100 mg/L) for 4 months demonstrated hypertensive effects; however, another 4-month study of rats exposed to 15 $mg/kg \cdot day^{-1}$ in drinking water found no effect. Human epidemiological studies with community drinking water containing from 2 to 10 mg/L barium did not provide definitive results (U.S. Environmental Protection Agency, 1989b; Agency for Toxic Substances and Disease Registry, 1992b).

The MCLG and MCL for barium are 5 mg/L, based on hypertension among humans (U.S. Environmental Protection Agency, 1991f).

Cadmium

Cadmium enters the environment from a variety of industrial applications, including mining and smelting operations, electroplating, and battery, pigment, and plasticizer production. Cadmium occurs as an impurity in zinc and may also enter consumers' tap water by galvanized pipe corrosion. Cadmium is also in food, with 27 μg/day in the average diet.

Cadmium acts as an emetic and can cause kidney dysfunction, hypertension, anemia, and altered liver function (U.S. Environmental Protection Agency, 1985a). It can build up in the kidney with time. Chronic occupational exposure has resulted in renal dysfunction and neuropsychological impairments. The mechanisms remain uncertain, but cadmium competes with calcium inside cells and across cell membranes. Cadmium has been shown to induce testicular and prostate tumors in laboratory animals injected subcutaneously (Agency for Toxic Substances and Disease Registry, 1993b). Lung tumor incidence is increased in people exposed to cadmium by inhalation.

As yet, USEPA considers cadmium unclassifiable as a human carcinogen regarding oral exposure because the observed human carcinogenicity occurs via inhalation. It is regulated based on its renal toxicity in humans, and, using an uncertainty factor of 10, an MCLG and an MCL of 5 μg/L have been adopted (U.S. Environmental Protection Agency, 1991e).

Chromium

Chromium occurs in drinking water in its +3 and +6 valence states, with +3 being more common. The valence is affected by the level of disinfection (increased proportion in the +6 valence), pH, dissolved oxygen, and presence of reducible organics. Primary sources in water are mining, wastes from electroplating operations, and garbage and fossil fuel combustion (U.S. Environmental Protection Agency, 1985a).

Chromium III is nutritionally essential, has low toxicity, and is poorly absorbed. Deficiency results in glucose intolerance, inability to use glucose, and other metabolic disorders. The NAS estimates a safe and adequate intake of 50 to 200 μg/day, which is the approximate range of daily dietary ingestion (National Research Council, 1989).

Chromium VI is toxic, producing liver and kidney damage, internal hemorrhage, and respiratory disorders, as well as causing cancer in humans and animals through inhalation exposure, but it has not been shown to be carcinogenic through ingestion exposure (U.S. Environmental Protection Agency, 1985a; U.S. Environmental Protection Agency, 1991e). Notably, chromium VI is reduced to III by reaction with salivary and gastric juices. For drinking water, USEPA considers chromium an unclassifiable human carcinogen.

Copper

Copper is commonly found in drinking water (U.S. Environmental Protection Agency, 1985a). Low levels (generally below 20 μg/L) can derive from rock weathering, and some industrial contamination also occurs, but the principal sources in water supplies are corrosion of brass and copper pipes and fixtures and the addition of copper salts during water treatment for algal control.

Copper is a nutritional requirement. Lack of sufficient copper leads to anemia, skeletal defects, nervous system degeneration, and reproductive abnormalities. The safe and adequate copper intake is 1.5 to 3 mg/day (National Research Council, 1989).

Copper doses in excess of nutritional requirements are excreted; however, at high doses, copper can cause acute effects, such as gastrointestinal (GI) disturbances, damage to the liver and renal systems, and anemia. A dose of 5.3 mg/day was the lowest at which GI tract irritation was seen. Exposure of mice via subcutaneous injection yielded tumors; however, oral exposure in several studies did not. Mutagenicity tests have been negative.

Copper is regulated under the special provisions of the Lead and Copper Rule (U.S. Environmental Protection Agency, 1991b). Under the Rule, if more than 10 percent of the residential tap samples have copper over the action level of 1.3 mg/L, then water purveyors must minimize corrosion. A secondary standard of 1.0 mg/L applies to water leaving the plant.

Fluoride

Fluoride occurs naturally in most soils and in many water supplies. For more than 40 years, fluoride has been added to supplies lacking sufficient natural quantities, for the purpose of reducing dental caries. However, acute overdosing in the 20 to 200 mg/L range, due to equipment failure in fluoridating systems, can result in nausea, diarrhea, abdominal pain, headache, and dizziness (Centers for Disease Control Division of Oral Health, 1992). Long-term health effects of fluoride are discussed in Chapter 15, "Water Fluoridation," but include dental and skeletal fluorosis. The MCL for fluoridating systems is 2.0 mg/L (U.S. Environmental Protection Agency, 1986), but the safe and effective target range established by the Public Health Service Ad Hoc Subcommittee on Fluoride (1991) is between 0.7 and 1.2 mg/L, depending on temperature and pH. The MCL and MCLG for naturally occurring fluoride is 4 mg/L.

Hardness

Hardness is generally defined as the sum of the polyvalent cations present in water and expressed as an equivalent quantity of calcium carbonate ($CaCO_3$). The most common such cations are calcium and magnesium. Although no distinctly defined levels exist for what constitutes a hard or soft water supply, water with less than 75 mg/L $CaCO_3$ is considered to be soft and above 150 mg/L $CaCO_3$ as hard.

An inverse relationship has been postulated between the incidence of cardiovascular disease and the amount of hardness in the water, or, conversely, a positive correlation with the degree of softness. Hypotheses outlined below suggest a protective effect from either the major or minor constituents of hard water, or, conversely, a harmful effect from elements more commonly found in soft water (National Academy of Sciences Safe Drinking Water Committee, 1980b).

Many investigators attribute a cardiovascular protective effect to the presence of calcium and magnesium (reviewed by Marx and Neutra, 1997; McCarron, 1998b). A moderate increase in calcium in the diet has been observed to lower levels of circulating organ cholesterol; this is speculated as a possible factor in relating water hardness and cardiovascular disease. Magnesium is theorized to protect against lipid deposits in arteries, to reduce cardiac irritability and damage, and may also have some anticoagulant properties that could protect against cardiovascular diseases by inhibiting blood clot formation.

A limited number of studies have been carried out that suggest that minor constituents often associated with hard water may exert a beneficial effect on the cardiovascular system. Candidate trace elements include vanadium, lithium, manganese, and chromium. On the other hand, other investigators suggest that certain trace metals found in higher concentrations in soft water, such as cadmium, lead, copper, and zinc, may be involved in the induction of cardiovascular disease.

For each of these hypotheses, whether drinking water provides enough of these elements to have any significant impact on the pathogenesis of cardiovascular disease is uncertain. Hard water generally supplies less than 10 percent of the total dietary intake for calcium and magnesium, for example. Water provides even smaller proportions of the total intake for the various suspect trace metals. Given the level of uncertainty in this area, USEPA currently has no national policy with respect to the hardness or softness of public water supplies. However, USEPA strongly supports corrosion-control measures (some of which add hardness) to reduce exposure to lead.

Iron

Commonly found in wells in many parts of North America, rust from iron or steel pipes can also increase the dissolved concentration in finished water, as well as total iron intake including particulates.

Because there is a nutritional requirement of 10 to 12 mg per day of iron for healthy adult men and 10 to 15 mg per day for women (National Research Council, 1989), it is unlikely that serious health problems will arise even from the maximum concentrations found in the surveys. However, in individuals genetically susceptible to hemochromatosis, too much iron can be accumulated in the body, resulting in liver, pancreatic, and heart dysfunction and failure after long-term high exposures (Motulsky, 1988). About 1 person in 200 is potentially at risk, though the actual incidence is much lower (Walker et al., 1998).

There is a secondary standard for iron of 0.3 mg/L, based on discoloration of laundry and a metallic taste that becomes noticeable in the 0.1 to 1.0 mg/L range (see the section entitled "Aesthetic Quality," later in this chapter).

Lead

Lead occurs in drinking water primarily from corrosion of lead pipe and solders and faucets constructed with leaded brass, especially in areas of soft or acidic water.

Health effects of lead are generally correlated with blood test levels. Infants and young children absorb ingested lead more readily than do older children and young adults (National Academy of Sciences Safe Drinking Water Committee, 1982). Maintaining high levels of the divalent ions, iron and calcium, in the diet reduces uptake. Lead exposure across a broad range of blood lead levels is associated with a continuum of pathophysiological effects, including interference with heme synthesis necessary for formation of red blood cells, anemia, kidney damage, impaired reproductive function, interference with vitamin D metabolism, impaired cognitive performance, delayed neurological and physical development, and elevations in blood pressure (U.S. Environmental Protection Agency, 1988b). USEPA has classified lead as a probable human carcinogen because some lead compounds cause renal tumors in rats (U.S. Environmental Protection Agency, 1985a).

In 1991, the CDC reduced the action level from 25 to 10 μg/dL in blood, based on a reassessment of neurodevelopmental toxicity (Centers for Disease Control, 1991). At this lower level, the lead from plumbing becomes more significant as a potential source.

Under the interim primary drinking water regulations, the MCL for lead was 50 μg/L. USEPA currently has an action level instead of an MCL (U.S. Environmental Protection Agency, 1991b). It requires that water purveyors sample residential taps for lead. If more than 10 percent of the residential tap samples are above the action level of 15 μg/L, there are treatment requirements. As of 1995, about 69 million people were served by 4167 community water systems exceeding the lead action level (U.S. Environmental Protection Agency, 1996b). The MCLG for lead is set at zero based on subtle effects at low blood lead levels, the overall USEPA goal of reducing total lead exposures, and probable carcinogenicity at very high doses.

Manganese

Manganese is ubiquitous in the environment and is often naturally present in significant amounts in groundwater. Man-made sources include discarded batteries, steel alloy production, and agricultural products. Manganese is an essential nutrient, fulfilling a catalytic role in various cellular enzymes. The National Research Council (1989) provisionally recommended from 2 to 5 mg/day as safe and adequate for adults, compared with a recommended 0.3 to 0.6 mg/day for infants. It has generally been regarded as nontoxic and naturally occurring. Manganese is removed from well water primarily due to aesthetic reasons. However, it has long been known that occupational inhalation exposure can lead to manganism, an irreversible, slow-onset neurotoxic disease.

Recently, a work group in USEPA developed an RfD based on a Greek study of manganism due to drinking water exposure (Valezquez and Du, 1994). A lowest-observed-adverse-effect level (LOAEL) of 2 mg/L and a no-observed-adverse-effect level (NOAEL) of 0.17 mg/L were identified. If validated, a daily allowable level of 0.2 mg/L would be lower than the NRC recommendation for overall diet.

Mercury

Mercury occurs in water primarily as an inorganic salt, and as organic (methyl) mercury in sediments and fish (U.S. Environmental Protection Agency, 1985a). Sources of mercury include the burning of fossil fuels, incineration of mercury-containing products, past use of mercury-containing pesticides, and leaching of organic mercury from antifungal outdoor paints, as well as natural origins.

Inorganic mercury is poorly absorbed in the adult GI tract, does not readily penetrate cells, and, therefore, is not as toxic as methyl mercury. (Calomel, mercurous chloride was used in the past as a laxative.) However, the absorption of inorganic mercury can be much higher in infants and young children. The most sensitive target organ of inorganic mercury in adult laboratory animals is the induction of an autoimmune disease of the kidney (Agency for Toxic Substances and Disease Registry, 1994d).

Organic forms, such as methyl mercury, are readily absorbed in the GI tract and easily enter the central nervous system (CNS), causing death and/or mental and motor dysfunctions. (Metallic mercury is also neurotoxic, but is not found at significant levels in water.) Organic mercury also easily crosses the placental barrier of the fetus. Larger saltwater fish and, in some locations, larger freshwater fish significantly bioaccumulate organic mercury. Fish contaminated with methyl mercury caused the well-known mercury poisonings near Japan's Minamata Bay, characterized by a large number of deaths, CNS disorders, and mental retardation. Since then, most epidemiological studies on children of women with a high consumption of saltwater fish have shown similar, though smaller, effects.

A recent study by the NTP found that inorganic mercury caused increased stomach tumors in male rats with equivocal evidence in female rats and male mice. In vitro inorganic and organic mercury give mixed results in mutagenicity and genotoxicity assays (Agency for Toxic Substances and Disease Registry, 1994d). Since these findings, the carcinogenicity of mercury has not been reevaluated by USEPA, but their significance has been questioned by the NTP (Schoeny, 1996).

The MCLG and MCL for inorganic mercury are 2 μg/L, based on induction of autoimmune kidney disease (U.S. Environmental Protection Agency, 1991e).

Molybdenum

Molybdenum is an essential trace element in the diet. The estimated safe average daily intake for molybdenum, 75 to 250 μg/day, matches the typical intake at 150 to

500 mg/day (National Research Council, 1989). Chronic exposure can result in weight loss, bone abnormalities, and male infertility. USEPA has not classified the carcinogenicity of molybdenum and no MCLG has been proposed (U.S. Environmental Protection Agency, 1998f).

Nickel

Nickel is common in drinking water. Most ingested nickel is excreted; however, some absorption from the GI tract does occur. Trace amounts are required by the body and about 200 to 500 μg/day (and perhaps as much as 2 mg/day) are provided by the average diet.

There is little useful data on chronic effects of overexposure except that nickel compounds are carcinogenic via inhalation and injection in laboratory animals. In humans, the incidence of respiratory tract cancers in nickel refinery workers is significantly higher (Agency for Toxic Substances and Disease Registry, 1993d). Nickel has not, however, been shown to be carcinogenic via oral exposure. Several studies suggest that it is not carcinogenic at 5 mg/L in drinking water given to rats and mice. Nickel chloride tested negatively in bacterial mutagenicity screening; however, both nickel chloride and nickel sulfate tested positive in mutagenicity and chromosomal aberration tests in mammalian cells (U.S. Environmental Protection Agency, 1985a; U.S. Environmental Protection Agency, 1990b).

For ingestion, the USEPA considers nickel unclassifiable regarding human carcinogenicity. The MCLG and MCL of 100 mg/L (U.S. Environmental Protection Agency, 1992c) are under review.

Nitrate and Nitrite

Nitrate is one of the major anions in natural waters, but concentrations can be greatly elevated due to leaching of nitrogen from farm fertilizer or from feed lots or from septic tanks. The mean concentration of nitrate nitrogen (NO_3-N, nitrate measured as nitrogen in testing) in a typical surface water supply would be around 0.2 to 2 mg/L; however, individual wells can have significantly higher concentrations. Adult daily dietary nitrate intake is approximately 20 mg, mostly from vegetables, like lettuce, celery, beets, and spinach (National Academy of Sciences Committee on Nitrite and Alternative Curing Agents in Food, 1981).

Nitrite does not typically occur in natural waters at significant levels, except under reducing conditions. It can also occur if water with sufficient ammonia is treated with permanganate. Sodium nitrite is widely used for cured meats, pickling, and beer. Rarely, buildings have been contaminated by faulty cross connections or procedures during boiler cleaning with nitrous acid.

Nitrite, or nitrate converted to nitrite in the body, causes two chemical reactions that can cause adverse health effects: induction of methemoglobinemia, especially in infants under one year of age, and the potential formation of carcinogenic nitrosamides and nitrosamines (National Academy of Sciences Safe Drinking Water Committee, 1977; U.S. Environmental Protection Agency, 1989b).

Methemoglobin, normally present at 1 to 3 percent in the blood, is the oxidized form of hemoglobin and cannot act as an oxygen carrier in the blood. Certain substances, such as nitrite ion, act as oxidizers (National Academy of Sciences Safe Drinking Water Committee, 1977). Nitrite is formed by reaction of nitrate with saliva, but in infants under one year of age the relatively alkaline conditions in the stomach allow bacteria there to form nitrite. Up to 100 percent of nitrate is reduced to nitrite in infants, compared with 10 percent in adults and children over one year of age. Furthermore, infants do not have the same capability as adults to reconvert methemoglobin back to hemoglobin. When the concentration of methemoglobin reaches 5 to 10 percent, the symptoms can include lethargy, shortness of breath, and a bluish skin color. Anoxia and death can occur at high concentrations of nitrites or nitrates.

Carcinogenic nitrosamines and nitrosamides are formed when nitrate or nitrite are administered with nitrosatable amines, such as the amino acids in proteins (International Agency for Research on Cancer, 1978). However, epidemiological studies, primarily on gastric cancer, have not yielded consistent results (Cantor, 1997). The carcinogenicity of nitrate and nitrite is currently under review.

The data on the role of nitrates in developmental effects, such as birth defects, are regarded as being inconclusive (U.S. Environmental Protection Agency, 1991e). Several epidemiologic case-control studies found an increased risk of developmental brain defects (Sever, 1995), but more studies are needed. If there is an association, maternal methemoglobinemia might be the critical risk factor.

The MCLGs and MCLs are 10 mg/L for nitrate measured as nitrogen (or 45 mg/L nitrate) and 1 mg/L for nitrite measured as nitrogen (U.S. Environmental Protection Agency, 1991e). In addition, the MCL for total nitrate-N and nitrite-N is 10 mg/L.

Selenium

Selenium is an essential dietary element, with most intake coming from food. The National Research Council of the National Academy of Sciences (1989) has recommended that the daily diet include between 55 and 75 μg selenium, the higher part of the range for adult males and pregnant or nursing females. Selenium is a key component of glutathione peroxidase, a vital thyroid enzyme (Corvilain et al., 1993). In humans, dermatitis, hair loss, abnormal nail formation and loss, diarrhea, liver degeneration, fatigue, peripheral nervous system abnormalities, and a garlic odor have been observed among people exposed to chronic, moderately high dietary selenium intakes (reviewed by Poirier, 1994 and by Patterson and Levander, 1997). The molecular form of dietary selenium was not determined. An RfD of 0.005 $mg/kg \cdot day^{-1}$ was calculated for total selenium intake, using an uncertainty factor of 3 to account for sensitive individuals (Poirer, 1994). Selenium reacts in vivo with other elements, protecting against heavy metal toxicity. Naturally occurring selenium compounds have not been shown to be carcinogenic in animals and selenium may inhibit tumor formation (U.S. Environmental Protection Agency, 1985a; Patterson and Levander, 1997), though an antitumor preventive function is controversial (Clark and Alberts, 1995). Selenium has not been found to be teratogenic in mammals (Poirer, 1994).

An MCLG and a new MCL of 50 μg/L (U.S. Environmental Protection Agency, 1991e) have superseded the MCL of 10 μg/L under the NIPDWR.

Sodium

Sodium is a naturally occurring constituent of drinking water. A survey of 2100 finished waters conducted between 1963 and 1966 by the U.S. Public Health Service found concentrations ranging from 0.4 to 1900 mg/L, with 42 percent having sodium more than 20 mg/L and 4 percent having more than 250 mg/L (White et al., 1967). This level can be increased at the tap by water softeners, which can add approximately 1 mg sodium for every 2 mg hardness removed (see Chapter 9).

There is strong evidence that there is an at-risk population of persons predisposed to high blood pressure (hypertension) from dietary sodium (reviewed by Ely, 1997; Luft and Weinberger, 1997; and Kotchen and McCarron, 1998). Others appear to be relatively unaffected (reviewed by McCarron, 1998a; Taubes, 1998). Coronary heart disease and stroke (Stamler, 1991) and certain other diseases are aggravated by high blood pressure. The long-term significance of sodium on hypertension in normal children is controversial (reviewed by Falkner and Michel, 1997, and by Simons-Morton and Obarzanek, 1997).

Food is the major source of sodium. Of a suggested maximum daily intake of 2400 mg (Krauss et al., 1996), drinking water, at a typical concentration of 20 mg/L, contributes less than 2 percent, assuming consumption of 2 L/day. Average adult intake is 10,000 mg/day (National Academy of Sciences Food and Nutrition Board, 1980). For persons requiring restrictions on salt intakes, sodium levels may be limited to as low as 500 mg/day (American Heart Association, 1969). The American Heart Association recommended a drinking water concentration of 20 mg/L, because dietary sodium restriction to less than 500 mg/day is difficult to achieve and maintain (National Academy of Sciences Safe Drinking Water Committee, 1977). This level has been adopted as guidance by USEPA, but an MCLG has not been proposed (U.S. Environmental Protection Agency, 1985a).

Sulfate

Sulfate is a naturally occurring anion. High concentrations of sulfate in drinking water may cause transitory diarrhea (U.S. Environmental Protection Agency, 1990b). A study of adults found that most experienced a laxative effect above 1000 mg/L, whereas medical case reports indicate that bottle-fed infants develop diarrhea at sulfate levels above 600 mg/L. Acute diarrhea can cause dehydration, particularly in infants and young children who may already have a microbial diarrheal condition. Adults living in areas having high sulfate concentrations in their drinking water easily adjust, with no ill effects.

USEPA had suggested a guidance level of 400 mg/L sulfate to protect infants, but it is under review (U.S. Environmental Protection Agency, 1992c). The current SMCL for sulfate is 250 mg/L based upon taste. The U.S. Army standard for soldiers drinking 5 L/day is 300 mg/L.

Zinc

Zinc commonly occurs in source waters and may leach into finished waters through corrosion of galvanized metal pipes.

The adult requirement for zinc is 15 mg per day for males and 12 mg per day for females (National Research Council, 1989). Drinking water typically contributes about 1 to 10 percent of this requirement. The equivalent of 40 mg/L zinc over a long period would cause muscular weakness and pain, irritability, and nausea (Cantilli, Abernathy, Donohue, 1994; Greger, 1994). Excess zinc also interferes with the absorption of other trace metals, such as copper and iron.

USEPA established an SMCL of 5 mg/L for zinc based upon taste (U.S. Environmental Protection Agency, 1979). Zinc was one of the seven constituents removed from the list of 83 contaminants to be regulated under the 1986 SDWA amendments because the available data indicate that zinc in drinking water does not pose a public health risk (U.S. Environmental Protection Agency, 1988b).

ORGANIC CONSTITUENTS

Organic compounds in water derive from three major sources: (1) the breakdown of naturally occurring organic materials, (2) domestic and commercial activities, and (3) reactions that occur during water treatment and transmission. Treatment additives may add small amounts of organic contaminants, such as monomers of polymer products. The first source predominates and is composed of humic materials from plants and algae, microorganisms and their metabolites, and high-molecular-weight aliphatic and aromatic hydrocarbons. These organics are typically benign, although some are nuisance constituents such as odoriferous metabolites. A few of the high-molecular-weight aliphatic and aromatic hydrocarbons may have adverse health effects. In addition, humics serve as precursors (reactants) in the formation of trihalomethanes (THMs) and other organohalogen oxidation products during disinfection.

Domestic and commercial activities contribute synthetic organic chemicals (SOCs) to wastewater discharges, agricultural runoff, urban runoff, and leachate from contaminated soils. Most of the organic contaminants identified in water supplies as having adverse health concerns are part of this group. They include pesticides (such as atrazine and aldicarb), solvents and metal degreasers (such as trichlorobenzene, tetrachloroethylene, trichloroethylene, and trichloroethane), and a family of compounds formerly in wide use, the polychlorinated biphenyls.

Organic contaminants formed during water disinfection include by-products such as trihalomethanes (e.g., chloroform) and haloacetic acids (e.g., di- and trichloroacetic acids). Other compounds such as acrylamide or epichlorohydrin are components of coagulants (e.g., polyacrylamide) that can leach out during treatment. During finished water transmission, undesirable components of pipes, coating, linings, and joint adhesives, such as polynuclear aromatic hydrocarbons (PAHs), epichlorohydrin, and solvents, have been shown to leach into water.

The term "synthetic organic chemicals" has become a regulatory rather than a chemical description. It has evolved to distinguish a group of mostly volatile organic chemicals (VOCs), regulated first under the 1986 amendments of the federal Safe Drinking Water Act, from "SOCs" regulated under Phase 2 and later regulations. However, some of the SOCs are also VOCs (e.g., ethylbenzene, styrene, toluene, and xylenes, and the fumigant pesticides). The bulk of SOCs are pesticides, but also include the polycyclic aromatic hydrocarbons, the polychlorinated biphenyls, and two water treatment polymers.

The predominance of naturally occurring organics in surface water is illustrated in Figure 2.2, showing the distribution of organic materials in the Mississippi River. The naturally occurring, high-molecular-weight constituents compose approximately 80 percent of the total, whereas anthropogenic organics, generally weighing less than 400 g/mol, comprise about 15 percent (the shaded area). The concentrations of the naturally occurring fraction are typically in the milligram-per-liter range, whereas the concentrations of the man-made organics are generally in the microgram- or nanogram-per-liter range.

Composite measures of organic constituents exist. Total organic carbon (TOC) is used to track the overall organic content of water. This is an important measure for surface water because it correlates with the production of disinfection by-products during chlorination. Total organic halogen (TOX) indicates the presence of halogenated organics. The TOX measure is useful because a large number of organics that are known or potential health hazards are halogenated. Both of these methods are technologically more simple and economically more attractive than measure-

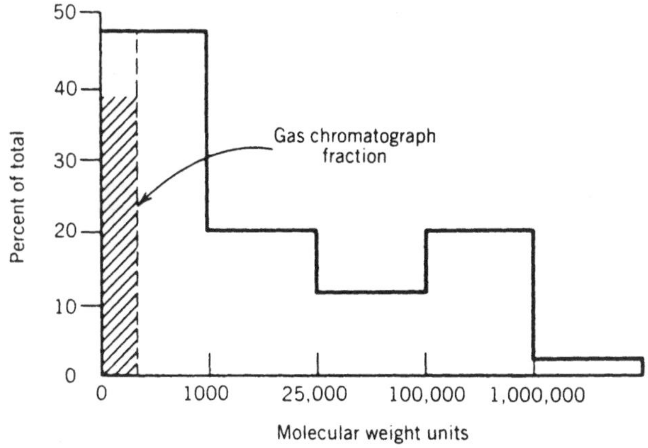

FIGURE 2.2 Organic size distribution, Mississippi River water. (*Source: Water Treatment Principles and Design,* James M. Montgomery Consulting Engineers, Inc., Copyright © 1984 John Wiley & Sons, Inc. Reprinted by permission of John Wiley & Sons, Inc.; after C. Anderson and W.J. Maler, "The Removal of Organic Matter from Water Supplies by Ion Exchange," Report for Office of Water Research and Technology, Washington, D.C. 1977.)

ment of individual compounds. They are useful in general comparisons of water supplies, in identifying pollution sources, and in helping to determine when additional, more specific analyses might be required. The TOC measure has become very common such that values from source and finished supplies throughout the United States can be compared on this basis. Figure 2.3 provides some illustrative values from a variety of natural waters.

Health aspects of organic constituents of interest in drinking water treatment are summarized later, grouped according to chemical characteristics. Drinking water regulations and health advisories for organic chemicals are listed in Table 2.4.

One new area of health concern that has developed during the 1990s is endocrine disruption caused by a wide variety of organic compounds. In particular, disruption of the development and reproductive function by chemicals that interfere with the function of thyroid and steroid hormones has affected animals in the wild and may be affecting humans (Colburn, vom Saal, and Soto, 1993; Porterfield, 1993; Toppari et al., 1996). Certain pesticides, certain dioxin and PCB congeners, and nonylphenol compounds (used in plastics and detergents) are among the chemicals of concern (Soto et al., 1995).

Volatile Organic Chemicals

The term "volatile organic chemical" refers to the characteristic evaporative abilities (or the vapor pressure) of these compounds, which can also apply when they are dissolved in water. Generally, the volatility of a chemical, whether pure or dissolved in water, decreases with increasing ionic polarity of the chemical. Three broad groups of VOCs have been found in drinking water. One group includes compounds found in petroleum products, especially aromatics like benzene, toluene, and xylenes. Typ-

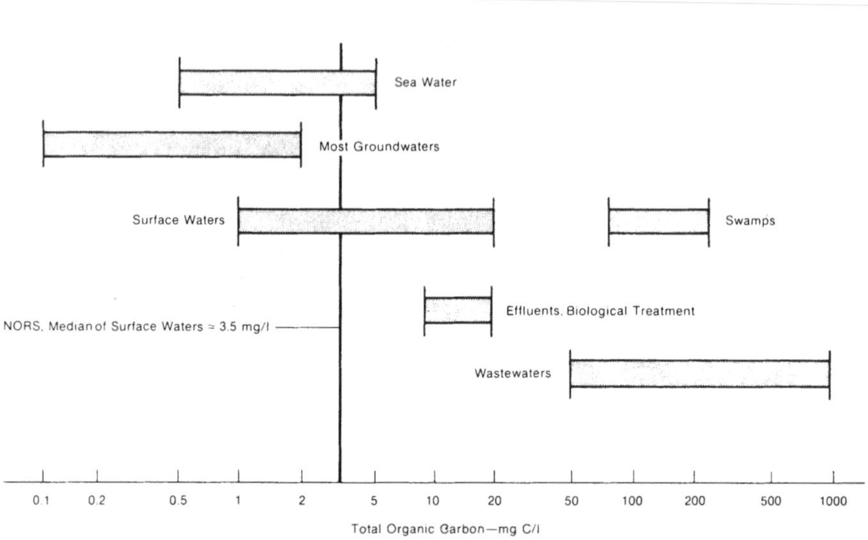

FIGURE 2.3 Ranges of TOC reported for a variety of natural waters. (*Source:* M. C. Kavanaugh, Coagulation for Improved Removal of Trihalomethane Precursors, *Journal AWWA,* vol. 70, no. 11, Nov. 1978, p. 613.)

ical sources are leaking fuel oil and gasoline tanks. Another group is the halogenated VOCs, used as solvents and degreasers. Their former use as septic tank cleaners also accounts for many instances of private well contamination. The third group includes some of the chlorinated organic disinfection by-products, particularly the trihalomethanes. Some of the trihalomethanes have also been found in groundwater due to industrial contamination.

Because of their volatility, the VOCs can be removed from water by air stripping methods. Generally, volatility is directly correlated with the Henry's Law coefficient (see Chapter 5) unless the chemical is soluble in water, such as methyl tert-butyl ether. As a rule of thumb, chemicals having coefficients greater than 1 atmosphere (atm) at or above room temperature can be removed from water by packed tower aeration (U.S. Environmental Protection Agency, 1989c), which can be interpreted as the definition of a VOC. In the same way, however, they easily volatilize from hot water used in the home and can be absorbed by inhalation. The air in bathrooms can represent a major source of exposure because of volatilization from shower and bathwater (McKone, 1987; Jo et al., 1990a,b; Maxwell et al., 1991). In addition, dermal absorption (across the skin) can occur in bathwater because the VOCs are lipophilic. [Because VOCs have low ionic polarity, they can easily cross the lipid (fatty) membrane that is the basic part of the cell surface.] The combination of dermal and inhalation absorption can represent as much or more exposure to VOCs as ingestion. By one estimate, they may account for up to 5 to 10 times more getting into the body than for ingestion (McKone, 1987).

The lipophilic characteristic also enables VOCs to cross into the brain and, at very high concentrations, can cause a reversible anesthetic effect, including dizziness, nausea, and cardiac depression. High exposure to most VOCs over many years causes brain damage and moderate exposure can alter kidney and liver function or cause damage. Vapor exposure at high concentration may produce irritation of the eyes, nose, and throat.

Benzene. Benzene is produced by petroleum refining, coal tar distillation, coal processing, and coal coking. It is also widely used as a chemical intermediate. Gasoline in the United States has contained anywhere from 0.5 to 5 percent benzene by volume, though currently the percentage is lower.

Chronic occupational exposure to benzene has been shown to be toxic to the hematopoietic (blood-forming) system. Effects include loss of various blood elements (anemia, thrombocytopenia, or leukopenia) and leukemia. It is considered a human carcinogen. By extrapolation from the occupational studies, a drinking water concentration of 10 μg/L benzene has been calculated to represent an excess lifetime cancer risk of 1/100,000 (U.S. Environmental Protection Agency, 1985b). Benzene-induced carcinogenesis (increased solid tumors and leukemias) have also been observed in exposed animals (U.S. Environmental Protection Agency, 1985b). Changes in chromosome number and chromosome breakage have been observed in humans and laboratory animals. One teratogenicity test showed positive results, but at doses just below lethal (Agency for Toxic Substances and Disease Registry, 1993a).

Carbon Tetrachloride. Carbon tetrachloride is used primarily in the manufacture of chlorofluoromethane, but has also been used as a grain fumigant, in fire extinguishers, as a solvent with many uses, and in cleaning agents (National Academy of Sciences Safe Drinking Water Committee, 1977). It has been found in community water systems with wells near industrialized areas.

There are few data on long-term health effects at low levels, but the well-known acute toxic effect in rats and humans is liver damage. Mutagenicity and genotoxicity testing using bacteria, mammalian cells, and short-term in vivo rat assays were mostly negative. Carcinogenicity bioassays produced hepatic carcinomas in mice, rats, and hamsters. No teratogenic effects have been demonstrated (Agency for Toxic Substances and Disease Registry, 1994a).

Dichlorobenzenes* (ortho-, meta-, para-, *or* o-, m-, p-*). Dichlorobenzenes are used as solvents, but *o*-dichlorobenzene is primarily used in the production of organic chemicals, pesticides, and dyes, and as a deodorant in industrial waste water treatment, whereas the water-soluble *p*-dichlorobenzene is used in very large amounts as an insecticidal fumigant, lavatory deodorant, and for mildew control. Dichlorobenzenes have been found in the milk of nursing mothers in several states (U.S. Environmental Protection Agency, 1985a).

The primary toxic effects of *o*- and *p*-dichlorobenzenes are on the blood, lung, kidney, and liver. Similar data are not available for the *meta*-isomer, although a few short-term assays suggest similar toxicity and the short-term assessments developed for the *ortho*-isomer are also applied to the *meta*-isomer. Some mutagenic activity has been observed for *o*-dichlorobenzene, but not *p*-dichlorobenzene in bacterial systems. Although there is no indication of carcinogenicity for *o*-dichlorobenzene, an NTP gavage bioassay on *p*-dichlorobenzene demonstrated treatment-related increases in the incidence of renal tubular cell adenocarcinomas in male rats and hepatocellular adenomas in male and female mice. Hepatoblastomas, usually rare, were also found in male mice. A previous long-term inhalation study revealed no treatment-related cancer increase (U.S. Environmental Protection Agency, 1987a,b).

1,2-Dichloroethane (Ethylene Dichloride). 1,2-Dichloroethane is or has been used in the manufacture of vinyl chloride, as a lead scavenger in gasoline, as an insecticidal fumigant, as a constituent of paint varnish and finish removers, as a metal degreaser, in soap and scouring compounds, in wetting agents, and in ore flotation. It is environmentally persistent and is moderately water soluble.

TABLE 2.4 USEPA Drinking Water Regulations and Health Advisories for Organic Contaminants

	Regulations					Health advisories								
						10-kg child			70-kg adult					
Chemicals	Status reg.	NIPDWR (mg/L)	MCLG (mg/L)	MCL (mg/L)	Status HA	1-day (mg/L)	10-day (mg/L)	Longer-term (mg/L)	Longer-term (mg/L)	RfD (mg/kg/day)	DWEL (mg/L)	Lifetime (mg/L)	10^{-4} cancer risk (mg/L)	USEPA cancer group
Acenaphthylene	—	—	—	—	—	—	—	—	—	0.06	—	—	—	—
Acifluorfen	—	—	—	—	F	2	0.2	0.1	0.4	0.013	0.4	—	0.1	B2
Acrylamide	P	—	0	TT[a]	F	1.5–	0.3	0.02	0.07	0.0002	0.007	—	0.001	B2
Acrylonitrile	T	—	0	—	D	—	—	—	—	—	—	—	0.006	B1[b]
Adipate (diethylhexyl)	F	—	0.4	0.4	—	20	20	20	60	0.6	20	0.4	3	C
Alachlor	F	—	0	0.002	F	0.1	0.1	—	—	0.01	0.4	—	0.04	B2
Aldicarb[c]	D	—	0.007	0.007[c]	D	—	—	—	—	0.001	0.035	0.007	—	D
Aldicarb sulfone[c]	D	—	0.007	0.007[c]	D	—	—	—	—	0.001	0.035	0.007	—	D
Aldicarb sulfoxide[c]	D	—	0.007	0.007[c]	D	—	—	—	—	0.001	0.035	0.007	—	D
Aldrin	—	—	—	—	D	0.0003	0.0003	0.0003	0.0003	0.00003	0.001	—	0.0002	B2
Ametryn	—	—	—	—	F	9	9	0.9	3	0.009	0.3	0.06	—	D
Ammonium sulfamate	—	—	—	—	F	20	20	20	80	0.28	8	2	—	D
Anthracene (PAH)[d]	—	—	—	—	—	—	—	—	—	0.3	—	—	—	D
Atrazine	F	—	0.003	0.003	F	0.1	0.1	0.05	0.2	0.035	0.2[b]	0.003[b]	—	C
Baygon	—	—	—	—	F	0.04	0.04	0.04	0.1	0.004	0.1	0.003	—	C
Bentazon	T	—	0.02	—	F	0.3	0.3	0.3	0.9	0.0025	0.09	0.02	—	D
Benz(*a*)anthracene (PAH)	—	—	—	—	—	—	—	—	—	—	—	—	—	B2
Benzene	F	—	0	0.005	F	0.2	0.2	—	—	—	—	—	0.1	A
Benzo(*a*)pyrene (PAH)	L	—	—	—	—	—	—	—	—	—	—	—	0.0002	B2[b]
Benzo(*b*)fluoranthene (PAH)	L	—	—	—	—	—	—	—	—	—	—	—	—	B2
Benzo(*g,h,i*)perylene (PAH)	L	—	—	—	—	—	—	—	—	—	—	—	—	D
Benzo(*k*)fluoranthene (PAH)	L	—	—	—	—	—	—	—	—	—	—	—	—	B2

bis-2-Chloroisopropyl ether	—	—	—	—	F	4	4	4	13	0.04	1	0.3	—	D
Bromacil	—	—	—	—	F	5	5	3	9	0.13	5	0.09	—	C
Bromochloromethane	—	—	—	—	F	0.1	0.1	0.1	0.5	0.013	0.05	0.01	—	
Bromodichloromethane (THM)[e]	F	0.1	0	0.08	D	6	6	4	13	0.02	0.7	—	0.06	B2
Bromoform (THM)[e]	F	0.1	0	0.08	D	5	2	2	6	0.02	0.7	—	0.4	B2
Bromomethane	T	—	—	—	F	0.1	0.1	0.1	0.5	0.001	0.04	0.01	—	D
Butylbenzylphthalate (BBP)	—	—	—	—	—	—	—	—	—	0.2	7	—	—	C
Butylate	—	—	—	—	F	2	2	1	4	0.05	2	0.35	—	D
Carbaryl	—	—	—	—	F	1	1	1	1	0.1	4	0.7	—	D
Carbofluran	P	—	0.04	0.04	F	0.05	0.05	0.05	0.02	0.005	0.2	0.04	—	E
Carbon Tetrachloride	F	—	0	0.005	F	4	0.2	0.07	0.3	0.0007	0.03	—	0.03	B2
Carboxin	—	—	—	—	F	1	1	1	4	0.1	4	0.7	—	D
Chloramben	—	—	—	—	F	3	3	0.2	0.5	0.015	0.5	0.1	—	D
Chlordane	F	—	0	0.002	F	0.06	0.06	—	—	0.00006	0.002	—	0.003	B2
Chlorodibromomethane (THM)[e]	F	0.1	0.6	0.08	D	6	6	2	8	0.02	0.7	0.06	—	C
Chloroform (THM)[e]	F	0.1	0	0.08	D	4	4	0.1	0.4	0.01	0.4	—	0.6	B2
Chloromethane	L	—	—	—	F	9	0.4	0.4	1	0.004	0.1	0.003	—	C
Chlorophenol (2-)	L	—	—	—	D	0.5	0.5	0.5	2.0	0.005	0.2	0.04	—	D
Chlorothalonil	—	—	—	—	F	0.2	0.2	0.2	0.5	0.015	0.5	—	0.15	B2
Chlorotoluene *o*-	L	—	—	—	F	2	2	2	7	0.02	0.7	0.1	—	D
Chlorotoluene *p*-	L	—	—	—	F	2	2	2	7	0.02	0.7	0.1	—	D
Chlorpyrifos	—	—	—	—	F	0.03	0.03	0.03	0.1	0.003	0.1	0.02	—	D
Chrysene (PAH)	—	—	—	—	—	—	—	—	—	—	—	—	—	B2
Cyanazine	T	0.001	—	—	D	0.1	0.1	0.02	0.07	0.002	0.07	0.001	—	C
2,4-D	F	0.1	0.07	0.07	F	1	0.3	0.1	0.4	0.01	0.4	0.07	—	D
Dacthal (DCPA)	L	—	—	—	F	80	80	5	20	0.5	20	4	—	D
Dalapon	F	—	0.2	0.2	F	3	3	0.3	0.9	0.026	0.9	0.2	—	D
Di[2-ethylhexyl]phalate	F	—	Zero	0.006	D	—	—	—	—	0.02	0.7	—	0.3	B2
Diazinon	—	—	—	—	F	0.02	0.02	0.005	0.02	0.00009	0.003	0.0006	—	E
Dibromoacetonitrile	L	—	—	—	D	2	2	2	8	0.02	0.8	0.02	—	C
Dibromochloropropane (DBCP)	F	—	0	0.0002	F	0.2	0.05	—	—	—	—	—	0.003	B2
Dibromomethane	—	—	—	—	—	—	—	—	—	—	—	—	—	D
Dibutyl phthalate	—	—	—	—	—	—	—	—	—	0.1	4	—	—	D
Dicamba	L	—	—	—	F	0.3	0.3	0.3	1	0.03	1	0.2	—	D
Dichloroacetic acid[e]	F	—	0	0.06	D	1	1	1	4	0.004	0.1	—	—	B2

TABLE 2.4 USEPA Drinking Water Regulations and Health Advisories for Organic Contaminants

	Regulations				Health advisories									
					10-kg child			70-kg adult						
Chemicals	Status reg.	NIPDWR (mg/L)	MCLG (mg/L)	MCL (mg/L)	Status HA	1-day (mg/L)	10-day (mg/L)	Longer-term (mg/L)	Longer-term (mg/L)	RfD (mg/kg/day)	DWEL (mg/L)	Lifetime (mg/L)	10^{-4} cancer risk (mg/L)	USEPA cancer group
Dichloroacetonitrile	L	—	—	—	D	1	1	0.8	3	0.008	0.3	0.006	—	C
Dichlorobenzene *p*-	F	—	0.075	0.075	F	10	10	10	40	0.1	4	0.075	—	C
Dichlorobenzene *o*-	F	—	0.6	0.6	F	9	9	9	30	0.09	3	0.6	—	D
Dichlorobenzene *m*-	—	—	—	—	F	9	9	9	30	0.09	3	0.6	—	D
Dichlorodifluoro-methane	L	—	—	—	F	40	40	9	30	0.2	5	1	—	D
Dichloroethane (1,2-)	F	—	0	0.005	F	0.7	0.7	0.7	2.6	—	—	—	0.04	B2
Dichloroethylene (1,1-)	F	—	0.007	0.007	F	2	1	1	4	0.009	0.4	0.007	—	C
Dichloroethylene (*cis*-1,2-)	F	—	0.07	0.07	F	4	1	1	1	0.01	0.4	0.07	—	D
Dichloroethylene (*trans*-1,2-)	F	—	0.1	0.1	F	20	2	2	6	0.02	0.6	0.1	—	D
Dichloromethane	F	—	0	0.005	F	10	2	—	—	0.06	2	—	0.5	B2
Dichlorophenol (2,4-)	—	—	—	—	D	0.03	0.03	0.03	0.1	0.003	0.1	0.02	—	D
Dichloropropane (1,2-)	F	—	0	0.005	F	—	0.09	—	—	—	—	—	0.06	B2
Dichloropropene (1,3-)	T	—	0	—	F	0.03	0.03	0.03	0.09	0.0003	0.01	—	0.02	B2
Dieldrin	L	—	—	—	F	0.0005	0.0005	0.0005	0.002	0.00005	0.002	—	0.0002	B2
Diethylphthalate (DEP)	—	—	—	—	D	—	—	—	—	0.8	30	5	—	D
Diethythexylphthalate (DEHP)	F	—	0	0.006	D	—	—	—	—	0.02	0.7	—	0.3	B2
Diisopropyl methylphosphonate	—	—	—	—	F	8	8	8	30	0.08	3	0.6	0.7	D
Dimethrin	—	—	—	—	F	10	10	10	40	0.3	10	2	—	D
Dimethylphthalate	L	—	—	—	—	—	—	—	—	—	—	—	—	D
Dinitrobenzene (1,3-)	—	—	—	—	F	0.04	0.04	0.04	0.14	0.0001	0.005	0.001	—	D
Dinitrotoluene (2,4-)	L	—	—	—	F	0.5	0.5	0.3	1	0.002	0.1	—	0.005	B2
Dinitrotoluene (2,6-)	L	—	—	—	F	0.4	0.4	0.4	1	0.001	0.04	—	0.005	B2
Dinoseb	F	—	0.007	0.007	F	0.3	0.3	0.01	0.04	0.001	0.04	0.007	—	D
Dioxane *p*-	—	—	—	—	F	4	0.4	—	—	—	—	—	0.7	B2
Diphenamid	—	—	—	—	F	0.3	0.3	0.3	1	0.03	1	0.2	—	D
Diphenylamine	—	—	—	—	F	1	1	0.3	1	0.03	1	0.2	—	D

Diquat	F	—	0.02	0.02	—	—	—	—	—	0.0022	0.08	0.02	—	D
Disulfoton	—	—	—	—	F	0.01	0.01	0.003	0.009	0.00004	0.001	0.0003	—	E
Dithiane (1,4-)	—	—	—	—	F	0.4	0.4	0.4	1	0.01	0.4	0.08	—	D
Diuron	—	—	—	—	F	1	1	0.3	0.9	0.002	0.07	0.01	—	D
Endothall	F	—	0.1	0.1	F	0.8	0.8	0.2	0.2	0.02	0.7	0.1	—	D
Endrin	F	—	0.002	0.002	F	0.02	0.02	0.003	0.01	0.0003	0.01	0.002	—	D
Epichlorohydrin	F	—	0	TT	F	0.1	0.1	0.07	0.07	0.002	0.07	—	0.4	B2
Ethylbenzene	F	—	0.7	0.7	F	30	3	1	3	0.1	3	0.7	—	D
Ethylene dibromide	F	—	0	0.00005	F	0.008	0.008	—	—	—	—	—	0.00004	B2
Ethylene glycol	—	—	—	—	F	20	6	6	20	2	40	7	—	D
Ethylene thiourea	L	—	—	—	F	0.3	0.3	0.1	0.4	0.00008	0.003	—	0.03	B2
Fenamiphos	—	—	—	—	F	0.009	0.009	0.005	0.02	0.00025	0.009	0.002	—	D
Fluometron	—	—	—	—	F	2	2	2	5	0.013	0.4	0.09	—	D
Fluorene (PAH)	—	—	—	—	—	—	—	—	—	0.04	—	—	—	D
Fluorotrichloromethane	L	—	—	—	F	7	7	3	10	0.3	10	2	—	D
Fonofos	—	—	—	—	F	0.02	0.02	0.02	0.07	0.002	0.07	0.01	—	D
Formaldehyde	D	—	—	—	D	10	5	5	20	0.15	5	1	—	B1[f]
Glyphosate	F	—	0.7	0.7	F	20	20	1	1	0.1	4	0.7	—	E
Heptachlor	F	—	0	0.0004	F	0.01	0.01	0.005	0.005	0.0005	0.02	—	0.0008	B2
Heptachlor epoxide	P	—	0	0.0002	F	0.01	—	0.0001	0.0001	0.00001	0.0004	—	0.0004	B2
Hexachlorobenzene	F	—	0	0.001	F	0.05	0.05	0.05	0.2	0.0008	0.03	—	0.002	B2
Hexachlorobutadiene	T	—	0.001	—	F	0.3	0.3	0.1	0.4	0.002	0.07	0.001	—	C
Hexachlorocyclo-pentadiene	F	—	0.05	0.05	—	—	—	—	—	0.007	0.2	—	—	—
Hexachloroethane	L	—	—	—	F	5	5	0.1	0.5	0.001	0.04	0.001	—	C
Hexane (*n*-)	—	—	—	—	F	10	4	4	10	—	—	—	—	D
Hexazinone	—	—	—	—	F	3	3	3	9	0.03[b]	1[b]	0.2[b]	—	D
HMX	—	—	—	—	F	5	5	5	20	0.05	2	0.4	—	D
Indeno(1,2,3-*c,d*)-pyrene (PAH)	—	—	—	—	D	—	—	—	—	—	—	—	—	B2
Isophorone	L	—	—	—	F	15	15	15	15	0.2	7	0.1	4	C
Isopropylmethyl-phosphonate	—	—	—	—	D	30	30	30	100	0.1	4.0	0.7	—	D
Isopropylbenzene	—	—	—	—	D	—	—	—	—	—	—	—	—	—
Lindane	F	—	0.0002	0.0002	F	1	1	0.03	0.1	0.0003	0.01	0.0002	—	C
Malathion	—	—	—	—	F	0.2	0.2	0.2	0.8	0.02	0.8	0.2	—	D
Maleic hydrazide	—	—	—	—	F	10	10	5	20	0.5	20	4	—	D
MCPA	—	—	—	—	F	0.1	0.1	0.1	0.4	0.0015	0.05	0.01	—	E
Methomyl	L	—	—	—	F	0.3	0.3	0.3	0.3	0.025	0.9	0.2	—	D
Methoxychlor	F	—	0.4	0.4	F	0.05	0.05	0.05	0.2	0.005	0.2	0.04	—	D
Methyl ethyl ketone	—	—	—	—	F	—	—	—	—	—	—	—	—	D

TABLE 2.4 USEPA Drinking Water Regulations and Health Advisories for Organic Contaminants

	Regulations				Health advisories									
					10-kg child			70-kg adult						
Chemicals	Status reg.	NIPDWR (mg/L)	MCLG (mg/L)	MCL (mg/L)	Status HA	1-day (mg/L)	10-day (mg/L)	Longer-term (mg/L)	Longer-term (mg/L)	RfD (mg/kg/day)	DWEL (mg/L)	Lifetime (mg/L)	10^{-4} cancer risk (mg/L)	USEPA cancer group
Methyl parathion	—	—	—	—	F	0.3	0.3	0.03	0.1	0.00025	0.009	0.002	—	D
Methyl *tert* butyl ether	L	—	—	—	D	24	24	3	12	0.03	1	0.2–0.02[g]	—	C[g]
Metolachlor	L	—	—	—	F	2	2	2	5	0.1	3.5	0.07	—	C
Metribuzin	L	—	—	—	F	5	5	0.3	0.5	0.013	0.5	0.1	—	D
Monochlorobenzene	P	—	0.1	0.1	F	2	2	2	7	0.02	0.7	0.1	—	D
Naphthalene	—	—	—	—	F	0.5	0.5	0.4	1	0.004	0.1	0.02	—	D
Nitroguanidine	—	—	—	—	F	10	10	10	40	0.1	4	0.7	—	D
Nitrophenol *p*-	—	—	—	—	F	0.8	0.8	0.8	3	0.008	0.3	0.06	—	D
Oxamyl (Vydate)	F	—	0.2	0.2	F	0.2	0.2	0.2	0.9	0.025	0.9	0.2	—	E
Paraquat	—	—	—	—	F	0.1	0.1	0.05	0.2	0.0045	0.2	0.03	—	E
Pentachlorophenol	F	—	0	0.001	F	1	0.3	0.3	1	0.03	1	—	0.03	D
Phenol	—	—	—	—	D	6	6	6	20	0.6	20	4	—	D
Picloram	F	—	0.5	0.5	F	20	20	0.7	2	0.07	2	0.5	—	D
Polychlorinated biphenyls (PCBs)	F	—	0	0.0005	P	—	—	—	—	—	—	—	0.0005	B2
Prometon	L	—	—	—	F	0.2	0.2	0.2	0.5	0.015[b]	0.5[b]	0.1[b]	—	D
Pronamide	—	—	—	—	F	0.8	0.8	0.8	3	0.075	3	0.05	—	C
Propachlor	—	—	—	—	F	0.5	0.5	0.1	0.5	0.013	0.5	0.09	—	D
Propazine	—	—	—	—	F	1	1	0.5	2	0.02	0.7	0.01	—	C
Propham	—	—	—	—	F	5	5	5	20	0.02	0.6	0.1	—	D
Pyrene (PAH)	—	—	—	—	—	—	—	—	—	0.03	—	—	—	D
RDX	—	—	—	—	F	0.1	0.1	0.1	0.4	0.003	0.1	0.002	0.03	C
Simazine	F	—	0.004	0.004	F	0.07	0.07	0.07	0.07	0.005	0.2	0.004	—	C
Styrene	F	—	0.1	0.1	F	20	2	2	7	0.2	7	0.1	—	C
2,4,5-T	L	—	—	—	F	0.8	0.8	0.8	1	0.01	0.35	0.07	—	D
2,3,7,8-TCDD (Dioxin)	F	—	0	3E-08[h]	F	1E-06	1E-07	1E-08	4E-08	1E-09	4E-08	—	2E-08	B2
Tebuthiuron	—	—	—	—	F	3	3	0.7	2	0.07	2	0.5	—	D
Terbacil	—	—	—	—	F	0.3	0.3	0.3	0.9	0.013	0.4	0.09	—	E
Terbufos	—	—	—	—	F	0.005	0.005	0.001	0.005	0.00013	0.005	0.0009	—	D
Tetrachloroethane (1,1,1,2-)	L	—	—	—	F	2	2	0.9	3	0.03	1	0.07	0.1	C
Tetrachloroethylene	F	—	0	0.005	F	2	2	1	5	0.01	0.5	—	0.07	—

Toluene	F	—	1	1	F	20	2	2	7	0.2	7	1	—	D
Toxaphene	F	—	0	0.003	F	—	—	—	—	0.1[b]	—	—	0.003	B2
2,4,5-TP	F	—	0.05	0.05	F	0.2	0.2	0.07	0.3	0.0075	0.3	0.05	—	D
Trichloroacetic acid[e]	F	—	0.3	0.06	D	4	4	4	13	0.1	4	0.3	—	C
Trichloroacetonitrile	L	—	—	—	D	0.05	0.05	—	—	—	—	—	—	—
Trichlorobenzene (1,2,4-)	F	—	0.07	0.07	F	0.1	0.1	0.1	0.5	0.001	0.04	0.07	—	D
Trichlorobenzene (1,3,5-)	—	—	—	—	F	0.6	0.6	0.6	2	0.006	0.2	0.04	—	D
Trichloroethane (1,1,1-)	F	—	0.2	0.2	F	100	40	40	100	0.035	1	0.2	—	D
Trichloroethane (1,1,2-)	F	—	0.003	0.005	F	0.6	0.6	0.4	1	0.004	0.1	0.003	—	C
Trichloroethylene	F	—	0	0.005	F	—	—	—	—	—	0.3	—	0.3	B2
Trichlorophenol (2,4,6-)	L	—	—	—	D	—	—	—	—	—	—	—	0.3	B2
Trichloropropane (1,2,3-)	—	—	—	—	F	0.6	0.6	0.6	2	0.006	0.2	0.04	0.5	B2
Trihalomethanes, total[e]	F	0.1	0	0.08	—	—	—	—	—	—	—	—	—	—
Trifluralin	L	—	—	—	F	0.08	0.08	0.08	0.3	0.0075	0.3	0.005	0.5	C
Trinitroglycerol	—	—	—	—	F	0.005	0.005	0.005	0.005	—	—	0.005	—	—
Trinitrotoluene	—	—	—	—	F	0.02	0.02	0.02	0.02	0.0005	0.02	0.002	0.1	C
Vinyl chloride	F	—	0	0.002	F	3	3	0.01	0.05	—	—	—	0.0015	A
Xylenes	F	—	10	10	F	40	40	40	100	2	60	10	—	D

[a] Treatment technique based on use of no more than 1 mg/L of polyacrylamide (as a flocculant) containing less than 0.05 percent monomer.

[b] Under review.

[c] Total aldicarb compounds cannot exceed 0.007 mg/L for MCLG/MCL because of similar mode of action.

[d] PAH means polycyclic aromatic hydrocarbon.

[e] Stage 1 Rule for disinfectants and disinfection by-products: Total THMs cannot exceed 0.08 mg/L and total haloacetic acids cannot exceed 0.06 mg/L.

[f] Carcinogenicity based on inhalation exposure.

[g] If the tentative cancer classification C is accepted, the Lifetime HA in 0.02 mg/L. If it is not classified, the Lifetime HA is 0.2 mg/L.

[h] The notation E means exponent base 10, e.g., 3E-8 = 3×10^{-4} or 0.00000003.

Note: See Table 2.3 for source, definition, and status codes.

Information based on:

U.S. Environmental Protection Agency. *Drinking Water Regulations and Health Advisories,* EPA 822-R-96-001. Washington, DC: Office of Water, 1996.

U.S. Environmental Protection Agency. *Integrated Risk Information System, April 1998.* Cincinnati, OH: Office of Research and Development, Office of Health and Environmental Assessment, 1998.

U.S. Environmental Protection Agency. "National Primary Drinking Water Regulations: Disinfectants and Disinfection Byproducts; Final Rule." *Federal Register,* 63, 1998h: 69389–69476.

Chronic human exposure to high vapor concentrations leads to injury to the liver, kidneys, and adrenals. Laboratory animals show the same responses to high vapor concentrations, whereas lower chronic doses lead to hepatic and renal changes. 1,2-Dichloroethane is mutagenic and genotoxic in most microbial and mammalian test systems, and administration by gavage to laboratory rats and mice caused carcinomas in the stomach, mammary glands, and lung. It has not been found to be teratogenic in mammalian laboratory animals (Agency for Toxic Substances and Disease Registry, 1994c).

1,1-Dichloroethylene (Vinylidene Chloride). 1,1-Dichloroethylene is used primarily as an intermediate in the synthesis of copolymers for food packaging films, adhesives, and coatings. It is formed by the hydrolysis of 1,1,1-trichloroethane. 1,1-dichloroethylene is moderately water soluble.

No definitive human health studies on 1,1-dichloroethylene have been conducted, but chronic effects in laboratory animals target the liver. With enzymatic activation 1,1-dichloroethylene was mutagenic in Ames testing, and in most in vitro mammalian assays (Agency for Toxic Substances and Disease Registry, 1994b; U.S. Environmental Protection Agency, 1998f). Carcinogenicity test results have been equivocal. One inhalation study with mice and rats appeared to produce malignant kidney tumors in rats and malignant mammary tumors and leukemia in mice. Although other inhalation and feeding studies have been negative, with the possible exception of mammary tumors, the methods and study design were generally not optimal for carcinogenicity (Agency for Toxic Substances and Disease Registry, 1994b).

1,2-Dichloroethylenes. 1,2-Dichloroethylenes exist as *cis*- and *trans*-isomers. Both are used, alone or in combination, as solvents and chemical intermediates. They are moderately water soluble.

For *trans*-1,2-dichloroethylene, a mouse feeding study provided a no-observed-adverse-effect level (NOAEL) of 17 $mg/kg \cdot day^{-1}$ for male mice (U.S. Environmental Protection Agency, 1989c). Toxic effects were minimal, but included an increased serum enzyme, indicative of toxicity and decreased thymus weight in females. Other studies with high exposure levels found liver and kidney toxicity (U.S. Environmental Protection Agency, 1985a). No genotoxic or mutagenic effects were observed for *trans*-1,2-dichloroethylene, though *cis*-1,2-dichloroethylene was positive in a few assays. There are no teratogenicity or carcinogenicity data.

Ethylbenzene. Ethylbenzene is used as a solvent and in the production of styrene. It is also one of many constituents of gasoline and fuel oil. It is slightly soluble in water.

Ethylbenzene is readily absorbed and would be expected to accumulate in the adipose tissue. It is not very toxic in acute exposure testing. Subacute and chronic testing show some liver and kidney pathologies. Ethylbenzene does not appear to be a mutagen. A recent study found kidney and testicular cancers in rats, while mice developed lung and liver cancers (Chan et al., 1998).

Methylene Chloride (Dichloromethane). Methylene chloride is widely used in the manufacture of paint removers, urethane foam, pesticides and other chemical production, degreasers, cleaners, pressurized spray products, and fire extinguishers. Methylene chloride is soluble at about 1 part to 50 parts of water and is persistent in anaerobic soils.

Observations on workers exposed to methylene chloride have not shown a link with cancer (Agency for Toxic Substances and Disease Registry, 1993c). Rats maintained on drinking water containing 0.13 g/L methylene chloride for 91 days showed no adverse effects. Methylene chloride was mutagenic in bacterial assays, but was mostly negative in mammalian cell assays. Carcinogenicity testing has also been mixed, but two studies found increased liver tumors, including one with methylene chloride in drinking water (U.S. Environmental Protection Agency, 1990b). An inhalation study found increased benign mammary tumors and an overall increase in a variety of cancers. There were no teratogenic effects observed (Agency for Toxic Substances and Disease Registry, 1993c).

Methyl* tert-*Butyl Ether (MTBE). Since MTBE was incorporated in the mid-1980s into gasoline mixtures as an antiknock replacement for aromatics and as an "oxygenator" to reduce carbon monoxide emissions, it has increasingly appeared in groundwater due to leaking underground storage tanks at gasoline stations. It is somewhat water soluble and its appearance typically marks the leading front of a plume.

In terms of noncarcinogenic effects, it has low oral toxicity, but at the gasoline pump and in the automobile, symptoms such as airway and eye irritation have been reported. In water, MTBE has a noticeable odor at 20 to 40 μg/L (U.S. Environmental Protection Agency, 1997a). Though MTBE is not mutagenic/genotoxic, exposure to high levels by inhalation (8000 ppm) or by ingestion (1000 mg/kg) was associated with the development of lymphoma and leukemia, as well as liver, renal, and testicular cancers in rodents (Burleigh-Flayer et al., 1992; Belpoggi et al., 1995).

The relevance of these cancers to human health is not clear, but "the weight of evidence suggests that MTBE is an animal carcinogen," and a 20 μg/L odor-based standard would give a 40,000- to 550,000-fold "margin of exposure" for cancer end points (U.S. Environmental Protection Agency, 1997a).

Tetrachloroethylene [Perchloroethylene (PCE)]. Tetrachloroethylene is used as a solvent in textile processing, metal degreasing, and dry cleaning, and as a precursor in the manufacture of fluorocarbons. It was also used as a septic tank cleaner. It is slightly soluble in water. Contamination in private wells has been as high as several parts per million (ppm).

Sublethal doses in laboratory animals cause no organ damage, and chronic feeding studies produced only increased liver weights, with some morphological changes. Limited data indicate that PCE is not teratogenic in rats or mice (National Academy of Sciences Safe Drinking Water Committee, 1983).

Several carcinogenicity assays have been conducted (U.S. Environmental Protection Agency, 1987a; U.S. Environmental Protection Agency, 1991e). A 1977 study reported hepatocellular carcinomas in mice, but not in rats, whereas other early studies failed to demonstrate carcinogenicity. The data from a more recent NTP study found increased leukemia and a rare kidney cancer. Perchloroethylene was mutagenic in a few of the assays. Based on the existing data, PCE had been assigned a B2 carcinogenicity classification, though this is currently being reviewed by USEPA. Recently, the International Agency for the Research of Cancer (1996) rated PCE as a probable human carcinogen, based on recent occupational and drinking water epidemiology.

Toluene (Methyl Benzene). Toluene was part of the aromatic fraction used in lead-free gasoline, as a starting material in the production of benzene and other chemicals, and as a solvent for paints, coatings, gums, glues, and resins. It is moderately soluble in water.

Most health effects data come from inhalation studies, but an NTP 13-week study of rats found increased liver weight, suggesting liver damage. Mutagenicity tests have been negative. Neither inhalation nor gavage assays of carcinogenicity have been positive.

1,1,1-Trichloroethane (Methyl Chloroform). 1,1,1-Trichloroethane is widely used as an industrial cleaner and degreaser, in adhesives and coatings, in drain cleaners, and as a solvent for aerosols (National Academy of Sciences Safe Drinking Water Committee, 1980b). Its water solubility is 4.4 g/L, and it has been widely identified in national drinking water surveys of community supplies in the United States at levels ranging as high as hundreds of parts per billion (U.S. Environmental Protection Agency, 1985b).

A chronic feeding study demonstrated diminished weight gain and decreased survival in mice and rats, but no histopathological changes. Both positive and negative results were reported in microbial mutagenicity testing. In an NCI carcinogenesis bioassay, an increase in liver carcinomas was observed in mice but reassessment indicated minimal carcinogenicity. No teratogenicity has been found in limited testing (Agency for Toxic Substances and Disease Registry, 1995).

The NAS found TCA to be one of the least toxic of the commonly used alkyl halocarbons (National Academy of Sciences Safe Drinking Water Committee, 1983).

Trichloroethylene [Trichloroethene (TCE)]. Trichloroethylene is used primarily in metal degreasing and textile processing. It is also used in dry cleaning, as a solvent, in organic synthesis, and in refrigerants and fumigants. Trichloroethylene is slightly soluble in water, but has been observed up to several ppm in some private wells.

Various chronic exposure tests conducted on rats and mice showed no or minimal liver damage. Limited teratogenicity testing has produced negative results (National Academy of Sciences Safe Drinking Water Committee, 1983). Mixed results have been obtained in mutagenicity testing. Two of six carcinogenicity studies showed significant increases in the incidence of liver tumors among male and female mice. Other studies have been flawed. USEPA has classified TCE as a probable human carcinogen (Group B2), based on sufficient animal evidence of carcinogenicity but inadequate human evidence (U.S. Environmental Protection Agency, 1987a), though the classification is currently being reviewed. Recently, the International Agency for the Research of Cancer rated TCE as a probable human carcinogen, based on recent occupational and environmental epidemiology (International Agency for the Research of Cancer, 1996).

Vinyl Chloride (Monochloroethene). Vinyl chloride is used in the production of PVC resins for building and construction industries. Its use in propellants and aerosols was banned in 1974. Vinyl chloride is slightly soluble in water.

The carcinogenicity of vinyl chloride is proven in humans. Occupational exposure by inhalation has produced the usually rare angiosarcomas of the liver, and possibly lung cancer, brain cancer, leukemia, and lymphoma. Liver damage was also common. Chronic assays in mice, rats, and hamsters exposed by inhalation or gavage have also demonstrated angiosarcomas, and a variety of carcinomas and adenocarcinomas, with most displaying a dose-response trend. Vinyl chloride is a mutagen, producing positive results in most microbial tests and mammalian cell assays. In addition, an increased number of chromosome aberrations have been noted in industrial workers (Agency for Toxic Substances and Disease Registry, 1993e).

Teratogenicity tests have been negative in rats and rabbits, but data for humans are inadequate (Agency for Toxic Substances and Disease Registry, 1993e). Occupational exposure of men to vinyl chloride may cause decreased testosterone levels and decreased sexual function.

Xylenes. Xylenes occur as three isomers: *ortho-*, *meta-*, and *para-*. The three are treated as one in the USEPA health effects evaluation. Xylenes are part of the aromatic fraction that has been used in aviation fuel and lead-free gasoline, and are used as solvents, and in the synthesis of many organic chemicals, pharmaceuticals, and vitamins. Xylenes are slightly soluble in water.

The primary chronic toxic effects of xylenes are kidney and liver damage (U.S. Environmental Protection Agency, 1989c; U.S. Environmental Protection Agency, 1991e). A long-term carcinogenicity bioassay conducted by the NTP found no increased incidence of neoplastic lesions in rats or mice of both sexes.

Pesticides

Pesticides are a wide range of compounds categorized primarily as insecticides, herbicides, and fungicides. The different categories are divided into chemical families. Chemicals within the same family typically result in related types of health effects, discussed below.

Pesticides display a range of solubilities in water, and many bind tightly to organic material in soil particles, slowing entry into surface or groundwater. Many of these compounds are chlorinated, increasing their persistence for agricultural uses, as well as in the general environment. Wells drawing from shallow aquifers have been shown to be susceptible to contamination by pesticides (see Chapter 4), but those that are more water soluble are able to affect deeper wells. In contrast, the poorly soluble organochlorines are rarely seen above trace levels in public water systems. The less soluble pesticides, when washed into surface waters, tend to bind to particulate organic matter that forms sediments or is removed by filtration processes at the treatment plant.

Pesticides can reach the point of major health concern after accidental spills or in wells located close to current or former farms, orchards, golf courses, and utility or rail right-of-way lands. (Occasionally, residential underground applications of a termiticide, like chlordane, have resulted in severe well contamination due to cracks in the well casing.) The National Pesticide Survey (U.S. Environmental Protection Agency, 1990a; U.S. Environmental Protection Agency, 1992d), conducted 1988 through 1990, found measurable pesticide in 10 percent of tested community water system wells, and in about 1 percent of wells tested the level was above the standard. State surveys have found similar results. In addition, triazine herbicides have frequently been observed at levels greater than their MCLs in rivers and reservoirs during periods of high runoff. Nevertheless, for the great majority of people, food represents a greater source of ingestion of pesticides than drinking water (MacIntosh et al., 1996).

Cancellation or severe restriction of use during the 1970s and 1980s (U.S. Environmental Protection Agency, 1994g) has reduced levels of some pesticides in the environment, including: the herbicide 2,4,5-TP (Silvex); fumigant insecticides, like dibromochloropropane (DBCP), 1,2-dichloropropane, and ethylene dibromide (EDB); and most of the organochlorine insecticides (chlordane, DDT, dieldrin, endrin, heptachlor, lindane, and toxaphene). The fumigants had been found frequently and still can be, as suggested by detections of DBCP in 3 of 540 wells in the National Pesticide Survey. Most surveys and compliance monitoring of the organochlorines in the last two decades have reported few detections (U.S. Environmental Protection Agency, 1985a; Parsons and Witt, 1988; U.S. Environmental Protection Agency, 1988d; U.S. Environmental Protection Agency, 1989c; U.S. Environmental Protection Agency, 1990a; U.S. Environmental Protection Agency, 1992d).

The health effects associated with pesticides vary according to chemical. They can be crudely differentiated into problems that occur acutely at high levels of exposure and chronically at low levels of exposure (Hodgson and Levi, 1996). The former can include liver and kidney damage; major interference with nervous, immune, and reproductive system functions; birth defects; and cancer risk. Problems that may occur with long-term, low-level exposure include birth defects, cancer risk, and mild interference with nervous and immune system function. Lesser effects on the nervous system are frequently expressed as "nonspecific" symptoms like dizziness, nausea, and tiredness. There may also be diarrheal symptoms. Importantly, not much is known about the effects of exposure to mixtures of pesticides.

Both cancer risk and birth defect risk estimates stem largely from studies in laboratory animals. However, epidemiologic studies have found increased incidence of soft-tissue sarcoma (STS) and lymphoid cancers among farmers (reviewed by Dich et al., 1997). There is increasing evidence that the chlorophenoxy herbicides, like 2,4-D, are associated with increased STS and lymphoma incidence. However, the carcinogenicity of these herbicides is probably related to the presence of dioxins produced in the past as a by-product (Kogevinas et al., 1997). The organochlorine insecticides have been linked to STS, non-Hodgkin's lymphoma, and leukemia. Organophosphate insecticides have been linked to increased non-Hodgkin's lymphoma and leukemia. Childhood cancers have also been linked to general pesticide exposure, particularly parental exposure during pregnancy and nursing (Hoar Zahm and Ward, 1998). Epidemiological studies of reproductive effects, including birth defects, were reviewed by Sever et al. (1997). Many of these studies suggested that maternal exposure was linked to reproductive outcome.

The environmental persistence associated with chlorinated pesticides, particularly those with more chlorination, is mirrored in the accumulation and persistence in fatty tissues in animals, including fish and humans. The chlorinated cyclodienes are notable for their storage in animal fat. Although the registration of DDT, and related DDD and DDE, for example, has been cancelled since 1972, they are still found in humans. Recently, DDT and its metabolites have been linked to breast cancer (reviewed by Laden and Hunter, 1998). Indeed, some pesticides have been found to affect the estrogenic endocrine system (Soto et al., 1995), possibly promoting cancer and reducing fertility. However, recent animal studies with low-dose mixtures of representative pesticides did not find a significant effect on reproduction (Heindel et al., 1994).

Carbamate (e.g., aldicarb, carbofuran, and oxamyl) and organophosphate (e.g., diazinon) insecticides inhibit the important neuromuscular enzyme, acetylcholinesterase. Biochemical similarities between insect and vertebrate nervous systems are the reason why these insecticides have the greatest potential to cause acute and medium-term health problems in sensitive individuals. In humans and most other vertebrates, acetylcholinesterase inhibition affects the skeletal muscles, causing weakness. It also affects the autonomic nervous system, which controls sweating, intestinal motility, salivation, tears, and constriction of the pupils. Inhibition of a related cholinesterase in blood can be used to monitor exposure. The mode of action of other insecticides, like the organochlorines, is not well understood, but appears to also involve the nervous system, including the mammalian nervous system at high exposure levels. In general, the levels found in public drinking water have probably not constituted an acute threat to the health of these individuals. The acute threat has declined as many of the more dangerous insecticides have been restricted or cancelled.

The herbicides, used to control weeds, are composed of the largest number of chemical families. In terms of water contamination, the important chemical families include the acetanilides, like alachlor and metolachlor; the triazines, like atrazine, cyanazine, prometon, and simazine; and the chlorophenoxy acids, like 2,4-D and mecoprop. Some of the herbicides may present cancer and birth defect risks. Cancelled herbicides include paraquat, diquat, and 2,4,5-T.

Although there are no fungicides regarded as significant health threats found at notable levels in water, a breakdown product of the ethylene bisdithiocarbamate (EBDC) chemical family (e.g., maneb, mancozeb, and zineb) is important: ethylene thiourea (ETU). It is water soluble and considered a probable human carcinogen, though it is not persistent. Uses of many of the EBDC fungicides were restricted in the late 1980s, but their use is still widespread.

Alachlor. Alachlor is a widely used agricultural herbicide frequently found in surface water and groundwater.

Although alachlor has low acute oral toxicity, chronic effects noted in laboratory feeding studies are hepatotoxicity, chronic nephritis, hemolytic anemia, and degeneration of certain components of the eye (U.S. Environmental Protection Agency, 1985a). Chronic feeding studies using mice have demonstrated that alachlor instigated lung tumors. Studies with rats have demonstrated stomach, thyroid, and nasal tumors. No teratogenic effects have been observed. One investigation of over 900 chemical workers at one plant found significantly elevated rates of chronic myeloid leukemia and colorectal cancer, especially among those highly exposed, but the number of cases was small (Leet et al., 1996)

Aldicarb, Aldicarb Sulfoxide, and Aldicarb Sulfone. Aldicarb is a carbamate, sold under the trade name, Temek, and was widely used in agriculture to control insects, mites, and nematodes. It is very soluble in water, but it is not persistent. Aldicarb sulfoxide and aldicarb sulfone are related forms, and are the forms of aldicarb most often detected in water.

Aldicarb's acute toxicity is the highest of any of the widely used pesticides. As a carbamate it acts on the nervous system through cholinesterase inhibition (see the introduction to this section). The effects end within hours after termination of exposure and may include nausea, diarrhea, sweating, muscle weakness, and constriction of the pupils of the eyes. Changes in immune response in humans and laboratory animals exposed to aldicarb in drinking water are considered inconclusive (U.S. Environmental Protection Agency, 1995a). Aldicarb has not been shown to be a mutagen or a carcinogen and there are few reproductive data.

Aldicarb and aldicarb sulfoxide were assigned a lifetime health advisory of 7 μg/L, based on a study of people exposed to pesticide-contaminated food and two small, controlled low-dose exposures of humans (U.S. Environmental Protection Agency, 1995a). Aldicarb sulfone was assigned the same lifetime health advisory based on cholinesterase inhibition in a one-year dog feeding study, supported by the human studies (U.S. Environmental Protection Agency, 1995a). Because the mechanism of neurotoxicity of these related compounds is the same, their presence in a mixture is considered additive. USEPA is reviewing its proposed MCL and MCLG.

Atrazine. Atrazine is a widely used agricultural herbicide. It is slightly soluble in water, but has been frequently detected in surface and groundwaters.

Two-year rat feeding/carcinogenicity studies, three long-term mouse studies, and a two-generation reproduction study with rats did not detect significant organ damage, whereas dogs fed atrazine showed cardiac pathologies at 150 and 1000 ppm mixed in food (U.S. Environmental Protection Agency, 1985a; U.S. Environmental Protection Agency, 1989c; U.S. Environmental Protection Agency, 1991e). An NOAEL from the reproduction study was developed into a human drinking water equivalent level (DWEL) of 0.2 mg/L. Long-term gavage and feeding studies in mice found no significant increase of cancer, whereas a feeding study in one strain of rats resulted in an increased number of mammary tumors in females (U.S. Environmental Protection Agency, 1985a; U.S. Environmental Protection Agency, 1989c; U.S. Environmental Protection Agency, 1991e). Atrazine appears to be mutagenic in animal assays, but not in microorganism assays (U.S. Environmental Protection Agency, 1988c). It was not teratogenic in laboratory animals.

Carbofuran. Because of its persistence and water solubility, carbofuran has often been found in groundwater.

As a carbamate pesticide, it has potential neurotoxic effects, as outlined in the introductory section. Noticeable effects can include headache and nausea, as well as changes in the autonomic nervous system that cause increased sweating, intestinal motility (diarrhea), salivation, tears, and constriction of the pupils (U.S. Environmental Protection Agency, 1989c; U.S. Environmental Protection Agency, 1991e). Inhibition of a related cholinesterase in blood can be used to monitor exposure. In laboratory animals testicular degeneration was also observed. Based on limited evidence, it does not appear to be carcinogenic.

Cyanazine. Cyanazine is one of the triazine herbicides, used for controlling grasses and broadleaf weeds in agriculture.

The kidney and liver were affected in laboratory animals eating cyanazine. One two-year feeding study in mice was negative for cancers, but rats had increased numbers of thyroid and adrenal tumors. However, this latter study is currently being reevaluated (U.S. Environmental Protection Agency, 1998f). There is little work on mutagenicity, though atrazine in the triazine family yielded mixed results. Teratogenic effects occurred in rabbits and rats, but there may have been maternal toxicity. There is no MCL, but an MCLG was proposed.

Dacthal. This preemergent herbicide, a dichloropropionic acid (or DCPA), has been widely detected in surface water and was the most frequently detected pesticide in groundwater in the 1988–1989 USEPA survey (U.S. Environmental Protection Agency, 1990a). Although it is chemically stable, it is metabolized by soil bacteria (a half-life of between two and three weeks).

Chronic feeding studies in rats have resulted in lung, liver, kidney, and thyroid pathology as well as alterations in thyroid hormone levels (U.S. Environmental Protection Agency, 1998f). There were no reproductive or teratological problems. Cancer data were equivocal. No information was available on the metabolites.

Dicamba. Dicamba is a dichlorobenzoic acid derivative used to control broadleaf weeds in grain, corn, and pastures, as well as brush in noncrop areas. Dicamba is highly mobile, but is not persistent, with a half-life in most soils of only a few weeks.

Long-term exposure to high levels of dicamba resulted in enlarged livers in rats. There was little apparent toxicity at lower levels in dogs or rats, no mutagenicity or carcinogenicity, and no significant teratogenicity. However, in rabbits there is significant maternal and fetal toxicity at 10 mg/kg. There is no MCL or MCLG, but the lifetime health advisory is 200 μg/L, whereas the longer-term advisory for young children is 300 μg/L (U.S. Environmental Protection Agency, 1998f).

2,4-D. 2,4-Dichlorophenoxyacetic acid, or 2,4-D, is a systemic herbicide used to control broadleaf weeds. Its primary current use is for lawns. It is very soluble (540 mg/L) and undergoes both chemical and biological degradation in the environment. In water, 2,4-D has a half-life of about one week, whereas in soil, 2,4-D persists for about six weeks (National Academy of Sciences Safe Drinking Water Committee, 1977).

A 90-day feeding study in rats resulted in hematologic (reduced hemoglobin and red blood cells), hepatic, and renal toxicity, though mice were less sensitive (U.S. Environmental Protection Agency, 1998f). There was no reproductive or teratogenic toxicity. Most microbial systems showed no mutagenic activity, and in vivo mammalian assays have also been negative. Carcinogenicity and teratogenicity data are inconclusive, and no dose-dependent tumor formation has been reported. There is considerable epidemiologic analysis, but the results have been mixed (Lilienfeld and Gallo, 1989).

Trace amounts of dioxins and related compounds (see 2,3,7,8-TCDD, later) created during production probably increased the toxicity of the chlorophenoxy herbicides. However, dioxins may not be at detectable levels in water, especially groundwater.

1,2-Dichloropropane. 1,2-Dichloropropane is a versatile compound used as a soil fumigant, a metal degreaser, a lead scavenger for antiknock fluids, and a component of dry-cleaning fluids (National Academy of Sciences Safe Drinking Water Committee, 1977).

Toxic effects target the liver with centrilobular necrosis, liver congestion, and hepatic fatty changes. It is mutagenic and an NTP carcinogenicity study reported a significant increase in nonneoplastic liver lesions and mammary gland adenocarcinomas in female rats but not in males when 1,2-dichloropropane was administered via gavage. When tests were run with mice, nonneoplastic liver lesions were observed in males, and hepatocellular adenomas were observed in both males and females. The results of these tests suggest 1,2-dichloropropane is carcinogenic in mice. Results for rats were equivocal (U.S. Environmental Protection Agency, 1985a). These health effects were similar for the related soil fumigant, 1,3-dichloropropene.

Ethylene Thiourea (ETU). Ethylene thiourea is a breakdown product of the ethylene bisdithiocarbamate (EDBC) fungicides, such as Zineb, Maneb, and Mancozeb. It is moderately water soluble, but can be metabolized by bacteria.

Liver and thyroid tumors were increased in mice and rats exposed to ETU, but the results of mutagenicity and genotoxicity testing were mixed (reviewed by Dearfield, 1994). There was also thyroid toxicity in adult rodents and developmental toxicity in exposed rodents. The developmental toxicity includes retarded CNS development that may be related to thyroid toxicity. The MCL and MCLG have not been set, but USEPA estimated that there is a 1-in-10,000 risk of cancer following a lifetime of exposure at 30 μg/L (U.S. Environmental Protection Agency, 1998f).

Mecoprop. This chlorophenoxy herbicide is persistent, soluble in water, and widely used as a lawn herbicide.

Rats exposed for 90 days at high levels developed enlarged kidneys and increased creatinine levels in females (U.S. Environmental Protection Agency, 1998f). Teratology studies showed decreased fetal growth and delayed ossification. There is no adequate carcinogenesis study.

Metolachlor. This acetanilide herbicide is used for preemergence weed control in agriculture. It is moderately soluble in water (up to 500 mg/L), is mobile in some soils, is persistent away from sunlight, and has frequently been found in surface water and groundwater.

Toxicity at high doses in long-term studies of rats includes liver and kidney enlargement and atrophy of seminal vesicle and testes in males. Reproductive and teratogenic effects were not seen. Evidence of increased liver tumors was seen in two of four oncogenicity studies, but mutagenicity tests were negative (U.S. Environmental Protection Agency, 1998f).

Metribuzin. Metribuzin is a triazine herbicide used in agriculture. It has been found frequently in surface water and groundwater due to its moderate persistence, solubility in water, and high mobility in soils.

Long-term exposure has produced thyroid enlargement in rats, kidney and liver enlargement in mice, but no apparent toxicity in dogs (U.S. Environmental Protection Agency, 1998f). No teratogenicity was observed. There is little evidence for mutagenicity or carcinogenicity.

Oxamyl. Oxamyl is a nonpersistent carbamate insecticide and nematocide in agriculture. It is highly soluble and mobile in soil, but is degraded by bacteria, with a half-life of one to five weeks in soil and one to two days in rivers.

The principal effect, particularly at low levels, is neurotoxicity. No histopathological changes were noted in organs, and no reproductive or developmental effects were seen (U.S. Environmental Protection Agency, 1990b; U.S. Environmental Protection Agency, 1992c). It is one of the few pesticides that is classified as a noncarcinogen.

Picloram. This rangelands and rights-of-way herbicide has been found in surface water and groundwater.

Liver toxicity is the most significant feature of chronic and subchronic feeding studies in rodents and dogs (U.S. Environmental Protection Agency, 1990b; U.S. Environmental Protection Agency, 1992c). One strain of rat also exhibited thyroid and parathyroid enlargement and testicular atrophy. No significant teratogenic effects were observed. Mutagenicity was negative or weak. The carcinogenicity of picloram is considered unclassifiable because benign hepatic nodules were increased in rats, but mice were negative.

Prometon. This triazine herbicide has often been found in surface water and groundwater.

It appears to have relatively low toxicity, but due to "low confidence" in the published studies, USEPA is conducting further investigations (U.S. Environmental Protection Agency, 1998f). There was little evidence of organ toxicity or teratogenicity and no data on mutagenicity and carcinogenicity.

Simazine. Simazine is a widely used agricultural triazine herbicide. It is nonvolatile and has a low solubility.

It appears to have low acute toxicity. Several recent studies have helped to characterize the long-term toxicity (U.S. Environmental Protection Agency, 1990b; U.S. Environmental Protection Agency, 1992c). One two-year rat feeding study found decreases in weight gain and hematologic parameters. A recent one-year dog feeding study also identified decreased hematological parameters. No adverse effects on female reproductive capacity were observed after a three-generation rat feeding study and no teratogenicity has been seen. Mutagenicity and genotoxicity tests yielded mixed results, but dose-related mammary carcinomas have been observed in rats, as has been noted for the related chemicals, atrazine and propazine. Pituitary adenomas were also found. Mice were unaffected.

2,3,7,8-TCDD (Dioxin). 2,3,7,8-Tetrachlorodibenzo-*p*-dioxin is not manufactured purposefully. It occurs as a contaminant of 2,4,5-trichlorophenol that is used in the production of several herbicides, such as 2,4,5-trichlorophenoxyacetic acid and Silvex. It also may be formed during pyrolysis of chlorinated phenols, chlorinated benzenes, and polychlorinated diphenyl esters. Dioxin is not mobile in soils and is not expected to be found in drinking water, but is accumulated in fatty tissues of fish.

Dioxin is readily absorbed and accumulates in adipose tissue. It is only slowly metabolized. Noncarcinogenic effects include thymic atrophy, weight loss, and liver damage (U.S. Environmental Protection Agency, 1990b). Mutagenicity test results have been conflicting. Dioxin is, however, a potent animal carcinogen. Oral exposure leads to hepatocellular carcinomas, thyroid carcinomas, and adrenal cortical adenomas. There are also developmental effects, including the immune system, the central nervous system, and the thyroid (U.S. Environmental Protection Agency, 1990b; Birnbaum, 1995).

Chemicals in Treatment Additives, Linings, and Coatings

Acrylamide. Acrylamide is used in manufacture of the polymer, polyacrylamide, from which unpolymerized monomer may leach. Polyacrylamide is used to enhance water and oil recovery from wells, as a flocculent and sludge-conditioning agent in drinking water and wastewater treatment, in food processing and papermaking, as a soil conditioner, and in permanent press fabrics. No monitoring data exist to detail its occurrence in water; however, its high solubility and the use of polyacrylamide within the drinking water industry make its presence likely (U.S. Environmental Protection Agency, 1985a).

Acrylamide's principal toxic effect is peripheral neuropathy (damage of the peripheral nerves, for example, in the arm or leg). Subchronic animal studies have also demonstrated atrophy of skeletal muscles in hind quarters, testicular atrophy, and weakness of the limbs. Case reports on humans indicate similar effects following exposure via the dermal, oral, or inhalation routes.

Recent evidence shows acrylamide given orally is carcinogenic in mice and rats (U.S. Environmental Protection Agency, 1991e). Mammary, central nervous system, thyroid, and uterine cancers were observed. An increase in lung adenomas was also observed in mice receiving oral exposure. USEPA calculated an upper 95 percent confidence estimate of lifetime cancer risk of 1×10^{-4}, assuming consumption of 1 L/day containing 1 μg/L of acrylamide. Although acrylamide was negative in the Ames mutagenicity test, it is a potent genotoxic agent (U.S. Environmental Protection Agency, 1991e).

Limitations on the amount of polyacrylamide polymer used as a flocculent to treat drinking water and on the amount of free acrylamide in the polymer have been adopted in lieu of an MCL (U.S. Environmental Protection Agency, 1991e).

Epichlorohydrin (ECH). Epichlorohydrin is a halogenated alkyl epoxide. It is used as a raw material in the manufacture of flocculents and epoxy and phenoxy resins, and as a solvent for resins, gums, cellulose, paints, and lacquers. The flocculents are sometimes used in food preparation and in potable water treatment, and the lacquers are sometimes used to coat the interiors of water tanks and pipes. Epichlorohydrin is quite soluble in water (6.6×10^4 mg/L) and has pronounced organoleptic properties. The threshold odor concentration is 0.5 to 1.0 mg/L, whereas the threshold for irritation of the mouth is at 100 μg/L (National Academy of Sciences Safe Drinking Water Committee, 1977).

Epichlorohydrin is rapidly absorbed. Occupational overexposure through inhalation has resulted in long-term liver damage (National Academy of Sciences Safe Drinking Water Committee, 1980b). Chronic exposure in laboratory animals can result in emphysema, bronchopneumonia, kidney tubule swelling, interstitial hemorrhage of the heart, and brain lesions (U.S. Environmental Protection Agency, 1985a). Epichlorohydrin or its metabolite, alpha-chlorohydrin, can produce reversible or nonreversible infertility in laboratory animals (National Academy of Sciences Safe Drinking Water Committee, 1980b). No teratogenic studies on ECH have been reported (U.S. Environmental Protection Agency, 1991e).

Epichlorohydrin is considered a potent mutagen, having produced positive results in a number of procaryotic systems and eucaryotic cell cultures. It has also been shown to be carcinogenic following oral and inhalation exposures. Laboratory rats given oral doses of epichlorohydrin developed dose-dependent increases in forestomach tumors (U.S. Environmental Protection Agency, 1985a).

Limitations on the amount of free ECH in flocculent polymer and the amount of polymer used to treat drinking water have been adopted in lieu of an MCL (U.S. Environmental Protection Agency, 1991e).

Polynuclear Aromatic Hydrocarbons (PAHs). Polynuclear (or polycyclic) aromatic hydrocarbons are a diverse class of compounds consisting of substituted and unsubstituted polycyclic and heterocyclic aromatic rings. They are formed as a result of incomplete combustion of organic compounds in the presence of insufficient oxygen. Benzo(*a*)pyrene is the most thoroughly studied because of its ubiquity and its known carcinogenicity in laboratory animals. It is formed in the pyrolysis of naturally occurring hydrocarbons and found as a constituent of coal, coal tar, petroleum, fuel combustion products, and cigarette smoke (National Academy of Sciences Safe Drinking Water Committee, 1982).

The PAHs are extremely insoluble in water (e.g., the solubility of benzo(*a*)pyrene is only 10 ng/L). Total PAH concentrations in surface waters have been reported to range from 0.14 to 2.5 μg/L. Because a large fraction of the PAHs are associated with particulates, water treatment can lower these concentrations to a reported range of 0.003 to 0.14 μg/L (National Academy of Sciences Safe Drinking Water Committee, 1982). Polynuclear aromatic hydrocarbons can leach from tar or asphalt linings of distribution pipelines.

Studies indicate an increased mortality from lung cancer in workers exposed to PAH-containing substances. In laboratory animals, subchronic and chronic doses of PAHs produce systemic toxicity, manifested by inhibition of body growth, and degeneration of the hematopoietic and lymphoid systems. The PAHs primarily affect organs with proliferating cells, such as the intestinal epithelium, bone marrow, lymphoid organs, and testes. The mechanism of toxicity with these organs may be the binding of PAH metabolites to DNA. Carcinogenic PAHs can also suppress the immune system, while noncarcinogenic PAHs do not have this effect. Most of the carcinogenic PAHs have also been demonstrated to be mutagenic (National Academy of Sciences Safe Drinking Water Committee, 1982).

DISINFECTANTS AND DISINFECTION BY-PRODUCTS

Chlorine has been the primary drinking water disinfectant in the United States for more than 90 years. The other major disinfectants used include chlorine dioxide, chloramines, and ozone. Other materials that can act as disinfectants include potassium permanganate, iodine, bromine, ferrate, silver, hydrogen peroxide, and ultraviolet (UV) light. Of these, potassium permanganate and UV light are used to a limited extent, whereas iodine and bromine are rarely used, and the others are even less established (U.S. Environmental Protection Agency, 1992d). A more complete discussion of disinfection is contained in Chapter 14.

The use of chlorine and other disinfectants such as ozone, although reducing the risk of waterborne disease, creates new potential risks, because compounds known as disinfection by-products (DBPs) are formed during the disinfection process. A wide variety of by-products have been identified. A number of these compounds have been shown to cause cancer and other toxic effects in animals under experimental conditions.

Epidemiological studies have been conducted to evaluate the association from exposure to chlorinated surface water with several adverse outcomes: cancer, cardiovascular disease, and adverse reproductive outcomes, including neural birth defects. Up to the mid-1990s, a number of studies had found small increases in bladder, colon, and rectal cancers, though others observed no association. In some cases, associations were linked with duration of exposure and volume of water consumed. Several recent well-done, well-designed studies have also observed a small elevated risk (Cantor, 1997). Based on a review of all the data, USEPA concluded that due to problems with characterization of exposures, the epidemiology data were inadequate to link cancer in humans with consumption of chlorinated water at this time (U.S. Environmental Protection Agency, 1998a). Other scientists, however, have concluded that the epidemiological data, when combined with the toxicological studies, are compelling and that consumption of chlorinated water potentially increases the cancer risk to individuals (U.S. Environmental Protection Agency, 1994f). In regard to cardiovascular disease, animal studies in the mid-1980s indicated a potential increase in the serum lipid levels in animals exposed to chlorinated water. Elevated serum lipid levels are indicators of cardiovascular disease. However, in a cross-sectional study in humans, comparing chlorinated and unchlorinated water supplies with varying water hardness, no adverse effects on serum lipid levels were found. Finally, recent studies have reported increased incidence of decreased birth weight, prematurity, intrauterine growth retardation, and neural tube defects with chlorinated water and, in some cases, trihalomethanes (U.S. Environmental Protection Agency, 1994a, 1998b, 1998f). As with the other adverse outcomes from the epidemiological studies, there is considerable debate in the scientific community on the significance of these findings.

USEPA finalized a rule in 1998 that pertains to all public water systems that disinfect and was intended to reduce risks from disinfectants and DBPs (U.S. Environmental Protection Agency, 1998h). This rule is referred to as EPA's Stage 1 D/DBP rule in the remainder of the chapter. Under this rule, public water systems will be required to achieve new limits for total trihalomethanes (chloroform, bromodichloromethane, dibromochloromethane, and bromoform), the sum concentration for five haloacetic acids (mono-, di-, and trichloroacetic acids and mono- and dibromoacetic acids), bromate, chlorite, chlorine, chlorine dioxide, chloral hydrate, and chloramines. Systems using settling and filtration will also be required to achieve percent reductions of total organic carbon, depending upon source water quality, prior to disinfection. Information on the disinfectants themselves and their inorganic by-products is given hereafter, followed by information on organic DBPs.

Disinfectants and Inorganic DBPs

Chlorine. At room temperature, chlorine is a greenish-yellow, poisonous gas. When added to water, chlorine combines with water to form hypochlorous acid that then ionizes to form hypochlorite ion, both of which can be measured as free chlorine. Because of their oxidizing characteristics and solubility, chlorine and hypochlorite are used in water treatment to disinfect drinking water, sewage and wastewater, swimming pools, and other types of water reservoirs. The relative amounts of hypochlorous acid and hypochlorite ion formed are dependent on pH and temperature. See Chapter 14, "Disinfection," for a complete discussion of chlorine chemistry. Maintenance of a chlorine residual throughout the distribution system is important for minimizing bacterial growth and for indicating (by the absence of a residual) water quality problems in the distribution system.

USEPA's Stage 1 D/DBP rule establishes a maximum residual disinfectant level goal (MRDLG) and maximum residual disinfectant level (MRDL) of 4.0 mg/L for residual free chlorine in water. The MRDLGs are analogous to MCLGs and are established at the level at which no known or anticipated adverse effects on the health of persons occur and which allow an adequate margin of safety. The MRDLGs are nonenforceable health goals based only on health effects and exposure information and do not reflect the benefit of the addition of the chemical for control of waterborne microbial contaminants. The MRDLs are enforceable standards, analogous to MCLs, which recognize the benefits of adding a disinfectant to water on a continuous basis and in addressing emergency situations such as distribution system pipe breaks. As with MCLs, USEPA has set the MRDLs as close to the MRDLGs as feasible (U.S. Environmental Protection Agency, 1994f, 1997c, 1998h).

No evidence of reproductive or developmental effects has been reported for chlorine. No systemic effects were observed in rodents following oral exposure to chlorine as hypochlorite in distilled water at levels up to 275 mg/L over a two-year period (U.S. Environmental Protection Agency, 1994c, 1997c). Chlorinated water has been shown to be mutagenic to bacterial strains and mammalian cells. However, assessment of the mutagenic potential of chlorine is confounded by the reactive nature of chlorine and by the presence of reaction products, which have been found to be mutagenic. Investigations with rodents to determine the potential carcinogenicity of chlorine have been negative.

Chloramines. Inorganic chloramines are formed in waters undergoing chlorination which contain ammonia. Hypochlorous acid reacts with ammonia to form monochloramine (NH_2Cl), dichloramine ($NHCl_2$), trichloramine or nitrogen trichloride (NCl_3), and other minor by-products. In most drinking waters, the predominant chloramine species is monochloramine. Chloramine is used as a disinfectant in drinking water to control taste and odor problems, limit the formation of chlorinated DBPs, and maintain a residual in the distribution system for controlling biofilm growth. The MRDLG and MRDL in USEPA's Stage 1 D/DBP rule are 4.0 mg/L, measured as chlorine residual in water (U.S. Environmental Protection Agency, 1994f, 1997c, 1998h).

In humans, observed adverse health effects associated with monochloramine in drinking water have been limited to hemodialysis patients. Chloramines in dialysis baths cause oxidation of hemoglobin to methemoglobin and denaturation of hemoglobin. Tests conducted on healthy human volunteers to evaluate the effects of monochloramine in drinking water at doses up to 24 mg/L (short term) and 5 mg/L (for 12 weeks) showed no effects (U.S. Environmental Protection Agency, 1994a, 1997c). Two lifetime rodent studies showed minimal toxicity effects, such as decreased body and organ weights in rodents, along with some effects to the liver (weight changes, hypertrophy, and chromatid pattern changes) which appear to be related to overall body weight changes caused by decreased water consumption due to the unpalatability of chloramines to the test animals (U.S. Environmental Protection Agency, 1994a, 1997c). One study examining the reproductive effects and another which examined developmental effects of chloramines concluded that there are no chemical-related effects due to chloramines (U.S. Environmental Protection Agency, 1994a, 1997c). Results on the mutagenicity of chloramines are inconclusive.

Chlorine Dioxide, Chlorite, and Chlorate. Chlorine dioxide is used as a disinfectant in drinking water treatment as well as an additive with chlorine to control tastes and odors in water treatment. It has also been used for bleaching pulp and paper, flour and oils, and for cleaning and tanning of leather. Chlorine dioxide is fairly unstable and rapidly dissociates into chlorite and chloride in water (U.S. Environmental Protection Agency, 1994d).

The primary concern with using chlorine dioxide has been the toxic effects attributed to residual chlorite and chlorate. Because chlorine dioxide converts to chlorite in vivo, health effects attributed to chlorine dioxide are assumed to be the same as for chlorite. Chlorine dioxide, chlorite, and chlorate have each been evaluated separately in humans and in laboratory animals.

Chlorine dioxide (ClO_2). Human volunteers given 250 mL of 40 mg/L ClO_2 experienced short-term (5 min) headaches, nausea, and light-headedness. In another study, subjects given increasing doses of ClO_2, from 0.1 to 24 mg/L, every 3rd day for a total of 16 days demonstrated no ill effects. Similarly, subjects given a dose of 5 mg/L daily for five weeks showed no effects (U.S. Environmental Protection Agency, 1994d). In subchronic and chronic studies, animals given chlorine dioxide–treated water exhibited osmotic fragility of red blood cells (1 $mg/kg \cdot day^{-1}$) and decreased thyroxine hormone levels (14 $mg/kg \cdot day^{-1}$), possibly due to altered iodine metabolism (U.S. Environmental Protection Agency, 1994d). Studies evaluating developmental or reproductive effects have described decreases in the number of implants and live fetuses per dam in female rats given chlorine dioxide in drinking water before mating and during pregnancy. Delayed neurodevelopment has been reported in rat pups exposed perinatally to chlorine dioxide– (14 $mg/kg \cdot day^{-1}$) treated water and was confirmed by a recent two-generation reproductive and developmental study by the Chemical Manufacturer Association with chlorite (Chemical Manufacturers Association, 1996; see the following chlorite discussion). Mutagenicity tests in mice and bacterial systems were negative (U.S. Environmental Protection Agency, 1994d, 1997c). No studies on the carcinogenic properties of orally ingested ClO_2 are available, but concentrates of chlorine dioxide–treated water did not increase the incidence of lung tumors in mice, nor was any initiating activity observed in mouse skin or rat liver bioassays (U.S. Environmental Protection Agency, 1994d). USEPA finalized an MRDLG and an MRDL of 0.8 mg/L for chlorine dioxide in the Stage 1 D/DBP rule based in part on the findings from the Chemical Manufacturers Association study (U.S. Environmental Protection Agency, 1998h).

Chlorite. Chlorite ion is relatively stable and degrades very slowly to chloride. Chlorite is used in the on-site production of chlorine dioxide and as a bleaching agent by itself, for pulp and paper, textiles, and straw. Chlorite is also used to manufacture waxes, shellacs, and varnishes (U.S. Environmental Protection Agency, 1994f).

Human volunteers given 1 L of water with increasing doses of sodium chlorite, from 0.01 to 2.4 mg/L, every 3rd day for a total of 16 days showed no effects. Another group given 500 mL of water containing 5 mg/L sodium chlorite for 12 weeks showed no significant effects. Adverse effects of chlorite on the hematological systems of laboratory animals are well documented. In a variety of reproductive effect tests, chlorite was associated with a decrease in the growth rate of rat pups between birth and weaning. Delayed neurodevelopment has been reported in rat pups exposed perinatally to chlorite (3 $mg/kg \cdot day^{-1}$) (U.S. Environmental Protection

Agency, 1994d, 1998d,g). A recent two-generation reproductive and developmental study conducted by the Chemical Manufacturers Association found neurodevelopmental effects at 6 mg/kg·day^{-1} (U.S. Environmental Protection Agency, 1998d,g). Mutagenicity testing of chlorite in mouse assays was negative. As with chlorine dioxide, no clear tumorigenic activity has been observed in animals given oral doses of chlorite (U.S. Environmental Protection Agency, 1994d). USEPA promulgated an MCLG of 0.8 mg/L and an MCL of 1.0 mg/L for chlorite in the Stage 1 D/DBP rule (U.S. Environmental Protection Agency, 1998g).

Chlorate. Chlorate may also be formed as a result of inefficient generation or generation of chlorine dioxide under very high or low pH conditions. Chlorate was once a registered herbicide to defoliate cotton plants during harvest and was used in leather tanning and in the manufacture of dyes, matches, and explosives (U.S. Environmental Protection Agency, 1994g).

Human tests conducted with an increasing dose range of 0.01 to 2.4 mg/L over 16 days and a 5 mg/L dose for 12 weeks showed no clinically significant effects. Because of its use as a weed killer, cases of chlorate poisoning have been reported. Methemoglobinemia, hemolysis, and renal failure occurred in people consuming large amounts. The lowest fatal dose reported was 15 g (U.S. Environmental Protection Agency, 1994d). Oral studies with chlorate demonstrate effects on hematological parameters and formation of methemoglobin, but at much higher doses than chlorite (between 157 and 256 mg/kg·day^{-1}). A study using African green monkeys given chlorate in drinking water at doses of 25 to 400 mg/L for 30 to 60 days did not result in any signs of red blood cell damage or oxidative stress. No carcinogenicity or reproductive effect tests have been conducted (U.S. Environmental Protection Agency, 1994d). Data were considered inadequate to develop an MCLG for chlorate in the USEPA Stage 1 D/DBP rule.

Bromine. Bromine has been used in swimming pools for disinfection and in cooling towers, but its use in drinking water has not been recommended. In water at neutral pH values, bromine exists primarily as hypobromous acid (HOBr), whereas below pH 6, Br_2, Br_3, bromine chloride, and other halide complexes occur. In the presence of ammonia or organic amines, bromoamines are formed. Most health effects information is on bromide salts because of their pharmaceutical use (National Academy of Sciences Safe Drinking Water Committee, 1980a).

Bromide occurs normally in blood at a range of 1.5 to 50 mg/L. Sedation occurs at a plasma concentration of about 960 mg/L, corresponding to a maintenance dose of 17 mg/kg·day^{-1}. Gastrointestinal disturbances can also occur at high doses. These effects have all been duplicated in laboratory animals. In addition, high doses in dogs were found to cause hyperplasia of the thyroid. No data have been developed on the mutagenicity, carcinogenicity, or teratogenicity of bromine (National Academy of Sciences Safe Drinking Water Committee, 1980b).

Iodine. In water, iodine can occur as iodine (I_2), hypoiodous acid (HOI), iodate (IO_3), or iodide (I^-). Iodoamines do not form to any appreciable extent. Iodine has been used to disinfect both drinking water and swimming pools. An iodine residual of about 1.0 mg/L is required for effective disinfection. Iodine is an essential trace element, required for synthesis of the thyroid hormone. The estimated adult requirement is 80 to 150 μg/day. Most intake of iodine is from food, especially seafood, and in the United States, table salt that has been supplemented with potassium iodide (National Academy of Sciences Safe Drinking Water Committee, 1987). Iodine is an irritant, with acute toxicity caused by irritation of the GI tract. A dose of 2 to 3 g may be fatal, although acute iodine poisoning is rare. No chronic data are available on iodine (Bull and Kopfler, 1990).

Ozone. Ozone is a very strong oxidant. It is moderately soluble in water and is typically used at a concentration of a few milligrams (O_3) per liter for drinking water disinfection. Over 1000 systems in Europe use ozone, and its use in the United States is increasing (over 100 systems). Ozone is unstable and is largely dissipated in the ozonation unit so that no ozone is present at the tap. Concerns about its safety are related to by-products such as bromate and formaldehyde, but not to the ozone itself (Bull and Kopfler, 1991).

Bromate. Bromate may be formed by the reaction of bromine with sodium carbonate. It is also formed in water following disinfection via ozonation of water containing bromide ion (U.S. Environmental Protection Agency, 1994f). Limited information exists concerning the concentration of bromate in drinking water. It was estimated that, if all surface water plants switched to ozone for predisinfection, the median nationwide distribution of bromate would be from 1 to 2 μg/L with the 90th to 95th percentiles in the range of 5 to 20 μg/L, based on bromide concentration (U.S. Environmental Protection Agency, 1994g).

Histopathological lesions in kidney tubules that coincided with decreased renal function were reported in rats exposed to 30 $mg/kg \cdot day^{-1}$ of bromate for 15 months. However, the study failed to provide dose response data and did not identify an NOAEL (U.S. Environmental Protection Agency, 1993, 1997c, 1998c). Bromate was shown to be mutagenic in several test systems (U.S. Environmental Protection Agency, 1993, 1997c, 1998c). Rats given sodium bromate orally over a 35-day period showed slightly reduced male epididymal sperm density at the highest dose given (26 $mg/kg \cdot day^{-1}$) (Wolfe and Kaiser, 1996). An increased incidence in kidney tumors was observed at doses as low as 60 mg/L. Thyroid tumor and peritoneal mesothelioma were detected at doses as low as 250 mg/L (U.S. Environmental Protection Agency, 1993, 1997c, 1998a). USEPA estimated the concentration of bromate in drinking water associated with excess cancer risk of 10^{-4} as 5 μg/L and classified bromate as a probable human carcinogen with an MCL of 0.01 mg/L (U.S. Environmental Protection Agency, 1994g).

Organic Disinfection By-Products

The use of oxidants for disinfection; for taste, odor, and color removal; and for decreasing coagulant demand also produces undesirable organic by-products. Surveys conducted since the mid-1970s have determined that the trihalomethanes and haloacetic acids, formed during drinking water chlorination, are the organic chemicals occurring the most consistently and at overall highest concentrations of any organic contaminant in treated drinking water. In addition to trihalomethanes and haloacetic acids, water chlorination can produce a variety of other compounds including haloacetonitriles, haloketones, haloaldehydes, chloropicrin, cyanogen chloride, MX, and chlorophenols. Exactly which compounds are formed, their formation pathways, and health effects are beginning to be better understood, but only 20 to 60 percent by weight of the halogenated material resulting from chlorination is accounted for by these compound classes (U.S. Environmental Protection Agency, 1995a).

Alternative disinfectants such as chloramines, chlorine dioxide, and ozone can also react with source water organics to yield organic by-products. The by-products are similar to those formed by chlorination but at lower concentrations (U.S. Environmental Protection Agency, 1994f). In addition, the presence of bromide ion during chlorination results in by-products with varying degrees of bromination and chlorination. In the United States, bromide levels in source waters range from undetectable to above 1 mg/L (Amy et al., 1993).

The concentrations of the chlorination by-products depend on several factors, with the total organic carbon (TOC) level or UV 254 (surrogates for the amount of precursor material) and the type and concentration of disinfectant being the most important. In general, the concentrations of chlorinated by-products are generally higher in surface waters than groundwaters because the level of TOC is generally higher in surface waters.

Discussion of the health effects of compounds that either have been found or are suspected to occur in the drinking water, and for which health effects data exist, follows.

Trihalomethanes (THMs). The THMs include chloroform, dibromochloromethane, dichlorobromomethane, and bromoform. All THMs are volatile and slightly water soluble. In most surface water systems chloroform is by far the most prevalent, but higher bromide concentrations can significantly shift these levels. The current MCL for THMs is 100 μg/L for total trihalomethanes and is based on a running annual average of quarterly samples. USEPA's Stage 1 D/DBP rule has an MCL of 80 μg/L for total trihalomethanes, based on a running annual average of quarterly samples. Health effects information on each THM follows.

Chloroform. Besides its formation during drinking water chlorination, chloroform is produced for use as a refrigerant and aerosol propellant; as a grain fumigant; and as a general solvent for adhesives, pesticides, fats, oils, rubbers, alkaloids, and resins (U.S. Environmental Protection Agency, 1994f).

Chloroform was used as an anesthetic and has been found to affect liver and kidney function in humans in both accidental and long-term occupational exposure situations. In experimental animals, chloroform causes changes in kidney, thyroid, liver, and serum enzyme levels. These responses are discernible in mammals from exposure to levels of chloroform ranging from 15 to 290 $mg/kg \cdot day^{-1}$ (U.S. Environmental Protection Agency, 1994e). Data concerning the developmental effects of chloroform indicate toxicity to the mother and fetus at high doses and suggest reproductive and developmental toxicity may occur (U.S. Environmental Protection Agency, 1994e, 1998e).

The results of a number of assays to determine the genotoxicity of chloroform are equivocally negative. Chloroform induced liver tumors in mice when administered by gavage in corn oil, but not via drinking water, and induced kidney tumors in male rats, regardless of the carrier vehicle (oil or drinking water) (U.S. Environmental Protection Agency, 1994e, 1998e). An expert panel formed by the International Life Sciences Institute viewed chloroform as a likely carcinogen to humans above a certain dose range, but considered it unlikely to be carcinogenic below a certain dose (International Life Science Institute, 1997). USEPA concluded in the Stage 1 D/DBP rule that the MCLG was zero.

Bromodichloromethane (BDCM). Only a small amount of BDCM is currently produced commercially in the United States for use as an intermediate for organic synthesis and as a laboratory reagent (U.S. Environmental Protection Agency, 1994f).

In mice and rats, BDCM caused changes in kidney, liver, and serum enzyme levels, and decreased body weight. These responses were discernible in rodents from exposure to levels of BDCM that ranged from 6 to 300 mg/kg. Bromodichloromethane significantly decreased maternal weight and increased relative kidney weights in rodents at 200 $mg/kg \cdot day^{-1}$. There were no increases in the incidence of fetotoxicity or external/visceral malformations. One study observed a decrease in sperm motility in rats given BDCM in drinking water for one year at an average dose of 39 $mg/kg \cdot day^{-1}$ (U.S. Environmental Protection Agency, 1994f, 1997c).

Mutagenicity studies reported mixed results in bacteria and yeasts, but BDCM was mutagenic in mouse lymphoma cells with metabolic activation. An increase in frequency of sister chromatid exchange was reported in most mammalian cells and the overall results suggest BDCM is genotoxic (U.S. Environmental Protection Agency, 1994e, 1997c). Bromodichloromethane in corn oil administered orally produced increased kidney tumors in male mice, liver tumors in female mice, and kidney and large intestine tumors in male and female rats (U.S. Environmental Protection Agency, 1994e, 1997c). USEPA estimated the concentration of BDCM associated with an excess cancer risk of 10^{-4} as 60 μg/L and classified BDCM as a probable human carcinogen with an MCLG of zero (U.S. Environmental Protection Agency, 1998h).

Dibromochloromethane (DBCM). Aside from its occurrence as a DBP, DBCM is used as a chemical intermediate in the manufacture of fire-extinguishing agents, aerosol propellants, refrigerants, and pesticides (U.S. Environmental Protection Agency, 1994f).

In mice and rats, DBCM causes changes in kidney, liver, and serum enzyme levels, and decreased body weight (U.S. Environmental Protection Agency, 1994e). In a chronic study, DBCM at doses of 40 and 50 mg/kg·day^{-1} in corn oil produced histopathological lesions in the liver of rats and mice, respectively (U.S. Environmental Protection Agency, 1994e, 1997c). A multigeneration reproductive study of mice treated with DBCM in drinking water showed maternal toxicity (weight loss, liver pathology) and fetal toxicity (U.S. Environmental Protection Agency, 1994e).

There was a significant increase in the incidence of benign and malignant liver tumors in female and male mice given DBCM at 100 mg/kg·day^{-1} in corn oil. Rats did not develop cancer (U.S. Environmental Protection Agency, 1994e). Several studies on the mutagenicity potential of DBCM have reported inconclusive results (U.S. Environmental Protection Agency, 1994e). USEPA has classified DBCM as a possible human carcinogen with an MCLG of 0.6 mg/L (U.S. Environmental Protection Agency, 1998h).

Bromoform. Bromoform occurs less frequently than the other THMs. Commercially, bromoform is used in pharmaceutical manufacturing, as an ingredient in fire-resistant chemicals and gauge fluid, and as a solvent for waxes, greases, and oils (U.S. Environmental Protection Agency, 1994f).

In mice and rats, bromoform caused changes in kidney, liver, and serum enzyme levels, decrease of body weight, and decreased operant response. These responses are discernible in mammals from exposure to levels of bromoform ranging from 50 to 250 mg/kg (U.S. Environmental Protection Agency, 1994e). A developmental study in rats showed no fetotoxicity or teratogenicity. One detailed reproductive toxicity study reported no apparent effects on fertility and reproduction (U.S. Environmental Protection Agency, 1994e, 1997c).

Studies on the in vitro genotoxicity of bromoform reported mixed results in bacteria, mutations in cultured mouse lymphoma cells, sister chromatid exchange in human lymphocytes, and sister chromatid exchange, chromosomal aberration, and micronucleus formation in mouse bone marrow cells (U.S. Environmental Protection Agency, 1994e). Bromoform administered orally in corn oil in rats produced an increased incidence of benign and malignant tumors of the intestinal colon or rectum (U.S. Environmental Protection Agency, 1994e). USEPA (1998h) estimated the concentration of bromoform associated with an excess cancer risk of 10^{-4} as 400 μg/L and classified bromoform as a probable human carcinogen with an MCLG of zero.

Haloacetic Acids. The haloacetic acids (HAAs) consist of nine highly water-soluble compounds and, as a group, are the next most prevalent DBPs after THMs. Dichloroacetic and trichloroacetic acid are the most prevalent HAAs. In high-bromide waters, the brominated species such as bromodichloroacetic acid and bromochloroacetic acid are more prevalent. The majority of the available human health data are for dichloroacetic acid and trichloroacetic acid. However, recent information suggests that some of the brominated HAAs may be of concern and warrant additional evaluation (International Life Sciences Institute, 1995).

Dichloroacetic acid (DCA). Dichloroacetic acid is used as a chemical intermediate, and an ingredient in pharmaceuticals and medicine. Dichloroacetic acid was used experimentally in the past to treat diabetes and hypercholesterolemia in human patients (U.S. Environmental Protection Agency, 1994f), and it is also a metabolite of trichloroethylene.

Humans treated with DCA for 6 to 7 days at 43 to 57 $mg/kg \cdot day^{-1}$ have experienced mild sedation, reduced blood glucose, reduced plasma lactate, reduced plasma cholesterol levels, and reduced triglyceride levels. A longer-term study in two young men receiving 50 mg/kg for 5 weeks up to 16 weeks, indicated that DCA significantly reduced serum cholesterol levels and blood glucose, and caused peripheral neuropathy (U.S. Environmental Protection Agency, 1994b, 1997c, 1998a). A dose-related increase of liver damage in mice that received DCA at 270 to 300 $mg/kg \cdot day^{-1}$ for 37 to 52 weeks has been reported. In another mouse study, DCA at 77 $mg/kg \cdot day^{-1}$ and above caused an increase in liver weight in mice. Dogs given DCA had toxic effects in liver, testis, and brain at the lowest dose tested, 12.5 $mg/kg \cdot day^{-1}$. Hindlimb paralysis was observed in three dogs in the high-dose group (U.S. Environmental Protection Agency, 1994b, 1997c).

Dichloroacetic acid appears to induce both reproductive and developmental toxicity (U.S. Environmental Protection Agency, 1994b, 1997c, 1998a). Damaged and atrophic sexual organs have been reported in male rats and dogs exposed to levels as low as 50 $mg/kg \cdot day^{-1}$. Malformation of the cardiovascular system was observed in rats exposed to 140 $mg/kg \cdot day^{-1}$ DCA from day 6 to 16 of pregnancy (U.S. Environmental Protection Agency, 1994b, 1997c, 1998a).

Several studies indicate that DCA induces liver tumors in both mice and rats exposed via drinking water. In one study with mice, exposure to DCA at 140 and 280 $mg/kg \cdot day^{-1}$ resulted in significant tumor formation. In male rats exposed to 4, 40, and 295 $mg/kg \cdot day^{-1}$, tumor prevalence increased only in the highest dose. However, at 40 $mg/kg \cdot day^{-1}$, there was an increase in the prevalence of proliferation of liver lesions that are likely to progress into malignant tumors (U.S. Environmental Protection Agency, 1994b, 1997c, 1998a). Although DCA has been found to be mutagenic, responses generally occur at relatively high doses (U.S. Environmental Protection Agency, 1994b, 1997c, 1998a). USEPA (1994f, 1998h) has classified DCA as a probable human carcinogen with an MCLG of zero.

Trichloroacetic acid (TCA). Trichloroacetic acid is used commercially in organic synthesis and in medicine to remove warts and damaged skin. It is used in the laboratory to precipitate proteins and as a reagent for synthetic medicinal products (U.S. Environmental Protection Agency, 1994f).

Male rats exposed to TCA in their drinking water at the highest dose (355 $mg/kg \cdot day^{-1}$) were found to have spleen weight reduction and increased liver and kidney weights. In another study, male rats exposed to TCA in their drinking water exhibited no significant changes in body weight, organ weight, or histopathology over the study duration (U.S. Environmental Protection Agency, 1994b, 1997c).

Several studies have shown that TCA, like DCA, can produce developmental malformations in rats, particularly in the cardiovascular system (U.S. Environmental Protection Agency, 1994b, 1997c). Teratogenic effects were observed at the lowest dose tested, 330 $mg/kg \cdot day^{-1}$ (U.S. Environmental Protection Agency, 1994b, 1997c).

Trichloroacetic acid was negative in Ames mutagenicity tests, but produced bone marrow chromosomal aberrations and sperm abnormalities in mice and induced single-strand DNA breaks in rats and mice exposed by gavage, and, overall, the available studies suggest TCA does not operate via mutagenic mechanisms (U.S. Environmental Protection Agency, 1994b, 1997c). Trichloroacetic acid has induced liver tumors in two tests with mice, one of 52 weeks' and another of 104 weeks' duration, but was not carcinogenic in male rats at doses as high as 364 mg/kg·day^{-1} (U.S. Environmental Protection Agency, 1994b, 1997c). In the 52-week study, a dose-related increase in the incidence of liver tumors was reported in male mice exposed to 164 and 329 mg/kg·day^{-1} for 52 weeks. In the 104-week study, there was a reported increase in the prevalence of liver tumors in male and female mice exposed to 583 mg/kg·day^{-1} TCA for 104 weeks, but not at 71 mg/kg·day^{-1} (U.S. Environmental Protection Agency, 1994b). USEPA (1998h) has classified TCA as a possible human carcinogen with an MCLG of 0.3 mg/L.

Haloacetaldehydes. The haloacetaldehydes include chloroacetaldehyde, dichloroacetaldehyde, and trichloroacetaldehyde monohydrate (chloral hydrate). Brominated analogues probably also occur. The health effects data for the haloacetaldehydes are limited.

Chloroacetaldehyde. Various studies report that chloroacetaldehyde is irritating and corrosive to lipids and membrane structures. It also causes a decrease in liver enzyme activity. Subchronic effects observed in rats subjected to intraperitoneal injections of 2.2 or 4.5 mg/kg chloroacetaldehyde for 30 days included a decrease in red blood cell components and an increase in white blood cell components. Also noted were organ-to-body weight ratio increases, with overall body weight decreases. Similar results were obtained in a longer-term (12 week), lower-dose (0.4 to 3.8 mg/kg) experiment, again using rats and intraperitoneal injections. This latter experiment also noted possibly premalignant necroses in the respiratory tract at the two highest doses (National Academy of Sciences Safe Drinking Water Committee, 1987; U.S. Environmental Protection Agency, 1994b). All chloroacetaldehydes were found to be mutagenic in bacteria, but no carcinogenicity has been observed in several studies (U.S. Environmental Protection Agency, 1994b).

Dichloroacetaldehyde. This chemical was found to be mutagenic in studies using *Salmonella typhimurium* strain TA100, but not in other strains (U.S. Environmental Protection Agency, 1994b).

Trichloroacetaldehyde (chloral hydrate, or CH). Chloral hydrate is used as a hypnotic or sedative drug ("knockout drops") in humans, including neonates. It is also used in the manufacture of DDT (U.S. Environmental Protection Agency, 1994f). The median concentration of chloral hydrate in 84 water systems was 2.1 μg/L.

Three 90-day studies with mice have been conducted for chloral hydrate using 16 and 160 mg/kg·day^{-1} (U.S. Environmental Protection Agency, 1994b). One study reported no treatment-related effects. The second study found a dose-related liver enlargement and biochemical microscopic changes in males. Females only displayed changes in liver biochemistry. The third study reported no effects in males, but exposure to the high dose resulted in decreased humoral immune function in females, though not cell-mediated immunity. Developmental effects (impaired passive avoidance learning tasks) were observed in mice pups from female mice receiving CH at 205 mg/kg·day^{-1} for three weeks prior to breeding and continuing until pups were weaned at 21 days of age (U.S. Environmental Protection Agency, 1994b). Chloral hydrate is weakly mutagenic, but caused chromosomal aberration in yeast and disrupted chromosomal rearrangement chromosomes during spermatogenesis (U.S. Environmental Protection Agency, 1994b, 1997c). Two cancer studies indicate that CH produces liver tumors in mice receiving 166 mg/kg·day^{-1} CH for 104 weeks (U.S. Environmental Protection Agency, 1994b).

USEPA estimated the concentrations of chloral hydrate in drinking water associated with an excess cancer risk of 10^{-4} as 40 μg/L and classified choral hydrate in Group C, possible human carcinogen (U.S. Environmental Protection Agency, 1994f).

Formaldehyde. Aldehydes are formed with most disinfectants. The highest concentrations have been observed with ozonation. The expected concentration range for the aldehydes in ozonated water is about 3 to 30 μg/L, with a mean of about 8 μg/L. Formation of aldehydes with chlorine and chloramines would be less (Bull and Kopfler, 1991).

Formaldehyde has been shown to be mutagenic and produce chromosomal alterations in a variety of test systems (Bull and Kopfler, 1990). Although it is regarded as a known carcinogen by inhalation, there are conflicting data regarding the oral carcinogenicity of formaldehyde. One study found an increased incidence of gastrointestinal tract tumors in rats. However, the incidence of tumors was small and no statistical analyses were provided. Two other rat studies with formaldehyde indicate carcinogenic effects (Bull and Kopfler, 1991).

Haloketones. The haloketones are minor constituents and include hex-, tetra-, tri-, di-, and monochloropropanones. Brominated analogues of these compounds would be expected if bromide is present in the water. Other than occurrence as DBPs, the haloketones are principally found as chemical intermediates in industrial processes. There is limited health effects information for the haloketones. Four haloketones (1,1,3-trichloroacetone, 1,1,3,3-tetrachloroacetone, pentachloroacetone, and hexachloroacetone) have been shown to be to be mutagenic using various strains of *Salmonella* (U.S. Environmental Protection Agency, 1994f). There is little indication that they are mutagenic in vivo (Bull and Kopfler, 1991). No information on the effects of long-term exposure to chlorinated ketones in drinking water is available.

Haloacetonitriles. The haloacetonitriles include chloroacetonitrile (CAN), dichloroacetonitrile (DCAN), trichloroacetonitrile (TCAN), bromochloroacetonitrile (BCAN), and dibromoacetonitrile (DBAN). These haloacetonitriles have limited commercial use, but are produced during chlorination of water containing organic matter. Commercially, the chlorinated acetonitriles are used as insecticides and fungicides (National Academy of Sciences Safe Drinking Water Committee, 1987; U.S. Environmental Protection Agency, 1991a).

Subchronic effects of haloacetonitriles (in corn oil) have been investigated using rats. Doses of 65 mg/kg of DCAN and DBAN for 90 days produced decreased weight gains, decreased organ weights and organ-to-body weight ratios of the liver, spleen, thymus, lungs, and kidneys. A decrease in blood cholesterol levels was also noted. Chloroacetonitrile, dichloroacetonitrile, and trichloroacetonitrile were all fetotoxic (decreased offspring weight and postnatal growth) when a dose of 55 mg/kg was administered to pregnant rats over 90 days (National Academy of Sciences Safe Drinking Water Committee, 1987; U.S. Environmental Protection Agency, 1991a). Studies on the mutagenicity of DBAN, BCAN, DCAN, and TCAN were positive in sister chromatid exchange and/or bacterial/microsome assays. Both DCAN and BCAN were shown to be mutagenic in *Salmonella.* In cultured human lymphoblasts, DNA strand breaks were produced by TCAN, BCAN, DBAN, and DCAN (U.S. Environmental Protection Agency, 1991a). Administration of TCAN and BCAN by gavage to mice at doses of 10 mg/kg produced a significant increase in the number of benign lung tumors, and dermal application caused skin cancer in mice. When tested for cancer-initiating activity, DBAN, DCAN, TCAN, and BCAN were all found to be negative (U.S. Environmental Protection Agency, 1991a). No EPA standards or goals have been proposed.

Chloropicrin. Chloropicrin, also known as trichloronitromethane or nitrochloroform, is used in organic synthesis, dyestuffs (crystal violet), fumigants, fungicides, insecticides, and tear gas. It is a strong irritant and toxicant when inhaled or ingested. The median concentration of chloropicrin is less than 1 μg/L.

No data are available for chloropicrin on adverse systemic effects, or effects on reproduction or the developing fetus. Chloropicrin does not appear to be carcinogenic or mutagenic, although there was a modest increase of stomach cancers in one study (National Academy of Sciences Safe Drinking Water Committee, 1987; U.S. Environmental Protection Agency, 1991a).

Cyanogen Chloride. Cyanogen chloride has been used in chemical synthesis, tear gas, and fumigant gases. Higher levels of cyanogen chloride were produced by systems that used chloramination for disinfection than by those that used chlorination. Median cyanogen chloride concentrations are about 1 μg/L in chlorinated drinking waters and about 2 μg/L in chloraminated waters (Bull and Kopfler, 1991).

Studies in various rodent and nonrodent species show that cyanogen chloride is acutely toxic via inhalation and oral administration. No data are available on the effects of cyanogen chloride on reproduction or the developing fetus or on its mutagenicity or carcinogenicity (U.S. Environmental Protection Agency, 1991a).

Chlorophenols. The chlorophenols include monochlorophenols (2-, 3-, 4-chlorophenol), dichlorophenols (2,4-, 2,6-dichlorophenol), and trichlorophenols (2,4,6-trichlorophenol). Chlorophenols are produced industrially as biocides and as intermediates in the synthesis of dyestuffs, pesticides, and herbicides. Chlorophenols appear to occur at less than 1 μg/L in drinking waters (Bull and Kopfler, 1991). Health effects data have been developed primarily for 2,4-dichlorophenol and 2,4,6-trichlorophenol (U.S. Environmental Protection Agency, 1989a).

2,4-Dichlorophenol. Like most of the other chlorophenols, 2,4-dichlorophenol has a low oral toxicity, up to 10,000 mg/kg, and no evidence of carcinogenic activity for rats or mice, except for a skin assay (National Academy of Sciences Safe Drinking Water Committee, 1987; U.S. Environmental Protection Agency, 1989a).

2,4,6-Trichlorophenol. Two short-term feeding studies have been conducted. A seven-week study using rats and mice showed a reduction in growth rate at levels greater than 500 mg/kg·day^{-1}. A 14-day study using rats given daily oral doses of 0 to 200 mg/kg detected a minimal effect on liver detoxification capabilities. 2,4,6-Trichlorophenol tested negative or equivocal in microbial mutagenicity assays. One oral carcinogenicity test with mice was equivocal, whereas in another, male rats developed lymphomas or leukemias, and mice developed liver cancers (National Academy of Sciences Safe Drinking Water Committee, 1982; U.S. Environmental Protection Agency, 1989a).

3-Chloro-4-(Dichloromethyl)-5Hydroxy-2(5H)-Furanone (MX). The organic disinfection by-product MX appears to be produced in water treated with chlorine and has been found in drinking water in several countries (Bull and Kopfler, 1991). The mean concentrations of MX found in Finland range from less than 0.004 μg/L to 0.067 μg/L, with a mean of 0.027 μg/L; in the United Kingdom, mean concentrations of MX range from 0.002 to 0.023 μg/L, with a mean of 0.007 μg/L.

One of the most potent mutagens in the Ames test is MX (Bull and Kopfler, 1991). Male and female rats given MX in drinking water developed cancer at multiple sites at all doses (Komulainen et al., 1997). No data are available on the effects of MX on reproduction.

RADIONUCLIDES

Radionuclides are unstable atoms that spontaneously disintegrate, releasing a portion of their nucleus as radioactivity in three forms:

1. Alpha radiation, consisting of large, positively charged particles made up of two protons and two neutrons (the same as a helium nucleus)
2. Beta radiation, consisting of electrons or positrons
3. Gamma radiation, consisting of electromagnetic, wave-type energy, such as X rays

Elements heavier than lead, such as radon, radium, thorium, and uranium isotopes, decay by the release of alpha, alpha and gamma, or beta and gamma emissions. Radionuclides lighter than lead generally decay by beta and gamma emissions.

The process of alpha and beta radiation results in the formation of a lighter isotope or element, while gamma ray emission by itself does not. (Isotopes are defined by the total number of neutrons and protons, whereas elements are defined by the total number of protons. For example, the uranium-238 (U-238) isotope is the 92nd element on the periodic chart because it has 92 protons, but it has a total of 238 neutrons and protons.) The isotope that decays is called the parent, and the new element is called the progeny, or daughter. Each naturally occurring isotope heavier than lead belongs to one of three series of stepwise decay pathways stemming from a long-lived isotope. For example, radium-226 and radon-222 are 2 of the 14 progenies of U-238. Eventually, the progenies in this series decay to the stable isotope lead-206. Radium-228 is 1 of 10 progenies of the thorium-232 series that decays eventually to the stable isotope lead-208. The parent of the third series is U-235.

Different isotopes decay at different rates. The half-life of an isotope is the time required for one-half of the atoms present to decay, and can range from billions of years (like U-238) to millionths of a second. Isotopes with longer half-lives have lower radioactivity. Isotopes with very short half-lives are not significant in that they do not survive transport through drinking water distribution systems; however, their progenies may be biologically significant radionuclides. Radioactivity is generally reported in units of curies (Ci), rads, or rems. One curie equals 3.7×10^{10} nuclear disintegrations per second (the radioactivity in 1 g of radium, by Marie Curie's definition). A common fraction is the picocurie (pCi), which equals 10^{-12} Ci. By comparison, 1 g of the relatively stable uranium-238 has 0.36×10^{-6} Ci of radioactivity.

Another factor distinguishing different isotopes is their source. Radioactivity in water can be naturally occurring (typical) or manmade (occasional). Naturally occurring radiation derives from elements in the earth's crust or from cosmic and X-ray bombardment within the atmosphere. Manmade radiation comes from three general sources: nuclear fission from weapons testing, radiopharmaceuticals, and nuclear fuel processing and use. About 200 manmade radionuclides occur or potentially occur in drinking water. However, only strontium-90 and tritium (a hydrogen isotope) have been detected on a consistent basis.

Based on occurrence in drinking water and health effects, the radionuclides of most concern are radium-226, radium-228, uranium-238, and radon-222 (U.S. Environmental Protection Agency, 1988a). Radium-224 and polonium-210 are also frequently found and should probably be included in this list. These are all naturally occurring isotopes. Radium-228, arising from the series originating in thorium-232, is a beta emitter, whose decay gives rise to a series of alpha-emitting daughters, whereas the others are all alpha emitters. Natural uranium actually includes U-234, U-235, plus U-238; however, U-238 makes up 99.27 percent of the composition. Radon is one of the noble gases, like helium or neon. Because it is a nonreactive gas, it quickly volatilizes from water and can significantly contribute to the level in indoor air. Radioactive thorium and lead isotopes occur only at low levels in groundwater.

Groundwater is the most important source of concern. In the United States, the highest levels of radium are found in the Upper Midwest and in the Piedmont region of the Southeast, particularly in smaller systems. Radon is found dissolved in groundwater, often near granitic formations containing uranium, but can quickly volatilize into a gas. The highest levels have been found in the northern half of the Appalachians and in the Upper Midwest and Rocky Mountain states.

Each type of radiation interacts differently with cells the human body. Alpha particles travel at speeds as high as 10 million meters per second, and when alpha-emitting radionuclides are ingested, the relatively massive alpha particles can be very damaging to nearby cells. Beta particles can travel close to the speed of light. Their smaller mass allows greater penetration, but creates less damage to organs and tissue. Gamma radiation has tremendous penetrating power, but has limited effect on the body at low levels. A rad quantifies the energy absorbed by tissue or matter, such that a rad of alpha particles creates more damage than a rad of beta particles. A rem quantifies radiation in terms of its dose effect, such that equal doses expressed in rems produce the same biological effect regardless of the type of radiation involved. Thus, an effective equivalent dose of 0.1 mrem/yr from any type of radiation corresponds to about 10^{-6} excess lifetime cancer risk level. Numerically, a rem is equal to the absorbed dose (in rads) times a quality factor (Q) that is specific to the type of radiation and reflects its biological effects (U.S. Environmental Protection Agency, 1991a). (In the international system, a becquerel is equal to one disintegration per second, or 22 pCi, and a sievert is equal to 100 rem.)

Humans receive an average annual dose of radiation of about 200 mrem from all sources. USEPA estimates that drinking water contributes about 0.1 to 3 percent of a person's annual dose. Local conditions can alter this considerably, however. For example, the level of 10,000 pCi/L of radon, found in some systems, corresponds to an annual effective dose equivalent of 100 mrem/yr (U.S. Environmental Protection Agency, 1991c).

Radioactivity can cause developmental and teratogenic effects, heritable genetic effects, and somatic effects including carcinogenesis. The carcinogenic effects of nuclear radiations (alpha, beta, and gamma) on the cell are thought to be ionization of cellular constituents leading to changes in the cellular DNA and followed by DNA-mediated cellular abnormalities. All radionuclides are considered to be human carcinogens, though their primary target organs and relative efficiencies differ.

Radium is accumulated in bone, and all radium isotopes (-224, -226, and -228) cause bone sarcomas. Radium-226 also induces head carcinomas, presumably because it decays into radon, which, as a gas, can enter the sinuses. Data about radium come from studies of radium watch dial painters, as well as persons medically treated with radium injections.

Direct association has been shown between inhalation of radon-222 by uranium miners and lung cancer, though there is also a small risk of other cancers (U.S. Environmental Protection Agency, 1991c). Evidence is strong that there is also significant lung cancer risk from residential indoor air (Lubin and Boice, 1997). Radon, an inert gas, can dissolve in groundwater following decay of radium-226. As an inert gas, it volatilizes out of water like the VOCs. It has been estimated that volatilization during showering releases between two-thirds and three-quarters of the dissolved radon and, as a result of typical water use (especially showering), 10,000 pCi/L in water results in an increase of 1 pCi/L of air in a typical house (U.S. Environmental Protection Agency, 1991c; Hopke et al., 1996). This compares with the USEPA residential indoor air guidance level of 4 pCi/L (of air), above which remedial measures are recommended, and the approximate level in outdoor air, 0.5 pCi/L. Inhalation of volatilized radon is the primary route of exposure from water. Estimates of total cancer risk at 4 pCi/L (of air) are in the 1 to 5 per 100 range among persons exposed over a lifetime and is increased by smoking (U.S. Environmental Protection Agency, 1991c). Total cancer risk from 100 pCi/L in water was estimated to be about 67 per million. The National Academy of Sciences Committee on the Risk Assessment of Exposure to Radon in Drinking Water (1998) recently estimated that 100 pCi/L of radon in water represents a cancer risk of about 56 per million, assuming that risk is linearly related to concentration. Because of its high concentration in some systems, radon-induced carcinogenicity is a concern.

Uranium-238 is a weak carcinogen because of its long half-life. It accumulates in the bones like radium. The USEPA classifies it as a human carcinogen, but it also has a demonstrated toxic effect on human kidneys leading to kidney inflammation and changes in urine composition.

Under the NIPDWR, MCLs were set for gross alpha emitters at 15 pCi/L, for radium-226 and -228 combined at 5 pCi/L, and for gross beta emitters at 4 mrem/year. The gross alpha standard is meant as a trigger for more testing, and is the only standard for alpha emitters like polonium-210 or radium-224.

Revised MCLs and new MCLGs have been proposed for radionuclides (U.S. Environmental Protection Agency, 1991c). Radium-226 and -228 would be regulated separately with MCLs of 20 pCi/L each, approximating an estimated lifetime risk of one cancer per 10,000 persons exposed to the MCL. Uranium would be regulated at 20 pCi/L based on its renal effects. The proposal of 300 pCi/L for radon may be modified because the background relative risk from radon in outdoor air (0.5 pCi/L) is considerably greater than would arise from indoor volatilization of 300 pCi/L of radon in water. Nevertheless, it was estimated that 150 pCi/L corresponds to a lifetime cancer risk of 1 in 10,000.

AESTHETIC QUALITY

In addition to health issues, consumer satisfaction and confidence are also important. Aesthetic components of drinking water quality include taste and odor, turbidity, color, mineralization, hardness, and staining. These problems can originate in source water, within the treatment plant, in distribution systems, and in consumer plumbing. Many are addressed in the National Interim Secondary Drinking Water Regulations (see Chapter 1). Aesthetic problems can be caused by some natural or commercial inorganic and organic chemicals, as well as by organisms.

Taste and Odor

Taste problems in water derive in part from salts [total dissolved solids (TDS)] and the presence of specific metals, such as iron, copper, manganese, and zinc. In general, waters with TDS less than 1200 mg/L are acceptable to consumers, although levels less than about 650 mg/L are preferable (Mallevialle and Suffet, 1987). Specific salts may be more significant in terms of taste, notably magnesium chloride and magnesium bicarbonate. The sulfate salts magnesium sulfate and calcium sulfate, on the other hand, have been found to be relatively inoffensive (Mallevialle and Suffet, 1987). Fluoride can also cause a distinct taste above about 2.4 mg/L. Testing of metals in drinking water (James M. Montgomery Consulting Engineers, Inc., 1985) showed the following taste thresholds (in mg/L):

Zinc	4–9
Copper	2–5
Iron	0.04–0.1
Manganese	4–30

(Taste and odor thresholds typically represent the level at which one or more individuals in a panel can sense the presence of the chemical.)

Objectionable tastes and odors may also occur in water contaminated with synthetic organics and/or as a result of water treatment or from coatings used inside tanks and pipes. Odor thresholds of some common solvents identified in groundwater are shown in Table 2.5.

Many consumers object to the taste of chlorine, which has a taste threshold of about 0.2 mg/L at neutral pH, whereas the disinfectant, monochloramine, has been found to have a taste threshold of 0.48 mg/L (Mallevialle and Suffet, 1987). Di- and trichloramine have stronger odors. Although chlorine can reduce many taste and odor problems by oxidizing the offending compound, it also reacts with organics to create taste and odor problems. Most notorious is the phenolic odor resulting from reactions between chlorine and phenols. The disinfection by-product, chloroform, has been detected by smell at 0.1 mg/L (Mallevialle and Suffet, 1987). The formation of aldehydes from chlorinated amino acids has also resulted in problem odors (Gittleman and Luttweiler, 1996). Other oxidants can also be used to remove problem chemicals.

TABLE 2.5 Odor Thresholds of Solvents

Solvent	Detection odor threshold mg/L
Benzene	0.19[1]
Carbon tetrachloride	0.2[2]–50[1]
1,4-Dichlorobenzene	0.0003[2]–0.03[1]
Methyl *tert*-butyl ether	0.015[3]
Tetrachloroethylene	0.3[2]
Toluene	1.0[3]
Trichloroethylene	0.5[2]–10[1]
Xylenes	0.5[1]

Sources:

[1] Verscheuren, K. *Handbook of Environmental Data on Organic Chemicals.* New York: Van Nostrand Reinhold, 1983.

[2] Van Gemert, L.J., and A.H. Nettenbreijer, eds. *Compilation of Odour Threshold Values in Air and Water.* Natl. Inst. for Wtr. Supply, Voorburg, Netherlands, and Centr. Inst. for Nutr. & Food Res. TNO, Zeist, Netherlands, 1977.

[3] Young, W.F., H. Horth, R. Crane, T. Ogden, and M. Arnott. "Taste and Odour Threshold Concentrations of Potential Potable Water Contaminants." *Water Res.,* 30, 1996: 331–340.

Decaying vegetation and metabolites of microbiota are probably the most universal sources of taste and odor problems in surface waters (Mallevialle and Suffet, 1987). The organisms most often linked to taste and odor problems are the filamentous bacteria actinomycetes, iron and sulfur bacteria, and blue-green algae, although other algal types, bacteria, fungi, and protozoans are also cited. The metabolites responsible for the tastes and odors are still being identified. Two highly studied metabolites of actinomycetes and blue-green algae are geosmin and methylisoborneol (MIB). These compounds are responsible for the common earthy, musty odors in water supplies and have been isolated from many genera of actinomycetes (e.g., *Actinomyces, Nocardia,* and *Streptomyces*) and of blue-green algae (e.g., *Anabaena* and *Oscillatoria*). Both geosmin and MIB can have odor threshold concentrations of less than 10 ng/L (Mallevialle and Suffet, 1987).

In groundwaters and in some distribution systems, a highly unpleasant odor of hydrogen sulfide may occur as the result of anaerobic bacterial action on sulfates (Mallevialle and Suffet, 1987). This rotten egg odor can be detected at less than 0.1 μg/L. The bacterium most often responsible for hydrogen sulfide production is *Desulfovibrio desulfuricans.* Other bacterially produced sulfur compounds creating swampy and fishy tastes and odors in distribution systems include dimethylpolysulfides and methylmercaptan. Various types of *Pseudomonas* bacteria can also produce obnoxious sulfur compounds.

Maintenance of a chlorine residual in the distribution system is an important measure in controlling growth of aesthetically significant microorganisms, many of which grow in biofilm on pipes (also, refer back to the section on opportunistic bacteria, in the section "Pathogenic Organisms" in this chapter). However, if nonpathogenic constituents, such as iron bacteria, nitrifying bacteria, or sulfur bacteria, have built up excessively in the distribution system, the chlorine demand can be significantly increased (American Water Works Association, 1995a). The subsequent decreased chlorine residual can allow increased growth of microorganisms in the distribution system, especially in slow-flow areas.

Methods for identification and treatment of taste and odor problems are discussed extensively by Mallevialle and Suffet (1987) and in Manual 7 of the American Water Works Association (1995).

Turbidity and Color

The appearance of a water can be a significant factor in consumer satisfaction. Low levels of color and turbidity are also important for many industries. Typical finished waters have color values ranging from 3 to 15 and turbidities under 1 NTU.

Sources of color in water can include natural metallic ions (iron and manganese), humic and fulvic acids from humus and peat materials, plankton, dissolved plant components, iron and sulfur bacteria, and industrial wastes. The added presence of turbidity increases the apparent, but not true color of water. Color removal is typically achieved by the processes of coagulation, flocculation, sedimentation (or floatation), and filtration.

Turbidity in water is caused by the presence of colloidal and suspended matter (such as clay, silt, finely divided organic and inorganic matter, plankton, and other microscopic organisms). The association of turbidity with infectious disease was described earlier. Iron bacteria can also be a source of turbidity. Suspended particles can be a significant health concern if heavy metals and hydrophobic chemicals (e.g., many pesticides) are adsorbed to them. Controlling turbidity as a component of treatment, as mandated by the surface water treatment rule (SWTR) (U.S. Environmental

Protection Agency, 1989b) and the interim enhanced SWTR, will require levels well below visual detection limits (U.S. Environmental Protection Agency, 1995b, 1998i).

Mineralization

Waters with high levels of salts, measured as TDS, may be less palatable to consumers, and, depending upon the specific salts present, may have a laxative effect on the transient consumer. High levels of sulfates are implicated in this latter aspect. Sulfate may also impart taste, at levels above 300 to 400 mg/L. Concentrations of chloride above 250 mg/L may give water a salty taste.

High salts can also adversely affect industrial cooling operations, boiler feed, and specific processes requiring softened or demineralized waters, such as food and beverage industries and electronics firms. Typically, high salts will force these industries to pretreat their water. High levels of chloride and sulfate can accelerate corrosion of metals in both industrial and consumer systems.

Removal of salts requires expensive treatment, such as demineralization by ion exchange, electrodialysis, or reverse osmosis, and is not usually done for drinking water. For excessively high-salt water, blending with lower-salt supplies may ameliorate the problem.

Hardness

Originally, the hardness of a water was understood to be a measure of the capacity of the water for precipitating soap. It is this aspect of hard water that is the most perceptible to consumers. The primary components of hardness are calcium and magnesium, although the ions of other polyvalent metals such as aluminum, iron, manganese, strontium, and zinc may contribute if present in sufficient concentrations. Hardness is expressed as an equivalent quantity of calcium carbonate ($CaCO_3$). Waters having less than 75 mg/L $CaCO_3$ are generally considered soft. Those having between 75 and 150 mg/L $CaCO_3$ are said to be moderately hard. Those having from 150 to 300 mg/L $CaCO_3$ are hard, and waters having more than 300 mg/L $CaCO_3$ are classified as very hard.

Calcium is of importance as a component of scale. The precipitation of $CaCO_3$ scale on cast-iron and steel pipes and on the lead-soldered joints of indoor copper plumbing helps inhibit corrosion, but the same precipitate in boilers and heat exchangers adversely affects heat transfer.

Staining

Staining of laundry and household fixtures can occur in waters with iron, manganese, or copper in solution. In oxygenated surface waters of neutral or near neutral pH (5 to 8), typical concentrations of total iron (mostly the ferric form) are around 0.05 to 0.2 mg/L. In groundwater, the occurrence of iron at concentrations of 1.0 to 10 mg/L is common. Higher concentrations (up to 50 mg/L of mostly the ferrous form) are possible in low-bicarbonated, low-oxygen waters. If water under these latter conditions is pumped from a well, red-brown ferric hydroxide precipitates on fixtures and laundry as soon as oxygen begins to dissolve in the water. The ferrous ion is oxidized to the ferric form by the oxygen. Manganese is often present with iron in groundwaters and may cause similar staining problems, yielding a red precipitate or, with the use of bleach, a dark brown to black staining precipitate. Excess copper in water may create blue stains.

SUMMARY

Health and aesthetic aspects of water quality are the driving force behind water quality regulations and water treatment practice. Because of the complexity of the studies summarized in this chapter, the reader is urged to review the cited literature, health advisories and criteria documents from USEPA, and the *Toxicologic Profile* series from the Agency for Toxic Substances and Disease Registry for more details on any contaminant of particular interest. Furthermore, because new information on waterborne disease-causing organisms and chemical contaminants is being discovered, review of literature since the publication of this chapter is recommended prior to using the information herein as the basis for decision making.

BIBLIOGRAPHY

Agency for Toxic Substances and Disease Registry. *Toxicologic Profile for Arsenic. TP/92-02.* Atlanta, GA: Agency for Toxic Substances and Disease Registry, 1992a.

Agency for Toxic Substances and Disease Registry. *Toxicologic Profile for Barium. TP/91-03.* Atlanta, GA: Agency for Toxic Substances and Disease Registry, 1992b.

Agency for Toxic Substances and Disease Registry. *Toxicologic Profile for Benzene. TP/92-03.* Atlanta, GA: Agency for Toxic Substances and Disease Registry, 1993a.

Agency for Toxic Substances and Disease Registry. *Toxicologic Profile for Cadmium. TP/92-06.* Atlanta, GA: Agency for Toxic Substances and Disease Registry, 1993b.

Agency for Toxic Substances and Disease Registry, Division of Toxicology. *Toxicologic Profile for Methylene Chloride. TP/92-13.* Atlanta, GA: Agency for Toxic Substances and Disease Registry, 1993c.

Agency for Toxic Substances and Disease Registry, Division of Toxicology. *Toxicologic Profile for Nickel. TP/92-14.* Atlanta, GA: Agency for Toxic Substances and Disease Registry, 1993d.

Agency for Toxic Substances and Disease Registry, Division of Toxicology. *Toxicologic Profile for Vinyl Chloride. TP/92-20.* Atlanta, GA: Agency for Toxic Substances and Disease Registry, 1993e.

Agency for Toxic Substances and Disease Registry, Division of Toxicology. *Toxicologic Profile for Carbon Tetrachloride. TP/93-02.* Atlanta, GA: Agency for Toxic Substances and Disease Registry, 1994a.

Agency for Toxic Substances and Disease Registry, Division of Toxicology. *Toxicologic Profile for 1,1-Dichloroethene. TP/93-07.* Atlanta, GA: Agency for Toxic Substances and Disease Registry, 1994b.

Agency for Toxic Substances and Disease Registry, Division of Toxicology. *Toxicologic Profile for 1,2-Dichloroethane. TP/93-06.* Atlanta, GA: Agency for Toxic Substances and Disease Registry, 1994c.

Agency for Toxic Substances and Disease Registry, Division of Toxicology. *Toxicologic Profile for Mercury. TP/93-04.* Atlanta, GA: Agency for Toxic Substances and Disease Registry, 1994d.

Agency for Toxic Substances and Disease Registry, Division of Toxicology. *Toxicologic Profile for 1,1,1-Trichloroethane.* Atlanta, GA: Agency for Toxic Substances and Disease Registry, 1995.

American Heart Association. *Sodium Restricted Diet—500 Milligrams (Revised).* Dallas, TX: American Heart Association, 1969.

American Public Health Association, American Water Works Association, Water Environment Federation. *Standard Methods for the Examination of Water and Wastewater (19th ed.).* (Method 2560). Washington, D.C.: American Public Health Assoc., 1995a.

American Public Health Association, American Water Works Association, Water Environment Federation. *Standard Methods for the Examination of Water and Wastewater (19th ed.).* (Method 9215). Washington, D.C.: American Public Health Assoc., 1995b.

American Public Health Association, American Water Works Association, Water Environment Federation. *Standard Methods for the Examination of Water and Wastewater (19th ed.).* (Method 2130). Washington, D.C.: American Public Health Assoc., 1995c.

American Society for Microbiology. "Microbiologists Consider Antiulcer, Anti-Stomach Cancer Strategies." *ASM News,* 60, 1994: 11–12.

American Water Works Association. *Problem Organisms in Water: Identification and Treatment. AWWA M7, 2nd ed.* Denver, CO: American Water Works Association, 1995a.

American Water Works Association. "Water Supply Suspected in B.C. Toxoplasmosis Outbreak." *Waterweek,* September 25, 1995b: 2.

Amy, G., M. Siddiqui, W. Zhai, J. DeBroux, and W. Odem. "Nation-wide Survey of Bromide Ion Concentrations in Drinking Water Sources." In *Annual Conference Proceedings.* Denver, CO: American Water Works Association, 1993.

Anna, C. H., R. R. Maronpot, M. A. Pereira, J. F. Foley, D. E. Malarkey, and M. W. Anderson. "Ras Proto-Oncogene Activation in Dichloroacetic Acid-, Trichloroethylene- and Tetrachloroethylene-induced Liver Tumors in B6C3F1 Mice." *Carcinogenesis,* 15, 1994: 2255–2261.

Armon, R. "Bacteriophage Monitoring in Drinking Water: Do They Fulfil The Index or Indicator Function?" *Water Sci. Tech.,* 27, 1993: 463–470.

Aschengrau, A., D. Ozonoff, C. Paulu, P. Coogan, R. Vezina, T. Heeren, and Y. Zhang. "Cancer Risk and Tetrachloroethylene (PCE) Contaminated Drinking Water in Massachusetts." *Arch. Environ. Health,* 48, 1993: 284–292.

Barker, J., M. R. W. Brown, P. J. Collier, I. Farrell, and P. Gilbert. "Relationship Between *Legionella pneumophila* and *Acanthamoeba polyphaga:* Physiological Status and Susceptibility to Chemical Inactivation." *Appl. Environ. Microbiol.,* 58, 1992: 2420–2425.

Bartlett, C. L. R. "Potable Water as Reservoir and Means of Transmission." In *Legionella: Proceedings of the 2nd International Symposium,* C. Thornsberry, A. Balows, J. C. Feeley, and W. Jakubowski, eds. American Society for Microbiology, Washington, D.C.: American Society for Microbiology, 1984.

Belpoggi, F., M. Soffritti, and C. Maltoni. "Methyl-tertiary-butyl ether (MTBE)—A Gasoline Additive—Causes Testicular and Lymphohaematopoietic Cancers in Rats." *Toxicol. Industr. Health,* 11, 1995: 119–149.

Benenson, M., E. T. Takafuji, S. M. Lemon, R. L. Greenup, and A. J. Sulzer. "Oocyst-transmitted toxoplasmosis associated with ingestion of contaminated water." *New Engl. Jour. Med.,* 307, 1982: 666–669.

Bennett, J .E. "Fungal infections." In *Harrison's Principles of Internal Medicine,* 13th ed., pp. 854–865, K. J. Isselbacher, E. Braunwald, J. D. Wilson, J. B. Martin, A. S. Fauci, and D. L. Kasper, eds. New York: McGraw-Hill, 1994.

Berman, D., and J. C. Hoff. "Inactivation of simian rotavirus SA11 by chlorine, chlorine dioxide, and monochloramine." *Appl. Environ. Microbiol.* 48, 1984: 317–323.

Bile, K., A. Isse, O. Mohamud, P. Allebeck, L. Nilsson, H. Norder, I. K. Mushahwar, and L. O. Magnus. "Contrasting roles of rivers and wells as sources of drinking water on attack and fatality rates in a hepatitis E epidemic in Somalia." *Amer. Jour. Trop. Med. Hyg.,* 51, 1994: 466–474.

Birnbaum, L. "Developmental Effects of Dioxins." *Environ. Health Perspectives,* 103(Suppl 7), 1995: 89–94.

Bisson, J. W., and V. Cabelli. "Clostridium perfringens as a water pollution indicator." *Jour. Water Poll. Control Fed.,* 52, 1980: 241–248.

Bitton, G. *Wastewater Microbiology,* p. 104. New York: Wiley-Liss, 1994.

Blaser, M. J., P. F. Smith, W-L. L. Wang, and J. C. Hoff. "Inactivation of Campylobacter jejuni by chlorine and monochloramine." *Appl. Environ. Microbiol.,* 51, 1986: 307–311.

Bonde, G. J. "Bacterial indication of water pollution." In *Advances in Aquatic Microbiology,* pp. 273–365, M. R. Droop and H. W. Jannasch, eds. London: Academic Press, 1977.

Bowden, F., V. Krause, J. Burrow, B. Currie, T. Heath, D. Fisher, M. Le Mire, S. Locarnini, C. Mansell, S. Nicholson, and B. Demediuk. "Hepatitis E in the Northern Territory: A Locally Acquired Case and Preliminary Evidence Suggesting Endemic Disease." *Communicable Diseases Intelligence,* 18(1), 1994: 2–3 (Australia).

Breiman, R. F. "Modes of Transmission in Epidemic and Nonepidemic *Legionella* Infection: Directions for Further Study." In *Legionella: Current Status and Emerging Perspectives,* pp. 30–35, Barbaree, J.M., R.F. Breiman, and A.P. Dufour, eds. Washington, D.C.: American Society for Microbiology, 1993.

Brock, T. D., and M. T. Madigan. In *Biology of Microorganisms* (*6th ed.*), ch. 11.4. Upper Saddle River, NJ: Prentice Hall, 1991a.

Brock, T. D., and M. T. Madigan. In *Biology of Microorganisms* (*6th ed.*), ch. 19.26. Upper Saddle River, NJ: Prentice Hall, 1991b.

Brock, T. D., and M. T. Madigan. In *Biology of Microorganisms* (*6th ed.*), ch. 21.4. Upper Saddle River, NJ: Prentice Hall, 1991c.

Bryan, R. T. "Microsporidiosis as an AIDS-related opportunistic infection." *Clin. Infect. Dis.,* 21(Suppl 1), 1995: 62–65.

Bull, R. J., and F. C. Kopfler. *Health Effects of Disinfectants and Disinfectant By-Products.* Denver, CO: AWWA Research Foundation, 1991.

Burleigh-Flayer, H. D., J. S. Chun, and W. J. Kintigh. *MTBE: Vapor Inhalation Oncogenicity Study in CD-1 Mice.* BRRC Report No. 91N0013A. Export, PA: Bushy Run Research Center, 1992.

Cabelli, V. J. "Clostridium perfringens as a water quality indicator." In *Bacterial Indicators/Health Hazard Associated with Water,* A.W. Hoadley and B.J. Dutka, eds. Philadelphia: American Society for Testing and Materials, 1977.

Cali, A. "General Microsporidium Features and Recent Findings on AIDS Isolates." *Jour. Protozool.,* 38, 1991: 625–630.

Canning, E. U., J. Lom, and I. Dykova. *The Microsporidia of Vertebrates,* ch. 4. San Diego, CA: Academic Press, 1986.

Cantilli, R., C. O. Abernathy, and J. M. Donohue. "Derivation of the Reference Dose for Zinc." In *Risk Assessment of Essential Elements,* W. Mertz, C.O. Abernathy, and S.S. Olin, eds. Washington, D.C.: International Life Sciences Institute Press, 1994.

Cantor, K. P. "Drinking Water and Cancer." *Cancer Causes Control,* 8, 1997: 292–308.

Centers for Disease Control. *Water-Related Disease Outbreaks: Annual Summary 1981.* Atlanta, GA: Centers for Disease Control, 1982.

Centers for Disease Control. *Water-Related Disease Outbreaks: Annual Summary 1984.* Atlanta, GA: Centers for Disease Control, 1985.

Centers for Disease Control. *Preventing Lead Poisoning in Young Children.* Atlanta, GA: U.S. Department of Health and Human Services, 1991.

Centers for Disease Control. "Primary Amebic Meningoencephalitis—North Carolina, 1991." *Morbidity and Mortality Weekly Rept.* 41, June 26, 1992, 437–440.

Centers for Disease Control, Division of Oral Health. *Fact Sheet. Unintentional High Fluoride Concentration: Acute Adverse Health Events. June 11.* Atlanta, GA: Centers for Disease Control National Center for Preventive Services, 1992.

Chan, P. C., J. K. Hasemani, J. Mahleri, and C. Aranyi. "Tumor Induction in F344/N Rats and B6C3F1 Mice Following Inhalation Exposure to Ethylbenzene." *Tox. Lett.*, 99, 1998: 23–32.

Chang, S-L. "Resistance of Pathogenic *Naegleria* to Some Common Physical and Chemical Agents." *Appl. Environ. Microbiol.,* 35, 1978: 368–375.

Chemical Manufacturer's Association. *Sodium Chlorite: Drinking Water Rat Two-generation Reproductive Toxicity Study.* Quintiles Report Reference CMA/17/96, 1996.

Chen, Y-S., and J.M. Vaughn. "Inactivation of Human and Simian Rotaviruses by Chlorine Dioxide." *Appl. Environ. Microbiol.* 56, 1990: 1363–1366.

Clark, L.C., and D.S. Alberts. "Selenium and Cancer: Risk or Protection?" *Jour. Nat. Cancer Inst.,* 87, 1995: 473–475.

Colburn, T., F.S. vom Saal, and A.M. Soto. "Developmental Effects of Endocrine-disrupting Chemicals in Wildlife and Humans." *Environ. Health Perspectives,* 101, 1993: 378–384.

Colbourne, J.S., P.J. Dennis, R.M. Trew, C. Berry, and G. Vesey. "*Legionella* and Public Water Supplies." *Water Sci. Tech.,* 20, 1988: 5–10.

Collins, M. "Algal Toxins." *Microbiol Rev.,* 42, 1978: 725–746.

Corvilain, B., B. Contempre, A.O. Longombe, et al. "Selenium and the Thyroid: How the Relationship was Established." *Am. Jour. Clin. Nutr.,* 57(Suppl. 2), 1993: 244S–248S.

Cover, T.L., and M.J. Blaser. "Helicobacter pylori: A Bacterial Cause of Gastritis, Peptic Ulcer Disease, and Gastric Cancer." *ASM News,* 61, 1995: 21–26.

Craun, G.F. "Statistics of Waterborne Outbreaks in the U.S. (1920–1980)." In *Waterborne Diseases in the United States,* pp. 73–159, G.F. Craun, ed. Boca Raton, FL: CRC Press, 1986.

Craun, G., D. Swerdlow, R. Tauxe, R. Clark, K. Fox, E. Geldreich, D. Reasoner, and E. Rice. "Prevention of Waterborne Cholera in the United States." *Jour. AWWA.* 83, 1991: 40–45.

Cubitt, W.D. "A Review of the Epidemiology and Diagnosis of Waterborne Viral Infections." *Water Sci. Tech.,* 24, 1991: 197–203.

Davis, J.P. *Cryptosporidiosis-Associated Mortality Following a Massive Waterborne Outbreak in Milwaukee, Wisconsin. Report of the Wisconsin Department of Health and Social Services.* Madison, WI: 1996.

Dearfield, K.L. "Ethylene thiourea (ETU). A review of the genetic toxicity studies." *Mutat. Res.,* 317, 1994: 111–132.

de Jonckheere, J., and H. van de Voorde. "Differences in Destruction of Cysts of Pathogenic and Nonpathogenic *Naegleria* and *Acanthamoeba* by chlorine." *Appl. Environ. Microbiol.,* 31, 1976: 294–297.

Dich, J., S. Hoar Zahm, A. Hanberg, and H.-O. Adami. "Pesticides and Cancer." *Cancer Causes and Control,* 8, 1997: 420–423.

Dreesman, G.R., and G.R. Reyes. "Hepatitis." In *Encyclopedia of Microbiology, Vol. 2,* J. Lederberg, ed. San Diego, CA: Academic Press, 1992.

Du Moulin, G.C., and K.D. Stottmeier. "Waterborne Mycobacteria: An Increasing Threat to Health." *ASM News,* 52, 1986: 525–529.

Du Moulin, G.C., K.D. Stottmeier, P.A. Pelletier, A.Y. Tsang, and J. Hedley-Whyte. "Concentration of *Mycobacterium avium* by Hospital Hot Water Systems." *J. Amer. Med. Assoc.,* 260, 1988: 1599–1601.

DuPont, H.L. "Enteropathogens." In *Encyclopedia of Microbiology, Vol. 2.* J. Lederberg, ed. San Diego, CA: Academic Press, 1992.

DuPont, H.L., C.L. Chappell, C.R. Sterling, P.C. Okhuysen, J.B. Rose, and W. Jakubowski. "The infectivity of Cryptosporidium parvum in healthy volunteers." *New Engl. Jour. Med.,* 332, 1995: 855–859.

Ely, D.L. "Overview of dietary sodium effects on and interactions with cardiovascular and neuroendocrine functions." *Am. Jour. Clin. Nutr.,* 65(Suppl 2), 1997: 594S–605S.

Ernest, J.A., B.L. Blagburn, and D.S. Lindsay. "Infection dynamics of Cryptosporidium parvum (Apicomplexa:Cryptosporiidae) in neonatal mice (Mus musculus)." *Jour. Parasit.,* 72(5), 1986: 796–798.

Falkner, B., and S. Michel. "Blood pressure response to sodium in children and adolescents." *Am. Jour. Clin. Nutr.,* 65(Suppl 2), 1997: 618S–621S.

Fiksdal, L., J. S. Maki, S. J. LaCroix, and J. T. Staley. "Survival and Detection of Bacteroides spp., Prospective Indicator Bacteria." *Appl. Environ. Microbiol.,* 49, 1985: 148–150.

Finch, G. R., E. K. Black, and L. L. Gyurek. "Ozone and chlorine inactivation of Cryptosporidium." In *Proceedings, AWWA Water Quality Technology Conference at New Orleans.* Denver, CO: American Water Works Association, 1995.

Fishback, J. L. "Toxoplasmosis." In *Encyclopedia of Microbiology, Vol. 4,* pp. 255–264, J. Lederberg, ed. San Diego, CA: Academic Press, 1992.

Fliermans, C. B., W. B. Cherry, L. H. Orrison, S. J. Smith, D. L. Tison, and D. H. Pope. "Ecological Distribution of *Legionella pneumophila.*" *Appl. Environ. Microbiol.,* 41, 1981: 9–16.

Frankova, E., and M. Horecka. "Filamentous Soil Fungi and Unidentified Bacteria in Drinking Water from Wells and Water Mains near Bratislava." *Microbiol. Res.,* 150, 1995: 311–313.

Fung, D. Y. C. "Foodborne Illness." In *Encyclopedia of Microbiology, Vol. 2,* pp. 209–218, J. Lederberg, ed. San Diego, CA: Academic Press, 1992.

Ganrot, P. O. "Metabolism and Possible Health Effects of Aluminum." *Environ. Health Perspectives,* 65, 1986: 363–441.

Gerba, C. P., J. B. Rose, C. N. Haas, and K. D. Crabtree. "Waterborne Rotavirus: a Risk Assessment." *Water Res.* 30(12), 1996: 2929–2940.

Gerba, C. P., J. B. Rose, and S. N. Singh. "Waterborne Gastroenteritis and Viral Hepatitis." *CRC Critical Reviews in Environmental Control,* 15(3), 1985: 213–226.

Girdwood, R. W. A. "Some Clinical Perspectives on Waterborne Parasitic Protozoa." In: *Proceedings, Protozoan Parasites and Water. Symposium held at York, England, in September 1994,* W. B. Betts, D. Casemore, C. Fricker, H. Smith, and J. Watkins, eds. London: Royal Society of Chemistry, 1995.

Gittleman, T. S., and P. Luitweiler. "Tastes and Odors in Treated Water from Algae, Urea, and Industrial Chemicals in Source Waters." In *Proceedings, AWWA Water Quality and Treatment Conference at Boston.* Denver, CO: American Water Works Association, 1996.

Gonzalez-de-la-Cuesta, N., M. Arias-Fernandez, E. Paniagua-Crespo, and M. Marti-Mallen. "Free-living Amoebae in Swimming Pool Waters from Galicia (Spain)." *Rev. Iber. Parasitol.,* 47, 1987: 207–210. Abstract in English only; text in Spanish.

Goodgame, R. W. "Understanding Intestinal Spore-forming Protozoa: Cryptosporidia, Microsporidia, Isospora, and Cyclospora." *Ann. Intern. Med.,* 1214, 1996: 429–441.

Graves, A. B., E. White, T. D. Koepsell, B. V. Reifler, G. van Belle, and E. B. Larson. "The Association between Aluminum-containing Products and Alzheimer's disease." *Jour. Clin. Epidemiol.,* 43, 1990: 35–44.

Gray, J. J., T. G. Wreghitt, W. D. Cubitt, and P. R. Elliot. "An Outbreak of Gastroenteritis in a Home for the Elderly Associated with Astrovirus Type 1 and Human Calicivirus." *Jour. Med. Virol.,* 23, 1987: 377–381.

Greger, J. L. "Zinc: An Overview from Deficiency to Toxicity." In: *Risk Assessment of Essential Elements.* W. Mertz, C. O. Abernathy, and S. S. Olin, eds. Washington, D.C.: International Life Sciences Institute Press, 1994.

Grohmann, G. S., et al. "Enteric Viruses and Diarrhea in HIV-Infected Patients." *New Engl. Jour. Med.,* 329, 1993: 14–20.

Havelaar, A. H. "Bacteriophages as Models of Human Enteric Viruses in the Environment." *ASM News,* 59, 1993: 614–619.

Havelaar, A. H., M. van Olphen, and Y. C. Drost. "F-specific RNA Bacteriophages are Adequate Model Organisms for Enteric Viruses in Fresh Water." *Appl. Environ. Microbiol.,* 59, 1993: 2956–2962.

Heindel, J. J., R. E. Chapin, D. K. Gulati, J. D. George, C. J. Price, M. C. Marr, C. B. Myers, L. H. Barnes, P. A. Fail, T. B. Grizzle, B. A. Schwetz, and R. S. H. Yang. "Assessment of the Reproductive and Developmental Toxicity of Pesticide/Fertilizer Mixtures Based on Confirmed Pesticide Contamination in California and Iowa Groundwater." *Fund. Appl. Toxicol.,* 22, 1994: 605–621.

Highsmith, A. K., and S. A. Crow. "Waterborne Diseases." In *Encyclopedia of Microbiology, Vol. 4,* pp. 377–384, J. Lederberg, ed., San Diego, CA: Academic Press, 1992.

Hoar Zahm, S., and M. H. Ward. "Pesticides and Childhood Cancer." *Environ. Health Perspectives,* 106(Suppl 3), 1998: 893–908.

Hopke, P. K., D. Vandana, B. Fitzgerald, T. Raunemaa, and K. Kuuspalo. *Critical Assessment of Radon Progeny Exposure While Showering in Radon-Bearing Water.* Denver, CO: AWWA Research Foundation, 1996.

Horan, T., D. Culver, W. Jarvis, G. Emori, S. Banerjee, W. Martone, and C. Thornsberry. "Pathogens Causing Noscomial Infections." *The Antimicrobic Newsletter,* 5, 1988: 65–67.

Huang, P., J. T. Weber, D. M. Sosin, E. G. Long, J. J. Murphy, F. Kocka, C. Peters, and C. Kallick. "The First Reported Outbreak of Diarrheal Illness Associated with Cyclospora in the United States." *Ann. Internal Med.,* 123, 1995: 409–414.

Hulten, K., et al. "*Helicobacter pylori* in the Drinking Water in Peru." *Gastroenterology,* 110, 1996: 1031–1035.

Hurst, C. J. "Presence of Enteric Viruses in Freshwater and Their Removal by the Conventional Drinking Water Treatment Process." *Bulletin of the World Health Organization* 69(1), 1991: 113–119.

Ibarra, H. V., S. G. Riedemann, F. G. Siegal, G. V. Reinhardt, C. A. Toledo, and G. Frosner. "Hepatitis E Virus in Chile (letter)." *Lancet,* 344, 1994: 1501.

Ing, D., G. I. Glass, C. W. LeBaron, and J. F. Lew. "Laboratory-based Surveillance for Rotavirus—United States, January 1989–May 1991." *CDC Surveillance Summaries, Morbidity and Mortality Weekly Report,* 41(SS-3):47–60, May 29, 1992.

International Agency for Research on Cancer. *IARC Monographs on the Evaluation of Carcinogenic Risks to Humans. Some N-Nitroso Compounds.* Lyon, France: World Health Organization, 1978.

International Agency for Research on Cancer. *IARC Monographs on the Evaluation of Carcinogenic Risks to Humans. Dry Cleaning. Some Chlorinated Solvents and Other Industrial Chemicals. Vol. 63.* Lyon, France: World Health Organization, 1996.

International Life Sciences Institute. *Disinfection By-Products in Drinking Water: Critical Issues in Health Effects Research.* Workshop Report, October 23–25, 1995. Washington, D.C.: International Life Sciences Institute, 1995.

International Life Sciences Institute. *An Evaluation of EPA's Proposed Guidelines for Carcinogen Risk Assessment Using Chloroform and Dichloroacetate as Case Studies: Report of an Expert Panel.* Washington, D.C.: International Life Sciences Institute, 1997.

Janoff, E. N., and L. B. Reller. "*Cryptosporidium* Species, a Protean Protozoan." *J. Clin. Microbiol.,* 25, 1987: 967–975.

Jo, W. K., C. P. Weisel, and P. J. Lioy. "Routes of Chloroform Exposure and Body Burden from Showering with Chlorinated Tap Water." *Risk Anal.,* 10, 1990: 575–580.

Jofre, J., E. Olle, F. Ribas, A. Vidal, and F. Lucena. "Potential Usefulness of Bacteriophages that infect Bacteroides fragilis as Model Organisms for Monitoring Virus Removal in Drinking Water Treatment Plants." *Appl. Environ. Microbiol.,* 61, 1995: 3227–3231.

Johnson, C. H., E. W. Rice, and D. J. Reasoner. "Inactivation of *Helicobacter pylori* by Chlorination." *Appl. Environ. Microbiol.,* 63, 1997: 4969–4970.

Kilvington, S. "Activity of Water Biocide Chemicals and Contact Lens Disinfectants on Pathogenic Free-living Amoebae." *Int. Biodeterior.,* 26, 1990: 127–138.

Kimmel, C. A. "Quantitative Approaches to Human Risk Assessment for Noncancer Health Effects." *Neurotoxicology,* 11, 1990: 189–198.

Klein, P. D., D. Y. Graham, A. Gaillour, A. R. Opekun, and E. O. Smith. "Water source as risk factor for *Helicobacter pylori* infection in Peruvian children." *Lancet,* 337, 1991: 1503–1506.

Knight, P. "Once Misidentified Human Parasite is a Cyclosporan." *ASM News,* 61, 1995: 520–522.

Koenraad, P. M. F. J., F. M. Rombouts, and S. H. W. Notermans. "Epidemiological Aspects of Thermophilic *Campylobacter* in Water-Related Environments: A Review." *Water Environment Research,* 69, 1997: 52–63.

Kogevinas, M., H. Becher, T. Benn, P. A. Bertazzi, P. Boffetta, H. Bas Bueno-de-Mewquita, D. Cogon, D. Colin, D. Flesch-Janys, M. Fingerhut, L. Green, T. Kauppinen, M. Littorin, E. Lynge, J. D. Matthews, M. Neuberger, N. Pearce, and R. Saracci. "Cancer Mortality in Workers Exposed to Phenoxy Herbicides, Chlorophenols, and Dioxins." *Am. Jour. Epidem.,* 145, 1997: 1061–1075.

Komulainen, H., V. Kosma, S. Vaittinen, T. Vartiainen, E. Korhonen, S. Lotojonen, R. Tuominen, and J. Tuomisto. "Carcinogenicity of the Drinking Water Mutagen 3-Chloro-4(dichoromethyl)-5-hydroxy-2(5H)-furanone in the Rat." *Journal of the National Cancer Institute,* 89(12), 1997: 848–856.

Korich, D. G., J. R. Mead, M. S. Madore, N. A. Sincliar, and C. R. Sterling. "Effects of ozone, chlorine dioxide, chlorine, and monochloramine on Cryptosporidium parvum oocyst viability." *Appl. Environ. Microbiol.,* 56, 1990: 1423–1428.

Kotchen, T. A., and D. A. McCarron. "Dietary electrolytes and blood pressure. A statement for healthcare professionals from the American Heart Association Nutrition Committee." *Circulation,* 98, 1998: 613–617.

Kramer, M. H., B. L. Herwaldt, G. F. Craun, R. L. Calderon, and D. D. Juranek. "Surveillance for Waterborne-Disease Outbreaks—United States, 1993–1994." *CDC Surveillance Summaries, Morbidity and Mortality Weekly Report,* 45(SS-1), 1996: 1–30.

Krauss, R. M., R. J. Deckelbaum, N. Ernst, E. Fisher, B. V. Howard, R. H. Knopp, et al. "Dietary Guidelines for Healthy American Adults. A Statement for Health Professionals from the Nutrition Committee, American Heart Association." *Circulation,* 94, 1996: 1795–1800.

Kühn, L., G. Allestam, G. Huys, P. Janssen, K. Kersters, K. Krovacek, and T. Stenström. "Diversity, Persistence, and Virulence of *Aeromonas* Strains Isolated from Drinking Water Distribution Systems in Sweden." *Appl. Environ. Microbiol.,* 63, 1997: 2708–2715.

Kurtz, J. B., and T. W. Lee. "Astroviruses: Human and Animal." In *Novel Diarrhoea Viruses.* New York: John Wiley & Sons, 1987.

Kwo, P. Y., G. G. Schlauder, H. A. Carpenter, P. J. Murphy, J. E. Rosenblatt, G. J. Dawson, E. E. Mast, K. Krawczynski, and V. Balan. "Acute Hepatitis E by a New Isolate Acquired in the United States." *Mayo Clin. Proc.,* 72, 1997: 1133–1136.

Laden, F., and D. J. Hunter. "Environmental Risk Factors and Female Breast Cancer." *Ann. Rev. Public Health,* 19, 1998: 101–123.

Latham, R., and W. Schaffner. "Hospital Epidemiology." In *Encyclopedia of Microbiology, Vol. 2,* J. Lederberg, ed. New York: Academic Press, 1992.

LeChevallier, M. W., T. M. Evans, and R. J. Seidler. "Effect of Turbidity on Chlorination Efficiency and Bacterial Persistence in Drinking Water." *Appl. Environ. Microbiol.,* 42, 1981: 159–167.

LeChevallier, M. W., and W. D. Norton. "Examining Relationships Between Particle Counts and Giardia, Cryptosporidium, and Turbidity." *Jour. AWWA,* 84, 1992: 54–60.

LeChevallier, M. W., and W. D. Norton. "*Giardia* and *Cryptosporidium* in Raw and Finished Water." *Jour. AWWA,* 87, 1995: 54–68.

LeChevallier, M. W., W. D. Norton, and R. G. Lee. "Evaluation of a method to detect *Giardia* and *Cryptosporidium* in water." In *Monitoring Water in the 1990's: Meeting New Challenges, ASTM STP 1102,* J. R. Hall and G. D. Glysson, eds. Philadelphia: American Society for Testing and Materials, 1991.

Leet, T., J. Acquavella, C. Lynch, M. Anne, N. S. Weiss, T. Vaughan, and H. Checkoway. "Cancer Incidence Among Alachlor Manufacturing Workers." *Am. Jour. Ind. Med.,* 30, 1996: 300–306.

Letterman, R. D. "What Turbidity Measurements Can Tell Us." *Opflow,* 20(8), 1994: 1, 3–5.

Lilienfeld, D. E., and M.A. Gallo. "2-D, 2,4,5-T, and 2,3,7,8-TCDD: An Overview." *Am. Jour. Epidem.,* 41, 1989: 28–58.

Lippy, E. C., and J. Erb. "Gastrointestinal Illness at Sewickley, Pa." *Jour. AWWA,* 68, 1996: 606–610.

Logsdon, G. S., F. B. Dewalle, and D. W. Hendricks. "Filtration as a Barrier to Passage of Cysts in Drinking Water." In *Giardia and Giardiasis,* S. L. Erlandsen and E. A. Meyer, eds. New York: Plenum Press, 1984.

Lubin, J. H., and J. D. Boice, Jr. "Lung Cancer Risk from Residential Radon: Meta-analysis of Eight Epidemiologic Studies." *Jour. Nat. Cancer Inst.,* 89, 1997: 49–57.

Luft, F. C., and M. H. Weinberger. "Heterogeneous Responses to Changes in Dietary Salt Intake: the Salt Sensitive Paradigm." *Am. Jour. Clin. Nutr.,* 65(Suppl 2), 1997: 612S–617S.

Lund, V. "Evaluation of *E. coli* as an Indicator for the Presence of *Campylobacter jejuni* and *Yersinia enterocolitica* in Chlorinated and Untreated Oligotrophic Lake Water." *Water Res.,* 30, 1996: 1528–1534.

Lye, D., G. S. Fout, S. R. Crout, R. Danielson, C. L. Thio, and C. M. Paszko-Kolva. "Survey of Ground, Surface, and Potable Waters for the Presence of *Legionella* Species by Environamp

PCR *Legionella* Kit, Culture, and Immunofluorescent Staining." *Water Research,* 31, 1997: 287–293.

MacIntosh, D. L., J. D. Spengler, H. Ozkaynak, L.-h. Tsai, and P. B. Ryan. "Dietary Exposures to Selected Metals and Pesticides." *Environ. Health Perspectives,* 104, 1996: 202–209.

MacKenzie, W. R., N. J. Hoxie, M. E. Proctor, M. S. Gradus, K. A. Blair, D. E. Peterson, J. J. Kazmierczak, D. G. Addiss, K. R. Fox, J. B. Rose, J. P. Davis. "A Massive Outbreak in Milwaukee of *Cryptosporidium* Infection Transmitted Through the Public Water Supply." *New Engl. J. Med.,* 331, 1994: 161–167.

Mallevialle, J., and F. H. Suffet, eds. *Identification and Treatment of Tastes and Odors in Drinking Water.* Denver, CO: American Water Works Association Research Foundation & Lyonnaise des Eaux, 1987.

Marx, A., and R. R. Neutra. "Magnesium in Drinking Water and Ischemic Heart Disease." *Epidem. Rev.,* 19, 1997: 258–272.

Mast, E. E., and K. Krawczynski. "Hepatitis E: An Overview." *Ann. Rev. Med.,* 47, 1996: 257–266.

Maxwell, N. I., D. E. Burmaster, and D. Ozonoff. "Trihalomethanes and Maximum Contaminant Levels: The Significance of Inhalation and Dermal Exposures to Chloroform in Household Water." *Reg. Tox. Pharm.,* 14, 1991: 297–312.

McCarron, D. "Diet and Blood Pressure—The Paradigm Shift." *Science,* 281, 1988a: 933–934.

McCarron, D. A. "Importance of Dietary Calcium in Hypertension." *J. Am. Coll. Nutr.,* 17, 1988b: 97–99.

McKone, T. E. "Human Exposure to Volatile Organic Compounds in Household Tap Water: The Indoor Inhalation Pathway." *Environ. Sci. Technol.,* 21, 1987: 1194–2101.

Melnick, J. L. "Enteroviruses." In *Encyclopedia of Microbiology, Vol. 2,* pp. 69–80, J. Lederberg, ed. New York: Academic Press, 1992.

Mitchell, R. *Water Pollution Microbiology.* New York: John Wiley & Sons, 1972.

James H. Montgomery Consulting Engineers, Inc. *Water Treatment Principles and Design.* New York: John Wiley & Sons, 1985.

Moore, A. C., B. L. Herwaldt, G. F. Craun, R. L. Calderon, A. K. Highsmith, and D. D. Juranek. "Surveillance for Waterborne-Disease Outbreaks—United States, 1991–1992." *CDC Surveillance Summaries, Morbidity and Mortality Weekly Report,* 42(SS-5), 1993: 1–22.

Moore, L. E., A. H. Smith, C. Hopenhayn-Rich, M. L. Biggs, D. A. Kalman, and M. T. Smith. "Micronuclei in exfoliated bladder cells among individuals chronically exposed to arsenic in drinking water." *Cancer Epidem. Biomarkers Prev.,* 6, 1997: 31–36.

Motulsky, A. G. "Hemochromatosis (Iron Storage Disease)." In *Cecil Textbook of Medicine,* J. B. Wyngaarden and L. H. Smith, eds. Philadelphia: W. B. Saunders Co., 1988.

Nagy, L. A., and B. H. Olson. "The Occurrence of Filamentous Fungi in Drinking Water Distribution Systems." *Can. J. Microbiol.,* 28, 1982: 667–671.

Nanda, S. K., I. H. Ansari, S. K. Acharya, S. Jameel, and S. K. Panda. "Protracted Viremia During Acute Sporadic Hepatitis E Virus Infection." *Gastroenterology,* 108, 1995: 225–230.

National Academy of Sciences Committee on Nitrite and Alternative Curing Agents in Food. *The Health Effects of Nitrate, Nitrite, and N-Nitroso Compounds.* Washington, D.C.: National Academy Press, 1981.

National Academy of Sciences Committee on the Risk Assessment of Exposure to Radon in Drinking Water. *Risk Assessment of Radon in Drinking Water.* Washington, D.C.: National Academy Press, 1998.

National Academy of Sciences Food and Nutrition Board. *Recommended Dietary Allowances,* 9th rev. Washington, D.C.: National Academy Press, 1980.

National Academy of Sciences Safe Drinking Water Committee. *Drinking Water and Health, Vol. 1.* Washington, D.C.: National Academy Press, 1977.

National Academy of Sciences Safe Drinking Water Committee. *Drinking Water and Health, Vol. 2.* Washington, D.C.: National Academy Press, 1980a.

National Academy of Sciences Safe Drinking Water Committee. *Drinking Water and Health, Vol. 3.* Washington, D.C.: National Academy Press, 1980b.

National Academy of Sciences Safe Drinking Water Committee. *Drinking Water and Health, Vol. 4.* Washington, D.C.: National Academy Press, 1982.

National Academy of Sciences Safe Drinking Water Committee. *Drinking Water and Health, Vol. 5.* Washington, D.C.: National Academy Press, 1983.

National Academy of Sciences Safe Drinking Water Committee. *Drinking Water and Health, Vol. 7.* Washington, D.C.: National Academy Press, 1987.

National Research Council. *Recommended Dietary Allowances, 10th edition.* Washington, D.C.: National Academy Press, 1989.

Nelson, M. A., I. M. Sanchez, R. J. Bull, and S. R. Sylvester. "Increased Expression of c-myc and c-H-ras in Dichloroacetate and Trichloroacetate-induced Liver Tumors in B6C3F1 mice." *Toxicology,* 64, 1990: 47–57.

Niemi, R. M., S. Knuth, and K. Lundstrom. "Actinomycetes and Fungi in Surface Waters and in Potable Water." *Appl. Environ. Microbiol.,* 43, 1982: 378–388.

Nieminski, E. C., and J. E. Ongerth. "Removing *Giardia* and *Cryptosporidium* by Conventional Treatment and Direct Filtration." *Jour. AWWA,* 87, 1995: 96–106.

Nomura, A., G. N. Stemmermann, P-H Chyou, I. Kato, G. I. Perez-Perez, and M. J. Blaser. "Helicobacter pylori Infection and Gastric Carcinoma Among Japanese Americans in Hawaii." *New Engl. Jour. Med.,* 325, 1991: 1132–1136.

North Atlantic Treaty Organization. *Drinking Water Microbiology,* pp. 193–197, EPA 570/9-84-006. Washington, DC: U.S. Environmental Protection Agency, Office of Drinking Water, 1984.

Oishi, I., K. Yamazaki, T. Kimoto, Y. Minekawa, E. Utagawa, S. Yamazaki, S. Inouye, G. S. Grohmann, S. S. Monroe, S. E. Stine, C. Carcamo, T. Ando, and R. I. Glass. "A Large Outbreak of Acute Gastroenteritis Associated with Astrovirus Among Students and Teachers in Osaka, Japan." *Jour. Infect. Dis.,* 170, 1994: 439–443.

Olivieri, V. P. "Measurement of Microbial Quality." In *Assessment of Microbiology and Turbidity Standards for Drinking Water,* EPA570/9-83-001, P.S. Berger and Y. Argamon, eds. Washington, D. C.: U.S. Environmental Protection Agency, Office of Drinking Water, 1983.

Ongerth, J. E., and H. H. Stibbs. "Identification of Cryptosporidium Oocysts in River Water." *Appl. Environ. Microbiol.,* 53, 1987: 672–676.

Ongerth, J. E., and J. P. Pecoraro. "Removing Cryptosporidium using multimedia filters." *Jour. AWWA,* 87(12), 1995: 83–89.

Parsonnet, J. "Gastrointestinal Microbiology." In *Encyclopedia of Microbiology,* Vol. 2, pp. 245–258, J. Lederberg, eds. New York: Academic Press, 1992.

Parsonnet, J., G. D. Friedman, D. P. Vandersteen, Y. Chang, J. H. Vogelman, N. Orentreich, and R. K. Sibley. "Helicobacter pylori Infection and the Risk of Gastric Carcinoma." *New Engl. Jour. Med.,* 325, 1991: 1127–1131.

Parsons, D. W., and J. M. Witt. *Pesticides in Groundwater in the United States of America. A Report of a 1988 Survey of State Lead Agencies,* EM 8406. Corvallis, OR: Oregon State University Extension Service, 1988.

Patterson, B. H., and O. A. Levander. "Naturally occurring selenium compounds incancer chemoprevention trials: a workshop summary." *Cancer Epidem. Biomarkers Prev.,* 6, 1997: 63–69.

Payment, P., and E. Franco. "Clostridium perfringens and Somatic Coliphages as Indicators of the Efficiency of Drinking Water Treatment for Viruses and Protozoan Cysts." *Appl. Environ. Microbiol.,* 59, 1993: 2418–2424.

Peeters, J. E., E. A. Mazas, W. J. Masschelein, I. V. Martinez de Maturana, and E. Debacker. "Effect of disinfection of drinking water with ozone or chlorine dioxide on survival of Cryptosporidium parvum oocysts." *Appl. Environ. Microbiol.,* 55, 1989: 1519–1522.

Pelletier, P. A., G. C. du Moulin, and K. D. Stottmeier. "Mycobacteria in Public Water Supplies: Comparative Resistance to Chlorine." *Microbiol. Sciences,* 5, 1988: 147–148.

Perera, F. P. "Molecular Epidemiology: Insights into Cancer Susceptibility, Risk Assessment, and Prevention." *Jour. Nat. Cancer Inst.,* 88, 1996: 496–509.

Peterson, W. L. "Helicobacter pylori and Peptic Ulcer Disease." *N. Eng. J. Med.,* 324, 1991: 1043–1048.

Peterson, D. A., T. R. Hurley, J. C. Hoff, and L. G. Wolfe. "Effect of Chlorine Treatment on Infectivity of Hepatitis A Virus." *Appl. Environ. Microbiol.,* 45, 1983: 223–227.

Pinto, R. M., F. X. Abad, R. Gajaardo, and A. Bosch. "Detection of infectious astroviruses in water." *Appl. Environ. Microbiol.,* 62, 1996: 1811–1813.

Pohland, A. E. "Mycotoxins in review." *Food Addit. Contam.,* 10, 1993: 17–28.

Poirier, K. "Summary of the Derivation of the Reference Dose for Selenium." In *Risk Assessment of Essential Elements,* W. Mertz, C. O. Abernathy, and S. S. Olin, eds. Washington, D.C.: International Life Sciences Institute Press, 1994.

Pontius F. W., K. G. Brown, and C.-J. Chen. "Health implications of arsenic in drinking water." *Jour. AWWA,* 86, 1994: 52–63.

Porterfield, S. P. "Vulnerability of the Developing Brain to Thyroid Abnormalities: Environmental Insults to the Thyroid System." *Environ. Health Perspectives,* 102(Suppl 2), 1994: 125–130.

Public Health Service Ad Hoc Subcommittee on Fluoride. *Review of Fluoride Benefits and Risks.* Washington, D.C.: Public Health Service, 1991.

Rabold, J. G., C. W. Hoge, D. R. Shlim, C. Kefford, R. Rajah, and P. Echeverria. "Cyclospora Outbreak Associated with Chlorinated Drinking Water." *Lancet,* 344, 1994: 1360–1361.

Rao, V.C., J. M. Symons, A. Ling, P. Wang, T. G. Metcalf, J. C. Hoff, and J. L. Melnick. "Removal of Hepatitis A Virus and Rotavirus by Drinking Water Treatment." *Jour. AWWA,* 80, 1988: 59–67.

Ray, R., R. Aggarwal, P. N. Salunke, N. N. Mehrotra, G. P. Talwar, and S. R. Naik. "Hepatitis E Virus Genome in Stools of Hepatitis Patients during Large Epidemic in North India." *Lancet,* 338, 1991: 783–784.

Reiber, S. H., and W. A. Kukull. *Aluminum, Drinking Water, and Alzheimer's Disease.* Denver, CO: AWWA Research Foundation, 1996.

Reid, J. "Arsenic Occurrence: USEPA Seeks a Clearer Picture." *Jour. AWWA.,* 86, 1994: 44–51.

Rendtorff, R. C. "The Experimental Transmission of *Giardia Lamblia* among Volunteer Subjects." In *Waterborne Transmission of Giardiasis,* pp. 64–81, EPA-600/9-79-001. W. Jakubowski and J. C. Hoff, eds. Cincinnati, OH: Environmental Protection Agency, 1979.

Rice, E. W., C. H. Johnson, R. M. Clark, K. R. Fox, D. J. Reasoner, M. E. Dunnigan, P. Panigrahi, J. A. Johnson, and J. G. Morris. "Vibrio cholerae 01 Can Assume a 'Rugose' Survival Form That Resists Killing by Chlorine, Yet Retains Virulence." *International J. Environ. Health Res.,* 3, 1993: 89–98.

Rice, E. W., C. H. Johnson, D. W. Naumovitz, M. M. Marshall, C. B. Plummer, and C. R. Sterling. "Chlorine Disinfection Studies of *Encephalitozoon* (*Septata*) *Intestinalis.*" Abstracts, 99th General Meeting, American Society for Microbiology, Q–130, 1999.

Rice, G., K. Fox, R. Miltner, D. Lytle, and C. Johnson. "Evaluating Plant Performance with Endospores." *Jour. AWWA,* 88, 1996: 122–130.

Rivera, F., A. Ortega, E. Lopez-Ochoterena, and M. E. Paz. "A Quantitative Morphological and Ecological Study of Protozoa Polluting Tap Water in Mexico City." *Trans. Amer. Micros. Soc.,* 98, 1979: 465–469.

Rogers, J., A. B. Dowsett, P. J. Dennis, J. V. Lee, and C. W. Keevil. "Influence of Plumbing Materials on Biofilm Formation and Growth of *Legionella pneumophila* in Potable Water Systems." *Appl. Environ. Microbiol.,* 60, 1994: 1842–1851.

Rose, J. B., C. P. Gerba, and W. Jakubowski. "Survey of Potable Water Supplies for Cryptosporidium and Giardia." *Environ. Sci. Technol.,* 25(8), 1991: 1393–1400.

Rosenzweig, W. D., H. A. Minnigh, and W. O. Pipes. "Chlorine Demand and Inactivation of Fungal Propagules." *Appl. Environ. Microbiol.,* 45, 1983: 182–186.

Rosenzweig, W. D., H. Minnigh, and W. O. Pipes. "Fungi in Potable Water Distribution Systems." *Jour. AWWA,* 78, 1986: 53–55.

Rusin, P., J. Rose, C. Haas, and C. Gerba. "Risk Assessment of Opportunistic Bacterial Pathogens in Drinking Water." In *Proceedings, AWWA Water Quality Technol. Conf.* (*1996, Boston*). Denver, CO: American Water Works Assoc., 1997. (On CD-ROM only.)

Saari, T. N., and G. P. Jansen. "Waterborne *Yersinia enterocolitica* in the Midwest United States." *Contr. Microbiol. Immunol.,* 5, 1979: 185–196.

Sawyer, T. K. "Free-living Pathogenic and Nonpathogenic Amoebae in Maryland Soils." *Appl. Environ. Microbiol.,* 55, 1989: 1074–1077.

Schiemann, D. A. "Isolation of *Yersinia Enterocolitica* from Surface and Well Waters in Ontario." *Can. J. Microbiol.,* 24, 1978: 1048–1052.

Schoeny, R. "Use of Genetic Toxicology Data in U.S. EPA Risk Assessment: The Mercury Study Report as an Example." *Environ Health Perspectives.,* 104(Suppl 3), 1996: 663–673.

Seal, D., F. Stapleton, and J. Dart. "Possible Environmental Sources of Acanthamoeba spp. in Contact Lens Wearers." *Br. Jour. Ophthalmology,* 76, 1992: 424–427.

Sever L. E. "Looking for causes of neural tube defects: where does the environment fit in?" *Envir. Health Perspectives,* 103(Suppl 6), 1995: 165–171.

Sever, L. E., T. E. Arbuckle, and A. Sweeney. "Reproductive and developmental effects of occupational pesticide exposure: the epidemiologic evidence." *Occ. Med. State Art Rev.,* 12, 1997: 305–325.

Shahamat, M., R. R. Colwell, and C. Paszko-Kolva. "Letter to Editor." *Jour. AWWA,* 84(10), 1992: 4.

Shahamat, M., U. Mai, C. Paszko-Kolva, and R. R. Colwell. "Use of Autoradiography to Assess Viability of Helicobacter pylori in Water." *Appl. Environ. Microbiol.,* 59, 1993: 1231–1235.

Shovlin, M. G., R. S. Yoo, D. R. Crapper-McLachlan, E. Cummings, J. M. Donohue, W. K. Hallman, Z. Khachaturian, J. Orme-Zavaleta, and S. Teefy. *Aluminum in Drinking Water and Alzheimer's Disease: A Resource Guide.* Denver, CO: American Water Works Association Research Foundation, 1993.

Simons-Morton, D. G., and E. Obarzanek. "Diet and blood pressure in children and adolescents." *Pediatr. Nephrol.,* 11, 1997: 244–249.

Smith, A. H., C. Rich-Hopenhayn, M. N. Bates, H. M. Goeden, I. Hertz-Picciotto, H. Duggan, R. Wood, M. Kosnett, and M. T. Smith. "Cancer Risks from Arsenic in Drinking Water." *Environ. Health Perspectives,* 97, 1992: 259–267.

Soave, R., and W. D. Johnson, Jr. "Cyclospora: Conquest of an Emerging Pathogen (comment)." *Lancet,* 345, 1995: 667–668.

Sobsey, M., and B. Olsen. "Microbial Agents of Waterborne disease." In *Assessment of Microbiology and Turbidity Standards for Drinking Water,* EPA570/9-83-001, P.S. Berger and Y. Argamon, eds. Washington, D.C.: U.S. Environmental Protection Agency, Office of Drinking Water, 1983.

Soto, A. M., C. Sonnenschein, K. L. Chung, M. F. Fernandez, N. Olea, and F. Olea Serrano. "The E-SCREEN Assay as a Tool to Identify Estrogens: An Update on Estrogenic Environmental Pollutants." *Environ. Health Perspectives,* 103(Suppl. 7), 1995: 113–122.

Stamler, R. "Implications of the INTERSALT study." *Hypertension,* 17(Suppl I), 1991: 16–20.

States, S. J., R. M. Wadowsky, J. M. Kuchta, R. S. Wolford, L. F. Conley, and R. B. Yee. "*Legionella* in Drinking Water." In *Drinking Water Microbiology,* pp. 340–367, G. A. McFeters, ed. New York: Springer-Verlag, 1990.

Sutton, L. D., W. W. Wilke, N. A. Lynch, and R. N. Jones. "Helicobacter pylori-containing Sewage Detected by an Automated Polymerase Chain Reaction Amplification Procedure." In *Proceedings, Annual Meeting of American Society for Microbiology,* Abstract C-395. American Society for Microbiology, 1995.

Taubes, G. "The (Political) Science of Salt." *Science,* 28, 1998: 898–907.

Thomas, D. L., P. O. Yarbough, D. Vlahov, S. A. Tsarev, K. E. Nelson, A. J. Saah, and R. H. Purcell. "Seroreactivity to hepatitis E virus in areas where the disease is not endemic." *J. Clin. Microbiol.,* 35, 1997: 1244–1247.

Thurman, E. T. "A Reconnaissance Study of Herbicides and Their Metabolites in Surface Water of the Midwestern United States Using Immunoassay and Gas Chromatography/Mass Spectrometry." *Jour. Envir. Sci. Technol.,* 26, 1992: 2440–2447.

Toppari, J., J. C. Larsen, P. Christiansen, A. Giwercman, P. Grandjean, L. J. Guillette, Jr., B. Jegou, T. K. Jensen, P. Jouannet, N. Keiding, H. Leffers, J. A. McLachlan, O. Meyer, J. Muller, W. Rajpert-De Meyts, T. Scheike, R. Sharpe, J. Sumpter, and N. E. Skakkebaek. "Male Reproductive Health and Environmental Xenoestrogens." *Environ. Health Perspectives,* 104(Suppl. 4), 1996: 741–803.

U.S. Environmental Protection Agency. "National Secondary Drinking Water Regulations." *Federal Register,* 44, 1979: 42198–42199.

U.S. Environmental Protection Agency. "National Primary and Secondary Drinking Water Regulations; Synthetic Organic Chemicals, Inorganic Chemicals and Microorganisms." *Federal Register,* 50, 1985a: 46936–47022.

U.S. Environmental Protection Agency. "National Primary Drinking Water Regulations; Volatile Synthetic Organic Chemicals." *Federal Register,* 50, 1985b: 46880–46901.

U.S. Environmental Protection Agency. "National Primary and Secondary Drinking Water Regulations; Fluoride." *Federal Register,* 51, 1986: 11396–11412.

U.S. Environmental Protection Agency. "National Primary Drinking Water Regulations; Synthetic Organic Chemicals; Monitoring for Unregulated Contaminants." *Federal Register,* 52, 1987a: 25690–25717.

U.S. Environmental Protection Agency. "Water Pollution Controls; National Primary Drinking Water Regulations; Volatile Synthetic Organic Chemicals; Para-Dichlorobenzene." *Federal Register,* 52, 1987b: 12876–12883.

U.S. Environmental Protection Agency. *Distribution Tables for the National Inorganics and Radionuclides Survey (NIRS) Results. Memorandum from Jon Longtin to Arthur Perler, February 23.* USEPA Water Docket, 1988a.

U.S. Environmental Protection Agency. "Drinking Water; Substitution of Contaminants and Drinking Water Priority List of Additional Substances which May Require Regulation under the Safe Drinking Water Act." *Federal Register,* 53, 1988b: 1892–1902.

U.S. Environmental Protection Agency. *Health Advisory for Atrazine.* Washington, D.C.: Office of Research and Development, 1988c.

U.S. Environmental Protection Agency. *Pesticides in Ground Water Data Base. 1988 Interim Report.* Washington, D.C.: Office of Pesticide Programs, 1988d.

U.S. Environmental Protection Agency. *Draft Drinking Water Health Criteria Document for Chlorophenols.* Washington, D.C.: Office of Drinking Water, 1989a.

U.S. Environmental Protection Agency. "Drinking Water; National Primary Drinking Water Regulations; Filtration, Disinfection; Turbidity, Giardia lamblia, Viruses, Legionella, and Heterotrophic Bacteria." *Federal Register,* 54, 1989b: 27486–27541.

U.S. Environmental Protection Agency. "National Primary and Secondary Drinking Water Regulations." *Federal Register,* 54, 1989c: 22062–22160.

U.S. Environmental Protection Agency. "Control of *Legionella* in Plumbing Systems." In *Reviews of Environmental Contamination and Toxicology, Vol. 107,* pp. 79–92, G.W. Ware, ed. New York: Springer-Verlag, 1989d.

U.S. Environmental Protection Agency. "Drinking Water; National Primary Drinking Water Regulations; Total Coliforms (Including Fecal Coliforms and E. coli)." *Federal Register,* 54, 1989e: 27544–27568.

U.S. Environmental Protection Agency. *National Pesticide Survey Phase I Report,* EPA570/9-90-015. Washington, D.C.: Government Printing Office, 1990a.

U.S. Environmental Protection Agency. "National Primary and Secondary Drinking Water Regulations; Synthetic Organic Chemicals and Inorganic Chemicals." *Federal Register,* 55, 1990b: 30370–30449.

U.S. Environmental Protection Agency. *Draft Drinking Water Health Criteria Document for Haloacetonitriles, Chloropicrin, and Cyanogen Chloride.* Washington, D.C.: Criteria and Standards Divisions, Office of Drinking Water, 1991a.

U.S. Environmental Protection Agency. "Maximum Contaminant Level Goals and National Primary Drinking Water Regulations for Lead and Copper." *Federal Register,* 56, 1991b: 26460–26564.

U.S. Environmental Protection Agency. "National Primary and Secondary Drinking Water Regulations; Radionuclides." *Federal Register,* 56, 1991c: 33050–33127.

U.S. Environmental Protection Agency. "National Primary Drinking Water Regulations; Synthetic Organic Chemicals." *Federal Register,* 56, 1991d: 30266–30281.

U.S. Environmental Protection Agency. "National Primary and Secondary Drinking Water Regulations; Synthetic Organic Chemicals and Inorganic Chemicals; Monitoring for Unregulated Contaminants; National Primary Drinking Water Regulations Implementation; National Secondary Drinking Water Regulations." *Federal Register,* 56, 1991e: 3526–3599.

U.S. Environmental Protection Agency. "National Primary Drinking Water Regulations—Monitoring for Synthetic Organic Chemicals; MCLGs and MCLs for Aldicarb, Aldicarb Sulfoxide, Aldicarb Sulfone, Pentachlorophenol, and Barium." *Federal Register,* 56, 1991f: 3600–3614.

U.S. Environmental Protection Agency. *Consensus Method for Determining Groundwaters under the Direct Influence of Surface Water Using Microscopic Particulate Analysis (MPA),* EPA 910/9-92-029. Port Orchard, WA: EPA Region 10 Environmental Services Division, 1992a.

U.S. Environmental Protection Agency. *Control of Biofilm Growth in Drinking Water Distribution Systems,* EPA625/R-92-001. Washington, DC: Office of Drinking Water, 1992b.

U.S. Environmental Protection Agency. "National Primary and Secondary Drinking Water Regulations—Synthetic Organic Chemicals and Inorganic Chemicals; National Primary Drinking Water Regulations Implementation." *Federal Register,* 57, 1992c: 31776–31849.

U.S. Environmental Protection Agency. *National Pesticide Survey Phase I Report,* EPA579/09-91-020. Washington, D.C.: Office of Water, Office of Pesticides and Toxic Substances, 1992d.

U.S. Environmental Protection Agency. *Occurrence Assessment for Disinfectants and Disinfection By-Products (Phase 6a) in Public Drinking Water.* Washington, D.C.: Office of Science and Technology, Office of Water, 1992e.

U.S. Environmental Protection Agency. *Draft Drinking Water Health Criteria Document for Bromate.* Washington, D.C.: Office of Science and Technology, Office of Water, 1993.

U.S. Environmental Protection Agency. *Draft Drinking Water Health Criteria Document for Chloramines.* Washington, D.C.: Office of Science and Technology, Office of Water, 1994a.

U.S. Environmental Protection Agency. *Draft Drinking Water Health Criteria Document for Chlorinated Acetic Acids/Alcohols/Aldehydes and Ketones.* Washington, D.C.: Office of Science and Technology, Office of Water, 1994b.

U.S. Environmental Protection Agency. *Draft Drinking Water Health Criteria Document for Chlorine, Hypochlorous Acid and Hypochlorite Ion.* Washington, D.C.: Office of Science and Technology, Office of Water, 1994c.

U.S. Environmental Protection Agency. *Final Draft Drinking Water Health Criteria Document for Chlorine Dioxide, Chlorite and Chlorate.* Washington, D.C.: Office of Science and Technology, Office of Water, 1994d.

U.S. Environmental Protection Agency. *Final Draft for the Drinking Water Criteria Document on Trihalomethanes.* Washington, D.C.: Health and Ecological Criteria Div., Office of Science and Technology, 1994e.

U.S. Environmental Protection Agency. "National Primary Drinking Water Regulations; Disfectants and Disinfection Byproducts; Proposed." *Federal Register,* 59, 1994f: 38668–38829.

U.S. Environmental Protection Agency. *Status of Pesticides in Reregistration and special review,* EPA738/R-94-008. Washington, D.C.: Government Printing Office, 1994g.

U.S. Environmental Protection Agency. *Aldicarb, Aldicarb Sulfoxide, and Aldicarb Sulfone.* Washington, D.C.: Health and Ecological Criteria Division, Office of Science and Technology, and Office of Water, 1995a.

U.S. Environmental Protection Agency. *Turbidity Criteria Document, Draft (September 1, 1985).* Washington, D.C.: U.S. Environmental Protection Agency (Office of Research and Development and Office of Drinking Water), 1995b.

U.S. Environmental Protection Agency. *Environmental Indicators of Water Quality in the United States. Fact Sheets,* EPA841/F-96-001. Washington, D.C.: Office of Water, 1996a.

U.S. Environmental Protection Agency. *Microscopic Particulate Analysis (MPA) for Filtration Plant Optimization,* EPA 910-R-96-001. Seattle, WA: Region 10, Office of Environmental Assessment, 1996b.

U.S. Environmental Protection Agency. *Proposed Guidelines for Cancer Risk Assessment,* EPA600/P-92-003C. Washington, D.C.: Office of Research and Development, 1996c.

U.S. Environmental Protection Agency. *Drinking Water Advisory: Consumer Acceptability Advice and Health Effects Analysis on Methyl Tertiary-Butyl Ether (MtBE).* Washington, D.C.: Office of Water, 1997a.

U.S. Environmental Protection Agency. "National Primary Drinking Water Regulations; Disinfectants and Disinfection Byproducts; Notice of Data Availability." *Federal Register,* 62, 1997b: 59388–59484.

U.S. Environmental Protection Agency. *Summary of New Health Effects Data on Drinking Water Disinfectants and Disinfectant Byproducts (D/DBPs) for the Notice of Data Availability (NODA),* October 10, 1997. Washington, D.C.: Office of Science and Technology, 1997c.

U.S. Environmental Protection Agency. *Dichloroacetic acid: Carcinogenicity Identification Characterization Summary,* March 1998a. Washington, D.C.: National Center for Environmental Assessment, Office of Research and Development.

U.S. Environmental Protection Agency. *EPA Panel Report and Recommendation for Conducting Epidemiological Research on Possible Reproductive and Developmental Effects of Exposure to Disinfected Drinking Water.* Washington, D.C.: Office of Research and Development, 1998b.

U.S. Environmental Protection Agency. *Health Risk Assessment/Characterization of the Drinking Water Disinfection Byproduct Bromate,* March 13, 1998. Washington, D.C.: Office of Science and Technology, Office of Water, 1998c.

U.S. Environmental Protection Agency. *Health Risk Assessment/Characterization of the Drinking Water Disinfection Byproduct Chlorine Dioxide and the Degradation Byproduct Chlorite.* March 13, 1998. Washington, D.C.: Office of Science and Technology, Office of Water, 1998d.

U.S. Environmental Protection Agency. *Health Risk Assessment/Characterization of the Drinking Water Disinfection Byproduct Chloroform,* March 13, 1998. Washington, D.C.: Office of Science and Technology, Office of Water, 1998e.

U.S. Environmental Protection Agency. *Integrated Risk Information System,* April. Cincinnati, OH: Office of Research and Development, Office of Health and Environmental Assessment, 1998f.

U.S. Environmental Protection Agency. "National Primary Drinking Water Regulations; Disinfectants and Disinfection Byproducts; Notice of Data Availability." *Federal Register,* 63, 1998g: 15674–15692.

U.S. Environmental Protection Agency. "National Primary Drinking Water Regulations: Disinfectants and Disinfection Byproducts; Final Rule." *Federal Register,* 63, 1998h: 69389–69476.

U.S. Environmental Protection Agency. "National Primary Drinking Water Regulations; Interim Enhanced Surface Water Treatment Rule; Final Rule." *Federal Register,* 63, 1998i: 69478–69521.

Uthus, E. O. "Estimation of Safe and Adequate Daily Intake for Arsenic." In *Risk Assessment of Essential Elements,* W. Mertz, C.O. Abernathy, and S.S. Olin, eds. Washington, D.C.: International Life Sciences Institute Press, 1994.

Velazquez, O., C. S. Harrison, C. Stetler, C. Avila, G. Ornelas, C. Alvarez, S. C. Hadler, D. W. Bradley, and J. Sepulveda. "Epidemic Transmission of Enterically Transmitted Non-A, Non-B Hepatitis in Mexico, 1986–1987." *Jour. Amer. Med. Assoc.,* 263, 1990: 3281–3285.

Valezquez, S. F., and J. T. Du. "Derivation of the Reference Dose for Managanese." In *Risk Assessment of Essential Elements,* W. Mertz, C.O. Abernathy, and S.S. Olin, eds. Washington, D.C.: International Life Sciences Institute Press, 1994.

Vaughn, J. M., Y. Chen, K. Lindburg, and D. Morales. "Inactivation of Human and Simian Rotaviruses by Ozone." *Appl. Environ. Microbiol.* 53, 1987: 2218–2221.

Vaughn, J. M., Y. Chen, and M. Z. Thomas. "Inactivation of of Human and Simian Rotaviruses by Chlorine." *Appl. Environ. Microbiol.,* 51, 1986: 391–394.

Walker, E. M., Jr., M. D. Wolfe, M. L. Norton, S. M. Walker, and M. M. Jones. "Hereditary hemochromatosis." *Ann. Clin. Lab. Sci.,* 28, 1998: 300–312.

Waller, T. "Sensitivity of Encephalitozoon cuniculi to Various Temperatures, disinfectants and drugs." *Lab. Anim.,* 13, 1979: 227–230.

Water Research Center. "Removal of Cryptosporidium during Water Treatment." In *Cryptosporidium, Giardia, and Other Encysting Parasites: Origins, Analytical Methodology, the Significance of Water vs. Other Vectors. Project DW-06.* UK Water Industry Research, Ltd., 1995.

Weber, R., R. Bryan, D. Schwartz, and R. Owen. "Human Microsporidial Infections." *Clin. Microbiol. Rev.,* 7, 1994: 426–461.

White, J. M., J. G. Wingo, L. M. Alligood, G. R. Cooper, J. Gutridge, W. Hydaker, R. T. Benack, J.W. Dening, and F.B. Taylor. "Sodium ion in drinking water. I. Properties, analysis, and occurrence." *Jour. Am. Dietetic Assoc.,* 50, 1967: 32–36.

Williams, N. "Phage transfer: a new player turns up in cholera infection." *Science,* 272, 1996: 1869–1870.

Wolfe, G., and L. Kaiser. *Final Report, Sodium Bromate: Short Term Reproductive and Developmental Toxicity Study when Administered to Sprague-Dawley Rats in the Drinking Water. Study No. NTP-PEST, 94007.* NTP/NIEHS No. NOI-ES-15323. 1996.

Young, W. F., H. Horth, R. Crane, T. Ogden, and M. Arnott. "Taste and Odour Threshold Concentrations of Potential Potable Water Contaminants." *Water Res.,* 30, 1996: 331–340.

Yu, V. L., J. J. Zuravleff, L. Gavlik, and M. H. Magnussen. "Lack of Evidence for Person-to-Person Transmission of Legionnaires' Disease." *J. Infect. Dis.* 147, 1983: 362.

CHAPTER 3

GUIDE TO SELECTION OF WATER TREATMENT PROCESSES

Gary Logsdon
Black & Veatch
Cincinnati, Ohio

Alan Hess
Black & Veatch
Philadelphia, Pennsylvania

Michael Horsley
Black & Veatch
Kansas City, Missouri

Water treatment process selection is a complex task. Circumstances are likely to be different for each water utility and perhaps may be different for each source used by one utility. Selection of one or more water treatment processes to be used at a given location is influenced by the necessity to meet regulatory quality goals, the desire of the utility and its customers to meet other water quality goals (such as aesthetics), and the need to provide water service at the lowest reasonable cost. Factors that should be included in decisions on water treatment processes include:

- Contaminant removal
- Source water quality
- Reliability
- Existing conditions
- Process flexibility
- Utility capabilities
- Costs
- Environmental compatibility

- Distribution system water quality
- Issues of process scale

This chapter begins with a brief discussion of alternatives to water treatment, followed by a review of the various factors that may influence the selection of a water treatment process. After these factors are covered, the chapter presents examples of water treatment process selection and explains the reasons for the choices made in the examples. The capabilities of commonly used treatment processes are presented in detail in the subsequent chapters of this book.

WATER SUPPLY APPROACHES

Use of the best source water quality that can be obtained economically is a concept that has been advocated by public health authorities for decades. The 1962 Public Health Service Drinking Water Standards (Public Health Service, 1969) stated, "The water supply should be obtained from the most desirable source which is feasible, and effort should be made to prevent or control pollution of that source. If the source is not adequately protected by natural means, the supply shall be adequately protected by treatment." The EPA's National Interim Primary Drinking Water Regulations (Environmental Protection Agency, 1976) stated, "Production of water that poses no threat to the consumer's health depends on continuous protection. Because of human frailties associated with protection, priority should be given to selection of the purest source." The fundamental concept of acquiring the best quality of source water that is economically feasible is an important factor in making decisions about source selection and treatment.

Alternative Sources

Water utilities and their engineers need to consider use of alternative sources when a new treatment plant or a major capacity expansion to an existing plant is being evaluated, or when a different and more costly approach to treatment is under study. When treatment costs are very high, development of a source of higher quality may be economically attractive. Among the options are:

- A different surface water source or a different groundwater source
- Groundwater instead of surface water
- Riverbank infiltration instead of direct surface water withdrawal

For medium or large water systems, switching to a different surface water source or groundwater source may be difficult because of the magnitude of the raw water demand. Small water systems with small demands may find it easier to obtain other sources within distances for which transmission of the water is economically feasible.

Alternatives to Treatment

In some instances, water utilities may be able to avoid investing large sums on treatment by choosing an alternative to treatment. One option that may be available to small water systems is to purchase water from another utility instead of treating water. This option might be selected when treatment requirements are made more stringent by regulations, or when capacity of the system has to be expanded to meet demand. This may be a particularly attractive choice when a nearby larger utility has excess capacity and can provide treated water of the quality needed.

Other alternatives to increased capacity for water treatment may occasionally be available. If the water utility needing to expand has not adopted universal metering for domestic water customers, the system demand might be significantly reduced if universal metering was put in place. Customers on flat rates may have little overall incentive, and no identifiable economic incentive, to be prudent in their use of water. If a system is unmetered, and the average per capita demand is substantially higher than demand in nearby metered systems, conserving the existing supply by spending money for meters may be a wiser investment than spending money for additional treatment facilities. When distribution systems have high rates of water loss, a program of leak detection and repair may result in increasing the amount of water available to consumers without an increase in production.

Examination of alternatives to treatment may in many instances reveal the existence of no practical or economically attractive alternatives to treatment of a presently used or a new water source. In such circumstances, modified, expanded, or new water treatment facilities will be necessary. Concepts on the selection of water treatment processes are presented in the remainder of this chapter. Treatment techniques, how they function, and their capabilities with regard to improving the quality of source water are discussed in following chapters of this book.

FACTORS INFLUENCING PROCESS SELECTION

Contaminant Removal

Contaminant removal is the principal purpose of treatment for many source waters, particularly surface waters. The quality of treated water must meet all current drinking water regulations. These regulations were reviewed after the passage of the 1996 Safe Drinking Water Act amendments by Pontius, who discussed not only the status of regulations but also the potential health effects and possible sources of regulated contaminants (Pontius, 1998). Furthermore, to the extent that future regulations can be predicted by careful analysis of proposed drinking water regulations, water treatment processes should be selected to enable the water utility to be in compliance with those future regulations when they become effective.

When water utility customers and water utility management place a strong emphasis on excellent water quality, the maximum contaminant levels (MCLs) of drinking water regulations may be viewed as an upper level of water contaminants that should be seldom or never approached, rather than as a guideline for finished water quality. Many water utilities choose to produce water that is much better in quality than water that would simply comply with the regulations. Such utilities may employ the same treatment processes that would be needed to provide the quality that complies with regulations, but operate those processes more effectively. Other utilities may employ additional treatment processes to attain the high finished water quality they seek.

Both surface waters and groundwaters may have aesthetic characteristics that are not acceptable to customers, even though MCLs are not violated. Utilities in some states may be required to provide treatment to improve the quality of water that has problems of taste, odor, color, hardness, high mineral content, iron, manganese, or other aesthetic problems resulting in noncompliance with secondary MCLs. Improvement of aesthetic quality is very important, however, because customer perceptions of water quality often are formed based on observable water quality factors, most of which are aesthetic. Water that has bad taste or odor or other aesthetic problems may be perceived as unsafe by customers. This can cause a loss of confidence in the utility by its customers, and might cause some persons to turn to an unsafe source of water in lieu of using a safe but aesthetically objectionable public water supply.

Much is known in general about the capabilities of various water treatment processes for removing both regulated contaminants and contaminants that cause aesthetic problems. A comprehensive review of drinking water treatment processes appropriate for removal of regulated contaminants was undertaken by the National Research Council (NRC) in the context of providing safe drinking water for small water systems (National Research Council, 1997), but many of the NRC's findings regarding treatment processes are applicable regardless of plant size. Information on the general effectiveness of treatment processes for removal of soluble contaminants is presented in Table 3.1. For removal of particulate contaminants, filtration and clarification (sedimentation or dissolved air flotation) processes are used. Site-specific information on process capabilities may be needed, however, before engineers select a process train for a plant, particularly when no previous treatment experience exists for the source water in question. Pilot plant studies may be an appropriate means of developing information on treatment processes and the water quality that can be attained by one or more process trains under evaluation. As soon as candidate treatment processes and treatment trains are identified, the potential need for a pilot plant study should be reviewed and the issue resolved. Carrying out a pilot study prior to process selection could take from 1 to 12 months for testing onsite and an additional 2 to 6 months for report preparation, but sometimes such a study holds the key to a cost-effective design and to ensuring that the quality goals will be met by the process train selected.

Information on the general capabilities of water treatment processes for removal of soluble contaminants is presented in Table 3.1. Much of the information in this table is drawn from the NRC report and from *Water Quality and Treatment,* Fourth Edition. Some soluble contaminants are more readily removed after oxidation, and this is indicated in the table. Not included in Table 3.1 are particulate contaminants and gases. Particulate contaminants are removed by the various filtration processes listed in the table, plus slow sand filtration, microfiltration, and ultrafiltration. In general, gaseous contaminants are treated by aeration or air stripping. Details of contaminant removal are presented in other chapters in this book.

The interaction of various processes on treated water quality must be considered in the regulatory context and in the broader context of water quality. Drinking water regulations generally have been written in a narrow context focusing on the contaminant or contaminants being regulated. Sometimes an approach to treatment for meeting a given MCL can cause problems of compliance with other regulations. For example, use of increased free chlorine residual might be an approach to meeting the CT requirement of the Surface Water Treatment Rule, but this could cause trihalomethanes (THMs) in the distribution system to exceed the MCL and possibly taste and odor problems. Maintaining a high pH in the distribution system might be helpful for meeting the requirements of the Lead and Copper Rule, but high pH increases the possibility of THM formation and decreases the efficacy of disinfection by free chlorine.

Some interactions between treatment processes are beneficial. Ozone can be used for a variety of purposes, including control of tastes and odors, disinfection, and oxidation of iron and manganese. Improved filter performance in terms of longer runs or improved particle removal or both can be an additional benefit of using ozone; however, ozonation by-products must be controlled to prevent biological regrowth problems from developing in the distribution system.

Source Water Quality

A comparison of source water quality and the desired finished water quality is essential for treatment process selection. With the knowledge of the changes in water quality that must be attained, the engineer can identify one or more treatment processes that would be capable of attaining the quality improvement. Depending on a water utility's past experience with a water source, the amount of data available on source water quality may range from almost nonexistent to fairly extensive. Learning about the source or origin of the raw water can be helpful for estimating the nature of possible quality problems and developing a monitoring program to define water quality. For surface waters, information about the watershed may reveal sources of contamination, either manmade or natural. Furthermore, an upstream or downstream user may possess data on source water quality. For groundwaters, knowledge of the specific aquifer from which the water is withdrawn could be very useful, especially if other nearby water utilities are using the same aquifer.

The capability of a water treatment plant to consistently deliver treated water quality meeting regulatory and water utility goals is strongly enhanced when the range of source water quality is always within the range of quality that the plant can successfully treat. Frequently, the source water database is limited. Water quality characteristics that may vary over a wide range, such as turbidity, can be studied by using probability plots. With such plots, estimates can be made of the source water turbidity that would be expected 90 or 99 percent of the time. When treatment processes such as slow sand, diatomaceous earth, or direct filtration are considered, careful study of the source water quality is needed to ensure that the high-quality source water required for successful operation of these processes will be available on a consistent basis. Source water quality problems can sometimes signal the need for a particular process, such as use of dissolved air flotation to treat algae-laden waters. When surface waters are treated, the multiple barrier concept for public health protection should be kept in mind. Sources subject to heavy fecal contamination from humans or from livestock (cattle, hogs, sheep, horses, or other animals capable of transmitting *Cryptosporidium*) will probably require multiple physical removal barriers [sedimentation or dissolved air flotation (DAF) followed by filtration].

Source water quality is an issue that can be used to eliminate a process from consideration, if the process has not been proven to be capable of successfully treating the range of source water quality that would be encountered at the site in question.

Reliability

Process reliability is an important consideration and in some cases could be a key aspect in deciding which process to select. Disinfection of surface water is mandatory, so this is an example of a treatment process that should be essentially fail-safe. The only acceptable action to take for a failure of disinfection in a plant treating surface water is to stop distributing water from the treatment works until the problem is corrected and proper disinfection is provided or until a "boil water" order can be put in place so the public will not drink undisinfected surface water. To avoid disin-

TABLE 3.1 General Effectiveness of Water Treatment Processes for Removal of Soluble Contaminants

						Membrane processes			Ion exchange		Adsorption		
Contaminant categories	Aeration and stripping	Coagulation, sedimentation or DAF,* filtration	Precoat filtration	Lime softening	Chemical oxidation and disinfection	Nanofiltration	Reverse osmosis	Electrodialysis/ ED reversal	Anion	Cation	Granular activated carbon	Powdered activated carbon	Activated alumina
Primary contaminants													
Inorganics													
Antimony							X†	X					
Arsenic (+3)		XO‡		XO			X	X	X				X
Arsenic (+5)		X		X			X	X	X				X
Barium				X			X	X		X			
Beryllium		X		X			X	X					
Cadmium		X		X			X	X		X			
Chromium (+3)		X		X			X	X		X			
Chromium (+6)							X	X	X				
Cyanide					X								
Fluoride				X			X	X					X
Lead§													
Mercury (inorganic)				X			X	X					
Nickel				X			X	X		X			
Nitrate							X	X	X				
Nitrite							X	X	X				
Selenium (+4)		X					X	X	X				X
Selenium (+6)							X	X	X				X
Thallium							X	X					X

Organic Contaminants												
Volatile organics	X										X	
Synthetic organics							X				X	X
Pesticides/Herbicides						X	X				X	X
Dissolved organic carbon		X				X	X				X	X
Radionuclides												
Radium (226 + 228)				X			X	X		X		
Uranium							X	X	X			
Secondary contaminants and constituents causing aesthetic problems												
Hardness				X		X	X	X		X		
Iron		XO	XO	X						X		
Manganese		XO	XO	X						X		
Total dissolved solids							X	X				
Chloride							X	X				
Sulfate						X	X	X				
Zinc				X			X	X		X		
Color		X			X	X	X				X	X
Taste and odor	X				X						X	X

* DAF, dissolved air flotation.
† X, appropriate process for this contaminant.
‡ XO, appropriate when oxidation used in conjunction with this process.
§ Lead is generally a product of corrosion and is controlled by corrosion control treatment rather than removed by water treatment processes.

fection failures and to minimize downtime in the event of an equipment failure, backup disinfection systems or spare parts must be kept on hand for dealing with emergencies. Process reliability would be a very important factor in evaluating alternative disinfection systems, as well as other processes whose failure could have immediate public health consequences.

Process reliability needs to be evaluated on a case-by-case basis, because factors that influence reliability in one situation may not apply at another situation. Factors that can influence reliability include:

Range of source water quality versus the range of quality the process can successfully treat

Rate of change of source water quality—slow and gradual or very rapid and severe

Level of operator training and experience

Staffing pattern—24 hours per day or intermittent, such as one shift per day

Mode of operation

- Continuous, or on-off each day
- Consistent rate of flow, or varying flow related to water system demand

Amount of instrumentation

Ability of the utility to maintain instruments in good working order and to keep them properly calibrated

Reliability of electric power supply

Capability to prevent or minimize source water deterioration over the long term

The concept of robustness is important to reliability. Robustness for water filtration plants was defined by Coffey et al. (1998) as ". . . the ability of a filtration system to provide excellent particle/pathogen removal under normal operating conditions and to deviate minimally from this performance during moderate to severe process upsets." Although the term "robustness" was not used, Renner and Hegg (1997) emphasize that changes in raw water quality should not impact the performance of sedimentation basins and filters in a self-assessment guide prepared for the Partnership for Safe Water. Drinking water literature has not focused on robustness through the years, but information does exist on processes that seem to resist upsets well and those that are less robust. For example, Kirmeyer (1979) showed that serious water quality deterioration occurred within about 15 minutes when coagulant chemical feed was lost in a direct filtration pilot plant treating low-turbidity water. In this episode, filtered water turbidity increased from 0.08 to 0.20 nephelometric turbidity unit (ntu), whereas chrysotile asbestos fibers increased from 0.1 million to 0.36 million fibers/L. After coagulant feed was restored, filtered water turbidity was reduced to 0.08 ntu, and the asbestos fiber count declined to 0.01 million fibers/L. Until more is published on robustness, engineering experience and judgment may be the best guide for considering this aspect of reliability.

Existing Conditions

The choice of processes to incorporate into a treatment train may be influenced strongly by the existing processes when a treatment plant is evaluated for upgrading or expanding. Site constraints may be crucial in process selection, especially in pretreatment when alternative clarification processes are available, some of which require only a small fraction of the space needed for a conventional settling basin. Hydraulic constraints can be important when retrofitting plants with ozone or granular activated carbon (GAC) adsorption. The extra head needed for some treatment processes could result in the necessity for booster pumping on-site to accommodate the hydraulic requirements of the process. This adds to the overall cost of the plant improvements and, in some cases, might result in a different process being selected. The availability of high head can influence process selection in some instances. Pressure filtration might be selected for treatment of groundwater after oxidation, for iron or manganese removal. In this situation, use of gravity filtration would involve breaking head and pumping after filtration, whereas with pressure filters it might be possible to pump directly from the well through the filters to storage.

Process Flexibility

The ability of a water treatment plant to accommodate changes in future regulations or changes in source water quality is quite important. In the present regulatory environment, water utilities must realize that more regulations are likely in the future. For some utilities, these future regulations may require additional treatment or more effective treatment, such as when a previously unregulated contaminant is present in the source water or a maximum contaminant level is lowered for a contaminant in the utility's source water. Some water treatment processes target a narrow range of contaminants and may not be readily adaptable to controlling other contaminants. For example, both microfiltration and diatomaceous earth filtration can provide excellent removal of particulate contaminants in the size range of protozoa. A surface water treatment plant employing either of those processes and treating a source water with an arsenic concentration of 0.03 to 0.04 mg/L (less than the present MCL, 0.05 mg/L) might not be able to meet a future arsenic MCL that was substantially lower than the present MCL. On the other hand, a surface water treatment plant employing coagulation and filtration might be able to attain sufficient arsenic removal to comply with a future lower MCL, depending on the arsenic concentration in the source water, the coagulant chemical and its dosage, and the pH of treatment. The coagulation and filtration treatment train in this example has more flexibility for dealing with a changing regulatory requirement.

Source water quality should be well established when a treatment plant is planned, so that good decisions on treatment processes can be made. Most treatment plants are built to last for several decades, and changes can occur in the quality of source waters with the passage of time. Long-term eutrophication of lakes can lead to increased algae blooms and to taste and odor problems. On the other hand, the positive changes in water quality in Lake Erie that have occurred since it was pronounced "dead" by some environmental advocates in the late 1960s have had some side effects. Some treatment plant operators believe that the water at present is more difficult to treat than it used to be. With the advent of zebra mussels and the elimination of some of the plankton in Lake Erie and other Great Lakes, the increased clarity has brought about the enhanced growth of benthic organisms in some places, with associated problems of taste and odor. Water quality problems of this nature generally cannot be foreseen when treatment processes are selected, and frequently cannot be prevented by the water utility. The defense against such problems is to incorporate process flexibility in a treatment plant, so that both present and unforeseen future quality problems can be addressed and finished water quality meeting the expectations of the utility and its customers can be produced for the long term.

Utility Capabilities

After treatment processes are selected, designed, and on-line, the water utility must be able to operate them successfully to attain the desired water quality. The issue of system size versus treatment complexity becomes important with smaller systems. If successful treatment plant operation requires more labor than a small system can afford, or if the level of technical skills exceeds that readily attainable in a community, treatment failure may occur. Availability and access to service and repair of equipment involves considerations of time and distance from service representatives, and this may be problematic for some small, very remote water utilities. Selected treatment processes need to be operable in the context for which they will be employed. System size is not the only determining factor in successful operation. Sometimes, management is not sufficiently progressive or does not realize the necessity of providing well-trained staff with modern tools and techniques to facilitate successful treatment plant operation. In this situation, utility management needs to be informed of the complexities and requirements for treatment processes before plans for treatment are adopted. Cleasby et al. (1989) reported that management attitudes about water quality were a key factor in attaining or failing to meet water quality goals. Introduction of relatively complex treatment processes at a water utility whose management is not supportive of actions that will be needed for successful operation is a recipe for trouble.

The adaptability of treatment to automation or enhanced supervisory control and data acquisition (SCADA) can be important for systems of all sizes. For large systems, automation or enhanced SCADA may be a way to keep operating costs in line by having a smaller but highly trained and talented operating staff. For small utilities, using automation or enhanced SCADA in conjunction with remote monitoring of processes may enable a small system to use a form of contract operation or circuit rider operation in which the highly trained specialist is not on-site all of the time but maintains close watch over the treatment processes through instrumentation and communications facilities.

Costs

Cost considerations usually are a key factor in process selection. Evaluation of costs for alternative process trains using principles of engineering economics might at first seem to be straightforward, but this may not be the case. When different treatment trains are evaluated, their capabilities are not likely to be identical, so the resulting treated-water quality from different trains likewise may not be identical. The basis for process comparison has to be decided upon in such situations. If a certain aspect of water quality improvement is beneficial but not really necessary, perhaps it is not sufficiently valuable to enter into cost considerations. For example, both diatomaceous earth filters and granular media filters with coagulation pretreatment can remove particulate matter, but the process train employing coagulation, flocculation, and sedimentation can remove more color and total organic carbon (TOC) from source water. For treatment of a water with low color and low TOC concentrations, the treatment for particulate contaminant removal may be sufficient, and the use of a lower-cost filtration process, such as diatomaceous earth filtration, might be favored. On the other hand, if additional water quality improvement is needed, then any process train under consideration must be able to attain that improvement.

Cost estimates should be made taking into consideration the entire life cycle cost of a process train. Both capital and operating and maintenance (O&M) costs must be included in the estimate. Estimating O&M costs can be difficult, and sometimes unforeseen major changes in the economy occur and invalidate earlier estimates. The very large increase in energy costs in the 1970s was not foreseen and caused some major reconsideration of operating practices and treatment process choices. Energy-intensive processes, such as reverse osmosis and recycling of calcium carbonate sludge to make lime at softening plants, were viewed as much less desirable after energy prices increased steeply in the mid-1970s. The need for repairs, for maintaining an inventory of spare parts or extra equipment, for operator staffing, and for routine maintenance activities must be included in cost determinations. Some water utilities have encountered high expenses for equipment upkeep and frequent replacement, negating the initial savings on the capital investment. Smaller utilities in particular must consider not only the amount of labor associated with the various treatment processes being considered, but also the skills required of that labor. For small utilities located in predominantly rural settings, far from large communities and far from sources of technical assistance on which to draw during times of crisis, the possibility of being able to attract and keep workers who can operate complex treatment equipment may become an important consideration. For some utilities, contract O&M arrangements or a circuit rider may be necessary for successful long-term operation.

Environmental Compatibility

Environmental compatibility issues cover a broad spectrum of concerns including residual waste management, the fraction of source water wasted in treatment processes, and energy requirements for treatment. The effect of water treatment extends beyond the treatment plant. The benefits of providing safe drinking water are very great, but caution must be taken that the treatment processes selected to provide that safe water do not create serious environmental problems. Making quantitative calculations about public health benefits and environmental damages attributed to alternative treatment processes is likely to involve much guesswork and only a limited amount of solid data, but the difficulty in making firm estimates about overall environmental effects should not discourage engineers and owners from considering these issues.

Residuals, or sludge and other by-products of water treatment, are commonly thought of when environmental compatibility is considered. Disposal of large volumes of water works sludge to surface waters is no longer permitted in most locations. Therefore, the residuals produced by coagulation, enhanced coagulation, and lime softening need to be dealt with in an environmentally acceptable manner. Disposal of brines from ion exchange or some membrane processes can present difficult issues in locations where brackish water or salt water is not nearby. Treatment of residuals can account for a significant portion of the total cost of water treatment; in some instances, concerns about residuals could influence process selection.

Water wastage is an issue that may be important in areas where water supplies are limited. Treatment employing membrane processes has some advantages over other approaches to filtration, but if the fraction of water rejected by a membrane process is excessive, then less water is available to satisfy the demand for treated water. Recycling of high-volume process waste streams, with or without additional treatment, is also practiced in many areas.

Energy usage by water utilities could become an environmental concern in the future. Water utilities currently use about 3 percent of the electricity used in the United States (Harmon et al., 1998). If global warming concerns increase in the future, and energy usage reductions are mandated in the United States, energy-intensive treatment processes may be viewed less favorably. The issues of global warming and energy usage are highly contentious in the United States. If evidence of actual global warming were to become scientifically and politically overwhelming, energy usage would become a more important factor in process selection, even though a majority of the energy used by water utilities is for pumping (Patton and Horsley, 1980). Developing estimates of future costs is very difficult. Those who consider the possible effect of future energy cost increases might look to the mid- to late 1970s, when the energy crisis and sharp increases in fuel prices occurred in the United States. A before-and-after comparison of the delivered prices of coagulant chemicals, sludge disposal costs, and electricity could be useful in an assessment of the vulnerability of a treatment plant employing coagulation versus vulnerability of a microfiltration plant to future energy price hikes.

Distribution System Water Quality

The influence of treatment processes on desired water quality in the distribution system is a factor to be considered in process evaluation, and includes:

- Chemical and microbiological stability of water leaving the treatment plant
- Prevention of internal corrosion and deposition
- Microbiological control in the distribution system
- Compatibility of the quality with water from other sources
- Minimization of formation of disinfection by-products in the distribution system

Regulatory requirements related to water distribution system monitoring are such that even if finished drinking water at the treatment plant meets MCLs, water quality deterioration in the distribution system could result in regulatory compliance problems.

Treatment processes should be selected to enhance water stability. For example, ozone's ability to break the molecular bonds of large organic molecules and form smaller organic molecules or molecular fragments can result in the formation of a more suitable food source for bacteria found in water, so use of ozone can promote growth of bacteria in water. If this growth takes place within a filter bed in the treatment plant, water with greater biological stability can be produced. On the other hand, if little or none of the organic matter were metabolized by bacteria in the filter bed, the organics would pass into the distribution system and could promote the growth of biofilms there. Distribution system biofilms can cause a variety of problems, including microbiological compliance violations, tastes and odors, excessive chlorine demand and free chlorine depletion, and corrosion of water mains.

If the pH and alkalinity of finished water are such that the water will not be stable over time, water quality in the distribution system may change sufficiently to cause corrosion problems, even though the water did not seem to be problematic at the treatment plant.

When multiple water sources are used by a single water utility, problems of water incompatibility can arise. These might be caused by the nature of the source waters, such as a water having high mineral content being mixed in a distribution system with a water of low mineral content. In addition, this situation could arise when a conventionally treated surface water and water treated by reverse osmosis are put into a common distribution system. Alternatively, water from different sources might be treated by different disinfection techniques. In general, it is considered inadvisable to mix chloraminated water and water disinfected with free chlorine in a distribution system. At the zone where the two different waters interact, the free chlorine can chemically react with the monochloramine, reducing the available free chlorine residual and forming dichloramine or nitrogen trichloride. Taste and odor complaints may also result from this practice.

Issues of Process Scale

Feasibility to scale processes up to very large sizes or to scale them down to very small sizes can be important in some cases. Complex treatment processes, such as coagulation and filtration of surface water or precipitative lime softening, can be scaled down physically, but the costs of equipment and the need for a highly trained operator may make the scaled-down process impractical. Processes that are practical and manageable at 10 mgd (38,000 m^3/day) or even 1 mgd (3,800 m^3/day) may be too complex at 0.01 mgd (38 m^3/day). On the other hand, processes that work very well for small water systems may not be practical for large systems. Membrane filtration has worked very well for small systems, but microfiltration plants in the size range of 100 to 500 mgd (3.8×10^5 to 1.9×10^6 m^3/day) would at this time entail a very large amount of piping and valving to interconnect large numbers of small modules. Processes that employ treatment modules (e.g., microfiltration) are expanded to larger sizes by joining together more modules. This can become problematic for a 100-fold size expansion. On the other hand, granular media filters can be expanded by designing the filter to have a large or small surface area. One single granular media filter bed could be as small as 4 ft^2 (0.37 m^2), or as large as over 1000 ft^2 (93 m^2), and filtration plants with capacities ranging from 27,000 gal/day (package plant) to 1 billion gal/day (100 m^3/day to 3.8×10^6 m^3/day) have been built.

EVALUATING PROCESS OPTIONS

When treatment of a new water source or expansion at an existing treatment plant is being considered, in most cases a number of options will be available. One task for project planners is to consider all reasonable options for treatment, and then gradually eliminate those that are not likely to be among the best choices, so that further efforts can be directed to identifying the process most appropriate for the given situation. A systematic approach for doing this is to develop a matrix table in which all treatment processes under consideration are listed on one axis, and the factors related to process selection are presented on the other axis. Each process is given a rating or ranking for each of the factors listed. Depending on the importance of some factors, a weighting system could be used to allow for greater influence of the more important aspects being considered.

For a surface water filtration plant, the following factors should be considered in a process evaluation report:

Meeting regulatory requirements

- Interim Enhanced Surface Water Treatment Rule
- Stage 1 Disinfectant/Disinfection By-Product Rule
- Expected Long-Term Enhanced Surface Water Treatment Rule
- Expected Stage 2 Disinfectant/Disinfection By-Product Rule

Process capability for treating variable raw water quality compared with expected raw water quality

Coping with spills in watershed

Staff experience with operating the process

Level of operator training needed

Process reliability/complexity

Process monitoring needs and capability of staff to manage the monitoring

Water industry experience with the process

Long-term viability

Customer acceptance

Compatibility with site's physical constraints

Compatibility with existing plant processes

Energy needs

Capital cost

Operation and maintenance cost

The factors that are considered during treatment process selection are not limited solely to engineering issues. Therefore, process evaluation and selection often involve not only consultants and water utility engineers, but also water utility managers and operators and perhaps others whose perspective must include an understanding of community issues and concerns. After a preliminary evaluation report is prepared, process selection may involve an extended meeting or workshop in which numerous interested parties participate and develop the ranking for the treatment processes in the matrix. Developing a consensus among those involved is an important step toward building broad public support for the water supply developments that are needed.

EXAMPLES OF TREATMENT PROCESS SELECTION

Hypothetical Examples

Surface Water Treatment. Surface water treatment can be accomplished by a variety of process trains, depending on source water quality. Some examples are given below, beginning with conventional treatment. All surface waters require disinfection, so regardless of the treatment train chosen to treat a surface water, that process train must include disinfection.

Disinfection Only with No Filtration. The number of water systems for which treatment of surface water consists only of disinfection is a small fraction of the total systems using surface water and is likely to decrease as a result of population growth and increasing difficulty associated with watershed ownership or control. Nevertheless, some systems, including some very large ones, now use this approach to water treatment.

The USEPA has addressed use of surface waters without filtration in the Surface Water Treatment Rule (SWTR) (EPA, 1989) and the Interim Enhanced Surface Water Treatment Rule (EPA, 1998). In 1989, USEPA established source water quality limits on fecal coliforms (equal to or less than 20 per 100 mL in at least 90 percent of samples for a six-month period), on total coliforms (equal to or less than 100 per 100 mL in at least 90 percent of samples for a six-month period), and on turbidity (not to exceed 5 ntu on any day unless the state determines that this is an unusual event) and required monitoring of source water prior to disinfection so data would be available to determine whether these conditions had been met. The SWTR stipulated that to avoid filtration a public water system must maintain a watershed control program that minimizes potential for source water contamination by viruses and by *Giardia* cysts. A watershed control program must:

- Characterize watershed ownership and hydrology
- Identify characteristics of the watershed and activities within the watershed that might have an adverse effect on water quality
- Provide for monitoring of activities that might have an adverse effect on source water quality

In 1998, USEPA promulgated additional criteria for avoiding filtration, requiring that the potential for contamination by *Cryptosporidium* also would have to be considered. Adequacy of the watershed control program would be based on:

- The comprehensiveness of the watershed review
- The effectiveness of the program for monitoring and controlling detrimental activities in the watershed
- The extent to which the water system has maximized its land ownership or controlled land use within the watershed, or both

Cryptosporidium oocysts, unlike *Giardia* cysts, are not susceptible to free chlorine and chloramine at residual concentrations and contact times commonly used by water systems that use unfiltered surface waters, so when surface waters are not filtered a very heavy reliance is placed on watershed protection to provide for public health protection from *Cryptosporidium.* Reliance on watershed protection will continue for systems without filtration until a substantial amount of information is developed on inactivation of *Cryptosporidium* by chemical disinfectants and by ultraviolet radiation, such that USEPA is able to establish criteria for effective disinfection of this pathogen. Even then, maintaining an effective watershed protection program will be the most crucial barrier against *Cryptosporidium* for systems that do not filter.

Conventional Treatment. Water treatment studies by George Fuller and his associates at Louisville in the 1890s established that effective pretreatment, including clarification, was necessary for effective filtration of turbid or muddy surface waters such as the Ohio River. In the decades following Fuller's work, a treatment train consisting of chemical feed, rapid mix, flocculation, sedimentation, and filtration came to be considered conventional treatment. Conventional treatment is the norm for water treatment plant process requirements in *Ten State Standards* (Great Lakes–Upper Mississippi River Board of Public Health and Environmental Managers, 1997). Disinfection is included in conventional treatment, with the point or points of addition of disinfectant varying at different treatment plants. A conventional treatment train is appropriate for source waters that are sometimes or always turbid, with turbidity exceeding 20 to 50 ntu for extended periods of time.

A modern hypothetical conventional filtration plant (Figure 3.1) for treatment of the Ohio River (depending upon its location on the river) would need to treat water having turbidity ranging from as low as about 10 ntu to a high of over 1000 ntu during floods. Coagulant dosages might be as low as 10 mg/L to over 100 mg/L during floods. Depending on the coagulant of choice, addition of alkalinity might be needed at some times. Rapid mixing would be followed by flocculation. Sedimentation might be accomplished in conventional long rectangular basins, or in basins aided by tube or plate settlers. Filtration would probably involve use of dual media (anthracite over sand). With the present emphasis on lowering disinfection by-product formation, chlorination would probably take place after sedimentation or after filtration. Total organic carbon concentrations on the Ohio generally are not so high as to require extraordinary measures for control of TOC. Process detention times would be shorter and filtration rates would be higher for a modern plant than for Fuller's designs for Ohio River plants, but his concept of clarification before filtration would still be employed because of the large amount of suspended matter that must be removed from the water for filtration to be practical and effective. Conventional treatment would be appropriate for many surface waters in the United States.

Conventional Treatment with Pretreatment. Some surface waters carry loads of sediment so high that water treatment plants employ a presedimentation step prior to the conventional treatment train. Earlier in the twentieth century, plain sedimentation with no chemical addition was practiced to remove a portion of the suspended solids before conventional treatment. Now, it is common to add some polymer or coagulant to enhance the first sedimentation step and reduce the load on the remainder of the plant. Thus, while the conventional treatment train can treat a wide range of source waters, some may be so challenging that even conventional treatment requires a form of pretreatment. Predisinfection using chloramines or chlorine dioxide may be used at some plants to decrease the concentrations of bacteria in the source water.

Processes for Source Waters of Very High Quality. For source waters having very low turbidities, low concentrations of TOC, and low concentrations of true color, some of the treatment steps employed in a conventional treatment plant may not be needed, or other filtration processes may be suitable. Treatment of very high-quality source waters can be accomplished by filtration without prior clarification

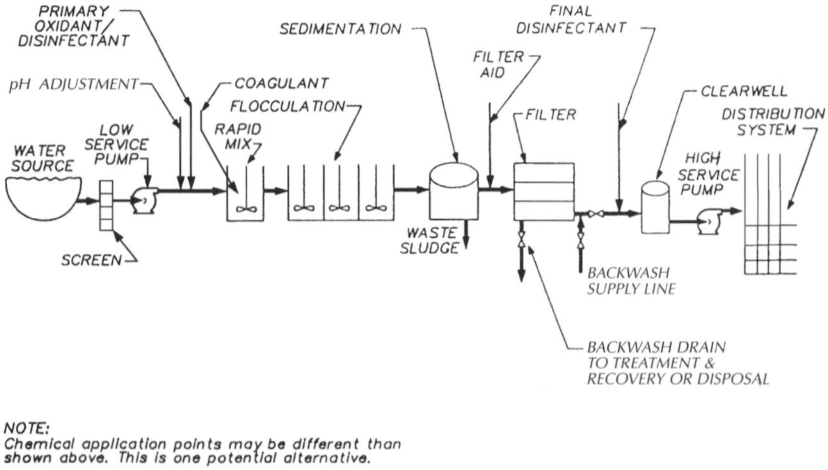

FIGURE 3.1 Conventional treatment, surface water.

using diatomaceous earth filtration, slow sand filtration, or by direct filtration, which deletes the sedimentation step from the conventional treatment train. Figure 3.2 is a process schematic diagram for direct filtration with an alternative for in-line filtration, in which flocculation is omitted. For waters not likely to form high concentrations of DBPs upon chlorination, free chlorine is a probable disinfectant.

Dissolved Air Flotation. For reservoirs and other surface waters with significant algal blooms, filtration processes lacking clarification can be quickly overwhelmed by filter-clogging algae. The processes suitable for low-turbidity source waters are not very successful when treatment of algal-laden water is necessary. The sedimentation basins employed in conventional treatment are not very successful for algae removal, though, because algae tend to float rather than to sink. The density of algae is close to that of water and when they produce oxygen, algae can create their own flotation devices. Therefore, a process that is better suited for algae removal is dissolved air flotation (DAF), in which the coagulated particulate matter, including algae if they are present, is floated to the top of a clarification tank. In DAF, the clarification process and the algae are working in the same direction. Like conventional treatment, DAF employs chemical feed, rapid mix, and flocculation, but then the DAF clarifier is substituted for the sedimentation basin. A DAF process scheme is shown in Figure 3.3. Waters having high concentrations of algae may also have high concentrations of disinfection by-products (DBP) precursors, so predisinfection with free chlorine could lead to DBP compliance problems. Chlorination just before or after filtration and use of alternative disinfectants, such as chloramines, may need to be considered.

Membrane Filtration. Membrane filtration covers a wide range of processes and can be used for various source water qualities, depending on the membrane process being used. Microfiltration, used for treatment of surface waters, can remove a wide range of particulate matter, including bacteria, protozoan cysts and oocysts, and particles that cause turbidity. Viruses, however, are so small that some tend to pass through the microfiltration membranes. Microfiltration is practical for application to a wider range of source water turbidities than slow sand filtration or diatomaceous earth (DE) filtration, but microfiltration can not handle the high turbidities

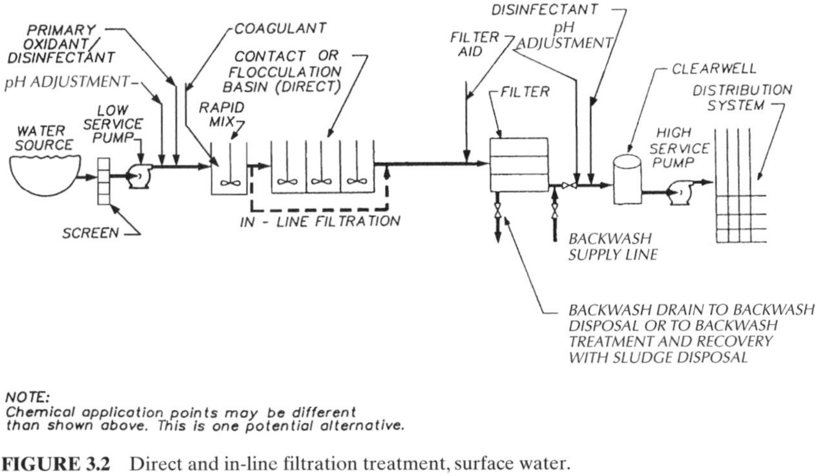

FIGURE 3.2 Direct and in-line filtration treatment, surface water.

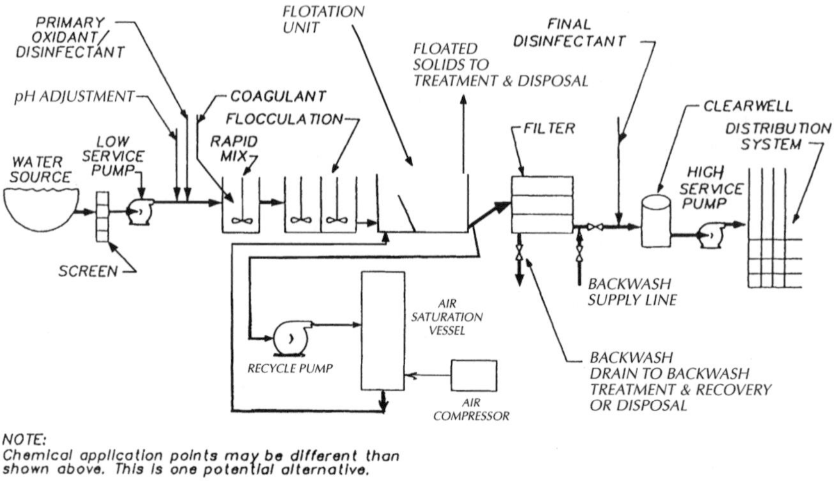

FIGURE 3.3 Dissolved air flotation/filtration treatment, surface water.

that are encountered in many conventional treatment plants. Microfiltration does not remove dissolved substances, so the disinfection process appropriate for water treated by this process will depend on the dissolved organic carbon (DOC) and precursor content of the source water. Advantages for membrane filtration include very high removal of *Giardia* cysts and *Cryptosporidium* oocysts, ease of automation, small footprint for a membrane plant, and the feasibility of installing capacity in small increments in a modular fashion rather than all at once in a major expansion, so that capital expenditures can be spread out over time. A microfiltration process train is shown in Figure 3.4.

Groundwater Treatment. Many groundwaters obtained from deep wells have very high quality with respect to turbidity and microbiological contaminants. If they do

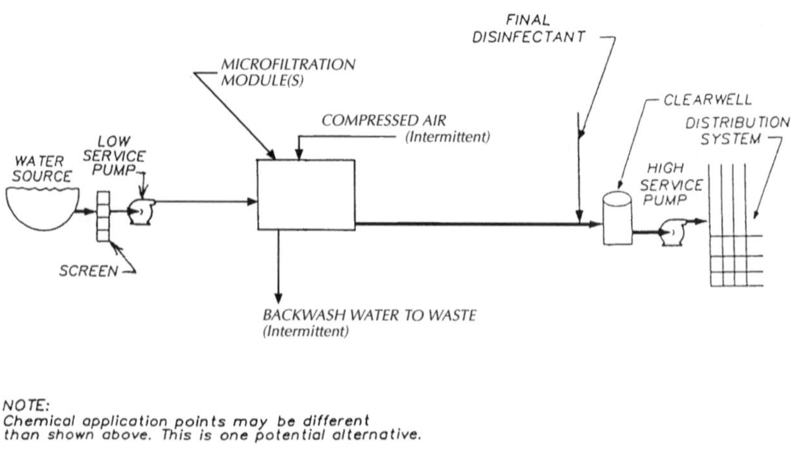

FIGURE 3.4 Microfiltration treatment, surface water.

not have mineral constituents requiring treatment, they may be suitable for consumption with disinfection as the only treatment. The minerals in groundwater in many cases result in the need or the desire for additional treatment.

Disinfection Only, or No Treatment. Some groundwaters meet microbiological quality standards and have a mineral content such that disinfection may be the only required treatment, and in some states disinfection may not be required. This may change when the Groundwater Rule is promulgated by USEPA. Circumstances favoring this situation are that the aquifer has no direct connection to surface water and the well has been properly constructed so the aquifer cannot be contaminated at the well site. For groundwaters of high quality, the most commonly used disinfectant is free chlorine.

Removal of Iron or Manganese, or Both, Plus Disinfection. If the minerals in the aquifer include iron or manganese, these inorganic constituents may be found in groundwater. For removal of iron and manganese, oxidation, precipitation, and filtration are commonly employed. Figure 3.5 shows processes for iron and manganese removal. Presence of organics in the source water can impair removal of iron and manganese by oxidation and filtration. Iron can be oxidized in many instances by aeration. Treatment at a pH of 8 or higher promotes a more rapid oxidation of iron by aeration, if natural organic matter (NOM) is not present in significant concentrations. Chlorine, potassium permanganate, chlorine dioxide, or ozone can be used to oxidize iron and manganese. Potassium permanganate is commonly used for manganese, which is more difficult to oxidize than iron. Greensand has been used in conjunction with potassium permanganate for iron and manganese removal in numerous treatment plants, especially for small- or medium-sized systems. Greensand can adsorb excess permanganate when it is overfed and later remove iron and manganese when permanganate is underfed, allowing operators to attain effective treatment without continuously matching the permanganate dosage to the iron and manganese content of the raw water. When chemical oxidants are used rather than aeration, pressure filters are sometimes used to accomplish iron or manganese removal without the need for repumping following treatment.

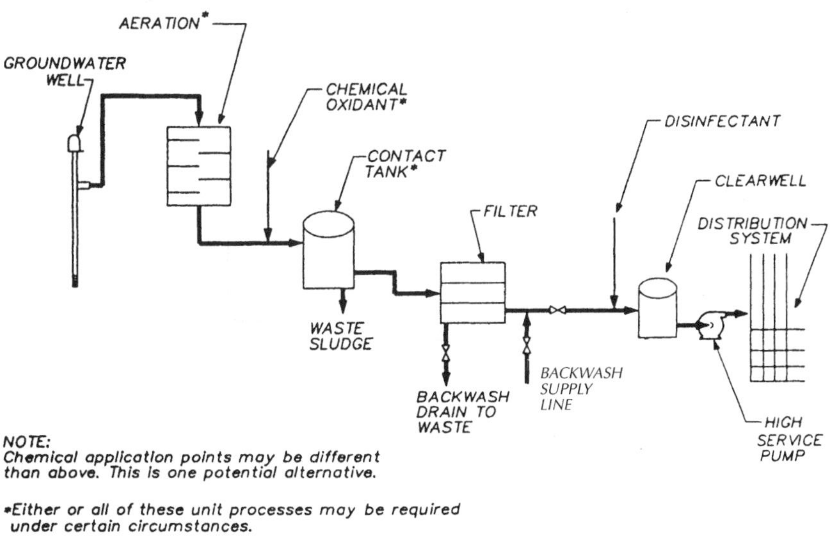

FIGURE 3.5 Iron and manganese treatment, groundwater.

Precipitative Lime Softening. Hard water contains excessive concentrations of calcium and magnesium. Both groundwater and surface water can be treated by precipitative lime softening to remove hardness. Treatment involves adding slaked lime or hydrated lime to water to raise the pH sufficiently to precipitate calcium or still higher to remove magnesium. If noncarbonate hardness is present, addition of soda ash may also be required for precipitation of calcium and magnesium. In precipitative lime softening the calcium carbonate and magnesium hydroxide precipitates are removed in a settling basin before the water is filtered. At softening plants that employ separate rapid mix, flocculation, and sedimentation processes, recirculating some of the lime sludge to the rapid mix step improves $CaCO_3$ precipitation and agglomeration of precipitated particles. Solids contact clarifiers combine the rapid mix, flocculation, and sedimentation steps in a single-process basin and generally are designed for higher rates of treatment than the long, rectangular settling basins. A two-stage softening process is shown in Figure 3.6. Solids contact clarifiers are an attractive alternative, especially for groundwater, because of the possibilities of lower capital cost and smaller space requirements, and are used more often than separate flocculation and sedimentation units. Use of solids contact clarifiers may reduce problems related to deposition of precipitates and scaling in channels and pipes connecting unit processes. When magnesium is removed, settled water has a high pH (10.6 to 11.0) and the pH must be reduced. Typically, this is accomplished by recarbonation (i.e., addition of carbon dioxide). Solids formed as a result of recarbonation can be removed by secondary mixing, flocculation, and sedimentation facilities. At some softening plants, carbon dioxide is added after the secondary settling to bring about further pH reduction and to stabilize the water.

Although two-stage recarbonation is more effective in optimizing hardness removal and controlling the stability of the softened water, a less expensive single-stage recarbonation process is sometimes used in excess lime treatment. Aeration sometimes is used before lime softening to remove carbon dioxide from groundwater, because lime reacts with carbon dioxide. The decision of whether to use aeration or simply to use more lime for carbon dioxide treatment can be aided by conducting an economic analysis of the cost of aeration versus the costs of the extra lime and the extra sludge produced.

Ion Exchange Processes. The most common ion exchange softening resin is a sodium cation exchange (zeolite) resin that exchanges sodium for divalent ions,

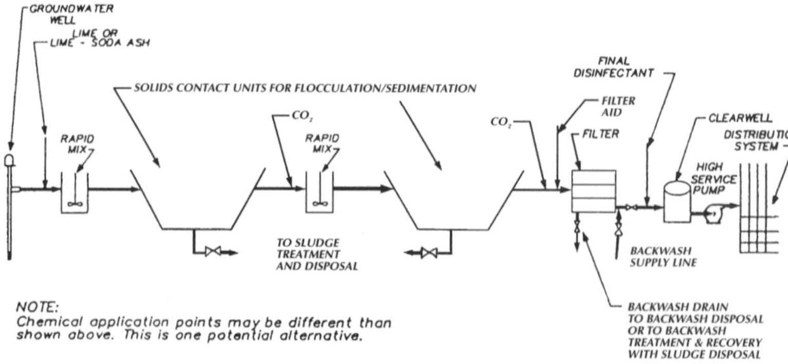

FIGURE 3.6 Two-stage excess lime softening treatment, groundwater.

including calcium, magnesium, and radium. When radium is present along with calcium or magnesium or both Ca and Mn, the hardness removal capacity of the resin is exhausted before the capacity for radium removal is reached, so hardness breaks through first. After the resin has reached its capacity for hardness removal, it is backwashed, regenerated with a sodium chloride solution, and rinsed with finished water. The regeneration step returns the resin to its sodium form so it can be used again for softening. A portion of the source water is typically bypassed around the softening vessel and blended with the softened water. This provides calcium ions to help stabilize the finished water.

Anion exchange resins are used in water treatment with equipment similar to that used for water softening with cation exchange resins. Anions such as nitrates and sulfates, along with other compounds, are removed with this process.

Ion exchange processes can be used for water softening and, in some instances, are used for removal of regulated contaminants such as nitrate or radium. Ion exchange is appropriate for water low in particulate matter, organics, iron, and manganese.

Pretreatment to remove iron and manganese should precede ion exchange if those inorganics are present. High concentrations of NOM can foul some ion exchange resins. Ion exchange, which is generally used in smaller plants, offers advantages over lime softening for water with varying hardness concentration and high noncarbonate hardness. Figure 3.7 is an ion exchange plant process diagram.

Case Studies

Dissolved Air Flotation and Filtration. The Greenville Water System (GWS) in Greenville, South Carolina, conducted studies on treatment of its two unfiltered source waters. Preliminary testing indicated that filter-clogging algae had the poten-

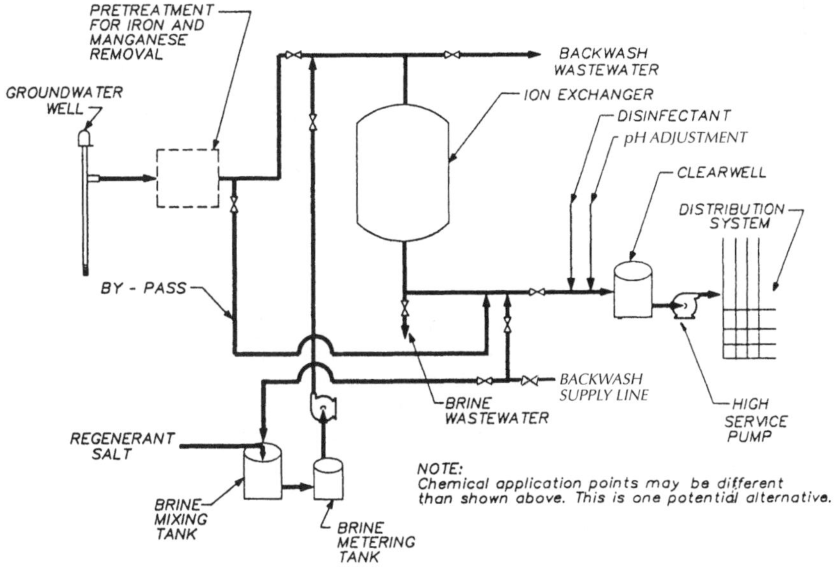

FIGURE 3.7 Ion exchange softening, groundwater.

tial to shorten filter runs if direct filtration were used to treat the low-turbidity source waters (Black & Veatch, 1987). Filter-clogging algae identified in the source waters included *Dinobrion, Asterionella,* and *Tabellaria.* Hutchison and Foley (1974) had indicated that a direct filtration plant in Ontario was troubled with short filter runs caused by filter-clogging diatoms including *Tabellaria.* As a result of that work and the observation of filter cloggers in the source waters at Greenville, caution was in order with regard to selection of direct filtration for treating North Saluda and Table Rock Reservoirs.

During follow-up pilot plant testing, treatment options included direct filtration with and without preozone and dissolved air flotation/filtration (DAF/filtration) without preozone (Ferguson et al., 1994). Both the direct filtration and the DAF/filtration process trains were able to provide excellent treated water quality. The filtered water turbidity goal of 0.10 ntu was met by both types of treatment. Manganese removal was effective for each treatment train. Total organic carbon removal was slightly greater by DAF/filtration, although the TOC concentration in the water from North Saluda Reservoir was quite low. Direct filtration without ozone yielded filter runs that were shorter than those of the DAF/filtration treatment train. Accordingly, when the comparison of alternatives was made the direct filtration option was evaluated at a filtration rate of 4 gpm/ft^2 (10 m/h), and DAF/filtration was evaluated at 6 gpm/ft^2 (15 m/h). Although the flotation process added cost to the treatment train, the superior filter performance provided for operation of filters at a higher rate, giving some savings over direct filtration. Additional savings would be realized by the lower water content of the residuals produced in DAF/filtration. The present worth costs for direct filtration and DAF/filtration were considered essentially equivalent. The similarity of costs resulted in the selection of DAF/filtration as the preferred treatment train (Black & Veatch, 1994) because of the capability of that treatment train to provide a higher level of treatment than direct filtration. The superior ability of DAF/filtration to remove particulate matter and algae and the presence of an additional barrier to prevent the passage of microbiological contaminants (such as *Cryptosporidium*) were important advantages resulting in the selection of DAF/filtration.

Direct Filtration. The Southern Nevada Water System provides water for the Las Vegas metropolitan area. The surface water source, Lake Mead, is treated at the Alfred Merritt Smith Water Treatment Facility (AMSWTF). Lake Mead has very low turbidity, with 1 ntu or lower being a typical value. The AMSWTF began as a 200 mgd (7.6×10^5 m^3/day) in-line filtration plant in 1971 (Spink and Monscvitz, 1974). Treatment consisted of prechlorination followed by addition of alum and polyelectrolyte in the rapid mix chamber. Filtration through dual media was accomplished at a rate of 5 gpm/ft^2 (12 m/h). Spink and Monscvitz reported use of alum dosages ranging from 3 to 15 mg/L. Turbidity of the treated water averaged under 0.10 ntu in 1972.

Several years later changes were studied, as a result of some problems that occurred during the first years of operation (Monscvitz et al., 1978). Aluminum was being carried over into the clear well, and plankton were found at times in the filtered water. Use of powdered activated carbon (PAC) during periods of reservoir destratification resulted in breakthrough of the PAC when the plant was operated at the normal 5 gpm/ft^2 (12 m/h) filtration rate. Process modifications were needed. Pilot plant studies indicated that improved filtration performance could be attained by addition of flocculation to the treatment train.

Following the pilot plant work described by Monscvitz et al. (1978), the AMSWTF was modified to include flocculation. Later, the filter media was changed from dual media to mixed media (anthracite, sand, and garnet), and the coagulant of choice was changed from alum to ferric chloride. The very low turbidity of the source water rendered filtered water turbidity monitoring of somewhat questionable value for the operating staff, so continuous, on-line particle-counting capability was installed. Plant operating decisions have been influenced strongly by particle count results, as filtered water turbidity is typically less than 0.10 ntu. As a result of rapid growth in the Las Vegas area, the plant was expanded to 400 mgd (1.5×10^6 m^3/day).

Recent pilot plant testing (Logsdon et al., 1996) demonstrated that substantially lower particle counts could be attained in filtered water when preozone was used as compared with water with no preozone treatment. In addition, particle counts in filtered water treated with preozone were lower than particle counts in filtered water treated with prechlorination. Ozone facilities were designed, and existing filters were uprated from 5 to 6 gpm/ft^2 (12 to 15 m/h). Additional filters were built, increasing total capacity of the plant to 600 mgd (2.3×10^6 m^3/day). The changes at the AMSWTF plant over the years have been made to improve filtration capability, to increase plant capacity, and to improve disinfection capabilities.

Microfiltration. The San Jose Water Company needed to replace a 5 mgd filtration plant to meet new requirements of California's Surface Water Filtration and Disinfection Rule. Yoo et al. (1995) explained that the new process needed to fit into a compact site and would have to cope with source water turbidity that could exceed 100 ntu during storms. Removal of *Giardia* cysts with a minimal disinfection contact time was a requirement.

Other considerations were the need for remote operation and a short (12 months) time frame for design and construction. Microfiltration was selected as the process that could satisfy all of these requirements. This plant was completed for a total capital expenditure of about $3.5 million 12 months after the purchase order was signed. Yoo et al. presented data on turbidity, showing that the microfiltration plant consistently produced very low filtered water turbidity from February through June 1994. Average source water turbidity was 97 ntu in February, when the peak filtered water turbidity was 0.13 ntu and the average was 0.05 ntu. Raw water turbidity ranged from 6 to 9 ntu during March, April, and May, and the maximum filtered water turbidity during those months was 0.06 ntu. Yoo et al. reported that the plant performance had exceeded the requirements of the SWTR and resulted in increased production at the site due to the capability of microfiltration to treat water having variable turbidity. Automation and use of a SCADA system have facilitated operation of the plant with minimal operator attention.

In a follow-up paper, Gere (1997) reported on operating costs at the Saratoga Water Treatment Plant, stating that noncapitalized expenditures (including power, labor, chemicals, membrane replacement, maintenance, and repairs) were $309 per million gallons ($82/1000 m^3) for 1995, the first full year of operation. Labor was the largest cost component, accounting for 31.6 percent of operating costs, based on 46 hours per week of scheduled work at the plant. This includes not only microfiltration process operation but also tasks such as cleaning the intake and manually removing debris as necessary. The second most significant portion of operating cost was electric power, which made up 28.6 percent of the cost. Electric power was purchased at an average cost of $0.103/kWh ($0.029/MJ) in 1995. Chemical costs were only 7 percent of the operating cost budget, and residuals management costs were less than 1 percent. Membrane replacement was considered to be an annual fixed cost based on an estimated membrane life of six years, and accounted for 22.5 percent of total operating costs. Gere concluded that microfiltration was highly reliable and cost-effective.

Slow Sand Filtration. Slow sand filtration is a process alternative that is attractive to many small water systems. Two examples provide interesting insights on this process.

Empire, Colorado, was a community of 450 persons when 110 cases of waterborne giardiasis occurred in 1981 (Seelaus, Hendricks, and Janonis, 1986). The water source, Mad Creek, had been treated by chlorination only. Mad Creek drains a meadow at an elevation of 9000 ft (2700 m), and the village is located at an elevation of 8600 ft (2580 m). Water from Mad Creek is usually cold or very cold, and the turbidity is generally 1 ntu. Slow sand filtration was selected for this small water system because research under way in Colorado (Bellamy et al., 1985) was demonstrating highly effective removal of *Giardia* cysts and coliform bacteria by slow sand filtration. The process was well suited to part-time operation that is generally necessary in small systems, and sufficient head was available that gravity flow from the source to the treatment plant and from the plant to Empire could be maintained. Electric power was not available at the site. Local materials and labor could be used in the construction.

Plant design was for 0.25 mgd (950 m^3/day) at a filtration rate of 0.10 gpm/ft^2 (0.24 m/h), with two filter beds of 27.5 × 30 ft (8.38 × 9.14 m). The filter bed was designed for 4 ft (1.2 m) of sand. A local sand having an effective size (D_{10}) of 0.21 mm and a uniformity coefficient of 2.67 was used. Delivered cost for all of the sand was $6270. A problem that can occur when using local sand is insufficient cleaning of the sand before placement in the filter. At Empire, when the filter was placed into service, the filtered water turbidity exceeded the raw water turbidity, indicating that fine particulate matter was being washed out of the filter bed. Filtered water turbidity declined from 11 ntu to 1 ntu within two weeks of operation. The filter was effective for removal of microbiological contaminants, as *Giardia* cysts were detected in the influent water during the first eight months of operation but not in the filtered water. Seelaus, Hendricks, and Janonis noted that the plant had low operating costs, with only a daily inspection trip by the operator to monitor head loss, rate of flow, and turbidity. About two hours per month were required for scraping and removing sand from the filter. Slow sand filtration was well suited for this application.

Camptonville, California, a community of about 260 persons, installed a slow sand filtration plant in 1991. Riesenberg et al. (1995) indicated that low capital and operating costs of slow sand filters, and the need to maintain gravity flow from the source through the plant to the community were reasons for selecting slow sand filtration. Use of ground water was considered too expensive because of problems with iron and manganese. Camptonville had experienced boil-water orders in 1973 and 1985, and later on in 1990 and 1991 while planning and construction for the project were under way. A noteworthy feature of this plant is the use of modular construction. The filter boxes were precast in the San Francisco Bay area and trucked to Camptonville. This enabled the utility to obtain higher-quality filter boxes at a lower price as compared with the option of constructing the filter boxes on-site and was an important consideration for the small water system. Modular construction also will allow incremental future expansion of the plant as needed. Total filter area for the plant is 1000 ft^2 (93 m^2). Maximum filtration rate is 0.10 gpm/ft^2 (0.24 m/h). Construction cost for the filter plant was $226,000, and costs of other facilities in the project brought the total construction cost to $532,000.

The authors reported that the total operation and maintenance time at the plant varied from 15 minutes to an hour per day, with total time for plant operation averaging 15 hours per month. Filter scraping requires about four hours of labor. Riesenberg et al. concluded that the facility provided the community with excellent quality water at a reasonable cost.

SUMMARY

At the beginning of a new century, the range of water treatment choices is expanding. New processes are being developed and brought into use, and processes that have been used for decades are being studied, refined, and improved. Engineers and water utilities today have many process options when water treatment plant expansions or new water treatment plants are being planned. Although the increased number of choices for water treatment processes will be beneficial for water utilities and for their customers, the availability of more options complicates the decision-making process and forces everyone involved to think more carefully before selecting a water treatment process. This situation will benefit water utilities and their customers in the long run, if choices are made wisely.

BIBLIOGRAPHY

Bellamy, W. D., G. P. Silverman, D. W. Hendricks, and G. S. Logsdon. "Removing *Giardia* Cysts with Slow Sand Filtration." *Jour AWWA,* 77(2), 1985: 52–60.

Black & Veatch. *Greenville Water System Treatability Studies.* 1987.

Black & Veatch. *Greenville Water System Preliminary Engineering Report—Phase* II. 1994.

Cleasby, J. L., A. H. Dharmarajah, G. L. Sindt, and E. R. Baumann. *Design and Operation Guidelines for Optimization of the High-Rate Filtration Process: Plant Survey Results,* pp. *x*, 89. Denver, CO: *AWWA Research Foundation and American Water Works Association,* 1989.

Coffey, B. M., S. Liang, J. F. Green, and P. M. Huck. "Quantifying Performance and Robustness of Filters during Non-Steady State and Perturbed Conditions." In *Proceedings of the 1998 AWWA Water Quality Technology Conference,* November 1–4, San Diego, CA. Denver, CO: AWWA, 1998. (CD-ROM.)

Ferguson, C., G. Logsdon, D. Curley, and M. Adkins. "Pilot Plant Evaluation of Dissolved Air Flotation and Direct Filtration." In *Proceedings 1994 AWWA Annual Conference; Water Quality,* pp. 417–435. Denver, CO: AWWA, 1994.

Gere, A. R. "Microfiltration Operating Costs." *Jour AWWA,* 89(10), 1997: 40–49.

Great Lakes–Upper Mississippi River Board of State Public Health and Environmental Managers. "Recommended Standards for Water Works." Albany, NY: Health Education Services, 1997.

Harmon, R., F. H. Abrew, J. A. Beecher, K. Carns, and T. Linville. "Roundtable: Energy Deregulation." *Jour AWWA* 90(4), 1998: 26, 28, 30, and 32.

Hutchison, W., and P. D. Foley. "Operational and Experimental Results of Direct Filtration." *Jour AWWA,* 66(2), 1974: 79–87.

Kirmeyer, G. J. *Seattle Tolt Water Supply Mixed Asbestiform Removal Study,* EPA-600/2-79-125. Cincinnati, OH: U.S. Environmental Protection Agency, 1979.

Logsdon, G., J. Monscvitz, D. Rexing, L. Sullivan, J. Hesby, and J. Russell. "Long Term Evaluation of Biological Filtration for THM Control." In *Proceedings AWWA Water Quality Technology Conference,* Boston, Massachusetts, November, 1996.

Monscvitz, J. T., D. J. Rexing, R. G. Williams, and J. Heckler. "Some Practical Experience in Direct Filtration." *Jour AWWA,* 70(10), 1978: 584–588.

National Research Council. "Safe Drinking Water from Every Tap." Washington, D.C.: National Academy Press, 1997.

Patton, J. L., and M. B. Horsley. "Curbing the Distribution Energy Appetite." *Jour AWWA,* 72(6), 1980: 314–320.

Pontius, F. W. "New Horizons in Federal Regulation." *Jour AWWA,* 90(3), 1998: 38–50.

Public Health Service. "Public Health Service Drinking Water Standards, Revised 1962." Public Health Service Publication No. 956, p. 2. Washington, D.C.: U.S. Government Printing Office, 1969.

Renner, R. C., and B. A. Hegg. *Self-Assessment Guide for Surface Water Treatment Plant Optimization,* p. 21. Denver, CO: AWWA Research Foundation, 1997.

Riesenberg, F., B. B. Walters, A. Steele, and R. A. Ryder. "Slow Sand Filters for a Small Water System." *Jour AWWA,* 87(11), 1995: 48–56.

Seelaus, T. J., D. W. Hendricks, and B. A. Janonis. "Design and Operation of a Slow Sand Filter." *Jour AWWA,* 78(12), 1986: 35–41.

Spink, C. M., and J. T. Monscvitz. "Design and Operation of a 200-mgd Direct-Filtration Facility." *Jour AWWA,* 66(2), 1974: 127–132.

U.S. Environmental Protection Agency. "National Primary Drinking Water Regulations; Interim Enhanced Surface Water Treatment; Final Rule." *Federal Register,* 63(241), December 16, 1998: 69478–69515.

U.S. Environmental Protection Agency. "Drinking Water; National Primary Drinking Water Regulations; Filtration, Disinfection; Turbidity, Giardia lamblia, Viruses, Legionella, and Heterotrophic Bacteria; Final Rule." *Federal Register,* 54(124), June 29, 1989: 27486–27541.

U.S. Environmental Protection Agency. "National Interim Primary Drinking Water Regulations," EPA-570/9-76-003, p. 25. Washington, D.C.: U.S. Government Printing Office, 1976.

Yoo, R. S., D. R. Brown, R. J. Pardini, and G. D. Bentson. "Microfiltration: A Case Study." *Jour AWWA,* 87(3), 1995: 38–49.

CHAPTER 4
SOURCE WATER QUALITY MANAGEMENT

GROUNDWATER

Stuart A. Smith, CGWP
Consulting Hydrogeologist, Smith-Comeskey Ground Water Science, Ada, Ohio

GROUNDWATER SOURCE QUALITY: RELATIONSHIP WITH SURFACE WATER AND REGULATION

The topic of groundwater source quality is a complicated one, with numerous influences, both natural and human in origin. Aside from frequently being more accessible (drilling a well near a facility is often more convenient than piping in surface water from a remote location), groundwater is typically chosen because its natural level of quality requires less treatment to ensure safe, potable consumption by humans. Although many groundwaters benefit from aesthetic treatment, and sometimes from treatment to remove constituents such as metals and arsenic that pose long-term health risks, in general groundwater sources compare favorably against surface water due to the reduced treatment needed. In the dangerous modern world, groundwater sources are far less vulnerable to terrorist attack than surface water reservoirs, just as they were recognized as being less vulnerable to nuclear fallout during the political atmosphere of the 1950s and 1960s.

Arguably, surface water and groundwater form a water resource continuum in any hydrologic setting (e.g., Winter et al., 1998), and from a groundwater-biased view, surface water bodies are often simply points where the water table rises above the surface. Thus impacts on one part of this continuum can affect the other over time. However, the hydraulic connections between these two resources can be obscure (at least superficially), and surface water and groundwater management strategies in the United States (regulatory issues are discussed in Chapter 1) tend to follow different paths.

Surface water management assumes that the source is impaired and unsafe for consumption without elaborate treatment (a point not always conceded where source surface waters are of high quality, such as in Portland, Oregon, and New York

City). By contrast, because the advantages of groundwater include a perceived reduced vulnerability and need for treatment, groundwater protection strategies, such as those in the United States, have traditionally focused on protection of that quality. This approach (and a procedural defeat for artificial separation of groundwater and surface water) is belatedly being reflected in the U.S. regulatory sphere in the change to source water assessment and protection instead of "wellhead protection" and "surface water rules" as separate emphases.

However, experience of the last several decades has shown that groundwater sources are not uniquely immune to contamination, and that once contaminated by chemical or radiological agents, they are almost always difficult to clean up. Recently, attention has refocused on the risk of pathogen transport to wells. A common conclusion of all groundwater contamination research is that prevention is far more effective than remediation or treatment in assuring the quality of groundwater supply sources. Prevention takes the form of management.

The intent of this chapter section is to introduce the reader to influences on groundwater quality, management options, and tools to improve understanding of the groundwater resource and thus improve its management and protection as it relates to water supply.

GENERAL OVERVIEW OF GROUNDWATER SOURCES AND IMPACTS ON THEIR QUALITY

The key to understanding and managing groundwater quality in water supply planning is to understand that both aquifer hydrologic characteristics and the causes and effects of groundwater contamination are complex and highly site-specific. Groundwater quality management is most effective when it can respond to the specifics of individual aquifers, wellfields, and even individual wells.

For example, in both porous media (sand/sand-and-gravel/sandstone) and fractured-rock aquifers, hydraulic conductivity and its derivative values can vary by orders of magnitude over short distances (meters to kilometers). Flow velocities can also vary by similar dimensions over meter differences horizontally and vertically. Pressure changes near a well may cause flow characteristics to vary significantly in a distance of one meter away from the well.

Changes in each of the just-mentioned hydrologic characteristics can affect groundwater quality by changing local constituent concentrations. Likewise, localized differences in formation geochemistry (e.g., organic content, iron, and other mineral transformations) affect water quality. A third factor is the influence of the aquifer microflora in a specific fracture or aquifer zone tapped by wells. Work in the last 20 years has revealed the extent and complexity of the microbial ecosystems that inhabit aquifers (e.g., Chapelle, 1993; Amy and Haldeman, 1997).

The extent of human impact also depends on (1) how potential contaminants are handled; (2) the physical-chemical characteristics of such materials if they are released to the ground; and (3) the hydrologic characteristics of the location where a release occurs. If the soil has a low hydraulic conductivity, contamination may be very limited, even if application (e.g., oil spills or herbicides) is relatively intense. On the other hand, if conductivity is high and there is a direct contact with an aquifer, a small release may have a large impact. A further human impact is the presence of abandoned wells or other underground workings that provide conduits through low-conductivity soils.

All of these factors are site-specific, but they can be understood and managed if identified. Thus, effective water supply management of source groundwater (and avoiding unnecessary treatment) depends on adequate local knowledge of the groundwater system being utilized.

Types of Aquifer Settings in North America and Their Quality Management Issues

North America has vast and widely distributed groundwater resources that occur in many types of aquifers (see Figure 4.1). In addition to this hydrogeologic variety, North America offers extreme variability in climate and degree of human development. From the standpoint of groundwater development, these range from areas of abundant groundwater and very low population density in western Canada to high population densities in urban centers in the dry southwestern United States and northern and central Mexico, where groundwater overdrafts are an important water management problem.

Large areas of the central and western United States (like large regions of the world) and Mexico struggle to effectively and equitably manage groundwater resources that could very easily become depleted by overuse. Prominent examples include the decades-long efforts to manage the Edwards aquifer in Texas, and the Ogallala aquifer, which spans the central United States east of the Front Range of the Rockies. In the long term, questions must be asked, such as how sustainable is the high-water-use, technological civilization in the desert U.S. West (established during a 200-year historically wet period)? What kind of population density is sustainable on

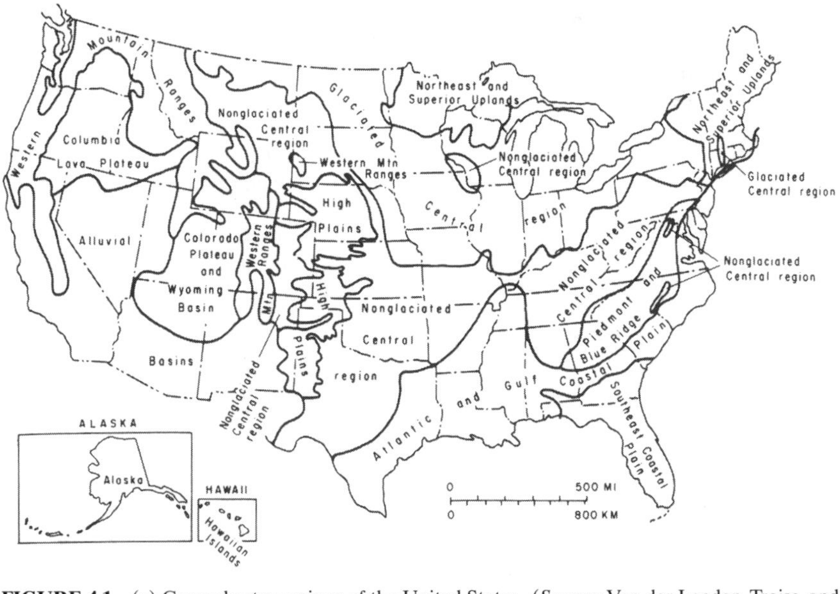

FIGURE 4.1 (*a*) Groundwater regions of the United States. (*Source:* Van der Leeden, Troise, and Todd. 1990. Originally from R. C. Heath. "Classification of ground-water regions of the United States." *Ground Water,* 20(4), 1982. Reprinted with the permission of the National Ground Water Association.)

FIGURE 4.1 (*Continued*) (*b*) Occurrence of aquifers in the United States.

[Abbreviations: (1) Aquifers: P, principal aquifer in region; I, important aquifer in region; M, minor aquifer in region; U, unimportant as an aquifer in region. (2) Rock terms: S, sand; Ss, sandstone; G, gravel; C, conglomerate; Sh, shale; Ls, limestone; Fm, formation; Gp, group.]

(1) Geologic Age and Rock Type	(2) Western Mountain Ranges	(3) Arid Basins	(4) Columbia Lava Plateau	(5) Colorado Plateau	(6) High Plains	(7) Unglaciated Central Region	(8) Glaciated Central Region	(9) Unglaciated Appalachian Region	(10) Glaciated Appalachian Region	(11) Atlantic and Gulf Coastal Plain	(12) Special Comments
Cenozoic											
Quaternary											
Alluvium and related deposits (primarily—recent and Pleistocene sediments and may include some of Pliocene age)	S and G deposits in valleys and along stream courses. Highly productive but not greatly developed—P to M.	S and G deposits in valleys and along stream courses. Highly developed with local depletion. Storage large but perennial recharge limited—P.	S and G deposits along streams, interbedded with basalt—I to M.	U	S and G along water courses. Sand dune deposits—P (in part).	S and G along water courses and in terrace deposits—I (limited).	S and G along water courses—M.	S and G along water courses and in terrace deposits. Not developed.		S and G along water courses and in terrace and littoral deposits, especially in the Mississippi and tributary valleys. Not highly developed in East and South. Some depletion in Gulf Coast—I.	The most widespread and important aquifers in the United States. Well over one-half of all ground-water pumped in United States is withdrawn from these aquifers. Many are easily available for artificial recharge and induced infiltration. Subject to saltwater
Glacial drift, especially outwash (Pleistocene)	S and G deposits in northern part of region—I.	S and G deposits especially in northern part of region and in some valleys—I.	S and G outwash, especially in Spo-kane area—I.	U	S and G outwash, much of it reworked (see above)—I.	S and G outwash especially along northern boundary of region—I.	S and G outwash, terrace deposits and lenses in till throughout region—P (in part).	S and G outwash in northern part. Not highly developed—M.	S and G outwash, terrace deposits and lenses in till. Locally highly	S and G outwash in Mississippi Valley (see above)—I.	

									devel-oped—I.		contamination in coastal areas.
Other Pleistocene sediments	Alluvial Fm and other basin deposits in the southern part—M to P (see Alluvium above).		U	U	Alluviated plains and valley fills—M to I.	U	U	U	U	Coquina, limestone, sand, and marl Fms in Florida—M.	
Tertiary Sediments, Pliocene	S and G in valley fill and terrace deposits. Not highly devel-oped—M.	Some S and G in valley fill—M.	U	U	Ogalalla Fm in High Plains. Extensive S and G with huge storage but little recharge locally. Much depletion—P (in part).	U	U	Absent	Absent	Dewitt Ss in Texas. Citronelle and LaFayette Fms in Gulf States—I.	
Miocene	Ellensburg Fm in Washington—I; elsewhere—U.	U	Ellensburg FM in Washington—I; elsewhere—U.	U	Arikaree Fm—M.	Arikaree Fm—M.	Flaxville and other terrace deposits. S and G in northwestern part—M.	Absent	Absent	New Jersey, Maryland, Delaware, Virginia—Cohansey and Calvert Fms—I. Delaware to North Carolina—St. Marys and Calvert Fms—I.	Aquifers in coastal areas subject to salt-water encroachment and contamination.

Continued

FIGURE 4.1 (*Continued*) (*b*) Occurrence of aquifers in the United States.

[Abbreviations: (1) Aquifers: P, principal aquifer in region; I, important aquifer in region; M, minor aquifer in region; U, unimportant as an aquifer in region. (2) Rock terms: S, sand; Ss, sandstone; G, gravel; C, conglomerate; Sh, shale; Ls, limestone; Fm, formation; Gp, group.]

(1) Geologic Age and Rock Type	(2) Western Mountain Ranges	(3) Arid Basins	(4) Columbia Lava Plateau	(5) Colorado Plateau	(6) High Plains	(7) Unglaciated Central Region	(8) Glaciated Central Region	(9) Unglaciated Appalachian Region	(10) Glaciated Appalachian Region	(11) Atlantic and Gulf Coastal Plain	(12) Special Comments
										Georgia and Florida—Tampa Ls, Alluvium Bluff Gp, and Tamiami Fm—I. Eastern Texas—Oakville and Catahoula Ss—I.	
Oligocene	U	U	U	U	Brule clay, locally—I; elsewhere—U.	U	U	Absent	Absent	Suwannee Fm, Byram Ls, and Vicksburg Gp—I.	
Eocene	Knight and Almy Fm in southwest Wyoming—M.	U	U	Knight and Almy Fm in southwest Wyoming, Chuska Ss, and Tohatchi Sh in northwest Arizona and northeast New Mexico—M.	U	Claibourne and Wilcox Gp in southern Illinois (?), Kentucky, and Missouri—M; elsewhere—U.	Absent	Absent	Absent	New Jersey, Maryland, Delaware, Virginia—Pamunkey Gp—I. North Carolina to Florida—Ocala Ls and Castle Hayne	Includes the principal formations (Ocala Ls, especially) of the great Floridan aquifer. Subject to saltwater contamina-

										Marl—P (in part). Florida—Avon Park Ls., South Carolina to Mexican border, Claibourne Gp, Wilcox Gp—I. Clayton Fm in Georgia—I.	tion in coastal areas but source of largest groundwater supply in southeastern United States.
Paleocene	U	U	U	U	Ft. Union Gp—M.	Ft. Union Gp—M.	Ft. Union Gp—M.	Absent	Absent	Absent	
Volcanic rocks, primarily basalt	U	Local flows—M.	Many interbedded basalt flows from Eocene to Pliocene—P.	Local flows—M.	Absent	Absent	Absent	Absent	Absent		

Source: Reprinted with permission from F. Van der Leeden, F. L. Troise, and D. K. Todd. *The Water Encyclopedia.* Boca Raton, FL: Lewis Publishers, 1990. Copyright CRC Press, Boca Raton, Florida.

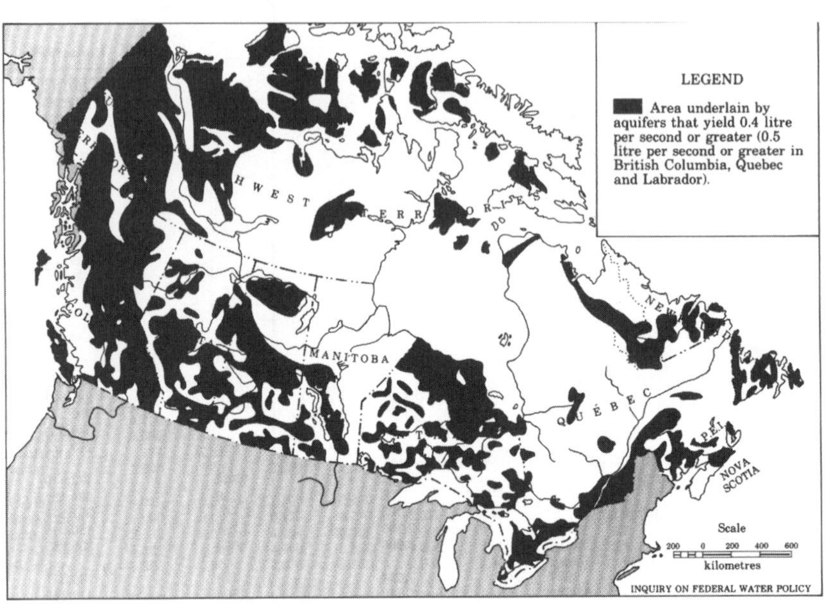

FIGURE 4.1 (*Continued*) (*c*) Groundwater potential in Canada. (*Source:* Van der Leeden, Troise, and Todd, 1990. Adapted from P. H. Pearse et al., 1985. *Currents of Change: Final Report—Inquiry on Federal Water Policy.* Ottawa, Canada. Reproduced with the permission of Environment Canada, 1999.)

Florida's abundant aquifers without water quality degradation? Do our current groundwater "sustainable yield" concepts bear scrutiny (Sophocleous, 1997)?

Groundwater management issues involve but extend beyond human water supply and also intersect with surface water quality and quantity issues. In the case of both the Edwards and the Ogallala aquifers, and the surficial aquifers of Florida, management has to consider the water needs of agricultural and human water supply, but equally so, the maintenance of wildlife habitats. The Edwards feeds prominent springs (supporting rare aquatic species), as well as culturally and ecologically important streams. The Everglades in Florida represent a water system incorporating both "surface" and "ground" water. The Ogallala involves formations too deep to be directly involved in surface water maintenance; however in the same region surficial aquifers along the North and South Platte, the Missouri, and other rivers in the western Mississippi watershed are critical to maintaining wildlife habitat. As with wetlands management strategies all over the United States, the Ogallala and High Plains management equation involves finding the optimal distribution of water withdrawal among groundwater and surface water resources. In all of these cases, quality is a factor. Where groundwater is overused, water quality and the quality of aquatic ecosystems are also commonly degraded.

What are the consequences? The technology to drill and pump wells essentially made settled rural and town life possible in the Great Plains of the United States and south-central Canada, as well as in very similar locations in Australia, where no useful surface water was available nearby. In many communities, stations, and farms in these areas, wells can be drilled to groundwater, often hundreds of meters deep. If groundwater sources become depleted or contaminated, many farms or communities in North Dakota or Queensland, for example, could not be maintained economically. With the years-long drought in Queensland, inhabitants recently were actually facing the possibility of abandoning their towns due to dwindling groundwater reserves.

Agriculture, especially in production of food for human consumption, has become highly dependent on irrigation. The food economy of the United States is highly dependent on the production of fresh vegetables from California, Florida, and Mexico, mostly sustained by irrigation. Israel, as a modern society established in a near-desert, is totally dependent upon irrigated agriculture. If large-scale, groundwater-dependent irrigation were to become impractical, major changes would be necessary in food production and distribution worldwide. In the case of the United States, water for irrigation reserved by prior appropriation can only be available for human water supply if these rights are purchased or abandoned. "Water farming" or purchasing prior agricultural water rights to groundwater reserves for urban water supply is an issue with many economic, cultural, and emotional aspects. If water farming reserves groundwater formerly allocated to agriculture for direct human use, how are fresh produce, cotton, or animal feed crops produced? Would all of this be shifted back to the wetter (but colder) eastern United States, raising costs and requiring increased expenditures of hydrocarbons for greenhouse heating?

Even within a relatively small and water-rich part of North America, great variety in water availability and quality can occur, illustrating the need for flexible and site-specific management. The situation in western Ohio and eastern Indiana (e.g., Lloyd and Lyke, 1995) is only one of many such examples. The region is underlain by an extensive carbonate-rock aquifer that is largely underutilized due to low population density, and relatively protected due to glacial clay till coverage. However, local over-drafting and contamination of this aquifer can and does occur. By contrast, carbonate rock in southern Ohio and Indiana (laid down under different depositional conditions) provides poor yields to wells. Within this area of unproductive rock, Dayton and Columbus, Ohio, and many smaller communities are underlain by large and productive glacial-outwash aquifers. These aquifers are at once both productive and vulnerable (sometimes being the only flat places to build factories, drill oil wells, etc.).

As in southern Ohio, in much of New England, municipal groundwater supplies can only be developed in relatively vulnerable (and highly developed) glacio-fluvial aquifers, although adequate household yields are possible from rock wells. However, population densities and intensity of land use is high. Overuse and vulnerability to contamination make management of these aquifers a critical environmental imperative for the region that is only now being addressed seriously.

PATTERNS OF PRIVATE AND PUBLIC GROUNDWATER SOURCE USES: QUALITY MANAGEMENT ISSUES

North America has a relatively high density of private groundwater supply use, particularly in the eastern United States (see Figure 4.2). Private wells have long been the principal mode of water supply for widely spaced rural homesteads and many small villages. This contrasts with France, for example, where public water service to rural properties is the rule.

Local variability in the quality and availability of groundwater influences pressure to develop or extend public water supply distribution systems in rural settings. Constructing individual water supply wells is always more cost-effective where groundwater is abundant and of suitable quality. Where natural groundwater quality

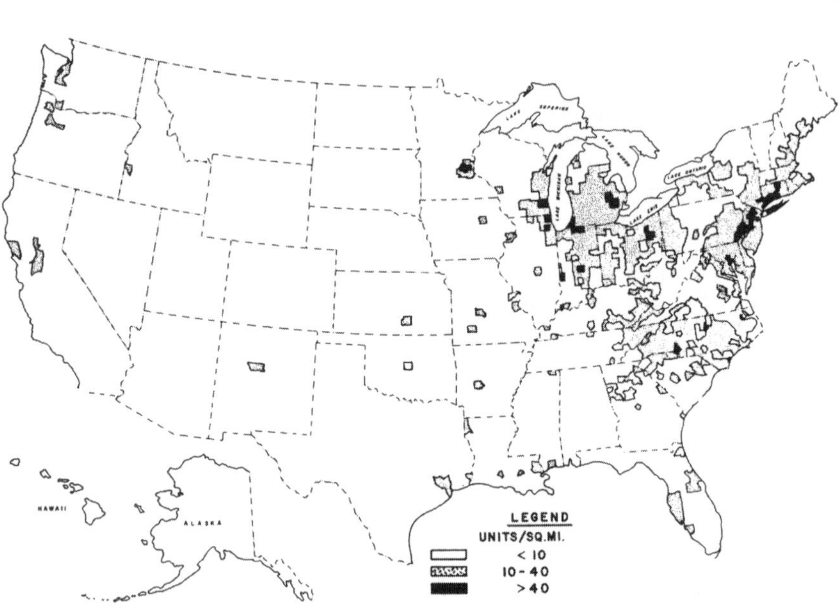

FIGURE 4.2 Density of housing units using on-site domestic water supply systems in the United States (by county). (*Source:* U.S. Environmental Protection Agency, Office of Water Supply, Office of Solid Waste Management Programs, 1977. *The Report to Congress: Waste Disposal Practices and Their Effects on Ground Water.* Reprinted in Van der Leeden, Troise, and Todd, 1990.)

is exceptionally poor or where supplies are insufficient, the more costly option of piping treated water from a centralized source is a solution to provide suitable water. In the United States, the development of systems to provide public water to rural residents has been promoted and funded by the federal government. Many areas have mixed individual private well and public piped water supply options.

Rural Groundwater Quality Management

Managing rural water quality is a significant challenge due to its site-specific nature and the typically inadequate financial and human resources available to address problems.

Microbial Health Risks. The most commonly detected problem of rural private wells is the occurrence of total-coliform (TC) bacteria positives. Statistically meaningful studies over the years have shown that a significant number of wells sampled are positive for total coliform bacteria. The most recent data available from large-scale studies (results from Midwest studies by the Centers for Disease Control and Prevention) show 41 percent TC positive and 11 percent fecal coliform positive (CDC, 1998) in the population of wells sampled. Such contamination is mostly due to well-construction deficiencies and deterioration (Exner et al., 1985; Smith, 1997; NGWA, 1998). It is rare that a large volume of an aquifer is contaminated by sewage waste, although such incidents have occurred (Ground Water Geology Section, 1961). A relatively new concern in the United States is the microbial impact of concentrated animal farm operations (CAFO) through faults in animal waste management. So far, this has been addressed as a surface water concern, but the potential exists for new phenomena such as the spread to nearby wells of antibiotic-resistant *Salmonella* sp. from chicken manure deposited on shallow bedrock.

Correlations made in CDC (1998) suggest that positive TC results increase with increasing well age, poor well condition and maintenance, shallow depth, and certain well types (e.g., dug), supporting the value of well construction standards and well maintenance (Smith, 1997). However, recent research in relatively undisturbed aquifers (e.g., Jones et al., 1989) raises the question of whether bacteria that are capable of positive growth in coliform media are native to aquifers, and not indicators of contamination from the outside.

Pesticides and Nitrates. Another problem is pesticide and nitrate infiltration from surface contamination, discussed in greater detail later in this chapter. This is a typically rural to suburban groundwater quality concern, since these chemicals are used in agricultural and horticultural activities. Pesticides and nitrates represent the main sources of chemical aquifer contamination in agricultural zones (Dupuy, 1997). The matter of pesticide and nitrate contamination is also a good illustration of the interactions among (and need to understand) use and application methods, hydrogeology, climatic factors, and site-specific circumstances in prevention and prevention in management. In contrast to the limited understanding of the microbial impacts of industrial-scale agriculture, nitrate impact management for facilities such as CAFO is relatively well understood. However, experience in the Netherlands, where rises in groundwater nitrate levels are attributed to manure spreading, provides a sobering example of the possibilities.

Public Water Supply Wells. Incidents of contamination are not unique to private wells. However, public wells are more likely to be isolated from contamination sources and better constructed. As is the case for coliform contamination, older, deteriorated, and poorly constructed wells (public or private) were more likely to be contaminated (EPA, 1990; USGS, 1997).

Nitrate and pesticide occurrence in wells is most likely where there is a high density of wells and on-site waste disposal systems (which are themselves often forgotten or neglected by their users). Situations can be particularly acute when combined with a vulnerable aquifer setting such as a shallow limestone. Numerous abandoned but unsealed or poorly sealed wells (Gass et al., 1977; Banks, 1984; Smith, 1994) may serve as conduits for contamination to move to the subsurface.

Natural Environmental Impacts on Groundwater Quality

Groundwater quality reflects the physical, chemical, and biological actions interacting with the water itself. As part of the hydrologic cycle, the water falls as precipitation and is influenced by soil and organisms at or near the surface. As groundwater in the saturated zone, it is affected by the nature of the formations (including their microbial ecologies) in which the water has been stored. The influence of the aquifer formation depends on how long the water has been held in the aquifer, and how rapidly stored water is recharged by fresh water from the surface.

Aquifer Formation Composition and Storage Effects

Groundwater movement is relatively slow, even in the most dynamic flow systems. Water moves in close contact with the minerals that make up the particles or fracture channels of the aquifer formation.

These minerals reflect the depositional environment of the formation. For example, limestone, laid down as carbonate-rich sediments, fossils, and fossil debris, consists of calcium carbonate with impurities. Dolomite rock is chemically altered limestone containing a high fraction of magnesium. Carbonate rock aquifers typically provide an alkaline, high total dissolved solids (TDS) water that also has high calcium and/or magnesium hardness, but as illustrated in Figure 4.3, quality can vary greatly among carbonate aquifers with somewhat varying histories (e.g., Cummings, 1989).

Sandstone (cemented quartz sands) and volcanic rock, by contrast, may be acidic (unless cemented by carbonates). The water may be relatively soft, especially with sodium minerals in the rock matrix, or low in TDS (but not always).

The chemical influences of glacial and alluvial sands and gravels tend to reflect the composition of the source rock of the formation material.

All natural aquifer formations have impurities that provide additional complexity to the water. Most aquifers contain iron, and many contain manganese and other minor metal constituents as well. Certain igneous and metamorphic rocks, and the shales, sandstones, and unconsolidated deposits derived from them, may have high heavy metal or radionuclide contents.

A tragic example is that of the arsenic-rich alluvial and deltaic formations of Bangladesh and the West Bengal state of India which were tapped for tubewells to replace poor-quality shallow groundwater and surface water sources (e.g., Bhattacharya et al., 1997; Mushtaq et al., 1997). These sediments were derived from As-rich continental source rocks. Formations laid down near geo-historical continental margins (which may be different from continental margins now) may contain significant amounts of evaporites, such as rock salt or gypsum, that contribute chlorides or sulfates.

Elevated radon and other radionuclides (e.g., radium) levels in groundwater closely match the geographical distribution of rock types rich in radioactive minerals (e.g., New England, the Canadian Shield region of the United States and Canada, shale areas of Ohio and the east, and sediments derived from them). See Figure 4.4. Sediments derived from metamorphic source rocks may contain elevated solid-phase radionuclides, for example the Kirkwood-Cohansey aquifer of New Jersey (USGS, 1998a). Phosphate deposits in Florida with high radioactive-isotope content may be added to this group.

Deep formations well isolated from surface hydrologic recharge usually contain briny fluids (usually geochemically distinct from shallower groundwater), which may migrate to freshwater aquifers along density gradients if there are pathways such as faults or boreholes open between the formations. Table 4.1 is a generalized summary of water qualities that may be expected from wells developed in various aquifers.

Table 4.2 shows variation in mean analytical values for selected parameters among aquifers in one area (Michigan), illustrating how difficult generalizations about water quality and source geochemistry can be.

Aquifer Biogeochemistry

Lithology (rock type and composition) sets the broad tone for natural groundwater quality, but the effects of microbial ecology in aquifers, typically driving redox reactions, provide additional complexity at the local scale. Many oxidation-reduction (redox) transformations of common metals can occur without the mediation of microorganisms, but are not likely to occur at ambient groundwater temperatures and measured bulk redox potentials. An example is the oxidation MnII to MnIV,

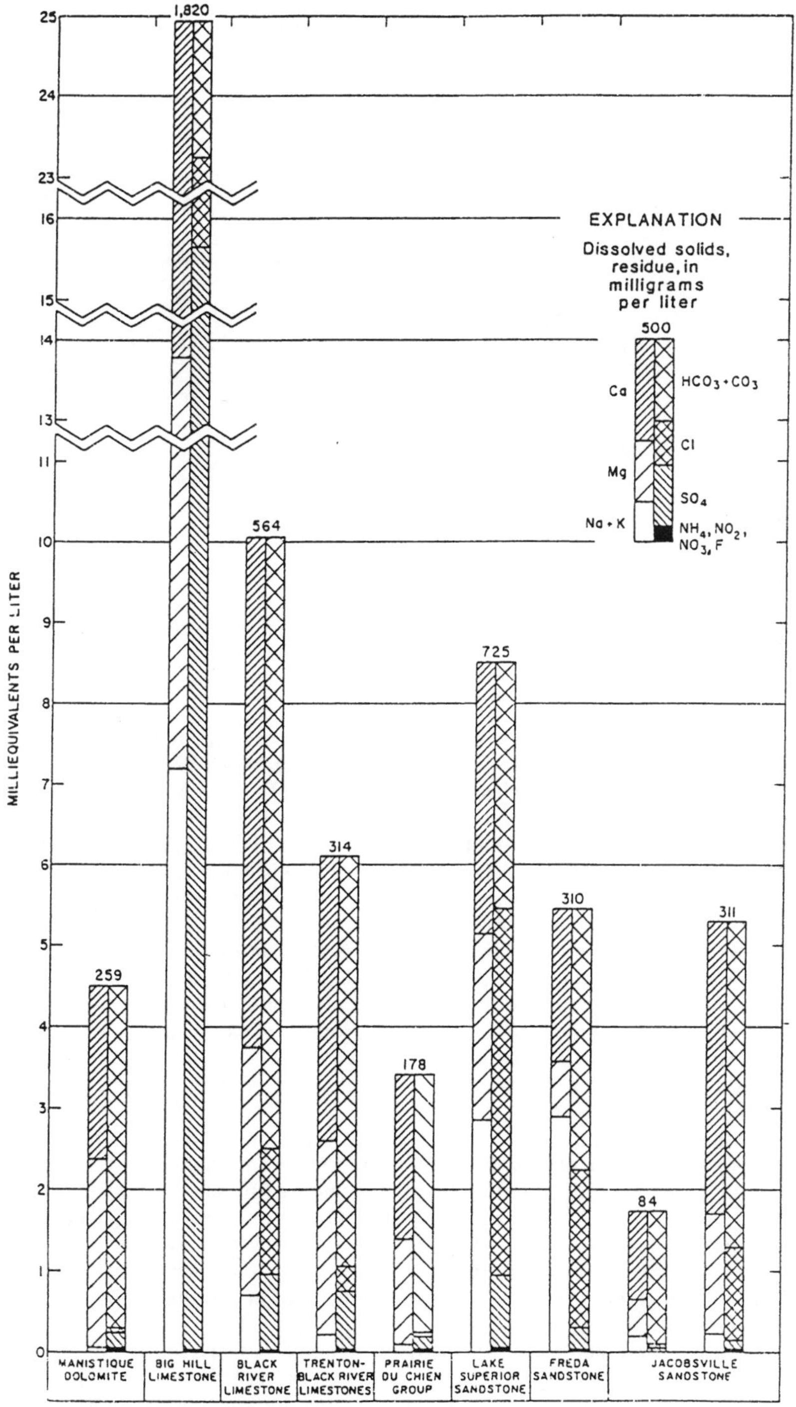

FIGURE 4.3 Chemical characteristics of water from bedrock (Silurian, Ordovician, Cambrian, and Precambrian ages). (*Source:* Cummings, 1989.)

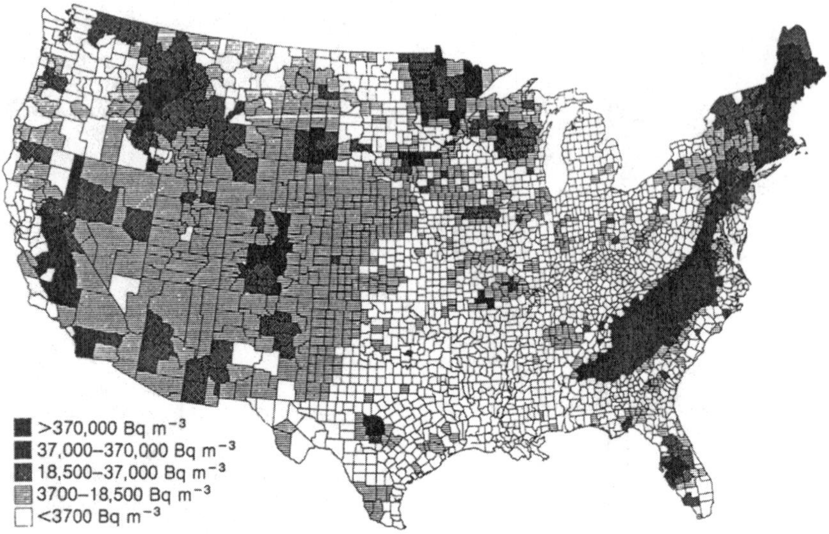

FIGURE 4.4 Distribution of radon in groundwater mapped by county, based on available data and aquifer type. (*Source:* J. Michel. 1987. In *Environmental Radon* (pp. 81–130). Ed. by C. R. Cothern and J. E. Smith. New York: Plenum Press.)

which occurs at approximately Eh +600 to 700 at pH and temperature ranges encountered in potable groundwater (Hem 1985). Figure 4.5 illustrates the Eh-pH ranges of stability for Mn species (gray areas are MnIV solids). Such a high redox potential is unlikely in groundwater, and in fact, MnIV is encountered in groundwater with much lower measured bulk Eh. Reductions of Fe, Mn, As, and a variety of other metals have been demonstrated to be microbially controlled. Both the oxidation and reduction of sulfur species largely involve bacteria except at extreme temperatures and pressures. Research on these activities to-date was summarized in Chapelle (1993) and Amy and Haldeman (1997) and continues actively.

The presence of organic carbon is an important factor influencing the activities of microorganisms. Both coal and hydrocarbons, which are extensively distributed in formations throughout the United States and around the world, are ready sources of organic carbon for bacteria. Acidic coal mine waters provide the preferred environment for iron-oxidizing bacteria, which, ironically, are chemolithotrophs and do not commonly utilize organic C.

Hydrocarbons, common in carbonate and other reduced sediments such as shales, provide the type of organic carbon compounds and the reductive conditions that promote sulfate reduction (and hydrogen sulfide formation) or methanogenesis by microorganisms. Jones et al. (1989) summarizes the occurrence and ecology of sulfate-reducing bacteria and methanogens in aquifers. Where ready sources of sulfate (e.g., gypsum) as well as short-chain hydrocarbons (e.g., acetate) are available, H_2S production can be intense. This is well illustrated in carbonate aquifers across the U.S. Great Lakes region.

Relatively high microbiological activity in aquifers can contribute significant amounts of carbon dioxide, producing gaseous waters and lower pH, thereby shifting the carbonate balance toward bicarbonate. This condition can in turn contribute to greater dissolution of Ca and Mg from rock and very high hardness. Figure 4.6 is a

TABLE 4.1 Representative Water Quality of Aquifer Types*

	Total dissolved solids	Carbonate hardness occurrence	Dominate dissolved mineral characteristics	Metal and radionuclide characteristics
Basalt	Very low <250 mg/L†	Very low <120 mg/L	Na, Cl, Si far from saturation	May have high and exotic metal and radionuclide concentrations
Carbonate	High >500	Very high >500	Ca, Mg carbonates (near saturation), major sulfate admixture, may be closely associated with evaporite deposits	High Fe, "dirtier" formations may have radionuclides at levels greater than U.S. MCLs
Igneous and metamorphic rocks	Low <250	Low to moderate <120–500	Mixed, depending on rock mineral origin	May have high and exotic metal and radionuclide concentrations
Surficial sands/gravels	Very low to moderate <250–500	Moderate 120–240. Glacial-fluvial deposits may be have high carbonate hardness.	Mixed, depending on rock mineral origin. Glacial-fluvial deposits may be more carbonate dominated.	Low Fe except where concentrated by biological action (e.g., Kirkwood-Cohansey aquifer, Pine Barrens, NJ), typically low metal contents in pumped groundwater.
Sandstone	Low to moderate	Moderate	Mixed, depending on rock mineral origin. Sandstones with carbonate cementation are closer to carbonate saturation. In the U.S. east, often intermixed with coal deposits.	Low to high depending on cementation material and sand origin, and degree of water flushing over time. May have radionuclides above U.S. MCLs depending on sand origin (sands are major sources of uranium deposits).
Shale	Moderate	Moderate		Often high radon and other radio nuclides and metals (typically poorly flushed).

* These summaries are highly generalized and may likely be different regionally or locally. Water quality should be investigated at the wellfield or aquifer scale.

† Numbers from Lehr et al. (1980) and considered general and representative.

TABLE 4.2 Selected Median Values of Physical-Chemical Parameters from Michigan Aquifers

Source	Ca hardness (total) in mg/L	Fe (dissolved) in μg/L	Boron (total) in μg/L	Chloride (dissolved) (mg/L)
Glacial sand (Pleistocene)	8	50	<20	1.5
Glacial gravel (Pleistocene)	27	270	40	1.8
Engadine dolomite	515	1290	80	1.2
Paleozoic dolomite	250	160	80	1.3
Sylvania sandstone	600	1800	110	47
Freda sandstone	101	16	640	11
Portage Lake volcanics	29	47	700	5.3

Source: Cummings (1989), Appendix A.

summary of the carbon system influencing aquifer conditions. Table 4.3 is a summary of physical-chemical activities often influenced by microorganisms in the subsurface.

When wells are installed in formations, redox gradients form due to both abiotic and biotic effects on metal species (Cullimore, 1993). These gradients can be very sharp. Iron can oxidize readily at the appropriate redox potential along these gradients, and microorganisms associated with iron precipitation are often found preferentially at Eh-pH environmental conditions as found in Figure 4.7 (Smith and Tuovinen, 1985). Manganese, where present and in solution as Mn(II), also oxidizes catalytically to Mn(IV) due to the activity of bacteria (Hatva et al., 1985; Vuorinen et al., 1988), a mechanism also involved in at least some Fe(II)-Fe(III) oxidation (Hatva et al., 1985). Hydrogen sulfide is also oxidized by various S-oxidizers to S^o and then to sulfate. The resulting oxidized products (solid S and sulfate salts) usually cause clogging in well intake screens, pumps, and piping, and discoloration and odors in the water.

Human Impacts

In addition to understanding and taking into consideration factors in natural groundwater quality, management of groundwater source quality is also concerned about human impacts. Within the overall scope of management, regulation is largely focused on human impacts. Adverse human impacts include providing routes for potential pathogens or chemicals to reach groundwater, and aquifer depletion. Other human activities can prevent or mitigate such adverse effects, and these are the agents of appropriate stewardship of the groundwater resource.

Health-Related Microbiological Occurrence in Groundwater

Groundwater is a preferred source of water supply due to natural removal of undesirable microorganisms and viruses, especially where minimal treatment is the highest priority (e.g., small systems and private wells).

Although bacteria are not uncommon in groundwater, attenuation by various means reduces total coliform and fecal indicator bacteria, commonly to undetectable levels. A persistent coliform bacteria presence in groundwater samples usually indicates short-circuited flow from a surface source [although some research (Jones et al., 1989) shows doubt about that assumption]. Pathogenic bacteria or protozoa are not known to be native to or commonly occur in groundwater except where direct and close sources of innoculant are present.

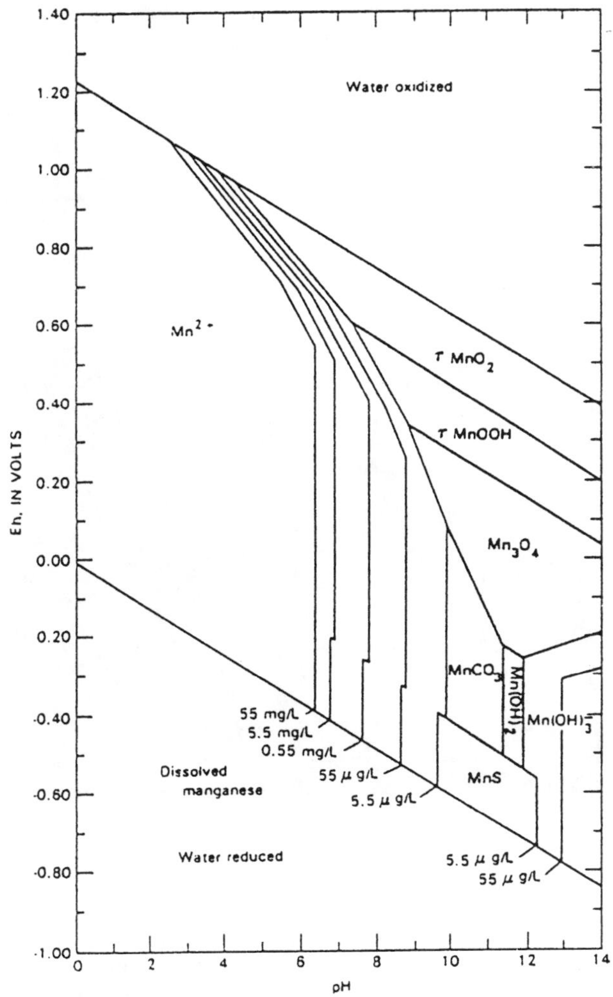

FIGURE 4.5 Fields of stability of manganese solids and equilibrium dissolved manganese activity as a function of Eh and pH at 25°C and 1 atmosphere pressure. Activity of sulfur species 96 mg/L as SO_4^{2-}, and carbon dioxide species 61 mg/L as HCO_3. (*Source:* Hem, 1985, p. 87.)

While viruses have been detected in groundwater, little has been known about their occurrence in aquifers. At the present time, intensive field research is limited, and conclusions that can be drawn from it are preliminary. A study sampling from 460 wells across the United States managed by the American Water Works Service Company (LeChevallier, 1997) showed that viral occurrence may be expected in many groundwater sources. Culturable viruses (cultured on tissue cells) were detected in samples from 7 percent of the sites (LeChevallier, 1997), illustrating how differences in methods affect detection statistics. Methods and their varying results are discussed in Abbaszadegan et al. (1998).

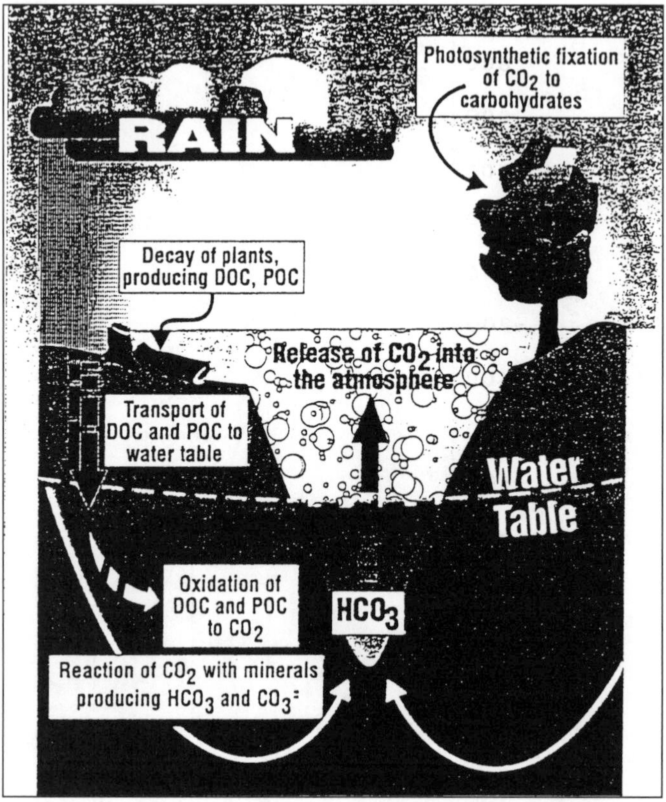

FIGURE 4.6 Carbon cycling in local flow systems. (*Source:* F. H. Chapelle. 1993. *Ground-Water Microbiology and Geochemistry.* Copyright © 1993 John Wiley & Sons. Reprinted by permission of John Wiley & Sons, Inc.)

Conclusions about viral numbers and role in human disease have been difficult to make, although viruses have been linked to enteric disease outbreaks from groundwater sources (e.g., Craun and Calderon, 1997). However, such outbreaks have to be obvious to be noticed. Assessment of the occurrence of viruses at less-than-outbreak levels has been hampered by limits in collection and analysis methods.

TABLE 4.3 Representative Microbially Influenced Chemical and Redox Activities in Aquifers

	Reaction activity*	Microbiological influence†
Calcite dissolution	$CaCO_3 + H_2CO_3 \rightarrow Ca^{2+} + 2\ HCO_3$	Respiration: DOC removed and CO_2 added.
Fe reduction	$Fe^{3+} + e = Fe^{2+}$	Dissimilatory use of ions such as Fe^{3+} or
SO_4^{2-} reduction	$SO_4^{2-} + 10H^+ + e = H_2S + 4H_2O$	SO_4^{2-} as final electron acceptors.
Methanogenesis	$CO_2 + 6H^+ + e = CH_4 + H_2O$	Dissimilatory reduction by archaebacteria.

Sources: * Freeze and Cherry (1979).
† Chapelle (1993).

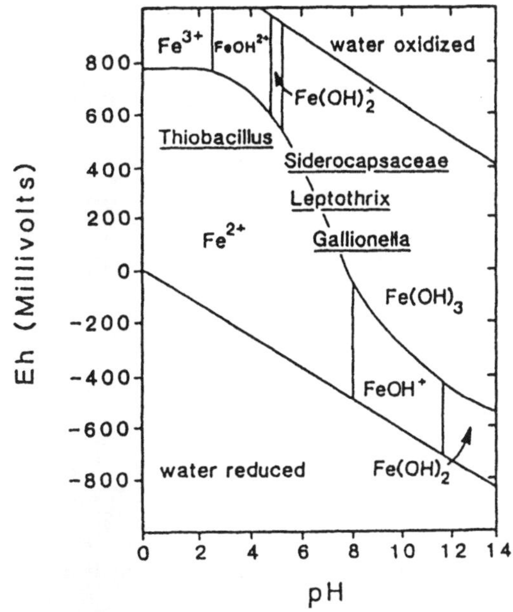

FIGURE 4.7 Eh-pH diagram for major iron species in relation to the occurrence of iron bacteria in the environment. (*Source:* S. A. Smith and O. H. Touvinen. "Environmental analysis of iron-precipitating bacteria in ground water and wells." *Ground Water Monitoring Review,* 5(4), 1985: 45–52.)

Transport and fate of microflora and viruses depends on a complex interaction of soil characteristics, hydrology, climate, water quality, and features of the organisms or viruses themselves. Gerba and Bitton (1984) provide a conceptual review of transport and attenuation of bacteria and viruses in soil and groundwater. Tables 4.4 and 4.5 summarize factors affecting enteric bacteria survival and viral movement in soils.

Groundwater has characteristics that may tend to favor the transport of viable viruses: absence of strong oxidants, cool temperatures, and frequently alkaline pH. However, reported transport distances have been on the meter scale. Yates (1997) reported viruses detected 402 m downgradient from a landfill on Long Island, New York, and three m below land-applied sludge (both potentially intense sources of viral inoculum). Bales et al. (1995) showed movement upwards of 14 m for bacteriophage PDF-1 in 24 days in a glacial sand. Attenuation factors clearly are at work at least in granular aquifers (Bales et al., 1995). Schijven and Rietveld (1997) report ≥2.6 to 2.7 log removal of enteroviruses and ≥4.7 to 4.8 log removal of reoviruses over 30 m bank filtration distances in the Netherlands.

While these studies provide evidence for mechanisms that provide impediment to viral transport, the degree of removal inactivation have been difficult to model (Schijven and Rietveld, 1997; Yates, 1997) and can be reversible (Bales et al., 1995).

All of these factors have relevance to water supply in determining what is necessary in risk assessment, and how to site and protect water supply wells. The Groundwater (Disinfection) Rule (GWDR), currently under development by the U.S. EPA

TABLE 4.4 Factors Affecting Survival of Enteric Bacteria in Soil

Factor	Comments
Moisture content	Greater survival time in moist soils and during times of high rainfall
Moisture-holding capacity	Survival time is less in sandy soils with lower water-holding capacity
Temperature	Longer survival at low temperatures; longer survival in winter than in summer
pH	Shorter survival time in acid soils (pH 3–5) than in alkaline soils
Sunlight	Shorter survival time at soil surface
Organic matter	Increased survival and possible regrowth when sufficient amounts of organic matter are present
Antagonism from soil microflora	Increased survival time in sterile soil

Source: C. P. Gerba and G. Bitton. "Microbial pollutants: Their survival and transport pattern to groundwater." Pp. 65–88 in *Groundwater Pollution Microbiology.* C. P. Gerba and G. Bitton, eds. New York: Wiley-Interscience, 1984. Copyright © 1984 John Wiley & Sons. Reprinted by permission of John Wiley & Sons, Inc.

(Macler, 1996; Macler and Pontius, 1997) and intended for promulgation early in the first decade of the twenty-first Century, will focus on minimizing risk of viral disease in groundwater-source public water systems. The mentioned research on viral occurrence, transport, and fate is intended to support the GWDR development effort, which is a regulatory process being developed on emerging and uncertain science.

TABLE 4.5 Factors That May Influence Virus Movement to Groundwater

Factor	Comments
Soil type	Fine-textured soils retain viruses more effectively than light-textured soils. Iron oxides increase the adsorptive capacity of soils. Muck soils are generally poor adsorbents.
pH	Generally, adsorption increases when pH decreases. However, the reported trends are not clear-cut due to complicating factors.
Cations	Adsorption increases in the presence of cations (cations help reduce repulsive forces on both virus and soil particles). Rainwater may desorb viruses from soil to its low conductivity.
Soluble organics	Generally compete with viruses for adsorption sites. No significant competition at concentrations found wastewater effluents. Humic and fulvic acid reduce virus adsorption to soils.
Virus type	Adsorption to soils varies with virus type and strain. Viruses may have different isoelectric points.
Flow rate	The higher the flow rate, the lower virus adsorption soils.
Saturated versus unsaturated flow	Virus movement is less under unsaturated flow conditions.

Source: C. P. Gerba and G. Bitton. "Microbial pollutants: Their survival and transport pattern to groundwater." Pp. 65–88 in *Groundwater Pollution Microbiology.* C. P. Gerba and G. Bitton, eds. New York: Wiley-Interscience, 1984. Copyright © 1984 John Wiley & Sons. Reprinted by permission of John Wiley & Sons, Inc.

Point-Source Chemical Contamination

Point-source contamination of groundwater is that resulting from defined sources of small areal extent such as single locations or properties. Eliminating it has been a major focus of public health regulation for many years in North America and elsewhere in the world. Leaking septic tanks or privies are point sources identified as public health risks generations ago. More recently, point sources of chemical contaminants such as underground storage tanks or waste pits have been the focus of environmental regulation.

Point sources are frequently identifiable as sources of contamination by field checking and testing. They are typically closely regulated by government agencies. Examples include municipal or hazardous waste landfills, underground storage tanks, or deep injection wells. They are readily eliminated as groundwater risks by such mitigating actions as repair, replacement, or sealing. Effectiveness of the "repair" can often be judged quickly by the result of greatly reduced or eliminated contamination. However, reduction or elimination of contamination after a point source is removed may also take many years. Point sources can also be addressed in public health regulation because it is often possible to assign blame for allowing the problem to occur.

Nonpoint Chemical Contamination

Surface-water nonpoint contamination is a well-known problem that is extensively discussed in the literature and elsewhere in this text. Nonpoint-source contaminants such as nitrate or pesticides also reach groundwater from regional use at the surface. An example might be elevated nitrate levels in shallow wells in an area of sandy soils where there is extensive use of fertilizers and septic tanks. This is a worldwide problem in areas of vulnerable soil situations.

In many regions, nitrate concentration levels in groundwater can reach and exceed water quality criteria (e.g., U.S. drinking water, 10 mg NO_3/L, E.U. ambient, 50 mg NO_3/L). The increasing use of mineral fertilizers in some regions and the intensive exploitation of the aquifers for crop irrigation have led to groundwater contamination by nitrates (e.g., Dupuy et al., 1997; USGS, 1997). The dynamics (long-term persistence) and extensiveness (regional contamination) of this contamination, as well as uncertainty about behavior in aquifers, make it a sensitive environmental issue.

The USGS (1997), summarizing current USGS study results, reports that over 300 studies of pesticide occurrences in groundwater and soils have been carried out during the past 30 years. Further data are being compiled in the National Water Quality Assessment Program (NAWQA). Provisional summary data and interpretation (subject to revision) are available from the USGS on the World Wide Web (e.g., USGS, 1998b), a mechanism that permits rapid, but fluid publication of such information. USGS studies have shown that:

1. Pesticides from every major chemical class have been detected in groundwater.
2. Pesticides are commonly present in low concentrations in groundwater beneath agricultural areas, but seldom exceed water quality standards.

Occurrence of pesticides in groundwater follows a pattern resembling that of coliform bacteria occurrence (Figure 4.8).

The USGS currently has come to a several conclusions on pesticide and nitrate contamination occurrences associated with cropland (USGS, 1997):

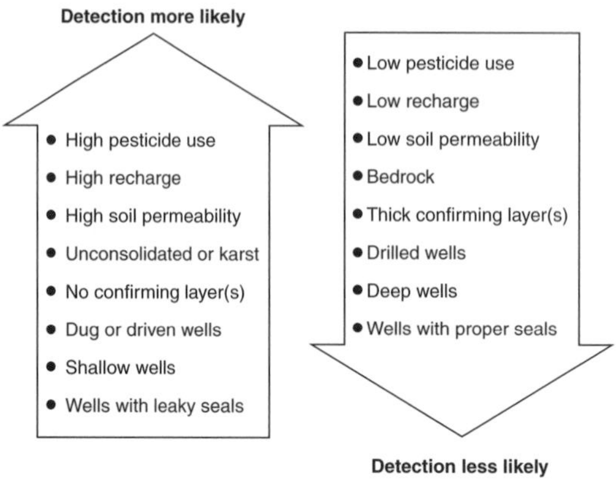

FIGURE 4.8 USGS arrow from //water.wr.usgs.gov/pnsp/images/gw7.gif.

1. Factors most strongly associated with increased likelihood of pesticide occurrence in wells are high pesticide use, high recharge by either precipitation or irrigation, and shallow, inadequately sealed, or older wells (consistent with results reported by EPA 1990) (Figures 4.9 and 4.10).
2. They are more likely to occur where groundwater is particularly vulnerable, such as shallow, unprotected sands, or highly solutioned (karstic) or fractured rock.
3. Pesticide contamination is generally more likely in shallow groundwater than in deep groundwater, and where well screens are located close to the water table, but such relations are not always clear-cut.
4. Frequencies of pesticide detection are almost always low in low-use areas, but vary widely in areas of high use.
5. Pesticide levels in groundwater show pronounced seasonal variability in agricultural areas [consistent with Dupuy (1997)], with maximum values often following spring applications. Temporal variations in pesticide concentrations decrease with increasing depth and are generally larger in unconsolidated deposits than in bedrock.

As with pesticides, the risk of groundwater contamination by nitrate is not the same everywhere (Nolan and Ruddy, 1996). Figure 4.11 shows four groups in order of increasing risk related to soil characteristics:

1. Poorly drained soils with low nitrogen input (white area on the map)
2. Well-drained soils with low nitrogen input (light gray area)
3. Poorly drained soils with high nitrogen input (medium gray area)
4. Well-drained soils with high nitrogen input (dark grayish area)

Well-drained soils can easily transmit water and nitrate to groundwater. In contrast, the other three groups have a lower risk of nitrate contamination because of poorly drained soils and/or low nitrogen input. Poorly drained soils transmit water and chemicals at a slower rate than well-drained soils (Nolan and Ruddy, 1996). Drains and ditches commonly are used to remove excess water from poorly drained agricul-

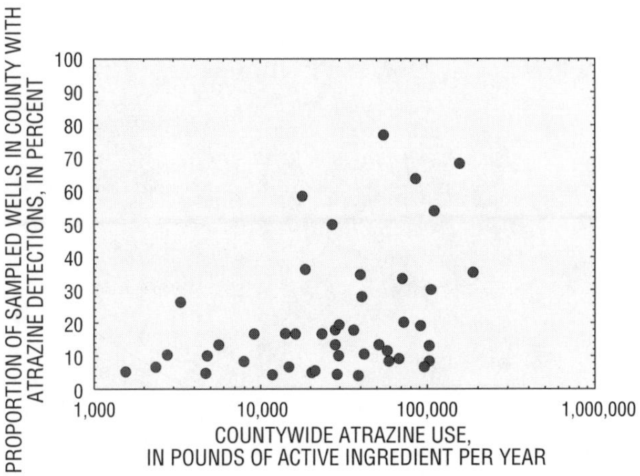

FIGURE 4.9 Proportion of sampled wells with atrazine detections in relation to countywide use. Wells were sampled as part of the National Alachlor Well-Water Survey. (*Source:* U.S. Geological Survey, Pesticides in Ground Water, USGS Fact Sheet FS-244-95, http://water.wr.usgs.gov/pnsp/gw/gw5.html.)

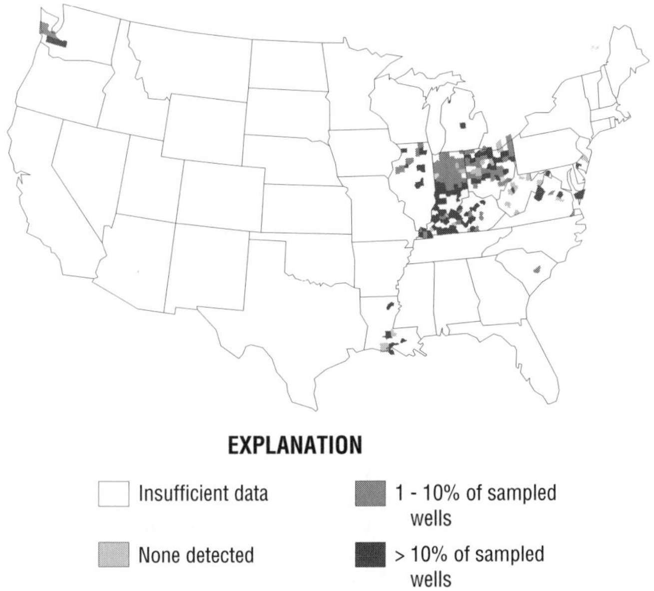

FIGURE 4.10 Frequency of triazine herbicide detection in counties with ten or more wells sampled during the Cooperative Private Well Testing Program. (*Source:* U.S. Geological Survey, Pesticides in Ground Water, USGS Fact Sheet FS-244-95, http://water.wr.usgs.gov/pnsp/gw/gw5.html.)

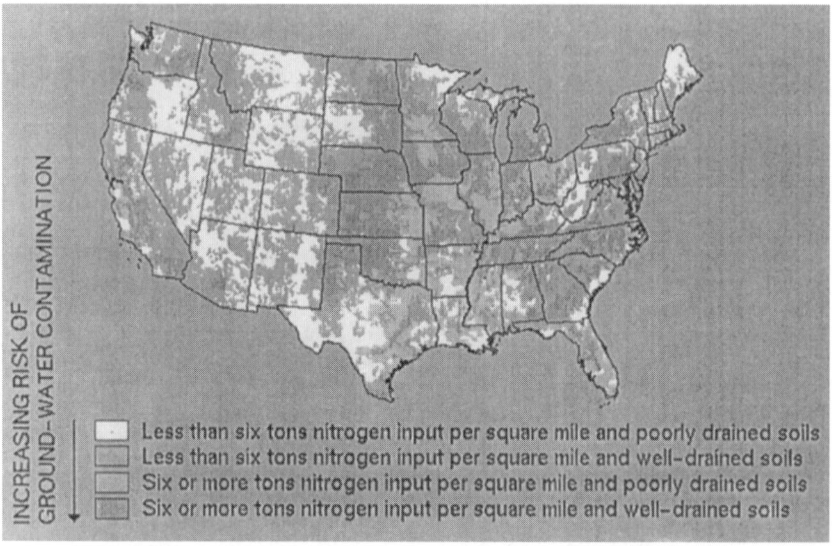

FIGURE 4.11 Areas in the United States most vulnerable to nitrate contamination of groundwater (shown in grayish blue on the map) generally have well-drained soils and high-nitrogen input from fertilizer, manure, and atmospheric deposition. High-risk areas occur primarily in the western, midwestern, and southeastern portions of the nation. (*Source:* U.S. Geological Survey Fact Sheet FS-092-96, http://wwwrvares.er.usgs.gov/nawqa/fs-092-96/fig1.html.)

tural fields, diverting nitrate to nearby streams (and thus making it a surface water problem).

The effects of application of fertilizers is illustrated by studies from the United Kingdom. The British Geological Survey (BGS) has been studying the problem of nitrates in groundwater for over 15 years. As is the case elsewhere in Europe, the concentrations of nitrate in groundwater in the principal British aquifers are generally rising, and some groundwater sources already exceed the EC Drinking Water Directive limit of 50 mg/L NO_3 (11.3 mg/L NO_3-N). The BGS suggests that many others are likely to exceed this limit if current agricultural practices continue unchanged.

Investigations at four grassland sites on the chalk of southern England between 1987 and 1989 (Figure 4.12) demonstrated that the majority of pore water nitrate concentrations under most of the grazed grassland sites investigated exceed the EC limit and that leaching from grazed grassland receiving more than about 100 kg N/ha/yr is likely to give rise to nitrate concentrations above the EC limit in groundwater recharge. Typical nitrate concentrations in the unsaturated zone beneath grazed grassland sites (Figure 4.13) are in the range of 10 to 100 mg/L NO_3-N, with marked peaks as high as 250 mg/L NO_3-N. This compares with values of less than 5 mg/L NO_3-N measured under lower-productivity grassland, and is also higher than levels observed under intensively cultivated land. Nitrate losses expressed as a percentage equivalent of the nitrogen input to the land range from 15 up to 45 percent, showing a broad correlation between the implied leaching loss and the quantity of nitrogen applied (BGS, 1997).

There is historically less information available on pesticide occurrence beneath nonagricultural land, such as residential areas and golf courses, despite chemical application rates that often exceed those for most crops. However, studies of golf courses on

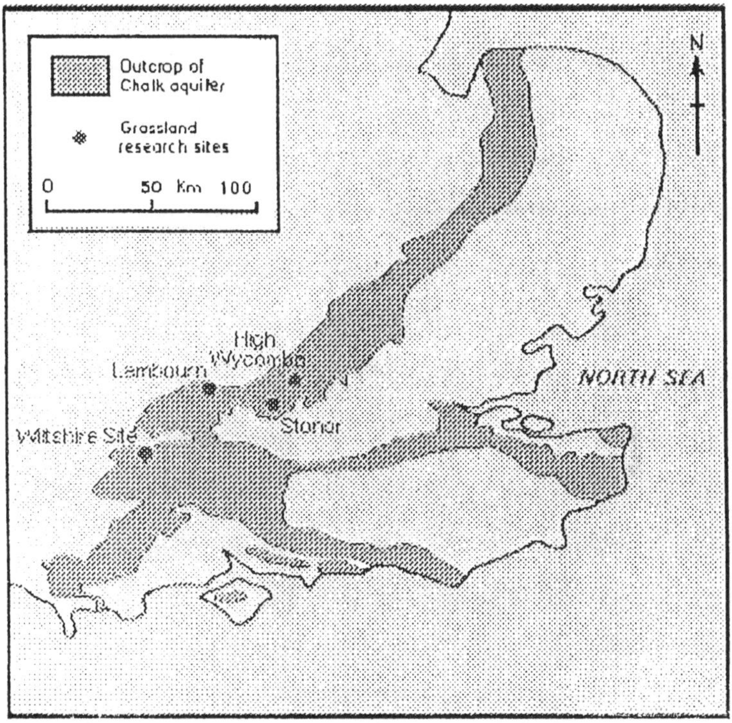

FIGURE 4.12 Outcrop of chalk aquifer in SE England and location of grassland investigation sites. (*Source:* BD/IPR/20-7, British Geological Survey. © NERC. All rights reserved. http://www.akw.ac.uk/bgs/w3/hydro/Nitrate.html.)

Cape Cod, Massachusetts, and in Florida (Cohen, 1996; Cohen et al., 1990; Swancar, 1996) shed some illumination. Swancar (1996) found pesticides in groundwater at seven of nine golf courses, with 45 percent at trace concentrations and 92 percent below MCLs or health advisory levels (HALs). In the Florida study, there was one trace diazinon detection, although this insecticide was banned for golf courses in the 1980s. This raises the question of the poorly known contribution of lawn application, a possible source of the diazinon occurrence in the Florida study (Cohen, 1996).

The nature of land use controls also has to be considered carefully. For example, protection zones around public supply wells are often used as a setback to reduce or prevent contamination of groundwater. Nitrate concentrations can be reduced by establishing zones in which agriculture would be restricted, perhaps by replacement of arable farming with grassland or woodland, with appropriate compensation to farmers. The BGS studies indicate that, where grassland is to be a land use option in plans to reduce nitrate concentrations in groundwater supply sources, restrictions on fertilizer applications to the grassland may be required.

The discharge of groundwater contaminants to surface water is another aspect of the relationship between groundwater and surface water. For example, sources of elevated TDS, salt, metals, or industrial chemicals may have their source in groundwater that contains or is contaminated with these constituents. The potential for close interaction of ground and surface waters is illustrated by detections of pesticides in sam-

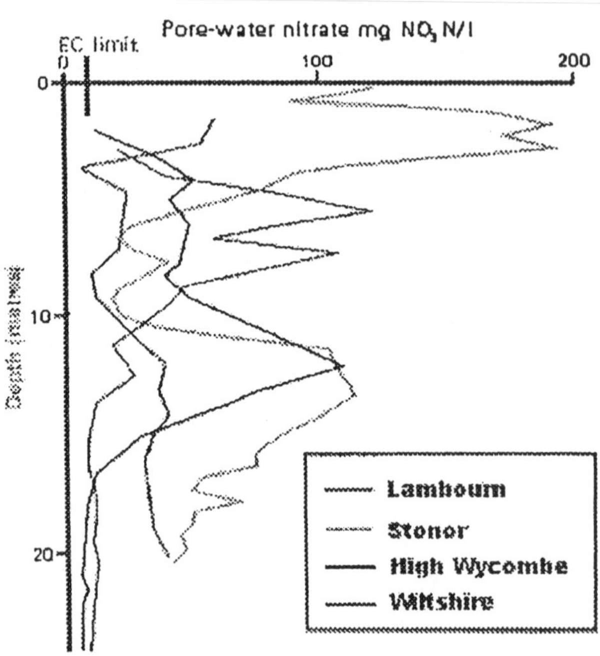

FIGURE 4.13 Unsaturated zone pore water nitrate profiles for selected grazed grassland sites. (*Source:* BD/IPR/20-7, British Geological Survey.)

ples (USGS, 1998b). On the one hand, high concentrations of pesticide contaminants in rivers may lead to contamination of shallow groundwaters in agricultural areas during periods of extensive seepage of river water into underlying alluvial aquifers. This is particularly the case following spring applications, when pesticide loads and river flows reach maximum levels. Conversely, pesticides in alluvial aquifers may flow into adjoining rivers during periods of low runoff. In many areas, "bank filtration" by alluvial aquifers has been found to be ineffective in removing pesticides from water drawn from pesticide-contaminated rivers into adjacent supply wells.

Industrial Agriculture: A Mixture of Point and Nonpoint Source Contamination

Agriculture has traditionally been a low-density activity on the land. Until recently, animal fecal wastes were thinly distributed over rangeland by grazing animals or spread mechanically. Problems, if they occurred, tended to be nonpoint types, such as runoff from inadequately protected fields flowing into surface waters.

In more recent times, hog, cattle, and poultry operations have become more concentrated. In large-animal (cattle and hog) CAFO operations, wastes are concentrated in lagoons in liquid form and then periodically pumped and spread or sprayed. This management method has the effect of creating a concentrated point source (the lagoon) in addition to nonpoint nitrate and bacteria land application. Effective groundwater protection against the risk posed by such a facility thus depends on assuring that the lagoon does not leak as well as on nonpoint runoff prevention.

Large-scale industrial chicken and turkey CAFO operations have a similar effect. Animal wastes must be dispersed in a manner that prevents surface and groundwater contamination. One strategy is to make the wastes available to area farmers for spreading. However, if nonpoint problems occur, the spreading farmer and not the generator may be at legal risk.

Egg and pullet operations may constitute a point source in addition to nonpoint bird waste problems. For example, egg washing for a 1-million-hen operation generates thousands of cubic meters of wastewater per day that must be properly treated or dispersed. In addition, large-scale bird and farrowing operations generate a certain percentage loss in dead animals, which must be properly composted on-site if they are not removed off-site. Animal disposal represents a serious risk of microbial contamination if performed improperly.

Effects of Excessive Withdrawal on Water Quality

In addition to water quality degradation by the introduction of wastes or chemicals to aquifers, excessive withdrawal may have localized or area-wide impacts on the quality of pumped groundwater. Examples include chronic seawater infiltration problems on Long Island and in New York, New Jersey, coastal Florida, and California. Another well-studied situation is that of pulling seawater into freshwater lenses on ocean islands, as in Hawaii and the Bahamas. Excessive pumping may also pull in undesirable water from other formations, such as upconing from brine-containing aquifers deeper in a sedimentary series.

This latter point introduces the topic of well or wellfield use in managing water quality. A particular aquifer may not be generally overtaxed or exhibit overuse-related water quality. However, individual wells or wellfields may cause water quality degradation locally if operated at unsustainable rates (which must be determined individually). The most common effect is excessive oxidation of Fe and Mn in a wellfield's area of influence.

Interrelationships Between Performance and Quality Decline

The primary negative effect of area-wide Fe and Mn oxidation is that Fe and Mn oxides can build up and reduce the hydraulic conductivity in the oxidized drawdown zone, causing further increases in drawdown. This process continues in a cause-effect cycle, and is part of the long-term aging process of some wellfields. Overall calculated aquifer transmissivity (flow capacity times aquifer thickness) may be noticeably reduced as drawdown increases. This results in increased well areas of influence, and increases the possibility of inducing poor-quality water or contaminants to flow toward affected wells. Migrating organic pollutants (e.g., petroleum products) can severely increase Fe and Mn available, as was discussed earlier.

The phenomenon of lifetime aquifer or wellfield aging is poorly understood. There does seem to be a pattern of increasing frequency and severity of well rehabilitation in wellfields observed over some decades. For example, in the Ohio River Valley, alluvial wellfields exhibit accelerated maintenance problems after about 20 years. This may be a near-well phenomenon, but it also occurs on the wellfield scale.

Groundwater Quality Management Options

Once the many factors that affect groundwater quality are understood, methods of management can be implemented or evaluated. Groundwater management is by nature localized. However, it can be planned simultaneously (1) in a general way at the regional or aquifer scale and then (2) more specifically at the wellfield scale.

The Big Picture: Regional and Aquifer Scale Management

Groundwater management may be considered on regional basin or aquifer scales. As is the case for large bodies of surface water such as lakes or rivers, aquifers and basins do not respect surface political jurisdictions (Figure 4.14), so "regional" may preferably have something other than artificial (such as state) boundaries. For example, on a system scale, the Atlantic and Gulf coastal plain aquifer systems each span several U.S. states. The Ogallala aquifer alone extends across several states. Managing such systems properly involves cooperation across major political jurisdictions (even international ones, e.g., at the U.S.-México and Washington–British Columbia frontiers, or more dramatically, in the Middle East, where nations in political conflict share aquifers).

On the other hand, specific aquifer units may be very localized, especially small glacial valley fill or alluvial fan aquifers. These also will not respect surface political boundaries. Entities such as towns or individual properties dependent upon an

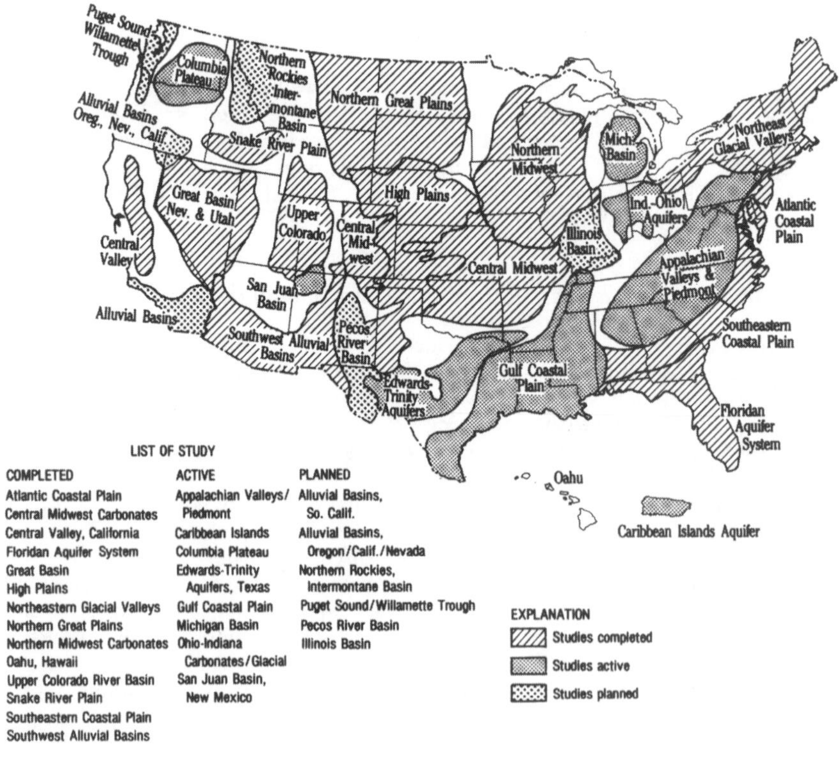

FIGURE 4.14 Regional aquifer studies in the United States. (*Source:* Van der Leeden, Troise, and Todd, 1990. Originally from Cardin, C. W., and others. 1986. Water Resources Division in the 1980s. U.S. Geological Survey Circular 1005.)

aquifer source may need to arrange with adjacent jurisdictions to protect areas outside their legal boundaries. Such cooperation at both the regional and local scale requires perhaps the greatest challenge in groundwater management: coordination of political and private entities for an intangible goal such as groundwater protection.

At both scales, the ideal situation is to have accurate and complete hydrogeologic information and mapping for planning for aquifer protection and management. Lithologic, depositional, depth, flow, and quality information should be available in groundwater resources maps. Where published by jurisdictions such as counties, specific hydrogeologic information should be included, as well as keys for linkage to adjacent maps, and references to sources. The advent of widespread availability of digital mapping information has made this somewhat easier.

However, in practice, one common problem with groundwater resources or contamination vulnerability mapping is the limited information available. With limited budgets, state or similar water resources agencies may not have the resources for highly detailed mapping, or may not verify file or regional information used. Local groundwater users should plan to conduct investigations to adapt general maps or assessments for the level of detail they need for local planning.

State/Provincial/Tribal and Federal Groundwater Management Strategies

Groundwater may be managed in a variety of ways, depending on political structure and tradition. In the Canadian, U.S., and Australian federal systems, for example, groundwater management and protection are accomplished at the subfederal (state, province, or tribal) level. Other nations make this a national government function or strictly a local matter.

In the case of the United States, federal legislation serves to protect groundwater and groundwater public water supply sources to some degree, but there is no single groundwater protection law. Both the Resource Conservation and Recovery Act (RCRA) and the Comprehensive Environmental Response, Compensation and Liability Act (CERCLA or "Superfund"), as well as other legislation, have groundwater protection as one goal. The Safe Drinking Water Act (SDWA) mandates source water (including wellhead) protection planning for public water supplies (and thus groundwater protection for systems using groundwater).

Water use is prioritized and allocated at the state and tribal level in the United States and Canada (Fetter, 1980). In the 17 U.S. western states, the doctrine of prior appropriation is used in surface water management. Under prior appropriation, use is "first-come, first served" with the junior rights holder holding lesser priority. Some states (e.g., New Mexico) also apply this doctrine to groundwater. Others, such as California, apply "correlative rights": the right to use groundwater underlying a property belongs to the landowner, who may use it as long as the use is "reasonable." In other states (Colorado for example) groundwater and surface water resources are integrated into basins, and are not treated separately.

Whatever the foundation of the water rights system, groundwater resources may be allocated based on criteria such as priority of need and water availability. In states such as North Dakota (and the Australian state of Queensland, for comparison), where water belongs to the state (or the Crown in the case of Australian states), water may be allocated on a volume basis. Landowners or public water purveyors or water districts have the right to use a certain volume per year. The allocation may apply to all water users, or some (such as private well owners) may be exempted. The allocation is usually based on available stored resources and available recharge expected. In times of drought, allocations may be more restricted.

In the more humid U.S. East, where rainfall exceeds evapotranspiration, two groundwater use doctrines prevail: English Rule, where a landowner has absolute right to groundwater under the property regardless of the value of the use, and American Rule, where neighbors have rights, too, and landowners may only use "reasonable" amounts of water so as not to harm a neighbor. American doctrine is increasingly applied throughout the eastern United States, mostly through individual legal decisions. However, like their western-state counterparts, some eastern and Midwestern states (Wisconsin for example) have adopted comprehensive, regional groundwater management plans rather than relying on case-by-case management.

The effectiveness of such plans is yet to be conclusively determined, since few have been in place for sufficient time to be evaluated properly. Most are still in the study stage [e.g., the Willamette River Basin in Oregon, (USGS, 1996)]. However, some are quite mature. Examples include the Edwards Aquifer Authority, the Water Resources Board system in North Dakota (North Dakota State Water Commission), and the Delaware River Basin Commission (Pennsylvania, New York, New Jersey, and Delaware). Management of artesian aquifer basins includes some notable successes. One of these is the Australian interstate project to limit hydrostatic head loss in the Great Artesian Basin in western Queensland, New South Wales, and the Northern Territories (Harth, 1993; Free et al., 1995), which has had great success in plugging uncontrolled artesian well flows, although desert springs and aquifer hydrostatic levels in general remain at risk.

Water rights for tribal reservations represents an interesting challenge for the water-short U.S. West. Native nations were forced out of the more water-rich areas of the United States settled by people of the Euro-American culture by the late 1830s, and settled on more arid land in the West. However, in a historic 1908 U.S. Supreme Court decision, it was ruled that the federal government must reserve water rights sufficient to make the tribal land productive. This amount was ruled to be that necessary to irrigate arable land. States could not reserve or allocate this water for other uses. While the 1908 case applied to surface water, in 1976 the Supreme Court extended the ruling to include groundwater.

One important issue in water rights adjudication is determining who has the rights to recharged or stored water. The priority of groundwater recharge (over other land uses) to replenish water supplies to historical levels was established by a California state ruling in 1969. Rights to use water equivalent to that injected to aquifer storage were also upheld in California. It is generally recognized in the United States that water committed to aquifer storage and recovery systems may be reclaimed by the entity pumping water into storage.

In terms of quality, states may apply specific environmental laws to protect groundwater, or extend nuisance laws or "reasonable use" doctrine for this purpose. In general, if a party causes harm to groundwater quality, or if it is determined that they may cause potential harm, the state may act. Federal laws are frequently applied by states with primacy in environmental enforcement. In nonprimacy states, or on tribal or territorial land, U.S. federal agencies may act directly. In Canada, either provincial or federal authorities may act in such circumstances.

States have the authority to enact legislation to protect groundwater from degradation due to overuse. Difficulties arise when states try to establish responsibility for changes in groundwater quality that develop due to problems other than direct contamination. In such cases, it may be more effective to take a regional approach to mitigating withdrawal-related quality problems. Examples include coastal states that establish recharge projects to protect large areas from seawater intrusion.

The Local Scale: Source Water (Including Wellhead and Aquifer) Protection

While state or regional activities may have defined regulatory authority and sometimes are effective in managing water use, local activities are more important in the practical management of groundwater quality problems. Sources of contamination are most typically located within meters to at most a few kilometers from the affected wells.

Recognizing this, the wellhead protection planning (WHPP) process established under the U.S. federal SDWA amendments of 1987 for public groundwater-using systems (and now part of the Source Water Protection Program under the 1996 amendments) is remarkably "federalist" in the original sense of the word, implying a high degree of local decision-making authority. A similar process is being implemented in Canada. In the United States, the existing WHPP and evolving SWAP processes permit broad state latitude in overall management. In Canada, provincial authority is presumed as in all resource management issues.

As a process, WHPP/SWAP depends on gathering local hydrogeological and potential pollution source risk information, and encourages local management and education initiatives. The entities involved in developing WHPP may range from state agencies to water conservancy districts to local water suppliers. The information necessary for the process is highly specific to the public water supply wellfields in question.

WHPP (as a subdivision of SWAP) applies to designated wellhead protection areas (WHPA). These are intended to be scientifically defined using hydrologic and geologic data gathering and analysis to establish valid groundwater time-of-travel zones or catchment areas around pumping wells. For many public water supply wellfields, the WHPA delineation process provides the most complete and accurate (and sometimes the only such) hydrogeologic information ever gathered for their management.

Both the WHPA delineation and the management planning processes of WHPP have the potential to go fundamentally wrong. Most WHPA delineation depends on hydrologic modeling of groundwater flow. Valid modeling depends on gathering data that represent the situation in the ground. Different groundwater models emphasize different parameters in their calculations. If data and calculations are incomplete or inaccurate, or the model invalid for the application, the shape of the WHPA may be wrong for management purposes. Both data and model validity have to be checked against the in-ground conditions by appropriate hydrogeologic analysis.

After the WHPA is delineated, risks to the wellfield within the zone must be assessed. Questions to be asked in the risk assessment process include: What constitutes risk? How are potential risks ranked? Are all significant real risks identified? What about historical impacts? Just what has transpired on that flat patch of grassy land near the river? What ever became of those oil wells or oil tanks? These kinds of questions add "industrial archaeologist" to the list of skills (including "engineer," "manager," and "hydrogeologist") that a water system may need for groundwater-source protection. Modern and active industrial activities are much less likely to have groundwater impacts than past activities because they work under current regulations. It becomes important to identify and assess whether further testing might be necessary at locations in question.

In the protection task, the water supplier actually takes action to protect the WHPA. It is easy to write and file a plan, but it is another matter to enact the plan. The WHPA often crosses jurisdictional boundaries. If a city files a WHPP for a WHPA that includes neighboring rural land, how does it induce the neighbors to take enforcement action to protect their wellfield? What authority is to be used in protecting this WHPA? Will enforcement or education be used?

Witten et al. (1995) identifies numerous options available. Among the most useful tools available for local governments are land-use regulations (depending on local-rule authority in individual states). Local entities may restrict land uses that could adversely affect groundwater quality. Industrial activities may be encouraged to relocate out of the WHPA, or permits required for certain activities such as handling hazardous materials. In practical terms, zoning restrictions must be balanced against excessive "takings" that may require compensation to the landowner. However, in some cases, paying compensation to remove risky operations may be a good investment.

An unfortunate unintended consequence of WHP programs in some jurisdictions has been the preclusion of groundwater use through denial of, or unreasonable restrictions on, permits for wells proximate to hypothetical, potential sources of contamination. Program administrators unable to control potential threats instead control groundwater use through existing permitting authority. The result is the "writing off" of certain groundwater resources that are not (and may never be) contaminated.

The Local Scale Beyond Wellhead Protection: Aquifer Protection

The WHPP process is specific to public water supplies. The U.S. SDWA gives the U.S. EPA no jurisdiction over private water supplies. Instead, private water supply protection is managed by state and local environmental health or water resources authorities. State and local jurisdictions typically regulate waste disposal and well placement and construction.

The evolving SWAP process is a broader approach than WHPP, and envisions a greater degree of state official participation in source water delineation, instead of the local lead in delineation under WHPP. In Ohio, for example, the state EPA will delineate source water zones, relieving the responsibilities of local officials, but certainly resulting in less detailed and less locally focused zones.

Beyond SWAP, federal point-source groundwater control legislation may play a role. Enforcement of landfill, injection well, or radioactive waste repository regulations may serve to protect the groundwater that supplies private wells.

U.S. federal law, for example, restricts certain chemicals, uses, storage, and disposal. However, there is little regulatory protection that is actually effective locally against nonpoint-source contamination. While specific regulatory authority may exist, identifying sources of hydrocarbons, nitrate, or ammonia in a well may be difficult, time-consuming, and expensive. Responding to something like an industrial agricultural operation risk to groundwater may be overwhelming to small populations of affected people.

To prevent pesticides and nitrates from reaching water supply wells vulnerable to nonpoint-source contamination, some process is necessary to identify and remove (1) sources of contamination and (2) preferential pathways for contamination to reach aquifers (such as abandoned wells). Examples of source-control strategies include switching from septic tank leachfield systems to more advanced systems or sewers in sensitive locations, or encouragement of on-site system maintenance, and best agricultural management practices to keep pesticides and fertilizer in the soil root zone, the intended point of application. Such an effort may require the formation of a county sewer district or active involvement of a county soil and water conservation district in regional groundwater planning.

Also on the list of "unenforced mandates" are abandoned wells. Almost all states have legislation that requires the proper sealing of abandoned wells, boreholes, and shafts of various kinds. However, the rules may be poorly or selectively enforced, for example on the "complaint system." Not all "wells" are covered by rules. For example, sealing of abandoned oil, gas, and water wells may be required, but not sealing of geotechnical boreholes. In oil and gas regions of the United States, millions of seismic shot holes are still unsealed. In the U.S. East, wells from as early as the 1870s remain effectively unplugged. Eger and Vargo (1989) describe one such problem and attempts to remediate it in Kentucky. Lost monitoring wells in hazardous sites (e.g., Welling and Foster, 1994) are also potential contamination routes to potable aquifers.

One example of an increasingly effective state program is that of Minnesota, where property sellers must provide information on any wells on a property, and are required to seal any that will not be maintained up to code as water wells. Mandatory sealing reports filed reached 12,000 per year in 1993 and have stayed high since (Herrick, 1997).

In the absence of effective governmental action, an effective groundwater protection mechanism may be a citizen-initiated aquifer protection effort. The usual incentive is an identifiable "enemy" such as a proposed landfill or large "chicken farm" industrial agriculture operation. Citizens usually have the option of challenging permits during public comment periods, or filing lawsuits to make a potential operation expensive to start up.

Problems with such "crisis-style" aquifer management include the excesses of expense and the hysteria associated with fear of water contamination and ignorance of hydrogeologic science. The developer may have resources and regulatory management experience that may hopelessly outclass the local opposition and stonewall efforts to control site activities. The proposed land use and environmental management plans may meet the letter of state permit requirements, so that there is no legal basis to deny operating permits. In response, local authorities sympathetic to their citizens may enact illegal spot zoning or discriminatory fire regulations. There are also times when the proposed land use will be so well controlled that, however distasteful it may seem, it will pose little real risk to groundwater quality.

A more effective approach is proactive groundwater protection planning before a crisis exists. This process may resemble wellhead protection planning, but has an overall aquifer protection goal. The British Columbia Ministry of Environment, Parks and Lands has published an eloquent and thorough treatise on groundwater management in Canada (BCMELP, 1996) that can serve as a model in any setting. Table 4.6 lists recommended features of an effective groundwater protection program.

One mechanism available in the United States is sole-source aquifer designation, established by the U.S. EPA under the SDWA. This process involves demonstrating to state authorities that a specific aquifer or aquifer zone is the sole practical water supply for a community. This community may consist of private individual well owners, public water supplies, or both. Such an application will involve hydrogeologic and water use mapping and shepherding the application through the process. The end result, if approved, is heightened levels of official protection and capacity to restrict land use.

Even in the case of well-planned aquifer protection entities, conflicting goals and political-legal conflicts pose challenges to effective management. The Edwards Aquifer Authority is a well-established aquifer management structure for cross-jurisdictional management of the carbonate Edwards aquifer in Texas. Conflict over water use priorities for the Edwards Aquifer have been very contentious (McCarl et al., 1993).

The most effective tool for both WHPP and aquifer protection is public education. For public education to be effective in protecting groundwater quality, people need to be made aware of both (1) the source of their water and (2) how their well

TABLE 4.6 Criteria for Effective Groundwater Protection Programs

Criterion	Scope
Goals and objectives	Protection programs should clearly define goals and objectives, reflect understanding of groundwater problems, have adequate legal authority, and have criteria for evaluating program success and the need for modifications.
Information	Programs should be based on information that permits resources and issues to be defined and preventive strategies to be evaluated.
Technical basis	Effective programs require a sound technical basis with which to link actions to results.
Source elimination and control	Long-term program goal should be to eliminate or reduce the sources of groundwater contamination.
Intergovernmental and interagency linkages	Comprehensive protection program must link actions at every level of government into coherent, coordinated action.
Effective implementation and adequate funding	Programs must have adequate legal authority, resources, and stable institutional structures to be effective.
Economic, social, political, and environmental impacts	A preventive program assumes that groundwater protection is the least costly strategy in the long run. Protective actions should be evaluated in terms of their economic, social, political, and environmental effects.
Public support and responsiveness	Programs must be responsive and credible to the public.

Source: U.S. Geological Survey, National Water Summary, 1986; modified from National Research Council, 1986. Reprinted in Van der Leeden, Troise, and Todd, 1990, p. 712.

use and maintenance and other activities affect the groundwater quality. For example, if septic tank and leachfield maintenance is a factor in reducing nitrate levels in an aquifer, people need to be convinced. Public awareness is also probably the most effective tool for effective abandoned-well sealing. If people become aware that their abandoned wells or their neighbors' represent a risk to their own well, more holes will be sealed.

One active and widely accepted process that encourages joint public and private efforts for groundwater protection at the local level is the Groundwater Foundation's Groundwater Guardian program (The Groundwater Foundation, Lincoln, Nebraska). In this process, a community sets goals for itself in groundwater protection, works to meet them, and reports to the Foundation annually. Public education is heavily emphasized. An oversight committee is needed to guide the process. The committee is expected to consist of diverse community interests. Reporting to the Foundation annually provides accountability and incentive to achieve goals.

Groundwater Quality Management: Technical Tools

In addition to public policy and regulation considerations, there is a range of technical tools available to assist in groundwater quality management. The best tool for protecting and managing groundwater is accurate and adequate information. To effectively protect a well, its owners and managers need to know what can be expected of it in volume and quality. From where does the water come? What is the area of influence of the wellfield? Is pumping going to draw in water from the landfill site or cause upconing of saline water? Are the well spacing and arrangement the best they can be to prevent overdrafting and maintenance problems? Answering such questions requires hydrogeologic analysis, combined with other planning data. Table 4.7 lists information needed for groundwater-management decisions.

The nature of the investigation tools and methods depends very much on the scale and criticality of the investigation. Is it a large-scale regional aquifer mapping study, a wellfield-scale contamination delineation, or a preventive time-of-travel (WHPA) study? What is already known? Can timesaving methods such as a geophysical investigation be used effectively? Will extensive drilling be needed to

TABLE 4.7 Major Components of an Information System Needed for Groundwater Management Decisions

- Hydrogeology
  - Soil and unsaturated zone characteristics
  - Aquifer characteristics
    - Depths involved
    - Flow patterns
    - Recharge characteristics
    - Transmissive and storage properties
    - Ambient water quality
    - Interaction with surface water
    - Boundary conditions
    - Mineralogy, including organic content
- Water extraction [withdrawals] and use patterns
  - Locations
  - Amounts
  - Purpose (domestic, industrial, agricultural)
  - Trends
- Potential contamination sources and characteristics
  - Point sources
    - Industrial and mining waste discharges
    - Commercial waste discharges
    - Hazardous material and waste storage
    - Domestic waste discharges
  - Nonpoint sources
    - Agricultural
    - Septic tanks
    - Land applications of waste
    - Urban runoff
    - Transportation spills (can also be considered a point source)
    - Pipelines (energy and waste) (can also be considered a point source)
- Population patterns
  - Demographic
  - Economic trends
  - Land-use patterns

Source: U.S. Geological Survey, National Water Summary, 1986; modified from National Research Council, 1986. Reprinted in Van der Leeden, Troise, Todd, p. 7.

define aquifer zones or dimensions? Will a review of file information suffice for the purpose at hand? An excellent summary of this decision-making process and the available tools is provided by Boulding (1995).

Analyzing the Hydrology and Water Quality of Aquifers and Wellfields

The value of any study depends on the acquisition of valid information. Sample collection procedures in groundwater exploration or monitoring studies must be chosen with the requirements of the analyses in mind. Sample acquisition planning must also take into consideration the three-dimensional nature of the resource and relative inaccessibility of points in the subsurface to be sampled, compared to surface water bodies.

Water samples may be collected as grab samples (as in those for regulatory monitoring) or flow-through samples (as for microscopic particulate analysis). In-situ electrodes may be used to examine the water quality of discrete well intervals. Core samples may also be analyzed to determine fluid characteristics of the aquifer matrix itself, which often differ greatly from that of pumped water from wells.

An ideal in sampling is to obtain a representative sample. A groundwater sample may be considered representative if it can be demonstrated that the sample came recently from the target aquifer and was handled to minimize change. Among solids collection methods (to assess the aquifer matrix itself), only rotary and drive cores can be considered fully representative. Other high-quality samples include certain dual-tube reverse circulation and rotasonic "cores" that may be considered "slightly altered" and quite useful for detailed analyses (Wright and Cunningham, 1994; ADITC, 1997). Drill cuttings may be used for initial exploration to identify gross contamination, for example by semivolatile hydrocarbons.

Analyzing samples depends on using equipment and procedures suitable for obtaining valid information from samples available. Gas monitoring instruments, such as photoionization detector (PID) tools, may be used to detect the presence of volatile organics from cores and cuttings, and are useful in compiling field profiles of contaminated zones and in selection of screening zones for monitoring points. Slurries can also be analyzed by photoionizing methods for the presence of organics. In addition to being used for scientific monitoring, core soil gas readings may be used to choose levels of personal protection.

Valid, representative fluid sampling also depends on the quality of the sampling point (Figure 4.15). Monitoring wells need to be selected or constructed to provide access to the formation at the three-dimensional coordinate where the sample needs to be taken. For example, a heavy solvent may sink to the bottom of an aquifer. Even when properly constructed, monitoring wells yield samples that represent the full screened length. Such samples may not be useful to pinpoint thin zones of contamination. Also, nonaqueous-phase organics may adsorb onto ("stick to") clays and not freely migrate to a well when it is pumped.

Where it can be used, the *direct-push sampling method* (Cordry, 1994; ADITC, 1997) is economical, highly portable, and causes less disturbance than drilling and monitoring well construction for exploration of potential contaminant plumes. Samples are taken at desired (often frequent) intervals to total desired depth, then the tool is withdrawn, and the process repeated at the next location. This is strictly a fluid-sampling program, but porosities can be estimated from inflow rates. The geometry and other characteristics of a contamination plume can usually be rapidly determined using direct-push or similar sampling methods, combined with field analysis of samples.

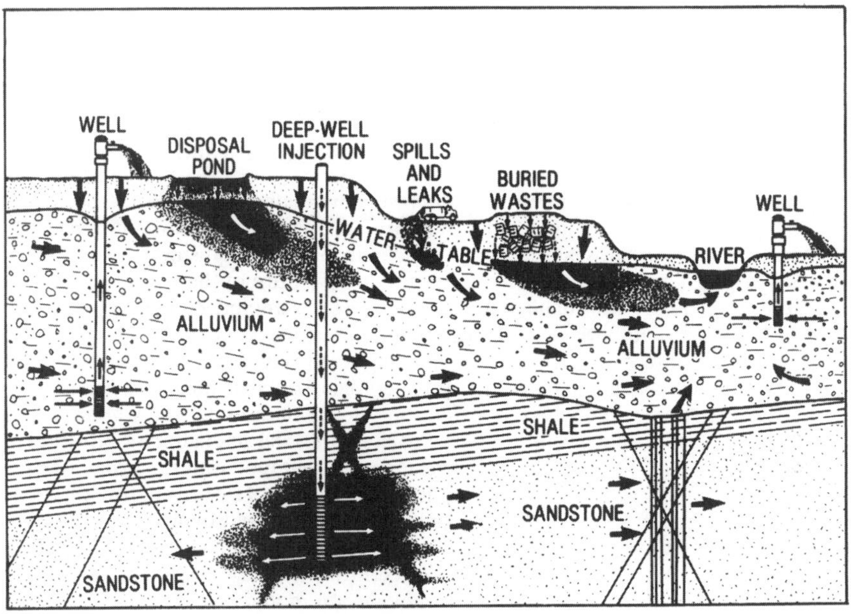

FIGURE 4.15 Waste disposal practices and contamination of groundwater (Movement of contaminants in unsaturated zone, alluvial aquifer, and bedrock shown by dark shading). (*Source:* Van der Leeden, Troise, and Todd, 1990. Originally from U.S. Geological Survey Water Fact Sheet. Toxic Waste, Ground-Water Contamination. 1983.)

Groundwater sampling from monitoring points is an area in which there is still considerable technical debate and methods improvements under way. Aside from direct-push methods, for example, techniques exist to isolate zones for pumping in monitoring wells, and to locate well intakes to provide more valid three-dimensional sampling opportunities.

While methods are still evolving (and readers should stay current on new developments), some requirements are likely to remain, despite improvements in technology. These include minimizing the volatilization of VOCs and using aseptic technique for bacteria. For details of environmental sampling methods for groundwater, refer to available general references such as Boulding (1995) and Nielsen (1991), but also consult with knowledgeable professionals.

Computer Modeling in Aquifer Flow Analyses and Wellhead Protection

Computer modeling software packages are mathematical tools used to simulate the physical aquifer condition and flow behavior in it. Calculations and algorithms in the software are used to determine flow conditions or solute transport. The validity of a particular model depends both on the validity of its internal assumptions and calculations, and on the quality of input data. Even very sophisticated modeling can produce erroneous results if the data used are invalid. All existing models require some simplification of the "real" condition, but poor data add to the distortion.

Computer modeling is an area of groundwater investigation where there is a high potential for method abuse and misuse. Modeling is used to assess groundwater flow directions, rates, and other crucial information such as time of travel. The use of modeling can take on a "Wizard of Oz" surreality, in which modeling wizards have to be trusted to apply the process correctly and the clients (groundwater users) do not understand how it is done.

Although not without limits (and occasional misuse in practice), models are very important in groundwater flow and solute transport analysis. They are the only tools available to provide backward and forward projections of currently identified trends in any practical human-scale time and reasonable cost. They also can be used to quickly visualize results if conditions (e.g., pumping patterns) change over time, and generally in groundwater management planning (e.g., Sun and Zheng, 1999). A very good conceptual summary of modeling techniques is provided by Boulding (1995).

Model types are proliferating and changing and capabilities expanding rapidly, taking advantage of advances in both hydrogeologic and computer science and technology. Any listing of models risks being rapidly out-of-date. The best approach to being conversant in the specific software package capabilities is to consult some of the many Internet, World Wide Web, and forum sites devoted to the topic, starting from a comprehensive site such as that maintained by the U.S. Geological Survey.

Generalizing, flow models in use may be classified in two broad groups: analytical and numerical types (with some hybrids). Analytical models (e.g., GPTRAC in U.S. EPA's WHPA package) are simpler, requiring less input time and computer power, and produce exact solutions, but they do not model often-crucial fine detail in complex aquifers.

Numerical models (e.g., MODFLOW, developed by the USGS) are much less simplifying than analytical models. Additionally, numerical models can provide a three-dimensional picture of the flow condition and can be used to simulate a system in fine detail. Numerical solutions are approximations, as opposed to exact solutions. However, numerical models can provide modeling of transient conditions, such as evolving pumping drawdown surfaces, whereas analytical models are limited to static, steady-state conditions.

Numerical modeling requires specific input to cells in a grid. This can be very time-consuming, although it is becoming less so with newer products, facilitated by user-friendly preprocessor packages. Memory and computing power are less of a problem with improving computer technology. For example, quite large and detailed multilayer numerical models may be constructed and run on conventional contemporary desktop computers, whereas mainframe systems were necessary only a few years ago.

Semianalytical models occupy an intermediate domain, using numerical methods to approximate complex analytical solutions. A major advantage of these model types is greater flexibility than analytical models provide, without significantly increasing the need for data.

Particle-path models (e.g., MODPATH) use the results of flow-model calculations to delineate time and direction of travel along a flow path. MODPATH, for example, is able to place discrete particles very flexibly to simulate travel in three dimensions. Such particle tracking is the basis for WHPA delineation modeling.

Solute transport modeling codes (e.g., SUTRA) are used to approximate (1) dispersion of a substance in groundwater, (2) retardation/degradation, and (3) chemical-reaction transport. Dispersion codes may be used if a conservative (poorly reactive) constituent such as nitrate or chloride is of first concern. Retardation/dispersion models can include a retardation or degradation factor. Chemical-reaction transport codes are the most complex (but not necessarily the most accurate) representations, incorporating geochemical transformation reactions such as redox changes. Solute transport modeling is needed if a constituent of concern is expected to be present and may travel faster or slower than the groundwater flow itself (a common condition).

Groundwater Quality Monitoring and Hydrogeologic Studies: Methods, Standards, and Issues

Decisions on groundwater testing usually require compromises based on geology, available methods, and the goals of the monitoring program. Managers of groundwater source supplies should be aware that methods and technology in groundwater testing are evolving rapidly. However, standard methods are available to help in guiding the process to make sure that good scientific practice is a primary element of any investigation. Scientists, engineers, and regulatory officials in North America, Australia, Europe, and elsewhere are in the process of adapting methods, standards, and regulations for the application of standards in groundwater sampling, hydrologic testing, and analysis. The American Society for Testing and Materials (ASTM) produces one extensive body of standards that can be used in project design and evaluation. Other sources include guidance manuals developed by U.S. states, and agencies and research arms of the federal governments of the United States, Canada, and Australia (among others). Example ASTM standards are listed in Table 4.8.

Other new standard guides are being formulated or drafted at the present time. In the groundwater science and engineering area, ASTM standards tend to be broad and flexible guides, rather than rigid "cookbook" standard practices. A common criticism of the consensus standards setting process, such as ASTM's, is that there is potential for abuse by activist parties who advocate their methods over alternative methods that have equivalent technical merit. Another is that publishing standards can pose stifling limits on the professional judgment necessary in groundwater science.

Because groundwater investigations are, of course, highly site-specific, generalized standards (even if highly flexible) will not cover every instance. As the technology and methods are still evolving, standards can become outdated in this type of environment. For these reasons, competent, professional hydrogeological guidance is always necessary for groundwater investigations.

Competent professionalism in hydrogeology is not as strictly defined as is the case for other professions such as engineering and law. Hydrogeologists may have

TABLE 4.8 Selected ASTM Standard Guides and Practices Relevant to Groundwater Studies

D5092	Practice for design and installation of groundwater monitoring wells in aquifers.
D5254	Practice for minimum set of data elements to identify a groundwater site.
D5299	Guide for decommissioning groundwater wells, vadose zone monitoring devices, boreholes and other devices for environmental activities.
D5521	Guide for development of groundwater monitoring wells in granular aquifers.
D5716	Test method to measure rate of well discharge by circular orifice weir.
D5737	Guide to methods for measuring well discharge.
D5786	Practice for (field procedure) for constant drawdown tests in flowing wells for determining hydraulic properties of aquifer systems.
D5787	Practice for monitoring well protection.

eclectic educational backgrounds, and not all jurisdictions license geologists or hydrogeologists. Education and practical experience are the most reliable criteria. Certification, such as that managed by the National Ground Water Association, American Institute of Professional Geologists, and the International Association of Hydrologists may be used as guides as well.

Water Quality Management to Optimize Natural Groundwater Quality

Wellfield planning can be used to optimize *natural groundwater quality,* that is, the quality of groundwater unaffected by human contamination. With adequate information on sources of poorer-quality water, well and wellfield operators can manage their pumping to keep undesirable water out of the system. Once the flow regime, wellfield behavior, risks, and other concerns have been analyzed, a water-quality optimization policy can be put into place.

As with surface water, land use and watershed management are critical to preventing nonpoint-source contamination. Risk assessment would follow the sole-source aquifer or WHPP model delineation. Control is a problem, but perhaps not an insurmountable one.

One example is a wellfield in a sand aquifer surrounded by brackish estuaries. Modeling of the system shows that overpumping of this wellfield will result in bringing in more saline, more highly colored water with a higher dissolved organic carbon content. Optimizing pumping would produce the highest yield from the aquifer that does not move the interface between better and poorer quality water. Making this determination may be accomplished by modeling the system, calibrating the model using valid local groundwater data. Jalink and Kap (1994) is an example of this kind of management process.

Another example is a wellfield in limestone with sulfide groundwater at deeper intervals. Excessive pumping draws in sulfide-containing groundwater, resulting in odor problems and black encrustations when it mixes with iron-containing water. Regular sampling of the wells may be used to detect sulfides. Pumping can then be calibrated to optimize output.

Aquifer storage and recovery (ASR) is the process of storing excess surface or treated water in aquifers in times of abundance, for withdrawal during times of scarcity, or to reduce impacts on limited aquifer resources. ASR can also be used as a groundwater quality management tool. Castro (1994) described a feasibility study conducted in the Atlantic Coastal Plain aquifer series near Myrtle Beach, South Carolina. Treated water was injected at 100 gpm into the aquifer zone, forming a fresh water zone around the screened portion of the injection wells, and a transition zone where injected water mixed with more highly mineralized formation water. The conclusion of the pilot study was that the mixed water would improve over time.

Remediation and Control of Contaminated Groundwater

Despite the availability of preventive measures, groundwater becomes contaminated. Groundwater users in some places have to contend with a legacy of ignorant or erroneous waste disposal and material handling practices of the past. Examples abound across the United States, for example, of contamination resulting from activities that occurred even early in the nineteenth century. A prominent example is the shallow, unconsolidated Kirkwood-Cohansey aquifer in New Jersey, with serious and varied contamination resulting from past industrial and agricultural activity, including pesticides and radium. Another is Woburn, Massachusetts, center of the famous legal proceedings concerning the source of contamination of two Woburn wells and responsibility for elevated numbers of leukemia and other cases. Woburn was an industrial center with land and water contamination (e.g., polyaromatic hydrocarbons, arsenic, and chromium) beyond the chlorinated organics attributed to the defendants in *Anderson v. W. R. Grace Co.* and related legal proceedings (Aurilio et al., 1995).

Since 1980 and the implementation of RCRA, practices have been changing in the United States. They are also changing in countries with similar environmental imperatives, including the United Kingdom, Canada, and Australia. Worldwide, however, major urban and rural groundwater users have ongoing, unconstrained groundwater contamination, often due to official neglect or economic poverty. Russia has horrendous problems resulting from oil pipeline leaks in Siberia that cannot be repaired. There is extensive radionuclide contamination of bedrock aquifers in the Ukraine. Water wells being used by poor neighborhoods dot the contaminated urban landscapes of Mexico City, Mexico; Buenos Aires, Argentina; and Lima, Peru.

In North America, groundwater contamination has proven to be a much less acute problem than it could have been, but more widespread and chronic in nature than originally thought. Low levels (usually below MCLs) of common chemical contaminants such as tetrachloroethylene (PCE), and fuel hydrocarbons and their additives (such as soluble, mobile, and recalcitrant methyl t-butyl ether, MtBE) are relatively common in more vulnerable aquifers.

Chemically contaminated groundwater has proven to be difficult to remediate. Most contaminants adhere to soil materials and do not readily flush into solution for removal. Halogenated dense nonaqueous phase liquids (DNAPL), which include very common degreasers, exhibit behavior in aquifers that is remarkably different from the water solute, and are notoriously difficult to "corral" for cleanup. Pump-and-treat remediation systems almost universally experience extensive maintenance problems due to the effects of microorganisms in iron- and carbon-rich contaminated groundwater (Smith, 1995).

Boulding (1995) presents a comprehensive review of the decision-making process in groundwater remediation planning and execution. A critical aspect of the solution to groundwater quality problems is defining an answer to "what is 'clean' "? The general approach in the United States is to require remediation to protect groundwater resources for drinking water use. CERCLA standards have been SDWA MCLs in the U.S. This is now undergoing review in the United States, and doctrine is being refined elsewhere, especially in Europe, which has widespread and serious groundwater contamination (Danahy, 1996).

In Europe, groundwater standards that require intervention and remediation are generally less stringent than drinking water standards (typically derived from WHO standards). For low to moderate groundwater contamination situations, the European approach may be more pragmatic than that of the United States (Danahy, 1996). In these situations, the European regulations do not require aquifer restoration to levels as stringent as drinking water standards. Instead, European regulations show a preference for providing the necessary water treatment prior to distribution of drinking water, resembling traditional U.S. surface water source regulation.

For groundwater cleanup standards, two levels exist in European practice (Danahy, 1996). "Target" or "reference" values are the expected natural or background concentrations of constituents. "Intervention values" are typically 10 to 1,000 times greater than target values, and represent levels which require remediation. At levels in between these limits, a study is required to find a technical solution.

Numerical values for various constituents vary from nation to nation in the European Union (EU), but the Netherlands 36-constituent list is a de facto standard list (where another official list does not exist). As in the United States (and as expected among independent nations), there is some inconsistency among EU jurisdictions. For example, in the United Kingdom and Germany, there is a groundwater source standard, but not a drinking standard, for benzene. Other European states also have a benzene drinking water standard.

North American practice seems to be evolving toward a policy of accepting that containment and cleanup to something like "intervention values" (rather than cleanup to SDWA MCL standards) is probably realistic for many situations. This philosophy is the basis for "brownfields" programs in a number of states, such as Pennsylvania and Ohio, which permit the use of less stringent alternative standards for the reclamation and reuse of existing properties with industrial histories if the risk to public health and safety is small. An example of a candidate property might be a closed refinery facility in an industrial area where local groundwater is not used. Brownfields programs were developed due to the recognition of one unintended consequence of older land cleanup standards: the abandonment of existing industrial properties and new construction on previously nonindustrial, often agricultural, "greenfield" properties.

Containment of contamination may take a variety of forms, including erecting physical barriers such as piling and sheet walls or establishing hydraulic barriers using pumping wells. An evolving procedure is the *biowall* concept in which a biologically active permeable treatment barrier is established in the groundwater flow path of a contaminant plume. A related procedure, *iron redox barriers,* establishes redox gradients and sorption capacity that immobilizes hydrocarbon plumes. Most projects include the concept of multiple barriers: for example, physical and hydraulic containment, in-situ saturated zone remediation, vadose (unsaturated) zone venting, and follow-up treatment. In some cases, natural attenuation (removal or position alteration of contaminants without active intervention) can be employed.

Making the difficult choice to settle for containment or partial remediation as opposed to full remediation to background values may depend on the human or environmental "value" of the contaminated aquifer. Contamination of a sole-source aquifer for a village water system would usually prompt the most complete cleanup possible. Likewise, contamination that would adversely affect a sensitive environmental area would also rate a very high level of cleanup. On the other hand, for a marginal aquifer under a refinery property in a city using surface water, the cost-benefit analysis would suggest containment.

The Case for Preventive Management Versus Reactive Remediation

In any case, by any standard, prevention, not reaction to contamination in the form of groundwater remediation, is the best approach to the potential for groundwater contamination. This is consistent with the ancient concept of stewardship, in which people are entrusted with something of value and expected to manage or improve it before passing it on. On strictly an accounting level, it is much easier and less costly to maintain an asset (vehicle, forest, river, or aquifer) than to repair or replace it after damage has taken place. With groundwater, the environmental economics of prevention are magnified due to the high cost and high uncertainty of remediation, relative to that for surface waters.

While management and protection approaches for groundwater are highly site-specific, any effective management process has several necessary features:

1. In the process of developing a groundwater source and treatment system, start with the best-quality water possible. The *quality* criterion should outweigh others such as *initial development cost* or *convenience.*
2. There should be a process to *keep* the groundwater source quality as high as possible, avoiding natural degradation and contamination from human sources. This is almost always the least-cost option (by orders of magnitude) compared to either remediation or developing alternative water supplies in the case of contamination.

Science, public involvement, and political will and cooperation are all necessary features of groundwater resource management. Developing high-quality groundwater sources and maintaining them in optimal condition requires knowledge of the resource, potential risks posed to its continued quality, and management options. Managers and regulators involved with groundwater protection have to display the will to provide the resources and to make the decisions necessary to manage the resource. The water-using public has to be involved in order for management of large areas with many human activities to be feasible. If these factors are all present, groundwater quality can be optimized.

BIBLIOGRAPHY

Abbaszadegan, M., P. W. Stewart, M. W. LeChevallier, and C. P. Gerba. *Application of PCR Technologies for Virus Detection in Groundwater.* Denver, Colorado: AWWA Research Foundation, 1998.

Amy, P. S., and D. L. Haldeman, eds. *The Microbiology of the Terrestrial Deep Subsurface.* Boca Raton, Florida: Lewis Publishers, 1997.

American Society for Testing and Materials, *Annual Book of Standards,* vol. 04.09, Soil and Rock (II): D 4943-latest; Geosynthetics. West Conshohocken, Pennsylvania, American Society for Testing and Materials, 1999.

Aurilio, A. C., J. L. Durant, H. F. Hemond, and M. L. Knox. "Sources and Distribution of Arsenic in the Aberjona Watershed, Eastern Massachusetts." *Water, Air, and Soil Pollution* 81, 1995: 265–282.

Australian Drilling Industry Training Committee. *Drilling: The Manual of Methods, Applications and Management.* Boca Raton, Florida: CRC Lewis Publishers, 1997.

Bales, R. C., S. Li, K. M. Maguire, M. T. Yahya, C. P. Gerba, and R. W. Harvey. "Virus and bacteria transport in a sandy aquifer, Cape Cod, Massachusetts." *Ground Water,* 33(4), 1995: 653–661.

Banks, R. S. "The imminent destruction of water resources of Lee County, Florida, and current management of the problem." *Proceedings of the NWWA Conference on Ground Water Management* (October 29–31, 1984, Orlando, Florida), pp. 432–446. National Water Well Association, Dublin, Ohio.

Bhattacharya, P., D. Chatterjee, and G. Jacks. "Occurrence of arsenic contaminated groundwater in alluvial aquifers from Delta Plains, Eastern India: Options for safe drinking water supply." *Int. Jour. Water Resources Management,* 13(1), 1997: 79–92.

BCMELP. A Groundwater Management Strategy for Canada, URL http://www.env.gov.bc.ca/wat/gws/gwis.html, British Columbia Ministry of Environment, Lands and Parks, Groundwater Section, Vancouver, BC, Canada. Note that URLs are somewhat mobile.

British Geological Survey. "Current research: Nitrates in groundwater, URL." http://www.nkw.ac.uk/bgs/, British Geological Survey, London, 1997.

Boulding, J. R. *Soil, Vadose Zone, and Ground-Water Contamination.* Boca Raton, Florida: CRC Press/Lewis Publishers, 1995.

Castro, J. E. *Proc. Second International Conference on Ground Water Ecology.* American Water Resources Assn., Herndon, Virginia, pp. 105–114.

CDC. *A Survey of the Quality of Drinking Water Drawn from Domestic Wells in Nine Midwest States.* Centers for Disease Control and Prevention National Center for Environmental Health, Atlanta, Georgia, 1990.

Chapelle, F. H. *Ground-Water Microbiology and Geochemistry.* New York: John Wiley & Sons, 1993.

Cohen, S. Z. "Ground Water Monitoring and Remediation." *Agricultural chemical news,* 16(3), 1996: 66–67.

Cohen, S. Z., S. Nickerson, R. Maxey, A. Duply, Jr., and J. A. Senita. "A ground water monitoring study for pesticides and nitrates associated with golf courses on Cape Cod." *Ground Water Monitoring Review,* 10(1), 1990: 160–173.

Cordry, K. "Practical guidelines for direct push sampling." Workshop Notebook, Eighth Annual Outdoor Action Conference, National Ground Water Association, Columbus, Ohio, 1994: 185–196.

Craun, G. F., and R. Calderon. "Microbial risks in groundwater systems epidemiology of waterborne outbreaks." *Under the Microscope: Examining Microbes in Groundwater,* pp. 9–20. Denver, Colorado: AWWA Research Foundation and AWWA, 1997.

Cullimore, D. R. *Practical Groundwater Microbiology.* Boca Raton, Florida: Lewis Publishers, 1993.

Cummings, T. R. *Natural Ground-Water Quality in Michigan, 1974–1989.* Open-File Report 89-259. Lansing, Michigan: U.S. Geological Survey and Michigan Department of Natural Resources, Geological Survey Division, 1989.

Danahy, T. "Sizing up global water quality standards." *International Ground Water Technology,* 2(4) 1996: 13–15.

Dupuy, A., R. Moumtaz, and O. Banton. "Contamination nitratée des eaux souterraines d'un bassin versant agricole hétérogene 2. Évolution des concentrations dans la nappe." (Groundwater pollution by nitrates in a heterogeneous agricultural watershed 2. Evolution of concentrations in the aquifer). *Rev. Sci. Eau,* 10(2) 1997: 185–198.

Eger, C. K., and J. S. Vargo. "Prevention: ground water contamination at the Martha oil field, Lawrence and Johnson Counties, Kentucky." *Proceedings of the Conference on Petroleum Hydrocarbons and Organic Chemicals in Ground Water: Prevention, Detection, and Restoration* (November 9–11, 1988, Houston, Texas), Vol. 1, pp. P3–P34. Dublin, Ohio: National Water Well Association, 1989.

EPA. *National Survey of Pesticides in Drinking Water Wells, Phase I Report.* EPA 570/9-90-015, Office of Water and Office of Pesticides and Toxic Substances, U.S. Environmental Protection Agency, Washington, DC., 1990.

Exner, M. E., C. W. Lindau, and R. F. Spaulding. "Ground water contamination and well construction in southeast Nebraska." *Ground Water* 23(1) 1985: 26–34.

Fetter, C. W. Applied Hydrogeology, Charles E. Merrill, Columbus, Ohio.

Free, D., R. Carruthers, and D. Stanfield. "Community Involvement in Developing a Management Strategy for an Overdeveloped Groundwater System." *Proceedings Murray Darling 1995 Workshop, Extended Abstracts,* 11–13 September 1995, Murray Darling Basin Commission.

Freeze, R. A., and J. A. Cherry. Groundwater. New Jersey: Prentice-Hall, Englewood Cliffs, 1979.

Gäss, T. E., J. H. Lehr, and H. W. Heiss, Jr. *Impact of Abandoned Wells on Ground Water,* EPA-600/3-77-095, U.S. Environmental Protection Agency. Ada, Oklahoma: R. S. Kerr Laboratories, 1977.

Gerba, C. P., and G. Bitton. "Microbial pollutants: Their survival and transport pattern to groundwater." In *Groundwater Pollution Microbiology,* G. Bitton and C. P. Gerba, eds., (New York: Wiley-Interscience, 1984) pp. 65–88.

Ground Water Geology Section. *Contamination of underground water in the Bellevue area.* Columbus, Ohio: Ohio Division of Water, 1961.

Harth G. "South Region's Role in the Great Artesian Basin Bore Rehabilitation program in Queensland." QDNR Groundwater Seminar, Toowoomba, Q., Australia, 1993.

Hatva, T., H. Seppänen, A. Vuorinen, and L. Carlson. "Removal of iron and manganese from groundwater by reinfiltration and slow sand filtration." *Aqua Fennica,* 15(2) 1985: 211–225.

Hem, J. D. *Study and Interpretation of the Chemical Characteristics of Natural Water,* Water Supply Paper 2254, 3rd ed. Reston, Virginia: U.S. Geological Survey, 1985.

Herrick, D. "Well abandonment—Minnesota style." *Water Well Journal* 51(6), 1997: 52–54.

Jalink, M. H., and A. Kap. *Proc. Second International Conference on Ground Water Ecology,* 115–124. American Water Resources Assn., Herndon, Virginia, 1994.

Jones, R. E., R. E. Beeman, and J. M. Suflita. "Anaerobic metabolic processes in the deep terrestrial subsurface." *Geomicrobiology Journal* 7(1/2) 1989: 117–130.

LeChevallier, M. W. "What do studies of public water system groundwater sources tell us? In *Under the Microscope: Examining Microbes in Groundwater.* (Denver, Colorado: AWWA Research Foundation and AWWA, 1997) 65–68.

Lehr, J. H., T. E. Gäss, and W. A. Pettyjohn. *Domestic Water Treatment.* New York: McGraw-Hill, 1980.

Lloyd, O. B. Jr., and W. L. Lyke. *Ground Water Atlas of the United States: Segment 10, Illinois, Indiana, Kentucky, Ohio and Tennessee,* Hydrologic Investigations Atlas 730-K. Reston, Virginia: U.S. Geological Survey, 1995.

Macler, B. A. "Developing the Ground Water Disinfection Rule." *Journal AWWA,* 88(3) 1996: 47–55.

Macler, B. A., and F. W. Pontius, "Update on the Ground Water Disinfection Rule," *Journal AWWA,* 89(1) 1997: 16, 18, 20, 115.

McCarl, B., W. Jordan, R. L. Williams, L. Jones, and C. Dillon. *Economic and Hydrologic Implications of Proposed Edwards Aquifer Management Plans* (TR 158), TWRI. College Station, Texas: Texas A&M University, 1993.

Mushtaq, A., A. Brandstetter, W. Walter, W. Blum, and W. E. H. Blum. "The arsenic calamity in Bangladesh." *Proc. 4th International conference on the Biogeochemistry of Trace Elements,* Berkeley, California, 1997: 263–264.

NGWA. *Manual of Water Well Construction Practices.* Westerville, Ohio: National Ground Water Association, 1998.

Nielsen, D. M. *Practical Handbook of Ground-Water Monitoring.* Boca Raton, Florida: CRC Press/Lewis Publishers, 1991.

Nolan, B. T. and B. C. Ruddy. *Nitrate in the Ground Waters of the United States—Assessing the Risk.* Fact Sheet FS-092-96, U.S. Geological Survey, Reston, Virginia, 1996.

Schijven, J. F., and L. C. Rietveld. "How do field observations compare with models of microbial removal." *Under the Microscope: Examining Microbes in Groundwater.* Denver, Colorado: AWWA Research Foundation and AWWA, 1997: 105–114.

Smith, S. A. *Well and Borehole Sealing: Importance, Materials, Methods and Recommendations for Decommissioning.* Dublin, Ohio: Ground Water Publishing Co., 1994.

———. Monitoring and Remediation Wells: Problem Prevention, Maintenance and Rehabilitation. Boca Raton, Florida: CRC/Lewis Publishers, 1995.

———. "Well construction, maintenance and abandonment: How they help in preventing contamination." *Under the Microscope: Examining Microbes in Groundwater.* Denver, Colorado: AWWA Research Foundation and AWWA, 1997: 115–126.

Smith, S. A., and O. H. Tuovinen. "Environmental analysis of iron-precipitating bacteria in ground water and wells." *Ground Water Monitoring Review* 5(4) 1985: 45–52.

Sophocleous, M. "Managing water resources systems: Why 'safe yield' is not sustainable." *Ground Water* 35(4) 1997: 561.

Sun, M., and C. Zheng. "Long-term groundwater management by a MODFLOW based dynamic optimization tool." *Jour. ARWA* 35(1) 1999: 99–111.

Swancar, A. *Water quality, pesticide occurrence, and effects of irrigation with reclaimed water at golf courses in Florida.* Water Resources Investigation Report 95-4250. Tallahassee, FL: U.S. Geological Survey, 1996.

USGS. *Summary of Willamette Basin Ground-Water Study,* URL http://wwworegon.we.usgs .gov/projs_dir/willgw/summary.html. Reston, Virginia: U.S. Geological Survey, 1996.

USGS. *Pesticides in Ground Water,* Fact Sheet FS-244-95, U.S. Geological Survey, Reston, Virginia, U.S. Geological Survey, 1997.

USGS. *Radium-226 and Radium-228 in Shallow Ground Water, Southern New Jersey,* Fact Sheet FS-062-98, West Trenton, New Jersey: U.S. Geological Survey, 1998a.

USGS. *Pesticides in Surface and Ground Water of the United States: Preliminary Results of the National Water Quality Assessment Program(NAWQA)*, July 1998, URL http://water.wr.usgs.gov/pnsp/allsum/ (provisional data—subject to revision). Reston, Virginia: U.S. Geological Survey, 1998b.

Van der Leeden, F., F. L. Troise, and D. K. Todd. *The Water Encyclopedia.* Boca Raton, Florida: Lewis Publishers, 1990.

Vuorinen, A., L. Carlson, H. Seppänen, and T. Hatva. "Chemical, mineralogical and microbiological factors affecting the precipitation of Fe and Mn from groundwater." *Water Science and Technology,* 20(3), 1998: 249.

Welling, W. B., and G. D. Foster. "Monitoring well inventory at the Love Canal inactive hazardous waste site and implications for monitoring well maintenance." *Proc. The Eighth National Outdoor Action Conference,* National Ground Water Assn., Dublin, Ohio, 1994: 701–713.

Winter, T. C., J. W. Harvey, O. L. Franke, and W. M. Alley. *Ground Water and Surface Water, A Single Resource,* Circular 1139. Denver, Colorado: U.S. Geological Survey, 1998.

Witten, J., S. Horsley, S. Jeer, and E. K. Flanagan. *A Guide to Wellhead Protection.* Chicago, Illinois: American Planning Association, 1995.

Wright, P. R., and W. L. Cunningham. "Use of rotasonic drilling for field investigation of ground water at Wright-Patterson Air Force Base, Ohio." *Proc. Eighth National Outdoor Action Conference,* National Ground Water Assn., Columbus, Ohio, 1994: 611–619.

Yates, M. V. "Fate and transport of microbial contaminants in the subsurface." *Under the Microscope: Examining Microbes in Groundwater,* AWWA Research Foundation and AWWA, Denver, Colorado, 1997: 161–170.

Other Recommended References

AWWARF. *Research Projects Relevant to Groundwater Utilities.* Denver, Colorado: AWWA Research Foundation, 1997.

Driscoll, D. *Groundwater and Wells.* St. Paul, Minnesota: Johnson Division, 1986.

Roscoe Moss Co. *Handbook of Ground Water Development.* New York: Wiley-Interscience, 1990.

AWWA. *Groundwater,* Manual M21, Denver, Colorado: American Water Works Association, 1989.

USGS. Ground Water Atlas of the United States, 1999. http://wwwcapp.er.usgs.gov/publicdocs/gwa/.

SURFACE WATER

John A. Hroncich
Senior Project Engineer
United Water Management and Services Company
Harrington Park, New Jersey

Source water management requires an understanding of the natural and human factors that impact water quality and the means to control, reduce, or eliminate those impacts where possible. Important quality concerns for suppliers using surface water sources are turbidity, eutrophication, and contamination from microbes, pesticides, and trihalomethane precursors (Chauret et al., 1995). Threats to water quality include urban development and agricultural activities.

The Hydrologic Cycle

The classic hydrologic cycle (illustrated in Figure 4.16) shows the relationship between surface and groundwater and the constant movement of water in the environment. Moisture on the earth's surface evaporates to form clouds, which deposit precipitation on land in the form of rain, snow, or hail. That moisture is absorbed by the soil and percolates into the ground. When the precipitation rate exceeds infiltration, overland flow occurs and streams, rivers and reservoirs receive the runoff. Water from surface sources will reach its final destination in the ocean, where the hydrologic cycle begins again with the evaporation process. The hydrologic cycle is affected by land features and ocean currents that determine changing weather patterns and deliver precipitation unevenly, with some areas getting ample rainfall and others getting little.

Sources of water contamination occur throughout the hydrologic cycle. Contaminants can be diluted, concentrated, or transported through the cycle and affect drinking water. The objective of source water quality management is to minimize (or eliminate, if possible) contaminant input within a watershed basin, the geographic area that drains to the source water intake.

Water sources can be affected by both *chronic* and *acute* water quality impacts. Chronic impacts may be subtle and persist over an extended period of time, resulting in a gradual deterioration of the source water. An example of a chronic impact is increased human activity within a watershed, producing increased nutrient levels in a lake or reservoir and accelerated eutrophication. An acute impact occurs as an incident, like an oil spill, which can be remedied quickly. The water quality manager needs to appreciate all source-water impacts in order to respond at the plant or on the watershed to minimize the impact on drinking water quality. The protection of water quality may require taking a treatment plant out of service for a time or installing additional treatments to meet the changing quality in the source water.

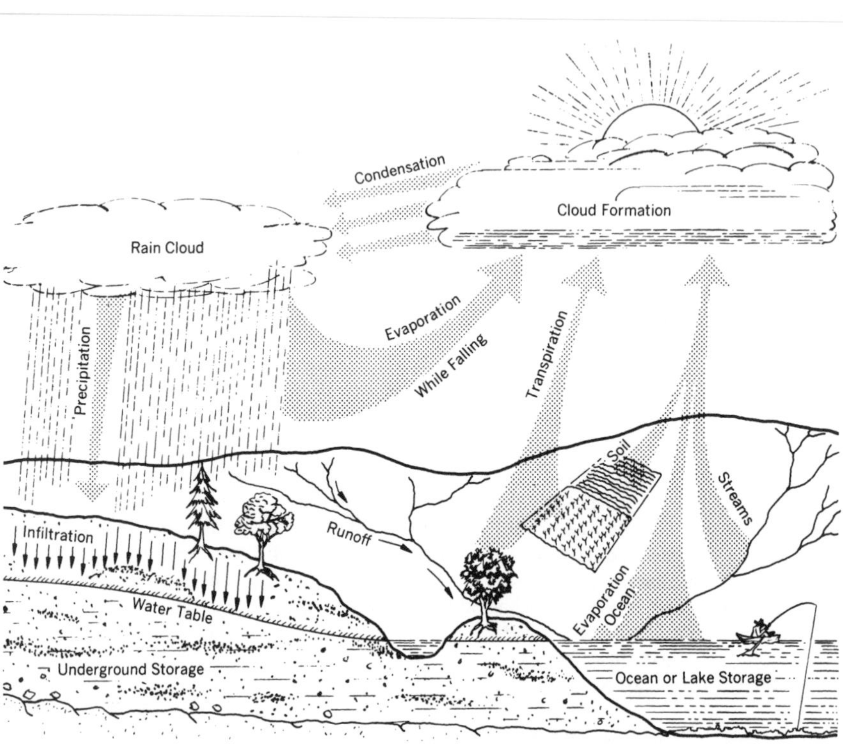

FIGURE 4.16 Hydrologic cycle.

The geology within a watershed basin influences source quality, as do seasonal flow variations and climate. Microbiological activity in lakes and reservoirs can affect water quality. A variety of human activities can introduce pathogens and increase nutrient levels contributing to eutrophication of lakes and reservoirs.

Factors Affecting Overall Source Water Quality

Factors that affect source water quality can be categorized as *natural* or *human* in origin, and as *point* and *nonpoint* impacts. Point sources emanate from a pipe or other definable point of discharge or release. Nonpoint sources are more diffuse over all or part of a watershed basin.

Typical natural factors that affect water quality include climate, watershed topography and geology, nutrients, fire, saltwater intrusion, and lake or reservoir density stratification. Typical human factors in the point-source category include wastewater discharges and hazardous waste facilities, spills, and releases. In the nonpoint source category, human factors include agricultural and urban runoff, land development, atmospheric deposition, and recreational activities. Understanding these factors in relation to their impact on water quality will allow for source water quality management decisions that will minimize water quality deterioration.

Water Quality Concerns Water quality parameters of concern include solids and turbidity, nutrients (including nitrogen and phosphorus) organic carbon, coliform bacteria, metals, oil, grease, sodium, toxic chemicals, algae, dissolved oxygen, and substances that degrade the aesthetic quality of water. Many of these are affected by multiple environmental factors involving both natural and human inputs.

High concentrations of solids may require high coagulant chemical dosages, which may in turn increase the amount of treatment residuals. Stricter filter performance requirements cause systems to achieve ever-lower finished water turbidity levels. Sources of solids include domestic sewage, urban and agricultural runoff, stream bank erosion, and construction activity.

Nutrients can increase microbial activity, which may cause increased taste and odor problems in the finished water. Nutrient sources include septic system and landfill leachate, wastewater plant discharges, lawn and road runoff, animal feedlots, agricultural runoff, and rainfall.

Total organic carbon (TOC) is a measure of the dissolved and particulate material related to the formation of disinfection by-products. Certain naturally occurring organic substances (particularly humic and fulvic acids) react with chlorine to form these by-products, some of which are regulated under the SDWA. *Natural organic matter* (NOM) consists of naturally occurring organic material derived from decaying organic matter and dead organisms. Other portions of TOC are derived from domestic and industrial activities that include wastewater discharge, agricultural and urban runoff, and leachate discharge.

Microbial contaminants are associated with health risks from pathogenic organisms. Coliform bacteria have been used for decades as surrogates for organisms that affect human health. Fecal coliform bacteria are indicators of warm-blooded animal fecal contamination. Sources include domestic sewage from wastewater discharges, leaking sewers, septic systems, urban runoff, animal farms and grazing areas, waterfowl droppings, and land application of animal wastes. Work done in Canada confirmed previous studies indicating that the protozoan parasites *Giardia lamblia* and *Cryptosporidium* were present in the raw surface water supplies in high percentages in the Ottawa River, downstream of population centers and agricultural activities. The study concluded that these protozoa were ubiquitous in surface waters—even in rivers considered to be pristine—and that human activities were a major source of these parasites (Chauret et al., 1995).

Metals may be a concern for both public health and aquatic life. Sources include industrial activities, wastewater, non-point-source runoff, and sometimes natural geologic sources. The SDWA regulates the metals antimony, arsenic, barium, beryllium, cadmium, chromium, copper, lead, mercury, selenium, and thallium. Renewed health concerns for arsenic is pushing the regulated standard lower and will affect water systems that have naturally occurring arsenic in their watershed. California water systems may be impacted, depending on the arsenic level ultimately regulated (Fruth et al., 1996). Mercury found in surface sources is typically from human activities and can range from being toxic to bioaccumulating in fish tissue. Mercury-contaminated sediment and soil will erode during storm events and carry the metal downstream.

Oil and grease interfere with biological treatment processes, causing maintenance problems and aesthetic concerns. Sources include runoff from kerosene, spilled lubricating and road oils, gas stations, industries, and food waste and cooking oil. Oil and grease concentrations are highly dependent on land use and have been found to range from 4.1 m/L in residential areas up to 15.3 mg/L in parking lots (Stenstrom et al., 1984). Deicing salts contribute to a short-term increase in sodium levels in surface water sources. Although sodium levels in drinking water may create health concerns for persons with high blood pressure and heart disease, drinking water is usually a very small contributor to the total sodium intake. Sodium is generally not considered a serious health problem or issue for the water industry and is a secondary SDWA standard. Studies show that deicing chemicals may pose a water quality problem from pollutants other than sodium and chloride that may include biological oxygen demand (BOD), chromium, copper, lead, nickel, and zinc in snow and snowmelt (Richards and Legrecque, 1973). Another study reported elevated levels of solids, phosphorus, lead, and zinc, attributed to the application of antiskid sand and deicing salts on Minnesota roads (Oberts, 1986).

Synthetic organic chemicals can affect both aquatic life and human health with an impact that can be acute or chronic. From a regulatory perspective, potential human carcinogenic affects from herbicides, pesticides, and polychlorinated biphenyls (PCBs) are the most common issues. Sources include agriculture, lawn care products, industrial sites, roads and parking lots, and wastewater.

Aesthetic issues include taste and odor, color, turbidity, and staining. Those parameters include:

- Taste and odor—industrial chemicals, algal metabolites, natural organic matter, urea, and other things that may react with chlorine in the treatment process
- Color—metals, natural organic matter, algae
- Turbidity—solids and algae
- Staining—metals

Algae may cause taste and odor in drinking water, as well as problems with the water treatment process. Metabolites like geosmin and methyl-isoborneol (MIB) can cause tastes and odor at low part-per-trillion levels. Other species may produce endotoxins and exotoxins that may be toxic to aquatic life, wildlife, and humans.

Dissolved oxygen deficiencies enable the release of iron and manganese into solution, causing possible water treatment problems. Taste and odor are also a concern in the absence of dissolved oxygen because of the potential production of hydrogen sulfide and other sulfur compounds.

Water suppliers need to recognize that a significant change in any of these parameters from watershed activities can result in a range of problems, from aesthetic to human health concerns. A summary of the water quality parameters, their possible source, and the potential effect on water supply can be found in Table 4.9.

Impacts on Surface Water Quality from Natural Factors

Climate Extreme wet and dry conditions affect water quality. Periods of heavy precipitation can resuspend bottom sediments and increase turbidity, microbial loading, color, metals, and other contaminants. Heavy precipitation can introduce an accumulation of naturally occurring organic compounds that form disinfection byproducts during the disinfection process. Dry climates or drought periods can result in stagnation in reservoirs and lakes and contribute to algae growth. Dry conditions can increase the impact of point-source discharges by reducing the effect of dilution and natural attenuation by the source water. Temperature can play an important role in affecting biological activity, oxygen saturation levels, and mass transfer rates. Seasonal precipitation has been found to increase concentrations of *Giardia* cysts and *Cryptosporidium* oocysts caused by the suspension of particulate matter from the river bottom and storm drains (Atherholt et al., 1998).

Watershed Characteristics Topography, vegetation, and wildlife play important roles in source water quality. During heavy runoff, steep slopes can introduce debris, sediment, and nutrients that may affect color, turbidity, and algae. Trihalomethane total organic carbon (THM) precursor sources include the biota of productive lakes and reservoirs, algae, animals, macrophytes, and sediments (Cooke and Carlson, 1989). Vegetation may have a beneficial effect on water quality by providing a natural filter for runoff of non-point-source contaminants. Wildlife carries and introduces pathogens like *Giardia lamblia* and *Cryptosporidium* that can have significant human health implications if not adequately treated. Studies suggest that migratory waterfowl can acquire oocysts from wilderness, agricultural, and recreational areas and then travel great distances to contaminate other aquatic and terrestrial locations (Fayer et al., 1997).

Geology The subsurface geology determines ground and surface water quality, including calcium and manganese contamination and radioactivity (uranium, radium, and radon) in some areas. Soils play a beneficial role in buffering acidic precipitation. Buffering capacity affects the biological activity in lakes and reservoirs, and treatment processes and corrosion rates in distribution systems. The weathering characteristics of the local geology will have an effect on the erosion rates. A cohesive soil will resist rainfall "splash erosion" more effectively than will loose soils. Soils that have shallow depths and low permeability are not well-suited for individual septic systems and can contribute to ground and surface water contamination in urban watersheds (Robbins et al., 1991).

Nutrients The natural life cycle of a lake or reservoir involves three stages or trophic levels where the distinguishing factors are nutrients and biological activity. The *oligotrophic stage* is associated with low nutrients and limited algal production and biological activity. The *mesotrophic stage* involves a moderate amount of nutrients and moderate biological activity. At the *eutrophic stage,* nutrient levels are high, along with microbiological activity. The eutrophic level is associated with depleted oxygen levels, high turbidity, color, and formation of disinfection by-product precursors. From an operational standpoint, treatment interference, filter clogging, and taste and odor problems can occur. The most common indicator of eutrophication is excessive algae in the water column. Algae blooms may cause oxygen depletion, creating a reducing environment in which minerals like iron and manganese solubilize, phosphorus may be released from bottom sediments, and nitrate and organic nitrogen may be converted to ammonia. All of these factors may cause problems for a treatment plant.

Stratification Seasonal density or thermal stratification varies for shallow (less than 20 feet) and deep (greater than 20 feet) lakes and reservoirs.

In shallow reservoirs, summer water temperatures and oxygen concentrations will depend on the amount of wind-induced mixing. As surface water temperatures rise in relation to bottom waters, stratified density layers will form in the water column. An oxygen deficiency will result at the sediment-water interface, creating anaerobic conditions that will solubilize nutrients and metals from bottom sediments. Taste, odors, color, and turbidity may increase. During the winter months, temperature and dissolved oxygen levels in the water column will remain somewhat uniform because water density is uniform and stratification is minimized. Cold winter temperatures form ice covers on the surface that may create anaerobic conditions if air and water exchange is minimized for an extended period of time.

Deep-water bodies experience thermal stratification and form three distinct layers of water below the surface. The top layer is called the *epilimnion;* the bottom layer is called the *hypolimnion;* and the layer between is called the *thermocline* or *meta-*

TABLE 4.9 Water Quality Concerns

Parameter	Sources	Effects on water supplier
Solids, Turbidity	Domestic sewage, urban and agricultural runoff, construction activity, mining	Hinder water treatment process. Reduce treatment effectiveness. Shield microorganism against disinfectants. Reduce reservoir capacity
Nutrients	Septic system leachate, wastewater plant discharge, lawn and road runoff, animal feedlots, agricultural lands, eroded landscapes, landfill leachate, rainfall (especially nitrogen)	Nitrates that may be toxic to infants and unborn fetuses. Accelerates eutrophication: High levels of algae Dissolved oxygen deficiencies Increase algae activity High color and turbidity Disinfection by-product formation Taste and odor problems
Natural organic matter (NOM)	Naturally occurring; wetlands in the watershed tend to increase concentrations	Influence nutrient availability Mobilize hydrophobic organics Disinfection by-product formation
Synthetic organic contaminants	Domestic and industrial activities, spills and leaks, wastewater discharges, agricultural and urban runoff, leachate, wastewater treatment and transmission	Adverse impacts on human health and aquatic life
Coliform bacteria	Domestic sewage from wastewater discharges, sewers, septic systems, urban runoff, animal farms and grazing, waterfowl droppings, land application of animal wastes	Fecal coliform are indicators of warm-blooded animal fecal contamination that pose a threat to human health with microbial pathogens like *giardia, cryptosporidium,* and viruses
Metals	Industrial activities and wastewater, runoff	Adverse effect to aquatic life and public health

Oil and grease	Runoff containing kerosene, lubricating and road oils from gas stations, industries, domestic, commercial and institutional sewage, food waste and cooking oil	Interfere with biological waste treatment causing maintenance problems Interfere with aquatic life Aesthetic impacts Human health associated with selected hydrocarbons
Sodium	Road deicing and salt storage	High blood pressure and heart disease
Toxics	Agriculture, lawn care, industrial sites, roads and parking lots, wastewater	Toxic to humans and aquatic life
Aesthetics	Taste and Odor: industrial chemicals, algae metabolites, NOM, urea Color: metals, NOM, algae, AOC, Turbidity: solids and algae Staining: Metals	Aesthetic problems Reduce public confidence in water supply safety
Algae	Wastewater plant discharges, septic systems, landfill leachate, urban and agricultural runoff, precipitation	Taste and odor Filter clogging Some algae species toxic to aquatic life
Dissolved oxygen	Organic matter, wastewater discharges, runoff, consumption by aerobic aquatic life and chemical substances	Water treatment problems Release of iron and manganese Taste and odor problems Ammonia

limnion. During the summer months the hypolimnion can develop because it becomes isolated from the epilimnion. This can create anaerobic conditions and associated water quality problems. During the winter months, oxygen deficiency in the hypolimnion layer is less likely to occur. A key concern for deep stratified reservoirs is the turnover experienced during seasonal water temperature changes. As water temperatures decline in the fall, cold surface water moves into the hypolimnion, resulting in a mixing of the water column and stirring up nutrients and anoxic water. This may result in color, turbidity, iron, manganese, ammonia, and taste and odor problems. Most surface water plant intakes can draw from various levels within the water column to reduce the impact of stratification on the treatment process.

Wildfire and Deforestation Under dry conditions and lightning, wildfires can destroy vegetative cover and increase the potential for erosion by reducing the natural filtering of runoff. Fires have a beneficial affect in rejuvenating forests with new vegetation. Vegetative cover has a number of beneficial affects on source water quality, including reducing erosion, filtering rainwater, and promoting the biological uptake of nutrients and contaminants (Robbins et al., 1991). Wildfires increase peak flows, sediment, turbidity, stream temperature, and nutrients (Tiedemann et al., 1979). Prescribed fires are used at times to reduce flammable brush and contain catastrophic wildfires in water supply watersheds (Berg, 1989).

Saltwater Intrusion Suppliers utilizing surface sources in coastal regions or in upstream tidal estuaries need to be aware of saltwater intrusion. A river supply has to deal with the migration of the freshwater-saltwater interface during low flow conditions.

Point-Source Impacts on Surface Water Quality from Human Factors

Point sources include wastewater and industrial discharges, hazardous waste facilities, mine drainage, spills, and accidental releases. Point-source discharges are typically those from an identifiable point like a pipe, lagoon, or vessel. A pipe discharge that collects storm runoff would not be considered a point discharge because the drainage comes from a wider geographical area that is typically considered a nonpoint source. Point discharges associated with a facility are usually regulated.

The impact of wastewater on the receiving stream depends on the stream's ability to assimilate pollutants. The assimilative capacity of a stream refers to its ability to self-purify naturally (Metcalf and Eddy, 1979). The TMDL (Total Maximum Daily Load) program under the Clean Water Act can provide some useful information on the pollutant load a stream can handle and meet surface water quality standards (USEPA, 1997).

Wastewater discharges are a major source of nutrients, bacteria, viruses, parasites, and chemical contamination. Design and operation of upstream wastewater treatment facilities are important in minimizing degradation of water quality for downstream water treatment plants. Most facilities are designed to provide secondary treatment (Robbins et al., 1991). Minimizing combined sewer overflows, pipe leaks or ruptures, and pump station failures throughout a watershed basin will reduce the stress that pollutants impose on a stream. Discharged treated wastewater with elevated levels of ammonia and nitrogen may support algae growth. Discharge of treated and untreated wastewater increases bacterial contamination and may cause dissolved oxygen depletion, fish kills, and negative impacts from nutrient input and organic and inorganic contamination.

Industrial discharges affect water quality through contaminant release via water, land disposal, and air emissions. Wastewater discharges upstream of the intake need to be identified and their potential risk to source water quality evaluated. Industrial discharges can introduce a contaminant to the source water that can bypass the treatment process and enter the distribution system. Facilities can introduce contaminants through a regulated discharge from an accidental spill. Outdated underground storage facilities with no secondary containment and monitoring program may be a concern. Untreated stack emissions may introduce airborne contaminants.

In 1977, the city of Cincinnati, Ohio, experienced an industrial carbon tetrachloride spill that was detected in the distribution system at 80 μg/L. Treatment processes were not able to fully remove the compound from the water. In response to this experience, the city moved to implement a comprehensive basin monitoring system that monitors for volatile solvents and petroleum in the Ohio River. The system has 17 monitoring stations from Pittsburgh, Pennsylvania, providing coverage along the Ohio River to Paducah, Kentucky. Three of the monitoring stations have automated process samplers that monitor continuously for volatile organic compounds (DeMarco, 1995).

Facilities that treat, store, and dispose of hazardous materials as defined by the Federal Resource Conservation and Recovery Act are required to take extensive precautions to prevent the release of hazardous contaminants. Inactive hazardous waste sites regulated under the Federal Comprehensive Environmental Response and Compensation Liability Act (CERCLA or Superfund) need to be considered as potential point sources.

Mining operations are associated with a number of water quality problems that include acid drainage, leaching and runoff of heavy metals, and sedimentation. Mine drainage becomes acidic in the presence of sulfur-bearing minerals, air exposure, and water that together form sulfuric acid. Contaminated drainage from mine spoils and tailings can acidify streams and cause dissolution of metals from the surrounding rock and soil and precipitate iron in streams that have a neutral pH (Robbins et al., 1991). Mining operations disturb the surface topography and remove vegetation, causing excessive erosion. Acid mine drainage may alter source water chemistry and carry dissolved iron, manganese, and other contaminants. Metals associated with mine drainage include zinc, lead, arsenic, copper, and aluminum (Davis and Bocgly, 1981). According to the USEPA Region VII, mining is the second-ranking activity (next to municipal wastewater facilities) in generating toxic metals (Chesters and Schierow, 1985).

Watersheds traversed by major highways and rail freight lines are vulnerable to spills from transportation accidents. Oil spills can generally be contained because their low-density allows them to float. Soluble materials may require neutralization, oxidation, precipitation, or adsorption measures to remove them from source water.

Non-Point-Source Impacts on Surface Water Quality from Human Factors

Many non-point-source impacts are related to land use for agricultural, development, recreation, and acid deposition. A non-point source is not easily definable, but it involves effluent from a wider geographic area than a point-source contaminant. Some non-point sources are regulated, especially those that are defined by a regulatory agency.

The most common agricultural non-point sources are sediment, dissolved solids, nutrients, bacteria, pathogenic organisms, and toxic materials (Christensen, 1983). Approximately four billion tons of sediment enter the waterways of the 48 contiguous states each year, of which 75 percent is from agricultural lands (National Research Council Committee on Agriculture and the Environment, 1974). Cropland alone (excluding other agricultural activities) accounts for 50 percent of the sediment load that enters U.S. waterways (USEPA, 1973). Over half of the nitrogen and 70 percent of the phosphorous loads are from agricultural sources in the 48 contiguous states (USEPA, 1975). Agricultural activities also involve the application of pesticides, herbicides, and fertilizers to improve crop yields. Some pesticides and herbicides have been banned from use, and those that are in use must be applied under controlled conditions. Applicators must be adequately trained and licensed. Although most pesticides and herbicides eventually decompose, many (particularly the triazine herbicides) may end up in the surface waters during spring runoff events at levels well above drinking water MCLs. Fertilizer use increases nutrients in the soil that, if not properly applied, can contribute to the eutrophication of the nearby surface source and exceedences of nitrate MCLs. Proper tilling techniques and vegetative strips can reduce soil erosion and sediment transport. Livestock contribute bacterial contamination and nutrients, while uncontrolled overgrazing of vegetative cover may increase erosion.

In the urban environment, runoff from highways, streets, and commercial areas can introduce numerous contaminants into a surface source, including nitrogen, phosphorus, suspended solids, coliform bacteria, heavy metals, and organic contaminants (Whipple et al., 1974). The predominant pollutants found in urban runoff are copper, lead, and zinc. Other urban contaminants include high levels of coliform bacteria, nitrogen, and phosphorus, and total suspended solids in sufficient quantities to accelerate eutrophication and exceed wastewater treatment discharges (USEPA, 1983).

Development of previously undisturbed land reduces the filtering capacity of natural vegetation and increases runoff from roofs, sidewalks, and streets. Whether the development is an office complex or single-family housing units, such human activity impacts surface water quality. Extensive landscaping may require the application of consumer-grade pesticides, herbicides, and fertilizers. Application of these chemicals does not require any specialized training or licensing, and the homeowners could apply excessive amounts, affecting nearby surface waters. (Although this is a common assumption, little concrete evidence is presently available.) Wastewater disposal in areas without sewers requires properly installed septic systems to ensure water quality protection. State or local governing authorities usually establish septic tank standards that recommend stream setbacks for leach fields and other technical parameters to ensure proper installation. Fuel storage for both the homeowner and commercial establishment needs to be considered as a possible contamination source since they are not regulated as stringently as larger underground storage tanks.

Recreational activities like swimming, boating, fishing, camping, and other motorized activities can impact water quality. Recreational activities in surface water sources have been undergoing much debate over the years. Local rural communities where water supply reservoirs are located prefer recreational use and consider that use compatible with management of those water supply sources (Robbins et al., 1991). The American Water Works Association (AWWA) has adopted a policy on recreational use of water supply reservoirs that would prohibit recreation where other nonwater supply sources are available. At those sources that are used both for water supply and recreation, the AWWA policy suggests that no body-contact activities such as swimming take place (AWWA, 1987). Body-contact recreation is a known source of fecal contamination that increases the levels of indicator and pathogenic organisms that include bacteria such as Shingella, enteric viruses like hepatitis and poliovirus, and the protozoa *Cryptosporidium* and *Giardia* (Stewart et al., 1997).

Motorized boating activities may contribute contaminating petroleum byproducts and additives to water supply reservoir. Fishing alone may not contribute to serious contamination, but one needs to consider if it is done from a boat or along the shoreline. Campers have been known to start accidental forest fires that defoliate forests and destroy vegetation. The watershed manager needs to measure the impacts and risks of these activities on the water supply and then make the appropriate management decision to minimize their impact on overall source water quality. Balancing recreational uses with water quality protection would require the need for optimization models that can generate alternatives. Three different recreational mixes that included white-water rafting, boating, and fishing were evaluated to determine the total recreational impact on the New River Gorge in West Virginia, a National River watershed (Flug et al., 1990).

A study conducted on two Maine lakes that were limnologically similar but had different recreational uses showed that the total coliform densities, when compared over a three-and-a-half-year period, were significantly higher in the recreational lake. The study contradicts a common perception that low levels of recreational use do not affect water quality (McMorran, 1997).

The concern from acid deposition comes from the ability of a water source to neutralize runoff from acidic precipitation and minimize the leaching and mobilization of contaminants in soil and rock (Perry, 1984). The contaminants of concern are mercury, aluminum, cadmium, lead, asbestos, and nitrates (Quinn and Bloomfield, 1985). Over half of the streams in the 27 eastern states in the United States that have areas sensitive to acid deposition are at risk or have already been altered by acid precipitation (OTA, 1984). Much of the research focus on acid precipitation is on the impact on the environment and aquatic ecosystem and little on source water quality.

QUALITY MANAGEMENT OF SURFACE WATER SOURCES

Regulatory Programs

The 1996 Safe Drinking Water Act Amendments (see Chapter 1) contain provisions for source water protection and support water quality managers in their efforts to protect surface and groundwater supplies. Existing federal, state and local regulations may have provisions to protect source waters from possible contamination. The Clean Water Act, for example, regulates the discharge of pollutants into source waters through the National Pollutant Discharge Elimination System. Publicly owned treatment works and industrial facilities that discharge treated wastewater are required to monitor their effluent and report results to the regulatory authority. The source water quality manager can use the regulations to identify those facilities that operate on the watershed and may provide input during permit renewal. Monitoring upstream and downstream of the facilities will provide additional information for water quality protection. Other established regulated programs include:

- Resource Conservation and Recovery Act (RCRA)
- Federal Insecticide, Fungicide, and Rodenticide Act (FIFRA)
- Comprehensive Environmental Response Compensation and Liability Act (CERCLA)
- Toxic Substances Control Act (TSCA)
- Clean Air Act (CAA)

Each regulation has a different focus. For example, the RCRA regulates active hazardous waste facilities, while the CERCLA regulates inactive or abandoned hazardous waste sites. Each state can make the federal regulation more stringent, depending on its individual priorities. For example, some New Jersey drinking water standards for volatile organic compounds are more stringent than the federal standard. Also, New Jersey regulates additional parameters that are not regulated by the federal SDWA. The watershed manager need not understand all these regulations, but needs to have a baseline working knowledge to enlist regulations in source protection.

The 1996 SDWA Amendments require states to conduct source water assessments and develop protection guidelines to protect public drinking water supplies. Source water protection managers need to become aware of the programs in the respective state where the water source is located to ensure that all needs are met and drinking water is protected. Source water assessments have to be completed within 2 years after the USEPA approves the state program, with an option to extend the deadline up to 18 months (USEPA, 1997).

Utility Management Programs

Source protection is the first barrier in reducing or eliminating contaminants that impact the water quality of the consumer. Increasing levels of contaminant input and source quality deterioration place added burdens and cost on the treatment plant. Public confidence can be affected if source water is impacted or is perceived as contaminated. Source water protection programs help to reduce public health risks associated with contaminants and pathogens at a time when current microbial analytical technology is developing. The following recommended procedures are a guide for watershed managers who are preparing or enhancing source water protection programs (Standish-Lee, 1995).

- Identify and characterize surface water sources
- Identify and characterize potential surface water impacts
- Determine the vulnerability of intake to contaminants
- Establish protection goals
- Develop protection strategies
- Implement the program
- Monitor and evaluate the program

1. Identify and Characterize Surface Water Sources. The outer boundary of the watershed with all the relevant water sources should be defined. The outlined area must include all upland streams and water bodies that drain to the treatment plant intake. The watershed boundary can be identified by physically traversing the watershed and utilizing topographic maps. When traversing the watershed, one would begin at the intake and proceed upstream along the drainage boundary, noting key features on a map or aerial photo. One may choose to first identify the watershed basin on topographic maps or aerial photos and then proceed to traverse the perimeter. The U.S. Geological Survey maintains topographic maps that can be utilized and are available in digital form for use in at computerized geographic information system (GIS). Other map sources include state agencies that regulate land use, mineral development, and water resources; local planning boards; and federal land management agencies like the U.S. Forest Service (Robbins et al., 1991).

Additional information that would be desirable to have on the watershed and surface water sources includes characteristics such as climate, topography, geology, soils, vegetation, wildlife, land use, and ownership. Groundwater recharge zones should be identified to determine the degree to which surface waters and groundwater sources affect each other. Understanding these factors allows the watershed manager to assess natural impacts on water quality.

2. Identify and Characterize Potential Surface Water Impacts. The next consideration is to determine potential source impacts. Those may include individual and municipal sewage disposal; urban, industrial, and agricultural runoff; farm animal populations; forests and recreation; solid and hazardous waste disposal; and vehicular traffic; to name a few. While conducting the inventory, the potential sources need to be identified with respect to their ability to degrade water quality and the extent of potential contamination, and ranked according to priority for control. The inventory can be divided into natural and human factors that affect water quality. Watershed activities and land uses that may be causing contamination need to be identified along with any monitoring data that may be available. Natural watershed basin features such as steep slopes, highly erosive and clayey soils, and wildlife and riparian areas need to be identified. Local authorities like the Soil Conservation Service, the USGS, and any regulatory authority responsible for land management may have supportive information on the local land use activities. Sources of contamination that have the potential to enter the watershed, such as air pollutants and transportation facilities, need to be included. An assessment needs to be made as to the future water quality impact from increased land use activities accompanying projected growth of the area. The inventory will need to be updated at least every five years, and more frequently if land use activities intensify. Chemical usage and recreational visits may need annual updating (Robbins et al., 1991).

Using a GIS model to conduct watershed assessments proved to be a powerful tool for the Metropolitan District Commission (MDC) that manages and protects the 297 square-mile watershed for the city of Boston water supply. The MDC is an independent agency that works closely with the Massachusetts Water Resources Authority that is responsible for treating and distributing water to over 2 million people. Together, the two agencies used GIS to study over 400 square miles of watershed to identify, map, and rank existing sources of potential pollution threats that included septic systems, recreational activities, storm water runoff, logging, petroleum storage, and natural impact, including erosion and animal populations. The study recommended ways of addressing current and future contamination sources and assisted in passing buffer zone legislation (USEPA et al., 1999).

A comprehensive inventory depends on the scale involved and level of treatment available. For example, it would be an onerous task for the city of New Orleans to gather and maintain such information on the entire Mississippi River basin, and it should not be expected to do so. One would have to determine the impacts that have the highest level of risk on a broader scale for a city like New Orleans and then prioritize those of greatest concern. Systems subject to vast watersheds will have to enlist the support of multistate agencies to determine the water quality impacts of concern.

3. Determine Vulnerability of Intake to Contaminants. Intake vulnerability to potential contaminants, from the sites inventoried, needs to be determined. Point discharges are easier to assess since there is a definable flow and pollutant load. As one moves to less definable non point sources, the affect on the water supply intake will become more difficult to assess because the frequency and intensity is based on runoff events. Accidental spills occur at random and are even more difficult to assess unless one can determine occurrence frequency based on past events (Robbins et al., 1991).

Intake vulnerability is based on the ability of the water treatment processes to remove contaminants and prevent them from passing to the distribution system or resulting in significant disinfection by product formation. Water quality monitoring, modeling, and on-site assessments are three ways to determine the effect of land use on source water quality.

4. Establish Protection Goals. The primary source protection objective is providing high-quality water to the consumer. Goals that support the primary objective can be based on the water quality parameters of concern and the characteristics of the watershed. Examples may include pollutant load reduction, protection from urban development, avoidance of treatment and disinfection changes, minimizing risks from hazardous chemicals, mitigating effects from natural disasters, land preservation, and enhancing fish and wildlife habitats. When establishing protection goals, interest groups and stakeholders need to be considered to mitigate obstacles from competing interests (Robbins et al., 1991).

5. Develop Protection Strategies. Protection strategies include land use controls and best management practices (BMP) in urban, agricultural and forest areas (see Table 4.10). In urban areas, BMPs can be *structural* and *nonstructural.* Structural BMPs involve the construction of physical structures that control water quality. An example of a structural BMP might be detention basins in a development to settle contaminants from storm runoff. A nonstructural BMP involves activities that a landowner may undertake to control the pollutant load. An example of a nonstructural BMP might be the use of an alternative herbicide. The challenge comes in choosing the control measures that maximize source water quality protection.

Agricultural BMPs may include judicious use of chemicals, grazing restrictions, animal waste management, contour farming, crop rotation, conservation tillage, terracing, and grassed waterways. Best management practices in forestry include haul road and access design and construction, postdisturbance erosion control, seasonal operating restrictions, slash disposal, and helicopter logging (Robbins et al., 1991).

Nonstructural practices in urban areas may include lot size minimums, cluster development, buffer zone setbacks, limits on impervious surfaces, land use prohibitions, wastewater restrictions, conservation easements, and revegetation. Some structural best management practices for urban areas include wet ponds, dry detention basins, infiltration controls, storm water diversions, oil-water separators, and constructed wetlands and grassed swales (Robbins et al., 1991).

Many strategies are subject to the whims of local politics and competing interests for land. One needs to work with stakeholders and convey the need for watershed protection to gain the support of those who oppose restrictions and BMPs. The associated BMPs benefit the utility, while the landowner may have to limit the use of the land and even build structures to reduce the pollutant load. One needs to remember that landowners do not have an inherent right to pollute a water source.

Two relatively innovative BMP strategies applied to the water body itself are biomanipulation with nutrient controls and bacterial inoculates/biostimulates. These strategies create competitive interactions between bacteria and algae for inorganic nitrogen and phosphorus, thereby reducing algae growth. These are largely experimental alternatives that work best under specific conditions and low hydraulic retention times or flushing rates (Coastal Environmental Services, 1996).

Land use controls include buffer zones, use restrictions, land acquisition, comprehensive planning, zoning, agreements, legal action, signage, public education, stakeholder participation, and regular inspections. Many of these strategies may take years to implement.

TABLE 4.10 Suggested BMP List

Agricultural	Forestry	Urban
	Nonstructural	
Tillage and cropland erosion control	Forestry preharvest	Land use planning and management
Pesticide and fertilizer application	Streamside management areas	Public acquisition of watershed land
Range and pasture management	Forest chemical management	Minimum lot size zoning restrictions
Contour farming and strip cropping	Fire management	Impervious surface restrictions
Confined feedlot management	Forest vegetation of disturbed areas	Buffer zones and setbacks
Cover cropping		Public information and education
Crop residue usage		Citizen advisory committees
Cropland irrigation management		Watershed sign posting
		Storm drain stenciling
		Illegal dumping and illicit connection controls
		Material exposure controls
		Material disposal and recycling
		Household hazardous pickup days
		Used motor oil collection
		Wastewater disposal restrictions
		Septic tank management
		Community wastewater systems control
		Sanitary sewer facilities planning and management
		Catch basin and street cleaning
		Construction site land stabilization
	Structural	
Animal waste management	Erosion and sediment controls	Detention/retention facilities
Terrace systems	Access roads	Wet detention ponds
Diversion systems	Skid trails	Extended detention ponds
Sediment basins	Stream crossings	Vegetated swales and strips
Filter strip and field borders.	Filter strip sediment controls	Constructed wetlands
		Infiltration ponds and trenches
		Drainage structure controls
		Inlet floatable controls
		Oil water separators
		Media filtration
		Erosion and sediment control
		Stream bank stabilization and riparian buffer restoration

Water body BMP: Direct use of the source water for recreational activities can be controlled by river and reservoir management restrictions on body contact recreation, motorboat engine restrictions, bird control and shoreline restoration.

Land acquisition can result in long-term tax and maintenance costs that may compete with or defer needed expenditures on treatment, and vice versa. The challenge is in finding an optimum balance or win-win partnership with landholders.

The watershed protection program manager must have the legal capacity to take action against polluters under existing environmental regulations. The public and certain stakeholders can be allies in protecting the watershed both from a public policy standpoint and as watershed guardians.

6. Implement the Program. The key to effective implementation of a protection program is getting support from upper management. When "selling" a watershed protection program, one needs to focus on the benefits, including cost savings or avoidance. The potential impacts have to be defined if certain aspects of the program are not implemented. Management support must come with the financial resources needed to support activities and personnel.

The degree to which a utility controls the watershed area is a key implementation factor. If the purveyor does not control all or a substantial portion of the watershed, then a relationship needs to be developed with those authorities who have control and will support watershed protection. All stakeholders that can exert indirect influence in watershed protection, such as environmental groups, can be identified and brought into the process. Stakeholder involvement and public relations tools can be developed to formalize communications. The program may have to be defended against legal challenges that could potentially weaken it.

7. Monitor and Evaluate the Program. Water quality monitoring is essential in determining the effectiveness of a watershed protection program. Routine monitoring data needs to be analyzed to determine what water quality changes are occurring. Monitoring the effectiveness of BMPs will provide an early warning signal to an impending problem. Special studies and sanitary surveys should be conducted periodically to better understand the watershed dynamics. By comparing the results of the evaluation to the goals and objectives of the program, the watershed manager can make adjustments.

Modeling

Models have broad use in the watershed protection applications, depending on what mechanism is under study. Reservoir loading and fate and transport models are commonly used to assess the impact of water quality from watershed activities. *Reservoir loading models* assess the trophic state of a lake or reservoir based on the nutrient inputs, algae production, transparency, and other relevant parameters. Larson and Mercier along with Vollenweider developed models predicting phosphorus concentrations in 1976. Carlson developed a trophic state model in 1977 that relates total phosphorus, chlorophyll, and transparency to trophic state graphically (North American Lake Management Society, 1990). Results are always debatable, and accuracy depends on how closely the model simulates real world conditions.

Fate and transport models are used to predict inputs of contaminants that affect water quality. A model conducted for the New York City watershed concluded that although dairy oocyst loads are minor in comparison to the wastewater load, oocyst-laden calf manure applications on the watershed could overtake the wastewater load and have important agricultural implications (Walker and Stedinger, 1999).

Watershed Management Technical Resources

The surface water quality portion of this chapter cannot be viewed as exhaustive because the existing amount of information and research on watershed management practices is far too voluminous to cover in such a short chapter. It can, however, be used as an introduction to watershed protection issues and a springboard to the vast information available. For a more in-depth understanding of watershed protection issues, the references at the end of this chapter are good resources to seek out and review. Other technical sources suggested for reference include: *Restoration and Management of Lakes and Reservoirs* (2nd edition) by G. Dennis Cooke, Eugene B. Welch, Spencer A. Peterson, and Peter R. Newroth (Lewis Publishers, Ann Arbor, Michigan); "Eutrophication: Causes, Consequences Correctives" (Symposium Proceedings, National Academy of Sciences, Washington, DC., 1969); and *Chemistry for Environmental Engineers* (3rd edition), by Clair N. Sawyer and Perry L. McCarty (McGraw-Hill Book Company, New York, 1978).

BIBLIOGRAPHY

American Water Works Association. "American Water Works Association Statement of Policy: Recreation Use of Domestic Water Supply Reservoirs" *Journal AWWA,* 79(12), 1987: 10.

Atherholt, T. B., M. W. LeChevallier, and J. S. Rosen. "Variation of Giardia Cyst and Cryptosporidium Oocyst Concentrations in a River Used for Potable Water." New Jersey Department of Environmental Protection Division of Science and Research, July 1998.

Berg, N. H. (ed). "Proceedings of the Symposium on Fire and Watershed Management." Berkeley, California: Pacific Southwest Forest and Range Experiment Station, 1989.

Chauret, Christian, Neil Armstrong, Jason Fisher, Ranu Sharma, Susan Springthorpe, and Syed Sattar. "Correlating Cryptosporidium and Giardia with Microbial Indicators." *Journal AWWA,* 87(11), November 1995: 76–84.

Chesters, G., and L. Schierow. "A Primer on Nonpoint Pollution." *Journal Soil and Water Conservation,* 40(1), 1985: 9.

Christensen, L. A. "Water Quality: A Multidisciplinary Perspective." *Water Resources Research: Problems and Potentials for Agricultural and Rural Communities,* T. L. Napier, ed. (Ankeny, Iowa: Soil Conservation Society of America, 1983).

Coastal Environmental Services. "The Development of a Management Plan for Oradell Reservoir, New Jersey." Project 95-1003-01, December 1996.

Cooke, G. Dennis, and Robert E. Carlson. "Reservoir Management for Water Quality and THM Precursor Control." American Water Works Association Research Foundation, 1989, p. 387.

Davis, E. C., and W. J. Boegly. "A Review of Water Quality Issues Associated with Coal Storage." *Journal Environmental Quality,* 10(2), 1981: 127.

DeMarco, J. "Case Study: What to do When You Can't Protect Your Source Water." AWWA Satellite Teleconference, Source Water Protection: An Ounce of Prevention, August 1995.

Fayer, Ronald, James M. Trout, Thaddeus K. Graczyk, C. Austin Farley, and Earl J. Lewis. "The Potential Role of Oysters and Waterfowl in the Complex Epidemiology of Cryptosporidium parvum." 1997 International Symposium on Waterborne Cryptosporidium Proceedings, March 2–5, 1997, 153–158.

Flug, Marshall, D. G. Fontane, and G. A. Ghoneim. "Modeling to Generate Recreational Alternatives." *American Society of Civil Engineers, Journal of Water Resources Planning and Management,* 116(5), September/October 1990: 625–638.

Fruth, Darrell A., Joseph A. Drago, Maria W. Tikkanen, Kenneth D. Reich, LeVal Lund, Susan B. Nielsen, and Lawrence Y. C. Leong. "A River Basin Based Method to Estimate Arsenic

Occurrence in California Surface Waters." *Proceedings 1996 Water Quality Technology Conference; Part I, Boston, Massachusetts, November 17–21, 1996.*

McMorran, Carl. "Observations on Recreation Uses and Total Coliform Densities in Two Surface Water Supplies." *Proceedings Annual Conference American Water Works Association,* Vol. B, 349–359. Atlanta, Georgia, June 15–19, 1997.

Metcalf & Eddy, Revised by George Tchobanoglous. *Wastewater Engineering: Treatment Disposal Reuse,* 2nd edition. New York: McGraw-Hill Book Company, 1979.

National Research Council Committee on Agriculture and the Environment. *Productive Agriculture and a Quality Environment.* Washington, DC: National Academy of Sciences, 1974.

North American Lake Management Society. *Lake and Reservoir Restoration Guidance Manual.* Washington, DC: U.S. Environmental Protection Agency Office of Water, 1990.

Oberts, G. L. "Pollutants Associated with Sand and Salt Applied to Roads in Minnesota." *Water Resources Bulletin,* 22(3) 1986: 479.

Office of Technology Assessment (OTA). *Acid Rain and Transported Air Pollutants: Implications for Public Policy.* OTA-0-204. Washington, DC: U.S. Congress Office of Technology Assessment, 1984.

Perry, J. A. "Current Research on the Effects of Acid Deposition." *Journal AWWA,* 76(3), 1984: 54.

Quinn, S. O., and N. Bloomfield (eds.). *Proceedings of a Workshop on Acidic Deposition, Trace Contaminants and Their Indirect Human Health Effects: Research Needs.* Corvallis, Oregon: EPA Environmental Research Laboratory, 1985.

Robbins, R. W., J. L. Glicker, and D. M. Bloem. "Effective Watershed Management for Source Water Supplies." American Water Works Association Research Foundation, 1991, 9–26.

Richards, J. L., and Associates, and Vezina Legrecque and Associates. *Snow Disposal Study for the National Capitol Area: Technical Discussion.* Ottawa, Ontario (Canada): Committee on Snow Disposal, 1973.

Standish-Lee, P. "Elements of a Source Water Protection Program." AWWA Satellite Teleconference, Source Water Protection: An Ounce of Prevention, August 1995.

Stenstrom, Michael K., Gary S. Silverman, and Taras A. Bursztynsky. "Oil and Grease in Urban Stormwaters." *Journal of Environmental Engineering,* 110(1), February 1984: 58–72.

Stewart, M., M. Yates, M. Anderson, C. Gerba, R. DeLeon, and R. Wolfe. "Modeling the Impact of Body-Contact Recreation on Cryptosporidium Levels in a Drinking Water Reservoir." *Proceedings 1997 International Symposium on Waterborne Cryptosporidium.* Newport Beach, California, March 2–5, 1997.

Tiedemann, A. R., C. E. Conrad, J. H. Dietrich, J. W. Hornbeck, W. F. Megahan, L. A. Viereck, and D. D. Wade. "Effects of Fire on Water." USDA Forest Service Gen. Tech. Report WO-10. Washington, DC: USDA Forest Service, 1979.

U.S. Environmental Protection Agency. *Methods and Practices for Controlling Pollution from Agricultural Nonpoint Sources.* EPA 430/9-73-015. Washington, DC: EPA Office of Water, 1973.

U.S. Environmental Protection Agency. *National Assessment of Water Pollution from Nonpoint Sources.* Washington, DC: EPA Office of Water, 1975.

U.S. Environmental Protection Agency. *Results of the Nationwide Urban Runoff Program, Vol. I—Final Report.* Washington, DC: Water Planning Division, 1983.

U.S. Environmental Protection Agency. *State Source Water Assessment and Protection Programs Guidance.* EPA 816-R-97-009. Washington, DC: EPA Office of Water, 1997.

U.S. Environmental Protection Agency and Association of Metropolitan Water Agencies. "Protecting Sources of Drinking Water: Selected Case Studies in Watershed Management." Office of Water, EPA 816-R-98-019, April 1999.

Walker, F. R., and J. R. Stedinger. "Fate and Transport Model of Cryptosporidium." *Journal of Environmental Engineering,* April 1999, 325.

Whipple, W., J. V. Hunter, and S. L. Yu. "Unrecorded Pollution from Urban Runoff." *Journal Water Pollution Control Federation,* 46(5), 1974: 873.

CHAPTER 5

AIR STRIPPING AND AERATION

David W. Hand, Ph.D.
David R. Hokanson, M.S.
John C. Crittenden, Ph.D., P.E., DEE
Department of Civil and Environmental Engineering
Michigan Technological University
Houghton, Michigan

Several different types of air stripping and aeration systems are widely used for a variety of water treatment applications. The most common types are diffused-air, surface aerator, spray, and packed-tower systems. Water treatment applications for these systems include the absorption of reactive gases for water stabilization and disinfection, precipitation of inorganic contaminants, and air stripping of volatile organic compounds (VOCs) and nuisance-causing dissolved gases. The diffused-aeration (or bubble) systems are primarily used for the absorption of reactive gases, such as oxygen (O_2), ozone (O_3), and chlorine (Cl_2). Oxygen is frequently used for the oxidation/precipitation of iron and manganese. Ozone is used for disinfection, color removal, and oxidation of total organic carbon (TOC). Chlorine is primarily used for disinfection and sometimes as a preoxidant for the oxidation of iron and manganese or for other purposes. Diffused-aeration systems have also been used for the stripping of odor-causing compounds and VOCs. Surface-aeration systems are primarily used for removal of VOCs. The packed-tower and spray nozzle systems are primarily used for the removal of NH_3, CO_2, H_2S, and VOCs. The packed-tower systems include countercurrent flow, cocurrent flow, and cross-flow configurations. Spray nozzle systems can include tower and fountain-type configurations.

A fundamental understanding of the theory of gas transfer is first discussed, followed by a description of the various unit operations, development of the design equations, and example design calculations.

THEORY OF GAS TRANSFER

To properly design and operate aeration and air stripping devices, a fundamental understanding of equilibrium partitioning of chemicals between air and water and of the mass transfer rate across the air-water interface is required. Equilibrium is the final state toward which the system is moving. The displacement of the system from equilibrium will dictate how much fluid (air) will be required for stripping or aeration and will define the driving force that governs mass transfer (i.e., the rate at which chemicals move from one phase to another), which in turn determines the vessel size required for stripping or aeration. Both equilibrium and mass transfer concepts are incorporated into mass balance equations to formulate the design equations. Consequently, these concepts are reviewed first.

Equilibrium

For most aeration and air stripping applications in water treatment, equilibrium partitioning of a gas or organic contaminant between air and water can be described by Henry's law. The Henry's law equilibrium description can be derived by considering the closed vessel shown in Figure 5.1. If the vessel contains both water and air and component A is in equilibrium with both phases at a constant temperature, equilibrium can be described by the following expression:

$$K_{eq} = \frac{\{A\}_{air}}{\{A\}_{aq}} \tag{5.1}$$

where: K_{eq} = equilibrium constant
$\{A\}_{air}$ = activity of component A in the gas phase
$\{A\}_{aq}$ = activity of component A in the aqueous phase

At a pressure of 1 atmosphere (atm), the gas behaves ideally and Eq. 5.1 reduces to:

$$H = K_{eq} = \frac{P_A}{\gamma_A[A]} \tag{5.2}$$

where: H = Henry's law constant (atm-L/mol) of component A
P_A = pressure A exerts in the gas phase (atm)
γ_A = the activity of A in the aqueous phase
$[A]$ = aqueous-phase molar concentration of A (mol/L)

(The presence of air does not affect the Henry's law constant for organic chemicals or gases because the components of air have a very low solubility in water.) For very low concentrations, Henry's law may be written as follows:

$$P_A = H[A] \tag{5.3}$$

When dissolved organic or inorganic species in the water do not affect the equilibrium partitioning, Eq. 5.3 is generally valid for concentrations less than 0.01 gmol/L, but it has been shown in some cases to be valid for concentrations as high as 0.1 gmol/L (Rogers, 1994). The units of H in Eq. 5.3 are atm·L/gmol, but H has other units. Figure 5.2 displays the three most commonly used unit measures of H. H is reported with the units of atmospheres (atm) when the gas-phase concentration of component A is expressed in atmospheres and the aqueous-phase concentration of component A is expressed in terms of mole fraction. H is reported in units of liters of water per

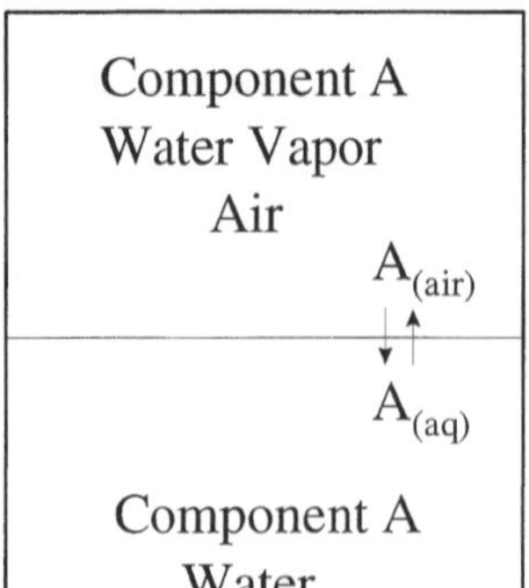

FIGURE 5.1 Schematic of equilibrium conditions for component A in air and water.

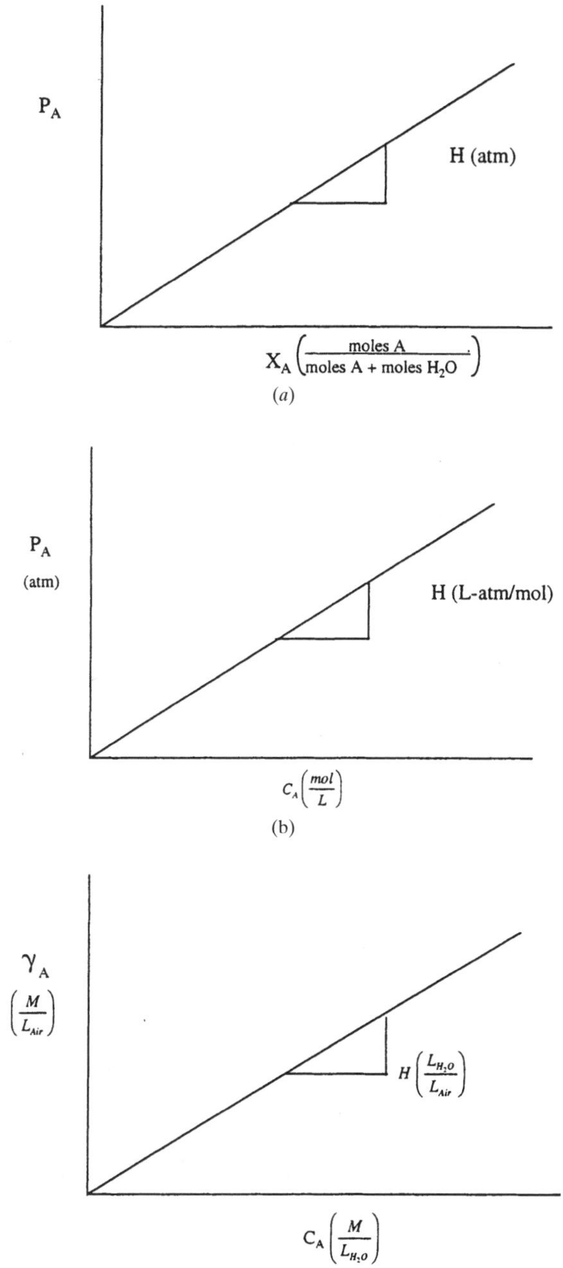

FIGURE 5.2 (*a*) Plot of *H* displayed in atm. (*b*) Plot of *H* displayed in liters of water·atm/mol. (*c*) Plot of *H* displayed in liters of water per liter of air.

atmospheres per gmole of gas (L·atm/gmol) when the gas-phase concentration of component A is expressed in atmospheres and the aqueous-phase concentration of component A is expressed in terms of moles of component A per liter of water. H is also reported in dimensionless units when both the gas and aqueous phases are expressed in the same concentration units. Because the reported units of H vary, it is necessary to convert from one system of units to another. Table 5.1 displays various unit conversions that can be used to convert from one system to another. Dimensionless units are very convenient for mass balances; consequently, dimensionless units are preferred and are used in this chapter.

When the Henry's law constant for a component of interest is not readily available, it can be estimated if the component's vapor pressure and aqueous solubility are known. There are two situations when it is possible to estimate the Henry's constant of component A: (1) when component A is perfectly miscible in the aqueous phase, and (2) when component A is immiscible in the aqueous phase. When component A is perfectly miscible and the mole fraction of A is equal to 1 ($x_{H_2O} = 0$), the pressure exerted by A is equal to the vapor pressure of pure A at a given temperature. In this case, H can be expressed as:

$$H = P_{v,A} \tag{5.4}$$

in which $P_{v,A}$ is the vapor pressure of pure component A at a given temperature (atm). For the case in which component A is immiscible in the aqueous phase, a third or separate phase of component A is formed within the aqueous phase once the sol-

TABLE 5.1 Unit Conversions for Henry's Law Constants

$H\left(\dfrac{L_{H_2O}}{L_{Air}}\right) = \dfrac{H\left(\dfrac{L\cdot atm}{mol}\right)}{RT}$
$H\left(\dfrac{L\cdot atm}{mol}\right) = H\left(\dfrac{L_{H_2O}}{L_{Air}}\right) \times RT$
$H\left(\dfrac{L_{H_2O}}{L_{Air}}\right) = \dfrac{H(atm)}{RT \times 55.6\,\dfrac{mol\ H_2O}{L_{H_2O}}}$
$H\left(\dfrac{L\cdot atm}{mol}\right) = \dfrac{H(atm)}{55.6\,\dfrac{mol\ H_2O}{L_{H_2O}}}$
$H(atm) = H\left(\dfrac{L\cdot atm}{mol}\right) \times 55.6\,\dfrac{mol\ H_2O}{L_{H_2O}}$
$H(atm) = H\left(\dfrac{L_{H_2O}}{L_{Air}}\right) \times RT \times 55.6\,\dfrac{mol\ H_2O}{L_{H_2O}}$
$R = 0.08205\,\dfrac{atm\cdot L}{mol\ °K} \quad T = °K$

ubility of A is exceeded. If this third phase contains water, then H cannot be determined from vapor pressure and solubility data because the partitioning in the third phase is unknown. However, if the third phase contains only pure component A, then the following expression can be used:

$$H = \frac{P_{y,A}}{C_{s,A}} \tag{5.5}$$

in which $C_{s,A}$ is the aqueous solubility of component A (mg/L). Equations 5.4 and 5.5 are estimation techniques that give H values within ±50 to 100 percent of experimentally reported values and should, therefore, only be used when measured values of the constants are not available.

Temperature, pressure, ionic strength, surfactants, and solution pH (for ionizable species such as NH_3 and CO_2) can influence the equilibrium partitioning between air and water. The impact of total system pressure on H is negligible because most aeration and stripping devices operate at atmospheric pressure.

For the range of temperatures encountered in water treatment, H tends to increase with increasing temperature because the aqueous solubility of the component decreases while its vapor pressure increases. Values of H for several organic compounds at different temperatures are summarized in Table 5.2. Table 5.3 displays a number of H values for gases at 20°C. Assuming the standard enthalpy change (ΔH^0) for the dissolution of a component in water is constant over the temperature range of interest, the change in H with temperature can be estimated using the following van't Hoff–type equation:

$$H_2 = H_1 \times \exp\left[\frac{-\Delta H^0}{R}\left(\frac{1}{T_2} - \frac{1}{T_1}\right)\right] \tag{5.6}$$

in which ΔH^0 is the standard enthalpy change in water due to the dissolution of a component in water (kcal/kmol), R is the universal gas constant (1.987 kcal/kmol·°K), H_1 is a known value of Henry's constant at temperature T_1 (°K), H_2 is the calculated Henry's law constant at the desired temperature T_2 (°K), and C is a constant. Eq. 5.6 can be simplified to Eq. 5.7 and values of ΔH^0 and C for selected compounds are summarized in Table 5.4.

$$H = C \times \exp\left(-\frac{\Delta H^0}{RT}\right) \tag{5.7}$$

Another common method of expressing the temperature dependence of H is to treat ΔH^0, R, and C as fitting parameters A and B using the following equation:

$$H = \exp\left(A - \frac{B}{T}\right) \tag{5.8}$$

Table 5.5 lists values of A and B for several compounds. These are valid for temperatures ranging from 283°K to 303°K (Ashworth et al., 1988).

Gases or VOCs in water supplies high in dissolved solids have higher volatility (or have a higher apparent Henry's law constant) than those with low dissolved solids. This results in a decrease in the solubility of the volatile component (i.e., a "salting-out effect"), which can be represented mathematically as an increase in the activity coefficient of component A, γ_A, in aqueous solution. γ_A will increase ($\gamma_A > 1$) with increasing ionic strength and this, in turn, causes the apparent Henry's law constant, H_{app}, to be greater than the thermodynamic value of H. The following equation can be used to calculate H_{app}:

TABLE 5.2 Henry's Law Constants in atm·m^3/gmole for 45 Organic Compounds*

Component	Henry's Law Constants, H				
	10°C	15°C	20°C	25°C	30°C
Nonane	0.400 (17.2)	0.496 (21.0)	0.332 (13.8)	0.414 (16.9)	0.465 (18.7)
n-hexane	0.238 (10.3)	0.413 (17.5)	0.883 (36.7)	0.768 (31.4)	1.56 (62.7)
2-methylpentane	0.697 (30.0)	0.694 (29.4)	0.633 (26.3)	0.825 (33.7)	0.848 (34.1)
Cyclohexane	0.103 (4.44)	0.126 (5.33)	0.140 (5.82)	0.177 (7.24)	0.223 (8.97)
Chlorobenzene	0.00244 (0.105)	0.00281 (0.119)	0.00341 (0.142)	0.00360 (0.147)	0.00473 (0.190)
1,2-dichlorobenzene	0.00163 (0.0702)	0.00143 (0.0605)	0.00168 (0.0699)	0.00157 (0.0642)	0.00237 (0.0953)
1,3-dichlorobenzene	0.00221 (0.0952)	0.00231 (0.0978)	0.00294 (0.122)	0.00285 (0.117)	0.00422 (0.170)
1,4-dichlorobenzene	0.00212 (0.0913)	0.00217 (0.0918)	0.00259 (0.108)	0.00317 (0.130)	0.00389 (0.156)
o-xylene	0.00285 (0.123)	0.00361 (0.153)	0.00474 (0.197)	0.00487 (0.199)	0.00626 (0.252)
p-xylene	0.00420 (0.181)	0.00483 (0.204)	0.00645 (0.268)	0.00744 (0.304)	0.00945 (0.380)
m-xylene	0.00411 (0.177)	0.00496 (0.210)	0.00598 (0.249)	0.00744 (0.304)	0.00887 (0.357)
Propylbenzene	0.00568 (0.245)	0.00731 (0.309)	0.00881 (0.366)	0.0108 (0.442)	0.0137 (0.551)
Ethylbenzene	0.00326 (0.140)	0.00451 (0.191)	0.00601 (0.250)	0.00788 (0.322)	0.0105 (0.422)
Toluene	0.00381 (0.164)	0.00492 (0.210)	0.00555 (0.231)	0.00642 (0.263)	0.00808 (0.325)
Benzene	0.00330 (0.142)	0.00388 (0.164)	0.00452 (0.188)	0.00528 (0.216)	0.00720 (0.290)
Methyl ethylbenzene	0.00351 (0.151)	0.00420 (0.178)	0.00503 (0.209)	0.00558 (0.228)	0.00770 (0.310)
1,1-dichloroethane	0.00368 (0.158)	0.00454 (0.192)	0.00563 (0.234)	0.00625 (0.256)	0.00776 (0.312)
1,2-dichloroethane	0.00117 (0.0504)	0.00130 (0.0550)	0.00147 (0.0612)	0.00141 (0.0577)	0.00174 (0.0700)
1,1,1-trichloroethane	0.00965 (0.416)	0.0115 (0.487)	0.0146 (0.607)	0.0174 (0.712)	0.0211 (0.849)
1,1,2-trichloroethane	0.000390 (0.0168)	0.000630 (0.0267)	0.000740 (0.0308)	0.000910 (0.0372)	0.00133 (0.0535)
cis-1,2-dichloroethylene	0.00270 (0.116)	0.00326 (0.138)	0.00360 (0.150)	0.00454 (0.186)	0.00575 (0.231)
trans-1,2-dichloroethylene	0.000590 (0.0254)	0.00705 (0.298)	0.00857 (0.356)	0.00945 (0.386)	0.0121 (0.469)
Tetrachloroethylene	0.00846 (0.364)	0.0111 (0.467)	0.0141 (0.587)	0.0171 (0.699)	0.0245 (0.985)
Trichloroethylene	0.00538 (0.237)	0.00667 (0.282)	0.00842 (0.350)	0.0102 (0.417)	0.0128 (0.515)

TABLE 5.2 Henry's Law Constants in atm·m³/gmole for 45 Organic Compounds* (*Continued*)

Component	Henry's Law Constants, *H*				
	10°C	15°C	20°C	25°C	30°C
Tetralin	0.000750 (0.0323)	0.00105 (0.0444)	0.00136 (0.0566)	0.00187 (0.0765)	0.00268 (0108)
Decalin	0.0700 (3.015)	0.0837 (3.54)	0.106 (4.41)	0.117 (4.79)	0.199 (8.00)
Vinyl chloride	0.0150 (0.646)	0.0168 (0.711)	0.0217 (0.903)	0.0265 (1.08)	0.028 (1.13)
Chloroethane	0.00759 (0.327)	0.00958 (0.405)	0.0110 (0.458)	0.0121 (0.495)	0.0143 (0.575)
Hexachloroethane	0.00593 (0.255)	0.00559 (0.237)	0.00591 (0.246)	0.00835 (0.342)	0.0103 (0.414)
Carbon tetrachloride	0.0148 (0.637)	0.0191 (0.808)	0.0232 (0.965)	0.0295 (1.21)	0.0378 (1.52)
1,3,5-trimethylbenzene	0.00403 (0.174)	0.00460 (0.195)	0.00571 (0.238)	0.00673 (0.275)	0.00963 (0.387)
Ethylene dibromide	0.000300 (0.0129)	0.000480 (0.0203)	0.000610 (0.0254)	0.000650 (0.0266)	0.000800 (0.0322)
1,1-dichloroethylene	0.0154 (0.663)	0.203 (8.59)	0.0218 (0.907)	0.0259 (1.06)	0.0318 (1.28)
Methylene chloride	0.00140 (0.0603)	0.00169 (0.0715)	0.00244 (0.102)	0.00296 (0.121)	0.00361 (0.145)
Chloroform	0.00172 (0.0741)	0.00233 (0.986)	0.00332 (0.138)	0.00421 (0.172)	0.00554 (0.223)
1,1,2,2-tetrachloroethane	0.000330 (0.0142)	0.000200 (0.00846)	0.000730 (0.0304)	0.000250 (0.0102)	0.000700 (0.0282)
1,2-dichloropropane	0.00122 (0.0525)	0.00126 (0.0533)	0.00190 (0.0790)	0.00357 (0.146)	0.00286 (0.115)
Dibromochloromethane	0.000380 (0.0164)	0.000450 (0.0190)	0.00103 (0.0428)	0.00118 (0.0483)	0.00152 (0.0611)
1,2,4-trichlorobenzene	0.00129 (0.0556)	0.00105 (0.0444)	0.00183 (0.0761)	0.00192 (0.0785)	0.00297 (0.119)
2,4-dimethylphenol	0.00829 (0.357)	0.00674 (0.285)	0.0101 (0.420)	0.00493 (0.202)	0.00375 (0.151)
1,1,2-trichlorotrifluoroethane	0.154 (6.63)	0.215 (9.10)	0.245 (10.2)	0.319 (13.0)	0.321 (12.9)
Methyl ethyl ketone	0.000280 (0.0121)	0.000390 (0.0165)	0.000190 (0.00790)	0.000130 (0.00532)	0.000110 (0.00443)
Methyl isobutyl ketone	0.000660 (0.0284)	0.000370 (0.0157)	0.000290 (0.0121)	0.000390 (0.0160)	0.000680 (0.0274)
Methyl cellosolve	0.0441 (1.90)	0.0363 (1.54)	0.116 (4.83)	0.0309 (1.26)	0.0381 (1.53)
Trichlorofluoromethane	0.0536 (2.31)	0.0680 (2.88)	0.0804 (3.34)	0.101 (4.13)	0.122 (4.91)

* Values in parentheses are *H* values in liters of water per liter of air.

Source: Reprinted from *Journal of Hazardous Materials,* Volume 18, Ashworth, Howe, Mullins, and Rogers, Air-Water Partitioning Coefficients of Organics in Dilute Aqueous Solutions, pp. 25–36, Copyright 1988, with permission from Elsevier Science.

TABLE 5.3 Henry's Law Constants at 20°C for Gases in Water Compounds*

Compound	H, (L_{H_2O}/L_{Air})
Ammonia	0.000574
Carbon dioxide	0.114
Chlorine	0.442
Chlorine dioxide	0.0408
Hydrogen sulfide	0.389
Methane	28.7
Oxygen	32.5
Ozone	3.77
Radon	1.69
Sulfur dioxide	0.0287

* J. M. Montgomery, Inc., 1985.

$$H_{\text{app}} = \gamma_A H = \frac{P_A}{[A]} \tag{5.9}$$

in which γ_A is a function of ionic strength and can be calculated using the following empirical equation:

$$\log_{10} \gamma_A = K_s \times I \tag{5.10}$$

where K_s is the Setschenow or "salting out" constant (L/mol), and I is the ionic strength of the water (mol/L). The magnitude of I can be estimated using the Lewis and Randall correlation:

$$I = \frac{1}{2} \sum_i (C_i Z_i^2) \tag{5.11}$$

in which C_i is the molar concentration of ionic species i (mol/L) and Z_i is the charge of species i. K_s values need to be determined by experimental methods because there is no general theory for predicting them. Table 5.6 displays the salting-out

TABLE 5.4 Temperature Correction Factors for H (in atm)*

Compound	$\Delta H \times 10^{-3}$	C
Oxygen	1.45	7.11
Methane	1.54	7.22
Carbon dioxide	2.07	6.73
Hydrogen sulfide	1.85	5.88
Carbon tetrachloride	4.05	10.06
Trichloroethylene	3.41	8.59
Benzene	3.68	8.68
Chloroform	4.00	9.10
Ozone	2.52	8.05
Ammonia	3.75	6.31
Sulfur dioxide	2.40	5.68
Chlorine	1.74	5.75

* J. M. Montgomery, Inc., 1985.

TABLE 5.5 Parameters for Calculating Henry's Law Constants (in atm m^3/gmole) as a Function of Temperature*

Component	*A*	*B*	r^2
Nonane	–0.1847	202.1	0.013
n-hexane	25.25	7530	0.917
2-methylpentane	2.959	957.2	0.497
Cyclohexane	9.141	3238	0.982
Chlorobenzene	3.469	2689	0.965
1,2-dichlorobenzene	–1.518	1422	0.464
1,3-dichlorobenzene	2.882	2564	0.850
1,4-dichlorobenzene	3.373	2720	0.941
o-xylene	5.541	3220	0.966
p-xylene	6.931	3520	0.989
m-xylene	6.280	3337	0.998
Propylbenzene	7.835	3681	0.997
Ethylbenzene	11.92	4994	0.999
Toluene	5.133	3024	0.982
Benzene	5.534	3194	0.968
Methyl ethylbenzene	5.557	3179	0.968
1,1-dichloroethane	5.484	3137	0.993
1,2-dichloroethane	–1.371	1522	0.878
1,1,1-trichloroethane	7.351	3399	0.998
1,1,2-trichloroethane	9.320	4843	0.968
cis-1,2-dichloroethylene	5.164	3143	0.974
trans-1,2-dichloroethylene	5.333	2964	0.985
Tetrachloroethylene	10.65	4368	0.987
Trichloroethylene	7.845	3702	0.998
Tetralin	11.83	5392	0.996
Decalin	11.85	4125	0.919
Vinyl chloride	6.138	2931	0.970
Chloroethane	4.265	2580	0.984
Hexachloroethane	3.744	2550	0.768
Carbon tetrachloride	9.739	3951	0.997
1,3,5-trimethylbenzene	7.241	3628	0.962
Ethylene dibromide	5.703	3876	0.928
1,1-dichloroethylene	6.123	2907	0.974
Methylene chloride	8.483	4268	0.988
Chloroform	11.41	5030	0.997
1,1,2,2-tetrachloroethane	1.726	2810	0.194
1,2-dichloropropane	9.843	4708	0.820
Dibromochloromethane	14.62	6373	0.914
1,2,4-trichlorobenzene	7.361	4028	0.819
2,4-dimethylphenol	–16.34	–3307	0.555
1,1,2-trichlorotrifluoroethane	9.649	3243	0.932
Methyl ethyl ketone	–26.32	–5214	0.797
Methyl isobutyl ketone	–7.157	160.6	0.002
Methyl cellosolve	–6.050	–873.8	0.023
Trichlorofluoromethane	9.480	3513	0.998

* Ashworth et al., 1988.

TABLE 5.6 Setschenow (or Salting-Out) Coefficients, K_s, at 20°C

Compound	K_s, (L·mol^{-1})	Reference
Tetrachloroethylene	0.213	Gossett, 1987
Trichloroethylene	0.186	Gossett, 1987
1,1,1-trichloroethane	0.193	Gossett, 1987
1,1-dichloroethane	0.145	Gossett, 1987
Chloroform	0.140	Gossett, 1987
Dichloromethane	0.107	Gossett, 1987
Benzene	0.195	Schwarzenbach et al., 1992
Toluene	0.208	Schwarzenbach et al., 1992
Naphthalene	0.220	Schwarzenbach et al., 1992

coefficients for several compounds at 20°C. For most water supplies, the ionic strength is less than 10 mM and the activity coefficient is equal to one. Significant increases in volatility and the apparent Henry's constant are only observed for very high ionic-strength waters, such as seawater.

pH does not affect the Henry's constant per se, but it does affect the distribution of species between ionized and unionized forms. This, in turn, influences the overall gas-liquid distribution of the compound because only the unionized species are volatile. For example, ammonia is present as ammonium ion at neutral pH and is not strippable. However, at high pH (greater than 10), ammonia is not an ion and may be stripped. To predict the effect of pH on solubility, one must consider the value of the appropriate acidity constant pKa. If the acid is uncharged, such as HCN or H_2S, and the pH is much less than the pKa (2 units lower), then equilibrium partitioning can be described using the Henry's law constant for the uncharged species. If the acid is charged, such as NH_4^+, and the pH is much higher than the pKa (2 units higher), or if the acid is uncharged, such as H_2S, and the pH is much less than the pKa (2 units lower), then the compound will be volatile and its equilibrium partitioning can be described using Henry's law constant for the uncharged species.

Surfactants can also affect the volatility of compounds. Most natural waters do not have high concentrations of surfactants; consequently, their influence does not affect the design of most stripping devices. However, if surfactants are present, they lower the volatility of compounds by several mechanisms. The most important factor is that they tend to collect at the air-water interface, diminishing the mole fraction of the compound at the interfacial area, thereby lowering its apparent Henry's law constant. In untreated wastewater for example, the solubility of oxygen can be lowered by 30 to 50 percent due to the presence of surfactants. Another effect for hydrophobic organics is the incorporation of the dissolved organics into micelles in solution (This would only occur above the critical micelle concentration) which, in turn, decreases the concentration of the organic compound at the air-water interface and lowers the compound's volatility.

Mass Transfer

The driving force for mass transfer between one phase and another derives from the displacement of the system from equilibrium. Figure 5.3 displays two situations in which mass transfer is occurring between the air and water. Figure 5.3(*a*) displays the situation in which mass is being transferred from the water to the air, and Figure 5.3(*b*) displays the situation in which mass is being transferred from the air to the water. Because the mechanisms and assumptions for mass transfer are essentially

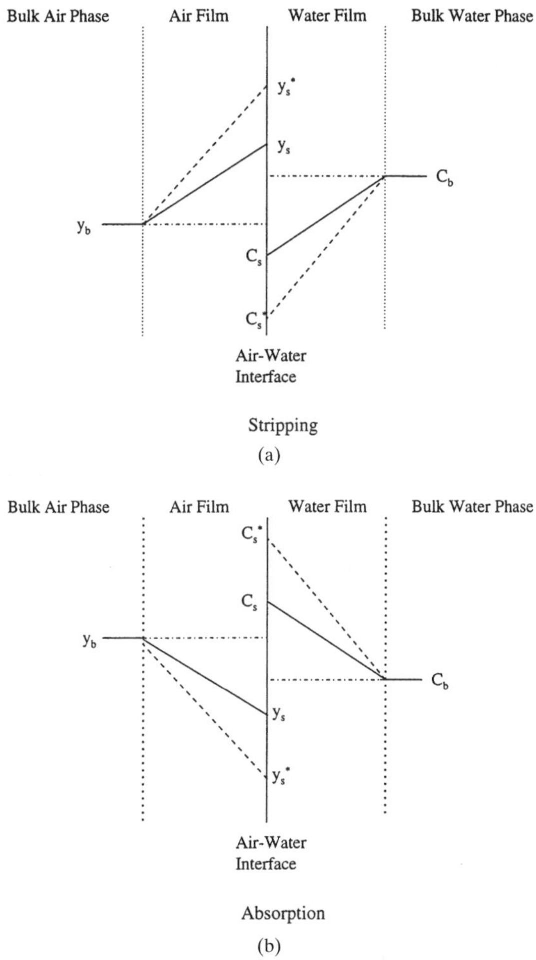

FIGURE 5.3 Diagram describing the equilibrium partitioning of a contaminant between the air and water phases using two-film theory.

the same for both cases, a detailed explanation of only one is warranted. Consider the case in which a volatile contaminant is being stripped from water to air [Fig. 5.3(*a*)]. The contaminant concentration in the water is high relative to the equilibrium concentration between the air and water. The tendency to achieve equilibrium is sufficient to cause diffusion of the aqueous-phase contaminant molecules from the bulk solution at some concentration, C_b (M/L^3), to the air-water interface where the aqueous phase concentration is C_s (M/L^3). Because C_b is larger than C_s, the difference between them provides the aqueous-phase driving force for stripping. Similarly, the contaminant concentration in the air at the air-water interface, y_s, is larger than the contaminant concentration in the bulk air, y_b, and diffusion causes the molecules to migrate from the air-water interface to the bulk air. The difference between y_s and y_b is the driving force for stripping in the gas phase.

Local equilibrium occurs at the air-water interface because, on a local scale of 10s of angstroms, random molecular movement causes the contaminant to dissolve in the aqueous phase and volatilize into the air. (These motions on a larger scale equalize the displacement of the system from equilibrium and cause diffusion.) Accordingly, Henry's law can be used to relate y_s to C_s (Lewis and Whitman, 1924).

$$y_s = HC_s \tag{5.12}$$

Fick's Law can be simplified to a linear driving force approximation, whereby the flux (the mass transferred per unit of time per unit of interfacial area) across the air-water interface is proportional to the concentration gradient. Mathematically, the flux of A, across the air-water interface, N_A, is given by:

$$N_A = k_l\,(C_b - C_s) = k_g\,(y_s - y_b) \tag{5.13}$$

in which k_l is the liquid-phase mass transfer coefficient that describes the rate at which contaminant A is transferred from the bulk aqueous phase to the air-water interface (L/t) and k_g is the gas-phase mass transfer coefficient that describes the rate at which contaminant A is transferred from the air-water interface to the bulk gas phase (L/t). Both k_l and k_g are sometimes called local mass transfer coefficients for the liquid and gas phases because they depend upon the conditions at or near the air-water interface in their particular phase. Because the interfacial concentrations y_s and C_s cannot be measured and are unknown, the flux cannot be determined from Eq. 5.13. Consequently, it is necessary to define another flux equation in terms of hypothetical concentrations that are easy to determine and that describe the displacement of the system from equilibrium. If it is hypothesized that all the resistance to mass transfer is on the liquid side, there is no concentration gradient on the gas side, and a hypothetical concentration C_s^* can be defined as shown in Figure 5.3(a)

$$y_b = HC_s \tag{5.14}$$

in which C_s^* is the aqueous-phase concentration of A at the air-water interface assuming no concentration gradient in the air phase (M/L^3); that is, that equilibrium exists between the bulk gas-phase concentration and the aqueous phase at the interface.

Alternatively, if it is hypothesized that all the resistance to mass transfer is on the gas side, there is no concentration gradient on the liquid side and a hypothetical concentration $y^*{}_s$ can be defined as shown in Figure 5.3(a).

$$y_s^* = H\,C_b \tag{5.15}$$

in which, y_s^* is the equilibrium gas-phase concentration of A at the air-water interface, assuming no concentration gradient in the liquid phase (M/L^3), and it may be calculated from the bulk liquid-phase concentration.

For stripping operations, mass balances are normally written on the liquid, and it is convenient to calculate the mass transfer rate using the hypothetical concentration, C_s^*, and an overall mass transfer coefficient, K_L, as shown in the following equation.

$$N_A = K_L\,(C_b - C_s^*) \tag{5.16}$$

Equating 5.13 and 5.16 results in:

$$N_A = k_l\,(C_b - C_s) = k_g\,(y_s - y_b) = K_L\,(C_b - C_s^*) \tag{5.17}$$

From Figure 5.3(a), one can obtain the following relationship:

$$(C_b - C_s^*) = (C_b - C_s) + (C_s - C_s^*) \tag{5.18}$$

Substituting $C_s = \frac{y_s}{H}$ and $C_s^* = \frac{y_b}{H}$ into Eq. 5.18 and combining with Eq. 5.17 yields:

$$\frac{N_A}{K_L} = \frac{N_A}{k_l} + \frac{N_A}{H\,k_g} \tag{5.19}$$

or

$$\frac{1}{K_L} = \frac{1}{k_l} + \frac{1}{H\,k_g} \tag{5.20}$$

Eq. 5.20 simply states that the overall resistance to mass transfer is equal to the sum of the mass transfer resistances for the liquid and gas phases. It describes the overall mass transfer rate without the use of the unknown local concentrations at the air-water interface.

When designing aeration and stripping processes, the rate of mass transfer is often expressed on a volumetric basis rather than an interfacial area basis. The flux term, N_A, is converted to a volumetric basis by multiplying by the specific interfacial area, *a*, which is defined as the interfacial area available for mass transfer divided by the system unit volume (L^2/L^3). Equation 5.20 can be expressed in terms of a volumetric mass transfer rate by multiplying N_A by a in Eq. 5.19 and rewritten as:

$$\frac{1}{K_L a} = \frac{1}{k_l a} + \frac{1}{H\,k_g a} \tag{5.21}$$

Although this modification is based on the liquid side, it includes the resistance to mass transfer in the gas phase and gives an exact representation of the mass flux across the air-water interface. A similar relationship may be developed for an overall mass transfer coefficient on the gas side, which is useful for scrubbing operations. The final equation is given as:

$$\frac{1}{K_G a} = \frac{H}{k_l a} + \frac{1}{k_g a} \tag{5.22}$$

in which K_G is the overall gas-phase mass transfer coefficient (L/t). The mass flux may be obtained from the gas-phase mass balance by multiplying K_G by the hypothetical driving force $(C_b - C_s^*)$.

Evaluating which phase controls the mass transfer rate is important in optimizing the design and operation of aeration and air stripping processes. For example, when the mass transfer rate is controlled by the liquid-phase resistance, increasing the mixing of the air will have little impact on the removal efficiency. Increasing the interfacial area will increase the removal efficiency. Equation 5.20 simply states that the overall resistance to mass transfer is equal to the sum of the resistance in liquid and gas phases and can be rewritten as:

$$R_T = R_L + R_G \tag{5.23}$$

in which R_T $(1/K_L)$ is the overall resistance to mass transfer (t/L), R_L $(1/k_l)$ is the liquid-phase resistance to mass transfer (t/L) and R_G $(1/k_g)$ is the gas-phase resistance to mass transfer (t/L). Equation 5.20 can also be rearranged as follows to evaluate which phase controls the rate of mass transfer:

$$\frac{R_L}{R_T} = \frac{\frac{1}{k_l}}{\frac{1}{k_l} + \frac{1}{Hk_g}} = \frac{1}{1 + \frac{1}{(k_g/k_l)H}} \tag{5.24}$$

or

$$\frac{R_L}{R_T}(\%) = \frac{100}{1 + \frac{1}{(k_g/k_l)H}} \tag{5.25}$$

Literature reported values of the ratio k_g/k_l range from 40 to 200 (Munz and Roberts, 1989) depending upon the type of aeration or stripping system. Assuming k_g/k_l is 100, Eq. 5.25 can be used to determine which phase controls the rate of mass transfer across the interface. Table 5.7 gives the phase that controls the rate of mass transfer for a number of compounds with a wide range of Henry's law constants. In general, the liquid phase controls the rate of mass transfer for compounds with H values greater than 0.05, and low liquid solubility. The gas phase controls mass transfer for compounds with H values less than 0.002, and high liquid solubility. For compounds with H values between 0.002 and 0.05 either phase may be preferred, and both the liquid and gas phases affect the rate of mass transfer.

UNIT OPERATIONS

Packed Towers

The primary applications of packed towers in water treatment are the stripping of VOCs, carbon dioxide, hydrogen sulfide, and ammonia. An example of a packed-tower system is displayed in Figure 5.4. Water is pumped to the top of the tower and through a liquid distributor where it then is allowed to flow by gravity over a packing material. At the same time, a blower is used to introduce fresh air into the bottom of the tower, and the air flows countercurrent to the water up through the void spaces between the wetted packing material. The packing material provides a large air-water interfacial area resulting in efficient transfer of the volatile contaminant from the water to the air. The contaminant-free, air-stripped water leaves the bottom of the tower while the air containing the contaminant exits the top of the tower for further treatment or exhaustion to the atmosphere. The packing material consists of individual pieces randomly dumped into the column. Figure 5.5 displays several different varieties of commercially available packing material. Table 5.8 lists some packing materials and their physical properties as reported by the manufacturers.

TABLE 5.7 Summary of Controlling Phase for Several Compounds

Compound	H at (25°C) ($L\ H_2O/L$ air)	R_L/R_T (%)	Controlling phase
Oxygen	29.29	100	Liquid
Carbon Tetrachloride	1.2	99	Liquid
Trichloroethene	0.40	98	Liquid
Chloroform	0.15	94	Liquid
Acetone	0.001	9	Gas

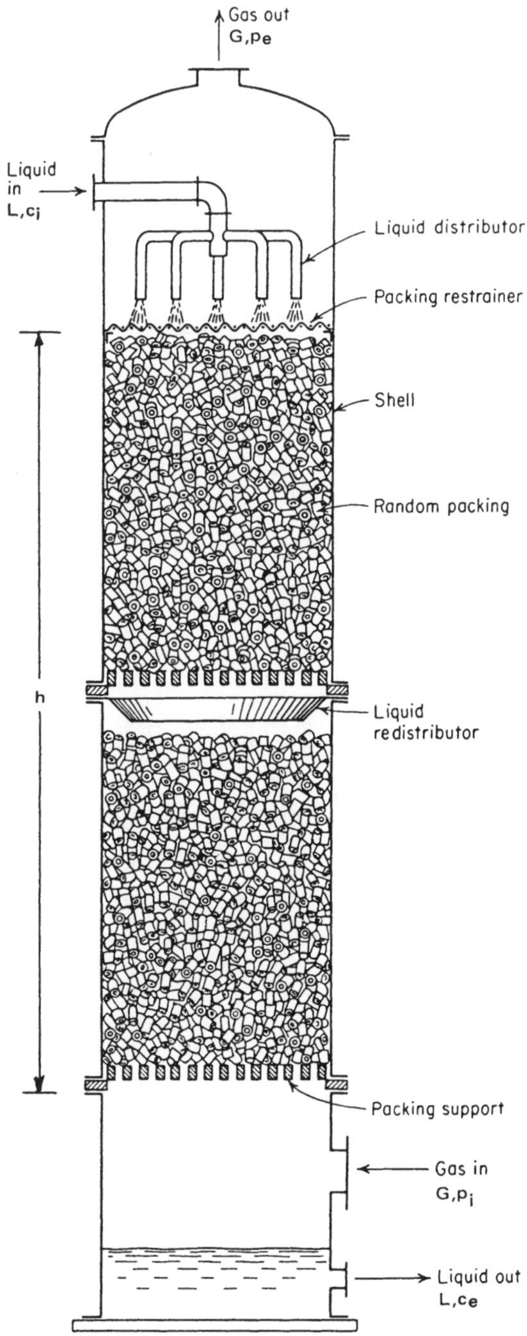

FIGURE 5.4 Schematic of a packed tower.

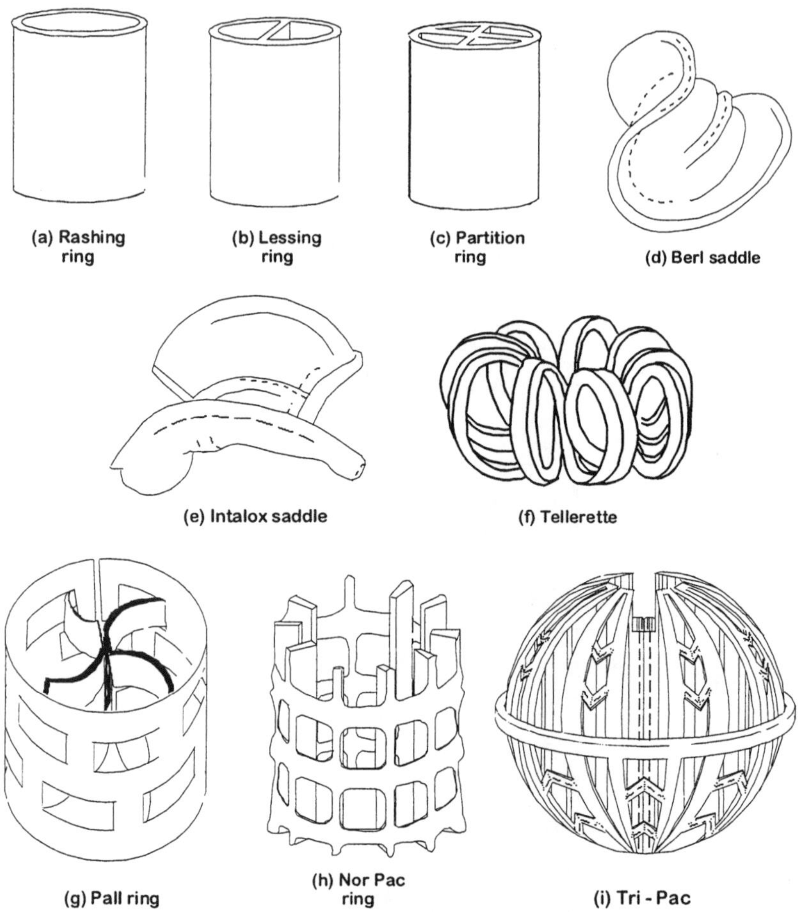

FIGURE 5.5 Example packing materials for air stripping towers.

Cross-flow, cascade, and cocurrent packed-tower systems have similar configurations, but the airflow is introduced differently. In the cross-flow system, the air flows across the tower packing at a 90° angle to the direction of the water flow. Methods for estimating the performance of the cross-flow system are available in the literature (Little and Selleck, 1991). In the cascade system, fresh air is introduced at various points along the tower and flows countercurrent to the water flow (Nirmalakhandan et al., 1991, 1992). Both the cross-flow and cascade systems provide larger airflow rates at lower gas pressure drops than those typically used in the conventional countercurrent system. The larger airflow rates will provide a greater driving force for stripping, leading to more efficient removal of contaminants with low volatility. The cross-flow and cascade systems are recommended for the packed-tower systems when very high air-to-water ratios are required to remove semi- and low-volatility contaminants from water. In the cocurrent system, both the air and water enter the top of the tower, flow through the packing, and exit at the bottom of the tower. The cocurrent system is rarely used in water treatment.

TABLE 5.8 Physical Characteristics of Packing Materials

Type of packing	Nominal diameter in.	Nominal diameter m	Packing factor	Specific surface area ft²/ft³	Specific surface area m²/m³	Surface tension lb-ft	Surface tension N/m
Plastic saddles[a]	3.0	0.0762	16.0	27.1	89.0	0.024	0.033
	2.0	0.0508	20.0	33.5	110.0	0.024	0.033
Plastic tripacks[b]	3.5	0.0889	14.0	42.0	138.0	0.024	0.033
	2.0	0.0508	15.0	47.8	157.0	0.024	0.033
Plastic pall rings[a]	2.0	0.0508	25.0	31.0	102.0	0.024	0.033
Plastic tellerettes[c]	2.0	0.0508	20.0	34.1	112.0	0.024	0.033
Nor-Pac[b]	1.5	0.0381	17.0	43.9	144.0	0.024	0.033
	2.0	0.0508	12.0	31.1	102.0	0.024	0.033
Flexring[d]	2.0	0.0508	24.0	35.1	115.0	0.024	0.033
	3.5	0.0889	20.0	28.0	92.0	0.024	0.033
IMPAC[e]	3.3	0.0838	15.0	64.9	213.0	0.024	0.033
	5.5	0.140	6.0	32.9	108.0	0.024	0.033
LANPAC[e]	2.3	0.0584	21.0	68.0	223.0	0.024	0.033
	3.5	0.0889	14.0	45.1	148.0	0.024	0.033

[a] Norton Co., Akron, Ohio.
[b] Jaeger Co., Houston, Texas.
[c] Ceilcote Co., Cleveland, Ohio.
[d] KOCH Co., Wichita, Kansas.
[e] LANTEC Co., Los Angeles, California.

Recently, sieve tray columns or low-profile air strippers have been used to remove VOCs from contaminated waters. Sieve tray columns operate as a countercurrent process. The columns are typically less than 10 ft (3 m) high and consist of several perforated trays placed in series along the column. Water enters at the top of the tower and flows horizontally across each tray. Inlet and outlet channels, or downcomers, are placed at the ends of each tray to allow the water to flow from tray to tray. At the same time, fresh air flows up from the bottom of the tower through the tray holes and the layer of water flowing across each tray. Large airflow rates are typically used, causing very small bubbles or frothing to occur upon air contact with the water. The frothing provides a high air-water surface area for mass transfer to occur. Because the water flows horizontally across each tray, the desired removal efficiency can be obtained by increasing the length or width of the trays instead of the height. The advantages of using a low-profile air stripper over a conventional packed-tower stripper are that (1) the low-profile air stripper is smaller and more compact, and (2) it is easier to perform periodic maintenance on a low-profile air stripper. The disadvantage is that for a given removal, the low-profile air stripper requires a significantly higher air flow than the conventional packed tower. Consequently, the operational costs will be greater. In addition, low-profile air strippers are limited to water flow rates less than 1000 gpm (0.0630 m^3/s).

Design Equations

Mathematical models describing countercurrent packed-tower process for water treatment are well established in the literature (e.g., Sherwood and Hollaway, 1940; Treybal, 1980; Kavanaugh and Trussel, 1980; Ball, Jones, and Kavanaugh, 1984; Singley et al., 1981; Umphres et al., 1983; Cummins and Westerick, 1983; McKinnon and Dyksen, 1984; Roberts et al., 1985; Gross and TerMaath, 1985; Roberts and Levy, 1985; Hand et al., 1986; Dzombak, Roy, and Fang, 1993; Hokanson, 1996). These models have been successfully used to describe the packed-tower air stripping process and, consequently, to size the towers for a given removal. Figure 5.6 displays a mathematical representation of the packed-tower process shown in Figure 5.4. The overall assumptions in the model development are: (1) plug-flow conditions prevail in the air and water streams; (2) the incoming air contains no contaminant; (3) the contaminant concentration in the influent water stream is constant; (4) air and water temperatures are constant and equal to the inlet water temperature; and (5) Henry's law describes the chemical equilibrium between the air and water phases. The first step in the development of the design equations is to establish the relationship between the air- and water-phase contaminant concentrations within the packed tower. Figure 5.7 compares the equilibrium line, as defined by Henry's law equation, with the operating line for packed-tower air stripping. The operating line relates the bulk air–phase contaminant concentration to the bulk water–phase contaminant concentration and is based on a mass balance around the bottom of the air stripping tower:

$$y_b(z) = \frac{Q}{\dot{V}} [C_b(z) - C_e] \tag{5.26}$$

in which y_b is the bulk air–phase concentration of the contaminant at any point along the packed-tower height (M/L^3), Q is the water flow rate to the tower (L^3/T), $\dot{V}$ is the volumetric airflow rate to the tower (L^3/T), C_b is the bulk water–phase concentration of the contaminant at any point along the packed-tower height (M/L^3), and C_e is the bulk water–phase contaminant concentration at the bottom of the tower (M/L^3). The driving force for stripping can be quantified by comparing the operating line with the equilibrium line. The bulk air–phase contaminant concentration corresponding to a given water-phase contaminant concentration is found by holding C constant and moving vertically until the operating line is intersected. This is the bulk air–phase contaminant concentration. The water-phase contaminant concentration, which would be in equilibrium with the air-phase contaminant concentration, is found by holding y constant until it intersects the equilibrium line. The difference between these two concentrations is the driving force for stripping and it must be positive for stripping to occur.

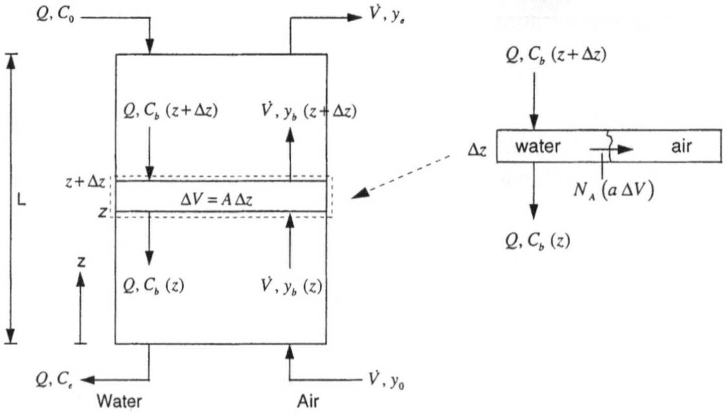

FIGURE 5.6 Schematic of a packed tower used in the development of the design equations.

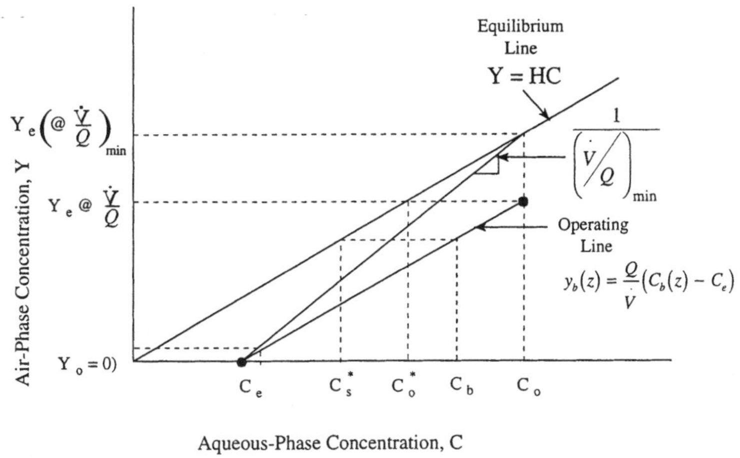

FIGURE 5.7 Operating line diagram for a packed tower.

Incorporating the above assumptions into an aqueous-phase mass balance around the differential element circumscribed by the dashed box in Figure 5.6 yields the following expression for removal of a contaminant within a packed tower:

$$\frac{dC_b}{dz} = \frac{K_L aA}{Q}(C_b - C_s^*) \tag{5.27}$$

in which C_s^* is the water-phase concentration of the contaminant in equilibrium with the bulk air–phase contaminant concentration at any point along the packed tower. The integral form of Eq. 5.27 can be expressed as:

$$\int_0^L dz = \frac{Q}{A\,K_L a}\int_{C_e}^{C_0}\frac{dC_b}{C_b - C_s^*} \tag{5.28}$$

Integration of Eq. 5.28 requires an expression for C_s^* in terms of C_b. From Figure 5.7, C_s^* can be expressed in terms of the bulk air–phase contaminant concentration using Henry's law and substituting into Eq. 5.26 to yield:

$$C_s^* = \frac{Q}{\dot{V}H}[C_b - C_e] \tag{5.29}$$

Substitution of Eq. 5.29 into Eq. 5.28 and integrating yields the following design equations for the countercurrent packed-tower process:

$$L = \frac{Q}{A\,K_L a}\left[\frac{C_0 - C_e}{C_0 - C_e - C_0^*}\right]\log_e\left[\frac{C_0 - C_0^*}{C_e}\right] \tag{5.30}$$

$$C_0^*(z = L) = \left(\frac{Q}{\dot{V}H}\right)(C_0 - C_e) \tag{5.31}$$

in which C_0^* is the aqueous-phase concentration of the contaminant in equilibrium with the exiting air-phase contaminant concentration, y_e (M/L^3). C_0^* is obtained by rearranging Eq. 5.26, assuming y_b equals y_e and C_b equals C_0, then substituting C_0^* for

y_e/H. Equations 5.30 and 5.31 are sometimes combined, and the quantity $(\dot{V}H)/(Q)$, defined as the stripping factor (R), is introduced.

The minimum air-to-water ratio required for stripping can be determined by considering the driving force for stripping at the top of the tower. Under the best scenario, the exiting air will be in equilibrium with the incoming water, which implies that C_0^* in Eq. 5.31 is equal to C_0, and upon substitution into Eq. 5.31, the following expression may be obtained.

$$\left(\frac{\dot{V}}{Q}\right)_{\min} = \frac{(C_0 - C_e)}{HC_0} \tag{5.32}$$

The minimum air-to-water ratio, $(\dot{V}/Q)_{\min}$, is the lowest air-to-water ratio that can be applied for a packed tower and meet a given contaminant's treatment objective, C_e. If the air-to-water ratio applied is less than the minimum air-to-water ratio, it will not be possible to design a packed tower capable of meeting the treatment objective, because equilibrium will be established in the tower before the treatment objective is reached. This can be explained using the stripping factor, R. R represents an equilibrium capacity parameter, and when it is greater than 1, there is sufficient capacity in the air to convey all the solute in the entering water, and complete removal by stripping is possible given a sufficiently long column. When $R < 1$, the system performance is limited by equilibrium, and the fractional removal is asymptotic to the value of R (i.e., $1 - C_e/C_0 \rightarrow R$ as $L \rightarrow \infty$, as shown by Eq. 5.32). When $R = 1$, the tower is operating at the minimum air-to-water ratio required for stripping.

With respect to the selection of the optimum air-to-water ratio, Hand et al. (1986) demonstrated that minimum tower volumes and power requirements were approximately achieved using 3.5 times the minimum air-to-water ratio for contaminants with Henry's law constants greater than 0.05 for high-percentage removals. This corresponds to a stripping factor of 3.5.

In packed-tower air stripping, the tower length is often defined as the product of the height of a transfer unit (or HTU) and the number of transfer units (or NTU). The following two equations define HTU and NTU:

$$HTU = \frac{Q}{A\, K_L a} \tag{5.33}$$

$$NTU = \left[\frac{C_0 - C_e}{C_0 - C_e - C_0^*}\right] \ln \left[\frac{C_0 - C_0^*}{C_e}\right] \tag{5.34}$$

or incorporating Eq. 5.31 and R into Eq. 5.30 yields:

$$NTU = \left[\frac{R}{R-1}\right] \ln \left[\frac{\frac{C_0}{C_e}(R-1) + 1}{R}\right] \tag{5.35}$$

Substituting Eq. 5.33 and Eq. 5.34 or 5.35 into Eq. 5.30 results in:

$$L = HTU \times NTU \tag{5.36}$$

The NTU is determined by the driving force for stripping and corresponds to the number of stages (or hypothetical, completely mixed tanks at equilibrium) required for stripping. Equation 5.34 shows that NTU depends upon R and VOC removal efficiency, and this relationship is displayed in Figure 5.8. For a given R, the removal efficiency increases with increasing NTU (or number of hypothetical, completely mixed tanks). In addition, for a given removal efficiency, increasing R or the air-to-water

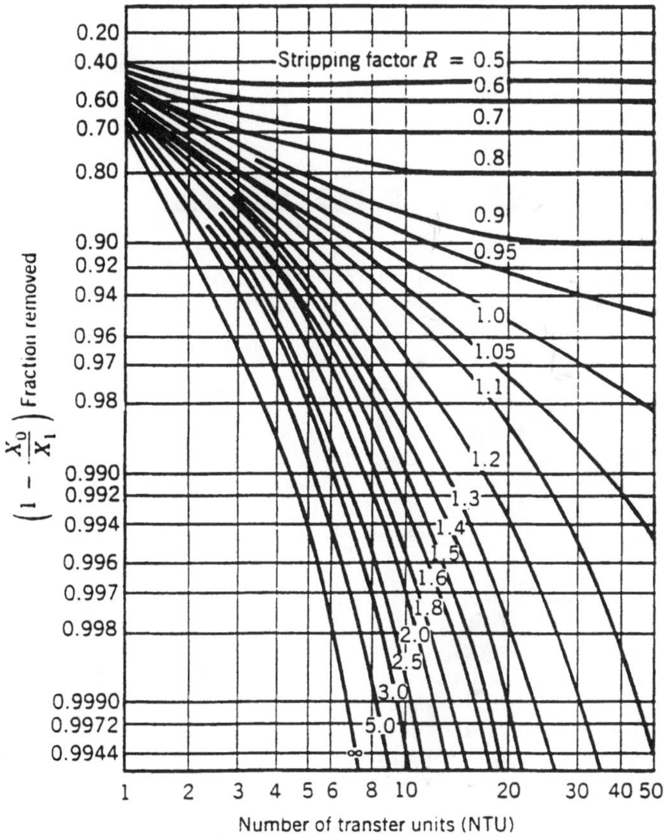

FIGURE 5.8 Dependence of removal efficiency on *NTU* on and stripping factor. (*Source:* Treybal, R. E. 1980. *Mass Transfer Operations.* New York: Chem. Engrg. Series, McGraw-Hill Book Co., 3rd. ed. Reproduced with permission of The McGraw-Hill Companies.)

ratio will decrease the number of *NTU*s required. The *HTU* is the height of one stage, as determined by the rate of mass transfer. To determine the packed-tower height from Eq. 5.30, the value of K_La and the tower diameter must be determined.

Determination of K_La

Values of K_La for packed towers can be determined by performing pilot plant studies or obtained from experimental values reported by packing manufacturers and previous reported field studies. When experimental K_La values are not available, they can be estimated from mass transfer correlations. A pilot plant study is the preferred way to determine K_La for a given VOC in a contaminated water. Pilot-scale packed towers range in size from 2 to 6 m (6.5 to 20 ft) in height and 0.3 to 0.61 m (1 to 4 ft) in diameter. The column diameter used will depend upon the desired packing size. It is generally recommended that column-diameter-to-packing-diameter ratios be greater than 10:1 (>15:1 is desired) to minimize error due to channeling of the water down the walls of the column (Treybal, 1980). It is also recommended the packed-column height-to-diameter ratio be greater than 1:1 to provide for proper liquid distribution (Roberts et al., 1995; Treybal, 1980). Pilot plant design guidelines such as these, as well as construction and operation of pilot-scale packed-tower units, is presented elsewhere (JMM Consulting Engineers, 1985).

Equation 5.36 is used in conjunction with pilot plant data to determine K_La for a given VOC. The value of K_La is based on VOC removal due to the packed-height portion of the tower. However, VOC removal also occurs as the water contacts the air above the packing at the top of the tower and at the bottom as the water falls into the clear well below the packing. This removal is sometimes referred to as *end effects* (Umphres et al., 1983). Consequently, an *NTU* correction factor for VOC removal due to end effects is used when determining K_La. The following equation can be used to calculate the *NTU* of the packing (Umphres et al., 1983):

$$NTU_{\text{packing}} = NTU_{\text{measured}} - NTU_{\text{end effects}} \tag{5.37}$$

where NTU_{measured} is the experimentally determined value from pilot data and $NTU_{\text{end effects}}$ is the *NTU* due to the end effects. Combining Eqs. 5.36 and 5.37 yields the following linear expression, which can be used to determine K_La:

$$NTU_{\text{measured}} = \frac{1}{HTU}(Z) + NTU_{\text{end effects}} \tag{5.38}$$

where Z is the distance from the top of the packing to a sample port location along the packed portion of the tower. For a given water and air loading rate, aqueous-phase concentration measurements are evaluated at the influent, effluent, and various sample port locations along the packed column. Equation 5.35 can be used to calculate NTU_{measured} at each sample port location or distance, Z, from the top of the packing where C_e is assumed to be the concentration measured at each sample port along the packed column. A plot is constructed of NTU_{measured} as a function of Z and should coincide with the linear expression given by Eq. 5.38. K_La is determined from the slope ($1/HTU$) and Eq. 5.33. Experimentally determined K_La values can be correlated as a function of water loading rate for several air-to-water ratios that would be expected during operation of the full-scale column. For a given water loading rate and air-to-water ratio of interest, a correlated K_La value can be determined and used with Eqs. 5.33 through 5.36 to determine the required full-scale packed-tower height.

A full-scale packed-tower height calculated using a K_La value determined from a pilot study is generally conservative. For a given packing size, K_La values generally increase as the tower diameter increases (Wallman and Cummings, 1986). This is caused by channeling of the water down the inside of the column walls, which is sometimes referred to as *wall effects.* The VOC removal rate is less along the walls of the column as compared with the removal within the packing because the air-water contact time, surface area, and mixing are less. As the tower diameter increases, the percentage of flow down the walls of the column decreases minimizing the wall effects. Consequently, the K_La value in a full-scale column will be larger than one in a smaller pilot column.

Table 5.9 displays numerous packed-tower field studies that reported experimentally determined K_La values for several VOCs and various contaminated water sources. The K_La values reported in these studies can be used, provided the operating conditions (temperature, water and air loading rate), and packing type and size are the same.

When K_La values are not readily available, they can be estimated from the following mass transfer correlations (Onda, Takeuchi, and Okumoto, 1968):

TABLE 5.9 Summary of Several Packed-Tower Air Stripping Pilot Plant Studies That Determined K_La Values for Several VOCs

Reference	Water matrix	VOCs
Umpheres et al., 1983	Sacramento-San Joaquin Delta water in Northern California	Chloroform Dibromochloromethane Bromodichloromethane Bromoform
Ball et al., 1984	Potomac tidal fresh estuary water mixed with nitrified effluent wastewater	Carbon tetrachloride Tetrachloroethylene Trichloroethylene Chloroform Bromoform
Byers and Morton, 1985	City of Tacoma, Washington, groundwater	1,1,2,2-tetrachloroethane *trans*-1,2-dichloroethylene Trichloroethylene Tetrachloroethylene
Roberts et al., 1985	Laboratory-Grade Organic Free Water	Oxygen Tetrachloroethylene Freon-12 1,1,1-trichloroethane Trichloroethylene Carbon tetrachloride
Bilello and Singley, 1986	North Miami Beach, Florida, groundwater and City of Gainsville, Florida, groundwater	Chloroform
Wallman and Cummins, 1986	Village of Brewster, New York, groundwater	*cis*-1,2-dichloroethylene Trichloroethylene Tetrachloroethylene
Lamarche and Droste, 1989	Gloucester, Ottawa, Ontario, groundwater	Chloroform Toluene 1,2-dichloroethane 1,1-dichloroethane Trichloroethylene Diethyl ether

$$a_w = a_t \left\{1 - \exp\left[-1.45\left(\frac{\sigma_c}{\sigma}\right)^{0.75}\left(\frac{L_m}{a_t\mu_l}\right)^{0.1}\left(\frac{L_m^{\ 2}\,a_t}{\rho_l^2\,g}\right)^{-0.05}\left(\frac{L_m^{\ 2}}{\rho_l a_t\,\sigma}\right)^{0.2}\right]\right\} \quad (5.39)$$

$$k_l = 0.0051\left(\frac{L_m}{a_w\mu_l}\right)^{2/3}\left(\frac{\mu_l}{\rho_l D_l}\right)^{-0.5}(a_t d_p)^{0.4}\left(\frac{\rho_l}{\mu_l\,g}\right)^{-1/3} \quad (5.40)$$

$$k_g = 5.23(a_t D_g)\left(\frac{G_m}{a_t\mu_g}\right)^{0.7}\left(\frac{\mu_g}{\rho_g D_g}\right)^{1/3}(a_t d_p)^{-2} \quad (5.41)$$

in which a_t is the total specific surface area of the packing material (L^2/L^3), a_w is the wetted surface area of the packing material, d_p is the nominal diameter of the packing material (L), D_g is the gas-phase diffusivity of the contaminant to be removed (L^2/t), D_l is the liquid-phase diffusivity of the contaminant to be removed (L^2/t), g is the gravitational constant (L/t^2), G_m is the air mass loading rate ($M/L^2 \cdot t^{-1}$), L_m is the water mass loading rate ($M/L^2 \cdot t^{-1}$), μ_L is the viscosity of water ($M/L \cdot t^{-1}$), μ_g is the viscosity of air ($M/L \cdot t^{-1}$), ρ_l is the density of water (M/L^3), ρ_g is the density of air (M/L^3), σ is the surface tension of water (M/t^2), and σ_c is the critical surface tension of the packing material (M/t^2).

Equation 5.40 is valid for water loading rates between 1.1 and 63 gpm/ft^2 (0.8 and 43 kg/m$^2 \cdot$ s^{-1}), and Eq. 5.41 is valid for air loading rates between 2.206 and 267.9 cfm/ft^2 (0.014 and 1.7 kg/m$^2 \cdot$ s^{-1}). To use these correlations, it is assumed that a_w is equivalent to a. Equations 5.40 and 5.41 were correlated for nominal packing sizes up to 2 in (0.0508 m). Roberts et al. (1985) and Cummins and Westrick (1983) showed that, for several VOCs, K_La values obtained from the Onda correlations compared favorably with pilot plant data. Lamarche and Droste (1989) evaluated available mass transfer models for packed-tower aeration (Onda, Takeuchi, and Okumoto, 1968; Sherwood and Holloway, 1940; and Shulman, Ullrich, and Wells, 1955) and determined that the Onda model gives the best predictions of the mass transfer coefficient.

Recent work by several researchers demonstrated that the use of Equations 5.39 through 5.41 in design calculations overestimates the magnitude of K_La (Lenzo et al., 1990; Thom and Byers, 1993; Djebbar and Narbaitz, 1995). This is especially true for situations in which large packing sizes (>1 in nominal diameter) are used. At the present time, there is no correlation that will predict K_La to within ±10 percent for larger packing sizes. For this reason, based on the literature cited above, it is recommended that a safety factor of 0.75 (i.e., K_La/Onda K_La) be applied for nominal packing sizes greater than 1 in to provide a conservative estimate of the packing height required.

Determination of the Tower Diameter

The diameter for a single packed tower can be determined using the generalized pressure drop curves shown in Figure 5.9 (Eckert, 1961). Once the packing factor for the media, the air-to-water ratio, and the air pressure drop per packing height have been specified, it is possible to determine the air loading rate and the tower diameter from Figure 5.9 or the following set of equations (Cummins and Westrick, 1983):

$$G_m = \sqrt{\frac{M\,\rho_g(\rho_l - \rho_g)}{C_f\,(\mu_l)^{0.1}}} \tag{5.42}$$

The empirical parameter M is defined by the following relationship:

$$\log_{10}(M) = a_0 + a_1\,E + a_2\,E^2 \tag{5.43}$$

The parameter E is defined as follows:

$$E = -\log_{10}\left[\left(\frac{\dot{V}}{Q}\right)\sqrt{(\rho_g/\rho_l) - (\rho_g/\rho_l)^2}\right] \tag{5.44}$$

The parameters a_0, a_1, and a_2 are defined by the following relationships:

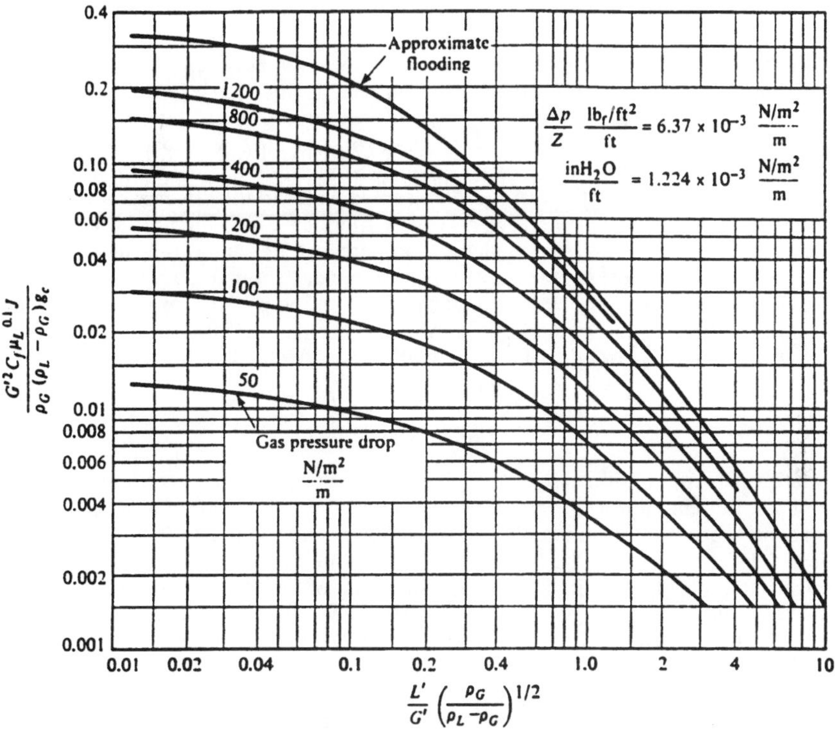

FIGURE 5.9 Flooding and pressure drop in random packed towers. (*Source:* Treybal, R. E. 1980. *Mass Transfer Operations.* New York: Chem. Engrg. Series, McGraw-Hill Book Co., 3rd. ed. Reproduced with permission of The McGraw-Hill Companies.)

$$a_0 = -6.6599 + 4.3077F - 1.3503F^2 + 0.15931F^3 \tag{5.45}$$

$$a_1 = 3.0945 - 4.3512F + 1.6240F^2 - 0.20855F^3 \tag{5.46}$$

$$a_2 = 1.7611 - 2.3394F + 0.89914F^2 - 0.11597F^3 \tag{5.47}$$

The parameter F that appears in Eqs. 5.41 through 5.43 is defined as:

$$F = \log_{10}(\Delta P/L) \tag{5.48}$$

The pressure drop correlation is valid for $\Delta P/L$ values between 50 and 1200 $N/m^2 \cdot m^{-1}$. Equations 5.42 through 5.48 were developed using a particular set of units. When applying them, it must be assured that values are supplied in the appropriate units: G_m is the air mass loading rate ($kg/m^2 \cdot s^{-1}$); M is a calculated empirical parameter; ρ_g is the air density (kg/m^3); ρ_l is the water density (kg/m^3); C_f is the packing factor (dimensionless); μ_l is the water viscosity ($kg/m \cdot s^{-1}$); a_0, a_1, and a_2 are calculated empirical parameters; E is a calculated empirical parameter; V/Q is the air-to-water ratio on a volumetric basis (m^3/m^3); F is a calculated empirical parameter; and $\Delta P/L$

is the air pressure drop gradient ($N/m^2 \cdot m^{-1}$). The tower diameter can be determined by dividing G_m by the air mass flow rate.

The practical operating range for packed towers is between abscissa values of 0.02 and 4.0 in Figure 5.9 (JMM Consulting Engineers, 1985). For abscissa values greater than 4, very large water loading rates can reduce the water-air contact area provided by the packing surface and inhibit proper airflow through the column, causing a decrease in VOC removal efficiency. Similarly, high airflow rates (abscissa values less than about 0.02) can cause entrained water in the tower as well as channeling of the air through the tower. For situations in which high airflow rates are required for high removal efficiencies (>95 percent), it is important to provide an even air inlet distribution at the bottom of the tower (Thom and Byers, 1993).

The total operating power for a single air stripping packed-tower system is the sum of the blower and pump brake power requirements. The blower brake power can be determined from the following relationship (Metcalf and Eddy, 1991):

$$P_{\text{blower}} = \left(\frac{G_m R_g T_{\text{air}}}{1000 n_a \, Eff_b}\right)\left[\left(\frac{P_{\text{in}}}{P_{\text{out}}}\right)^{n_a} - 1\right] \tag{5.49}$$

where Eff_b is the blower net efficiency, expressed as a decimal, which accounts for both the fan and motor on the blower; G_m is the mass flow rate of air (M/T); n_a is a constant used in determining blower brake power and is equal to 0.283 for air; P_{in} is the inlet air pressure in the packed tower (bottom of tower) ($M/L \cdot T^{-2} \cdot L^{-2}$); P_{out} is the outlet air pressure in the packed tower (top of the tower) and is usually equal to the ambient pressure ($M/L \cdot T^{-2} \cdot L^{-2}$); R_g is the universal gas constant; and T_{air} is the absolute air temperature, which is typically assumed equal to T.

The term P_{in} refers to the pressure at the bottom of the tower, which is the inlet for the airstream. P_{in} is calculated as the sum of the ambient pressure and the pressure drop caused by the packing media, demister, packing support plate, duct work, inlet, and outlet of the tower. The equation used to find P_{in} is:

$$P_{\text{in}} = P_{\text{ambient}} + \left[\left(\frac{\Delta P}{L}\right)L\right] + \Delta P_{\text{losses}} \tag{5.50}$$

ΔP_{losses} is determined by (Hand et al., 1986):

$$\Delta P_{\text{losses}} = \left(\frac{\dot{V}}{A}\right)^2 k_p \tag{5.51}$$

in which $\dot{V}$ is the volumetric airflow rate (L^3/T), A is the tower cross-sectional area (L^2), and k_p is a constant equal to 275 N s^2/m^4 and is used to estimate the air pressure drop for losses other than that caused by the tower packing (M/L^3).

Eq. 5.51 represents the air pressure drop through the demister, packing support plate, duct work, inlet, and outlet of the tower. It is assumed that turbulent flow conditions prevail and most of the losses occur in the tower (i.e., in the packing support plate and the demister).

The pump power requirement can be determined from the following equation:

$$P_{\text{pump}} = \frac{\rho_l Q H g}{1000 \, Eff_p} \tag{5.52}$$

in which Eff_p is the pump efficiency and H is the vertical distance from the pump to the liquid distributor at the top of the tower (L). Eq. 5.52 accounts only for the pressure loss resulting from the head required to pump the water to the top of the tower.

Design Procedure

The procedure used to solve a particular packed-tower design problem will depend upon the type of design problem. The most common design problems are modifications to existing towers and developing design criteria for new ones. Consequently, the design engineer may use a different approach for each problem. For example, one problem may be to determine ways of improving process efficiency of an existing tower by increasing the airflow rate, inserting a more efficient packing type, or increasing the packed-tower height. Or a designer may need to design a new packed-tower system using standard vendor column sizes. Problems such as these require many tedious calculations, which can easily be performed using commercially available software. Software tools such as ASDC (Dzombak et al., 1993), AIRSTRIP (Haarhoff, 1988), and ASAP (Hokanson, 1996) can be used to evaluate the impact of process variables on the process performance. These tools contain the design equations and correlations presented above, graphical user interfaces for ease in software application, and they provide databases for many commercially available packing types and physical properties of several VOCs, which have been encountered in surface and groundwaters.

Presented herein is a step-by-step design procedure for sizing a packed tower and important considerations that should be addressed during the design and operation phases. This procedure is only one of several methods (Kavanaugh and Trussell, 1980; J. M. Montgomery, Inc., 1985) that has been successfully used to design a packed tower. This procedure was presented in detail elsewhere (Hand et al., 1986) and will simply be highlighted herein. The design parameters for packed-tower air stripping are: (1) the air-to-water ratio, (2) the gas pressure drop, and (3) the type of packing material. Once the physical properties of the contaminant(s) of interest, the influent concentration(s), treatment objective(s), water, and air properties are known, these design parameters can be selected to give the lowest capital, and operation and maintenance costs. As discussed above, the optimum air-to-water ratio for contaminants with Henry's law constants greater than 0.05, which will minimize the tower volume and total power consumption, is around 3.5 times the minimum air-to-water ratio. Therefore, the optimum air-to water ratio can be estimated by multiplying the minimum air-to-water ratio calculated from Eq. 5.32 by 3.5. A low gas pressure drop should be chosen to minimize the blower power consumption. In most cases, a gas pressure drop of 50 to 100 $N/m^2 \cdot m^{-1}$ is chosen. To determine the optimum gas pressure drop that will result in a least-cost design, a detailed cost analysis would have to be performed to determine the true costs associated with the tower volume and power requirement. A number of researchers have performed detailed cost analyses, and their results show that using a low gas pressure drop of around 50 to 100 $N/m^2 \cdot m^{-1}$ and an air-to-water ratio of around 3.5 times the minimum yields the lowest total annual treatment cost (Cummins, 1985; Dzombak, Roy, and Fang, 1993). An additional advantage of operating at a low gas pressure drop is that if the blower is sized conservatively, the airflow rate can be increased to meet a greater removal demand without major changes in the process operation. However, the required height using this approach may sometimes be too large for a particular application. In this case, the air-to-water ratio can be increased by increasing the airflow rate to obtain a smaller tower height for a given removal.

The criteria for choosing the type and size of packing will depend upon the water flow rate and the desired degree of operational flexibility of the design. For small water flow rates, it is recommended that nominal diameter packing of 2 in (0.0508 m) or less be used to minimize channeling or short circuiting of the water down the wall of the tower. Minimizing the impact of channeling requires that the ratio of tower diameter to nominal packing diameter be greater than 12 to 15. If a design is needed with a high degree of operational flexibility, a larger, less efficient packing type should be chosen for the initial design of the tower. Tower designs with larger packing sizes will result in larger tower volumes because the larger packing sizes usually have a lower specific area than the smaller ones. However, if the treatment objective becomes more stringent or other less strippable compounds appear in the influent, the less efficient, larger packing can be replaced with smaller, more efficient packing to provide for the needed removal. Building in this type of operational flexibility is much cheaper than adding additional height to the existing tower (if possible) or adding another tower in series. Spending additional capital on providing operational flexibility in the initial design can ultimately reduce future costs and prevent headaches.

In most situations in water treatment, multiple contaminants are usually present in the water and the packed tower must be designed to remove all the contaminants to some specified treatment level. For a given tower design, the following equation can be used to determine the effluent concentrations of multiple contaminants for a given tower design:

$$C_e = \frac{C_0(R-1)}{\left\{R\exp\left[\frac{L\,K_{\mathrm{L}}a(R-1)}{(Q/A)R}\right]-1\right\}} \tag{5.53}$$

However, at the design stage, the limiting contaminant that controls the design must first be determined. In general, the contaminant with the lowest Henry's constant is used to determine the required air-to-water ratio, and the contaminant with the highest removal efficiency is used to determine the required packing height. The following example problem illustrates a typical design procedure for sizing a packed tower.

EXAMPLE 5.1 ***Sample Packed-Tower Aeration Calculation for TCE***
Design a packed tower to reduce the trichloroethylene (TCE) concentration from 100 μg/L to 5.0 μg/L. Listed below are the operating conditions, TCE, physical and chemical properties of air and water, packing type, and packing parameters. For a design water flow of 0.1577 m^3/s (2500 gpm), determine the tower diameter, packed-tower height, and power consumption requirements.

Operating Conditions

Property	Value
Temperature (T, T_{air}), C	10
Pressure (P, P_{ambient}), atm	1

TCE, Water, and Air Characteristics

Property	Value
TCE molecular weight	131.39
Henry's constant (H), dimensionless	0.230
Influent TCE concentration (C_{i}), μg/L	100
Effluent TCE concentration (C_e), μg/L	5.0
Molarvolume @ boiling point (V_{b}), cm^3/gmol	107.1
TCE normal boiling pt temperature, °C	87.0
Density of water (ρ_1), kg/m^3	999.75
Viscosity of water (μ_1), $kg/m \cdot s^{-1}$	1.31×10^{-3}
Density of air (ρ_g), kg/m^3	1.25
Viscosity of air (μ_g), $kg/m \cdot s^{-1}$	1.72×10^{-5}
Surface tension of water (σ), N/m	0.0742

Packing Characteristics	
Packing Type: 3.5-Inch Nominal Diameter Tripacks	
Property	Value
Nominal diameter (d_p), m	0.0889
Nominal surface area (a_t), m^2/m^3	125
Critical surface tension (σ_c), N/m	0.0330

Constants	
Property	Value
Acceleration due to Gravity (g), m/s^2	9.81
Blower efficiency (Eff_b), %	35
Pump efficiency (Eff_p), %	80

SOLUTION:

1. Calculate the minimum air-to-water ratio $(\dot{V}/Q)_{min}$, using Eq. 5.32.

$$\dot{V}/Q)_{min} = \frac{C_0 - C_e}{H\,C_0} = \frac{100\,\frac{\mu g}{L} - 5.0\,\frac{\mu g}{L}}{0.230 \times 100\,\frac{\mu g}{L}} = 4.13$$

2. Calculate a reasonable $\dot{V}/Q$ that is some multiple i.e. assume 3.5 times of $(\dot{V}/Q)_{min}$.

$$(\dot{V}/Q)_{mult} = (\dot{V}/Q)_{min} \times 3.5 = 4.13 \times 3.5 \approx 15.0$$

3. Calculate the air flow rate, $\dot{V}$.

$$\dot{V} = \frac{\dot{V}}{Q} \times Q = 15.0 \times 0.1577\,\frac{m^3}{s} = 2.37\,\frac{m^3}{s}$$

4. Calculate the tower diameter, D.

a. Choose a low value of $\Delta P/L = 50$ N/m$^2 \cdot$ m^{-1}.

b. Calculate F, a_0, a_1, a_2, E, M and G_m using Eqs. 5.42 through 5.48. The tower area can then be determined from either G_m or L_m.

$$F = \log_{10}(\Delta P/L) = \log_{10}\left(50\,\frac{N}{m^2 \cdot m}\right) = 1.699$$

$$a_0 = -6.6599 + 4.3077F - 1.3503F^2 + 0.15931F^3$$

$$= -6.6599 + 4.3077(1.699) - 1.3503\,(1.699)^2 + 0.15931(1.699)^3$$

$$= -2.45758$$

$$a_1 = 3.0945 - 4.3512F + 1.6240F^2 - 0.20855F^3$$

$$= 3.0945 - 4.3512(1.699) + 1.6240(1.699)^2 - 0.20855(1.699)^3$$

$$= -0.63315$$

$$a_2 = 1.7611 - 2.3394F + 0.89914F^2 - 0.11597F^3$$
$$= 1.7611 - 2.3394(1.699) + 0.89914(1.699)^2 - 0.11597(1.699)^3$$
$$= -0.18684$$

$$E = -\log_{10}\left[\left(\frac{\dot{V}}{Q}\right)\sqrt{(\rho_g/\rho_l) - (\rho_g/\rho_l)^2}\right]$$

$$= -\log_{10}\left[15.0\sqrt{\left(\frac{1.25\ \dfrac{\text{kg}}{\text{m}^3}}{999.75\ \dfrac{\text{kg}}{\text{m}^3}}\right) - \left(\frac{1.25\ \dfrac{\text{kg}}{\text{m}^3}}{999.75\ \dfrac{\text{kg}}{\text{m}^3}}\right)^2}\right]$$

$$= 0.2756$$

$$\log_{10}(M) = a_0 + a_1 E + a_2 E^2$$
$$= -2.45762 + -0.63313(0.2756) + -0.18683(0.2756)^2$$
$$= -2.6463 \Rightarrow$$
$$M = 0.0022578$$

c. Calculate air mass loading rate, G_m, using Eq. 5.42.

$$G_m = \sqrt{\frac{M\rho_g(\rho_l - \rho_g)}{C_f(\mu_l)^{0.1}}}$$

$$= \sqrt{\frac{2.2578 \times 10^{-3} \times 1.25\ \dfrac{\text{kg}}{\text{m}^3}\ (999.75\ \dfrac{\text{kg}}{\text{m}^3} - 1.25\ \dfrac{\text{kg}}{\text{m}^3})}{12(1.31 \times 10^{-3}\ (\text{kg/m}\cdot\text{s}^{-1})^{0.1}}}$$

$$= 0.675\ \text{kg/m}^2\cdot\text{s}^{-1}$$

d. Calculate water mass loading rate, L_m.

$$L_m = \frac{G_m}{\left(\dfrac{\dot{V}}{Q}\right)\left(\dfrac{\rho_g}{\rho_l}\right)} = \frac{0.675\ \dfrac{\text{kg}}{\text{m}^2/\text{s}}}{\left(15.0\ \dfrac{\text{m}^3}{\text{m}^3}\right)\left(\dfrac{1.25\ \text{kg/m}^3}{999.7\ 5\ \text{kg/m}^3}\right)} = 36.0\ \text{kg/m}^2\cdot\text{s}^{-1}$$

e. Calculate tower cross-sectional area, A.

$$A = \frac{Q\rho_l}{L_m} = \frac{0.1577\ \dfrac{\text{m}^3}{\text{s}} \times 999.75\ \dfrac{\text{kg}}{\text{m}^3}}{36.0\ \text{kg/m}^2\cdot\text{s}^{-1}} = 4.38\ \text{m}^2$$

f. Calculate tower diameter, D.

$$D = \sqrt{\frac{4A}{\pi}} = \sqrt{\frac{4(4.38\ \text{m}^2)}{\pi}} = 2.36\ \text{m}\ (7.75\ \text{ft})$$

Standard packed-column sizes are typically available in diameter increments of 0.3048 m (1 ft). Consequently, a tower diameter of 2.44 m (8.0 ft) is chosen for this design. The following design parameters were adjusted for a tower diameter of 2.44 m.

$$A = 4.67\ \text{m}^2 \quad L_\text{m} = 33.76\ \frac{\text{kg}}{\text{m}^2\cdot\text{s}} \quad G_\text{m} = 0.634\ \frac{\text{kg}}{\text{m}^2\cdot\text{s}} \quad \frac{\Delta p}{L} = 43\ \frac{\text{Pa}}{\text{m}}$$

5. Calculate tower length, L.

a. Calculate the TCE aqueous phase concentration in equilibrium with the exiting air, C^*_0, using Eq. 5.31.

$$C^*_0 = \left(\frac{1}{(\dot{V}/Q)\times H}\right)(C_0 - C_e) = \left(\frac{1}{(15.0)\times 0.230}\right)\left(100\ \frac{\mu\text{g}}{\text{L}} - 5.0\ \frac{\mu\text{g}}{\text{L}}\right)$$

$$= 27.54\ \frac{\mu\text{g}}{\text{L}}$$

b. Calculate the specific surface area available for mass transfer, a_w, using Eq. 5.39.

$$a = a_\text{w}$$

$$a_\text{w} = a_\text{t}\left\{1 - \exp\left[-1.45\left(\frac{\sigma_\text{c}}{\sigma}\right)^{0.75}\left(\frac{L_\text{m}}{a_\text{t}\mu_\text{l}}\right)^{0.1}\left(\frac{L_\text{m}^{\,2}a_\text{t}}{\rho_\text{l}^2 g}\right)^{-0.05}\left(\frac{L_\text{m}^{\,2}}{\rho_\text{l}a_\text{t}\sigma}\right)^{0.2}\right]\right\}$$

$$= 125\ \frac{\text{m}^2}{\text{m}^3}\left\{1 - \exp\left[-1.45\left(\frac{0.0330\frac{\text{N}}{\text{m}}}{0.0742\frac{\text{N}}{\text{m}}}\right)^{0.75}\left(\frac{33.76\ \frac{\text{kg}}{\text{m}^2\cdot\text{s}}}{125\ \frac{\text{m}^2}{\text{m}^3}\times 1.31\times 10^{-3}\ \frac{\text{kg}}{\text{m}\cdot\text{s}}}\right)^{0.1}\times\left(\frac{\left(33.76\ \frac{\text{kg}}{\text{m}^2\cdot\text{s}}\right)^2\times 125\ \frac{\text{m}^2}{\text{m}^3}}{\left(999.75\ \frac{\text{kg}}{\text{m}^3}\right)^2\times 9.81\ \frac{\text{m}}{\text{s}^2}}\right)^{-0.05}\times\left(\frac{\left(33.76\ \frac{\text{kg}}{\text{m}^2\cdot\text{s}}\right)^2}{999.75\ \frac{\text{kg}}{\text{m}^3}\times 125\ \frac{\text{m}^2}{\text{m}^3}\times 0.0742\ \frac{\text{N}}{\text{m}}}\right)^{0.2}\right]\right\}$$

$$= 83.10\ \frac{\text{m}^2}{\text{m}^3}$$

c. Calculate the TCE diffusivity in water, D_l using the following correlation (Hayduk and Laudie, 1974).

$$D_\text{l} = \frac{13.26\times 10^{-5}}{\mu_\text{w}^{1.14}\times V_\text{b}^{0.589}} = \frac{13.26\times 10^{-5}}{(1.31\ cp)^{1.14}\times\left(107.1\ \frac{\text{cm}^3}{\text{gmol}}\right)^{0.589}\times\frac{100\ \text{cm}^2}{1\ \text{m}^2}}$$

$$= 6.21\times 10^{-10}\ \frac{\text{m}^2}{\text{s}}$$

μ_w = viscosity of water (centipoise)

V_b = molar volume, cm^3/gmol

d. Calculate the liquid phase mass transfer coefficient, k_l, using Eq. 5.40.

$$k_l = 0.0051\left(\frac{L_m}{a_w\mu_l}\right)^{2/3}\left(\frac{\mu_l}{\rho_l D_l}\right)^{-0.5}(a_t d_p)^{0.4}\left(\frac{\rho_l}{\mu_l g}\right)^{-1/3}$$

$$= 0.0051\left(\frac{33.76\left(\frac{\text{kg}}{\text{m}^2}\cdot\text{s}^{-1}\right)}{83.10\,\frac{\text{m}^2}{\text{m}^3}\times 1.31\times 10^{-3}\left(\frac{\text{kg}}{\text{m}}\cdot\text{s}^{-1}\right)}\right)^{2/3}\left(\frac{1.31\times 10^{-3}\left(\frac{\text{kg}}{\text{m}}\cdot\text{s}^{-1}\right)}{999.75\,\frac{\text{kg}}{\text{m}^3}\times 6.21\times 10^{-10}\,\frac{\text{m}^2}{\text{s}}}\right)^{-0.5}$$

$$\times\left(125\,\frac{\text{m}^2}{\text{m}^3}\times 0.0889\text{ m}\right)^{0.4}\left(\frac{999.75\,\frac{\text{kg}}{\text{m}^3}}{1.31\times 10^{-3}\,(\text{kg/m}\cdot\text{s}^{-1})\times 9.81\,\frac{\text{m}}{\text{s}^2}}\right)^{-1/3}$$

$$= 3.12\times 10^{-4}\,\frac{\text{m}}{\text{s}}$$

e. Calculate the gas-phase diffusivity, D_g, using the following correlation by Hirschfelder-Bird-Spotz (1949) (loc cit. Treybal, 1980).

i. To calculate the gas diffusivity, the following parameters must be determined.

$$r_A = 1.18(V_b)^{1/3} = 1.18\times\left(0.1071\,\frac{\text{m}^3}{\text{kgmol}}\right)^{1/3} = 0.561\text{ nm (TCE)}$$

$$\varepsilon_A/K = 1.21(T_b)_A = 1.21\times(87+273.15)°\text{K} = 435.6°\text{K (TCE)}$$

in which, r_A is the radius of molecule A, nm (TCE); V_b is the molar volume of molecule A at the boiling point (TCE); ε_A is the force constant of molecule A, Joule (TCE); K is the Boltzmann's Constant, Joule/°K; T_{bA} is the boiling temperature of molecule A, °K (TCE); r_B is the characteristic radius of the gas medium, nm, and is equal to 0.3711 nm for air.

Calculate the molecular separation at collision, r_{AB}, for TCE in air.

$$r_{AB} = (r_A + r_B)/2 = \frac{0.561\text{nm} + 0.3711\text{nm}}{2} = 0.4661\text{nm (TCE in air)}$$

Calculate the energy of molecular attraction between molecule A (TCE) and molecule B (air).

$$\frac{\varepsilon_{AB}}{K} = \sqrt{\frac{\varepsilon_A}{K}\times\frac{\varepsilon_B}{K}} = \sqrt{435.6\times 78.6} = 185.0\text{ (TCE in air)}$$

Calculate the collision function for diffusion of molecule A (TCE) in B (air). [The collision function was presented by Treybal (1980) and correlated by Cummins and Westrick (1983).]

$$EE = \log_{10}(KT/\varepsilon_{AB}) = \log_{10}\left(\frac{283.15°\text{K}}{185.0}\right) = 0.185$$

$$NN = \begin{pmatrix} -0.14329 - 0.48343(EE) + 0.1939(EE)^2 + 0.13612(EE)^3 \\ -0.20578(EE)^4 + 0.083899(EE)^5 - 0.011491(EE)^6 \end{pmatrix}$$

$$= \begin{pmatrix} -0.14329 - 0.48343(0.185) + 0.1939(0.185)^2 \\ +0.13612(0.185)^3 - 0.20578(0.185)^4 + 0.083899(0.185)^5 \\ -0.011491(0.185)^6 \end{pmatrix}$$

$$= -0.225$$

$$f(KT/\varepsilon_{AB}) = 10^{NN} = 10^{-0.225} = 0.595$$

ii. Calculate gas-phase diffusivity, D_g, as follows:

$$D_g = \left\{ \frac{10^{-4}\left(1.084 - 0.249\sqrt{\frac{1}{M_A} + \frac{1}{M_B}}\right)(T^{1.5})\sqrt{\frac{1}{M_A} + \frac{1}{M_B}}}{P_t(r_{AB})^2 f\left(\frac{KT}{\varepsilon_{AB}}\right)} \right\} \frac{\text{m}^2}{\text{s}}$$

$$= \left\{ \frac{10^{-4}\left(1.084 - 0.249\sqrt{\frac{1}{131.39} + \frac{1}{28.95}}\right)(283.15^{1.5})\sqrt{\frac{1}{131.39} + \frac{1}{28.95}}}{101325\,\frac{\text{N}}{\text{m}^2}(0.4661\text{nm})^2\, 0.595} \right\} \frac{\text{m}^2}{\text{s}}$$

$$= 7.71 \times 10^{-6}\, \frac{\text{m}^2}{\text{s}}$$

f. Calculate gas-phase TCE mass transfer coefficient, k_g, using Eq. 5.41.

$$k_g = 5.23(a_t D_g)\left(\frac{G_m}{a_t \mu_g}\right)^{0.7} \left(\frac{\mu_g}{\rho_g D_g}\right)^{1/3} (a_t d_p)^{-2}$$

$$= 5.23\left(125\,\frac{\text{m}^2}{\text{m}^3} \times 7.71 \times 10^{-6}\,\frac{\text{m}^2}{\text{s}}\right)\left(\frac{0.634\ (\text{kg/m}\cdot\text{s}^{-1})}{125\,\frac{\text{m}^2}{\text{m}^3} \times 1.72 \times 10^{-5}\ (\text{kg/m}\cdot\text{s}^{-1})}\right)^{0.7}$$

$$\times \left(\frac{1.72 \times 10^{-5}\ (\text{kg/m}\cdot\text{s}^{-1})}{1.25\,\frac{\text{kg}}{\text{m}^3} \times 7.71 \times 10^{-6}\,\frac{\text{m}^2}{\text{s}}}\right)^{1/3} \left(125\,\frac{\text{m}^2}{\text{m}^3} \times 0.0889\text{m}\right)^{-2}$$

$$= 2.65 \times 10^{-3}\,\frac{\text{m}}{\text{s}}$$

g. Calculate $K_L a$ using Eq. 5.21.

$$\frac{1}{K_L a} = \frac{1}{k_l a_w} + \frac{1}{k_g a_w H}$$

$$= \frac{1}{3.12 \times 10^{-4}\,\frac{\text{m}}{\text{s}} \times 83.1\,\frac{\text{m}^2}{\text{m}^3}} + \frac{1}{2.65 \times 10^{-3}\,\frac{\text{m}}{\text{s}} \times 83.1\,\frac{\text{m}^2}{\text{m}^3} \times 0.230}$$

$$K_L a = 0.0171\ \text{s}^{-1}$$

h. Calculate K_La assuming a safety factor of 0.75.

$$K_La = K_La \times (SF)_{K_La} = 0.0171\text{s}^{-1} \times 0.75$$
$$= 0.0128\text{s}^{-1}$$

i. Calculate tower length, *L*, using Eq. 5.30.

$$L = \frac{Q}{A\,K_La}\left[\frac{C_o - C_{TO}}{C_o - C_{TO} - C^*_0}\right]\ln\left[\frac{C_0 - C^*_0}{C_{TO}}\right]$$

$$= \frac{0.1577\,\dfrac{\text{m}^3}{\text{s}}}{4.67\text{ m}^2 \times 0.0171\text{s}^{-1}}\left[\frac{100\,\dfrac{\mu\text{g}}{\text{L}} - 5.0\,\dfrac{\mu\text{g}}{\text{L}}}{100\,\dfrac{\mu\text{g}}{\text{L}} - 5.0\,\dfrac{\mu\text{g}}{\text{L}} - 27.54\,\dfrac{\mu\text{g}}{\text{L}}}\right]\ln\left[\frac{100\,\dfrac{\mu\text{g}}{\text{L}} - 27.54\,\dfrac{\mu\text{g}}{\text{L}}}{5.0\,\dfrac{\mu\text{g}}{\text{L}}}\right]$$

$$= 9.94\text{ m (32.6 ft)}$$

Typical packed-tower heights usually do not exceed about 9.0 m (30 ft) in height. Consequently, for this design, the air-to-water ratio can be increased by increasing the airflow rate to achieve the same treatment objective but with a smaller tower height. For example, if the air-to-water ratio is increased to 30, a packed-tower height of 7.62 m (25 ft) is required for the same removal.

6. Calculate power requirements.

a. Calculate blower power requirements.

i. Calculate the air mass flow rate, G_m, from the volumetric air flow rate.

$$G_{me} = \dot{V} \times \rho_g = 2.37\,\frac{\text{m}^3}{\text{s}} \times 1.25\,\frac{\text{kg}}{\text{m}^3} = 2.96\,\frac{\text{kg}}{\text{s}}$$

ii. Calculate the pressure drop through the demister, the packing support plate, duct work, inlet and outlet, ΔP_{losses}, using Eq. 5.51.

$$\Delta P_{losses} = \left(\frac{\dot{V}}{A}\right)^2 k_p = \left(\frac{2.37\text{ m}^3/\text{s}}{4.67\text{ m}^2}\right)^2 \times 275\text{ N}\cdot\text{s}^2/\text{m}^4 = 70.83\text{ N/m}^2$$

k_p = a constant used to estimate gas pressure drop in a tower for losses other than that in the packing,

$$275\,\frac{\text{N}\cdot\text{s}^2}{\text{m}^4}\text{ (Hand et al., 1986).}$$

iii. Calculate the inlet pressure to the packed tower, P_{in}, using Eq. 5.50.

$$P_{in} = P_{ambient} + \left[\left(\frac{\Delta P}{L}\right)L\right] + \Delta P_{losses}$$

$$= 101325\,\frac{\text{N}}{\text{m}^2} + \left[\left(43\,\frac{\text{N/m}^2}{\text{m}}\right)9.94\text{ m}\right] + 70.83\,\frac{\text{N}}{\text{m}^2}$$

$$= 101823\,\frac{\text{N}}{\text{m}^2}$$

iv. Calculate the blower brake power, P_{blower}, using Eq. 5.49.

$$P_{blower} = \left(\frac{G_{me}R_g T_{air}}{1000 n_a Eff_b}\right)\left[\left(\frac{P_{in}}{P_{out}}\right)^{n_a} - 1\right]$$

$$= \left(\frac{2.96\,\frac{kg}{s} \times 286.7\,\frac{J}{kg \cdot K} \times 283.15\,K}{1000\,\frac{W}{kW}\,(0.283)(0.35)}\right)\left[\left(\frac{101823\,\frac{N}{m^2}}{101325\,\frac{N}{m^2}}\right)^{0.283} - 1\right]\left(\frac{1W}{1\frac{J}{s}}\right)$$

$$= 3.37\ kW$$

b. Calculate pump power requirements, P_{pump}, using Eq. 5.52.

$$P_{pump} = \frac{\rho_l Q\, L\, g}{1000\ Eff_p}$$

$$= \left(\frac{999.75\,\frac{kg}{m^3} \times 0.1577\,\frac{m^3}{s} \times 9.94\,m \times 9.81\,\frac{m}{s^2}}{1000\,\frac{W}{kW} \times 0.80}\right)\left(\frac{1W}{1\,\frac{kg\,m}{s^2}\cdot\frac{m}{s}}\right)$$

$$= 19.2\ kW$$

Eff_p = pump efficiency

c. Calculate total power requirements, P_{total}.

$$P_{total} = P_{blower} + P_{pump} = 3.37\ kW + 19.2\ kW = 22.6\ kW$$

Impact of Dissolved Solids on Tower Performance

One major concern with packed-tower air stripping is the potential for precipitation of inorganic compounds onto packing media causing a steady decrease in the tower void volume and a steady increase in tower pressure drop, eventually leading to possible plugging of the tower. This is especially true for some groundwaters and hypolimnetic waters from stratified lakes and reservoirs that contain considerable amounts of free carbon dioxide. It is not uncommon to encounter groundwaters with 30 to 50 mg/L of carbon dioxide. In an air stripping tower, carbon dioxide together with VOCs can be removed from the water. Because carbon dioxide is an acidic gas, its removal tends to raise the pH of the water. As the pH increases, the alkalinity forms change, with the result that carbon dioxide can also be extracted both from bicarbonates and from carbonates in accordance with the following equilibrium equations:

$$2HCO_3^- \Leftrightarrow CO_3^= + H_2CO_3 \tag{5.54}$$

$$CO_3^= + H_2O \Leftrightarrow 2OH^- + CO_2 \tag{5.55}$$

Thus, the removal of carbon dioxide by air stripping tends to cause a shift in the forms of alkalinity present from bicarbonate to carbonate, and from carbonate to hydroxide.

In natural waters containing appreciable amounts of calcium ion, calcium carbonate precipitates when the carbonate-ion concentration becomes so great that the calcium carbonate solubility product is exceeded:

$$Ca^{++} + CO_3^{=} \Leftrightarrow CaCO_3(s) \tag{5.56}$$

Air contains about 0.035 percent by volume of carbon dioxide. The Henry's law constant for carbon dioxide at 25°C is about 61.1 (1500 mg/L·atm^{-1}); therefore, the equilibrium concentration of carbon dioxide with air is about 0.45 mg/L. Air stripping can reduce the free carbon dioxide concentration to its equilibrium concentration with air. In most situations, theoretical precipitation calculations can be performed to determine the maximum amount and rate of precipitation that will cause plugging of the tower. However, the free carbon dioxide concentration is usually reduced to that above its equilibrium concentration with air, and the time taken to plug the air stripping tower will be much longer than that predicted from the maximum theoretical precipitation. The actual rate of precipitation can only be determined using pilot plant testing. The control of precipitation using acid or scale inhibitors can be very expensive, and in some cases it can be a major expense. Consequently, this needs to be carefully considered.

Air Stripping Off-Gas Control Using Adsorption

Air pollution resulting from VOC removal in air stripping processes has prompted many regulatory agencies to set emissions standards. Off-gas treatment technologies, such as gas-phase granular activated carbon (GAC) adsorption, thermal incineration, and catalytic incineration, are capable of reducing, and in some cases eliminating, the discharge of VOCs from air strippers. Thermal and catalytic incineration are used in cases when the VOC concentrations are very high or the VOCs are very weakly adsorbing. Volatile organic chemical concentrations are usually dilute in most water treatment applications. Consequently, GAC fixed-bed adsorption is the most widely used process because it is cost-effective for treating dilute off-gas streams, and many of the common VOC contaminants (e.g., trichloroethene, tetrachloroethene, toluene, xylenes) are amenable to adsorption onto GAC (Crittenden et al., 1988). Presented below is a short discussion of the important parameters that need to be considered for fixed-bed adsorber design and operation for VOC off-gas control from air strippers.

The design principles and application of adsorption for water treatment are presented in Chapter 13. The fundamental principles for gas-phase adsorption are very similar to those for liquid-phase adsorption with the exception that the transport fluid is air instead of water. For example, the same equilibrium and mass transfer concepts used in design of liquid-phase systems are used in gas-phase adsorber design. The main differences between these two adsorption processes are: the impact of water vapor on VOC adsorption must be considered in gas-phase adsorption; the impact of NOM on the adsorption of organic contaminants is negligible and does not need to be considered; gas-phase adsorbents have properties (e.g., particle size, pore size distribution, hydrophobicity) that are designed to be more amenable to gas-phase adsorption; and the adsorbent reactivation techniques may be different depending upon the type of adsorbent that is used.

A schematic of a typical GAC off-gas control system for an air stripper is displayed in Figure 5.10. The VOC-laden air stripper off-gas contains water vapor that

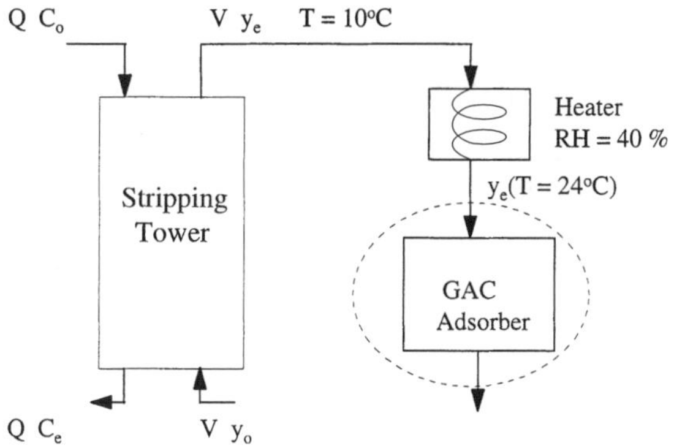

FIGURE 5.10 Schematic of air stripping off-gas control using a GAC adsorber.

has a relative humidity (RH) of 100 percent. Water vapor can have a large influence on the adsorption capacity of adsorbents such as GAC. Figure 5.11 displays water vapor isotherms for several GAC adsorbents. For most of these GAC adsorbents, the adsorption of water is significant when the RH is greater than about 40 percent. At high-RH values, a phenomena called *capillary condensation* takes place when the water vapor will begin to condense in the micropores of the adsorbent. The condensed water will then compete with the VOCs for adsorption sites, and the VOC capacity of the adsorbent will be significantly reduced. However, for RH values less than about 40 percent, the amount of water adsorbed onto the GAC is small because capillary condensation of the water vapor is negligible and consequently, its impact on VOC adsorption capacity can be neglected (Crittenden et al., 1988). For most gas-phase GAC adsorbents, it is recommended the off-gas RH be reduced to less than 40 percent prior to the adsorption process. A water vapor isotherm performed on a particular adsorbent of interest should be used to determine the RH value that will provide the most cost-effective design. As shown in Figure 5.11, heating the off-gas is usually the most common method used to reduce the RH when GAC is used. Recently, synthetic adsorbents have been created that are hydrophobic at RH values approaching 100 percent. However, the high cost of these adsorbents may make them economically unfeasible in most applications.

Typical design parameters for gas-phase GAC fixed-bed adsorbers consist of empty-bed contact times (EBCTs) from 1.5 to 5.0 s and air loading rates from 0.25 to 0.50 m/s. The GAC used in gas-phase applications is usually a microporous carbon with nominal particle sizes ranging from 4×6 to 6×16 U.S. standard mesh size. Gas-phase GAC particles are typically larger than those for liquid-phase applications to minimize the air pressure drop through the fixed bed. The air pressure drop in gas-phase fixed-bed adsorbers ranges from 0.2 to 5.0 kPa/m packed-bed depth.

In gas-phase adsorption, the mass transfer zone (MTZ) of an organic contaminant is very short and occupies a very small fraction of the fixed bed. For moderate to strongly adsorbing VOCs (e.g., trichloroethylene, toluene, tetrachloroethylene), MTZ lengths range from 0.025 to 0.043 m (Crittenden et al., 1988). For a typical packed-bed depth of 1.0 m, the MTZ would occupy less than 4.3 percent of the bed. Consequently, for preliminary design calculations, equilibrium models can be used to determine the GAC usage rate for organic contaminants. Based on the prelimi-

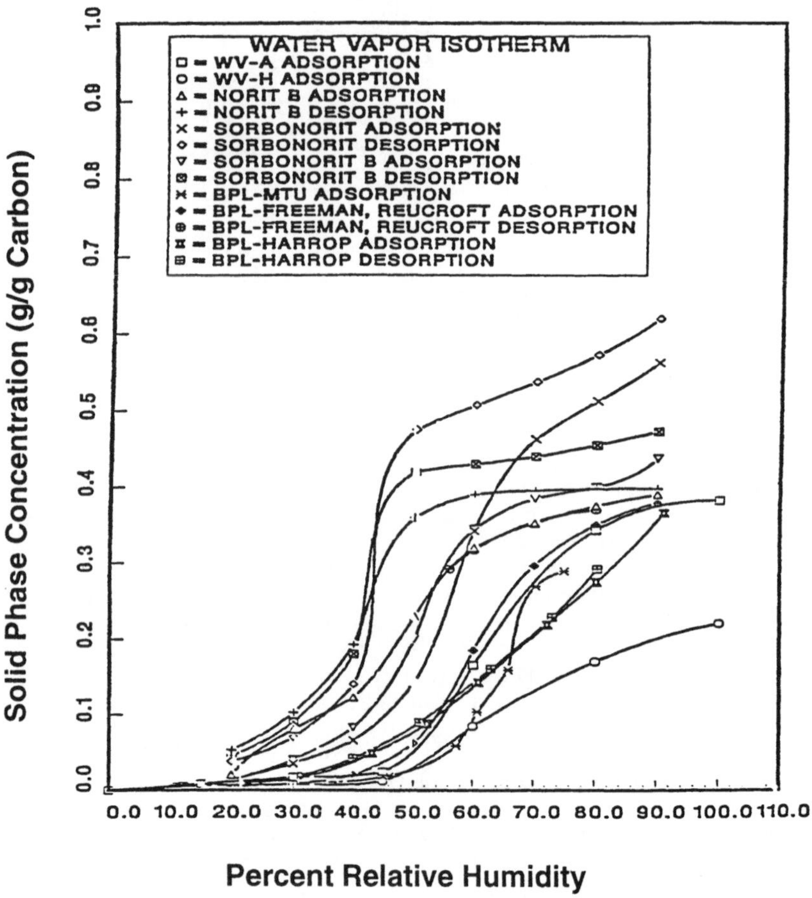

FIGURE 5.11 Water vapor isotherms for various granular activated carbons (Tang et al., 1987).

nary design calculations, pilot studies can be performed and used to evaluate the most cost-effective design and operation.

The Dubinin-Raduskevich (D-R) equation was shown to correlate the isotherms of several gas-phase VOCs (Tang et al., 1987). Based on the work of several researchers (Reucroft et al., 1971; Rasmuson, 1984, Crittenden at al., 1988), the following form of the D-R equation can be used to estimate the single-solute gas-phase capacity for a VOC:

$$q = \rho_L W = \rho_L W_0 \exp\left[\frac{-B\varepsilon^2}{\alpha^2}\right] \tag{5.57}$$

$$\varepsilon = RT \ln\left(\frac{P_s}{P}\right) \tag{5.58}$$

TABLE 5.10 Typical W_0 and B Values for Several Commercial Gas-Phase Adsorbents

Adsorbent type	W_0 (cm^3/gm)	B (cm^6/cal^2)	Reference
Calgon BPL (6 × 16 mesh)	0.515	2.99×10^{-5}	Crittenden et al., 1988
Calgon BPL (4 × 6 mesh)	0.460	3.22×10^{-5}	Crittenden et al., 1988
CECA GAC—410G	0.503	2.42×10^{-5}	CECA Inc., 1994

where q is the solid-phase VOC concentration in equilibrium with P (M/M), W is the adsorption space or pore volume occupied by the adsorbate (L^3/M), W_0 is the maximum adsorption space (L^3/M), B is the microporosity constant ($M^2/L^6 \cdot t^{-2} \cdot M^{-1} \cdot L^{-2}$), ε is the adsorption potential ($M/L^2 \cdot M^{-1} \cdot t^{-2}$), T is the temperature (°K), P_s is the saturation vapor pressure of the VOC at T°K (M/L^2), P is the partial pressure of the VOC in the gas (M/L^2), ρ_L is the liquid density of the VOC (M/L^3), and α is the polarizability of the VOC (L^{3}M) and may be calculated from the Lorentz-Lorentz equation if it is not known:

$$\alpha = \frac{(\eta^2 - 1)M}{(\eta^2 + 2)\rho_l} \tag{5.59}$$

where η is the refractive index of the VOC and M is the molecular weight of the VOC. Equations 5.57 and 5.58 were correlated with isotherm data for several compounds with dipole moments less than 2 debyes (2×10^{-18} esu cm) (Reucroft et al., 1971; Crittenden et al., 1988); W_o and B depend only on the nature of the adsorbent. When isotherm data is plotted as W versus $(\varepsilon/\alpha)^2$ for a given adsorbent, the data conform to essentially one line for different VOCs and temperatures (Crittenden et al., 1988). Table 5.10 summarizes W_o and B values for three gas-phase adsorbents. The D-R values are based on toluene, which was used as the reference compound. Two important limitations of the D-R equation (Eqs. 5.57 and 5.58) are (1) it is only valid for relative pressures, P/P_s, less than 0.2 because capillary condensation of the VOC occurs, and (2) it is only valid for relative humidities (RH) values less than about 50 percent. For relative pressures greater than 0.2, the adsorbent capacity can be calculated by assuming the adsorbent micropore volume is filled with the condensed VOC. For cases in which multiple VOCs are present in an air stripper off-gas, equilibrium column model (ECM) calculations can be used to estimate the adsorbent usage rate for each VOC (Cortright, 1986; Crittenden et al., 1987). When the GAC capacity is exhausted, it is usually sent back to the manufacturer for reactivation and reuse.

EXAMPLE 5.2 ***Air Stripper Off-Gas Control Using Adsorption Calculation*** Perform preliminary design calculations for a gas-phase adsorber that will remove TCE from the air stripper off-gas in Example 5.1. Determine the GAC usage rate that will be required to meet a nondetectable limit for TCE and the bed life for the following conditions: an EBCT of 2.5 s and a loading rate of 0.5 m/s. A summary of the air stripper off-gas and TCE properties required to perform the calculations follows.

Air Stripper data:

$$\dot{V} = Q \times \frac{\dot{V}}{Q} = 0.1577 \frac{\text{m}^3}{\text{s}} \times 30 \frac{\text{m}^3 \text{ air/s}}{\text{m}^3 \text{ water/s}} = 4.73 \frac{\text{m}^3}{\text{s}}$$

$$C_0\,(\text{TCE}) = 100 \frac{\mu\text{g}}{\text{L}} \quad C_e(\text{TCE}) = 5.0 \frac{\mu\text{g}}{\text{L}}$$

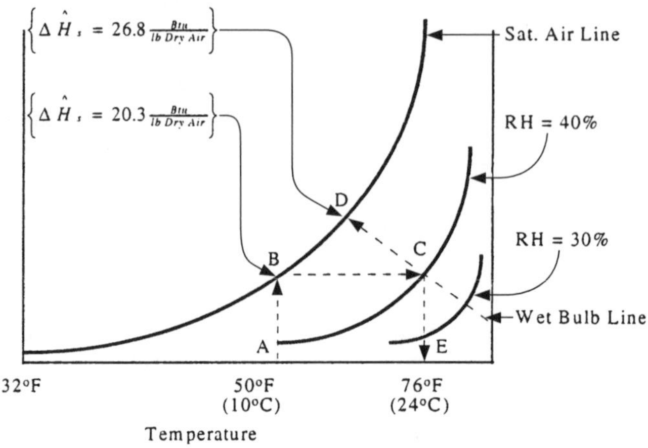

FIGURE 5.12 Determining the gas stream temperature after heating.

Off-gas temperature assumed to be 10°C and RH = 100 percent

$$y_e\left(\text{TCE concentration in off-gas at 10°C}\right) = \frac{C_o - C_e}{\dot{V}/Q}$$

$$= \frac{(100.0 - 5.0)\ \dfrac{\mu g}{L}}{30} = 3.2\ \frac{\mu g}{\text{L Air}}$$

1. The first step is to determine the off-gas heating requirements necessary to reduce the impact of RH on the adsorption of TCE. Determine the energy and electrical requirements necessary to reduce the RH from 100 percent to 40 percent, the temperature of the off-gas leaving the heater and the TCE concentration leaving the heater.

a. From a psychrometric chart, the change in enthalpy from 100 percent RH to 40 percent RH can be determined as illustrated in the following. In addition, the temperature of the gas stream after heating can also be determined as illustrated in Figure 5.12.

The following steps can be used in conjunction with a psychrometric chart to determine the heating requirements and the temperature required for a RH of 40 percent.

Step 1. Locate the inlet temperature on the *x*-axis of the chart at point A.

Step 2. Follow the vertical line straight up from point A until the 100 percent RH line is intersected at point B. At point B, the enthalpy at saturation can be determined to be 20.3 Btu/lb dry air.

Step 3. From point B, travel horizontally until the 40 percent RH line is intersected at point C. Then travel along the wet bulb line up to point D. At point D, the enthalpy at saturation can be determined to be 26.8 Btu/lb dry air.

Step 4. The temperature of the heated air can be determined by traveling from point C vertically downward to the *x*-axis and obtaining the temperature at point E, which is 76°F (or about 24°C).

Step 5. The following calculations illustrate the conversion of the heat required to the energy required.

$$\{\text{Required heat (Btu/day)}\} = \left(4.73\ \frac{\text{m}^3}{\text{s}}\right)\left(\frac{\text{kg mol Air}}{22.4\ \text{m}^3}\right)\left(\frac{28.97\ \text{kg Air}}{\text{kg mol}}\right)$$

$$\left(\frac{86{,}400\ \text{s}}{\text{day}}\right)\left(\frac{2.203\ \text{lb}}{\text{kg}}\right)\left(\frac{26.8 - 20.3\ \text{Btu}}{\text{lb}}\right)$$

$$= 7.57 \times 10^6\ \frac{\text{Btu}}{\text{day}}$$

$$\{\text{Required energy (kW)}\} = \left(7.57 \times 10^6\ \frac{\text{Btu}}{\text{day}}\right)\left(\frac{\text{day}}{86{,}400\ \text{s}}\right)\left(\frac{1\ \text{kW}}{0.9478\ \frac{\text{Btu}}{\text{s}}}\right) = 92.4\ \text{kW}$$

$$\{\text{Electrical requirements (kw} \cdot \text{h}^{-1}\text{/day)}\} = (92.4\ \text{kW})\left(\frac{24\ \text{hr}}{\text{day}}\right) = 2{,}218\ \frac{\text{kWh}}{\text{day}}$$

b. From step 4 shown above, the off-gas temperature leaving the heater and entering the adsorber is 24°C (76.4°F).

c. The TCE concentration in the heated off-gas will be decreased by a small amount due to heating the off-gas. The ideal gas law at constant pressure of 760 mm Hg can be used to adjust for this change as shown in the following:

$$y_e(24°\ \text{C}) = 3.2\ \frac{\mu\text{g}}{\text{L Air}}\ (10°\ \text{C})\left(\frac{273 + 10}{273 + 24}\right) = 3.05\ \frac{\mu\text{g}}{\text{L Air}}$$

d. Calculate the best possible GAC usage rate assuming no mass transfer resistance. Assume that the adsorbent that will be used in the design is Calgon BPL 4 × 6 mesh GAC (D-R parameters: $B = 3.22 \times 10^{-5}$ cm^6/cal^2, $W_o = 0.460$ cm^3/gm). Performing a mass balance on the GAC adsorber, the following expression can be obtained for calculating the GAC usage rate:

$$\frac{\hat{M}}{\dot{V}\, t_e} = \frac{y_e}{q_e} = \text{GAC usage rate}$$

where $\hat{M}$ is the mass of GAC in the fixed bed, t_e is the time it takes to exhaust GAC fixed-bed, and q_e is the adsorbed phase concentration on TCE in equilibrium with y_e. q_e can be calculated using equation 5.57.

If equations 5.57 and 5.58 are combined and divided by $M \times 10^{-6}$, the following expression can be used to calculate q_e.

$$q_e = \left(\frac{W_o \rho_l}{M \times 10^{-6}}\right) \exp\left[\frac{-B}{\alpha^2}\left\{RT \ln\left(\frac{P_s}{P}\right)\right\}^2\right]$$

From Lange's *Handbook of Chemistry,* the Antoine's equation is used to calculate P_s.

$$P_s = 10^{(A - B/T(°\text{C}) + C)}$$

For TCE: $A = 7.72, B = 1742, C = 273$

$$P_s(T = 24°C) = 10^{(7.72 - 1742/24 + 273)} = 71.56 \text{ mm Hg}$$

y_e can be converted in terms of P using the following expression:

$$P(24°C) = y_e RT = \left(3.05 \frac{\mu g}{L}\right)\left(0.08205 \frac{L - atm}{mol - °K}\right)(297° K)$$

$$\left(\frac{mol}{131.39 \text{ gm}}\right)\left(\frac{gm}{10^6 \mu g}\right)\left(\frac{760 \text{ mm Hg}}{1 \text{ atm}}\right) = 4.3 \times 10^{-4} \text{ mm Hg}$$

Equation 5.59 is used to calculate the polarizability of TCE. From the *CRC*, 71st Edition, η(TCE) = 1.4773.

$$\alpha(TCE) = \frac{((1.4773)^2 - 1)131.39 \dfrac{gm}{mol}}{((1.4773)^2 + 2)1.464 \dfrac{gm}{cm^3}} = 25.37 \frac{cm^3}{mol}$$

$$q_e = \left(\frac{(0.460 \dfrac{cm^3}{gm})(1.464 \dfrac{gm}{cm^3})}{131.39 \times 10^{-6} \dfrac{gm}{\mu \text{ mol}}}\right)$$

$$\exp\left[\frac{-3.22 \times 10^{-5} \dfrac{cm^6}{cal^2}}{25.37^2 \dfrac{cm^6}{mol^2}}\left\{\left(1.987 \frac{cal}{mol - °K}\right)(297° K) \ln\left(\frac{71.56 \text{ mm Hg}}{4.3 \times 10^{-4} \text{ mm Hg}}\right)\right\}^2\right]$$

$$q_e = 413 \frac{\mu mol \text{ TCE adsorbed}}{gm \text{ GAC}}$$

$$\{\text{GAC usage rate}\} = \frac{y_e}{q_e} = \frac{\left(3.05 \dfrac{\mu g}{L \text{ Air}}\right)\left(\dfrac{1 \text{ kg}}{1000 \text{ gm}}\right)}{\left(413 \dfrac{\mu mol}{gm}\right)\left(131.39 \dfrac{\mu g}{\mu mol}\right)\left(\dfrac{m^3}{1000 \text{ L}}\right)}$$

$$= 5.62 \times 10^{-5} \frac{kg \text{ GAC}}{m^3 \text{ air treated}}$$

$$\{\text{GAC usage rate}\} = 17{,}793 \frac{m^3 \text{ Air treated}}{kg \text{ GAC}}$$

e. Calculate the mass of GAC required and the bed life of the GAC adsorber assuming an EBCT = 2.5 s and $v_s = 0.5$ m/s.

{Mass of GAC in 2.5 s EBCT bed}

$$= \text{EBCT} \, \dot{V} \rho_F = (2.5 \text{ s})\left(4.73 \frac{m^3}{s}\right)\left(531 \frac{kg}{m^3}\right) = 6{,}279 \text{ kg}$$

{Volume of air treated for a 2.5 s EBCT}

$$= \frac{\text{(Mass of GAC for 2.5 s EBCT)}}{\text{(TCE usage rate)}} = \frac{6{,}279 \text{ kg GAC}}{\left(\dfrac{\text{kg GAC}}{17{,}793 \text{ m}^3 \text{ air}}\right)} = 1.12 \times 10^8 \text{ m}^3 \text{ air}$$

$$\{\text{GAC bed life}\} = \frac{(1.12 \times 10^8 \text{ m}^3 \text{ air})}{\left(4.73 \dfrac{\text{m}^3 \text{ air}}{\text{s}}\right)\left(\dfrac{86{,}400\text{s}}{\text{d}}\right)} = 274 \text{ days}$$

DIFFUSED OR BUBBLE AERATION

The diffused or bubble aeration process consists of contacting gas bubbles with water for the purposes of transferring gas to the water (e.g., O_3, CO_2, O_2) or removing VOCs from the water by stripping. The process can be carried out in a clear well or in special rectangular concrete tanks typically 9 to 15 ft in depth. Figure 5.13 displays different types of diffused aeration systems. The most commonly used diffuser system consists of a matrix of perforated tubes (or membranes) or porous plates arranged near the bottom of the tank to provide maximum gas-to-water contact. Various types

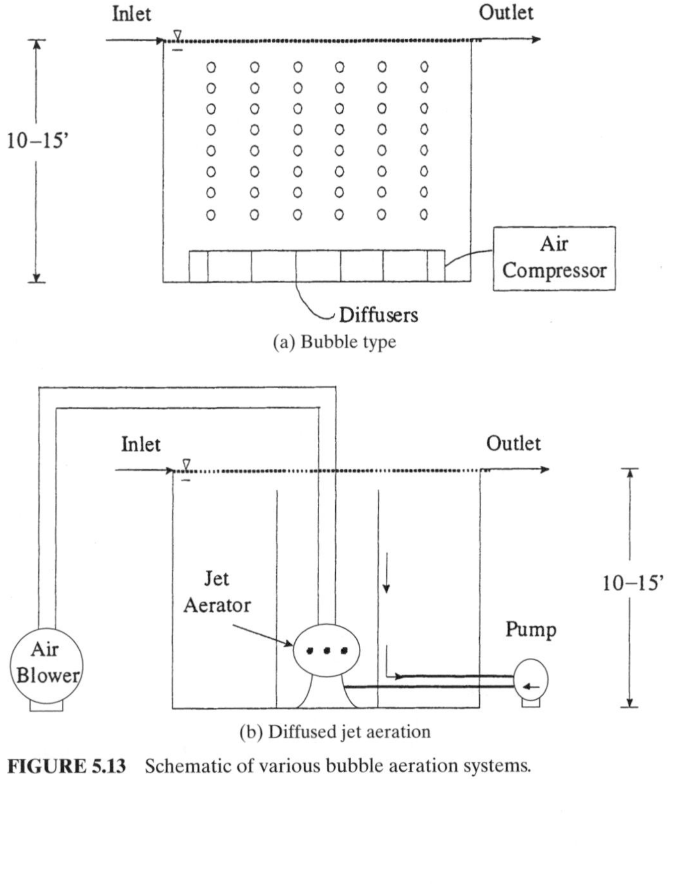

FIGURE 5.13 Schematic of various bubble aeration systems.

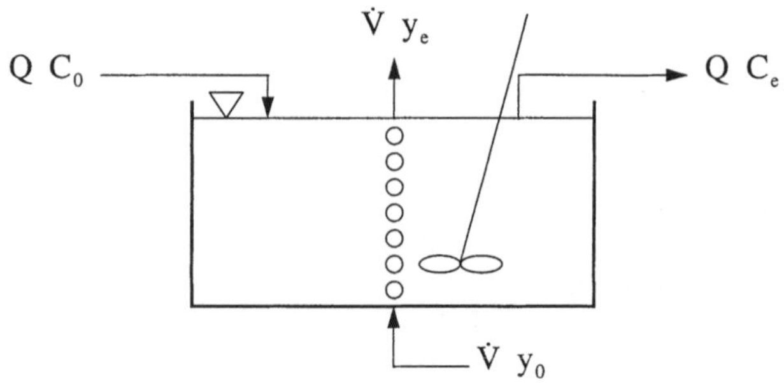

FIGURE 5.14 Schematic of a completely mixed bubble aeration tank for development of the design equations.

of diffusers and diffuser system layouts are presented in the Environmental Protection Agency's technology transfer design manual on fine-pore aeration systems (loc. cit., EPA/625/1-89/023). Jet aerator devices are also used to provide good air-to-water contact (Mandt and Bathija, 1978). These aerators consist of jets that discharge fine gas bubbles and provide enhanced mixing for increased absorption efficiency.

Design Equations

Figure 5.14 shows a schematic of a bubble aeration system for a single tank. Model development for air stripping using bubble aeration has been described in the literature (Matter-Müller, Gujer, and Giger, 1981; Munz and Roberts, 1982; Roberts, Munz, and Dändliker, 1984). The development of the process design equations for bubble aeration incorporates the following assumptions: (1) the liquid phase is completely mixed, (2) the gas phase is plug flow, (3) the process is at steady state, and (4) the inlet VOC gas concentration is zero. A liquid-phase mass balance around the tank shown in Figure 5.14, incorporating the above assumptions, results in the following relationship:

$$QC_0 - QC_e - \dot{V}(y_o - y_e) = 0 \tag{5.60}$$

Applying the assumptions that the inlet VOC gas concentration is zero (i.e., $y_o = 0$) and the exiting air is in equilibrium with the bulk liquid (i.e., $y_e = HC_e$), Eq. 5.60 can be solved for $(\dot{V}/Q)$, which is for the minimum air-to-water ratio for bubble aeration:

$$\left(\frac{\dot{V}}{Q}\right)_{\min} = \frac{C_0 - C_e}{HC_e} \tag{5.61}$$

The minimum air-to-water ratio represents the smallest air-to-water ratio that can be applied to a bubble aeration tank to meet the treatment objective, C_e.

Figure 5.15 shows a schematic of a gas bubble within a bubble aeration tank. Eliminating the assumption of equilibrium between the bulk liquid and the exiting air and performing a mass balance around the gas bubble (assuming plug flow conditions prevail in the gas phase) result in an expression for y_e in Eq. 5.60 (Hokanson, 1996; Matter-Müller et al., 1981):

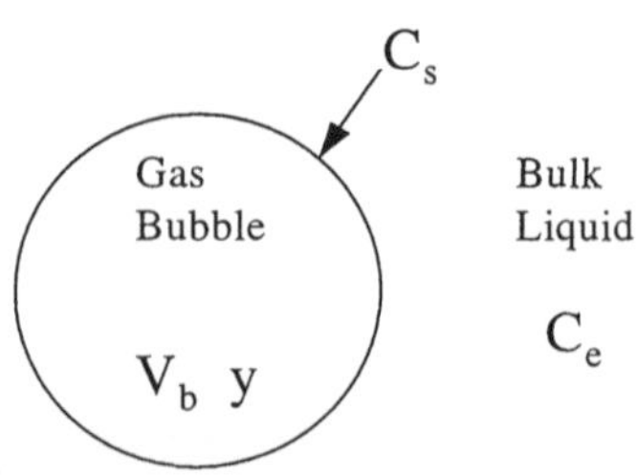

FIGURE 5.15 Schematic of a single bubble inside a bubble aeration tank.

$$y_e = HC_e\,(1 - e^{-\phi}) \tag{5.62}$$

in which ϕ is equal to the Stanton number, defined by:

$$\phi = \frac{K_L aV}{H\dot{V}} \tag{5.63}$$

in which K_La is the overall mass transfer coefficient for bubble aeration (T^{-1}), V is the volume of a bubble aeration tank (L^3), H is the Henry's law constant (dimensionless), and $\dot{V}$ is the air flow rate (L^3/T).

Substitution of Eq. 5.62 into Eq. 5.60 results in the design equation for bubble aeration:

$$QC_0 - QC_e - \dot{V}HC_e\,(1 - e^{-\phi}) = 0 \tag{5.64}$$

To design a new bubble aeration tank, Eq. 5.64 can be rearranged and solved for V (note that C_e in Eq. 5.64 represents the treatment objective):

$$V = -\frac{H\dot{V}}{K_L a} \ln\left\{1 - \left[\left(\frac{C_0}{C_e} - 1\right)\left(\frac{1}{\frac{\dot{V}}{Q}H}\right)\right]\right\} \tag{5.65}$$

Recall that the air-to-water ratio, $\dot{v}/Q$ in Eq. 5.65, must be greater than the minimum air-to-water ratio, $(\dot{v}/Q)_{\min}$, defined in Eq. 5.61, for the bubble aeration tank to be able to meet the treatment objective. To estimate the effluent concentrations of various compounds for an existing bubble aeration tank with volume, V, it is necessary to solve Eq. 5.65 for C_e:

$$C_e = \frac{C_0}{1 + \frac{\dot{V}}{Q}H(1 - e^{-\phi})} \tag{5.66}$$

In either case (new tank or existing tank), the gas-phase effluent concentration from the tank, y_e, can be determined from:

$$y_e = \frac{C_0 - C_e}{\frac{\dot{v}}{Q}} \tag{5.67}$$

Process performance can often be improved significantly (i.e., lower total volume requirements) by configuring the bubble aeration system as tanks in series. Figure 5.16 shows a schematic of a bubble aeration system operated in a tanks-in-series configuration. Sometimes this is easy to accomplish by placing baffles in the tank. The process design equations that follow are for the general case of n tanks in series. If it is assumed that all the tanks are equally sized and all the assumptions for a single-tank system are valid here, a mass balance around tank 1 yields the following expression:

$$C_1 = \frac{C_0}{1 + \frac{\dot{V}}{Q}H(1 - e^{-\phi})} \tag{5.68}$$

in which C_1 is the effluent liquid-phase concentration from tank 1. A mass balance around tank 2 yields the following expression:

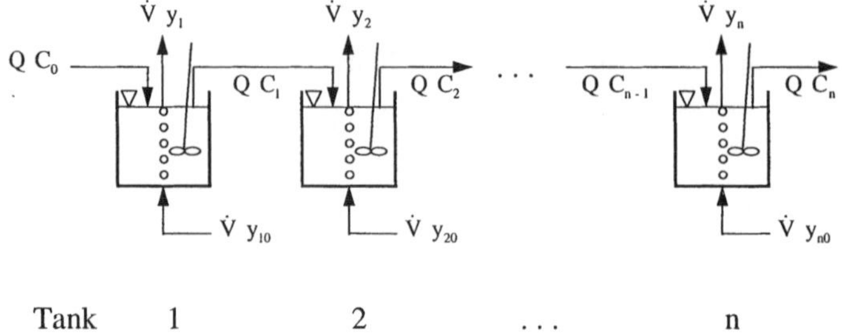

FIGURE 5.16 Schematic of a tanks-in-series bubble aeration system.

$$C_2 = \frac{C_1}{1 + \frac{\dot{V}}{Q}H(1 - e^{-\phi})} = \frac{C_0}{\left[1 + \frac{\dot{V}}{Q}H(1 - e^{-\phi})\right]^2} \tag{5.69}$$

in which C_2 is the effluent liquid-phase concentration from tank 2. Similarly, a mass balance written around tank n results in:

$$C_n = \frac{C_{n-1}}{1 + \frac{\dot{V}}{Q}H(1 - e^{-\phi})} = \frac{C_0}{\left[1 + \frac{\dot{V}}{Q}H(1 - e^{-\phi})\right]^n} \tag{5.70}$$

in which C_n is the effluent liquid-phase concentration from tank n. Equation 5.70 is the general expression for the effluent concentration for a bubble aeration system consisting of tanks in series. For the case of designing new tanks-in-series systems, both the influent and desired effluent concentrations are known; consequently, Eq. 5.70 can be rearranged to determine the volume of each individual tank, V_n, using the following equation:

$$V_n = -\frac{H\dot{V}}{K_L a} \ln\left\{1 - \left[\left(\frac{C_0}{C_e}\right)^{1/n} - 1\right]\left(\frac{1}{\frac{\dot{v}}{Q}H}\right)\right\} \tag{5.71}$$

As for the single-tank case, the air-to-water ratio supplied to each tank must exceed the minimum air-to-water ratio so that the system will be able to meet its treatment objective. The minimum air-to-water ratio for a tanks-in-series configuration can be determined as follows:

$$\left(\frac{\dot{v}}{Q}\right)_{\min} = \left[\left(\frac{C_0}{C_e}\right)^{1/n} - 1\right]\frac{1}{H} \tag{5.72}$$

It is sometimes necessary to determine the gas-phase concentration leaving the tanks. The gas-phase effluent concentration from each tank in series may be calculated from the following:

$$y_n = \frac{C_{n-1} - C_n}{\frac{\dot{v}}{Q}} \tag{5.73}$$

Equations that describe the transfer of compounds from the gas to the water phase are similar to the equations used to describe stripping in a bubble aeration tank, a process known as absorption. Figure 5.14, the bubble aeration schematic shown previously, is also relevant to this case. The development of the process design equations for absorption using bubble aeration incorporates the same assumptions as for stripping, except that the inlet liquid concentration, rather than the inlet gas-phase concentration, is zero. A mass balance around the tank shown in Figure 5.14 results in the following:

$$-QC_e + \dot{V}y_0 - \dot{V}y_e = 0 \tag{5.74}$$

For sizing a tank, and substituting Eq. 5.62 into Eq. 5.74 and solving it for V, the following expression can be derived:

$$V = -\frac{H \cdot \dot{V}}{K_{\mathrm{L}}a} \ln\left(1 - \left[\frac{1}{H\frac{\dot{v}}{Q}\left(\frac{y_0}{y_e} - 1\right)}\right]\right) \tag{5.75}$$

As for the stripping case, there is a minimum water-to-air ratio that must be exceeded for absorption to be viable:

$$\frac{Q}{\dot{V}} \geq \left(\frac{Q}{\dot{V}}\right)_{\min} = H\left(\frac{y_0}{y_e} - 1\right) \tag{5.76}$$

To estimate the gas-phase effluent concentrations of various compounds for an existing bubble aeration tank with volume, V, it is necessary to solve Eq. 5.74 for C_e, substitute the result into Eq. 5.75, and solve for y_e:

$$y_e = \frac{y_0}{1 + \left[\frac{1}{H\dot{v}/Q(1 - e^{-\phi})}\right]} \tag{5.77}$$

For bubble and surface aeration, it is difficult to determine $K_{\mathrm{L}}a$ for a compound from Eq. 5.21 because the quantities k_{l} and k_{g} vary greatly from system to system and are difficult to measure. The general equation for estimating the $K_{\mathrm{L}}a$ of a compound in bubble or surface aeration is found by expanding Eq. 5.21 based on oxygen as a reference compound:

$$K_{\mathrm{L}}a_i = K_{\mathrm{L}}a_{\mathrm{O}_2}\left(\frac{D_{\mathrm{l},i}}{D_{\mathrm{l,O}_2}}\right)^m \left[1 + \frac{1}{H_i\left(\frac{k_{\mathrm{g}}}{k_{\mathrm{l}}}\right)}\right]^{-1} \tag{5.78}$$

in which $K_{\mathrm{L}}a_{\mathrm{O}_2}$ is the overall mass transfer coefficient for oxygen (1/T), $D_{\mathrm{l},i}$ is the liquid diffusivity of component i(L^2/T), and m is the diffusivity exponent (dimensionless). $k_{\mathrm{g}}/k_{\mathrm{l}}$ is the ratio of gas phase to liquid phase mass transfer coefficient (dimensionless), which tends to be relatively constant for a given type of aeration system. $D_{\mathrm{l},i}$ can be found from the Hayduk and Laudie correlation (1974) for small molecules in water. $D_{\mathrm{l,O}_2}$ is found from a correlation presented in Holmén and Liss (1984):

$$D_{\mathrm{l,O}_2} = [10^{(3.15 + (-831.0)/T)}](1.0 \times 10^{-9}) \tag{5.79}$$

in which T is the temperature in °K. The recommended value for the diffusivity exponent, m, in bubble aeration is 0.6 (Holmén and Liss, 1984). Munz and Roberts (1989) point out that a constant value of $k_{\mathrm{g}}/k_{\mathrm{l}} \approx 100$ to 150 is widely used in air-water mass transfer. Therefore, a value of $k_{\mathrm{g}}/k_{\mathrm{l}} = 100$ is recommended as a conservative estimate in bubble aeration. $K_La_{O_2}$ can be found from clean-water oxygen transfer test data (Brown and Baillod, 1982; Baillod et al., 1986):

$$K_La_{O_2} = K_La^*_{20}\,(1.024)^{T-20} \tag{5.80}$$

in which T is the temperature in °C, $K_La^*_{20}$ is the true mass transfer coefficient of oxygen at 20°C and $K_La^*_{20}$ can be determined from the following series of relationships:

$$K_La^*_{20} = \frac{K_La_{20}}{1 - \dfrac{K_La_{20}}{2\,\phi_O}} \tag{5.81}$$

$$K_La_{20} = \frac{\text{SOTR}}{V\,C^*_\infty} \tag{5.82}$$

$$C^*_\infty = C^*_s\left[\frac{P_b - P_v + \gamma_w d_e}{P_s - P_v}\right] \tag{5.83}$$

$$\phi_O = \frac{M_O \rho_a Q_a}{M_a H_O (P_b + \gamma_w d_e)} \tag{5.84}$$

in which ϕ_O is the oxygenation coefficient (1/T); C^*_∞ is the dissolved oxygen saturation concentration attained at infinite time (M/L^3); SOTR is the standard oxygen transfer rate (M/T); V is the water volume in the tank (L^3); C^*_s is the tabulated value of dissolved oxygen saturation concentration at 20°C (M/L^3); P_b is the barometric pressure ($M/L \cdot T^{-2} \cdot L^{-2}$) P_s is the barometric pressure under standard conditions ($M/L \cdot T^{-2} \cdot L^{-2}$); P_v is the vapor pressure of water at 20°C ($M/L \cdot T^{-2} \cdot L^{-2}$); γ_w is the weight density of water (M/L^3); d_e is the effective saturation depth (L), which is equal to one-third times the water depth; M_O is the molecular weight of oxygen (amu); M_a is the molecular weight of air (amu); Q_a is the volumetric flow rate of air (L^3/t); and H_O is Henry's constant for oxygen (M/L^3)($M/L \cdot T^{-2} \cdot L^{-2}$). These correlations are valid only for clean waters and should not be used for wastewaters.

The total operating power for bubble aeration is equal to the blower brake power requirement for each tank times the number of tanks times the number of blowers per tank. The blower brake power is calculated from Eq. 5.49. The term P_{in} refers to the pressure at the top of the tank, which represents the sum of the ambient pressure and the head required to raise the water to the inlet of the tank. The equation used to find P_{in} is:

$$P_{in,b} = P_{ambient} + \rho_l\, H_b g \tag{5.85}$$

in which the pressure head, H_b, is assumed equal to the water depth in the tank, d_b.

EXAMPLE 5.3 ***Bubble Aeration Sample Calculation***
A portion of an existing clear well is to be modified to include a bubble aeration system capable of reducing total trihalomethane (THM) concentrations from 160 μg/L to less than 50 μg/L. The total volume of clear well is 3785 m^3 (1 mgal) (depth = 5.0 m, length = 30 m, width = 25 m). The average plant flow rate is 0.2191 m^3/s (5 mgd), and the water temperature is 20°C. The influent concentrations (C_O), molecular weights (MW), Henry's constants (H), and molar volumes at the normal boiling point (V_b) for the compounds are summarized in the following. Tank

manufacturer information on clean-water oxygen transfer data necessary to determine $K_La_{O_2}$ is also presented hereafter.

Determine the retention time, airflow rate, and power requirements necessary to meet the treatment objective for a single tank. Also, assume that the clear well can be modified to perform like three tanks in series. Assume a blower efficiency of 30 percent.

Properties of the Compounds in the System

Compound	C_O (μg/L)	MW (g/mol)	H (—)	V_b (cm³/mol)
Chloroform ($CHCl_3$)	120.0	119.38	0.143	87.0
Bromodichloromethane ($BrCl_2CH$)	30.0	163.82	0.087	94.5
Dibromochloromethane (Br_2ClCH)	10.0	208.27	0.033	102.0

Manufacturer's Clean-Water Oxygen Transfer Test Data for a Tank

Property	Value
SOTR (kg O_2/day)	1500
Volumetric airflow rate (m³/hr)	1700
Tank volume (m³)	500
Tank depth (m)	3

SOLUTION:

1. Determine the retention time for a single tank.

$$\tau = \frac{V}{Q} = \frac{3785 \text{ m}^3}{0.2191 \dfrac{\text{m}^3}{\text{s}}} = 172755 \text{ s} = 287 \text{ min} = 4.79 \text{ h}$$

2. Find the oxygen mass transfer rate coefficient, $K_La_{O_2}$.

a. In the calculation of $K_La_{O_2}$, the following quantities, in addition to the clean-water oxygen test data that was given, are needed:

C_s^*	Tabulated value of dissolved oxygen surface saturation concentration at 20°C (M/L^3) = 9.09 mg/L (Metcalf and Eddy, 1991)
P_b	Barometric pressure ($M/L \cdot T^{-2} \cdot L^{-2}$) = 1 atm
P_s	Barometric pressure under standard conditions ($M/L \cdot T^{-2} \cdot L^{-2}$) = 1 atm
P_v	Vapor pressure of water at 20°C ($M/L \cdot T^{-2} \cdot L^{-2}$) = 2.34 kN/m² (Metcalf and Eddy, 1991)
γ_w	Weight density of water = 62.4 lb/ft³
d_e	Effective saturation depth (L) = 1/3 × water depth = 1.67 m
M_O	Molecular weight of oxygen (g/mol) = 32.0
ρ_a	Density of air at 20°C = 1240mg/L
M_a	Molecular weight of air (g/mol) = 28.0
H_O	Henry's constant for oxygen (M/L^3)/($M/L \cdot T^{-2} \cdot L^{-2}$) = 50 mg/L · atm^{-1}

Using these quantities along with those provided in the problem statement, $K_La_{O_2}$ is calculated as follows:

b. Calculate C^*_∞ using Eq. 5.83.

$$C^*_\infty = C^*_s\left[\frac{P_b - P_v + \gamma_w d_e}{P_s - P_v}\right]$$

$$= 9.09\,\frac{\text{mg}}{\text{L}}\left[\frac{1\text{ atm} - \dfrac{2340\text{ Pa}}{101325.0\text{ Pa/atm}} + \dfrac{62.4\text{ lb/ft}^3 \times 1.67\text{ m} \times 3.281\text{ ft/m}}{144\text{ in}^2/\text{ft}^2 \times 14.696\text{ psi/atm}}}{1\text{ atm} - \dfrac{2340\text{ Pa}}{101325.0\text{ Pa/atm}}}\right]$$

$$= 10.59\,\frac{\text{mg}}{\text{L}}$$

c. Calculate K_La_{20} using Eq. 5.82.

$$K_La_{20} = \frac{\text{SOTR}}{VC_\infty{}^*} = \frac{1500\,\dfrac{\text{kg}}{\text{d}} \cdot \left(10^6\,\dfrac{\text{mg}}{\text{kg}}\right)\left(\dfrac{1\text{d}}{86400\text{ s}}\right)}{500\text{ m}^3 \cdot 10.59\,\dfrac{\text{mg}}{\text{L}} \cdot 1000\,\dfrac{\text{L}}{\text{m}^3}} = 3.28 \times 10^{-3}\,\frac{1}{\text{s}}$$

d. Calculate ϕ_O from Eq. 5.84.

$$\phi_O = \frac{M_O \rho_a Q_a}{M_a H_O V_{\text{test}}(P_b + \gamma_w d_e)}$$

$$= \frac{32 \times 1240\,\dfrac{\text{mg}}{\text{L}} \times \dfrac{1700\text{ m}^3/\text{hr}}{3600\text{ s/hr}}}{28.95 \cdot 50\,\dfrac{\text{mg}}{\text{L} \cdot \text{atm}} \cdot 500\text{ m}^3\left(1\text{ atm} + \dfrac{62.4\text{ lb/ft}^3 \times 1.67\text{ m} \times 3.281\text{ ft/m}}{144\text{ in}^2/\text{ft}^2 \times 14.696\text{ psi/atm}}\right)}$$

$$= 0.0301\,\frac{1}{\text{s}}$$

e. Calculate $K_La_{20}{}^*$ using Eq. 5.81.

$$K_La_{20}{}^* = \frac{K_La_{20}}{1 - \dfrac{K_La_{20}}{2\phi_O}} = \frac{3.28 \times 10^{-3}\text{ 1/s}}{1 - \dfrac{3.28 \times 10^{-3}\text{1/s}}{2(0.0301\text{1/s})}} = 3.5 \times 10^{-3}\text{ 1/s}$$

f. Calculate $K_La_{O_2}$ using Eq. 5.80.

$$K_La_{O_2} = K_La_{20}{}^*\,(1.024)^{T-20} = 3.5 \times 10^{-3}\,\frac{1}{\text{s}}(1.024)^{20-20} = 3.5 \times 10^{-3}\,\frac{1}{\text{s}}$$

3. Calculate the liquid diffusivities of oxygen and THMs.

a. Calculate the liquid diffusivity of oxygen using the correlation of Holmen and Liss (1984).

$$D_l\,(O_2) = 10^{3.15 + -831/293.15} \times 1E - 09 = 2.07E - 09\,\frac{\text{m}^2}{\text{s}}$$

b. Calculate the diffusivities of the THMs using the correlation of Hayduk and Laudie (1974).

$$D_l(Cl_3CH) = \frac{13.26 \times 10^{-5}}{(1.02\ cp)^{1.14}(83.3)^{0.589}} = 9.6 \times 10^{-6}\ \frac{cm^2}{s} = 9.6 \times 10^{-10}\ \frac{m^2}{s}$$

$$D_l(BrCl_2CH) = 9.234 \times 10^{-10}\ \frac{m^2}{s}$$

$$D_l(Br_2ClCH) = 8.9 \times 10^{-10}\ \frac{m^2}{s}$$

4. Calculate the overall mass transfer rate of the THMs.

$$K_L a_{Cl_3CH} = 4.075 \times 10^{-3}\ \frac{1}{s}\left(\frac{9.6 \times 10^{-10}\ \frac{m^2}{s}}{2.07 \times 10^{-9}\ \frac{m^2}{s}}\right)^{0.6}\left(1 + \frac{1}{0.143 \times 100}\right)^{-1}$$

$$= 2.402 \times 10^{-3}\ \frac{1}{s}$$

$$K_L a_{BrCl_2CH} = 2.252 \times 10^{-3}\ \frac{1}{s}$$

$$K_L a_{Br_2ClCH} = 1.88 \times 10^{-3}\ \frac{1}{s}$$

5. Calculate the minimum air-to-water ratio using Eq. 5.61. Because $CHCl_3$ is the largest THM present, assume a $C_e(CHCl_3)$ of 30 μg/L.

$$\left(\frac{\dot{V}}{Q}\right)_{min(CHCl_3)} = \frac{C_O - C_e}{HC_e} = \frac{(120 - 30)\frac{\mu g}{L}}{0.143 \times 30\frac{\mu g}{L}} = 21$$

6. Calculate the airflow rate.

$$\dot{V} = \left(\frac{\dot{V}}{Q}\right)_{min} \times Q = 21 \times 0.2191\ \frac{m^3}{s} = 4.6\frac{m^3}{s}$$

7. Calculate the Stanton number, ϕ, for each THM using Eq. 5.63.

$$\phi_{Cl_3CH} = \frac{2.402 \times 10^{-3}\ \frac{1}{s} \times 3785\ m^3}{4.6\ \frac{m^3}{s} \times 0.143} = 13.8$$

$$\phi_{BrCl_2CH} = 21.3$$

$$\phi_{Br_2ClCH} = 46.9$$

8. Calculate the THM effluent concentration, C_e, from the tank, using Eq. 5.66.

$$C_{e_{CHC_3}} = 30\ \frac{\mu g}{L}$$

$$C_{e_{BrCl_2CH}} = \frac{30\frac{\mu g}{L}}{1 + 21 \times 0.087 \times (1 - e^{-21.3})} = 10.6\ \frac{\mu g}{L}$$

$$C_{e_{Br_2ClCH}} = \frac{10\frac{\mu g}{L}}{1 + 21 \times 0.033 \times (1 - e^{-46.9})} = 5.9\ \frac{\mu g}{L}$$

$$\text{Total effluent THM conc.} = 30 + 10.6 + 5.9 = 46.5\ \frac{\mu g}{L}$$

The assumed $CHCl_3$ effluent concentration of 30 μg/L is slightly conservative for calculation of the minimum air-to-water ratio required for stripping. This value could be increased slightly and still meet the THM treatment objective of 50 μg/L.

9. Calculate the total blower power required to meet the treatment objective.

a. Calculate the air mass flow rate, G_{me}.

$$G_{me} = 4.6\ \frac{m^3}{s} \times 1.24\ \frac{kg}{m^3} = 5.07\ \frac{kg}{s}$$

b. Calculate the inlet air pressure to the aeration basin, $P_{in,b}$, using Eq. 5.85.

$$P_{in,b} = 101325\ \frac{N}{m^2} + 998.26\ \frac{kg}{m^3} \times 5\ m \times 9.81\ \frac{m}{s^2} = 150290\ \frac{N}{m^2}$$

c. Calculate the total blower power, P_{blower}, using Eq. 5.49.

$$P_{blower} = \left[\frac{5.07\ \frac{kg}{s} \times 286.7\ \frac{J}{kg \cdot K} \times 293.15\ K}{100\ \frac{W}{kW} \times 0.283 \times 0.30}\right] \times \left[\left(\frac{150290\ \frac{N}{m^2}}{101325\ \frac{N}{m^2}}\right)^{0.283} - 1\right] \times \left(\frac{1\ W}{1\ \frac{J}{s}}\right)$$

$$= 592\ kW$$

For comparison purposes, assume a portion of the clear well can be modified to provide a three-tanks-in-series system with a total retention time of 1 h.

1. Calculate the total volume if the total retention time is 1 h:

$$V_n = Q\tau_n = 0.2191\ \frac{m^3}{s} \times 1\ h \times \frac{3600\ s}{1\ h} = 789\ m^3$$

2. Determine the volume per tank, V.

$$V = \frac{V_n}{\text{No. of tanks}} = \frac{789\ m^3}{3\ \text{tanks}} = 266\ m^3$$

Assume the same tank depth as the single tank and mass transfer coefficients do not change for tanks in series. The removal efficiency after the third tank can be determined as shown in the following calculations.

3. Determine the minimum air-to-water ratio for the three-tanks-in-series system using Eq. 5.72. Assume a $CHCl_3$ effluent concentration of 30 μg/L.

$$\left(\frac{\dot{v}}{Q}\right)_{\min(CHCl_3)} = \left[\left(\frac{C_O}{C_e}\right)^{1/n} - 1\right]\frac{1}{H} = \left[\left(\frac{120\frac{\mu g}{L}}{30\frac{\mu g}{L}}\right)^{1/3} - 1\right]\frac{1}{0.143} = 4.11$$

4. Determine the airflow rate for each tank.

$$\dot{V} = \left(\frac{\dot{v}}{Q}\right)_{\min(CHCl^3)} \times Q = 4.11 \times 0.2191\ \frac{m^3}{s} = 0.901\ \frac{m^3}{s}$$

5. Calculate the Stanton number, ϕ, for each THM.

$$\phi_{Cl_3CH} = \frac{2.402 \times 10^{-3}\ \frac{1}{s} \times 266\ m^3}{0.901\frac{m^3}{s} \times 0.143} = 4.96$$

$$\phi_{BrCl_2CH} = 7.64$$

$$\phi_{Br_2ClCH} = 16.8$$

6. Determine the total effluent THM concentration from the tank, C_3.

$$C_{3_{Cl_3CH}} = \frac{120\ \frac{\mu g}{L}}{[1 + 4.11 \times 0.143 \times (1 - e^{-5.827})]^3} = 30.2\ \frac{\mu g}{L}$$

$$C_{3_{BrCl_2CH}} = 12.0\ \frac{\mu g}{L}$$

$$C_{3_{Br_2ClCH}} = 6.83\ \frac{\mu g}{L}$$

$$\text{Total effluent THM concentration} = 30.2\ \frac{\mu g}{L} + 12.0\ \frac{\mu g}{L} + 6.83\ \frac{\mu g}{L} = 49.03\ \frac{\mu g}{L}$$

The effluent concentration for the three-tanks-in-series design meets the total THM treatment objective of 50 μg/L.

7. Determine the required blower power for the three tanks in series.

a. Find the air mass flow rate, G_{me}.

$$G_{me} = .901\frac{m^3}{x} \times 1.24\ \frac{kg}{m^3} = 1.12\ \frac{kg}{s}$$

b. Calculate the inlet air pressure to the aeration basin, $P_{in,b}$.

$$P_{in,b} = 101325\ \frac{N}{m^2} + 998.26\ \frac{kg}{m^3} \times 5\ m \times 9.81\ \frac{m}{s^2} = 150290\ \frac{N}{m^2}$$

c. Calculate the total blower power, P_{blower}.

$$P_{\text{blower}} = \left[\frac{1.12\dfrac{\text{kg}}{\text{s}} \times 286.7\dfrac{\text{J}}{\text{kg}\cdot\text{K}} \times 293.15\ \text{K}}{1000\dfrac{\text{W}}{\text{kW}} \times 0.283 \times 0.30}\right]$$

$$\times \left[\left(\frac{150290\dfrac{N}{\text{m}^2}}{101325\dfrac{N}{\text{m}^2}}\right)^{0.283} - 1\right] \times \left(\frac{1\ \text{W}}{1\dfrac{\text{J}}{\text{s}}}\right)$$

$$= 131\ \text{kW per tank}$$

The total blower power required for three tanks in series with a retention time of 1 h is 393 kW. The power consumption for the single tank is 592 kW, which is 34 percent larger than the three-tank design. From an operational point of view, the three tanks in series would probably be more cost-effective. However, a total cost analysis would have to be performed to determine which design is the least cost-effective one.

EXAMPLE 5.4 ***Sample Ozone Absorption Problem***

A diffused ozone system with a retention time of 15 min is set up to apply ozone to a surface water at a dosage of 3.0 mg/L. Air is used to generate the ozone at a generation rate of 0.02 mg O_3/mg O_2. The water flow rate is 5000 gpm. The temperature of the system (water and air) is 23°C and the pressure is 1 atm. Properties of the ozone are provided below. The tank manufacturer has provided the clean-water oxygen transfer test data needed to estimate $K_L a_{O_2}$. This information follows.

Assuming that the decay rate of ozone in the water is negligible, determine the effluent concentration of the ozone in the exiting gas (y_e) and in the water. (Note that one can account for the decay rate by incorporating a first-order reaction term in Eq. 5.60 and solving).

Properties of Ozone

Property	Value
Molecular weight (mg/mmol)	48.0
Henry's constant (—)	3.77
Liquid diffusivity (m^2/s)	1.7×10^{-9}

Manufacturer's Clean-Water Oxygen Transfer Test Data for a Tank

Property	Value
SOTR (kg O_2/day)	1500
Volumetric Airflow rate (m^3/hr)	1700
Test tank volume (m^3)	500
Tank depth (m)	3

SOLUTION:

1. Calculate the airflow rate, $\dot{V}$.

$$\dot{V} = \frac{Q \cdot \text{dosage}}{\rho_a \cdot (\%O_3 \text{ in air})}$$

$$= \frac{\left(5000 \frac{\text{gal}}{\text{min}} \cdot \frac{1 \text{ min}}{60 \text{ s}} \cdot \frac{1000 \text{ L}}{264.17 \text{ gal}}\right) \cdot 3 \frac{\text{mg } O_3}{\text{L}}}{1.19 \frac{\text{kg air}}{\text{m}^3 \text{ air}} \cdot \left[\left(0.02 \frac{\text{mg } O_3}{\text{mg } O_2}\right)\left(0.21 \frac{\text{mg } O_2}{\text{mg air}}\right)\right] \cdot \left(10^6 \frac{\text{mg air}}{\text{kg air}}\right)} = 0.189 \frac{\text{m}^3}{\text{s}}$$

2. Find the volume of the tank, V.

$$V = \tau \cdot Q = (15 \text{ min})\left(5000 \frac{\text{gal}}{\text{min}} \cdot \frac{1 \text{ m}^3}{264.17 \text{ gal}}\right) = 284 \text{ m}^3$$

3. Find the oxygen mass transfer coefficient, $K_L a_{O_2}$. See steps 2a through 2e in previous problem.

$$K_L a_{O_2} = K_L a_{20}^* (1.024)^{T-20} = 0.00408 \frac{1}{\text{s}} (1.024)^{23-20} = 0.00438 \frac{1}{\text{s}}$$

4. Find the liquid diffusivities of oxygen and ozone. Calculate the liquid diffusivity of O_2 using the correlation of Holmen and Liss (1984).

$$D_l(O_2) = 10^{(3.15 - 831/296° \text{ K})} \times \left(1.0 \times 10^{-9} \frac{\text{m}^2}{\text{s}}\right) = 2.20 \times 10^{-9} \frac{\text{m}^2}{\text{s}}$$

5. Find the overall mass transfer coefficient for ozone, $K_L a_{O_3}$.

$$K_L a_{O_3} = K_L a_{O_2} \left(\frac{D_{l,O_3}}{D_{l,O_2}}\right)^{0.6} \left[1 + \frac{1}{H_i(100)}\right]^{-1}$$

$$= 0.00438 \frac{1}{\text{s}} \left(\frac{1.7 \times 10^{-9}}{2.2 \times 10^{-9}}\right)^{0.6} \left[1 + \frac{1}{3.77(100)}\right]^{-1} = 3.7 \times 10^{-3} \frac{1}{\text{s}}$$

6. Calculate y_e.

$$y_e = \frac{y_0}{1 + \left[\frac{1}{H(\dot{V}/Q)(1 - e^{-\varphi})}\right]}$$

$$= \frac{3 \frac{\text{mg}}{\text{L}}}{1 + \left\{1/\left[3.77\left(\frac{0.189}{0.315}\right)\left(1 - e^{[-0.0037(1/\text{s})284\text{m}^3]/[3.770 \cdot 189(\text{m}^3/\text{s})]}\right)\right]\right\}} = 1.75 \frac{\text{mg}}{\text{L}}$$

7. Calculate C_e assuming no ozone reaction.

$$C_e = \frac{\dot{V}}{Q}(y_0 - y_e) = \left(\frac{0.189}{0.315}\right)(3 - 1.90) \frac{\text{mg}}{\text{L}} = 0.66 \frac{\text{mg}}{\text{L}}$$

SURFACE AERATION

Surface aeration has been primarily used for oxygen absorption and the stripping of gases and volatile contaminants when the required removals are less than about 90 percent. Surface aeration devices consist of the brush type or turbine type, as shown in Figure 5.17. The brush-type aerator consists of several brushes attached to a rotary drum, which is half-submerged in water in the center of the tank. As the drum rotates, it disperses the water into the surrounding air providing reasonable contact between the air and water for mass transfer to take place. The turbine-type aerator consists of a submerged propeller system located in the center of the tank and surrounded by draft tubs. As the submerged propeller rotates it draws water from outside the draft tubs through the inner section and into the air creating contact between the air and water. These types of systems have been extensively used in the aeration of wastewater and their design and operation have been well documented (*Design of Municipal Wastewater Treatment Plants,* WEF-ASCE, Vol. 1, 1992).

Design Equations

Figure 5.18 displays a schematic of a surface aeration system for a single tank. Model development for air stripping using surface aeration has been described in the literature (Matter-Müller, Gujer, and Giger, 1981; Roberts and Dändliker, 1983; Roberts,

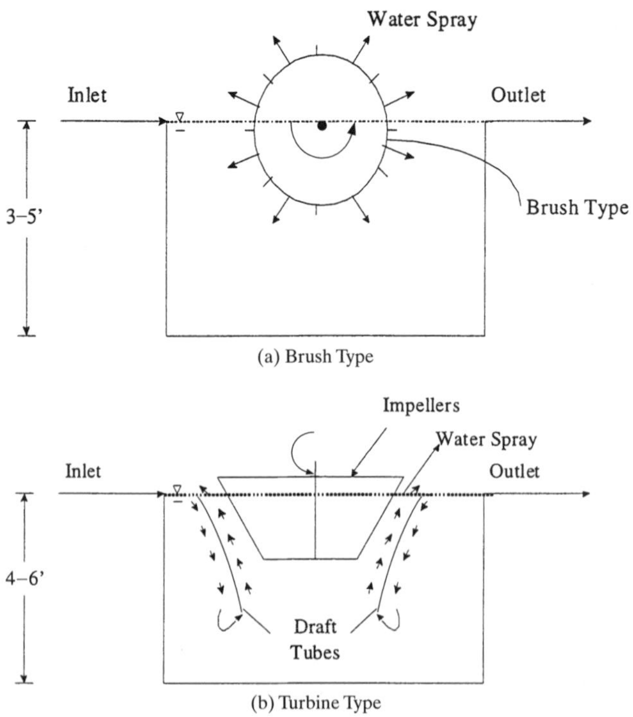

FIGURE 5.17 Schematic of various surface aeration devices.

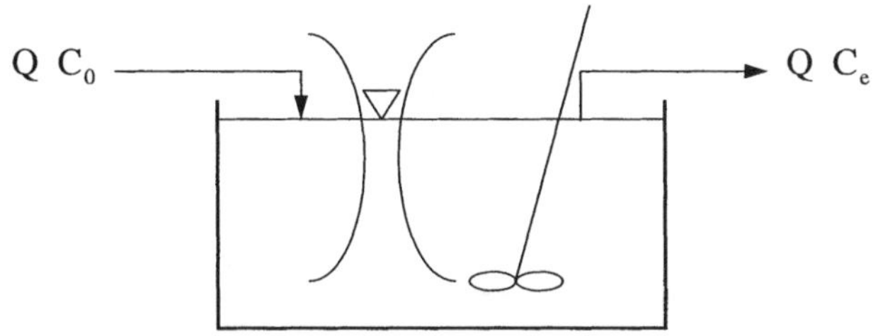

FIGURE 5.18 Schematic of a mechanical surface aerator used for the development of the design equation for a single tank.

Munz, and Dändliker, 1984; Roberts and Levy, 1985; Munz and Roberts, 1989). The development of the process design equations incorporate the following assumptions: (1) the tank is completely mixed in the liquid phase, (2) the gas-phase concentration of the contaminant in the tank equals zero, (3) the process is at steady state, and (4) Henry's constant applies. Incorporating these assumptions into an overall mass balance around the tank shown in Figure 5.18 yields the following expression (Hokanson, 1996):

$$QC_0 - QC_e - K_L a C_e V = 0 \tag{5.86}$$

Solving Eq. 5.86 for V results in the minimum tank volume required for a single tank in order to meet the treatment objective, C_e:

$$V = \frac{Q(C_0 - C_e)}{K_L a C_e} \tag{5.87}$$

Alternatively, Eq. 5.87 may be expressed in terms of the hydraulic retention time, τ, where τ is equal to V/Q. Substituting this relationship into Eq. 5.87 gives:

$$\tau = \frac{(C_0 - C_e)}{K_L a C_e} \tag{5.88}$$

If the removal of a given contaminant is required for an existing tank, Eq. 5.86 or Eq. 5.87 can be solved for C_e to yield:

$$C_e = \frac{C_0}{1 + K_L a\left(\dfrac{V}{Q}\right)} = \frac{C_0}{1 + K_L a \tau} \tag{5.89}$$

The process performance can be improved significantly with less total volume requirement by configuring the surface aeration system as tanks in series. Figure 5.19 shows a schematic of a surface aeration system operated with a tanks in series configuration. If it is assumed that all the tanks are equally sized and the assumptions for a single-tank system are also valid here, a mass balance around tank 1 yields the following expression:

$$C_1 = \frac{C_0}{1 + K_L a\left(\dfrac{V}{Q}\right)} \tag{5.90}$$

in which C_1 is the effluent concentration from tank 1.

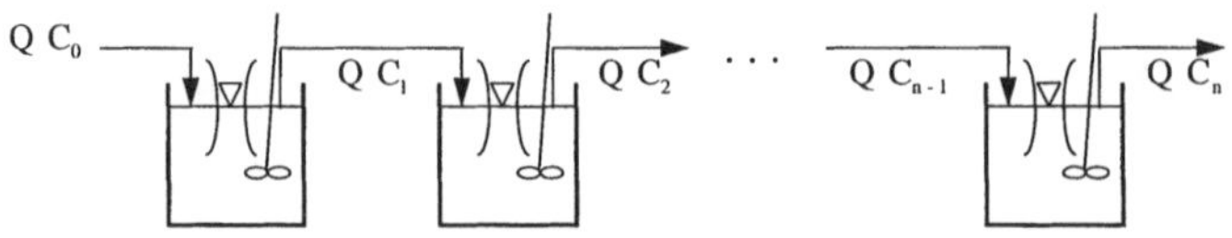

FIGURE 5.19 Schematic of a tanks-in-series surface aeration system.

A mass balance around tank 2 yields the following expression:

$$C_2 = \frac{C_1}{1 + K_{\mathrm{L}}a\left(\dfrac{V}{Q}\right)} = \frac{C_0}{\left[1 + K_{\mathrm{L}}a\left(\dfrac{V}{Q}\right)\right]^2} \tag{5.91}$$

in which C_2 is the effluent concentration from tank 2.

Similarly, if a mass balance is written around tank n, this results in:

$$C_n = \frac{C_{n-1}}{1 + K_{\mathrm{L}}a\left(\dfrac{V}{Q}\right)} = \frac{C_0}{\left[1 + K_{\mathrm{L}}a\left(\dfrac{V}{Q}\right)\right]^n} \tag{5.92}$$

in which C_n is the effluent contaminant concentration from tank n, and C_{n-1} is the effluent concentration from tank $n - 1$. Equation 5.92 is the general expression for the effluent concentration for a surface aeration system consisting of tanks in series.

Solving Eq. 5.92 in terms of the volume for each individual tank yields:

$$V = \frac{Q}{K_{\mathrm{L}}a}\left[\left(\frac{C_0}{C_n}\right)^{1/n} - 1\right] \tag{5.93}$$

Substitution of τQ for V into Eq. 5.93 yields the hydraulic retention time, τ, for each tank in series.

$$\tau = \frac{1}{K_{\mathrm{L}}a}\left[\left(\frac{C_0}{C_n}\right)^{1/n} - 1\right] \tag{5.94}$$

The total volume for all tanks, V_n, can be obtained by multiplying both sides of Eq. 5.93 by n:

$$V_n = V \times n = \frac{nQ}{K_{\mathrm{L}}a}\left[\left(\frac{C_0}{C_n}\right)^{1/n} - 1\right] \tag{5.95}$$

Likewise, the total hydraulic retention time, τ_n, for all tanks can be obtained by multiplying both sides of Eq. 5.94 by n:

$$\tau_n = \tau \times n = \frac{n}{K_{\mathrm{L}}a}\left[\left(\frac{C_0}{C_n}\right)^{1/n} - 1\right] \tag{5.96}$$

Because the quantities V, V_n, τ, and τ_n are not independent, calculation of one of these variables from the preceding equations allows for determination of the other three parameters because the flow rate, Q, and the number of tanks are specified.

The K_La for surface aeration can be calculated using Eq. 5.78. A value of 0.5 is recommended for the diffusivity exponent, m, and a value of 40 is recommended for the ratio k_g/k_l (Munz and Roberts, 1989). The overall mass transfer coefficient for oxygen can be estimated using the following correlation (Roberts and Dändliker, 1983):

$$K_La_{O_2} = 2.9 \times 10^{-5}\left(\frac{P}{V}\right)^{0.95} \tag{5.97}$$

in which P/V is the power input per unit tank volume (W/m^3). Eq. 5.97 is valid for P/V values between 10 and 200 W/m^3 and it is assumed the reactor is completely mixed. The liquid diffusivity of oxygen can be calculated using the correlation of Hayduk and Laudie (1974).

The total power required for surface aeration is estimated based on the power supplied to the aerators. The following equation can be used to calculate the power requirements:

$$P_{\text{surface}} = \left(\frac{P}{V}\right)\frac{V \times n_{\text{tanks}}}{1000\ Eff_s} \tag{5.98}$$

in which Eff_s is the surface aerator motor efficiency.

EXAMPLE 5.5 ***Surface Aeration Sample Calculation***
An existing clear well is to be modified to include a surface aeration system capable of reducing total trihalomethane (THM) concentrations from 160 μg/L to less than 50 μg/L. The total volume of the clear well is 3785 m^3 (1 mgal) (depth = 5.0 m, length = 30 m, width = 25 m). The average plant flow rate is 0.2191 m^3/s (5 mgd) and the water temperature is 20°C. The following list summarizes the influent concentrations (C_0), molecular weights (MW), Henry's constants (H), and molar volumes at the normal boiling point (V_b) for the THMs.

Determine the power input per unit tank volume and the total power input for aeration of the whole clear well. Assuming the clear well can be modified into individual tanks in series, determine the power input per unit tank volume and the total power input for aeration assuming a total retention time of 1 h. Assume the blower efficiency is 30 percent.

THM Chemical and Physical Properties

Compound	C_0 (μg/L)	MW (g/mol)	H (—)	V_b* (cm^3/mol)
Chloroform ($CHCl_3$)	120.0	119.38	0.143	87.0
Bromodichloromethane ($BrCl_2CH$)	30.0	163.82	0.087	94.5
Dibromochloromethane (Br_2ClCH)	10.0	208.27	0.033	102.0

* Determined using Le Bas method.

SOLUTION:

1. Determine the overall mass transfer coefficients for each THM by first assuming a power input per unit volume of 15 W/m^3.

$$T_{\text{water}} = 20°\text{C} = 293°\ \text{K} \quad \mu_{\text{water}} = 1.02\ \text{cp}$$

   a. Calculate the liquid diffusivity of O_2 using the correlation of Holmen and Liss (1984).

$$D_l(O_2) = 10^{(3.15 - 831/293°\text{ K})} \times \left(1.0 \times 10^{-9}\ \frac{\text{m}^2}{\text{s}}\right) = 2.06 \times 10^{-9}\ \frac{\text{m}^2}{\text{s}}$$

b. Calculate the liquid diffusivity of each THM using the correlation of Hayduk and Laudie (1974).

$$D_l(\text{CHCL}_3) = \frac{13.26 \times 10^{-5}}{(1.02\ \text{cp})^{1.14}\,(87.0)^{.589}} = 9.34 \times 10^{-6}\ \frac{\text{cm}^2}{\text{s}} = 9.34 \times 10^{-10}\ \frac{\text{m}^2}{\text{s}}$$

$$D_l(\text{BrCl}_2\text{CH}) = 8.9 \times 10^{-10}\ \frac{\text{m}^2}{\text{s}}$$

$$D_l(\text{Br}_2\text{ClCH}) = 8.52 \times 10^{-10}\ \frac{\text{m}^2}{\text{s}}$$

c. Calculate $K_L a\ (O_2)$ using Eq. 5.97

$$K_L a(\text{O}_2) = 2.9 \times 10^{-5} \times \left(\frac{P}{V}\right)^{0.95} = 2.9 \times 10^{-5} \times \left(15\ \frac{\text{W}}{\text{m}^3}\right)^{0.95} = 3.8 \times 10^{-4}\ \text{s}^{-1}$$

d. Calculate the $K_L a$ for each THM using Eq. 5.96.

$$K_L a(\text{CHCl}_3) = 3.8 \times 10^{-4}\ \text{s}^{-1} \left(\frac{\dfrac{9.34 \times 10^{-10}\ \frac{\text{m}^2}{2}}{}}{2.06 \times 10^{-9}\ \dfrac{\text{m}^2}{\text{s}}}\right)^{0.6} \left[1 + \frac{1}{(0.143)(40)}\right]^{-1}$$

$$= 2.06 \times 10^{-4}\ \text{s}^{-1}$$

$$K_L a(\text{BrCl}_2\text{CH}) = 1.78 \times 10^{-4}\ \text{s}^{-1}$$

$$K_L a(\text{Br}_2\text{ClCH}) = 1.27 \times 10^{-4}\ \text{s}^{-1}$$

2. Calculate the surface aeration removal efficiency using Eq. 5.90.
Volume of clear well = 1 mgd = 3786 m^3
Average flow = 5 mgd = 0.2191 m^3/s

$$C_e(\text{CHCl}_3) = \frac{120\ \dfrac{\mu\text{g}}{\text{L}}}{1 + (2.06 \times 10^{-4}\ \text{s}^{-1})\left(\dfrac{3{,}786\ \text{m}^3}{0.2191\ \text{m}^3/\text{s}}\right)} = 26.3\ \frac{\mu\text{g}}{\text{L}}$$

$$C_e(\text{BrCl}_2\text{CH}) = 7.4\ \frac{\mu\text{g}}{\text{L}}$$

$$C_e(\text{Br}_2\text{ClCH}) = 3.1\ \frac{\mu\text{g}}{\text{L}}$$

$$\text{Total effluent THM conc.} = 26.3\ \frac{\mu\text{g}}{\text{L}} + 7.4\ \frac{\mu\text{g}}{\text{L}} + 3.1\ \frac{\mu\text{g}}{\text{L}} = 36.8\ \frac{\mu\text{g}}{\text{L}}$$

For a P/V of 15 W/m^3, the THM effluent concentration is less than the treatment objective of 50 μg/L. The above calculation can be repeated using a slightly lower P/V value but the solution is conservative.

3. Calculate the total power requirements for 1 tank using Eq. 5.98.

$$P = \left(15\ \frac{\text{W}}{\text{m}^3}\right)\left(\frac{3{,}786\text{m}^3}{0.50}\right)\left(\frac{\text{kW}}{1000\ \text{W}}\right) = 114\ \text{kW}$$

4. Calculate the P/V requirement for the three-tanks-in-series system using Eq. 5.98 and assuming a total retention time of 45 min.

$$\text{Assume } \frac{P}{V} = 15\ \frac{\text{W}}{\text{m}^3}$$

$$C_e(\text{CHCl}_3) = \frac{120\ \dfrac{\mu\text{g}}{\text{L}}}{\left[1 + 2.06 \times 10^{-4}(15\ \text{min})\left(\dfrac{60\ \text{s}}{\text{min}}\right)\right]^3} = 72.0\ \frac{\mu\text{g}}{\text{L}}$$

Because C_e should be similar to the value for the single tank to meet the treatment objective, the P/V value needs to be increased.

Assume $P/V = 60\ \text{W/m}^3$, recalculate the K_La values for O_2 and the THMs and the C_e values.

$$K_La(\text{O}_2) = 1.42 \times 10^{-3}\ \text{s}^{-1} \quad K_La(\text{CHCl}_3) = 7.64 \times 10^{-4}\ \text{s}^{-1} \quad C_e(\text{CHCl}_3) = 25.0\ \frac{\mu\text{g}}{\text{L}}$$

$$K_La(\text{BrCl}_2\text{CH}) = 6.67 \times 10^{-4}\ \text{s}^{-1} \quad C_e(\text{BrCl}_2\text{CH}) = 7.3\ \frac{\mu\text{g}}{\text{L}}$$

$$K_La(\text{Br}_2\text{ClCH}) = 4.76 \times 10^{-4}\ \text{s}^{-1} \quad C_e(\text{Br}_2\text{ClCH}) = 3.4\ \frac{\mu\text{g}}{\text{L}}$$

$$\text{Total effluent THM conc.} = 25.0\ \frac{\mu\text{g}}{\text{L}} + 7.3\ \frac{\mu\text{g}}{\text{L}} + 3.4\ \frac{\mu\text{g}}{\text{L}} = 35.7\ \frac{\mu\text{g}}{\text{L}}$$

It appears that a P/V of $60\ \text{W/m}^3$ will be sufficient to meet the treatment objective.

5. Calculate the total power requirements for the three-tanks-in-series system using Eq. 5.98.

{Total volume of the three-tank system}

$$= \left(5 \times 10^6\ \frac{\text{gal}}{\text{day}}\right)\left(\frac{1\ \text{day}}{1440\ \text{min}}\right)(45\ \text{min})\left(\frac{\text{ft}^3}{7.48\ \text{gal}}\right)\left(\frac{0.3048\ \text{m}}{\text{ft}}\right)^3 = 592\ \text{m}^3$$

{Total power requirement for the three-tank system}

$$= \left(60\ \frac{\text{W}}{\text{m}^3}\right)\left(\frac{592\ \text{m}^3}{0.5}\right)\left(\frac{1\ \text{kW}}{1000\ \text{W}}\right) = 71\ \text{kW}$$

A modification of a small portion of the clear well to incorporate a three-tanks-in-series system as opposed to aeration of the whole clear well could reduce the power costs by about 50 percent.

SPRAY AERATORS

Spray aerators have been used in water treatment for many years to oxygenate groundwater for the purpose of iron and manganese removal and the air stripping of gases (i.e., carbon dioxide, hydrogen sulfide) and volatile organic compounds

(VOCs). Effective iron oxidation by aeration usually requires at least 1 h of retention time after aeration. Manganese oxidation by aeration is very slow and not practical for waters with pH values below 9.5. Manganese removal usually requires a stronger oxidant. Carbon dioxide and hydrogen sulfide removals have ranged from 50 to 90 percent depending upon the pH of the water. Chapter 12 discusses in more detail the chemistry of metals and their oxidation. Volatile organic chemical removals have been as high as 90 percent depending upon the Henry's law constant.

Spray aerator systems consist of a series of fixed nozzles on a pipe grid. The grids can be placed in towers, commonly known as spray towers (or fountains), that spray onto the surface of raw water reservoirs. Pressurized nozzles disperse fine water droplets into the surrounding air, creating a large air-water surface for mass transfer. Two types of pressurized spray nozzles, hollow- and full-cone, are commonly used in water treatment. Full-cone nozzles deliver a uniform spray pattern of droplets. The hollow-cone nozzle delivers a circular spray pattern with most of the droplets concentrated at the circumference. The hollow-cone nozzle is generally preferred over the full-cone type because it provides smaller droplets for better mass transfer even though it has a larger pressure drop requirement. Hollow-cone spray droplets are around 5 mm and are prone to plugging. It is recommended that in-line strainers be installed in the spray nozzle manifold to prevent plugging.

The fountain-type spray aerators have been more widely used in water treatment because they can be easily adapted to existing water treatment systems. The design approach and application of the fountain type is presented below. The design approach and application for spray towers is shown elsewhere (J. M. Montgomery, Inc., 1st edition, 1985).

Design Equations

Figure 5.20 displays a schematic of a single fountain-type spray aerator. A design equation can be derived for estimating the removal of a volatile contaminant by per-

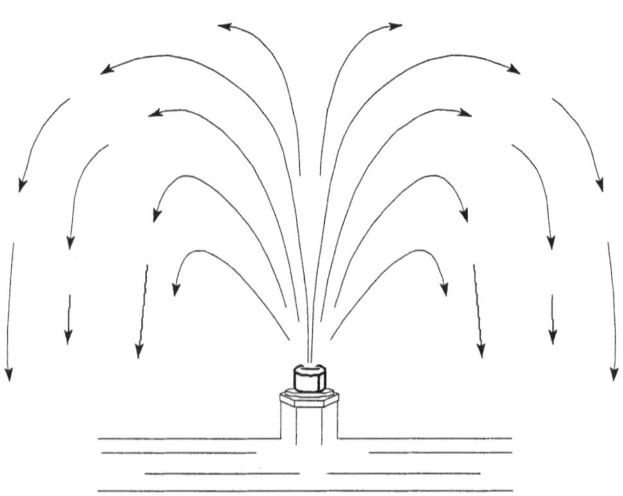

FIGURE 5.20 Schematic of a single-fountain spray aerator.

forming a mass balance on a water drop while it is exposed to the air. If it is assumed that all the water drops are the same size and have the same exposure time in the air, the following mass balance can be performed:

$$\left\{\begin{array}{l}\text{Mass lost from}\\ \text{the water drop}\\ \text{per unit time}\end{array}\quad = \quad \begin{array}{l}\text{Mass transferred across}\\ \text{the air-water interface}\\ \text{of a water drop}\\ \text{per unit time}\end{array}\right\} \tag{5.99}$$

$$V_d \frac{dC}{dt} = K_L a[C(t) - C_s(t)]V_d \tag{5.100}$$

in which V_d is equal to volume of the drop (L^3). In spray aeration, it is assumed that the gas-phase contaminant concentration in the open air is zero. As a result of this assumption, C_s is also assumed to be zero because it is in equilibrium with the contaminant gas-phase concentration. Rearranging Eq. 5.100 and integrating over the time the drop is exposed to the air yields the following expression for the final contaminant concentration of the water drop after being exposed to the air:

$$C_e = C_0 e^{-K_L a t} \tag{5.101}$$

in which C_0 is the initial contaminant liquid-phase concentration (M/L^3) of the drop before being exposed to the air, and t is the time of contact between the water drops and the air and is dependent upon the water drop exiting velocity and trajectory. The t can be estimated from the following equations:

$$t = \frac{2v_d \sin \alpha}{g} \tag{5.102}$$

in which α is the angle of spray measured from the horizontal (degrees), g is the acceleration due to gravity (L/t^2) and v_d is the exit velocity of the water drop from the nozzle (L/t):

$$v_d = C_v \sqrt{2\,gh} \tag{5.103}$$

in which C_v is the velocity coefficient of the nozzle (–) and h is the total head of the nozzle (L). C_v varies from 0.4 to 0.95 and can be obtained from the nozzle manufacturers. The flow rate Q through a given nozzle can be estimated from the following equation:

$$Q = C_d A \sqrt{2\,gh} \tag{5.104}$$

in which A is the area of the nozzle opening (L^2) and C_d is the coefficient of discharge from the nozzle (dimensionless).

The overall mass transfer coefficient, K_L, can be estimated from the following correlations (Higbie, 1935; Jury, 1967):

$$\text{For } \frac{2\,D_l t^{1/2}}{d_p} < 0.22 \quad K_L = 2\left(\frac{D_l}{\pi t}\right)^{1/2} \tag{5.105}$$

$$\text{For } \frac{2\,D_l t^{1/2}}{d_p} > 0.22 \quad K_L = \frac{10\,D_l}{d_p} \tag{5.106}$$

in which d_p is the sauter mean diameter (SMD) of the water drop (in) (SMD is obtained by dividing the total volume of the spray by the total surface area), D_l = contaminant liquid diffusivity (cm^2/s), and t is the contact time of the water drop with the air (s). The interfacial surface area available for mass transfer for the water drop can be obtained from the following equation:

$$a = \frac{6}{d_p} \tag{5.107}$$

EXAMPLE 5.6 ***Spray Aeration Sample Calculation***
Well water contains 1.0 mg/L of total sulfide at a pH of 6.2. The well pumps 0.0631 m^3/s (1000 gpm), and the pump has the capacity to deliver an additional 28 m (40 psi) of head. Determine the number of nozzles required and the expected H_2S removal efficiency. It is assumed that the water has sufficient buffer capacity to maintain the pH at 6.2 during aeration. The following information was obtained from the nozzle manufacturer: SMD = 0.10 cm, $\alpha = 90°$, $C_v = 0.45$, $C_d = 0.1$, nozzle diameter = 1.25 cm.

1. Determine the number of nozzles required.

$$A_{\text{nozzle}} = \frac{\pi}{4}\left(\frac{1.25\ \text{cm}}{100\ \frac{\text{cm}}{\text{m}}}\right)^2 = 1.23 \times 10^{-4}\ \text{m}^2$$

The flow rate through one nozzle, Q_{nozzle}, can be calculated using Eq. 5.104:

$$Q_{\text{nozzle}} = C_d A_{\text{nozzle}} \sqrt{2\,gh} = 0.25(1.23 \times 10^{-4}\ \text{m}^2)\sqrt{2(9.81\ \frac{\text{m}}{\text{s}^2})(28\ \text{m})}$$

$$= 7.21 \times 10^{-4}\ \frac{\text{m}^3}{\text{s}}$$

The number of nozzles can be calculated by dividing the total flow by the flow through each nozzle.

$$\{\text{Number of nozzles required}\} = \frac{Q}{Q_{\text{nozzle}}} = \frac{0.0631\ \frac{\text{m}^3}{\text{s}}}{7.21 \times 10^{-4}\ \frac{\text{m}^3}{\text{s}}} = 88$$

2. Determine the H_2S removal efficiency. Eq. 5.103 is used to calculate the velocity of the water exiting the nozzle.

$$v_d = C_v \sqrt{2\,gh} = 0.45\sqrt{2(9.81\ \frac{\text{m}}{\text{s}^2})(28\ \text{m})} = 10.6\ \frac{\text{m}}{\text{s}}$$

The contact time of the water drop with the air can be calculated using Eq. 5.102.

$$t = \frac{2\,v_d \sin\alpha}{g} = \frac{2(10.6\ \frac{\text{m}}{\text{s}})\sin(90°)}{9.81\ \frac{\text{m}}{\text{s}^2}} = 2.16\ \text{s}$$

The overall liquid-phase mass transfer coefficient is calculated using Eq. 5.105.

$$\frac{2D_l t^{1/2}}{d_p} = \frac{2(1.6 \times 10^{-5}\ \frac{\text{cm}^2}{\text{s}})(2.14\ \text{s})^{1/2}}{0.10\ \text{cm}} = 4.68 \times 10^{-4}$$

$$4.68 \times 10^{-4} << 0.22$$

$$K_L = 2\left(\frac{D_l}{\pi t}\right)^{1/2} = 2\left(\frac{1.5 \times 10^{-5}\ \frac{\text{cm}^2}{\text{s}}}{\pi(2.14\ \text{s})}\right)^{1/2} = 0.0031\ \frac{\text{cm}}{\text{s}}$$

The interfacial area for mass transfer is calculated using Eq. 5.107.

$$a = \frac{6}{d_p} = \frac{6}{0.1\ \text{cm}} = 60\ \text{cm}^{-1}$$

Calculate $K_L a$:

$$K_L a = \left(0.0031\ \frac{\text{cm}}{\text{s}}\right)(60\ \text{cm}^{-1}) = 0.19\ \text{s}^{-1}$$

The H_2S reacts with water to form both HS^- and $S^=$ as shown by the following equations:

$$H_2S + H_2O \Leftrightarrow H_3O^+ + HS^- \qquad K_1 = 1.0 \times 10^{-6.99}\ \frac{\text{mol}}{\text{L}}\ (25°\text{C})$$

$$HS^- + H_2O \Leftrightarrow H_3O^+ + S^{2-} \qquad K_2 = 1.0 \times 10^{-18.2}\ \frac{\text{mol}}{\text{L}}\ (25°\text{C})$$

In the gas side, the only sulfide species is hydrogen sulfide, H_2S. In the liquid side, sulfide can exist in two species: H_2S or HS^-. Therefore, the following general relationship applies on the liquid side:

$$C_S = C_{H_2S} + C_{HS^-}$$

in which C_S is the total sulfide concentration in the liquid-phase (M/L^3) and is equal to the sum of C_{H_2S} and C_{HS^-}. Conveniently, C_{H_2S} can be related to C_S by making use of the ionization fraction, α_0, which is defined as:

$$\alpha_0 = \frac{C_{H_2S}}{C_S} = \left(1 + \frac{K_{al}}{[H^+]} + \frac{K_{a1}\,K_{a2}}{[H^+]^2}\right)^{-1}$$

If Eq. 5.104 is rewritten in terms of total sulfide and integrated using the same assumptions, the following expression to the effluent liquid-phase concentration of hydrogen sulfide can be written as:

$$C_{Se} = C_{So}e^{(-\alpha_0 K_L a_{H_2S} t)}$$

in which C_{Se} is the effluent liquid-phase concentration of total sulfide after spray aeration, C_{Si} is the influent total liquid-phase sulfide concentration, and $K_L a_{H^2S}$ is the overall mass transfer coefficient for hydrogen sulfide. The first ionization constant, α_0, is calculated as:

$$\alpha_0 = \frac{C_{H_2S}}{C_S} = \left(1 + \frac{10^{-6.99}}{[10^{-6.2}]} + \frac{(10^{-6.99})(10^{-18.2})}{[10^{-6.2}]^2}\right)^{-1} = 0.860$$

The influent H_2S concentration is 1.0 mg/L, then the total initial sulfide concentration is:

$$C_{So} = \frac{C_{(H_2S)_o}}{\alpha_o} = \frac{1.0\,\frac{mg}{L}}{0.860} = 1.16\,\frac{mg}{L}$$

The effluent total liquid-phase sulfide concentration after stripping is calculated as:

$$C_{Se} = 1.16\,\frac{mg}{L}\,e^{(-0.860 \times 0.19\,s^{-1} \times 2.14\,s)} = 0.818\,\frac{mg}{L}$$

The effluent liquid-phase hydrogen sulfide concentration is:

$$C_{e,H_2S} = \alpha_o \times C_{Se} = (0.860) \times \left(0.818\,\frac{mg}{L}\right) = 0.703\,\frac{mg}{L}$$

The H_2S removal due to stripping is calculated as:

$$\{H_2S \text{ removal due to stripping}\} = \frac{1.0\,\frac{mg}{L} - 0.703\,\frac{mg}{L}}{0.10\,\frac{mg}{L}} \times 100 = 29.7 \text{ percent}$$

BIBLIOGRAPHY

Ashworth, R. A., G. B. Howe, M. E. Mullins, and T. N. Rogers. "Air-Water Partitioning Coefficients of Organics in Dilute Aqueous Solutions." *Journal of Hazardous Materials,* 18, 1988: 25–36.

Cummins, M. D. *Field Evaluation of Packed Column Air Stripping: Bastrap, Louisiana, Interim Report.* Cincinnati, OH: U.S. Environmental Protection Agency, Office of Drinking Water, Technical Support Division, 1985.

Baillod, C. R., W. L. Paulson, J. J. McKeown, and H. J. Campbell, Jr. "Accuracy and Precision of Plant Scale and Shop Clean Water Oxygen Transfer Tests." *Journal WPCF,* 58, 1986: 4.

Ball, W. P., M. D. Jones, and M. C. Kavanaugh. "Mass Transfer of Volatile Organic Compounds in Packed Towers." *Jour WPCF,* 56, 1984: 127.

Brown, L. C., and C. R. Baillod. "Modeling and Interpreting Oxygen Transfer Data." *J. Environ. Eng. Div. Proc. Am. Soc. Civ. Eng.,* 108 (EE4), 1982: 607.

Cortright, R. D. *Gas-Phase Adsorption of Volatile Organic Compounds from Air Stripping Off-Gas onto Granulated Activated Carbon.* Houghton, MI: Michigan Technological University, 1986.

Crittenden, J. C., R. D. Cortright, B. Rick, S. Tang, and D. L. Perram. "Using GAC to Remove Air Stripper Off-Gas." *Jour. AWWA,* 80(5), 1988: 73–84.

Cummins, M. D., and J. J. Westrick. Proceedings ASCE Environmental Engineering Conference, Boulder, Colorado, July 1983, pp. 442–449.

Djebbar, Y., and R. M. Narbaitz. "Mass Transfer Correlations for Air Stripping Towers." *Environmental Progress,* 14(3), 1995: 137–145.

Dzombak, D. A., S. B. Roy, and H. J. Fang. "Air-Stripper Design and Costing Computer Program." *Jour AWWA,* 85(10), 1993: 63–72.

Eckert, J. S. "Design Techniques for Sizing Packed Towers." *Chem. Eng. Prog.,* 57(9), 1961.

Gross, R. L., and S. G. TerMaath. "Packed Tower Aeration Strips Trichloroethene from Groundwater." *Environmental Progress,* 4(2), 1985.

Hand, D. W., J. C. Crittenden, J. L. Gehin, and B. L. Lykins, Jr. "Design and Evaluation of an Air-Stripping Tower for Removing VOCs from Groundwater." *Jour. AWWA,* 78(9), 1986: 87.

Haarhoff, J. AIRSTRIP. AIRSTRIP, Inc., Ames IA. 1988.

Hayduk, W., and H. Laudie. "Prediction of Diffusion Coefficients for Non-electrolytes in Dilute Aqueous Solutions." *Jour. AIChE,* 28, 1974: 611.

Higbie, R. "The Rate of Adsorption of a Pure Gas into a Still Liquid During Short Periods of Exposure." *Trans. Am. Inst. Chem. Engrs.,* 31, 1935: 365.

Hirschfelder, J. O., R. B. Bird, and E. L. Spotz. *Trans. ASME,* 71, 1949:921. *Chem. Rev.,* 44, 1949: 205.

Hokanson, D. R. *Development of Software Design Tools for Physical Property Estimation, Aeration, and Adsorption* (Master's Thesis). Houghton, MI: Michigan Technological University, 1996.

Holmén, Kim, and Peter Liss. "Models for Air-Water Gas Transfer: An Experimental Investigation." *Tellus,* 36B, 1984: 92–100.

Jury, S. H. *A.I.Ch.E.J.,* 13, 1967: 1924.

Kavanaugh, M. C., and R. R. Trussell. "Design of Aeration Towers to Strip Volatile Contaminants from Drinking Water." *Jour AWWA,* 72(12), 1980: 684.

Lamarche, P., and R. L. Droste. "Air-Stripping Mass Transfer Correlations for Volatile Organics." *Jour AWWA,* 81(1), 1989: 78–89.

Lenzo, F. C., T. J. Frielinghaus, and A. W. Zienkiewicz. "The Application of the Onda Correlation to Packed Column Air Stripper Design: Theory Versus Reality." In *Proceedings of the AWWA Conference,* Cincinnati, Ohio. Denver, CO: AWWA, June 1990.

Lewis, W. K., and W. E. Whitman. "Principles of Gas Absorption." *Ind. Engng. Chem.,* 16, 1924: 1215–1220.

Little, J. C., and R. E. Selleck. "Evaluating the Performance of Two Plastic Packings in a Crossflow Aeration Tower." *Jour. AWWA,* 83(6), June 1991: 88.

Mandt, M. G., and P. R. Bathija. "Jet Fluid Gas/Liquid Contacting and Mixing." *AIChE Symposium Series,* 73(167), 1987: 15.

Matter-Müller, C., W. Gujer, and W. Giger. "Transfer of Volatile Substances from Water to the Atmosphere." *Water Research,* 15, 1981: 1271–1279.

McKinnon, R. J., and J. E. Dyksen. "Removal of Organics from Groundwater Using Aeration Plus Carbon Adsorption." *Jour AWWA,* 76(5), 1984: 42.

Metcalf and Eddy, Inc. *Wastewater Engineering,* 3rd Ed. New York: McGraw-Hill, 1991.

Montgomery, J. M., Inc. *Water Treatment Principles and Design.* New York: Wiley-Interscience, 1985.

Munz, C., and P. V. Roberts. "Mass Transfer and Phase Equilibria in a Bubble Column." *Proc. AWWA Annual Conference,* 1982: 633–640.

Munz, C., and P. V. Roberts. "Gas- and Liquid-Phase Mass Transfer Resistances of Organic Compounds during Mechanical Surface Aeration." *Water Research,* 23(5), 1989: 589–601.

Nirmalakhandan, N., W. Jang, and R. E. Speece. "Evaluation of Cascade Air Stripping—Pilot-Scale and Prototype Studies." *A.S.C.E. Jour Environmental Engineering,* 117(6), 1991: 788–798.

Nirmalakhandan, N., W. Jang, and R. E. Speece. "Removal of 1,2 Dibromo-3-Chloropropane by Countercurrent Cascade Air Stripping." *A.S.C.E. Jour Environmental Engineering,* 118(2), 1992: 226–237.

Onda, K., H. Takeuchi, and Y. Okumoto. "Mass Transfer Coefficients between Gas and Liquid Phases in Packed Columns." *Jour. Chem. Engrg. Japan,* 1(1), 1968: 56–62.

Rasmuson, A. C. "Adsorption Equilibria on Activated Carbon of Mixtures of Solvent Vapors." In *Fundamentals of Adsorption, Proceedings of the Engineering Foundation Conference,* A. Meyers and G. Belfort, eds. New York: Engineering Foundation, 1984.

Reucroft, P. J., W. H. Simpson, and L. A. Jonas. "Sorption Properties of Activated Carbon." *Jour. Phys. Chem.* 75(23), 1971: 3526.

Roberts, P. V., and P. G. Dändliker. "Mass Transfer of Volatile Organic Contaminants from Aqueous Solution to the Atmosphere during Surface Aeration." *Environ. Sci. Technol.,* 17(8), 1983: 484–489.

Roberts, P. V., C. Munz, and P. Dändliker. "Modeling Volatile Organic Solute Removal by Surface and Bubble Aeration." *Jour WPCF,* 56(2), 1984: 157–163.

Roberts, P. V., G. D. Hopkins, C. Munz, and A. H. Riojas. "Evaluating Two-Resistance Models for Air Stripping of Volatile Organic Contaminants in a Countercurrent, Packed Column." *Environ. Sci. Technol.,* 19, 1985: 161–173.

Roberts, P. V., and J. A. Levy. "Energy Requirements for Air Stripping Trihalomethanes." *Jour AWWA,* 77(4), 1985: 138.

Rogers, T. N. *Predicting Environmental Physical Properties from Chemical Structure Using a Modified UNIFAC Model,* Ph.D. Dissertation. Houghton, MI: Michigan Technological University, 1994.

Sherwood, T. K., and F. A. Hollaway. "Performance of Packed Towers—Liquid Film Data for Several Packings." *Trans. Amer. Inst. Chem. Engrs.,* 36, 1940: 39.

Shulman, H. L., C. F. Ullrich, and N. Wells. "Performance of Packed Columns. I. Total, Static, and Operating Holdups." *Amer. Inst. Chem. Engrs. Jour.,* 1(2), June 1955: 247.

Singley, et al. *Trace Organics Removal by Air Stripping.* Denver, CO: Supplementary Rept. to AWWARF, 1981.

Tang, S. R, et al. *Description of Adsorption Equilibria for Volatile Organic Chemicals in Air Stripping Tower Off-Gas.* Denver, CO: AWWARF Report, 1987.

Thom, J. E., and W. D. Byers. "Limitations and Practical Use of a Mass Transfer Model for Predicting Air Stripper Performance." *Environmental Progress,* 12(1), 1993: 61–66.

Treybal, R. E. *Mass Transfer Operations,* 3rd Ed., Chem. Engrg. Series. New York: McGraw-Hill, 1980.

Umphres, M. D., C. H. Tate, M. C. Kavanaugh, and R. R. Trussell. "A Study of THM Removal by Packed Tower Aeration." *Jour AWWA,* 75(8), 1983: 414.

Wallman, H., and M. D. Cummins. "Design Scale-Up Suitability for Air-Stripping Columns." *Jour. Public Works,* 153(10), 1986: 73–78.

CHAPTER 6

COAGULATION AND FLOCCULATION

Raymond D. Letterman, Ph.D., P.E.
Professor, Department of Civil and Environmental Engineering, Syracuse University, Syracuse, New York

Appiah Amirtharajah, Ph.D., P.E.
Professor, School of Civil and Environmental Engineering, Georgia Institute of Technology, Atlanta, Georgia

Charles R. O'Melia, Ph.D., P.E.
Abel Wolman Professor, Department of Geography and Environmental Engineering, The Johns Hopkins University, Baltimore, Maryland

Coagulation is a process for increasing the tendency of small particles in an aqueous suspension to attach to one another and to attach to surfaces such as the grains in a filter bed. It is also used to effect the removal of certain soluble materials by adsorption or precipitation. The coagulation process typically includes promoting the interaction of particles to form larger aggregates. It is an essential component of conventional water treatment systems in which the processes of coagulation, sedimentation, filtration, and disinfection are combined to clarify the water and remove and inactivate microbiological contaminants such as viruses, bacteria, and the cysts and oocysts of pathogenic protozoa. Although the removal of microbiological contaminants continues to be an important reason for using coagulation, a newer objective, the removal of natural organic material (NOM) to reduce the formation of disinfection by-products, is growing in importance.

Aluminum and ferric iron salts have long been used to remove color caused by NOM. These organic substances are present in all surface waters and in many groundwaters. They can be leached from soil, diffused from wetland sediments, and released by plankton and bacteria. Natural organic material adsorbs on natural particles and acts as a particle-stabilizing agent in surface water. It may be associated with toxic metals and synthetic organic chemicals (SOCs). Natural organic material includes precursor compounds that form health-related by-products when chlorine and other chemical disinfectants are used for disinfection and oxidation. For these reasons, considerable attention is being directed at the removal of NOM by coagulation in water treatment, even when color removal is not the principle objective. A treatment technique requirement in the U.S. Environmental Protection Agency's (USEPA's) Stage 1 Disinfection By-Products Rule requires NOM removal in conventional treatment systems by the practice of *enhanced coagulation.*

Coagulation has been an important component of high-rate filtration plants in the United States since the 1880s. Alum and iron (III) salts have been employed as coagulant chemicals since the beginning, with alum having the most widespread use. In the 1930s, Baylis perfected activated silica as a "coagulant aid." This material, formed on site, is an anionic polymer or a small, negatively charged colloid. Synthetic organic polymers were introduced in the 1960s, with cationic polymers having the greatest use. Natural starches were employed before the synthetic compounds. Polymers have helped change pretreatment and filtration practice, including the use of multimedia filters and filters with deep, uniform grain-size media, high-rate filtration, direct filtration (rapid mixing, flocculation, and filtration, but no sedimentation), and in-line filtration (rapid mixing and filtration only).

Coagulants are also being used to enhance the performance of membrane microfiltration systems (Wiesner et al., 1989) and in pretreatment that prolongs the bed life of granular activated carbon (GAC) contactors (Nowack and Cannon, 1996). The development of new chemicals, advances in floc removal process and filter design, and particle removal performance standards and goals have stimulated substantial diversity in the design and operation of the coagulation process, and change can be expected to continue into the future.

In evaluating high-rate filtration plants that were producing high-quality filtered water, Cleasby et al. (1989) concluded, "Chemical pretreatment prior to filtration is more critical to success than the physical facilities at the plant." Their report recommends that plant staff use a well-defined coagulant chemical control strategy that considers variable raw-water quality. There is no question that high-rate (rapid sand) filtration plants are coagulant-based systems that work only as well as the coagulants that are used.

DEFINITIONS

Coagulation is a complex process, involving many reactions and mass transfer steps. As practiced in water treatment the process is essentially three separate and sequential steps: coagulant formation, particle destabilization, and interparticle collisions. Coagulant formation, particle destabilization, and coagulant-NOM interaction typically occur during and immediately after chemical dispersal in rapid mixing; interparticle collisions that cause aggregate (floc) formation begin during rapid mixing but usually occur predominantly in the flocculation process. For example, using the aluminum sulfate salt known as alum $[Al_2(SO_4)_3 \cdot 14H_2O]$ in coagulation involves formation of an assortment of chemical species, called aluminum hydrolysis products, that cause coagulation. These species are formed during and after the time the alum is mixed with the water to be treated. Coagulants are sometimes formed (or partially formed) prior to their addition to the rapid-mixing units. Examples include activated silica and synthetic organic polymers, and the more recently introduced prehydrolyzed metal salts, such as polyaluminum chloride (PACl) and polyiron chloride (PICl).

The terminology of coagulation has not been standardized. However, in most of the water treatment literature, coagulation refers to all the reactions and mechanisms that result in particle aggregation in the water being treated, including in situ coagulant formation (where applicable), particle destabilization, and physical interparticle contacts. The physical process of producing interparticle contacts is termed *flocculation.*

These definitions of coagulation and flocculation are based on the terminology used by early practitioners, such as Camp (1955). However, in the colloid science literature, LaMer (1964) considered only chemical mechanisms in particle destabilization and used the terms coagulation and flocculation to distinguish between two of them. LaMer defined destabilization by simple salts such as NaCl (a so-called indifferent electrolyte) as "coagulation." Destabilization of particles by adsorption of large organic polymers and the subsequent formation of particle-polymer-particle bridges was termed "flocculation."

The water treatment literature sometimes makes a distinction between the terms "coagulant" and "flocculant." When this distinction is made, a coagulant is a chemical used to initially destabilize the suspension and is typically added in the rapid-mix process. In most cases, a flocculant is used after the addition of a coagulant; its purpose is to enhance floc formation and to increase the strength of the floc structure. It is sometimes called a "coagulant aid." Flocculants are often used to increase filter performance (they may be called "filter aids" in this context) and to increase the efficiency of a sludge dewatering process. In any case, depending on how and where it is used and at what dosage, a coagulant is sometimes a flocculant and vice versa. In this chapter, no distinction is made between coagulants and flocculants. The term "coagulant" is used exclusively.

Coagulants and Treatment Waste

The type and amount of coagulant or coagulants used in a water treatment facility can have a significant effect on the type and amount of residue produced by the plant. The amount of residue (weight and volume) impacts the cost of treatment and the overall environmental significance of the plant. Because, in most water treatment facilities, coagulation is the process that generates the bulk of the residual materials, their handling and disposal processes and costs must be considered in the selection and use of coagulants. The use of enhanced coagulation is an important example of this, because the higher coagulant dosages may produce residuals that are much more difficult to dewater. Water treatment plant waste handling, treatment, and disposal are covered in Chapter 16.

CONTAMINANTS

Natural Organic Material

Humic substances are typically the major component of NOM in water supplies. They are derived from soil and are also produced within natural water and sediments by chemical and biological processes such as the decomposition of vegetation. Humic substances are anionic polyelectrolytes of low to moderate molecular weight; their charge is primarily caused by carboxyl and phenolic groups; they have both aromatic and aliphatic components and can be surface active; they are refractive and can persist for centuries or longer. Humic substances are defined operationally by the methods used to extract them from water or soil. Typically, they are divided into the more soluble fulvic acids (FAs) and the less soluble humic acids (HAs), with FAs predominating in most waters (Christman, 1983).

The concentration of NOM in water is typically expressed using the amount of organic carbon. Organic carbon that passes through a 0.45 μm pore-size membrane filter is defined as dissolved organic carbon (DOC), and the amount that does not is known as particulate organic carbon (POC). Total organic carbon (TOC) is the sum of DOC and POC. Most groundwaters have a DOC of less than 2 mg C/L, whereas the DOC of lakes ranges from 2 mg C/L or less (oligotrophic lakes) to 10 mg C/L (eutrophic lakes) (Thurman, 1985). The DOC of small, upland streams will typically fall in the range 1 to 3 mg C/L; the DOC of major rivers ranges from 2 to 10 mg C/L. The highest DOC concentrations (10 to 60 mg C/L) are found in wetlands (bogs, marshes, and swamps). The DOC concentration in upland lakes has been shown to be directly related to the percentage of the total watershed area that is near-shore wetlands (Driscoll et al., 1994). The median raw water TOC concentration for U.S. plants treating surface water is approximately 4 mg C/L (Krasner, 1996).

Disinfection By-Products. The amount of by-products formed by disinfectant chemicals such as chlorine is proportional to the amount of organic carbon in the water. A number of relationships between organic carbon and disinfection by-product concentration have been presented in the literature. For example, Chapra, Canale, and Amy (1997) used data from groundwater, agricultural drains, and surface waters (rivers, lakes, and reservoirs) to show a highly significant correlation ($r^2 = 0.936$, $n = 133$) between the TOC and the trihalomethane formation potential (THMFP). The relationship is given by

$$\text{THMFP} = 43.78\ \text{TOC}^{1.248} \tag{6.1}$$

where THMFP is in μg/L and TOC is in mg C/L. The data gathered by Chapra et al. (1997) are consistent with the frequent observation that high-TOC waters (with their higher fraction of humic acids) yield a greater amount of THMs per amount of TOC than do low-TOC waters. The yield was 20 to 50 μg THMFP/mg C for low-TOC waters and 50 to 100 μg THMFP/mg C for high-TOC waters. Disinfection by-product formation is covered in detail in Chapter 12.

Specific Ultraviolet Light Absorbance (SUVA). Organic compounds that are aromatic in structure or that have conjugated double bonds absorb light in the ultraviolet wavelength range. The higher molecular weight fraction of NOM (the fraction that tends to be removed by coagulation and that has the greater yield of disinfection by-products) absorbs UV light, and consequently, UV light absorbance (typically at a wavelength of 254 nm) can be used as a simple surrogate measure for DOC. Also, the ratio of the UV absorbance to the DOC concentration (called the specific UV absorbance, or SUVA) can be used as an indicator of the molecular weight distribution of the NOM in the water. Based on the absorbance (at 254 nm), expressed as the reciprocal of the light path length in meters, divided by the DOC concentration in mg C/L, the units of SUVA are L/mg C·m^{-1}. Waters with a low humic acid fraction (generally low-DOC waters) tend to have SUVAs that are less than 2 L/mg C·m^{-1}, whereas waters with a high humic acid fraction have SUVAs between 3 and 5 L/mg C·m^{-1}. A higher SUVA value means that the DOC of the water will tend to control the coagulant dosage and relatively high removals of DOC can be expected (50 to 80 percent). When the SUVA is less than 3 L/mg C·m^{-1}, the effect of the DOC on the coagulant dosage may be negligible, and relatively low removal percentages (20 to 50 percent) are likely (Edzwald and Van Benschoten, 1990).

USEPA's Enhanced Coagulation Requirement. The USEPA's 1998 Stage 1 Disinfection By-Products Rule (DBPR) requires the use of an NOM removal strategy called "enhanced coagulation" to limit the formation of all DBPs. The requirement applies to conventional water treatment facilities that treat surface water or groundwater that is under the influence of surface water. The amount of TOC a plant must remove is based on the raw water TOC and alkalinity.

Enhanced coagulation ties the TOC removal requirement to the raw water alkalinity to avoid forcing a utility to add high dosages of hydrolyzing metal salt (HMS) coagulants to reduce the pH to between 5 and 6, a range where HMS coagulants frequently appear to be most efficient. It also recognizes that higher TOC removal is usually possible when the raw water TOC concentration is relatively high and the fraction of the NOM that is more readily removed by HMS coagulants is typically greater. The matrix in Table 6.1 gives Stage 1 DBPR's required TOC removal percentages.

The application of Table 6.1 is illustrated by the following example. A plant's source water has a TOC of 3.5 mg C/L and an alkalinity of 85 mg/L as $CaCO_3$. According to the table, the required TOC removal is 25.0 percent. The TOC of the water before the application of chlorine would have to be less than 2.6 mg C/L, calculated using the relationship $2.6 = 3.5 \times (1 - 0.25)$.

The regulatory negotiators who formulated the enhanced coagulation requirement were concerned that coagulant chemical costs might be excessive for utilities that treat water with a high fraction of NOM that is not amenable to removal by coagulants. For these plants, the removal requirements of Table 6.1 may be infeasible. The Rule allows them to use a jar test procedure to determine an appropriate, alternative TOC removal requirement (White et al., 1997).

The alternative TOC removal requirement is determined by performing jar tests on at least a quarterly basis for one year. In these tests, alum or ferric coagulants are added in 10 mg/L increments until the pH is lowered to a target pH value that varies with the source water alkalinity. For the alkalinity ranges of 0 to 60, more than 60 to 120, more than 120 to 240, and more than 240 mg/L as $CaCO_3$, the target pH values are 5.5, 6.3, 7.0, and 7.5, respectively. When the jar test is complete, the residual TOC concentration is plotted versus the coagulant dosage (in mg coagulant/L) and the alternative TOC percentage is found at the point called the "point of diminishing returns," or PODR. The PODR is the coagulant dosage where the slope of the TOC-coagulant dosage plot changes from greater than 0.3 mg C/10 mg coagulant to less than 0.3 mg C/10 mg coagulant. If the plot does not yield a PODR, then the water is considered to be not amenable to enhanced coagulation and the primary agency may grant the system a waiver from the enhanced coagulation requirement. Details of the jar test procedure are given in the USEPA's *Guidance Manual for Enhanced Coagulation and Enhanced Precipitative Softening* (USEPA, 1999).

TABLE 6.1 Required Percent Removals of Total Organic Carbon by Enhanced Coagulation in the 1998 Stage 1 Disinfection By-Products Rule

	Source water alkalinity (mg/L as $CaCO_3$)		
Source water total organic carbon (mg C/L)	0–60	>60–120	>120
>2.0–4.0	35	25	15
>4.0–8.0	45	35	25
>8.0	50	40	30

The Stage 1 DBPR provides alternative compliance criteria for the enhanced coagulation, treatment technique requirement. The six criteria are listed below:

1. The system's source water TOC is less than 2.0 mg C/L.
2. The system's treated-water TOC is less than 2.0 mg C/L.
3. The system's source water TOC is less than 4.0 mg C/L, the source water alkalinity is more than 60 mg/L as $CaCO_3$, and the system is achieving TTHM less than 40 μg/L and HAA5 (haloacetic acids) less than 30 μg/L.
4. The system's TTHM is less than 40 μg/L, HAA5 is less than 30 μg/L, and only chlorine is used for primary disinfection and maintaining a distribution system residual.
5. The system's source water SUVA prior to any treatment is less than or equal to 2.0 $L/(mg{\cdot}m^{-1})$.
6. The system's treated-water SUVA is less than or equal to 2.0 $L/(mg{\cdot}m^{-1})$.

The measurements used to test compliance with criteria 1, 2, 5, and 6 are made monthly and a running annual average is calculated quarterly. Compliance with criteria 3 and 4 is based on monthly measurements of TOC, alkalinity, quarterly measurements of TTHMs, and HAA5, and the running annual average is calculated quarterly.

Particles

Particles in natural water vary widely in origin, concentration, size, and surface chemistry. Some are derived from land-based or atmospheric sources (e.g., clays and other products of weathering, silts, pathogenic microorganisms, asbestos fibers, and other terrestrial detritus and waste discharge constituents), and some are produced by chemical and biological processes within the water source (e.g., algae, precipitates of $CaCO_3$, $FeOOH$, MnO_2, and the organic exudates of aquatic organisms). Certain toxic metals and SOCs are associated with solid particles, so coagulation for particle aggregation can be important in the removal of soluble, health-related pollutants.

Particle size may vary by several orders of magnitude, from a few tens of nanometers (e.g., viruses and high-molecular-weight NOM) to a few hundred micrometers (e.g., zooplankton). All can be effectively removed by properly designed and operated coagulation, floc separation, and filtration facilities. The very important cysts and oocysts of pathogenic protozoa (e.g., *Giardia* and *Cryptosporidium*) are ovoid particles with overall dimensions in the 4 to 12 μm (micrometer) range. A comparison of the size spectra of waterborne particles and filter pores is shown in Figure 6.1.

The smallest particles, those with one dimension less than 1 μm, are usually called "colloidal," and those that are larger than this limit are said to be "suspended." The operational definition of "dissolved" and "suspended" impurities is frequently established by a 0.45 μm pore-size membrane filter, but colloidal particles can be smaller than this dimension. The effect of gravity on the transport of colloidal particles tends to be negligible compared with the diffusional motion caused by interaction with the fluid (Brownian motion) and, compared with suspended particles, colloidal particles have significantly more external surface area per unit mass.

Measuring Particle Concentration. The principal methods for measuring the performance of particle removal processes in water treatment systems are turbidity and

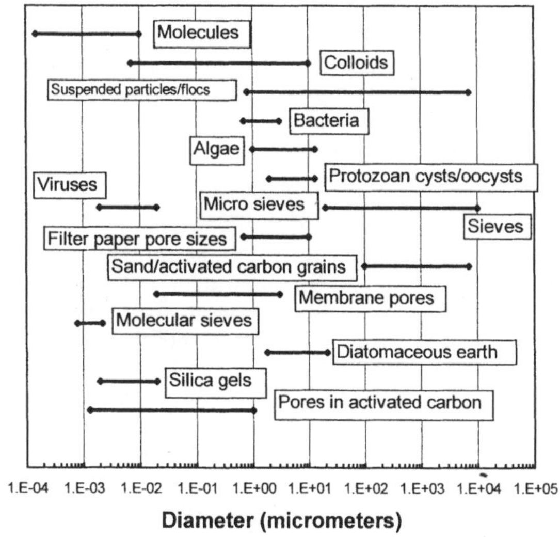

FIGURE 6.1 Size spectrum of waterborne particles and filter pores (from Stumm and Morgan, 1981).

particle counting. Both techniques have limitations, and, consequently, a single method may not provide all the information needed to successfully monitor and control process performance.

Turbidity is measured using an instrument called a turbidimeter, or nephelometer, that detects the intensity of light scattered at one or more angles to an incident beam of light. Light scattering by particles is a complex process and the angular distribution of scattered light depends on a number of conditions including the wavelength of the incident light and the particle's size, shape, and composition (Sethi et al., 1996). Consequently, it is difficult to correlate the turbidity with the amount, number, or mass concentration of particles in suspensions.

When the turbidity measurement is used for regulatory purposes, it should theoretically be possible to take a given suspension and measure its turbidity at any water treatment facility and obtain an unbiased result that is reasonably close to the average turbidity measured at all other facilities. To achieve reasonable agreement, three factors must be considered: the design of the instruments, the material used to calibrate the instrument, and the technique used to make the measurement. Given this need, turbidimeter design, calibration, and operation criteria have been developed using the consensus process by a number of organizations, including *Standard Methods for the Examination of Water and Wastewater* (Section 2130), the American Society for Testing Materials (ASTM, Method D 1889), the International Standards Organization (ISO 7027-1948E), and the United States Environmental Protection Agency (Method 180.1). However, standardization is a difficult and imperfect process, and it has been shown (Hart et al., 1992) that instruments designed and calibrated using the criteria of these standards can give significantly different responses.

Turbidity measurements were first used to maintain the aesthetic quality of treated water. In 1974, after the passage of the Safe Drinking Water Act, the USEPA lowered the limit for filtered water to one nephelometric turbidity unit (1 NTU) with the explanation that particles causing turbidity can interfere with the disinfection process by enmeshing and, therefore, protecting microbiological contaminants from chemical disinfectants such as chlorine. Today, the turbidity measurement is also used to assess filter performance. It is viewed as an important indicator of the extent to which disinfectant-resistant pathogens have been removed by the filtration process. Filtered-water turbidity criteria must be met before the protozoan cyst/oocyst and virus removal credits allowed by the Surface Water Treatment Rule of the 1986 Amendments of the Surface Water Treatment Rule can be applied (Pontius, 1990).

Particle-counting instruments are becoming widely used in the drinking water industry, especially for monitoring and controlling filtration process performance. Plants use them to detect early filter breakthrough and to maintain plant performance at a high level. On-line devices that continuously measure particle concentrations in preselected size ranges at various points in the treatment system are especially important. Batch sampling devices are also used. Two types of particle-counting/sizing sensors are important in water treatment applications: light-blocking (light-obscuration) devices and light-scattering devices (Hargesheimer et al., 1992; Lewis et al., 1992). At the present time, instruments with light-blocking-type sensors are more common.

The types of particle-counting instruments used in water treatment plants have limitations (Hargesheimer et al., 1992). Most do not detect particles smaller than about 1 μm, and therefore, they must be used in conjunction with turbidimeters that do detect these smaller particles. Differences in the optical characteristics of the sensors make achieving direct count and size agreement between instruments difficult. There are no industrywide standards for sensor resolution or for particle counting and sizing accuracy. For a given particle suspension, it is not possible to make similar sensors yield identical particle counts and sizes. Until this is feasible, the regulatory use of particle count measurements will be limited.

STABILITY OF PARTICLE SUSPENSIONS

In water treatment, the coagulation process is used to increase the rate or kinetics of particle aggregation and floc formation. The objective is to transform a stable suspension [i.e., one that is resistant to aggregation (or attachment to a filter grain)] into an unstable one. Particles that may have been in lake water for months or years as stable, discrete units can be aggregated in an hour or less following successful destabilization. The design and operation of the coagulation process requires proper control of both particle destabilization and the subsequent aggregation process.

As particles in a suspension approach one another, or as a particle in a flowing fluid approaches a stationary surface such as a filter grain, forces arise that tend to keep the surfaces apart. Also, there are forces that tend to pull the interacting surfaces together. The most well-known repulsive force is caused by the interaction of the electrical double layers of the surfaces ("electrostatic" stabilization). The most important attractive force is called the London–van der Waals force. It arises from spontaneous electrical and magnetic polarizations that create a fluctuating electromagnetic field within the particles and in the space between them. These two types of forces, repulsive and attractive, form the basis of the Derjaguin, Landau, Verwey, and Overbeek (DLVO) theory of colloid stability. Other forces include those associated with the hydration of ions at the surfaces (a repulsive force) and the presence of adsorbed polymers, which can cause either repulsion ("steric" interaction) or attraction ("polymer bridging"). As particles approach one another on a collision course, the fluid between them must move out of the way. The repulsive force caused by this displacement of fluid is called hydrodynamic retardation.

Electrostatic Stabilization

Origins of Surface Charge. Most particles in water, mineral and organic, have electrically charged surfaces, and the sign of the charge is usually negative (Niehof and Loeb, 1972; Hunter and Liss, 1979). Three important processes for producing this charge are considered in the following discussion. First, surface groups on the solid may react with water and accept or donate protons. For an oxide surface such as silica, the surface site might be indicated by the symbol ≡SiOH and the surface site ionization reactions by

$$\equiv SiOH_2^+ \Leftrightarrow \equiv SiOH + H^+ \tag{6.2a}$$

$$\equiv SiOH \Leftrightarrow \equiv SiO^- + H^+ \tag{6.2b}$$

An organic surface can contain carboxyl (COO^-) and amino (NH_3^+) groups that become charged through ionization reactions as follows:

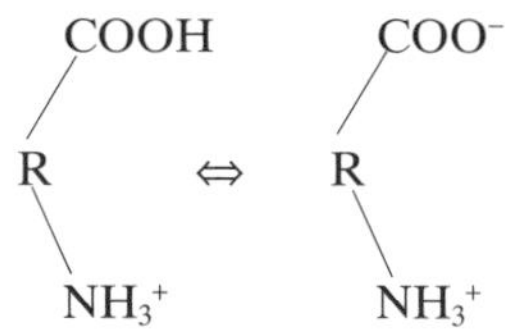

(6.3a)

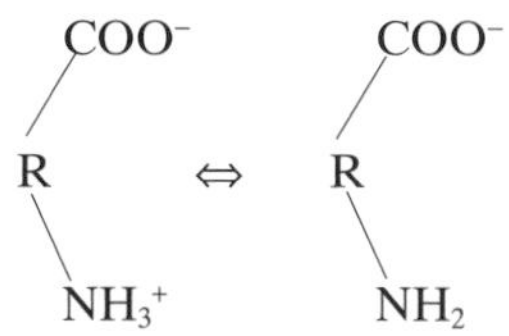

(6.3b)

In these reactions, the surface charge on a solid particle depends upon the concentration of protons ($[H^+]$) or the pH ($= -\log [H^+]$) in the solution. As the pH increases (i.e., $[H^+]$ decreases), Eqs. 6.2 and 6.3 shift to the right and the surface charge becomes increasingly negative. Silica is negatively charged in water with a pH above 2; proteins contain a mixture of carboxyl and amino groups and usually have a negative charge at a pH above about 4. The adsorption of NOM onto particles can be responsible for site behavior like that shown previously.

Second, surface groups can react in water with solutes other than protons. Again, using the ≡SiOH surface groups of silica,

$$\equiv SiOH + Ca^{2+} \Leftrightarrow \equiv SiOCa^+ + H^+ \tag{6.4}$$

$$\equiv SiOH + HPO_4^{2-} \Leftrightarrow \equiv SiOPO_3H^- + OH^- \tag{6.5}$$

These surface complex formation reactions involve specific chemical reactions between chemical groups on the solid surface (e.g., silanol groups) and adsorbable solutes (e.g., calcium and phosphate ions). Surface charge is again related to solution chemistry.

Third, a surface charge may arise because of imperfections within the structure of the particle; this is called isomorphic replacement, or substitution. It is responsible for a substantial part of the charge on many clay minerals. Clays are layered structures and in these structures sheets of silica tetrahedra are typically cross-linked with sheets of alumina octahedra. The silica tetrahedra have an average composition of SiO_2 and may be depicted as shown in Figure 6.2(*a*). If an Al atom is substituted for an Si atom during the formation of this lattice, a negatively charged framework results [Figure 6.2(*b*)].

Similarly, a divalent cation such as Mg(II) or Fe(II) may substitute for an Al(III) atom in the aluminum oxide octahedral network, also producing a negative charge. The sign and magnitude of the charge produced by such isomorphic replacements are independent of the characteristics of the aqueous phase after the crystal is formed.

The Electrical Double Layer. In a colloidal suspension, there can be no net imbalance in the overall electrical charge; the primary charge on the particle must be counterbalanced in the system. Figure 6.3 shows schematically a negatively charged particle with the counterbalancing cloud of ions (the "diffuse layer") around the particle. Because the particle is negatively charged, excess ions of opposite charge (positive) accumulate in this interfacial region. Ions of opposite charge accumulating in the interfacial region, together with the primary charge, form an electrical double layer. The diffuse layer results from electrostatic attraction of ions of opposite charge to the particle ("counterions"), electrostatic repulsion of ions of the same charge as the particle ("coions"), and thermal or molecular diffusion that acts against the concentration gradients produced by the electrostatic effects. The formation of diffuse layers is shown in Figures 6.3 and 6.4(*a*).

Because of the primary charge, an electrostatic potential (voltage) exists between the surface of the particle and the bulk of the solution. This electric potential can be pictured as the electric pressure that must be applied to bring a unit charge having

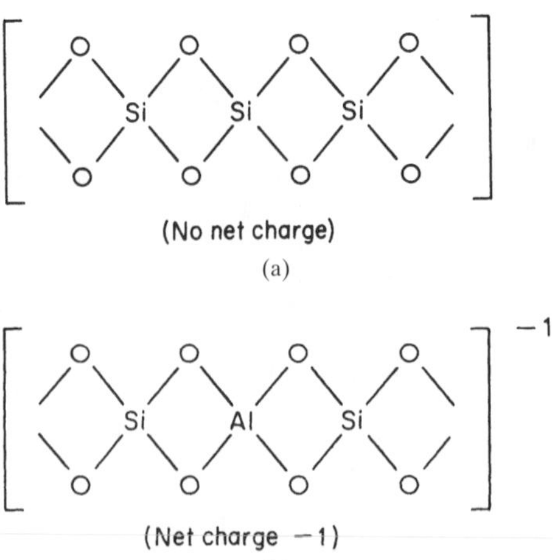

FIGURE 6.2 SiO_2 structure. (*a*) With no net charge. (*b*) With –1 net charge.

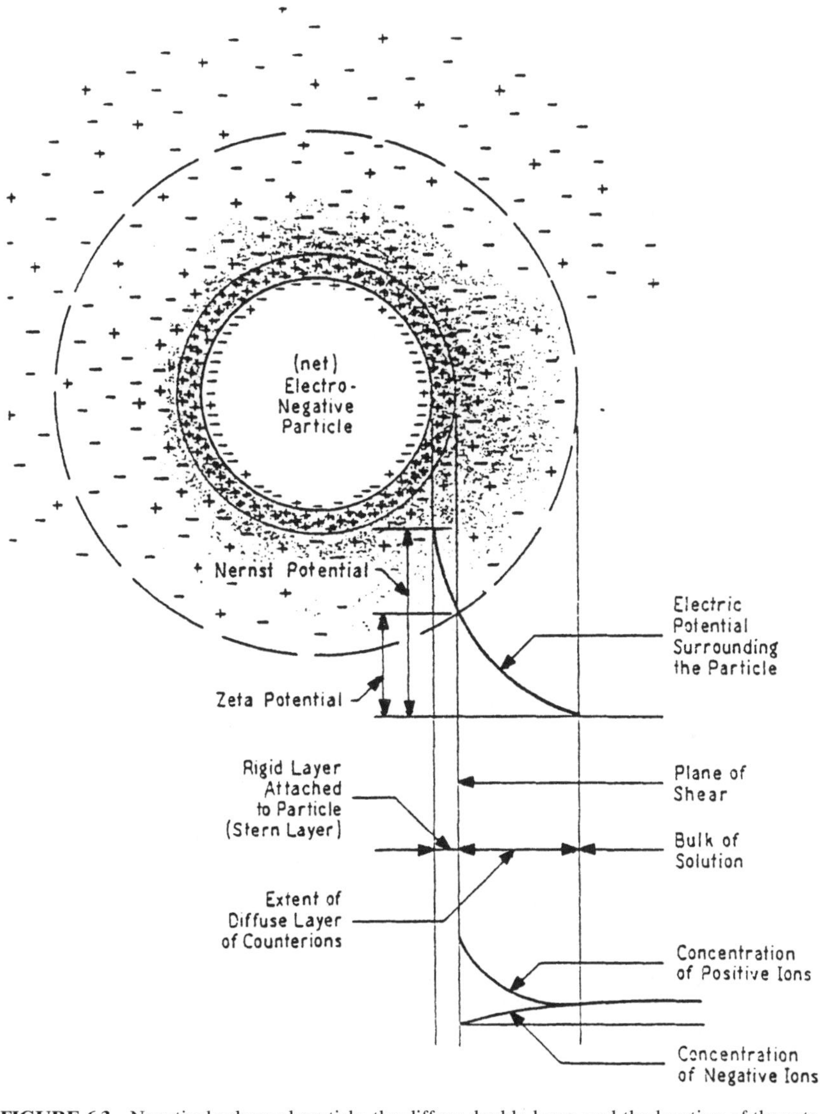

FIGURE 6.3 Negatively charged particle, the diffuse double layer, and the location of the zeta potential.

the same sign as the primary charge from the bulk of the solution to a given distance from the particle surface, shown schematically in Figure 6.4(*b*). The potential has a maximum value at the particle surface and decreases with distance from the surface. This decrease is affected by the characteristics of the diffuse layer, and, thus, by the type and concentration of ions in the bulk solution.

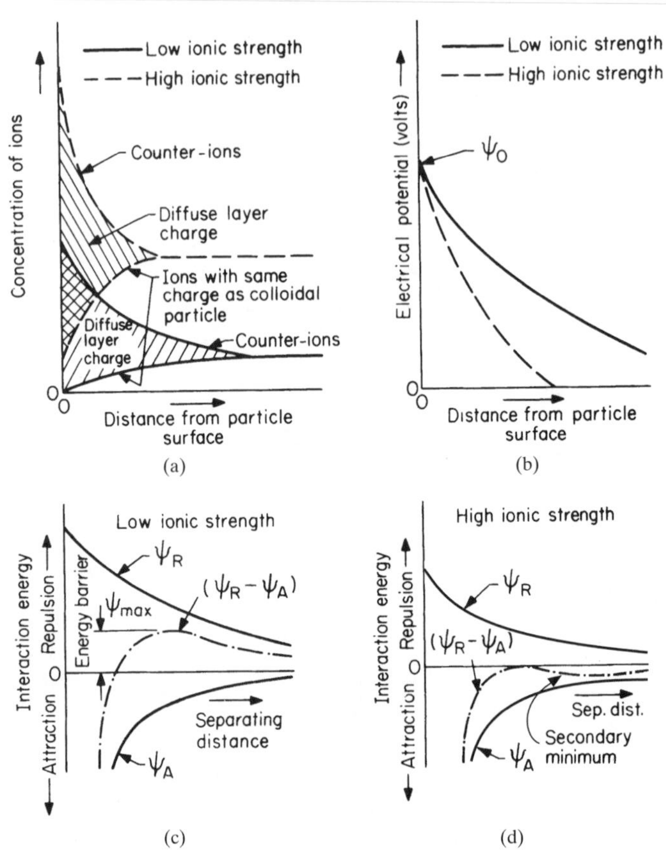

FIGURE 6.4 Schematic representations of (*a*) the diffuse double layer; (*b*) the diffuse layer potential; (*c* and *d*) two cases of particle-particle interaction energies in electrostatically stabilized colloidal systems.

Dynamic Aspects of the Electrical Double Layer. The interaction of two particles can be evaluated in terms of potential energy (i.e., the amount of energy needed to bring two particles from infinite separation up to a given separation distance). If the potential energy is positive, the overall interaction is unfavorable (repulsive) because energy must be provided to the system. If the potential energy is negative, the net effect is attractive.

When particles are forced to move in a fluid or when the fluid is forced to move past a stationary particle, some of the charge that balances the surface charge moves with the fluid and some of it does not. Thus, the electrical potential at the plane of shear (which is the boundary between the charge that remains with the particle and the charge that does not) is less than the electrostatic potential at the surface of the particle. The exact location of the slipping plane is not known, and, therefore, it is difficult to relate the potential measured at the plane of shear (by electrokinetic techniques such as electrophoresis or streaming current) to the surface potential. Lyklema (1978) locates the slipping plane at the outer border of the Stern layer as shown in Figure 6.3.

The mathematical treatment of the double layer, shown schematically in Figures 6.3 and 6.4, was developed independently by Gouy and Chapman. The result is called the Gouy-Chapman model; details of it have been presented by Verwey and Overbeek (1948). In this model, electrostatic or coulombic attraction and diffusion are the interacting processes responsible for the formation of the diffuse layer. Ions are treated as point (dimensionless) charges and have no other physical or chemical characteristics.

When two similar colloidal particles approach each other, their diffuse layers begin to overlap and to interact. This comingling of charge creates a repulsive potential energy, Ψ_R, that increases in magnitude as the distance separating the particles decreases. Figure 6.4 (*c* and *d*) plots the repulsive energy of interaction as a function of the distance between the interacting surfaces.

Attractive forces (van der Waals forces) exist between all types of particles, no matter how dissimilar they may be. These forces arise from interactions among dipoles, either permanent or induced, in the atoms composing the interacting surfaces and the water. The magnitude of the attractive energy of interaction, Ψ_A, decreases with increasing separation distance. Schematic curves of the van der Waals attractive potential energy of interaction are shown in Figure 6.3 (*c* and *d*). Unlike electrostatic repulsive forces, the van der Waals attractive forces are essentially independent of the composition of the solution; they depend upon the kind and number of atoms in the colloidal particles and the continuous phase (water).

Summation of Ψ_R and Ψ_A yields the net interaction energy between two colloidal particles. This sum with the proper signs is $(\Psi_R - \Psi_A)$ and is shown schematically as a function of separation distance in Figure 6.4 (*c* and *d*). The force acting on the particles is the derivative $d(\Psi_R - \Psi_A)/ds$, where s is the separation distance. When electrostatic repulsion dominates during particle-particle interactions, the suspension is said to be electrostatically stabilized and to undergo only "slow" coagulation. Electrostatic stabilization is fundamental to the current understanding of coagulation in water treatment. For additional insight into electrostatic stabilization, the mathematical treatment of Verwey and Overbeek (1948) and a summary by Lyklema (1978) should be consulted. Valuable summaries of the methods, mathematics, and meanings of models for electrostatic stabilization are also contained in texts by Morel (1983), Stumm and Morgan (1980), and Elimelech et al. (1995).

The Gouy-Chapman model provides a good qualitative picture of the origins of electrostatic stabilization. It does not provide a quantitative, predictive tool for such important characteristics as coagulant requirements and coagulation rates. Its principal drawback lies in the characterization of all ions as point charges. This allows for physical description of electrostatics, but omits description of chemical interactions. For example, the sodium ion (Na^+), the dodecylamine ion ($C_{12}H_{25}NH_3^+$), and the proton (H^+) have identical charges but are not identical coagulants.

Quantitative models for the surface chemistry of oxides and other types of surfaces with ionizable surface sites are available. Some have been incorporated into widely used computer programs such as MINEQL, allowing self-consistent calculations to be made simultaneously for surface and solution. The most popular are the surface complexation models, which describe the formation of charge, potential, and the adsorption of ions at the particle-water interface. The fundamental concept upon which all surface complexation models are based is that adsorption takes place at defined coordination sites (surface hydroxyl groups present in finite number) and that adsorption reactions can be described quantitatively by mass action expressions (Goldberg, Davis, and Hem, 1996). The various models use different assumptions for the electrostatic interaction terms. For example, some use a simple linear relationship between charge and potential at the surface [the "constant capacitance model" of Schindler and Stumm (Stumm, 1992)], and others use more elaborate relationships such as the triple-layer model of Davis and coworkers (1978). All of the surface complexation models reduce to a similar set of simultaneous equations that are typically solved numerically: (1) mass action equations for all surface reactions, (2) mass balance equations for surface hydroxyl groups, (3) equations for calculation of surface charge, (4) a set of equations that describes the charge and potential relationships of the electrical double-layer. Attempts have been made, with limited success, to use these models to predict the effects of solution and suspension variables on flocculation efficiency when aluminum salt coagulants are used (Letterman and Iyer, 1985) and on the rate of coagulation of iron oxide particles (Liang and Morgan, 1988).

Secondary Minimum Aggregation. In certain systems, the kinetic energy and electrical double-layer characteristics of the interacting particles may be such that aggregation takes place in what is called a secondary minimum [see Figure 6.4 (*d*)]. Under this condition, the particles are held in proximity (at least momentarily) by van der Waals attraction but remain nanometers apart due to the repulsive force associated with the interacting electrical double layers. If the kinetic energy of the interacting particles is increased, the repulsive force may not be great enough to limit the close approach of the particles, and aggregation will tend to occur in the primary minimum, with the surfaces of the particles close together. In this case, the separation distance in Figure 6.4 (*c*) is reduced until the quantity $(\Psi_R - \Psi_A)$ becomes less than Ψ_{max}.

Steric Stabilization

Steric stabilization can result from the adsorption of polymers at solid-water interfaces. Large polymers can form adsorbed segments on a solid surface with loops and tails extending into the solution (Lyklema, 1978) as illustrated in Figure 6.5. Adsorbed polymers can be either stabilizing or destabilizing, depending on the relative amounts of polymer and solid particles, the affinities of the polymer for the solid and for water, electrolyte type and concentration, and other factors. A stabilizing polymer may contain two types of groups, one of which has a high affinity for the

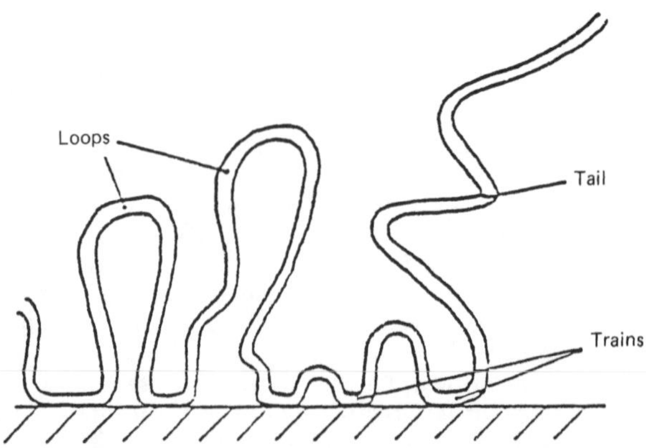

FIGURE 6.5 Illustration of adsorbed polymer configuration with loops, trains, and tails (from Lyklema, 1978).

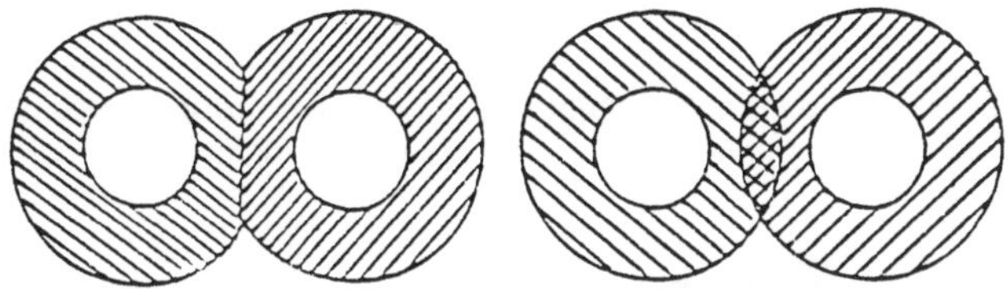

FIGURE 6.6 Two possible repulsive interactions of adsorbed polymer layers in sterically stabilized colloidal systems (Gregory, 1978). Shaded areas are zones occupied by polymers. Zone thickness relative to particle size is arbitrary.

solid surface and a second, more hydrophilic group, that extends into the solution (Gregory, 1978). The configuration of such an interfacial region is difficult to characterize either theoretically or experimentally. This, in turn, prevents quantitative formulation of the interaction forces between two such interfacial regions during a particle-particle encounter in coagulation. Some useful qualitative descriptions can, however, be made.

Gregory (1978) summarized two processes that can produce a repulsive force when two polymer-coated surfaces interact at close distances. These are illustrated in Figure 6.6. First, the adsorbed layers can each be compressed by the collision, reducing the volume available for the adsorbed molecules. This reduction in volume restricts the movement of the polymers (a reduction in entropy) and causes a repulsion between the particles. Second, and more frequently, the adsorbed layers may interpenetrate on collision, increasing the concentration of polymer segments in the mixed region. If the extended polymer segments are strongly hydrophilic, they can prefer the solvent to reaction with other polymer segments. An overlap or mixing then leads to repulsion. These two processes are separate from and in addition to the effects of polymer adsorption on the charge of the particles and the van der Waals interaction between particles. Charged polymers (polyelectrolytes) can alter particle charge; organic polymers can also reduce the van der Waals attractive interaction energy. Steric stabilization is widely used in the manufacture of industrial colloids, such as paints and waxes. Natural organic materials such as humic substances are ubiquitous in water supplies. They are anionic polyelectrolytes, absorb at interfaces, can be surface active, and may contribute to particle stability by steric effects (O'Melia, 1995).

COAGULANTS

Introduction

Coagulants are widely used in water treatment for a number of purposes. Their principal use is to destabilize particulate suspensions and enhance the rate of floc formation. Hydrolyzing metal salt coagulants are also used to form flocculent precipitates that adsorb NOMs and certain inorganic materials, such as phosphates, arsenic compounds, and fluoride.

For many years most water treatment plants that required a coagulant used the hydrolyzing metal salt, aluminum sulfate (alum). Organic polymers (polyelectrolytes) came into widespread use in the 1960s. In the last 10 years, the number of coagulant products used in water treatment has grown substantially. Today, the list includes ferric iron salts and prehydrolyzed metal salts plus an assortment of chemical mixtures and products supplemented with additives. In some countries, organic compounds derived from natural materials are used as coagulants.

Destabilization Mechanisms

Destabilization is the process in which the particles in a stable suspension are modified to increase their tendency to attach to one another (or to a stationary surface like a filter grain or deposit). The aggregation of particles in a suspension after destabilization requires that they be transported toward one another. In filtration they must be transported to the stationary filter surface. Transport processes in suspensions are discussed in this chapter in the section on flocculation. Transport in filtration is discussed in Chapter 8.

Double-Layer Compression. The classical method of colloid destabilization is double-layer compression. To affect double-layer compression, a simple electrolyte such as NaCl is added to the suspension. The ions that are opposite in sign to the net charge on the surface of the particles enter the diffuse layer surrounding the particle. If enough of these counterions is added, the diffuse layer is compressed, reducing the energy required to move two particles of like surface charge into close contact. Destabilization by double-layer compression is not a practical method for water treatment, because the salt concentrations required for destabilization may approach that of seawater and, in any case, the rate of particle aggregation would still be relatively slow in all but the most concentrated suspensions. Double-layer compression, however, is an important destabilization mechanism in certain natural systems [e.g., estuaries (O'Melia, 1995)].

Surface Charge Neutralization. Destabilization by surface charge neutralization involves reducing the net surface charge of the particles in the suspension. As the net surface charge is decreased the thickness of the diffuse layer surrounding the particles is reduced and the energy required to move the particles into contact is minimized.

Two processes are used to accomplish surface charge neutralization. In the first, coagulant compounds that carry a charge opposite in sign to the net surface charge of the particles are adsorbed on the particle surface. (In some cases, the coagulant is a very small particle that deposits on the particle surface.) The coagulants used to accomplish this usually have a strong tendency to adsorb on (attach to) surfaces. Examples include the synthetic and natural organic polyelectrolytes and some of the hydrolysis products formed from hydrolyzing metal salt coagulants. The tendency for these compounds to adsorb is usually attributable to both poor coagulant-solvent interaction and a chemical affinity of the coagulant, or chemical groups on the coagulant, for the particle surface. Most of the coagulants that are used for charge neutralization can adsorb on the surface to the point that the net surface charge is reversed and, in some cases, increased to the point that the suspension is restabilized.

Adjustment of the chemistry of the solution can be used to destabilize some common types of suspensions by reducing the net surface charge of the particle surfaces. For example, when most of the surface charge is caused by the ionization of surface sites (see Eq. 6.1), pH adjustment with acid or base may lead to destabilization. For some surfaces, such as positively charged oxides and hydroxides, the adsorption of simple multivalent anions (such as sulfate and phosphate) or complex polyvalent organic compounds (such as humic materials), will reduce the positive charge and destabilize the suspension.

Heterocoagulation is a destabilization mechanism that is similar to the process of surface charge neutralization by the adsorption of oppositely charged soluble species. However, in this case, the process involves one particle depositing on another of opposite charge. For example, large particles with a high negative surface charge may contact smaller particles with a relatively low positive charge. Because the particles have opposite surface charge, electrostatic attraction enhances particle-particle interaction. As the stabilizing negative charge of the larger particles is reduced by the deposited positive particles, the suspension of larger particles is destabilized.

Adsorption and Interparticle Bridging. Destabilization by bridging occurs when segments of a high-molecular-weight polymer adsorb on more than one particle, thereby linking the particles together. When a polymer molecule comes into contact with a colloidal particle, some of the reactive groups on the polymer adsorb on the particle surface and other portions extend into the solution. If a second particle with open surface is able to adsorb the extended molecule, then the polymer will have formed an interparticle bridge. The polymer molecule must be long enough to extend beyond the electrical double layer (to minimize double-layer repulsion when the particles approach) and the attaching particle must have available surface. The adsorption of excess polymer may lead to restabilization of the suspension. Ions such as calcium are known to affect the bridging process, apparently by linking sites on interacting polymer chains (Black et al., 1965; Lyklema, 1978; Dentel, 1991).

Hydrolyzing Metal Salt (HMS) Coagulants

The most widely used coagulants in water treatment are sulfate or chloride salts that contain the metal ions Al^{3+} or Fe^{3+}. In aqueous solutions, these small, highly positive ions form such strong bonds with the oxygen atoms of six surrounding water molecules that the oxygen-hydrogen atom association in the water molecules is weakened, and hydrogen atoms tend to be released to the solution (see Figure 6.7). This process is known as *hydrolysis* and the resulting aluminum and ferric hydroxide

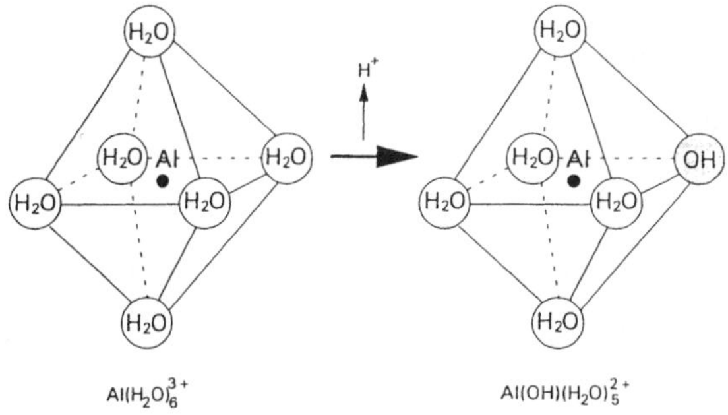

FIGURE 6.7 Deprotonation of the aquo aluminum ion, initial step in aluminum hydrolysis (from Letterman, 1991).

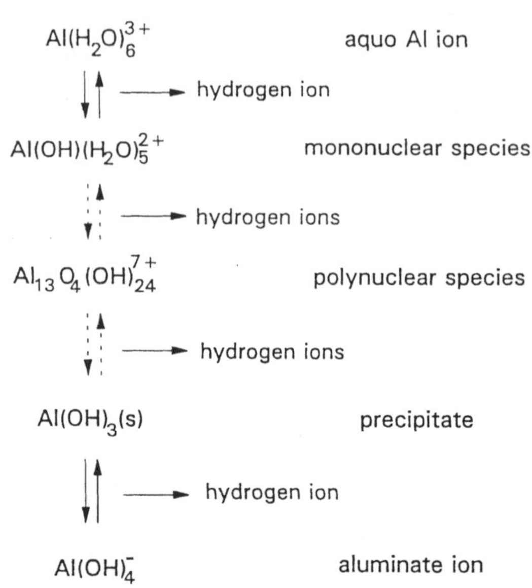

FIGURE 6.8 Aluminum hydrolysis products (from Letterman, 1991).

species are called *hydrolysis products.* The water molecules are frequently omitted when writing the chemical formula for these species, but their role is important in determining species behavior.

The chemistry of aluminum and iron hydrolysis reactions and products is complex and not completely understood. As hydrolysis proceeds, and if there is sufficient total metal ion in the system, simple mononuclear products can form complex polynuclear species, which in turn can form microcrystals and the metal hydroxide precipitate (see Figure 6.8). Hydrolysis products can adsorb (and continue to hydrolyze) on many types of particulate surfaces. Hydrolysis products tend to react with the higher-molecular-weight fraction of NOM. Soluble hydrolysis products may bind with NOM functional groups and small, positively charged microcrystalline metal hydroxide particles can be stabilized by the adsorption of a negatively charged coating of NOM (Kodama and Schnitzer, 1980; Bertsch and Parker, 1996).

The solubility of the metal hydroxide precipitate is one factor that must be considered in maximizing coagulant performance and in minimizing the amounts of residual Al and Fe in treated water. At low pH, the dissolution of the metal-hydroxide precipitate produces positively charged, soluble hydrolysis products and the aquo-metal ion (Al^{3+} and Fe^{3+})[1]. At high pH, the negatively charged, soluble hydrolysis products $Al(OH)_4^-$ and $Fe(OH)_4^-$ are formed. These species are tetrahedral rather than octahedral, so no further deprotonation can occur. The minimum-solubility pH of aluminum hydroxide precipitate at 25°C is approximately 6.3 (Figure 6.9); for ferric hydroxide it is about 8 (Figure 6.10). The pH of minimum solubility increases with decreasing temperature. At 4°C, the pH of minimum solubility of freshly precipitated aluminum hydroxide is approximately 6.8 (Van Benschoten and Edzwald, 1990).

[1] From this point on, water molecules are omitted from the formulas for hydrolysis products.

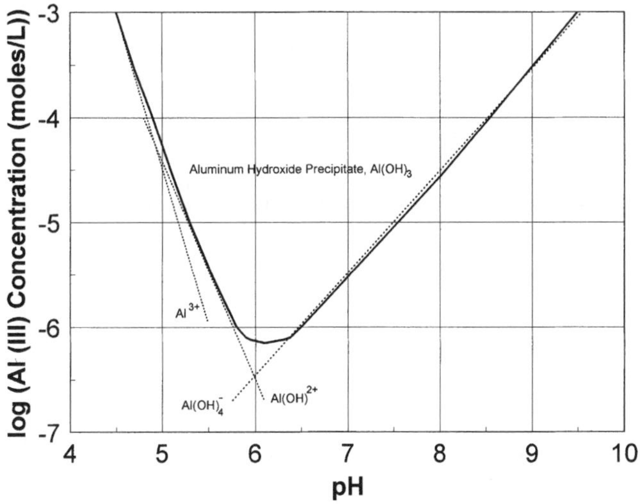

FIGURE 6.9 Solubility diagram for amorphous, freshly precipitated aluminum hydroxide. Water molecules are omitted in species notation.

The solubility diagrams of Figures 6.9 and 6.10 were plotted using the formation constants ($\beta_{x,y}$) and solubility constants (K_{sp}) listed in Table 6.2. The ionic strength was assumed to be 0.001 M. The formation constants in Table 6.2 are for reactions and mass action expressions of the form

$$x\,M^{3+} + y\,H_2O \Leftrightarrow M_x(OH)_y^{(3x-y)+} + y\,H^+ \tag{6.6}$$

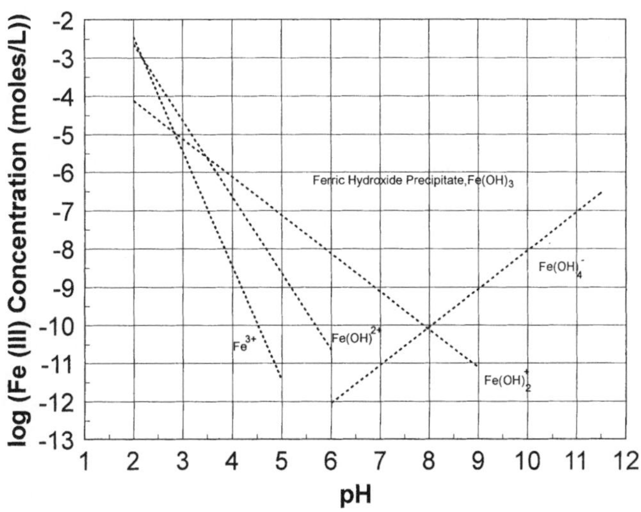

FIGURE 6.10 Solubility diagram for amorphous, freshly precipitated ferric hydroxide. Water molecules are omitted in species notation.

and

$$\beta_{x,y} = \frac{[M_x(OH)_y^{(3x-y)+}][H^+]^y}{[M^{3+}]^x} \tag{6.7}$$

The solubility constants are for reactions and solubility product relationships of the form

$$M(OH)_3(s) + 3\,H^+ \Leftrightarrow M^{3+} + 3\,H_2O \tag{6.8}$$

and

$$K_{sp} = \frac{[M^{3+}]}{[H^+]^3} \tag{6.9}$$

A number of researchers have assumed that the important mononuclear Al hydrolysis products include $Al(OH)^{2+}$, $Al(OH)_2^+$, $Al(OH)_3^0$, and $Al(OH)_4^-$ (Bertsch and Parker, 1996). Others, including Letterman and Driscoll (1994), Hayden and Rubin (1974), and VanBenschoten and Edzwald (1990), have reported that the effect of pH on the solubility of freshly precipitated aluminum hydroxide can be accurately predicted using just the aquo-Al species (Al^{3+}) and two mononuclear hydrolysis products [$Al(OH)^{2+}$, $Al(OH)_4^-$]. This assumption was used to plot Figure 6.9.

The dotted lines plotted in Figure 6.9 give the concentrations of each Al species when it is in equilibrium with the $Al(OH)_3$ precipitate. The equation of each line was derived by combining Equations 6.6 and 6.8 and using the appropriate equilibrium constants corrected for ionic strength. For example, the equation for log [$Al(OH)^{2+}$] when the ionic strength $I = 0.001$ M is

$$\log[Al(OH)^{2+}] = \log\beta_{1,1} + \log K_{sp} + 2\log H^+ = 5.57 - 2pH \tag{6.10}$$

Values of log $\beta_{1,1}$ and log K_{sp} for $I = 0.001$ M are listed in Table 6.2.

As an Al salt solution is titrated with strong base, hydrolysis of the Al increases, forming a varying combination of hydrolysis products. The distribution of the Al(III) species at equilibrium depends on the pH and the total Al concentration (Al_T). The effect of pH on the fractional distribution of Al hydrolysis products was calculated

TABLE 6.2 Formation Constants and Solubility Constants Used to Plot Solubility Diagrams for Amorphous Aluminum Hydroxide and Amorphous Ferric Hydroxide Precipitates, $T = 25°C$ and $I = 0$*

Hydrolysis product	$p\beta_{x,y}$ and pK_{sp}	Reference
	Amorphous Aluminum Hydroxide	
$Al(OH)^{2+}$	4.99 (5.07)	Ball, Nordstrom, and Jenne, 1980
$Al(OH)_4^-$	23.00 (23.12)	Ball, Nordstrom, and Jenne, 1980
$Al(OH)_3(s)$	−10.50 (10.64)	Ball, Nordstrom, and Jenne, 1980
	Amorphous Ferric Hydroxide	
$Fe(OH)^{2+}$	2.2	Stumm and Morgan, 1981
$Fe(OH)_2^+$	5.7	Stumm and Morgan, 1981
$Fe(OH)_4^-$	21.6	Stumm and Morgan, 1981
$Fe(OH)_3(s)$	−3.6	Stumm and Morgan, 1981

* The values in parentheses have been corrected for ionic strength, $I = 0.001$ M.

using Equations 6.7 and 6.9 and the Al hydrolysis products and equilibrium constants ($I = 0.001$ M) listed in Table 6.2. The results are plotted in Figures 6.11 (*a* and *b*) for total Al concentrations (Al_T) of 1×10^{-6} M [0.03 mg Al/L or 0.33 mg alum ($Al_2(SO_4)_3\ 14H_2O$)/L] and 3×10^{-4} M [8.1 mg Al/L or 89 mg alum/L], respectively.

According to Figure 6.11(*a*) for $Al_T = 1 \times 10^{-6}$ M, a small amount of $Al(OH)_3$ precipitate forms between pH = 5.8 and 6.5. At pH = 6.1, where the amount of precipitate is maximum, $Al(OH)_4^-$ and $Al(OH)_2^+$ are the principal hydrolysis products, representing over 70 percent of the total Al in the system. For the higher Al_T (3×10^{-4} M) of Figure 6.11(*b*), $Al(OH)_3$ precipitate begins to form at pH = 4.7 and by pH = 9.0 all of it has dissolved to form $Al(OH)_4^-$. Over most of this pH interval (from pH 5.5 to 8.5), the precipitate is the predominant hydrolysis product.

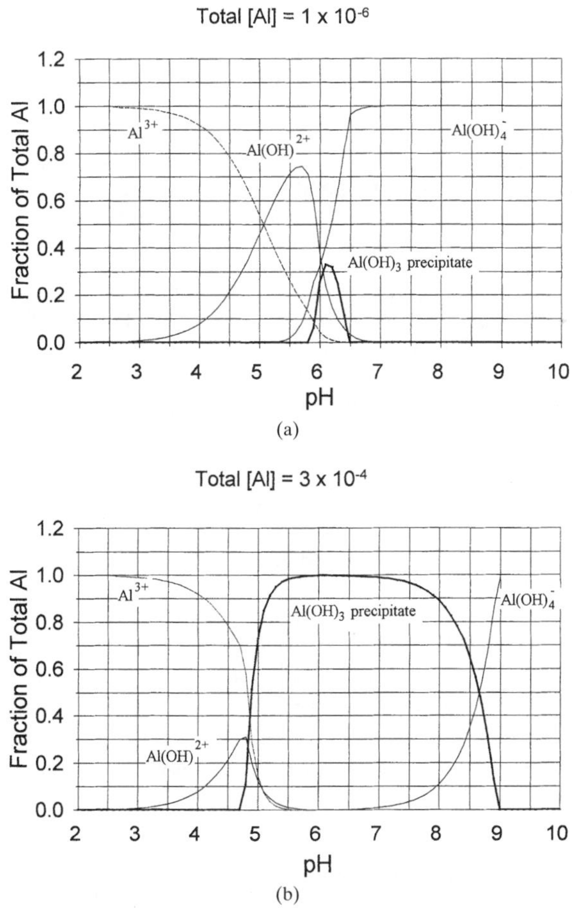

FIGURE 6.11 Effect of pH and total aluminum concentration on the speciation of Al(III): (*a*) $Al_T = 1 \times 10^{-6}$ M, and (*b*) $Al_T = 3 \times 10^{-4}$ M.

TABLE 6.3 Examples of Polynuclear Al(III) Species

Polynuclear species	[OH]/[Al]	Reference
$Al_8(OH)_{20}^{4+}$	2.5	Hayden and Rubin (1974)
$Al_{13}O_4(OH)_{24}^{7+}$	2.46	Bertsch (1986)
$Al_{14}(OH)_{32}^{10+}$	2.29	Turner (1976)

The literature contains convincing evidence that polynuclear hydrolysis products are formed under certain conditions. Table 6.3 lists three examples of the many polynuclear forms that have been suggested. Most have molar ratios of OH to Al in the 2.3 to 2.5 range. Nuclear magnetic resonance (NMR) spectroscopy (Bertsch and Parker, 1996) has provided strong evidence for the existence of $Al_{13}O_4(OH)_{24}^{7+}$. Bertsch and Parker conclude that polynuclear species are metastable intermediates in solutions that reach or exceed a critical degree of supersaturation with respect to a solid phase such as gibbsite (Al_2O_3) or amorphous $Al(OH)_3$.

In polynuclear (and microcrystalline) hydrolysis products most of the metal atoms are interconnected by double-hydroxide bonds. Double-hydroxide bonds are not easily broken. Therefore, when complex metal hydroxide species are transferred from the solution in which they were formed to one in which they could not be formed and are not stable, the transition to the new set of species is typically slow. For example, if a concentrated solution of polynuclear species is added to a suspension as a coagulant, it is likely that a significant amount of time (days to months) will be required for a new stable distribution of hydrolysis products to form in the system; that is, the polynuclear species will tend to persist. Consequently, particle or NOM removal diagrams prepared using preformed polynuclear species (Dempsey, Ganho, and O'Melia, 1984) may show effective coagulation in areas where the corresponding solubility diagram based on equilibrium conditions [e.g., Figure 6.9 for $Al(OH)_3$] does not show the formation of destabilizing hydrolysis products.

Bertsch and Parker (1996) have argued that it is not appropriate to explain polynuclear species formation and stability using complex formation reactions (such as Eq. 6.6). These reactions use electrolyte theory and conventions to describe single-ion activities. It is not correct to use electrolyte theory to describe the activity of large, polyvalent, polynuclear species, and, consequently, it is not thermodynamically correct to include polynuclear species in the calculations used to plot solubility diagrams such as Figure 6.9 and 6.10.

Polynuclear species are typically detected at high total Al concentrations ($>1 \times 10^{-3}$ M) and at pH values between 7 (on the right) and the acid boundary of the $Al(OH)_3$ precipitate region of Figure 6.9 (on the left). Dispersed polynuclear species cannot be detected in the presence of significant concentrations of destabilizing multivalent anions, such as sulfate, or in solutions with high ionic strengths (de Hek, Stol, and deBruyn, 1976).

Types of HMS Coagulants Used in Water Treatment

The essential HMS product groups are described below. The list is not comprehensive because new types of products are constantly entering the water treatment market in this highly competitive business.

Simple Metal Salts. The simple HMS coagulants are aluminum sulfate (alum), ferric sulfate, and ferric chloride. These are sold as dry crystalline solids and as concentrated (~2 M) aqueous solutions. Alum is still the predominant HMS coagulant; however, the iron salts are growing in importance.

Prehydrolyzed Metal Salts. As noted above, when HMSs are added to and diluted in the water to be treated, the hydrolysis reaction produces hydrogen ions that react with alkalinity species in the solution. If some of this acid is neutralized with base when the coagulant is manufactured, the resulting product is a prehydrolyzed metal salt coagulant solution. The degree to which the hydrogen ions produced by hydrolysis are preneutralized is called the *basicity*. The basicity is given by

$$\text{Basicity}(\%) = \left(\frac{100}{3}\right) \times \frac{[\text{OH}]}{[M]} \tag{6.11}$$

where $[\text{OH}]/[M]$ is the weighted average of the molar ratio of the bound hydroxide to metal ion for all the metal hydrolysis products in the undiluted coagulant solution. For example, if a hypothetical product solution contained just one hydrolysis product, the polynuclear species $Al_{13}O_4(OH)_{24}{}^{7+}$, [OH]/[Al] would be effectively equal to $2.46 = [24 + (2 \times 4)]/13$ and the basicity would be 82 percent $= (100/3) \times 2.46$. For commercial coagulant solutions, the basicity varies from ~10 (low prehydrolysis) to around 83 percent. As the basicity is increased beyond about 75 percent, it becomes increasingly difficult to keep the metal hydroxide precipitate from forming in the product solution during shipping and extended storage. AWWA Standard B408-98, *Liquid Polyaluminum Chloride,* provides a laboratory method for determining the basicity of product solutions.

To avoid forming a precipitate in prehydrolyzed product solutions, the higher basicity products are usually made with chloride as the dominant anion; multivalent anions such as sulfate ($SO_4^{=}$) tend to destabilize positively charged polynuclear and microcrystalline hydrolysis products. Prehydrolyzed metal salt coagulants made with aluminum chloride are called *polyaluminum chloride,* or sometimes, *polyaluminum hydroxychloride,* or simply *PACl.* Use of the term *polyaluminum* (or *polyiron* for Fe salt products) is based on the assumption that the product solution contains significant amounts of polynuclear metal hydrolysis products, which tends to be true only for the higher basicity (70-plus percent) solutions. Some investigators (Bottero et al., 1987) believe that the principal hydrolysis product in PACl solutions with basicities greater than about 75 percent is $Al_{13}O_4(OH)_{24}{}^{7+}$. Prehydrolyzed iron solutions exist but are still a relatively uncommon commercial product. The significance of the basicity of iron salt coagulant solutions has been discussed by Tang and Stumm (1987).

Metal Salts Plus Strong Acid. Several coagulant manufacturers prepare coagulant solutions that contain the metal salt (e.g., alum) and an amount of strong acid, typically sulfuric acid. The typical acid-supplemented alum product (also called *acidulated alum* or *acid alum*) contains 5 percent to 20 percent (weight basis) 93 percent sulfuric acid. Iron salt solutions are available that contain supplemental sulfuric acid. For a given amount of metal ion, Al or Fe, added to the water, strong acid-fortified products react with more alkalinity and depress the pH to a greater extent than the nonfortified metal salt solutions.

Metal Salts Plus Additives. Metal salt coagulant solutions are available with special additives including phosphoric acid, sodium silicate, and calcium salts. Alum with phosphoric acid has some of the characteristics of acid-supplemented alum, but $AlPO_4$ precipitate is formed when the solution is added to the water. Metal salt solutions are also sold premixed with polyelectrolyte coagulant compounds such as epichlorohydrin dimethylamine (epiDMA) and polydiallyl dimethylammonium chloride (polyDADMAC).

Sodium Aluminate. Sodium aluminate ($NaAlO_2$), a common chemical in papermaking, is sometimes used as a coagulant in water treatment, typically in combination with alum to treat waters with low alkalinity. The chemical properties of sodium aluminate make it more difficult to handle than alum and the other metal salt products and this—and cost—have limited its use. It is a basic aluminum salt; when added to water and aluminum hydroxide precipitate forms, the alkalinity and pH of the solution tend to increase.

$$AlO_2^- + 2H_2O \Leftrightarrow Al(OH)_3(s) + OH^-$$

A sodium aluminate dosage of 1 mg Al/L increases the alkalinity 0.037 milliequivalents per liter (meq/L) or 1.9 mg/L as $CaCO_3$.

Impurities in HMS Coagulant Solutions

Most of the impurities in HMS coagulants are derived from the raw materials used to make them. For example, alum is usually made by digesting an aluminum source in sulfuric acid. Typical aluminum sources are bauxitic or high-aluminum clays, aluminum trihydrate, and high-purity bauxite. Impurities in the aluminum source tend to appear in the alum product. The most significant contaminant in aluminum salt coagulants is iron. Standard alum solution (4.2 percent Al) may contain 1000 ppm Fe. Small amounts of heavy metals such as chromium and lead can be found in standard-grade alum solution (Lind, 1995).

Iron salt coagulants are manufactured by dissolving various iron sources (iron ores and scrap iron) in sulfuric or hydrochloric acid or by reprocessing materials such as acidic iron salt solutions from iron mills and foundries. Ferric chloride is made from reprocessed titanium dioxide liquors. Like alum, the iron salt coagulants typically contain metal contaminants, usually Mn, Cu, V, Zn, Pb, and Cd. The amount varies with the source of the product so checking metal concentration is a prudent procedure.

In most cases, the low amounts of heavy metal contaminants in HMS coagulants will not have a significant effect on metal concentrations in the treated water. The metals may already be in an insoluble form or they are likely to precipitate or adsorb on the floc when the coagulant is added to the water. The metals may, however, increase the heavy metal content of treatment residue.

AWWA standards for HMS coagulants (e.g., AWWA B403-98, *Aluminum Sulfate—Liquid, Ground, or Lump*) include only a general statement about limits on impurity levels in the products. The statement says that the contaminants should not be present in quantities that will cause "deleterious or injurious effects" on human health when the product is used properly. For alum products the only specific impurity limit is for iron.

Aluminum in drinking water has been implicated as a contributing factor in Alzheimer's disease. However, to date, researchers have been unable to verify or refute these claims. An AWWA Executive Committee paper (Anonymous, 1997) on aluminum salt coagulants states that research results are not sufficiently consistent or accurate to support concerns about aluminum in general or aluminum in drinking water as causal agents for Alzheimer's disease. The amount of residual Al in treated water can be minimized by optimizing filtration to maximize the removal of particulate matter and by keeping the pH during coagulation and flocculation between 6 and 6.5 [i.e., near the pH of the minimum solubility of aluminum hydroxide (Letterman and Driscoll, 1993)].

Acidity of Hydrolyzing Metal Coagulants

When an acidic HMS coagulant solution, such as alum or PACl, is diluted in the water to be treated, the hydrolysis reaction produces hydrogen ions that decrease the alkalinity of the solution and tend to lower the pH, for example:

$$H^+ + HCO_3^- \rightarrow H_2CO_3\ (CO_2 + H_2O) \tag{6.12}$$

The strong acid content or acidity of commercial coagulant solutions depends on the basicity of prehydrolyzed products (Eq. 6.10) and the acid content of acid-supplemented products. Letterman, Chappell, and Mates (1996) have shown that the effective acidity can be estimated using the following expression:

$$\text{Effective acidity (meq/mg metal)} = \frac{300 - 3B}{100\ (AW)} + \frac{A}{52.7\ (MC)} \tag{6.13}$$

where

B = basicity of the prehydrolyzed products (percent)
A = weight percent of 93 percent sulfuric acid solution in acid-supplemented products
AW = atomic weight of the metal; Al = 27 and Fe = 55.9
MC = metal concentration in the product solution (weight percent)

Equation 6.13 is based on the assumption that the predominant hydrolysis product present after coagulant addition is the metal hydroxide precipitate, $Al(OH)_3$ or $Fe(OH)_3$, with a molar ratio of hydroxide to metal of 3. Because it is very unlikely that a manufacturer would add supplemental sulfuric acid to a prehydrolyzed product solution, it can be assumed that when $B > 0$, then $A = 0$, and, conversely, when $A > 0$, then $B = 0$. Example values of coagulant solution acidities are listed in Table 6.4 for four commercial HMS products.

The effective acidity of a coagulant product can be used to determine the relationship between the coagulant dosage and the pH after flocculation and floc separation.

TABLE 6.4 Calculated Effective Acidities of Selected Commercial Coagulant Products*

Type of coagulant solution	Basicity (%)	Sulfuric acid content (%, 93% sulfuric acid)	Metal concentration in product solution (weight %)	Calculated acidity (meq/mg M)‡
Aluminum sulfate (alum)	0	0	4.3	0.111
Polyaluminum chloride (PACl)	75	0	12.3	0.028
Acid-supplemented alum	NA†	10	2.95	0.168
Ferric sulfate	7	0	12	0.051

* The solution attributes used in the calculations are from the manufacturer's product data sheets.
† NA = not applicable.
‡ M = aluminum or iron.

In one approach to establishing this relationship, the water to be treated is slowly titrated with a coagulant solution of known acidity while the pH is measured. The titration results are plotted as pH versus the coagulant dosage in meq/L. The coagulant dosage in meq/L is equal to the dosage in mg metal/L times the acidity in meq/mg metal. If the coagulant dosage is initially recorded as volume of coagulant solution per volume of water, then the conversion to mass concentration units is given by

Coagulant dosage

$$\text{(mg metal/L)} = \text{volumetric dosage } (\mu\text{L/L}) \times SG \times \rho_w \times MC/100 \qquad (6.14)$$

where SG is the specific gravity of the coagulant solution, ρ_w is the density of water (1 mg/μL), and MC is the metal concentration (as weight percent) in the coagulant solution. The specific gravity and metal concentration can be found on coagulant product data sheets. The relationship between the amount of coagulant added and the pH of the solution can vary with the time allowed for reaction. It is also affected by the transport of carbon dioxide between the solution and the atmosphere. Consequently, the relationship obtained from laboratory measurements may not always agree exactly with what is seen in the treatment plant.

Example. Figure 6.12 is an example titration curve for a water with an initial alkalinity of 3.7 meq/L (185 mg/L as $CaCO_3$) and an initial pH of 8.75. The titrant was alum with an acidity of 0.111 meq/mg Al. In this example, the objective is to use this titration curve to determine the final pH when the coagulant is polyaluminum chloride (75 percent basicity and effective acidity of 0.028 meq/mg Al), and the dosage is 8 mg Al/L (47.5 μL of commercial PACl solution per liter of suspension). The effective acidity for this PACl product is used with the dosage to determine the amount of acid added by the coagulant solution (i.e., 8 mg Al/L × 0.028 meq/mg Al = 0.22 meq/L). According to Figure 6.12, at a coagulant dosage of 0.22 meq/L, the final pH will be approximately 7.8.

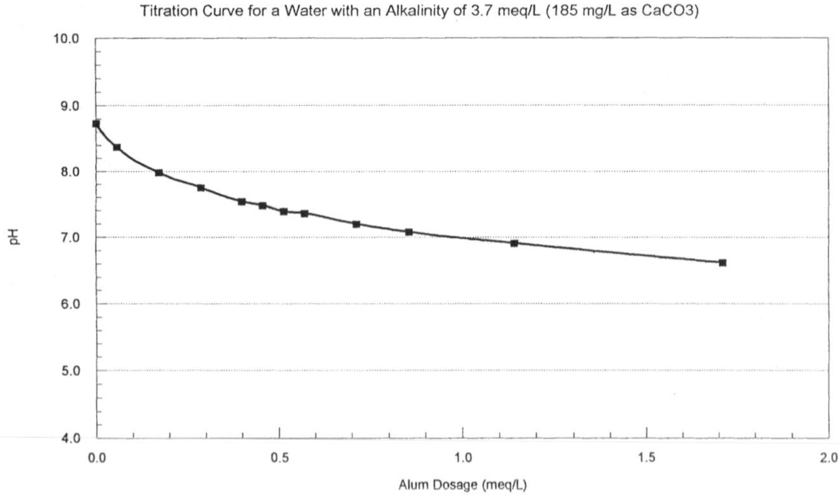

FIGURE 6.12 Titration of a bicarbonate solution, initial alkalinity of 3.7 meq/L (185 mg/L as $CaCO_3$), with alum [$Al_2(SO_4)_3 \cdot 14H_2O$].

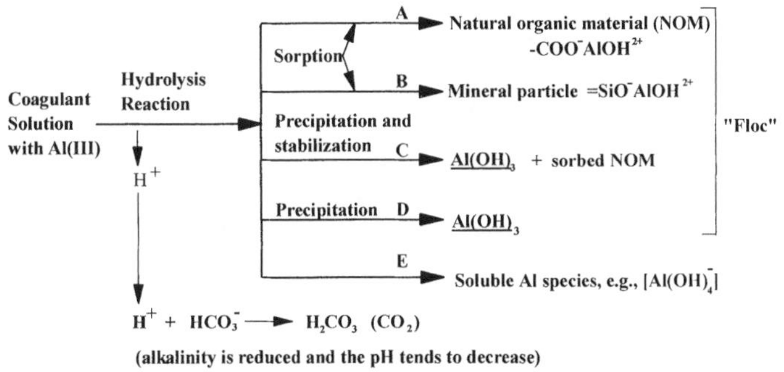

FIGURE 6.13 Reaction pathways that hydrolysis products may follow when an HMS coagulant is added to water with particles or NOM.

The final pH can also be calculated using the equilibrium chemistry relationships that describe the inorganic carbon system[2] (Snoeyink and Jenkins, 1980). First, the initial inorganic carbon concentration (C_T) of the water is calculated using the initial pH and initial alkalinity (8.75 and 3.70 meq/L, respectively). For this example, assuming a solution temperature of 25°C and an ionic strength of 4×10^{-3} moles/L, $C_T = 3.6 \times 10^{-3}$ moles/L. The final alkalinity is then determined using the initial alkalinity of the water (3.70 meq/L) and the acidity and dosage of the PACl coagulant solution of Table 6.4 (0.028 meq/mg Al and 8 mg Al/L, respectively). The final alkalinity is equal to 3.48 meq/L = 3.70 meq/L − (8 mg Al/L × 0.028 meq/mg Al). Finally, the equilibrium chemistry relationships are used again with the final alkalinity (3.48 meq/L), the initial C_T (3.6×10^{-3} moles/L), and the temperature and ionic strength to calculate a final pH of 7.75.

Using the initial C_T to calculate the final pH requires the assumption that the system does not exchange significant amounts of carbon dioxide with the atmosphere during flocculation and floc separation. This has been shown to be a reasonable assumption for typical full-scale treatment plants (Shankar and Letterman, 1996).

Action of Hydrolyzing Metal Coagulants

Figure 6.13 lists reaction pathways that the hydrolysis products may follow when an HMS coagulant is added to a water that contains particles or NOM. If the dosage of the coagulant is sufficient and chemical and physical conditions are conducive, the presence of hydrolysis products may lead to the formation of particle aggregates (flocs). The processes associated with each pathway are discussed below.

Pathways A and B. Pathways A and B represent the reaction of the metal hydrolysis products with metal binding sites on the surfaces of contaminant particles (minerals, microorganisms, and so forth) and may include the process by which soluble

[2] The inorganic carbon relationships can only be used if bases other than carbonates and OH^- are insignificant compared with the total amount of alkalinity in the system. For example, if silicates and organic acids are present in significant amounts, the relationships discussed in this example may not give accurate results.

hydrolysis products bind with sites on soluble natural organic compounds (VanBenschoten and Edzwald, 1990). Dempsey (1989) and Bottero and Bersillon (1989) have argued that soluble hydrolysis products binding with soluble NOM is not a significant step in the removal of NOM by HMS coagulants and that the adsorption of the NOM on precipitate particles is the controlling process for all conditions.

If the binding sites are abundant relative to the concentration of hydrolysis products and the sites have an affinity for Al and Fe hydrolysis products, then essentially all the hydrolysis products formed will react irreversibly with the surface. The hydrolysis products that follow these reaction pathways will not be available for the formation of the metal hydroxide precipitate.

In the case of negatively charged particles such as clays and microorganisms, the positively charged, site-bound metal hydrolysis products may neutralize the surface charge of the particles and the suspension will be destabilized. This is usually called *coagulation by charge neutralization.*

For example, for a silica particle with negative surface sites ($\equiv SiO^-$), the hydrolysis-adsorption reaction might be

$$Al^{3+} + \equiv SiO^- \rightarrow \equiv SiO^- \, Al(OH)_2^+ + 2H^+ \tag{6.15}$$

In the scheme of Eq. 6.15, the overall charge of the silica site with bound positive aluminum hydrolysis product is +1. A silica particle after treatment with an aluminum salt, and with equal numbers of $\equiv SiO^-$ and $\equiv SiO^- Al(OH)_2^+$ sites, would have a net surface charge of zero and, according to the principles of colloid stability discussed above, the particles in the suspension would be destabilized by the mechanism of surface charge neutralization.

In some treatment plants, typically direct filtration systems, where large, settleable flocs are not required or even desirable, the target for coagulant addition is frequently the dosage that yields charge neutralization. The amount of positive charge required for charge neutralization is proportional to the site density on the particle surface and the total surface area concentration of the suspension. The pH of the solution and the presence of site-binding cations and anions that compete with the hydrolysis products for sites will also be factors. When destabilization by charge neutralization is used with a dilute suspension, the rate of flocculation is usually much lower than the rate obtained when the voluminous metal hydroxide precipitate is formed.

Pathways C and D. Metal hydroxide precipitate and soluble metal hydrolysis products begin to form after the coagulant demand of Pathways A and B has been essentially satisfied. Edzwald and VanBenschoten (1990) argue that the tendency to follow one reaction pathway over the other depends on the relative affinity of the hydrolysis products for surface sites and metal-binding species in solution as well as on the ratio of their effective concentrations. When the amount of sites that bind metal hydrolysis products is relatively low, reaction with the binding sites and formation of the metal hydroxide precipitate in the bulk solution may occur simultaneously. It is likely that the quality and intensity of mixing used to disperse the coagulant solution in the water are factors that determine the predominant reaction pathway (Clark and Srivastava, 1993; David and Clark, 1991).

When the metal hydroxide begins to precipitate in the presence of NOM, the more hydrophobic, higher-molecular-weight fraction of the NOM tends to adsorb and coat the colloidal microcrystals as they form (Pathway C). This process creates the significant coagulant demand of waters that contain adsorbable NOM. When the density of adsorbed NOM is relatively high, the microcrystal particles will have a net negative surface charge and will remain stable and dispersed by a process called *peptization* or *steric stabilization.*

As the dosage of the HMS is increased, the surface area concentration of the microcrystals increases and the amount of NOM adsorbed per unit area of particle surface decreases. Eventually, at a higher HMS dosage, there is not enough negatively charged NOM on the surfaces to stabilize the positively charged precipitate particles, and the suspension becomes unstable, forming flocs that consist of the metal hydroxide precipitate and sorbed NOM. The dosage of HMS required to reach this point has been shown by a number of investigators to be proportional to the initial concentration of NOM in the water (Narkis and Rebhun, 1977; Dempsey, 1987; Edzwald, 1993). There is evidence that it is inversely proportional to the initial positive charge on the metal hydroxide precipitate. Differences in initial hydroxide surface charge may explain why, at a given pH, ferric hydroxide generally has a higher coagulant demand for NOM removal than aluminum hydroxide. Edwards (1997) has used a NOM adsorption mechanism as the basis for a model that predicts the efficiency of NOM removal by HMS coagulants.

At coagulant dosages that exceed the coagulant demand, the formation of an unstable (flocculent) precipitate leads to the relatively rapid formation of visible floc. This process is usually called *sweep flocculation,* or *enmeshment.* In most cases, the rate of flocculation increases in proportion to the volume concentration of precipitate in the suspension. Floc formation by enmeshment involves both the interaction of particles of metal hydroxide precipitate and contact with contaminant particles. The metal-binding sites on the contaminant particles that interact with the precipitate may be occupied by simple hydrolysis products, and NOM, if present, may be adsorbed on the precipitate surfaces.

It is possible, especially at low pH (<6.5) and with dilute solutions, for the microcrystalline precipitate to become stabilized and dispersed by its high positive surface charge. When this charge reversal occurs, removal of the colloidal precipitate by floc separation processes (sedimentation, flotation, and coarse-media filtration) becomes essentially impossible. Multivalent anions, such as sulfate and phosphate, as well as NOM, tend to adsorb on positive hydroxide surfaces, reducing the net charge on the surface and destabilizing the precipitate suspension. Consequently, when the water to be treated contains significant amounts of multivalent anions, or if significant amounts are added via the coagulant solution (e.g., sulfate from alum and sulfuric acid), then the formation of a stable, precipitate suspension at low pH may not occur (Letterman and Vanderbrook, 1983).

In the treatment of some dilute waters at low pH, destabilization by charge neutralization may occur at a relatively low coagulant dosage. As the dosage is increased and a stable colloidal precipitate is formed, the suspension may become restabilized by the high positive charge of sorbed hydrolysis products. Settleable flocs will not be formed under these conditions. Dentel's (Dentel, 1988; Dentel, 1991) precipitation-charge neutralization model of coagulation with HMS coagulants assumes that destabilization of the suspension begins after the solubility limit of the metal hydroxide has been exceeded. Small, positively charged units of the precipitate deposit on the negative particle surfaces by heterocoagulation, and destabilization occurs by charge neutralization. The model does not account for the uptake of soluble hydrolysis species by the particle surfaces before the precipitate begins to form but could be modified to do so.

Pathway E. Pathway E in Figure 6.13 represents the process by which the HMS forms soluble metal hydrolysis species. For a given coagulant dosage and coagulant demand, the amount of flocculent metal hydroxide precipitate formed depends on the final pH of the solution (as well as reaction time, temperature, and other factors). If the final pH is close to the pH of minimum solubility of the metal hydroxide precipitate, the amount of precipitate will be maximized and the amount of soluble "residual" metal ion in the water will be minimized. The pH of minimum solubility is approximately 6 for aluminum hydroxide and 8 for ferric hydroxide (see Figures 6.9 and 6.10). The final pH is determined by the initial alkalinity (and pH) of the water and the amount of hydrogen ions added to or formed within the system by the coagulant solution. The amount of hydrogen ions is quantified by the coagulant solution acidity (i.e., the "effective acidity" discussed earlier). As noted previously, the effective acidity (Eq. 6.13) is a key parameter in predicting the relationship between the coagulant dosage and the final pH of the solution.

Interpreting Coagulation Results

The reaction pathways and processes described above act in combination to determine the performance of HMS coagulants. The key elements are the coagulant dosage, the alkalinity and pH of the raw water, the initial NOM and particulate concentrations, and the metal hydroxide speciation, including the precipitate surface characteristics and solubility. The complexity of these factors makes it difficult to accurately predict coagulation performance for a given raw water. However, analysis of the coagulation process in operation, or using laboratory simulations, allows the process to be understood qualitatively. Once this is done, it is possible to identify the governing mechanisms and thereby allow improved process design and operation for each specific water and coagulant. The laboratory simulations, typically done as batch tests, known as jar tests, allow a wide range of conditions to be explored. The following examples show how different coagulation phenomena can be observed and interpreted in jar tests.

The results presented in the following examples include the electrophoretic mobility (EM) measurement. The EM measurement is discussed later in this chapter. It is made by measuring the velocity of the particle in an applied electric field and is related to the zeta potential and charge near the particle surface. It is one of the electrokinetic measurements that can be used to relate (with some assumptions) the magnitude of particle surface charge to suspension stability. For many suspensions, as the EM varies from zero in either the positive or negative direction, the tendency for the particles in the suspension to aggregate decreases.

Coagulation of Particulates with Controlled pH and Negligible NOM. The data plotted in Figure 6.14 were obtained in jar test experiments in which aluminum nitrate was used at constant pH (pH = 6) to destabilize a silica suspension. The nitrate salt was used so the sulfate ion concentration could be varied independently of the aluminum concentration. The point at which the EM passes through zero corresponds to the point at which the residual turbidity, measured after flocculation and sedimentation, passes through a minimum. This observation suggests that destabilization occurred by charge neutralization. Particle aggregation at aluminum concentrations greater than 10^{-5} molar (0.27 mg Al/L) was negligible because, at the pH of these experiments and in the absence of an adsorbing/destabilizing anion such as sulfate, the positively charged particles and the colloidal hydrolysis products are stable, and the enmeshment process cannot occur.

Figure 6.15 is a plot of EM and residual turbidity data obtained under the same conditions as Figure 6.14, except that the solution contained a sulfate concentration of 3×10^{-3} M (288 mg $SO_4^{=}$/L). Sulfate was added as the sodium salt during the preparation of the suspension.

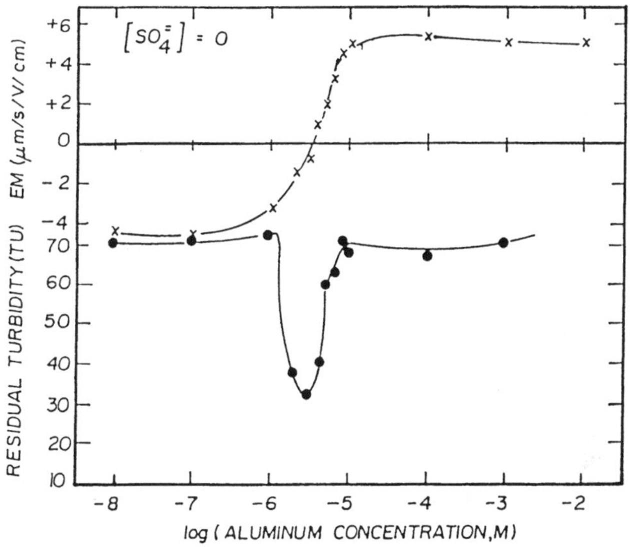

FIGURE 6.14 Effect of the aluminum concentration on particle EM and the residual turbidity after flocculation and sedimentation. The aluminum was added as aluminum nitrate. No sulfate was present in the system and pH = 6.

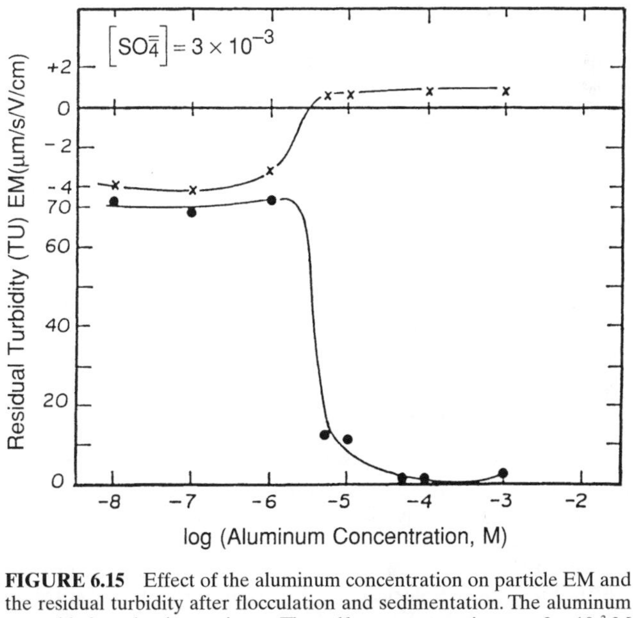

FIGURE 6.15 Effect of the aluminum concentration on particle EM and the residual turbidity after flocculation and sedimentation. The aluminum was added as aluminum nitrate. The sulfate concentration was 3×10^{-3} M (288 mg $SO_4^{=}$/L) and pH = 6.

According to Figures 6.14 and 6.15, the amount of Al that must be added to accomplish charge neutralization (EM = 0) is approximately 3×10^{-6} M (0.081 mg Al/L) whether sulfate is present or absent. This finding agrees with Meng's observation (Meng, 1989) that the sulfate ion has a negligible tendency to react with the site-bound aluminum hydrolysis products on a silica surface.

As shown in Figure 6.14, with high aluminum concentrations and no sulfate, the silica particle suspension is restabilized by the high positive surface charge associated with the coating of site-bound aluminum. The EM under these conditions is greater than +5 $(\mu m \cdot s^{-1})/(V \cdot cm^{-1})$. The negative sulfate ion destabilizes the dispersed positive hydrolysis products (by adsorbing in the Stern layer and reducing the net surface charge) and, at aluminum concentrations somewhat greater than the amount needed for charge neutralization ($\approx 1 \times 10^{-5}$ M or 0.27 mg Al/L), enmeshment occurs. The EM of the sulfate-destabilized, precipitate-coated particles is about +1 $(\mu m \cdot s^{-1})/(V \cdot cm^{-1})$ (Figure 6.15).

In most applications of the jar test, the sulfate concentration will not be controlled at a constant value as in this example. In fact, if alum is used as a coagulant, the sulfate added with the alum will increase the sulfate concentration in direct proportion to the alum dose. At low doses, Figure 6.14 will be applicable, but at high doses, the effects of sulfate illustrated in Figure 6.15 (most important, the absence of charge reversal) will apply.

The residual turbidity at the higher Al concentrations in Figure 6.15 is much less than the minimum (32 NTU) shown in Figure 6.14 at the charge neutralization point. Although both systems are destabilized under these conditions, the rate of particle aggregation is much greater when the voluminous precipitate is present under enmeshment conditions, as opposed to when the particles have only the site-bound aluminum. As shown in Figure 6.15 for [Al] > 1×10^{-5} (0.27 mgAl/L), as the aluminum concentration is increased, more precipitate is formed, and particle removal by sedimentation increases.

Combination of the Various Factors. The coagulation examples previously described become even more complex when the effects of variable alkalinity, the presence of NOM, and metal hydroxide solubility are combined. In the following example, these factors are shown to complicate the coagulation process, but results can be logically analyzed in terms of the governing physical-chemical interactions.

Figure 6.16 is a plot of aluminum concentration (in mg/L) versus pH. The figure includes two titration curves, A and B, superimposed on four curves showing the amount of freshly precipitated aluminum hydroxide (0.1, 1, 10, and 100 mg Al/L). For example, on or above the curve labeled >10 mg Al/L, the amount of aluminum hydroxide precipitate is equal to or greater than 10 mg Al/L. The curves showing the amount of aluminum hydroxide precipitate were drawn using the formation and solubility constants listed in Table 6.2.

In plotting the titration curves of Figure 6.16, it was assumed that the coagulant is standard commercial alum with an effective acidity of 0.111 meq/mg Al. For titration Curve A, the initial alkalinity and initial pH are 3.5 meq/L (175 mg/L as $CaCO_3$) and 8.7; for Curve B, they are 0.25 meq/L (12.5 mg/L as $CaCO_3$) and 7.5. In this example, the water being treated is assumed to contain a moderate amount of NOM (approximately 5 mg C/L) with a relatively high SUVA but a negligible amount of solid particulate matter.

Figure 6.17 (*a–d*) shows the organic carbon concentration, turbidity, soluble Al concentration, and filtrate pH plotted as a function of the coagulant dosage for the conditions of titration Curve A of Figure 6.16. The results are discussed below for each of the four dosage intervals, I, II, III, and IV, indicated in the figures.

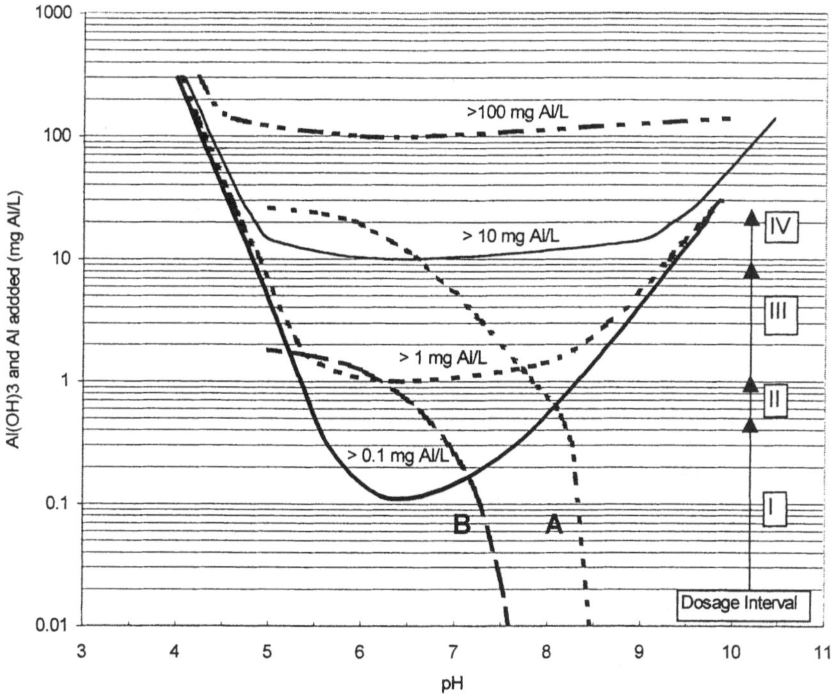

FIGURE 6.16 Titration curves superimposed on curves showing the amount of aluminum hydroxide precipitate formed. Curve A is for an initial alkalinity of 3.5 meq/L (175 mg/L as $CaCO_3$); Curve B is for an initial alkalinity of 0.25 meq/L (12.5 mg/L as $CaCO_3$). *Note:* 1 mg Al/L is equal to 12.8 mg $Al_2(SO_4)_3 \cdot 14H_2O$/L and 2.9 mg $Al(OH)_3$/L.

Dosage intervals I and II. Alum dosages less than 1.0 mg Al/L have a negligible effect on the removal of organic carbon and turbidity by filtration [Figure 6.17 (*a* and *b*)]. In dosage interval I (dosage < 0.5 mg Al/L), the solubility product of $Al(OH)_3$ has not been exceeded and there are no filterable particles with Al binding surface sites; consequently, the soluble Al concentration in the filtrate (although very low) is essentially equal to the amount of Al added. In the next interval, 0.5 to 1.0 mg Al/L, the solubility product has been exceeded and particulate aluminum hydroxide has begun to form. This "precipitate," however, is a colloidal suspension of microcrystalline aluminum hydroxide particles stabilized by a negative coating of NOM. Because the colloidal particles are not filterable, the filtrate turbidity begins to increase [Figure 6.17(*b*)] and the removal of organic carbon is negligible [Figure 6.17(*a*)]. Across dosage interval II the total Al concentration (soluble plus microcrystalline forms) in the filtrate [Figure 6.17(*c*)] is equal to the amount of Al added; however, as microcrystals form, the amount of soluble Al in the filtrate decreases.

Dosage interval III. As the dosage enters the third interval, 1.0 to 10 mg Al/L, and the surface area concentration of the microcrystalline precipitate increases, the density of NOM adsorbed on the surfaces of the particles becomes less than the amount needed to stabilize the particles and the suspension begins to flocculate. At this point, the organic carbon concentration [Figure 6.17(*a*)], the turbidity [Figure 6.17(*b*)], and total Al concentration [Figure 6.17(*c*)] in the filtrate decrease. If the

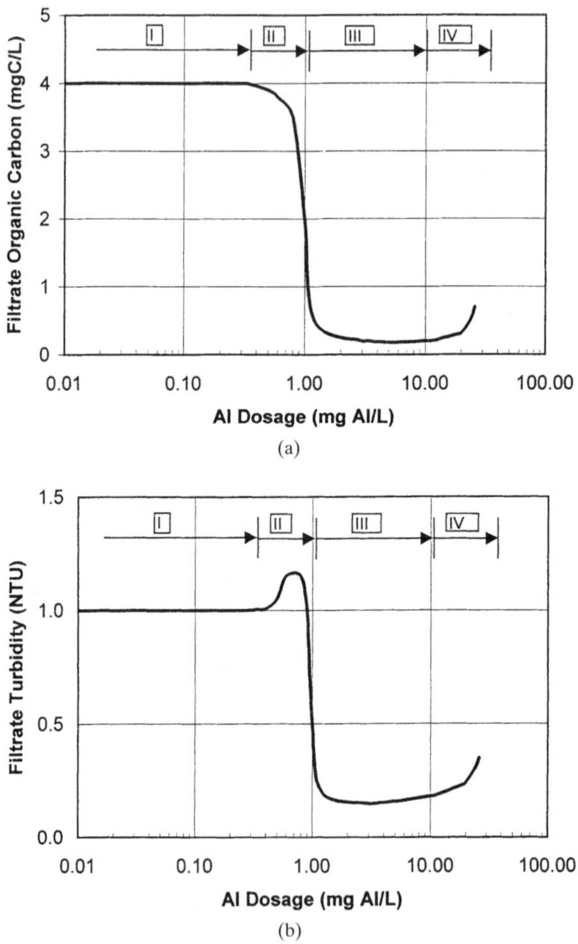

FIGURE 6.17 Effect of the coagulant dosage on (*a*) organic carbon concentration and (*b*) turbidity for titration Curve A in Figure 6.16. The dosage intervals of Figure 6.16 are shown at the top of the figures.

NOM is mostly high molecular weight, hydrophobic species with a strong tendency to adsorb on the metal hydroxide surface, the filtrate organic carbon concentration will decrease abruptly. If the NOM is made up of species with a broad distribution of molecular weights with varying affinity for the hydroxide surfaces, the transition will be more gradual and the filtrate organic carbon concentration at high coagulant dosages will not be as low. The dosage that initiates significant organic carbon removal is typically proportional to the initial organic carbon concentration (in this example, it is about 1.0 mg Al/L).

The pH decreases as the dosage increases across interval III [Figure 6.17(*d*)]. However, because the pH remains greater than 6, the pH of minimum solubility of $Al(OH)_3$, the precipitate becomes less soluble and the filtrate Al concentration decreases [Figure 6.17(*c*)].

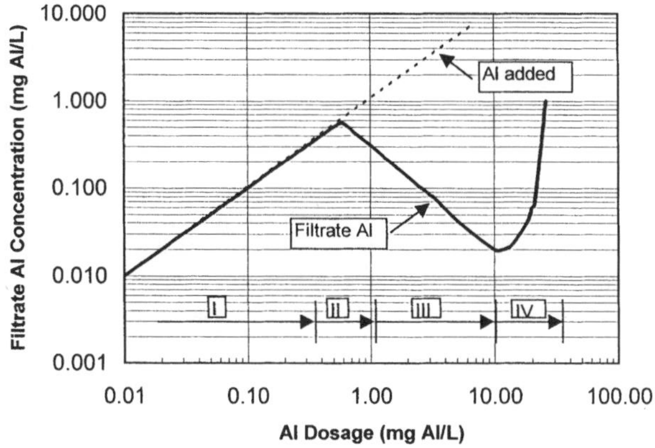

(c)

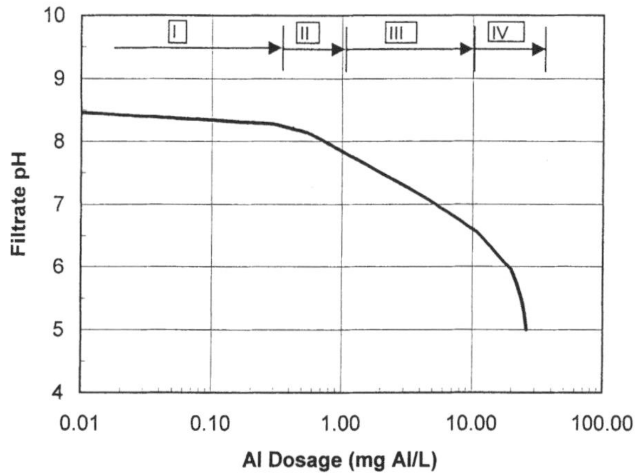

(d)

FIGURE 6.17 (*Continued*) Effect of the coagulant dosage on (*c*) soluble aluminum concentration, and (*d*) filtrate pH for titration Curve A in Figure 6.16. The dosage intervals are shown by the roman numerals in the figures.

Dosage interval IV. The fourth and last dosage interval of Figure 6.17(*a–d*) is from 10 to 25 mg Al/L. In this interval, the pH passes through the point of minimum $Al(OH)_3$ solubility, and the filtrate Al concentration reaches a minimum and then begins to increase [Figure 6.17(*c*)]. Also, as the pH decreases [Figure 6.17(*d*)], the charge on the surface of the precipitate becomes increasingly positive, and this change tends to stabilize the component particles and may make them disperse. These changes in precipitate solubility and stability may cause NOM and turbidity removal to become less efficient [Figure 6.17 (*a* and *b*)] across interval IV.

The calculations used to plot the curves shown in Figure 6.17 (*c* and *d*) assume that chemical equilibrium is reached at each point in the titration with alum. Dosage interval IV is the interval where this assumption is most likely to be inadequate, especially if the actual titration with alum was done quickly and aluminum hydroxide, formed at a higher pH, is not allowed to dissolve to its new, lower, equilibrium amount before the titration proceeds. When this is the case, the equilibrium filtrate aluminum concentrations plotted in dosage interval IV of Figure 6.17(*c*) and the equilibrium pH values of interval IV of Figure 6.17(*d*) will both be high compared with measured values.

Water with Low Initial Alkalinity. Curve B of Figure 6.16 is an example alum titration curve for a low initial alkalinity water (0.25 meq/L or 12.5 mg/L as $CaCO_3$). The titration curve passes through the lower part of the $Al(OH)_3$ precipitation region. At the pH of minimum solubility of $Al(OH)_3$, the amount of precipitate formed is less than the amount needed to meet the coagulant demand for the initial NOM concentration. Consequently, there is essentially no removal of NOM; that is, very few of the NOM-stabilized microcrystals are removed by filtration, and the filtrate total Al concentration is equal to the Al concentration added to the water. For a low initial alkalinity condition such as this, improved coagulant performance can be achieved by supplementing the initial alkalinity using a base [e.g., NaOH, $Ca(OH)_2$, or Na_2CO_3] and/or by using a coagulant solution with a lower effective acidity (e.g., a PACl product with high basicity).

Estimating the Required Coagulant Product Acidity Using Coagulation Diagrams. The effective acidity of HMS coagulant products can be used in conjunction with coagulation diagrams to select the appropriate coagulant product and coagulant dosage for a given application. A coagulation diagram is a figure in which coagulant metal concentration-pH coordinates are used to outline regions of coagulant performance. The regions are usually drawn on top of the thermodynamic stability diagram for the metal hydroxide precipitate (e.g., Figures 6.9 and 6.10). A coagulation diagram has practical utility only if the coagulation data used to prepare it were obtained with the appropriate solution chemistry and temperature, particle type and concentration (or NOM type and concentration), and type of coagulant (Al or Fe). The floc separation method (sedimentation, membrane filtration, flotation, or other method) should also be the same.

Figure 6.18 is a coagulation diagram (Amirtharajah and Mills, 1982) that is appropriate for alum and low-basicity (≤50 percent) PACl products treating dilute suspensions of particulate matter with a negligible concentrations of NOM. Floc separation is by sedimentation. In this example, the objective is to achieve "optimum sweep coagulation." According to the diagram, the minimum Al concentration and pH to reach this condition are 8×10^{-5} M (2.16 mg Al/L or 24 mg $Al_2(SO_4)_3 \cdot 14H_2O$/L) and 7.5, respectively. The titration curve (Figure 6.12) that was used in conjunction with an earlier example is applied in this case. This titration curve is for a water with an initial alkalinity of 3.7 meq/L (185 mg/L as $CaCO_3$). According to Figure 6.12, to reach pH = 7.5, a coagulant dosage of 0.43 meq/L is required. Because the Al concentration needed to reach the performance objective is 2.16 mg Al/L, the coagulant product selected must have an effective acidity of 0.2 meq/mg Al {0.2 meq/mg Al = [0.43 meq/L/(2.16 mg Al/L)]}. Because commercial alum solutions have an effective acidity of about 0.1 meq/mg Al, the results show that an acid-supplemented product should be considered. According to Equation 6.13, an acid-supplemented alum with approximately 15 percent of 93 percent sulfuric acid would have an effective acidity of approximately 0.2 meq/mg Al. Alternatively, strong acid could be used to reduce the initial alkalinity before the coagulant solution is added.

NOM Removal. The jar test results shown in Figures 6.19 through 6.22 are from a study of an unfiltered community water supply in upstate New York. The raw water sample had a pH of 7.6, an alkalinity of 1 meq/L (50 mg/L as $CaCO_3$), a total dissolved solids concentration of 190 mg/L, a turbidity of 0.9 NTU, and Pt-Co color of 50 units.

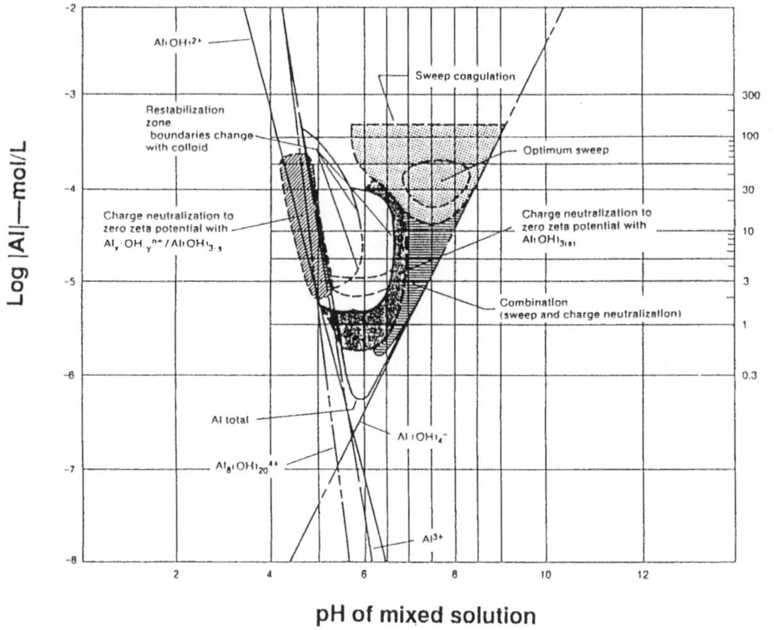

FIGURE 6.18 Alum coagulation diagram from Amirtharajah and Mills (1982).

The TOC concentration was approximately 9 mg C/L. The "alum dose" plotted in the figures is reagent-grade aluminum sulfate in milligrams of $Al_2(SO_4)_3{\cdot}18H_2O$ per liter.

The effect of the coagulant concentration on the removal of NOM was determined by filtering an aliquot of supernatant using a 0.45-μm-pore-size membrane filter. Sedimentation was not effective because of the low raw water turbidity (0.9 NTU) and, therefore, low floc density.

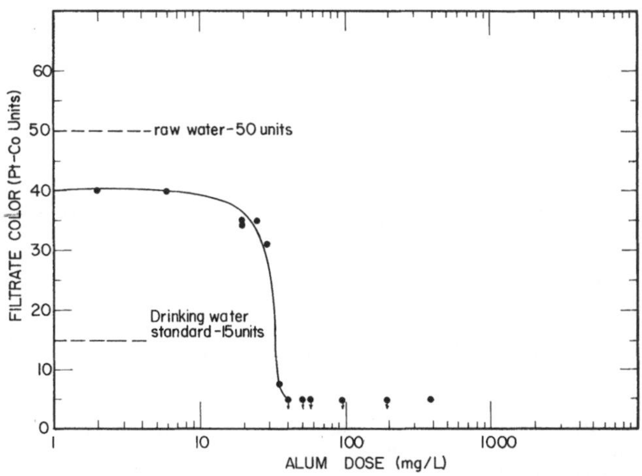

FIGURE 6.19 Jar test results—effect of alum dosage on filtrate color.

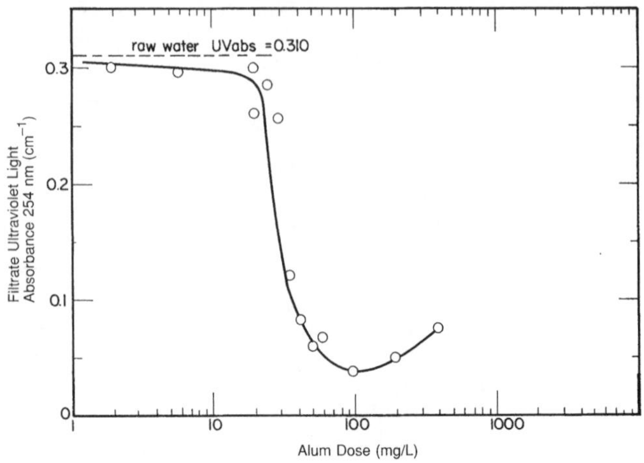

FIGURE 6.20 Jar test results—effect of alum dosage on the spectrophotometric absorbance of ultraviolet light by filtrate samples.

Figure 6.19 shows the effect of the alum dose on the filtrate color (in Pt-Co units). The color was reduced 20 percent by membrane filtration without added aluminum salt. Effective removal of color required approximately 40 mg $Al_2(SO_4)_3 \cdot 18H_2O/L$ (3.2 mg Al/L). The residual color in the filtrate could not be accurately determined when the value was less than about 5 Pt-Co units.

As discussed above, the ratio of the UV absorbance and the dissolved organic carbon concentration (the SUVA) is a useful measure of the fraction of the NOM that can be removed by HMS coagulants. In Figure 6.20, the UV light absorbance of fil-

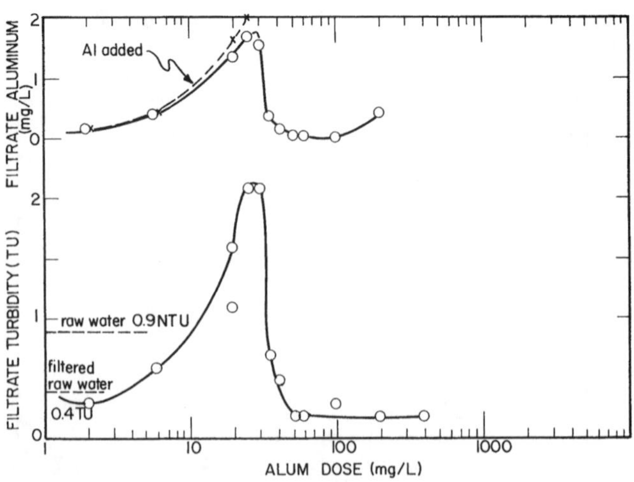

FIGURE 6.21 Jar test results—effect of the alum dosage on filtrate total aluminum concentration and turbidity. Dotted line in the upper graph is the amount of Al added.

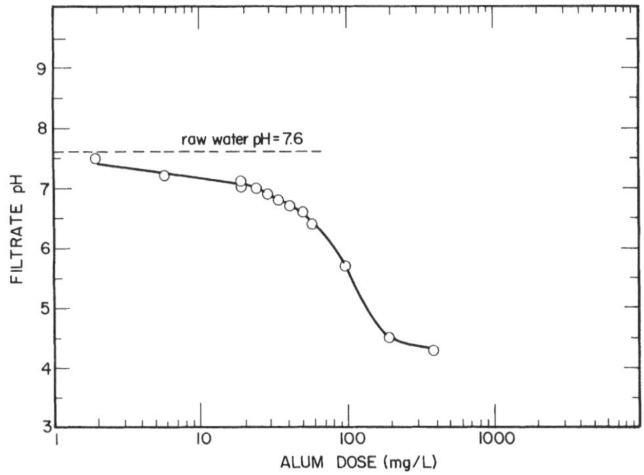

FIGURE 6.22 Jar test results—effect of the alum dosage on the filtrate pH.

trate samples is plotted versus the alum dose. Effective removal of UV light-absorbing organic compounds required an alum dose of at least 30 mg $Al_2(SO_4)_3 \cdot 18H_2O$/L (2.4 mg Al/L). Maximum removal was obtained at approximately 100 mg $Al_2(SO_4)_3 \cdot 18H_2O$/L. About 12 percent of the UV light-absorbing compounds could not be removed by coagulation (at any alum dosage) and filtration. This low unremovable fraction is consistent with the high raw water SUVA of 3.5 L/mg C·m^{-1}

The filtrate UV absorbance curve in Figure 6.20 shows a relatively sharp decrease in the NOM concentration as the alum dosage is increased. This is characteristic of NOM that is relatively high in molecular weight and homogeneous in composition. If this system had contained a lower-molecular-weight NOM with a broader-molecular-weight distribution (typical of low, <5 mg C/L, NOM concentration waters) the decrease in the residual NOM with increasing coagulant dosage would have been more gradual and the maximum removal of NOM at high coagulant dosages would not have been as great. The effect of the NOM characteristics on coagulant performance has been discussed by Randtke (Randtke, 1987, 1988).

Figure 6.21 shows the effect of the alum dose on the filtrate turbidity and aluminum concentration. For alum dosages in the interval 2 to 25 $Al_2(SO_4)_3 \cdot 18H_2O$/L (0.2 to 2 mg Al/L), most of the aluminum added passed through the membrane filter and was measured in the filtrate. The increasing filtrate turbidity shown in the lower plot (and the high filtrate color and UV absorbance measurements in Figures 6.19 and 6.20) suggest that the aluminum was present in a colloidal, light-scattering form. At an alum dose of approximately 30 mg/L (2.4 mg Al/L), the filtrate turbidity and aluminum concentration decreased abruptly, indicating that at this point the suspension of partially coated aluminum hydroxide particles had become unstable and filterable flocs had begun to form.

The solution's pH decreased with increasing alum dose, from about 7 at 10 mg $Al_2(SO_4)_3 \cdot 18H_2O$/L (0.8 mg Al/L) to 4.3 at 200 mg $Al_2(SO_4)_3 \cdot 18H_2O$/L (16.2 mg Al/L) (Figure 6.22). As aluminum hydrolysis released hydrogen ions, they depleted the limited alkalinity of the system, and the pH was depressed.

According to the filtrate aluminum plot of Figure 6.21, as the alum dosage was increased beyond about 100 mg $Al_2(SO_4)_3 \cdot 18H_2O$/L (8.1 mg Al/L), the filtrate aluminum concentration increased. This behavior can be attributed, at least in part, to the solubility of aluminum hydroxide at the low pH values indicated at high alum doses in Figure 6.22. Also, at low pH, colloidal aluminum hydrolysis products may acquire a high positive charge and pass through the membrane filter.

Sedimentation, in this example, is not likely to be an effective separation process for pretreatment. Because a rather significant minimum alum dose [30 to 40 mg $Al_2(SO_4)_3 \cdot 18H_2O/L$] is needed to obtain effective removal of NOM, direct filtration may not be a feasible alternative. (The high-volume concentration of the precipitate in the influent suspension may limit filter run length to values that are too low for economical filter operation; see Chapter 8.) The results suggest that pretreatment separation processes, such as dissolved air flotation and coarse bed filtration, should be considered.

Polyelectrolyte Coagulants

The polyelectrolyte coagulants used in water treatment are high-molecular-weight, synthetic organic compounds called *polymers* that have a strong tendency to adsorb on the surfaces of most particles in an aqueous suspension. Polyelectrolyte polymers consist of subunits called *monomers.* A polymer's total number and types of monomer units can be varied in manufacture; consequently, a wide variety of polymers can be and has been produced. The fact that polymer chains may be linear, branched, or cross-linked adds to their complexity.

Types of Polyelectrolytes. The monomer units in a polymer may have positively or negatively charged sites. In some cases, charged sites are formed by ionization reactions, with the overall charge on the molecule a function of the solution's pH and ionic strength. Polymers with a preponderance of negative sites are called *anionic;* those with predominantly positive sites are called *cationic.* Polymers with no charged sites or a very low tendency to develop them in aqueous solution are known as *nonionic polymers.* (Although not polyelectrolytes by the strictest chemical definition, the term is still used in describing nonionic polymers.) *Ampholyte polymers* have both positive and negative sites.

Usage. Estimates indicate (AWWA Coagulation and Filtration Committee, 1982) that over half the water treatment plants in the United States use one or more polyelectrolytes to improve treatment efficiency. Cleasby et al. (1989) found that of 23 treatment plants with very high-quality filtered water, 20 were using one or more polymeric flocculants and/or filter aids. The Cleasby report states that these additives are essentially required at high-rate (rapid sand) filtration plants.

Impurities in Polyelectrolyte Products. Polyelectrolyte formulations contain contaminants from the manufacturing process, such as residual monomers and other reactants and reaction by-products, that have potential human health significance (Letterman and Pero, 1988). Because of concern about certain contaminants, Switzerland and Japan do not permit the use of polyelectrolytes in drinking water treatment, and West Germany, France, and Canada have stringent limits on application rates. Some states in the United States require plants that use polyacrylamide and epichlorohydrin/dimethylamine (epi/DMA) compounds to notify them of monomer content.

For many years, the U.S. Public Health Service and the USEPA evaluated the health significance of polyelectrolyte products and maintained a list of "accepted" products that many states used in their regulatory/review activities. The list included a maximum dosage for each accepted product. In the 1980s, the USEPA sponsored the development and initiation of a new system based on a voluntary standard prepared by the National Sanitation Foundation (NSF Standard 60, 1988).

A large number of polymer products are listed for use by the USEPA and NSF, but this does not mean that hundreds of different compounds are being used in potable water treatment. In fact, Letterman and Pero (1988) suggest that fewer than 12 compounds make up the products on these lists. It appears that only polyDADMAC and epi/DMA are widely used and that many products are simply dilutions of the same compound.

Types of Polyelectrolytes Used in Water Treatment. Polyelectrolytes used to treat drinking water are often categorized as either *primary coagulant polymers* or *flocculent* (coagulant aid) *polymers.* Primary coagulant polymers are generally cationic and, with few exceptions, have relatively low molecular weight (<500,000). Primary coagulant polymers seem to enhance the coagulation and deposition (filtration) of negatively charged particles by adsorption and particle surface-charge neutralization. If one adds too much polymer to a suspension, each particle's overall surface charge may become positive. This occurrence, known as *restabilization,* can adversely affect coagulation and filtration.

Primary coagulant polymers may not be as effective as aluminum and iron salts to treat dilute inorganic suspensions and water with significant amounts of color-causing organics when aggregate removal is by, for example, sedimentation. Enmeshment by the voluminous metal hydroxide precipitates can be used to increase the volume concentration of the suspension and increase the rate of particle aggregation to efficient levels. Enmeshment is not usually a significant factor with polyelectrolytes and, although a dilute suspension can be effectively destabilized by a primary coagulant, the rate of particle aggregation may be too low to produce large, settleable aggregates in a reasonable period of time.

On the other hand, when the pretreatment objective is to produce small, high-density aggregates for direct filtration (i.e., with no intervening aggregate removal step like sedimentation), a primary coagulant used alone is often an effective alternative. One must be careful, however, never to exceed the product dosage limit recognized by the review authority for health protection.

Cationic Polyelectrolytes. The two most widely used cationic polyelectrolytes in water treatment are polydiallyldimethyl ammonium chloride (polyDADMAC) and epichlorohydrin dimethylamine (epiDMA). Both compounds are sold as aqueous solutions, and both are in a class called *quaternary amines.* Each monomer in a quaternary amine contains a nitrogen atom bound to four carbon atoms and carries a positive charge. This positive charge, unlike the charge of an ionizable group, is not affected to a significant extent by a dilute solution's chemistry (especially the pH).

Figure 6.23 shows the results of a jar test experiment in which a silica suspension was treated with a cationic polyelectrolyte (polyDADMAC). Turbidity was maximally removed at a polyelectrolyte-product solution concentration of 10^{-4} g/L (0.1 mg/L). The product solution in this case contained about 20 percent by mass polyelectrolyte. The range for all the types of products sold as solutions is roughly 10 to 60 percent.

The electrophoretic mobility at the dosage of maximum effectiveness (minimum residual turbidity) in Figure 6.23 is approximately zero. Maximum particle removal at the point of zero EM is consistent with a charge neutralization mechanism. Above a product dosage of about 1 mg/L (for this silica particle size and concentration), the adsorption of polyelectrolyte causes the EM to become positive. The high positive

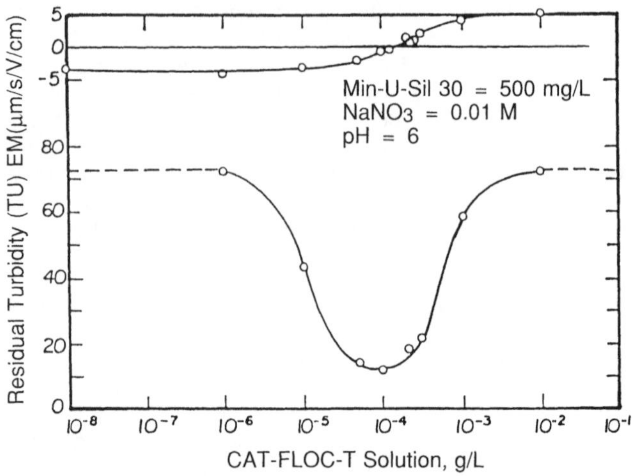

FIGURE 6.23 Effect of polyelectrolyte concentration (PolyDADMAC) on particle EM and residual turbidity after flocculation and sedimentation.

charge on the particles causes restabilization of the suspension, with turbidity removal negligible.

Anionic and Nonionic Polyelectrolytes. Flocculent polymers are generally anionic or nonionic and have relatively high molecular weight, perhaps 10 times or more that of the typical primary coagulant polymer. They are often added after the flocculation process to increase the size and strength of particle aggregates. Because of their high molecular weight and, therefore, appreciable length, the polymers are able to bridge or interconnect particles in the agglomerates.

Low dosages of high-molecular-weight nonionic polymers (0.005 to 0.05 mg/L) are often applied just before granular-bed filtration. This procedure may improve overall effluent quality and reduce the tendency for perturbations, such as an abrupt change in the filtration rate, to have an adverse effect on performance. The ripening period of granular-bed filters has been decreased by adding low dosages of nonionic polymers to all or part of the backwash water (Yapijakis, 1982; Harris, 1972).

Most anionic and nonionic polymers are made with the monomer acrylamide and are known as *polyacrylamide polymers.* Surveys of the water supply industry (AWWA Coagulation and Filtration Committee, 1982; Dentel et al., 1989) indicate that many treatment plants use this type of polymer and, as noted previously, Cleasby et al. (1989) suggest that their use is almost essential to high performance in high-rate filtration. Dentel et al. (1989) have prepared a manual that guides the testing needed to determine a feasible type and cost-effective dosage of flocculent polymers used in water treatment.

Polymeric Coagulants from Natural Organic Compounds. Polymeric coagulants made from natural organic compounds are sometimes used for water treatment. This practice is most common in developing countries where these compounds may be cheaper than hydrolyzing metal salt coagulants. An example is the water extract of the seeds of the tropical shade plant *Moringa oleifera* (Jahn, 1988; Sutherland et al., 1989). The extract contains a dimeric cationic protein with a molecular weight of approximately 13 kilodaltons. Another example is chitin, from crustacean shells.

Activated Silica

The first application of activated silica was described by Baylis (1937). Its use today is limited. The material was prepared by acidifying a sodium silicate solution and it was used during cold winter months in Chicago as a flocculant with alum to improve the efficiency of flocculation and sedimentation. Stumm et al. (1967) and O'Melia (1985) have described the mode of action of activated silica and the factors that affect its performance. Many factors, both kinetic and thermodynamic, can be important.

In preparing activated silica, a concentrated sodium silicate solution (1.5 to 1.78 percent SiO_2) is acidified to make it oversaturated with respect to the precipitation of amorphous silica. Acidification can be carried out with a variety of acids including sulfuric acid, hydrochloric acid, alum, and carbon dioxide. The precipitation begins with polymerization reactions in which monomers condense to form Si-O-Si bonds (Stumm et al., 1967). These polysilicates react further by cross-linking and aggregation to form negatively charged silica sols. Continued reaction can produce a silica gel; however, the reactions are typically terminated or reversed by diluting the acidified solution to near or just out of the insolubility limit. A wide variety of activated silicas containing species ranging from anionic polysilicates to colloidal SiO_2 precipitates can be prepared by varying the concentration and basicity of the initial stock solution, the method and extent of acidification, the aging time, the extent of dilution, and the temperature. These activated silicas, although thermodynamically unstable, can remain active as coagulants for periods of up to a few weeks.

The dosage of activated silica used in treatment is typically expressed as mg SiO_2/L. It has been observed that for a given pH, the optimum dosage is proportional to the alum dosage. Concentrations in the range 7 to 11 percent of the alum dosage are typical. Overdosing and restabilization of the suspension, apparently by coating the floc particles with negative silica polymers, is possible. The dosage and effectiveness tend to vary with the chemical composition of the water. Activated silica can be used in lieu of polyelectrolyte flocculants and filter aids. The trade-off is lower material costs versus a greater need for operator time and attention.

Combinations of Coagulants

In a 1982 national survey of treatment plants (AWWA Coagulation and Filtration Committee, 1982), a majority of respondents stated that they use cationic polymers in combination with hydrolyzable metal coagulants such as alum. The benefits cited included reduced sludge handling and disposal problems, reduced consumption of alkalinity, and improved performance of direct filtration systems.

Results obtained in jar test studies with silica suspensions (Letterman and Sricharoenchaikit, 1982) suggest that under a condition where the positively charged, aluminum hydrolysis products cause destabilization by charge neutralization, a simple trade-off relationship exists between the amounts of Al and cationic polyelectrolyte required to achieve maximum turbidity removal. In experiments in which aluminum hydroxide precipitate was formed and enmeshment of the silica particles occurred, the adsorbed polyelectrolyte became covered by the precipitate, and the stability of the suspension was controlled by the precipitate coating. A number of companies are marketing coagulant solutions that are a mixture of two active ingredients (e.g., alum or prehydrolyzed aluminum salt with a cationic polyelectrolyte).

Hubel and Edzwald (1987) studied the use of combinations of cationic polyelectrolyte and alum for the removal of NOM from low-turbidity, low-alkalinity water. Low-molecular-weight cationic polyelectrolytes used alone were less effective than alum used alone. Without alum the dosage of cationic polyelectrolyte required to achieve charge neutralization and effective removal was proportional to the initial NOM concentration. This dosage was very high for all but the lowest initial NOM concentrations and tended to exceed the health-based product dosage limit.

The use of high-molecular-weight polyelectrolytes as flocculants in water that had already been treated with alum improved turbidity removal and reduced the sludge volume but did not improve NOM removal over what was obtained with alum alone (Hubel and Edzwald, 1987). The use of high-charge-density cationic polyelectrolytes as primary coagulants with alum as a supplemental coagulant achieved high NOM removal and reduced the amount of solids produced by the process.

Ozone and Coagulation

Ozone is gaining increasing acceptance as a primary disinfectant in North American water-treatment practice. In addition to being an effective disinfectant and oxidant, ozone can cause particle destabilization and flocculation. In the new Los Angeles Aqueduct treatment plant, preozonation has been shown to improve filtration efficiency and to reduce the need for conventional coagulants and long flocculation periods (Prendiville, 1986; Monk et al., 1985).

The mechanism(s) by which ozone causes particle destabilization is not well understood. It is possible that the ozone oxidizes metal ions such as Fe(III), which then form insoluble precipitates. Humic materials, both free and sorbed, may be oxidized and made more polar. It has also been suggested that ozone treatment causes the desorption of stabilizing, natural organic material from the surfaces of mineral colloids (Edwards and Benjamin, 1992; Chandrakanth and Amy, 1996; Paralkar and Edzwald, 1996).

THE FLOCCULATION PROCESS

Purpose of Flocculation

The purpose of the flocculation process is to promote the interaction of particles and form aggregates that can be efficiently removed in subsequent separation processes such as sedimentation, flotation, and coarse bed filtration. For efficient flocculation to occur, the suspension must be destabilized. This is usually accomplished by the addition of a coagulant.

Transport Mechanisms

A number of mechanisms can cause relative motion and collisions between particles in a destabilized suspension, including Brownian motion, velocity gradients in laminar flow, unequal settling velocities, and turbulent diffusion. Flocculation rate equations have been derived for each of these mechanisms by assuming that the aggregation process is a second-order rate process in which the rate of collision (N_{ij}) between i- and j-size particles is proportional to the product of the concentrations of the two colliding units (n_i and n_j). The general form of these relationships is:

$$N_{ij} = \alpha_{ij} k_{ij} n_i n_j \tag{6.16}$$

where k_{ij} is a second-order rate constant that depends on the transport mechanism and a number of factors including particle size; α_{ij} is a flocculation rate correction factor that is usually called a *collision efficiency factor* in the water treatment literature. It is equal to the inverse of the stability ratio, W, of the colloid chemistry literature, and its magnitude ranges from 0 to slightly greater than 1.

The α_{ij} term in Eq. 6.16 corrects for the simplifying assumption that transport is not influenced by the short-range forces that affect particle motion when two particles move close together. The important short-range forces are double-layer repulsion, van der Waals attraction, and hydrodynamic retardation. The first two were discussed earlier in the chapter; hydrodynamic retardation is caused by the viscous flow of fluid from between the particles as they start to collide. Hydrodynamic retardation and double-layer repulsion tend to slow particle motion and inhibit collisions, and van der Waals attraction tends to promote them. Equations for predicting the magnitude of α_{ij} are usually from studies in which numerical solutions of the complete equations of motion for interacting particles are compared with solutions obtained with the simplified transport equations that do not consider the short-range forces.

Brownian Diffusion. Small particles suspended in a fluid move about in a random way due to continuous bombardment by the surrounding water molecules. The intensity of this motion is a function of the thermal energy of the fluid, k_BT, where k_B is Boltzmann's constant and T is the absolute temperature. The process is called Brownian diffusion, and the particle interaction it causes is Brownian, or perikinetic, flocculation. The flocculation rate constant (k_{ij} of Eq. 6.16) for perikinetic flocculation in an infinite stagnant fluid, k_p, is given by

$$k_p = \left(\frac{2}{3}\right) \frac{k_BT}{\mu} \frac{(d_1 + d_2)^2}{(d_1\, d_2)} \tag{6.17}$$

According to Equation 6.17, the rate of perikinetic flocculation is proportional to the absolute temperature, and, for particles of equal diameter ($d_1 = d_2$), is independent of the particle size. For transport by Brownian diffusion, the flocculation rate correction factor of Equation 6.16, α_{ij}, is α_p. It accounts for the retarding effect of hydrodynamic forces and double-layer repulsion on particle-particle interaction. The magnitude of α_p can be estimated using expressions derived by Kim and Rajagopolan (1982). These expressions use the following parameters: the height of the potential barrier in the net energy of interaction curve [Ψ_{max} in Figure 6.4(c)], $A/(6k_BT)$, d_1, and d_2. The quantity A is the Hamaker constant.[3]

In systems with significant mixing and convection, Brownian diffusion becomes relatively unimportant in the bulk fluid. However, where small particles contact large floc particles, Brownian diffusion can still control transport of the small particles across a layer of fluid at the floc surface. The rate of transport to the floc surface is a function of the floc size, and the intensity of the turbulent fluid motion as well as the Brownian diffusivity, D, where $D = k_BT/(3\pi\mu d)$; d is the diameter of the diffusing particle.

Transport in Laminar Shear. When particles are suspended in a laminar flow field, a particle located at a point with high fluid velocity tends to move faster than

[3] To use Kim and Rajagopolan's (1982) equations to estimate α_p, the depth of the secondary minimum in the net energy of interaction curve [Figure 6.4(d)] must be low (i.e., less than $0.5k_BT$).

one at a point with low velocity. If the particles are close enough together, their different velocities will eventually cause them to come into contact. This process is called *orthokinetic flocculation,* and its flocculation rate constant, $k_{ij} = k_o$, is given by

$$k_o = \left[\frac{(d_1 + d_2)^3}{6}\right]\left(\frac{du}{dz}\right) \tag{6.18}$$

where du/dz is the magnitude of the velocity gradient.

The flocculation rate correction factor in Equation 6.16 for orthokinetic flocculation ($\alpha_{ij} = \alpha_o$) has been evaluated by van de Ven and Mason (1977). They presented an approximate expression for the factor when double-layer repulsion is negligible ($\Psi_{max} \rightarrow 0$) and van der Waals attraction and hydrodynamic retardation are the controlling short-range forces. The two interacting particles are assumed to have the same diameter, d. The expression is

$$\alpha_o = 0.8\left[\frac{A}{(4.5\,\pi\,\mu\,d^3\,(du/dz))}\right]^{0.18} \tag{6.19}$$

According to Equation 6.19, hydrodynamic retardation reduces the collision rate in orthokinetic encounters even when double-layer repulsion is negligible. The effect is greater for larger particles and higher shear rates.

Differential Settling. Flocculation by differential settling occurs when particles have unequal settling velocities and their alignment in the vertical direction makes them tend to collide when one overtakes the other. The driving force for this mechanism is gravity and the flocculation rate constant, $k_{ij} = k_d$, is given by

$$k_d = \frac{\pi\,g(s-1)}{72\,\nu}\,(d_1 + d_2)^3\,(d_1 - d_2) \tag{6.20}$$

where s is the specific gravity of the particles and ν is the kinematic viscosity. This expression is based on several assumptions; the particles are spherical and have the same density and their settling velocities are predicted by Stokes' law. According to Eq. 6.20, the rate of flocculation by differential settling is maximized when both particles are large and dense and the difference in their sizes is great. The effect of the short-range forces on the rate of flocculation by differential settling has been evaluated by Han and Lawler (1991).

Turbulent Transport. In turbulent flow, the fluctuating motion of the fluid forms eddies that vary in size. The largest eddies are of comparable size to the vessel or impeller. The kinetic energy in large-scale eddies is transferred to smaller and smaller eddies. Eventually, below a certain length scale, the energy is dissipated as heat. An eddy size, known as the Kolmogoroff microscale, separates the inertial size range, where energy is transferred with very little dissipation, from the viscous subrange, where the energy is dissipated as heat.

Within the eddies of turbulent flow are time-varying velocity gradients. These velocity gradients cause relative motion of entrained particles and this relative motion, as in laminar flow, causes flocculation.

The flocculation rate constant for collisions between two neutrally buoyant particles in homogeneous and isotropic turbulent fluid motion, $k_{ij} = k_t$, is given by

$$k_t = \left[\frac{(d_1 + d_2)^3}{6.18}\right]\left(\frac{\varepsilon}{\nu}\right)^{1/2} \tag{6.21}$$

where ε is the local rate of turbulent energy dissipation per unit mass of fluid. Equation 6.21 was derived by Saffman and Turner (1956) using the assumption that the diameter of the particles is less than the Kolmogoroff microscale of the turbulent motion, $\eta = (\nu^3/\varepsilon)^{1/4}$. According to Equation 6.21, the rate at which particles collide by turbulent diffusion increases with particle size and with the rate of energy dissipation in the fluid.

Spielman (1977) has stated that the magnitude of the flocculation rate correction factor for the interaction of equal-size particles in turbulent flow, $\alpha_{ij} = \alpha_t$, can be approximated using Eq. 6.19, the expression derived by van de Ven and Mason (1976) for particle interactions in laminar shear. Spielman substituted the quantity $(\varepsilon/\nu)^{1/2}$ for the laminar velocity gradient (du/dz) and obtained

$$\alpha_t = 0.8 \left\{ \frac{A}{[4.5\,\pi\,\mu\,d^3\,(\varepsilon/\nu)^{1/2}]} \right\}^{0.18} \tag{6.22}$$

This expression suggests that α_t decreases with increasing particle diameter and increasing mixing intensity.

Delichatsios and Probstein (1975) derived a kinetic model for flocculation in isotropic turbulent flow for two conditions. In the first the radius of the collision sphere, $(d_1 + d_2)/2$, is smaller than the Kolmogoroff microscale and in the second it is greater. For the first condition, the equation for the flocculation rate constant is similar in form to Eq. 6.21 with a constant of 1/19.6 instead of 1/6.18. For the second condition, $(d_1 + d_2)/2$ larger than the microscale, the expression is

$$k_{\mathrm{t,d-p}} = (0.427)\,(d_1 + d_2)^{7/3}\,\varepsilon^{1/3} \tag{6.23}$$

This equation suggests that for larger floc particles, the rate of particle interaction is proportional to ε to the 1/3 power instead of the 1/2 power of Eq. 6.21. Also, in contrast to Eq. 6.21, the rate of interaction is independent of the viscosity of the fluid.

G Value Concept. In the 1940s, Camp and Stein (1943) used Smoluchowski's (1917) equation for flocculation in uniform laminar shear to derive a widely used flocculation rate equation for turbulent flow. They calculated that for turbulent fluid motion a root-mean-square (rms) velocity gradient can be used in place of the laminar velocity gradient in the laminar shear flocculation rate equation (Eq. 6.18). Their rms velocity gradient, or *G value,* is given by

$$G = \left(\frac{P}{V\mu} \right)^{1/2} \tag{6.24}$$

where P is the power input to the fluid (through, for example, the blades of a rotating impeller), V is the volume of water in the vessel, and μ is the absolute viscosity of the water. The flocculation rate constant in Camp and Stein's modified Smoluchowski equation for orthokinetic flocculation is

$$k_{\mathrm{o,c-s}} = \frac{(d_1 + d_2)^3}{6}\,G \tag{6.25}$$

Equation 6.25 becomes very similar to the more rigorous expression (Eq. 6.21) derived by Saffman and Turner (1956) if it is assumed that the G value is equal to $(\varepsilon/\nu)^{1/2}$ where ε, as noted above, is the local rate of energy dissipation per unit mass of fluid.

The limitations of the G value concept have been discussed by a number of authors including Cleasby (1984), McConnachie (1991), Hanson and Cleasby (1990), and Han and Lawler (1992). Clark (1985) has argued that inappropriate assumptions were made in the derivation of the Camp and Stein G value and, therefore, it is not a valid representation of the average velocity gradient in a turbulent fluid. The theoretical validity of the G value of Eq. 6.24 as a design parameter and reactor scale-up tool is questionable because the rate of flocculation is affected not only by the intensity of the fluid motion but also by how the kinetic energy of the fluid is distributed over the size of the eddies in the turbulent flow field. For a given mixing intensity (either overall power input to the fluid or local rate of energy dissipation), the distribution of energy is determined by the size and geometry of the impeller and vessel system.

Use of the G value concept in flocculation is affected by three other considerations. First, the local rate of energy dissipation in a mechanically mixed vessel varies widely with location in the vessel, from several hundred times the overall energy input rate per unit mass of fluid ($P/V\rho$) near the rotating impeller, to a fraction of this overall value away from the impeller (Schwartzberg and Treybal, 1968; McConnachie, 1991). Consequently, the local rate of particle aggregation (and possibly the local rate of floc breakup) will tend to vary widely with location in the vessel, increasing the tendency for floc structure and density to change through breakup and reformation as the flocs are moved around the vessel by the fluid (Clark and Flora, 1991).

Second, some of the energy supplied to the rotating impeller shaft is dissipated directly as heat at the surfaces of the impeller and vessel walls and does not produce turbulent fluid motion. The weighted average (by fluid mass) of ε for all elements of fluid in the vessel will tend to be less than $P/(V\rho)$, the overall energy input per unit mass of fluid in the vessel; the efficiency with which the motion of the impeller increases the kinetic energy of the turbulent flow will vary with the geometry of the mixing system.

Third, according to theoretical calculations by Han and Lawler (1992), fluid shear, and therefore the G value, are controlling factors in flocculation kinetics only when the interacting particles are larger than 1 μm and about the same size. When one particle is large and the other is small (<1 μm), Brownian diffusion tends to control the relative motion between the particles and, although mixing intensity is still important, the form of the relationship between flocculation rate and the overall energy dissipation rate in the fluid is likely to be different than what is predicted by Eq. 6.21.

Our understanding of turbulent flow in mechanically agitated vessels is incomplete but growing rapidly. Unfortunately, sensitive methods for measuring flocculator performance, especially the performance of full-scale units, are not available. Until this capability improves, it seems reasonable to use the Camp and Stein G value for flocculator design and scale-up.

Flocculation Kinetics in Batch and Plug Flow Reactors

Flocculation in a well-mixed batch reactor is described by the following expression for the rate of change in the concentration of k-size particles. The expression assumes that k-size particles are formed by the collision of i- and j-size particles (i.e., $k = i + j$).

$$\frac{dn_k}{dt} = \frac{1}{2} \sum_{\substack{i+j \to k \\ i=1}}^{i=k-1} \alpha_{ij} k_{ij} n_i n_j - n_k \sum_{k=1}^{\infty} \alpha_{ik} k_{ik} n_i \tag{6.26}$$

The term on the left, dn_k/dt, is the rate of change in the concentration of k-size particles in the reactor. The first term on the right is for the rate of formation of k-size particles by the collision of i- and j-size particles and the second term on the right accounts for the loss of k-size particles by their collision with any other particle. This expression only applies to irreversible aggregation; it does not account for the breakup or erosion of aggregates. For continuous-particle size distributions, an integral form of this expression can be written (Elimelech et al., 1995).

The use of Eq. 6.26 is difficult. One of the most significant problems is assigning values to the rate constants, k_{ij} and k_{ik}, and the flocculation rate correction factors, α_{ij} and α_{ik}. These depend on the nature of the particles and on the transport processes that cause the particle interaction (Elimelech et al., 1995). Relationships must be established for how aggregate shape and density vary with size. The simplest assumption is that the particles coalesce like oil droplets, but this is obviously very approximate for the interaction of solid particles.

Most flocculators are mechanically mixed, continuous-flow reactors, and the fluid motion is turbulent; particle interaction through laminar velocity gradients and unequal settling velocities tend to be insignificant. The next section describes relationships that predict the performance of completely mixed, continuous-flow flocculation reactors in which turbulent fluid motion is the predominant particle transport mechanism and where simplifying assumptions have been made about the size distributions of the particles.

Ideal Continuous-Flow Flocculation Reactors

Argaman and Kaufman (1970) have shown that the performance of a continuous-flow flocculation reactor can be predicted using relationships based on the completely mixed, tanks-in-series model (Levenspiel, 1972). It was assumed that each completely mixed (CFSTR) compartment in the reactor contains a bimodal particle size distribution, where the smaller particles in the distribution are from the influent, and are called primary particles, and the larger particles are the aggregates or flocs. The flocs grow by the attachment of primary particles. Because each compartment is assumed to be completely mixed, at steady state the effluent characteristics (a mixture of flocs and primary particles) are the same as the vessel contents (see Figure 6.24).

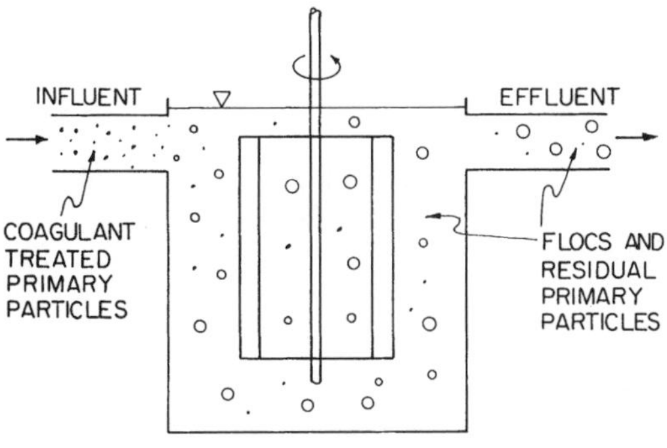

FIGURE 6.24 Schematic diagram of a compartment in an ideal completely mixed flocculation reactor containing a bimodal particle size distribution.

The following equations are based on a simplified version of Argaman and Kaufman's (1970) model in which floc disruption is ignored. The rate at which primary particles contact the larger floc particles in one of the completely mixed compartments is calculated using Eqs. 6.16 and 6.21. It is assumed that the flocs have a spherical shape and are much larger than the primary particles (i.e., $d_2 >> d_1$). By substituting $\Phi = n_2 \pi d_2^3/6$, where n_2 and d_2 are the floc number concentration and diameter, and Φ is the floc volume fraction (volume of floc per volume of suspension), Eqs. 6.16 and 6.21 yield

$$N_{12} = \alpha_{12} \frac{0.97}{\pi} \Phi \left(\frac{\varepsilon}{\nu}\right)^{1/2} n_1 \tag{6.27}$$

Equation 6.27 is a first-order expression in terms of n_1. A mass balance on primary particles in the completely mixed reactor compartment can be combined with this expression and the assumption that $\alpha_{12} = 1$ to result in

$$\frac{n_1}{n_{10}} = \frac{1}{1 + k\bar{t}} \tag{6.28}$$

where the flocculation rate constant, k, is

$$k = \left(\frac{0.97}{\pi}\right) \Phi\, G \tag{6.29}$$

$\bar{t}$ is the mean residence time of the fluid in the reactor compartment, and n_1 and n_{10} are the effluent and influent primary particle number concentrations, respectively. It is assumed that ε is equal to $P/(V\rho_w)$, the overall energy input rate per unit mass of fluid; V is the volume of water in the reactor; and ρ_w is the density of water.

For destabilization by charge neutralization, the floc volume concentration in the reactor is determined by the volume of primary particles that has been incorporated in the flocs and by the density of the flocs:

$$\Phi = V_1 n_{10} \left(1 - \frac{n_1}{n_{10}}\right) \frac{\rho_p - \rho_w}{(\rho_f - \rho_w)} \tag{6.30}$$

where V_1 is the volume of a primary particle, and ρ_f and ρ_p are the floc and primary particle densities, respectively. The quantity $V_1 n_{10}$ is equal to C_0/ρ_p, where C_0 is the mass concentration of primary particles in the influent to the first compartment. According to Eqs. 6.29 through 6.31, flocculator performance (n_1/n_{10}) is a function of n_{10}, ρ_f, ρ_p, the dimensionless product $G \times \bar{t}$, and the amount of primary particles that attach to the flocs in the reactor. These equations are approximate due to the simplifying assumptions that have been made; however, they illustrate the importance of mixing intensity, residence time, and influent particle concentration in affecting the performance of continuous-flow flocculation reactors.

It is frequently noted that raw water with a relatively high concentration of particulate matter is "easier to treat" than water with a low concentration. There are also proprietary treatment systems that use a buoyant coarse media in the flocculator (Schulz et al., 1994) or the addition of supplemental suspensions such as silica particles to enhance the rate of flocculation for what would otherwise be dilute and slowly flocculating suspensions. One of the reasons for this observation and the efficacy of the proprietary systems is that, for a given value of $G \times \bar{t}$, as the influent particle concentration increases, the floc volume concentration increases and, consequently, the aggregation rate and the fractional conversion of primary particles to flocs become greater.

TABLE 6.5 Equations Used to Calculate the Performance of a Compartmentalized, Continuous-Flow Flocculation Reactor*

$\frac{n_{1m}}{n_{10}} = \prod_{i=1}^{m} \frac{1}{1 + k_j \bar{t}_j}$	(6.31)
$k_i = \left(\frac{0.97}{\pi}\right) \Phi_i \, G_i$	(6.32)
$G_i = (\varepsilon/\nu)_i^{1/2}$	(6.33)
$\Phi_i = \frac{C_0}{\rho_p}\left(1 - \frac{n_{1i}}{n_{10}}\right) \frac{\rho_p - \rho_w}{(\rho_f - \rho_w)}$	(6.34)

* In these equations, the subscript i refers to the ith compartment in a series of m compartments. n_{10} is the number concentration of particles in the influent to the first compartment, and n_{1m} is the concentration in the effluent of the mth compartment. C_0 is the mass concentration of primary particles in the influent to the first compartment. The overall mean fluid residence time for the reactor is $T = \sum \bar{t}_i$. It is assumed that α for primary particle-floc interaction is equal to 1.

The typical continuous-flow flocculator has multiple compartments and the mean fluid detention time and G value may be different for each compartment. As primary particles contact the flocs in each compartment, the floc volume concentration and the flocculation rate constant increase. The equations listed in Table 6.5 are based on Eqs. 6.28 through 6.30 and were used to calculate values of the quantity n_{1m}/n_{10} as a function of the dimensionless product $G \times T$ for three values of the influent particle mass concentration. (n_{10} is the number concentration of primary particles in the influent to the first compartment, n_{1m} is the number concentration in the effluent of the mth compartment, and $T = \sum \bar{t}_i$.) For this example it was assumed that the flocculator has three equal volume compartments ($m = 3$), each with the same value of $\bar{t}_i$ and the same mixing intensity ($G_i = G$); the primary particle and floc densities are 2.5 and 1.02 g/cm^3, respectively. The results are plotted in Figure 6.25.

According to Figure 6.25, flocculator performance increases with increasing $G \times T$ and increasing influent particle concentration.[4] For $C_o = 5$ mg/L, as $G \times T$ increases, the effect of flocculation on the primary particle concentration in the flocculator effluent is insignificant until $G \times T$ is approximately 50,000. From $G \times T = 50{,}000$ to 200,000, the quantity n_{1m}/n_{10} decreases from slightly less than 1 to approximately 0.02. For $C_o = 5$ mg/L, $n_{1m}/n_{10} = 0.5$ corresponds to $G \times T = 72{,}000$. For $C_o = 100$ mg/L, significant primary particle removal begins at $G \times T = 2{,}800$; and $n_{1m}/n_{10} = 0.5$ corresponds to $G \times T = 3{,}600$. When C_o is equal to 50 mg/L, $n_{1m}/n_{10} = 0.5$ corresponds to $G \times T = 7{,}200$. This general relationship between the influent primary particle concentration and the primary particle removal efficiency exists because the flocculation rate constant for a given flocculator compartment (k_i) depends on the magnitude of the floc volume concentration in that compartment (Φ_i), and Φ_i depends on the pri-

[4] In *Water Treatment Plant Design* (AWWA, 1998), it is stated that the range of $G \times T$ values used in modern systems is from 24,000 to 84,000, and the typical value of T is 20 min.

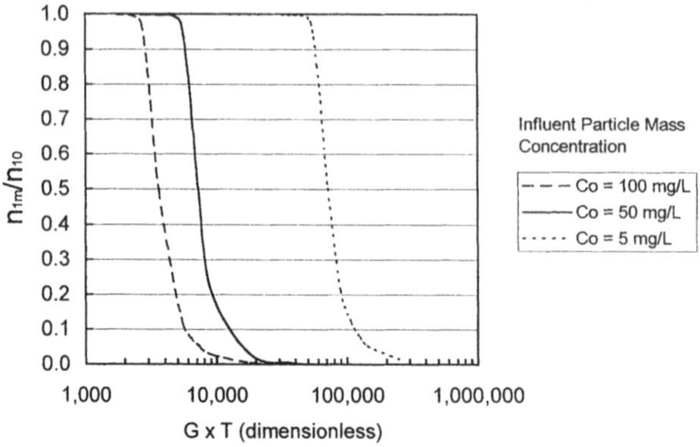

FIGURE 6.25 Effect of the dimensionless parameter $G \times T$ on the performance of a flocculation reactor with three equal compartments for several influent primary particle concentrations. The primary particle and floc density are 2.5 and 1.02 g/cm^3 and the G value in each compartment is 25 s^{-1}.

mary particle concentration in the flocculator influent (see Eqs. 5.2 and 5.4, Table 6.5). This dependency has been observed in practice (Hudson, 1981).

The equations in Table 6.5 can be used to show that the relationship between flocculator performance (n_{1m}/n_{10}) and $G \times T$ depends on the number of equal-volume compartments (m) in the flocculation reactor. For example, for $C_o = 100$ mg/L and $m = 3$, $n_{1m}/n_{10} = 0.5$ is obtained at $G \times T = 3600$. For a single compartment ($m = 1$), this value of $G \times T$ decreases to 2200, and for $m = 5$, it increases to 5300. Similar trends are calculated at other values of C_o. In general, a flocculation reactor with a single, CFSTR compartment requires the lowest value of $G \times T$ to reach a given level of performance. This is because the performance of each compartment depends on the product $G \times T/m \times \Phi_i$ (see Eqs. 6.31, 6.32, and 6.34 of Table 6.5). In the initial compartments of a multiple-compartment flocculator, the magnitude of Φ_i is lower than in a single-compartment reactor, and, therefore, a larger value of $G \times T/m$ is required to achieve a given level of overall performance.

The curves in Figure 6.26 were plotted using the equations in Table 6.5 to illustrate the effect of flocculator compartmentalization and the influent primary particle concentration on flocculator performance for a given value of $G \times T$. The primary particle and floc density were 2.5 and 1.02 g/cm^3, and the G value in each compartment was 25 s^{-1}. The product $G \times T$ was constant and equal to 90,000. For the lowest influent primary particle concentration, $C_o = 10$ mg/L, $m = 5$ yields maximum performance. The fraction of primary particles remaining is approximately 0.12 when $m = 1$, and 0.017 when $m = 5$. For $C_o = 20$ and 100 mg/L and $m < 10$, flocculator performance increases with increasing compartmentalization. For a given number of compartments, the fraction of primary particles remaining in the flocculator effluent decreases with increasing influent primary particle concentration.

Floc Disaggregation. Floc disaggregation can apparently influence flocculator performance, especially when the mixing intensity is high ($G > 100$ s^{-1}) and the flocs have grown to a significant size. Investigators have observed with batch reactors, using a high intensity and relatively long duration of mixing, that the flocs tend to approach a constant size distribution. One explanation for this behavior is that with

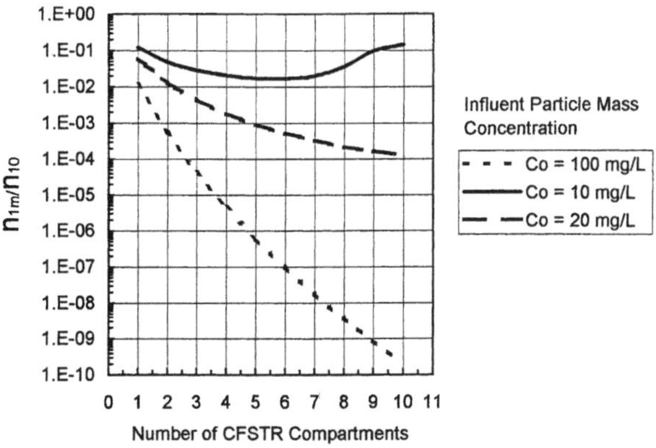

FIGURE 6.26 Effect of influent primary particle concentration and the number of completely mixed compartments on flocculator performance, $G \times T = 90{,}000$. T is the overall mean fluid detention time. The primary particle and floc density are 2.5 and 1.02 g/cm^3, and the G value in each compartment is 25 s^{-1}.

time, the surfaces of the flocs and primary particles change (chemically and physically), and the floc suspension becomes restabilized. In another explanation, it is assumed that a dynamic equilibrium is eventually reached between the rates of particle aggregation and disaggregation.

According to a number of investigations conducted in the 1970s (Spielman, 1978; Argaman and Kaufman, 1970; Parker et al., 1972), the principal mechanisms of disaggregation or floc breakup are:

1. Surface erosion of primary particles from the floc
2. Fracture of the floc to form smaller, daughter aggregates

Argaman and Kaufman (1970) considered both particle aggregation and floc erosion in deriving an expression similar to Eq. 6.21 for predicting the performance of a single-compartment, completely mixed, continuous-flow flocculator. The following first-order rate expression was used to describe the disappearance and formation of primary particles by aggregation and erosion mechanisms:

$$r = -k_a GC + k_b G^2 \tag{6.35}$$

where C is the mass concentration of primary particles, k_a is the agglomeration rate constant, and k_b is the floc breakup (disaggregation) coefficient. For a single-compartment, completely mixed flocculator reactor, Eq. 6.35 yields the following flocculator performance equation:

$$\frac{n_{11}}{n_{10}} = \frac{1 + k_b\, G^2\, \bar{t}}{1 + k_a\, G\, \bar{t}} \tag{6.36}$$

The effect of including the erosion term in the rate expression is illustrated by Figure 6.27, where flocculator performance curves are shown for $k_a = 5 \times 10^{-5}$ and $k_b = 1.0 \times 10^{-7}$ s and $k_b = 0$ s. With $k_b = 0$ s, the plotted curve is essentially the same as that obtained using Equations 6.28, 6.29 and 6.30. According to Figure 6.27, including floc

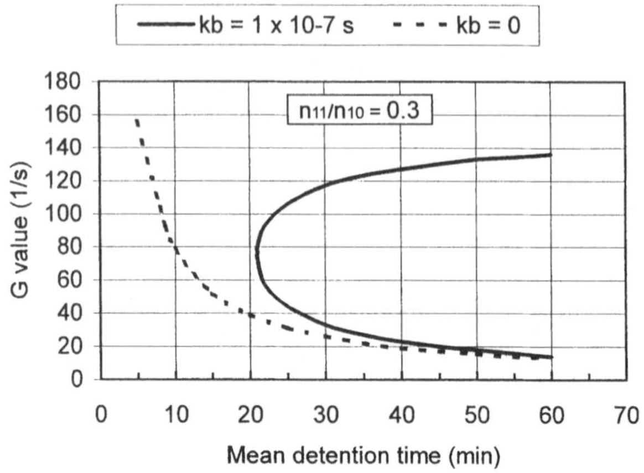

FIGURE 6.27 Flocculator performance curves illustrating the significance of the floc disaggregation term in Eq. 6.36 (G versus T for $n_{11}/n_{10} = 0.3$; $k_b = 1.0 \times 10^{-7}$ s; and $k_b = 0$ s).

disaggregation in the continuous-flow flocculator performance equation yields an optimum mixing intensity, a result that differs from that obtained using Eq. 6.28. Also, when floc erosion is included, constant $G \times T$ does not yield constant performance.

Equation 6.35 is appropriate only for completely mixed reactors, where each compartment is populated by a floc suspension of constant size distribution. It is not appropriate to use it (or Eq. 6.27) to derive an expression for flocculation in a plug flow reactor where the floc size distribution is not constant with time.

Floc Size and Density. Experiments have shown that the floc size distribution is a function of the intensity of the turbulence and the structural characteristics of the flocs. A number of investigators (Parker et al., 1972; Mühle and Domasch, 1991; Tambo and François, 1991) have suggested the following simple relationship between a characteristic floc size, d_f (e.g., the "maximum" diameter), and G for a steady-state condition in which the rate of disaggregation equals the aggregation rate, where the exponent b is a positive integer. According to this expression, the steady-state floc size decreases as the mixing intensity increases:

$$d_f = \frac{P}{G^b} \tag{6.37}$$

The coefficient P in Equation 6.37 has been related to the "strength" of the floc (Parker et al., 1972; Tambo and François, 1991). According to Parker et al., the magnitude of the exponent b depends on the size of the flocs relative to the characteristic size of the turbulent eddies that cause the erosion of primary particles or floc breakage. For flocs that are large relative to the Kolmogoroff microscale of the turbulence, $b = 2$, and for flocs smaller than this scale, $b = 1$. For the floc breakage mechanism, in both turbulence scale regimes, $b = 0.5$. Some experimental investigations (Argaman and Kaufman, 1970; Lagvankar and Gemmell, 1968) have reported the value of the exponent b to be 1.0 (i.e., the characteristic size of the flocs is inversely proportional to G). Expressions similar in form to Eq. 6.37 have been derived using theoretical approaches for floc disruption in turbulent flow (Tambo and François, 1991; Mühle, 1993). Tambo and François (1991) summarize the literature on floc size and breakup mechanisms. The removal of flocs in most separation processes, such as sedimentation, is determined by floc density and floc density is related to floc size and composition.

Using a computer simulation of the aggregation process, Vold (1963) determined the following inverse relationship between floc buoyant density ($\rho_f - \rho_w$) and diameter, d_f (of a circle with equal projected area):

$$\rho_f - \rho_w = B\, d_f^{-0.7} \tag{6.38}$$

This relationship, which shows that the floc density tends to decrease as the floc size increases, is in general agreement with experimental results (Lagvankar and Gemmell, 1968). Equations 6.28 and 6.29 considered together suggest that as the mixing intensity used in the flocculator is increased, the steady-state, characteristic floc size will decrease (Eq. 6.37), and the floc buoyant density (Equation 6.38) will increase. Tambo and François (1991) have reviewed studies on the relationship between floc size, structure, and density.

According to Eqs. 6.28 through 6.30, increasing the floc bouyant density tends to have an adverse affect on flocculator performance. For example, for a primary particle density of 1.5 g/cm^3 and with the floc buoyant density increasing from 2×10^{-3} to 1×10^{-2} g/cm^3, the fraction of primary particles remaining in the continuous-flow flocculator effluent increases from 1.6×10^{-5} to 2×10^{-3}. This result is obtained because as the buoyant density increases, Φ, and the flocculation rate constant, k in Eq. 6.28, decrease. It suggests that the beneficial effect of increasing the mixing intensity on flocculator performance may be offset to some extent by the tendency for higher mixing intensities to produce smaller (see Eq. 6.37), less voluminous (i.e., higher-density) flocs (see Eq. 6.38).

Fractal geometry has been used to characterize the physical characteristics of particle aggregates (Meakin, 1988). As a fractal object, the mass of an aggregate (M) is related to its size (for example, its diameter R) by

$$M = R^F \tag{6.39}$$

where the exponent F is called the fractal dimension (or mass fractal dimension). If a particle (e.g., a spherical oil droplet) is formed by coalescence, then F is equal to 3. In general, when aggregates of irregular shape are formed from solid primary particles, the magnitude of F can be considerably less than 3. Its magnitude decreases as the aggregate structure becomes more open and irregular. The exponent in the relationship between buoyant density and aggregate size (e.g., Eq. 6.38) is related to the fractal dimension by the equation $F = 3 +$ exponent. The value of F for Eq. 6.39 is $3 - 0.7 = 2.3$. Fractal dimensions in the range 1.6 to 2 are typical for simulation results and model systems (Elimelech et al., 1995).

Floc Density—Significance of Enmeshment by Hydroxide Precipitates. When alum or a ferric iron salt is used at a concentration in the enmeshment range, the metal hydroxide precipitate can significantly influence the floc volume concentration and buoyant density as well as flocculator performance. This effect is especially significant at low influent primary particle concentrations. Letterman and Iyer (1985) determined that, for alum and an enmeshment condition, Φ is given by

$$\Phi = \frac{S}{\rho_p} + 120\,[Al_p] \tag{6.40}$$

where S is the mass concentration of primary particles of density ρ_p that are enmeshed in the precipitate, and $[Al_p]$ is the molar concentration of Al in the precipitate. The coefficient, 120 L/mole Al, was determined by measuring the buoyant density of flocs as a function of the ratio of Al precipitate to primary particle mass.

For aluminum salt coagulants, the buoyant density of the flocs is determined by the ratio of the primary particle mass concentration and the alum dosage. An empirical relationship, derived by Letterman and Iyer (1981), using floc density measurements by Lagvankar and Gemmell (1968) is:

$$\rho_f - \rho_w = 4.5 \times 10^{-3}\left[1 + \frac{S}{Q}\right]^{0.8} \tag{6.41}$$

where Q is the alum ($Al_2(SO_4)_3{\cdot}18\ H_2O$) dosage. Both S and Q are in mg/L and $\rho_f - \rho_w$ is in g/cm^3. When $S/Q << 1$, $\rho_f - \rho_w$ is a minimum and approximately equal to 4.5×10^{-3} g/cm^3, the buoyant density of the aluminum hydroxide precipitate. For $S/Q >> 1$, the flocs' buoyant density is roughly proportional to S/Q, the ratio of floc-incorporated primary particles and amount of metal hydroxide precipitate. Because the primary particles in Lagvankar and Gemmell's experiments were kaolin clay platelets, Eq. 6.41 pertains only when the primary particles are of this type. There is evidence that the chemical makeup of the solution affects the structure and density of the metal hydroxide precipitate and this factor must also be considered.

According to Equation 6.41, the amount of higher-density particulate material that is enmeshed in the lower-density precipitate matrix has a significant effect on the density of the flocs. Decreasing the amount of coagulant relative to the concentration of particulate matter (increasing S/Q) increases the density of the flocs. However, as indicated by Eq. 6.40, reducing the amount of coagulant also reduces the floc volume concentration and, according to Eqs. 6.28 to 6.30, this general reduction can decrease the performance of the flocculator. Of course, as the coagulant concentration becomes less than the amount needed for suspension destabilization, the tendency for particles to aggregate and form flocs will become negligible.

The effect of metal hydroxide precipitate on the performance of a flocculation reactor is especially significant when the influent particle concentration is low. The following example assumes that the flocculator is a single compartment. Equations 6.28 through 6.30 were used with $C_o = S = 5$ mg/L, $G \times \bar{t} = 1 \times 10^5$, $\rho_p = 2.5$ g/cm^3, and $Q = 20$ mg/L. For aggregation in the absence of precipitate ($Q = 0$), it was assumed that the flocs have a density of 1.01 g/cm^3, a reasonable density for particles destabilized with a polyelectrolyte, and Eq. 6.30 was used in place of Equation 6.40 to calculate Φ.

For a low influent particle concentration ($C_o = S = 5$ mg/L), the calculated Φ (Eq. 6.40) for flocs formed in the presence of aluminum hydroxide precipitate is 7.2×10^{-3} L/L and significantly less, 3×10^{-4} L/L, for flocs formed without a precipitate matrix (Eq. 6.30). Consequently, the fraction of primary particles remaining in the flocculator effluent is 0.005 with Al and significantly higher, 0.10, without Al.

When C_o is increased to 100 mg/L and all other parameters are held constant, the fraction of primary particles remaining is approximately 0.005 for both conditions. The higher influent primary particle concentration yields a floc volume concentration (6×10^{-3} L/L), which does not require augmentation with precipitate to obtain a high level of primary particle aggregation in a simple, single-compartment flocculator.

Rapid Mixing

Rapid, or flash, mixing, is a high-intensity mixing step used before the flocculation process to disperse the coagulant(s) and to initiate the particle aggregation process.

In the case of hydrolyzing metal salts, the primary purpose of the rapid mix is to quickly disperse the salt so that contact between the simpler hydrolysis products and the particles occurs before the metal hydroxide precipitate has formed. Rapid dispersal before precipitation helps ensure that the coagulant is distributed uniformly among the particles. This process is poorly understood, but probably depends on factors such as the concentration of salt in the coagulant feed solution, the coagulant dosage, the concentration and size distribution of the particulate matter, the temperature and ionic constituents of the solution, and the turbulent flow conditions (overall energy input and the flow and kinetic energy spectrum of the turbulent motion) in the rapid mixing device. Amirtharajah and O'Melia (1990) have reviewed how some of these factors might affect rapid-mix unit performance. The AWWA Research Foundation publishes another useful book, *Mixing in Coagulation and Flocculation* (AWWA Research Foundation, 1991).

The significance of the rapid-mix step when polyelectrolyte coagulants are used is probably similar to that for hydrolyzing metals except that the reaction between the coagulant and the solution is not as important. Polyelectrolytes rapidly and irreversibly adsorb on the particulate surfaces. Therefore, in the absence of intense mixing at the point of coagulant addition, it is logical to assume that some particles might adsorb more polymer than others. If the overdosed particles became surrounded by other particles that have little or no adsorbed polymer, it is possible that the aggregation process would slow or stop before sufficiently large flocs were formed.

Rapid mixing is also the start of the flocculation process. By adding coagulant, the particles become destabilized and the high-intensity mixing leads to rapid aggregation. Particle disaggregation may become important as the aggregates grow. Evidence (AWWA Research Foundation, 1991) suggests that a steady-state size distribution of relatively small aggregates may characterize the suspension leaving the rapid-mix process. Furthermore, there is limited evidence that mixing at high intensity for too long can be detrimental to subsequent process performance, possibly because aggregates that are eroded or broken have a reduced tendency to reform with time because of changes in surfaces' chemical or physical properties.

Effect of Temperature on Coagulation and Flocculation

The temperature of the water can have a significant effect on coagulation and flocculation (Hanson and Cleasby, 1990; Kang and Cleasby, 1995). In general, the rate of floc formation and efficiency of primary particle removal decrease as the temperature decreases. The negative effect of temperature tends to be greatest with dilute suspensions. Temperature affects the solubility of the metal hydroxide precipitate and the rate of formation of the metal hydrolysis products. In general, $Fe(OH)_3$ and $Al(OH)_3$ decrease in solubility with decreasing temperature and the pH of minimum solubility (see Figures 6.9 and 6.10) increases slightly. The rate of hydrolysis and metal hydroxide precipitation and the rate of hydrolysis product dissolution or reequilibration (as in the disappearance of a polynuclear species when the product solution is diluted in the water to be treated) decrease with decreasing temperature. At lower temperatures, polynuclear species will tend to persist for a longer period of time. Temperature also alters the distribution of kinetic energy over the scale of fluid motion in a turbulent flow field. Kang and Cleasby (1990) concluded that chemical factors are more significant than fluid motion effects. Their use of constant *p*OH instead of constant pH yielded some improvement in performance at low temperature but not to the level of primary particle removal observed at room temperature.

Morris and Knocke (1984), on the other hand, have presented evidence that physical factors rather than chemical kinetics are behind the effect of temperature on coagulation and flocculation performance. In their work, they observed a significant effect of temperature on the size distribution of aluminum hydroxide floc.

Electrokinetic Measurements

Electrokinetic measurements are used to monitor the effect of coagulants and changes in solution chemistry on particle surface chemistry and the stability of particle and precipitate suspensions. Two types of electrokinetic measurements, electrophoretic mobility (zeta-potential) and streaming current, are used in water treatment practice to control the addition of coagulants and monitor the conditions that affect coagulant performance.

Electrophoretic Mobility Measurements. The engineering literature contains numerous references to the use of electrophoretic mobility (EM), or zeta-potential (ZP), measurements as a coagulation process control technique. Many of these papers describe attempts to correlate ranges of EM values or changes in the EM, such as sign reversal, with the efficiency of particle removal by flocculation followed by sedimentation and filtration.

Electrophoresis is an electrokinetic effect and is, therefore, explained by the same fundamental principles as other electrokinetic phenomena such as streaming current (or streaming potential), sedimentation potential, and electroosmosis. In electrophoresis, particles suspended in a liquid are induced to move by the application of an electric field across the system. This technique has been used by colloid chemists for many years to determine the net electric charge or near-surface (zeta) potential of particles with respect to the bulk of the suspending phase (Hunter, 1981; Anonymous, 1992).

The use of EM measurements as a method to control the application of coagulant chemicals in solid-liquid separation systems has been known for many years. However, although there have been convincing advocates for the use of EM measurements for coagulation process control, inconsistent and difficult-to-interpret results and the time-consuming nature of the EM determination appear to have limited the widespread use of EM measurements in routine treatment plant operation.

A number of different techniques are used to determine particle EM. The most important of these in water treatment applications has been microelectrophoresis. In microelectrophoresis, the suspension is contained in a small-diameter glass or plastic tube that has, in most cases, a round or rectangular cross-section. An electric field is applied across the contents of the tube in the axial direction using a stable, constant-voltage power supply and inert (e.g., platinized platinum) electrodes inserted in sealed-fluid reservoirs at the end of the tube. When the voltage is applied, the particles tend to migrate in the axial direction. The EM is determined by measuring the average velocity of particle migration and dividing this by the voltage gradient across the cell. Unless special procedures are used, the measured velocity must be corrected for the particle movement caused by electroosmotic flow in the electrophoresis cell. The voltage gradient is determined by dividing the applied voltage by the effective length of the electrophoresis cell.

Streaming Current Measurements. The original streaming current detector (SCD) was patented by Gerdes (1966) in 1965 and variants of this device have been used in water treatment plants for coagulant dosage control for a number of years. The electronic output from an SCD has been shown by Dentel et al. (1988) to be proportional to the net charge on the surfaces of the particles that have been treated with the coagulant chemical.

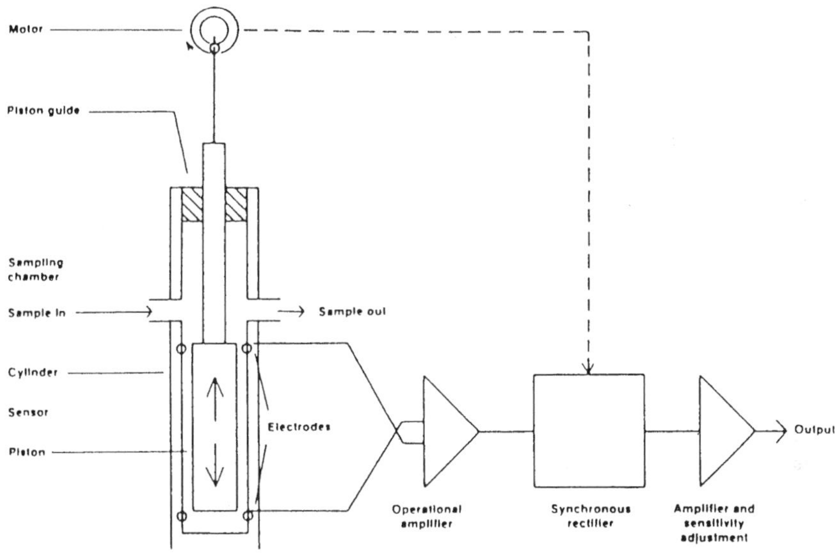

FIGURE 6.28 Cross-section of a widely used type of streaming current detector.

The cross-section of a widely used type of SCD is shown in Figure 6.28. The purpose of the sample chamber is to contain the coagulant-treated suspension. Within the chamber is a reciprocating piston that moves vertically inside a cylinder. The movement of the piston causes water to flow inside the annular space between the cylinder and piston and to be pumped through the sample chamber. The SC is detected by electrodes attached inside the cylinder.

The streaming current is determined by the charged particles that attach to the walls of the stationary cylinder and the moving piston. As water flows back and forth cyclically in the annulus, the ions in the electrical double layer next to the charged particles are transported with the flow. This displacement of electrical charge by the movement of the fluid past the stationary particles creates the sinusoidal current that tends to flow between the electrodes that are in contact with the solution within the cylinder. The magnitude of the current depends on the amount of charge on the attached particles. The current is amplified, rectified, and time-smoothed by circuitry in the instrument. The processed signal is called the streaming current.

Interpreting EM and SC Measurements. The rate of particle interaction for orthokinetic flocculation is determined (approximately) by the product of the collision efficiency factor α and the volume concentration of the suspension Φ (see Eq. 6.27). Theoretical relationships show that α is related to the EM (or SC) of the particles in the suspension (Anonymous, 1992). As the EM or SC approaches zero, α should approach a maximum value that is equal to or slightly greater than one. As discussed previously, the magnitude of Φ depends on the volume concentration of the particles and precipitate in the suspension.

Electrophoretic mobility and SC measurements effectively predict the rate of flocculation when Φ is relatively constant and the rate is influenced only on the magnitude of α (see Figures 6.16, 6.17, and 6.23). When the addition of a coagulant, typically an HMS coagulant, affects both α and Φ, the interpretation of EM and SC measurements can be difficult.

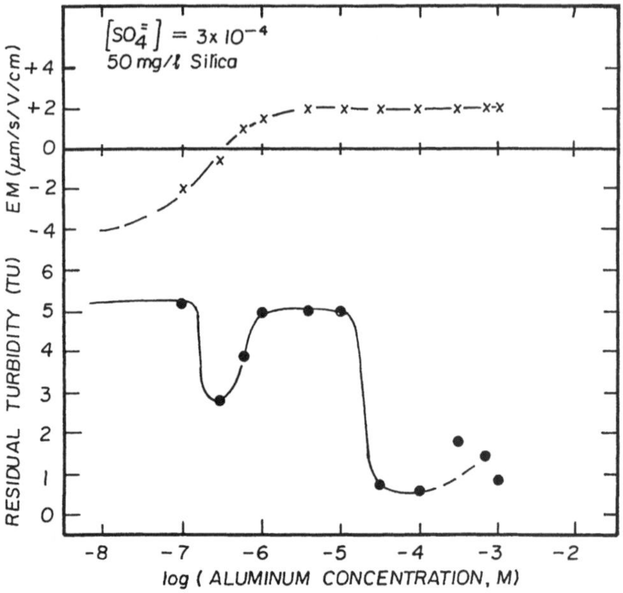

FIGURE 6.29 Jar test data showing the effect of the aluminum nitrate dosage on the EM and residual turbidity after flocculation and sedimentation for 50 mg/L silica, pH = 6, and $[SO_4^=] = 3 \times 10^{-4}$.

The data plotted in Figure 6.29 were obtained using a silica concentration of 50 mg/L and a sulfate concentration of 3×10^{-4} M. Because the silica concentration is less than the amount used to obtain the results plotted in Figures 6.14 and 6.15, the amount of aluminum needed for charge neutralization ($\approx 2 \times 10^{-7}$ M or 5.4×10^{-3} mg Al/L) is less than that shown in the previous figures.

Figure 6.29 illustrates an important aspect of hydrolyzing metal coagulants that can make the application of EM measurements for dosage control difficult. Note that between aluminum concentrations of about 10^{-6} and 10^{-5} (0.027 and 0.27 mg Al/L), the residual turbidity is high, suggesting that restabilization has occurred. However, as the aluminum concentration is increased beyond 2×10^{-5} M (0.54 mg Al/L), the residual turbidity decreases to values lower than the minimum observed at the point of charge neutralization (EM = 0). The EM measurement remains at +2 and gives no indication that flocculation efficiency should improve with increasing aluminum concentration beyond the point of charge neutralization.

The results presented in Figure 6.29 can be explained by a consideration of the effect of the aluminum concentration on the rate of flocculation and on the magnitude of the product $\alpha\Phi$. Apparently, under the experimental conditions used to obtain the results plotted in Figure 6.29 and at aluminum concentrations between 10^{-6} (0.027 mg Al/L) and 10^{-5} (0.27 mg Al/L), the magnitude of α and the product $\alpha\Phi$ are not high enough to yield efficient flocculation. However, as the aluminum concentration is increased beyond 10^{-5} (0.27 mg Al/L), the magnitude of Φ increases until the modest value of α is compensated for and efficient flocculation is obtained.

The streaming current measurement has been found to be a useful technique for controlling the HMS coagulant dosage in the removal of NOM. According to Dempsey (1994), the streaming current detector response is sensitive to conditions that lead to effective NOM removal. However, Dentel (1994) has noted that although streaming current appears to have promise for monitoring the enhanced coagulation process, problems may occur with low-turbidity, high-NOM-concentration waters, especially at high pH.

BIBLIOGRAPHY

American Society for Testing Materials. *Standard Test Method for Turbidity of Water,* Method D 1889-88a. Philadelphia: ASTM, 1988.

American Water Works Association. *Water Treatment Plant Design,* 3rd ed., Denver, CO. New York: McGraw-Hill, 1998.

American Water Works Association, Coagulation and Filtration Committee. "Survey of polyelectrolyte use in the United States. *Jour. AWWA,* 74, 1982: 600–607.

Amirtharajah, A., and Mills, K. M. "Rapid-mix design for mechanisms of alum coagulation." *Jour. AWWA,* 74(4), 1982: 210–216.

Amirtharajah, A., and C. R. O'Melia. "Coagulation processes: Destabilization, mixing, and flocculation. In *Water Quality and Treatment,* ch. 6, F. W. Pontius, ed. New York: McGraw-Hill, 1990.

Argaman, Y., and W. F. Kaufman. "Turbulence and flocculation." *Jour. Sanitary Eng. Div., ASCE,* 96, 1970: 223.

AWWA Research Foundation. *Mixing in Coagulation and Flocculation,* pp. 256–281, A. Amirtharajah, M. M. Clark, and R. R. Trussell, eds. Denver, CO: American Water Works Association Research Foundation, 1991.

Anonymous. *Use of Aluminum Salts in Drinking Water Treatment,* AWWA Executive Committee White Paper, approved April 11, 1997, 4 pp. Denver, CO: AWWA, 1997.

Anonymous. *Operational Control of Coagulation and Filtration Processes,* Am. Waterworks Assoc, 1st ed., Manual M37. Denver, CO: AWWA, 1992.

Ball, J. W., D. K. Nordstrom, and E. A. Jenne. *Additional and Revised Thermochemical Data for WATEQ-2 Computerized Model for Trace and Major Element Speciation and Mineral Equilibrium of Natural Waters,* U.S. Geological Survey Water Resource Investigations. Menlo Park, CA: U.S. Geological Survey, 1980.

Baylis, J. R. "Silicates as aids to coagulation." *Jour. AWWA,* 29(9), 1937: 1355.

Bertsch, P. M., and D. R. Parker. "Aqueous Polynuclear Aluminum Species." In *The Environmental Chemistry of Aluminum,* 2nd ed., ch. 4, p. 117, G. Sposito, ed. Boca Raton, FL: CRC Press, 1996.

Bertsch, P. M., G. W. Thomas, and R. I. Barnhisel. "Characterization of hydroxy aluminum solutions by aluminum-27 nuclear magnetic resonance spectroscopy." *Soil Sci. Soc. Am. J.,* 50, 1986: 825.

Black, A. P., et al. "Destabilization of dilute clay suspensions with labeled polymers." *Jour. AWWA,* 57(12), 1965: 1547.

Bottero, J. Y., and J. L. Bersillon. "Aluminum and iron(III) chemistry: Some implications for organic substance removal." In *Aquatic Humic Substances: Influence on Fate and Treatment of Pollutants,* ch. 26, pp. 425–442, I. Suffet and P. MacCarthy, eds. Washington, D.C.: Am. Chem. Soc., 1989.

Bottero, J. Y., M. Axelos, D. Tchourbar, J. M. Cases, J. J. Fripiat, and F. Fiessinger. "Mechanism of formation of aluminum trihydroxide from Keggin Al_{13} polymers." *J. Colloid Interface Sci.,* 117, 1987: 47.

Camp, T. R. "Flocculation and flocculation basins." *Am. Soc. Civil Eng. Trans.,* 120, 1955: 1–16.

Camp, T. R., and P. C. Stein. "Velocity gradients and internal work in fluid motion." *Jour. of the Society of Civil Engineering,* 30, 1943: 219–237.

Chandrakanth, M. S., and G. L. Amy. "Effects of ozone on the colloidal stability and aggregation of particles coated with natural organic matter." *Environ. Sci. Technol.,* 30(2), 1996: 431–443.

Chapra, S. C., R. P. Canale, and G. L. Amy. "Empirical models for disinfection by-products in lakes and reservoirs." *Jour. of Environ. Engineering,* ASCE, 123(7), 1997: 714–715.

Clark, M. M. "Critique of Camp and Stein's rms velocity gradient." *Jour. of Environ. Engineering,* ASCE, 111(6), 1985: 741–764.

Clark, M. M., and J. R. V. Flora. "Floc restructuring in varied turbulent mixing." *Jour. of Colloid and Interface Sci.,* 147(2), 1991: 407–421.

Clark, M. M., and R. M. Srivastava. "Mixing and aluminum precipitation." *Environ. Sci. Technol.,* 27(10), 1993: 2181–2189.

Cleasby, J. L. "Is velocity gradient a valid turbulent flocculation parameter?" *Jour. of Environ. Engineering,* ASCE, 110(5), 1984: 875–895.

Cleasby, J. L., A. H. Dharmarajah, G. L. Sindt, and E. R. Baumann. *Design and Operations Guidelines for Optimization of the High-Rate Filtration Process: Plant Survey Results,* Final project report. Denver, CO: AWWA and AWWA Research Foundation, 1989.

David, D., and M. M. Clark. "Micromixing models and application to aluminum neutralization precipitation reactions." In *Mixing and Coagulation and Flocculation,* ch. 5, A. Amirtharajah, M. M. Clark, and R. R. Trussell, eds. Denver, CO: American Water Works Association Research Foundation, 1991.

Davis, J. A., and J. O. Leckie. "Surface ionization and complexation at the oxide/water interface. II. Surface properties of amorphous iron oxyhydroxide and adsorption of metal ions." *Jour. Colloid Interface Sci.,* 67, 1978: 90–107.

de Hek, H., R. J. Stol, and P. L. DeBruyn. "Hydrolysis-precipitation studies of aluminum (III) solutions. Part 3: The role of the sulfate ion." *Jour. Colloid Interface Sci.,* 64, 1978: 72.

Delichatsios, M. A., and R. F. Probstein. "Coagulation in turbulent flow: theory and experiment." *Jour. Colloid Interface Sci.,* 51, 1975: 394.

Dempsey, B. A. "Chemistry of coagulants." In *AWWA Seminar Proceedings, Influence of Coagulation on the Selection, Operation, and Performance of Water Treatment Facilities.* AWWA Annual Conference, June 14, 1987, Kansas City, MO 1987.

Dempsey, B. A., R. M. Ganho, and C. R. O'Melia. "The coagulation of humic substances by means of aluminum salts." *Jour. AWWA,* 76(4), 1984: 141–150.

Dempsey, B. A. "Reactions between fulvic acid and aluminum: Effects on the coagulation process." In *Aquatic Humic Substances: Influence on Fate and Treatment of Pollutants,* ch. 25, pp. 409–424, I. Suffet and P. MacCarthy, eds. Washington D.C.: Am. Chem. Soc., 1989.

Dempsey, B. A. "Enhanced coagulation using charge neutralization." In *Proceedings of the AWWA Enhanced Coagulation Research Workshop,* December 4–6, 1994, Charleston, SC. Denver, CO: AWWA, 1994.

Dentel, S. K. "Use of the streaming current detector in enhanced coagulation processes." In *Proceedings of the AWWA Enhanced Coagulation Research Workshop,* December 4–6, 1994, Charleston, SC. Denver, CO: AWWA, 1994.

Dentel, S. K. "Application of the precipitation-charge neutralization model of coagulation." *Environ. Sci. Technol.,* 22(7), 1988: 825.

Dentel, S. K., B. M. Gucciardi, T. A. Bober, P. V. Shetty, and J. J. Resta. *Procedures Manual for Polymer Selection in Water Treatment Plants,* Publication 90553. Denver, CO: American Water Works Association Research Foundation, 1989.

Dentel, S. K. "Coagulant control in water treatment." *Critical Reviews in Environmental Control,* 21(1), 1991: 41–135.

Dentel, S. K., and K. M. Kingery. *An Evaluation of Streaming Current Detectors,* Publication 90536. Denver, CO: American Water Works Association Research Foundation, 1988.

Driscoll, C. T., C. Yan, C. L. Schofield, R. Munson, and J. Holsapple. "The mercury cycle and fish in the Adirondack Lakes." *Environ. Sci. Technol.,* 28(3), 1994: 136–143.

Edwards, M. "Predicting DOC removal during enhanced coagulation." *Jour. AWWA,* 85(5), 1997: 78–89.

Edwards, M., and M. M. Benjamin. "Effect of preozonation on coagulant-NOM interactions." *Jour. AWWA,* 84(8), 1992: 63–72.

Edzwald, J. K., and J. E. Van Benschoten. "Aluminum coagulation of natural organic material." In *Chemical Water and Wastewater Treatment,* H. H. Hahn and R. Klute, eds. Berlin: Springer-Verlag, 1990.

Edzwald, J. K. "Coagulation in drinking water treatment: Particles, organics and coagulants." *Wat. Sci. Tech.,* 27(11), 1993: 21–35.

Elimelech, M., J. Gregory, X. Jia, and R. Williams. *Particle Deposition and Aggregation.* Oxford: Butterworth-Heinemann Ltd., 1995.

Christman, R. F. *Aquatic and Terrestrial Humic Materials.* Ann Arbor, MI: Ann Arbor Science, 1983.

Gerdes, W. F. "A new instrument—the streaming current detector." 12th Natl. ISA Analysis Instrument Symp., 1966.

Goldberg, S., J. A. Davis, and J. D. Hem. "The surface chemistry of aluminum oxides and hydroxides. In *The Environmental Chemistry of Aluminum,* ch. 7, 2nd ed., p. 271, G. Sposito, ed. Boca Raton, FL: CRC Press, 1996.

Gregory, J. "Effects of polymers on colloid stability." In *The Scientific Basis of Flocculation,* K. J. Ives, ed. The Netherlands: Sijthoff and Noordhoff, 1978.

Han, M., and Lawler, D. F. "Interaction of two settling spheres; settling rates and collision efficiencies." *Journal of Hydraulics Division, Am. Soc. Civil Engrs.,* 17(10), 1991: 1269–1289.

Han, M., and D. F. Lawler. "The (relative) insignificance of G in flocculation." *Jour. AWWA,* 84(10), 1992: 79–91.

Hanson, A. T., and J. L. Cleasby. "The effect of temperature on turbulent flocculation: Fluid dynamics and chemistry." *Jour. AWWA,* 82(11), 1990: 56–73.

Hargesheimer, E. E., C. M. Lewis, and C. M. Yentsch. *Evaluation of Particle Counting as a Measure of Treatment Plant Performance,* Final project report, Am. Water Works Assoc. Denver, CO: AWWA, 1992.

Harris, W. L. "Use of polyelectrolytes as aids during backwash of filters." In *Proceedings of the AWWA Seminar on Polyelectrolytes—Aids to Better Water Quality.* Chicago, IL, June 4, 1972. Denver, CO: AWWA, 1972.

Hart, V., C. E. Johnson, and R. D. Letterman. "An analysis of low-level turbidity measurements." *Jour. AWWA,* 84(12), 1992: 40–45.

Hayden, P. L., and A. J. Rubin. "Systematic investigation of the hydrolysis and precipitation of aluminum (III)." In *Aqueous-Environmental Chemistry of Metals,* A. J. Rubin, ed. Ann Arbor, MI: Ann Arbor Science, 1974.

Hudson, H. E., Jr. *Water Clarification Processes: Practical Design and Evaluation.* New York: Van Nostrand Reinhold, 1981.

Hunter, R. J. *Zeta Potential in Colloid Science.* London: Academic Press, 1981.

Hunter, R. J., and P. S. Liss. "The surface charge of suspended particles in estuarine and coastal waters." *Nature,* 282, 1979: 823.

International Standards Organization. *Water Quality—Determination of Turbidity,* Ref. No. ISO 7027-1984(E). Zurich: International Organization for Standardization, 1984.

Jahn, S. A. A. "Using Moringa seeds as coagulants in developing countries." *Jour. AWWA,* 90, 1988: 43–50.

Kang, L.-S., and J. L. Cleasby. "Temperature effects on flocculation kinetics using Fe(III) coagulant." *Jour. of Env. Eng,* ASCE, 121(12), 1995: 893–910.

Kim, J. S., and R. Rajagopolan. "A comprehensive equation for the rate of adsorption of colloidal particles and for stability ratio." *Colloids and Surfaces,* 4, 1982: 17–31.

Kodama, H., and M. Schnitzer. "Effect of fulvic acid on the crystallization of aluminum hydroxides." *Geoderma,* 24, 1980: 195.

Krasner, S. W. "The regulatory role of TOC measurements in the ICR and DBP rules." In *Proceedings of the AWWA Water Quality Technology Conference,* November 17–21, Boston, MA, 1996.

Lagvankar, A. L., and R. S. Gemmell. "A size-density relationship for flocs." *Jour. AWWA,* 60(9), 1968: 1040.

LaMer, V. K. "Coagulation symposium introduction." *Jour. Colloid Sci.,* 19, 1964: 291.

Letterman, R. D. *Filtration Strategies to Meet the Surface Water Treatment Rule.* Denver, CO: American Water Works Association, 1991.

Letterman, R. D., R. Chappell, and B. Mates. "Effect of pH and alkalinity on the removal of NOM with Al and Fe salt coagulants." In *Proceedings of the AWWA Water Quality Technology Conference,* November 17–21, Boston, MA, 1996.

Letterman, R. D., and S. G. Vanderbrook. "Effect of solution chemistry on coagulation with hydrolyzed Al III." *Water Research,* 17(1), 1983: 195–204.

Letterman, R. D., and D. Iyer. "Modeling the effects of adsorbed hydrolyzed aluminum and solution chemistry on flocculation kinetics." *Environ. Sci. and Technol.,* 19(8), 1985: 673–681.

Letterman, R. D., and C. T. Driscoll. *Control of Residual Aluminum in Filtered Water,* Final Report to American Water Works Association Research Foundation, Denver, CO, 135 pp. Denver, CO: American Water Works Assoc., 1993.

Letterman, R. D., and R. W. Pero. "Contaminants in polyelectrolyte coagulant products used in water treatment." *Jour. AWWA,* 82(11), 1990: 87–97.

Letterman, R. D., and P. Sricharoenchaikit. "Interaction of hydrolyzable metal and cationic polyelectrolyte coagulants. *Jour. Of Environmental Engineering,* ASCE, 108(5), 1982: 883–898.

Levenspiel, O. *Chemical Reaction Engineering,* 2nd ed. New York: John Wiley & Sons, 1972.

Lewis, C. M., E. E. Hargesheimer, and C. M. Yentsch. "Selecting counters for process monitoring." *JAWWA,* 84(12), 1992: 46–53.

Liang, L., and J. J. Morgan. *Effects of Surface Chemistry on Kinetics of Coagulation of Submicron Iron Oxide Particles* (α-Fe_2O_3) *in Water,* Report No. AC5-88 of the W. M Keck Laboratory of Environmental Engineering Science. Pasedena, CA: California Institute of Technology, 1988.

Lind, C. "A coagulant road map." *Public Works,* March 1995: 36–38.

Lyklema, J. "Surface chemistry of colloids in connection with stability." In *The Scientific Basis of Flocculation,* K. J. Ives, ed. The Netherlands: Sijthoff and Noordhoff, 1978.

McConnachie, G. L. "Turbulence intensity of mixing in relation to flocculation." *Journal of Environmental Engineering,* ASCE, 117(6), 1991: 731–750.

Meakin, P. "Fractal aggregates." *Adv. Colloid Interface Sci.,* 28, 1988: 249–331.

Meng, X-G. Ph.D. Dissertation, Department of Civil and Environmental Engineering, Syracuse University, Syracuse, New York, 1989.

Monk, R. D. G., et al. "Prepurchasing ozone equipment." *Jour. AWWA,* 77, 1985: 49–54.

Morel, F. M. M. *Principles of Aquatic Chemistry.* New York: Wiley-Interscience, 1983.

Morris, J. K., and W. R. Knocke. "Temperature effects on the use of metal-ion coagulants for water treatment." *Jour. AWWA,* 76(3), 1984: 74.

Mühle, K., and K. Domasch. "Stability of particle aggregates in flocculation with polymers." *Chem. Eng. Progress,* 29, 1991: 1–8.

Mühle, K. "Floc stability in laminar and turbulent flow." In *Coagulation and Flocculation, Theory and Applications,* pp. 355–390, B. Dobiáš, ed. New York: Marcel Dekker, 1993.

Narkis, N., and M. Rebhun. "Stoichiometric relationship between humic and fulvic acids and flocculants." *Jour. AWWA,* 69(6), 1977: 325.

Niehof, R. A., and G. I. Loeb. "The surface charge of particulate matter in sea water." *Limnology Oceanography,* 17, 1992: 7.

Nowack, K. O., and F. S. Cannon. "Ferric chloride plus GAC for removing TOC." In *Proceedings of the AWWA Water Quality Technology Conference,* November 17–21, Boston, MA, 1996.

NSF Standard 60, NSF International, Ann Arbor, MI, 1988.

O'Melia, C. R. "From algae to aquifers: Solid-liquid separation in aquatic systems." In ACS Advances in Chemistry Series No. 244, *Aquatic Chemistry: Interfacial and Interspecies Processes,* C. P. Huang, C. R. O'Melia, and J. J. Morgan, eds. Washington, D.C.: American Chemical Soc., 1995.

O'Melia, C. R. "Polymeric inorganic flocculants." In *Flocculation, Sedimentation, and Consolidation,* M. Moudgil and P. Semasundaran, eds. New York: Engineering Foundation, 1985.

Paralkar, A., and J. K. Edzwald. "Effect of ozone on EOM and coagulation." *Jour. AWWA,* 88(4), 1996: 143–154.

Parker, D. S., W. J. Kaufmann, and D. Jenkins. "Floc breakup in turbulent flocculation processes." *Jour. Sanitary Engineering Div.,* ASCE, 98, 1972: 79.

Pontius, F. W. "Complying with the new drinking water regulations." *Jour. AWWA,* 82(2), 1990: 32–52.

Prendiville, P. W. *Ozone Sci. Eng.,* 8, 1986: 77–93.

Randtke, S. J. "Organic contaminant removal by coagulation and related process combinations." *Jour. AWWA,* 80(5), 1988: 40.

Randtke, S. J. "The influence of coagulation on organic contaminant removal and activated carbon adsorption." In *Proceedings of the Am. Water Works Assoc. Seminar, Influence of coagulation on the selection, operation, and performance of water treatment facilities,* AWWA Annual Conference, June 14, Kansas City, MO, 1987.

Saffman, P. G., and J. S. Turner. "On the collision of drops in turbulent clouds." *Jour. Fluid Mech.,* 1, 1956: 16.

Schulz, C. R., P. C. Singer, R. Gandley, and J. E. Nix. "Evaluating buoyant coarse media flocculation." *Jour. AWWA,* 86(8), 1994: 51–63.

Schwartzberg, H. G., and R. E. Treybal. "Fluid and particle motion in turbulent stirred tanks." *Ind. Eng. Chem. Fundamentals,* 7, 1968: 1–12.

Sethi, V., P. Patniak, Clark R. M. Biswas, and E. W. Rice. "Evaluation of optical detection methods for waterborne suspensions." *Jour. AWWA,* 89(2), 1997: 98–112.

Shankar, S., and R. D. Letterman. "Modeling pH in water treatment plants—The effect of carbon dioxide transport on pH profiles." Poster presented at the AWWA Annual Conf., June 23–27, Toronto, ON, 1996.

Smoluchowski, M. "Versuch einer mathematischen Theorie der Koagulationskinetic kolloider Losunger." *Zeitschrift Physicalische Chemie,* 92, 1917: 129.

Snoeyink, V. L., and D. Jenkins. *Water Chemistry.* New York: John Wiley & Sons, 1980.

Spielman, L. A. "Hydrodynamic aspects of flocculation." In *The Scientific Basis of Flocculation,* K. J. Ives, ed. The Netherlands: Sijthoff and Noordhoff, 1978.

Sricharoenchaikit, P., and R. D. Letterman. "Effect of Al(III) and sulfate ion on flocculation kinetics." *Jour. of Environ. Engineering,* ASCE, 113(5), 1987: 1120–1138.

Standard Methods for the Examination of Water and Wastewater, Section 2130, 18th ed., A. E. Greenberg, L. S. Clesceri, and A. D. Eaton, eds. Washington, D.C.: American Public Health Assoc., American Water Works Assoc., and Water Environment Fed., 1992.

Stumm, W. *Chemistry of the Solid-Water Interface.* New York: John Wiley and Sons, 1992.

Stumm, W., H. Huper, and R. L. Champlin. "Formulation of polysilicates as determined by coagulation effects." *Environ. Sci. Technol.,* 1, 1967: 221.

Stumm, W., and J. J. Morgan. *Aquatic Chemistry,* 2nd ed. New York: John Wiley and Sons, 1981.

Sutherland, J. P., G. K. Folkard, and W. D. Grant. "Seeds of Moringa species as naturally occurring flocculants for water treatment." *Sci. Tech. and Dev.,* 7, 1989: 191–197.

Tambo, N., and R. J. François. "Mixing, breakup, and floc characteristics." In *Mixing in Coagulation and Flocculation,* pp. 256–281. A. Amirtharajah, M. M. Clark, and R. R. Trussell, eds. American Water Works Association Research Foundation, Denver, CO: 1991.

Tang, H.-X., and W. Stumm. "The coagulating behaviors of Fe(III) polymeric species—I." *Wat. Res.,* 21(1), 1987: 115–121.

Tang, H.-X., and W. Stumm. "The coagulating behaviors of Fe(III) polymeric species—II." *Wat. Res.,* 21(1), 1987: 123–128.

Thurman, E. M. *Developments in biochemistry: Organic geochemistry of natural waters.* Dordrecht, Germany: Nijhoff and Junk Publ., 1985.

Turner, R. C. "A second series of polynuclear hydroxyaluminum cation, its formation and some of its properties." *Can. J. Chem.,* 54, 1976: 1528.

United States Environmental Protection Agency. *Determination of Turbidity by Nephelometry,* Revision 2.0, Method 180.1. Cincinnati, OH: Environmental Monitoring Systems Laboratory, 1993.

United States Environmental Protection Agency. *Guidance Manual for Enhanced Coagulation and Enhanced Precipitative Softening.* Preliminary draft prepared by Malcolm Pirnie, Inc., White Plains, NY, for the USEPA, Regulation Management Branch, Washington, D.C., 1994.

van de Ven, T. G. M., and S. G. Mason. "The microrheology of colloidal suspensions. VII: Orthokinetic doublet formation of spheres." *Colloid Polymer Science,* 255, 1977: 468.

VanBenschoten, J. E., and J. K. Edzwald. "Chemical aspects of coagulation using aluminum salts—I. Hydrolytic reactions of alum and polyaluminum chloride." *Wat. Res.,* 24(12), 1990, 1519–1526.

Verwey, E. J. W., and J. Th. G. Overbeek. *Theory of the Stability of Lyophobic Colloids.* Amsterdam: Elsevier, 1948.

Vold, M. J. "Computer simulation of floc formation in a colloidal suspension." *Jour. Colloid Sci.,* 18, 1963: 684–693.

White, M. C., J. D. Thompson, G. W. Harrington, and P. C. Singer. "Evaluating criteria for enhanced coagulation." *Jour. AWWA,* 89(5), 1997: 64–77.

Wiesner, M. R., M. M. Clark, and J. Mallevialle. "Membrane filtration of coagulated suspensions." *Jour. of Environ. Engineering,* ASCE, 115, 1989: 20–40.

Yapajakis, C. "Polymer in backwash serves dual purpose." *Jour. AWWA,* 74(8), 1982: 426–428.

CHAPTER 7
SEDIMENTATION AND FLOTATION

Ross Gregory
WRc Swindon
Frankland Road, Blagrove
Swindon, Wiltshire
England

Thomas F. Zabel
WRc Medmenham
Medmenham, Oxfordshire
England

James K. Edzwald
Professor, Department of Civil and
Environmental Engineering
University of Massachusetts
Amherst, Massachusetts
U.S.A.

Sedimentation and flotation are solid-liquid separation processes used in water treatment mostly to lower the solids concentration, or load, on granular filters. As a result, filters can be operated more easily and cost effectively to produce acceptable-quality filtered water. Many sedimentation and flotation processes and variants of them exist, and each has advantages and disadvantages. The most appropriate process for a particular application will depend on the water to be treated as well as local circumstances and requirements.

HISTORY OF SEDIMENTATION

Early History

Sedimentation for the improvement of water quality has been practiced, if unwittingly, since the day humans collected and stored water in jars and other containers.

Water stored undisturbed and then poured or ladled out with little agitation will improve in quality, and this technique is used to this day.

As societies developed, reservoirs and storage tanks were constructed. Although constructed for strategic purposes, reservoirs and storage tanks did improve water quality. Various examples are known that predate the Christian era. Ancient surface water impounding tanks of Aden were possibly constructed as early as 600 B.C. and rainwater cisterns of ancient Carthage about 150 B.C. (Ellms, 1928). The castellae and piscinae of the Roman aqueduct system performed the function of settling tanks, even though they were not originally intended for that purpose.

Modern Sedimentation

The art of sedimentation progressed little until the industrial age and its increased need for water. Storage reservoirs developed into settling reservoirs. Perhaps the largest reservoirs constructed for this purpose were in the United States at Cincinnati, Ohio, where two excavated reservoirs held approximately 1480 ML (392 million gallons) and were designed to be operated by a fill-and-draw method, though they never were used in this way (Ellms, 1928). The development of settling basins led to the construction of rectangular masonry settling tanks that assured more even flow distribution and easier sludge removal. With the introduction of coagulation and its production of voluminous sludge, mechanical sludge removal was introduced.

Attempts to make rectangular tanks more cost-effective led to the construction of multilayer tanks. Very large diameter [60-m (200-ft)] circular tanks also were constructed at an early stage in the development of modern water treatment. Other industries, such as wastewater treatment, mineral processing, sugar refining, and water softening, required forms of sedimentation with specific characteristics, and various designs of settling tanks particular to certain industries were developed. Subsequently, wider application of successful industrial designs were sought. From this, circular radial-flow tanks emerged, as well as a variety of proprietary designs of solids-contact units with mechanical equipment for premixing and recirculation.

The inclined plate settler also has industrial origins (Barham, Matherne, and Keller, 1956) (Figure 7.1), although the theory of inclined settling dates back to experiments using blood in the 1920s and 1930s (Nakamura and Kuroda, 1937; Kinosita, 1949). Closely spaced inclined plate systems for water treatment have their origins in Sweden in the 1950s, resulting from a search for high-rate treatment processes compact enough to be economically housed against winter weather. Inclined

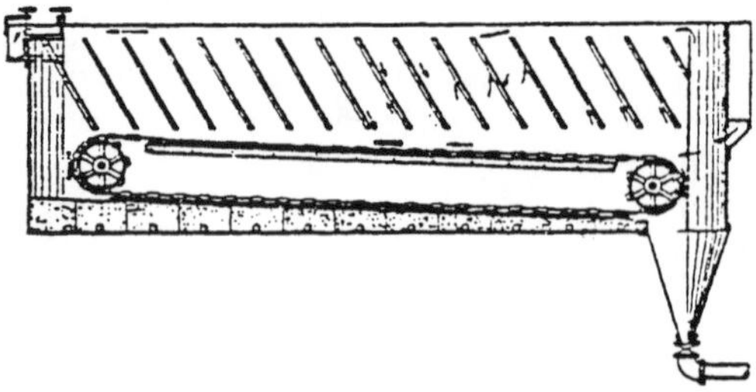

FIGURE 7.1 Early patent for inclined settling. (*Source:* Barham et al., 1956.)

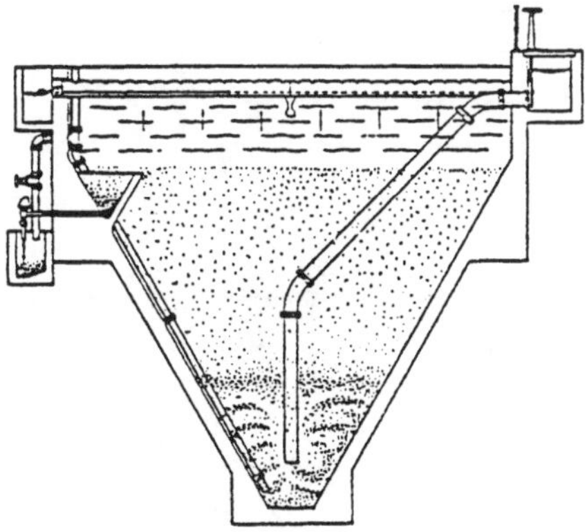

FIGURE 7.2 The pyramidal Candy floc-blanket tank. (*Source:* by PWT.)

tube systems were spawned in the United States in the 1960s. The most recent developments have involved combining inclined settling with ballasting of floc to reduce plant footprint further (de Dianous, Pujol, and Druoton, 1990).

Floc-Blanket Sedimentation and Other Innovations

The floc-blanket process for water treatment emerged from India about 1932 as the pyramidal Candy sedimentation tank (Figure 7.2). A tank of similar shape was used by Imhoff in 1906 for wastewater treatment (Kalbskopf, 1970). The Spaulding Precipitator soon followed in 1935 (Figure 7.3) (Hartung, 1951). Other designs that were mainly solids contact clarifiers rather than true floc-blanket tanks were also introduced.

The Candy tank can be expensive to construct because of its large sloping sides, so less costly structures for accommodating floc blankets were conceived. The aim was to decrease the hopper component of tanks as much as possible, yet to provide good flow distribution to produce a stable floc blanket. Development from 1945 progressed from tanks with multiple hoppers or troughs to the present flat-bottom tanks. Efficient flow distribution in flat-bottom tanks is achieved with either candelabra or lateral inlet distribution systems (Figures 7.4 and 7.5).

An innovation in the 1970s was the inclusion of widely spaced inclined plates in the floc-blanket region (Figure 7.5). Other developments that also have led to increased surface loadings include the use of polyelectrolytes, ballasting of floc with disposable or recycled solids, and improvements in blanket-level control. The principal centers for these developments have been in the United Kingdom, France, and Hungary.

SEDIMENTATION THEORY

The particle-fluid separation processes of interest to water engineers and scientists are difficult to describe by a theoretical analysis, mainly because the particles

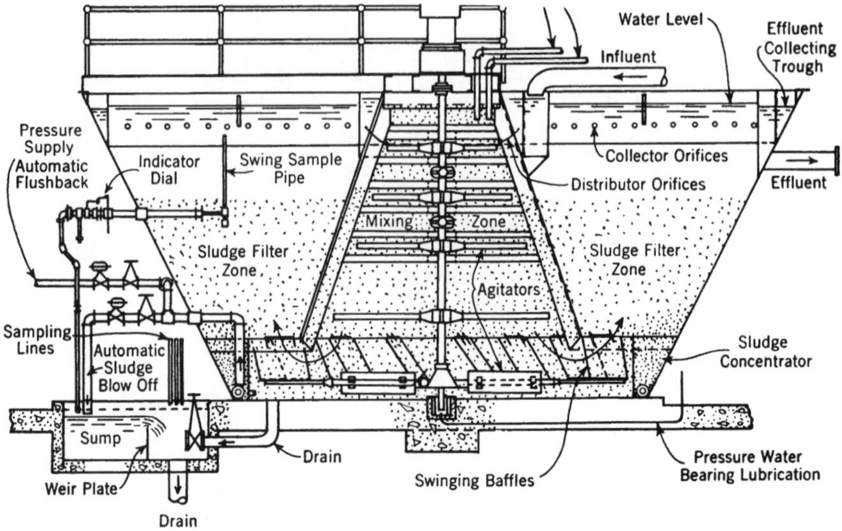

FIGURE 7.3 The Spaulding Precipitator solids contact clarifier. (*Source:* Hartung, 1951.)

involved are not regular in shape, density, or size. Consideration of the theory of ideal systems is, however, a useful guide to interpreting observed behavior in more complex cases.

The various regimes in settling of particles are commonly referred to as *Types 1 to 4*. The general term *settling* is used to describe all types of particles falling through

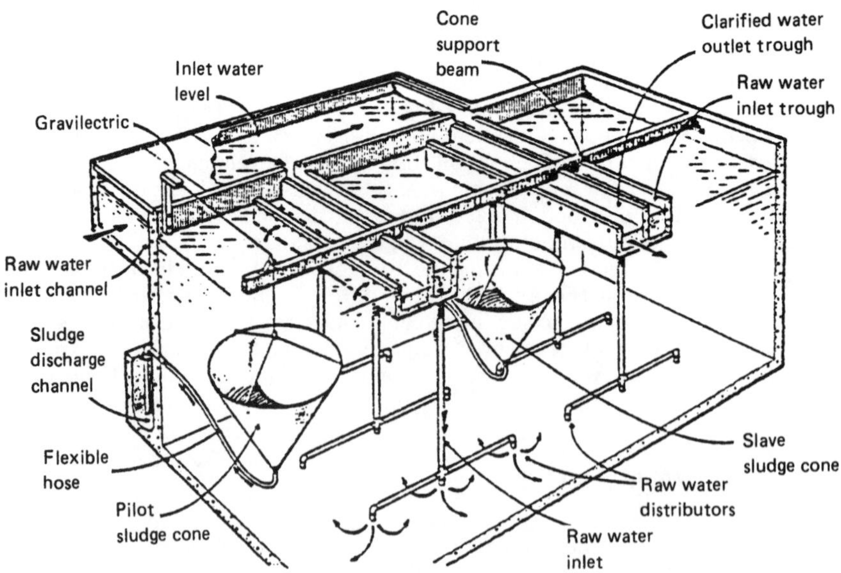

FIGURE 7.4 The flat-bottom clarifier with candelabra flow distribution. (*Source:* by PWT.)

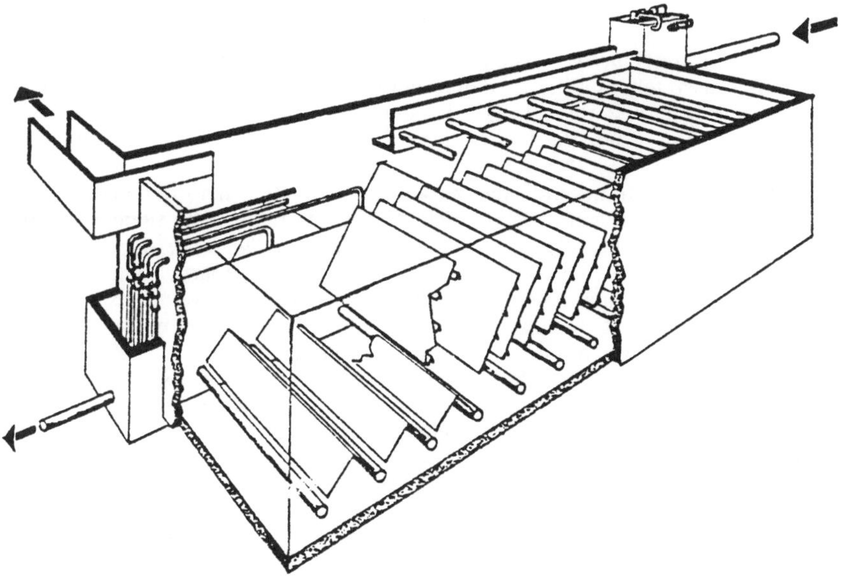

FIGURE 7.5 The Superpulsator flat-bottom clarifier with lateral-flow distribution. (*Source:* Courtesy of Infilco Degremont, Inc., Richmond, VA.)

a liquid under the force of gravity and settling phenomena in which the particles or aggregates are suspended by hydrodynamic forces only. When particles or aggregates rest on one another, the term *subsidence* applies. The following definitions of the settling regimes are commonly used in the United States and are compatible with a comprehensive analysis of hindered settling and flux theory:

Type 1. Settling of discrete particles in low concentration, with flocculation and other interparticle effects being negligible.

Type 2. Settling of particles in low concentration but with coalescence or flocculation. As coalescence occurs, particle masses increase and particles settle more rapidly.

Type 3. Hindered, or zone, settling in which particle concentration causes interparticle effects, which might include flocculation, to the extent that the rate of settling is a function of solids concentration. Zones of different concentrations may develop from segregation of particles with different settling velocities. Two regimes exist—a and b—with the concentration being less and greater than that at maximum flux, respectively. In the latter case, the concentration has reached the point that most particles make regular physical contact with adjacent particles and effectively form a loose structure. As the height of this zone develops, this structure tends to form layers of different concentration, with the lower layers establishing permanent physical contact, until a state of compression is reached in the bottom layer.

Type 4. Compression settling or subsidence develops under the layers of zone settling. The rate of compression is dependent on time and the force caused by the weight of solids above.

Settling of Discrete Particles (Type 1)

Terminal Settling Velocity. When the concentration of particles is small, each particle settles discretely, as if it were alone, unhindered by the presence of other particles. Starting from rest, the velocity of a single particle settling under gravity in a liquid will increase, where the density of the particle is greater than the density of the liquid.

Acceleration continues until the resistance to flow through the liquid, or drag, equals the effective weight of the particle. Thereafter, the settling velocity remains essentially constant. This velocity is called the *terminal settling velocity,* v_t. The terminal settling velocity depends on various factors relating to the particle and the liquid.

For most theoretical and practical computations of settling velocities, the shape of particles is assumed to be spherical. The size of particles that are not spherical can be expressed in terms of a sphere of equivalent volume.

The general equation for the terminal settling velocity of a single particle is derived by equating the forces upon the particle. These forces are the drag f_d, buoyancy f_b, and an external source such as gravity f_g. Hence,

$$f_d = f_g - f_b \tag{7.1}$$

The drag force on a particle traveling in a resistant fluid is (Prandtl and Tietjens, 1957):

$$f_d = \frac{C_D v^2 \rho A}{2} \tag{7.2}$$

where C_D = drag coefficient
v = settling velocity
ρ = mass density of liquid
A = projected area of particle in direction of flow

Any consistent, dimensionally homogeneous units may be used in Eq. 7.2 and all subsequent rational equations.

At constant (i.e., terminal settling velocity) v_t

$$f_g - f_b = Vg(\rho_p - \rho) \tag{7.3}$$

where V is the effective volume of the particle, g is the gravitational constant of acceleration, and ρ_p is the density of the particle. When Eqs. 7.2 and 7.3 are substituted in Eq. 7.1

$$\frac{C_D v_t^2 \rho A}{2} = Vg(\rho_p - \rho) \tag{7.4a}$$

rearranging:

$$v_t = \left[\frac{2g(\rho_p - \rho)V}{C_D \rho A}\right]^{1/2} \tag{7.4b}$$

When the particle is solid and spherical,

$$v_t = \left[\frac{4g(\rho_p - \rho)d}{3C_D \rho}\right]^{1/2} \tag{7.5}$$

where d is the diameter of the sphere.

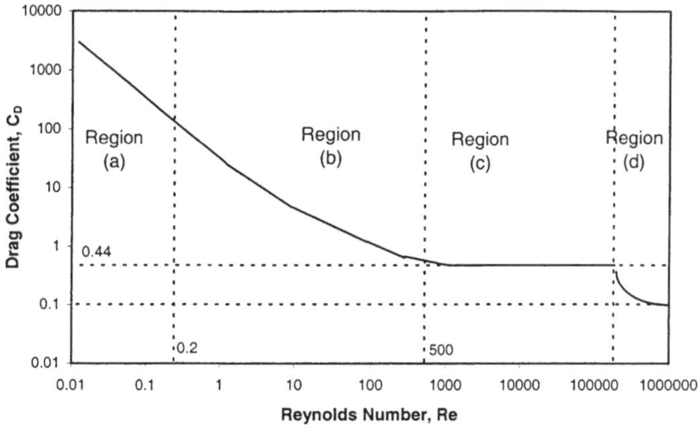

FIGURE 7.6 Variation of drag coefficient, C_D, with Reynolds number, Re, for single-particle sedimentation.

The value of v_t is the difference in velocity between the particle and the liquid and is essentially independent of horizontal or vertical movement of the liquid, although in real situations there are secondary forces caused by velocity gradients, and so on. Therefore, the relationship also applies to a dense stationary particle with liquid flowing upward past it or a buoyant particle with liquid flowing downward.

Calculation of v_t for a given system is difficult because the drag coefficient, C_D, depends on the nature of the flow around the particle. This relationship can be described using the Reynolds number, Re (based on particle diameter), as illustrated schematically in Figure 7.6, where

$$\text{Re} = \frac{\rho v d}{\mu} \tag{7.6}$$

and μ is the absolute (dynamic) liquid viscosity, and v is the velocity of the particle relative to the liquid.

The value of C_D decreases as the value of Re increases, but at a rate depending on the value of Re, such that for spheres only:

Region (a): 10^{-4} < Re < 0.2. In this region of small Re value, the laminar flow region, the equation of the relationship approximates to

$$C_D = 24/\text{Re} \tag{7.7}$$

This, substituted in Eq. 7.1, gives Stokes' equation for laminar flow conditions:

$$v_t = \frac{g(\rho_p - \rho)d^2}{18\mu} \tag{7.8}$$

Region (b): 0.2 < Re < 500 to 1000. This transition zone is the most difficult to represent, and various proposals have been made. Perhaps the most recognized representation of this zone for spheres is that promoted by Fair, Geyer, and Okun (1971):

$$C_D = \frac{24}{\text{Re}} + \frac{3}{\text{Re}^{1/2}} + 0.34 \tag{7.9}$$

For many particles found in natural waters, the density and diameter yield Re values within this region.

Region (c): 500 to 1000 < Re < 2 × 10⁵. In this region of turbulent flow, the value of C_D is almost constant at 0.44. Substitution in Eq. 7.5 results in Newton's equation:

$$v_t = 1.74\left[\frac{(\rho_p - \rho)gd}{\rho}\right]^{1/2} \tag{7.10}$$

Region (d): Re > 2 × 10⁵. The drag force decreases considerably with the development of turbulent flow at the surface of the particle called *boundary-layer turbulence,* such that the value of C_D becomes equal to 0.10. This region is unlikely to be encountered in sedimentation in water treatment.

Effect of Particle Shape. Equation 7.4b shows how particle shape affects velocity. The effect of a nonspherical shape is to increase the value of C_D at a given value of Re.

As a result, the settling velocity of a nonspherical particle is less than that of a sphere having the same volume and density. Sometimes, a simple shape factor, Θ, is determined, for example, in Eq. 7.7:

$$C_D = \frac{24\Theta}{\text{Re}} \tag{7.11}$$

Typical values found for Θ for rigid particles are (Degremont, 1991):

Sand	2.0
Coal	2.25
Gypsum	4.0
Graphite flakes	22

Details on the settling behavior of spheres and nonspherical particles can be found in standard texts (e.g., Coulson and Richardson, 1978).

Flocculation. A shape factor value is difficult to determine for floc particles because their size and shape are interlinked with the mechanics of their formation and disruption in any set of flow conditions. When particles flocculate, a loose and irregular structure is formed, which is likely to have a relatively large value shape factor. Additionally, while the effective particle size increases in flocculation, the effective particle density decreases in accordance with a fractal dimension (Lagvankar and Gemmel, 1968; Tambo and Watanabe, 1979) (see Chapter 6).

Flocculation is a process of aggregation and attrition. Aggregation can occur by Brownian diffusion, differential settling, and velocity gradients caused by fluid shear, namely flocculation. Attrition is caused mainly by excessive velocity gradients (see Chapter 6).

The theory of flocculation detailed in Chapter 6 recognizes the role of velocity gradient (G) and time (t) as well as particle volumetric concentration Φ. For dilute suspensions, optimum flocculation conditions are generally considered only in terms of G and t:

$$Gt = \text{constant} \tag{7.12}$$

Camp (1955) identified optimum Gt values between 10^4 and 10^5 for flocculation prior to horizontal flow settlers. In the case of floc-blanket clarifiers (there being no prior flocculators), the value of G is usually less than in flocculators, and the value of Gt is about 20,000 (Gregory, 1979). This tends to be less than that usually considered necessary for flocculation prior to inclined settling or dissolved air flotation.

In concentrated suspensions, such as with hindered settling, the greater particle concentration (e.g., volumetric concentration, Φ) contributes to flocculation by enhancing the probability of particle collisions, and increasing the velocity gradient that can be expressed in terms of the head loss across the suspension. Consequently, optimum flocculation conditions for concentrated suspensions may be better represented by Fair, Geyer, and Okun (1971); Ives (1968); and Vostrcil (1971):

$$Gt\Phi = \text{constant} \tag{7.13}$$

The value of the constant at maximum flux is likely to be about 4000, when Θ is measured as the fractional volume occupied by floc, with little benefit to be gained from a larger value (Gregory, 1979; Vostrcil, 1971).

Measurement of the volumetric concentration of floc particle suspensions is a problem because of variations in particle size, shape, and other factors. A simple settlement test is the easiest method of producing a measurement (for concentrations encountered in floc blanket settling) in a standard way (e.g., half-hour settlement in a graduated cylinder) (Gregory, 1979). A graduated cylinder (e.g., 100 mL or 1 L) is filled to the top mark with the suspension to be measured: the half-hour settled-solids volume is the volume occupied by the settled suspension measured after 30 min, and it is expressed as a fraction of the total volume of the whole sample.

The process of flocculation continues during conditions intended to allow settlement. Assuming collision between flocculant particles takes place only between particles settling at different velocities at Stokes' velocities, then the collision frequencies N_{ij} between particles of size d_i and d_j of concentrations n_i and n_j is given by Amirtharajah and O'Melia (1990):

$$(N_{ij})_{\mathrm{d}} = \frac{\pi g(s-1)}{72\nu}(d_i - d_j)^3(d_i - d_j)n_i n_j \tag{7.14}$$

where s is the specific gravity of particles and ν is the kinematic viscosity.

Settlement in Tanks. In an ideal upflow settling tank, the particles retained are those whose terminal settling velocity exceeds the liquid upflow velocity:

$$v_{\mathrm{t}} \geq \frac{Q}{A} \tag{7.15}$$

where Q is the inlet flow rate to the tank, and A is the cross-sectional area of the tank.

In a horizontal-flow rectangular tank, the settling of a particle has both vertical and horizontal components, as shown in Figure 7.7:

$$L = \frac{tQ}{HW} \tag{7.16}$$

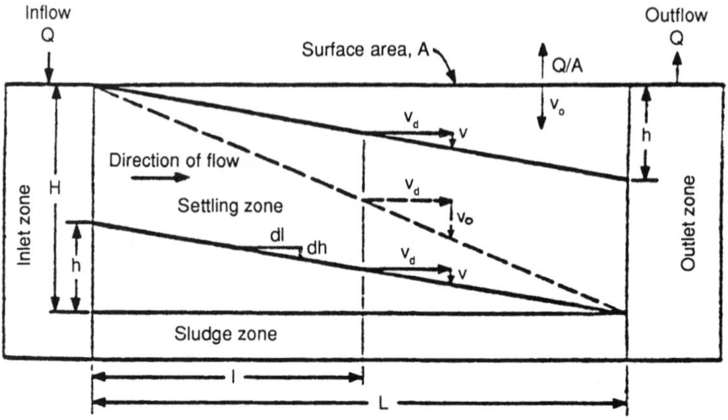

FIGURE 7.7 Horizontal and vertical components of settling velocity. (*Source:* Fair, Geyer, and Okun, 1971.)

where L = horizontal distance traveled
t = time of travel
H = depth of water
W = width of tank

and for the vertical distance traveled, h:

$$h = vt \tag{7.17a}$$

Hence, the settling time for a particle that has entered the tank at a given level, h, is

$$t = \frac{h}{v} \tag{7.17b}$$

Substitution of this in Eq. 7.16 gives the length of tank required for settlement to occur under ideal flow conditions:

$$L = \frac{hQ}{vHW} \tag{7.18a}$$

or

$$v = \frac{hQ}{HLW} \tag{7.18b}$$

If all particles with a settling velocity of v are allowed to settle, then h equals H, and, consequently, this special case then defines the surface-loading or overflow rate of the ideal tank, v^*:

$$v^* = \frac{Q}{L^*W} \tag{7.19a}$$

$$v^* = \frac{Q}{A^*} \tag{7.19b}$$

where L^* is the length of tank over which settlement ideally takes place, and A^* is the plan area of tank, with horizontal flow, over which the settlement ideally takes place.

All particles with a settling velocity greater than v^* are removed. Particles with a settling velocity less than v^* are removed in proportion to the ratio $v{:}v^*$.

Particles with a settling velocity v' less than v^* need a tank of length, L', greater than L^* for total settlement, such that

$$\frac{v'}{v^*} = \frac{L^*}{L'} \tag{7.20}$$

This ratio defines the proportion of particles with a settling velocity of v' that settle in a length L^*. Equation 7.19a states that settling efficiency depends on the area available for settling. The same result applies to circular tanks.

Equation 7.19a shows that settling efficiency for the ideal condition is independent of depth H and dependent on only the tank plan area. This principle is sometimes referred to as *Hazen's law.* In contrast, retention time, t, is dependent on water depth, H, as given by

$$t = \frac{AH}{Q} \tag{7.21}$$

In reality, depth is important because it can affect flow stability if it is large and scouring if it is small.

Predicting Settling Efficiency (Types 1 and 2). For discrete particle settling (Type 1), variation in particle size and density produce a distribution of settling velocities. The settling velocity distribution may be determined from data collected from a column settling test. The settling column test produces information on x_1 (fraction of particles with settling velocities less than or equal to v_1) and v_1. This data is used to produce a settling-velocity analysis curve (Figure 7.8) (Metcalf and Eddy, 1991).

Equation 7.20 defines the proportion of particles with settling velocity v smaller than v^*, which will be removed in a given time. If x^* is the proportion of particles having settling velocities less than or equal to v^*, the total proportion of particles that could be removed in settling is defined by Thirumurthi (1969):

$$F_t = (1 - x^*) + \int_0^{x} \frac{v}{v^*}\, dx \tag{7.22}$$

This can be solved using a version of Figure 7.8.

In flocculant settling (Type 2), flocculation occurs as the particles settle. To evaluate the effect of flocculation as a function of basin depth requires a column test with sampling ports at various depths (Zanoni and Blomquist, 1975). The settling column should be as deep as the basin being designed. A set of samples is taken every 20 min or so for at least 2 h. The suspended-solids concentration is determined in each sample and expressed as a percentage difference, removal R, of the original concentration. These results are plotted against time and depth, and curves of equal percentage removal are drawn. Figure 7.9 is an example for flocculated-particle Type 2 settling, with increase in settling velocity as settlement progresses. An effective settling rate for the quiescent conditions of the column can be defined as the ratio of the effective depth divided by the time required to obtain a given percentage of removal.

For Figure 7.9, any combination of depth h_n and time t_n on one of the isopercentage lines will establish a settlement velocity v_n:

$$v_n = \frac{h_n}{t_n} \tag{7.23}$$

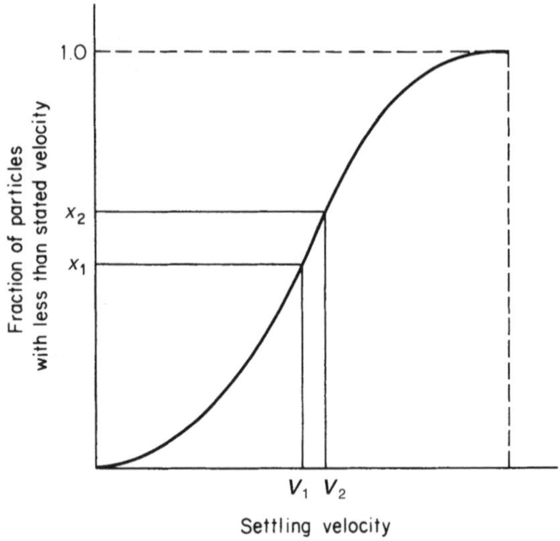

FIGURE 7.8 Settling-velocity analysis curve for discrete particles. (*Source:* Camp, 1936; Metcalf and Eddy, Engineers, 1991. *Wastewater Engineering,* 3rd ed. New York: McGraw-Hill. Reproduced by permission of the McGraw-Hill Companies.)

Thus, all particles with a settling velocity equal to or greater than v_n will be removed. Particles with a velocity v less than v_n are assumed to be removed in the proportion v/v_n.

For the same time t_n, this would be the same as taking the depth ratios, h/h_n, for the same reason as in Eq. 7.20. Then the overall removal of particles to a depth h_n is given by

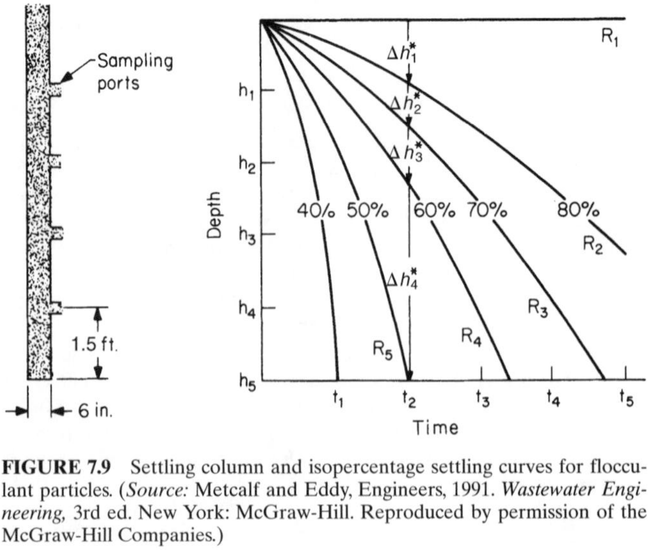

FIGURE 7.9 Settling column and isopercentage settling curves for flocculant particles. (*Source:* Metcalf and Eddy, Engineers, 1991. *Wastewater Engineering,* 3rd ed. New York: McGraw-Hill. Reproduced by permission of the McGraw-Hill Companies.)

$$\Delta R = \frac{1}{h_n}\int_0^{h_n} R \cdot dh \tag{7.24}$$

This might be solved as [Metcalf and Eddy (1991)]:

$$\Delta R = \sum_0^n \left[\frac{\Delta h_n}{h_5} \times \frac{R_n + R_{n+1}}{2}\right] \tag{7.25a}$$

Then with respect to Fig 7.9, this can be written as

$$\Delta R = \left[\frac{\Delta h_1}{h_5} \times \frac{R_1 + R_2}{2}\right] + \left[\frac{\Delta h_2}{h_5} \times \frac{R_2 + R_3}{2}\right] + \left[\frac{\Delta h_3}{h_5} \times \frac{R_3 + R_4}{2}\right] + \left[\frac{\Delta h_4}{h_5} \times \frac{R_4 + R_5}{2}\right] \tag{7.25b}$$

Although of limited practical value, Agrawal and Bewtra (1985) and Ali San (1989) have suggested improvements for this computation that overcome weaknesses in the assumptions.

In practice, to design a full-scale settling tank to achieve comparable removal, the settling rate from the column test should be multiplied by a factor of 0.65 to 0.85, and the detention time should be multiplied by a factor of 1.25 to 1.5 (Metcalf and Eddy, 1991).

Hindered Settling (Types 3a and 3b)

The following addresses Type 3 settling relevant to clarification. Types 3 and 4 settling as relevant to thickening are addressed in Chapter 16.

Particle Interaction. At high particle concentrations, individual particle behavior is influenced, or hindered, by the presence of other particles, and the flow characteristics of the bulk suspension can be affected. With increased particle concentration, the free area between particles is reduced causing greater interparticle fluid velocities and alteration of flow patterns around particles. Consequently, the average settling velocity of the particles in a concentrated suspension is generally less than that of a discrete particle of similar size.

When particles in a suspension are not uniform in size, shape, or density, individual particles will have different settling velocities. Particles with a settling velocity less than the suspension increase the effective viscosity. Smaller particles tend to be dragged down by the motion of larger particles. Flocculation may increase the effective particle size when particles are close together (i.e., flocculation due to differential settling, Eq. 7.14).

Solids Flux. The settling velocity of the suspension, v_s, depends on particle concentration in the suspension. The product of velocity and mass concentration, C, is solids mass flux F_M, the mass of solids passing a unit area per unit of time:

$$F_M = v_s C \tag{7.26a}$$

The equivalent relationship holds for solids volumetric concentration, Φ, to define solids volume flux, F_V:

$$F_V = v_s \Phi \tag{7.26b}$$

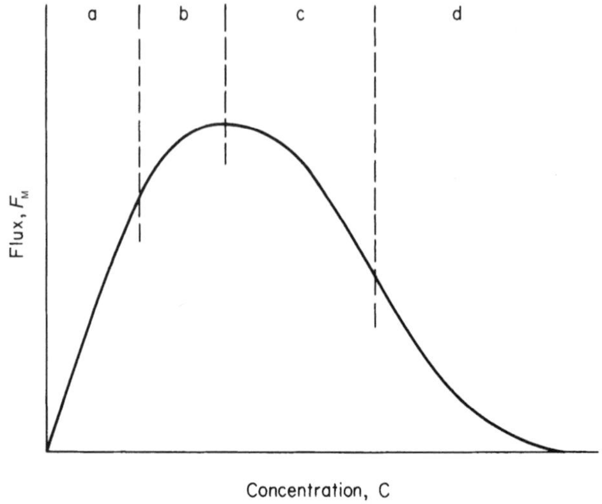

FIGURE 7.10 Typical relationship between flux and concentration for batch settlement.

The relationship between F_M and C is shown in Figure 7.10 and is complex because v_s is affected by concentration. The relationship can be divided into four regions.

Region (a): Type 1 and 2 settling. Unhindered settling occurs such that the flux increases in proportion to the concentration. A suspension of particles with different settling velocities has a diffuse interface with the clear liquid above.

Region (b): Type 3a settling. With increase in concentration, hindered-flow settling increasingly takes effect, and ultimately a maximum value of flux is reached. At about maximum flux, the diffuse interface of the suspension becomes distinct with the clear liquid above when all particles become part of the suspension and settle with the same velocity.

Region (c): Type 3b settling. Further increase in concentration reduces flux because of the reduction in settling velocity. In this region, the suspension settles homogeneously.

Region (d): Type 4 settling. Associated with the point of inflection in the flux-concentration curve, the concentration reaches the point where thickening can be regarded to start leading ultimately to compression settling.

Equations for Hindered Settling. The behavior of suspensions in regions (b) and (c) has attracted considerable theoretical and empirical analysis and is most important in understanding floc-blanket clarification. The simplest and most convenient relationship is represented by the general equation (Gregory, 1979)

$$v_s = v_0 \exp(-q \Phi) \tag{7.27}$$

where q = constant representative of the suspension
v_0 = settling velocity of suspension for concentration extrapolated to zero
Φ = volume concentration of the suspension

Other empirical relationships have been proposed. The most widely accepted and tested relationship was initially developed for particles larger than 0.1 mm diameter in rigid particle-fluidized systems. This relationship has been shown to be applicable to settling and is known as the *Richardson and Zaki equation* (Coulson and Richardson, 1978):

$$v_s = v_t E^n \tag{7.28}$$

where E = porosity of the suspension (i.e., volume of fluid per volume of suspension, $E = 1 - \Phi$)
n = power value dependent on the Reynolds number of the particle
v_t = terminal settling velocity of particles in unhindered flow (i.e., absence of effect by presence of other particles)

For rigid particles, this equation is valid for porosity from about 0.6 (occurring at around minimum fluidization velocity) to about 0.95. The Reynolds number determines the value of n (Coulson and Richardson, 1978). For a suspension with uniform-size spherical particles, $n = 4.8$ when Re is less than 0.2. As the value of Re increases, n decreases until Re is greater than 500 when n equals 2.4.

When Eq. 7.28 is used for flocculent suspensions (Gregory, 1979) correction factors must be included to adjust for effective volume to account for particle distortion and compression. If particle volume concentration is measured, for example, by the half-hour settlement test, then because such a test as this is only a relative measurement providing a measure of the apparent concentration, then such adjustments are necessary.

$$v_s = v_t k_1 (1 - k_2 \Phi^*)^r \tag{7.29}$$

where k_1, k_2 = constants representative of the system
Φ^* = apparent solids volumetric concentration
r = power value dependent on the system

Equation 7.28 can be substituted in Eq. 7.26 for flux with 1 – E substituted for Φ (Coulson and Richardson, 1978):

$$F_M = v_t E^n (1 - E) \tag{7.30}$$

Differentiating this equation with respect to E gives:

$$\frac{dF_M}{dE} = v_t n E^{n-1} - v_t (n+1) E^n \tag{7.31}$$

The flux F_M has a maximum value when dF_M/dE equals zero and E equals E^+ (the porosity at maximum flux). Hence, dividing Eq. 7.31 by $v_t\, E^{n-1}$ and equating to zero produces

$$0 = n - (n+1)\, E^+ \tag{7.32a}$$

or

$$n = \frac{E^+}{1 - E^+} \tag{7.32b}$$

This means that the porosity E^+, or the volume concentration, at maximum flux, Φ^+, is an important parameter in describing the settling rates of suspensions. In the case of rigid uniform spheres, if n ranges from 2.4 to 4.6, the maximum flux should occur at a volumetric concentration between 0.29 and 0.18. In practice, the range of values generally found for suspensions of aluminum and iron flocs for optimal coagulant dose and coagulation pH, when concentration is measured as the half-hour settled volume, is 0.16 to 0.20 (Gregory, 1979; Gregory, Head, and Graham, 1996), in which case n ranges from 4.0 to 5.26.

If Eq. 7.31 is differentiated also, then

$$\frac{d^2F_M}{dE^2} = v_t[n(n-1)E^{n-2} - (n+1)nE^{n-1}] \tag{7.33}$$

and when $d^2F_M/dE^2 = 0$ for real values of E, a point of inflection will exist, given by

$$0 = n - 1 - (n+1)\,E \tag{7.34a}$$

such that

$$E = \frac{n-1}{n+1} \tag{7.34b}$$

For rigid uniform spheres, if n ranges from 2.4 to 4.6, the point of inflection occurs at a concentration between 0.59 and 0.35. For nonrigid, irregular-shaped, and multi-sized particles, the situation is more complex.

The point of inflection is associated with the transition from Type 3 to Type 4 settling. Type 3 and 4 settling in the context of thickening are considered in Chapter 16.

Prediction of Settling Rate. The hindered settling rate can be predicted for suspensions of rigid and uniform spheres using Eqs. 7.5 and 7.28. For suspensions of nonuniform and flocculent particles, however, settling rate has to be measured. This is most simply done using a settling column; a 1-L measuring cylinder is usually adequate. The procedure is to fill the cylinder to the top measuring mark with the sample and record at frequent intervals the level of the interface between the suspension and the clear-water zone. The interface is only likely to be distinct enough for this purpose if the concentration of the sample is greater than that at maximum flux. The results are plotted to produce the typical settling curve (Figure 7.11). The slope of the curve over the constant-settling-rate period is the estimate of the Type 3 settling rate for quiescent conditions. If the concentration of the sample was greater than that at the inflection point in the mass flux curve, the transition from region (c) to (d) in Figure 7.10, then a period of constant settling rate and the compression point (CP) will not be found as represented by line A.

The compression point signifies the point at which all the suspension has passed into the Type 4 settling or compression regime. Up to that time, a zone of solids in the compression regime has been accumulating at the bottom of the suspension with its upper interface moving upward. The compression point, thus, is where that interface reaches the top of the settling suspension.

Fluidization

When liquid is moving up through a uniform stationary bed of particles at a low flow rate, the flow behavior is similar to when the flow is down through the bed. When the upward flow of liquid is great enough to cause a drag force on particles equal to the apparent weight (actual weight less buoyancy) of the particles, the particles rearrange to offer less resistance to flow and bed expansion occurs. This process continues as the liquid velocity is increased until the bed has assumed the least stable form of packing. If the upward liquid velocity is increased further, individual particles sep-

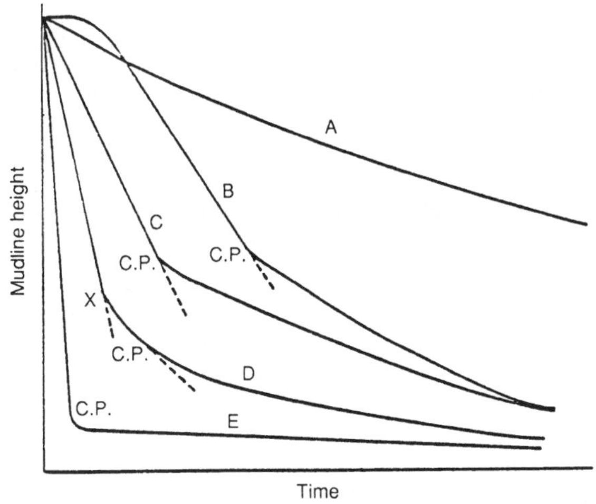

A Concentrated, flocculated or unflocculated pulp
B Intermediate flocculated or unflocculated pulp
C As (B) but showing no induction period
D Dilute, flocculated pulp
E Dilute, unflocculated pulp
C.P. Compression point

FIGURE 7.11 Typical batch-settling curves. (*Source:* Pearse, 1977.)

arate from one another and become freely supported in the liquid. The bed is then said to be *fluidized.*

For rigid and generally uniform particles, such as with filter sand, about 10 percent bed expansion occurs before fluidization commences. The less uniform the size and density of the particles, the less distinct is the point of fluidization. A fluidized bed is characterized by regular expansion of the bed as liquid velocity increases from the minimum fluidization velocity until particles are in unhindered suspension (i.e., Type 1 settling).

Fluidization is hydrodynamically similar to hindered, or zone Type 3, settling. In a fluidized bed, particles undergo no net movement and are maintained in suspension by the upward flow of the liquid. In hindered settling, particles move downward, and in the simple case of batch settling, no net flow of liquid occurs. The Richardson and Zaki equation, Eq. 7.28, has been found to be applicable to both fluidization and hindered settling (Coulson and Richardson, 1978) as have other relationships.

In water treatment, floc-blanket clarification is more a fluidized bed rather than a hindered settling process. Extensive floc-blanket data (Gregory, 1979) with Φ^* determined as the half-hour settled-solids volume, such that Φ^+ tended to be in the range 0.16 to 0.20, allowed Eq. 7.29 to be simplified to

$$v_s = v_0 (1 - 2.5\Phi^*) \tag{7.35a}$$

or

$$v_0 = \frac{v_s}{(1 - 2.5\Phi^*)} \tag{7.35b}$$

The data that allowed this simplification was obtained with alum coagulation (optimal coagulant dose and coagulation pH) of a high-alkalinity, organic-rich river water, but the value for k_2 (in Eq. 7.29) of 2.5 should hold for other types of water producing similar quality floc. The value of k_2 can be estimated as the ratio of the concentration at the compression point to the half-hour settled concentration. Values predicted for v_0 by Eq. 7.35b are less, about one-half to one-third, than those likely to be estimated by Stokes' equation for v_t, assuming spherical particles (Gregory, 1979).

The theory of hindered settling and fluidization of particles of mixed sizes and different densities is more complex and is still being developed. In some situations, two or more phases can occur at a given velocity, each phase with a different concentration. This has been observed with floc blankets to the extent that an early but temporary deterioration in performance occurs with increase in upflow (Gregory and Hyde, 1975; Setterfield, 1983). An increase in upflow leads to intermixing of the phases, with further increase in upflow limited by the characteristics of the combined phase. The theory has been used to explain and predict the occurrence of intermixing and segregation in multimedia filter beds during and after backwash (Patwardan and Tien, 1985; Epstein and LeClair, 1985).

EXAMPLE PROBLEM 7.1 Predict the maximum volume flux conditions for floc-blanket sedimentation.

SOLUTION For a floc blanket that can be operated over a range of upflow rates, collect samples of blanket at different upflow rates. For these samples, measure the half-hour settled volume. Example results are listed below:

Upflow rate (m/h)	1.6	1.95	2.5	3.05	3.65	4.2	4.7	5.15
Half-hour floc volume (%)	31	29	25	22	19	16	13	10
Blanket flux = upflow rate × half-hour floc volume (%m/h)	49.6	56.6	62.5	67.1	69.4	67.2	61.1	51.5

These results predict that maximum flux occurs at an upflow rate of 3.7 m/h. If flux is plotted against upflow rate and against half-hour floc volume, then the maximum flux is located at 3.4 m/h for a half-hour floc volume of 20 percent, as shown in Figure 7.12.

The above results can be fitted to Eq. 7.35a:

$$v_s = v_0\,(1 - 2.5\Phi^*)$$

$$3.44 = v_0\,(1 - 2.5 \times 0.2)$$

$$v_0 = 3.44/0.5 = 6.9 \text{ m/h}$$

This means that at the maximum flux, the theoretical terminal settling velocity of the blanket is 6.9 m/h. The maximum operating rate for a floc blanket in a stable tank is about 70 percent of this rate, or 4.8 m/h.

Inclined (Tube and Plate) Settling

The efficiency of discrete particle settling in horizontal liquid flow depends on the area available for settling. Hence, efficiency can be improved by increasing the area. Some tanks have multiple floors to achieve this. A successful alternative is to use lightweight structures with closely spaced inclined surfaces.

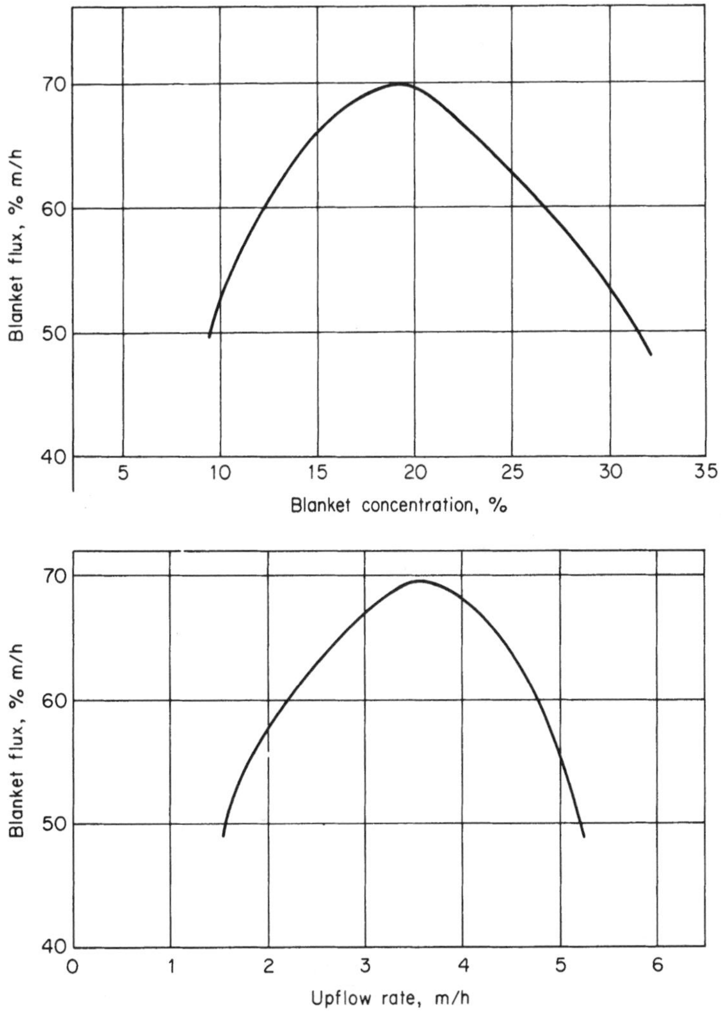

FIGURE 7.12 Relationship between blanket flux, blanket concentration, and upflow rate for Example Problem 7.1.

Inclined settling systems (Figure 7.13) are constructed for use in one of three ways with respect to the direction of liquid flow relative to the direction of particle settlement: countercurrent, cocurrent, and cross-flow. Comprehensive theoretical analyses of the various flow geometries have been made by Yao (1970). Yao's analysis is based on flow conditions in the channels between the inclined surfaces being laminar. In practice, the Reynolds number must be less than 800 when calculated using the mean velocity v_θ between and parallel to the inclined surfaces and hydraulic diameter of the channel d_H:

$$d_H = \frac{4A_H}{P} \tag{7.36}$$

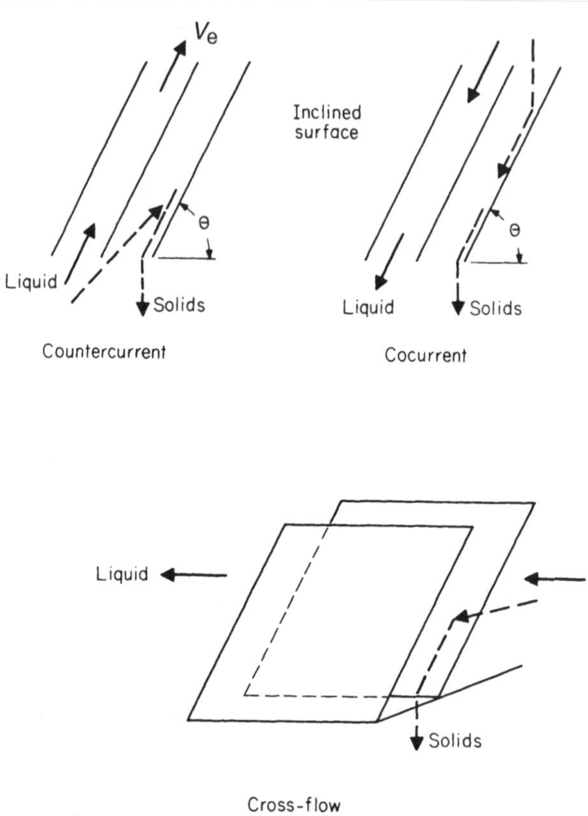

FIGURE 7.13 Basic flow geometries for inclined settling systems.

where A_H is the cross-sectional area of channel-to-liquid flow, and P is the perimeter of A_H, such that Eq. 7.6 becomes:

$$\mathrm{Re} = \frac{\rho v_\theta d_H}{\mu} \tag{7.37}$$

Countercurrent Settling. The time, t, for a particle to settle the vertical distance between two parallel inclined surfaces is:

$$t = \frac{w}{v \cos \theta} \tag{7.38}$$

where w is the perpendicular distance between surfaces, and θ is the angle of surface inclination from the horizontal. The length of surface, L_p, needed to provide this time, if the liquid velocity between the surfaces is v_θ, is

$$L_p = \frac{w(v_\theta - v \sin \theta)}{v \cos \theta} \tag{7.39a}$$

By rearranging this equation, all particles with a settling velocity, v, and greater are removed if

$$v \geq \frac{v_\theta w}{L_p \cos\theta + w \sin\theta} \tag{7.39b}$$

When many plates or tubes are used

$$v_\theta = \frac{Q}{Nwb} \tag{7.40}$$

where N is the number of channels made by $N + 1$ plates or tubes, and b is the dimension of the surface at right angles to w and Q.

Cocurrent Settling. In cocurrent settling, the time for a particle to settle the vertical distance between two surfaces is the same as for countercurrent settling. The length of surface needed, however, has to be based on downward, and not upward, liquid flow:

$$L_p = w\frac{(v_\theta + v \sin\theta)}{v \cos\theta} \tag{7.41a}$$

Consequently, the condition for removal of particles is given by

$$v \geq \frac{v_\theta w}{L_p \cos\theta - w \sin\theta} \tag{7.41b}$$

Cross-Flow Settling. The time for a particle to settle the vertical distance between two surfaces is again given by Eq. 7.39. The liquid flow is horizontal and does not interact with the vertical settling velocity of a particle. Hence

$$L_p = \frac{v_\theta w}{v \cos\theta} \tag{7.42a}$$

and

$$v \geq \frac{v_\theta w}{L_p \cos\theta} \tag{7.42b}$$

Other Flow Geometries. The above three analyses apply only for parallel surface systems. To simplify the analysis for other geometries, Yao (1970) suggested a parameter, S_c, defined as

$$S_c = v\frac{(\sin\theta + L_r \cos\theta)}{v_\theta} \tag{7.43}$$

where $L_r = L/w$ is the relative length of the settler.

When v^* is the special case that all particles with velocity, v, or greater are removed, then for parallel surfaces, S_c is equal to 1.0. However, the value for circular tubes is 4/3 and for square conduits is 11/8 (Yao, 1973). Identical values of S_c for different systems may not mean identical behavior.

The design overflow rate is also defined by v^* in Eq. 7.43, and Yao has shown by integration of the differential equation for a particle trajectory that the overflow rate for an inclined settler is given by

$$v^* = \frac{k_3 \, K \, v_\theta}{L_r} \tag{7.44}$$

where k_3 is a constant equal to 8.64×10^2 m^3/day · m^{-2} and

$$K = \frac{S_C \, L_r}{\sin\theta + L_r \cos\theta} \tag{7.45a}$$

For given values of overflow rate and surface spacing and when $\theta = 0$, Eq. 7.45a becomes

$$\frac{S_C}{L} = \text{constant} \tag{7.45b}$$

Equation 7.45b indicates that the larger the value of S_C the longer the surface length must be to achieve the required theoretical performance. In practice, compromises must be made between theory and the hydrodynamic problems of flow distribution and stability that each different geometry poses.

EXAMPLE PROBLEM 7.2 A tank has been fitted with 2.0 m (6.6 ft) square inclined plates spaced 50 mm (2.0 in) apart. The angle of inclination of the plates can be altered from 5° to 85°. The inlet to and outlet from the tank can be fitted in any way so that the tank can be used for either countercurrent, cocurrent, or cross-flow sedimentation. If no allowances need to be made for hydraulic problems due to flow distribution and so on, then which is the best arrangement to use?

SOLUTION Equation 7.39b for countercurrent flow, Eq. 7.41b for cocurrent flow, and Eq. 7.42b for cross-flow sedimentation are compared. As an example, the calculation for countercurrent flow at 85° is

$$\frac{v}{v_\theta} = \frac{50}{2000 \cos 85 + 50 \sin 85} = \frac{50}{174.3 + 49.8} = 0.223$$

The smallest value of v is required. Thus, for the range:

Angle (θ)	5	15	30	45	60	75	85
Countercurrent (v/v_θ)	0.025	0.026	0.028	0.035	0.048	0.088	0.223
Cocurrent (v/v_θ)	0.025	0.026	0.029	0.036	0.052	0.106	0.402
Cross-flow (v/v_θ)	0.025	0.026	0.029	0.035	0.050	0.096	0.287

From the above, little difference exists between the three settling arrangements for an angle of less than 60°. For angles greater than 60°, countercurrent flow allows settlement of particles with the smallest settling velocity.

Floc-Blanket Clarification

A simple floc-blanket tank has a vertical parallel-walled upper section with a flat or hopper-shaped base. Water that has been dosed with an appropriate quantity of a suitable coagulant, and pH adjusted if needed, is fed downward into the base. The resultant expanding upward flow allows flocculation to occur, and large floc particles remain in suspension within the tank. Particles in suspension accumulate slowly at first, but then at an increasing rate due to enhanced flocculation and other effects, eventually reaching a maximum accumulation rate limited by the particle characteristics and the upflow velocity of the water. When this maximum rate is reached, a floc blanket can be said to exist.

As floc particles accumulate, the volume occupied by the suspension in the floc blanket increases and its upper surface rises. The level of the floc blanket surface is controlled by removing solids from the blanket to keep a zone of clear water or supernatant liquid between the blanket and the decanting troughs, launders, or weirs.

A floc blanket is thus a fluidized bed of floc particles even though the process can be regarded as a form of hindered settling. However, true hindered settling exists only in the upper section of sludge hoppers used for removing accumulated floc for blanket-level control. Thickening takes place in the lower section of the sludge hoppers. Excess floc removed from the floc blanket becomes a residue stream and may be thickened to form sludge (see Chapter 16).

Mechanism of Clarification. Settling, entrainment, and particle elutriation occur above and at the surface of a blanket. The mechanism of clarification within a floc blanket is more complex, however, and involves flocculation, entrapment, and sedimentation. In practice, the mean retention time of the water within a blanket is in excess of the requirements for floc growth to control the efficiency of the process (i.e., the opportunity for the small particles to become parts of larger and more easily retained floc is substantial, so other factors cause particles to pass through a blanket).

Physical removal by interception and agglomeration, similar to surface capture in deep-bed filtration, occurs throughout a floc blanket. Probably the most important process is mechanical entrapment and straining, in which rising small particles cannot pass through the voids between larger particles that comprise the bulk of the blanket. (The mechanisms are not the same as in filtration through a fixed bed of sand, because all the particles are in fluid suspension.) The efficiency of entrapment is affected by the spacing of the larger suspended floc particles, which, in turn, is related to floc quality (shape, density, and so on) and water velocity. When suspension destabilization, coagulation, is not optimal, then flocculation will be poor and will result in a greater number of smaller particles that can pass through the floc blanket. (See Chapter 6 for material on coagulation and particle destabilization.)

Performance Prediction. Within a floc blanket, the relationship between floc concentration and upflow velocity of the water is represented by Eqs. 7.26 through 7.29 for hindered settling and fluidization. Unsuccessful attempts have been made to establish a simple theory for predicting solids removal (Cretu, 1968; Shogo, 1971). The relationships between settled-water quality and floc concentration in the blanket, upflow velocity, and flux (Figure 7.14) are of practical importance for understanding and controlling plant performance (Gregory, 1979). Recently, however, floc-blanket clarification has been modeled successfully (Hart, 1996; Gregory, Head, and Graham, 1996; Head, Hart, and Graham, 1997).

The modeling by Head and associates has been successfully tested in dynamic simulations of pilot and full-scale plants. The modeling is based on the theories and work of various researchers, including Gould (1967) and Gregory (1979). Although the model accommodates the principle that the removal rate of primary particles is

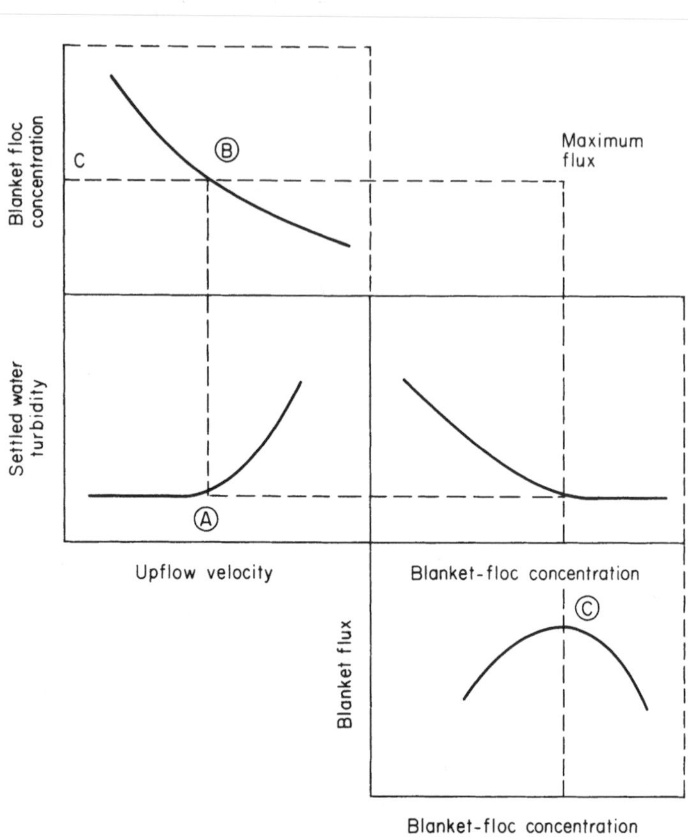

FIGURE 7.14 Typical relationships between settled-water quality and blanket concentration, upflow velocity and blanket flux. (*Source:* Gregory, 1979.)

dependent on blanket concentration, removal is simulated on the basis that the blanket region of a clarifier is a continuous stirred-tank reactor (CSTR). To take into account the possibility of poor coagulation, the model can assume a nonremovable fraction of solids.

The relationships in Figure 7.14 show that settled-water quality deteriorates rapidly (point A) as the floc concentration (point B) is decreased below the concentration at maximum flux (point C). Conversely, little improvement in settled-water quality is likely to be gained by increasing floc concentration to be greater than that at maximum flux (to the left of points A and B). This is because for concentrations greater than that at maximum flux, interparticle distances are small enough for entrapment to dominate the clarification process.

As the concentration decreases below that at maximum flux, interparticle distances increase, especially between the larger particles, and their motion becomes more intense. Some of the larger particles might not survive the higher shear rates that develop. Thus, smaller particles may avoid entrapment and escape from the floc blanket. Consequently, the maximum flux condition represents possible optimum operating and design conditions. Maximum flux conditions and performance depend on various factors, which account for differences between waters and particle surface and coagulation chemistry, and are described later.

Effects of Upflow Velocity. The surface loading for floc-blanket clarification is expressed as the upflow velocity, or overflow rate. For some floc-blanket systems, the performance curve is quartic, reflecting an early or premature deterioration in water quality of limited magnitude with increase in upflow velocity (Figure 7.15). This deterioration is associated with segregation of particles, or zoning, in the blanket at low surface loading (Gregory, 1979) because of the wide range in particle settling velocities. This has been observed not only in the treatment waters with a high silt content but also with the use of powdered carbon (Setterfield, 1983) and in precipitation softening using iron coagulation (Gregory and Hyde, 1975). As surface loading is increased, remixing occurs at the peak of the "temporary" deterioration as the lower-lying particles are brought into greater expansion.

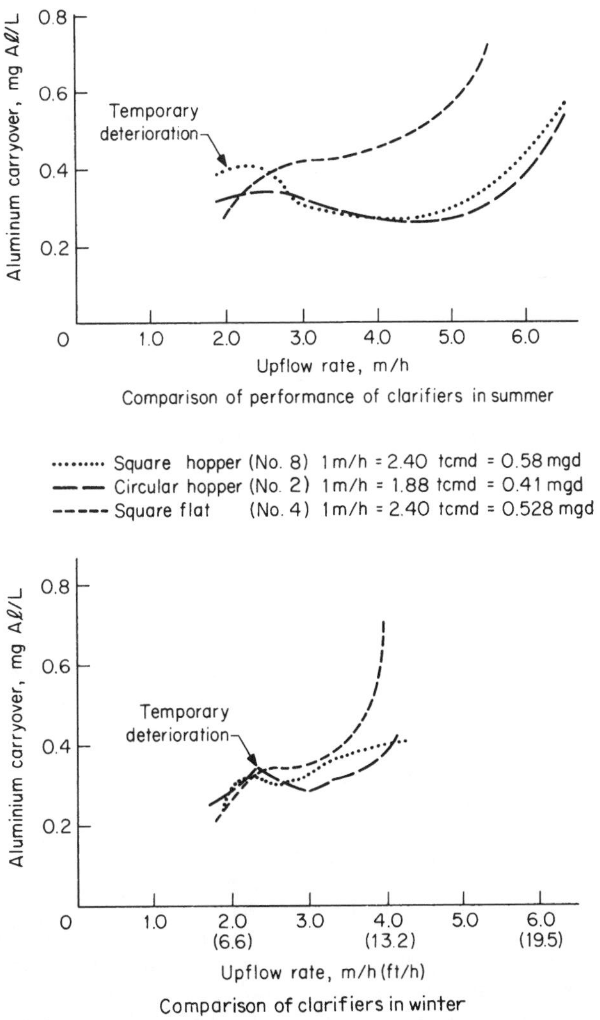

FIGURE 7.15 Floc-blanket performance curves showing "temporary" deterioration in settled-water quality. (*Source:* Setterfield, 1983.)

For floc-blanket clarification, the "corner" of the performance curve (point A in Figure 7.14) is associated with the point of maximum flux. The limit to upflow velocity for reliable operation has been expressed by some (Bond, 1965; Tambo et al., 1969; Gregory, 1979) in simple terms of terminal velocity. Bond (1965) noted that the blanket surface remained clearly defined up to a velocity of about half of the zero-concentration settling rate, v_0, that slight boiling occurred above $0.55v_0$, and that clarification deteriorated noticeably at about $0.65v_0$. Tambo et al. (1969) found that a floc blanket is stable for velocities less than about $0.7v_0$ and very unstable at velocities greater than $0.8v_0$. Gregory (1979) found that the velocity at maximum flux was about $0.5v_0$, as given by Eq. 7.35, when the maximum-flux half-hour settled-solids volume is 20 percent. Gregory also observed that for velocities less than at maximum flux, the blanket interface was sharp. However, as upflow velocity increased beyond this, the blanket surface became more diffuse to the extent that a blanket was very difficult to sustain for a velocity greater than $0.75v_0$. Hence, the best guideline for optimum operation is to use the velocity that creates a blanket concentration at which the blanket surface becomes diffuse.

The half-hour settled-solids volume at maximum flux has been found to be in the range of 16 to 20 percent, when coagulant dose and coagulation pH are selected from jar tests to minimize metal-ion concentration and turbidity. This has been observed for both alum and iron coagulation with and without using polyelectrolyte flocculant aid (Gregory, Head, and Graham, 1996). The actual value depends on the quality of the water as well as the choice of coagulation chemistry, because these govern such characteristics as floc strength, size, and density. The lowest blanket solids volume concentration that a discernible blanket can be found to exist with is about 10 to 12 percent. If upflow rate is increased to cause further dilution, then the blanket effectively becomes "washed out."

Fluid Mixing and Residence Time Distribution in Sedimentation Tanks

Ideal Flow Conditions. The simplest flow condition is plug flow, when all liquid advances with equal velocity. Conditions only approximate to this when turbulence is small and uniform throughout the liquid. In laminar, nonturbulent flow conditions, a uniform velocity gradient exists, with velocity zero at the wall and maximum at the center of the channel through which the liquid flows, and, therefore, plug flow cannot exist.

Major departures from plug and laminar flow conditions in sedimentation tanks are associated with currents caused by poor flow distribution and collection, wind, or rising bubbles, and density differences caused by temperature or concentration. Currents caused by these factors result in short-circuiting of flow and bulk mixing, and reduce the performance of the process predicted by ideal theory. The extent of departure from ideal plug flow performance can be assessed by residence time distribution analysis with the help of tracer studies and modeling with computational fluid dynamics (CFD).

Residence Time. The theoretical mean residence time of a process is the volume of the process from the point of dosing or end of the previous process through to the point at which separation efficiency is measured or the outlet of the process, divided by the flow-through rate. For horizontal flow sedimentation, the volume of the entire tank is important in assessing the effect on sedimentation efficiency. For inclined settlers the volume within the inclined surfaces, and for floc blankets the volume of the blanket itself, are most important. Depth can be used to reflect the volume of a floc blanket.

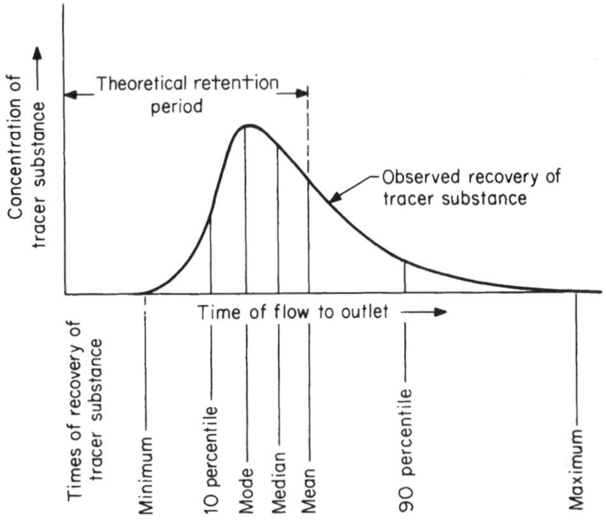

FIGURE 7.16 Typical plot of a flow-through curve for a tracer dosed as a slug. (*Source: Elements of Water Supply and Wastewater Disposal,* 2nd ed., G. M. Fair, J. C. Geyer, and D. A. Okun. Copyright © 1971 John Wiley & Sons, Inc. Reprinted by permission of John Wiley & Sons, Inc.)

The mean residence time of a process is effectively the length, in the direction of flow, divided by the average flow velocity. Thus, the mean residence time reflects the average velocity, or overflow rate, and will relate to sedimentation efficiency accordingly.

Flow-through curves (fluid residence time distributions) (Figure 7.16) are a graphical depiction of the distribution of fluid element residence times. These can be analyzed to produce estimates of efficiency of flow distribution and volumetric utilization of a tank. Consequently, the extent to which sedimentation efficiency might be improved by improving flow conditions can be estimated. Clements and Khattab (1968) have shown with model studies that sedimentation efficiency is correlated with the proportion of plug flow.

Tracer Tests. In simplest form, residence time distribution analysis is carried out by injecting a tracer into the liquid entering the process and monitoring the concentration of the tracer in the liquid leaving the process. The results are plotted to produce a flow-through curve; the likely outcome when the tracer is introduced as a slug is illustrated by Figure 7.16. Alternative tracer test methods and the ways in which results can be analyzed are described in standard texts (e.g., Levenspiel, 1962). The performance indices commonly used in analyses of flow-through curves for sedimentation and other tanks are listed in Table 7.1 (Rebhun and Argaman, 1965; Marske and Boyle, 1973; Hart and Gupta, 1979). A more analytical approach is to produce what is called the $F(t)$ curve (Wolf and Resnick, 1963; Rebhun and Argaman, 1965; Hudson, 1975). $F(t)$ represents the fraction of total tracer added that has arrived at the sampling point, in response to continuous addition of the tracer during the test (as opposed to pulse addition), and is usually plotted against t/T, in which t is the time from injection of tracer, and T is the theoretical residence time or hydraulic detention time. Consequently, mathematical modeling predicts:

TABLE 7.1 Performance Indices Commonly Used in Analyses of Flow-Through Curves

Indices	Definition
T	Theoretical retention time
t_i	Time interval for initial indication of tracer in effluent
t_p	Time to reach peak concentration (mode time)
t_g	Time to reach centroid of curve
t_{10}, t_{50}, t_{90}	Time of 10, 50, and 90 percent of tracer to have appeared in effluent (t_{50} = mean time)
t_{90}/t_{10}	Morril dispersion index, indicates degree of mixing; as value increases, degree of mixing increases
Ideally, the following indices approach values of 1.0 under perfect plug-flow conditions:	
t_i/T	Index of short-circuiting
t_p/T	Index of modal retention time
t_g/T	Index of average retention time
t_{50}/T	Index of mean retention time

$$1 - F(t) = \left[\exp\frac{-1}{(1-p)(1-m)}\right]\left(\frac{t}{T} - p^{(1-m)}\right) \tag{7.46}$$

where p = fraction of active flow volume acting as plug flow
$1 - p$ = fraction of active flow volume acting as mixed flow
m = fraction of total basin volume that is dead space

Rebhun and Argaman (1965) compared using the $F(t)$ curve with using the flow-through curve, and they concluded comparable results are produced. Plotting the $F(t)$ curve provides, however, a quantitative measure, and its results have a clear physical meaning. [Flow-through analysis has been described in the context of measuring the performance of ozone contact tanks in the *SWTR Guidance Manual* (AWWA, 1990).]

The chemical engineering dispersion index, δ, as it applies to tracer studies, was introduced by Levenspiel (1962) and Thirumurthi (1969). The dispersion index, δ, is calculated from the variance of the dye dispersion curve, such that ideal plug flow conditions are indicated when the value of δ approaches zero. Marske and Boyle (1973) compared the dispersion index with other indices, as listed in Table 7.1. They found that the dispersion index has the strongest statistical probability to accurately describe the hydraulic performance of a contact basin. Of the conventional parameters, the Morril index, t_{90}/t_{10}, is the best approximation of δ.

Some of the above hydraulic characteristics were determined in Japan for a range of flow rate for five types of circular and rectangular tanks and compared (Kawamura, 1991). The results showed that a rectangular tank with an inlet diffuser wall was generally the superior design for minimizing short-circuiting and maximizing plug flow.

OPERATIONAL AND DESIGN CONSIDERATIONS FOR SEDIMENTATION

Types of Sedimentation Tanks

Horizontal-Flow Tanks

Rectangular Tanks. With rectangular horizontal-flow tanks, the water to be settled flows in one end, and the treated water flows out at the other end. The inlet flow arrangement must provide a flow distribution that maximizes the opportunity for particles to settle. If flocculation has been carried out to maximize floc particle size, then the flow at the inlet should not disrupt the flocs. This requires minimizing the head loss between the distribution channel and the main body of the tank. A certain amount of head loss is necessary, however, to achieve flow distribution. One option is to attach the final stage of flocculation to the head of the sedimentation tank to assist flow distribution.

The length and cross-sectional shape of the tank must not encourage the development of counterproductive circulatory flow patterns and scour. Outlet flow arrangements also must ensure appropriate flow patterns. The principal differences between tanks relate to inlet and outlet arrangements; length, width, and depth ratios; and the method of sludge removal. For horizontal-flow tanks with a small length-width ratio, the end effects dominate efficiency. Inlet and outlet flow distribution substantially affect overall flow patterns and residence time distribution. When the depth is greater than the width, the length-depth ratio is more important than the length-width ratio.

A length-width ratio of 20 or more ideally is needed to approach plug flow (Hamlin and Wahab, 1970; Marske and Boyle, 1973) and maximum efficiency for horizontal flow and, presumably, inclined settlers, as shown in Figure 7.17, by determination of the reactor dispersion index, δ. Such a high-value ratio may not be economically acceptable, and a lower ratio, possibly as low as about 5, may give acceptable efficiency if the flow distribution is good. The length-width ratio can be increased by installing longitudinal baffles or division walls.

Increasing the length-width ratio also has the effect of increasing the value of the *Froude number:*

$$\mathrm{Fr} = \frac{2v^2}{d_{\mathrm{H}} g} \tag{7.47}$$

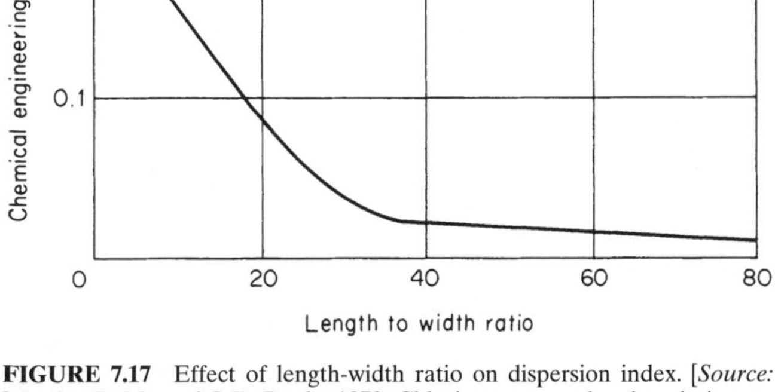

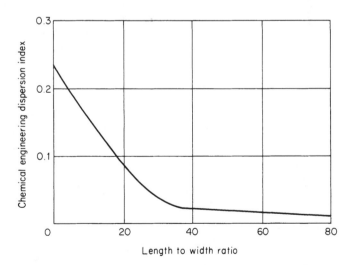

FIGURE 7.17 Effect of length-width ratio on dispersion index. [*Source:* Marske, D. M., and J. D. Boyle. 1973. Chlorine contact chamber design—a field evaluation. *Water and Sewage Works, 120: 70.* Reprinted with permission from *Water and Sewage Works* (1973).]

The value of Fr increases with the length-width ratio because of the increase in velocity v, and decrease in hydraulic diameter d_H. Camp (1936) has shown that the increase in value of Fr is associated with improved flow stability.

Poor flow distribution may produce currents or high flow velocities near the bottom of a tank. This may cause scour, or resuspension, of particles from the layer of settled sludge. Scour may cause transportation of solids along the bottom of the tank to the outlet end. An adequate tank depth can help to limit scour, and, consequently, depths less than 2.4 m (8 ft) are rarely encountered (Gemmell, 1971). To avoid scour, the ratio of length to depth or surface area to cross-sectional area must be kept less than 18 (Kalbskopf, 1970).

Initially, sludge was removed from tanks by simple manual and hydraulic methods. To avoid interruption in operation and to reduce manpower, mechanically aided sludge removal methods were introduced. In the most common method, mechanical scrapers push the sludge to a hopper at the inlet end of the tank. Periodically, the hopper is emptied hydraulically.

If sludge is not removed regularly from horizontal-flow tanks, allowance must be made to the tank depth for sludge accumulation so that the sedimentation efficiency remains unaffected. Sludge can be allowed to accumulate until settled-water quality starts to be impaired. The tank floor should slope toward the inlet, because the bulk of solids generally settle closer to the inlet end.

The frequency of sludge removal depends upon the rate of sludge accumulation. This can be estimated by mass balance calculations. Sometimes, the decomposition of organic matter in the sludge necessitates more frequent sludge removal. Decomposition can be controlled by prechlorination, if this practice is acceptable; otherwise, decomposition may produce gas bubbles that disturb the settled sludge and create disruptive flow patterns.

Frequent sludge removal is best carried out with mechanical sludge scrapers that sweep the sludge to a hopper at the inlet end of the tank. Frequent removal results in easy maintenance of tank volumetric efficiency, and better output efficiency with continuous operation.

Multistory Tanks. Multistory, or tray, tanks are a result of recognizing the importance of settling area to settling efficiency. Two basic flow arrangements are possible with multistory tanks. The trays may be coupled in parallel with flow divided between them (Figure 7.18), or coupled in series with flow passing from one to the next. A few of the latter reverse-flow tanks exist in the United States with two levels.

The Little Falls Water Filtration Plant of the Passaic Valley Water Commission, Clifton, New Jersey, uses tanks with two layers of reverse-flow (four levels in total) coupled in series. Coagulated water enters the lower pass and returns on the level above. Clarified water is removed using submerged launders. Sludge collectors move in the direction of the flow, scraping settled material to sludge hoppers at the far end of the first pass. Each collector flight is trapped at the effluent end on the return pass so that collected material drops down into the path of the influent to the bottom pass.

Multistory tanks are attractive where land value is high. Difficulties with these tanks include a limited width of construction for unsupported floors, flow distribution, sludge removal, and maintenance of submerged machinery. Successful installations in the United States show that these difficulties can be overcome in a satisfactory manner. However, tanks with reverse flow (180° turn) tend to be the least efficient (Kawamura, 1991).

Circular Tanks. Circular tank flow is usually from a central feedwell radially outward to peripheral weirs (Figure 7.19). The tank floor is usually slightly conical to a central sludge well. The floor is swept by a sludge scraper that directs the sludge toward the central wall. Circular tanks incorporate central feedwells that are needed

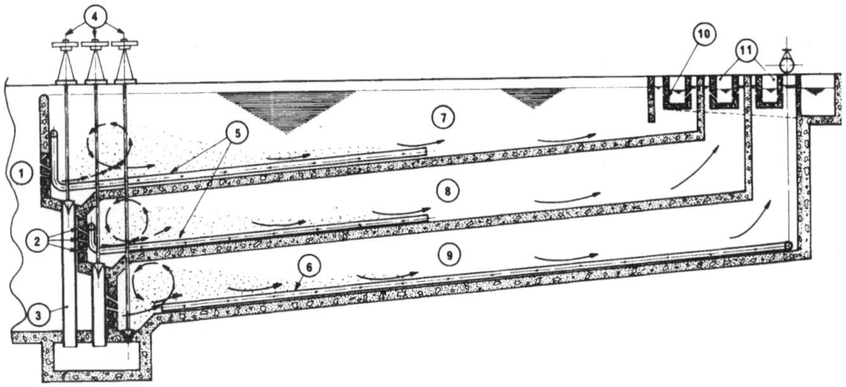

① Flocculation tank
② SPLIT-ROLL inlets
③ Upper floor sludge drain
④ Manually controlled TOP VALVE
⑤ COMBCET: Sludge suction system (optional)
⑥ WATERINSE: Water flushing system (optional)
⑦ Upper clarifying compartment
⑧ Intermediate clarifying compartment
⑨ Lower clarifying compartment
⑩ DUCK-LIPS: Effluent collectors
⑪ Clarified water collecting channels

FIGURE 7.18 Multistory horizontal tank with parallel flow on three levels. (*Source:* Courtesy of OTV, Paris, France, and Kubota Construction Co., Ltd., Tokyo, Japan.)

to assist flow distribution but also are often designed for flocculation and so incorporate some form of agitation. If this agitation is excessive, it carries through to the outer settling zone and affects sedimentation efficiency (Parker et al., 1996).

Radial outward flow is theoretically attractive because of the progressively decreasing velocity. The circumference allows a substantial outlet weir length and hence a relatively low weir loading. Weirs must be adjustable, or installed with great accuracy, to avoid differential flow around a tank. Circular tanks are convenient for constructing in either steel or concrete, although they might be less efficient in the use of land than rectangular tanks. Sludge removal problems tend to be minimal.

The settling efficiency might be less than expected because of the problem of achieving good flow distribution from a central point to a large area. The principal differences between circular tanks are associated with floor profile and sludge scraping equipment.

Inclined (Plate and Tube) Settlers. Individual or prefabricated modules of inclined plate or tube settlers can be constructed of appropriate materials. The advantages of prefabricated modules include efficient use of material, accuracy of separation distances, lightweight construction, and structural rigidity. Inclined surfaces may be contained within a suitably shaped tank for countercurrent, cocurrent, or cross-flow sedimentation. Adequate flocculation is a prerequisite for inclined settling if coagulation is carried out. The tank containing the settler system also can incorporate the flocculation stage and preliminary sludge thickening (Figure 7.20). Particle removal can be enhanced by ballasting the floc, as in the *Actiflo* process (de Dianous, Pujol, and Druoton, 1990).

The angle of inclination of the tubes or plates depends upon the application, the tendency for self-cleaning, and the flow characteristics of the sludge on the inclined surface. If the angle of inclination of the inclined surfaces is great enough, typically

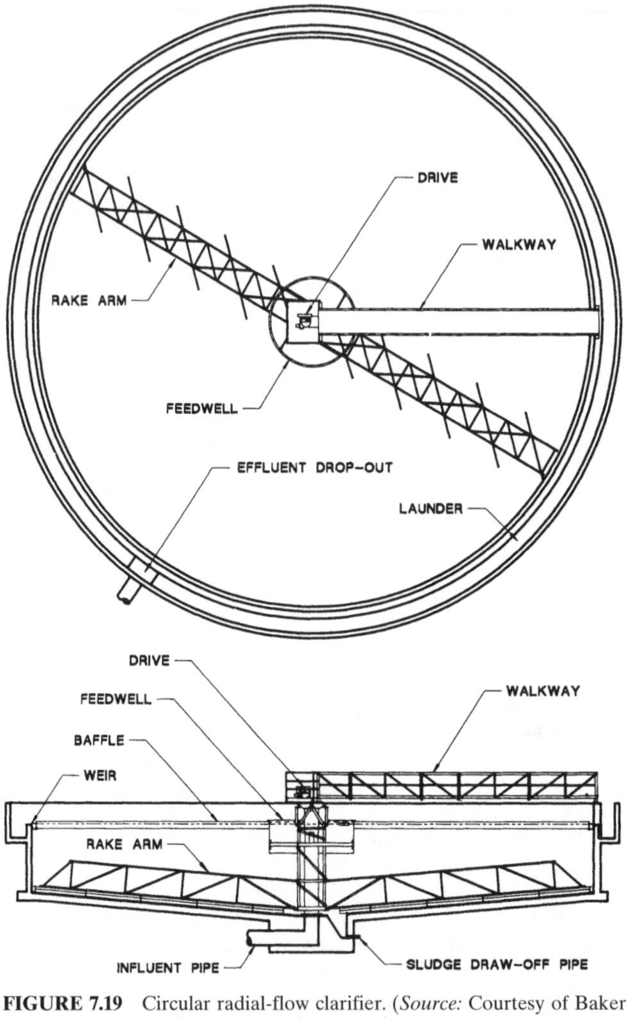

FIGURE 7.19 Circular radial-flow clarifier. (*Source:* Courtesy of Baker Process Equipment Co., Salt Lake City, Utah.)

more than 50° to 60° (Yao, 1973), self-cleaning of surfaces occurs. Demir (1995) found for inclined plates fitted at the end of a pilot horizontal-flow settler the optimum angle is about 50°, with this becoming more pronounced as surface loading rate increases. When the angle of inclination is small, the output of the settler must be interrupted periodically for cleaning. This is because the small distance between inclined surfaces allows little space for sludge accumulation. An angle of as little as 7° is used when sludge removal is achieved by periodic backflushing, possibly in conjunction with filter backwashing. The typical separation distance between inclined surfaces for unhindered settling is 50 mm (2 in) with an inclined length of 1 to 2 m (3 to 6 ft).

The main objective in inclined settler development has been to obtain settling efficiencies close to theoretical. Considerable attention must be given to providing

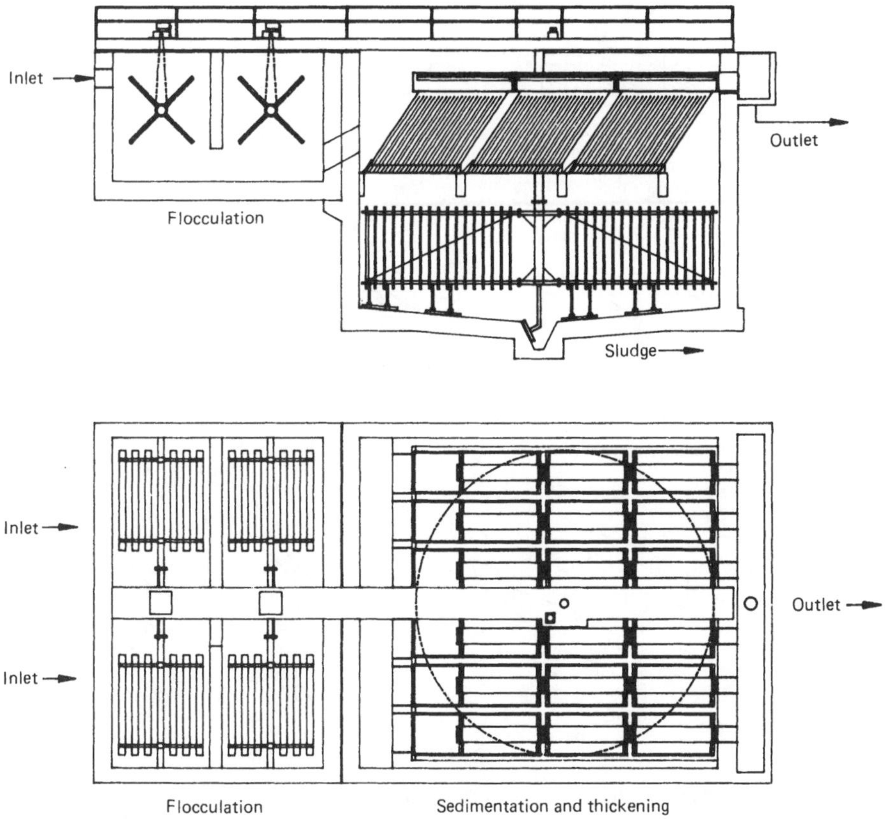

FIGURE 7.20 Inclined-plate settler with preflocculation and combined thickening. (*Source:* Courtesy of US Filter–Zimpro Products, Rothschild, Wis.)

equal flow distribution to each channel, producing good flow distribution within each channel, and collecting settled sludge while preventing its resuspension.

With inclined settlers, the velocity along the axis of the channels defines the flow regime. In practice, the efficiency is usually related either (1) to the surface loading based on the plan area occupied by the settling system, (2) to the upflow velocity, or (3) to the loading based on the total area available for settlement.

Countercurrent Settlers. In countercurrent inclined settlers, the suspension is fed below the settling modules, and the flow is up the channels formed by the inclined surfaces (Figure 7.13). Solids settle onto the lower surface in each channel. If the angle of inclination is great enough, the solids move down the surface counter to the flow of the liquid; otherwise, periodic interruption of flow, possibly with flushing, is necessary for cleaning.

Tube settlers are used mostly in the countercurrent settling mode. Tube modules have been constructed with various configurations (Figure 7.21), including square tubes between vertical sheets, alternating inclination between adjacent vertical sandwiches, chevron-shaped tubes between vertical sheets, and hexagonal tubes.

Countercurrent modular systems are suitable for installing in existing horizontal-flow tanks and some solids contact clarifiers to achieve upgrading and uprating.

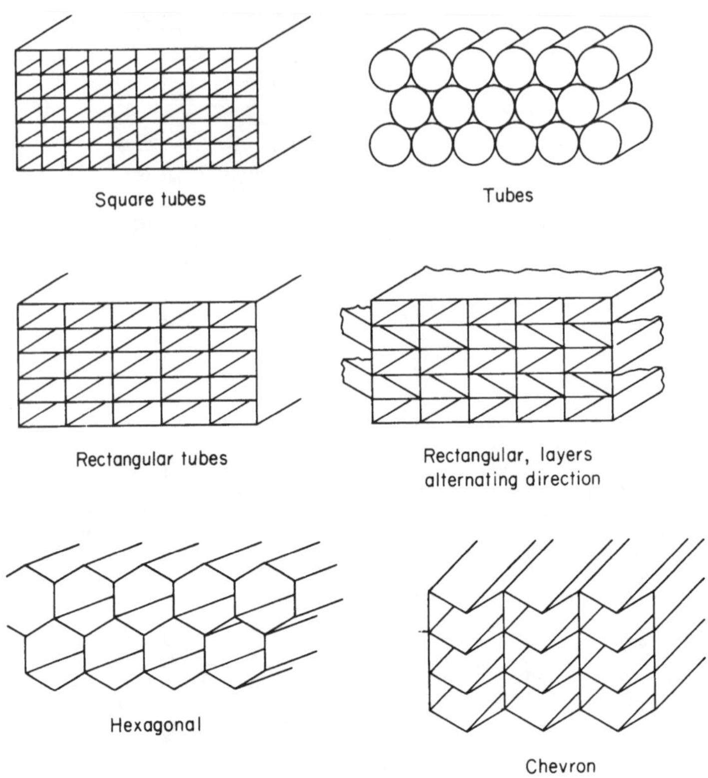

FIGURE 7.21 Various formats for tube modules.

Closely spaced inclined surface systems are not cost-effective in floc-blanket clarifiers, although widely spaced [0.3 m (1 ft)], inclined plates are. Tube modules may aid uprating by acting in part as baffles that improve flow uniformity.

Cocurrent Settlers. In cocurrent sedimentation, the suspension is fed above the inclined surfaces and the flow is down through the channels (Figure 7.13). Settled solids on the lower surface move down the surface in the same direction as the liquid above. Special attention must be given to collecting settled liquid from the lower end of the upper surface of each channel to prevent resuspension of settled solids.

Cross-Flow Settlers. In cross-flow sedimentation, the suspension flows horizontally between the inclined surfaces, and the settled solids move downward (Figure 7.13). In this case, resuspension of settled solids is usually less of a problem than in countercurrent and cocurrent settling. This might not be true in some systems in which the direction of inclination alternates (Figure 7.22) (Gomella, 1974). Alternating inclination can allow efficient use of tank volume and results in rigidity of modular construction. Development and application of cross-flow systems has occurred mainly in Japan.

Solids Contact Clarifiers. Solids contact clarifiers are generally circular in shape and contain equipment for mixing, flow recirculation, and sludge scraping. There is a wide variety of these tanks, and most are of proprietary design. Hartung (1951) has

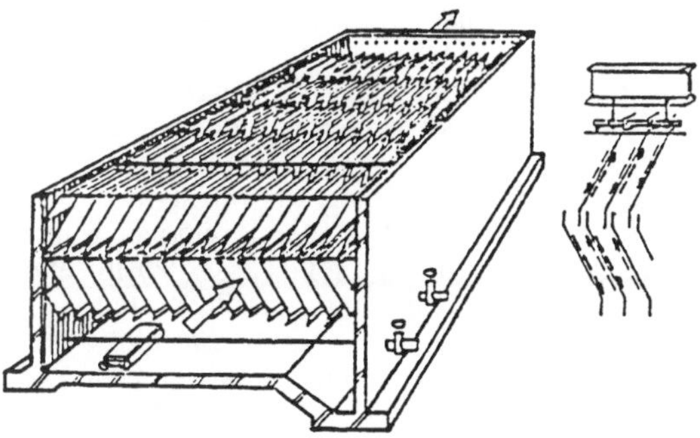

FIGURE 7.22 Alternating cross-flow lamella settler. [*Source:* Gomella, Co. Clarification avant filtration; ses progres recents (Rapport General 1). Int'l Water Supply Assoc. Int'l Conf., 1974.]

presented a review of eight different designs. Contact clarifiers are of two types: premix (like the example in Figure 7.3) and premix-recirculation (like that in Figure 7.23).

In the simpler premix system, water is fed into a central preliminary mixing zone that is mechanically agitated. This premix zone is contained within a shroud that acts as the inner wall of the outer annular settling zone. Chemicals can be dosed into the premix zone. Water flows from the premix zone under the shroud to the base of the settling zone.

In the premix-recirculation system, water is drawn out of the top of the premix zone and fed to the middle of the settling zone. The recirculation rate can exceed the actual flow of untreated water to the tank such that the excess flow in the settling

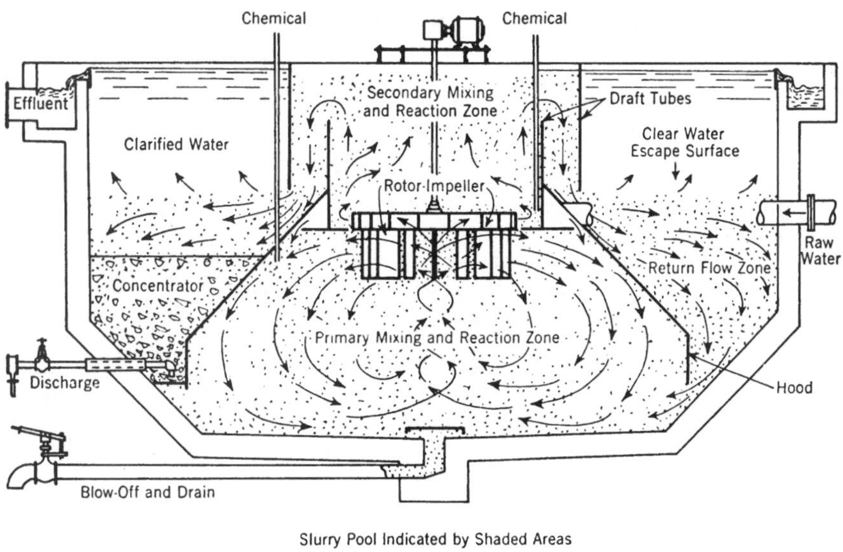

FIGURE 7.23 The Accelator solids contact clarifier. (*Source:* Hartung, 1951.)

zone is drawn downward and under the shroud back into the premix zone. This movement recirculates solids, which can assist flocculation in the premix zone.

Mechanical equipment associated with solids contact clarifiers must be adjusted or tuned to the throughput. Excessive stirring motion in the premix zone can be counterproductive, whereas too little stirring can result in poor radial-flow distribution under the shroud as well as poor chemical mixing and flocculation. In solids contact units, sludge settles to the tank floor and is removed with mechanical equipment. Clarifiers with recirculation to keep solids in suspension allow excess solids to accumulate in sludge pockets, or concentrators, as illustrated in Figure 7.23. Appropriate operation of these pockets contributes to controlling the concentration of solids in suspension and influences the sedimentation efficiency of the clarifier.

Floc-Blanket Clarifiers. Both types of solids contact tanks, premix and premix-recirculation, can function as floc-blanket clarifiers if stable and distinct floc blankets can be established and easily maintained in the settling zone. Only a few designs of solids contact clarifiers have been developed with this objective. Usually, the volume and concentration of solids in circulation in contact units is not great enough to maintain a blanket in the outer separation zone.

Hopper-Bottomed Tanks. The first designed floc-blanket tanks had a single hopper bottom, square or circular in cross-section. In these units, coagulant-dosed water was fed down into the apex of the hopper. The hopper shape assists with even flow distribution from a single-point inlet to a large upflow area. The expanding upward flow allows floc growth to occur, large particles to remain in suspension, and a floc blanket to form. The pressure loss through the floc blanket, although relatively small, helps to create homogeneous upward flow.

A single hopper, conical or pyramidal, only occupies 33 percent of available volume relative to its footprint. In addition, it is expensive to construct, and its size is limited by constructional constraints. Consequently, alternative forms of hopper tanks have been developed to overcome these drawbacks yet retain the hydraulic advantage of hoppers. These include tanks with multiple hoppers, a wedge or trough, a circular wedge (premix-type of clarifier, Figure 7.3), and multiple troughs.

As a floc blanket increases in depth, settled-water quality improves, but with diminishing return (Figure 7.24) (Miller, West, and Robinson, 1966). The blanket depth defines the quantity of solids in suspension. If *effective depth* is defined as the total volume of blanket divided by the area of its upper interface with supernatant, then the effective depth of a hopper is roughly one-third the actual depth. As a result, flat-bottomed tanks have an actual depth that is much less than that of hopper tanks with the same effective depth. Effective depths of blankets are typically in the range 2.5 to 3 m (8 to 10 ft).

The quantity of solids in suspension, established by the blanket depth and concentration, affects sedimentation efficiency because of the effect on flocculation. In addition, the head loss assists flow distribution, ensuring a more stable blanket and, thereby, greater blanket concentration.

A very stable floc blanket can be operated with little depth of supernatant water, less than 0.3 m (1 ft), without significant carryover (Figure 7.25) (Miller, West, and Robinson, 1966). In practice, the feasibility of this depends on weir spacing and likely disturbance by wind. Generally, blankets are operated in a manner likely to produce a blanket that has a diffuse surface, a tendency to exhibit some unstable boiling, and poor level control methods. Thus, a supernatant water depth of at least 1 m (3 ft) may be necessary to minimize carryover, especially after increases in upflow velocity. Supernatant water depths of 2 m (6 ft) are commonly provided, but this may be unnecessary if there is good blanket-level control and if care is taken when increasing the upflow rate (Hart, 1996).

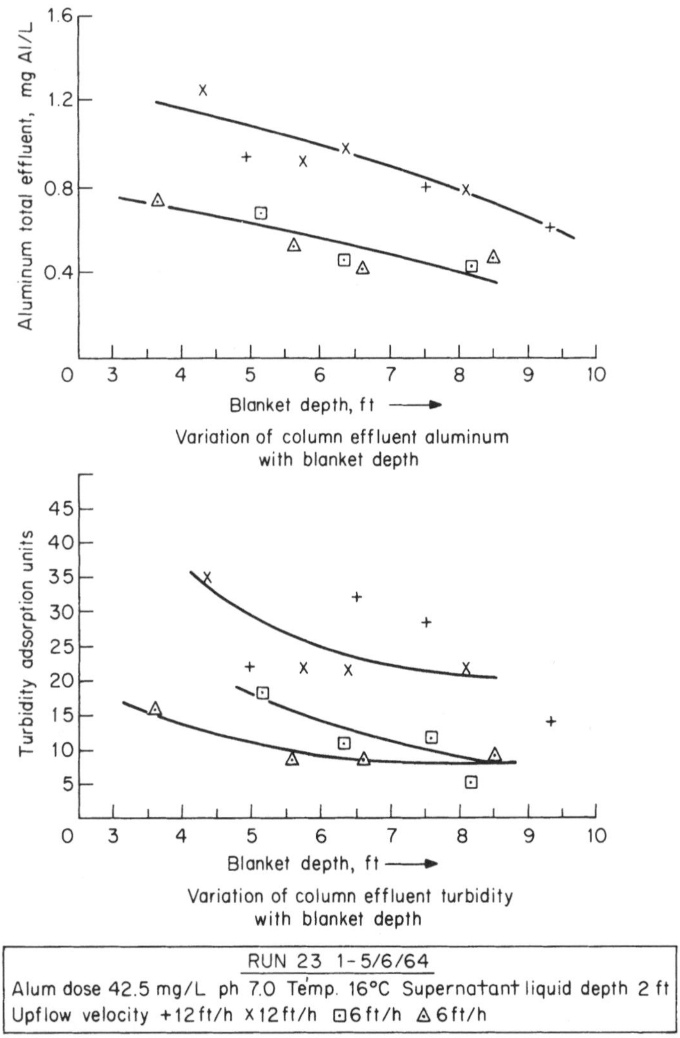

FIGURE 7.24 Change in settled-water quality as depth of blanket increases. (*Source:* Miller, West, and Robinson, 1966.)

Control of the blanket surface is easily achieved using a slurry weir or sludge hopper with sills set at an appropriately high level. Sludge hoppers can be emptied frequently by automatically timed valves. One proprietary suspended sludge hopper system utilizes a strain gauge to initiate drainage of sludge from suspended canvas cones. Another method is to monitor the magnitude of the turbidity in a sample drawn continuously from a suitable point.

Sludge hoppers, pockets, and cones must be sized to allow efficient removal of sludge (Pieronne, 1996). They must be large enough to allow in situ preliminary thickening, even when the sludge removal rate has to be high. High rates might

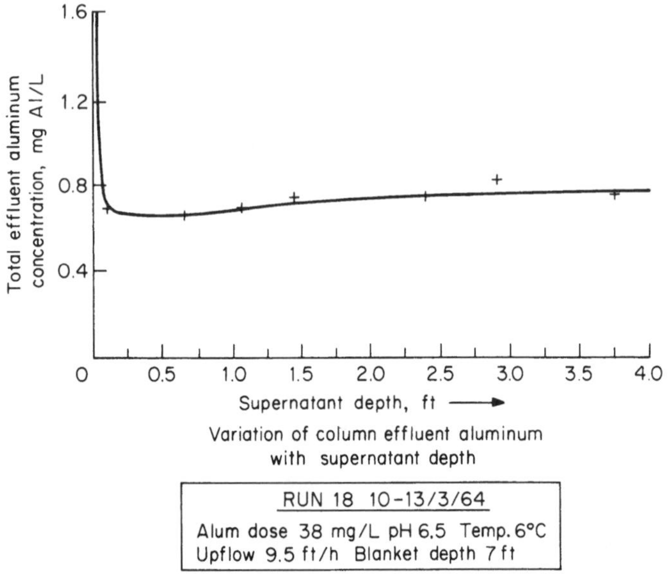

FIGURE 7.25 Change in settled-water quality as depth of supernatant increases above a floc blanket. (*Source:* Miller, West, and Robinson, 1966.)

occur following a substantial increase in flow-through rate during periods of high chemical doses. Dynamic simulation modeling by Hart (1996); Gregory, Head, and Graham (1996); and Head, Hart, and Graham (1997) has shown the importance of correct sizing of sludge hoppers and their operation, especially for handling increases in flow through the clarifiers. If sizing and operation are inadequate, then loss of blanket level control and poor settled-water quality will occur when the flow is increased.

EXAMPLE PROBLEM 7.3 What area will be needed for removing floc to control the blanket level in a floc-blanket clarifier using the conditions for Example Problem 7.1?

SOLUTION Assume the aluminum coagulant dose to the water is 3.2 mg Al/L. The aluminum concentration in the floc blanket at maximum flux (when the floc volume concentration is 20 percent) is 110 mg Al/L. The concentration of aluminum in the blanket is proportional to the floc volume. From Example Problem 7.1, the conditions are the following:

Upflow rate (m/h)	1.6	1.95	2.5	3.05	3.65	4.2	4.7	5.15
Half-hour floc volume (%)	31	29	25	22	19	16	13	10

The proportion of the floc-blanket tank area needed for removing floc is determined by a mass balance:

$$\begin{gathered}(\text{aluminum dose}) \times (\text{total volumetric flow rate to tank}) \\ = (\text{blanket aluminum concentration}) \\ \times (\text{volumetric settlement rate into removal area})\end{gathered}$$

But:

(total volumetric flow rate to tank)
= (upflow rate) × (total tank upflow area)

and:

(volumetric settlement rate into removal area)
= (upflow rate) × (area for removal)

Thus:

(area for removal) × (blanket aluminum concentration)
= (total tank upflow area) × (aluminum dose)

This means that the proportion of tank area needed for removing floc is the ratio of aluminum concentration in the blanket to that being dosed. For example, at the upflow rate of 1.95 m/h:

$$\text{concentration of aluminum in the blanket} = 110 \times 29/20$$

Therefore

$$\text{proportion of area needed} = 3.2/110 \times 20/29 \times 1/100\% = 2.0\%$$

Hence

Upflow rate (m/h)	1.6	1.95	2.5	3.05	3.65	4.2	4.7	5.15
Proportion of area needed (%)	1.9	2.0	2.3	2.6	3.1	3.6	4.5	5.8

For a different aluminum dose, the area will need to be accordingly proportionally greater or less. In practice, a greater area will be required to cope with the short-term need to remove excess floc at a high rate to prevent the blanket from reaching the launders when the upflow rate is increased quickly.

Flat-Bottomed Tanks. It is simpler and cheaper to build a floc-blanket tank with a flat bottom; therefore, few tanks are now built with hoppers. In flat-bottomed tanks, good flow distribution is achieved using either multiple-downward, inverted-candelabra feed pipes, or laterals across the floor (Figures 7.4 and 7.5).

An inverted-candelabra system can ensure good distribution for a wide range of flows but may obstruct installation of inclined settling systems. The opposite can apply to a lateral distribution system. In the proprietary Pulsator design, reliability of flow distribution is ensured by periodically pulsing the flow to the laterals.

Inclined Settling with Floc Blankets. Tube modules with the typical spacing of 50 mm (2 in) between inclined surfaces are not cost-effective in floc-blanket tanks (Gregory, 1979). With the blanket surface below the tube modules, the settled-water quality is no better than from a stable and efficient tank without modules.

With the blanket surface within the modules, the floc concentration in the blanket increases by about 50 percent, but no commensurate improvement in settled-water quality occurs. The failure of closely spaced inclined surfaces to increase hindered settling rates relates to the proximity of the surfaces and a circulatory motion at the blanket surface that counteracts the entrapment mechanism of the blanket (Gregory, 1979).

The problem with closely spaced surfaces diminishes with more widely spaced inclined surfaces. An effective spacing is about 0.3 m (1 ft), but no optimization studies are known to have been published. Large (2.9 m) plates, however, have been shown to be preferable to shorter (1.5 m) plates (Casey, O'Donnel, and Purcell, 1984). The combined action of suppressing currents and inclined settling with widely spaced surfaces can result in about a 50 percent greater throughput than with a good floc blanket without inclined surfaces. The proprietary Superpulsator tank is the Pulsator design with widely spaced inclined surfaces.

Ballasted-Floc Systems. Floc produced by coagulating clay-bearing water generally settles faster than floc produced by coagulating water containing little mineral turbidity. Consequently, mineral turbidity added purposely to increase floc density can be useful. Bentonite is the usual choice, and fly ash has been considered in eastern Europe. The advantages of ballasting floc also arise with powdered activated carbon, dosed for taste and odor control or pesticide removal (Standen et al., 1995), and when precipitation softening is carried out in association with iron coagulation. Sometimes fine sand has been used as the ballasting agent.

A process based on fine sand ballasting was developed in Hungary in the 1950s and 1960s. In this process, sand is recycled for economy (Figure 7.26). The process has found favor in France, and is sometimes known by the proprietary name of Cyclofloc (Sibony, 1981) or Simtafier (Webster et al., 1977). Recovered sand is conditioned with polyelectrolyte and added to the untreated water before the metal-ion coagulant is added. A second polyelectrolyte might be used as a flocculant aid prior to floc-blanket settling. Sand is recovered by pumping the sludge through small hydrocyclones. The Densadag system (Degremont, 1991) and the Actiflo system (de Dianous, Pujol, and Druoton, 1990) combine sand ballasting with inclined settling. The Fluorapide system combines floc-blanket settling and sand ballasting with inclined settling (Sibony, 1981).

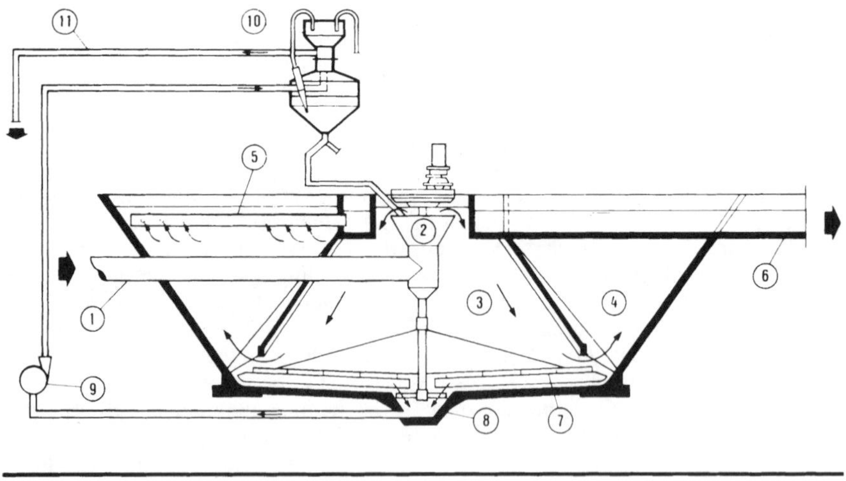

1 *Sewage inlet*
2 *Bell-shaped inlet work*
3 *Reaction zone*
4 *Clarification zone*
5 *Recovery trough*
6 *Clarified water outlet*
7 *Sludge scraper*
8 *Sludge and microsand collection pump*
9 *Recycling pump*
10 *Sludge and microsand hopper*
11 *Sludge discharge*

FIGURE 7.26 The Cyclofloc clarification system. (*Source:* Courtesy of OTV, Paris, France, and Kubota Construction Co., Tokyo, Japan.)

Powdered magnetite, recovered by magnetic drum filters, also has been suggested as a ballasting agent. A process from Australia (Dixon, 1984), with the proprietary name Sirofloc, is based on recycling magnetite without coagulation, but the magnetite is chemically conditioned with sodium hydroxide such that a metal-ion coagulant is not needed. Sirofloc is almost efficient enough for subsequent filtration to be unnecessary. High-rate (20 m/h) filtration may be needed to remove residual magnetite and manganese that Sirofloc cannot normally remove (Gregory, Maloney, and Stockley, 1988).

Other Factors Influencing Sedimentation Efficiency

Flow-Through Rate

Surface Loading. The surface loading of a sedimentation tank is expressed as the flow rate per unit of surface area of that part of the tank in which settling is meant to happen. The performance of all settling processes is influenced by surface loading. Settled-water quality deteriorates when surface loading is increased (Figure 7.15). The reasons for this deterioration are various, but (particularly for discrete particle-settling processes) changes in flow rate may affect flocculation performance as well as sedimentation efficiency.

The point at which the performance curve (Figure 7.15) crosses the limit of acceptable quality defines the maximum reliable surface loading. Cost effectiveness requires the highest surface loadings possible; however, the efficiency will then be more sensitive to major variations in surface loading and water quality.

The reliable surface loading for a given type of settling process depends on a wide range of factors. For example, when treating a highly colored, low-alkalinity water in winter by floc-blanket clarification, the loading might be only 1 m/h (3 ft/h), whereas when treating a minerally turbid, high-alkalinity water in summer with a flocculent aid, the loading could be greater than 7 m/h (23 ft/h). (Note, 1 ft/h is equivalent to 0.125 gpm/ft^2, hence 23 ft/h is equivalent to 2.9 gpm/ft^2.)

Size and Shape of Tank

Size and Number. The number of tanks can affect the flexibility of plant operation. When operating close to the limit of acceptable quality, isolation of one tank is likely to impose a reduction in total plant production. No fewer than two tanks should be used for reliability, whereas three will provide greater operational flexibility.

The size of tanks might be limited by constructional constraints, and, consequently, this will dictate the minimum number of tanks. Although factors that affect sedimentation efficiency, such as loading, velocity, and various dimension ratios, can be maintained with any size tank, the performance of extra-large tanks may become unacceptable and limit tank size. For example, tanks with large surface area will tend to be more vulnerable to environmental effects, such as wind-induced circulation.

When considering the number of tanks, the performance of a floc blanket is dependent on the upflow velocity, blanket-floc concentration, effective blanket depth, and dosing and delay time conditions. Thus, provided that the size and shape of floc-blanket tanks do not affect these factors, any such size or shape might not be expected to affect sedimentation efficiency. The efficiency is affected, however, because differences occur between tanks in their hydrodynamic and hydraulic conditions for flocculation and flow distribution.

The length-width ratio is not relevant to floc-blanket clarifiers, although baffles have been shown to be useful (Gregory and Hyde, 1975). With inclined settlers, the length of the flow path affects the sedimentation efficiency (Yao, 1973). When lengths are short, end effects may limit efficiency. Additionally, problems of flow distribution may mean that inclined plate settlers with narrow plates are more efficient than those with wide plates.

Depth. According to Hazen's law (Eq. 7.19), for discrete particle settling, the settling efficiency depends on the tank area and is independent of the depth. However, in practice, increasing tank depth promotes settling efficiency and is interrelated with width in horizontal and inclined settlers. A minimum depth may be needed to limit scour as mentioned previously. Depth also defines the spacing between inclined surfaces and, therefore, number of layers and hence inclined settler efficiency.

Flow Arrangements

Inlet and Outlet. The purpose of an inlet is to distribute the incoming water uniformly over the cross-section of a tank for a wide range of flow rates. The outlet arrangement is as important as the inlet in ensuring good flow distribution. Numerous investigations (Fair and Geyer, 1954; Thirumurthi, 1969) have shown how inlets and outlets in horizontal-flow tanks influence residence time distribution and settling efficiency. Tests with model tanks (Price and Clements, 1974; Kawamura, 1981) have shown that symmetry is desirable, and obstacles (such as by support walls and piers) can disturb uniformity. Complex arrangements can be as good as simple ones, and a uniformly fed submerged weir can give good results without a baffle.

For all types of settlers, flow velocities in approach channels or pipe work that are low enough to allow premature settlement must be avoided. Conversely, flows high enough to disrupt floc must also be avoided.

In floc-blanket tanks, the injection velocity from the inlet pipes governs the input energy, which can affect blanket stability and "boiling." Boiling causes direct carryover and a reduction in blanket concentration, resulting in deterioration of settled-water quality.

Outlet weir, or launder length, and loading affect outlet flow distribution. Therefore, they must be kept clean and unobstructed, but the effectiveness of long launders is questionable (Kawamura and Lang, 1986).

Baffling. Baffles are useful in horizontal-flow tanks at the inlet and outlet to assist flow distribution, longitudinally to increase length-width ratio, and as vanes to assist changes in horizontal-flow direction. Kawamura (1981) did numerous model tests with diffuser walls and arrived at a number of design guides relating to the position of walls and the free area in the walls.

In hopper-bottomed floc-blanket tanks, baffles might help inlet flow distribution (Gould, 1967; Hale, 1971), but the centering of the inlet should be checked first (Gregory, 1979). A matrix of vertical baffles in the floc blanket is an alternative method for improving tank performance if flow distribution is poor (Gregory and Hyde, 1975).

Vertical baffles have been installed in some solids contact clarifiers so that the central mixer does not cause the fluid to rotate in the settling zone. Fluid rotation will lift the floc up the outer side of the settling zone. A proprietary tube-module baffle system (Envirotech Corp.) has been used to improve flow distribution in radial-flow tanks. Baffles at the water surface can help to counteract wind disturbance.

Particulate and Water Quality

Seasonal Water Quality. Sedimentation efficiency may vary with seasonal changes in temperature, alkalinity, and similar parameters, as well as with changes in the nature of color and turbidity being coagulated.

Temperature affects efficiency by influencing the rate of chemical reactions, solubilities, the viscosity of water, and hence particle-settling velocities. Temperature can also be a surrogate for change in other parameters that occur on a similar seasonal basis. Changes in alkalinity, color, turbidity, and orthophosphate concentration for eutrophic waters affect coagulation reactions and the properties and rate of settling of resulting floc particles.

With floc-blanket clarification, for example, the reliable upflow velocity in the summer can be more than twice that in the winter (Figure 7.15) (Gregory, 1979; Setterfield, 1983). This is important when plant size or reliable plant output is defined.

EXAMPLE PROBLEM 7.4 What will be the floc-blanket upflow rate at maximum flux for different temperatures, given the conditions defined in Example Problem 7.1?

SOLUTION The Stoke's settling velocity of particles is inversely proportional to liquid viscosity, Eq. 7.8:

$$v_{\text{temp }1} = v_{\text{temp }2} \times (\mu_{\text{temp }2}/\mu_{\text{temp }1})$$

The viscosity of water for different temperatures is as follows:

Temperature (°C)	5	10	15	20	25
Viscosity (mN·s/m²)	1.52	1.31	1.14	1.01	0.893

If the results given in Example Problem 7.1 were obtained at the temperature of 10°C, then assuming no change in water chemistry and other factors that might affect the floc, the maximum flux upflow rate at 20°C can be estimated, for example as

$$v_{20°\text{C}} = 3.44 \times 1.31/1.01 = 4.46 \text{ m/h}$$

Hence, for the above temperature range:

Maximum flux upflow rate (m/h)	2.96	3.44	3.95	4.46	5.05

This means that the upflow rate for reliable operation in the summer could be more than 1.5 times that in the winter for this temperature range.

Coagulation. Particle-settling velocity is affected by various particle characteristics, principally size, shape, and density. The choice of coagulation, chemistry, and application efficiency affects floc particle characteristics. The choice of coagulant should be dictated by chemical considerations (see Chapter 6). Optimum chemical doses for sedimentation can be predicted reliably by jar tests. Jar tests also reliably predict color removal and soluble metal-ion concentrations, but not turbidity removal because of the difference in hydrodynamic conditions in the batch jar test and the continuous settling process.

Adequate control of dosing and mixing of the chemicals should be provided. If a change in coagulant dose or pH improves the sedimentation efficiency, the current chemical doses may be incorrect or the mixing or flocculation inadequate. The difference in efficiency between identical tanks supposedly receiving the same water can be because of differences in chemical doses if made separately to each tank, poor chemical mixing before flow splitting, or different flows through the tanks.

Flocculation. The efficiency of horizontal- and inclined-flow sedimentation is dependent on prior flocculation if coagulation is carried out. The efficiency of solids contact units depends on the quality of preliminary mixing. Mixing assists with chemical dispersion, flocculation, solids resuspension, and flow distribution. Because

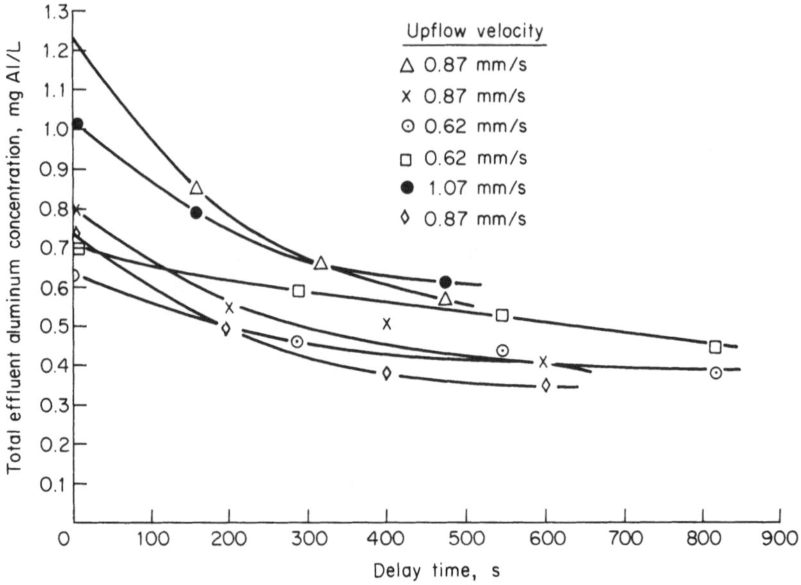

FIGURE 7.27 Change in settled-water quality with increase in delay time between chemical dosing and inlet to a floc blanket. (*Source:* Yadav and West, 1975.)

these can conflict with each other, mixing rates should be adjusted when changes are made to the flow-through rate. This is rarely done, however, because it is usually too difficult.

The extent of flocculation is dependent on velocity gradient and time (see Chapter 6). The time delay between chemical dosing and inlet to a floc blanket, to allow preliminary flocculation in connecting pipe work and channels, affects the floc-blanket clarification efficiency (Figure 7.27) (Miller and West, 1968; Yadav and West, 1975). The delay times between dosing the coagulant and dosing a polyelectrolyte, and between dosing the polyelectrolyte and the inlet to the tank, also affect the efficiency (Figure 7.28). The reasons for needing such delay times are not entirely understood, but relate to efficiency of mixing, rate of hydrolysis and precipitation, and the rate of formation of primary and subsequent floc particles. Useful delay times are likely to be in the range ½ to 8 min, similar to those needed for direct filtration. The greatest times tend to be needed when coagulating very cold water with aluminum sulfate. High-rate flocculation for inclined settling or dissolved-air flotation is similarly affected, such that flocculation times of from 5 to 10 minutes may be effective when water is warm, but much longer times may be needed when water is very cold, especially with aluminum sulfate. Water temperature has less effect on flocculation with polyaluminium chlorides and ferric salts (see Chapter 6).

Delay time is not a substitute for initial rapid chemical dispersion. Many operators have learned that simple improvements to flow conditions and dispersion arrangements at the point of dosing can produce substantial improvements in chemical utilization and sedimentation efficiency.

Polyelectrolytes and other agents. Flocculent aids improve sedimentation efficiency by increasing floc strength and size. Jar tests can be used to select suitable polyelectrolytes and determine a trial dosing range. Optimal doses must be carefully

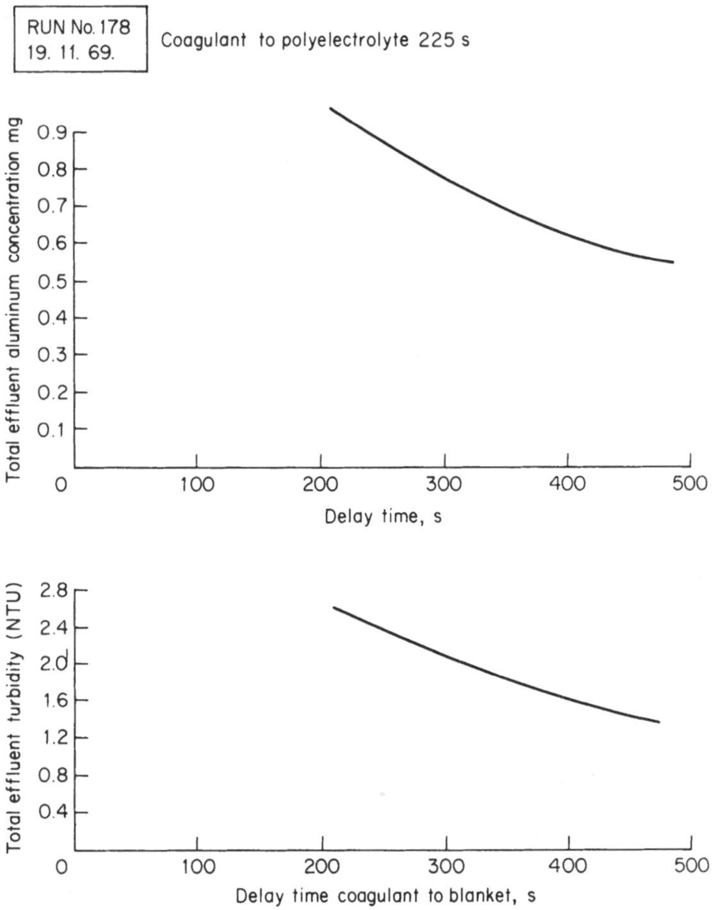

FIGURE 7.28 Change in settled-water quality with increase in delay time between dosing a polyelectrolyte and inlet to a floc blanket. (*Source:* Yadav and West, 1975.)

determined by full-scale plant tests. An excessive dose producing good settled water may be detrimental to filters. If the apparent optimal dose of polyelectrolyte on the plant is much greater than that indicated by jar tests, then mixing and flocculation conditions must be checked.

Dosing of flocculent aids increases particle-settling velocities in floc-blanket clarification, resulting in an increase in the blanket floc concentration. A diminishing return also exists in increasing the dose of flocculent aid and is associated with the blanket floc concentration increasing with an increase in flocculent dose and becoming greater than at maximum flux (Figure 7.29) (Gregory 1979). Thus the polyelectrolyte dose need be only enough to ensure that the floc concentration is greater than at maximum flux and that the blanket surface is distinct and not diffuse, and certainly not so great as to take the floc concentration into the thickening regime. Additives that increase floc density and settling velocity, such as floc ballasting with bentonite or fine sand, have an effect on sedimentation similar to flocculent aids.

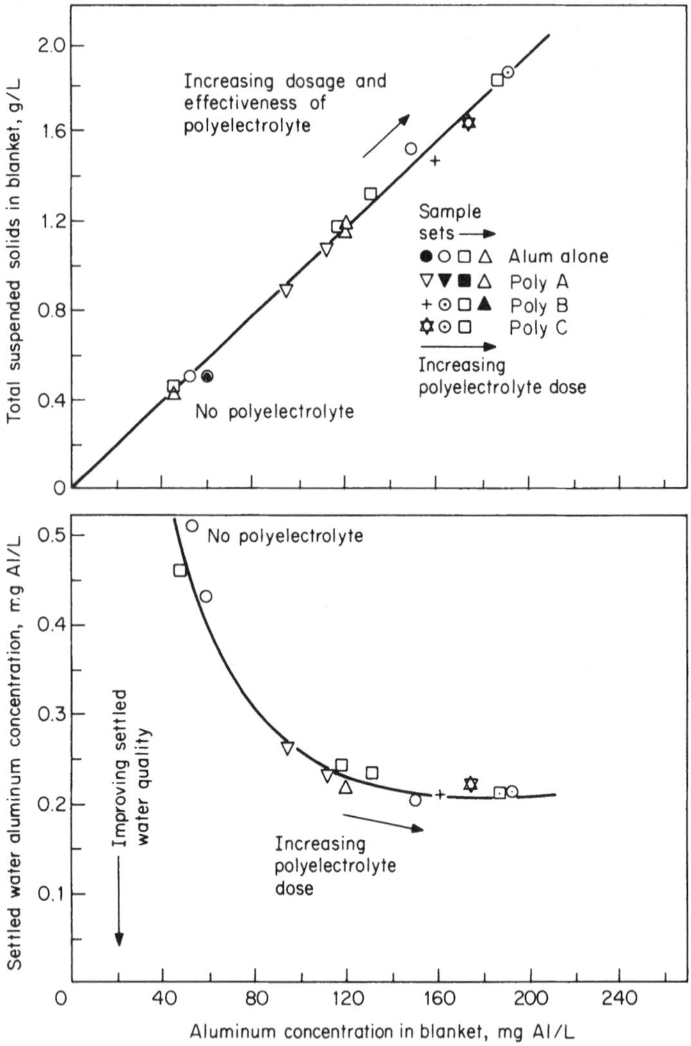

FIGURE 7.29 Effect of polyelectrolyte type and dose on blanket solids concentration and settled-water quality. (*Source:* Gregory, 1979.)

Climate and Density Currents. Wind can induce undesirable circulation in horizontal-flow tanks, with cross-flow winds reducing sedimentation efficiency the most (Price and Clements, 1974). Therefore, wind effects should be minimized by constructing tanks to align with prevailing winds. Strong winds can disrupt floc-blanket stability. This can be counteracted by covering tanks with roofs, placing floating covers between the launders, constructing windbreaks around each tank, aligning the launders across the prevailing wind, installing scumboards or baffles at the water surface across the prevailing wind, or installing fully submerged baffles in the supernatant water. Wind combined with low temperatures can cause severe ice formation and affect outlet flow. Total enclosure may be needed to prevent ice formation.

High solar radiation can cause rapid diurnal changes in water temperature and density. Rapid density changes because of temperature, solids concentration, or salinity can induce density currents that can cause severe short circuiting in horizontal tanks (Hudson, 1981) and inclined settlers (White, Baskerville, and Day, 1974). Hudson (1981) has reviewed methods of minimizing or preventing induced currents. He considered the methods to fall into several categories: (1) use of a surface weir or launder takeoff over a large part of the settling basin, (2) improved inlet arrangement, (3) schemes to provide increased basin drag and friction, and (4) slurry recirculation. Substantial diurnal variation in temperature also can result in release of dissolved gases, especially in biologically active waters. The resulting rising columns of bubbles can disrupt floc blankets.

DISSOLVED-AIR FLOTATION

History of Flotation

Over 2000 years ago, the ancient Greeks used a flotation process to separate the desired minerals from the gangue, the waste material (Gaudin, 1957). Crushed ore was dusted onto a water surface, and mineral particles were retained at the surface by surface tension while the gangue settled. In 1860, Haynes patented a process in which oil was used for the separation of the mineral from the gangue (Kitchener, 1984). The mineral floated with the oil when the mixture was stirred in water.

In 1905, Salman, Picard, and Ballot developed the *froth flotation* process by agitating finely divided ore in water with entrained air. A small amount of oil was added, sufficient enough to bestow good floatability to the sulfide grains (Kitchener, 1984). The air bubbles, together with the desired mineral, collected as foam at the surface while the gangue settled. The first froth flotation equipment was developed by T. Hoover in 1910 (Kitchener, 1984), and except for size, it was not much different than the equipment used today. Later, in 1914, Callow introduced air bubbles through submerged porous diffusers. This process is called *foam flotation*. The two processes, froth and foam flotation, are generally known as dispersed-air flotation and are widely used in the mineral industry.

Elmore suggested in 1904 the use of electrolysis to produce gas bubbles for flotation. This process, although not commercially used at that time, has been developed into electrolytic flotation (Bratby, 1976). Elmore also invented the dissolved-air (vacuum) flotation (DAF) process, whereby air bubbles are produced by applying a vacuum to the liquid, which releases the air in the form of minute bubbles (Kitchener, 1984).

The original patent for the dissolved-air pressure flotation process was issued in 1924 to Peterson and Sveen for the recovery of fibers and white water in the paper industry (Lundgren, 1976). In pressure dissolved-air flotation, the air bubbles are produced by releasing the pressure of a water stream saturated with air above atmospheric pressure.

Initially, DAF was used mainly in applications in which the material to be removed, such as fat, oil, fibers, and grease, had a density less than that of water. In the late 1960s, however, the process also became acceptable for wastewater and potable water treatment applications.

Dissolved-air flotation has been applied extensively for wastewater sludge thickening. In water treatment, it has become accepted as an alternative to sedimentation, particularly in the Scandinavian countries and the United Kingdom, where more than 50 (Dahlquist, 1997) and 90 (Gregory, 1997) plants, respectively, are in operation or under construction. There is an increasing interest in the United States with some 12 plants in operation and at least another 10 under design or construction (Nickols and Crossley, 1997). There are at least 8 plants constructed or in hand in Canada (Adkins, 1997). Flotation is employed mainly for the treatment of nutrient-rich reservoir waters that may contain heavy algal blooms, and for low-turbidity, low-alkalinity, colored waters (Zabel and Melbourne, 1980). These types of water are difficult to treat by sedimentation, because the floc produced by chemical treatment has a low settling velocity. In the United States, most of the DAF plants built or planned so far are for treating supplies that were previously unfiltered. These supplies have high, raw water quality of low turbidity and low color or TOC.

Types of Flotation Processes

Flotation can be described as a gravity separation process in which gas bubbles attach to solid particles to cause the apparent density of the bubble-solid agglomerates to be less than that of the water, thereby allowing the agglomerate to float to the surface. Different methods of producing gas bubbles give rise to different types of flotation processes. These are electrolytic flotation, dispersed-air flotation, and dissolved-air flotation (Lundgren, 1976).

Electrolytic Flotation. The basis of electrolytic flotation, or electroflotation, is the generation of bubbles of hydrogen and oxygen in a dilute aqueous solution by passing a DC current between two electrodes (Barrett, 1975). The bubble size generated in electroflotation is very small, and the surface loading is therefore restricted to less than 4 m/h (13.3 ft/h). The application of electrolytic flotation has been restricted mainly to sludge thickening and small wastewater treatment plants in the range 10 to 20 m^3/h (50,000 to 100,000 gpd).

Dispersed-Air Flotation. Two different dispersed-air flotation systems are used: foam flotation and froth flotation (Sherfold, 1984). Dispersed-air flotation is generally unsuitable for water treatment as the bubble size tends to be large and either high turbulence or undesirable chemicals are used to produce the air bubbles required for flotation. However, the French Ozoflot system applies ozone-rich air for treating waters high in algae using diffusers to achieve dispersion.

Dissolved-Air Flotation. In dissolved-air flotation, bubbles are produced by the reduction in pressure of a water stream saturated with air. The three main types of DAF are vacuum flotation (Zabel and Melbourne, 1980), microflotation (Hemming, Cottrell, and Oldfelt, 1977), and pressure flotation (Barrett, 1975). Of these three, pressure flotation is currently the most widely used. In pressure flotation, air is dissolved in water under pressure. Three basic pressure DAF processes can be used: full-flow, split-flow, and recycle-flow pressure flotation (Zabel and Melbourne, 1980). For water treatment applications requiring the removal of fragile floc, recycle-flow pressure flotation is the most appropriate system (Figure 7.30). In this process, the whole influent flows either initially through the flocculation tank or directly to the flotation tank if separate flocculation is not required. Part of the clarified effluent is recycled, pressurized, and saturated with air. The pressurized recycle water is introduced to the flotation tank through a pressure-release device and is mixed with the flocculated water. In the pressure-release device, the pressure is reduced to

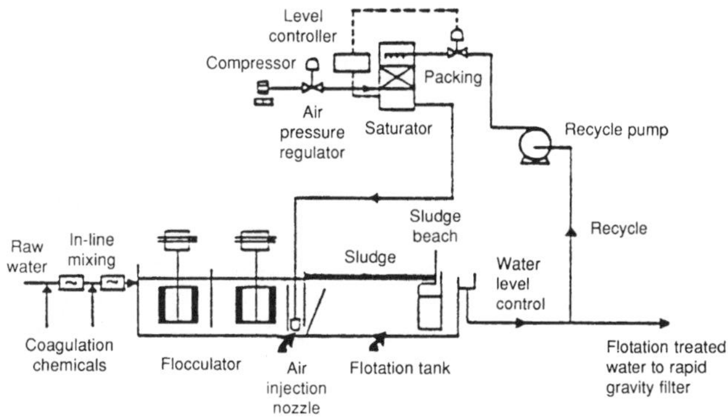

FIGURE 7.30 Schematic diagram of a flotation plant for potable water treatment. (*Source:* T. F. Zabel and J. D. Melbourne, "Flotation," in W. M. Lewis, ed. *Developments in Water Treatment,* vol. 1. Elsevier, Essex, England, 1980.)

atmospheric pressure, releasing the air in the form of fine bubbles (10 to 100 μm in diameter). The air bubbles attach themselves to the flocs, and the aggregates float to the surface. The floated material (the float) is removed from the surface, and the clarified water is taken from the bottom of the flotation tank.

THEORY OF DISSOLVED-AIR FLOTATION

To achieve efficient clarification by DAF, particles and natural color present in the water must be coagulated and flocculated effectively prior to the introduction of the microbubbles to form the bubble-floc aggregates.

Mechanism of Flotation

Floatable bubble-floc agglomerates might form by any of three distinct mechanisms: entrapment of bubbles within a condensing network of floc particles, growth of bubbles from nuclei within the floc, and attachment of bubbles to floc during collision. Work by Kitchener and Gochin (1981) has shown that all three mechanisms can occur but that the principal mechanism in DAF for potable water treatment is the attachment mechanism. Their work also indicates that the organic content of surface waters is usually high enough to render the floc surface sufficiently hydrophobic for bubble attachment. Only in waters very low in dissolved organic matter will flotation efficiency be reduced.

Flotation Models

Dissolved-air flotation plants have been designed largely on experience. Models that describe the primary variables affecting flotation can be used to improve DAF design and operation and to guide pilot plant studies for collection of design criteria. Dissolved-air flotation tanks have two zones and functions (Figure 7.31). The pur-

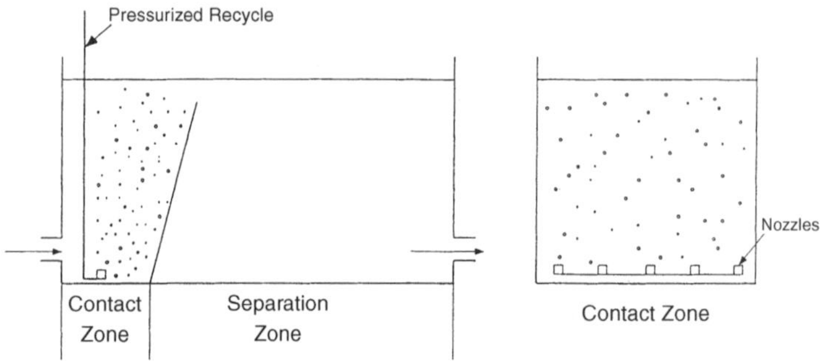

FIGURE 7.31 Schematic of a flotation tank showing contact and separation zones. (*Source:* Edzwald, J. K., J. P. Malley, Jr., and C. Yu. "A conceptual model for dissolved air flotation in water treatment," *Water Supply* 8, 1990: 141–150. Reprinted with permission from Blackwell Science Ltd.)

pose of the contact zone is to provide contact opportunities between floc particles and bubbles leading to attachment and formation of particle-bubble aggregates with densities less than water. In the second zone, separation, or clarification, occurs as particle-bubble aggregates rise to the surface. Modeling of the clarification zone is similar to sedimentation in that Stokes' law–type expressions are used to describe the rise velocity of the particle-bubble aggregates.

Two modeling approaches have been used to describe collisions among bubbles and particles in the contact zone. Tambo and coworkers (Tambo and Matsui, 1986; Fukushi, Tambo, and Matsui, 1995) use a heterogeneous flocculation model that describes collisions among bubbles and particles. Edzwald and coworkers (Edzwald, Malley, and Yu, 1990; Edzwald, 1995) consider the white water of air bubbles in the contact zone to be much like a filter blanket and model the bubbles as collectors of particles.

Heterogeneous Flocculation-Based Model. Tambo and coworkers model the kinetics of collisions among particles and particles with attached air bubbles using a heterogeneous flocculation-based model similar to particle flocculation kinetic models (see Chapter 6). They write population balance kinetic equations describing the change in floc particles with attached air bubbles due to collisions (Eq. 7.48) and the change in the number of particles without attached bubbles (Eq. 7.49).

$$\frac{dN_{f,i}}{dt} = -kN_b(\alpha_{f,i}N_{f,i} - \alpha_{f,i-1}N_{f,i-1}) \quad (i = 1 - m_f) \tag{7.48}$$

$$\frac{dN_{f,0}}{dt} = -k\alpha_f N_b N_{f,0} \quad (i = 0) \tag{7.49}$$

where $N_{f,i}$ = number concentration of floc particles with i attached air bubbles
$N_{f,0}$ = number concentration of floc particles without attached air bubbles
N_b = number concentration of air bubbles
α_f = collision attachment factor
α_0 = collision attachment factor with no attached air bubbles
m_f = maximum number of attached air bubbles to floc particles
k = kinetic coefficient dependent on the mass transfer collision mechanism

Collisions among bubbles and floc particles are modeled by considering turbulent bulk flow conditions causing fluid shear in the contact zone. This collision mechanism is described by:

$$k = aG(d_b + d_f)^3 = a\left(\frac{\varepsilon}{\mu}\right)^{1/2}(d_b + d_f)^3 \tag{7.50}$$

where a = a constant
G = mean velocity gradient
ε = mean energy dissipation rate
μ = absolute water viscosity
d_b = bubble diameter
d_f = floc diameter

Because the bubble size is fixed in the range of from 10 to 100 μm (for example, a mean of 40 μm), then Eq. 7.50 shows that the collision rate increases with floc size. This is an important modeling result.

White Water Collector Model Summary. The continuously rising cloud or blanket of bubbles (white water) in the contact zone is much like a filter bed and is conceptually analogous. Collisions between particles and the rising blanket of bubbles of the white water are modeled by considering the mass transport of particles to a collector surface (air bubble). The mass transport equations are expressed in terms of the collection efficiency of a single bubble. The single collector efficiency equations are then incorporated into an overall equation describing particle loss in the contact zone accounting for the number concentration of bubbles.

The model is based on the single collector efficiency concept (also used to model water filtration, see Chapter 8) to describe mass transport of particles from the bulk water to bubble surfaces in which the bubbles act as collectors of particles. A basic assumption is that bubbles are formed rapidly, almost instantaneously, in the contact zone. This is a good assumption based on the large pressure difference across the nozzles, or valves, used to inject pressurized recycle water. Other assumptions are detailed in the papers of Edzwald (Edzwald, Malley, and Yu, 1990; Edzwald, 1995). Air bubbles and water move upward through the contact zone (Figure 7.31). An air bubble has a net upward velocity relative to the water equal to the bubble rise velocity; therefore, streamlines of flow move downward around the bubble equal to its rise rate.

Single collector collision efficiency equations describing mass transport of particles to a bubble by Brownian diffusion (η_D), interception as the bubble rises (η_I), differential settling of particles onto a bubble (η_S), and inertia (η_{IN}) follow:

$$\eta_D = 6.18\left[\frac{k_B T^\circ}{g\rho_w}\right]^{2/3}\left[\frac{1}{d_p}\right]^{2/3}\left[\frac{1}{d_b}\right]^2 \tag{7.51}$$

$$\eta_I = \frac{3}{2}\left[\frac{d_p}{d_b}\right]^2 \tag{7.52}$$

$$\eta_S = \left[\frac{(\rho_p - \rho_w)}{\rho_w}\right]\left[\frac{d_p}{d_b}\right]^2 \tag{7.53}$$

$$\eta_{IN} = \left[\frac{g\rho_p\rho_w d_b d_p^{\,2}}{324\mu^2}\right] \tag{7.54}$$

where k_B = Boltzmann's constant
T° = absolute temperature
d_p = particle diameter
ρ_p = particle density
μ = water viscosity
ρ_w = water density
v_b = bubble rise velocity (relative fluid motion around bubble)
g = gravitational constant of acceleration

Collisions by inertia are not significant for particles and bubbles less than 100 μm, so they are ignored here. The total single collector collision efficiency (η_T) is the sum of Eqs. 7.51 to 7.53:

$$\eta_T = \eta_D + \eta_I + \eta_S \qquad (7.55)$$

Only a fraction of the collisions between particles and collectors lead to attachment, so a particle-bubble attachment efficiency term (α_{pb}) is used to compute the removal efficiency (R) of particles by a single bubble:

$$R = \alpha_{pb}\,\eta_T\,(100\%) \qquad (7.56)$$

The following particle removal rate equation is obtained by extending the removal of a single bubble to the white water blanket in the contact zone containing a bubble number concentration of N_b:

$$\frac{dN_p}{dt} = -(\alpha_{pb}\eta_T)N_p(A_b v_b N_b) \qquad (7.57)$$

where N_p = particle number concentration
A_b = projected area of the bubble
v_b = bubble rise velocity (relative fluid motion around bubble)

Equation 7.57 may be expressed in terms of the bubble volume concentration (Φ_b) and bubble diameter (d_b):

$$\frac{dN_{p,i}}{dt} = -\left(\frac{3}{2}\right)\left(\frac{\alpha_{pb}\eta_T N_p v_b \Phi_b}{d_b}\right) \qquad (7.58)$$

The fraction of particles leaving the contact zone not attached to air bubbles is described with Eq. 7.59.

$$\frac{N_{p,i}}{N_{p,e}} = \exp\left(\frac{-3\alpha_{pb}\eta_T v_b \Phi_b t_{cz}}{2d_b}\right) \qquad (7.59)$$

where $N_{p,i}$ = the influent particle concentration to the contact zone
$N_{p,e}$ = the effluent particle concentration leaving the contact zone
t_{cz} = the detention time of the contact zone assuming plug flow

White Water Collector Model—Discussion. Equation 7.59 shows that the contact zone efficiency depends on the variables in the right-hand side of the equation. Two of the variables (α_{pb} and η_T) are influenced by pretreatment, because η_T depends on floc size (d_p). Several variables (η_T, Φ_b, v_b and d_b) are affected by conditions in the contact zone. Table 7.2 summarizes these variables and their dependence.

TABLE 7.2 Contact Zone Model Variables

Variable	Dependence	Comments
	Pretreatment variables	
α_{pb} (particle-bubble attachment efficiency)	1. Particle-bubble charge interactions 2. Hydrophilic nature of particles	1. Favorable flotation requires reduction in particle charge and hydrophobic particles 2. α_{pb} approaches 1 with optimum coagulation
d_p (mean size of floc in the contact zone)	1. Flocculation time 2. Flocculation mixing intensity	Affects mass transport, η_T
	Flotation variables	
η_T (total single collector collision efficiency)	Mass transport collision efficiency of particles to bubbles from Brownian diffusion, interception, and settling	1. Minimum η_T for floc particles of ≈1 μm 2. Flocculation should produce floc particles of 10s of μm
d_b (mean bubble diameter)	1. Controlled by pressure difference across the injection system 2. Type of nozzle or injection valve	1. Desire microbubbles, smaller bubbles better performance; range is normally 10–100 μm, median of 40 μm 2. η_T is proportional to d_b^{-2}; rate of particle collection is proportional to d_b^{-1}
v_b (bubble rise velocity)	1. Size of bubbles 2. Water temperature	1. Affects η_T 2. As bubbles collect particles, see v_{pb} (Eq. 7.65)
Φ_b (bubble volume concentration)	1. Recycle ratio 2. Saturator pressure	1. Increasing Φ_b, increases collision opportunities, more bubbles or collectors 2. Increasing Φ_b, more bubble volume to reduce particle-bubble aggregate density

Effect of coagulation. α_{pb} depends on coagulation (type, dosage, and pH). It is empirically determined, and has values ranging from near zero (for poor attachment) to 1 (where all bubble contacts with particles lead to attachment, termed *favorable attachment*). Two conditions are necessary for favorable attachment: charge neutralization of particles and production of hydrophobic particles. Coagulant chemicals are used to obtain favorable attachment conditions. Most likely values for α_{pb} of 0.1 to 0.5 are obtained in practice with good coagulation chemistry. η_T is the single collector collision efficiency factor accounting for mass transport collisions between particles and bubbles. These collisions occur in the contact zone, so it is identified above as a flotation tank variable. However, it depends on the size of floc particles (d_p), which is affected by flocculation.

Equations 7.51 to 7.53 are instructive. They show that η_T is dependent on bubble size according to d_b^{-2} for all cases, and its dependence on particle size is $d_p^{-2/3}$ for

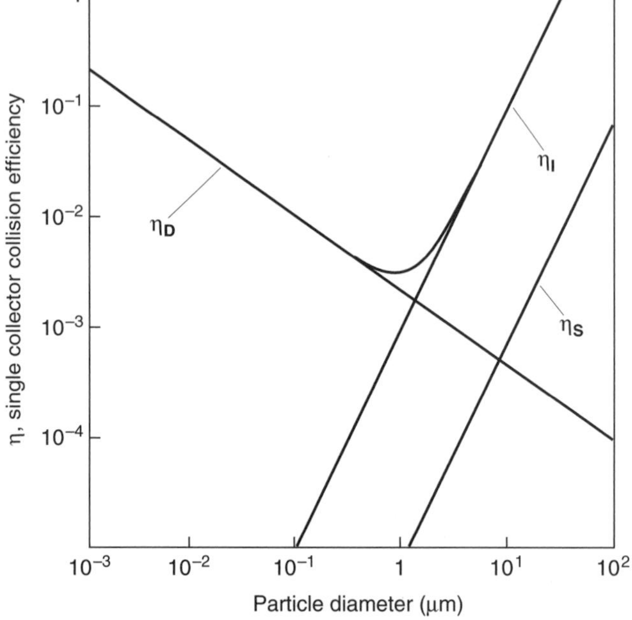

FIGURE 7.32 Single collector collision efficiency versus particle size for particle density of 1.01 g/cm^3, bubble diameter of 40 μm, and 25°C (η_D is Brownian diffusion, η_I is interception, η_S is sedimentation). (*Source:* Edzwald, 1995.)

Brownian diffusion and $d_p^{\,2}$ for interception and settling. The effect of particle size on η_T is shown in Figure 7.32 for a bubble diameter of 40 μm. Flocculation reduces particle densities, so floc densities are usually not much greater than water in the range of 1.001 to 1.01 g/cm^3 (Chapter 6). Here, a conservative value of 1.01 g/cm^3 was used in the calculations for the contact zone in Figure 7.32. The model results depicted in the figure show that collisions due to settling of particles onto air bubbles (η_S) is not an important mechanism. The model also shows that η_T has a minimum value for floc particles with a size of about 1 μm with increasing values for smaller and larger particles. η_T is controlled by Brownian diffusion (η_D) for particle diameters smaller than 1 μm, and by interception (η_I) for particles larger than 1 μm.

Effect of bubble size. Overall, the particle removal rate by bubbles in the contact zone depends on the inverse of the bubble diameter as shown by Eq. 7.59. Smaller bubbles improve flotation performance, a primary reason why DAF is more effective than dispersed-air flotation processes. The bubble size is controlled primarily by the pressure difference across the nozzles, or recycle injection device (Table 7.2). A distinctive characteristic of DAF is the white water due to formation of microbubbles in the 10 to 100 μm size range.

The rise velocity (v_b) depends on bubble size and water temperature (Table 7.2). To achieve maximum collision and attachment between air bubbles and floc parti-

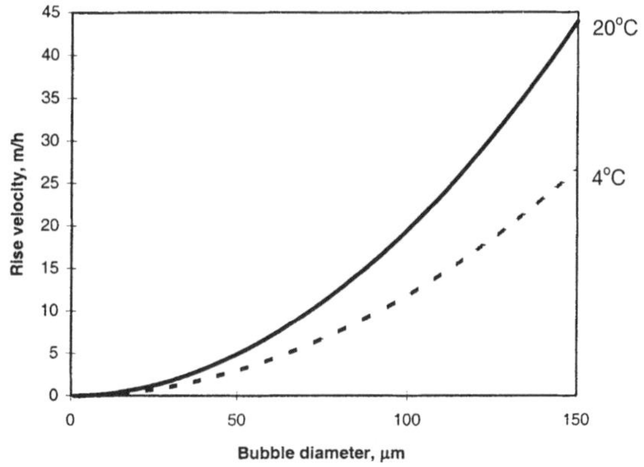

FIGURE 7.33 Rate of rise of air bubbles in water as a function of bubble size and temperature.

cles, the bubbles must rise under laminar flow conditions. This avoids shedding of floc such as can occur in the turbulent regime. The maximum bubble diameter for laminar flow is 130 μm (Turner, 1975). For bubble sizes less than 130 μm, Stokes' law for laminar flow conditions, can be used to calculate the rise rate:

$$v_b = \frac{g(\rho_w - \rho_b)d_b^2}{18\mu} \tag{7.60}$$

where ρ_b is the density of the gas bubble.

The relationship between rise rate and bubble diameter for single bubbles is given in Figure 7.33, calculated from Eq. 7.60 for 4° and 20°C (using viscosity and density values used in Example 7.6).

Air Dissolution and Release. Over the temperature and pressure ranges used in DAF (0 to 30°C and 200 to 800 kPa), air is mostly a mixture of nitrogen and oxygen. Assuming this mixture obeys Henry's law:

$$p = K_H\, C_{air} \tag{7.61}$$

where p = pressure of the air (kPa)
C_{air} = concentration of air dissolved in water (mg/L)
K_H = Henry's law constant ($kPa/mg \cdot L^{-1}$)

In a continuous saturation system, the gas phase above the water does not have the same composition as air, because oxygen is more soluble in water than nitrogen. In order that the same quantities of oxygen and nitrogen leave the saturator in the pressurized water as enter it in the compressed air, the nitrogen content of the gas in the saturator will rise, creating a nitrogen-rich atmosphere. This results in a reduction of about 9 percent in the mass of gas that can be dissolved. In assessing the performance of a continuously operating saturation system, the 100 percent saturation level should be taken as that achievable, assuming a nitrogen-rich atmosphere, as per Figure 7.34 (prepared after Haarhoff and Rykaart, 1995). (The diagram of total dis-

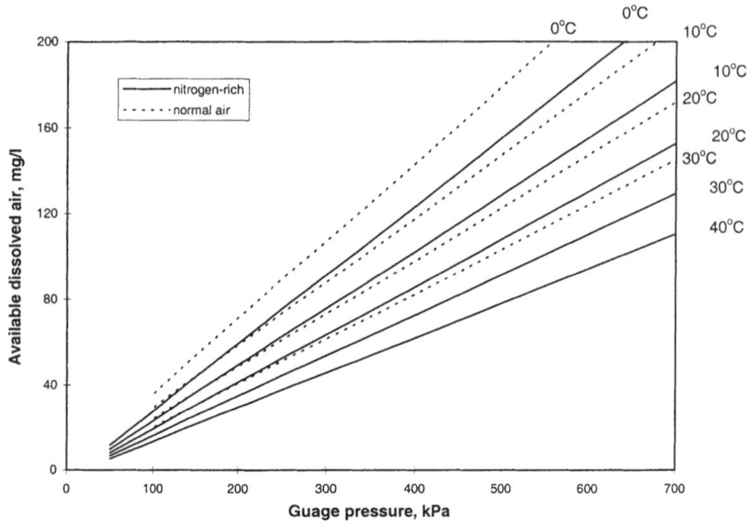

FIGURE 7.34 Available dissolved air in water as a function of gauge pressure and temperature, for normal and nitrogen-rich atmospheres.

solved air versus absolute pressure is almost identical to Figure 7.34, for the same atmospheric pressure.)

Although a saturator might be very efficient, the amount of air actually released on injection of the recycle into the flocculated water stream is dependent also on saturator pressure and nozzle design (Steinbach and Haarhoff, 1997). It would appear that greater air precipitation efficiencies are associated with greater dissolved-air concentrations in the recycle (due to high saturator efficiency and pressure) and nozzle dimensions and integral impinging surfaces.

The quantity of air released is controlled by the saturator pressure and the recycle flow. Given that the saturator pressure does not vary much for a particular plant, the more important operating variable in controlling the released air is the recycle flow.

The density of floc produced in water treatment is very similar to that of water, and only very small air bubbles are required to float floc to the surface. The smaller the air bubbles that can be produced, the larger the number of bubbles produced per unit volume of gas released. The presence of a large number of bubbles increases the chance of bubble-floc attachment in very dilute floc suspensions that are typical in water treatment applications. The smaller the bubble size (and the lower the water temperature), the slower the rise rate of the bubble (Figure 7.33). Consequently, a larger flotation tank is required to allow bubbles to reach the surface (Figure 7.35). In practice, the bubble size produced in DAF ranges from 10 to 120 μm, with a mean size of approximately 40 μm (Rees, Rodman, and Zabel, 1979b).

Bubble Volume and Number Concentration. The mass concentration (C_b) of air released is obtained from the following equation.

$$C_b = \frac{(C_r - C_{fl})}{1 + r} r \qquad (7.62)$$

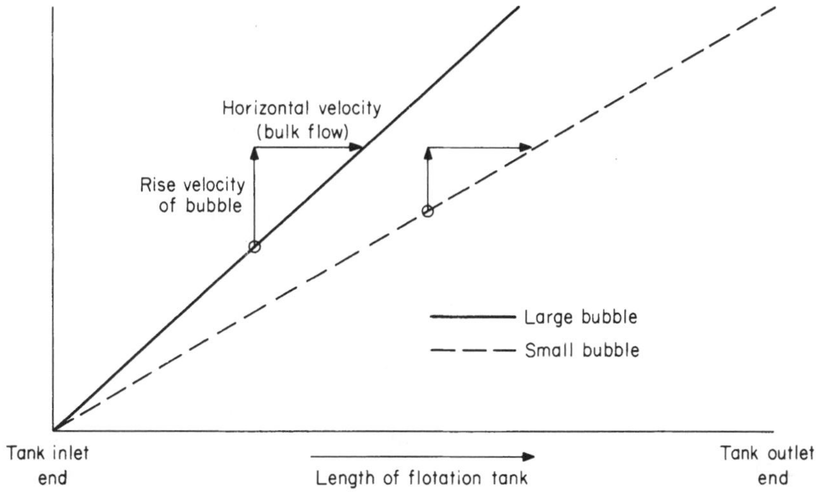

FIGURE 7.35 Effect of bubble size on flotation tank size.

where C_r = the mass concentration of air in the recycle flow (mg/L)
C_{fl} = the mass concentration of air in the floc tank effluent (mg/L)
r = is the recycle ratio expressed on a fractional basis

C_r is computed from a Henry's law expression (Eq. 7.61) for a saturator pressure of interest. Equation 7.62 assumes that the air in the water leaving the flocculation tank is at saturation. The air concentration in the contact zone is expressed fundamentally in terms of either the bubble number concentration (N_b), as indicated in Eq. 7.57, or the bubble volume concentration (Φ_b), as expressed in Eqs. 7.58 and 7.59 and shown in Table 7.2. Once C_b is obtained, the fundamental variables of bubble volume concentration (Φ_b) and bubble number concentration (N_b) are calculated from the following equations.

$$\Phi_b = \frac{C_b}{\rho_{air}} \tag{7.63}$$

$$N_b = \frac{6\Phi_b}{\pi d_b^3} \tag{7.64}$$

where ρ_{air} = the density of air saturated with water vapor (kg/m^3)
π = the mathematical constant (3.14)
d_b = the mean bubble diameter (μm)

Separation Zone. The floc particle-bubble aggregate rise velocity can be calculated from a Stokes' law expression

$$v_{pb} = \frac{g(\rho_w - \rho_{pb})d_{pb}^2}{18\mu} \tag{7.65a}$$

where v_{pb} = the rise velocity of the floc particle-bubble aggregate (m/h).

For transitional flow (1<Re<50) and considering the drag on floc-particle aggregates with sphericity of 0.8 (Tambo and Watanabe, 1979), the modified Stokes' rise velocity is

$$v_{pb} = \frac{16.7g^{0.8}(\rho_w - \rho_{pb})^{0.8} d_{pb}^{1.4}}{\rho_w^{0.2}\mu^{0.6}} \tag{7.65b}$$

The floc particle-bubble aggregate density is calculated from:

$$\rho_{pb} = \left[\frac{\rho_p d_p^3 + N_{ab} \cdot \rho_b d_b^3}{d_p^3 + N_{ab} \cdot d_b^3}\right] \tag{7.66}$$

where ρ_{pb} = the floc particle-bubble aggregate density (kg/m^3)
N_{ab} = the number of attached air bubbles to floc particle

The equivalent spherical diameter (d_{pb}) of the floc particle-bubble aggregate is calculated from

$$d_{pb} = [d_p^3 + N_{ab} \cdot d_b^3]^{1/3} \tag{7.67}$$

Schers and van Dijk (1992) and Liers et al. (1996) have used Edzwald's white water collector model and coupled it to a separation zone efficiency to compute the overall DAF tank efficiency. The separation zone efficiency is simply determined by computing the fraction of particle-bubble aggregates with rise velocities equal to or greater than the separation zone hydraulic loading. Both groups did a model sensitivity analysis and concluded the following:

1. The hydraulic residence time in the contact zone should be at least 90 s.
2. The contact zone should be designed to ensure plug flow.
3. The contact zone efficiency will exceed 95 percent for good coagulation chemistry ($\alpha_{pb} > 0.1$) and if the bubble volume concentration (Φ_b) is at least 5000 ppm [5 L air/m^3 water treated (i.e., about 6 g air/m^3)].

Edzwald et al. (1992) made a similar conclusion about the minimum bubble volume concentration and the significance of this parameter in DAF.

Flotation performance will improve as N_b increases because of greater collision opportunities among bubbles and particles, as shown by Eq. 7.57. This is also shown in terms of the bubble volume concentration (Φ_b) in Eqs. 7.58 and 7.59. The attached air bubbles also provide another role of lowering the floc particle density. The larger the bubble volume attached to floc, the lower the density yielding high-rise velocities of floc particle-bubble aggregates.

The above equations are instructive as illustrated in the following examples:

EXAMPLE PROBLEM 7.5 Consider a DAF plant operating at 10 percent recycle and at a saturator pressure of 483 kPa (70 psig). The flocculated water enters the contact zone of the flotation tank with a floc particle concentration (N_p) of 5000 particles/mL and a floc volume concentration (Φ_p) of 5 ppm. Compute the air mass concentration (c_b), bubble volume concentration (Φ_b), and number concentrations (N_b) in the contact zone of the DAF tank, and compare the concentrations of bubbles to floc particles. For these calculations, use a water temperature of 20°C; at this temperature, ρ_{air} is 1.19 kg/m^3. Assume the flocculated water has no oxygen deficit, so the air concentration is 24 mg/L.

SOLUTION

a. *Mass of air in DAF tank.* First, the air dissolved in the water through the saturator is determined. From Figure 7.34, the mass of air in the recycle water would be approximately 125 mg/L for 584 kPa (483 kPa gauge pressure plus 101 kPa atmospheric). Next, a mass balance for air is made for the contact zone of the DAF tank using Eq. 7.62.

$$C_b = \frac{(C_r - C_{fl})r}{1 + r} = \frac{(125 - 24)0.10}{1.10} \text{ mg/L}$$

$$= 9.2 \text{ mg/L}$$

b. *Bubble volume concentration.* Now, Φ_b is calculated from Eq. 7.63:

$$\Phi_b = \frac{C_b}{\rho_{air}} = \frac{9.2 \text{ mg/L}}{1.19 \text{ kg/m}^3} = 7.7\left(\frac{\text{mg}}{\text{kg}}\right)\left(\frac{\text{m}^3}{\text{L}}\right) = 7.7\left(\frac{\text{mg}}{\text{kg}} \times \frac{\text{kg}}{10^6 \text{ mg}}\right)\left(\frac{\text{m}^3}{\text{L}} \times \frac{10^3\text{L}}{\text{m}^3}\right)$$

$$= \frac{7700 \text{ m}^3 \text{ of air}}{10^6 \text{ m}^3 \text{ of water}} = 7700 \text{ parts per million (ppm)}$$

c. *Bubble number concentration.* N_b is calculated from Eq. 7.64 using a mean bubble diameter of 40 μm (40×10^{-6} m).

$$N_b = \frac{6\Phi_b}{\pi d_b^{\,3}} = \left[\frac{6 \times 7700 \times 10^{-6}}{\pi(40 \times 10^{-6})^3}\right] = 7.22 \times 10^{11}\left(\frac{\text{bubbles}}{\text{m}^3}\right) \times \left(\frac{\text{m}^3}{10^6 \text{ mL}}\right)$$

$$= 7.22 \times 10^5 \left(\frac{\text{bubbles}}{\text{mL}}\right)$$

d. *Ratios of concentrations.*

$$\frac{N_b}{N_p} = \frac{7.72 \times 10^5 \text{ bubbles/mL}}{5000 \text{ particles/mL}} \cong 150$$

$$\frac{\Phi_b}{\Phi_p} = \frac{7700 \text{ ppm}}{5 \text{ ppm}} \cong 1500$$

These calculations illustrate that there is a much larger number of bubbles compared with particles ensuring opportunities for bubble collisions and attachment to particles. They also show that the bubble volume is quite large relative to the particle volume, ensuring a reduction in floc particle-bubble density and large rise velocities of particle-bubble aggregates. Furthermore, the calculations assumed 100 percent saturator efficiency in dissolving air and no loss of air between saturator and injection into the DAF tank. In practice, the actual mass of air calculated in part **a** may be from 70 to 90 percent of this value. However, the bubble numbers and volumes would be very large, as illustrated in the example.

Generally, particle number concentrations following flocculation for most drinking water applications range from 1,000 to 10,000 particles/mL, whereas bubble concentrations for DAF systems operating at saturator pressures of 450 to 500 kPa and recycle ratios of from 6 to 12 percent are in the range of 1×10^5 to 2×10^5 bubbles/mL. This yields bubble number to particle number ratios of at least 10 to 1 and as high as 200 to 1.

Likewise, particle volume concentrations following flocculation generally range from 1 to 10 ppm. Bubble volume concentrations range from about 3500 to 8000 ppm. This yields ratios of bubble volume to particle volume of at least 350 to 1 and as high as 8000 to 1. Even a highly turbid water (100 mg/L suspended solids) coagulated with

100 mg/L of alum would have a particle volume concentration of no greater than 100 ppm; therefore, the bubble-volume-to-particle-volume ratio is at least 35 to 1.

EXAMPLE PROBLEM 7.6 Calculate (a) the settling velocity of a 10 μm floc particle with a particle density of 1100 kg/m³ for summer (20°C) and winter (4°C) water temperatures and (b) the rise velocity of this floc attached to 1 air bubble of 40 μm for the same two temperature conditions.

SOLUTION

a. *Floc settling velocity.* To solve the problem requires water density and absolute viscosity for the two water temperatures:

Temperature (°C)	Density (kg/m³)	Absolute Viscosity (N·s/m²)
4	999.97	1.6527×10^{-3}
20	998.20	1.0019×10^{-3}

1. For 20°C, from Stokes' law for settling of particles, Eq. 7.8:

$$v_p = \frac{g(\rho_p - \rho_w)d_p^2}{18\mu} = \frac{9.8(1100 - 998.2)(10 \times 10^{-6})^2}{18 \times 1.0019 \times 10^{-3}} \frac{\text{m}}{\text{s}} \times \frac{3600 \text{ s}}{\text{h}}$$

$$= 0.02 \text{ m/h}$$

2. For 4°C:

$$v_p = 0.012 \text{ m/h}$$

b. *Rise velocity of 40 μm air bubble attached to floc particle of 10 μm.* To solve the problem requires water density and viscosity constants (presented above) and the density of air: air densities (saturated with water vapor) at 4° and 20°C are 1.27 and 1.19 kg/m³, respectively.

1. For 20°C:

To compute the rise velocity, the density of the air bubble with attached floc particle and the equivalent diameter of the floc particle-bubble aggregate must be calculated. Equations 7.66 and 7.67 are used.

$$\rho_{pb} = \left[\frac{\rho_p d_p^3 + N_{ab}(\rho_b d_b^3)}{d_p^3 + N_{ab}(d_b^3)}\right] = \left[\frac{1100(10)^3 + 1(1.19)(40)^3}{10^3 + 40^3}\right] \frac{\text{kg}}{\text{m}^3}$$

$$= 18.1 \text{ kg/m}^3$$

$$d_{pb} = [d_p^3 + N_{ab}(d_b)^3]^{1/3} = [10^3 + 1(40)^3]^{1/3} \mu\text{m}$$

$$= 40.2 \ \mu\text{m}$$

Note that the equivalent spherical diameter of the floc particle-bubble aggregate is not much larger than the bubble. This is because $d_b >> d_p$. From Eq. 7.65a, the particle-bubble agglomerate rise velocity is determined:

$$v_{pb} = \frac{g(\rho_w - \rho_{pb})d_{pb}^2}{18\mu} = \frac{9.8(998.2 - 18.1^2)(40.2 \times 10^{-6})^2}{18 \times 1.0019 \times 10^{-3}} \frac{\text{m}}{\text{s}} \times \frac{3600 \text{ s}}{\text{h}}$$

$$= 3.1 \text{ m/h}$$

2. For 4°C, the results are as follows:

$$\rho_{pb} = 18.2 \text{ kg/m}^3$$

$$d_{pb} = 40.2 \ \mu\text{m}$$

$$v_{pb} = 1.9 \text{ m/h}$$

These rise velocities are much greater than particle-settling velocities and show in a simplistic way why DAF tanks can be designed with greater hydraulic loading rates than horizontal-flow settling tanks.

OPERATIONAL AND DESIGN CONSIDERATIONS FOR FLOTATION

Types of Flotation Tanks

Circular Tanks. Circular tanks are used mainly in small flotation plants treating wastewater or for sludge thickening applications that require no preflocculation prior to flotation (Figure 7.36).

For potable water treatment, a preflocculation stage is typically required prior to flotation to flocculate the impurities present in water to form agglomerates suitable for removal by flotation. For circular tanks the transfer of the flocs to the flotation tank without breakage creates problems. In larger plants, the flocculated water must be introduced close to the bottom of the center section of the flotation tank to

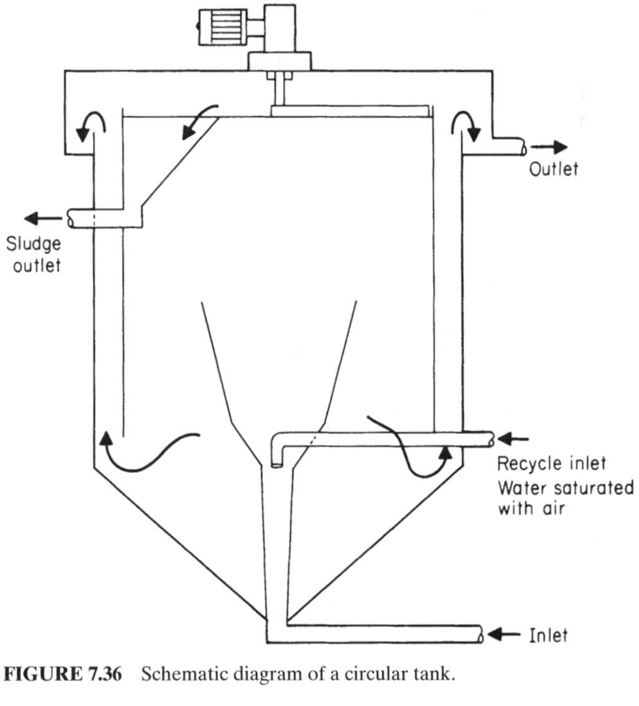

FIGURE 7.36 Schematic diagram of a circular tank.

achieve even distribution. As a result, most large flotation plants for water treatment use rectangular tanks. However, small and large flotation package plants have been built where flocculation and flotation are contained within the same circular tank.

Rectangular Tanks. Rectangular flotation tanks offer advantages in terms of scale-up, simple design, easy introduction of flocculated water, easy float removal, and a relatively small area requirement. Flotation tanks are typically designed with a depth of approximately 1.5 m (5 ft) and overflow rates of between 8 and 12 m/h [26 and 40 ft/h (3 and 5 gpm/ft^2)] depending on the type of water treated. Tanks are equipped at the inlet with an inclined baffle (60° to the horizontal) to direct the bubble-floc agglomerates toward the surface and to reduce velocity extremes in the incoming water, to ensure minimum disturbance of the float layer accumulating on the water surface (Figure 7.30). The gap between the top of the baffle and the water surface should be designed to achieve a horizontal water velocity similar to the velocity in the top section of the inclined baffle area. However, the size and nature of design features such as these continue to be challenged as investigational tools and computational fluid dynamic modeling are developed (Fawcett, 1997; O'Neill, 1997).

Maximum tank size is determined by hydraulic conditions and the design of the sludge removal system. Tanks with surface areas in excess of 80 m^2 (860 ft^2) are in operation. The nominal retention time in the flotation tank is between 5 and 15 min, depending on surface loading and tank depth (Zabel, 1985). The flotation tank must be covered because both rain and wind can cause breakup of floated solids and because freezing of the float can cause problems. Treated water should be withdrawn, preferably via a full-width weir, to maintain uniform hydraulic conditions and to minimize changes in water level because of variations in flow through the plant.

Countercurrent Flotation. The basic design of DAF has some analogy to horizontal-flow sedimentation, with flocculated water entering one end of a tank where it is mixed with dissolved air, followed by separation in a horizontal dimension. Radical reappraisal of the concept of DAF concluded that because flotation is mainly in the vertical dimension, then flow through the flotation tank should be the same. This led to the COCODAF® design, which is sometimes combined with filtration (Figure 7.37). (Both applications are proprietary processes of Thames Water licensed to Paterson Candy Limited.) These processes were developed with the aid of computational fluid dynamic analysis to overcome some of the hydraulic problems (Eades et al., 1997). In this design, the DAF tank bears more resemblance to an upflow floc-blanket clarifier in contrast to a normal DAF tank having similarities with a horizontal-flow sedimentation tank. In the Coco-DAF design, the flocculated water is introduced above the recycle, uniformly across the area of the tank, so that the water flows down through the rising bubble cloud to the treated water outlet, or to the filter bed as in Coco-DAFF.

Combined Flotation and Filtration. The combination of flotation and filtration was pioneered in Sweden (Figure 7.38) (Zabel and Melbourne, 1980), but experience with this process has also been reported from other countries (Krofta and Wang, 1984; Van Vuuren, de Wet, and Cillie, 1984). A rapid gravity sand or anthracite-sand filter is incorporated in the lower section of the flotation tank. This concept has the advantage of providing an extremely compact plant. The flotation rate of the plant is, however, limited by the filtration rate that can be achieved. It is also possible to apply the concept to existing filters, of appropriate dimensions, by retrofitting the DAF equipment.

The tank depth of a flotation-filtration plant tends to be deeper, approximately 2.5 m (8.3 ft), to accommodate the filter bed and underdrain system, compared with

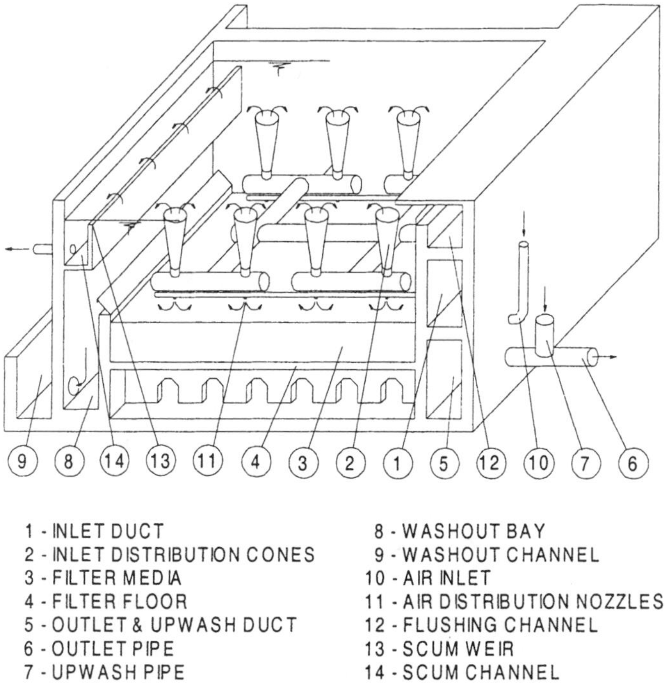

FIGURE 7.37 Typical arrangement of a COCODAF® unit. (*Source:* Courtesy of Thames Water and Paterson Candy Ltd, Isleworth, U.K.)

1.5 m (5 ft) for a separate flotation unit. In addition, the flow to the plant and any coagulant dosing has to be stopped periodically to facilitate cleaning of the filter that is backwashed in the normal way by air scour and water wash. The compactness of this system makes it particularly suitable for package plants. The reported performance of the package plants in terms of treated water quality is comparable with standard flotation plants followed by rapid gravity filtration (Zabel and Melbourne, 1980).

Air Saturation Systems

Approximately 50 percent of the power costs of the flotation process is for pumping the recycle against the saturator pressure. As a result, optimization of the recycle system design is important in minimizing operating costs.

Various methods are employed for dissolving air under pressure in the recycle stream. These include sparging the air into the water in a pressure vessel (or saturator), trickling the water over a packed bed, spraying the water into an unpacked saturator, entraining the air with eductors, and injecting the air into the suction pipe of the recycle pump (Figure 7.39) (Zabel and Melbourne, 1980).

Introducing the air to the recycled water either on the suction side of the recycle pump or through eductors before entering the saturation vessel leads to substantially higher pumping costs compared with using a separate compressed-air supply. Saturation levels of between 60 and 80 percent achieved by introducing the air on

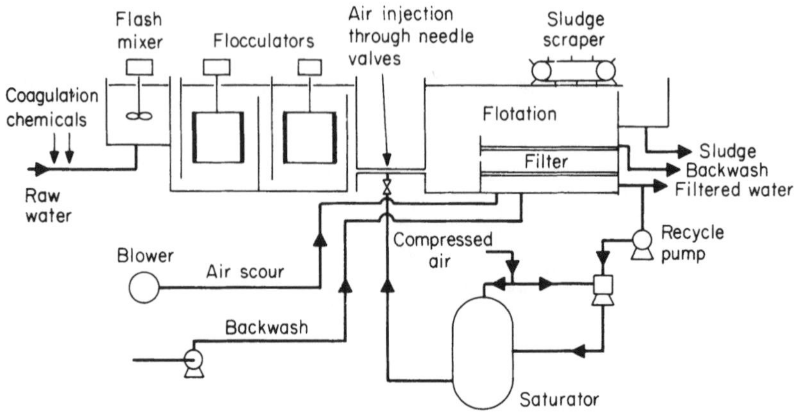

FIGURE 7.38 Schematic diagram of combined flotation-filtration plant.

the suction side of the pump could be increased to 90 percent by providing a turbine mixer in the saturation vessel (Bratby and Marais, 1977).

Tests have shown that the packed saturator system has the lowest ratio of operating cost to saturation level achieved (Vrablik, 1959; Rees, Rodman, and Zabel, 1980). A possible disadvantage of using packings is the danger of blockage caused by biological growth or other precipitates. A problem with iron precipitates has been encountered in a few plants treating drinking water.

Extensive research has been carried out on optimizing the design of packed saturators (Rees, Rodman, and Zabel, 1980). It has been shown that the saturator can be operated over the range of hydraulic loading 300 to 2000 $m^3/m^2 \cdot d^{-1}$ [36 to 274 ft/h (5 to 30 gpm/ft^2)] without any decrease in saturation efficiency. A packing depth of 0.8 m (2.6 ft) of 25-mm (1-in) polypropylene rings is sufficient to achieve 100 percent saturation (Figure 7.40) (Rees, Rodman, and Zabel, 1980). This packing depth, however, was substantially higher than the 0.3 m (1 ft) reported elsewhere (Bratby and Marais, 1977).

In terms of operating costs, an unpacked saturator must be operated at a pressure of about 200 kPa (30 psi) above that of a packed saturator to supply the same amount of air to the flotation tank (Zabel and Hyde, 1977). Water saturated with air under pressure is corrosive. If mild steel is used for construction of the saturation vessel, a corrosion-resistant lining should be provided. Plastic pipes should be used for connecting pipe work between the saturator and flotation tank.

Factors Influencing Dissolved-Air Flotation Efficiency

Coagulation. In the contact zone of the flotation tank, collisions occur between bubbles and particles. To achieve favorable conditions for attachment of bubbles with particles requires good coagulation pretreatment. The chemicals used as coagulants have two roles in flotation: (1) to produce floc particles with reduced surface charge, and (2) to produce floc particles that are hydrophobic. The coagulants that are effective for flotation are the same as for sedimentation and direct filtration. Selection depends not on flotation, but on costs, local preference and availability, and raw water quality {particularly on alkalinity, organic content [DOC, natural organic matter (NOM), color], and temperature}.

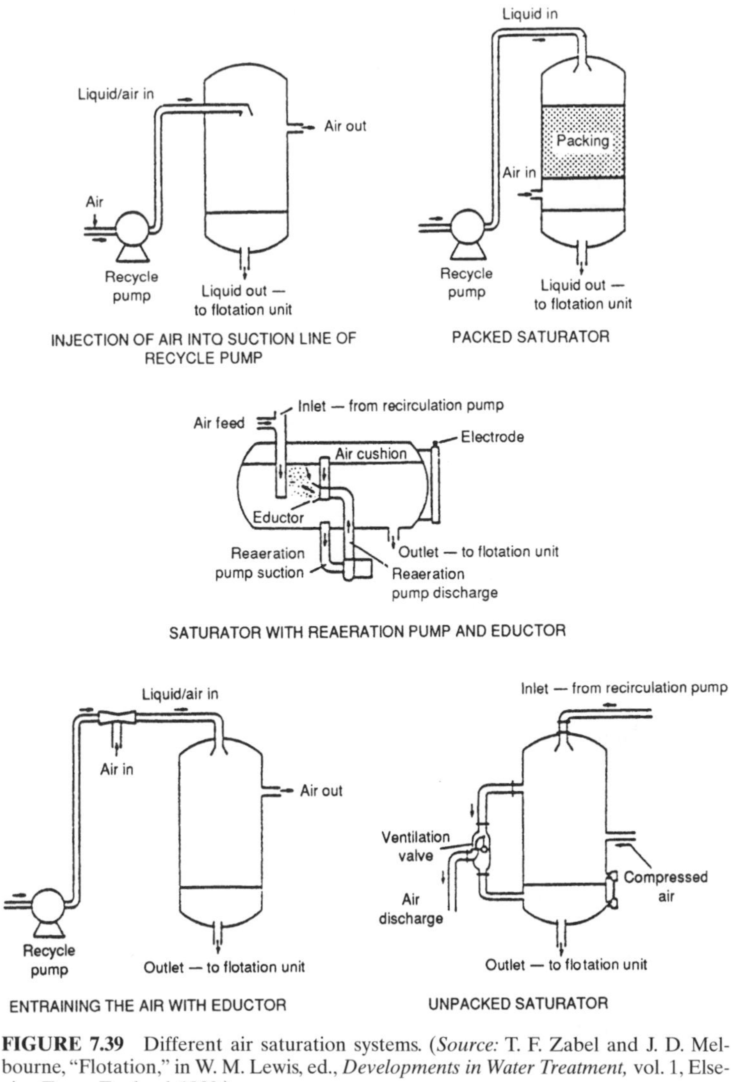

FIGURE 7.39 Different air saturation systems. (*Source:* T. F. Zabel and J. D. Melbourne, "Flotation," in W. M. Lewis, ed., *Developments in Water Treatment,* vol. 1, Elsevier, Essex, England, 1980.)

Figure 7.41 illustrates the effect of coagulant dose on floc particle charge and on flotation efficiency for the removals of DOC and turbidity. The floc charge is presented as particle electrophoretic mobility (EPM) measurements. The figure shows clearly that underdosing of coagulant is associated with negative particle charge and little difference in turbidities before and after flotation. Minimum turbidity after flotation is associated with a particle charge of about zero and with good removal of organic matter. In practice, measurements of particle charge and hydrophobicity are not needed to determine optimum coagulation conditions. Optimum coagulant dose

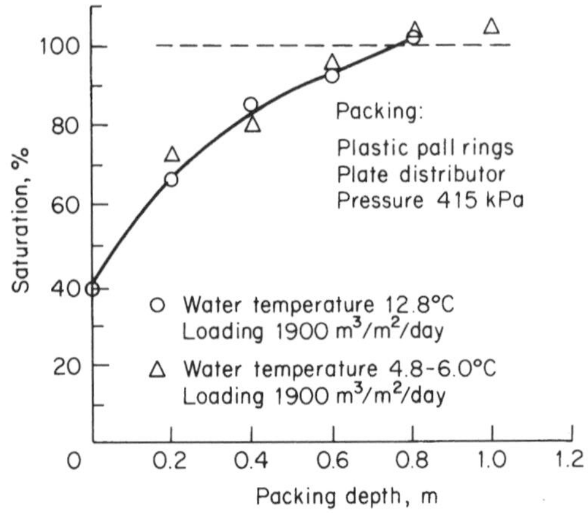

FIGURE 7.40 Effect of packing depth on air dissolution. (*Source:* Rees, Rodman, and Zabel, 1980.)

and coagulation pH may be determined using standard sedimentation or flotation jar test apparatus (see Chapter 6).

The destabilization of particles with coagulants is fundamentally the same for flotation and sedimentation. Optimum coagulant doses for flotation and sedimentation generally are the same. An exception is that lower coagulant doses may be appropriate in treating high-quality waters, whereas sedimentation processes based on discrete particle behavior may require coagulant doses high enough to establish sweep floc conditions for effective flocculation (see Chapter 6).

Another important difference is that flotation does not normally require addition of high-molecular-weight polyelectrolyte flocculants. The most difficult type of water to treat by flotation, and sedimentation, has low alkalinity, turbidity, and DOC, especially when alum is used and the water is very cold. If alum is used, control of coagulation pH must take account of water temperature effects on aluminum solubility and optimum coagulation.

For best treated water quality, chemicals must be thoroughly and rapidly mixed with the raw water. Some form of in-line mixing is preferable to the use of a flash mixer. If both a coagulant and a pH adjustment chemical are required, good mixing of the first chemical with the source water must be completed before the addition of the second chemical. This is particularly important when treating soft waters with high DOC. The order of chemical addition can be important if stress conditions of high flow, low water temperature, or poor raw water quality occur.

Flocculation. Before coagulated impurities can be removed successfully by flotation, flocculation into larger agglomerates (floc) is required. Flocculation time, degree of agitation, and the means of providing agitation affect flotation performance. Flocculation for flotation has a different objective than for sedimentation. Although unhindered sedimentation needs floc particles of 100 μm and larger, flotation is effective with smaller floc particles (Edzwald et al., 1992; Valade et al., 1996).

Flocculation Time. The flocculator usually consists of a tank subdivided by partial baffles into two equal-sized compartments in series, each agitated by slow-moving paddle or propeller units. Additional compartments can be used, but this is usually

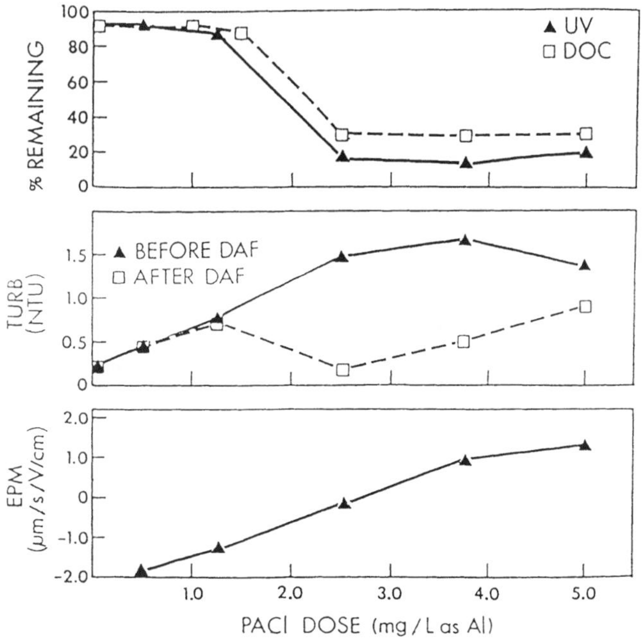

FIGURE 7.41 Effect of polyaluminium chloride coagulant dose on flotation performance at pH 5.5. (*Source:* Edzwald and Malley, 1990.)

not cost-effective. The flocculation time required differs with the type of raw water being treated, coagulant used, and water temperature. For turbid and warm waters, a flocculation time of less than 10 min may be viable, whereas for soft, colored water using aluminum sulfate when the water is close to freezing, a flocculation time of 20 min or more may be required. From a design view, plants could be designed for 10 min flocculation if winter flows are low enough to allow longer flocculation time.

An increase in flocculation time that produces larger floc particles (d_p) improves flotation efficiency as shown in Figure 7.32 and summarized in Table 7.2; however, floc sizes need only be 10s of micrometers. Valade et al. (1996) have shown good filtration performance with short flocculation times (see Table 7.3).

Degree of Agitation. Besides the flocculation time, the degree of agitation is also very important for efficient flocculation. Agitation is usually provided by a slow-moving, four-blade gate paddle in each flocculator compartment, although different paddle designs are being used. To avoid excess shear, which prevents adequate floc growth, the tip speed of the paddles should not exceed 0.5 m/s (1.6 ft/s). More recently, propeller units have been applied with apparent effectiveness in flotation (Franklin, 1997) and filter performance (Valade et al., 1996).

The degree of agitation can be expressed by the *mean velocity gradient, G* (see Chapter 6). Tests have indicated that the optimum mean velocity gradient for flotation is about 70 s^{-1}, independent of the type of surface water treated (Rees, Rodman, and Zabel, 1979a). This compares with an optimum G value for horizontal sedimentation of between 10 and 50 s^{-1}. Flotation performance, using either gate or propeller units for flocculation, was affected little by G value in the range 30 to 70 s^{-1} (Valade

TABLE 7.3 Comparison of the Effect of 5 and 20 min Flocculation on Flotation and Filtered Water Turbidity and Particle Numbers

	Floc time*			
	5 min		20 min	
Sample	Turbidity (ntu)	Part. #/mL (2–200 μm)	Turbidity (ntu)	Part. #/mL (2–200 μm)
Raw wat.	1.0	2040	0.9	4010
Floc. effl.	6.7	18480	6.9	14010
DAF effl.	1.0	1620	0.8	1530
Filter effl.	0.01	6	0.04	12

* Conditions: *5 min floc.:* ferric sulfate, 5.7 mg/L as Fe; water temp., 2.1°C; recycle, 6%. *20 min floc.:* ferric sulfate, 5.1 mg/L as Fe; water temp., 5.8°C; recycle, 9%. Common conditions: 3-stage flocculation *G,* 70 s_{-1}; DAF loading, 8 m/h (3.3 gpm/ft^2); dual media filters loading, 10 m/h (4.1 gpm/ft^2).

Source: From work by Valade et al., 1996.

et al., 1996). However, there may need to be a minimum degree of agitation in flocculators to prevent settlement of floc in the flocculators or short circuiting through the tanks. It is also important to produce relatively small and compact flocs to avoid floc breakup when the air is introduced.

Values of *Gt* between 40,000 and 60,000 are usually considered necessary for efficient flotation. However, the move toward short flocculation times is showing that *Gt* values less than 20,000 can be effective (Valade et al., 1996).

Hydraulic Flocculation. An alternative approach to mechanical flocculation is the use of hydraulic flocculation in which the energy required for flocculation is provided by the water flowing through the flocculator, which can be a baffled tank. Tests have shown that one-half the flocculation time, with a higher *G* value (150 s^{-1}), was required for hydraulic flocculation compared with mechanical flocculation ($G = 70\ s^{-1}$) (Figure 7.42) (Rodman, 1982). The difference in *G* value required is probably due to the more uniform velocity distribution in the hydraulic floccula-

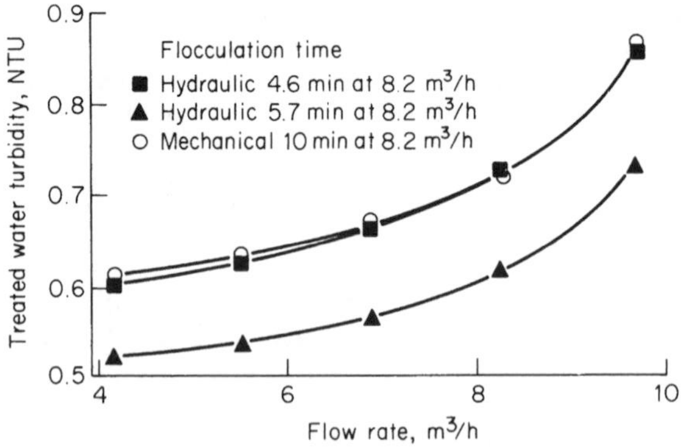

FIGURE 7.42 Comparison of hydraulic and mechanical flocculation for flotation. (*Source:* Rodman, D. J. 1982. *Investigation into Hydraulic Flocculation with Special Emphasis on Algal Removal.* Master's thesis. Water Research Centre, Stevenage, U.K.)

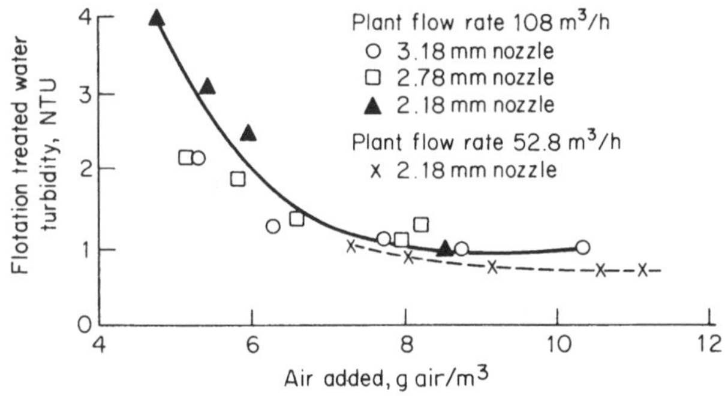

FIGURE 7.43 Effect of quantity of air added and nozzle size on flotation-treated water quality. (*Source:* A. J. Rees, D. J. Rodman, and T. F. Zabel, *Water Clarification by Flotation,* 5, TR 114, Water Research Centre, Medmenham, U.K., 1979.)

tors, thus avoiding excess shear and floc breakup. However, from these results, the *Gt* value for efficient flotation is independent of whether hydraulic or mechanical flocculation is employed.

Quantity of Air Required for Flotation. The quantity of air supplied to the flotation tank can be varied by altering either the saturator pressure or the amount of recycle or both. If a fixed orifice is used for controlling the injection of recycle, an increase in saturator pressure is associated with a small increase in recycle rate. Thus, different nozzle sizes require different combinations of flow and pressure to deliver the same amount of air.

Experiments varying the recycle rate by using different nozzle sizes and saturator pressures have shown that treated water quality is dependent on only the total amount of air supplied, not the pressure and recycle rate employed (Figure 7.43) (Rees, Rodman, and Zabel, 1979a). This confirmed work also by Bratby and Marais (1975).

The quantity of air required for treatment of surface waters depends only on the total quantity of water treated and is independent of the suspended solids present, unless the suspended solids concentration is very high (>1000 mg/L). The air-solids ratio required for surface water treatment is approximately 380 mL of air per gram of solids for a solids concentration in the raw water of 20 mg/L. That is much higher than for activated sludge thickening, for which 15 to 30 mL of air per gram of solids is required (Maddock, 1977). The large air-solids ratio required for treatment of surface water with low turbidity is probably necessary to ensure adequate collision between floc particles and air bubbles, to facilitate attachment before separation.

With a packed saturator, an operating pressure of between 350 and 420 kPa (50 and 60 psi) and a recycle rate of between 7 and 8 percent, corresponding to about 8 to 10 g air/m^3 water treated, were found adequate for optimum performance (Rees, Rodman, and Zabel, 1979a). Subsequent experience has shown this generally to be the case.

Fundamentally, volume concentrations of bubbles are more important than mass concentrations. It shows that larger volumes lead to greater collision rates with particles and produce densities of particle-bubble aggregates much less than water (see Example Problems 7.5 and 7.6).

Air-Release Devices. Different types of pressure-release devices are employed, ranging from proprietary nozzles and needle valves to simple gate valves. To achieve effective air release, the pressure should be reduced suddenly and highly turbulent conditions should exist in the device. The velocity of the recycle stream leaving the pressure-reduction device should be low enough to prevent floc breakup. As water passes through the pressure-release devices at high velocities and as air is released, erosion and cavitation can occur, and so the devices should be made from stainless steel. For larger plants a number of these devices, usually at a spacing of approximately 0.3 m (1 ft), are used to obtain good mixing and distribution between the flocculated water and the air bubbles. Good mixing of the recycle stream containing the released air bubbles with the flocculated water stream is also essential to facilitate contact between bubbles and floc.

Releasing the pressure of the recycle stream close to the point of injection into the flocculated water is important to minimize coalescence of air bubbles, or premature release of air in the recycle prior to injection, which could result in a loss of bubbles available for flotation.

An air injection nozzle was patented and developed (Brit. Pat. 1976) that consists of two orifice plates, to reduce the pressure and to create turbulence, and a shroud section, to decrease the velocity of the stream of recycled water before it is mixed with the flocculated water. The size of the first orifice plate, which is the smaller, controls the amount of recycled water added to the flotation tank. More than 95 percent of the bubbles produced by the nozzle were in the size range 10 to 120 μm, with a mean size of about 40 μm. A comparison of the proprietary nozzle and a needle valve showed that the nozzle produced smaller air bubbles; however, both devices achieved similar flotation-treated water quality (Rees, Rodman, and Zabel, 1979a).

Because the production and size of air bubbles is regarded as important, considerable further attention has been given to understanding and improving nozzle design (Van Craenenbroeck et al., 1993; van Puffelen et al., 1995; Heinänan et al., 1995; Rykaart and Haarhoff, 1995; Offringa, 1995; Franklin, 1997). These investigations have demonstrated that orifice diameter, impingement distance, and other features are important in efficient production of bubbles of the preferred mean size and size range. Franklin found that small improvements in flotation efficiency could be obtained through changes to nozzle design and bubble size, whereas Van Craenenbroeck et al. found no effect. In general, the size and number of bubbles produced is primarily a function of the energy available, which is not otherwise dissipated in turbulence and friction: the higher the nozzle back-pressure the smaller the bubble size (Jackson, 1994; Rykaart and Haarhoff, 1995). Also, the number of nucleation sites is greatly increased when certain impurities, especially surface-active agents, are present (Jackson, 1994).

Float Removal. The sludge that accumulates on the flotation tank surface, called *float,* can be removed either continuously or intermittently by flooding or mechanical scraping. Flooding involves raising the water level in the flotation tank sufficiently by closing the treated-water outlet or lowering the outlet weir to allow the float and water to flow into the float collection trough. The flooding method has the advantages of low equipment costs and minimal effect on the treated-water quality, but at the expense of high water wastage (up to 2 percent of plant throughput) and very low sludge solids concentration (less than 0.2 percent). Therefore, with this float removal method, one advantage of flotation—the production of a sludge with a high solids concentration—is lost.

The most widely used mechanical float removal devices are of two types: (1) part- or full-length scrapers usually with rubber or brush blades that travel over the tank surface and push the float over the beach into the collection channel, and (2) beach scrapers that consist of a number of rubber blades rotating over the beach.

As float is removed from the beach, float from the remainder of the flotation tank surface flows toward the beach. The beach scrapers, especially if operated continuously, have the advantage of reducing the danger of float breakup during the removal process because the float is minimally disturbed. Beach scrapers are also of simpler construction compared with full-length scrapers. They have the disadvantage, however, of producing relatively thin sludges (1 to 3 percent), because thicker sludge will not flow toward the beach. If a part- or full-length scraper is used, selecting the correct frequency of float removal and travel speed of the scraper is important to minimize deterioration in treated-water quality because of float breakup.

Effect of Air-Solids Ratio on Float. In water treatment, the aim is to produce water of a good quality, and thickness of the float is of secondary importance. In general, the thicker the float, which means the longer the float is allowed to accumulate on the tank surface, the more severe will be the deterioration in treated-water quality during the float removal process. The variations in air-solids ratio, produced by varying the amount of air added to the system, have no influence on the float concentration produced.

Influence of Source Water on Float Characteristics. The characteristics of float obtained from the treatment of different source waters vary considerably. Therefore, the most appropriate float-removal system must be selected for the source water being treated. For example, experience has shown that for cases in which a low-alkalinity, highly colored water was treated, the float started to break up after only 30 min of accumulation. The most appropriate device for such an application is a full-length scraper operating at a removal frequency and blade spacing that does not allow the float to remain on the flotation tank surface for longer than 30 min. The optimum scraper speed for this application, in terms of treated-water quality and float solids concentration, was 30 m/h (0.028 ft/s), producing a sludge of 1 percent solids concentration.

Conversely, float produced from turbid river water or stored algal-laden waters is very stable. Accumulation of float for more than 24 h does not result in float breakup or deterioration in treated-water quality. Beach scrapers and part- and full-length scrapers have been used successfully for these applications, producing solids concentrations in excess of 3 percent with little deterioration in treated-water quality, provided the float is not allowed to accumulate for too long. These floats were suitable for filter pressing, producing cake solids concentration of between 16 and 23 percent without polyelectrolyte addition.

The float stability is independent of the primary coagulant used, for optimum coagulant dose and coagulation pH. The addition of flocculant aid can be found useful when using aluminum sulfate in very cold waters, with the flocculant also helping in weak floc situations to reduce float breakup and increase filter run length.

For optimum operation in terms of treated-water quality and float solids concentration, beach scrapers should be operated continuously, the water level in the flotation tank should be adjusted close to the lower edge of the beach, and a thin, continuous float layer of about 10 mm (0.4 in) should be maintained on the surface of the flotation tank. This operation produces a float concentration, depending on the source water treated, of between 1 and 3 percent with minimum deterioration of treated-water quality.

Equipment costs for float removal systems are significant and can be as much as 10 to 20 percent of the total plant cost. Selection of the most appropriate and cost-effective removal system for a particular application is important.

Performance of Dissolved-Air Flotation Plants

Extensive studies with both pilot- and full-scale plants have been conducted on the performance of the DAF process treating different source waters and reported in detail during the past 20 years (Rees, Rodman, and Zabel, 1979b; Zabel and Melbourne, 1980; Wilkinson, Bolas, and Adkins, 1981; Krofta and Wang, 1984, 1985; Drajo, 1984; Edzwald et al., 1995; Ives and Bernhardt, 1995; Valade et al., 1996; CIWEM, 1997). The types of source water investigated include colored (low-alkalinity) water, mineral-bearing (high-alkalinity) water, algal-bearing water, and low-turbidity, low-color water.

Treatment of Lowland Mineral-Bearing (High-Alkalinity) River Water. Under optimum operating conditions (Rees, Rodman, and Zabel, 1979a), flotation reduced source water turbidities of up to 100 NTU to levels rarely exceeding 3 NTU at a design flow rate of 12 m/h (39.6 ft/h). When the source water turbidity exceeded 60 NTU, treated-water quality was improved significantly by reducing the flow rate through the plant by about 10 to 20 percent. Color was reduced from as much as 70 to less than 5 color units (CU), and residual coagulant concentrations before filtration were in the range of from 0.25 to 0.75 mg Al/L.

A floc-blanket sedimentation plant operated at an upflow rate of 2 m/h (6.5 ft/h) produced similar treated-water quality to that of the flotation plant during low-turbidity periods but better quality (by 1 to 2 NTU) when source water turbidity was greater than 100 NTU. Selection of the correct coagulant dosage and coagulation pH was critical during flood conditions. Because of the short residence time in the flotation plant, changes in source water quality had to be followed closely to maintain optimum coagulation conditions.

Directly abstracted river water can be treated successfully by flotation. However, if the source water turbidity varies rapidly with high turbidity peaks (>100 NTU) and large doses of coagulant are needed, then sedimentation tends to be the more appropriate treatment process.

Treatment of Colored (Low-Alkalinity) Stored Water. Table 7.4 shows a comparison of the water quality achieved by flotation, sedimentation, and filtration treating a colored (low-alkalinity) stored water (Rees, Rodman, and Zabel, 1979a). The flotation plant was operated at 12 m/h (39.6 ft/h) upflow rate.

The floc-blanket sedimentation plant could only be operated at less than 1 m/h (3.24 ft/h), however, even with the addition of polyelectrolyte. The floc produced by coagulation of these waters is very light and has low settling velocities. The quality of the waters treated by the two processes was quite similar. Initially, the residual coagulant concentration of the sedimentation-treated water was usually lower by about 0.2 mg Fe/L. By increasing the flocculation time from 12 to 16 min, however, the residual coagulant concentration in the flotation-treated water was reduced to that of the sedimentation-treated water.

Another advantage of flotation was that the plant consistently produced good treated-water quality even at temperatures below 4°C (39°F), which occurred frequently during the winter months. At these low temperatures, the floc blanket in the sedimentation tank tended to become unstable, resulting in a deterioration in treated-water quality.

Treatment of Algal-Bearing (High-Alkalinity) Stored Water. Recent emphasis on source water storage in water resources management has led to the construction of source water storage reservoirs. Severe algal problems have been experienced in some storage reservoirs containing nutrient-rich waters, which in turn has led to prob-

TABLE 7.4 Comparison of Qualities Achieved with Flotation,* Sedimentation,† and Filtration (Colored Low-Alkalinity Water) Following Iron Coagulation (Rees, Rodman & Zabel 1979a, Zabel & Melbourne 1980)

Type of Water	Turbidity (NTU)	Dose (mgFe/L)	Color (PtCo)	pH	Iron (mg/L)	Manganese (mg/L)	Aluminum (mg/L)
Source	3.2	—	45	6.2	0.70	0.11	0.23
Flotation-treated‡	0.72	8.5	2	4.8	0.58	0.16	0.01
Flotation-filtered	0.19	—	0	9.0	0.01	<0.02	0.01
Sedimentation-treated	0.50	6.0§	0	5.05	0.36	0.14	0.10
Sedimentation-filtered	0.29	—	0	10.5	0.01	<0.02	0.10

* Upflow rate: 12 m/h (39 ft/h).
† Upflow rate: 1 m/h (3.25 ft/h).
‡ Improved flotation-treated water quality similar to that achieved with sedimentation was obtained by increasing the flocculation time from 12 to 16 min.
§ Plus 0.8 mg polyelectrolyte/L required to maintain the floc blanket.
Source: Rees, Rodman, and Zabel, 1979a; Zabel and Melbourne, 1981.

lems in existing sedimentation treatment plants operating at less than 2 m/h (6.5 ft/h). Algae tend to float and are, therefore, difficult to remove by sedimentation.

Table 7.5 shows the removal efficiency of flotation and sedimentation for different algal species (Rees, Rodman, and Zabel, 1979a). At times the algal counts in the flotation-treated water were lower than those in the sedimentation and filtered water (Figure 7.44). Efficient coagulation and flocculation is essential for effective algal removal. Only 10 to 20 percent algal removal was obtained when the flotation plant was operated without coagulation.

Figure 7.45 shows a comparison of algae removal rates achieved by three different coagulants at their optimum pH for minimum coagulant residuals. Aluminum sulfate gave the best removal. For polyaluminum chloride (PACl), an equivalent dosage in terms of aluminum was required to achieve an algae removal similar to that of aluminum sulfate. The poorest treated-water quality was obtained with chlorinated ferrous sulfate. Tests have shown that algae removal is improved by lowering the pH. The poorer algae removal achieved with chlorinated ferrous sulfate might have been a result of the higher coagulation pH required for minimum coagulant

TABLE 7.5 Comparison of Algae Removal Efficiency of Flotation* and Sedimentation† Using Chlorinated Ferrous Sulphate as Coagulant

Algae type	Source water (cells/mL)	Sedimentation-Treated Water (cells/mL)	Flotation-Treated‡ Water (cells/mL)
Aphanizomenon	179,000	23,000	2,800
Microcystis§	102,000	24,000	2,000
Srephanodiscus	53,000	21,900	9,100
Chlorella	23,000	3,600	2,200

* Upflow rate: 12 m/h (39 ft/h).
† Upflow rate: 2 m/h (6.5 ft/h).
‡ Before filtration.
§ Aluminum sulphate used as coagulant.
Source: Rees, Rodman, and Zabel, 1979a; Zabel and Melbourne, 1980.

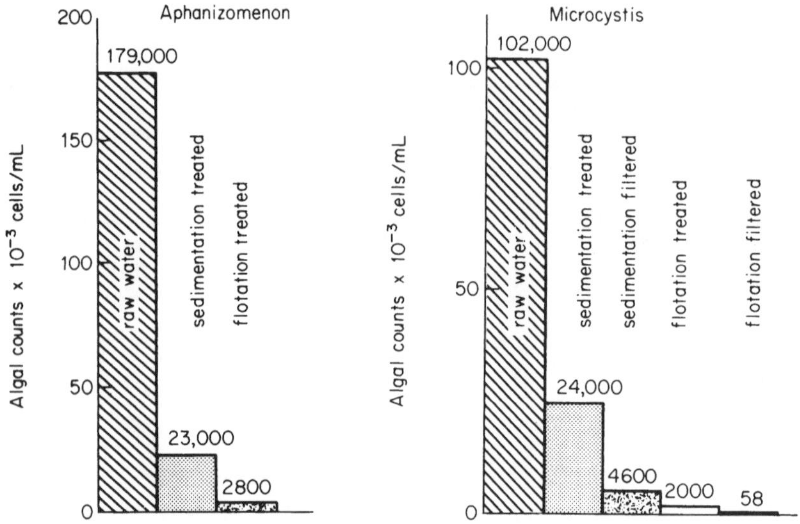

FIGURE 7.44 Removal of algae by flotation [12 m/h (39 ft/h)], floc-blanket sedimentation [2 m/h (6.5 ft/h)], and filtration. (*Source:* A. J. Rees, D. J. Rodman, and T. F. Zabel, *Water Clarification by Flotation,* 5, TR 114, Water Research Centre, Medmenham, U.K., 1979.)

Coagulant		Raw water algal count cells/m, microcystis	Raw water turbidity NTU
Aluminum sulphate	○	40,000	3.5 – 4.1
Aluminum sulphate	●	120,000	5.5 – 5.9
PACl	■	120,000	5.5 – 5.9
Chlorinated ferrous sulphate	▲	48,000	3.6

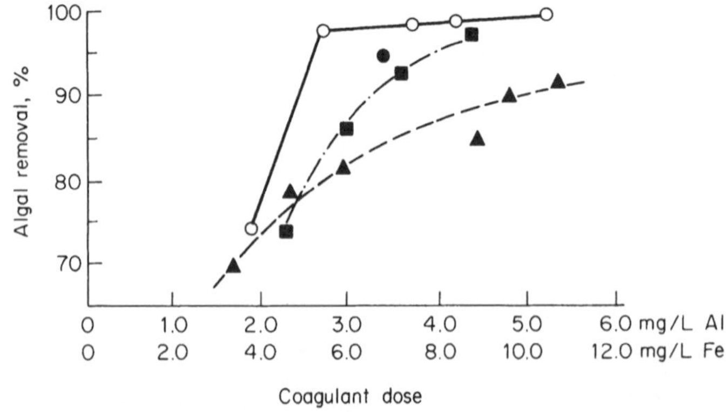

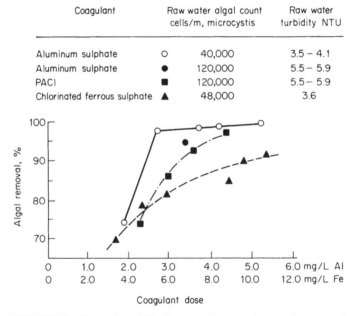

FIGURE 7.45 Comparison of effectiveness of three coagulants on algae removal by flotation. (*Source:* A. J. Rees, D. J. Rodman, and T. F. Zabel, *Water Clarification by Flotation,* 5, TR 114, Water Research Centre, Medmenham, U.K., 1979.)

residual in the treated water (pH 8.3 to 8.7 for ferrous sulfate, compared with pH 6.8 to 7.2 for aluminum sulfate).

Treatment of Low-Turbidity, Low-Color Waters. In the United States, there are a number of DAF plants being designed and built to treat waters of high quality (e.g., Boston, Massachusetts; Bridgeport, Connecticut; Greenville, South Carolina; and Cambridge, Massachusetts). Raw water turbidities are about 1 NTU, color is typically 15 CU or less, and TOC is 3 mg/L or less.

They are being designed at DAF loadings of 10 to 20 m/h (4 to 8 gpm/ft^2) and producing DAF effluent turbidities below 1 NTU and often in the range 0.2 to 0.5 NTU. This allows integration with filtration at high rates of 15 to 20 m/h. Several U.S. cities have limited land and space, so the reduced footprint is a considerable advantage. The design of a 400 mgd plant for Boston also includes flocculation tank detention time of 10 min, which further reduces the plant footprint and costs (Shawcross et al., 1997).

Filtration of Flotation-Treated Water

Tests comparing the performance of rapid gravity sand filters fed with water treated by flotation and by floc-blanket clarification showed that these waters have similar turbidities and residual coagulant concentrations. The presence of air bubbles in flotation-treated water has no influence on filter performance (Rees, Rodman, and Zabel, 1979b).

APPLICATIONS

Factors Influencing Choice of Process

A comparison of the various types of processes and DAF sedimentation for removing floc prior to normal deep-bed filtration is summarized in Table 7.6. The various types of coarse-bed filtration that could also be used have most of the same disadvantages. The factors influencing choice—those principally relating to cost, source water quality, compactness of plant, rapid start-up, and sludge removal—are discussed below.

Solids Loading. Solids removal prior to filtration is needed when the solids concentration is too great for the filtration to cope alone. Pretreatment is needed when the coagulant dose required for very soft waters exceeds about 1 mg Al or Fe per liter (12 mg alum/L), or greater than about 2 mg Al/L for hard waters (Gregory, 1991).

The concentration of solids in a floc blanket using alum coagulation is equivalent to about 100 mg Al/L or 200 mg Fe/L if iron coagulant is used. Floc-blanket sedimentation becomes inappropriate for coagulant doses greater than about 7.5 mg Al/L or 15 mg Fe/L because blanket-level control requirements become unreasonable. Other types of sedimentation then will be more appropriate.

Floc-blanket sedimentation can be used with partial lime softening in conjunction with iron coagulation. For simple precipitation softening, a solids contact clarifier, which is not too complex and is easily maintained, is likely to be the best choice. Consideration should be given to using pellet reactors, however, especially when softening groundwaters. When the raw water bears heavy silts, horizontal and inclined settlers are most appropriate. They will, however, need good arrangements for discharging the sediment with robust sludge-scraping equipment. When silt con-

TABLE 7.6 Comparison of Sedimentation and Dissolved-Air Flotation

	Horizontal flow (rectangular)	Radial flow (circular)	Upflow, US-type clarifier	Upflow floc blanket	Floc blanket (widely spaced inclined plates)	Ballasted floc	Inclined plates (unhindered)	Dissolved-air flotation
Sedimentation regime	Unhindered	Unhindered	Unhindered or hindered	Hindered	Hindered	Hindered or unhindered	Unhindered	Unhindered (hindered by air)
Theoretical removal of particles with $v_t < v_t^*$	Partial, depending on entrance position	Partial, depending on entrance position	Partial, depending on whether blanket formed	No removal	No removal	No removal	Partial, depending on entrance position	Depends on entrance position
Appropriate for heavy silts and lime precipitate as well as light floc	Yes	Yes	Yes, if bottom scraped	Flat tanks limited to light floc, hopper tanks can take wider range of settling velocities		For heavy silts and light floc	Yes	No
Appropriate for eutrophic (algal) waters	No	No	No	No	No	No	No	Yes
Ease of start-up and on/off operation	Easy	Easy	Easy for unhindered	Slow—may take several days to form a new blanket		Quick, but some skill needed	Quick, but some skill needed	Quick and easy
Skill level of operation	Low	Low	Medium	Medium	Medium	High	Medium-high	Medium-high
Retention time	High	High	Medium	Medium-low	Low	Low	Low	Low
Short circuiting	High	High	Medium	Medium-low	Medium-low	Medium-low	Medium-low	Low
Coagulation and flocculation effectiveness	Little difference if adequately designed and properly operated with rapid dispersion of chemicals							
Polymer usage	Low-medium	Low-medium	Low-medium	Low-medium	Low-medium	Medium-high	Medium-high	None-low
Relative capital cost	High-medium	High-medium	Medium	Medium	Medium-low	Medium-low	Medium-low	Medium-low
Relative operating cost	Medium	Medium-low	Medium-low	Low	Low	Medium	Medium	Medium

centrations are very high, sedimentation prior to coagulation may be appropriate. When silt concentrations are low but algae pose a problem, and for colored (low-alkalinity) waters, flotation is likely to be the best choice for pretreatment. Flotation might also be a better choice for other reasons, as mentioned below.

Costs. The comparison of costs of different sedimentation processes and comparison of sedimentation with flotation must include both capital and operating costs. This is necessary because of the substantial differences in the distribution of costs in the various processes, as summarized in Table 7.6. Floc-blanket sedimentation has a relatively high capital cost but low operating cost. In contrast, flotation has a relatively low capital cost but a high operating cost because of its energy consumption.

The chemical cost of basic coagulation should be the same for the various processes. Although no advantage occurs from using polyelectrolyte as a flocculation aid with flotation, its use can be very cost-effective for any form of sedimentation because loading rates can be increased. High loading rates and, hence, lower capital costs are also possible with ballasted floc, but additional operating costs of the ballasting agent, conditioning chemicals, and energy for agent recovery and recirculation are incurred.

The relatively high energy cost of flotation, although still possibly less than the cost of chemicals or labor, is largely incurred in pumping the recycle through the saturator to achieve the required level of saturation with air, and through the nozzles with enough backpressure to prevent premature release of air downstream of the saturators. Most of the remaining energy cost is required for flocculation. The cost of flocculation for discrete particle-settling systems, horizontal flow, and inclined sedimentation is basically the same. Because a large portion of the operating cost of DAF is in the recycle, it is important that total recycle flow, or pressure, can be reduced for low plant flows. This will need to be done in conjunction with reducing the number of nozzles so as to maintain nozzle backpressure and prevent premature release of air in the recycle.

The large differences between processes in fixed and variable costs mean that dissolved-air flotation is especially attractive for plants with low utilization, less than full flow, or infrequent use, because of its relatively lower capital cost and the reduced variable costs. Other factors besides cost can affect the choice of process, however. Such factors include rate of start-up, operational flexibility, ease of removing algae or color, coping with high mineral loadings, and producing the best treated-water quality.

Dissolved-air flotation is likely to be less expensive than floc-blanket sedimentation, for a fully utilized plant, only if the sedimentation process cannot be operated at rates greater than about 2 to 3 m/h (6 to 9 ft/h) (Gregory, 1977). This rule is easy to derive because the rating of flotation is relatively constant and independent of the type of water to be treated. As the expected plant utilization decreases, the dissolved-air flotation becomes relatively cheaper. The same should apply to ballasted sedimentation. Inclined-plate sedimentation should have a similar economic advantage, whereas the other types of unhindered sedimentation and coarse-bed filtration are likely to be better for full-utilization applications.

Sometimes the solids concentration in the water, after initial chemical treatment, is low enough to be within the capacity of filtration to operate without prior sedimentation or flotation and still achieve the desired objective. Just as each form of filtration has a limited capacity to remove solids, so does each form of sedimentation and flotation. This limited ability to remove solids efficiently may limit the surface loading and, therefore, cost-effectiveness of the process. However, some of the big new plants in North America are exploiting DAF cost-effectively in treating high-quality raw waters: DAF effluent is so good, even at high rates, that high-rate filtration is also possible, allowing very compact plants to be built.

With the typical applications of flotation, solids loading and air requirements are independent of each other. If the influent solids concentration increases, a greater recycle rate or pressure may be required to achieve efficient flotation, however, resulting in a greater operating cost. The greater the solids concentration the more likely that horizontal-flow sedimentation will be selected. The extreme situation is equivalent to thickening and then presedimentation before coagulation may be justified.

For certain types of source water, flotation does produce better water for filtration and, therefore, better filtrate quality than sedimentation. This may be because of better use of flocculation (this is applicable also to inclined settling), and the mechanism of flotation that makes it especially effective for algae removal. The aeration in flotation also may be attractive because it might increase dissolved-oxygen concentration or even cause some desorption or stripping of volatile contaminants.

Compactness. Inclined settling, ballasted-floc settlement, dissolved-air flotation, and coarse-bed filtration are regarded as "high-rate" processes because of the relatively high surface loadings possible. This label also implies they are compact processes, and therefore, they occupy less area.

Compactness is important where land is at a premium. These processes are also likely to need shallower and smaller tanks (except coarse-bed filtration), which is important in coping with preparing foundations in difficult sites. Both of these points are relevant where a plant has to be housed in cold climates or for other environmental or strategic reasons. The compactness of the flotation-filtration plant (flotation carried out over the filter) has made it especially attractive for package plants.

Rapid Start-Up. A floc blanket can take several days to form when starting with an empty tank, especially when the solids concentration is low. In contrast, high-rate inclined settlers and dissolved-air flotation can produce good quality water within 45 min from start-up. Such rapid start-up is useful where daily continuous operation is not needed. The disadvantage of high-rate processes is that they are more sensitive to failure of chemical dosing and flocculation and changes in inlet water quality. Rapid start-up plants are suitable for unstaffed sites using automatic shutdown and call-out alarms as a means of avoiding relying on difficult automatic quality control.

Sludge Removal. The ability to accumulate sludge, prior to removal, requires space and this depends on the process:

1. In horizontal-flow sedimentation, the whole plan area of the floor of the tank is available for sludge accumulation.
2. In dissolved-air flotation, the whole plan area of the surface of the flotation tank is available, but thickness of the floated sludge and ease of its removal are more important.
3. In solids contact clarifiers with scrapers, the whole plan area of the floor is available.
4. Floc-blanket clarifiers have only a small area for excess floc removal as sludge. However, when it is designed well, this will be adequate to ensure some thickening before discharge.
5. Some inclined settler systems are combined with thickeners.

Some systems, such as dissolved-air flotation and those with combined thickeners, can produce sludge concentrated enough for dewatering without additional thickening (see Chapter 16). Although systems that produce the more concentrated sludges are likely to incur greater operating costs for the scraping and pumping, this is likely to be more than offset by a saving in subsequent sludge treatment processes.

Emerging Technology

Developments in sedimentation processes and dissolved-air flotation, as currently known, have probably reached the point of diminishing return. Application lags behind knowledge, including making those plants that already exist work better. This will be done in part by paying more attention to the chemical engineering of reagent dosing, mixing, and flocculation rather than the clarification process itself. In the case of flotation, more attention also is likely to be given to the transfer of flocculated water to the flotation tank and its mixing with air in the contact zone. The use of computational fluid dynamic analysis (CFD) might lead to radical changes in tank design. More attention also is likely to be given to selecting the best chemistry and process for a particular application and to the design of the chosen process with regard to materials of construction, control equipment, and compactness.

Ballasted-floc systems, particularly those using recycled ballast, do not appear to have received as much attention in the past as they might deserve. This might change in the future. The economics of ballasted-floc systems may be like flotation, by having a lower capital cost but greater operating cost than normal sedimentation processes. Just as flotation has its niche for application, then there will be a niche for ballasted-floc systems. An interesting process of this type is the Sirofloc process, which recycles magnetite. An early claim for this process was that it can be so efficient that subsequent filtration might not be necessary, though in practice the four plants in the United Kingdom include filtration to satisfy the barrier principle for disinfection, to scavenge for residual magnetite, and at two of the plants to remove manganese (Home et al., 1992).

Flotation is generally rated at 8 to 12 m/h (24 to 36 ft/h or 3.3 to 5 gpm/ft^2). However, based on pilot trials, plans exist in the United States for high-rate flotation (19.5 m/h or 8 gpm/ft^2) in combination with high-rate filtration (29.3 m/h or 12 gpm/ft^2) to treat waters with low or moderate turbidities of 1 to 10 NTU (Shawcross et al., 1997; Adkins, 1997). Flotation has been a target for the application of the modern design and investigational tool of CFD (Fawcett, 1997; Eades et al., 1997; Ta and Brignall, 1997) and is helping to show and provide confidence that high-rate flotation, up to 25 m/h, should be viable with attention to appropriate design features. Once CFD has been developed for three-phase conditions, it might possibly lead to further understanding and exploitation of flotation.

Carryover of air from DAF onto the filters is sometimes regarded as a problem. Dahlquist (1997) has reported that an inclined-plate system on the DAF tank floor, through which the treated water must pass to the tank outlet, not only helps prevent carryover of small air bubbles but also allows DAF to be operated at two to three times greater rates.

No clear benefit has been demonstrated for the aeration that water receives in flotation, but no doubt there is some, even though it may be small. However, an example of combined ozonation and flotation has been reported as the Flottazone process (Baron et al., 1997). In this process, the recycle is saturated with ozone-rich air. This differs from the Ozoflot process, in which the flocculated water passes over porous plate diffusers releasing ozone-rich air. More examples of combined ozonation and flotation might occur should it be found appropriate to use ozonation in the treatment of waters for which flotation is also applicable.

Sometimes filtration has been used as an alternative to sedimentation. Various types of filtration have been used for this purpose including upflow filtration, continuous or moving-bed filters, and buoyant-media filters (see Chapter 8 on filtration). Some of these options will find more frequent application for clarification.

A wide variety of options has emerged for treatment prior to final filtration. Individual circumstances will dictate the choice. Furthermore, no single process will likely take a major share of new applications because costs of alternatives are generally the same, unless substantial technical advantages exist as has occurred for flotation.

NOMENCLATURE

a	constant
A	area
A_b	projected area of a bubble
A_H	cross-sectional area of channel-to-liquid flow
A^*	plan area of tank with horizontal flow for ideal settlement
b	breadth
C	mass concentration
C_b	mass concentration of bubbles
C_D	drag coefficient
C_{fl}	mass concentration of air in flotation tank effluent
C_r	mass concentration of air in recycle
d	diameter
d_b	diameter of bubble
d_f	diameter of floc
d_H	hydraulic diameter
d_p	diameter of particle
f_b	buoyancy
f_d	drag force
f_g	external force such as gravity
F	fraction
Fr	Froude number
F_M	solids mass flux
F_V	solids volume flux
F_t	fraction at time t
g	gravitational constant of acceleration
G	velocity gradient
h	vertical distance $\leq H$
h_n	height of isopercentage line at time t_n
H	height or depth
k_B	Boltzmann's constant
k, k_1, etc.	coefficients
K	constant from Yao's inclined settling equation
K_H	Henry's law constant
L	horizontal distance or length
L_p	length of surface for inclined settlement
L_r	Yao's relative length of settler
L^*	length of tank for ideal settlement
L'	greater than or equal to L^*
m	fraction as dead space
m_f	maximum number of attached air bubbles to floc particles
n	power index dependent on Re

n_i, n_j	number concentrations
N	number of channels, compartments, or stages, or number concentration
p	fraction as plug flow, or partial pressure
P	perimeter
q	constant
Q	volumetric flow rate
r	power index, or recycle ratio
R	removal efficiency
R_n	removal efficiency (of isopercentage) relating to Δh_n at t_n
Re	Reynold's Number
s	particle specific gravity = ρ_p/ρ_w
S_c	Yao's inclined-flow geometry number
t	detention time
t_n	time to reach height, h_n
t_p	time to reach tracer peak concentration in effluent
t_{10}, t_{90}	time required for 10 or 90 percent of tracer to emerge in outflow
t_{50}	mean detention time of tracer
T	theoretical retention time
$T°$	absolute temperature
v	particle or liquid velocity
v_b	bubble rise velocity
v_d	horizontal velocity in Figure 7.7
v_n	velocity relating to h_n and t_n
v_0	settling velocity of suspension for concentration extrapolated to zero
v_s	settling velocity of suspension
v_t	terminal settling velocity
v_θ	velocity at angle θ
v^*	overflow rate of the ideal settling tank
v'	less than or equal to v^*
V	volume
w	perpendicular spacing
W	width
x	mole or mass fraction
x_1	fraction of particles with settling velocities less than or equal to v_1
x^*	mass fraction of particles with settling velocity v^*
α	collision attachment factor
δ	dispersion index
ε	mean energy dissipation rate
E	porosity
E^+	porosity at maximum flux
Φ	particle volume concentration
Φ_b	bubble volume concentration
Φ^*	apparent volume concentration
Φ^+	particle volume concentration at maximum flux
N_{ij}	collision frequency between particles of concentrations n_i and n_j
η	collector collision efficiency
μ	absolute viscosity
ν	kinematic viscosity = μ/ρ
π	the geometric constant pi = 3.14
θ	angle of inclination
Θ	shape factor
ρ	density
ρ_b	density of gas bubble
ρ_p	density of particle
ρ_w	density of water

BIBLIOGRAPHY

Adkins, M. F. "Dissolved air flotation and the Canadian experience." In *Dissolved Air Flotation,* Proc. Int. Conf., London, U.K. CIWEM, 1997.

Agrawal, S. K., and K. Bewtra. "Modified approach to evaluate column test data." *J. Envl. Engng.,* 111 (2), 1985: 231–234.

Ali San, H. "Analytical approach for evaluation of settling column data." *J. Envl. Engng.,* 115(2), 1989: 455–461.

Amirtharajah, A., and C. R. O'Melia. "Coagulation processes: Destabilization, mixing and flocculation." In *Water Quality & Treatment,* 4th Ed., ch. 6. New York: McGraw-Hill, 1990.

American Water Works Association. *Guidance Manual for Compliance with the Filtration and Disinfection Requirements for Public Water Systems Using Surface Water Sources.* Denver, CO: Am. Water Wks. Assoc., 1990.

Barham, W. L., J. L. Matherne, and A. G. Keller. *Clarification, Sedimentation and Thickening Equipment—A Patent Review,* Bulletin No. 54, Engrg. Expt. Station. Baton Rouge, LA: Louisiana State University, 1956.

Baron, J., N. Martin Ionesco, and G. Bacquet. "Combining flotation and ozonation—the Flottazone process." In *Dissolved Air Flotation,* Proc. Int. Conf., London, U.K. CIWEM, 1997.

Barrett, F. "Electroflotation—Development and application." *Water Pollution Control,* 74, 1975: 59.

Bond, A. W. "Water-solids separation in an upflow." *Instn. Engrs. Aust., Civ. Engrg. Trans.,* 7, 1965: 141.

Bratby, J. *Dissolved-Air Flotation in Water and Waste Treatment.* Doctoral dissertation, University of Cape Town, Cape Town, South Africa, 1976.

Bratby, J., and G. vR. Marais. "Dissolved-air flotation—An evaluation of inter-relationships between process variables and their optimization for design." *Water SA,* 1 (2), 1975: 57–69.

Bratby, J., and C. V. R. Marais. *Solid-Liquid Separation Equipment Scale Up,* ch. 5, D. B. Purchas, ed. Croydon, London: Uplands Press Ltd., 1977.

Brit. Pat. Spec. Nos. 1.444.026 and 1.444.027. Published 28 July 1976.

Camp, T. R. "A study of the rational design of settling tanks." *Sewage Wks. Jour.,* 8, 1936: 742.

Camp, T. R. "Flocculation and flocculation basins." *Trans. ASCE,* 120, 1955: 1.

Casey, J. J., K. O'Donnel, and P. J. Purcell. "Uprating sludge blanket clarifiers using inclined plates." *Aqua,* 2, 1984: 91.

CIWEM. *Dissolved Air Flotation.* Proc. Int. Conf., London, U.K.: Chart. Instn. Water & Envtl. Management, 1997.

Clements, M. S., and A. F. M. Khattab. "Research into time ratio in radial flow sedimentation tanks." *Proc. Inst. Civ. Engrs., U.K.,* 40, 1968: 471.

Coulson, J. M., and J. F. Richardson. *Chemical Engineering,* 3rd ed., vol. 2. Oxford, U.K.: Pergamon Press, 1978.

Van Craenenbroeck, W., J. Van den Bogaert and J. Ceulemans. "The use of dissolved air flotation for the removal of algae—The Antwerp experience." *Water Supply,* 11(3/4), 1993: 123–133.

Cretu, G. "Contribution to the theory of water treatment using a sludge blanket." *Hydrotechnia Gospodarirea Apelor, Meterologia, Romania,* 13, 1968: 634.

Dahlquist, J. "The state of DAF development and applications to water treatment in Scandinavia." In *Dissolved Air Flotation,* Proc. Int. Conf., London, U.K. CIWEM, 1997.

Degremont. *Water Treatment Handbook,* 6th ed., vols. I, II. Degremont, France: 1991.

Demir, A. "Determination of settling efficiency and optimum plate angle for plated settling tanks." *Wat. Res.,* 29 (2), 1995: 611–616.

de Dianous, F., E. Pujol, and J. C. Druoton. "Industrial application of weighted flocculation: Development of the Actiflo clarification process." In *Proc 4th Gothenburg Symp., Oct. Chem-*

ical Water and Wastewater Treatment, Madrid, pp. 127–137, H. H. Hahn and R. Klute, eds. Berlin: Springer Verlag, 1990.

Dixon, D. R. "Colour and turbidity removal with reusable magnetite particles—VII." *Water Research, U.K.,* 18, 1984: 529.

Drajo, J. A. "Clarification of Sacramento Delta water in a large scale dissolved air flotation pilot plant." In *Proc. AWWA Ann. Conf.,* Dallas, Texas. 1984.

Eades, A., D. Jordan, and S. Scheidler. "Counter-current dissolved air flotation filtration COCO-DAFF." In *Dissolved Air Flotation,* Proc. Int. Conf., London, U.K., April 1997. CIWEM, 1997.

Edzwald, J. K. "Principles and applications of dissolved air flotation. Flotation processes in water and sludge treatment." *Wat. Sci. Tech.,* 31 (3–4), 1995: 1–23.

Edzwald, J. K., and J. P. Malley, Jr. *Removal of Humic Substances and Algae by Dissolved Air Flotation,* EPA/600/2-89-032. Cincinnati, OH: U.S. Environmental Protection Agency, 1990.

Edzwald, J. K., J. P. Malley, Jr., and C. Yu. "A conceptual model for dissolved air flotation in water treatment." *Water Supply,* 8, 1990: 141–150.

Edzwald, J. K., J. P. Walsh, G. S. Kaminski, and H. J. Dunn. "Flocculation and air requirements for dissolved air flotation." *Jour. AWWA,* 84 (3), 1992: 92–100.

Ellms, J. W. *Water Purification.* New York: McGraw-Hill, 1928.

Envirotech Corp. Eimco Modular Energy Dissipating Clarifier Feedwells, Form No. MED 121-10-72-3M. Brisbane, CA: Envirotech Corporation.

Epstein, N., and B. P. LeClair. "Liquid fluidization of binary particle mixtures—II." *Chem. Engrg. Sci.,* 40, 1985: 1517.

Fair, G. M., and J. C. Geyer. *Water Supply and Wastewater Disposal.* New York: John Wiley and Sons, 1954.

Fair, G. M., J. C. Geyer, and D. A. Okun. *Elements of Water Supply and Waste Water Disposal,* 2nd ed. New York: John Wiley and Sons, 1971.

Fawcett, N. S. J. "The hydraulics of flotation tanks: computational modelling." In *Dissolved Air Flotation,* Proc. Int. Conf., London U.K. CIWEM, 1997.

Franklin, B. "Ten years experience of dissolved air flotation in Yorkshire Water." In *Dissolved Air Flotation,* Proc. Int. Conf., London, U.K. CIWEM, 1997.

Fukushi, K., N. Tambo, and Y. Matsui. "A kinetic model for dissolved air flotation in water and wastewater treatment." *Wat. Sci. Tech.,* 31 (3–4), 1995: 37–48.

Gaudin, A. M. *Flotation,* 2nd ed. New York: McGraw-Hill, 1957.

Gemmell, R. S. "Mixing and sedimentation." In *Water Quality and Treatment,* 3rd ed., American Water Works Association, Inc. New York: McGraw-Hill, 1971.

Gomella, C. "Clarification avant filtration ses progres recents," (Rapport General 1). Intl. Water Supply Assoc. Intl. Conf., 1974.

Gould, B. W. "Low cost clarifier improvement." *Aust. Civ. Engrg. and Constn.,* 8, 1967: 49.

Gregory, R. "A cost comparison between dissolved air flotation and alternative clarification processes." In *Papers and Proceedings of the Conference on Flotation for Water and Waste Treatment,* J. D. Melbourne and T. F. Zabel, eds. Swindon, U.K.: Water Research Centre, 1977.

Gregory, R. *Controlling Coagulant Residuals: Direct Filtration Operation and Performance,* UM1273. Swindon, U.K.: Water Research Centre, 1991.

Gregory, R. *Floc Blanket Clarification,* TR 111. Swindon, U.K.: Water Research Centre, 1979.

Gregory, R. "Summary of general developments in DAF for water treatment since 1976." In *Dissolved Air Flotation.* London, U.K.: CIWEM, 1997.

Gregory, R., R. Head, and N. J. D. Graham. "Blanket solids concentration in floc blanket clarifiers." In *Proc. Gothenburg Symposium,* Edinburgh. 1996.

Gregory, R., and M. Hyde. *The Effects of Baffles in Floc Blanket Clarifiers, TR7.* Swindon, U.K.: Water Research Centre, 1975.

Gregory, R., R. J. Maloney, and M. Stockley. "Water treatment using magnetite: A study of a Sirofloc pilot plant. *Jour. Instn. Water & Envir. Management,* 2(5), 1988: 532.

Haarhoff, J., and E. M. Rykaart. "Rational design of packed saturators." *Water Sci. Tech.,* 31(3–4), 1995: 179–190.

Hale, P. E. *Floc Blanket Clarification of Water,* doctoral dissertation. London, U.K.: London University, 1971.

Hamlin, M. J., and A. H. Abdul Wahab. "Settling characteristics of sewage in density currents." *Water Research, U.K.,* 4, 1970: 609.

Hart, F. L., and S. K. Gupta. "Hydraulic analysis of model treatment units." *Proc. Amer. Soc. Civ. Engrs.,* 104(EE4), 1979: 785.

Hart, J. *Application of Process Simulation in Water Treatment.* London, U.K.: Association Generale Hygienistes Techniciens Municipaux 76th Congress, 1996.

Hartung, H. O. "Committee Report: Capacity and loadings of suspended solids contact units." *Jour. AWWA,* 43(4), April 1951: 263.

Head, R., J. Hart, and N. Graham. "Simulating the effect of blanket characteristics on the floc blanket clarification process." *Wat. Sci. Tech.,* 36(4), 1997: 77–84.

Heinänan, J., P. Jokela, and T. Ala-Peijari. "Use of dissolved air flotation in potable water treatment in Finland. Flotation processes in water and sludge treatment." *Wat. Sci. Tech.,* 31(3–4), 1995: 225–238.

Hemming, M. L., W. R. T. Cottrell, and S. Oldfelt. "Experience in the treatment of domestic sewage by the micro-flotation process." In *Papers and Proceedings of the Water Research Centre Conference on Flotation for Water and Waste Treatment,* J. D. Melbourne and T. F. Zabel, eds. Medmenham, U.K.: Water Research Centre, 1977.

Home, G. P., M. Stockley, and G. Shaw. "The Sirofloc process at Redmires Water Treatment Works." *Jour. Instn. Water & Env. Man.,* 6(1), 1992: 10–19.

Hudson, H. E. "Residence times in pretreatment." *Jour. AWWA,* 67(1), January 1975: 45.

Hudson, H. E. "Density considerations in sedimentation." In *Water Clarification Processes Practical Design and Evaluation.* New York: Van Nostrand Reinhold, 1981.

Ives, K. J. "Theory of operation of sludge blanket clarifiers." *Proc. Instn. Civ. Engrs., U.K.,* 39, 1968: 243.

Ives, K., and H. J. Bernhardt, eds. "Flotation processes in water and sludge treatment." *Wat. Sci. Tech.,* 31(3–4), 1995.

Jackson, M. L. "Energy effects in bubble nucleation." *Ind. Eng. Chem. Res.,* 33(4), 1994: 929–933.

Kalbskopf, K. H. "European practices in sedimentation." In *Water Quality Improvement by Physical and Chemical Processes, Water Resources Symposium No. 3,* E. F. Gloyna and W. W. Eckenfelder, eds. Austin, TX: University of Texas Press, 1970.

Kawamura, S. "Hydraulic scale-model simulation of the sedimentation process." *Jour. AWWA,* 73(7), 1981: 372.

Kawamura, S. *Integrated Design of Water Treatment Facilities.* New York: John Wiley & Sons, 1991.

Kawamura, S., and J. Lang. "Re-evaluation of launders in rectangular sedimentation basins." *Journal Water Pollution Control Federation,* 58, 1986: 1124.

Kinosita, K. "Sedimentation in tilted vessels." *Jour. Colloid Science,* 4, 1949: 525.

Kitchener, J. A. "The froth flotation process: Past, present and future—in brief." In *The Scientific Basis of Flotation,* K. J. Ives, ed. NATO ASI Series. The Hague, Netherlands: Martinus Nijhoff Publishers, 1984.

Kitchener, J. A., and R. J. Gochin. "The mechanism of dissolved-air flotation for potable water: Basic analysis and a proposal." *Water Research, U.K.,* 15, 1981: 585.

Krofta, M., and L. K. Wang. "Development of innovative flotation-filtration systems for water treatment. Part A: First full-scale sand float plant in US." In *Proceedings Water Reuse Symposium III: Future of Water Reuse,* San Diego, California. 1984.

Krofta, M., and L. K. Wang. "Application of dissolved air flotation to the Lennox Massachusetts water supply: water purification by flotation." *Journal New England Water Works Association,* 99(3), 1985: 249.

Lagvankar, A. L., and R. S. Gemmel. "A size-density relationship for flocs." *Jour. AWWA,* 60(9), 1968: 1040.

Levenspiel, O. *Chemical Reaction Engineering.* New York: John Wiley and Sons, 1962.

Liers, S., J. Baeyens, and J. Mochtar. "Modeling dissolved air flotation." *Water Environment Research,* 68(6), 1996: 1061.

Lundgren, H. "Theory and practice of dissolved-air flotation." *Journal Filtration and Separation,* 13(1), 1976: 24.

Maddock, J. L. "Research experience in the thickening of activated sludge by dissolved-air flotation." In *Papers and Proceedings of the Conference on Flotation for Water and Waste Treatment,* J. D. Melbourne and T. F. Zabel, eds. Medmenham, U.K.: Water Research Centre, 1977.

Marske, D. M., and J. D. Boyle. "Chlorine contact chamber design—A field evaluation." *Water and Sewage Works,* 120, 1973: 70.

Metcalf & Eddy Engineers. *Wastewater Engineering,* 3rd ed. New York: McGraw-Hill, 1991.

Miller, D. G., and J. T. West. "Pilot plant studies of floc blanket clarification." *Jour. AWWA,* 60(2), February 1968: 154.

Miller, D. G., J. T. West, and M. Robinson. "Floc blanket clarification—1." *Water and Water Engrg.,* 70, June 1966: 240.

Nakamura, H., and K. Kuroda. "La cause de l'acceleration de la vitesse de sedimentation des suspensions dans les recipients inclines." *Keijo Jour. Med., Japan,* 8, 1937: 265.

Nickols, D., and I. A. Crossley. "The current status of dissolved air flotation in the USA." In *Dissolved Air Flotation,* Proc. Int. Conf., London, U.K. CIWEM, 1997.

Offringa, G. "Dissolved air flotation in Southern Africa, flotation processes in water and sludge treatment." *Wat. Sci. Tech.,* 31(3–4), 1995: 159–172.

O'Neill, S. "Physical modelling study of the dissolved air flotation process." In *Dissolved Air Flotation,* Proc. Int. Conf., London, U.K. CIWEM, 1997.

Parker, D., et al. "Design and operations experience with flocculator-clarifiers in large plants." *Wat. Sci. Tech.,* 33(12), 1996: 163–170. (Also in *WQ International,* Nov/Dec 1996: 32–36.)

Patwardan, V. S., and Tien Chi. "Sedimentation and liquid fluidization of solid particles of different sizes and densities." *Chem. Engng. Sci.,* 40, 1985: 1051.

Pieronne, P. *Design of an Optimized Water Treatment Plant, in Particular Sludge Extraction, by Means of a Pilot Study.* London, U.K.: Assoc. Gen. des Hygienistes & Technicians Mun. 76th Congress, 29–31 May, 1996.

Prandtl, L., and O. G. Tietjens. *Applied Hydro and Aeromechanics.* New York: Dover, 1957.

Price, G. A., and M. S. Clements. "Some lessons from model and full-scale tests in rectangular sedimentation tanks." *Water Pollution Control, U.K.,* 73, 1974: 102.

van Puffelen, J., P. J. Buijs, P. N. A. M. Nuhn, and W. A. M. Hijnen. "Dissolved air flotation in potable water treatment: The Dutch experience." *Wat. Sci. Tech.,* 31(3–4), 1995: 149–157.

Rebhun, M., and Y. Argaman. "Evaluation of hydraulic efficiency of sedimentation basins." *Proc. Amer. Soc. Civ. Engrs.,* 91(SA5), 1965: 37.

Rees, A. J., D. J. Rodman, and T. F. Zabel. *Water Clarification by Flotation—5,* TR 114. Medmenham, U.K.: Water Research Centre, 1979a.

Rees, A. J., D. J. Rodman, and T. F. Zabel. "Dissolved-air flotation for solid-liquid separation." *Journal of Separation Process Technology,* 2, 1979b: 1.

Rees, A. J., D. J. Rodman, and T. F. Zabel. *Evaluation of Dissolved-Air Flotation Saturator Performance,* TR 143. Medmenham, U.K.: Water Research Centre, 1980.

Rodman, D. J. *Investigation into Hydraulic Flocculation with Special Emphasis on Algal Removal.* Master's thesis. Stevenage, U.K.: Water Research Centre, 1982.

Rykaart, E. M., and J. Haarhoff. "Behaviour of air injection nozzles in dissolved air flotation, in flotation processes in water and sludge treatment." *Wat. Sci. Tech.,* 31(3–4), 1995: 25–36.

Schers, G. J., and J. C. van Dijk. "Dissolved-air flotation: Theory and practice." In *Chemical Water and Wastewater Treatment II,* pp. 223–246, E. Klute and H. H. Hahn, eds. New York: Springer-Verlag, 1992.

Setterfield, G. H. "Water treatment trials at Burham." *Effluent and Water Treat. Jour.,* 23, 1983: 18.

Shawcross, J., T. Tran, D. Nickols, and C. R. Ashe. "Pushing the envelope: Dissolved air flotation at ultra-high rate." In *Dissolved Air Flotation,* Proc. Int. Conf., London, U.K. CIWEM, 1997.

Sherfold, H. L. "Flotation in mineral processing." In *The Scientific Basis of Flotation,* NATO ASI Series, K. J. Ives, ed. The Hague, Netherlands: Martinus Nijhoff Publishers, 1984.

Shogo, T. "Slurry-blanket type suspended solids contact clarifiers: Part 5." *Kogyo Yoshui, Japan,* 153, 1971: 19.

Sibony, J. "Clarification with microsand seeding—A state of the art." *Water Research, U.K.,* 15, 1981: 1281.

Standen, G., P. J. Inole, K. J. Shek, and R. A. Irwin. "Optimization of clarifier performance for pesticide removal." *3rd Int. Conf. Water and Waste Water Treatment,* Harrogate, U.K., M. White, ed. BHR Group Conf. Series, Publication No. 17, 1995: 13–33.

Steinbach, S., and J. Haarhof. "Air precipitation efficiency and its effect on the measurement of saturator efficiency." In *Dissolved Air Flotation,* Proc. Int. Conf., London, U.K. CIWEM, 1997.

Ta, C. T., and W. J. Brignal. "Application of single phase computational fluid dynamics techniques to dissolved air flotation tank studies." In *Dissolved Air Flotation,* Proc. Int. Conf., London, U.K. CIWEM, 1997.

Tambo, N., et al. "Behaviour of floc blankets in an upflow clarifier." *Jour. Jap. Water Wks. Assoc.,* 44, 1969: 7.

Tambo, N., and Y. Matsui. "A kinetic study of dissolved air flotation." *World Congress of Chem. Engr., (Tokyo),* 1986: 200–203.

Tambo, N., and Y. Watanabe. "Physical characteristics of floc—I: The floc density function and aluminium floc." *Water Research,* 13, 1979: 409.

Thirumurthi, D. A. "Breakthrough in the tracer studies of sedimentation tanks." *Journal Water Pollution Control Federation,* 41[11(pt.2)], 1969: R405.

Turner, M. T. "The use of dissolved air flotation for the thickening of waste activated sludge." *Effluent and Water Treatment Journal,* May 1975.

Valade, M. T., J. K. Edzwald, J. E. Tobiason, J. Dahlquist, T. Hedberg, and T. Amato. "Pretreatment effects on particle removal by flotation and filtration." *JAWWA,* 88(12), 1996: 35–47.

Vostrcil, J. "The effect of organic flocculants on water treatment and decontamination of water by floc blanket." Prace a Studie, Sesit 129. Prague: Water Research Inst., 1971.

Vrablik, E. R. "Fundamental principles of dissolved-air flotation of industrial wastes." In *Proceedings of the 14th Industrial Waste Conference,* Purdue University, Lafayette, Indiana. 1959.

Van Vuuren, L. R., F. J. de Wet, and F. J. Cillie. *Dissolved-Air Flotation-Filtration Studies in South Africa,* pp. 10–12. Denver, CO: AWWA Research Foundation Research News, 1984.

Webster, J. A., H. C. Webster, and J. P. Fairley. *Aspects of Design for Efficient Plant Operation for Water Treatment Processes Using Coagulants.* Paper presented to Scottish Section, Instn. Water Engrs. and Scientists, U.K. 1977.

White, M. J. D., R. C. Baskerville, and M. C. Day. *Increasing the Capacity of Sedimentation Tanks by Means of Sloping Plates.* Paper presented to Inst. Water Pollution Control, East Midlands Branch, U.K., (November).(62) reproduced as: *The Application of Inclined Tubes or Plates to Sedimentation Tanks in Waste-Water Treatment,* Notes on Water Pollution No. 68, March 1975. U.K.: Water Research Centre, 1974.

Wilkinson, P. D., P. M. Bolas, and M. F. Adkins. *British Experience with Flotation Process at Bewl Bridge Treatment Works.* Denver, CO: AWWA Research Foundation Water Quality Research News, 1981.

Wolf, D., and W. Resnick. "Residence time distribution in real systems." *Ind. and Engrg. Chem. Fund.,* 2, 1963: 287.

Yadav, N. P., and J. T. West. *The Effect of Delay Time on Floc Blanket Efficiency,* TR9. Medmenham, U.K.: Water Research Centre, 1975.

Yao, K. M. "Theoretical study of high-rate sedimentation." *Journal Water Pollution Control Federation,* 42, 1970: 218.

Yao, K. M. "Design of high-rate settlers." *Proc. Amer. Soc. Civ. Engrs.,* 99(EE5), 1973: 621.

Zabel, T. F. "The advantages of dissolved-air flotation for water treatment." *Journal American Water Works Association,* 77(5), May 1985: 42.

Zabel, T. F., and R. A. Hyde. "Factors influencing dissolved-air flotation as applied to water clarification." In *Papers and Proceedings of Water Research Centre Conference on Flotation for Water and Waste Treatment,* J. D. Melbourne & T. F. Zabel, eds. Medmenham, U.K.: Water Research Centre, 1977.

Zabel, T. F., and J. D. Melbourne. "Flotation." In *Developments in Water Treatment,* vol. 1, W. M. Lewis, ed. London, U.K.: Applied Science Publishers Ltd., 1980.

Zanoni, A. E., and M. W. Blomquist. "Column settling tests for flocculant suspensions." *Proc. Am. Soc. Civ. Engrs.,* 101(EE3), 1975: 309.

CHAPTER 8
GRANULAR BED AND PRECOAT FILTRATION

John L. Cleasby, Ph.D., P.E.
Professor Emeritus
Department of Civil and Construction Engineering
Iowa State University
Ames, Iowa

Gary S. Logsdon, D.Sc., P.E.
Director, Water Process Research
Black and Veatch, Engineers-Architects
Cincinnati, Ohio

AN OVERVIEW OF POTABLE WATER FILTRATION

The filtration processes discussed in this chapter are used primarily to remove particulate material from water. Filtration is one of the unit processes used in the production of potable water. Particulates removed may be those already present in the source water or those generated during treatment processes. Examples of particulates include clay and silt particles; microorganisms (bacteria, viruses, and protozoan cysts); colloidal and precipitated humic substances and other natural organic particulates from the decay of vegetation; precipitates of aluminum or iron used in coagulation; calcium carbonate and magnesium hydroxide precipitates from lime softening; and iron and manganese precipitates.

Types of Filters

A number of different types of filters are used in potable water filtration, and they may be described by various classification schemes. The granular bed and precoat filters discussed herein are comprised of porous granular material. In recent years, interest has grown in the use of membrane filtration in place of, or in addition to, granular bed filtration. Membrane processes are discussed in Chapter 11.

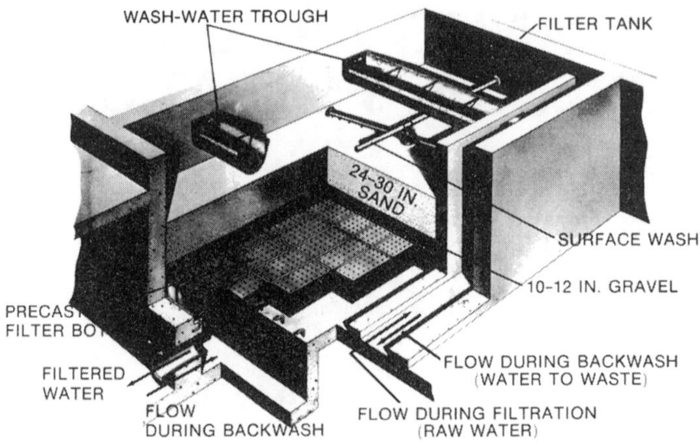

FIGURE 8.1 A rapid sand filtration system. (*Source:* Courtesy of F. B. Leopold Company.)

One classification scheme for granular bed filters is based on the type of media used. These filters commonly use a substantial depth of sand or anthracite coal or granular activated carbon or combinations thereof. A typical granular bed filter is shown in Figure 8.1. In contrast, precoat filters use a thin layer of very fine medium such as diatomaceous earth (DE) that is disposed of after each filter cycle—typically about a day in duration. Recovery, cleaning, and reuse of the medium is possible but not common. A typical precoat filter is shown in Figure 8.2, with the circular flat-plate septa that support the precoat.

Filters also may be described by the hydraulic arrangement employed to pass water through the medium. *Gravity filters* are open to the atmosphere, and flow through the medium is achieved by gravity, such as shown in Figure 8.1. *Pressure filters* utilize a pressure vessel to contain the filter medium. Water is delivered to the vessel under pressure and leaves the vessel at slightly reduced pressure. The two systems are merely two ways to provide a hydraulic gradient across the filter.

Filters may also be described by the rate of filtration, that is, the flow rate per unit area. Granular bed filters can be operated at various rates, for example, rapid granular bed filters provide higher filtration rates than do slow sand filters, which favor surface removal of particulates at the top of the sand bed.

Finally, filtration can be classified as *depth filtration* if the solids are removed within the granular material, or *cake filtration* if the solids are removed on the entering face of the granular material. Rapid granular bed filters are of the former type, while precoat and membrane filters are of the latter type. Slow sand filters utilize both cake and depth mechanisms, as will be explained later.

After a period of operation referred to as a *filter cycle,* the filter becomes clogged with removed particulates and must be cleaned. Rapid filters are cleaned by backwashing, using an upward, high-rate flow of water. Slow sand filters are cleaned by scraping off the dirty layer from the surface.

Thus, a filter can be fully described by an appropriate choice of adjectives. For example, a rapid, gravity, dual-media filter would describe a deep bed comprised of two media (usually anthracite coal on top of sand) operated at high enough rates to

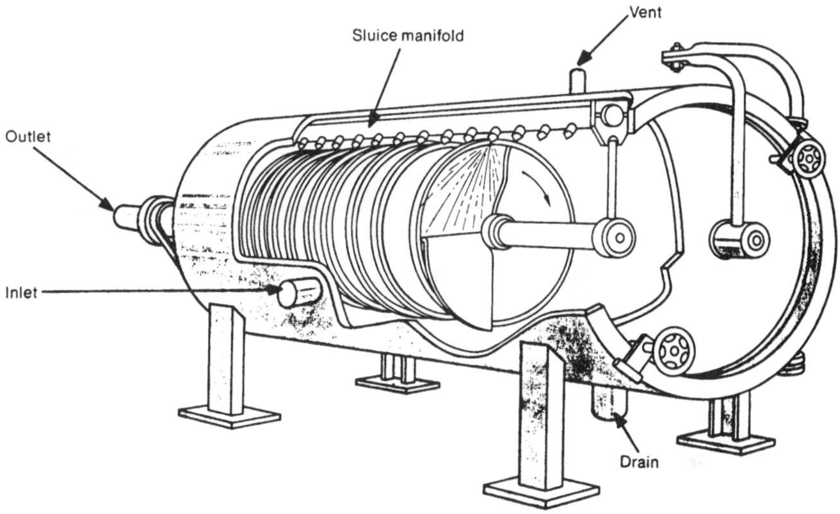

FIGURE 8.2 Precoat filter of rotating leaf type (sluice type during backwash). (*Source:* Courtesy of Manville Products Corporation.)

encourage depth removal of particulates within the bed, and operated by gravity in an open tank.

Dominant Mechanisms, Performance, and Applications

Cake filtration of particulates involves physical removal by straining at the surface. In addition, for the slow sand filter the surface cake of accumulated particulates includes a variety of living and dead micro- and macroorganisms. The biological metabolism of the organisms causes some alteration in the chemical composition of the water, and the development of this dirty layer (or *schmutzdecke*) enhances removal of particulates as well. As the filter cake develops, the cake itself assumes a dominant role in particulate removal. Because of this, filtrate turbidity improves as the filter run progresses, and deterioration of the filtrate turbidity is normally not observed at the end of the filter cycle. Because the mechanism of cake filtration is largely physical straining, chemical pretreatments such as coagulation and sedimentation are not generally provided. To obtain reasonable filter cycles, however, the source water must be of quite good quality (which will be defined later).

In contrast, depth filtration involves a variety of complex mechanisms to achieve particulate removal. Particles to be removed are generally much smaller than the size of the interstices formed between filter grains. Transport mechanisms are needed to carry the small particles into contact with the surface of the individual filter grains, and then attachment mechanisms hold the particles to the surfaces. These mechanisms will be discussed in more detail later.

Chemical pretreatment is essential to particulate removal in depth filtration. It serves to flocculate the colloidal-sized particulates into larger particles, which enhances their partial removal in pretreatment processes (such as sedimentation, flotation, or coarse-bed filtration, all located ahead of the filter) and/or enhances the transport mechanisms in filtration. In addition, chemical pretreatment enhances the attachment forces retaining the particles in the filter. The focal point of particulate removal in depth filtration moves progressively deeper into the bed as the cycle progresses, and if the cycle continues long enough, deterioration of the filtrate may be observed.

The provision of pretreatment makes the depth filtration process more versatile in meeting a variety of source water conditions. With appropriate coagulation, flocculation, and solids separation ahead of depth filtration, source waters of high turbidity or color can be treated successfully. Better quality source waters may be treated by coagulation, flocculation, and depth filtration, a process referred to as *direct filtration,* or by *in-line filtration,* which utilizes only coagulation and very limited flocculation before depth filtration.

In some cases, biological metabolism will result in partial removal of biodegradable organic matter in depth filters, if biological growth is allowed to develop in the filter medium. This is especially important if ozonation precedes filtration. Chapter 13 includes information on biological filtration.

Regulatory Requirements for Filtration

The U.S. Environmental Protection Agency's (USEPA) Surface Water Treatment Rule (SWTR), promulgated on June 29, 1989 (Federal Register 40 CFR Parts 140 and 141, p. 27486–27568), requires community water systems to disinfect all surface waters and requires filtration for most surface water sources. The Surface Water Treatment Rule imposed stricter turbidity limits on filtration processes and made them specific to the type of process used (see Table 8.1). The total extent of inactivation and physical removal must be at least 3-log (99.9 percent) for *Giardia* cysts and 4-log (99.99 percent) for viruses. The supplementary information published with the rule presented recommended minimum levels of disinfection and assumed log removals to be credited to the following defined filtration processes: conventional filtration, direct filtration, diatomaceous earth filtration, and slow sand filtration. Conventional filtration and direct filtration were defined in the SWTR to include chemical coagulation; flocculation; and in the case of conventional filtration, sedimentation; ahead of the filtration process. Slow sand filtration was defined as filtration of water, without chemical coagulation, through a bed of sand at rates of up to 0.16 gpm/ft^2 (0.40 m/h). Filtration processes that do not function on the principles of

TABLE 8.1 SWTR Assumed Log Removals and Turbidity Requirements

Filtration process	Log removals*		Turbidity requirement
	Giardia	Virus	
Conventional	2.5	2.0	= or < 0.5 ntu in 95% of samples each month and never > 5 ntu
Direct	2.0	1.0	= or < 0.5 ntu in 95% of samples each month and never > 5 ntu
Slow sand	2.0	2.0	= or < 1 ntu in 95% of samples each month** and never > 5 ntu
Diatomaceous earth	2.0	1.0	= or < 1 ntu in 95% of samples each month and never > 5 ntu

* From Table IV-2 in Supplementary Information, p. 27511.

** Special provision was made for slow sand filters to exceed 1 ntu in some cases, providing effective disinfection was maintained.

the processes defined in the SWTR are called *alternative filtration processes,* and the log removal for *Giardia* cysts or viruses that can be allowed for alternative processes must be determined for each alternative process.

Removal of Microorganisms by Granular Bed and Precoat Filtration

In North America, waterborne disease outbreaks caused by *Giardia lamblia* and *Cryptosporidium parvum,* pathogenic protozoa with high resistance to disinfectants, have resulted in numerous studies of pilot plants and full-scale treatment plants to evaluate removal of microorganisms. Many of these studies focused on filtration process trains involving coagulation, but some investigations of the efficacy of slow sand filtration and diatomaceous earth filtration have also been carried out.

Pilot plant studies (Logsdon et al., 1985; Al-Ani et al., 1986) established the need for attaining effective coagulation and filtered water turbidity in the range of 0.1 to 0.2 ntu for effective removal of *Giardia* cysts. Additional research by Nieminski and Ongerth (1995) on *Giardia* and *Cryptosporidium* confirmed the necessity of attaining low filtered water turbidity and the importance of maintaining proper coagulation chemistry. They evaluated a small (4900 m^3/d) plant for removal of protozoan cysts and concluded that a properly operated treatment plant producing finished water turbidity of 0.1 to 0.2 ntu, using either the direct filtration mode or the conventional treatment mode, could achieve 3-log removal of *Giardia* cysts and about 2.6-log removal of *Cryptosporidium.* However, when they corrected the test results for recovery efficiency in both the influent and the effluent samples, they reported 3.7- to 4.0-log removals for *Giardia* and *Cryptosporidium* in both conventional treatment and direct filtration. The similarity of results for direct filtration and conventional treatment is different from the outcome reported by Patania et al. (1995), who observed 1.4- to 1.8-log additional removal for *Cryptosporidium* and 0.2- to 1.8-log additional removal for *Giardia* when sedimentation was included in the treatment train, in comparison with in-line filtration treatment employing only coagulation and filtration. Nieminski and Ongerth (1995) reported that oocyst removals calculated on the basis of the oocyst concentration in the seed suspension being fed to the raw water were 1-log higher than oocyst removals calculated on the basis of the concentration of oocysts actually measured in the seeded influent water, which is the more rational method of evaluation.

LeChevallier et al. (1991) examined raw and filtered water samples from 66 surface water treatment plants in 14 states and 1 Canadian province. Log removals of *Giardia* and *Cryptosporidium* ranged from less than 0.5 to greater than 3.0. Log removals averaged slightly above 2.0 for each organism. Some of these plants were practicing disinfection (usually chlorination) during clarification, so the *Giardia* results may have been influenced somewhat by disinfection. They reported that production of low turbidity water (<0.5 ntu) did not ensure that the treated effluents were free of cysts or oocysts. Treatment plants evaluated by LeChevallier et al. very probably would have removed cysts and oocysts more effectively if they had been attaining lower filtered water turbidity. In pilot plant testing carried out by Patania et al. (1995), when filtered water turbidity was equal to or less than 0.1 ntu, removal of cysts and oocysts was greater by as much as 1.0-log, as compared to removal when filtered water turbidity was between 0.1 and 0.3 ntu.

Payment (1997) reported on biweekly monitoring of a conventional water treatment plant (coagulation, flocculation, sedimentation, and filtration) for over 12 months. Clarification through filtration before any use of disinfectant attained the following results:

- 3.0-log or greater removal of human enteric viruses in 20 of 30 samples
- 3.0-log or greater removal of coliphage in 20 of 32 samples
- 4.0-log or greater removal of *Clostridium perfringens* in 20 of 33 samples
- 3.0-log or greater removal of *Giardia* cysts in 24 of 32 samples

Giardia cysts were detected in the filtered water in only 1 of 32 samples. Removals of *Cryptosporidium* oocysts appeared to be lower than those for *Giardia,* but only 2 of 16 raw water samples contained sufficient densities of *Cryptosporidium* oocysts to permit calculation of 3-log removal, based on the detection limit for *Cryptosporidium* in filtered water. During the study, 98 percent of the monthly average turbidity values for individual filters were equal to or less than 0.20 ntu, and 80 percent were equal to or less than 0.10 ntu. These results show that well-run conventional treatment plants present a formidable barrier to the passage of pathogens, through effective sedimentation (or other solids separation) followed by filtration.

Slow sand filtration is effective for removal of *Giardia* and *Cryptosporidium.* Pilot plant studies gave *Giardia* cyst removals of over 3-log at filtration rates of up to 0.16 gpm/ft^2 (0.40 m/h) (Bellamy et al., February 1985; Bellamy, Hendricks, and Logsdon, 1985). Field testing at a small slow sand filter (area 37 m^2) operated at 0.03 gpm/ft^2 (0.08 m/h) (Pyper, 1985) gave 3.7- to 4.0-log *Giardia* removal at temperatures of 7.5 to 21°C, but at temperatures below 1°C, removals ranged from 1.2- to 3-log. Schuler, Ghosh, and Boutros (1988) obtained 3.7 or higher log removal for *Cryptosporidium* in a pilot filter operated at 0.11 gpm/ft^2 (0.26 m/h), using 0.27 mm effective size sand.

Diatomaceous earth filtration is very effective for removal of pathogens in the size range of *Giardia* cysts and *Cryptosporidium* oocysts. The removal mechanism involved is straining, and when an appropriate grade of diatomaceous earth is selected, the pore structure of the filter cake physically blocks the passage of cysts and oocysts into filtered water. Pilot plant studies using a 0.1 m^2 test filter have demonstrated *Giardia* cyst removals ranging from 2- to 4-log, typically at filtration rates of 1.0 to 1.5 gpm/ft^2 (2.4 to 3.7 m/h) (Lange et al., 1986). Schuler and Ghosh (1990), utilizing three common grades of diatomaceous earth in pilot filtration studies, reported 100 percent removal of *Giardia muris* and greater than 3-log removal of *Cryptosporidium* oocysts, all at 2 gpm/ft^2 (4.9 m/h) and without the use of coagulants. Use of alum or cationic polymer to coat the media did not enhance the cyst removal, but did aid in the removal of turbidity and coliform bacteria. Principe et al. (1994) reported 3.7-log *Giardia* cyst removal by diatomaceous earth filtration (unspecified grade of DE) at 3 gpm/ft^2 (7.3 m/h).

Membrane filtration has been found to be very effective in removing protozoan cysts and oocysts, (See Chapter 11 for information on membrane filtration.)

In summary, filtration options for the removal of pathogenic microorganisms include rapid filtration, slow sand filtration, DE filtration, and membrane filtration. The most widely used filtration process, however, is coagulation and rapid filtration, and generally the process train includes flocculation and sedimentation (conventional treatment). For conventional treatment or for direct or in-line filtration, effective removal of microorganisms requires careful control of coagulation chemistry and operation of filters to consistently attain very low filtered water turbidity. Based on pilot plant and full-scale studies, removal of microorganisms can be maximized by the following:

- Filtered water turbidity should be 0.10 ntu or lower.
- The duration of higher turbidity at the beginning of a filter run should be minimized (less than 1 hour).

- Filtered water turbidity above 0.2 ntu should be considered turbidity breakthrough, signaling the need to backwash filters.

The U.S. Environmental Protection Agency's Surface Water Treatment Rule, effective since June 1993, mandates disinfection of all surface waters used for community water systems and also requires filtration for most surface waters. In response to waterborne disease outbreaks caused by *Cryptosporidium* (Kramer et al., 1996), this rule may be made more stringent in the future.

FILTER MEDIA

Types of Media

The common types of media used in granular bed filters are silica sand, anthracite coal, and garnet or ilmenite. These may be used alone or in dual- or triple-media combinations. Garnet and ilmenite are naturally occurring, high-density minerals and are described further in the next paragraph. American Water Works Association Standard B100-96 (AWWA, 1996) provides standard requirements for properties, sampling, testing, placement, and packaging of these filter materials. Other types of media have also been used in some cases. For example, granular activated carbon (GAC) has been used for reducing taste and odor in granular beds that serve both for filtration and adsorption, that is, as filter adsorbers (Graese, Snoeyink, and Lee, 1987). Granular activated carbon is also being used after filtration for adsorption of organic compounds. American Water Works Association Standard B604-90 (AWWA, 1991) provides standard requirements for physical properties, sampling, testing, and packaging of GAC.

Garnet is somewhat of a generic term referring to several different minerals, mostly almandite, andradite, and grossularite, which are silicates of iron, aluminum, and calcium mixtures. *Ilmenite* is an iron titanium ore, which invariably is associated with hematite and magnetite, both iron oxides. Garnet specific gravities range from 3.6 to 4.2, while those for ilmenite range from 4.2 to 4.6.

Precoat filters use diatomaceous earth or perlite as a filter medium. Standard requirements for precoat filter media are presented in AWWA Standard B101-94 (AWWA, 1995b). Diatomaceous earth (DE or diatomite) is composed of the fossilized skeletons of microscopic diatoms that grew in fresh or marine waters. Deposits of this material from ancient lakes or oceans are mined and then processed by flux calcining, milling, and air classification into various size grades for assorted filtration applications. The grades used in potable water filtration have a mean pore size of the cake from about 5 to 17 μm.

A less common medium for precoat filtration is *perlite,* which comes from glassy volcanic rock. It is a siliceous rock containing 2 to 3 percent water. When heated, the rock expands to form a mass of glass bubbles. It is crushed and classified into several size grades.

Important Granular Media Properties

A number of properties of a filter medium are important in affecting filtration performance and also in defining the medium. These properties include size, shape, density, and hardness. The porosity of the granular bed formed by the grains is also important.

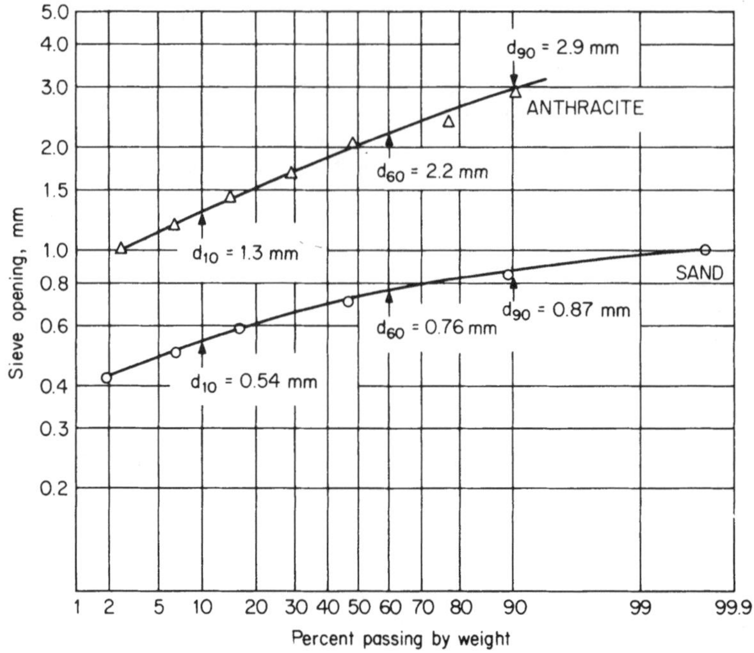

FIGURE 8.3 Typical sieve analysis of two filter media.

Grain Size and Size Distribution. Grain size has an important effect on filtration efficiency and on backwashing requirements for a filter medium. It is determined by sieve analysis using American Society for Testing and Materials (ASTM) Standard Test C136-92, Sieve Analysis of Fine and Coarse Aggregates (ASTM, 1993). A log-probability plot of a typical sieve analysis is presented in Figure 8.3. Sieve analyses of most filter materials plot in nearly a linear manner on log-probability paper.

In the United States, the size gradation of a filter medium is described by the effective size (ES) and the uniformity coefficient (UC). The ES is that size for which 10 percent of the grains are smaller by weight. It is read from the sieve analysis curve at the 10 percent passing point on the curve, and is often abbreviated by d_{10}. The UC is a measure of the size range of the media. It is the ratio of the d_{60}/d_{10} sizes that are read from the sieve analysis curve, d_{60} being the size for which 60 percent of the grains are smaller by weight. In some other countries, the lower and upper size of the media are specified with some maximum percentage allowance above and below the specified sizes.

Values of d_{10}, d_{60}, and d_{90} can be read from an actual sieve analysis curve such as shown in Figure 8.3. If such a curve is not available and if a linear log-probability plot is assumed, they can be interrelated by the following equation derived from the geometry of such linear plots for different UC:

$$d_{90} = d_{10}(10^{1.67 \log \mathrm{UC}}) \qquad (8.1)$$

This relationship is useful because the d_{90} size is recommended for calculation of the required backwash rate for a filter medium.

Grain Shape and Roundness. The shape and roundness of the filter grains are important because they affect the backwash flow requirements for the medium, the fixed bed porosity, the head loss for flow through the medium, the filtration efficiency, and the ease of sieving.

Different measures of grain shape have evolved in the geological and chemical engineering literature, leading to considerable confusion in terminology (Krumbein, 1941; McCabe and Smith, 1976). The chemical engineering literature defines the sphericity (ψ) as the ratio of the surface area of an equal volume sphere (diameter of d_{eq}) to the surface area of the grain (McCabe and Smith, 1976). The equivalent spherical diameter can be determined by a tedious count, weigh, and calculation procedure (Cleasby and Fan, 1981). In the absence of such data, the mean size for any fraction observed from the sieve analysis plot (e.g., Figure 8.3) can be used as an acceptable approximation of the equivalent spherical diameter.

The chemical engineering definition will be used in the following discussion. The sphericity of filter media by this definition can be determined indirectly by measuring pressure drop for flow of water or air through a bed of uniformly sized grains. The Kozeny or Ergun equation for flow through porous media (presented later) is used to calculate ψ after determining all other parameters of the equation (Cleasby and Fan, 1981).

Grain Density or Specific Gravity. Grain density, mass per unit grain volume, is important because it affects the backwash flow requirements for a filter medium. Grains of higher density but of the same diameter require higher wash rates to achieve fluidization.

Grain density is determined from the specific gravity following ASTM Standard Test C128-93, Specific Gravity and Absorption of Fine Aggregate (ASTM, 1993). This ASTM test uses a displacement technique to determine the specific gravity. Two alternative tests are detailed in the ASTM standard. The procedure for "bulk specific gravity, saturated surface dry" would be best from a theoretical standpoint for fluidization calculations. However, starting with a reproducible saturated surface dry condition is difficult. Therefore, the "apparent specific gravity" that starts with an oven-dry sample is more reproducible and is an acceptable alternative for fluidization calculations. For porous materials such as anthracite coal or GAC, the sample should be soaked to fill the pores with water before final measurements are made.

Grain Hardness. The hardness of filter grains is important to the durability of the grains during long-term service as a filter medium. Hardness is usually described by the MOH hardness number, which involves a scale of hardness based on the ability of various minerals to scratch or be scratched by one another. A sequence of minerals of specified hardnesses is listed by Trefethen (1959).

The two materials of known MOH hardness that can and cannot scratch the filter medium are used to estimate the hardness of the medium. This is a rather crude test and difficult to apply to small filter grains. Of the filter media listed earlier, only anthracite coal and GAC have low hardness worthy of concern. Silica sand, garnet, and ilmenite are very hard, and their hardness need not be of concern. A minimum MOH hardness of 2.7 or 3 is often specified for anthracite coal filter medium, although accurately measuring partial values closer than 0.5 is doubtful.

Two standard mechanical abrasion tests are presented in the AWWA Standard for Granular Activated Carbon (Standard B604-90, AWWA, 1991) to evaluate the abrasion resistance of GAC. Although GAC is more friable than anthracite, the progressive reduction that takes place in its grain size due to backwashing and regeneration operations is not reported to be a serious problem in actual practice (Graese, Snoeyink, and Lee, 1987).

Fixed-Bed Porosity. Fixed-bed porosity is the ratio of void volume to total bed volume, expressed as a decimal fraction or as a percentage. It is important because it affects the backwash flow rate required, the fixed-bed head loss, and the solids-holding capacity of the medium. Fixed-bed porosity is affected by the grain sphericity; angular grains (i.e., those with lower sphericity) have higher fixed-bed porosity (Cleasby and Fan, 1981), as is evident in Table 8.2. For the low UC commonly specified for filter media, UC has no effect on porosity. However, for natural materials with very high UC, the nesting of small grains within the voids of the large grains can reduce the average bed porosity.

Fixed-bed porosity is determined by placing a sample of known mass and density in a transparent tube of known internal diameter. The depth of the filter medium in the tube is used to calculate the bed volume. The grain volume is the total mass of medium in the column divided by the density. The void volume is thus the bed volume minus the grain volume. The fixed-bed porosity is substantially affected by the extent of compaction of the medium placed in the column and by the column diameter. The loose-bed porosity can be measured in a column of water. If the bed is agitated by inversion and then allowed to settle freely in the water with no compaction, the highest porosity (i.e., the loose-bed porosity) will be obtained. It may be as much as 5 percent greater than the porosity measured after gentle compaction of the bed. Materials of lower sphericity show greater change in porosity between the loose bed and compacted bed conditions.

The porosity of the granular filter medium is higher near the wall of the filter. This can be important in small pilot scale filter columns because it causes the average bed porosity to be higher than in full-scale filters, and affects the particulate removal and head loss behavior during filtration studies. It is common to make the diameter of such filter columns at least 50 times the grain size of the coarser filter grains to be studied (e.g., d_{90}) to minimize such wall effects.

Sieve Analysis Considerations

The standard procedure for conducting a sieve analysis of a filter medium is detailed in ASTM Standard Test C136-92 (ASTM, 1993). This standard does not specify a sieving time or a specific mechanical apparatus for shaking the nest of sieves. Rather, it specifies that sieving should be continued "for a sufficient period and in such manner that, after completion, not more than 1 weight percent of the residue on any individual sieve will pass that sieve during one minute of hand sieving" conducted in a manner described in the ASTM standard. With softer materials such as anthracite coal or GAC, abrasion of the material may occur when attempting to meet the 1 percent passing test.

TABLE 8.2 Typical Properties of Common Filter Media for Granular-Bed Filters (Cleasby and Fan, 1981; Dharmarajah and Cleasby, 1986; Cleasby and Woods, 1975)

	Silica sand	Anthracite coal	Granular activated carbon	Garnet	Ilmenite
Grain density, ρ_s, Kg/m^3	2650	1450–1730	1300–1500*	3600–4200	4200–4600
Loose-bed porosity ϵ_o	0.42–0.47	0.56–0.60	0.50	0.45–0.55	**
Sphericity ψ	0.7–0.8	0.46–0.60	0.75	0.60	**

* For virgin carbon, pores filled with water, density increase when organics are adsorbed.
** Not available.

When sieving a 100-g sample of hard materials such as sand, on 8-in sieves, and using a Ro-Tap type of sieving machine, it is common to require three sieving periods of 5 min each to satisfy the 1 percent passing test. The Ro-Tap machine imparts both a rotary shaking and a vertical hammering motion to the nest of sieves. With some other sieving machines, the ASTM requirement will not be achieved even after three 5-min periods of sieving.

When sieving anthracite coal, the sample should be reduced to 50 g because of its lower density. The Ro-Tap machine should be used and the time fixed at 5 min. This will not meet the 1 percent passing test, but prolonged sieving may cause continued degradation of the anthracite, yielding a more erroneous result.

Because of sieving and sampling difficulties, and because of the tolerance allowed in manufacture of the sieves themselves (ASTM Standard Test E11-87, Wire-cloth Sieves for Testing Purposes, 1993), a reasonable tolerance should be allowed in the specified size when specifying filter media. Otherwise, producers of filter material may not be able to meet the specification, or a premium price will be charged. A tolerance of 10 percent plus or minus is suggested. For example, if a sand of 0.5 mm ES is desired, the specification should read 0.45 to 0.55 mm ES. If an anthracite coal of 1.0 mm ES is desired, the specification should read 0.9 to 1.1 mm ES.

Typical Properties of Granular Filter Media

With the prior understanding of the importance of various properties of granular filtering materials, Table 8.2 is presented to illustrate typical measured values for some properties. The large difference in grain densities evident in Table 8.2 allows the construction of dual- and triple-media filters, with coarse grains of low-density material on top and finer grains of higher density beneath. Alluvial sands have the highest sphericity, and crushed materials such as anthracite, ilmenite, and some garnet have lower sphericity. Some anthracites contain an excessive amount of flat, elongated jagged grains, resulting in lower sphericity. The loose-bed porosity is inversely related to the sphericity (i.e., the lower the sphericity the higher the loose-bed porosity). An approximate empirical relationship between sphericity and loose-bed porosity was used in developing a predictive model for fluidization that will be presented later (Cleasby and Fan, 1981).

HYDRAULICS OF FLOW THROUGH POROUS MEDIA

Head Loss for Fixed-Bed Flow

The head loss (i.e., pressure drop) that occurs when clean water flows through a clean filter medium can be calculated from well-known equations. The flow through a clean filter of ordinary grain size (i.e., 0.5 mm to 1.0 mm) at ordinary filtration velocities (2 to 5 gpm/ft^2; 4.9 to 12.2 m/h) would be in the laminar range of flow depicted by the Kozeny equation (Fair, Geyer, and Okun, 1968) that is dimensionally homogeneous (i.e., any consistent units may be used that are dimensionally homogeneous*):

* Units will not be presented for all dimensionally homogeneous equations in this chapter.

$$\frac{h}{L} = \frac{k\mu}{\rho g}\frac{(1-\varepsilon)^2}{\varepsilon^3}\left(\frac{a}{v}\right)^2 V \tag{8.2}$$

where h = head loss in depth of bed, L
g = acceleration of gravity
ε = porosity
a/v = grain surface area per unit of grain volume = specific surface (S_v) = $6/d$ for spheres and $6/\psi d_{eq}$ for irregular grains
d_{eq} = grain diameter of sphere of equal volume,
V = superficial velocity above the bed = flow rate/bed area (i.e., the filtration rate)
μ = absolute viscosity of fluid
ρ = mass density of fluid
k = the dimensionless Kozeny constant commonly found close to 5 under most filtration conditions (Fair, Geyer, and Okun, 1968)

The Kozeny equation is generally acceptable for most filtration calculations because the Reynolds number Re based on superficial velocity is usually less than 3 under these conditions, and Camp (1964) has reported strictly laminar flow up to Re of about 6:

$$\text{Re} = d_{eq}\, V\rho/\mu \tag{8.3}$$

The Kozeny equation can be derived from the fundamental Darcy-Weisbach equation for flow through circular pipes:

$$h = f\frac{LU^2}{D(2g)} \tag{8.4}$$

where f = the friction factor, a function of pipe Reynolds number
D = the pipe diameter
U = the mean flow velocity in the pipe

The derivation is achieved by considering flow through porous media analogous to flow through a group of capillary tubes of hydraulic radius r (Fair, Geyer, and Okun, 1968). The hydraulic radius is approximated by the ratio of the volume of water in the interstices per unit bed volume divided by the grain surface area per unit bed volume. If N is the number of grains per unit bed volume, v is the volume per grain, and a is the surface area per grain, then the bed volume = $Nv/(1-\varepsilon)$, and the interstitial volume = $Nv\varepsilon/(1-\varepsilon)$, and the surface area per unit bed volume is the product of N times a leading to $r = \varepsilon v/(1-\varepsilon)a$. The following additional substitutions are made: $D = 4r$; U = interstitial velocity = V/ε; $f = 64/\text{Re}'$ for laminar flow; and $\text{Re}' = 4(V/\varepsilon)r\,\rho/\mu$ is the Reynolds number based on interstitial velocity.

For larger filter media, where higher velocities are used in some applications, or for velocities approaching fluidization (as in backwashing considerations), the flow may be in the transitional flow regime, where the Kozeny equation is no longer adequate. Therefore, the Ergun equation (Ergun, 1952a), Equation 8.5, should be used because it is adequate for the full range of laminar, transitional, and inertial flow through packed beds (Re from 1 to 2,000). The Ergun equation includes a second term for inertial head loss.

$$\frac{h}{L} = \frac{4.17\mu}{\rho g}\frac{(1-\varepsilon)^2}{\varepsilon^3}\left(\frac{a}{v}\right)^2 V + k_2\frac{(1-\varepsilon)}{\varepsilon^3}\left(\frac{a}{v}\right)\frac{V^2}{g} \tag{8.5}$$

Note that the first term of the Ergun equation is the viscous energy loss that is proportional to V. The second term is the kinetic energy loss that is proportional to V^2. Comparing the Ergun and Kozeny equations, the first term of the Ergun equation (viscous energy loss) is identical with the Kozeny equation, except for the numerical constant. The value of the constant in the second term, k_2, was originally reported to be 0.29 for solids of known specific surface (Ergun, 1952a). In a later paper, however, Ergun reported a k_2 value of 0.48 for crushed porous solids (Ergun, 1952b), a value supported by later unpublished studies at Iowa State University. The second term in the equation becomes dominant at higher flow velocities because it is a square function of V. The Kozeny equation, however, is more convenient to use and is quite acceptable up to Re = 6.

As is evident from equations 8.2 or 8.5, the head loss for a clean bed depends on the flow rate, grain size, porosity, sphericity, and water viscosity. As filtration progresses and solids are deposited within the void spaces of the medium, the porosity decreases, and sphericity is altered. Head loss is very dependent upon porosity, and reduction in porosity causes the head loss to increase.

The ability to calculate head loss through a clean fixed bed is important in filter design because provision for this head must be made in the head loss provided in the plant. In addition, of course, head must be provided in the plant design for the increase in head loss resulting from the removal of particulates from the influent water on top of and within the filter bed. This is referred to as the *clogging head loss.* The clogging head loss to be provided is usually based on prior experience for similar waters and treatment processes or on pilot studies.

EXAMPLE 8.1 Calculate the head loss for the 3-ft- (0.91-m) deep bed of filter sand shown in Figure 8.3 at a filtration rate of 6 gpm/ft² (14.6 m/h) and a water temperature of 20°C, using a grain sphericity of 0.75 and a porosity of 0.42 estimated from Table 8.2. Sphericity and porosity can be assumed to be constant for the full bed depth.

SOLUTION Because the sand covers a range of sizes and will be stratified during backwashing, divide the bed into five equal segments and use the middle sieve opening size for the diameter term in the solution. Solving in SI units:

Kozeny Equation
$$\frac{h}{L} = \frac{k\mu}{\rho g} \frac{(1-\varepsilon)^2}{\varepsilon^3} \left(\frac{a}{v}\right)^2 V \tag{8.2}$$

$$\frac{a}{v} = \frac{6}{\psi d} = \frac{6}{0.75d}$$

$$\frac{\mu}{\rho} = \upsilon = 1.004\text{E} - 6\ \text{m}^2/\text{s at } 20°\text{C}$$

$g = 9.81$ m/s²
k = Kozeny's constant, which is typically 5 for common filter media
$V = 6$ gpm/ft² $= 4.08\text{E} - 3$ m/s

$$\frac{h}{L} = 5 \cdot \frac{1.004\text{E} - 6}{9.81} \cdot \frac{(1-0.42)^2}{0.42^3} \cdot \frac{6^2}{0.75^2\text{d}^2} 4.08\text{E} - 3 = \frac{6.0671\text{E} - 7}{\text{d}^2}$$

From Figure 8.3, select mid-diameters and calculate h/L for each selected mid-diameter.

Size	Middiameter (mx1000)	$\frac{h}{L}$	Layer depth (ft)	h (ft)
d_{10}	0.54	2.08	0.6	1.25
d_{30}	0.66	1.39	0.6	0.83
d_{50}	0.73	1.14	0.6	0.68
d_{70}	0.80	0.95	0.6	0.57
d_{90}	0.87	0.80	0.6	0.48
	Average	1.27	Total	3.81

Alternatively, because each layer was the same depth, the average h/L can be used to calculate the head loss for 3 ft (0.91 m) of bed of sand; $L = 3$ ft; average $h/L = 1.27$; $h = 1.27 \times 3 = 3.81$ ft (1.16 m) of water.

Head Loss for a Fluidized Bed

The American practice of filter backwashing has for many years been above the minimum fluidization velocity of the filter media. Therefore, some fluidization fundamentals are essential to proper understanding of this backwashing practice.

Fluidization can best be described as the upward flow of a fluid (gas or liquid) through a granular bed at sufficient velocity to suspend the grains in the fluid. During upward flow, the energy loss (pressure drop) across the fixed bed will be a linear function of flow rate at low superficial velocities when flow is laminar. For coarser or heavier grains, it may become an exponential function at higher flow rates if the Re enters the transitional regime, Re > 6. As the flow rate is increased further, the resistance equals the gravitational force, and the particles become suspended in the fluid. Any further increase in flow rate causes the bed to expand and accommodate to the increased flow while effectively maintaining a constant pressure drop (equal to the buoyant weight of the media). Two typical curves for real filter media fluidized by water are shown in Figure 8.4.

The pressure drop Δp after fluidization is equal to the buoyant weight of the grains and can be calculated from the following equation:

$$\Delta p = h\rho g = L\,(\rho_s - \rho)g\,(1 - \varepsilon) \tag{8.6}$$

in which ρ_s = mass density of the grains and other terms are as defined before.

Point of Incipient Fluidization

The point of incipient fluidization, or minimum fluidizing velocity, V_{mf}, is the superficial fluid velocity required for the onset of fluidization. It can be defined by the intersection of the fixed-bed and fluidized-bed head loss curves, the points labeled V_{mf} on Figure 8.4.

The calculation of minimum fluidization velocity is important in determining minimum backwash flow rate requirements. The rational approach to the calculation is based upon the fixed-bed head loss being equal to the constant head loss of the fluidized bed at the point of incipient fluidization. Thus, the Ergun equation (Equation 8.5) can be equated to the constant head loss equation (Equation 8.6) and solved for the velocity, that is, V_{mf}. The accuracy of the result is very dependent upon using realistic values for sphericity, ψ, and fixed-bed porosity, ε. Such data for actual media may not be available, making the result uncertain. By substituting an approximate relation between ψ and ε_{mf} into the aforementioned equation (Ergun equation = constant

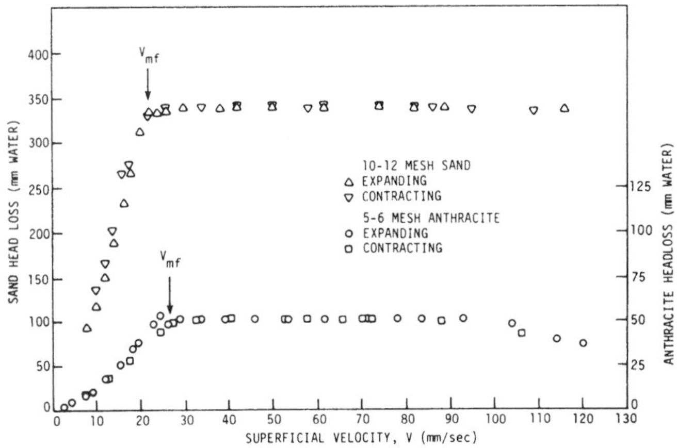

FIGURE 8.4 Head loss versus superficial velocity for 10–12 mesh sand at 25°C, L_o = 37.9 cm, ϵ_o = 0.446, and for 5–6 mesh anthracite at 25°C, L_o = 19.8 cm, ϵ_o = 0.581. (*Source:* J. L. Cleasby and K. S. Fan, "Predicting Fluidization and Expansion of Filter Media," *J. Environ. Eng.* 107(3): 455. Copyright © 1981 American Society of Civil Engineers. Reproduced by permission of ASCE.)

head loss equation), Wen and Yu (1966) were able to eliminate both ψ and ε_{mf} from the calculation of V_{mf}. The resulting equation is:

$$V_{mf} = \frac{\mu}{\rho d_{eq}}(33.7^2 + 0.0408\ \text{Ga})^{0.5} - \frac{33.7\mu}{\rho d_{eq}} \tag{8.7}$$

where Ga is the dimensionless Galileo number:

$$\text{Ga} = d_{eq}^3 \frac{\rho(\rho_s - \rho)g}{\mu^2} \tag{8.8}$$

For a bed containing a gradation in particle sizes, the minimum fluidization velocity is not the same for all particles. Smaller grains become fluidized at a lower superficial velocity than do larger grains. Therefore, a gradual change from the fixed-bed to the totally fluidized state occurs. In applying Equation 8.7 to a real bed with grains graded in size, calculating V_{mf} for the coarser grains in the bed is necessary to ensure that the entire bed is fluidized. The d_{90} sieve size would be a practical diameter choice in this calculation. Generally, the d_{eq} is not conveniently available, and the d_{90} diameter from the sieve analysis may be used as an acceptable approximation.

Furthermore, the minimum backwash rate selected must be higher than V_{mf} for the d_{90} sieve size (V_{mf90}) to allow free movement of these coarse grains during backwashing. A backwash rate equal to 1.3 V_{mf90} has been suggested to ensure adequate free movement of the grains (Cleasby and Fan, 1981). However, a rate closer to V_{mf} may be better to avoid movement of graded gravel support layers.

For a filter medium that has a very wide range of grain sizes, the use of V_{mf} for the d_{90} size during backwashing could possibly expand the finest grains so much that they could be lost to overflow. This is never a problem with the uniformity coefficients commonly specified for filter media (UC less than 1.7, often less than 1.5). The expansion of the entire bed during backwashing can be calculated as shown later under Backwashing.

EXAMPLE 8.2 Calculate the minimum fluidization velocity for the anthracite shown in Figure 8.3, at a water temperature of 20°C, using an anthracite density of 1.6 g/cm^3 (1600 Kg/m^3) estimated from Table 8.2.

SOLUTION Calculate the fluidization velocity (V_{mf}) for the d_{90} size of the anthracite as suggested in the text. The d_{90} size from Figure 8.3 = 0.29 cm (2.9E – 3 m).

Wen and Yu, Equation 8.7:

$$V_{mf} = \frac{\mu}{\rho d}(33.7^2 + 0.0408\ \text{Ga})^{0.5} - \frac{33.7\mu}{\rho d}$$

Galileo number Equation 8.8

$$\text{Ga} = d_{eq}^3 \frac{\rho(\rho_s - \rho)g}{\mu^2}$$

Solving in SI units:

$$\mu = 1.002\text{E} - 3\ \text{Ns/m}^2$$

$$\rho = 998.207\ \text{Kg/m}^3$$

$$\mu/\rho = \upsilon = 1.004\text{E} - 6\ \text{m}^2\text{/s}$$

$$g = 9.81\ \text{m/s}^2$$

$$d_{eq}\ (\text{use } d_{90}) = 2.9\text{E} - 3\ \text{m}$$

$$\text{Ga} = \frac{(2.9\text{E} - 3)^3 \cdot 998\ (1600 - 998)\ 9.81}{(1.002\text{E} - 3)^2} = 143{,}170 \text{ dimensionless}$$

$$V_{mf} = \frac{1.002\text{E} - 3\ (33.7^2 + .0408 \cdot 143170)^{0.5}}{998 \cdot (2.9\text{E} - 3)} - \frac{33.7 \cdot (1.002\text{E} - 3)}{998 \cdot (2.9\text{E} - 3)}$$

$$= 0.01725\ \text{m/s}$$

The recommended backwash rate would be up to 30 percent higher as discussed in text.

$$\text{Backwash rate} = 1.3 \cdot 0.01725 = 0.0223\ \text{m/s}\ (80\ \text{m/h})$$

$$\frac{0.0223\ \text{m}^3}{\text{m}^2\text{s}}\ \frac{1000\ L}{\text{m}^3}\ \frac{\text{gal}}{3.785\ L}\ \frac{0.0929\ \text{m}^2}{\text{ft}^2}\ \frac{60\text{s}}{\text{min}} = 32.8\ \text{gpm/ft}^2$$

This is a higher than normal backwash rate because of the very coarse anthracite grain size, the rather warm water, and the choice of a 30 percent factor above V_{mf}. Without that factor applied, the rate would be 25.2 gpm/ft^2.

RAPID GRANULAR BED FILTRATION

General Description

Rapid granular bed filtration, formerly known as "rapid sand filtration," usually consists of passage of pretreated water through a granular bed at rates of between 2 and

10 gpm/ft^2 (5 to 25 m/h). Flow is typically downward through the bed, although some use of upflow filters is reported in Latin America, Russia, and the Netherlands. Both gravity and pressure filters are used, although some restrictions are imposed against the use of pressure filters on surface waters or other polluted source waters or following lime soda softening (Great Lakes Upper Mississippi River Board of State Sanitary Engineers, 1992).

During operation, solids are removed from the water and accumulate within the voids and on the top surface of the filter medium. This clogging results in a gradual increase in head loss (i.e., clogging head loss) if the flow rate is to be sustained. The total head loss may approach the maximum head loss provided in the plant, sometimes called the *available head loss.* After a period of operation, the rapid filter is cleaned by backwashing with an upward flow of water, usually assisted by some auxiliary scouring system. The operating time between backwashes is referred to as a *filter cycle* or a *filter run.* The head loss at the end of the filter run is called the *terminal head loss.*

The need for backwash is indicated by one of the following three criteria, whichever occurs first:

1. The head loss across the filter increases to the available limit or to a lower established limit (usually 8 to 10 ft, i.e., 2.4 to 3.0 m) of water.
2. The filtrate begins to deteriorate in quality or reaches some set upper limit.
3. Some maximum time limit (usually 3 or 4 days) has been reached.

Typical filter cycles range from about 12 hours to 96 hours, although some plants operate with longer cycles. Setting an upper time limit for the cycle is desirable because of concern with bacterial growth in the filter, and because of concern that compaction of the solids accumulated in the filter will make backwashing difficult.

Pretreatment of surface waters by chemical coagulation (see Chapter 6) is essential to achieve efficient removal of particulates in rapid filters. In addition, filter-aiding polymers may be added to the water just ahead of filtration to strengthen the attachment of the particles to the filter media. Groundwaters treated for iron and manganese removal by oxidation, precipitation, and filtration generally do not need other chemical pretreatment.

Filter Media for Rapid Filters

Common filter materials used in rapid filters are sand, crushed anthracite coal, GAC, and garnet or ilmenite. Typical configurations of filter media are shown in Figure 8.5. The most commonly used of these configurations are the conventional sand and dual-media filters, but a substantial number of triple-media filters have been installed in the United States. Granular activated carbon replaces sand or anthracite in filter-adsorbers. It can be used alone or in dual- or triple-media configurations. The first three configurations in Figure 8.5 are backwashed with full fluidization and expansion of the bed. Fluidization results in stratification of the finer grains of each medium near the top of that layer of medium.

The single-medium deep-bed filter using coarse sand or anthracite coal [Figure 8.5(4)] differs from the conventional sand filter in two ways. First, because the medium is coarser, a deeper bed is required to achieve comparable removal of particulates. Second, because excessive wash rates would be required to fluidize the coarse medium, it is washed without fluidization by the concurrent upflow of air and water. The air-water wash causes mixing of the medium, and little or no stratification by size occurs.

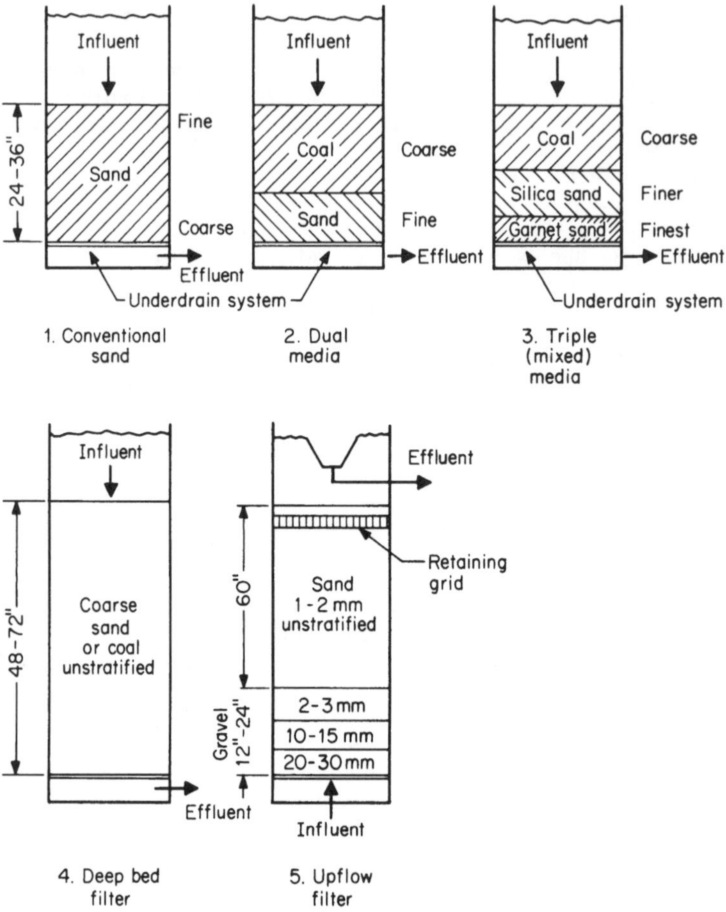

FIGURE 8.5 Schematic diagrams of filter configurations for rapid filtration. Media 1, 2, and 3 are washed with fluidization, whereas 4 and 5 are washed with air-plus-water without fluidization.

An upflow filter is used in some wastewater filtration plants and in a few potable-water treatment plants in other countries. It may include a restraining grid to resist uplift, as shown in Figure 8.5(5), or it may be operated with a deeper sand layer and to a limited terminal head loss so that the mass of the sand itself acts to resist uplift. The upflow filter is backwashed using air and water together during part of the backwash cycle. Adoption of the upflow filter as the only filtration step in potable water treatment is doubtful because of the potential for contamination of the filtered water caused by both the dirty backwash water and the filtered water exiting above the filter medium. However, upflow filtration exists in preengineered, package, potable water treatment plants in the United States, where it is used in the pretreatment process in place of sedimentation or other solids separation processes, ahead of downflow filtration.

Typical grain sizes used in rapid filters are presented in Table 8.3 for various potable water applications. The UC of the filter medium is usually specified to be less than 1.65 or 1.7. Use of a lower UC is beneficial for coarser filter media sizes that are to be backwashed with fluidization, however, because this will minimize the d_{90} size and thereby reduce the required backwash flow rate. But the lower the specified UC, the more costly the filter medium, because a greater portion of the raw material falls outside the specified size range. Therefore, the lowest practical UC is about 1.4. Anthracite coal that will meet this UC is commercially available.

There is growing interest in the use of deeper beds of filtering materials, either mono or dual media, especially for direct filtration applications. For example, deep beds of dual media were used in pilot studies of cyst removal by Patania et al. (1995) in Seattle. These studies included in-line filtration using dual media with 80 in (2.0 m) of 1.25-mm ES anthracite over 10 in (0.25 m) of 0.6-mm ES sand. The media were selected to meet stringent turbidity and particle removal goals and operated with optimized chemical pretreatment. Pilot studies for proposed plants for Sydney, Australia, included deep-bed mono- and dual-media configurations (Murray and Roddy, 1993). The dual media studied included anthracite 39 to 118 in (1.0 to 3.0 m) in depth, with 1.7- to 2.5-mm ES over 6 to 12 in (0.15 to 0.30 m) of crushed sand with 0.65- to 1.0-mm ES. The coarser dual media performed best in production per cycle and in filtrate quality. One reason expressed for the interest in deep beds was the possible future conversion of the filters to GAC with empty-bed contact times of at least 7.5 minutes.

In addition to the configurations of filtering materials shown in Figure 8.5, other proprietary media are being used in some applications. The following examples are discussed in more detail in the section titled "Other Filters" in this chapter. A buoyant crushed plastic medium is being used in an upflow mode as a contact flocculator and pretreatment filter ahead of a downflow triple-media bed (Benjes, Edlund, and Gilbert, 1985). Several manufacturers are marketing traveling backwash filters in which the filter is divided into cells and utilizes a shallow layer of fine sand, usually about 12 in. in depth (Medlar, 1974).

TABLE 8.3 Typical Grain Sizes for Different Applications

	Effective size, mm	Total depth, m
A. *Common U.S. Practice After Coagulation and Settling*		
1. Sand alone	0.45–0.55	0.6–0.7
2. Dual media	0.9–1.1	0.6–0.9
Add anthracite (0.1 to 0.7 of bed)		
3. Triple media	0.2–0.3	0.7–1.0
Add garnet (0.1 m)		
B. *U.S. Practice for Direct Filtration*		
Practice not well established. With seasonal diatom blooms, use coarser top size.		
Dual-media or deep mono-medium, 1.5-mm ES.		
C. *U.S. Practice for Fe and Mn Filtration*		
1. Dual media similar to A-2 above		
2. Single medium	<0.8	0.6–0.9
D. *Coarse Single-Medium Filters Washed with Air and Water Simultaneously*		
1. For coagulated and settled water	0.9–1.0	0.9–1.2
2. For direct filtration	1.4–1.6	1–2
3. For Fe and Mn removal	1–2	1.5–3

The Concept of Equivalent Depth of Filter Media

When considering the use of deeper beds of coarser media, the provision of a deeper bed is sometimes used so that the ratio of bed depth to grain diameter L/d is held constant. Hopefully, this will result in equal filtrate quality when filtering the same influent suspension at the same filtration rate. The concept is supported by the experiments of Ives and Sholji (1965). However, these authors also analyzed the work of other investigators and suggested the relationship should be L/d^{β}, with β values from 1.5 to 1.67.

Nevertheless, the use of L/d has grown in recent years. For graded beds, the effective size d_e has been suggested for the diameter term (Montgomery 1985, p. 538). For dual and multimedia filters, the sum of the L/d_e for each layer is calculated (i.e., the weighted average L/d_e for the bed). Some typical values for L/d_e for common bed configurations, quoted directly from Kawamura (1991, p 211), are as follows:

- $L/d_e \geq 1000$ for ordinary fine sand and dual-media beds
- $L/d_e \geq 1250$ for triple-media (anthracite, sand, garnet) beds
- $L/d_e \geq 1250$ for a deep, monomedium beds (1.5 mm> d_e >1.0 mm)
- $L/d_e = 1250–1500$ for very coarse, deep, monomedium beds (2.0 mm> d_e > 1.5mm)

For a common dual-media filter with 2.0 ft (0.6 m) of anthracite of 1.0-mm ES over 1.0 ft (0.3 m) of sand of 0.5-mm ES, the total L/d_e would be 600/1.0 plus 300/0.5 = 1200, meeting Kawamura's criterion. Of course, this is a simplistic concept for a complex process, but it does assist in selecting alternative filter media for pilot scale evaluations.

Monomedium Versus Multimedia Filters

The media utilized in rapid filters evolved from fine sand, monomedium filters about 2 to 3 ft (0.5 to 0.9 m) deep to dual media and later to triple (mixed) media of about the same depth. Still later, success at Los Angeles had led to strong interest in deep-bed (4 to 6 ft, 1.2 to 1.8 m), coarse, monomedium filters. The rationale for this evolution is discussed in this section.

Early research on rapid sand filters with sand about 0.4- to 0.5-mm ES demonstrated that most of the solids were removed in the top few inches of the sand, and that the full bed depth was not being well-utilized. The dual-media filter bed, consisting of a layer of coarser anthracite coal on top of a layer of finer silica sand, was therefore developed to encourage penetration of solids into the bed.

The use of dual-media filters is now widespread in the United States. It is not a new idea. Camp (1961) reported using dual media for swimming pool filters beginning about 1940, and later in municipal treatment plant filters. Baylis (1960) described early work in the mid-1930s at the Chicago Experimental Filtration Plant, where a 3-in layer (7.5 cm) of 1.5-mm ES anthracite over a layer of 0.5-mm ES sand greatly reduced the rate of head loss development in treatment of Lake Michigan water.

The benefit of dual media in reducing the rate of head loss development—thus lengthening the filter run—is well-proven by a number of later studies (Conley and Pitman, 1960a; Conley, 1961; Tuepker and Buescher, 1968). However, the presumed benefit to the quality of the filtrate is not well-demonstrated.

Based on their experiences, Conley and Pitman (1960b) concluded that the alum dosage should be adjusted to achieve low levels of uncoagulated matter in the filtrate (low turbidity) early in the filter run (after 1 h), and that the filter-aiding polymer should be adjusted to the minimum level required to prevent terminal breakthrough of alum floc near the end of the run. At the same time, to prevent excessive head loss development, the dosage of polymer should not be higher than necessary.

Further research comparing dual media with a fine sand medium was reported by Robeck, Dostal, and Woodward (1964). They compared three filter media during filtration of alum-coagulated surface water. These comparisons were made by running filters in parallel, so the benefits of dual media were more conclusively demonstrated. The rate of head loss development for the dual-media filter was about one half of that for the sand medium, but the effluent turbidity was essentially the same prior to breakthrough, which was observed under some weak flocculation conditions.

The evidence clearly demonstrated lower head loss for a dual-media filter, as compared to a traditional fine sand filter. For a typical dual media with anthracite ES that is about double the sand ES, the head loss development rate should be about one-half the rate of the fine sand filter when both are operated at the same filtration rate on the same influent water.

The benefits gained by the use of dual-media filters led to the development of the triple-media filter, in which an even finer layer of high-density media (garnet or ilmenite) is added as a bottom layer (Conley, 1965; Conley, 1972). The bottom layer of finer material should improve the filtrate quality in some cases, especially at higher filtration rates. There is growing evidence to support that expectation.

The triple-media filter is sometimes referred to as a *mixed-media* filter because in the original development, the sizes and uniformity coefficients of the three layers were selected to encourage substantial intermixing between the adjacent layers. This was done to come closer to the presumed ideal configuration of "coarse to fine" filtration. The original patents have now expired, and other specifications for triple media are being used with various degrees of intermixing.

The initial clean-bed head loss will be higher for the triple-media filter due to the added layer of fine garnet or ilmenite. Thus, for a plant with a particular total available filter system head loss, the clogging head loss available to sustain the run is reduced, which may shorten the run compared to the run obtained with a dual-media filter (Robeck 1966).

Some comparisons of triple media versus dual media have demonstrated triple media to be superior in filtered water quality. On Lake Superior water at Duluth, a mixed-media (triple-media) filter was superior to dual media in amphibole fiber removal (Logsdon and Symons, 1977; Peterson, Schleppenbach, and Zaudtke, 1980). Twenty-nine out of 32 samples of filtrate were below or near detection level for the dual-media filter, and 18 out of 18 for the mixed-media filter. Mixed media was recommended for the plant. Mixed media was also reported to be superior in resisting the detrimental effects of flow disturbances on filtrate quality (Logsdon, 1979).

In contrast, Kirmeyer (1979) reported pilot studies for the Seattle water supply in which two mixed-media and two dual-media filters were compared. No difference in filtrate quality was observed in either turbidity or asbestos fiber content with filtration rates of 5.5 to 10 gpm/ft^2 (13.4 to 24.4 m/h). Some differences in production per unit head loss were observed, favoring dual media at lower rates and mixed media at higher rates.

A number of laboratory studies have compared triple media versus single-medium (fine sand) filters, (Diaper and Ives, 1965; Rimer, 1968; Oeben, Haines, and Ives, 1968; Mohanka, 1969). All of these studies have clearly shown the head loss benefit gained by filtering in the direction of coarse grains to fine grains. Three of the studies also clearly showed benefits to the filtrate quality for the triple media (Diaper and Ives, 1965; Rimer, 1968; Oeben, Haines, and Ives, 1968).

Cleasby et al. (1992) summarized twelve unpublished pilot and plant scale studies conducted for utilities, comparing dual, triple-media, and deep monomedium filters in high-rate filtration of surface waters. The studies supported the superiority of triple-media filters over dual media (with comparable depths of anthracite and sand) in producing the best-quality filtered water. However, higher initial head losses and better solids capture resulted in shorter filter cycles for triple-media filters, other factors being the same. The studies comparing deep-bed, coarse, monomedium filters with dual- or triple-media filters were inconclusive with regard to quality of filtrate. However, the deep-bed, coarse, monomedium filters achieved longer filter cycles and greater water production per cycle than traditionally sized dual- or triple-media filters.

Use of GAC in Rapid Filtration

Granular activated carbon is being used in filter-adsorbers that serve both for filtration of particles and for adsorption of organic compounds, as well as in biological filters and post-filter-adsorbers. (See Chapter 13 for more detail.) The principal application of filter-adsorbers to date is for removal of taste and odor, where full-scale experience shows successful removal for periods of from 1 to 5 years before the GAC must be regenerated (Graese, Snoeyink, and Lee, 1987). High concentrations of competing organic compounds, however, can reduce this duration. Most existing GAC filter-adsorbers are retrofitted rapid filters where GAC has replaced part or all of the sand in rapid sand filters, or replaced anthracite in dual- or triple-media filters. Several filter adsorbers, however, have been initially constructed with GAC.

Granular activated carbon used in retrofitted filters is typically 15 to 30 in (0.38 to 0.76 m) of 12 × 40 mesh GAC (ES 0.55 to 0.65 mm) or 8 × 30 mesh GAC (ES 0.80 to 1.0 mm) placed over several inches of sand (ES 0.35 to 0.60 mm) in a dual-media configuration. These GAC materials have a higher UC (≤2.4) than traditionally used filter materials (≤1.6). The high UC can contribute to more fine grains in the upper layers and shorter filter cycles, but it also results in less size intermixing during backwashing, which is beneficial to adsorption.

The use of retrofitted filters as filter-adsorbers is primarily for removal of trihalomethane (THM) precursors. The removal of less strongly adsorbed compounds such as trihalomethanes and volatile organic compounds is limited. The short empty-bed contact times typical of such filter-adsorbers (about 9 min) would result in frequent regeneration or replacement of the GAC if used for THM adsorption.

Some filter-adsorbers designed for GAC initially use up to 48 in (1.2 m) of coarser GAC (ES-1.3 mm) with lower uniformity coefficients (1.4). This provision should result in longer filter cycles, and longer periods between GAC regenerations.

Granular activated carbon is being used successfully on both monomedium and dual-media filter-adsorbers. It is as effective in its filtration function as conventional (sand or dual) filter media, provided an appropriately sized medium has been selected (Graese, Snoeyink, and Lee, 1987). Particulates removed in the filter-adsorber, however, do impede the adsorption function somewhat, compared to post-filter-adsorbers with the same contact time (see Chapter 13).

The layer of fine sand in the dual-media filter-adsorber may be essential where very-low-turbidity filtered water is the goal. The use of sand and GAC in dual-media beds, however, causes problems during regeneration because of the difficulty of removing sand-free GAC from the filter-adsorber, and of separating the sand after removal. Sand causes difficulties in regeneration furnaces (Graese, Snoeyink, and Lee, 1987).

Bacteria proliferate in GAC beds, making periodic backwashing essential. Backwashing should be minimized, however, because of its possible effect on adsorption efficiency. Application of chlorine to the influent of the adsorbers does not prevent bacterial growth and can cause other detrimental effects (see Chapter 13). Post-adsorption disinfection is essential.

Granular activated carbon is softer than anthracite, but the attrition of GAC media in full-scale plants has not been excessive. Losses of from 1 to 6 percent per year have been reported (Graese, Snoeyink, and Lee, 1987), which are not higher than typical losses of anthracite. A modest reduction in grain size of GAC has been measured at some plants. Granular activated carbon has a lower density than anthracite, posing some concerns about losses during backwashing (a topic that will be discussed later).

Rates of Filtration

Fuller (1898) is commonly credited with establishing a standard rate of filtration of 2 gpm/ft^2 (5 m/h) for chemically pretreated surface waters. This filtration rate was considered practically inviolable for the first half of the twentieth century in the United States. Fuller observed, however, that with properly pretreated water, higher rates gave practically the same water quality. Of equal importance, Fuller acknowledged that without adequate chemical pretreatment, no assurance of acceptable filtered water existed even at filtration rates of 2 gpm/ft^2 (5 m/h).

In the 1950s and 1960s, many plant-scale studies were conducted comparing filter performance at different filtration rates (Brown, 1955; Baylis, 1956; Hudson, 1962). These studies were generally conducted as utilities were considering uprating existing filters, or building new plants with filtration rates higher than the traditional 2 gpm/ft^2 (5 m/h).

The results of such studies—as illustrated in Tables 8.4 and 8.5, and other similar studies (Cleasby and Baumann, 1962; Conley and Hsuing, 1969)—demonstrated that higher filtration rates do result in somewhat poorer filtrate quality, as both theory and intuition would predict. However, at the time, the turbidity standard was 1 Jackson Turbidity Unit, and it was presumed that chlorine disinfection would handle any pathogens that happened to pass through the filters. Nevertheless, there was a gradual acceptance of filtration rates of up to 4 gpm/ft^2 (9.8 m/h), with conventional pretreat-

TABLE 8.4 Full-Scale Results at Three Filtration Rates (Brown, 1955)*

	Filter no.		
Item	12 (2 gpm/ft^2) (4.9 m/h)	13 (3 gpm/ft^2) (7.3 m/h)	14 (4 gpm/ft^2) (9.8 m/h)
Length of run, h	135.2	116.7	81.3
Wash water, %	1.21	0.89	0.99
Turbidity, ppm	0.34	0.38	0.43
Bacteria, colonies/mL	0.32	0.42	0.36
Coliform organisms	Negative	Negative	Negative

* The total amount of water passing through the individual filters during the 3-year tests period is not known, because they were unmetered. It may be assumed, however, that the quantities were proportional to the rates, since the accuracy of the standard venturi controllers was checked before and during the tests. The filters were operated continuously during the trial period, except when backwashing.

TABLE 8.5 Full-Scale Filtrate Quality Data on Several Filtration Rates as Indicated by Solids Captured on Cotton Plug Filters (Baylis, 1956)

Year	Average of all filters	Filtration rate, gpm/ft² (m/h) 2 (4.9)	4 (9.8)	4.5 (11.0)	5 (12.2)
		Ash, ppm*			
1949	0.047		0.055	0.059	0.066
1950	0.037	0.028	0.048	0.045	0.060
1951	0.042	0.039	0.067	0.064	0.084
1952	0.077	0.039	0.058	0.084	0.084
1953	0.060	0.057	0.067	0.089	0.087
1954	0.058	0.056	0.073	0.090	0.082
Avg.	0.054	0.044	0.059	0.072	0.077

* Ash remaining after ignition of cotton plug filter.

ment and without the use of filter-aiding polymers to increase filtration efficiency (King et al., 1975).

Pioneering work at even higher filtration rates, assisted by filter-aiding polymers, included the following studies. Robeck, Dostal, and Woodward (1964) compared performance of pilot filters at 2 to 6 gpm/ft² (5 to 15 m/h), filtering alum-coagulated surface waters through single and dual-media filters. They concluded that with proper coagulation ahead of the filters, the effluent turbidity, coliform bacteria, polio virus, and powdered carbon removal was as good at 6 gpm/ft² (15 m/h) as at 4 or 2 gpm/ft² (10 or 5 m/h). Pretreatment included activated silica when necessary to aid flocculation and a polyelectrolyte as a filter aid (referred to as a coagulant aid in the original article). The benefit of using filter-aiding polymers in retarding terminal breakthrough is shown in Figure 8.6.

Conley and Pitman (1960a) showed the detrimental effect of high filtration rates of up to 15 gpm/ft² (37 m/h) in the treatment of Columbia River water, using alum coagulation followed by short detention flocculation and sedimentation before filtration. A proper dose of nonionic polymer added to the water as it entered the filters, however, resulted in the same filtrate quality from 2 to 35 gpm/ft² (5 to 85 m/h). [Note that the turbidity unit being reported in Conley's studies was later acknowledged (Conley, 1961) to be equivalent to about 50 Jackson Turbidity Units.]

However, the more recent concerns about *Giardia* and *Cryptosporidium* have emphasized the need to achieve filtered water turbidities at or below 0.10 ntu, and have emphasized log reduction of cysts or cyst-sized particles to ensure the absence of protozoan pathogens. Some results of such studies were presented in the Introduction section of this chapter, and more follow.

When chemical pretreatment was optimized for turbidity and particle removal, resulting in filtered water turbidity below 0.10 ntu, Patania et al. (1995) found no difference in *Giardia* cyst and *Cryptosporidium* oocyst removal in pilot studies at Contra Costa, California, at filtration rates of 3 and 6 gpm/ft² (7 to 15 m/h), and at Seattle at 5 and 8 gpm/ft² (12 to 20 m/h). Conventional pretreatment preceded filtration at Contra Costa, whereas in-line filtration was used at Seattle. Both used deep-bed dual-media filters and dual coagulants (alum plus cationic polymer). Filter-aiding polymer was also used in the Seattle pilot plant.

Design and operation guidelines for optimization of high-rate filtration plants were reported by Cleasby et al. (1989, 1992). The reports were based on a survey of

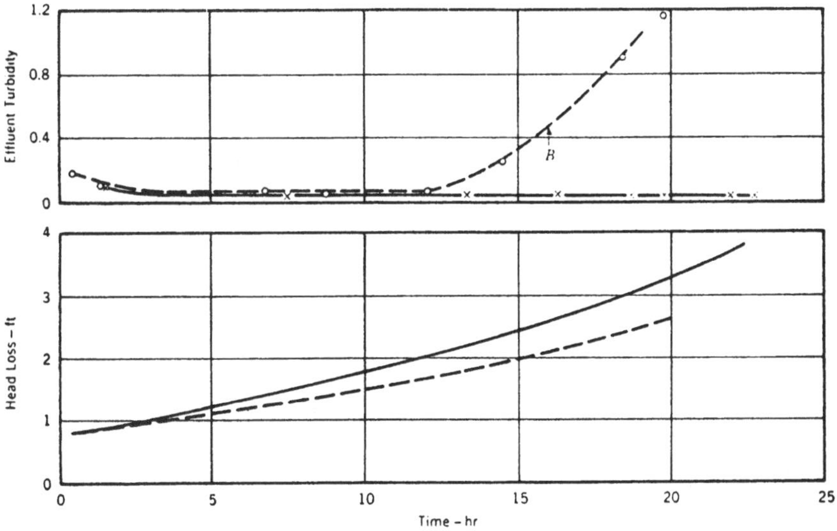

FIGURE 8.6 Effect of polyelectrolyte on length of run. The data shown were obtained under the following operating conditions: raw-water turbidity, 10 units; alum dose, 75 mg/L; filtration rate 2 gpm/ft^2; settling tank effluent turbidity, 8 units; and activated carbon, 2 mg/L. In both graphs, the dashed curve is for the filter influent with no polyelectrolyte added; and the solid curve is for 0.08 mg/L polyelectrolyte added. In the upper graph, Point B shows the time of filter breakthrough, 16 h. The length of run with polyelectrolyte added was more than 22 h. (*Source:* G. G. Robeck; K. A. Dostal; and R. L. Woodward, "Studies of Modification in Water Filtration," *Jour. AWWA,* 56(2), February 1964: 198.)

21 surface water treatment plants with consistent operational success in producing filtered water turbidity below 0.2 ntu at filtration rates at or above 4 gpm/ft^2 (10 m/h). These plants were characterized by management support of a low-turbidity goal, optimal chemical pretreatment, the use of polymeric flocculation and/or filter-aiding chemicals, the use of dual- or tri-media filters, continuous monitoring of each filter effluent turbidity, and good operator training.

The turbidity and particle count results of this study were summarized by Bellamy et al. (1993) as presented in Tables 8.6 and 8.7. Table 8.6 shows that when source water turbidities were above 5 ntu, the log reductions from source water to finished water turbidity agreed well with log reductions in total particle count and cyst-sized particle count. Log reductions in turbidity and particle counts were 2-log or higher in all of the plants in Table 8.6. However, with lower source water turbidity (Table 8.7), the agreement between log reductions in turbidity and particle count was not good because of the inability to measure turbidities below about 0.05 ntu. In spite of this, the log reduction in cyst-sized particles was near 2-log except in 4 of the 11 plants treating such waters.

The use of unusually high filtration rates was reported at the Contra Costa County Water District plant. By precoating the dual-media filters with a small dose of polymer during the backwash operation, Harris (1970) reported successful operation at 10 gpm/ft^2 (24 m/h). Harris also reported that the initial period of poorer water quality was eliminated by this precoating operation. The Contra Costa County Water District plant is now authorized by the State of California to operate at 10 gpm/ft^2 (24 m/h).

TABLE 8.6 Turbidity and Particle Count Data for Plants Treating Source Waters with Turbidity Above About 5 ntu (Bellamy et al., 1993)

	Turbidity		Total particle count*		Cyst-sized particles	
					Log reduction,†‡	
City (plant)	Source ntu	Log reduction†	Number/mL	Log reduction†	3.7-μm size	7.8-μm size
Conventional plants						
Glendale, Ariz. (Cholla)	4.4	2.0	69,000	2.1	2.0	2.2
Phoenix, Ariz. (Val Vista)	6.9	2.0	99,000	2.4	2.4	2.4
Contra Costa, Calif. (Bollman)	9.0	2.9	170,000	2.8	2.8	2.8
Winnetka, Ill.	49.0	3.7	500,000	3.0	3.1	3.2
Corvallis, Ore. (Taylor)	4.9	1.8	71,000	2.0	2.0	2.2
Merrifield, Va. (Corbalis)	5.7	2.3	51,000	2.4	2.4	2.4
Two-stage clarification and lime softening						
Johnson County, Kan. (Hanson)	9.2	1.9	66,000	1.6	1.5	1.9

* Particle counter range—1 μm to 60 μm

† Source water to filter effluent

‡ Particle range for mean diameter of 3.7 μm was 3.06 to 4.43 μm; range for mean diameter of 7.8 μm was 6.43 to 9.33 μm.

TABLE 8.7 Turbidity and Particle Count Data for Plants Treating Source Waters with Turbidity Less Than 5 ntu (Bellamy et al., 1993)

	Turbidity		Total particle count*		Cyst-sized particles	
					Log reduction,[†‡]	
City (plant)	Source ntu	Log reduction[†]	Number/mL	Log reduction[†]	3.7-μm size	7.8-μm size
Conventional plants						
Tuscaloosa, Ala. (Ed Love)	1.2	1.6	17,000	1.5	1.4	1.5
Colo. Springs, Colo. (Pine Valley)	0.96	1.5	52,000	2.5	2.4	2.5
Loveland, Colo. (Chasteens Grove)	1.5	1.7	11,000	2.0	2.1	2.2
Duluth, Minn. (Lakewood)	0.4	1.2	4,000	2.0	2.1	1.9
Rapid mix, detention, filtration plants						
Onondaga, N.Y. (Otisco Lake)	2.4	1.3	73,000	1.5	1.4	1.8
Eugene, Ore. (Haydon Bridge)	1.5	1.5	29,000	2.0	2.1	2.0
Lake Oswego, Ore.	2.4	1.5	46,000	1.5	1.5	1.4
Direct filtration plants						
Phoenix, Ariz. (Union Hills)	1.2	0.8	29,000	1.4	1.4	1.3
Los Angeles, Calif. (LA Aqueduct)	3.6	1.7	55,000	1.9	2.0	2.1
Oceanside, Calif. (Wiese)	2.1	1.5	38,000	1.9	2.0	1.8
Las Vegas, Nev. (Smith)	0.21	0.5	3,200	1.0	1.2	1.3

* Particle counter range—1 μm to 60 μm

[†] Source water to filter effluent

[‡] Particle range for mean diameter of 3.7 μm was 3.06 to 4.43 μm; range for mean diameter of 7.8 μm was 6.43 to 9.33 μm.

Another example of an unusually high design filtration rate is the 13.5 gpm/ft^2 (33 m/h) selected for the Los Angeles direct filtration plant. This filtration rate was selected and approved by the state after more than five years of extensive pilot scale studies (Trussell et al., 1980; McBride and Stolarik, 1987). The plant uses ozone, ferric chloride, and cationic polymer in pretreatment, and deep-bed (6 ft, 1.8 m) coarse anthracite (1.5 mm ES, 1.4 UC), monomedium filters. It has operated successfully at the design rate, producing filtered water below 0.1 ntu on a consistent basis.

Both of these are large treatment plants, treating high-quality surface water, with competent management and operation. Thus, filtration rates of these magnitudes probably will not become common. With poorer source waters and in smaller plants with less careful surveillance or operation, lower filtration rates at or below 4 gpm/ft^2 (9.8 m/h) will continue to be the prudent choice.

Nevertheless, the success at Los Angeles has stimulated great interest in deep-bed monomedium filtration. Other similar plants are now in service, but not a such high filtration rates. For example, the Modesto, California, plant uses the same filter medium as in the Los Angeles plant, but the design filtration rate is 6 gpm/ft^2 (15 m/h) (Short, Gilton, and Henderson, 1996).

Acceptance of filtration rates above 2 gpm/ft^2 (5 m/h) varies among state regulatory agencies, however, and plant or pilot-scale demonstrations are still required by some state agencies before acceptance.

The use of higher filtration rates shortens the filter cycle in proportion roughly inversely to the rate. This problem is minimized by the use of dual- or triple-media filters or deep-bed, coarser, monomedium filters. These filters provide for better penetration of solids into the filter medium, and thus better utilization of the medium for solids storage during the filter cycle.

Acceptable Run Length and Production per Run

Higher filtration rates have two conflicting impacts on plant production. First, the clean-bed head loss is increased. Thus, for a given amount of available head loss in the plant hydraulic profile, less clogging head loss is available to sustain the filter cycle. This will shorten the filter cycle length if it is terminated based on reaching maximum available head loss. Second, the head loss increases at a faster rate per hour at higher filtration rates because solids are captured at a faster rate. However, the production per unit time also increases, so that production per day is increased substantially unless the filter cycles get too short. Therefore, the common importance attached to the length of filter run can be misleading. Rather than emphasizing the effect of filtration rate on length of run, one should emphasize the effect of filtration rate on filtrate quality, net plant production, and production efficiency.

Figure 8.7 illustrates the effect of unit filter run volume on net water production at three filtration rates (Trussell et al., 1980). *Unit filter run volume* is the actual throughput of a filter during one filter run; it also can be called *gross production per filter run.* The figure is based on the assumption that 100 gal/ft^2 (4 m^3/m^2) is used for each backwash operation, and 30 min downtime is required per backwash. Similar figures can be constructed with other assumptions. The following example illustrates the preparation of the figure.

- Assume four cycles per day per filter.
- Use a run time per cycle = 360 min/cycle – 30 min/wash = 330 min/cycle.
- Assume desired maximum filtration rate = 5 gpm/ft^2.

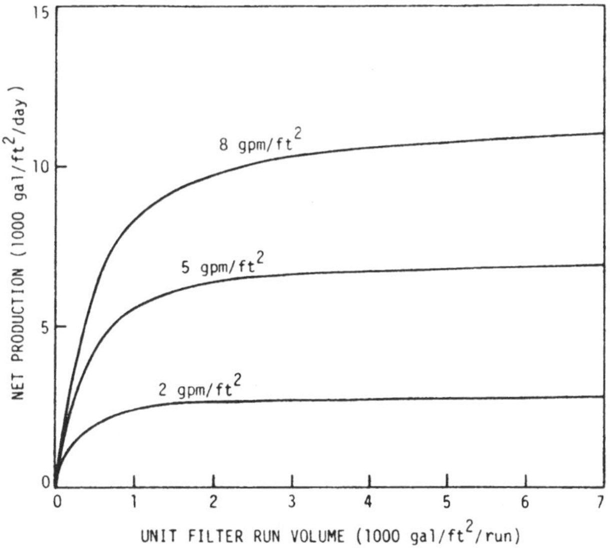

FIGURE 8.7 Gross production per square foot per run versus net production, assuming 30 min downtime per backwash and 100 gal/ft^2 per backwash. (*Source:* R. R. Trussell, A. R. Trussell, J. S. Lang, and C. H. Tate, "Recent Development in Filtration System Design," *Jour. AWWA,* 72(12) December 1980: 705.)

- Unit filter run volume 5 × 330 = 1,650 gal/ft^2/cycle.
- Gross volume filtered/ft^2/day = 1,650 × 4 cycles/day = 6,600 gal/ft^2/day.
- Net water production/ft^2/day = 6,600 – backwash volume = 6,600 – 100 × 4 cycles/day = 6,200 gal/ft^2/day.
- Production efficiency expressed as percent of gross volume filtered = (6,200/6,600) 100 = 94 percent.

In general, allowing for backwash volumes of 100 to 200 gal/ft^2/wash (4 to 8 m^3/m^2/wash) and a downtime of 30 min/wash, production efficiency will remain high if the unit filter run volume exceeds about 5,000 gal/ft^2/run (200 m^3/m^2/run). The calculations just presented are the same whether the backwash water is recovered by recycling it through the plant or not recovered. Above 5,000 gal/ft^2/run (200 m ^{3}/m ^{2}/run) unit filter run volume, the net production increases almost linearly with rate. Thus, the key issue when considering the impact of high filtration rates on production is not run length alone, but rather the gross production per run (run length times filtration rate) and the filtrate quality. A unit filter run volume of 5,000 gal/ft^2/run means a run length of 2,500 min (42 h) at 2 gpm/ft^2 (5 m/h) or 1,000 min (16.7 h) at 5 gpm/ft^2 (12 m/h). So a 42 h run at 2 gpm/ft^2 (5 m/h) is no better than 16.7 h at 5 gpm/ft^2 (12 m/h) because both would yield 98 percent production efficiency.

In summary, high rate filtration does have a definite effect on development of head loss and length of filter run that must be anticipated. The benefits of greater net production/ft^2/day at higher filtration rates, however, more than justify that consideration because of lower capital cost for new plants, or increased production from existing plants.

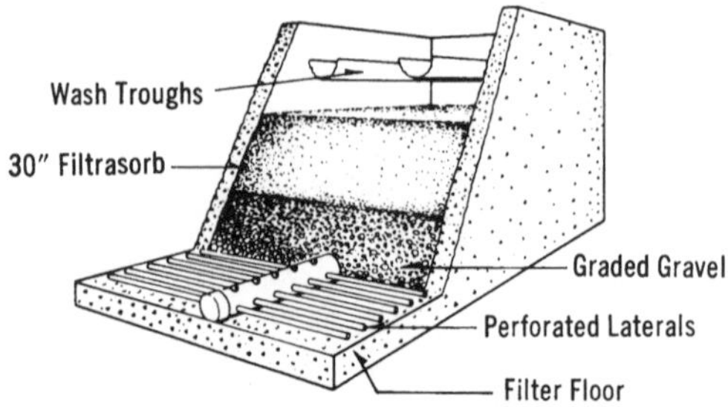

FIGURE 8.8 Rapid gravity filter with manifold and lateral underdrain system. (*Source:* After C. P. Hoover, Water Supply & Treatment, National Lime Assoc.)

Underdrain and Support Gravel

The underdrain system serves to support the filter medium, collect filtered water (in downflow filters), and distribute backwash water and air scour if air is used. A wide variety of underdrain systems are in use, but they can be grouped into three major types.

The first and oldest type is the *manifold-lateral system,* in which perforated pipe laterals are located at frequent intervals along a manifold (Figure 8.8). Perforations in the laterals are ¼ to ½ in (6 to 13 mm) in diameter, located on 3- to 12-in (8- to 30-cm) spacing.

The second type is the *fabricated self-supporting underdrain system* that is grouted to the filter floor as in Figure 8.1. One example is a vitrified clay block underdrain now marketed by several companies. Recently, a similar plastic block underdrain has been marketed that is capable of delivering backwash air or water, or air and water simultaneously. Top openings on these types of underdrains are about ¼ in (6 mm) in diameter.

The third type of underdrain is the *false-floor underdrain with nozzles* (Figure 8.9). A false-floor slab (or a steel plate in pressure filters) is located 1 to 2 ft (0.3 to 0.6 m) above the bottom of the filter, thus providing an underdrain plenum below the false floor. Nozzles to collect the filtrate and distribute the backwash water are located at 5- to 8-in centers (13 to 20 cm). Nozzles may have coarse openings of about ¼ in (6 mm), or they may have very fine openings that are sufficiently small to retain the filter medium. The nozzles may be equipped with a stem protruding about 6 to 9 in (15 to 23 cm) into the underdrain plenum. One or two small holes or slots in the stem serve to distribute air, either alone or in combination with water.

Most filters use underdrain systems with openings larger than the filter medium to be supported. As a result, several layers of graded gravel between the underdrain openings and the filter medium are necessary to prevent the medium from leaking downward into the underdrain system. The conventional gravel system begins with coarse-sized gravel at the bottom, with progressively finer-sized gravel layers up to the filter medium. Typical rules of thumb for the gradations of adjacent layers, reported by Huisman (1974) in Europe, are as follows:

- Each gravel layer should be as uniform as possible, preferably with the 10 and 90 percent passing diameters not more than a factor of $\sqrt{2}$ apart. In the United

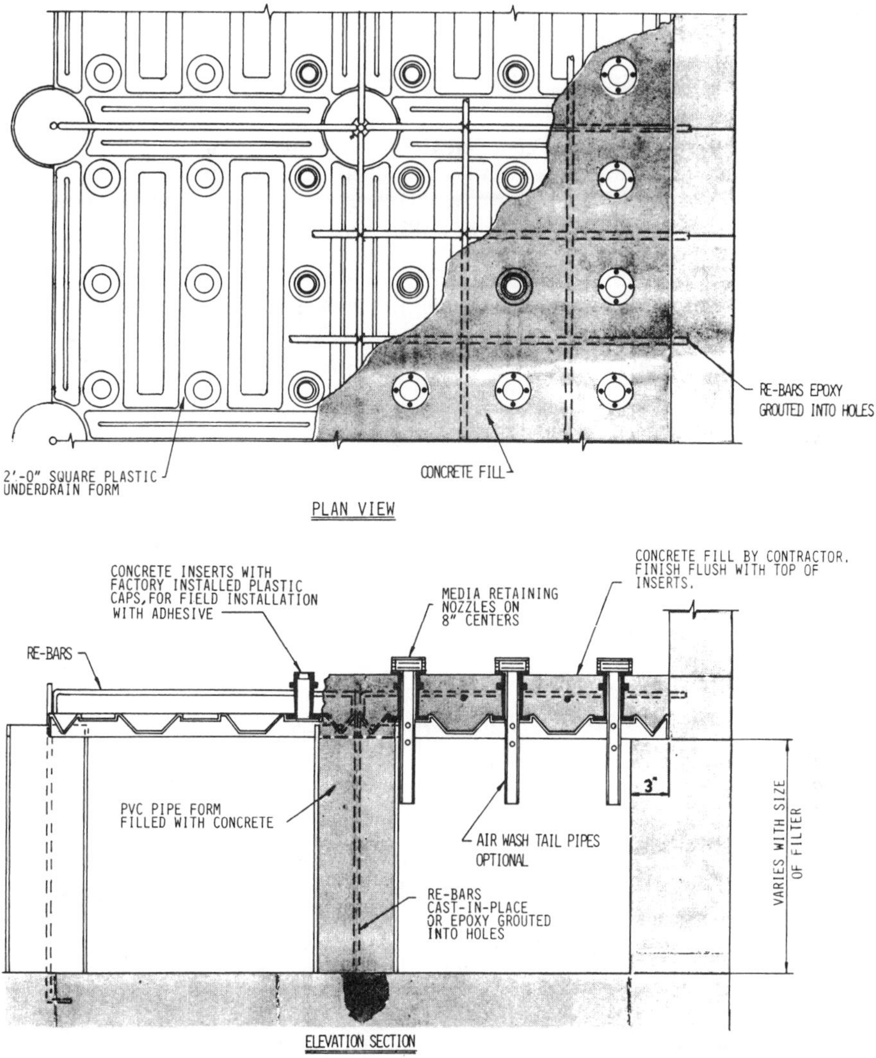

FIGURE 8.9 Nozzle underdrain system consisting of a monolithic, cast-in-place, concrete slab on a permanent plastic underdrain form with nozzles capable of air and water distribution. (*Source:* Multicrete II™, Courtesy of General Filter Co., Ames, Iowa.)

States, a sieve size ratio of 2 for the passing and retaining sieves is commonly accepted (American Water Works Association 1996).

- Bottom layer lower grain size limit should be 2 to 3 times the orifice diameter of the underdrain system.
- Top layer lower grain size limit should be 4 to 4.5 times the ES of the media to be retained.

- From layer to layer, the upper grain size limit of the coarser layer should be 4 times the lower grain size limit of the adjacent finer layer.
- Each layer should be at least 3 in (7 cm) thick, or 3 times the upper grain size limit of the layer, whichever is greater.

The U.S. interpretation of these guidelines is that the upper and lower grain size limits refer to the sieve sizes that pass and retain the gravel. No more than 8 percent by dry weight shall be finer than the lower grain size limit, and a minimum of 92 percent shall be finer than the upper grain size limit (American Water Works Association 1996, Standard B100-96).

When air scour is to be delivered through the supporting gravel, great danger exists that the gravel will be disrupted during the backwash cycle, especially if air and water are used simultaneously. Two solutions are being used. One is to use a nozzle underdrain with small enough openings so that gravel is not required. The other is to use a double-reverse graded gravel system in which the layers are graded from coarse on the bottom to fine in the middle and back up to coarse on the top. This concept—originally used by Baylis to solve gravel migration problems in large filters in Chicago (American Water Works Association 1971)—is now being used in air and water backwash filters supplied by several companies.

In addition to the systems just described, various new proprietary underdrain systems are being marketed that eliminate the need for supporting graded-gravel layers. The long-term success of these new systems remains to be demonstrated. Since filter failures are often attributed to underdrain deficiencies, adoption of unproven systems is a matter of concern.

THEORY OF RAPID FILTRATION AND MODELING

Mechanisms of Filtration

The mechanisms involved in removal of suspended solids during rapid filtration are complex. Many workers have discussed the various factors that may play an important role in particle removal. A notable early example of such discussions is by O'Melia and Stumm (1967).

During rapid granular bed filtration, particle removal is primarily within the filter bed and is referred to as *depth filtration.* The efficiency of depth removal depends on a number of mechanisms. Some solids may be removed by the simple mechanical process of physical screening if the particle is larger than the smallest opening through which the water flows. Removal of other solids, particularly the smaller ones, involves two steps. First, a transport mechanism must bring the small particle from the bulk of the fluid within the interstices close to the surfaces of the grains. Transport mechanisms may include gravitational settling, Brownian diffusion, interception, and hydrodynamics that are affected by such physical characteristics as size and shape of the filter grains, filtration rate, fluid temperature, and the density, size, and shape of the suspended particles.

Second, as particles approach the surface of a grain, or previously deposited solids on the grain, short-range surface forces must be favorable for attachment to occur. If the particles have been destabilized sufficiently so that electrostatic repulsive forces are minimized, the interaction of these forces and short-range attractive Van der Waal's forces results in a net attractive force, and attachment of the particle to the grain surface (or to previous deposits) can occur. The collision and attachment process is comparable to coagulation of destabilized particles (see Chapter 6).

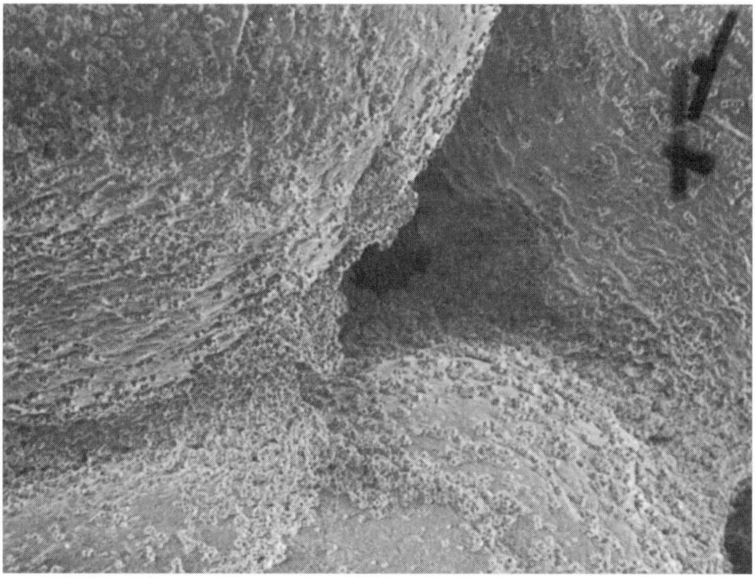

FIGURE 8.10 Scanning electron micrograph of sand grains with collected 5–10 μm silicate particles. Influent suspension was 200 mg/L, coagulated with cationic polymer. Photograph taken after 3 h of filtration at depth of 46 cm, sand grains about 1 mm in size. (*Source:* Courtesy of R. B. Robinson, A. M. Saatci, and E. R. Baumann, Iowa State University. Engineering Research Institute reprint 80106, Particle removal in deep-bed filters observed with the scanning electron microscope, January 1980.)

Figure 8.10 is a scanning electron micrograph of small particles (5 to 10 μm) retained on three filter sand grains about 0.5 to 1.0 mm.

Mathematical Descriptions of Filtration

Filtration models can be grouped in two distinct categories: *phenomenological* (macroscopic) models and *fundamental* (microscopic) models. Both types of models attempt to predict particle removal in the filter, but only the fundamental models consider the mechanisms of particle transport and attachment. Each of these types of models will be described briefly here. More comprehensive presentations are available elsewhere (Herzig, LeClerc, and LeGoff, 1970; Tien, 1989; Elimelech et al., 1995).

Fundamental (Microscopic) Models. Particle removal depends upon a *transport step* and an *attachment step* as discussed earlier. Some researchers have concentrated on developing models for the various possible transport mechanisms that can bring the small particles close to the grain surface so that collision and capture is possible. Flow through the filter is usually laminar, and the small particles are carried along the flow stream lines unless a transport mechanism causes particle transport across the stream lines. The particles to be removed are usually considerably smaller than the size of the grains of the filter media. For example, typical particles range from about 0.1 to 50 μm, while typical filter grains may range from about 500 to 2000 μm.

Yao, Habibian, and O'Melia (1971) presented the following models to describe the three important mechanisms of the filtration process, namely, *sedimentation, interception,* and *diffusion.* These models, which are extensions of work done in air filtration, are useful in indicating the various parameters that are important in transport of particles of different size and density in granular bed filters of different grain size, water velocity, and temperature. Each mechanism is expressed in terms of a theoretical single spherical collector efficiency η, which is the ratio of the number of successful collisions to the total number of potential collisions in the projected cross-sectional area of the collector.

The larger or heavier particles can be transported to the collector by sedimentation. The single collector efficiency for sedimentation transport η_s, is the ratio of the Stokes's settling velocity V_s to the filtration rate (i.e., approach velocity of the flow, V):

$$\eta_s = \frac{V_s}{V} = \frac{(\rho_s - \rho)\, g d_p^2}{18\mu V} \tag{8.9}$$

where d_p is the particle diameter and other terms as defined earlier.

If the stream line carrying a particle passes within $d_p/2$ of the collector surface, the particle can be removed by interception. The single collector model for interception is:

$$\eta_I = \frac{3}{2}\left(\frac{d_p}{d_c}\right)^2 \tag{8.10}$$

where d_c is the collector diameter.

Very small colloidal-sized particles, less than about 1 μm in size, will be moved in a random pattern away from their stream lines by Brownian diffusion. The single collector model for diffusive transport, η_D, which incorporates Einstein's equation for the diffusion coefficient of suspended particles, is:

$$\eta_D = 0.9\left(\frac{KT}{\mu d_p d_c V}\right)^{2/3} \tag{8.11}$$

where K is Boltzman's constant, T is absolute temperature, and the other terms are as defined before.

The overall collector efficiency is the sum of the three individual collector efficiencies:

$$\eta_o = \eta_s + \eta_I + \eta_D \tag{8.12}$$

If the particles are not adequately destabilized, the removal efficiency of the collector will be less than η_o. An empirical collision efficiency factor α is applied to account for this, so that the actual single collector removal efficiency is:

$$\eta = \alpha\, \eta_o \tag{8.13}$$

With adequate destabilization, $\alpha = 1$ since repulsive forces are eliminated.

These relationships predict that collection of particles will be hindered by higher filtration rates (V^{-1} or $V^{-2/3}$) (except for interception which is independent of velocity); by larger grain sizes (d_c^{-2} or $d_c^{-2/3}$); or by colder waters of higher viscosity (μ^{-1} or $\mu^{-2/3}$). Capture is also hindered by smaller and lighter particles (d_p^2 and $\rho_s - \rho$) unless they are small enough for diffusive transport ($d_p^{-2/3}$). These observations based on theoretical models are in general agreement with actual observations of filtration and with intuition. Higher rates of filtration, larger grain sizes, and colder waters yield poorer filtration. Because real suspensions contain a range of particle sizes, all three mechanisms may be effective in removing some of the particles, so the impact of particle size on efficiency is less easily observed.

This fundamental modeling approach has been extended to determine the rate of particle deposition from the analysis of particle trajectories in the flow field. This approach, known as *trajectory analysis,* has been developed most extensively by Tien and his coworkers (Tien, 1989), and the work is also presented by Elimelech et al. (1995). The trajectory of a particle over a given collector is determined by the various forces acting on the particle. Trajectory analysis is applied to particles not significantly affected by Brownian motion.

The rigorous use of trajectory analysis to predict particle capture requires correction for hydrodynamic interaction as the particle approaches a grain surface. These interactions retard the rate of deposition of particles. Based on the results of trajectory analysis, Rajagopalin and Tien (1976) obtained an approximate expression for η_o from the observed dependence of η_o on various dimensionless transport-related parameters. That expression, as modified and presented by Elimelech et al. (1995), is as follows:

$$\eta_o = 4.0A_s^{1/3}\left(\frac{D_\infty}{Vd_c}\right) + A_sN_{LO}^{1/8}R^{15/8} + 3.38 \times 10^{-3}A_sN_G^{1.2}R^{-0} \tag{8.14}$$

where the following new symbols are defined, with other symbols as defined before:

A_s = a dimensionless flow parameter to account for the effect of neighboring collectors in the flow field around a single collector = 38 for a sphere-in-cell model and a typical porosity of 0.4 (Elimelech et al., 1995)

D_∞ = the bulk particle diffusion coefficient, calculated from the Stokes-Einstein equation = $KT/3\Pi d_p\mu$

N_{LO} = the dimensionless van der Waals number = $4A/9\Pi\mu d_p^2V$, with A being the Hamaker constant of the interacting media

R = a dimensionless ratio of particle to collector size = d_p/d_c

N_G = a dimensionless gravitational force number equal to the right-hand side of Equation 8.9.

Equation 8.14 can be used to calculate single collector removal efficiencies for particles of different sizes after selecting appropriate values for the other constants and variables. The single collector efficiency can then be related to the removal efficiency of a clean bed of uniform spheres with the following equation (Yao Habibian and O'Melia, 1971):

$$\ln\left(\frac{C}{C_o}\right) = -\frac{3}{2}\,\frac{(1-\varepsilon)\,\alpha\eta_oL}{d_c} \tag{8.15}$$

where C = effluent particle concentration

C_o = influent particle concentration

L = bed depth, with other symbols as defined before

The log removal efficiency is the negative log of C/C_o. Elimelech et al. (1995) used Equations 8.14 and 8.15 to illustrate the effect of particle size on removal efficiency (Figure 8.11) and the effect on log removal efficiency of filtration rate (Figure 8.12), grain size (Figure 8.13), and bed depth (Figure 8.14). It should be emphasized that these theoretical curves are for a clean bed of uniformly sized spheres and that these equations can not be used to predict changes in efficiency with time as the bed becomes clogged with removed particulates. These figures

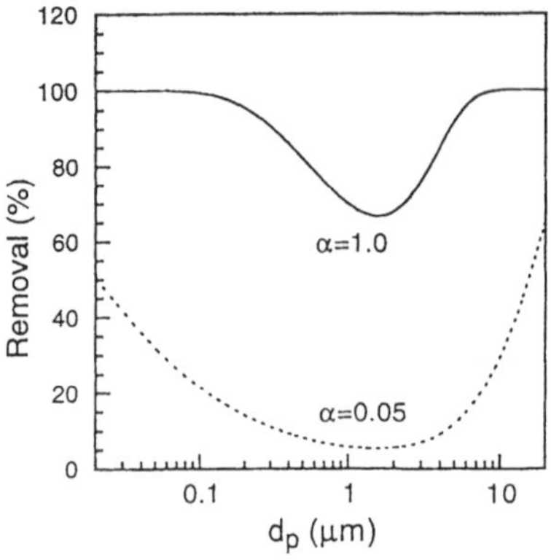

FIGURE 8.11 Removal efficiency of particles as a function of particle size, calculated for filtration at two different attachment (collision) efficiencies. Parameters used in the calculations: $V = 0.14$ cm/s, $L = 60$ cm, $\epsilon = 0.4$, $d_c = 0.5$ mm, $T = 293$K, $A = 10E-20$ J, $\rho_p = 1.05$ g/cm^3. (*Source:* Elimelech et al., reproduced from *Particle Deposition and Aggregation Measurement, Modelling and Simulation,* 1995, Butterworth-Heinemann Ltd., Oxford, UK.)

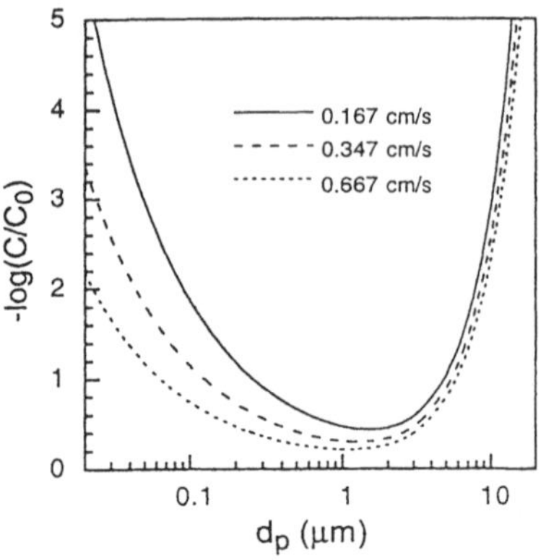

FIGURE 8.12 Log removal of particles as a function of particle diameter, calculated for three different filtration rates (approach velocities). Other parameters used in calculation same as in Figure 8.11, and $\alpha = 1.0$. (*Source:* Elimelech et al., reproduced from *Particle Deposition and Aggregation,* 1995, by permission of the publisher, Butterworth-Heinemann.)

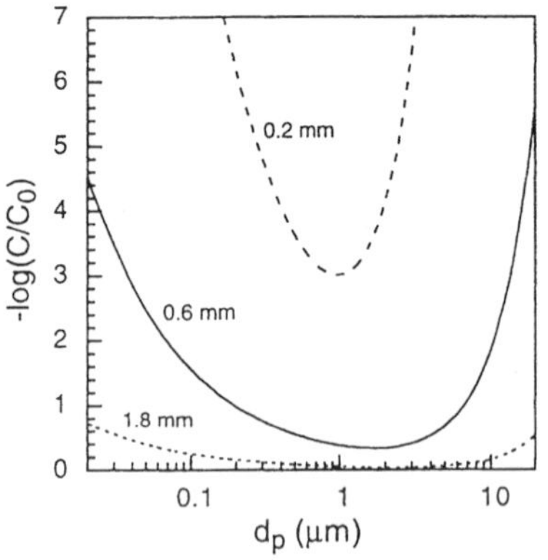

FIGURE 8.13 Log removal of particles as a function of particle diameter, calculated for three different grain diameters. Other parameters used in calculation same as in Figure 8.11, and $\alpha = 1.0$. (*Source:* Elimelech et al., reproduced from *Particle Deposition and Aggregation,* 1995, by permission of the publisher, Butterworth-Heinemann.)

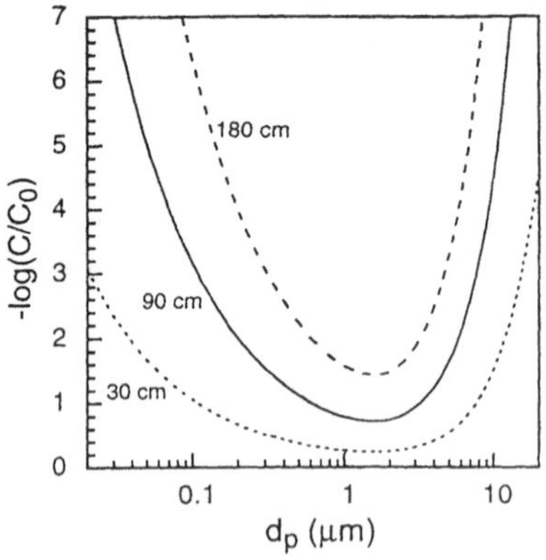

FIGURE 8.14 Log removal of particles as a function of particle diameter, calculated for three different filter bed depths. Other parameters used in calculation same as in Figure 8.11, and $\alpha = 1.0$. (*Source:* Elimelech et al., reproduced from *Particle Deposition and Aggregation,* 1995, by permission of the publisher, Butterworth-Heinemann.)

show that the lowest efficiency is evident for particles about 1 μm in size, with better removal for larger-size particles (by gravity and interception) and better removal for smaller-size particles (by diffusion), as previously reported by Yao, Habibian, and O'Melia (1971).

Phenomenological (Macroscopic) Models. Phenomenological models make no attempt to consider the mechanisms of particle transport or attachment. Rather, a mass balance equation and an empirical rate expression are combined to describe the rate of particle deposition as a function of time of filtration and depth in the filter medium.

The mass balance equation merely states that the mass or volume of the particles removed from suspension in the filter must result in an equal mass or volume of accumulated solids in the pores. Mathematically;

$$V\left(\frac{\delta C}{\delta z}\right)_t + \left(\frac{\delta(\sigma + \varepsilon C)}{\delta \theta}\right)_z = 0 \tag{8.16}$$

where z = depth into the bed in the direction of flow
θ = corrected time as in the following equation

$$t - \int_o^z \frac{dz}{V/\varepsilon} \tag{8.17}$$

σ = specific deposit (volume of deposited solids/filter volume)
ε = porosity
C = concentration of particles in suspension (volume of particles/suspension liquid volume)

The σ and C terms can also be expressed in mass/volume units. In most cases, the change of the concentration of solids in the pores (εC) is trivial compared to the change in the amount of deposited solids, so the equation is often simplified by omitting the (εC) term resulting in the form originally presented by Iwasaki (1937).

The rate expression describes a first-order removal, with depth proportional to the local particle concentration in the fluid ($\delta C/\delta z = -\lambda C$) as originally presented by Iwasaki (1937), with the proportionality constant, λ, being called the *filter coefficient.* Combining this expression with the simplified form of Equation 8.16 (i.e., without the εC term) results in Equation 8.18 as presented by Tien (1989).

$$\frac{\delta \sigma}{\delta \theta} = \lambda V C \tag{8.18}$$

where λ varies with the amount of specific deposit (σ) starting with λ_o for the clean bed.

Various empirical rate expressions have been proposed to relate λ to λ_o as some function of σ. Ten such expressions have been summarized by Tien (1989, p. 25). Two examples will illustrate the evolution of such equations. In 1967, Heertges and Lerk (1967) proposed the following:

$$\frac{\delta \sigma}{\delta t} = k_1 (\varepsilon_o - \sigma) VC \tag{8.19}$$

where ε_o = clean-bed porosity
k_1 = empirical attachment coefficient (same as λ_o)

This equation states that the rate of particle deposition inside a filter lamina at a given time is proportional to the particle flux VC and the volume remaining available for deposition ($\varepsilon_o - \sigma$).

Adin and Rebhun proposed a rate expression in 1977 that includes a second term to allow for detachment of already deposited particles (Adin and Rebhun, 1977), i.e.,

$$\frac{\delta\sigma}{\delta t} = k_1 VC(F - \sigma) - k_2 \sigma J \tag{8.20}$$

where k_2 = empirical detachment coefficient
F = theoretical filter capacity (i.e., the amount of deposit that would clog the pores completely)
J = hydraulic gradient, and other terms as defined previously.

The first term is almost identical to the previous expression, but a new filter capacity term F is proposed rather than the clean-bed porosity ε_o. The second term states that the probability of detachment depends on the product of the amount of material deposited already σ and the hydraulic gradient J.

The mass balance equation and one rate expression must be solved simultaneously in order to find C and σ as a function of time of filtration in the cycle and depth in the filter. Various numerical or analytical solutions have been utilized to solve the equations (Heertges and Lerk, 1967; Adin and Rebhun, 1977; Saatci and Oulman, 1980; Tien, 1989). One of the analytical solutions (Saatci and Oulman, 1980) was developed based on the close similarity between the Langmuir equation for adsorption and the filtration rate expression.

The use of these phenomenological models depends upon the collection of pilot or full-scale filtration data to generate the appropriate attachment and, in some cases, detachment coefficients for the particular suspension being filtered. This need, plus the complex nature of the solutions, has limited the use of such equations in routine filter design. Nevertheless, the use of such models with the appropriate empirical constants is acceptable in predicting filtrate quality, specific deposit, and head loss as a function time and depth.

The two rate expressions presented (Equations 8.19 and 8.20) draw attention to a long-standing question in filtration research, namely, whether detachment of deposited solids occurs in a filter operating at a constant filtration rate, or new influent particles merely bypass previously clogged layers. Evidence indicates that detachment does occur as a result of impingement of newly arriving particles (Clough and Ives, 1986). This evidence was collected using an industrial endoscope inserted into the filter to magnify, observe, and record the deposition process on videotape. The observations were made while filtering kaolin clay on sand filters but without the use of coagulant. Similar unpublished observations (Ives, 1997) were made while filtering a 100 mg/L kaolin suspension that had been coagulated and flocculated with 25 mg/L of alum ($Al_2(SO_4)_3 16H_2O$). Detachment was observed during constant-rate filtration, caused by the impact of arriving particles on unstable deposits, and by hydraulic shear stresses without the influence of arriving particles. Further studies of detachment were presented by Ginn, Amirtharajah, and Karr (1992), who indicated that detached aggregates in the filter effluent could distort log removal results for cyst-sized particles obtained by particle counting.

Effect of Filtration Variables on Performance

Laboratory and pilot scale studies conducted under controlled conditions have evaluated the effects of filtration rate, grain size, and viscosity (Cleasby and Baumann, 1962; Ives and Sholji, 1965; Hsiung and Cleasby, 1968). Ives and Sholji (1965) determined empirically that the filter coefficient λ was inversely proportional to the filtration rate, grain size, and the square of the viscosity. Comparison of these observations with Equations 8.9, 8.10, and 8.11 shows that the filtration rate exponent agrees with Equation 8.9. While the grain size and viscosity exponents do not agree precisely with the three equations, they are qualitatively in agreement.

RAPID FILTER PERFORMANCE

General Pattern of Effluent Quality

The quality of filtered water is poorer at the beginning of the filtration cycle and may also deteriorate near the end of the cycle if the cycle is prolonged for a sufficient time period. The particles removed in the filter are held in equilibrium with the hydraulic shearing forces that tend to tear them away and carry them deeper into, or through, the filter. As deposits build up, the velocities through the more nearly clogged upper layers of the filter increase, and these layers become less effective in removal. The burden of removal passes deeper and deeper into the filter (Eliassen, 1941; Stanley, 1955; Ling, 1955). Ultimately, inadequate clean bed depth is available to provide the desired effluent quality, and the filter run must be terminated because of deterioration of the filtrate quality. This deterioration is called *breakthrough,* but it should be emphasized that some particles pass through the filter continuously throughout the filter cycle.

If the filtration rate on a filter that contains deposited solids is suddenly increased, the hydraulic shearing forces also suddenly increase. This disturbs the equilibrium existing between the attachment forces holding the deposited solids and the hydraulic shearing forces, and some solids will be dislodged to pass out with the filtrate. This aspect of filtration will be discussed in detail later.

Cleasby and Baumann (1962) presented detailed data on the filtered water quality early in the filter run, observed during filtration of precipitated iron (Figure 8.15). The filtrate was observed to deteriorate for a few minutes and then improve over about 30 min before reaching the best level of the entire run. Thereafter, as the filter cycle progressed, deterioration of filtrate quality occurred at the highest filtration rate of 6 gpm/ft^2 (15 m/h) (Figure 8.16). Deterioration of filtrate later in the filter run did not occur, however, at 2 and 4 gpm/ft^2 (5 and 10 m/h), nor in a constant head loss run where the filtration rate declined from 6 gpm/ft^2 (15 m/h) at the beginning of the run to about 4 gpm/ft^2 (10 m/h) at the end of the run (Figure 8.16).

The initial peak occurred at approximately the theoretical detention time of the filter plus the detention time of the effluent appurtenances to the point of turbidity monitoring. Amirtharajah and Wetstein (1980) have used these data and their own data to formulate a "two-peak" conceptual model for the initial degradation and improvement period. The first of the two peaks results from the residual backwash water remaining within the filter media at the end of the backwash, and the second peak results from backwash water above the media at the end of the backwash.

Similar observations of water quality during the filter cycle have been presented by Robeck, Dostal, and Woodward (1964) during surface water treatment, shown in Figures 8.17 and 8.18. During strong floc conditions, the filtrate quality was nearly constant after the initial improvement period (Figure 8.17). During weak floc conditions, however, the terminal turbidity breakthrough was rather abrupt (Figure 8.18) and occurred at low head loss. Figures 8.17 and 8.18 represent the two most typical patterns

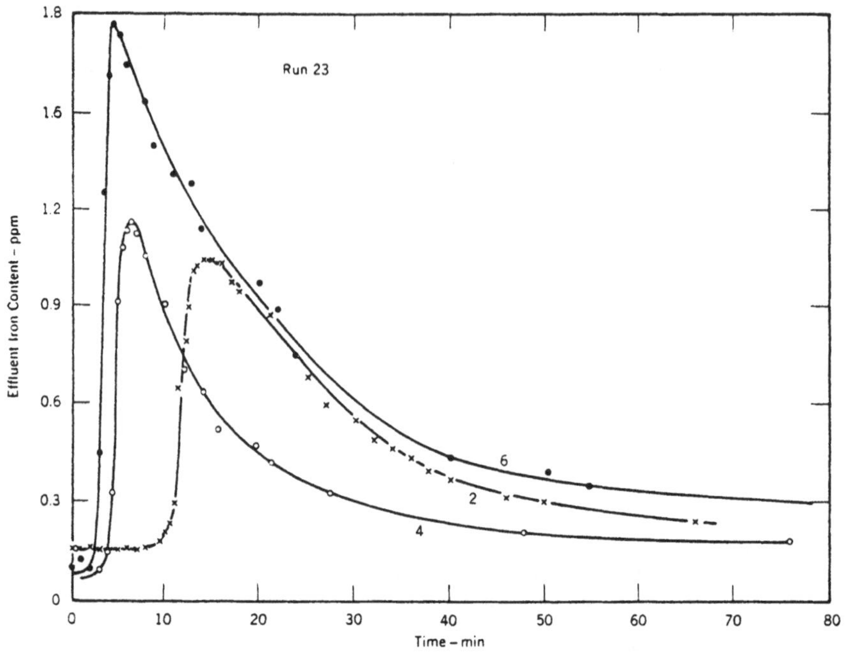

FIGURE 8.15 Initial effluent improvement. After water in filter has been displaced, turbidity drops to minimum. Numbers by curves indicate filtration rate in gallons per minute per square foot. (*Source:* J. L. Cleasby and E. R. Baumann, "Selection of Sand Filtration Rates," *Jour. AWWA,* 54(5) May 1962: 579.)

of filtrate quality observed. In some cases, however, the filtrate quality may improve throughout the run, and no terminal breakthrough is observed (Cleasby, 1969b).

The Value of Filtering-to-Waste

The initial water quality degradation period also has been demonstrated in studies using *Giardia* cysts (Logsdon et al., 1981). *Giardia muris* was used as a model for the human pathogen, *Giardia lamblia. G. muris* was spiked into a low turbidity surface water, coagulated with alum alone or alum and nonionic polymer, flocculated, and filtered through granular media filters. Initial concentration of cysts in the filtrate were from 10 to 25 times higher than those following the initial improvement period, even though the turbidity improved less than 0.1 ntu during the period.

In conventional water treatment practice, passage of turbidity during the initial degradation and improvement (i.e., "ripening") period is small when averaged over the entire filter run, and compared to the total turbidity passed during the run. For many years, little attention was paid to the impact of the initial period on the average filtrate quality, on the assumption that disinfection would control any pathogenic microorganisms that passed through the filters. Consequently, based on this assumption, the early practice of filtering-to-waste at the beginning of the filter cycle was largely abandoned. However, there is renewed interest in filtering-to-waste because of

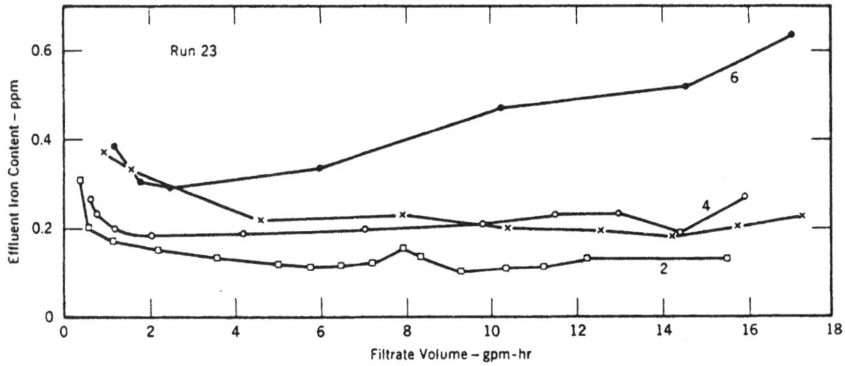

FIGURE 8.16 Effluent iron content and filtrate volume. Observations for curves were made after initial improvement period. Numbers by curves indicate filtration rate. Curve along "X" marks indicates constant pressure. Rate of this run started at 6 gpm/ft² and decreased to 4 gpm/ft² as head loss increased. (*Source:* J. L. Cleasby and E. R. Baumann, "Selection of Sand Filtration Rates," *Jour. AWWA,* 54(5), May 1962: 579.)

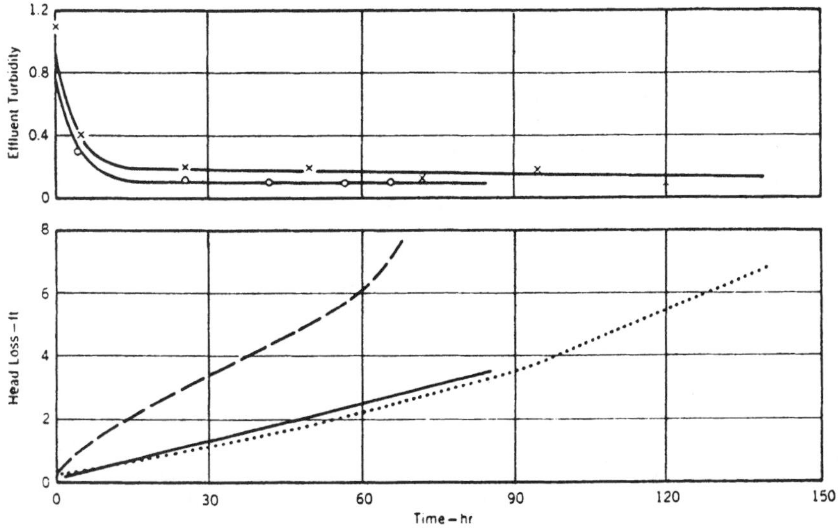

FIGURE 8.17 Effect of filter media on length of run with strong floc (summer conditions). The data shown were obtained under the following operating conditions: raw-water turbidity, 80 units; settling tank effluent turbidity, 2 units; alum dose, 75 mg/L; filtration rate, 2 gpm/ft². In the upper graph, the curve determined by the "X" points is for the medium consisting of both coal and sand; and the curve determined by the circular points is for both coal alone and sand alone. In the lower graph, the dashed curve is for sand; the solid, coal; and the dotted, coal and sand combined. (*Source:* G. G. Robeck, K. A. Dostal, and R. L. Woodward, "Studies of Modification in Water Filtration," *Jour. AWWA,* 56(2), February 1964: 198.)

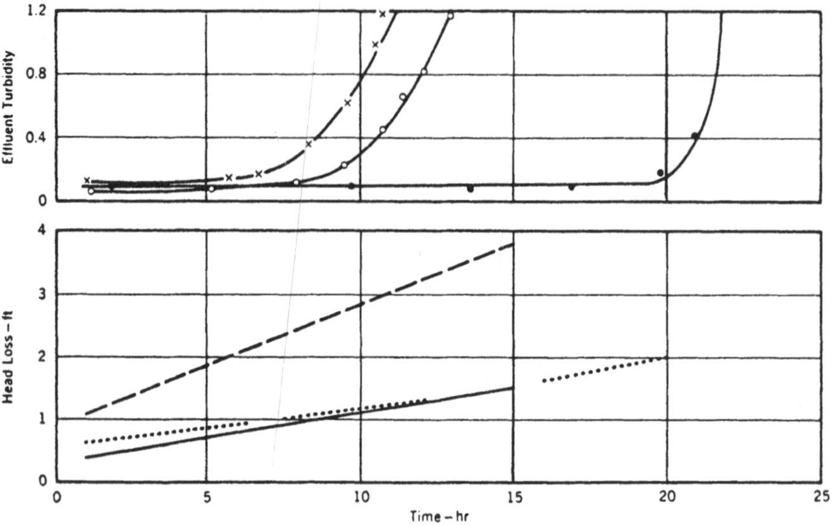

FIGURE 8.18 Effect of filter media on length of run with weak floc. The data shown were obtained under the following operating conditions: source-water turbidity, approximately 20 units; alum dose, 100 mg/L; activated carbon, 2 mg/L; filtration rate, 2 gpm/ft²; settling tank effluent turbidity, 15 units. In the upper graph, the curve determined by the "X" points is for coal; that by open circles, sand; and that by solid circles, coal and sand. In the lower graph, the dashed curve is for sand; the dotted, coal and sand; and the solid, coal. (*Source:* G. G. Robeck, K. A. Dostal, and R. L. Woodward, "Studies of Modification in Water Filtration," *Jour. AWWA,* 56(2), February 1964: 198.)

increasing outbreaks of *Giardiasis* and *Cryptosporidiosis* caused by protozoan cysts that are very resistant to chlorine disinfection (Hibler et al., 1987; Kramer et al., 1996). Provision of the "filter-to-waste period" is now strongly encouraged because of the low infective dose for these protozoan pathogens (Kirner, Littler, and Angelo, 1978; Kramer et al., 1996). (See Chapters 2 and 18.)

The Value of Continuous Turbidity Monitoring and Particle Counting

From the foregoing discussion of effluent quality patterns, the value of providing continuous monitoring of the filtrate quality of each filter becomes apparent. It is useful to observe the length of the initial improvement period to guide the duration of filtering-to-waste. It is also useful to detect the onset of terminal breakthrough. The cycle could then be terminated at the onset of breakthrough, even if the head loss has not reached the normal maximum available at the plant. This would be especially important if the passage of *Giardia* cysts, *Cryptosporidium* oocysts, viruses, or asbestos fibers are of concern, because increases in turbidity indicate simultaneous increases in other particulates, often of larger relative magnitude than the increases in turbidity. For example, Figures 8.19 and 8.20 indicate increases in viruses and asbestos fibers, coinciding with breakthrough in turbidity (Robeck, Dostal, and Woodward, 1964; Logsdon, Symons, and Sorg, 1981). In Figure 8.20 the fiber count increases much more than the turbidity. Similar evidence in Figure 8.21 shows the change in concentration of *Giardia* cysts during a turbidity breakthrough and fol-

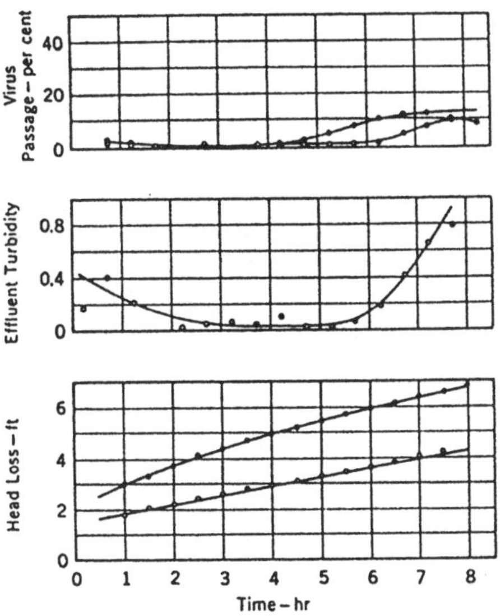

FIGURE 8.19 Virus and floc breakthrough at high rate on coarse sand or coal. The curves with solid circles represent sand filters. Those with open circles, coal plus sand. The filtration rate was 6 gpm/ft^2, with blended water containing an alum dose of 10 mg/L. The virus load was 8400 PFU/mL; turbidity load, 10 Jackson units. (*Source:* G. G. Robeck, K. A. Dostal, and R. L. Woodward, "Studies of Modification in Water Filtration," *Jour. AWWA,* 56(2), February 1964: 198.)

lowing the backwash operation (Logsdon et al., 1985). In this case a threefold change in turbidity corresponds to a 30- to 40-fold change in concentration of cysts.

Large numbers of very small particles can exist in filtered water with turbidity less than 1 ntu. This has been demonstrated for asbestos fibers (Logsdon and Symons 1977), bacteria (O'Connor et al., 1984), *Giardia* cysts (Logsdon et al., 1981) and *Cryptosporidium* oocysts (Patania et al., 1995). Because small changes in filtered water turbidity can be accompanied by large increases in microscopic particles, water utilities are increasingly applying on-line particle counters to monitor filtered water quality. Particle counters can detect increases in particle concentration that may not be associated with increases in filtrate turbidity, and so can provide plant operators with early warning of incipient turbidity breakthrough. This is especially true for low turbidity waters below 0.1 ntu because the turbidity measurement is reaching its lower limit of detection. The use of particle counting can improve monitoring and control of the filtration process.

The Importance of Adequate Pretreatment

Adequate and continuous pretreatment is absolutely essential to producing good filtrate. Any interruption of good pretreatment results in almost immediate deteriora-

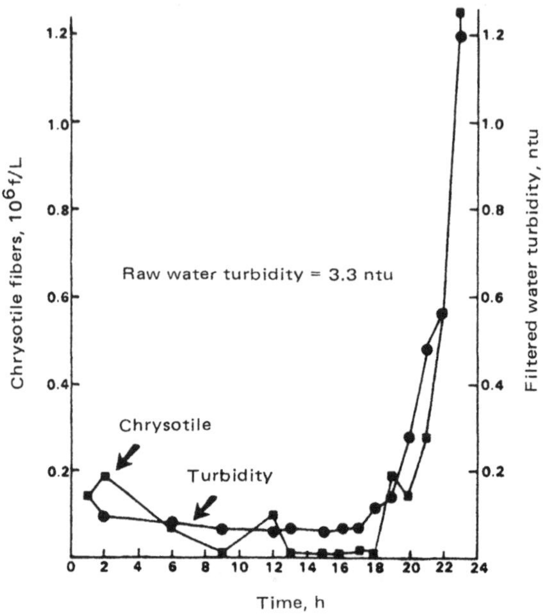

FIGURE 8.20 Finished water chrysotile counts and turbidity versus time-run #120, Seattle pilot plant. (*Source:* G. S. Logsdon, J. M. Symons, and T. J. Sorg, "Monitoring Water Filters for Asbestos Removal," *J. Environ. Eng.* 107(6):1297. Copyright © 1981 American Society of Civil Engineers. Reproduced by permission of ASCE.)

tion of filtrate quality. Fuller's famous work (1898) recognized this as evidenced by this quotation from his 1898 report:

"In all cases experience showed that for successful filtration the coagulation of the water as it enters the filter must be practically complete."

Robeck, Dostal, and Woodward (1964) had a good illustration of the effect of a loss of adequate pretreatment from a study at Gaffney, South Carolina (Figure 8.22). Two filters at 2 and 5 gpm/ft^2 (5 and 25 m/h) were operating in parallel and producing about equal turbidity in the filtrate. However, when the quality of the source water suddenly worsened and coagulation was thereby upset, the effluent of both filters deteriorated sharply. One other paper demonstrated similar findings in a direct filtration pilot study as shown in Figure 8.23 (Trussel et al., 1980). In this case, the coagulant feed was discontinued for 30 min, and filtrate turbidity rose sharply until the chemical feed was resumed.

Thus, all precautions must be taken to ensure that the chemical dosage is adequate and that the feed is reliably maintained. This is even more critical in direct filtration because of the short detention time ahead of the filters.

Detrimental Effects of Rate Increases and Restarting Dirty Filters

If the rate of filtration on a dirty filter is suddenly increased, this disturbs the equilibrium that previously existed between the attachment forces holding the solids in

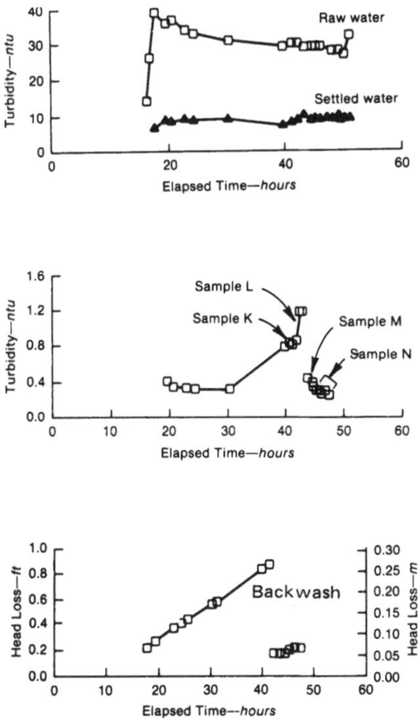

FIGURE 8.21 Filter performance before and after a backwash operation during conventional treatment using alum coagulation of a surface water. Turbidity breakthrough at low head loss is reflected in higher cyst concentrations, which were reduced after the backwash. With continuous feed of *Giardia muris* at 11,400 cysts/L, the effluent cyst concentrations per L were $K = 440$, $L = 240$, $M = 8.7$, and $N = 14.5$. (*Source:* G. S. Logsdon, V. C. Thurman, E. S. Frindt, and J. G. Stoecker, "Evaluating Sedimentation and Various Filter Media for Removal of Giardia Cysts," *Jour. AWWA,* 77(2), February 1985: 61.)

the filter and the hydraulic shearing forces tending to dislodge those solids. The result is a temporary flushing of solids deeper into the filter and into the filtrate.

Evidence of this phenomenon was presented by Cleasby, Williamson, and Baumann in 1963, showing that the amount of material flushed through the filter was greater for sudden increases in filtration rate than for gradual increases in rate (Figures 8.24 and 8.25). The amount of material released was greater for large increases than for small increases, but the amount was not affected by the duration of the maximum imposed rate of filtration. Different types of suspended solids encountered at different water plants exhibited different sensitivities to increases in filtration rate. These observations are important in many water plant operational decisions.

The detrimental impact of sudden rate increases on dirty filters also is evident in the work of Tuepker and Buescher (1968) and in the studies of DiBernardo and

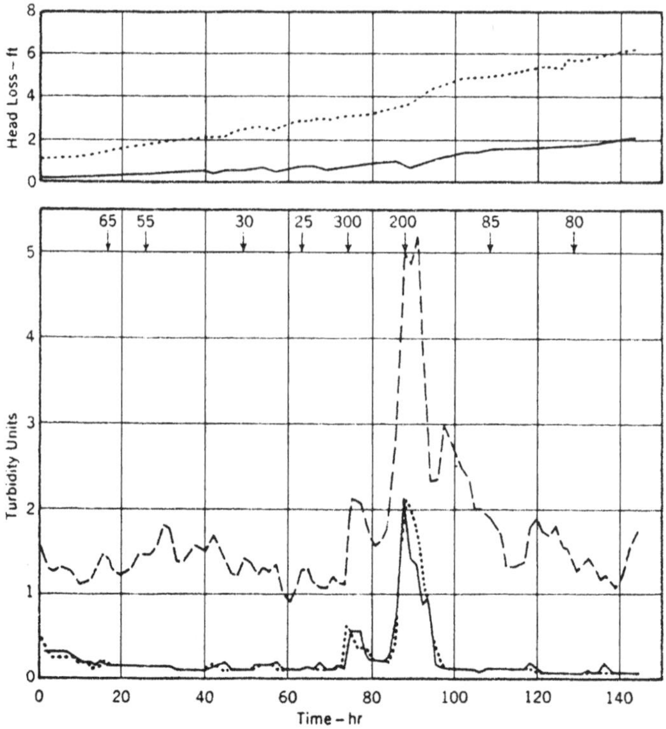

FIGURE 8.22 Turbidity values at filtration rates of 2 and 5 gpm/ft^2. In the upper graph, the dotted curve is for 5 gpm/ft^2; the solid, 2 gpm/ft^2. In the lower graph, the numbers at the top are the raw-water turbidity values at the time shown during the run; the dashed curve is for filter influent; the dotted for filter effluent at 2 gpm/ft^2; and the solid for filter effluent at 5 gpm/ft^2. (*Source:* G. G. Robeck, K. A. Dostal, and R. L. Woodward, "Studies of Modification in Water Filtration," *Jour. AWWA,* 56(2), February 1964: 198.)

Cleasby (1980). Filtration rate increases on dirty filters should be avoided or made gradually (over 10 min).

Studies by Logsdon et al. (1981) showed similar effects when the filtration rate was suddenly increased from 4.5 to 11 gpm/ft^2 (11 to 27 m/h). Turbidity in the effluent rose sharply and then rapidly declined. Concentrations of *G. muris* cyst followed the turbidity trends. "A fourfold increase in turbidity was accompanied by a 25-fold increase in the cyst concentration in the filtered water." The same study demonstrated the detrimental effect of loss of coagulant feed, and of extending the filter run into the period of terminal breakthrough. In both instances, large increases in concentration of cysts were observed in the effluent.

Similar disturbances sometimes occur when starting up a dirty filter after brief idle periods, as evident in Figure 8.26. In the investigation of the causes of a *Cryptosporidiosis* outbreak in Georgia, Logsdon, Mason, and Stanley (1990) reported that restarting dirty filters resulted in filtered water turbidities of 0.5 to 5 ntu that persisted for 1 to 1.5 hours. In contrast, filters operated continuously after backwashing had low turbidities of 0.1 to 0.2 ntu. As a result, it was recommended that dirty filters should not be restarted.

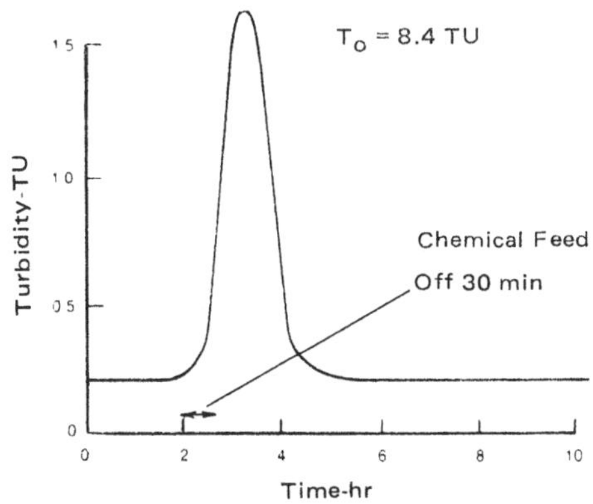

FIGURE 8.23 Influence of loss of chemical feed on filter performance. Direct filtration; no flocculation; 15 m/h (6 gpm/ft^2); influent turbidity, 8.4 TU; chemical feed, 2 mg/L polymer and 2 mg/L alum (shut down for 30 min 2 h into operation). (*Source:* R. R. Trussell, A. R. Trussell, J. S. Lang, and C. H. Tate, "Recent Development in Filtration System Design," *Jour. AWWA,* 72(12), December 1980: 705.)

Filters with automatic rate controllers sometimes exceed the target rate on startup and disturb previously deposited solids. Thus, systems that provide no mechanical effluent rate manipulation are attractive (Dibernardo and Cleasby, 1980; Cleasby, 1969a). Nonmechanical filter control systems will be discussed later.

The Detrimental Effects of Negative Head

Some filter arrangements and operating practices can result in pressures below atmospheric pressure (i.e., negative head) in the filter medium during a filter cycle. This can occur in gravity filters when the total head loss down to any point in the filter medium exceeds the static head (i.e., water depth) down to that point. Negative head is more likely to occur if gravity filters are operated with low submergence of the medium, and if the filter effluent exits to the clearwell at an elevation below the top elevation of the filter medium.

Negative head is undesirable because dissolved gases in the influent water may be released in the zone of negative pressure, causing gas bubbles to accumulate (a process called *air binding*) during the filter cycle. Such accumulations of gas lead to more rapid development of head loss and poorer quality of filtrate because the flow velocity is higher around the gas pockets. Negative head can be completely avoided by terminating the filter run before the total head loss reaches the submergence depth of the medium, or by causing the effluent to exit at or above the surface of the filter medium.

Head Loss Development During a Filter Run

The rate of increase in head loss during the filter run is roughly proportional to the rate of solids capture by the filter. Assuming essentially complete capture of incoming solids, the head loss will develop in proportion to the filtration rate, *V,* and the

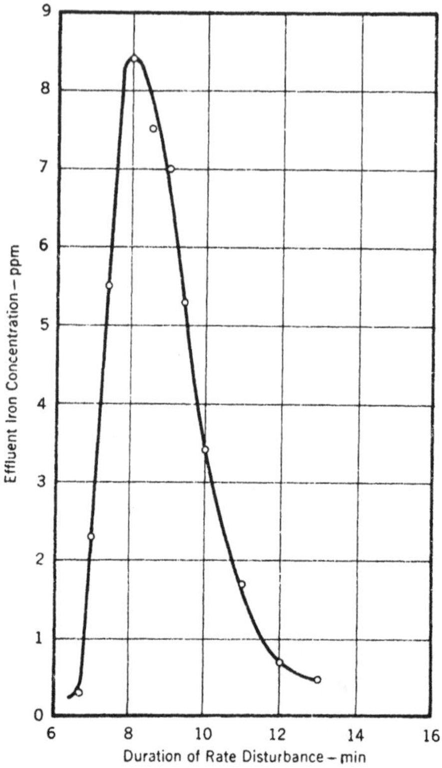

FIGURE 8.24 Effect of typical rate disturbance on effluent quality. The iron concentration in the filter effluent builds up rapidly after the disturbance has been initiated. The curve represents Run 9a, which had an instantaneous rate change from 2 to 2.5 gpm/ft^2. (*Source:* J. L. Cleasby, M. M. Williamson, and E. R. Baumann, "Effect of Rate Changes on Filtered Water Quality," *Jour. AWWA*, 55(7), July 1963: 869.)

influent suspended solids concentration, C_o. The rate of head loss development is reduced if the solids capture occurs over a greater depth of the medium, rather than in a thin upper layer of the medium. A coarser grain size encourages greater penetration of solids into the bed, and possibly different deposit morphology, thereby reducing the rate of head loss development per unit mass of solids captured.

The most common head loss pattern encountered in rapid filtration is linear (or nearly linear) with respect to volume of filtrate, as shown in Figure 8.27. This linear head loss would be typical for most alum or iron coagulated waters.

An accelerating head loss pattern per unit filtrate volume is less commonly observed. This pattern is caused by partial capture of solids in a surface cake and/or by very heavy solids removal within a thin layer of the medium at the top (Cleasby and Baumann, 1962). In this case, increasing the filtration rate reduces the exponential tendency by encouraging greater penetration of solids into the medium, and increases the production to a given head loss. Quality of filtrate also must be observed, however, if such increases in filtration rate are being considered.

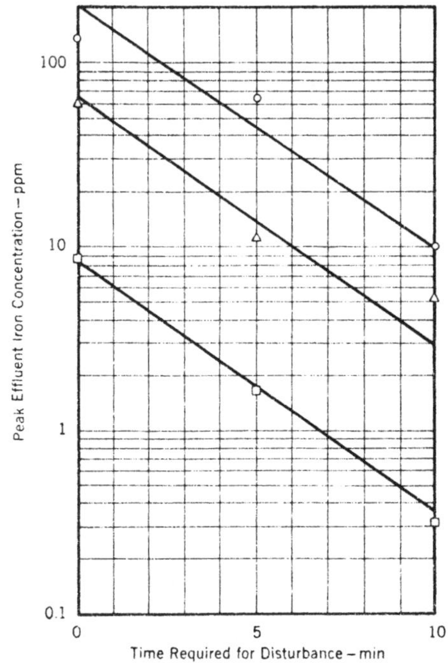

FIGURE 8.25 Peak concentration against disturbance time. The curves indicate a first-order relationship between peak concentration and time required to make the disturbance. At a base rate of 2 gpm/ft^2, circles indicate 100 percent increase; triangles, 50 percent increase; and squares, 25 percent increase. (*Source:* J. L. Cleasby, M. M. Williamson, and E. R. Baumann, "Effect of Rate Changes on Filtered Water Quality," *Jour. AWWA,* 55(7), July 1963: 869.)

In the absence of plant- or pilot-scale data to assist in predicting development of head loss, experimental values of mass of solids captured per unit filter area per unit increase in head loss are sometimes used for prediction of clogging head loss. This is an admittedly simplistic concept. The values used depend on the density of the solids, the ES of the medium where flow enters the filter, and the filtration rate. Typical values range from 0.035 to 0.35 lb/ft^2/ft (555 to 5550 g/m^2/m) (Montgomery, 1985), more commonly less than 0.1 lb/ft^2/ft (1580 g/m^2/m) for flocculent solids using typical potable water filter media and filtration rates. This approach can only be applied when a linear head loss versus time pattern is expected.

DIRECT FILTRATION

Process Description, Advantages, and Disadvantages

Direct filtration is a surface water treatment process that includes addition of coagulant, rapid mixing, flocculation, and filtration. In some cases, the flocculation tank is

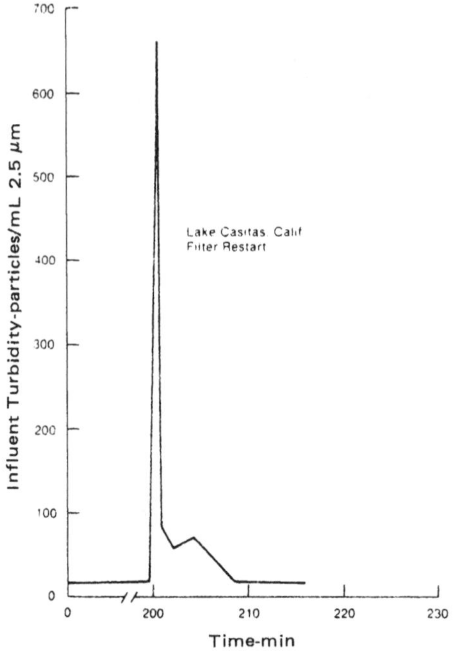

FIGURE 8.26 Influence of stop-start on filter performance. Dual media; 15 m/h (6 gpm/ft^2); influent particles (≥2.5 μm), 2000 particulates/mL; polymer, 1 mg/L; alum, 4 mg/L. (*Source:* R. R. Trussell, A. R. Trussell, J. S. Lang, and C. H. Tate, "Recent Development in Filtration System Design," *Jour. AWWA,* 72(12), December 1980: 705.)

omitted and the process is referred to as *in-line filtration,* with flocculation occurring in the filter itself.

The use of direct filtration for good quality, low-color, surface waters is increasing because it has several advantages over conventional treatment for such waters. Capital costs are lower because no sedimentation tank is required. Lower coagulant dosages are generally used in direct filtration with the goal of forming a pinpoint-sized floc that is filterable, rather than a large settleable floc. Therefore, direct filtration results in lower chemical costs compared to conventional treatment, and lower sludge production, resulting in lower costs for sludge treatment and disposal. Direct filtration also results in lower operation and maintenance costs because the sedimentation tank (and sometimes the flocculation tank) need not be powered or maintained.

Several disadvantages to direct filtration exist, however. It cannot handle waters that are high in turbidity and/or color. Less response time is available for the operator to respond to changes in source water quality, and less detention time is available for controlling seasonal tastes and odors.

Instrumentation is very important for automatic control of a direct filtration plant. Monitoring of source and finished water quality are needed to alert the operator to changes, and a fail-safe plant shutdown mode is desirable in the event the finished water does not meet treatment goals. If taste and odor episodes are expected,

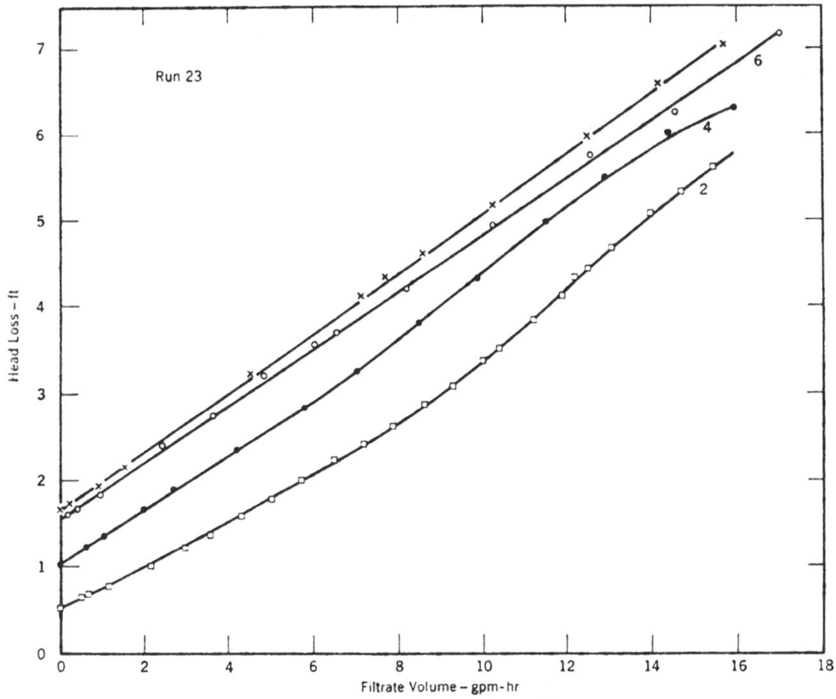

FIGURE 8.27 Total head loss and filtrate volume. No optimum-rate tendency appears. Uncontrolled run (along "X" marks) started at 6 gpm/ft^2 and was allowed to decrease in rate as head loss increased. (*Source:* J. L. Cleasby, and E. R. Baumann, "Selection of Sand Filtration Rates," *Jour. AWWA*, 54(5), May 1962: 579.)

a pretreatment or contact basin may be used to provide the needed contact time with powdered activated carbon or oxidizing chemicals (Conley and Pitman, 1960b).

Appropriate Source Waters for Direct Filtration

The desire to use direct filtration has prompted a number of efforts to define acceptable source waters for the process (Culp, 1977). The key issues in determining appropriate waters for direct filtration are the type and dosage of coagulants needed to achieve desired quality of filtrate as demonstrated by pilot or full-scale observations. Hutchison (1976) suggested that if alum is the primary coagulant, 12 mg/L of alum ($Al_2(SO_4)_3 14H_2O$) on a continuous basis would yield 16- to 20-h filter runs at 5 gpm/ft^2 (12 m/h). This was considered sufficiently long to limit backwash water volume to 4 percent of the product. Waters with up to 25 color units were considered as suitable candidates for direct filtration, with partial substitution of polymer for alum. Hutchison also suggested that diatom levels from 1000 to 2000 areal standard units per mL (asu/mL) required use of coarser anthracite and more frequent use of polymer to prevent breakthrough. One asu is 20 × 20 μm or 400 (μm)^2. Anthracite with 1.5-mm ES could handle diatoms of 2500 asu/mL and produce 12-hour runs at 5 gpm/ft^2 (12 m/h).

Wagner and Hudson (1982) suggested the use of jar tests and filtration through standard laboratory filter paper as a screening technique to select appropriate waters for direct filtration. Waters requiring more than 15 mg/L of alum (as filter alum) to produce an acceptable quality of filtrate were doubtful candidates for direct filtration. Waters requiring only 6 to 7 mg/L of alum and a small dose of polymer were considered favorable candidates. If the results of this bench-scale screening test are favorable, then pilot plant tests are needed to determine full-scale design parameters.

An AWWA report on direct filtration (Committee Report, 1980) defined a water meeting the following criteria as a "perfect candidate" for direct filtration:

Color	< 40 color units
Turbidity	< 5 ntu
Algae	< 2000 asu/mL
Iron	< 0.3 mg/L
Manganese	< 0.05 mg/L

However, the proposed "enhanced coagulation" approach to controlling DBP precursor materials tends to make these criteria obsolete (see Chapter 6).

Cleasby et al. (1984) considered the AWWA committee guidelines acceptable except for turbidity, which was considered too low. During seasons with low algae, they suggested turbidity limits of 12 ntu when using alum alone, or 16 ntu when using cationic polymer alone. During seasons with high algae, they suggested limits of 7 ntu using alum alone and 11 ntu using cationic polymer alone. Best performance was achieved during seasons with low algae with alum dosages between 5 and 10 mg/L. During seasons with high algae, dosages up to 20 mg/L were attempted with only moderate success, and with runs as short as 12 h at 3 gpm/ft^2 (7.3 m/h).

Chemical Pretreatment for Direct Filtration

The selection of the coagulant dosage for direct filtration is best determined by full-length filter cycles using pilot or full-scale filters. Jar tests are often misleading because the goal in direct filtration is to form small pinpoint floc that are barely visible but that will filter effectively. Thus, the usual criteria used to judge jar test observations, such as formation of a large floc or a clear supernatant liquor, after settling, are not appropriate for direct filtration. The jar test and filter paper filtration technique (Wagner and Hudson, 1982) is somewhat better, but it provides no information on terminal breakthrough behavior or rate of generation of head loss. Substitution of a miniature granular bed filter for the filter paper (Bower, Bowers, and Newkirk, 1982) suffers from similar weaknesses. Kreissl, Robeck, and Sommerville (1968) demonstrated that a clean filter could be operated at very high rates to simulate the hydraulic shear that would exist in the more clogged layers of the filter late in the filter run. This would predict whether the chemical dosage would be successful through the full length of the filter run. In spite of these various techniques of using bench or pilot tests to select chemical dosages, the most common approach involves confirmation of the selected dosage with pilot scale or full-scale filter runs of full duration at the design filtration rate.

When using alum or iron salts alone, the optimum dosage of coagulant for direct filtration is the lowest dosage that will achieve filtrate quality goals. An optimum dosage is normally not observed based on filtrate quality alone. Rather, as the dosage is increased, the filtrate quality gets marginally better—but the head loss develops at a higher rate and early breakthrough is encouraged (Cleasby et al., 1984). One advantage of alum or iron coagulants is that the dosage is less sensitive to the quality of the source water.

When using cationic polymer alone as a primary coagulant, a distinct optimum dose may be found that produces the best-quality filtrate. In pilot studies, one method of detecting whether a particular dosage is above or below the optimum is to shut off the polymer feed for a few minutes. If the filtrate improves momentarily, the prior dose is excessive. If the filtrate deteriorates immediately, the prior dose was at or below optimum (Cleasby et al., 1984).

A number of studies have reported successful use of alum and cationic polymers simultaneously. Typical dosages are 2 to 10 mg/L of alum plus about 0.2 to 2 mg/L of cationic polymer (Cleasby et al., 1984; McBride et al., 1977; Tate, Lang, and Hutchinson, 1977; Monscvitz et al., 1978). These dosages are drawn from pilot and plant-scale experience.

The impact of flocculation on direct filtration has been the subject of several studies that have been summarized by Cleasby et al. (1984). These studies can assist the operator in deciding whether flocculation should be provided in direct filtration. Comparing operation with and without flocculation, the provision of flocculation was generally found to (1) improve filtrate quality before breakthrough, (2) shorten the initial improvement period of the filter cycle, and (3) reduce the rate of head loss development, but (4) result in earlier breakthrough. Even though head loss is reduced, the length of the filter run may be shortened because of early breakthrough.

The benefits or detriments of providing flocculation are best determined on a case-by-case basis by pilot or full-scale studies. In some cases, flocculation has been found to be of no benefit (Al-Ani et al., 1986). Nevertheless, current practice usually includes a short flocculation period in direct-filtration plants, typically about 10 min of high-energy flocculation at a velocity gradient (G) of up to 100 sec^{-1}, although the range of flocculation detention times in existing plants has been very great, ranging from no flocculation up to 60 min of flocculation (Letterman and Logsdon, 1976).

In-line filtration is appropriate for source waters that are consistently very low in turbidity and color. If flocculation were provided for such waters, long times of flocculation would be required because of the poor opportunity for collision of particles. For example, flocculation was added to the in-line filtration plant at Las Vegas when the capacity was increased from 200 mgd to 400 mgd (Monscvitz et al., 1978), but 30 min of detention was needed for this excellent source water. Therefore, waters that are low in turbidity and color are more economically treated by the addition of coagulant and an extended rapid mix of three to five min to achieve only the initial stages of particle aggregation before filtration. Few in-line plants are being built, however, because flocculation adds flexibility to the plant operation. A bypass around the flocculator can be provided for contingencies.

Filter Details for Direct Filtration

Two summaries of direct filtration (Committee Report, 1980; Letterman and Logsdon, 1976) indicated filtration rates of full-scale operating plants between 1 and 6 gpm/ft^2 (2.4 and 15 m/h). Direct-filtration pilot studies at three locations showed, however, that effluent turbidity was nearly constant over the range from 2 to 12 gpm/ft^2 (5 to 29 m/h) and in one case as high as 18 gpm/ft^2 (44 m/h) (Trussel et al., 1980). In these studies both alum and cationic polymer were used for chemical pretreatment. Dual media and deep beds of coarse sand were compared.

An unusual direct-filtration plant is operated by the Los Angeles Department of Water and Power to treat water from the Los Angeles Aqueduct. The plant design was based on extensive pilot studies that led to the choice of deep beds of anthracite (6 ft, 1.8 m) with an ES of 1.5 mm and UC < 1.4. The design filtration rate is 13.5 gpm/ft^2 (33 m/h). Pretreatment includes ozone to assist coagulation with ferric chloride and cationic polymer, and flocculation is provided before filtration. The filters are backwashed by air alone first at 4 cfm/ft^2 (1.2 m/min), followed by air plus water at 10 gpm/ft^2 (24 m/h). The air is terminated before overflow occurs to prevent loss of filter medium, and the water wash is continued alone at higher rate during overflow (McBride and Stolarik, 1987). Some of the pilot studies were reported by Trussell et al. (1980). Certainly, this facility represents a marked departure from current direct filtration practice.

FLOW CONTROL IN FILTRATION

Why Control Is Needed

The need for some method of filter flow control has been recognized since the early days of potable water filtration. If no attempt is made to control flow, the filters may not share the total plant flow in a reasonably equal manner, or sudden changes in flow rate may occur, both of which could cause filtrate quality to suffer. For that reason, a number of papers presenting alternative flow control systems have been published (Aultman, 1959; Baylis, 1959b; Hudson, 1959; Cleasby, 1969a; Arboleda, 1974; Arboleda-Valencia, 1985; Cleasby and DiBernardo, 1980; Cleasby, 1981; Committee Report, 1984).

Rate Control Systems for Gravity Filters

Filter control systems currently in use can be divided into two categories: *mechanical* control systems, and *nonmechanical* control systems (that achieve control by the inherent hydraulics of the operating filters). Control systems can also be classified as *equal rate* systems or *declining rate* systems, as was done by Browning (1987). Five control systems are shown in Figure 8.28, using Browning's classification system and nomenclature:

Equal-Rate Systems

- Variable-level influent flow splitting
- Proportional-level influent flow splitting
- Proportional-level equal rate

Declining-Rate Systems

- Variable-level declining rate
- Proportional-level declining rate

It should be noted that none of the systems are called "constant-rate filtration" systems. True constant-rate filtration can only occur if the total plant flow rate is constant and the number of filters in service is constant. This is seldom the case. Plant

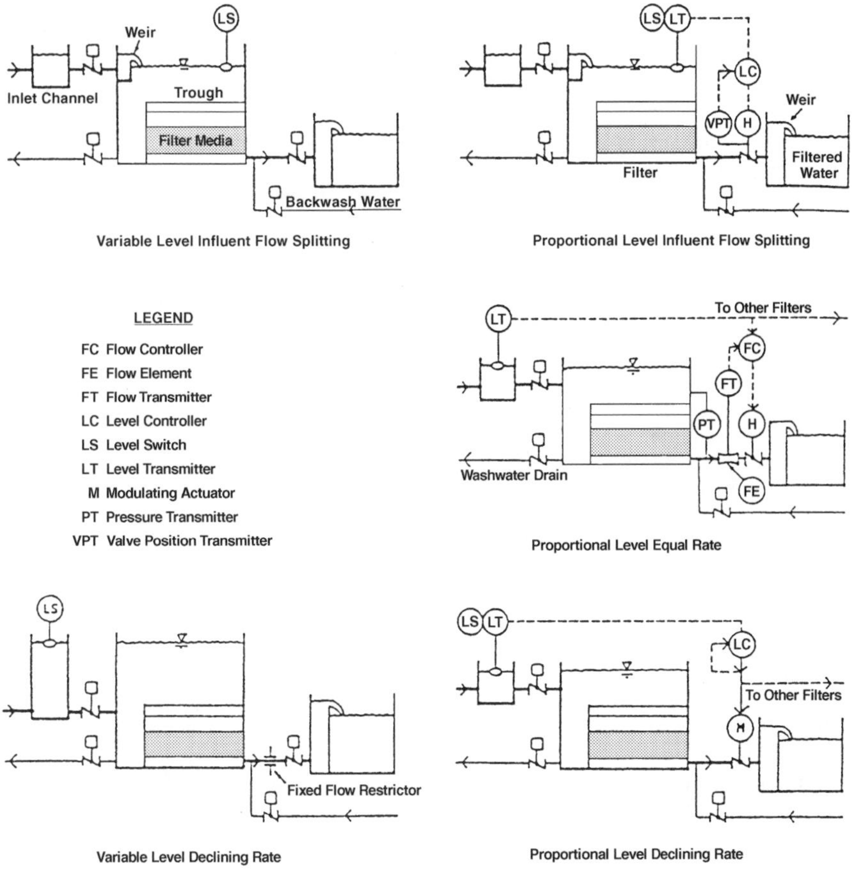

FIGURE 8.28 Schematic plant diagrams for the most common control systems for gravity filters, showing typical control elements. (*Source:* From Browning, 1987, with permission of the Australian Water and Wastewater Association.)

flow rates are varied to meet variations in demand. Removing one filter from service for backwashing forces the other filters to pick up the load, thereby increasing the rate of filtration on the filters remaining in service. Thus, in most water plants, constant-rate filtration is not possible.

The key features of each of these systems are described briefly in the following paragraphs.

Variable-Level Influent Flow Splitting. This is an equal-rate, nonmechanical system in which each filter receives an equal (or near-equal) portion of the total plant flow. This is achieved by splitting the flow by means of an inlet weir box or orifice on each filter inlet above the maximum water level of the filter. The filter effluent discharges to the clearwell at a level above the surface of the filter medium. As solids accumulate in the filter medium, the water level rises in the filter box to provide the head required to drive the flow through the filter medium. The water level in each filter box is different and depends on the extent to which the filter medium is clogged. When the water level in a filter box rises to some selected maximum level, that filter must be backwashed.

No instrumentation is required for flow rate or head loss measurement on the individual filters. This is the simplest system for operators because the head loss is clearly evident merely from observing the water level in the filter box, and there is no control equipment or instrumentation to maintain.

Proportional-Level Influent Flow Splitting. This is an equal-rate mechanical system in which flow control occurs by inlet flow splitting as in the prior system. A flow-splitting weir or orifice at the inlet of each filter is located above the operating water level of the filter. However, each filter requires a level transmitter, a level controller, a modulating valve, and a head-loss instrument. The modulating valve acts to control the water level in the associated filter, not as a flow-rate controller. The degree of opening of the effluent modulating valves is proportional to the water level in the associated filter, being closed when the level is at the bottom of a preselected water level band, and open when the water level is at the top of the band.

Head loss is monitored for backwash initiation. No flow-measurement instrumentation is required for the individual filters.

Proportional-Level Equal Rate. This has been the most common equal-rate mechanical control system. It splits the total plant flow equally among the operating filters, and maintains the water level in the influent channel within a preselected band. Each filter requires a flow meter, a flow controller, a modulating control valve, and a head-loss instrument. A single level transmitter in the inlet channel, serving a bank of several filters, sends an equal signal to the flow controller in each filter that is proportional to the water level within the band, and represents the required flow. The flow controllers adjust the modulating valves to make measured flow match required flow.

When everything is working properly, flow rate through each filter will be equal. However, equipment malfunction often results in unequal flow that often goes undetected. Head loss is monitored for backwash initiation.

Variable-Level Declining Rate. This is a declining-rate nonmechanical system. Flow enters the filter below the normal water level in each filter and discharges to the clearwell above the level of the filter medium. Because the inlet to the filters is below the normal water level in each filter, all filters connected by a common inlet channel or pipe operate at approximately the same water level and thus have the same total head loss available to the effluent weir level at any instant. Therefore, the cleanest filter operates at the highest filtration rate, and the dirtiest filter operates at the lowest filtration rate.

As solids accumulate in the filter media, the water level rises in all connected filters to provide the head required to drive the flow through the filter media. Water level is monitored at a single location for backwash initiation. In some plants, backwash is initiated when the filtration rate of the dirtiest filter drops to a preselected level, or on a time-interval basis. The filtration rate of each filter declines in a stepwise fashion after each backwash. As a clean (backwashed) filter is returned to service, it assumes the highest rate of filtration.

A fixed-flow restrictor (orifice plate or fixed-position valve) is provided for each filter to limit the starting rate on a clean filter. No instrumentation for flow rate or head loss measurement is required on individual filters. However, operating personnel prefer to have some flow-rate indication to help them understand the system and for diagnostic purposes. Less total filter system head loss must be provided because head consumed in turbulent head losses early in the filter run is reallocated to clogging head loss as the filtration rate declines (DiBernardo and Cleasby, 1980; Cornwell et al., 1991).

Proportional-Level Declining Rate. Browning (1987) suggested one additional flow-control system combining the features of declining rate with proportional level control from the level in the filter inlet channel. As in the preceding system, submerged filter inlets result in a common water level in all filters. However, in this system a modulating valve in the effluent of each filter is controlled by the water level in the inlet channel. All valves receive the same set point, and are fully closed when water level in the channel is at the bottom of the band and are at their maximum utilized opening when the water level is at the top of the band. This system functions like a variable declining rate system, but with filter water level controlled in a preselected band width. Browning claimed that the depth of the filter box could be reduced with this system, as compared with the previous declining-rate system.

Common Elements of Filter Control Systems

Some common elements of the various rate-control systems should be recognized. The total filter system head provided to operate the gravity filter is the vertical distance from the water level (hydraulic grade line) in the filter inlet conduit to the water level at the downstream control point. The downstream control point can be an overflow weir to the clearwell or an upturned elbow in the clearwell, or the level in the clearwell if it is higher.

If the downstream control point is located above the top surface of the medium, no possibility of having pressures below atmospheric pressure (i.e., negative head) exists anywhere within the filter medium or underdrain system because the static water level is above the surface of the medium. Also, no possibility of accidentally allowing the filter water level to drop below the top surface of the medium (i.e., partially dewatering the filter medium) exists. These advantages are not obtained without cost, however, because all of the needed head to operate the filter must be placed above the downstream control level, and this means a deeper filter box.

The three equal-rate systems have the common characteristic that the entire filter system operating head provided, which is essentially constant, is consumed throughout the filter run. The head that is not utilized by the dirty filter medium is wasted either in the modulating control valve of the mechanical systems, or in free fall into the filter box for the influent flow-splitting system. The declining-rate systems have a lower total filter system head requirement in the plant hydraulic profile because the total head is reallocated as the filter cycle proceeds as described earlier.

The two nonmechanical systems (i.e., without proportional level control) have the inherent advantage that sudden changes in filtration rate cannot be imposed on the filter. If the total plant flow is increased or a filter is removed from service, the filters can only pick up the load by changing water level to generate the head needed to accommodate the increased flow. This changing of water level takes time, so that the change in filtration occurs slowly and smoothly without mechanical devices.

Slow, smooth changes in filtration rate are also possible in properly designed proportional-level mechanical control systems. The rate-of-change of the modulating control valve position should be proportional to the divergence of the measured variable (flow rate or level) from the desired value. The proportional level control band should be substantial. The drive for the control valve should be an electric motor rather than a pneumatic or hydraulic valve drive, because the latter two drive types are prone to sticking, overshooting, and hunting problems as they age. No industry standards for filter control systems are available, so that some systems do not have all of the attributes listed above.

Choice of the Appropriate Control System

With the foregoing information at hand, some generalizations about the five systems are possible that may affect the choice of a control system to meet a specific utility's desired goals and needs. A particular system should be used only when conditions are appropriate for that system.

If full automation of the plant is desired, proportional-level systems will be favored, although the two nonmechanical systems can be partially automated. If, on the other hand, minimizing mechanical equipment and instrumentation is desired, nonmechanical systems will be favored. (This will be especially important when equipment and repair parts must be imported to a developing country with limited foreign exchange resources.) In small nonautomated plants with four or less filters and with unskilled operators, the variable-level influent flow-splitting system is ideal, because of its simplicity of operation. In larger, nonautomated plants, the variable-level declining-rate method may be favored, because it requires less total filter system head to generate equal filter cycle lengths, and because it produces better filtrate quality in some cases (DiBernardo and Cleasby, 1980). But this is not assured in all cases (Hilmoe and Cleasby, 1986; Cornwell et al., 1991).

In all of the control systems just discussed, there is legitimate concern about the passage of protozoan cysts during the ripening phase at the beginning of the filter cycle. As discussed elsewhere, provision for a filter-to-waste mode is the best available solution.

The design of variable-level declining-rate filters—including the selection of the required total filter system head loss, and the prevention of excessive starting filtration rate for a clean declining rate filter—has been presented by Cleasby (1993).

BACKWASHING OF RAPID FILTERS

Effective backwashing of rapid filters is essential to long-term successful service of the system. The goal of the backwashing operation is to keep the filter acceptably clean while avoiding problems such as mudballs and filter cracks (to be discussed later) in the filter medium.

Alternative Methods of Backwashing

The backwashing system is the most frequent source of filter failure. Therefore, selecting the type of backwashing system and ensuring the proper design, construction, and operation of that system are key elements in the success of a water treatment plant. Two prominent systems of backwashing currently in use are compared in Table 8.8.

Upflow Wash with Full Fluidization. The traditional backwash system in the United States uses an upflow water wash with full bed fluidization. Backwash water is introduced into the bottom of the bed through the underdrain system. It should be

TABLE 8.8 Comparison of Two Backwash Alternatives for Granular Bed Filters

	Backwash method	
	With fluidization	Without fluidization
Applications	1. Fine sand 2. Dual media 3. Triple media	Coarse monomedium Sand or Anthracite
Routines used	1. Water wash + surface wash 2. Water wash + air scour Air first Water second No air during overflow	Air scour + water wash simultaneously during overflow (See text for precautions) Finish with water wash only.
Fluidization	Yes, during water wash	No
Bed expansion	15 to 30 percent	Nil
Wash troughs	Usually used	Usually not used
Horizontal water travel to overflow	Up to 3 ft (0.9 m)	Up to 13 ft (4 m)
Vertical height to overflow	2.5 to 3 ft (0.76 to 0.91 m)	2 ft (0.5m)

turned on gradually over at least a 30-s time interval to avoid disturbing the gravel layers or subjecting the underdrain to sudden momentary pressure increases. The filter medium gradually assumes a fluidized state as the backwash flow rate is increased and the bed expands. The backwash flow is continued with full fluidization until the waste wash water is reasonably clear: a turbidity level of about 10 ntu is sufficiently clear. Then the supply valve is shut off. Shutoff is not as crucial as the opening phase because no danger to the underdrain or gravel exists. However, a slow shutoff will result in a greater degree of restratification.

According to Baylis (1959c), backwash by water fluidization alone is a weak washing method. The reason for that weakness, discussed by Amirtharajah in a later paper (1978), is attributed to a lack of any abrasion between the grains in a fluidized bed. For that reason, backwashing is usually assisted by an auxiliary scouring system, such as surface wash or air scour. The contrasts between the auxiliary scouring systems are presented in Table 8.9 and in the following sections.

Surface Wash Plus Fluidized Bed Backwash. Surface wash has been used extensively and successfully to improve the effectiveness of fluidized bed backwashing. Surface wash systems inject jets of water from orifices located about 1 to 2 in (2.5 to 5 cm) above the surface of the fixed bed. Surface wash jets are operated for 1 to 2 min before the upflow wash and usually are continued during most of the upflow wash, during which time they become immersed in the fluidized filter medium. Surface wash is terminated 2 or 3 min before the end of the upflow wash.

Surface wash is accomplished either with a grid of fixed, vertical pipes located above the granular medium, or with rotary water distribution arms containing orifices or nozzles that supply high-pressure jets of water. Orifice sizes are typically $\frac{3}{32}$ to $\frac{1}{8}$-in in diameter (2 to 3 mm) and are directed downward 15° to 45° below the horizontal. Operating pressures are typically 50 to 75 psig (350 to 520 kPa). Fixed-nozzle systems deliver 2 to 4 gpm/ft^2 (4.9 to 9.8 m/h) and rotary systems deliver 0.5 to 1 gpm/ft^2 (1.2 to 2.4 m/h).

TABLE 8.9 Contrasts Between Backwash Alternatives

	With fluidization			Without fluidization
	Without auxiliary scour	With surface wash auxiliary	With air scour auxiliary	Simultaneous air + water backwash
Wash effectiveness	Weak	Fair	Fair	Good
Solids transport to overflow	Fair	Fair	Fair	Good
Compatible with fine media	Yes	Yes	Yes	No
Compatible with dual & triple media	Yes	Yes	Yes	No
Compatible with graded support gravel	Yes	Yes	Yes	No
Potential for media loss to overflow	Nil	Yes, mainly for coal	Yes, unless used properly	Major, unless used properly

Surface wash has a number of advantages and disadvantages (Cleasby et al., 1977). The advantages include the following:

- It is relatively simple, requiring only a source of high-pressure water in conjunction with a system of distribution nozzles.
- It is accessible for maintenance and repair, because it is located above the surface of the fixed bed.

Some disadvantages of surface wash systems are:

- Rotary-type washers sometimes stick temporarily and fail to rotate as intended.
- If mudballs do form in the bed and reach sufficient size and density, they can sink into the fluidized bed and no longer come under the action of the surface wash jets.
- Fixed-nozzle surface-wash systems obstruct convenient access to the filter surface for maintenance and repair.

Air-Scour-Assisted Backwash. Air-scour systems supply air to the full filter area from orifices located under the filter medium. Air scour is used to improve the effectiveness of the backwashing operation. If air scour is used during overflow, there is substantial danger of losing some of the medium. Therefore, the system must be properly designed and operated to avoid such loss.

The air scour operating routines are different for finer filter media and for coarse sand media, as shown in Table 8.8. Operating sequences are listed below.

Air scour alone before water backwash. This system can be used for fine sand, dual media, and triple media:

- Lower the water level about 6 in (15 cm) below the edge of the backwash overflow.
- Turn on air scour alone for 1 to 2 min.
- Turn off air scour.
- Turn on water wash at a low rate to expel most of the air from the bed before overflow occurs.
- Increase the water wash rate to fluidize and restratify the bed and wash until clean. Some additional air will be expelled.

Simultaneous air scour and water backwash during rising level but before overflow. This system can be used for fine sand, dual media, triple media, and coarse monomedium anthracite:

- Lower the water level to just above the surface of the filter medium.
- Turn on air scour for 1 to 2 minutes.
- Add low-rate water wash at below half the minimum fluidization velocity as water level rises.
- Shut off air scour about 6 in (15 cm) below overflow level while water wash continues. Most air will be expelled before overflow.
- After overflow occurs, increase the water wash rate to fluidize and restratify the bed, and wash until clean. Some additional air will be expelled.

Simultaneous air scour and water backwash during overflow. This system can be used for coarse (1.0 mm ES or larger) monomedium sand or anthracite. However, for anthracite, special baffled overflow troughs are essential to prevent loss of anthracite.

- Turn on air scour.
- Add water wash at below half the minimum fluidization velocity and wash with simultaneous air and water for about 10 minutes during overflow.
- Turn off air scour and continue water wash until overflow is clean. The water wash rate is sometimes increased to hasten the cleanup but is usually kept below fluidization velocity.

For the simultaneous wash routine just outlined, the wash is very effective even though the bed is never fluidized. A slow transport of the grains occurs, caused by the simultaneous air and water flow that causes abrasion between the grains. This abrasion, plus high interstitial water velocities, results in an effective backwash.

Several precautions must be taken to prevent the loss of sand or anthracite when using simultaneous air and water wash during overflow. The water and airflow rates are varied appropriately for the size of the filter medium. For coarse sand, backwash troughs are not generally used, and the dirty washwater exits over a horizontal concrete wall or central gullet, with the vertical distance from the surface of the sand to the washwater overflow being at least 1.6 ft (0.5 m). The top edge of the overflow wall(s) is sloped downward 45° toward the filter bed, so that any sand grains that fall on the sloping wall during backwashing will reenter the filter bed (Degremont 1973). Alternatively, if backwash troughs are used, specially shaped baffles can be located around each trough to prevent the loss of filter medium (Dehab and Young, 1977). Such baffled troughs are especially important when using anthracite medium.

Air-Scour Delivery Systems. Air scour may be introduced to the filter through a pipe system that is completely separate from the backwash water system, or it may be through the use of a common system of nozzles (strainers) that distribute both the air and water, either sequentially or simultaneously. In either method of distribution, if the air is introduced below graded support gravel, concern exists over the movement of the finer gravel by the air—especially by air and water used concurrently by intention or accident. This concern has led to the use of media-retaining strainers (Figure 8.9) in some filters that eliminate the need for graded support gravel in the filter.

Backwash Water and Air-Scour Flow Rates. Typical flow rates used in backwash practice are summarized in Table 8.10. Hewitt and Amirtharajah (1984) did an experimental study of the particular combinations of air and subfluidization water flow that caused the formation and collapse of air pockets within the bed, a condition they called "collapse-pulsing." This condition was presumed to create the best abrasion between the grains and the optimum condition for air plus subfluidization water backwashing. An empirical equation relating airflow rate, fluidization velocity, and backwash water flow rate was presented. In a companion paper, Amirtharajah (1984) developed a theoretical equation for collapse-pulsing using concepts from soil mechanics and porous media hydraulics. The resulting equation was:

$$0.45Q_a^2 + 100\left(\frac{V}{V_{mf}}\right) = 41.9 \tag{8.21}$$

in which Q_a is the airflow rate in standard cubic feet per minute per square foot and V/V_{mf} is the ratio of superficial water velocity divided by the minimum fluidization velocity based on the d_{60} grain size of the medium.

The empirical and the theoretical equations give almost the same results over the range of airflow rates from 2 through 6 scfm/ft^2 (37 through 110 m/h). For a given airflow rate, both of the equations predict somewhat higher water flow rates than the typical values presented in Table 8.10. The tabulated values should be used until future research reconciles these differences.

Some of the advantages of auxiliary air-scour in contrast with surface-wash systems, are the following (Cleasby et al., 1977):

- It covers the full area of rectangular filters and is adaptable to any filter dimensions.
- It agitates the entire filter depth. Therefore, it can agitate the interfaces in dual and multimedia beds, and can reach mudballs that have sunk deep into the filter.

TABLE 8.10 Typical Water and Air-Scour Flow Rates for Backwash Systems Employing Air Scour

Filter Medium	Backwash Sequence	Air Rate scfm/sf (m/h)	Water Rate* gpm/sf (m/h)
Fine sand 0.5mm ES	Air first	2–3 (37–55)	
	Water second		15 (37)
Fine dual and triple media 1.0 mm ES anthracite	Air first	3–4 (55–73)	
	Water second		15–20 (37–49)
Coarse dual media 1.5 mm ES anthracite	Air first	4–5 (73–91)	
	Air + water on rising level	4–5 (73–91)	10 (24)
	Water third		25 (61)
Coarse sand 1.0 mm ES	Air + water 1st simultaneously	3–4 (55–73)	6–7 (15–17)
	Water second		Same or double rate
Coarse sand 2 mm ES	Air + water 1st simultaneously	6–8 (110–146)	10–12 (24–29)
	Water second		Same or double rate
Coarse anthracite 1.5 mm ES	Air + water 1st simultaneously	3–5 (55–91)	8–10 (20–24)
	Water second		Same or double rate

* Water rates for dual and triple media vary with water temperature and should fluidize the bed to achieve restratification of the media. See Eq. (8.7).

Some disadvantages of the air-scour auxiliary are:

- The need for a separate air blower and piping system.
- The potential for loss of media, especially if air and water are used simultaneously during overflow.
- A greater possibility of moving the supporting gravel exists if air is delivered through the gravel concurrently with water. Special gravel designs are required as described earlier.

Backwash Troughs and Wash Water Required

If backwash troughs are used, they must be of adequate size and of appropriate spacing to take away the dirty backwash water without surcharging. Spacing is dictated by maximum horizontal travel of the water (see Table 8.8). At maximum backwash rate, there should be a free fall of the water into the troughs, even at the upstream end. The top edges of the troughs should all be at the same elevation, and all should be level so that the flow is withdrawn evenly over the whole bed.

The volume of wash water required to wash a filter will depend on the depth of the filter medium and the vertical distance from the fixed-bed surface to the overflow level. The greater the depth of medium, the greater the water volume needed to flush the dirt out of the medium and into space above the medium. The greater the vertical distance from the surface of the medium to the wash-water overflow level, the greater the volume needed to wash the dirt out of the filter box. Typically, it will take about four displacement volumes for the void volume in the bed plus the water volume above the bed to reach the overflow level. Typical wash water volumes used per unit area per wash range from 100 to 200 gal/ft^2/wash (4 to 8 m^3/m^2/wash).

In filters with backwash troughs and washed with full-bed fluidization, the expanded media surface should be lower than the bottom of the troughs. If the expanded media rises up between the troughs, the effective vertical flow velocity is increased and the danger of carrying the filter medium into the troughs is increased.

The wash-water volume required per wash to clean the filter could be reduced by decreasing the vertical distance from the fixed bed surface up to the top edge of the troughs. The temptation to do this must be resisted, however, because the danger of loss of filter medium would be increased. Loss of filter medium is greater when anthracite is used. In this case, to minimize the loss of anthracite, the vertical distance to the edge of the troughs should be increased above the traditional 2.5 to 3 ft (0.75 to 0.9 m), to 3.5 to 4 ft (1.1 to 1.2 m). This, in turn, will require a greater volume of wash water to complete the backwash.

Expansion of Filter Medium During Backwashing

When using upflow wash with full fluidization, the filter bed expands about 15 to 30 percent above its fixed-bed depth. The degree of expansion is affected by many variables associated with the filter medium and the water. Filter medium variables include the size and size gradation, and grain shape and density. Water variables include viscosity and density. The ability to predict expansion is important, for example, in determining whether the expanded medium will rise too high, possibly above the bottom of the troughs.

The following model for predicting expanded bed porosity during backwashing was developed by extending a Reynolds number versus porosity function that had been used for fixed beds into the expanded bed region (Dharmarajah and Cleasby, 1986). The modified Reynolds number Re_1 uses interstitial velocity V/ε for the characteristic velocity and a term approximating the mean hydraulic radius of the flow channel, $\varepsilon/[S_v(1-\varepsilon)]$, for the characteristic length.

$$Re_1 = \frac{V}{\varepsilon}\,\frac{\varepsilon}{S_v(1-\varepsilon)}\cdot\frac{\rho}{\mu} = \frac{V\rho}{S_v(1-\varepsilon)\mu} \tag{8.22}$$

where S_v = specific surface of the grains ($6/d$ for spheres and $6/\psi d_{eq}$ for non-spheres). The porosity function used previously to correlate fixed-bed pressure drop data to Re_1 was modified by combining it with the constant head-loss equation for a fluidized bed (Equation 8.6) resulting in a new dimensionless porosity function for fluidized beds (A1):

$$\mathrm{A1} = \frac{\varepsilon^3}{(1-\varepsilon)^2}\cdot\frac{\rho(\rho_s-\rho)g}{S_v^3\mu^2} \tag{8.23}$$

Using the data for many different sizes and types of filter media, log A1 was correlated with log Re_1 using a stepwise regression analysis and resulting in the following expansion correlation (Dharmarajah and Cleasby, 1986):

$$\log \mathrm{A1} = 0.56543 + 1.09348 \log \mathrm{Re}_1 + 0.17971 (\log \mathrm{Re}_1)^2$$
$$-0.00392 (\log \mathrm{Re}_1)^4 - 1.5 (\log \psi)^2 \tag{8.24}$$

The equation can be used to predict the expanded porosity (ε) of a filter medium of any uniform size (i.e., S_v = specific surface) at any desired backwash rate (V). Because both A1 and Re_1 are functions of ε, the solution is trial and error and is best solved by computer. When applying the equation to a real filter medium with size gradations, the bed must be divided into several segments of approximately uniform size using the sieve analysis data, and the expanded porosity of each size segment must be calculated. The expanded depth of each segment then can be calculated from the following equation that is based on the total grain volume remaining unchanged as the bed expands (Cleasby and Fan, 1981):

$$\frac{L}{L_o} = \frac{1-\varepsilon_o}{1-\varepsilon} \tag{8.25}$$

in which L/L_o is the ratio of expanded depth (L) to fixed-bed depth L_o, ε is the expanded porosity, and ε_o is the fixed loose bed porosity. Typical values of ε_o were presented earlier in Table 8.2.

EXAMPLE 8.3 Calculate the expansion for the d_{50} size anthracite shown in Figure 8.3 at 20°C and at the backwash rate of 32.8 gpm/ft² calculated in Example Problem No. 2. Use a sphericity of 0.55 and a density of 1.6 estimated from Table 8.1.

SOLUTION IN SI UNITS

$$\mathrm{Re}_1 = \frac{V\rho}{S_v\,(1-\varepsilon)\mu} \tag{8.22}$$

$V = 32.8\ \mathrm{gpm/ft^2} = 0.0223$ m/s as before
$\rho = 998.207\ \mathrm{kg/m^3}$
$\mu = 1.002\mathrm{E}-3\ \mathrm{Ns/m^2}$

$\mu/\rho = \upsilon = 1.004\text{E} - 6 \text{ m}^2/\text{s}$
$d_{50} = 2.0\text{E} - 3 \text{ m}$
$S_v = 6/\psi d = 6/0.55 \cdot 2.0\text{E} - 3 = 5454 \text{ m}^{-1}$

$$\text{Re}_1 = \frac{0.0223}{5454\,(1 - \varepsilon)\,1.004\text{E} - 6} = \frac{4.073}{(1 - \varepsilon)} \text{ (dimensionless)}$$

ε is the desired expanded porosity

$$\text{A1} = \frac{\varepsilon^3}{(1 - \varepsilon)^2} \frac{\rho(\rho_s - \rho)g}{(S_v)^3\mu^2} \text{ (dimensionless)} \tag{8.23}$$

$$= \frac{\varepsilon^3}{(1 - \varepsilon)^2} \cdot \frac{998(1600 - 998)9.81}{(5454)^3(1.002\text{E} - 3)^2} = \frac{36.184\varepsilon^3}{(1 - \varepsilon)^2}$$

Insert these Re_1 and A1 values into Equation 8.24 as follows:

$$\log\left[36.184 \frac{\varepsilon^3}{(1 - \varepsilon)^2}\right] = 0.56543 + 1.09348 \log \frac{4.073}{(1 - \varepsilon)} +$$

$$0.17971 \left[\log \frac{4.073}{(1 - \varepsilon)}\right]^2 - 0.00392 \left[\log \frac{4.073}{(1 - \varepsilon)}\right]^4 - 1.5\,(\log 0.55)^2$$

Note that the only unknown in the foregoing equation is ε, but it appears on both sides of the equation. A trial and error solution is necessary.

Equation 8.24 solution yields $\varepsilon = 0.616$.

The expansion of this size anthracite can be calculated from Equation 8.25, assuming an initial porosity of 0.56 from Table 8.2.

$$\frac{L}{L_o} = \frac{1 - \varepsilon_o}{1 - \varepsilon} \tag{8.25}$$

$$\frac{L}{L_o} = \frac{1 - 0.56}{1 - 0.616} = 1.15$$

Therefore, this mid-size material would be expanded 15 percent. To obtain the expansion of the entire bed, the bed must be divided into about five sizes and the expanded depth of each layer calculated and totaled. To obtain the expansion of dual or triple media beds, the expansion of each layer is calculated and totalled without regard to any intermixing at the interfaces between layers.

Stratification and Intermixing During Backwashing

The related phenomena of stratification and intermixing are important issues in filter construction and backwashing. These will be discussed in the following subsections:

Stratification and Skimming. In the case of a single-medium filter such as a sand filter, during backwash with fluidization the grains tend to stratify by size, with the finer grains resting on top and the coarser grains setting on the bottom. The tendency to stratify at a given backwash rate (above fluidization velocity) is driven by bulk density differences between the fluidized grains of different sizes. Smaller grains expand more and have a lower bulk density (grains plus fluidizing water) and thus rise to the top of the bed. The concept of bulk density is discussed further in the next section, "Intermixing of Adjacent Layers." This stratification is upset to a varying extent by nonuniform upflow of the backwash water that creates localized regions of above-average upflow velocity. Larger grains are transported upward rapidly into the upper bed, while in adjacent regions the sand will be moving downward carrying finer grains down into the bed. These regions of excessive upflow are referred to as *sand boils,* or *jet action* in water filtration literature, and as *gulf streaming* in some fluidization literature.

The tendency to stratify is used beneficially during construction of a filter to remove unwanted fine grains from the bed. The filter is washed above fluidization velocity and the fine grains accumulate at the upper surface. After the backwash is completed, the fine grains are skimmed from the surface to avoid leaving a blinding layer of fines on top.

Better stratification is achieved at lower upflow rates, just barely above the minimum fluidization velocity of the bed. So in preparing for skimming, the bed should first be fully fluidized and then the upflow wash rate slowly reduced over several minutes to bring to the surface as many of the fine grains as possible. The fluidization and skimming process may be repeated two or three times for maximum effectiveness.

During the installation of dual- or triple-media filter beds, skimming each layer as it is completed is common. The same concepts and procedures just described are equally appropriate for each layer. If the very fine grains are not removed from the lower layers, they may rise high in the next upper layer, partially negating the desired benefit of the coarse upper layer (Cleasby et al., 1984).

Intermixing of Adjacent Layers. Intermixing will tend to occur between adjacent layers of dual- and triple-media filters. For example, the upper finer sand grains of a dual media bed move up into the lower coarser grains of the anthracite bed that lies above.

The tendency to intermix between two adjacent layers of filter media can be estimated by comparing the bulk density of the two media calculated independently at any particular backwash flow rate and temperature. The bulk density is the mixed density of the grains and fluidizing water calculated as follows (Cleasby and Woods, 1975):

$$\rho_b = (1 - \varepsilon)\, \rho_s + \rho\varepsilon \tag{8.20}$$

in which ρ_b is the bulk density. The first term on the right-hand side is the density contributed by the solid fraction, and the second term is the density contributed by the water fraction.

The tendency to intermix increases with the backwash flow rate because the bulk densities tend to converge at higher flow rates. Intermixing actually begins before the bulk densities become equal because of uneven flow distribution and mixing and circulation patterns that exist in the fluidized layers. Mixing was observed to occur when the bulk densities converged to within 3 to 8 lb/ft^3 (50 to 130 kg/m^3) (Cleasby and Woods, 1975). Similarly, as the backwash rate is decreased slowly, the bulk densities of the adjacent layers diverge, and intermixing decreases.

The bulk density model would predict that inversion of layers should occur at very high wash rates, and such inversion was observed experimentally (Cleasby and Woods, 1975). Wash rates were, however, far higher than rates used in practice.

The tendency for sand and anthracite to intermix during backwashing is much less than for silica and garnet sand in the usual backwash flow range. This is because the differences in bulk density are greater for the usual sand and anthracite sizes used in current practice.

Backwash of GAC Filter-Adsorbers

The properties of GAC are sufficiently different from conventional anthracite or sand media to require some special precautions when backwashing GAC filter-adsorbers (Graese, Snoeyink, and Lee, 1987). The lower density of GAC means lower backwash rates are required to fluidize the common size gradations currently being used. Lower wash rates mean lower hydraulic shear and less effective upflow wash. Mudballs have been a common problem in GAC filter-adsorbers. Therefore, surface wash or auxiliary air scour are essential for GAC backwashing.

The higher UC of typical gradations of GAC being used results in a greater percent expansion and better size stratification if the full bed is fluidized. There is greater potential for loss of GAC into the backwash troughs. In retrofitting existing filters into GAC filter-adsorbers, raising the backwash troughs may be necessary to reduce the loss of GAC.

Granular activated carbon is abrasive and corrosive to many metals. Therefore, all metals in contact with GAC should be resistant to abrasion and corrosion. Metals can be coated with corrosion-resistant substances.

Special precautions are important when preparing a new GAC filter-adsorber for service. The filter box should be disinfected before installing the GAC. After installation of the GAC, it must be submerged and soaked for at least a day to allow water to penetrate the pores of the GAC. Initial backwashing must be done at a very low rate to be sure that all air is out of the bed to avoid loss of GAC. After all air has been removed, the backwash rate can be increased to wash out undesirable fines from the bed. Release of additional air and fines may be observed during the first week of normal backwashing operation. See Chapter 13 for further information on the backwashing of GAC adsorbers.

Backwash Water Recovery

Recovery of dirty backwash water has been very common for a number of reasons. It represents a rather large volume of water with low solids content. Typically, the volume is 1 to 5 percent of total plant production. Therefore, its recovery represents a savings in water resources and in the chemicals that were expended to treat it initially. Discharge of water plant wastes is usually prohibited under current pollution regulations, although some agencies are relaxing requirements for economic reasons and in recognition of the minimal environmental impact of these wastes (Vicory and Weaver, 1984).

However, most regulatory agencies require proper handling of wastes in new treatment plants and in plants being enlarged or upgraded. Further discussion on this subject is presented in Chapter 16.

The concern about possible recycling of *Giardia* cysts and *Cryptosporidium* oocysts led to a plant-scale study of the impact of waste stream recycling on filtered water quality (Cornwell and Lee, 1994). The authors concluded that proper management of the waste streams can render them acceptable for recycling in many situations. Backwash water and sedimentation basin sludge should be subjected to further sedimentation at a low overflow rate or by adding a polymer to remove cysts before recycling the supernatant. Recycled streams should be equalized and blended over the full 24-hour operating period, or the plants operating period if it is less than 24 hours. Recycle streams should be regularly monitored for cysts or by particle counting to judge the effectiveness of removing cysts from the waste stream prior to recycle.

PROBLEMS IN RAPID FILTERS

Dirty Filter Media and Mineral Deposits

Dirty filters result from inadequate backwashing of the filter, including the absence or improper operation of auxiliary scour systems. Typical manifestations of inadequate backwashing include filter cracks and mudballs. Inadequate cleaning leaves a thin layer of compressible matter around each grain of the medium. As pressure drop across the filter medium increases during the subsequent filter run, the grains are squeezed together and cracks form in the surface of the medium, usually along the walls first. In severe cases, cracks may develop at other locations as well.

The heavier deposits of solids near the top surface of the medium break into pieces during backwashing, resulting in spherical-shaped accretions referred to as *mudballs*. Mudballs are composed of the filter grains and the solids removed by the filter and range from pea size to 1 to 2 in (2.5 to 5 cm) or more. The formation of mudballs is accentuated by the use of polymers as coagulant aids or as filter aids that form stronger attachments between the filter grains and the removed solids. If the mudballs are small and of low density, they may float on the surface of the fluidized medium. Larger or heavier ones may sink into the filter, to the bottom, or in dual-media filters, to the coal-sand interface. Subsurface accumulations can lead to solidified inactive regions of the filter bed that are not remedied in normal backwashing. This, in turn, increases the filtration rate through the remaining active portions of the bed, with potential detriment to the quality of the filtrate as well as shorter filter cycles.

Surface mudballs can be removed manually by (1) lifting them from the surface with a large strainer while the backwash water is running at a low rate; (2) breaking them up with rakes; or (3) breaking them up with a high-pressure hose jet. Subsurface mudballs and agglomerations can be reduced by (1) probing the bed with lances delivering jets of high-pressure water; (2) pumping the medium through an ejector; (3) washing the filter and using chemical additions described below; or (4) digging out the hard spots and removing, cleaning, or replacing the medium.

Mineral deposits can develop on the grains of the filter medium, causing them to change in size, shape, and density. Calcium carbonate deposits are common in many lime-soda ash softening plants if recarbonation is not sufficient to deliver a stable water to the filters. Such deposits also may occur in surface water plants if lime is added ahead of filtration for corrosion control. The deposition of calcium carbonate can be minimized by adding a low dosage of polyphosphate ahead of the filters. Iron oxides and manganese oxides are common deposits on the filter medium of iron and manganese removal plants. Aluminum oxide or iron oxide deposits can occur in surface water plants that use alum or iron salts for coagulation.

All of these mineral deposits are subject to later leaching into the filtered water, if the pH is changed in a direction to increase solubility. The increased size of the grains caused by mineral deposits results in an increase in the bed depth and can impair filtration efficiency. Thus, cleaning the deposits from the grains or replacing the medium may become necessary.

Dirty filters can usually be corrected by proper use of auxiliary scour devices such as surface washers or air scour. Where auxiliary scour is not successful, use of chemicals may in some cases be necessary to attempt to clean the filter medium in place. Various chemicals have been used, including chlorine, copper sulfate, acids, and alkalies (Babbitt, Doland, and Cleasby, 1962). Chlorine may be used where the material to be removed includes living and dead organisms or their metabolites. Copper sulfate is effective in killing algae growing on filter walls or medium. Carrying a chlorine residual in the filter influent aids in control of microorganisms in the filter. Acidifying the water to a pH of about 4 aids in dissolving deposits of calcium carbonate and the oxides of iron, aluminum, and manganese. Sulfuric acid and hydrochloric acid have been used, but care must be taken to prevent high local concentrations from damaging concrete filter walls. Citric acid has been used in the Saint Louis, Missouri, softening plant, and diluted, food-grade, glacial acetic acid at the lime softening plant at Decatur, Illinois (Mayhugh et al., 1996). These organic acids are less dangerous to handle and may offer public relations advantages. Caustic soda has been used for removing deposits resulting from the use of alum and for deposits of organic material. It is used at the rate of 1 to 3 lb/ft^2 (5 to 15 kg/m^2). Chemical solutions are generally left in contact with the filter medium for 1 to 2 days, and the filter is then thoroughly washed and returned to service.

If a filter can be removed from service for an extended period, deposits of calcium carbonate can be removed by feeding, at a low flow rate, a water acidified with carbon dioxide. The flow rate is selected so that the water leaves the filter with the CO_2 fully consumed at a pH of about 8.

Movement of Gravel During Backwashing

Many filters have layers of graded gravel to support the filter medium. The conventional gravel layers are graded from coarse on the bottom to fine at the top, according to appropriate size guidelines presented earlier. With this conventional gravel arrangement, some cases of mounding of the gravel caused by lateral movement of the fine gravel have been reported. In severe cases, the finest gravel layer may be completely removed from some areas of the filter bed. This can lead to leakage of the filter medium into the underdrain system, and ultimately to the need to clean the underdrain and rebuild the filter gravel and medium. Mechanisms contributing to movement of gravel have been exhaustively discussed by Baylis (American Water Works Association, 1971; Baylis, 1959a) and those discussions related to fluidized bed backwashing of sand beds are summarized here.

As backwashing begins, sand grains do not move apart quickly and uniformly throughout the bed. Time is required for the sand to equilibrate at its expanded spacing in the upward flow of wash water. If the backwash is turned on suddenly, it lifts the sand bed bodily above the gravel layer, forming an open space between the sand and gravel. The sand bed then breaks at one or more points, as shown in Figure 8.29, causing sand boils and subsequent upsetting of the supporting gravel layers. This then requires frequent rebuilding of the graded gravel.

Opening the backwash valve slowly is essential. The time from start to full backwash flow should be at least 30 s and perhaps longer, and should be restricted by devices built into the plant. The most destructive filter-washing blunder is to turn the backwash water on quickly when the bed has been drained. Gross disturbance of the gravel results, and rupture of the filter underdrain system has resulted in many cases.

The upward flow from the gravel is never completely uniform. As a result, in some parts of the expanded sand bed the water and sand travel upward at rates higher than the average backwash velocity, and at other places the sand actually travels downward. This leads to jet action at the sand-gravel junction, as illustrated in Figure 8.30. In some cases the jet velocity is so great that it will move grains of small gravel several diameters larger than the sand grains adjacent to the gravel. The problem of gravel movement is greater at wash rates above 15 gpm/ft^2 (37 m/h).

The importance of sand in generating the movement of gravel is illustrated by the observation that an upward flow through a graded gravel bed at backwash rates exceeding 25 gpm/ft^2 (61 m/h) will not move any of the gravel particles unless sand

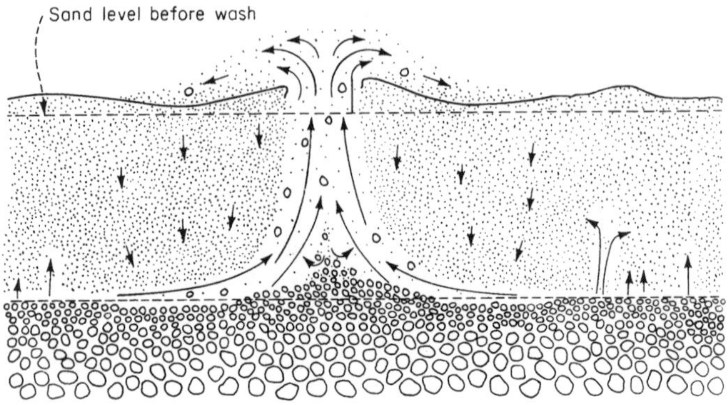

FIGURE 8.29 Sand boil at beginning of wash.

is present on top of the gravel (Baylis, 1959a). The presence of the sand greatly increases the mobility of the gravel.

Movement of the gravel can also occur when air scour and water backwash are used simultaneously. Provisions being used to avoid movement of gravel are the use of double reverse graded gravel, or the use of media-retaining nozzle underdrains that require no gravel, as discussed earlier.

Three methods are used to determine whether the upper surface of the gravel has moved: (1) pushing a ¼-in metal rod gently down to the gravel surface of a newly backwashed filter while the filter is at rest; (2) lowering a probe with a flat plate on the end through the expanded filter medium during the fluidized bed backwash; or (3) draining the filter to the gravel, inserting a thin-walled pipe of about 1 in (2.5 cm)

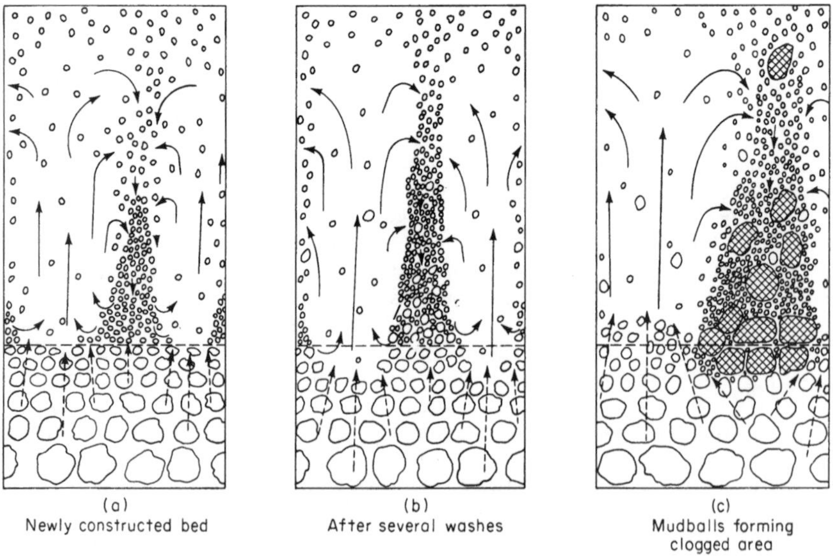

FIGURE 8.30 Jet action at the sand-gravel interface.

diameter into the medium, and removing the media in segments until the top layer of gravel is reached in successive probings. Each method has advantages and disadvantages. The first is quick, but distinguishing the fine gravel from the bottom filter medium is difficult. The second method requires a long period of fluidized backwash. The third method is the most positive but also the most time-consuming and labor-intensive. Numerous measurements made on a gridiron pattern will reveal the contour of the top surface of the gravel.

Underdrain Failures

The filter underdrain is a vitally important component of the rapid filter. It serves to collect the filtered water, support the filter gravel (if used) and filter medium, and to distribute backwash water.

The most serious failures of rapid filters are usually related to problems with the underdrain. Proper design and construction of the underdrain are essential. Any underdrain system, if designed improperly, can lead to uneven distribution of the backwash water. This, in turn, can lead to problems such as dirty areas in the filter and migration of gravel, if gravel is used to support the medium.

False-floor underdrains with nozzles have failed for several reasons. If the nozzles have fine openings to retain the fine filter medium, the openings may become clogged with debris from construction of the underdrain or from solids in the backwash water, or because of mineral deposits. If steel filter vessels are allowed to corrode, the rust can clog backwash nozzles. If the clogging becomes excessive, the backwash flow rate may be diminished and the pressure drop across the floor causes uplift during backwashing that may become excessive. The problem can be repaired by removing the filter medium and cleaning the nozzles. If not corrected, excessive pressure drop can result in an uplift structural failure of the false floor. An open pressure-relief pipe is often provided that projects an appropriate length above backwashing water level. Excessive pressure in the underdrain is evident when this pipe starts to flow during the backwash operation.

Some modern plastic nozzles are not strong enough to accept the abuse of construction, during or after installation. They may be weakened during construction, and the nozzle heads may break off during future backwashing operations. The result is leakage of medium into the underdrain plenum.

The manifold and lateral type of underdrain is not subject to uplift failure. It can fail to function properly, however, if the distribution orifices are reduced in size because of mineral deposits, or enlarged in size because of corrosion or erosion.

Vitrified clay block underdrains have proven quite successful when properly installed and if precautions are taken to prevent air from entering the backwash system. The manufacturer's recommendations for installation and grouting must be followed explicitly to obtain a good installation. Careless workmanship can lead to grout in the water-carrying channels. This can lead to maldistribution of washwater, and in severe cases, to excessive backwash pressures and uplift failure. Some plastic block underdrains eliminate the danger of grout intrusion by the use of o-ring gaskets or other positive exclusion systems.

PRESSURE GRANULAR BED FILTERS

Pressure filters are sometimes used for rapid filtration. The filter medium is contained in a steel pressure vessel. The water to be filtered enters the filter under pres-

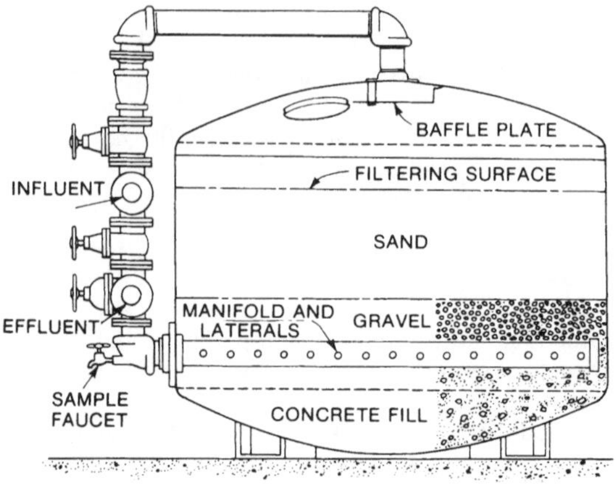

FIGURE 8.31 Cross section of typical pressure filter.

sure and leaves at slightly reduced pressure because of the head loss encountered in the filter medium, underdrain, and piping connections.

Description of Pressure Filters

The pressure vessel may be a cylindrical tank with a vertical axis such as shown in Figure 8.31, or it may be a horizontal-axis cylindrical tank. The horizontal cylindrical configuration has the disadvantage that the width of the filter medium is not constant from top to bottom, usually being wider at the top. This leaves dead areas along the walls that do not receive adequate fluidization during backwashing and therefore, may not be washed effectively.

In some horizontal-axis cylindrical tanks, the filter is divided into multiple (four or five) cells by vertical bulkheads. Cost advantages are gained by this configuration because only a single pressure vessel is needed. One filter cell can be backwashed by the production of the other cells that remain in service. This requires a filtration rate sufficiently high so that the total production of the operating cells is sufficient to fluidize the medium of the single cell being backwashed.

Comparison of Pressure and Gravity Filtration

While the outward appearance of pressure and gravity filters are quite different, the filtration process is the same. The same mechanisms for capturing particles are functioning in both. The same filter medium, the same filtration rates, and the same terminal head loss should be utilized if comparable filtrate quality is desired.

The use of higher filtration rates and high terminal head loss in a pressure filter is tempting because the influent is under pressure and more potential head loss is available. This temptation should be resisted, however, unless no detriment to the quality of the filtrate can be demonstrated on a case-by-case basis.

One advantage gained by pressure filtration is that water leaves the filter under a positive gauge pressure, and therefore no negative pressure can ever exist in the filter medium. The potential problems associated with negative pressure discussed earlier are therefore avoided.

Operation of Pressure Filters

Because of the similarities between pressure and gravity filters, the operating principles are identical. For example, appropriate pretreatment is equally important to pressure and gravity filters; patterns of filtrate quality will be the same; the impact of sudden rate increases will be just as detrimental; and the importance of proper backwashing is equally important.

The operation of a pressure filter is similar in most respects to the operation of a gravity filter. Proper backwashing of a pressure filter is more difficult, however, because the filter medium is not conveniently visible to the operator during the backwash operation. The various analyses that can be conveniently made by visual observation for gravity filters are difficult or impossible to perform with pressure filters. These include:

1. Presence of filter cracks before the backwash, or mudballs after the backwash
2. Uniformity of backwash water distribution
3. Uniformity of rate of cleanup of the wash water over the full filter areas
4. Proper functioning of the auxiliary scour device such as surface wash or air scour
5. Elevation and appearance of the top surface of the filter medium after the backwash
6. Whether the medium is fully fluidized during the backwash (if that is the intended washing method), and the extent of expansion of the filter medium
7. Extent of loss of the filter medium

Because of these difficulties, pressure filters have been implicated in waterborne disease outbreaks that can be partially attributed to poor condition of the filter medium. (Kirner, Littler, and Angelo, 1978; Lippy, 1978). Prior concerns about the reliability of pressure filters and other concerns have caused some state regulatory agencies to exclude the use of pressure filters in the treatment of surface waters or other polluted source waters and lime softened waters (Great Lakes Upper Mississippi River Board of State Sanitary Engineers, 1992).

Rate Control for Pressure Filters

Rate control for pressure filters is equally as important as it is for gravity filters because the same filtration mechanisms are functioning in both cases. Fewer options are available for pressure filters, however, because the pressure filter operates full of water under pressure. As a result, options involving changing water levels are not available, and influent gravity flow splitting is not available. Therefore, the benefits of slow rate changes by allowing water levels to change are not available to pressure filters.

The usual arrangement for a bank of pressure filters is based on the assumption that flow through the system will self-equalize. The total filter area of the filter bank is sized appropriately to meet local regulatory requirements based on feed pump capacity. Individual filter flow controllers and flow meters are not provided. The system is designed symmetrically so that flow is distributed equally to each filter in the bank. Thus, after backwashing all of the filters in short succession, if one filter is passing more flow than it should, it is assumed that that filter will clog more quickly than the other filters, and that the flow will be reduced because of the clogging resistance. Thus, the system is considered self-equalizing during operation. However, if the multiple filters are backwashed at random intervals during a filter cycle, the cleanest filter will provide the highest flow and the dirtiest filter the lowest flow. The extent of these differences will not be known if individual flow meters are not provided.

In spite of the common practice just described, it is desirable to have a flow meter on each pressure filter in a bank of several filters. Flow metering is useful to the plant operators for diagnostic purposes to observe if something is wrong with one filter, causing the flows to not be equally split. This could happen if one filter is not cleaning up properly during backwashing, perhaps due to a clogged underdrain. Or, one filter may flow at an excessive rate due to the loss of some or all of its filtering medium.

Applications of Pressure Filters

Pressure filters tend to be used in small water systems. Many pressure filters are used in industrial water and wastewater filtration applications. They also are used widely in swimming pool filtration.

Advantages of pressure filters over gravity filters include the following:

- The filtrate, which is under pressure, can be delivered to the point of use without repumping. In some treatment plants, source water can be pumped from the source through the treatment plant and directly to the point of use by the source water pumps.
- A treatment plant equipped with pressure filters is somewhat easier to automate.
- Some groundwaters containing iron can be treated by pressure aeration and/or chemical oxidation and then filtered directly on pressure filters. This approach has received considerable application for small communities. (Not all iron-bearing waters respond successfully to this method of treatment, however, and prior pilot testing is usually required.)

SLOW SAND FILTERS

Description and History

The slow sand filter is a sand filter operated at very low filtration rates without the use of coagulation in pretreatment. The grain size of the sand is somewhat smaller than that used in a rapid filter, and this, plus the low filtration rate, results in the solids being removed almost entirely in a thin layer on the top of the sand bed. This layer, composed of dirt and living and dead micro- and macroorganisms from the water (i.e., the *schmutzdecke* or *dirty skin*), becomes the dominant filter medium as the filter cycle progresses. When the head loss becomes excessive, the filter is cleaned by draining it below the sand surface and then physically removing the dirty layer along with ½ to 2 in (13 to 50 cm) of sand. Typical cycle lengths may vary from 1 to 6 months (or longer) depending on source water quality and the filtration rate.

The slow sand filter was developed in England in the early nineteenth century and continues to be used successfully there, notably on River Thames water, which serves

London. In the United States, this treatment process has been used to treat high-quality upland surface waters in New York and New England since the early 1900s.

Renewed interest in the slow sand filter occurred because numerous outbreaks of giardiasis occurred in the latter 1970s and early 1980s in communities using unfiltered or poorly filtered water. The slow sand filter is a simple technology requiring no knowledge of coagulation chemistry and is quite attractive for small installations treating high-quality surface waters. The possibilities for applying this treatment process for small water systems led to research efforts to demonstrate the efficacy of the slow sand filter in removal of *Giardia lamblia* and to its application in numerous communities in New England and the Pacific Northwest.

Mechanisms of Filtration and Performance

Removal mechanisms in a slow sand filter are both physical and biological and were discussed in detail by Haarhoff and Cleasby (1991). During the use of the filter, an ecosystem becomes established at the top of and within the filter bed. The effectiveness of the filter improves over the first few cycles as the microflora develop and then remains high thereafter. In addition, living organisms in the filter cause reductions in concentrations of organic constituents and can promote chemical transformations such as the oxidation of ammonia to nitrate.

During slow sand filter operation a biologically active layer (*schmutzdecke*) builds up on the top of the filter medium and assists filtration. As the water enters the *schmutzdecke,* biological action breaks down some organic matter, and inert suspended particles may be physically strained out of the water. The water then enters the top layer of sand, where more physical straining and biological action occur, and particles become attached onto the surfaces of the sand grains. The ecosystem in a slow sand filter typically includes bacteria; protozoa such as rhizopods and ciliates; rotifers; copepods; and aquatic worms. In general, smaller organisms are found at or near the top of the filter and are preyed upon by larger organisms that range deeper into the filter bed, as shown in Figure 8.32. Algae also may be found at or near the surface of the filter in open filters.

The removal of microorganisms in filtration by predatory action is probably unique to slow sand filtration. The role of ciliate protozoa as bacterial predators was investigated in laboratory studies (Lloyd, 1996) of a pilot filter having a bed of sterilized sand. Removal of bacteria in filtration tests without ciliate protozoa present were about 1-log lower than removals when ciliate protozoa were inoculated into the sand bed (between 2- to 3-log). Lloyd pointed out that the improvement resulting from the presence of ciliate protozoa was impressive because attaining the last few percent of removal becomes exponentially more difficult.

The *schmutzdecke* that develops on the surface of the sand bed removes significant amounts of particulate matter, and an initial improvement period (or ripening period) frequently can be observed at the beginning of each cycle after the *schmutzdecke* has been removed (Cleasby et al., 1984; Cullen and Letterman, 1985; Cleasby, Hilmoe, and Dimitracopoulos, 1984). The initial improvement period was observed to vary from 6 h to 2 weeks (Cullen and Letterman, 1985), although most improvement periods have been reported to be less than 2 days (Cleasby et al., 1984; Cullen and Letterman, 1985). A filtering-to-waste period of 2 days was recommended where cysts of *Giardia lamblia* are of concern (Cleasby et al., 1984; Cleasby, Hilmoe, and Dimitracopoulos, 1984), but Bellamy et al. (1985) found no effect of scraping on *Giardia* cyst removal after 41 weeks of filter operation, when the sand bed was biologically mature.

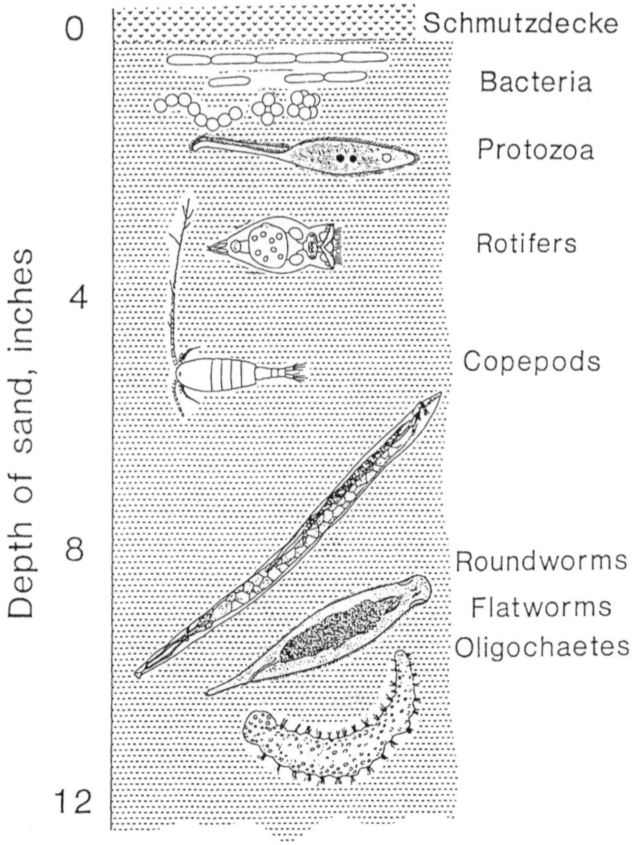

FIGURE 8.32 Typical slow sand filter biota at different depths. (*Source:* American Public Health Association, American Water Works Association, and Water Environment Federation. 1995. *Standard Methods for the Examination of Water and Wastewater,* 19th ed. Washington, D.C.: APHA.)

In addition to the initial improvement after scraping, a distinct improvement in the performance of a filter with new sand occurs as it matures over several cycles, as indicated by the removal of coliform bacteria, particles, and *Giardia* cysts (Cleasby, Hilmoe, and Dimitracopoulos, 1984; Bellamy et al., 1985; Fox et al., 1984). In a midwestern pilot study, after the first four filter cycles, average reduction of turbidity was 97.8 percent or better; removal of cyst-sized particles (7 to 12 μm) was 96.8 percent or better; removal of total particles (1 to 60 μm) was 98.1 percent or better; removal of total coliform bacteria was 99.4 percent or better; and removal of chlorophyll-*a* was 95 percent or better (Cleasby et al., 1984).

Slow sand filters are effective for controlling a wide variety of microorganisms. Recent work in England, Germany, the Netherlands, and by the World Health Organization has been cited to demonstrate the high efficiency of slow sand filters for removal of bacteria, viruses, and organic and inorganic pollutants (Bellamy et al., 1985). These authors found that a new sand bed was able to remove 85 percent of source water coliform bacteria and 98 percent of *Giardia* cysts. As the sand bed matured biologically, removal of coliform bacteria exceeded 99 percent and removal of *Giardia* was virtually 100 percent. Pyper (1985) evaluated a small municipal slow sand filter (area 37 m^2) that was not being used to produce drinking water, and spiked the influent water with *Giardia* cysts and total coliform bacteria. This full-scale filter, operated at 0.03 gpm/ft^2 (0.08 m/hr), resulted in 3.7- to 4.0-log removal of *Giardia* at temperatures of 7.5 to 21°C, but at temperatures below 1°C, removals ranged from 1.2- to 3-log. Removal of *Cryptosporidium* by slow sand filtration was studied by Schuler, Ghosh, and coworkers at Pennsylvania State University. The pilot slow sand filter achieved log removals of 3.7 or higher when the filter was operated at 0.11 gpm/ft^2 (0.26 m/hr), using 0.27-mm ES sand (Schuler, Ghosh, and Boutros, 1988). Their filtration testing was conducted at ambient temperatures for Shingletown Reservoir water, which ranged from 4.5 to 16.5°C (Ghosh, Schuler, and Gopalan, 1989).

Removal of turbidity often parallels removal of other particulates and is generally excellent, with filtrate turbidity at full-scale plants usually below 0.5 ntu (Cullen and Letterman, 1985), but exceptions have been noted. A low percentage of turbidity was removed in one pilot study of a lake water containing a very fine clay from mountain runoff, and 1 ntu filtered water was seldom produced (Bellamy et al., 1985; Bellamy, Hendricks, and Logsdon, 1985).

Because the slow sand filter is partially a biological process, prechlorination might be expected to hinder performance, but conflicting results have been reported. No detrimental effect of prechlorination was seen in a comparison of three full-scale plants with prechlorination, when compared to four plants without prechlorination (Cullen and Letterman, 1985). In one pilot-scale study, a slow sand filter that received superchlorinated water with about 6 to 12 mg/L free chlorine and with free chlorine in the filtrate performed better than a filter without prechlorination in removal of bacteria and in filter run length (Baumann, Willrich, and Ludwig, 1963). On the other hand, Bellamy, Hendricks, and Logsdon (1985) reported that heavy chlorination of a slow sand filter was quite detrimental to filter performance.

Appropriate Waters and Pretreatment for Slow Sand Filters

Slow sand filters are not successful in treating clay-bearing river waters typical of most of the United States, because the clay penetrates too deeply into the filter and cannot be removed in the normal surface scraping operation. In addition, they are not very effective for removing color, typically achieving only 25 percent removal. Early references suggested that appropriate waters should have less than 30 ppm turbidity and 20 ppm color (Cleasby et al., 1984). The units for these measures have changed since those early reports, and therefore, translating those values to current units is difficult.

A 1984 survey of 27 existing full-scale slow sand filter plants in the United States found 74 percent using lakes or reservoirs as their source water, 22 percent using rivers or streams, and 4 percent using groundwater. The mean turbidity of the source water was about 2 ntu, with the peak about 10 ntu (Slezak and Sims, 1984). Mean lengths of filter runs varied from 42 days in the spring to 60 days in the winter. Wide variations were reported, with two reports of cycles up to one year.

A survey of 7 New York plants (Cullen and Letterman, 1985) including 3 of the plants in the prior reference (Slezak and Sims, 1984) revealed source waters with generally less than 3 ntu turbidity. One plant with 6 to 11 ntu was not meeting the existing (1984) 1-ntu finished water turbidity maximum contaminant level (MCL). Average run lengths varied from 1 to 6 months.

Waters with periodic algal blooms may have short filter runs during such periods. Some measure of plankton should be included in source water guidelines. Cleasby et al. (1984) suggested that acceptable source waters should have turbidity less than 5 ntu and chlorophyll-*a* less than 5 mg/m^3 (μg/L).

In northern climates where freezing will occur on uncovered filters, the run length is of crucial importance. If the filter run terminates in midwinter, the layer of ice will prevent a normal scraping operation. Therefore, if any possibility of runs shorter than the period of ice cover on the filter exists, the filter must be covered.

There is considerable worldwide interest in the use of pretreatment systems to extend the application of slow sand filters to somewhat poorer quality source waters. Various upflow, downflow, and horizontal flow roughing filters have been studied for this purpose (Logsdon, 1991). It is hoped that such devices will make slow sand filtration more widely useful, especially in developing countries where simple technology is of utmost importance.

Preozonation has been evaluated for enhancement of color removal because slow sand filters are not very effective for removal of color and trihalomethane formation potential (THMFP) precursors. Yordanov et al. (1996) reported that preozonation and slow sand filtration reduced color an average of 58 percent at the 16 mgd (60 ML/d) Invercannie Water Treatment Works in Scotland. Color in the source water ranged between 9 and 74 Hazen units and averaged 35 Hazen units. (Hazen units are based on the platinum-cobalt standard and are the same as Pt-Co color units.) The raw water total organic carbon (TOC) concentration, which averaged 5 mg/L, was reduced 28 percent by preozonation and slow sand filtration. The filters at this plant operated at a rate of 0.04 gpm/ft^2 (0.1 m/hr), and ozone was applied at an ozone/TOC dose ratio that generally was less than 0.5.

Preozonation facilities were added to North West Water's Oswestry Treatment Works, a 54-mgd (205 ML/d) slow sand filter plant treating colored upland water in the United Kingdom in 1994 (Cable and Jones, 1996). This plant had one bank of filters receiving water not treated by ozone, while three banks of filters received ozonated water, so a full-scale comparison of the effects of preozonation was possible. Trihalomethane formation potential was reduced an average of 28 percent by slow sand filtration alone, while reduction by preozonation and slow sand filtration averaged 50 percent. The raw water had low pH (about 6), and color typically ranged from 10 to 35 Hazen units. Removal of color averaged 19 percent without ozone and 52 percent with ozone.

In response to the need to effectively treat water for removal of trace levels of pesticides, Thames Water Utilities, Ltd., which serves the London metropolitan area and the River Thames watershed, has developed a slow sand filter incorporating a layer of GAC at mid-depth. The GAC Sandwich™ (Bauer et al., 1996) filter was piloted and then tested in a full-scale filter at the Ashford Common Works. The 33,600 sf (3121 m^2) test filter consisted of a bottom layer of 12 in (300 mm) of filter sand, 6 in (150 mm) of ChemViron (Calgon) F400 GAC, and finally a top layer of 18 in (450 mm) of sand. A control filter of the same size was emptied and resanded with 36 in (900 mm) of clean sand. Both filters were put into service in July 1991 and operated at 0.04 to 0.12 gpm/ft^2 (0.1 to 0.3 m/hr). Long-term testing has shown no difference between the two filters for turbidity and particle removal (Bauer et al., 1996).

The 209-mgd (790 ML/d) Hampton Water Treatment Works was converted to GAC Sandwich™ filters in 19 of 25 beds, which resulted in compliance with European Community standards for pesticides, whereas compliance had not been attained in the prior year without the GAC Sandwich™ filters. At Hampton, removal of THMFP averaged 3.6 percent in May 1995 in sand filters, versus 28 to 42 percent for GAC Sandwich™ filters. Chlorine demand was reduced 12 percent in a control slow sand filter, whereas the filters containing both sand and GAC reduced chlorine demand 33 to 44 percent. Bauer et al. (1996) reported that after two years of use and treatment of more than 36,000 bed volumes, the GAC Sandwich™ filters were still able to produce water of significantly higher quality than conventional slow sand filters, with respect to chlorine demand, THMFP, and TOC.

Physical Details

Early slow sand filter practice has been summarized by Cleasby et al. (1984). Sand effective sizes ranged from 0.15 to 0.40 mm (0.3 mm most common), sand uniformity coefficients from 1.5 to 3.6 (2 most common), and initial bed depths from 1.5 to 5.0 ft (0.46 to 1.52 m), with 3 ft (0.9 m) most common. The sand was supported on graded gravel 6 to 36 in deep (0.15 to 0.91 m), with 18 to 24 in (0.45 to 0.60 m) the common range.

Filtration rates ranged from 0.016 to 0.16 gpm/ft^2 (0.04 to 0.40 m/h), with 0.03 to 0.05 gpm/ft^2 (0.07 to 0.12 m/h) most common on source waters that received no prior pretreatment. Flow rates higher than 0.12 gpm/ft^2 (0.29 m/h) were used only following some pretreatment step to lengthen the filter cycle, such as sedimentation or plain rapid filtration without coagulants.

The available head loss for operating the filter ranged from 2.5 to 14 ft (0.76 to 4.3 m) but was most commonly from 3 to 5 ft (0.9 to 1.5 m). A survey of slow sand filtration plants in operation in the United States in 1984 indicated that the effective size and bed depth of the sand and the filtration rates fell within the ranges just noted (Slezak and Sims, 1984). An experimental filter that was intentionally equipped with an unsieved local sand with an ES of 0.18 mm and a UC of 4.4 performed satisfactorily, indicating the possibility of reducing costs by using less rigid specifications (Slezak and Sims, 1984).

Pyper and Logsdon (1991) reviewed slow sand filter design criteria recommended in Ten State Standards; criteria recommended for European filters; and criteria recommended by Visscher (1990) for filters in developing countries. This review considered Visscher's recommendations to be more pragmatic and appropriate for small water systems in the United States. Visscher recommended filtration rates of 0.04 to 0.08 gpm/ft^2 (0.1 to 0.2 m/hr), sand effective size of 0.15 to 0.30 mm, uniformity coefficient less than 5 but less than 3 preferred, and initial bed depth of 31 to 35 inches (0.8 to 0.9 m), with 12 to 20 inches (0.3 to 0.5 m) of support media beneath the sand.

With the interest in slow sand filtration, the AWWA Research Foundation sponsored development of a design manual for slow sand filters (Hendricks et al., 1991) because this material was generally not covered in university civil engineering curricula. The manual is intended to provide design guidance for engineers who recommend and design slow sand filters for small communities.

Operation and Cleaning of Slow Sand Filters

Slow sand filters are cleaned by removing the *schmutzdecke* along with a small amount of sand, an operation known as *scraping*. The sand that is removed is usually cleaned hydraulically and stockpiled for later replacement in the filter. The scraping operation can be repeated several times until the depth of the filter bed has decreased to about 16 to 20 in (0.4 to 0.5 m), at which time the depth should be replenished—an operation referred to as *resanding* the filter.

The sand in small plants is usually skimmed by hand using broad shovels. Scraping and resanding are labor-intensive operations, and consideration should be given to reducing the manual labor involved. A 1985 survey of seven full-scale plants by Cullen and Letterman reported labor requirements ranging from 4 to 42 h/1000 ft^2 (4 to 46 h/100 m^2) of filter area for one scraping. By the use of efficient motorized buggies or hydraulic transport, and by limiting removal to a 1 in. (2.5 cm) depth, 5 h/1000 ft^2 (5.4 h/100 m^2) was considered adequate (Cullen and Letterman, 1985). Resanding with 6 to 12 in. (15 to 30 cm) of sand requires approximately 50 h/1000 ft^2 (54 h/100 m^2).

One small plant reported use of asphalt rakes to scrape the dirty surface into windrows that were then shoveled into buckets for transport (Seelaus, Hendricks, and Janonis, 1986). Using this technique, only about 0.2 in. (5 mm) was scraped from the filter. This was accomplished by two workers in about 30 min for an 825 ft^2 (77 m^2) filter, about 1.2 h/1000 ft^2 (1.3 h/100 m^2).

Scraping operations for the large slow sand filters serving Salem, Oregon, are conducted mechanically with a minimum of manual shoveling. A self-propelled scraping machine picks up the sand and delivers it by a system of belt conveyors to a truck that is driven alongside the scraper. The truck carries the sand out of the filter to a used sand stockpile. Thus labor is limited mainly to the drivers of the machines.

The longer a filter is drained for the scraping operation, the longer will be the initial improvement period during the subsequent run (Cullen and Letterman, 1985). Therefore, cleaning should be accomplished quickly.

Some plants try to extend the periods between scrapings by raking the surface one to five times between each scraping. The run time gained with each successive raking diminishes, however, and when the scraping is finally required, up to 6 in (15 cm) of sand must be removed to reach clean sand (Cleasby et al., 1984).

An innovative filter-cleaning technique has been developed by the operators of the slow sand filtration plant at West Hartford, Connecticut (Collins, Eighmy, and Malley, 1991). When terminal head loss is reached, the supernatant water is drained down to a depth of about 12 in. (30 cm) over the sand. A rubber-tired tractor equipped with a comb-tooth harrow is then used to stir the top 12 in. (30 cm) of the sand bed. At the same time, supernatant water is drawn off at one side of the filter bed, creating a horizontal flow across the surface of the bed. This flow carries away the dirt and *schmutzdecke* to waste. When the supernatant water level drops to about 3 in. (8 cm), harrowing is stopped and the filter is backfilled to return the supernatant water level to about 12 in. (30 cm). Harrowing is again resumed. This cycle of harrowing and backfilling is continued until the entire filter bed has been cleaned. The authors reported that the run lengths typically were 4 to 8 weeks and that sand has been removed from the filter beds for cleaning at about 8- to 10-year intervals.

Application to Small Systems

Slow sand filters are especially well-adapted for use by small systems because of their simplicity of operation. These filters are quite suitable for the part-time operation affordable at many small systems because coagulation chemistry is not involved. Duties for an operator would include checking head loss, checking and adjusting filter production to provide for system demand, and measuring and recording filtered water turbidity and disinfectant residual.

Tanner and Ongerth (1990) reviewed the performance of slow sand filters operated by small water systems in northern Idaho. During their one-year study, the plants were sampled weekly for measurement of turbidity, total coliforms, fecal coliforms, and heterotrophic plate count bacteria. The authors concluded that if a slow sand filter is designed and operated according to accepted standards, it should provide effective water treatment on a consistent basis. Of special relevance to small systems, they noted that plant operators should be educated about proper operating techniques for slow sand filters. They indicated that drinking water regulatory personnel should provide training to operators and perform sanitary surveys to help assure that slow sand filters are operated and maintained acceptably.

Leland and Damewood (1990) presented the case study of a slow sand filter project in Westfir, Oregon (population 310), from pilot testing through design, construction, and initial operations for almost two years. The authors presented diagrams of a readily constructed, modestly priced slow sand pilot filter, as well as a recommended schedule for collecting data during pilot testing of a filter. The relationship of raw and filtered water turbidity during the first year of full-scale operation was quite different from that for raw and filtered water total coliform bacteria. Whereas coliform removal rapidly improved, and nearly all total coliform samples were less than 1/100 mL after the first couple of months, filtered water turbidity exceeded raw water turbidity for nearly a year. This was attributed to a higher-than-expected content of fines in the filter sand, which points out the potential for water quality problems caused by using inadequately cleaned local sand as the filter medium. Even though a temporary turbidity problem was encountered, this project was a success.

An interesting development for small systems is the use of precast concrete filter boxes (Riesenberg et al., 1995) rather than constructed-in-place concrete filter boxes. The precast boxes were reported to provide better quality control during fabrication at a reduced cost, as compared to filters constructed in place. This type of construction would lend itself well to a concept of modular slow sand filter units and might decrease the time needed for constructing the plant.

Use of slow sand filters by small systems with good quality source waters is expected to increase in the future.

PRECOAT FILTRATION

Precoat filtration is a U.S. Environmental Protection Agency accepted filtration technique for potable water treatment. When the Surface Water Treatment Rule went into effect in 1989, it provoked a renewed interest in this filtration process. Precoat filters use a thin layer of very fine material such as diatomaceous earth as a filter medium. In precoat filtration, the water to be filtered is passed through a uniform layer of the filter medium that has been deposited (precoated) on a *septum,* a permeable material that supports the filter medium. The septum is supported by a rigid structure termed a *filter element.* As the water passes through the filter medium and septum, suspended particles about 2 μm and larger are captured and removed (American Water Works Association, 1995a). The majority of particles removed by the filter are strained (i.e., trapped) at the surface of the layer of filter medium, with some being trapped within the layer. As the filter cycle proceeds, additional filter medium, called *body feed,* is regularly metered into the influent water flow in proportion to the solids being removed. Without the regular addition of body feed, the head loss across the precoat layer would increase rapidly. Instead, the dirt particles intermingle with the body feed particles so that permeability of the cake is maintained as the thickness of the cake gradually increases. By maintaining permeability of the cake in this way, the length of the filter cycle is extended.

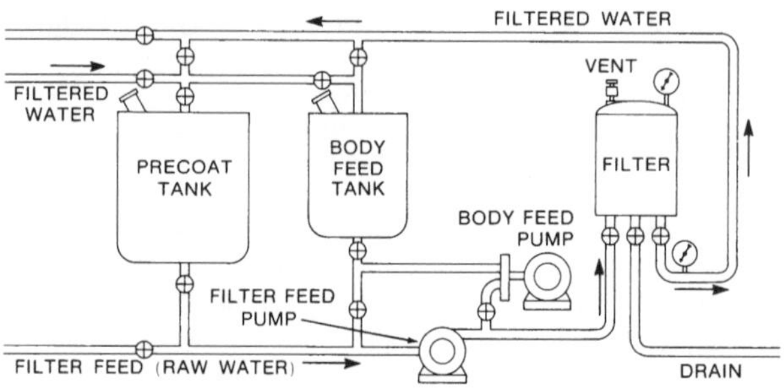

FIGURE 8.33 Typical precoat filtration system. (Courtesy of Manville Filtration & Minerals Division.)

Ultimately, a gradually increasing pressure drop through the filter system reaches a point where continued filtration is impractical. The forward filtration process is stopped; the filter medium and collected dirt are washed off the septum; a new precoat of filter medium is applied; and filtration continues. A typical flow schematic diagram is shown in Figure 8.33.

Precoat filters accomplish particle removal by physically straining the solids out of the water. The thickness of the initial layer of precoat filter medium is normally about ⅛ in. (3 mm), and the water passageways through this layer are so small and numerous that even very fine particles are retained.

Five or six grades of filter medium (sometimes called *filter aid*) are commonly used in potable water treatment. They offer a range of performance with respect to clarity and flow characteristics. With an appropriate selection from among these grades, particles as small as 1 μm can be removed by the precoat filter cake. This includes most surface water impurities. Where colloidal matter or other finely dispersed particles are present, however, filtration alone may not be adequate to produce 1 ntu water as required by the Surface Water Treatment Rule based on earlier results obtained by Logsdon et al. (1981).

Applications and Performance

Precoat filters are widely used in industrial filtration applications and in swimming pool filtration. They also have also been used in municipal potable water treatment, primarily in the direct, in-line filtration of high-quality surface waters (turbidity 10 ntu or less and acceptable color), and in the filtration of iron and manganese from groundwaters after appropriate pretreatment to precipitate these contaminants.

Since 1949, more than 170 potable water treatment plants utilizing precoat filtration with diatomaceous earth or other filter media have been designed, constructed, and operated in the United States (American Water Works Association, 1995a). About 90 percent utilize surface water supplies and 10 percent groundwater supplies. The largest existing plant is a 20-mgd system in San Gabriel, California. A proposed 450-mgd (1700 ML/d) precoat filtration plant for the Croton supply of New York City has been studied on a pilot scale and on a 3-mgd (11.3 ML/d) prototype scale (Principe et al., 1994). The proposed process would utilize preozone, BAC, and diatomaceous earth filtration at 3 gpm/ft^2 (7.3 m/h). Diatomaceous earth would be recovered and reused for body feed application.

Where the source water and other conditions are suitable, precoat filtration can offer a number of benefits to the user, including the following:

1. Capital cost savings may be possible because of smaller land and plant building requirements.
2. Treatment costs may be slightly less than conventional coagulation/sedimentation/granular media filtration when filterable solids are low (AWWA, 1995a; Bryant and Yapijakis, 1977; Ris, Cooper, and Goodard, 1984), although sedimentation would not usually be needed for such high-quality source waters.
3. The process is entirely a physical/mechanical operation and can attain high log removals of *Giardia* cysts and *Cryptosporidium* oocysts (as described in the Introduction section of this chapter) without operator expertise in water chemistry relating to coagulation.
4. The waste residuals are easily dewatered and in some cases may be reclaimed for other uses, including soil conditioning and land reclamation. Research is under way to determine the feasibility of reusing filter medium as body feed.
5. Acceptable finished water clarity is achieved as soon as precoating is complete and filtration starts. A filter-to-waste period is generally not necessary to bring turbidity of the finished water within acceptable limits.
6. Terminal turbidity breakthrough is not generally observed because precoat filtration is dominantly a surface filtration process.

The disadvantages of precoat filtration are:

1. A continued cost is associated with purchase and disposal of the filter medium, which is usually discarded at the end of each filter cycle.
2. Precoat filtration is less cost-effective for waters that require pretreatment for algae, color, and taste and odor problems or DBP precursor removal. Waters containing only larger plankton such as diatoms can sometimes be treated economically by microstraining prior to the precoat filtration.
3. Proper design, construction, and operation are absolutely essential to prevent the filter cake from dropping off of the septa or cracking during operation, which could result in the system failing to remove the target particulates.

Precoat filtration has been shown to be very effective in the removal of *Giardia* cysts over a broad range of operating conditions typical of potable water filtration (Logsdon et al., 1981; Lange et al., 1986). The need for proper operation and maintenance is emphasized, however, and a precoat rate of 0.2 lb/ft^2 (1 kg/m^2) is recommended. Other information on removal of cysts was presented earlier in the Introduction section of this chapter.

The capability of precoat filteration to remove cysts and the fact that chemical coagulation is not required makes this option attractive to very small communities facing filtration because of concern about cysts in source waters. Operators skilled in coagulation are less likely to be available in small installations, but operators possessing mechanical skills are required.

The removal of small particulates such as bacteria and viruses depends on the grade of filter aid used and other operating conditions (Lange et al., 1986). Capture of smaller particles such as bacteria, viruses, and asbestos fibers can be enhanced by using aluminum or iron coagulants or cationic polymers to coat the filter medium (Lange et al., 1986; Brown, Malina, and Moore, 1974; Baumann, 1975; Burns, Baumann, and Oulman, 1970).

The Filter Element and Septum

The filter element is a vital element of a precoat filter because it supports the septum and the filter medium. The attributes of good filter elements and septa are detailed in the *AWWA Precoat Filtration Manual* (American Water Works Association, 1995a). Filter elements may be either *flat* or *tubular* (Figures 8.34 and 8.35). Flat elements, often referred to as *leaves,* may be rectangular, or round. Tubular elements are available in several different cross-sectional shapes, but are generally round.

Overlaying the filter element is the septum material, consisting of either tightly woven stainless steel wire mesh or a tightly fitted synthetic cloth bag, usually made from monofilament polypropylene weave. The precoat layer forms on and is supported by the septum. For this to occur, the size of the clear openings in the septum must be small enough for the filter medium particles to form and maintain stable bridges across the openings. Generally, a clear opening of 0.005 in (about 125 μm) or less in one direction is desirable.

The septum must be firmly supported so that it does not yield, flex, or become distorted as the differential pressure drop increases during the filter cycle, which could result in temporary cracks in the filter medium.

The Filter Vessel

Two basic types of precoat filter vessels that contain multiple filter elements and septa are available. They are the *vacuum filter* and the *pressure filter.*

Vacuum Filters. The vessel containing vacuum filter elements and their septa is an open tank at atmospheric pressure. A filter discharge pump or a vacuum discharge leg downstream of the filter creates a suction. This suction enables the available atmospheric pressure to move water through the precoat and filter medium cake as the cake builds up. The open filter tanks permit easy observation of the condition of the septa and elements and general condition of the filter during filtration and cleaning.

The maximum available differential pressure across the vacuum filter is limited to the net positive suction head of the filter pump or the vacuum leg. This limitation influences the length of filter cycles.

The effect of any entrained air or dissolved gases coming out of solution because of the decrease in pressure also must be considered with vacuum filters. Gas will have an adverse effect on the filter cake, tending to disrupt the integrity of the filter medium on the septum. One advantage of the vacuum filter is that the condition of the cake can be easily observed at the open water surface. This is not possible in a pressure filter.

Pressure Filters. A filter feed pump or influent gravity flow produces higher than atmospheric pressure on the inlet (upstream) side of a pressure filter, forcing liquid through the filter medium cake (Figures 8.2 and 8.35). Large pressure drops across the filter are theoretically possible (limited only by the strength of the filter shell and the filter elements and septa), but the maximum economical differential pressure drop is generally limited to 30 to 40 psi (206 to 276 kPa). Typically, a higher pressure drop across a pressure filter will yield longer cycles than with vacuum filters, which in turn will yield greater removal of suspended solids per pound of filter medium. However, increased pumping costs for differential pressures much over 30 to 40 psi (206 to 276 kPa) usually offset savings in costs of filter medium.

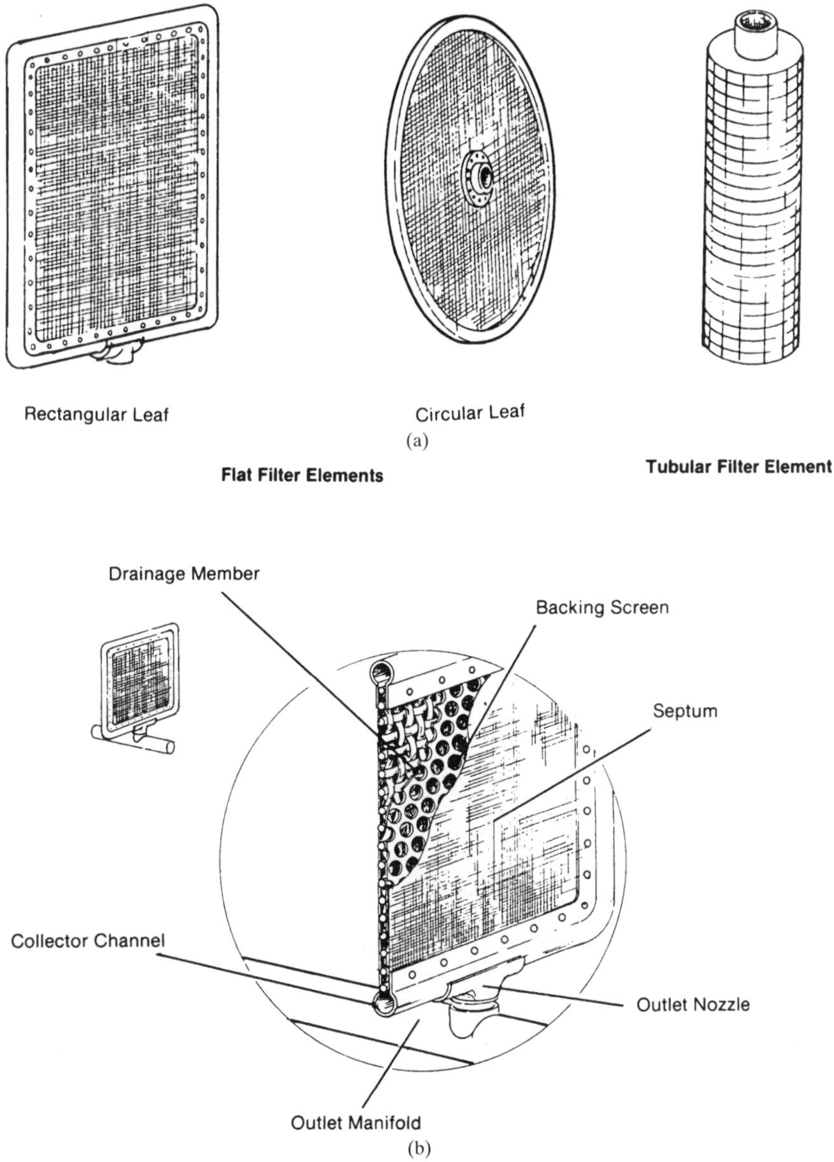

FIGURE 8.34 Types of filter elements for precoat filters, and construction details of a flat-leaf element.

Filter Media

Diatomaceous earth and perlite are used as filter media for precoat filtration of water. Both materials are available in various grades to allow creation of a filter cake with the desired pore size. AWWA B101-94, Standard for Precoat Filter Materials, covers quality control, density, permeability, and particle size distribution of precoat

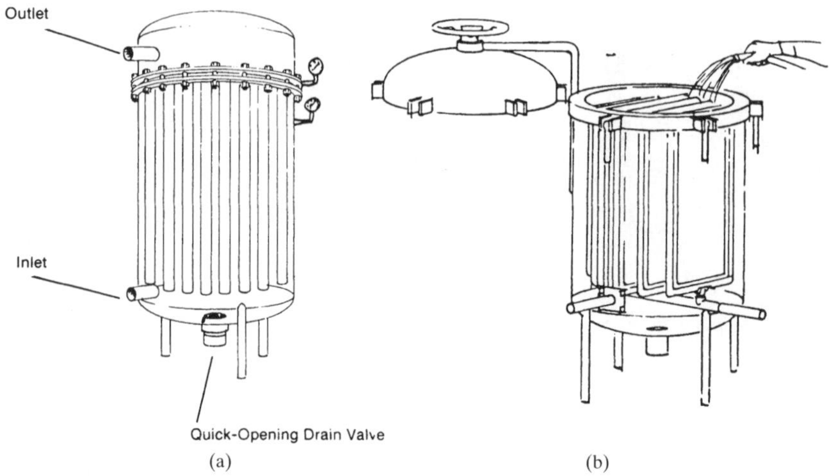

FIGURE 8.35 Typical pressure tubular filter and leaf filter for precoat filtration.

filter media (American Water Works Association, 1995b). Pilot testing can determine the two or three most suitable grades of filter medium that will produce the desired clarity, simplifying the task of optimizing a final selection when working with a full-scale plant.

Filter Operation

Precoating. Successful precoating requires the uniform application of precoat filter aid to the entire surface of the clean septa. This is accomplished by recirculating a concentrated slurry of clean water and filter medium (generally 12 percent or greater) through the filter at 1.0 to 1.5 gpm/ft^2 (2.4 to 3.7 m/h) until most of the medium has been deposited on the septa and the recirculating water turbidity is lowered to the desired treated water quality.

Because particles of the medium are much smaller than the clear openings in the septum, their retention and the formation of a stable precoat take place as a result of bridging. As the particles crowd together when passing through the openings, they jam and interlock, forming bridges over the openings (Figure 8.36). As bridges form, additional particles are caught and the filter septum is coated. A typical precoating system is shown in Figure 8.33. The thickness of the precoat layer is typically $\frac{1}{16}$ to $\frac{1}{8}$ in. (1.5 to 3 mm).

Body-feeding. The amount of body feed to be added to the source water is determined by the nature and amount of the solids to be removed. Pilot testing during representative source water quality periods will generally indicate the type and range of solids that will be encountered and the amount and type of body feed needed to provide an incompressible and permeable cake.

Typical body feed ratios of 1 to 10 mg/L of diatomite for each 1 mg/L of suspended solids are required, depending on the type of solids being filtered. Details of selecting the appropriate ratio will be presented later. Compressible solids such as alum or iron coagulation solids require the highest ratios. Proper control of the body

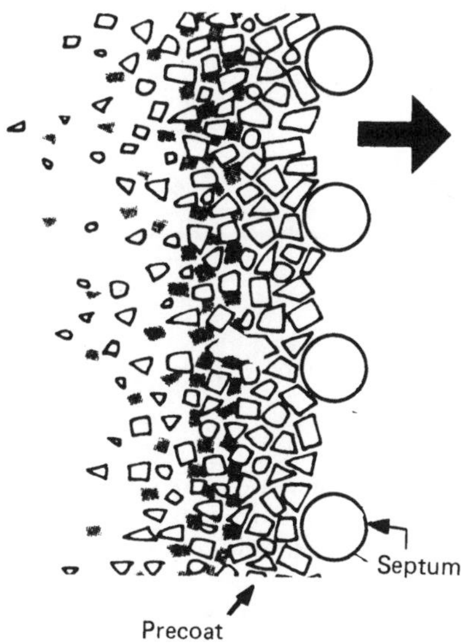

FIGURE 8.36 Illustration of precoat formation, with bridges of precoat between the elements of the septum material. (*Source:* Courtesy of Manville Filtration & Minerals Division.)

feed dosage is the most important factor contributing to economical operation of a precoat filtration plant.

Body feed equipment can be classified as *dry* or *wet* systems. The wet, or *slurry,* feeders are the most common type. The attributes of good precoat and body feed equipment have been presented in the *AWWA Precoat Filtration Manual* (American Water Works Association, 1995a).

Spent Cake Removal. At the end of each filter cycle, the spent filter cake is removed from the filter in preparation for the precoat that will begin the next cycle. If spent solids are not fully removed from the filter vessel, the material could be resuspended and deposited on the filter septa. The resuspended dirty material could foul the septa, although the effect would usually develop gradually so that the operator would become aware of it only after a number of cycles.

The most reliable means of determining the cleanliness of the septa involves visual observation of the bare septa, followed by observation of the uniformity and completeness of precoat. When the septa cannot be fully inspected, a higher than normal differential pressure immediately after precoating (at the start of the filtration cycle) would suggest that the septa are becoming fouled.

Techniques for removing the spent cake vary with the different kinds of filter vessels and filter elements. Spent cake removed from a precoat filter is a mixture of filter medium and the materials removed from the source water. This waste matter is usually removed from the filter in slurry form. Although some systems may be able to dispose of the slurry into a sanitary sewer system, most plants must dewater the waste material and make separate provisions for the solids and liquid wastes. Methods of handling wastes are covered further in Chapter 16, Water Treatment Plant Waste Management.

Theoretical Aspects of Precoat Filtration

Understanding the probable effect of flow rate, terminal head loss, and body feed rate on the operation of the filter is necessary to obtain maximum economy in the design and operation of a precoat filter. The most important of these items affecting filter operation is the prudent use of body feed during a filter run.

Effect of Concentration of Body Feed. The results of a typical series of filter runs conducted at a constant rate of filtration and to a constant terminal pressure drop, but with varying amounts of body feed, are shown in Figure 8.37. The total volume of filtrate is plotted against the total head loss in feet of water. The initial head loss is caused by the clean precoat layer, septum, and filter element. With no body feed

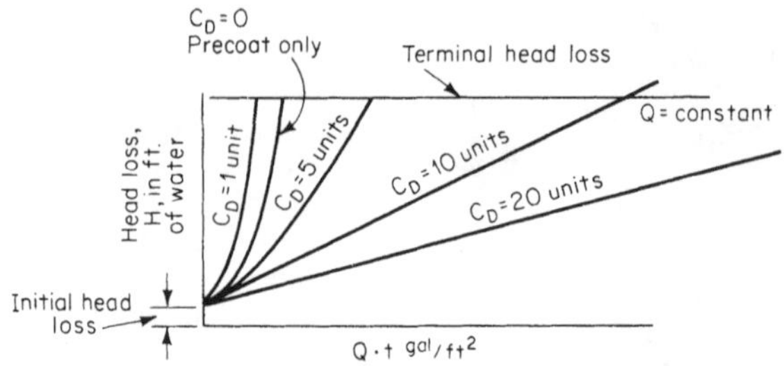

FIGURE 8.37 Effect of varying the concentration of body feed C_D on the relationship of head loss to volume of filtrate for flat septa.

or insufficient amounts of body feed, the head loss will increase more rapidly as the run continues. This is because suspended solids that are removed on the surface of the precoat layer soon form a compressible, impermeable layer of solids that will not readily permit the passage of water.

With increased amounts of body feed, the mixture of body feed and solids removed from the water will be less compressible and of higher porosity and permeability. As a result, the head loss will develop more slowly (e.g., $C_D = 5$ units in Figure 8.37). With sufficient body feed, the cake will be incompressible, and head loss will develop in a linear manner when using flat septa (e.g., $C_D = 10$ units in Figure 8.37). With still higher rates of body feed, the slope of the straight-line plot will get flatter. The most economical use of body feed generally occurs at the lowest concentration that generates a linear head loss curve when using flat septa.

The linear head loss curves shown in Figure 8.37 for higher body feed concentrations only occur with flat septa at constant total flow rates. With tubular septa, as the body feed layer gets thicker, the area of filtration increases and the effective filtration rate decreases for the outer layers of the cake. This causes head loss to increase more slowly as the cycle progresses (Dillingham, Cleasby, and Baumann, 1967a).

Effect of Filtration Rate. Consider the linear head loss curve for a flat septa at a constant flow rate and with adequate body feed as shown in Figure 8.37. The total head loss at any time is the sum of the initial head loss (caused by precoat layer, filter element, and septum) plus the head loss of the body feed layer. The flow through the entire filter cake at the usual filtration velocities is laminar. Therefore, the following can be stated with regard to head loss development.

1. Head loss for a particular instant (i.e., cake thickness) is proportional to the filtration rate Q because of laminar flow conditions.
2. Head loss across the body feed layer is proportional to the thickness of the layer. This, in turn, is proportional to the volume of filtrate V' = filtration rate Q times running time t at constant body feed concentration.
3. Thus, combining 1 and 2 above, the head loss in the body feed layer at any instant is proportional to Q^2t.

Using the above concepts, the rate of head loss development at two different filtration rates Q_1 and Q_2 can be compared:

$$h_1 \propto Q_1^2 t_1 \quad \text{and} \quad h_2 \propto Q_2^2 t_2 \tag{8.26}$$

Dividing these two expressions and for the case of the same terminal head loss, $h_1 = h_2$:

$$\frac{t_1}{t_2} = \frac{Q_2^2}{Q_1^2} \tag{8.27}$$

Similarly,

$$V_1' = Q_1 t_1 \quad \text{and} \quad V_2' = Q_2 t_2 \tag{8.28}$$

Dividing these two equations and inserting the previous relation for t_1/t_2:

$$\frac{V_1'}{V_2'} \simeq \frac{Q_2}{Q_1} \tag{8.29}$$

According to these expressions, for a given terminal head loss, the length of filter run is inversely proportional to the square of the filtration rate, and the filtrate volume is inversely proportional to the filtration rate. (These relationships are approximate because the impact of the precoat head loss has been ignored because it is usually trivial compared to the body feed head loss.) Equations 8.27 and 8.29 help explain why the most economical operation of precoat filters usually occurs at fairly low filtration rates, 1 gpm/ft^2 (2.4 m/h) being most common, but with 0.5 to 3 gpm/ft^2 (1.2 to 7.3 m/h) used in some plants.

Mathematical Model for Precoat Filtration. A rigorous mathematical model has been presented for head loss development for a precoat filter operating at constant total flow rate. The model describes the head loss for either flat or tubular septa operating with adequate body feed to achieve an incompressible cake (Dillingham, Cleasby, and Baumann, 1967a; Weber, 1972). To use the model, empirical filter cake resistance indices must be determined for the precoat and the body feed layers by pilot filter operation. The body feed resistance index can be correlated empirically to the ratio of source water suspended solids concentration to the body feed concentration (Dillingham, Cleasby, and Baumann, 1967b). The model can be used in an optimization program to select the particular combination of filtration rate, terminal head loss, and body feed concentration for best overall economy (Dillingham, Cleasby, and Baumann, 1966).

OTHER FILTERS

A variety of proprietary filtration systems have been developed by manufacturers. Several examples of such filters are described in the following subsections. In some cases, the patents for these filters have expired, and more than one manufacturer may now be producing similar products.

Low-Head Continuous Backwash Filters

The low-head, continuous backwash filter has been used on some supplies (Medlar, 1974). Several manufacturers now offer this type of filter. The filter usually consists of a shallow bed of sand about 1 ft (0.3 m) deep, with an ES of 0.4 to 0.5 mm. The bed is divided into multiple compartments, with the filtrate flowing to a common effluent channel. A traveling-bridge backwash system, equipped with a backwash pump and a backwash collection system, washes each cell of the filter in succession as it traverses the length of the filter. The backwash system is capable of traversing the filter intermittently or continuously.

Because the filter is washed frequently at low head loss, the solids are removed primarily near the top surface of the fine sand. This facilitates removal of the solids during the short backwash period, which lasts only about 15 s for each cell.

A primary advantage of this filter is that it is backwashed frequently at low head loss and therefore does not require as deep a filter box as a conventional gravity filter. This type of filter also eliminates the need for a filter gallery, a large backwash pump, and the associated large backwash piping. Thus the capital cost is lower than for a conventional gravity filter plant.

Some concerns regarding the low-head continuous backwash system deserve consideration, however. The backwash is very brief and is unassisted by auxiliary scour. Therefore, if the filter is blinded by a short-term influent flow containing high levels of solids, cleanup of the filter by the normal backwash routine may be difficult. The flow rate and filtrate quality from individual cells are not monitored. During the backwashing operation, the cells that are washed first immediately return to service while the traveling bridge continues to backwash the remaining cells. Therefore, the cells backwashed first operate at higher rates of filtration for a brief period preceding completion of the entire filter wash—leading to concern over the passage of solids through the newly cleaned cells during this period of higher filtration rate. The impact of the initial improvement period on the filtrate quality is also of concern because the backwashes occur more frequently. Therefore, a greater portion of the total operating time occurs during the initial improvement period.

In response to these concerns, this type of filter is now being marketed with options for deeper medium and with filter-to-waste capability. It is also being preceded by coagulation and sedimentation to enhance its range of application and reliability.

Two-Stage Filtration Systems

Two-stage filtration systems were introduced to extend the application of direct filtration to somewhat poorer quality surface waters. These systems have received considerable use in package plant installations for smaller systems, but they are not limited to small systems. Multiple modules can be utilized in larger systems.

These systems generally utilize a first-stage coarse-medium bed ahead of a conventional multimedia downflow filter. Coagulating chemicals are added ahead of the coarse-medium bed, which serves both as a contact flocculator and as a roughing filter or clarifier to prepare the water for the multimedia filter. Additional coagulants or filter-aiding polymers may be added ahead of the second-stage filter.

In one system of this type, the first stage consists of a bed of plastic chips with density less than water, which float up against a retaining screen in an open tank (Benjes, Edlund, and Gilbert, 1985). Flow is upward through this bed, and it is backwashed periodically with simultaneous air and water. The lower density of the air-water mixture results in the plastic sinking away from the screen so it can be washed more effectively in a semifluidized state. Another system uses a deep bed of coarse sand as the first-stage contact "clarifier," which is operated downflow in a pressure tank and washed periodically upflow using air and water (MacNeill and MacNeill, 1985). Each of these and similar systems coming on the market possesses possible advantages and liabilities. Some systems use a first-stage upflow roughing filter, but there is danger of the medium being lifted as head loss develops. As new systems come on the market, state regulatory agencies usually require demonstation testing before granting approval for full-scale installations. A careful evaluation of the track record of any new system is a prudent precaution before adoption.

Bag and Cartridge Filters

Bag filters and cartridge filters are technology developed for small to very small systems and for point-of-use filtration applications. These filters are basically pressure vessels containing a woven bag or a cartridge with a wound-filament filter. Water passes through the bag or the wound-filament cartridge, and particles large enough to be trapped in the pores of the bag or cartridge are removed.

Bag filters and cartridge filters should be applied only for treatment of high-quality source waters. Chemical coagulation is not employed, and these filters do not remove dissolved substances such as true color or soluble contaminants. Viruses and bacteria can pass through most of these filters. Source waters should have little virus or bacteria contamination, as disinfection will be relied upon to control these microorganisms. High turbidity can clog these filters prematurely, or excessive turbidity can pass through. Algae also can clog bag filters or cartridge filters prematurely.

Cartridge filters are usually rated by their manufacturers as to the nominal particle size to be retained, with the smallest being about 0.2 μm and the largest going up to about 10 μm. Some pore sizes may be smaller or larger than the nominal size, so care must be used in the selection of these filters for removal of *Giardia* cysts and *Cryptosporidium* oocysts. The importance of selecting a proper pore size is complicated by the capability of cysts and oocysts to deform somewhat and to squeeze through pores seemingly small enough to prevent their passage.

As water is filtered through a bag or cartridge filter, eventually the pressure drop becomes so great that terminating the filter run is necessary. When this is done, the filter is not backwashed, but rather the bag or cartridge is thrown away and replaced by a clean one. Filter run length has an important influence on bag filtration and cartridge filtration economics, as frequent bag or cartridge replacement will drive up operating costs.

Monitoring and operating requirements for bag filters and cartridge filters are uncomplicated. Turbidity of filtered water should be monitored daily, and the operator should monitor head loss through the filter and total gallons of water filtered so as to be able to estimate when replacement of the existing bag or cartridge is necessary. Operators need to exercise care in changing filter bags or cartridges when they have become clogged and terminal head loss is attained. The manufacturer's instructions should be followed carefully so that the clean cartridges or bags are not damaged during installation.

Disinfection is an essential process in a treatment train involving a bag or cartridge filter, so extra care is needed for this process. The entire SWTR requirement for virus removal and inactivation needs to be met by disinfection. With the use of free chlorine, this would not be difficult, in most situations. The very short residence time of disinfectants in the filters, though, means that disinfectant contact time may be needed in storage after filtration.

EMERGING TECHNOLOGY

The desire to reduce overall costs or to treat some waters more effectively is driving the development of new technologies involving filtration.

The poor settleability of algal-laden or colored surface waters after coagulation and flocculation has led to the development of proprietary systems incorporating flotation and filtration in the same tank. (See Chapter 7, Sedimentation and Flotation, for further details.)

There is growing use of membrane filters in place of or in addition to granular bed filters. Membrane filters are discussed in Chapter 11.

Various granular filters that use moving media during the filtration process are also on the market. Their primary application is in wastewater filtration, but some consideration is being given to potable water applications.

WASTE DISPOSAL

Wastes generated in the filtering operations covered in this chapter include waste backwash water from rapid filters and precoat filters, and dirty sand from scraping of slow sand filters.

Backwash water in rapid filtration plants is often recovered and recycled through the plant, as discussed earlier. In conventional plants, it may be recycled without sedimentation, in which case the backwash solids end up in the sedimentation tank waste solids. In direct and in-line filtration plants, the solids must be separated before recycle because this is the only waste solids discharge from the plant. The separated solids may be generated in a slurry that must be handled as a waste stream.

Spent cake removed from precoat filters at the end of each filter cycle is a mixture of filter media and organic and inorganic solids removed from the source water. It is usually transported as a slurry to a solids separation step. After separation of the solids from the slurry, the water may be recycled or wasted. The solids are quite dewaterable because of the high content of body feed and precoat media, and may be further dewatered for land disposal.

Dirty sand removed from slow sand filters is a mixture of sand, biological organisms from the *schmutzdecke,* and other organic and inorganic solids captured from the source water. The dirty sand is usually cleaned hydraulically and stockpiled for later replacement in the filter. The hydraulic cleaning generates a slurry of all solids except the sand. This slurry must be handled as a waste stream. If the sand is not recovered, the dirty sand can be disposed of as a solid waste. It is preferable to clean and reuse the sand because it meets the desired size specification and is free of dust after washing.

Methods of handling the waste streams or solids from water plants are covered further in Chapter 16, Water Treatment Plant Waste Management.

BIBLIOGRAPHY

Adin, A., and M. Rebhun. "A Model to Predict Concentration and Head Loss Profiles in Filtration." *Journal AWWA,* 69(8), August 1977: 444.

Al-Ani, M. Y., D. W. Hendricks, G. S. Logsdon, and C. P. Hibler. "Removing Giardia Cysts from Low Turbidity Waters by Rapid Rate Filtration." *Journal AWWA,* 78(5), 1986: 66–73.

American Public Health Association, American Water Works Association, and Water Environment Foundation. *Standard Methods for the Examination of Water and Wastewater,* 19th ed. Washington, DC: APHA, 1995.

American Society for Testing and Materials. *Concrete and Aggregates,* 1993 Annual Book of ASTM Standards, Vol. 04.02. Philadelphia, Pennsylvania: ASTM, 1993.

American Water Works Association. *Water Quality and Treatment,* 3rd ed. New York: McGraw Hill 1971.

American Water Works Association. *AWWA Standard for Filtering Material, Standard B100-96.* Denver, Colorado: AWWA, 1996.

American Water Works Association. *AWWA Standard for Granular Activated Carbon, Standard B604-90.* Denver, Colorado: AWWA, 1991.

American Water Works Association. *Precoat Filtration,* Manual M30. Denver, Colorado: AWWA, 1995a.

American Water Works Association. *AWWA Standard for Precoat Filter Materials, Standard B101-94.* Denver, Colorado: AWWA, 1995b.

Amirtharajah, A. "Optimum Backwashing of Sand Filters." *Jour. Envir. Engr. Div.—ASCE,* 104(EE5), October 1978: 917.

Amirtharajah, A. "Fundamentals and Theory of Air Scour." *Jour. Envir. Engr. Div.—ASCE,* 110(3), June 1984: 573.

Amirtharajah, A., and D. P. Wetstein. "Initial Degradation of Effluent Quality During Filtration." *Journal AWWA,* 72(9), September 1980: 518.

Arboleda, J. V. "Hydraulic Control Systems of Constant and Declining Flow Rate in Filtration." *Journal AWWA,* 66(2), February 1974: 87.

Arboleda-Valencia, J. "Hydraulic Behavior of Declining-Rate Filtration." *Journal AWWA,* 77(12), December 1985: 67.

Aultman, W. W. "Valve Operating Devices and Rate-of-Flow Controllers." *Journal AWWA,* 51(11), November 1959: 1467.

Babbitt, H. E., J. J. Doland, and J. L. Cleasby. *Water Supply Engineering,* 6th ed. New York: McGraw-Hill, 1962.

Bauer, M. J., J. S. Colbourne, D. M. Foster, N. V. Goodman, and A. J. Rachwal. "GAC Enhanced Slow Sand Filtration (GAC Sandwich™)." In: *Advances in Slow Sand And Alternative Biological Filtration,* N. Graham and R. Collins, ed., 223–232. New York: John Wiley & Sons, 1996.

Baumann, E. R., T. Willrich, and D. D. Ludwig. "For Purer Water Supply, Consider Prechlorination." *Agricultural Engineering,* 44(3), 1963: 138. (The above is a condensation of: "Effect of Prechlorination on Filtration and Disinfection." Paper No. 62–208, 7 pp., Amer. Soc. of Agr. Engrs., June 19, 1962.)

Baumann, E. R. "Diatomite Filters for Removal of Asbestos Fibers." In *Proc. AWWA Ann. Conf.,* Minneapolis, Minnesota, 1975.

Baylis, J. R. "Seven Years of High-Rate Filtration." *Journal AWWA,* 48(5), May 1956: 585.

Baylis, J. R. "Nature and Effect of Filter Backwashing." *Journal AWWA,* 51(1), January 1959a: 131.

Baylis, J. R. "Variable Rate Filtration." *Pure Water,* May 1959b.

Baylis, J. R. "Review of Filter Design and Methods of Washing." *Journal AWWA,* 51(11), November 1959c: 1433.

Baylis, J. R. "Discussion of Conley and Pitman." *Journal AWWA,* 52(2) February 1960: 214.

Bellamy, W. D., J. L. Cleasby, G. S. Logsdon, and M. J. Allen. "Assessing Treatment Plant Performance." *Journal AWWA,* 85(12), December 1993, 34–58.

Bellamy, W. D., D. W. Hendricks, and G. S. Logsdon. "Slow Sand Filtration: Influences of Selected Process Variables." *Journal AWWA,* 77(12), December 1985: 62–66.

Bellamy, W. D., G. P. Silverman, D. W. Hendricks, and G. S. Logsdon. "Removing *Giardia* cysts with Slow Sand Filtration." *Journal AWWA,* 77(2), February 1985: 52–60.

Benjes, H. H., C. E. Edlund, and P. T. Gilbert. "Adsorption Clarifier™ Applied to Low Turbidity Surface Supplies." In *Proc. AWWA Ann. Conf.,* Washington, D.C., June 1985.

Bowers, D. A., A. E. Bowers, and D. D. Newkirk. "Development and Evaluation of a Coagulant Control Test Apparatus for Direct Filtration." In *Proc. AWWA Water Qual. Tech. Conf.,* Nashville, Tennessee, 1982.

Brown, W. G. "High-rate filtration experience at Durham, N.C." *Journal AWWA,* 47(3), March 1955: 243.

Brown, T. S., J. F. Malina, Jr., and B. D. Moore. "Virus Removal by Diatomaceous Earth Filtration" (Part 2). *Journal AWWA,* 66(12), December 1974: 735.

Browning, R. C. "New Method and Revised Nomenclature for Granular Filter Flow Control." *Proc. Australian Water and Wastewater Association* (pp. 202–209), Adelaide, Australia, March 23–27, 1987.

Bryant, A., and C. Yapijakis. "Ozonation-Diatomite Filtration Removes Color and Turbidity." *Water and Sewage Works* (October 1977) Part 1, 124:96; Part 2, 124:94.

Burns, D. E., E. R. Baumann, and C. S. Oulman. "Particulate Removal on Coated Filter Media." *Journal AWWA,* 62(2), February 1970: 121.

Cable, C. J., and R. G. Jones. "Assessing the Effectiveness of Ozonation Followed by Slow Sand Filtration in Removing THM Precursor Material from an Upland Water." In: *Advances in Slow Sand And Alternative Biological Filtration,* N. Graham and R. Collins, ed., pp. 29–38. New York: John Wiley & Sons, 1996.

Camp, T. R. "Discussion of Conley—Experiences with Anthracite Filters." *Journal AWWA,* 53(12), December 1961: 1478.

Camp, T. R. "Theory of Water Filtration." *Journal San. Eng. Div.—ASCE,* 90(SA4), 1964:1.

Cleasby, J. L. "Filter Rate Control Without Rate Controllers." *Journal AWWA,* 61(4), April 1969a: 181.

Cleasby, J. L. "Approaches to a Filterability Index for Granular Filters." *Journal AWWA,* 61(8), August 1969b: 372.

Cleasby, J. L. September 1981. "Declining Rate Filtration." *Journal AWWA,* 73(9), September 1981: 484.

Cleasby, J. L. "Status of Declining Rate Filtration Design." *Water Science Technology,* 27(10), 1993: 151.

Cleasby, J. L., A. Arboleda, D. E. Burns, P. W. Prendiville, and E. S. Savage. "Backwashing of Granular Filters." (AWWA Filtration Subcommittee Report). *Journal AWWA,* 69(2), February 1977: 115.

Cleasby, J. L., and E. R. Baumann. "Selection of Sand Filtration Rates." *Journal AWWA,* 54(5), May 1962: 579.

Cleasby, J. L., A. H. Dharmarajah, G. L. Sindt, and E. R. Baumann. *Design and Operation Guidelines for Optimization of the High-Rate Filtration Process: Plant Survey Results.* Denver, Colorado: AWWA Research Foundation, 1989.

Cleasby, J. L., and L. DiBernardo. "Hydraulic Considerations in Declining-Rate Filtration Plants." *Journal Envir. Engr. Div.—ASCE,* 106(EE6), December 1980: 1043.

Cleasby, J. L., and K. S. Fan. "Predicting Fluidization and Expansion of Filter Media." *Journal Envir. Engrg. Div.—ASCE,* 107(EE3), June 1981: 455.

Cleasby, J. L., D. J. Hilmoe, C. J. Dimitracopoulos, and L. M. Diaz-Bossio. "Effective Filtration Methods for Small Water Supplies." *Final Report: USEPA Cooperative Agreement CR808837-01-0, NTIS No. PB84-187-905,* 1984.

Cleasby, J. L., D. J. Hilmoe, and C. J. Dimitracopoulos. "Slow Sand and Direct In-Line Filtration of a Surface Water." *Journal AWWA,* 76(12), December 1984: 44.

Cleasby, J. L., G. L. Sindt, D. A. Watson, and E. R. Baumann. *Design and Operation Guidelines for Optimization of the High-Rate Filtration Process: Plant Demonstration Studies.* Denver, Colorado: AWWA Research Foundation, 1992.

Cleasby, J. L., M. M. Williamson, and E. R. Baumann. "Effect of Rate Changes on Filtered Water Quality." *Journal AWWA,* 55(7), July 1963: 869.

Cleasby, J. L., and C. W. Woods. "Intermixing of dual and multi-media granular filters." *Journal AWWA,* 67(4), April 1975: 197.

Clough, G., and K. J. Ives. "Deep Bed Filtration Mechanisms Observed with Fiber Optic Endoscopes and CCTV." In: *Proceedings 4th World Filtration Congress, Part II,* Ostend, Belgium, 1986. Published by The Royal Flemish Society of Engineers, K.VIV, Jan van Rijswijcklaan 58, B-2018 Antwerpen (Belgium).

Collins, M. R., T. T. Eighmy, and J. P. Malley, Jr. "Evaluating Modifications to Slow Sand Filters," *Journal AWWA,* 83(9), 1991, 62–70.

Committee Report. "The Status of Direct Filtration." *Journal AWWA,* 72(7), July 1980: 405.

Committee Report. "Comparison of Alternative Systems for Controlling Flow through Filters." *Journal AWWA,* 76(1), January 1984: 91.

Conley, W. R. "Experiences with Anthracite Sand Filters." *Journal AWWA,* 53(12), December 1961: 1473.

Conley, W. R. "Integration of the Clarification Process." *Journal AWWA,* 57(10), October 1965: 1333.

Conley, W. R. "High Rate Filtration." *Journal AWWA,* 64(3), March 1972: 203.

Conley, W. R., and K. Y. Hsuing. "Design and Application of Multimedia Filters." *Journal AWWA,* 61(2), February 1969: 97.

Conley, W. R., and R. W. Pitman. "Test Program for Filter Evaluation at Hanford." *Journal AWWA,* 52(2), February 1960a: 205.

Conley, W. R., and R. W. Pitman. "Innovations in Water Clarification." *Journal AWWA,* 52(10), October 1960b: 1319.

Cornwell, D. A., M. M. Bishop, T. R. Bishop, and N. E. McTigue. *Full scale Evaluation of Declining and Constant Rate Filtration.* Denver, Colorado: AWWA Research Foundation, 1991.

Cornwell, D. A. and R. G. Lee. "Waste Stream Recycling: Its Effect on Water Quality." *Journal AWWA,* 86(11), November 1994: 50.

Cullen, T. R., and R. D. Letterman. "The Effect of Slow Sand Filter Maintenance on Water Quality." *Journal AWWA,* 77(12), December 1985: 48.

Culp, R. L. "Direct Filtration." *Journal AWWA,* 68(6), July 1977: 375.

Degremont, G. *Water Treatment Handbook.* Caxton Hill, Hertford, England: Stephen Austin & Sons Ltd., 1973.

Dehab, M. F., and J. C. Young. "Unstratified-Bed Filtration of Wastewater." *Jour. Envir. Engr. Div.—ASCE,* 103(1), February 1977: 21.

Dharmarajah, A. H., and J. L. Cleasby. "Predicting the Expansion of Filter Media." *Journal AWWA,* 78(12), December 1986: 66.

Diaper, E. W. J., and K. J. Ives. "Filtration Through Size-Graded Media." *Jour. San. Eng. Div.—ASCE,* 91(SA3), June 1965: 89.

DiBernardo, L., and J. L. Cleasby. "Declining Rate Versus Constant Rate Filtration." *Journal of Environmental Engineering,* 106(6), 1980: 1023.

Dillingham, J. H., J. L. Cleasby, and E. R. Baumann. "Optimum Design and Operation of Diatomite Filtration Plants." *Journal AWWA,* 58(6), June 1966: 657.

Dillingham, J. H., J. L. Cleasby, and E. R. Baumann. "Diatomite Filtration Equations for Various Septa." *Jour. San. Engr. Div.—ASCE,* 93(SA1), February 1967a: 41.

Dillingham, J. H., J. L. Cleasby, and E. R. Baumann. "Prediction of Diatomite Filter Cake Resistance." *Jour. San. Engr. Div.—ASCE,* 93(SA1), February 1967b: 57.

Eliassen, R. "Clogging of Rapid Sand Filters." *Journal AWWA,* 33(5), May 1941: 926.

Elimelech, M., J. Gregory, X. Jia, and R. A. Williams. *Particle Deposition and Aggregation, Measurement, Modelling and Simulation.* Oxford, UK: Butterworth- Heinemann, Ltd., 1995.

Ergun, S. "Fluid Flow Through Packed Columns." *Chemical Engineering Progress,* 48(2), February 1952a: 89.

Ergun, S. "Determination of geometric surface area of crushed porous solids." *Analytical Chemistry,* 24, 1952b: 388.

Fair, G. M., J. C. Geyer, and D. A. Okun. *Water and Wastewater Engineering* (Vol. 2). New York: John Wiley & Sons, Inc., 1968.

Fox, K. R., R. J. Miltner, G. S. Logsdon, D. L. Dicks, and L. F. Drolet. "Pilot Plant Studies of Slow-Rate Filtration." *Journal AWWA,* 76(12), December 1984: 62.

Fuller, G. W. *The Purification of the Ohio River Water at Louisville, Kentucky.* New York: D. Van Nostrand Co., 1898.

Gainey, P. L., and T. H. Lord. *Microbiology of Water and Sewage.* Englewood Cliffs, New Jersey: Prentice-Hall, Inc. 1952.

Ginn Jr., T. M., A. Amirtharajah, and P. R. Karr. "Effects of Particle Detachment in Granular-Media Filtration." *Journal AWWA,* 84(2), 1992: 66–76.

Ghosh, M. M., P. F. Schuler, and P. Gopalan. "Field Study of *Giardia* and *Cryptosporidium* Removal from Pennsylvania Surface Waters by Slow Sand and Diatomaceous Earth Filtration". Final Report for Department of Environmental Resources, Commonwealth of Pennsylvania. University Park, Pennsylvania: Environmental Resources Research Institute, Pennsylvania State University, 1989.

Graese, S. L., V. L. Snoeyink, and R. G. Lee. *GAC Filter-Adsorbers.* Denver, Colorado: American Water Works Association Research Foundation, 1987.

Great Lakes Upper Mississippi River Board of State Public Health and Environmental Managers. *Recommended Standards for Water Works* (1992 edition). Published by Health Education Service, P.O. Box 7283, Albany, New York 12224.

Haarhoff, J., and J. L. Cleasby. "Biological and Physical Mechanisms in Slow Sand Filtration." In: *Slow Sand Filtration.* G. S. Logsdon, ed., pp. 69–100. New York, NY: Task Committee on Slow Sand Filtration, American Society of Civil Engineers, 1991.

Harris, W. L. "High-Rate Filter Efficiency." *Journal AWWA,* 62(8), August 1970: 515.

Heertges, P. M. and C. E. Lerk. "The Functioning of Deep Filters: Part I, The Filtration of Flocculated Suspensions." *Trans. Instn. of Chem. Eng.,* 45(T138), 1967.

Hendricks, D., J. M. Barrett, J. Bryck, M. R. Collins, B. A. Janonis, and G. S. Logsdon. *Manual of Design for Slow Sand Filtration.* Denver, Colorado: AWWA Research Foundation, 1991.

Herzig, J. P., D. M. LeClerc, and P. LeGoff. "Flow of Suspensions Through Porous Media." *Industrial & Engineering Chemistry,* 62(5), May 1970: 8.

Hewitt, S. R., and A. Amirtharajah. "Air Dynamics Through Filter Media During Air Scour." *Jour. Envir. Engr. Div.—ASCE,* 110(3), June 1984: 591.

Hibler, C. P., C. M. Hancock, L. M. Perger, J. G. Wegrzyn, and K. D. Swabby. *Inactivation of Giardia Cysts with Chlorine at 0.5°C to 5°C.* Denver, Colorado: American Water Works Association Research Foundation, 1987.

Hilmoe, D. J., and J. L. Cleasby. "Comparing Constant Rate and Declining Rate Filtration of a Surface Water." *Journal AWWA,* 78(12), December 1986: 26.

Hsiung, K. Y., and J. L. Cleasby. "Prediction of Filter Performance." *Jour. San. Engrg. Div.—ASCE,* 94(SA6), December 1968: 1043.

Hudson, H. E., Jr. "Filter Design—Declining Rate Filtration." *Journal AWWA,* 51(11), November 1959: 1455.

Hudson, H. E., Jr. "High Quality Water Production and Viral Disease." *Journal AWWA,* 54(10), October 1962: 1265.

Huisman, L. *Rapid Filtration* (Part 1), Delft, The Netherlands: Delft University of Technology, 1974.

Hutchison, W. R. "High-Rate Direct Filtration." *Journal AWWA,* 68(6), June 1976: 292.

Iwasaki, T. "Some Notes on Sand Filtration." *Journal AWWA,* 29(10), December 1937: 1591–1602.

Ives, K. J., and I. Sholji. "Research on Variables Affecting Filtration." *Jour. San. Engrg. Div.—ASCE,* 91(SA4), August 1965: 1.

Ives, K. J. Private communication. 1997.

Kawamura, S. *Integrated Design of Water Treatment Facilities.* New York: John Wiley and Sons, 1991.

King, P. H., R. L. Johnson, C. W. Randall, and G. W. Rehberger. "High-Rate Water Treatment: The State of the Art." *Journal Envir. Engr. Div.—ASCE,* 101(EE4), August 1975: 479.

Kirmeyer, G. J. *Seattle Tolt Water Supply Mixed Asbestiform Removal Study.* EPA 600/2-79-125 (August 1979).

Kirner, J. C., J. D. Littler, and L. A. Angelo. "A Waterborne Outbreak of Giardiasis in Camus, Wash." *Journal AWWA,* 70(1), January 1978: 41.

Kramer, M. H., B. L. Herwaldt, G. F. Craun, R. L. Calderon, and D. D. Juranek. "Waterborne Diseases 1993 and 1994." *Journal AWWA,* 88(3), March 1996: 66–80.

Kreissl, J. F., G. G. Robeck, and G. A. Sommerville. "Use of Pilot Filters to Predict Optimum Chemical Feeds." *Journal AWWA,* 60(3), March 1968: 299.

Krumbein, W. C. "Measurement and Geological Significance of Shape and Roundness of Sedimentary Particles." *Journal Sedimentary Petrology,* 11(2), August 1941: 64.

Lange, K. P., W. D. Bellamy, D. W. Hendricks, and G. S. Logsdon. "Diatomaceous Earth Filtration of *Giardia* Cysts and Other Substances." *Journal AWWA,* 78(1), January 1986: 76–84.

LeChevallier, M. W., W. D. Norton, R. G. Lee, and J. B. Rose. *Giardia and Cryptosporidium in Water Supplies.* Denver, Colorado: AWWA Research Foundation and AWWA, 1991.

Leland, D. E., and M. Damewood III. "Slow Sand Filtration in Small Systems in Oregon." *Journal AWWA,* 82(6), 1990: 50–59.

Letterman, R. D., and G. S. Logsdon. "Survey of Direct Filtration Practice." In *Proc. AWWA Ann. Conf.,* New Orleans, Louisiana, June 1976.

Ling, J. T. "A Study of Filtration Through Uniform Sand Filters." In *Proc. ASCE San. Engr. Div.* (Paper 751), 1955.

Lippy, E. C. "Tracing a Giardiasis Outbreak at Berlin, N. H." *Journal AWWA,* 70(9), September 1978: 512.

Lloyd, B. J. "The Significance of Protozoal Predation and Adsorption for the Removal of Bacteria by Slow Sand Filtration." In: *Advances in Slow Sand And Alternative Biological Filtration,* N. Graham and R. Collins, eds., pp. 129–138. New York: John Wiley & Sons, 1996.

Logsdon, G. S. (ed.). *Slow Sand Filtration. A Report of the Task Committee on Slow Sand Filtration.* New York: American Society of Civil Engineers, 1991.

Logsdon, G. S. *Water Filtration for Asbestos Fiber Removal.* EPA 600/2-79-206, December 1979.

Logsdon, G. S., L. Mason, and J. B. Stanley, Jr. "Troubleshooting an Existing Treatment Plant." *Journal of New England Water Works Association,* 104(1), March 1990: 43–56.

Logsdon, G. S. and J. M. Symons. "Removal of Asbestiform Fibers by Water Filtration." *Journal AWWA,* 69(9), September 1977: 499.

Logsdon, G. S., J. M. Symons, and T. J. Sorg. "Monitoring Water Filters for Asbestos Removal." *Jour. Envir. Engrg. Div.—ASCE,* 107(6), December 1981: 1297.

Logsdon, G. S., J. M. Symons, R. L. Hoye, Jr., and M. M. Arozarena. "Removal of *Giardia* Cysts and Cyst Models by Filtration." *Journal AWWA,* 73(2), February 1981: 111.

Logsdon, G. S., V. C. Thurman, E. S. Frindt, and J. G. Stoecker. "Evaluating Sedimentation and Various Filter Media for Removal of Giardia Cysts." *Journal AWWA,* 77(2), February 1985: 61–66.

MacNeill, J. S., Jr., and A. MacNeill. *Feasibility Study of Alternative Technology for Small Community Water Supply.* Project Summary EPA-600/S2-84-191. Cincinnati, Ohio: USEPA Water Engineering Research Laboratory, March 1985.

Mayhugh, J. R., Sr., J. A. Smith, D. B. Elder, and G. S. Logsdon. "Filter Media Rehabilitation at a Lime-Softening Plant." *Journal AWWA,* 88(8), August 1996: 64–69.

McBride, D. G., R. C. Siemak, C. H. Tate, and R. R. Trussell. "Pilot Plant Investigations for Treatment of Owens River Water." In *Proc. AWWA Ann. Conf.,* Anaheim, California, June 1977.

McBride, D. G., and G. F. Stolarik. "Pilot to Full-Scale: Ozone and Deep Bed Filtration at the Los Angeles Aqueduct Filtration Plant." *AWWA Seminar Proceedings, Coagulation and Filtration: Pilot to Full Scale.* No. 20017. Annual Conference, Kansas City, Missouri, June 14, 1987.

McCabe, W. L., and J. C. Smith. *Unit Operations of Chemical Engineering,* (3rd ed.). New York: McGraw Hill, 1976.

Medlar, S. "This Filter Cleans Itself." *The American City,* 89(6), June 1974: 63.

Mohanka, S. S. "Multilayer Filtration." *Journal AWWA,* 61(10), October 1969: 504.

Monscvitz, J. T., D. J. Rexing, R. G. Williams, and J. Heckler. "Some Practical Experience in Direct Filtration." *Journal AWWA,* 70(10), October 1978: 584.

Montgomery, James M. Consulting Engineers. *Water Treatment Principles and Design.* New York: John Wiley & Sons, 1985.

Murray, B. A., and S. J. Roddy. "Treatment of Sydney's Water Supplies." *Water,* 20(2), April 1993: 17–20, 38.

Nieminski, E. C., and J. E. Ongerth. "Removing Giardia and Cryptosporidium by Conventional Treatment and Direct Filtration." *Journal AWWA,* 87(9), 1995: 96–106.

Oeben, R. W., H. P. Haines, and K. J. Ives. "Comparison of Normal and Reverse Graded Filtration." *Journal AWWA,* 60(4), April 1968: 429.

O'Connor, J. T., B. J. Brazos, W. C. Ford, L. L. Dusenberg, and B. Summerford. "Chemical and Microbiological Evaluation of Drinking Water Systems in Missouri." In *Proc. AWWA Water Qual. Tech. Conf.,* American Water Works Association, Denver, Colorado, 1984.

O'Melia, C. R., and W. Stumm. "Theory of Water Filtration." *Journal AWWA,* 59(11), November 1967: 1393.

Patania, N. L., J. G. Jacangelo, L. Cummings, A. Wilczak, K. Riley, and J. Oppenheimer. *Optimization of Filtration for Cyst Removal.* Denver, Colorado: AWWA Research Foundation, 1995.

Payment, P. *A Prospective Epidemiological Study of the Gastrointestinal Health Effects due to the Consumption of Drinking Water.* Denver, Colorado: AWWA Research Foundation, 1998.

Peterson, D. L., F. X. Schleppenbach, and T. M. Zaudtke. "Studies of Asbestos Removal by Direct Filtration of Lake Superior Water." *Journal AWWA,* 72(3), March 1980: 155.

Principe, M. R., D. Mastronardi, D. Bailey, and G. Fulton. "New York City's First Water Filtration Plant," pp 147–173. *Proceedings AWWA Annual Conference,* New York, New York, 1994.

Pyper, G. R. *Slow Sand filter and Package Treatment Plant Evaluation: Operating Costs and Removal of Bacteria, Giardia, and Trihalomethanes.* USEPA/600/2-85/052, Cincinnati, Ohio: U.S. Environmental Protection Agency, 1985.

Pyper, G. R., and G. S. Logsdon. "Slow Sand Filter Design." In: *Slow Sand Filtration,* G. S. Logsdon, ed., pp. 122–149. New York: Task Committee on Slow Sand Filtration, American Society of Civil Engineers, 1991. New York, NY.

Rajagopalan, R., and C. Tien. "Trajectory Analysis of Deep Bed Filtration with the Sphere-in-cell Porous Media Model." *American Inst. Chem. Eng. J.,* 22, 1976: 523–533.

Riesenberg, F., B. B. Walters, A. Steele, and R. A. Ryder. "Slow Sand Filters for a Small Water System." *Journal AWWA,* 87(11), 1995: 48–56.

Rimer, A. E. "Filtration Through a Trimedia Filter." *Jour. San. Engr. Div.—ASCE,* 94(SA3), June 1968: 521.

Ris, J. L., I. A. Cooper, and W. R. Goodard. "Pilot Testing and Predesign of Two Water Treatment Processes for Removal of *Giardia lamblia* in Palisade, Colorado." *Proc. AWWA Ann. Conf.,* Dallas, Texas, June 1984.

Robeck, G. G. "Discussion of Conley." *Journal AWWA,* 58(1), January 1966: 94.

Robeck, G. G., K. A. Dostal, and R. L. Woodward. "Studies of Modification in Water Filtration." *Journal AWWA,* 56(2), February 1964: 198.

Saatci, A. A., and C. S. Oulman. "The Bed Depth Service Time Design Method for Deep Bed Filtration." *Journal AWWA,* 72(9), September 1980: 524.

Schuler, P. F., M. M. Ghosh, and S. N. Boutros. "Comparing the Removal of Giardia and Cryptosporidium Using Slow Sand and Diatomaceous Earth Filtration." In *Proceedings AWWA Annual Conference,* Denver, Colorado, June 19–23, 1988.

Schuler, P. F., and M. M. Ghosh. "Diatomaceous Earth Filtration of Cysts and Other Particulates Using Chemical Additives." *Journal AWWA,* 82(12), 1990: 67–75.

Seelaus, T. J., D. W. Hendricks, and B. A. Janonis. "Design and Operation of a Slow Sand Filter." *Journal AWWA,* 78(12), December 1986: 35.

Short, A. C., M. B. Gilton, and R. E. Henderson. "Blending Problems Overcome During the Design and Start Up of the Modesto Domestic Water Project." Proceedings AWWA Annual Conference (Water Quality, Vol. D, paper 56–7B, pp. 217–235), Toronto, Ontario, June 23–27, 1996.

Slezak, L. A., and R. C. Sims. "The Application and Effectiveness of Slow Sand Filtration in the United States." *Journal AWWA,* 76(12), December 1984: 38.

Stanley, D. R. "Sand Filtration Studies with Radio-Tracers." *Proc. Amer. Soc. Civ. Eng.,* 81, 1955: 592.

Tanner, S. A., and J. E. Ongerth. "Evaluating the Performance of Slow Sand Filters in Northern Idaho." *Journal AWWA,* 82(12), 1990: 51–61.

Tate, C. H., J. S. Lang, and H. L. Hutchinson. "Pilot Plant Tests of Direct Filtration." *Journal AWWA,* 69(7), July 1977: 379.

Tate, C. H., and R. R. Trussell. "Use of Particle Counting in Developing Plant Design Criteria." *Journal AWWA,* 70(12), December 1978: 691.

Tien, C. *Granular Filtration of Aerosols and Hydrosols.* Stoneham, Massachusetts: Butterworth Publishers, 1989.

Trefethen, J. M. *Geology for Engineers,* 2nd ed. Princeton, New Jersey: D. Van Nostrand Co., 1959.

Trussell, R. R., A. R. Trussell, J. S. Lang, and C. H. Tate. "Recent Development in Filtration System Design." *Journal AWWA,* 72(12), December 1980: 705.

Tuepker, J. L., and C. A. Buescher, Jr. "Operation and Maintenance of Rapid Sand Mixed-Media Filters in a Lime Softening Plant." *Journal AWWA,* 60(12), December 1968: 1377.

Vicory, A. H., and L. Weaver. "Controlling Discharges of Water Plant Wastes to the Ohio River." *Journal AWWA,* 76(4), April 1984: 122.

Visscher, J. T. "Slow Sand Filtration: Design, Operation, and Maintenance." *Journal AWWA,* 82(6), 1990: 67–71.

Wagner, E. G., and H. E. Hudson. "Low-Dosage, High-Rate Direct Filtration." *Journal AWWA,* 74(5), May 1982: 256.

Weber, W. J. *Physicochemical Processes for Water Quality Control.* New York: Wiley-Interscience, 1972.

Wen, C. Y., and Y. H. Yu. "Mechanics of Fluidization." *Chemical Engineering Progress Symposium Series 62,* vol. 62, New York, 1966.

Yao, K. M., M. T. Habibian, and C. O'Melia. "Water and Waste Water Filtration: Concepts and Applications." *Environmental Science & Technology,* 5(11), November 1971: 1105.

Yordanov, R. V., A. J. Lamb, M. A. L. Melvin, and J. Littlejohn. "Biomass Characteristics of Slow Sand Filters Receiving Ozonated Water." In: *Advances in Slow Sand And Alternative Biological Filtration,* N. Graham and R. Collins, eds., pp. 107–118. New York: John Wiley & Sons, 1996.

CHAPTER 9
ION EXCHANGE AND INORGANIC ADSORPTION

Dennis A. Clifford, Ph.D., P.E., DEE
Professor and Chairman
Department of Civil and Environmental Engineering
University of Houston
Houston, Texas

INTRODUCTION AND THEORY OF ION EXCHANGE

Contaminant cations such as calcium, magnesium, barium, strontium, and radium, and anions such as fluoride, nitrate, fulvates, humates, arsenate, selenate, chromate, and anionic complexes of uranium can be removed from water by using ion exchange with resins or by adsorption onto hydrous metal oxides such as activated alumina (AAl) granules or coagulated Fe(II), Fe(III), Al(III), and Mn(IV) surfaces. This chapter deals only with the theory and practice of ion exchange with resins and adsorption with activated alumina (AAl). The reader interested in cation and anion adsorption onto hydrous metal oxides in general is referred to Schindler's and Stumm's publications on the solid-water interface (Schindler, 1981; Stumm, 1992) as a starting point.

Ion exchange with synthetic resins and adsorption onto activated alumina are water treatment processes in which a presaturant ion on the solid phase, the *adsorbent,* is exchanged for an unwanted ion in the water. In order to accomplish the exchange reaction, a packed bed of ion-exchange resin beads or alumina granules is used. Source water is continually passed through the bed in a downflow or upflow mode until the adsorbent is exhausted, as evidenced by the appearance (*breakthrough*) of the unwanted contaminant at an unacceptable concentration in the effluent.

The most useful ion-exchange reactions are reversible. In the simplest cases, the exhausted bed is regenerated using an excess of the presaturant ion. Ideally, no permanent structural change takes place during the exhaustion/regeneration cycle. (Resins do swell and shrink, however, and alumina is partially dissolved during regeneration.) When the reactions are reversible, the medium can be reused many times before it must be replaced because of irreversible fouling or, in the case of alumina, excessive attrition. In a typical water supply application, from 300 to as many as 300,000 *bed volumes* (BV) of contaminated water may be treated before exhaustion. Regeneration typically requires from 1 to 5 bed volumes of regenerant, followed by 2 to 20 bed volumes of rinse water. These wastewaters generally amount to less than 2 percent of the product water; nevertheless, their ultimate disposal is a major consideration in modern design practice. Disposal of the spent media may also present a problem if it contains a toxic or radioactive substance such as arsenic or radium.

Uses of Ion Exchange in Water Treatment

By far the largest application of ion exchange to drinking water treatment is in the area of softening, that is, the removal of calcium, magnesium, and other polyvalent cations in exchange for sodium. The ion-exchange softening process is applicable to both individual home use and municipal treatment. It can be applied for whole-house (*point-of-entry* or POE) softening or for softening only the water that enters the hot water heater. Radium and barium are ions more preferred by the resin than calcium and magnesium; thus the former are also effectively removed during ion-exchange softening. Resins beds containing chloride-form anion exchange resins can be used for nitrate, arsenate, chromate, selenate, *dissolved organic carbon* (DOC), and uranium removal, and more applications of these processes will be seen in the future. Activated alumina is being used to remove fluoride and arsenate from drinking water, particularly high *total dissolved solids* (TDS) waters, at point-of-use (POU), (POE), and municipal scales.

The choice between ion exchange or alumina adsorption (to remove arsenic from water, for example) is largely determined by (a) the background water quality—including TDS level, competing ions, alkalinity, and contaminant concentration—and (b) the resin or alumina affinity for the contaminant ion in comparison with the competing ions. The affinity sequence determines the run length, chromatographic peaking (if any), and process costs. As previously mentioned, process selection will be affected by spent regenerant and spent medium disposal requirements, and regenerant reuse possibilities, particularly if hazardous materials are involved. Each of these requirements is dealt with in some detail in the upcoming design sections for the specific processes summarized in Table 9.1.

Past and Future of Ion Exchange

Natural zeolites (i.e., crystalline aluminosilicates) were the first ion exchangers used to soften water on a commercial scale. Later, zeolites were completely replaced by synthetic resins because of the latters' faster exchange rates, longer life, and higher capacity. Aside from softening, the use of ion exchange for removal of specific contaminants from municipal water supplies has been limited. This is primarily because of the expense involved in removing what is perceived as only a minimal health risk resulting from contaminants such as fluoride, nitrate, or chromate. The production of pure and ultrapure water by ion-exchange demineralization (IXDM) is the largest use of ion exchange resins on a commercial scale. The complete removal of contaminants, which occurs in demineralization (DM) processes, is not necessary for drinking water treatment, however. Furthermore, costs associated with these treatments

TABLE 9.1 Advantages and Disadvantages of Packed-Bed Inorganic Contaminant Removal Processes

Ion exchange
Advantages • Operates on demand. • Relatively insensitive to flow variations, short contact time required. • Relatively insensitive to trace-level contaminant concentration. • Essentially zero level of effluent contaminant possible. • Large variety of specific resins available. • Beneficial selectivity reversal commonly occurs upon regeneration. • In some applications, spent regenerant may be reused without contaminant removal.
Disadvantages • Potential for chromatographic effluent peaking when using single beds. • Variable effluent quality with respect to background ions when using single beds. • Usually not feasible at high levels of sulfate or total dissolved solids. • Large volume/mass of regenerant must be used and disposed of.
Activated alumina adsorption
Advantages • Operates on demand. • Relatively insensitive to total dissolved solids and sulfate levels. • Low effluent contaminant level possible. • Highly selective for fluoride and arsenic.
Disadvantages • Both acid and base are required for regeneration. • Relatively sensitive to trace-level contaminant concentration. • Media tend to dissolve, producing fine particles. • Slow adsorption kinetics and relatively long contact time required. • Significant volume/mass of spent regenerant to neutralize and dispose of.

are high compared with those of the alternative membrane processes (i.e., reverse osmosis and electrodialysis) for desalting water (see Chapter 11).

Adherence to governmentally mandated *maximum contaminant levels* (MCLs) for *inorganic contaminants* (IOCs) will mean more use of ion exchange and alumina for small-community water treatment operations to remove barium, arsenic, nitrate, fluoride, uranium, and other IOCs. An AWWA survey (1985) indicates that 400 communities exceeded the 10 mg/L nitrate-N MCL, 400 exceeded the 4.0 mg/L fluoride MCL (USEPA, 1985), and 200 exceeded the 2.0 mg/L secondary limit on barium. Regarding radiological contaminants, an estimated 1,500 communities exceed the proposed 20 μg/L MCL for uranium (USEPA, 1991), and many others may exceed the MCL goal for radon (Rn) contamination when it is established. In most of these cases, new contaminant-free sources cannot readily be developed, and a treatment system will eventually be installed.

ION EXCHANGE MATERIALS AND REACTIONS

An ion exchange resin consists of a crosslinked polymer matrix to which charged functional groups are attached by covalent bonding. The usual matrix is polystyrene crosslinked for structural stability with 3 to 8 percent divinylbenzene. The common functional groups fall into four categories: strongly acidic (e.g., sulfonate, —SO_3^-); weakly acidic (e.g., carboxylate, —COO^-); strongly basic (e.g., quaternary amine, —$N^+(CH_3)_3$); and weakly basic (e.g., tertiary amine—$N(CH_3)_2$).

A schematic presentation of the resin matrix, crosslinking, and functionality is shown in Figure 9.1. The figure is a schematic three-dimensional bead (sphere) made up of many polystyrene polymer chains held together by divinylbenzene crosslinking. The negatively charged ion exchange sites (—SO_3^-) or (—COO^-) are fixed to the resin backbone or *matrix*, as it is called. Mobile positively charged counterions (positive charges in Figure 9.1) are associated by electrostatic attraction with each negative ion exchange site. The resin exchange *capacity* is measured as the number of fixed charge sites per unit volume or weight of resin. *Functionality* is the term used to identify the chemical composition of the fixed-charge site, for example sulfonate (—SO_3^-) or carboxylate (—COO^-). *Porosity* (e.g., microporous, gel, or macroporous) is the resin characterization referring to the degree of openness of the polymer structure. An actual resin bead is much tighter than implied by the schematic, which is shown as fairly open for purposes of illustration only. The water

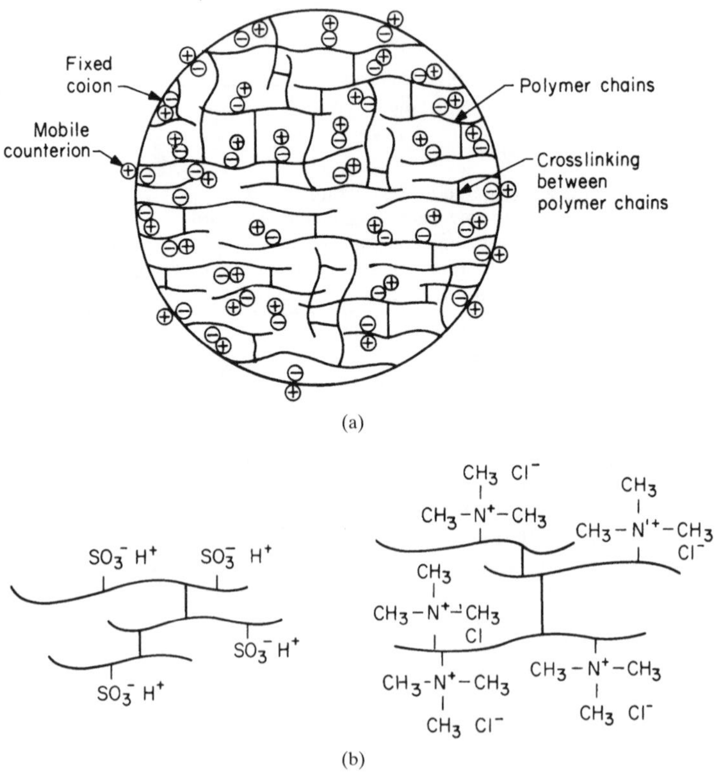

FIGURE 9.1 (*a*) Organic cation-exchanger bead comprising polystyrene polymer cross-linked with divinylbenzene with fixed coions (minus charges) of negative charge balanced by mobile positively charged counterions (plus charges). (*b*) Strong-acid cation exchanger (*left*) in the hydrogen form and strong-base anion exchanger (*right*) in the chloride form.

(40 to 60 percent by weight) present in a typical resin bead is not shown. This resin-bound water is an extremely important characteristic of ion exchangers because it strongly influences both the exchange kinetics and thermodynamics.

Strong- and Weak-Acid Cation Exchangers

Strong acid cation (SAC) exchangers operate over a very wide pH range because the sulfonate group, being strongly acidic, is ionized throughout the pH range (1 to 14). Three typical SAC exchange reactions follow. In Equation 9.1, the neutral salt $CaCl_2$, representing noncarbonate hardness, is said to be *split* by the resin, and hydrogen ions are exchanged for calcium, even though the equilibrium liquid phase is acidic because of HCl production. Equations 9.2 and 9.3 are the standard ion exchange softening reactions in which sodium ions are exchanged for the hardness ions Ca^{2+}, Mg^{2+}, Fe^{2+}, Ba^{2+}, Sr^{2+}, and/or Mn^{2+}, either as noncarbonate hardness (Equation 9.2) or carbonate hardness (Equation 9.3). In all these reactions, R denotes the resin matrix, and the overbar indicates the solid (resin) phase.

$$2\,\overline{RSO_3^-H^+} + CaCl_2 \Leftrightarrow \overline{(RSO_3^-)_2Ca^{2+}} + 2HCl \tag{9.1}$$

$$2\,\overline{RSO_3^-Na^+} + CaCl_2 \Leftrightarrow \overline{(RSO_3^-)_2Ca^{2+}} + 2NaCl \tag{9.2}$$

$$2\,\overline{RSO_3^-Na^+} + Ca(HCO_3)_2 \Leftrightarrow \overline{(RSO_3^-)_2Ca^{2+}} + 2NaHCO_3 \tag{9.3}$$

Regeneration of the spent resin is accomplished using an excess of concentrated (0.5 to 3.0 M) HCl or NaCl, and amounts to the reversal of Equations 9.1 through 9.3.

Weak acid cation (WAC) resins can exchange ions only in the neutral to alkaline pH range because the functional group, typically carboxylate ($pK_a = 4.8$), is not ionized at low pH. Thus, WAC resins can be used for carbonate hardness removal (Equation 9.4) but fail to remove noncarbonate hardness, as is evident in Equation 9.5.

$$2\,\overline{RCOOH} + Ca(HCO_3)_2 \Rightarrow \overline{(RCOO^-)_2Ca^{2+}} + H_2CO_3 \tag{9.4}$$

$$2\,\overline{RCOOH} + CaCl_2 \Leftarrow \overline{(RCOO^-)_2Ca^{2+}} + 2HCl \tag{9.5}$$

If Equation 9.5 were to continue to the right, the HCl produced would be so completely ionized that it would *protonate* (i.e., add a hydrogen ion to the resin's weakly acidic carboxylate functional group, and prevent exchange of H^+ ions for Ca^{2+} ions). Another way of expressing the fact that Equation 9.5 does not proceed to the right is to say that WAC resins will not *split neutral salts* (i.e., they cannot remove noncarbonate hardness). This is not the case in Equation 9.4, in which the basic salt, $Ca(HCO_3)_2$, is split because a very weak acid, H_2CO_3 ($pK_1 = 6.3$), is produced.

In summary, SAC resins split basic and neutral salts (remove carbonate and noncarbonate hardness), whereas WAC resins split only basic salts (remove only carbonate hardness). Nevertheless, WAC resins have some distinct advantages for softening, namely TDS reduction, no increase in sodium, and very efficient regeneration resulting from the carboxylate's high affinity for the regenerant H^+ ion.

Strong- and Weak-Base Anion Exchangers

The use of *strong-base anion* (SBA) exchange resins for nitrate removal is a fairly recent application of ion exchange for drinking water treatment (Clifford and W. J. Weber, 1978; Guter, 1981), although they have been used in water demineralization for decades. In anion exchange reactions with SBA resins, the quaternary amine functional group (—$N^+[CH_3]_3$) is so strongly basic that it is ionized, and therefore useful as an ion exchanger over the pH range of 0 to 13. This is shown in Equations 9.6 and 9.7, in which nitrate is removed from water by using hydroxide or chloride-form SBA resins. (Note that R_4N^+ is another way to write the quaternary exchange site, —$N^+(CH_3)_3$)

$$\overline{R_4N^+OH^-} + NaNO_3 \Leftrightarrow \overline{R_4N^+NO_3^-} + NaOH \tag{9.6}$$

$$\overline{R_4N^+Cl^-} + NaNO_3 \Leftrightarrow \overline{R_4N^+NO_3^-} + NaCl \tag{9.7}$$

In Equation 9.6 the caustic (NaOH) produced is completely ionized, but the quaternary ammonium functional group has such a small affinity for OH^- ions that the reaction proceeds to the right. Equation 9.7 is a simple ion exchange reaction without a pH change. Fortunately, all SBA resins have a much higher affinity for nitrate than chloride (Clifford and W. J. Weber, 1978), and Equation 9.7 proceeds to the right at near-neutral pH values.

Weak-base anion (WBA) exchange resins are useful only in the acidic pH region where the primary, secondary, or tertiary amine functional groups (Lewis bases) are protonated and thus can act as positively charged exchange sites for anions. In Equation 9.8 chloride is, in effect, being adsorbed by the WBA resin as hydrochloric acid, and the TDS level of the solution is being reduced. In this case, a positively charged Lewis acid-base adduct (R_3NH^+) is formed, which can act as an anion exchange site. As long as the solution in contact with the resin remains acidic (just how acidic depends on basicity of the R_3N:, sometimes pH $\leq$ 6 is adequate), ion exchange can take place as is indicated in Equation 9.9—the exchange of chloride for nitrate by a WBA resin in acidic solution. If the solution is neutral or basic, no adsorption or exchange can take place, as indicated by Equation 9.10. In all these reactions, R represents either the resin matrix or a functional group such as —CH_3 or —C_2H_5, and overbars represent the resin phase.

$$\overline{R_3N:} + HCl \Leftrightarrow \overline{R_3NH^+Cl^-} \tag{9.8}$$

$$\overline{R_3NH^+Cl^-} + HNO_3 \Leftrightarrow \overline{R_3NH^+NO_3^-} + HCl \tag{9.9}$$

$$\overline{R_3N:} + NaNO_3 \Rightarrow \text{no reaction} \tag{9.10}$$

Although no common uses of WBA resins are known for drinking water treatment, useful ones are possible (Clifford and W. J. Weber, 1978). Furthermore, when activated alumina is used for fluoride and arsenic removal, it acts as if it were a weak-base anion exchanger, and the same general rules regarding pH behavior can be applied. Another advantage of weak-base resins in water supply applications is the ease with which they can be regenerated with bases. Even weak bases such as lime ($Ca[OH]_2$) can be used, and regardless of the base used, only a small stoichiometric excess (less than 20 percent) is normally required for complete regeneration.

Activated Alumina Adsorption

Packed beds of activated alumina can be used to remove fluoride, arsenic, selenium, silica, and humic materials from water. Coagulated Fe(II) and Fe(III) oxides (McNeill and Edwards, 1995; Scott, Green et al., 1995) and iron oxides coated onto sands (Benjamin, Sletten et al., 1996) can also be employed to remove these anions, but these processes are not covered in this chapter. The mechanism, which is one of exchange of contaminant anions for surface hydroxides on the alumina, is generally called *adsorption,* although *ligand exchange* is a more appropriate term for the highly specific surface reactions involved (Stumm, 1992).

The typical activated aluminas used in water treatment are 28- × 48-mesh (0.3- to 0.6-mm-diameter) mixtures of amorphous and gamma aluminum oxide (γ-Al_2O_3) prepared by low-temperature (300 to 600°C) dehydration of precipitated $Al(OH)_3$. These highly porous materials have surface areas of 50 to 300 m^2/g. Using the model of an hydroxylated alumina surface subject to protonation and deprotonation, the following ligand exchange reaction (Equation 9.11) can be written for fluoride adsorption in acid solution (alumina exhaustion) in which ≡Al represents the alumina surface and an overbar denotes the solid phase.

$$\overline{\equiv\text{Al–OH}} + H^+ + F^- \Rightarrow \overline{\equiv\text{Al–}F^-} + \text{HOH} \qquad (9.11)$$

The equation for fluoride desorption by hydroxide (alumina regeneration) is presented in Equation 9.12.

$$\overline{\equiv\text{Al–F}} + OH^- \Rightarrow \overline{\equiv\text{Al–OH}} + F^- \qquad (9.12)$$

Another common application for alumina is arsenic removal, and reactions similar to Equations 9.11 and 9.12 apply for exhaustion and regeneration when $H_2AsO_4^-$ is substituted for F^-.

Activated alumina processes are sensitive to pH, and anions are best adsorbed below pH 8.2, a typical *zero point of charge* (ZPC), below which the alumina surface has a net positive charge, and excess protons are available to fuel Equation 9.11. Above the ZPC, alumina is predominantly a cation exchanger, but its use for cation exchange is relatively rare in water treatment. An exception is encountered in the removal of radium by plain and treated activated alumina (Clifford, Vijjeswarapu et al., 1988; Garg and Clifford, 1992).

Ligand exchange as indicated in Equations 9.11 and 9.12 occurs chemically at the internal and external surfaces of activated alumina. A more useful model for process design, however, is one that assumes that the adsorption of fluoride or arsenic onto alumina at the optimum pH of 5.5 to 6.0 is analogous to weak-base anion exchange. For example, the uptake of F^- or $H_2AsO_4^-$, requires the protonation of the alumina surface, and that is accomplished by preacidification with HCl or H_2SO_4, and reducing the feed water pH into the 5.5 to 6.0 region. The positive charge caused by excess surface protons may then be viewed as being balanced by exchanging anions (i.e., ligands such as hydroxide, fluoride, and arsenate). To reverse the adsorption process and remove the adsorbed fluoride or arsenate, an excess of strong base (e.g., NaOH) must be applied. The following series of reactions (9.13–9.17) is presented as a model of the adsorption/regeneration cycle that is useful for design purposes.

The first step in the cycle is acidification, in which neutral (water-washed) alumina (Alumina·HOH) is treated with acid (e.g., HCl), and protonated (acidic) alumina is formed as follows:

$$\overline{\text{Alumina·HOH}} + \text{HCl} \Rightarrow \overline{\text{Alumina·HCl}} + \text{HOH} \qquad (9.13)$$

When HCl-acidified alumina is contacted with fluoride ions, they strongly displace the chloride ions providing that the alumina surface remains acidic (pH 5.5 to 6.0). This displacement of chloride by fluoride, analogous to ion exchange, is shown as

$$\overline{\text{Alumina·HCl}} + \text{HF} \Rightarrow \overline{\text{Alumina·HF}} + \text{HCl} \qquad (9.14)$$

To regenerate the fluoride-contaminated adsorbent, a dilute solution of 0.25 to 0.5 N NaOH alkali is used. Because alumina is both a cation and an anion exchanger, Na^+ is exchanged for H^+, which immediately combines with OH^- to form HOH in the alkaline regenerant solution. The regeneration reaction of fluoride-spent alumina is

$$\overline{\text{Alumina·HF}} + 2\text{NaOH} \Rightarrow \overline{\text{Alumina·NaOH}} + \text{NaF} + \text{HOH} \qquad (9.15)$$

Recent experiments have suggested that Equation 9.15 can be carried out using fresh or recycled NaOH from a previous regeneration. This suggestion is based on the field studies of Clifford and Ghurye (1998) in which arsenic-spent alumina was regenerated with equally good results using fresh or once-used 1.0 M NaOH. The spent regenerant, fortified with NaOH to maintain its hydroxide concentration at 1.0 M, probably could have been used many times, but the optimum number of spent-regenerant reuse cycles was not determined in the field study.

To restore the *fluoride removal capacity,* the basic alumina is acidified by contacting it with an excess of dilute acid, typically 0.5 N HCl or H_2SO_4:

$$\overline{\text{Alumina·NaOH}} + 2\text{HCl} \Rightarrow \overline{\text{Alumina·HCl}} + \text{NaCl} + \text{HOH} \qquad (9.16)$$

The acidic alumina, alumina·HCl, is now ready for another fluoride (or arsenate or selenite) ligand-exchange cycle as summarized by Equation 9.14. Alternatively, the feed water may be acidified prior to contact with the basic alumina, thereby combining acidification and adsorption into one step as summarized by Equation 9.17:

$$\overline{\text{Alumina·NaOH}} + \text{NaF} + 2\text{HCl} \Rightarrow \overline{\text{Alumina·HF}} + 2\text{NaCl} + \text{HOH} \qquad (9.17)$$

The modeling of the alumina adsorption-regeneration cycle as being analogous to weak-base anion exchange fails in regard to regeneration efficiency, which is excellent for weak-base resins but quite poor on alumina. This is caused by the need for excess acid and base to partially overcome the poor kinetics of the semicrystalline alumina, which exhibits very low solid-phase diffusion coefficients compared with resins that are well-hydrated, flexible gels offering little resistance to the movement of hydrated ions. A further reason for poor regeneration efficiency on alumina is that alumina is amphoteric and reacts with (consumes) excess acid and base to produce soluble forms ($Al(H_2O)_6^{3+}$, $Al(H_2O)_2(OH)_4^-$) of aluminum. Resins are totally inert in this regard (i.e., they are not dissolved by regenerants).

Special-Purpose Resins

Resins are practically without limit in their variety because polymer matrices, functional groups, and capacity and porosity are controllable during manufacture. Thus, numerous special-purpose resins have been made for water-treatment applications. For example, bacterial growth can be a major problem with anion resins in some water supply applications because the positively charged resins tend to "adsorb" the negatively-charged bacteria that metabolize the adsorbed organic material—negatively charged humate and fulvate anions. To correct this problem special resins have been invented, which contain bacteriostatic long-chain *quaternary amine functional groups* ("quats") on the resin surface. These immobilized quats kill bacteria on contact with the resin surface (Janauer, Gerba et al., 1981).

The strong attraction of polyvalent humate and fulvate anions (*natural organic matter,* [NOM]) for anion resins has been used as the basis for removal of these *total organic carbon* (TOC) compounds from water by using special highly porous resins. Both weak- and strong-base macroporous anion exchangers have been manufactured to remove these large anions from water. The extremely porous resins originally thought to be necessary for adsorption of the large organic anions tended to be structurally weak and break down easily. More recently, however, it has been shown that both gel and standard macroporous resins, which are highly crosslinked and physically very strong, can be used to remove NOM (Fu and Symons, 1990). Regeneration of resins used to remove NOM is often a problem because of the strong attraction of the aromatic portion of the anions for the aromatic resin matrix. This problem has at least been partially solved using acrylic-matrix SBA resins. More details on the use of ion exchange resins to remove NOM appears later in this chapter.

Resins with chelating functional groups such as imino-diacetate (Calmon, 1979), amino-phosphonate, and ethyleneamine (Matejka and Zirkova, 1997) have been manufactured that have extremely high affinities for hardness ions and troublesome metals such as Cu^{2+}, Zn^{2+}, Cr^{3+}, Pb^{2+}, and Ni^{2+}. These resins are used in special applications such as trace-metal removal and metals-recovery operations (Brooks, Brooks et al., 1991). The simplified structures of these resins are shown in Figure 9.2. Table 9.2 summarizes the features of some of the special ion exchangers available commercially from a variety of sources (Purolite, 1995).

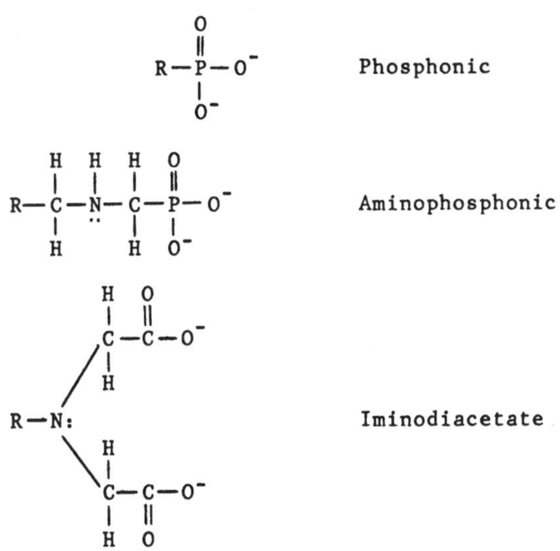

FIGURE 9.2 Structure of highly selective cation exchangers for metals removal.

ION EXCHANGE EQUILIBRIUM

Selectivity Coefficients and Separation Factors

Ion exchange resins do not prefer all ions equally. This variability in preference is often expressed semiquantitatively as a position in a selectivity sequence or, quantitatively, as a separation factor, α_{ij}, or a selectivity coefficient, K_{ij}, for binary exchange. The selectivity, in turn, determines the run length to breakthrough for the contaminant ion; the higher the selectivity, the longer the run length. Consider, for example, Equation 9.18, the simple exchange of Cl^- for NO_3^- on an anion exchanger

TABLE 9.2 Special Ion Exchangers—Commercially Available

Type of resin	Functional group	Typical application
Chelating	Thio-uronium	Selective removal of metals, especially mercury.
Chelating	Imino-diacetic	Selective removal of polyvalent ions, especially transition metals.
Chelating	Amino-phosphonic	Decalcification of brine and removal of metals from wastewaters
Silver impregnated, SAC	Sulfonic	Softening resin with bacteriostatic properties
NSS, Nitrate-over-sulfate selective (sulfate rejecting), hydrophobic, SBA	Triethyl and tripropyl quaternary amines	Nitrate removal in high sulfate waters
Iodine releasing	Quaternary amine SBA in triiodide form, —$R_4N^+I_3^-$	Disinfection by iodine release into product water

Source: Purolite, 1995.

whose equilibrium constant is expressed numerically in Equation 9.19 and graphically in Figure 9.3*a:*

$$\overline{Cl^-} + NO_3^- \Rightarrow \overline{NO_3^-} + Cl^- \tag{9.18}$$

$$K = \frac{\{\overline{NO_3^-}\}\{Cl^-\}}{\{\overline{Cl^-}\}\{NO_3^-\}} \tag{9.19}$$

In Equations 9.18 to 9.20, overbars denote the resin phase, and the matrix designation R has been removed for simplicity; K is the thermodynamic equilibrium constant, and braces denote ionic activity. Concentrations are used in practice because they are measured more easily than activities. In this case, Equation 9.20 based on concentration, the selectivity coefficient $K_{N/Cl}$ describes the exchange. Note that $K_{N/Cl}$ includes activity coefficient terms that are functions of ionic strength and, thus, is not a true constant (i.e., it varies somewhat with different ionic strengths).

$$K_{N/Cl} = \frac{[\overline{NO_3^-}][Cl^-]}{[\overline{Cl^-}][NO_3^-]} = \frac{q_N C_{Cl}}{q_{Cl} C_N} \tag{9.20}$$

where [ ] = concentration, mol/L
q_N = resin phase equivalent concentration (normality) of nitrate, eq/L
C_N = aqueous phase equivalent concentration (normality), eq/L

The binary separation factor $\alpha_{N/Cl}$, used throughout the literature on separation practice, is a most useful description of the exchange equilibria because of its simplicity and intuitive nature:

$$\alpha_{ij} = \frac{\text{distribution of ion } i \text{ between phases}}{\text{distribution of ion } j \text{ between phases}} = \frac{y_i / x_i}{y_j / x_j} \tag{9.21}$$

$$\alpha_{N/Cl} = \frac{(y_N/x_N)}{(y_{Cl}/x_{Cl})} = \frac{y_N\, x_{Cl}}{x_N\, y_{Cl}} = \frac{(q_N/q)\,(C_{Cl}/C)}{(C_N/C)\,(q_{Cl}/q)} \tag{9.22}$$

where y_i = equivalent fraction of ion i in resin, q_i/q
y_N = equivalent fraction of nitrate in resin, q_N/q
x_i = equivalent fraction of ion i in water, C_N/C
x_N = equivalent fraction of nitrate in water, C_N/C
q_N = concentration of nitrate on resin, eq/L
q = total exchange capacity of resin, eq/L
C_N = nitrate concentration in water, eq/L
C = total ionic concentration of water, eq/L

Equations 9.20 and 9.22 show that for homovalent exchange (i.e., monovalent/monovalent and divalent/divalent exchange), the separation factor α_{ij} and the selectivity coefficient K_{ij} are equal. This is expressed for nitrate/chloride exchange as

$$K_{N/Cl} = \alpha_{N/Cl} = \frac{q_N\, C_{Cl}}{C_N\, q_{Cl}} \tag{9.23}$$

For exchanging ions of unequal valence (i.e., heterovalent exchange), the separation factor is not equivalent to the selectivity coefficient. Consider, for example, the case of sodium ion-exchange softening as represented by Equation 9.24, the simplified form of Equation 9.2:

$$2\,\overline{Na^+} + Ca^{2+} = \overline{Ca^{2+}} + 2\,Na^+ \tag{9.24}$$

$$K_{Ca/Na} = \frac{q_{Ca}\, C_{Na}^2}{C_{Ca}\, q_{Na}^2} \tag{9.25}$$

Using a combination of Equations 9.21 and 9.25,

$$\alpha_{\text{divalent/monovalent}} \quad \text{or} \quad \alpha_{Ca/Na} = K_{Ca/Na}\,\frac{q\, y_{Na}}{C\, x_{Na}} \tag{9.26}$$

The implication from these equations is that the intuitive separation factor for divalent/monovalent exchange depends inversely on solution concentration C and directly on the distribution ratio y_{Na}/x_{Na} between the resin and the water, with q constant. The higher the solution concentration C, the lower the divalent/monovalent separation factor [i.e., selectivity tends to reverse in favor of the monovalent ion as ionic strength—I (which is a function of C)—increases]. This reversal of selectivity is discussed in detail in the following paragraphs

Selectivity Sequences

A *selectivity sequence* describes the order in which ions are preferred by a particular resin or by a solid porous oxide surface such as AlOOH (activated alumina granules or hydrated aluminum oxide precipitate), FeOOH (hydrous iron oxide), or MnOOH (hydrous manganese oxide). Although special-purpose resins (such as chelating resins) can have unique selectivity sequences, the commercially available cation and anion resins exhibit similar selectivity sequences. These are presented in Table 9.3, where the most-preferred ions (i.e., those with the highest separation factors) are listed at the top of the table and the least-preferred ions are at the bottom. For example, the $\alpha_{Ca/Na}$ value of 1.9 means that at equal concentration in the aqueous phase, calcium is preferred by the resin 1.9/1.0 over sodium (on the basis of equivalents, not moles). Weak acid cation resins with carboxylic functional groups exhibit the same selectivity sequence as SAC resins except that hydrogen is the most pre-

TABLE 9.3 Relative Affinities of Ions for Resins*

Strong acid cation resins†		Strong base anion resins‡	
Cation, i	α_{i/Na^+}	Anion, i	α_{i/Cl^-} §
Ra^{2+}	13.0	$UO_2(CO_3)_3^{4-}$	3200
Ba^{2+}	5.8	ClO_4^- ¶	150
Pb^{2+}	5.0	CrO_4^{2-}	100
Sr^{2+}	4.8	SeO_4^{2-}	17
Cu^{2+}	2.6	SO_4^{2-}	9.1
Ca^{2+}	1.9	$HAsO_4^{2-}$	4.5
Zn^{2+}	1.8	HSO_4^-	4.1
Fe^{2+}	1.7	NO_3^-	3.2
Mg^{2+}	1.67	Br^-	2.3
K^+	1.67	SeO_3^{2-}	1.3
Mn^{2+}	1.6	HSO_3^-	1.2
NH_4^+	1.3	NO_2^-	1.1
Na^+	1.0	Cl^-	1.0
H^+	0.67	BrO_3^-	0.9
		HCO_3^-	0.27
		CH_3COO^-	0.14
		F^-	0.07

* Above values are approximate separation factors for 0.005–0.010 *N* solution (TDS = 250–500 mg/L as $CaCO_3$).

† SAC resin is polystyrene divinylbenzene matrix with sulfonate functional groups.

‡ SBA resin is polystyrene divinylbenzene matrix with $-N^+(CH_3)_3$ functional groups (i.e., a Type 1 resin).

§ Separation factors are approximate and are based on various literature sources and on experiments performed at the University of Houston.

¶ ClO_4^-/Cl^- separation factor is for polystyrene SBA resins; on polyacrylic SBA resins, the ClO_4^-/Cl^- separation factor is approximately 5.0.

ferred cation, and the magnitude of the separation factors differ from those in Table 9.3. Similarly, WBA resins and SBA resins exhibit the same selectivity sequence, except that hydroxide is most preferred by WBA resins, and the WBA separation factors differ in magnitude but have the same trend as those in Table 9.3.

Some general rules govern selectivity sequences. In dilute solution (e.g., in the TDS range of natural waters) the resin prefers the ion with the highest charge and lowest degree of hydration.

Selectivity is affected by the nature of the ion. Hydrophobic ions (e.g., nitrate and chromate) prefer hydrophobic resins (i.e., highly crosslinked macroporous resins without polar matrices and/or functional groups), whereas hydrophilic ions (e.g., bicarbonate and acetate) prefer moderately crosslinked (gel) resins with polar matrices and/or functional groups. Divalent ions, (e.g., sulfate and calcium) prefer resins with closely spaced exchange sites, where their need for two charges can be satisfied (Clifford and W. J. Weber, 1983; Sengupta and Clifford, 1986; Subramonian and Clifford, 1988; Horng and Clifford, 1997).

Activated alumina operated in the acidic to neutral pH range for anion adsorption has a selectivity sequence that differs markedly from anion exchange resins. Fortunately, some of the ions such as fluoride, which is least preferred by resins (and therefore not amenable to removal by resins) are highly preferred by the alumina. Based on research by the author, his coworkers, and other investigators (Trussell and

Trussell et al., 1980; Singh and Clifford, 1981; Rosenblum and Clifford, 1984; Schmitt and Pietrzyk, 1985), activated alumina operated in the pH range of 5.5 to 8.5 prefers anions in the following order:

$$OH- > H_2AsO_4^-, Si(OH)_3O^- > F^- > HSeO_3^- > SO_4^{2-} > CrO_4^{2-}$$
$$>> HCO_3^- > Cl^- > NO_3^- > Br^- > I^- \quad (9.27)$$

Humic- and fulvic-acid anions are more preferred than sulfate, but because of their widely differing molecular weights and structures, and the different pore-size distributions of commercial aluminas, no exact position in the selectivity sequence can be assigned. Reliable separation factors for ions in the above selectivity sequence (such as fluoride, arsenate, silicate, and biselenite) are not available in the literature, but this is not particularly detrimental to the design effort because alumina has an extreme preference for these ions. For example, when fluoride or arsenate is removed from water, the presence of the usual competing ions—bicarbonate and chloride—is nearly irrelevant in establishing run length to contaminant ion breakthrough (Singh and Clifford, 1981; Rosenblum and Clifford, 1984). Sulfate does, however, offer some small but measurable competition for adsorption sites. The problem with the extremely preferred ions is that they are difficult to remove from the alumina during regeneration, which necessitates the use of hazardous, chemically strong (e.g., NaOH and H_2SO_4), and potentially destructive (of the medium) regenerants.

Isotherm Plots

The values of α_{ij} and K_{ij} can be determined from a constant-temperature, equilibrium plot of resin-phase concentration versus aqueous-phase concentration (i.e., the ion-exchange isotherm). Favorable and unfavorable isotherms are depicted in Figure 9.3*a* and *b*, where each curve depicts a constant separation factor, $\alpha_{NO3/Cl}$ for Figure 9.3*a* and $\alpha_{HCO3/Cl}$ for Figure 9.3*b*.

A "favorable" isotherm (convex to *x*-axis) means that species *i* (NO_3^- in Figure 9.3*a*), which is plotted on each axis, is preferred to species *j* (Cl^- in Figure 9.3*a*), the hidden or exchanging species. An "unfavorable" isotherm (concave to the *x*-axis) indicates that species *i* (HCO_3^- in Figure 9.3*b*) is less preferred than *j* (Cl^- in Figure

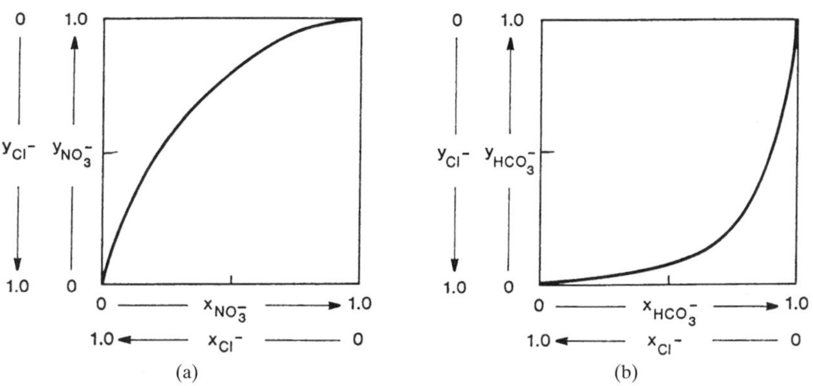

FIGURE 9.3 (*a*) Favorable isotherm for nitrate-chloride exchange according to reaction (9.18) with constant separation factor $\alpha_{NO^3/Cl} > 1.0$. (*b*) Unfavorable isotherm for bicarbonate-chloride exchange with constant separation factor $\alpha_{HCO_3/Cl} < 1.0$.

9.3*b*). During column exhaustion processes, favorable isotherms result in sharp breakthroughs when i is in the feed and j is on the resin, whereas unfavorable isotherms lead to gradual breakthroughs under these conditions. (This is discussed in detail later, under the heading "Column Processes and Calculations" where Figure 9.8 is explained.) In viewing these binary isotherms, note that

$$x_i + x_j = 1.0 \tag{9.28}$$

$$C_i + C_j = C \tag{9.29}$$

$$y_i + y_j = 1.0 \tag{9.30}$$

$$q_i + q_j = q \tag{9.31}$$

Therefore, the concentration or equivalent fraction of either ion can be directly obtained from the plot, which in Figure 9.3*a* and *b* is a "unit" isotherm because equivalent fractions (x_i, y_j) rather than concentrations have been plotted in the range 0.0 to 1.0. Figure 9.3*a* represents the favorable isotherm for nitrate-chloride exchange, and Figure 9.3*b* the unfavorable isotherm for bicarbonate-chloride exchange.

For nonconstant separation factors (e.g., the divalent/monovalent [Ca^{2+}/Na^+] exchange case described by Equations 9.24 and 9.26, a separate isotherm exists for every total solution concentration C. As the solution concentration or TDS level decreases, the resin exhibits a greater preference for the divalent ion, as evidenced by a progressively higher and more convex isotherm. The phenomenon can be explained by solution theory: As the solution concentration increases, the aqueous phase becomes more ordered. This results in polyvalent ion activity coefficients that are significantly less than 1.0 (i.e., the tendency for polyvalent ions to escape from the water into the resin is greatly diminished, leading to a reduction in the height and convexity of the isotherm). This phenomenon of diminishing preference for higher-valent ions with increasing ionic strength I of the solution has been labeled *electroselectivity* and can eventually lead to selectivity reversal, whereupon the isotherm becomes concave (Helfferich, 1962). This trend is shown in Figure 9.4, where the sulfate-chloride isotherm is favorable in 0.06 N solution and unfavorable in 0.6 N solution.

The exact ionic strength at which electroselectivity reversal occurs is dependent on the ionic makeup of the solution, and highly dependent on the resin structure (Boari, Liberti et al., 1974) and its inherent affinity for polyvalent ions. Electroselectivity reversal is very beneficial to the sodium ion exchange softening process in that it causes the divalent hardness ions to be highly preferred in dilute solution ($I \leq 0.020\ M$) during resin exhaustion and highly nonpreferred (i.e., easily rejected) during regeneration with relatively concentrated (0.25 to 2.0 M) salt solution.

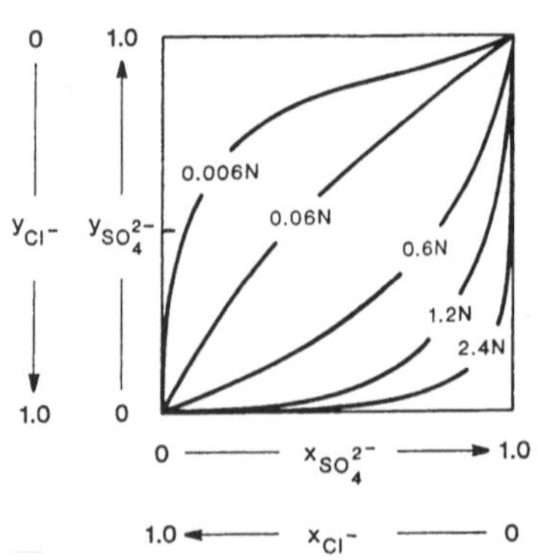

FIGURE 9.4 Electroselectivity of a typical type 1 strong-base anion-exchange resin used for divalent-monovalent (SO_4^{2-}/Cl^-) anion exchange.

EXAMPLE 9.1 The following solved example problem briefly describes the experimental technique necessary to obtain isotherm data and illustrates the calculations required to construct a nitrate-chloride isotherm for a strong-base anion exchange resin. By using the isotherm data or the plot, the individual and average separation factors α_{ij} can be calculated. Only minor changes are necessary to apply the technique to weak-base resins or to cation resins. For example, acids (HCl and HNO_3) rather than sodium salts would be used for equilibration of weak-base resins.

To obtain the data for this example, weighted amounts of air-dried chloride-form resin of known exchange capacity were placed in capped bottles containing 100 mL of 0.005 *N* (5.0 meq/L) $NaNO_3$ and equilibrated by tumbling for 16 hours. Following equilibration, the resins were settled, and the nitrate and chloride concentrations of the supernatant water were determined for each bottle. The nitrate/chloride equilibrium data are in Table 9.4. The total capacity q of the resin is 3.63 meq/g. Note that the units of resin capacity used here are meq/g rather than eq/L, because for precise laboratory work a mass rather than volume of resin must be used.

SOLUTION

1. Verify that, within the expected limits of experimental error, the total concentration C of the aqueous phase at equilibrium is 0.005 *N*. Large deviations from this value usually indicate that concentrated salts were absorbed in the resin and leached out during the equilibration procedure. This problem can be avoided by extensive prewashing of the resin with the same normality of salt, in this case 0.005 *N* NaCl, as is used for equilibration.
2. Calculate the equivalent fractions, x_N and x_{Cl}, of nitrate and chloride in the water at equilibrium.
3. Using the known total capacity of the resin, q_{Cl}, calculate the milliequivalents (meq) of chloride remaining on the resin at equilibrium by subtracting the meq of chloride found in the water.
4. Calculate the meq of nitrate on the resin, q_N, by assuming that all the nitrate removed from solution is taken up by the resin.
5. Calculate the equivalent fractions, y_N and y_{Cl}, of nitrate and chloride in the resin phase at equilibrium.
6. Calculate the separation factor, α_{ij}, which is equal to the selectivity coefficient, K_{ij}, for homovalent exchange.
7. Repeat steps (1) through (6) for all equilibrium data points, and plot the isotherm.

Solution (with the Equilibrium Data Point for 0.2 g Resin as an Example):

1. $C = C_N + C_{Cl} = 1.17 + 3.78 = 4.95$ meq/L (This is well within the expected ±5 percent limits of experimental error; (5.00 – 4.95)/5.00 = 1.0 percent error)

TABLE 9.4 Example Data for Plot of Nitrate/Chloride Isotherm

g resin / 100 mL	C_N meq/L	C_{Cl} meq/L	*C meq/L*	x_N	x_{Cl}	y_N	y_{Cl}	α_{ij}
0.020	4.24	0.722	*4.96*	*0.854*	*0.146*	*0.980*	*0.020*	*8.6*
0.040	3.56	1.32	*4.88*	*0.730*	*0.270*	*0.920*	*0.091*	*4.25*
0.100	2.18	2.77	*4.98*	*0.440*	*0.560*	*0.760*	*0.240*	*4.12*
0.200	1.17	3.78	*4.95*	*0.236*	*0.764*	*0.523*	*0.477*	*3.55*
0.400	0.53	4.36	*4.89*	*0.108*	*0.892*	*0.300*	*0.700*	*3.59*
1.20	0.185	4.49	*4.68*	*0.040*	*0.960*	*0.110*	*0.890*	*2.99*

The first three columns represent experimental data. The remaining italicized columns were obtained by calculation as described in the example.

2.
$$x_N = \frac{C_N}{C} = \frac{1.17 \text{ meq/L}}{4.95 \text{ meq/L}} = 0.236$$

$$x_{Cl} = \frac{C_{Cl}}{C} = \frac{3.78 \text{ meq/L}}{4.95 \text{ meq/L}} = 0.764$$

Checking: $x_N + x_{Cl} = 0.236 + 0.764 = 1.00$

3. Calculate chloride remaining on the resin at equilibrium, q_{Cl}:

$$q_{Cl} = q_{Cl,\text{initial}} - \text{chloride lost to water per gram of resin}$$

$$q_{Cl,\text{initial}} = q = 3.63 \text{ meq/g}$$

$$q_{Cl} = 3.63 \text{ meq/g} - 3.78 \text{ meq/L}\left(\frac{0.100 \text{ L}}{0.200 \text{ g}}\right) = 1.74 \text{ meq/g}$$

4. Calculate nitrate on resin at equilibrium

$$q_N = q_{N,\text{initial}} + \text{nitrate lost from water per gram of resin}$$

$$q_N = 0 + [(5.00 - 1.17) \text{ meq/L}]\left(\frac{0.100 \text{ L}}{0.200 \text{ g}}\right) = 1.91 \text{ meq/g}$$

Checking: $q_N + q_{Cl} = 1.74 + 1.91 = 3.65$ meq/g (within 5 percent of 3.63)

5. Calculate the resin-phase equivalent fractions, y_N and y_{Cl}, at equilibrium.

$$y_N = \frac{1.91 \text{ meq/g}}{3.65 \text{ meq/g}} = 0.523$$

$$y_{Cl} = \frac{1.74 \text{ meq/g}}{3.65 \text{ meq/g}} = 0.477$$

6. Calculate the separation factor, α_{ij}.

$$\alpha_{ij} = \frac{y_N\, x_{Cl}}{x_N\, y_{Cl}} = \frac{0.523 \times 0.764}{0.236 \times 0.477} = 3.55$$

Note: Each data point will have an associated α_{ij} value. These α_{ij} values can be averaged, but it is preferable to plot the isotherm data, construct the best-fit curve, and use the curve at $x_N = 0.5$ to obtain an average α_{ij} value. The bad data points will be evident in the plot and can be ignored. Due to mathematical sensitivity, resin inhomogeneity, and imprecise experimental data, the calculated α_{ij} values are not constant, as can be seen in Table 9.4. The α_{ij} values at the ends of the isotherm are particularly nonrepresentative.

7. Plot the isotherm of y_N versus x_N. The nitrate versus chloride isotherm plot should appear similar to that in Figure 9.3*a*.

ION EXCHANGE AND ADSORPTION KINETICS

Pure Ion Exchange Rates

As is usual with interphase mass transfer involving solid particles, resin kinetics is governed by liquid- and solid-phase resistances to mass transfer. The liquid-phase resistance, modeled as the stagnant thin film, can be minimized by providing turbulence around the particle such as that resulting from fluid velocity in packed beds or mechanical mixing in batch operations. The speed of "pure" ion-exchange reactions [i.e., reactions not involving (a) WAC resins in the RCOOH form or (b) free-base forms of weak-base resins] can be attributed to the inherently low mass-transfer resistance of the resin phase that is caused by its well-hydrated gelular nature. Resin beads typically contain 40 to 60 percent water within their boundaries, and this water can be considered as a continuous extension of the aqueous phase within the flexible polymer network. This pseudo-continuous aqueous phase in conjunction with the flexibility of the resin phase can result in rapid kinetics for "pure ion-exchange" reactions (i.e., ion exchange of typical inorganic ions using fully hydrated strong resins). Reactions involving the acid or base forms of weak resins, reactions involving large ions, and reactions of chelating resins are not considered "pure ion exchange"; these reactions are generally not rapid.

Alumina and SBA Resins Compared

Unlike adsorption onto *granular activated carbon* (GAC) or activated alumina, requiring on the order of hours to days to reach equilibrium, pure ion exchange using resins is a rapid process at near-ambient temperature. For example, the half-time to equilibrium for adsorption of arsenate onto granular 28- × 48-mesh (0.29- to 59-mm-dia) activated alumina was found to be approximately 2 days (Rosenblum and Clifford, 1984), while the half-time to equilibrium during the exchange of arsenate for chloride on a strong-base resin was only 5 min (Horng, 1983; Horng and Clifford, 1997). Similarly, the exchange of sodium for calcium on a SAC resin is essentially complete within 5 min (Kunin, 1972).

Rates Involving Tight Resin Forms

In contrast, ion exchange with WAC and WBA resins can be very slow because of the tight, nonswollen nature of the acid form $\overline{(RCOOH)}$ of WAC resins or free-base forms (e.g., $\overline{R_3N:}$) of WBA resins. In reactions involving these tight forms, the average solid-phase diffusion coefficients change drastically during the course of the exchange, which is often described using the progressive-shell, shrinking-core model (Helfferich, 1965; Helfferich, 1966) depicted in Figure 9.5. In these reactions, which are effectively neutralization reactions, either the shell or the core can be the swollen (more hydrated) portion, and a rather sharp line of demarcation exists between the tight and swollen zones. Consider, for example, the practical case of softening with WAC resins in the H^+ form (Equation 9.4). As the reaction proceeds, the hydrated, calcium-form shell comprising $\overline{(RCOO^-)_2Ca^{2+}}$ expands inward and replaces the shrinking, poorly hydrated core of $\overline{RCOOH}$. The entire process is reversed upon regeneration with acid, and the tight shell of $\overline{RCOOH}$ thickens as it proceeds inward and replaces the porous, disappearing core of $\overline{(RCOO^-)_2Ca^{2+}}$.

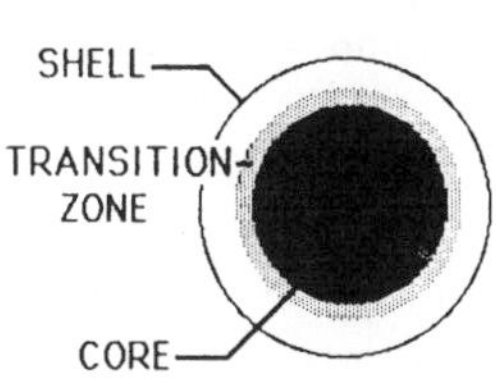

FIGURE 9.5 Progressive-shell model of ion exchange with weak resins.

In some cases, "pure ion exchange" with weak resins is possible, however, and proceeds as rapidly as pure ion exchange with strong resins. For example, the "pure" exchange of sodium for calcium on a WAC resin (Equation 9.32) does not involve conversion of the resin RCOOH in contrast with Equation 9.4 and would take place in a matter of minutes as with SAC resins (Equation 9.3).

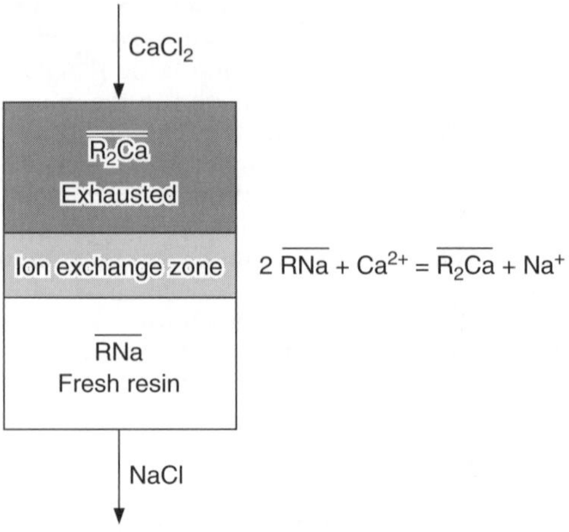

FIGURE 9.6 Resin concentration profile for binary ion exchange of sodium for calcium.

$$2\,\overline{RCOO^-Na^+} + Ca(HCO_3)_2 = \overline{(RCOO^-)_2Ca^{2+}} + 2NaHCO_3 \qquad (9.32)$$

Although weak resins involving $\overline{RCOOH}$ and $\overline{R_3N}$; may require several hours to attain equilibrium in a typical batch exchange, they may still be used effectively in column processes where the contact time between the water and the resin is only 1 to 5 min. There are two reasons for the column advantage: (1) an overwhelming amount of unspent resin is present relative to the amount of water in the column; and (2) the resin is typically exposed to the feed water for periods in excess of 24 h before it is exhausted. Prior to exhaustion, the overwhelming ratio of resin exchange sites present in the column to exchanging ions present in the column water nearly guarantees that an ion will be removed by the resin before the water carrying the ion exits the column. The actual contaminant removal takes place in the "adsorption" or "ion-exchange" or "mass transfer" zone (see Figure 9.6) which characterized the breakthrough curve of interest.

In summary, ion exchange of small inorganic ions using strong resins is fundamentally a fast, interphase transfer process because strong resins are well-hydrated gels exhibiting large solid-phase diffusion coefficients and little resistance to mass transfer. This is not the case with weak resins in the acid (RCOOH) or free-base (R_3N:) forms, nor is it true for alumina, because these media offer considerably more solid-phase diffusion resistance. Irrespective of fast- or slow-batch kinetics, all these media can be effectively used in column processes for contaminant removal from water, because columns exhibit enormous contaminant-removal capacity and are exhausted over a period of many hours to many days. Leakage of contaminants, will, however, be much more significant with media that exhibit relatively slow mass transfer rates.

COLUMN PROCESSES AND CALCULATIONS

Binary Ion Exchange

Ion-exchange and adsorption column operations do not result in a fixed percentage of removal of contaminant with time, which would result, for example, in a steady-state coagulation process. These column processes exhibit a variable degree of contaminant

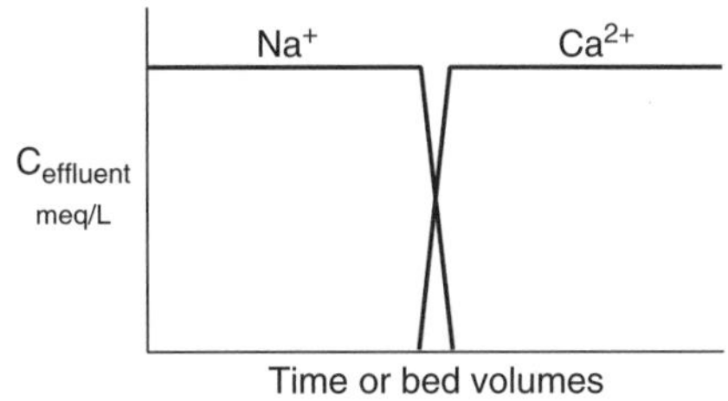

FIGURE 9.7 Effluent concentration histories (breakthrough curves) for the softening reaction in Figure 9.6.

removal and gradual or sharp contaminant breakthroughs similar to (but generally much more complicated than) the breakthrough of turbidity through a granular filter. First, we consider the hypothetical case of pure binary ion exchange before proceeding to the practical drinking water treatment case of multicomponent ion exchange.

If pure calcium chloride solution is softened by continuously passing it through a bed of resin in the sodium form, ion exchange (Equation 9.2) immediately occurs in the uppermost differential segment of the bed (at its inlet). Here all the resin is converted to the calcium form in the moving ion-exchange zone, where mass transfer between the liquid and solid phases occurs. These processes are depicted in Figure 9.6.

The resin phase experiences a calcium wave front that progresses through the column until it reaches the outlet. At this point, no more sodium-form resin exists to take up calcium, and calcium "breaks through" into the effluent, as shown in Figure 9.7. In this pure binary ion-exchange case, the effluent calcium concentration can never exceed that of the influent; this is, however, generally not true for multicomponent ion exchange, as we will show later. The sharpness of the calcium breakthrough curve depends on both equilibrium (i.e., selectivity) and kinetic (i.e., mass transfer) considerations. Imperfect (i.e., noninstantaneous) interphase mass transfer of sodium and calcium, coupled with flow channeling and axial dispersion, always act to reduce the sharpness of the breakthrough curve and result in a broadening of the ion-exchange zone. This is equivalent to saying that nonequilibrium (noninstantaneous) mass transfer produces a diffuse calcium wave and a somewhat gradual calcium breakthrough.

A breakthrough curve can be gradual even if mass transfer is instantaneous, and flow channeling and axial dispersion are absent, because the first consideration in determination of the shape is the resin's affinity (an equilibrium consideration) for the exchanging ions. Mass transfer is the second consideration. If the exchange isotherm is favorable, as is the case here (i.e., calcium is preferred to sodium), then a perfectly sharp (square-wave) theoretical breakthrough curve results. If the ion-exchange isotherm is unfavorable, as is the case for the reverse reaction of sodium chloride fed to a calcium-form resin, then a gradual breakthrough curve results even for instantaneous (equilibrium) mass transfer. These two basic types of breakthrough curves, sketched in Figure 9.8, result from the solution of mass balance equations assuming instantaneous equilibrium and constant adsorbent capacity.

Multicomponent Ion Exchange

The breakthrough curves encountered in water supply applications are much more complicated than those in Figures 9.7 and 9.8. The greater complexity is caused by the multicomponent nature of the exchange reactions when treating natural water. Some ideal resin concentration profiles and breakthrough curves for hardness

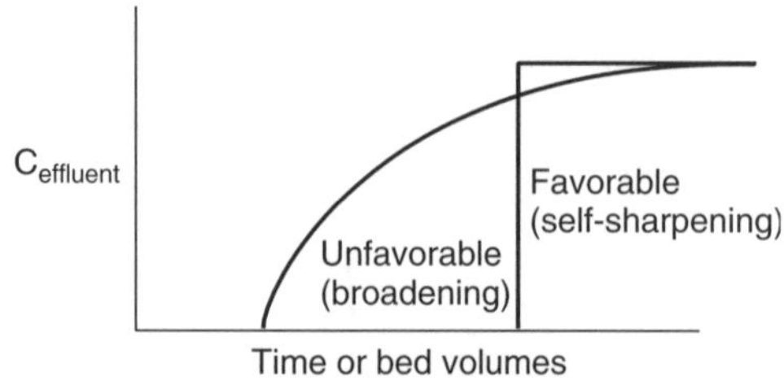

FIGURE 9.8 Theoretical breakthrough curves for equilibrium ion exchange with no mass transfer limitations. An unfavorable isotherm (Figure 9.3*b*) results in a broadening wave front (breakthrough), while a favorable isotherm (Figure 9.3*a*) results in a self-sharpening wave front.

removal by ion-exchange softening and for nitrate removal by chloride-form anion exchange are sketched in Fig. 9.9*a* and *b*. The important determinants of the shapes of these breakthrough curves are (1) the feed water composition, (2) the resin capacity, and (3) the resin's affinity for each of the ions as quantified by the separation factor, α_{ij}, or the selectivity coefficient, K_{ij}. The *order* of elution of ions from the resin, however, is determined solely by the selectivity sequence, which is the ordering of the components from i = 1 – *n*, where 1 is the most-preferred and *n* is the least-preferred species. Finally, before continuing with our discussion of multicomponent ion-exchange column behavior, we must remind ourselves of Equation 9.26, which shows that the α_{ij} values for di- and higher-valent ions, and thus the order of elution of ions, will be determined by the total ionic concentration *C* of the feed water.

In carrying out the cation-or anion-exchange reactions, ions in addition to the target ion (e.g., calcium or nitrate) are removed by the resin. All the ions are concentrated, in order of preference, in bands or zones in the resin column, as shown in the resin concentration profiles of Figures 9.9*a* and 9.9*b*. As these resin boundaries (wave fronts) move through the column, the breakthrough curves shown in the figures result. These are based on theory (Helfferich and Klein, 1970) but have been verified in the actual breakthrough curves published by Clifford (1982 and 1995), Snoeyink et al. (1987), and Guter (1995).

Some useful rules can be applied to effluent histories in multicomponent ion-exchange (and adsorption) systems (Helfferich and Klein, 1970; Clifford, 1982; Clifford, 1991):

1. Ions higher in the selectivity sequence than the presaturant ion tend to have long runs and sharp breakthroughs (like all those except HCO_3^- in Figure 9.9*b*); those less preferred than the presaturant ion will always have early, gradual breakthroughs, as typified by HCO_3^-.
2. The most-preferred species (radium in the case of softening, and sulfate in the case of nitrate removal) are last to exit the column, and their effluent concentrations never exceed their influent concentrations.
3. The species exit the column in reverse preferential order, with the less preferred ions (smaller separation factors with respect to the most-preferred species) leaving first.
4. The less-preferred species will be concentrated in the column and will at some time exit the column in concentrations exceeding their influent concentrations (chromatographic peaking). This is a potentially dangerous situation, depending on the toxicity of the ion in question. Good examples of chromatographic peak-

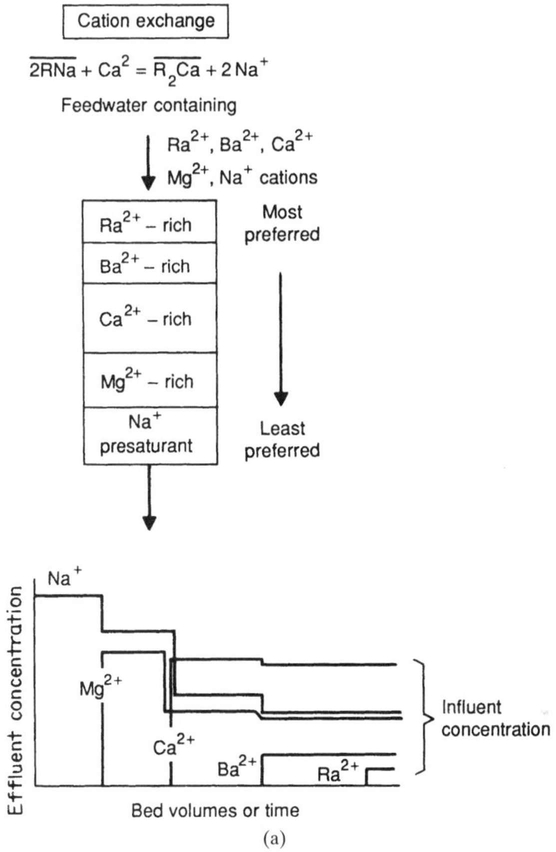

FIGURE 9.9 (*a*) Ideal resin concentration profile (*above*) and breakthrough curves (*below*) for typical softening and radium removal. Note that the column was run far beyond hardness breakthrough and slightly beyond radium breakthrough. The most preferred ion is Ra^{2+}, followed by $Ba^{2+} > Ca^{2+} > Mg^{2+} > Na^+$.

ing (i.e., effluent concentration greater than influent concentration) are visible in Figure 9.9*a* and *b*. A magnesium peak is shown in Figure 9.9*a*, and bicarbonate and nitrate peaks in Figure 9.9*b*.

5. When all the breakthrough fronts have exited the column, the entire resin bed is in equilibrium with the feed water. When this happens, the column is exhausted, and the effluent and influent ion concentrations are equal.
6. The effluent concentration of the presaturant ion (Na^+ in Figure 9.9*a*, and Cl^- in Figure 9.9*b*) decreases in steps as each new ion breaks through, because the total ionic concentration of the water (C, meq/L) must remain constant during simple ion exchange.

One way to eliminate the troublesome chromatographic peaking of toxic ions such as nitrate and arsenate is by inverting the selectivity sequence so that the toxic contam-

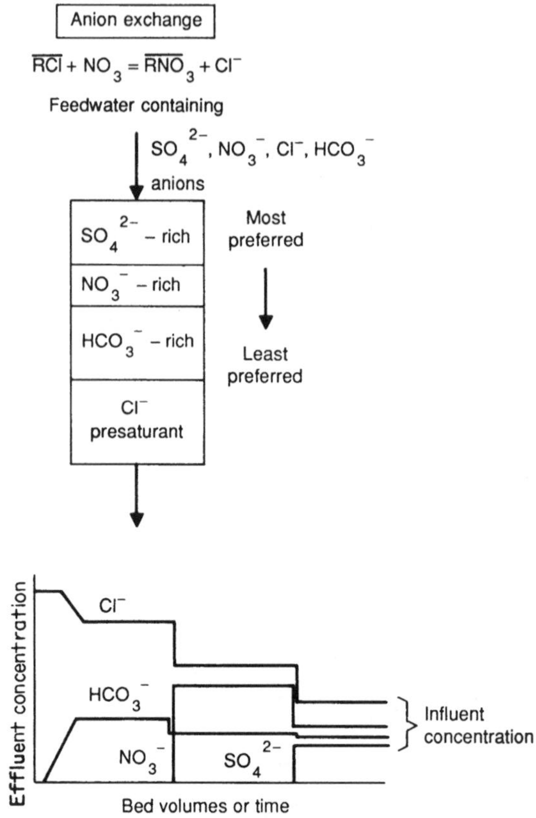

(b)

FIGURE 9.9 *(Continued)* (*b*) Ideal resin concentration profile (*above*) and breakthrough curves (*below*) for nitrate removal by chloride-form anion exchange with a strong-base resin. Note that the column was run far beyond nitrate breakthrough and somewhat beyond sulfate breakthrough. The most preferred ion is SO_4^{2-}, followed by $NO_3^- > Cl^- > HCO_3^-$.

inant is the ion most preferred by the resin. This requires the preparation of special-purpose resins. This has been done in the case of nitrate removal and will be discussed later under that heading. Potential peaking problems still remain with other inorganic contaminants, notably arsenic [As(V)] and selenium [Se(IV)] (Clifford, 1991). An alternative means of eliminating or minimizing peaking is to operate several columns in parallel, as will be discussed in the section on "Multicolumn Processes."

Breakthrough Detection and Run Termination

Clearly an ion-exchange column run must be stopped before a toxic contaminant is "dumped" during chromatographic peaking. Even without peaking, violation of the

MCL will occur at breakthrough, when the contaminant feed concentration exceeds the MCL. Effective detection and prevention of a high effluent concentration of contaminant depend on the frequency of sampling and analysis. Generally, continuous on-line analysis of the contaminant (e.g., nitrate or arsenate) is too sophisticated for small communities, where most of the inorganic contaminant problems exist (AWWA, 1985). On-line conductivity detection, the standard means of effluent quality determination in ion-exchange demineralization processes, is not easily applied to the detection of contaminant breakthrough in single-contaminant processes such as radium, barium, nitrate, or arsenate removal. This is because of the high and continuously varying conductivity of the effluents from cation or anion beds operated on typical water supplies. Nevertheless conductivity should not be ruled out completely, because even though the changes may be small, as the various ions exit the column a precise measurement may be possible in selected applications.

On-line pH measurement is a proven, reliable technique that can sometimes be applied as a surrogate for contaminant breakthrough. For example, pH change can be used to signal the exhaustion of a weak-acid resin (—RCOOH) used for carbonate hardness removal. When exhausted, the WAC resin ceases to produce acidic carbon dioxide, and the pH quickly rises to that of the feed water. This pH increase is, however, far ahead of the barium or radium breakthrough. The pH can sometimes be used as an indicator of nitrate breakthrough, as discussed in the section on nitrate removal.

The usual method of terminating an ion-exchange column run is to establish the relevant breakthrough curve by sampling and analysis and then use these data to terminate future runs based on the metered volume of throughput with an appropriate safety factor. If a breakthrough detector such as a pH or conductivity probe is applied, the sample line to the instrument can be located ahead (e.g., 6 to 12 in.) of the bed outlet to provide advance warning of breakthrough.

Typical Service Cycle for a Single Column

Ion-exchange and adsorption columns operate on similar service cycles consisting of six steps: (1) exhaustion, (2) backwash, (3) regeneration, (4) slow rinse, (5) fast rinse, and (6) return to service. (Backwash may not be required after every exhaustion.) A simple single-column process schematic is shown in Figure 9.10, which includes an optional bypass for a portion of the feed water. Bypass blending will be a common procedure for drinking water treatment applications because ion-exchange resins can usually produce a contaminant-free effluent that is purer than that required by law. Therefore, to minimize treatment costs, part of the contaminated feed water, typically 10 to 50 percent, will be bypassed around the process and blended with the effluent to produce a product water approaching some fraction (e.g., 70 percent) of the MCL acceptable to the regulatory agency. An alternative means of providing efficient column utilization when significant contaminant leakage is allowed is to operate several columns in parallel as discussed in the section on "Multicolumn Processes."

Partial Regeneration and Regenerant Reuse

Yet another means of optimizing column utilization and minimizing process costs is to use the technique of partial regeneration. This involves the use of only a fraction (e.g., 25 to 50 percent) of the regenerant required for "complete" (e.g., 90 to 100 per-

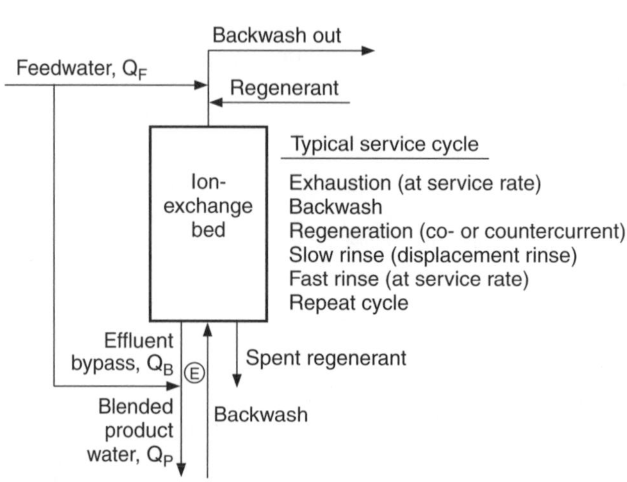

FIGURE 9.10 Schematic and service cycle of a single-column ion-exchange process.

cent) removal of the contaminant from the exhausted resin. The result is often, but not always, a large leakage of contaminant on the next exhaustion run, caused by the relatively high level of contaminant remaining on the resin. Such large leakage can often be tolerated without exceeding the MCL. Partial regeneration is particularly useful in nitrate removal, as will be discussed in detail later. Generally, either bypass blending or partial regeneration will be used; simultaneous use of both processes is possible, but creates significant process control problems.

Reuse of spent regenerant is another means of reducing costs and minimizing waste disposal requirements. In order for a spent regenerant to be reused, the target contaminant ion must either be removed from the regenerant before reusing it, or the resin must have a strong preference in favor of the regenerant ion as compared to the contaminant ion, which accumulates in the recycle brine. The recent literature suggests that spent brine reuse is possible in more applications than were previously thought possible. Removing nitrate from the recycle brine by means of biological denitrification was the approach used by the author and his colleagues (Clifford and Liu, 1993b; Liu and Clifford, 1996) for their nitrate ion-exchange process with brine-reuse. In their pilot-scale experiments, a denitrified 0.5 M Cl^- brine was reused 38 times without disposal. Clifford, Ghurye, et al. (1998) also determined that spent arsenic-contaminated brine could be reused more than 20 times by simply maintaining the Cl^- concentration at 1.0 M without removing the arsenic. Kim and Symons (1991) showed that DOC anions could be removed from drinking water by strong-base-anion exchange with regenerant reuse. No deterioration of DOC removal was noted during 9 exhaustion-regeneration cycles with spent brine (a mixture of NaCl and NaOH) reuse when the Cl^- and OH^- levels were maintained at 2.0 and 0.5 *M*, respectively. Further information on these processes is provided in the "Applications" section of this chapter.

Reusing the entire spent-regenerant solution is not necessary. In the case where there is a long tail on the contaminant elution curve, the first few bed volumes of regenerant are discarded, and only the least-contaminated portions are reused. In this case a two-step roughing-polishing regeneration can be utilized. The roughing regeneration is completed with the partially contaminated spent regenerant, and the polishing step is carried out with fresh regenerant. The spent regenerant from the polishing step is then used for the next roughing regeneration.

Regenerant reuse techniques are relatively new to the ion-exchange field and are yet to be proved in full-scale long-term use for water supply applications. Although possessing the advantages of conserving regenerants and reducing the volume of waste discharges regenerant reuse can also result in some significant disadvantages, including (1) increased process complexity; (2) increased contaminant leakage; (3) progressive loss of capacity caused by incomplete regeneration and fouling; (4) the need to store and handle spent regenerants; and (5) buildup (concentration) of trace contaminants as the number of regenerant reuse cycles increases.

Multicolumn Processes

Ion-exchange or adsorption columns can be connected (1) in series to improve product purity and regenerant usage, or (2) in parallel to increase throughput, minimize peaking, and smooth out product water quality variations. If designed properly, multiple-column systems can be operated in parallel, series, or parallel-series modes.

Columns in Series. A series roughing-polishing sequence is shown in Figure 9.11. In such a process, a completely exhausted roughing column is regenerated when the partially exhausted polishing column effluent exceeds the MCL. This unregenerated polishing column becomes the new roughing column, and the old roughing column, now freshly regenerated, becomes the new polishing column. Often three columns are used. While two are in service, the third is being regenerated. A multicolumn system consisting of three or more columns operated in this manner is referred to as a "merry-go-round system," which should not be confused with "carousel system" (AST, 1995), which is usually operated as columns in parallel.

Columns in Parallel. In addition to bypass blending (see Figure 9.10), an alternative means of allowing a predetermined amount of contaminant leakage in the product water is to employ multiple columns in parallel operated at different stages of exhaustion. For example, with three columns operated in parallel, the first one could be run beyond MCL breakthrough while the second and third columns would not have achieved breakthrough. Thus, even after contaminant breakthrough in the first column, the average concentration of the three blended efflluents would be below the MCL. Multiple-parallel-column operation will give a more consistent product water quality and can also prevent, or at least smooth out, chromatographic peaks from serious overruns. During normal operation of multiple-parallel-column systems, some columns are being exhausted while others are being rinsed, regenerated, or are in standby mode. A recently described carousel-ion-exchange process

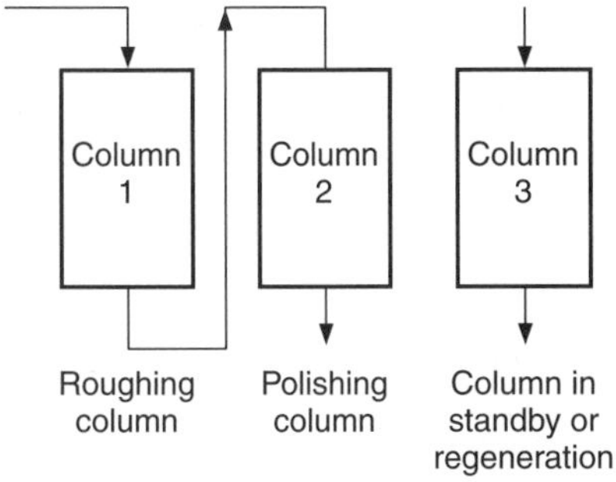

FIGURE 9.11 Two-column roughing-polishing system operated in a merry-go-round fashion. After exhaustion of column 1, it will be taken out of service, and the flow sequence will be column 2 and then column 3. Following exhaustion of column 2 and regeneration of column 1, the roughing-polishing sequence will be column 3 then column 1.

typically uses 10 to 20 parallel columns in the exhaustion zone and produces a very consistent product water quality (AST, 1995).

Process Differences: Resins Versus Alumina

The design of a process for activated alumina exhaustion and regeneration is similar to that for ion-exchange resins, but with some significant exceptions. First, contaminant leakage is inherently greater with alumina, adsorption zones are longer, and breakthrough curves are more gradual, because alumina adsorption processes are much slower than ion exchange with strong resins. Second, effluent chromatographic peaking of the contaminant (fluoride, arsenic, or selenium) is rarely seen during alumina adsorption because these contaminants are usually the most-preferred ions in the feed water. (An exception is the peaking of arsenic, which was observed in Albuquerque when treating pH 8.5 groundwater with activated alumina. The sharp arsenic peaking observed at breakthrough was thought to be caused either by hydroxide or silicate, which may be more preferred than arsenate [Clifford, Ghurye et al., 1998].) Finally, complex, two-step base-acid regeneration is required to rinse out the excess base and return the alumina to a useful form.

DESIGN CONSIDERATIONS

Resin Characteristics

Several hundred different resins are available from U.S. and European manufacturers. Of these, resins based on the polystyrene divinylbenzene matrix see the widest use. Representative ranges of properties of these resins are shown in Table 9.5 for the two major categories of resins used in water treatment. Ion-exchange capacity is expressed in milliequivalents per milliliter (wet-volume capacity) because resins are purchased and installed on a volumetric basis (meq/mL $\times$ 21.8 = kgrain $CaCO_3/ft^3$). A wet-volume capacity of 1.0 meq/mL means that the resin contains 6.023×10^{20} exchange sites per milliliter of wet resin, including voids. The dry-weight capacity in milliequivalents per gram of dry resin is more precise, and is often used in scientific research.

The *operating capacity* is a measure of the actual performance of a resin under a defined set of conditions including, for example, feed water composition, service rate, and degree of regeneration. The operating capacity is always less than the advertised exchange capacity because of incomplete regeneration and early contaminant leakage, which causes early run termination. Some example operating capacities during softening are given in Table 9.6, where the operating capacity for softening is seen to be a function of the amount of regenerant used.

Bed Size and Flow Rates

A resin bed depth of 30 in (76 cm) is usually considered the minimum, and beds as deep as 12 ft (3.67 m) are not uncommon. The *empty-bed contact time* (EBCT) chosen determines the volume of resin required and is usually in the range of 1.5 to 7.5 min. The reciprocal of EBCT is the *service flow rate* (SFR) or *exhaustion rate,* and its accepted range is 1 to 5 gpm/ft^3. These relationships are expressed as

TABLE 9.5 Properties of Styrene-Divinylbenzyl, Gel-Type Strong-Acid Cation and Strong-Base Anion Resins

Parameter	Strong-acid cation resin	Type I, strong-base anion resin
Screen size, U.S. mesh	−16 + 50	−16 + 50
Shipping weight, lb/ft^3 (kg/m^3)	53, (850)	44, (700)
Moisture content, %	45 – 48	43 – 49
pH range	0 – 14	0 – 14
Maximum operating temp. °F (°C)	280, (140)	OH^- form 140, (60) Cl^- form 212, (100)
Turbidity tolerance, NTU	5	5
Iron tolerance, mg/L as Fe	5	0.1
Chlorine tolerance, mg/L Cl_2	1.0	0.1
Backwash rate, gal/min ft^2 (m/hr)	5 – 8, (12 – 20)	2 – 3, (4.9 – 7.4)
Backwash period, min	5 – 15	5 – 20
Expansion volume, %	50	50 – 75
Regenerant and concentration[1]	NaCl, 3 – 12%	NaCl, 1.5 – 12%
Regenerant dose, lb/ft^3 (kg/m^3)	5 – 20, (80 – 320)	5 – 20, (80 – 320)
Regenerant rate, gal/min ft^3 (min/BV)	0.5, (15)	0.5, (15)
Rinse volume, gal/ft^3 (BV)	15 – 35, (2 – 5)	15 – 75, (2 – 10)
Exchange capacity, kgr $CaCO_3$/ft^3 (meq/mL)[2]	39 – 41, (1.8 – 2.0)	22 – 28, (1 – 1.3)
Operating capacity, kgr $CaCO_3$/ft^3 (meq/mL)[3]	20 – 30 (0.9 – 1.4)	12 – 16 (0.4 – 0.8)
Service rate, gal/min ft^3 (BV/hr)	1 – 5, (8 – 40)	1 – 5, (8 – 40)

[1] Other regenerants such as H_2SO_4, HCl and $CaCl_2$ can also be used for SAC resins while NaOH, KOH and $CaCl_2$ can be used for SBA regeneration.

[2] Kilograins $CaCO_3$/ft^3 are the units commonly reported in resin manufacturer literature. To convert kgr $CaCO_3$/ft^3 to meq/mL, multiply by 0.0458.

[3] Operating capacity depends on method of regeneration, particularly on the amount of regenerant applied. See Table 9.6 for SAC resins.

TABLE 9.6 Softening Capacity as a Function of Regeneration Level

Regeneration level		Hardness capacity		Regeneration efficiency	
lb NaCl / ft^3 resin	kg NaCl / m^3 resin	kgr $CaCO_3$ / ft^3 resin	eq $CaCO_3$ / L resin	lb NaCl / kgr $CaCO_3$	eq NaCl / eq $CaCO_3$
4	64	17	0.78	0.24	1.40
6	96	20	0.92	0.30	1.78
8	128	22	1.00	0.36	2.19
10	160	25	1.15	0.40	2.38
15	240	27	1.24	0.56	3.30
20	320	29	1.33	0.69	4.11
infinite	infinite	45	2.06	infinite	infinite

These operating capacity data are based on the performance of Amberlite IR-120 SAC resin. Other manufacturers' resins are comparable. Values given are independent of EBCT and bed depth providing the minimum criteria (EBCT = 1.0 min, bed depth = 2.5 ft) are met.

$$EBCT = \frac{V_R}{Q_F} = \text{average fluid detention time in an empty bed} \quad (9.33)$$

$$\text{Service flow rate} = SFR = \frac{1}{EBCT} = \frac{Q_F}{V_R} \quad (9.34)$$

where Q_F = volumetric flow rate, gal/min, (L/min)
V_R = resin bed volume including voids, ft^3, (m^3)

Fixed-Bed Columns

Ion-exchange columns are usually steel pressure vessels constructed so as to provide (1) a good feed and regenerant distribution system; (2) an appropriate bed support, including provision for backwash water distribution; and (3) enough free space above the resin bed to allow for expected bed expansion during backwashing. Additionally, the vessel must be lined so as to avoid corrosion problems resulting from concentrated salt solutions and, in some cases, acids and bases used for regeneration or resin cleaning. There must be minimal dead space below the resin bed, where regenerants and cleaning solutions might collect and subsequently bleed into the effluent during the service cycle.

COCURRENT VERSUS COUNTERCURRENT REGENERATION

Historically, downflow exhaustion followed by downflow (cocurrent) regeneration has been the usual mode of operation for ion-exchange columns. However, the recent trend, especially in Europe, is to use upflow (countercurrent) regeneration for the purpose of minimizing the leakage of contaminant ions on subsequent exhaustions of ion exchange demineralizers. Theoretically, countercurrent regeneration is better because it exposes the bottom (exit) of the bed to a continuous flow of fresh regenerant, and leaves the resin near the outlet of the bed in a well-regenerated condition. The author's research on nitrate (Clifford, Lin et al., 1987) and arsenate removal (Clifford, Ghurye et al., 1998) has, however, called into question the conventional wisdom that countercurrent is always better than cocurrent regeneration. It has been found that cocurrent downflow regeneration is superior to countercurrent upflow regeneration for contaminants such as arsenate and nitrate, which are concentrated at the bed outlet at the end of a run. The proposed reason for the superiority of downflow regeneration in these situations is that the contaminant is not forced back through the entire resin bed during regeneration. The forcing of the contaminant back through the bed tends to leave relatively more contaminant in the resin. This will be discussed in more detail in the sections on nitrate and arsenic removal.

Spent Brine Reuse

In order for a spent regenerant to be reused, the target contaminant ion must either be removed from the regenerant before reusing it, or the resin must have a strong preference in favor of the regenerant ion as compared to the target ion, which accumulates in the recycle brine. The recent literature suggests that spent brine reuse is possible in more applications than were previously thought possible. Removing nitrate from the recycle brine by means of biological denitrification is the approach used by Van der Hoek, Van der Van et al. (1988) and Clifford and Liu (1993a) for their innovative nitrate ion-exchange processes. In the latter's pilot-scale experiments, their denitrified 0.5 *M* Cl^- brine was reused 38 times without disposal. Clifford, Ghurye, et al. (1998) also determined that spent arsenic-contaminated brine could be reused more than 20 times by simply maintaining the Cl^- concentration at 1.0 *M* without removing the arsenic. Kim and Symons (1991) showed that DOC anions could be removed from drinking water by strong-base-anion exchange with regenerant reuse. No deterioration of DOC removal was noted during 9 exhaustion-regeneration cycles with spent brine (a mixture of NaCl and NaOH) reuse when the Cl^- and OH^- levels were restored to 2.0 and 0.5 *M*, respectively, after each regeneration. Further information on regenerant reuse processes is provided in the sections on nitrate, arsenic, and organics removal.

APPLICATIONS OF ION EXCHANGE AND ADSORPTION

Sodium Ion-Exchange Softening

As already mentioned, softening water by exchanging sodium for calcium and magnesium using SAC resin (see Equation 9.2) is the major application of ion-exchange technology for the treatment of drinking water. Prior to the advent of synthetic resins, zeolites (i.e., inorganic crystalline aluminosilicate ion exchangers in the sodium form) were utilized as the exchangers. The story of one major application of ion-exchange softening at the Weymouth plant of the Metropolitan Water District of Southern California is well-told by A. E. Bowers in *The Quest for Pure* Water (Bowers, 1980). In that application, which included 400 Mgd (1.5×10^6 m^3/d) of softening capacity, softening by ion exchange eventually supplanted excess lime-soda ash softening because of better economics, fewer precipitation problems, and the requirement for a high alkalinity level in the product water to reduce corrosion. One advantage of the lime soda ash softening process is that it reduces the TDS level of the water by removing calcium and magnesium bicarbonates as $CaCO_3(s)$ and $Mg(OH)_2(s)$. The concomitant removal of alkalinity is, however, sometimes detrimental, thus favoring ion-exchange softening that deals only with cation exchange while leaving the anions intact.

As with most ion-exchange softening plants, the zeolite medium at the Weymouth plant was exchanged for resin in the early 1950s, shortly after polystyrene SAC resins were introduced. The SAC softening resins used today are basically the same as these early polystyrene resins. Their main features are high chemical and physical stability, even in the presence of chlorine; uniformity in size and composition; high exchange capacity; rapid exchange kinetics; a high degree of reversibility; and long life. A historical comparison between the life of the zeolites and that of the resins indicated that zeolites could process a maximum of 1.6×10^6 gal H_2O/ft^3 zeolite (214,000 BV) before replacement, whereas the resins could process up to 20×10^6 gal H_2O/ft^3 resin (2,700,000 BV) before they needed replacement. The softeners designed and installed for resins at this plant in 1966 were 28×56 ft (8.5×17 m) reinforced-concrete basins filled to a depth of 2.5 ft (0.76 m), with each containing 4000 ft^3 (113 m^3) of resin.

The Weymouth plant utilized ion-exchange softening for over 30 years. Softening ceased in 1975 when the source water hardness was reduced by blending. At that time, the 9-year-old resin in the newest softeners was still good enough to be resold. Other interesting design features of this plant included disposal of waste brine to a waste water treatment plant through a 20-mi-long (32 km) pipe flowing at 15 ft^3/s (0.43 m^3/s), and the upflow exhaustion at 6 gpm/ft^2 (3.1-min EBCT in a 2.5-ft-deep bed) followed by downflow regeneration.

EXAMPLE 9.2 ***Softening Design Example***

This typical design example illustrates how to establish the ion-exchange resin volume, column dimensions, and regeneration requirements for a typical water softener.

DESIGN PROBLEM A groundwater is to be partially softened from 275 down to 150 mg per liter of $CaCO_3$ hardness. Ion exchange has been selected instead of lime softening because of its simplicity and the ease of cycling on and off to meet the water demand. The well pumping capacity is 1.0 mgd (700 gpm), and the system must be sized to meet this maximum flow rate. The source water contains only traces of iron; therefore, potential clogging problems because of suspended solids are not significant. In applications where raw water suspended solids would foul the resins, filtration pretreatment with dual- or multi-media filters would be required.

OUTLINE OF SOLUTION

1. Select a resin and a regeneration level, using the resin manufacturer's literature.
2. Calculate the allowable fraction, f_B, of bypass source water.
3. Choose the service flow rate (SFR, gpm/ft^3) or EBCT (min).
4. Calculate the run length, t_H and the bed volumes V_F that can be treated prior to hardness breakthrough.
5. Calculate the volume of resin V_R required.
6. Determine the minimum out-of-service time, in hours, during regeneration.
7. Choose the number of columns in the system.
8. Dimension the columns.
9. Calculate the volume and composition of wastewater.

CALCULATIONS

1. *Selection of resin and resin capacity.* Once the resin and its regeneration level have been specified, the ion-exchange operating capacity is fixed based on experimental data of the type found in Table 9.6. The data are for a polystyrene SAC resin subjected to cocurrent regeneration using 10 percent (1.7 *N*) NaCl. If a regeneration level of 15 lb NaCl/ft^3 resin is chosen, the resulting hardness capacity prior to breakthrough is 27 kgr of hardness as $CaCO_3/ft^3$ resin (i.e., 1.24 meq/mL resin).
2. *Calculation of bypass water allowance.* Assume that the water passing through the resin has zero hardness. (Actually, hardness leakage during exhaustion will be detectable but usually less than 5 mg/L as $CaCO_3$.) The bypass flow is calculated by writing a hardness balance at blending point, point E in Figure 9.10, where the column effluent is blended with the source water bypass.

   Mass balance on hardness at point E:

$$Q_B C_B + Q_F C_E = Q_P C_P \tag{9.35}$$

Balance on flow at point E:

$$Q_B + Q_F = Q_P \tag{9.36}$$

where Q_B = bypass flow rate
Q_F = column feed and effluent flow rate
Q_P = blended product flow rate (i.e., total flow rate)
C_B = concentration of hardness in bypass raw water, 275 mg/L as $CaCO_3$
C_E = concentration of hardness in column effluent, assumed to be zero, mg/L
C_P = chosen concentration of hardness in blended product water, 150 mg/L as $CaCO_3$

The solution to these equations is easily obtained in terms of the fraction bypassed, f_B:

$$f_B = \frac{Q_B}{Q_P} = \frac{C_P}{C_B} = 0.55 \tag{9.37}$$

The fraction, f_F, which must be treated by ion exchange is:

$$f_F = 1 - f_B = 0.45 \tag{9.38}$$

3. *Choosing the exhaustion flow rate.* The generally acceptable range of SFR for ion exchange is 1 to 5 gal/min ft^3. Choosing a value of 2.5 gal/min ft^3 results in an EBCT of 3.0 min and an approach velocity, v_o, of 6.25 gal/min ft^2 if the resin bed is 2.5 ft (30 in) deep.

$$\text{EBCT} = \frac{1 \text{ min ft}^3}{2.5 \text{ gal}} \times \frac{7.48 \text{ gal}}{1 \text{ ft}^3} = 3 \text{ min} \tag{9.39}$$

$$v_o = \frac{\text{depth}}{\text{detention time}} = \frac{\text{depth}}{\text{EBCT}} = \text{SFR} \times \text{depth} \tag{9.40}$$

$$v_o = 2.5 \text{ (gal/min ft}^3) \times 2.5 \text{ (ft)} = 6.25 \text{ gal/min ft}^2 \tag{9.41}$$

4. *Calculation of run length.* The exhaustion time to hardness breakthrough t_H and the bed volumes BV_H to hardness breakthrough are calculated from a mass balance on hardness, assuming again that the resin effluent contains zero hardness. Expressed in words, this mass balance is:

equivalents of hardness *removed from the water* during the run = equivalents of hardness *accumulated on the resin* during the run

$$Q_F C_F t_H = V_F C_F = q_H V_R \tag{9.42}$$

where q_H = hardness capacity of resin at selected regeneration level, eq/L (kgr/ft^3)
V_R = volume of resin bed including voids, L
$Q_F t_H = V_F$, volume of water fed to column during time t_H, L

Then:

$$\frac{V_F}{V_R} = \text{BV}_H = \frac{q_H}{C_F} \tag{9.43}$$

Based on the hardness capacity in Table 9.6, the bed volumes to hardness breakthrough BV_H following a regeneration at 15 lb NaCl/ft^3 is:

$$BV_H = \frac{1.24 \text{ equiv } CaCO_3}{\text{L resin}} \times \frac{1 \text{ L } H_2O}{275 \text{ mg } CaCO_3} \times \frac{50{,}000 \text{ mg } CaCO_3}{\text{equiv } CaCO_3}$$

$$BV_H = 225 \text{ Volumes of } H_2O \text{ treated/volume of resin} \tag{9.44}$$

The time t_H to hardness breakthrough is related to the bed volumes to breakthrough BV_H and the EBCT:

$$t_H = \text{EBCT} \times BV_H \tag{9.45}$$

$$t_H = 3.0 \frac{\text{min}}{\text{BV}} \times 225 \text{ BV} \times \frac{1 \text{ hr.}}{60 \text{ min.}} = 11.2 \text{ h} \tag{9.46}$$

If the EBCT is decreased by increasing the flow rate through the bed (i.e., SFR), then the run time is proportionately shortened even though the total amount of water treated V_F remains constant.

5. *Calculation of resin volume* V_R. The most important parameter chosen was the service flow rate (SFR) because it directly specified the necessary resin volume V_R according to the following relationships based on Equation 9.34:

$$V_R = \frac{Q_F}{\text{SFR}} = Q_F\,(\text{EBCT}) \tag{9.47}$$

Numerically, for a column feed flow Q_F of 45 percent (the amount not bypassed) of 1.0 mgd

$$V_R = \frac{0.45 \times 10^6 \text{ gal}}{\text{day}} \times 3.0 \text{ min} \times \frac{1 \text{ day}}{1440 \text{ min}} \times \frac{1 \text{ ft}^3}{7.48 \text{ gal}}$$

$$V_R = 125 \text{ ft}^3 \tag{9.48}$$

6 and 7. *Calculation of the number of columns and the minimum out-of-service time for regeneration.* For a reasonable system design, two columns are required—one in operation and one in regeneration or standby. A single-column design with product water storage is possible, but provides no margin of safety in case the column has to be serviced. Even with two columns, the out-of-service time t_{OS} for the column being regenerated should not exceed the exhaustion run time to hardness breakthrough t_H:

$$t_{OS} \leq t_{BW} + t_R + t_{SR} + t_{FR} \tag{9.49}$$

where t_{BW} = time for backwashing, 5 to 15 min
t_R = time for regeneration, 30 to 60 min
t_{SR} = time for slow rinse, 10 to 30 min
t_{FR} = time for fast rinse, 5 to 15 min

A conservative out-of-service time would be the sum of the maximum times for backwashing, regeneration, and rinsing (i.e., 2 h). This causes no problem with regard to continuous operation because the exhaustion time is more than 11 h.

8. *Calculation of column dimensions.* The resin depth h was specified earlier as 2.5 ft; thus the column height, after we allow for 100 percent resin bed expansion during backwashing, is 5.0 ft. The bed diameter D is then:

$$D = \sqrt{\frac{4\,V_R}{\pi h}} = 8 \text{ ft} \tag{9.50}$$

The resulting ratio of resin bed depth to column diameter is 2.5:8 or 0.3:1. This is within the acceptable range of 0.2:1 to 2:1 if proper flow distribution is provided. Increasing the resin depth to 4 ft increases the column height to 8 ft and reduces its diameter to 6.3 ft. Clearly, a variety of depths and diameters is possible. Before specifying these, the designer should check with equipment manufacturers because softening units in this capacity range are available as predesigned packages.

Important: Another alternative would be to use three or more columns, with two or more in service and one or more in standby. This offers a more flexible design. For a three-column system with two in service and one in standby, the resin volume of the in-service units would be 125/2 = 62.5 ft^3 each (i.e., the flow would be split between two 62.5-ft^3 resin beds operating in parallel). Regeneration would be staggered such that only one column would undergo regeneration at any time. *An important advantage of operating columns in parallel with staggered regeneration is that product water quality is less variable compared with single-column operation.* This can be a major consideration when the contaminant leakage or peaking is relatively high during a portion of the run; when this happens, combining the high leakage from one column with the low leakage from another produces an average leakage—presumably less than the MCL—over the duration of the run. Also, operating multiple columns in parallel with staggered regeneration is appropriate when nontargeted contaminants are removed for a portion of the run and then are subject to peaking before the target contaminant run is complete. A good example is the removal of the target-contaminant arsenic (which exceeds its MCL in the feed) in the presence of the nontarget contaminant, nitrate (which is present, but does not exceed its MCL in the feed). Generally, the arsenic run length would be 400 to 1200 BV, whereas nitrate would typically break through before 400 BV and peak at 1.2 to 3 times its feedwater concentration. If the nitrate peaking causes the nitrate-N to exceed its MCL, then averaging the product water from two or more columns in parallel will be necessary to keep nitrate below its MCL while still allowing a long run length for arsenic. Carousel systems generally operate up to 20 columns in parallel, which protects against peaking of all contaminants and provides a consistent (averaged) product water quality.

9. *Calculation of volume and composition of wastewater.* The spent regenerant solution comprises the regenerant and the slow-rinse (displacement-rinse) volumes. These waste solutions must be accumulated for eventual disposal, as detailed in Chapter 16. The actual wastewater volume per regeneration will depend on the size of the resin bed (i.e., whether one, two, or three beds are chosen for the design). The following calculations are in terms of bed volumes which can be converted to fluid volume once the column design has been fixed. The chosen regeneration level, 15 lb NaCl/ft^3 resin, is easily converted to BV of regenerant required. Given that a 10 percent NaCl solution has a specific gravity of 1.07, the regenerant volume applied is

$$15\,\frac{\text{lb NaCl}}{\text{ft}^3\text{ resin}} \times \frac{1\text{ lb soln}}{0.1\text{ lb NaCl}} \times \frac{\text{ft}^3\text{ soln}}{1.07 \times 62.4\text{ lb soln}} = 2.25\text{ BV} \tag{9.51}$$

Following the salt addition, a displacement (slow) rinse of 1 to 2 BV is applied. The total regenerant wastewater volume is made up of the spent regenerant (2.25 BV) and the displacement rinse (2.0 BV):

$$\text{Regenerant wastewater} = 2.25 + 2.0 = 4.25\text{ BV}$$

In this example, the wastewater volume amounts to approximately 1.9 percent [(4.25/225) 100] of the treated water, or 0.9 percent (1.9 × 0.45) of the blended product water. Choosing 1.0 L of resin as a convenient bed size (for calculation purposes), we set 1.0 BV = 1.0 L; thus, the regenerant wastewater volume from a 1.0-L bed is 4.25 L. For our example bed containing 125 ft^3 of resin, the wastewater volume is 531 ft^3 (4000 gal).

If the small quantity of ions in the water used to make up the regenerant solution is disregarded, the waste brine concentration (eq/L) can be calculated as follows:

$$\text{Total ionic concentration of wastewater} = \text{Hardness concentration} + \text{excess NaCl concentration} \tag{9.52}$$

$$\text{Wastewater hardness concentration} = \frac{\text{equiv of hardness removed}}{\text{regenerant wastewater volume}} \tag{9.53}$$

From Table 9.6, we see that 1.24 equivalents of hardness are removed from the feedwater per liter of resin at a regeneration level of 15 lb NaCl/ft^3 resin. During steady-state operation, this is the amount of hardness removed from the resin with each regeneration. Therefore

$$\text{Waste hardness concentration} = \frac{1.24\ \text{equiv}}{4.25\ \text{L}} = 0.29\ \frac{\text{equiv}}{\text{L}} \tag{9.54}$$

This amounts to 14,500 mg $CaCO_3$ hardness/L, which can be further broken down into the separate Ca and Mg concentrations using the known ratio of Ca to Mg in the raw water.

The excess NaCl concentration is also calculated from Table 9.6 and the following relationship:

$$\text{excess NaCl concentration} = \frac{\text{equiv NaCl applied} - \text{equiv hardness removed}}{\text{regenerant wastewater volume}}$$

$$\text{excess NaCl concentration} = \frac{(1.24)\ 3.3 - 1.24}{4.25} = 0.67\ \frac{\text{equiv NaCl}}{\text{L}} \tag{9.55}$$

This excess NaCl concentration corresponds to 39,300 mg NaCl/L. The total cation composition of the wastewater is 0.96 eq/L, made up of 0.29 eq/L hardness and 0.67 eq/L sodium. These, in addition to chloride, are the major constituents of the wastewater. Other minor contaminant cations removed from the source water will be present in the regenerant wastewater. These can include Ba^{2+}, Sr^{2+}, Ra^{2+}, Fe^{2+}, Mn^{2+}, and others. The minor anionic contaminants in the wastewater will be bicarbonate and sulfate from the source water used for regeneration and rinsing.

Brine Disposal from Softening Plants

The usual method for disposing of spent regenerant brine is through metering into a sanitary sewer. In coastal locations, direct discharge into the ocean is a possibility. Other alternatives are properly lined evaporation ponds in arid regions or brine disposal wells in areas where such wells are permitted or already in existence. The uncontrolled discharge (batch dumping) of the spent regenerant brine into surface waters or sanitary sewers is never recommended because of the potential damage to biota from localized high salinity. For the interested reader the subject of waste disposal is covered in detail in Chapter 16.

Hydrogen Ion-Exchange Softening

Softening water without the addition of sodium is sometimes desirable. In this case, hydrogen can be exchanged for hardness ions using either strong- or weak-base cation exchangers. Hydrogen-form strong-acid resins (Equation 9.1) are seldom used in this application because of the acidity of the product water, the inefficiency of regeneration with acid, and the problems of excess acid disposal. Hydrogen-form weak-acid cation (WAC) exchangers are sometimes used for sodium-free softening. Only the removal of temporary hardness is possible, and this proceeds according to Equation 9.4, resulting in partial softening, dealkalization, and TDS reduction. Regeneration is accomplished with strong acids such as HCl or H_2SO_4, and proceeds according to Equation 9.5 backwards (from right to left).

The partially softened, alkalinity-free column effluent must be stripped of CO_2 and blended with raw water to yield a noncorrosive product water. Alternatively, the pH of the column effluent can be raised by adding NaOH or $Ca(OH)_2$ following CO_2 stripping. This, however, costs more and results in the addition of either sodium or hardness to the product water.

BARIUM REMOVAL BY ION EXCHANGE

During all types of ion-exchange softening, barium is removed in preference to calcium and magnesium. This is shown graphically in Figure 9.9*a* where, in theory, barium breaks through long after hardness. This is true even though barium-contaminated groundwater will always contain much higher levels of calcium and magnesium. Snoeyink, Cairns-Chambers et al. (1987) summarized their considerable research on barium removal using hydrogen- and sodium-form SAC and WAC resins operated to barium breakthrough. They found that the main problem with using SAC resin for barium removal is the difficulty of removing barium from the exhausted resin. Barium accumulates on the resin and reduces the exchange capacity of subsequent runs if sufficient NaCl is not used for regeneration.

When using WAC resins in the hydrogen form for barium removal, the same considerations apply as with WAC softening. Divalent cations are preferentially removed; cation removal is equivalent to alkalinity; partial desalting occurs; CO_2 must be stripped from the column effluent, and bypass blending will be required. The advantage of WAC resins is that barium is easily removed during regeneration with a small excess (typically 20 percent) of HCl or H_2SO_4. In summary, using a hydrogen-form WAC resin for barium, or combined hardness and barium removal, produces a better-quality product water with less wastewater volume compared to a sodium-form resins. The WAC process is, however, more complex and more expensive because of chemical costs, the need for acid-resistant materials of construction, wastewater neutralization, and the need to strip CO_2 from the product water.

RADIUM REMOVAL BY ION EXCHANGE

Radium-226 and radium-228 are natural groundwater contaminants occurring at ultratrace levels. Their combined maximum contaminant level (MCL) is limited to 5.0 picoCuries/L, which corresponds to 11.1 Ra-226 disintegrations/min L or 0.185 Becquerel/L. For Ra-226 alone, this corresponds to 5×10^{-9} mg Ra/L. According to a 1985 survey (Cothern and Lappenbush, 1984), more than 550 community water supplies in the USA exceed the current radium MCL, but very few supplies will exceed the proposed separate MCLs of 20 pCi/L for Ra-226 and Ra-228 (USEPA, 1991). Cation exchange with either sodium- or hydrogen-form resins is a very effective means of radium removal because radium is more preferred than all the common cations found in water. As with hardness and barium removal, radium can effectively be removed using sodium-form SAC or hydrogen-form WAC resins.

Radium Removal During Softening

During the normal sodium ion-exchange softening process, radium is completely (>95 percent) removed; thus, softening is an effective technique for meeting the radium MCL. Recently completed pilot studies on groundwater containing 18 pCi/L total radium and 275 mg/L total hardness in Lemont, Illinois, have resulted in the following conclusions regarding sodium ion exchange softening for radium removal (Subramonian, Clifford et al., 1990):

1. On the first exhaustion run to radium breakthrough, hardness breakthrough occurred at 300 BV while radium did not break through until 1200 BV.
2. On the second and subsequent exhaustion runs following salt regeneration at 15 lb NaCl/ft^3 resin, radium broke through simultaneously with hardness at 300 BV. This long, first run to radium breakthrough, followed by shorter subsequent runs, was not an anomaly but was repeated with other SAC resins.
3. When the resin bed was operated in the normal fashion (i.e., exhausted downflow and never run beyond hardness breakthrough), no radium was removed during the first three cocurrent regenerations at 15 lb NaCl/ft^3 resin. And five exhaustion-regeneration cycles were required to reach a steady state where radium sorption during exhaustion equaled radium desorption during regeneration.
4. Weak-acid cation resins had a lower relative affinity for radium than SAC resins. Macroporous SAC resins had the highest affinity for radium, but were more difficult to regenerate because of this high affinity.
5. Radium never broke through before hardness in any of the 80 experimental runs with five different SAC and WAC resins. Furthermore, the radium concentration of the effluent never exceeded that of the influent (i.e., chromatographic peaking of radium never occurred).
6. Radium was very difficult to remove from exhausted resins, presumably because it is a very large, poorly hydrated ion that seeks the relatively inaccessible hydrophobic regions of the resin phase. Increasing the regeneration time beyond the usual 15 to 30 min helped only slightly to remove the radium from the spent resin. In this regard, radium behaved differently from barium (Myers, Snoeyink et al., 1985), which was much more efficiently eluted at 30 min compared with 15 min EBCT for the regenerant.

Calcium-Form Resins for Radium Removal

Calcium-form SAC resins can be used for radium and barium removal (Myers, Snoeyink et al., 1985; Subramonian, Clifford et al., 1990) when concomitant softening is not necessary. In this process, 1 to 2 M $CaCl_2$ is used as the regenerant at a level of 14.2 lb $CaCl_2$/ft^3, and counterflow regeneration for an extended time (60 min) is the preferred mode of operation. With counterflow regeneration, care must be taken not to mix the exhausted resin bed prior to or during regeneration. In theory, the run time to radium breakthrough is independent of the form (sodium or calcium) of the resin. This theoretical constant run length irrespective of the resin's initial condition was verified in the Lemont pilot study for the first exhaustion of a calcium-form resin, which also ran for 2500 BV before radium reached the MCL (Subramonian, Clifford et al., 1990). The lengths of subsequent runs of calcium-regenerated resin were, however, very much a function of the regeneration conditions. The best counterflow $CaCl_2$ regeneration resulted in a typical run length of 500 BV to radium breakthrough with a continuous radium leakage prior to breakthrough of 3 pCi/L when the feed was 18 pCi/L. Cocurrent $CaCl_2$ regeneration resulted in immediate radium leakage (8 to 10 pCi/L) that continually decreased until radium breakthrough. Therefore, cocurrent $CaCl_2$ regeneration is not an acceptable way to operate this radium-removal process. When calcium-form resins are utilized for radium removal, only countercurrent regeneration should be employed. Furthermore, extreme care must be taken to keep traces of radium-contaminated resin away from the column exit during exhaustion.

Dealing with Radium-Contaminated Brines

If one is consistent with existing practices regarding the disposal of radium-contaminated brines, disposal into the local sanitary sewer should be allowed. For example, in many midwest communities where water is being softened on both a residential and a municipal scale, radium removal is also taking place and the radium-contaminated brines are being disposed of in the usual fashion (i.e., by metering into the local sanitary sewer). To determine the radium concentration in waste-softener brines, a simple calculation can be done. In a hypothetical situation in which the raw water contains as much as 20 pCi/L of radium and the softening run length is 300 BV, the 5 BV of spent regenerant contains an average of 1200 pCi/L of radium. This is a concentration factor of 60 (300/5) and is typical of softeners in general.

If necessary, a spent-regenerant brine solution could be decontaminated prior to disposal by passing it through a radium-specific adsorbent such as the Dow "complexer" or $BaSO_4$-loaded activated alumina (Clifford, 1990). (Note: The Dow *radium-selective complexer* [RSC] is not really a complexer but a $BaSO_4$-impregnated SAC resin [Dow patent] that adsorbs radium onto the $BaSO_4$ even in the presence of a high concentration of competing ions such as Ca^{2+}, Mg^{2+}, and Na^{+}.) This brine-decontamination process has been tested on a small municipal scale (Mangelson, 1988) and found to work well. The RSC was loaded to a level of 2.7 nCi/cm^3 and was still decontaminating spent brine after 1 year of operation. Unfortunately, disposal of a radium-containing solid at a level of 2.7 nCi/cm^3 (2.7×10^6 pCi/L) is potentially a far more serious problem than disposal of the original radium-contaminated brine. Partially because of the disposal problem, Dow abandoned production of the RSC even though it had been proved technically very effective.

NITRATE REMOVAL BY ION EXCHANGE

Ion exchange of chloride for nitrate is currently the simplest and lowest-cost method for removing nitrate from contaminated groundwater to be used for drinking. Only a few applications of the process (shown schematically in Figure 9.10) now exist, and these are restricted to small-community and noncommunity water supplies. As additional communities are forced into compliance, however, more applications of nitrate removal by ion exchange will be seen. This prediction is based on an AWWA survey of inorganic contaminants (1985), which reported that nitrate concentration in excess of 10 mg/L NO_3-N MCL was the most common reason compelling the shutdown of small-community water supply wells.

The anion-exchange process for nitrate removal is similar to cation-exchange softening except that (1) anions rather than cations are being exchanged; (2) nitrate is a monovalent ion, whereas calcium is divalent; and (3) nitrate, unlike calcium, is not the most preferred common ion involved in the multicomponent ion-exchange process with a typical nitrate-contaminated groundwater. The latter two exceptions lead to some significant differences between softening and nitrate removal.

Chloride-form SBA exchange resins are used for nitrate removal according to Equation 9.7. Excess NaCl at a concentration of 1.5 to 12 percent (0.25 to 2.0 *M*) is used for regeneration to produce a reversal of that reaction. The apparently simple process is not without complications, however, as is detailed in the following paragraphs.

Effects of Water Quality on Nitrate Removal

The source water quality and, in particular, the sulfate content influence the bed volumes that can be treated prior to nitrate breakthrough, which is shown along with that for chloride and bicarbonate in Figure 9.12. The effect that increasing sulfate concentration has on nitrate breakthrough is shown in Figure 9.13, constructed from data obtained by spiking Glendale, Arizona, water with sodium sulfate (Clifford, Lin et al., 1987). As sulfate increased from the natural value of 43 mg/L (0.9 meq/L) to 310 mg/L (6.5 meq/L) (i.e., an increase of 5.6 meq/L), the experimental run length to nitrate breakthrough decreased 55 percent, from 400 to 180 BV. Similar response to increasing sulfate is expected for all conventional Type 1 and Type 2 strong-base resins, disregarding the special nitrate-selective resins.

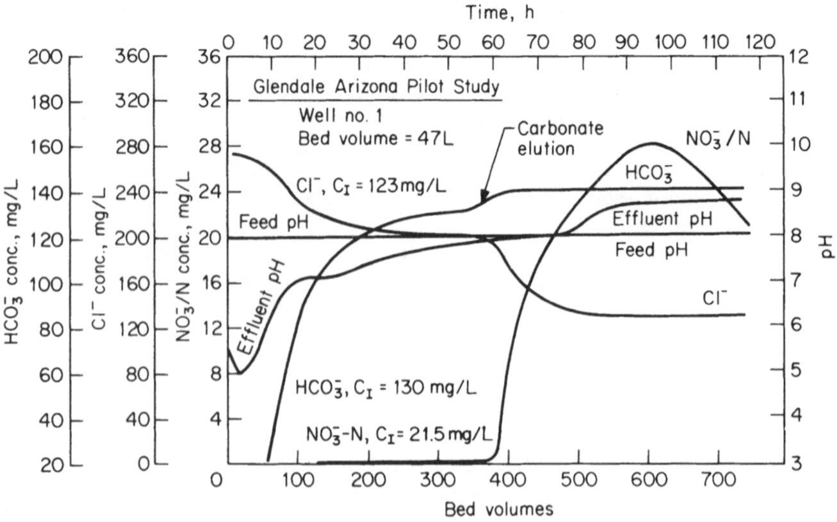

FIGURE 9.12 Breakthrough curves for nitrate and other anions following complete regeneration of type 2 gel SBA resin in chloride form. C_I = influent concentration.

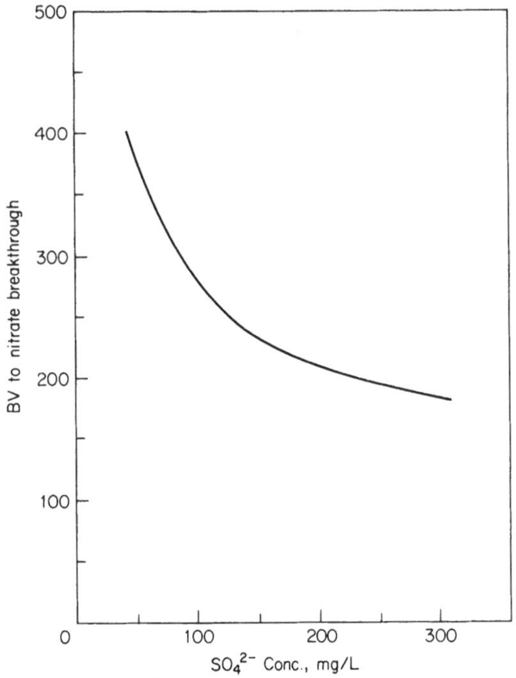

FIGURE 9.13 Effect of sulfate concentration on nitrate breakthrough for source and sulfate-spiked Glendale water, polystyrene SBA resin.

The detrimental effect of sulfate shown in Figure 9.13 is well-predicted by multi-component chromatography calculations (Clifford, 1982). Such calculations can be used to estimate the effects of additional chloride and bicarbonate on nitrate breakthrough for typical Type 1 and 2 SBA resins. When the chloride concentration in the Glendale water was hypothetically increased by 5.6 meq/L (200 mg/L), the calculated nitrate breakthrough occurred at 265 BV—a reduction of 34 percent. Adding 5.6 meq/L (340 mg/L) of bicarbonate only reduced the nitrate run length by 15 percent, to 340 BV. These reductions in nitrate run length were in accord with qualitative predictions based on the selectivity sequence (i.e., sulfate > nitrate > chloride > bicarbonate).

Because all commercially available SBA resins prefer sulfate to nitrate at the TDS levels and ionic strengths of typical groundwater, chromatographic peaking of nitrate occurs following its breakthrough. This peaking, when it does occur, means that effluent nitrate concentrations will exceed the source water nitrate level. No easy way is available to calculate whether a peak will occur or what the magnitude of the expected nitrate peak will be. Peak magnitude depends primarily on the TDS level; the specific composition of the source water, including its sulfate, nitrate, and alkalinity concentrations; and the type of SBA resin used. Computer estimates of the magnitude of the nitrate peak can be made using the EMCT Windows (Tirupanangadu, 1996) and IX Windows computer Programs (Guter 1998).

As the TDS level of the source water increases, selectivity reversal can occur, with the result that nitrate is preferred to sulfate. In this case, which was reported by

Guter in his McFarland, California, nitrate-removal pilot studies (Guter, 1981), sulfate broke through prior to nitrate, and no nitrate peaking occurred.

In other laboratory and pilot studies (Clifford and W. J. Weber, 1978; Clifford, Lin et al., 1987), nitrate peaking was clearly observed with both simulated groundwater and actual groundwater. The theoretical nitrate breakthrough curve for the standard Type 1 and Type 2 SBA resins is depicted in Figure 9.9*b,* where idealized nitrate peaking is shown. Figure 9.12 depicts the pH, nitrate, and bicarbonate breakthrough curves observed in the Glendale nitrate-removal pilot studies (Clifford, Lin et al., 1987). In this groundwater, which is considered somewhat typical of nitrate-contaminated groundwater, the effluent nitrate concentration peaked at 1.3 times the feed water value.

Detecting Nitrate Breakthrough

When the usual Type 1 and Type 2 SBA resins are used to remove nitrate from a typical groundwater containing sulfate and having an ionic strength $I \leq 0.010\ M$, chromatographic peaking of nitrate can occur. Even if it does not occur, the column run must still be terminated prior to the breakthrough of nitrate. If the source water concentration is reasonably constant, this termination can be initiated by a flowmeter signal when a predetermined volume of feed has passed through the column. Terminating a run at a predetermined volume of treated water cannot always be recommended, because of the variable nature of nitrate contamination in some types of groundwater. For example, in a 6-month period in McFarland (Guter, 1981), the nitrate content of one well varied from 5 to 25 mg/L NO_3–N. Such extreme variations did not occur during the 15-month pilot study in Glendale (Clifford, Lin et al., 1987), where the nitrate content of the study well (which was pumped occasionally to fill a large tank for the pilot study) only varied from 18 to 25 mg/L NO_3-N.

In the Glendale pilot study, a pH change of about 1.0 pH unit accompanied the nitrate breakthrough. This pH wave is visible in Figure 9.12, where we see that if the run had been terminated when the effluent pH equaled the feed pH, the nitrate peak would have been avoided. The observed pH increase resulted from the simultaneous elution of carbonate and nitrate. The carbonate elution was a fortunate coincidence, and it allowed pH or differential (feed versus effluent) pH to be used to anticipate nitrate breakthrough. Such pH waves are not unique to the Glendale water but are expected when significant (>1.0 meq/L) concentrations of sulfate and bicarbonate are present and when ionic strength $\leq 0.01\ M$. In practice, the pH detector should be located slightly upstream of the column effluent to provide a safety factor. Another way to avoid nitrate peaking is to employ multiple-parallel columns as described in the "Multicolumn Operations" section.

Choice of Resin for Nitrate Removal

Both laboratory and field studies of nitrate removal by chloride-form anion exchange have shown that no significant performance differences exist between the standard commercially available SBA resins. Both Type 1 and Type 2 polystyrene divinylbenzene SBA resins have been used successfully. Guter (1981) reported on the performance of a Type 1 resin in a full-scale application, and Clifford, Lin et al. (1987) made extensive use of a Type 2 SBA resin during a 15-month pilot-scale study in Glendale. When even potential chromatographic peaking of nitrate must be avoided, however, a special nitrate-selective resin is necessary. Liu and Clifford (1996) also reported that nitrate-selective resins were required with their brine denitrification and reuse process when the feed water contained 200 mg/L sulfate. Guter (1981) described several such special resins that were nitrate-selective with respect to sulfate based on their increased charge separation distance and hydrophobicity. The nitrate-sulfate selectivities of these resins were in accord with the predictions of Clifford and Weber (1978; 1983), who found that (1) the sulfate preference of a resin was significantly reduced by increasing the distance between charged ion-exchange sites, and (2) the nitrate preference of a resin was improved by increasing matrix and functional group hydrophobicity.

The nitrate-over-sulfate selective (NSS) resins reported on by Guter are available from several major ion-exchange resin manufacturers (Rhom and Haas, 1994; Purolite, 1995; Sybron, 1995). Basically, NSS resins are similar to the standard Type 1 resins, which contain trimethyl amine functionality ($RN[CH_3]_3$), but the NSS resins have ethyl, propyl, or butyl groups substituted for the methyl groups to increase the charge separation distance and resin hydrophobicity. Although the tripropyl NSS resins have the highest relative nitrate preference, the less expensive tri-ethyl resins proved to be adequate for those applications that require a nitrate-selective resin (Liu and Clifford, 1996).

Regeneration of Nitrate-Laden Resin

Regardless of the regeneration method used, all previous studies have demonstrated that sulfate is much easier to elute than nitrate. Typical sulfate and nitrate elution curves are shown in Figure 9.14. For a complete regeneration, divalent sulfate, the

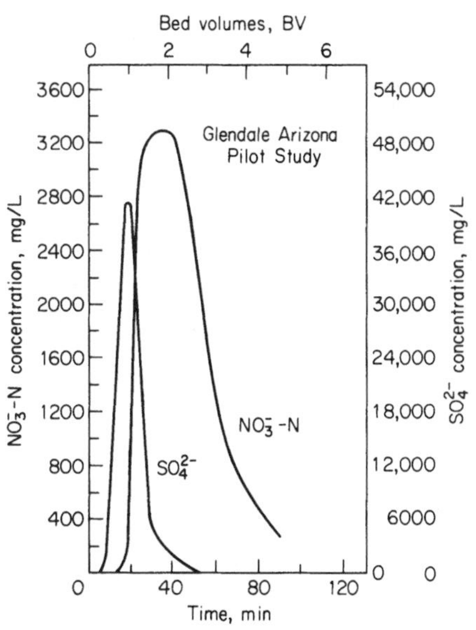

FIGURE 9.14 Elution of sulfate and nitrate during complete regeneration of type 2 gel SBA resin following exhaustion run shown in Figure 9.12. Cocurrent downflow regeneration using 12 percent NaCl (2 *N*) at EBCT = 19 min. Essentially 100 percent recovery of sulfate and nitrate observed.

most preferred ion during exhaustion, is the first to be stripped from the resin during regeneration because of the selectivity reversal that occurs in high-ionic-strength (0.25 to 3 *M*) salt solution.

Regeneration of the nitrate-laden resin was studied extensively in Glendale by using the complete and partial regeneration techniques. For complete regeneration (i.e., the removal of more than 95 percent of the sorbed nitrate), the more dilute the regenerant, the more efficient it was. For example, 0.5 *M* NaCl required 4.0 equivalent of chloride per equivalent of resin, whereas 1.7 *M* NaCl (10 percent NaCl) required 170 percent of this value. A further advantage of dilute regenerants is their amenability to biological denitrification of the spent brine, because the spent-regenerant denitrification rate has been found to decrease significantly with increasing NaCl concentration (Van der Hoek, Van der Van et al., 1988). An obvious disadvantage of dilute regenerants is that they produce more wastewater and require longer regeneration times.

Complete regeneration is not very efficient in terms of the amount of salt required, but it does result in minimal leakage of nitrate during the subsequent exhaustion run. But because significant nitrate leakage (up to 7 mg NO_3-N/L) is permissible based on the MCL of 10 mg NO_3-N/L, complete regeneration is not required. Thus, nitrate removal is an ideal application for incomplete or partial regeneration (i.e., the removal of 50 to 60 percent of the sorbed contaminant).

Partial regeneration levels of 0.64 to 2.0 equivalent per equivalent of resin (3 to 10 lb NaCl per ft^3 of resin) were studied at Glendale, with the result that 1.0 eq Cl^-/eq resin was acceptable if 7.0 mg NO_3-N/L could be tolerated in the effluent prior to the nitrate breakthrough that occurred earlier (350 BV) than in the complete regeneration case (410 BV). Guter (1981) found that regeneration levels as low as 0.64 equivalent Cl^- per equivalent of resin (3 lb NaCl per ft^3 of resin) were acceptable in his pilot studies in McFarland. The acceptable level must be determined for each application, and it depends primarily on the feed water quality and the allowable nitrate leakage. Regardless of the level of salt used for partial regeneration, complete mixing of the resin bed is mandatory following regeneration. If mixing is not accomplished, excessive nitrate leakage will occur during the first 80 to 100 BV of the exhaustion cycle. The required mixing can be achieved mechanically or by introducing air into the backwash, but conventional (uniform) backwashing, even for extended periods, will not provide adequate mixing.

In the Glendale pilot study, a comparison of the optimum complete and partial regeneration methods indicated that partial regeneration would use 37 percent less salt. Bypass blending could not be used with partial regeneration because the nitrate leakage was 7 mg NO_3-N/L.

Disposal of Nitrate-Contaminated Brine

Because of its eutrophication potential, nitrate-contaminated brine cannot be disposed of into rivers or lakes even if it is slowly metered into the receiving water. Furthermore, the high sodium concentration prevents disposal of spent regenerant onto land where its nitrogen content could serve as a fertilizer. To eliminate the high-sodium problem, calcium chloride could be used in place of NaCl, but this would be much more expensive and could result in the precipitation of $CaCO_3$(s) during regeneration. Potassium chloride (KCl) is also feasible as a regenerant, but currently (1999) it is much more expensive than NaCl.

In one municipal-scale application of nitrate removal by chloride ion exchange (Guter, 1981), the spent regenerant was simply metered into the sanitary sewer for subsequent biological treatment. This is probably only possible in those applications where nitrate removal is being carried out in a fraction (e.g., less than 25 percent) of the wells in the supply system. If all the wells were being treated, the waste salt and varying salinity would surely have a negative effect on the biological treatment process.

In the Glendale studies, direct reuse of the spent brine (without removing the nitrate) was attempted but without success. Electroselectivity reversal does not occur with monovalent ions like nitrate, and the high affinity of the resin for the nitrate in the brine resulted in excessive nitrate on the resin following reuse. The nitrate remaining on the resin was subsequently driven off by sulfate in the feed water, causing early nitrate leakage during the next exhaustion.

Biological Denitrification and Reuse of Spent Brine

Removing the nitrate from the spent brine prior to its reuse is possible via biological denitrification. Bench and pilot-scale studies of this process have been reported by Van der Hoek et. al (1988), who found that biological denitrification and reuse was feasible below about 15,000 mg NaCl/L. Clifford and Liu (1993*a* and *b*) developed a *sequencing batch reactor* (SBR) denitrification process for biological denitrification of spent nitrate ion exchange brine. Their bench-scale experiments examined the feasibility of using an SBR to denitrify 0.5 *M* (3 percent) NaCl brine containing up to 835 mg NO_3-N/L. After acclimation, the denitrification rate in 0.5 *M* NaCl was only 10 percent lower than in the no-salt control reactor. The effect of mass ratio R of methanol to nitrate-nitrogen on denitrification rate and residual TOC in the denitrified SBR effluent was studied in the range of 2.2 to 3.2. At the optimum methanol-to-nitrate-nitrogen ratio of 2.7, the time for >95 percent denitrification was 8 h. In one set of runs, actual spent regenerant was reused 15 times; each time it was denitrified in the SBR, filtered, and compensated with NaCl before reuse. Denitrification and reuse of the brine resulted in a 50 percent reduction in NaCl consumption and a 90 percent reduction in spent brine discharge. Their bench-scale research indicated that a denitrifying batch reactor provided simple operation, reliable effluent quality, and compatibility with the inherent batch operation of the nitrate ion-exchange process. Later Liu and Clifford (1996) reported on pilot-scale tests of the SBR denitrification brine reuse process (see Figure 9.15) for nitrate removal in McFarland, California, on a high-sulfate (140 to 220 mg/L) water. In those tests, which evaluated three different resins and partial and complete regeneration with and without brine reuse, the brine was denitrified and reused successfully up to 38 times. Because of sulfate buildup to 16,000 mg/L in the recycle brine, a nitrate-selective resin (triethyl or tripropyl amine) was preferred to minimize nitrate leakage and maximize the number of brine reuse cycles. Even though the biologically denitrified brine was repeatedly in contact with microorganisms, methanol, and high TOC, the product water quality in terms of heterotrophic plate count, volatile organics, TOC, and methanol was not significantly different than product water from a conventional nitrate ion exchange process. The pilot study of the brine denitrification-reuse process verified that NaCl consumption could be reduced up to 75 percent compared with the optimum partial regeneration process without brine reuse. Depending on the number of reuse cycles, the brine discharge volume could be reduced up to 95 percent compared with operation without brine reuse.

The brine denitrification-reuse process, although more complex than processes using direct disposal of waste regenerants, should be considered where wastewater disposal is a major consideration.

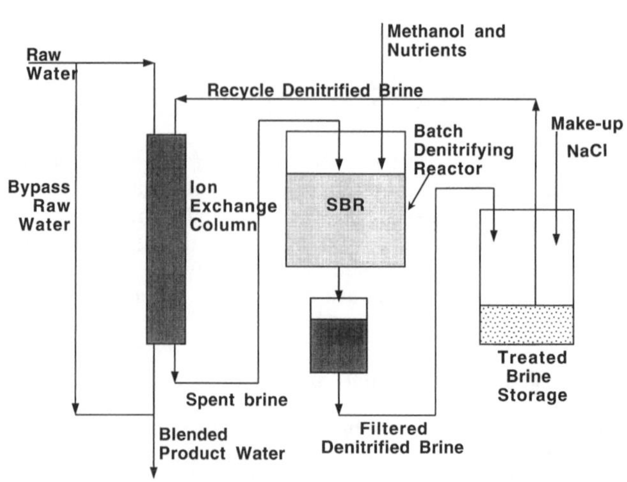

FIGURE 9.15 Ion-exchange/biological denitrification process with sequencing batch reactor (SBR).

FLUORIDE REMOVAL BY ACTIVATED ALUMINA

Activated alumina, a semicrystalline porous inorganic adsorbent, is an excellent medium for fluoride removal. Alumina is far superior to synthetic organic anion-exchange resins because fluoride is one of the ions most preferred by alumina, whereas with resins, fluoride is the least preferred of the common anions. (Compare the position of fluoride in Equation 9.27, the activated alumina anion selectivity sequence, with Table 9.3, containing the relative affinity of anions for resins.) The usefulness of alumina as a fluoride adsorbent has been known for more than 50 years, and municipal defluoridation of public water supplies using packed beds of activated alumina has been practiced since the 1940s (Maier, 1953). More recently, POE and POU fluoride-removal systems utilizing activated alumina either have come into common use (Water Quality Research Council, 1987) or at least are being considered for individual-home treatment (Bellen, Anderson et al., 1986).

The fundamentals of fluoride adsorption onto activated alumina were covered in the section "Activated Alumina Adsorption," where attention was drawn to Equations 9.11 through 9.17. The following discussion focuses on the important design considerations for municipal defluoridation. Although pH adjustment is not ordinarily performed prior to POE and POU defluoridation, the discussion may also prove useful to designers of these systems because the important factors, including pH, governing fluoride capacity are explained.

Alumina Defluoridation System Design

A typical fluoride-removal plant utilizing activated alumina consists of two or more adsorption beds operated alternately or simultaneously. The source water pH is adjusted to 5.5 to 6.0 and passed downflow through a 3- to 5-ft-deep bed of fine (28 × 48 mesh) medium that adsorbs the fluoride. The fluoride breakthrough curve (some typical examples are shown in Figure 9.16) is not sharp compared with the usual ion-exchange resin breakthrough curves. The effluent fluoride concentration is con-

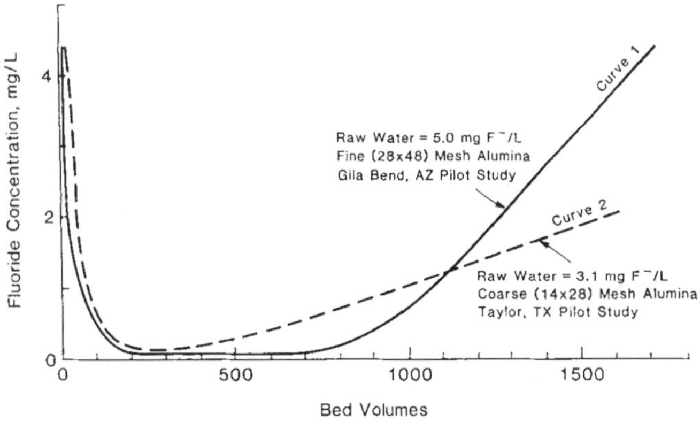

FIGURE 9.16 Typical fluoride breakthrough curves for activated alumina operated at a feed pH of 5.5 and EBCT of 5 min.

tinuously increasing, thereby making the use of the bypass blending technique shown in Figure 9.10 difficult to implement. Instead, a product water storage tank is sometimes provided to equalize the column effluent fluoride concentration and to maximize the column run length.

A target level of fluoride, usually in the range of 1 to 3 mg/L, must be chosen for the process effluent before the operating procedures and economics can be established. The MCL for fluoride is 4 mg/L, and the secondary maximum contaminant level (SMCL) is 2 mg/L (see Chapter 15). When the source water contains 3 to 6 mg/L fluoride and a maximum fluoride effluent (breakthrough) concentration of 1.4 mg/L is chosen, a run length of approximately 1000 to 1300 BV can be expected with an alumina similar to Alcoa F-1. Some typical data from pilot-scale defluoridation processes are presented in Table 9.7 for source water fluoride concentrations of 2 to 5 mg/L, and some detailed design recommendations are given in Table 9.8.

Following exhaustion, the medium is backwashed and then subjected to a two-step regeneration with base followed by acid. Because many types of high-fluoride groundwater are located in hot, arid climates, the spent-regenerant brines are neutralized and sent to a lined evaporation pond for interim disposal. The ultimate disposal of high-fluoride salt residues is a problem that still remains unsolved.

TABLE 9.7 Fluoride Capacity of F-1 Activated Alumina Columns at pH 5.5 to 6.0—Field Results

Feedwater fluoride mg/L	Feedwater TDS mg/L	Alumina mesh size[1]	Run length BV to 1.4 mg F^-/L	Fluoride capacity g/m^3
2.0[2]	810	28 × 48	2300	3700
3.0[3]	1350	14 × 28	1200	3000
5.0[4]	1210	28 × 48	1150	4600

[1] 28 × 48 mesh is "fine" alumina, the recommended media 14 × 28 mesh is a "coarse" alumina that can also be used.

[2] San Ysidro, New Mexico; avg. of 3 pilot-scale runs (Clifford and Lin, 1991).

[3] Taylor, Texas; avg. of 3 pilot-scale runs

[4] Gila Bend, Arizona; typical pilot-scale run (Rubel and Woosley, 1979).

TABLE 9.8 Process Design Criteria for Fluoride Removal by Activated Alumina

Parameter	Typical value or range
Fluoride concentration	3–6 mg/L
Media[1]	Alcoa F-1 activated alumina
Media size[2]	28 × 48 mesh, (0.29 to 0.59 mm)
Media depth	3–5 ft
Fluoride capacity	3000–5000 g/m^3 (1300–2200 grains/ft^3)
Bed volumes to 1.4 mg F^-/L[3]	1000–1500 (pH = 5.5–6.0)
Exhaustion flow rate	1.5 gal/min ft^3 (EBCT = 5 min)
Exhaustion flow velocity	4–8 gal/min ft^2
Backwash flow rate	8–9 gal/min ft^2
Backwash time	5–10 minutes using raw water
	NaOH regeneration
Volume of regenerant	5 BV
Regenerant flow rate[4]	0.5 gal/min ft^3 (EBCT = 15 min)
Regenerant concentration[5]	1% NaOH (0.25 N)
Total regenerant contact time	75 min
Displacement rinse volume	2 BV
Displacement rinse rate	0.5 gal/min ft^3
	H_2SO_4 neutralization
Acid concentration[6]	2.0% (0.4 N H_2SO_4)
Acid volume[7]	Sufficient to neutralize the bed to pH 5.5, typically 1.5 BV
Displacement rinse volume	2 BV
Displacement rinse rate	0.5 gal/min ft^3

[1] Mention of trade names does not imply endorsement.
[2] Coarse 14 × 28 mesh alumina has also been used successfully.
[3] Capacity and BV to fluoride breakthrough depend to some extent on the fluoride level in the raw water, sulfate level in the pH adjusted feedwater, and the severity of regeneration.
[4] Cocurrent (downflow) or countercurrent regeneration may be utilized.
[5] Higher regenerant NaOH concentrations (e.g. 4%) have also been used successfully.
[6] Lower and higher H_2SO_4 concentrations have been used successfully. HCl can also be used, and it may be preferred to H_2SO_4 because chloride does not compete with fluoride for adsorption sites.
[7] Alternatively, the neutralization and acidification steps can be combined. In this case the initial feedwater pH is lowered to approximately 2.5 to produce a bed effluent pH slowly dropping from a high initial value (>13) down to a continuous effluent pH in the range of 5.5–6.0.

Factors Influencing the Fluoride Capacity of Alumina

The fluoride capacity of alumina is very sensitive to pH, as shown in Figure 9.17, based on laboratory equilibrium data obtained by using minicolumns of granular alumina exposed for up to 30 days to pH-adjusted fluoride solutions in deionized water. The fluoride capacities for Figure 9.17 can be considered the "maximum attainable (equilibrium) capacities" for F-1 alumina because potentially competing anions such as sulfate, silicate, arsenate, and selenite were minimized or entirely absent in the test solutions (Singh and Clifford, 1981). Pilot-scale field studies on typical fluoride-contaminated water have resulted in the observation that single columns of fine (28 × 48 mesh) alumina operated to 1.4 mg fluoride/L breakthrough and 5-min EBCT can attain approximately 50 percent of the capacities shown in Figure 9.17 (Rubel and Woosley, 1979; Clifford and Lin, 1991). Using coarse-mesh alumina under the same conditions yields about 37 percent of the maximum attainable (equilibrium) capacity. These percentages (37 percent for coarse medium and 50 percent for fine medium) can be used in conjunction with Figure 9.17 to estimate the

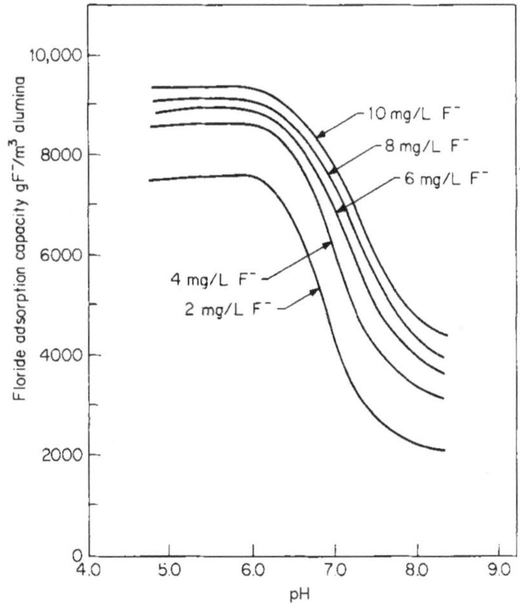

FIGURE 9.17 Effects of pH on equilibrium fluoride adsorption capacity on F-1 alumina. Distilled water solutions of NaF, pH-adjusted with Na_2CO_3 and H_2SO_4.

single-column fluoride capacities attainable in actual single-column installations when the natural or adjusted source water pH is known.

The inability of a pilot- or full-scale continuously operated alumina column to rapidly and closely approach fluoride equilibrium is primarily related to solid-phase mass transfer limitations. These limitations cause the fluoride breakthrough curve to be gradual and the run to be terminated long before the alumina is saturated. This is a classic case in which the process economics can be improved by operating two columns in series in a roughing-polishing sequence, as depicted in Figure 9.11. The two-column sequence will allow the roughing column to attain a near-saturation fluoride loading prior to regeneration. The obvious disadvantages of such systems are greater complexity and higher capital costs.

In POE and POU alumina systems, a closer approach to equilibrium may be expected because of the intermittent operation of the alumina bed. In these installations, the solid-phase fluoride concentration gradient has a chance to relax completely (approaching equilibrium) during the many, prolonged "off" periods experienced by the column. This results in a high concentration gradient between the liquid and the surface of the solid and much improved fluoride removal when the column is restarted.

Another reason for the less-than-equilibrium fluoride capacities of actual columns is the effect of competing anions such as silicate and sulfate present in the source water. During a laboratory study (Singh and Clifford, 1981) of the effect of competing ions on fluoride capacity at optimum pH, bicarbonate and chloride ions did not compete measurably with fluoride for adsorption sites. Sulfate did, however, cause a significant reduction in fluoride adsorption. For example, 50 mg sulfate/L lowered the fluoride capacity by about 16 percent. The maximum reduction in fluoride capacity because of sulfate was about 33 percent, and it occurred at approximately

500 mg sulfate/L. Beyond that, no further fluoride capacity reduction occurred, probably because the alumina contains a high proportion of fluoride-specific sites that are not influenced by sulfate.

Previously unpublished results of fluoride-removal pilot studies in Taylor, Texas, indicated that silicate ($Si(OH)_3O^-$) adsorption onto alumina occurs at about pH 7 and above and can cause a serious reduction in the fluoride capacity of a column operated in the pH range of 7 to 9. This reduction is particularly detrimental during cyclic operation because silicate is only partially removed during a normal regeneration to remove fluoride. The presence of silica in the water had no observable effect on fluoride capacity for columns operated in the optimum (5.5 to 6.0) pH range.

Regeneration of Fluoride-Spent Alumina Columns

Prior to regeneration to elute fluoride, the exhausted column must be backwashed to remove entrained particles, break up clumps of alumina, and reclassify the medium to eliminate packing and channeling. Following a 5- to 10-min backwash, regeneration is accomplished in two steps—the first utilizing NaOH and the second utilizing H_2SO_4. Because fluoride leakage is such a serious problem, the most efficient regeneration is accomplished by using countercurrent flow. Nevertheless, cocurrent regeneration is more commonly used because of its simplicity. Rubel (Rubel and Woosley, 1979) recommends a two-stage, upflow-then-downflow application of base, followed by downflow acid neutralization. A slow rinse follows both the base-addition and acid-addition steps, to displace the regenerant and conserve chemicals. The rinsing process is completed by using a fast rinse applied at the usual exhaustion flow rate. The total amount of wastewater produced amounts to about 4 percent of the blended product water and is made up of 8-BV backwash, 5-BV NaOH, 2-BV NaOH displacement rinse, 1.5-BV acid, 2-BV acid displacement rinse, and 30 to 50 BV of fast rinse water. Further details pertaining to regeneration can be found in Table 9.8.

Essentially complete (>95 percent) removal of fluoride from the alumina is possible by using the regeneration techniques described; however, in addition to the complexity of regeneration, its efficiency is poor. For example, an alumina bed employed to remove fluoride from a typical 800 to 1200 mg TDS/L groundwater containing 4 to 6 mg fluoride/L would contain an average fluoride loading of 4600 g F^-/m^3 (2000 grains F^-/ft^3). Removing this fluoride would typically require 5 BV of 1 percent NaOH (30 gal of 1 percent NaOH/ft^3). This equates to 19 percent base regeneration efficiency, or 5.2 hydroxide ions required for each fluoride ion removed from the alumina. When the amount of acid required (1.5 BV of 0.4 N H_2SO_4) is added into the efficiency calculation, an additional 2.5 hydrogen ions are required for each fluoride ion removed and the overall efficiency drops to 13 percent, or a total of 7.7 regenerant ions required for each fluoride ion removed from the water. In spite of the complexity of the process and the inefficiency of regeneration on an ion-for-ion basis, fluoride removal by activated alumina is probably the cheapest alternative compared with other acceptable methods, such as reverse osmosis and electrodialysis (USEPA, 1985). To the author's knowledge, no study has reported on (1) the influence of natural organic matter (NOM) adsorption on the alumina capacity for fluoride or (2) the influence of activated alumina in the aluminum content of the treated water. Because the affinity of alumina for NOM anions (fulvates and humates) is significant, one would expect a decrease in fluoride capacity with increasing NOM concentration. Because aluminum hydroxide is widely used as the primary coagulant for surface water, there has not been much concern about the aluminum from activated alumina, which is basically low-temperature dehydrated aluminum hydroxide. Nevertheless, aluminum in treated water from activated alumina systems should be studied, especially in low-pH (5.5 to 7.0) systems.

ARSENIC REMOVAL BY RESINS AND ALUMINA

As with other toxic inorganic contaminants, arsenic is almost exclusively a groundwater problem. Although it can exist in both organic and inorganic forms, only inorganic arsenic in the +III or +V oxidation state has been found to be significant where potable water supplies are concerned (Andreae, 1978).

The current (1996) MCL for total arsenic is 0.05 mg/L (see Chapter 2), but a more stringent MCL, probably in the range of 2 to 20 μg/L, is likely by 2001 (Pontius, 1995). Depending on the redox condition of the groundwater, either arsenite (As(III)) or arsenate (As(V)) forms will be predominant. The pH of the water is also very important in determining the arsenic speciation. The primary arsenate (As(V)) species found in groundwater in the pH range of 6 to 9 are monovalent $H_2AsO_4^-$ and divalent $HAsO_4^{2-}$. These anions result from the dissociation of arsenic acid (H_3AsO_4), which exhibits pK_a values of 2.2, 7.0, and 11.5. Uncharged arsenious acid (H_3AsO_3) is the predominant species of trivalent arsenic found in natural water. Only at pH values above its pK_a of 9.2 does the monovalent arsenite anion ($H_2AsO_3^-$) predominate.

Both the redox potential and pH are important with regard to arsenic removal from groundwater using the most economical alternatives—anion exchange and activated alumina adsorption (Frey, Edwards et al., 1997). This is because both processes require an anionic species to effectively remove arsenic. Theoretically, at least, As(III) is not removed by either process. Arsenic(V), however, is removed by both processes. Which process to choose is primarily dependent on the sulfate and TDS levels in the source water. Because of competition by background ions for exchange sites, anion exchange is not economically attractive at high TDS (>500 mg/L) or sulfate (>150 mg/L) levels, whereas alumina adsorption is so specific for arsenate that it is not greatly affected by these variables. Depending on the local conditions, hazardous chemicals handling and/or brine disposal may also be important considerations.

In the removal of As(V) by anion exchange, pH adjustment is not required, and ordinary sodium chloride can be used to achieve essentially complete elution of arsenic from spent resin at steady-state operation, which is achieved when the mass of arsenic eluted during regeneration is equal to that removed during exhaustion. Activated alumina, however, requires both sodium hydroxide and sulfuric acid and typically recovers only 75 percent of the adsorbed arsenic. A significant potential disadvantage of ion exchange is that chromatographic peaking of arsenic is possible depending on the sulfate level in the feed water (i.e., the higher the sulfate, the greater the potential for arsenic peaking). This potential problem can be solved by operating multiple columns in parallel or by careful monitoring of the flow or chemical quality of the effluent. For example, if the raw water quality is reasonably constant with respect to sulfate, it is only necessary to totalize the flow through the bed and regenerate on a fixed total volume throughput, based on some test runs to determine the point of arsenic breakthrough. On the other hand, when the column feed water quality is subject to significant variation, especially with regard to sulfate, it may be necessary to frequently measure either sulfate or arsenic in the effluent to determine when to stop the run and regenerate the resin. To anticipate arsenic breakthrough, the samples for sulfate or arsenic analysis would be taken upstream from the column effluent.

Naturally contaminated arsenic-bearing types of groundwater have been reported to have relatively high pH and high alkalinity (National Research Council, 1977). These water types, if they are also low in sulfate, as was the case in Hanford, California (Clifford and Lin, 1986), are very amenable to treatment by chloride-form anion exchange providing As(III) is oxidized (see below) to As(V). High-sulfate, high-TDS water such as the water supply encountered in San Ysidro, New Mexico (Clifford and Lin, 1991), is not amenable to ion exchange, but can be effectively treated by activated alumina adsorption following oxidation of As(III) to As(V) and pH reduction to 5.5 to 6.0.

Oxidation of As(III) to As(V)

To achieve effective removal of arsenic from groundwater by means of columns or membranes, arsenite must be oxidized to arsenate. Figure 9.18 illustrates the importance of oxidative pretreatment ahead of activated alumina columns operated at the optimum pH of 6.0. We see that arsenic breakthrough curve 1, representing a source water containing 100 μg As(III)/L, reaches 50 μg As/L after only 300 BV. But curve 4, representing the same water but containing 100 μg As(V)/L, does not reach 50 μg As/L until 23,400 BV. This is an 80-fold improvement in performance due simply to converting As(III) to As(V). The intermediate plots, curves 2 and 3, also illustrate the need for oxidation, but are not as dramatic. In these curves, actual field data from pilot studies in San Ysidro, New Mexico, and Hanford, California, were used. Some unplanned oxidation of As(III) to As(V) within the alumina column occurred during the New Mexico study (curve 2) and resulted in better-than-expected performance of the alumina. The performance of the alumina on chlorinated Hanford, California, groundwater, curve 3, containing nearly 100 percent As(V) was somewhat poorer than expected and was probably caused by fouling of the alumina by the negatively charged black "mica" particles in this water. For properly oxidized source water in a municipal arsenic-removal application, a breakthrough curve somewhere between curves 3 and 4 would be expected. Further data on alumina column performance as a function of oxidation state and pH is presented in Table 9.9.

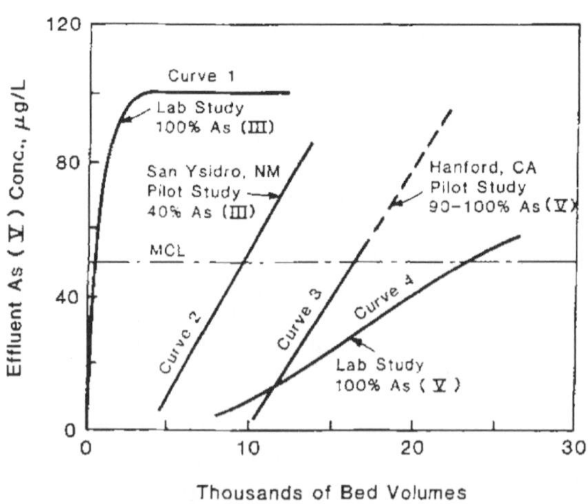

FIGURE 9.18 Arsenic (total) breakthrough curves for fine (28 × 48) mesh F-1 activated alumina used in laboratory and pilot-scale columns. Feedwater pH = 6.0, arsenic (total) concentration = 88 to 100 ppb, and As(III) or As(V) percentages are indicated. EBCT = 3 to 5 min.

TABLE 9.9 Summary Arsenic (Total) Capacities of 28 × 48 Mesh F-1 Alumina Columns

Column feedwater	As(III) μg/L	As(V) μg/L	As(total) μ/L	pH	BV to 50 μg As/L	Arsenic (total) capacity[1] g/m^3	Comment
Synthetic groundwater	100	0	100	6	300	20	pure As(III)
Hanford, CA groundwater	80	10	90	6	700	60	90% As(III)
San Ysidro, NM groundwater[2]	31	57	88	6	9000	575	40% As(III)
Chlorinated Fallon, NV groundwater[3]	0	110	100	5.5	13100	1280	100% As(V)
Chlorinated Hanford, CA groundwater	0	98	98	6	16000	1410	100% As(V)
Synthetic groundwater	0	100	100	6	23000	1920	100% As(V)
Performance at Unadjusted pH[4]							
Chlorinated Fallon, NV groundwater	0	110	110	9	800	42	100% As(V)
Hanford, CA groundwater	80	10	90	8.6	800	61	90% As(III)
Chlorinated Hanford, CA groundwater	0	98	98	8.8	900	83	100% As(V)

[1] To convert g As/m^3 alumina to g As/kg alumina, use an F-1 Alumina density of 0.85 g/cm^3 or 850 kg/m^3.

[2] Unplanned oxidation of As(III) to As(V) occurred on the activated alumina column; thus the run was longer than expected.

[3] Measurement of As(III) was not performed. 100% As(V) was presumed due to the presence of free Cl_2 in the column feedwater.

[4] Normally POE and POU systems would be operated in this manner (i.e., no pH adjustment of the raw water). Although unadjusted pH runs are much shorter than runs at pH 6, the unadjusted pH process is feasible for POE and POU applications.

Here, at relatively high pH (8.6), it did not matter whether As(III) was oxidized, because about the same mediocre performance was obtained (i.e., 800 to 900 BV). At pH 8.6, the adsorption or ligand-exchange capacity of alumina was severely reduced by competition from hydroxide ions. This led to poor As(V) uptake compared with adsorption at the optimum pH of 6.0, where the hydroxide competition was insignificant. The As(III) uptake at pH 8.6 was, however, slightly improved compared with that at pH 6.0 because of the increase in the fraction of charged arsenite anion, $H_3AsO_2^-$, present at the higher pH.

In a laboratory study (Frank and Clifford, 1986), greater than 95 percent oxidation of As(III) to As(V) was observed in the presence of 1 mg/L free chlorine in the 6 to 10 pH range in less than 5 sec. In that same study, pure oxygen could not oxidize As(III) to As(V) in 1 h, but complete oxidation of As(III) to As(V) occurred during 2 months of ambient-temperature storage of synthetic groundwater. Inadvertent, unpredictable, microbially assisted oxidation of As(III) to As(V) has been observed in all the University of Houston studies with As(III) (Rosenblum and Clifford, 1984; Clifford and Lin, 1986; Frank and Clifford, 1986; Clifford and Lin, 1991). Arsenic reduction can occur, too. When As(V)-contaminated groundwater was transported from Albuquerque, New Mexico, to Houston, Texas, and stored for a few days, a significant portion of the As(V) had been reduced to As(III) (Tong 1997). Therefore, As(III) and As(V) should be considered as interconvertible under appropriate conditions.

Alumina System Design for Arsenic Removal

The design of an activated alumina system for arsenic removal is similar to that already described for fluoride removal. A few differences, however, do exist:

1. Because the arsenic concentration (typically ≤0.06 mg/L) is so low compared with the fluoride concentration (typically ≥4 mg/L), a continuous arsenic run may last as long as 30 to 90 days.
2. Because of the difficulty in removing arsenic, the regenerant NaOH concentration may be as high as 4 percent (1.0 *N*), compared with the 1 percent NaOH used for fluoride removal.
3. Prechlorination or an alternative form of oxidation will probably be required to ensure the presence of As(V) as opposed to As(III).
4. A spent-regenerant treatment system involving the coprecipitation and sludge disposal of arsenates with $Fe(OH)_3(s)$ or $Al(OH)_3(s)$ will probably also be required because of the greater toxicity of arsenic compared with fluoride.

Factors Influencing the Arsenic Capacity of Alumina

Activated alumina adsorption of arsenate, a ligand exchange process, is highly dependent on pH; at alkaline pHs where hydroxide competition is significant, arsenate adsorption is poor. As the pH is lowered arsenate adsorption increases dramatically until about pH 5. Based on a limited number of full-scale and pilot-scale column studies, Rubel recommends pH 5.5 for both fluoride (Rubel and Woosley, 1979) and arsenic (Rubel and Hathaway, 1985); and a combination of equilibrium isotherm studies and pilot-scale column studies indicated that the optimum pH is closer to 6.0 for both fluoride (Singh and Clifford, 1981) and arsenic (Rosenblum and Clifford, 1984; Simms and Azizian, 1997; Clifford, Ghurye et al., 1998). The equilibrium isotherm data for arsenic in Figure 9.19 shows the arsenic capacity

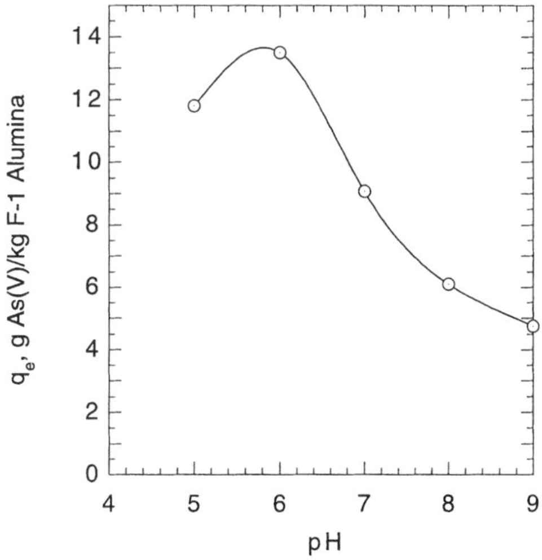

FIGURE 9.19 Effect of pH on equilibrium As(V) capacity of alumina. C_e = 1 mg/L, TDS = 1000 mg/L, SO_4^{2-} = 240–370 mg/L, hardness = 0. (*Source:* Rosenblum and Clifford, 1984.)

declining rapidly as pH increases above about pH 5.7. The isotherms summarized in Figure 9.19 were developed using minicolumn equilibrium experiments with Alcoa F-1 alumina in synthetic background water containing approximately 1000 mg TDS/L and the competing anions sulfate (368 mg/L) and chloride (266 mg/L) at pH 5 and 6, and sulfate (240 mg/L), chloride (178 mg/L), and bicarbonate (305 mg/L) at pH 7 to 9. Simms and Azizian found a similar trend in arsenic capacity reduction with increasing pH in their Severn-Trent Water Authority pilot studies in the UK (Simms and Azizian, 1997), where the groundwater contained about 22 μg/L As(V). When measuring arsenic adsorption capacity to 10 μg/L arsenic breakthrough using pilot-scale columns of Alcan AA-400G alumina, they found the following capacities: pH 6.0 (2.64 g As/kg AAl), pH 6.5 (1.32 g/kg), pH 7.0 (0.44 g/kg), and pH 7.5 (0.19 g/kg). Fortunately, the large amount of arsenic adsorbed at the lower pH did not desorb when the acid-feed system was shut down and the feed water pH was allowed to rise. In spite of a 14-times-higher capacity at pH 6.0, Simms and Azizian chose to operate the scaled-up process at pH 7.5 to avoid having to lower the pH and neutralize much of the alkalinity before treatment and then raise the pH again after treatment.

As with fluoride adsorption, the As(V) capacity of alumina is significantly reduced in the presence of sulfate ions, but nearly unaffected by high concentrations of chloride, which suggests that HCl rather than H_2SO_4 would be preferred for pH adjustment. The influence of these ions is demonstrated in the 25°C equilibrium isotherms presented in Figure 9.20. In actual field operation, however, the alumina column capacity at an arsenic MCL of 10 or 50 μg/L was far less than the batch-equilibrium values as shown in Table 9.10. The four column capacities shown were obtained from single columns operated at pH 5.5 to 6.0 until the breakthrough of 50 or 10 μg As(V)/L. The effects of competing anions and the nonequilibrium mass transfer limitations already discussed (under the topic of fluoride removal) are reasons for the low column capacities observed. An additional reason is the possible fouling of the porous alumina by particulate and colloidal constituents such as col-

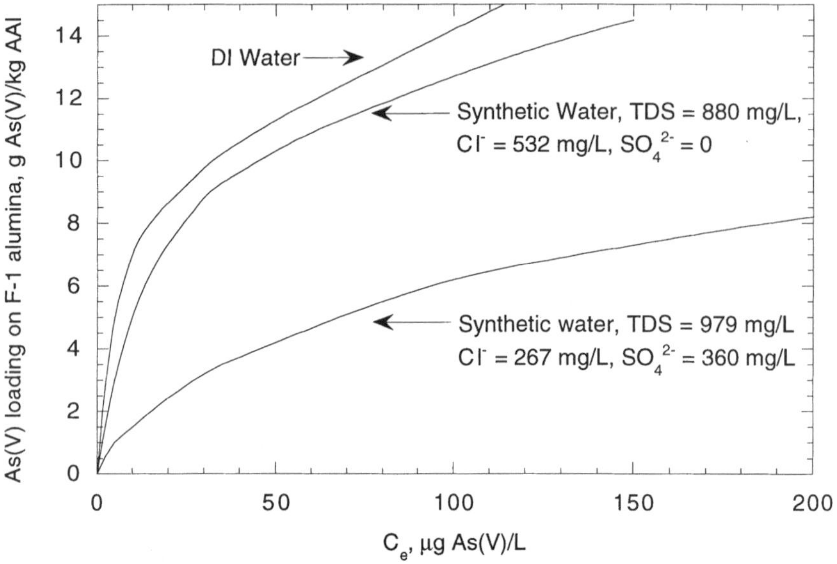

FIGURE 9.20 Equilibrium As(V) adsorption onto F-1 activated alumina at pH 6.0 for deionized water and high TDS synthetic waters at 25°C.

loidal silica and mica. The latter was identified as a problem in the Hanford study (Clifford and Lin, 1986). (Note: Natural organic matter and biogrowths are also potential alumina foulants; these were not expected or measured in the deep Hanford groundwater.) The exceptionally long run length (110,500 BV) and high As(V) capacity (2.64 g/kg) in the Severn-Trent UK study are difficult to explain in light of the performance of the same alumina (Alcan AA-400G) in Albuquerque, where only 15,600 BV and 0.41 g/kg As(V) capacity were observed. Based on the higher sulfate concentration and shorter (3 min) EBCT, the UK water should have produced a shorter run length and lower capacity. Neither water contained significant iron, which might have aided arsenic removal. The only significant difference was the 6-fold higher hardness level in the UK water, which suggests that calcium and/or magnesium may be involved in the uptake of arsenate by alumina. Clearly, more research is needed in order to better predict the effect of water quality on the removal of arsenic by alumina.

Alumina particle size and empty-bed contact time can significantly influence arsenic removal by alumina. Clifford and Lin (1991) and Simms and Azizian (1997) reported that finer particles of alumina (28 × 48 mesh, 0.6 to 3 mm) have higher arsenic capacity, lower arsenic leakage, and longer run length than larger alumina particles (14 × 28 mesh, 1.18 to 0.6 mm). During the same study, Simms and Azizian found that arsenic run length was linearly proportional to EBCT in the range of 3 min (9,000 BV) to 12 min (14,000 BV) when operating with 14 × 28 mesh Alcan AA-400G alumina at pH 7.5. To minimize bed size and alumina inventory, however, they preferred to operate in the 3 to 6 min EBCT range. In the recent Albuquerque arsenic studies, a similar relationship between EBCT and run length was observed when using the finer 28 × 48 mesh alumina and operating at pH 6.0. At 5-min EBCT, the run length was 6,400 BV, while at 10-min EBCT the run length was 8,800 BV.

TABLE 9.10 Comparison of Alumina Capacities for Arsenic(V) at pH 5.5–6.0

Groundwater	Deionized water[1]	Synthetic water[2]	Fallon, NV[3]	Hanford, CA[4]	Severn-Trent, UK[5]	Albuquerque, NM[6]
Reference	Rosenblum and Clifford, 1984	Rosenblum and Clifford, 1984	Hathaway and Rubel, 1987	Clifford and Lin, 1986	Simms and Azizian, 1998	Clifford and Ghurye, 1998
Alumina	F-1	F-1	F-1	F-1	Alcan AA-400G	Alcan AA-400G
Influent As(V) μg/L	NA	NA	110	98	23	22
EBCT, min	7-day batch equilibration	7-day batch equilibration	5.0	5.0	3.0	5.0
TDS, mg/L	<5	979	535	213	350 (est.)	328
Hardness, mg/L $CaCO_3$	0	0	5	10	331	53
Sulfate, mg/L	0	360	96	5	117	70
MCL[7]	NA	NA	50	50	10	10
BV to MCL	NA	NA	14,450	16,000	110,500	15,600
Capacity[8]	11.3 @ 50 μg/L	4.2 @ 50 μg/L	1.50	1.66	2.64	0.41
g As/kg AAl	7.0 @ 10 μg/L	1.5 @ 10 μg/L				

[1–6] The first two columns represent 7-day batch equilibration tests (C_e = 50 μg/L or 10 μg/L) that presumably yield maximum As(V) capacities for comparison with the field column studies in columns (3)–(6). The DI water and synthetic water contained no hardness.

[7] Fallon and Hanford tests (1984–86) assumed an arsenic MCL of 50 μg/L, whereas the Severn-Trent and Albuquerque tests (1998) assumed an MCL of 10 μg/L.

[8] A batch equilibrium capacity of 11.3 @ 50 μg/L means 11.3 g As/kg AAl at C_e = 50 μg As/L

NA = Not Applicable

Regeneration of Arsenic-Spent Alumina

Compared with adsorbed fluoride, arsenic is much more difficult to remove from alumina. For this reason, a higher concentration, 4 percent NaOH, should be used for the base-regeneration step. The usual acid concentration, 2 percent H_2SO_4, can be used, however, for the acid-neutralization step. Even with additional caustic, only 50 to 70 percent of the absorbed arsenic was eluted from the column during regeneration, and the arsenic capacity of the alumina appeared to deteriorate about 10 to 15 percent on each subsequent run during the University of Houston field studies in San Ysidro and Albuquerque (Clifford and Lin, 1986; Clifford and Lin, 1991). In spite of incomplete removal of arsenic from spent alumina during regeneration and greater arsenic leakage from regenerated alumina, the process appears to be feasible based on more recent field studies in Albuquerque, New Mexico (Clifford, Ghurye et al., 1998), and the Severn Trent Water Authority in the United Kingdom (Simms and Azizian, 1997). After four regenerations, the column capacity for arsenic at feed pH 7.5 and 6 min EBCT stabilized at about 80 percent of the virgin run capacity during the UK studies. The arsenic breakthrough curves for six sequential runs following base-acid regenerations are shown in Figure 9.21.

Because of difficulties associated with regenerating alumina and disposing of the arsenic-contaminated residue, the engineer should consider POU or POE treatment without pH adjustment. Refer again to Table 9.9 and to the work of Simms and Azizian (1997) who chose natural pH 7.5 even for a large-scale system with on-site regeneration. Point-of-use systems without preoxidation or pH reduction, used intermittently, should achieve 1000-BV throughput prior to exhaustion. Exhausted medium would simply be thrown away, not regenerated. Although not verified, the spent medium would probably pass the standard Toxicity Characteristic Leaching Procedure (TCLP) or extraction procedure (EP) toxicity tests as a nonhazardous

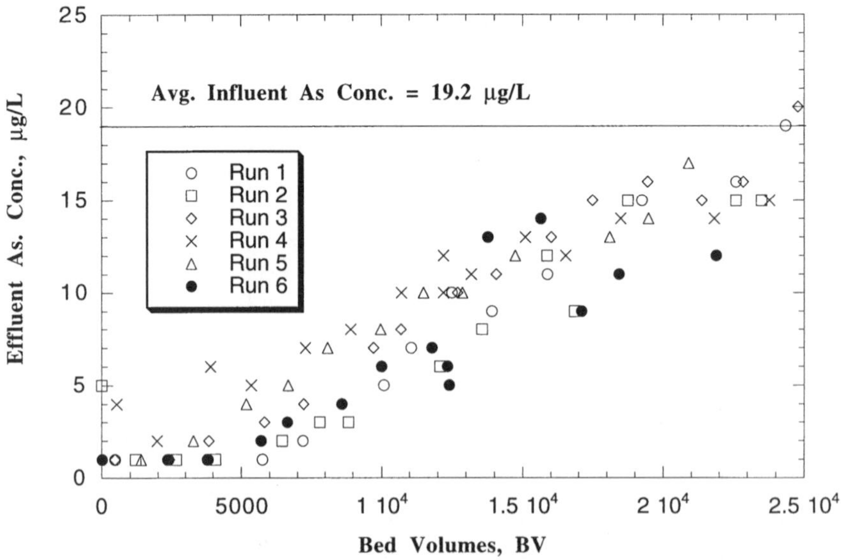

FIGURE 9.21 Arsenic breakthrough curves for six pilot-scale column runs at influent pH 7.5, NaOH regeneration with acid rinse. [*Source:* data from Severn Trent, UK pilot study (Simms and Azizian, 1997).]

waste. The reason is that the arsenic loading is very low and the tests are performed at pH 5, which is near the optimum pH for arsenic adsorption onto alumina. Furthermore, arsenic-laden $Al(OH)_3(s)$ sludges from spent-regenerant treatment are known to pass the EP toxicity test, and these sludges have very similar chemistry to that of activated alumina containing adsorbed arsenic.

Arsenic Removal from Spent Alumina Regenerants

During normal regeneration and acidification of spent alumina, enough aluminum dissolves to make precipitation of $Al(OH)_3(s)$ a feasible treatment step for the removal of arsenic from the spent-regenerant wastewater. When lowering the pH to approximately 6.5 with HCl or H_2SO_4, the As(V) quantitatively coprecipitates with the resulting $Al(OH)_3(s)$. Following dewatering, the dried arsenic-contaminated sludge should easily pass the EP toxicity test when 5 mg As/L (100 times the drinking water MCL) is allowed in the leachate. Hathaway and Rubel (1987) and Clifford and Lin (1991) used this procedure to treat spent alumina regenerant and produced leachates containing 0.036 and 0.6 mg As/L, respectively. The latter sludge contained some As(III), which caused the higher arsenic concentration in the leachate. The leachate arsenic concentration can be minimized by oxidizing the sludge (e.g., with chlorine) to ensure the presence of As(V) as opposed to As(III).

ARSENIC REMOVAL BY ION EXCHANGE

If the source water contains <500 mg/L TDS and <150 mg/L sulfate, ion exchange may be the arsenic-removal process of choice (Clifford and Lin, 1986; Clifford, Ghurye et al., 1997; Ghurye, Clifford et al., 1998; Clifford, Ghurye et al., 1999). Preoxidation to convert As(III) to As(V) (Frank and Clifford, 1986) is necessary, but raw water pH adjustment generally is not required. The chlorinated and filtered raw water is passed downflow through a 2.5- to 5-ft-deep bed of chloride-form strong-base anion-exchange resin, and the chloride-arsenate ion-exchange reaction (Equation 9.56) takes place in the near-neutral pH range. Regeneration, according to Equation 9.57, is not difficult because a divalent ion (arsenate) is being replaced by a monovalent ion (chloride) in high ionic strength solution where electroselectivity favors monovalent ion uptake by the resin. Regeneration returns the resin to the chloride form, ready for another exhaustion cycle:

$$2\,\overline{RCl} + HAsO_4^{2-} = \overline{R_2HAsO_4} + 2\,Cl^- \tag{9.56}$$

$$\overline{R_2HAsO_4} + 2NaCl = 2\,\overline{RCl} + Na_2HAsO_4 \tag{9.57}$$

Although chloride-arsenic ion exchange is simple in concept, several issues must be addressed when implementing the process for drinking water treatment. Among the important factors are the following: (1) choice of SBA resin, (2) effect of multiple contaminants such as arsenic and nitrate, (3) arsenic leakage, (4) effect of sulfate concentration, (5) optimum EBCT, (6) regenerant strength (percent NaCl), (7) regenerant level (lbs $NaCl/ft^3$ resin), (8) spent-brine reuse, and (9) spent-brine treatment. The author and his coworkers have completed three major lab and field studies for arsenic removal from groundwater that have dealt with these issues. The field locations were Hanford, California (Clifford and Lin, 1986), McFarland, California (Ghurye, Clifford et al., 1998), and Albuquerque, New Mexico (Clifford, Ghurye et al., 1999). The earliest study, Hanford, proved the feasibility of arsenic ion exchange for low-sulfate waters, but arsenic leakage below 2 μg/L was not reported because 2 μg/L was then the limit of detection. The more recent McFarland and Albuquerque studies focused on low (<2 μg/L) arsenic leakage because they were undertaken during the time when an arsenic MCL in the range of 0.002 to 0.005 mg/L was anticipated (Pontius, 1995). An additional emphasis of the later studies was regenerant optimization, including brine conservation and reuse.

Breakthrough Curves in As(V) Ion Exchange

Prechlorination followed by ion exchange proved to be a very effective means of As(V) removal in Hanford. A typical set of breakthrough curves for pH, arsenic, and sulfate is shown in Figure 9.22. Here, a commercially available SBA-exchange resin (Dowex 11) could effectively treat 4200 BV (16.3-day run length at a 5.6 min EBCT) of water before the arsenic level reached 0.05 mg/L. However, in spite of its very low concentration, sulfate was still the most preferred anion, and it eventually drove arsenate off the column and caused arsenic to peak at 0.136 mg As(V)/L—a value that was 160 percent of its feed concentration. (As previously mentioned, in normal operation the run would have been stopped at a predetermined volume of throughput, before the peak occurred.)

In contrast with nitrate-chloride ion exchange, a significant pH change did not occur simultaneously with As(V) breakthrough in Figure 9.22. The effluent pH began at a relatively low value (5.6) and rose to the influent value of 8.7 at about 1300 BV (5 days), while As(V) did not reach 0.02 mg/L (a potential run-termination point based on an MCL of 0.05 mg/L) until about 3500 BV. In this or a similar situation, a column run could simply be terminated at a predetermined length of, say, 3500 BV, which is far short of the point at which the effluent As(V) concentration

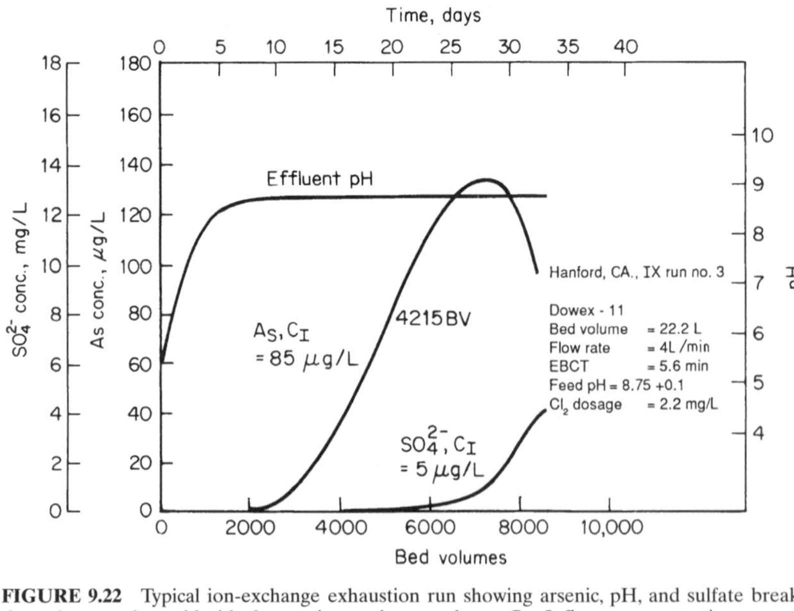

FIGURE 9.22 Typical ion-exchange exhaustion run showing arsenic, pH, and sulfate breakthrough curves from chloride-form anion-exchange column. C_I = Influent concentration.

reaches its influent value of 0.085 mg/L (5200 BV). With an MCL in the range of 0.002 to 0.020 mg/L, the predicted run length would probably be much shorter and would require additional pilot-scale tests.

Effect of Process Variables on As(V) Removal by Ion Exchange

Choice of Resins for Arsenic Removal. Five different commercially available SBA exchange resins, including polystyrene type 1, were tested for As(V) removal performance in Hanford, California (Clifford and Lin, 1986). After the exchange capacity was taken into account, the resins differed very little in their ability to remove As(V). Dowex 11, an improved-porosity polystyrene type 1 resin, and Ionac ASB-2, a polystyrene type 2 resin, seemed to have a slight edge over the others in terms of BV to arsenic breakthrough, so they were studied extensively. In contrast with subsequent field studies in McFarland (Ghurye, Clifford et al., 1998) and Albuquerque (Clifford, Ghurye et al., 1999), where the run length remained stable after many regenerations, run length deteriorated 3 to 5 percent following each regeneration in the original Hanford study. The reason for the decrease was not established but was thought to have been caused by the mica fouling observed on the exhausted resin. Most but not all of the black coating was removed upon NaCl regeneration. Presumably this sort of fouling could be eliminated by multimedia filtration ahead of the anion resin, as it is never a good idea to use a resin bed as a particulate filter.

Combined Arsenic and Nitrate Removal. Combined arsenic and nitrate removal was studied in McFarland, where the chlorinated water contained 13 μg/L arsenic, 10 mg/L NO_3–N, and 40 mg/L sulfate. Here, the triethyl (IRA 996) and tributyl (Ionac SR-6) nitrate-over-sulfate-selective (NSS) SBA resins were compared with two conventional sulfate-selective (CSS) resins. The results of a typical exhaustion cycle are summarized in Figure 9.23, which presents the arsenic breakthrough curves for all four resins. Clearly the conventional SBA resins are far superior to the nitrate-selective resins on the basis of run length to arsenic breakthrough. Also evident is the arsenic peaking, which occurs with all of the resins, because arsenate is not the most-preferred ion by any of the resins. Table 9.11, which gives more details of the runs depicted in Figure 9.23, shows that, because nitrate is the most-preferred ion, it does not peak with the nitrate-selective resins, but does peak with the conventional sulfate-selective resins.

To summarize, arsenic is removed from water as divalent $HAsO_4^{2-}$, which has poor affinity for monovalent- (nitrate-) selective resins. Therefore, when removing both nitrate and arsenic, nitrate-selective resins should be avoided, and the runs should be stopped before nitrate breakthrough to avoid peaking of nitrate and, possibly, arsenate. It does little good to lower the feed pH to produce the monovalent $H_2AsO_4^-$, because $H_2AsO_4^-$ is much lower in the selectivity sequence than is divalent $HAsO_4^{2-}$.

Effect of Arsenic Concentration on Run Length. Because arsenic is a trace species and not the most-preferred ion, its concentration doesn't significantly influence its major breakthrough, according to multicomponent chromatography theory (Helfferich and Klein, 1970). Nevertheless, higher concentration produces higher leakage and shorter run length. To test these predictions, the As(V) concentration was deliberately increased 14-fold, from 0.085 to 1.2 mg/L, by spiking the feed water during one run at Hanford (Clifford and Lin, 1986). The general trend predicted from theory did occur; the run length to arsenic breakthrough decreased only 46 percent, which was not linearly proportional to the 14-fold increase in concentration. The shorter run was primarily a result of greater arsenic leakage at the 14-fold higher influent arsenic concentration.

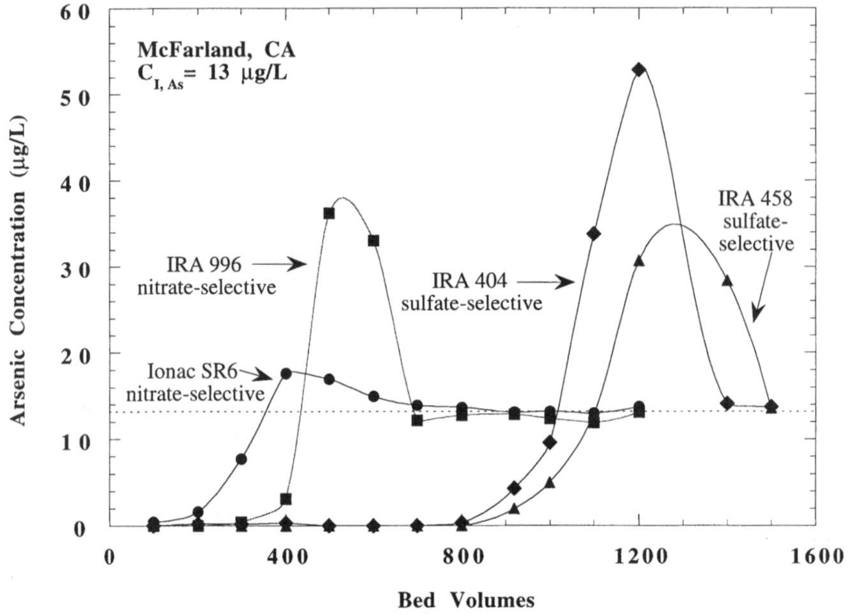

FIGURE 9.23 Comparison of arsenic (V) breakthrough curves for conventional and nitrate-selective resins.

Arsenic Leakage During Exhaustion. The 1995–1996 ion exchange field tests in McFarland, Hanford, and Albuquerque proved that ion exchange could easily meet a 0.002 mg/L arsenic MCL when treating a 10 to 50 μg/L arsenic-contaminated groundwater containing up to 220 mg/L sulfate. Average arsenic leakages during runs with EBCTs in the range of 1.5 to 3 min were 0.2 to 0.8 μg/L for the three SBA resins tested: ASB-2 (polystyrene type 2), IRA 458 (polyacrylic type 1), and IRA 404 (polystyrene type 1). Runs with 3-min EBCT produced essentially the same arsenic leakage as 1.5-min EBCT runs. In all cases, cocurrent downflow regenerations with 1 *M* (6 percent) NaCl at 20 lbs NaCl/ft^3 were employed. Subsequent experiments with regeneration levels as low as 5 lbs NaCl/ft^3 resulted in similar leakage values.

TABLE 9.11 Performance of Nitrate and Sulfate-Selective Resins for Combined Arsenic and Nitrate Removal

Parameter	Ionac SR-6	IRA 996	IRA 404	IRA 458
Selective for	Nitrate, NSS	Nitrate, NSS	Sulfate, CSS	Sulfate, CSS
Matrix	STY-DVB	STY-DVB	STY-DVB	Acrylic
Functionality	Tributyl	Triethyl	Trimethyl	Trimethyl
As breakthrough, BV[1]	200	300	800	900
Nitrate breakthrough, BV	680	750	560	610
As peaking factor[2]	1.4	3.7	2.7	4.0
Nitrate peaking factor[2]	1.0	0.9	2.5	2.2

[1] Arsenic breakthrough to 2 μg/L.
[2] Peaking factor is the ratio of peak concentration/influent concentration.

Arsenic leakage from SBA resin columns dramatically increased in the presence of particulate iron (III), which strongly adsorbed As(V) and was not well filtered by the resin. This fact was discovered during some tests in McFarland using water that was visibly contaminated with particulate iron(III). Serious arsenic leakage (up to 6 μg/L) tracked the iron contamination in the ion-exchange column effluent. The problem was solved by adding a multimedia prefilter ahead of the ion-exchange column; the prefilter eliminated the particulate iron in the arsenic-contaminated feed to the SBA column, and arsenic leakage returned to normal.

Effect of Sulfate on Arsenic Run Length. As was observed during nitrate removal, increasing sulfate concentration leads to shorter ion-exchange runs, because sulfate is preferred to nitrate and arsenate when using the recommended conventional sulfate selective (CSS) resins. Table 9.12 compares arsenic run lengths for ASB-2 resin using various waters with different TDS values and sulfate concentrations. Even with 100 mg/L sulfate in the feed water, a run length of 480 BV was observed with the spiked McFarland water. The 250 BV run length at 220 mg/L sulfate is probably too short for economical full-scale operation, which is why <150 mg/L sulfate is suggested as one criterion for selecting ion exchange for arsenic removal.

Regeneration of Arsenic-Spent Resins

During the original (1984) Hanford studies, arsenic recoveries upon downflow (cocurrent) regeneration were essentially complete. Three BV of 1.0 *N* NaCl (11 lb/ft^3 NaCl resin) was more than adequate to elute all the adsorbed arsenic, which was even easier to elute than bicarbonate—a very nonpreferred ion. One reason why arsenic elutes so readily is that it is a divalent ion ($HAsO_4^{2-}$) and is thus subject to a selectivity reversal in the high-ionic-strength (>1 *M*) environment of the regenerant solution. This ease of regeneration is a strong point in favor of ion exchange, as compared with alumina for arsenic removal in low-TDS, low-sulfate waters.

In confirmation of similar findings from the nitrate studies in Glendale, Arizona (Clifford, Lin et al., 1987), the 1984 Hanford experiments (Clifford and Lin, 1986) showed that dilute regenerants (0.25–0.5 *N*) were more efficient than concentrated ones for eluting arsenic. The greater efficiency of dilute regenerants was further verified in the Albuquerque field study (Clifford, Ghurye et al., 1999) where 0.5 *N* NaCl outperformed 1.0 and 2.0 *N* regenerants in downflow regeneration experiments on similar aliquots of exhausted resin. When analyzing the data, however, it appeared that the regenerant flow velocity was an important factor that had not

TABLE 9.12 Arsenic Run Length as a Function of Influent Sulfate Concentration[1]

Location	As, μg/L	TDS, mg/L	Sulfate, mg/L	Run length, BV[2]
Hanford, CA	50	213	5	1500
McFarland, unspiked	13	170	40	1030[3]
Albuquerque	26	328	82	640
McFarland, SO_4^{2-}-spiked	13	259	100	490
McFarland, SO_4^{2-}-spiked	13	436	220	250

[1] Run lengths for ASB-2 type 2 SBA resin regenerated with 20 lbs NaCl/ft^3. When regenerated with 10 lbs NaCl/ft^3, run lengths are decreased by about 25%.

[2] Based on run termination at effluent arsenic concentration of 2 μg/L.

[3] Extrapolated value based on comparison with IRA 404 performance in McFarland.

previously been considered. At the low superficial flow velocities (volumetric flow/cross-sectional area <2 cm/min) associated with the higher regenerant concentrations, regenerant channeling occurred, which resulted in poor arsenic recoveries. The resulting rule of thumb was that the regenerant *superficial linear velocity* (SLV) should exceed 2 cm/min with ASB-2 and similar resins (16 × 50 U.S. mesh). With SLVs >2 cm/min, there wasn't much difference between the performance of 0.5 and 1.0 *N* regenerants at regeneration levels of 1 to 2 eq Cl^-/equiv resin (5 to 10 lb $NaCl/ft^3$).

Resin mixing following regeneration was tried in Albuquerque and found to be successful at reducing the average arsenic leakage during exhaustion to 0.19 μg/L from 0.85 μg/L without mixing. In these experiments, the resin was removed from the regenerated column, physically mixed, and returned to the column before the next exhaustion. In actual practice, resin mixing has been achieved in nitrate removal applications by upflowing gas (air or nitrogen) in a column containing resin and water (Liu and Clifford, 1996) or by an uneven flow of backwash water as reported by Lauch and Guter (1986).

Downflow versus Upflow Regeneration. It is conventional wisdom that countercurrent upflow regeneration is better than downflow to minimize contaminant leakage in cyclic ion-exchange operations. Normally this is true—and especially so in demineralizer operation, where minimum leakage is critical. However, with arsenic ion exchange, it is not necessary to use counterflow (upflow) regeneration to minimize leakage. When upflow was compared with downflow regeneration in the Albuquerque field study (Clifford, Ghurye et al., 1999), conventional downflow regeneration performed better (i.e., it gave lower arsenic leakage during subsequent exhaustions). The reason for the superiority of cocurrent downflow regeneration is that, at exhaustion, most of the arsenic on the resin is concentrated in a thin zone near the column outlet. Experimental results indicate that it is preferable to elute the arsenic by the shortest route (i.e., cocurrent regeneration), as opposed to flushing it back through the entire resin bed, where some apparently remains during countercurrent regeneration.

Reuse of Spent Regenerant. The major finding of the Albuquerque arsenic study was that spent regenerant could be reused without treatment to remove arsenic. In the series of exhaustion-regeneration cycles shown in Figure 9.24, the ASB-2 column was exhausted and regenerated 18 times, using recycled regenerant that had been compensated each cycle with NaCl to maintain the chloride concentration at 1 *N*. In spite of the fact that the arsenic concentration in the reuse brine rose to 19,000 μg/L, sulfate reached 151 g/L (3.1 *N*), and bicarbonate reached 24.4 g/L (0.39 *N*), the arsenic leakage during exhaustion was typically <0.4 μg/L, and run lengths were consistently in the 400 to 450 BV range. After eighteen cycles of reuse, attempts were made to continue reusing the brine without chloride compensation in the hope that the high bicarbonate level in the brine would be sufficient to elute the arsenic. These attempts failed as the arsenic leakage rose to 10 μg/L after only four runs without chloride addition. When chloride addition was resumed with the same recycle brine, the arsenic leakage dropped back to 0.4 μg/L, but the run lengths were a little shorter (380 to 400 BV). Based on the Albuquerque field study results, it appears that spent brine can be reused at least 20 times, and possibly more. Eventually, the spent brine will have to be wasted, but it should be treated to remove arsenic before disposal. This can be accomplished by iron (III) or aluminum (III) precipitation as discussed in the text that follows.

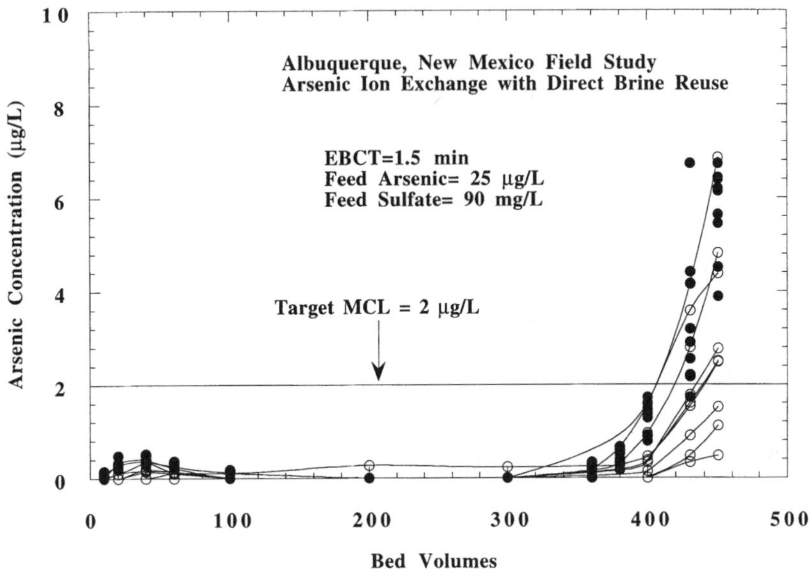

FIGURE 9.24 Arsenic breakthrough curves for 18 exhaustion regeneration cycles, reusing the same brine with chloride concentration maintained at 1 *M*.

Treatment of Spent Ion-Exchange Regenerant

Arsenic(V) can be precipitated from the waste brine by using iron (III) or aluminum salts such as $FeCl_3{\cdot}6H_2O$ or $Al_2(SO_4)_3{\cdot}18H_2O$ or lime ($Ca(OH)_2$). The stoichiometric reactions are:

$$3\ Na_2HAsO_4 + 3\ H_2O + 2\ FeCl_3 = Fe(OH)_3(s) + Fe(H_2AsO_4)_3(s) + 6\ NaCl \quad (9.58)$$

$$2\ Na_2HAsO_4 + NaHCO_3 + 4\ Ca(OH)_2 = CaCO_3(s) + Ca_3(AsO_4)_2(s) + 3\ H_2O + 5\ NaOH \quad (9.59)$$

$Al_2(SO_4)_3{\cdot}18H_2O$ may be substituted for $FeCl_3$ in reaction 9.58 to produce $Al(H_2AsO_4)_3(s)$.

Iron or aluminum is required in greater than stoichiometric amount in actual practice. A wastewater of pH 9.9 resulting from a 1.0 *N* NaCl regeneration was treated to remove As(V) in Hanford, California. For the wastewater, which contained 90 mg As(V)/L and 50,000 mg TDS/L, about 12 times the stoichiometric amount of iron (adjusted final pH 8.5) or aluminum (adjusted final pH 6.5) was required to reduce the As(V) level from 90 to less than 5 mg/L. To reduce As(V) to less than 1.5 mg/L required a 20-fold stoichiometric dosage of either metal salt. The iron floc was heavier and settled better than the alum floc. When subjected to the EP toxicity test to determine As(V) leachability from the air-dried sludges, both sludges passed the test with about 1.5 mg As(V)/L in the leachate when the EP MCL was 5 mg As/L (Clifford and Lin, 1986).

Similar iron(III)-precipitation-of-arsenic experiments were carried out in the Albuquerque field study (Clifford, Ghurye et al., 1999). In these tests, however, the spent brine initially contained only 3.5 mg/L arsenic. The arsenic removals achieved were found to be strongly pH-dependent. When using a molar Fe/As ratio of 20/1, the arsenic removals observed were 98.7 and 99.5 percent at final pHs of 6.4 and 5.5, respectively. Eventually, TCLP tests will have to be run on precipitated arsenic sludges to determine their suitability for land disposal. If the arsenic MCL is lowered, it is possible that the acceptable TCLP leachate arsenic concentration, which is now set at 100 times the MCL for drinking water, will also be lowered.

SELENIUM REMOVAL BY RESINS AND ALUMINA

Selenium contamination of potable water supplies at concentrations above the current MCL of 0.05 mg/L is rare. For example, a nationwide survey of potable water supplies conducted by the AWWA in 1982 and supplemented by U.S. Environmental Protection Agency data found only 44 supplies with selenium in excess of 0.03 mg/L. All were groundwater supplies.

When selenium is found in groundwater, it is generally a result of natural, not artificial, contamination. At typical groundwater pH values (7.0 to 9.5), only the anionic forms of selenious (Se(IV)) and selenic (Se(VI)) acid are found. The dissociation equations and constants for the two acids are presented in Equations 9.60 through 9.63.

Se(IV), Selenious Acid Dissociation Equilibria

$$H_2SeO_3 = H^+ + HSeO_3^- \quad pK_1 = 2.55 \tag{9.60}$$

$$HSeO_3^- = H^+ + SeO_3^{2-} \quad pK_2 = 8.15 \tag{9.61}$$

Se(VI), Selenic Acid Dissociation Equilibria

$$H_2SeO_4 = H^+ + HSeO_4^- \quad pK_1 = -3.0 \tag{9.62}$$

$$HSeO_4^- = H^+ + SeO_4^{2-} \quad pK_2 = 1.66 \tag{9.63}$$

Under oxidizing conditions, Se(VI) will predominate, and divalent selenate (SeO_4^{2-}), an anion with chemical behavior similar to that of sulfate, will be found. Under reducing conditions Se(IV) will predominate, and at pH values below 8.15 the monovalent biselenite anion ($HSeO_3^-$) will be the dominant form; above pH 8.15, the divalent selenite anion (SeO_3^{2-}), will dominate.

Depending on the form of selenium present, either ion exchange or activated alumina adsorption will be chosen for treatment. Selenite (Se(IV)) forms are best removed by alumina, and SBA exchange is the process of choice for selenate (Se(VI)). The justification for this recommendation is based on the relative affinity values for anions on alumina (Equation 9.27) and anion resins (Table 9.3).

Oxidation of Selenite (Se(IV)) to Selenate (Se(VI))

Selenate is so strongly preferred by anion resins that oxidizing selenite to selenate prior to treatment is the best approach. Although Se(IV) is considerably more difficult to oxidize than As(III), laboratory research has demonstrated that this oxidation can be readily accomplished with free chlorine. In synthetic groundwater containing sulfate, chloride, and bicarbonate at pH 8.3, the reaction is first-order in both Se(IV) and free chlorine concentrations (Boegel and Clifford, 1986).

The selenite oxidation rate was optimum between pH 6.5 and 8.0. In this range, 60 percent of the Se(IV) was converted to Se(VI) within 5 min at a free chlorine concentration of 2 mg/L. At pH 9, only 15 percent of the Se(IV) could be converted in 5 min with 2 mg free chlorine per liter. In these studies, pure oxygen was ineffective (no measurable oxidation in 1 h), while H_2O_2 and $KMnO_4$ were not nearly as effective as free chlorine. TOC was not present in the Se(IV) oxidation experiments, and it is expected to slow the oxidation rate by also reacting with the free chlorine.

Ion Exchange for Selenate Removal

To date, only laboratory studies with synthetic water have been done to test the possibility of using chloride-form SBA exchange for selenate removal. Even with a high TDS level (700 mg/L), high sulfate level (192 mg/L), and source water containing 0.1 mg Se(VI)/L, an acceptable run length (275 BV) to 0.03 mg Se(VI)/L was obtained (Boegel and Clifford, 1986). As expected, selenate eluted after sulfate and was not subject to chromatographic peaking. (Chromatographic peaking of Se(IV) will occur, however, if it is present with Se(VI).) No attempts were made to regenerate the resin, but this should not be a problem because SeO_4^{2-} is a divalent ion subject to selectivity reversal, as are sulfate, arsenate, and chromate, all strongly adsorbed divalent anions that are readily eluted during regeneration with NaCl.

Activated Alumina for Selenite Removal

Trussell et al. (1980) demonstrated that alumina adsorption is very effective for removal of Se(IV) (i.e., $HSeO_3^-$) in the optimum pH range of 5 to 6. They also found, not unexpectedly, that alumina is a relatively poor medium for the removal of selenate (Se(VI)) because of strong competition from sulfate. Yaun et al. (1983) also reported that alumina was good for biselenite ($HSeO_3^-$) removal and found nearly the same optimum pH range—3 to 8. Because biselenite adsorption onto alumina is approximately as strong as fluoride adsorption, complete regenerability of the spent alumina by using the typical base-acid sequence described for fluoride regeneration is expected. These laboratory studies demonstrate that the feasibility and cost of an activated alumina process for selenium removal will depend on the oxidation state of the selenium. Activated alumina adsorption could be an excellent process for water in which Se(IV) is the only selenium species, but it becomes less attractive as the fraction of Se(VI) increases. It should be noted that in petrochemical wastewaters, selenocyanate ($SeCN^-$) may also be present in significant amount relative to selenite and selenate. Selenocyanate is a highly preferred anion, which is readily removed by strong-base anion exchange, but is very difficult to remove from the exhausted resin.

In water containing both Se(IV) and Se(VI), oxidation with chlorine followed by ion exchange is more feasible than microbial reduction (or some as-yet undetermined chemical reduction means) followed by activated alumina adsorption. This is especially obvious given that the product water for distribution usually has to be disinfected with chlorine.

CHROMATE REMOVAL BY ANION EXCHANGE

The two common oxidation states for chromium in natural water are Cr(III) and Cr(VI). Trivalent chromium occurs as a hydrated cation Cr^{3+} at low pH, but it readily hydrolyzes to form insoluble $Cr(OH)_3(s)$ and soluble cationic and neutral hydroxide complexes such as $Cr_3(OH)_4^+$, $Cr(OH)_2^+$, and $Cr(OH)_3$, which are significant in the 7 to 9 pH range (Baes and Mesmer, 1976). Because of its relative insolubility under typical groundwater conditions, Cr(III) is not a significant groundwater contaminant, whereas hexavalent chromium (i.e., Cr(VI) from both natural and artificial sources) is found in groundwater (Robertson, 1975). The current (1999) MCL for total chromium in drinking water is 0.1 mg/L (USEPA, 1992). Chromate (CrO_4^{2-}), the divalent anion of chromic acid (H_2CrO_4; $pK_1 = 0.08$, $pK_2 = 6.5$) is the predominant form of Cr(VI) in the 7 to 9 pH range. Theoretically, anion exchange with synthetic resins is an ideal process for chromate removal because, with the exception of uranyl carbonate and perchlorate, chromate is the most preferred of the common anions (see Table 9.3). In practice, ion exchange works well, as demonstrated by the author's previously unpublished research with chloride-form SBA exchange to remove chromate from a well in Scottsdale, Arizona. The Scottsdale well used for the field research contained 0.042 mg Cr(VI)/L present as the CrO_4^{2-} anion. Other water contaminants present at the milligram/liter level were bicarbonate (244), chloride (24), sulfate (9), magnesium (28), calcium (19), and sodium (38). The pH was 7.6, TDS by evaporation was 276 mg/L, and silica was 32 mg/L as SiO_2.

Effect of Resin Matrix on Chromate Removal

Although all anion-exchange resins strongly prefer chromate, the resin matrix exhibits a significant influence on run length to chromate breakthrough. The extreme affinity of chromate for all anion resins and the effects of resin matrix and porosity are demonstrated in Table 9.13. The longest run (32,000 BV and 98 days) was achieved with a macroporous polystyrene resin. With regard to CrO_4^{2-} affinity, macroporous resins are better than the gel and improved-porosity types, and hydrophobic polystyrene resins are better than the hydrophilic polyacrylic type.

In ion-exchange processes, the resin with the highest affinity for the contaminant is not necessarily the resin of choice, after regenerability has been taken into account. In fact, the resin with the highest affinity and longest virgin run length is ordinarily the hardest to regenerate. This was the case with chromate sorption onto macroporous polystyrene resin (IRA 900) observed in the Scottsdale study.

Regenerability of Chromate-Spent Resin

Various concentrations and amounts of salt (NaCl) along with mixtures of salt and caustic (NaOH), typical of chromate-removal processes for cooling water treatment, were experimented with to elute chromate from spent resin in Scottsdale. In the initial experiments, it was found that NaOH was not necessary to achieve a good regeneration; thus most regenerations were done with NaCl alone. Chromate was surprisingly easy to remove from the resin, as seen in Table 9.14, which compares the regenerability of three resins using 1 *N* NaCl. The macroporous polystyrene resin (IRA 900), which had the highest chromate affinity, was more difficult to regenerate than both the polystyrene gel resin (Dowex 11) and the macroporous polyacrylic resin (IRA 958).

In spite of the extreme affinity of chromate for resins, only 3 to 4 equivalent of Cl^- per equivalent of resin (11 to 15 lb $NaCl/ft^3$ resin) was required for "complete" cocurrent (downflow) regeneration. The range of maximum chromate recovery for all the 1 *N* NaCl regenerations was 67 to 89 percent. The always-less-than-100 percent recovery was attributed to Cr(VI) reduction to Cr(III) with subsequent precipitation of

TABLE 9.13 Effects of Resin Matrix and Porosity on Run Length to Chromate Breakthrough[1]

			Run length to 0.010 mg Cr(VI)/L	
Resin	Matrix/Porosity[2]	Capacity meq/mL	BV	days
IRA 900	Polystyrene DVB/MR	1.1	32,000	98
Dowex 11	Polystyrene DVB/iso	1.2	20,700	68
IRA 958	Polyacrylic DVB/MR	0.8	14,600	44

[1] Data from Scottsdale Pilot Study. Feed pH = 7.6, SO_4^{2-} = 9 mg/L, TDS = 276, Cr(VI) = 0.042 mg/L.
[2] DVB = Divinylbenzene, MR = Macroporous, iso = isoporous or unique porosity gel resin designed for organics removal.

greenish $Cr(OH)_3(s)$, observed in the top one-fifth of the resin bed. This $Cr(OH)_3(s)$ was not completely removed during backwashing and regeneration. The seriousness of the Cr(III) precipitation problem was not extensively studied in Scottsdale. Due to the long runs, only one repeat run (with Dowex 11) was made. Following regeneration, the second run with Dowex 11 was 20,700 BV (i.e., slightly longer than the initial 19,700 BV run). Thus, no significant reduction was observed due to resin fouling with $Cr(OH)_3(s)$. This is not surprising in light of the fact that, at chromate breakthrough, only 3 percent of the resin exchange sites were in the chromate form.

Effect of Chromate Concentration

To assess the influence of chromate level on run length, the chromate concentration of the Scottsdale water was increased from 0.042 mg/L to 0.195 mg/L by spiking. As predicted from *equilibrium multicomponent chromatography theory* (EMCT), the run length was not decreased proportionately (Helfferich and Klein, 1970). With IRA 900 resin and raw water containing 0.042 mg Cr/L, the initial run length was 32,000 BV. This decreased only 32 percent when the chromate concentration was increased 360 percent to 0.195 mg/L. Actually, the EMCT predicts much less of a concentration influence on breakthrough for a trace species such as chromate; but because of mass transfer limitations and nonattainment of equilibrium, higher chromate leakage occurs as its concentration in the raw water increases.

Chromate Removal from Spent Regenerant

Chromate in the saline spent-regenerant solution can be reduced to Cr(III) and precipitated as $Cr(OH)_3(s)$ by using a reductant such as ferrous sulfate or acidic sodium sulfite. This spent-regenerant treatment process has been studied extensively on a

TABLE 9.14 Regeneration of Resins Used for Chromate Removal[1]

Resin	Regenerant (1)	eq chloride / eq resin	lbs NaCl / ft^3 resin	Percent Cr(VI) recovered
Dowex 11	3.7 BV 1 *N* NaCl	3.08	11.3	89
IRA 958	2.6 BV 1 *N* NaCl	3.25	11.9	87
IRA 900	4.0 BV 1 *N* NaCl	3.40	14.6	67

[1] Data from Scottsdale pilot study.

laboratory scale (Siegel and Clifford, 1988). The Cr(VI) reduction precipitation reactions are:

$$HCrO_4^- + 3\,Fe^{2+} + 7\,H^+ \rightarrow Cr^{3+} + 3\,Fe^{3+} + 4\,H_2O \tag{9.64}$$

$$2\,HCrO_4^- + 3\,H_2SO_3 + 2\,H^+ \rightarrow 2\,Cr^{3+} + 3\,SO_4^{2-} + 5\,H_2O \tag{9.65}$$

When acidified sodium sulfite is the reductant, it is necessary to reduce the pH to below 1.5 to achieve a reasonable reaction rate. The entire process involves sulfite addition at 1.3 times the stoichiometric amount, pH reduction, 30- to 60-min reaction time, pH increase to 8.3 to precipitate $Cr(OH)_3(s)$, and sludge settling.

For Cr(VI) reduction and precipitation with the ferrous ion, the optimum pH range is 5 to 7. Outside this range, significant amounts of soluble chromium are found in the supernatant after settling. Reduction with the ferrous ion comprises ferrous sulfate addition, followed by slow mixing while pH is controlled within the 5 to 7 range, and then settling.

For both reduction processes, the high salinity of the spent regenerant improves the floc formation rate and increases the floc size. When properly operated, a total chromium concentration less than 0.10 mg/L is expected in the filtered regenerant following Cr(VI) reduction and $Cr(OH)_3(s)$ precipitation. Additional research needs to be done to establish the feasibility of reusing the regenerant following chromate removal. Cost estimates comparing sodium sulfite to ferrous sulfate reduction were made for 0.1- and 4-Mgd chromate-removal plants. When treatment chemicals and land disposal of sludge were taken into consideration, the ferrous sulfate reduction was significantly less costly than the sulfite reduction in spite of the five-fold greater sludge volume resulting from $Fe(OH)_3(s)$ in addition to the unavoidable $Cr(OH)_3(s)$.

Chromate-Removal Process Recommendations

Based on the results of the Scottsdale study, the recommended ion exchange process for chromate removal from groundwater comprises chloride-form strong-base anion exchange with a polystyrene gel or isoporous resin. Following chromate breakthrough at about 0.01 mg/L, the spent resin should be backwashed before cocurrent regeneration with 4 BV of 1 *N* NaCl (15 lb NaCl per ft^3 of resin), followed by the appropriate slow and fast rinses. Chromic hydroxide ($Cr(OH)_3(s)$) should be precipitated from the spent regenerant brine with ferrous sulfate. Consideration should be given to reuse of treated regenerant. Further study is needed to examine the feasibility of reusing the spent regenerant without treatment to remove chromate. The potential for direct regenerant reuse is good because chromate is a divalent ion similar in electroselectivity reversal behavior to divalent As(V), which doesn't seriously influence regenerant reuse even when it builds up in the recycling regenerant (Clifford, Ghurye et al., 1999).

COLOR AND ORGANICS REMOVAL BY RESINS

Naturally occurring *dissolved organic carbon* (DOC) consisting primarily of humate and fulvate anions has been recognized for more than 40 years as a common foulant of the anion exchangers used in demineralization processes (McGarvey and Reents, 1954; Frisch and Kunin, 1957; Wilson, 1959). The symptoms of organic fouling are product water quality deterioration, resin darkening, brown-colored regenerant, exchange capacity loss, and increased frequency of regeneration. Early on it was found that occasionally regenerating the fouled SBA resin with a mixture of NaCl and NaOH would reverse most of the fouling and lead to acceptable run lengths and resin life (Wilson, 1959). In spite of an early history of problems with DOC in surface waters treated by SBA resins, it was found that DOC could be removed from groundwater and surface water by using resins, especially macroporous anion exchangers in the chloride form. Published research on the ion-exchange process for DOC removal is scarce, and much remains to be done to optimize the process; however, the few reported studies to date have suggested that ion exchange is a viable process for DOC and disinfection by-product (DBP) control in drinking water supplies.

History of DOC Removal by Ion Exchange

One of the earliest reports on the use of resins to treat drinking water to remove DOC was a successful two-year pilot study of color removal on the Merrimack River in Massachusetts during 1962–1964 (Coogan, 1968). Duolite A-7–a hydrophilic matrix, granular weak-base phenol formaldehyde anion resin–regenerated with NaOH was tested extensively. Although A-7 is an anion exchanger at low pH, it had virtually no ion-exchange capacity at the pH of the influent water (6.4 to 6.9). Thus, the removal of color was by adsorption of anion-cation pairs rather than by ion exchange. The main findings of the study were as follows:

1. With APHA color values of 50 to 100 on the raw water, diatomaceous earth prefiltration to remove about 10 percent of the presumably larger color bodies was required ahead of the resin bed.
2. EBCTs in the 3.75 to 7.5 min range were acceptable, with the 3.75 min time giving only a small decrease in run length, which was 2,000 to 2,500 BV for 7.5 min.
3. Typically, 80 to 100 percent color removal was attained at the start of a run, decreasing to 50 to 75 percent at the end.
4. NaOH at a level of 4 lbs/ft^3 was adequate to regenerate the resin during more than 200 exhaustion-regeneration cycles in which the resin capacity for DOC remained relatively constant.
5. Iron removals were about 25 percent across the diatomaceous earth filter and 95 percent through the resin bed, which led to the suggestion that soluble iron was being removed by complexation with the adsorbed color bodies.

This study and others reported by Abrams (1982) led researchers to try A-7 and similar weak-base resins for adsorption rather than ion exchange of DOC. Later, it was found that strong-base anion resins regenerated with NaOH or a mixture of NaCl and NaOH could also be used in cyclic operation to control organics in surface- and groundwaters.

Baker, Davies et al. (1977) reported that the substitution of the acrylic-matrix SBA resin IRA-458 for a conventional polystyrene type 2 resin in a demineralizer system solved the organic fouling problems that had plagued the system for years. The acrylic matrix resin performed better because it was more efficiently regenerated than the original polystyrene resin. The authors explained that there is minimal van der Waals attraction between the hydrophobic organic molecules and the hydrophilic resin matrix, which makes regeneration easier and more complete. In 1977, when Baker's paper was published, the acrylic resin had been placed into service at 125 sites worldwide because of its superior regenerability and resistance to organic fouling.

Kolle (1984) described a DOC removal process utilizing anion exchange for treatment of a highly colored Hannover, West Germany, groundwater. The

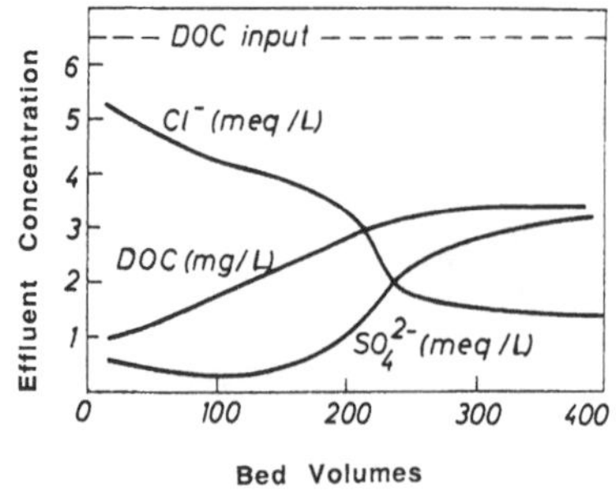

FIGURE 9.25 Typical breakthrough curves for DOC, sulfate, and chloride from a macroporous SBA exchanger removing humate and fulvate anions from a Hannover, West Germany, groundwater. Column is operated to 5000 BV before regeneration. Between 400 and 5000 BV the DOC removal is approximately 50 percent. Influent sulfate = 130 mg/L.

2500-m^3/d (15.8-mgd) full-scale process used a macroporous type 1 polystyrene SBA resin (Lewatit MP 500 A) in four single beds operated in parallel, with EBCT = 1.2 min. Regeneration was accomplished with 2 BV of a mixture of 1.7 *N* (10 percent) NaCl, and 0.5 *N* (2 percent) NaOH. A striking feature of the process was the reuse of regenerant 7 times, with the NaCl and NaOH concentrations adjusted after each use. The DOC was not removed from the recycled regenerant, and it built up to 25,000 mg/L prior to disposal, which was triggered when the sulfate concentration increased to the chloride level. This reuse method allowed for a 20,000:1 ratio of product water to spent regenerant.

Typical breakthrough curves for DOC, chloride, and sulfate from the early portion of the run in the Hannover process are shown in Figure 9.25. After sulfate breakthrough at 250 BV, the DOC removal remained at about 50 percent until 5,000 BV—the point of run termination. Although far from complete, the 50 percent DOC removal was sufficient to increase the bacterial doubling time in the product water from 5 to 12 h and to drastically reduce the chlorine consumption with its attendant organic halide formation problems. The overall treatment efficiency was reduced only 10 percent because of fouling and resin losses over a period of 2 years of full-scale operation.

Brattebo, Odegaard et al. (1987) described a similar DOC removal process designed to treat a highly colored Norwegian surface water containing 3 to 6 mg/L DOC, primarily in the >10K molecular weight (MW) range. During their pilot studies, they verified the superiority of macroporous MP 500 A type 1 SBA resin compared with a limited number of other polystyrene resins of macroporous and gel porosity. They also effectively reused the 1.7 *N* NaCl/0.5 *N* NaOH regenerant as many as 8 times by wasting a fraction of the first eluate containing most of the DOC and making up the remainder with fresh NaCl and NaOH. For water containing up to 4 mg DOC/L, they recommended the ion-exchange method, implemented by using a four-column merry-go-round ion-exchange system (see Figure 9.11) in which the resting, freshly regenerated fourth column automatically becomes the final polisher when the effluent DOC reaches a predetermined level. At this time, the lead column is taken out of service and regenerated. They also determined that the EBCT for the series of columns was the most important design parameter, and they suggested a range of 5 to 10 min for this surface water, which is considerably longer than the 1.2 min used in the full-scale Hannover groundwater system. The differences in optimum EBCTs are probably due to differences in the nature (e.g., size, aromaticity, and hydrophobicity) of the different surface and groundwater organics involved.

The same Norwegian research group (Brattebo, Odegaard, et al., 1987) tested the multibed anion-exchange process on a small water supply using a full-scale, 3 m^3/h (13 gpm) plant and reported the following conclusions:

1. NaOH was a necessary constituent of the regenerant, because with NaCl alone the DOC recovery was less than 80 percent.

2. Reusing up to 75 percent of the spent regenerant did not reduce the DOC adsorption capacity upon subsequent runs.
3. Mass transfer of DOC was particle-diffusion limited such that a reduction in temperature from 15 to 5°C required a 20 percent increase in EBCT.
4. For small water supplies with relatively low (<4 mg/L) DOC, the anion-exchange process was estimated to be less costly to build and operate than a conventional coagulation process.

Fundamentals of DOC Ion Exchange

In spite of the full-scale implementation of anion exchange for DOC removal, no fundamental study of the exhaustion-regeneration process had been undertaken prior to the work of Fu and Symons (1990). Their objectives were to (1) establish the relative importance of adsorption compared with ion exchange for removal of aquatic humic substances; (2) determine the influences of resin properties on DOC uptake; and (3) describe the chromatographic behavior of the multicomponent DOC mixture during columnar ion exchange. Their study utilized two macroporous resins, five gel resins, and two macroporous polymeric adsorbents to remove DOC from <1K, 1 to 5K, 5 to 10K, and >10K MW fractions of Lake Houston surface water. The percentages of DOC in the various raw-water MW fractions were 52, 9, 28, and 11, respectively.

Their ion-exchange isotherm experiments indicated that, within the limits of experimental error, each milliequivalent of DOC in the 1 to 5K fraction sorbed was balanced by an equal amount of chloride ion released. See Figure 9.26. Overall, their results showed that the sorption mechanism for DOC removal was ion exchange with only a small percentage of the DOC in the <1K and >10K fractions being removed by adsorption. This may seem to but does not contradict the results of Coogan et. al. (1968), who observed that fulvic acid anions were removed primarily by adsorption on *weak-base* resins such as A-7. Each fulvate anion taken up by the strong-base anion resin was balanced by a fixed positively charged quaternary amine site, whereas, on the A-7 weak-base resin, the anions were balanced by mobile cations such as Fe^{2+}, Ca^{2+}, Mg^{2+}, or Na^{+}.

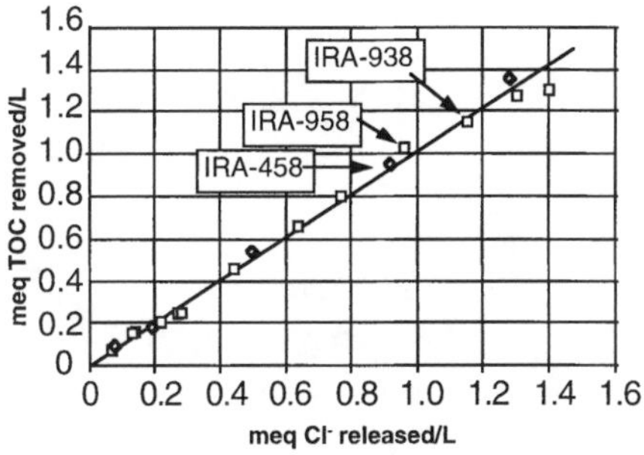

FIGURE 9.26 Proof of ion-exchange mechanism using chloride ion exchange with 1K–5K TOC fraction at pH = 7.5. [*Source:* P. L.-K. Fu and J. M. Symons (1990). "Removing aquatic organic substances by anion exchange resins." *Journal AWWA,* 82(10), 1990: 70–77.]

Using equilibrium isotherm tests, Fu and Symons (1990) found that the resin matrix and porosity exhibited significant influences on DOC removal. Generally, for DOC MW ≥1K, the hydrophilic polyacrylic resins were preferred for DOC removal compared with the relatively hydrophobic polystyrene resins. And, for all MW fractions greater than 1K, macroporous resins performed better than the gel types (i.e., the greater the resin porosity, the better the DOC removal). At <1K MW, the resin porosity did not influence the DOC removal, which was, however, influenced by the resin matrix. In this low MW range, hydrophobic polystyrene resins performed somewhat better than the hydrophilic polyacrylic resins. Thus, the resin of choice for removing natural organic matter will depend on the actual water supply being treated. They also found that the resin's ability to sorb DOC ≥1K was positively correlated with its tendency to swell in water (i.e., its hydrophilicity), which is another way of explaining why acrylic resins were preferred by Baker, Davies et al. (1977) over polystyrene resins for DOC removal.

To establish the effect of partial DOC removal by coagulation prior to the ion-exchange process, Fu and Symons measured the DOC in the various fractions before and after clarification. They found that 70 to 90 percent of the DOC > 1K was removed by coagulation, but only 35 percent removal was observed for the <1K fraction. Thus, coagulation performed in a similar manner to ion exchange for removing DOC, and neither process was particularly good for the <1K fraction. Using the best performer from the isotherm tests, a macroporous acrylic resin (IRA 958), they studied the chromatographic behavior of the <1K and the 5 to 10K fractions. Column study results suggested that the DOC could be divided into three categories: (1) not removed, (2) less-preferred than sulfate, and (3) more-preferred than sulfate. Figure 9.27 shows the TOC effluent history and the sulfate breakthrough point during a test with the <10K TOC fraction. With only 6 mg/L sulfate and 18 mg/L chloride in the dialyzed feed water, the run length to sulfate breakthrough, 3000 BV, was exceptionally long compared with typical drinking waters. Prior to sulfate breakthrough, TOC removal was about 75 percent. At sulfate breakthrough, a TOC peak appeared, which was the TOC less-preferred-than-sulfate being driven off the column by the advancing sulfate wave front. After sulfate breakthrough, the TOC more preferred than sulfate was being removed for a total TOC removal of about 50 to 60 percent until 9000 BV. Then TOC removal dropped gradually to zero at about 12,000 BV.

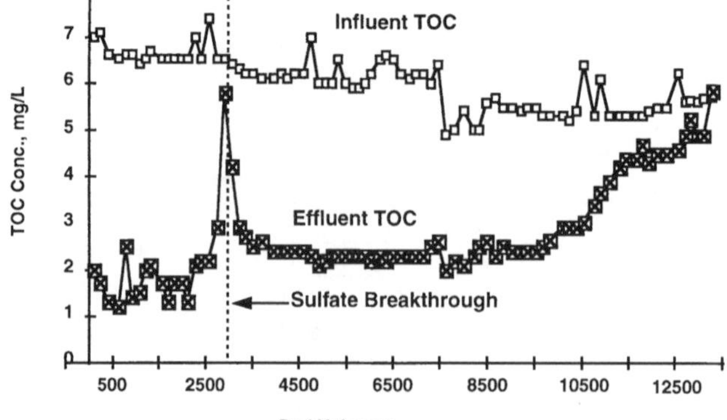

FIGURE 9.27 TOC effluent history for a bench-scale test with IRA 958 resin fed with <1K TOC fraction in low TDS water. [*Source:* P. L.-K. Fu and J. M. Symons (1990). "Removing aquatic organic substances by anion exchange resins." *Journal AWWA,* 82(10), 1990: 70–77.]

These column results were similar to those of Coogan et al. (1968) and Kolle et al. (1984) in that TOC removal started out at 70 to 80 percent and dropped to a long period of about 50 percent removal following sulfate breakthrough.

Kim and Symons (1991) focused their research on the field performance of the macroporous polyacrylic SBA resin (IRA 958) when treating a coagulated, filtered surface water at a City of Houston water plant. Their objectives were to verify the lab studies of Fu and Symons (1990), to determine the influence of regeneration and regenerant reuse, and to evaluate the control of THM percursors with a combination of ion exchange and GAC treatment. Following conventional treatment, the water fed to their experimental minicolumns was cartridge-filtered (1 and 0.45 μm), dechlorinated (Na_2SO_3), and pH adjusted (to 6.8). Figure 9.28 illustrates the virgin resin breakthrough curves for the sulfate and TOC (which was DOC for these filtered waters). As with the previously described nitrate and arsenate removal processes and all the column experiments with DOC removal, the background sulfate concentration determined the performance of the system. When sulfate broke through at about 500 BV, it caused the TOC that was less preferred than sulfate to peak at about 120 percent of the influent TOC concentration (4.0 mg/L). When the effluent histories of the various TOC fractions were studied, it was found that, with the exception of the <0.5K fraction, all peaked at essentially the same time, indicating that all fractions, except the <0.5K fraction, contained some TOC less-preferred-than-sulfate. The <0.5K fraction was only about 25 percent removed throughout the run, and was not influenced by sulfate, which suggests that either it was (1) more preferred than sulfate, (2) sorbed rather than ion exchanged, or (3) a combination of (1) and (2). The TOC effluent histories also indicate that, generally, the higher the MW the better is the TOC removal before and after the sulfate breakthrough. For example, the initial TOC removals of the >10K versus the 1 to 5K fraction were 80 and 60 percent, respectively. After sulfate breakthrough, the respective removals were 50 and 25 percent. Running beyond sulfate breakthrough wasn't very effective with this coagulated water because only about 25 percent overall TOC removal was achieved from 1300 to 2700 BV.

The Kim and Symons (1991) TOC regeneration study determined that the most effective TOC regenerant was a mixture of 2 *N* NaCl with 0.5 *N* NaOH, which is in agreement with the 1.7 *N* NaCl/0.5N NaOH mixture used by Kolle (1984) and Brattebo, Odegaard et al. (1987) in their larger-scale studies. This 2 *N* NaCl with 0.5 *N* NaOH mixture was used in the regenerant reuse study, which showed that the IRA

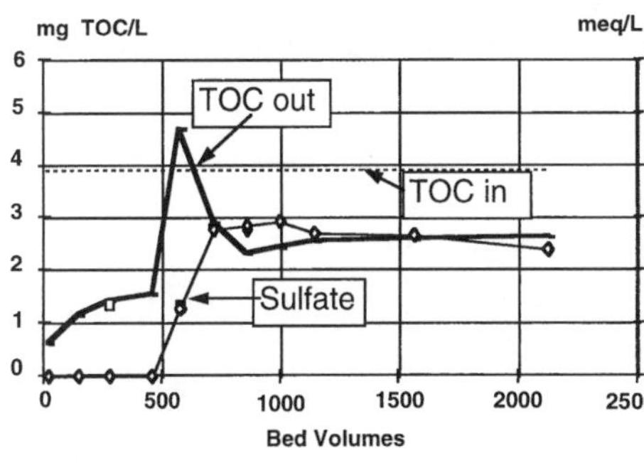

FIGURE 9.28 TOC and sulfate breakthrough curves from virgin IRA 958 resin during pilot tests with treated Houston water. pH = 6.8; EBCT = 10 min. [*Source:* P. H.-S. Kim and J. M. Symons (1991). "Using anion exchange resins to remove THM precursors." *Journal AWWA,* 83(12), 1991: 61–68.]

958 resin could be exhausted and regenerated nine times with the same regenerant, which had been compensated with salt and caustic after each regeneration to maintain the original chloride and hydroxide concentrations. Although the TOC removal during regeneration following the eighth reuse cycle was only 13 percent, the overall performance of the resin for TOC removal up to sulfate breakthrough was essentially unchanged. Kim and Symons suggested that the TOC not removed during regeneration was building up in the deeper pores of the resin where it didn't interfere with the reversible ion exchange, which appeared to be taking place closer to the resin surface. More study is needed in relation to the detrimental effects of TOC buildup after many exhaustion/regeneration cycles.

When Kim and Symons (1991) tested the sequence of ion exchange (IRA 958) followed by GAC (Calgon F-400) for THMFP control, they found that it was much superior to GAC alone. When regenerating the resin at sulfate breakthrough, the IX/GAC process consistently achieved 95 percent trihalomethane formation potential (THMFP) removal to 5000 BV, whereas IX alone achieved only 50 percent removal after 200 BV, and GAC alone achieved only 75 percent removal after about 2000 BV.

In summary, experimental evidence shows that naturally occurring organic matter in groundwater and surface water can be removed by adsorption onto weak-base resins and is also amenable to removal by using macro- and microporous chloride-form strong-base anion resins with both polystyrene and polyacrylic matrices. With SBA resins, the sulfate concentration of the raw water determines run length to the first DOC breakthrough, which peaks above the influent DOC concentration. After this, DOC removal may be continued if 50 percent and lower removals can be tolerated. Filtration should always be employed ahead of ion exchange to protect against resin fouling. When coagulation precedes ion exchange, the DOC remaining is difficult to remove because the two processes are similar in their DOC removal profiles.

URANIUM REMOVAL BY ANION EXCHANGE

The U.S. EPA has recently proposed an MCL of 20 μg/L and a Maximum Contaminant Level Goal (MCLG) of zero pCi/L for uranium (USEPA, 1991) due to health concerns; it is a known kidney chemotoxin and a suspected human carcinogen (Cothern and Lappenbusch, 1983). Based on the results from a national survey, it was estimated that 1,500 community water supplies in the United States could be out of compliance and would have to reduce uranium concentrations to the proposed level (USEPA, 1991).

Uranium Chemistry and Speciation

Uranium has four oxidation states: III, IV, V and VI; however, only the IV and VI states are stable. U(III) oxidizes readily to form U(IV) in air, and U(V) disproportionates to U(IV) and U(VI), which is fairly stable in very acidic aqueous solutions.

The high affinity of uranium for oxygen results in the formation of uranium's most common oxygen-containing compound, uranyl ion UO_2^{2+}, which is the most stable state of uranium in aerated aqueous solution under acidic condition (pH < 5.0). The uranyl ion (UO_2^{2+}) in water readily combines with anionic ligands such as CO_3^{2-}, F^-, Cl^-, NO_3^-, SO_4^{2-}, and HPO_4^{2-} to form stable complexes without itself being changed. When F^-, $H_2PO_4^-$ or HPO_4^{2-} are present in oxidized water, these ligands preferentially complex the UO_2^{2+} at pHs below about 8. (Langmuir, 1978).

Carbonate is the most significant uranium ligand in natural water. In the near-neutral pH range, UO_2^{2+} combines with bicarbonate and carbonate anions in aqueous solution to form stable, charged, and uncharged uranyl carbonates. Figure 9.29 demonstrates the distribution of typical uranium complexes as a function of pH in aerobic groundwater at a CO_2 partial pressure of 10^{-2} atm (Langmuir, 1978). As shown in Figure 9.29, under mildly acidic conditions of pH 5.0 to 6.5, the principal species is $(UO_2CO_3)^0$, and under neutral and slight alkaline conditions, the principal species are $UO_2(CO_3)_2^{2-}$ and $UO_2(CO_3)_3^{4-}$. Uranyl hydroxy complexes such as UO_2OH^+ and $(UO_2)_3(OH)_5^+$ are also formed, but generally in small percentages unless at high temperature or in carbonate-depleted water at pH > 10 (Mo, 1980).

History of Ion Exchange for Uranium Removal

Both cation and anion exchange resins have been used for uranium removal, but because of their requirements for low feed water pH and their relatively unsatisfactory uranium removal efficiency, cation resins are probably not practical for drinking water treatment (Lee and Bondietti, 1983).

Anion exchange exhibited much better performance than cation exchange for removing uranium from contaminated water. Sorg (1988) summarized the results of EPA-funded studies for uranium removal by anion exchange, which he suggested would be the cost-effective method for treating small-community water supplies. In these bench, pilot, and full-scale uranium ion-exchange studies, it was apparent that strong-base anion resins exhibited an enormous capacity for the uranyl carbonate complexes—$UO_2(CO_3)_2^{2-}$ and $UO_2(CO_3)_3^{4-}$. For example, during field studies in Colorado and New Mexico, columns receiving 22 to 104 μg/L uranium treated 8,000 to 60,000 bed volumes of groundwater without exceeding 1.0 μg/L uranium, and in associated studies, two columns containing SBA resins (Dowex 21K and Ionac 641) were operated to 17,400 BV and 31,300 BV, and still removed 95 percent and 90 percent, respectively, of the total uranium in the feed water (Hathaway, 1983). Recent uranium-removal field studies by Zhang and Clifford (Zhang and Clifford, 1994; Clifford and Zhang, 1995) have confirmed this ear-

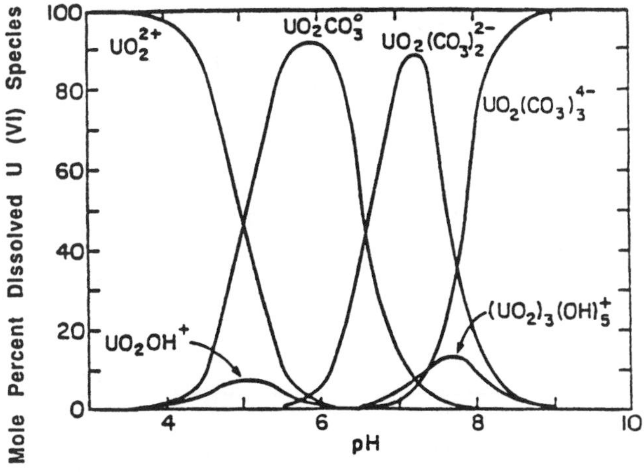

FIGURE 9.29 Distribution of uranyl-carbonate and uranyl hydroxide complexes versus pH for $P_{CO_2} = 10^{-2}$ atm, 25°C, and $C_{T,U} = 2.4$ μg/L.

lier work and shown that chloride-form ion exchange is an excellent process for removing uranium from groundwater. In their Chimney Hill, Texas, field study they reported that the type 1 macroporous polystyrene SBA resin, Ionac A-642, exhibited enormous capacity for uranium removal because of its exceptionally high affinity for $UO_2(CO_3)_3^{4-}$. (Similar resins are available from other suppliers and should perform in a similar manner.) A pilot-scale SBA column was operated continuously for 478 days for a total throughput of 302,000 BV at pH 7.6 to 8.2. The feed water contained 120 µg/L of uranium and 25 pCi/L of radium in a background water quality of 310 mg/L TDS, 150 mg/L alkalinity, 47 mg/L chloride, and <1 mg/L sulfate. Even after 300,000 BV, the effluent uranium concentration was still under 6 µg/L, which corresponds with a removal efficiency of 95 percent. See Figure 9.30. Before 260,000 BV, the uranium leakage was under 1 µg/L, i.e., more than 99 percent removal. Compared with typical granular activated carbon (GAC) columns, which showed a capacity of a few thousand BV throughput before >20 µg/L uranium leakage, the capacity of anion resin for uranium was enormous in this low-sulfate water. After 300,000 BV, the uranium concentration averaged over the whole bed was 30 g/L resin, which corresponds to 2.21 lbs U_3O_8/ft^3 anion resin, or 63 percent of resin sites in uranium form based on a resin capacity of 1.0 meq/mL. Although this ultimate uranium capacity is truly impressive, drinking water treatment columns are not generally operated for 200 to 500 days to uranium breakthrough, because of problems with resin fouling and excessive pressure drop. Cyclic run lengths in the range of 30,000 to 50,000 BV would be more appropriate for uranium removal from drinking water.

Effect of pH and Competing Ions on Uranium Removal

Even though SBA resins exhibit enormous capacity for removing uranium from contaminated ground waters, the removal efficiency and resin capacity are some-

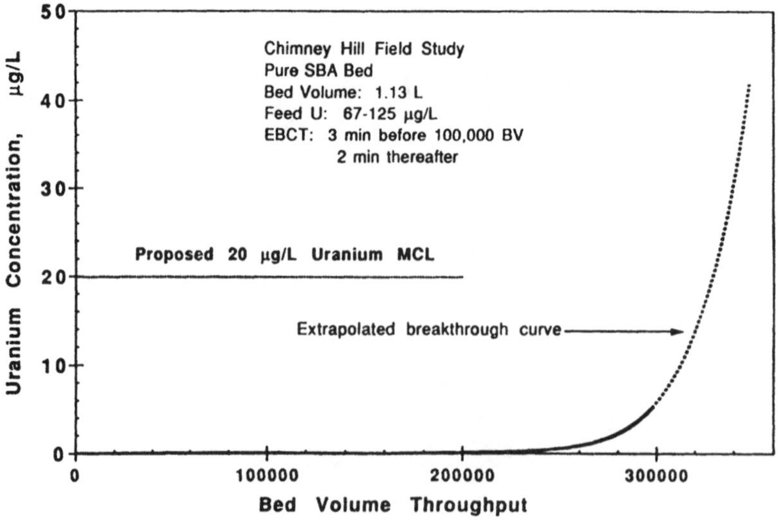

FIGURE 9.30 Effluent uranium levels during virgin exhaustion of a single bed of Ionac A-642 SBA resin at pH 8.0.

what influenced by the usual parameters including feed pH, and the feed concentrations of uranium, sulfate, chloride, and possibly bicarbonate.

The simple ion-exchange reactions expected between uranium compounds and SBA resins in the 6.5 to 9.5 range, where $UO_2(CO_3)_2^{2-}$ and $UO_2(CO_3)_3^{4-}$ are the predominant form, can be expressed as follows:

$$2R_4N^+Cl^- + UO_2(CO_3)_2^{2-} \Leftrightarrow (R_4N^+)_2UO_2(CO_3)_2^{2-} + 2Cl^- \qquad 9.66$$

$$4R_4N^+Cl^- + UO_2(CO_3)_3^{4-} \Leftrightarrow (R_4N^+)_4UO_2(CO_3)_3^{4-} + 4Cl^- \qquad 9.67$$

where R_4N^+ represents an anion exchange site on the resin. Because the resin highly prefers the tetravalent $UO_2(CO_3)_3^{4-}$ complex, there is a tendency for this complex *to actually be formed within the resin* from neutral and divalent uranyl carbonate complexes. This is similar to the in-resin formation of carbonate from bicarbonate as described by Horng and Clifford (Horng, 1983; Horng and Clifford, 1997). The in-resin formation of $UO_2(CO_3)_3^{4-}$, as further explained in the next subsection, helps explain why the performance of SBA resin is so good at feed water pH 5.8, where the predominant form of uranium in water is the neutral species $UO_2CO_3^0$.

Effect of pH. Based on the changes in uranium speciation with pH that are shown in Fig. 9.29, pH variations in the 5.8 to 8 range are expected to have a significant impact on the resin performance when removing uranium. A substantial decrease in the resin's capacity for uranium is expected at pH's below 7 because the charge on the uranyl carbonate complex decreases with decreasing pH, as does the carbonate concentration in the feed water. However, when pH was decreased by HCl addition from 7.8 to 5.8, uranium removal efficiency was unchanged for 270,000 BV during the Chimney Hill field study (Zhang and Clifford, 1994; Clifford and Zhang, 1995). When the pH was further decreased to 4.3, however, serious uranium leakage (about 50 percent of the feed U concentration) occurred from the very beginning of a run with virgin chloride-form resin.

This unexpectedly excellent performance at pH 5.8 probably resulted from the formation of uranyl tricarbonate complexes within the resin, whereas these complexes were virtually absent from the aqueous phase. (See Figure 9.29, where at pH 5.8, the dominant uranium species in the aqueous system in the absence of anion resin is the zero-charged $UO_2CO_3^0$ complex.)

It is theorized that the direct formation of charged $UO_2(CO_3)_3^{4-}$ from uncharged $UO_2CO_3^0$ occurs within the ion exchanger because of the resin's high affinity for polyvalent anions and the high stability constant of the tricarbonate uranium complex. The suggested exchange reactions are:

$$4R_4N^+Cl^- + 4HCO_3^- \Leftrightarrow 4R_4N^+HCO_3^- + 4Cl^- \qquad 9.68$$

$$4R_4N^+HCO_3^- + UO_2CO_3^0 \Leftrightarrow (R_4N^+)_4UO_2(CO_3)_3^{4-} + 2H_2CO_3(aq) \qquad 9.69$$

Reaction 9.68 is the conversion of four resin sites to the bicarbonate form by simple ion exchange with chloride. In Reaction 9.69, the bicarbonate-form ($R_4N^+HCO_3^-$) resin sites are converted to the highly preferred uranyl tricarbonate $(R_4N^+)_4UO_2(CO_3)_3^{4-}$ form with the production of carbonic acid ($H_2CO_3(aq)$). In the Chimney Hill Studies, the resulting pH lowering by the $H_2CO_3(aq)$ produced was insignificant because of the very low concentration (0.0005 mM) of $UO_2CO_3^0$ in the feed water, compared with the total carbonate concentration (C_{T,CO_3}) of the carbonate buffer system—3.00 mM in the groundwater.

Adding Reactions 9.68 and 9.69 yields the net ion exchange reaction:

$$4R_4N^+Cl^- + UO_2CO_3^0 + 4HCO_3^- \Leftrightarrow (R_4N^+)_4UO_2(CO_3)_3^{4-} + 2H_2CO_3(aq) + 4Cl^- \qquad 9.70$$

Reaction 9.70 is, effectively, the uptake of the neutral $UO_2CO_3^0$ molecule from an acidic (pH 5.8) solution by a chloride-form SBA resin.

Effects of Uranium, Sulfate, and Chloride on Run Length. Uranium, sulfate, and chloride concentrations in the feed water have significant effects on uranium breakthrough. Increasing the concentration of these components results in decreasing run length. However, they affect the runs in different degrees.

Uranium feed concentration was expected to have a big influence on column capacity; however, because of the extremely long runs and the potential regulatory problems associated with uranium spiking of raw water during the Chimney Hill field tests, it was not practical to vary the uranium concentration in the test water to determine its influence on BV to uranium breakthrough. Thus, computer predictions of uranium breakthrough were made using the *Equilibrium Multicomponent Chromatography Program with Constant Separation Factors* (EMCT/CSF) Program (Horng, 1983; Clifford, 1993; Tirupanangadu, 1996). This program is a user-friendly implementation of the equilibrium multicomponent chromatography theory (EMCT) of Helfferich and Klein (1970), which has been used by the author (Clifford, 1982) for prediction of column-effluent concentrations during selected-ion separations.

Using this program, the bed volume throughputs to uranium breakthrough were calculated at various uranium concentrations in the Chimney Hill water. The results from the computer prediction are presented in the Figure 9.31 curve labeled "Chloride Water." This curve predicts that for a sulfate-free water like the Chimney Hill feed water, an increase in the uranium concentration to 240 μg/L would result in a decreased run length of 203,000 BV. Decreasing the feed water uranium concentration to 20 μg/L would increase the run length to 815,000 BV. This high sensitivity of run length to uranium concentration was due to the fact that (1) uranium was a trace species with an exceptionally high affinity for the resin, (2) uranium complexes occu-

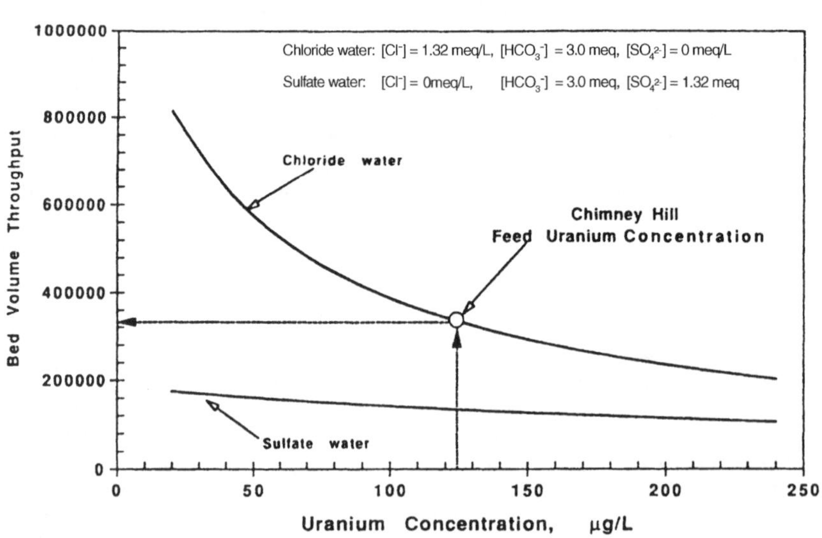

FIGURE 9.31 Calculated bed volume throughput from a SBA resin column before uranium breakthrough at different concentrations of uranium, chloride, and sulfate.

pied a significant fraction of the resin sites at exhaustion, and (3) there was no sulfate in the feed water.

Among the common anions in groundwaters, sulfate will exhibit the biggest impact on run length. The EMCT/CSF program was also used to predict the influence of sulfate on uranium run length at variable uranium concentration. In this simulation, sulfate replaced chloride in the Chimney Hill water, which was initially free of sulfate. The line labeled "Sulfate Water" in Figure 9.31 demonstrates that replacing chloride with sulfate greatly affects the SBA resin's capacity for uranium removal. At the natural uranium concentration of 120 μg/L, when the sulfate concentration increases from 0 to 1.32 meq/L (64 mg/L), run length drops 60 percent from the experimental 340,000 BV down to a predicted 135,000 BV. This decrease at higher sulfate concentration is attributed to the fact that, compared with chloride, sulfate ion has a high affinity for the SBA resin sites. (See Table 9.3.) Therefore, water containing significant concentrations of sulfate will produce significantly shorter runs than will low-sulfate waters.

Regenerability of Uranium-Spent SBA Resins

The Chimney Hill field experiments demonstrated that, in spite of the high affinity of SBA resins for uranyl carbonate complexes, the uranium-spent resins are not difficult to regenerate because of the electroselectivity reversal that takes place in the presence of 1 to 6 *M* chloride regenerant solutions. Furthermore, it is not necessary to completely remove the uranium from the resin during regeneration because the residual uranium does not leak into the effluent on subsequent runs (Zhang and Clifford, 1994). What is necessary for a practical cyclic exhaustion/regeneration process is that a steady state be reached in which the mass of uranium eluted during regeneration is equal to the mass sorbed during exhaustion. Such a steady state can be attained in uranium removal by chloride-form SBA exchange.

Effect of NaCl Concentration and Regeneration Level. Uranium recovery efficiency is strongly dependent on the regenerant NaCl concentration (Zhang and Clifford, 1994). Increasing NaCl concentration results in increased uranium recovery at a fixed regeneration level. This was established in the Chimney Hill field test, where NaCl concentrations ranging from 0.8 to 4.0 *N* were applied to columns partially exhausted to 30,000 or 41,000 BV. Although the NaCl concentration varied, the total amount of NaCl applied was maintained constant at 4.0, or 6.0 eq NaCl/eq. resin during two sets of experiments. At a regeneration level of 6.0 eq NaCl/eq. resin (21.6 lbs NaCl/ft^3), the recovery efficiencies for 4.0, 3.0, 2.0, 1.33, 1.0, and 0.8 *N* NaCl resin were 91, 86, 75, 67, 54 and 47 percent, respectively, as shown in Figure 9.32. The uranium recovery did not appear to level off at 4.0 *N*, which suggests that even higher NaCl concentrations [e.g., saturated NaCl (6.15 *N* @ 20°C)] might be tried.

Increasing the regeneration level from 4 to 6 eq Cl^-/eq resin did improve recovery but not by very much at any NaCl concentration. (See Figure 9.32.) Thus, regeneration levels lower than 4 eq Cl^-/eq resin should be considered, because this would conserve salt, and as pointed out previously, it is not necessary to remove all or even most of the uranium from the exhausted resin as it does not seriously leak on subsequent exhaustions.

Other observations made during the regeneration experiments in Chimney Hill are as follows: (1) Uranium was mostly desorbed in the first one or two BV of regenerant when 2 to 4 *N* NaCl was used. (2) Addition of NaOH to the NaCl solution greatly reduced the recovery of uranium. When NaOH represented one-third of the

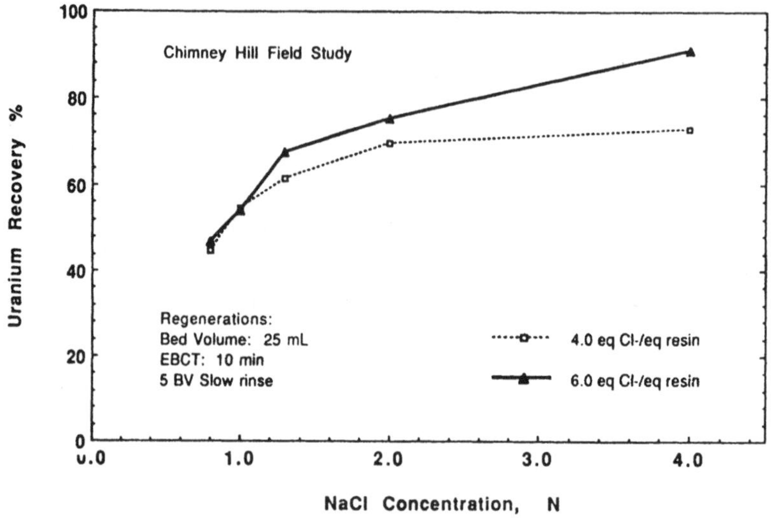

FIGURE 9.32 Effect of equivalents and normality of NaCl on percent uranium recovery during cocurrent regeneration of SBA resin exhausted to 41,000 BV.

regenerant equivalents at 4.0 *N* regenerant concentration, uranium recovery efficiency dropped from 91 to 15 percent. Reductions in regeneration efficiency were also observed when Na_2CO_3 and $NaHCO_3$ were added to NaCl. Uranium recovery efficiency was not affected by the degree of exhaustion when the anion resin was far away from complete exhaustion. Columns that were exhausted to 500 BV exhibited the same uranium recovery percentages (75 to 78 percent) during regeneration as did columns exhausted to 30,000 and 40,000 BV, whereas complete exhaustion was approximately 300,000 BV.

Combined Radium and Uranium Removal. In the Chimney Hill field study conducted by the author and his colleagues (Clifford and Zhang, 1994), a common NaCl-regenerated strong-acid-cation resin water softener modified by the addition of approximately 10 percent strong-base anion (SBA) resin was used for the simultaneous removal of radium (25 pCi/L) and uranium (120 μg/L) from groundwater. Although a high regeneration level (37 lb NaCl/ft^3) was used in the field tests, typical softener regeneration levels in the range of 10 to 20 lb NaCl/ft^3 are expected to be adequate for a cyclic radium and uranium removal process. However, to effectively remove uranium it is important to keep the regenerant NaCl concentration as high as possible.

Intimately mixing the SAC and SBA resins was preferred to stratifying the lighter anion resin over the cation resin in the modified softener. Stratified SBA/SAC beds, especially those containing 25 percent or more anion resin, produced greater radium and uranium leakages after cocurrent regeneration than did mixed beds. In actual practice, backwashing the softener will tend to stratify the anion resin over the cation resin. This less-desirable stratified condition may, however, be adequate for combined radium and uranium removal when the stratified bed contains ≤10 percent anion resin.

As a rule of thumb, an anion bed EBCT ≥ 0.2 min and an anion resin bed depth greater than 1 in were recommended for the stratified SBA resin layer to produce

<10 μg/L uranium leakage with 120 μg/L in the feed water. Extensive tests with the stratified and mixed beds showed that radium (and possibly the uranium leakage) in stratified-bed softeners was attributed to precipitation of $CaCO_3$(s) during regeneration of the cation resin from carbonate ions taken up and eluted from the anion resin. Minimizing the amount of anion resin and mixing it with the cation resin minimized the $CaCO_3$(s) precipitation and the subsequent radium and uranium leakage. Countercurrent regeneration is also expected to eliminate excessive radium and uranium leakage.

Although 10 percent anion resin appeared to be a near optimum amount, as little as 2.5 percent anion resin removed 50 percent of uranium from the Chimney Hill groundwater at conventional flow rates (EBCT = 3 min based on combined cation and anion resin volume). Potassium chloride proved to be superior to NaCl for radium elution from the softener in cyclic operation (Clifford and Zhang, 1994). The use of KCl is recommended for the additional health reason that potassium is preferable to sodium as an exchanging ion.

PERCHLORATE REMOVAL BY ANION EXCHANGE

As this chapter was in the final stages of completion, the perchlorate problem in groundwater had become a national issue, and the California Department of Health Services had established an *action level* for perchlorate in drinking water of 18 μg/L, which was based on the potential for perchlorate to inhibit the uptake of iodine by the thyroid gland. Perchlorate at levels up to several hundred μg/L had been found in groundwaters in California and other states as a result of contamination from ammonium perchlorate in rocket fuel. Two approaches to removing perchlorate are being researched intensively: biological destruction and ion exchange. Ion exchange was chosen for study because of literature reports that perchlorate has a very high affinity for the common polystyrene SBA resins. Perchlorate ion exchange is similar to nitrate removal by anion exchange, except that perchlorate has a much higher affinity for resins than does nitrate. The following brief introduction to perchlorate anion exchange is based on unpublished work done at the University of Houston and on published work related to the San Gabriel Watermaster study of perchlorate removal by ion exchange.

Factors Affecting Perchlorate Resin Selection

Equilibrium batch studies of ClO_4^-/Cl^- binary ion exchange demonstrated (1) the extremely high affinity of perchlorate for polystyrene SBA resins, (2) the even higher affinity of perchlorate for nitrate-selective resins, and (3) the relatively low affinity of perchlorate for polyacrylic resins. The approximate $\alpha_{ClO4/Cl}$ values for polystyrene and polyacrylic resins are 150 and 5, respectively; see Table 9.3. These affinity values are expected, based on the hydrophobic character of the perchlorate ion and the hydrophobic nature of polystyrene resins, compared with the relatively hydrophilic character of polyacrylic resins such as Rohm and Haas IRA 458 and Purolite A-850. Bench-scale resin testing at the University of Houston also verified the reported difficulty regenerating the perchlorate-spent resins, especially the polystyrene and nitrate-selective resins.

As with nitrate removal, the choice of a resin for perchlorate removal depends on water quality, especially sulfate, and the presence of other contaminants such as

TABLE 9.15 Summary of Processes for Removing Inorganic Cations

Contaminant and its MCL	Usual form at pH 7–8	Removal options	Typical BV to MCL*	Pretreatment required	Typical regenerants and % recovery of sorbed contaminant†	Effect of TDS on BV	Effect of Hardness on BV	Notes
Hardness (no MCL)	Ca^{2+}	Na^+ IX softening with SAC resin	200 to 700	Iron removal	6 to 12% NaCl 80 to 100% recovery	V. signif. reduction		1, 2,3
	Mg^{2+}	H^+ IX softening with WAC resin	200 to 500	None	1 to 2 N HCl or H_2SO_4	V. signif. reduction		4, 5,6
Barium 2.0 mg/L	Ba^{2+}	Na^+ IX softening with SAC resin	200 to 700	Iron removal	1 to 12% NaCl, 50 to 100% recovery	V. signif. reduction	V. signif. reduction	1, 2, 3, 7, 8
Radium 5 pCi/L (old MCL)	Ra^{2+}	Na^+ IX softening with SAC resin	200 to 700	Iron removal	6 to 12% NaCl, 0 to 100% recovery	V. signif. reduction	V. signif. reduction	1, 2, 3 9, 10, 11, 12, 20
1999 Proposed Ra-226, 20 pCi/L; Ra-228, 20 pCi/L		Ca^{2+} IX with SAC resin	300 to 1500	Iron removal	10 to 15% $CaCl_2$, 50 to 100% recovery	Slight reduction	Slight reduction	12,13, 14, 15, 16, 20
		Dow Radium Selective Complexer	20,000 to 50,000	Iron removal	Not regenerable	Slight reduction	Slight reduction	12, 17, 18, 19
		$BaSO_4$-impregnated alumina	20,000 to 50,000	Iron removal	Not regenerable	Slight reduction	Slight reduction	12, 17, 18, 19
		Activated alumina	1000 to 3000	None	2% HCl followed by 1% NaOH, 70 to 100% recovery	None	Slight reduction	12, 20

* Generally, run length depends on raw water contaminant concentration, allowable effluent concentration, competing ions, leakage, and the actual resin or adsorbent used.

† Percent recovery of contaminants is shown for the first few regenerations of a virgin alumina or resin. For most processes, the percent recovery approaches 100% at steady-state cyclic (exhaustion-regeneration) operation. The steady-state run lengths are always shorter than the virgin run when removing highly preferred contaminants.

[1] Hardness capacity depends on regeneration level (lbs $NaCl/ft^3$ resin). KCl is a better regenerant but it costs more.

[2] Typically, spent regenerant is disposed of to sanitary sewer.

[3] Iron removal by oxidation and filtration should be considered if total iron concentration exceeds 0.3 mg/L.

[4] Only carbonate hardness can be removed using WAC resins.

[5] Carbon dioxide produced in the IX reaction must be removed from product water; and pH adjustment will be necessary.

[6] TDS reduction occurs as a result of removing both the hardness and alkalinity.

[7] Barium tends to build up on the resin and breakthrough with hardness when insufficient regenerant is used.

[8] If necessary, $BaSO_4$ can be precipitated from the spent regenerant by adding sulfate.

[9] With a virgin resin, radium breaks through long after hardness but in cyclic operation they eventually elute simultaneously.

[10] Even though radium accumulates on the resin, no leakage of radium occurs before hardness breakthrough.

[11] Current disposal practices allow the discharge of radium-contaminated spent regenerant to the sanitary sewer.

[12] Radon-222 is continuously generated from the radium-226 on the resin. Radon peaks can occur after idle periods.

[13] Immediate serious radium leakage occurs if extensive countercurrent regeneration is not used.

[14] $CaCl_2$ is much more expensive than NaCl as a regenerant.

[15] When using $CaCl_2$ for regeneration, sodium is not added to the product water.

[16] No softening is produced in the calcium exchange process, and magnesium is exchanged for calcium.

[17] Radon generation is more serious with the RSC and $BaSO_4$-alumina because of the large amount of radium on the media.

[18] Disposal of the spent media is a serious problem because it is considered a low-level radioactive waste.

[19] The RSC and $BaSO_4$-alumina may not be commercially available because of disposal problems.

[20] Spent regenerant disposal may be a problem if sanitary sewer disposal is not allowed.

TABLE 9.16 Summary of Processes for Removing Inorganic Anions

Contaminant and its MCL	Usual form at pH 7–8	Removal options	Typical BV to MCL*	Pretreatment required	Typical regenerants and % recovery of sorbed contaminant†	Effect of TDS on run BV	Effect of SO_4^{2-} on run BV	Notes
Fluoride 4.0 mg/L	F^-	Activated alumina	1,000 to 2,500	pH = 5.5 to 6.0	1% NaOH followed by 2% H_2SO_4 90–100% recovery	None	Slight reduction	0
Nitrate-N 10 mg/L	NO_3^-	Anion exchange (complete regeneration)	300 to 600	Usually none	0.25 to 2.0 N NaCl (1.5 to 12% NaCl) 90–100% recovery	V. signif. reduction	V. signif. reduction	1,2,4
		Anion exchange (partial regeneration)	200 to 500	Usually none	0.5 to 2.0 N NaCl (3.0 to 12% NaCl) 50% recovery	V. signif. reduction	V. signif. reduction	1,3,4
Arsenic 0.05 mg/L (old MCL)	$HAsO_4^-$ As(V)	Activated alumina adsorption	5,000 to 25,000	pH = 5.5 to 6.0 oxidize	4% NaOH followed by 2% H2SO4 70% recovery	None	Slight reduction	0, 5, 6
0.002–0.020 mg/L (1999 proposed)	$HAsO_4^{2-}$ As(V)	Anion exchange	500 to 5,000	Oxidize and prefilter to remove iron	1.0 N NaCl >95% recovery	V. signif. reduction	V. signif. reduction	1, 5, 7, 8
Selenium 0.05 mg/L	$HSeO_3$ Se(IV)	Activated alumina adsorption	1,000 to 2,500	pH = 5.5 to 6.0	1% NaOH followed by 2% H_2SO_4	None	Slight reduction	0
	SeO_4^{2-} Se(VI)	Anion exchange	300 to 1500	None	1.0 N NaCl 90–100% recovery	V. signif. reduction	Signif. reduction	9, 10
Chromium 0.10 mg/L	CrO_4^{2-} Cr(VI)	Anion exchange	10,000 to 50,000	None	1.0 N NaCl 60–90% recovery	Slight reduction	Slight reduction	11, 12, 13
Color and DOC	Fulvates and etc.	Anion exchange	400 to 5,000	Prefiltration for turbidity	Mixture of 2.0 N NaCl with 0.5 N NaOH	V. signif.	V. signif.	13, 14, 15

Uranium 20 μg/L (proposed)	$UO_2(CO_3)_3^{4-}$	Anion exch.	30,000 to 300,000	Prefiltration for turbidity	1–6 N NaCl 70–90% recovery	Slight	Signif.	16
Perchlorate	ClO_4^-	Anion exchange	500 to 6000	Prefiltration for turbidity	0.5–3 N NaCl 0–90% recovery	Slight reduction	V. signif. reduction	13, 17

* Generally, run length depends on raw water contaminant concentration, allowable effluent concentration, competing ions, leakage, and the actual resin or adsorbent used. Run lengths are averages for typical water supplies.

† Percent recovery of contaminants is shown for the first few regenerations of a virgin alumina or resin. For most processes, the percent recovery approaches 100% at steady-state cyclic (exhaustion-regeneration) operation. The steady-state run lengths are always shorter than the virgin run when removing highly preferred contaminants.

[0] Operation at natural feed pH is also possible to simplify the process. Runs are much shorter at feed pHs above 6.5.

[1] Chromatographic peaking of contaminant is possible after breakthrough. Use multiple parallel columns to avoid this.

[2] No significant leakage of nitrate occurs prior to breakthrough if complete regeneration is used.

[3] Continuous, significant (>5 mg/L) leakage of nitrate occurs following partial regeneration during all runs.

[4] Resin must be mixed mechanically following regeneration to avoid excessive early nitrate leakage. Batch denitrification of spent brine can be employed to remove nitrate and reuse regenerant up to 40 times if Cl^- conc. is maintained above 0.5 M.

[5] As(III) in the form of uncharged arsenious acid(H_3AsO_3) must be oxidized to As(V) prior to adsorption or ion exchange. Arsenic MCLs in the range of 0.002 to 0.020 mg/L are being considered (1999).

[6] Spent regenerant can be reused several times if OH^- conc. is maintained at about 1 M. Eventually, arsenic (V) can be coprecipitated from regenerant by lowering pH (to 6–8) to precipitate $Al(OH)_3(s)$.

[7] Spent regenerant can be reused up to 20 times if Cl– conc. is maintained above about 0.5 M. Eventually, arsenic (V) can be coprecipitated from regenerant by lowering pH (to 5–6) and adding Fe(III) to precipitate $Fe(OH)_3(s)$.

[8] Chloride-form anion exchange can be the process of choice for low sulfate (<120 mg/L) and low TDS (<500 mg/L) water.

[9] Se(IV) can typically be oxidized to Se(VI) by 1 to 2 mg/L free chlorine in 30 to 60 minutes at pH 6.5 to 8.5.

[10] Selenite, Se(IV) should be absent if ion exchange is used, because it peaks before sulfate and selenate, Se(VI), breakthrough.

[11] Cr(III) can be precipitated from the spent regenerant after reduction with ferrous sulfate or acidic sodium sulfite.

[12] Macroporous resins and polystyrene resins have a higher preference for chromate than gel and acrylic resins.

[13] Resins with higher contaminant affinity are relatively more difficult to regenerate.

[14] No MCL for DOC. Major DOC breakthrough occurs just ahead of sulfate breakthrough. Typical removal of DOC = 60–80% before sulfate breakthrough and 40–60% after sulfate breakthrough. Highly colored waters should be prefiltered.

[15] Reuse of spent regenerant for DOC removal by IX is possible if chloride and hydroxide concentrations are maintained.

[16] All SBA resins have an extremely high affinity of uranyl carbonate complexes. For cyclic operations, run lengths in the range of 30,000–50,000 BV are suggested. Higher regenerant NaCl concentrations are better.

[17] Polystyrene resins have a much higher affinity for perchlorate compared with polyacrylic resins. Perchlorate is very difficult to remove from resins during regeneration. Counter-flow regeneration is recommended to minimize leakage.

arsenate and nitrate. High sulfate will shorten the runs significantly, and nitrate and arsenate will generally exit before perchlorate and be subject to chromatographic peaking. With polyacrylic resins, the selectivity sequence is sulfate > perchlorate > arsenate > nitrate > chloride > bicarbonate. For polystyrene resins, the sequence is perchlorate > sulfate > arsenate > nitrate > chloride > bicarbonate. Recall that the ions exit the column in the reverse order of selectivity, with the least-preferred ions leaving first.

Perchlorate Ion Exchange Process Considerations

The major process design consideration in perchlorate removal is the regenerability of the resin. In the San Gabriel bench- and pilot-scale perchlorate removal study (Najm, Trussell et al., 1999), polystyrene resins produced bench-scale virgin run lengths of 6,000 to 10,000 BV with a spiked feed water containing 200 μg/L perchlorate, 9 mg/L NO_3–N, 55 mg/L sulfate, 40 mg/L chloride, 200 mg/L bicarbonate, and 300 mg/L TDS when using 18 μg/L as the perchlorate MCL. Unfortunately, with a strong 15 lb NaCl/ft^3 cocurrent downflow regeneration, the run lengths were shortened to less than 1,000 BV after regeneration, and the perchlorate leakage was always greater than 10 μg/L. The polyacrylic resin produced only 600 BV on the virgin run, but also produced 600 BV on subsequent runs with the same regeneration. Unfortunately, the leakage was 30 μg/L from the incompletely regenerated polyacrylic resin. They concluded that countercurrent regeneration with at least 30 lb NaCl/ft^3 would be required for a process with low perchlorate leakage and long (>600 BV) run length.

Following the bench-scale tests, Najm et. al. pilot-tested two polyacrylic and two polystyrene SBA resins on the San Gabriel. Using an EBCT of 1.5 min and 0.5 *N* NaCl (3 percent) countercurrent regeneration with 30 lb NaCl/ft^3, they determined that all resins gave more than 600 BV run lengths to 10 μg/L perchlorate breakthrough, but that early nitrate breakthrough was a problem with the polyacrylic resins. The best performance in terms of low leakage (≤4 μg/L) and long run length (750 BV) over 31 cycles was obtained with a type 2 polystyrene resin (Ionac ASB-2). These investigators used dilute (0.5 *N,* 3 percent) NaCl because they intend to develop an ion-exchange process with brine reuse. To achieve this, attempts are being made to (1) develop a robust biological culture for perchlorate destruction in the regenerant brine, and (2) combine the ion-exchange and biological perchlorate reduction steps into a complete process with brine reuse similar to the previously developed nitrate ion-exchange process with biological denitrification and brine reuse (Liu and Clifford, 1996). Other investigators are attempting to use elevated-temperature physical-chemical processes to catalytically reduce the perchlorate in the brine prior to its reuse.

WASTE DISPOSAL

To the extent possible, waste disposal considerations have been covered in the discussions of each contaminant removal process. The focus has been on the design of processes that minimize regenerant use and on the reuse of the spent regenerant to maximize the ratio of product water to spent regenerant for final disposal. Further explanations of how to deal with process residues are given in Chapter 16.

SUMMARY

In the first part of the chapter, the fundamentals of ion-exchange and adsorption processes were explained with the goal of demonstrating how these principles influence process design for inorganic contaminant removal. In the second part, ion-exchange and adsorption processes were described in detail for the removal of hardness, barium, radium, nitrate, fluoride, arsenic, selenium, chromate, DOC, uranium, and perchlorate. The selection of a process for removal of a given contaminant can be confusing because of the many variables to be considered (e.g., contaminant speciation, resins, adsorbents, competing ions, foulants, regenerants, and column flow patterns). Summary Tables 9.15 and 9.16 have been added to aid the reader in choosing a process. In Table 9.15 the important cation-removal process alternatives have been summarized, while in Table 9.16 anion removal is covered. The reader should refer to the discussion of fundamentals and process alternatives to answer questions arising from the tables.

BIBLIOGRAPHY

Abrams, I. M. "Organic fouling of ion exchange resins." In *Physicochemical Methods for Water and Waste Water Treatment* by L. Pawlowski (Amsterdam: Elsevier Publishing Co., 1982), 213–224.

Advanced Separation Technologies. *Continuous adsorption ion exchange systems.* Lakeland, Florida: Calgon Corporation, 1995.

American Water Works Association. "An AWWA survey of inorganic contaminants in water supplies." *Jour. AWWA,* 77(5), 1985: 67.

Andreae, M. O. "Distribution and speciation of Arsenic in natural waters and some marine algea." *Deep Sea Research* 25(4), 1978: 391–402.

Baes, C. F., and R. E. Mesmer. *The hydrolysis of cations.* New York: Wiley-Interscience, 1976.

Baker, B., V. R. Davies, et al. "Use of Acrylic strong base anion resins in treatment of organic bearing waters." 38th Annual Intl. water conference, Pittsburgh, Pennsylvania, 1977.

Bellen, G. E., M. Anderson, et al. *Point-of-use treatment of control organic and inorganic contaminant in drinking water.* Cincinnati, Ohio: U.S. EPA, 1986.

Benjamin, M. M., R. S. Sletten, et al. "Sorption and filtration of metals using iron-oxide-coated sand." *Water Research,* 30, 1996: 2609–2620.

Boari, G., L. Liberti, et al. "Exchange equilibria on anion resins." *Desalination,* 15, 1974: 145–166.

Boegel, J. V., and D. A. Clifford. *Selenium oxidation and removal by ion exchange.* Cincinnati, Ohio: U.S. EPA, 1986.

Bowers, A. E. "Ion exchange softening." In *The quest for pure water* (Denver, Colorado: AWWA, 1980), 2.

Brattebo, H., H. Odegaard, et al. "Ion exchange for the removal of humic acids in water treatment." *Water Res.* 21(9), 1987: 1045.

Brooks, C. S., P. L. Brooks, et al. *Metal recovery from industrial waste.* Chelsea, Michigan: Lewis Publishers, 1991.

Calmon, C. "Specific ion exchangers." In *Ion exchange for pollution control* by C. Calmon and H. Gold (Boca Roton, Florida: CRC Press, 1979), 2.

Clifford, D. A. "Multicomponent ion exchange calculations for selected ion separations." *Industrial and Engineering Chemistry-Fundamentals* 21(2), 1982: 141–53.

———. "Removal of radium from drinking water." In *Radon, Uranium and Radium in Drinking Water.* (Chelsea, Michigan: Lewis Publishers, 1990), 225–247.

———. "Chromatographic peaking of toxic contaminants during water treatment by ion exchange." *Proceedings of 1991 International Conference on Ion Exchange, Tokyo, Japan,* Japan Association of Ion Exchange.

———. "Computer prediction of ion exchange." *Jour. AWWA,* 85(4), 1993: 20.

———. "Computer prediction of arsenic ion exchange." *Jour. AWWA,* 87(4), 1995.

Clifford, D. A., G. Ghurye, et al. *Arsenic ion-exchange process with reuse of spent brine.* Dallas, Texas: American Water Works Association, 1998.

———. *Final report: phases 1 and 2 city of Albuquerque arsenic study, field studies on arsenic removal in Albuquerque, New Mexico using the University of Houston/USEPA mobile drinking water treatment research facility.* Houston, Texas: University of Houston, 1997.

———. *Final report: Phase 3 city of Albuquerque arsenic study, field studies on arsenic removal in Albuquerque, New Mexico using the University of Houston/USEPA mobile drinking water treatment research facility.* Houston, Texas: University of Houston, 1998.

———. "Development of an arsenic ion exchange process with direct reuse of spent brine." *Jour. AWWA,* submitted August, 1999.

Clifford, D. A., and C. C. Lin. *Arsenic removal from groundwater in Hanford, California—a summary report.* Houston, Texas: University of Houston, 1986.

———. *Arsenic(III) and arsenic(V) removal from drinking water in San Ysidro, New Mexico.* Cincinnati, Ohio: U.S. EPA, 1991.

Clifford, D. A., C. C. Lin, et al. *Nitrate removal from drinking water in Glendale, Arizona.* Cincinnati, Ohio: U.S. EPA, 1987.

Clifford, D. A., and X. Liu "Biological denitrification of spent regenerant brine using a sequencing batch reactor." *Water Research* 27(9), 1993*a:* 1477–84.

———. "Nitrate removal using ion exchange with batch denitrification of spent regenerant brine." *Jour. AWWA,* 85(4), 1993*b:* 135–143.

Clifford, D. A., W. Vijjeswarapu, et al. "Evaluating various adsorbents and membranes for removing Radium from groundwater." *Jour. AWWA,* 80(7), 1988: 94.

Clifford, D. A., and W. J. Weber. *Nitrate removal from water supplies by ion exchange.* Cincinnati, Ohio: USEPA, 1978.

———. "The determinants of divalent/monovalent selectivity in anion exchangers." *Reactive Polymers* 1, 1983: 77–89.

Clifford, D. A., and Z. Zhang. "Combined Uranium and Radium removal by ion exchange." *Jour. AWWA,* 86(4), 1994: 214–227.

———. "Removing uranium and radium from ground water by ion exchange resins." In *Ion Exchange Technology: Recent Advances in Pollution Control* by A. K. Sengupta. (Lancaster, Pennsylvania: Technomic Publishing Company, 1995), 1–59.

Coogan, G. J. "Color removal from surface waters by use of resins-development report." *Jour. AWWA,* 82(1), 1968: 1–4.

Cothern, C. R., and W. L. Lappenbusch. "Occurrence of uranium in drinking water in the United States." *Health Physics* 45(1), 1983: 89–100.

———. "Compliance data for the occurrence of Radium and gross alpha particle activity in drinking water supplies in the United States." *Health Physics* 46(3), 1984: 503.

Frank, P., and D. A. Clifford. *Arsenic (III) oxidation and removal from drinking water.* Cincinnati, Ohio: USEPA, 1986.

Frey, M. M., M. A. Edwards, et al. *National Compliance assessment and costs for the regulation of arsenic in drinking water.* Denver Colorado: American Water Works Association, 1997.

Frisch, N. W. and R. Kunin. "Long term operating characteristics of anion exchange resins." *Ind. and Eng. Chem.* 49(9), 1957: 1365–1372.

Fu, P. L.-K., and J. M. Symons. "Removing aquatic organic substances by anion exchange resins." *Jour. AWWA,* 82(10), 1990: 70–77.

Garg, D., and D. A. Clifford. *Removing radium from water by plain and treated activated alumina.* Cincinnati, Ohio: USEPA, 1992.

Ghurye, G. L., D. A. Clifford, et al. "Combined Arsenic and Nitrate removal by ion exchange." *Jour. AWWA,* 1998.

Guter, G. A. *IX Windows Pro.* Bakersfield, California: Cathedral Peak Software, 1998.

———. *Removal of nitrate from contaminated water supplies for public use.* Cincinnati, Ohio: U.S. EPA, 1981.

———. "Nitrate removal from contaminated groundwater by anion exchange." In *Ion Exchange Technology,* A. K. Sengupta, ed. Lancaster, PA: Technomic Publishing Company, 1995.

Hathaway, S., and F. Rubel Jr. "Removing arsenic from drinking water." *Jour. AWWA,* 79(8), 1987: 61–65.

Hathaway, S. W. "Uranium removal process." *Proceedings American Water Works Association Annual Conference, Las Vegas, Nevada, 1983.* American Water Works Association.

Helfferich, F. "Ion exchange kinetics V: ion exchange accompanied by reactions." *J. Phys. Chem,* 69, 1965: 1178–1187.

Helfferich, F. "Ion exchange kinetics." In *Ion exchange: a series of advances* by J. A. Marinsky (New York: Marcel Dekker, 1966), 1.

Helfferich, F., and G. Klein. *Multicomponent chromography: theory of interference.* New York: Marcel Dekker, 1970.

Helfferich, F. G. *Ion Exchange.* New York: McGraw-Hill, 1962.

Horng, L. L. Reaction mechanisms and chromatographic behavior of Polyprotic acid anions in multicomponent ion exchange. Ph.D. dissertation, University of Houston, Houston, Texas: 1983.

Horng, L. L., and D. A. Clifford. "The behavior of polyprotic anions in ion exchange resins." *Reactive and Functional Polymers,* 35(1/2), 1997: 41–54.

Janauer, G. E., C. P. Gerba, et al. "Insoluble polymer contact disinfectants: an alternative approach to water disinfection." In *Chemistry in Water Reuse* by W. J. Cooper (Ann Arbor, Michigan: Ann Arbor Science, 1981).

Kim, P. H.-S., and J. M. Symons. "Using anion exchange resins to remove THM precursors." *Journal AWWA,* 83(12), 1991: 61–68.

Kolle, W. *Humic acid removal with macroreticular ion exchange resins at Hannover.* Washington, DC., USEPA, 1984.

Kunin, R. *Ion exchange resins.* Huntington, New York: Robert E. Krieger Publishing Co, 1972.

Langmuir, D. "Uranium solution-mineral equilibria at low temperatures with applications to sedimentary ore deposits." *Geochimica et Cosmochimica,* 42, 1978: 547–569.

Lauch, R. P., and G. A. Guter. "Ion exchange for removal of nitrate from well water." *Jour. AWWA,* 1986: 83.

Lee, S. Y., and E. A. Bondietti. "Removing uranium from drinking water by metal hydroxides and anion-exchange resins." *Jour. AWWA,* 75(10), 1983: 537.

Liu, C. X., and D. A. Clifford. "Ion exchange with denitrified brine reuse." *Jour. AWWA,* 88, 1996: 88–99.

Maier, F. J. "Defluouridation of municipal water supplies." *Jour. AWWA,* 45(8), 1953: 879.

Mangelson, K. A. *Radium removal for a small community water supply system.* Cincinnati, Ohio: USEPA, 1988.

Matejka, Z., and Z. Zirkova. "The sorption of heavy-metal cations from EDTA complexes on acrylamide resins having oligo(ethyleneamine) moieties." *Reactive and Functional Polymers* 35(1/2), 1997: 81–88.

McGarvey, F. X., and A. C. Reents. "Get rid of fouling in ion exchangers." *Chemical Engineer,* 61, 1954: 205.

McNeill, L. S., and M. Edwards. "Soluble arsenic removal at water treatment plants." *Jour. AWWA,* 87(4), 1995: 105–113.

Mo, T. J. *Chemistry of uranium in aqueous environments.* U.S. EPA Office of Radiation Programs—Criteria and Standards Division—Radioactive Waste Standard Branch, 1980.

Myers, A. G., V. L. Snoeyink, et al. "Removing barium and radium through calcium cation exchange." *Jour. AWWA,* 77(5), 1985: 60.

Najm, I., R. R. Trussell, L. Boulos, B. Gallagher, R. Bowcock, and D. Clifford. "Evaluating Ion Exchange Technology for Perchlorate Removal," *Proc. 1999 AWWA Annual Conference, Water Research.* Denver, Colo.: American Water Works Assoc.

National Research Council. "Arsenic." *Drinking water and health,* National Academy of Science, 1977: 1.

Pontius, F. W. "An update of the federal drinking water regulations." *Jour. AWWA,* 87(2), 1995: 48–58.

Purolite Inc. *Purolite resin products description.* Bala Cynwyd, Pennsylvania: Purolite Inc, 1995.

Rhom and Haas. *Amberlite resin product descriptions.* Philadelphia, Pennsylvania: Rhom and Haas Co, 1994.

Robertson, F. N. "Hexavalent Chromium in the ground water in Paradise Valley, Arizona." *Groundwater* 13(6), 1975: 516–527.

Rosenblum, E. R., and D. A. Clifford. *The equilibrium arsenic capacity of activated alumina.* Cincinnati, Ohio: USEPA, 1984.

Rubel, F., Jr., and S. W. Hathaway. *Pilot study for the removal of arsenic from drinking water at the Fallon, Nevada, Naval Air Station.* Cincinnati, Ohio: USEPA, 1985.

Rubel, F., Jr., and R. D. Woosley. "The removal of excess Fluoride from drinking water by activated Alumina." *Jour. AWWA,* 71(1), 1979: 45.

Schindler, P. W. "Surface complexes at oxide-water interfaces." In *Adsorption of Inorganics at Solid-Liquid Interfaces* by M. A. Anderson and A. J. Rubin (Ann Arbor, Michigan: Ann Arbor Science, 1981).

Schmitt, G. L., and D. J. Pietrzyk. "Liquid chromatographic separation of inorganic anions on an alumina column." *Annual Chemistry,* 57, 1985: 2247.

Scott, K. N., J. F. Green, et al. "Arsenic removal by coagulation." *Jour. AWWA,* 87(4), 1995: 114–126.

Sengupta, A. K., and D. A. Clifford. "Chromate ion-exchange mechanism for cooling water." *Ind. Eng. Chem. Fund.* 25(2), 1986: 249–258.

Siegel, S. K., and D. A. Clifford. *Removal of chromium from ion exchange regenerant solution.* Cincinnati, Ohio: USEPA, 1988.

Simms, J., and F. Azizian. "Pilot-Plant Trials on Removal of Arsenic from Potable Water using Activated Alumina." *AWWA Water Quality Technology Conference,* Denver, Colorado, November 1997, American Water Works Association.

Singh, G., and D. A. Clifford. *The equilibrium flouride capacity of activated alumina.* Cincinnati, Ohio: USEPA, 1981.

Snoeyink, V. L., C. Cairns-Chambers, et al. "Strong-acid ion exchange for removing barium, radium, and hardness." *Jour. AWWA* 79(8), 1987: 66.

Sorg, T. J. "Methods for removing Uranium from drinking water." *Jour. AWWA,* 80(7), 1988: 105.

Stumm, W. *Chemistry of the solid-water interface.* New York: John Wiley & Sons, 1992.

Subramonian, S., and D. A. Clifford. "Monovalent/Divalent selectivity and the charge separation concept." *Reactive Polymers,* 9, 1988: 195–209.

Subramonian, S., D. A. Clifford, et al. "Evaluating ion exchange for the removal of Radium from groundwater." *Jour. AWWA,* 1990: 61–70.

Sybron, I. *Ionac resin product descriptions.* Philadelphia, Pennsylvania: Sybron Inc, 1995.

Tirupanangadu, M. "A visual basic application for multicomponent chromatography in ion exchange columns." M.S. thesis. Houston, Texas: University of Houston, 1996, 179.

Tong, J. "Development of an ion(III)-coagulation-microfiltration process for arsenic removal from ground water." M.S. thesis (University of Houston, Houston, Texas: 1997), 93.

Trussell, R. R., A. Trussell, et al. *Selenium removal from groundwater using activated Alumina.* Cincinnati, Ohio: USEPA, 1980.

USEPA. 40 CRF Parts 141, 142, and 143. National primary drinking water regulations; fluoride; final rule and proposed rule. *Federal Register.* 50, 1985: 47159-163.

USEPA. CFR parts 141 and 142. National primary drinking water regulations; radionuclides; advanced notice of proposed rulemaking. *Federal Register.* 56, 1991: 138.

USEPA. SOCs and IOCs final rule. *Federal Register.* 56, 1992: 3526.

Van der Hoek, J. P., P. J. M. Van der Van, et al. "Combined ion exchange/biological denitrification of Nitrate removal from ground water under different process conditions." *Water Res.,* 22(6), 1988: 679–684.

Water Quality Research Council. EPA approves point-of-use/point-of-entry use by public water system. *Point of Use,* 2, 1987.

Wilson, A. L. "Organic fouling of strongly basic anion-exchange resins." *Applied Chemistry* 9, 1959: 352.

Yuan, J. R., M. M. Ghosh, et al. "Adsorption of Arsenic and Selenium on activated Alumina." *National conference on environmental engineering, New York, 1983,* American Society of Civil Engineers.

Zhang, Z., and D. A. Clifford. "Exhaustion and regeneration of resins for Uranium removal." *Jour. AWWA,* 86(4), 1994: 228–241.

CHAPTER 10
CHEMICAL PRECIPITATION

Larry D. Benefield, Ph.D.
Professor
Department of Civil Engineering
Auburn University, Alabama

Joe M. Morgan, Ph.D.
Associate Professor
Department of Civil Engineering
Auburn University, Alabama

Chemical precipitation is an effective treatment process for the removal of many contaminants. Coagulation with alum, ferric sulfate, or ferrous sulfate and lime softening both involve chemical precipitation. The removability of substances from water by precipitation depends primarily on the solubility of the various complexes formed in water. For example, heavy metals are found as cations in water and many will form both hydroxide and carbonate solid forms. These solids have low solubility limits in water. Thus, as a result of the formation of insoluble hydroxides and carbonates, the metals will be precipitated out of solution.

Although coagulation with alum, ferric sulfate, or ferrous sulfate involves chemical precipitation, extensive coverage of coagulation is given in Chapter 6 and will not be repeated here. The discussion of the application of chemical precipitation in water treatment presented in this chapter will emphasize the reduction in the concentration of calcium and magnesium (water softening) and the reduction in the concentration of iron and manganese. Attention will also be given to the removal of heavy metals, radionuclides, and organic materials in the latter part of the chapter.

FUNDAMENTALS OF CHEMICAL PRECIPITATION

Chemical precipitation is one of the most commonly used processes in water treatment. Still, experience with this process has produced a wide range of treatment efficiencies. Reasons for such variability will be explored in this chapter by considering precipitation theory and translating this into problems encountered in actual practice.

Solubility Equilibria

A chemical reaction is said to have reached equilibrium when the rate of the forward reaction is equal to the rate of the reverse reaction so that no further net chemical change occurs. A general chemical reaction that has reached equilibrium is commonly expressed as

$$aA + bB \rightleftharpoons cC + dD \tag{10.1}$$

The equilibrium constant K_{eq} for this reaction is defined as

$$K_{Eq.} = \frac{(C)^c(D)^d}{(A)^a(B)^b} \tag{10.2}$$

where the equilibrium activities of the chemical species A, B, C, and D are denoted by (A), (B), (C), and (D) and the stoichiometric coefficients are represented as *a, b, c,* and *d*. For dilute solutions, molar concentration is normally used to approximate activity of aqueous species while partial pressure measured in atmospheres is used for gases. By convention, the activities of solid materials, such as precipitates, and solvents, such as water, are taken as unity. Remember, however, that the equilibrium constant expression corresponding to Equation 10.1 must be written in terms of activities if one is interested in describing the equilibrium in a completely rigorous manner.

The state of solubility equilibrium is a special case of Equation 10.1 that may be attained either by formation of a precipitate from the solution phase or from partial dissolution of a solid phase. The precipitation process is observed when the concentrations of ions of a sparingly soluble compound are increased beyond a certain value. When this occurs, a solid that may settle is formed. Such a process may be described by the reaction

$$A^+ + B^- \rightleftharpoons AB(s) \tag{10.3}$$

where (s) denotes the solid form. The omission of "(*s*)" implies the species is in the aqueous form.

Precipitation formation is both a physical and chemical process. The physical part of the process is composed in two phases: nucleation and crystal growth. Nucleation begins with a supersaturated solution (i.e., a solution that contains a greater concentration of dissolved ions than can exist under equilibrium conditions). Under such conditions, a condensation of ions will occur, forming very small (invisible) particles. The extent of supersaturation required for nucleation to occur varies. The process, however, can be enhanced by the presence of preformed nuclei that are introduced, for example, through the return of settled precipitate sludge, back to the process.

Crystal growth follows nucleation as ions diffuse from the surrounding solution to the surfaces of the solid particles. This process continues until the condition of supersaturation has been relieved and equilibrium is established. When equilibrium is achieved, a saturated solution will have been formed. By definition, this is a solution in which undissolved solute is in equilibrium with solution.

No compound is totally insoluble. Thus, every compound can be made to form a saturated solution. Consider the following dissolution reaction occurring in an aqueous suspension of the sparingly soluble salt:

$$AB(s) \rightleftharpoons AB \tag{10.4}$$

The aqueous, undissociated molecule that is formed then dissociates to give a cation and anion:

$$AB \rightleftharpoons A^+ + B^- \tag{10.5}$$

The equilibrium constant expressions for Equations 10.4 and 10.5 may be manipulated to give Equation 10.6, where the product of the activities of the two ionic species is designed as the thermodynamic activity product K_{ap}:

$$K_{ap} = (A^+)\,(B^-) \tag{10.6}$$

The concentration of a chemical species, not activity, is of interest in water treatment. Because dilute solutions are typically encountered, this parameter may be employed without introducing significant error into calculations. Hence, in this chapter all relationships will be written in terms of analytical concentration rather than activity. Following this convention, Equation 10.6 becomes

$$K_{sp} = [A^+]\,[B^-] \tag{10.7}$$

This is the classical solubility product expression for the dissolution of a slightly soluble compound where the brackets denote molar concentration. The equilibrium constant is called the *solubility product constant.* The more general form of the solubility product expression is derived from the dissolution reaction

$$A_xB_y(s) \rightleftharpoons xA^{y+} + yB^{x-} \tag{10.8}$$

and has the form

$$K_{sp} = [A^{y+}]^x[B^{x-}]^y \tag{10.9}$$

The value of the solubility product constant gives some indication of the solubility of a particular compound. For example, a compound that is highly insoluble will have a very small solubility product constant. Solubility product constants for solutions at or near room temperature are listed in Table 10.1.

Equation 10.9 applies to the equilibrium condition between ion and solid. If the actual concentrations of the ions in solution are such that the ion product $[A^{y+}]^x \cdot [B^{x-}]^y$ is less than the K_{sp} value, no precipitation will occur and any quantitative information that can be derived from Equation 10.9 will apply only where equilibrium conditions exist. Furthermore, if the actual concentrations of ions in solution are so great that the ion product is greater than the K_{sp} value, precipitation will occur (assuming nucleation occurs). Still, however, no quantitative information can be derived directly from Equation 10.9.

If an ion of a sparingly soluble salt is present in solution in a defined concentration, it can be precipitated by the other ion common to the salt, if the concentration of the second ion is increased to the point that the ion product exceeds the value of the solubility product constant. Such an influence is called the *common-ion effect.* Furthermore, precipitating two different compounds is possible if two different ions share a common third ion and the concentration of the third ion is increased so that the solubility product constants for both sparingly soluble salts are exceeded. This type of precipitation is normally possible only when the K_{sp} values of the two compounds do not differ significantly.

The common-ion effect is an example of LeChâtelier's principle, which states that if stress is applied to a system in equilibrium, the system will act to relieve the stress and restore equilibrium, but under a new set of equilibrium conditions. For example, if a salt containing the cation A (e.g., AC) is added to a saturated solution of AB, AB(s) would precipitate until the ion product $[A^+]\,[B^-]$ had a value equal to the solubility product constant. The new equilibrium concentration of A^+, however, would be greater than the old equilibrium concentration, while the new equilibrium concentration of B^- would be lower than the old equilibrium concentration. The follow-

TABLE 10.1 Solubility Product Constants for Solutions at or near Room Temperature

Substance	Formula	K_{sp}*
Aluminum hydroxide	$Al(OH)_3$	2×10^{-32}
Barium arsenate	$Ba_3(AsO_4)_2$	7.7×10^{-51}
Barium carbonate	$BaCO_3$	8.1×10^{-9}
Barium chromate	$BaCrO_4$	2.4×10^{-10}
Barium fluoride	BaF_2	1.7×10^{-6}
Barium iodate	$Ba(IO_3)_2 2H_2O$	1.5×10^{-9}
Barium oxalate	$BaC_2O_4H_2O$	2.3×10^{-8}
Barium sulfate	$BaSO_4$	1.08×10^{-10}
Beryllium hydroxide	$Be(OH)_2$	7×10^{-22}
Bismuth iodide	BiI_3	8.1×10^{-19}
Bismuth phosphate	$BiPO_4$	1.3×10^{-23}
Bismuth sulfide	Bi_2S_3	1×10^{-97}
Cadmium arsenate	$Cd_3(AsO_4)_2$	2.2×10^{-33}
Cadmium hydroxide	$Cd(OH)_2$	5.9×10^{-15}
Cadmium oxalate	$CdC_2O_4 3H_2O$	1.5×10^{-8}
Cadmium sulfide	CdS	7.8×10^{-27}
Calcium arsenate	$Ca_3(AsO_4)_2$	6.8×10^{-19}
Calcium carbonate	$CaCO_3$	8.7×10^{-9}
Calcium fluoride	CaF_2	4.0×10^{-11}
Calcium hydroxide	$Ca(OH)_2$	5.5×10^{-6}
Calcium iodate	$Ca(IO_3)_2 6H_2O$	6.4×10^{-7}
Calcium oxalate	$CaC_2O_4H_2O$	2.6×10^{-9}
Calcium phosphate	$Ca_3(PO_4)_2$	2.0×10^{-29}
Calcium sulfate	$CaSO_4$	1.9×10^{-4}
Cerium(III) hydroxide	$Ce(OH)_3$	2×10^{-20}
Cerium(III) iodate	$Ce(IO_3)_3$	3.2×10^{-10}
Cerium(III) oxalate	$Ce_2(C_2O_4)_3 9H_2O$	3×10^{-29}
Chromium(II) hydroxide	$Cr(OH)_2$	1.0×10^{-17}
Chromium(III) hydroxide	$Cr(OH)_3$	6×10^{-31}
Cobalt(II) hydroxide	$Co(OH)_2$	2×10^{-16}
Cobalt(III) hydroxide	$Co(OH)_3$	1×10^{-43}
Copper(II) arsenate	$Cu_3(AsO_4)_2$	7.6×10^{-76}
Copper(I) bromide	$CuBr$	5.2×10^{-9}
Copper(I) chloride	$CuCl$	1.2×10^{-6}
Copper(I) iodide	CuI	5.1×10^{-12}
Copper(II) iodate	$Cu(IO_3)_2$	7.4×10^{-8}
Copper(I) sulfide	Cu_2S	2×10^{-47}
Copper(II) sulfide	CuS	9×10^{-36}
Copper(I) thiocyanate	$CuSCN$	4.8×10^{-15}
Iron(III) arsenate	$FeAsO_4$	5.7×10^{-21}
Iron(II) carbonate	$FeCO_3$	3.5×10^{-11}
Iron(II) hydroxide	$Fe(OH)_2$	8×10^{-16}
Iron(III) hydroxide	$Fe(OH)_3$	4×10^{-38}
Lead arsenate	$Pb_3(AsO_4)_2$	4.1×10^{-36}
Lead bromide	$PbBr_2$	3.9×10^{-5}
Lead carbonate	$PbCO_3$	3.3×10^{-14}
Lead chloride	$PbCl_2$	1.6×10^{-5}
Lead chromate	$PbCrO_4$	1.8×10^{-14}
Lead fluoride	PbF_2	3.7×10^{-8}
Lead iodate	$Pb(IO_3)_2$	2.6×10^{-13}
Lead iodide	PbI_2	7.1×10^{-9}
Lead oxalate	PbC_2O_4	4.8×10^{-10}
Lead sulfate	$PbSO_4$	1.6×10^{-8}
Lead sulfide	PbS	8×10^{-28}
Magnesium ammonium phosphate	$MgNH_4PO_4$	2.5×10^{-13}
Magnesium arsenate	$Mg_3(AsO_4)_2$	2.1×10^{-20}
Magnesium carbonate	$MgCO_3 3H_2O$	1×10^{-5}

TABLE 10.1 Solubility Product Constants for Solutions at or near Room Temperature (*Continued*)

Substance	Formula	K_{sp}*
Magnesium fluoride	MgF_2	6.5×10^{-9}
Magnesium hydroxide	$Mg(OH)_2$	1.2×10^{-11}
Magnesium oxalate	$MgC_2O_42H_2O$	1×10^{-8}
Manganese(II) hydroxide	$Mn(OH)_2$	1.9×10^{-13}
Mercury(I) bromide	Hg_2Br_2	5.8×10^{-23}
Mercury(I) chloride	Hg_2Cl_2	1.3×10^{-18}
Mercury(I) iodide	Hg_2I_2	4.5×10^{-29}
Mercury(I) sulfate	Hg_2SO_4	7.4×10^{-7}
Mercury(II) sulfide	HgS	4×10^{-53}
Mercury(I) thiocyanate	$Hg_2(SCN)_2$	3.0×10^{-20}
Nickel arsenate	$Ni_3(AsO_4)_2$	3.1×10^{-26}
Nickel carbonate	$NiCO_3$	6.6×10^{-9}
Nickel hydroxide	$Ni(OH)_2$	6.5×10^{-18}
Nickel sulfide	NiS	3×10^{-19}
Silver arsenate	Ag_3AsO_4	1×10^{-22}
Silver bromate	$AgBrO_3$	5.77×10^{-5}
Silver bromide	$AgBr$	5.25×10^{-13}
Silver carbonate	Ag_2CO_3	8.1×10^{-12}
Silver chloride	$AgCl$	1.78×10^{-10}
Silver chromate	Ag_2CrO_4	2.45×10^{-12}
Silver cyanide	$Ag[Ag(CN)_2]$	5.0×10^{-12}
Silver iodate	$AgIO_3$	3.02×10^{-8}
Silver iodide	AgI	8.31×10^{-17}
Silver oxalate	$Ag_2C_2O_4$	3.5×10^{-11}
Silver oxide	Ag_2O	2.6×10^{-8}
Silver phosphate	Ag_3PO_4	1.3×10^{-20}
Silver sulfate	Ag_2SO_4	1.6×10^{-5}
Silver sulfide	Ag_2S	2×10^{-49}
Silver thiocyanate	$AgSCN$	1.00×10^{-12}
Strontium carbonate	$SrCO_3$	1.1×10^{-10}
Strontium chromate	$SrCrO_4$	3.6×10^{-5}
Strontium fluoride	SrF_2	2.8×10^{-9}
Strontium iodate	$Sr(IO_3)_2$	3.3×10^{-7}
Strontium oxalate	$SrC_2O_4H_2O$	1.6×10^{-7}
Strontium sulfate	$SrSO_4$	3.8×10^{-7}
Thallium(I) bromate	$TlBrO_3$	8.5×10^{-5}
Thallium(I) bromide	$TlBr$	3.4×10^{-6}
Thallium(I) chloride	$TlCl$	1.7×10^{-4}
Thallium(I) chromate	Tl_2CrO_4	9.8×10^{-13}
Thallium(I) iodate	$TlIO_3$	3.1×10^{-6}
Thallium(I) iodide	TlI	6.5×10^{-8}
Thallium(I) sulfide	Tl_2S	5×10^{-21}
Tin(II) sulfide	SnS	1×10^{-25}
Titanium(III) hydroxide	$Ti(OH)_3$	1×10^{-40}
Zinc arsenate	$Zn_3(AsO_4)_2$	1.3×10^{-28}
Zinc carbonate	$ZnCO_3$	1.4×10^{-11}
Zinc ferrocyanide	$Zn_2Fe(CN)_6$	4.1×10^{-16}
Zinc hydroxide	$Zn(OH)_2$	1.2×10^{-17}
Zinc oxalate	$ZnC_2O_42H_2O$	2.8×10^{-8}
Zinc phosphate	$Zn_3(PO_4)_2$	9.1×10^{-33}
Zinc sulfide	ZnS	1×10^{-21}

* The solubility of many metals is altered by carbonate complexation. Solubility predictions without consideration for complexation can be highly inaccurate.

Source: Robert B. Fischer and Dennis G. Peters, *Chemical Equilibrium.*

ing example problem is presented to illustrate calculations involving the common-ion effect.

EXAMPLE PROBLEM 10.1 Determine the residual magnesium concentration that exists in a saturated magnesium hydroxide solution if enough sodium hydroxide has been added to the solution to increase the equilibrium pH to 11.0.

SOLUTION

1. Write the appropriate chemical reaction.

$$Mg(OH)_2(s) \rightleftharpoons Mg^{2+} + 2OH^-$$

From Table 10.1 the solubility product constant for this reaction is 1.2×10^{-11}.

2. Determine the hydroxide ion concentration.

$$K_w = [H^+]\,[OH^-] = 10^{-14} \text{ at } 25°\text{ C}$$

Because

$$[H^+] = 10^{-pH} = 10^{-11} \text{ mol/L}$$

we know that

$$[OH^-] = \frac{10^{-14}}{10^{-11}} = 10^{-3} \text{ mol/L}$$

3. Establish the solubility product constant expression and solve for the magnesium ion concentration

$$K_{sp} = [Mg^{2+}]\,[OH^-]^2$$

$$[Mg^{2+}] = \frac{1.2 \times 10^{-11}}{(10^{-3})^2}$$

$$= 1.2 \times 10^{-5} \text{ mol/L or } 0.29 \text{ mg/L}$$

Since hardness ion concentrations are frequently expressed as $CaCO_3$, multiply the concentration by the ratio of the equivalent weights.

$$0.29 \times \frac{50}{12.2} = 1.2 \text{ mg/L as } CaCO_3$$

Metal Removal by Chemical Precipitation

Consider the following equilibrium reaction involving metal solubility:

$$MA_x(s) \rightleftharpoons M^{x+} + xA^- \tag{10.10}$$

$$K_{sp} = [Me^{x+}]\,[A^-]^x \tag{10.11}$$

Equation 10.11, the solubility product expression for Equation 10.10, indicates that the equilibrium concentration (in precipitation processes this is referred to as the *residual concentration*) of the metal in solution is solely dependent upon the concentration of A^-. When A^- is the hydroxide ion the residual metal concentration is a function of pH such that

$$\log\,[M^{x+}] = \log K_{sp} - x \log K_w - X\text{pH} \tag{10.12}$$

This relationship is shown as line A in Figure 10.1, where $K_{sp} = 10^{-10}$, $K_w = 10^{-14}$, and $X = 2$ (assumed values). The solubility of most metal hydroxides is not accurately described by Equation 10.12, however, because they exist in solution as a series of complexes formed with hydroxide and other ions. Each complex is in equilibrium with the solid phase and their sum gives the total residual metal concentration. For the case of only hydroxide species and a divalent metal, the total residual metal concentration is given by Equation 10.13.

$$M_{T1} = M^{2+} + M(OH)^{+} + M(OH)_2^0 + M(OH)_3^- + \cdots \tag{10.13}$$

For this situation, the total residual metal concentration is a complex function of the pH as illustrated by line B in Figure 10.1. Line B shows that the lowest residual metal concentration will occur at some optimum pH value and the residual concentration will increase when the pH is either lowered or raised from this optimum value.

Nilsson (1971) computed the logarithm of the total residual metal concentration as a function of pH for several pure metal hydroxides (see Figure 10.2). Bold lines show those areas where the total residual metal concentration is greater than 1 mg/L. If the rise in pH occurs by adding NaOH, the total residual Cr(III) and total residual Zn(II) will rise again when the pH values rise above approximately 8 and 9, respectively, because of an increase in the concentration of the negatively charged hydroxide complexes. If the rise in pH occurs by adding lime, then a rise in the residual concentration does not occur, because the solubilities of calcium zincate and calcium chromite are relatively low.

Numeric estimations on metal removal by precipitation as metal hydroxide should always be treated carefully because oversimplification of theoretical solubility data can lead to error of several orders of magnitude. Many possible reasons exist for such

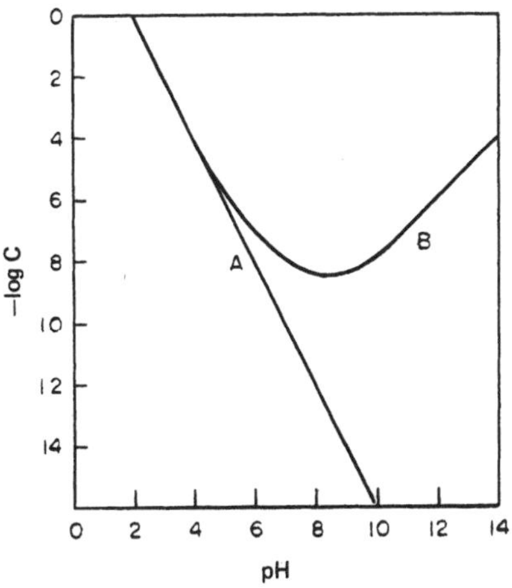

FIGURE 10.1 Theoretical solubility of hypothetical metal hydroxide, with and without complex formation. *A* = without complex formation, *B* = with complex formation. (*Source:* J. W. Patterson and R. A. Minear, "Physical-Chemical Methods of Heavy Metal Removal," in P. A. Kenkel (ed.), *Heavy Metals in the Aquatic Environment,* Pergamon Press, Oxford, 1975.)

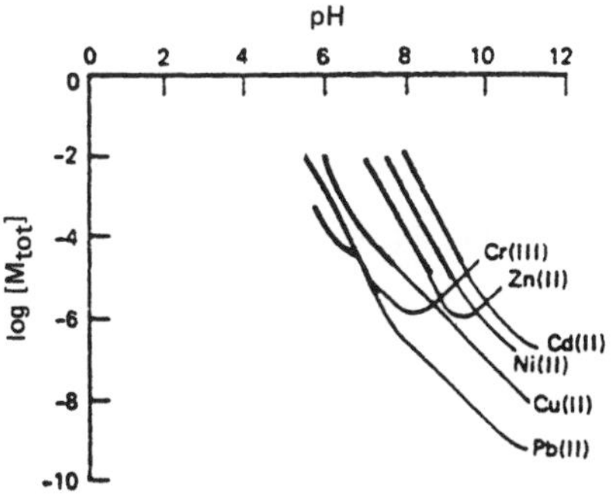

FIGURE 10.2 The solubility of pure metal hydroxides as a function of pH. Heavy portions of lines show where concentrations are greater than 1 mg/L. *Note:* If NaOH is used for pH adjustment, Cr(III) and Zn(II) will exhibit amphoteric characteristics. (*Source:* Reprinted with permission from *Water Research,* Vol. 5, R. Nilsson, "Removal of Metals by Chemical Treatment of Municipal Wastewater." Copyright 1971. Pergamon Press.)

discrepancies. For example, changes in the ionic strength of a water can result in significant differences between calculated and observed residual metal concentrations when molar concentrations rather than activities are used in the computations (high ionic strength will result in a higher-than-predicted solubility). The presence of organic and inorganic species other than hydroxide, which are capable of forming soluble species with metal ions, will increase the total residual metal concentration. Two inorganic complexing agents that result in very high residual metal concentrations are cyanide and ammonia. Small amounts of carbonate will significantly change the solubility chemistry of some metal hydroxide precipitation systems. As a result, deviations between theory and practice should be expected because precipitating metal hydroxides in practice is virtually impossible without at least some carbonate present.

Temperature variations can explain deviations between calculated and observed values if actual process temperatures are significantly different from the value at which the equilibrium constant was evaluated. Kinetics may also be an important consideration because under process conditions the reaction between the soluble and solid species may be too slow to allow equilibrium to become established within the hydraulic retention time provided. Furthermore, many solids may initially precipitate in an amorphous form but convert to a more insoluble and more stable crystalline structure after some time period has passed.

Formation of precipitates other than the hydroxide may result in a total residual metal concentration lower than the calculated value. For example, the solubility of cadmium carbonate is approximately two orders of magnitude less than that of the hydroxide. Effects of coprecipitation on flocculating agents added to aid in settling the precipitate may also play a significant role in reducing the residual metal concentration. Nilsson (1971) found that when precipitation with aluminum sulfate was employed, the actual total residual concentrations of zinc, cadmium, and nickel were much lower than the calculated values because the metals were coprecipitated with aluminum hydroxide.

In summary, the solubility behavior of most slightly soluble salts is very complex because of competing acid-base equilibria, complex ion formation, and hydrolysis. Still, many precipitation processes in water treatment can be adequately described when these reactions are ignored. This will be the approach taken in this chapter. A more detailed discussion on solubility equilibria may be found in Stumm and Morgan (1981); Snoeyink and Jenkins (1980); and Benefield, Judkins and Weand (1982).

Carbonic Acid Equilibria

The pH of most natural waters is generally assumed to be controlled by the carbonic acid system. The applicable equilibrium reactions are

$$CO_2 + H_2O \rightleftharpoons (H_2CO_3) \rightleftharpoons H^+ + HCO_3^- \tag{10.14}$$

$$HCO_3^- \rightleftharpoons H^+ + CO_3^{2-} \tag{10.15}$$

Because only a small fraction of the total CO_2 dissolved in water is hydrolyzed to H_2CO_3, summing the concentrations of dissolved CO_2 and H_2CO_3 to define a new concentration term, $H_2CO_3^*$, is convenient. Equilibrium constant expressions for Equations 10.14 and 10.15 have the form

$$K_1 = \frac{[H^\pm]\,[HCO_3^-]}{[H_2CO_3{*}]} \tag{10.16}$$

$$K_2 = \frac{[H^\pm]\,[CO_3^{2-}]}{[HCO_3^-]} \tag{10.17}$$

where K_1 and K_2 represent the equilibrium constants for the first and second dissociation of carbonic acid, respectively. Rossum and Merrill (1983) have presented the following equations to describe the relationships between temperature and K_1 and K_2:

$$K_1 = 10^{14.8435 - 3404.71/T - 0.032786T} \tag{10.18}$$

$$K_2 = 10^{6.498 - 2909.39/T - 0.02379T} \tag{10.19}$$

where T represents the solution temperature in degrees Kelvin (i.e., °C + 273).

The total carbonic species concentration in solution is usually represented by C_T and defined in terms of a mass balance expression.

$$C_T = [H_2CO_3^*] + [HCO_3^-] + [CO_3^{2-}] \tag{10.20}$$

The distribution of the various carbonic species can be established in terms of the total carbonic species concentration by defining a set of ionization fractions, α, where

$$\alpha_0 = \frac{[H_2CO_3^*]}{C_T} \tag{10.21}$$

$$\alpha_1 = \frac{[HCO_3^-]}{C_T} \tag{10.22}$$

$$\alpha_2 = \frac{[CO_3^{2-}]}{C_T} \tag{10.23}$$

Through a series of algebraic manipulations (Snoeyink and Jenkins, 1980)

$$\alpha_0 = \frac{1}{1 + K_1/[H^+] + K_1K_2/[H^+]^2} \tag{10.24}$$

$$\alpha_1 = \frac{1}{[H^+]/K_1 + 1 + K_2/[H^+]} \tag{10.25}$$

$$\alpha_2 = \frac{1}{[H^+]^2/(K_1K_2) + [H^+]/K_2 + 1} \tag{10.26}$$

The effect of pH on the species distribution for the carbonic acid system is shown in Figure 10.3. Because the pH of most natural waters is in the neutral range, the alkalinity (assuming that alkalinity results mainly from the carbonic acid system) is in the form of bicarbonate alkalinity.

Calcium Carbonate and Magnesium Hydroxide Equilibria

The solubility equilibrium for $CaCO_3$ is described by Equation 10.27:

$$CaCO_3(s) \rightleftharpoons Ca^{2+} + CO_3^{2-} \tag{10.27}$$

The addition of $Ca(OH)_2$ to a water increases the hydroxyl ion concentration and elevates the pH that, according to Figure 10.3, shifts the equilibrium of the carbonic acid system in favor of the carbonate ion, CO_3^{2-}. This increases the concentration of the CO_3^{2-} ion and, according to LeChâtelier's principle, shifts the equilibrium described by Equation 10.27 to the left (common-ion effect). Such a response results

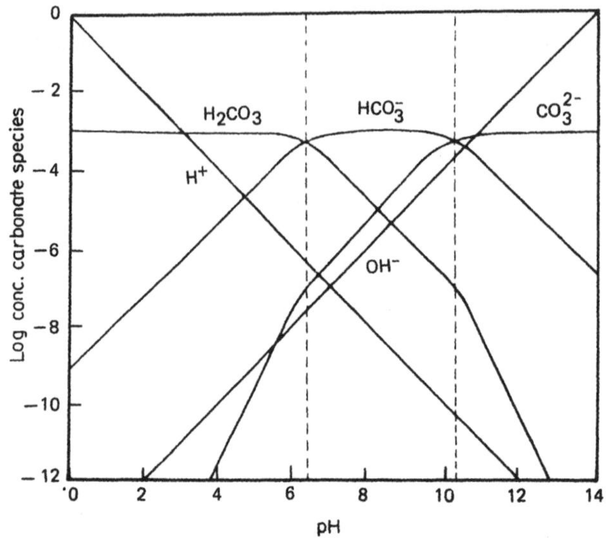

FIGURE 10.3 Concentration distribution diagram for carbonic acid. (*Source: Handbook of Water Resources and Pollution Control.* H. W. Gehm and J. I. Bregman, eds. Van Nostrand Reinhold Co., New York, 1976.)

in the precipitation of $CaCO_3(s)$ and a corresponding decrease in the soluble calcium concentration.

The solubility equilibrium for $Mg(OH)_2$ is described by

$$Mg(OH)_2(s) \rightleftharpoons Mg^{2+} + 2OH^- \tag{10.28}$$

According to LeChâtelier's principle, the addition of hydroxyl ions shifts the equilibrium described by Equation 10.28 to the left (common-ion effect), resulting in the precipitation of $Mg(OH)_2$ and a corresponding decrease in the soluble magnesium concentration.

The solubility product expressions for Equations 10.27 and 10.28 have the forms

$$K_{sp} = [Ca^{2+}]\,[CO_3^{2-}] \tag{10.29}$$

$$K_{sp} = [Mg^{2+}]\,[OH^-] \tag{10.30}$$

The effects of temperature on the solubility product constants for calcium carbonate and magnesium hydroxide is given by the empirical equations (Rossum and Merrill, 1983; Faust and McWhorter, 1976; Lowenthal and Marais, 1976)

$$\text{Calcium carbonate: } K_{sp} = 10^{[13.870 - 3059/T - 0.04035T]} \tag{10.31}$$

$$\text{Magnesium hydroxide: } K_{sp} = 10^{[-0.0175t - 9.97]} \tag{10.32}$$

where T and t are the solution temperature in °K and °C, respectively. The K_{sp} for calcium carbonate presented in Equation 10.31 is based on the classical 1942 constant of Larson and Buswell. A modern constant has been introduced by Plummer and Busenberg (1982); see also APHA, AWWA, and WEF (1989).

Complex ion formation reactions that contribute to the total soluble calcium and magnesium concentrations are listed in Table 10.2. These reactions can be used to

TABLE 10.2 Complex Ion Formation Reactions of Calcium and Magnesium Ions*

Reaction	Equilibrium constant	Temperature correction T, K
1. Calcium		
a. $Ca^{2+} + OH^- \rightleftharpoons CaOH^+$	$K_3 = [CaOH^+]/[Ca^{2+}][OH^-]$	$pK_3 = -1.299 - 260.388\ 1/T - 1/298.15$
b. $Ca^{2+} + HCO_3^- \rightleftharpoons CaHCO_3^+$	$K_4 = [CaHCO_3^+]/[Ca^{2+}][HCO_3^-]$	$pK_4 = 2.95 - 0.0133T$
c. $Ca^{2+} + CO_3^{2-} \rightleftharpoons CaCO_3^0$	$K_5 = [CaCO_3^0]/[Ca^{2+}][CO_3^{2-}]$	$pK_5 = 27.393 - 4114/T - 0.05617T$
d. $Ca^{2+} + SO_4^{2-} \rightleftharpoons CaSO_4^0$	$K_6 = [CaSO_4^0]/[Ca^{2+}][SO_4^{2-}]$	$pK_6 = 691.70/T$
2. Magnesium		
a. $Mg^{2+} + OH^- \rightleftharpoons MgOH^+$	$K_7 = [MgOH^+]/[Mg^{2+}][OH^-]$	$pK_7 = -0.684 - 0.0051T$
b. $Mg^{2+} + HCO_3^- \rightleftharpoons MgHCO_3^+$	$K_8 = [MgHCO_3^+]/[Mg^{2+}][HCO_3^-]$	$pK_8 = -2.319 + 0.011056T - (2.29812 \times 10^{-5})T$
c. $Mg^{2+} + CO_3^{2-} \rightleftharpoons MgCO_3^0$	$K_9 = [MgCO_3^0]/[Mg^{2+}][CO_3^{2-}]$	$pK_9 = -0.991 - 0.00667T$
d. $Mg^{2+} + SO_4^{2-} \rightleftharpoons MgSO_4^0$	$K_{10} = [MgSO_4^0]/[Mg^{2+}][SO_4^{2-}]$	$pK_{10} = 707.07/T$

* Temperature corrections are from Truesdell and Jones (1973).

determine the effect of complex ion formation on calcium carbonate and magnesium hydroxide solubility by writing mass balance relationships for total residual calcium and total residual magnesium that consider these species. Such relationships have the form

$$[Ca]_1 = [Ca^{2+}] + [CaOH^+] + [CaHCO_3^+] + [CaCO_3^0] + [CaSO_4^0] \qquad (10.33)$$

that reduces to

$$[Ca]_T = \frac{K_{sp}}{\alpha_2 C_T}\left(1 + \frac{K_w K_3}{[H^+]} + K_4\alpha_1 C_T + K_5\alpha_2 C_T + K_6[SO_4^{2-}]\right) \qquad (10.33a)$$

and

$$[Mg]_T = [Mg^{2+}] + [MgOH^+] + [MgHCO_3^+] + [MgCO_3^0] + [MgSO_4^0] \qquad (10.34)$$

which reduces to

$$[Mg]_1 = \frac{K_{sp}[H^+]^2}{(K_w)^2}\left(1 + \frac{K_2 K_7}{[H^+]} + K_8\alpha_1 C_T + K_g\alpha_2 C_T K_{10}[SO_4^{2-}]\right) \qquad (10.34a)$$

where

$$K_w = 10^{[6.0486 - 4471.33/T - 0.017053(T)]} \qquad (10.35)$$

Figures 10.4 and 10.5 illustrate the effect of complex ion formation on calcium carbonate and magnesium hydroxide, respectively. For convenience, a solution temperature of 25°C and a sulfate ion concentration of zero was assumed. The results show that the equilibrium carbonic species concentration has virtually no effect on the total residual magnesium concentration (Figure 10.5) but significantly affects the total residual calcium concentration (Figure 10.4).

Cadena et al. (1974) indicate that at 25°C the $CaCO_3^0$ species accounts for 13.5 mg/L of soluble calcium expressed as $CaCO_3$. Their work is based in part on the following relationship for the variation in the dissociation constant for $CaCO_3^0$ with temperature:

$$\log K = \frac{2280}{T} - 12.10 \qquad (10.36)$$

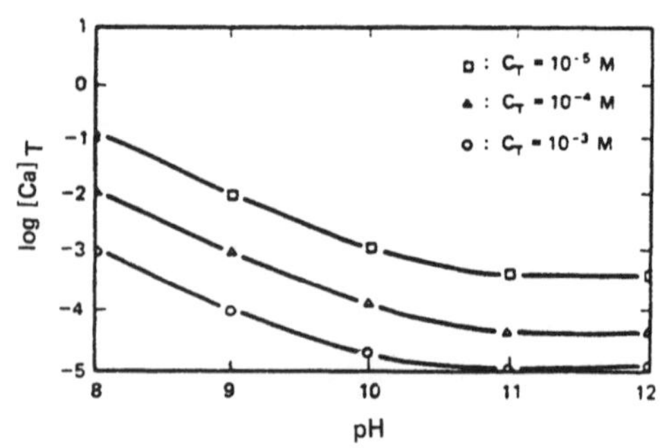

FIGURE 10.4 Relationship between total soluble calcium, pH, and equilibrium total carbonic species concentration. (*Source:* L. D. Benefield, J. F. Judkins, and B. L. Weand, *Process Chemistry for Water and Wastewater.* Copyright 1982, pp. 124, 292. Reprinted by permission of Prentice-Hall, Inc., Englewood Cliffs, New Jersey.)

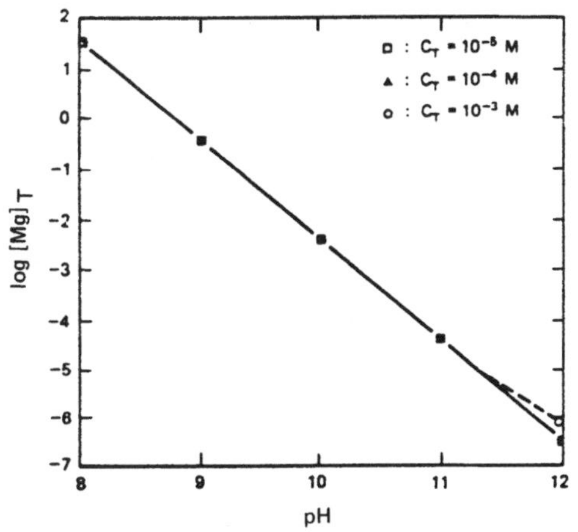

FIGURE 10.5 Relationship between total soluble magnesium, pH, and final equilibrium total carbonic species concentration. (*Source:* L. D. Benefield, J. F. Judkins, and B. L. Weand, *Process Chemistry for Water and Wastewater.* Copyright 1982, pp. 124, 292. Reprinted by permission of Prentice-Hall, Inc., Englewood Cliffs, New Jersey.)

where T represents the temperature in degrees Kelvin. The concentration of $CaCO_3^0$ may be estimated by dividing the solubility product expression for calcium carbonate by the equilibrium constant expression for $CaCO_3^0$. This gives

$$[CaCO_3^0] = \frac{K_{sp}}{K_{CaCO_3^0}} \tag{10.37}$$

A graphical representation of the variation in the $CaCO_3^0$ concentration with temperature is presented in Figure 10.6. Trussell et al. (1977) do not consider the $CaCO_3^0$ species to be important. These workers indicate that the concentration of $CaCO_3^0$ in a saturated solution of calcium carbonate will be about 0.17 mg/L as $CaCO_3$ rather than 13.5 mg/L. Experimental evidence by Pisigan and Singley (1985) supports this. They found that the concentration of $CaCO_3^0$ is insignificant in fresh water in the pH range of 6.20 to 9.20.

For a detailed explanation of the calcium carbonate system and the ion pairs $CaHCO_3^+$ and $CaCO_3^0$, the reader is directed to the rigorous work of Plummer and Busenberg (1982).

WATER SOFTENING BY CHEMICAL PRECIPITATION

Hardness in natural waters is caused by the presence of any polyvalent metallic cation. Principle cations causing hardness in water and the major anions associated with them are presented in Table 10.3. Because the most prevalent of these species are the divalent cations calcium and magnesium, *total hardness* is typically defined as

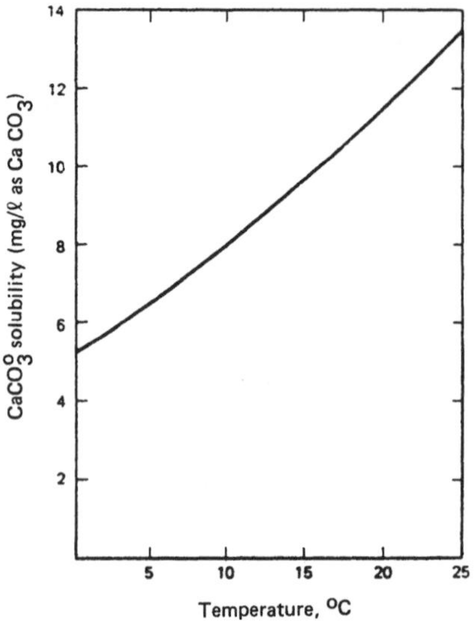

FIGURE 10.6 Variation in solubility of $CaCO_3$ complex ion with temperature. (*Source:* D. T. Merrill, "Chemical Conditioning for Water Softening and Corrosion Control," *Proc. 5th Envir. Engr. Conf.,* Montana State University, June 1976.)

the sum of the concentration of these two elements and is usually expressed in terms of mg/L as $CaCO_3$. Within the United States, significant regional variation in the total hardness of both surface and groundwaters occurs. Approximate hardness values of municipal water supplies are depicted in Figure 10.7.

The hardness of water is that property that causes it to form curds (Ca or Mg oleate) when soap is used with it. Some waters are very hard, and the consumption of soap by these waters is commensurately high. Other adverse effects such as bathtub rings, deterioration of fabrics, and, in some cases, stains, also occur. Many of these problems have been alleviated by the development of detergents and soaps that do not react with hardness.

Public acceptance of hardness varies from community to community, consumer sensitivity being related to the degree to which the consumer is accustomed. Because of this variation in consumer acceptance, finished water hardness produced by different utility softening plants will range from 50 mg/L to 150 mg/L as $CaCO_3$. According to the hardness classification scale presented by Sawyer and McCarty (1967; see Table 10.4), this hardness range covers the scale from soft water to hard water.

Hardness is classified in two ways. These classes are (with respect to the metallic ions and with respect to the anions associated with the metallic ions):

1. *Total hardness:* Total hardness represents the sum of multivalent metallic cations that are normally considered to be only calcium and magnesium. Generally, chemical analyses are performed to determine the total hardness and calcium hardness present in the water. Magnesium hardness is then computed as the difference between total hardness and calcium hardness.

TABLE 10.3 Principal Cations Causing Hardness in Water and the Major Associated Anions

Principal cations causing hardness	Anions
Ca^{2+}	HCO_3^-
Mg^{2+}	SO_4^{2-}
Sr^{2+}	Cl^-
Fe^{2+}	NO_3^-
Mn^{2+}	SiO_3^{2-}

Source: Sawyer and McCarty, 1967.

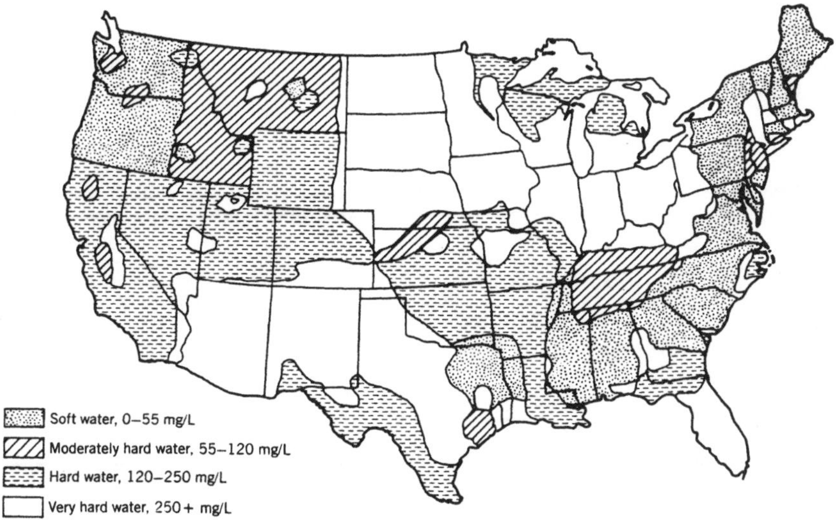

FIGURE 10.7 Distribution of hard water in the United States. The areas shown define approximate hardness values for municipal water supplies. (*Source:* Ciaccio, L., ed. *Water and Water Pollution Handbook*. Marcel Dekker, Inc., New York, 1971.)

2. *Carbonate and noncarbonate hardness:* Carbonate hardness is caused by cations from the dissolution of calcium or magnesium carbonate and bicarbonate in the water. Carbonate hardness is hardness that is chemically equivalent to the alkalinity where most of the alkalinity in natural waters is caused by the bicarbonate and carbonate ions. Noncarbonate hardness is caused by cations from calcium and magnesium compounds of sulfate, chloride, or silicate that are dissolved in the water. Noncarbonate hardness is equal to the total hardness minus the carbonate hardness. Thus, when the total hardness exceeds the carbonate and bicarbonate alkalinity, the hardness equivalent to the alkalinity is carbonate hardness and the amount in excess of carbonate hardness is noncarbonate hardness. When the total hardness is equal to or less than the carbonate and bicarbonate alkalinity, then the total hardness is equivalent to the carbonate hardness and the noncarbonate hardness is zero. Example Problem 2 illustrates the carbonate hardness and noncarbonate hardness classification.

TABLE 10.4 Hardness Classification Scale

Hardness range, mg/L as $CaCO_3$	Hardness description
0–75	Soft
75–150	Moderately hard
150–300	Hard
>300	Very hard

Source: Sawyer and McCarty, 1967.

EXAMPLE PROBLEM 10.2 A groundwater has the following analysis: calcium 75 mg/L, magnesium 40 mg/L, sodium 10 mg/L, bicarbonate 300 mg/L, chloride 10 mg/L, and sulfate 109 mg/L. Compute the total hardness, carbonate hardness, and noncarbonate hardness all expressed as mg/L $CaCO_3$.

SOLUTION

1. Construct a computation table, and convert all concentrations to mg/L $CaCO_3$.
   a. The species concentration in meq/L is calculated from the relationship

$$[\text{meq/L of species}] = \frac{\text{mg/L of species}}{\text{equivalent weight of species}}$$

   b. The species concentration expressed as mg/L $CaCO_3$ is computed from the relationship

$$[\text{mg/L } CaCO_3] = \text{mg/L of species } (50/\text{equivalent weight of species})$$

Chemical specie	Concentration, mg/L	Equivalent weight	Concentration, meq/L	Concentration mg/L $CaCO_3$
Ca^{2+}	75	20.0	3.7	187
Mg^{2+}	40	12.2	3.3	164
Na^+	10	23.0	0.4	22
			7.4	373
HCO_3^-	300	61.0	4.9	246
Cl^-	10	35.5	0.3	14
SO_4^{2-}	109	48.0	2.2	113
			7.4	373

2. Draw a bar diagram of the raw water indicating the relative proportions of the chemical species important to the softening process. Cations are placed above the anions on the diagram.

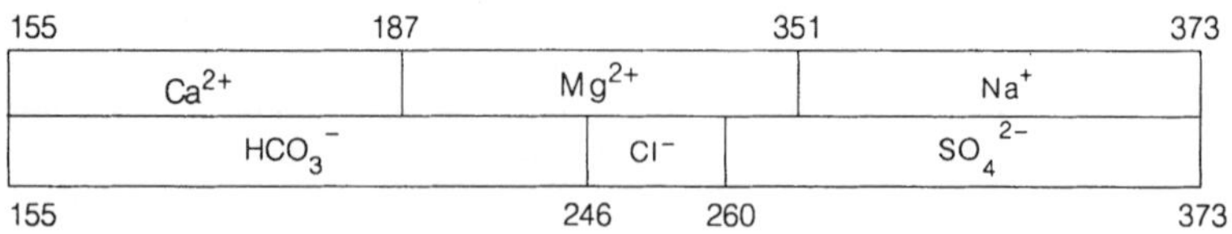

3. Calculate the hardness distribution for this water.
   Total hardness = 187 + 164 = 351 mg/L as $CaCO_3$
   Alkalinity = Bicarbonate alkalinity = 246 mg/L as $CaCO_3$
   Carbonate hardness = Alkalinity = 246 mg/L as $CaCO_3$
   Noncarbonate hardness = 351 – 246 = 105 mg/L as $CaCO_3$

Process Chemistry

During precipitation softening, calcium is removed from water in the form of $CaCO_3(s)$ precipitate while magnesium is removed as $Mg(OH)_2(s)$ precipitate. The concentrations of the various carbonic species and the system pH play important roles in the precipitation of these two solids.

Carbonate hardness can be removed by adding hydroxide ions and elevating the solution pH so that the bicarbonate ions are converted to the carbonate form (pH above 10). Before the solution pH can be changed significantly, however, the free carbon dioxide or carbonic acid must be neutralized. The increase in the carbonate concentration from the conversion of bicarbonate to carbonate causes the calcium and carbonate ion product ($[Ca^{2+}][CO_3^{2-}]$) to exceed the solubility product constant for $CaCO_3(s)$, and precipitation occurs. The result is that the concentration of calcium ions, originally treated as if they were associated with the bicarbonate anions, is reduced to a low value. The remaining calcium (noncarbonate hardness), however, is not removed by a simple pH adjustment. Rather, carbonate, usually sodium carbonate (soda ash), from an external source must be added to precipitate this calcium. Carbonate and noncarbonate magnesium hardness are removed by increasing the hydroxide ion concentration until the magnesium and hydroxide ion product ($[Mg^{2+}][OH^-]^2$) exceeds the solubility product constant for $Mg(OH)_2(s)$ and precipitation occurs.

In the lime-soda ash softening process, lime is added to provide the hydroxide ions required to elevate the pH while sodium carbonate is added to provide an external source of carbonate ions. The least expensive form of lime is quicklime (CaO), which must be hydrated or slaked to $Ca(OH)_2$ before application. Reactions of the lime-soda ash softening process are:

$$H_2CO_3 + Ca(OH)_2 \rightarrow CaCO_3(s) + 2H_2O \tag{10.38}$$

$$Ca^{2+} + 2HCO_3^- + Ca(OH)_2 \rightarrow 2CaCO_3(s) + 2H_2O \tag{10.39}$$

$$Ca^{2+} + \begin{bmatrix} SO_4^{2-} \\ 2Cl^- \end{bmatrix} + Na_2CO_3 \rightarrow CaCO_3(s) + 2Na^+ + \begin{bmatrix} SO_4^{2-} \\ 2Cl^- \end{bmatrix} \tag{10.40}$$

$$Mg^{2+} + 2HCO_3^- + 2Ca(OH)_2 \rightarrow 2CaCO_3(s) + Mg(OH)_2(s) + 2H_2O \tag{10.41}$$

$$Mg^{2+} + \begin{bmatrix} SO_4^{2-} \\ 2Cl^- \end{bmatrix} + Ca(OH)_2 \rightarrow Mg(OH)_2(s) + Ca^+ + \begin{bmatrix} SO_4^{2-} \\ 2Cl^- \end{bmatrix} \tag{10.42}$$

$$Ca^{2+} + \begin{bmatrix} SO_4^{2-} \\ 2Cl^- \end{bmatrix} + Na_2CO_3 \rightarrow CaCO_3(s) + 2Na^+ + \begin{bmatrix} SO_4^{2-} \\ 2Cl^- \end{bmatrix} \tag{10.43}$$

Equation 10.38 represents the neutralization reaction between free carbon dioxide or carbonic acid and lime that must be satisfied before the pH can be elevated significantly. Although no net change in water hardness occurs as a result of Equation 10.38, this reaction must be considered because a lime demand is created. If both carbonic acid and lime are expressed in terms of calcium carbonate, stoichiometric coefficient ratios suggest that for each mg/L of carbonic acid (expressed as $CaCO_3$) present, 1 mg/L of lime (expressed as $CaCO_3$) will be required for neutralization.

The removal of calcium carbonate hardness is reflected in Equation 10.39. This reaction shows that for each molecule of calcium bicarbonate present, two carbonate ions can be formed by elevating the pH. One of the carbonate ions can be assumed to react with one of the calcium ions originally present as calcium bicarbonate, while the other carbonate ion can be assumed to react with the calcium ion released from the lime molecule added to elevate the pH. In both cases calcium carbonate will precipitate. If both the calcium bicarbonate and the lime are expressed in terms of $CaCO_3$, stoichiometric coefficient ratios show that for each mg/L of calcium bicarbonate (calcium carbonate hardness) present, 1 mg/L of lime (expressed as $CaCO_3$) will be required for its removal.

Equation 10.40 represents the removal of calcium noncarbonate hardness. If the calcium noncarbonate hardness is expressed in terms of $CaCO_3$, stoichiometric coefficient ratios suggest that for each mg/L of calcium noncarbonate hardness present, 1 mg/L of sodium carbonate (expressed as $CaCO_3$) will be required for its removal.

Equation 10.41 is somewhat similar to Equation 10.39, in that it represents the removal of carbonate hardness, except in this case it is magnesium carbonate hardness. By elevating the pH, two carbonate ions can be formed from each magnesium bicarbonate molecule. Because no calcium is considered to be present in this reaction, enough calcium ion must be added in the form of lime to precipitate the carbonate ion as calcium carbonate before the hydroxide ion concentration can be increased to the level required for magnesium removal. The magnesium is precipitated as magnesium hydroxide. If magnesium bicarbonate and lime are expressed in terms of $CaCO_3$, stoichiometric coefficient ratios state that for each mg/L of magnesium carbonate hardness present, 2 mg/L of lime (expressed as $CaCO_3$) will be required for its removal.

Equation 10.42 represents the removal of magnesium noncarbonate hardness. If the magnesium noncarbonate hardness and lime are expressed in terms of $CaCO_3$, stoichiometric coefficient ratios state that for each mg/L of magnesium noncarbonate hardness present, 1 mg/L of lime (expressed as $CaCO_3$) will be required for its removal. In this reaction, however, note that no net change in the hardness level occurs because for every magnesium ion removed a calcium ion is added. Thus, to complete the hardness removal process, sodium carbonate must be added to precipitate this calcium. This is illustrated in Equation 10.43, which is identical to Equation 10.40.

Based on Equations 10.39 to 10.43, the chemical requirements for lime-soda ash softening can be summarized as follows if all constituents are expressed as equivalent $CaCO_3$: 1 mg/L of lime as $CaCO_3$ will be required for each mg/L of carbonic acid (expressed as $CaCO_3$) present; 1 mg/L of lime as $CaCO_3$ will be required for each mg/L of calcium carbonate hardness present; 1 mg/L of soda ash as $CaCO_3$ will be required for each mg/L of calcium noncarbonate hardness present; 2 mg/L of lime as $CaCO_3$ will be required for each mg/L of magnesium carbonate hardness present; 1 mg/L of lime as $CaCO_3$ and 1 mg/L of soda ash as $CaCO_3$ will be required for each mg/L of magnesium noncarbonate hardness present. To achieve removal of magnesium in the form of $Mg(OH)_2(s)$, the solution pH must be raised to a value greater than 10.5 [see Figure 10.5, which shows the solubility of $Mg(OH)_2$ as a function of pH]. This will require a lime dosage greater than the stoichiometric requirement.

Chemical Dose Calculations for Lime-Soda Ash Softening

Calculations Based on Stoichiometry. The characteristics of the source water will establish the type of treatment process necessary for softening. Four process types are listed by Humenick (1977). Each process name is derived from the type and amount of chemical added. These processes are:

1. *Single-stage lime process:* Source water has high calcium, low magnesium carbonate hardness (less than 40 mg/L as $CaCO_3$). No noncarbonate hardness.
2. *Excess lime process:* Source water has high calcium, high magnesium carbonate hardness. No noncarbonate hardness. May be a one- or two-stage process.
3. *Single-stage lime-soda ash process:* Source water has high calcium, low magnesium carbonate hardness (less than 40 mg/L as $CaCO_3$). Some calcium noncarbonate hardness.

4. *Excess lime-soda ash process:* Source water has high calcium, high magnesium carbonate hardness and some noncarbonate hardness. It may be a one- or two-stage process.

Example problems 3 through 6 illustrate chemical dose calculations and hardness distribution determinations for each type of process. (*Hoover's Water Supply and Treatment,* revised in 1995 by Nicholas G. Pizzi and the National Lime Association, is also an excellent reference for additional examples. Chapter 8, "Removal of Hardness and Scale-Forming Substances," from the 1998 *Chemistry of Water Treatment* by Faust and Aly should also be consulted if additional information is required.

EXAMPLE PROBLEM 10.3 ***Straight Lime Softening***

A groundwater was analyzed and found to have the following composition (all concentrations are as $CaCO_3$):

$pH = 7.0$
$Ca^{2+} = 210$ mg/L
$Mg^{2+} = 15$ mg/L
Alk. = 260 mg/L
Temp. = 10°C

Estimate the lime dose required to soften the water.

SOLUTION

1. Estimate the carbonic acid concentration.
   a. Determine the bicarbonate concentration in mol/L by assuming that at pH = 7.0, all alkalinity is in the bicarbonate form.

$$[HCO_3^-] = 260\ [61/50]\ [1/1{,}000]\ [1/61]$$
$$= 5.2 \times 10^{-3}\ \text{mol/L}$$

   b. Compute the dissociation constants for carbonic acid at 10°C using Equations 10.18 and 10.19.

$$K_1 = 10^{14.8435 - 3404.71/283 - 0.032786(283)}$$
$$= 3.47 \times 10^{-7}$$
$$K_2 = 10^{6.498 - 2909.39/283 - 0.02379(283)}$$
$$= 3.1 \times 10^{-11}$$

   c. Compute α_1 from Equation 10.25.

$$\alpha_1 = \frac{1}{1.0 \times 10^{-7}/3.47 \times 10^{-7} + 1 + 3.1 \times 10^{-11}/1.0 \times 10^{-7}}$$

   d. Determine the total carbonic species concentration from Equation 10.22.

$$C_T = 5.2 \times 10^{-3}/0.77 = 6.75 \times 10^{-3}\ \text{mol/L}$$

   e. Compute the carbonic acid concentration from a rearrangement of Eq. 10.20 while neglecting the carbonate term, because it will be insignificant at a pH of 7.0.

$$[H_2CO_3^*] = C_T - [HCO_3^-] = 6.75 \times 10^{-3} - 5.2 \times 10^{-3}$$
$$= 1.55 \times 10^{-3}\ \text{mol/L}$$

or

$$[H_2CO_3^*] = 155 \text{ mg/L as } CaCO_3$$

2. Draw a bar diagram of the untreated water.

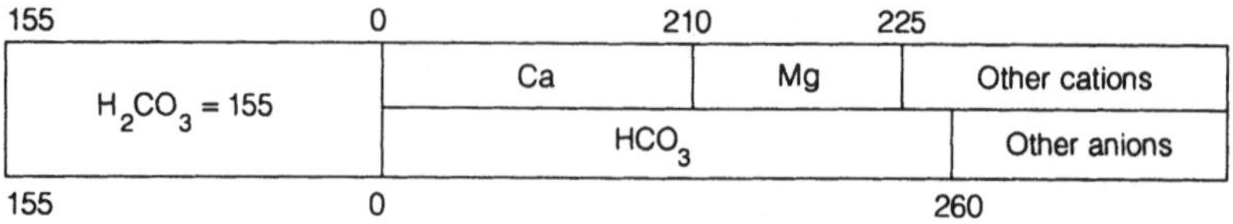

3. Establish the hardness distributed based on the measured concentrations of alkalinity, calcium, and magnesium.
Total hardness = 210 + 15 = 225 mg/L
Calcium carbonate hardness = 210 mg/L
Magnesium carbonate hardness = 15 mg/L
Note: Generally no need for magnesium removal exists when the concentration is less than 40 mg/L as $CaCO_3$.

4. Estimate the lime dose requirement by applying the following relationship for the straight lime process:
Lime dose for straight lime process = carbonic acid concentration + calcium carbonate hardness

$$= 155 + 210 = 365 \text{ mg/L as } CaCO_3$$

or

$$\text{Lime dose} = 365 \times 37/50 = 270 \text{ mg/L as } Ca(OH)_2$$

This calculation assumes that the lime is 100 percent pure. If the actual purity is less than 100 percent, the lime dose must be increased accordingly.

5. Estimate the hardness of the finished water. The final hardness of the water is all the Mg^{2+} in the untreated water plus the practical limit of $CaCO_3$ removal. Although calcium carbonate has a finite solubility, the theoretical solubility equilibrium concentrations are seldom reached because of factors such as insufficient detention time in the softening reactor, the interaction of Ca^{2+}, CO_3^{2-}, and OH^- with soluble anionic or cationic impurities to precipitate insoluble salts in a separate phase from $CaCO_3$, and inadequate particle size for effective solids removal. For most situations the practical lower limit of calcium achievable is between 30 and 50 mg/L as $CaCO_3$. Sometimes a 5 to 10 percent excess of the stoichiometric lime is added to accelerate the precipitation reactions. In such cases the excess should be added to the lime dose established in Step 4.

EXAMPLE PROBLEM 10.4 *Excess Lime Softening*

A water was analyzed and found to have the following composition, with all concentrations as $CaCO_3$:

pH = 7.0
Ca^{2+} = 180 mg/L
Mg^{2+} = 60 mg/L
Alk = 260 mg/L
Temp. = 10°C

Estimate the lime dose required to soften the water.

SOLUTION

1. Estimate the carbonic acid concentration. From step 1, Example Problem 3, the carbonic acid concentration is 155 mg/L as $CaCO_3$.
2. Draw a bar diagram of the untreated water.

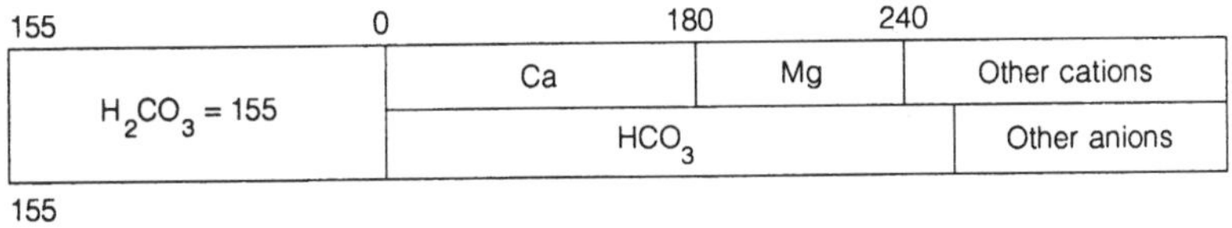

3. Establish the hardness distribution based on the measured concentrations of alkalinity, calcium, and magnesium:
   Total hardness = 180 + 60 = 240 mg/L
   Calcium carbonate hardness = 180 mg/L
   Magnesium carbonate hardness = 60 mg/L
   Note: In determining the required chemical dose for this process, sufficient lime must be added to convert all bicarbonate alkalinity to carbonate alkalinity, to precipitate magnesium as magnesium hydroxide, and to account for the excess lime requirement.
4. Estimate the lime dose requirements by applying the following relationship for the excess lime process:
   Lime dose for excess lime process = carbonic acid concentration + total alkalinity + magnesium hardness + 60 mg/L excess lime

$$= 155 + 260 + 60 + 60$$

$$= 535 \text{ mg/L as } CaCO_3$$

   or

$$\text{Lime dose} = 535 \times 37/50 = 396 \text{ mg/L as } Ca(OH)_2$$

   A high hydroxide ion concentration is required to drive the magnesium hydroxide precipitation reaction to completion. This is normally achieved when the pH is elevated above 11.0. To ensure that the required pH is established, 60 mg/L as $CaCO_3$ of excess lime is added.
5. Estimate the hardness of the finished water. See Step 5, Example Problem 3 for explanation. Normally the practical lower limit of calcium achievable is between 30 and 50 mg/L as $CaCO_3$ while the practical limit of magnesium achievable is between 10 and 20 mg/L as $CaCO_3$ with an excess of lime of 60 mg/L as $CaCO_3$. In this case, however, the finished water calcium concentration will be slightly higher than the normal range because of the excess lime added.

EXAMPLE PROBLEM 10.5 ***Straight Lime–Soda Ash Process***

A water was analyzed and found to have the following composition where all concentrations are as $CaCO_3$:

pH = 7.0
Ca^{2+} = 280 mg/L
Mg^{2+} = 10 mg/L
Alk = 260 mg/L
Temp. = 10°C

Estimate the lime and soda ash dosage required to soften the water.

SOLUTION

1. Estimate the carbonic acid concentration. From step 1, Example Problem 3 the carbonic acid concentration is 155 mg/L as $CaCO_3$.
2. Draw a bar diagram of the untreated water.

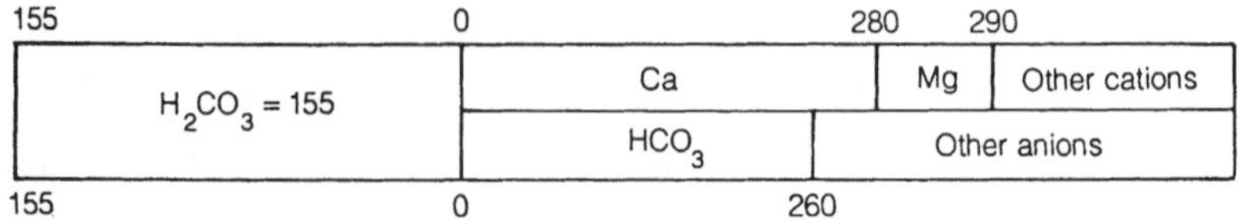

3. Establish the hardness distribution based on the measured concentrations of alkalinity, calcium, and magnesium.
   Total hardness = 280 + 10 = 290 mg/L
   Calcium carbonate hardness = 260 mg/L
   Calcium noncarbonate hardness = 280 – 260 = 20 mg/L
   Magnesium carbonate hardness = 0 mg/L
   Magnesium noncarbonate hardness = 10 mg/L
4. Estimate the lime and soda ash requirements by applying the following relationships for the straight lime-soda ash process:

Lime dose for straight lime–soda ash process

$$= \text{Carbonic acid concentration} + \text{Calcium carbonate hardness}$$

$$= 155 + 260$$

$$= 415 \text{ mg/L as } CaCO_3$$

or

$$\text{Lime dose} = 415 \times 37/50 = 307 \text{ mg/L as } Ca(OH)_2$$

and

$$\text{Lime dose for straight lime–soda ash process} = \text{calcium noncarbonate hardness}$$

$$= 20 \text{ mg/L as } CaCO_3$$

$$\text{Lime dose} = 20 \times 53/50 = 21 \text{ mg/L as } Na_2CO_3$$

5. Estimate the hardness of the finished water. See Step 5, Example Problem 3 for explanation. The final hardness of the water is all the Mg^{2+} in the untreated water plus the practical limit of calcium achievable, which is between 30 and 50 mg/L as $CaCO_3$.

EXAMPLE PROBLEM 10.6 ***Excess Lime–Soda Ash Process***

A water is analyzed and found to have the following composition, where all concentrations are as $CaCO_3$:

pH = 7.0
Ca^{2+} = 280 mg/L
Mg^{2+} = 80 mg/L
Alk = 260 mg/L
Temp. = 10°C

Estimate the lime and soda ash dosage required to soften the water.

SOLUTION

1. Estimate the carbonic acid concentration. From step 1, Example Problem 3 the carbonic acid concentration is 155 mg/L as $CaCO_3$.
2. Draw a bar diagram of the untreated water.

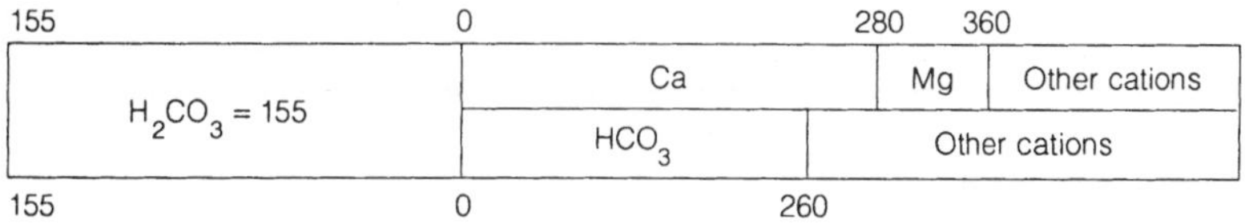

3. Establish the hardness distribution based on the measured concentrations of alkalinity, calcium, and magnesium.
   Total hardness = 280 + 80 = 360 mg/L
   Calcium carbonate hardness = 260 mg/L
   Calcium noncarbonate hardness = 280 – 260 = 20 mg/L
   Magnesium carbonate hardness = 0 mg/L
   Magnesium noncarbonate hardness = 80 mg/L
4. Estimate the lime and soda ash requirements by applying the following relationships for the excess lime-soda ash process:
   Lime dose for excess lime-soda ash process = carbonic acid concentration + calcium carbonate hardness + 2 magnesium carbonate hardness + magnesium noncarbonate hardness + 60 mg/L excess lime

$$= 155 + 260 + (2)(0) + 80 + 60$$

$$= 555 \text{ mg/L as } CaCO_3$$

or

$$\text{Lime dose} = 555 \times 37/50 = 411 \text{ mg/L as } Ca(OH)_2$$

and

Soda ash dose for excess lime-soda ash process

= calcium noncarbonate hardness + magnesium noncarbonate hardness

$$= 20 + 80$$

$$= 100 \text{ mg/L as } CaCO_3$$

or

$$\text{Soda ash dose} = 100 \times 53/50 = 106 \text{ mg/L as } Na_2CO_3$$

5. Estimate the hardness of the finished water. See Step 5, Example Problem 3 for explanation. The practical limit of calcium achievable is between 30 and 50 mg/L as $CaCO_3$, while the practical limit of magnesium achievable is between 10 and 20 mg/L as $CaCO_3$ with an excess lime of 60 mg/L as $CaCO_3$. Although excess lime was added, no excess soda ash was added to remove these extra calcium ions.

Calculations Based on Caldwell-Lawrence Diagrams. An alternative to the stoichiometric approach is the solution of simultaneous equilibria equations to estimate the dosage of chemicals in lime-soda ash softening. A series of diagrams have been developed that allow such calculations with relative ease. These diagrams are called *Caldwell-Lawrence (C-L) diagrams.* Only a brief discussion of the principles of these diagrams and their application will be presented in this chapter. The interested reader is referred to the publication *Corrosion Control by Deposition of $CaCO_3$ Films* (AWWA, 1978) for an excellent introduction to the use of C-L diagrams. Detailed discussions on the application of C-L diagrams in the solution of lime-soda ash softening problems have been presented by Merrill (1978) and Benefield, Judkins, and Weand (1982). Also available from AWWA is a computer software application for working with Caldwell-Lawrence diagrams, The Rothberg, Tamburini, and Winsor Model for Corrosion Control, and Process Chemistry.

A C-L diagram is a graphical representation of saturation equilibrium for $CaCO_3$ (Figure 10.8). Any point on the diagram indicates the pH, soluble calcium concentration, and alkalinity required for $CaCO_3$ saturation. The coordinate system for the diagram is defined as follows:

$$\text{Ordinate} = \text{acidity} \tag{10.44}$$

$$\text{Abscissa} = C_2 = \text{Alk} - \text{Ca}. \tag{10.45}$$

where acidity = acidity concentration expressed as mg/L $CaCO_3$
Alk = alkalinity concentration as mg/L $CaCO_3$
Ca = calcium concentration as mg/L $CaCO_3$

When C-L diagrams are employed to estimate chemical dosages for water softening, it is necessary to use both the direction format diagram and the Mg-pH nomograph located on each diagram. The general steps involved in solving water softening problems with C-L diagrams are as follows:

1. Measure the pH, alkalinity, soluble calcium concentration, and soluble magnesium concentration of the water to be treated.
2. Evaluate the equilibrium state with respect to $CaCO_3$ precipitation of the untreated water. This is done by locating the point of intersection of the measured pH and alkalinity lines. Determine the value of the calcium line that passes through that point. Compare that value to the measured calcium value. If the measured value is greater, the water is oversaturated with respect to $CaCO_3$. If the measured value is less than the value obtained from the C-L diagram, the water is undersaturated with respect to $CaCO_3$.
3. To use the direction format diagram, the water must be saturated with $CaCO_3$. The procedure for establishing this point for waters that are not saturated is as follows:
   - ***a.*** *Source water oversaturated:* Locate the point of $CaCO_3$ saturation by allowing $CaCO_3$ to precipitate until equilibrium is established. This point is located at the point of intersection of horizontal line through the ordinate value given by $[\text{acidity}]_{\text{initial}}$ and a vertical line through the abscissa value $C_2 = [\text{Alk}]_{\text{initial}} - [\text{Ca}]_{\text{initial}}$.
   - ***b.*** *Source water undersaturated:* Locate the point of $CaCO_3$ saturation by allowing recycled $CaCO_3$ particles to dissolve until equilibrium is established. This point is located by the same procedure followed in Step 3a.
4. Establish the pH required to produce the desired residual soluble magnesium concentration. This is accomplished by simply noting the pH associated with the desired concentration on the Mg-pH nomograph.
5. On a C-L diagram, $Mg(OH)_2$ precipitation produces the same response as the addition of a strong acid. This response is indicated on the direction format dia-

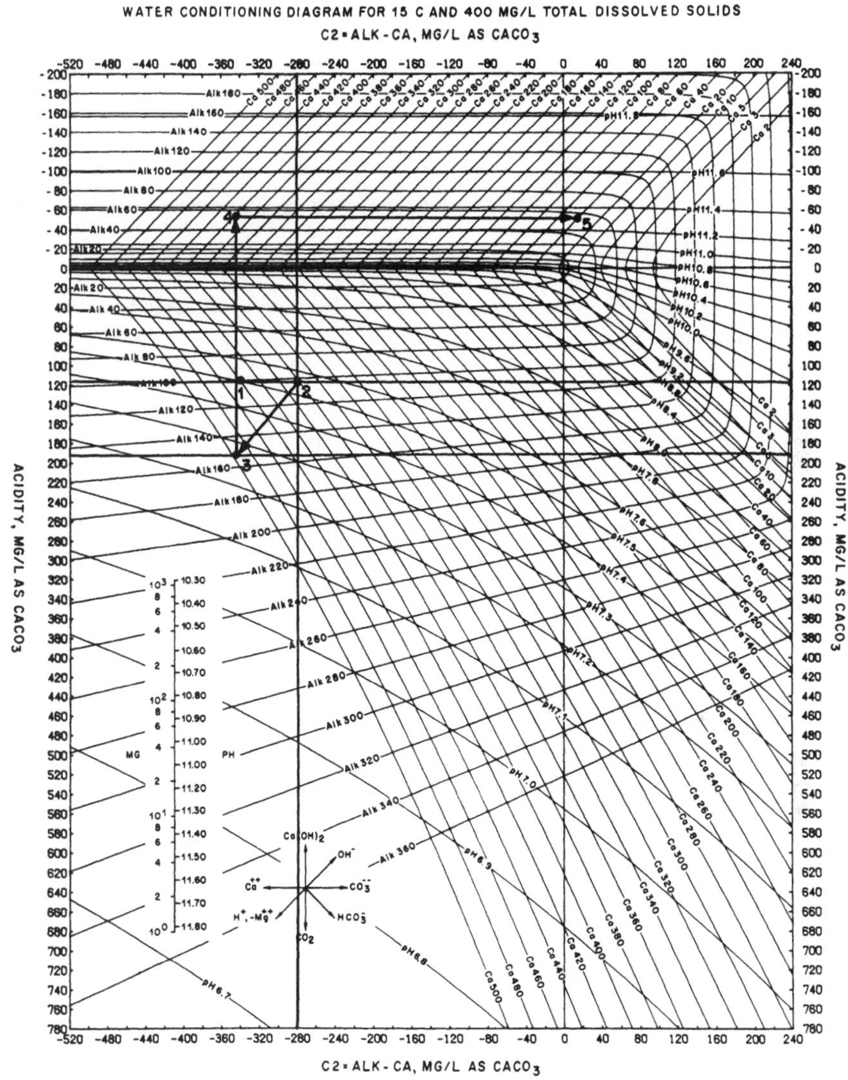

FIGURE 10.8 Water-conditioning diagram for 15°C and 400 mg/L TDS. (*Source: Corrosion Control by Deposition of* $CaCO_3$ *Films,* AWWA, Denver, 1978.)

gram as downward and to the left at 45°. When using a C-L diagram for softening calculations, the effect of $Mg(OH)_2$ precipitation should be accounted for before the chemical dose is computed. The starting point for the chemical dose calculation is located as follows:

a. Compute the change in the magnesium concentration as a result of $Mg(OH)_2$ precipitation:

$$\Delta Mg = [Mg]_{initial} - [Mg]_{desired} \tag{10.46}$$

b. From the initial saturation point construct a vector downward and to the left at 45°. The magnitude of this vector should be such that the horizontal and vertical projections have a magnitude equal to Mg.

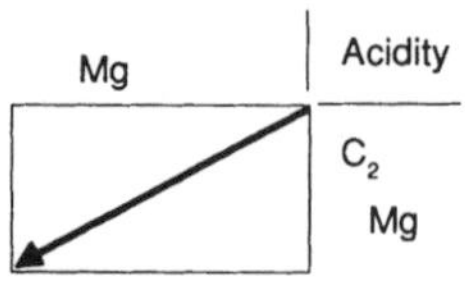

6. From the point located in Step 5b, construct a vector whose head intersects the pH line established in Step 4. The direction of this vector will depend on the chemical selected for the softening process. If lime is selected, the direction format diagram shows that the vector will move vertically. On the other hand, if sodium hydroxide is added the diagram shows that the vector will move upward and to the right at 45°. The required $Ca(OH)_2$ dose is given by the magnitude of the projection of the lime addition vector on the ordinate, and the required NaOH dose is given by either the magnitude of the horizontal projection on the C_2 axis or by the magnitude of the vertical projection on the ordinate of the sodium hydroxide vector.
7. Evaluate the calcium line that passes through the point located at the intersection of the pH line established in Step 4 and the chemical addition vector established in Step 6. If this line indicates that the soluble calcium concentration is too high, soda ash addition is necessary. The required soda ash dose is determined as follows:
   a. Locate the calcium line representing the desired residual calcium concentration.
   b. Construct a vector from the point established in Step 6 to intersect the desired calcium line. The direction format diagram indicates that the direction of this vector is horizontal and to the right.
   c. The required soda ash dose is given by the magnitude of the projection of the soda ash vector on the C_2 axis. The saturation state defined by the intersection of the desired calcium line and the soda ash vector describes the characteristics of the softened water.

The use of a C-L diagram for computing the required chemical dose for water softening is illustrated in Example Problem 7.

EXAMPLE PROBLEM 10.7 A source water has the following characteristics:

$pH = 7.5$
$Ca^{2+} = 380$ mg/L as $CaCO_3$
$Mg^{2+} = 80$ mg/L as $CaCO_3$
Alk = 100 mg/L as $CaCO_3$
temp. = 15°C
$I = 0.01$

where I represents ionic strength of the water. Determine the chemical dose requirement for excess lime-soda ash softening if the finished water is to contain 40 mg/L calcium and 10 mg/L magnesium (both as $CaCO_3$).

SOLUTION

1. Evaluate the equilibrium state of the untreated water.
   a. Locate the intersection of the initial pH and initial alkalinity lines (shown as point 1 in Figure 10.8).

b. The calcium line that passes through point 1 is 440 mg/L as $CaCO_3$.
Because this represents a concentration greater than 380 mg/L as $CaCO_3$, the water is undersaturated with respect to $CaCO_3$.

2. Compute the initial acidity of the untreated water. Construct a horizontal line through Point 1 and read the acidity value at the point where that line intersects the ordinate: acidity = 117 mg/L as $CaCO_3$.

3. Compute the C_2 value for the untreated water: C_2 = ([Alk] – [Ca]) = (100 – 380) = –280 mg/L as $CaCO_3$

4. Locate the system equilibrium point at the intersection of a horizontal line through acidity = 117 and a vertical line through C_2 = –280 (shown as point 2 in Figure 10.8).

5. Establish the pH required to produce the desired residual soluble magnesium concentration. This is obtained from the Mg-pH nomograph, which shows that a pH of 11.32 is required to reduce the soluble magnesium concentration to 10 mg/L as $CaCO_3$.

6. Compute the change in the magnesium concentration as a result of $Mg(OH)_2$ precipitation:

$$\Delta Mg = [Mg]_{initial} - [Mg]_{desired}$$

$$= 80 - 10$$

$$= 70 \text{ mg/L as } CaCO_3$$

7. Construct a downward vector from Point 2*b,* that will account for the effects of $Mg(OH)_2$ precipitation.

a. Draw a horizontal line through the acidity value of (117 + 70 initial acidity + ΔMg) = 187 mg as $CaCO_3$.

b. Beginning at Point 2, construct a vector downward and to the left at 45° until it intersects the horizontal line through acidity = 187 mg/L as $CaCO_3$ (shown as Point 3 in Figure 10.8).

8. Construct a vertical vector beginning at Point 3 to intersect the pH = 11.32 line (shown as Point 4 in Figure 10.8). The lime dose is equal to the magnitude of the projection of this vector onto the ordinate (from point 3 up to *O* on the ordinate is 187 units, and from *O* up to point 4 on ordinate is 50 units):

$$\text{Lime dose} = 187 + 50 = 237 \text{ mg/L as } CaCO_3$$

9. Construct a horizontal vector beginning at Point 4 to intersect the Ca = 40 line (shown as Point 5 in Figure 10.8). The soda ash dose is equal to the magnitude of the projection of this vector onto the C_2 axis (from Point 4 to *O* on the abscissa is 350 units, and from *O* to Point 5 on the abscissa is 18 units):

$$\text{Soda ash dose} = 350 + 18 = 368 \text{ mg/L as } CaCO_3$$

Note: Chemical dosages computed in Steps 8 and 9 are lower than would be calculated by the stoichiometric approach, because the C-L diagram assumes that equilibrium is achieved, which actually does not happen in real plants.

Recarbonation

Depending on the softening process utilized (straight lime, excess lime, straight lime-soda ash, or excess lime-soda ash), the treated water will usually have a pH of 10 or greater. It is necessary to lower the pH and stabilize such water to prevent the deposition of hard carbonate scale on filter sand and distribution piping. Recarbonation is the process most commonly employed to adjust the pH. In this process carbon dioxide (CO_2) is added to the water in sufficient quantity to lower the pH to within the range of 8.4 to 8.6.

When low magnesium waters are softened, no excess lime will be added. After softening, the water will be supersaturated with calcium carbonate and have a pH between 10.0 and 10.6. When carbon dioxide is added to this water, the carbonate ions will be converted to bicarbonate ions according to the following reaction:

$$Ca^{2+} + CO_3^{2-} + CO_2 + H_2O \rightleftharpoons Ca^{2+} + 2HCO_3^- \tag{10.47}$$

When high-magnesium waters are softened, excess lime will be added to raise the pH above 11 to precipitate magnesium hydroxide. For this situation enough carbon dioxide must be added to neutralize the excess hydroxide ions as well as to convert the carbonate ions to bicarbonate ions. To achieve the first requirement (i.e., neutralize the excess hydroxide ions), carbon dioxide is added to lower the pH to between 10.0 and 10.5. In this pH range calcium carbonate is formed as shown by Equation 10.48, while magnesium hydroxide that did not precipitate, as well as that which did not settle, is converted to magnesium carbonate as shown by Equation 10.49:

$$Ca^{2+} + 2OH^- + CO_2 \rightleftharpoons CaCO_3(s) + H_2O \tag{10.48}$$

$$Mg^{2+} + 2OH^- + CO_2 \rightleftharpoons Mg^{2+} + CO_3^{2-} + H_2O \tag{10.49}$$

Additional carbon dioxide is required to lower the pH to between 8.4 and 8.6. Here the previously formed calcium carbonate redissolves and the carbonate ions are converted to bicarbonate ions as described by Equations 10.50 and 10.51.

$$CaCO_3(s) + H_2O + CO_2 \rightleftharpoons Ca^{2+} + 2HCO_3^- \tag{10.50}$$

$$Mg^{2+} + CO_3^{2-} + CO_2 + H_2O \rightleftharpoons Mg^{2+} + 2HCO_3^- \tag{10.51}$$

Process Description. Two types of recarbonation processes are used in conjunction with the four types of softening processes previously discussed. For treatment of low-magnesium waters where excess lime addition is not required, single-stage recarbonation is used. A typical plant arrangement for single-stage softening with recarbonation is shown in Figure 10.9*a*. In this process, lime is mixed with the source water in a rapid-mix chamber, resulting in a pH of 10.2 to 10.5. If noncarbonate hardness removal is required, soda ash is added along with the lime. After rapid mixing, the water is slow mixed for 40 min to 1 h to allow the particles to agglomerate. After agglomeration the water passes to a sedimentation basin for 2 to 3 h where most of the suspended material is removed. Following sedimentation the water, carrying some particles still in suspension, moves to the recarbonation reactor. Here carbon dioxide is added to reduce the pH to 8.5 to 9.0. Any particles remaining in suspension after recarbonation are removed during the filtration step.

For treatment of high-magnesium waters where excess lime is required, two-stage recarbonation is sometimes used. A typical plant arrangement for two-stage softening with recarbonation is shown in Figure 10.9*b*. In this process excess lime is added in the first stage to raise the pH to 11.0 or higher for optimum magnesium removal. Following first-stage treatment carbon dioxide is added to reduce the pH to 10.0 to 10.6, the optimum value for calcium carbonate precipitation. If noncarbonate hardness removal is required, soda ash is added in the second stage. During second-stage

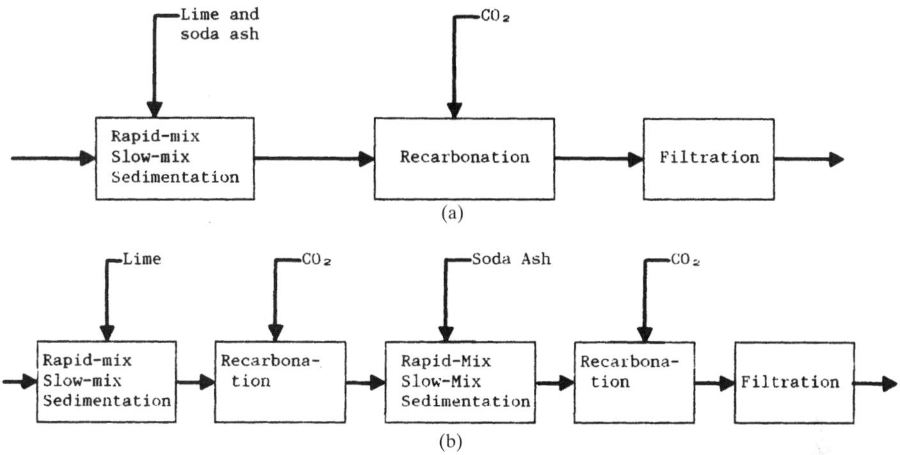

FIGURE 10.9 (*a*) Typical plant arrangement for single-stage softening with recarbonation. (*b*) Typical plant arrangement for two-stage softening with recarbonation.

treatment carbon dioxide is added to reduce the pH to 8.4 to 8.6. Because of the capital cost savings realized through the elimination of one set of settling basins and recarbonation units, single-stage recarbonation is usually the method of choice for high-magnesium waters. Still, certain advantages to the use of two-stage recarbonation exist. These include a lower operating cost because of the lower requirement for carbon dioxide dosages and a better finished water quality. The water produced by two-stage softening and recarbonation is softer and lower in alkalinity than water from a single-stage softening and recarbonation process. In most situations, however, the latter advantage is not important because water hardness concentrations between 80 and 120 mg/L as $CaCO_3$ are normally acceptable for municipal use.

Dose Calculations for Recarbonation. The quantity of gas required for recarbonation varies with the quantity of water treated, the amounts of carbonate and hydroxide alkalinity in the water, and the degree to which recarbonation is to be performed. Example Problem 8 illustrates the stoichiometric approach to estimating carbon dioxide requirements.

EXAMPLE PROBLEM 10.8 Estimate the carbon dioxide requirements for the following treatment situations:

1. Example Problem 3 with single-stage recarbonation
2. Example Problem 4 with single-stage recarbonation
3. Example Problem 5 with single-stage recarbonation
4. Example Problem 6 with two-stage recarbonation

SOLUTION

1. *Example Problem 3:* Estimate the carbon dioxide dose using the following relationship for single-stage recarbonation for straight lime softening:

Carbon dioxide requirement = estimated carbonate alkalinity of softened water where

Estimated carbonate alkalinity of softened water
= source water alkalinity – source water calcium hardness
– estimated residual calcium hardness of settled softened water

Therefore, assuming the residual calcium hardness in the settled softened water is 50 mg/L as $CaCO_3$,

$$\text{Carbon dioxide requirement} = 260 - (210 - 50) = 100 \text{ mg/L as } CaCO_3$$

or

$$\text{Carbon dioxide requirement} = 100 \times 22/50 = 44 \text{ mg/L as } CO_2$$

2. *Example Problem 4:* Estimate the carbon dioxide dose using the following relationship for single-stage recarbonation for excess lime softening:

Carbon dioxide requirement
= estimated carbonate alkalinity of softened water + 2 excess lime dose
+ estimated residual magnesium hardness of settled softened water

where

Estimated carbonate alkalinity of softened water
= source water alkalinity – source water total hardness – excess lime dose
+ estimated residual calcium hardness of settled softened water

Therefore, assuming the residual calcium hardness and residual magnesium hardness in the settled softened water are 30 and 20 mg/L as $CaCO_3$, respectively,

$$\text{Carbon dioxide requirement} = 260 - 240 + 60 - 50 + 2(60) + 20$$

$$= 150 \text{ mg/L as } CaCO_3$$

or

$$\text{Carbon dioxide requirement} = 150 \times 22/50 = 66 \text{ mg/L as } CO_2$$

3. *Example Problem 5:* Estimate the carbon dioxide dose using the following relationship for single-stage recarbonation for straight lime-soda ash softening:

Carbon dioxide requirement = estimated carbonate alkalinity of softened water

where

Estimated carbonate alkalinity of softened water
= source water alkalinity + soda ash dose – source water calcium hardness
– estimated residual calcium hardness of settled softened water

Therefore, assuming the residual calcium hardness in the settled softened water is 45 mg/L as $CaCO_3$,

$$\text{Carbon dioxide requirement} = 260 + 20 - 280 - 45$$

$$= 45 \text{ mg/L as } CaCO_3$$

or

$$\text{Carbon dioxide requirement} = 45 \times 22/50 = 20 \text{ mg/L as } CO_2$$

4. *Example Problem 6:* Estimate the carbon dioxide dose using the following relationship for two-stage recarbonation for excess lime-soda ash softening:
a. First-stage:

Carbon dioxide requirement = Estimated hydroxide alkalinity of softened water

where

Estimated hydroxide alkalinity of softened water = excess lime dose
+ estimated residual magnesium hardness of settled softened water

Therefore, assuming the residual magnesium hardness in the settled softened water is 15 mg/L as $CaCO_3$,

$$\text{Carbon dioxide requirement} = 60 + 15$$
$$= 75 \times 22/50 = 33 \text{ mg/L as } CO_2$$

b. Second-stage:

Carbon dioxide requirement = estimated residual alkalinity of softened water

where

Estimated carbonate alkalinity of softened water
= source water alkalinity + soda ash dose – source water total hardness
– estimated residual total hardness of settled softened water

Because C_T is increased after first-stage recarbonation, the calcium carbonate solubility during second-stage softening will be lowered. Hence, the residual calcium hardness after second-stage softening will be less than that normally observed in a single-stage process. In this problem a residual calcium hardness of 30 mg/L as $CaCO_3$ will be assumed. The residual magnesium hardness is the same as that assumed for first-stage treatment (i.e., 15 mg/L as $CaCO_3$).

$$\text{Carbon dioxide requirement} = 260 + 100 - 360 - 45$$

or

$$\text{Carbon dioxide requirement} = 45 \times 22/50 = 20 \text{ mg/L as } CO_2$$

Caldwell-Lawrence diagrams may also be used to compute the required carbon dioxide dose for both two-stage and single-stage softening. Example Problem 9 illustrates the procedure for making such calculations.

EXAMPLE PROBLEM 10.9 Compute the chemical doses required to neutralize the softened water described by point 1 in Figure 10.10, using two-stage recarbonation and single-stage recarbonation. Assume that the pH must be reduced to 8.7 during neutralization.

SOLUTION Two-stage recarbonation:

1. Locate the point of minimum calcium concentration.
a. Starting at Point 1, Figure 10.10, construct a vector vertically downward until the head of the vector intersects a horizontal line constructed through the ordi-

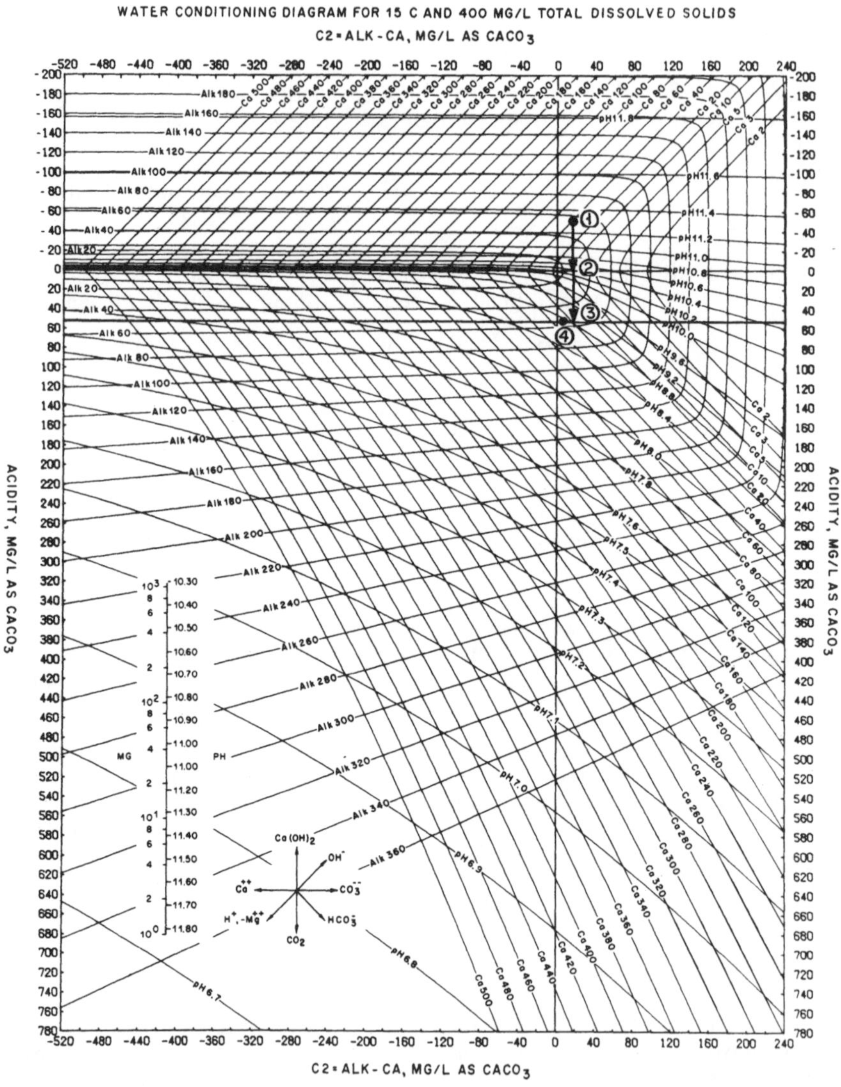

FIGURE 10.10 Example Problem 10.9 water-conditioning diagram for 15°C and 400 mg/L total dissolved solids. (*Source: Corrosion Control by Deposition of $CaCO_3$ Films,* AWWA, Denver, 1978.)

nate value $C_1 = 0$. This is the point of minimum calcium concentration and is shown as Point 2 in Figure 10.10. Note that if the vector had been extended beyond this point, it would have begun to intersect calcium lines of increasing concentration. First-stage recarbonation should be terminated at this point.

b. Determine the Ca, alkalinity, and pH for the settled water (i.e., determine the water characteristics at Point 2 in Figure 10.10).

The Ca and pH values can be determined directly from the C-L diagram. Distinguishing between the alkalinity lines in this region of the diagram is very difficult in many cases. For those situations alkalinity can be computed from the C_2 value associated with the point of minimum calcium concentration.

$$C_2 = \text{Alk} - \text{Ca}$$

or

$$\text{Alk} = C_2 + \text{Ca}$$

For point 2, pH = 10.4, Ca = 10, and $C_2 = 18$ (Point 3). Then,

$$\text{ALK} = 18 + 10 = 28$$

c. Compute the CO_2 requirement for first-stage recarbonation. The CO_2 requirement is given by the magnitude of the projection of the vector onto the ordinate:

$$\text{First-stage } CO_2 \text{ dose} = 50 \text{ mg/L as } CaCO_3$$

d. Determine the CO_2 requirement for second-stage recarbonation. The direction format diagram cannot be applied beyond point 2 because the addition of CO_2, after the $CaCO_3$ particles precipitated during first-stage recarbonation have been removed, will produce an undersaturated condition. To circumvent this problem, compute the acidity of the final state at a pH of 8.7, assuming CO_2 addition does not affect alkalinity as long as $CaCO_3$ does not precipitate. The following equation can be used for this calculation: Final acidity (mg/L as $CaCO_3$) =

$$\frac{([\text{Alk}] - K_w/[H^+] + [H^+])(1 + [H^+]/K_1)}{(1 + K_2/[H^+])} + [H^+] - \frac{K_w}{[H^+]} \qquad (10.52)$$

where K_w, K_1, and K_2 = equilibrium constants for the ionization of H_2O, $H_2CO_3^*$, and HCO_3^-, respectively
[Alk] = alkalinity of water after first-stage recarbonation (mg/L as $CaCO_3$)
$[H^+]$ = hydrogen ion concentration at the final state, in this case
pH = 8.7 (mg/L as $CaCO_3$)

According to this equation,

$$\text{Acidity} = 28 \text{ mg/L as } CaCO_3$$

The CO_2 required to reduce the pH to 8.7 is given by the difference between the acidity of the final state (computed) and the acidity at Point 2; see Point 4.

$$\text{Second-stage } CO_2 \text{ dose} = 28 - 0 = 28 \text{ mg/L as } CaCO_3$$

e. Calculate the total CO_2 requirement for first-stage and second-stage recarbonation:

$$\text{Total } CO_2 \text{ requirement} = 50 + 28 = 78 \text{ mg/L as } CaCO_3 \text{ (mg/L as } CaCO_3)$$

2. Single-stage Recarbonation

a. Locate the point describing the limiting saturated state that can be attained by CO_2 addition. Starting at Point 1 in Figure 10.10, construct a vector vertically downward until the head of the vector intersects calcium line 40 on the opposite side of the acidity = 0 line (shown as Point 3 in Figure 10.10). The sequence of events occurring between Points 1 and 3 can be visualized as $CaCO_3$ precipitation between Points 1 and 2 and dissolution of the precipitated $CaCO_3$ between Points 2 and 3. The direction format diagram cannot be used to show the reaction path past Point 3 because any further CO_2 addition produces an undersaturated condition.

b. Evaluate the Ca, alkalinity, and pH for the saturated conditions at point 3. These values are read directly from the C-L diagram:

$$\text{Ca} = 40$$

$$\text{Alk} = 58 \text{ (point 5)}$$

$$\text{pH} = 8.8$$

c. Compute the CO_2 dose requirement.

(1) Remembering that alkalinity was assumed to not change with CO_2 as long as $CaCO_3$ does not precipitate, locate the final state at the intersection of the Alk = 58 and pH = 8.7 lines (shown as Point 4 in Figure 10.10).

(2) Construct a horizontal line through Point 5 and determine the acidity of the final state.

$$\text{Acidity} = 58 \text{ mg/L as } CaCO_3 \text{ (Point 6)}$$

(3) The CO_2 requirement is given by the change in acidity between the initial and final states (in this case between point 7 and point 6):

$$CO_2 \text{ requirement} = 50 + 58 = 108 \text{ mg/L as } CaCO_3 \text{ (mg/L as } CaCO_3)$$

The use of C-L diagrams for computing carbon dioxide dosages does not recognize the demand for unsettled $CaCO_3$ floc. This demand may, at times, be quite high.

Split Treatment

Several types of processes fit the split treatment category (Montgomery 1985). This category includes processes where two or more streams are treated separately and are then combined. Three such processes are split treatment with excess lime, parallel softening and coagulation, and blended lime-softened stream with another stream, either ion exchange or reverse osmosis treated water. In this chapter, only the split treatment with excess lime process will be discussed.

For certain types of water, split treatment with excess lime softening is less costly (because of reduced chemical costs) than conventional excess lime treatment. Experience has shown that split-treatment softening should be considered when the magnesium content of the water to be treated is high and the noncarbonate hardness content is insignificant. Larson, et al. (1959) have suggested that the magnesium concentration of finished water should not exceed 40 mg/L as $CaCO_3$ if water heater fouling is to be prevented. Because total hardness levels of 60 to 120 mg/L as $CaCO_3$ are often acceptable, the calcium concentration of finished water can be allowed to vary between 20 and 80 mg/L as $CaCO_3$.

Split-treatment excess lime softening is a treatment technique where the source water flow is divided into two streams. One of the streams receives excess lime treatment for calcium carbonate hardness and magnesium hardness removal. This reduces magnesium hardness in this stream to its practical solubility limit near 10 mg/L as $CaCO_3$. The idea is to treat enough of the total water volume so that

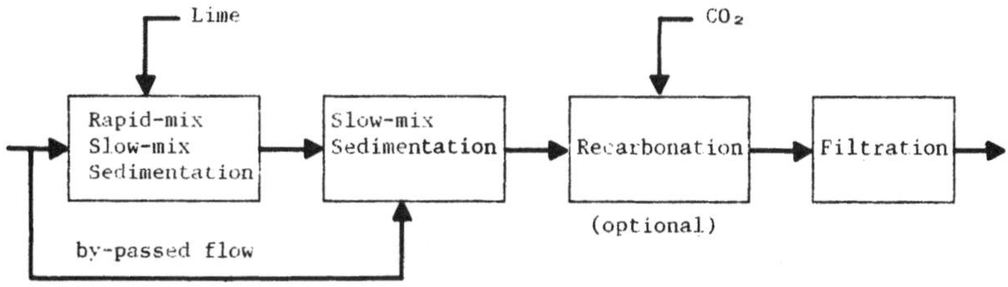

FIGURE 10.11 Flow schematic for split-treatment excess lime softening.

when the treated and untreated streams are mixed, the magnesium concentration will be less than 40 mg/L as $CaCO_3$. Dissolved carbon dioxide in the bypassed stream is used to neutralize the hydroxide alkalinity produced by the excess lime dose. The blended waters are normally allowed to settle before filtration. Because of neutralization by the free carbon dioxide contained in the untreated water, however, frequently recarbonation is not required. A flow diagram for a typical split-treatment excess lime softening process is presented in Figure 10.11.

Once the desired finished water magnesium concentration is established, the bypass flow fraction can be determined using the formula

$$X = \frac{(\text{Mg})_e - (\text{Mg})_t}{(\text{Mg})_r - (\text{Mg})_t} \tag{10.53}$$

where X = Fraction of total flow to be bypassed
$(\text{Mg})_e$ = Desired magnesium concentration in plant effluent (mg/L as $CaCO_3$)
$(\text{Mg})_t$ = Magnesium concentration in stream from lime treatment process (mg/L as $CaCO_3$)*
$(\text{Mg})_r$ = Magnesium concentration in source water (mg/L as $CaCO_3$)

The upper limit for the fraction of water to bypass can be determined by noting that the lower limit for the magnesium concentration that can be achieved in the treated water is zero. Hence, setting $(\text{Mg})_t = 0$ and solving for X give

$$(X)_{\text{max}} = \frac{(\text{Mg})_e}{(\text{Mg})_r} \tag{10.54}$$

Example Problem 10.10 illustrates the use of C-L diagrams to determine the chemical requirements for split-treatment excess lime softening.

EXAMPLE PROBLEM 10.10 A groundwater was analyzed and found to have the following composition:

pH = 7.1
Ca^{2+} = 200 mg/L as $CaCO_3$
Mg^{2+} = 80 mg/L as $CaCO_3$
Alk = 300 mg/L as $CaCO_3$
Temp. = 15°C

* This is normally assumed to be 10 mg/L as $CaCO_3$, which is the practical solubility limit of $Mg(OH)_2$(s).

Determine the lime dose required to soften this water if split-treatment excess lime softening is used. The magnesium concentration in the finished water should not exceed 40 mg/L as $CaCO_3$, and the total hardness in the effluent should not exceed 120 mg/L as $CaCO_3$.

SOLUTION

1. Determine the maximum fraction of water that can be bypassed from Equation 10.54:

$$(X)_{max} = 40/80 = 0.5$$

Because this is the maximum fraction that can be bypassed and still reach the desired effluent Mg^{2+} concentration, 0.4 of the flow will actually be bypassed because Mg^{2+} is generally not removed to 0 mg/L by chemical precipitation.

2. Compute the required magnesium concentration in the treated stream when 0.4 of the flow is bypassed:

$$(Mg)_2 = \frac{40 - (80)(0.4)}{1 - 0.4} = 13 \text{ mg/L as } CaCO_3$$

3. Evaluate the equilibrium state of the untreated water.

a. Locate the intersection of initial pH and initial alkalinity lines (shown as Point 1 in Figure 10.12).

b. The calcium line that passes through Point 1 is 370 mg/L as $CaCO_3$. Because this represents a concentration greater than 200, the water is unsaturated with respect to $CaCO_3$.

4. Compute the initial acidity of the untreated water. Construct a horizontal line through Point 1, and read the acidity value at the point where this line intersects the ordinate:

$$\text{Acidity} = 415 \text{ mg/L as } CaCO_3 \text{ (Point 1}a\text{)}$$

5. Compute the C_2 value for the untreated water:

$$C_2 = [\text{Alk}] - [\text{Ca}] = 300 - 200 = 100 \text{ mg/L as } CaCO_3$$

6. Locate the system equilibrium point at the intersection of a horizontal line through acidity = 415 mg/L as $CaCO_3$ and a vertical line through C_2 = 100 mg/L as $CaCO_3$ (shown as Point 2 in Figure 10.12).

7. Establish the pH required to produce the desired residual soluble magnesium concentration. This is obtained from the Mg-pH nomograph, which shows that a pH of 11.25 is required to reduce the soluble magnesium concentration to 13 mg/L as $CaCO_3$.

8. Compute the change in the magnesium concentration as a result of $Mg(OH)_2$ precipitation:

$$\Delta Mg = [Mg]_{initial} - [Mg]_{desired}$$
$$= 80 - 13 = 67 \text{ mg/L as } CaCO_3$$

9. Construct a downward vector from Point 2 that will account for the effects of $Mg(OH)_2$ precipitation.

a. Draw a horizontal line through the acidity value of (415 + 67) = 482 mg/L as $CaCO_3$.

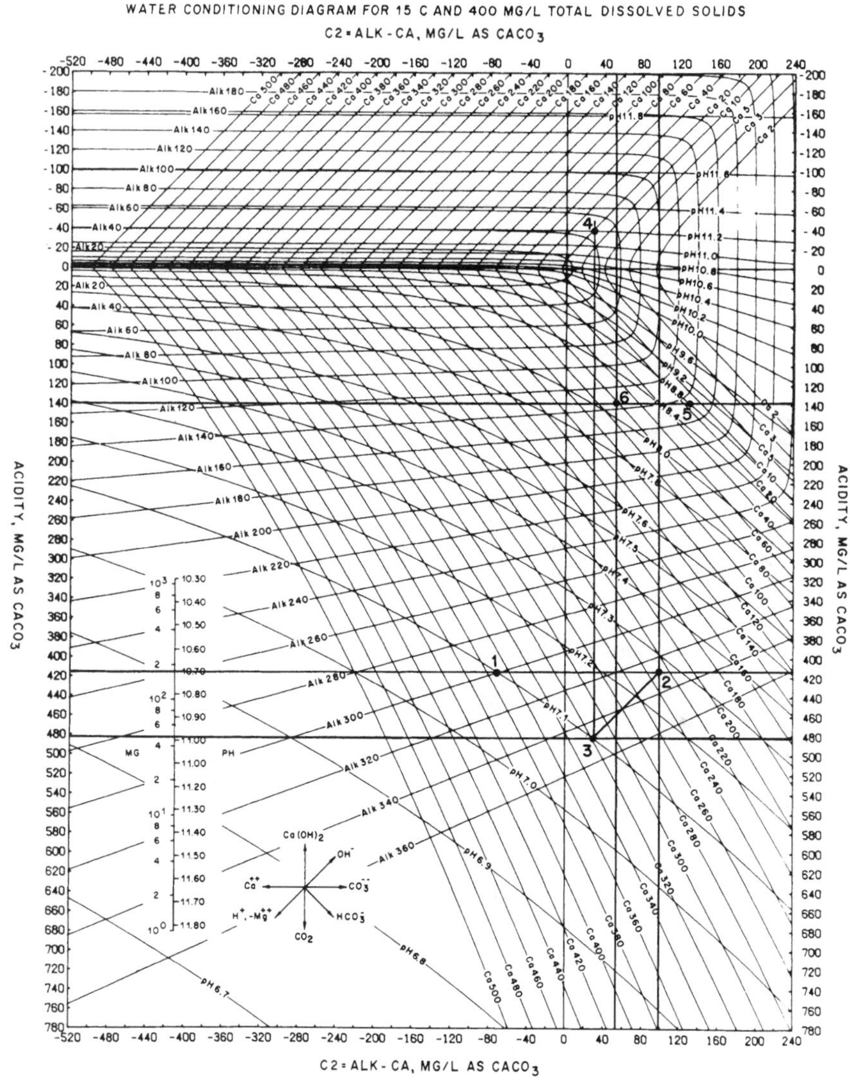

FIGURE 10.12 Example Problem 10.10 water-conditioning diagram for 15°C and 400 mg/L total dissolved solids. (*Source: Corrosion Control by Deposition of $CaCO_3$ Films,* AWWA, Denver, 1978.)

b. Construct a vector, beginning at Point 2, downward and to the left at 45° until it intersects the horizontal line through acidity = 482 mg/L as $CaCO_3$ (shown as Point 3 in Figure 10.12).

10. Construct a vertical vector, beginning at Point 3, to intersect the pH = 11.25 line (shown as Point 4 in Figure 10.12). The lime dose is equal to the magnitude of the projection of this vector (3.4) onto the acidity axis:

$$\text{Lime dose} = 482 + 43 = 525 \text{ mg/L as } CaCO_3$$

11. Combine the treated and bypassed water, and evaluate the equilibrium state of the mixture.

a. Assume that the $CaCO_3$ is infinitely soluble and compute the Ca, alkalinity, and acidity of the mixed streams.

$$[Ca]_{mix} = [Ca]_{pt\,4}(0.6) + [Ca]_{untreated}(0.4)$$
$$= (25)(0.6) + (200)(0.4)$$
$$= 95 \text{ mg/L as } CaCO_3$$
$$[Alk]_{mix} = [Alk]_{pt\,4}\,(0.6) + [Alk]_{untreated}\,(0.4)$$
$$= (50)(0.6) + (300)(0.4)$$
$$= 150 \text{ mg/L as } CaCO_3$$
$$[Acidity]_{mix} = [Acidity]_{pt\,4}\,(0.6) + [Acidity]_{untreated}\,(0.4)$$
$$= (-43)(0.6) + (415)(0.4)$$
$$= 140 \text{ mg/L as } CaCO_3$$

b. Construct a horizontal line through the ordinate value given by an acidity of 140 mg/L as $CaCO_3$ (Point 4). Locate the intersection of this line and the alkalinity = 150 mg/L as $CaCO_3$ (shown as Point 5 in Figure 10.12).

c. The calcium line that passes through Point 5 is 15 mg/L as $CaCO_3$. Because this represents a concentration less than 95 mg/L as $CaCO_3$, the mixture is oversaturated with respect to $CaCO_3$.

12. Determine the final equilibrium state of the system.

a. Remove the condition of infinite solubility and allow $CaCO_3$ to precipitate. Recall two things: Acidity remains constant during precipitation, and because equivalent amounts of alkalinity and Ca are removed, [Alk] – [Ca] = C_2 will remain constant during $CaCO_3$ precipitation.

b. Construct a vertical line through the C_2 value of [150] – [95] = 55 mg/L as $CaCO_3$ (Point 5). Locate the intersection of this line and the horizontal line through Point 5 (shown as Point 6 in Figure 10.12). The characteristics of the final equilibrium state at point 6 are pH = 8.3, alkalinity = 137 mg/L as $CaCO_3$, and Ca = 82 mg/L as $CaCO_3$. Thus, the theoretical final hardness of the water is 122 mg/L as $CaCO_3$ (Ca = 82 mg/L as $CaCO_3$ and Mg = 40 mg/L as $CaCO_3$). Particle carryover will increase this value, but it will probably still be less than 120 mg/L. A final pH of 8.3 was achieved without the need for recarbonation. Note that these values will differ from those calculated by the stoichiometric approach, because the C-L diagram assumes that equilibrium is achieved, which actually does not happen in real plants.

Softening by Means of Pellet Reactors

A pellet softener is basically a fluidized bed of grains on which the crystallization of $CaCO_3$ takes place. One of the most commonly used pellet softener systems is the pellet reactor (see Figure 10.13). This system was developed in 1938 by Zentner in Czechoslovakia (Dept. of the Air Force, 1984). It consists of an inverted conical tank in which the softening reactions take place in the presence of a suspended bed of fine sand 0.1 to 0.2 mm in diameter that acts as a catalyst. Source water and chemicals enter tangentially at the bottom of the cone and mix immediately. The treated water then rises through the reactor in a swirling motion. The magnitude of the upward velocity is sufficient to keep the sand fluidized. The contact time between the treated water and the sand grains is 8 to 10 min. During this period, precipitated hardness particles attach to the surface of the sand grains so that the grains increase in diameter. Because a bed of large grains has a small reactive surface, some of the grains are removed regularly at the bottom of the reactor and replaced by smaller-diameter seeding grains. During treatment the pH should be kept low enough to prevent the precipitation of magnesium hydroxide. This material does not adhere well to the surface of the sand grains and, as a result, will pass out of the reactor, creating a high solids loading on the filter.

Advantages of the pellet reactor are its small size, low installation cost, and rapid treatment. Because removing magnesium in these systems is difficult, however, they should not be considered when the water to be treated has a high magnesium content.

Pellet reactors have been used for softening in The Netherlands for many years. Their experience has resulted in a high degree of standardization so that pellet reactors are now considered an established technique in that country (Graveland et al., 1983). Over the past few years these units have also been installed at a number of locations in the United States.

Process Considerations in Water Softening

Although water softening is often considered as a treatment process whose use is limited to groundwater, surface water supplies require softening in a number of locations. If split treatment is used with surface water, however, some considerations must be given to problems such as taste and odor in the bypass water. When softening surface water, preceding the softening process with coagulation may be benefi-

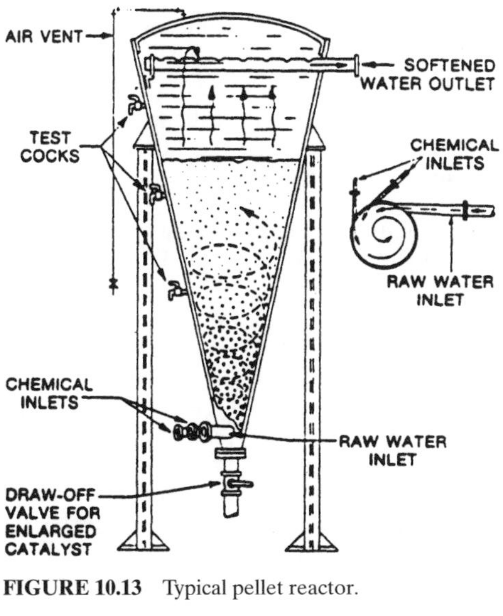

FIGURE 10.13 Typical pellet reactor.

cial if the water has a high turbidity, contains organic colloids (such materials have been found to inhibit the growth of $CaCO_3$ crystals), or if recalcining of the sludge is practiced. Even when softening groundwater, which is relatively free from turbidity, treatment efficiency is often increased when coagulants are added to enhance agglomeration of the collodial size calcium and magnesium precipitates into particles that are rapidly removed by sedimentation. Although small amounts of metal coagulants such as aluminum salts and iron salts are often used for this purpose, polymeric coagulants offer an attractive alternative because: (a) a smaller volume of sludge is generated than with metal coagulants, (b) they are effective over a much broader pH range than metal coagulants, and (c) sludges produced with polymeric coagulants tend to dewater easier than sludges produced with metal coagulants (Reh, 1978).

Flow diagrams for two full-scale lime softening plants in Illinois that use metal coagulants to enhance treatment efficiency are presented in Figure 10.14. In this figure, aeration is shown in the scheme for treatment plant B. This operation is necessary to reduce the CO_2 content in waters where it is high, and, thereby, reduce the lime requirement for carbonic acid neutralization.

The discussion on softening presented in this chapter has been limited to lime–soda ash softening. In many cases, however, caustic soda (NaOH) may be substituted for both these chemicals. Four factors must be considered when deciding whether to use lime-soda ash or caustic soda for a particular application. These factors are:

1. *Cost.* Total chemical cost will generally be less when lime and soda ash are used. Caustic soda is most competitive for either very low or very high alkalinity waters.
2. *Total dissolved solids (TDS).* A greater increase in TDS occurs in the finished water when caustic soda is used, whereas TDS will frequently decrease when lime is used. Furthermore, the use of caustic soda may increase the sodium concentration to a level high enough to become a health concern for some.
3. *Sludge production.* Generally less sludge is produced when caustic soda is used than when lime and soda ash are used. When lime and soda ash are used, sludge production increases with increasing alkalinity, while the quantity of sludge production is independent of alkalinity when caustic soda is employed for softening a water of a particular hardness.
4. *Chemical stability.* Storing and feeding caustic soda is easier than for lime. Caustic soda does not deteriorate during storage, but hydrated lime may adsorb CO_2 and water from the air and form $CaCO_3$ while quicklime may slake in storage.

In the final analysis, constraints for the use of caustic soda will probably be chemical costs and the guideline for sodium content of the finished water.

Design Considerations. In lime-soda ash or caustic soda softening plants, the softening process may be carried out by a unit sequence of rapid-mix, flocculation, and sedimentation (see Figure 10.15) or in a solids-contact softener where rapid-mix, flocculation, and sedimentation occur in a single unit. The process begins with rapid dispersion of the chemicals into the flow stream, followed at once by violent agitation. This is referred to as *rapid mixing* (sometimes called *initial* or *flash mixing*). Where coagulation of colloids is to occur simultaneously with softening, the initial introduction of chemicals into the water is very often a key factor in determining the amount to be used. Introduction of the destabilizing chemicals into the inlet pipeline by jet injection appears to be the most effective means of achieving rapid dispersion

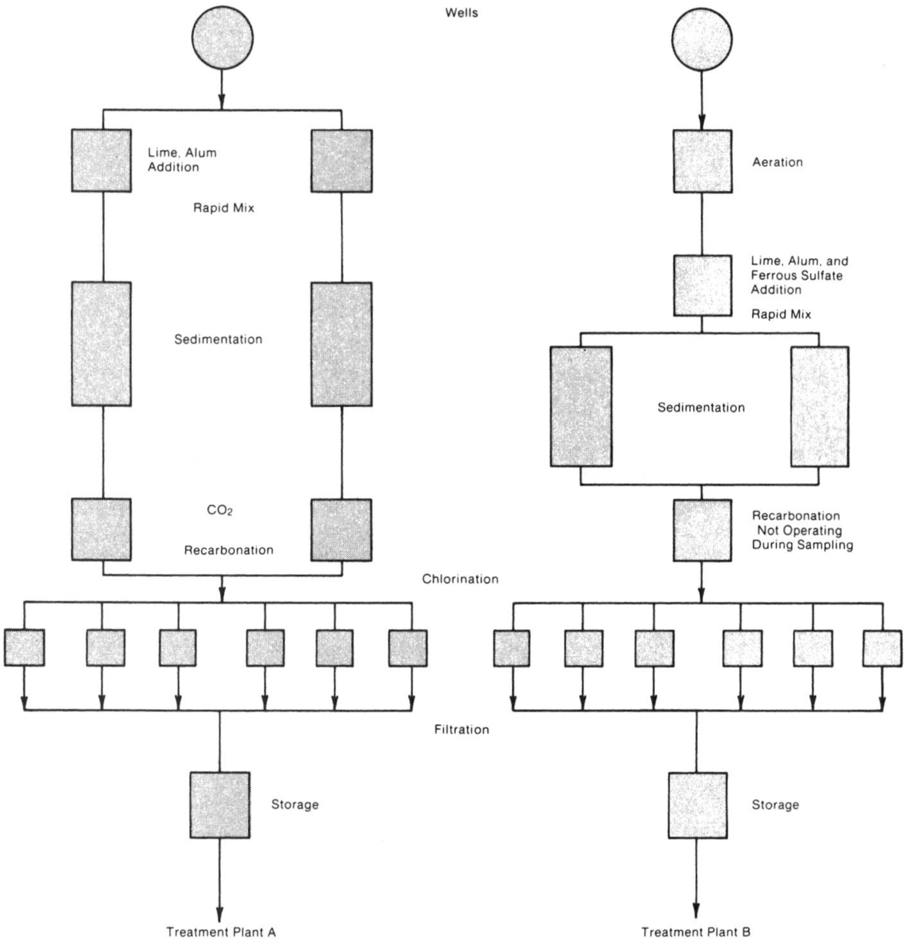

FIGURE 10.14 Flow diagram of full-scale lime softening treatment plant. (*Source:* T. J. Sorg and G. S. Logsdon, "Treatment Technology to Meet the Interim Primary Drinking Water Regulations for Inorganics: Part 5," *J. AWWA,* vol. 72, no. 7, July 1980, p. 411.)

(Montgomery, 1985). When designing the rapid-mix step remember that lime dissolves rather slowly in water. In those situations where coagulants will be added to enhance the effectiveness of the softening process, lime is added and dispersed before the addition of the coagulant.

The purpose of the flocculation step is to provide the retention time and contact opportunities required for the chemical precipitate to grow to a size large enough to be removed by gravity settling. During this step the water is gently mixed (sometimes called the slow-mix step). For large flows, rectangular basins with horizontal and vertical paddle flocculators are commonly used, while smaller installations use turbine mixers or end-around flow channels. In the past, most flocculation basins consisted of single, uniformly mixed tanks. Both the degree of flocculation and the subsequent clarification step, however, have been found to improve if flocculation

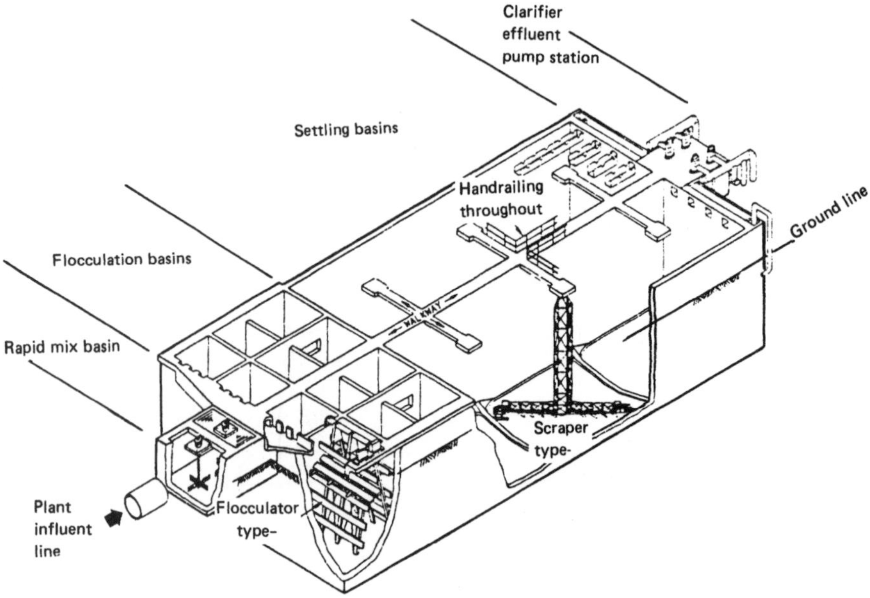

FIGURE 10.15 Chemical clarifier, Orange County Water District. (*Source:* R. L. Culp, G. M. Wesner, and G. L. Culp, *Handbook of Advanced Wastewater Treatment,* 2d ed., Van Nostrand Reinhold, 1978.)

basins are compartmentalized. Current design favors the use of basins with three or more compartments, with the velocity gradient G decreasing from compartment to compartment. A minimum of three stages for conventional softening, with the G value decreasing from 50 s^{-1} to 10 s^{-1} has been recommended (Montgomery, 1985). In groundwater lime softening, however, the recirculation of sludge is more important to the kinetics of crystal formation than compartmentalization.

The detention time in a flocculation basin is also an important parameter because it determines the amount of time that particles are exposed to the velocity gradient and thus is a measure of contact opportunity in the basin. A minimum retention time of 30 min is recommended for conventional water softening (Montgomery, 1985). Sludge return to the head of the flocculator will reduce chemical requirements as well as provide floc nuclei for precipitate growth (Ryder, 1977). The estimated portion of returned sludge is 10 to 25 percent of the source water flow.

Sedimentation is the next step following flocculation. Settling rates for chemical precipitates are a function of particle size and density. To ensure efficient removal of the precipitates formed in the softening process, a retention time of 1.5 to 3 h is normally provided in the sedimentation basin. Sedimentation basins are constructed in several configurations that include horizontal flow units (both rectangular and circular shaped), inclined flow units (tube settlers mounted in rectangular or circular shaped basins), and upflow flow units (units that incorporate chemical mixing, flocculation, and sedimentation in a single tank). The upflow clarifier, specifically the sludge-blanket clarifier, is often used in water softening operations. A schematic of this type unit is shown in Figure 10.16. This system combines mixing and sludge recirculation. The recirculated settled sludge provides additional particles that increase the probability of particle contact (and nuclei for crystal growth) and forms

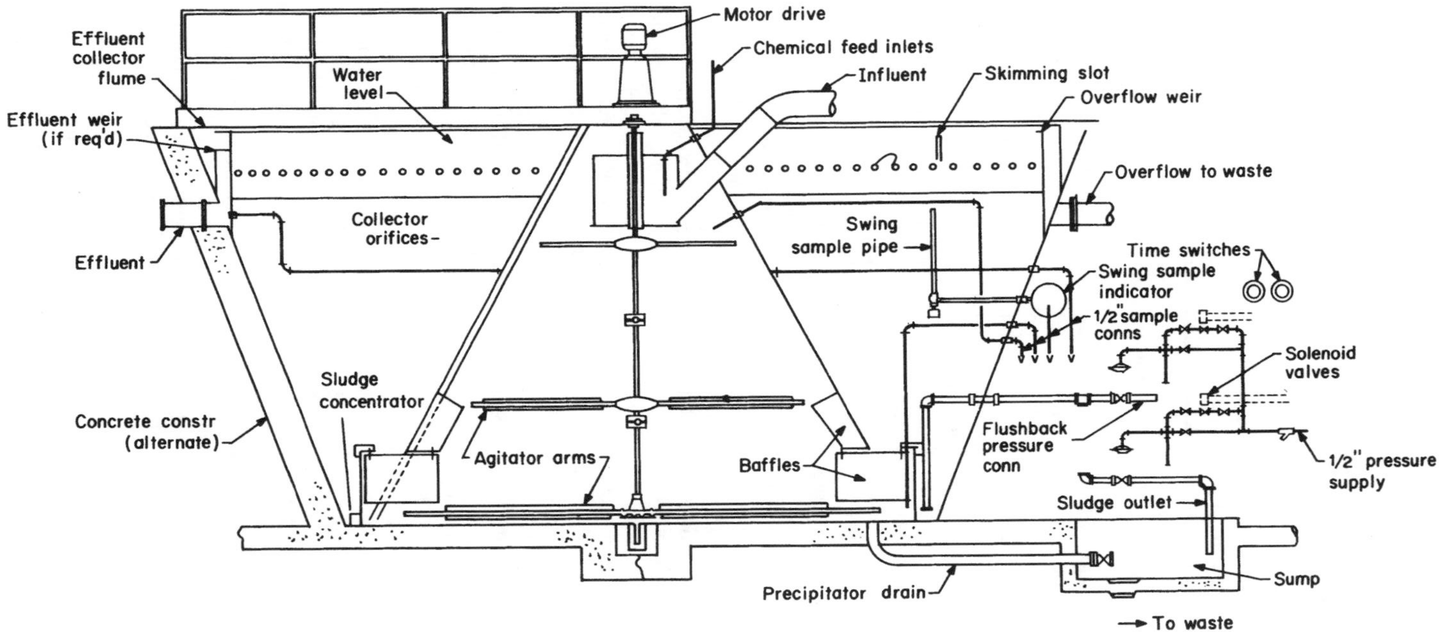

FIGURE 10.16 Vertical sludge blanket clarifier. (*Source:* Reproduced, with permission, from R. L. Sanks, ed., *Water Treatment Plant Design,* Butterworth Publishers, Stoneham, Mass., 1978.)

a dense sludge blanket. The sludge blanket concentrates, traps, and settles out suspended particles and floc before they are discharged over the effluent weir.

Water softening produces a very unstable water of high causticity. Such waters have objectionable tastes, cause filter-sand encrustation, and cause scale formation on pipes and valves. To eliminate these problems, the pH of the water is usually reduced by adding carbon dioxide. In the past CO_2 was produced on site by the combustion of oil or gas in either underwater or external burners. Liquid CO_2 is, however, now readily available in bulk quantities. Transfer efficiencies for liquid CO_2 are near 100 percent and its use significantly reduces the operation and maintenance problems normally encountered with the recarbonation process.

Chemical Feeders and Mixers. The amounts of chemicals added to water must be carefully controlled to ensure uniform treatment. Certain chemicals, such as lime, soda ash, and most nonionic and anionic polymers, are available only in dry form. Because most chemicals must be in solution form before mixing with the water to be treated, the use of these chemicals normally requires two distinct operations: (1) the preliminary preparation of the chemical, and (2) feeding of the prepared chemical.

Preliminary preparation of dry chemical requires that a specific volume or weight of chemical be measured and dissolved in water. Although both dry feeders and saturators can be used for this purpose, dry feeders are normally selected because of their accuracy and ease of operation. Two types of dry feeders are available: (1) the volumetric feeder that meters the chemical by volume per unit time, and (2) the gravimetric feeder that meters the chemical by weight per unit time. Volumetric dry feeders are simpler, less expensive, and less accurate than gravimetric feeders. The types of dry feeding mechanisms, which may be controlled either volumetrically or gravimetrically, are rotating disc (suitable for feed rates less than 10 lb/h), oscillating (suitable for feed rates between 10 and 100 lb/h), rotary gate (suitable for feed rates between 200 and 500 lb/h), belt (suitable for feed rates between 500 and 20,000 lb/h), and screw (generally volumetric and suitable for feed rates ranging from 10 to 24,000 lb/h). Unless the savings in chemicals resulting in the greater accuracy of the gravimetric control are warranted, the smaller feeders are generally volumetric.

The dry chemical discharged from the feeder falls into a tank, where it is dissolved in water to form the feed solution. This solution is fed to the source water mixing point. Where gravity delivery cannot be achieved, piston or diaphragm applicators are generally employed to deliver the feed solution under pressure.

Bulk handling of dry chemicals usually will be more economical than bag handling and batch preparation. Chemical storage facilities should be planned with sufficient capacity for at least a month of storage. Soda ash is noncorrosive and relatively safe to handle and may be stored in containers like barrels or drums. Quicklime and hydrated lime are also noncorrosive, but care must be taken to keep the storage units airtight and watertight. In most medium to large softening plants quicklime is generally more economical to use. Before quicklime can be applied, however, it must be slaked. In a typical lime slaker, quicklime is combined with water in a paddle-agitated compartment to form a paste at near-boiling temperature. In some plants, the paste is pumped with a progressive cavity pump without dilution to avoid deposits in the line that will result if it is diluted.

Residues from Lime–Soda Ash Softening. The residue from water softening plants are predominantly calcium carbonate or a mixture of calcium carbonate and magnesium hydroxide. Calcium carbonate sludges are generally dense, stable, and inert materials that dry well. The sludge solids content is typically near 5 percent, although a range between 2 and 30 percent has been observed. The sludge pH is normally greater than 10.5.

Theoretically, the type of hardness removed and the chemicals used in the softening process establish the amount of sludge produced (see Table 10.5). The amount of sludge resulting from only carbonate hardness removal can be estimated from Eq. 10.55 (Committee Report, 1981):

$$\Delta S = 86.4Q(2.0\text{Ca} + 2.6\text{Mg}) \tag{10.55}$$

where ΔS = Dry weight of sludge solids formed, kgd
Q = Source water flow, m^3/s
Ca = Calcium carbonate hardness removed, mg/L as $CaCO_3$
Mg = Magnesium carbonate hardness removed, mg/L as $CaCO_3$

In those situations where coagulants are added to increase the efficiency of the softening process Eq. 10.55 must be modified to account for the additional solids generated (Committee Report, 1981):

$$\Delta S = 86.4Q(2.0\text{Ca} + 2.6\text{Mg} + 0.44\text{Al} + 1.9\text{Fe} + \text{SS} + \text{A}) \tag{10.56}$$

where Al = Alum dose as 17.1 percent Al_2O_3, mg/L
Fe = Iron dose as Fe, mg/L
SS = Suspended solids concentration in source water, mg/L
A = Additional chemicals such as polymer, clay, or activated carbon, mg/L

The composition of the sludge significantly affects its dewatering characteristics. Relatively pure calcium carbonate sludge can be easily dewatered to a solids content up to 50 or 60 percent. Increasing magnesium hydroxide content of the sludge, however, causes it to be more difficult to handle and dewater. Normally a sludge with a Ca:Mg ratio less than 2 will be difficult to dewater, while a sludge with a Ca:Mg ratio greater than 5 will be fairly easy to dewater (see Figure 10.17) (Calkins and Novak, 1973). The presence of lime in the sludge will also have an adverse effect on dewatering. Normally this is a result of poor slaking or incomplete dissolution.

Various methods are in use today to concentrate and dewater softening sludges: gravity thickening, dewatering lagoons, sand drying beds, centrifugation, vacuum filtration, pressure filtration, belt filtration, sludge pelletization, recalcination, and land application. These methods may be used separately or in combinations to accomplish a specific management objective (see Chapter 16).

When the sludge is predominantly calcium carbonate, solids concentrations greater than 30 percent have been achieved in gravity thickeners at loading rates of 40 lb/ft^2-day (Committee Report, 1981). Solids concentrations greater than 50 percent have been reported for dewatering lagoons and sand drying beds. Even higher solids concentrations can be achieved with centrifugation, vacuum filtration, and pressure filtration.

TABLE 10.5 Theoretical Solids Production, mg Dry Solids/mg Hardness Removed as $CaCO_3$

	Carbonate hardness		Noncarbonate hardness	
Treatment chemical	Calcium	Magnesium	Calcium	Magnesium
Lime and soda ash	2.0	2.6	1.0	1.6
Sodium hydroxide	1.0	0.6	1.0	0.6

Source: Committee Report, 1981.

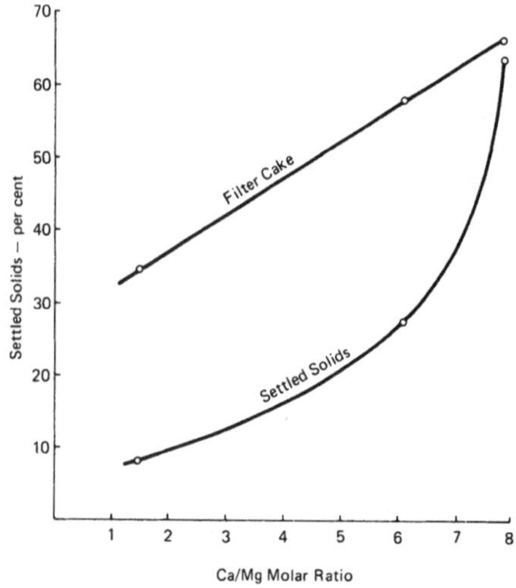

FIGURE 10.17 Effect of Ca:Mg ratio on sludge solids concentration for lime sludges. (*Source:* R. J. Calkins and J. T. Novak, "Characterization of Chemical Sludges," *J. AWWA,* vol. 65, no. 6, June 1973, p. 423.)

Sludge pelletization occurs during the suspended-bed (pellet reactor) softening process. The precipitated hardness is withdrawn as sand-like granules, which are much easier to dispose of than sludge from a conventional softening process. When the pelletized sludge leaves the reactor it is near 60 percent solids by weight. The entrained water can be readily drained to produce a residue containing greater than 90 percent solids. In The Netherlands, several methods of using the pellets have been found, including treatment of aggressive groundwater, neutralization of acid wastewater, and utilization for road construction, cement manufacturing, and metal industries (Graveland et al., 1983).

Calcium carbonate sludge from lime water softening operations can be converted to calcium oxide (quicklime) by recalcination (burning the calcium carbonate sludge in a furnace at about 1,850°F). (Magnesium hydroxide, clay, and other inorganic materials are not completely oxidized and remain in the quicklime as impurities, ultimately leading to reduced lime recovery.) Recalcination will not only produce lime that can be used in the softening process, but it will also dramatically reduce the volume of sludge requiring disposal, as well as producing carbon dioxide that can be used in the recarbonation process. The basic reaction is heat

$$CaCO_3 \rightarrow CaO + CO_2 \tag{10.57}$$

The high energy demand is the greatest disadvantage to the use of this process.

Traditionally, water treatment plant wastes have been disposed of by discharge to rivers and lakes, either directly or by way of a storm sewer. Federal law no longer allows this to occur. Alternative methods of ultimate lime sludge disposal include discharge to sanitary sewers, landfills, and drying lagoons, and land spreading in agricultural areas. Water treatment plant waste management is discussed in Chapter 16.

REMOVAL OF NOMINAL ORGANIC MATERIAL

Removal of nominal organic material (NOM) is significant to the drinking water community in that color, total organic carbon (TOC), and disinfection by-products (DBPs) are NOM subsets and controlled by water treatment due to regulatory and/or aesthetic constraints. Color, TOC and DBPs are partially removed by softening. The removal of color and DBPs can be related to TOC removal. TOC is measured as mg/L C and is a direct measure of NOM. Not all NOM or TOC produces color or regulated DBPs; hence TOC is a more universal measure of organic material in drinking water. Most if not all of the TOC removed during lime softening is in the form of nonpurgeable dissolved organic carbon (NPDOC). TOC can be in a suspended or gaseous form in some drinking water sources, however these TOC forms are either easily removed during drinking water treatment or not DBP precursors, which prior to disinfection are in the form of NPDOC. TOC is commonly used to describe NOM in drinking water treatment, but readers should realize that usually the TOC is in the NPDOC form.

Investigators have found that softening removed TOC but was less effective for TOC removal than coagulation; that the addition of coagulants during softening enhanced TOC removal; and that chemical structure affected TOC removal. A survey of water treatment plants participating in the information collection rule (ICR) found that 30 and 40 percent of TOC was removed during lime softening in the 2–4 mg/L and 4–8 mg/L TOC groups respectively. They suggested additional TOC removal should not be required by regulation after 0.2 meq/L Mg removal, 0.8–1.2 meq/L alkalinity removal, or if major changes of existing facilities would be required to accommodate the more slowly settling $Mg(OH)_2$ floc or the additional sludge (Clark and Lawler, 1996). Increasing doses of ferric sulfate to 9.5 mg/L Fe^{+3} was observed to increase TOC removal to 75 percent as softening pH increased to 10.3 (Quinn et al., 1992). Bench scale jar testing using waters from nine utilities found that TOC removal was correlated with increasing TOC concentration, hydrophobic TOC fraction, and the magnesium removed during softening. A significant relationship between TOC removed and magnesium removed was observed. (Thompson et al., 1997). Softening of Mississippi River water was found to remove less TOC than coagulation, although higher molecular weight hydrophobic organic solutes were removed by both processes. (Semmens and Staples, 1986) Liao and Randtke (1986) suggested that co-precipitation was the primary mechanism for removal of organic solutes during softening, and organic removal was limited to anionic compounds which could absorb onto $CaCO_3$ solids. Polymeric electrolytes containing acidic oxygen-containers such as carboxyl, phenol, and sulfuryl groups were not expected to be removed during lime softening unless they polymerized with or contained phosphorous-containing functional groups that could interact with calcium or calcium carbonate solids.

Calcium and Magnesium Precipitation

The removal of calcium and magnesium has been described in the sections discussing calcium carbonate and magnesium hydroxide equilibrium, and softening. The phenomena described in these sections can be related to the removal of TOC by relating TOC removal to a pH domain where $CaCO_3$ precipitation occurs (<pH 10.3) and to a pH domain where $Mg(OH)_2$ precipitation occurs (>pH 10.8). During lime softening, calcium removal due to $CaCO_3$ precipitation increases with pH to pH

10.3. At pH 10.3 nearly all of the calcium or carbonate alkalinity has been precipitated as $CaCO_3$ because of equilibrium (K_2, K_{sp}). Removal of calcium hardness is typically optimized at pH 10.3 in lime softening. Past pH 10.3, there is not enough carbonate alkalinity to precipitate the calcium solubilized from lime. Some slight additional calcium removal will be realized in a caustic softening process, but typically the vast majority of $CaCO_3$ precipitation is complete at pH 10.3. Because of $Mg(OH)_2$ equilibrium, adequate magnesium removal is typically not achieved until pH ≥ 10.8. The exact pH for optimized $CaCO_3$ and desired $Mg(OH)_2$ precipitation may differ slightly from 10.3 and 10.8 due to calcium and magnesium interactions with other solutes. However, $CaCO_3$ and $Mg(OH)_2$ precipitation occurs in different pH ranges and can be related to TOC removal.

USEPA and AWWARF (Taylor, 1984; Taylor, 1986; Randtke, 1999) have investigated the removal of color, TOC and DBP precursors. The raw water quality of three different sources investigated by Randtke (Randtke, 1999) is shown in Table 10.6. These waters vary from a soft water with low magnesium content and low TOC concentration (Lawrence, Kansas) to a hard water with high magnesium content and high TOC concentration (Grand Forks, North Dakota) with intermediate conditions in Kansas City, Missouri. TOC varies directly with both calcium and magnesium hardness for these three waters. The removal of TOC and initial total hardness (ITH) for varying pH during lime softening of these three waters is shown in Figure 10.18. ITH removed was determined by deducting the calcium and magnesium removal from the initial hardness until calcium removal was maximized (≈pH 10.3). Past pH 10.3 ITH removed was determined by deducting the magnesium removed and the maximum calcium removal from the ITH. This allowed the calcium and magnesium removed to be related to the TOC removed as shown in Figure 10.18.

TOC removal increases with pH for each of these waters. Prior to pH 10.3 the TOC removal varies from approximately 20 to 30 percent. ITH reduction is approximately 50 percent at pH 10.3, is due to CaCO3 precipitation, and occurs simultaneously with 20 to 30 percent TOC reductions. Past pH 10.3 TOC reduction is increased by approximately 25 percent and is associated with approximately 30 percent reduction of ITH, which is due to $Mg(OH)_2$ precipitation. TOC removal due to $CaCO_3$ precipitation was limited to 30 percent. TOC removal was increased to 55 percent when $Mg(OH)_2$ was precipitated and indicates that removal of magnesium hardness in a softening process will increase TOC removal. This has also been observed by Taylor (Taylor, 1983; Taylor, 1987).

The log of TOC removed versus log of Mg and Log of Ca removed is shown in Figure 10.19. Calcium data was taken prior to pH 10.3 so that calcium removed would not be offset by calcium from excess lime. The slope of the magnesium data sets varies from 0.32 to 0.40 and indicates an equilibrium relationship between TOC

TABLE 10.6 Raw Water Characteristics

Parameter		Lawrence KS	Kansas City MO	Grand Forks ND
TOC	mg/L	3.7	5.15	14.02
Ca Hardness	meq/L	100	158	178
Mg Hardness	meq/L	29	74	132
Total Hardness	meq/L	129	232	310
pH		8.3	8.1	8.2
Alkalinity	meq/L	112	170	218
Turbidity	NTU	5.4	180	25

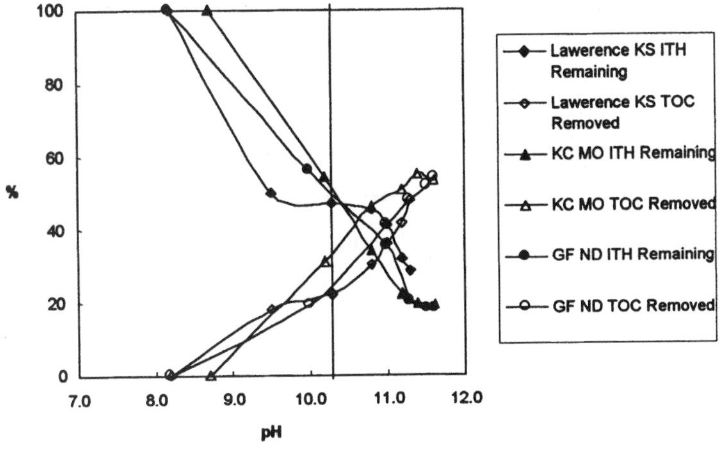

FIGURE 10.18 Percent TOC removed and initial TH remaining versus softening pH.

and magnesium removal. The arithmetic slopes of the TOC removed/metal removed indicates that the capacity of magnesium for TOC removal is also an order of magnitude greater than the calcium capacity for TOC removal.

If magnesium is removed by lime softening, the excess calcium from the lime required to go from pH 10.3 to 10.8 or greater must be removed by primary recarbonation. The $CaCO_3$ precipitated in primary recarbonation must either be removed by sedimentation or by filtration. If sedimentation is used, then a settling area equal to the settling area for the lime softening process is required. If additional settling is not provided, then the $CaCO_3$ solids formed in primary recarbonation are passed to the filters.

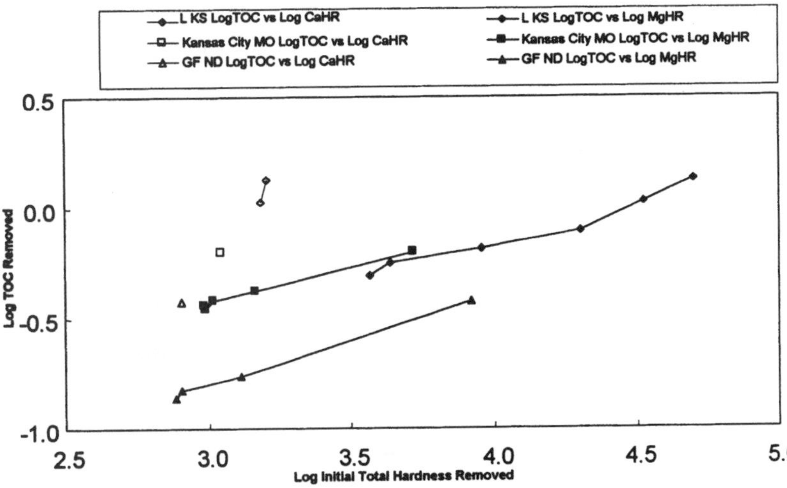

FIGURE 10.19 Log TOC removed versus log Ca and log Mg removed.

Iron and Aluminum Enhancement

Iron and aluminum salts have been added during lime softening as a settling aid and to increase color removal. Randtke investigated the addition of alum and ferric sulfate for TOC removal at Grand Forks, North Dakota (Randtke, 1996). As shown in Figure 10.20, TOC removal was increased from approximately 30 to 35 percent by the addition of 10-mg/L alum or ferric sulfate. These results show that addition of small amounts of alum or ferric sulfate can improve TOC removal by approximately 10 percent and indicate that the TOC removal by in-situ addition is limited to approximately 10 percent for this water. Taylor found that the addition of 20 mg/L of alum could increase TOC removal by approximately 10 percent during magnesium precipitation at Melbourne, Florida, but that further alum addition removed no more TOC and resulted in soluble aluminum in the finished water. Advantages of iron and alum in-situ addition with a softening process are a slight increase in removal of TOC and other organic solutes. Disadvantages of iron and aluminum in-situ addition are increased sludge volume and the possibility of aluminum postprecipitation in the distribution system.

Color and THMFP

Percent removal of THMFP, color and TOC is shown in Figure 10.21 for Lake Washington, Florida. Lake Washington is a surface water source with very high natural color, TOC, and formation potential. The relationship between TOC removal and pH is similar to what was observed previously for the Kansas, Missouri, and North Dakota waters. Moreover, increasing color and THMFP removal are also observed past pH 10.3 for the Lake Washington source. THMFP removal is similar to TOC removal; however even greater color removal is observed past pH 10.3. Consequently enhanced color and THMFP removal as well as enhanced TOC removal was attained at higher softening pHs.

Softening and coagulation have been used as serial unit operations in water treatment. Taylor observed at Acme Improvement District (AID), Florida, and Olga, Florida, that TOC removal by alum coagulation and lime softening to pH 10.3 removed no more TOC than alum coagulation alone, and that the sequence of softening and coagulation has no effect on TOC removal (Taylor, 1983; Taylor, 1987). Randtke observed similar results with ferric sulfate coagulation and softening of Austin, Texas, waters.

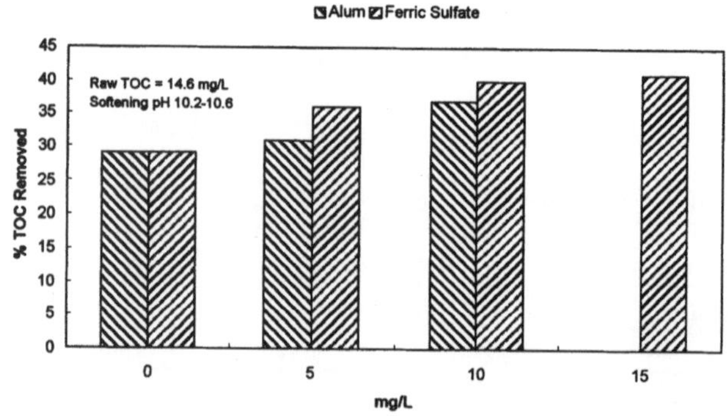

FIGURE 10.20 TOC removal by lime softening and in-situ addition for Grand Forks, North Dakota.

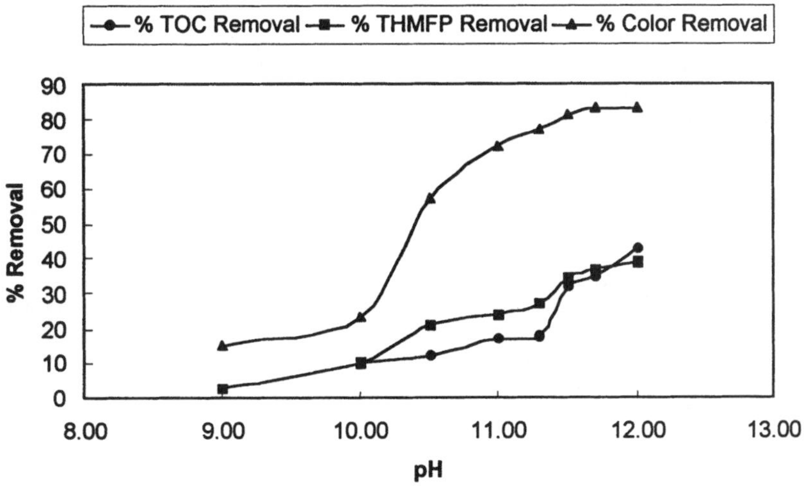

FIGURE 10.21 TOC, THMFP, and color removal versus pH for Lake Washington, Florida.

Summary

Removal of organic solutes in softening processes is unique to a given water source. However, some generalizations can be made regarding softening:

- *Calcium Carbonate Precipitation:* Generally removes from 10 to 30 percent of the color, TOC, and DBP precursors. Has the least capacity for organic removal of solids generally precipitated in precipitative softening.
- *Magnesium Hydroxide Precipitation:* Generally removes from 30 to 60 percent of the TOC and DBP precursors, and 50 to 80 percent of the color. Requires primary recarbonation to remove excess calcium if lime is used, produces excess magnesium and calcium sludge, and requires either additional sedimentation basins or solids loading on filters if excess calcium is removed.
- *Iron and Aluminum Augmentation:* Generally removes an additional 5 to 15 percent of the color, TOC, and DBP precursors in either calcium or magnesium precipitation. Will cause excess sludge formation. Aluminum may be passed through the process and postprecipitate in the distribution system.
- *Sequential Treatment:* Softening in series with coagulation will remove additional color, TOC, and DBP precursors relative to only softening. Softening in series with coagulation will not remove additional color, TOC, and DBP precursors relative to coagulation alone.

REMOVAL OF OTHER CONTAMINANTS BY PRECIPITATION

Although this chapter has focused on removal of calcium, magnesium, and organic contaminants, heavy metals, radionuclides, organics, and viruses may also be removed from water using some type of chemical precipitation process. Probably the most popular method of removing toxic heavy metals from water is precipitation of the metal hydroxide. This process normally involves the addition of caustic soda or lime to adjust the solution pH to the point of maximum insolubility. Figure 10.2 gives a graphical representation of the experimentally determined solubilities of several metal hydroxides of interest in water treatment. This figure also illustrates the amphoteric nature of certain metal hydroxides (i.e., those metal hydroxides that act as both acids and bases and will redissolve in excessively acid or alkaline solutions). Because heavy metal contamination of drinking water supplies has not been a frequent problem, few studies have been conducted that consider the removal of a specific heavy metal from drinking water by precipitation treatment. Sorg et al. (1977) have, however, discussed the application of various treatment technologies for removal of inorganics. Table 10.7 summarizes the effectiveness of chemical coagulation and lime softening processes for the removal of inorganic contaminants from drinking water.

Depending on the contaminant and its concentration, either or both precipitation and coprecipitation will play a major role in removal during chemical coagulation or lime softening. In most cases, however, coprecipitation results in the removal of soluble metal ions during coagulation and lime treatment. Four types of coprecipitation exist:

1. *Inclusion.* This process involves mechanical entrapment of a portion of the solution surrounding the growing particle. This type of coprecipitation is normally significant only for large crystals.

TABLE 10.7 Effectiveness of Chemical Coagulation and Lime Softening Processes for Inorganic Contaminant Removal

Contaminant	Method	Removal, %
Arsenic		
As^{3+}	Oxidation to As^{5+} required	>90
As^{5+}	Ferric sulfate coagulation, pH 6–8	>90
	Alum coagulation, pH 6–7	>90
	Lime softening, pH 11	>90
Barium	Lime softening, pH 10–11	>80
Cadmium*	Ferric sulfate coagulation, pH > 8	>90
	Lime softening, pH > 8.5	>95
Chromium*		
Cr^{3+}	Ferric sulfate coagulation, pH 6–9	>95
	Alum coagulation, pH 7–9	>90
	Lime softening, pH > 10.5	>95
Cr^{6+}	Ferrous sulfate coagulation, pH 6.5–9 (pH may have to be adjusted after coagulation to allow reduction to Cr^{3+})	>95
Lead*	Ferric sulfate coagulation, pH 6–9	>95
	Alum coagulation, pH 6–9	>95
	Lime softening, pH 7–8.5	>95
Mercury,* inorganic	Ferric sulfate coagulation, pH 7–8	>60
Selenium* Se^{4+}	Ferric sulfate coagulation, pH 6–7	70–80
Silver*	Ferric sulfate coagulation, pH 7–9	70–80
	Alum coagulation, pH 6–8	70–80
	Lime softening, pH 7–9	70–90

* No full-scale experience.

Source: Sorg, Love, and Logsdon, 1977.

2. *Adsorption.* This type of coprecipitation involves the attachment of an impurity onto the surface of a particle of precipitate. This type of coprecipitation is generally not important if the particle size is large when the precipitation is complete because large particles have surface areas that are very small in proportion to the amount of precipitate they contain. Adsorption may, however, be a major means of contaminant removal if the particles are small.
3. *Occlusion.* In this form, a contaminant is entrapped in the interior of a particle of precipitate. This type of coprecipitation occurs by adsorption of the contaminant onto the surface of a growing particle, followed by further growth of the particle to enclose the adsorbed contaminant.
4. *Solid-solution formation.* In this type of occlusion, a particle of precipitate becomes contaminated with a different type of particle that precipitates under similar conditions and is formed from ions whose sizes are nearly equal to those of the ions of the original precipitate.

During coagulation, some metals will coprecipitate with either iron or aluminum hydroxide. Iron coagulants seem to perform better than aluminum coagulants, primarily because iron hydroxide is insoluble over a wider pH range and is less soluble than aluminum hydroxide. Iron coagulants not only form a stronger and heavier floc, but coprecipitation of iron-metal complexes appears to be a significant factor. Mukai et al. (1979) found that the apparent solubility of Cd^{2+}, Cu^{2+}, and Zn^{2+} could be dramatically reduced through the addition of Fe^{3+}. Removal efficiencies for various heavy metals for a lime-treated secondary effluent have been presented by Culp et al. (1978). In all cases studied, the residual metal concentration was found to be less than 0.1 mg/L.

Because radium occurs naturally and is sometimes found in drinking water, this element is a radionuclide of interest in water treatment. Radium can be effectively removed from water by lime softening. Removal efficiency is a function of pH, however, and if a high degree of treatment is required, the pH during softening should be elevated above 10.8 (see Figure 10.22) (Brinck et al., 1978).

Most natural waters contain a certain amount of organic matter known as humic substances that, when present in high concentrations, impart a yellowish-brown color to water. Rook (1976) identified this material as precursors for trihalomethanes in chlorinated water. Many water supplies are treated with alum coagulation or lime softening, and although these processes are not intended to remove organic contaminants, they are generally effective in removing a significant amount

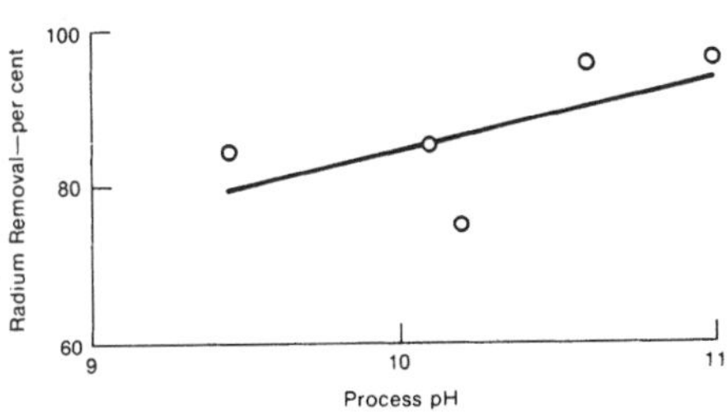

FIGURE 10.22 Radium removal versus softening pH. (*Source:* W. L. Brinck et al. "Radium-Removal Efficiencies in Water-Treatment Processes," *J. AWWA*, vol. 70, no. 1, January 1978, p. 31.)

of organic material. Hall and Packham (1965) found that organic color and clay turbidity were removed by entirely different mechanisms in the coagulation process. These workers suggest that the removal of organic color with alum is a chemical process in which a partially hydrolyzed aluminum ion of empirical formula $Al(OH)_{2.5}$ interacts with ionic groups on the humic acid colloid. Such a response results in the precipitation of an insoluble humate or fulvate. Edzwald (1978) found good removal of humic substances using alum with high molecular weight polymers in the pH range of 4.5 to 6.5. Reductions in humic acid of 90 percent or greater were obtained at a pH of 6 using the following dosages: 10 mg/L alum, 0.5 mg/L of cationic polymer and 5 mg/L humic acid, 10 mg/L alum, 1 mg/L of anionic polymer and 5 mg/L humic acid, and 10 mg/L alum, 1 mg/L of nonionic polymer and 5 mg/L humic acid. Soluble organic contaminants may also be removed from water by lime softening. Randtke et al. (1982) discuss the removal of soluble organic contaminants from wastewater by lime precipitation. Their data indicate a chemical oxygen demand removal range between 24 and 70 percent. Johnson and Randtke (1983) have presented data (see Table 10.8) that illustrates the importance of the point of chlorination on the removal of nonpurgeable organic chlorine and total organic carbon (TOC) by lime precipitation. These data suggest that prechlorination with free chlorine can have a detrimental effect on the removal of TOC by lime precipitation. Liao and Randtke (1985) found that lime softening could remove a significant fraction of fulvic acid from groundwater. Conditions favoring a high removal efficiency were a high pH, a high calcium concentration, and a low carbonate concentration. The results of this study are summarized in Table 10.9. Weber and Godellah (1985) have presented Figures 10.23 and 10.24 that show the effect of alum coagulation and lime softening on TOC removal from a humic acid solution, a fulvic acid solution, and Huron river water. These data indicate that at high alum dosages more than 80 percent of the humic and fulvic acids are removed, while TOC removal from the Huron River sample only slightly exceeded 50 percent. A similar trend was noted for lime softening.

TABLE 10.8 Removal of Nonpurgeable Organic Chlorine and Total Organic Carbon from Three Water Sources by Lime Softening

Sample	Treatment*	Removal of NPOCl		Removal of TOC	
		μg/L	%	mg/L	%
River water	Chlorination only	131.5		1.55	
	Prechlorination, softening	95.0	28	1.52	2
	Softening, postchlorination	84.8	36	1.34	14
	Prechlorination, hydrolysis	110.8	13		
Groundwater	Chlorination only	368.5		3.10	
	Prechlorination softening	253.4	31	2.48	20
	Softening, postchlorination	231.6	37	2.11	32
	Prechlorination, hydrolysis	301.4	18		
Secondary effluent	Chlorination only	171.8		6.41	
	Prechlorination, softening	128.1	25	5.34	17
	Softening, postchlorination	135.7	21	5.00	22
	Prechlorination, hydrolysis	134.5	22		

* Prechlorination was at point 1.48 h before softening. Hydrolysis was effected by softening the samples at pH 11.0 and then acidifying them to pH 2.0 after sedimentation (without solids separation), thereby dissolving the precipitated solids back into solution.

Source: Johnson and Randtke, 1983.

TABLE 10.9 Effects of Operational Changes on the Removal of Groundwater Fulvic Acid and Water Hardness

Process variable	TOC removed, %	Residual hardness, mg/L as $CaCO_3$
pH		
9	28	25
11*	35	16
12	44	12
Chemical addition		
Calcium-rich	41	9
Carbonate-rich	29	7
Sludge recycling		
1:1 (old:fresh)	31	9
2:1 (old:fresh)	23	8
Additives		
Ca:Mg = 6:2	72	14
Ca:Mg = 6:1	60	13
Ca:P = 6:0.2	43	52
Ca:P = 6:0.1	35	50
Two-stage chemical addition	53	7

* Standard condition: $[CO_3^{2-}] = [CO_3^{2-}] = 6$ mM, single stage, pH = 11, no sludge recycling, no additives present, and TOC = 3 mg/L.
Source: Liao and Randtke, 1985.

Sinsabaugh et al. (1986) investigated the effects of charge, solubility, and molecular size on the removal of dissolved organic carbon by ferric sulfate coagulation and settling. They found that molecular size, independent of any charge or solubility correlation, was the most significant factor. In every charge and solubility category studied, removal efficiency declined monotonically with molecular weight.

Wolf et al. (1974) conducted a large-scale pilot study of virus removal by both lime and alum. For an Al:P ratio of 7:1, they observed bacterial virus removals as high as 99.845 percent for coagulation-sedimentation and 99.985 percent for coagulation-sedimentation-filtration processes. At lower alum dosages a marked decrease in virus removal occurred. They also found that treating with lime to elevate the solution pH above 11.0 resulted in excellent virus removal, but the actual percentages were not quantified. Rao et al. (1988) studied the influences of water softening on virus removal. They found that during calcium hardness removal at pH 9.6, rotavirus was not as effectively removed as poliovirus and hepatitis-A virus. Greater than 90 percent of rotavirus, however, was removed during Mg^{2+} hardness removal at pH 10.8 at 37°C. During total hardness removal at pH 11, all viruses were efficiently removed.

Logsdon et al. (1994) supplemented and expanded on this work with a study funded by AWWARF. The purpose of this study was to develop information on the inactivation of viruses and *Giardia* under laboratory conditions representative of those found at lime softening plants and to evaluate the removal and inactivation of bacteria and viruses, as well as the removal of *Giardia* cysts, at water treatment plants. A survey of lime softening plants in the United States was carried out to determine the appropriate condition for conducting bench-scale inactivation experiments using lime softening. Data was collected on softening pH, detention time, disinfectant used, and residual disinfectant at different steps in the softening process. Inactivation of bacteria, *Giardia,* and viruses during lime softening was evaluated

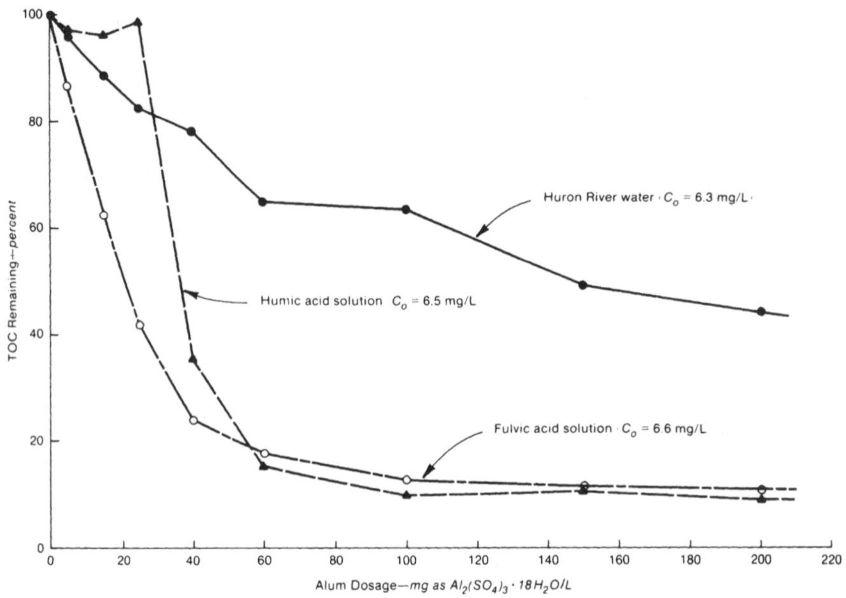

FIGURE 10.23 Removal of TOC by alum coagulation. (*Source:* W. J. Weber, Jr., and A. M. Jodellah, "Removing Humic Substances by Chemical Treatment and Adsorption," *J. AWWA,* vol. 77, no. 4, April 1985, p. 132.)

showing the effects of softening alone with no disinfectant and also with either free chlorine or monochloramine. Finally, a field sampling and analysis phase was carried out to evaluate the physical removal of *Giardia* and the removal of inactivation of bacteria and viruses at lime softening treatment plants.

Additional work continues in this subject area. Andrews et al. (1993) conducted related research on behalf of the city of Edmonton, Alberta. Johnson (1989) and Battigelli and Sobsey (1993) are examples of additional investigations.

FUTURE TRENDS IN SOFTENING

With continuing technological advances in membrane process manufacturing, it is anticipated that lime softening may gradually be replaced by membrane processes for some applications. This option appears to be particularly applicable for small installations that may achieve significant cost savings through reduced operator attention and remote monitoring of membrane processes. In addition, membrane processes produce residuals containing only the constituents removed from the source water, potentially making those processes attractive to regulatory agencies concerned with residuals disposal to source waters. Membrane processes may also remove regulated contaminants in conjunction with softening.

Lime softening uses large amounts of chemicals (primarily lime) and produces large amounts of residuals in comparison to coagulation treatment alone. Because of

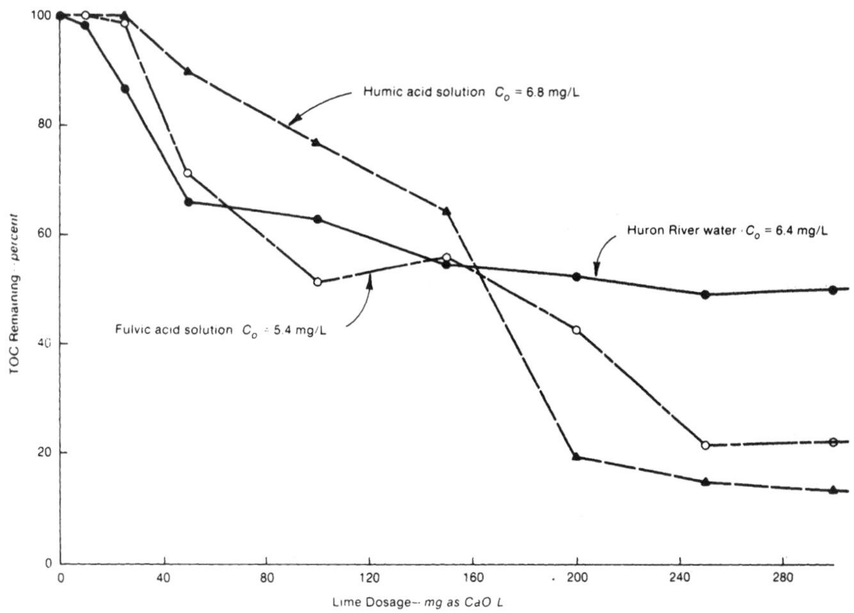

FIGURE 10.24 Removal of TOC by lime softening. (*Source:* W. J. Weber, Jr., and A. M. Jodellah, "Removing Humic Substances by Chemical Treatment and Adsorption," *J. AWWA,* vol. 77, no. 4, April 1985, p. 132.)

the cost of lime, other chemicals, and residuals disposal, some utilities have considered reducing their standards for hardness removal. Investigation is expected to continue on the question of the level of hardness removal that is beneficial and cost-effective for the consumers of a particular utility.

BIBLIOGRAPHY

American Water Works Association. *Corrosion Control by Deposition of $CaCO_3$ Films.* Denver, Colorado: AWWA, 1978.

American Water Works Association. *The Rothberg, Tamburini, and Winsor Model for Corrosion Control and Process Chemistry,* version 3.0. Denver, Colorado: AWWA, 1997.

Andrews, R. C., M. Ferguson, T. Lee, and J. Reske. "Evaluation and Optimization of Conventional Disinfectants Using the CT Concept." *Proceedings of the Fifth National Conference on Drinking Water.* Winnipeg, Manitoba, September 13–15, 1992. Denver, Colorado: American Water Works Association, 1993.

APHA, AWWA, WPCF. *Standard Methods for the Examination of Water and Wastewater* (17th ed.). Washington, D.C.: APHA, 1989.

Benefield, L. D., J. F. Judkins, and B. L. Weand. *Process Chemistry for Water and Wastewater Treatment.* Englewood Cliffs, New Jersey: Prentice-Hall, 1982.

Brinck, W. L., R. J. Schliekelman, D. L. Bennett, C. R. Bell, and I. M. Markwood. "Radium-Removal Efficiencies in Water-Treatment Processes." *Jour. AWWA,* 70(1), January 1978: 31.

Cadena, F., W. S. Midkiff, and G. A. O'Conner. "The Calcium Carbonate Ion-Pair as a Limit to Hardness Removal." *Jour. AWWA,* 66(9), September 1974: 524.

Calkins, R. J., and J. T. Novak. "Characterization of Chemical Sludges." *Jour. AWWA,* 65(6), June 1973: 423.

Clark, S. G., and D. F. Lawler. "Enhanced Softening: Calcium, Magnesium and TOC Removal by Geography." *Proceeding of AWWA WQTC,* November 1996.

Committee Report. "Lime Softening Sludge Treatment and Disposal." *Jour. AWWA,* 73(11), November 1981: 600.

Culp, R. L., G. M. Wesner, and G. L. Culp. *Handbook of Advanced Wastewater Treatment.* New York: Van Nostrand Reinhold Company, 1978.

Department of the Air Force. *Maintenance and Operation of Water Supply, Treatment, and Distribution Systems.* Department of the Air Force Regulation AFR 91.26, 1984.

Edzwald, J. K. "Coagulation of Humic Substances." *AICHE Symposium Series* No. 190, 75:54, 1978.

Engelbrecht, R. S., J. T. O'Conner, and M. Gosh. "Significance and Removal of Iron in Water Supplies." Fourth Annual Environmental Engineering & Water Resources Conference, Vanderbilt University, Nashville, Tennessee, 1965.

Faust, S. D., and J. G. McWhorter. "Water Chemistry." In *Handbook of Water Resources and Pollution Control,* H. W. Gehm and J. I. Bregman, eds. New York: Van Nostrand Reinhold Company, 1976.

Faust, S. D., and O. M. Aly. *Chemistry of Water Treatment* (2nd ed.). Chelsea, Michigan: Ann Arbor Press, 1998.

Graveland, A., J. C. Van Dyk, P. J. deMoel, and J. H. C. M. Oomen. "Developments in Water Softening by Means of Pellet Reactors." *Jour. AWWA,* 75(12), December 1983: 619.

Hall, E. S., and R. F. Packham. "Coagulation of Organic Color with Hydrolyzing Coagulants." *Jour. AWWA,* 57, 1965: 1149.

Humenick, Michael J. *Water and Wastewater Treatment.* New York: Marcel Dekker, Inc., 1977.

James M. Montgomery Consulting Engineers, Inc. *Water Treatment Principles and Design.* New York: John Wiley & Sons, 1985.

Johnson, D. E., and S. J. Randtke. "Removing Nonvolatile Organic Chlorine and its Precursors by Coagulation and Softening." *Jour. AWWA,* 75(5), May 1983: 249.

Johnson, S. L. "A Virus Study Using Chloramine Disinfection." *Proceedings Water Quality Technology Conference; Advances in Water Analysis and Treatment.* St. Louis, Missouri, November 13–17, 1988. Denver, Colorado: American Water Works Association, 1989.

Larson, T. E., and A. M. Buswell. "Calcium Carbonate Saturation Index and Alkalinity Interpretations." *Jour. AWWA,* 34(11), November 1942: 1667.

Larson, T. E., R. W. Lane, and C. H. Neff. "Stabilization of Magnesium Hydroxide in the Solids-Contact Process." *Jour. AWWA,* 51(12), December 1959: 1551

Liao, M. Y., and S. J. Randtke. "Predicting Removal of Soluble Organic Contaminants by Lime Softening." *Water Research* 20(1), 1986: 27–35.

Liao, M. Y., and S. J. Randtke. "Removing Fulvic Acid by Lime Softening." *Jour. AWWA,* 77(8), August 1985: 78.

Logsdon, G. S., M. M. Frey, T. D. Stefanich, S. L. Johnson, D. E. Feely, J. B. Rose, and M. Sobsey. *The Removal and Disinfection Efficiency of Lime Softening Processes for Giardia and Viruses.* Denver, Colorado: AWWA Research Foundation and American Water Works Association, 1994.

Lowenthal, R. E., and G. V. R. Marais. *Carbonate Chemistry of Aquatic Systems: Theory and Application.* Ann Arbor, Michigan: Ann Arbor Science Publishers, 1976.

Merrill, D. T. "Chemical Conditioning for Water Softening and Corrosion Control." *Proceedings, Fifth Environmental Engineers' Conference,* Montana State University, June 16–18, 1976.

Merrill, D. T. "Chemical Conditioning for Water Softening and Corrosion Control." In *Water Treatment Plant Design,* Robert L. Sanks, ed. Ann Arbor, Michigan: Ann Arbor Science, 1978.

Mukai, S., T. Wakamatsu, and Y. Nakahiro. "Study on the Removal of Heavy Metal Ions in Wastewater by the Precipitation-Flotation Method." *Recent Developments in Separation Science.* 67, 1979.

Nilsson, R. "Removal of Metals by Chemical Treatment of Municipal Wastewater." *Water Research* 5(51), 1971.

O'Connell, R. T. "Suspended Solids Removal." In *Water Treatment Plant Design,* Robert L. Sanks, ed. Ann Arbor, Michigan: Ann Arbor Science, 1978.

Pisigan, R. A., and J. E. Singley. "Calculating the pH of Calcium Carbonate Saturation." *Jour. AWWA,* 77(10), October 1985: 83.

Pizzi, N. G., and National Lime Association. *Hoover's Water Supply and Treatment.* Dubuque, Iowa: Kendall/Hunt Publishing, 1995.

Plummer & Busenberg. "Geochim. Cosmochim." *Acta.* 46, 1982: 1011.

Quinn, S. R., S. A. Hasham, and N. I. Ansari. "TOC Removal by Coagulation and Softening." *Journal of Environmental Engineering* 118(3), May/June 1992: 432–436.

Randtke, S. J. et al. *Precursor Removal by Coagulation and Softening.* Denver, Colorado: AWWA Research Foundation and American Water Works Association, 1999.

Randtke, S. J., C. E. Thiel, M. Y. Liao, and C. N. Yamaya. "Removing Soluble Organic Contaminants by Lime-Softening." *Jour. AWWA,* 74(4), April 1982: 192.

Rao, V. C., J. M. Symons, A. Ling, P. Wang, T. G. Metcalf, J. C. Hoff, and J. L. Melnick. "Removal of Hepatitis A Virus and Rotavirus in Drinking Water Treatment Processes." *Jour. AWWA,* (80)2, February 1988: 59.

Reh, C. W. "Lime-Soda Softening Processes." In *Water Treatment Plant Design,* Robert L. Sanks, ed. Ann Arbor, Michigan: Ann Arbor Science, 1978.

Rook, J. J. "Haloforms in Drinking Water." *Jour. AWWA,* 68(3), March 1976: 168.

Rossum, J. R., and D. T. Merrill. "An Evaluation of the Calcium Carbonate Saturation Indexes." *Jour. AWWA,* 75(2), February 1983: 95.

Ryder, R. A. "State of the Art in Water Treatment Design, Instrumentation and Analysis." *Jour. AWWA,* 69(11), November 1977: 612.

Sawyer, C. N., and P. L. McCarty. *Chemistry for Sanitary Engineers* (2nd ed.). New York: McGraw-Hill, Inc., 1967.

Semmens, M. J., and A. B. Staples. "The Nature of Organics Removal During Treatment of Mississippi River Water." *Jour. AWWA,* 78(2), February 1986: 76–81.

Sinsabaugh, R. L., R. C. Hoehn, W. R. Knocke, and A. E. Linkins. "Removal of Dissolved Organic Carbon by Coagulation with Iron Sulfate." *Jour. AWWA,* 78(5), May 1986: 74.

Snoeyink, V. L., and D. Jenkins. *Water Chemistry.* New York: John Wiley and Sons, 1980.

Sorg, T. J. and G. S. Logsdon. "Treatment Technology to Meet the Interim Primary Drinking Water Regulations for Inorganics: Part 5." *Jour. AWWA,* 72(7), July 1980: 411.

Sorg, T. J., O. T. Love, Jr., and G. Logsdon. *Manual of Treatment Techniques for Meeting the Interim Primary Drinking Water Regulations.* USEPA Report 600/8.77.005, MERL, Cincinnati, Ohio, 1977.

Stumm, W., and J. J. Morgan. *Aquatic Chemistry.* New York: Wiley-Interscience, 1981.

Taylor, J. S., B. R. Snyder, B. Ciliax, C. Ferraro, A. Fisher, P. Muller, and D. Thompson. *Trihalomethane Precursor Removal by the Magnesium Carbonate Process.* EPA/600/S2-84/090. Cincinnati, Ohio: Water Engineering Research Laboratory, 1984.

Taylor, J. S., D. Thompson, B. R. Snyder, J. Less, and L. Mulford. *Cost and Performance Evaluation of In-Plant Trihalomethane Control Techniques.* EPA/600/S2-85/138. Cincinnati, OH: Water Engineering Research Laboratory, 1986.

Thompson, J. D., M. C. White, G. W. Harrington, and P. C. Singer. "Enhanced Softening: factors influencing DBP Precursor Removal." *Jour. AWWA,* 89(6), June 1997: 94–105.

Truesdell, A. H., and B. F. Jones. *WATES, A Computer Program for Calculating Chemical Equilibria of Natural Waters.* PB 220464 Washington, D.C.: NTIS, U.S. Department of Commerce, 1973.

Trussell, R. R., L. L. Russell, and J. F. Thomas. "The Langelier Index." In *Water Quality in the Distribution System,* Fifth Annual AWWA Water Quality Technology Conference, Kansas City, Missouri, 1977.

Weber, W. J., Jr., and A. M. Godellah. "Removing Humic Substances by Chemical Treatment and Adsorption." *Jour. AWWA,* 77(4), April 1985: 132.

Wolf, H. W., R. S. Safferman, A. R. Mixson, and C. E. Stringer. "Virus Inactivation During Tertiary Treatment." *Jour. AWWA,* 66(9), September 1974: 526.

CHAPTER 11
MEMBRANES

J. S. Taylor, Ph.D., P.E.
Alex Alexander Professor of Engineering
Civil and Environmental Engineering Department
University of Central Florida
Orlando, Florida

Mark Wiesner, Ph.D.
Professor of Engineering
Environmental Sciences and Engineering Department
Rice University
Houston, Texas

Membranes represent an important new set of processes for drinking water treatment. Their tremendous potential results from universal treatment capabilities and competitive cost. There are very few drinking water contaminants that cannot be removed economically by membrane processes, and several applications have been described in textbooks on water treatment (Weber, 1972; Belfort, 1984; Nalco, 1988). However, membrane processes with the greatest immediate application to potable water treatment are reverse osmosis (RO), nanofiltration (NF), electrodialysis (ED), ultrafiltration (UF), and microfiltration (MF).

Reverse osmosis is primarily used to remove salts from brackish water or seawater, although RO is also capable of very high rejection of synthetic organic compounds (SOCs). Nanofiltration, the most recently developed membrane process, is used to soften fresh waters and remove disinfection by-product (DBP) precursors. Electrodialysis is used to demineralize brackish water and seawater and to soften fresh water. Ultrafiltration and microfiltration are used to remove turbidity, pathogens, and particles from fresh waters. In the broadest sense, a membrane—the common element of all these processes—could be defined as any barrier to the flow of suspended, colloidal, or dissolved species in any solvent. The applicable size ranges for membrane processes are shown in Figure 11.1.

Typically, the cost of membrane treatment increases as the size of the solute removed decreases. The ionic range in Figure 11.1 encompasses potable water solutes such as sodium, chloride, total hardness, most total dissolved solids, and smaller DBP precursors. The macromolecular range includes large and small colloids, bacteria, viruses, and color. The fine particle range includes larger turbidity-producing particles, most total suspended solids, cysts, and larger bacteria. The

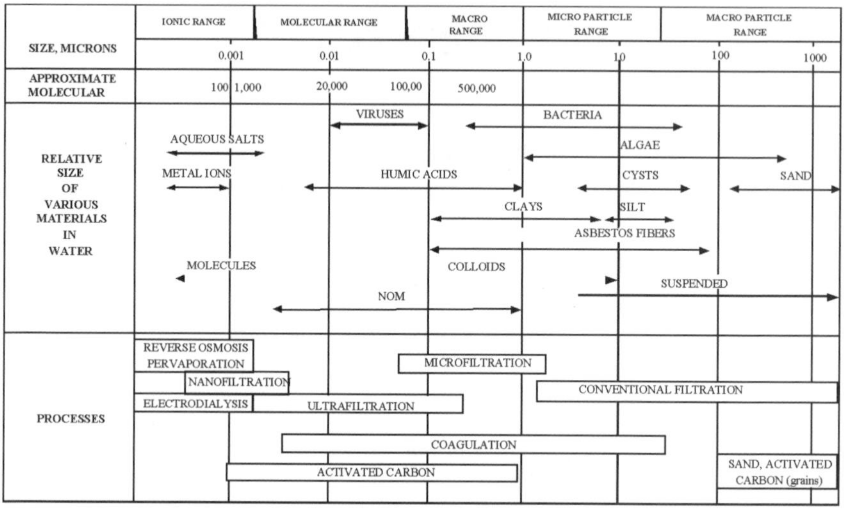

FIGURE 11.1 Size ranges of membrane processes and contaminants.

membrane processes normally used in the ionic range remove macromolecules and fine particles, but because of operational problems, they are not as cost effective as membranes with larger pores.

Drinking water contaminants are presented as biological, inorganic, and organic contaminants, as well as radionuclides, particulates, and other groupings. The ability of membranes to remove these contaminants can be inferred from Table 11.1 and Figure 11.1, both of which show effective size ranges of contaminants that can be partially or completely removed by each membrane process. If sieving or size exclusion is the mechanism of solute rejection, complete removal can be achieved by a membrane system free of defects. All membrane processes can reject contaminants such as turbidity or pathogens, but UF and MF are the most cost-effective processes for control of large particles. Smaller contaminants are removed via size exclusion, charge repulsion, or diffusion mechanisms in membrane processes, leaving some residual. Although such a residual may be below the detection level, it is present.

Contaminants larger than the maximum pore size of the membrane are completely removed by sieving in a diffusion-controlled process. Contaminant rejection by diffusion-controlled membrane processes increases as species charge and size increase. Consequently, satisfactory removal of metals, total dissolved solids (TDS), biota, radionuclides, and disinfection by-product precursors can be attained. No commercially available membrane effectively removes dissolved and uncharged species such as hydrogen sulfide (H_2S) and small, uncharged organic contaminants. Other aqueous contaminants should be treatable by membrane processes, although, once again, the cost of this treatment generally increases as the size of the removed contaminant decreases. Many membrane manufacturers specify the molecular weight cutoff (MWC) values for membranes. The MWC represents a nominal molecular weight of a known species that would always be rejected in a fixed percentage, using a specific membrane under specific test conditions. (However, MWC testing conditions vary among manufacturers, which limits the use of MWC for membrane specification.)

Although many factors affect solute separation by these processes, a general understanding of drinking water applications can be achieved by associating mini-

TABLE 11.1 Potable Water Contaminants Classified by Effective Membrane Size Range

		Size range		
		Ionic	Macromolecular/Colloidal	Fine particle
Biological				
	Viruses		X	
	Bacteria		X	
	Helemitis			X
	Algae			X
	Protozoa			X
	Cysts			X
	Fungi			X
Inorganics				
	Metals	X		
	Chlorides	X		
	Fluoride	X		
	Sulfate	X		
	Nitrate	X		
	Cyanide	X		
Organics				
	Priority pollutants	X		
	Surfactants		X	
	NOM		X	
	THM precursors	X		
	DPB precursors	X		
Radionuclides				
Particulates				
	Turbidity			X
	TSS			X
Other				
	Color		X	
	TDS	X		

NOM, natural organic matter; THM, trihalomethane; TSS, total suspended solids; TDS, total dissolved solids.

mum size of solute rejection with membrane process and regulated contaminants (Taylor et al., 1989). One correct interpretation of Figure 11.1 is to assume that each membrane process has the capability of rejecting solutes larger than the size shown in the exclusion column. As shown in Table 11.1, regulated drinking water solutes can be simplified to the categories of pathogens, organic solutes, and inorganic solutes. Pathogens can be subdivided into cysts, bacteria, and viruses. Organics can be subdivided into DBPs and SOCs. Inorganic parameters are total dissolved solids, total hardness, and heavy metals, among others.

Electrodialysis and electrodialysis reversal (EDR) processes are capable of removing the smallest charged contaminant ions to 0.0001 μm. Consequently, ED and EDR are limited to treatment of ionic contaminants and are ineffective for pathogen and organics removal in most cases. Contaminant rejection by RO and NF occurs by both diffusion and sieving. They can remove all pathogens and many organic contaminants by sieving; by diffusion, they can achieve almost total removal of ionic contaminants. The RO and NF processes have the broadest span of treatment capabilities. Commercial UF membranes can achieve greater than 6-log removal of all currently known pathogens from drinking water. Commercial MF membranes can achieve greater than 6-log removal of cysts. Consequently, these processes effectively remove turbidity and microbiological contaminants, making them ideal for treating the majority of drinking water sources in the United States.

Both pressure-driven membranes (RO, NF, UF, and MF) and coagulation, sedimentation, and filtration (CSF) processes remove particles and pathogens from drinking water. A pressure-driven membrane process accomplishes this goal by presenting a static barrier that rejects any pathogen too large to pass through the membrane's pores. A CSF process removes pathogens through a dynamic process. Coagulation encourages formation of nuclei, which agglomerate during flocculation into large particles that can be removed by sedimentation/filtration. Effective removal requires that pathogens collide and remain with flocculated particles. Without effective coagulation, even large pathogens like cysts can pass through a CSF system.

Cysts, bacteria, and viruses can be grouped into ranges of 10–1 μ, 1–0.1 μ, and less than 0.1 μ. The nominal pore size of an MF membrane is less than those of cysts and bacteria, but this process alone cannot accomplish virus removal (Jacangelo et al., 1991; Taylor, Reiss, and Robert, 1999). Ultrafiltration has been shown to achieve more than 6-log removal of all pathogens. Reverse osmosis and nanofiltration have been shown to achieve 4-log pathogen removal (Taylor, Reiss, and Robert, 1999). Direct comparisons of conventional treatment (CSF) with membrane processes is difficult due to lack of literature; however, many state regulatory requirements allow 3-log removal of *Giardia* by CSF processes. No process is capable of absolute pathogen removal; however, the literature indicates a significant advantage for membrane processes over conventional CSF treatment for pathogen rejection.

Regulatory Environment for Membrane Processes

Existing regulations have been and will be modified to include more stringent control of chemical and biological toxins. The Safe Drinking Water Act (SDWA) amendments that were modified in 1996 still require the United States Environmental Protection Agency (USEPA) to create new drinking water regulations. The regulatory changes will continue to create a need for new drinking water technology to meet these challenges.

The surface water treatment rule (SWTR) has been in effect since 1993 and was developed to reduce the potential for pathogenic contamination of drinking water. The SWTR requires that all surface waters and waters under the direct influence of surface waters achieve a minimum of a 3-log reduction of *Giardia* cysts and a 4-log reduction of enteric viruses. In addition, disinfection is required by the SWTR to meet a CT (mg/L disinfectant × disinfectant contact time) standard. The required CT is dependent on disinfectant used, pathogen, and water quality parameters. The enhanced SWTR scheduled to be promulgated in 2000 will link the required log removal of pathogens to the source water quality. The USEPA has stated that well-operated coagulation, sedimentation, and filtration systems are capable of 2.5-log removal of *Giardia* and 3-log removal of viruses. However, pilot studies have shown that some membrane processes can consistently achieve greater than 6- to 7-log pathogen removal (Jacangelo et al., 1992). The current proposal for the groundwater disinfection rule (GWDR), planned for finalization in 2000, requires most groundwater source utilities to maintain disinfectant residuals. Based on pore size alone, MF should be capable of removing bacteria and cysts, while UF, NF, and RO should remove all pathogens—even viruses. However, all of the membrane processes must be followed by disinfection to ensure distribution system water quality.

Contrary to the focus on chemical contamination during the 1980s, improving methods for detection of biological pathogens have shown that contamination by waterborne disease-causing agents may well be the more serious problem. Specific contamination by *Salmonella, Legionella, E. coli,* and *Cryptosporidium* have recently been documented in the United States, and indications are that many incidents of waterborne disease go unidentified and unreported. Consequently, the need to use the best treatment technology and ensure distribution system water quality has been emphasized in recent regulatory action. The Total Coliform Rule, in effect since 1991, requires a maximum concentration of zero for total coliform, fecal coliform, and *E. coli.* These species can be totally rejected by any pressure-driven membrane process, which provides the maximum protection by a treatment process. In addition, NF and RO can remove a significant amount of disinfectant demand, enhancing the stability of the residual in the distribution system. Membranes are among the best options to meet existing and future drinking water regulations regarding pathogens.

Inorganic compounds (IOCs), synthetic organic compounds (SOCs), and volatile organic compounds (VOCs) are regulated by amendments to the SDWA. Five phases of IOC, SOC, and VOC regulations are in place, and the final phases will eventually be enacted. The radionuclides rule, which covers radon gas, uranium, radium, and some alpha emitters, is expected to be finalized in 2000. Sulfate and arsenic rules are proposed, and the regulations are expected to be finalized by 2000 and 2001, respectively.

The IOC, SOC, and VOC regulations involve hundreds of specific compounds, which have varying susceptibility to removal by membrane processes. Some generalizations can be made, though. No commercially available membrane process is being used for VOC removal at this time. Research has shown that membranes are feasible for VOC removal, but this process has not been developed by suppliers. VOCs are uncharged, have low molecular weights (MWs), and pass through membranes with water. Like any dissolved gas, the VOC concentrations in the feed, permeate, and concentrate streams are essentially identical in conventional membrane processes. However, gas separation membranes are used commercially and may be adapted to drinking water treatment in the future.

The EDR, MF, and UF processes by themselves generally are not capable of rejecting SOC or IOC contaminants. The RO and NF processes can achieve significant SOC and IOC rejection because the exclusion limit of these membranes is so small that many SOCs cannot pass or are diffusion controlled (Duranceau, Taylor, and Mulford, 1992). It would be incorrect to infer that all SOCs or IOCs can be rejected by RO or NF, but these processes have shown promise to remove such contaminants. The RO and NF processes rely primarily on both sieving and diffusion mechanisms to reject SOCs and IOCs from drinking water. Rejection increases as the molecular weight and charge of the contaminant increase. While EDR, RO, and NF remove arsenic and sulfates, UF and MF do not; UF and MF could remove SOCs if powdered activated carbon (PAC) were added prior to membrane treatment. EDR should be considered as a viable means of removing any charged solute from drinking water.

The lead and copper rule (LCR), finalized since 1991, is intended to control the concentrations of these metals in drinking water. Corrosion in the distribution system emphasizes the importance of finished water chemistry. Corrosion is significantly affected by the sulfate, sodium, chloride, and bicarbonate ion concentrations in the finished water. The UF and MF processes do not affect corrosion because they remove no ions. The RO and NF processes reject sodium, sulfate, chloride, calcium, and bicarbonate ions, potentially producing a corrosive finished water in the absence of buffering. However, if a base is added to the RO or NF permeate stream prior to aeration, the carbon dioxide will be converted to bicarbonate, and alkalinity recovery will be accomplished. If CaO is used as the base, the finished water will contain very little ion content except for calcium and bicarbonate. Generally, the least corrosive water would be a stable water with from 1 to 3 meq/L of calcium and bicarbonate alkalinity, which is easily produced by alkalinity recovery, as discussed later in the chapter.

The information collection rule (ICR) was put into effect in mid-1996. In addition to extensive water quality monitoring, it requires large utilities using surface water with total organic carbon (TOC) greater than 4 mg/L and groundwater with finished water greater than 2 mg/L to conduct bench or pilot studies using either granular activated carbon (GAC) or membrane processes. The primary intent is to gain information on a national level on the ability of GAC and membrane processes to remove DBP precursors. The information collected under the ICR will be used to formulate future regulations such as the Enhanced Surface Water Treatment Rule (ESWTR) and the disinfection by-products rule (DBPR). The DBPR has two phases, with phase 1 finalized in 1998 and phase 2 to be finalized in 2002. The existing DBP maximum contaminant level (MCL) of 100 μg/L for THMs applies to utilities serving more than 10,000 customers. Phase 1 of the DBPR will change the MCL to 80 μg/L for THMs and 60 μg/L for haloacetic acids (HAAs) and will apply to all utilities. If enacted, phase 2 of the DBPR will reduce the MCLs to 40 and 30 μg/L THMs and HAAs, respectively. Membrane processes can effectively control DBP formation by removing precursors in the form of natural organic matter, usually measured as TOC. Not all TOC solutes are DBP precursors, but all organic DBP precursors are TOC. By removing the precursors, a utility can maintain a free chlorine residual in the distribution system and reduce DBPs below the proposed phase 2 MCLs.

Since EDR does not remove any uncharged species, and since the vast majority of organic DBP precursors are uncharged, EDR is not effective for organic DBP control. Pores of UF and MF membranes are too large to reject significant amounts of DBP precursors, but these processes can be used to control DBP and TOC if they are preceded by coagulation. They replace conventional sedimentation and filtration and are limited to the 50 to 75 percent TOC reductions achieved by coagulation (Taylor et al., 1986). The RO and NF processes can achieve more than 90 percent TOC removal. By rejecting disinfectant demand, these processes greatly reduce the disinfectant dose, while achieving maximum distribution system integrity because the residual lasts longer. A final major advantage of membranes is the removal of contaminants without producing any oxidation by-products. All oxidants, including chloramines, produce organic by-products that may be controlled by future regulations. In the not too distant future, TOC may be regulated, further encouraging use of membranes. A summary of membrane process applications and drinking water regulations is shown in Table 11.2.

The growth of drinking water regulations for both chemical and biological species has created treatment applications that can be met by membrane processes. Membranes can be effectively used for total removal of pathogens, for high removals of inorganic and organic contaminants, and for maintaining the highest possible distribution system integrity. There are very few instances where membranes cannot be utilized to meet or exceed all drinking water regulations.

CLASSIFICATIONS AND CONFIGURATIONS OF MEMBRANE PROCESSES

Some membrane processes rely on pressure as the driving force to transport fluid across the membranes. They can be classified by the types of materials they reject and

TABLE 11.2 Summary of Membrane Process Applications for Drinking Water Regulations

	Membrane process				
Rule	EDR	RO	NF	UF	MF
SWTR/ESWTR	no	yes	yes	yes	yes
CR*	no	yes	yes	yes	yes
LCR	yes	yes	yes	no	no
IOC	yes	yes	yes	no	no
SOC	no	yes	yes	yes (+PAC)	yes (+PAC)
Radionuclides	yes (no radon)	yes (no radon)	yes (no radon)	no	no
DBPR	no	yes	yes	yes (+ coagulation)	yes (+ coagulation)
GWDR	no	yes (expected)	yes (expected)	yes (expected)	yes (expected)
Arsenic	yes	yes	yes	yes (+ coagulation)	yes (+ coagulation)
Sulfates	yes	yes	yes	no	no

* CR, Coliform Rule.

the mechanisms by which rejection occurs. The progression of microfiltration to ultrafiltration to nanofiltration to reverse osmosis corresponds to a decreasing minimum size of components rejected by membranes as well as increasing transmembrane pressures required to transport fluid across the membranes and decreasing recoveries. Although a continuum of mechanisms likely contributes to separation in these pressure-driven processes, very clear differences in separation methods distinguish reverse osmosis from microfiltration. These mechanisms are discussed later in the text.

While pressure-driven processes like RO pass water through the membranes, electrodialysis involves the passage of the solute rather than the solvent through the membrane. As a consequence, both the mechanism of separation and the physical characteristics of membranes in ED differ substantially from those in pressure-driven processes. The ED membranes are fundamentally porous sheets of ion-exchange resin with a relatively low permeability for water.

Membranes are classified by solute exclusion size, which is sometimes referred to as *pore size.* A reverse osmosis or hyperfiltration membrane rejects solutes as small as 0.0001 μm, which is in the ionic or molecular size range. A nanofiltration membrane rejects solutes as small as 0.001 μm, which is also in the ionic and molecular size range. Solute mass transport in these processes is diffusion controlled. Ultrafiltration and microfiltration membranes have a minimum solute rejection size of 0.01 and 0.10 μm, respectively. These membranes reject colloidal particles, bacteria, and suspended solids by size exclusion, and contaminant rejection is not diffusion controlled. Pressure drives the transport of water (the solvent) through these membranes. Electrodialysis relies on charge for solute separation and pulls ions through ED membranes, so it is unaffected by pore size.

Membranes can be classified by molecular weight cutoffs, solute and solvent permeability, solute and solvent solubility in the membrane film, active film material, active film thickness, surface charge, and active film surface. The molecular weight cutoff is the degree of exclusion of a known solute, as determined for a given set of test conditions in the laboratory. Typical known solutes used for determination of molecular weight cutoff are sodium chloride, magnesium sulfate, dextrose, and some dyes. Solute mass transport through a diffusion-controlled membrane is influenced by solute type and aqueous environment, so attempts to characterize them benefit from the use of additional organic solutes such as aromatic and aliphatic compounds of known molecular weight and structure.

Diffusion-controlled solute and solvent permeability are best described by mass transfer coefficients (MTC), although percent rejection is sometimes used to describe solute mass transfer for a given environment. The solvent and solute mass transfer coefficients, like molecular weight cutoffs, are measured in the laboratory with standardized test conditions and test cells. Continuous flow test cells that operate at less than 1 percent recovery can be used to minimize concentration-polarization so that linearly determined mass transfer coefficients can be used to characterize membranes. Such test cells can also be used to determine the changes in solvent mass transfer coefficients over time for different films (membranes) to determine a potential correlation between films and fouling.

Solute and solvent solubility in the membrane film are very important characteristics of diffusion-controlled membrane processes. Ideally the active film would have no solubility for the contaminants or foulants and complete solubility for water. The solute mass transfer coefficient increases as the solubility of the contaminant in the film increases. Hence, the lower the contaminant solubility in the film, the lower the contaminant concentration in the permeate. If needed, the film material can be modified to effect solubility by inclusion of functional groups, or a different film material can be substituted.

The thickness and chemical structure of the film are important membrane characteristics and are difficult to measure, although techniques do exist. Plasma etching has been used to measure film thickness. A plasma is introduced to initiate a chemical reaction with the top layer of the film, producing gases. The rate of gas production during etching indicates surface morphology and thickness of a nonporous film. Electron spectroscopy and X-ray photoelectron spectrometry can be used to determine the thickness of the top layer.

Membrane characterization also gives desirable information about the potential of particular membranes to reject pesticides. This information could be used to modify existing membranes or develop new membranes that would effectively reject pesticides from drinking water sources.

Classification by Material

The development of a variety of membrane materials, both synthetic and modified natural polymers coupled with various manufactured forms, has played an integral part in the development of industrial-scale membrane separation applications. In any such process, several important membrane characteristics must be considered. Membrane selectivity, permeability, mechanical stability, chemical resistance, and thermal stability are among these critical factors, which are highly dependent upon the type of material and the process control variables applied during manufacturing (Rautenbach and Albrecht, 1989).

Among the many raw materials used for membrane manufacturing, most basic types involve various forms of modified natural cellulose acetate materials and a variety of synthetic materials. To name a few, these synthetic materials are primarily composed of polyamides, polysulfone, vinyl polymers, polyfuran, polybenzimidazole, polycarbonate, polyolefins, and polyhydantoin. The diverse chemistries of the specific polymers and the associated production kinetics yield the individual attributes that provide for the myriad of membrane selectivity and productivity combinations (Rautenbach and Albrecht, 1989).

Porous materials produced by precipitation from a homogeneous polymer solution are termed *phase inversion membranes.* They incorporate both symmetrical (homogeneous) and asymmetrical structures. The basic production process consists of five fundamental steps. A homogeneous polymer solution must first be produced. The polymer film is then cast, followed by partial evaporation of the solvent from the polymer film. Immersion of the polymer film in a precipitation solution then allows the solvent to be exchanged for the precipitation agent. Imperfections in the precipitated membrane film are restructured by treatment in a heated bath solution. Variations in environmental conditions for each of these steps can generate an assorted range of membrane structures that affect system performance. Typical membrane structural profiles can range from well-defined cavities in the shape of fingers to pores arranged in a dense sponge structure (Figure 11.2). These configurations result from the membrane structure both at the membrane surface and within the support structure itself (Rautenbach and Albrecht, 1989).

Both symmetric and asymmetric membranes can be produced by the phase inversion process. The difference between these two membrane classifications is the environmental conditions in which the membranes are produced and their resulting structural profiles. In the production of symmetrical membranes, homogeneous conditions for material formation throughout the membrane matrix lead to a uniform polymer structure. Conversely, structural properties vary throughout asymmetric membranes. Production of an asymmetric membrane forms a dense surface layer of submicron thickness that gives the membrane its selectivity properties. This active layer is in turn supported by a porous support structure. The combination yields membrane materials that are both selective and mechanically stable, with enhanced productivity. Given the reduction in thickness of the active layer, hydraulic losses across the membrane are significantly less than those associated with symmetric membranes of comparable thickness. Consequently, development of the asymmetric membrane structure is an additional component integral to the commercial success of both RO and UF. An actual scanning electron microscope (SEM) image of one asymmetric membrane is provided in Figure 11.3.

Another classification of phase inversion membranes is the composite membrane. In such a membrane, the materials composing the active surface are different

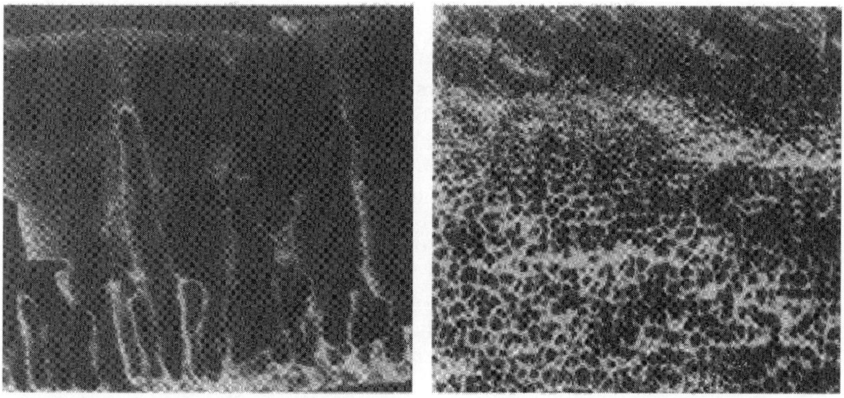

(a) (b)

FIGURE 11.2 SEM photograph of two common membrane structures: (*a*) finger structure; (*b*) sponge structure. (*Source:* R. Rautenbach and R. Albrecht, *Membrane Processes.* Copyright © 1989 John Wiley & Sons. Reprinted by permission of John Wiley & Sons Limited.)

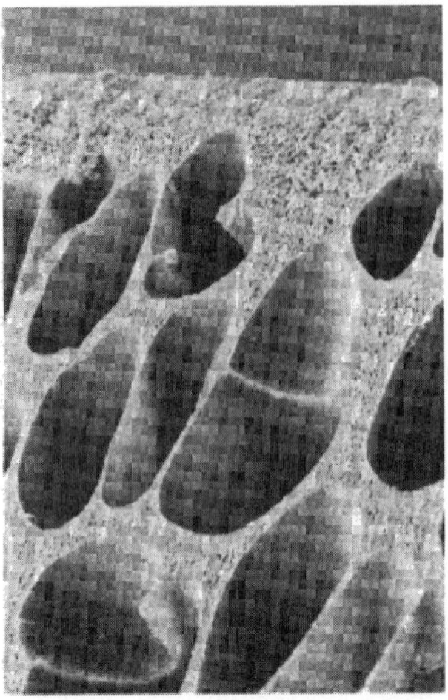

FIGURE 11.3 Polyhydantoin asymmetric membrane. (*Source:* R. Rautenbach and R. Albrecht, *Membrane Processes.* Copyright © 1989 John Wiley & Sons. Reprinted by permission of John Wiley & Sons Limited.)

from those of the support material. These membranes are produced by lamination of the active surface layer onto the support layer (e.g., polysulphone). This class of membranes is generally regarded as an improvement in membrane material design in that a specific active surface layer can be matched with a support layer of optimum porosity. This combination enhances membrane productivity while retaining the desired rejection properties offered by the dense active surface layer. Figure 11.4 shows the characteristic layers in a composite membrane. Interestingly, the support layer itself is an asymmetric membrane.

The active layer or membrane film is a polymer or combination of polymers forming a composite layer of varying thin films or a single thin-film layer. These polymers are generally in the form of straight chain compounds such as cellulose acetate or aromatic compounds such as a polyamide or polyimide. Several different interactions occur between the polymers that form the active layer of the membrane and, consequently, between the membrane and solutes that pass through it. The three types of secondary forces are dipole forces, dispersion forces, and hydrogen bonding forces. Covalent and ionic forces are primary forces, with stronger effects than those of the secondary forces in the active membrane film. Average values of primary and secondary forces in membrane polymers are 400 kJ/mole for covalent and ionic forces, 40 kJ/mole for hydrogen bonding forces, 20 kJ/mole for dipole forces, and 2 kJ/mole for dispersion forces. These membrane forces can interact with corresponding forces associated with solutes, possibly promoted by functional groups on pesticides.

Hyperfiltration (reverse osmosis at 800 to 1200 psi), reverse osmosis, and nanofiltration membrane films have active layers of cellulose compounds, aliphatic or aromatic polyamides, and thin-film composites. Cellulose triacetate is used as the active film for many desalination applications. Cellulose derivatives have good properties for membranes, since their crystalline and hydrophilic properties enhance durability and their capacity to transport water. Cellulose membranes are subject to chemical degradation by hydrolysis and biological degradation by oxidation. They must be operated at ambient temperatures from pH 4.0 to 6.5 with a biocide to avoid degradation.

Polyamide membranes are also very effective films. Aromatic polyamides are generally preferred over aliphatic polyamides because of their mechanical, thermal, chemical, and hydrolytic stability and their permaselective properties. The aliphatic polyamides are porous and therefore not permaselective. They may be used for sieving applications such as ultrafiltration and microfiltration processes.

The development of the cross-linked fully aromatic polyamide thin-film composite membrane in the late 1970s represented a major advance in membrane technology. Thin-film composites provided very thin active films that required much less

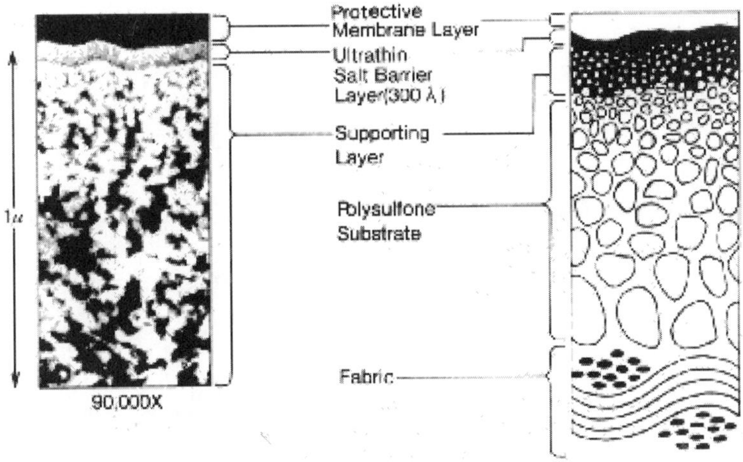

FIGURE 11.4 Composite membrane structure (PEC 1000 Toray). (*Source:* R. Rautenbach and R. Albrecht, *Membrane Processes.* Copyright © 1989 John Wiley & Sons. Reprinted by permission of John Wiley & Sons Limited.)

energy to induce fluid passage than previous materials, making them more economical to use on a large scale. As previously mentioned, thin-film composites layer asymmetric films to form membranes with several different characteristics. Both hydrophilic and hydrophobic films are laid in a composite film by cross-linking different polymers. The thickness of the nonporous layer is typically less than 1 μm. Cross-linking to the porous film provides needed support for the nonporous film. Removal of macromolecules and ionic species is achieved by a nonporous film in a pressure-driven membrane process.

Classification by Geometry

Reverse osmosis and nanofiltration membranes are made from different materials and in different configurations. Both the material and configuration of such a membrane affect its mass transport or performance. As the effects of materials and configurations on solute rejection by membranes is largely unknown, the following paragraphs briefly discuss membrane materials and configurations.

RO/NF Configuration. The RO and NF membranes for drinking water treatment have either spiral wound (SW) or hollow fine fiber (HFF) configurations. The SW configuration is the most common for production of drinking water. The HFF configuration is used extensively for desalination of seawater in the Middle East. The geometry of an SW membrane is subject to fewer dead areas than that of an HFF membrane, it can be cleaned more thoroughly, and it is less subject to fouling. The ratio of surface area to volume is higher for an HFF element than for an SW element. However, dominant fouling mechanisms in seawater may differ from those in brackish waters.

An HFF membrane element consists of a tube that contains a bundle of HFFs, as shown in Figure 11.5. A group of HFFs folded in a U form are sealed into a tube to create a bundle. A 4-in DuPont B-10 membrane, for example, contains 650,000 HFF,

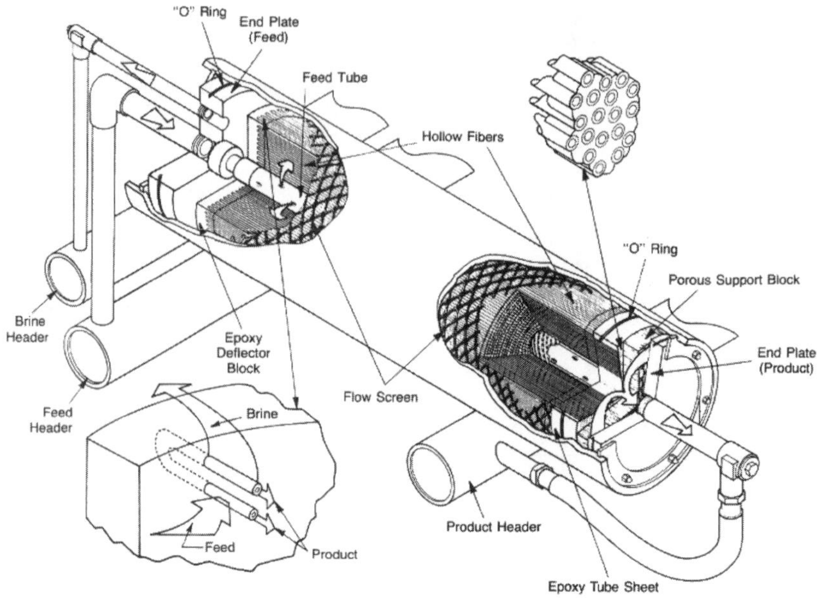

FIGURE 11.5 Diagram of a hollow fine fiber reverse osmosis membrane. (*Source:* Courtesy of DuPont Permasep®.)

each approximately 4 ft (3.28 m) long with 1500 ft^2 (139 m^2) of surface area. The inside and outside diameters of the fibers are 1.33E-4 ft (4.1E-5 m) and 2.94E-4 ft (11.0 E-5m), respectively. As the feed stream passes along the outside surface of the hollow fine fibers, the permeate or purified stream passes from the outside to the inside. The feed stream enters the hollow fine fiber element from a feed tube in the center of the element. The feed stream flows radially from the center feed tube to the brine collection channel at the outside of the element. The highest feed stream velocity occurs next to the feed stream tube, and the lowest velocity occurs at the brine collection tube. The recovery from a hollow fine fiber element ranges from 10 to 50 percent and is typically higher than that from an SW element. The radial feed stream velocity along the outside surface of the HFF varies from approximately 0.01 to 0.001 ft/s (0.003 to 0.0003 m/s), which gives a Reynolds number ranging from 100 to 500 for transport through the membrane. There are other sizes of hollow fine fiber elements but the feed stream velocities, recoveries, and Reynolds numbers are similar to those of the described B-10 element. The feed flow is in the laminar region and is most likely to produce chemical or colloidal fouling near the brine collector. Also, fouling from the collection of filtered particles may occur near the feed tube. The physical configuration of the hollow fine fiber element is prone to fouling by particulates removed by sieving, and the bundle of HFFs is difficult to clean.

Spiral wound elements are manufactured using flat-sheet membranes, as opposed to bundles of fibers, as shown in Figure 11.6. A typical SW element consists of envelopes attached to a center tube that collects the permeate stream. Designs of SW elements differ among manufacturers; however, the following description is applicable to Filmtec, Desal, Hydranautics, and Fluid Systems SW membranes.

An envelope is formed by folding one flat sheet over a permeate stream spacer. The sheet itself consists of at least two layers, a nonporous active membrane film and

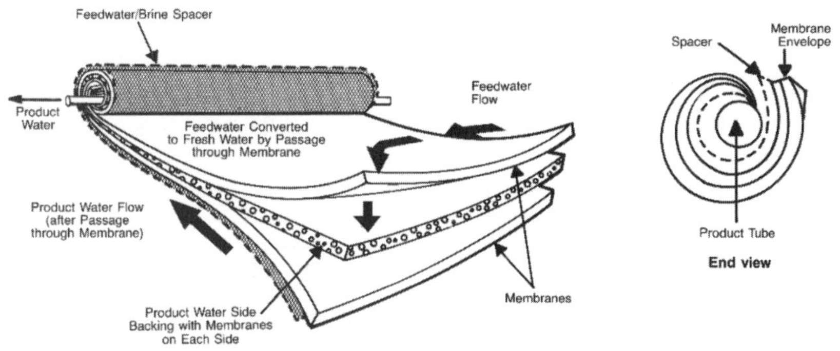

FIGURE 11.6 Diagram of a spiral wound reverse osmosis membrane. (*Source:* Courtesy of DuPont Permasep®.)

a porous membrane support. The active layer is on the outside of the fold. The envelope is glued along three open sides and near the fold, completely enclosing the permeate spacer. The glue line on the fold end is a short distance away from the fold, because the fold end is attached to the center collection tube. The glue line at the fold end stops the flow of the feed stream and allows the remaining pressure in the permeate stream to drive it through the membrane into the center collection tube. A feed stream spacer is attached to each envelope prior to establishing the fold-end glue line. Several envelopes and feed stream spacers are attached to the center collection tube and wrapped in a spiral around it. An epoxy shell or tape wraps are applied around the envelopes, completing the SW element.

The feed stream enters the end of the SW element in the channel created by the feed stream spacer. The feed stream can flow either in a path parallel to the center collection tube or through the active membrane film and membrane supports into a channel created by the permeate stream spacers. The permeate stream follows a spiral path into the center collection tube and is taken away as product water in a drinking water application. As with hollow fine fiber membranes, the feed stream becomes progressively more concentrated as it passes to a succeeding element. A 4-in. Filmtec NF70 membrane contains four envelopes with approximately 90 ft^2 (8.33 m^2) of surface area in a sheet that measures 3 ft (0.91 m) × 3.75 ft (1.14 m). The total element is 3.33 ft (1.01 m) long, but the feed stream path along the active membrane film is approximately 3 ft (0.91 m).

The recovery in an SW element varies from approximately 5 to 20 percent. The manufacturer-specified maximum feed and concentrate stream flows in a 4-in (1.57-cm) element are approximately 16 gpm (4.2E-3 m^3/min) and 3 gpm (7.9E-4 m^3/min). Neglecting the effect of the feed stream spacer, the Reynolds number typically ranges from 100 to 1000. The feed stream spacer creates additional turbulence and increases the Reynolds number. The physical configuration of the SW element produces a more turbulent feed stream than that in an HFF element and leaves the membrane more easily accessible to cleaning agents. The highest and lowest feed stream velocities occur at the entrance and exits of the element, respectively. The feed flow is in the laminar region and is most likely to produce chemical or colloidal fouling in the last elements in series. Fouling from particle deposition occurs mainly in the first elements in series.

MF-UF Configuration. The flow in microfiltration or ultrafiltration processes can run from inside out or outside in, as shown in Figure 11.7. These processes typically

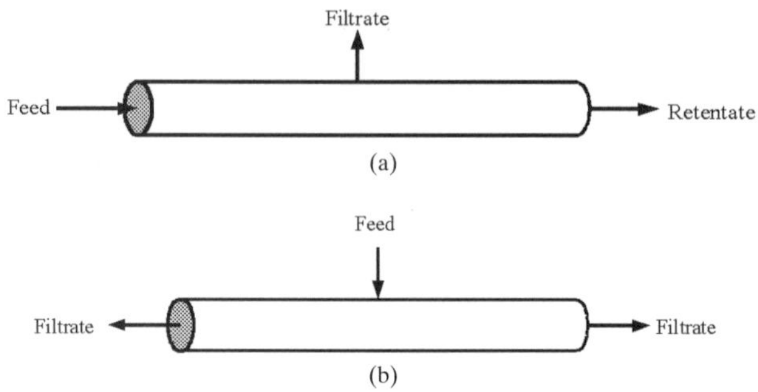

FIGURE 11.7 Individual fiber flow pattern for (*a*) inside-out and (*b*) outside-in flow.

use hollow fibers larger in diameter than the HFFs previously described. The inside and outside diameters of the HFF are 1.33E-4 ft (4.1E-5 m) and 2.94E-4 ft (11.0 E-5m), whereas the inner and outer diameters of an HF fiber are 3.08E-4 ft (1E-3M) and 6.16E-3 ft (2E-3), respectively. Some UF or MF membranes have been produced in SW configurations for industrial applications. In a conventional UF or MF process, the driving force to produce filtrate can work in two ways; positive pressure moves fluid through the fibers, usually at a rating lower than 35 psi, and negative pressure moves fluid through fibers under vacuum pressure. Combining the two different flow regimes and the two driving forces allows four different configurations:

Inside-out flow with positive pressure

Outside-in flow with positive pressure

Outside-in flow with negative pressure

Inside-out flow with negative pressure

Both inside-out and outside-in flow patterns can be further characterized as either dead-end or cross-flow operations. In a dead-end operation, the entire feed flow passes through the membrane. In one pass, the dead-end mode of operation is analogous to conventional coarse filtration in that the retained particles accumulate and form a type of cake layer at the membrane surface. In a cross-flow operation, only a portion of the flow passes through the membranes, and only a portion of the retained solutes accumulate at the membrane surface. The remaining flow (retentate) is recycled on the feed side. The cross-flow regime also incorporates a tangential flow that shears the cake and minimizes the accumulation of solids on the membrane surface. In all cases, a frequent backwash (every 15 min to 1 h or more frequently) removes the cake formed on the membrane surface.

Classification by Driving Force

Membrane processes can be classified based on the driving forces that induce transport of materials across the membranes. Examples of driving forces and corresponding membrane processes are listed in Table 11.3.

TABLE 11.3 Categorization of Membrane Processes by Driving Force

Driving force	Examples of membrane processes
Temperature gradient	Thermoosmosis
Concentration gradient	Dialysis, pervaporation, osmosis
Pressure gradient	RO, NF, UF, MF, piezodialysis
Electrical potential	Electrodialysis, electroosmosis

Although interest in industrial applications for pervaporation is growing, industrial-scale applications of membrane processes for environmental quality control have so far been dominated by the pressure-driven processes such as RO and by electrodialysis and electrodialysis reversal, which employ electrical potential as the driving force.

Electrodialysis

Electrodialysis is a membrane process driven by electric potential for removing charged species (ions) from an aqueous stream. Electrodialysis is an electrochemical separation process in which ions are transferred through ion exchange membranes by means of a direct current (DC) voltage. In a simple electrolytic cell, negatively charged ions (anions) are drawn toward the positively charged electrode, or anode, and positively charged ions (cations) are drawn toward the negatively charged electrode, or cathode.

The ED process differs fundamentally from the pressure-driven membrane processes by transport mechanisms and effect. An electrochemical separation process removes ions from a process stream, whereas a pressure-driven separation process removes water from the process stream. Both membrane processes can remove ions, but electrochemical separation processes cannot remove pathogens, as pressure-driven processes can, giving them no role in disinfection. Electrochemical separation processes are often less costly than NF/RO, however, because they achieve higher recoveries through lower salt rejection and mass transport advantages. A basic diagram of a batch electrodialysis process is shown in Figure 11.8.

ED Cell. The basic ED cell consists of alternating anion-permeable and cation-permeable membranes, which provide a basis for separation of ions under DC voltage. Redox reactions occur during ED. Water and chlorine are oxidized at the cathode, and water is reduced at the anode:

$$H_2O \rightarrow 4H^+ + 4e^- + O_2 \uparrow$$

$$2Cl^- \rightarrow Cl_2 + 2e^- \uparrow$$

$$H_2O + 2e^- \rightarrow 2OH^- + H_2 \uparrow$$

These by-products of the ED cell are removed as electrode waste products and treated by aeration and neutralization (reduction), if necessary.

A simplified diagram of a complete cell for NaCl removal is shown in Figure 11.9. Sodium ions pass through the cation-transfer membrane, and chloride ions pass

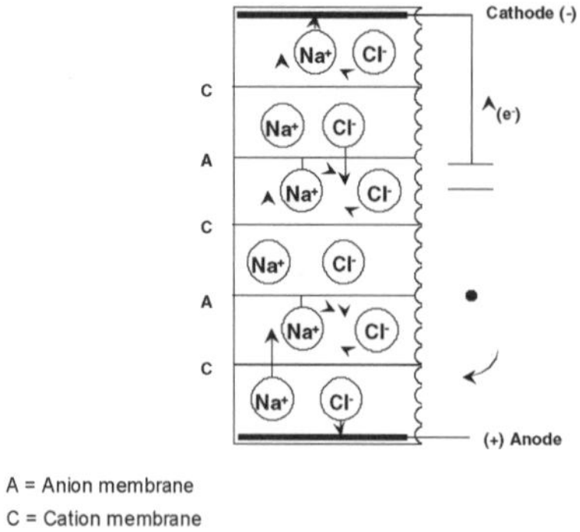

FIGURE 11.8 Diagram of an electrodialysis batch process under applied DC voltage. (Courtesy of Ionics, Inc.)

through the anion-transfer membrane. The sodium and chloride ions are trapped in the brine channel by the alternating ion-exchange membranes, as only cations can pass the cation-permeable membranes and only anions can pass the anion-permeable membranes. The alternating ion-exchange membranes produce a demineralized stream and a concentrate stream, as shown in Figure 11.9. Water flows across these membranes, not through them, as it does in pressure-driven processes.

As shown in Figure 11.10, an ion-exchange membrane has a polymeric support structure with fixed sites and water-filled passages that reject common ions and pass counterions through the membrane. In Figure 11.10, a fixed, negatively charged, sulfonated functional group and a fixed, positively charged, tertiary amine are shown in the anionic and cationic membranes, respectively. These membranes are (1) impermeable to water, (2) electrically conductive, and (3) ion selective. The anion membrane is composed of a cast anion exchange resin with a fixed negative charge in sheet form. The fixed negative charge assists cations in passing and keeps anions from passing through the anion membrane. The cation membrane has a fixed positive charge and utilizes the same mechanism to repel cations and pass anions.

A spacer is placed between the anion-exchange and cation-exchange membranes to form a basic cell pair, as shown in Figure 11.11. The cell pair consists of an anion-exchange membrane, a concentrating spacer, a cation-exchange membrane, and a demineralizing spacer. The spacers are assembled with cross straps to promote turbulence and reduce polarization at the membrane surfaces. Several hundred cell pairs are grouped into a common arrangement known as a *membrane stack* or *module,* as shown in Figure 11.12. These streams either pass to succeeding cells for additional treatment or are discharged from the ED cell as concentrate waste and demineralized product.

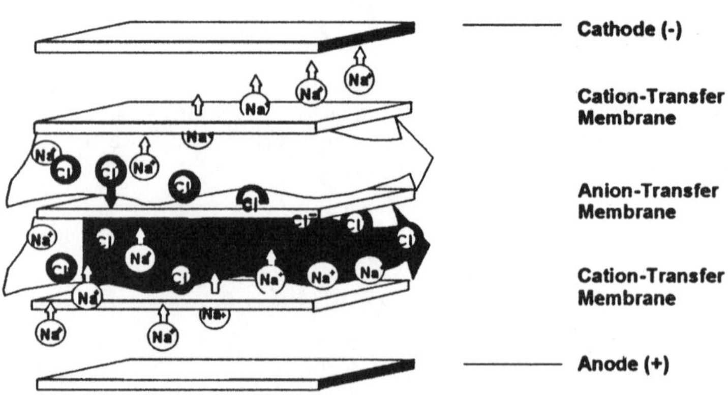

FIGURE 11.9 Simplified diagram of an electrodialytic cell. (Courtesy of Ionics, Inc.)

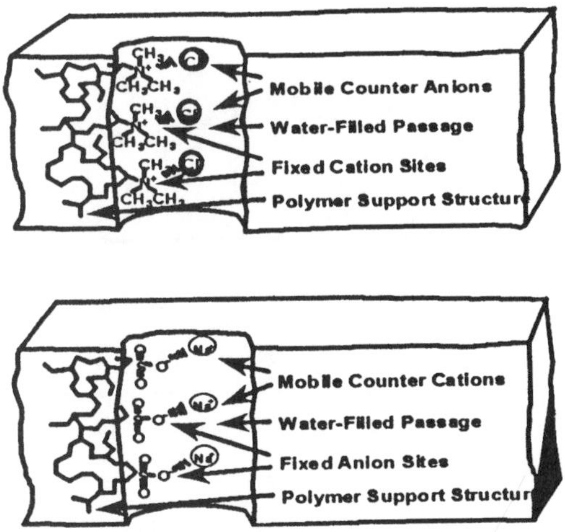

FIGURE 11.10 Ion exchange cast membrane sheets showing a fixed charge and mechanisms of mass transport in the electrodialysis process. (Courtesy of Ionics, Inc.)

The concentrate waste stream does not contain the redox products in the electrode waste streams, although these streams may be combined in some applications.

ED Processes. Electrodialysis processes must work within polarization and scaling limitations like those in pressure-driven, diffusion-controlled membrane processes, because of ion concentrations at the membrane surfaces. However, polarization problems arise in ED processes due to the lack of ions at the surface. Ions arrive at a membrane surface by electrical transport, diffusion, and convection, and they are transported through the membrane electrically. As ions are transported through the membrane, the demineralized stream—membrane interface becomes depleted of ions, causing an exponential increase in electrical resistance and current density. This process continues until polarization occurs, disassociating water into protons and hydroxide ions. This point is identified as the limiting current density. These prob-

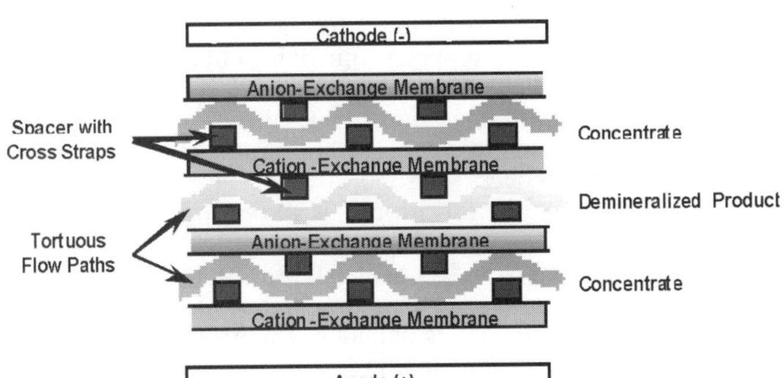

FIGURE 11.11 Basic electrodialysis cell pair. (Courtesy of Ionics, Inc.)

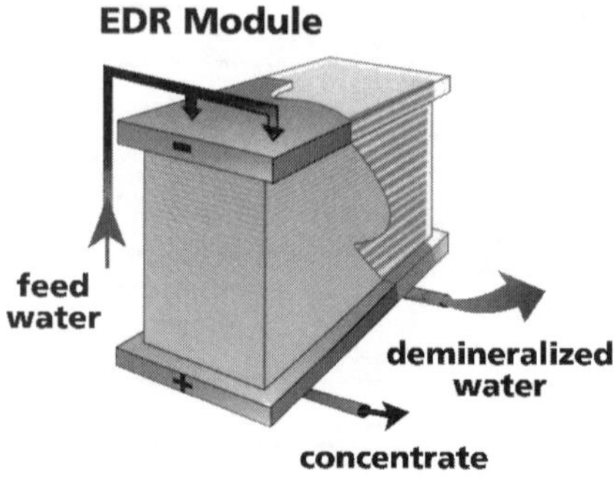

FIGURE 11.12 ED or EDR module. (Courtesy of Ionics, Inc.)

lems have been reduced in ED applications by reversing module polarity every 15 to 20 min during plant applications. This process, known as electrodialysis reversal, has the advantages of (1) reducing scaling potential, (2) breaking up fresh scale, (3) reducing microbiological growth on membrane surfaces, (4) reducing membrane cleaning frequency, and (5) cleaning electrodes with acid formed during anodic operation. The EDR processes reportedly achieve slightly higher recoveries than NF or RO processes, which can become an economic advantage when removal of un-ionized solutes such as disinfection by-product formation potential (DBPFP) or pathogens is not an issue.

The rate of ion removal in an electrochemical separation process is controlled by the feed water characteristics, design parameters, and equipment selection. The water quality and temperature of the feed water determine the system recovery and rate of mass transfer. Ion removal increases as temperature increases and as charge increases. System recovery is typically limited by precipitation of the least soluble salt, as in RO or NF systems. Design parameters include current, voltage, current efficiency, polarization, and finished water quality. As current or current density increases, the removal of ions also increases. Voltage is a function of the current, temperature, and ionic water quality. Current efficiency is determined by the efficiency of salt transfer. A 100 percent efficient process would transfer 1-gram equivalent of salt for every 26.8 ampere-hours. Polarization is controlled by limiting the allowable current density to 70 percent of the limiting current density. Finished water quality standards are set by design, within limits of back-diffusion. A high concentration gradient between the demineralized and concentrate streams leads to diffusion of ions from the concentrate to the demineralized stream. This effect, termed *leakage,* is controlled by limiting the concentration ratio of the concentrate stream to the demineralized stream to less than 150.

An example of a three-stage EDR process is shown in Figure 11.13. The system recovery from this process is 90 percent, and the overall TDS rejection from the feed stream is 75 percent. Contrary to an RO or an NF process, the product water is taken only from the final stage, as ions are continually removed from the product stream until the desired TDS concentration is obtained. The feed stream to the EDR process is pretreated with acid and/or an inhibitor, as is an RO or an NF feed stream, to prevent scaling at the desired recovery. Recycling of the concentrate stream to the feed stream is not shown in Figure 11.13 but is often used in an EDR process to (1) reduce equipment requirements for a given recovery, (2) reduce concentrate

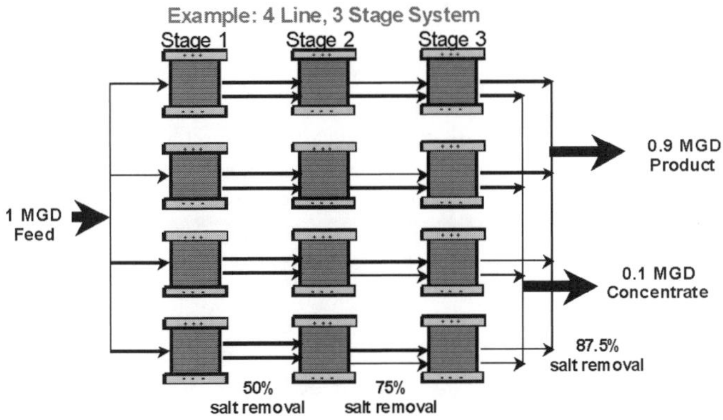

FIGURE 11.13 Four-line, three-stage EDR process. (Courtesy of Ionics, Inc.)

stream discharge, (3) reduce pretreatment, and (4) increase the transfer of ions to the concentrate stream.

MEMBRANE PROPERTIES AND REJECTION CHARACTERISTICS

Characterization of membrane properties is desirable because of the potential of relating these characteristics to solute rejection and membrane fouling studies. This information could be used to modify existing membranes or develop new membranes that would effectively reject inorganic and organic solutes from drinking water sources without fouling the membranes.

Fundamental research for mass transfer, membrane fouling, and solute-solvent interactions has been conducted or is in progress using several different techniques. Currently, feasible techniques are capable of measuring membrane surface charge, pore size, thickness, roughness, surface energy, and surface atomic composition. These techniques are presented in Table 11.4. These seven have actually been conducted in membrane research; others are in the R&D stage.

Dissolved Solute-Membrane Interactions

Dissolved inorganic and organic solutes markedly influence electrokinetic properties of RO and NF membranes through various solute-membrane interactions. Among such interactions are adsorption of inorganic and organic solutes on the membrane surface and/or within the membrane pores. These solute-membrane interactions have a significant impact on membrane rejection and fouling behavior. In the following text, the effect of solution composition (i.e., dissolved solutes) on membrane rejection and fouling is documented based on recent publications.

Influence of Dissolved Solutes on Electrokinetic Properties of Membranes

The effect of dissolved solutes on electrokinetic properties of RO and NF membranes has been investigated by the streaming potential technique, which estimates

TABLE 11.4 Techniques to Characterize Membrane Surfaces

Techniques	Application on membrane research	References
Streaming potential	Charge on membrane surface to determine electrostatic interactions.	Elimelech, Chen, and Waypa, 1994
Contact angle	Surface energy to distinguish hydrophilic and hydrophobic membranes	Oldani and Schock, 1989
Attenuated total reflectance-Fourier transform infrared spectroscopy (ATR-FTIR)	Kinetics study on membrane-solute-solvent interface	Ridgway and Flemming, 1996
Secondary ion mass spectrometry (SIMS)	Chemical analysis for sorption and fouling for clean and fouled membrane	Spevack and Deslandes, 1996
X-ray photoelectron spectroscopy (XPS)	Chemical analysis for sorption and fouling for clean and fouled membrane	Jucker and Clark, 1994
Scanning probe microscopy	Topological information to characterize roughness and pore size for clean and fouled membrane	Fritsche, et al., 1992
Scanning electron microscopy (SEM)	High-resolution image to characterize roughness and pore size for clean and fouled membrane	Fritsche, et al., 1992

charge on membrane surface and in membrane pores (AWWA committee report, 1998). Streaming potential measurements by Childress and Elimelech (1996) demonstrated that typical commercial RO and NF membranes are amphoteric, and the isoelectric point ranges from pH 3 to 5. These authors also reported that, above the isoelectric point, a membrane becomes more negatively charged with increasing pH. This pH dependence is attributed to deprotonation of membrane surface functional groups originated from manufacturing processes.

Membrane charge is also greatly influenced by divalent cations. Hong and Elimelech (1997) showed that membrane surface charge becomes less negative with increasing divalent cation concentration. The decrease in the negative charge of the membrane was attributed to charge neutralization and effective screening of the membrane surface charge by divalent cations. Specific adsorption of divalent cations to membrane surface functional groups may also, in part, have been responsible for the decrease in the negative charge of the membrane.

Dissolved organic solutes such as NOM readily adsorb to the membrane surface and/or pores, affecting the membrane charge. It has been observed that the membrane surface becomes more negatively charged in the presence of humic acids (Childress and Elimelech, 1996; Hong and Elimelech, 1997). The increase in the negative charge is ascribed to adsorption of the humic acids to the membrane surface. The adsorbed humic substances mask inherent membrane charge and dominate the surface charge of the membrane.

Impacts on Membrane Rejection

Charge repulsion is an important rejection mechanism of RO and NF membranes. Thus, changes in electrokinetic properties of the membrane due to solute-membrane interactions have a significant influence on membrane rejection characteristics (Braghetta, 1995). Hong and Elimelech (1997) observed a decline in organic removal by NF membranes with decreasing pH. This observation can be explained by the reduced electrostatic repulsion between organic matter and membrane surfaces due to the lower membrane surface charge at pH less than 4.

Impacts on Membrane Fouling. The degree of membrane fouling is determined by interplay between several chemical and physical interactions. Electrostatic repulsion, one of the critical interactions, is affected by solution composition (Song and Elimelech, 1995; Zhu and Elimelech, 1997). A recent NF study by Hong and Elimelech (1997) reported that NOM fouling increases with increasing electrolyte concentration and decreasing solution pH. In particular, at fixed solution ionic strength and pH, the presence of divalent cations markedly increases NOM fouling. Divalent cations substantially reduce the electrostatic repulsion between NOM and the membrane surface, resulting in a substantial increase in NOM deposition on the membrane. Jucker and Clark (1994) also recognized the effect of divalent cations on NOM adsorption to membrane surfaces.

Organic Solute Removal

The removal of disinfection by-product precursors or nonpurgeable dissolved organic carbon (NPDOC) by RO or NF has been studied extensively (Taylor et al., 1986; Taylor, Thompson, and Carswell, 1987; 1989*a,b;* Jones and Taylor, 1992). Nanofiltration membranes have been shown to control trihalomethane formation potential (THMFP) in highly organic (>10 mg/L NPDOC) potable water sources (Taylor et al., 1986). These efforts have often been necessitated by inadequate efficiency of DBP removal using conventional coagulation and softening treatment processes.

An early investigation using diffusion-controlled membrane processes had shown that membrane material would achieve the same solute rejection (McCarty and Aieta, 1983). In addition to membrane chemistry, characterization of raw water organic fractions was found to be an important consideration for appropriate membrane selection (Fouroozi, 1980; Conlon and Click, 1984; Taylor, 1989*a,b;* Tan and Amy, 1991). In research conducted by Jones and Taylor (1992), over 90 percent of the NPDOC was removed for both surface water and groundwater supplies by membranes for which a manufacturer reported an MWC of 300 to 500 Daltons. The THMs and HAAs in the permeate averaged 15 and 4 μg/L, respectively, and represented more than a 90 percent DBPFP reduction. However, lower MWC membranes removed very little additional DBPFP.

Effective DBPFP removal has also been demonstrated for highly organic potable water sources where limited removal of inorganic contaminants was desired. Research published by Taylor (1989*a,b*) and Spangenburg et al. (1997) have shown high DBP removal capability while simultaneously meeting treatment objectives for hardness or alkalinity concentrations in the treated permeate. It should also be noted that energy requirements for membranes offering comparable DBPFP removal efficiency have been shown to vary as much as 50 percent. This has been demonstrated in single-stage membrane pilot systems operated in parallel using different film chemistries and membrane manufacturers (Tan and Amy, 1991). Research attention has also been focused upon the rejection of specific THM species using membranes. The relative percentage of brominated THMs increased in permeates from membrane processes with increasing MWC (Laine, Clark, and Malleviale, 1990). While NF was found to control brominated DBPs, advanced pretreatment was necessary to sustain production when using the bromide spiked surface water source. Many NF membranes do not remove bromides effectively; hence, higher ratios of brominated DBPs are formed as the Br:NOM ratio increases in the permeate.

High DBPFP reduction has been demonstrated for groundwaters and surface waters alike. However, due to increased fouling potential, surface water is generally more difficult to treat by membrane processes than highly organic groundwater. Consequently, advanced pretreatment is frequently required to counter significant membrane production losses in these applications. Long-term investigation of THMFP control at the Flagler Beach, Florida, water treatment plant (a groundwater site) and at the Punta Gorda, Florida, water treatment plant (a surface water site) provided for comparison of groundwater and surface water membrane treatment. Although consistent control of permeate THMFP was achieved, severe water flux loss was experienced in the system treating surface water. Membrane cleanings were conducted on 20 occasions in an attempt to sustain a water flux of 10 gsfd. The operating system at the groundwater site required prefiltration and acidification in order to sustain a flux of over 15 gsfd with only a semiannual cleaning frequency (Taylor et al., 1989*a*). In research reported by European investigators, NOM adsorption onto a membrane surface has been verified by XPS analysis (Heimstra and Nederlof, 1997). As a result, an alternative membrane was studied that offered similar organic removal and sustained operation capability (three-month cleaning frequency).

The referenced DBP investigations are summarized in Table 11.5. In all of these investigations, the removal of naturally occurring dissolved organic carbon (NPDOC) or DBPFP by membranes was found to be virtually independent of operating condition (i.e., flux or recovery), suggesting that natural organics removal is generally sieve controlled. Similar findings using nanofiltration and reverse osmosis to remove trace-level pesticides are reported by Duranceau and Taylor (1990). If the DBPFPs are assumed to be uncharged NPDOC, which is likely, then the removal of uncharged organics such as pesticides and naturally occurring dissolved organic compounds may be sieving controlled. Removal of organic compounds by sieving may be increased if they can be placed in a state of increased size or steric hindrance. Some NOM compounds are diffusion controlled and affected by variations in flux and recovery. However, differences in permeate NOM concentrations due to diffusion-controlled variables (flux and recovery) are in the tenths of mg/L and unlikely to have any effect on DBP MCL compliance.

Suppliers of drinking water are subject to stringent government regulations for potable water quality regarding allowable pesticide and herbicide (i.e., SOCs) concentrations. In particular, European standards require less than 0.1 μg/L for any one particular pesticide or herbicide and no greater than 0.5 μg/L for total pesticides and herbicides in drinking water. Many investigators have shown that RO and NF are effective techniques for pesticide and herbicide removal. However, specific mechanisms underlying SOC rejection are largely unknown. Some general statements can be made about SOC rejection in membranes. Rejection has been observed to increase as SOC molecular weight and charge increased and membrane polarity increased. In Table 11.6, results and significant findings from published accounts of pesticide and SOC removal are summarized.

Pathogen Removal

The removal of *Giardia* by microfiltration and ultrafiltration has been well documented in the literature (Jacangelo et al., 1991; Coffey, 1993). In these studies, a removal higher than 4 logs was reported, with ultrafiltration (Figure 11.14) and

TABLE 11.5 Summary of DBP Precursor Studies with Membrane Processes

Citation	Water source	Pretreatment	Membrane technology	Feed water THMFP (μg/L)	Treated THMFP (μg/L)	Percent THMFP removal
Taylor, Thompson, and Carswell, 1987	Ground	Antiscalant, prefiltration	NF	961	28–32	97
			NF	961	31–39	96–97
			UF	961	326–947	2–66
Amy, Alleman, and Cluff, 1990	Surface	Prefiltration	NF	157–182	55–84	49–70
	Ground	Prefiltration	NF	176–472	6–95	78–98
Parker, 1991	Surface	None	MF	60–630	40–420	20
		Coagulation	MF	70–80	30–40	40–60
Tan & Amy, 1991	Ground	Prefiltration	NF	259	39	85
Duranceau, Taylor, and Mulford, 1992	Ground	pH adjustment Prefiltration	NF	120	6	95
Laine, Clarke, and Mallevialle, 1993	Surface	Prefiltration	UF	40–460	NA	<10
		Prefiltration	NF	40–460	NA	30–90
		UF	NF	40–460	NA	90

TABLE 11.6 Literature Summary for SOC and Pesticide Removal

Membrane	SOC and pesticides	Rejection (%)	Significant findings	Researchers
CA	Sodium alkyl benzene sulphonate	99.9	Flat-sheet application	Ironside and Sourirajan, 1967
CA	DDT TDE BHC Lindane	99.9 99.5 52.0 79.0	Flat-sheet application	Hindin, Bennett, and Narayanan, 1969
TFC (aromatic polyamide)	Heptachlor Lidane DDT Aalathion Parathion Atrazine Captan	99.5 99.5 99.5 98.0 98.0 72.0 99.0	(1) Application of TFC membrane (2) Organic solutes were rejected higher in TFC membrane than in CA membrane	Chian, Bruce, and Fang, 1975
RO (Toray PEC100, UOP TFC 4600, FimTec Ft30 Desal DSI)	Acetic acid Benzene Propylene oxide Ethylene di-chloride Formal-dehyde 24-D P-chlorobenzo tri-fluoride	11–72 0–99 28–70 0–93 8–95 83–99 77–99	Pilot application	Whittaker and Szaplonczay, 1985; Whittaker and Clark, 1985
TFC Nylon Amide CA	Alachlor	98.5 84.6 71.4	(1) Pilot application (2) TFC > PA > CA	Miltner, Fronk, and Speth, 1987
TFC	Ethylene dibromide (EDB) Alachlor Metolachlor	350 100 100	(1) Pilot application (2) TFC > PA > CA	Fronk, 1987
NF (NF-70)	Ethylene dibromide Dibromochloropropane Heptachlor Methoxychlor Chlordane Alachlor	0 28 100 100 100 100	(1) Pilot application (2) SOC rejections were dependent on charge and MW.	Duranceau, Taylor, and Mulford, 1992
NF (NF-70, PVD1, PZ, SU-610)	Simazine Atrazine Diuron Bentazone DNOC Dinoseb	66–94 78–99 45–92 97–100 38–98 80–100	(1) Pilot application (2) Pesticide removal	Hofman et al., 1993
NF (CA-50, BQ-01, Desal 5-DK, NTC-20, NTC-60, PVD-1, NTR-7250)	Uncharged SOC Simazine Atrazine Terbutylazine Diuron Metazachlorine Charged SOC TCA Mecoprop	 0–80 5–90 12–96 5–90 20–95 61–95 60–90	(1) Flat-sheet and pilot applications (2) Charged SOC were rejected higher in charged membranes.	Berg and Gimbel, 1997

TABLE 11.6 Literature Summary for SOC and Pesticide Removal (*Continued*)

Membrane	SOC and pesticides	Rejection (%)	Significant findings	Researchers
RO and NF (20 membranes)	Simazine	14–95	(1) Flat-sheet application	Chen and Taylor, 1997
	Atrazine	41–99	(2) PA > CA	
	Cyanazine	33–99	(3) Inorganic and organic solutes didn't affect pesticide rejection.	
	Bentazone	0–99		
	Diuron	15–83		
	DNOC	9–95		
	Pirimicarb	48–97		
	Metamitron	12–98		
	Metribuzin	45–97		
	MCPA	0–99		
	Mecoprop	78–99		
	Vinclozolin	64–95		

microfiltration membranes having provided absolute removal of this protozoan cyst. In these cases, the level of removal was limited only by the concentration of the organism in the feed water. A more recent study reported that at bench scale, all the tested membranes (three MF and three UF) except one MF membrane (for which the membrane seal ring was found to be defective) removed *Cryptosporidium* and *Giardia* below the detection limit (1 cyst/L; Jacangelo, Adham, and Laine, 1995). These results were confirmed at pilot scale. Removal effectiveness ranging from 6 to 7 logs was limited only by the influent concentration of the *Cryptosporidium* and *Giardia*. Polymeric MF and UF membranes behave as significant barriers to protozoan cysts as long as the membranes and all system components remain intact. However, no process should be regarded as an absolute barrier to all pathogens.

Coffey observed a 0.32 mean log removal for heterotrophic plate count (HPC) in an MF process. The permeate contained an average of 182 CFU/mL for a mean of 1480 CFU/mL in the feed water. In the same study, no *E. coli* or *Giardia muris* were found in the filtrate for a feed concentration of 9.8×10^5 to 2.67×10^6 and 2.75×10^4, respectively. The low HPC removal was due to regrowth on the permeate channel

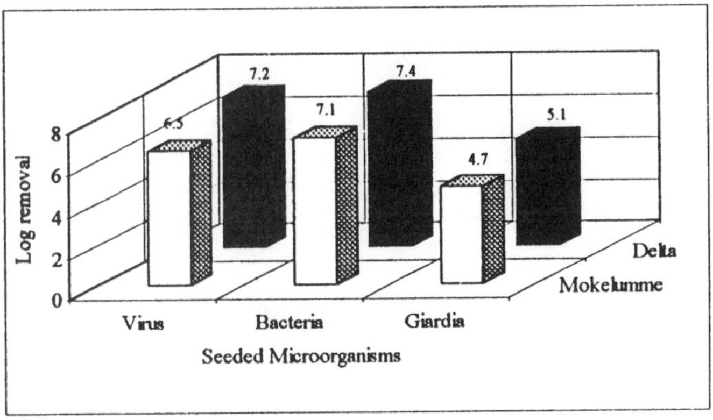

FIGURE 11.14 Removal of seeded microorganisms by UF of Mokelumme and delta waters.

and the low HPC in the feed stream. These results are consistent with the results published by Yoo et al. (1995): an intact 0.2 μm MF filtration process appears to provide a significant barrier to *Cryptosporidium* and *Giardia.* A microfiltration pilot has been used at the Fishing Creek water supply for Frederick, Maryland, removing all the total coliform in the feed water, which ranged from 10 to 1000 per mL (Olivieri et al., 1991). At Manitowoc, Wisconsin, no total coliform, fecal coliform, or *E. coli* were detected in the MF filtrate. The feed contained 0 to 140 total coliform, 0 to 10 fecal coliform per mL, and 0 to 6 *E. coli* per mL (Kothari et al., 1997). For the Saratago Water Treatment Plant, a microfiltration process reached more than 6-log removal of *Giardia* and *Cryptosporidium* (Yoo et al., 1995). Between 1.7 and 2.9 log removal of MS2 virus in a California surface water with a 0.2-μm pore size hollow fiber microfiltration membrane was observed. The range of MS2 virus number in the feed water was from 1.3×10^6 to 3×10^7, and the filtrate was found to contain 2.2×10^4 to 3.4×10^5 MS2 viruses (Coffey, 1993). Total removal of the MS2 virus was found for these feed concentrations using a hollow fiber UF membrane with a 0.01-μm nominal pore size (Jacangelo et al., 1991). Membrane processes, like any process, will eventually fail, and pathogen passage is more likely to be detected as feed concentration increases. However, membrane processes offer greater bacteria and protozoa removal than other water treatment processes intended for the removal of suspended or dissolved matter.

Membrane Integrity

Membrane integrity is an essential issue in implementation and regulatory approval of membrane processes. The high efficacy of MF and UF processes for removal of turbidity and pathogens can be compromised if the membranes or some system components are damaged. Monitoring membrane system integrity is important to avoid any water quality degradation. As shown in Table 11.7, different membrane integrity tests are divided into direct and indirect monitoring methods. This section presents only the direct monitoring methods.

Air-Pressure Testing. This five-step test is used by U.S. Filter-Memtec to detect a compromised module. First, isolate modules by closing all their inlets and outlets. Second, drain membrane lumens via the filtrate exhaust line. Third, pressurize the filtrate side to 15 lb/in^2 with the feed side open to ambient pressure. Fourth, turn off the membrane test pressure to the filtrate side, and watch for any decay in filtrate pressure. Fifth, exhaust the pressure to the filtrate exhaust line. Usually the pressure is monitored after 2 and 4 min. A pressure drop between the readings at 2 and 4 min exceeding 0.4 lb/in^2 may indicate breaks in one or more fibers.

To take the broken fiber out of service, maintenance personnel remove the element end cap and insert a pin into the broken fiber, which is identified by escaping air bubbles. If no broken fiber is found, then an O-ring or seal is defective. The air-pressure test method cannot be used for air-permeable membranes.

TABLE 11.7 Different Membrane Integrity Tests

Indirect monitoring methods	Direct monitoring methods
Particle counting	Air-pressure testing
Particle monitoring	Bubble-point testing
Turbidity monitoring	Sonic sensors

Bubble-Point Testing. The bubble-point test is used to detect a compromised Aquasource UF module. The module is taken off line, and the water outside the fiber is drained. Air pressure of 2 bars (29.4 lb/in^2) is applied on the external side of the fibers. A low-concentration surfactant solution is applied at the surface of the module end to assist in identifying any air bubbles coming through damaged fibers (Adham, Jacangelo, and Laine, 1995). A pressure decay higher than 50 mbar (0.7 lb/in^2) in 5 min indicates damage to one or more fibers.

Sonic Sensor Method. An in-line sonic sensor is also used to detect defects in an Aquasource UF module. A sensor located on the module is monitored for hydraulic noise in a certain frequency range, which indicates membrane integrity loss. When membrane integrity is compromised, hydraulic noise increases due to a rise in turbulence in the module.

MASS TRANSPORT AND SEPARATION

For an isothermal process, the total driving force ($F_{\text{tot},i}$) for the transport of component i across the membrane can be expressed as the sum of the concentration potential, electrical potential, and a pressure gradient:

$$F_{\text{tot},i} = \frac{RT}{\delta_m}\frac{\Delta c_i}{c_i} + \frac{z_i\mathcal{F}}{\delta_m}\Delta\varphi + \frac{V_i}{\delta_m}\Delta p \tag{11.1}$$

Where R = universal gas constant
T = temperature
Δc_i = concentration gradient
c_i = bulk stream concentration
δ_m = membrane film thickness
$\Delta\varphi$ = potential gradient
V_i = molar volume
Δp = pressure gradient

Different driving forces typically come into play in transporting solutes, colloids, larger particles, and water across the membrane. Summing flux values for individual components of a raw water (e.g., Na^+, Cl^-, and H_2O) gives J_i, a weighted sum of a minimum number of driving forces F_k, affecting all of the components. This sum is mathematically described by the Onsager relationship (Onsager, 1931) as:

$$J_i = \sum_{k=1}^{n} L_{ik}F_k \tag{11.2}$$

Each phenomenological coefficient, L_{ik}, relates the flux of component i to force F_k. When one driving force dominates mass transport across the membrane, Eq. 11.1 can be simplified in conjunction with the Onsager relationship to obtain expressions for the transport of water or solutes in a pressure-driven or electrically driven membrane process. Thus, these two expressions are the basis for describing a variety of membrane processes. For example, in electrodialysis the last term in Eq. 11.1 is negligible, and expressions can be derived for the transport of solute across the electrodialysis membrane. This issue will be taken up in a later consideration of principles of separation in electrodialysis. First, however, some simplifications allow descriptions of performance for pressure-driven membrane processes such as microfiltration, ultrafiltration, nanofiltration, and reverse osmosis.

Mass Transport Considerations in Pressure-Driven Membrane Processes

If pressure is the only driving force, as in UF and MF, Eqs. 11.1 and 11.2 reduce to a Darcy-type expression for the flux of water across the membrane:

$$J = \frac{\Delta p}{\mu R_m} \tag{11.3}$$

where Δp is the pressure drop across the membrane (the transmembrane pressure drop or TMP), μ is the absolute viscosity (of the water), and R_m is the hydraulic resistance of the clean membrane with dimensions of reciprocal length. In this case, there is a single phenomenological coefficient L_{11}, which by comparison of Eqs. 11.2 and 11.3, is seen to be equal to $(\mu R_m)^{-1}$. This expression is similar in form to the Kedem-Katchalsky equation commonly derived from irreversible thermodynamics to describe solute transport across reverse osmosis membranes. However, in this case, the flux of solute and the buildup of an osmotic pressure across the membrane must be accounted for. If the flux of permeate J is much greater than the flux of solute J_s, then such a derivation leads to the following result:

$$J \cong \left(\frac{L_v}{V_w}\right)(\Delta p - \sigma\Delta\Pi) \tag{11.4}$$

where L_v is a phenomenological coefficient, V_w is the molar volume of water, σ is the reflection coefficient (derived as the ratio of two phenomenological coefficients), and $\Delta\Pi$ is the difference in osmotic pressure across the membrane. Empirically, permeate flux is calculated from Eq. 11.4 as a function of the transmembrane pressure, the osmotic pressure, and two empirical constants corresponding to (L_v/V_w) and σ. By analogy, Eq. 11.3 may be modified directly to account for the reduction in the net transmembrane pressure due to the effects of osmotic pressure:

$$J = \frac{(\Delta p - \sigma_k\Delta\Pi)}{\mu R_m} \tag{11.5}$$

Thus, permeate flux across a clean membrane is not predicted to occur until the transmembrane pressure Δp exceeds the difference in osmotic pressure across the membrane. The osmotic pressure of a solute is inversely proportional to its molecular weight. Larger macromolecules, colloids, and particles produce very little osmotic pressure. As a result, the correction for osmotic pressure is negligible for pressure-driven processes such as MF or UF that reject only these larger species.

Osmotic pressure is one of many phenomena that may reduce permeate flux as the result of the rejection of materials by the membrane. Reductions in permeate flux due to the accumulation of materials on, in, or near the membrane are referred to as *membrane fouling.* As water moves across the membrane, it draws solutes and particles toward the membrane. If these materials do not pass through the membrane, they may begin to accumulate on or near its surface, leading to the formation of additional layers of material through which water must pass. Particles may form cakes on the membrane surface, and macromolecules may form gel layers. Virtually all species achieve higher concentrations near the membrane surface in a flowing concentration boundary layer referred to as the *concentration-polarization layer.* Concentration-polarization is often a precursor to cake or gel formation. In addition, materials may precipitate or adsorb on or in the membrane, leading to reductions in permeate flux that may be difficult to reverse. Elevated concentrations near the membrane resulting from the rejection of and subsequent concentration-polarization of these materials tend to exacerbate precipitative or adsorptive membrane fouling.

Permeate flux decline can be described mathematically by generalizing Eq. 11.5 to the case where resistance to permeate flux is produced by both the membrane and by materials accumulated near, on, and in the membrane. These layers are assumed to act in series to present additional resistance to permeation, designated as R_c and R_{cp}, respectively. These components of resistance vary as a function of the composition of materials rejected and thickness of each layer. The resulting expression for permeate flux is

$$J = \frac{(\Delta p - \sigma_k \Delta \Pi)}{\mu(R_m(t) + R_c(\delta_c(t), \ldots) + R_{cp}(k, J))} \tag{11.6}$$

where δ_c is the thickness of the cake (or gel) layer, and k is the mass transport coefficient of the material in the concentration-polarization layer. The resistance terms are all functions of time and are related to the hydrodynamics of the membrane system and the feed water quality. Adsorption or precipitation of materials within the membrane matrix as well as compaction of the membrane may lead to an increase in the membrane resistance $R_m(t)$ over time. The resistance of the cake R_c can be expressed as the product of the specific resistance of the material that forms the cake $\hat{R}_c$ and the cake thickness δ_c. By the Kozeny equation, the specific resistance of an incompressible cake composed of uniform particles can be calculated as:

$$\hat{R}_c = \frac{180(1 - \varepsilon_c)^2}{d_p^2 \varepsilon_c^3} \tag{11.7}$$

where ε_c is the porosity of the cake, and d_p is the diameter of particles deposited. This expression predicts that resistance to permeation by a deposited cake should increase as the particles composing the cake decrease in size. The resistance produced by RO and NF membranes is likely to be large in comparison with the resistance of deposited colloidal materials or cakes. However, gel layers of macromolecular materials may produce significant resistance. Cake resistance may also be small in comparison with the resistance of a UF or MF membrane if the particles deposited in the cake are large compared with the effective pore size of the membrane. For feed streams containing large particles, permeate flux may be relatively independent of the concentration of particles. The morphology of the cake appears to be an important variable in particle filtration, and cake porosity appears to vary as a function of the hydrodynamics of the membrane module as well as the size distribution of the particles. By comparison with the resistance produced by cake or gel layers, resistance from the concentration-polarization layer is typically small.

Equations 11.3 through 11.6 at first appear to predict that permeate flux increases indefinitely with increasing TMP. However, it is frequently observed that as pressure increases, a maximum permeate flux is eventually attained and permeate flux becomes pressure-independent (Figure 11.15). This maximum is usually interpreted as a mass transport–limited flux. Under conditions of mass transport–limited flux, an increase in pressure that would otherwise increase the flow of permeate across the membrane is instantaneously balanced by an increased accumulation of permeate-limiting materials near the membrane. In other words, increases in the resistance terms in the denominator of Eqs. 11.4 through 11.6 offset the increase in Δp in the numerator.

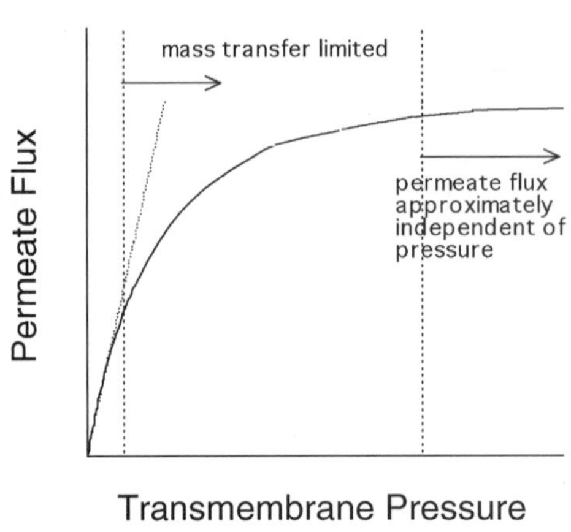

FIGURE 11.15 Transition from pressure-dependent to mass transfer–limited permeate flux.

For a given TMP or under conditions of mass transfer–limited permeate flux, the accumulation of materials near the membrane can be envisioned as a balance between advection of materials toward the membrane due to permeation and back-diffusion that occurs as a concentration gradient builds up near the membrane. Assuming, therefore, a constant permeation rate and concentration-polarization layer with axial distance, mass balance on the concentration-polarization layer yields

$$-D\frac{\partial c}{\partial y} = Jc \tag{11.8}$$

where c is the concentration of the species subject to concentration-polarization, D is the diffusivity of this species, and y is the distance with the boundary layer such that $c = c_{mem}$ at $y = 0$ and $c = c_{bulk}$ at $y = \delta_{cp}$. Integrating this expression over the thickness of the concentration-polarization layer δ_{cp} results in the following expression for the concentration of solute at the membrane surface c_{mem}:

$$c_{mem} = c_{bulk} \exp\left(\frac{\delta_{cp}}{D} J\right) \tag{11.9}$$

For constant operating conditions, the exponential term in Eq. 11.9 can be taken as a constant, referred to as the *polarization factor* (PF). Thus, concentrations of rejected materials at the membrane surface exceed the local bulk concentration by a factor of PF, sometimes estimated as an exponential function of the recovery r such that PF = exp(Kr). The semiempirical constant K typically takes on values of 0.6 to 0.9 for commercially available RO modules.

When permeate flux is mass transfer–limited, Eq. 11.9 can be rearranged to describe the limiting permeate flux J as a function of the bulk concentration c_{bulk}, the limiting concentration at the membrane c_{mem}, the diffusion coefficient for the solute, and the concentration-polarization layer thickness:

$$J = \frac{D}{\delta_{cp}} \ln \frac{c_{mem}}{c_{bulk}} \tag{11.10}$$

The ratio of diffusivity to concentration-polarization layer thickness in this "film theory" model defines a mass transfer coefficient k:

$$k = \frac{D}{\delta_{cp}} \tag{11.11}$$

The thickness of the concentration-polarization layer is a function of the hydrodynamics of the membrane module. For example, for a module with tangential flow across the membrane surface, a higher cross-flow velocity tends to decrease the thickness of the concentration-polarization layer if Brownian diffusion is the only mechanism of back-transport ($D = D_B$). In this case, the mass transfer coefficient can be calculated directly using correlations for the Sherwood number Sh of the form:

$$\text{Sh} = \frac{kd_h}{D_B} = A(\text{Re})^{\alpha}\,(\text{Sc})^{\beta}\left(\frac{d_h}{L}\right)^{\omega} \tag{11.12}$$

where $$\text{Re} = \frac{u_{\text{ave}}d_h}{\nu}$$

$$\text{Sc} = \frac{\nu}{D_B}$$

ν = kinematic viscosity
u_{ave} = average cross-flow velocity
d_h = hydraulic diameter of the membrane element (e.g., diameter of the hollow fiber)
A, α, β, and ω = adjustable parameters

The Graetz-Leveque correlation, which is valid for laminar flow when the velocity field is fully developed and the concentration boundary layer is not fully developed, is often used to estimate the mass transfer coefficient:

$$\text{Sh} = 1.86\,(\text{Re})^{0.33}\,(\text{Sc})^{0.33}\left(\frac{d_h}{L}\right)^{0.33} \tag{11.13}$$

The Linton and Sherwood correlation can be used to calculate the mass transfer coefficient in turbulent flow, which may occur in tubular membranes:

$$\text{Sh} = 0.023\,(\text{Re})^{0.83}\,(\text{Sc})^{0.33} \tag{11.14}$$

Concentration-Polarization and Precipitative Fouling. As one consequence of concentration-polarization in RO and NF membranes, salts may more readily precipitate as a scale on membranes. The average concentrations of scale-forming species rejected by NF or RO membranes in the bulk flow inevitably increase as water permeating through the membrane is removed from the salt-bearing solution. Concentration-polarization further elevates rejected salt concentrations near the membrane and exacerbates the tendency to form a scale.

Consider a sparingly soluble salt s consisting of anion A with charge z_A and cation B with charge z_B in equilibrium with precipitated solid phase. For example, metals such as calcium, magnesium, or iron most commonly precipitate as hydroxide, carbonate, or sulfate scales. In the absence of interactions with other salts, the activities of A and B, a_A and a_B, in equilibrium with the precipitated phase are related to the solubility product K_{sp} for that precipitate. This relation can be expressed as a function of the concentrations of A and B as follows:

$$K_{sp} = \gamma_A^x[A^{y-}]^x \gamma_B^y[B^{x+}]^y \tag{11.15}$$

where $\gamma_{A,B}$ are the free ion activity coefficients of A and B, $[A]$ and $[B]$ and x and y are, respectively, the molal concentrations in solution and the stoichiometric coefficients for the precipitation reaction of A and B. For example, if calcium and sulfate are present in sufficient quantities to initiate precipitation of $CaSO_4$, then at equilibrium at a temperature of 20°C, the product of the activities of calcium and sulfate in solution will be equal to the K_{sp} for $CaSO_4$; approximately 1.9×10^{-4}. The mean activity coefficients can be estimated as function of ionic strength I as,

$$\log \gamma_{A,B} = -0.509 z_A z_B \sqrt{I} \tag{11.16}$$

For dilute solutions typical of most natural waters, the activity coefficients are approximately equal to 1. However, as the feed water concentration increases, the activity coefficients may decrease sufficiently to make concentration a poor approximation of activity. The presence of other electrolytes may further decrease ion activities through effects on ionic strength and ion pairing. The average (bulk) concentration of the rejected salts (for example, the cation B) in the concentrate stream exiting a membrane module c_r increases over the feed concentration c_f as the recovery r and global rejection R increase:

$$c_r = c_f \frac{1 - r(1 - R)}{(1 - r)} \tag{11.17}$$

The salt may precipitate when the ratio of the product of ion activities in the concentrate (the right-hand side of Eq. 11.17 after substituting $[A]_r$ and $[B]_r$ calculated from the feed concentration) to the solubility product K_{sp} is greater than 1. However, concentration-polarization further increases the ion product at the membrane surface since $c_{mem} = c_{bulk} * \text{PF} = c_r * \text{PF}$. Thus, taking the case of $x = y = 1$ (calcium sulfate, for example) the theoretical conditions to avoid scale formation are given by:

$$(\text{PF})^2 \left\{ \frac{1 - r(1 - R)}{(1 - r)} \right\}^2 [B]_f [A]_f < K_{sp} \tag{11.18}$$

Transport of Colloids and Particles. The preceding discussion clearly suggests that the diffusivity of contaminants in water plays an important role in determining the permeate flux. However, Brownian diffusion alone does not adequately describe the transport of particulate and colloidal species near the membrane. The transport of colloidal and particulate species is of greatest significance in MF and UF systems, which are designed to remove these species. In a flowing suspension, particles may collide with one another, producing random rotary and translational motions that, on the average, result in a net particle migration from regions of high concentration and shear (near the membrane) to regions of lower concentration and shear (Davis, 1992). Building on the work of Eckstein and coworkers (Eckstein et al., 1977), Leighton and Acrivos (1987) proposed the following expression for the shear-induced diffusion coefficient. The equation was reported as valid up to particle volume fractions (the concentration of particle volume per volume of water) of $\phi = 0.5$:

$$D_{sh} = a_p^2 \dot{\gamma} \hat{D}_{sh}(\phi) \tag{11.19}$$

where a_p is the particle radius, γ is the shear rate, and D_{sh} is a dimensionless function of ϕ estimated for suspensions of rigid spheres as:

$$\hat{D}_{sh} = 0.33\phi^2(1 + 0.5e^{8.8\phi}) \tag{11.20}$$

Brownian diffusion is more important for smaller species such as solutes and small macromolecules, while shear-induced diffusion is increasingly important for larger particles. Contaminants encountered in water treatment typically span several orders of magnitude in hydrodynamic size, requiring consideration of both Brownian and shear-induced diffusion in estimating concentration polarization and permeate flux. The sum of Brownian and shear-induced diffusivity D can be rewritten as (Sethi and Wiesner, 1997),

$$D = \frac{kT}{6\pi\mu_o r_p} + \frac{\tau_{\text{wall}}}{\mu_o \eta(\phi)} r_p^2 \tilde{D}_{sh}(\phi) \tag{11.21}$$

where τ_{wall} is the shear stress at the membrane wall, μ_o is the absolute viscosity of the water at low-particle volume fractions, and $\eta(\phi)$ is the relative viscosity, which varies with volume fraction. Simultaneous consideration of Brownian and shear-induced diffusivity yields a minimum in particle back-transport for particles in the size range of several tenths of a micron and a corresponding minimum in permeate flux. The sum of the Brownian and shear-induced diffusion coefficients exhibits a minimum for particles approximately 10^{-1} μm in size for flow conditions typical of hollow fiber or SW modules (Figure 11.16). Consequently, the relatively low mass transfer coefficients of species in this unfavorable size range are predicted to produce minimum permeate flux, while species either smaller or larger than this intermediate size are predicted to produce higher permeate fluxes. Until recently, only indirect experimental evidence supported this hypothesized minimum in back-transport as interpreted from permeate flux data (Fane, 1984; Wiesner, Clark, and Mallevialle, 1989; Lahoussine-Turcaud et al., 1990). However, a recent report gave the first direct

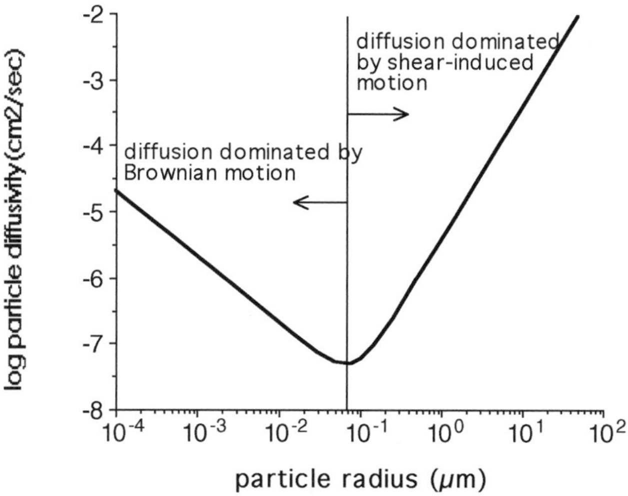

FIGURE 11.16 Brownian and shear-induced diffusivity as a function of particle size for condition typical of hollow fiber UF membranes.

experimental confirmation of a minimum back-transport based on measurements of particle residence time distributions (Chellam and Wiesner, 1997).

Diffusive transport is significant near the membrane where boundary layers may develop. Differential transport of smaller species in the bulk flow (outside the boundary layers) is generally negligible. However, in laminar cross-flow filtration, the transport of larger particles from the bulk into the boundary layer may be affected by interactions between particles and the fluid flow relatively far from the membrane surface. In particular, inertial lift arising from nonlinear interactions of particles with the surrounding flow field may offset the drag force on particles associated with the flow of permeate across the membrane. This effect may occur where the particle Reynolds number becomes important (as it does with large particle sizes or fast cross-flow velocities). For fast laminar flows (Re >> 1), as is typical in membrane filtration, the maximum value of the lift velocity in a clean channel is given by:

$$v_{lo} = \frac{0.036\ \rho a_p^3 \dot{\gamma}_o^2}{\mu_o} \tag{11.22}$$

where $\dot{\gamma}_o$ is the shear rate at the membrane (in the absence of a cake layer), and ρ is the fluid density. Particle transport into the boundary layer is favored when the magnitude of the permeation velocity exceeds that of the inertial lift velocity (Belfort, Davis, and Zydney, 1994).

In addition to these processes governing the transport of particulate species to and from boundary layers, rolling or sliding of particles along the membrane surface due to convection must also be considered. In one approach (e.g., Leonard and Vassilieff, 1984; Davis and Birdsell, 1987), particles deposited as a cake are considered, as a continuum may, under sufficient shear, begin to flow tangentially along the membrane surface toward the filter exit. An alternative approach (e.g., Stamatakis and Tien, 1993) considers the force and torque balances on a single particle on the membrane or cake surface. These balances are used to develop criteria for an estimate of whether the particle will adhere to the surface.

Mechanisms of Separation. Pressure-driven membrane processes may differ from one another substantially in the mechanisms by which they achieve separation, also referred to as *rejection.* In MF and UF, size exclusion is the primary mechanism of separation. Rejection of contaminants by RO membranes is largely a function of the relative chemical affinity of the contaminant for the membrane vis-à-vis water. The NF and UF membranes often represent intermediate cases that achieve rejection by combinations of mechanisms.

Rejection is usually defined as 1 minus the ratio of the permeate and feed stream concentration. This *global* rejection R is calculated as:

$$R = 1 - \left(\frac{c_p}{c_f}\right) \tag{11.23}$$

where c_p and c_f are the permeate and feed concentrations. Occasionally, as in bench-scale test modules and associated scale-up, analysis focuses on the inherent *local* rejection R_{local} of a given membrane for a specific contaminant. The evaluation must account for local changes in rejection that occur with increases in the bulk concentration of a material as flow proceeds along the membrane (x coordinate) due to the permeation of water through the membrane. In addition, rejected materials tend to accumulate near the membrane compared with the bulk flow due to concentration-polarization. Thus, the effective rejection of materials by a membrane may be very different from that calculated based on the average feed and permeate concentrations. Local rejection is a function of the concentrations directly on either side of the membrane at a specific location and is defined as:

$$R_{\text{local}}(x) = 1 - \left(\frac{c_p(x)}{c_{\text{mem}}(x)} \right) \tag{11.24}$$

where c_{mem} is the concentration at the membrane surface calculated from the bulk concentration at a given location within the module such that $c_{\text{mem}}(x) = (\text{PF})c_{\text{bulk}}(x)$. If a mass balance is performed over a membrane module operating in tangential flow, the following expression is derived relating the global rejection to the apparent rejection:

$$R = 1 - \left(\frac{c_p}{c_f} \right) = 1 - \frac{1 - (1 - r)^{(1 - R_{\text{local}})\text{PF}}}{r} \tag{11.25}$$

where r is the single-pass recovery of the module, and the local rejection is assumed constant along the membrane. This latter assumption is intimately related to the mechanism of rejection.

Mechanical Sieving at the Membrane Surface. Membranes intended specifically to remove or separate particles and colloidal materials, such as MF and UF membranes, reject materials largely by mechanical sieving. However, electrostatic interactions, dispersion forces, and hydrophobic bonding may significantly affect the rejection of materials with dimensions similar to the pore sizes in UF or MF membranes. These considerations tend to be more important for separations involving larger macromolecules such as humic and fulvic acids, where charge, adsorptive affinity to the membrane, and hydrodynamic size are interrelated.

A quantitative analysis of mechanical sieving (or steric rejection) was first presented by J. D. Ferry in 1936 (Ferry, 1936). Assuming pores of cylindrical geometry and spherical particles, he derived an expression for the fraction of particles p passing through a pore of radius R. For particles with radii equal to or greater than the pore radius, perfect rejection is predicted. For particles much smaller than the pore, no significant rejection occurs. Only particles with radii slightly smaller than pore radius show nontrivial rejection behavior. This expression was subsequently modified (Zeman and Wales, 1981) to account for the lag velocity of particles with respect to the fluid permeating through the membrane pore. The rejection of particles by a membrane $(1 - p)$ can be estimated using this expression, as a function of the nondimensionalized particle radius $\lambda = r_p/r_{\text{pore}}$, as

$$p = \begin{cases} (1 - \lambda)^2[2 - (1 - \lambda)^2]G & \lambda \le 1 \\ 1 & \lambda > 1 \end{cases} \tag{11.26}$$

where G is the lag coefficient, empirically estimated by Zeman and Wales (1981) as:

$$G = \exp(-0.7146\, \lambda^2) \tag{11.27}$$

Rigorously derived, $(1 - p)$ corresponds to the local rejection of the membrane R_{local}. Measurements of apparent rejection can be used to calculate a value p^* that theoretically corresponds to the product of the polarization factor PF and the particle passage p. Removal of materials in deposited cake or gel layers may further alter the apparent rejection of the membrane. Further modifications can account for electrostatic repulsion and dispersion forces near the pore wall (Matsuura and Sourirajan,

1983). A more rigorous treatment would incorporate drag on particles transported through membrane pores by advection and diffusion (Deen, 1987), and the diffusion and convection of flexible macromolecules through cylindrical pores of molecular dimensions (Davidson and Deen, 1988). One approach extends the physical sieving model for particle removal to describe the rejection of macromolecular compounds such as humic materials by substituting each molecule's hydrodynamic radius for particle radius. Empirical constants in an expression for the molecule's hydrodynamic radius are then manipulated to fit a given data set. For larger molecular weight materials, hydrodynamic radius can often be correlated with the compound's molecular weight, $\overline{M}$. These correlations are typically of the form:

$$a_p = Z_1(\overline{M})^{Z_2} \tag{11.28}$$

where Z_1 and Z_2 are empirical constants. The constant Z_2 contains information on molecule geometry. For a perfectly spherical molecule, Z_2 takes on its theoretical minimum value of 1/3. The theoretical maximum value of 1 for Z_2 corresponds to a linear molecule. Z_2 may take on values outside its theoretical range when Z_1 and Z_2 are treated as fitting constants within the context of the sieving model.

Thus, rejection of organic compounds is predicted to increase with molecular weight. Such behavior has been observed in processes removing larger molecular weight fractions of natural organic matter by UF and NF membranes (e.g., Taylor et al., 1988; Taylor, Thompson, and Carswell, 1987).

While the membrane itself determines the initial removal of materials and serves as an absolute barrier, the rejection characteristics of the membrane system may change over time as materials are deposited on the membrane surface. This concept is exploited in the application of "dynamic" membranes, which formed effective barriers from materials deposited on a porous support that are periodically removed and replenished. Mathematical treatment of this process has borrowed concepts from packed-bed filtration. Suspended colloidal materials transported toward the cake with the permeating water may be removed at the cake surface by simple sieving. Particles entering the cake may be removed by previously deposited particles that form a type of micropacked bed. Removal within the cake may be due to simple interception of mobile colloids, if the streamline they follow brings them in contact with a previously deposited (immobile) colloid. Gravity and Brownian motion may cause particles to deviate from their streamlines, helping to bring particles in contact with the immobile particles composing the cake. The previously deposited particles are termed *collectors* in the packed-bed filtration literature.

Solute Transport and Rejection. The solution diffusion model proposed by Lonsdale, Merten, and Riley (1965; Merten, 1966) describes solute and solvent transport across membranes in terms of the relative affinities of these components for the membrane and their diffusive transport within the membrane "phase." Rejection of a given solute is therefore a function of the degree to which it partitions into the membrane and its diffusivity within the membrane. The driving forces for transport are the differences in chemical potential across the membrane due to differences in concentration and pressure. A basic diagram representing osmosis and reverse osmosis cells is shown in Figure 11.17. In the osmosis cell, water flows through the membrane from a low to high TDS concentration due to osmotic pressure, and ions flow in a reverse path due to diffusion. Reverse osmosis, represented in the adjacent cell, is achieved by applying a pressure greater than the osmotic pressure to the high-TDS water. The transport of water occurs by convection and is dependent on pressure, while transport of solutes is independent of pressure. Consequently, the

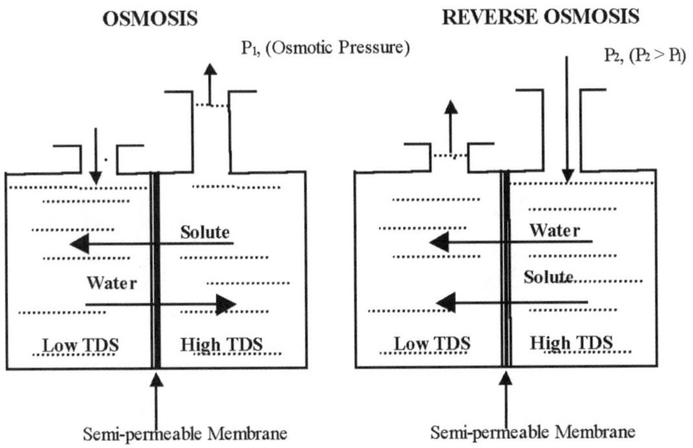

FIGURE 11.17 Diagram of reverse osmosis mass transport.

concentration of the water passing through the RO membrane decreases as pressure increases.

Thus far, the focus has been on the transport of water across membranes and the role of materials in water in potentially reducing the permeate flux. However, expressions for the flux of solutes across membranes can also be derived from very general principles beginning with Eqs. 11.1 and 11.2. The ratio of solute to water flux J_s/J yields the permeate concentration c_p and, therefore, the rejection as calculated by Eq. 11.23. Models for solute rejection tend to differ in the degree of mechanistic detail they offer or the specific mechanisms they invoke. However, where solutes can enter the membrane matrix, the underlying assumption is usually that separation of solutes occurs due to the relative rates of transport across the membrane rather than by size exclusion, as is typically assumed in the rejection of macromolecules, colloids, and particles. An important implication of this distinction is that rejection of solutes depends on the feed (or reject) concentration, while rejection due to size exclusion is independent of feed concentration. The flux of water is given as:

$$J_W = -\frac{D_W c_{W,\text{mem}}}{RT}\frac{d\mu_W}{dz} \approx -\frac{D_W c_{W,\text{mem}}}{RT}\frac{\Delta\mu_W}{\delta_m} \tag{11.29}$$

Note that the form of Eq. 11.29 gives the permeate flux in proportion to a driving force (chemical potential), concentration, and mobility (as expressed by diffusivity). Substituting the Lewis equation for osmotic pressure, the following equation is derived for water flux:

$$J_W = -K_W \frac{(\Delta p - \Delta\Pi_W)}{\delta_m} \tag{11.30}$$

where the specific hydraulic permeability K_W is a function of the diffusivity D_W and concentration $C_{W,\text{mem}}$ of the water in the membrane, and the molar volume of water V_W such that $K_W = (D_W C_{W,\text{mem}} V_W/RT)$. Comparison of this result with Eq. 11.4 shows a similar result to that obtained from irreversible thermodynamics. Indeed, the models are identical, given the underlying interpretations of the phenomenological coefficients.

If water and solute transport are not coupled ($\sigma = 1$). The flux of solute J_s is given as:

$$J_s = -D_s k_s \frac{(c_{\text{mem}} - c_p)}{l} \tag{11.31}$$

where k_s is the distribution coefficient that describes the relative affinity of the solvent for the membrane. Similar to the result for permeate flux, this equation expresses the solute flux as a function of a driving force (concentration difference across the membrane), a mobility (diffusivity), and a concentration. Solute flux is predicted to be independent of the transmembrane pressure drop Δp. Therefore, increases in Δp result in an increase in J_W and in local rejection, since:

$$R_{\text{local}}(x) = 1 - \left(\frac{J_s(x)/J_W(x)}{c_{\text{mem}}(x)}\right) \tag{11.32}$$

This model has been generalized (Sherwood, Brian, and Fisher, 1967) to include advective transport through membrane pores as well as diffusion. This modification leads to the following expression for local rejection by a membrane:

$$R_{\text{local}} = \left[1 + \left(\frac{D_s K_d}{K_W}\right)\left(\frac{1}{\Delta p - \Delta \Pi}\right) + \left(\frac{K_{\text{pf}}}{K_W}\right)\left(\frac{\Delta p}{\Delta p - \Delta \Pi}\right)\right]^{-1} \tag{11.33}$$

where K_{pf} is an additional fitting constant describing pore flow through the membrane. Equation 11.33 predicts increasing local (and therefore global) rejection with increasing pressure, but only to an asymptotically limiting value. In practice, higher operating pressures have been observed to increase global rejection as described by the following empirical equation (Sourirajan, 1970):

$$R = \frac{z_1 \Delta p}{(z_2 \Delta p + 1)} \tag{11.34}$$

where z_1 and z_2 are empirically determined constants.

One variation on such a generalized solution-diffusion model, the preferential sorption-capillary flow model, involves a mechanistic interpretation of transport within the membrane matrix based on differential influence on a layer of water by interfacial processes near polymer surfaces. This layer may exist due to (1) repulsion of ions by membrane materials of low dielectric constant, (2) the excessive energy required to strip ions of their hydration spheres as they move from the bulk water to the water-membrane interface, or (3) the hydrophilic nature of the membrane surface, which actively coordinates and orders water near the membrane surface. Transport of the solute across the membrane within this layer is impeded due to the rise in free energies required to move the solute through a progressively more ordered fluid. The solute is assumed to move through the membrane by diffusion, advection, or both (Sourirajan, 1970). Ions with smaller hydrated radii may able to diffuse through the adsorbed water film or may be convected through pores.

Charges on membrane surfaces can play a significant role in the rejection of ionic solutes. Although these effects are most important in electrically driven membrane separations, they may also play a role in pressure-driven processes. Charged functional groups on membranes attract ions of opposite charge (counterions). This effect in combination with a deficit of like-charged ions (co-ions) in the membrane results in a so-called Donnan potential. Although co-ions are largely unable to enter the membrane, water can pass through under pressure. The accumulation of co-ions in the membrane concentrate is accompanied by an accumulation of counterions as well, due to the need to preserve electroneutrality in the solution. By this mechanism, membrane rejection is predicted to increase with increasing membrane charge and with co-ion valence (Dresner, 1972). Similarly, for membranes that are semipermeable to some ionic species, the requirement for electroneutrality in the concentrates and permeates may lead to significant modification of rejection characteristics in the presence of mixed solutes. For example, as an NF membrane rejects calcium, higher rejections of monovalent anions such as nitrate or chloride may also occur. Varying the concentration of competing anions in the concentrate can affect anion rejection by NF membranes. For example, increasing sulfate concentration increases sulfate rejection and decreases nitrate and chloride relative to levels in single-solute solutions (Rautenbach and Groeschl, 1990).

Mass Transport Considerations in Electrodialysis

In electrodialysis, the driving force for separation is a gradient in electrical potential. An ED system applies an electrical potential across charged ion-exchange membranes, and cations or anions are differentially transported across the membrane and removed from the feed flow. A stack of cation- and anion-selective membranes can remove ionic species in applications such as desalination of brackish water.

In the absence of a pressure drop across the membrane, the driving force for transport of ions across an ion selective membrane becomes:

$$F_{\text{tot},i} = \frac{RT}{\delta_m}\frac{\Delta c_i}{c_i} + \frac{z_i\mathcal{F}}{\delta_m}\Delta\varphi \tag{11.35}$$

Assuming ion flux proportional to this driving force and to ion concentration and mobility, the following equation is derived for cation transport across a cation-exchange membrane:

$$J_+ = -U_+c_+\left(\frac{RT}{\delta_m}\frac{\Delta c_+}{c_+} + \frac{z_i\mathcal{F}}{\delta_m}\Delta\varphi\right) \tag{11.36}$$

The fraction of current I carried by the flux of cations FJ_+ is defined as the cation transport number t_+:

$$t_+ = \frac{\mathcal{F}J_+}{I} \tag{11.37}$$

The flux of cations through a cation-exchange membrane under the driving force of an electrical potential can then be rewritten as:

$$J_+ = \frac{t_+I}{z\mathcal{F}} \tag{11.38}$$

Increases in current are therefore predicted to induce increases in cation transport (or similarly anion transport). The electrical potential E required to produce a given current is calculated from Ohm's law:

$$E = IR \tag{11.39}$$

where R is the resistance of the total membrane stack. The resistance of the membrane stack can be considered as the sum of n cell pairs, each with a resistance R_{cell} such that

$$R = R_{\text{cell}}\, n \tag{11.40}$$

Resistance within each cell pair can be expressed, in turn, as the sum of resistances due to the anion-exchange membrane R_{an}, the cation exchange membrane R_{cat}, the concentrate stream R_{con}, and the dialysate stream R_{dial}:

$$R_{\text{cell}} = R_{\text{an}} + R_{\text{cat}} + R_{\text{con}} + R_{\text{dial}} \tag{11.41}$$

Unfortunately, the increase in ion flux with current is limited by a theoretical maximum current required to transport all available ions. This maximum is a function of the ion concentration at the membrane surface and the concentration in bulk flows of concentrate and dialysate. Differential transport of cations across the membrane leads to the formation of boundary or concentration-polarization layers near either side of the membrane. On the side of the membrane where cations are depleted (the dialysate), concentrations are lower than the bulk concentration. On the concentrate side of the membrane, cation concentration is greater in the boundary layer than in the bulk flow. The flux of cations through the boundary layers can be calculated as a function of a transport number $t_{+,\text{bl}}$ for flux due to the applied potential and the concentration gradient. At equilibrium, cation flux through the membrane must equal the flux of cations through the boundary layer:

$$J_+ = \frac{t_+ I}{z\mathcal{F}} = \frac{t_{+,\text{bl}} I}{z\mathcal{F}} - D\frac{dc_+}{dy} \tag{11.42}$$

Integration of Eq. 11.42 with the boundary conditions $c_+ = c_{\text{bulk}}$ at $y = \delta_{\text{cp}}$ and $c_+ = c_{+,\text{mem}}$ at $y = 0$ yields the following expression:

$$c_{+,\text{mem}} = c_{+,\text{bulk}} \pm \frac{(t_+ - t_{+,\text{bl}}) I \delta_{\text{cp}}}{Dz\mathcal{F}} \tag{11.43}$$

The plus sign applies to calculations of concentrations on the concentrate side of the membrane, and the minus sign applies to calculations of the depletion of cations in the boundary layer near the membrane on the dialysate side. Analogous results are obtained for transport across anion-exchange membranes. The depletion of ions near the membrane leads to an increase in electrical resistance across the dialysate compartment R_{dial}, which increases power consumption and reduces efficiency. Rearranging Eq. 11.43 yields the following expression for the current density in the concentration-polarization layer on the dialysate side of the membrane:

$$I = \frac{Dz\mathcal{F}(c_{+,\text{bulk}} - c_{+,\text{mem}})}{(t_+ - t_{+,\text{bl}})\delta_{\text{cp}}} \tag{11.44}$$

The limiting case (maximum current) occurs when the current density is high enough to drive the cation concentration (or similarly the anion concentration) to zero on the dialysate side of the membrane. This limiting current is then given as:

$$I_{\text{lim}} = \frac{Dz\mathcal{F} c_{+,\text{bulk}}}{(t_+ - t_{+,\text{bl}})\delta_{\text{cp}}} \tag{11.45}$$

In a situation analogous to concentration-polarization in pressure-driven processes, the thickness of the polarization layer is a function of the flow across the membrane.

The cross-flow can therefore be manipulated to a limited extent to optimize performance.

Temperature Effects on Flux

The flux of water through a pressure-driven membrane is affected by temperature. The following section presents three techniques to compensate for the effect of temperature on flux.

Theoretical Normalized Flux Equation. The Hagen-Poiseuille equation (Eq. 11.46) measures flow through pores of a membrane as a function of viscosity, pore diameter, porosity, applied pressure, and membrane thickness (Cheryan, 1988). This equation indicates that water flux is inversely proportional to solvent viscosity. The viscosity of water decreases as temperature increases, which increases flux.

$$F_W = \frac{\Delta P}{R_W} = \frac{\varepsilon r^2 \Delta P}{8\delta\,\mu} = \frac{1}{\mu}\,\frac{\varepsilon r^2 \Delta P}{8\delta} \qquad (11.46)$$

where ε = porosity
r = pore radius
μ = solvent viscosity
δ = membrane thickness

If the porosity, pore radius, viscosity, and thickness are assumed to be constant, Eq. 11.46 can be used to develop a temperature correction factor. The normalized temperature of 25°C is a worldwide reporting standard (PEM, 1982). To derive the normalized flux equation, θ in Eq. 11.47 is determined graphically or by linear regression.

$$\text{TCF} = \frac{F_{T^\circ\text{C}}}{F_{25^\circ\text{C}}} = \theta^{(T-25)} \qquad (11.47)$$

where TCF = temperature correction factor
T = temperature (°C)

Applying the Hagen-Poiseuille equation, Eq. 11.47 can be written as:

$$\log \frac{F_{T^\circ\text{C}}}{F_{25^\circ\text{C}}} = \log \frac{\mu_{25^\circ\text{C}}}{\mu_{T^\circ\text{C}}} = (T-25)\log\theta \qquad (11.48)$$

Log θ then can be solved by regressing or plotting log $\mu_{25^\circ\text{C}}/\mu_{T^\circ\text{C}}$ versus $(T-25)$. The result for this theoretical derivation is shown in Eq. 11.49. Membrane suppliers have developed temperature correction equations for their products that account for temperature changes of water and membrane. Equation 11.49 accounts only for the viscosity change with temperature.

$$\text{TCF} = \frac{F_{T^\circ\text{C}}}{F_{25^\circ\text{C}}} = 1.026^{(T-25)} \qquad (11.49)$$

Temperature Correction Factor Based on Operational Data. The theory developed so far assumes no effect of temperature on the membrane. Another method derives a TCF by developing a statistical relationship of the mass transfer coefficient and temperature using operational data. This technique accounts for temperature effects on both water and the membrane; however, it does assume the membrane is effectively cleaned.

The same equation form as shown in Eq. 11.47 is used.

$$\text{TCF} = \frac{F_{T°\text{C}}}{F_{25°\text{C}}} = \theta^{(T-25)} \tag{11.47}$$

Flux is proportional to the mass transfer coefficient. Taking the logarithm of both sides gives:

$$\log \frac{F_{T°\text{C}}}{F_{25°\text{C}}} = \log \frac{\text{MTC}_{T°\text{C}}}{\text{MTC}_{25°\text{C}}} = (T-25)\log\theta \tag{11.50}$$

Again, plotting log $\text{MTC}_{T°\text{C}}/\text{MTC}_{25°\text{C}}$ of operational data versus $(T - 25)$ gives log θ as the slope, which is determined by linear regression. Manufacturers commonly develop TCFs for their membranes. For example, a TCF of 1.04 was developed for a Memtec MF product using operational data from Manitowoc, Wisconsin (Kothari et al., 1997).

DuPont Equation. The TCF for DuPont B-10 membrane is commonly used for normalized flux and is presented in Eq. 11.51 (PEM, 1982). This figure of 1.03 is essentially equal to the theoretical θ of 1.026.

$$\text{TCF} = \frac{F_{T°\text{C}}}{F_{25°\text{C}}} = 1.03^{(T-25)} \tag{11.51}$$

PROCESS DESIGN CRITERIA

This section introduces priorities for design and associated material. A conventional RO/NF treatment system includes pretreatment, membrane filtration, and posttreatment, as shown in Figure 11.18. Common terminology for membrane processes appears in Table 11.8.

As can be seen in Figure 11.18, the overall process can be rationally separated into advanced pretreatment, conventional pretreatment, membrane filtration, and posttreatment subsections. Any raw water stream used as a feed stream to a membrane process must undergo either conventional or advanced pretreatment. Con-

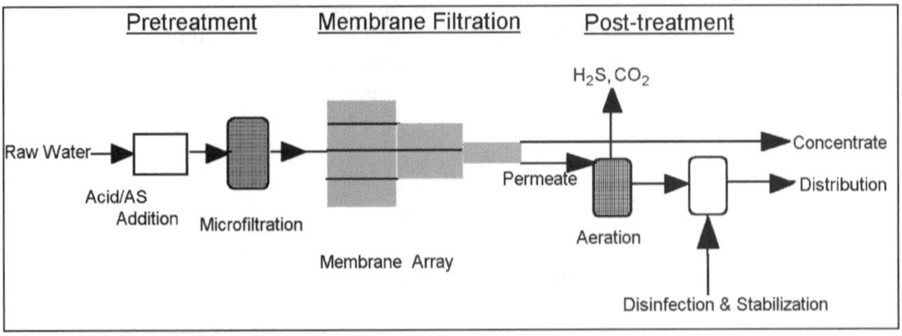

FIGURE 11.18 Conventional RO/NF membrane process.

TABLE 11.8 Membrane Terminology

Raw	Input stream to the membrane process
Conventional RO/NF process	Treatment system consisting of acid or antiscalant addition for scaling control (pretreatment), RO/NF membrane filtration and aeration, and chlorination, plus corrosion control for posttreatment
Feed	Input stream to the membrane array
Concentrate, reject, retentate, residual stream	Membrane output stream that contains higher TDS than the feed stream
Brine	Concentrate stream containing total dissolved solids greater than 36,000 mg/L
Permeate or product	Membrane output stream that contains lower TDS than the feed stream
Membrane element	Single membrane unit containing a bound group of SW or HFF membranes to provide a nominal surface area
Pressure vessel	Single tube with several membrane elements in series
Stage or bank	Parallel pressure vessels
Array or train	Multiple interconnected stages in series
High recovery array	Array where the concentrate stream from one array becomes feed to a succeeding array to increase recovery
System arrays	Several arrays that produce the required plant flow
Rejection	Percentage solute concentration reduction of the permeate stream relative to the feed stream
Solute	Dissolved solids in raw, feed, permeate, and concentrate streams
Solvent	Liquid, usually water, containing dissolved solids
Flux	Mass ($mL^{-2}t^{-1}$) or volume (Lt^{-1}) rate of transfer through a membrane surface
Scaling	Precipitation of solids in an element due to solute concentration in the feed stream
Fouling	Deposition of solid material from the feed stream in a membrane element
Mass transfer coefficient (MTC)	Mass or volume unit transfer through a membrane based on driving force

ventional pretreatment includes acid or antiscalant addition to prevent precipitation of salts during membrane filtration. Advanced pretreatment is required before conventional pretreatment when the raw water contains excessive fouling materials. Membrane filtration is passage of the pretreated water through an active RO/NF membrane surface with apparent pore sizes in the range of 0.001 to 0.0001 μm. Posttreatment includes many unit operations common to drinking water treatment such as aeration, disinfection, and corrosion control.

In a typical membrane process, one stream enters the membrane element, and two streams exit. The entering stream is the feed stream. The streams exiting the membrane are referred to as *concentrate* and *permeate streams*. A portion of the concentrate stream is sometimes recycled to the feed stream to increase cross-flow velocity and recovery.

Fouling Indexes

Membrane fouling is an important consideration in the design and operation of a membrane system. Cleaning frequency, pretreatment requirements, operating conditions, cost, and performance are affected by membrane fouling. Fouling indexes are estimates of the fouling and pretreatment requirements for RO or NF systems. The silt density index (SDI) and the modified fouling index (MFI) are the most common fouling indexes, which are defined using the basic resistance model as quantitative indicators of water quality and potential for membrane fouling.

Currently, fouling indexes are determined from simple membrane tests. These values resemble mass transfer coefficients for membranes used to produce drinking water. The water must be passed through a 0.45-μm Millipore filter with a 47-mm internal diameter at 30 psig to determine any index. The time required to complete data collection for such a test varies from 15 min to 2 h depending on the character of the water. Although similar data are collected for each index, significant differences separate them. Because differences in filtration equipment can affect test results, only a Millipore filter apparatus can be used to generate accurate SDIs and MFIs.

Silt Density Index. The SDI, the most widely used fouling index, is calculated as shown in Eq. 11.52. The Millipore test apparatus is used to determine three time intervals for calculation of the SDI. The first two intervals are the times to collect initial and final 500-mL samples. The third time interval—5, 10, or 15 min—is the time between the collection of the initial and final samples. The 15-min interval is used unless the water has such extreme fouling potential that the filter plugs before that period expires. The interval between the initial and final sample collection is decreased until a final 500-ml sample can be collected.

$$\mathrm{SDI} = \frac{100(1 - (t_i/t_f))}{t} \tag{11.52}$$

where t_i = time to collect initial 500 ml of sample
t_f = time to collect final 500 ml of sample
t = total running time of the test

The SDI value is a static measurement of resistance determined by samples taken at the beginning and the end of the test. The SDI does not measure the rate of change of resistance during the test, and it is the least sensitive of the fouling indexes. The SDI is not dynamic, is not measured in a cross-flow mode, does not use the same material or pore size as a membrane element, measures only static resistance, and is not reflective of a continuously operated membrane process.

Modified Fouling Index. The MFI is determined using the same equipment and procedure as for the SDI, except that the volume is recorded every 30 s over a 15-min filtration period (Schippers and Verdouw, 1980). The development of the MFI is consistent with Darcy's law, in that the thickness of the cake layer formed on the membrane surface is directly proportional to the filtrate volume. The total resistance is the sum of the filter and cake resistance. The MFI is derived in Eqs. 11.53 to 11.55 and is defined graphically as the slope of a curve relating inverse flow to cumulative volume (Figure 11.19).

$$\frac{dV}{dt} = \frac{\Delta P}{\mu} \frac{A}{(R_f + R_k)} \tag{11.53}$$

$$t = \frac{\mu V R_f}{\Delta P A} + \frac{\mu V^2 I}{2 \Delta P A^2} \tag{11.54}$$

$$\frac{1}{Q} = a + \mathrm{MFI} \times V \tag{11.55}$$

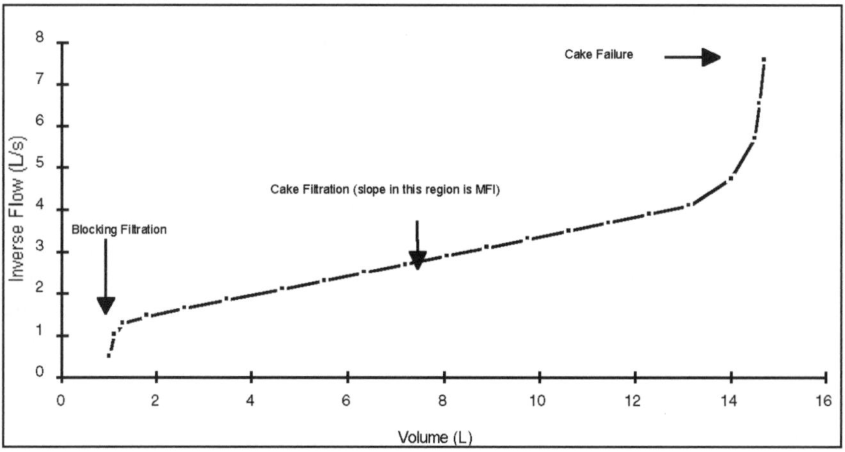

FIGURE 11.19 Typical MFI curve.

where R_f = resistance of the filter (L^{-1})
R_k = resistance of the cake (L^{-1})
I = measure of fouling potential (L^2t)
Q = average flow (v/t^{-1})
a, b = constants
μ = absolute viscosity ($ML^{-1}t^{-1}$)

Based on the work of Schippers and Verdouw, Morris (1990) determined an instantaneous MFI by calculating the ratio of flow over volume in 30-s increments to increase sensitivity. An MFI plot is shown in Figure 11.19. Typically the formation, buildup, and compaction or failure of the cake can be seen in three distinct regions on an MFI plot. The regions corresponding to blocking filtration and cake filtration represent productive operation, whereas compaction indicates the end of a productive cycle.

Index Guidelines. Some approximations for index values required prior to conventional membrane treatment are given in Table 11.9. The indexes apply to the raw or advanced pretreated water before conventional pretreatment unit operations. These numbers are only approximations and do not eliminate the need for pilot studies. Pretreatment requirements cannot be determined for most installations without pilot studies unless actual plant operating data can be obtained on very sim-

TABLE 11.9 Fouling Index Approximations for RO/NF

Fouling index	Range (s/L^2)	Application
MFI	0–2	Reverse osmosis
	0–10	Nanofiltration
SDI	0–2	Reverse osmosis
	0–3	Nanofiltration

Source: Morris, 1990; Sung, 1994.

ilar waters. Pilot studies have been omitted in the design of some reverse osmosis plants for treatment of raw waters with TOC and SDI values less than 1.

Pretreatment

As mentioned, membrane operations require some feed water pretreatment. First, however, it is important to realize that pretreatment is specific for individual processes and feed waters. Needs differ from application to application and site to site.

Pretreatment is the first step in controlling membrane fouling, and it can be quite involved. In its simplest form, pretreatment involves microstraining with no chemical addition. However, when a surface water is treated, the pretreatment procedure may be much more involved and include pH adjustment, chlorination, addition of coagulants (e.g., alum, polyelectrolytes), sedimentation, clarification, dechlorination (e.g., addition of sodium bisulphite), adsorption onto activated carbon, addition of complexing agents (e.g., EDTA, SHMP) and final polishing. Several important factors must be considered in contemplating pretreatment:

- Material of membrane construction (asymmetric cellulose or noncellulose membranes, thin-film ether, or amide composite membranes)
- Module configuration (spiral wrap, hollow fine fiber, tubular)
- Feed water quality
- Recovery ratio
- Final water quality

Figure 11.20 gives some indication of substances known to affect membrane performance. Concentrations and/or presence of these components in the feed water must be controlled (Taylor and Jacobs, 1996).

Turbidity levels stipulated by membrane manufacturers are normally attained by conventional clarification techniques such as coagulation followed by sedimentation and sand filtration. In seawaters that are rich in suspended nutrients and for treating surface waters with high TOCs, ultrafiltration has been advocated as a viable pretreatment option capable of reducing the suspended solids concentration to acceptable standards to feed finer membrane processes (Strohwald and Jacobs, 1992; Metcalf et al., 1992; Taylor et al., 1989*a*). The minimum pretreatment processes for RO or NF consist of antiscalant and/or acid addition and cartridge microfiltration. These pretreatment processes help to control scaling and to protect the membrane elements. Such precautions are required for conventional reverse osmosis or nanofiltration systems.

Advanced Pretreatment. Advanced pretreatment operations precede scaling control and cartridge microfiltration. These unit operations might include coagulation, oxidation followed by greensand filtration, groundwater recharge, continuous microfiltration, and GAC filtration. Any other unit operations upstream of conventional pretreatment would also be advanced pretreatment by definition. In some pretreatment unit operations, such as alum coagulation, the feed water is saturated with a salt, such as alum hydroxide. The solubility of such salts must be accounted for in the feed water stream to avoid precipitation in the membrane. Several other types of pretreatment may form part of RO or NF membrane systems.

Membrane fouling is not a well-understood phenomenon, and many different pretreatment processes can be used to control different types of fouling mechanisms. For example, to prevent problems with biological fouling, a bactericide that is

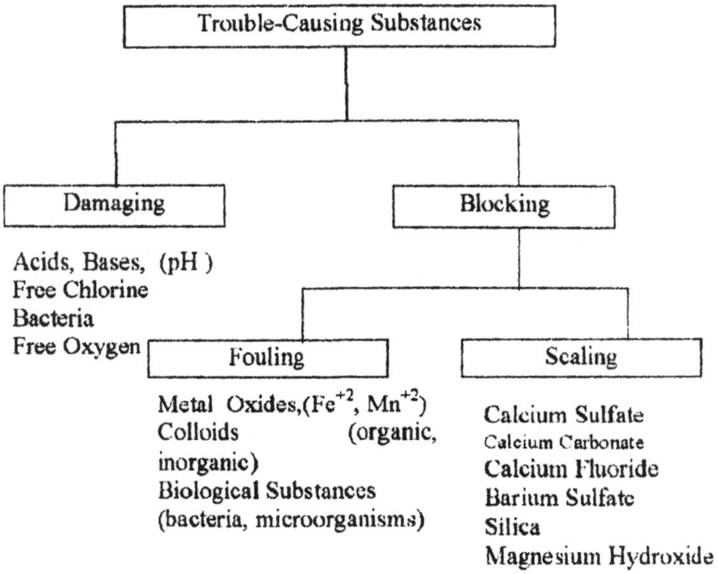

FIGURE 11.20 Substances potentially harmful to membranes. (*Source:* R. Rautenbach and R. Albrecht. *Membrane Processes.* Copyright © 1989 John Wiley & Sons. Reprinted by permission of John Wiley & Sons Limited.)

not harmful to the membranes might be introduced. To treat for biodegradable foulants, biologically activated carbon would be a feasible option. Microfiltration or ultrafiltration would also be useful for removing some foulants. Many processes could be combined with membranes for specific applications.

The most serious fouling problems often involve treatment of highly organic surface waters by NF or RO for TOC or DBP control. Such applications may require cleaning frequencies of less than one month, a period not typical for conventional RO or NF processes. However membrane applications for organic control are not typical, and new methods will have to be developed with time. One method for controlling fouling is to highly automate the membrane cleaning process. Currently, conventional RO plants treat nonfouling groundwaters, averaging cleaning frequencies of six months or more. Consequently, the cleaning operation is not automated, and it disrupts normal operation. Development of nondamaging membrane cleaning agents and automation of the cleaning process offer potential answers to fouling control.

Scaling Control. Scaling control is essential in RO/NF membrane filtration. Control of precipitation or scaling within the membrane element involves identifying the limiting salt and determining an acid and/or inhibitor treatment that will prevent precipitation at the desired recovery.

EXAMPLE 11.1 ***Limiting Salt.***
The required dosage of antiscalant or acid is determined by the limiting salt. A diffusion-controlled membrane process naturally concentrates salts on the feed side of the membrane. If excessive water is passed through the membrane, this concentration process continues until a salt precipitates and scaling occurs. Scaling reduces

membrane productivity and limits recovery to the allowable recovery just before the limiting salt precipitates. The limiting salt can be determined from the solubility products of potential limiting salts and the actual feed stream water quality, as shown Eq. 11.56. Ionic strength must also be considered in these calculations, because the natural concentration of the feed stream during the membrane process increases the ionic strength, allowable solubility, and recovery. Calcium carbonate scaling is commonly controlled by sulfuric acid addition; however, sulfate salts are often the limiting salt. Commercially available antiscalants can prevent scaling by complexing metal ions and preventing precipitation. Specific antiscalant formulas are proprietary, but phosphates, silicates, and other materials commonly used for corrosion control also help with scaling control. Antiscalants are not toxic, but they can add significant nutrients for biogrowth in the feed stream and in the concentrate stream after discharge. Equilibrium constants for commercial antiscalants are not available, which prevents direct calculation. However, some manufacturers provide computer programs for estimating the required antiscalant doses for given recovery, water quality, and membrane parameters. General formulas for the solubility products and ionic strength approximations are given in Eqs. 11.57 and 11.58.

$$A_nB_m \Leftrightarrow nA^{+J} + mB^{-K}$$

$$K_{sp} = (A^{+J})^n(B^{-K})^m \tag{11.56}$$

$$K_{sp} = \left(a\frac{A^{+J}}{x}\right)^n\left(b\frac{B^{-K}}{x}\right)^m$$

where x = fraction remaining
$1 - x$ = fraction recovered
K_{sp} = solubility product
a = fraction of cation retained
b = fraction of anion retained

$$u = 0.5\Sigma C_iZ_i^2 \cong (2.5 \times 10^{-5})(\text{TDS}) \tag{11.57}$$

where u = ionic strength
C = moles/L
Z = ion charge
TDS = mg/L

$$\log \gamma \cong -0.5Z^2 \frac{\sqrt{u}}{1+\sqrt{u}} \tag{11.58}$$

where γ = activity coefficient

A limiting salt calculation is shown in Table 11.10. Rejection of ions varies by operating conditions, membrane type, molecular weight, and charge. The solubility product has been modified by a and b coefficients, as shown in Eq. 11.56, for consideration of less-than-complete rejection. In the example, 0.90 rejection has been assumed for divalent ions. A more exact estimate of fraction rejected can be obtained from the manufacturer's literature or pilot studies. A raw-water quality is given with the solubility products of limiting salts at 25°C. The limiting salt is identified by determining which salt allows the least recovery or minimizes $1 - x$. The solubility levels of all possible salts are considered here. The first calculations demonstrate supersaturation of both $CaCO_3$ and $BaSO_4$. (This conclusion is suggested by the negative decimal fraction for r; a positive r would indicate the fraction

TABLE 11.10 Limiting Salt Example

Feed-water quality used for example				Salts to consider	
Parameter	Concentration	Parameter	Concentration	Salt	pK_{sp}@25°C
pH	8.0	Cl^-	730 mg/L	$CaCO_3$	8.3
Na^+	695 mg/L	NO^{3-}	0 mg/L	$Ca_3(PO)_4$	6.8
K^+	8 mg/L	F^-	1.1 mg/L	CaF_2	10.3
Ca^{+2}	8 mg/L	SO_4^{-2}	79 mg/L	$CaSO_4$	4.7
Mg^{+2}	2 mg/L	O-PO_4	0.7 mg/L	$BaSO_4$	11.7
Fe^{+2}	0.5 mg/L	HCO_3^-	631 mg/L	$SrSO_4$	6.2
Mn^{+2}	0.02 mg/L	SiO_2	24 mg/L	SiO_2	2.7
Cu^{+2}	0 mg/L	TDS	2200		
Ba^{+2}	0.04 mg/L				

For $CaCO_3 \rightarrow Ca^{2+} + CO_3^{2-}$ $K_{sp} = [Ca^{2+}][CO_3^{2-}] = 10^{-8.3}$

$$Ca^{2+} = 8 \text{ mg/L} = 0.0002 \text{ m/L}$$

$$CO_3^{2-} = HCO_3^- \frac{\alpha_2}{\alpha_1} = HCO_3^- \frac{K_2}{H^+} = 0.00005 \text{ M/L} = 3 \text{ mg/L}$$

$$\text{Then} \left[\frac{0.0002}{X}\right]\left[\frac{0.00005}{X}\right] = 10^{-8.3} \qquad X = 1.41 \text{ L}$$

$$R = 1 - X = -0.41 \text{ L}$$

For $BaSO_4 \rightarrow Ba^{2+} + SO_4^{2-}$ $K_{sp} = [Ba^{2+}][SO_4^{2-}] = 10^{-9.7}$

if $Ba^{2+} = 0.04 \text{ mg/L} = 3 \times 10^{-7}$ M/L, $SO_4^{2-} = 79 \text{ mg/L} = 0.0008$ M/L

$$\text{Then} \left[\frac{3 \times 10^{-7}}{X}\right]\left[\frac{0.0008}{X}\right] = 10^{-9.7} \qquad X = 1.09 \text{ L}$$

$$R = 1 - X = -0.09 \text{ L}$$

of the feed stream that could be recovered.) Although recovery seems impossible, when ionic strength is considered, the allowable recovery rises to 60 percent. $CaCO_3$ precipitation will be controlled by acid addition.

The effect of ionic strength is shown as follows for the $BaSO_4$ example, approximated for brevity. Recovery increases because the increasing TDS boosts the ionic strength, which increases the solubility of the limiting salt at equilibrium. Iteration using the adjusted solubility product allows convergence of the recovery calculation, as illustrated in the example. The recovery is observed to increase from 0 to 72 percent when ionic strength is used following iteration.

Assuming 80 percent rejection of TDS and 60 percent recovery, the following approximation can begin:

$$\text{TDS} = (0.8/0.4)2200 = 4400 \text{ mg/L}$$

$$\mu = (2.5 \times 10^{-5})\text{TDS} = 0.11$$

$$\log \gamma = -0.5(2)^2 \frac{\sqrt{0.11}}{1 + \sqrt{0.11}} = -0.50$$

$$pK_c = pK_{sp} - (m)p\gamma - (n)p\gamma = 9.7 - (1)0.50 - (1)0.50 = 8.7$$

Recalculate X from:

$$K_c = \left(a\frac{[A^{+m}]}{X}\right)n\left(b\frac{[B^{-n}]}{X}\right)m = \left[(0.9)\,\frac{3\times 10^{-7}}{X}\right]\left[(0.9)\,\frac{0.0008}{X}\right] = 10^{-8.7}$$

$$X = 0.31\text{L} \qquad R = 1 - X = 0.69\text{ L or } 69\%$$

Iterate calculations from beginning but use 0.31 for X in place of 0.4 to determine new TDS concentration in concentrate stream. After two iterations of recalculating TDS and K_c, the recovery converges at 72 percent.

EXAMPLE 11.2 ***Acid Addition.***
Acid is usually added to allow $CaCO_3$ recovery to the limiting salt recovery, as illustrated in the example. The equilibrium constants for the carbonate system, alkalinity, pH, calcium, and recovery are required to make these calculations. In the example, pH 7.18 was required to avoid scaling by $CaCO_3$. Addition of 162 mg/L H_2SO_4 produces this pH. The additional sulfate from acid addition should be considered for calculating the limiting salt recovery.

Acid addition to control $CaCO_3$ based on ionic strength:
Example:
Given: $Ca^{2+} = 8$ mg/L, $HCO_3^- = 631$ mg/L, pH = 8.0, R = 75%, $\mu = 0.12$
Find: H_2SO_4 dose to prevent scaling of calcium carbonate
Solution: *Determine pH required on the feed side of membrane*

For $CaCO_3 \rightarrow Ca^{2+} + CO_3^{2-}$ $\quad K_{sp} = [Ca^{2+}][CO_3^{2-}] = 10^{-8.3}$

$$\log\gamma = -0.5(1)^2\frac{\sqrt{0.12}}{1+\sqrt{0.12}} = -0.13 \qquad p\gamma_{+/-1} = 0.13\; p\gamma_{+/-2} = 0.51$$

$$pK_c = pK_{sp} - (m+n)p\gamma = 8.3 - (1)0.51 - (1)0.51 = 7.31$$

$$K_c = (a\frac{[A^{+m}]}{X})^n(b\frac{[B^{-n}]}{X})^m = [(0.9)\frac{Ca^{2+}}{X}][(0.9)\frac{CO_3^{2-}}{X}] = 10^{-7.31}$$

$$Ca^{2+} = (8/40000) = 0.0002\text{ M/L}$$

for $HCO_3^- \rightarrow H^+ + CO_3^{2-}$ $\quad K_2 = \dfrac{[H^+][CO_3^{2-}]}{[HCO_3^-]} = 10^{-10.3}$

$$K_2 = \frac{[H^+][\gamma_2 CO_3^{2-}]}{[\gamma_1 HCO_3^-]}, pK_{c2} = pK_2 + p\gamma_1 - p\gamma_2 = 10.3 + 0.13 - 0.51 = 9.92$$

$$CO_3^{2-} = \frac{[K_{c2}][HCO_3^-]}{[H^+]} = \frac{[10^{-9.92}][(631/61000)]}{[H^+]} = \frac{10^{-11.9}}{[H^+]}$$

$$K_c = [(0.9/0.25)Ca^{2+}][(0.9/0.25)CO_3^{2-}]$$

$$= (0.9/0.25)^2[0.0002][10^{-11.9}/[H^+]] = 10^{-7.31}$$

$[H^+] = 10^{-7.18}$ or pH = 7.18 not to scale calcium carbonate

The estimated pH to prevent scaling may be improved either by iteration or by inserting $\alpha_1 C_t$ for HCO_3^- before solving. However, the technique is an approximation and provides satisfactory results as presented.

This example calculation has determined the pH that will not scale calcium carbonate at 75 percent recovery. The example assumed 90 percent rejection of calcium and carbonate ions. This calculation is approximate, but it does illustrate the effect of ionic strength and the control of calcium carbonate solubility by adjusting pH. Control of calcium carbonate solubility is required in almost every RO/NF plant that treats groundwater.

Once the required pH has been determined for calcium carbonate scaling, the required acid dose can be calculated.

Determine dose of H_2SO_4 to achieve pH:

$$K_1 = \frac{(H^+)(\gamma HCO_3^-)}{(H_2CO_3)} = 10^{-6.3} \qquad pK_{c1} = pK_1 - p\gamma$$

$$\log \gamma = -0.5(1)^2 \frac{\sqrt{0.12}}{1+\sqrt{0.12}} = -0.13 \qquad p\gamma = 0.13$$

$$pK_{c1} = pK_1 - p\gamma = 6.3 - 0.13 = 6.17$$

$$K_{c1} = \frac{(H^+)(HCO_3^-)}{[H_2CO_3]} = 10^{-6.17} = \frac{[10^{-7.18}][(0.9/0.25)(631/61000) - X]}{[0 + X]}$$

where X is the H^+ needed to react with HCO_3^- $10^{-6.17}$ $X = -10^{-7.18}X + 10^{-8.61}$ $X = 0.0033$ M/L$H^+ = (0.0033 * 98000)/2 = 162$ mg/LH_2SO_4

This is the H_2SO_4 that needs to be added in the last element of the last stage. Consequently this concentration is reduced by (0.25)(162) = 40.5mg/L to determine feed stream H_2SO_4 addition.

Manufacturers' Programs. Antiscalants can also be used to determine allowable recovery. As noted, manufacturers such as Hydranautics, Desal, Fluid Systems, Dow-Filmtec, DuPont, and Trisep provide computer programs to help with estimating recovery. Most antiscalants complex metal cations, making them unavailable for precipitation. Polyacrylic acid (PAA), one common antiscalant, interferes with nucleation and crystal growth. Unfortunately, the equilibrium constants are not available, and the predicted recoveries cannot be verified. However, antiscalants certainly work, as they are used worldwide in membrane plants to increase recovery. Many membrane plants combine sulfuric acid and antiscalant pretreatment. Table 11.11 shows sample output from a program to calculate antiscalant dosage. Even though the $BaSO_4$ is 608 percent saturated, a dose of 4.2 mg/L is indicated to control scaling. This program, like many others, predicts the saturation percentages of selected salts and the doses required to control scaling, but it does not give the exact requirements for a particular membrane. That judgment reflects a design choice, as all scaling could possibly be controlled by antiscalant addition. Antiscalants are typically more costly than acid based on mass, but they cost less than acid in use. Antiscalants commonly lose effectiveness if they react preferentially with metals other than the limiting salt metals. High iron concentrations sometimes greatly decrease the effectiveness of certain antiscalants. These factors and others point out the need for pilot studies.

Cartridge Microfiltration. Cartridge microfilters typically used for RO/NF pretreatment are sieving filters with pore diameters of 5 to 20 μm. Membranes in these elements have larger pores than the continuous MF membranes discussed previously in this chapter. The pressure drop through a cartridge microfilter does not usually exceed 5 lb/in^2 in nanofiltration applications and 10 lb/in^2 in reverse osmosis applications. Cartridge microfiltration upstream of a nanofiltration or reverse osmosis process protects only against particulate foulants. The technology offers no pro-

TABLE 11.11 Sample Antiscalant Dosage Calculation

Flocon 100 dose projection

Prepared by: Date: 3/7/94

Feed water analysis Cations			Recovery = 75% Anions	
Ca	8		HCO3	583.5
Mg	2		SO4	202.2
Na	695		Cl	730
K	8		F	1.1
Ba	.04		NO3	0
Sr	0		PO4	.7
Fe	.05		CO3	.72
Silica	24		CO2	45.12
Temperature	25		Pressure	150

The above feed water analysis reflects the addition of sulfuric acid.

H_2SO_4 (100%) added (ppm)	45.09
Ionic strength, feed.	035
Total cation as $CACO_3$	1554
Total anion as $CACO_3$	1723
TDS (feed)	2267
TDS (brine)	9070
pH, raw water	8
pH, FEED	7.2
pH, brine	7.72

Caution Anion/cation balance exceeds 5% error.
A more accurate water analysis is desirable.

Flocon 100 dose for $CaCO_3$ control (ppm)	Minimum = 2.5	Recommended = 3.8
Flocon 100 dose for BaSO4 control (ppm)	Minimum = 3	Recommended = 4.2

Degree of saturation of potential scales, %

$CaCO_3$ (LSI)	$CaSO_4$	$SrSO_4$	$BaSO_4$
.97	2	0	767

To ensure operation in a scale-free position Flocon 100 should be dosed consistent with the greatest scale potential in the system as indicated by the highest dosage given above.

Ion product, $CaCO_3$ = 3.833975E-08
Ion product, $CaSO_4$ = 6.721338E-06
Ion product, $SrSO_4$ = 0
Ion product, $BaSO_4$ = 11.805316E-06
Ion product, CaF_2 = 4.28174E-11

tection against scaling. Also, cartridge microfiltration alone cannot remove foulants from a feed stream with a high turbidity or suspended solids concentration. Generally, the size distribution of the foulants removed in this way includes particles with diameters smaller than the cartridge microfilter pore sizes. Consequently, a maximum value for the feed stream fouling indexes is specified prior to cartridge microfiltration. A cartridge microfilter in a reverse osmosis or nanofiltration process should be thought of as a means of protecting the membrane elements against periodic upsets from failure of solids removal processes or sand entrainment from well pumping.

Array Models

This section describes different models and design techniques used to size arrays of membrane elements. A linear solution diffusion model, a film theory solution diffusion model, and a coupling model are described. Single and multistaged array equations for predicting permeate water quality are developed, and a simple design example is presented.

Linear Solution Diffusion Model. Many different theories and models attempt to describe mass transfer in diffusion-controlled membrane processes. However, a few basic principles or theories are used to develop most of these models: convection, diffusion, film theory, and electroneutrality. These principles or theories could be used to group models into linear diffusion models, exponential diffusion models, and coupling models. Most of the modeling efforts have been developed using very small test cells. Also, they have not incorporated product recovery, limiting their practical use. The basic equations used to develop these models are shown in Eqs. 11.59 through 11.63. The standard membrane element configuration is shown in Figure 11.21.

$$J = k_W (\Delta P - \Delta\Pi) = \frac{Q_p}{A} \tag{11.59}$$

$$J_i = k_i \Delta C = \frac{Q_p C_p}{A} \tag{11.60}$$

$$r = \frac{Q_p}{Q_f} \tag{11.61}$$

$$Q_f = Q_c + Q_p \tag{11.62}$$

$$Q_f C_f = Q_c C_c + Q_p C_p \tag{11.63}$$

where J = water flux (L^3/L^2t)
J_i = solute flux (M/L^2t)
k_W = solvent mass transfer coefficient (L^2t/M)
k_i = solute mass transfer coefficient (L/t)
ΔP = pressure gradient (L), $(P_f + P_c)/2 - P_p$

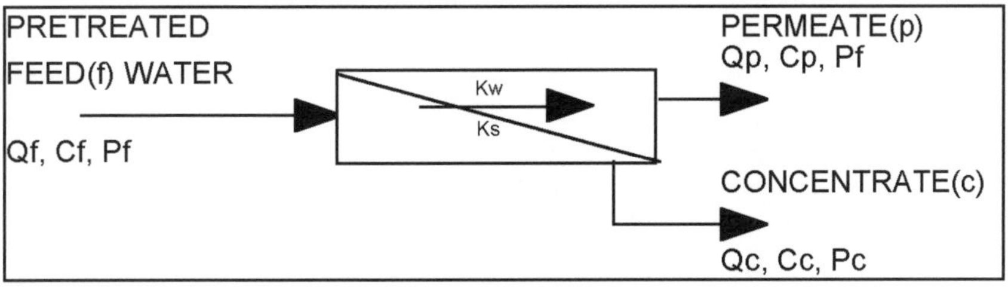

FIGURE 11.21 Basic diagram of mass transport in a membrane.

$\Delta\Pi$ = osmotic pressure (L) $(\Pi_f + \Pi_c)/2 - \Pi_p$
ΔC = concentration gradient (M/L^3), $(C_f + C_c)/2 - C_p$
Q_f = feed stream flow (L^3/t)
Q_c = concentrate stream flow (L^3/t)
Q_p = permeate stream flow (L^3/t)
C_f = feed stream solute concentration (M/L^3)
C_c = concentrate stream solute concentration (M/L^3)
C_p = permeate stream solute concentration (M/L^3)
r = recovery
A = membrane area (L^2)
Z = combined mass transfer term

If ΔC is defined as the difference of the average feed and brine stream concentrations and the permeate stream concentration, then Eq. 11.64 can be derived from Eqs. 11.59 and 11.63. This model can be described as a linear homogeneous solution-diffusion model in that it predicts diffusion-controlled solute flow and pressure (convection)-controlled solvent flow. Equation 11.64 can be simplified by including a Z term, which incorporates the effects of the mass transfer coefficients, pressure, and recovery into a single term.

$$C_p = \frac{k_i C_f}{k_W(\Delta P - \Delta\Pi)(2 - 2r/2 - r) + k_i} = Z_i C_f \tag{11.64}$$

This is the simplest model, but it does incorporate the effects of five independent variables on permeate water quality, as shown in Figure 11.22. If pressure increases and all other variables are held constant, then permeate concentration will decrease. If recovery is increased and all other variables are held constant, then permeate concentration increases. These effects may be hard to realize if an existing membrane array is considered. It is impossible in such an environment to increase pressure without increasing recovery. However, array designs can increase pressure without varying recovery. Decreasing feed stream concentration may identify pretreatment as an option for decreasing the permeate stream concentration. Different membranes may have different mass transfer characteristics. Using a membrane with a lower molecular weight cutoff would probably decrease the permeate concentration, although the solvent and solute MTCs must be considered before such a result can be expected.

Film Theory Model. The linear model for array design can be modified by the incorporation of film theory, which assumes that the solute concentration exponentially increases from the center of the feed stream channel toward the surface of the membrane and diffuses back into the bulk stream. Mathematically, this is shown in Eq. 11.65, and the back-diffusion constant is introduced in Eq. 11.66.

$$J_i = -D_s \frac{dC}{dx} + C_i J \tag{11.65}$$

where D_s = diffusivity
C_i = concentration from the bulk to the membrane interface
x = path length or film thickness

$$\left[\frac{C_s - C_p}{C_b - C_p}\right] = e^{(J/k_b)} \tag{11.66}$$

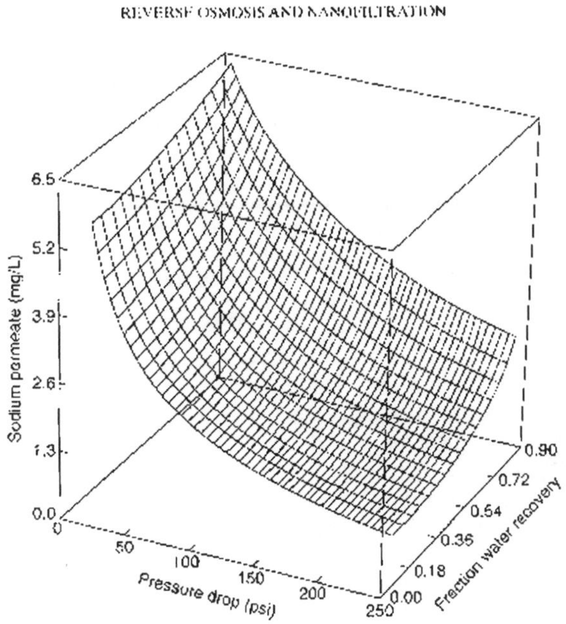

FIGURE 11.22 Sodium permeate concentration for varying pressure and recovery.

where C_p = solute concentration in the membrane permeate
C_b = solute concentration in the membrane bulk
C_s = solute concentration at the membrane surface
J = water flux through the membrane
$k_b = D_s/x$ = diffusion coefficient from the surface to the bulk

Incorporation of Eq. 11.66 into Eq. 11.59 through 11.63 results in Eq. 11.67, which is a development of the homogeneous solution diffusion model using concentration-polarization. This model predicts a higher concentration at the membrane surface than in the bulk of the feed stream. Such an effect is documented in the literature (Sung, 1993; Hofman et al., 1995), and the model in Eq. 11.67 accounts for this phenomenon. The back-diffusion coefficient k_b represents solute diffusion from the membrane surface to bulk in the feed stream, which is different than the mass transfer coefficient k_W, which represents solute diffusion through the membrane to the permeate stream.

$$C_p = \frac{C_f k_i e^{(J/k_b)}}{k_W(\Delta P - \Delta\pi)(2 - 2r/2 - r) + k_i e^{(J/k_b)}} \tag{11.67}$$

If solutions containing ions are assumed to be electrically neutral due to the presence of counterions, then solutes do not pass through a membrane in a charged state but in a coupled state. Such electroneutrality is observed in diffusion-controlled membrane processes when the permeate streams become more concentrated than the feed streams. Relatively high concentrations of sulfates have been observed to increase rejection of calcium and decrease rejection of chlorides. One interpretation of such observations suggests that the strong calcium sulfate couple is retained in the feed stream and the weaker sodium chloride couple is forced to pass through the membrane to maintain equilibrium.

Coupling. The coupling effect on mass transfer through membranes has been modeled using a statistical modification of free energy for single-solute systems, as shown in Eq. 11.68 (Rangarajan, Matsuura, and Sourirajan, 1976; Sung, 1993). The free energy term is assumed to be different in the bulk solution and at the membrane surface because of a difference of ion concentration. Consequently, the energy required to bring the ions to the surface, shown in Eq. 11.69, is the difference of the free energy in bulk and surface solutions. The $\Delta\Delta G$ values for each ion are determined by experiment.

$$\frac{1}{\Delta G} = -\frac{1}{E}r - \frac{\Delta}{E} \tag{11.68}$$

where ΔG = free energy of coupled ion
E = solution-dependent constant
r = coupled ion radius
Δ = statistical constant

$$\Delta\Delta G = \Delta G_I - \Delta G_B \tag{11.69}$$

where $\Delta\Delta G$ = difference of coupled ion free energy at interface I and bulk B.

Membrane-specific solute mass transfer coefficients for single-solute systems have been determined by experiment. They are related to free energy as indicated by Eq. 11.70. Once the membrane-specific constant ln (C^*) has been determined for a reference solute (e.g., NaCl), it is possible to determine k_i for any given solute in a single-solute system. Once k_i is known, the mass transport of any single solute in a diffusion-controlled membrane can be predicted.

$$\ln (k_i) = \ln (C^*) + \Sigma\left(-\frac{\Delta\Delta G}{RT}\right) \tag{11.70}$$

where C^* = membrane-specific constant
k_i = mass transfer coefficient
R = gas constant
T = temperature

Coupling can be used to model mass transport in a multisolute system. As with the single-solute system, electroneutrality is assumed in all phases (bulk, surface, permeate, film). A coupling model can incorporate either a homogeneous or an exponential solution-diffusion model with the free energy model. The model requires the mass transfer coefficients for water and a reference couple, the back-transfer coefficient and the $\Delta\Delta G$ values for all ions to predict flux, and the concentration of the permeate stream. The mass transfer models shown in Eqs. 11.64 and 11.65 can both be used as coupling models as long as the k_i term is determined using $\Delta\Delta G$ or feed stream solute composition. However this term is membrane-specific and must be determined for a given product. No model accounts for the effects of the membrane surface on a fundamental basis.

The determination of $\Delta\Delta G$ is based on free energy and is therefore affected by feed stream composition. The determination of free energy for any given reaction is a good method of determining the likelihood that the reaction will proceed, but this method relies on consideration of all the free energies for the major chemical interactions. As the solutes and free energies are not known for many of the constituents in natural water, a basic free energy model has not been developed for practical membrane design.

Array Modeling. The design process for membrane element arrays will be illustrated using the linear model for simplicity and to take advantage of a membrane manufacturer's computer program. The mass transfer coefficients for the linear model can be developed from field criteria, and the model can easily be used to predict permeate water quality for given changes in operation. Variations in sodium permeate concentration are shown in Figure 11.22 for varying pressures and recoveries. Reverse osmosis and nanofiltration processes allow considerable flexibility in permeate water quality. Designers can vary flux and recovery to achieve water quality goals. Anyone involved in the design or operation of a membrane facility should be aware of the coupling phenomena and concentration-polarization. The mass transfer coefficients for the film theory model are more difficult to develop, but they have been shown to increase accuracy of pilot plant data. However, the best design criteria would come from site-specific pilot studies. Pilot studies may not be required for systems to desalinate some brackish waters with very low organic concentration and no potential foulants, but many physical, chemical, and biological factors not initially recognized can cause serious operational problems. The only way to avoid these types of unrealized problems in design is to conduct a pilot study.

Modeling a Linear Array. Equations 11.64 and 11.67 are very useful tools for determining the effect of ΔP, C_f, r, $\Delta\Pi$, k_W, and k_i on C_p without experimentation if the six variables are known. The first four can be accurately assumed by the user for a particular feed water and membrane operating conditions. However, k_W and k_i should be measured experimentally for a given membrane and source; they can be taken from the literature if reports describe a given membrane and a similar source. In Eq. 11.71, Z_i states a modified MTC, which represents the effect of solvent mass transfer, solute mass transfer, recovery and pressure on permeate stream concentration for any single membrane element. In this way, the expression simplifies Eq. 11.63. If Eq. 11.66 were used, the modified mass transfer coefficient would be represented by Z_{cpi}, which would include consideration of back-diffusion.

$$C_{p,i} = \frac{k_{i,i}C_{f,i}}{k_{W,i}\,\Delta P_i(2 - 2r_i/2 - r_i) + k_{i,i}} = Z_iC_{f,i} \tag{11.71}$$

where Z_i = modified mass transfer coefficient

$\Delta P_i = \Delta P - \Delta\Pi$ = net average driving pressure including hydraulic losses

The subscript i in Eq. 11.71 denotes any stage in a multistage membrane array. Membrane systems are configured in arrays that consist of stages. Consequently, Eq. 11.71 can be expanded to describe an entire membrane system. The configuration for a typical two-stage membrane system is shown in Figure 11.23. The first stage consists of two pressure vessels, each of which typically contains six membrane elements. The second stage consists of one pressure vessel, also typically with six elements. The combination of these two stages is a 2–1 array. The model can be used to predict the system permeate concentration or the effect of any of the six independent variables on permeate concentration if MTCs, pressures, and recoveries for each stage are known.

As noted in Figure 11.23, the stage 2 feed stream flow and concentration always equals the stage 1 concentrate stream flow and concentration in a high-recovery system for potable water treatment. The interstage flow and solute concentration in a multiple-array membrane system can be related by an interstage mass balance, as shown in Eqs. 11.72 through 11.75. The resulting X_i term is used in Eq. 11.75 to deter-

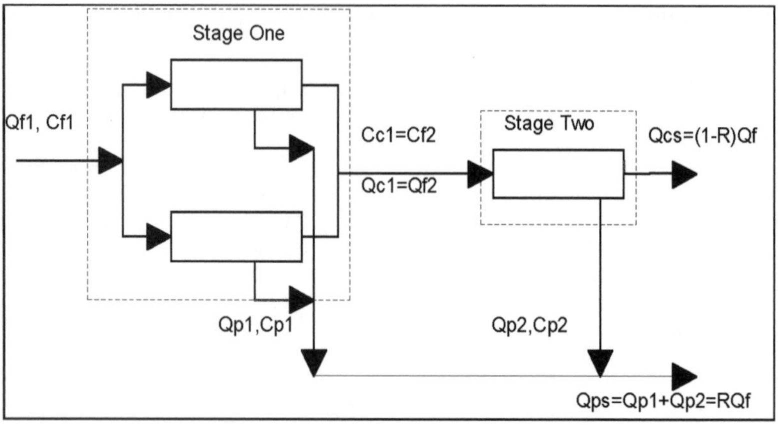

FIGURE 11.23 One-array (two-stage) membrane system.

mine the feed stream concentration for any stage from the initial feed stream concentration. It can be considered a concentration factor of stage *i*.

$$C_{C,i} = \frac{Q_{f,i}C_{f,i} - Q_{p,i}C_{p,i}}{Q_{c,i}} = \frac{C_{f,i} - r_iC_{p,i}}{1 - r_i} \tag{11.72}$$

$$C_{c,i} = C_{f,i+1} = C_{f,i}\left(\frac{1 - Z_ir_i}{1 - r_i}\right) \tag{11.73}$$

$$C_{c,i} = X_iC_{f,i} \tag{11.74}$$

$$X_i = \frac{1 - Z_ir_i}{1 - r_i} \qquad \text{Define } X_0 \equiv 1 \tag{11.75}$$

Consequently, an interstage equation for permeate concentration can be developed by interrelating all stages in any diffusion-controlled membrane system. Equation 11.75 interrelates all stages in a simple, flow-weighted concentration equation that represents the total solute mass from the permeate stream in each stage, shown in the numerator of Eq. 11.75, and the total volume of solvent from the permeate stream in each stage, shown in the denominator. Equation 11.75 is less complex than an expression incorporating the full Z_i and X_j terms, as shown in Eq. 11.76. This equation can be used to describe membrane system performance with as much versatility as Eq. 11.64 or Eq. 11.67, which describes a single stage or element. The subscripts *j* and *i* denote the stage number in a multistage membrane array, as shown in Eq. 11.77.

$$C_{p,\text{system}} = \frac{C_f \sum_{i=1}^{n} \left[A_ik_{i,i}\Delta P_i \frac{k_{i,i}}{k_{W,i}\Delta P_i(2 - 2r_i/2 - r_i) + k_{i,i}} \prod_{j=0}^{i-1} \frac{1 - k_{i,j}r_j/k_{W,j}\Delta P_j(2 - 2r_j/2 - r_j) + k_{i,j}}{1 - r_j} \right]}{\sum_{i=1}^{n} A_ik_{W,i}\Delta P_i} \tag{11.76}$$

$$C_{p,\text{system}} = \frac{C_f \sum_{i=1}^{n} \left(A_ik_{i,i}\Delta P_iZ_i \prod_{j=0}^{i-1} X_j \right)}{\sum_{i=1}^{n} A_ik_{i,i}\Delta P_i} \tag{11.77}$$

In Eq. 11.77, the term $\prod_{j=0}^{i-1} X_j$ is the product of the concentration factors of all stages before stage *i*.

The concentration factor is therefore mathematically defined as 1 before stage 1 ($X_o = 1$) (Duranceau, 1992). The exponent J/k_b used in the film theory model, as shown in Eq. 11.67, can be approximated by a constant times the ratio of the flow through the membrane to the flow across the membrane, as shown in Eq. 11.78. This expression approximates the compensation for the effects of multisolute solutions in a high-recovery or plant application.

$$\frac{J}{k_b} = k\left(\frac{r_e}{1 - r_e}\right) \tag{11.78}$$

where k_b = statistically derived coefficient
r_e = average membrane recovery per element

The system permeate equation can also be modified to incorporate the film theory model, as shown in Eq. 11.79 and 11.80.

$$C_p = \frac{k_i e^{(J/k_b)} C_f}{k_W(\Delta P - \Delta\Pi)(2 - 2r/2 - r) + k_i e^{(J/k_b)}} = Z_{i,cp} C_f \tag{11.79}$$

where $Z_{i,cp}$ = modified film theory feed stream mass transfer coefficient

$$C_{p,\text{system}} = \frac{C_f \sum_{i=1}^{n} \left(A_i k_{W,i} \Delta P_i Z_{i,cp} \prod_{j=0}^{i-1} X_{j,cp}\right)}{\sum_{i=1}^{n} A_i k_{W,i} \Delta P_i} \tag{11.80}$$

where $X_{j,cp}$ = modified film theory concentrate stream concentration factor

Practical Array Design

Membranes are typically incorporated in elements, as mentioned early in this chapter. Unfortunately, the terms in Table 11.8 are not universally accepted, and meanings vary. For example, a membrane "array" can be defined as stage 1 or as the total of stages 1, 2, and 3. Care must be taken to ensure that parties discussing membrane systems are using the same terminology.

Simple Sizing. This section gives an example assuming a simplistic 4-2-1 array to illustrate design concepts for membrane configurations. Basic assumptions in this design are 8-in- (20-cm-) diameter membrane elements with 350-ft^2 (32.5-m^2) surface areas operating at a flux of 15 gsfd (0.7 m^3/m^2/day) and a 75 percent recovery. One element will produce 5250 gpd. As seven elements are in each pressure vessel and seven pressure vessels are in one array, an array will produce approximately 250,000 gpd. Four such arrays would provide 1 mgd of permeate and 0.33 mgd of concentrate. Note that osmotic pressure, not illustrated, would continuously decrease the recovery in each succeeding element.

EXAMPLE 11.3

Assumptions:

1. Membrane element flux = 15 gal/ft^2 day (7×10^{-6} m/s)
2. Design for a 4-2-1 array

3. Each pressure vessel can contain 6 to 8 elements.
4. Element surface area = 350 ft² (32.5 m²)
5. Recovery = 75 percent

Water 1 element can produce:

$$Q_p = JA = (15\ \text{gal/ft}^2\text{/day})(350\ \text{ft}^2\text{/element}) = 5250\ \text{gal/day/element}\ (19.87\ \text{m}^3\text{/day/element})$$

Number of elements on 1 array:

$$(7\ \text{PV/array})(7\ \text{elements/PV}) = 49\ \text{elements/array}$$

Gallons per day produced per array:

$$(5250\ \text{gal/day/element})(49\ \text{elements/array}) = 257{,}250\ \text{gal/train/day}\ (973.69\ \text{m}^3\text{/train/day})$$

*Trains needed to supply 1 mgd (3786 m³/*day):

$$(10^6\ \text{gal/day})(\text{array/day/257,250 gal}) = 3.9\ \text{arrays}$$

Total number of elements needed:

$$(4\ \text{arrays})(49\ \text{elements/array}) = 196\ \text{elements}$$

Flow in and out of 1 array:

$$Q_f = Q_p/r = 1\ \text{mgd}/0.75 = 1.33\ \text{mgd}(5048\ \text{m}^3\text{/d})$$

$$\text{flow in per array} = 1.33\ \text{mgd}/4 = 0.33\ \text{mgd} = 231\ \text{gpm}\ (0.88\ \text{m}^3\text{/min})$$

$$\text{flow out per array} = 1\ \text{mgd}/4 = 0.25\ \text{mgd} = 174\ \text{gpm}\ (0.66\ \text{m}^3\text{/min})\ \text{permeate}$$

$$\text{flow out per array} = 0.33\text{–}0.25 = 0.08\ \text{mgd} = 57\ \text{gpm}\ (0.22\ \text{m}^3\text{/min})\ \text{permeate}$$

The film theory model can be used to estimate permeate water quality. In the following example, MTCs were determined from a pilot study or given by the membrane manufacturer. The concentration-polarization factor is also given. The example calculation shows the chloride concentrate in the permeate stream from each stage and the array (Hofman et al., 1993).

EXAMPLE 11.4

Given: A 2–1 array with 6 elements/pv, element surface area = 30 m² (323 ft²)
5-bar driving force, $k_W = 1.46 \times 10^{-6}$ m/s – bar (0.2135 gsfd/lb/in²),
$k_i = 1.87 \times 10^{-6}$ m/s (0.53 ft/day)
$k' = 0.75$, $r = 0.5$ stage, $r_e = 0.083$/element
Using the array film theory model:

$$C_{p,\text{system}} = \frac{C_f \sum_{i=1}^{n} \left(A_i k_{W,i} \Delta P_i Z_{i,cp} \prod_{j=0}^{i-1} X_{j-1,cp} \right)}{\sum_{i=1}^{n} A_i k_{W,i} \Delta P_i}$$

$$\sum_{i=1}^{2} A_i k_{W,i} \Delta P_i = \text{stage 1 (2 vessels/stage)(6 elements/p. vessel)(30 m}^2\text{/element)}$$

$$(1.46 \times 10^{-6}\ \text{m/s/bar})(5\ \text{bar}) + \text{stage 2 (1 vessel/stage)}$$

$$(6 \text{ elements/p. vessel})$$
$$(30 \text{ m}^2/\text{element})(1.46 \times 10^{-6} \text{ m/s/bar})(5 \text{ bar})$$
$$= 2.628 \times 10^{-3} \text{ m}^3/\text{s} + 1.324 \times 10^{-3} \text{ m}^3/\text{s}$$
$$= 3.942 \times 10^{-3} \text{ m}^3/\text{s} \ (89{,}999 \text{ gal/day})$$

$$Z_{i,cp} = \frac{k_{S,i} e^{k(r_e/1 - r_e)}}{k_{W,i}\Delta P_i(2 - 2r_i/2 - r_i) + e^{k(r_e/1 - r_e)k_{S,i}}} = Z_1 = Z_2$$

$$= \frac{((1.87 \times 10^{-6} \text{ m/s})e^{0.75(0.083/1 - 0.083)}}{\left((1.46 \times 10^{-6} \text{ m/s bar})(5 \text{ bar})\left(\frac{2 - 2(0.5)}{2 - 0.5}\right) + (1.87 \times 10^{-6} \text{ m/s})e^{0.75(0.083/1 - 0.083)}\right)}$$

$$= 0.29 \prod_{j=0}^{1} X_j = (1)\left(\frac{1 - Z_1 R_1}{1 - R_1}\right) = (1)\left(\frac{1 - (0.29)(0.50)}{1 - 0.5}\right) = 1.71$$

$C_{p,\text{system}}$

$$= \frac{100 \text{ mg/L } (2.628 \times 10^{-3} \text{ m}^3/\text{s})(0.29) + 100 \text{ mg/L } (1.314 \times 10^{-3} \text{ m}^3/\text{s})(0.29)(1.71)}{(5.4 \times 10^{-3} \text{ m}^3/\text{s})}$$

$$= 100 \text{ mg/L } (0.67)(0.29) + 171 \text{ mg/L } (0.33)(0.29)$$
$$= 29.0 \text{ mg/L } (0.67) + 49.6 \text{ mg/L } (0.33)$$
$$= 19.4 \text{ mg/L} + 16.4 \text{ mg/L} = 35.8 \text{ mg/L}$$

Some comments will clarify this model application. The five-bar feed stream pressure gradient through each stage has been assumed to include the effects of osmotic pressure and hydraulic losses. These effects can be estimated by assuming that 1 mg/L TDS produces 0.7 millibars of osmotic pressure, a 0.20-bar pressure loss across each element, and a 0.34-bar stage entrance loss. The hydraulic effects on pressure gradient are accounted for during design and typically controlled by high-pressure pumps and flow restrictors. The first-stage and second-stage Z terms are identical, because stage recovery, pressure gradient, and membranes are identical. The Z_1 term is used to calculate the X_2 term, as the stage 1 concentrate stream is the stage 2 feed stream. The denominator of Eq. 11.80 is the permeate flow produced by the two-stage system and is shown as the first calculation in the example application. The final calculation gives the system permeate Cl^- concentration, estimated at 35.8 mg/L. The stage 1 and 2 Cl^- concentrations are 29.0 and 49.6 mg/L, shown in the calculation of the flow-weighted system concentration. Since recovery, pressure, and membranes are identical from stage 1 to stage 2, the second-stage permeate stream Cl^- concentration will always be higher than that in the first stage, because the feed stream Cl^- concentration is higher in stage 2 (171 mg/L) than in stage 1 (100 mg/L).

Manufacturers' Programs. Manufacturers like Dow-Filmtec, Hydranautics, Fluid System, and TriSep provide computer programs for developing RO plant design criteria. These programs do not yield final design specifications, and no manufacturer accepts responsibility for their use by anyone not specifically authorized. These programs are only tools for developing and testing various system configurations, and their output should not be regarded as completed designs. The programs do provide a means of estimating water production and quality for given parameters. Table 11.12 gives an example of output from a Dow-Filmtec computer program that

TABLE 11.12 Sample Design Program Output

FILMTEC RO SYSTEM ANALYSIS, SEPT XX VERSION

PREPARED FOR: AWWARF
PREPARED BY: J. M. H.
DATE: 8-2-XX

FEED:	204.71	1933 MG/L	25 DEG C
RECOVERY:	85%		

ARRAY:	1	2	3
NO. OF PV:	4	2	1
ELEMENT:	SW8040	SW8040	SW8040
NO. EL/PV:	7	7	7
EL. TOTAL:	28	14	7

FOULING FACTOR: 0.85

	FEED	REJECT	AVERAGE
PRESSURE (PSIG)	355.0	248.1	311.2
OSMOTIC PRESSURE (PSIG)	18.7	122.9	51.2

NDP (MEAN) = 260.1 PSIG
AVERAGE PERMEATE FLUX = 17.3 GFD PERMEATE FLOW = 174.05

GPM

			PERMEATE		FEED		P
ARRAY	EL. NO.	RECOVERY	GPD	MG/L	GPM	MG/L	(PSIG)
1	1	0.089	6531	11	51.2	1933	350
	2	0.094	6337	12	46.6	2120	343
	3	0.101	6155	14	42.2	2339	336
	4	0.109	5978	17	38.0	2601	331
	5	0.119	5801	20	33.8	2918	327
	6	0.131	5614	24	211.8	3310	323
	7	0.145	5404	29	25.9	3805	320
2	1	0.081	5150	33	44.3	4445	313
	2	0.084	4938	38	40.7	4833	307
	3	0.088	4728	44	37.3	5274	302
	4	0.092	4514	51	34.0	5780	298
	5	0.097	4292	61	30.8	6362	294
	6	0.101	4058	73	27.9	7036	288
	7	0.106	3807	90	25.0	7819	288
3	1	0.055	3549	101	44.8	8731	281
	2	0.054	3322	115	42.3	9234	275
	3	0.054	3107	131	40.0	9759	269
	4	0.053	2897	147	37.9	10307	264
	5	0.052	2694	165	35.9	10877	260
	6	0.051	2496	185	34.0	11467	255
	7	0.050	2304	209	32.3	12073	252

ARRAY:	TOTAL	1	2	3
REJECT GPM:		88.5	44.8	30.7
REJECT MG/L:		4445	8731	12692
PERM GPD:	250627	167283	62975	20369
PERM MG/L:	37	18	54	145

TABLE 11.12 Sample Design Program Output (*Continued*)

PERMEATE, MG/L AS ION				
NH_4	0.0	0.0	0.0	0.0
K	0.3	0.1	0.5	1.3
Na	8.5	3.9	12.3	34.0
Mg	0.6	0.3	0.8	2.1
Ca	4.1	2.1	6.0	15.3
HCO_3	0.8	0.4	1.2	3.3
NO_3	1.5	0.7	2.2	6.5
Cl	111.7	11.4	28.5	76.7
F	0.0	0.0	0.0	0.0
SO_4	1.7	0.9	2.5	6.2
SiO_2	0.0	0.0	0.0	0.0

FEED/REJECT, MG/L AS ION				
NH_4	0.0	0.0	0.0	0.0
K	10.0	22.9	44.8	64.9
Na	343.0	788.0	1545.1	2242.2
Mg	40.0	92.1	181.2	263.9
Ca	300.0	690.9	13511.5	19711.6
HCO_3	28.7	65.8	128.9	186.8
NO_3	15.0	33.8	64.6	91.5
Cl	1056.0	24211.2	4772.5	6938.9
F	0.0	0.0	0.0	0.0
SO_4	1311.9	322.4	634.6	924.6
SiO_2	0.0	0.0	0.0	0.0

TO BALANCE: 4.9 MG/L SODIUM AND 0.0 MG/L CHLORIDE ADDED TO FEED
FEED WATER IS WELL OR SOFTENED WATER (BW) SDI <3

SCALING CALCULATIONS

	FEED	ACIDIFIED FEED	REJECT
HCO3 (MG/L)	193.8	28.7	186.8
CO2 (MG/L)	10.0	1211.2	1211.2
SO4 (MG/L)	10.0	1311.9	924.6
Ca (MG/L)	300.0	300.0	19711.6
TDS (MG/L)	1967.8	1932.6	12692.4
pH	7.50	5.56	6.37
LSI	0.63	–2.14	0.26
IONIC STRENGTH (MOLAL)	0.043	0.044	0.289
IP CaSO4 (SQ. MOLAR)	0.78E-06	0.11E-04	0.48E-03

IP CaSO4 AT SATURATION (SQ. MOLAR) = 0.57E-03
SULFURIC ACID DOSING (MG/L, 100%) = 132.6
TEMPERATURE (DEGREE C) = 25.0
RECOVERY (PERCENT) = 85.0
ESTIMATED PERMEATE pH IS 4.0
SULFURIC ACID CONSUMPTION (100% CONCENTRATION) = 147.8 KG/DAY

was used to develop an initial design of an RO array. Feed flow rate, TDS concentration, feed temperature, recovery, array type (i.e., 4-2-1 array), number of elements per pressure vessel, and element type are required program inputs. The program outputs pressure, flux, flow, and water quality for all process streams by stage and element for each array. Scaling calculations are provided for calcium sulfate.

Posttreatment

Posttreatment consists of several different unit operations for RO and NF membrane systems. The steps chosen and their sequence depend on the designer's preferences and water quality goals. The primary posttreatment unit operations are aeration, disinfection, and stabilization. Additional posttreatment operations of concern are removal of hydrogen sulfide, if present, and alkalinity recovery. A systems view of posttreatment can help a designer to realize important goals. The membrane process removes essentially all pathogens and the majority of the DBP precursors, salts, and other solutes in the feed stream. Solute removal eliminates carbonate alkalinity, but all dissolved gases including carbon dioxide and hydrogen sulfide pass through the membrane. The designer must produce a finished water after posttreatment with an appropriate alkalinity profile and disinfection without significant sulfur turbidity. The posttreatment example in Table 11.13 uses the permeate stream water quality illustrated in the previous membrane section without hydrogen sulfide. The water quality resulting from each posttreatment unit operation is illustrated in this table.

These changes in water quality will be discussed in the following sections. The sequence of unit operations assumed here is disinfection followed by alkalinity recovery and then aeration/stabilization. Other situations may call for different sequences of posttreatment steps, but this sequence assumes no problem with hydrogen sulfide, and it offers the advantage of minimizing equipment, as disinfection and aeration/stabilization are conducted simultaneously in separate unit operations.

Disinfection. If chlorine is added to the process stream before aeration, stabilization occurs during aeration. Almost no chlorine demand remains following a reverse osmosis or nanofiltration process. The chlorine converts some alkalinity that passes through the membrane to carbon dioxide. The pH following chlorination can be determined using pK_1 for the carbonate system and the alpha for OCl^-. This equation is applicable only when HCO_3^- is present. Once HCO_3^- is neutralized during chlorination, pH can be determined by summing the protons from the HCl added past the point of neutralization to the protons at neutralization.

$$Cl_2 + H_2O \Rightarrow HOCl + HCl$$

$$HCl \Rightarrow H^+ + Cl^-$$

$$HOCl \Leftrightarrow H^+ + OCl^- \; pK = 7.4$$

$$pH = pK_{H_2CO_3} + \log \left| \frac{HCO_3^- - (1 + \alpha_{OCl^-})C_{T_{Cl_2}}}{H_2CO_3 + (1 + \alpha_{OCl^-})C_{T_{Cl_2}}} \right|$$

TABLE 11.13 Posttreatment Water Quality Changes by Unit Operation

	Permeate	Disinfection	Alkalinity recovery	Aeration/stabilization
pH	4	3.9	6.3	11.0
H_2CO_3	129 mg/L	130 mg/L	67 mg/L	0.6 mg/L
HCO_3^-	0.8	0	62	62
H_2S	0	0	0	0
SO_4^{-2}	1.7	1.7	4.7	4.7
Cl^-	111.7	23.7	23.7	23.7
Ca^{+2}	4.1	4.1	4.1	4.1
TDS	34	37	79	79
DO	0	0	0	8.9
Cl_2	0	3	3	3

Chlorine addition to water will produce equal moles of hypochlorous acid and hydrochloric acid. The hypochlorous acid will partially ionize to hypochlorite ions and protons. The hydrochloric acid will completely ionize, producing protons and chloride ions. One mole of protons will be produced for every mole of hydrochloric acid and every mole of hypochlorite ion produced. Consequently, the complete proton production during chlorination would be canceled by the addition of OH^-, as shown here. An iterative process can solve the pH during chlorination. Typical chlorine doses following a reverse osmosis or nanofiltration process range from 3 to 10 mg/L. The following example uses the water quality illustrated in Table 11.12. Since essentially no alkalinity is available at pH 4, the pH is determined from the strong acid, HCl, produced from chlorination.

EXAMPLE 11.5

Given: 0.8 mg/L $HCO_3^- = 10^{-4.9}$ mol/L HCO_3^-

pH = 4.0

$Cl_2 = 3.0$ mg/L $Cl_2 = 10^{-4.37}$ mol/L H^+

After chlorination all HCO_3^- alkalinity will be converted to H_2CO_3 and additional H^+ will depress the pH.

$$\text{Total } H^+ = 10^{-4.0} + (10^{-4.37} - 10^{-4.9}) = 10^{-3.9}$$

$$\text{pH} = 3.9$$

Alkalinity Recovery. Alkalinity recovery becomes a consideration during scaling control. CO_2 converted from HCO_3^- during pretreatment or posttreatment will be available in a closed system. Consequently, the desired carbonate alkalinity in the finished water can be attained by CO_2 conversion before aeration, given presence of adequate CO_2. Normally, 1 to 3 meq/L of bicarbonate alkalinity is considered desirable for corrosion control. Since carbon dioxide passes unhindered through a membrane, the desired amount of alkalinity can be recovered in the permeate by acidifying the desired amount of HCO_3^-, passing it through the membrane, and adding the desired amount of base to convert the carbon dioxide back to its original bicarbonate form. The reactions are shown as follows.

$$HCO_3^- + H^+ \Rightarrow H_2CO_3$$

$$H_2CO_3 + OH^- \Rightarrow HCO_3^- + H_2O$$

The following example calculation illustrates alkalinity recovery using the water quality achieved following disinfection. In the example, the pH before alkalinity recovery is past the point of alkalinity neutralization. Consequently, additional base must be added to reach the point of alkalinity neutralization before alkalinity recovery can begin.

EXAMPLE 11.6 ***Alkalinity Recovery***

Assuming pH 3.9, 0.0 mg/L HCO_3^-, 130 mg/L $H_2CO_3 = 10^{-2.68}$ M, for permeate stream after chlorination

$$\text{Base for chlorination} \left| \frac{3 \text{ mg/L}}{71 \text{ mg/mmol } Cl_2} \right| \frac{1 \text{ mol } H^+}{1 \text{ mol } Cl_2} \left| \frac{1 \text{ mol NaOH}}{1 \text{ mol } H^+} \right| \frac{40 \text{ mg}}{\text{meq NaOH}} \right|$$

$$\approx 2 \frac{\text{mg NaOH}}{\text{L}}$$

Now pH 4.0, 0.8 mg/L HCO_3^-, 129 mg/L $H_2CO_3 = 10^{-2.68}$ M, for permeate conditions before chlorination

$$\text{Alkalinity recovery} \left| \frac{62\text{ mg/L}}{62\text{ mg/mmol }H_2CO_3} \right| \frac{1\text{ meq }H_2CO_3}{1\text{ meq NaOH}} \left| \frac{40\text{ mg}}{\text{meq NaOH}} \right|$$

$$= 40\frac{\text{mg NaOH}}{\text{L}}$$

$$\text{Finished alkalinity} \left| \frac{61.8\text{ mg/L}}{61\text{ mg/mmol }HCO_3} \right| \frac{1\text{ meq }HCO_3}{1\text{ mmol }HCO_3^-} \left| \frac{50\text{ mg }CaCO_3}{\text{meq}} \right|$$

$$= 50.6\frac{\text{mg }CaCO_3}{\text{L}}$$

$$pH = pK - \log\left[\frac{HCO_3}{H_2CO_3}\right] = 6.3 - \log\left[\frac{10^{-3}M}{(10^{-2.68} - 10^{-3})M}\right] = 6.28 \approx 6.3$$

Aeration and Stabilization. If calcium and bicarbonate are present, the pH following aeration is controlled by $CaCO_3$ buffering and can be estimated by assuming $CaCO_3$ equilibrium.

$$pH_s = pK_2 + pK_{sp} + pCa + pHCO_3^-$$

EXAMPLE 11.7 ***Aeration***

$$CaCO_3 = Ca^{+2} + CO_3^{-2}\ K_{sp} = 10^{-8.3}$$

$$HCO_3^- = H^+ + CO_3^{-2}\ K_2 = 10^{-10.3}$$

$$pH = pK_2 - pK_{sp} + p[Ca^{+2}] + p[Alk]$$

$$[Ca^{+2}] = \left| \frac{4.1\text{ mg/L}}{40\text{ mg/mmol Ca}} \right| = 0.103\text{ mM} = 10^{-3.99}\text{ M}$$

$$[HCO_3^-] = \left| \frac{61.8\text{ mg/L}}{61000\text{ mg/mol}} \right| = 10^{-2.99}\text{ M}$$

$$pH_s = pK_2 - pK_{sp} + p[Ca^{+2}] + p[Alk] = 10.3 - 8.3 + 3.99 + 2.99 = 9.0$$

WASTE DISPOSAL

Waste disposal is a more significant problem for any membrane plant than for a conventional water treatment plant because concentrate disposal is highly regulated by government agencies. A membrane plant can collect significant amounts of waste with high concentrations of many constituents including TDS and organics.

The quality and quantity of concentrate streams from RO/NF/ED plants can be easily estimated using simple recovery and rejection calculations. A 10-mgd plant operating at 90 percent recovery will produce a concentrate stream of 1.11 mgd, which is determined by dividing the permeate flow, 10 mgd, by the recovery, 0.11. The quality of the concentrate stream can be determined by dividing the decimal fraction solute rejection into the feed stream recovery. The same plant producing 10 mgd at 90 percent recovery and rejecting 80 percent of 100 mg/L of the feed stream chlorides would produce 20 mg/L of chlorides in the permeate stream and 820 mg/L

TABLE 11.14 Concentration Disposal Technique Distribution by Plants and Flow

Technique/plant size (mgd)	<.3	.3–1	1–3	>3	Total
Surface	34	12	9	11	66
Land application*	14	2	0	1	17
Sewer	18	8	3	3	32
Deep well injection	3	1	5	5	14
Evaporation pond	8	0	0	0	8
Total	77	23	17	20	137

* None planned in future.

of chlorides in the concentrate stream. Rejection and recovery are sometimes confused. The calculations are illustrated as follows.

$$Q_c = \frac{Q_p(1 - R)}{R} = \frac{10 \text{ mgd } (1.0 - 0.9)}{0.9} = \frac{1.0}{0.9} = 1.11 \text{ mgd}$$

$$C_p = (1 - \text{Rej.})(C_f) = (1 - 0.8)(100 \text{ Cl}^- \text{ mg/L}) = 20 \text{ Cl}^- \text{ mg/L}$$

$$C_c = \frac{Q_f C_f - Q_p C_p}{Q_c} = \frac{(1 \text{ L})(100 \text{ Cl}^- \text{ mg/L}) - (0.9 \text{ L})(20 \text{ Cl}^- \text{ mg/L})}{0.1 \text{ L}} = 820 \text{ Cl}^- \text{ mg/L}$$

Table 11.14 lists techniques used for concentrate disposal in the United States by plant size and number (Mickley, Hamilton, and Truesdall, 1993). Mickley identified five basic techniques used by 137 plants surveyed in an AWWARF study of concentrate disposal. These techniques show surface discharge as the most common technique, accounting for nearly half of the techniques surveyed. Land application was not planned for any proposed new plant. However, land application of NF concentrates is possible in some locations because of their low TDS concentrations, typically 1000 to 2000 mg/L, relative to RO concentrates. Sewer discharge, usually an option only for very small plants, was the second most common technique for plants under 0.3 mgd. Deep well injection was more common in Florida than in any other states because of differences in regulatory environments.

Concentrate discharge is regulated through programs in the Clean Water Act (CWA) and the Safe Drinking Water Act (SDWA). Surface waters fall under the National Pollution Discharge Elimination System (NPDES). Groundwaters fall under the Underground Injection Control program. Regulatory environments for concentrate disposal are controlled by states or by USEPA, depending on primacy. Although these programs have existed since the 1970s, the regulatory environment is becoming more stringent. Bioassay testing is becoming more common in Florida and California and can be used to require additional treatment or stop planned disposal of concentrate streams.

BIBLIOGRAPHY

Adham, S. S., J. G. Jacangelo, and J.-M. Laine. "Assessing integrity." *Jour. AWWA,* 87(3), 1995: 62–75.

Amy, G., B. C. Alleman, and C. B. Cluff. "Removal of dissolved organic matter by nanofiltration." *Journal of Environmental Engineering,* 116(1), 1990: 200–205.

AWWA committee report. "Membrane Processes." *Jour. AWWA,* 90(6), 1998: 91–105.

Belfort, G. *Synthetic Membrane Processes, Fundamentals and Water Applications.* New York: Academic Press, 1984.

Belfort, G., D. R. Davis, and A. L. Zydney. "The behavior of suspensions and macro-molecular solution in crossflow microfiltration." *Journal of Membrane Science,* 96, 1994: 1–58.

Berg, P., and P. Gimbel. "Rejection of trace organics by nanofiltration." In *Proc. AWWA Membrane Technology Conference,* New Orleans, LA, 1997.

Braghetta, A. "The influence of solution chemistry operating conditions on nanofiltration of charged and uncharged organic macromolecules." Ph.D. dissertation, University of North Carolina, Chapel Hill, 1995.

Chen, S., and J. S. Taylor. "Flat sheet testing for pesticide removal by varying RO/NF membrane." In *Proc. AWWA Membrane Technology Conference,* New Orleans, LA, 1997.

Chellam, S., and M. R. Wiesner. "Particle back-transport and permeate flux behavior in crossflow membrane filters." *Environmental Science and Technology* 31(3), 1997: 819–824.

Cheryan, M. *Ultrafiltration Handbook.* Lancaster, PA: Technomic Publishing, 1988.

Chian, E. S. K., W. N. Bruce, and H. H. P. Fang. "Removal of pesticides by reverse osmosis." *Journal of Environmental Science and Technology,* 9(1), 1975: 52–59.

Childress, A. E., and M. Elimelech. "Effect of solution chemistry on the surface charge of polymeric reverse osmosis and nanofiltration membranes." *Journal of Membrane Science,* 119(10), 1996: 253–268.

Coffey, B. M. "Evaluation of MF for metropolitan's small domestic water system." In *Proc. AWWA Membrane Technology Conference,* Baltimore, MD, 1993.

Conlon, W. J., and J. D. Click. "Surface water treatment with ultrafiltration." In *Proc. 58th FSAWWA, FPCA and FW&PCOA Annual Technical Conference,* 1984.

Davidson, M. G., and W. M. Deen. "Hydrodynamic theory for the hindered transport of flexible macromolecules in porous membranes." *Journal of Membrane Science,* 35, 1988: 167–192.

Davis, R. H. "Modeling of fouling in crossflow microfiltration membranes." *Separation and Purification Methods,* 21(2), 1992: 75–126.

Davis, R. H., and S. A. Birdsell. "Hydrodynamic model and experiments with crossflow microfiltration." *Chemical Engineering Communication,* 49, 1987: 217–234.

Deen, W. M. "Hindered transport of large molecules in liquid-filled pores." *AICHE J.,* 33, 1987: 1409–1425.

Dresner, L. "Some remarks on the integration of the extended Wernst-Planck equation in hyperfiltration of multicomponent solution." *Desalination,* 10, 1972: 27.

Duranceau, S. J., and J. S. Taylor. "Investigation and modeling of membrane mass transfer." In *Proc. the National Water Improvement Supply Association,* 1990.

Duranceau, S. J., J. S. Taylor, and L. A. Mulford. "SOC removal in a membrane softening process." *Jour. AWWA,* 84, 1992: 68–78.

Eckstein, E. C., et al. "Self diffusion of particles in shear flow of a suspension." *Journal of Fluid Mechanics,* 79(1), 1977: 191–208.

Electrodialysis-Electrodialysis Reversal Technology. Watertown, MA: Ionics Incorporated, September 1997.

Elimelech, M., W. Chen, and J. Waypa. "Measuring the zeta (electrokinetic)-potential of reverse osmosis membranes by a streaming potential analyzer." *Desalination,* 95, 1994: 269–286.

Fane, A. G. "Ultrafiltration of suspensions." *Journal of Membrane Science,* 20, 1984: 249–259.

Ferry, J. D. "Statistical evaluation of sieve constants in ultrafiltration." *Journal of General Physiology,* 20, 1936: 95–104.

Fourozzi, J. "Nominal molecular weight distribution of color, TOC, TTHm precursors, and acid strength in a highly organic portable water source." Master's thesis, University of Central Florida, Orlando, FL, 1980.

Fritsche, et al. "The structure and morphology of the skin of polyethersulfone ultrafiltration membranes: A comparative atomic force microscope and scanning electro microscope study." *Journal of Applied Polymer Science,* 45, 1992: 1945.

Fronk, C. A. "Removal of low molecular weight organic contaminants from drinking water using reverse osmosis membranes." In *Proc. Annual Conference and Exhibition of the American Water Works Association,* Kansas City, MO, 1987.

Heimstra, P., and M. Nederlof. "Reduction of color and hardness of groundwater with nanofiltration—Is pilot-plant operation really important?" In *Proc. AWWA Membrane Technology Conference,* New Orleans, LA, 1997.

Hindin, E., P. J. Bennett, and S. S. Narayanan. "Organic compounds removed by reverse osmosis." *Water and Sewage Works,* 116(12), 1969: 466–470.

Hofman, J. A. M. H., Th. H. M. Noij, J. C. Kruithof, and J. C. Schippers. "Removal of pesticides by nanofiltration." In *Proc. AWWA Membrane Technology Conference,* Reno, NV, 1993.

Hofman, J. A. M. H., J. S. Taylor, J. C. Schippers, S. J. Duranceau, and J. C. Kruithof. "Vereenvoudigd model voor de beschrijving van de weking van nano-en hyperfiltratie-installaties." *H_2O Journal, Europe,* 1, Jan. 1995.

Hong, S., and M. Elimelech. "Chemical and physical aspects of natural organic matter (NOM) fouling of nanofiltration membranes." *Journal of Membrane Science,* 132(9), 1997: 159–181.

Ironside, R., and S. Sourirajan. "The reverse osmosis membrane separation technique for water pollution control." *Water Research,* 1(2), 1967: 179–180.

Jacangelo, J. G., J.-M. Laine, K. E. Carns, E. W. Cummings, and J. Malevialle. "Low-pressure membrane filtration for removing giardia and microbial indicators." *Jour. AWWA,* 83(9), 1991: 97–106.

Jacangelo, J. G., J.-M. Laine, W. Booe, and J. Malevialle. "Low pressure membrane filtration for particle removal," AWWARF report, 1992.

Jacangelo, J. G., S. S. Adham, and J.-M. Laine. "Cryptosporidium, giardia and MS2 virus removal by MF and UF." *Jour. AWWA,* 87(9), 1995: 107–121.

Jones, P. A., and J. S. Taylor. "DBP control by nanofiltration, cost and performance." *Jour. AWWA,* 84, 1992: 104–116.

Jucker, C., and M. M. Clark. "Adsorption of aquatic humic substances on hydrophobic ultrafiltration membranes." *Journal of Membrane Science,* 97, 1994: 37.

Kothari, N., W. A. Lovins, C. Robert, S. Chen, K. Kopp, and J. S. Taylor. "Pilot scale microfiltration at Manitowoc." In *Proc. AWWA Membrane Technology Conference,* New Orleans, LA, 1997.

Lahoussine-Turcaud, V., M. R. Wiesner, and J. Y. Bottero. "Coagulation pretreatment for ultrafiltration of a surface water." *Jour. AWWA,* 82(12), 1990: 76–81.

Laine, J.-M., M. M. Clark, and J. Malleviale. "Ultrafiltration of lake water: Effects of pretreatment on the partitioning of organics, THMFP, and flux." *Jour. AWWA,* 82(12), 1990: 82–87.

Leighton, D., and A. Acrivos. "The shear-induced migration of particles in concentrated suspensions." *Journal of Fluid Mechanics,* 181, 1987: 415–439.

Leonard, E. F., and C. S. Vassilief. "Deposition of rejected matter in membrane separation processes." *Chem. Eng. Commun.,* 30(3–5), 1984: 209–217.

Lonsdale, H. K., U. Merten, and R. L. Riley. "Transport properties of cellulose acetate osmotic membrane." *J. Applied Polymer Science,* 9, 1965: 1341.

Matsuura, T., and S. Sourirajan. *Journal of Colloid and Interface Science,* 95, 1983: 10.

McCarty, P. L., and E. M. Aieta. "Chemical indicators and surrogate parameters for water treatment." In *Annual Conference Proceedings,* pp. 625–650, Las Vegas, NV, 1983.

Merten, U. *Desalination by Reverse Osmosis.* Cambridge, MA: MIT Press, 1966.

Metcalf, P. J., et al. "Water science and technology." In *Proc. IAWPRC Conference,* Cape Town, South Africa, 1992.

Mickley, M., R. Hamilton, and J. Truesdall. *Membrane Concentrate Disposal.* Denver, CO: AWWA Research Foundation, 1993.

Miltner, R. J., C. A. Fronk, and T. F. Speth. "Removal of alachlor from drinking water." In *Proc. National Conference on Environmental Engineering,* pp. 204–211, Orlando, FL, 1987.

Morris, K. M. "Predicting fouling in membrane separation processes." Master's thesis, University of Central Florida, Orlando, FL, 1990.

Nalco Chemical Company. *Nalco Water Handbook,* F. K. Kemmer, ed. New York: McGraw-Hill, 1988.

Oldani, M., and G. Schock. "Characterization of ultrafiltration membranes by infrared spectroscopy, ESCA, and contact angle measurements." *Journal of Membrane Science,* 3, 1989: 243.

Olivieri, V., et al. "Continuous microfiltration of surface water." In *Proc. Technologies Conference in the Water Industry,* Orlando, FL, 1991.

Onsager, L. *Physi. Rev.,* 37, 1931: 405.

Parker, D. Y., Jr. "Removal of trihalomethane precursors by microfiltration." Essay submitted to the Johns Hopkins University, Baltimore, MD, 1991.

PEM Products Engineering Manual. Wilmington, DE: E. I. DuPont de Nemours & Co., 1982.

Rautenbach, R., and R. Albrecht. *Membrane Processes.* New York: John Wiley & Sons, 1989.

Ridgway, H. F., and H.-C. Flemming. "Membrane biofouling." In *Water Treatment Membrane Processes.* New York: McGraw-Hill, 1996.

Rangarajan, R., E. C. Matsuura, and S. Sourirajan. "Free energy parameter for reverse osmosis separation of some inorganic ions and ion pairs in aqueous solutions." *I & EC Process Design and Development,* 15(4), 1976: 529–534.

Rautenbach, R., and A. Groeschl. "Separation potential of nanofiltration membranes." *Desalination,* 77(1–3), 1990: 73–84.

Schippers, J. C., and J. Verdouw. "The modified fouling index: A method of determining the fouling characteristics of water." *Desalination,* 32, 1980: 137–148.

Sethi, S., and M. R. Wiesner. "Modeling the transient permeate flux in crossflow membrane filtration incorporating multiple transport mechanisms." *Journal of Membrane Science,* 136(1–2), 1997: 191–205.

Sethi, S., and M. R. Wiesner. "Performance and cost modeling of ultrafiltration." *Journal of Environmental Engineering,* 121(12), 1995: 874–883.

Sherwood, T. K., P. L. T. Brian, and R. E. Fisher. Desalination by reverse osmosis. *Ind. Eng. Chem. Fundam.,* 6(1), 1967: 2–12.

Song, L., and M. Elimelech. "Particle deposition onto a permeable surface in laminar flow." *Journal of Colloid and Interface Science,* 173(7), 1995: 165–180.

Sourirajan, S. *Reverse Osmosis.* New York: Academic Press, 1970.

Spevack, P., and Y. Deslandes. "TOF-SIMS analysis of adsorbate-membrane interactions, 1. Adsorption of dehydroabietic acid on PVDE." *Applied Surface Science,* 99, 1996: 41.

Spangenburg, C., et al. "Selection, evaluation and optimization of organic selective membranes for color and DBP precursor removal." In *Proc. AWWA Membrane Technology Conference,* New Orleans, LA, 1997.

Stamatakis, K., and C. Tien. "A simple model of crossflow filtration based on particle adhesion." *AIChE J,* 39(8), 1993: 1292–1302.

Strohwald, N. K. H., and E. P. Jacobs. "An investigation into UF systems in the pretreatment of sea water for RO desalination." *Water Science and Technology,* 25(10), 1992: 69–78.

Sung, L. "Modeling mass transfer in nanofiltration." Ph.D. dissertation, University of Central Florida, Orlando, FL, 1993.

Tan, L., and G. L. Amy, "Comparing ozonation and membrane separation for color removal and disinfection by-product control." *Jour. AWWA,* 83, 1991: 74–79.

Taylor, J. S., C. Reiss, and R. Robert. "Integrated membrane systems state of the art." In *Proc. NWSIA Conference on Microfiltration,* San Diego, CA, 1999.

Taylor, J. S., and E. P. Jacobs. "Nanofiltration and reverse osmosis." In *Water Treatment Membrane Processes.* New York: McGraw-Hill, 1996.

Taylor, J. S., et al. "Cost and performance evaluation of in-plant trihalomethane control techniques," EPA Final Report No. 600/2-85-168. Washington, DC: Environmental Protection Agency, 1986.

Taylor, J. S., et al. "Comparison of membrane processes at ground and surface water sites." In *Proc. Annual Conference AWWA,* Orlando, FL, 1988.

Taylor, J. S., L. A. Mulford, W. M. Barrett, S. J. Duranceau, and D. K. Smith. *Cost and Performance of Membrane Processes for Organic Control on Small Systems.*" Cincinnati, OH: U.S. Environmental Protection Agency Water Engineering Research Laboratory, 1989*a.*

Taylor, J. S., L. A. Mulford, S. J. Duranceau, and W. M. Barrett. "Cost and performance of a membrane pilot plant." *Jour. AWWA,* 81(11), 1989*b:* 52–60.

Taylor, J. S., D. Thompson, and J. K. Carswell. "Removal of THM precursors by membrane processes from a groundwater source." *Jour. AWWA,* August 1987.

Taylor, J. S., et al. *Assessment of Potable Water Membrane Application and Research Needs.* Denver, CO: AWWA Research Foundation Report, 1989.

Weber, W. J. *Physicochemical Process for Water Quality Control.* New York: John Wiley & Sons, 1972.

Wiesner, M. R., M. M. Clark, and J. Mallevialle. "Membrane filtration of coagulated suspensions." *Journal of Environmental Engineering,* 115(1), 1989: 20–40.

Whittaker, H., and T. Szaplonczay. "Testing of reverse osmosis on chemical solutions." In *Proc. 2nd Annual Tech. Seminar on Chemical Spills,* Toronto, Canada, 1985.

Whittaker, H., and R. Clark. "Cleanup of PCB contaminated groundwater by reverse osmosis." In *Proc. 2nd Annual Tech. Seminar on Chemical Spills,* Toronto, Canada, 1985.

Yoo, S., et al. "Microfiltration: A case study." *Jour. AWWA,* March 1995: 38–49.

Zeman, L., and M. Wales. "Polymer solute rejection by ultrafiltration membranes." In *Synthetic Membranes, Vol. II Hyper and Ultrafiltration Uses,* A. F. Turbak, ed. Washington, DC: American Chemical Society, 1981.

Zhu, X., and M. Elimelech. "Colloidal fouling of reverse osmosis membranes: Measurements and fouling mechanisms." *Environmental Science and Technology,* 31, 1997: 3654–3662.

CHAPTER 12
CHEMICAL OXIDATION[1]

Philip C. Singer, Ph.D.
Professor, Department of Environmental Sciences and Engineering
University of North Carolina
Chapel Hill, North Carolina

David A. Reckhow, Ph.D.
Professor of Civil and Environmental Engineering
University of Massachusetts
Amherst, Massachusetts

Chemical oxidation processes play several important roles in the treatment of drinking water. Chemical oxidants are used for the oxidation of reduced inorganic species, such as ferrous iron, Fe(II); manganous manganese, Mn(II); and sulfide, S(-II); and hazardous synthetic organic compounds such as trichloroethylene (TCE) and atrazine. Oxidants can also be used to destroy taste- and odor-causing compounds and eliminate color. In addition, in some cases, they may improve the performance of, or reduce the required amount of, coagulants.

Because many oxidants also have biocidal properties, they can be used to control nuisance aquatic growths, such as algae, in pretreatment basins, and may be used as primary disinfectants to meet CT (disinfectant concentration times contact time) requirements (see Chapter 14). These oxidants are often added at the head of the treatment plant, prior to or at the rapid mix basin, but they can also be employed after clarification, prior to filtration, after a substantial portion of the oxidant demand has been removed.

The most common chemical oxidants used in water treatment are chlorine, ozone, chlorine dioxide, and permanganate. Ozone is sometimes used in conjunction with hydrogen peroxide or ultraviolet irradiation to produce radicals that have powerful oxidative properties. Mixed oxidant technologies are also available.

Free chlorine has traditionally been the oxidant (and disinfectant) of choice in the United States, but concerns about the formation of potentially harmful halogenated disinfection by-products (DBPs) produced by reactions between free chlorine and natural organic material (NOM), exacerbated in some cases by the presence of bromide, have caused many water systems to adopt alternative chemical oxidants (and disinfectants) to lower halogenated DBP formation. These other oxidants may also react with NOM and bromide to various degrees, depending upon

[1] Acknowledgment: We would like to thank Dr. William H. Glaze of the University of North Carolina at Chapel Hill who wrote the earlier version of this chapter, which provided a starting point for the current material.

the properties of the oxidant, to form oxidation by-products, some of which also have adverse public health effects or result in downstream operational problems in the treatment plant or distribution system.

This chapter reviews thermodynamic and kinetic principles associated with the use of chemical oxidants in general, the types and properties of the chemical oxidants used in water treatment, specific applications of oxidation processes for the treatment of drinking water, and the formation and control of oxidation and disinfection by-products. Comparisons among the different oxidant choices are presented where information is available.

PRINCIPLES OF OXIDATION

Thermodynamic Considerations

Thermodynamics establishes the bounds or constraints for oxidation reactions. Chemical kinetics fills in much of the detail. In many cases there are simply no other available data than the thermodynamic enthalpies and entropies of reaction. Despite its limitations, the domain of thermodynamics is where one must begin the task of characterizing and understanding oxidation reactions. In this section, the most basic thermodynamic concepts relating to oxidation reactions will be presented. For a more comprehensive treatment of the subject, there are many excellent textbooks that can be consulted (e.g., Stumm and Morgan, 1996; Pankow, 1991).

Electrochemical Potentials. Oxidation reactions are often viewed as reactions involving the exchange of electrons. Since acids are frequently defined as proton donors and bases as proton acceptors, one can think of oxidants as electron acceptors and reductants as electron donors. In fact, it's not quite this simple. Many oxidants actually donate an electron-poor element or chemical group, rather than simply accept a lone electron. Nevertheless, it's useful to treat all oxidation reactions as simple electron transfers for the purpose of balancing equations and performing thermodynamic calculations.

Thermodynamic principles can be used to determine if specific oxidation reactions are possible. This generally involves the calculation of some form of reaction potential. Although in most cases oxidation equilibria lie very far to one side or the other, it is sometimes instructive to calculate equilibrium concentrations of the reactants and products.

The first step is to identify the species being reduced and those being oxidized. Appropriate half-cell reactions and their standard half-cell potentials (E^o_{red} and E^o_{ox}, respectively) are available in tables of thermodynamic constants (a few are listed in Tables 12.1 and 12.2). These may be combined to get the overall standard cell potential E^o_{net} (Eq. 12.1).

$$E^o_{\text{net}} = E^o_{\text{ox}} + E^o_{\text{red}} \tag{12.1}$$

Much as a pK_a describes the tendency of an acid to give up a hydrogen ion, an electrochemical potential E describes the tendency of an oxidant to take up an electron, or a reductant to give one up. The standard-state Gibbs Free Energy of reaction ΔG^o is related to the standard electrochemical cell potential by Faraday's constant F and the number of electrons transferred n:

$$\Delta G^o = -nFE^o_{\text{net}} \tag{12.2}$$

For a one-electron transfer reaction, this becomes:

$$\Delta G^o\ (K\ \text{cal}) = -23E^o_{\text{net}}\ (\text{volts}) \tag{12.3}$$

TABLE 12.1 Standard Half-Cell Potentials for Chemical Oxidants Used in Water Treatment

Oxidant	Reduction half-reaction	E^o_{red}, volts
Ozone	$\frac{1}{2}O_3(aq) + H^+ + e^- \rightarrow \frac{1}{2}O_2(aq) + \frac{1}{2}H_2O$	2.08
Hydroxyl radical	$OH + H^+ + e^- \rightarrow H_2O$	2.85
Hydrogen peroxide	$\frac{1}{2}H_2O_2 + H^+ + e^- \rightarrow H_2O$	1.78
Permanganate	$\frac{1}{3}MnO_4^- + \frac{4}{3}H^+ + e^- \rightarrow \frac{1}{3}MnO_2(s) + \frac{2}{3}H_2O$	1.68
Chlorine dioxide	$ClO_2 + e^- \rightarrow ClO_2^-$	0.95
Hypochlorous acid	$\frac{1}{2}HOCl + \frac{1}{2}H^+ + e^- \rightarrow \frac{1}{2}Cl^- + \frac{1}{2}H_2O$	1.48
Hypochlorite ion	$\frac{1}{2}OCl^- + H^+ + e^- \rightarrow \frac{1}{2}Cl^- \frac{1}{2}H_2O$	1.64
Hypobromous acid	$\frac{1}{2}HOBr + \frac{1}{2}H^+ + e^- \rightarrow \frac{1}{2}Br^- + \frac{1}{2}H_2O$	1.33
Monochloramine	$\frac{1}{2}NH_2Cl + H^+ + e^- \rightarrow \frac{1}{2}Cl^- + \frac{1}{2}NH_4^+$	1.40
Dichloramine	$\frac{1}{4}NHCl_2 + \frac{3}{4}H^+ + e^- \rightarrow \frac{1}{2}Cl^- + \frac{1}{4}NH_4^+$	1.34
Oxygen	$\frac{1}{4}O_2(aq) + H^+ + e^- \rightarrow \frac{1}{2}H_2O$	1.23

Sources: Lide (1995); American Water Works Assoc. (1990); Stumm and Morgan (1996).

Classical thermodynamics indicates that reactions with a negative Gibbs Free Energy (or a positive E^o) will spontaneously proceed in the direction as written (i.e., from left to right), and those with a positive value (or negative E^o) will proceed in the reverse direction.

Consider a generic oxidation reaction:

$$aA_{ox} + bB_{red} \rightarrow aA_{red} + bB_{ox} \tag{12.4}$$

where substance A picks up one electron from substance B. In order to determine which substance is being reduced and which is being oxidized, one must calculate and compare oxidation states of the reactant atoms and product atoms.

The equilibrium constant K for this reaction defines the concentration quotient for the reactants and products at equilibrium:

$$K = \frac{[A_{red}]^a[B_{ox}]^b}{[A_{ox}]^a[B_{red}]^b} \tag{12.5}$$

The overall standard cell potential is then directly related to this equilibrium constant by:

$$E^o_{net} = \frac{RT}{nF} \ln K \tag{12.6}$$

TABLE 12.2 Standard Half-Cell Potentials for Some Oxidation Reactions That Can Occur During Drinking Water Treatment

Oxidation half-reaction	E^o_{ox}, volts
$\frac{1}{2}Br^- + \frac{1}{2}H_2O \rightarrow \frac{1}{2}HOBr + \frac{1}{2}H^+ + e^-$	−1.33
$\frac{1}{2}Mn^{+2} + H_2O \rightarrow \frac{1}{2}MnO_2(s) + 2H^+ + e^-$	−1.21
$Fe^{+2} + 3H_2O \rightarrow Fe(OH)_3(s) + 3H^+ + e^-$	−1.01
$\frac{1}{8}NH_4^+ + \frac{3}{8}H_2O \rightarrow \frac{1}{8}NO_3^- + 1\frac{1}{4}H^+ + e^-$	−0.88
$\frac{1}{2}NO_2^- + \frac{1}{2}H_2O \rightarrow \frac{1}{2}NO_3^- + H^+ + e^-$	−0.84
$\frac{1}{8}H_2S + \frac{1}{2}H_2O \rightarrow \frac{1}{8}SO_4^{-2} + 1\frac{1}{4}H^+ + e^-$	−0.30
$\frac{1}{2}H_2S \rightarrow \frac{1}{2}S(s) + H^+ + e^-$	−0.14
$\frac{1}{2}HCOO^- \rightarrow \frac{1}{2}CO_2(g) + \frac{1}{2}H^+ + e^-$	+0.29

Sources: Lide (1995); American Water Works Assoc. (1990); Stumm and Morgan (1996).

and for a one-electron-transfer reaction at 25°C, this simplifies to:

$$\log K = \frac{1}{0.059} E^{\circ}_{\text{net}} \tag{12.7}$$

Oxidation-Reduction Reactions

Oxidation State. Oxidation state is characterized by an oxidation number, which is the charge one would expect for an atom if it were to dissociate from the surrounding molecule or ion (assigning any shared electrons to the more electronegative atom). Oxidation number may be either a positive or a negative number—usually an integer between –VII and +VII, although in their elemental forms, for example, S(*s*), $O_2(aq)$, atoms have an oxidation number of zero. This concept is useful in balancing chemical equations and performing certain calculations. The rules for calculating oxidation number are described in most textbooks on general chemistry.

Balancing Equations. The first step in working with oxidation reactions is to identify the role of the reacting species. At least one reactant must be the oxidizing agent (i.e., containing an atom or atoms that become reduced), and at least one must be a reducing agent (i.e., containing an atom or atoms that become oxidized). The second step is to balance the gain of electrons from the oxidizing agent with the loss of electrons from the reducing agent. Next, oxygen atoms are balanced by adding water molecules to one side or another, and hydrogens are balanced with H^+ ions. For a more detailed treatment on calculations using oxidation reactions, the reader is referred to a general textbook on aquatic chemistry (e.g., Stumm and Morgan, 1996; Pankow, 1991).

As an example consider the oxidation of manganese by ozone (Eq. 12.8). The substance being oxidized is manganese (i.e., the reducing agent) and the one doing the oxidizing (i.e., being itself reduced) is ozone.

$$\text{Mn} + \text{O}_3 \rightarrow \text{products} \tag{12.8}$$

Next, the products formed need to be evaluated. It might be known from experience that reduced soluble manganese (i.e., Mn^{+2}) can be oxidized in water to the relatively insoluble manganese dioxide. It might also be known that ozone ultimately forms hydroxide and oxygen after it becomes reduced.

$$\text{Mn}^{+2} + \text{O}_3 \rightarrow \text{MnO}_2 + \text{O}_2 + \text{OH}^- \tag{12.9}$$

The next step is to determine the oxidation state of all atoms involved (Eq. 12.10).

$$\overbrace{\text{Mn}}^{+\text{II}}{}^{+2} + \overbrace{\text{O}}^{0}{}_3 \rightarrow \overbrace{\text{Mn}}^{+\text{IV}}\overbrace{\text{O}}^{-\text{II}}{}_2 + \overbrace{\text{O}}^{0}{}_2 + \overbrace{\text{O}}^{-\text{II}}\overbrace{\text{H}}^{+\text{I}}{}^- \tag{12.10}$$

From this analysis, it is clear that manganese is oxidized from +II to +IV, which involves a loss of two electrons per atom. On the other side of the ledger, the ozone undergoes a gain of two electrons per molecule, as one of the three oxygen atoms goes from an oxidation state of 0 to –II. The two half-reactions can be written as single electron transfers. These half-reactions are balanced by adding water molecules and H^+ ions to balance oxygen and hydrogen, respectively.

$$\tfrac{1}{2}\text{Mn}^{+2} + \text{H}_2\text{O} \rightarrow \tfrac{1}{2}\text{MnO}_2 + 2\text{H}^+ + e^- \tag{12.11}$$

By convention, when hydroxide appears in a half-reaction, additional H^+ ions are added until all of the hydroxide is converted to water. This is done to the reduction half-reaction.

$$\tfrac{1}{2}\text{O}_3 + \text{H}^+ + e^- \rightarrow + \tfrac{1}{2}\text{O}_2 + \tfrac{1}{2}\text{H}_2\text{O} \tag{12.12}$$

From this point, it is a simple matter of combining the equations and canceling out terms or portions of terms that appear on both sides. At the same time, the standard electrode potentials can be combined to get the overall potential.

$$\tfrac{1}{2}Mn^{+2} + H_2O \rightarrow \tfrac{1}{2}MnO_2(S) + 2H^+ + e^- \qquad -1.21\ V\ (E^o_{ox})$$

$$\tfrac{1}{2}O_3(aq) + H^+ + e^- \rightarrow \tfrac{1}{2}O_2(aq) + \tfrac{1}{2}H_2O \qquad +2.04\ V\ (E^o_{red})$$

$$\tfrac{1}{2}O_3(aq) + \tfrac{1}{2}Mn^{+2} + \tfrac{1}{2}H_2O \rightarrow \tfrac{1}{2}O_2(aq) + \tfrac{1}{2}MnO_2(S) + H^+ \qquad +0.83\ V\ (E^o_{net}) \tag{12.13}$$

Immediately, it is seen that this reaction will proceed toward the right (the E^o_{net} is positive). But how far to the right will it go? To answer this, Eq. 12.7 is rearranged to get

$$K = e^{16.95E^o_{net}} \tag{12.14}$$

So for this reaction

$$K = e^{16.95*0.83} = 1.29 \times 10^6 \tag{12.15}$$

and using the concentration quotient from the reaction stoichiometry,

$$1.29 \times 10^6 = \frac{[O_2(aq)]^{0.5}[MnO_2(s)]^{0.5}[H^+]}{[O_3(aq)]^{0.5}[Mn^{+2}]^{0.5}[H_2O]^{0.5}} \tag{12.16}$$

Because the activity of solvents (i.e., water) and solid phases are, by convention, equal to 1,

$$1.29 \times 10^6 = \frac{[O_2(aq)]^{0.5}[H^+]}{[O_3(aq)]^{0.5}[Mn^{+2}]^{0.5}} \tag{12.17}$$

Furthermore, if the pH is 7.0 and a dissolved oxygen concentration of 10 mg/L and an ozone concentration of 0.5 mg/L is maintained in the contactor, an equilibrium Mn^{+2} concentration of 1.8×10^{-25} M or about 10^{-27} mg/L can be calculated. Thermodynamic principles therefore indicate that this reaction essentially goes to completion.

Now, knowing that the Mn^{+2} should react essentially completely to form manganese dioxide, it might be desirable to determine if ozone can possibly oxidize the manganese dioxide to a higher oxidation state, that is, to permanganate. To examine this, the preceding ozone equation must first be combined with the reverse of the permanganate reduction equation (from Table 12.1).

$$\tfrac{1}{2}O_3(aq) + H^+ + e^- \rightarrow \tfrac{1}{2}O_2(aq) + \tfrac{1}{2}H_2O$$

$$\tfrac{1}{3}MnO_2 + \tfrac{2}{3}H_2O \rightarrow \tfrac{1}{3}MnO_4^- + \tfrac{4}{3}H^+ + e^-$$

$$\tfrac{1}{2}O_3(aq) + \tfrac{1}{3}MnO_2 + \tfrac{1}{3}H_2O \rightarrow \tfrac{1}{3}MnO_4^- + \tfrac{1}{3}H^+ + \tfrac{1}{2}O_2(aq) \tag{12.18}$$

This allows the net potential to be calculated:

$$E^o_{net} = E^o_{ox} + E^o_{red} = (-1.68V) + (+2.04V) = +0.36V \tag{12.19}$$

Again, this is a favorable reaction. The equilibrium constant is:

$$\log K = \frac{1}{0.059} E^o_{net} = \frac{1}{0.059}(+0.36V) = 6.1$$

$$K = 1.26 \times 10^6 \tag{12.20}$$

The equilibrium quotient can now be formulated directly from the balanced equation. Note that neither manganese dioxide (MnO_2) nor water (H_2O) appears in this quotient. This is because both are presumed present at unit activity. Manganese dioxide is a solid and as long as it remains in the system, it is considered to be in a pure, undiluted state. The same may be said for water. As long as the solutes remain dilute, the concentration of water is at its maximum and remains constant.

$$K = \frac{[MnO_4^-]^{0.33}[H^+]^{0.33}[O_2]^{0.5}}{[O_3]^{0.5}} = 10^{6.1} \quad (12.21)$$

So under typical conditions where the pH is near neutrality (i.e., $[H^+] = 10^{-7}$), dissolved oxygen is near saturation (i.e., $[O_2(aq)] = 3 \times 10^{-4}$ M), and the ozone residual is 0.25 mg/L (i.e., $[O_3(aq)] = 5 \times 10^{-6}$ M), the expected equilibrium permanganate concentration should be:

$$K = \frac{[MnO_4^-]^{0.33}[10^{-7}]^{0.33}[3 \times 10^{-4}]^{0.5}}{[5 \times 10^{-6}]^{0.5}} = 10^{6.1} \quad (12.22)$$

and solving for permanganate

$$[MnO_4^-]^{0.33} = 3.5 \times 10^7$$

$$[MnO_4^-] = 327 \quad (12.23)$$

Obviously, one cannot have 327 mol/L of permanganate. Nevertheless, the system will be forced in this direction so that all of the manganese dioxide would be converted to permanganate. Once the manganese dioxide is gone, the reaction must stop.

As already mentioned, the preceding thermodynamic analysis is quantitatively accurate when all reactions are at equilibrium. However, this is rarely the case. Many oxidation reactions are quite slow or, in some cases, kinetically unfavored, and the actual concentrations of reactants and products observed during water treatment are far from those predicted by classical thermodynamics. For this reason, oxidation chemistry must rely heavily on kinetics.

Kinetics and Mechanism

Reaction Kinetics. Thermodynamics indicates whether a reaction will proceed as written. However, it will not indicate whether this reaction will produce significant change within milliseconds or thousands of years. For this, chemical kinetics must be considered. As an example, consider the reaction between hypochlorous acid and bromide ion.

$$HOCl + Br^- = HOBr + Cl^- \quad (12.24)$$

In order for a molecule of hypochlorous acid and a molecule of bromide to combine to form products, the two molecules must come into contact with each other (contact meaning approach within a certain distance so that bonding forces can play a role). The probability that a single $HOCl{:}Br^-$ molecular encounter will occur within any fixed time period is directly proportional to the number of molecules of each type in the system. It will also depend on the rate of movement of each of the reactant molecules. As a consequence, the rate of formation of products—for example,

HOBr—will be dependent on a number of factors, including the concentration of hypochlorous acid and the concentration of bromide in the reacting solution. This is the kinetic law of mass action, which is expressed mathematically in Eq. 12.25.

$$\frac{d[\text{HOBr}]}{dt} = k_f[\text{HOCl}][\text{Br}^-] \tag{12.25}$$

The reactants and products are expressed in molar units of concentration and k_f is called the *forward reaction rate constant.* The units for k_f are liters/mole per unit time. The reaction rate constant is going to be a function of such things as the rate of movement of the molecules and the probability of HOBr formation, given that a collision between hypochlorous acid and bromide has already occurred. Because the concentrations of HOCl and Br^- that appear in Eq. 12.25 are raised to the first power, it is said that this rate law is first order in both reactants. The overall order of the reaction is the sum of the individual orders (i.e., second order in this case).

In a more general sense, Eq. 12.26 is the rate law for any elementary reaction of the type described by Eq. 12.27.

$$-\frac{d[\text{A}]}{dt} = k_{fa}[\text{A}]^a[\text{B}]^b \tag{12.26}$$

$$a\text{A} + b\text{B} \rightarrow c\text{C} + d\text{D} \tag{12.27}$$

where the capital letters represent chemical species participating in the reaction and the small letters are the stoichiometric coefficients (i.e., the numbers of each molecule or ion required for the reaction). The overall order describes the extent of dependence of the reaction rate on reactant concentrations. For the reaction in Eq. 12.27, it is equal to $(a + b)$. The order with respect to species A is a, and the order with respect to species B is b. Thus, the reaction in Eq. 12.27 is first order in both reactants and second order overall.

Chemical reactions may be either elementary or nonelementary. Elementary reactions are those reactions that occur exactly as they are written, without any intermediate steps. These reactions almost always involve just one or two reactants. The number of molecules or ions involved in elementary reactions is called the *molecularity* of the reaction. Thus, for all elementary reactions, the overall order equals the molecularity. Nonelementary reactions involve a series of two or more elementary reactions. Many complex environmental reactions are nonelementary. In general, reactions with an overall reaction order greater than 2 or reactions with some noninteger reaction order are nonelementary.

Reaction rate constants for the various oxidants with similar solutes are often positively correlated. In other words, a compound favored for oxidation by one oxidant is generally favored by others as well. Those that are relatively resistant to oxidation by one will likewise be unreactive with others. A good case study is the extensive research done on the oxidation of phenolic compounds, as presented by Tratnyek and Hoigne (1994). These data highlight the similarities between the chemical structure of a reactant and its reactivity with various oxidants. Chemists have used such relationships to develop quantitative structure-activity relationships (QSARs). The Hammett equations are one of the most widely used QSARs (see Brezonik [1994] for more detail on this subject).

Temperature Dependence. As mentioned previously, the reaction rate constant k is a function of temperature. The Arrhenius equation (Eq. 12.28) is the classic model that describes this relationship:

$$k = k_o e^{-E_a/RT} \tag{12.28}$$

where k_o is called the frequency factor or the preexponential factor, E_a is the activation energy, R is the universal gas constant (199 cal/°K-mole), and T is the temperature in °K. The values for k_o and E_a may be either found in the literature or determined from experimental measurements.

Types of Reactions. To this point, considerations have addressed whether or not a certain oxidation reaction can occur, and perhaps how fast it can occur. However, it is sometimes quite useful to know how the reaction occurs on a molecular scale. In other words, by what mechanism or pathway does it go from reactants to products? For example, the problem of disinfection by-products is one of chemical pathways. There is no inherent problem with oxidizing natural organic matter using chlorine. However, when that reaction occurs through addition and substitution reactions (see below) rather than simple electron transfer reactions, chlorinated organic byproducts such as the trihalomethanes (THMs) are obtained.

Oxidation reactions can generally be categorized as those involving electron transfer and those involving transfer of atoms and groups of atoms. They may also be characterized as reactions involving species with paired electrons (ionic) and those involving unpaired electrons (radical). Aqueous chlorine presents a wide array of ionic reactions (e.g., see Morris [1975]) that will serve as illustrative examples for this discussion.

Table 12.3 presents a summary of the major types of ionic reactions occurring in drinking water. Hypochlorous and hypobromous acid can be added to olefinic bonds (i.e., carbon-carbon double bonds), forming halohydrins. This is an electrophilic reaction where the initial attack is by the halogen atom (on the positive side of the HOX dipole). The most stable configuration places the halogen on the carbon with the most hydrogen atoms (producing the most stable carbonium ion: Markovnikov's rule). The other carbon becomes a carbonium ion, which subsequently reacts with the HO portion of the HOX species or with water.

Activated ionic substitution can occur with both aromatic and aliphatic compounds. As with the addition reactions, this type of reaction will also lead to the formation of organohalide compounds. Aromatic substitution reactions occur readily when an electron-donating substituent is bound to the ring. Functional groups on the aromatic ring, such as OH and NH_2, can be thought of as creating a partial negative charge on the ortho and para positions (second closest and farthest carbon atoms from the functional group, respectively). The halogen end of the HOX molecule attacks one of these carbons. Next, there is a loss of the OH end of the molecule, and displacement of the H atom from the carbon under attack. Substitution on aliphatic species is also a multistep reaction, as exemplified by the haloform reaction (see next section on pathways).

When substitution (or transfer) of a halogen occurs onto a nitrogen atom, a relatively reactive N-halo organic compound results. These compounds retain some of the oxidizing capabilities of hypohalous acid and, consequently, the reactions are not considered to cause an oxidant demand. The rates of substitution reactions with nitrogenous organic compounds generally increase as the basicity of the nitrogen atom increases.

Oxidation reactions with halogens are characterized by the formation of the inorganic halide ion, and an oxidized (nonhalogenated) form of the reacting compound. With organic compounds, it is quite common to observe addition of an oxygen atom. For example, oxidation reactions transform unsaturated hydrocarbons into alcohols, then to aldehydes and ketones, and finally to carboxylic acids. Some oxidations do not result in a net transfer of atoms. For these electron transfer reactions, it is common to form free radical intermediates. When this happens, chain reactions can occur, sometimes leading to the types of reactions listed in Table 12.4.

TABLE 12.3 Major Types of Ionic Reactions

Reaction type	Example
1. Addition to an olefinic bond	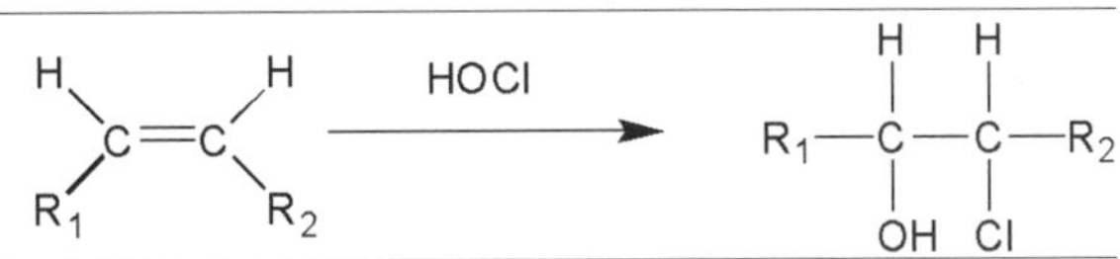
2. Activated aromatic substitution	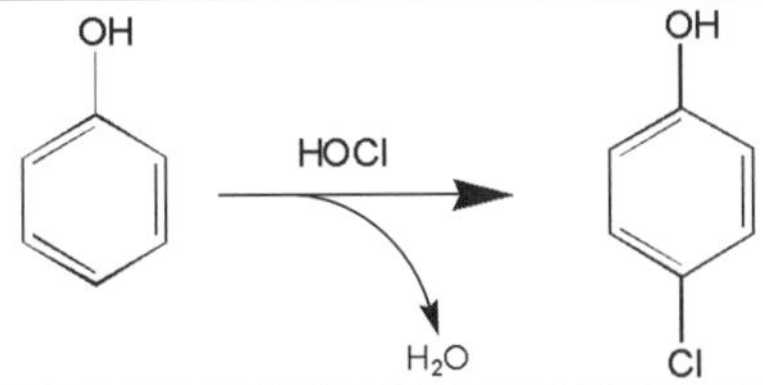
3. Substitution onto nitrogen	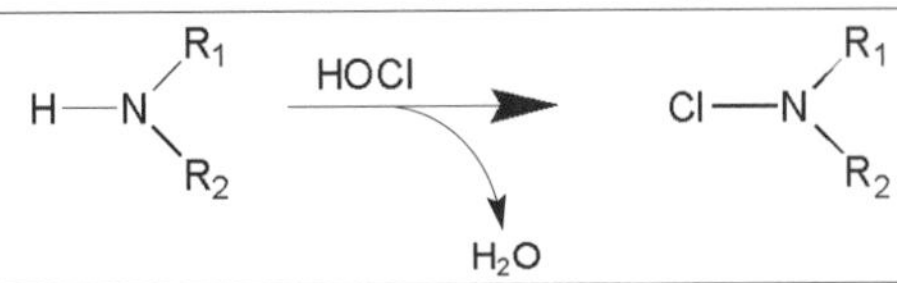
4. Oxidation with oxygen transfer	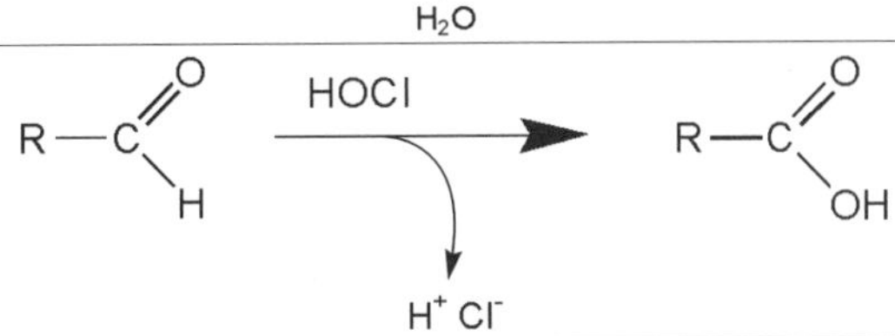
5. Oxidation with electron transfer	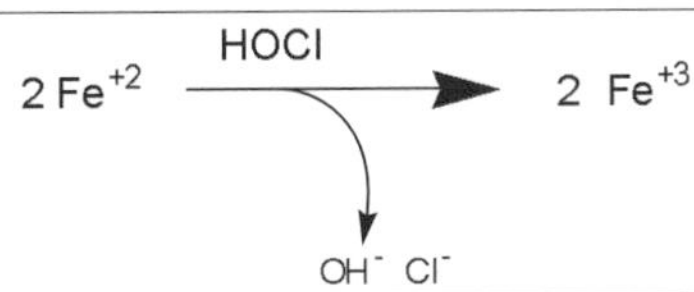

In addition to these ionic reactions, there are several reactions involving free radical species that can occur following addition of drinking water oxidants. These types of reactions are commonly encountered with ozone, chlorine dioxide, and especially the advanced oxidation processes (see below). For example, addition of ozone will always lead to some decomposition and subsequent formation of hydroxyl radicals (•OH). These reactive species engage in reactions that generally lead to the formation of new free radical species. The most common types are addition reactions, hydrogen abstractions, and single electron transfers.

Catalysis. Many types of oxidation reactions are strongly affected by the presence of catalysts. These are compounds that alter reaction rates without being formed or consumed in the reaction. They typically participate in some key, rate-limiting step and are regenerated during some later step. Catalysts generally provide an alternative pathway to a reaction with a lower activation energy.

Probably the most important catalytic processes involve the participation of acids and bases. Specific acid and specific base catalysis involve H^+ and OH^-, respectively. General acid and base catalysis involves any electron acceptor (e.g., a proton donor) and electron donor (e.g., a proton acceptor), respectively. A good example of general base catalysis is the classic haloform reaction (Figure 12.1). Here the rate-limiting

TABLE 12.4 Major Types of Radical Reactions

Reaction type	Example
6. Radical addition reaction	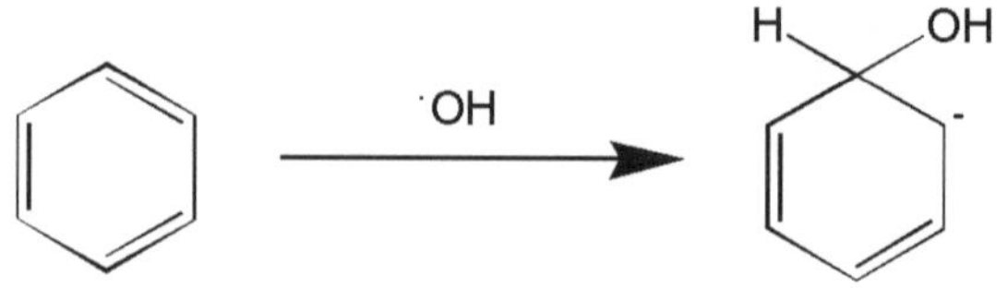
7. Hydrogen abstraction reaction	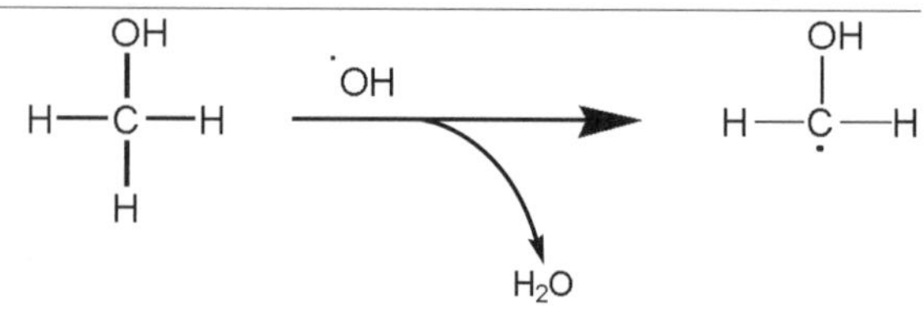
8. Radical oxidation reaction with single electron transfer	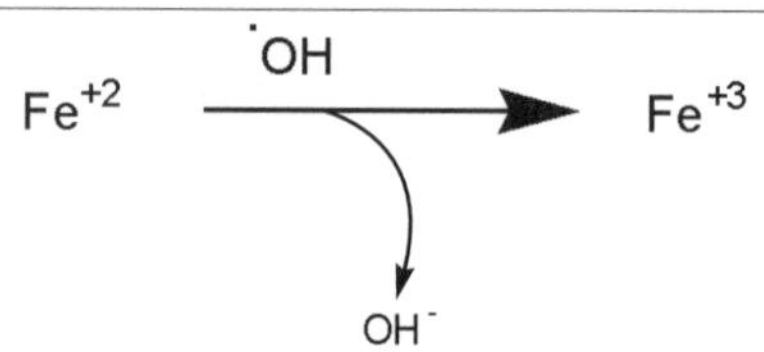

step is the loss of a proton giving the enol. While it is shown in Figure 12.1 as occurring by reaction with hydroxide, any strong base will participate. Also the base (or hydroxide) consumed in the first step is regenerated in the second.

Another type of catalysis that is important in oxidation processes comes from the initiation of a free radical chain reaction. Examples include the decomposition of ozone by hydroxide and the decomposition of chlorine by iron (e.g., see Brezonik [1994]). In either case, the original oxidant will not react appreciably with recalcitrant compounds such as oxalate. However, in the presence of sufficient catalyst, decomposition is initiated, leading to a series of chain propagation reactions whereby oxalate can be easily converted to carbon dioxide.

Reaction Pathways. Oxidation reactions in drinking waters can be very complex. They may begin with one of the mechanisms discussed here, but then may be followed by a wide range of nonoxidation processes, such as elimination reactions, hydrolysis reactions, radical chain reactions, and rearrangement reactions.

The formation of trihalomethanes may occur through many different reaction mechanisms. One of the most widely discussed is the haloform reaction (Figure 12.1), which involves the stepwise chlorine substitution of the enolate form of a methyl ketone. This classic reaction begins with a base-catalyzed halogenation ultimately leading to a carboxylic acid and chloroform. It is base-catalyzed because the species that reacts with hypochlorous acid is the enol form of the methyl ketone. This undergoes electrophilic substitution, forming a monohalogenated intermediate. The presence of halogens on this carbon speeds subsequent enolization, which leads to complete halogenation of the α-carbon. The resulting intermediate (a trihalogenated acetyl compound) is subject to base-catalyzed hydrolysis, giving a trihalomethane and a carboxylic acid. If hypochlorous acid is the only halogenating species, chloroform is the result.

Many early studies with acetone (propanone) indicated that the rate-limiting step was the initial enolization. Once the enolate was formed, the molecule quickly

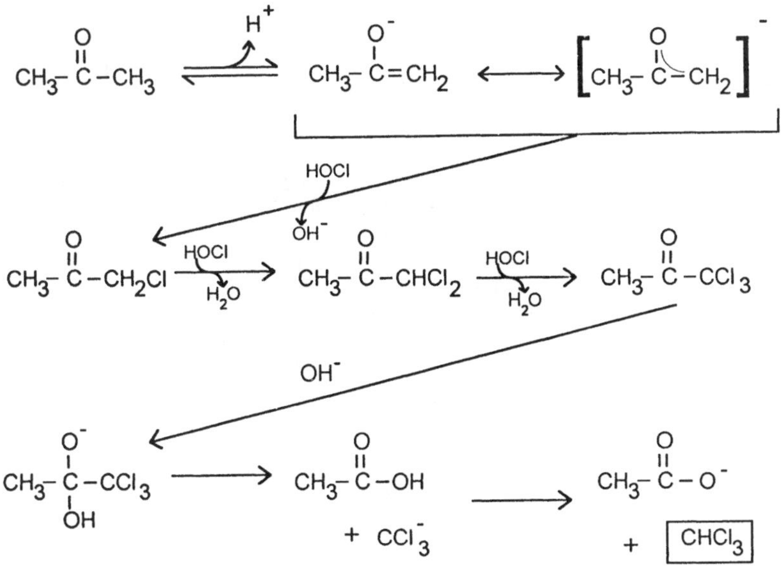

FIGURE 12.1 The haloform reaction.

proceeded through the entire reaction pathway. Thus, the reaction rate expression often cited in the chemical literature was

$$\frac{d[\mathrm{CH_3COCH_3}]}{dt} = -k[\mathrm{CH_3COCH_3}] \qquad (12.29)$$

However, under drinking water conditions (i.e., neutral pH, low chlorine residual), other steps may be rate limiting. This changes the rate law and complicates attempts to characterize it. In addition, a competing pathway exists that leads to the formation of dichloroacetic acid. Trichloropropanone may undergo further base-catalyzed chlorine substitution to form pentachloropropanone (Reckhow and Singer, 1985). This compound will rapidly hydrolyze to form chloroform and dichloroacetic acid. The rate law proposed for the loss of trichloropropanone is

$$-\frac{d[\mathrm{CH_3COCCl_3}]}{dt} = \left\{\frac{k_1 k_2[\mathrm{HOCl}][\mathrm{P}_T]}{k_{-1}[\mathrm{P}_T] + k_2[\mathrm{HOCl}]} + k_3[\mathrm{OCl^-}] + k_4[\mathrm{OH^-}]\right\}[\mathrm{CH_3COCCl_3}] \qquad (12.30)$$

This complicated rate law reflects catalysis by hypochlorite (OCl^-), hydroxide (OH^-), and phosphate (P_T).

OXIDANTS USED IN WATER TREATMENT

Chlorine

Chlorine is the most widely used oxidant (and disinfectant; see Chapter 14) in water treatment practice. Chlorine is available in gaseous form, as Cl_2; as a concentrated aqueous solution, sodium hypochlorite, NaOCl (e.g., bleach); or as a solid, calcium hypochlorite, $Ca(OCl)_2$.

When chlorine gas is added to water, the chlorine rapidly disproportionates to form hypochlorous acid (HOCl) and the chloride ion (Cl^-):

$$Cl_2 + H_2O \rightarrow HOCl + H^+ + Cl^- \tag{12.31}$$

The equilibrium constant for Eq. 12.31 is 5×10^{-4} at 25°C, indicating that the reaction goes relatively far to the right, as shown. The residual concentration of molecular Cl_2 in solution will represent only a small fraction of the total chlorine concentration, except at very low pH conditions or at high chloride concentrations. For example, in water at pH 2 with a 10^{-3} M chloride concentration, only 2 percent of the total chlorine will exist as molecular Cl_2. The chloride produced by Eq. 12.31 is essentially inert with respect to its oxidizing and disinfecting properties.

Hypochlorous acid is a weak acid ($pK_a = 7.5$ at 25°C), which partially dissociates to hypochlorite ion (OCl^-):

$$HOCl = H^+ + OCl^- \tag{12.32}$$

The ratio of hypochlorous acid to hypochlorite may be calculated from the expression:

$$\log\left(\frac{[HOCl]}{[OCl^-]}\right) = 7.5 - pH \tag{12.33}$$

The sum of the three species, Cl_2, HOCl, and OCl^-, is commonly referred to as free available chlorine (FAC), and the concentrations of the individual species and their sum are usually expressed in units of mg/L as Cl_2.

Equations 12.31 and 12.32 show that the species of chlorine present in water will depend on the total concentration of chlorine, pH, and temperature. Figure 12.2 is a diagram of the relative amounts of the three species as a function of pH. At 25°C, hypochlorous acid is the predominant species between pH 1 and pH 7.5, and hypochlorite ion is predominant at pH values greater than 7.5. The concentrations of the two species are equal at pH 7.5 (the pK_a value). The distribution shifts somewhat with changing temperature because the equilibrium constants for Eqs. 12.31 and 12.32 are temperature dependent.

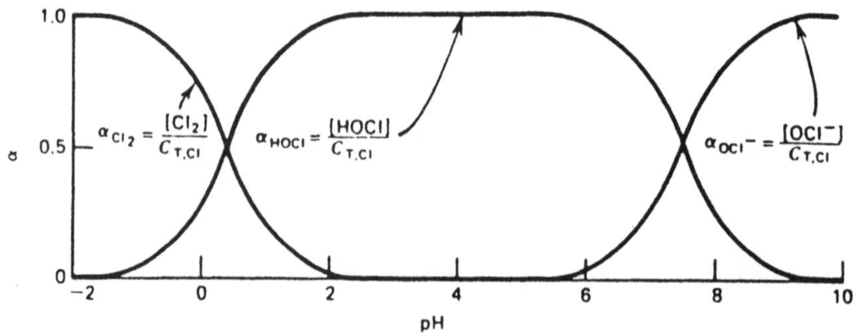

FIGURE 12.2 Distribution diagram for molecular chlorine, hypochlorous acid, and hypochlorite ion in water as a function of pH ($[Cl^-] = 10^{-3}M$). (*Source:* Snoeyink and Jenkins, 1980.)

If chlorine is added to water as liquid sodium hypochlorite, the following reactions occur:

$$NaOCl \rightarrow Na^+ + OCl^- \tag{12.34}$$

$$OCl^- + H_2O = HOCl + OH^- \tag{12.35}$$

If calcium hypochlorite granules are dissolved in water, they also form the hypochlorite ion, which, like the hypochlorite formed from the addition of NaOCl, reacts with water to form hypochlorous acid:

$$Ca(OCl)_2 \rightarrow Ca^{2+} + 2OCl^- \tag{12.36}$$

$$OCl^- + H_2O = HOCl + OH^- \tag{12.37}$$

Just as in the case of the species formed by the addition of chlorine gas, the relative distribution of HOCl and OCl^- that result from the addition of sodium hypochlorite or calcium hypochlorite will be determined by pH, temperature, and the total chlorine concentration. Again, this distribution will be in accordance with Eqs. 12.31 and 12.32 and Figure 12.2. No matter in what form the chlorine is added to water, hypochlorite, hypochlorous acid, and molecular chlorine will be formed as described by the chemical equilibria represented by the previous reactions. The only difference is that chlorine gas produces an acidic reaction which lowers the pH of the solution, while sodium hypochlorite and calcium hypochlorite are both bases which will raise the pH of the water. The extent of the pH change will depend on the alkalinity of the water.

As shown in Table 12.1, hypochlorous acid and hypochlorite ion are both strong oxidizing agents, but HOCl is the stronger of the two. Hence, oxidation reactions of chlorine tend to be more effective at low pH values. This is also true of the disinfecting properties of chlorine (see Chapter 14).

Molecular chlorine is typically provided for water treatment applications in pressurized tanks so that the chlorine exists as a liquid under pressure. The chlorine is then added to water by reducing the pressure in the tank and releasing the chlorine as a gas. Various types of gas-feeding equipment are available to introduce the gaseous chlorine to a sidestream of water, after which the resulting concentrated sidestream solution of chlorine is blended with the main flow through a diffuser at the desired point of application.

Alternatively, if liquid sodium hypochlorite solution is used, it is usually added from a concentrated NaOCl bulk solution directly to the main plant flow with the help of a liquid metering pump. Solid granules of calcium hypochlorite can be added to the water directly or a concentrated solution of calcium hypochlorite can be prepared, from which the hypochlorite is metered into the main flow, as in the case of sodium hypochlorite.

Historically, gaseous chlorine in pressurized tanks has been the most common method of applying chlorine in municipal water treatment practice, but, more recently, because of concerns about transport and handling of hazardous chemicals, the use of liquid sodium hypochlorite is becoming more widely practiced despite its higher cost. A potential concern involving the use of sodium hypochlorite is its stability. The hypochlorite tends to degrade over time, particularly when it is stored at high temperatures and/or exposed to sunlight. One of the degradation products is chlorate (ClO_3^-), a potential health concern (see subsequent discussion of chlorine dioxide). Recent reports also indicate the possible presence of bromate, another inorganic disinfection by-product usually associated with ozonation, in sodium hypochlorite feedstocks at levels which may exceed regulatory limits when the hypochlorite is fed at the typical doses used in practice.

Free chlorine can also be generated electrolytically on-site from brine (sodium chloride, NaCl) solutions. The molecular chlorine generated from the electrolysis reaction can be dissolved in sodium hydroxide (NaOH), which is also a by-product of the electrolysis reaction, to produce a concentrated sodium hypochlorite solution:

$$Cl_2 + NaOH \rightarrow NaOCl + H^+ + Cl^- \tag{12.38}$$

Again, while the use of electrolysis cells to produce chlorine gas on-site tends to be more costly than purchasing chlorine gas directly in pressurized tanks, the hazards associated with the transport and handling of pressurized chlorine containers is avoided.

Reactions of Chlorine with Organic Compounds. Chlorine reacts with organic material by a combination of oxidation and substitution reactions. For example, chlorine reacts with amino acids to produce nonhalogenated oxidation by-products, such as aldehydes and organic acids. Conversely, chlorine reacts with phenol to produce chlorinated phenolic compounds, such as ortho- and parachlorophenol. Formation of the latter contributes to an enhancement of taste and odor in chlorinated water containing phenol (see below). In the case of aromatic compounds, the presence of electron-donating substituents on the aromatic ring facilitates both substitution reactions and oxidative ring cleavage reactions.

Chlorine reacts with natural organic material (e.g., humic substances) to produce a variety of chlorine-substituted halogenated disinfection by-products, such as chloroform, di- and trichloroacetic acid, and chloropicrin. In view of the importance of these reactions to water quality and treatment, the subject of disinfection by-product formation and control is presented in a separate section.

Reactions of Chlorine with Ammonia (Formation of Chloramines). Chloramines are formed by the reaction of aqueous chlorine with ammonia. Chloramine formation may be done purposefully by adding ammonia to a water containing free available chlorine if it is desirable to form monochloramine for maintenance of a stable disinfectant residual in the distribution system (see Chapter 14). Conversely, chloramine formation may occur during the course of water treatment when chlorine is added to a source water containing ammonia. The mixture that results from reaction between free chlorine and ammonia may contain monochloramine (NH_2Cl), dichloramine ($NHCl_2$), and trichloramine or nitrogen trichloride (NCl_3):

$$NH_3 + HOCl = NH_2Cl + H_2O \tag{12.39}$$

$$NH_2Cl + HOCl = NHCl_2 + H_2O \tag{12.40}$$

$$NHCl_2 + HOCl = NCl_3 + H_2O \tag{12.41}$$

The sum of the concentrations of the three chloramine species is typically referred to as combined chlorine and is often expressed in units of mg/L as Cl_2. Combined chlorine is analytically distinguishable from free chlorine (*Standard Methods,* 1995). The sum of the free and combined chlorine concentrations is referred to as total chlorine.

The relative amounts of the three species formed will depend primarily on pH, temperature, and the ratio of chlorine to ammonia. At a chlorine:ammonia molar ratio less than 1:1 or a weight ratio (Cl_2:N) less than 5:1 and a pH of about 8, conditions that are common during the chloramination of drinking water, the principal chloramine species formed is monochloramine. At higher chlorine:ammonia ratios and lower pH values, dichloramine formation becomes important and, at mildly acidic pH values in the range of 6 to 6.5, nitrogen trichloride can be formed. Because dichloramine is unstable at pH 8 and decomposes essentially to nitrogen gas and chloride, the addition of excess chlorine to an ammonia-containing water results in the following overall reaction:

$$2NH_3 + 3HOCl \rightarrow N_2(gas) + 3H^+ + 3Cl^- + 3H_2O \qquad (12.42)$$

This equation illustrates the so-called breakpoint phenomenon whereby chlorine applied in sufficient doses (at a Cl_2:N molar ratio greater than 1.5 or a weight ratio greater than 7.5) will oxidize ammonia, resulting in the formation of a free chlorine residual. This equation can be used to estimate the chlorine demand of ammonia, although in practice the molar ratio tends to be closer to 2:1 (10:1 by weight).

If it is desirable to produce monochloramine to serve as a stable disinfectant residual in the distribution system, this is usually done by adding more than 0.2 mg/L ammonia as N for each mg/L of free Cl_2 in order to ensure that the molar ratio of Cl_2:N is less than 1 so that only monochloramine is formed (Eq. 12.39). This is usually done because monochloramine, being a weaker oxidant, is a more stable and persistent species than free chlorine. This contributes to its desirability as a secondary disinfectant in water distribution systems.

From the standpoint of its oxidation potential, monochloramine is too weak to oxidize reduced iron and manganese and most taste- and odor-causing organics, and it is too weak to eliminate natural organic color. While monochloramine will oxidize natural organic material to some degree and produce some halogenated organic material, it does not generally produce trihalomethanes (see below).

Part of the oxidation capacity of the chloramines derives from the hydrolysis reactions represented by the reverse reactions of Eqs. 12.39 to 12.41. The free chlorine liberated, although often in small amounts, may be responsible in part for the formation of some halogenated by-products associated with chloramines. Likewise, the free ammonia liberated may contribute to nitrification problems in water distribution systems.

In general, the principal value of chloramines in water treatment is not as an oxidant but as a secondary disinfectant (see Chapter 14).

Reactions of Chlorine with Bromide. In drinking waters containing bromide, chlorine will oxidize the bromide to produce hypobromous acid (HOBr):

$$HOCl + Br^- \rightarrow HOBr + Cl^- \qquad (12.43)$$

This reaction is very fast, with a second-order rate constant of 2.95×10^3 M^{-1} sec^{-1}. Hence, bromide represents another source of chlorine demand, and the presence of bromide contributes to the depletion of free chlorine.

The resulting hypobromous acid, like hypochlorous acid, is a weak acid ($pK_a = 8.7$ at 25°C) and will dissociate to some degree to form the hypobromite ion (OBr^-), depending upon pH:

$$HOBr \rightarrow OBr^- + H^+ \qquad (12.44)$$

As noted in Table 12.1, hypobromous acid is a somewhat weaker oxidant than hypochlorous acid. The formation of hyprobromous acid is important, however, because hypobromous acid is capable of reacting with natural organic material to produce undesirable brominated disinfection by-products, such as bromoform ($CHBr_3$)

and dibromoacetic acid. The presence of bromide and its subsequent oxidation to hyprobromous acid is what contributes, along with hypochlorous acid, to the formation of mixed brominated and chlorinated DBPs such as bromodichloromethane ($CHBrCl_2$) and bromochloroacetic acid (see below). Because of their relative acidity constants, more hypobromous acid remains in the undissociated HOBr form than hypochlorous acid under most pH conditions encountered in practice. Accordingly, because the undissociated forms of these species are stronger oxidants than their deprotonated counterparts, hypobromous acid behaves as a stronger substituting agent than hypochlorous acid, which results in the formation of greater amounts of halogenated disinfection by-products in bromide-containing waters.

Chlorine Dioxide

Chlorine dioxide (ClO_2) is a greenish-yellow gas at room temperature and has an odor similar to that of chlorine. It is unstable at high concentrations and can explode upon exposure to heat, light, electrical sparks, or shocks. Accordingly, it is never shipped in bulk, but instead is generated on-site. Aqueous solutions are usually prepared from the gaseous chlorine dioxide generated, as chlorine dioxide is very soluble in water. It does not hydrolyze in water like chlorine does and remains in its molecular form as ClO_2. It is much more volatile than chlorine, and can easily be stripped from aqueous solution if not properly handled.

Chlorine dioxide is usually generated by reacting sodium chlorite ($NaClO_2$) with either gaseous chlorine (Cl_2) or hypochlorous acid (HOCl) under acidic conditions, in accordance with the following reactions:

$$2NaClO_2 + Cl_2(g) \rightarrow 2ClO_2(g) + 2Na^+ + 2Cl^- \tag{12.45}$$

$$2NaClO_2 + HOCl \rightarrow 2ClO_2(g) + 2Na^+ + Cl^- + OH^- \tag{12.46}$$

In order to drive the reaction to completion and avoid the presence of unreacted chlorite (ClO_2^-) in the product stream, some generators use excess chlorine, although several variations in generator design have been developed which allow for the sodium chlorite to be completely converted to chlorine dioxide without using excess chlorine. Additionally, to overcome the basicity of sodium chlorite and the hydroxide produced by Eq. 12.46, acid is sometimes added along with the hypochlorous acid to maintain the optimal pH for chlorine dioxide generation. Typically, pH values in the range of 3.5 to 5.5 are preferred; more acidic pH values lead to the formation of chlorate (ClO_3^-):

$$NaClO_2 + Cl_2 + H_2O \rightarrow ClO_3^- + 2Cl^- + 2H^+ + Na^+ \tag{12.47}$$

Chlorine dioxide may also be prepared by acidification of a sodium chlorite solution:

$$5NaClO_2 + 4H^+ \rightarrow 4ClO_2(g) + 5Na^+ + Cl^- + 2H_2O \tag{12.48}$$

Once generated, chlorine dioxide can be dissolved in water and is stable in the absence of light and elevated temperatures. In the presence of the latter, or at high pH values, it disproportionates to form both chlorite and chlorate:

$$2ClO_2 + 2OH^- \rightarrow ClO_2^- + ClO_3^- + H_2O \tag{12.49}$$

both of which are undesirable in drinking water (see below).

The primary application of chlorine dioxide in the past has been for taste and odor control, although it is also an effective oxidant for reduced iron and manganese, and is a good primary disinfectant (see Chapter 14). One of its principal advantages is that it does not react with ammonia. Hence, much lower doses of chlorine dioxide are required for most oxidative applications compared to chlorine dosage requirements. Another advantage is that chlorine dioxide does not enter into substitution reactions with NOM to the same degree that free chlorine does and, accordingly, does not form trihalomethanes, haloacetic acids, or most other commonly observed halogenated disinfection by-products that result from chlorination. Richardson et al. (1994) and others have identified a number of disinfection/oxidation by-products of chlorine dioxide treatment—for example, aldehydes, carboxylic acids, and some halogenated compounds—but most of the latter were present at extremely low concentrations. Because it does not tend to form halogenated DBPs to any significant degree, chlorine dioxide is enjoying renewed interest as a water treatment oxidant. An additional benefit is that chlorine dioxide reacts only very slowly with bromide (Hoigne and Bader, 1994). Hence, brominated by-products, either organic or inorganic (e.g., bromate), are not a concern following treatment with chlorine dioxide.

Hoigne and Bader (1994) and Tratnyek and Hoigne (1994) provide a listing and discussion of the rate constants for reactions between chlorine dioxide and a variety of organic and inorganic compounds, a number of which are of relevance to the treatment of drinking water.

Chlorine dioxide typically reacts with most reducing agents (e.g., ferrous iron, natural organic material) through a one-electron transfer (see Table 12.1):

$$ClO_2 + e^- \rightarrow ClO_2^- \tag{12.50}$$

As a result, chlorite is considered to be the principal oxidation by-product of chlorine dioxide usage. Most researchers (e.g. Werdehoff and Singer, 1987) have reported that approximately 50 to 70 percent (by mass) of the chlorine dioxide applied during the course of drinking water treatment ends up as chlorite.

Chlorite has been demonstrated to exhibit a number of potential adverse health effects based on studies with laboratory animals. Because of this, the Disinfectants/Disinfection By-Products Rule (USEPA, 1998) establishes a maximum contaminant level (MCL) of 1.0 mg/L for chlorite at representative locations in the distribution system and a maximum residual disinfectant level (MRDL) of 0.8 mg/L for chlorine dioxide at the point of entry to the distribution system. No regulations exist for chlorate at this time. Given the observations that chlorine dioxide is rapidly consumed during the course of water treatment and that up to 70 percent of the applied chlorine dioxide is reduced to chlorite, the practical upper limit for chlorine dioxide doses would be approximately 1.4 to 1.5 mg/L unless chlorite is removed (see below).

A second concern associated with residual chlorite in the distribution system is that it reacts with free chlorine, producing low levels of chlorine dioxide and/or chlorate:

$$HOCl + 2ClO_2^- \rightarrow 2ClO_2 + Cl^- + OH^- \tag{12.51}$$

$$HOCl + ClO_2^- \rightarrow ClO_3^- + Cl^- + H^+ \tag{12.52}$$

If present in tap water, the chlorine dioxide, being volatile, can be released into the home or office environment when customers open their taps. This can lead to offensive chlorinous odors or, if new carpeting has recently been installed, the escaping chlorine dioxide can react with organic compounds released from the carpeting to produce other offensive odors.

Chlorite, whether it is present as a result of incomplete oxidation of sodium chlorite in the chlorine dioxide generator (Eqs. 12.45 and 12.46) or by chemical reduction of the chlorine dioxide during the course of treatment (Eq. 12.50), can be removed from water by the application of ferrous iron salts (Iatrou and Knocke, 1992) or reduced sulfur compounds. No practical method for the removal of chlorate is available; hence, its formation during chlorine dioxide generation should be minimized.

Ozone and Advanced Oxidation Processes

Ozone is an unstable gas that must be generated on-site. A simplified representation of the chemistry involved in the formation of ozone (O_3) is as follows:

$$O_2 + \text{energy} \rightarrow O + O \tag{12.53}$$

$$O + O_2 \rightarrow O_3 \tag{12.54}$$

The energy required to produce nascent or elemental oxygen (O) from molecular oxygen (O_2) is usually supplied by an electric discharge with a peak voltage from 8 to 20 kV, depending on the apparatus used. Dry, refrigerated, particle-free air, oxygen, or oxygen-enriched air is passed through a narrow gap between two electrodes and a high-energy discharge is generated across the gap between the electrodes. This corona or cold plasma discharge is induced by an alternating current that creates a voltage cycle between the two electrodes. The yield of ozone will depend on the voltage, the frequency, the design of the ozone generator, and the type and quality of the feed gas used. Ozone streams containing up to 14 percent ozone by volume can be produced. Current ozone generators are available as low-frequency (50 to 60 Hz), medium-frequency (400 to 1000 Hz), and high-frequency (2000 to 3000 Hz) systems.

Once generated, the ozone-enriched air or oxygen gas is passed through a gas absorption device to transfer ozone into solution. This can be achieved through either a countercurrent multistage bubble contactor, an in-line gas injection system, or other such gas transfer devices.

Ozone is very unstable in aqueous solution. It is very reactive with a number of common constituents in drinking water (e.g., NOM), and it also undergoes a spontaneous decomposition process, sometimes referred to as auto-decomposition. The auto-decomposition of ozone is a complex chain reaction process involving several free radical species. Decomposition may be initiated by a number of different water constituents, such as hydroxide ion (e.g., high pH values), natural organic material, and ferrous iron, or it may be initiated by the addition of hydrogen peroxide or by irradiation with ultraviolet light. The reactions shown in Eqs. 12.55 through 12.60 illustrate the auto-decomposition scheme when hydroxide ion is the initiator.

$$OH^- + O_3 \rightarrow HO_2 + O_2^- \tag{12.55}$$

$$HO_2 = H^+ + O_2^- \tag{12.56}$$

$$O_2^- + O_3 \rightarrow O_2 + O_3^- \tag{12.57}$$

$$O_3^- + H^+ \rightarrow HO_3 \tag{12.58}$$

$$HO_3 \rightarrow O_2 + OH \tag{12.59}$$

$$OH + O_3 \rightarrow HO_2 + O_2 \tag{12.60}$$

These reactions constitute a chain mechanism because the hydroperoxyl radical (HO_2) and the superoxide ion (O_2^-) produced by the initiation reaction (Eq. 12.55) generate new chain reactions that further contribute to ozone decomposition. In pure water, the chain may be very long; that is, hundreds of ozone molecules may be decomposed by a single initiation step. In natural waters, the lifetime of ozone depends on several variables, including pH, temperature, total organic carbon (TOC) concentration, and bicarbonate and carbonate concentrations. Bicarbonate and carbonate increase the lifetime of ozone by reacting with the hydroxyl radical (OH)

$$OH + HCO_3^- \rightarrow OH^- + HCO_3 \qquad (12.61)$$

$$OH + CO_3^{-2} \rightarrow OH^- + CO_3^- \qquad (12.62)$$

and thereby decelarating the chain mechanism shown in Eqs. 12.56 to 12.60.

The bicarbonate and carbonate radicals (HCO_3 and CO_3^-, respectively) are relatively unreactive intermediates that cannot propagate the chain. Thus, waters high in bicarbonate and carbonate alkalinity will retain an ozone residual for longer periods of time than low-alkalinity waters. This is especially important when ozone is used as a disinfectant (see Chapter 14). Additionally, the radical scavenging activity of the carbonate species increases with increasing pH because carbonate is a more effective scavenger than bicarbonate. This partially offsets the more rapid rate of the hydroxide-induced initiation reaction (Eq. 12.55) at higher pH values (higher hydroxide ion concentrations).

The hydroxyl radical (OH), one of the intermediates produced by the decomposition of ozone, is one of the strongest chemical oxidants known and is capable of rapidly reacting with a myriad of organic and inorganic compounds (see below). Accordingly, the oxidative properties of ozone depend significantly on the oxidative characteristics of this free radical species.

Aieta et al. (1988) have presented an illustrative schematic of the behavior of ozone in aqueous solution, based on the fundamental work of Hoigne and his coworkers (e.g., Hoigne and Bader, 1976; Staehelin and Hoigne, 1982). Figure 12.3 shows that ozone reacts by two distinct types of pathways: a direct pathway involving molecular ozone (O_3) and an indirect pathway originating with the decomposition of ozone to produce the hydroxyl free radical (OH). Direct reactions involving molecular ozone are very selective; ozone reacts very rapidly with some species—for

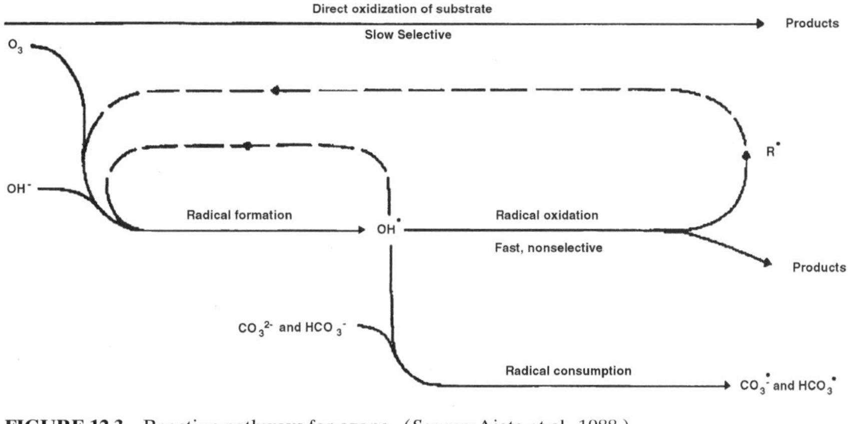

FIGURE 12.3 Reaction pathways for ozone. (*Source:* Aieta et al., 1988.)

example, phenol and mercaptans—but very slowly with other species—for example, benzene and tetrachloroethylene (PCE). Conversely, the OH radical is nonselective in its behavior, reacting rapidly with a large number of species. Additionally, the OH radical reacts rapidly with molecular ozone (see Eq. 12.60 and Figure 12.3), thereby contributing to the autocatalytic rate of ozone decomposition. Hydroxyl radical scavengers, such as the bicarbonate and carbonate ion, react with the hydroxyl radical (see also Eqs. 12.61 and 12.62), removing it from the cycle and, in so doing, decelerating the kinetics of ozone decomposition, which promotes the stability of molecular ozone in solution, as already noted.

Hydroxyl free radicals can be produced through a number of other pathways in addition to the hydroxide-induced ozone decomposition chain previously described. For example, the addition of both hydrogen peroxide and ozone to water accelerates the decomposition of ozone and enhances production of the hydroxyl radical. The hydrogen peroxide (H_2O_2) dissociates into the hydroperoxide ion (HO_2^-):

$$H_2O_2 = HO_2^- + H^+ \tag{12.63}$$

The hydroperoxide ion then reacts with molecular ozone to produce the superoxide ion (O_2^-) and the hydroxyl radical, plus molecular oxygen:

$$HO_2^- + O_3 \rightarrow OH + O_2^- + O_2 \tag{12.64}$$

The hydroxyl radical and the superoxide ion then participate in the ozone decomposition cycle depicted by Eqs. 12.57 to 12.60, leading to an accelerated production of more hydroxyl free radicals. Because reactions 12.63 and 12.64 tend to be appreciably faster than reaction 12.55 under most conditions, the conjunctive use of O_3 and H_2O_2 tends to be a more effective method of generating the highly reactive, nonselective hydroxyl radicals for chemical oxidation reactions.

Another method for generating hydroxyl radicals is by ultraviolet (UV) irradiation of hydrogen peroxide. The UV irradiation provides the energy to split the hydrogen peroxide into two hydroxyl radicals:

$$H_2O_2 + UV \rightarrow 2\ OH \tag{12.65}$$

This process tends to be slower than reactions 12.63 and 12.64 and is therefore a less effective method of generating hydroxyl radicals.

Ultraviolet irradiation of waters containing dissolved molecular ozone leads to the formation of hydrogen peroxide. The hydrogen peroxide then reacts with molecular ozone in the same manner as it does through Eqs. 12.63 and 12.64, which produces the hydroxyl radical and other species that enter the ozone decomposition cycle depicted by Eqs. 12.57 to 12.60.

These reactions, all of which involve the accelerated production of the hydroxyl free radical, are termed *advanced oxidation processes* (AOPs). Often, when applying ozone to water, it is difficult to distinguish between reactions that are attributable to molecular ozone and those that are attributable to the hydroxyl radical. It is believed that hydroxyl radical reactions are at the heart of all reactions involving molecular ozone, except perhaps for disinfection reactions, reactions with some solutes that have very high reaction rate constants, and reactions that take place in the presence of high concentrations of hydroxyl radical scavengers.

Kinetics of Ozone Reactions. The kinetics of reactions of molecular ozone and hydroxyl radicals have been studied extensively. The most comprehensive listing of reaction rate constants is that provided by Hoigne and coworkers (e.g., Hoigne and

Bader, 1983*a,b;* Hoigne et al., 1985). As previously noted, some solutes react very quickly with molecular ozone. For example, the oxidation of sulfide by molecular ozone is extremely rapid, with a second-order rate constant on the order of 10^9 M^{-1} sec^{-1}. The second-order rate constants for phenol and naphthalene oxidation by molecular ozone are 1300 and 3000 M^{-1} sec^{-1}, respectively, also relatively high. Conversely, the second-order rate constants for benzene and tetrachloroethylene oxidation are 2 and <0.1 M^{-1} sec^{-1}, respectively. The oxidation of trichloroethylene and atrazine, two common contaminants of drinking water, are also relatively low, being in the range of 10 to 20 M^{-1} sec^{-1}. Also, ammonia reacts very slowly with molecular ozone in the pH range of interest in drinking water treatment. Amines and amino acids react quickly when the amino group is not protonated.

In general, compounds that dissociate in water react with molecular ozone at rates that depend on the individual species present. For example, the phenolate anion reacts much faster than its protonated counterpart phenol. The same is true for formate relative to formic acid. Hence, the rates of ozonation may increase or decrease as pH changes, depending on the specific compound being oxidized. Also, while benzene itself is relatively unreactive toward molecular ozone, derivatives of benzene tend to be much more reactive, especially with the addition of electron-withdrawing substituents on the aromatic ring.

In the case of the hydroxyl radical, the magnitude of the second-order reaction rate constants with various solutes is much more uniform, tending to range from 10^7 to 10^{10} M^{-1} sec^{-1}. It is for this reason that the hydroxyl radical is often called nonselective with respect to its reactivity, in contrast to molecular ozone, which is highly selective in that its rate constants with various solutes range over more than 12 orders of magnitude. Atrazine, trichloroethylene, and tetrachloroethylene, all relatively inert toward molecular ozone, are oxidized at appreciable rates by the hydroxyl radical, and therefore by AOPs. Because of the large magnitude of the rate constants for oxidation reactions involving the hydroxyl radical, the rate-determining step in these reactions is the rate at which the radicals are generated either by ozone auto-decomposition or by the various AOPs previously discussed. Hence, in many cases (e.g., atrazine or trichloroethylene), the oxidation of these solutes is enhanced at elevated pH values because of the more rapid generation of hydroxyl radicals by hydroxide-induced ozone decomposition.

Formation of Biodegradable Organic Material. When ozone reacts with organic contaminants in water, including natural organic material, it partially oxidizes them to lower molecular weight, more polar species, including a variety of aldehydes and organic acids (see below). These oxidation by-products, while not believed to be harmful in themselves, tend to be biodegradable and may contribute to biofouling problems in the water distribution system if not properly controlled. Often, ozonation is followed by a biologically active filtration process to remove these biodegradable organic materials. Ozonation by-products are discussed in greater detail later.

Reactions of Ozone with Bromide. While ozone will not react with natural organic material to directly produce halogenated DBPs, it may result in the formation of brominated DBPs in waters containing bromide. Ozone is capable of oxidizing bromide to hypobromous acid in the same manner that chlorine does:

$$O_3 + Br^- + H^+ \rightarrow HOBr + O_2 \qquad (12.66)$$

This is a very fast reaction, with a reaction rate constant of 160 M^{-1} sec^{-1}. The hypobromous acid can then react with natural organic material to produce brominated

DBPs, such as bromoform, dibromoacetic acid, or bromopicrin. The extent of formation of these DBPs depends upon the bromide ion and TOC concentration of the water and its pH. In general, the concentration of these halogenated DBPs produced indirectly by ozonation is at least an order of magnitude lower than those formed by chlorination. This is discussed in greater detail later.

A particular concern associated with the ozonation of bromide-containing waters is the formation of bromate (BrO_3^-). Bromate is a possible human carcinogen and is currently regulated at 10 μg/L, the practical quantitation level (PQL) for bromate, although it is expected that this level will decrease in the near future as more sensitive analytical methodologies are developed. Bromate is produced by a number of possible pathways involving molecular ozone and the hydroxyl radical. One such pathway (unbalanced stoichiometrically) proceeds through the formation of hypobromite (OBr^-) and bromite (BrO_2^-):

$$O_3 + Br^- \rightarrow OBr^- \tag{12.67}$$

$$OBr^- + O_3 \rightarrow BrO_2^- \tag{12.68}$$

$$BrO_2^- + O_3 \rightarrow BrO_3^- \tag{12.69}$$

In the pH range of most natural waters, a large portion of the OBr^- is protonated as HOBr (the pK_a for hypobromous acid is 8.7 at 25°C), so that the rate of formation of bromite by reaction 12.68 is slowed considerably with decreasing pH. If the rate of formation of hydroxyl radicals in the water is high, bromate may also be formed through the hypobromite radical (BrO) as follows:

$$HOBr/OBr^- + OH \rightarrow BrO \tag{12.70}$$

$$BrO + O_3 \rightarrow BrO_3^- \tag{12.71}$$

Techniques for controlling bromate formation most often involve ozonation at slightly acidic pH values, multistage ozonation in which the ozone is added at several different application points, and the use of ammonia to tie up the hypobromous acid produced:

$$NH_3 + HOBr \rightarrow NH_2Br + H_2O \tag{12.72}$$

Additionally, once formed, bromate can be removed by chemical reduction using reduced sulfur compounds, such as bisulfite (HSO_3^-) or ferrous iron. Granular activated carbon is also capable of adsorbing bromate, albeit to a limited degree, and medium-pressure ultraviolet irradiation decomposes bromate to bromide. Bromate can also be reduced under anaerobic conditions.

Potassium Permanganate

Potassium permanganate ($KMnO_4$) has been used as a water treatment oxidant for decades. It is commercially available in crystalline form and either it is fed into solution directly using a dry chemical feeder or a concentrated solution is prepared on-site from which the desired dose is metered into the water.

Permanganate contains manganese in the +VII oxidation state. Under most treatment applications, oxidation by permanganate involves a three-electron transfer, with the permanganate (MnO_4^-) being reduced to insoluble manganese dioxide, $MnO_2(s)$:

$$MnO_4^- + 4H^+ + 3e^- \rightarrow MnO_2(s) + 2H_2O \tag{12.73}$$

The manganese dioxide produced is a black precipitate that, if not properly removed by a suitable solid-liquid separation process, will create black particulate deposits in the distribution system and on household plumbing fixtures. Most often, removal of $MnO_2(s)$ is achieved by conventional clarification or filtration processes. Because most operators are fearful of seeing pink water (reflecting unreacted permanganate) coming through their filters, permanganate is commonly added at the head of the treatment plant, as close to the intake as possible. This allows sufficient time for the permanganate to perform its oxidative function and to be reduced completely to solid manganese dioxide prior to filtration.

The kinetics of oxidation reactions involving permanganate tend to be more rapid with increasing pH values. Hence, in some cases, addition of a base prior to filtration may be desirable to hasten the reduction of permanganate.

The manganese dioxide that results from permanganate reduction (Eq. 12.73) may have some beneficial attributes. $MnO_2(s)$ is an effective adsorbent for ferrous iron (Fe^{2+}), manganous manganese (Mn^{2+}), radium (Ra^{2+}), and other trace inorganic cationic species. Accordingly, some additional removal of these contaminants occurs as a result of permanganate treatment beyond that achieved simply by oxidation. In fact, the adsorptive behavior of $MnO_2(s)$ is the principle underlying the historic manganese greensand process, in which the filter media is coated with manganese dioxide, which subsequently serves as an adsorbent for Fe^{2+}, Mn^{2+}, and Ra^{2+} in the filter influent. The filter backwash water is treated with permanganate, or low doses of permanganate are applied to the filter influent to oxidize the adsorbed metals, thereby creating additional adsorption sites.

Solid manganese dioxide is also capable of adsorbing natural organic material that serves as DBP precursors. This benefit is particularly pronounced in hard waters (Singer and Colthurst, 1982), presumably because of the bridging action of calcium and magnesium.

Mixed Oxidants

The electrolysis of brine has been used since the nineteenth century to produce chlorine on an industrial scale. This technology has recently been modified and adopted to electrochemically produce a mixture of free chlorine and other powerful oxidants and disinfectants for drinking water applications for small, rural water supplies. Both liquid- and gas-phase generators are available.

The underlying principle is based on fundamental electrochemical theory. At the anode of the electrochemical cell, chloride is oxidized to chlorine, which subsequently hydrolyzes to hypochlorous acid:

$$2Cl^- \rightarrow Cl_2 + 2e^- \qquad (12.74)$$

$$Cl_2 + H_2O \rightarrow HOCl + Cl^- + H^+ \qquad (12.75)$$

The pH of the anodic stream tends to be on the order of 3 to 5. At the cathode, the water is reduced to hydrogen gas, producing a strongly alkaline solution with a pH of about 10 to 11:

$$2H_2O + 2e^- \rightarrow H_2 + 2OH^- \qquad (12.76)$$

The anodic and cathodic streams are separated by a semipermeable barrier.

While free chlorine is the primary oxidant produced by these mixed oxidant generators, other reactions occur at the anode, purportedly resulting in the formation of ozone, chlorine dioxide, hydrogen peroxide, other short-lived reactive oxidant species, and a number of inorganic by-products such as chlorite, chlorate, and bromate. However, the composition of the oxidant streams from these mixed oxidant generators has not been fully characterized. The output from one of these generators has been shown to produce on the order of 200 to 400 mg/L of free available chlorine (Dowd, 1994; Gordon, 1998), but no ozone, chlorine dioxide, or hydrogen peroxide was detected in the anodic stream (Gordon, 1998). These findings appear to contradict those of Venczel et al. (1997), who found that the oxidant product had disinfecting properties distinctly different than those of free chlorine alone. More research is needed to characterize these product streams before widespread use of these generators is recommended.

APPLICATIONS OF OXIDATION PROCESSES

Application of Oxidants to Water Treatment Practice

This section describes typical applications of oxidants and the role they play in overall water treatment practice. These oxidants are most often used for the oxidation of reduced iron and manganese, destruction of taste- and odor-causing organic contaminants, elimination of color, and the destruction of synthetic organic chemicals of public health concern. Additionally, many of these oxidants act as coagulant aids and are also employed as part of an overall program for the control of potentially harmful disinfection by-products. The formation and control of oxidation and disinfection by-products are discussed in a separate section. Many of these oxidants are also powerful disinfectants and therefore serve the dual purposes of oxidation and disinfection. Disinfection is discussed in detail in Chapter 14.

Control of Iron and Manganese. Iron and manganese are relatively soluble under reducing conditions, for example, in groundwaters, stagnant surface waters, and in the hypolimnetic waters of eutrophic lakes, reservoirs, and impoundments. Correspondingly, they are quite insoluble under oxidizing conditions, for example, in flowing streams and in the epilimnetic waters of lakes or impoundments, or in hypolimnetic waters that have been subject to hypolimnetic aeration to maintain oxidizing conditions. The reduced forms of iron and manganese—ferrous iron, Fe(II), and manganous manganese, Mn(II)—may occur as the free metal ions, Fe^{2+} and Mn^{2+}, which is often the case in most groundwaters, or they may be found complexed to various degrees with natural organic material, as is often the case in surface waters and highly colored groundwaters. During and immediately following lake overturn—that is, when the iron- and manganese-rich hypolimnetic water is mixed with the remainder of the lake water—dissolved iron and manganese levels in the upper portions of the lake can increase appreciably.

The primary concern with elevated levels of dissolved iron and manganese in water is that when they become oxidized to insoluble ferric hydroxide, $Fe(OH)_3(s)$, and manganese dioxide, $MnO_2(s)$, they precipitate and cause reddish-orange or black deposits, respectively, to appear on plumbing fixtures and to create stains during laundering operations:

$$Fe^{2+} + \tfrac{1}{4}O_2 + \tfrac{5}{2}H_2O \rightarrow Fe(OH)_3(s) + 2H^+ \qquad (12.77)$$

$$Mn^{2+} + \tfrac{1}{2}O_2 + H_2O \rightarrow MnO_2(s) + 2H^+ \qquad (12.78)$$

Dissolved Fe(II) and Mn(II) are usually removed from water by oxidizing them under engineered conditions to their insoluble forms through the addition of an oxidant and then removing the precipitated ferric hydroxide and manganese dioxide by sedimentation and filtration. The oxidants used most often for this purpose are oxygen, chlorine, permanganate, chlorine dioxide, and ozone.

The kinetics of oxidation of Fe(II) by oxygen are relatively rapid at pH values above 7, provided that the ferrous iron is not complexed by organic material. This is often the case in the removal of iron from groundwater. The rate expression for the oxygenation of ferrous iron is:

$$-\frac{d[\text{Fe(II)}]}{dt} = k[\text{Fe(II)}][\text{OH}^-]^2 P_{\text{O}_2} \tag{12.79}$$

The kinetics are first order with respect to the concentration of ferrous iron and the partial pressure of oxygen (P_{O_2}) and second order with respect to the hydroxide ion concentration (Stumm and Morgan, 1996); the latter illustrates the strong pH dependency of Fe(II) oxidation by oxygen. If the ferrous iron is complexed by organic material, the rate of oxidation by oxygen can be very slow and a stronger oxidant, such as chlorine or permanganate, may be needed.

In the case of Mn(II), the kinetics of oxidation by oxygen have been shown to conform to the following rate expression:

$$-\frac{d[\text{Mn(II)}]}{dt} = k_o[\text{Mn(II)}] + k[\text{Mn(II)}][\text{MnO}_2(\text{s})]) \tag{12.80}$$

where $k = k_1[OH^-]^2 P_{O_2}$. This is an autocatalytic reaction whereby the by-product of the oxidation reaction, manganese dioxide or $MnO_2(s)$, catalyzes the oxidation of reduced manganese (Stumm and Morgan, 1996). Because the initial oxidation step represented by the first term in Eq. 12.80 is relatively slow at pH values below 9, a strong chemical oxidant, such as chlorine or permanganate, is often used in place of oxygen. Alternatively, because the reaction is catalyzed by manganese dioxide, and manganese dioxide strongly adsorbs Mn^{2+}, a common practice for manganese oxidation and removal is to allow the filter media to be coated by manganese dioxide. Reduced manganese then adsorbs to the manganese dioxide coating and can be oxidized within the filter bed using chlorine or permanganate. This sequence of reactions can be represented as:

$$\text{Mn}^{2+} + \text{MnO}_2(\text{s}) \rightarrow \text{MnO}_2(\text{s}) - \text{Mn}^{2+} \tag{12.81}$$

$$\text{MnO}_2(\text{s}) - \text{Mn}^{2+} + \text{HOCl} + \text{H}_2\text{O} \rightarrow 2\,\text{MnO}_2(\text{s}) + \text{Cl}^- + 3\text{H}^+ \tag{12.82}$$

Commercially available glauconite (greensand) can also be used as a filter medium in the same manner. It has an affinity for the adsorption of manganese.

Last, it should be noted that, because ozone is such a strong oxidant, it is capable of oxidizing reduced manganese to permanganate (see illustrative calculations at the beginning of this chapter):

$$\text{Mn}^{2+} + \tfrac{5}{2}\text{O}_3 + \tfrac{3}{2}\text{H}_2\text{O} \rightarrow \text{MnO}_4^- + \tfrac{5}{2}\text{O}_2 + 3\text{H}^+ \tag{12.83}$$

As a result, it is not uncommon to observe pink water if excess ozone is added to a water containing reduced manganese.

Destruction of Tastes and Odors. Tastes and odors occur in water from a variety of sources, most notably algae, actinomycetes, organic and inorganic sulfides (e.g., mercaptans and hydrogen sulfide, respectively), and industrial contaminants, such as phenols (Mallevialle and Suffet, 1987). The presence of blue-green algae, in particular, is associated with a variety of specific chemical compounds that produce unpleasant odors at ng/L levels, such as methylisoborneol (MIB) and geosmin, in waters drawn from impoundments. Because of the algal origin of these odorous compounds, they tend to occur on a seasonal basis.

Chemical oxidation and activated carbon adsorption (see Chapter 13) are the two most common methods of eliminating taste and odor from drinking water. Free chlorine is an effective chemical oxidant for the destruction of odors associated with reduced sulfur compounds, but is less effective in destroying phenolic compounds and MIB and geosmin. In the case of phenols, low doses of chlorine can produce chlorophenols by substitution reactions of the type shown earlier in this chapter. Some of these chlorophenols have stronger odors than the parent phenolic compounds themselves (Burttschell et al., 1959). Hence, the application of chlorine intensifies the odor. In some cases, the chlorinous odor associated with free chlorine itself simply masks the odor associated with the odor-causing contaminant.

Ozone, chlorine dioxide, and permanganate have all been used for oxidizing and eliminating odorous compounds in water, but their success varies, and all are limited to some extent for different types of odor-producing compounds and in different types of waters (Glaze et al., 1990). Generally, all are effective for oxidizing phenolic compounds. Ozone is the most effective agent for oxidizing MIB and geosmin, although when coupled with hydrogen peroxide or UV irradiation to produce the hydroxyl radical (see above), the resulting advanced oxidation processes tend to be more effective in destroying MIB and geosmin than ozone alone.

It should be noted that aldehydes are often formed from the reactions of ozone with natural organic material (see below). These aldehydes will often impart a fruity odor to the water.

Also, the application of chlorine dioxide has been shown, in some cases, to produce kerosene-like and cat-urine-like odors. This has been attributed to reactions between chlorite (the principal inorganic by-product of chlorine dioxide treatment), free chlorine (which oxidizes the residual chlorite to chlorine dioxide, which is volatile), and volatile organic chemicals released from new carpeting (Hoehn et al., 1990). When chloramines are used as a secondary disinfectant in place of free chlorine, they are not strong enough to oxidize chlorite to chlorine dioxide, and therefore the problem goes away.

Color Removal. Color in water tends to be associated with the presence of polyaromatic compounds arising from natural vegetative decay processes. These compounds are often referred to as *humic acids* or *humic substances.* They impart a yellowish hue to the water, which can be quantified by the measurement of absorbance at 430 nm or by comparison to platinum-cobalt standards (*Standard Methods,* 1995). The nature of these substances and the molecular basis of the color vary with the source water. Color can also be correlated to the absorbance of UV radiation at 254 nm, a parameter that is often used to quantify the presence of natural organic material that contributes to the formation of disinfection by-products (see below). The humic substances are frequently complexed with trace metals in the water, a phenomenon which contributes to the appearance of color.

The most common processes for the removal of natural organic color from water are chemical oxidation and coagulation (see Chapter 6). In the case of chemical oxidation, the oxidizing agent attacks the chromophoric portion of the molecules that are responsible for the absorption of visible light; these are usually the carbon-carbon double bonds, the groups that bind trace metals, and the polyaromatic structures themselves. Chlorine had been used historically for color removal, but it became apparent that the reactions between chlorine and the color-causing organic materials are the ones that lead to the formation of halogenated disinfection by-products that are of public health concern (see below). Ozone and chlorine dioxide are also effective chemical oxidants for color removal, but these oxidants also produce by-products that can be problematic, such as biodegradable organic material that can lead to microbial regrowth in the distribution system if not properly controlled.

Accordingly, it may be more desirable to remove the color-causing organic materials by coagulation processes, thereby lowering the oxidant demand of the water, and use chemical oxidation to polish any residual color that may remain after clarification and/or filtration.

Oxidation of Synthetic Organic Chemicals. Strong chemical oxidants are capable of reacting with synthetic organic chemicals, such as pesticides and chlorinated solvents, that may be present in water supplies. These reactions are highly dependent on the nature of the organic compound, the specific oxidant in question, other constituents in the water, pH, and temperature. Some organic compounds are relatively easy to oxidize, while others are much more resistant. Additionally, for any particular oxidant, the products of the reaction vary, depending upon the oxidant, the ratio of the concentration of the oxidant to that of the contaminant, and general water quality characteristics.

For example, the oxidation of phenol by low doses of chlorine leads to the formation of mono-, di-, and trichlorophenols (Burttschell et al., 1959). At higher doses, the aromatic ring is cleaved to generate halogen-containing products. In the case of ozone, oxidation of phenol at low ozone doses results in the formation of catechol and o-quinone before the ring is cleaved at higher doses to produce a variety of low-molecular-weight carboxylic acids (Singer and Gurol, 1981). With chlorine dioxide, oxidation of phenol produces hydroxophenols, quinones, and several chlorophenols and chloroquinones at low oxidant doses (Wajon, Rosenblatt, and Burrows, 1982; Stevens, 1982). Subsequent ring cleavage at higher chlorine dioxide doses leads to the formation of carboxylic acids, as is the case with ozone. With chlorine dioxide, it is not clear whether the chlorinated species formed evolved from reactive chlorine produced during the reaction between chlorine dioxide and phenol or from the presence of free chlorine as a contaminant in the chlorine dioxide feed solution.

As previously noted, Hoigne and coworkers (1983*a,b,* 1985) present rate constants for reactions of ozone with a variety of inorganic and organic compounds in water. Their summary attests to the selectivity of molecular ozone for various types of contaminants. Phenol and naphthalene are rapidly oxidized by molecular ozone, whereas atrazine and trichloroethylene are only poorly oxidized by ozone. On the other hand, atrazine and trichloroethylene are rapidly oxidized by the hydroxyl radical generated from the decomposition of ozone. Hence, conditions that favor the formation of hydroxyl radicals are conducive to elimination of these harmful contaminants. Such conditions include ozonation at elevated pH values or the use of ozone in conjunction with UV irradiation or hydrogen peroxide addition (AOPs). Conversely, conditions that favor the scavenging of hydroxyl radicals interfere with the effectiveness of these AOPs. This is true for high-alkalinity waters that contain high concentrations of bicarbonate and carbonate, both of which are hydroxyl radical scavengers (see above). Additionally, because the micropollutants are usually present at relatively low (microgram per liter) concentrations compared to natural organic material, which is present at milligram-per-liter concentrations, these advanced oxidation processes tend to be less effective in waters that contain high concentrations of NOM.

Chlorine dioxide is also a highly selective chemical oxidant. While little work has been done with pesticides and chlorinated solvents, Hoigne and Bader (1994) present a summary of reaction rate constants for a variety of organic and inorganic species. Their report shows that olefins, aromatic hydrocarbons, primary and secondary amines, aldehydes, ketones, and carbohydrates are not very reactive with chlorine dioxide under water treatment conditions, whereas tertiary amines, thiols, and deprotonated phenols had high reaction rate constants. Again, in waters with high concentrations of NOM, the oxidant demand of the NOM must be overcome before appreciable elimination of the target micropollutant can be achieved.

Oxidation as an Aid to Coagulation and Flocculation. The application of oxidants to raw waters drawn from surface supplies has been observed to assist in subsequent coagulation and flocculation processes. While this has been reported to a limited degree for chlorine and chlorine dioxide, the use of ozone for this purpose has been widely touted (Langlais, Reckhow, and Brink, 1991). The phenomenon appears to be associated with the fact that most particles in raw drinking water are negatively charged due to the adsorption of natural organic material to the particle surface, and that natural organic material in bulk solution exerts an appreciable coagulant demand. Accordingly, mechanisms cited for the coagulation and flocculation benefits of oxidants include oxidation of adsorbed organics to more polar forms which desorb from particles, thereby making the particles less stable and more amenable to aggregation; alteration of the configuration of the adsorbed organics so that they more effectively bind coagulants such as Al(III) and Fe(III) at the particle surface; oxidation of organics in bulk solution to form carboxylic acid functional groups that bind metals such as calcium, resulting in direct precipitation of the organic material; oxidation of organics, causing them to release bound metals such as Fe(III), which can then assist in particle coagulation by conventional mechanisms (Reckhow, Singer, and Trussell, 1986). While the actual mechanisms are not entirely clear, there are a number of waters where these phenomena have been observed, although there are many other waters where no such benefits have been found.

Control of Biological Growth in Treatment Plants. One of the many benefits of chlorine is that it controls nuisance aquatic growths in water treatment plants. This includes growths in flocculation and sedimentation basins, and on filter media. Maintenance of a free chlorine residual throughout the treatment plant minimizes the potential growth of microorganisms, especially in open basins exposed to sunlight, which encourages algal growth. Ozone, chlorine dioxide, and permanganate also exhibit biocidal attributes, but, because of the doses usually employed and the relative chemical instability of these oxidants, it is not common to maintain a residual with these oxidants (disinfectants).

In fact, because ozone and chlorine dioxide (and probably permanganate) oxidize organics to produce low-molecular-weight biodegradable organic compounds, it can be anticipated that oxidative treatment will increase the potential for microbial growth in pretreatment basins and on filter media in the absence of a disinfectant residual. Accordingly, such growths must be controlled in order to minimize subsequent biofouling problems. This can include the use of free or combined chlorine following the application of the oxidant, periodic "shock" treatment of the basins or filters with free chlorine, or engineering and operating the filters in a biological mode.

Summary. It is clear that chemical oxidants provide a variety of tangible benefits in drinking water treatment practice. Some are easy to define and quantify. Others are more diffuse and tend to be more difficult to delineate and assess. Ultimately, however, chemical oxidation is a multiobjective process that serves many useful purposes and helps to improve the overall quality of drinking water. The specific benefits of the process and of each candidate chemical oxidant need to be evaluated on a case-by-case basis.

FORMATION OF DISINFECTION (AND OXIDATION) BY-PRODUCTS

General Considerations

Disinfection (and oxidation) by-products are chemical compounds produced as an unintended result of drinking water disinfection and oxidation. The compounds of greatest concern contain chlorine and bromine atoms. Some of these compounds have been found to be carcinogenic or to cause adverse reproductive or developmental effects in animal studies (see Chapter 2). Others have been shown to be mutagenic and hepatotoxic. As a result, the U.S. Environmental Protection Agency promulgated rules in 1979 and 1998 regulating DBP concentrations in finished drinking waters (see Chapter 1).

Water treatment oxidants/disinfectants derive their effectiveness from their general chemical reactivity. The same attributes which give disinfectants the ability to react with cell membranes, nucleic materials, and cellular proteins also lead to reactions with abiotic dissolved organic matter. In fact, most of the oxidant/disinfectant demand can be attributed to reactions with such abiotic molecules in water.

Most organic matter in surface and groundwater is of natural origin. Some of this natural organic matter is highly reactive with a wide range of oxidants. The reaction products include reduced forms of the oxidants (e.g., chloride, hydroxide, chlorite) and oxidized forms of the organic or inorganic reactants (e.g., the organic disinfection by-products; see Figure 12.4).

The sites of disinfectant attack on NOM are often carbon-carbon double bonds. The organic by-products formed are more highly oxidized, often containing more

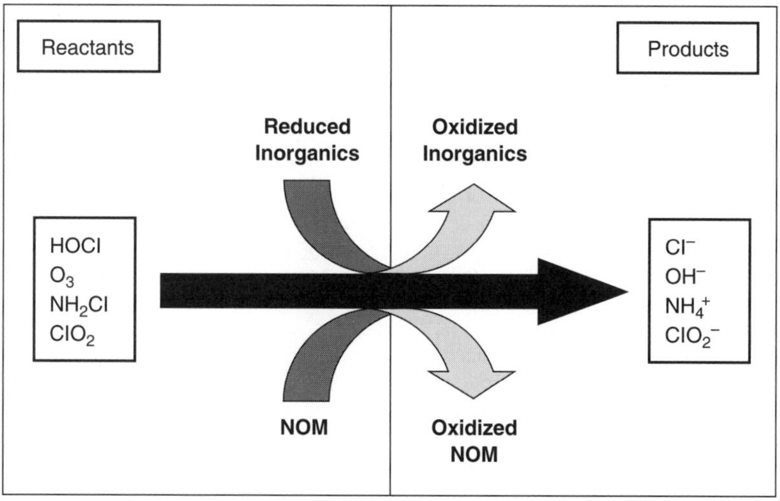

FIGURE 12.4 Schematic illustration of reactions of various oxidants with natural organic material and reduced inorganic substances.

oxygen atoms. As the extent of reaction increases, the organic matter becomes more fragmented, and the specific by-products are simpler in structure. General oxidation by-products include the C1-C3 acids, diacids, aldehydes, and ketones (e.g., Griffini and Iozzelli, 1996). Specific examples include oxalic acid, acetic acid, and formic acid.

A few of the disinfectants are capable of producing by-products that have halogen atoms (i.e., chlorine, bromine, and iodine) incorporated into their structure. Aqueous chlorine and bromine will do this to the greatest extent, followed by chloramines and ozone. In the latter case, high concentrations of bromide are required for substantial halogen incorporation. The organic halide by-products can be measured as a group by the total organic halide (TOX, or more accurately, dissolved organic halide, DOX) analysis. Because NOM contains very low levels of TOX, this analysis presents an opportunity to easily measure a large and diverse group of compounds that are indisputably DBPs. It also targets a subset of the total DBPs that are viewed as the compounds of greatest concern.

Aqueous chlorine and ozone are capable of oxidizing naturally occurring bromide to form active bromine (see above). The latter will react with NOM to form brominated organic compounds (e.g., bromoform) and, in the presence of free chlorine, mixed bromo-chloro-organics. These halogenated by-products can all be measured as TOX.

Identity of Disinfection By-products

Since the discovery of THMs in the early 1970s, hundreds of specific compounds have been identified as drinking water disinfection by-products. Many of these are summarized in Table 12.5. More detailed listings of individual compounds can be found in the review by Richardson (1998).

Each of the four disinfectants presented in this table has its own unique chemistry. For example, ozone seems to be the only disinfectant that produces measurable quantities of bromate. Nevertheless, many by-product classes and specific compounds are common to two or more of the major disinfectants. This is illustrated by the simple aliphatic carboxylic acids (e.g., acetic acid), which are universal by-products, regardless of the disinfectant/oxidant. Itoh and Matsuoka (1996) found that all oxidants produce carbonyls (e.g., formaldehyde), with ozone and chlorine dioxide producing the most, and chlorine and chloramines only slightly behind. Some halogenated compounds, such as dihaloacetic acids, may be produced by all four disinfectants, but the amounts produced range over several orders of magnitude, depending on the disinfectant, the disinfectant dose, and the bromide level. For this reason, an attempt has been made in Table 12.5 to classify by-product abundance based on an order-of-magnitude scale (very high > 100 μg/L, high = 10–100 μg/L, medium = 1–10 μg/L, low < 1 μg/L) as assessed for an average water under typical treatment conditions.

Many worthwhile studies have been conducted with solutions of isolated NOM (e.g., aquatic fulvic acids). Since these studies are often conducted under extreme conditions designed to maximize DBP formation, no attempt has been made to render a judgment on the likely concentration level based on such studies. Also useful, but further removed from practice, are the model compound studies (designated "Models" in Table 12.5). Entries that are not labeled with "NOM" or "Model" refer to expected occurrence levels in typical drinking water.

Chlorination By-products. The chlorination by-products include a wide range of halogenated and nonhalogenated organic compounds. Regulatory agencies have

Table 12.5 Chemical By-products of the Four Major Disinfectants

By-product class	Examples	Chlorine	Chloramines	Chlorine dioxide	Ozone
	Compounds with O-X bonds				
Oxychlorines	Chlorate, chlorite			Very high[34,35]	
Oxybromines	Bromate, hypobromate				Medium[50–52,56]
	Compounds with C-X bonds				
Trihalomethanes	Chloroform, bromodichloromethane, chlorodibromomethane	High			
	Bromoform	Medium			Medium[48,51]
Other haloalkanes	1,2-dibromoethane, 1,2-dibromopropane	Low[14,15]			
Halohydrins	3-bromo-2-methyl-2-butanol	(NOM[20])	(Models[3,23])	(Models[37])	Medium[66]
	9-chloro-10-hydroxyl methyl stearate				
Haloacids (saturated)	Dichloroacetic acid, trichloroacetic acid	High	Medium[17,18]	(NOM[29])	
	Monochloroacetic acid, bromochloroacetic acid, bromodichloroacetic acid	Medium			
	Monobromoacetic acid, dibromoacetic acid, tribromoacetic acid	Low			Medium[49]
	6,6,-dichlorohexanoic acid		(NOM[17,18])		
Haloacids (unsaturated)	3,3-dichloropropenoic acid	(NOM[15])	(NOM[17,18])		
Halodiacids	2,2-dichlorobutanedioic acid	(NOM[15,16])	(NOM[17,18])		
	2,3-dichlorobutenedioic acid				
Halohydroxy-acids	3,3,3-trichloro-2-hydroxypropanoic acid	(NOM[15])	(NOM[17,18])		
	2-chloro-4-hydroxybutanoic acid				
	2,3-dichloro-3,3-dihydroxy propanoic acid				
	4-chloro-4-hydroxypentenoic acid				
Haloketones	1,1,1-trichloropropanone	Medium[1,2,3]			
	Chloropropanone		(NOM[17,18])		
	Bromopropanone				Unknown[58]
	1,1,3,3-tetrachloropropranone			Unknown[28]	
	1,1,1-trichloro-2-butanone, pentachloro-3-buten-2-one	(NOM[4])			

Table 12.5 Chemical By-products of the Four Major Disinfectants (*Continued*)

By-product class	Examples	Chlorine	Chloramines	Chlorine dioxide	Ozone
Haloaldehydes	Chloral	Medium			
	Chloroacetaldehyde, dichloroacetaldehyde		(NOM[17,18])		
	Dichloropropanal, 3-chlorobutanal	(NOM[4])	(NOM[17,18])		
	2,3,3-trichloropropenal				
Haloketoacids	2,3-dichloro-4-oxopentenoic acid	(NOM[15])			
	2,5-dichloro-4-bromo-3-oxopentanoic acid				
Halonitriles	Dichloroacetonitrile, trichloroacetonitrile	Medium[1,2,5]			
	Dibromoacetonitrile				
Cyanogen halides	Cyanogen chloride	Low	Medium		
	Cyanogen bromide		Low		
C-chloro amines			(Models[3,21])		
Halophenols		(NOM[1,6])			
Chloroaromatic acids	5-chloro-2-methoxybenzoic acid	(NOM[7,8])			
	Dichloromethoxybenzoic acid				
Halothiophenes	Tetrachlorothiophene	(NOM[1])			
Chlorinated PAHs		Unknown[9]			
MX and related compounds	MX, EMX, red-MX, ox-EMX	Low[11–13]	(NOM[17,18,27])		
	2,2,4-trichlorocyclopentene-1,3-dione	Low[10]			
Halonitromethanes	Chloropicrin	Medium			
	Bromopicrin	Low			Medium
	Compounds with N-X bonds				
N-chloro-amino acids	N-chloroglycine	(Models)			
N-chloro-amines		(Models[26])	(Models[22,24,25])		

Compounds without halogens					
Aliphatic	Formic acid, acetic acid, butyric acid, pentanoic acid	High		High[28,29,38,40]	High[43,56]
Monoacids	Hexadecanoic acid				Low[58]
Aliphatic diacids (saturated)	Oxalic acid	High		High[32,40]	Very high[56]
	Succinic acid, glutaric acid, adipic acid			(NOM[29])	Unknown[46]
Aliphatic diacids (unsaturated)	Butenedioic acid	(NOM[15,16])			
	2-tert-butylmaleic acid, 2-ethy-3-methylmaleic acid			Unknown[28,32]	
Aromatic acids	Benzoic acid, 3,5-dimethylbenzoic acid	(NOM[15,16])		Unknown[28]	Unknown[44–46]
Other aromatics	p-benzoquinone			(Models[33])	
	Hydroxy-PAHs			(Models[36])	
	3-ethyl styrene, 4-ethyl styrene			Unknown[28]	
	Naphthalene, 1-methylnaphthalene			Unknown[28]	
Aldehydes	Formaldehyde, acetaldehyde, propanal		(Models[1])	Unknown[30]	High
	Glyoxal, methylglyoxal				High
	Benzaldehyde, ethylbenzaldehyde				Unknown[46,56]
Ketones	Acetone, propyl ethyl ketone				Medium[41,53]
	Dioxopentane, 1,2-dioxobutane				Unknown[41]
	Acetophenone, 4-phenyl-2-butanone				Unknown[46]
	2-hexenal, 6-methyl-5-hepten-2-one				Low[58]
	2,3,4-trimethylcyclopent-2-en-1-one			Unknown[28]	
	2,6,6,-trimethyl-2-cyclohexene-1,4-dione				
Ketoacids	Pyruvic acid, glyoxalic acid, ketomalonic acid				High
	Oxobutanoic acid, 4-oxo-2-butenoic acid				Unknown[41,46]
	Ketosuccinic acid, ketoglutaric acid				Unknown[41,59]
	Dioxopropanoic acid, dioxopentanoic acid				Unknown[41]
Hydroxy acids	Hydroxymalonic acid				(NOM[44])
Hydroxy carbonyls	Hydroxyacetaldehyde				(Models[57])
Furans	Methylfurancarboxylic acid			(NOM[29])	

Table 12.5 Chemical By-products of the Four Major Disinfectants (*Continued*)

By-product class	Examples	Chlorine	Chloramines	Chlorine dioxide	Ozone
Epoxides			(Models[23])	(Models[31,39])	(Models[42,54])
Organic peroxides					(Models)
Nitriles			(Models[1,19])		
Misc.	5-methoxy-α-prone				(NOM[44])

Notes: Data marked "(NOM)" and "(Models)" are from studies using solutions of natural organic matter extracts and model compounds, respectively. All other data are from treated drinking waters and unaltered natural waters. Concentrations are classified as follows: Very high (>100 μg/L), High (10–100 μg/L), Medium (1–10 μg/L), Low (<1 μg/L), Unknown (not quantified).

Sources:

1. Le Cloirec and Martin, 1985
2. Brass et al., 1977
3. Minisci and Galli, 1965
4. Smeds, Holmbom, and Tikkanen, 1990
5. Kanniganti, 1990
6. Shank and Whittaker, 1988
7. Backlund, Kronberg, and Tikkanen, 1988
8. Kronberg and Vartiainen, 1988
9. Burttschell et al., 1959
10. Fielding and Horth, 1986
11. Franzén and Kronberg, 1994
12. Peters, de Leer, and Versteegh, 1994
13. Franzén and Kronberg, 1994
14. de Leer et al., 1985
15. Oliver, 1983
16. Kanniganti et al., 1992
17. Kanniganti, 1990
18. Kanniganti et al., 1992
19. Hausler and Hausler, 1930
20. Havlicek et al., 1979
21. Neale, 1964
22. Scully, 1986
23. Carlson and Caple, 1977
24. Jensen and Johnson, 1989
25. Crochet and Kovacic, 1973
26. Kringstad et al., 1985
27. Backlund, Kronberg, and Tikkanen, 1988
28. Richardson, Thurston, and Collette, 1994
29. Colclough, 1981
30. Stevens et al., 1978
31. Legube et al., 1981
32. Masschelein, 1979
33. Wajon, Rosenblatt, and Burrows, 1982
34. Steinbergs, 1986
35. Werdehoff and Singer, 1987
36. Luikkonen et al., 1983
37. Ghanbari, Wheeler, and Kirk, 1983
38. Somsen, 1960
39. Carlson and Caple, 1977
40. Griffini and Iozzelli, 1996
41. Le Lacheur et al., 1993
42. Carlson and Caple, 1977
43. Lawrence, 1977
44. Benga, 1980
45. Paramisigamani et al., 1983
46. Glaze, 1986
47. Edwards, 1990
48. Cooper et al., 1986
49. Daniel et al., 1989
50. Haag and Hoigne, 1983
51. Glaze, Weinberg, and Cavanagh, 1993
52. Krasner et al., 1993
53. Fawell and Watts, 1984
54. Chen, Junk, and Svec, 1979
55. Lawrence et al., 1980
56. Griffini and Iozzelli, 1996
57. Le Lacheur and Glaze, 1996
58. Richardson et al., 1996
59. Hwang et al., 1997
60. Cavanagh et al., 1992

focused on the halogenated compounds, especially the trihalomethanes and haloacetic acids (e.g., see Singer [1994] for historical background on U.S. regulatory actions). These are small, highly substituted end products of the reaction of chlorine with organic matter. In waters with low bromide levels, the fully chlorine-substituted forms will predominate (e.g., chloroform, di- and trichloroacetic acid). Waters with moderate to high levels of bromide will form elevated levels of the bromine-containing analogs (e.g., bromodichloromethane, bromodichloroacetic acid).

Nearly all DBP studies in the 1970s focused on the trihalomethanes. Due to their volatility, chemical stability, and high halogen/carbon ratio, these compounds could be easily analyzed with a minimum of analytical equipment and expertise. Consequently, these were the first by-products to be found in finished drinking waters (Rook, 1974; Bellar et al., 1974), the first to be the subject of an established analytical method, and the first to be included in a large survey of public water supplies (Symons et al., 1975). In a matter of just a few years, it was recognized that the THMs were always present whenever chlorine was used as a disinfectant.

The discovery of haloacetic acids (HAAs) in chlorinated waters (Miller and Uden, 1983; Christman et al., 1983) and subsequent occurrence studies trailed the THMs by several years. One early survey (Krasner et al., 1989) showed the HAAs, like the THMs, to be ubiquitous in chlorinated waters, although present at slightly lower levels. More recent data have supported this; however, the lack of available standards for all of the HAAs in these earlier studies may have resulted in underestimation of their concentrations (Singer, 1994). Figure 12.5 presents a summary of some DBP yields presented on a per-mg-carbon basis. This is probably a valid basis of comparison, because chlorine is generally added in excess when used as a secondary disinfectant (i.e., it is added so that the dose exceeds the long-term demand). Thus the carbonaceous precursors are the primary limiting reactants (see discussion below). This figure presents two contrasting situations. The fulvic acid data show yields expected under the extreme conditions of a formation potential (FP) test

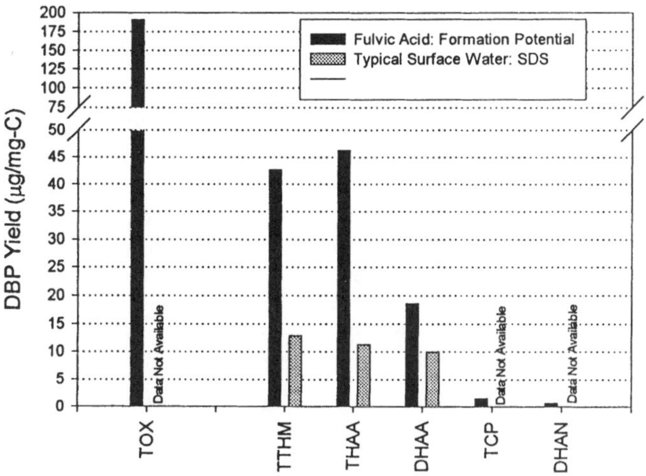

FIGURE 12.5 Specific halogenated DBP formation based on literature values (TOX: total organic halides, THAA: trihaloacetic acids, DHAA: dihaloacetic acids, TCP: trichloropropanone, DHAN: dihaloacetonitile). (Fulvic acid: from Reckhow et al., 1990; surface water: calculated from MW model [Montgomery Watson, 1993].)

where the water has organic material that is entirely humic in character. The second set of results represents a surface water with less humic content and chlorinated under conditions more typical of a drinking water system (simulated distribution system [SDS] test). Most specific DBP yields should fall within 50 percent of the SDS values in Figure 12.5, although some systems may approach the FP values.

Many other halogenated by-products have been widely reported in chlorinated drinking waters. The most intensively studied of these nonregulated compounds are the halopropanones, the haloacetonitriles, chloropicrin, and chloral hydrate. This group owes its large industry-wide database to the analytical method which it shares with the THMs (i.e., like the THMs, they are all volatile neutral compounds that respond well to analysis by gas chromatography with electron capture detection). The haloacetonitriles are thought to be products of the chlorination of amino acids and proteinaceous material (Shiraishi et al., 1985). Nitrogenous structures in humic substances will also form haloacetonitriles via cyano acid intermediates (Backlund, Kronberg, and Tikkanen, 1988). The halopropanones (e.g., 1,1,1-trichloropropanone) and chloral hydrate (a haloaldehyde) are halogenated analogs of some common ozonation by-products. They are commonly found at elevated concentrations in waters that had been previously ozonated (Reckhow, Singer, and Trussell, 1986; McKnight and Reckhow, 1992).

About 50 percent of the TOX produced upon chlorination can be attributed to the major by-products already discussed. The remainder is largely unknown. A similar mass balance on mutagenic activity reveals that more than 50 percent can be accounted for among the known by-products. Much of the identified mutagenicity is found in a single compound, MX (Backlund, Kronberg, and Tikkanen, 1988; Meier, 1988). This is a chlorinated furanone that is produced in small quantities by free chlorine and chloramines. Relatively little is known about its occurrence, because it is so costly to measure.

Because the TOX represents only a small fraction of the total chlorine demand (typically less than 10 percent), it can be concluded that most of the organic chlorination by-products do not contain chlorine. A number of these nonhalogenated by-products have been identified (Table 12.5). Most are aliphatic mono- and di-acids, and benzenepolycarboxylic acids. Because these compounds are probably not of health concern, they have not been extensively studied. It should be noted that many of these compounds are readily biodegradable, and therefore contribute to a water's biodegradable organic carbon (BDOC) content, sometimes measured as assimilable organic carbon (AOC).

Chloramine By-products. Monochloramine is less reactive than free chlorine with NOM and most model compounds. Identifiable by-products include dichloroacetic acid (DCAA), cyanogen chloride, and very small amounts of chloroform and trichloroacetic acid (TCAA; see Table 12.5). While it's not clear that TCAA and the THMs are true by-products of chloramines (i.e., they may be formed due to the presence of a small free chlorine residual), it does seem that DCAA and cyanogen chloride are true by-products.

Backlund and coworkers (Backlund, Kronberg, and Tikkanen, 1988) found that monochloramine also formed MX from humic materials, although the amount measured was less than 25 percent of that formed during chlorination. Shank and Whittaker (1988) have proposed that small amounts of the known carcinogen, hydrazine ($H_2N\text{-}NH_2$), may be formed when drinking waters are chloraminated. However, they have not been able to detect this compound in drinking water distribution systems.

Jensen et al. (1985) found that at high monochloramine/carbon ratios (~10 mg/mg), about 5 percent of the oxidant demand went toward TOX formation. This is identical to the results for chlorine at an equally high chlorine/carbon ratio. Most of the chlorinated organic by-products remained tied up in high-molecular-weight compounds. This contrasts with free chlorine, as monochloramine is a much weaker oxidant and is much less likely to create the smaller fragments that can be analyzed by gas chromatography.

Monochloramine is known to transfer active chlorine to the nitrogen of amines and amino acids forming organic chloramines (Scully, 1986). This reaction also occurs with free chlorine. Model compound studies have shown that monochloramine can also add chlorine to activated aliphatic carbon-carbon double bonds (Johnson and Jensen, 1986). The adjacent carbon may become substituted with an amine group or with oxygen. Other types of reactions involve the simple addition of amine or chloramine to unsaturated organic molecules. Under certain conditions chlorine substitution onto activated aromatic rings has been observed. For example, monochloramine will slowly form chlorophenols from phenol (Burttschell et al., 1959). Preformed chloramines will also add chlorine to phloroacetophenone, a very highly activated aromatic compound, to produce chloroform (Topudurti and Haas, 1991). However, the rate of reaction for this compound is low and the molar yield (i.e., moles DBP formed per mole precursor) is only 3 percent as compared to 400 percent for free chlorine.

Chlorine Dioxide By-products. Chlorine dioxide undergoes a wide variety of oxidation reactions with organic matter to form oxidized organics and chlorite (see Table 12.5 and Eq. 12.50). The concentrations of chlorite account for 50 to 70 percent of the chlorine dioxide consumed (Rav Acha et al., 1984; Werdehoff and Singer, 1987). Chlorite may also be formed, along with chlorate (ClO_3^-), by the disproportionation of chlorine dioxide (see Eq. 12.49). All three of the oxidized chlorine species (chlorine dioxide, chlorite, and chlorate) are considered to have adverse health effects, and their presence in finished water is a source of concern.

Chlorine dioxide can also undergo a limited number of chlorine substitution reactions. Reactions with specific model compounds demonstrate the formation of chlorinated aromatic compounds and chlorinated aliphatics. Trihalomethanes, however, have not been detected as reaction products. As with the other chlorine-containing oxidants, chlorine addition/substitution products are favored at low oxidant-to-carbon ratios and oxidation reactions are favored at high ratios. Studies using drinking waters and NOM have shown that small amounts of TOX form upon treatment with typical levels of chlorine dioxide. This may be due to the formation of HOCl when chlorine dioxide reacts with NOM, and subsequent reaction with other organics (Werdehoff and Singer, 1987). The relatively small amount of HOCl formed in this manner probably leads to sparsely halogenated macromolecular TOX, which would account for the lack of identifiable organo-halide by-products.

Ozonation By-products. Ozone can lead to the formation of brominated by-products when applied to waters with moderate to high bromide levels. This is a direct result of ozone's ability to oxidize bromide to hypobromous acid and related species (see Eq. 12.66). Some of this oxidized bromine will continue to react to form bromate ion. Much of the remaining hypobromous acid will react with NOM, forming brominated organic compounds. These by-products encompass the same general classes reported for the halogenated by-products of chlorine—that is, THMs (bromoform), HAAs (dibromoacetic acid), HANs (haloacetonitriles, such as dibromoacetonitrile). One recent study showed that 7 percent of the raw water bromide becomes incorporated as TOX (or total organic bromide, TOBr) following ozonation under conditions typical of drinking water treatment (Song, 1997).

Most ozonation by-products are not halogenated. The majority of ozone by-products are similar to the general oxidation products reported for other disinfectants. For example, parallel field studies with ozone and chlorine dioxide found both to produce about the same level of low-molecular-weight-carboxylic acids (Griffini and Iozzelli, 1996). Nevertheless, a number of studies suggest that ozone produces higher levels of simple aldehydes and keto-acids (or aldo-acids) than the other major disinfectants. Figure 12.6 presents some dose-specific yield data for these major by-products. Other oxidation by-products attributed to ozone include the hydroxy-acids, aromatic acids, and hydroxyaromatics (see Table 12.5). Organic peroxides and epoxides are also expected ozonation by-products, although their detection in treated drinking waters has proved to be a challenge.

Because of the strong link between ozonation by-products and biodegradable organic matter (BOM) in water, attempts have been made to attribute this BOM to specific compounds. Krasner et al. (1996) have shown that as much as 40 percent of the assimilable organic carbon and 20 percent of the biodegradable dissolved organic carbon can be assigned to the known major ozonation by-products. However, this total still represents less than 5 percent of the dissolved organic carbon (DOC) compound.

Factors Influencing By-product Formation

Time. Reaction time is among the most important factors determining DBP concentrations under conditions where a disinfectant residual persists. The major halogenated DBPs (e.g., THMs and HAAs) are chemically stable. They accumulate in disinfected waters and their concentrations will increase with reaction time for as

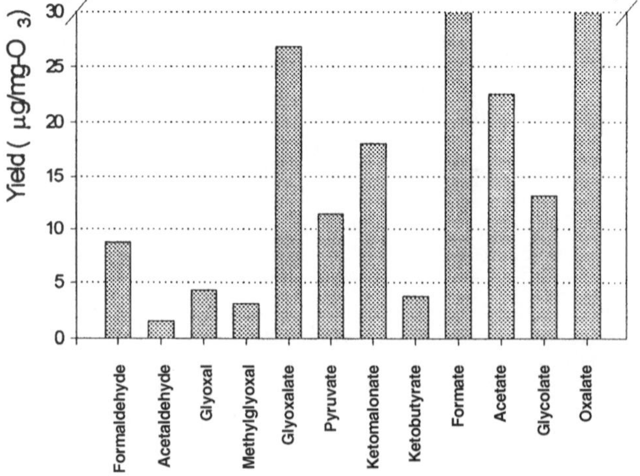

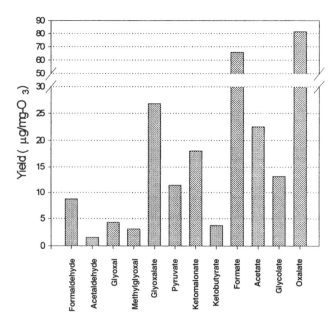

FIGURE 12.6 Yields of major ozonation by-products normalized per mg of ozone applied, as reported in the literature (after Reckhow, 1999).

long as a disinfectant residual exists (Figure 12.7). However, there are cases where HAA concentrations drop to near zero after long residence times in real drinking water distribution systems. This phenomenon is generally attributed to biodegradation, and it does not appear to occur with the THMs (the latter appears to be biodegradable only under anaerobic conditions).

Laboratory tests have shown that HAAs form more rapidly than THMs. Studies have also shown the brominated analogs to form more rapidly than the purely chlorinated compounds. This causes the HAA/THM or the TTHM/chloroform ratio to be high in the early stages of the reaction and to slowly drop with reaction time. These observations are supported by data collected from full-scale treatment plants and distribution systems.

In contrast to the case for chlorine, ozonation by-products form quickly and show little long-term increase. This is due to the rapid dissipation of ozone residuals. Once dissolved ozone is gone, by-product formation can only continue by means of hydrolysis reactions, which represent a relatively minor contribution to the total by-product concentration.

Many halogenated DBPs are chemically unstable and are subject to hydrolysis or further oxidation. For these compounds, there is a reaction time during which the concentration reaches a maximum (see subsequent discussion on pH effects). Some DBPs, like dichloroacetonitrile, decompose slowly and reach a maximum concentration only after reaction times on the order of days. Others are more reactive and decay to undetectable levels within minutes to hours (e.g., 1,1-dichloropropanone).

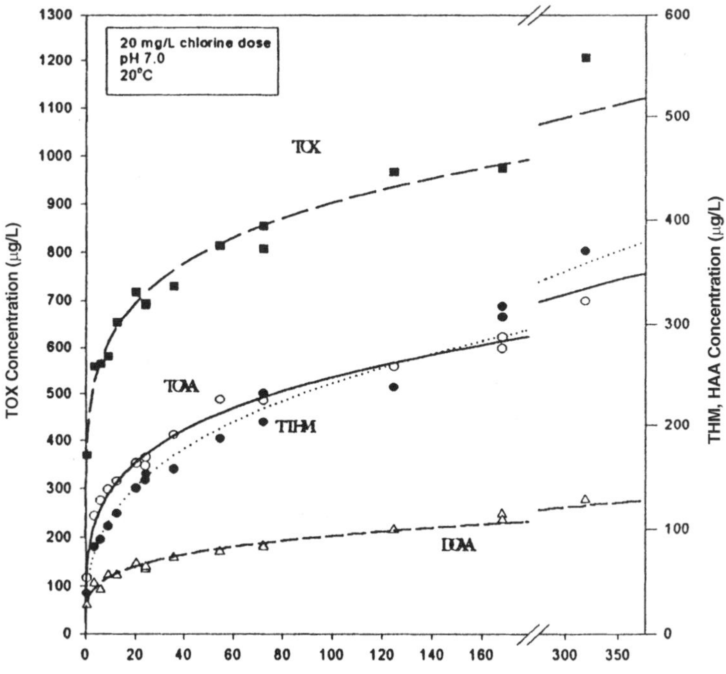

FIGURE 12.7 Effect of reaction time on the major chlorination by-products (from Reckhow and Singer, 1984).

Disinfectant Dose. Disinfectant dose has a variable impact on DBP formation. Small changes in the dose used for residual disinfection often have minor effects on DBP formation. This is because these systems have an excess of disinfectant, and they are therefore precursor-limited. When the residual drops below about 0.3 mg/L, DBP formation becomes disinfectant-limited. Under these circumstances, changes in disinfectant dose will have a large effect. Figure 12.8 shows that when 3 and 5 mg/L of chlorine are added to Connecticut River water, it produces THMs in direct proportion to those doses. However, when an excess is added (>5 mg/L), a residual persists and the THM formation levels off. The other two waters in this figure show only the latter behavior, because they had lower TOC values and their chlorine demand was less than the minimum dose tested (3 mg/L).

As a general rule, disinfectant dose plays a greater role in DBP formation during primary disinfection than during secondary disinfection. This is because primary disinfectants are usually added in amounts well below the long-term demand. Therefore, the disinfectant is the limiting reactant, not the organic precursors. Figure 12.8 shows a near linear relationship between ozone dose and glyoxalic acid formation. When the ozone is applied after coagulation and filtration, the start of a plateau appears. This is a reflection of the removal of ozone-demanding organics and a closer approach to precursor-limiting conditions.

The relationship between disinfectant dose and DBP formation can be illustrated with a simple kinetic model. Figure 12.9 shows the results of a kinetic simulation for the simple second-order reaction as follows:

$$A + B \xrightarrow{k} C$$

where A represents the NOM precursor material, B is the disinfectant, C is the particular DBP, and k is the second-order rate constant. If the initial precursor level (A_o) is held constant, and a rate constant and reaction time are arbitrarily chosen,

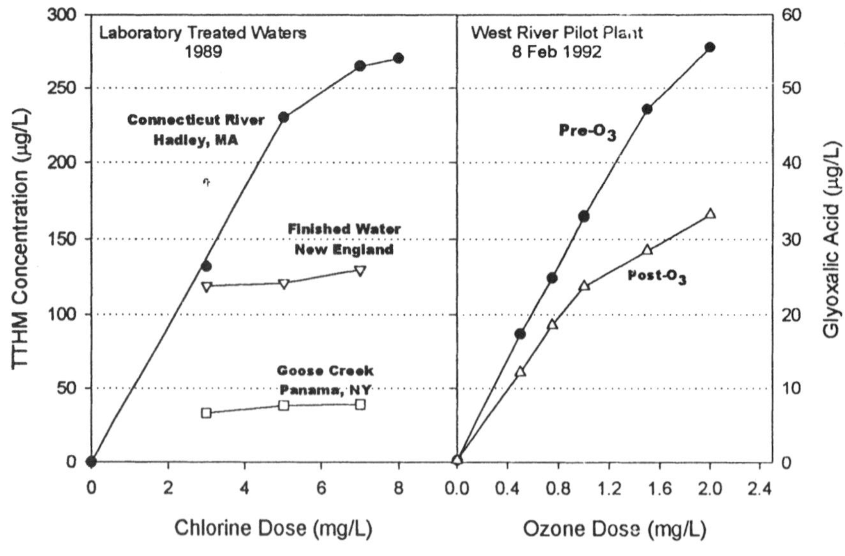

FIGURE 12.8 Observations on effect of disinfectant dose on DBP formation. Chlorination conditions: pH 7, 20°C, 3-day reaction time. (Figures redrawn from Coombs, 1990; Reckhow et al., 1993.)

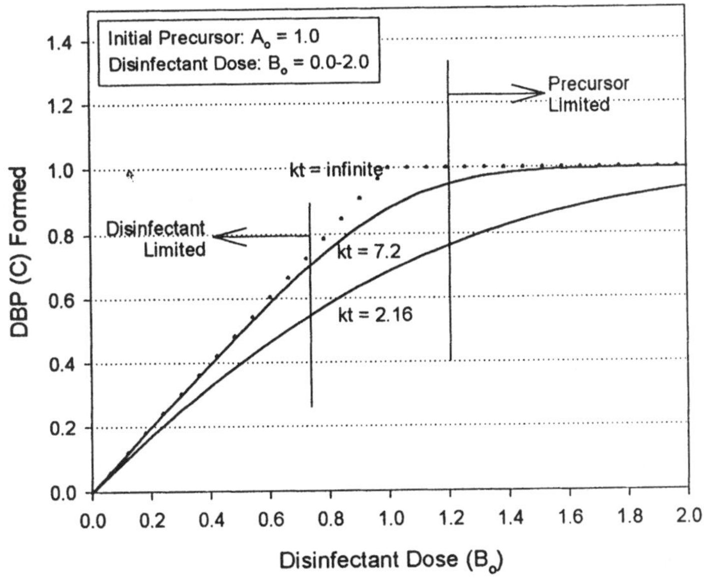

FIGURE 12.9 Theoretical second-order kinetic plot showing effect of reactant dose on product formation.

the DBP formation can be calculated as a function of the disinfectant dose. The three curves in Figure 12.9 are the result of this calculation for three different combinations of *k* and reaction time *t*. As the product of *k* and *t* gets large, the DBP formation curve approaches two straight lines (one where DBP formation increases linearly with dose, and one where DBP formation is independent of dose). The upward sloping line corresponds to the disinfectant-limiting case. The horizontal line corresponds to the precursor-limiting case. This model is consistent with the experimental observations in Figure 12.8.

When using chloramines, the chlorine-to-nitrogen ratio must also be a consideration. As the Cl_2/N ratio increases from 2 to 5 (weight basis), the water's exposure to a transient free chlorine residual increases substantially. This results in a shift in the DBP character (yields and types) from that typical of pure chloramines to that of free chlorine. Model compound studies (DeLaat et al., 1982) have shown that highly reactive precursors (e.g., resorcinol) will experience this shift at lower Cl_2/N ratios than the more slowly reacting compounds (e.g., acetone). It's likely that the same could be said for waters with highly reactive NOM versus less reactive waters.

pH. The overall reaction between chlorine and NOM is relatively insensitive to pH. However, the concentration of TOX and specific halogenated by-products is strongly influenced by pH (e.g., Fleischacker and Randtke, 1983; Reckhow and Singer, 1984). Nearly all by-products (THAA, TCP, DHAN, etc.) decrease in concentration with increasing pH. One important exception is the THMs. Although pH can influence chlorination reactions in many ways, it is probably base-catalyzed hydrolysis mechanisms which have the greatest overall effect. Many DBPs (e.g., 1,1,1-trichloropropanone) are decomposed by base-catalyzed hydrolysis. These compounds are less prevalent in waters with high pH, and they tend to drop in concentration at long residence times (Type 1, Table 12.6). The THMs increase at high pH because many hydrolysis reactions actually lead to THM formation (Type 2). Other by-products are themselves unaffected by base hydrolysis (e.g., the HAAs), but their formation pathways may be altered at high pH, resulting in less compound produced at high pH (Type 3).

None of the major ozonation by-products are subject to base hydrolysis. Instead, pH plays a role by altering the rate of decomposition of ozone to hydroxyl radicals (see above). As pH increases, ozone decomposition accelerates, and this is thought to be responsible for a decrease in the classical by-products of ozonation (e.g., aldehydes; Schechter and Singer 1995). However, the chemistry of ozone and hydroxyl radical reactions with NOM are not well understood, and there may be cases where some carbonyl by-products increase with pH (e.g., Itoh and Matsuoka, 1996).

Bromate is formed in ozonated waters from a series of reactions between ozone or hydroxyl radicals and naturally occurring bromide (see above). A key intermediate in this mechanism is hypobromite ion. At lower pHs, more of the hypobromite becomes protonated, forming the less reactive hypobromous acid. This causes the overall formation of bromate to decrease as pH is decreased. On the other hand, lower pHs favor the formation of TOX (presumably TOBr) during ozonation (Song, 1997). This is probably due to suppressed decomposition of ozone, changes in the speciation of the oxidized bromine, and base-catalyzed hydrolysis of brominated by-products.

Temperature. The rate of formation of DBPs generally increases with increasing temperature. Laboratory and full-scale data suggest that chloroform formation is more sensitive to temperature than DCAA formation. The relationship is not as clear for TCAA formation. At high temperatures, accelerated depletion of residual chlorine slows DBP formation, even though the rate constants for DBP-producing reactions might increase. This may be especially true for TCAA, as its formation is more sensitive to chlorine residual than is chloroform or DCAA formation.

High temperatures may also accelerate DBP degradation processes. Biodegradation of HAAs would be expected to proceed more quickly at high temperature. Reactive DBPs would undergo abiotic reactions with bases or active chlorine at a faster rate as temperature increases. For this reason, increases in temperature may actually cause a decrease in the concentration of certain DBPs.

Precursor Material. It is well established that halogenated DBP formation is, by a first approximation, proportional to the TOC concentration at the point of chlorina-

Table 12.6 Impact of Base-Catalyzed Hydrolysis on DBP Concentration Profiles

Type	Characteristics	pH of max. concentration	Drop at high residence times*	Examples
1	DBP susceptible to base hydrolysis	Low	Yes	Haloketones, chloral hydrate
2	DBP formed as a result of base hydrolysis	High	No	THMs
3	DBP not susceptible to hydrolysis, but precursors are	Low	No	THAAs

* Not considering possible biodegradation.

tion. This is why, in waters with high DBP precursor concentrations, shifting the point of chlorination from before coagulation/settling to after yields substantial reductions in DBP formation. Furthermore, a 40 percent reduction in TOC across conventional treatment often translates to a 50 percent or more reduction in some DBPs when this switch is made. This is partly attributed to preferential removal of precursor organics by coagulation and the accompanying shift in the nature of the NOM at the point of chlorination.

Chlorination studies with NOM extracts and whole waters have suggested that the activated aromatic content is an important determinant in its tendency to form major chlorination by-products (Reckhow et al., 1993). This probably explains why UV absorbance is such a good predictor of a water's tendency to form THMs and HAAs, as UV absorbance by NOM is generally attributed to activated aromatic chromophores.

Bromide. Bromide is readily oxidized by aqueous chlorine and ozone to form hypobromous acid (Eqs. 12.43 and 12.66, respectively). This reacts in concert with other oxidants to form brominated DBPs. As bromide levels increase, formation of the brominated DBPs increases. When chlorine is the disinfectant, families of mixed bromo-chloro DBPs will result, such as the THMs. Figure 12.10 shows how increasing bromide concentration causes shifts in the THM speciation from chloroform to progressively more brominated members of the group. A similar effect has been observed for the HAAs, halopicrins, halopropanones, haloacetaldehydes, and haloacetonitriles (e.g., Trehy and Bieber, 1981; Pourmoghaddas and Dressman, 1993; Xie and Reckhow, 1993; Cowman and Singer, 1996). Most laboratory studies have also shown that the molar formation of THMs increases slightly with bromide concentration. Since the brominated forms are heavier than their chlorinated analogs, the mass-based THM level (e.g., μg/L) increases even more sharply with increasing bromide level.

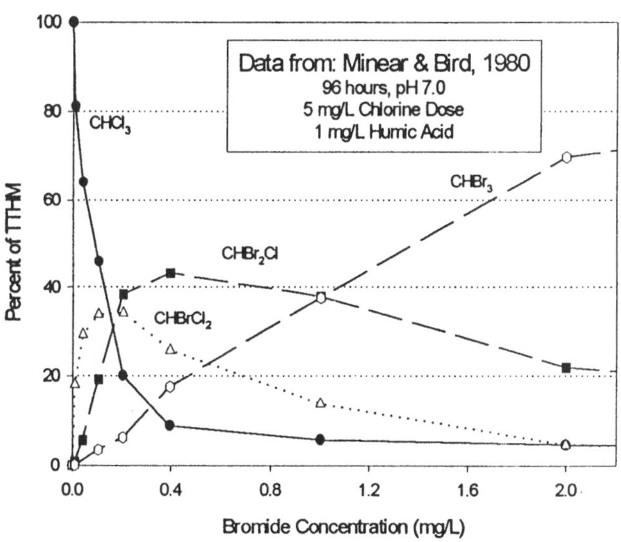

FIGURE 12.10 The effect of bromide concentration on THM speciation (redrawn from Minear and Bird, 1980).

The formation of nonhalogenated DBPs is probably insensitive to varying bromide levels. One study of ozonation by-products showed only very minor effects of bromide on aldehyde levels (Schechter and Singer, 1995).

Seasonal Effects. DBP concentrations are strongly influenced by seasons. The underlying factors are temperature and seasonally induced changes in water quality (NOM characteristics, bromide, pH). In general, DBP concentrations are highest in summer. This is attributed to high temperatures, which accelerate DBP-forming reactions. Furthermore, disinfectant demand reactions are faster, requiring higher disinfectant doses in order to maintain target residuals. Competing with this is the fact that chemical degradation is faster at high temperatures and biodegradation may play a larger role.

With some sources, water quality changes may also be important. For example, the San Francisco Bay delta, which provides raw drinking water to a large portion of the population in central and southern California, experiences significant seasonal variations in bromide concentrations. During the spring season, high freshwater flows from the Sierras keep salt water, with its concomitant high bromide content, out of the delta so that chlorinated DBPs constitute most of the DBP content of the treated drinking water. On the other hand, during the summer and early fall when freshwater flows are low, more saltwater intrusion into the delta occurs and bromine-containing species are the dominant DBPs formed.

Control of Oxidation/Disinfection By-products

There are three general approaches to controlling DBP concentrations: (1) minimize DBP formation by reducing the organic precursor material at the point of disinfection; (2) minimize DBP formation by reducing the disinfectant dose, changing the nature of the disinfectant, or optimizing the conditions of disinfection; and (3) removal of DBPs after their formation. Most efforts have focused on the first approach, as represented by the interest in "enhanced coagulation." However, DBP control strategies should also consider process changes in accord with the second approach. The third approach (DBP removal) is most appropriate for control of biodegradable ozonation by-products.

Minimizing Organic Precursors. The objective of this particular strategy is to minimize the quantity and reactivity of the organic precursors at the point of disinfection. This can be done by either reducing the precursor content of the raw water, improving precursor removal through the plant, shifting the point of disinfection to a later stage of treatment, or some combination of the three. It has been proposed that watershed management practices that help reduce primary productivity in impoundments will also result in reduced THM precursor levels (Karimi and Singer, 1991; Chapra et al., 1997). Reductions in organic precursors and bromide levels may also be achieved through careful source water blending. Once they have entered the plant, DBP precursors may be removed by coagulation, adsorption, oxidation, or membrane separation. Coagulation with aluminum and ferric salts has proven to be a very effective means of removing hydrophobic organic precursors (see Chapter 6). Adsorption with granular activated carbon (GAC) can also be quite effective (see Chapter 9). Other approaches that have been shown to remove TOC are also likely to remove DBP precursors (e.g., membrane processes, Chapter 11; ion exchange, Chapter 13).

Modifying Disinfection. Various elements of the disinfection step can sometimes be changed to help control DBP formation. One obvious approach is to minimize the disinfectant dose to the extent possible within the constraints of good disinfection practice. For example, primary and secondary disinfectant doses may be reduced in some plants, while relying on booster disinfection out in the distribution system to take care of areas susceptible to low residuals. pH adjustment can also be used to favorably affect DBP formation. Base addition for corrosion control might

Table 12.7 General Advantages and Disadvantages of Common Water Treatment Oxidants

Oxidant	Advantages	Disadvantages
Chlorine	Strong oxidant Strong disinfectant Produces persistent residual Inexpensive Easy to use Long history of use	Produces halogenated DBPs May contribute to taste and odor problems
Chloramines	Does not produce THMs Produces persistent residual Controls microbial growth Easy to use Long history of use	Weak oxidant Weak disinfectant Produces unidentified TOX Can lead to nitrification problems in distribution system May contribute to taste and odor problems
Ozone	Very strong oxidant Very strong disinfectant Effective for taste and odor Does not produce halogenated DBPs except in bromide-rich waters May aid in coagulation and flocculation	Does not produce a persistent disinfectant residual Relatively costly Produces bromate in bromide-rich waters Produces biodegradable organic material that must be controlled
Advanced oxidation processes	Very strong oxidant Most effective for taste and odor control Does not produce halogenated DBPs	Limited disinfection properties Relatively costly Produces biodegradable organic material that must be controlled
Chlorine dioxide	Strong oxidant Strong disinfectant Effective for certain types of taste and odor Does not produce halogenated DBPs Does not react with ammonia	Produces chlorite as an inorganic DBP May produce undesirable odors Difficult to maintain a persistent disinfection residual
Permanganate	Easy to feed Effective for Fe and Mn oxidation Does not produce halogenated DBPs Effective for certain types of taste and odor	Produces manganese dioxide which must be removed Can lead to pink water if dosage not carefully controlled Limited disinfection capabilities
Oxygen	Easy to feed Does not produce halogenated by-products	Relatively weak oxidant for most water treatment applications except Fe(II) and sulfide oxidation

be delayed until after disinfection with chlorine to help depress formation of THMs. A widely used approach is to substitute one disinfectant for another, "alternative" disinfectant. Common examples include the use of chloramines as a secondary disinfectant in place of free chlorine, or the use of preozonation as a primary disinfectant in place of prechlorination.

Removing By-products. In general, DBPs are hydrophilic and low in molecular weight, characteristics which make them difficult to remove by most physicochemical processes. For example, the removal of THMs in disinfected water requires either air stripping or GAC adsorption with frequent regeneration. On the other hand, some DBPs are biodegradable and may therefore be removed by biologically active filtration. The most common example of this is the removal of aldehydes, acids, and keto-acids following ozonation. There is also evidence that HAAs can be removed in this way.

SUMMARY

Based upon the information presented in this chapter, Table 12.7 summarizes the relative advantages and disadvantages of each of the common chemical oxidants used in water treatment practice. It must be cautioned that the summary gives generalizations only. These generalizations do not always hold in that the relative effectiveness of each of the oxidants may vary depending upon local source water characteristics, treatment objectives, and operational compatibility.

BIBLIOGRAPHY

Aieta, E. M., K. M. Reagan, J. S. Lang, L. McReynolds, J. W. Kang, and W. H. Glaze. "Advanced oxidation processes for treating groundwater contaminated with TCE and PCE: Pilot-scale evaluations." *Jour. AWWA,* 80(5), 1988: 64.

American Water Works Association. *Water Quality and Treatment,* 4th ed. New York: McGraw-Hill, 1990.

Backlund, P., L. Kronberg, and L. Tikkanen. "Formation of Ames mutagenicity and of the strong bacterial mutagen 3-chloro-4-(dichloromethyl)-5-hydroxy-2-(5H)-furanone and other halogenated compounds during disinfection of humic water." *Chemosphere,* 17(7), 1988: 1329–1336.

Bellar, T. A., J. J. Lichtenberg, and R. C. Kroner. "The occurrence of organohalides in chlorinated drinking water," *Jour. AWWA,* 66, 1974: 703.

Benga, J. "Oxidation of humic acid by ozone or chlorine dioxide." Ph.D. dissertation, Miami University, Ohio, 1980.

Brass, H. J., M. A. Feige, T. Halloran, J. W. Mello, D. Munch, and R. F. Thomas. "The national organic monitoring survey: Samplings and analyses for purgeable organic compounds." In *Drinking Water Quality Enhancement Through Source Protection,* R. B. Pojasek, ed., pp. 393–416. Ann Arbor, MI: Ann Arbor Science Publ., 1977.

Brezonik, P. L. *Chemical Kinetics and Process Dynamics in Aquatic Systems.* Boca Raton: Lewis Publishers, 1994.

Burttschell, R. H., A. A. Rosen, F. M. Middleton, and M. B. Ettinger. "Chlorine derivatives of phenol and phenol causing taste and odor." *Jour. AWWA,* 51(2), 1959: 205–214.

Carlson, R., and R. Caple. "Chemical biological implications of using chlorine and ozone for disinfection," EPA-600/3-77-066. Washington, DC: U.S. Environmental Protection Agency, Ecological Research Series, 1977.

Cavanagh, J. E., H. S. Weinberg, A. Gold, R. Sangaiah, D. Marbury, W. H. Glaze, T. W. Collette, S. D. Richardson, and A. D. Thruston, Jr. "Ozonation byproducts: Identification of bromohydrins from the ozonation of natural waters with enhanced bromide levels." *Environmental Science and Technology,* 26(8), 1992: 1658–1662.

Chapra, S. C., R. P. Canale, and G. L. Amy. "Empirical models for disinfection by-products in lakes and reservoirs." *Journal of Environmental Engineering,* 123(7), 1997: 714–715.

Chen, P. N., G. A. Junk, and H. J. Svec. "Reactions of organic pollutants. I. Ozonation of acenaphthylene and acenaphthene." *Environmental Science and Technology,* 13(4), 1979: 451–454.

Christman, R. F., D. L. Norwood, D. S. Millington, et al. "Identity and yields of major halogenated products of aquatic fulvic acid chlorination." *Environmental Science and Technology,* 17(10), 1983: 625.

Colclough, C. A. "Organic reaction products of chlorine dioxide and natural aquatic fulvic acids." M.S. report, University of North Carolina at Chapel Hill, 1981.

Colthurst, J. M., and P. C. Singer. "Removing trihalomethane precursors by permanganate oxidation and manganese dioxide adsorption." *Jour. AWWA,* 74(2), 1982: 78.

Coombs, K. K. "Development of a rapid trihalomethane formation potential test." M.S. thesis, University of Massachusetts, Amherst, 1990.

Cooper, W. J., G. L. Amy, C. A. Moore, and R. G. Zika. "Bromoform formation in ozonated groundwater containing bromide and humic substances." *Ozone: Science & Engineering,* 8(1), 1986: 63–76.

Cowman, G. A., and P. C. Singer. "Effect of bromide ion on haloacetic acid speciation resulting from chlorination and chloramination of aquatic humic substances." *Env. Sci. Tech.,* 30(1), 1996: 16.

Crochet, R. A., and P. Kovacic. "Conversion of o-hydroxyaldehydes and ketones into o-hydroxyanilides by monochloramine." *Journal of the Chemical Society, Chemical Communications 1973,* 19, 1973: 716–717.

Daniel, P. A., P. F. Meyerhofer, M. Lanier, and J. Marchand. "Impact of ozonation on formation of brominated organics." In *Ozone in Water Treatment: Proc. 9th Ozone World Congress,* L. J. Bollyky, ed., pp. 348–360, Zurich: International Ozone Association, 1989.

DeLaat, J., N. Merlet, and M. Dore. "Chlorination of organic compounds: chlorine demand and reactivity in relationship to trihalomethane formation." *Water Res.,* 16(10), 1982: 1437–1450.

de Leer, E. W. B., J. S. S. Damste, C. Erkelens, and L. de Galan. "Identification of intermediates leading to chloroform and C-4 diacids in the chlorination of humic acid." *Environmental Science and Technology,* 19, 1985: 512–522.

Dowd, M. T. "Assessment of THM formation with MIOX," M.S. report. Department of Environmental Science Engineering, University of North Carolina, 1994.

Edwards, M. "Transformation of natural organic matter, effect of organic matter-coagulation interactions, and ozone-induced particle destabilization." Ph.D. thesis, University of Washington at Seattle, 1990.

Fawell, J. K., and C. D. Watts. "The nature and significance of organic by-products of ozonation—A review." In *Seminar on Ozone in UK Water Treatment.* London: The Institution of Water Engineers and Scientists, 1984.

Fielding, M., and H. Horth. *Water Supply,* 4, 1986: 103–126.

Fleischacker, S. J., and S. J. Randtke. "Formation of organic chlorine in public water supplies." *Jour. AWWA,* 75(3), 1983: 132.

Franzén, R., and L. Kronberg. *Environmental Science and Technology,* 28(12), 1994: 2222–2227.

Ghanbari, H. A., W. B. Wheeler, and J. R. Kirk. "Reactions of chlorine and chlorine dioxide with free fatty acids, fatty acid esters, and triglycerides." In *Water Chlorination: Environmental Impact and Health Effects, Volume 4, Book 1, Chemistry and Water Treatment, Proc. Fourth Conference on Water Chlorination,* R. L. Jolley, W. A. Brungs, J. A. Cotruvo, R. B. Cumming, J. S. Mattice, and V. A. Jacobs, eds., pp. 167–177. Pacific Grove, CA, 1983.

Glaze, W. H. "Reaction products of ozone: A review." *Environmental Health Perspectives* 69(11), 1986: 151–157.

Glaze, W. H., R. Schep, W. Chauncey, E. C. Ruth, J. S. Zarnoch, E. M. Aieta, C. H. Tate, and M. J. McGuire. "Evaluating oxidants for the removal of model taste and odor compounds from a municipal water supply." *Jour. AWWA,* 82(5), 1990: 79.

Glaze, W. H., H. S. Weinberg, and J. E. Cavanagh. "Evaluating the formation of brominated DBPs during ozonation." *Jour. AWWA,* 85(1), 1993: 96–103.

Gordon, G. *Electrochemical Mixed Oxidant Treatment: Chemical Detail of Electrolyzed Salt Brine Technology.* Cincinnati, OH: U.S. Environmental Protection Agency, 1998.

Griffini, O., and P. Iozzelli. "The influence of H_2O_2 in ozonation treatment: Experience of the water supply service of Florence, Italy." *Ozone: Science and Engineering,* 18(2), 1996: 117–126.

Haag, W. R., and J. Hoigne. "Ozonation of bromide-containing water: Kinetics of formation of hypobromous acid and bromate." *Environmental Science and Technology,* 17(5), 1983: 261–267.

Hausler, C. R., and M. L. Hausler. "Research on chloramines. I. Orthochlorobenzalchlorimine and anisalchlorimine." *Journal of the American Chemical Society,* 52(5), 1930: 2050–2054.

Havlicek, S. C., J. H. Reuter, R. S. Ingols, J. D. Lupton, M. Ghosal, J. W. Ralls, I. El-Barbary, L. W. Strattan, J. H. Cotruvo, and C. Trichilo. *Abstracts of Papers, 177th National Meeting of the American Chemical Society.* American Chemical Society, 1979.

Hoehn, R. C., A. M. Dietrich, W. S. Farmer, M. P. Orr, R. G. Ramon, E. M. Aieta, D. W. Wood, and G. Gordon. "Household odors associated with the use of chlorine dioxide." *Jour. AWWA,* 82(4), 1990: 162.

Hoigne, J., and H. Bader. "Kinetics of reactions of chlorine dioxide (OClO) in water. I. Rate constants for inorganic and organic compounds." *Water Res.,* 28(1), 1994: 45.

Hoigne, J., and H. Bader. "Rate constants of reactions of ozone with organic and inorganic compounds in water. I. Non-dissociating organic compounds. *Water Res.,* 17(2), 1983*a:* 173.

Hoigne, J., and H. Bader. "Rate constants of reactions of ozone with organic and inorganic compounds in water. II. Dissociating organic compounds. *Water Res.,* 17(2), 1983*b:* 185.

Hoigne, J., and H. Bader. "The role of hydroxyl radical reactions in ozonation processes in aqueous solutions." *Water Res.,* 10, 1976: 377.

Hoigne, J., H. Bader, W. R. Haag, and J. Staehelin. "Rate constants of reactions of ozone with organic and inorganic compounds in water. III. Inorganic compounds and radicals. *Water Res.,* 19(8), 1985: 993.

Hwang, C. J., M. J. Sclimenti, L. Trinh, and S. W. Krasner. "Consolidated method for the determination of aldehydes and oxo-acids in ozonated drinking water." *Proc. 1996 Water Quality Technology Conference,* Denver, CO: AWWA, 1997.

Iatrou, A., and W. R. Knocke. "Removing chlorite by addition of ferrous iron," *Jour. AWWA,* 84(11), 1992: 63.

Itoh, S., and Y. Matsuoka. "Contributions of disinfection by-products to activity inducing chromosomal aberrations of drinking water." *Water Res.,* 30(6), 1996: 1403–1410.

Jensen, J. N., and J. D. Johnson. "Specificity of the DPD and amperometric titration methods for free available chlorine: A review." *Jour. AWWA,* 81(12), 1989: 59–64.

Jensen, J. N., J. D. Johnson, J. St. Aubin, and R. F. Christman. "Effect of monochloramine on isolated fulvic acid." *Organic Geochemistry,* 8(1), 1985: 71–76.

Johnson, J. D., and J. N. Jensen. "THM and TOX formation: routes, rates, and precursors." *Jour. AWWA,* 78(4), 1986: 156–162.

Kanniganti, R. "Characterization and gas chromatography/mass spectrometry analysis of mutagenic extracts of aqueous monochloraminated fulvic acid." M.S. report, University of North Carolina at Chapel Hill, 1990.

Kanniganti, R., J. D. Johnson, L. M. Ball, and M. J. Charles. "Identification of compounds in mutagenic extracts of aqueous monochloraminated fulvic acid." *Environmental Science and Technology,* 26(10), 1992: 1998–2004.

Karimi, A. A., and P. C. Singer. "Trihalomethane formation in open reservoirs." *Jour. AWWA,* 83(3), 1991: 84–88.

Krasner, S. W., J. P. Croue, J. Buffle, and E. M. Perude. "Three approaches for characterizing NOM." *Jour. AWWA,* 88(6), 1996: 66.

Krasner, S. W., W. H. Glaze, H. S. Weinberg, P. A. Daniel, and I. N. Najm. "Formation and control of bromate during ozonation of waters containing bromide." *Jour. AWWA,* 85(1), 1993: 73–81.

Krasner, S. W., M. J. McGuire, J. J. Jacangelo, et al. "The occurrence of disinfection by-products in US drinking water." *Jour. AWWA,* 81(8), 1989: 41.

Kringstad, K. P., F. de Sousa, and L. M. Strömberg. "Studies on the chlorination of chlorolignins and humic acid." *Environmental Science and Technology,* 19(5), 1985: 427–431.

Kronberg, L., and T. Vartiainen. *Mutation Research* 206, 1988: 177–182.

Langlais, B., D. A. Reckhow, and D. R. Brink. *Ozone in Water Treatment.* AWWA Research Foundation, Chelsea, MI: Lewis Publishers, 1991.

Lawrence, J. "The oxidation of some haloform precursors with ozone." In *Proc. 3rd Ozone World Congress,* International Ozone Institute, 1977.

Lawrence, J., H. Tosine, F. I. Onuska, and M. E. Comba. "The ozonation of natural waters: Product identification." *Ozone: Science and Engineering,* 2(1), 1980: 55–64.

Le Cloirec, C., and G. Martin. "Evolution of amino acids in water treatment plants and the effect of chlorination on amino acids." In *Water Chlorination: Environmental Impact and Health Effects,* R. L. Jolley, R. J. Bull, W. P. David, S. Katz, M. H. Roberts, Jr., and V. A. Jacobs, eds., pp. 821–834. Chelsea, MI: Lewis Publishers, 1985.

Legube, B., B. Langlais, B. Sohm, and M. Dore. "Identification of ozonation products of aromatic hydrocarbon micropollutants: Effect on chlorination and biological filtration." *Ozone: Science and Engineering,* 3(1), 1981: 33–48.

Le Lacheur, R. M., and W. H. Glaze. "Reactions of ozone and hydroxyl radicals with serine." *Environmental Science and Technology,* 30(4), 1996: 1072–1080.

Le Lacheur, R. M., L. B. Sonnenberg, P. C. Singer, R. F. Christman, and M. J. Charles. "Identification of carbonyl compounds in enviromental samples." *Environmental Science and Technology,* 27(13), 1993: 2745–2753.

Lide, D. R., ed. *CRC Handbook of Chemistry and Physics,* 76th ed. Boca Raton, Fla.: CRC Press, 1995.

Liukkonen, R. J., S. Lin, A. R. Oyler, M. T. Lukasewycz, D. A. Cox, Z.-J. Yu, and R. M. Carlson. "Product distribution and relative rates of reaction of aqueous chlorine and chlorine dioxide with polynuclear aromatic hydrocarbons." In *Water Chlorination: Environmental Impact and Health Effects, Volume 4, Book 1, Chemistry and Water Treatment, Proc. Fourth Conference on Water Chlorination,* R. L. Jolley, W. A. Brungs, J. A. Cotruvo, R. B. Cumming, J. S. Mattice, and V. A. Jacobs, eds., pp. 151–165. Pacific Grove, CA, 1983.

Mallevialle, J., and I. H. Suffet. Identification and treatment of tastes and odors in drinking water. Denver, CO: AWWA Research Foundation, 1987.

Masschelein, W. J. *Chlorine Dioxide: Chemistry and Environmental Impact of Oxychlorine Compounds.* Ann Arbor, MI: Ann Arbor Science Publishers, 1979.

McKnight, A., and D. A. Reckhow. "Reactions of ozonation by-products with chlorine and chloramines." *Proc. 1992 Annual Conference,* Vancouver, British Columbia, Canada. Denver, CO: AWWA, 1992: 399–409.

Meier, J. R. "Genotoxic activities of organic chemicals in drinking water." *Mutation Res.,* 196, 1988: 211–245.

Miller, J. W., and P. C. Uden. "Characterization of nonvolatile aqueous chlorination products of humic substances." *Environmental Science and Technology,* 17(3), 1983:150.

Minear, R. A., and J. C. Bird. "Trihalomethanes: Impact of bromide ion concentration on yield, species, distribution, rate of formation, and influence of other variables." In *Water Chlorination: Environmental Impact and Health Effects, Volume 3, Proc. Third Conference on Water Chlorination: Enviromental Impact and Health Effects,* Robert L. Jolley, William A. Brungs, Robert B. Cummings, eds., pp. 151–160. Colorado Springs, CO, 1980.

Minisci, J. R., and R. Galli. *Tetrahedron Letters,* 1965(8), 1965: 433.

Montgomery Watson Consulting Engineers. *Mathematical Modeling of the Formation of THMs and HAAs in Chlorinated Natural Waters, Final Project Report.* Denver, CO: American Water Works Association and Water Industry Technical Action Fund, 1993.

Morris, J. C. "Formation of halogenated organics by chlorination of water supplies," EPA-600/1-75-002. Washington, D.C.: U.S. Environmental Protection Agency, 1975.

Neale, R. "The chemistry of ion radicals: The free-radical addition of N-chloramines to olefinic and acetylenic hydrocarbons." *Journal of the American Chemical Society,* 86, 1964: 5340–5342.

Oliver, B. G. "Dihaloacetonitriles in drinking water: Algae and fulvic acid as precursors." *Environmental Science and Technology,* 17(2), 1983: 80–83.

Pankow, J. F. *Aquatic Chemistry Concepts.* Chelsea, MI: Lewis Publishers, 1991.

Paramisigamani, V., M. Malaiyandi, F. M. Benoit, R. Helleur, and S. Ramaswamy. "Identification of ozonated and/or chlorinated residues of fulvic acids." In *Proc. 6th Ozone World Congress,* p. 88. Vienna, VA: International Ozone Association, 1983.

Peters, R. J. B., E. W. B. de Leer, and J. F. M. Versteegh. "Identification of halogenated compounds produced by chlorination of humic acid in the presence of bromide." *Journal of Chromatography,* A 686, 1994: 253–261.

Pourmoghaddas, H., and R. C. Dressman. "Determination of nine haloacetic acids in finished drinking water. *Proc. 1992 Water Quality Technology Conference,* Toronto, Ontario, Canada. Denver, CO: AWWA, 1993: 447–464.

Rav-Acha, C., A. Serri, E. Choshen, and B. Limoni. "Disinfection of drinking water rich in bromide with chlorine and chlorine dioxide, while minimizing the formation of undesirable byproducts." *Water Science and Technology,* 17, 1984: 611.

Reckhow, D. A. "Control of disinfection by-product formation using ozone." In P. C. Singer, ed. *Formation and Control of Disinfection By-Products in Drinking Water.* Denver, CO: American Water Works Association, 1999: 179–294.

Reckhow, D. A., and P. C. Singer. "The removal of organic halide precursors by preozonation and alum coagulation." *Jour. AWWA* 76(4), 1984: 151–157.

Reckhow, D. A., and P. C. Singer. "Mechanisms of organic halide formation during fulvic acid chlorination and implications with respect to preozonation." *Water Chlorination: Environmental Impact and Health Effects,* vol. 5, R. L. Jolley, R. J. Bull, W. P. David, S. Katz, M. H. Roberts, Jr., and V. A. Jacobs, eds., pp. 1229–1257. Chelsea, MI: Lewis Publishers, 1985.

Reckhow, D. A., and P. C. Singer, "Chlorination by-products in drinking waters: from formation potentials to finished water concentrations." *Jour. AWWA* 82(4), 1990: 173–180.

Reckhow, D. A., P. C. Singer, and R. L. Malcolm. "Chlorination of humic materials: by-product formation and chemical interpretations." *Enviromental Science and Technology,* 24(11), 1990: 1655–1664.

Reckhow, D. A., P. C. Singer, and R. R. Trussell. "Ozone as a coagulant aid." In *Ozonation: Recent Advances and Research Needs,* pp. 17–46. Denver, CO: AWWA, 1986.

Reckhow, D. A., Y. Xie, R. McEnroe, P. Byrnes, J. E. Tobiason, and M. S. Switzenbaum. "The use of chemical surrogates for assimilable organic carbon." *Proc. 1993 Annual Conference,* San Antonio, TX, June 1993. Denver, CO: AWWA, 1993: 251–278.

Richardson, S. D. "Identification of drinking water disinfection by-products." In *Encyclopedia of Environmental Analysis and Remediation,* vol. 3. R. A. Meyers, ed., New York: John Wiley & Sons, 1998: 1398–1421.

Richardson, S. D., A. D. Thruston, T. W. Collette, et al., "Multispectral identification of chlorine dioxide disinfection byproducts in drinking water." *Env. Sci. Tech.,* 28(4), 1994: 592.

Rook, J. J. "Formation of haloforms during chlorination of natural waters." *Water Treatment and Examination* 23, 1994: 234.

Schechter, D. S., and P. C. Singer. "Formation of aldehydes during ozonation." *Ozone Science and Engineering,* 1995: 53.

Scully, F. E., Jr. "N-chloro compounds: Occurrence and potential interference in residual analysis." In *Proc. Water Quality Technology Conference: Advances in Water Analysis and Treatment,* pp. 611–622. Houston, TX, 1986.

Shank, R. C., and C. Whittaker. "Formation of enotixic hydrazine by the chloramination of drinking water, Technical Completion Report, Project No. W-690. California: University of California Water Resources Center, 1988.

Shiraishi, H., N. H. Polkington, A. Otsuke, and K. Fuwa. "Occurrence of chlorinated polynuclear aromatic hydrocarbons in tap water." *Environmental Science and Technology* 19(7), 1985: 585–590.

Singer, P. C., and M. D. Gurol. "Dynamics of the ozonation of phenol: I. Experimental observations." *Water Research,* 17, 1983: 1163–1171.

Singer, P.C. "Control of disinfection by-products in drinking water." *Journal of Environmental Engineering* 120(4), 1994: 727–744.

Snoeyink, V. L., and D. Jenkins. *Water Chemistry.* New York: John Wiley & Sons, 1980.

Somsen, R. A. "Oxidation of some simple organic molecules with aqueous chlorine dioxide solution: II. reaction products." *Journal of the Technical Association of the Pulp and Paper Industry.* 43(2), 1960: 157–160.

Song, R., P. Westerhoff, R. Minear, and G. L. Amy. "Bromate minimization during ozonation." *Jour. AWWA* 89(6), 1997: 69–78.

Staehelin, J., and J. Hoigne. "Decomposition of ozone in water: Rate of initiation by hydroxide ions and hydrogen peroxide." *Env. Sci. Tech.,* 16(10), 1982: 676.

Standard Methods for the Examination of Water and Wastewater, APHA, AWWA, WEF, 19th ed., 1995.

Steinbergs, C. Z. "Removal of by-products of chlorine and chlorine dioxide at a hemodialysis center." *Jour. AWWA,* 78(6), 1986: 94–98.

Stevens, A. A. "Reaction Products of Chlorine Dioxide." *Env. Health Perspectives,* 46, 1982: 101.

Stevens, A. A., C. J. Slocum, D. R. Seeger, and G. G. Robeck. "Chlorination of organics in drinking water." In *Water Chlorination: Environmental Impact and Health Effects, Volume 1, Proc. Conference on the Environmental Impact of Water Chlorination,* R. L. Jolley, ed., pp. 77–104. Oak Ridge, TN, 1978.

Stumm, W., and J. J. Morgan. *Aquatic Chemistry,* 3rd ed. New York: Wiley Interscience, 1996.

Symons, J. M., T. A. Bellar, J. K. Carswell, et al. "National organics reconnaissance survey for halogenated organics in drinking water," *Jour. AWWA,* 67, 1975: 634.

Topudurti, K. V., and C. N. Haas. "THM formation by the transfer of active chlorine from monochloramine to phloroacetophenone." *Jour. AWWA,* 83(5), 1991: 62–66.

Tratnyek, P. G., and J. Hoigne. "Kinetics of reactions of chlorine dioxide (OClO) in water. II. Quantitative structure-activity relationships for phenolic compounds." *Water Res.,* 28(1), 1994: 57.

Trehy, M. L., and T. I. Bieber. "Detection, identification, and quantitative analysis of dihaloacetonitriles in chlorinated natural waters." In *Advances in the Identification and Analysis of Organic Pollutants in Water,* vol. 2, L. H. Keith, ed. Ann Arbor, MI: Ann Arbor Science, 1981: 941–975.

US Environmental Protection Agency. "Disinfectants and disinfection by-products: final rule." *Federal Register,* 63(241), 1998: 69478.

Venczel, L. V., M. Arrowood, M. Hurd, and M. D. Sobsey. "Inactivation of Cryptosporidium parvum oocysts and Clostridium perfringens spores by a mixed-oxidant disinfectant and by free chlorine." *Appl. Environ. Microbiol.,* 63(4), 1997: 1598.

Wajon, J. E., D. H. Rosenblatt, and E. P. Burrows. "Oxidation of phenol and hydroquinone by chlorine dioxide." *Env. Sci. Tech.,* 16, 1982: 396.

Werdehoff, K. S., and P. C. Singer. "Chlorine dioxide effects on THMFP, TOXFP, and the formation of inorganic by-products." *Jour. AWWA,* 79(9), 1987: 107.

Xie, Y., and D. A. Reckhow. "Formation and control of trihaloacetaldehydes in drinking water." *Proc. 1992 Water Quality Technology Conference,* Toronto, Ontario, Canada, Denver, CO: AWWA, 1993: 1499–1511.

CHAPTER 13
ADSORPTION OF ORGANIC COMPOUNDS

Vernon L. Snoeyink, Ph.D.
Ivan Racheff Professor of Environmental Engineering
Department of Civil and Environmental Engineering
University of Illinois at Urbana-Champaign
Urbana, Illinois

R. Scott Summers, Ph.D.
Professor of Environmental Engineering
Civil, Environmental, and Architectural Engineering
University of Colorado
Boulder, Colorado

Adsorption of a substance involves its accumulation at the interface between two phases, such as a liquid and a solid or a gas and a solid. The molecule that accumulates, or adsorbs, at the interface is called an *adsorbate,* and the solid on which adsorption occurs is the *adsorbent.* Adsorbents of interest in water treatment include activated carbon; ion exchange resins; adsorbent resins; metal oxides, hydroxides, and carbonates; activated alumina; clays; and other solids that are suspended in or in contact with water.

Adsorption plays an important role in the improvement of water quality. Activated carbon, for example, can be used to adsorb specific organic molecules that cause taste and odor, mutagenicity, and toxicity, as well as natural organic matter (NOM) that causes color and that can react with chlorine to form disinfection by-products (DBPs). NOM is a complex mixture of compounds such as fulvic and humic acids, hydrophilic acids, and carbohydrates. The aluminum hydroxide and ferric hydroxide solids that form during coagulation will also adsorb NOM. Adsorption of NOM on anion exchange resins may reduce their capacity for anions (see Chapter 9), but ion exchange resins and adsorbent resins are available that can be used for efficient removal of selected organic compounds. Calcium carbonate and magnesium hydroxide solids formed in the lime softening process have some adsorption capacity, and pesticides adsorbed on clay particles can be removed by coagulation and filtration (Chapters 6 and 8).

The removal of organic compounds by adsorption on activated carbon is very important in water purification and therefore is the primary focus of this chapter. A study conducted by two committees of the AWWA showed that approximately 25 percent of 645 United States utilities, including the 500 largest, used powdered activated carbon (PAC) in 1977 (American Water Works Association, 1977). In 1986, 29 percent of the 600 largest utilities reported using PAC (American Water Works Association, 1986), predominantly for odor control. More attention is being given now to granular activated carbon (GAC) as an alternative to PAC. GAC is used in columns or beds that permit higher adsorptive capacities to be achieved and easier process control than is possible with PAC. The higher cost for GAC can often be offset by better efficiency, especially when organic matter must be removed on a continuous basis. GAC should be seriously considered for water supplies when odorous compounds or synthetic organic chemicals of health concern are frequently present, when a barrier is needed to prevent organic compounds from spills from entering finished water, or in some situations that require DBP control. GAC has excellent adsorption capacity for many undesirable substances and it can be removed from the columns for reactivation when necessary. The number of drinking water plants using GAC, principally for odor control, increased from 65 in 1977 (American Water Works Association, 1977) to 135 in 1986 (Fisher, 1986); in 1996, there were approximately 300 plants treating surface water and several hundred more treating contaminated groundwater. The promulgated as well as proposed DBP regulations will drive many utilities to consider GAC for removal of organic compounds in the next 10 years. GAC is also used as a support medium for bacteria in processes to biologically stabilize drinking water before distribution.

This chapter also covers the use of ion exchange and adsorbent resins for the removal of organic compounds. Removal of inorganic ions by ion exchange resins and activated alumina is discussed in Chapter 9.

ADSORPTION THEORY

Adsorption Equilibrium

Adsorption of molecules can be represented as a chemical reaction:

$$A + B \Leftrightarrow A \cdot B$$

where A is the adsorbate, B is the adsorbent, and $A \cdot B$ is the adsorbed compound. Adsorbates are held on the surface by various types of chemical forces such as hydrogen bonds, dipole-dipole interactions, and van der Waals forces. If the reaction is reversible, as it is for many compounds adsorbed to activated carbon, molecules continue to accumulate on the surface until the rate of the forward reaction (adsorption) equals the rate of the reverse reaction (desorption). When this condition exists, equilibrium has been reached and no further accumulation will occur.

Isotherm Equations. One of the most important characteristics of an adsorbent is the quantity of adsorbate it can accumulate. The constant-temperature equilibrium relationship between the quantity of adsorbate per unit of adsorbent q_e and its equilibrium solution concentration C_e is called the *adsorption isotherm.* Several equations or models are available that describe this function (Sontheimer, Crittenden, and Summers, 1988), but only the more common equations for single-solute adsorption, the Freundlich and the Langmuir equations, are presented here.

The Freundlich equation is an empirical equation that is very useful because it accurately describes much adsorption data. This equation has the form

$$q_e = KC_e^{1/n} \tag{13.1}$$

and can be linearized as follows:

$$\log q_e = \log K + \frac{1}{n} \log C_e \tag{13.2}$$

The parameters q_e (with units of mass adsorbate/mass adsorbent, or mole adsorbate/mass adsorbent) and C_e (with units of mass/volume, or moles/volume) are the equilibrium surface and solution concentrations, respectively. The terms K and $1/n$ are constants for a given system; $1/n$ is unitless, and the units of K are determined by the units of q_e and C_e. Although the Freundlich equation was developed to empirically fit adsorption data, a theory of adsorption that leads to the Freundlich equation was later developed by Halsey and Taylor (1947).

The parameter K in the Freundlich equation is related primarily to the capacity of the adsorbent for the adsorbate, and $1/n$ is a function of the strength of adsorption. For fixed values of C_e and $1/n$, the larger the value of K, the larger the capacity q_e. For fixed values of K and C_e, the smaller the value of $1/n$, the stronger is the adsorption bond. As $1/n$ becomes very small, the capacity tends to be independent of C_e and the isotherm plot approaches the horizontal level; the value of q_e then is essentially constant, and the isotherm is termed *irreversible.* If the value of $1/n$ is large, the adsorption bond is weak, and the value of q_e changes markedly with small changes in C_e.

The Freundlich equation cannot apply to all values of C_e, however. As C_e increases, for example, q_e increases (in accordance with Equation 13.1) only until the adsorbent approaches saturation. At saturation, q_e is a constant, independent of further increases in C_e, and the Freundlich equation no longer applies. Also, no assurance exists that adsorption data will conform to the Freundlich equation over all concentrations less than saturation, so care must be exercised in extending the equation to concentration ranges that have not been tested.

The Langmuir equation,

$$q_e = \frac{q_{max} b C_e}{1 + bC_e} \tag{13.3}$$

where b and q_{max} are constants and q_e and C_e are as defined earlier, has a firm theoretical basis (Langmuir, 1918). The constant q_{max} corresponds to the surface concentration at monolayer coverage and represents the maximum value of q_e that can be achieved as C_e is increased. The constant b is related to the energy of adsorption and increases as the strength of the adsorption bond increases. The Langmuir equation often does not describe adsorption data as accurately as the Freundlich equation. The experimentally determined values of q_{max} and b often are not constant over the concentration range of interest, possibly because of the heterogeneous nature of the adsorbent surface (a homogeneous surface was assumed in the model development), lateral interactions between adsorbed molecules (all interaction was neglected in the model development), and other factors.

Factors Affecting Adsorption Equilibria. Important adsorbent characteristics that affect isotherms include surface area, pore size distribution, and surface chem-

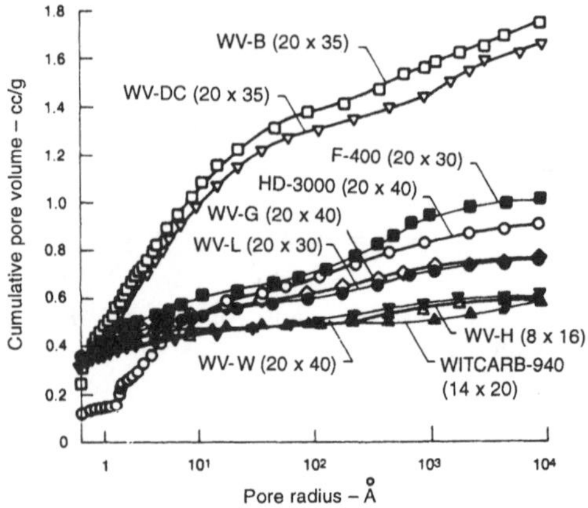

FIGURE 13.1 Pore size distributions for different activated carbons. (*Source:* Lee, Snoeyink, and Crittenden, 1981.)

istry. The maximum amount of adsorption is proportional to the amount of surface area within pores that is accessible to the adsorbate. Surface areas range from a few hundred to more than 1500 m^2/g, but not all of the area is accessible to aqueous adsorbates. The range of pore size distributions in an arbitrary selection of GACs is shown in Figure 13.1. A relatively large volume of micropores (pores less than 2 nm diameter d) (Sontheimer, Crittenden, and Summers, 1988) generally corresponds to a large surface area and a large adsorption capacity for small molecules, whereas a large volume of mesopores ($2 < d < 50$ nm) and macropores ($d > 50$ nm) is usually directly correlated to capacity for large molecules. The fulvic acid isotherms in Figure 13.2 are for the same activated carbons whose pore size distributions are shown in Figure 13.1. Note that the activated carbons that have a relatively small volume of macropores also have a relatively low capacity for the large fulvic acid molecule. Lee et al. (1981) showed that the quantity of humic substances of a given size that was adsorbed was correlated with pore volume within pores of a given size. The relative positions of the isotherms for the activated carbons in Figure 13.1 might be entirely different than those in Figure 13.2 if the adsorbate were a small molecule, such as a phenol, which can enter pores much smaller than those accessible to fulvic acid. Summers and Roberts (1988b) showed that if the amount adsorbed was normalized for the available surface area, the differences in adsorption capacity of different carbons for a humic acid could be attributed to the surface chemistry of the carbon.

The surface chemistry of activated carbon and adsorbate properties also can affect adsorption (Coughlin and Ezra, 1968; Gasser and Kipling, 1959; Kipling and Shooter, 1966; Snoeyink and Weber, 1972; Snoeyink et al., 1974). Several researchers (Coughlin and Ezra, 1968; Gasser and Kipling, 1959; Kipling and Shooter, 1966) demonstrated that extensive oxidation of carbon surfaces led to large decreases in the amounts of phenol, nitrobenzene, benzene, and benzenesulfonate that could be adsorbed. Oxidation of the activated carbon surface with aqueous chlorine was also found to increase the number of oxygen surface functional groups and correspondingly to decrease the adsorption capacity for phenol (Snoeyink et al., 1974). Thus, oxygenating a carbon surface decreases its affinity for simple aromatic compounds.

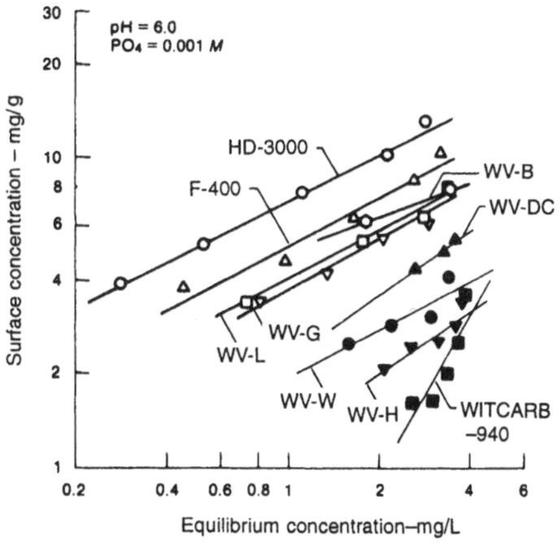

FIGURE 13.2 Adsorption isotherms for peat fulvic acid. (*Source:* Lee, Snoeyink, and Crittenden, 1981.)

The tendency of a molecule to adsorb is a function of its affinity for water as compared to its affinity for the adsorbent. Adsorption onto GAC from water, for example, generally increases as the adsorbate's solubility decreases (Weber, 1972). As a molecule becomes larger through the addition of hydrophobic groups such as —CH_2—, its solubility decreases and its extent of adsorption increases as long as the molecule can gain entrance to the pores. When an increase in size causes the molecule to be excluded from some pores, however, adsorption capacity may decrease as solubility decreases. As molecular size increases, the rate of diffusion within the activated carbon particle decreases, especially as molecular size approaches the particle's pore diameter.

The affinity of weak organic acids or bases for activated carbon is an important function of pH. When pH is in a range at which the molecule is in the neutral form, adsorption capacity is relatively high. When pH is in a range at which the species is ionized, however, the affinity for water increases and activated carbon capacity accordingly decreases. Phenol that has been adsorbed on activated carbon at pH below 8, where phenol is neutral, can be desorbed if the pH is increased to 10 or above, where the molecule is anionic (Fox, Keller, and Pinamont, 1973). If adsorption occurs on resins by means of the ion exchange mechanism, the specific affinity of the ionic adsorbate for charged functional groups may also cause good removal.

The inorganic composition of water also can have an important effect on the extent of NOM adsorption, as shown in Figure 13.3 for fulvic acids (Randtke and Jepsen, 1982). After 70 days, a small GAC column was nearly saturated with fulvic acid. Addition of $CaCl_2$ at this point resulted in a large increase in adsorbability of fulvic acid, as reflected in the reduced column effluent concentration. After 140 days, elimination of the $CaCl_2$ resulted in desorption of much of the fulvic acid. Calcium ion apparently associates (complexes) with the fulvic acid anion to make fulvic acid more adsorbable (Randtke and Jepsen, 1982; Weber, Voice, and Jodellah, 1983). Presumably many other divalent ions can act in similar fashion, but calcium is of special interest because of its relatively high concentration in many natural waters. Similar effects are expected for other anionic adsorbates, but salts are not expected to have much effect on the adsorption of neutral adsorbates (Snoeyink, Weber, and Mark, 1969).

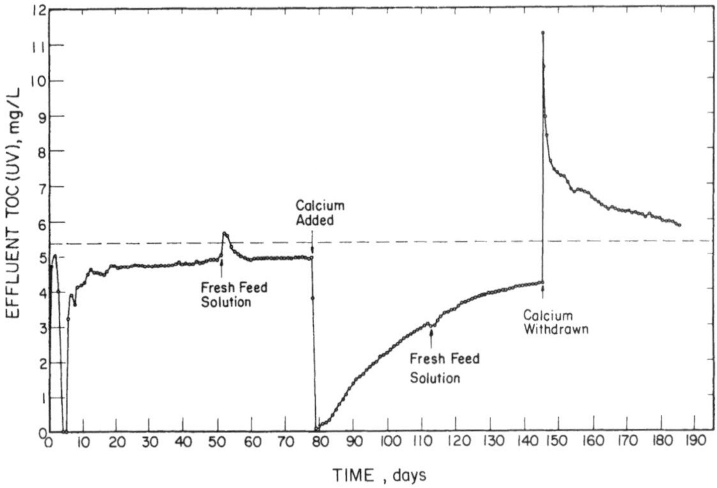

FIGURE 13.3 Effects of calcium chloride addition and withdrawal on column performance (pH = 8.3; TOC = 5.37 mg/L, peat fulvic acid buffer = 1.0 m*M* $NaHCO_3$). (*Source:* Randtke and Jepsen, 1982.)

Inorganic substances such as iron, manganese, and calcium salts or precipitates may interfere with adsorption if they deposit on the adsorbent. Pretreatment to remove these substances, or to eliminate the supersaturation, may be necessary if they are present in large amounts.

Adsorption isotherms may be determined for heterogeneous mixtures of compounds using group parameters such as total organic carbon (TOC), dissolved organic carbon (DOC), chemical oxygen demand (COD), dissolved organic halogen (DOX), UV absorbance, and fluorescence as a measure of the total concentration of substances present. Because the compounds within a mixture can vary widely in their affinity for an adsorbent, the shape of the isotherm will depend on the relative amounts of compounds in the mixture. For example, isotherms with the shape shown in Figure 13.4 are expected if some of the compounds are nonadsorbable and some are more strongly adsorbable than the rest (Randtke and Snoeyink, 1983). The strongly adsorbable compounds can be removed with small doses of adsorbent and yield large values of q_e. In contrast, the weakly adsorbable compounds can only be removed with large doses of adsorbent that yield relatively low values of q_e. The nonadsorbable compounds produce a vertical isotherm at low C_e values. In contrast to single-solute isotherms, the isotherm for a heterogeneous mixture of compounds will be a function of initial concentration and the fraction of the mixture that is adsorbed. The relative adsorbabilities of compounds within a mixture have an important effect on the performance of adsorption columns. The nonadsorbable fraction cannot be removed regardless of the column design, whereas the strongly adsorbable fraction may cause the effluent concentration to slowly approach the influent concentration.

Competitive Adsorption in Bisolute Systems. Competitive adsorption is important in drinking water treatment because most compounds to be adsorbed exist in solution with other adsorbable compounds. The quantity of activated carbon or other adsorbent required to remove a certain amount of a compound of interest

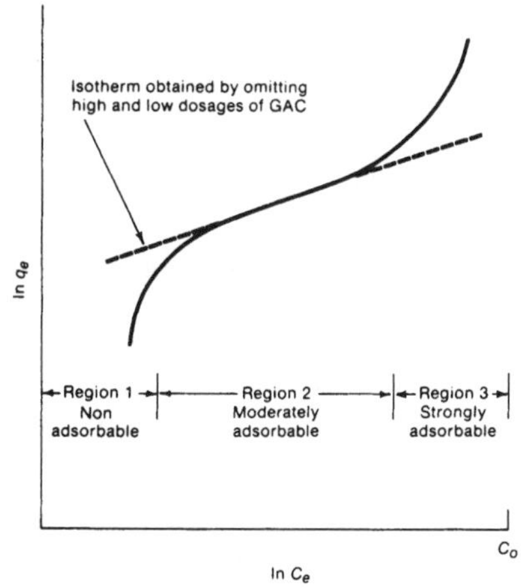

FIGURE 13.4 Nonlinear isotherm for a heterogeneous mixture of organic compounds. (*Source:* Randtke and Snoeyink, 1983.)

from a mixture of adsorbable compounds is greater than if adsorption occurs without competition, because part of the adsorbent's surface is utilized by the competing substances.

The extent of competition on activated carbon depends upon the strength of adsorption of the competing molecules, the concentrations of these molecules, and the type of activated carbon. Some examples illustrate the possible magnitude of the competitive effect. Jain and Snoeyink (1973) showed that as *p*-bromophenol (PBP) equilibrium concentration increased from 10^{-4} to 10^{-3} *M* (17 to 173 mg/L), the amount of *p*-nitrophenol (PNP) adsorbed at an equilibrium concentration of 3.5×10^{-5} *M* (~ 5 mg/L) decreased by about 30 percent.

Displacement of previously adsorbed compounds by competition can result in a column effluent concentration of a compound that is greater than the influent concentration, as shown in Figure 13.5. A dimethylphenol (DMP) concentration about 50 percent greater than the influent resulted when dichlorophenol (DCP) was introduced to the influent of a column

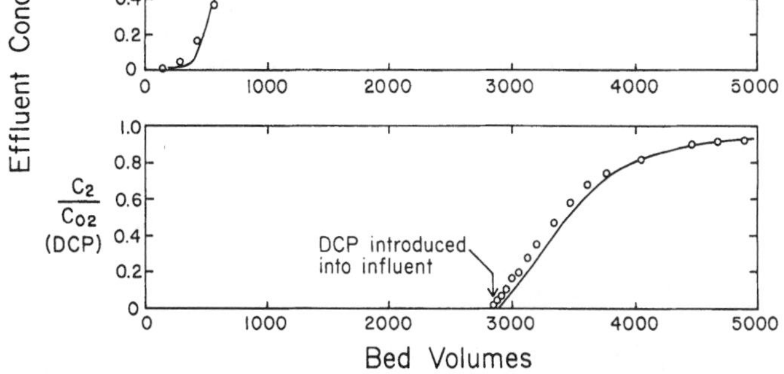

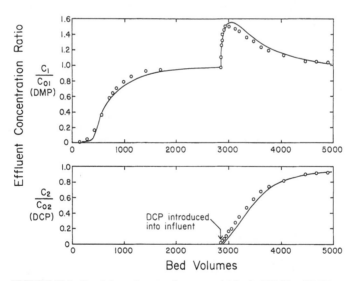

FIGURE 13.5 Breakthrough curves for sequential feed of DMP and DCP to a GAC adsorber (C_0 = 0.990 mmol/L, C_{02} = 1.02 mmol/L, EBCT = 25.4 s). (*Source:* W. E. Thacker, J. C. Crittenden, and V. L. Snoeyink, 1984. "Modeling of Adsorber Performance: Variable Influent Concentration and Comparison of Adsorbents," *Journal Water Pollution Control Federation* 56: 243. Copyright © Water Environment Federation, reprinted with permission.)

saturated with DMP (Thacker, Crittenden, and Snoeyink, 1984). Similar occurrences have been observed in full-scale GAC systems. Effluent concentrations in excess of influent concentrations can be prevented through careful operation. Crittenden et al. (1980) showed that the magnitude of the displacement decreased when the value of $C_{\text{eff}}/C_{\text{inf}}$ was lowered at the time the second compound was introduced. Thus, a reasonable strategy to prevent the occurrence of an undesirable compound at a concentration greater than the influent is (1) to monitor the column for that compound and (2) to replace the activated carbon before complete saturation at the influent concentration occurs (i.e., before $C_{\text{eff}} = C_{\text{inf}}$).

A number of isotherm models have been used to describe competitive adsorption. A common model for describing adsorption equilibrium in multiadsorbate systems is the Langmuir model for competitive adsorption, which was first developed by Butler and Ockrent (1930) and which is presented in the fourth edition of this book (Snoeyink, 1990). This model is based on the same assumptions as the Langmuir model for single adsorbates. Jain and Snoeyink (1973) modified this model to account for a fraction of the adsorption taking place without competition. This can happen if the adsorbates have different sizes and only the smaller adsorbate can enter the smaller pores (Pelekani and Snoeyink, 1999), or if some of the surface functional groups adsorb one compound but not the other. Other models that can be used to describe and predict competitive effects are the Freundlich-type isotherm of Sheindorf, Rebhun, and Sheintuck (1981) and the ideal adsorbed solution theory of Radke and Prausnitz (1972) described in the next section. The latter has proven to be applicable to a large number of situations.

Competitive Adsorption in Natural Waters. Adsorption of organic compounds at trace concentrations from natural waters is an important problem in water purification. Essentially all synthetic organic chemicals that must be removed in water treatment by adsorption must compete with natural or background organic matter for adsorption sites. The heterogeneous mixture of compounds in natural waters adsorbs on activated carbon and reduces the number of sites available for the trace compounds, either by direct competition for adsorption sites or by pore blockage (Pelekani and Snoeyink, 1999). The amount of competition and the capacity for the trace compound depend on the nature of the background organic matter and its concentration, as well as the characteristics of the activated carbon. Also important is the concentration of the trace compound, because this concentration affects how much of this compound can adsorb on the carbon. For example, Figure 13.6 shows that the adsorption capacity of 2-methylisoborneol (MIB), an important earthy/

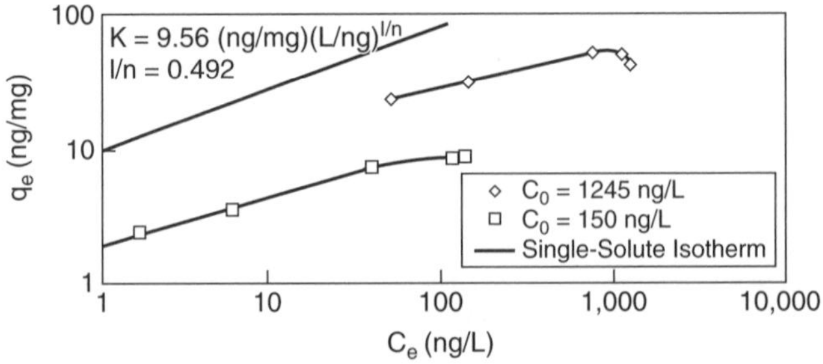

FIGURE 13.6 Effect of initial concentration on MIB capacity in Lake Michigan water. (*Source:* Gillogly et al., 1998b.)

musty odor compound, is lower in natural water than in distilled water, and that this capacity is further reduced as initial concentration decreases (Gillogly et al., 1998b).

It is important to have a procedure to predict capacity as a function of initial concentration, because the capacity of activated carbon depends in such an important way on initial concentration and because the concentrations of trace organic compounds vary widely in natural waters. The ideal adsorbed solution theory (IAST) can be used for this purpose. The following two equations, based on the IAST as developed by Radke and Prausnitz (1972) and modified by Crittenden et al. (1985) to include the Freundlich equilibrium expression, describe equilibrium in a two-solute system,

$$C_{1,0} - q_1 C_c - \frac{q_1}{q_1 + q_2}\left(\frac{n_1 q_1 + n_2 q_2}{n_1 K_1}\right)^{n_1} = 0 \tag{13.4}$$

$$C_{2,0} - q_2 C_c - \frac{q_2}{q_1 + q_2}\left(\frac{n_1 q_1 + n_2 q_2}{n_2 K_2}\right)^{n_2} = 0 \tag{13.5}$$

where q_1 and q_2 = equilibrium solid phase concentrations of compounds 1 and 2
$q_1 = (C_{1,0} - C_{1,e})/C_c$ and $q_2 = (C_{2,0} - C_{2,e})/C_c$
$C_{1,0}$ and $C_{2,0}$ = initial liquid phase concentrations of compounds 1 and 2
$C_{1,e}$ and $C_{2,e}$ = equilibrium concentrations of compounds 1 and 2
K_1 and K_2 = single-solute Freundlich parameters for compounds 1 and 2
$1/n_1$ and $1/n_2$ = single-solute Freundlich exponents for compounds 1 and 2
C_c = carbon dose

These equations show the relationship between the initial concentration of each adsorbate, the amount of adsorbed compound per unit weight of carbon, and the carbon dose. The Freundlich parameters are derived from single-solute tests in organic-free water.

In natural waters, the organic matter present is a complex mixture of many different compounds; representing each of these compounds, even if they could be identified, would be computationally prohibitive. Several researchers have modeled NOM adsorption by defining several fictive components that represent groups of compounds with similar adsorption characteristics, as expressed by Freundlich K and n values (Sontheimer, Crittenden, and Summers, 1988). Extending Equations 13.4 and 13.5 to N components yields

$$C_{i,0} - C_c q_i - \left(\frac{q_i}{\sum_{j=1}^{N} q_j}\right)\left[\frac{\sum_{i=1}^{N} n_j q_j}{n_i K_i}\right]^{n_i} = 0 \tag{13.6}$$

where N = number of components in the solution
$C_{i,0}$ = initial liquid-phase concentration of compound i
C_c = carbon dose
q_i = equilibrium solid-phase concentration of compound i
n_i and K_i = single-solute Freundlich parameters for compound i

These equations can be solved simultaneously to determine the concentrations for each component assumed to be in solution.

Crittenden et al. (1985) used this fictive component approach to describe the adsorption of a target compound in the presence of NOM. With a single-solute isotherm of the target compound and experimental results from isotherms measured

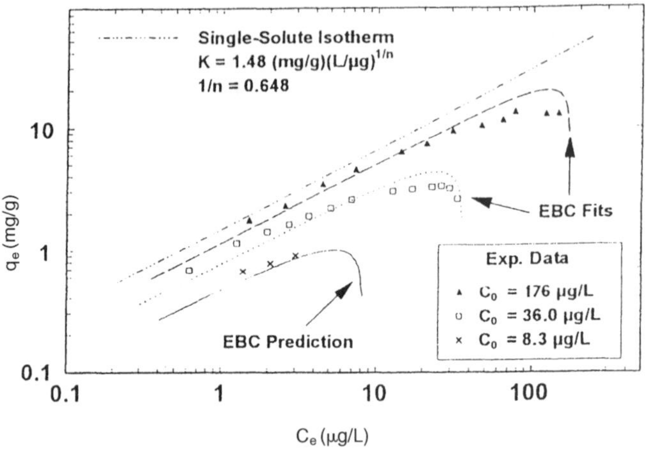

FIGURE 13.7 EBC model results for atrazine isotherms in Illinois groundwater. (*Source:* Reprinted with permission from D. R. U. Knappe et al. 1998. "Predicting the capacity of powdered activated carbon for trace organic compounds in natural waters." *Environmental Science & Technology,* 32: 1694–1698. Copyright 1998 American Chemical Society.)

using the natural water, parameters for each of the fictive components were found through a best-fit search procedure. These results were then applied to describe the adsorption of other compounds in that water.

The IAST was applied to the problem of trace organic adsorption in natural waters by Najm, Snoeyink, and Richard (1991) using a procedure that was subsequently modified by Qi et al. (1994) and Knappe et al. (1998). These researchers assumed that the background organic matter that competed with the trace compound could be represented as a single compound, called the equivalent background compound (EBC). This approach involved the determination of the single-solute isotherm for the trace compound, and an isotherm in natural water for the trace compound at two different initial concentrations. A search routine was used to find the Freundlich parameters K and $1/n$ and the initial concentration C_0 for the EBC that gave the observed amount of competition. For example, Figure 13.7 shows isotherms determined for atrazine in organic-free water and in Illinois groundwater at initial concentrations of 176 and 36 μg/L (Knappe et al., 1998). These data were used to determine the following EBC characteristics:

$$K_{\mathrm{EBC}} > 1.0 \times 10^6\ (\mu\text{mole/g})(\text{L}/\mu\text{mole})^{1/n},\ 1/n_{\mathrm{EBC}} = 0.648,\ C_{0,\mathrm{EBC}} = 0.870\ \mu\text{mole/L}$$

The K value for the EBC was arbitrary above 1.0×10^6 (μmole/g)(L/μmole)$^{1/n}$. These EBC parameters are specific for the type of carbon, the type and concentration of background organic matter, and the type of synthetic organic chemical (SOC). They can be used in Equations 13.4 and 13.5 together with the initial concentration of the trace compound and its single-solute Freundlich parameters to calculate the surface coverage of trace compound as a function of carbon dose C_e. Given the surface coverage q, the initial concentration C_0, and the carbon dose C_c, the equilibrium concentration of the trace compound can be calculated from the equation $q = (C_0 - C_e)/C_c$. This approach was used to determine the predicted isotherm for atrazine at an initial concentration of 8.3 μg/L, shown in Figure 13.7, which compares very well with the measured data. It is interesting to note that at an equilibrium concentration of 1 μg/L, there is a 63 percent reduction in capacity for atrazine as the initial concentration is reduced from 176 to 8.3 μg/L.

An important modification of the EBC model was developed by Knappe et al. (1998), who found that the amount of adsorption on a unit mass of activated carbon was directly proportional to the initial concentration of that trace compound in an adsorption test if (1) the Freundlich exponents for the trace compound $1/n_1$ and the EBC $1/n_2$ both fall between 0.1 and 1, as is generally the case, and (2) the solid-phase concentration of the background organic matter is much in excess of that of the trace compound, which also is often true. When these two approximations hold, the IAST model simplifies to

$$q_1 = C_{1,0}\left[C_c + \frac{1}{q_2}\left(\frac{n_2 q_2}{n_1 K_1}\right)^{n_1}\right]^{-1} \tag{13.7}$$

This equation shows that the surface loading of trace compound q_1 is directly proportional to its initial concentration. The equation can also be manipulated to show that the percent removal or percent remaining of a compound for any carbon dose, when the two preceding assumptions are valid, is a constant. The implication of this result is that only one isotherm need be determined for low concentrations of trace compound in a natural water. This isotherm can be plotted as percent remaining versus carbon dose, as shown in Figures 13.8 and 13.9 for the same data shown in Figures 13.6 and 13.7, respectively. In Figure 13.9, the atrazine data for $C_0 \leq 36$ μg/L plot on a single percent remaining versus carbon dose line, and the $C_0 = 176$ μg/L line is only slightly higher. All the MIB data in Figure 13.6 plot on one line in Figure 13.8.

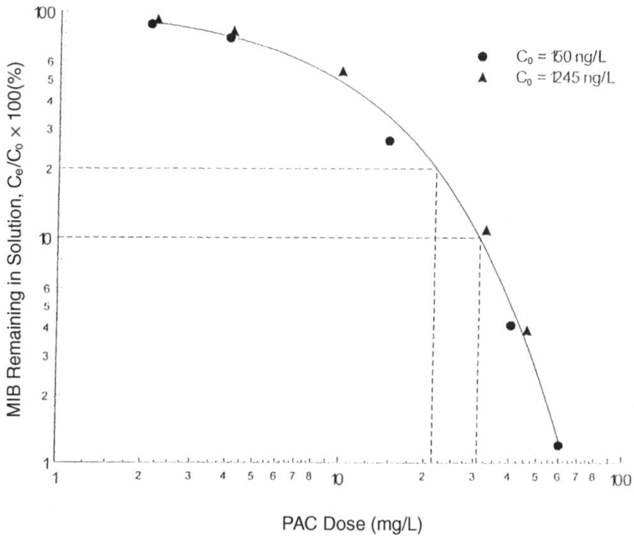

FIGURE 13.8 Percent MIB remaining as a function of PAC dose. (*Source:* Gillogly et al., 1998b.)

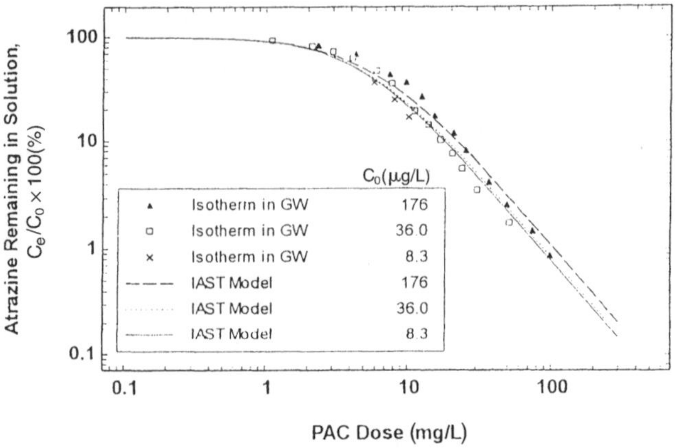

FIGURE 13.9 Removal efficiency as a function of PAC dose for the adsorption of atrazine from Illinois groundwater (GW). (*Source:* Reprinted with permission from D. R. U. Knappe et al., 1998. "Predicting the capacity of powdered activated carbon for trace organic compounds in natural waters." *Environmental Science & Technology,* 32: 1694–1698. Copyright 1998 American Chemical Society.)

Graham et al. (1999) independently derived a different form of Eq. 13.7 and applied it using the EBC approach to adsorption of MIB and geosmin from four natural waters. The same EBC parameters [$K = 1.35$ (μg/mg)(L/μg)$^{1/n}$ and $1/n = 0.20$] were found to be applicable to both MIB and geosmin and, at different initial concentrations (15 to 51 μg/L), to all four natural waters. For three of the four natural waters, the EBC initial concentration was about 0.45 percent of the TOC initial concentration.

These results allow the following general procedure to be used to determine adsorption capacity for a trace compound in natural water:

1. Determine one adsorption isotherm for the trace compound in natural water at a sufficiently low initial concentration.
2. Plot the data on a log-log percent remaining versus carbon dose plot.
3. Use this isotherm plot to determine the carbon dose required for any desired percent removal for any initial concentration that satisfies the assumptions.

An important question now is what concentration is sufficiently low that the assumptions made in developing the percent remaining versus carbon dose plot are valid. Research to date has shown that MIB (Gillogly et al., 1998b) and geosmin (Graham et al., 1999) concentrations less than 1000 ng/L, and atrazine concentrations less than about 50 μg/L[32], will give a satisfactory plot. However, it is necessary to expand this database for other trace compounds.

Desorption. Adsorption of many compounds is reversible, which means that they can desorb. Desorption may be caused by displacement by other compounds, as discussed previously, or by a decrease in influent concentration. Both phenomena may occur in some situations. An analysis of desorption by Thacker, Snoeyink, and Crittenden (1983) showed that the quantity of adsorbate that can desorb in response to a decrease in influent concentration increased as (1) the diffusion coefficient of the adsorbate increased, (2) the amount of compound adsorbed increased, (3) the strength of adsorption decreased (e.g., as the Langmuir *b* value decreased or the Freundlich $1/n$ value increased), and (4) the activated carbon particle size decreased. Volatile organic compounds are especially susceptible to displacement because they are weakly adsorbed and diffuse rapidly. Summers and Roberts (1988a, b) have shown that NOM only partially desorbs and for the desorbing fraction, the desorption diffusivity is lower than that during adsorption.

Adsorption Kinetics

Transport Mechanisms. Removal of organic compounds by physical adsorption on porous adsorbents involves a number of steps, each of which can affect the rate of removal:

1. *Bulk solution transport* Adsorbates must be transported from bulk solution to the boundary layer of water surrounding the adsorbent particle. The transport occurs through diffusion if the adsorbent is suspended in quiescent water such as a sedimentation basin, or through turbulent mixing such as during turbulent flow through a packed bed of GAC, or when PAC is being mixed in a rapid mix unit or flocculator.
2. *External (film) resistance to transport* Adsorbates must be transported by molecular diffusion through the stationary layer of water (hydrodynamic boundary layer) that surrounds adsorbent particles when water is flowing past them. The distance of transport, and thus the time for this step, is determined by the flow rate past the particle. The higher the flow rate, the shorter the distance.
3. *Internal (pore) transport* After passing through the hydrodynamic boundary layer, adsorbates must be transported through the adsorbent's pores to available adsorption sites. Intraparticle transport may occur by molecular diffusion through the solution in the pores (pore diffusion), or by diffusion along the adsorbent surface (surface diffusion) after adsorption takes place.
4. *Adsorption* After transport to an available site, an adsorption bond is formed between the adsorbate and adsorbent. This step is very rapid for physical adsorption (Adamson, 1982) and as a result one of the preceding diffusion steps will control the rate at which molecules are removed from solution. If adsorption is accompanied by a chemical reaction that changes the nature of the molecule, the chemical reaction may be slower than the diffusion step and thereby control the rate of compound removal.

The transport steps occur in series, so the slowest step, called the *rate-limiting step,* will control the rate of removal. In turbulent flow reactors, a combination of film diffusion and pore diffusion very often controls the rate of removal for some of the types of molecules to be removed from drinking water. Initially, film diffusion may control the rate of removal, and after some adsorbate accumulates within the pore, pore transport may control the rate of removal. The mathematical models of the adsorption process, therefore, usually include both steps.

Both molecular size and adsorbent particle size have important effects on the rate of adsorption. Diffusion coefficients, in particular, decrease as molecular size increases, and thus longer times are required to remove the large-molecular-weight humic substances than are needed for the low-molecular-weight phenols, for example. Adsorbent particle size is also important because it determines the time required for transport within the pore to available adsorption sites. If the rate of adsorbate uptake is controlled by intraparticle diffusion, and the effective diffusion coefficient is constant, the time to reach equilibrium is directly proportional to the diameter of the particle squared. Calculations by Randtke and Snoeyink (1983) for activated carbon illustrate these points. For the low-molecular-weight dimethylphenol, nearly 8 days is estimated for near-equilibrium ($C_{\text{final}} = 1.01\ C_e$) of 2.4-mm-diameter activated carbon, but only about 25 min is required for 44-µm-diameter activated carbon. For very large-molecular-weight (approximately 50,000) humic acid, the 2.4-mm-diameter particle is expected to take much longer than a year to equilibrate, but only 2 days is required for the 44-µm-diameter particle. Calculations for 10,000-molecular-weight fulvic acid showed only about 25 percent saturation of 2.4-mm-diameter particles after 40 days of contact. Thus, the smaller the particle, the faster equilibrium is achieved in both column and complete-mix adsorption systems.

Some conclusions that can be drawn from these observations are that (1) granular carbon should be pulverized for isotherm measurement, especially when the capacity of large molecular weight compounds is to be determined (pulverizing does not affect the total surface available for adsorption if all of the GAC is pulverized) (Randtke and Snoeyink, 1983); (2) the smallest activated carbon, consistent with other process constraints such as head loss and loss during reactivation, should be chosen for the best kinetics; and (3) depending on the compound, all the capacity of large activated carbon particles in a column may not be used because the time interval between activated carbon replacements is not sufficient for equilibrium to be achieved.

Mass Transfer Zone and Breakthrough Curves for Packed Bed Reactors. The region of an adsorption column in which adsorption is taking place, the mass transfer zone (MTZ), is shown in Figure 13.10*a*. The activated carbon behind the MTZ has been completely saturated with adsorbate at $C_e = C_0$, and the amount adsorbed per unit mass of GAC is $(q_e)_0$. The activated carbon in front of the MTZ has not been exposed to adsorbate, so solution concentration and adsorbed concentration are both zero. Within the MTZ, the degree of saturation with adsorbate varies from 100 percent ($q = [q_e]_0$) to zero. The length of the MTZ, L_{MTZ}, depends upon the rate of adsorption and the solution flow rate. Anything that causes a higher rate of adsorption, such as a smaller carbon particle size, higher temperature, a larger diffusion coefficient of adsorbate, and/or greater strength of adsorption of adsorbate (i.e., a larger Freundlich K value), will decrease the length of the MTZ. In some circumstances, L_{MTZ} will be reduced sufficiently that it can be assumed to be zero, yielding the ideal plug-flow behavior, as shown in Figure 13.10*b*. If L_{MTZ} is negligible, analysis of the adsorption process is greatly simplified.

The breakthrough concentration C_B for a column is defined as the maximum acceptable effluent concentration. When the effluent concentration reaches this value, the GAC must be replaced. The critical depth of a column L_{critical} is the depth that leads to the immediate appearance of an effluent concentration equal to C_B when the column is started up. For the situation in which C_B is defined as the minimum detectable concentration, the critical depth of an activated carbon column is equal to the length of the MTZ. The length of the MTZ is fixed for a given set of conditions, but L_{critical} varies with C_B. The critical depth, the flow rate Q, and the area of the column A, can be used to calculate the minimum empty bed contact time (EBCT) (EBCT = Q/V, where V is the bulk volume of GAC in the contactor):

$$\frac{L_{\text{critical}}}{Q/A} = \text{EBCT}_{\text{min}} \tag{13.8}$$

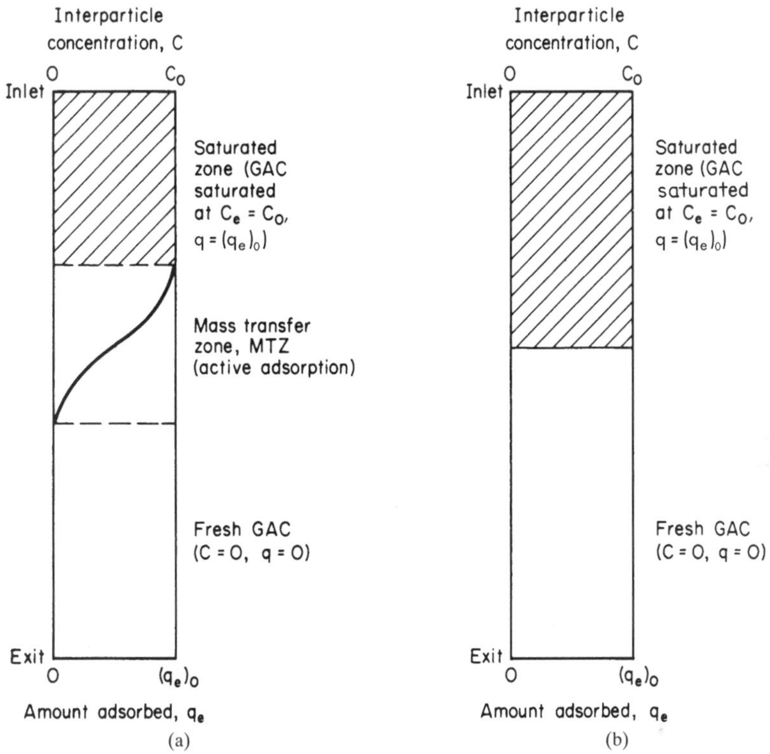

FIGURE 13.10 Adsorption column MTZ. (*a*) Column with MTZ. (*b*) Column without MTZ.

When C_B is greater than the minimum detectable concentration, the critical depth is less than L_{MTZ} and its value can be determined as shown in a later section (see Figure 13.14 and related discussion).

The breakthrough curve is a plot of the column effluent concentration as a function of either the volume treated, the time of treatment, or the number of bed volumes (BV) treated (BV = V_B/V, where V_B is the volume treated). The number of bed volumes is a particularly useful parameter because the data from columns of different sizes and with different flow rates are normalized. A breakthrough curve for a single adsorbable compound is shown in Figure 13.11. The shape of the curve is affected by the same factors that affect the length of the MTZ, and in the same way. Anything that causes the rate of adsorption to increase will increase the sharpness of the curve, while increasing the flow rate will cause the curve to "spread out" over a larger volume of water treated. The breakthrough curve will be vertical if $L_{MTZ} = 0$, as shown in Figure 13.10*b*. As shown in Figure 13.11, the breakthrough capacity, defined as the mass of adsorbate removed by the adsorber at breakthrough, and the degree of column utilization, defined as the mass adsorbed at breakthrough/mass adsorbed at complete saturation at the influent concentration, both increase as the rate of adsorption increases.

The breakthrough curve can be used to determine the activated carbon usage rate (CUR), the mass of activated carbon required per unit volume of water treated:

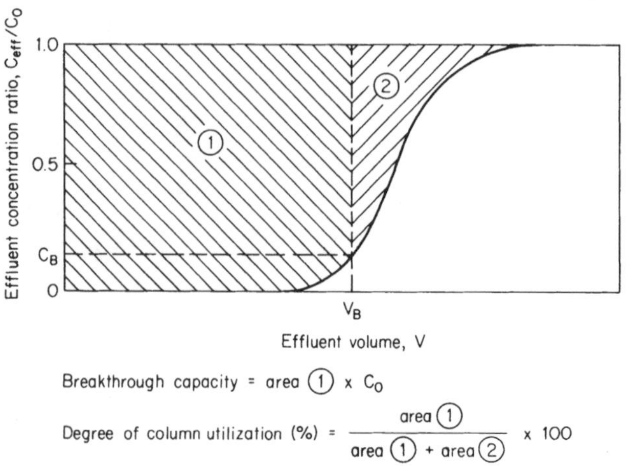

FIGURE 13.11 Adsorption column breakthrough curve.

$$\text{CUR}\left(\frac{\text{mass}}{\text{volume}}\right) = \frac{\text{mass of GAC in column}}{\text{volume treated to breakthrough } V_B} \tag{13.9}$$

Breakthrough curves are strongly affected by the presence of nonadsorbable compounds, the biodegradation of compounds in a biologically active column, slow adsorption of a fraction of the molecules present, and the critical depth of the column relative to the length of the column. Immediate breakthrough of adsorbable compounds occurs if the L_{MTZ} is greater than the activated carbon bed depth (compare curves A and B in Figure 13.12). Nonadsorbable compounds immediately appear in the column effluent, even when the carbon depth is greater than the L_{MTZ} (compare curves B and C in Figure 13.12). Removal of adsorbable, biodegradable compounds by microbiological degradation in a column results in continual removal, even after the carbon is saturated with adsorbable compounds (see curve D in Figure 13.12). If a fraction of compounds adsorbs slowly, the upper part of the breakthrough curve will be similar to that produced by biodegradation but will slowly approach $C_{eff}/C_0 = 1$. Breakthrough curves shown in later sections (see Figures 13.15 and 13.22, for example) also illustrate some of these effects.

GAC ADSORPTION SYSTEMS

Characteristics of GAC

Physical Properties. A wide variety of raw materials can be used to make activated carbon (Hassler, 1974), but wood, peat, lignite, subbituminous coal, and bituminous coal are the substances predominately used for drinking water treatment carbons in the United States. Both the physical and chemical manufacturing processes involve carbonization (conversion of the raw material to a char) and activation (oxidation to develop the internal pore structure). With physical activation, carbonization, or *pyrolysis,* is usually performed in the absence of air at temperatures less than 700°C, while activation is carried out with oxidizing gases such as

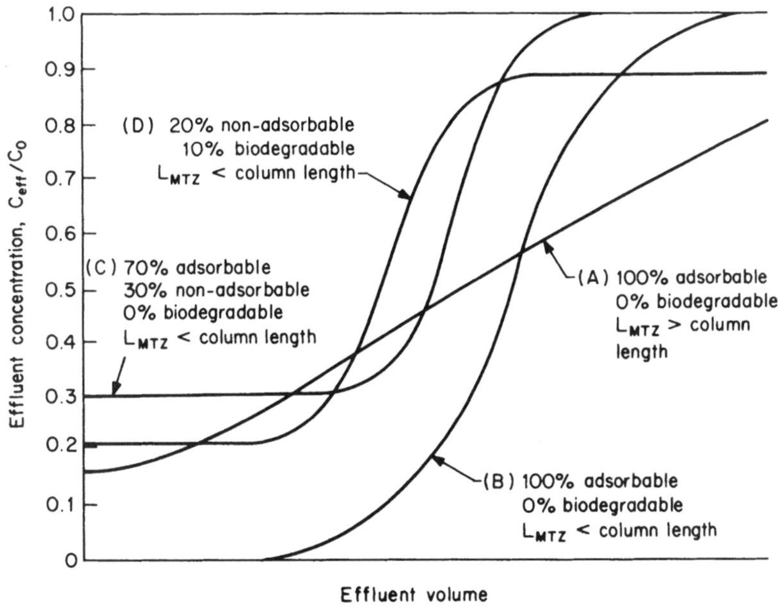

FIGURE 13.12 Effect of biodegradation and the presence of nonadsorbable compounds on the shapes of breakthrough curves.

steam and CO_2 at temperatures of 800 to 900°C. Chemical activation combines carbonization and activation steps. Patents describing carbonization and activation procedures are given by Yehaskel (1978).

Various characteristics of activated carbon affect its performance.* The particle shape of crushed activated carbon is irregular, but extruded activated carbons have a smooth cylindrical shape. Particle shape affects the filtration and backwash properties of GAC beds. Particle size is an important parameter because of its effect on rate of adsorption, as discussed previously. Particle size distribution refers to the relative amounts of different-size particles that are part of a given sample, or lot, of carbon, and has an important impact on the filtration properties of GAC in GAC columns that are used both as filters to remove particles and as adsorbers (i.e., filter-adsorbers) (Graese, Snoeyink, and Lee, 1987). Commonly available activated carbon sizes are 12×40 and 8×30 US Standard mesh, which range in apparent diameter from 1.68 to 0.42 mm and from 2.38 to 0.59 mm, respectively. The uniformity coefficient (see Chapter 8) is often quite large, typically on the order of 1.9, to promote stratification during backwashing. Commercially available activated carbons usually have a small percentage of material smaller than the smallest sieve and larger than the largest sieve, which significantly affects the uniformity coefficient. Extruded carbon particles all have the same diameter, but vary in length. There is no method comparable to the sieve analysis procedure to characterize the distribution of lengths, however.

* Note: Descriptions of the analytical procedures for testing activated carbon are given in ASTM Standards (American Society for Testing Materials, 1988) available from ASTM, 1916 Race St., Philadelphia, PA 19103, as well as in AWWA Standards B604-96 and B600-66, available from AWWA, 6666 West Quincy Ave., Denver, CO 80235.

The *apparent density** of activated carbon is the mass of nonstratified dry activated carbon per unit volume of activated carbon, including the volume of voids between grains. Typical values for GAC are 350 to 500 kg/m^3 (25 to 31 lb/ft^3). Distinguishing between the apparent density and the *bed density, backwashed and drained* (i.e., stratified, free of water) is important, however. The former is a characteristic of activated carbon as shipped. The latter is about 10 percent less than the apparent density and is typical of activated carbon during normal operation unless it becomes destratified during backwashing. The latter is an important parameter because it determines how much activated carbon must be purchased to fill a filter of given size.

The *particle density wetted in water* is the mass of solid activated carbon plus the mass of water required to fill the internal pores per unit volume of particle. Its value for GAC typically ranges from 1300 to 1500 kg/m^3 (90 to 105 lb/ft^3) and it determines the extent of fluidization and expansion of a particle of given size during backwash.

Particle hardness is important because it affects the amount of attrition during backwash, transport, and reactivation. In general, the harder the activated carbon, the less the attrition for a given amount of friction or impact between particles. Activated carbon hardness is generally characterized by an experimentally determined hardness or abrasion number, using a test such as the ASTM Ball Pan Hardness Test that measures the resistance to particle degradation upon agitation of a mixture of activated carbon and steel balls (American Society for Testing Materials, 1988). The relationship between the amount of attrition that can be expected when activated carbon is handled in a certain way and the hardness number has not been determined, however.

Adsorption Properties. A number of parameters are used to describe the adsorption capacity of activated carbon (Sontheimer, Crittenden, and Summers, 1988). The *molasses number* or *decolorizing index* is related to the ability of activated carbon to adsorb large-molecular-weight color bodies from molasses solution, and generally correlates well with the ability of the activated carbon to adsorb other large adsorbates. The *iodine number* (American Society for Testing Materials, 1988) measures the amount of iodine that will adsorb under a specified set of conditions, and generally correlates well with the surface area available for small molecules. Other numbers have been developed for specific applications, such as the *carbon tetrachloride activity,* the *methylene blue number,* and the *phenol adsorption value.* The values of these numbers give useful information about the abilities of various activated carbons to adsorb different types of organics. However, isotherm data for the specific compounds to be removed in a given application, if available, are much better indicators of performance.

Two of the more important characteristics of an activated carbon are its pore size distribution (discussed previously) and surface area. The manufacturer provides typical data that usually include the *BET surface area.* This parameter is determined by measuring the adsorption isotherm for nitrogen gas molecules and then analyzing the data using the Brunauer-Emmett-Teller (BET) isotherm equation (Adamson, 1982) to determine the amount of nitrogen required to form a complete monolayer of nitrogen molecules on the carbon surface. Multiplying the surface area occupied per nitrogen molecule (0.162 nm^2/molecule of N_2) by the number of molecules in the monolayer yields the BET surface area. Because nitrogen is a small molecule, it can

* This definition is based on ASTM Standard D2854-83 (American Society for Testing Materials, 1988). It conflicts with ASTM Standard C128-84 for apparent specific gravity, as used in Chapter 8, which does not include the volume of interparticle voids.

enter pores that are unavailable to larger adsorbates. As a result, all of the BET surface area may not be available for adsorbates in drinking water.

Tabulations of single-solute isotherm constants are very useful when only rough estimates of adsorption capacity are needed to determine whether a more intensive analysis of the adsorption process is warranted. The Freundlich isotherm constants of Dobbs and Cohen (1980), as tabulated by Faust and Aly (1983), are reproduced in Table 13.1 for this purpose.* Values from Speth and Miltner (1998) have also been listed. Additional values can be found in Sontheimer, Crittenden, and Summers (1988). These data can be used to judge relative adsorption efficiency. The K values of isotherms that have nearly the same values of $1/n$ show the relative capacity of adsorption. For example, if a GAC column is satisfactorily removing 2-chlorophenol [$K = 51$ (mg/g)(L/mg)$^{1/n}$ and $1/n = 0.41$], the removal of compounds with larger values of K and approximately the same concentration will very likely be better. (An exception might occur if the organic compounds adsorb to particles that pass through the adsorber.) If the $1/n$ values are much different, however, the capacity of activated carbon for each compound of interest should be calculated at the equilibrium concentration of interest using Equation 13.1, because the relative adsorbability will depend on the equilibrium concentration. The use of isotherm values to estimate adsorber life and PAC usage rate will be discussed later.

GAC Contactors

GAC contactors can be classified by the following characteristics: (1) driving force—gravity versus pressure; (2) flow direction—downflow versus upflow; (3) configuration—parallel versus series; and (4) position—filter-adsorber versus postfilter-adsorber.

GAC may be used in pressure or gravity contactors. Pressure filters enclose the GAC and can be operated over a wide range of flow rates because of the wide variations in pressure drop that can be used. An advantage of these filters is that they can be prefabricated and shipped to the site. A disadvantage is that the GAC cannot be visually observed with ease. Gravity contactors are better suited for use when wide variations in flow rate are not desirable because of the need to remove turbidity, when large pressure drops are undesirable because of their impact on operation costs, and when visual observation is needed to monitor the condition of the GAC. For many systems the decision between pressure or gravity contactors is made on the basis of cost. Medium-size and large systems normally use gravity contactors.

Water may be applied to GAC either upflow or downflow, and upflow columns may be either packed bed or expanded bed. Downflow columns are the most common and seem best suited for drinking water treatment. McCarty, Argo, and Reinhard (1979) found that carbon fines were produced during packed-bed upflow operation and not during downflow operation. The pulsed-bed contactor can also be used to decrease carbon usage rate from that of a single contactor. The flow is applied upward through the column; the spent GAC, a fraction of the total amount present, is periodically removed from the bottom of the column and an equal amount of fresh GAC is applied to the top.

* Dobbs and Cohen (1980) and Speth and Miltner (1998) should be consulted to determine the type of activated carbon and the experimental conditions that were used. The data of Dobbs and Cohen were not determined in a way that would ensure that equilibrium was achieved for all adsorbates, but are suitable to show relative absorbability of compounds and to make rough estimates of activated carbon life. If precise values are needed, new isotherms should be determined using the water to be treated.

TABLE 13.1 Freundlich Adsorption Isotherm Parameters for Organic Compounds

Compound	K (mg/g)(L/mg)$^{1/n}$	$1/n$	Reference
PCB	14,100	1.03	*
Bis(2-ethylhexyl phthalate)	11,300	1.5	†
Heptachlor	9,320	0.92	*
Heptachlor epoxide	2,120	0.75	*
Butylbenzyl phthalate	1,520	1.26	†
Hexachlorocyclopentadiene	1,400	0.504	*
Dichloroacetonitrile	1,300	0.232	*
Toxaphene	950	0.74	*
Endosulfan sulfate	686	0.81	†
Endrin	666	0.8	†
Fluoranthene	664	0.61	†
Aldrin	651	0.92	†
PCB-1232	630	0.73	†
β-Endosulfan	615	0.83	†
Dieldrin	606	0.51	†
p-Dichlorobenzene	588	0.691	*
1,3,5-Trichlorobenzene	586	0.324	*
Alachlor	482	0.257	*
m-Dichlorobenzene	458	0.63	*
m-Dichlorobenzene	458	0.63	*
Hexachlorobenzene	450	0.6	†
Pentachlorophenol	443	0.339	*
Pentachlorophenol	436	0.34	*
Oxamyl	416	0.793	*
Anthracene	376	0.7	†
p-Chlorotoluene	376	0.34	*
4-Nitrobiphenyl	370	0.27	†
m-Xylene	343	0.614	*
Styrene	334	0.479	*
Fluorene	330	0.28	†
DDT	322	0.5	†
2-Acetylaminofluorene	318	0.12	†
o-Chlorotoluene	316	0.378	*
α-BHC	303	0.43	†
Anethole	300	0.42	†
3,3-Dichlorobenzidine	300	0.2	†
Lindane	299	0.433	*
Atrazine	289	0.291	*
γ-BHC (lindane)	285	0.43	*
2,4-Dinitrotoluene	284	0.157	*
2-Chloronaphthalene	280	0.46	†
Carbofuran	275	0.408	*
Phenylmercuric acetate	270	0.44	†
o-Dichlorobenzene	263	0.378	*
Hexachlorobutadiene	258	0.45	†
p-Nonylphenol	250	0.37	†
4-Dimethylaminoazobenzene	249	0.24	†
Cyanazine	244	0.126	*
PCB-1221	242	0.7	†
Acifluorofen	236	0.198	*
Metolachlor	233	0.125	*

TABLE 13.1 Freundlich Adsorption Isotherm Parameters for Organic Compounds (*Continued*)

Compound	K (mg/g)(L/mg)$^{1/n}$	$1/n$	Reference
DDE	232	0.37	†
Acridine yellow	230	0.12	†
p-Xylene	226	0.418	*
Benzidine dihydrochloride	220	0.37	†
β-BHC	220	0.49	†
n-Butylphthalate	220	0.45	†
n-Nitrosodiphenylamine	220	0.37	†
Dibromochloropropane (DBCP)	220	0.501	*
Silvex	215	0.38	*
Phenanthrene	215	0.44	†
Bromobenzene	213	0.364	*
Dimethylphenylcarbinol	210	0.34	†
Dinoseb	209	0.279	*
4-Aminobiphenyl	200	0.26	†
β-Naphthol	200	0.26	†
p-Xylene	200	0.42	*
Glyphosate	199	0.119	*
α-Endosulfan	194	0.5	†
Chlordane	190	0.33	*
Acenaphthene	190	0.36	†
4,4′-Methylene-bis-(2-chloroaniline)	190	0.64	†
Metribuzin	185	0.193	*
Benzol |*k*| fluoranthene	181	0.57	†
Acridine orange	180	0.29	†
α-Naphthol	180	0.32	†
o-Xylene	174	0.47	*
4,6-Dinitro-*o*-cresol	169	0.27	†
Ethyl benzene	163	0.415	*
α-Naphthylamine	160	0.34	†
1,1,1-Trichloropropanone	159	0.11	*
2,4-Dichlorophenol	157	0.15	†
1,2,4-Trichlorobenzene	157	0.31	†
2,4,6-Trichlorophenol	155	0.4	†
β-Naphthylamine	150	0.3	†
Simazine	150	0.227	*
2,4-Dinitrotoluene	146	0.31	†
2,6-Dinitrotoluene	145	0.32	†
4-Bromophenyl phenyl ether	144	0.68	†
Tetrachlorethene	143	0.516	*
p-Nitroaniline	140	0.27	†
1,1-Diphenylhydrazine	135	0.16	†
Aldicarb	133	0.402	*
Naphthalene	132	0.42	†
1-Chloro-2-nitrobenzene	130	0.46	†
p-Chlorometacresol	124	0.16	†
1,4-Dichlorobenzene	121	0.47	†
Benzothiazole	120	0.27	†
Diphenylamine	120	0.31	†
Guanine	120	0.4	†
1,3-Dichlorobenzene	118	0.45	†

TABLE 13.1 Freundlich Adsorption Isotherm Parameters for Organic Compounds (*Continued*)

Compound	K (mg/g)(L/mg)$^{1/n}$	1/*n*	Reference
Acenaphthylene	115	0.37	†
Methoxychlor	115	0.36	*
4-Chlorophenyl phenyl ether	111	0.26	†
Diethyl phthalate	110	0.27	†
Chlorobenzene	101	0.348	*
Chlorobenzene	101	0.348	*
2-Nitrophenol	99	0.34	†
Dimethyl phthalate	97	0.41	†
Hexachloroethane	97	0.38	†
Toluene	97	0.429	*
Bromoform	92	0.655	*
Dicamba	91	0.147	*
Chloropicrin	88	0.155	*
Pichloram	81	0.18	*
2,4-Dimethylphenol	78	0.44	†
4-Nitrophenol	76	0.25	†
Acetophenone	74	0.44	†
1,2,3,4-Tetrahydronaphthalene	74	0.81	†
1,2,3-Trichloropropane	74	0.613	*
Ethylene thiourea	73	0.669	*
Adenine	71	0.38	†
Dibenzo |*a,h*| anthracene	69	0.75	†
1,1,1,2-Tetrachlorethane	69	0.604	*
Nitrobenzene	68	0.43	†
2,4-D	67	0.27	*
Isophorone	63	0.271	*
Methyl isobutyl ketone	61	0.279	*
3,4-Benzofluoranthene	57	0.37	†
Trichloroethene	56	0.482	*
2,4,5-Trichlorophenoxy acetic acid	55	0.21	*
Trichloroacetic acid	52	0.216	*
2-Chlorophenol	51	0.41	†
o-Anisidine	50	0.34	†
Benzene	50	0.533	*
Dibromochloromethane	47	0.636	*
5-Bromouracil	44	0.47	†
Dichloroacetic acid	40	0.462	*
1,1-Dichloropropene	35	0.374	*
Methomyl	35	0.29	*
Benzo |*a*| pyrene	34	0.44	†
2,4-Dinitrophenol	33	0.61	†
1,1,2-Trichloroethane	33	0.652	*
Isophorone	32	0.39	†
1,3-Dichloropropane	28	0.497	*
Thymine	27	0.51	†
Chloral hydrate	27	0.051	*
5-Chlorouracil	25	0.58	†
N-Nitrosodi-*n*-propylamine	24	0.26	†
Bis(2-Chloroisopropyl) ether	24	0.57	†
Carbon tetrachloride	23	0.594	*

TABLE 13.1 Freundlich Adsorption Isotherm Parameters for Organic Compounds (*Continued*)

Compound	K (mg/g)(L/mg)$^{1/n}$	1/*n*	Reference
Dalapon	23	0.224	*
1,2-Dibromoethane	23	0.471	*
1,2-Dibromoethene (EDB)	22	0.46	*
Endothall	22	0.329	*
Bromodichloromethane	22	0.655	*
Phenol	21	0.54	†
1,2-Dichloropropane	19	0.597	*
Methyl ethyl ketone	19	0.295	*
1,1-Dichloroethene	16	0.515	*
trans-1,2-Dichloroethene	14	0.452	*
1,1,1-Trichloroethane	13	0.531	*
Diquat	12	0.242	*
cis-1,2-Dichloroethene	12	0.587	*
Bis(2-Chloroethoxy) methane	11	0.65	†
Uracil	11	0.63	†
Benzo |*g,h,i*| perylene	11	0.37	†
1,1,2,2-Tetrachloroethane	11	0.37	†
Chloroform	9.4	0.669	*
Dibromomethane	9	0.701	*
1,2-Dichloropropene	8.2	0.46	†
1,1-Dichloroethane	8	0.706	*
Cyclohexanone	6.2	0.75	†
tert-Butyl-methyl ether	6	0.479	*
Trichlorofluoromethane	5.6	0.24	†
5-Fluorouracil	5.5	1	†
1,2-Dichloroethane	5	5.33	*
2-Chlorothethyl vinyl ether	3.9	0.8	†
Methylene chloride	1.6	0.801	*
Acrylonitrile	1.4	0.51	†
Acrolein	1.2	0.65	†
Cytosine	1.1	1.6	†
Ethylenediaminetetraacetic acid	0.86	1.5	†
Benzoic acid	0.76	1.8	†
Chloroethane	0.59	0.95	†
N-Dimethylnitrosamine	6.8×10^{-5}	6.6	†

* Speth and Miltner, 1998.
† Dobbs and Cohen, 1980/Faust and Aly, 1983.

Single-stage contactors are often used for small groundwater systems, but if more than one contactor is required, lower activated carbon usage rates can be achieved by arranging the contactors either in series or in parallel as shown in Figure 13.13, possibly yielding a lower-cost system. GAC in a single-stage contactor must be removed about the time the MTZ begins to exit the column (Figure 13.11). At this point only a portion of the activated carbon is saturated at the influent concentration, so the activated carbon usage rate may be relatively high. Alternatively, columns may be arranged in series so that the MTZ is entirely contained within the downstream columns after the lead column has been saturated with the influent concentration. When the activated carbon is replaced in the lead column, the flow is redirected so that it goes through the freshest activated carbon last. Thus, the activated carbon "moves" countercurrent to the flow of

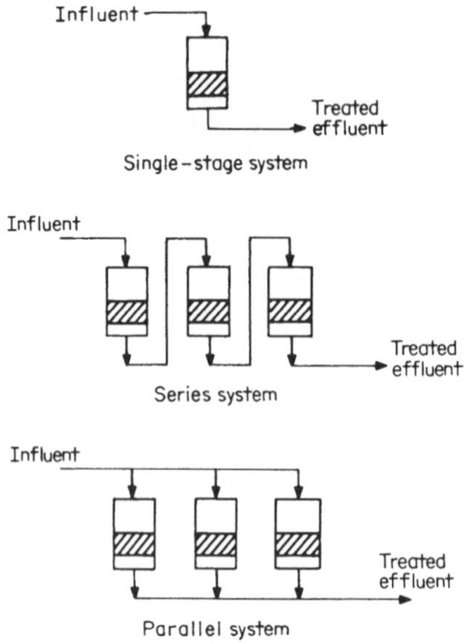

FIGURE 13.13 Adsorber systems.

water, and lower activated carbon usage rates are achieved than with single-stage contactors. Series configuration is best utilized when the effluent criterion is very low compared to the influent concentration (Wiesner, Rook, and Fiessinger, 1987). The increased cost of plumbing counters the cost benefit of reduced activated carbon usage rate, however, especially when more than two columns must be used in series.

When parallel-flow activated carbon adsorbers are operated in staggered mode, they can also be used to decrease the activated carbon usage rate from that which is possible with a single-stage contactor (Westrick and Cohen, 1976; Roberts and Summers, 1982). Because the effluent from each of the units is blended, each unit can be operated until it is producing a water with an effluent concentration in excess of the treated water goal. Only the composite flow must meet the effluent quality goal.

Other flow arrangements can be used to produce lower activated carbon usage rates. A parallel-series arrangement of gravity filters is used in North Holland (Schultink, 1982), and Sontheimer and Hubele (1987) report that a similar arrangement using pressure filters with two layers of GAC was employed at Pforzheim, West Germany. Each of the two layers can be backwashed and replaced independently, and the order of flow through the layers can be reversed. A 35 percent lower activated carbon usage rate for this system for removing halogenated hydrocarbons from groundwater was reported compared to a single-stage system.

GAC contactors can also be classified by their position in the treatment train. The filter-adsorber employs GAC to remove particles as well as dissolved organic compounds. These contactors may be constructed simply by removing all or a portion of the granular media from a rapid filter and replacing it with GAC. Alternatively, a new filter box and underdrain system for the GAC may be designed and constructed. Graese, Snoeyink, and Lee (1987a) discuss these types of filters in detail. The postfilter adsorber is preceded by a granular media filter, and thus has as its only objective the removal of dissolved organic compounds. Backwashing of these adsorbers is unnecessary for particle removal, but if extensive biological growth occurs, backwashing may be required as often as once per week, especially if immediately preceded by ozonation (Fiessinger, 1983; Sontheimer, 1983).

PERFORMANCE OF GAC SYSTEMS

Factors Affecting Organic Compound Removal Efficiency

Adsorbate and GAC properties both have important effects on adsorption that have been discussed in earlier sections. Additional factors that must be considered in the design of full-scale systems are presented here.

GAC Particle Size. The effect of particle size on the rate of approach to equilibrium in isotherm determination was discussed previously. It has a similar effect on the rate of adsorption in columns. If the rate of adsorption is controlled by intraparticle diffusion, the time to reach equilibrium with a given solution concentration in a column approximates that for a batch test. With all other factors constant, decreasing particle size will decrease the time required to achieve equilibrium and will decrease the length of the MTZ in a column. Thus, to improve adsorption efficiency and to minimize the size of column required, the particle size selected for a contactor should be as small as possible.

The rate of head loss buildup caused by particle removal may limit the size of GAC that can be used in adsorbers. The smaller the GAC, the higher the initial head loss and rate of head loss buildup; thus, cost of energy and availability of head have an important influence on the GAC size selected for a design. Additionally, if a filter-adsorber is constructed by replacing media in an existing rapid filter, turbidity removal efficiency generally increases as the GAC size decreases. If the medium is too small, however, the rate of head loss buildup because of particle accumulation becomes excessive, and the net water production will be too small for cost-effective operation. The filters also become difficult to clean by backwashing.

The commercial sizes of GAC are typically characterized by a relatively large uniformity coefficient of up to 1.9. This large coefficient causes the bed to restratify more easily after backwashing. This large uniformity coefficient also requires that greater percent expansion of the adsorber be used during backwash in order to expand the bottom media (Graese, Snoeyink, and Lee, 1987a). Some GAC filters use GAC with a small uniformity coefficient (~1.3) in deep beds to improve depth removal of turbidity and to increase net water production (Graese, Snoeyink, and Lee, 1987b). Mixing of the media in these filters undoubtedly is more than in filters with large uniformity coefficients, so they should not be used in applications where desorption from the mixed GAC will require early GAC replacement.

Common practice is to use 12 × 40 US Standard mesh (1.68 × 0.42 mm) or similar activated carbon in postfilter adsorbers, because backwashing is rarely required due to head loss buildup. This carbon size also is commonly used in filter-adsorbers when it is the only filter medium and the filter depth is less than about 75 cm (30 in.). Deeper beds that will be used to remove turbidity commonly employ 8 × 30 US Standard mesh (2.38 × 0.60 mm) or larger activated carbon to promote longer filter runs. The option of using custom-sized GAC to obtain a better media design for a particular application also is available.

Contact Time, Bed Depth, and Hydraulic Loading Rate. The most important GAC adsorber design parameter that affects performance is the contact time, most commonly described by the EBCT. For a given situation, a critical depth of GAC and a corresponding minimum EBCT (Equation 13.8) exist that must be exceeded to contain the MTZ and minimize or eliminate immediate breakthrough. As the EBCT increases, the bed life or service time (expressed in bed volumes of product water to breakthrough) will increase until a maximum value is reached. Correspondingly, the activated carbon usage rate will decrease to a minimum value. For example, Figure 13.14 shows that the operating time or service time of a column will increase with increasing depth, although the increase is not always linear with depth. These curves are commonly called bed depth–service time curves and may be used to determine the critical depth as shown in Figure 13.14. Figure 13.14 also shows that the percentage of activated carbon in a column exhausted at MTZ breakthrough increases as depth or EBCT increases. The mass of organic matter adsorbed per unit mass of activated carbon increases as percent exhaustion increases and, correspondingly, the

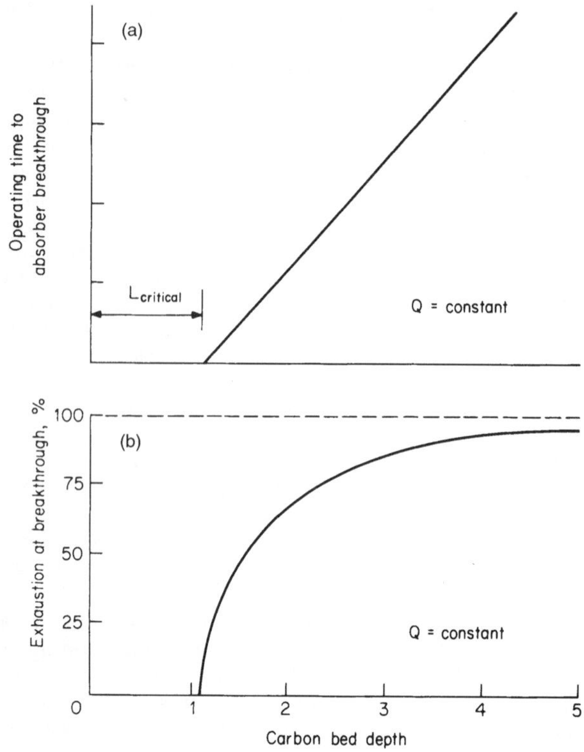

FIGURE 13.14 Bed depth–service time and percent exhaustion at breakthrough versus depth.

number of bed volumes of water that can be processed before MTZ breakthrough also increases to a maximum value.

Increasing EBCT or bed depth at a constant hydraulic application rate can impact treatment costs (Kornegay, 1979; Lee et al., 1983; Wiesner, Rook, and Fiessinger, 1987). As contactor size increases, fixed costs increase because of the greater cost of larger systems. Operating costs decrease because of decreasing carbon usage rate and replacement frequency. Thus an optimum depth or EBCT can be achieved.

Pilot data that show the effect of increasing EBCT on bed life for volatile organic chemicals (VOCs) have been given by Love and Eilers (1982). In general, these data showed that the carbon usage rate for *cis*-1,2-dichloroethylene, 1,1,1-trichloroethane, and carbon tetrachloride decreased substantially as the EBCT was increased from 6 to 12 min, but little change was noted when the EBCT was increased to 18 min. Summers et al. (1997) reported that for three of the four waters examined for TOC and DBP precursor control, no significant increase in the bed volumes treated or carbon usage rate occurred when the EBCT was increased from 10 to 15 to 20 min. Collecting pilot plant data for a range of EBCT values is important if the lowest activated carbon usage for a given application is to be determined.

Other factors must be considered when selecting a hydraulic application rate. As application rate increases, the thickness of the hydrodynamic boundary layer decreases, although this is not a major effect for most applications because film transport often plays a relatively small role in affecting rate of uptake of many compounds from aqueous solution. Cover and Pieroni (1969) review data that show that adsorbers with the same EBCT, but with different hydraulic loading rates, give essentially the same performance in terms of number of bed volumes processed to breakthrough provided the depth is considerably greater than the critical depth L_{MTZ}. Head loss will increase with increasing hydraulic rate, so energy costs must be considered. Also, if the GAC is being used to remove particles as well as dissolved organics, the effect of application rate on removal of turbidity must also be considered.

The EBCTs in use today range from a few minutes for some filter-adsorbers (Graese, Snoeyink, and Lee, 1987a) to more than 4 h for the removal of high concentrations of some specific contaminants (Snoeyink, 1983). Hydraulic application rates vary from 1 to 30 m/h (0.4 to 12 gpm/ft^2), with a typical value being 7 to 10 m/h (3 to 4 gpm/ft^2).

Backwashing. Backwashing of GAC filter-adsorbers is essential to remove solids, to maintain the desired hydraulic properties of the bed, and possibly to control biological growth, and is often necessary for postfilter adsorbers. Backwashing should be minimized, however, because of its possible effect on adsorption efficiency. Mixing of the bed may take place during backwashing; if it does, GAC with adsorbed molecules near the top of the bed will move deeper into the bed where desorption is possible. Molecules that are easily reversibly adsorbed, such as carbon tetrachloride and other VOCs, may be partially desorbed in this new position, leading to a spreading out of the MTZ and to early breakthrough (Wiesner, Rook, and Fiessinger, 1987). Desorption will not occur if the molecules are irreversibly adsorbed, or if they are removed by a destructive mechanism, such as biodegradation, instead of adsorption. The large uniformity coefficient of most commercial activated carbons promotes restratification after backwash, but if the underdrain system does not properly distribute the washwater, or if the backwash is not carried out in a manner that aids restratification, substantial mixing of the activated carbon can occur with each backwash (Graese, Snoeyink, and Lee, 1987a). Hong and Summers (1994) have shown that backwashing had little impact on the time to 50 percent breakthrough for the TOC and THM precursors of four waters. For two of the waters, some increase in TOC after backwashing was detected, but the maximum amount was only 0.2 mg/L.

Biological Activity. Biological activity on GAC has several beneficial aspects. Specific compounds, such as phenol (Chudyk and Snoeyink, 1984), the odor-causing compounds geosmin and MIB (Namkung and Rittman, 1987; Silvey, 1964), *p*-nitrophenol and salicylic acid (DeLaat, Bouanga, and Dore, 1985), ammonia (Bablon, Ventresque, and Ben Aim, 1988), trichlorobenzene (Summers et al., 1989), bromate (Kirisits and Snoeyink, 1999), and probably many more compounds (Rittmann and Huck, 1989), can be removed by biological oxidation rather than adsorption. Some evidence for biological removal of chlorinated benzenes and aromatic hydrocarbons was also found in reclamation water (McCarty, Argo, and Reinhard, 1979). Additionally, some portion of the DOC in natural waters can be biologically oxidized on activated carbon as shown in Figure 13.15. Sontheimer and Hubele (1987) found a small amount of biological oxidation if the water was not preozonated, but application of 1.1 mg O_3/mg DOC resulted in removal of 35 to 40 percent of the influent DOC by biological oxidation. Biodegradable compounds may be removed by microbes, without prior

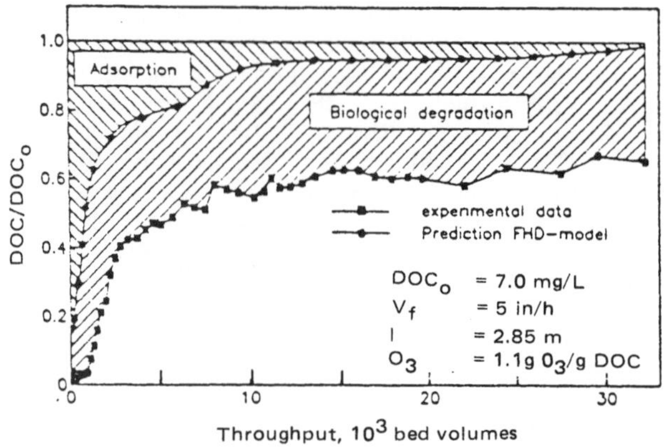

FIGURE 13.15 DOC removal by adsorption and biodegradation during GAC filtration of an ozonated humic acid solution. (*Source:* Sontheimer and Hubele, 1987.)

adsorption to the GAC, if a biofilm capable of degrading such compounds is developed before they are applied. Adsorbable biodegradable compounds may be adsorbed first if the biofilm is not developed when the compounds enter the column, and then desorbed and degraded as the biofilm develops. The use of GAC in biofilters has recently been reviewed by Servais et al. (1998).

Preozonation does not always produce increased amounts of biological oxidation. Glaze et al. (1982) studied a GAC system with a 24-min EBCT; one set of columns was preceded by ozonation at 0.5 to 0.6 mg O_3/mg TOC and another parallel system was not preceded by ozone. Typical results showed about 40 to 50 percent TOC removal by GAC for the 20- to 44-wk period of operation in both systems, thus indicating little effect of the ozone. About 0.6 to 0.8 mg/L of the approximately 2 mg/L TOC being removed by each column could be attributed to biological activity. It is possible that no significant effect of the ozone was noted because the dose is on the lower end of the 0.5- to 1.0-mg O_3/mg TOC dose recommended for promoting biological oxidation (Sontheimer and Hubele, 1987), or because a large fraction of the organic matter was biodegradable before it reacted with ozone.

An important observation made by Glaze et al. (1982) at Shreveport was that biological activity was high during the summer months when the water temperature was in the 25 to 35°C range, but decreased to a relatively insignificant amount as water temperature dropped to 8 to 12°C during the winter months. Thus, biological oxidation may vary significantly throughout the year if low temperatures are expected (Servais et al., 1998).

Sontheimer and Hubele (1987) give typical process parameters for an ozone-GAC system for the type of water they studied (Table 13.2). The 100-g DOC/m^3-day biological oxidation compares well with the 75- to 150-g TOC/m^3-day found by Glaze et al. (1982).

TOC removal by biologically active GAC systems seems to be a conservative indicator for DBP precursor removal (Miltner et al., 1996; Wang, Summers, and Miltner, 1995). Table 13.3 shows the removal of TOC, DBP precursors as measured by the formation potential (FP) for THMs and total organic halide (TOX), and assimilable organic carbon (AOC), a measure of the biodegradable fraction of NOM, by biofiltration of ozonated and settled Ohio River water. All filters had a total depth

TABLE 13.2 Process Parameters for Activated Carbon Following Ozone

Parameter	Value
Ozone dosage	0.5 to 1.0 g ozone/g DOC
Biological degradation	~100 g DOC/(m^3-day)
Oxygen demand for DOC oxidation	~200 g oxygen/(m^3-day)
EBCT	15 to 30 min

Source: Sontheimer and Hubele, 1987.

of 0.76 m (30 in.), including a bottom layer of 0.20 m (8 in.) of sand, yielding an overall EBCT of 9.2 min. The steady-state removals reported in the table were taken between 5 and 11 mo of operation. The GAC columns outperformed the inert media columns, and the micro- and mesoporous GACs seem to perform best in the biological mode. However, Servais et al. (1998) report that the macroporous GAC may be better suited for cold waters. They also report that EBCTs of 5 min were enough to remove the rapidly biodegradable fraction, except for cold waters where longer EBCTs would be required.

An important beneficial effect of ozone-GAC systems was noted by van der Gaag, Kruithof, and Puijker (1985) when treating coagulated and settled Rhine River water. The mutagenicity of chlorinated effluent from an ozone-GAC system was significantly lower than the mutagenicity of chlorinated nonozonated GAC effluent. The authors caution that this response may not be the same for all waters. Sontheimer and Hubele (1987) reference similar results from research in Israel. There the ozone-GAC effluent showed a much lower response in cell tissue tests, thus indicating a lower mutagenicity.

Biologically active GAC generally does cause the concentration of microorganisms in a GAC column effluent to be higher than in the influent. A thorough analysis of this is given by Symons et al. (1981). They report data from Beaver Falls, Pennsylvania, that show both coliform and standard plate counts were higher in GAC effluent than influent when the water temperature was greater than about 10°C, even though 1 to 2 mg/L of chlorine residual was present in the influent to the bed. Apparently the GAC reduced the chlorine and allowed the bacteria to regrow. When the water temperature was below about 10°C, no regrowth was noticed. Other data from Philadelphia showed that coliform organisms such as *Citrobacter freundii, Enterobacter cloacae,* and *Klebsiella pneumoniae* were found in the GAC filters. In all cases, postdisinfection produced water meeting United States Environmental Protection Agency regulations.

TABLE 13.3 Impact of Media Type on Performance—Ozonated/Settled Ohio River Water

	Removal (%)				
Parameter	Anthracite—sand	Sand	GAC—microporous	GAC—mesoporous	GAC—macroporous
TOC	16	20	29	27	21
AOC-NOX	39	43	51	47	42
THM FP	23	23	40	34	27
TOX FP	28	25	52	44	31

Source: Wang, Summers, and Miltner, 1995.

Even though bacteria grow readily in GAC filters, and high microorganism counts are observed in effluents from these processes, the disinfectant demand to achieve microorganism kill is much reduced by GAC filtration. An exception to this occurs if activated carbon particles penetrate the underdrain and provide a habitat for microbes that protects them from being killed by disinfectant (LeChevallier et al., 1984).

Zooplankton can grow in GAC filters if the filters are biologically active. Organisms such as *oligochaetes* and *rotatoria* have been reported to increase as water was processed through GAC during the summer months at Rotterdam and North Holland (van der Kooij, 1983), probably because they use bacteria for food. Similar observations have been made in West Germany (Sontheimer, 1983) and France (Fiessinger, 1983). Sontheimer and Fiessinger recommend backwashing with air scour once every five days or so to wash the eggs of these organisms out of the adsorber before they can hatch. In contrast, microscreening is used to remove the organisms from the GAC-filtered water at North Holland.

Control of Microbial Growth. Biologically active carbon must be controlled to avoid undesirable effects. Anaerobic conditions may develop, with attendant odor problems, if the system is not kept aerobic. This may happen if large concentrations of ammonia enter the filter (each mg/L of NH_3 requires about 3.8 mg/L of dissolved oxygen if it is converted to NO_3^-), if insufficient dissolved oxygen is in the water, or if the bed is allowed to stand idle for a period of time.

Control is also possible through proper design. Figure 13.16 shows the distribution of microorganisms in a biologically active GAC filter that is treating nonchlorinated water (Sontheimer, Crittenden, and Summers, 1988; Topalian, 1987). Much larger numbers of organisms are on the activated carbon at the entrance to the bed than at greater depths. This distribution is consistent with larger amounts of adsorbed compounds in the upper level (Sontheimer and Hubele, 1987). Growth in the upper part of a deep bed has the opportunity to be removed deeper within the bed when it sloughs off. van der Kooij (1983) noted that whereas increasing the EBCT by increasing the bed depth can bring about sharp reductions in numbers of organisms in GAC filtrate, increasing EBCT by decreasing flow rate did not decrease, but sometimes increased, the plate count. Available data on the effect of backwashing on effluent organism concentration are inconclusive (van der Kooij, 1983).

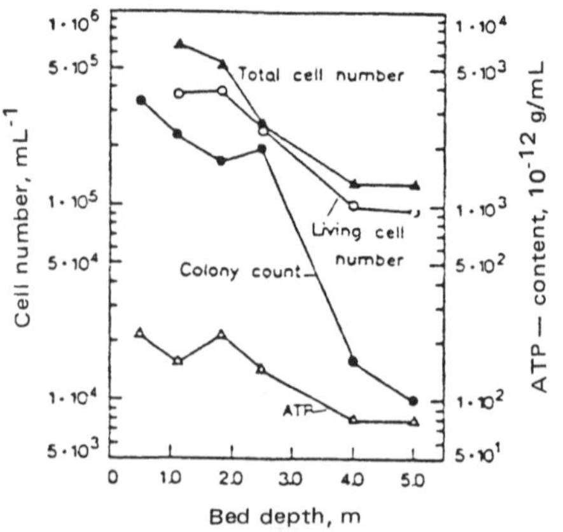

FIGURE 13.16 Microbiological parameters in a GAC filter. (*Source:* Sontheimer and Hubele, 1987.)

Application of chlorine to GAC adsorbers does not prevent growth, and increases the concentration of adsorbed chloroorganics. The potential exists for chlorine to make the activated carbon become more friable and break up more easily, especially during backwash, because chlorine destroys some of the activated carbon when it is reduced (Snoeyink and Suidan, 1975). Also, the potential exists for formation of unique organic compounds through the catalytic action of the activated carbon surface, as shown later. Thus, application of chlorine to GAC filters is not recommended.

Pretreatment for GAC Systems. Pretreatment can have a significant impact on the performance of activated carbon systems. The influent concentration of organics may be lowered, the species of compounds may be changed, thus changing adsorbability and biodegradability, and the inorganic composition may be changed in a manner that affects absorbability and the tendency of the activated carbon to become fouled. The removal of NOM by coagulation, sedimentation, and filtration reduces the quantity of organics that must be removed by adsorption. Lower-cost operation of GAC systems for TOC and DBP precursor removal may then be possible (see Chapter 10 for a discussion of organic compound removal by coagulation).

Summers et al. (1994) and Hooper et al. (1996a), as shown in Figure 13.17, have summarized the impact of the initial TOC concentration TOC_0 on the run time to a 50 percent TOC breakthrough, measured as bed volumes BV_{50} for a wide range of source waters. The relationship can be expressed as

$$BV_{50} = 18{,}000/TOC_0$$

Twenty-eight case studies of GAC bench, pilot, and full-scale contactors from 21 different source waters were evaluated. All systems utilized bituminous-coal-based GAC, and the influent pH was between 7 and 8 for the river, lake, and groundwater examined. The relationship was verified by five GAC runs of the same isolated NOM, diluted to different concentrations (hollow symbols in Figure 13.17). Part of the variability of the data around the regression line is likely due to differences in the adsorbability of the NOM caused by pretreatment and differences between sources.

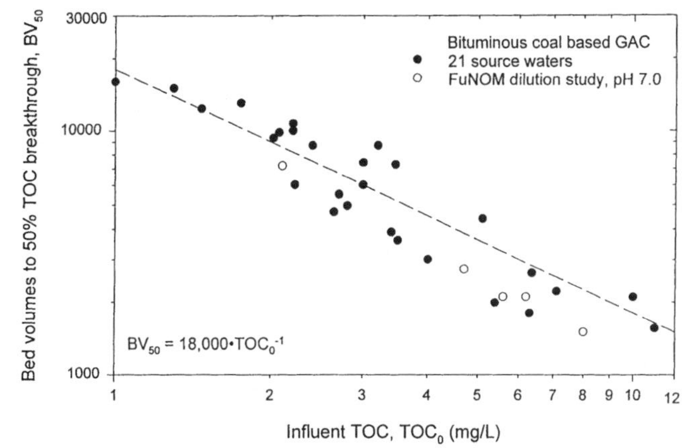

FIGURE 13.17 Correlation between influent TOC concentration and bed volumes to 50 percent TOC breakthrough. (*Source:* Hooper et al., 1996a.)

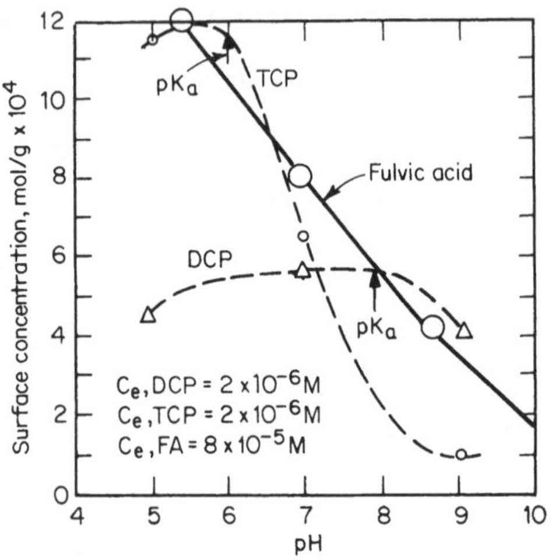

FIGURE 13.18 Effect of pH on TCP, DCP, and soil fulvic acid (FA) adsorption. (TCP and DCP data taken from Murin and Snoeyink, 1979; FA data, expressed in moles C, taken from McCreary and Snoeyink, 1980.)

Adsorption of organic acids and bases by GAC is generally affected by solution pH, so pretreatment steps that affect pH have an important effect on adsorption. In general, both the undissociated and ionized forms of an adsorbate can be adsorbed on GAC, with the undissociated form being more strongly adsorbed. The ionic form has a much higher affinity for polar water molecules and thus tends to remain in solution. Typical data are shown in Figure 13.18 for three weak organic acids: 2,4,6-trichlorophenol, 2,4-dichlorophenol, and soil fulvic acid (McCreary and Snoeyink, 1980; Murin and Snoeyink, 1979). Different types of humic substances may have both different capacities and capacity dependencies on pH.

Several researchers have shown the impact of influent pH on the adsorption of TOC. Unfortunately, some of the work has been done with different initial TOC concentrations, and the increased performance attributed to low pH may be because of the lower TOC_0 (Figure 13.17). A relationship between the relative adsorption capacity for TOC and pH is shown in Figure 13.19 for 13 different source waters and a bituminous-coal-based GAC (Hooper, Summers, and Hong, 1996). Within the pH range shown, a decrease in the pH of one unit yielded a 6 percent increase in adsorption capacity.

Several investigators have reported better GAC performance for TOC control after coagulation or after increasing the coagulant dose to achieve enhanced coagulation. Hooper and coworkers (Hooper, Summers, and Hong, 1996; Hooper et al., 1996a,b) have shown that the increase in GAC run time after enhanced coagulation can be attributed to the lower pH and lower initial TOC concentration associated with the coagulated water.

Reactions of chlorine, or other oxidative pretreatment chemicals such as oxygen, ozone, chlorine dioxide, and permanganate, with GAC or with organic compounds in aqueous solution or on the GAC surface can alter the adsorption performance. For example, ozone can react with humic substances to produce more polar inter-

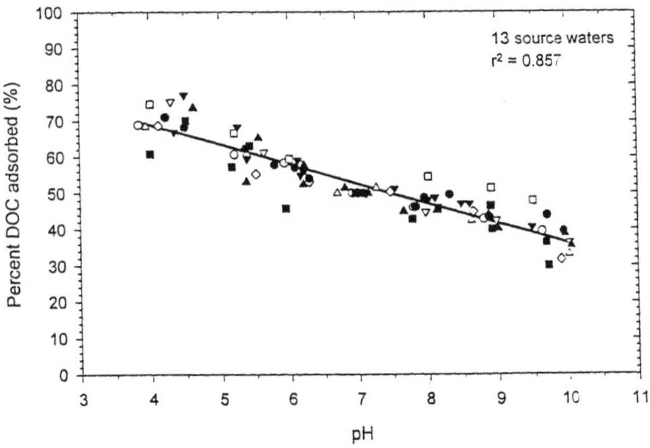

FIGURE 13.19 Impact of pH on percent DOC adsorbed for several waters. (*Source:* Hooper, Summers, and Hong, 1996.)

mediates that are less adsorbable on GAC (Chen, Snoeyink, and Fiessinger, 1987) but usually more biodegradable (Sontheimer and Hubele, 1987). If TOC is more biodegradable, increased removals by microbiological activity in a GAC contactor are expected.

However, if biological treatment is not effective, then the weakly adsorbing compounds can have a negative impact on the overall GAC performance. Solarik et al. (1996) systematically evaluated the impact of ozonation and biotreatment on subsequent GAC performance for five waters, as illustrated in Figure 13.20 for Ohio River water. They showed that ozonation and biotreatment decreased the humic and intermediate molecular size fractions, which are the most strongly adsorbing fractions. They found that the early part of the breakthrough was dominated by the relative increase in the weakly adsorbing fraction, which in some cases led to earlier breakthrough, while the overall lower influent TOC concentration dominated the latter portion of the breakthrough curve and in some cases led to longer run times.

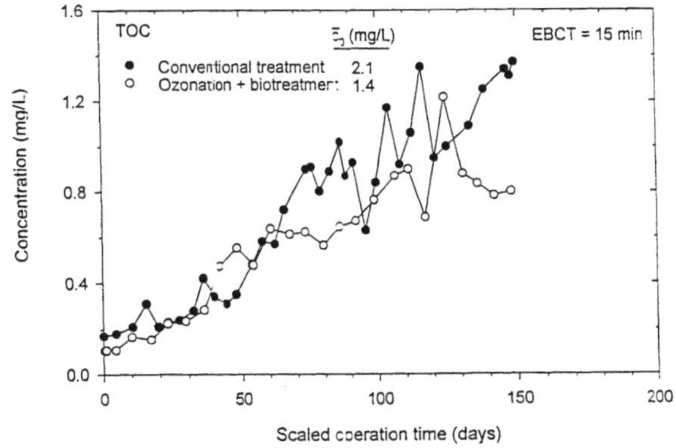

FIGURE 13.20 Effect of pretreatment on TOC breakthrough for Ohio River water. (*Source:* Solarik et al., 1996.)

Chlorine-containing disinfectants (HOCl, ClO_2, or NH_2Cl) may react both with activated carbon and adsorbed compounds. Unusual products not characteristic of solution reactions may be formed when activated carbon is present. For example, the HOCl reaction with adsorbed 2,4-dichlorophenol (2,4-DCP) resulted in a series of hydroxylated PCBs at HOCl concentrations normally encountered in drinking water treatment practice (see Table 13.4). A similar product mixture was also obtained when the GAC was first treated with HOCl and then 2,4-DCP was adsorbed. The HOCl-activated carbon surface caused similar compounds to form as adsorption took place. Furthermore, some of these products may desorb from the activated carbon column. Additional data are given on this effect by Voudrias, Larson, and Snoeyink (1985). The reaction products that have been found to date have been formed only in laboratory systems. Further research is needed to show whether they will also form in field installations in the presence of humic substances. Because such compounds might form, and because their health effects are unknown, the application of chlorine-containing disinfectants to GAC adsorbers needs to be eliminated where possible.

Vidic and Suidan (1991) showed that dissolved oxygen also reacted with adsorbed compounds such as phenols, converting them to products that allowed more of the target adsorbate to be removed from solution. The capacity for these compounds was thus much greater if oxygen was present than if the solution was free of oxygen.

Pretreatment to prevent fouling of GAC is also important. Application of water that is supersaturated with salts such as calcium carbonate will lead to blockage of the activated carbon pores and possibly to complete coverage of the particle. Iron and manganese precipitates may also interfere with adsorption. If ammonia concen-

TABLE 13.4 Reaction Products from HOCl-2,4-Dichlorophenol-GAC Reaction

Compound

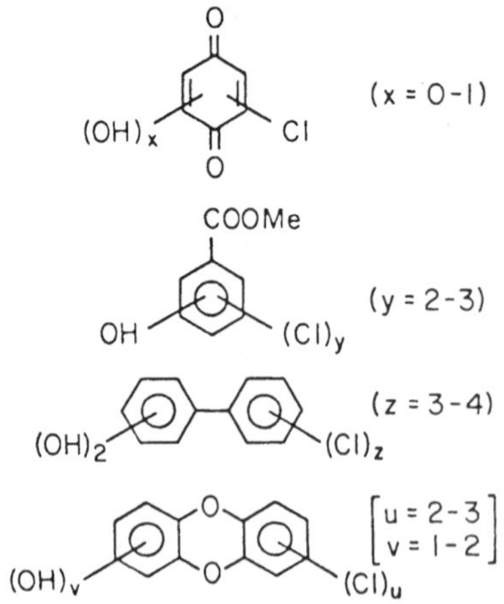

Source: Reprinted with permission from E. A. Voudrias, R. A. Larson, and V. L. Snoeyink. "Effects of activated carbon on the reactions of free chlorine with phenols." *Environmental Science and Technology,* 1985:441. Copyright 1985 American Chemical Society.

tration is very high, pretreatment for ammonia removal may also be necessary to prevent depletion of dissolved oxygen in biologically active beds.

Reactions of Inorganic Compounds with Activated Carbon

Activated carbon in water treatment may inadvertently contact oxidants such as oxygen, aqueous chlorine, chlorine dioxide, and permanganate and react with them. Virgin activated carbon has been shown by Prober, Pyeha, and Kidon (1975) to react with 10 to 40 mg aqueous O_2 per gram of carbon over a time span of 1700 h, and Chudyk and Snoeyink (1981) found that 3 to 4 mg O_2 per gram of carbon reacted over a time span of 130 h. Some of this oxygen is converted to surface oxides (Prober, Pyeha, and Kidon, 1975).

Free Chlorine-Activated Carbon Reactions. The well-known reactions of HOCl and OCl^- with activated carbon are as follows:

$$HOCl + C^* \rightarrow C^*O + H^+ + Cl^- \tag{13.10}$$

$$OCl^- + C^* \rightarrow C^*O + Cl^- \tag{13.11}$$

where C* and C*O represent the carbon surface and a surface oxide, respectively. These reactions proceed rapidly, with the one in Equation 13.10 ($pH < 7.5$) being faster than that in Equation 13.11 ($pH > 7.5$) (Suidan, Snoeyink, and Schmitz, 1976, 1977a,b). Reactions of free chlorine with activated carbon will result in the production of organic by-products. The TOX on the surface increases as extent of reaction increases, and some of these compounds may be found in the column effluent if the reaction proceeds for a very long time (Dielmann, 1981).

Combined Chlorine-Activated Carbon Reactions. Bauer and Snoeyink (1973) hypothesized that the following reactions were taking place between monochloramine, NH_2Cl, and activated carbon:

$$NH_2Cl + H_2O + C^* \rightarrow NH_3(aq) + C^*O + H^+ + Cl^- \tag{13.12}$$

$$2\,NH_2Cl + C^*O \rightarrow N_2(g) + 2H^+ + 2Cl^- + H_2O + C^* \tag{13.13}$$

Initially, all the NH_2Cl was converted to NH_3 and Cl^- in accordance with Equation 13.12, but after a period of reaction, some of the NH_2Cl was converted to $N_2(g)$ and HCl in accordance with Equation 13.13. The rate of reaction was much slower than the reactions of either free chlorine or dichloramine with activated carbon (Kim and Snoeyink, 1980). Additional studies of this reaction were made by Komorita and Snoeyink (1985), who showed that the rate of reaction was high initially and then reached a plateau value. The higher initial rate is useful in design of some GAC dechlorination systems.

Dichloramine ($NHCl_2$) reacts very rapidly with activated carbon according to the following reaction (Bauer and Snoeyink, 1973; Kim, Snoeyink, and Schmitz, 1978):

$$2NHCl_2 + H_2O + C^* \rightarrow N_2(g) + C^*O + 4H^+ + 4Cl^- \tag{13.14}$$

When excess ammonia was present, Kim, Snoeyink, and Schmitz (1978) found evidence of the parallel reaction,

$$NH_4^+ + 3NHCl_2 \rightarrow 2N_2(g) + 7H^+ + 6Cl^- \tag{13.15}$$

Bromate, Chlorine Dioxide, Chlorite, and Chlorate-Activated Carbon Reactions. Other halogen oxides react with activated carbon. For example, Miller, Snoeyink, and Harrell (1995) showed that BrO_3^- was converted to Br^- by virgin activated carbon. ClO_2 reacts rapidly with activated carbon, but the nature of the reaction changes as pH changes. At pH 3.5, Cl^-, ClO_2^-, and ClO_3^- are found in the effluent of a column receiving only ClO_2, but Cl^- is the predominant product. At pH 7.9, the same end products are formed, but ClO_2^- is now the predominant species (Chen, Larson, and Snoeyink, 1982). When ClO_2^- solutions at pH 7 were applied to virgin GAC, ClO_2^- readily reacted, presumably in accordance with the reaction (Voudrias et al., 1983)

$$ClO_2^- + C^* \rightarrow C^*O_2 + Cl^- \tag{13.16}$$

The reaction capacity of the fresh carbon was saturated after 80 to 90 mg ClO_2^- reacted per gram of GAC. The chlorate ion was not reduced by activated carbon, but was adsorbed to a slight extent (~0.03 mg ClO_3^-/g), presumably by an ion exchange mechanism (Dielmann, 1981).

Removal of Other Inorganic Ions. Huang (1978) reviewed the removal of inorganic ions by activated carbon. A number of ions can be removed from water by GAC, but the capacity for most substances is quite low. The gold-cyanide complex, for example, is adsorbed on GAC in a widely used method for recovering gold in the mining industry. Huang reviewed data that show some removal of cadmium (II) at higher pH, and that removal can be increased slightly by complexing before adsorption with chelating agents. The removal of chromium involves adsorption of Cr(III) or Cr(VI), and under some conditions Cr(VI) is chemically reduced to Cr(III) by the activated carbon. Mercury adsorption is best at low pH. From 0.1 to 0.5 mmole Cu/g at equilibrium concentrations of 8 mmol/L have also been observed to adsorb. Activated carbon will also catalyze the oxidation of Fe(II) to Fe(III).

Adsorption Efficiency of Full-Scale Systems

Taste and Odor Removal. Many types of taste and odor problems are encountered in drinking water. Troublesome compounds may result from biological growth or industrial activities. They may be produced in the water supply, in the water treatment plant from reactions with treatment chemicals, in distribution systems, and in consumers' plumbing systems (American Water Works Association, 1987). Activated carbon—both PAC and GAC—has an excellent history of success in removing taste and odor compounds from raw water.

GAC filter-adsorber systems are reported to effectively remove odor from source water for typically one to five years (Graese, Snoeyink, and Lee, 1987b). The bed life is dependent upon the intensity and frequency of appearance of taste and odor compounds, the presence of organics that compete for adsorption sites, and the concentration of these compounds that is acceptable in the treated water. Case history information is difficult to apply at different utilities because the intensity and frequency of appearance usually is not well documented and the acceptable level of taste and odor varies from community to community.

Regina, Saskatchewan, is an interesting example of the effect of high-intensity tastes and odors on GAC performance (Gammie and Giesbrecht, 1986; Snoeyink and Knappe, 1994). The water treatment plant has postfilter-adsorbers with an EBCT of 15 min and a bed depth of 10 ft. The GAC was reactivated once per year. The influent water had a DOC concentration of 2 to 3 mg/L and typical threshold odor numbers (TONs) of 5 to 15 with spikes of 40 to 60. The effluent goal was a TON of 1.

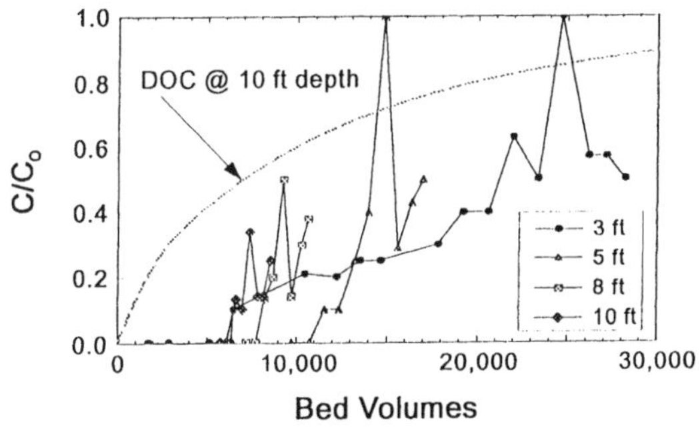

(a) TON Breakthrough Curves

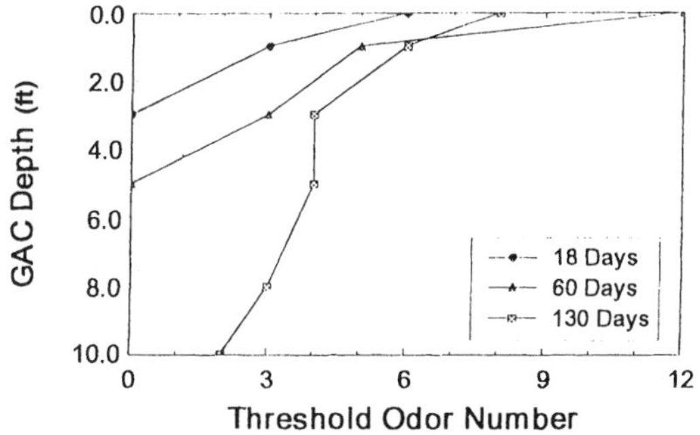

(b) TON Mass Transfer Zone

FIGURE 13.21 TON adsorption at Regina. (*Source:* Snoeyink and Knappe, 1994.)

Figure 13.21*a* depicts TON breakthrough curves for different depths within the GAC adsorber as a function of the number of bed volumes treated. The figure shows that the optimal filter depth was 5 ft. At greater filter depths, increasing the bed depth reduced the bed volumes treated before breakthrough. This observation is indicative of preloading* by NOM, which reduces the accessible adsorption sites for taste- and odor-causing substances. By 8000 bed volumes, 50 percent of the DOC had broken through. Also, a bed depth of 3 ft appeared to be too short to effectively reduce TON.

Figure 13.21*b* shows the progression of the TON mass transfer zone through the GAC bed. After 18 days of operation, the TON reached a value of 3 at a GAC depth of 1 ft, and after 60 days the same TON level was found at a bed depth of 3 ft. After 130 days of operation (13,000 bed volumes), the entire GAC bed had become ineffective for reaching the effluent TON goal.

* *Preloading* means the adsorption of NOM or similar organic matter in the lower section of the adsorber, thereby reducing the adsorption capacity for the target compound, before the target compound reaches that section.

Some useful observations have been made using data from full-scale GAC systems. Background organic matter, for example, usually breaks through much earlier than the taste and odor compounds (Love et al., 1973; Robeck, 1975), as can be observed in Figure 13.21*a*.

The experience at Stockton East Water District, Stockton, CA, showed that factors other than adsorption may complicate the removal of MIB and geosmin (Thomas, 1986). GAC filters were operated over a period of approximately 2 yr when MIB and geosmin in the adsorber influent were often in the 5- to 20-ng/L range. Extraction of GAC taken from cores of the filter after about 2 yr of operation showed no measurable MIB, and geosmin levels of about 0.1 to 0.2 μg/g GAC in the upper part of the filter (detection limits were 0.01 μg/g). These measurements were consistent with the concept that MIB, and possibly some geosmin, were removed by biodegradation, because both compounds are biodegradable (Namkung and Rittmann, 1987; Silvey, 1964).

One hypothesis for an MIB and geosmin removal mechanism is that adsorption capacity is needed to prevent these molecules from passing through the GAC adsorber. After the initial accumulation on the GAC, a biofilm develops that is capable of biodegrading these compounds. They then desorb and diffuse to the biofilm, where they are biologically oxidized. The ability of GAC to remove odor decreases over time because adsorption sites are gradually taken up by natural organics and thus are not available for MIB and geosmin. GAC replacement is necessary when adsorption capacity has been used up by these competitors.

GAC adsorbers with an 11-min EBCT were used to control a sulfide odor problem at the Goleta Water District in California (Lawrence, 1968). The odor was attributed to hydrogen polysulfides in the groundwater that were not removed by aeration and that GAC was only partially effective in removing (Monsitz and Auinesworth, 1970). Influent TON varied from 6 to 1000, while effluent TON ranged from odor free to 35 (Love et al., 1973). These beds were in service for 2 yr before the activated carbon was replaced.

Total Organic Carbon and DBP Precursors. The type and adsorbability of NOM in water vary widely from location to location. Sufficient differences in adsorbability exist, with respect to the quantity of TOC adsorbed and the competitive effect of the TOC on adsorption of trace organic compounds, that adsorption tests should be done on the water in question if the results are needed for design. However, Figure 13.17 can be used to provide estimates of the range of adsorbability for conventionally treated waters.

Roberts and Summers (1982) presented an excellent summary of full-scale plant performance for TOC removal. They evaluated removals from 47 different plants, including some wastewater reclamation plants. The ranges of design conditions are given in Table 13.5, and typical breakthrough curves are given in Figure 13.22. The

TABLE 13.5 Design Conditions for GAC Adsorbers

Parameter	Median	Range	Typical range
EBCT, min	10	3–34	5–24
Depth of bed, m	1	0.2–8	0.5–4
Hydraulic loading, m/h	6	1.9–20	2.6–17
Influent TOC, mg/L	3.5	1–16	2–6

Data from 47 plants were analyzed. m/h × 0.42 = gpm/ft^2.
Source: Roberts and Summers, 1982.

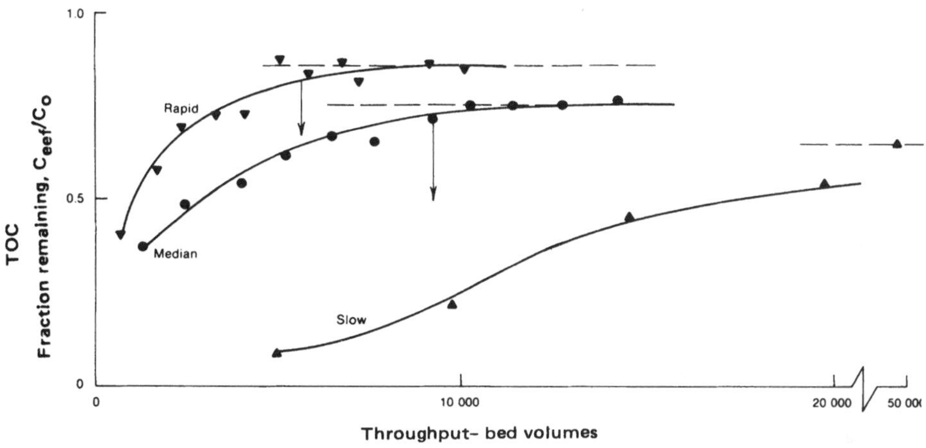

FIGURE 13.22 Representative TOC breakthrough curves. (*Source:* Roberts and Summers, 1982.)

breakthrough curves reach an effluent concentration plateau about 10 to 25 percent below the influent concentration. Roberts and Summers found that the amount of organic matter remaining in the effluent immediately after start-up decreased as the EBCT increased to about 20 min, and that the time to reach a steady-state effluent concentration increased as the EBCT increased.

Graese, Snoeyink, and Lee (1987a) summarized TOC removal data for a criterion of 50 percent removal and showed that adsorbers with EBCT values of less than 10 min had a life of less than 30 days, and that service time increased as EBCT increased. To obtain the lowest-cost adsorption system for TOC removal, EBCTs in the range of 10 to 20 min should be closely examined.

The performance for GAC removal of DBP precursors usually parallels that for the removal of TOC (Hooper et al., 1996a; Lykins, Clark, and Adams, 1988). Symons et al. (1981) have summarized much of the research prior to 1981. Hooper et al. (1996a), Solarik et al. (1995a,b), and Summers et al. (1994) assessed the use of TOC and UV absorbance as indicators of DBP precursor breakthrough for seven source waters and the following DBPs: THMs, haloacetic acids (HAAs), TOX, and chloral hydrate (CH). The precursors were assessed using the uniform formation conditions (UFC) test (Hooper et al., 1996a). In all but one water, TOC was a good conservative surrogate in that it broke through prior to the DPB precursors, as illustrated in Figure 13.23. In one case, THM precursors broke through a filter adsorber with an EBCT of 6.3 min prior to TOC. Organics measured at a UV absorbance wavelength of 254 nm tended to break through after the DBP precursor. TOC removal can vary seasonally because the adsorbability and initial concentration may change with season (Solarik et al., 1995b).

The bromide concentration of the water has an important bearing on the use of GAC for DBP control. Symons et al. (1981) observed that formation of brominated THMs occurs more rapidly in GAC adsorber product water. Graveland, Kruithof, and Nuhn (1981) made a similar observation, and noted a shift to the formation of more highly brominated forms of THMs in GAC effluent. This stems from the rapid formation of bromine-substituted THMs compared to chloroform when precursor concentration is low during the first part of the breakthrough curve, because the DBP formation reactions are precursor limited (Summers et al., 1993).

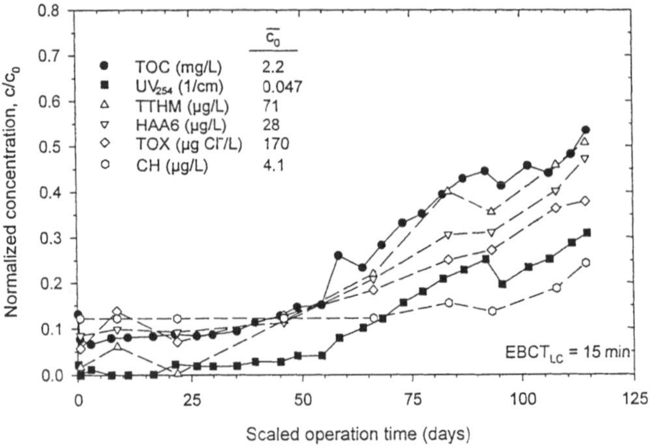

FIGURE 13.23 Comparison of RSSCT, UV_{254}, and UFC-DBP breakthrough for Salt River Project water. (*Source:* Hooper et al., 1996.)

A pilot study that evaluated the removal of organic carbon by GAC and BAC at the pilot scale was conducted in Newport News, VA (Snoeyink and Knappe, 1994). The design of the pilot test consisted of two columns in series for both the GAC and the BAC tests, with each column having an EBCT of 15 min. The influent to both the GAC and the BAC columns consisted of coagulated, settled, and filtered water; in addition, the BAC influent was ozonated. The TOC concentration of the influent was 3 mg/L, which gave a distribution system THMFP of 60 µg/L before ozonation. The duration of the study was 107 days for the GAC columns and 83 days for the BAC columns. Figure 13.24 depicts the THMFP breakthrough curves at different EBCTs. In both cases, EBCT had little impact on the breakthrough curves, with the exception of the GAC breakthrough curve for an EBCT of 5 min. (The 5-min curve was considered unreliable because of the difficulty in maintaining a constant depth of GAC above the first sampling port.) A comparison of GAC and BAC curves in Figure 13.24 shows that both the GAC and the BAC treatment trains resulted in breakthrough curves that were beginning to plateau at a dimensionless THMFP concentration of about 0.7, an indication that biological removal of organic carbon occurred both for the ozonated and the nonozonated influent. Additional run time is needed to fully establish the effect of ozone on the plateau. However, the shape of the breakthrough curves suggests that biological activity developed earlier in the BAC columns than in the GAC columns.

These results highlight some important issues that must be considered when dealing with biologically active filters. It must be determined whether the plateau region meets the treatment objective and whether the plateau region is a function of the ozone dose that was applied to the filter influent and EBCT. If the plateau region meets the treatment objective, very long run times can be achieved with BAC filters. Long pilot test runs are required to establish the plateau value as a function of EBCT. The life of the filter as a function of ozone dose and EBCT also needs to be examined if the treatment objective is exceeded before the plateau is reached. As shown in the Newport News pilot study, the life of the filter would have been greater for the BAC than for the GAC for a treatment objective that was reached before the plateau, because the BAC breakthrough curve was not as steep.

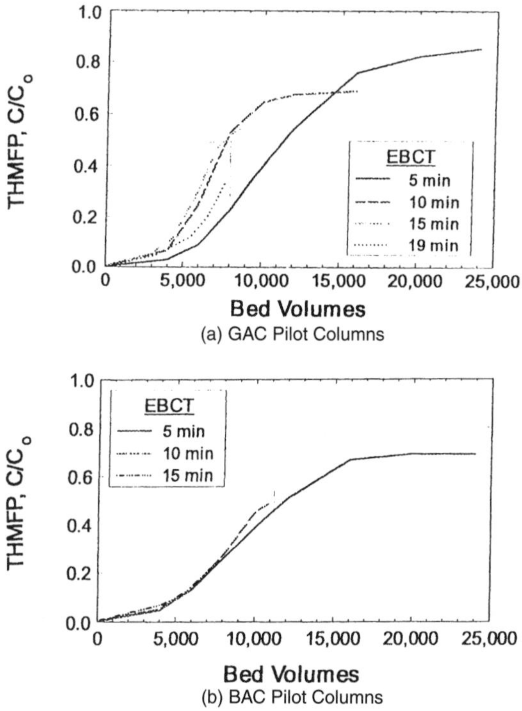

FIGURE 13.24 THMFB breakthrough curves. (*Source:* Snoeyink and Knappe, 1994.)

Removal of VOCs and SOCs. VOCs include compounds such as tetrachloroethylene and trichloroethylene, which are adsorbed relatively strongly, and 1,1,1-trichloroethane, 1,2-dichloroethane, and chloroform, which are adsorbed relatively weakly. The isotherm data in Table 13.1 show the relative adsorbability of these compounds. GAC adsorbers can be used to remove VOCs directly from contaminated water (Hess, 1981; Love and Eilers, 1982; Snoeyink, 1983) or from the off-gases from air-stripping towers (Crittenden et al., 1988). The best type of system will depend upon the type of VOC to be removed and air emission standards in effect.

VOCs are commonly found in contaminated groundwaters, and although this type of water is generally low in TOC (Solarik et al., 1995a; Zimmer, Brauch, and Sontheimer, 1989), large competitive effects are still noted. For example, Zimmer, Brauch, and Sontheimer (1989) studied the adsorption of three VOCs—tetrachloroethylene, trichloroethylene, and 1,1,1-trichloroethane—from distilled water and from groundwater in field installations. The full-scale field installations had EBCTs ranging from 12 to 15 min. The groundwater contained 0.4 to 2 mg/L of DOC and was applied at a rate of 10 to 15 m/h (4 to 6 gpm/ft). As the data in Table 13.6 show, the amount of water containing 50 μg/L of tetrachloroethylene that could be processed in the field was about 8 percent of the distilled water value. The capacity for trichlolorethylene and 1,1,1-trichloroethane was reduced to 22 percent and 24 percent, respectively, of the distilled water value. Apparently, natural organic compounds "foul" much of the GAC surface.

TABLE 13.6 GAC Bed Life With and Without Competition from Natural Organic Matter

Compound	Single-solute isotherm constants		Bed volumes to saturation, without competition, $C_0 = 50$ μg/L†	Observed capacity of full-scale adsorbers in practice*		
	K	1/n		K′	1/n′	Bed volumes to breakthrough, $C_0 = \sim 50$ μg/L, $C_{\text{breakthrough}} = \sim 5$ μg/L
Tetrachloroethylene	219	0.42	620,000	20.4	0.46	51,000
Trichloroethylene	78	0.46	197,000	27.4	0.61	44,000
1,1,1-Trichloroethane	23	0.60	38,000	62	1.4	9,400

* Zimmer, Brauch, and Sontheimer, 1989. An adsorber isotherm of the form $q = K'C_0^{1/n'}$, where C_0 is the adsorber influent concentration, K' and $1/n'$ are constants, and q is the amount adsorbed in mg/g at the time of adsorber breakthrough, was used to describe the data. These data are based on the performance of several full-scale plants in Germany (Zimmer, Brauch, and Sontheimer, 1989).

† Units on K and K' are $(\text{mg/g})(\text{L/mg})^{1/n}$. This number is calculated using the assumption that all influent compound is adsorbed until saturation is reached. 52,500 bed volumes can be processed in one year if the EBCT is 10 min.

Hess (1988) used pilot plant data (Love and Eilers, 1982) to predict the effect of influent concentration and type of VOC on GAC life (Figure 13.25), and trichloroethylene influent concentration and effluent concentration at breakthrough on GAC bed life (Figure 13.26). The figures show that GAC adsorber life decreases as influent concentration increases and as the required effluent concentration decreases. Each water that is treated will have different background organic matter; thus, Figures 13.25 and 13.26 should be taken only as estimates of performance. Other data for Mississippi River water at Jefferson Parish, Louisiana, at low VOC concentration show that a bed life of approximately 200 days (30,000 BV) can

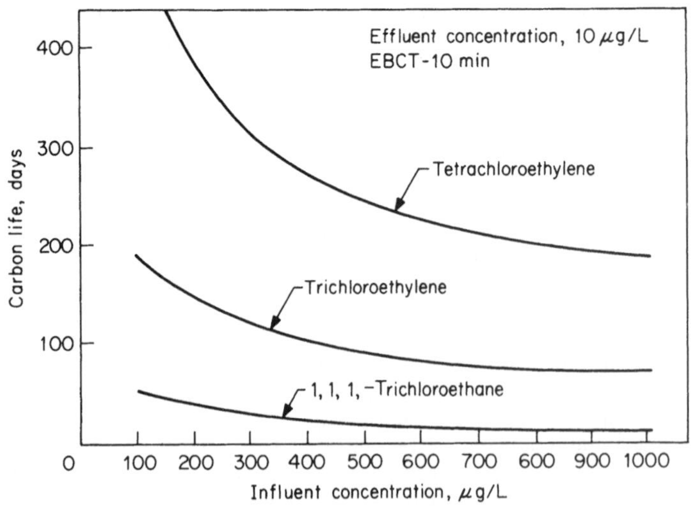

FIGURE 13.25 Effect of contaminant type and influent concentration on carbon life. (Note: 100 days = 14,400 bed volumes if EBCT = 10 min.) (*Source:* Hess, 1981.)

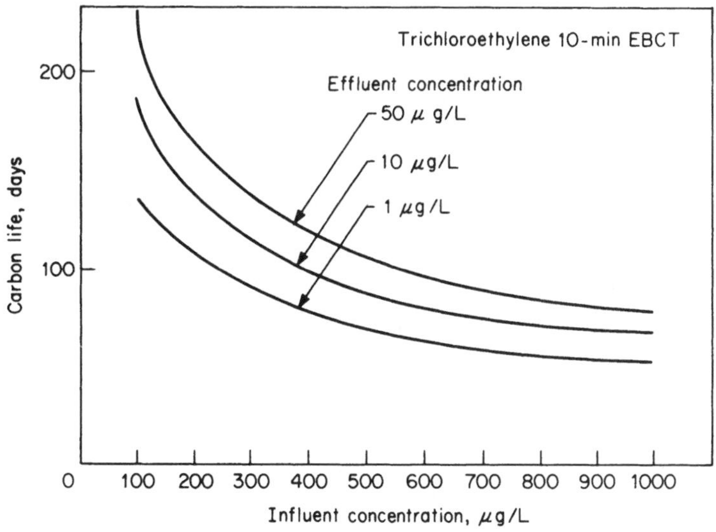

FIGURE 13.26 Effect of influent and effluent contaminant concentrations on carbon life. (Note: 100 days = 14,400 bed volumes if EBCT = 10 min.) (*Source:* Hess, 1981.)

be achieved for trichloroethylene for an influent concentration of 14 μg/L and an effluent concentration of 7 μg/L, and 400 days of life (60,000 BV) can be achieved for tetrachloroethylene for an influent concentration of 22 μg/L and an effluent concentration of 11 μg/L. In each case, the EBCT was 10 min (Qi et al., 1992; Snoeyink and Knappe, 1994).

Crittenden et al. (1988) developed mass transfer models to predict the removal of VOCs in air stripper off-gas by GAC, and investigated the regeneration of this GAC with steam. They found that adsorption efficiency decreased as relative humidity increased, and that heating of the gas stream to reduce the relative humidity to 40 to 50 percent was beneficial. Air stripping followed by GAC treatment to purify the off-gas was a good alternative to treatment of the water by GAC alone for several VOC treatment applications, even though steam regeneration of spent gas-phase GAC was not feasible. This advantage results because the natural organics that interfere with adsorption from the aqueous phase are not present in the stripper off-gas.

Symons et al. (1981) summarized most of the data available in 1981 on trihalomethane adsorption. Consistent with the size of the Freundlich K values listed in Table 13.1, the brominated THMs are adsorbed much better than chloroform. Graese, Snoeyink, and Lee (1987a) presented data from Cincinnati and Miami for an 80 percent chloroform removal criterion that showed that about 5000 to 6000 bed volumes of water could be processed before breakthrough. If the EBCT was less than 8 to 10 min, however, fewer bed volumes could be processed because of the short length of the MTZ. The performance will vary as the breakthrough criterion and the composition of the water change.

Pesticides are common synthetic organic chemicals that require removal. GAC beds were evaluated for the removal of selected pesticides spiked into river water (Robeck et al., 1965). One activated carbon column was exhausted for removal of background organic matter as measured by carbon chloroform extract and COD. Pesticide-spiked, sand-filtered river water was applied to the exhausted activated carbon column, and the effluent from it was applied to a second, fresh column. With concentrations of dieldrin as high as 4.3 μg/L, the effluent from the first column was reduced to 0.3 μg/L. Further reduction in the second column reached concentrations as low as 0.05 μg/L and often below the detection limit of 0.01 μg/L.

At Jefferson Parish, Louisiana (Koffskey and Brodtmann, 1981), 18 chlorinated hydrocarbon insecticides at a combined concentration of 18 to 88 ng/L were removed to less than about 5 ng/L for 1 yr in an adsorber with an EBCT of 20 min. Alachlor at concentrations of 13 to 593 ng/L was reduced to the limits of detectability during the same period. Atrazine was present at 30 to 560 ng/L and was removed essentially 100 percent by the postfilter-adsorber (24-min EBCT), but concentrations of 10 to 20 ng/L were often found in the effluent of the filter-adsorber (14-min EBCT). Other analyses at Jefferson Parish showed that phthalates, *n*-alkanes, and substituted benzenes at the ng/L level were not removed by GAC.

Baldauf and Henkel (1988) ran a pilot test using a single-stage adsorber 2.5 m deep. Samples were taken at different depths throughout the filter corresponding to EBCTs between 3 and 15 min as a function of time. The influent to the GAC column was a groundwater with a DOC concentration of 2.3 mg/L and an average atrazine concentration of 2.2 μg/L. Figure 13.27*a* shows the breakthrough curves as a function of the volume of water treated, and Figure 13.27*b* shows the same data as a function of the number of bed volumes treated. Figure 13.27*a* shows that atrazine reached the 0.1 μg/L MCL of the European Union (EU) very quickly at EBCTs of

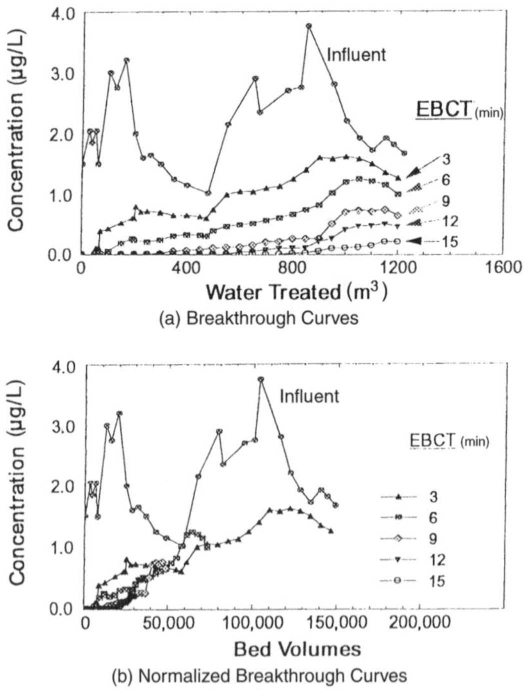

FIGURE 13.27 Atrazine breakthrough curves from groundwater. [*Source:* Baldauf, G., and M. Henkel, "Entfernung von Pestiziden bei der Trinkwasser-aufbereitung." In *20. Bericht der Arbeitsgemeinschaft Wasserwerke Bodensee-Rhein,* Teil 7, 127 (1988).]

3 and 6 min, while the longer EBCTs showed a much more gradual breakthrough. Figure 13.27*b* shows similar breakthrough behavior at each EBCT after the data are normalized on the basis of number of bed volumes treated, but the early breakthrough for EBCTs of 3 and 6 min is still apparent. The CUR was in the range of 65 to 82 g/m^3 for the 3- and 6-min EBCTs, decreasing to approximately 21 g/m^3 for EBCTs of 9 min and longer (Snoeyink and Knappe, 1994). At the shorter EBCTs of 3 and 6 min, it appears that the MTZ was not fully contained within the bed, leading to the immediate appearance of atrazine in the effluent.

Figure 13.28 shows normalized atrazine breakthrough curves obtained from pilot studies that investigated different EBCTs and preloading times (time of service before test) (Knappe, 1997). For comparative purposes, the breakthrough curves for virgin GAC are included. Influent atrazine concentrations for the three pilot tests were about 4 μg/L. The data obtained with a GAC that was preloaded for 5 mo (December 1991 sample) demonstrated a great decrease in atrazine removal efficiency as the EBCT decreased, while an EBCT of 8.6 min resulted in a performance similar to that of virgin GAC and an EBCT of 2.3 min led to a removal efficiency nearly as low as that observed for a GAC preloaded for 20 mo (March 1993 sample). In contrast, the GAC preloaded for 20 mo showed a uniformly low atrazine removal efficiency that was not a function of EBCT. The reduction in atrazine capacity due to NOM preloading primarily governed the observed breakthrough behavior.

Polynuclear aromatic hydrocarbons are effectively removed by GAC. Tap water spiked with 50 μg/L of naphthalene was passed through a GAC column at Cincinnati, OH (Robeck, 1975). After 7 mo of operation at 5 m/h (2 gpm/ft^2), the nonvolatile total organic carbon (NVTOC) front for 50 percent removal had penetrated the first 51 cm (1.7 ft) of the bed, while the 50 percent removal point for naphthalene was only about 5 cm (0.2 ft) down the column. Riverbank filtration followed by activated carbon treatment reduced the concentration of polynuclear aromatic hydrocarbons in water by about 99 percent, whereas rapid sand filtration followed by ozonation or chlorination was not effective (Andelman, 1973), probably because of the absence of GAC.

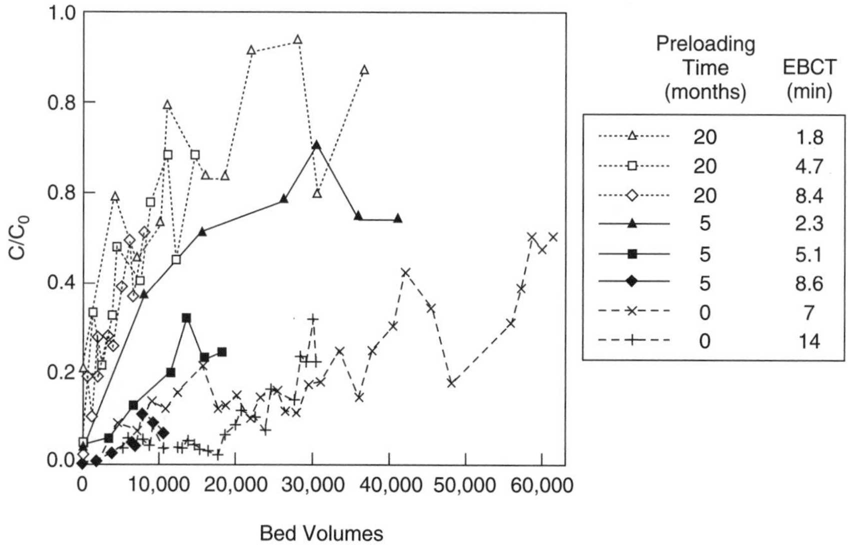

FIGURE 13.28 Effect of EBCT on atrazine removal efficiency at different preloading times.

Performance of Filter-Adsorbers versus Postfilter Adsorbers. Filter-adsorbers, which have resulted from the retrofitting of inert media (e.g., sand) beds, have proven to be effective for taste and odor removal for typically 1 to 5 yr. However, their use for removing weakly adsorbed compounds, including THMs and some VOCs, is limited (Graese, Snoeyink, and Lee, 1987a) because of the short EBCTs typically found in such systems. The typical EBCT of operating sand replacement systems is about 9 min at actual flow rates, but much less at design flow rates. Control of weakly adsorbing organic compounds with such short EBCTs would require replacement of GAC after only a few weeks or months of operation. Replacement at a high frequency would be operationally cumbersome and costly.

Contributing factors to lower organic loadings and earlier breakthrough for filter-adsorbers may be the reduced stratification resulting from mixing that occurs during frequent backwashing, and interference by trapped floc. A thorough economic analysis is required to determine whether the costs associated with a higher carbon usage rate in the filter-adsorber, if any, would be justified by the reduced capital costs of the filter-adsorber.

Chowdhury et al. (1996), Hong and Summers (1994), Summers et al. (1994, 1995, 1997), and Wulfeck and Summers (1994) have shown that filter-adsorbers operated in parallel may be a cost-effective approach for controlling DBP formation when the influent TOC concentrations are not high. For several waters, these researchers have shown that the carbon usage rate was not a function of EBCT in the range of 10 to 20 min. Based on the relationship shown in Figure 13.17, a water with an influent TOC concentration of 2.0 mg/L and an EBCT of 10 min would lead to an effluent TOC concentration of 1.0 mg/L after 60 days of operation. If the filter-adsorber effluents are blended, the replacement interval would be about 4 to 5 mo. Regular backwashing of filter-adsorbers has been shown to have little impact on the run time to 50 percent TOC breakthrough (Hong and Summers, 1994; Summers et al., 1997).

Turbidity Removal by Filter-Adsorbers*

A review of the available data on the efficiency of GAC for turbidity reduction compared to sand or coal showed that GAC is as good as or better than sand or coal (Graese, Snoeyink, and Lee, 1987a). However, when turbidity loadings were high, a more rapid rate of head loss buildup was experienced because of the larger uniformity coefficient of GAC. When the turbidity loading was low, the difference in uniformity coefficient seemed to make little difference in performance.

A sand layer is often retained in the filter below the GAC if turbidity is to be removed. The efficiency of turbidity removal by GAC indicates this is not always necessary for good performance, although it may be useful if the decision is made to use a large-diameter GAC or if very low levels of turbidity are desired. A sand layer below the GAC is a nuisance when GAC must be removed, because the sand often mixes with the GAC. It interferes with resale of the reactivated product if the reactivated carbon is to be reused, and it may interfere with reactivation itself. Additional sand may be required after GAC removal to keep the same amount of sand under the GAC. Gravel under the GAC can cause similar problems, but gravel is necessary to prevent the GAC from entering many types of underdrains.

Nozzle underdrains can be used that do not require either sand or gravel, and they can be used to provide both air and water for backwashing. These appear to

* This section includes only those aspects unique to the use of GAC. Chapter 8 discusses this topic in more detail.

have significant advantages over the types that require gravel; however, a pressure-relief device on the washwater influent line should be provided to prevent pressure damage to underdrains when and if the nozzles clog (Richard, 1986). Monitoring of the plenum pressure during backwashing can give advance warning of nozzle clogging problems.

The backwash system must be properly designed to prevent accumulation of mudballs in the GAC. Mudballs accumulate if the GAC is not properly cleaned, and GAC is difficult to clean because it is lighter than sand and anthracite. The density of GAC wetted in water is in the range of 1.3 to 1.5 g/cm^3; thus, providing sufficient energy into the bed to clean it is more difficult if water wash alone is used. Provision of good surface scour or air scour is important for good performance.

GAC PERFORMANCE ESTIMATION

Isotherm Determination and Application

Useful information can be obtained from the adsorption isotherm. It has been used in previous sections of this chapter to show the variation in adsorbability of different types of organic compounds, the differences between activated carbons, and, in some cases, the magnitude of competitive effects. Because the adsorption capacity of a GAC is a major determinant of the activated carbon's performance in columns, the isotherm can be used (1) to compare candidate activated carbons, (2) as a quality control measure for purchased activated carbons (some variation from lot to lot of a given manufacturer's activated carbon is expected), (3) as an input function to the mathematical methods that can be used to predict performance (discussed later), and (4) to evaluate the residual adsorption capacity of GAC in a contactor that has been in use and is thus partially exhausted. Important limitations to the use of isotherms must be recognized if they are to be properly used, however. The purpose of this section is to demonstrate how isotherms can be used to estimate performance, and to note several factors of importance in isotherm determination.

Types of Adsorbates. The determination procedures must take into account the adsorbate characteristics. The simplest procedure can be followed if a single non-volatile adsorbate is present in an aqueous solution with no background organics that might compete for adsorption sites. This isotherm is not a function of initial concentration, and no precautions need be taken to avoid volatilization. The test is conducted by agitating known volumes of the solution with accurately weighed portions of GAC until equilibrium is achieved. The time required for equilibrium can be estimated using the information in Randtke and Snoeyink (1983) (see discussion in an earlier section of this chapter), and can be determined by measuring solution concentration versus time. When equilibrium is reached, the data are analyzed using the mass balance,

$$q_e = \frac{(C_0 - C_e)V}{m} \tag{13.17}$$

where q_e = amount adsorbed per unit mass of carbon (mol/g or mass/g)
C_0 = initial concentration (mol/L or mass/L)
C_e = equilibrium concentration (mol/L or mass/L)
V = solution volume (L)
m = mass of carbon (g)

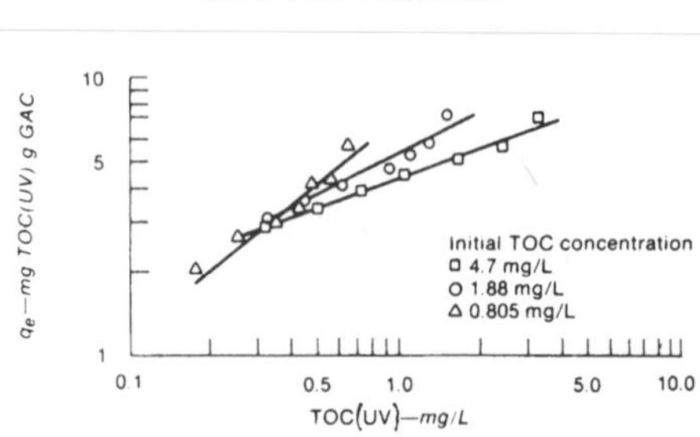

FIGURE 13.29 Effect of initial TOC concentration on the adsorptive capacity of carbon A (14 × 20 mesh pulverized prior to adsorption) for peat fulvic acid. (*Source:* Randtke and Snoeyink, 1983.)

The data can be plotted to determine the Freundlich parameters (plot log q_e versus log C_e according to Equation 13.2) or other parameters as appropriate.

Adsorption from a mixture of adsorbable organic compounds is more complicated because of the effect of the different adsorbabilities of compounds in the mixture. Figure 13.4 shows a typical isotherm for a heterogeneous mixture of compounds. The small fraction of strongly adsorbable compounds is often overlooked when determining an isotherm because of the analytical difficulty of obtaining good data for C_e close to C_0 (i.e., for very small doses of activated carbon). This fraction is important because it is responsible, at least in part, for the slow approach of effluent concentration to influent concentration in column operation. The small, but important, nonadsorbable fraction is often not detected if a sufficiently large dose of carbon is not added to the solution.

Initial concentration affects the adsorption of a complex mixture of organic matter, such as a fulvic acid, as is shown in Figure 13.29 (Randtke and Snoeyink, 1983). The difference in the isotherms is caused primarily by the different mass concentration of strongly adsorbable compounds in each of the solutions. The same type of difference is expected if the amount of a single compound that is adsorbed in the presence of different concentrations of background organics is measured. The isotherm of the single compound will be a function of the initial concentration of background organics if competition occurs between them.

Other adsorbate-related factors must be carefully controlled during isotherm determination. Biological activity in isotherm bottles must be controlled either if the NOM is biodegradable, such as after ozonation of a natural water, or if the target trace compound is biodegradable. Biological growth can be minimized if the isotherm is determined at low temperature, for example less than 5°C. If the adsorbate is volatile, equilibrium must take place in a headspace-free vessel, and mixing of the system must take place by continual inversion of the test bottle. Control of pH is important for compounds whose adsorption is a function of pH, and other salts in solution can be important for ionic substances or for substances that compete with organic ions for adsorption sites. The reader is referred to Randtke and Snoeyink (1983) for a more detailed discussion of these factors.

Activated Carbon. The major concerns related to activated carbon are (1) obtaining a representative sample for testing and (2) preparing the sample to obtain a rapid approach to equilibrium without destroying its representative nature. The adsorption capacity of some commercial activated carbons is a function of particle size because of the manufacturing process that is used (Randtke and Snoeyink, 1983); thus, selecting a sample for testing with the same particle size distribution as the bulk activated carbon is important. Activated carbon sizes may separate during shipment, so special equipment or techniques are required to obtain the representative sample. By using equipment such as a sample reducer and a sample splitter, representative samples from large quantities (bags) of GAC can be obtained. The sample reducer reduces sample sizes greater than 4.5 kg (10 lb) to representative samples that are smaller at a 16:1 ratio. Further reduction to a size that can be used for laboratory tests is possible using the sample splitter, or *riffle*. The particle size distribution and quality of the final sample are then representative of those of the original large quantity of GAC. Using such equipment to establish that a shipment of GAC meets specification is important. A simpler alternative that yields a good approximation of the characteristics of a batch is to sieve the sample and use only those particles in the dominant size range.

The time to reach equilibrium is an important function of particle size, but the adsorption capacity of a sample of activated carbon is not altered by crushing it to a smaller particle size (Randtke and Snoeyink, 1983). Grinding a granular activated carbon to 325 US Standard mesh (0.044 mm) or less reduces the time to reach equilibrium to 2 days for a large humic acid molecule, and much shorter times for smaller molecules. Avoiding significant biological activity should be possible if the sample is not seeded with microorganisms and the test time is 2 days or less. Prefiltration of the solution with a filter membrane having 0.45-μm pore diameter will remove most bacteria from the sample. Postfiltration of the sample through filter paper having 0.45-μm pore diameter will be necessary to remove the activated carbon before testing for the concentration of residual organics. Filter paper must be selected that does not contribute organic compounds that interfere with the quantitative analysis of the test compounds.

Isotherms can be used to obtain a rough estimate of activated carbon loading and bed life. These are useful in determining the applicability of GAC. Assuming that (1) all of the GAC in an adsorber will reach equilibrium with the adsorber influent concentration and (2) the capacity obtained by extrapolating the isotherm data to the initial concentration is a good value, the bed life, in bed volumes of water, can be calculated.* For example, if $(q_e)_0$ is the mass adsorbed (mg/g) when $C_e = C_0$, the bed life Y, the volume of water that can be treated per unit volume of carbon, can be calculated:

$$Y = \frac{(q_e)_0(\text{mg/g GAC})}{(C_0 - C_1)(\text{mg/L})} \cdot \rho_{\text{GAC}}(\text{g/L}) \tag{13.18}$$

where C_0 is the influent concentration, C_1 is the effluent concentration that represents an average for the entire column run, and ρ_{GAC} is the apparent density of the GAC. C_1 is zero for a strongly adsorbed compound that has a sharp breakthrough curve, and is the concentration of nonadsorbable compounds when such substances are present. An estimate of the activated carbon usage rate, the rate at which activated carbon is spent as defined in Equation 13.9, is given by

$$\text{CUR(g/L)} = \frac{(C_0 - C_1)(\text{mg/L})}{(q_e)_0(\text{mg/g})} \tag{13.19}$$

* These assumptions are consistent with the assumption that the length of the MTZ, L_{MTZ}, is negligible.

EXAMPLE 13.1 Estimate the bed life and carbon usage rate for a GAC adsorber that is to treat water containing 1 mg/L of toluene from solution. The density of the carbon $\rho_{GAC} = 500$ g/L.

1. From Table 13.1, $K = 100\ (\text{mg/g})(\text{L/mg})^{1/n}$ and $1/n = 0.45$ for toluene.
2. Applying the Freundlich equation (Equation 13.1),

$$(q_e)_0 = KC^{1/n}$$
$$= 100\ (\text{mg/g})(\text{L/mg})^{0.45}(1\ \text{mg/L})^{0.45}$$
$$= 100\ \text{mg/g}$$

3. Applying Equation 13.21, assuming that $C_1 = 0$, gives

$$Y = \frac{(q_e)_0(\text{mg/g})}{(C_0 - C_1)(\text{mg/L})} \cdot \rho_{GAC}(\text{g/L})$$
$$= \frac{100\ \text{mg/g}}{1\ \text{mg/L}} \cdot 500\ \text{g/L}$$
$$= 50{,}000\ \text{L H}_2\text{O/L GAC} = \text{bed life}$$

If the EBCT is 15 min, 4 bed volumes per hour or 96 bed volumes per day can be processed. The bed life in days is then 50,000 bed volumes/96 bed volumes per day = 521 days.

4. Applying Equation 13.22 to obtain CUR gives

$$\text{CUR} = \frac{(C_0 - C_1)(\text{mg/L})}{(q_e)_0(\text{mg/g})}$$
$$= \frac{1\ \text{mg/L}}{100\ \text{mg/g}}$$
$$\text{CUR} = 0.01\ \text{g GAC/L H}_2\text{O}$$

Note: the bed life will be reduced, and the CUR will be increased, under practical conditions where other organics are in the water that compete for adsorption sites.

Several limitations affect the use of isotherm data to estimate activated carbon usage rate. This approach is only valid for columns in series, or for very long columns for which the assumption that all the activated carbon in the column is in equilibrium with the influent concentration is valid. Furthermore, it provides no indication of the effect of biological activity. Finally, when competition does occur, its impact in a batch test often is not the same as is experienced in a column, because the molecules will separate in a column according to their strength of adsorption. The NOM that adsorbs in the upper region of a column will be dominated by the strongly adsorbing fraction of organic matter, especially in the initial portion of a column run, whereas the more weakly adsorbing fraction of NOM will selectively accumulate in the lower reaches of the column. In a batch test, all carbon particles are exposed equally to each fraction of the NOM. Competitive effects of NOM on trace organics may be much greater in a column for this reason (Zimmer, Brauch, and Sontheimer, 1989).

One approach to better estimate the adsorption capacity in the field based on isotherm values is to increase the distilled water carbon usage rates based upon the adsorbability of the compound. Ford et al. (1989) proposed that a natural water correction factor be applied to the distilled water carbon use rate. From their data, the following relationship between natural water CURs (CUR_{NW}) taken from full- and pilot-scale adsorbers and those from isotherms in distilled water (CUR_{DW}) can be developed:

$$CUR_{NW} = 0.7\ CUR_{DW}^{0.5} \tag{13.20}$$

where the CUR is in lb/1000 gal and all values of CUR_{DW} are below 1 lb/1000 gal. While there is a large amount of scatter in the more than 75 data points evaluated, the square root nature of the relationship shows that the more strongly adsorbing compounds—that is, low CURs—will be more impacted by the presence of NOM. For example, a weakly adsorbing compound with a high CUR_{DW} of 0.1 lb/1000 gal will have a CUR in natural water of 0.22 lb/1000 gal, a 2.2-fold increase. By contrast, a strongly adsorbing compound with a low CUR_{DW} of 0.001 lb/1000 gal will have a CUR in natural waters of 0.022 lb/1000 gal, a 22-fold increase.

Small-Scale Column Tests and Applications

Various types of small-scale column tests have been developed to obtain data that can be used to estimate the performance of large contactors. Rosene et al. (1980) developed the high-pressure minicolumn (HPMC) technique that uses a high-pressure liquid chromatography column loaded to a depth of 2 to 2.5 cm (0.8 to 1 in.) with about 50 mg of activated carbon. The activated carbon is crushed until it passes through a 0.149-mm (100 mesh) sieve. This column is used with a headspace-free reservoir and a flow rate of 2 to 3 mL/min to determine the capacity of activated carbon for volatile compounds. The headspace-free reservoir is not required if the adsorbable molecules are not volatile. Bilello and Beaudet (1983) modified the HPMC by using only the 230 × 325 mesh fraction (0.063 × 0.044 mm) of crushed GAC, which led to lower pressure drop in the column. Prefiltration of samples containing many particles was sometimes necessary to prevent excessive head loss.

The HPMC technique has several advantages. It allows rapid determination of GAC capacity under conditions that are closer to those experienced in full-scale installations, and is especially advantageous for volatile compounds because it is easier to avoid their loss. It has the potential for being used in a way that will give good information on the extent of competition, although more research is needed to determine how these tests should be run to duplicate the competition caused by NOM in full-scale systems. Thus, HPMC can be used to screen potentially applicable carbons. It also allows a rapid determination of the order of elution of compounds expected from a carbon column. Its disadvantages include the large pressure drop, the need for a high-pressure pump, and the difficulty of getting good kinetic information that can be scaled up to accurately predict the performance of large adsorbers. Other types of small columns can also be used to obtain good kinetic data, which, together with equilibrium data from isotherms or HPMC tests, can be used in mathematical models to predict the performance of larger columns (Liang and Weber, 1985).

Crittenden, Berrigan, and Hand (1986) and Crittenden et al. (1987, 1991) developed an alternative procedure called the rapid small-scale column test (RSSCT) for designing small columns such that the data obtained could be scaled up to predict the breakthrough behavior of larger columns. In their first studies (Crittenden, Berrigan, and Hand, 1986), these researchers used a mathematical model to develop the relationship between the breakthrough curves of large-scale and small-scale columns. Based on the assumption that the internal diffusion coefficient was constant and not a function of GAC particle diameter, they found

$$\frac{EBCT_{SC}}{EBCT_{LC}} = \left[\frac{d_{SC}}{d_{LC}}\right]^2 = \frac{t_{SC}}{t_{LC}} \tag{13.21}$$

where d is the diameter of the GAC particle, the subscript SC refers to the small column, and the subscript LC refers to the large column. The fraction t_{SC}/t_{LC} is the time required to conduct a small-column test divided by the time necessary to conduct a large-column test. This constant diffusivity design equation is valid for various combinations of pore and surface diffusion controlled adsorption. It predicts that a column with 0.1-mm-diameter GAC, and an EBCT 0.01 times as large as that of a column with 1-mm-diameter GAC, will produce the same breakthrough curve. Furthermore, the time required to process a given number of bed volumes of water through the small column is 0.01 times the time required to process the same number of bed volumes through the large column.

Subsequent studies by Crittenden et al. (1987) showed that the internal diffusion coefficient could not always be assumed to be constant with respect to particle diameter. These researchers developed a second relationship, termed the *proportional diffusivity design,* based on the assumption that the internal diffusion coefficient varied linearly with particle diameter. The following equation resulted:

$$\frac{EBCT_{SC}}{EBCT_{LC}} = \frac{d_{SC}}{d_{LC}} = \frac{t_{SC}}{t_{LC}} \tag{13.22}$$

where each of the terms has the same definition as in Equation 13.21. Considerably longer tests are required if Equation 13.22 applies instead of Equation 13.21. For example, if the particle diameter ratio is 0.1:1.0, the small-scale test will require one-tenth the time of the full-scale test if its EBCT is one-tenth as large. Thus, the time savings is not as great as can be achieved if the diffusion coefficient is constant with respect to particle diameter. An example of an RSSCT column is shown in Figure 13.30. Particle sizes commonly used range from 0.05 to 0.2 mm, and column diameters range from 5 to 11 mm (Crittenden et al., 1991). Using these particle and column

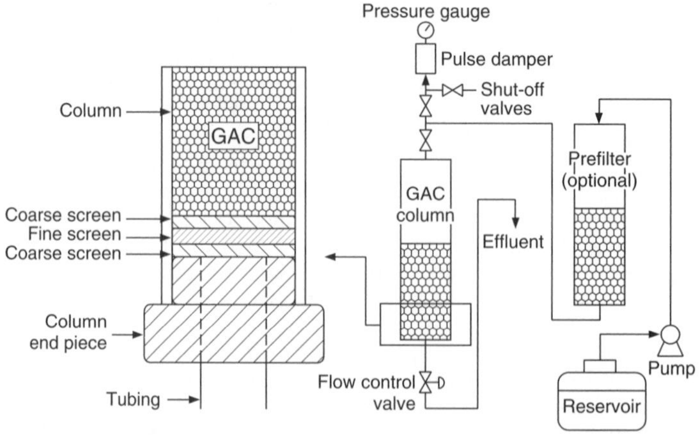

FIGURE 13.30 RSSCT setup. (*Source:* Summers, Hooper, and Hong, 1996.)

sizes yields RSSCT flow rates of 50 to 150 ml/min for the constant diffusivity design and 5 to 20 ml/min for proportional diffusivity design.

Unfortunately, the a priori selection of the design equation, either constant (Equation 13.21) or proportional (Equation 13.22) diffusivity, is not possible for specific organic compounds such as VOCs, SOCs, or pesticides. At least one pilot-plant column run with the water to be tested is needed to verify the RSSCT design for these compounds (Crittenden et al., 1991). However, the proportional diffusivity design approach in Equation 13.22 seems to work in all cases for NOM. Summers, Hooper, and Hong (1996) and Summers et al. (1995) have recently reviewed the use of proportional diffusivity design for nine raw water sources and four GAC types. The 30 verification runs showed that the RSSCT predicted the pilot- or full-scale breakthrough behavior for TOC, UV absorbance, and precursors of THMs, HAAs, TOX, and CH. An RSSCT design and operation manual has been recently developed for the treatment studies requirement by USEPA's Information Collection Rule (Summers, Hooper, and Hong, 1996).

Blending of the effluent from parallel GAC contactors can be simulated with the RSSCT by collecting all of the RSSCT column effluent into one container and sampling that composite container with time (Chowdhury et al., 1996; Roberts and Summers, 1982). The resulting integrated breakthrough curve can be used to determine how long a single contactor can be run until the blended effluent exceeds a target concentration. This analysis assumes that a large number of contactors will be used, and that the carbon will be replaced at equally spaced intervals (see the following discussion on analysis of parallel contactors).

The RSSCT technique is very useful because it significantly reduces the time required to obtain typical performance data. It does have limitations, however. Much of the variability in effluent concentration is caused by changes in influent concentration. A test that lasts from a fraction of a day to at most a few days cannot account for the changes expected over a much longer time interval. In addition, the short-term test cannot determine the effect of biological activity in the full-scale column. It also cannot be used to determine the remaining capacity of carbon that has been in service for a period of time, because unlike virgin carbon, used carbon cannot be ground to a smaller size without affecting its capacity.

Pilot Plant Testing

Pilot testing is useful in most situations. Isotherm or small-scale column tests, or the experience of others (e.g., Figure 13.17), may provide good reason to anticipate that activated carbon will be a technically and economically feasible solution to a water quality problem. The treatability of each water is somewhat unique with respect to day-to-day and season-to-season variation in contaminants, background organics, other water quality parameters that can affect adsorption, and the type and level of pretreatment. Thus, questions remain about how activated carbon will perform on a specific water even though results of small-scale tests are available. The pilot study should (1) show the effectiveness of the carbon, (2) permit a determination of the best design parameters to use for the full-scale system, and (3) establish the best operating procedure to use. A good estimate of the cost of the full-scale system can then be made. Although pilot testing is expensive, the excessive cost of operating an improperly designed system because of insufficient information may more than justify the cost of a pilot study to obtain that information.

Pilot studies are not always necessary. Available information may be sufficient to determine whether activated carbon should be used instead of an alternative process, or whether a particular design will achieve the desired removal efficiency. The type of GAC may be chosen based on the experience of others with similar water supplies. Use of GAC for taste and odor removal from many water supplies provides an example of this (see previous discussion of taste and odor removal performance). Also, the cost of a pilot study may not be justified by the potential savings that would be made possible by an optimized design. GAC adsorbers for removal of VOCs from small supplies may be designed on the basis of existing information, possibly supplemented with isotherm or small-scale column test information, with an appropriate safety factor to compensate for the lack of a pilot test.

Pilot Plant Design and Operation. The pilot plant should be designed with the same pretreatment that will be used in the full-scale system. It should be operated at the location where the full-scale adsorber is to be constructed so the water quality for pilot and full-scale operations will be the same. Data that are often useful for designing a pilot plant are:

1. A detailed chemical analysis of the water including the concentrations of specific organics, TOC, DBP precursors, taste and odor compounds, turbidity, and inorganics that might affect adsorption and removal by biodegradation, such as dissolved oxygen, pH, hardness, and alkalinity. Preferably, these analyses will be done at different times so that the range of concentrations expected at the full-scale plant can be estimated accurately. The analyses should also show what pretreatment is necessary, if any.
2. Isotherm tests or small-scale column tests, preferably on water samples taken at different times and coupled with an analysis of the experience of others, are needed to obtain estimates of activated carbon usage rates for the different candidate activated carbons. The best activated carbon for pilot testing can then be selected. Activated carbon cost and head loss during operation should be included in this consideration. A preliminary sizing of the full-scale plant should be done to establish flow rate, depth, and EBCT. The pilot plant then should be designed and operated to show whether these factors are optimum or whether they should be modified for best performance.

The GAC influent should be representative of that anticipated for the future full-scale system. If this influent is not available, the pilot plant should include the necessary pretreatment processes. For postfilter-adsorbers, three or four GAC columns in series, such as shown in Figure 13.31, provide maximum operating flexibility. Alternatively, one or two long columns with several taps can be used so that samples can be taken at different depths. For filter-adsorbers, the GAC and other media, including underdrain media, would be contained in one column. The materials of construction may range from plastics such as clear acrylic and polyethylene to the more expensive stainless steel, polytetrafluoroethylene (PTFE) or glass. The study of Kreft et al. (1981) showed no leaching of VOCs or TOC from, or adsorption of VOCs and TOC on, clear acrylic columns and polyethylene tubing in a pilot plant. These researchers did find that phthalates leached from transparent vinyl plastic tubing, so this material should not be used where phthalates will interfere. The possibility that some SOCs may be adversely affected by some materials of construction cannot be ruled out, however. Stainless steel, PTFE, and glass can be used if avoiding any question of possible organics added by plastic is necessary.

The total depth of the activated carbon should be sufficient to allow determination of the optimum EBCT (or optimum depth at a given flow rate). A total depth

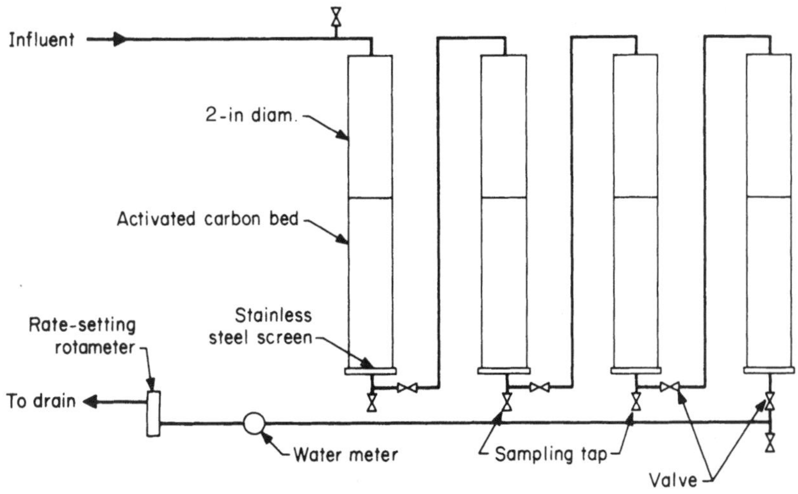

FIGURE 13.31 Downflow pilot carbon columns.

greater than the depth likely to be used in the full-scale plant, 3 to 4 m (10 to 13 ft), should be used. The ratio of column diameter to average particle diameter should be at least 30:1 (Summers, Hooper, and Hong, 1996), or it should be demonstrated that short-circuiting does not occur if smaller ratios are used. Hydraulic loading rates of 5 to 12.5 m/h (2 to 5 gpm/ft^2) are usually used (Kornegay, 1987). Within this range, the assumption can be made that at a given EBCT, activated carbon usage rate is not a function of flow rate. If possible, the duration of the test should be long enough to show the impact of seasonal variations in quality of the source water. The concentrations of all substances that might have a significant effect on the primary contaminant should be monitored. The data obtained can be used to calculate total annual costs (capital plus operating costs) for alternative designs to determine the most economical design. A pilot plant design and operation manual has been recently developed for the DBP precursor treatment studies requirement of USEPA's Information Collection Rule (Summers, Hooper, and Hong, 1996).

Analysis of Column Data

The GAC effluent data collected at different EBCT values should be plotted as breakthrough curves to show the effect of EBCT on effluent quality. The effluent concentration C_{eff}, or the effluent concentration divided by the influent concentration C_{eff}/C_0, can be plotted against time of operation, volume of water processed (see Figure 13.32), or bed volumes of water processed (see Figure 13.33). Because flow rate may vary with time, the latter two abscissa parameters are the most useful and most commonly used. Because most full-scale plants use time, a double abscissa is suggested with time being one independent variable. Using the number of bed volumes as the abscissa has the important advantage of normalizing the data collected at different depths and at different flow rates. The number of bed volumes processed must be calculated using only the volume of carbon above the sample port, however.

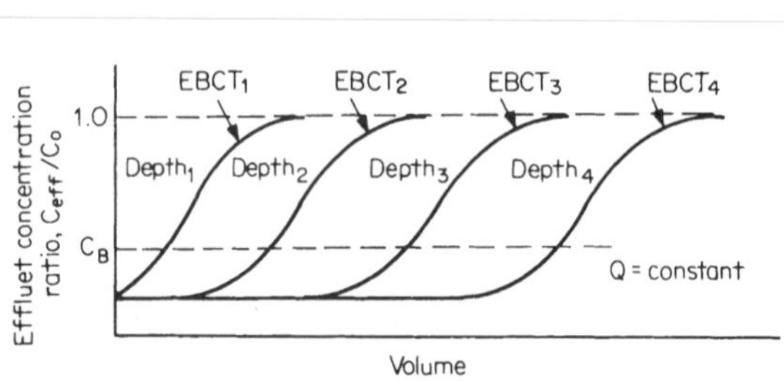

FIGURE 13.32 Breakthrough curves as a function of total volume of water processed.

As EBCT values increase, the C_{eff}/C_0 versus bed volume breakthrough curves eventually will superimpose as shown in Figure 13.33; when this occurs, the activated carbon usage rate will be a minimum and a function only of the breakthrough criterion that is used. The CUR, as originally defined in Equation 13.9, can be calculated from the breakthrough curve:

$$\text{CUR(g/L)} = \frac{\rho_{\text{GAC}}\text{(g/L)}}{\text{Bed Volumes to Breakthrough}} \tag{13.23}$$

where ρ_{GAC} is the apparent density of the GAC.

The breakthrough curves should be carefully analyzed to determine whether the number of bed volumes to breakthrough increases to a maximum value and then begins to decrease as EBCT increases. Arbuckle (1980) observed that GAC at the end of an adsorber train adsorbed less trace compound per unit mass than GAC at the front of the adsorber train. Similar effects were observed for VOC adsorption from groundwater (Baldauf and Zimmer, 1986; Zimmer, Brauch, and Sontheimer, 1989) and for adsorption of chlorinated organics from Rhine River water (Summers et al., 1989). The explanation given for this phenomenon was that large molecules of background organics adsorbed first in the lower reaches of the adsorber and either blocked pores or occupied adsorption sites so that adsorption of trace compounds was decreased. The effect, called the *premature exhaustion* or *preloading* effect, results in higher activated carbon usage rates as EBCT is increased beyond a certain value. The implication of this phenomenon is that a single EBCT exists that will give the lowest activated carbon usage rate.

Bed depth–service time plots (see Figure 13.14) also are useful as an aid to interpreting breakthrough curve data. For a constant flow rate, the length of time to breakthrough (ordinate) is plotted as a function of GAC depth (abscissa). Data from

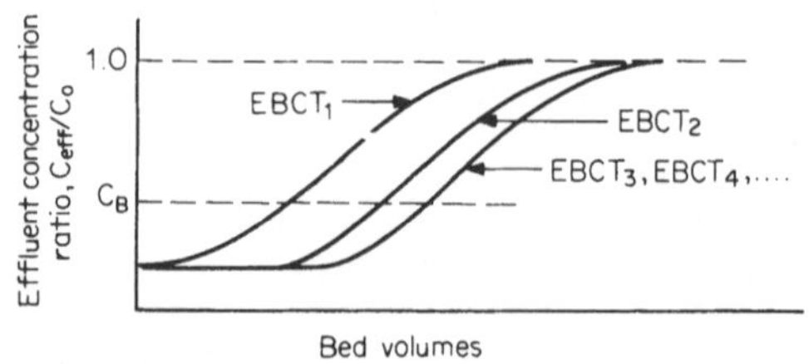

FIGURE 13.33 Breakthrough curves as a function of the number of bed volumes processed.

plots similar to those in Figure 13.32 are used, except that time of operation instead of volume is used as the abscissa. This allows a determination of the $L_{critical}$ (see Figure 13.14) for a certain breakthrough concentration.

Parallel Column Analysis. One of the objectives of pilot- or small-scale column testing should be to determine the activated carbon usage rate if GAC columns are to be used in parallel. Pilot column data can be analyzed using the approach of Roberts and Summers (1982), who showed that the fraction, or concentration, of organic matter remaining in the composite effluent of parallel adsorbers, $\bar{f}$, was given by:

$$\bar{f} = \frac{1}{n}\sum_{i=1} f_i \tag{13.24}$$

where f_i is the fraction, or concentration, of organic matter remaining in the effluent from the ith adsorber, and n is the number of adsorbers, each of equal capacity, in parallel. Values of f_i can be determined from a single breakthrough curve such as is shown in Figure 13.34*a,* assuming that replacement of GAC in each adsorber will take place at equal intervals. Given that θ_n is the number of bed volumes processed through each adsorber in the parallel system at the time of replacement, the abscissa of the breakthrough curve for the individual contactor from 0 to θ_n is divided into

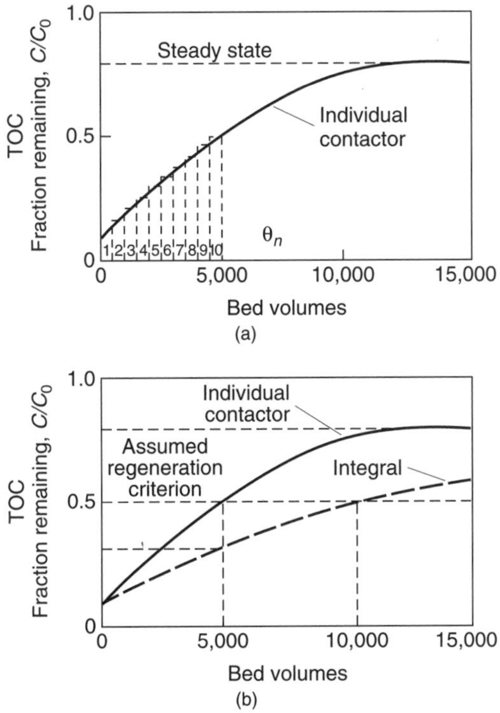

FIGURE 13.34 Integral breakthrough curve for characterizing performance of multiple parallel contactor operation. (*Source:* Roberts and Summers, 1982.)

θ_n/n equal increments. For a given value of θ_n, the value of f_i for each increment can be read from the figure and Equation 13.24 can be used to calculate the concentration of the blended water. For example, if 10 adsorbers are used in parallel, determination of $\bar{f}$ for several different values of θ_n using Equation 13.24 leads to the integral curve shown in Figure 13.34*b*. If the 10 contactors are operated in parallel, each adsorber can process 10,000 bed volumes of throughput if the effluent TOC criterion is 50 percent of the influent, compared to 5,000 bed volumes if only a single contactor were used or if all 10 contactors were operated in parallel but were replaced at the same time.

Data from an individual GAC pilot contactor, plotted as shown in Figures 13.33 and 13.34*a*, are needed to develop the integral breakthrough curve shown in Figure 13.34*b*. The integral curve is developed assuming (1) that each contactor will be the same size, (2) that each contactor will be operated for the same number of bed volumes θ_n before replacement, and (3) that only one contactor is replaced at a time at intervals of θ_n/n bed volumes for n contactors in parallel, that is, in staggered operation. For a given value of n, $\bar{f}$ versus θ is calculated by applying Equation 13.24 for different assumed values of θ_n. (The procedure is illustrated in Figure 13.34 and related discussion for $\theta_n = 5000$ bed volumes and an n of 10.) The number of bed volumes to breakthrough can then be determined from the integral curve. Separate integral curves should be calculated for each possible value of n. An economic analysis can then be made for each parallel arrangement to determine the best choice.

EXAMPLE 13.2 An ozone/GAC pilot plant was used to remove THMFP from a shallow reservoir supply after coagulation, sedimentation, and filtration. The GAC columns had a diameter of 4 in (ID). Samples were taken from ports located 20 in (#1), 44 in (#2), and 64 in (#3) from the inlet. A flow rate of 0.24 gpm gave corresponding EBCT values of 4.6, 10.1, and 14.8 min. The data (except for points at large volumes processed at ports 2 and 3) are shown in Figure 13.35. Calculate the GAC replacement frequency if four GAC adsorbers, each with 5-min EBCTs, are to be operated in parallel. The THMFP concentration of the blended GAC effluent is to be 25 μg/L. The C_0 for THMFP is 60 μg/L.

1. Plot the breakthrough curve (shown in Figure 13.35) as a function of the bed volumes of water treated. To convert volume of water processed to bed volumes,

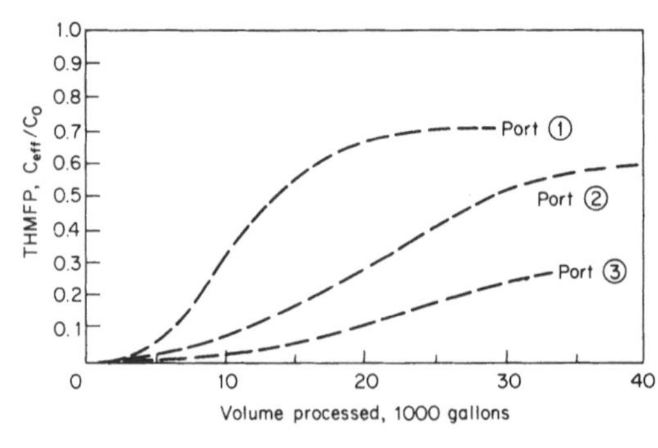

FIGURE 13.35 Breakthrough curve for samples.

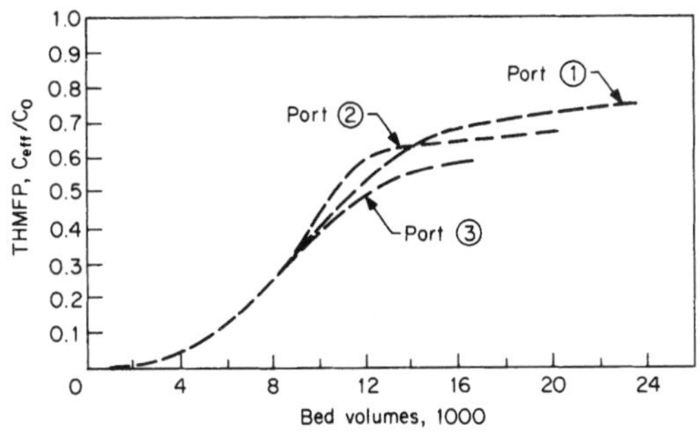

FIGURE 13.36 Breakthrough curve for full-scale system.

divide the volume processed by the volume of GAC above the port. The volume of GAC above each port is 1.09, 2.39, and 3.48 gal for ports 1, 2, and 3, respectively. For example, the point (vol = 20,000 gal, $C_{eff}/C_0 = 0.28$) for port 2 in Figure 13.35 corresponds to

$$\mathrm{BV} = \left(\frac{20{,}000}{2.39} = 8{,}400\right) \qquad \left(\frac{C_{eff}}{C_0} = 0.28 \text{ in Fig. 13.36}\right)$$

2. Use Figure 13.36 to obtain a good estimate of the breakthrough curve expected for a full-scale system. More emphasis is given to data from ports 2 and 3 because the depth of beds for these ports should be much greater than the critical depth. The EBCT for port 3 is also about 15 min, and should have about the same amount of biological degradation as expected for the full-scale system. Use the composite of the three curves as the resulting breakthrough curve, which is shown in Figure 13.37.
3. Determine the expected breakthrough curve when four columns in parallel are blended (integral curve) by applying Equation 13.24, as discussed previously.

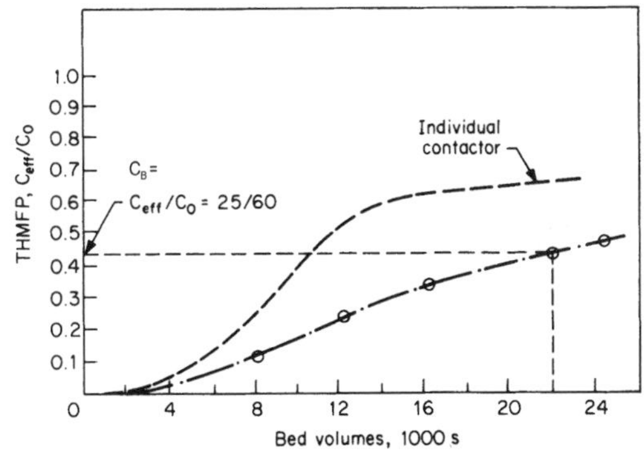

FIGURE 13.37 Composite breakthrough curve.

Assumed values of θ_4	Corresponding values of f_i (from Fig. 13.37)	$\bar{f}$
8000 BV	$f_1(2000\text{ BV}) = 0.02$	
$\left(\frac{8000}{4} = 2000\right)$	$f_2(4000\text{ BV}) = 0.05$	
	$f_3(6000\text{ BV}) = 0.13$	$\frac{1}{n}\sum f_i = \frac{0.43}{4} = 0.11$
	$f_4(8000\text{ BV}) = 0.23$	
	$\sum f_i = 0.43$	
12,000 BV	$f_1(3000) = 0.03$	
$\left(\frac{12{,}000}{4} = 3000\right)$	$f_2(6000) = 0.13$	
	$f_3(9000) = 0.35$	$\frac{1.03}{4} = 0.26$
	$f_4(12{,}000) = 0.52$	
	$\sum f_i = 1.03$	
16,000 BV		0.36
20,000 BV		0.48

4. Use the integral curve to determine the number of bed volumes that can be processed for a blended effluent concentration, C_B, of 25 μg/L ($C_{eff}/C_0 = 25/60 = 0.42$). As shown in Figure 13.37, each column can be operated for 22,000 bed volumes. At this time, the effluent concentration of the single contactor will be 40 μg/L ($C_{eff}/C_0 = 0.67$), but because it is blended with effluent from the other three columns, which have lower effluent concentrations, the blended water will remain below 25 μg/L.

Given the EBCT of 15 min (or 96 BV/day), each adsorber can be operated for 22,000 BV/(96 BV/day) = 230 days before GAC replacement. For a single contactor, the 25-μg/L effluent standard was exceeded at 11,000 BV or 115 days; thus blending yields a run time that is twice that of a single contactor. Because carbon must be replaced at equal intervals, and the life of each adsorber is 230 days, 1 of the 4 adsorbers will be replaced every 230/4 = 59 days.

Series Column Analysis. The advantage of series operation of columns was discussed earlier. The pilot test should be carefully designed to show the reduction in activated carbon usage rate that can be achieved by such operation.

The simplest case is experienced if only a single compound with no competing adsorbates is present in the water to be treated. The breakthrough curves for each column in a two-column series are shown in Figure 13.38*a*. Each column is equal in size and thus has the same EBCT. The run can be terminated when the effluent concentration in the second column reaches C_B, because moving column 2 to position 1 and placing a fresh GAC column in position 2 should give the same breakthrough curves shown in Figure 13.38*a* between V_1 and V_2. The activated carbon usage rate (g/L) is then the mass of GAC in one column (g) divided by the total volume of water processed between replacements $V_2 - V_1$ (L). The number of bed volumes processed is ($V_2 - V_1$) divided by the volume of GAC in one column.

The situation is more complex if competing organics are present with the contaminant, or if a mixture of compounds is to be removed. The run must begin with fresh GAC in each position, and then operated until the C_{eff} in the second column

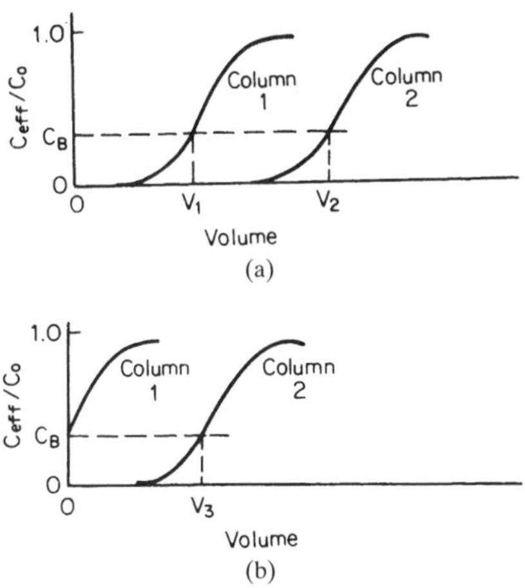

FIGURE 13.38 Breakthrough curves for two columns in series. (*a*) Fresh GAC in each column at start of test. (*b*) Column 1 was in the second position until CB was reached and then moved to the first position. Fresh GAC was in column 2 at start of test.

reaches C_B. The data should be plotted as shown in Figure 13.38*a*. Column 2 then must be moved to position 1, a fresh GAC column must be placed in position 2, and operation must continue until C_{eff} for the second column again reaches the breakthrough concentration. These data should be plotted as shown in Figure 13.38*b*. However, volume V_3 probably will not be the same as volume $V_2 - V_1$ because of the effects of competition. The activated carbon usage rate (g/L) for this case now becomes the mass of GAC in one column (g) divided by V_3(L).

The plots in Figure 13.38 show that column 1 is completely saturated when column 2 reaches C_B (point A). If this does not happen, a lower activated carbon usage rate can probably be achieved with more than two columns in series. The pilot test to determine this carbon usage rate, assuming competing organics are present, must be run until the column initially in the last position rotates into the first position, followed by continued operation until this sequence of columns reaches breakthrough.

Mathematical Models

Several useful mathematical models of the adsorption process are available (Sontheimer, Crittenden, and Summers, 1988; Weber and Smith, 1987). These models have led to a better understanding of the adsorption process, which is useful in determining high-priority areas for future study, and potential cost-effective ways to design and operate activated carbon systems. At this time (1999), the models do not permit the making of long-term, accurate predictions of trace compound removal from a mixture of adsorbable organics, however. The purpose of this section is to summarize what can be done with models and to refer the reader to other sources of information on the subject.

The homogeneous surface diffusion model (HSDM) and modifications of it have been widely used to predict performance of adsorption systems. Crittenden and Weber (1978a, b) extensively developed this model and it subsequently has been employed to accurately predict adsorption and desorption for single-solute and bisolute systems with similar-size adsorbates (Thacker, Crittenden, and Snoeyink, 1984; Qi et al, 1994). Hand, Crittenden, and Thacker (1984) developed a simplified approach to solving the model. A major difficulty in extending the use of the model is the accurate description of the competitive interactions between adsorbates with different characteristics, especially size. The nature of the competition between trace compounds and natural organics also is not well understood and is difficult to predict based on laboratory tests.

Some success with the prediction of TOC adsorption (and other group parameters) from heterogeneous mixtures of organics has been achieved. Lee, Snoeyink, and

Crittenden (1981) treated humic substances as a single compound in the HSDM with some success. Summers and Roberts (1984) developed a solution to this same problem using the assumptions that DOC could be treated as a single compound and that it had a linear isotherm. Their model has an analytical solution, and Summers and Roberts were able to obtain good agreement between predictions and the performance of operating systems. Sontheimer, Crittenden, and Summers (1988) developed the approach of simulating NOM solutions with many adsorbates by a hypothetical solution with only a few fictive components. The concentration and adsorption properties of the fictive components are assigned to give a composite adsorption isotherm that is the same as the isotherm of the mixture. The HSDM combined with a competitive adsorption equilibrium model can then be used to predict the performance of various types of adsorbers for adsorption of the mixture of fictive compounds.

Liang and Weber (1985), Weber and Pirbazari (1982), and Weber and Smith (1987) have dealt with the problem of one or more target compounds in the presence of complex mixtures of organics by incorporating the effects of the background organics in the kinetic and equilibrium constants for the target compound(s). The target compound(s) are then modeled as though only they are present in the water.

A good application of mathematical models is to use them to predict small-scale column and pilot plant results, and to determine factors that should be tested in pilot studies.

PAC ADSORPTION SYSTEMS

Comparison to GAC

The primary characteristic of PAC that differentiates it from GAC is particle size. Commercially available PACs typically show 65 to 90 percent passing a number 325 mesh (44 μm) sieve. For comparison, Kruithof et al. (1983) give the particle size distributions of two powdered activated carbons available in the Netherlands that show 23 to 40 mass percent smaller than 10 μm diameter, and 10 to 18 percent larger than 74 μm. Graham and Summers (1996) measured the particle size distribution of a commonly used PAC and found a log-normal distribution with a mean size of 24 μm, with about 30 percent smaller than 10 μm and about 5 percent larger than 74 μm. Particle size distribution is important because the smaller PAC particles in the absence of aluminum or ferric floc adsorb organic compounds more rapidly than large particles (Najm et al., 1990), although studies to show the effect of particle size in the presence of floc have not yet been done. PAC is made from a wide variety of materials, including wood, lignite, and coal. Its apparent density ranges from 0.36 to 0.74 g/cc (23 to 46 lg/ft^3), and is dependent on the type of materials and the manufacturing process. The iodine number, molasses number, and phenol number are often used to characterize PAC. The AWWA standard for PAC specifies a minimum iodine number of 500, for example (American Water Works Association, 1996), but in general, commercial GAC products have higher iodine numbers.

The primary advantages of using PAC are the low capital investment costs and the ability to change the PAC dose as the water quality changes. The latter advantage is especially important for systems that do not require an adsorbent for much of the year. The disadvantages, according to Sontheimer (1976), are the high operating costs if high PAC doses are required for long periods of time, the inability to regenerate, the low TOC removal, the increased difficulty of sludge disposal, and the difficulty of completely removing the PAC particles from the water.

Application of PAC

Points of Addition. PAC can be purchased and stored in bags, and fed as a powder using dry feed machines, or it can be bought in large quantities and fed as a slurry using metering pumps (American Water Works Association, 1971). Storage in bags requires specially designed and operated facilities to avoid generation of nuisance conditions from PAC dust. A survey by Graham et al. (1995, 1997) in 1995 of 95 utilities using PAC for taste and odor control indicated that points of PAC addition in conventional plants during treatment include presedimentation (16 percent), rapid mix (49 percent), flocculation (10 percent), sedimentation (7 percent), and filter influent (10 percent). Twenty-three percent of the plants had the ability to apply PAC at multiple points. Another point of addition that should be considered, although it is not commonly used (7 percent), is a continuous-flow slurry contactor that precedes the rapid mix. The PAC can be intensely mixed with the water, enabling rapid adsorption onto the small PAC particles, and then incorporated into the floc in the rapid mix for subsequent removal by sedimentation and filtration. Table 13.7 summarizes some of the important advantages and disadvantages of each of these points.

Important criteria for selecting the point of addition include (1) the provision of good mixing, or good contact between the PAC and all the water being treated, (2) sufficient time of contact for adsorption of the contaminant, (3) minimal interference of treatment chemicals with adsorption on PAC, and (4) no degradation of finished water quality. The PAC must be added in a way that ensures its contact with all of the flow.

Sufficient time of contact is necessary, and the time required is an important function of the characteristics and concentration of the molecule to be adsorbed (Meijers and van der Leer, 1983). Figure 13.39 shows the dramatic effect of PAC particle size on the rate of adsorption of trichlorophenol (TCP) from a groundwater (Adham et

TABLE 13.7 Advantages and Disadvantages of Different Points of PAC Addition

Point of Addition	Advantages	Disadvantages
Intake	Generally long contact time; good mixing	Some possibility (slight) of adsorbing compounds that otherwise would be removed by coagulation, thus increasing carbon usage rate.
Rapid mix	Good mixing during rapid mix and flocculation; reasonable contact time	Some chance of reduced adsorption rate because of coagulant interference. Contact time may be too short to reach equilibrium, thus requiring increased PAC dose.
Filter inlet	Sufficient time to reach full adsorption capacity of PAC	PAC may penetrate the filter and deposit in clear well or distribution system, or cause consumer "gray water" complaints. Requires good filter aid. PAC dose is limited by the rate of head loss buildup and reductions in filter run length.
Slurry contactor preceding rapid mix	Excellent mixing; no interference from coagulants	New basin and mixer may have to be installed.

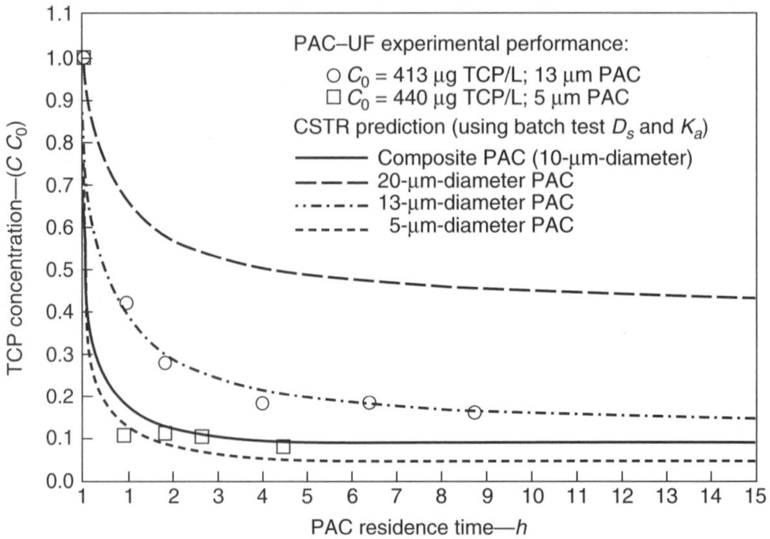

FIGURE 13.39 CSTR model prediction versus PAC-UF performance for the adsorption of TCP from groundwater using all PAC size fractions (PAC dose = 15 mg PAC/L). (*Source:* Adham et al., 1993.)

al., 1993). PAC 5 μm in diameter adsorbs TCP much more rapidly than PAC 20 μm in diameter. Up to 10 h is required for the plateau in the kinetic curve to be reached. Some of the difference in rate in the curves shown in Figure 13.39 is attributable to the lower capacity of the 20-μm-diameter carbon, but the other fractions have the same capacity per unit mass of PAC (Adham et al., 1993). Adsorption kinetics and equilibrium capacity are dependent on the type of adsorbate, the type of PAC used, and the competition from background organic matter. If insufficient time is allowed for equilibration, an increased PAC dose must be used to compensate.

Incorporation of PAC into floc particles is one factor that may reduce the rate of adsorption when PAC is added at the rapid mix (Gauntlett and Packham, 1973; Sontheimer, 1976). Gauntlett and Packham (1973) conducted jar tests that showed a lower removal rate of chlorophenol when PAC was added at the same time as alum compared to removal when PAC was added in the absence of alum. They also found that addition of PAC a short time after addition of alum gave a higher rate of removal than when PAC was added together with alum. They reasoned that PAC added after the alum adhered to the outer surface of the alum floc, thus avoiding rate interference. Najm et al. (1989) found little reduction, however, in the rate of TCP adsorption on PAC because of incorporation of PAC particles into coagulant floc when groundwater was treated with alum. Similarly, Graham and Summers (1996) found that alum had only a small impact on the kinetics of trichloroethene adsorption both in organic-free water and with NOM.

Addition at the intake has the advantage of providing extra contact time and generally good mixing in the intake line. Adsorption of compounds that would otherwise be removed by coagulation, flocculation, and sedimentation might increase the required PAC dosage, but this has not been shown to be an important factor in studies to date.

Addition of PAC just before the filter is advantageous because the PAC can be kept in contact with the water longer, thereby better using its capacity. The average PAC residence time is equal to one-half of the time between two successive backwashings, assuming PAC is continuously added to the filter influent. Also, all of the PAC could be added at once, just after backwash. The addition must be performed carefully to avoid having PAC escape the filter and penetrate the distribution system, however. The maximum dosage of PAC for this point of addition is limited by the ability of the filter to retain the PAC, and by the head loss in the filter, which is expected to increase as PAC dosage increases. Selected polyelectrolytes may be added to retain the PAC, but careful monitoring of the filter to ensure that PAC does not penetrate it is required. The millipore filter test involving filtration of a fixed quantity of filter effluent is a good method to determine if PAC is penetrating the filter. Comparison of the color of the filter pad with the color of filter pads that have been used to remove known quantities of PAC from water allows a good estimate to be made of the concentration of PAC that is escaping.

Sontheimer (1976) suggested adding a separate reactor between the sedimentation basin and the filter to increase the time of contact, and noted that this procedure would have the advantage of having eliminated competing organics to the maximum possible extent by coagulation-sedimentation before PAC addition. A major disadvantage is that the PAC would have to be removed from the water by another coagulation and filtration step, so the cost of this approach would be high.

Careful attention must be paid to the interaction of PAC with water treatment chemicals. Activated carbon is an efficient chemical reducing agent that will chemically reduce substances such as free and combined chlorine, chlorine dioxide, ozone, and permanganate, thereby increasing the demand for these substances and the cost of treatment. Reaction of activated carbon with chlorine will reduce the adsorption capacity of the activated carbon for selected compounds (Gillogly et al., 1998a; McGuire and Suffet, 1984; Snoeyink and Suidan, 1975). As much as 50 percent of MIB capacity was lost when 12 mg/L PAC was reacted with 3 mg/L of free chlorine (Gillogly et al., 1998a). Lalezary-Craig et al. (1988) found a reduction in the ability of PAC to adsorb both geosmin and MIB when PAC was applied to water containing free chlorine and monochloramine; the effect of the monochloramine appeared to be greater than that of free chlorine. Furthermore, as chlorine is destroyed by PAC, more chlorine must be added to achieve the desired degree of disinfection.

Addition of PAC to a water that is supersaturated with $CaCO_3$ or other precipitates, or an increase of pH to cause supersaturation just after PAC is added, such as in lime softening, may lead to coating of the particle with precipitates and to a corresponding decrease in adsorption efficiency. Greene, Snoeyink, and Pogge (1994) found that PAC added to a lime-softening basin was much less effective at removing atrazine than PAC added after the softening process. Also, adsorption at high pH is often poorer than at low pH because many organic contaminants are weak acids that ionize at high pH.

Various techniques of applying PAC help to improve its ability to adsorb slowly diffusing compounds such as DBP precursors. For example, addition of PAC to solids contact clarifiers has the potential for improved adsorption efficiency because the carbon can be kept in contact with the water for a longer time (Hoehn et al., 1987; Lettinga, Beverloo, and van Lier, 1978). Richard (1986) noted that the PAC dose for detergent removal could be reduced by 25 to 40 percent if the activated carbon were added to the influent of a floc blanket clarifier instead of to a conventional system. Najm, Snoeyink, and Richard (1993) found that trichlorophenol reached equilibrium with PAC in a floc blanket reactor, and thus that lower PAC doses could be used than if PAC was added to the rapid mix followed by flocculation and sedimentation. More research is needed to optimize this process, especially for TOC removal, and to develop coagulation procedures to increase adsorption kinetics, such as using polyelectrolyte alone instead of alum or ferric salts (Lettinga, Beverloo, and van Lier, 1978; Meijers and van der Leer, 1983).

In the Roberts-Haberer process (Haberer and Normann, 1979; Hoehn et al., 1984), buoyant polystyrene spheres 1 to 3 mm in diameter are coated with PAC. The spheres are held in a reactor by means of a screen, and the water to be treated is passed upflow through the media. After saturation, the PAC is removed from the beds by backwashing the media with a high flow rate. New PAC can then be applied to the beds. A primary advantage of the process is the ability to retain the PAC until its adsorption capacity is utilized, without incorporation of the PAC in floc as is necessary in most other applications. Recovery of the PAC for regeneration may be possible. In addition, the Roberts-Haberer filter may be used as a roughing filter to remove suspended solids ahead of the rapid filter, thereby reducing the solids load to the rapid filter. Studies are needed of the adsorption efficiency for specific organics and of the ability of the polystyrene beads to retain the PAC during filtration, however.

PAC can also be used in conjunction with cross-flow microfiltration (MF) or ultrafiltration (UF) membrane systems. These membranes are effective for removal of particles and microorganisms (see Chapter 11), but pores in the membrane are relatively large, and dissolved organic matter readily passes through. Removal of organic matter such as pesticides, taste- and odor-causing compounds, and DBP precursors requires pretreatment of the water with an adsorbent. PAC particles are large enough so that PAC can be applied just before the membrane. A flow diagram of the PAC/UF system is shown in Figure 13.40. This system is highly effective for removal of dissolved organic carbon (Adham et al., 1993; Anselme et al., 1997) and atrazine (Campos et al., 1998b). Anselme et al. (1997) report that several full-scale plants in France are now operating such systems.

A more recent innovation involves the two-stage application of PAC in the floc blanket reactor (FBR)/PAC/UF process (Campos et al., 1998a; Schimmoler et al., 1995). The flow diagram for this process is shown in Figure 13.41. Fresh PAC is continuously applied just in advance of the UF process, where it tends to come to equilibrium with the concentration of organic matter in the UF process effluent. As the PAC and other particles accumulate on the membrane, head loss gradually increases until the membrane must be backwashed. The backwash effluent is recycled to the FBR, where the PAC is trapped in the floc. The PAC now tends to equilibrate with the concentration of organic matter in the effluent of the FBR. The solids residence time in the FBR is typically 12 to 72 h, so there is sufficient time for extensive

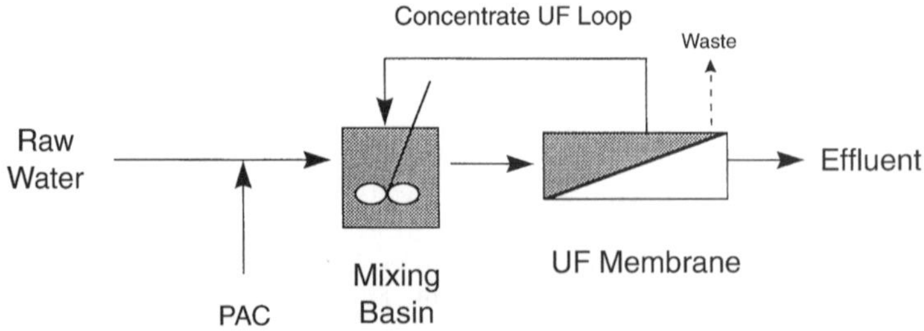

FIGURE 13.40 Schematic of a PAC-UF system.

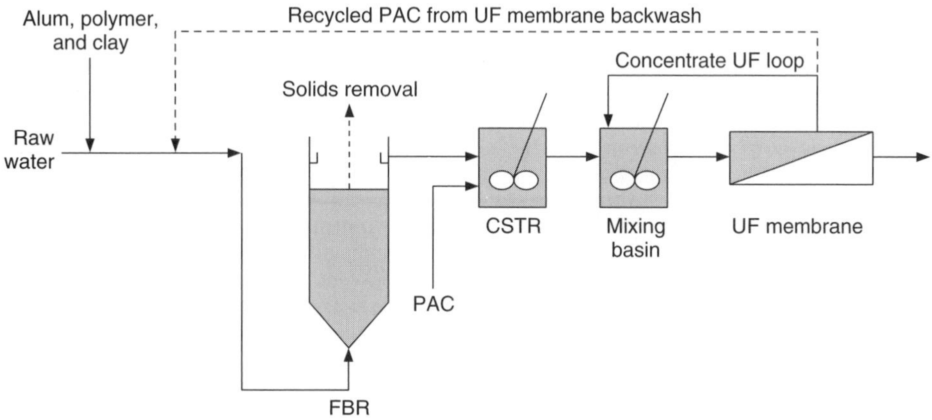

FIGURE 13.41 Schematic of an FBR-PAC-UF system.

adsorption to occur. The FBR effluent concentration is much higher than the UF effluent concentration, so the amount of adsorption per unit mass of PAC is significantly increased. The benefit of a counter current, two-stage process such as this depends on the type of organic matter being removed.

PAC PERFORMANCE

PAC Performance Estimation

Estimation of PAC Dose. Jar tests are often used to determine an estimate of the PAC dose required to achieve the desired removal in conventional plants. Jar tests can be used if water is available with a typical concentration of the contaminant in question. Use of a mixing rate that does not allow the PAC to settle in the jar is important (Meijers and van der Leer, 1983); the chemical doses, rapid mix, and settling time should correspond as nearly as possible to those that will be encountered in the full-scale plant. The percent removal or final concentration of contaminant in the jar test can be plotted against dose to determine the required amount to be added. The time of sampling should correspond to the estimated time of contact of the PAC in the plant, keeping in mind that well-coagulated PAC may settle out soon after the PAC enters the sedimentation basin. This dosage will only be an estimate of the full-scale plant dosage because of the many differences between contact in a jar test and in the plant. Therefore, the dosage may have to be modified based on full-scale performance of the PAC. Some bench-scale or pilot testing also may be necessary to determine the best filter aid polymer to prevent PAC from penetrating the rapid filter.

Often it is of interest to study the expected performance of PAC for many conditions, various contact times, several types of PAC, and various water quality conditions. Two cases are presented in the following example to illustrate the procedures for estimating performance.

EXAMPLE 13.3

CASE I

Conditions PAC is to be applied in such a way that 4 h of contact time will be achieved. Atrazine ranging from 3 to 20 μg/L must be reduced to a plant goal of 1 μg/L. The best PAC must be selected.

Approach A 4-h atrazine isotherm should be determined for each PAC to be evaluated. The water sample for the isotherm tests should have the pH, mineral composition, and background organic matter concentration expected in the full-scale plant on a consistent basis. An initial concentration of 50 μg/L or less of atrazine should be used for the test.

Data analysis The data should be analyzed by plotting percent remaining versus PAC dose as shown in Figure 13.42. One curve will be determined for each carbon. This curve can be used to determine the required PAC dose for each initial concentration expected. For example, if the initial concentration is 20 μg/L, the required percent removal is

$$\frac{20-1}{20} \times 100 = 95\%$$

Assuming that Figure 13.42 is valid for this carbon and water, the required PAC dose to achieve 95 percent removal (5 percent remaining) is 30 mg/L. Repeating this procedure for each carbon under consideration, and knowing the cost of each carbon, allows the selection of the most cost-effective carbon to achieve the required 95 percent removal. The required percent removal will decrease as the influent concentration decreases, given that the desired effluent concentration remains 1 μg/L.

CASE II

Conditions MIB is present at 20 ng/L and must be reduced by 75 percent to the threshold odor level of 5 ng/L. However, several points of addition must be evaluated, each of which has a different contact time, and the adsorption process does not necessarily reach equilibrium.

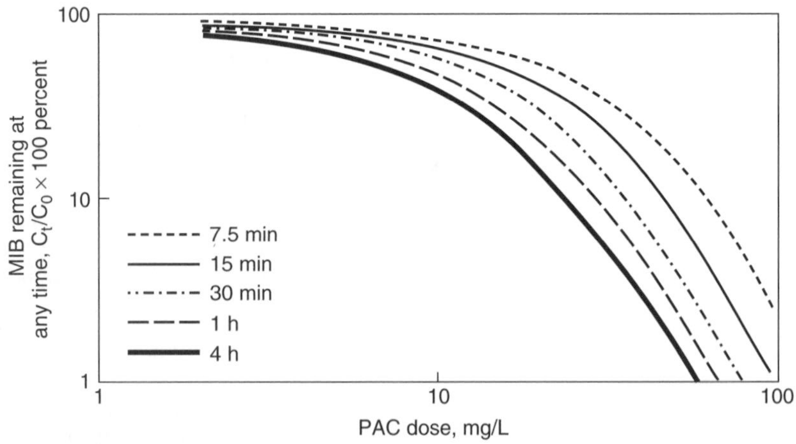

FIGURE 13.42 HSDM fit and prediction of batch kinetic test data for MIB adsorption with PAC E, based on kinetic test plateau values. (*Source:* Gillogly et al., 1998b.)

Approach Gillogly et al. (1998b) showed that the percent adsorbate remaining versus carbon dose plot *at any time* is independent of initial concentration if the diffusion coefficient is not a function of test conditions. Thus, the percent removal at each desired time can be determined and plotted as shown in Figure 13.42. The jar test procedure for MIB adsorption has been described by Gillogly et al. (1998b); these tests should be conducted for each desired contact time. Assuming the same PAC and the same water with the same background organic matter as the water studied by Gillogly et al. (1998b), Figure 13.42 can be used to determine a carbon dose of 16 mg/L if 4 h of contact time is available, and 30 mg/L if only 15 min of contact time is available.

Experience with PAC

The literature provides many reports of successful removal of taste and odor using PAC; a few are summarized here. Twenty-five percent of the 683 United States water plants surveyed (including the 500 largest utilities) in 1976 reported that they used PAC (American Water Works Association, 1977), and similar data have been reported for 1984 (Fisher, 1986). The predominant reason for the use of PAC is taste and odor control, but it also can be used to remove nonodorous organics. PAC doses range from a few to more than 100 mg/L, but are typically less than 25 to 50 mg/L (Degremont, 1979; Graham et al., 1997). TON removal is most commonly used to measure its effectiveness. Reports of the effectiveness of PAC are complicated because an odor may be caused by the combination of the effects of more than one compound, or a compound may mask the odor of another compound and the odor of the second may appear after the first compound has been adsorbed. Furthermore, odors may form after treatment through chemical or biological reactions in the distribution system (American Water Works Association, 1987), thus making PAC appear ineffective. A reliable catalog of odors that can be removed well by PAC does not exist, probably because of the complicating factors just mentioned. As analytical capabilities improve and the ability to qualitatively and quantitatively analyze compounds that cause odor advances, however, development of such a catalog may be possible.

PAC performance for removal of THMs and VOCs has not been very good (Singley et al., 1979; Symons et al., 1981), consistent with theoretical calculations. The single-solute isotherm data in Table 13.1 show small Freundlich K values for substances such as chloroform, benzene, carbon tetrachloride, and trichloroethane. In general, if the K value for a compound is less than 30 to 50 mg/g(L/mg)$^{1/n}$, PAC doses greater than approximately 50 mg/L are expected to achieve removals of 90 percent.

The rates of removal of compounds with Freundlich K values greater than about 30 to 50 (mg/g)(L/mg)$^{1/n}$ are much better than those for the lower molecular-weight VOCs (Aly and Faust, 1965; Burttschell et al., 1959; El-Dib, Moursy, and Badawy, 1978; Singley et al., 1977). Organics may adsorb to particles (American Water Works Association, 1979), however, and strongly adsorbing compounds may not be removed well by PAC if such particles are not removed by clarification processes. Pesticides and herbicides are generally quite well removed (Aly and Faust, 1965; American Water Works Association, 1979; El-Dib and Aly, 1977; Robeck et al., 1965).

TOC and THMFP removals by PAC in conventional plants have not been very high, possibly because removal is limited by slow kinetics of adsorption and equilibrium capacity (Hoehn et al., 1987; Lange and Kawczynski, 1978; Love et al., 1976). Typical TOC isotherms for natural waters show adsorption capacities in the range of 10 to 100 mg/g. If the equilibrium capacity is 50 mg/g, a PAC dose of 1 mg/L can remove a maximum of 0.05 mg/L TOC. Further, the time for adsorption in conventional water treatment plants is often insufficient for equilibrium to be achieved.

THERMAL REACTIVATION OF GAC

Reactivation Principles

The reactivation process can be described as consisting of four basic steps (van Vliet, 1985): (1) drying, including loss of highly volatile adsorbates, at temperatures up to 200°C, (2) vaporization of volatile adsorbates and decomposition of unstable adsorbates to form volatile fragments, at temperatures of 200 to 500°C, (3) pyrolysis of nonvolatile adsorbates and adsorbate fragments to form a carbonaceous residue or char on the activated carbon surface, at temperatures of about 500 to 700°C, and (4) oxidation of the pyrolyzed residue using steam or carbon dioxide as the oxidizing agent, at temperatures above about 700°C.

The drying step eliminates the 40 to 50 percent water associated with the spent GAC (Schuliger and MacCrum, 1973). The proportions of adsorbates that are volatilized (steps 1 and 2) or converted to char (step 3) depend upon the nature of the adsorbate and the strength of the adsorbate-activated carbon bond (Juntgen, 1978). Thus, the type of activated carbon also plays an important role. Suzuki et al. (1978) related the amount volatilized to the boiling point of the compound (the lower the boiling point, the larger the amount of volatilization) and its degree of aromaticity (the more aromatic, the smaller the amount of volatilization). Molecules such as lignin, phenolic compounds, and humic substances produced the largest amount of char. Tipnis and Harriott (1986) studied a series of phenolic compounds, for example, and showed that 20 to 50 percent was converted to char. van Vliet (1985) cautions that pyrolysis should not take place above 850°C because a char with a more graphitized structure similar to that of the base activated carbon will be formed. The stronger oxidation conditions required to remove the graphitized char will also result in extensive attack and loss of the activated carbon.

Oxidation to gasify the char is critical to returning the GAC as closely as possible to virgin conditions. The objective is to remove the char, with minimal loss of the activated carbon, in a manner that does not alter the structure and adsorption properties of the activated carbon. The reactions used are similar to those employed to produce activated carbon. Steam and carbon dioxide are the oxidizing agents:

$$H_2O(g) + C(s) \rightarrow CO(g) + H_2(g) \tag{13.25}$$

$$CO_2(g) + C(s) \rightarrow 2\,CO(g) \tag{13.26}$$

CO can then react with steam:

$$CO(g) + H_2O(g) \rightarrow CO_2(g) + H_2(g) \tag{13.27}$$

The reactions in Equations 13.25 and 13.26 are endothermic and thus require heat in order to proceed. This property makes the reactions easy to control by controlling the heat input rate (Juhola, 1977a). In contrast, the reaction between oxygen and carbon,

$$\frac{3}{2}O_2(g) + 2C(s) \rightarrow CO_2(g) + CO(g) \tag{13.28}$$

is exothermic, or heat-producing, and is thus self-promoting and difficult to control (Loven, 1973; van Vliet, 1985). Thus, oxygen is more likely to attack the base-activated carbon structure, altering the pore volume distribution and increasing activated carbon loss, and should not be used under normal reactivation conditions. Juhola (1977a) noted that carbon dioxide promoted oxidation of the GAC exterior and led to larger volumes of large pores than did steam, indicating that carbon dioxide was the stronger oxidizing agent.

Factors affecting reactivation product quality are primarily those that affect the oxidation step. Important factors are the type of oxidizing gas, the time and temperature of oxidation, the amount and type of adsorbate on the carbon, the type and quantity of inorganic substances that accumulate on the carbon, and the type of activated carbon.

The activated carbon should be relatively homogeneous and have a lower reactivity with steam and carbon dioxide than the char deposit (Juntgen, 1976). This difference in reactivity should make possible the selection of reactivation conditions that minimize oxidation of the activated carbon while removing the deposit. Juntgen (1976) found that the reactivity of several activated carbons varied widely depending on the raw material used to make the activated carbon. For example, activated carbon made from wood was far more reactive than activated carbon made under the same conditions from hard coal. Juhola (1970) reported that activated carbons with a large volume of large pores, such as lignite activated carbon, usually reactivated at lower temperatures than activated carbons with smaller pores. The larger pores made desorbing the adsorbate molecules easier and also allowed oxidizing gases to reach the deposits of carbonized adsorbate.

Reactivation studies indicate a trade-off between the time required for reactivation and the temperature of reactivation: the higher the temperature the shorter the reactivation time (Juhola, 1970). A temperature of 925°C was near optimum for one activated carbon that was used to remove organics from secondary effluent (Juhola, 1970). At lower temperatures, the recovery of small pores that were needed for this application was not as good as at higher temperatures; also, the oxidation rate was very slow at a temperature of 700°C. At temperatures above 925°C, the carbon structure itself was very reactive and activated carbon losses were expected to be high, although they were not measured. The optimum temperature for reactivation has been shown to depend upon the type of activated carbon (Juntgen, 1976). van Vliet and Venter (1984) thoroughly studied the reactivation of GAC from a wastewater reclamation plant and found optimum reactivation conditions to include a temperature and time of reactivation range of 800°C for 10 min to 850°C for 5 min. Longer times and higher temperatures resulted in higher losses and a less desirable pore structure, while shorter times and lower temperatures gave incompletely reactivated GAC.

Schuliger and MacCrum (1973) showed that the type of adsorbed organic compound significantly affects the time of reactivation. Reactivation times from 5 to 125 min were required, with a typical range being 20 to 60 min for a variety of applications. The required time increased as the quantity of adsorbed organics increased.

In the application of activated carbon for purification of surface water and possibly groundwater, the need to remove adsorbed humic substances will probably control the reactivation step even though the primary purpose of applying the activated carbon is to remove trace organic compounds. Because humic substances are the predominant organic fraction in most surface waters and they readily adsorb (McCreary and Snoeyink, 1977), the quantities of adsorbed organic compounds present at trace concentrations will be small by comparison. Thus, the quantity of pyrolyzed deposit will probably depend on the manner in which humic substances are converted to char. Suzuki et al. (1978) found that 26 percent of one humic acid was converted to char during pyrolysis. Humic substances vary widely in molecular weight; the larger molecules will probably adsorb in the larger activated carbon pores, while the smaller molecules and most trace organics will likely adsorb in the smaller pores. Good removal of organics from all pores will then be necessary if the reactivated carbon is to function well.

Certain inorganic substances, if present in the water being treated, may form a precipitate on the activated carbon surface or may accumulate on the activated carbon surface because of their association with the organic molecules being adsorbed. Juhola (1970) noticed that there was an increase in ash content of several percent after a few reactivations of a bituminous carbon that had been used to remove organics from municipal secondary effluent, and that recovering the small pores during reactivation was difficult. When the inorganic materials were removed with an HCl wash prior to the reactivation step, the small pore recovery was very good. Smisek and Cerny (1970) indicated that oxides and carbonates of metals such as sodium, potassium, iron, and copper catalyze the reactivation step. Umehara et al. (1983) found that the Na_2SO_4 produced by reactivation of a carbon loaded with the adsorbed detergent sodium dodecyl benzene sulfonate had a significant catalytic effect on the rate of oxidation of char. Thus, the optimum temperature and other reactivation conditions are likely to change as ash builds up on the carbon surface. Poggenburg et al. (1979), reporting on a study of treatment of Rhine River water in Germany, noted no increase in ash content over three reactivations of a particular activated carbon. Knappe et al. (1992) and Cannon et al. (1993) found that calcium accumulated on many water treatment plant carbons. In agreement with others, they showed that calcium catalyzed the reaction of reactivated gases with GAC and made the recovery of micropores difficult.

Reactivation Furnaces

Common types of furnaces are the rotary kiln, the multiple hearth, and the fluidized bed furnace (Sontheimer, Crittenden, and Summers, 1988). The multiple hearth furnace is the most commonly used. Scrubbers are used to remove particles from the off-gases, and afterburners, which are maintained at 750 to 1000°C by burning fuel with excess oxygen, are used to destroy organic compounds that are volatilized but not combusted to CO_2 during the reactivation step.

Reactivation Performance. Loss of mass during reactivation is very important because of its impact on the cost of using activated carbon. DeMarco et al. (1983) reported total losses (transport plus loss in the furnace) of 16 to 19 percent in an experimental system at Cincinnati, while Koffskey and Lykins (1987) observed 9 percent loss (7 percent in the furnace and 2 percent during transport) at Jefferson Parish, LA. Proper design of the transport system is very important because of the high losses that can take place in a poorly designed system (Schuliger, Riley, and Wagner, 1987). Schuliger, Riley, and Wagner note that losses are a function of the transport system design, the type of adsorbate, the loading of adsorbate on the carbon, and the type of furnace.

Along with the loss in mass may come a change in pore size distribution. Often small pores less than 2 μm in diameter are lost while larger pores are created (Hutchins, 1973). Pore size distribution has a significant effect on both rate of uptake and capacity for a particular adsorbate (Juhola, 1977b; Poggenburg et al., 1979). Modifying reactivation conditions to produce the best pore size distribution for a particular adsorbate should be possible, but research is needed to show how this can be done and to show the desired pore size for a given application (Sontheimer, Crittenden, and Summers, 1988).

It has been observed that the particle diameter becomes smaller during reactivation, with the disadvantage of creating higher head loss in fixed-bed adsorbers but with the advantage of allowing the rate of uptake of molecules to occur more rapidly. The decrease in particle diameter should be minimized as much as possible, because this decrease normally correlates with loss in mass. This change in particle size, however, apparently does not affect the rate of reactivation of the particles (Juhola, 1970). The hardness of activated carbon will decrease upon reactivation because of the destruction of part of the structure of the activated carbon. As the activated carbon becomes less hard, the losses in the furnace and during handling will increase.

Generally, density, molasses number or decolorization index, iodine number, and similar parameters are used to measure the quality of the activated carbon. An improved procedure involves the use of adsorbate molecules in the water to be treated to test the quality of the product. Poggenburg et al. (1979) used this approach to evaluate activated carbons for removal of organics found in Rhine River water. Their procedure involved extensive use of adsorption isotherms and adsorption rate tests and modeling of the performance of full-scale adsorbers. Changes in activated carbon quality produced by reactivation then can be expressed as changes in the length of time to breakthrough and the amount of material adsorbed at breakthrough for a column of a given design. Because the rate of adsorption depends upon pore size distribution and adsorbate size, as well as other factors, using the molecule in the water to be treated as a test substance will yield a meaningful comparison of the activated carbon quality for the application in question.

Kornegay (1979) analyzed the cost of reactivation as a function of furnace size. He showed rapidly increasing unit costs as activated carbon usage became smaller than 5000 to 6000 lb/day. Kornegay noted that on-site thermal reactivation will not be economical if usage is smaller than 500 to 1000 lb/day, while Hess (1981) proposed a lower economical limit of about 2000 lb/day. Transport of spent GAC to off-site thermal reactivation facilities may be appropriate for intermediate values of activated carbon usage rate. The cost of reactivation would then include transportation to and from the site, the cost of makeup activated carbon, and (presumably) profit for the owner of the furnace. This approach to reactivation is common in Europe, where transportation distances are relatively short, the incidence of GAC usage for surface water is high, and the number of on-site furnaces is small. Further, assurance would have to be given that only water treatment plant activated carbon would be returned to a plant. Inclusion of regenerated activated carbon that had been used for municipal or industrial waste treatment would be unacceptable.

Reactivation By-products

Activated carbon must be regenerated in a way that does not pollute the atmosphere. For this reason, scrubbers and afterburners are used to minimize particulate and gaseous emissions, respectively, from operating facilities. Off-gas control is important in controlling dioxins and furans that are produced during reactivation. Lykins, Clark, and Cleverly (1988) studied operating reactivation systems at Cincinnati and Jefferson Parish and found these compounds in various effluent streams before treatment of these effluent streams, even though the compounds were not on GAC entering the furnace. Small amounts of the dioxins and furans do pass through the stack gas control equipment, but Lykins, Clark, and Cleverly found the concentrations to be so low that the estimated cancer risk was negligible.

ADSORPTION OF ORGANIC MATTER ON RESINS

Many types of synthetic resins that differ both in the matrix that supports the functional groups and in the functional groups themselves have been used for removal of organic compounds. (See Chapter 9 for a discussion of the removal of inorganic substances.) Some of the more common matrices and functional groups are shown in Figure 13.43 (Kim, Snoeyink, and Saunders, 1976). A few resins called *adsorbent resins,* such as the styrene divinylbenzene (SDVB) resin and the phenol-formaldehyde (PF) resin, have been made without functional groups in addition to those that are part of

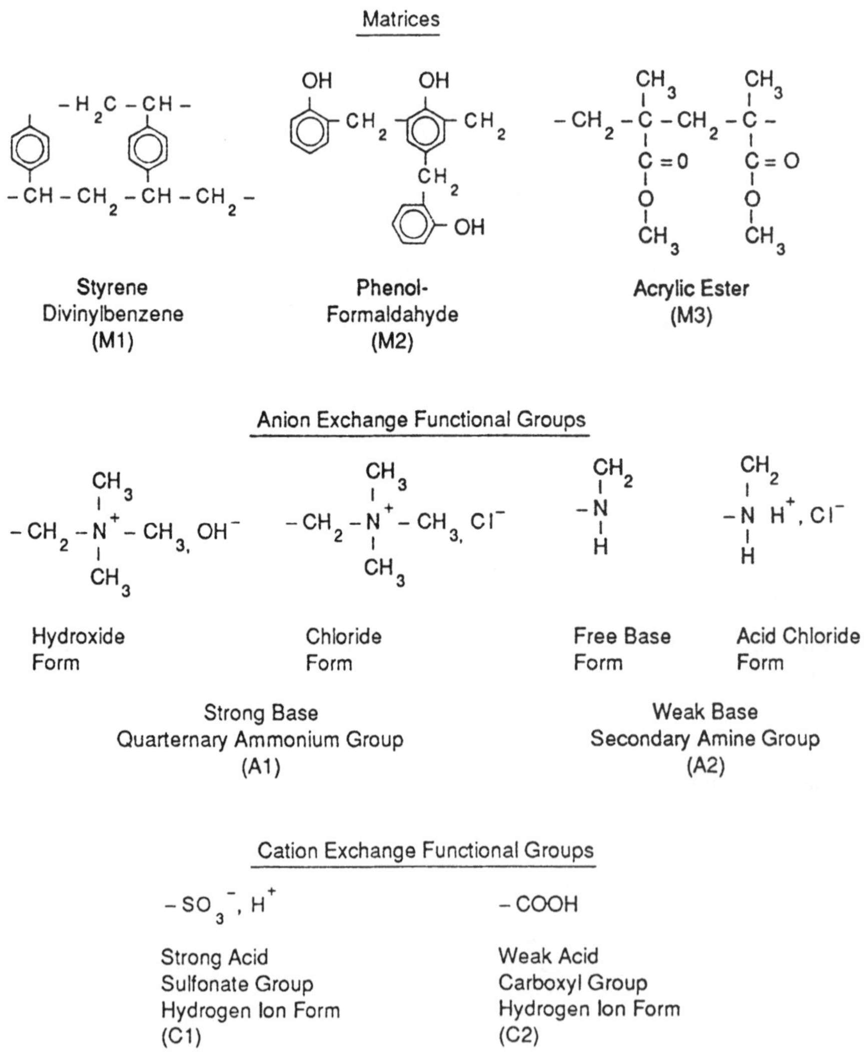

FIGURE 13.43 Matrices and functional groups of resins commonly used for water purification. (*Source:* Kim, Snoeyink, and Saunders, 1976.)

the matrix. More typically, resins have functional groups that make it possible for substances to be taken up by ion exchange or by specific interaction with the functional group. The strong base ion exchange resins, for example, frequently have the quaternary ammonium functional group (Figure 13.43, A1), whereas the weak base resins generally have an amine functional group (Figure 13.43, A2). The strong acid and weak acid functional groups can remove substances from water by cation exchange; typical functional groups are the sulfonate group (Figure 13.43, C1) and the carboxyl group (Figure 13.43, C2). In addition, the degree of cross-linking between the polymers that constitute the matrix can be varied, and thus resins with different pore size distributions can be prepared.

Kolle (1976) found good removal of TOC during groundwater treatment at Hannover, Germany, using a macroporous strong base SDVB resin. The resin was regenerated with an NaCl-NaOH solution (100 and 20 g/L, respectively), and the solution could be reused several times before disposal. The volume of spent regenerant used could be limited to 1 unit per 25,000 units of product water. Brattebo, Odegaard, and Halle (1987) used the same resin and regenerant for a surface water in Norway and concluded that an average regenerant reuse of 78 percent would not degrade the resin. The strong base SDVB resin was also used in England on Thames River water and resulted in much poorer removals of TOC than could be achieved with activated carbon (Gauntlett, 1975). The amount of alkylbenzene sulfonate (ABS) adsorbed during the test and the capacity for trace concentrations of dichlorophenol were also low.

Fu and Symons (1988) concentrated the organics in Lake Houston, Texas, water by reverse osmosis and investigated their removal as a function of molecular weight by anion exchange. The researchers found the removal to occur predominantly through ion exchange between the organic anion and the chloride ion. They found the acrylic strong base exchanger to have the higher capacity and generally to have the better performance.

Koechling et al. (1997), Summers, DiCarlo, and Palepu (1990), and Summers et al. (1993, 1999) have evaluated the use of macroporous strong base resins alone and as a pretreatment to GAC for the removal of NOM and THM precursors. They found that GAC alone often outperformed the resin alone; the combined resin-GAC system at the same EBCT as the GAC alone yielded better performance. This was attributed to the removal by resins of compounds that are weakly adsorbed to the GAC (Koechling et al., 1997).

The performance of weak base resins under different conditions of adsorption differs markedly from that of the strong base resins. One of the major differences is the pH dependence of adsorption. Adsorption of phenol on a weak base PF resin, for example, takes place on the functional group in the free base form that predominates at pH 5 and above; as pH decreases below pH 5, acid is adsorbed by the resin and competes with phenol, thereby decreasing phenol capacity. As pH increases above approximately 8 to 9, capacity for phenol decreases because the OH groups of the PF matrix ionize, and phenol itself ionizes. The maximum adsorption for phenol occurs near neutral pH. On the other hand, ABS is negatively charged and is taken up by the acid form of the resin (Figure 13.43, A2) by an exchange mechanism. The capacity of weak base resins for ABS is very good at low pH, but they do not adsorb very well on the free base form of the resin.

An advantage of weak base resins compared to strong base resins is that the former can usually be regenerated more easily. They have a lower affinity for anionic organics, and as a result the organics can be removed by sodium hydroxide solution or salt solution more easily. Strong acid (Figure 13.43, C1) and weak acid (Figure 13.43, C2) cation exchangers have found little application in removal of organic materials from water, presumably because most of the organics in water are either negatively charged or neutral.

The SDVB resin with no functional groups (Figure 13.43, M1) has received much attention as an adsorbent. It has been used to concentrate pesticides and related compounds from raw and finished waters in a study of contamination of water supplies (Junk et al., 1976), and it has been shown to be particularly advantageous for removal of chlorinated pesticides from an industrial wastewater (Kennedy, 1973). Reactivation was readily accomplished with acetone or isopropanol, and the spent regenerant could be reclaimed for reuse. This resin also adsorbs phenols, although its capacity is somewhat less than that of weak base resins (Kumagai and Kaufman, 1968). The SDVB resin did not remove TOC from Thames River water to an appreciable extent (Gauntlett, 1975).

The Water Research Center in England evaluated synthetic resins in comparison to activated carbon for application in water treatment plants using coagulated and filtered Thames River water (Gauntlett, 1975). The Center eliminated the SDVB resins without functional groups because of very low capacity and selected three strong base resins and one weak base resin for additional testing. The results showed that the activated carbon removed TOC more effectively than any of the resins, and that the strong base SDVB and acrylic matrix resins performed somewhat better than the weak base PF resin. Activated carbon removed the trace concentrations of the pesticide BHC very effectively; the weak base resin removed it nearly as well, but the strong base resin removed very little.

Jayes and Abrams (1968) reported on the results of the application of a PF weak base resin for removing color at the Lawrence, MA test facility. The resin was generally quite effective for removing this material, although when the river water was extremely high in color a substantial amount of leakage through the column occurred. Reactivation was accomplished with 4 lb of sodium hydroxide per ft^3. On the basis of this test, a resin life in excess of 200 cycles of adsorption-reactivation was estimated.

Commercial humic acid was adsorbed very well on weak base PF resins and strong base resins; the capacities achieved were comparable to those for GAC (Boening, Beckmann, and Snoeyink, 1980). The capacity of the weak base PF resin decreased when the pH was either decreased from 8.3 to 5.5 or increased from 8.3 to 9.5. At low pH, the amine functional groups adsorb acid, thereby reducing the amount of organic material that can be adsorbed at those sites. As the pH is increased above 8.3, the OH functional groups on the PF resin matrix ionize, thus giving the resin a negative charge. Because of the negative charge on humic acid, which increases as pH is increased, adsorption capacity on the negatively charged resin is low. The strong base resin did not show a great variation in capacity with changes in pH over the range of 5.5 to 9.5. The weak base PF resin had essentially no capacity for the earthy-musty odor compound MIB (Chudyk et al., 1979). The SDVB adsorbent resin did adsorb MIB, but its capacity was lower than those of low-activity bituminous carbons.

A general conclusion concerning resins is that they are not applicable as a general adsorbent for drinking water treatment. They are more selective than activated carbon and do not meet the criterion that an adsorbent must be able to remove a wide variety of compounds. They may be applicable in specific situations that require only a particular type of contaminant removal.

BIBLIOGRAPHY

Adamson, A. W. *Physical Chemistry of Surfaces* (4th ed.) John Wiley & Sons, New York, 1982.

Adham, S. S. et al. *Jour. AWWA,* 85(12):59, 1993.

Aly, O. M., and S. D. Faust. *Jour. AWWA,* 57(2):221, Feb. 1965.

American Society for Testing Materials, "Refractories; Carbon and Graphite Products; Activated Carbon." *Annual Book of ASTM Standards,* Vol. 15.01. Philadelphia, 1988.

American Water Works Association, *Water Quality and Treatment* (3rd ed.). McGraw-Hill, New York, 1971.

American Water Works Association Committee Report. *Jour. AWWA,* 69(5):267, May 1977.

American Water Works Association, *AWWA Standard for Powdered Activated Carbon, B600-96,* AWWA, Denver, CO, 1996.

American Water Works Association Committee Report. *Jour. AWWA,* 71(10):588, 1979.

American Water Works Association, *1984 Utility Operating Data,* AWWA Denver, CO, 1986.

American Water Works Association, Research Foundation–Lyonnaise des Eaux. *Identification and Treatment of Tastes and Odors in Drinking Water* (J. Mallevialle and I. H. Suffet, eds.). AWWA Research Foundation, Denver, CO, 1987.

Andelman, J. B. Chapter in *Proc. 15th Water Quality Conf.,* Univ. of Illinois Bull., 70:122, Urbana, IL, June 4, 1973.

Anselme, C., et al. "Drinking Water Production by UF and PAC Adsorption." *Proc. AWWA Specialty Conference on Membrane Technology,* New Orleans, LA, 1997.

Arbuckle, W. B. Chapter in *Activated Carbon Adsorption,* vol. 2 (J. M. McGuire and I. H. Suffet, eds.). Ann Arbor Science Publishers, Ann Arbor, MI, 1980.

Bablon, G., C. Ventresque, and R. Ben Aim. *Jour. AWWA,* 80(12):47, 1988.

Baldauf, G., and M. Henkel. "Entfernung von Pestiziden bei der Trinkwasser-aufbereitung." In *20. Bericht der Arbeitsgemeinschaft Wasserwerke Bodensee-Rhein,* Teil 7, 127, 1988.

Baldauf, G., and G. Zimmer. *Vom Wasser,* 66:21, 1986.

Bauer, R. C., and V. L. Snoeyink. *J. Water Pollution Cont. Fed.,* 45:2290, 1973.

Bilello, L. J., and B. A. Beaudet. Chapter in *Treatment of Water by Granular Activated Carbon,* Vol. 1 (M. J. McGuire and I. H. Suffet, eds.). American Chemical Society, Washington, D.C., 1983.

Boening, P. H., D. D. Beckmann, and V. L. Snoeyink. *Jour. AWWA,* 72(1):54, Jan. 1980.

Brattebo, H., H. Odegaard, and O. Halle. *Water Res.,* 21(9):1045, 1987.

Burttschell, R. H. et al. *Jour. AWWA,* 51(2):205, 1959.

Butler, J. A. V., and C. Ockrent. *J. Phys. Chem.,* 34:2841, 1930.

Campos, C. et al. "Adsorption in the Floc Blanket Reactor-Powdered Activated Carbon-Ultrafiltration System." *Proc. AWWA Ann. Conf.,* Dallas, Texas, June 1998a.

Campos, C. et al. *Desalination* 117(1):265, 1998b.

Cannon, F. S. et al. *Jour. AWWA,* 85:3, 76–89, 1993.

Chen, A. S. C., R. A. Larson, and V. L. Snoeyink. *Environ. Sci. Technol.,* 16:268, 1982.

Chen, A. S. C., V. L. Snoeyink, and F. Fiessinger. *Environ. Sci. Technol.,* 21:83, 1987.

Chowdhury, Z. K. et al. "NOM Removal by GAC Adsorption: Implications of Blending." *Proc. AWWA Ann. Conf.,* Toronto, Ontario, Canada, 1996.

Chudyk, W. A., and V. L. Snoeyink. *The Removal of Low Levels of Phenol by Activated Carbon in the Presence of Biological Activity.* Univ. of Illinois Wtr. Res. Ctr. Rept. No. 154, 1981.

Chudyk, W. A., and V. L. Snoeyink. *Environ. Sci. Technol.,* 18:1, 1984.

Chudyk, W. A. et al. *Jour. AWWA,* 71:529, 1979.

Coughlin, R. W., and F. Ezra. *Environ. Sci. Technol.,* 2:291, 1968.

Cover, A. E., and L. J. Pieroni. *Evaluation of the Literature on the Use of GAC for Tertiary Waste Treatment.* Report No. TWRC-11, U.S. Dept. of the Interior, FWPCA, Cincinnati, OH, 1969.

Crittenden, J. C., J. K. Berrigan, and D. W. Hand. *J. Water Pollution Cont. Fed.,* 58:312, 1986.

Crittenden, J. C., and W. J. Weber Jr. *J. Env. Eng. Div., ASCE,* 104:1175, 1978a.

Crittenden, J. C., and W. J. Weber Jr. *J. Env. Eng. Div., ASCE,* 104:433, 1978b.

Crittenden, J. C. et al. *J. Water Pollution Cont. Fed.,* 52:2780, 1980.

Crittenden, J. C. et al. *Environ. Sci. Technol.,* 19(11):1037, 1985.

Crittenden, J. C. et al. *J. Env. Engr. Div., ASCE,* 113(2):243, 1987.

Crittenden, J. C. et al. *Jour. AWWA,* 80(5):73, May 1988.

Crittenden, J. C. et al. *Jour. AWWA,* 83(1):77, 1991.

Degremont, *Water Treatment Handbook,* 5th ed., Degremont, Paris, 1979.

DeLaat, J., F. Bouanga, and M. Dore. Chapter in *Organic Micropollutants in Drinking Water and Health* (H. A. M. de Kruif and H. J. Kool, eds.). Elsevier, New York, 1985.

DeMarco, J. et al. "Experiences in Operating a Full-Scale Granular Activated Carbon System with On-Site Reactivation." In *Treatment of Water by Granular Activated Carbon* (M. J. McGuire and I. H. Suffet, eds.). American Chemical Society, Washington, DC, 1983.

Dielmann, L. M. J. III. *The Reaction of Aqueous Hypochlorite, Chlorite and Hypochlorous Acid with GAC.* MS thesis, University of Illinois, Urbana, IL, 1981.

Dobbs, R. A., and J. M. Cohen. *Carbon Adsorption Isotherms for Toxic Organics.* EPA-600/8-80-023, USEPA, Cincinnati, OH, 1980.

El-Dib, M. A., and O. A. Aly. *Water Res.,* 11:617, 1977.

El-Dib, M. A., A. S. Moursy, and M. I. Badawy. *Water Res.,* 12:1131, 1978.

Faust, S. D., and O. M. Aly. *Chemistry of Water Treatment.* Ann Arbor Science, Boca Raton, FL, 1983.

Fiessinger, F., personal communication, 1983.

Fisher, J. L., Calgon Carbon Corp., personal communication, Nov. 1986.

Ford, R. et al. "Developing Carbon Usage Rate Estimates for Synthetic Organic Chemicals." *Proc. AWWA Ann. Conf.,* Los Angeles, CA, 1989.

Fox, R. D., R. T. Keller, and C. J. Pinamont. *Recondition and Reuse of Organically Contaminated Waste Brines.* EPA-R2-73-200, USEPA, Washington, DC, 1973.

Fu, P. L. K., and J. Symons. "Removal of Aquatic Organics by Anion Exchange Resins." *AWWA Ann. Conf.,* Orlando, FL, June 1988.

Gammie, L., and G. Giesbrecht. "Full-Scale Operation of GAC Contactors at Regina/Moose Jaw, Saskatchewan." *Proc AWWA Annual Conf.,* Denver, CO, 1986.

Gasser, C. G., and J. J. Kipling. *Proc. Conf. on Carbon,* 4th, 55, 1959.

Gauntlett, R. B. *A Comparison Between Ion-Exchange Resins and Activated Carbon for the Removal of Organics from Water.* Water Research Center Technical Report TR 10, Medmenham, UK, 1975.

Gauntlett, R. B., and R. F. Packham. "The Use of Activated Carbon in Water Treatment." *Proc. Conf. on Activated Carbon in Water Treatment,* Univ. of Reading, Wtr. Res. Assn., Medmenham, UK, April 1973.

Gillogly, T. E. T. et al. *Jour. AWWA,* 90(2):107, 1998a.

Gillogly, T. E. T. et al. *Jour. AWWA,* 90(1):98, 1998b.

Glaze, W. H. et al. *Evaluation of Biological Activated Carbon for Removal of Trihalomethane Precursors.* Report to USEPA, Cincinnati, OH, 1982.

Graese, S. L., V. L. Snoeyink, and R. G. Lee. *Jour. AWWA,* 79(12):64, Dec. 1987a.

Graese, S. L., V. L. Snoeyink, and R. G. Lee. *GAC Filter Adsorbers.* AWWA Research Foundation, Denver, CO, 1987b.

Graham, M., and R. S. Summers. "The Role of Floc Formation and Its Presence on Adsorption by Powdered Activated Carbon." *Proc. AWWA Ann. Conf.,* Toronto, Ontario, Canada, 1996.

Graham, M. et al. *Optimization of PAC Application for MIB and Geosmin Control.* AWWA Research Foundation, Denver, CO, 1997.

Graham, M. et al. "Modeling Equilibrium Adsorption of 2-Methylisoborneol and Geosmin in Natural Waters." *Water Res.,* in press, 1999.

Graham, M. R. et al. "Control of Taste-and-Odor Problems With Powdered Activated Carbon." *Proc. AWWA Water Quality Technol. Conf.,* New Orleans, LA, 1995.

Graveland, A., J. C. Kruithof, and P. A. N. M. Nuhn. "Production of Volatile Halogenated Compounds by Chlorination after Carbon Filtration." Paper presented at the American Chemical Society Meeting, Atlanta, GA, April 1981.

Greene, B. G., V. L. Snoeyink, and F. W. Pogge. *Adsorption of Pesticides by Powdered Activated Carbon.* AWWA Research Foundation, Denver, CO, 1994.

Haberer, K., and S. Normann. *Vom Wasser,* 49:331, 1979.

Halsey, G., and H. S. Taylor. *J. Chem. Phys.,* 15:624, 1947.

Hand, D. W., J. C. Crittenden, and W. E. Thacker. *J. Env. Eng. Div., ASCE,* 110:440, 1984.

Hassler, J. W. *Activated Carbon.* Chemical Publishing Co., Inc., New York, 1974.

Hess, A. F. "GAC Treatment Designs and Costs for Controlling Volatile Organic Compounds in Ground Water." Presented at the American Chemical Society Meeting, Atlanta, GA, 1981.

Hoehn, R. C. et al. "A Pilot-Scale Evaluation of the Roberts-Haberer Process for Removing Trihalomethane Precursors from Surface Water with Activated Carbon." *Proc. AWWA Ann. Conf.,* Dallas, TX, June 1984.

Hoehn, R. C. et al. "THM-Precursor Control with Powdered Activated Carbon in a Pulsed-Bed, Solids Contact Clarifier." *Proc. AWWA Ann. Conf.,* Kansas City, MO, June 1987.

Hong, S., and R. S. Summers. "Impact of Backwashing and Desorption on GAC Breakthrough of Natural Organic Matter," *Proc. AWWA Ann. Conf.,* New York, NY, 1994.

Hooper S., R. S. Summers, and S. Hong. "A Systematic Evaluation of the Role of Influent TOC and pH on GAC Performance after Enhanced Coagulation." *Proc. AWWA Water Quality Technol. Conf.,* Boston, MA, 1996.

Hooper, S. M. et al. "GAC Performance for DBP Control: Effect of Influent Concentration, Seasonal Variation, and Pretreatment." *Proc. AWWA Annual Conference,* Toronto, Ontario, Canada, 1996a.

Hooper, S. M. et al. *Jour. AWWA,* 88(8):107, 1996b.

Huang, C. P. "Chemical Interactions between Inorganics and Activated Carbon." In *Carbon Adsorption Handbook* (Cheremisinoff and Ellerbusch, eds.). Ann Arbor Science, Boca Raton, FL, 1978.

Hutchins, R. A. *Chem. Engr. Prog.,* 69:48, 1973.

Jain, J. S., and V. L. Snoeyink. *J. Water Pollution Cont. Fed.,* 45:2463, 1973.

Jayes, D. A., and I. M. Abrams. *J. NEWWA,* 82:15, 1968.

Juhola, A. J. *Optimization of the Regeneration Procedures for Granular Activated Carbon.* USEPA Report No. 17020 DAO, 1970.

Juhola, A. J. *Kemia-Kemi,* 11, 1977a.

Juhola, A. J. *Kemia-Kemi,* 12, 1977b.

Junk, G. A. et al. *Jour. AWWA,* 68(4):218, 1976.

Juntgen, H. "Phenomena of Activated Carbon Regeneration." Translation of *Reports on Special Problems of Water Technology,* vol. 9, *Adsorption.* USEPA EPA-600/9-76-030, 1976.

Kennedy, C. *Environ. Sci. & Technol.,* 7:138, 1973.

Kim, B. R., and V. L. Snoeyink. *Jour. AWWA,* 72(8):488, Aug. 1980.

Kim, B. R., V. L. Snoeyink, and F. M. Saunders. "Adsorption of Organic Compounds by Synthetic Resins," *J. Water Pollution Cont. Fed.,* 48:120, 1976.

Kim, B. R., V. L. Snoeyink, and R. A. Schmitz. *J. Water Pollution Cont. Fed.,* 50:122, 1978.

Kipling, J. J., and P. V. Shooter. *J. Colloid and Interface Sci.,* 21:238, 1966.

Kirisits, M. J., and V. L. Snoeyink. "The Reduction of Bromate in a Biologically Active Carbon Filter." *Jour. AWWA,* in press, 1999.

Knappe, D. R. U. *Predicting the Removal of Atrazine by Powdered and Granular Activated Carbon,* PhD thesis, Univ. of Illinois, Urbana, IL, 1997.

Knappe, D. R. U. et al. *Jour. AWWA,* 84(8): 73–80, 1992.

Knappe, D. R. U. et al. *Environ. Sci. Technol.,* 32:1694, 1998.

Koechling, M. et al. "Combination of Anion Exchange Resin and GAC: An Alternative Method of DBP Precursor Removal." *Proc. AWWA Ann. Conf.,* Atlanta, GA, 1997.

Koffskey, W. E., and N. V. Brodtmann. *Organic Contaminant Removal in Lower Mississippi Drinking Water by Granular Activated Carbon.* USEPA, Cincinnati, OH, 1981.

Koffskey, W. E., and B. W. Lykins, Jr. "Experiences with GAC Filtration and On-Site Reactivation at Jefferson Parish, La." *AWWA Ann. Conf.,* Kansas City, MO, June 1987.

Kolle, W. "Use of Macroporous Ion Exchangers for Drinking Water Purification." translation of *Reports on Special Problems of Water Technology* (H. Sontheimer, ed.). EPA Report No. 600/9-76-030 USEPA, Cincinnati, OH, 1976.

Komorita, J. D., and V. L. Snoeyink. *Jour. AWWA,* 77(1):62, 1985.

Kornegay, B. H. "Control of Synthetic Organic Chemicals by Activated Carbon—Theory, Application, and Regeneration Alternatives." Presented at the Seminar on Control of Organic Chemicals in Drinking Water, USEPA, Feb. 13–14, 1979.

Kornegay, B. H. "Determining Granular Activated Carbon Process Design Parameters." *AWWA Ann. Conf.,* Kansas City, MO, June 1987.

Kreft, P. et al. *Jour. AWWA,* 73(10):558, Oct. 1981.

Kruithof, J. C. et al. Chapter in *Activated Carbon in Drinking Water Technology.* AWWA Research Foundation, Denver, CO, 1983.

Kumagai, J. S., and W. J. Kaufman. *Removal of Organic Contaminants, Phenol Sorption by Activated Carbon and Selected Macroporous Resins.* SERL Report No. 68-8, San. Eng. Res. Lab., Univ. of California, Berkeley, CA, July 1968.

Lalezary-Craig, S. et al. *Jour. AWWA,* 80(3):73, March 1988.

Lange, A. L., and E. Kawczynski. "THM Studies, Contra Costa County Water District." *Proc. Water Treatment Forum VII,* Calif.-Nev. Sect. AWWA, Palo Alto, CA, April 1978.

Langmuir, I. *J. Am. Chem. Soc.,* 1931, 1918.

Lawrence, C. H. *Wtr. Waste Engrg.,* 5:46, 1968.

LeChevallier, M. W. et al. *Appl. Environ. Microb.,* 48(5):918, Nov. 1984.

Lee, M. C., V. L. Snoeyink, and J. C. Crittenden. "Activated Carbon Adsorption of Humic Substances," *Jour. AWWA,* 73(8):440, Aug. 1981.

Lee, M. C. et al. *J. Env. Eng. Div., ASCE,* 109(3):631, 1983.

Lettinga, G., W. A. Beverloo, and W. C. van Lier. *Prog. Water Tech.,* 10:5, 1978.

Liang, S., and W. J. Weber, Jr. *Chem. Engr. Commun.,* 35:49, 1985.

Love, O. T., and R. G. Eilers. *Jour. AWWA,* 74(8):413, Aug. 1982.

Love, O. T. Jr. et al. Chapter in *Proc. Conf. on Activated Carbon in Water Treatment.* Univ. of Reading, Water. Res. Assn., Medmenham, UK, 279, April 1973.

Love, O. T. et al. *Treatment for the Removal of Trihalomethanes in Drinking Water.* USEPA, Drinking Water Research Division, Cincinnati, OH, 1976.

Loven, A. W. *Chem. Engr. Prog.,* 69:56, 1973.

Lykins, B. W. Jr., R. M. Clark, and J. Q. Adams. *Jour. AWWA,* 80(5):85, May 1988.

Lykins, B. W. Jr., R. M. Clark, and D. H. Cleverly. *J. Env. Engr. Div. ASCE,* 114(2):300, 1988.

McCarty, P. L., D. Argo, and M. Reinhard. *Jour. AWWA,* 71(11):683, Nov. 1979.

McCreary, J. J., and V. L. Snoeyink. *Jour. AWWA,* 69(8):437, Aug. 1977.

McCreary, J. J., and V. L. Snoeyink. *Water Res.,* 14:151, 1980.

McGuire, M. J., and I. H. Suffet. *J. Env. Eng. Div., ASCE,* 110(3):629, June 1984.

Meijers, J. A. P., and R. C. van der Leer. Chapter in *Activated Carbon in Drinking Water Technology.* AWWA Research Foundation, Denver, CO, 1983.

Miller, J. A., V. L. Snoeyink, and S. Harrell. Chapter in *Disinfection By-Products in Water Treatment* (R. Minear and G. Amy, eds.). Lewis Publishers, Ann Arbor, MI, 1995.

Miltner, R. et al. "A Comparative Evaluation of Biological Filters." *Proc. AWWA Water Quality Technol. Conf.,* Boston, MA, 1996.

Monsitz, J. T., and L. D. Ainesworth. *Public Works,* 101:113, 1970.

Murin, C. J., and V. L. Snoeyink. *Environ. Sci. Technol.,* 13:305, 1979.

Najm, I. N., V. L. Snoeyink, and Y. Richard. *Jour. AWWA,* 83(8):57, 1991.

Najm, I. N., V. L. Snoeyink, and Y. Richard. *Water Res.,* 27(4):551, 1993.

Najm, I. N. et al. "PAC in Floc Blanket Reactors." *Proc. AWWA Ann. Conf.,* Los Angeles, CA, June 1989.

Najm, I. N. et al. *Jour. AWWA,* 82:65, Jan. 1990.

Namkung, E., and B. E. Rittmann. *Jour. AWWA,* 79(7):109, July 1987.

Pelekani, C., and V. L. Snoeyink. "Competitive Adsorption in Natural Water: Role of Activated Carbon Pore Size." *Water Res.,* in press, 1999.

Poggenburg, W. et al. "Untersuchungen zur Optimierung der Aktivkohleanwendung bei der Trinkwasseraufbereitung am Rhein Unter besonderer Berucksichtigung der Regeneration nach thermischen Verfahren." Heft 12, *Veroffentlichugen des Bereichs und Lehrstuhls fur Wasserchemie.* Universitat Karlsruhe, Germany, 1979.

Prober, R., J. J. Pyeha, and W. E. Kidon. *AIChEJ.,* 21:1200, 1975.

Qi, S. et al. *Jour. AWWA,* 84(9):113, 1992.

Qi, S. et al. *J. Env. Eng.,* 120(1):202, 1994.

Radke, C. J., and J. M. Prausnitz. *AIChEJ.,* 18:761, 1972.

Randtke, S. J., and C. P. Jepsen. "Effects of Salts on Activated Carbon Adsorption of Fulvic Acids," *Jour. AWWA,* 74(2):84, Feb. 1982.

Randtke, S. J., and V. L. Snoeyink. "Evaluating GAC Adsorptive Capacity," *Jour. AWWA,* 75(8):406, Aug. 1983.

Richard, Y., personal communication, 1986.

Rittmann, B. E., and P. M. Huck. "Biological Treatment of Public Water Supplies." *CRC Crit. Rev. Environ. Control,* 19:2, June 1989.

Robeck, G. G. *Evaluation of Activated Carbon.* Water Supply Research Laboratory, National Environmental Research Center, Cincinnati, OH, March 3, 1975.

Robeck, G. G. et al. *Jour. AWWA,* 57(2):181, Feb. 1965.

Roberts, P. V., and R. S. Summers. "Performances of Granular Activated Carbon for Total Organic Carbon Removal," *Jour. AWWA,* 74(2):113, Feb. 1982.

Rosene, M. R. et al. Chapter in *Activated Carbon Adsorption from the Aqueous Phase,* vol. I (M. J. McGuire and I. H. Suffet, eds.). Ann Arbor Science, Boca Raton, FL, 1980.

Schimmoler, L. J., et al. "Performance of a Floc Blanket Reactor-PAC-UF System for DOC Removal." *Proc. AWWA Membrane Conf.,* Reno, NV, 1995.

Schuliger, W. G., and J. M. MacCrum. "GAC Reactivation System Design and Operating Conditions." AIChE Mtg., Detroit, MI, June 5, 1973.

Schuliger, W. G., G. N. Riley, and N. J. Wagner. "Thermal Reactivation of GAC: A Proven Technology." Paper presented at the AWWA Ann. Conf., Kansas City, MO, June 1987.

Schultink, B. "Provincial Waterworks of North Holland," personal communication, 1982.

Servais, P. et al. "BOM in Water Treatment," in *Biodegradable Organic Matter in Drinking Water* (M. Prevost, ed.). 1998.

Sheindorf, C., M. Rebhun, and M. Sheintuch. *J. Colloid Interface Sci.,* 79(1):136, 1981.

Silvey, J. K. G. *Jour. AWWA,* 56(1):60, 1964.

Singley, J. E. et al. *Minimizing Trihalomethane Formation in a Softening Plant.* USEPA, Water Supply Research Division, Municipal Environmental Research Laboratory, Cincinnati, OH, 1977.

Singley, J. E. et al. "Use of PAC for Removal of Specific Organic Compounds." *Sem. Proc.: Controlling Organics in Drinking Water,* AWWA Ann. Conf., San Francisco, CA, 1979.

Smisek, M., and S. Cerny. *Active Carbon.* Elsevier, New York, 1970.

Snoeyink, V. L. Chapter in *Occurrence and Removal of Volatile Organic Chemicals from Drinking Water.* AWWA Research Foundation/KIWA Report, AWWA Research Foundation, Denver, CO, 1983.

Snoeyink, V. L. "Adsorption of Organic Compounds." In *Water Quality and Treatment,* 4th ed. McGraw-Hill, New York, 1990.

Snoeyink, V. L., and D. R. U. Knappe. "Analysis of Pilot and Full Scale Granular Activated Carbon Performance Data." *Proc. AWWA Ann. Conf.,* New York, 1994.

Snoeyink, V. L., and M. T. Suidan. Chapter in *Disinfection: Water and Wastewater* (J. D. Johnson, ed.). Ann Arbor Science Publishers, Ann Arbor, MI, 1975.

Snoeyink, V. L., and W. J. Weber, Jr. Chapter in *Progress in Surface and Membrane Science* (J. F. Danielli, ed.) Pergamon Press, 1972.

Snoeyink, V. L., W. J. Weber, Jr., and H. B. Mark, Jr. *Environ. Sci. Technol.,* 3:1918, 1969.

Snoeyink, V. L. et al. Chapter in *Chemistry of Water Supply, Treatment and Distribution* (A. Rubin, ed.). Ann Arbor Science, Boca Raton, FL, 1974.

Solarik, G. et al. "Predicting and Characterizing the Removal of DBP Precursors by GAC." *Proc. AWWA Ann. Conf.,* Anaheim, CA, 1995a.

Solarik, G. et al. "DBP Precursor Treatment Studies at Little Falls, New Jersey, to Meet the ICR Requirements." *Proc. AWWA Water Quality Technol. Conf.,* New Orleans, LA, 1995b.

Solarik, G. et al. "The Impact of Ozonation and Biotreatment on GAC Performance for NOM Removal and DBP Control." *Proc. AWWA Ann. Conf.,* Toronto, Ontario, Canada, 1996.

Sontheimer, H. "The Use of Powdered Activated Carbon." *Translation of Reports of Special Problems of Water Technology,* vol. 9, *Adsorption.* Report EPA-600/9-76-030, USEPA, Cincinnati, OH, Dec. 1976.

Sontheimer, H., personal communication, 1983.

Sontheimer, H., J. C. Crittenden, and R. S. Summers. *Activated Carbon for Water Treatment,* 2nd ed. DVGW-Forschungstelle am Engler-Bunte-Institut der Universitat Karlsruhe, Karlsruhe, Germany, 1988.

Sontheimer, H., and C. Hubele. "The Use of Ozone and Granulated Activated Carbon in Drinking Water Treatment." In *Treatment of Drinking Water for Organic Contaminants* (P. M. Huck and P. Toft, eds.). Pergamon Press, Oxford, 1987.

Speth, T. F., and R. J. Miltner. *Jour. AWWA* 82:72, Feb. 1990, and 90:171, April 1998.

Suidan, M. T., V. L. Snoeyink, and R. A. Schmitz. *Water—1976: I. Physical Chemical Wastewater Treatment, AIChE Symposium Series,* 73:18, 1976.

Suidan, M. T., V. L. Snoeyink, and R. A. Schmitz. *Environ. Sci. Technol.,* 11:785, 1977a.

Suidan, M. T., V. L. Snoeyink, and R. A. Schmitz. *J. Environ. Eng. Div., ASCE.,* 103:677, 1977b.

Summers, R. S., D. DiCarlo, and S. S. Palepu. "GAC Adsorption in the Presence of Background Organic Matter: Pretreatment Approaches and Attenuation of Shock Loadings." *Proc. AWWA Ann. Conf.,* Cincinnati, OH, 1990.

Summers, R. S., S. Hooper, and S. Hong. "GAC Precursor Removal Studies." In *ICR Manual for Bench- and Pilot-Scale Treatment Studies.* EPA 814-B-96-003, USEPA, Office of Water, 1996.

Summers, R. S., and P. V. Roberts. *J. Env. Eng. Div., ASCE,* 110(1):73, 1984.

Summers, R. S., and P. V. Roberts. *J. Colloid and Interface Sci.,* 122(2):367–381, 1988a.

Summers, R. S., and P. V. Roberts. *J. Colloid Interface Sci.,* 122(2):382, 1988b.

Summers, R. S. et al. *Jour. AWWA,* 81(5):66, 1989.

Summers, R. S. et al. *Jour. AWWA,* 85(1):88, 1993.

Summers, R. S. et al. "Adsorption of Natural Organic Matter and Disinfection By-Product Precursors." *Proc. AWWA Ann. Conf.,* New York, 1994.

Summers, R. S. et al. *Jour. AWWA* 87(8):69, 1995.

Summers, R. S. et al. *DBP Precursor Control with GAC Adsorption.* AWWA Research Foundation Report, Denver, CO, 1997.

Summers, R. S. et al. "Combined Use of Ion Exchange and Granular Activated Carbon for the Control of Organic Matter and Disinfection By-Products," *Water Res.,* in press, 1999.

Suzuki, M. et al. *Chem. Eng. Sci.,* 33:271, 1978.

Symons, J. M. et al. *Treatment Techniques for Controlling Trihalomethanes in Drinking Water.* EPA-60012-81-156, USEPA, Cincinnati, OH, 1981.

Thacker, W. E., J. C. Crittenden, and V. L. Snoeyink. "Modeling of Adsorber Performance: Variable Influent Concentration and Comparison of Adsorbents," *J. Water Pollution Cont. Fed.,* 56:243, 1984.

Thacker, W. E., V. L. Snoeyink, and J. C. Crittenden. *Jour. AWWA,* 75(3):144, March 1983.

Thomas, J., personal communication, 1986.

Tipnis, P. R., and P. Harriott. *Chem. Eng. Commun.,* 46:11, 1986.

Topalian, P., referenced by H. Sontheimer and C. Hubele, "The Use of Ozone and Granular Activated Carbon in Drinking Water Treatment." In *Treatment of Drinking Water for Organic Contaminants* (P. M. Huck and P. Toft, eds.) Pergamon Press, Oxford, 1987.

Umehara, T., P. Harriott, and J. M. Smith. *AIChE,* 29(5):737, 1983.

van der Gaag, M. A., J. C. Kruithof, and L. M. Puijker. Chapter in *Organic Micropollutants in Drinking Water and Health,* (H. A. M. de Kruif and H. J. Kool, eds.). Elsevier, New York, 1985.

van der Kooij, D. "Biological Processes in Carbon Filters." In *Activated Carbon in Drinking Water Technology,* KIWA/AWWA Research Foundation Report, Denver, CO, 1983.

van Vliet, B. M. "Regeneration Principles." *Proc. Symp. on Design and Operation of Plants for the Recovery of Gold by Activated Carbon,* South African Inst. of Mining and Metallurgy, Johannesburg, South Africa, Oct. 14–18, 1985.

van Vliet, B. M., and L. Venter. *Water Sci. Technol.,* 17:1029, 1984.

Vidic, R. D., and M. T. Suidan. *Environ. Sci. Technol.,* 25:1612, 1991.

Voudrias, E. A., R. A. Larson, and V. L. Snoeyink. *Environ. Sci. Technol.,* 19:441, 1985.

Voudrias, E. A. et al. *Water Res.,* 17(9):1107, 1983.

Wang, J. Z., R. S. Summers, and R. J. Miltner. *Jour. AWWA,* 87(12):55, 1995.

Weber, W. J. Jr. *Physicochemical Processes.* Wiley-Interscience, New York, 1972.

Weber, W. J. Jr., and M. Pirbazari. *Jour. AWWA,* 74(4):203, 1982.

Weber, W. J. Jr., and E. H. Smith. *Environ. Sci. Technol.,* 21(11):1040, 1987.

Weber, W. J. Jr., T. C. Voice, and A. Jodellah. *Jour. AWWA,* 75(12):612, Dec. 1983.

Westrick, J. J., and J. M. Cohen. *J. Water Pollution Cont. Fed.,* 48:323, 1976.

Wiesner, M. R., J. J. Rook, and F. Fiessinger. *Jour. AWWA,* 79(12):39, Dec. 1987.

Wulfeck, W. M. Jr., and R. S. Summers. "Control of DBP Formation Using Retrofitted GAC Filter-Adsorbers and Ozonation." *Proc. AWWA Water Quality Technol. Conf.,* San Francisco, CA, 1994.

Yehaskel, A. *Activated Carbon Manufacture and Regeneration.* Noyes Data Corp., Park Ridge, NJ, 1978.

Zimmer, G., H. J. Brauch, and H. Sontheimer. Chapter in *Aquatic Humic Substances: Influence on Fate and Treatment of Pollutants* (I. Suffet and P. MacCarthy, eds.). American Chemical Society, Washington, D.C., 1989.

CHAPTER 14
DISINFECTION

Charles N. Haas, Ph.D.
LD Betz Professor of Environmental Engineering
Drexel University
Philadelphia, Pennsylvania

Disinfection is a process designed for the deliberate reduction of the number of pathogenic microorganisms. While other water treatment processes, such as filtration or coagulation-flocculation-sedimentation, may achieve pathogen reduction, this is not generally their primary goal. A variety of chemical or physical agents may be used to carry out disinfection. The concept of disinfection preceded the recognition of bacteria as the causative agent of disease. Averill (1832), for example, proposed chlorine disinfection of human wastes as a prophylaxis against epidemics. Chemical addition during water treatment for disinfection became accepted only after litigation on its efficacy (Race, 1918). The prophylactic benefits of water disinfection soon became apparent, particularly with respect to the reduction of typhoid and cholera.

While significant progress is being made in controlling the classic waterborne diseases, newly recognized agents have added to the challenge. These include viruses (Melnick et al., 1978; Mosley, 1966), certain bacteria (*Campylobacter,* Palmer et al., 1983; *Yersinia,* Brennhovd et al., 1992; Reasoner, 1991; or *Mycobacteria,* Geldreich, 1971; Iivanainen et al., 1993; Reasoner, 1991; for example), and protozoans (*Giardia,* Brown et al., 1992; Le Chevallier et al., 1991; Miller et al., 1978; Reasoner, 1991; Renton et al., 1996; Rose et al., 1991; *Cryptosporidium,* Bridgman et al., 1995; Centers for Disease Control and Prevention, 1995; Gallaher et al., 1989; Goldstein et al., 1996; Hayes et al., 1989; Le Chevallier et al., 1991; Leland et al., 1993; Mac Kenzie et al., 1994; Miller, 1992; Reasoner, 1991; Richardson et al., 1991; Rose et al., 1991; Rush et al., 1990; Smith, 1992). Occasional outbreaks of drinking-water-associated hepatitis have also occurred (Nasser, 1994; Rosenberg et al., 1980). In addition, new viral agents are continually being found to be capable of waterborne transmission.

The state of disinfection practice in the United States in the late 1980s was summarized in a survey of the AWWA Disinfection Committee (Haas et al., 1992). Most water utilities continue to rely on chlorine or hypochlorite as their primary disinfection chemicals (Table 14.1), although increasing numbers are using ammonia (for pre- or postammoniation) or chlorine dioxide or ozone. With the increasing concern for removing and inactivating some of the more resistant pathogens, such as *Giardia* and *Cryptosporidium,* while minimizing disinfection by-products, options other than

TABLE 14.1 Water Utility Disinfection Practices According to 1989 AWWA Survey ($N = 267$)

	No ammonia	Ammonia
Chlorine alone		
Gas	67.42%	19.85%
Hypochlorite	5.99%	0.75%
Chlorine + ClO_2	3.37%	1.50%
Ozone	0.37%	
Other	0.75%	

Source: Haas et al., 1992.

traditional chlorination are gaining popularity. This chapter will cover the use of chlorine, as well as the major alternative agents, for the purpose of disinfection.

HISTORY OF DISINFECTION

Chlorine

Chlorine gas was first prepared by Scheele in 1774, but chlorine was not regarded as a chemical element until 1808 (Belohlav and McBee, 1966). Early uses of chlorine included the use of Javelle water (chlorine gas dissolved in an alkaline potassium solution) in France for waste treatment in 1825 (Baker, 1926) and its use as a prophylactic agent during the European cholera epidemic of 1831 (Belohlav and McBee, 1966).

Disinfection of water by chlorine first occurred in 1908 at Bubbly Creek (Chicago) and the Jersey City Water Company. Within two years, chlorine was introduced as a disinfectant at New York City (Croton), Montreal, Milwaukee, Cleveland, Nashville, Baltimore, and Cincinnati, as well as other smaller treatment plants. Frequently, dramatic reductions in typhoid accompanied the introduction of this process (Hooker, 1913). By 1918, over 1000 cities, treating more than 3 billion gal/day (1.1×10^7 m^3/day) of water, were employing chlorine as a disinfectant (Race, 1918).

Chloramination, the addition of both chlorine and ammonia either sequentially or simultaneously, was first employed in Ottawa, Canada, and Denver, Colorado, in 1917. Both of these early applications employed prereaction of the two chemicals prior to their addition to the full flow of water. Somewhat later, preammoniation (the addition of ammonia prior to chlorine) was developed. In both cases, the process was advocated for its ability to prolong the stability of residual disinfectant during distribution and for its diminished propensity to produce chlorophenolic taste and odor substances. Shortages of ammonia during World War II, and recognition of the superiority of free chlorine as a disinfectant, reduced the popularity of the chloramination process. Recent concerns about organic by-products of chlorination, however, have increased the popularity of chloramination (Wolfe et al., 1984).

Chlorine Dioxide

Chlorine dioxide was first produced from the reaction of potassium chlorate and hydrochloric acid by Davy in 1811 (Miller et al., 1978). However, not until the industrial-scale preparation of sodium chlorite, from which chlorine dioxide may more readily be generated, did its widespread use occur (Rapson, 1966).

Chlorine dioxide has been used widely as a bleaching agent in pulp and paper manufacture (Rapson, 1966). Despite early investigations on the use of chlorine dioxide as an oxidant and disinfectant (Aston and Synan, 1948), however, its ascendancy in both water and wastewater treatment has been slow. As recently as 1971 (Morris, 1971), it was stated that "... ClO_2 has never been used extensively for water disinfection."

By 1977, 84 potable water treatment plants in the United States were identified as using chlorine dioxide treatment, although only one of these relied upon it as a primary disinfectant (Miller et al., 1978). In Europe, chlorine dioxide was being used as either an oxidant or disinfectant in almost 500 potable water treatment plants (Miller et al., 1978).

Ozone

Ozone was discovered in 1783 by Van Marum, and named by Schonbein in 1840. In 1857, the first electric discharge ozone generation device was constructed by Siemens, with the first commercial application of this device occurring in 1893 (Water Pollution Control Federation, 1984).

Ozone was first applied as a potable water disinfectant in 1893 at Oudshoorn, Netherlands. In 1906, Nice, France, installed ozone as a treatment process, and this plant represents the oldest ozonation installation in continuous operation (Rice et al., 1981). In the United States, ozone was first employed for taste and odor control at New York City's Jerome Park Reservoir in 1906. In 1987, five water treatment facilities in the United States were using ozone oxidation primarily for taste and odor control or trihalomethane precursor removal (Glaze, 1987). Since the 1993 Milwaukee *Cryptosporidium* outbreak, there has been an upsurge in interest in ozone as a disinfectant.

UV Radiation

The biocidal effects of ultraviolet radiation (UV) have been known since it was established that short-wavelength UV was responsible for microbial decay often associated with sunlight (Downes and Blount, 1877). By the early 1940s, design guidelines for UV disinfection were proposed (Huff et al., 1965). UV has been accepted for treating potable water on passenger ships (Huff et al., 1965). Historically, however, it has met with little enthusiasm in public water supply applications because of the lack of a residual following application. In wastewater treatment, in contrast, over 600 plants in the United States are either using, currently designing, or constructing UV disinfection facilities (Scheible et al., 1992).

Other Agents

A variety of other agents may be used to effect inactivation of microorganisms. These include heat, extremes in pH, metals (silver, copper), surfactants, permanganate, and electron beam irradiation. Heat is useful only in emergencies as in "boil water" orders, and is uneconomical. An alkaline pH (during high lime softening) may provide some microbial inactivation, but is not usually sufficient as a sole disinfectant. Potassium permanganate has been reported to achieve some disinfecting effects; however, the magnitudes have not been well characterized. High-energy electrons for disinfection of wastewaters and sludges have also been studied (Farooq et al., 1993); however, their feasibility in drinking water is uncertain. In this chapter, therefore, primary consideration will be given to chlorine compounds, ozone, chlorine dioxide, and UV.

Regulatory Issues for Disinfection Processes

SWTR and GDR Requirements. Amendments to the Safe Drinking Water Act require that all surface water suppliers in the United States filter and/or disinfect to protect the health of their customers. The filtration and disinfection treatment requirements for public water systems using surface water sources or groundwater under the direct influence of surface water are included in what is called the Surface Water Treatment Rule (SWTR, June 1989).

The SWTR requires that all surface water treatment facilities provide filtration and disinfection that achieves at least (1) a 99.9 percent (3-log) removal-inactivation of *Giardia lamblia* cysts and (2) a 99.99 percent (4-log) removal-inactivation of enteric viruses. The SWTR assumes that for effective filtration, a conventional treatment plant achieves 2.5-log removal of *Giardia* and a 2-log removal of viruses. Disinfection is required for the remainder of the removal-inactivation. The amount of disinfection credit to be awarded is determined with the CT concept, CT being defined as the residual disinfectant concentration (*C*, mg/L) multiplied by the contact time (*T*, min) between the point of disinfectant application and the point of residual measurement. The SWTR Guidance Manual provides tables of CT values for several disinfectants, which indicate the specific disinfection or CT credit awarded for a calculated value of CT. A large safety factor is incorporated into the CT values included in the Guidance Manual tables. In addition to relying on the CT tables to calculate disinfection credit, the SWTR allows utilities to demonstrate the effectiveness of their disinfection systems through pilot-scale studies, which may be prohibitively expensive for smaller operations. The SWTR is being revised to take into account knowledge developed since the mid-1980s, and the anticipated formal promulgation of the Enhanced Surface Water Treatment Rule (ESWTR) will further affect the level of required disinfection. A more complete discussion of the SWTR is included in Chapter 1.

Furthermore, under the Safe Drinking Water Act, EPA is required to promulgate rules for the disinfection of groundwaters. While the regulatory development of the anticipated Groundwater Disinfection Rule is currently pending, this is expected to require a level of disinfection either by chemical agents or by virtue of aquifer passage of all groundwaters being used in community water supply systems.

Disinfection By-product Requirements. Along with disinfection requirements, since 1974 there have been explicit regulations on disinfection by-products—first with respect to trihalomethanes, and more recently with respect to haloacetic acids, bromate, and other possible by-products. The combination of the requirement to achieve disinfection along with the requirement to minimize disinfection by-products has led to an increasing spectrum of options being considered.

DISINFECTANTS AND THEORY OF DISINFECTION

Basic Chemistry

Chlorine and Chlorine Compounds. Chlorine may be used as a disinfectant in the form of compressed gas under pressure that is dissolved in water at the point of application, solutions of sodium hypochlorite, or solid calcium hypochlorite. The three forms are chemically equivalent because of the rapid equilibrium that exists between dissolved molecular gas and the dissociation products of hypochlorite compounds.

Elemental chlorine (Cl_2) is a dense gas that, when subject to pressures in excess of its vapor pressure, condenses into a liquid with the release of heat and with a reduction in specific volume of approximately 450-fold. Hence, commercial shipments of chlorine are made in pressurized tanks to reduce shipment volume. When chlorine is to be dispensed as a gas, supplying thermal energy to vaporize the compressed liquid chlorine is necessary.

The relative amount of chlorine present in chlorine gas, or hypochlorite salts, is expressed in terms of *available chlorine.* The concentration of hypochlorite (or any other oxidizing disinfectant) may be expressed as available chlorine by determining the electrochemical equivalent amount of Cl_2 to that compound. Equation 14.1 shows that 1 mole of elemental chlorine is capable of reacting with two electrons to form inert chloride:

$$Cl_2 + 2\ e^- = 2\ Cl^- \tag{14.1}$$

Equation 14.2 shows that 1 mole of hypochlorite (OCl^-) may react with two electrons to form chloride:

$$OCl^- + 2\ e^- + 2H^+ = Cl^- + H_2O \tag{14.2}$$

Hence, 1 mole of hypochlorite is electrochemically equivalent to 1 mole of elemental chlorine, and may be said to contain 70.91 g of available chlorine (identical to the molecular weight of Cl_2).

Calcium hypochlorite ($Ca(OCl)_2$) and sodium hypochlorite (NaOCl) contain 2 and 1 moles of hypochlorite per mole of chemical, respectively, and, as a result, 141.8 and 70.91 g available chlorine per mole, respectively. The molecular weights of $Ca(OCl)_2$ and NaOCl are, 143 and 74.5, respectively, so that pure preparations of the two compounds contain 99.2 and 95.8 weight percent available chlorine; hence, they are effective means of supplying chlorine for disinfection purposes.

Calcium hypochlorite is available commercially as a dry solid. In this form, it is subject to a loss in strength of approximately 0.013 percent per day (Laubusch, 1963). Calcium hypochlorite is also available in a tablet form for use in automatic feed equipment at low-flow treatment plants.

Sodium hypochlorite is available in 1 to 16 weight percent solutions. Higher-concentration solutions are not practical because chemical stability rapidly diminishes with increasing strength. At ambient temperatures, the half-life of sodium hypochlorite solutions varies between 60 and 1700 days, respectively, for solutions of 18 and 3 percent available chlorine (Baker, 1969; Laubusch, 1963).

It should be noted that the loss of strength in sodium hypochlorite solutions may also result in the formation of by-products that may be undesirable. Thermodynamically, the autodecomposition of hypochlorite to chlorate is highly favored by the following overall process (Bolyard et al., 1992):

$$3\ ClO^- \rightarrow 2\ Cl^- + ClO_3^- \tag{14.3}$$

Measurements of sodium hypochlorite disinfectant solutions at water utilities have revealed that the mass concentration of chlorate is from 1.7 to 220 percent of the mass concentration of free available chlorine (Bolyard et al., 1992, 1993). The concentration of chlorate present in these stock solutions is kinetically controlled and may be related to the solution strength, age, temperature, pH, and presence of metal catalysts (Gordon et al., 1993, 1995).

When a chlorine-containing compound is added to a water containing insignificant quantities of kjeldahl nitrogen, organic material, and other chlorine-demanding substances, a rapid equilibrium is established among the various chemical species in solution. The term *free available chlorine* is used to refer to the sum of the concentrations of molecular chlorine (Cl_2), hypochlorous acid (HOCl), and hypochlorite ion (OCl^-), each expressed as *available chlorine.*

The dissolution of gaseous chlorine to form dissolved molecular chlorine is expressible as a phase equilibrium, and may be described by Henry's law:

$$Cl_2(g) = Cl_2(aq) \qquad H(\text{mol/L-atm}) = [Cl_2(aq)]/P_{Cl_2} \tag{14.4}$$

where quantities within square brackets represent molar concentrations, P_{Cl_2} is the gas phase partial pressure of chlorine in atmospheres, and H is the Henry's law constant, estimated from the following equation (Downs and Adams, 1973):

$$H = 4.805 \times 10^{-6} \exp (2818.48/T) \text{ (mol/L-atm)} \tag{14.5}$$

Dissolved aqueous chlorine reacts with water to form hypochlorous acid, chloride ions, and protons as indicated by Equation 14.6.

$$Cl_2(aq) + H_2O = H^+ + HOCl + Cl^-$$

$$K_H = \frac{[H^+][HOCl][Cl^-]}{[Cl_2(aq)]}$$

$$= 2.581 \exp\left(-\frac{2581.93}{T}\right) (\text{mol}^2/\text{L}^2) \tag{14.6}$$

This reaction typically reaches completion in 100 ms (Aieta and Roberts, 1985; Morris, 1946) and involves elementary reactions between dissolved molecular chlorine and hydroxyl ions. The extent of chlorine hydrolysis, or disproportionation (because the valence of chlorine changes from 0 on the left to +1 and –1 on the right), as described by Equation 14.6, decreases with decreasing pH and increasing salinity; hence, the solubility of gaseous chlorine may be increased by the addition of alkali or by the use of fresh, rather than brackish, water.

Hypochlorous acid is a weak acid and may dissociate according to Equation 14.7:

$$HOCl = OCl^- + H^+$$

$$K_a = [OCl^-][H^+]/[HOCl] \tag{14.7}$$

The pK_a of hypochlorous acid at room temperature is approximately 7.6 (Brigano et al., 1978). Morris (1966) has provided a correlating equation for K_a as a function of temperature:

$$\ln(K_a) = 23.184 - 0.0583\ T - 6908/T \tag{14.8}$$

where T is specified in degrees Kelvin (K = °C + 273). Figure 14.1 illustrates the effect of pH on the distribution of free chlorine between OCl^- and HOCl.

One practical consequence of the reactions described by Equations 14.4 through 14.8 is that the chlorine vapor pressure over a solution depends on solution pH, decreasing as pH increases (because of the increased formation of nonvolatile hypochlorite acid). Therefore, the addition of an alkaline material such as lime or

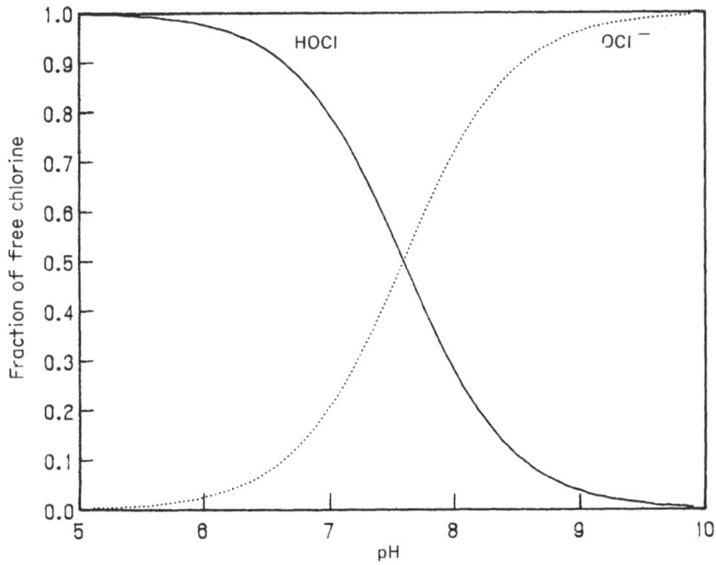

FIGURE 14.1 Effect of pH on relative amount of hypochlorous acid and hypochlorite ion at 20°C.

sodium bicarbonate will reduce the volatility of chlorine from accidental spills or leaks and thus minimize danger to exposed personnel.

The acid-base properties of gaseous chlorine, or the hypochlorite salts, will also result in a loss or gain, respectively, of alkalinity, and a reduction or increase, respectively, in pH. For each mole of free chlorine (i.e., 1 mole of Cl_2, or of NaOCl or 0.5 mole of $Ca(OCl)_2$), there will be a change of one equivalent of alkalinity (increase for sodium and calcium hypochlorite, and decrease for chlorine gas).

EXAMPLE 14.1 The solution produced by a gas chlorinator contains 3500 mg/L available chlorine at a pH of 3. What is the equilibrium vapor pressure of this solution at 20°C (given that the value of the hydrolysis constant K_H is 4.5×10^4 at this temperature)?

1. The pH is sufficiently low that the dissociation of hypochlorous acid to form hypochlorite can be ignored. Therefore, a balance over chlorine species yields:

$$[Cl_2] + [HOCl] = (3500 \times 10^{-3})/71$$

2. The factor of 71 reflects the fact that 1 mole of either dissolved chlorine or hypochlorous acid contains 71 g of available chlorine.
3. The hydrolysis equilibrium constant can be used to develop an additional equation:

$$4.5 \times 10^4 = [H^+][Cl^-][HOCl]/[Cl_2]$$

or, because the pH is given,

$$4.5 \times 10^7 = [Cl^-][HOCl]/[Cl_2]$$

4. Because chlorine gas was used to generate the dissolved free chlorine, the disproportionation reaction requires that for each mole of HOCl produced, 1 mole of Cl^- must have been produced. If the initial concentration of chloride (in the feedwater to the chlorinator) was minimal, then a third equation results:

$$[Cl^-] = [HOCl]$$

5. These three equations can be manipulated to produce a quadratic equation in the unknown $[Cl_2]^{1/2}$:

$$[Cl_2] + 6708[Cl_2]^{1/2} - 0.05 = 0$$

6. The single positive root is the only physically meaningful one, hence:

$$[Cl_2]^{1/2} = 7.45 \times 10^{-6}$$

or

$$[Cl_2] = 5.55 \times 10^{-11}$$

7. The Henry's law constant can be computed from Equation 14.5 as 0.072 moles/L-atm, and therefore the partial pressure of chlorine gas is found:

$$P_{Cl_2} = 5.55 \times 10^{-11}/0.072 = 7.7 \times 10^{-10} \text{ atm}$$
$$= (0.77 \text{ ppb})$$

8. The OSHA permissible exposure limit (PEL) is reported as 1 ppm (ACGIH, 1994). Therefore, this level is of no apparent health concern to the workers.

Chlorine Dioxide. Chlorine dioxide (ClO_2) is a neutral compound of chlorine in the +IV oxidation state. It has a boiling point of 11°C at atmospheric pressure. The liquid is denser than water and the gas is denser than air (Noack and Doeff, 1979).

Chemically, chlorine dioxide is a stable free radical that, at high concentrations, reacts violently with reducing agents. It is explosive, with the lower explosive limit in air variously reported as 10 percent (Downs and Adams, 1973; Masschelein, 1979b) or 39 percent (Noack and Doeff, 1979). As a result, virtually all applications of chlorine dioxide require the synthesis of the gaseous compound in a dilute stream (either gaseous or liquid) on location as needed.

The solubility of gaseous chlorine dioxide in water may be described by Henry's law, and a fit of the available solubility data (Battino, 1984) results in the following relationship for the Henry's law constant (in units of atm^{-1}):

$$\ln(H) = \text{mole fraction dissolved } ClO_2(aq)/P_{ClO_2}$$
$$= 58.84621 + (47.9133/T) - 11.0593 \ln(T) \quad (14.9)$$

Under alkaline conditions, the following disproportionation into chlorite (ClO_2^-) and chlorate (ClO_3^-) occurs (Gordon et al., 1972):

$$2\,ClO_2 + 2\,OH^- = H_2O + ClO_3^- + ClO_2^- \quad (14.10)$$

In the absence of catalysis by carbonate, the reaction (Equation 14.10) is governed by parallel first- and second-order kinetics (Gordon et al., 1972; Granstrom and Lee, 1957). The half-life of aqueous chlorine dioxide solutions decreases substantially with increasing concentration and with pH values above 9. Even at neutral pH values, however, in the absence of carbonate at room temperature, the half-life of chlorine dioxide solutions of 0.01, 0.001, and 0.0001 mol/L is 0.5, 4, and 14 h, respectively. Hence, the storage of stock solutions of chlorine dioxide for even a few hours is impractical.

The simple disproportionation reaction to chlorate and chlorite is insufficient to explain the decay of chlorine dioxide in water free of extraneous reductants. Equation 14.10 predicts that the molar ratio of chlorate to chlorite formed should be 1:1. Medir and Giralt (1982), however, found that the molar ratio of chlorate to chlorite to chloride to oxygen produced was 5:3:1:0.75, and that the addition of chloride enhanced the rate of decomposition and resulted in the predicted 1:1 molar ratio of chlorite to chlorate. Thus, the oxidation of chloride by chlorate, and the possible formation of intermediate free chlorine, may be of significance in the decay of chlorine dioxide in demand-free systems (Gordon et al., 1972).

The concentration of chlorine dioxide in solution is generally expressed in terms of g/L as chlorine by multiplying the molarity of chlorine dioxide by the number of electrons transferred per mole of chlorine dioxide reacted and then multiplying this by 35.5 g Cl_2 per electron mole. Conventionally, the five-electron reduction (Equation 14.11) is used to carry out this conversion.

$$ClO_2 + 5e^- + 4H^+ = Cl^- + 2\,H_2O \qquad (14.11)$$

Note, however, that the typical reaction of chlorine dioxide in water, being reduced to chlorite, is a one-electron reduction as follows:

$$ClO_2 + e^- = ClO_2^- \qquad (14.11a)$$

Hence, according to Equation 14.11, 1 mole of chlorine dioxide contains 67.5 g of mass, and is equivalent to 177.5 (=5 × 35.5) g Cl_2. Therefore, 1 g of chlorine dioxide contains 2.63 g as chlorine. In examining any study on chlorine dioxide, due care with regard to units of expression of disinfectant concentration is warranted.

Ozone. Ozone is a colorless gas produced from the action of electric fields on oxygen. It is highly unstable in the gas phase; in clean vessels at room temperature the half-life in air is 20 to 100 h (Manley and Niegowski, 1967).

The solubility of ozone in water can be described by a temperature- and pH-dependent Henry's law constant. The following provisional relationship (H in atm^{-1}) has been suggested (Roy, 1979):

$$H = 3.84 \times 10^7\,[OH^-]\,\exp(-2428/T) \qquad (14.12)$$

Practical ozone generation systems have maximum gaseous ozone concentrations of about 50 g/m^3; thus, the maximum practical solubility of ozone in water is about 40 mg/L (Stover et al., 1986). Upon dissolution in water, ozone can react with water itself, with hydroxyl ions, or with dissolved chemical constituents, as well as serving as a disinfecting agent. Details of these reactions will be discussed later in this chapter and in Chapter 12.

DISINFECTANT DEMAND REACTIONS

Chlorine

Reactions with Ammonia. In the presence of certain dissolved constituents in water, each of the disinfectants may react and transform to less biocidal chemical forms. In the case of chlorine, these principally involve reactions with ammonia and amino nitrogen compounds. In the presence of ammonium ion, free chlorine reacts in a stepwise manner to form chloramines. This process is depicted in Equations 14.13 through 14.15:

$$NH_4^+ + HOCl = NH_2Cl + H_2O + H^+ \tag{14.13}$$

$$NH_2Cl + HOCl = NHCl_2 + H_2O \tag{14.14}$$

$$NHCl_2 + HOCl = NCl_3 + H_2O \tag{14.15}$$

These compounds, monochloramine (NH_2Cl), dichloramine ($NHCl_2$), and trichloramine (NCl_3), each contribute to the total (or combined) chlorine residual in a water. The terms *total available chlorine* and *total oxidants* refer, respectively, to the sum of free chlorine compounds and reactive chloramines, or total oxidating agents. Under normal conditions of water treatment, if any excess ammonia is present, at equilibrium the amount of free chlorine will be much less than 1 percent of total residual chlorine. Each chlorine atom associated with a chloramine molecule is capable of undergoing a two-electron reduction to chloride; hence, each mole of monochloramine contains 71 g available chlorine; each mole of dichloramine contains 2×71 or 142 g; and each mole of trichloramine contains 3×71 or 223 g of available chlorine. Inasmuch as the molecular weights of mono-, di-, and trichloramine are 51.6, 86, and 110.5, respectively, the chloramines contain, respectively, 1.38, 1.65, and 2.02 g available chlorine per gram. The efficiency of the various combined chlorine forms as disinfectants differs, however, and thus the concentration of available chlorine does not completely characterize process performance. On an approximate basis, for example, for coliforms, the biocidal potency of $HOCl{:}OCl^-{:}NH_2Cl{:}NHCl_2$ is approximately 1:0.0125:0.005:0.0166; and for viruses and cysts, the combined chlorine forms are considerably less effective (Chang, 1971). As Equation 14.12 indicates, the formation of monochloramine is accompanied by the loss of a proton, because chlorination reduces the affinity of the nitrogen moiety for protons (Weil and Morris, 1949a).

The significance of chlorine speciation on disinfection efficiency was graphically demonstrated by Weber et al. (1940) as shown in Figure 14.2. As the dose of chlorine is increased, the total chlorine residual (i.e., remaining in the system after 30 min) increases until a dose of approximately 50 mg/L, whereupon residual chlorine decreases to a very low value, and subsequently increases linearly with dose indefinitely. The "hump and dip" behavior is paralleled by the sensitivity of microorganisms to the available chlorine residual indicated by the time required for 99 percent inactivation of *Bacillus metiens* spores. At the three points indicated, the total available chlorine is approximately identical at 22 to 24 mg/L, yet a 32-fold difference in microbial sensitivity occurred.

The explanation for this behavior is the "breakpoint" reaction between free chlorine and ammonia (Figure 14.3). At doses below the hump in the chlorine residual curve (zone 1), only combined chlorine is detectable. At doses between the hump and the dip in the curve, an oxidative destruction of combined residual chlorine accompanied by the loss of nitrogen occurs (zone 2) (Taras, 1950). One possible reaction during breakpoint is:

$$2\ NH_3 + 3\ HOCl = N_2 + 3\ H^+ + 3\ Cl^- + 3\ H_2O \tag{14.16}$$

This reaction also may be used as a means to remove ammonia nitrogen from water or wastewaters (Pressley et al., 1972). Finally, after the ammonia nitrogen has

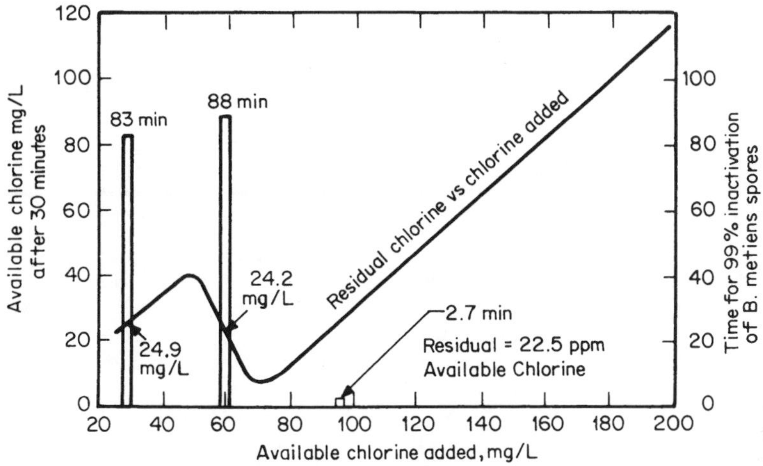

FIGURE 14.2 Effect of increased chloride dosage on residual chlorine and germicidal efficiency; pH 7.0, 20°C, NH_3 10 mg/L. (*Source:* Adapted from Weber et al., 1940.)

been completely oxidized, the residual remaining consists almost exclusively of free chlorine (zone 3). The minimum in the chlorine residual–versus–dose curve (in this case Cl_2:NH_4^+ – N weight ratio of 7.6/1) is called the *breakpoint* and denotes the amount of chlorine that must be added to a water before a stable free residual can be obtained.

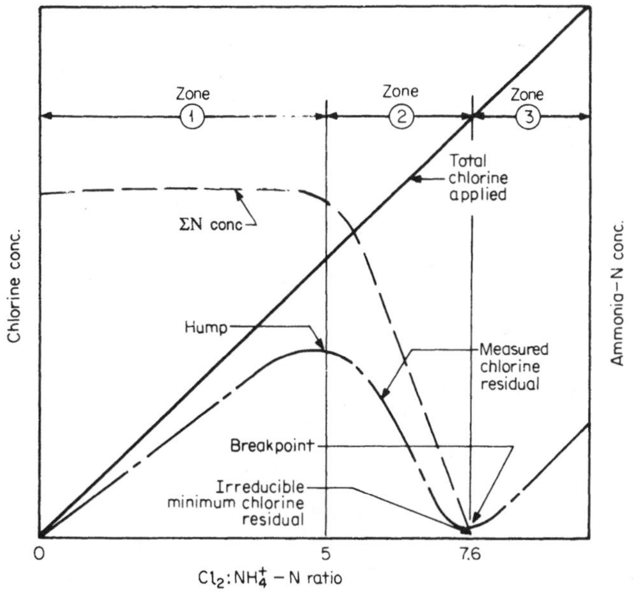

FIGURE 14.3 Schematic idealization of breakpoint curve. (*Source:* Adapted from G. C. White, *Disinfection of Wastewater and Water for Reuse,* Van Nostrand Reinhold, New York. Copyright 1978.)

In their investigations of the chlorination of drinking water, Griffin and Chamberlin (1941a,b) observed that:

1. The classical hump and dip curve is only seen at water pHs between 6.5 and 8.5.
2. The molar ratio between chlorine and ammonia nitrogen dose at the breakpoint under ideal conditions is 2:1, corresponding to a mass dose ratio (Cl_2:$NH_4^+ - N$) of 10:1.
3. In practice, mass dose ratios of 15:1 may be needed to reach breakpoint.

The breakpoint reaction may also affect the pH of a water. If sodium hypochlorite is used as the source of active chlorine, as breakpoint occurs, the pH decreases due to an apparent release of protons during the breakpoint process (Equation 14.16). If gaseous chlorine is used, this effect is reinforced by the release of protons by hydrolysis of gaseous chlorine according to Equations 14.6 and 14.7 (McKee, 1960).

The oxidation of ammonia nitrogen by chlorine to gaseous nitrogen at the breakpoint would theoretically require 1.5 mol of chlorine (Cl_2) per mole of nitrogen oxidized according to Equation 14.16. The observed stoichiometric molar ratio between chlorine added and ammonia nitrogen consumed at breakpoint is typically about 2:1, suggesting that more oxidized nitrogen compounds are produced at breakpoint rather than N_2 gas. Experimental evidence (Saunier and Selleck, 1979) indicates that the principal additional oxidized product may be nitrate formed via Equation 14.17:

$$NH_4^+ + 4\,HOCl = NO_3^- + 4\,Cl^- + 6\,H^+ + H_2O \qquad (14.17)$$

Depending upon the relative amount of nitrate formed in comparison to nitrogen at breakpoint, between 1.5 and 4.0 mol of available chlorine may be required, which is consistent with the available data.

Below the breakpoint, inorganic chloramines decompose by direct reactions with several compounds. For example, monochloramine may react with bromide ions to form monobromamine (Trofe, 1980). If trichloramine is formed, as would be the case for applied chlorine doses in excess of that required for breakpoint, it may decompose either directly to form nitrogen gas and hypochlorous acid or by reaction with ammonia to form monochloramine and dichloramine (Saguinsin and Morris, 1975). In distilled water, the half-life of monochloramine is approximately 100 h (Kinman and Layton, 1976). Even in this simple circumstance, however, the decomposition products have not been completely characterized. Valentine (1986) found that the decomposition of pure solutions of monochloramine produces an unidentified product that absorbs UV light at 243 nm and is capable of being oxidized or reduced.

Where the pH is below 9.0 (so that the dissociation of ammonium ion is negligible), the amount of combined chlorine in dichloramine relative to monochloramine after the reactions in Equations 14.13 and 14.14 have attained equilibrium is given by the following relationship (McKee, 1960):

$$A = \frac{BZ}{1 - \sqrt{1 - BZ(2 - Z)}} - 1 \qquad (14.18)$$

In Equation 14.18, A is the ratio of available chlorine in the form of dichloramine to available chlorine in the form of monochloramine, Z is the ratio of moles of chlorine (as Cl_2) added per mole of ammonia nitrogen present, and B is defined by Equation 14.19:

$$B = 1 - 4\,K_{eq}[H^+] \qquad (14.19)$$

The equilibrium constant in Equation 14.19 refers to the direct interconversion between dichloramine and monochloramine as follows:

$$H^+ + 2\ NH_2Cl = NH_4^+ + NHCl_2$$

$$K_{eq} = [NH_4^+][NHCl_2]/[H^+][NH_2Cl]^2 \qquad (14.20)$$

At 25°C, K_{eq} has a value of 6.7×10^5 L/mol (Gray et al., 1978). From these relationships, determination of the equilibrium ratio of dichloramine to monochloramine as a function of pH and applied chlorine dose ratio is possible (assuming no dissipative reactions other than those involving the inorganic chloramines). As pH decreases and the Cl:N dose ratio increases, the relative amount of dichloramine also increases (Figure 14.4). As the Cl:N molar dose ratio increases, the relative amount of dichloramine also increases. As the Cl:N molar dose ratio increases beyond unity, the amount of dichloramine relative to monochloramine rapidly increases as well. For the conversion from dichloramine to trichloramine, the equilibrium constant given at 0.5 M ionic strength and 25°C indicates that the amount of trichloramine to be found in equilibrium with di- and monochloramine at molar dose ratios of up to 2.0 is negligible (Gray et al., 1978). This agrees with experimental measurement of the individual combined chlorine species as a function of approach to breakpoint (White, 1972).

These findings, coupled with the routine observation of the breakpoint at molar doses at or below 2:1 (Cl_2-to-N weight ratios below 10:1), indicate that trichloramine is not an important species in the breakpoint reaction. Rather, the breakpoint reaction leading to oxidation of ammonia nitrogen and reduction of combined chlorine is initiated with the formation of dichloramine.

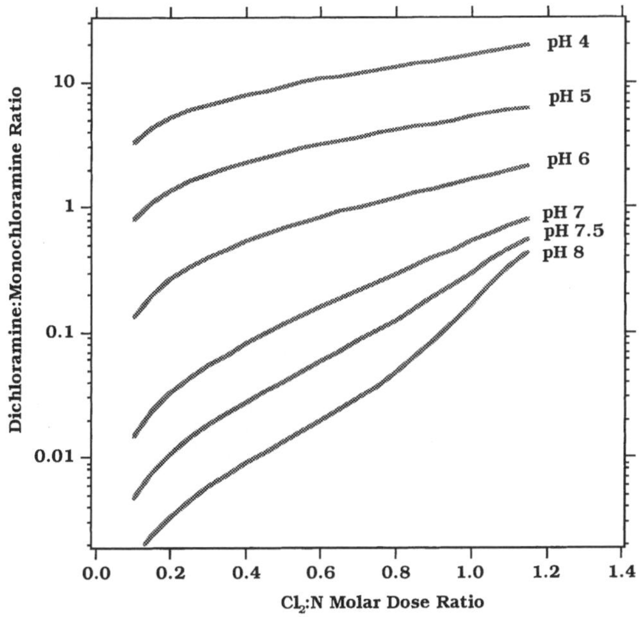

FIGURE 14.4 Effect of pH and Cl_2:NH_4^+ molar ratio on dichloramine-to-monochloramine ratio (25°C).

The kinetics of formation of chloramine species have been investigated by various researchers since initial attempts by Weil and Morris (1949b). The formation of monochloramine is a first-order process with respect to both hypochlorous acid and un-ionized ammonia. However, determining whether this, or a process involving hypochlorite ions reacting with ammonium cations, is the actual mechanism of reaction is not possible solely through kinetic arguments. If the neutral species are selected as the reactants, then the rate of formation of monochloramine (r) may be described by (Morris and Isaac, 1983):

$$r\ (\text{mol/L-s}) = 6.6 \times 10^8 \exp(-1510/T)\ [\text{HOCl}][\text{NH}_3] \tag{14.21}$$

Because hypochlorous acid dissociates into hypochlorite with a pK_a of approximately 7.4 and ammonia is able to associate with a proton to form the ammonium cation, with the pK_a for the latter of approximately 9.3, for a constant chlorine:nitrogen dose ratio, the maximum rate of monochloramine formation occurs at a pH where the product $HOCl \times NH_3$ is maximized, which is at the midpoint of the two pK values or 8.4. At this optimum pH and the usual temperatures encountered in practice, the formation of monochloramine attains equilibrium in seconds to 1 min; however, at either a higher or lower pH, the speed of the reaction slows.

A number of the other reactions in the chlorine-ammonia system may be kinetically limited. These have recently been reviewed; Table 14.2 is a compilation of the known reaction kinetics involving chlorine, ammonia, and intermediate species.

The reaction of NH_2Cl with HOCl to form $NHCl_2$ is catalyzed by a number of acidic species that may be present in water (Valentine and Jafvert, 1988). Possibly, a number of the other reactions in Table 14.2 can also be catalyzed in a similar manner; however, insufficient data are available to evaluate this possibility.

When free chlorine is contacted with a water containing ammonia, the initial velocity of monochloramine formation is substantially greater than the velocity of the subsequent formation of dichloramine. Hence, relative to equilibrium levels, an initial accumulation of monochloramine will occur if large dose ratios are employed, until the dichloramine formation process can be driven (Palin, 1983).

TABLE 14.2 Summary of Chlorine Reaction Kinetics

Reaction	Forward rate expression	Reverse rate expression
$NH_3 + HOCl \Leftrightarrow NH_2Cl + H_2O$	$6.6 \times 10^8 \exp\left(-\frac{1510}{T}\right)$	$1.38 \times 10^8 \exp\left(-\frac{8800}{T}\right)$
$NH_2Cl + HOCl \Leftrightarrow NHCl_2 + H_2O$	$3 \times 10^5 \exp\left(-\frac{2010}{T}\right)$	7.6×10^{-7} L/mol-s*
$NHCl_2 + HOCl \Leftrightarrow NCl_3 + H_2O$	$2 \times 10^5 \exp\left(-\frac{3420}{T}\right)$	$5.1 \times 10^3 \exp\left(-\frac{5530}{T}\right)$
$2NH_2Cl \Leftrightarrow NHCl_2 + NH_3$	$80 \exp\left(-\frac{2160}{T}\right)$	24.0 L/mol-s*

Rates are in units of L/mol-s.
Concentrations are in mol/L.
Reactions are elementary and water is at unit activity.
* Rate constant at 25°C.
Source: Morris and Isaac, 1983.

The kinetic evolution of the chlorine-ammonia speciation process in batch systems is described by a series of coupled ordinary differential equations. While these are highly nonlinear, various authors have applied numerical integration techniques for their solutions and, below the breakpoint, have found reasonable concordance between model predictions and experimental measurements (Haag and Lietzke, 1980; Isaac et al., 1985; Saunier and Selleck, 1979; Valentine and Jafvert, 1988).

The breakpoint process involves a complex series of elementary reactions, of which Equations 14.16 and 14.17 are the net results. Saunier and Selleck (1979) proposed that hydroxylamine (NH_2OH) and NOH may be intermediates in this reaction. However, sufficient evaluation of their proposed kinetic scheme for the breakpoint process has not yet been achieved to justify its use for design applications.

EXAMPLE 14.2 A water supply is to be postammoniated. If the water has a pH of 7.0, a free chlorine residual of 1.0 mg/L, and a temperature of 25°C, how much ammonia should be added such that the ratio of dichloramine to monochloramine is 0.1 (assume that, upon the addition of ammonia, none of the residual dissipates)?

1. From Equations 14.17 and 14.18, the following is determined:

$$B = 1 - 4\,K_{eq}\,(10^{-7}) = 1 - 4\,(6.7 \times 10^5)(10^{-7})$$
$$= 0.732$$

2. From Equation 14.16, noting that the problem condition specifies $A = 0.1$, the following equation is to be solved:

$$0.1 = -1 + \frac{0.732\,Z}{1 - \sqrt{1 - 0.732(2 - Z)Z}}$$

This can be rearranged into a quadratic equation

$$-0.289\,Z^2 + 0.134\,Z = 0$$

3. The single nonzero root gives $Z = 0.463$, which is molar ratio of chlorine (as Cl_2) to ammonia nitrogen. Because chlorine has a molecular weight of 70, 1 mg/L of free chlorine has a molarity of 1.43×10^{-5}. Therefore, 3.09×10^{-5} molarity of ammonia is required, or (multiplying by the atomic weight of nitrogen, 14) a concentration of 0.43 mg/L as N of ammonia must be added.

Reactions with Organic Matter. Morris (1967) has determined that organic amines react with free chlorine to form organic monochloramines. The rate laws for these reactions follow patterns similar to the inorganic monochloramine formation process, except that the rate constants are generally less. In addition, the rate constants for this process correlate with the relative basicity of the amine reactant. Organic chloramines may also be formed by the direct reaction between monochloramine and the organic amine, and this is apparently the most significant mechanism of organic *N*-chloramine formation at higher concentrations such as might exist at the point of application of chlorine to a water (Isaac and Morris, 1980). Pure solutions of amino acids and some proteins yield breakpoint curves identical in shape to those of ammonia solutions (Baker, 1947; Wright, 1936).

Free chlorine reacts with organic constituents to produce chlorinated organic byproducts. Murphy (1975) noted that phenols, amines, aldehydes, ketones, and pyrrole groups are readily susceptible to chlorination. Granstrom and Lee (1957) found that phenol could be chlorinated by free chlorine to form chlorophenols of various degrees of substitution. The kinetics of this process depend upon both phenolate ions and hypochlorous acid. If excess ammonia was present, however, the formation of chlorophenols was substantially inhibited.

More recently, DeLaat (1982) determined that polyhydric phenols are substantially more reactive than simple ketones in the production of chloroform, and that the rates of these processes are first order with respect to the phenol concentration and the free chlorine concentration. More significantly, the reactivity of these compounds was observed to be greater than the reactivity of ammonia with hypochlorous acid. Therefore, even if subbreakpoint chlorination is practiced, some chloroform may be formed rapidly prior to the conversion of free to combined chlorine. Chapter 12 presents additional discussion of the formation of trihalomethanes and other disinfection by-products that can arise from reactions with naturally occurring dissolved humic substances.

The reactivity of the chlorine species with compounds responsible for taste and odor depends on the predominant form of chlorine present. In field tests, Krasner (1986) determined that free chlorine, but not combined chlorine, could remove tastes and odors associated with organic sulfur compounds.

Reactions with Other Inorganic Compounds. The rates of reaction between free chlorine residuals and other inorganic compounds likely to be present in water are summarized in Table 14.3 (Wojtowicz, 1979). These reactions are generally first order in both the oxidizing agent (hypochlorous acid or hypochlorite anion) and the reducing agent.

Nitrites present in partially nitrified waters react with free chlorine via a complex, pH-dependent mechanism (Cachaza, 1976). While combined chlorine residuals were generally thought to be unreactive with nitrite, Valentine (1985) has found that the rate of decay of monochloramine in the presence of nitrite was far greater than would be predicted based on reaction of the equilibrium free chlorine, implicating a direct reaction between NH_2Cl and NO_2—N.

Overall Chlorine Demand Kinetics. Chlorine demand is defined as the difference between the applied chlorine dose and the chlorine residual measured at a particular time. The rate of exertion of chlorine demand in complex aqueous solutions has been the subject of numerous studies. The most systematic work has been that of

TABLE 14.3 Summary of Kinetics of HOCl and OCl^- Reduction by Miscellaneous Reducing Agents after Wojtowicz (1979)

Oxidizing agent	Reducing agent	Oxidation product	Log *k*, L/m-s, 25°C
OCl^-	IO^-	IO_4^-	−5.04
OCl^-	OCl^-	ClO_2^-	−7.63
OCl^-	ClO_2^-	ClO_3^-	−5.48
OCl^-	SO_3	SO_4^{2-}	3.93
HOCl	NO_2^-	NO_3^-	0.82
HOCl	$HCOO^-$	H_2CO	−1.38
HOCl	Br^-	BrO^-	3.47
HOCl	OCN^-	HCO_3^-, N_2	−0.55
HOCl	$HC_2O_4^-$	CO_2	1.20
HOCl	I^-	IO^-	8.52

Source: Feng, 1966.

Taras (1950), who chlorinated pure solutions of various organic compounds and found that chlorine demand kinetics could be described by Equation 14.22:

$$D = kt^n \tag{14.22}$$

where t is the time in hours, D is the chlorine demand, and k and n are empirical constants. In subsequent work, Feben and Taras (1950, 1951) found that chlorine demand exertion of waters blended with wastewater could be correlated to Equation 14.22, with the value of n correlated to the 1-h chlorine demand.

Haas and Karra (1984) developed Equation 14.23 to describe chlorine demand exertion kinetics.

$$D = C_o\{1 - [x \exp(-k_1 t) + (1 - x) \exp(-k_2 t)]\} \tag{14.23}$$

where x is an empirical parameter, typically 0.4 to 0.6, k_1 and k_2 are rate constants, typically 1.0 min^{-1} and 0.003 min^{-1}, respectively, and C_o is the chlorine dose in mg/L.

Dugan et al. (1995) developed a Monod (Langmuir Hinshelwood) model for describing free chlorine decay in drinking water in the absence of ammonia. It describes chlorine decay as a reaction with total organic carbon (TOC) in water according to the following differential equation:

$$\frac{dC}{dt} = -\frac{k(\text{TOC})C}{K(\text{TOC}) + C} \tag{14.24}$$

where TOC (assumed constant) is in mg C/L and C is the free chlorine concentration in mg Cl_2/L. Equation 14.24 can be integrated to the following implicit equation for chlorine concentration at time t (C_t):

$$C_t = K\,(\text{TOC}) \ln\left(\frac{C_0}{C_t}\right) - k\,(\text{TOC})\,t + C_0 \tag{14.25}$$

Furthermore, k and K were correlated with the initial chlorine dose (C_0) and TOC concentration. In tests conducted on a variety of waters, the constants were found to be given by the following equations (it should be noted that the pH and temperature were fixed at 8 and 20°C, and the waters were of relatively low ionic strength, so that the applicability of these relationships under other conditions is unclear):

$$K = -0.85\left(\frac{C_0}{\text{TOC}}\right)$$

$$k = 0.030 - 0.0060\left(\frac{C_0}{\text{TOC}}\right) \tag{14.26}$$

Dechlorination. When the chlorine residual in a treated water must be lowered prior to distribution, the chlorinated water can be dosed with a substance that reacts with or accelerates the rate of decomposition of the residual chlorine. Compounds that may perform this function include thiosulfate, hydrogen peroxide, ammonia, sulfite/bisulfite/sulfur dioxide, and activated carbon; however, only the latter two materials have been widely used for this purpose in water treatment (Snoeyink and Suidan, 1975).

Chlorine Dioxide

The reaction of chlorine dioxide with material present in waters containing chlorine dioxide demand appears to be less significant than in the case of chlorine. Rather, the dominant causes of loss of chlorine dioxide during disinfection may be the direct reactions with water and interconversions to chlorite and chloride, as outlined in Equations 14.10 and 14.11. At mg/L concentrations, ammonia nitrogen, peptone, urea, and glucose have insignificant chlorine dioxide demand in 1 h (Ingolls and Ridenour, 1948; Sikorowska, 1961). However, a variety of inorganic and biological materials will react (Werderhoff and Singer, 1987).

Masschelein (1979a) concluded that only the following organic-ClO_2 reactions are of significance to water applications:

1. Oxidation of tertiary amines to secondary amines and aldehydes
2. Oxidation of ketones, aldehydes, and (to a lesser extent) alcohols to acids
3. Oxidation of phenols
4. Oxidation of sulfhydryl-containing amino acids

Wajon et al. (1982) found a reaction stoichiometry of 2 mol of chlorine dioxide consumed per mole of phenol (or hydroquinone) consumed. Products formed included chlorophenols, aliphatic organic acids, benzoquinone, and (in the case of phenol) hydroquinone. The mechanism appeared to include the possible formation of hypochlorous acid as an active intermediate, and the rate of this process was found to be base catalyzed and first order in each of the reactants.

In general, chlorine dioxide itself has been found to produce fewer organic byproducts with naturally occurring dissolved organic material, although some nonpurgeable organic halogenated compounds are formed (Rav-Acha, 1984). In practice, however, chlorine dioxide may be generated in a manner in which chlorine is present as an impurity. Therefore, the reactions of such a stream may also include those discussed earlier regarding chlorine reactions. The inorganic byproducts consist of chloride, chlorate, and chlorite; specific ratios may depend on the precise application conditions (Noack and Doeff, 1981; Werderhoff and Singer, 1987).

Ozone

Upon addition to water, ozone reacts with hydroxide ions to form hydroxyl radicals and organic radicals. These radicals cause increased decomposition of ozone, and also are responsible for nonselective (compared to the direct ozone reaction) oxidation of a variety of organic materials. Carbonate, and possibly other ions, may act as radical scavengers and slow this process (Hoigne and Bader, 1975, 1976).

Gurol and Singer (1982) determined that ozone decomposition kinetics in various aqueous solutions are second order in ozone concentration and base promoted. Some systematic difference between various buffer systems employed does occur, with borate giving higher decomposition rates than phosphate, and phosphate at higher ionic strength giving lower decomposition rates than phosphate at lower ionic strength (1 versus 0.1 *M*). This effect was suggested as being caused by phosphate being a radical scavenger (and by radical decomposition being important at higher pH values).

As a result of these decomposition processes, the half-life of ozone in water, even in the absence of other reactive constituents, is quite short, on the order of seconds

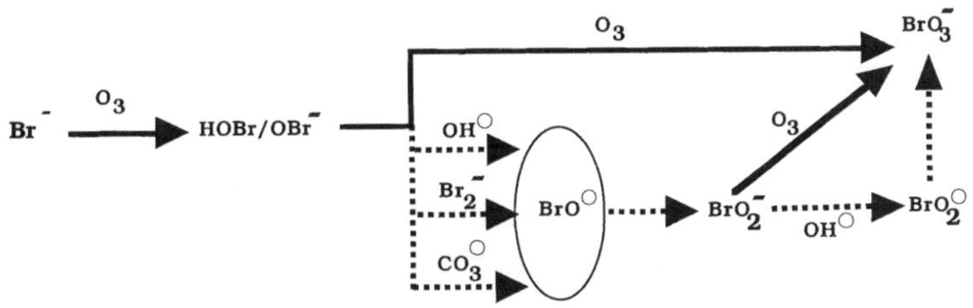

FIGURE 14.5 Schematic of bromate formation pathways. Solid lines: direct ozone reactions. Dashed lines: radical reactions (*Source:* Reprinted with permission from von Gunten, U., and J. Hoigne, 1994. Bromate formation during ozonation of bromide-containing waters: interaction of ozone and hydroxyl radical reactions. *Environmental Science and Technology* 28(7): 1234–1242. Copyright 1994 American Chemical Society.)

to minutes. Water chemistry may exert a strong influence on the rate and extent of ozone demand in a given application. Reactions of ozone in aqueous solution are discussed further in Chapter 12.

Bromide reacts with ozone under aqueous conditions typical of drinking water disinfection. Products of the reaction may be hypobromous acid, hypobromite, and/or bromate. Higher concentrations of bromide can reduce the rate of ozone decomposition. Under alkaline conditions, this may be influenced by trace metal catalysts and organic sinks for radicals and oxidized bromide species (Cooper et al., 1985).

Ozone will react with cyanides at a very fast rate. The mechanism involves reaction of the cyanide ion (to form unknown products), and the process is inhibited by iron complexes but catalyzed by copper complexes of cyanide (Gurol et al., 1985).

The reaction of ozone with bromide may proceed to the further product of bromate (BrO_3^-) by a complex process that involves direct reaction as well as hydroxyl radical mediation (von Gunten and Hoigne, 1994). The overall process is summarized in Figure 14.5. The formation of bromate by ozonation is highly important in view of the potential carcinogenicity of bromate in disinfected waters (Bull and Kopfler, 1991).

Demand for UV

For UV disinfection, the "dose" may be described in terms of the emitted lamp power in the germicidal range per unit volume of fluid under irradiation, for example, W/m^3. This can also be expressed as an integral over the disinfection reactor volume of the surface intensity (in W/m^2, for example) (Severin et al., 1983a, 1984b).

With ultraviolet light disinfection systems, the equivalent of demand results from dissolved and suspended materials, such as proteins, humic material, and iron compounds, that absorb radiation and thus shield microorganisms. Huff et al. (1965) found that intensity monitoring within the reactor itself could be used to correct for such effects.

One particular problem unique to physical systems such as UV is the need to assure complete mixing in the transverse direction so that all microorganisms may come equally close to the UV source. Cortelyou (1954) analyzed this effect for batch UV reactors, and the analysis was extended to flow-through reactors by Haas and Sakellaropoulous (1979). This phenomenon results in the desirability to achieve turbulent flow conditions in a UV reactor.

ASSESSMENT OF MICROBIAL QUALITY (INDICATORS)

The microbial quality of a source water, or the efficacy of a treatment system for removing microorganisms, can be assessed either by direct monitoring of pathogens or by the use of an indicator system. Because pathogens are a highly diverse group, generally requiring a highly specialized (and often insensitive and expensive) analytical technique for each pathogen, the use of indicator organisms is a more popular technique.

An indicator group of organisms can be used either to assess source water contamination or degree of treatment; however, the same indicator group is often used to assess both properties. This places severe constraints on the group of indicator organisms chosen. Bonde (1966) has proposed that an ideal indicator must:

1. Be present whenever the pathogens concerned are present
2. Be present only when the presence of pathogens is an imminent danger, that is, be unable to proliferate to any greater extent in the aqueous environment
3. Occur in much greater numbers than pathogens
4. Be more resistant to disinfectants and to the aqueous environment than pathogens
5. Grow readily on relatively simple media
6. Yield characteristic and simple reactions enabling, as far as possible, an unambiguous identification of the group
7. Be randomly distributed in the sample to be examined, or be able to be uniformly distributed by simple homogenization procedures
8. Grow widely independent of other organisms present when inoculated in artificial media, that is, not be seriously inhibited in growth by the presence of other bacteria

The use of coliforms as indicator organisms stems from the pioneering work of Phelps (1909). The basic rationale was that coliforms and enteric bacterial pathogens originate from a common source—namely human fecal contamination. Subsequent work by Butterfield et al. (1943, 1946), Kabler (1951), and Wattie and Butterfield (1944) confirmed that these organisms were at least as resistant to free or combined chlorine as enteric bacterial pathogens.

The coliform group is a heterogeneous conglomerate of microorganisms, including forms native to mammalian gastrointestinal tracts as well as a number of exclusively soil forms. The common fermentation tube (FT) and membrane filter (MF) procedures are subtly different in the organisms they enumerate. Classically, coliforms have been defined as "Gram-negative, non-sporeforming bacteria which [sic] ferment lactose at 35–37°C, with the production of acid and gas" (APHA, AWWA, and WPCF, 1989). The FT procedure, however, ignores anaerogenic and lactose-negative coliforms, and the MF procedure ignores non-lactose-fermenting strains (Clark and Pagel, 1977).

Furthermore, interferences can selectively reduce coliforms as measured by one or the other method. Allen (1977), for example, found that high concentrations (>500 to 1000/mL) of standard plate count (SPC) organisms appeared to reduce the recovery of coliforms by the MF technique when compared to the FT technique.

The fecal coliform group of organisms is that subset of coliforms that are capable of growing at elevated temperature (44.5°C). The original rationale for development of this test was to provide a more selective indicator group, excluding mesophilic coliforms primarily indigenous to soils. Total coliforms, however, continue to be the basic U.S. microbiological standard for drinking water because the absence of coliforms ensures the absence of fecal coliforms, which is a conservative standard.

While coliforms, either fecal or total, may be reasonably good indicators of fecal contamination of a water supply, reservations were expressed as early as 1922 (Anonymous, 1922) about the relative resistance of coliforms to chlorine vis-à-vis pathogenic bacteria and the resulting adequacy of the coliform test as an indicator of disinfection efficiency. In more recent work, coliforms have been found to be more sensitive to disinfection by one or more forms of chlorine than various human enteric viruses (Grabow et al., 1983; Kelly and Sanderson, 1958) and the protozoan pathogens *Naegleria* (Rubin et al., 1983), *Giardia* (Jarroll, 1981; Korich et al., 1990; Leahy, 1985; Rice et al., 1982), and *Cryptosporidium* (Kovich et al., 1990). In addition, viruses (Scarpino et al., 1977) and protozoan cysts (Leahy, 1985) have been found to be more resistant to ClO_2 inactivation than coliforms. Farooq (1976) has determined that coliforms are more resistant to ozone than viruses. Rice and Hoff (1981) found that *Giardia lamblia* cysts survived exposure to UV doses sufficient to effect over 99.99 percent inactivation of *E. coli.* Human enteric viruses have been isolated in full-scale water treatment plants practicing conventional treatment, and meeting turbidity and coliform standards in the presence of free residual chlorine (Payment et al., 1985; Rose et al., 1986).

As a result of the problems with the coliform group of organisms, a number of workers have investigated alternative indicator systems with greater resistance to disinfectants than coliforms. Among the most successful of these are the acid-fast bacteria and yeasts studied by Engelbrecht et al. (1977, 1979) and Haas et al. (1983a, b; 1985a, b). In addition, work using endotoxins (Haas and Morrison, 1981), *Clostridia* (Cabelli, 1977; Payment and Franco, 1993), and bacteriophage (Abad et al., 1994; Grabow, 1968; Grabow et al., 1983; Payment and Franco, 1993) has been carried out. In addition, to some degree, heterotrophic plate count (HPC) organisms may provide a conservative indicator of treatment efficiency. Despite these studies, however, in U.S. practice, no alternative to the total coliform group of organisms has yet found widespread application.

PATHOGENS OF CONCERN

A variety of pathogenic organisms capable of transmission by the fecal-oral route may be found in raw wastewaters. Waterborne outbreaks of shigellosis, salmonellosis, and various viral agents have been reported, in many cases associated with sewage-contaminated water supplies (Blostein, 1991; Drenchen and Bert, 1994; Haas, 1986; Herwaldt et al., 1991, 1992; Levine et al., 1990; Reeve et al., 1989; Rosenberg et al., 1976, 1980). Among the bacteria, *Salmonella, Shigella,* and *Vibrio cholerae* organisms are the classical agents of concern (Mosley, 1966). In more recent times, concern has expanded to other agents that have been found in wastewater—viruses and protozoa.

Among the viruses, enteroviruses (ECHO virus, Coxsackievirus), rotavirus, reovirus, adenovirus, and parvovirus have been isolated from wastewater (Melnick et al., 1978). New viruses that are suspected of waterborne transmission have been identified at the rate of about one organism per year (Gerba, personal communication). Among the more important of these newly identified agents may be Norwalk virus and calicivirus.

Over the past 15 years, significant concerns have increased over the risk from pathogenic protozoa in drinking water, particularly *Giardia* and *Cryptosporidium*

(Gallaher et al., 1989; Goldstein et al., 1996; LeChevallier et al., 1991; Leland et al., 1993; Richardson et al., 1991; Rose et al., 1991; Smith, 1992). The SWTR arose, to a significant extent, from concerns over *Giardia* (Regli et al., 1988). Revisions to drinking water regulations presently under discussion are concerned with assuring an adequate degree of protection from *Cryptosporidium.*

DISINFECTION KINETICS

The information needed for the design of a disinfection system includes knowledge of the rate of inactivation of the target, or indicator, organism(s) by the disinfectant. In particular, the effect of disinfectant concentration on the rate of this process will determine the most efficient combination of contact time (i.e., basin volume at a given design flow rate) and the dose to employ.

Chick's Law and Elaborations

The major precepts of disinfection kinetics were enunciated by Chick (1908), who recognized the close similarity between microbial inactivation by chemical disinfectants and chemical reactions. A good overview of the principles of kinetic modeling of disinfection has been presented by Gyurek and Finch (1998). Disinfection is analogous to a bimolecular chemical reaction, with the reactants being the microorganism and the disinfectant, and can be characterized by a rate law as are chemical reactions:

$$r = -kN \tag{14.27}$$

where r is the inactivation rate (organisms killed/volume-time) and N is the concentration of viable organisms. In a batch system, this results in an exponential decay in organisms, because the rate of inactivation equals dN/dt, assuming that the rate constant k is actually constant (e.g., the disinfectant concentration is constant).

Watson (1908) proposed Equation 14.28 to relate the rate constant of inactivation k to the disinfectant concentration C:

$$k = k'C^n \tag{14.28}$$

where n is termed the coefficient of dilution and k' is presumed independent of disinfectant concentration, and, by virtue of Equation 14.27, microorganism concentration.

From the Chick-Watson law, when C, n, and k' are constant (i.e., no demand, constant concentration), the preceding rate law may be integrated so that in a thoroughly mixed batch system,

$$\ln(N/N_0) = -k'C^n t \tag{14.29}$$

where N and N_0 are, respectively, the concentrations of viable microorganisms at time t and time 0. When disinfectant composition changes with time, or when a configuration other than a batch (or plug flow) system is used, the appropriate rate laws characterizing disinfectant transformation (Haas and Karra, 1984b) along with the applicable mass balances must be used to obtain the relationship between microbial inactivation and concentration and time.

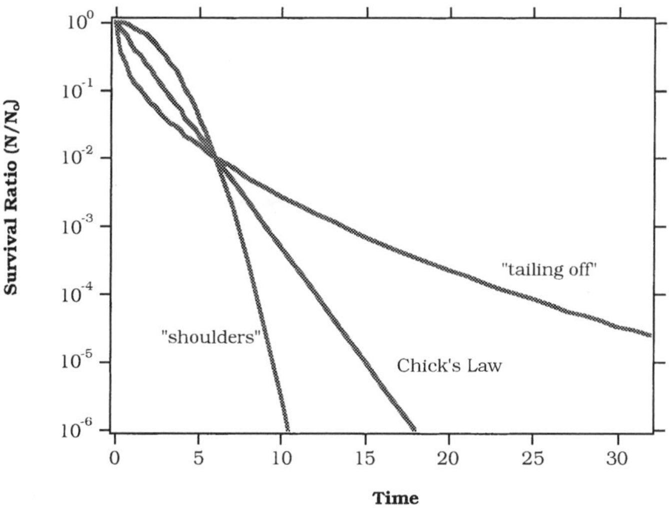

FIGURE 14.6 Chick's law and its deviations.

Inactivation of microorganisms in batch experiments, even when disinfectant concentration is kept constant, does not always follow the exponential decay pattern predicted by Equation 14.29. Indeed, two common types of deviations are noted (Figure 14.6). In addition to the linear Chick's law decay, the presence of "shoulders" or time lags until the onset of disinfection is often observed. Also, some microorganisms and disinfectants exhibit a "tailing" in which the rate of inactivation progressively decreases. In some cases, a combination of both of these behaviors is seen.

Even if deviations from Chick-Watson behavior are observed, plotting combinations of disinfectant concentration and time to produce a fixed percent inactivation is generally possible. Such plots tend to follow the relationship $C^n t$ = constant, where the constant is a function of the type of organism, pH, temperature, form of disinfectant, and extent of inactivation. Such plots are linear on a log-log scale (Figure 14.7). If the value of n is greater than 1, a proportionate change in disinfectant concentration produces a greater effect than a proportionate change in time. In many cases (Hoff, 1986), the Chick-Watson law n value is close to 1.0, and hence a fixed value of the product of concentration and time (CT product) results in a fixed degree of inactivation (at a given temperature, pH, etc.).

In the chemical disinfection of a water, the concentration of disinfectant may change with time, and particularly during the initial moments of contact the chemical form(s) of halogens such as chlorine undergo rapid transformations from the free to the combined forms. Because C would thus not be a constant, typically disinfection results obtained in batch systems exhibit tailing, the degree of which may depend on the demand and the concentration of reactive constituents (such as ammonia) in the system (Olivieri et al., 1971). Determination of the disinfectant residual (and its chemical forms) is more critical than the disinfectant dose in these systems.

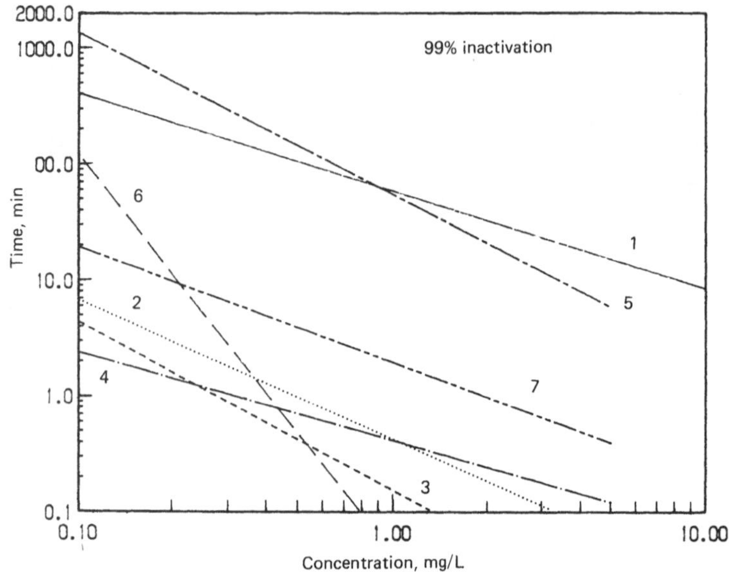

FIGURE 14.7 Concentration-time relationships for 99 percent inactivation of various microorganisms by various disinfectants. (1) *Giardia lamblia;* free chlorine, 5°C (*Source:* Hoff and Akin, 1986). (2) *E. coli;* free chlorine, 2 to 5°C, pH 8.5 (*Source:* Haas and Karra, 1984a). (3) *E. coli;* free chlorine, 20 to 25°C, pH 8.5 (*Source:* Haas and Karra, 1984a). (4) Poliovirus 1 (Mahoney); free chlorine, 2°C, pH 6 (*Source:* Haas and Karra, 1984a). (5) *E. coli;* combined chlorine, 3 to 5°C, pH 7 (*Source:* Haas and Karra, 1984a). (6) Poliovirus 1 (Mahoney); ozone, 20°C, pH 7.2 (*Source:* Roy et al., 1981a). (7) *Giardia muris;* ozone, 5°C, pH 7 (*Source:* Wickramanayake et al., 1985).

In the chlorine system, for example, knowing the rate laws for inactivation by individual separate species and the dynamics of chlorine species interconversions as described previously enables an overall model for chlorine inactivation to be formulated (Haas and Karra, 1984b). In doing this computation, the individual rates are usually assumed to be additive (Fair et al., 1948), although this assumption has not yet been experimentally verified.

The presence of shoulders in inactivation curves is often seen in organisms that form clumps. This means that more than one cell must be inactivated to achieve inactivation of a colony or plaque-forming unit. For example, Rubin et al. (1983) found that cysts of *Naegleria gruberii* in demand-free water showed shoulder-type inactivation to free chlorine. Similarly, when cells of *E. coli* were agglutinated, they displayed shoulder-type inactivation, which was absent in unagglutinated cultures (Carlson et al., 1975). Severin et al. (1984a) found shoulder-type inactivation curves in the case of *E. coli* with preformed chloramines (i.e., solutions of ammonia and chlorine prereacted to form combined chlorine prior to addition of microorganisms), *Candida parapsilosis* (a yeast organism proposed as a possible disinfection-resistant indicator) with both preformed chloramines, and free chlorine and poliovirus with iodine.

Shoulder inactivation curves may be explained by a multitarget model (Hiatt, 1964), by a series event model (Severin et al., 1984a), or by a diffusional model

(Haas, 1980). Tailing inactivation curves may be explained either by a vitalistic hypothesis in which individuals in a population are nonidentical, and their inherent resistance is distributed in a permanent (time-independent) manner, or by a mechanistic concept (Cerf, 1977). In the latter case, four particular mechanisms have been advanced leading to tailing:

1. Conversion to resistant form during inactivation (hardening)
2. Existence of genetic variants of differing sensitivity
3. Protection of a subpopulation, or variations in received dose of disinfectant
4. Clumping of a subpopulation

The hardening process and resultant tailing have received wide attention, following discoveries of apparent hardening in the formaldehyde inactivation of poliovirus prepared for the Salk vaccine (Nathanson and Langmuir, 1963). Gard (1960) has proposed an empirical rate law for this behavior, which has been used by Selleck et al. (1978) in the analysis of wastewater chlorination kinetics. Tailing behavior has been found for viral and coliform inactivation by ozone (Katzenelson et al., 1974) and for coliform inactivation by free chlorine (Haas and Morrison, 1981; Olivieri et al., 1971).

Hom (1972) developed a flexible but highly empirical kinetic formulation for the inactivation rate based on modifying Equations 14.27 and 14.28 to the following form:

$$r = -k'mN\,t^{m-1}C^n \tag{14.30}$$

This equation is difficult to use as a rate model since it contains time as an explicit variable. A formulation leading to the classical Hom integrated relationship can be written as (Haas and Joffe, 1994):

$$r = -mN(kC^n)^{1/m}\left[-\ln\left(\frac{N}{N_0}\right)\right]^{(1-1/m)} \tag{14.31}$$

Upon integration, if C is constant, this results in the following relationship:

$$\ln\left(\frac{N}{N_0}\right) = -k'C^n t^m \tag{14.32}$$

Depending upon the value of m, both shoulders and tailing may be depicted by Equation 14.32. In early work, Fair et al. (1948) used a model of the form of Equation (14.32) with $m = 2$ to analyze *E. coli* inactivation by free and combined chlorine.

EXAMPLE 14.3 A certain water supply has operational problems due to high levels of HPC organisms. To maintain adequate system water quality, a decision has been made to keep the concentration of HPC organisms below 10/mL at the entry point to the distribution system (i.e., following disinfection). Disinfection using free residual chlorine is practiced. As part of the laboratory investigation to develop design criteria for this system, the inactivation of the HPC organisms is determined in batch reactors (beakers). The pH and temperature are held constant at the expected final water conditions. Using water with an initial HPC of 1000/mL, the following data are taken:

Cl_2 residual, mg/L	Contact time, min	HPC remaining, number/mL
0.5	10	40
0.5	20	4
0.5	30	1
1	5	35
1	10	4
1	15	1
1.5	2	98
1.5	5	10
1.5	10	1

From this information, determine the best fit using the Hom inactivation model, and compute the necessary chlorine residual that will achieve HPC < 10/mL (from an initial concentration of 1000/mL) at a contact time of 10 min.

1. This problem can be solved using a maximum likelihood technique to fit the Hom model to the data. Regression methods, however, can be used in two different ways—multiple linear regression and nonlinear regression (Haas and Heller, 1989; Haas et al., 1988). In a batch system, without chlorine demand, the Hom model becomes

$$\ln\left(\frac{N}{N_0}\right) = -kC^n t^m$$

This can be rearranged as

$$\ln\left[-\ln\left(\frac{N}{N_0}\right)\right] = \ln(k) + n\ln(C) + m\ln(t)$$

2. A multiple linear regression using $\ln[-\ln(N/N_0)]$ as the dependent variable and $\ln(C)$ and $\ln(t)$ as the independent variables produces an intercept (equal to $\ln(k)$) and slopes equal to n and m. This computation can be handled by common spreadsheet programs as well as statistical packages. Transformation of the data given produces the following values of the dependent and independent variables:

$\ln(-\ln(N/N_0))$	$\ln(t)$	$\ln(C)$
1.169	2.303	–0.693
1.709	2.996	–0.693
1.933	3.401	–0.693
1.210	1.609	0
1.709	2.303	0
1.933	2.708	0
0.843	0.693	0.405
1.527	1.609	0.405
1.933	2.303	0.405

The result is

$$k = 1.11$$

$$n = 0.68$$

$$m = 0.70$$

$$\text{Correlation coefficient} = 0.997$$

3. The final estimation equation then becomes:

$$\ln\left(\frac{N}{N_0}\right) = -1.11\ C^{0.68}t^{0.70}$$

4. Inserting the design specifications gives the following results:

$$\ln\left(\frac{10}{1000}\right) = -1.11\ C^{0.68}10^{0.70}$$

$$C^{0.6833} = 0.83$$

$$C = 0.77\ \frac{\text{mg}}{\text{L}}$$

Hence, if a 10-min contact time is accepted as a worst-case condition, and assuming good contactor hydraulics, the maximum chlorine residual required to achieve the design inactivation is 0.77 mg/L. From this information, and the chlorine demand of the water, the capacity of a chlorine feed system can be computed.

Another class of models can be obtained by assuming that inactivation is other than first order in surviving microbial concentrations. Depending upon the order chosen, either tailing or shoulders can be produced. For example, Roy and coworkers (Roy, 1979; Roy et al., 1981a, b), using continuously stirred tank reactor studies on inactivation of Poliovirus 1 with ozone in demand-free systems, developed the following rate law:

$$r = -kCN^{0.69} \tag{14.33}$$

A similar model was used by Benarde et al. (1967) to analyze *E. coli* inactivation by chlorine dioxide, and for analysis of various organisms in ozone contacting (Zhou and Smith, 1994, 1995).

Disinfection, like all other rate processes, is temperature dependent. This dependency may be quantified by the Arrhenius relationship:

$$k = k_o \exp(-E/RT) \tag{14.34}$$

where k is a rate constant characterizing the reaction (such as the Chick-Watson k' value), T is the absolute temperature, R is the ideal gas constant, k_o is called the *frequency factor,* and E, with units of energy/mole, is called the *activation energy.* E is always positive, and as it increases, the effect of temperature becomes more pronounced. The values of E and k_o may be determined from rates of inactivation obtained as a function of temperature. As E increases, the effect of temperature on the rate increases. For example, an E of 10 kcal/mol doubles the rate between 10 and 20°C. In contrast, activation energies for breaking hydrogen bonds are 3 to 7 kcal/mol (Bailey and Ollis, 1986). Activation energies less than this range suggest physical (e.g., transport) limitations rather than chemical reactions.

If the disinfectant concentration is not constant, then the rate laws for inactivation must be combined with those for disinfectant demand. For example, in the case of the Hom model with first-order decay in a batch system, the following approach has been developed (Haas and Joffe, 1994). The time course of disinfectant concentration is given by:

$$C = C_0 \exp(-k^*t) \tag{14.35}$$

where C_0 is the initial concentration and k^* is the demand constant. This equation is substituted into the Hom rate expression and then into a batch mass balance to yield:

$$\frac{dN}{dt} = -mN[k(C_0\exp(-k^*t))^n]^{1/m}\left[-\ln\left(\frac{N}{N_0}\right)\right]^{(1-1/m)} \tag{14.36}$$

This has an analytical solution in terms of an incomplete gamma function, which under conditions generally present in disinfection can be approximated as:

$$\ln\left(\frac{N}{N_0}\right) = -kC_0^{n}t^{m}\left[\frac{1-\exp\left(-\frac{nk^*t}{m}\right)}{\left(\frac{nk^*t}{m}\right)}\right]^{m} \tag{14.37}$$

For other disinfection decay rate laws, the solution may be obtained by numerical integration (Finch et al., 1993).

In general, microbial inactivation kinetics have been determined in batch systems. Real contactors, however, are continuous and may have nonideal flow patterns with *backmixing* or short-circuiting. If the residence time distribution of a continuous reactor is known, for example from tracer experiments, then it is possible to develop reasonable estimates of inactivation efficiency in these systems (Haas, 1988; Haas et al., 1995; Lawler and Singer, 1993; Stover et al., 1986; Trussell and Chao, 1977).

The residence time distribution is determined from a pulse or step tracer experiment. In a pulse experiment, a virtually instantaneous "slug" of tracer is introduced, while in a step experiment a virtually instantaneous step change in concentration is effected. Aspects of these experiments in disinfection systems have been described by several recent authors (Bishop et al., 1993; Boulos et al., 1996; Teefy and Singer, 1990). From this experiment, descriptive characteristics of the residence time distribution, such as the mean residence time and dispersion, may be obtained; in addition, the cumulative residence time distribution (*F curve*) and its density function (*E curve*) are also directly obtained. The E curve, written as a function of time, $E(t)$, is the probability density function that gives the fraction of the fluid elements leaving the reactor that were in the reactor a period of time between t and $t + dt$.

Using the E curve and the batch kinetic rate expression, the composition of the effluent may be obtained. The approach assumes a completely segregated system (Dankwerts, 1958; Levenspiel, 1972), and estimates the survival ratio in a disinfection reactor by the following equation:

$$\ln\left(\frac{N}{N_0}\right)_{\text{continuous}} = \int_0^\infty S(t)E(t)dt \tag{14.38}$$

where $\ln(N/N_0)_{\text{continuous}}$ = predicted continuous survival ratio

$S(t)$ = predicted batch survival at time t, given influent concentration as initial condition (N/N_0)

$E(t)$ = normalized density function for the residence time distribution

The complete segregation model is one extreme of micromixedness behavior. Prior work has suggested that, while there is a difference between the predictions from the perfect segregation assumption and the perfect micromixed assumption, the numerical differences are less than those associated with variations in the residence time distribution itself (Haas, 1988). The integration in Equation 14.38 may be most effectively performed if the E curve is fitted to a parametric model, such as the tanks-in-series model or axial dispersion model (Haas et al., 1997; Nauman and Buffham, 1983) given, respectively, by:

$$E(t) = \frac{1}{t\Gamma(w)}\left(\frac{wt}{\theta}\right)^{w}\exp\left(-\frac{wt}{\theta}\right) \tag{14.39}$$

and

$$E(t) = \sqrt{\frac{\theta}{2\pi t^3 \nu}} \exp\left[-\frac{(t-\theta)^2}{2\theta t \nu}\right] \tag{14.40}$$

where w is the number of equal-volume tanks in series and ν is the dimensionless variance, which is related to the Peclet number—the reciprocal of the dimensionless dispersion number—by the following relationship:

$$\nu = \frac{2}{Pe} - \frac{2}{Pe^2}(1 - e^{-Pe}) \tag{14.41}$$

The axial dispersion model also has the integrated solution that gives F, the cumulative fraction of fluid that spends a residence time $\leq t$ in the system as

$$F(t) = \Phi\left[\left(\frac{t}{\theta} - 1\right)\sqrt{\frac{\theta}{t\nu}}\right] + \exp\left(\frac{2}{\nu}\right)\Phi\left[-\left(\frac{t}{\theta} + 1\right)\sqrt{\frac{\theta}{t\nu}}\right] \tag{14.42}$$

where $\Phi(z)$ is the standard normal-probability integral (i.e., area under the normal distribution from negative infinity to z).

Application of this method is illustrated by the following example.

EXAMPLE 14.4 Batch studies of inactivation of *Giardia* in water by chlorine have shown the applicability of the Hom model with first-order decay. The following are the kinetic parameters:

$$k^* = 0.03$$

$$k = 0.1$$

$$m = 1.2$$

$$n = 0.9$$

A contactor is to be designed with a mean residence time of 30 min and a Peclet number of 50. If a chlorine dose of 2 mg/L is to be used, what is the anticipated survival ratio?

1. First, from Equation 14.41, the dimensionless variance is computed as:

$$\nu = \frac{2}{50} - \frac{2}{50^2}(1 - e^{-50}) = 0.0392$$

2. Now, the trapezoidal rule (Chapra and Canale, 1988) can be used to evaluate the integral as shown in the following table.

Time, min	$S(t)$	$E(t)$	$S(t) \times E(t)$	Trapezoid area
0	1	0	0	
5	0.3000	8.189*E*-24	2.457*E*-24	6.141*E*-24
10	0.0750	1.435*E*-08	1.076*E*-09	2.690*E*-09
15	0.0192	3.228*E*-04	6.206*E*-06	1.552*E*-05
20	0.0053	0.0147	7.808*E*-05	2.107*E*-04
25	0.0016	0.0577	9.234*E*-05	4.261*E*-04
30	5.308*E*-04	0.0672	3.565*E*-05	3.200*E*-04
35	1.934*E*-04	0.0393	7.609*E*-06	1.081*E*-04
40	7.713*E*-05	0.0151	1.162*E*-06	2.193*E*-05
45	3.349*E*-05	0.0044	1.461*E*-07	3.271*E*-06
50	1.574*E*-05	0.0010	1.637*E*-08	4.062*E*-07
55	7.954*E*-06	2.158*E*-04	1.716*E*-09	4.522*E*-08
60	4.299*E*-06	4.035*E*-05	1.735*E*-10	4.724*E*-09

Time intervals are selected—the finer the time spacing, the more precise the evaluation of the integral. More sophisticated integration approaches, such as Simpson's rule, or higher-order methods may also be chosen to improve precision (Chapra and Canale, 1988). The second and third columns are the computed values of the survival (S) and residence time density function (E) at the particular time (t). The fourth column is the product of S and E. The final column represents the contribution of the rectangle to the area. If t_0 and t_1 represent the prior row and the current row, then the values in the final column are given by:

$$\left[\frac{S(t_1)E(t_1) + S(t_0)E(t_0)}{2}\right](t_1 - t_0)$$

The sum of the final column is 0.0011061—indicating that there is just under 3 logs removal, that is, 0.11 percent survival.

Note that times beyond 60 min were not examined, since at these higher times, as shown in the right column, the contribution to the integral is very small.

THE CT APPROACH IN REGULATION

For regulatory purposes, under the SWTR, the adequacy of disinfection is judged using the product of the final residual concentration of disinfectant (in mg/L) and the contact time (in minutes). The contact time is evaluated as that which is exceeded by 90 percent of the fluid (this is designated as the t_{10}—denoting that 10 percent of the fluid has a smaller residence time in the system). The t_{10} for disinfection systems may be evaluated from tracer studies, or by use of default multipliers of the theoretical hydraulic residence time (V/Q). Critical values of the CT to achieve varying levels of disinfection, incorporating a margin of safety, have been published under EPA guidance documents (Malcolm Pirnie and HDR Engineering, 1991). These are indicated in Table 14.4 for ozone, chlorine dioxide, and chloramines. In the case of free chlorine, the CT values are functions of temperature, pH, and also concentration (under the regulations, measured at the end of the contact chamber). Space does not permit a recapitulation of the full CT tables for free chlorine—the values at a few conditions are given in Table 14.5.

The use of the CT approach is illustrated in the following example.

EXAMPLE 14.5 The disinfection contactor described in Example 14.4 is to be used to achieve a 3-log inactivation of *Giardia.* Compute the required residual necessary to achieve this result from the CT tables if the pH is 7 and the temperature is 20°C.

TABLE 14.4 CT Values for 99.9 Percent Reduction of *Giardia lamblia* with Ozone, Chlorine Dioxide, and Chloramines

		Temperature, °C					
Disinfectant	pH	1	5	10	15	20	25
Ozone	6–9	2.9	1.9	1.4	0.95	0.72	0.48
Chlorine dioxide	6–9	63	26	23	19	15	11
Chloramines	6–9	3800	2200	1850	1500	1100	750

Source: Malcolm Pirnie and HDR Engineering, 1991.

TABLE 14.5 CT Values for 99.9 Percent Reduction of *Giardia lamblia* with Free Chlorine Under Selected Conditions

Temperature, °C pH	5 7	5 9	20 7
C (mg/L)			
0.4	139	279	52
0.6	143	291	54
0.8	146	301	55
1	149	312	56
1.2	152	320	57
1.4	155	329	58
1.6	158	337	59
1.8	162	345	61
2	165	353	62
2.2	169	361	63
2.4	172	368	65
2.6	175	375	66
2.8	178	382	67
3	182	389	68

Source: Malcolm Pirnie and HDR Engineering, 1991.

1. First, the t_{10} must be computed. Since the contactor is an axial dispersion contactor, equation 14.42 can be used. We therefore need to solve the following equation for the time at which 10 percent of the fluid has exited:

$$0.1 = \Phi\left[\left(\frac{t_{10}}{30} - 1\right)\sqrt{\frac{30}{t_{10}(0.0392)}}\right] + \exp\left(\frac{2}{(0.0392)}\right)\Phi\left[-\left(\frac{t_{10}}{30} + 1\right)\sqrt{\frac{30}{t_{10}(0.0392)}}\right]$$

This equation must be solved by trial and error—however, this can readily be done on a spreadsheet (in which the cumulative normal distribution function is available). Doing this, it is found that $t_{10} = 25.3$ min.

2. We now need to examine the last column of Table 14.5 to determine what chlorine residual would satisfy the requirement at this t_{10}. The following table illustrates the computation:

C	CT	$t = (CT/C)$
2.4	65	27.08
2.6	66	25.38
2.8	67	23.93

3. Finally, an inverse interpolation is needed to obtain the C that yields the desired t, which is performed as follows:

$$C = 2.6 + (2.8 - 2.6)\left(\frac{25.3 - 25.38}{23.93 - 25.38}\right)$$

$$= 2.61 \text{ mg/L}$$

Therefore, the final residual chlorine should exceed 2.61 mg/L.

UV PROCESSES

In the application of kinetic models to the analysis of UV disinfection processes, the concentration of disinfectant is replaced by the incident light intensity (in units of energy per unit area). This may be determined either by direct measurement in the actual disinfection reactor or by modeling of physical aspects of light transmission (Stover et al., 1986; Water Pollution Control Federation, 1984). The potential absorbance of effective light by dissolved components and scattering by suspended solids may both contribute to demand for disinfectant.

MODE OF ACTION OF DISINFECTANTS

Chlorine

Since Nissen (1890), free chlorine at low pH has been known to be more biocidal than free chlorine at high pH. Holwerda (1928) proposed that hypochlorous acid was the specific agent responsible for inactivation, and thus the pH effect. Fair et al. (1948) determined that the pH dependency of free chlorine potency correlated quantitatively with the dissociation constant of hypochlorous acid. Chang (1944) determined that the association of chlorine with cysts of *Entamoeba hystolytica* was greater at low pH than at high pH. Friberg (1957), Friberg and Hammarstrom (1956), and Haas and Engelbrecht (1980a), using radioactive free chlorine, found similar results applied with respect to bacteria, and also found that the microbial binding of chlorine could be described by typical chemical isotherms. With respect to viruses, this association of chlorine parallels the biocidal efficacy of hypochlorous acid, hypochlorite, and monochloramine (Dennis et al., 1979a, b; Olivieri et al., 1980).

Once taken into the environment of the living organism, chlorine may enter into a number of reactions with critical components causing inactivation. In bacteria, respiratory, transport, and nucleic acid activity are all adversely affected (Haas and Engelbrecht, 1980a, b; Venkobachar et al., 1975, 1977). In bacteriophage f2, the mode of inactivation appears to be disruption of the viral nucleic acid (Dennis et al., 1979b). With poliovirus, however, the protein coat, and not the nucleic acid, appears to be the critical site for inactivation by free chlorine (Fujioka et al., 1985; Tenno et al., 1980).

The rate of inactivation of bacteria by monochloramine is greater than could be attributed to the equilibrium-free chlorine present in solution. This argues strongly for a direct inactivation reaction of combined chlorine (Haas and Karra, 1984). Although organic chloramines are generally measured as combined or total chlorine by conventional methods, they are of substantially lower effectiveness as disinfectants than inorganic chloramines (Feng, 1966; Wolfe and Olson, 1985).

In general, the rate of inactivation of microorganisms by various disinfectants increases with increasing temperature. This may be characterized by an activation energy, a Q_{10} (factor of increase for every 10°C temperature increase), or a temperature multiplier.

Surprisingly, Scarpino et al. (1972) reported that viruses were more sensitive to free chlorine at high pH than at low pH. A variety of subsequent authors confirmed these findings with viruses and with bacteria (Berg et al., 1989; Haas, 1981; Haas et al., 1986, 1990). Hypochlorite can form neutral ion pairs with sodium, potassium, and lithium, and (particularly at ionic strengths approaching 0.1 M) these can increase disinfection efficiency by free chlorine at high pH (Haas et al., 1986). More recently, calcium enhancement of chlorine inactivation of coliforms has been reported (Haas and Anotai, 1996).

Chlorine Dioxide

The dependence of inactivation efficiency on pH is weaker for chlorine dioxide than for chlorine, and more inconsistent. Benarde et al. (1965), working with *E. coli,* and Scarpino et al. (1979), working with Poliovirus 1, found that the degree of inactivation by chlorine dioxide increases as pH increases. However, for amoebic cysts, as pH increases, the efficiency of inactivation by chlorine dioxide decreases (Chen et al., 1985). The physiological mode of inactivation of bacteria by chlorine dioxide has been attributed to a disruption of protein synthesis (Benarde et al., 1967). In the case of viruses, chlorine dioxide preferentially inactivated capsid functions, rather than nucleic acids (Noss et al., 1985; Olivieri et al., 1985).

Benarde et al. (1967) computed the activation energy for the inactivation of *E. coli* by chlorine dioxide at pH 6.5 as 12 kcal/M. An identical number was computed for the disinfection of Poliovirus 1 by chlorine dioxide at pH 7 (Scarpino et al., 1979).

Ozone

Understanding of the mode of inactivation of microorganisms by ozone remains hindered by difficulties in measuring low concentrations of dissolved ozone. The effect of pH on ozone inactivation of microorganisms appears to be predominantly associated with changing the stability of residual ozone, although additional work is needed. Farooq (1976) found little effect of pH on the ability of dissolved ozone residuals to inactivate acid-fast bacteria. Roy (1979) found a slight diminution of the virucidal efficacy of ozone residuals as pH decreased; however, Vaughn, as cited in Hoff (1986) and Hoff and Akin (1986), noted the opposite effect. The principal action of ozone as a disinfectant occurs via the dissolved ozone, rather than physical contact with ozone gas bubbles (Dahi and Lund, 1980; Farooq, 1976).

Bacterial cells lacking certain DNA polymerase gene activity were found to be more sensitive to inactivation by ozone than wild-type strains, strongly implicating physicochemical damage to DNA as a mechanism of inactivation by ozone (Hamelin and Chung, 1978). For poliovirus, the primary mode of inactivation by ozone also appears to be nucleic acid damage (Roy et al., 1981b).

Activation energies for ozone inactivation of *Giardia* and *Naegleria* cysts were reported by Wickramanayake et al. (1985). At pH 7, the activation energies were 9.7 and 16.7 kcal/M. For Poliovirus 1, Roy et al. (1981a) estimated an activation energy of 3.6 kcal/M at pH 7.2. If the latter is correct, its low value suggests that ozone inactivates virus by a diffusional rather than a reaction-controlled process.

UV Light

The mode of inactivation of microorganisms by ultraviolet radiation is quite well characterized. Specific deleterious changes in nucleic acid arise upon exposure to UV radiation (Jagger, 1967). These may be repaired by light-activated as well as dark repair enzymes in vegetative microorganisms. The phenomenon of photoreactivation of UV-disinfected microorganisms has been demonstrated in municipal effluents (Scheible and Bassell, 1981). However, the operation of these repair processes in microorganisms discharged to actual distribution systems or receiving waters is not clear.

Severin et al. (1983b) have shown that the series event model for inactivation describes the kinetics of UV disinfection quite well. Kinetic parameters for inactivation are shown in Table 14.6.

Activation energies for UV inactivation using this model were lower than for chemical disinfection (indicating the relative insensitivity to temperature) (Severin et al., 1983b). These are considerably lower than the activation energies for chemical disinfectants in accord with the apparent (purely physical) mechanism of UV inactivation. This also indicates, as a practical matter, that the effect of temperature on UV performance is much less than for chemical agents such as ozone or chlorine.

The pH dependency of UV inactivation has not been characterized in controlled systems. However, since the mechanism of UV inactivation appears to be purely physical, it is not anticipated that pH would dramatically alter the efficiency of UV disinfection. Insofar as pH may affect the light absorption characteristics of humic materials, an indirect effect of pH (by changing the extent of demand) on inactivation efficiency may exist.

Influence of Physical Factors on Disinfection Efficiency

The apparent increase in microbial resistance by clumping has already been discussed.

Solids Association. Microorganisms can also be partially protected against the action of disinfectants by adsorption to or enmeshment in nonviable solid particles present in a water. Stagg et al. (1978) and Hejkal et al. (1979) found that fecal material protected poliovirus against inactivation by combined chlorine. Boardman and Sproul (1977) found that kaolinite, alum flocs, and lime sludge increased the resistance of Poliovirus 1 to free chlorine in demand free systems.

For chlorine dioxide, bentonite turbidity protects poliovirus against the action of chlorine dioxide (Brigano et al., 1978). Although there was some protection from ozone inactivation afforded to coliform bacteria and viruses by fecal matter and by cell debris, at ozone doses usually employed in disinfection, it was still possible to achieve more than 99.9 percent inactivation within 30 s (Sproul et al., 1978).

Influence of Physiological Factors on Disinfection Efficiency

The physiological state of microorganisms, especially vegetative bacteria, may influence their susceptibility to disinfectants. Milbauer and Grossowicz (1959b) found that coliforms grown under minimal conditions were more resistant than cells grown under enriched conditions. Similarly, Berg et al. (1985) found that chemostat-grown

TABLE 14.6 Kinetics of UV Inactivation

	k (cm^2/(mW-s))	Number of events	Activation energy (kcal/mole)
E. coli	1.538	9	0.554
Candida parapsilosis	0.891	15	0.562
f2 virus	0.0724	1	1.023

Source: Severin et al., 1983.

cells produced at high growth rates were more sensitive to disinfectants than cells harvested from low growth rates. A simple subculturing of aquatic strains of *Flavobacterium* has been found to increase sensitivity to chlorine disinfection (Wolfe and Olson, 1985).

Postexposure conditions can also influence apparent microbial response to disinfectants (Milbauer and Grossowicz, 1959a). In three New England water treatment plants and distribution systems sampled for total coliforms using both standard MF techniques and media to recover sublethally injured organisms (m-T7 agar medium), from 8 to 38 times as many coliforms were recovered on the latter medium than on the former medium (McFeters et al., 1986). This sublethal injury does not adversely affect pathogenicity to mice (Singh et al., 1986).

With viruses, the phenomenon of multiplicity reactivation can occur when individual viruses inactivated by different specific events are combined in a single host cell to produce a competent and infectious unit. This has been demonstrated to occur in enteric viruses inactivated by chlorine (Young and Sharp, 1979).

The survivors of disinfection can exhibit inheritable increased resistance to subsequent exposure. This was first demonstrated for poliovirus exposed to chlorine (Bates et al., 1978). However, demonstration of this phenomenon in bacteria has not been consistent (Haas and Morrison, 1981; Leyval, 1984).

DISINFECTANT RESIDUALS FOR POSTTREATMENT PROTECTION

One factor that may be important in evaluating the relative merits of alternative disinfectants is their ability to maintain microbial quality in a water distribution system. With respect to chlorine, it has been suggested that free chlorine residuals may serve to protect the distribution system against regrowth, or at least as a sentinel for the presence of contamination (Snead et al., 1980). However, other studies have noted the lack of correlation between distribution system water quality and the form or concentration of chlorine residual (Haas et al., 1983b). Similarly, LeChevallier et al. (1990) have reported that microbial slimes grown in tap water may be more sensitive to inactivation by combined chlorine than to free chlorine transported by the overlying water. It must be recognized that, regardless of the disinfectant chosen, the water distribution system can never be regarded as biologically sterile. As shown by Means et al. (1986a), shifts in the dominant form of disinfectant (e.g., from free chlorine to monochloramine) can result in shifts in the taxonomic distribution of microorganisms that inhabit the distribution system.

It must be noted that there is a difference in practice between U.S. systems and many systems in Europe (Haas, 1999; Hydes, 1999; Trussell, 1999). In the United States, most utilities strive to maintain minimum chlorine residuals in the distribution system (and in fact there is a strong incentive to do so under regulatory requirements), with minimum residuals generally exceeding 0.2 mg/L. In a number of countries in Europe, the philosophical approach to maintaining distribution system water quality relies on nutrient control (principally degradable organic matter) rather than on disinfectant residuals.

The ability of chlorine dioxide residuals to maintain distribution system microbial water quality has not been well studied. With respect to both ozone and UV, the absence of a residual may necessitate the addition of a second disinfectant if a residual in the distribution system is desired. For further discussion on this point, see Chapter 18.

APPLICATION OF TECHNOLOGIES

With the increasing and conflicting objectives that must be met by disinfection processes, water utilities are moving toward the use of multiple disinfectants. In general, disinfectants may be applied at three locations within treatment. Application of a disinfectant prior to coagulation is generally termed *preoxidation* or *predisinfection.* While some inactivation may occur due to predisinfection, generally this is minor due to substantial disinfectant demand. Application of a disinfectant subsequent to sedimentation, either before or after filtration, but prior to a contactor, clear well, or hydraulic device with substantial contact, is termed *primary disinfection.* The majority of inactivation by disinfection processes is expected to occur due to this application. If another application occurs following primary disinfection (or if ammonia is added to convert a free residual to a combined residual), this is termed *secondary disinfection.* Generally the objective of secondary disinfection is to allow the penetration of an active disinfectant residual into the distribution system.

Chlorination

Chlorine may be obtained for disinfection in three forms, as well as generated onsite. For very small water treatment plants, solid calcium hypochlorite ($Ca(OCl)_2$) can be used. This can be applied as a dry powder, or in proprietary tablet dispensers. Calcium hypochlorite is more expensive than the other chemical forms, and particularly in hard waters, its use can lead to scale formation.

Generally, on a per unit mass basis of active chlorine, the least expensive form at large usage rates is liquified chlorine gas. The use of liquified chlorine gas carries with it certain risks associated with accidental leakage of the gas. As a result, a number of utilities have elected to use the somewhat more expensive sodium hypochlorite (NaOCl) as a source of disinfectant.

Upon addition to a water, chlorine gas will reduce the pH and alkalinity, while sodium hypochlorite will raise the pH and alkalinity. In a poorly buffered water, the addition of a pH control agent may thus be necessary to control the distribution system water aggressiveness.

Chlorine and hypochlorites have been produced from the electrolysis of brines and saline solutions since the early 20th century (Rideal, 1908). This remains an attractive option for remote treatment plants near a cheap source of brine. The basic principle is the use of a direct current electrical field to effect the oxidation of chloride ion with the simultaneous and physically separated reduction usually of water to gaseous hydrogen.

In actual practice, it is necessary to operate electrolytic chlorine-generating units at voltages as high as 3.85 volts in order to provide reasonable rates of generation. At these overvoltages, however, additional oxidations such as chlorate formation, ohmic heating, and incomplete separation of hydrogen from oxidized products with subsequent dissipative reaction combine to produce system inefficiencies. For typical electrolytic generating units, current efficiencies of 97 percent may be obtained along with energy efficiencies of 58 percent (Downs and Adams, 1973). These efficiencies are related to the physical configuration of the electrolysis cells, brine concentration, and desired degree of conversion to available chlorine (Bennett, 1978; Michalek and Leitz, 1972).

Chlorine can be applied at a variety of points within treatment. Table 14.7 from the 1989 AWWA Committee Survey illustrates the frequency with which chlorine is applied at various locations.

TABLE 14.7 Points of Application of Chlorine or Hypochlorite

Point of application	Frequency of plants ($N = 268$)
Before coagulation	18.66%
After coagulation	5.97%
After sedimentation and before filtration	31.34%
After filtration	47.01%
Within the distribution system	15.67%

Percentages sum to more than 100 percent due to multiple points of application.
Source: Haas et al., 1992.

Source Water Chlorination/Preoxidation. *Prechlorination,* the addition of chlorine at an early point within treatment, is designed to minimize operational problems associated with biological slime formation on filters, pipes, and tanks, and also release of potential taste and odor problems from such slimes. In addition, prechlorination can be used for the oxidation of hydrogen sulfide or reduced iron and manganese. Probably the most common point of addition of chlorine for prechlorination is the rapid mix basin (where flocculant is added).

However, due to present concerns for minimizing the formation of chlorine byproducts, the use of prechlorination is being supplanted by the use of other chemical oxidants (e.g., ozone, permanganate) for the control of biological fouling, odor, or reduced iron or manganese.

Postchlorination. *Postchlorination,* or *terminal disinfection,* is the primary application for microbial reduction. It has been most common to add chlorine for these purposes either immediately before the clear well or immediately before the sand filter. In the latter case, the filter itself serves, in effect, as a contact chamber for disinfection.

In general, the use of specific contact chambers subsequent to the addition of chlorine to a water has been uncommon. Instead, the clear well, or finished water reservoir, serves the dual function of providing contact to ensure adequate time for microbial inactivation prior to distribution. The distribution system itself, from the entry point until the first consumer's tap, provides additional contact time (see Table 14.8).

The hydraulic characteristics of most finished water reservoirs, however, are not compatible with the ideal characteristics of chlorine contact chambers. The latter are most desirably plug flow, while the former most usually have a large degree of dispersion.

TABLE 14.8 Residual and Contact Time to First Customer: 1989 AWWA Disinfection Committee Survey of Utilities ($N = 178$)

	Contact time to first customer, min						
Residual, mg/L	0	1–9	10–29	30–60	75–240	>240	Total
0–0.35	4.49%	0.56%	0.56%	0.56%	1.12%	0.56%	7.87%
0.4–0.95	10.11%	3.93%	3.93%	3.93%	3.37%	2.81%	28.09%
1–1.5	14.61%	5.62%	2.25%	4.49%	1.69%	5.62%	34.27%
1.6–2	3.93%	1.69%	1.69%	3.93%	3.37%	1.69%	16.29%
>2	6.74%	0.56%	0.56%	0.56%	2.81%	2.25%	13.48%
Total	39.89%	12.36%	8.99%	13.48%	12.36%	12.92%	100.00%

Superchlorination/Dechlorination. In the process of superchlorination/dechlorination, which has generally been employed for treatment of a poor-quality water (with high ammonia nitrogen concentrations, or perhaps severe taste and odor problems), chlorine is added beyond the breakpoint. This oxidizes the ammonia nitrogen present. Generally, the residual chlorine obtained at this point is higher than may be desired for distribution. The chlorine residual may be decreased by the application of a dechlorinating agent (sulfur compounds or activated carbon).

A modern application of chlorination-dechlorination in water treatment may be judicious where both high degrees of microbial inactivation and low levels of by-product formation are desired. It may be possible, for example, to hold a water with free chlorine for a period (sufficient to ensure disinfection, but not so long as to produce substantial by-products), then to partially (or completely) dechlorinate the water to minimize the production of organic by-products.

Chloramination. Chloramination, the simultaneous application of chlorine and ammonia or the application of ammonia prior to the application of chlorine, resulting in a stable combined residual, has been a long-standing practice at many utilities (Table 14.9). As noted in Table 14.1, approximately 20 percent of U.S. utilities use ammonia addition in conjunction with chlorine or hypochlorite.

Jefferson Parish, Louisiana, has been using simultaneous addition of ammonia and chlorine in its disinfection process for over 35 years (Brodtmann and Russo, 1979). To further reduce total trihalomethane formation, the point of disinfectant addition was changed to immediately upstream of the filters (rather than upstream of clarifiers). This change maintained satisfactory distribution system water quality, using a chloramine residual of 1.6 mg/L exiting the filters.

From a survey of utilities (Trussell and Kreft, 1984), it was found that:

1. Seventy percent use anhydrous ammonia, 20 percent aquo ammonia, 10 percent ammonium sulfate.

TABLE 14.9 Utilities with Long Experience of Chloramine Use

City	Approximate start of chloramination
Denver	1914
Portland (OR)	1924
St. Louis	1934
Boston	1944
Indianapolis	1954
Minneapolis	1954
Dallas	1959
Kansas City (MO)	1964
Milwaukee	1964
Jefferson Parish (LA)	1964
Philadelphia	1969
Houston	1982
Miami (FL)	1982
Orleans Parish (LA)	1982
San Diego	1982

Source: Trussell and Kreft, 1984.

2. Most utilities use 3:1 to 4:1 chlorine-to-ammonia feed ratios. Excess ammonia is generally used to make monochloramine predominant; however, some utilities use higher ratios to form more effective dichloramine.
3. There is no clear consensus as to the point of ammonia application (i.e., pre- or postammoniation).

In current practice, chloramination has been regaining popularity as a means to minimize organic by-product formation.

A major concern that has been identified in chloramination arises during the transition from free chlorination to chloramination (Means et al., 1986a). Data on distribution system water quality were collected at the Metropolitan Water District of Southern California before and after the distribution system was changed from a free chlorine residual to a combined (monochloramine) chlorine residual. There was no effect on coliform counts; however, the plate counts on m-R2A medium increased dramatically (with some increase also observed using pour plates/TGEA medium). In one of the reservoirs, a precipitous drop in chlorine residual, associated with nitrification in the reservoir and growth of microorganisms, occurred following the switchover. This was postulated to occur as a reaction between nitrites and monochloramine.

It is important for water utilities to alert hospitals and kidney dialysis centers to switchovers from free residual chlorination to chloramination. Birrell et al. (1978) and Eaton et al. (1973) reported cases of chloramine-induced hemolytic anemia in such centers.

Influence of Relative Point of Addition of Chlorine and Ammonia. Details of chloramination practice can dramatically influence process performance. Options as to preversus postammoniation or pH, in particular, must be considered.

Prereacted chloramine residuals are more effective bactericides (*E. coli,* demand free batch tests) at pH 6 than at pH 8, and at high Cl_2-to-N ratios (5:1) rather than low ratios (down to 2:1). Concurrent addition of ammonia and chlorine was as effective as preammoniation (and at pH 6 was nearly as effective as free residual chlorination). Both concurrent addition and preammoniation were more effective than prereaction of the chlorine, except at pH 8, where all three modes behaved in a similar manner (Ward et al., 1984).

In pilot plant studies, Means et al. (1986b) found that concurrent and sequential methods (chlorine at rapid mix and ammonia at end of flocculation tank) gave better performance at removing m-SPC bacteria than preammoniation (but poorer than free chlorination). Concurrent addition gave about as low a TTHM value as preammoniation.

In the presence of concentrations of organic nitrogen similar to ammonia, prereacted chloramines may give better performance than dynamically formed chloramines from preammoniation due to the favorable competition for chlorine by many organic N compounds (and their low biocidal potency) relative to inorganic nitrogen (Wolfe and Olson, 1985).

Characterizing and Improving Contact Tank Hydraulics. The presence of backmixing—deviations from ideal plug flow behavior—will reduce the disinfection efficiency of chlorine contactors. In serpentine contactors, the degree of backmixing (i.e., the dispersion) can be estimated from the geometry of the tank, in particular the length-to-width ratio (Stover et al., 1986).

When a finished water reservoir is used as a contact tank, severe backmixing and also stratification may occur (Boulos et al., 1996; Grayman et al., 1996). Various types of baffles may be used to counteract these tendencies, as shown in Figure 14.8 (Grayman et al., 1996). In addition, both hydraulic scale models and computational

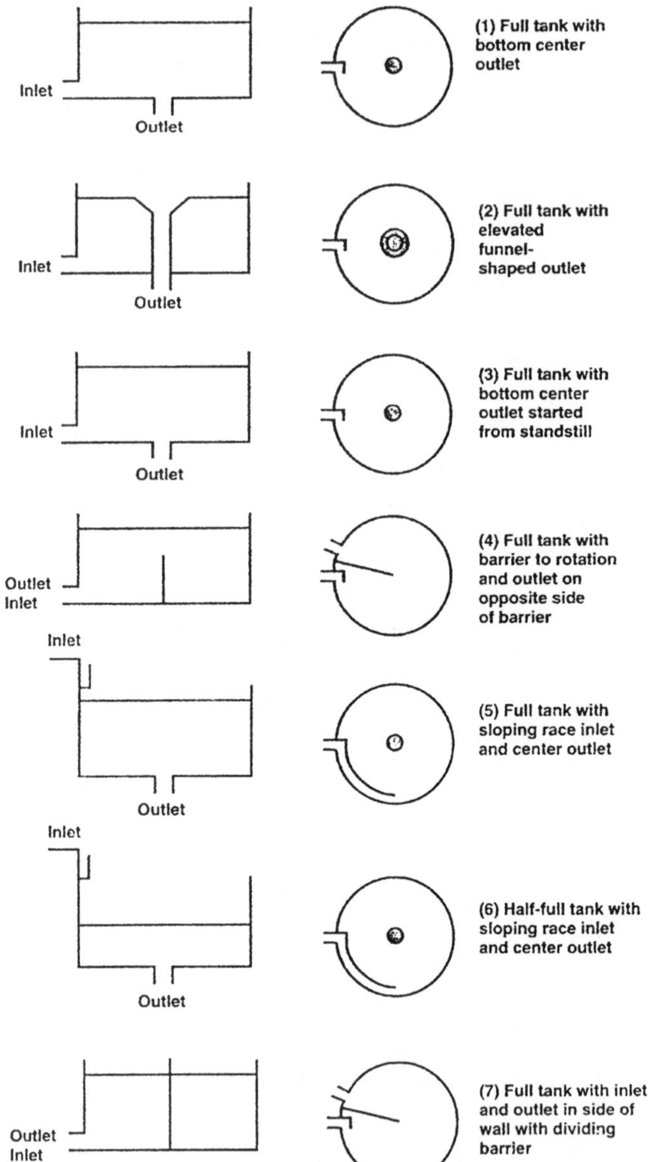

FIGURE 14.8 Schematic of baffling arrangements for reservoirs. (*Source:* Grayman et al., 1996.)

fluid dynamic models may be useful in assessing potential improvements in flow patterns from alternative baffling arrangements (Boulos et al., 1996; Grayman et al., 1996; Taras, 1980). Similarly, chlorine contact chambers may be upgraded in terms of hydraulic performance (increasing t_{10} and decreasing dispersion) by the installation of baffles (Bishop et al., 1993).

Chlorine Dioxide

It is necessary to generate chlorine dioxide on a continuous basis for use as a disinfectant. Although a few European potable water treatment plants have been reported to use the acid-chlorite generation process (Miller et al., 1978), the most common synthesis route for disinfectant ClO_2 generation is the chlorine-chlorite process.

Chemistry of Generation. Theoretically, chlorine dioxide may be produced by either the oxidation of a lower-valence compound or reduction of a more oxidized compound of chlorine. Chlorites (ClO_2^-) or chlorous acid ($HClO_2$) may be oxidized by chlorine or persulfate to chlorine dioxide, or may undergo autooxidation (disproportionation) to chlorine dioxide in solutions acidified with either mineral or organic acids. Chlorates (ClO_3^-) may be reduced either by use of chlorides, sulfuric acid, sulfur dioxide, or oxalic acid, or electrochemically to form chlorine dioxide (Masschelein, 1979a).

For practical purposes in water treatment, chlorine dioxide is generated exclusively from chlorite inasmuch as the reductive processes using chlorate as a starting material are capital intensive and competitive only at larger capacities (Masschelein, 1979b).

In the acid-chlorite process, sodium chlorite and hydrochloric acid react according to Equation 14.43:

$$5\ NaClO_2 + 4\ HCl = 4\ ClO_2 + 5\ NaCl + 2\ H_2O \qquad (14.43)$$

The resulting chlorine dioxide may be evolved as a gas or removed in solution. Mechanistically, this process occurs via a series of coupled reactions, some of which may involve the in situ formation of chlorine, catalysis by chloride, and the oxidation of chlorite by chlorine (Gordon et al., 1972; Masschelein, 1979a; Noack and Doeff, 1979). In addition, the yield of the reaction as well as the rate of the process are improved by low pH values in which formation of both gaseous chlorine and chlorous acid is favored. Under these favorable conditions, the reaction proceeds in the order of minutes; however, to achieve these conditions, excess hydrochloric acid is required.

During the acid-chlorite reaction, the following side reactions result in chlorine production:

$$5\ ClO_2^- + 5\ H^+ = 3\ ClO_3^- + Cl_2 + 3\ H^+ + H_2O \qquad (14.44)$$

$$4\ ClO_2^- + 4\ H^+ = 2\ Cl_2 + 3\ O_2 + 2\ H_2O \qquad (14.45)$$

$$4\ HClO_2 = 2\ ClO_2 + HClO_3 + HCl + H_2O \qquad (14.46)$$

if equal weights of the reactants are added, than close to 100 percent of the conversion of chlorite may occur; this means a final pH below 0.5 (Masschelein, 1979b).

Alternatively, chlorine dioxide may be produced by the oxidation of chlorite with chlorine gas according to Equation 14.47:

$$2\ NaClO_2 + Cl_2 = NaCl + 2\ ClO_2 \qquad (14.47)$$

As in the previous case, low pH accelerates the rate of this process, as does excess chlorine gas. However, if chlorine gas is used in stoichiometric excess, the resultant product may contain a mixture of unconsumed chlorine as well as chlorine dioxide.

The rate of the direct reaction between dissolved Cl_2 and chlorite has been measured (Aieta and Roberts, 1985), with a forward second-order rate constant given by the following:

$$k_f = 1.31 \times 10^{11} \exp(-4800/T) \qquad 1/\text{M-s} \qquad (14.48)$$

In the chlorine-chlorite process, sodium chlorite is supplied as either a solid powder or a concentrated solution. A solution of chlorine gas in water is produced by a chlorinator-ejector system of design similar to that used in chlorination. The chlorine-water solution and a solution of sodium chlorite are simultaneously fed into a reactor vessel packed with Raschig rings to promote mixing (Miller et al., 1978). Equation 14.46 shows that 1 mole of chlorine is required for 2 moles of sodium chlorite—or 0.78 parts of Cl_2 per part of $NaClO_2$ by weight. However, for this reaction to proceed to completion, it is necessary to reduce the pH below that provided by the typically acidic chlorine-water solution produced by an ejector. At 1:1 feed ratios by weight, only 60 percent of the chlorite typically reacts (Miller et al., 1978).

To provide greater yields, several options exist. First, it is possible to produce chlorine-water solutions in excess of 3500 mg/L using pressurized injection of gas. In this case, however, there will be an excess of unreacted chlorine in the product solution, and the resulting disinfectant will consist of a mixture of chlorine and chlorine dioxide. The second option consists of addition of acid to the chlorine-chlorite solution. For example, a 0.1 M HCl/M chloride addition enabled the production of a disinfectant solution of 95 percent purity in terms of chlorine dioxide, and achieved a 90 percent conversion of chlorite to chlorine dioxide (Jordan, 1980). A third process, developed by CIFEC (Paris, France), involves recirculation of the chlorinator ejector discharge water back to the ejector inlet to produce a strong (5000–6000 mg/L) chlorine solution, typically at pH below 3.0, and in this manner to increase the efficiency of chlorite conversion (Miller et al., 1978). It has been reported that this last option is capable of producing 95 to 99 percent pure solutions of chlorine dioxide. The intricacies of chlorine dioxide reactions and by-products necessitate careful process monitoring during operation of the generator (Lauer et al., 1986).

EXAMPLE 14.6 A water utility has a chlorination capacity of 1000 tons/day (454 kg/day) and is considering a switchover to chlorine dioxide to be generated using the chlorine-chlorite process. If the existing chlorination equipment is to be used, what is the maximum production capacity of chlorine dioxide, and how much sodium chlorite must be used under these conditions? Assume ideal stoichiometry and no excess chlorite or chlorine requirements.

1. From Equation 14.47, 2 moles of sodium chlorite ($NaClO_2$) react with 1 mole of chlorine to produce 2 moles of chlorine dioxide. Because chlorine has a molecular weight of 70, the current chlorinators have a capacity of 454,000/70 = 6486 mol/day of chlorine. Therefore, 12,971 mol/day of sodium chlorite are required, and the result would be an equal number of moles of chlorine dioxide. The molecular weights are:

   Chlorine dioxide $35 + 2(16) = 67$
   Sodium chlorite $23 + 35 + 2(16) = 89$

   Therefore, the sodium chlorite required is $12{,}971(89) = 1.15 \times 10^6$ g/day (2,540 lb/day). The chlorine dioxide produced would be $12{,}971(67) = 0.87 \times 10^6$ g/day (1,914 lb/day).

Application. The use of chlorine dioxide is limited by two factors. First, the maximum residual that does not cause adverse taste and odor problems is 0.4 to 0.5 mg/l as ClO_2 (Masschelein, 1979b). Second, the chlorite produced by reduction of chlorine dioxide as demand is exerted has been found to cause certain types of anemia, and therefore the maximum chlorine dioxide dose must be 1 mg/l to minimize this effect.

Augenstein (1974) suggested that chlorine dioxide residuals have moderate stability during distribution. However, it is not clear whether the analytical methods used in that study could adequately differentiate chlorine dioxide from its reaction products.

Insofar as chlorine dioxide is produced free of chlorine (in the acid-chlorite process, or in "optimized" chlorine-chlorite processes), the reactions with organic material to produce chlorinated by-products appear less significant than with chlorine (Aieta and Berg, 1986; Lykins and Griese, 1986).

Ozonation

Generation. The use of ozone in water treatment in the United States has been confined to preoxidation (Glaze, 1987). Additional information on the use of ozone as an oxidant is provided in Chapter 12.

Due to its instability, ozone is produced from gas-phase electrolytic oxidation of oxygen, either using very dry air or pure oxygen. The ozone-enriched gaseous phase is then contacted with the water to be treated in a bubble contactor (either diffused air or turbine mixed) or in a countercurrent tower contactor. Due to the cost of ozone, it is highly desirable to maximize the efficiency of transfer of ozone from gas to liquid.

Most ozone generators used in water treatment use one of two designs (Glaze, 1987). The most common for large plants is a bank of glass tube generators as shown in Figure 14.9. Small plants may use this type of generator on a smaller scale, or a plate-type generator in which the ozone is generated between ceramic plates. Cooling the tubes or plates increases the efficiency of ozone production. Ozone generators may use pure oxygen, oxygen-enriched air, or air as the feed gas. If air is used, the most economical operation gives a product stream that contains about 2 percent ozone by weight. Enhancement of the amount of oxygen in the stream increases the economical yield of ozone; for example, pure oxygen can generate a stream containing 5 to 7 percent ozone economically. In any case, the gas stream must have a very

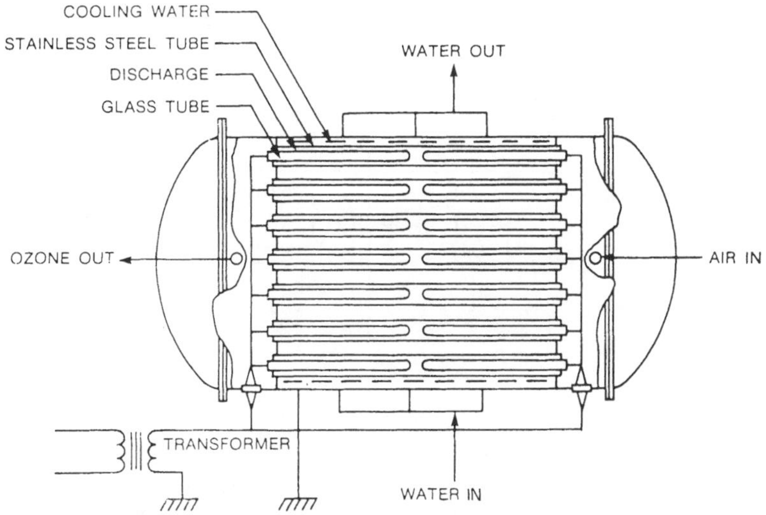

FIGURE 14.9 Large-scale tube-type generator for production of ozone from air or oxygen by cold plasma discharge.

low dew point (–50°C) and must be free of organic vapors. Figure 14.10 shows a flow diagram for a plant that utilizes oxygen enriched air.

Ozone generator designs that utilize different electrode configurations, such as surface discharge models, are also available. Also, ozone may be generated by irradiation of air with high-energy ultraviolet radiation (with wavelengths less than 200 nm). Photochemical generators are not yet capable of producing as much ozone as plasma generators, but may be particularly useful for small-scale applications such as swimming pool disinfection.

Contactors. After generation, ozone is piped to a contactor where it is transferred into the water. The most common type of contactor is the countercurrent sparged tank with diffuser (Figure 14.11). In this reactor, ozone-containing gas forms small bubbles as it is passed through a porous stone at the bottom of the tank. As the bubbles rise through the tank, ozone is transferred from the gas phase into water according to the rate equation

$$\text{Rate of transfer } [\text{mol/(m}^3)(\text{s})] = K_L a\,(C^* - C) \tag{14.49}$$

where C (mol/m^3) is the prevailing concentration of ozone in the liquid, C^* is the concentration at saturation, and $K_L a$ is the overall transfer coefficient (s^{-1}). The value of C^* depends on the percentage of ozone in the gas and may be calculated from the equation

$$C^* = \frac{P_{\text{gas}}}{H} \tag{14.50}$$

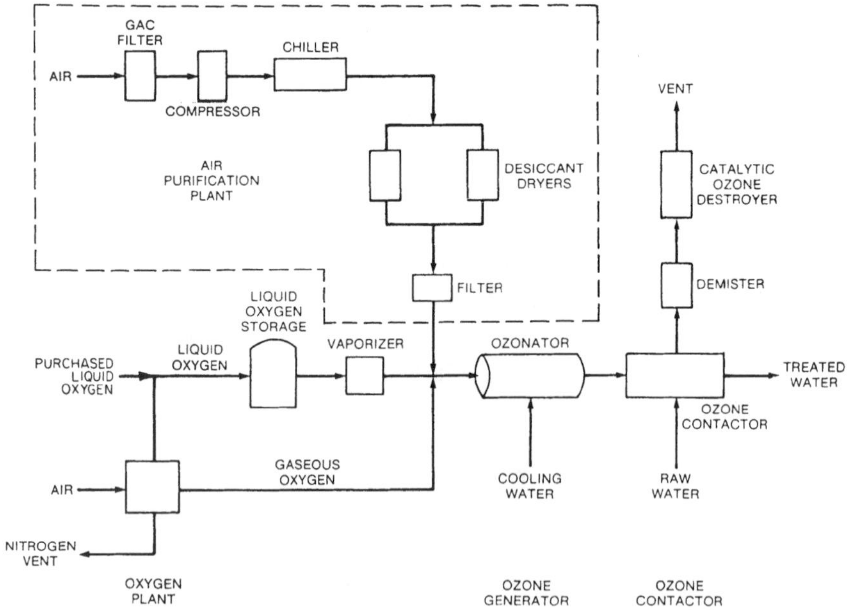

FIGURE 14.10 Flow diagrams for air and oxygen purification for ozone production from oxygen-enriched air. The air purification unit may be omitted when pure oxygen is used, or it may be used without oxygen enrichment.

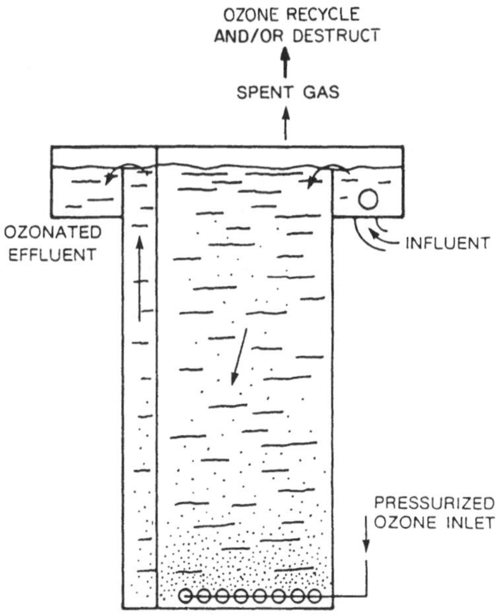

FIGURE 14.11 Diagram of typical countercurrent sparged column for transfer of ozone from gas to liquid phase.

where P_{gas} is the partial pressure of ozone in the gas phase (atm) and H is the Henry's law constant for ozone (0.082 atm-m^3/mol at 25°C). At a partial pressure of ozone of 0.02 atm (corresponding to 3 weight percent of ozone in air), Equation 14.36 predicts that the concentration of ozone in water at saturation will be 0.24 mol/m^3 (12 mg/L). More details on theory and application of liquid to gas transfer may be found in Chapter 5.

Ozone contactors such as that shown in Figure 14.11 may be sized to transfer ozone on a small or very large scale. Alternately, other designs such as sparged stirred tank reactors and venturi injectors may be used. If injected under a large head of water pressure, ozone may be transferred at higher rates because the saturation concentration of ozone, the C^* term in Equation 14.49, is higher.

Ozone contactors must have systems for the collection of ozone off-gas. Ozone is toxic and must be kept within OSHA allowable limits (ACGIH, 1994) within the treatment plant and surrounding areas. In some regions of the United States, ozone discharge from the treatment plants may be regulated. As a consequence of these requirements, thermal and catalytic ozone destroyers are routinely used in plants employing ozonation.

Ozone in water is so highly corrosive that only certain construction materials may be used in plants that generate and utilize ozone for water treatment. Metal in contact with ozone should be 304 stainless steel, and gasket materials should be some form of inert polymer such as a fluorocarbon. Concrete is a typical material for construction of basins, but joints must be caulked with an inert material.

Unfortunately, the characteristics that promote efficient gas-liquid mass transfer, particularly the desirability of intense agitation, lead to hydraulic characteristics that decrease disinfection efficiency (short-circuiting). Therefore, laboratory results on ozone inactivation of microorganisms are poor predictors of field performance, unless detailed aspects of field-scale hydraulic features are considered.

In one pilot study of surface water treated with granular activated carbon (GAC) with no other pretreatment (Morin et al., 1975), it was found that a contactor with 57 s of contact and an ozone dosage of 1.45 mg/l (with a transferred dose of 1.13 mg/l) resulted in complete viral inactivation (>7 logs). Satisfactory coliform results (<2/100 ml) were obtained at a dose of 1.29 mg/l (transferred dose of 1.04 mg/l). An economic analysis showed that the GAC-ozone process was cheaper than a process employing full conventional treatment with chlorination.

The contact time required for ozone inactivation of microorganisms at commonly employed ozone doses (several mg/L) is typically quite short—seconds to several minutes—rather than the characteristically longer disinfection times commonly used for chlorine or chloramines.

The lack of a persistent residual following ozone treatment has generally necessitated application of an additional terminal disinfection process such as chloramination.

Ozone Contactor Hydraulics. The role of hydraulics in ozone contactors has two influences on performance. It influences ozone transfer efficiency, and also, by virtue of the influence of dispersion on performance, it influences the effectiveness with which a given combination of residual and time can be effective. The complex interactions involved are illustrated in several papers (Lev and Regli, 1992; Martin et al., 1992; Roustan et al., 1991). The designer is faced with a trade-off in which multiple stages, each approximating a completely mixed system, are used to transfer ozone. As the number of stages increases, the overall system hydraulics and also potentially the transfer efficiency may improve. However, this also would significantly increase construction and operating costs for the ozonation units.

UV Radiation

UV radiation can be effectively produced by the use of mercury vapor (or more recently, antimony vapor) lamps. While to date there has been limited experience with UV disinfection in water treatment, there is a substantial operational data base associated with UV disinfection of wastewater effluents (Stover et al., 1986; Water Pollution Control Federation, 1984).

There are a variety of physical configurations in which UV has been employed. In two of these, the UV lamp(s) are surrounded by quartz sheaths and the jacketed lamps are immersed in the flowing water. The flow may be in a closed or open vessel, and may be either parallel or perpendicular to the axes of the lamps. In the third configuration, the water flows through Teflon tubes (which are relatively transparent to UV radiation) surrounded by UV lamps.

There are a number of newer UV technologies emerging for disinfection application. These include (O'Brien et al., 1996):

- Use of medium-pressure UV lamps with higher-intensity emissions and potential for longer lamp lifetimes
- High-frequency pulsed UV disinfection

For design purposes, it is necessary to ensure that there is turbulence (to allow all elements of fluid to come sufficiently close to the lamp surfaces) and yet to minimize the degree of transverse mixing (short-circuiting). The latter must be explicitly considered in design calculations (Stover et al., 1986).

The contact times for UV disinfection systems can be quite short, generally under 1 min. Therefore, the space required for UV disinfection units is relatively small. However, since there is no residual, some additional terminal disinfection process would usually be required. The potential for UV reactions to produce organic byproducts is minor since the intensities required for UV disinfection are less than those needed to cause photochemical effects. Operationally, it is essential to employ an effective cleaning program to periodically removal biological and chemical fouling materials from lamp jacket or teflon tube surfaces (Stover et al., 1986; Water Pollution Control Federation, 1984). In wastewater disinfection systems, lamp sizing is typically (to within a factor of 2) 1 kW per million gallons (3780 m^3) peak design flow (Scheible, 1993).

Role of Hydraulics. The role of hydraulics in UV disinfection has two effects on performance. In addition to a negative influence of backmixing, due to the phenomenon of dispersion noted earlier, the movement of microorganisms relative to lamp surfaces may dramatically influence the integrated exposure to radiation during contacting. The effect of turbulence to decrease the potential for stratification (defined as travel of a given microorganism at a fixed distance from lamp surfaces) is desirable to increase UV disinfection efficiency (Haas and Sakellaropoulous, 1979).

There is an interaction between the residence time distribution and the distribution of intensities of exposure that microorganisms experience in transiting a UV disinfection unit. The use of computational hydraulic models may assist in providing a more precise prediction of the effect of flow rate and water quality parameters (absorbency) on the inactivation expected in full-scale UV units (Chiu et al., 1999; Lyn et al., 1999).

USE OF MULTIPLE DISINFECTANTS

In view of the reexamination of disinfection practices that has occurred over the past 20 years, a number of case studies of multiple disinfectants have been described. These might include a relatively reactive primary disinfectant (such as O_3 or ClO_2) followed by a secondary disinfectant that is used to maintain a chemical residual in the distribution system.

For example, at the Louisville Water Company (Hubbs et al., 1980), there was a plant-scale examination of a process in which ClO_2 is added between the coagulation basin effluent and the softening basin influent along with excess ammonia to convert the free chlorine to combined chlorine (predominantly monochloramine). Some additional chlorine is added after filtration to provide a monochloramine residual in the distribution system. There was some regrowth in the filters; however, bacterial quality in the distribution system was not altered, and TTHMs were reduced from 30 ppb (free chlorination) to less than 5 ppb.

The use of multiple disinfectants may be advantageous in terms of achieving disinfection efficiency, since possible synergistic effects may occur—although further study is necessary to delineate the exact extent of such phenomena. For example, the exposure of *E. coli* to mixtures of free chlorine and chloramine resulted in substantially greater inactivation than would be predicted by their individual effectiveness (Kouame and Haas, 1991). Similarly, the following combinations of disinfectants may offer some level of inactivation of the highly resistant *Cryptosporidium* oocysts (Finch et al., 1998):

- Free chlorine followed by monochloramine
- Ozone followed by either free or combined chlorine
- Chlorine dioxide followed by either free or combined chlorine

RELATIVE COMPARISONS

The dominant disinfection technology in the United States is presently free residual chlorination. However, chloramination, chlorination with partial dechlorination, and chlorine dioxide treatment are viable alternative primary disinfectants. In addition, ozone, or possibly UV, when supplemented with a chemical that can produce a lasting residual in the distribution system, have application.

Table 14.10 summarizes important aspects of the major technologies and their technical pros and cons.

DIAGNOSTICS AND TROUBLESHOOTING

Since real-time measurements of microorganisms surviving disinfection processes are not presently possible, the operation of disinfection processes relies on indirect assessment of performance. With chlorine contacting systems, control is most frequently based on the use of residual measurements on a batch or continuous basis. A variety of control schemes are available. The feed of chlorine may be controlled to automatically provide a relatively constant residual. The presence of excessive chlorine demand may be symptomatic of a process upset or malfunction (Stover et al., 1986).

With ozone systems, it is possible to use either off-gas ozone monitors or measurement of dissolved residual for control of ozonation. However, the degree of

TABLE 14.10 Applicability of Alternative Disinfection Techniques

Consideration	Cl_2	O_3	ClO_2	UV
Equipment reliability	Good	Good	Good	Fair to good
Relative complexity of technology	Simple	Complex	Moderate	Moderate
Safety concerns	Yes	Moderate	Yes	Moderate
Bactericidal	Good	Good	Good	Good
Virucidal	Moderate	Good	Moderate	Good
Efficacy against protozoa	Fair	Moderate	Fair	Fair to moderate
By-products of possible health concern	Yes	Some	Some	None known
Persistent residual	Long	None	Moderate	None
Reacts with ammonia	Yes	No	No	No
pH dependent	Yes	Slight	Slight	No
Process control	Well developed	Developing	Developing	Developing
Intensiveness of operations and maintenance	Low	High	Moderate	High

automation in control systems is not as readily developed as in the case of chlorination systems.

For UV disinfection, light absorption may be monitored continually, and lamps may be controlled (by either turning on or off sets of lamps, or varying power) to produce a given delivered energy intensity. This can also allow for compensation of fouling of lamp and jacket surfaces with operations.

BIBLIOGRAPHY

Abad, F. X., R. M. Pinto, J. M. Diez, and A. Bosch. "Disinfection of Human Enteric Viruses in Water by Copper and Silver in Combination with Low Levels of Chlorine." *Applied and Environmental Microbiology,* 60(7):2377–2383, 1994.

ACGIH. "Threshold Limit Values for Chemical Substances and Physical Agents and Biological Exposure Indices." American Conference of Governmental Industrial Hygienists, Cincinnati, OH, 1994.

Aieta, E., and J. Berg. "A Review of Chlorine Dioxide in Drinking Water Treatment." *Journal AWWA,* 78(6):62, 1986.

Aieta, E., and P. Roberts. "The Chemistry of Oxo-Chlorine Compounds Relevent to Chlorine Dioxide Generation," In *Water Chlorination: Environmental Impact and Health Effects,* vol. 5 (R. Jolley, ed.). Boca Raton, FL: Lewis Publishers, 1985.

Allen, M. Presented at the American Society for Microbiology Annual Conference, 1977.

Anonymous. "Is Chlorination Effective Against all Waterborne Disease?" *Journal of the American Medical Association,* 78:283, 1922.

APHA, AWWA, and WPCF. *Standard Methods for the Examination of Water and Wastewater* (17th ed.). American Public Health Assoc. Washington, DC, 1989.

Aston, R., and J. Synan. "Chlorine Dioxide as a Bactericide in Waterworks Operation." *Journal of the New England Water Works Association,* 62:80, 1948.

Augenstein, H. "Use of Chlorine Dioxide to Disinfect Water." *Journal AWWA,* 66(12):716, 1974.

Averill, C. *Facts Regarding the Disinfecting Powers of Chlorine: Letter to Hon. J. I. Degraff, Mayor of the City of Schenectady NY.* SS Riggs Printer, Schenectady, NY, 1832.

Bailey, J. E., and D. F. Ollis. *Biochemical Engineering Fundamentals* (2nd ed.). McGraw-Hill, New York, 1986.

Baker, J. "Use of Chlorine in the Treatment of Sewage." *Surveyor,* 69:241, 1926.

Baker, R. "Studies on the Reaction Between Sodium Hypochlorite and Proteins. I. Physicochemical Study of the Course of the Reaction." *Biochemical Journal,* 41:337, 1947.

Baker, R. J. "Characteristics of Chlorine Compounds." *Journal of the Water Pollution Control Federation,* 41:482, 1969.

Bates, R., S. Sutherland, and P. Shaffer. "Development of Resistant Poliovirus by Repetitive Sublethal Exposure to Chlorine," In *Water Chlorination: Environmental Impact and Health Effects* (R. Jolley, ed.). Butterworths, Stoneham, MA, 1978.

Battino, R. *Chlorine Dioxide Solubilities.* Pergamon, New York, 1984.

Belohlav, L. R., and E. T. McBee. "Discovery and Early Work," In *Chlorine: Its Manufacture, Properties and Uses: ACS Monograph No. 154* (J. Sconce, ed.). Van Nostrand Reinhold, New York, 1966.

Benarde, M. A. et al. "Efficiency of Chlorine Dioxide as a Bactericide." *Applied Microbiology,* 13(5):776, 1965.

Benarde, M. A. et al. "Kinetics and Mechanism of Bacterial Disinfection by Chlorine Dioxide." *Applied Microbiology,* 15(2):257, 1967.

Bennett, J. "On Site Generation of Hypochlorite Solutions by Electrolysis of Seawater." *AIChE Symposium Series,* 74(178):265, 1978.

Berg, G., H. Sanjaghsaz, and S. Wangwongwatana. "Potentiation of the Virucidal Effectiveness of Free Chlorine by Substances in Drinking Water." *Applied and Environmental Microbiology,* 55(2):390–393, 1989.

Berg, J. D. et al. "Disinfection Resistance of *Legionella pneumophila, Escherichia coli* Grown in Continuous and Batch Culture." In *Water Chlorination: Environmental Impact and Health Effects* (R. Jolley ed.). Lewis Publishers, Chelsea, MI, 1985.

Birrell, D. J. et al. "An Epidemic of Chloramine Induced Anemia in a Hemodialysis Unit." *Medical Journal of Australia,* 2:288, 1978.

Bishop, M., J. Morgan, B. Cornwell, and D. Jamison. "Improving the Disinfection Detention Time of a Water Plant Clearwell." *Journal of AWWA,* 85(3):68–75, 1993.

Blostein, J. "Shigellosis from Swimming in a Park Pond in Michigan." *Public Health Reports,* 106(3):317–321, 1991.

Boardman, G. D., and O. J. Sproul. Presented at the American Water Works Association, 97th Annual Conference, Anaheim, CA, 1977.

Bolyard, M., P. S. Fair, and D. P. Hautman. "Occurrence of Chlorate in Hypochlorite Solutions Used for Drinking Water Disinfection." *Environmental Science and Technology,* 26(8):1663–1665, 1992.

Bolyard, M., P. S. Fair, and D. Hautman. "Sources of Chlorate Ion in US Drinking Water." *Journal AWWA,* 85(9):81–88, 1993.

Bonde, G. "Bacteriological Methods for Estimation of Water Pollution." *Health Laboratory Science,* 3(2):124, 1966.

Boulos, P., W. Grayman, R. Bowcock, J. Clapp, L. Rossman, R. Clark, R. Deininger, and A. Dhingra. "Hydraulic Mixing and Free Chlorine Residual in Reservoirs." *Journal AWWA,* 88(7):48–59, 1996.

Brennhovd, O., G. Kapperud, and G. Langeland. "Survey of Thermotolerant *Campylobacter* spp and *Yersinia* spp in 3 Surface Water Sources in Norway." *International Journal of Food Microbiology,* 15(3–4):327–338, 1992.

Bridgman, S. A., R. M. P. Robertson, and P. R. Hunter. "Outbreak of cryptosporidiosis associated with a disinfected ground water supply." *Epidemiology and Infection,* 115(3):555–566, 1995.

Brigano, F. A. O. et al. Presented at the American Society for Microbiology Annual Conference, 1978.

Brodtmann, N., and P. Russo. "The Use of Chloramine for Reduction of Trihalomethanes and Disinfection of Drinking Water." *Journal AWWA,* 71(1):40, 1979.

Brown, T. J., J. C. Hastie, P. J. Kelly, R. Vanduivenboden, J. Ainge, N. Jones, N. K. Walker, D. G. Till, H. Sillars, and F. Lemmon. "Presence and Distribution of Giardia cysts in New-Zealand Waters." *New Zealand Journal of Marine and Freshwater Research,* 26(2):279–282, 1992.

Bull, R. J., and F. C. Kopfler. *Health Effects of Disinfectants and Disinfection Byproducts.* AWWA Research Foundation and AWWA, Denver, CO, 1991.

Butterfield, C. T. et al. "Influence of pH and Temperature on the Survival of Coliforms and Enteric Pathogens When Exposed to Free Chlorine." *US Public Health Reports.* 58:1837, 1943.

Butterfield, C. T. et al. "Influence of pH and Temperature on the Survival of Coliforms and Enteric Pathogens When Exposed to Chloramine." *US Public Health Service Reports,* 1946.

Cabelli, V. "*Clostridium perfringens* as a Water Quality Indicator," In *Bacterial Indicators/Health Hazards Associated with Water* (A. Hoadley and B. Dutka, eds.). ASTM, Philadelphia, PA, 1977.

Cachaza, J. "Kinetics of Oxidation of Nitrite by Hypochlorite in Aqueous Basic Solution." *Canadian Journal of Chemistry,* 54:3401, 1976.

Carlson, S. et al. "Water Disinfection by Means of Chlorine: Killing of Aggregate Bacteria." *Zbl. Bakteriol. Hygiene. I. Abt. Orig. B,* 161:233, 1975.

Centers for Disease Control and Prevention. "Assessing the Public Health Threat Associated with Waterborne Cryptosporidiosis: Report of a Workshop." *Morbidity and Mortality Weekly Reports,* 44(RR6):1–19, 1995.

Cerf, O. "Tailing of Survival Curves of Bacterial Spores." *Applied Bacteriology,* 42:1, 1977.

Chang, S. "Destruction of Microorganisms." *Journal AWWA,* 36(11):1192, 1944.

Chang, S. "Modern Concept of Disinfection." *ASCE Journal of the Sanitary Engineering Division,* 97(689), 1971.

Chapra, S. C., and R. P. Canale. *Numerical Methods for Engineers* (2nd ed.). McGraw-Hill, New York, 1988.

Chen, Y., O. Sproul, and A. Rubin. "Inactivation of *Naegleria gruberi* Cysts by Chlorine Dioxide." *Water Research,* 19(6):783, 1985.

Chick, H. "An Investigation of the Laws of Disinfection." *Journal of Hygiene,* 8:92–157, 1908.

Chiu, K., D. A. Lyn, P. Savoye, and E. R. Blatchley, III. "Integrated UV Disinfection Model Based on Particle Tracking." *Journal of Environmental Engineering,* 125(1):7–16, 1999.

Clark, J., and J. Pagel. "Pollution Indicator Bacteria Associated with Municipal Raw and Drinking Water Supplies." *Canadian Journal of Microbiology,* 23:465, 1977.

Cooper, W., R. Zika, and M. Steinhauer. "The Effect of Bromide Ion in Water Treatment. II. A Literature Review of Ozone and Bromide Interactions and the Formation of Organic Bromine Compounds." *Ozone Science and Engineering,* 7:313, 1985.

Cortelyou, J. R. et al. "The Effects of Ultraviolet Irradiation on Large Populations of Certain Waterborne Bacteria in Motion." *Applied Microbiology,* 2:227, 1954.

Dahi, E., and E. Lund. "Steady State Disinfection of Water by Ozone and Sonozone." *Ozone Science and Engineering* 2(1):13, 1980.

Danckwerts, P. V. "The Effect of Incomplete Mixing on Homogeneous Reactions." *Chemical Engineering Science,* 8(2):93–102, 1958.

De Laat, J. "Chloration de Composés Organisés: Demande en Chlore et Reactivité vis-à-vis de la Formation des Trihalomethe Incidence de l'Azote Ammoniacal." *Water Research,* 16:143–147, 1982.

Dennis, W. H. et al. "The Reaction of Nucleotides with Aqueous Hypochlorous Acid." *Water Research,* 13:357, 1979a.

Dennis, W. H. et al. "Mechanism of Disinfection: Incorporation of ^{36}Cl into f2 Virus." *Water Research,* 13:363, 1979b.

Downes, A., and T. Blount. "Research on the Effect of Light upon Bacteria and Other Organisms." *Proceedings of the Royal Society of London,* 26:488, 1877.

Downs, A., and C. Adams. *The Chemistry of Chlorine, Bromine, Iodine and Astatine.* Pergamon, Oxford, 1973.

Drenchen, A., and M. Bert. "A Gastroenteritis Illness Outbreak Associated with Swimming in a Campground Lake." *Journal of Environmental Health.* 57(2):7–10, 1994.

Dugan, N., R. Summers, R. Miltner, and H. Shukairy. Presented at the AWWA Water Quality Technology Conference, New Orleans, 1995.

Eaton, J. W. et al. "Chlorinated Urban Water: A Cause of Dialysis Induced Hemolytic Anemia." *Science,* 181:463, 1973.

Engelbrecht, R. S., C. N. Haas, J. A. Shular, D. L. Dunn, D. Roy, A. Lalchandani, B. F. Severin, and S. Farooq. *Acid-Fast Bacteria and Yeasts as Indicators of Disinfection Efficiency.* EPA-600/2-79-091. US Environmental Protection Agency, 1979.

Engelbrecht, R. S., B. F. Severin, M. T. Masarik, S. Farooq, S. H. Lee, C. N. Haas, and A. Lalchandani. *New Microbial Indicators of Disinfection Efficiency.* EPA-600/2-77-052. US Environmental Protection Agency, 1977.

Fair, G. M., J. C. Morris, S. L. Chang, I. Weil, and R. P. Burden. "The Behavior of Chlorine as a Water Disinfectant." *Journal of the American Water Works Association,* 40:1051–1061, 1948.

Farooq, S. PhD Dissertation. University of Illinois at Urbana-Champaign, 1976.

Farooq, S., C. N. Kurucz, T. D. Waite, and W. J. Cooper. "Disinfection of Wastewaters: High Energy Electron vs. Gamma Irradiation." *Water Research,* 27(7):1177–1184, 1993.

Feben, D., and M. J. Taras. "Chlorine Demand Constants of Detroit's Water Supply." *Journal AWWA,* 42:453–461, 1950.

Feben, D., and M. J. Taras. "Studies on Chlorine Demand Constants." *Journal AWWA,* 43:922–932, 1951.

Feng, T. "Behavior of Organic Chloramines in Disinfection." *Journal of the Water Pollution Control Federation,* 38:614, 1966.

Finch, G. R., E. K. Black, C. W. Labatiuk, L. Gyurek, and M. Belosevic. "Comparison of *Giardia lamblia* and *Giardia muris* Cyst Inactivation by Ozone." *Applied and Environmental Microbiology,* 59(11):3674–3680, 1993.

Finch, G. R., L. L. Gyurek, L. R. J. Liyanage, and M. Belosevic. *Effects of Various Disinfection Methods on the Inactivation of Cryptosporidium.* AWWA Research Foundation and American Water Works Association, Denver, CO, 1998.

Friberg, L. "Further Qualitative Studies on the Reaction of Chlorine with Bacteria in Water Disinfection." *Acta Pathologica et Microbiologica Scandinavica,* 40:67, 1957.

Friberg, L., and E. Hammarstrom. "The Action of Free Available Chlorine on Bacteria and Bacterial Viruses." *Acta Pathologica et Microbiologica Scandinavica,* 38:127, 1956.

Fujioka, R., K. Tenno, and P. Loh. "Mechanism of Chloramine Inactivation of Poliovirus: A Concern for Regulators." In *Water Chlorination: Environmental Impact and Health Effects* (R. Jolley, ed.). Lewis Publishers, Chelsea, MI, 1985.

Gallaher, M. M., J. L. Herndon, L. J. Nims, C. R. Sterling, D. J. Grabowski, and H. F. Hill. "Cryptosporidiosis and Surface Water." *American Journal of Public Health,* 79(1):39–42, 1989.

Gard, S. "Theoretical Considerations in the Inactivation of Viruses by Chemical Means." *Annals of the New York Academy of Sciences,* 83:638, 1960.

Geldreich, E. "Waterborne Pathogens," In *Water Pollution Microbiology* (R. Mitchell, ed.). Wiley-Interscience, New York, 1971.

Glaze, W. "Drinking Water Treatment with Ozone." *Environmental Science and Technology,* 21:224, 1987.

Goldstein, S., D. Juranek, O. Ravenholt, A. Hightower, D. Martin, J. Mesnick, S. Griffiths, A. Bryant, R. Reich, and B. Herwaldt. "Cryptosporidiosis: An Outbreak Associated with Drinking Water Despite State of the Art Water Treatment." *Annals of Internal Medicine,* 124(5):459–468, 1996.

Gordon, G., L. C. Adam, and B. P. Bubnis. "Minimizing Chlorate Ion Formation." *Journal of the American Water Works Association,* 87(6):97–106, 1995.

Gordon, G., L. C. Adam, B. P. Bubnis, B. Hoyt, S. J. Gillette, and A. Wilczak. "Controlling the Formation of Chlorate Ion in Liquid Hypochlorite Feedstocks." *Journal AWWA,* 85(9):89–97, 1993.

Gordon, G. et al. "The Chemistry of Chlorine Dioxide." *Progress in Inorganic Chemistry,* 15:201, 1972.

Grabow, W. "The Virology of Waste Water Treatment." *Water Research,* 2:675, 1968.

Grabow, W. O. K. et al. "Inactivation of Hepatitis A Virus and Indicator Organisms in Water by Free Chlorine Residuals." *Applied and Environmental Microbiology,* 46:619, 1983.

Granstrom, M. L., and G. F. Lee. "Rates and Mechanisms of Reactions Involving Oxychlorine Compounds." *Public Works,* 88(12):90–92, 1957.

Gray, E., D. Margerum, and R. Huffman. "Chloramine Equilibria and the Kinetics of Disproportionation in Aqueous Solution," In *Organometals and Metalloids, Occurrence and Fate in the Environment,* vol. Symposium #82 F. E. Brenchman and J. M. Bellama (eds.). American Chemical Society, Washington DC, p. 264–277, 1978.

Grayman, W., R. Deininger, A. Green, P. Boulos, R. Bowcock, and C. Godwin. "Water Quality and Mixing Models for Tanks and Reservoirs." *Journal AWWA* 88(7):60–73, 1996.

Griffin, A., and N. Chamberlin. "Relation of Ammonia Nitrogen to Breakpoint Chlorination." *American Journal of Public Health,* 31:803, 1941a.

Griffin, A., and N. Chamberlin. "Some Chemical Aspects of Breakpoint Chlorination." *Journal of the New England Water Works Association,* 55:3, 1941b.

Gurol, M., and P. Singer. "Kinetics of Ozone Decomposition: A Dynamic Approach." *Environmental Science and Technology,* 16(7):377, 1982.

Gurol, M. D. et al. "Oxidation of Cyanides in Industrial Wastewaters by Ozone." *Environmental Progress,* 4(1):46, 1985.

Gyurek, L. L., and G. R. Finch. "Modeling Water Treatment Chemical Disinfection Kinetics." *Journal of Environmental Engineering,* 124(9):783–793, 1998.

Haag, W. R., and M. H. Lietzke. "A kinetic model for predicting the concentrations of active halogen species in chlorinated saline cooling waters." *Water Chlorination: Environ. Impact Health Eff.,* 3:415–426, 1980.

Haas, C., and J. Anotai. Presented at the Disinfecting Wastewater for Discharge and Reuse Conference, Portland, OR, 1996.

Haas, C., and G. Sakellaropoulous. Presented at the ASCE Environmental Engineering Specialty Conference, San Francisco, 1979.

Haas, C. N. "A Mechanistic Kinetic Model for Chlorine Disinfection." *Environmental Science and Technology,* 14:339–340, 1980.

Haas, C. N. "Sodium Alterations of Chlorine Equilibria: Quantitative Description." *Environmental Science and Technology,* 15:1243–1244, 1981.

Haas, C. N. "Wastewater Disinfection and Infectious Disease Risks." *CRC Critical Reviews in Environmental Control,* 17(1):1–20, 1986.

Haas, C. N. "Micromixing and Dispersion in Chlorine Contact Chambers." *Environmental Technology Letters,* 9(1):35–44, 1988.

Haas, C. N. "Benefits of Using a Disinfectant Residual." *Journal AWWA,* 91(1):65–69, 1999.

Haas, C. N., and R. S. Engelbrecht. "Chlorine Dynamics During Inactivation of Coliforms, Acid-Fast Bacteria and Yeasts." *Water Research,* 14:1749–1757, 1980a.

Haas, C. N., and R. S. Engelbrecht. "Physiological Alterations of Vegetative Microorganisms Resulting From Aqueous Chlorination." *Journal of the Water Pollution Control Federation,* 52:1976–1989, 1980b.

Haas, C. N., M. S. Heath, J. Jacangelo, J. Joffe, U. Anmangandla, J. C. Hornberger, and J. Glicker. *Development and Validation of Rational Design Methods of Disinfection.* American Water Works Association Research Foundation, 1995.

Haas, C. N., and B. Heller. "Statistics of Microbial Disinfection." *Water Science and Technology,* 21(3):197–201, 1989.

Haas, C. N., and J. Joffe. "Disinfection Under Dynamic Conditions: Modification of Hom's Model for Decay." *Environmental Science and Technology,* 28(7):1367–1369, 1994.

Haas, C. N., J. Joffe, M. Heath, and J. Jacangelo. "Continuous Flow Residence Time Distribution Function Characterization." *Journal of Environmental Engineering,* 123(2):107–114, 1997.

Haas, C. N., M. G. Karalius, D. M. Brncich, and M. A. Zapkin. "Alteration of Chemical and Disinfectant Properties of Hypochlorite by Sodium Potassium and Lithium." *Environmental Science and Technology,* 20:822–826, 1986.

Haas, C. N., and S. B. Karra. "Kinetics of Microbial Inactivation By Chlorine. I. Review of Results in Demand-Free Systems." *Water Research,* 18:1443–1449, 1984a.

Haas, C. N., and S. B. Karra. "Kinetics of Microbial Inactivation By Chlorine. II. Kinetics in the Presence of Chlorine Demand." *Water Research,* 18:1451–1454, 1984b.

Haas, C. N., and S. B. Karra. "Kinetics of Wastewater Chlorine Demand Exertion." *Journal of the Water Pollution Control Federation,* 56:170–173, 1984c.

Haas, C. N., M. A. Meyer, and M. S. Paller. "The Ecology of Acid-Fast Organisms in Water Supply Treatment and Distribution Systems." *Journal AWWA,* 75:139–144, 1983a.

Haas, C. N., M. A. Meyer, and M. S. Paller. "Microbial Alterations in Water Distribution Systems and Their Relationship to Physical-Chemical Characteristics." *Journal AWWA,* 75:475–481, 1983b.

Haas, C. N., M. A. Meyer, M. S. Paller, and M. A. Zapkin. "The Utility of Endotoxins as a Surrogate Indicator in Potable Water Microbiology." *Water Research,* 17:803–807, 1983c.

Haas, C. N., and E. C. Morrison. "Repeated Exposure of *Escherichia coli* to Free Chlorine: Production of Strains Possessing Altered Sensitivity." *Water Air and Soil Pollution,* 16:233–242, 1981.

Haas, C. N., B. F. Severin, D. Roy, R. S. Engelbrecht, and A. Lalchandani. "Removal of New Indicators by Coagulation-Flocculation and Sand Filtration." *Journal AWWA,* 77:67–71, 1985a.

Haas, C. N., B. F. Severin, D. Roy, R. S. Engelbrecht, A. Lalchandani, and S. Farooq. "Field Observations on the Occurrence of New Indicators of Disinfection Efficiency." *Water Research,* 19:323–329, 1985b.

Haas, C. N., J. G. Sheerin, and C. Lue-Hing. "Effects of Ceasing Disinfection on a Receiving Water." *Journal of the Water Pollution Control Federation,* Boca Raton, FL. 60:667–673, 1988.

Haas, C. N., C. D. Trivedi, and J. R. O'Donnell. "Further Studies of Hypochlorite Ion Pair Chemistry and Disinfection Efficiency," In *Water Chlorination: Environmental Impact and Health Effects, vol. 6* (R. L. Jolley, ed.). Lewis Publishers, p. 729–740, 1990.

Haas, C. N. et al. "Survey of Water Utility Disinfection Practices." *Journal AWWA,* 84(9):121–128, 1992.

Hamelin, C., and Y. Chung. "Role of the POL, REC and DNA Gene Products in the Repair of Lesions Produced in *E. coli* by Ozone." *Studia (Berlin),* 68:229, 1978.

Hayes, E. B., T. D. Matte, T. R. O'Brien, T. W. McKinley, G. S. Logsdon, J. B. Rose, B. L. Ungar, D. M. Word, P. F. Pinsky, and M. L. Cummings. "Large community outbreak of cryptosporidiosis due to contamination of a filtered public water supply." *New England Journal of Medicine,* 320(21):1372–1376, 1989.

Hejkal, T. W. et al. "Survival of Poliovirus Within Organic Solids During Chlorination." *Applied and Environmental Microbiology,* 38:114, 1979.

Herwaldt, B. L., G. F. Craun, S. L. Stokes, and D. D. Juranek. "Waterborne Disease Outbreaks, 1989–1990." *Morbidity and Mortality Weekly Report,* 40(SS-3):1–21, 1991.

Herwaldt, B., G. F. Craun, S. L. Stokes, and D. D. Juranek. "Outbreaks of Waterborne Disease in the United States: 1989–90." *Journal AWWA,* 84(4), 129–135, 1992.

Hiatt, C. "Kinetics of the Inactivation of Viruses." *Bacteriological Reviews,* 28:150, 1964.

Hoff, J. *Inactivation of Microbial Agents by Chemical Disinfectants.* EPA/600/2-86/067. USEPA, 1986.

Hoff, J., and E. Akin. "Microbial Resistance to Disinfectants: Mechanisms and Significance." *Environmental Health Significance,* 69:7–13, 1986.

Hoigne, J., and H. Bader. "Ozonation of Water: Role of Hydroxyl Radicals as Oxidizing Intermediates." *Science,* 190:782, 1975.

Hoigne, J., and H. Bader. "The Role of Hydroxyl Radical Reactions in Ozonation Processes in Aqueous Solutions." *Water Research,* 10(377), 1976.

Holwerda, K. "Mededeelingen van den Dienst der Volksgezondheid in Ned-Indie." 17:251, 1928.

Hom, L. W. "Kinetics of Chlorine Disinfection of an Ecosystem." *Journal of the Sanitary Engineering Division, ASCE,* 98(SA1):183–194, 1972.

Hooker, A. *Chloride of Lime in Sanitation.* Wiley, New York, 1913.

Hubbs, S., M. Goers, and J. Siria. "Plant-Scale Examination and Control of a Chlorine Dioxide Chloramination Process at the Louisville Water Company," In *Water Chlorination: Environmental Impact and Health Effects* (R. Jolley, ed.). Butterworths, Stoneham, MA, 1980.

Huff, C. B. et al. "Study of Ultraviolet Disinfection of Water and Factors in Treatment Efficiency." *Public Health Reports,* 80(8):695–705, 1965.

Hydes, O. "European Regulations on Residual Disinfection." *Journal of the American Water Works Association,* 91(1):70–74, 1999.

Iivanainen, E. K., P. J. Martikainen, P. K. Vaananen, and M. L. Katila. "Environmental Factors Affecting the Occurrence of Mycobacteria in Brook Waters." *Applied and Environmental Microbiology,* 59:398–404, 1993.

Ingols, R., and G. Ridenour. "Chemical Properties of Chlorine Dioxide in Water Treatment." *Journal of the American Water Works Association,* 40(11):1207, 1948.

Isaac, R., and J. Morris. "Rates of Transfer of Active Chlorine Between Nitrogenous Substances," In *Water Chlorination: Environmental Impact and Health Effects* (R. Jolley, ed.). Butterworth, Stoneham, MA, 1980.

Isaac, R. A. et al. "Subbreakpoint Modeling of the HOBr-NH3-OrgN Reactions." *Water Chlorination: Environmental Impact and Health Effects,* 5:985–998, 1985.

Jagger, J. *Introduction to Research in Ultraviolet Photobiology.* Prentice Hall, Inc., Englewood Cliffs, NJ, 1967.

Jarroll, E. L., A. K. Bingham, and E. A. Meyer. "Effect of Chlorine on *Giardia lamblia* Cyst Viability." *Applied and Environmental Microbiology,* 41(2):483–487, 1981.

Jordan, R. "Improved Method Generates More Chlorine Dioxide." *Water and Sewage Works,* 127(10):44, 1980.

Kabler, P. "Relative Resistance of Coliform Organisms and Enteric Pathogens in the Disinfection of Water with Chlorine." *Journal of the American Water Works Association,* 43(7):553, 1951.

Katzenelson, E., B. Kletter, and H. Shuval. "Inactivation Kinetics of Viruses and Bacteria in Water by Use of Ozone." *Journal of the American Water Works Association,* 66(12):725, 1974.

Kelly, S., and W. Sanderson. "The Effect of Chlorine in Water on Enteric Viruses." *American Journal of Public Health,* 48:1323, 1958.

Kinman, R. N., and R. F. Layton. "New Method for Water Disinfection." *Journal AWWA,* (6):298–302, 1976.

Korich, D. G., J. R. Mead, M. S. Madore, N. A. Sinclair, and C. R. Sterling. "Effects of ozone, chlorine dioxide, chlorine, and monochloramine on *Cryptosporidium parvum* oocyst viability." *Applied and Environmental Microbiology,* 56(5):1423–1428, 1990.

Kouame, Y., and C. N. Haas. "Inactivation of *E. coli* by Combined Action of Free Chlorine and Monochloramine." *Water Research,* 25:1027–1032, 1991.

Krasner, S. Presented at the AWWA Annual Conference, Denver, CO, 1986.

Laubusch, E. "Sulfur Dioxide." *Public Works,* 94(8):117, 1963.

Lauer, W. C. et al. "Experience with Chlorine Dioxide at Denver's Reuse Plant." *Journal of the American Water Works Association,* 78(6):79, 1986.

Lawler, D. F., and P. C. Singer. "Analyzing Disinfection Kinetics and Reactor Design: A Conceptual Approach Versus the SWTR." *Journal AWWA,* 85(11):67–76, 1993.

Leahy, J. MS Thesis. Ohio State University, 1985.

LeChevallier, M. W., C. D. Lowrey, and R. G. Lee. "Disinfecting Biofilms in a Model Distribution System." *Journal AWWA,* 82(6):87–99, 1990.

LeChevallier, M. W., W. D. Norton, and R. G. Lee. "*Giardia* and *Cryptosporidium* spp. in Filtered Drinking Water Supplies." *Applied and Environmental Microbiology,* 57:2617–2621, 1991.

Leland, D., J. McAnulty, W. Keene, and G. Stevens. "A Cryptosporidiosis Outbreak in a Filtered-Water Supply." *Journal AWWA,* 85(6):34–42, 1993.

Lev, O., and S. Regli. "Evaluation of Ozone Disinfection Systems—Characteristic Concentration-C." *Journal of Environmental Engineering, ASCE,* 118(4):477–494, 1992.

Levenspiel, O. *Chemical Reaction Engineering* (2nd ed.). John Wiley & Sons, New York, 1972.

Levine, W. C., W. T. Stephenson, and G. F. Craun. "Waterborne Disease Outbreaks, 1986–1988." *Mmwr Cdc Surveill Summ.* 39(1):1–13, 1990.

Leyval, C. *Environmental Technology Letters,* 5(8):359, 1984.

Lykins, B., and M. Griese. "Using Chlorine Dioxide for Trihalomethane Control." *Journal AWWA,* 78(6):88, 1986.

Lyn, D. A., K. Chiu, and E. R. Blatchley, III. "Numerical Modeling of Flow and Disinfection in UV Disinfection Channels." *Journal of Environmental Engineering,* 125(1):17–26, 1999.

Mac Kenzie, W. R., N. J. Hoxie, M. E. Proctor, M. S. Gradus, K. A. Blair, D. E. Peterson, J. J. Kazmierczak, K. R. Fox, D. G. Addias, J. B. Rose, and J. P. Davis. "Massive Waterborne Outbreak of *Cryptosporidium* Infection Associated with a Filtered Public Water Supply, Milwaukee, Wisconsin, March and April 1993." *New England Journal of Medicine,* 331(3):161–167, 1994.

Malcolm Pirnie and HDR Engineering. *Guidance Manual for Compliance with the Filtration and Disinfection Requirements for Public Water Systems Using Surface Water Sources.* American Water Works Association, USA, 1991.

Manley, T. C., and S. Niegowski. In *Kirk-Othmer Encyclopedia of Chemical Technology* (2nd ed.), vol. 14. R. E. Kirk et al., eds., Wiley, New York, 1967.

Martin, N., M. Benezet-Toulze, C. Laplace, M. Faivre, and B. Langlais. "Design and Efficiency of Ozone Contactors for Disinfection." *Ozone Science and Engineering,* 14(5):391–405, 1992.

Masschelein, W. *Chlorine Dioxide.* Ann Arbor Science, Ann Arbor, MI, 1979a.

Masschelein, W. *Use of Chlorine Dioxide for the Treatment of Drinking Water, Oxidation Techniques in Drinking Water Treatment.* EPA 570/9-79-020. USEPA, Cincinnati, OH, 1979b.

McKee, J. "Chemical and Colicidal Effects of Halogens in Sewage." *Journal of the Water Pollution Control Federation,* 32:795, 1960.

McFeters, G. A. et al. "Injured Coliforms in Drinking Water." *Applied and Environmental Microbiology,* 51:1–5, 1986.

Means, E., K. Scott, M. Lee, and R. Wolfe. Presented at the American Water Works Association Annual Conference, Denver, CO, 1986a.

Means, E., T. Tanaka, D. Otsuka, and M. McGuire. "Effect of Chlorine and Ammonia Application Points on Bactericidal Efficiency." *Journal AWWA,* 78(1):62, 1986b.

Medir, M., and F. Giralt. "Stability of Chlorine Dioxide in Aqueous Solution." *Water Research,* 16:1379, 1982.

Melnick, J. L., C. P. Gerba, and C. Wallis. "Viruses in Water." *Bulletin of the World Health Organization,* 56(4):499–508, 1978.

Michalek, S., and F. Leitz. "On Site Generation of Hypochlorite." *Journal of the Water Pollution Control Federation,* 44(9):1697, 1972.

Milbauer, R., and N. Grosswicz. "Reactivation of Chlorine-Inactivated *Escherichia coli.*" *Applied Microbiology,* 7:67, 1959a.

Milbauer, R., and N. Grosswicz. "Effect of Growth Conditions on Chlorine Sensitivity of *Escherichia coli.*" *Applied Microbiology,* 7:71–74, 1959b.

Miller, D. G. "A Survey of Cryptosporidium Oocysts in Surface and Groundwaters in the UK." *Journal of the Institution of Water and Environmental Management,* 6:697–703, 1992.

Miller, G. W. et al. *An Assessment of Ozone and Chlorine Dioxide for Treatment of Municipal Water Supplies.* EPA 600/8-78-018. USEPA, 1978.

Morin, R. A. et al. "Ozone Disinfection Pilot Plant Studies at Laconia, New Hampshire." *Journal of the New England Water Works Association:* 206, 1975.

Morris, J. "The Acid Ionization Constant of HOCl from 5°C to 35°C." *Journal of Physical Chemistry,* 70:3798, 1966.

Morris, J. "Kinetics of Reactions Between Aqueous Chlorine and Nitrogen Compounds." In *Principles and Applications of Water Chemistry* (S. Faust, ed.). Wiley, New York, 1967.

Morris, J. C. "The Mechanism of the Hydrolysis of Chlorine." *Journal of the American Chemical Society,* 68:1692–1694, 1946.

Morris, J. C. "Chlorination and Disinfection—State of the Art." *Journal AWWA,* 63(12):769, 1971.

Morris, J. C., and R. A. Isaac. "A Critical Review of Kinetics and Thermodynamic Constants for Aqueous Chlorine–Ammonia Systems," In *Water Chlorination: Environmental Impact and Health Effects,* vol. 4(#1) (R. L. Jolley et al. (eds.)). Ann Arbor Science Publishers, Ann Arbor, MI, p. 49–63, 1983.

Mosley, J. "Transmission of Viral Diseases by Drinking Water," In *Transmission of Viruses by the Water Route,* (G. Berg, ed.). Wiley, New York, 1966.

Murphy, K. "Effect of Chlorination Practice on Soluble Organics." *Water Research,* 9:389, 1975.

Nasser, A. M. "Prevalence and Fate of Hepatitis A in Water." *Critical Reviews in Environmental Science and Technology,* 24(4):281, 1994.

Nathanson, N., and A. D. Langmuir. "The Cutter Incident: Poliomyelitis Following Formaldehyde-Inactivated Poliovirus Vaccination in the United States During the Spring of 1955. II. Relationship of Poliomyelitis to Cutter Vaccine." *American Journal of Epidemiology,* 78:29–60, 1963.

Nauman, E. B., and B. A. Buffham. *Mixing in Continuous Flow Systems.* John Wiley & Sons, New York, 1983.

Nissen, F. *Zeitsch. für Hygiene.* 8:62, 1890.

Noack, M., and R. Doeff. "Chlorine Dioxide, Chlorous Acid and Chlorites," In *Kirk-Othmer Encyclopedia of Chemical Technology* (3rd ed.), vol. 5 (H. Mark, ed.). John Wiley, New York, 1979.

Noack, M., and R. Doeff. "Reactions of Chlorine, ClO_2 and Mixtures Thereof with Humic Acid," In *Water Chlorination, Environmental Impact and Health Effects,* vol. 2 (R. Jolley et al., ed.). Ann Arbor Science, Ann Arbor, MI, 1981.

Noss, C., W. Dennis, and V. Olivieri. "Reactivity of Chlorine Dioxide with Nucleic Acids and Proteins," In *Water Chlorination: Environmental Impact and Health Effects* (R. Jolley, ed.). Lewis Publishers, Chelsea, MI, 1985.

O'Brien, W. J., G. L. Hunter, J. J. Rosson, R. A. Hulsey, and K. E. Carns. Presented at the Disinfecting Wastewater for Discharge and Reuse, Portland, OR, 1996.

Olivieri, V. et al. "Inactivation of Virus in Sewage." *ASCE Journal of the Sanitary Engineering Division,* 97(5):661, 1971.

Olivieri, V. P. et al. "Reaction of Chlorine and Chloramines with Nucleic Acids Under Disinfection Conditions," In *Water Chlorination: Environmental Impact and Health Effects* (R. Jolley, ed.). Butterworths, Stoneham, MA, 1980.

Olivieri, V. P. et al. "Mode of Action of Chlorine Dioxide on Selected Viruses," In *Water Chlorination: Environmental Impact and Health Effects* (R. Jolley, ed.). Lewis Publishers, Chelsea, MI, 1985.

Palin, A. T. *Chemistry and Control of Modern Chlorination.* La Motte Chemical Products Co., Chestertown, MD, 1983.

Palmer, S. R. et al. "Waterborne Outbreak of *Campylobacter* Gastroenteritis." *Lancet,* 8319:287–90, 1983.

Payment, P., and E. Franco. "*Clostridium perfringens* and Somatic Coliphages as Indicators of the Efficiency of Drinking Water Treatment for Viruses and Protozoan Cysts." *Applied and Environmental Microbiology,* 59(8):2418–2424, 1993.

Payment, P., M. Trudel, and R. Plante. "Elimination of Viruses and Indicator Bacteria at Each Step of Treatment During Preparation of Drinking Water at Seven Water Treatment Plants." *Applied Environmental Microbiology,* 49:1418, 1985.

Phelps, E. "The Disinfection of Sewage and Sewage Filter Effluents." *USGS Water Supply Paper,* 229:1909.

Pressley, T., F. Dolloff, and S. Roan. "Ammonia Nitrogen Removal by Breakpoint Chlorination." *Environmental Science and Technology,* 6:622, 1972.

Race, J. *Chlorination of Water.* Wiley, New York, 1918.

Rapson, W. H. "From Laboratory Curiosity to Heavy Chemical." *Chemistry Canada,* 18(1):2531, 1966.

Rav-Acha, C. "The Reactions of Chlorine Dioxide with Aquatic Organic Materials and Their Health Effects." *Water Research,* 18:1329, 1984.

Reasoner, D. J. Presented at the Water Quality Technology Conference, American Water Works Association, Orlando, Florida, 1991.

Reeve, G. D. L., J. Martin, R. E. Pappas, R. E. Thompson, and K. D. Greene, "An Outbreak of Shigellosis Associated with the Consumption of Raw Oysters." *New England Journal of Medicine,* 321(4):224–7, 1989.

Regli, S., A. Amirtharajah, B. Borup, C. Hibler, J. Hoff, and R. Tobin. "Panel Discussion on the Implications of Regulatory Changes for Water Treatment in the United States." *Advances in Giardia Research,* 275, 1988.

Renton, J. I., W. Moorehead, and A. Ross. "Longitudinal Studies of *Giardia* Contamination in Two Community Drinking Water Supplies: Cyst Levels, Parasite Viability and Health Impact." *Applied and Environmental Microbiology,* 62(1):47–54, 1996.

Rice, E. W., and J. C. Hoff. "Inactivation of *Giardia lamblia* Cysts by Ultraviolet Irradiation." *Applied and Environmental Microbiology,* 42(3):546–547, 1981.

Rice, E. W., J. C. Hoff, and F. W. Schaefer, III. "Inactivation of *Giardia* Cysts by Chlorine." *Applied and Environmental Microbiology,* 43(1):250–251, 1982.

Rice, R. G., C. Robson, G. Miller, and A. Hill. "Uses of Ozone in Drinking Water Treatment." *Journal of the American Water Works Association,* 73(1):44, 1981.

Richardson, A. J., R. A. Frankenberg, A. C. Buck, J. B. Selkon, J. S. Colbourne, J. W. Parsons, and W. R. Mayon "An Outbreak of Waterborne Cryptosporidiosis in Swindon and Oxfordshire." *Epidemiology and Infection,* 107(3):485–495, 1991.

Rideal, S. "Application of Electrolytic Chlorine to Sewage Purification and Deodorization in the Dry Chlorine Process." *Transactions of the Faraday Society,* 4:179, 1908.

Rose, J. B., C. P. Gerba, and W. Jakubowski. "Survey of Potable Water Supplies for *Cryptosporidium* and *Giardia.*" *Environmental Science and Technology,* 25:1393–1400, 1991.

Rose, J. B. et al. "Isolating Viruses from Finished Water." *Journal of the American Water Works Association,* 78(1):56, 1986.

Rosenberg, M. L., K. K. Hazlet, J. Schaefer, J. G. Wells, and R. C. Pruneda. "Shigellosis from Swimming." *Journal of the American Medical Association,* 236(16):1849–1852, 1976.

Rosenberg, M. L. et al. "The Risk of Acquiring Hepatitis from Sewage Contaminated Water." *American Journal of Epidemiology,* 112:17, 1980.

Roth, J. *Ozone Solubilities.* Pergamon, New York, 1984.

Roustan, M., Z. Stambolieva, J. Duget, O. Wable, and J. Mallevialle. "Influence of Hydrodynamics on *Giardia* Cyst Inactivation by Ozone. Study by Kinetics and by 'CT' Approach." *Ozone Science and Engineering,* 13(4):451–462, 1991.

Roy, D. PhD Thesis. University of Illinois at Urbana-Champaign, 1979.

Roy, D., R. S. Engelbrecht, and E. S. K. Chian. "Kinetics of Enteroviral Inactivation by Ozone." *Journal of the Environmental Engineering Division, ASCE,* 107(EE5):887–899, 1981a.

Roy, D. et al. "Mechanism for Enteroviral Inactivation by Ozone." *Applied and Environmental Microbiology,* 41(3):718, 1981b.

Rubin, A. J. et al. "Disinfection of Amoebic Cysts in Water with Free Chlorine." *Journal of the Water Pollution Control Federation,* 55(9):1174, 1983.

Rush, B. A., P. A. Chapman, and R. W. Ineson. "A probable waterborne outbreak of cryptosporidiosis in the Sheffield area." *Journal of Medical Microbiology,* 32(4):239–242, 1990.

Saguinsin, J. L. S., and J. C. Morris. "The Chemistry of Aqueous Nitrogen Trichloride," In *Disinfection: Water and Wastewater* (J. Johnson, ed.). Ann Arbor Science, Ann Arbor, MI, p. 277–279, 1975.

Saunier, B. M., and R. E. Selleck. "The Kinetics of Breakpoint Chlorination in Continuous Flow Systems." *Journal of the American Water Works Association,* 71:164–172, 1979.

Scarpino, P. V. et al. "A Comparative Study of the Inactivation of Viruses in Water by Chlorine." *Water Research,* 6:959, 1972.

Scarpino, P. V., et al. Presented at the Water Quality Technology Conference, American Water Works Association, Kansas City, MO, 1977.

Scarpino, P. V. et al. *Effect of Particulates on Disinfection of Enteroviruses and Coliform Bacteria in Water by Chlorine Dioxide.* EPA-600/2-79-054. USEPA, 1979.

Scheible, O., and C. Bassell. *Ultraviolet Disinfection of a Secondary Wastewater Treatment Plant Effluent.* EPA-600/1-81-139. USEPA, 1981.

Scheible, O., A. Gupta, and D. Scannel. *Ultraviolet Disinfection Technology Assessment,* EPA 832-R-92-004. USEPA, Office of Water, 1992.

Scheible, O. K. Presented at the Planning, Design and Operations of Effluent Disinfection Systems, Whippany, NJ, 1993.

Selleck, R., H. Sollins, and G. White. "Kinetics of Bacterial Deactivation with Chlorine." *ASCE Journal of the Environmental Engineering Division,* 104:1197, 1978.

Severin, B. F., M. T. Suidan, and R. S. Engelbrecht. "Kinetic Modeling of UV Disinfection of Water." *Water Research,* 17(11):1669–1676, 1983a.

Severin, B. F., M. T. Suidan, and R. S. Engelbrecht. "Series-Event Kinetic Model for Chemical Disinfection." *Journal of Environmental Engineering,* 110(2):430–439, 1984a.

Severin, B. F. et al. "Effects of Temperature on Ultraviolet Light Disinfection." *Environmental Science and Technology,* 17:717, 1983b.

Severin, B. F. et al. "Inactivation Kinetics in a Flow-through UV Reactor." *Journal of the Water Pollution Control Federation,* 56(2):164–169, 1984b.

Sikorowska, C. "Influence of Pollutions on Chlorine Dioxide Demand of Water." *Gig. Woda Tech. Sanita,* 35(12):4645, 1961.

Singh, A., R. Yeager, and G. McFeters. "Assessment of in vivo Revival, Growth and Pathogenicity of *Escherichia coli* Strains after Copper and Chlorine Induced Injury." *Applied and Environmental Microbiology,* 52(4):832, 1986.

Smith, H. V. "*Cryptosporidium* and Water—A Review." *Journal of the Institution of Water and Environmental Management,* 6(4):443–451, 1992.

Snead, M. C. et al. *Benefits of Maintaining a Chlorine Residual in Water Supply Systems.* EPA-600/2-80-010. USEPA, 1980.

Snoeyink, V., and M. Suidan. "Dechlorination by Activated Carbons and Other Dechlorinating Agents." In *Disinfection: Water and Wastewater* (J. Johnson, ed.). Ann Arbor Science, Ann Arbor, MI, 1975.

Sproul, O. J. et al. Presented at the American Water Works Association Annual Conference, Atlantic City, NJ, 1978.

Stagg, C. et al. "Chlorination of Solids Associated Coliphages." *Progress in Water Technology,* 10(12):3817, 1978.

Stefan, H. G., T. R. Johnson, H. L. McConnell, C. T. Anderson, and D. R. Martenson. "Hydraulic Modeling of Mixing in Wastewater Dechlorination Basin." *Journal of Environmental Engineering (ASCE),* 116(3):524–541, 1990.

Stover, E. L., C. N. Haas, K. L. Rakness, and O. K. Scheible. *Design Manual: Municipal Wastewater Disinfection.* USEPA, Cincinnati, OH, 1986.

Taras, M. J. "Preliminary Studies on the Chlorine Demand of Specific Chemical Compounds." *Journal AWWA* 42:462–473, 1950.

Teefy, S. M., and P. C. Singer. "Performance and Analysis of Tracer Tests to Determine Compliance of a Disinfection Scheme with the SWTR." *Journal AWWA,* 82(12):88–98, 1990.

Tenno, K., R. Fujioka, and P. Loh. "The Mechanism of Poliovirus Inactivation by Hypochlorous Acid," In *Water Chlorination: Environmental Impact and Health Effects* (R. Jolley, ed.). Butterworths, Stoneham, MA, 1980.

Trofe, T. W. "Kinetics of Monochloramine Decomposition in the Presence of Bromine." *Environmental Science and Technology,* 14:544, 1980.

Trussell, R., and P. Kreft. Presented at the AWWA Seminar Proceedings: Chloramination for THM Control, Dallas, Texas, 1984.

Trussell, R. R. "Safeguarding Distribution System Integrity." *Journal of the American Water Works Association,* 91(1):46–54, 1999.

Trussell, R. R., and J.-L. Chao. "Rational Design of Chlorine Contact Facilities." *Journal of the Water Pollution Control Federation,* 49(7):659–667, 1977.

Valentine, R. "Disappearance of Monochloramine in the Presence of Nitrite," In *Water Chlorination: Environmental Impact and Health Effects,* vol. 5 (R. Jolley, ed.). Lewis Publishers, Chelsea, MI, 1985.

Valentine, R. "A Spectrophotometric Study of the Formation of an Unidentified Monochloramine Decomposition Product." *Water Research,* 20(8):1067, 1986.

Valentine, R. L., and C. T. Jafvert. "General Acid Catalysis of Monochloramine Disproportionation." *Environmental Science and Technology,* 22:691, 1988.

Venkobachar, C. et al. "Mechanism of Disinfection." *Water Research,* 9:119, 1975.

Venkobachar, C. et al. "Mechanism of Disinfection: Effect of Chlorine on Cell Membrane Functions." *Water Research,* 11:727, 1977.

von Gunten, U., and J. Hoigne. "Bromate Formation During Ozonation of Bromide-Containing Waters: Interaction of Ozone and Hydroxyl Radical Reactions." *Environmental Science and Technology,* 28(7):1234–1242, 1994.

Wajon, J. "Oxidation of Phenol and Hydroquinone by Chlorine Dioxide." *Environmental Science and Technology,* 16:396, 1982.

Ward, N., R. Wolfe, and B. Olson. "Disinfectant Activity of Inorganic Chloramines with Pure Culture Bacteria: Effect of pH, Application Technique and Chlorine to Nitrogen Ratio." *Applied and Environmental Microbiology,* 48:508, 1984.

Water Pollution Control Federation. *Wastewater Disinfection, Manual of Practice FD-10.* Water Pollution Control Federation, Alexandria, VA, 1984.

Watson, H. E. "A Note on the Variation of the Rate of Disinfection with Change in the Concentration of the Disinfectant." *Journal of Hygiene,* 8:536–542, 1908.

Wattie, E., and C. Butterfield. "Relative Resistance of *Escherichia coli* and *Eberthella typhosa* to Chlorine and Chloramines." *US Public Health Service Reports,* 59:1661, 1944.

Weber, G. R. et al. "Effect of Ammonia on the Germicidal Efficiency of Chlorine in Neutral Solutions." *Journal AWWA,* 32:1904, 1940.

Weil, I., and J. Morris. "Equilibrium Studies on N-Chloro Compounds." *Journal of the American Chemical Society,* 71:3, 1949a.

Weil, I., and J. C. Morris. "Kinetic Studies on the Chloramines. I. The Rates of Formation of Monochloramine, N-Chloromethylamine and N-Chlorodimethylamine." *Journal of the American Chemical Society,* 71:1664–1671, 1949b.

Werderhoff, K., and P. Singer. "Chlorine Dioxide Effects on THMFP, TOXFP and the Formation of Inorganic Byproducts." *Journal AWWA,* 79(9):107–113, 1987.

White, G. C. *Handbook of Chlorination.* Van Nostrand, New York, 1972.

White, G. C. *Disinfection of Wastewater and Water for Reuse.* Van Nostrand Reinhold, New York, 1978.

Wickramanayake, G. B., A. J. Rubin, and O. J. Sproul. "Effects of Ozone and Storage Temperature on *Giardia* Cysts." *Journal of the American Water Works Association,* 77(8):74–77, 1985.

Wojtowicz, J. A. "Chlorine Monoxide, Hypochlorous Acid, and Hypochlorites," In *Kirk-Othmer Encyclopedia of Chemical Technology* (3d ed.). vol. 5 (R. E. Kirk et al., eds.). Wiley, New York, p. 580–611, 1979.

Wolfe, R., and B. Olson. "Inability of Laboratory Models to Accurately Predict Field Performance of Disinfectants," In *Water Chlorination: Environmental Impact and Health Effects* (R. Jolley, ed.). Lewis Publishers, Chelsea, MI, 1985.

Wolfe, R., N. Ward, and B. Olson. "Inorganic Chloramines as Drinking Water Disinfectants: A Review." *Journal of the American Water Works Association,* 76(5):74, 1984.

Wright, N. "The Action of Hypochlorites on Amino Acids and Proteins: The Effects of Acidity and Alkalinity." *Biochemical Journal,* 30:1661, 1936.

Young, D., and D. Sharp. "Partial Reactivation of Chlorine Treated Enterovirus." *Applied and Environmental Microbiology,* 37(4):766, 1979.

Zhou, H., and D. Smith. "Kinetics of Ozone Disinfection in Completely Mixed System." *Journal of Environmental Engineering,* 120(4):841–858, 1994.

Zhou, H., and D. W. Smith. "Evaluation of Parameter Estimation Methods for Ozone Disinfection Kinetics." *Water Research,* 29(2):679–686, 1995.

CHAPTER 15
WATER FLUORIDATION

Thomas G. Reeves, P.E.
National Fluoridation Engineer
U.S. Public Health Service
Centers for Disease Control and Prevention
Atlanta, Georgia

Fluoridation of public water supplies has been practiced since 1945. Few public health measures have been accorded greater clinical and laboratory research, epidemiological study, clinical trials, and public attention than water fluoridation. This chapter will present an overview of the history of fluoridation, as well as the public health and engineering aspects.

Fluoridation is the deliberate adjustment of the fluoride concentration of a public water supply in accordance with scientific and medical guidelines. Fluoride, a natural trace element, is present in small but widely varying amounts in practically all soils, water supplies, plants, and animals, and is a normal constituent of all diets (Hodges and Smith, 1965). The highest concentrations in mammals are found in bones and teeth. Virtually all public water supplies in the United States contain at least trace amounts of fluoride from natural sources.

HISTORY

The study of the relationship between fluoride in drinking water and dental health has an interesting and intriguing history. The series of studies that led to a demonstration that fluoridated water had caries-inhibiting properties was one of the most extensive programs carried out in the epidemiology of chronic disease. It began in 1901, when a U.S. Public Health Service (USPHS) physician stationed in Naples, Italy wrote that black teeth observed in emigrants from a nearby region were popularly believed to have been caused by using water charged with volcanic fumes. It was later determined that the water supply contained an extremely high amount of fluoride and that everyone drinking it was afflicted with discolored (or "mottled") teeth, a condition referred to as *dental fluorosis.*

In its mildest form, dental fluorosis is characterized by very slight opaque, whitish areas on some posterior teeth. As the defect becomes more severe, discoloration is more widespread, and changes in color range from shades of gray to black. In the most severe cases, gross calcification defects occur, resulting in pitting of the enamel. In some of the latter cases, teeth are subject to such severe attrition that they wear down to the gum line, and complete dentures must be obtained.

In 1916 Dr. Frederick S. McKay, a practicing dentist, reported that many of his patients in Colorado Springs, Colorado had this defect (McClure, 1970). After further study, McKay concluded that the condition was caused by an undetermined substance in the drinking water. McKay recommended that the water supply of Oakley, Idaho be changed because of the high incidence of such dental defects among the children there. The supply was changed in 1925 to a nearby spring that had been used by a few other children whose teeth were not discolored.

The cause of dental fluorosis was discovered almost simultaneously in the late 1920s by two different groups of scientists working independently with different tools and methods. A. W. Petrey, a chemist and head of the testing division of the laboratory at the Aluminum Company of America at Pittsburgh, noticed the calcium fluoride band in a spectroscopic examination for aluminum in a water sample from Bauxite, Arkansas. The chief chemist of these laboratories, H. V. Churchill, reported in 1931 that similar examinations of water samples from areas where dental fluorosis was endemic invariably showed the presence of fluoride (McNeil, 1957).

In 1931 a paper by Churchill describing the relationship between fluoride and dental fluorosis appeared in the *Journal of the American Water Works Association* (Churchill, 1931). Churchill reported that endemic regions of mottling had waters containing 2 mg/L or more fluoride, while areas without mottling had water supplies with less than 1.0 mg/L. Also in 1931, H. V. Smith, M. C. Smith, and E. M. Lantz at the University of Arizona helped establish the cause of mottling by duplicating the condition in rats by feeding concentrated, naturally fluoridated water and comparing the results with the mottling observed when a diet high in fluorides was used (McNeil, 1957). The Smiths also confirmed Churchill's findings by reporting that water sources from areas with no endemic mottling contained less than 0.72 mg/L fluoride (Smith et al., 1931).

The work by Churchill and the University of Arizona investigators was followed by epidemiological studies by H. T. Dean of the USPHS. Dean confirmed that many localities have water supplies containing fluoride. Areas with the largest number of such supplies containing the highest levels of fluorides include those states running from North Dakota to Texas, those along the Mexican border, and Illinois, Indiana, Ohio, and Virginia. Similar supplies have been found in the British West Indies, China, Holland, Italy, Mexico, North Africa, South America, Spain, and India.

Through observation of thousands of children in communities with varying fluoride levels, Dean developed what he termed a *mottled enamel index*—a numerical method for measuring the severity of fluorosis (Dean et al., 1936). Using this index, Dean established the fluoride level below which the use of such water contributed no significant discoloration. This level in the latitude of Chicago was about 1.0 mg/L.

Many investigators, including McKay, observed during the 1920s that less decay occurred in children whose teeth were afflicted with mottling. To confirm this, Dean examined 7257 children in 21 cities with water supplies containing varying fluoride levels. Results of this study, some of which are shown in Figure 15.1, revealed a remarkable relationship between waterborne fluorides, fluorosis, and caries incidence (McClure, 1943). Three conclusions were drawn from Dean's study:

1. When the fluoride concentration exceeds about 1.5 mg/L, any further increase does not significantly decrease the decayed, missing, and filled (DMF) tooth incidence, but does increase the occurrence and severity of mottling.

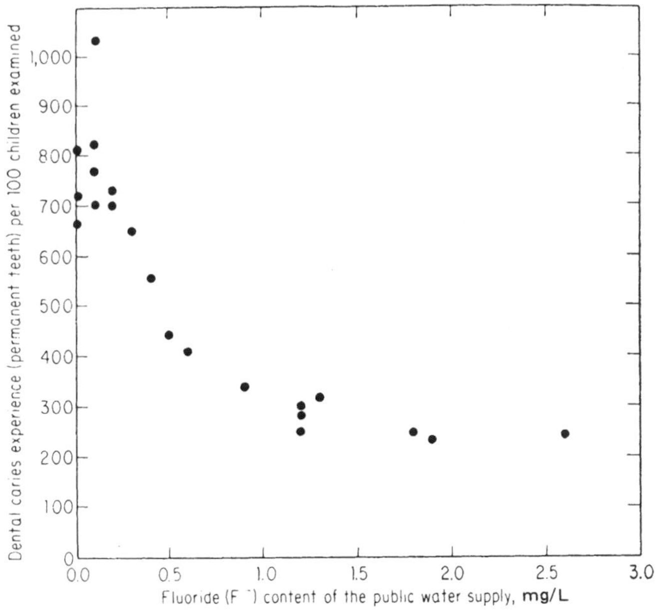

FIGURE 15.1 Relation between amount of dental caries (permanent teeth) observed in 7257 selected 12- to 14-year-old white schoolchildren in 21 cities in four states and the fluoride content of public water supplies. (*Source:* McClure, 1943.)

2. At a fluoride concentration of about 1.0 mg/L, the optimum occurs—maximum reduction in caries with no aesthetically significant mottling. At this level, DMF tooth rates were reduced by 60 percent among the 12- to 14-year-old children.
3. At fluoride concentrations below 1.0 mg/L, some benefits occur, but caries reduction is not as great and decreases as the fluoride level decreases until, as zero fluoride is approached, no observable improvement occurs.

Studies on fluoride were interrupted by World War II, but in 1945 and 1947, four classic studies were initiated with the intent to demonstrate conclusively the benefits of adding fluoride to community drinking water (Dean et al., 1950; Ast et al., 1956; Brown and Poplove, 1963; Hill et al., 1961). Fluoridation began in January 1945 in Grand Rapids, Michigan; in May 1945 in Newburgh, New York; in June 1945 in Brantford, Ontario; and in February 1947 in Evanston, Illinois. When caries rates in these localities were compared to those in a nonfluoridated "control city," a 50 to 65 percent reduction in dental caries was found in the fluoridated cities without evidence of any adverse effects. These initial studies established fluoridation as a practical and effective public health measure that would prevent dental caries.

Once the safety and effectiveness of fluoridation had been established, engineering aspects needed to be developed before community water fluoridation could be implemented. In the 1950s and 1960s, Franz J. Maier, a sanitary engineer, and Ervin Bellack, a chemist, both with the USPHS, made major contributions to the engineering aspects of water fluoridation. Maier and Bellack helped determine which chemicals were the most practical to use in water fluoridation, the best mechanical equipment to use, and the best process controls. Bellack also contributed to major advances in fluoride testing. In 1963, Maier published the first comprehensive book on the technical aspects of fluoridation (Maier, 1963) and, in 1972, Bellack, then with the U.S. Environmental Protection Agency (USEPA), published an engineering manual (Bellack, 1972) that was replaced in 1986 by the manual entitled *Water Fluoridation—A Training Manual for Engineers and Technicians* (Reeves, 1986).

Over the past 40 years, fluoride and fluoridation have been the subject of numerous studies undertaken by the USPHS, state health departments, and nongovernmental research organizations. Since 1970, over 3700 such studies have been conducted (Michigan Department of Public Health, 1979; Newbrun, 1989; Ripa, 1993). These studies have strongly supported the beneficial effect of water fluoridation.

Water fluoridation has been supported for approximately 50 years as a public policy by the U.S. Government. Both the U.S. Environmental Protection Agency and the U.S. Public Health Service continue to recommend water fluoridation as an effective means to help prevent dental caries (USPHS, 1992, 1995).

High levels of fluoride in drinking water have been found to cause adverse health effects. As a result, USEPA has established regulatory limits on the fluoride content of drinking water. On the basis of a detailed review of health effects studies on fluoride, (National Research Council, 1993) USEPA set a maximum contaminant level of 4 mg/L in water systems to prevent crippling skeletal fluorosis (Federal Register, National Primary and Secondary Drinking Water Regulations, 1986). A secondary level of 2 mg/L was established by USEPA to protect against objectionable dental fluorosis. These limits, to be reviewed by USEPA every three years, or as new health data becomes available, primarily impact systems that have naturally high fluoride levels.

RECENT STUDIES

Two different studies reviewed the complete health effects of water fluoridation. The *Review of Fluoride—Benefits and Risks,* a report of the Ad Hoc Subcommittee on Fluoride of the Committee to Coordinate Environmental Health and Related Programs, was published in February 1991 (Public Health Service, 1991). This report is a comprehensive review and evaluation of the public health benefits and risks of fluoride in drinking water and other sources. It was prompted by a study of the National Toxicology Program that found "equivocal evidence" of carcinogenicity based on the occurrence of a small number of malignant bone tumors in male rats. This report recommends the continued use of fluoride to prevent dental caries and the continued support of optimal fluoridation of drinking water. It also recommends scientific conferences to determine an optimal level of fluoride exposure from all sources combined, not only from drinking water. The CDC in Atlanta is involved with initiating these recommendations.

A second study, *Health Effects of Ingested Fluoride,* was the report of the Subcommittee of Health Effects of Ingested Fluoride of the National Research Council (National Research Council, 1993). Members of the National Research Council are drawn from the councils of the National Academy of Sciences, the National Academy of Engineering, and the Institute of Medicine. The report was the result of a request by the U.S. Environmental Protection Agency (EPA) to review the health effects of ingested fluoride and determine whether EPA's maximum contaminant level (MCL) of 4 mg/L was appropriate. The subcommittee conducted a detailed examination of data for the intake, metabolism, and disposition of fluoride; dental fluorosis; bone strength and risk of bone fracture; effects on the renal, gastrointestinal, and immune systems; reproductive effects in animals; genotoxicity; and carcinogenicity in animals and humans. There was no evidence that fluoride affected hypersensitive individuals, but the report suggested that more research is needed in this area. The report concluded that EPA's current MCL of 4 mg/L for fluoride in drinking water was appropriate. The subcommittee did recommend additional research in various areas, as it felt the fluoride data was incomplete.

Concerns have recently been expressed that increases in the prevalence of mild or very mild fluorosis are occurring in communities with negligible and optimal water fluoride concentrations, because of increased total fluoride consumption from various sources (Leverett, 1986). It is true, as Corbin (1989) has written, that it is "virtually impossible to find 'non-fluoride' communities due to the many opportunities for alternative exposures to fluoride," i.e., fluoridated toothpaste, mouth rinses, beverages, and others. Ripa (1993) states that "communities in the U.S. still may be classified as being optimal fluoridated or fluoride-deficient based upon the concentration of fluoride in the drinking water. However, because fluoride is ubiquitous in food and dental health products, practically no American today is unexposed to fluoride." The goal in water fluoridation, as it always has been, is to obtain the lowest rate of tooth decay with the least amount of dental fluorosis.

PRESENT STATUS OF FLUORIDATION

The Centers for Disease Control and Prevention (CDC) estimated that approximately 145 million Americans, or about 62 percent of those served by public water supplies, consumed fluoridated water daily as of January 1, 1993 (CDC, 1993). Some 10.5 million of these people are served by naturally fluoridated supplies. Over the years the USPHS has been setting health objectives to be met for each decade. The fluoridation objective for the year 2000 states: "to increase by at least 75% the proportion of people served by community water systems providing optimal levels of fluoride." Twenty-one states and the District of Columbia provide fluoridated water to 75 percent or more of their populations. As of 1996, nine states require fluoridation, at least for cities above a minimum population. The increase in the U.S. population served by fluoridated drinking water systems is shown in Figure 15.2 (CDC, 1985b).

Data about water fluoridation in other countries is not readily available. But it is known that other countries have large populations consuming fluoridated water: Australia (67 percent), Canada (40 percent), Ireland (62 percent), Israel (42 percent), Malaysia (60 percent), New Zealand (64 percent), and the United Kingdom (10 percent) (Hammer, 1996). The city-states of Hong Kong and Singapore are totally fluoridated. Israel is planning to double the population drinking fluoridated water by the year 2000.

Some progress has been made toward achieving community fluoridation in Central and South America, especially in Brazil. Brazil requires fluoridation for all communities with populations over 50,000. There also are strong fluoridation efforts in Argentina and Chile. The Pan American Health Organization (a branch of the World Health Organization) has been very active in the promotion of fluoridation in Latin America (World Health Organization, 1984). Japan and South Korea are planning major efforts in water fluoridation.

Although community water fluoridation has been shown to be both safe and the most cost-effective method of preventing dental caries, a small percentage of the

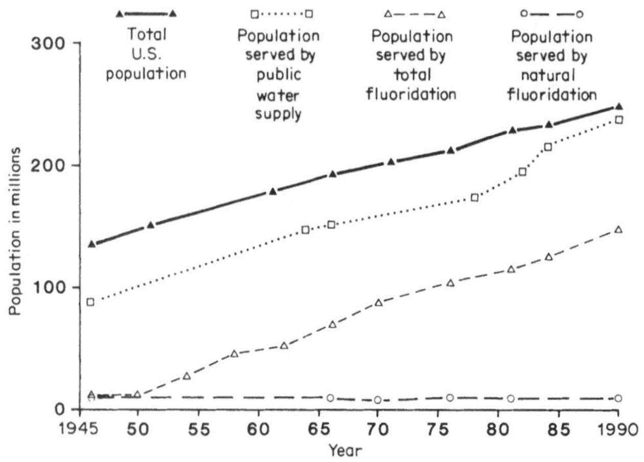

FIGURE 15.2 Fluoridation growth in the United States, 1945–1984. (*Source:* CDC, 1985b.)

population continues to oppose its introduction into community water systems. When fluoridation is being considered for adoption by a community, persons opposed to fluoridation often attempt to refute the benefits, safety, and efficacy of this effective public health measure. Charges against fluoridation and the corresponding truths have been discussed elsewhere (AWWA, 1988). Assistance in responding to false charges against fluoridation may be obtained from the Oral Health Program of the CDC, Atlanta, Georgia. The National Institute of Dental Research in Bethesda, Maryland, a branch of the National Institutes of Health, and the American Dental Association are other sources of information concerning water fluoridation.

Alternative means of providing the benefits of fluoride besides the fluoridation of municipal water supply systems are available. The six most important methods are listed in Table 15.1.

Public health authorities, however, typically recommend that these methods only be considered in situations where municipal water fluoridation is not possible. Municipal water fluoridation is the most cost-effective means available for reducing the incidence of caries in a community. Topical fluoride methods can be used in conjunction with water fluoridation (optimally fluoridated water in community or school water systems or naturally fluoridated water). Systemic fluoride methods are sufficient alone in preventing tooth decay, and other methods should not be used in conjunction with them. The cost and effectiveness of alternatives to municipal water fluoridation are shown in Table 15.2 (Gish, 1978).

TABLE 15.1 Alternative Means of Fluoride Supply

Topical methods	Systemic methods
Fluoride gels	Fluoride tablets
Fluoride mouth rinses	Fluoride drops
Fluoride dentifrices	Salt fluoridation

TABLE 15.2 Comparison of Effectiveness of Different Types of Fluoride Applications

Procedure	Cavities prevented per $100,000 spent	Cost/cavities prevented
Water fluoridation		
Municipal	500,000	$0.20
School	111,100	0.90
Topical fluorides		
Supervised application of paste or rinse in school	55,500	1.82
Professional application of topical fluoride	25,600	3.90
Systemic fluorides		
Supervised distribution of fluoride tablets in school	16,542	6.06
Individually prescribed fluoride tablets or drops	10,000	10.00

Source: Gish (1978).

THEORY

Causes of Dental Caries

Tooth decay is a complex process, and all factors involved are not entirely understood. It is usually characterized by loss of tooth structure (enamel, dentin, and cementum) as a result of destruction of these tissues by acids. Evidence indicates that acids are produced by the action of oral bacteria and enzymes on sugars and carbohydrates entering the mouth. This takes place beneath the plaque, an invisible film composed of gummy masses of microorganisms that adhere to the teeth. Oral bacteria are capable of converting some of the simpler sugars into acids, and the bacteria and enzymes acting in combination are capable of converting carbohydrates and more complex sugars into acids. The production of acids is a result of the natural existence of bacteria and enzymes in the mouth.

Until the middle or late 20th century, a very high prevalence of dental caries existed in all developed countries (Burt et al., 1992). Until water fluoridation became widespread, almost 98 out of 100 Americans experienced some tooth decay by the time they reach adulthood. The highest tooth decay activity is found in schoolchildren. Tooth decay begins in early childhood, reaches a peak in adolescence, and diminishes during adulthood (National Institute of Dental Research, 1981).

Dental Benefits of Fluoride in Drinking Water

When water containing fluoride is consumed, some fluoride (about 50 percent) is retained by fluids in the mouth and is incorporated onto the teeth by surface uptake (topical effect). The rest (about 50 percent) enters the stomach, where it is rapidly adsorbed by diffusion through the stomach walls and intestine. Fluoride enters the blood plasma and is rapidly distributed to all parts of the body, including the teeth (systemic effect). Because of the systemic effect, the fluoride ion is able to pass freely through all cell walls and is available to all organs and tissues of the body. Distributed in this fashion, the fluoride ion is available to all skeletal structures of the body in which it may be retained and stored in proportions that generally increase with age and intake.

Bones, teeth, and other parts of the skeleton tend to attract and retain fluoride. Soft tissues do not retain fluoride. Fluoride is a "bone seeker," with about 96 percent of the fluoride found in the body deposited in the skeleton. Because teeth are part of the skeletal system, incorporation of fluoride in teeth is basically similar to that in other bones. It is most rapid during the time of the child's formation and growth. The fluoride ion is actually incorporated into the apatite crystal of the tooth enamel. During formation of the tooth, the fluoride ion, F^-, replaces the hydroxyl ion, OH^-, in the crystal lattice. Thus, the enamel of the tooth is greatly strengthened so that it is more resistant to bacterial acids and inhibits the growth of certain kinds of bacteria that produce acids (Keyes, 1969; Whitford, 1996). In addition, fluoride appears to actually aid in the remineralization of teeth (Mellberg and Mallon, 1984; Silverstone, 1984). Erupted teeth differ from other parts of the skeleton in that once they are formed, with the exception of the dentin (inner part of the tooth) and the root, cellular activity virtually ceases. As a result, very little change occurs in the fluoride level in teeth after they are formed. Children must drink the proper amount of fluoridated water during early development of permanent teeth, preferably before they start school, in order to realize full benefits.

The relationship between dental caries, dental fluorosis, and fluoride level is shown in the classic chart in Figure 15.3 (Dunning, 1986). The beneficial effect of optimally fluoridated water ingested during the years of tooth development has been amply demonstrated. At the optimal concentration in potable water, fluoride will reduce dental caries from 20 to 40 percent among children who ingest this water from birth (Newbrun, 1989). Evidence that water fluoridation is effective in preventing caries has been repeatedly demonstrated, starting with the initial community trials in the United States and Canada in the 1940s. In recent years, however, the *relative* impact of water fluoridation appears to have diminished as other sources of fluoride supplementation (toothpastes, food, etc.) have increased. Continuation of benefits into adult life is inevitable. Stronger teeth result in fewer caries that require fewer and less extensive fillings, fewer extractions, and fewer artificial teeth. Fluoridated water helps prevent cavities on exposed roots as a result of receding gums in adults who develop periodontal disease. Early evidence indicated that higher levels of fluoride would strengthen bones of older people, thereby reducing the incidence of bone fractures (Jowsey et al., 1972; Riggs et al., 1982). However, this has now been shown to be untrue (National Research Council, 1993).

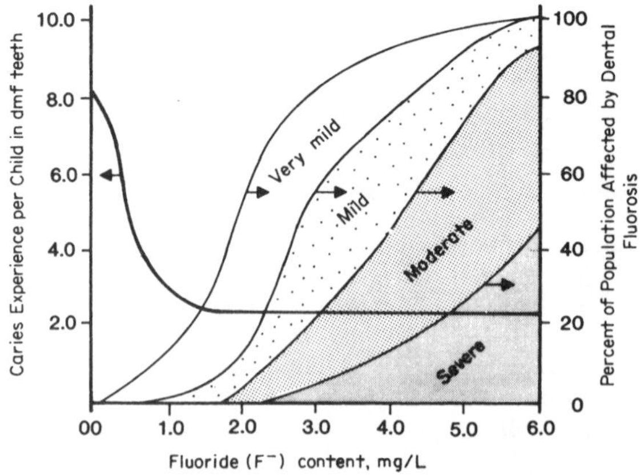

FIGURE 15.3 Dental caries and dental fluorosis in relation to fluoride in public water supplies. (*Source:* Reprinted by permission of the publisher from *Principles of Dental Public Health* by J. M. Dunning, Cambridge, MA: Harvard University Press, copyright © 1962 by the President and Fellows of Harvard College.)

Various studies in fluoridated communities over the last 30 years have shown a dramatic increase in the number of teenagers who are completely caries free. Teenagers without lifetime exposure realize benefits from fluoridation, and benefits increase for those with lifetime exposure. Conservative estimates indicate that 20 percent of the teenagers in a fluoridated community will be caries free (CDC, 1978). This is about six times as many as are caries free in a fluoride-deficient community.

Fluoridation can substantially reduce costs associated with restorative dentistry. According to the CDC, for every dollar spent (1992) on water fluoridation, a potential $80 in dental bills may be saved (CDC, 1992). In 1992, the CDC estimated the cost of fluoridation to be about the cost of a candy bar per person per year.

OPERATIONAL AND DESIGN CONSIDERATIONS

Optimal Fluoride Levels

The optimal fluoride level in water is the level that produces the greatest protection against caries with the least risk of fluorosis. Initially, this figure was obtained after examining the teeth of thousands of children living in various places with differing fluoride levels. It was not based on any direct or accurate knowledge of how much water children drank at various times and at different places. Early in such investigations, variation in the optimal figure was observed depending on the local air temperature, which had a direct bearing on the amount of water children consumed at different ages. Studies in California and Arizona, where temperatures are considerably above the average of other parts of the United States, showed a definitely lower optimal fluoride level. This was demonstrated by observing the prevalence of dental fluorosis in places with various natural fluoride levels in their public water supplies and by estimating the actual quantities of water ingested by children of various age groups and weights.

The optimal fluoride level for a water system is usually established by the appropriate state regulatory agency. Optimal fluoride concentrations and control ranges recommended by the USPHS and CDC may be used as guidelines if state limits have not been established. These levels, shown in Table 15.3 (Reeves, 1986), are based on the annual average of the maximum daily air temperature in the particular school or community.

Many water supplies contain fluorides from natural sources. For these systems, the question of the practicability of supplementing natural fluoride with enough fluoride to bring the concentration up to the optimal level must be addressed. Usually, addition of the small amount of fluoride needed to reach the optimal level can be shown to be economically justified based on the resulting benefits to the community.

Fluoride Chemicals and Chemistry

Fluorine, a gaseous halogen, is the thirteenth most abundant element found in the earth's crust. It is a pale yellow noxious gas that is highly reactive. It is the most electronegative of all elements and cannot be oxidized to a positive state. Fluorine is not found in a free state in nature, as it is a gas. It is always found in combination with chemical radicals or other elements as fluoride compounds. When dissolved in water, these compounds dissociate into ions. All fluoride chemicals commonly used in water fluoridation dissociate to a great degree.

TABLE 15.3 Optimal Fluoride Levels Recommended by the U.S. Public Health Service, Centers for Disease Control

Annual average of maximum daily air temperatures*, °F	Recommended fluoride concentration		Recommended control range			
			Community systems, mg/L		School systems, mg/L	
	Community, mg/L	School†, mg/L	0.1 below	0.5 above	20% low	20% high
50.0–53.7	1.2	5.4	1.1	1.7	4.3	6.5
53.8–58.3	1.1	5.0	1.0	1.6	4.0	6.0
58.4–63.8	1.0	4.5	0.9	1.5	3.6	5.4
63.9–70.6	0.9	4.1	0.8	1.4	3.3	4.9
70.7–79.2	0.8	3.6	0.7	1.3	2.9	4.3
79.3–90.5	0.7	3.2	0.6	1.2	2.6	3.8

* Based on temperature data obtained for a minimum of 5 years.
† Based on 4.5 times the optimum fluoride level for communities.
Source: Reeves (1986).

Fluoride can be found in a solid form in fluoride-containing minerals such as fluorspar, cryolite, and apatite. Fluorspar is a mineral containing from 30 to 98 percent calcium fluoride (CaF_2). Cryolite (Na_3AlF_6) is a compound of aluminum, sodium, and fluoride. Apatite is a mixture of calcium compounds that includes calcium phosphates, calcium fluorides, and calcium carbonates. Trace amounts of sulfates are usually present as impurities. Apatite contains from 3 to 7 percent fluoride and is the main source of fluoride used in water fluoridation at the present time. Apatite is also the raw material for phosphate fertilizers. Cryolite is not a major source of fluoride in the United States.

Fluoride is widely distributed in the lithosphere and hydrosphere. Because of the dissolving power of water and the movement of water in the hydrologic cycle, fluoride is found naturally in all waters. High concentrations of fluoride are not common in surface water, but may occur in groundwaters, hot springs, and geothermal fluids.

Fluoride forms compounds with every element except helium, neon, and argon. Polyvalent cations such as aluminum, iron, silicon, and magnesium form stable complexes with the fluoride ion. The extent to which complex formation takes place depends on several factors, including the complex stability constant, pH, concentrations of fluoride, and complexing species (Eichenberger, 1982).

Sodium fluoride, sodium fluorosilicate, and fluorosilicic acid are the three most commonly used fluoride chemicals in the United States. Standards for these chemicals are published by AWWA for use by the water industry (AWWA, 1994a–c). All chemicals used for fluoridation should be comparable in quality to the requirements of these standards.

From time to time, shortages of fluoride chemicals have occurred. Generally, most of these are not shortages at the manufacturer's plant, but temporary shortages at the local distributor level. Local shortages are usually eliminated quickly. In the past, shortages at the manufacturing level, especially of fluorosilicic acid and sodium silicofluoride, have occurred.

Fluorosilicic acid and sodium fluorosilicate (and most sodium fluoride) are byproducts of the manufacture of phosphoric acid, the main ingredient of phosphate fertilizer. Sales of fertilizer will have a direct effect on the volume of fluoride chemicals produced. In the past, slow sales of fertilizer have resulted in a temporary shortage of these two chemicals. Shortages have been relatively mild because the number of fluoridated communities was smaller and lower volumes of sodium fluorosilicate and fluorosilicic acid were needed than at the present time. Shortages occurred in 1955 to 1956, in the summer of 1969, in the spring and summer of 1974, in the summer of 1982, and in the early part of 1986, with the last one being the most severe (Reeves, 1985).

During production of fluoride chemicals, trace amounts of impurities may be introduced into the chemical, especially arsenic, lead, and/or zinc. Normally, impurities are at levels far below those that would necessitate the establishment of maximum impurity limits.

Sodium Fluoride. Sodium fluoride (NaF) is a white, odorless material available either as a powder or as crystals of various sizes. It has a molecular weight of 42.00, a specific gravity of 2.79, and a practically constant solubility of 4.0 g/100 mL (4 percent) in water at temperatures generally encountered in water treatment practice. When added to water, sodium fluoride dissociates into sodium and fluoride ions:

$$NaF \Leftrightarrow Na^+ + F^- \tag{15.1}$$

The pH of a sodium fluoride solution varies with the type and amount of impurities present. Solutions prepared from common grades of sodium fluoride have a pH near neutrality (approximately 7.6). Sodium fluoride is available in purities ranging from 97 to over 98 percent, with impurities consisting of water, free acid or alkali, sodium silicofluoride, sulfites, and iron, plus traces of other substances. Approximately 8.6 kg (19 lb) of sodium fluoride will add 1 mg/L of fluoride to 1.0 mil gal (3.8 ML) of water.

Sodium fluoride is used in the manufacture of vitrified enamel and glasses, as a steel degassing agent, in electroplating, in welding fluxes, in heat-treating salt compounds, in sterilizing equipment in breweries and distilleries, in paste and mucilage, as a wood preservative, and in the manufacture of coated paper. In the past it was also used as a rodenticide. It is no longer used as such and is not included on the Environmental Protection Agency's list of registered rodenticides.

Sodium Fluorosilicate. Sodium fluorosilicate (Na_2SiF_6) is a white, odorless crystalline material with a molecular weight of 188.06 and a specific gravity of 2.679. Its solubility varies from 0.44 g/100 mL of water at 0°C to 2.45 g/100 mL at 100°C.

When sodium fluorosilicate is dissolved in water, virtually 100 percent dissociation occurs rapidly:

$$Na_2SiF_6 \Leftrightarrow 2Na^+ + SiF_6^= \tag{15.2}$$

Fluorosilicate ions ($SiF_6^=$) may react in two ways. Most common is hydrolysis of $SiF_6^=$ releasing fluoride ions and silica (SiO_2):

$$SiF_6^= + 2H_2O \Leftrightarrow 4H^+ + 6\,F^- + SiO_2 \downarrow \tag{15.3}$$

Silica, the main ingredient in glass, is very insoluble in water. Alternatively, $SiF_6^=$ dissociates very slowly, releasing fluoride ions and the gas silicon tetrafluoride (SiF_4):

$$SiF_6^= \Leftrightarrow 2F^- + SiF_4 \uparrow \tag{15.4}$$

Silicon tetrafluoride reacts quickly with water to form silicic acid or silica:

$$SiF_4 \uparrow + 3H_2O \Leftrightarrow 4HF + H_2SiO_3 \tag{15.5a}$$

$$SiF_4 \uparrow + 2H_2O \Leftrightarrow 4HF + SiO_2 \downarrow \tag{15.5b}$$

$$HF \Leftrightarrow H^+ + F^- \tag{15.5c}$$

Solutions of sodium fluorosilicate are acidic, with saturated solutions usually exhibiting a pH of between 3 and 4 (approximately 3.6). Sodium fluorosilicate is available in purities of 98 percent or higher. Principal impurities are water, chlorides, and silica. Approximately 6.3 kg (14 lb) of sodium fluorosilicate will add 1 mg/L of fluoride to 1.0 mil gal (3.8 ML) of water.

Sodium fluorosilicate has some other industrial uses including laundry scouring (neutralizing industrial caustic soaps), the manufacture of opal glass, and mothproofing woolens. It has been used in the past as a rodenticide, but like sodium fluoride, it is no longer used in this way. The U.S. Environmental Protection Agency does not list it as a registered rodenticide. As in the case of sodium fluoride, the principal hazard associated with handling sodium fluorosilicate is dust.

Fluorosilicic Acid. Fluorosilicic acid, also known as hexafluorosilicic, silicofluoric, or hydrofluorosilicic acid (H_2SiF_6), has a molecular weight of 144.08 and is available commercially as a 20 to 35 percent aqueous solution. It is a straw-colored, transparent, fuming, corrosive liquid having a pungent odor and an irritating action on the skin. Solutions of 20 to 35 percent fluorosilicic acid have a low pH (1.2), and at a concentration of 1 mg/L of fluoride in poorly buffered potable waters, a slight depression of pH can occur. If the alkalinity of the drinking water is less than 5 mg/L as $CaCO_3$ and the pH of the water is 6.8–7.0, then adding 1 mg/L of fluoride could lower the pH of the drinking water to 6.2–6.4.

Fluorosilicic acid dissociates in solution virtually 100 percent. Its chemistry is very similar to that of Na_2SiF_6:

$$H_2SiF_6 \Leftrightarrow 2HF + SiF_4 \uparrow \tag{15.6a}$$

$$SiF_4 \uparrow + 2H_2O \Leftrightarrow 4HF + SiO_2 \downarrow \tag{15.6b}$$

$$SiF_4 \uparrow + 3H_2O \Leftrightarrow 4HF + H_2SiO_3 \tag{15.6c}$$

$$HF \Leftrightarrow H^+ + F^- \tag{15.6d}$$

Fluorosilicic acid should be handled with great care because of its low pH and the fact that it will cause a "delayed burn" on skin tissue. Fluorosilicic acid (23 percent) will freeze at approximately 4°F (–15.5°C). Approximately 20.8 kg (46 lb) of 23 percent acid are required to add 1 mg/L of fluoride to 1.0 mil gal (3.8 ML) of water.

Hydrofluoric acid and silicon tetrafluoride are common impurities in fluorosilicic acid that result from production processes. Hydrofluoric acid is an extremely corrosive material. Its presence in fluorosilicic acid, whether due to intentional addition (i.e., "fortified" acid) or normal production processes, demands careful handling. Unlike chlorine fumes, fluorosilicic acid fumes are lighter than air and will rise instead of settling to the floor. (Silicon tetrafluoride [SiF_6] is a gas that is heavier than air but is not toxic.) Fluorosilicic acid seems to have a special affinity for electrical switches, contacts, and control panels, as well as concrete.

Other Fluoride Chemicals. Ammonium silicofluoride, magnesium silicofluoride, potassium fluoride, hydrofluoric acid, and calcium fluoride (fluorspar) are being used or have been used for water fluoridation. Each material has properties that make it desirable in a specific application, but each also has undesirable characteristics. None of these chemicals has widespread application in the United States. Calcium fluoride, however, is sometimes used in South America.

Ammonium silicofluoride has the peculiar advantage of supplying all or part of the ammonium ion necessary for the production of chloramine, when this form of disinfectant is preferred to chlorine in a particular situation. It has the disadvantage of hindering disinfection if there are short contact times. Also, it is more expensive than sodium fluorosilicate.

Magnesium silicofluoride and potassium fluoride have the advantage of extremely high solubility, of particular importance in such applications as school fluoridation, where infrequent refills of solution containers are desired. In addition, potassium fluoride is quite compatible with potassium hypochlorite, so a mixture of the two solutions (in the same container) can be used for simultaneous fluoridation and chlorination. They cannot be fed in dry form. Also, they are both (especially potassium fluoride) more expensive than sodium silicofluoride. Magnesium silicofluoride is widely used in Europe as a concrete curing compound and thus is mass produced, but it is still more expensive than sodium silicofluoride. Potassium fluoride is one of the main ingredients in the manufacture of nerve gas.

Calcium fluoride (fluorspar) is the least expensive of all the compounds used in water fluoridation, but it is also the most insoluble. It has been successfully fed by first dissolving it in alum solution and then utilizing the resultant solution to supply both the alum needed for coagulation and the fluoride ion. Some attempts have been made to feed fluorspar directly in the form of ultrafine powder, on the premise that the powder would eventually dissolve or at least remain in suspension until consumed. These attempts have not been very successful.

APPLICATION

Fluoride Feed Systems

Three methods of feeding fluoride are common in community water supply systems. They are:

1. Dry chemical feeder with a dry fluoride compound
2. Chemical solution feeder with a liquid fluoride compound or with a prepared solution of a dry chemical
3. Fluoride saturator

The first two methods are also commonly used to feed other water treatment chemicals. The saturator is a unique method for feeding fluoride.

Selection of the best fluoridation system for a situation must be based on several factors, including population served or water usage rate, chemical availability, cost, and operating personnel available (AWWA, 1988; Reeves, 1986). Although many options will be possible, some general limitations are imposed by the size and type of facility. In general, very large systems will use the first two methods, whereas smaller systems will use either an acid feeder or the saturator.

Manuals describing considerations and alternatives involved in selecting the optimal fluoridation system are available (AWWA, 1988; Reeves, 1986). Factors important in the selection, installation, and operation of a fluoride feed system are the type of equipment used, the fluoride injection point, safety, and waste disposal.

Types of Equipment. Fluoride chemicals are added to water as liquids, but they may be measured in either liquid or solid form. Solid chemicals must be dissolved into solution before feeding. This is usually accomplished by using a dry chemical feeder that delivers a predetermined quantity of chemical in a given time interval. Two types of dry feeders exist. Each has a different method of controlling the rate of delivery. A volumetric dry feeder delivers a measured *volume* of dry chemical per unit of time. A gravimetric dry feeder delivers a measured *weight* of chemical per unit of time.

Many water treatment plants that treat surface water utilize dry feeders to add other treatment chemicals and so use dry feeders for fluorides to maintain consistency with other equipment. Dry feeders are used almost exclusively to feed sodium fluorosilicate because of the high cost of sodium fluoride.

The saturator feed system is unique to fluoridation and is based on the principle that a saturated fluoride solution (4%) will result if water is allowed to trickle through a bed containing a large amount of sodium fluoride. A small pump is used to feed the saturated solution into the water being treated. Although saturated solutions of sodium fluoride can be manually prepared, automatic feed systems are preferred.

Selection of Fluoridation Systems. While there is no specific type of fluoridation system that is solely applicable to a specific situation, there are some general limitations imposed by the size and type of water facility. For example, a large metropolitan water plant would hardly be likely to consider a fluoridation installation involving the manual preparation of sodium fluoride solution, nor would a small facility consisting of one unattended well consider the use of a gravimetric dry feeder installation.

Prior to the actual design of a fluoridation system, a decision must be made on the type of chemical to be used. This will largely determine the type of fluoridated water system that will be designed.

To determine the type of fluoride chemical to use and thus the type of fluoridation system to be designed, the following items must be considered:

1. Chemical availability
2. Water usage
3. Type of existing facilities
   a. Compatibility with proposed system
   b. Space available
   c. Number of treatment sites required (fluoride injection points)
4. Characteristics of the water
   a. Natural fluoride and optimal fluoride levels
   b. Type of flow (variable or steady state)
   c. Pressure (discharge)
5. Estimated overall cost
   a. Capital (initial) cost
   b. Operation and maintenance costs
   c. Chemical costs
6. Operator preference and skill
7. State rules, regulations, and preference

The selection of the chemical is a judgment made after considering some or all of the above items. There is no exact or perfect solution, and different people will make somewhat different judgments. While in many cases the facts clearly favor one chemical, sometimes they will not; therefore, well-informed, knowledgeable persons could come to different conclusions. The following example illustrates how fluoride chemical selection might be made and the chemical equipment selected.

EXAMPLE 15.1 The Town of Pelion, Iowa (population 525) has decided to fluoridate its water supply system. The water system also serves a large rural school (population 2,000). Pelion's water system consists of two city wells that are not attended on a full-time basis. The average daily production rate is 0.2 MGD. The optimal fluoride level for this community's water system is 0.8 mg/L. All three fluoride chemicals are readily available from a nearby chemical supplier.

Well No. 1 has a maximum pumping rate (capacity) of 290 gpm (417,600 gpd) and a discharge pressure of 65 psi. Well No. 2 has a capacity of 250 gpm (360,000 gpd) and a discharge pressure of 60 psi. The natural fluoride level in the water from both wells is 0.1 mg/L.

The wells are located approximately 1 mile apart. Both wellhouses are large and contain equipment for feeding chlorine, polyphosphate, and soda ash. Also, the wellhouses contain electricity and the necessary piping.

$$\text{Sodium fluoride feed rate} = \frac{0.7\ \text{mg/L} \times 0.2\ \text{MGD} \times 8.34\ \text{lb/gal}}{0.45 \times 0.98}$$

Sodium fluoride feed rate = 2.65 lbs/day
NaF needed/yr = 2.65 lbs/day × 365 days/yr = 967.3 lb/yr

Amount of fluorosilicic acid solution (H_2SiF_6, 23% solution) needed:

Fluorosilicic acid solution feed rate (lb/day)

$$= \frac{\text{dosage (mg/L)} \times \text{capacity (MGD)} \times 8.34\ \text{lb/gal}}{\text{AFI} \times \text{solution strength}}$$

AFI = atomic weight of fluoride in H_2SiF_6 divided by the molecular weight of H_2SiF_6 = grams of fluoride per gram of H_2SiF_6 = (6 × 19)/144.1 = 0.79

Solution strength = 23 grams of H_2SiF_6 per 100 grams of solution = 0.23 grams H_2SiF_6 per gram solution

$$\text{Solution feed rate} = \frac{0.7\ \text{mg/L} \times 0.2\ \text{MGD} \times 8.34\ \text{lb/gal}}{0.79 \times 0.23} = 6.4\ \text{lb/day}$$

Solution (23%) needed/yr = 6.4 lb/day × 365 days/yr = 2336 lb/year
Comparing the cost of the two chemicals:

Chemical Costs Comparison

Chemical (item)	Cost, ¢/lb	Chemical used, lb/yr	Chemical cost, $/yr	Difference, $
NaF	90	967	870	–167
Acid	30	2336	703	0

1. As the town has relatively small unattended wells, the use of sodium fluorosilicate for dry feeders should be ruled out immediately. Thus, the choice is between using sodium fluoride and a saturator and using fluorosilicic acid in carboys and a metering pump.

A saturator will require slightly more space, but that is not a problem here. Both the acid system and the saturator system are compatible with the water system. There will be two fluoride injection points, one at each well, because of the location of the wells. There is a steady flow and adequate pressure at each well.

While the chemicals are readily available, there will be a difference in cost of both chemicals and equipment. These costs can be estimated, compared, and evaluated:

Data:

Average daily production rate = 0.2 MGD
Fluoride dosage (mg/L) = optimal fluoride level (mg/L) – natural fluoride level (mg/L)
Fluoride dosage = 0.8 mg/L – 0.1 mg/L = 0.7 mg/L

Amount of sodium fluoride (NaF with saturator) needed:

Sodium fluoride feed rate (lbs/day)

$$= \frac{\text{Fluoride dosage (mg/L)} \times \text{capacity (MGD)} \times 8.34\ \text{lb/gal}}{\text{AFI} \times \text{chemical purity of the sodium fluoride}}$$

AFI = fluoride content of sodium fluoride = gram atomic weight of fluoride (19) divided by the molecular weight of sodium fluoride (42) = 19/42 = 0.45

2. There is approximately a $167 difference in yearly chemical costs between using fluorosilicic acid and sodium fluoride. Also, there is a difference of approximately $2,200 in capital costs. As the equipment for the acid installation and the yearly chemical costs are cheaper, a judgment can be made that the acid system is preferred. (The fact that the Pelion water system also serves a large rural school is not a factor that will influence the selection of the fluoride chemical. It is generally best to base the decision on which chemical to use on cost.) Thus, in this problem, the decision is to use hydrofluorosilicic acid.

As the project develops and specific kinds of equipment are selected, rough designs are made, and additional information is gathered, the estimated costs may become very inaccurate. If this happens, another cost comparison should be made to ensure that it is still more economical to use the acid.

The type of feeder chosen for a particular fluoride installation is determined by cost (primarily), availability, service reputation of the manufacturer or sales representative, and, again, personal preference. Once it has been decided which fluoride chemical to use, the choice of the fluoride feeder will be limited. If fluorosilicic acid is to be used, then a metering pump will be required. If a saturator (with sodium fluoride) is to be used, a metering pump is necessary. If sodium fluoride (as a dry chemical) or sodium fluorosilicate is to be used, then a dry feeder is required. Only the specific model of each general type of feeder will need to be determined after the chemical has been selected.

Fluoride Injection Point. Ideally, the fluoride injection point should be at a location through which all water to be treated passes. In a treatment plant, this could be a channel where other water treatment chemicals are added, a main coming from the filters, or the clear well. If a combination of facilities exists, such as a treatment plant for surface water plus supplemental wells, a point where all water from all sources passes must be selected. If no common point exists, a separate fluoride feeding installation is needed for each facility.

Another consideration in selecting the fluoride injection point is the possibility of fluoride losses through reaction with and adsorption on other treatment chemicals. Whenever possible, fluoride should be added after filtration to avoid substantial losses that can occur, particularly with heavy alum dosages or when magnesium is present and the lime-soda ash softening process is being used. A fluoride loss of up to 30 percent can result if the alum dosage rate is 100 mg/L (Bellack, 1984). If aluminum or iron salt coagulants are used and a fluoride compound is added before the metal hydroxide precipitate is removed, soluble aluminum and iron complexes can be formed, especially when the coagulation pH is less than about 6.5. In some situations, addition of fluoride before filtration may be necessary, such as in cases where the clear well is inaccessible.

When other chemicals are being fed, the question of chemical compatibility must be considered. The fluoride injection point should be as far away as possible from the injection points for chemicals that contain calcium, in order to minimize loss of fluoride by local precipitation. For example, if lime is being added to the main leading from the filters for pH control, fluoride can be added to the same main but at another point, or it can be added at the clear well. If lime is added to the clear well, fluoride should be added to the opposite side. If injection point separation is not possible, an in-line mixer must be used to prevent local precipitation of calcium fluoride and to ensure that the added fluoride dissolves.

In a single-well system, the well pump discharge can be used as the fluoride injection point. If more than one well pump is used, the line leading to the distribution system can be used as the injection point. In a surface water treatment plant or softening plant, the ideal location of the fluoride injection point is in the line from the filters to the clear well. This location provides for maximum mixing. Sometimes the clear well is located directly below the filters, and discharging any chemicals directly to the clear well is difficult. In this situation the fluoride injection point must be at another location, such as in the main line to the distribution system or before the filters. All fluoride injection points should have an antisiphon device included.

Safety Considerations. Manuals describing operational hazards and safety practices for fluoride chemical feed systems are available (AWWA, 1983, 1988; Reeves, 1986). Treatment plant operators must use proper handling techniques to avoid overexposure to fluoride chemicals. Dusts are a particular problem when sodium fluoride and sodium fluorosilicate are used. The use of personal protective equipment (PPE) should be required when any fluoride chemical is handled or when maintenance on fluoridation equipment is performed.

Recommended Emergency Procedures for Fluoride Overfeeds. When a community fluoridates its drinking water, a potential exists for a fluoride overfeed. Most overfeeds do not pose an immediate health risk; however, some fluoride levels can be high enough to cause immediate health problems. All overfeeds should be corrected immediately because some have the potential to cause serious long-term health effects.

Specific actions should be taken when equipment malfunctions or when an adverse event occurs in a community public water supply system that causes a fluoride chemical overfeed. The CDC publishes recommended actions for handling fluoride overfeed events in community water systems (CDC, 1995).

BIBLIOGRAPHY

American Dental Association Council on Access, Prevention and Interprofessional Relations, *JADA,* vol. 126, Chicago, IL, p. 19-S, June 1995.

Ast, D. B., D. J. Smith, B. Wachs, H. C. Hodges, H. E. Hilleboe, E. R. Schesinger, H. C. Chase, K. T. Cantwell, and D. E. Overton. "Newburgh-Kingston Caries-Fluorine Study: Final Report." *J. Am. Dental Assoc.,* 52: 290, 1956.

AWWA B701, *Standard for Sodium Fluoride.* AWWA, Denver, CO, 1994a.

AWWA B702, *Standard for Sodium Fluorosilicate.* AWWA, Denver, CO, 1994b.

AWWA B703, *Standard for Fluorosilicic Acid.* AWWA, Denver, CO, 1994c.

AWWA, *Safety Practice for Water Utilities,* AWWA Manual M3. AWWA, Denver, CO, 1983.

AWWA, *Water Fluoridation Principles and Practices* (3rd ed.), AWWA Manual M4. AWWA, Denver, CO, 1988.

Bellack, E. *Fluoridation Engineering Manual.* U.S. Environmental Protection Agency, Washington, DC, 1972; reprinted September 1984.

Brown, H. K., and M. Poplove. "Brantford-Sarnia-Statford, Fluoridation Caries Study, Final Survey, 1963." *J. Can. Dental Assoc.,* 31 (8): 505, 1965.

Burt, B. A., S. A. Eklund, and D. W. Lewis. *Dentistry, Dental Practice and the Community* (4th ed.). Saunders, Philadelphia, 1992.

CDC. *FL-98 Caries-Free Teenagers Increase with Fluoridation.* United States Department of Health and Human Services, Public Health Service, Centers For Disease Control, Atlanta, 1978.

CDC. NIOSH Recommendations for Occupational Safety and Health Standards. MMWR 1985; 34 (Supplement): 175, 1985a.

CDC. "Dental Caries and Community Water Fluoridation Trends—U.S." *Morbidity and Mortality Weekly Report,* 34 (6): 77, 1985b.

CDC. "Public Health Focus: Fluoridation of Community Water Systems." *Morbidity and Mortality Weekly Report,* 41:372–375, 381, 1992.

CDC. *Fluoridation Census 1991.* U.S. Department of Health and Human Services, Public Health Service, CDC, Atlanta, 1993.

CDC. Engineering and Administrative Recommendations for Water Fluoridation, 1995 MMWR, vol. 44, No. RR-13, 1995.

Churchill, H. V. "The Occurrence of Fluorides on Some Waters of the United States." *Jour. AWWA,* 23 (9): 1399, 1931.

Corbin, S. B. "Fluoridation Then and Now." *Am. J. Public Health,* 79, 1989: 561–563.

Dean, H. T. "Chronic Endemic Dental Fluorosis (Mottled Enamel)." *JAMA,* 107: 1269, 1936.

Dean, H. T., F. A. Arnold, J. Phillip, and J. W. Knutson. *Studies on Mass Control of Dental Caries through Fluoridation of Public Water Supply.* Public Health Reports 65, Grand Rapids-Muskegon, MI, 1950.

Dunning, J. M. *Principles of Dental Public Health* (4th ed.). Harvard University Press, Cambridge, MA, p. 399, 1986.

Eichenberger, B. A., and K. Y. Chen. "Origin and Nature of Selected Inorganic Constituents in Natural Waters," In *Water Analysis,* vol. 1: *Inorganic Species,* Part 1, (R. A. Minear and L. H. Keith, eds.). Academic Press, New York, 1982.

Gish, C. "Relative Efficiency of Methods of Caries Prevention in Dental Public Health," In *Proc. Workshop on Preventive Methods in Dental Public Health,* University of Michigan, Ann Arbor, MI, June 1978.

Hammer, C. T. *The Status of Fluoridation in the State of Israel,* unpublished report, February–May 1996.

Hill, I. N., J. R. Blayney, and W. Wolf. "Evanston Fluoridation Study—Twelve Years Later." *Dental Prog.,* 1: 95, 1961.

Hodges, H. C., and F. A. Smith. In *Fluorine Chemistry,* vol. 4 (J. H. Simons, ed.). Academic Press, New York, 1965.

Jowsey, J., L. B. Riggs, P. J. Kelly, and D. L. Hoffman. "Effect of Combined Therapy with Sodium Fluoride, Vitamin D, and Calcium in Osteoporosis." *Am. J. Med.,* 53: 43, 1972.

Keyes, P. H. "Present and Future Measures for Dental Caries Control." *J. Am. Dental Assoc.,* 79: 1395, 1969.

Leverett, D. "Prevalence of Dental Fluorosis in Fluoridated and Nonfluoridated Communities—A Preliminary Investigation." *J. Pub. Health Dentist.,* 46: 4, 1986.

Maier, F. J. *Manual of Water Fluoridation Practice.* McGraw-Hill, New York, 1963.

McClure, F. J. "Ingestion of Fluoride and Dental Caries—Quantitative Relations based on Food and Water Requirements of Children 1–12 Years Old." *Am. J. Diseases Children,* 66: 362, 1943.

McClure, F. S. *Water Fluoridation, The Search and the Victory.* National Institutes of Health, Bethesda, MD, 1970.

McNeil, D. R. *The Fight for Fluoridation.* Oxford University Press, New York, 1957.

Mellberg, J. R., and D. E. Mallon. "Acceleration of Remineralization, in vitro, by Sodium Monofluorophosphate and Sodium Fluoride." *J. Dental Res.,* 63(9): 1130, 1984.

Michigan Department of Public Health. *Michigan Department of Public Health Policy Statement on Fluoridation of Community Water Supplies and Synopsis of Fundamentals of Relation of Fluorides and Fluoridation to Public Health,* 1979.

National Institute of Dental Research, National Caries Program. *The Prevalence of Dental Caries in United States Children, 1979–1980. The National Dental Caries Prevalence Survey,* National Institutes of Health, Bethesda, MD, December 1981.

National Institute for Occupational Safety and Health. *Pocket Guide to Chemical Hazards.* U.S. Department of Human Services, Public Health Service, CDC; DHHS(NIOSH) publication no. 94-116, 1994.

"National Primary and Secondary Drinking Water Regulations; Fluoride; Final Rule." *Federal Register,* 51 (April 2): 11396, 1986.

National Research Council. *Health Effects of Ingested Fluoride.* National Academy of Sciences, National Academy Press, Washington, DC, 1993.

Newburn, E. "Effectiveness of Water Fluoridation." *J. Public Health Dentistry,* 49 (5): special issue, 1989.

Occupational Safety and Health Administration. "Respiratory Protective Devices: Final Rules and Notice." *Federal Register* 60:30336–30402, 1995.

Public Health Service. *Review of Fluoride: Benefits and Risks—Report of the Ad Hoc Subcommittee on Fluoride of the Committee to Coordinate Environmental Health and Related Programs.* U.S. Department of Health and Human Services, Public Health Service, Washington, DC, 1991.

Reeves, T. G. "The Availability of Fluoride Chemical Supplies." *JADA,* 110 (April): 513–515, 1985.

Reeves, T. G. *Water Fluoridation: A Manual for Engineers and Technicians.* U.S. Department of Health and Human Services, Public Health Service, CDC, Atlanta, 1986.

Reeves, T. G. *Water Fluoridation: A Manual for Water Plant Operators.* U.S. Department of Health and Human Services, Public Health Service, CDC, Atlanta, 1994.

Riggs, L. B., E. Seeman, S. F. Hodgson, D. R. Taves, and W. M. O'Fallon. "Effects of the Fluoride/Calcium Regimen of Vertebral Fracture Occurrence in Postmenopausal Osteoporosis." *N. Engl. J. Med.,* 306 (8): 446, 1982.

Ripa, L. W. "A Half-Century of Community Water Fluoridation in the United States: Review and Commentary." *J Public Health Dent.* 53: 17–44, 1993.

Silverstone, L. M. "The Significance of Remineralization in Caries Prevention." *J. Can. Dental Assoc.,* 50 (2): 157, 1984.

Smith, M. C., E. M. Lantz, and H. V. Smith. "The Cause of Mottled Enamel, a Defect of Human Teeth." *University of Arizona Agricultural Experiment Station Bulletin,* No. 32, 1931.

United States Environmental Protection Agency (USEPA), Letter of Support, 1986.

United States Public Health Service (USPHS), Policy Statement, 1992.

United States Public Health Service (USPHS), Surgeon General Statement, 1995.

Whitford, G. M. *The Metabolism and Toxicity of Fluoride* (2nd ed.). Karger, New York, 1996.

World Health Organization. *Fluorides and Oral Health* (WHO Technical Report Series: 846). World Health Organization, Geneva, 1994.

World Health Organization. *Fluorine and Fluorides.* Environmental Health Criteria 36 (WHO Technical Report Series: 846). World Health Organization, Geneva, 1984.

CHAPTER 16

WATER TREATMENT PLANT RESIDUALS MANAGEMENT

David A. Cornwell, Ph.D., P.E.
President
Environmental Engineering & Technology, Inc.

Water treatment plants typically produce some type of waste stream. The quality and characteristics of these waste streams are related to the main treatment process. Furthermore, waste streams can impact the finished water quality of the treatment process itself. This is especially true when the waste is stored internal to the process or recycled.

Despite the strong linkage between the treatment process and its waste streams, however, water treatment plant waste management has historically been treated as a stand-alone management issue. Whatever the treatment process produced was dealt with in a technically appropriate manner. With increasing costs associated with managing waste streams, it has become prudent to consider the waste stream quality and characteristics as part of the overall evaluation and design of the main water treatment process. Water treatment processes produce unique waste streams, each of which has different associated waste handling costs. The waste streams must be viewed as part of the overall process to be optimized when determining the most economical method for meeting a specific set of finished water quality goals. As an example, it is now recognized that storing solids in a sedimentation basin is not desirable from a water quality perspective. Including solids storage considerations in the overall design of the water treatment process will then influence the decision regarding the type of sedimentation basin to install. Similarly, some filter media combinations produce more spent filter backwash waste than do others. In determining the main treatment process components, waste streams should be considered in the overall decision tree, not viewed as an issue that is handled separately.

It is interesting to note the goals of waste treatment as described in the last edition of this book (1990)—"What must be removed? Where will it be disposed? What treatment is necessary to prepare it for disposal?" Recently, however, a new set of issues other than just disposal of the waste has become important. Instead of disposal, the first approach to end use is now beneficial utilization and solids treatment. Systems are often geared to preparing a material that can be used. Minimization of the liquid volumes of waste produced is also increasingly important. Water quality issues associated with storing residuals in the process train or associated with recycling the water back to the treatment plant have become important in planning a waste management system.

The original term used to describe all water treatment plant wastes was *sludge.* In fact, sludge is really only the solid or liquid-solid component of some types of waste streams. The term *residuals* is now used to describe all water treatment plant process wastes, either liquid, solid, or gaseous.

Hydrolyzing metal salts or synthetic organic polymers are added in the water treatment process to coagulate suspended and dissolved contaminants and yield relatively clean water suitable for filtration. Most of these coagulants and the impurities they remove settle to the bottom of the settling basin where they become part of the waste stream. These residuals are referred to as *alum, iron,* or *polymeric sludge* (even though they may be made up largely of water), being named after the primary coagulant used. These residuals account for approximately 70 percent of the water plant solids generated. Similar solids, called *lime sludge,* are produced in treatment plants where water softening is practiced, and these lime or lime/soda ash plants account for an additional 25 percent of the industry's solids production. It is therefore apparent that most of the waste generation where solids are produced involves water treatment plants using coagulation or softening processes. The above wastes are solid/liquid wastes in that the liquid waste (water) contains suspended solids (and, as indicated above, are referred to as sludges). Other solid/liquid wastes produced in the water industry include wastes from iron or manganese removal plants, spent GAC, spent precoat filter media, wastes from slow sand filter plants, and spent filter backwash water (SFBW).

The water industry also produces liquid-phase residuals, referred to as such because the liquid phase (water) contains primarily dissolved solids which are within the liquid phase itself. These wastes are often called *brines* or *concentrates* and include spent brine from ion exchange regeneration, reject water from membrane systems (although microfilter and ultrafilter membranes will produce a concentrate containing suspended solids, they are included in this category), reject water from electrodialysis plants, and spent regenerant from specific adsorption media such as activated alumina. Gas-phase residuals are produced as off-gases from air stripping systems, and off-gas from ozone contactors. The major types of treatment plant residuals are shown in Table 16.1. This chapter primarily addresses characterization, handling, and ultimate utilization of sludges. Some introduction to the handling of liquid-phase residuals is included. More information on handling other residuals can be found in AWWA (1996) and Cornwell, Bishop, Gould, and Vandermeyden (1987). A good presentation on softening pellets is in Cornwell and Koppers (1990).

QUANTITY OF SOLID/LIQUID RESIDUALS GENERATED

Most conventional coagulation plants produce two major residuals—residuals from the sedimentation basin (commonly referred to as *sludge*) and residuals from backwashing a filter (referred to as *spent filter backwash water*).

The quantity of these solid/liquid residuals generated from water treatment plants depends upon the raw water quality, dosage of chemicals used, performance of the treatment process, method of sludge removal, efficiency of sedimentation, and backwash frequency.

One of the most difficult tasks facing the utility or engineer in planning and designing a residuals treatment process is determining the amount of material (volume and solids) to be handled. The solids quantity is usually determined as an

TABLE 16.1 Major Water Treatment Plant Wastes

Solid/liquid residuals
Alum sludges
Iron sludges
Polymeric sludges
Softening sludges
Spent filter backwash water
Spent GAC or discharge from carbon systems
Slow sand filter wastes
Wastes from iron and manganese removal plants
Spent precoat filter media
Softening pellets
Liquid-phase residuals
Ion-exchange regenerant brine
Waste regenerant from activated alumina
Membrane concentrate
GAC transport water
Gas-phase residuals
Air stripping off-gases
Ozone off-gases

annual average for a given design year and is a function of flow demand projections. Sometimes overlooked, but very important, is information on seasonal or monthly variations. It is not unusual for order of magnitude differences in sludge production to exist for different months of the year.

The amount of alum (or iron) sludge generated can be calculated fairly closely by considering the reactions of alum or iron in the coagulation process. Using an empirical relation to account for the sludge contribution from turbidity will improve the estimate, and the contribution from other sources can be added as required.

When alum is added to water as aluminum sulfate, the reaction with respect to sludge production is typically represented by the simplified equation that includes three waters of hydration in the product (Cornwell et al., 1987; Cornwell and Koppers, 1990; AWWA, 1996).

$$Al_2(SO_4)_3 \cdot 14H_2O + 6HCO_3^- = 2Al(OH)_3 \cdot 3H_2O + 6CO_2 + 11H_2O + 3SO_4^{-2} \tag{16.1}$$

If inadequate alkalinity is present, lime or sodium hydroxide is normally added to maintain the proper pH. The three waters of hydration satisfy the covalent bonding number of six for aluminum. Not including the waters of hydration in the reaction will tend to underestimate the amount of solids that are produced. This chemically bound water increases the sludge quantity, increases the sludge volume, and also makes it more difficult to dewater because the chemically bound water cannot be removed by normal mechanical methods. Commercial alum has a molecular weight of 594 and contains two moles of aluminum, each with a molecular weight of 27. Therefore, alum is about 9.1 percent aluminum (54/594). The resulting aluminum hydroxide species ($Al(OH)_3 \cdot 3H_2O$) has a molecular weight of 132, and therefore, 1 mg/L of aluminum will produce 4.89 mg of solids (132/27), or 1 mg/L of alum added to water will produce approximately 0.44 mg/L of inorganic aluminum solids (0.091×4.89). Suspended solids present in the raw water produce an equivalent weight of sludge solids because they are nonreactive. It can be assumed that other additives, such as polymer and powdered activated carbon, produce sludge on a one-to-one basis. The amount of sludge produced in an alum coagulation plant for the removal of suspended solids is then:

$$S = 8.34\ Q\ (0.44\text{Al} + SS + A) \tag{16.2}$$

where S = sludge produced (lb/day)
Q = plant flow (mgd)
Al = dry alum dose (mg/L) (as 17.1 percent Al_2O_3)
SS = raw water suspended solids (mg/L)
A = additional chemicals added, such as polymer, clay, or powdered activated carbon (mg/L)
Note: To convert from lb/day to kg/day, multiply by 0.45.

If iron is used as the coagulant, then the equivalent product of equation 16.1 is $Fe(OH)_3 \cdot 3\ H_2O$ with a molecular weight of 161. The solids production equation becomes:

$$S = 8.34\ Q\ (2.9\ \text{Fe} + SS + A) \tag{16.3}$$

where the iron dose is expressed as mg/L of Fe^{3+} added or produced via Fe^{2+} oxidation (note that significant Fe^{2+} in the raw water will also produce sludge at a factor of 2.9 if it is oxidized). For iron coagulants, the solids production is best expressed as a function of iron because iron coagulant is purchased in may different forms. It should not be interpreted from equations 16.2 and 16.3 that iron produces several times the amount of sludge that alum produces. The units for the coagulant are significantly different for the two equations. In reality, 1 mole of coagulating equivalent of iron produces about 20 to 25 percent more dry-weight sludge than 1 mole of aluminum, based on the ratio of molecular weights of the product. When iron is purchased as ferric chloride ($FeCl_3$), the coagulant dose is usually reported as equivalent dry weight of chemical without waters of hydration (although this should be confirmed with the manufacturer) and, thus, the coagulant has a molecular weight of 162.3. This results in the production of 1.0 mg of solids produced for each milligram of ferric chloride ($FeCl_3$) added. Ferric sulfate is reported by different manufacturers with different waters of hydration and the individual products need to be referenced.

EXAMPLE 16.1 A treatment plant in the mid-Atlantic area of the United States had an average raw water turbidity of 4.5 ntu for the period 1991 through 1994 and used an average ferric chloride dose of 11.5 mg/L (as $FeCl_3$). After conducting a correlation study between SS and turbidity, they found the b value, to convert from turbidity (TU) to suspended solids (SS), to be 1.4 at an average flow of 198 mgd. What was the annual sludge production?

SOLUTION The solution can be found using equation 16.3. In this equation, the coagulant dose is expressed as Fe. Therefore, convert $FeCl_3$ dose to Fe dose by:

$$\frac{MW\ \text{Fe}}{MW\ \text{FeCl}_3} = \frac{56}{161}\ (11.5\ \text{mg/L}) = 4\ \text{mg/L as Fe}$$

and the solids production

$$S = 8.34\ Q\ (2.9\ \text{Fe} + b\text{TU})$$

$$S = 8.34\ (198)\ [2.9(4) + 1.4\ (4.5)]$$

$$= 29{,}558\ \text{lb/day}$$

or 149 lb/MG (million gallons)

Polyaluminum chloride (PACl) is the third major coagulant used. Care especially needs to be used when converting PACl dose to solids production, because each manufacturer may use different strengths and utilities report these doses differently. Some utilities report PACl dose as a neat solution, some as Al_2O_3, and some as PACl product. A "typical" manufactured PACl liquid contains about 30 to 35 percent PACl and around 10 percent Al_2O_3. One manufactured product contains 33.3 percent PACl and 10.3 percent Al_2O_3 or, in this case, the PACl itself contains 30.9 percent Al_2O_3. This is equivalent to 16.4 percent aluminum, and therefore 1 mg of PACl (as PACl) will produce 0.8 mg of solid product (0.164×4.89).

The above equations can then be used to track yearly or even daily variation changes in sludge dry weight produced. One difficulty in applying the relationships is that most plants do not routinely analyze raw water suspended solids concentrations. The logical correlation is to equate a turbidity unit to a suspended solids unit. Unfortunately, the relationship is generally not 1 to 1:

$$SS \text{ (mg/L)} = b \cdot \text{TU} \tag{16.4}$$

The value of b for low-color, predominately turbidity removal plants can vary from 0.7 to 2.2 (Cornwell et al., 1987). It may vary seasonally for the same raw water supply. A utility can therefore either continually measure suspended solids, or it may be possible to develop a correlation between turbidity and suspended solids. Figure 16.1 shows one such correlation for a low-color raw water source (Cornwell, 1981). Ideally, this correlation should be done weekly until information is learned as to seasonal variations in the suspended solids/turbidity relationship. Afterward, a monthly correlation may be sufficient.

Another complication exists for raw water sources that contain a significant amount of total organic carbon (TOC). Total organic carbon can be a large contributor to the sludge production. Values of b for low-turbidity, high-TOC raw waters can be as high as 20, but unless turbidity and TOC vary together, a correlation between suspended solids and turbidity will not exist.

Figure 16.2 shows the relationship between calculated and measured solids production done by the City of Philadelphia (EE&T, 1996). The City used iron as the coagulant during this time period. A correlation was developed that showed the ratio of suspended solids to turbidity was 1.4. The calculated quantities using equation 16.3 were within 5 percent of the measured quantities. Through careful calibration and measurements, a complete solids mass balance can be prepared, as was done by the City of Philadelphia and shown in Figure 16.3.

Through similar theoretical considerations, a general equation has been developed (Cornwell et al., 1987; AWWA, 1981) for plants that use a lime softening process for carbonate hardness removal with or without the use of alum, iron, or polymer. The equation is:

$$S = 8.34\, Q\, [2.0 \text{ Ca} + 2.6 \text{ Mg} + 0.44 \text{ Al} + 2.9 \text{ Fe} + SS + A] \tag{16.5}$$

where S = sludge production (lb/day)
Ca = calcium hardness removed as $CaCO_3$ (mg/L)
Mg = magnesium hardness removed as $Mg(OH)_2$ (mg/L)
Fe = iron dose as Fe (mg/L)
Al = dry alum dose (mg/L) (as 17.1 percent Al_2O_3)
Q = plant flow (mgd)
SS = raw water suspended solids (mg/L)
A = other additives (mg/L)

The preceding equations or prediction procedures allow estimation of the dry weight of sludge produced. For sludge productions from noncarbonate hardness

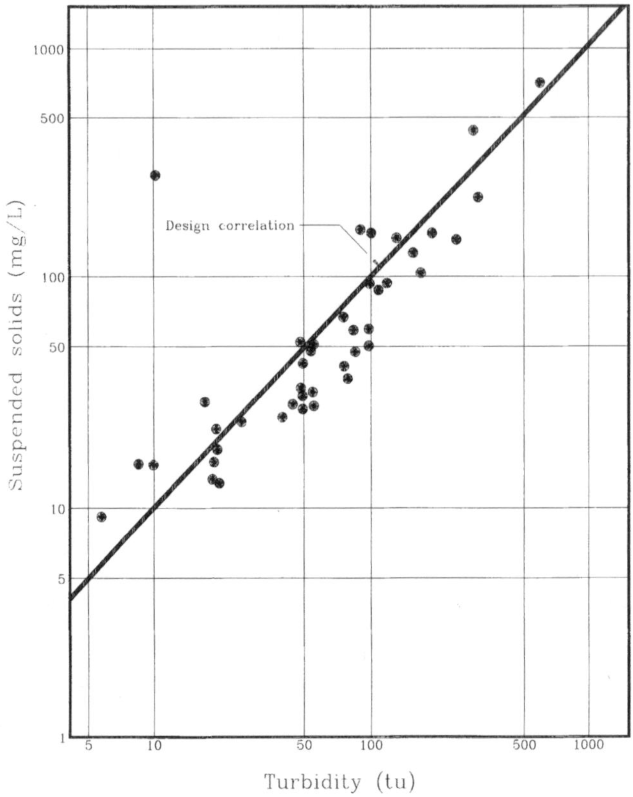

FIGURE 16.1 Suspended solids versus turbidity. (*Source:* Cornwell, 1981.)

removal or when sodium hydroxide is used, see AWWA (1981). These equations do not predict the volume of sludge that will be produced.

Volumes and suspended solids concentrations of sludges leaving the sedimentation basins or clarifiers are a function of raw water quality, treatment, and the sludge removal method. When basins are cleaned only periodically by manual procedures accumulating sludges tend to compact and thicken at the bottom. There is often a stratification of solids with the heavier particles settling to the bottom and the hydroxide, or lighter, particles at the top. However, the actual volume produced will depend largely on the amount of water used to flush the solids out of the basin during cleaning. With increasing finished water quality standards, there will be a trend to remove the solids as quickly as possible, generally with continuous collection equipment. In this case, the solids concentrations will be lower because compaction height and time have been less. Solids concentrations using continuous collection equipment for sludges produced with alum or iron coagulants and for low- to moderate-turbidity raw waters will be about 0.1 to 1.0 percent leaving the sedimentation basin. Some of the upflow clarifier devices will produce sludge at a concentration below 0.1 percent, whereas some of the sludge blanket clarifiers can produce sludge at over 2 percent solids concentration. The higher the ratio of coagulant-to-raw-water-solids, the lower the solids concentration and the higher

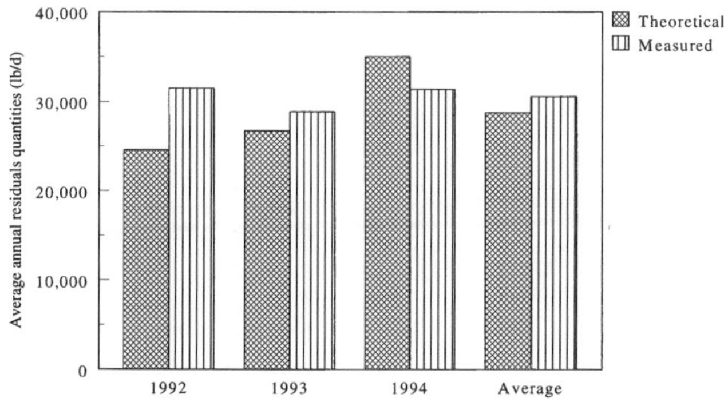

FIGURE 16.2 Baxter WTP theoretical versus measured residuals quantities. (*Source:* EE&T, 1996.)

the sludge volume. Coagulant sludges from highly turbid raw waters may be in the 2 to 4 percent solids concentration range and occasionally higher. Sludge volumes from sedimentation basins tend to be 0.1 to 3 percent of the raw water flow, with one survey (Cornwell and Susan, 1979) finding an average of 0.6 percent. Softening sludges will concentrate higher, usually as a function of the $CaCO_3$:$Mg(OH)_2$ ratio and the type of clarifier. Conventional sedimentation basins may only produce solids concentrations of 2 to 4 percent, whereas sludge blanket clarifiers can produce solids concentrations of up to 30 percent. Sludge volumes will correspondingly vary considerably, from 0.5 to 5 percent of the water plant flow.

Spent filter backwash water is characterized by its large water volume, high instantaneous flow rate, and low solids concentration. Filters can be backwashed at anywhere from 15 to 30 gpm/ft^2, depending upon the media size and water temperature, and the backwash time may be 15 to 20 min. Backwash water volumes are in the range of 3 to 10 percent of plant production. Accurate plant records often exist on the amount of backwash water used. The percentage of plant production used for backwashing can be computed from the ratios of the unit run volumes. For example, a filter producing water at 4 gpm/ft^2 with a 24-h run time has a unit production of 5760 gal/(run ft^2). If it is backwashed at 20 gpm/ft^2 for 20 min, the unit volume of backwash water is 400 gal/(run ft^2), for a ratio of about 7 percent backwash water compared with production water. Spent filter backwash water will typically contain 10 to 20 percent of the total solids production and have suspended solids concentrations of 30 to 300 mg/L depending upon the applied turbidity to the filters and the ratio of backwash water to production.

PHYSICAL AND CHEMICAL CHARACTERISTICS OF SOLID/LIQUID RESIDUALS

Physical characterization of water plant wastes is primarily directed at solid/liquid waste streams of various percent suspended solids concentrations. *Solid/liquid wastes* are terms used to describe free-flowing liquids that are predominantly water all the way up to mixtures that are predominantly solids and behave like a soil texture. Therefore, whenever referring to the physical properties of sludge, it is important to know the suspended solids concentration of the solid/liquid mixture to assess

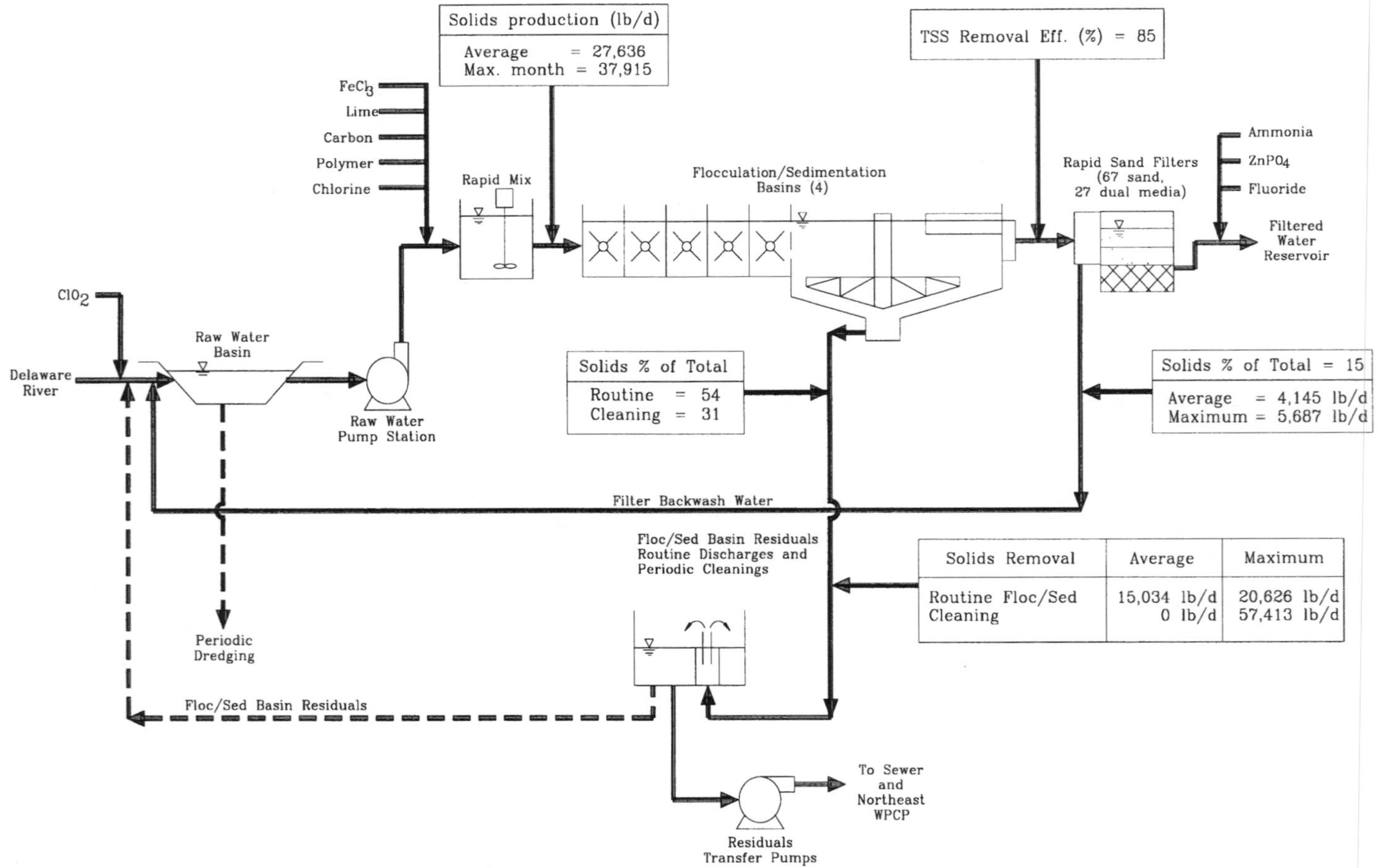

FIGURE 16.3 Baxter WTP baseline residuals distribution. (*Source:* EE&T, 1996.)

the physical state. Cornwell and Wang (Cornwell et al., 1992) used the Atterberg limit test to classify a sludge's physical state. The Atterberg test was originally developed to describe quantitatively the effect of varying the water content on the consistency of fine-ground soils. The test consists of measuring five limits; however, the liquid limit and the plastic limit are the most applicable to sludges. Figure 16.4 shows the relative location of the liquid and plastic limits. The plastic limit identifies the solids concentration at which a sludge transitions from a semisolid to a plastic stage (the plastic state ranges from soft butter to stiff putty). The liquid limit is the solids concentration below which the sludge exhibits viscous behavior; the consistency could be described as ranging from soft butter to a pea soup–type slurry. Coagulant sludges that were tested had liquid limits in the 15 to 20 percent solids concentrations range. Solids concentrations below but near this range would result in a material that still had free water associated with it but may not flow. Generally, a coagulant sludge is still free flowing up to an 8 to 10 percent solids concentration. The plastic limit for coagulant sludges was found to be anywhere from 40 to 60 percent solids concentration.

Knocke and Wakeland (1983) divided the physical properties of sludge into *macroproperties* and *microproperties*. Macroproperties include parameters such as specific resistance, settling rates, and solids concentrations. The indices described above could be considered macroproperties. Microproperties included particle size distribution and density. Vandermeyden et al. (1997) studied the micro- and macroproperties of about 80 water plant sludges. Coagulant sludges had a median particle

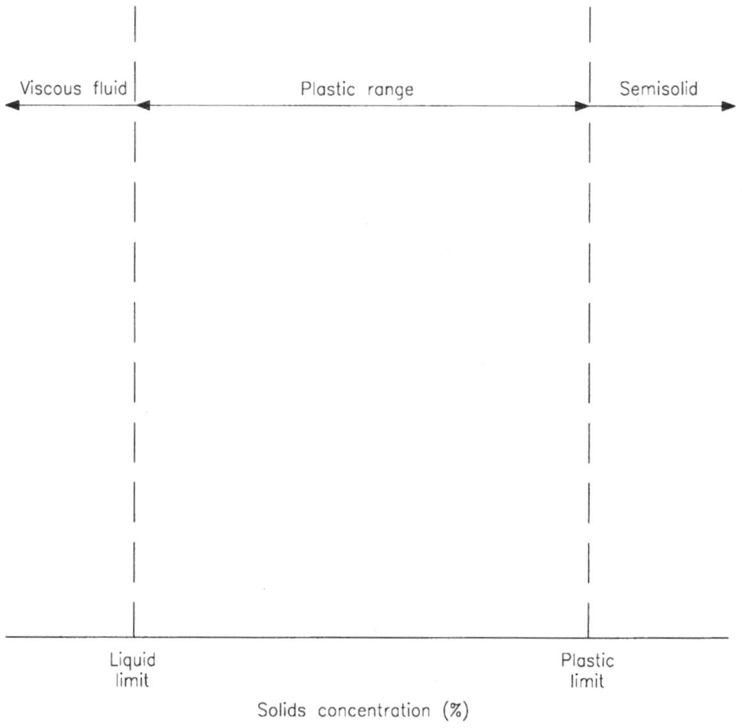

FIGURE 16.4 Relative location of liquid and plastic limits.

diameter of 0.005 mm with a range of approximately 0.001 to 0.03 mm, as shown in Figure 16.5. Lime sludge had a similar range, but the median diameter was 0.012 mm. They also measured the specific gravity of the solid material in the sludge mixtures, as shown in Figure 16.6. The coagulant residuals had an average specific gravity of 2.32 and the lime residuals averaged 2.50. Koppers (Cornwell and Koppers, 1990) reported the dry density of coagulant sludges to be about 2.5. Knocke et al. (1993) found densities for coagulant sludges to range from 2.45 to 2.86, lime sludges to be 2.47, and polymer sludges to be 1.60.

Vandermeyden et al. (1997) also evaluated drainage properties of the 80 residuals using the capillary suction time (CST), specific resistance (SR), and time to filter (TTF) tests. The CST test is a fast and relatively simple test that is performed to

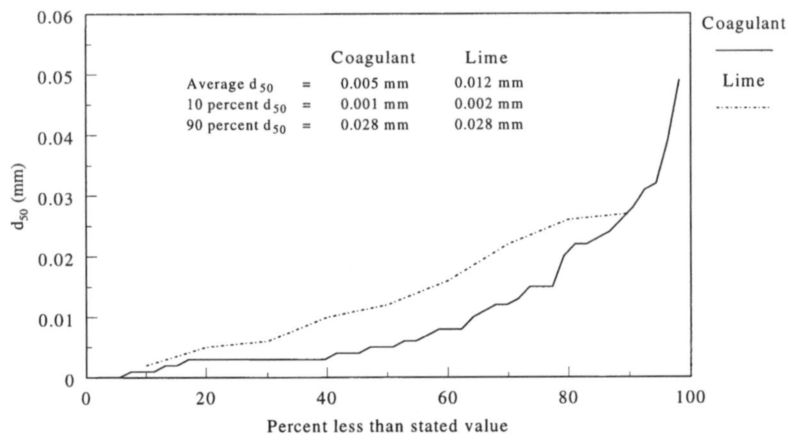

FIGURE 16.5 Average particle diameter for coagulant and lime residuals. (*Source:* Vandermeyden et al., 1997.)

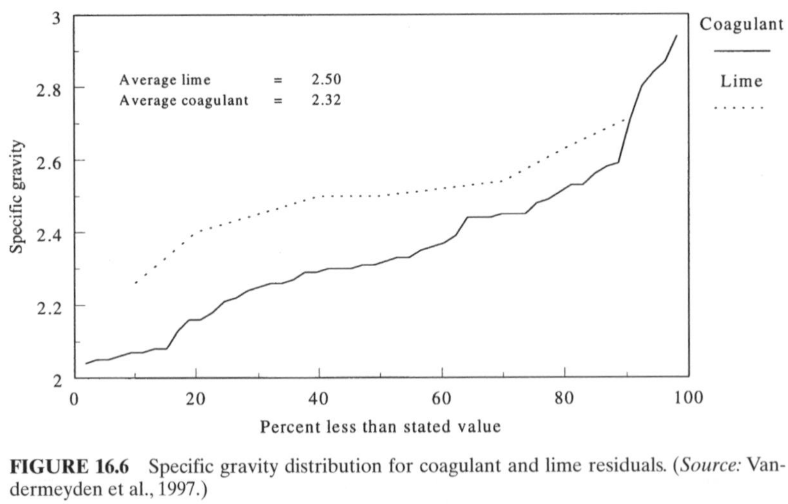

FIGURE 16.6 Specific gravity distribution for coagulant and lime residuals. (*Source:* Vandermeyden et al., 1997.)

determine the rate of free water release from a residual sample. The test is especially useful for comparing the drainage characteristics of different residuals and for optimizing polymer conditioning of residuals. The test consists of measuring the time in seconds for free water to travel 1 cm when a 5- to 7-mL sample of residuals is placed in a special cylinder on top of Whatman No. 17 chromatography paper. As the free water drains from the residuals through the chromatography paper, it passes by an electronic sensor that activates a timer. The timer stops when the free water reaches a second electronic sensor, 1 cm away. The time, in seconds, recorded by the instrument is the CST time.

The resistance to fluid flow exerted by a cake of unit weight of dry solids per unit area is defined as the specific resistance (SR). To evaluate SR, a sample of residuals is subjected to a vacuum using a Buchner funnel apparatus. Typically, 100-mL portions of residuals are added to the Buchner funnel, which is lined with a paper filter, and a vacuum is applied to the filter apparatus. The volume of filtrate generated at various times is recorded. This procedure is continued until enough water has been drawn out to produce cracking in the cake on the filter paper, and subsequent loss of vacuum.

A simplification of the SR test is the TTF test. This test is set up with the same Buchner funnel apparatus as the SR test, but is much simpler to run. The only data collected is the amount of time it takes for one-half of the sample volume to filter. The result is expressed in seconds. Details of the procedures for all these tests are in Vandermeyden et al. (1997).

Vesilind (1988) presented a model for what occurs during the performance of a CST test. He proposed that the rate at which water is released from the sludge material into the chromatography paper is a function of two distinct and separate processes. The first is absorption associated with the test instrument, and the second is water release associated with the sludge material. The absorption associated with the test instrument can be quantified as a function of the test apparatus and the chromatography paper. In terms of the test apparatus, the flow of free water from the solids is a function of the bottom diameter area of the stainless steel reservoir, the permeability of the chromatography paper used, and viscosity. The values and effects of each of these parameters can be evaluated and determined through simple measurements conducted on the test instrument. Viscosity is a function of temperature, so its value must be determined for each test conducted. Because in all likelihood these tests will all be conducted on the same equipment, an instrument constant can be evaluated.

This instrument constant accounts for the change in diameter between the first and second sets of electrodes used to measure the CST. It also quantifies the permeability of the chromatography paper and the effects on dewatering associated with the reservoir.

Vesilind further proposed that the water released from the sludge material is a function of solids concentration and viscosity. It has long been recognized that solids concentration has an effect on CST. The sludge concentration is directly proportional to the filterability constant. The filterability constant can be determined as follows:

$$\chi = 10^{-6}\,\Phi\left[\frac{\mu SS}{\text{CST}}\right] \tag{16.6}$$

where χ = filterability constant [$kg^2/(s^2\,m^4)$]
Φ = dimensionless instrument constant
μ = viscosity [centipoise (cP)]
SS = solids concentration (mg/L)

Equation 16.6 suggests that the filterability constant is a fundamental measure of sludge dewaterability. Due to the cumbersome nature of the SR test, it is often preferable to use an easily determined value, such as CST, which allows calculation of a measure of dewaterability that is independent of solids concentration.

If the filterability index and specific resistance are both measures of dewaterability, then they should plot linearly. Figure 16.7 (Vandermeyden et al., 1997) shows a plot of the inverse of the filterability constant and SR. As predicted by Vesilind, a strong correlation exists between the filterability index and the specific resistance.

Both the CST and the TTF tests measure the rate of water release from the sludge, and therefore one would expect a relationship to exist between the two tests. A plot of such a relationship is shown in Figure 16.8. The strong correlation suggests that either test could be successfully used to evaluate sludge drainage characteristics.

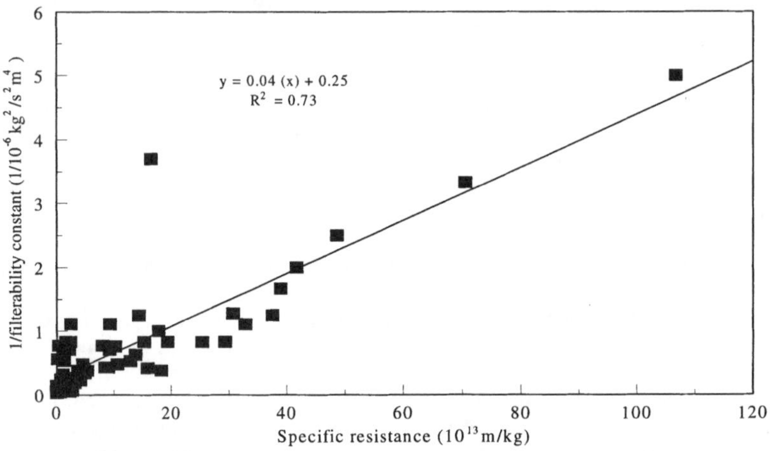

FIGURE 16.7 Inverse filterability constant versus specific resistance. (*Source:* Vandermeyden et al., 1997.)

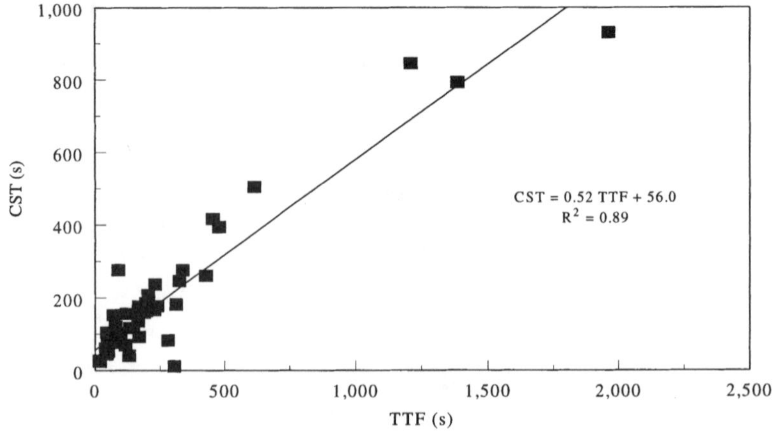

FIGURE 16.8 CST versus TTF for coagulant residuals. (*Source:* Vandermeyden et al., 1997.)

Table 16.2 shows the summary statistics for each of the measured drainage tests (Vandermeyden et al., 1997). Interestingly, the general trend from all the tests is similar. For example, lime sludges had the lowest CST, TTF, and SR. The tests would predict the ease of dewatering order to be lime, ferric, alum, and PACl.

Two additional physical factors, *compaction density* and the *shear strength,* are important when a dry sludge is to be disposed of in a landfill or monofill, or utilized as backfill or a soil substitute. Compaction tests can be used to determine the achievable dry density for a particular sludge. This value can be useful when estimating volumes required for a landfill or the truck volumes required to haul a certain weight of sludge. The first work completed on using compaction density values for water plant sludge characterization was done in Europe in the mid-1980s, as reported by Koppers (Cornwell and Koppers, 1990). Koppers used moisture-density curves to show optimum compaction for a particular sludge. At low solids concentrations, the density is essentially the same as water (62 lb/ft^3, 1,000 kg/m^3), whereas he reported highly dewatered coagulant sludges to have a density of 100 lb/ft^3 (1700 kg/m^3). In work by Cornwell (Cornwell et al., 1992), coagulant sludges at about an 80 percent solids concentration (highly dewatered) were reported to have a density of 110 to 125 lb/ft^3 (1850 to 2100 kg/m^3) and, up to a 20 percent solids concentration, the densities were about 60 lb/ft^3 (1000 kg/m^3), essentially the same as water.

The shear strength of sludges relates to the overall ability of the sludge to support itself and external loadings, such as vehicle traffic or earth-moving equipment. Measuring shear strengths of water plant sludges is complicated by the fact that the shear strength will vary with sludge age and with disturbance. Novak and Calkins (1973) first used the shear value of sludges as a way to measure their handleability. They reported that to achieve a shear value of 0.7 psi (4.7 kN/m^2), where they believed the sludge could be "handled," required a solids concentration for an alum, iron, and lime sludge of 30, 40, and 55 percent, respectively. Koppers reported shear values for coagulant sludges at 20 to 30 percent solids concentration to be 0.3 to 0.6 psi (2 to 4 kN/m^2). They felt that appropriate landfilling would require a shear strength of 1.4 psi (10 kN/m^2) or above, which would require above a 35 percent solids concentration. Figure 16.9 shows an example of shear strength as a function of solids concentrations for three different water plant sludges (Cornwell et al., 1992). The techniques used by Cornwell and Wang (Cornwell et al., 1992) resulted in similar conclusions as Koppers, although they predicted that a shear value of over 4 psi (28 kN/m^2) would be needed for monofilling sludge to support earth-moving equipment, which would require a solids concentration for coagulant sludges of between 30 and 50 percent. There was a wide variation in results for the three sludges tested.

Chemical characterization of solid/liquid water plant residuals is primarily concerned with determining total metal concentrations, leachable metals, and nutrient

TABLE 16.2 Summary Statistics for Drainage Parameters

	CST(s)			TTF(s)			SR(10^{13} m/kg)		
Sludge type	*n*	Mean	Std. dev.	*n*	Mean	Std. dev.	*n*	Mean	Std. dev.
Alum	38	194.1	195.4	38	319.5	412.6	38	15.8	21.3
Ferric	9	103.0	64.5	9	104.7	79.5	9	6.4	8.1
PACl	5	289.8	258.8	5	410.9	562.5	5	13.8	11.0
Lime	9	70.0	34.5	9	34.3	20.4	9	0.5	0.82

Source: Vandermeyden et al., 1997.

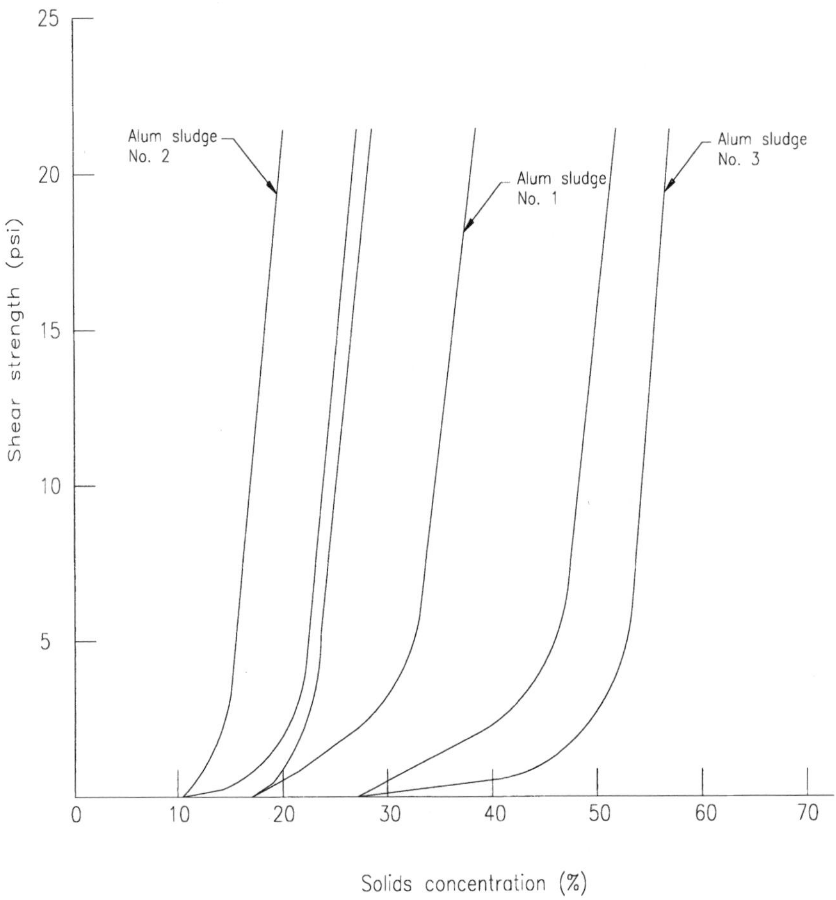

FIGURE 16.9 Shear strength versus solids concentration, cone penetration, and triaxial compression tests. (*Source:* Cornwell et al., 1992.)

levels. Several publications (Schmitt and Hall, 1975; Cornwell et al., 1987; AWWA, 1996; Cornwell et al., 1992; Cornwell and Koppers, 1990) have presented values for total metal concentrations of coagulant sludges; one such example is shown in Table 16.3. The source of the metals can be the raw water as well as the coagulant itself. Metals that are often found in coagulant sludges include aluminum, arsenic, occasionally cadmium, chromium, copper, iron, lead, manganese, nickel, and zinc.

Extraction tests are designated by USEPA to be one of the methods to determine if a waste is hazardous as per Subtitle C of RCRA. The toxicity characteristic leach procedure (TCLP) is used to determine if a waste is toxic, and therefore classified as hazardous. The presence in the extract from the TCLP test of any one of a number of contaminants above a specified regulatory level constitutes failure of the test and results in the waste being classified as hazardous. There are no reported coagulant or lime sludges that have failed the TCLP test. In fact, it is rare to even find detectable levels of the regulated contaminants in the extract from a TCLP test on a coagulant

TABLE 16.3 Example Total Metal Analysis for Coagulant Sludges

Metal	Alum sludge 1 (mg/kg dry weight)	Alum sludge 2 (mg/kg dry weight)	Alum sludge 3 (mg/kg dry weight)
Aluminum	107,000	123,000	28,600
Arsenic	25.0	32.0	9.2
Barium	30	<30	230
Cadmium	1	1	2
Chromium	120	130	50
Copper	168	16	52
Iron	48,500	15,200	79,500
Lead	11	9	40
Manganese	1,180	233	4,800
Mercury	0.1	<0.1	0.2
Nickel	24	23	131
Selenium	<2	<2	<2
Silver	<2	<2	<2
Zinc	91.7	393	781

Source: Cornwell et al., 1992.

or lime sludge. Due to the pH of lime sludges, the TCLP test is not used, and the sludges are classified as nonhazardous according to this procedure.

Leaching tests provide another method that can characterize the release of metals from a sludge. Cornwell et al. (1992) conducted leaching tests in a lysimeter, in which the solids were subjected to rainfall of volume equivalent to about 12 years of normal precipitation and the leachate was analyzed for metal concentrations. Figure 16.10 shows an example of the leaching of arsenic from three sludges and compares the leach levels with in-stream water quality standards and to the drinking water MCL. Only arsenic, copper, iron, manganese, and zinc leached from all three sludges. A slight degree of leaching of nickel and cadmium was exhibited by the ferric sludge only. Selenium leached from two sludges, but only during the first week. Aluminum

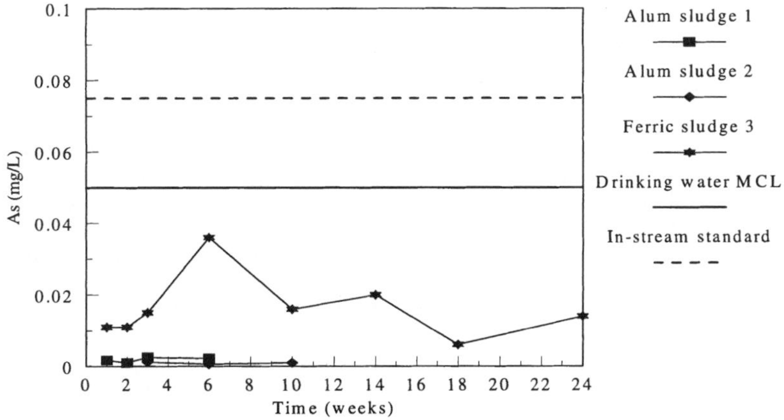

FIGURE 16.10 Arsenic leaching in relation to drinking water and in-stream water quality from an alum sludge. (*Source:* Cornwell et al., 1992.)

did not leach from any of the sludges, and none of the primary drinking water MCLs for metals monitored were exceeded in any of the leachate samples analyzed.

Water plant coagulant sludges are generally considered to be lower in nutrient content than biosolids, particularly nitrogen and phosphorus. However, some residuals can have as much as 3 percent total nitrogen or phosphate content, which can have some fertilizer value. In fact, in some cases nitrogen can be the limiting constituent in determining the maximum amount of residuals that can be applied to a specific land (EE&T, 1992).

Although specific inorganic concentrations are not reported in the literature for lime sludges, it is certain that they will be present to the extent that they are removed from the raw water. Some removal from the raw water of most of inorganic contaminants will take place during lime softening, with high removals for compounds such as lead, cadmium, and arsenic (at a high softening pH).

Many groundwaters being treated for hardness removal also contain background concentrations of naturally occurring radium. Radium is a naturally occurring daughter product of U^{238}. Decay of U^{238} over millions of years passes through a series of elements eventually producing radium. The parent elements of radium are generally insoluble in water so that radium is often the first radioactive element which is found in drinking water supplies. The major threat to human health from radium daughters comes from breathing air containing radon and its very short lived daughters, which can accumulate as solids in the lungs. This exposes the lungs and other internal organs to continuous radiation. In addition, the sludge can directly expose humans to gamma radiation from the decay of radium. However, for the most part, the safe handling and disposal of lime sludges containing radium involves the prevention of radon exposure.

Measurement of radioactive components is expressed in curies or picocuries, pCi (10^{-12} Ci). A curie is the official unit of radioactivity, defined as exactly 3.70×10^{10} disintegrations per second. Snoeyink (1984) reported data for radium concentrations in various lime softening sludges. The sedimentation basin sludge concentrations of Ra^{226} ranged from 1,000 to 11,000 pCi/L of sludge, Ra^{228} varied from 200 to 12,000 pCi/L. Because the radium is associated with the sludge solids, its concentration in the liquid stream is a function of the solids concentration. The concentration per gram of solids is 10 to 20 pCi/g for Ra^{226} and 1 to 11 pCi/g for Ra^{228}. Backwash water concentrations for Ra^{226} ranged from 6 to 50 pCi/L.

THICKENING SOLID/LIQUID RESIDUALS

After removal from a clarifier or sedimentation basin, sludges can be thickened in a gravity concentration tank. Thickening can be economically attractive in that it reduces the sludge volume and produces a more concentrated sludge for further treatment in the dewatering process, or for perhaps hauling to a land application site. Some dewatering systems will perform more efficiently with higher solids concentrations. Thickening tanks can also serve as equalization facilities to provide a uniform feed to the dewatering step. Although there are a few types of thickeners available on the market, the water industry almost exclusively uses gravitational thickening.

Sludge thickening is performed primarily for reduction in the volume of sludge. The relationship between the volume of sludge and the solids concentration is expressed as:

$$V = \frac{M}{rsP} \tag{16.7}$$

where V = volume of sludge (m^3)
M = mass of dry solids (kg)
r = density of water = 10^3 kg/m^3 (at 5°C)
s = specific gravity of the sludge mixture
P = percent solids expressed as a decimal (weight/weight)

An approximation, assuming the specific gravity of the sludge does not change and that there is 100 percent solids capture, for determining volume reduction based on percent solids is expressed as:

$$\frac{V_2}{V_1} = \frac{P_1}{P_2} \tag{16.8}$$

This is a quick and useful equation because the specific gravity of the sludge is not always known, but generally does not change within the limits of most thickening operations. Therefore, for thickening a 1 percent solids concentration sludge to 10 percent solids concentration a volume reduction of approximately 90 percent is achieved.

EXAMPLE 16.2 An alum sludge is produced from a sedimentation basin at a 0.3 percent solids concentration. The plant produces 100 kg of sludge per day. What is the volume of sludge produced? What would be the volume if a thickener were used to increase the solids concentration to 2 percent?

To determine the volume from Eq. 16.7, the specific gravity of the sludge must be known. It was reported earlier that the specific gravity for low solids concentrations is essentially 1.0. For alum sludges, this seems to hold up to about a 20 percent solids concentration, well above the performance of thickeners. Therefore, the specific gravity of 1.0 can be used in Eq. 16.7 and Eq. 16.8 applies.

$$V = \frac{M}{rsP} = \frac{100}{10^3\,(1)\,(0.003)} = 33.3\ \text{m}^3$$

The volume after thickening would be:

$$V_2 = \frac{P_1}{P_2}\,(V_1) = \frac{0.3}{2.0}\,(33.3)$$

$$V_2 = 5\ \text{m}^3$$

Gravity sludge thickeners are generally circular settling basins with either a scraper mechanism in the bottom (see Figure 16.11), or equipped with sludge hoppers (Figure 16.12). They may be operated as continuous flow or as batch "fill-and-draw" thickeners. For continuous flow thickeners, the sludge normally enters the thickener near the center of the basin and is distributed radially. The settled water exits the thickener over a peripheral weir, or trough, and the thickened sludge is drawn from the basin. For tanks equipped with a scraper mechanism, the scraper is located at the thickener bottom and rotates slowly. This movement directs the sludge to the draw-off pipe near the bottom, center of the basin. The slow rotation of the scraper mechanism also prevents bridging of the sludge solids. The basin's bottom is sloped to the center to facilitate collection of the thickened sludge.

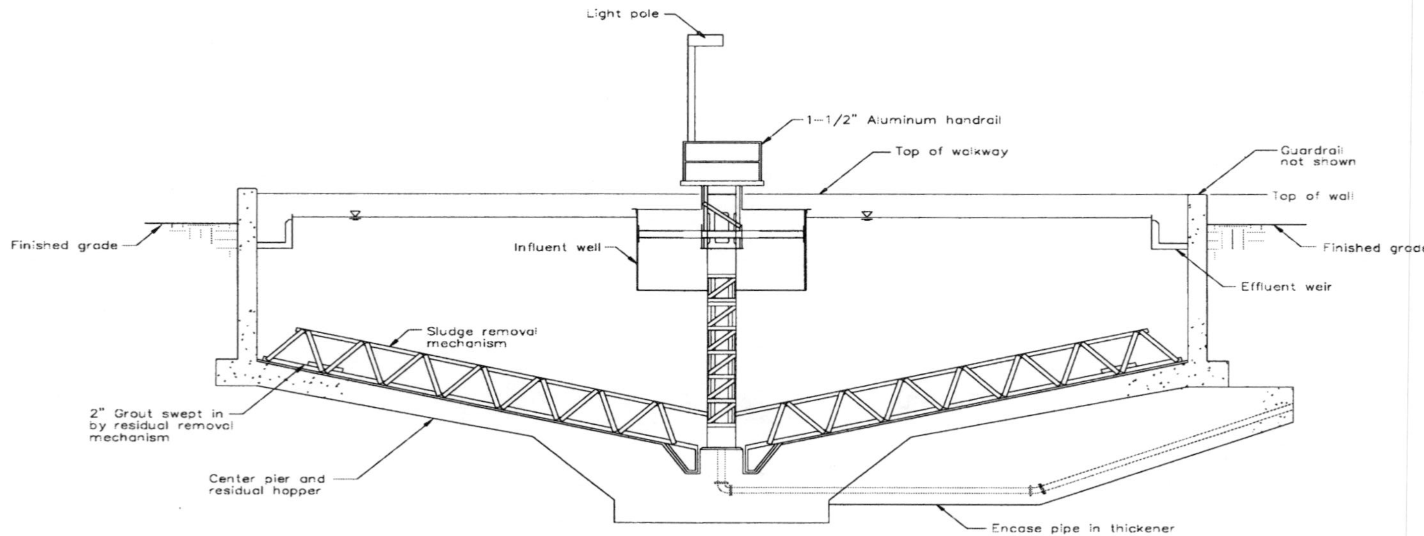

FIGURE 16.11 Continuous flow gravity thickener.

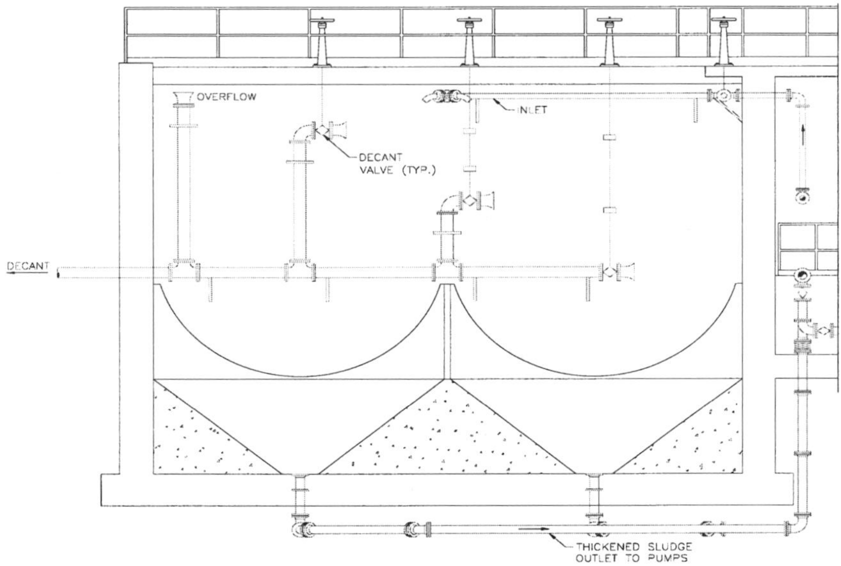

FIGURE 16.12 Batch-thickening tank schematic.

Batch fill-and-draw thickening tanks are often equipped with bottom hoppers as was shown in Figure 16.12. In these tanks, sludge flows into the tank, from either a periodic or continuous removal of sludge from the sedimentation basin, until the thickening tank is full. The sludge is allowed to quiescently settle and a telescoping decant pipe is used to remove supernatant. The decant pipe may be continually lowered as the solids settle until the desired solids concentration is reached or the sludge will not thicken further. The thickened sludge is then pumped out of the bottom hoppers to further treatment or disposal.

Design of batch or continuous flow thickeners is usually accomplished based on previous experience of similar full-scale installations or on laboratory or pilot settling tests. Thickeners are used not only to increase the concentration of the sludge, but also to remove solids from the liquid phase so that the supernatant can be discharged to a receiving stream or recycled to the head of the treatment plant. Quality requirements for discharge are primarily concerned with meeting a set turbidity or suspended solids level. Considerations for quality suitable for recycle include the effective removal of parasitic cysts (see the section on recycle). For designs based on thickening the residuals, the thickener surface area is often solids limited, and a flux rate is used for determining the thickening area in units such as kg/(m^2 h) or lb/(day ft^3). However, units designed to meet a given supernatant quality are often hydraulically limited such that sizing is based on the solids settling velocity, m/h or gpm/ft^2. Although there is some available literature to set flux rates for coagulant or calcium carbonate sludges, site-specific study is needed to determine thickening requirements and appropriate settling velocities when supernatant quality is the criterion. Settling velocity studies can be conducted similarly to those for designing sedimentation basins (see Chapter 7). AWWA (1996) reports that a typical loading rate for alum sludge thickening is 4 lb/(ft^2 day) [20 kg/(m^2 day)], whereas later in this chapter, it is reported that for cyst removal from sludge streams, a hydraulic loading of

0.05 gpm/ft^2 (0.12 m/hr) may be needed if polymer is not used. For these hydraulic and flux values, the thickener would be hydraulically limited rather than solids limited.

The common settling test used in the laboratory is conducted in a transparent cylinder filled with sludge and mixed to evenly distribute the solids. At time zero, the mixing is stopped and the solids are allowed to settle. Water plant sludges from clarifiers and sedimentation basins will generally settle as a blanket with a well-defined interface. By recording the height of the interface with time, a plot such as Figure 16.13 can be created. The free settling velocity is then determined as the slope of the straight-line portion of the plot.

In considering the size of the test system, the cylinder diameter is probably the most critical factor. Vesilind (1979) has evaluated the effects of various cylinder diameters. At low suspended solids concentrations (<0.4 percent), the smaller cylinders tended to underestimate the settling velocity, which would result in a more conservative design. However, at suspended solids concentrations over 0.5 percent, the smaller cylinders overestimated the settling velocity. The results are probably site specific, but their work does show the importance of selecting the proper size cylinder for pilot studies. Vesilind made four recommendations for conducting thickening tests:

1. The cylinder diameter should be as large as possible; 8 in is a practical compromise.
2. The initial height should be the same as the prototype thickener depth. When this is not practical, 3 ft should be considered minimum.
3. The cylinder should be filled from the bottom.
4. The sample should be stirred throughout the test, but very slowly—0.5 rpm is a reasonable speed for an 8 in cylinder. This slow stirring will help the test results of small cylinders better approach that of full-scale.

After completing a settling test as described by Figure 16.13, the test is repeated with several different initial suspended solids concentrations, resulting in plot A of Figure 16.14. The solids flux, F, is then computed as:

$$F = vC_i \tag{16.9}$$

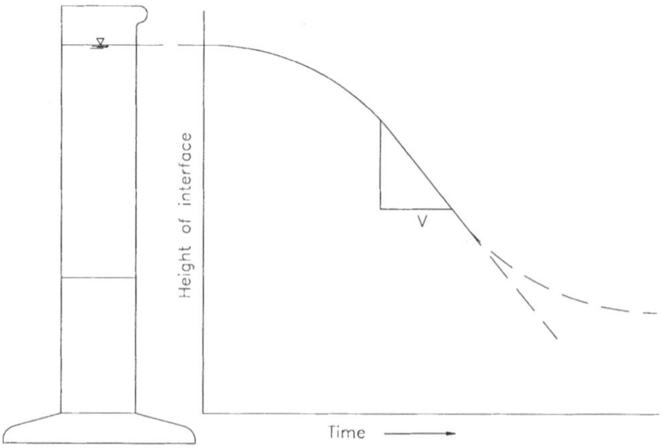

FIGURE 16.13 Thickening test in a cylinder with resulting interface-height-versus-time curve.

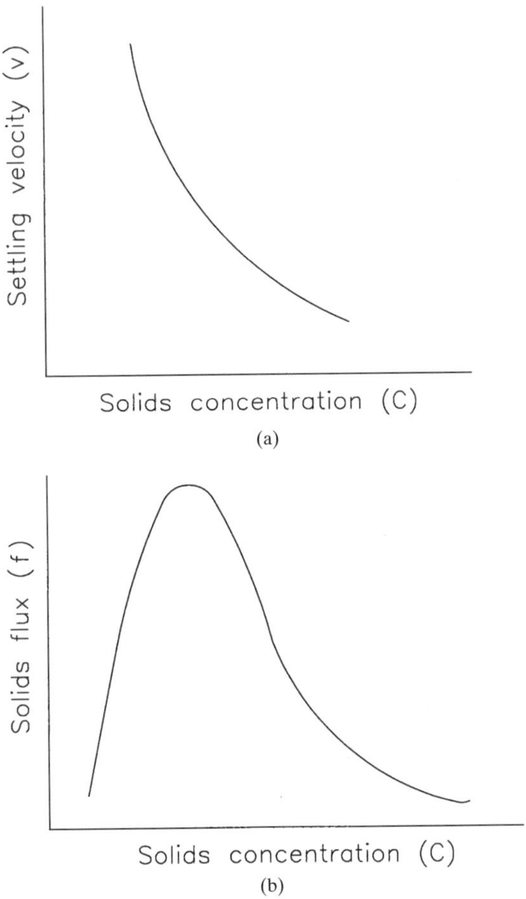

FIGURE 16.14 Thickening test in a cylinder with resulting batch-thickening test plots. *Note:* The results of several batch-thickening tests ploted as (*a*) interface velocity versus initial solids concentration, and (*b*) solids flux versus initial solids concentration.

where F = solids flux [kg/(m^2 h)]
v = settling velocity (m/h)
C_i = initial suspended solids concentration (kg/m^3)

The flux curve will generally take the shape as shown in plot B of Figure 16.14. The flux is zero at zero suspended solids concentration, and the settling velocity approaches zero as the solids concentration increases, thereby driving the flux to zero and representing the maximum possible solids concentration to be achieved.

For batch fill-and-draw tanks, the curves of plot A of Figure 16.14 can be directly used to estimate the settling time required and predict the thickened solids concentrations. Similarly, the curves of plot B of Figure 16.14 could be used to develop possible flux rates for the anticipated range of influent solids concentrations.

For continuous flow thickeners, the solids move to the bottom of the tank not only due to the batch sedimentation discussed above, but also due to the velocity created by the underflow of sludge being removed from the thickener. The flux due to the sludge withdrawal is:

$$F_u = v_u\, C_i \tag{16.10}$$

where F_u = flux due to underdrain withdrawal of sludge
v_u = downward velocity caused by sludge removal
C_i = solids concentration at a given layer in the thickener

The flux due to settling has already been defined and can be labeled for the continuous flow thickener as:

$$F_B = v_i C_i \tag{16.11}$$

where F_B = flux due to solids settling
v_i = settling velocity of solids concentration, C_i
C_i = solids concentration at a given layer in the thickener

Therefore, the total flux, F, is:

$$F = v_i C_i + v_u C_i \tag{16.12}$$

The total flux is plotted in Figure 16.15. The minimum point of this flux curve occurs at the solids concentration layer which restricts performance, and thus the flux cannot be higher. Several methods are available to find this maximum flux, most of the methods having been described by Vesilind (1979). A widely used method was developed by Coe and Clevenger (1916).

Coe and Clevenger used mass and liquid balance equations to develop the following expression for thickener area:

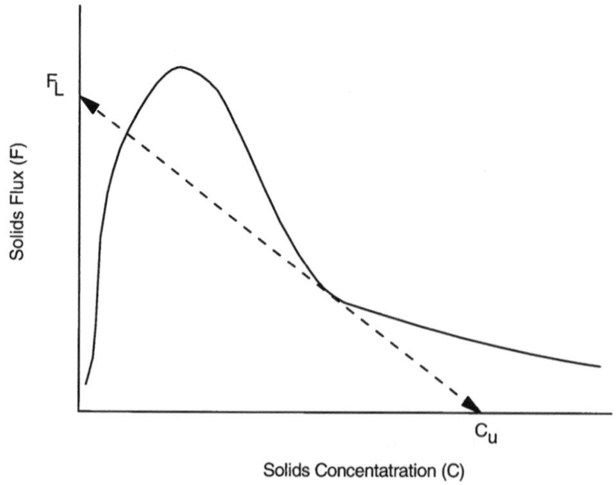

FIGURE 16.15 Solids flux curve for continuous thickener with Yoshioko solution. (*Source:* Visiland, 1979.)

$$A = \frac{Q_0 C_0}{v_i}\left(\frac{1}{C_i} - \frac{1}{C_u}\right) \tag{16.13}$$

where A = thickener area (m^2)
Q_0 = feed flow rate (m^3/sec)
C_0 = influent sludge solids concentration (kg/m^3)
v_i = settling velocity of solids, C_i (m/sec)
C_i = design solids concentration of given layer within the thickener (kg/m^3)
C_u = desired underflow concentration (kg/m^3)

It is necessary to find a number of values of v_i and C_i, and conduct a series of area calculations. The largest area calculated should be used as the minimum design thickener area. A graphical solution was introduced by Yoshioko et al. (1957), as shown in Figure 16.15. A line is drawn tangent to the minimum point in the flux curve beginning at the desired underflow solids concentration. This line will intersect the limiting flux and allow calculations of the required area.

EXAMPLE 16.3 A thickener is to be sized to accomplish both meeting required discharge limits for the supernatant of 10 mg/L suspended solids and to thicken the underflow solids to 2 percent suspended solids from an initial suspended solids concentration of 0.1 percent. The initial sludge flow is 500 gpm.

To properly size this thickener, two types of pilot study are required. The first study would determine the effective solids settling velocity to achieve a supernatant suspended solids of 10 mg/L. This testing could be done using techniques described in AWWA M37 (1992) and would result in determining a settling velocity, for example, of 0.4 cm/min. The area required for appropriate settling to meet the discharge limit would be found from this value.

$$(0.4 \text{ cm/min})\left(0.03\,\frac{\text{ft}}{\text{cm}}\right)\left(\frac{\text{ft}^2}{\text{ft}^2}\right)(7.5 \text{ gal/ft}^3)$$

$$= 0.10 \text{ gpm/ft}^2$$

The area is found as

$$\frac{500 \text{ gpm}}{0.10 \text{ gpm/ft}^2} = 5000 \text{ ft}^2$$

The area must also be found to accomplish the required thickening. A series of settling tests would be conducted resulting in the construction of a curve such as Figure 16.15, resulting in the determination of the allowed flux, for instance, of 4 lb/(ft^2 day). The pounds of solids loaded to the thickener on an equivalent day basis is:

$$\left(\frac{0.1}{100}\right)\left(\frac{500 \text{ gal}}{\text{min}}\right)\left(\frac{8.34 \text{ lb}}{\text{gal}}\right)\left(\frac{1440 \text{ min}}{\text{day}}\right)$$

$$= 6004 \text{ lb/day}$$

and the required area is

$$\frac{6004 \text{ lb/day}}{4 \text{ lb/(ft}^2 \text{ day)}} = 1500 \text{ ft}^2$$

In this case, the limiting area would be controlled by the discharge requirements for supernatant quality and would be 5000 ft^2.

To avoid the need to conduct several settling tests at different solids concentrations as the procedure above requires, researchers (Kynch, 1952; Talmage and Fitch, 1955) have developed simpler test procedures to determine the flux rates. They have formulated techniques that make it possible to develop a complete velocity-concentration profile from only one settling test. Unfortunately, experiments have proven that these methods do not apply to highly compressible materials and thus are not applicable in designing water plant sludge thickeners (Vesilind, 1979).

NONMECHANICAL DEWATERING OF SOLID/LIQUID RESIDUALS

Nonmechanical dewatering, as the name implies, is the dewatering of water treatment plant residuals through means that do not require the use of mechanical devices such as centrifuges or filter presses. Nonmechanical dewatering is used in locations where land is available and where it can be both economical and efficient for dewatering water treatment plant residuals.

A variety of means are employed to accomplish nonmechanical dewatering. The most basic of these is separation of solids and free water through sedimentation followed by natural air drying of the residuals. A second method allows free water to be percolated through sand and into an underdrain system, while additional solids concentration increases are achieved through evaporation. In northern climates, a third system is utilized whereby water treatment plant residuals are subjected to freezing and thawing, which dramatically reduces the residuals volume and correspondingly increases the solids concentration.

Sand Drying Beds

Sand drying beds were initially developed for dewatering municipal wastewater biosolids, but they have since been used to dewater residuals from water treatment plants. Drainage (percolation), decanting, and evaporation are the primary mechanisms for dewatering. Following residuals application to the drying bed, free water is allowed to drain from the residuals into a sand bottom from which it is transported via an underdrain system consisting of a series of lateral collection pipes. This process continues until the sand is clogged with fine particles or until all the free water has been drained, which may require several days. Secondary free water can be removed by decanting once a supernatant layer has formed. Decanting can also be utilized to remove rainwater that would otherwise hinder the overall drying process. Water remaining after initial drainage and decanting is removed by evaporation over a period of time necessary to achieve the desired final solids concentration.

Several variations on sand drying beds are currently in use, and Rolan (1980) proposed the following classification categories:

1. *Sand drying beds.* These are conventional rectangular beds with sidewalls and a layer of sand or gravel with underdrain piping. These are built with or without the provisions for mechanical removal of the dried residuals and with or without a roof or greenhouse-type covering.

2. *Paved rectangular beds.* These have a center sand drainage strip, with or without heating pipes, buried in the paved section and with or without covering to prevent rain incursion. The paved bottom beds are referred to as solar drying beds.
3. *Drying beds with a wedge-wire septum.* These incorporate provisions for an initial flood with a thin layer of water followed by introduction of liquid residuals on top of the water layer, controlled formation of cake, and provisions for mechanical cleaning.

The layout and construction of sand drying beds is very site specific—topography, available land, and operational constraints must all be considered. Topography plays a key role in how beds are laid out on a site, and operational constraints (such as residuals pumping distance) must also be considered when siting a bed location. Materials used in construction are typically cast-in-place concrete or concrete block when the beds are constructed at grade, or earthen sides with a liner when the beds are constructed below grade.

Underdrain systems for sand drying beds are used to collect water that has percolated down through the sand and gravel and transmit it to a point of discharge. Some states require an impervious clay or liner below the underdrain piping. When the beds are equipped for decanting, the flow from the decant mechanism is often tied to the underdrains so that a combined effluent is produced. Underdrains are typically constructed of vitrified clay or plastic piping and many configurations exist, but the most common is collecting drainage with laterals and conveying the flow to a header pipe. Figure 16.16 shows a section view of a typical sand drying bed.

Several sand drying beds are typically used at a given site, which offers some advantages from an operations point of view. Chief among these is the ability to rotate bed use, so that as one sand drying bed is loaded and the residuals begin to dry, another bed is cleaned and readied for a new application of residuals. Cleaning of the sand drying bed can be accomplished with mechanical equipment if concrete support runners are properly installed in the bed. Front-end loaders and Vac-haul trucks have been used successfully by utilities operating sand drying beds.

The time required to evaporate the water remaining after percolation and decanting is the controlling factor in determining the bed size required. Therefore, maximizing the removal of water prior to evaporation will maximize the bed yield. The solids that remain after drainage and decanting are referred to as the drained solids. The value of the drained solids concentration is dependent upon the initial solids concentration, the use of conditioning agents such as polymer, the applied depth and the efficiency of the water removal system. A series of pilot tests can be conducted to determine the combination of loading depth, initial solids concentration, and polymer use that maximizes the drained solids concentration. Vandermeyden et al. (1997) developed the following unit specific equation to determine the bed yield once the drained solids concentration is known:

$$Y = (0.624)\,\frac{SS_{\mathrm{f}}\,SS_{\mathrm{d}}\,E}{SS_{\mathrm{f}} - SS_{\mathrm{d}}} \tag{16.14}$$

where 0.624 = unit conversion assuming a specific gravity of 1 (the conversion factor is 1.2 for a yield in kg/(m^2 year) and evaporation in cm/month)
Y = sand bed yield [lb/(ft^2 year)]
SS_{f} = desired final solids concentration (percent)
SS_{d} = drained solids concentration (percent)
E = net pan evaporation (in/month)

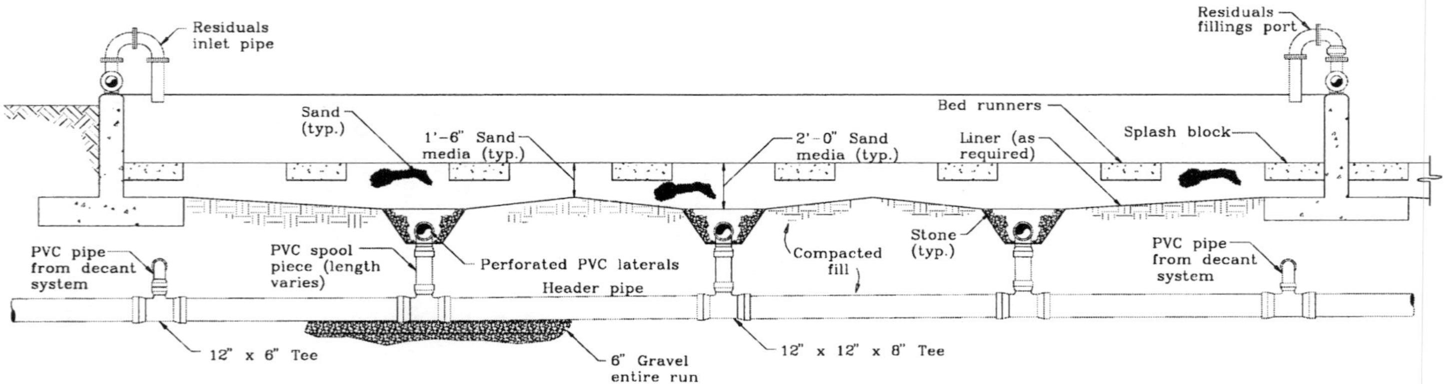

FIGURE 16.16 Typical sand drying bed section.

The bed area required should be determined based on monthly solids production and evaporation numbers rather than annual averages to account for production and climatic variables (Vandermeyden et al., 1997). Sand bed yields are very site specific but tend to be in the 10 to 30 lb/(ft^2 year) [50 to 150 kg/(m^2 year)] range for coagulant sludges, with individual bed loadings in the 1 to 4 lb/ft^2 range (5 to 20 kg/m^2). Polymer conditioning in the 2 lb/ton (1 g/kg) range can significantly increase the drained solids concentrations and, therefore, yields for coagulant sludges. Loading rates for lime residuals vary anywhere from 4 to 20 lb/ft^2 (20 to 100 kg/m^2). Polymer has little effect on improving the yield for lime sludges.

Vandermeyden also found that when using net pan evaporation to predict the evaporation rate, the evaporation was underestimated. In three different field trials, a 25 percent dry cake solids concentration was reached in about two-thirds of the time predicted by pan evaporation data. Thus, under the test conditions studied, pan evaporation would result in a conservative sizing of the sand drying beds. At higher solids concentrations (>30 percent), however, pan evaporation overestimated the actual field evaporation.

Solar Drying Beds

Solar drying beds are similar to sand drying beds in terms of shape and operation; however, they are constructed with sealed bottoms, and have sometimes been referred to as paved drying beds. These beds have little or no provisions for water to be removed through drainage; all residuals thickening and drying is accomplished through decant of free water and evaporation. A principal advantage of this type of drying bed is low maintenance costs and ease of cleaning. No sand replacement costs are associated with this type of drying bed, and because the bottoms of these beds are sealed, neither initial underdrain costs nor underdrain repair costs are incurred. Also, because the entire solar bed bottom is often paved or concrete, cleaning with front-end loaders can be done quickly and efficiently. Because solar beds rely primarily on evaporation, they typically have lower solids loading rates than sand drying beds. Most solar beds are located in the southern and southwestern parts of the country where evaporation rates are high.

Dewatering Lagoons

Dewatering lagoons are very similar to sand drying beds except that they operate at much higher initial loadings, and therefore have longer drying times between cleaning. The dewatering lagoons are equipped with a decant structure and underdrains, similar to a sand bed. For a dewatering lagoon, the lagoon is filled over a long time period (3 to 12 months) and then allowed to dry for a long period while another lagoon is filled. Dewatering lagoons can have an advantage over sand drying beds in reducing peaks, because the loading is often spread over several months. Because dewatering lagoons use a much higher loading rate, the drainage volume as a percent of total applied volume would generally be lower than a sand drying bed. The main difficulty in sizing a dewatering lagoon is in predicting the drained solids concentration [SS(d)] after the loading is complete. Plugging of the sand media on the bottom of the dewatering lagoon with multiple loadings is difficult to predict and would require a carefully planned pilot test with dewatering columns, or even pilot-scale dewatering lagoons would be necessary to accurately size and design the system. The bottom of the lagoon would have a higher solids concentration than the top of the lagoon and a net average solids concentration must be estimated. During the evaporation phase the bottom layers often do not dry out. Some utilities have found that tilling the sludge during the evaporative cycle helps to expose all of the residuals to drying.

A lagoon can be sized by selecting a desired fill cycle (total pounds of sludge loaded to the lagoon) and estimating the drained solids concentration. The lagoon area can be calculated for varying drained solids concentrations in percent as:

$$\text{Dewatering lagoon area (ft}^2\text{)} = \frac{\text{(lb sludge)}}{\text{(Depth)}\,(SS(\text{d}))\,(0.624)} \tag{16.15}$$

Freeze-Thaw Beds

The mechanism for dewatering by freezing is the separation of solid and liquid fractions during ice crystal formation. Because of the highly organized structure of ice, it cannot accommodate other atoms or molecules without severe local strain (Martel, 1987). An ice crystal grows by adding water molecules to its structure. If a growing crystal comes in contact with other atoms or impurities (i.e., solids), it rejects them in favor of water (Chalmers, 1959). The water selectively freezes, forcing the solids to come in contact with each other. This mechanical process of separation of the water and solids is critical to the freeze-conditioning process. As contact is made between the particles, the double layer is reduced and agglomeration of the particles occurs. It is uncertain whether the solids are coagulated by compression or dehydration (Cornwell and Koppers, 1990). In alum or iron sludge, water moves from the sludge into the ice crystal, leaving the crystalline form of aluminum (or ferric) hydroxide. Cornwell (1981) indicated that this process takes place in two stages. The first stage reduces the sludge volume by selectively freezing the free water molecules. In the second stage, the solids are dehydrated when they become frozen. However, Halde (1979) has argued that the solids are coagulated by the compressive forces of expanding ice. In any event, the end result is a separation of the water and solids, with an agglomerating of the solids such that when the ice is thawed and the water drained, the solids remain agglomerated.

When sludge is frozen and subsequently thawed, the resulting volume reduction and increased solids concentration is appreciable. Typically, the volume reduction is well over 70 percent and solids concentrations may reach as high as 80 percent when freeze-thaw is followed by evaporation. Freeze-thaw followed by evaporation dramatically converts the residuals from a fine particle suspension to granular particles. The granular particles often resemble coffee grounds in both size and appearance, and they do not break apart even after vigorous agitation. If the freeze mixture is placed on a porous medium, the water drains away easily upon thawing (Martel and Diener, 1991). As one might expect, freeze-thaw beds are operated most effectively in northern climates, with a range of effective operation beginning at approximately 40° north latitude and extending northward.

Some water treatment plants in cold climates already take advantage of this process by modifying the operation of their lagoons or drying beds. One technique is to decant a lagoon down to the residuals interface and allow it to freeze over the winter months. Martel and Diener (1991) report that this technique is not always successful because the residuals do not freeze to the bottom. Another technique is to pump a shallow layer (20 to 45 cm) of residuals from a storage lagoon into drying beds or ponds that are then allowed to freeze in the winter. This technique works well because the residuals usually freeze completely, but it requires a considerable amount of land and storage volume.

Combination sand drying beds and freeze-thaw beds can also be utilized. In this case, the design must consider the evaporative condition for the drying bed cycle and the freezing and thawing conditions for the freeze-thaw cycle.

Combining the concepts presented in the literature and observations of residuals freezing operations on drying beds and lagoons, a unit operation called a *residuals freezing bed* was developed (Martel, 1989). To maximize residuals dewatering by natural freeze-thaw, a freezing bed includes the following features:

1. It is designed to apply residuals in several thin layers rather than a single thick layer. Each layer is applied as soon as the previous layer has frozen, thereby maximizing the total depth of residuals that can be applied.
2. The bed is covered to prevent snow and rain from entering it. This feature is critical if the bed is to utilize all of the available freezing time in the winter. An open freezing bed would have less capacity because snow accumulations on the surface would slow down the freezing rate. Also, snow removal would be practically impossible if a large snowfall occurred soon after residuals were applied. In this case, the operator would have to delay snow removal until the frozen residuals were thick enough to support snow removal equipment. A covering would also prevent rainfall from rewetting the thawed residuals.
3. The sides of the bed housing are left open to allow free air circulation. However, a half-wall or louvered wall is recommended to prevent drifting snow from entering the bed. Also, the roof is made to be transparent so that solar radiation can help thaw and dry the residuals in the spring. Incoming solar radiation in the winter is expected to be negligible because of the sun's low azimuth and the likelihood of snow on the roof.

A conceptual sketch of the freezing bed is shown in Figure 16.17. Essentially, it consists of an in-ground containment structure that is waterproofed to prevent groundwater infiltration. A ramp is provided at one end to allow vehicle access for residuals removal and to distribute the incoming residuals evenly within the bed. The opposite end of the bed is equipped with an overflow gate or drain valves to draw off supernatant during thaw. The bottom of the bed is underdrained with wedge-wire screen or sand to allow drainage of the filtrate. Both overflow and filtrate are collected in a sump and pumped back to the plant (Martel, 1989).

The freezing bed must be complemented with appropriate residuals storage in a separate lagoon or tankage.

The area required for a freeze-thaw bed is generally determined by the depth of residual that can be frozen. For some climates with long freezing periods, the depth that can be thawed can be the controlling depth. Martel (1989) developed equations to allow the calculation of these two depths. The freezing depth can be found by:

$$D(z) = \frac{t(\mathrm{f})\,(T_\mathrm{f} - T)}{\rho_\mathrm{f}\, F\left(\dfrac{1}{h} - \dfrac{d(z)}{2K}\right)} \tag{16.16}$$

where $D(z)$ = total depth of sludge that can be frozen (m)
T_f = freezing point temperature = 0°C
T = average ambient temperature (°C)
$t(\mathrm{f})$ = the freezing time (hours)
ρ_f = density of frozen sludge = 917 kg/m^3
F = latent heat of fusion = 93 W·h/kg

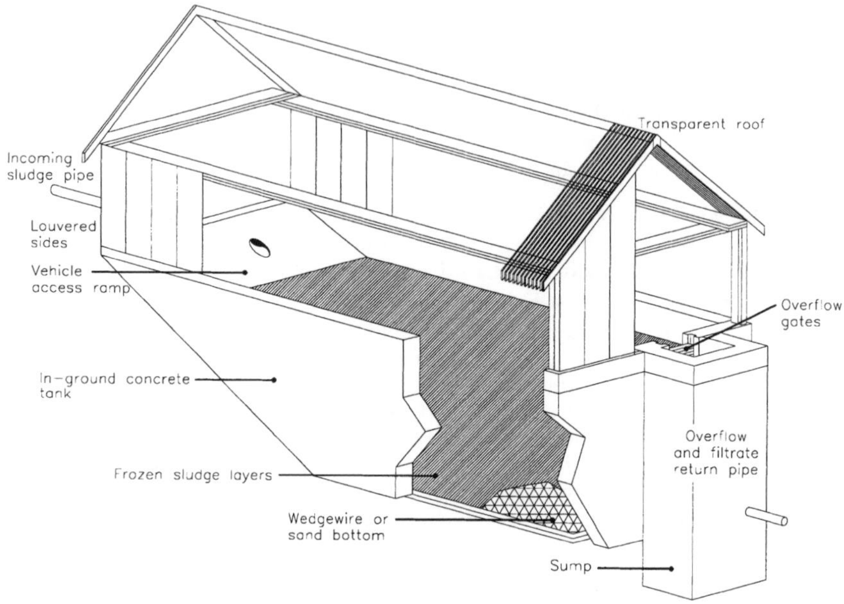

FIGURE 16.17 Conceptual sketch of residuals freezing bed. (*Source:* Martel, C. James. Development and design of sludge freezing beds. *J. Environ. Eng.* 115(4): 799–808. Copyright © 1989 American Society of Civil Engineers. Reproduced by permission of ASCE.)

h = convection coefficient = 7.5 W/m^2°C
$d(z)$ = the thickness of the sludge layer (m)
K = conductivity coefficient = 2.21 W/m°C

Because many of these parameters are known or assumed constants, the equation was reduced by Vandermeyden et al. (1997) to:

$$D(z) = \frac{-t(\mathrm{f})T}{11{,}371 + 19{,}294\,(d(z))} \tag{16.17}$$

Equation 16.17 would be used when the design calls for multiple layers of sludge to be frozen, each layer of thickness $d(z)$. In this case, the utility personnel would apply the layer to the bed, and as soon as one layer had frozen, then another would be applied. In the above equation, $t(\mathrm{f})$ and T can be found (for U.S. locations) from the NOAA through records that are generally on file at the National Climatic Data Center in Asheville, North Carolina (www.ncdc.noaa.gov).

For the case of a one-time bed loading, $D(z)$ would be set equal to $d(z)$. Solving this equation requires use of the quadratic rule which results in the following expression (Vandermeyden et al., 1997):

$$D(z) = \frac{-11{,}371 + \sqrt{1.3 \times 10^8 - 7.7 \times 10^4\; Tt(\mathrm{f})}}{3.9 \times 10^4} \tag{16.18}$$

Mechanically assisted freeze-thaw has also found some limited application in Europe (Cornwell and Koppers, 1990). The mechanical freeze systems generally consist of two chambers and a refrigeration unit. The two chambers, which contain heat exchangers, are operated alternately—while sludge is frozen in one container, frozen sludge is being thawed in the other chamber using heat generated from the freezing process. The limited application of mechanical freezing is primarily due to the high energy consumption and difficulties in obtaining a uniformly frozen material.

MECHANICAL DEWATERING OF SOLID/LIQUID RESIDUALS

Several mechanical devices are available to dewater water treatment plant residuals. However, only five general types are in use—centrifuges, plate-and-frame filter presses, diaphragm filter presses, belt filter presses, and vacuum filters. Vacuum filters have only found use on lime sludges, whereas the other four are used on coagulant as well as lime sludges. Centrifuges, belt presses, and vacuum filters are all considered low-pressure systems, whereas the two types of filter presses can both operate at higher pressures and thus produce a higher solids concentration cake. On coagulant sludges, centrifuges and belt presses will produce a sludge cake in the 15 to 25 percent solids concentration range. Diaphragm and plate-and-frame presses can produce a cake of 30 to 45 percent solids concentration, although there is limited operating experience with the diaphragm presses. All of the devices tend to produce cakes in the 55 to 65 percent range on lime sludges produced by softening groundwater. The choice of equipment can be first narrowed down by the required final cake concentrations, which for coagulant sludges especially will separate the low-pressure devices from the high-pressure devices. However, even when a high cake solids concentration is required, some utilities have found it economical to use a lower-pressure device followed by air drying to produce the desired final cake solids concentration. Ultimately, pilot studies comparing performance of the different devices on a site-specific situation are generally used to select the best equipment. Scale up from pilot results to full-scale then becomes an important design issue.

Vacuum Filtration

Many types of vacuum filters exist. Each is subject to the same limitation; that is, the maximum theoretical pressure differential that can be applied is atmospheric, 14.7 psi (103 kPa). In practice, a differential pressure of about 10 psi (70 kPa) is achieved.

The equipment itself consists of a horizontal cylindrical drum that rotates partially submerged in a vat of sludge that, to assist dewatering, is usually conditioned by either a coagulant or a body feed such as fly ash (Figure 16.18). The drum surface is covered by a filtering medium that is fine enough to retain a thin cake of sludge solids as it is formed. The filtering medium usually consists of a fabric mesh. The drum surface is divided into sections around its circumference. Each section is sealed from its adjacent section and the ends of the drum. A vacuum is applied to the appropriate zone and subsequently to each section of the drum. From 10 to 40 percent of the drum surface is submerged in a vat containing the sludge slurry. The submerged area is the cake-forming zone. When the vacuum is applied to this zone, it causes filtrate to pass through, leaving a cake formed on the cloth. The next zone, the cake-drying zone, represents from 40 to 60 percent of the drum surface. In this zone, moisture is removed from the cake under vacuum. The zone terminates at the point at which the vacuum is shut off. Finally, the sludge cake enters the cake discharge zone, where it is removed from the medium. This is accomplished by the filter cloth

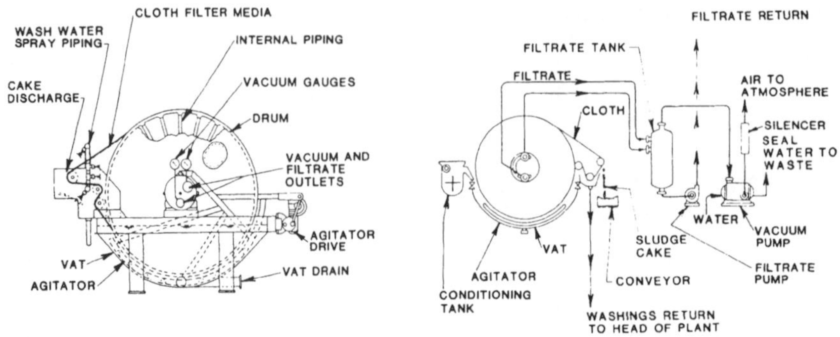

FIGURE 16.18 Typical vacuum filter schematic. (*Source:* Innocenti, 1988.)

belt leaving the drum surface and passing over a small-diameter discharge roll that facilitates cake discharge. No vacuum is applied to this zone.

The best way to describe the performance of a vacuum filter press is the filter yield. The filter yield is defined as the mass of dry cake solids discharged from the filter media per hour per square foot of filter. Filter yield can be expressed as the product of filtrate production per unit area per unit time multiplied by the parameter, *w,* mass of cake deposited per unit volume of filtrate. Based on the specific resistance equation, the following can be used to find the filter yield [see Vesilind (1979) for detailed development].

$$Y = \left(\frac{2PwD}{u\ SR\ t_c} \right)^{0.5} \tag{16.19}$$

where Y = filter cake yield [kg/(m^2 s)]
P = pressure (N/m^2)
w = feed solids concentration (kg/m^3)
D = drum submergence (fraction of the drum circumference below the sludge surface in the pan), dimensionless
u = viscosity (Ns/m^2)
SR = specific resistance (m/kg)
t_c = cycle time [time for a complete revolution of the drum] (s)

All variables in this expression are easily determined by the characteristics of the sludge and the equipment operational setup, except the variable *SR.* Although this equation is theoretically correct, it has not proven accurate enough for design from lab-scale specific resistance tests. However, the equation can be useful for scale-up where the specific resistance is determined in pilot testing and then the sizing parameters can be appropriately scaled.

The first use of a vacuum filter in water treatment was the application of a belt vacuum filter to a lime sludge in Minot, North Dakota. A relatively high magnesium content lime sludge was successfully dewatered. In the late 1960s, Boca Raton, Florida, tested and installed a belt vacuum filter on a calcium carbonate sludge. A 1 to 4 percent solids concentration sludge from the softening reactors was concentrated to 28 to 32 percent by a thickener. The vacuum filter could be loaded at rates of 60 lb/(ft^2 h) resulting in a final cake concentration of 65 percent. The cake was further air dried prior to disposal.

Numerous applications have since been installed with similar results. The two primary factors affecting performance are the solids feed concentration and the magnesium hydroxide content relative to the calcium carbonate content. Yields of 10 to 20 lb/(ft^2 h) [50 to 100 kg/(m^2 year)] are typical for lime sludges with high magnesium hydroxide contents with yields of 40 to 90 lb/(ft^2 h) [200 to 450 kg/(m^2 h)] can be obtained in primarily calcium carbonate sludges.

Belt Filter Press

Belt filter presses use a combination of gravity draining and mechanical pressure to dewater sludges. A typical belt filter press consists of a chemical conditioning stage, a gravity drainage stage, and a compression dewatering stage (see Figure 16.19).

The dewatering process starts after the feed sludge has been properly conditioned, usually with polymer. The slurry enters the gravity drainage stage, where it is evenly distributed onto a moving porous belt. Readily drainable water passes through the belt as the slurry travels over the full length of the dewatering stage. Typically, 1 or 2 minutes are necessary to allow for the filtrate separation in the drainage stage.

Following gravity drainage, the partially dewatered sludge enters the compression dewatering stage. Here, the sludge is "sandwiched" between two porous cloth media belts which travel in an S-shape path over numerous rollers. Both belts operate under a specific tension, which induces dewatering pressure onto the sludge. The S-shape path the sludge follows creates shear forces, which assist in the dewatering process. The compressive and shear forces acting on the sludge increase over the length of this dewatering stage. The final sludge cake is removed from the belts by blades.

Proper sludge conditioning is considered critical for obtaining acceptable dewatering results. A typical sludge conditioning unit consists of chemical conditioner storage, metering pumps, mixing equipment (chemical and chemical/sludge), controls, and process piping.

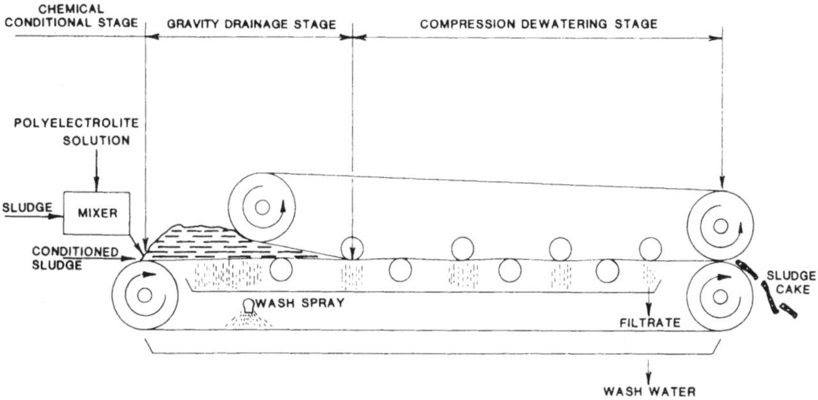

FIGURE 16.19 The three basic stages of belt press. (*Source:* Innocenti, 1988.)

In general, polymer is used for chemical conditioning at a dose of 2 to 5 lb/ton (1 to 2.5 g/kg). To achieve proper sludge conditioning, the polymer is first diluted to between 0.25 and 0.50 percent by weight before it is applied to the feed sludge. Next, the sludge and the polymer are thoroughly mixed. The required mixing time depends on sludge characteristics and type of polymer used.

The design and selection of a belt filter press is often based on the throughput of the machine (i.e., the rate the sludge can be dewatered by the press). The throughput capacity either can be limited by the water in the sludge (hydraulically) or can be solids limited. A belt filter press having a particular type of belt at a particular width has a maximum loading capacity for a particular sludge. Generally, the solids loading is considered the most critical factor and the throughput is expressed in terms of solids loading. The loading units are usually similar to a yield except expressed as belt width—mass/width/time. Loading rates for coagulant sludges are around 100 lb/(ft h) [150 kg/(m h)].

Centrifuges

Centrifugation of sludge is basically a shallow-depth settling process enhanced by applying centrifugal force. The basic physical principle of centrifugal force is that a moving body tends to continue in the same direction; if that body is forced to change directions, it resists the change and exerts a force against whatever is resisting it. In the case of centrifugal force, the force applied by the body is radially outward from the axis of rotation.

Centrifugation enhances settlement of the solids. In conventional settling tanks, the solids are acted on by the force of acceleration due to gravity (g). In centrifugation, the applied force is $r\omega^2$, where r is the distance of the particle from the axis of rotation and ω is the rotational speed. In modern centrifuges, $r\omega^2$ may be 1500 to 4000 times the value of g.

The comparison of $r\omega^2$ to g has led to efforts by many to develop equations for centrifugation by substituting $r\omega^2$ for g. However, this substitution relates to discrete particles only and does not account for hindered settling and the effect of scrolling (moving the solids out of the bowl). These deficiencies in the theory of centrifugation limit the use of sedimentation theory as a basis for the design of centrifuges, and evaluation and design need to be based on pilot studies.

The major type of centrifuge used for the dewatering of water plant sludge is the scroll-discharge, solid-bowl decanter centrifuge. The solid-bowl centrifuge (also called scroll or decanter centrifuge) is a horizontal unit that utilizes a scroll conveyor inside the centrifuge bowl (see Figure 16.20). The unit is fed continuously with the solids settling against the bowl wall. The scroll rotates at a slightly different speed than the bowl and conveys the dewatered sludge to the small end of the centrifuge where it is discharged. The water is directed from the central axis of the centrifuge toward the centrifuge's large end where it is discharged. The water exits through adjustable weirs (level rings), which also control the pool depth.

The best way to evaluate centrifugation is pilot tests on prototype equipment. Tests should be conducted on a centrifuge exactly like the type to be used in full-scale except smaller. Tests should be conducted for operational parameters of concern such as: feed flow rate, feed suspended solids concentration, polymer conditioning, bowl speed, pool depth, and scroll speed. Scale-up considering only liquid loading is often referred to as the *sigma concept*. The sigma concept is based on Stokes' Law description of the settling of discrete particles under the influence of

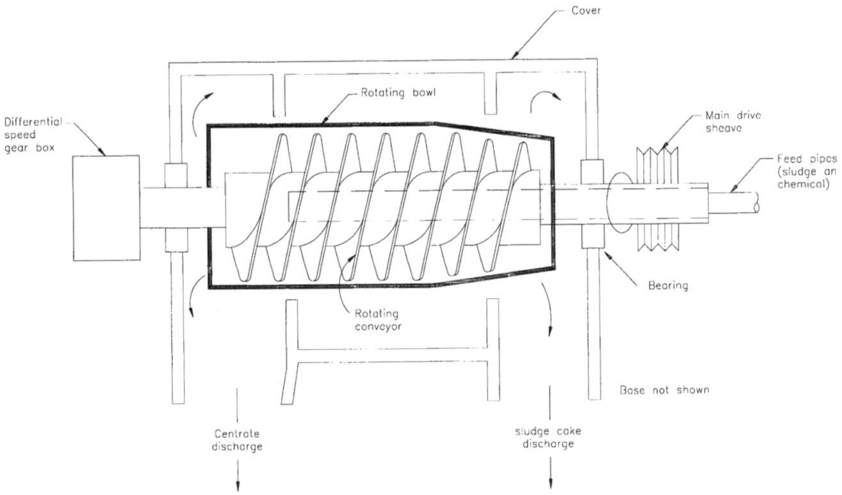

FIGURE 16.20 Schematic of horizontal scroll centrifuge. (*Source:* EE&T, Inc., 1996.)

gravity. Gravity is replaced by the centrifugal acceleration and the expression is integrated over the depth of the water pool. One then ends up with a term for the allowable flow through the centrifuge:

$$Q = \left(\frac{V \omega^2}{g \ln \frac{r_2}{r_1}} \right) \left(\frac{g (P_p - P) d^2}{18u} \right) \tag{16.20}$$

where V = volume of sludge and water in the pool
ω = radial velocity of centrifuge (rad/s)
r_2 = radius from centerline of centrifuge to bowl
r_1 = radius from centerline of centrifuge to pool level
P_p = particle density
P = fluid density
d = particle diameter
u = viscosity

Note that the left-hand term consists of machine variables and the right-hand term is sludge variables. Therefore, in scale-up, if it is assumed that the sludge is the same for full-scale as in the pilot studies,

$$Q_2 = \frac{Q_1 \Sigma_2}{\Sigma_1} \tag{16.21}$$

where Q_2 is the allowable flow in the full-scale centrifuge based on the optimal flow (Q_1) obtained in the pilot plant, and

$$\Sigma = \frac{V \omega^2}{g \ln \frac{r_2}{r_1}} \tag{16.22}$$

which are variables obtainable for the particular size pilot and full-scale centrifuge.

An analysis of the solids loading limitation is known as the *beta concept,* and is expressed as:

$$Q_{s2} = \frac{Q_{s1}\,\beta_2}{\beta_1} \tag{16.23}$$

where Q_s is the solids throughput in units, such as kg/hr, and:

$$\beta = \Delta WSNDz\pi \tag{16.24}$$

where ΔW = bowl/conveyor differential speed
S = pitch of blades
N = number of leads
D = total bowl diameter
z = pool depth

Again, all beta terms are made up of machine variables and set by the manufacturer for the unit of interest.

In scale-up, the limiting conditioning should be calculated using both the sigma and beta concepts.

EXAMPLE 16.4 A centrifuge pilot study is typically conducted to evaluate performance as well as to allow for scale-up to select the appropriate size for the final installation. Figure 16.21 shows example results from such testing with a pilot centrifuge. Recovery refers to the solids capture and a desired value is over 95 percent. If, for this example, a 24 percent cake was desired, then the pilot centrifuge performed adequately at about 40 gpm using 4 to 5 lb/ton of polymer. The following table shows the pertinent information for the pilot centrifuge, as well as for an example full-scale unit. Using the sigma concept, what is the capacity of the full-scale unit?

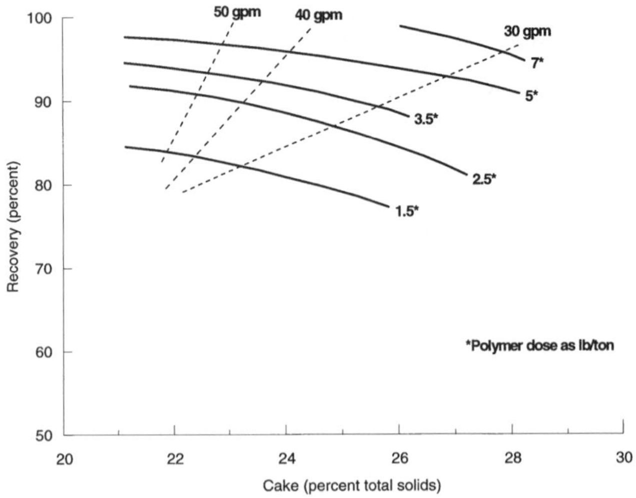

FIGURE 16.21 Centrifuge pilot performance data.

	Pilot unit	Full-scale
Clarification length (in)	31	91.6
Bowl radius (in)	8.4	14.5
Radius to pool (in)	5.1	9.5
rpm	3250	3150

To calculate the sigma values, Eq. 16.22 needs to be used. The volume of sludge or water in the pool, *V*, can be found as:

$$V = \pi L (r_2^2 - r_1^2)$$

where *L* is the clarification length. Therefore, for the pilot unit:

$$V = \pi (31)(8.4^2 - 5.1^2) = 4339 \text{ in}^3$$

and ω is found as

$$\omega = (\text{rpm}) \left(\frac{\text{min}}{60 \text{ s}}\right) \left(2\pi \frac{\text{rad}}{\text{revolution}}\right)$$

$$= 3250 \left(\frac{1}{60}\right) (2\pi) = 340 \text{ rad/s}$$

Using Eq. 16.22:

$$\Sigma_1 = \frac{(4{,}339)\,(340)^2}{(32.2)\,(12)\,\ln\,(8.4/5.1)}$$

$$\Sigma_1 = 2{,}601{,}462 \text{ in}^3 = 1{,}505 \text{ ft}^3$$

Similarly, Σ_2 – 13,320 ft^3, such that the sigma ratio, or scale-up factor, is 13,320/1,505 = 8.85 and, from Eq. 16.21, the design flow for the full-scale centrifuge would be 40(8.85) = 354 gpm.

Filter Presses

The filter press is another process option available for dewatering sludge, and it generally results in the production of the highest final cake concentration of any of the mechanical dewatering devices. Early filter presses were frequently used in Europe for dewatering thin slurries, such as china clays and wastewater sludges. Their practical use for water treatment residues began around 1965. Experiments commenced in England in 1956, but were disappointing until the advent of the use of polymers as conditioners. The first known use in the United States was at the Atlanta Waterworks and at the Little Falls Treatment Plant of the Passaic Valley Water Commission.

At the commencement of a filter cycle, sludge is forced into contact with the cloth, which retains the solid matter while passing the liquid filtrate. Very quickly the cloth becomes coated with a cake of sludge solids, and all future filtering occurs through this cake, which increases in depth as succeeding layers build up. The type of cloth does not affect the rate of filtration after the first few minutes, and it can be ignored from a theoretical point of view.

Filter presses (Figure 16.22) are very heavy, cumbersome pieces of equipment demanding costly foundations and relatively large buildings. Apart from minor refinements, for several decades filter press design changed little until the advent of the diaphragm filter. The original design of the plate and frame filter (Figure 16.23) consisted of a series of frames into which sludge is passed under high pressure, up to 225 psi (1570 kPa), to dewater the sludge against the outer cloth-covered plate. The depth of the cake was consequently fixed, being governed by the distance between filter plates. Different sizes of plates were manufactured to give cakes of, for example, ¾, 1, or 1½ in (19, 25, or 38 mm) depth. Such filters have been used to dewater sludge in an acceptable manner for many years.

A considerable change in design resulted with the introduction of diaphragm filters (Figure 16.24). The advantage of this system is that the thickness of the cake is infinitely variable within the limits of the machine dimensions. Sludge is filtered through a cloth for a fixed period of time, perhaps 20 min, at which stage the sludge supply is cut off and water or compressed air is applied behind an expandable diaphragm that further squeezes water out of the sludge. The cake is dislodged by shaking or by rotating the cloth, depending on the manufacturer's design, and falls into a hopper for disposal. Hanging cakes, where the cake refuses to leave the cloth, an unhappy feature of the older plate and frame press, is consequently eliminated. Diaphragm presses also have the advantage that conditioning of the sludge, although desirable in some circumstances, is not always necessary. As a result, although output is somewhat reduced, a considerable amount of capital and operations costs are eliminated. A diaphragm press is currently installed at the Moores Bridges Water Treatment Plant in Norfolk, Virginia, and was designed for a peak output of 138,000 lb (62,600 kg) of solids per week.

FIGURE 16.22 Conventional filter press. (*Source:* Courtesy of US Filter–Zimpro Products, Rothschild, WI.)

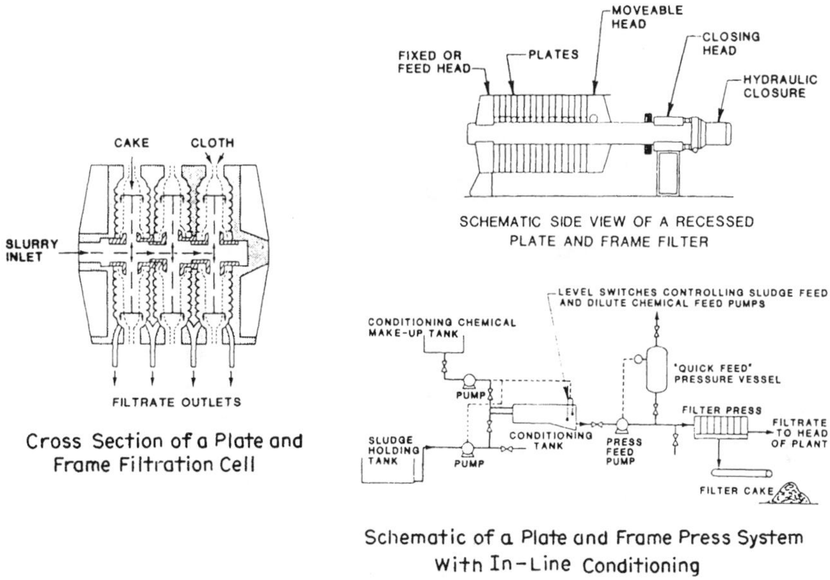

Cross Section of a Plate and Frame Filtration Cell

Schematic of a Plate and Frame Press System With In-Line Conditioning

FIGURE 16.23 Plate and frame filter press schematic. (*Source:* Innocenti, 1988.)

RECYCLE

It is common for many utilities to recycle various streams back to the beginning of the treatment process. Recycling is most often done due to water conservation measures, to improve plant operation, or due to the lack of being able to discharge the liquids to a watercourse or sewer. The liquid volume associated with sedimentation underflow and spent filter backwash water can be 3 to 10 percent of the plant production. In water-scarce areas, the water is too valuable a resource to discharge, and returning it to the treatment process is a necessary water conservation measure. Lime softening plants find that recycling solids can provide nucleation sites for calcium carbonate precipitation, thereby making more efficient use of lime by not having to overfeed chemical to start the precipitation process. Discharge into watercourses is regulated by states under the National Pollutant Discharge Elimination System (NPDES). In some areas, an NPDES permit cannot be obtained or a sewer is not available making recycling the only option.

Recycling can be back to a terminal reservoir, to the beginning of the treatment train, or in some cases to an intermediate point in the treatment process. Possible streams that could be recycled include the following:

1. Filter-to-waste
2. Spent filter backwash water
   a. With the solids from filtration
   b. Without the solids from filtration (after settling)
3. Clarifier or sedimentation basin sludge
4. Sludge thickener overflow (supernatant)
5. Sludge lagoon overflows

FIGURE 16.24 Diaphragm filter press. (*Source:* Courtesy of US Filter–JWI Products, Holland, MI.)

6. Dewatering operation liquid wastes
  a. Pressate from belt or filter press
  b. Centrate from centrifuge
  c. Leachate from sand drying bed

Recycling of these wastes may upset the treatment process and affect the quality of the finished water. The impacts could be caused by the solids in some of the recycle streams, or by constituents in the recycle streams. The principal constituents that could be in recycle streams and be of water quality concern include:

- Microbiological contaminants, including *Giardia* and *Cryptosporidium* parasites
- Total organic carbon
- Disinfection by-products
- Turbidity and suspended solids
- Metals, such as manganese, aluminum, and iron
- Taste- and odor-causing compounds

Giardia and *Cryptosporidium* cysts and oocysts can be present in spent filter backwash water and sedimentation basin sludges at relatively high concentrations (Cornwell and Lee, 1993). For example, one plant studied had *Giardia* and *Cryptosporidium* concentrations of more than 150 cysts/L in the spent filter backwash water, as compared with 0.2 to 3 cysts/L in the raw water.

Laboratory- and full-scale confirmation in that research showed that sedimentation was effective in reducing particles (and cyst levels) in the spent filter backwash prior to recycle. However, very low overflow rates [less than 0.05 gpm/ft^2 (0.12 m/h)] were required to achieve 70 to 80 percent particle removal in the cyst size range. A nonionic polymer was effective in increasing particle removals to more than 90 percent at overflow rates of 0.2 to 0.3 gpm/ft^2 (0.5 to 0.75 m/h).

Nearly all coagulant sludges contain high concentrations of manganese. Quantities of dissolved manganese in the water surrounding sludge can be in the range of 1 to 7 mg/L, and upon storage of the solids, the release of manganese to the surrounding water can reach 20 to 30 mg/L. As sludge accumulates in manually cleaned sedimentation basins, the manganese levels in the clarified water may gradually increase. Therefore, some manganese will be released to sludge thickener overflows and recycled to the plant or will be released in manually cleaned sedimentation basins to the clarified water. Normally, the manganese concentrations are low unless large spikes of waste streams are recycled. However, if accumulated sludge is allowed to occupy a significant portion of the thickener or manually cleaned basin, or if a hydraulic upset occurs, a situation could develop in which large concentrations of manganese are recycled or released from the manually cleaned sedimentation basin.

In the research by Cornwell and Lee (1993), it was generally found that if the solids were removed from the waste streams prior to recycle, total trihalomethane formation potential (TTHMFP) in the recycle streams was no higher than in the raw waters. The same was found for total organic carbon (TOC). However, without solids removal, TTHMFP and TOC levels can be quite high in the waste streams. The recycle streams may contain preformed trihalomethane (THM), and therefore the THM concentration leaving the plant with recycle is sometimes found to be higher than that without recycle. This increase could impact a utility's distribution system THM average.

Waste streams can be treated prior to recycle and the type of treatment depends upon the types of contaminants to be removed and the degree of removal required. Generally, sedimentation is sufficient to remove 80 percent of the solids leaving a relatively clear supernatant to be recycled. Polymer or coagulant assistance can increase the removal efficiency of particles. The supernatant stream could also be disinfected prior to recycle using ozone, chlorine dioxide, or perhaps potassium permanganate. If particulate removal above that achievable by sedimentation is required, filters could be added or membrane technology employed. If recycle of streams is necessitated by conservation or discharge issues, it appears that an appropriate technology can be selected to allow for recycle without impacting finished water; however, more research and site-specific investigations are needed to determine appropriate treatment for each situation.

MEMBRANE AND ION EXCHANGE RESIDUALS

The water industry produces liquid wastes primarily from ion exchange or membrane processes. If the wastes contain a high total dissolved solids (TDS) content, they are often referred to as brine wastes. This would apply to ion exchange processes or to electrodialysis (ED) or membrane processes used to remove salt from seawater or brackish groundwater. Membrane processes that produce a low-TDS waste are referred to as concentrate. Ion exchange has been used for a number of years as a softening process. Most large plants that have used ion exchange have been located near coastal areas so that brine wastes were discharged to the ocean. Many of these large plants have been abandoned due to corrosion and high maintenance costs. However, ion exchange is still practiced by small treatment systems. Ion exchange columns are generally backwashed, or regenerated, at a rate of 5 to 6 gpm/ft^2 for about 10 min. The frequency of regeneration depends upon the hardness level and type of resin being used.

The amount of concentrate or brine produced by membrane processes will vary by application and membrane type. Table 16.4 shows a range of concentrate production for various types of membranes.

Conventional methods of concentrate disposal involve discharge to surface bodies of water, spray irrigation combined with a dilution stream, deep well injection, drain fields, boreholes, or into the sewer. Concentrate disposal methods must be evaluated in light of geographic, environmental, and regulatory impacts. Table 16.5 summarizes concerns and requirements associated with conventional concentrate disposal methods (AWWA, 1996).

ULTIMATE DISPOSAL AND UTILIZATION OF SOLIDS

The treatment and disposal of water treatment plant residuals is rapidly becoming an integral part of operating water treatment plant facilities as local, state, and federal regulations require more stringent standards for traditionally practiced residuals management programs. The discharge of untreated residuals to most surface waters is severely restricted under the National Pollutant Discharge Elimination System (NPDES) of the Clean Water Act. Discharge of residuals to sanitary sewers is becoming more restrictive through wastewater pretreatment standards, the availability of wastewater treatment plant capacity, the capability of digesters to handle the mostly inorganic residuals, and the wastewater plants' effluent standards. Sanitary landfill disposal of dewatered liquid/solid residuals has become very costly, and the residuals consume valuable space in a disposal system that is already subject to shortages in the future. As a result, beneficial use programs for residuals are increasingly being considered by utilities, not only as a cost-effective alternative but also as a publicly acceptable management practice.

The desired residuals management goal for any water utility is to operate an economically efficient and environmentally attractive residuals management plan by developing one or more long-term agreements, which will allow for the proper utilization of residuals in a beneficial application. The beneficial use markets that have exhibited the greatest potential for success with coagulant sludges are the following:

TABLE 16.4 Membrane Concentrate Generation

Membrane process	Percent recovery of feed water	Percent disposal as concentrate
UF	80 to 90	10 to 20
NF	80 to 95	5 to 20
Brackish water RO	50 to 85	15 to 30
Seawater RO	20 to 40	60 to 80
ED	80 to 90	10 to 20

Source: AWWA, 1996.

TABLE 16.5 Concerns and Requirements Associated with Conventional Disposal Methods

Disposal method	Regulatory concerns	Other requirements
Disposal to surface water	Receiving stream limitations Radionuclides Odors (hydrogen sulfide) Low dissolved oxygen levels Sulfide toxicity Low pH	Mixing zone Possible pretreatment Multiple-port diffusers Modeling of receiving stream
Deep well injection	Confining layer Upcoming to USDWs Injection well integrity Corrosivity	Well liner Monitoring well Periodic integrity test Water quality of concentrate must be compatible with the water quality in the injection zone
Spray irrigation	Groundwater protection	Monitoring wells Possible pretreatment Backup disposal method Need for irrigation water Availability of blend waters
Drainfield or borehole	Groundwater protection	Monitoring wells Proper soil conditions and/or rock permeability
Sanitary sewer collection systems	Effect on local wastewater treatment plant performance (toxicity to biomass or inhibited settleability in clarifiers)	None

Source: AWWA, 1996.

Commercial products. The use of sludges in making commercial products has been successful in such areas as turf farming and topsoil blending. In both cases, the sludge is a substitute for natural soil material and offers economical benefits to the commercial user. Brick manufacturing is a potential market for coagulant sludges and experiences by the Santa Clara Valley Water District (Migneault, 1988) and the City of Durham, North Carolina (Rolan, 1976), indicate that this can be a workable program.

Co-use with biosolids. Incorporating sludge in biosolids management programs, such as land application and composting, can be beneficial. Blended products tend to have lower metal concentrations making the product more marketable. Also, for utilities that operate both a water and wastewater facility, permitting, record keeping, and monitoring requirements are reduced when the sludge and biosolids are managed under one program.

Land application. Land application of sludge to agricultural or forested land is a feasible beneficial use alternative. It is not widely practiced because the need has only been here quite recently and residuals must frequently compete with biosolids for appropriate sites.

Application of coagulant sludges to turf farms is a management option that benefits a farmer's harvest and supplements the farmed areas with a new "soil" base. Turf grass has a relatively low nutrient demand but requires significant moisture levels, particularly for initial growth phases. Dewatered sludge applied to a turf farm at the beginning of the seeding process can provide excellent water retention capabilities. The preferred method of application is with a manure-type spreader. The application rate is a function of the type of grass, supplemental fertilization, and residuals quality. A 1-in application depth of residuals at a 20 percent solids concentration is equal to a solids loading rate of 17 dry tons per acre or a mass loading rate of 1.7 percent. Greenhouse and leaching experiments undertaken by the University of Buffalo (Van Benschoten, 1991) showed the success of using residuals in grass growing. Sludge mixed with native soils at 25, 50, 75, and 100 percent loadings clearly enhanced turf grass growth in the greenhouse studies. As shown in Figure 16.25, measured turf grass growth in sludge/soil mixtures was up to three to four times the growth observed in native soils. This improvement in yield allows a turf farmer to bring his fields to harvest sooner and, therefore, provides a marketable turf product sooner than experienced under current harvesting conditions. Increases in labile phosphorus (as determined by the Bray-1 phosphorus tests) followed the recorded increases in grass mass, indicating a beneficial increase in soil phosphorus available to plants.

Commercial producers of topsoil utilize a variety of raw soil products to develop a marketable product for nurseries, homeowners, professional landscapers, and so forth. In this process, the raw soils are screened and blended with some organic material before being sold as a product. Sludge can be blended during the topsoil production process to increase the nutrient value and water retention capabilities. To a topsoil producer, the sludge is a raw material that can be obtained at little or no cost and increases the profit margin on the final product. The amount of sludge added to the topsoil may be 10 percent or less and is a function of consistency, quality, and availability. The acceptable quality of the sludge is determined by individual topsoil producers. An example of metal limits established by one commercial topsoil producer is shown in Table 16.6 (Vandermeyden and Cornwell, 1993).

Blending sludge with topsoil can be a mutually beneficial operation for the utility and the producers. Key elements for long-term success are reasonably consistent solids quality, reliable delivery of dewatered residuals, and strong contractual arrangement between all parties.

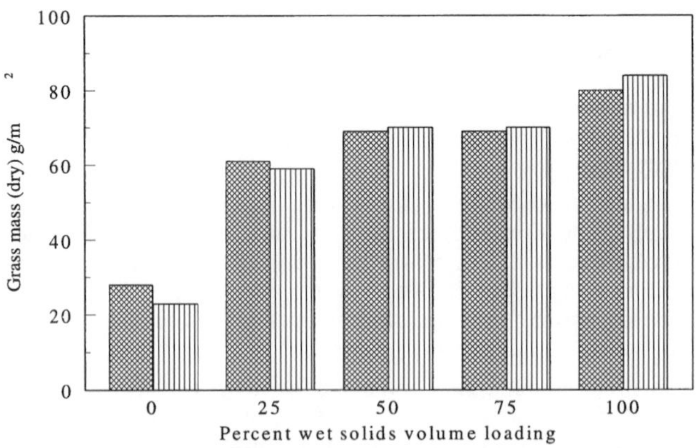

FIGURE 16.25 University of Buffalo greenhouse study: Grass mass versus percent residuals loading. *Note:* Each bar shows the sum of three harvests and represents a unique test. (*Source:* Van Benschoten, J. E., J. N. Jensen, and A. R. Griffin. 1991. *Land Application of Water Treatment Plant Sludge.* Prepared for the Erie County, NY, Water Authority.)

TABLE 16.6 Topsoil Blending: Example of Metal Content Limits (ppm)

Parameter	Preapproved	Requires review	Not accepted
Cadmium	<2.0	2.0–25.0	>25.0
Chromium	<1000	—	>1000
Copper	<100	100–1000	>1000
Lead	<200	200–400	>400
Mercury	<0.3	0.3–10.0	>10.0
Nickel	<200	—	>200
Zinc	<300	300–1200	>1200

Source: Vandermeyden and Cornwell, 1993.

Sludge can be mixed with biosolids and, subsequently, be a coproduct in the overall end-use management strategy of the biosolids. For a utility that operates both water and wastewater facilities, this type of solids management program has certain benefits including: (1) it avoids separate permitting and monitoring of the residuals, (2) it provides a beneficial and often cost-effective end-use avenue for the residuals, and (3) it reduces most of the metal concentrations in the biosolids product because of the diluting effect of the residuals. Even in situations in which the water and wastewater facilities are owned and operated by separate entities, the residuals can enhance the quality of the biosolids product and provide a source of revenue. Residuals can be incorporated with biosolids in a variety of methods, including:

- Discharge of liquid residuals to the sanitary sewer
- Discharge of liquid residuals at the wastewater plant influent
- Discharge of liquid residuals at the wastewater plant solids handling stage
- Blending of dewatered residuals with dewatered biosolids

Water plant residuals containing high concentrations of inorganic compounds or suspended solids can significantly impact primary settler overflow quality, digester space, and digester efficiency. Residuals should also not degrade the end-use biosolids product quality, such as lowering nutrient values or increasing/introducing higher metal concentrations. This is particularly true for land application and composting processes.

The land application of residuals can be an environmentally safe and cost-effective method for residuals management. Due to the variable characteristics of different raw water sources, allowable residuals application rates must be considered according to the composition of each individual residual. Heavy metals and the potential for phosphorus binding are the primary concerns associated with land application. Alum sludge, when added to soil, has the capability to absorb inorganic phosphorus, which prohibits plants from extracting P from the soil for plant growth. However, with good soil management and crop selection, P depletion can be prevented with proper loading rates and P fertilization. The aluminum concentration of alum sludge can range from 5 to 15 percent of the total dry solids mass, which is 50 to 100 percent higher than the concentration of the aluminum in most soils (Elliott and Dempsey, 1991). Elliott and Dempsey (1991) determined that for a 10-ton-per-acre sludge loading rate with an Al concentration of 30 percent, the soil Al level was reported to only increase by 0.3 percent. Aluminum phytotoxicity is dependent on Al solubility and is not a problem when the pH range in the soils is maintained between pH 6 to 6.5 (Elliott and Dempsey, 1991). Cornwell et al. (1992) has found that aluminum does not leach from the sludges but remains complexed within the solids matrix.

The addition of alum sludge to soils could also change the soils, physical structure, or bulk density. A soil with a high-bulk density is more compact, which is unfavorable to plant growth because root penetration is restricted. A low-bulk density (less compacted soil) has more pore space for air and water, which is beneficial for plant growth (Tisdale and Nelson, 1975). Rengasamy et al. (1980) found that soil mixed with alum residuals increased soil aggregation and moisture retention and, as a result, increased the dry yield of maize. Bugbee and Frink (1985) demonstrated that improvements in aeration and moisture retention, promoted by the addition of alum residuals, were made to offset the phosphorus deficiency in lettuce.

The impact of land application on various crops has been investigated by Virginia Polytechnic Institute and State University. Novak (1993) studied the impact of alum and PACl residuals on corn when applied at 1.3 to 2.5 percent by dry weight (13 to 25 tons per acre). Crop yields from the treated plots were not statistically different from the untreated plots. Mutter et al. (1994) studied the impact of PACl residuals on wheat when applied at 2, 4, and 8 percent. No negative effects on wheat grain or biomass yield were observed at these loading rates. Although soil aluminum levels were significantly increased at these loading rates, after two crop rotations the soil Al concentrations were found to be similar to background Al levels. The wheat leaf tissues were slightly increased in Al concentration as compared with the control. This increase in Al concentration, however, was not found to be significant.

Lime sludges have been land-applied for over 40 years. In many farming regions, the application of nitrogen fertilizers causes a reduction in soil pH. Farmers normally apply sufficient quantities of lime to obtain the desired soil pH. Lime sludges are high in $CaCO_3$ and provide the same or better neutralizing value as commercially available limestone (AWWA, 1981). The Ohio Department of Health (AWWA, 1981) reported that liming materials typically available to farmers have a total neutralizing power (TNP) of 60 to 90. The department evaluated sludges from seven utilities and found the TNP of lime sludges to range between 92 to 100, or better than that of commercially available materials. In Illinois, a calcium carbonate equivalent performed on several sludges indicated that the softening sludges were superior to agricultural limestones available locally.

The sludge can be applied to farm lands by either spraying liquid sludge (<15 percent solids concentration) from a tank truck or by spreading and tilling dewatered lime sludge from a hopper-bed truck with a spinner device for spreading the sludge.

Transportation costs and farmer acceptance appear to be major drawbacks to more widespread use of land application of lime sludges. Consideration must also be given to the problem that lime sludge is produced continuously at the water plant, but is needed only seasonally by the farmer.

Special considerations apply to lime sludges containing radium. Illinois, for example, requires that the sludge be mixed with the soil so as to not increase the radium by more than 0.1 pCi/g. The maximum allowable application rate can be found by the following (Cornwell, 1987):

$$AR = 1390\, Sd\, \frac{\text{Ra}}{R} \tag{16.25}$$

where AR = maximum allowed lime sludge application, tons/acre, dry weight (Note: tons/acre times 0.23 equals kg/m^2)
S = specific gravity of soil (for example, =2)
d = depth of sludge/soil mixing (ft)

Ra = allowed radium increases (pCi/g)
R = radium in sludge (pCi/g dry weight)

Landfilling of water plant solids is the disposal choice when beneficial options cannot be obtained. Landfilling of water plant residuals has three basic categories:

- Sludge monofills
- Codisposal in a municipal or industrial landfill
- Hazardous waste landfills

Because water plant solids are usually not hazardous, the last category is not used. Codisposal into a municipal or industrial waste landfill is a matter of contacting the available landfills in the area and determining these requirements and their tipping fee. Generally landfills will require that the sludge be tested to prove it is nonhazardous and it must be of a high enough solids concentration that there is no free water. Due to the costs of codisposal landfills, some utilities are building monofills. Monofills are disposal sites specifically constructed to hold water plant residuals. The two major types of sludge monofills are trench filling and area filling.

In trench landfills, sludge is placed entirely below the original ground surface. The trench depth is dependent typically on the depth to groundwater and bedrock to maintain sufficient soil buffers between the sludge and substrata. Trench depth is a function of sidewall stability and equipment limitations as well. The trenching method encompasses both narrow- and wide-trench-type disposal areas, which range in width from several to 50 ft (15.2 m).

Narrow trenches are generally employed for sludges with low solids concentrations which could not support any type of heavy equipment (see section of sludge physical characteristics). Wide trenches are used for sludges when solids concentrations are sufficient to achieve the necessary shear strength to support heavy equipment. Trenching is a convenient way to operate a landfill because trucks can unload sludge from firm ground above the trench while a hydraulic excavator located outside the trench or a tracked dozer operating inside the trench places and compacts the sludge. Trenches are also relatively quick and easy to construct, thus minimizing construction costs.

The planning and design of sludge disposal trenches involves determination of the following parameters to predict the required acreage needed for a long-term disposal plan:

- Cover thickness
- Excavation depth
- Length
- Orientation
- Sludge fill depth
- Spacing
- Width

Final cover thickness depends on the trench width and type of equipment (land-based or sludge-based) to be employed in final cover operations. It should be noted that a daily cover for odor control is typically not required for water plant sludges. Factors influencing excavation depth include location of groundwater and bedrock, soil permeability, soil cation exchange capacity, equipment limitations, and sidewall stability. Trench length is limited by sludge solids content and ground slopes;

trenches must be discontinued or dikes implemented to contain sludge with a low solids content in a sloping area. For optimal land utilization, trenches should be oriented parallel to one another. Sidewall stability, in addition to ultimately controlling excavation depth, determines trench spacing. In general, for every foot of trench depth, 1 to 1.5 ft (0.3 to 0.5 m) of spacing should be provided between trenches (USEPA, 1978). Spacing should not hinder vehicular access or preclude stockpiling of trench spoil. Sludge solids content and equipment limitations are both important considerations in appropriate trench width determinations.

Unlike the trench landfilling technique, in which sludge is placed below ground, in the area filling technique sludge is generally placed above the original surface. Area filling may be accomplished in one of three ways:

1. Area mound, where sludge is mixed with soil such that it becomes stable enough to be stacked in mounds
2. Area layer, in which sludge is spread evenly in layers over a large tract of land
3. Diked containment, where earthen dikes are constructed above ground to form a containment structure into which the sludge can be disposed

The requirement that sludge must be capable of supporting equipment due to the lack of sidewall containment necessitates reasonably good sludge stability and bearing capacity. These characteristics are typically achieved through good dewatering, dewatering followed by air drying, freeze-thaw dewatering, or mixing sludge with bulking agents (Cornwell et al., 1992). A combination of these methods can also be employed. Areas with high water tables and those with bedrock close to the surface are particularly amenable to area fill methods of sludge monofilling.

The area fill mound technique is a disposal method used for wastewater sludges (USEPA, 1978). For wastewater sludge, a soil bulking agent is generally mixed with the sludge to enhance stability and increase bearing capacity to the degree required based on the number of lifts to be constructed and the weight of the equipment. After being piled in mounds approximately 6 ft high, the sludge/soil mixture is covered with at least 3 ft of soil cover material (more if additional lifts are to be piled on top of the first mound). This disposal method exhibits good land utilization and reasonably high application rates. On the negative side, the tendency of mounds to slump, particularly under high rainfall conditions, and the resulting need for readjustment thereof induces higher manpower and equipment requirements. This monofilling method has also been used for water plant sludges. The need for bulking agents with water plant sludges, however, should be evaluated based on factors such as the sludge shear strength and the size of the monofill (Cornwell et al., 1992).

In the area fill layer disposal method, sludge is spread in 6 to 12 in (15.2 to 30.5 cm) layers. This, in turn, allows additional air drying of the sludge to achieve higher solids concentration and shear strengths. This method would seem favorable for coagulant sludges, which are typically difficult to dewater. Applying the layering method eliminates the need for a separate air drying area outside the monofill, provided the monofill cell is large enough. The area fill layering technique usually results in relatively stable fill areas when completed and, therefore, requires less extensive equipment and manpower efforts for maintenance than the area fill mound technique.

In the diked disposal method, earthen dikes are constructed above ground to form a containment structure into which the sludge can be disposed. Containment areas are sometimes placed at the top of a slope, which provides a portion of the containment structure itself. Access roads are constructed on top of the dikes so that sludge haul trucks can unload sludge directly into the disposal cell.

The sludge can be disposed inside the diked containment in either a layering or a mounding technique although the layer technique seems to be preferred. Access should be provided into the disposal cell itself for trucked equipment and trucks delivering the sludge.

The sludge physical data presented earlier showed shear strength values at various solids concentrations. The importance of these data allows the determination of the minimum solids concentration required, which exhibits adequate shear strength to (1) create stable side slopes to prevent slope failure and (2) support heavy earthmoving equipment utilized in the monofill operations. Once the required minimum solids concentration is determined, the monofill size can be established based on sludge generation rates, density, and disposal depth. Both slope stability and bearing capacity analysis must be performed to determine which requirement will govern the monofill design. Details of determining these factors has been presented by Cornwell et al. (1992).

Figure 16.26 presents a schematic diagram that highlights the essential planning considerations for disposal of water plant sludge in a monofill. The diagram shows the determining factors and the effects of each decision in the planning process as indicated on the figure. Sludge treatment requirements and project economics are affected not only by the quantity of sludge produced, but also by the sludge's physical and chemical characteristics.

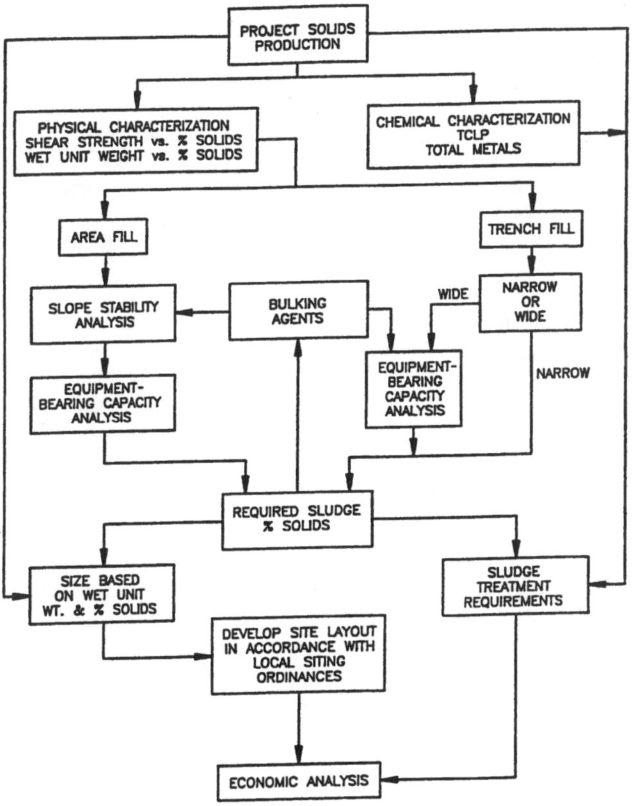

FIGURE 16.26 Considerations for sludge monofill design.

BIBLIOGRAPHY

AWWA. *Management of Water Treatment Plant Residuals.* Denver, CO: American Water Works Association, 1996.

AWWA M37. *Operational Control of Coagulation and Filtration Process.* Denver, CO: AWWA, 1992.

AWWA "Committee report: Lime softening sludge treatment and disposal." *Jour. AWWA,* 73(11), 1981: 600.

Bugbee, G. J., and C. R. Frink. "Alum sludge as soil amendment: Effects on soil properties and plant growth." *Connecticut Agricultural Experiment Station,* Bulletin 827, 1985.

Chalmers, B. "How water freezes." *Sci. American,* 200, 1959: 114–122.

Coe, H. S., and G. H. Clevenger. "Methods for determining the capacities of slime settling tanks." *Trans. AIME,* 55, 1916: 356.

Cornwell, D. A., and R. G. Lee. *Recycle Stream Effects on Water Treatment.* Denver, CO: AWWARF, 1993.

Cornwell, D. A., Carel Vandermeyden, Gail Dillow, and Mark Wang. *Landfilling of Water Treatment Plant Coagulant Sludges.* Denver, CO: AWWARF, 1992.

Cornwell, D. A., and H. Koppers. *Slib, Schlamm, Sludge.* Denver, CO: AWWARF, 1990.

Cornwell, D. A., M. M. Bishop, R. G. Gould, and C. Vandermeyden. *Water Treatment Plant Waste Management.* Denver, CO: AWWARF, 1987.

Cornwell, D. A. "Management of Water Treatment Plant Sludges." *Sludge and Its Ultimate Disposal.* Ann Arbor, MI: Ann Arbor Science, 1981.

Cornwell, D. A., and J. Susan. "Characteristics of acid-treated alum sludges." *Jour. AWWA,* 71(10), 1979: 604.

EE&T. *Residuals Management Options.* Prepared for the Philadelphia Water Department, Philadelphia, PA, 1996.

EE&T, Inc. *Alternative Evaluation and Preliminary Design Studies.* Prepared for Charlotte-Mecklenburg Utility Department, Charlotte, NC, 1992.

Elliott, H. A., and B. A. Dempsey. "Agronomic effects of land application of water treatment sludges." *Jour. AWWA,* 83(4), 1991: 126–131.

Halde, R. "Concentration of impurities by progressive freezing." *Wtr. Res.,* 15, 1979: 575–580.

Innocenti, P. "Techniques for handling water treatment sludge." *Upflow,* 14(2), 1988: 1.

Knocke, W. R., C. M. Dishman, and G. F. Miller. "Measurement of chemical sludge floc density and implications related to sludge dewatering." *Water Environ. Res.,* 65, 1993: 735.

Knocke, W. R., and D. L. Wakeland. "Fundamental characteristics of water treatment plant sludges." *Jour. AWWA,* 75(10), 1983: 516.

Kynch, G. "A theory of sedimentation." *Trans Faraday Society,* 48, 1952: 166.

Martel, C. James, and Carel J. Diener. "A pilot-scale study of alum sludge dewatering in a freezing bed." *Jour. AWWA,* 83(12), 1991: 51–55.

Martel, C. James. *Development and Design of Sludge Freezing Beds,* Doctoral Disertation. Fort Collins, CO: Colorado State University, 1987.

Martel, C. James. "Development and design of sludge freezing beds." *JEED,* 115(4), 1989: 799–808.

Migneault, W. H. "Potential for brickmaking. Freshwater utility recycles its sludge." *BioCycle,* 4, 1988: 63.

Mutter, R., et al. *An Assessment of Cropland Application of Alum Sludge.* Master's Thesis, Blacksburg, Va: VPI & SU, 1994.

Novak, J. *Demonstration of Cropland Application of Alum Sludges.* Denver, CO: AWWARF, 1993.

Novak, J. T., and D. C. Calkins. "Sludge dewatering and its physical properties." *Jour. AWWA,* 67, 1973: 42–45.

Rengasamy, P., J. M. Oades, and T. W. Hancock. "Improvement of soil structure and plant growth by addition of alum sludge." *Communications in Soil Science and Plant Analysis,* 11(6), 1980: 533–545.

Rolan, A. T. "Determination of design loading for sand drying beds." *Jour. North Carolina Section, AWWA and North Carolina WPCA,* L5(1), 1980: 25.

Rolan, A. T. "Evaluation of water plant sludge disposal methods. Proc. Annual Meeting." *Jour. North Carolina Section,* 12(1), 1976: 56.

Schmitt, C. R., and J. E. Hall. "Analytical characterization of water treatment plant sludges." *Jour. AWWA,* 67(1), 1975: 40.

Snoeyink, V. L. "Characteristics and Handling of Wastes from Groundwater Treatment Systems." Sunday Seminar on Experience with Groundwater Contamination. Presented at the AWWA National Conference. 1984.

Talmage, W., and B. Fitch. "Determining Thickener Unit Areas." *Ind. & Eng. Chem.,* 47, 1955: 38.

Tisdale, S. L., and W. L. Nelson. *Soil Fertility and Fertilizers,* 3rd ed., pp. 105–189. New York: Macmillian, 1975.

USEPA. *Technology Transfer Process Design Manual—Municipal Sludge Landfills.* EPA-62511-78-010. Cincinnati, OH: USEPA, 1978.

Van Benschoten, J. E., J. N. Jensen, and A. R. Griffin. *Land Application of Water Treatment Plant Sludge.* Prepared for the Erie County Water Authority, 1991.

Vandermeyden, C., D. A. Cornwell, and K. Schenkelberg. *Nonmechanical Dewatering of Water Plant Residuals.* Denver, CO: AWWARF, 1997.

Vandermeyden, C., and D. A. Cornwell. "A U.S. perspective of beneficial use programs for water treatment and plant residuals." In *Proc. of WEF Biosolids Conference.* Washington, D.C.: Water Environment Foundation, 1993.

Vesilind, P. A. *Treatment and Disposal of Wastewater Sludges.* Ann Arbor, MI: Ann Arbor Science, 1979.

Vesilind, P. A. "Capillary suction time as fundamental measure of sludge dewaterability." *JWPCF,* 60(2), 1988: 215–219.

Yoshioko, N., et al. "Continuous thickening of homogeneous flocculated slurries." *Chem. Eng. (Tokyo),* 21, 1957.

CHAPTER 17

INTERNAL CORROSION AND DEPOSITION CONTROL[1]

Michael R. Schock, Chemist
U.S. Environmental Protection Agency
Water Supply and Water Resources Division
Cincinnati, Ohio

Corrosion is one of the most important problems in the drinking water industry. It can affect public health, public acceptance of a water supply, and the cost of providing safe water. Deterioration of materials resulting from corrosion can necessitate huge yearly expenditures of resources for repairs, replacement, and system. Many times the problem is not given the attention it needs until expensive changes or repairs are required.

Corrosion tends to increase the concentrations of certain metals in tap water. Two potentially toxic metals (lead and cadmium) are attributable almost entirely to leaching caused by corrosion. Three other metals—copper, iron, and zinc—cause staining of fixtures, or metallic taste, or both. Low levels of tin and antimony can be caused by the corrosion of lead-free solders (Herrera, Ferguson, and Benjamin, 1982; Subramanian, Connor, and Meranger, 1991; Subramanian, Sastri, and Connor, 1994). Nickel has sometimes been mentioned as a potential contaminant from the plating of decorative plumbing fixtures. The promulgation of the Lead and Copper Rule by the U.S. Environmental Protection Agency (USEPA) in 1991 has created an emphasis on corrosion control in distribution systems, as well as domestic, public, and institutional plumbing systems (*Federal Register,* 1991a,b, 1994a).

Corrosion products attached to pipe surfaces or accumulated as sediments in the distribution system can shield microorganisms from disinfectants (see Chapter 18). These organisms can reproduce and cause problems such as bad tastes, odors, slimes, and additional corrosion. Several researchers have recently promoted corrosion control within the distribution system as an effective way to maintain water quality and adequate disinfection (Rompré et al., 1996; Schreppel, Frederickson, and Geiss, 1997; Camper, 1997; Kiéné, Lu, and Lévy, 1996; Norton et al., 1995; Olson, 1996).

[1] The views expressed in this paper are those of the author and do not necessarily reflect the views or policies of the U.S. Environmental Protection Agency.

The term *corrosion* is also commonly applied to the dissolution of cement-based materials, and the leaching of their free lime component. The most common manifestation of this problem is the increase in pH, which can be detrimental to disinfection and the aesthetic quality of the water, as well as reducing the effectiveness of phosphate corrosion inhibitor chemicals intended to control the corrosion of metals. The release of asbestos fibers is of regulatory concern, and in extreme cases, the chemical attack on the pipe by the water may cause a reduction of structural integrity and, ultimately, failure.

Even when a water system passes all regulatory requirements, the release of corrosion by-products by miles of distribution system and domestic piping, and the application of corrosion inhibitor chemicals containing metals such as zinc, can be significant sources of metal loading of wastewater treatment plants. This contamination source can affect their ability to meet discharge or sludge disposal limits. Phosphate-based corrosion inhibitors can provide unwanted nutrients to wastewater plants and can cause violations of wastewater or other discharge regulations, or water quality problems in ecosystems receiving the water.

Corrosion-caused problems that add to the cost of water include the following:

1. Increased pumping costs caused by tuberculation and hydraulic friction
2. Loss of water and water pressure caused by leaks
3. Water damage to the dwelling, requiring that pipes and fittings be replaced
4. Replacing hot water heaters
5. Customer complaints of "colored water," "stains," or "bad taste," for which the response may be expensive both in terms of money and public relations
6. Increased wastewater and sludge treatment and disposal costs
7. Increased dosage of chlorine to maintain a distribution system residual

Corrosion is the deterioration of a substance or its properties because of a reaction with its environment. In the waterworks industry, the "substance" that deteriorates may be a metal pipe or fixture, the cement mortar in a pipe lining, or an asbestos-cement (A-C) pipe. For internal corrosion, the "environment" of concern is water.

All waters are corrosive to some degree. A water's corrosive tendency will depend on its physical and chemical characteristics. Also, the nature of the material with which the water comes in contact is important. For example, water corrosive to galvanized iron pipe may be relatively noncorrosive to copper pipe in the same system. Corrosion inhibitors added to the water may protect a particular material, but may either have no effect or may be detrimental to other materials.

Physical and chemical interactions between pipe materials and water may cause corrosion. An example of a physical interaction is the erosion or wearing away of a pipe elbow from high flow velocity in the pipe. An example of a chemical interaction is the oxidation or rusting of an iron pipe. Biological growths in a distribution system (Chapter 18) can also cause corrosion by providing an environment in which physical and chemical interactions can occur. The actual mechanisms of corrosion in a water distribution system are usually a complex and interrelated combination of these physical, chemical, and biological processes. They depend greatly on the materials themselves, and the chemical properties of the water. The purpose of this chapter is to provide an introduction to the concepts involved in corrosion and deposition phenomena in potable waters.

Each material that can corrode has a body of literature devoted to it. Detail on the form of corrosion of each metal or piping material and specific corrosion inhibition practices that might be employed can be found in a comprehensive text (AWWARF-TZW, 1996; Trussell, 1985; Snoeyink and Kuch, 1985) and water treatment journal articles. Table 17.1, modified slightly from the original source (Singley, Beaudet, and Markey, 1984; AWWA, 1986), briefly relates various types of materials to corrosion resistance and the potential contaminants added to the water. In general, plastic plumbing materials are more corrosion-resistant, but they are not without their own potential problems.

CORROSION, PASSIVATION, AND IMMUNITY

Electrochemical Reactions

Metal species can be released into water either from the simple dissolution of existing scale materials, or actual electrochemical corrosion followed by dissolution. In some cases, scale materials formed from corrosion by-products may be eroded from the pipe surfaces. Almost all mineral salts dissolve in water to some extent, from insignificant traces to high concentrations in seawater. This section will provide a general overview of some aspects of the electrochemistry of metallic corrosion as it applies in the context of drinking water treatment. However, many specialized texts on electrochemistry and electrochemical corrosion are available (Piron, 1991; Ailor, 1970; Bockris and Reddy, 1973; Butler and Ison, 1966; NACE, 1984; Pourbaix, 1966, 1973; Thompson, 1970) and should be consulted by readers who are interested in a comprehensive examination of the subject.

For corrosion of any type to occur, all of the components of an electrochemical cell must be present. These include an *anode,* a *cathode,* a *connection* between the anode and cathode for electron transport, and an *electrolyte solution* that will conduct ions between the anode and cathode. The anode and cathode are sites on the metal that have different electrical potential. Differences in potential may arise because metals are not completely homogeneous. If any one of these components is absent, a corrosion cell does not exist and corrosion will not occur.

Oxidation and dissolution of the metal takes place at the anode. The electrons generated by the anodic reaction migrate to the cathode, where they are discharged to a suitable electron acceptor, such as oxygen. The positive ions generated at the anode will tend to migrate to the cathode, and the negative ions generated at the cathode will tend to migrate to the anode. Migration occurs as a response to the concentration gradients and to maintain an electrically neutral solution.

At the phase boundary of a metal in an electrolyte solution an electrical potential difference exists between the solution and the metal surface. This potential is the result of the tendency of the metal to reach chemical equilibrium with the electrolyte solution. This oxidation reaction, representing a loss of electrons by the metal, can be written as

$$Me \Leftrightarrow Me^{z+} + ze^{-} \tag{17.1}$$

Equation 17.1 indicates that the metal corrodes, or dissolves, as the reaction goes to the right. This reaction will proceed until the metal is in equilibrium with the electrolyte containing ions of this metal.

The current that results from the oxidation of the metal is called the *anodic current.* In the reverse reaction, the metal ions are chemically reduced by combining with electrons. The current resulting from the reduction (the reaction, Eq. 17.1, going

TABLE 17.1 Corrosion Properties of Materials Frequently Used in Water Distribution Systems*

Plumbing material	Corrosion resistance	Primary contaminants from pipe
Copper	Good overall corrosion resistance; subject to corrosive attack from high flow velocities, soft water, chlorine, dissolved oxygen, low pH, and high inorganic carbon levels (alkalinities). May be prone to "pitting" failures.	Copper and possibly iron, zinc, tin, antimony, arsenic, cadmium, and lead from associated pipes and solder.
Lead	Corrodes in soft water with pH < 8, and in hard waters with high inorganic carbon levels (alkalinities) and pH below ~7.5 or above ~8.5.	Lead.
Mild steel	Subject to uniform corrosion; affected primarily by high high dissolved oxygen and chlorine levels, and poorly buffered waters.	Iron, resulting in turbidity and red-water complaints.
Cast or ductile iron (unlined)	Can be subject to surface erosion by aggressive waters and tuberculation in poorly buffered waters.	Iron, resulting in turbidity and red-water complaints.
Galvanized iron or steel	Subject to galvanic corrosion of zinc by aggressive waters, especially of low hardness; corrosion is accelerated by contact with copper materials; corrosion is accelerated at higher temperatures as in hot-water systems; corrosion is affected by the workmanship of the pipe and galvanized coating.	Zinc and iron; cadmium and lead (impurities in galvanizing process).
Asbestos-cement, concrete, cement linings	Good corrosion resistance; immune to electrolysis; aggressive (soft) waters can leach calcium from cement; polyphosphate sequestering agents can deplete the calcium and substantially soften the pipe.	Asbestos fibers; increase in pH, aluminum, and calcium.
Plastic	Resistant to corrosion.	Some pipes contain metals in plasticizers, notably lead.
Brass	Good overall resistance; different types of brass respond differently to water chemistry; subject to dezincification by waters of pH > 8.3 with high ratio of chloride to carbonate hardness. Conditions causing mechanical failure may not directly correspond to those promoting contaminant leaching.	Lead, copper, zinc.

* ***Source:*** Adapted from Singley, J. E., B. A. Beaudet, and P. H. Markey, "Corrosion manual for internal corrosion of water distribution systems," U.S. Environmental Protection Agency, EPA/570/9-84-001. Prepared for Office of Drinking Water by Environmental Science and Engineering, Inc., Gainesville, FL, 1984.

to the left) is called the *cathodic current.* At equilibrium, the forward reaction proceeds at the same rate as the reverse reaction, and the anodic current is equal to the cathodic current. Thus, no net corrosion is occurring at equilibrium. The velocity of an electrochemical reaction, unlike that of a normal chemical reaction, is strongly influenced by the potential itself.

Corrosion results from the flow of electric current between electrodes (anodic and cathodic areas) on the metal surface. These areas may be microscopic and in very close proximity, causing general uniform corrosion. Alternatively, they may be large and somewhat remote from one another, causing pitting, with or without tuberculation. Electrode areas may be induced by various conditions, some because of the characteristics of the metal and some because of the character of the water at the boundary surface. Especially significant are variations in the composition of the metal or the water from point to point on the contact surface. Impurities in the metal, sediment accumulations, adherent bacterial slimes, and accumulations of the products of corrosion are all related either directly or indirectly to the development of electrode areas for corrosion circuits.

Figure 17.1 shows an example of corrosion reactions taking place on a fresh pipe surface with proximate anodic and cathodic areas (Snoeyink and Jenkins, 1980). In almost all forms of pipe corrosion, the metal goes into solution at the anodic areas. As the metal dissolves, a movement of electrons occurs and the metal develops an electric potential. Electrons liberated from the anodic areas flow through the metal to the cathodic areas where they become involved in another chemical reaction, and the metal develops another electric potential. The focus of corrosion control by water treatment methods is usually attempting to retard either or both of the primary electrode reactions.

The Nernst Equation

The Nernst equation is a relationship that allows the driving force of the reaction to be computed from the difference in free energy levels of corrosion cell components. The free energy difference under such conditions depends on the electrochemical potential, which, in turn, is a function of the type of metal and the solid- and aqueous-phase reaction products. Electrons (electricity) will then flow from certain areas of a metal surface to other areas through the metal. A metal may go into solution as an ion, or may react in water with another element or molecule to form a complex, an ion pair, or insoluble compound.

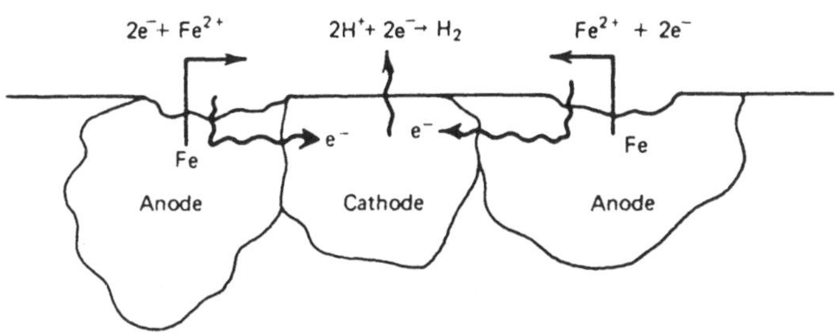

FIGURE 17.1 Adjoining anodes and cathodes during the corrosion of iron in acidic solution. (*Source: Water Chemistry,* V. L. Snoeyink and D. Jenkins. Copyright © 1980, John Wiley & Sons, Inc. Reprinted by permission of John Wiley & Sons, Inc.)

The equilibrium potential of a single electrode can be calculated by using the Nernst equation for the general reaction (Eq. 17.1):

$$E_{Me/Me^{z+}} = E^0_{Me/Me^{z+}} - \frac{RT}{zF} \ln \{Me^{z+}\} \qquad (17.2)$$

where $E_{Me/Me^{z+}}$ = the potential (volts)
$E^0_{Me/Me^{z+}}$ = the standard potential (volts), a constant that can be obtained from tables of standard reduction potentials
{ } denote activity of the ion Me^{z+}
R = the ideal gas constant (about 0.001987 Kcal/deg·mol^{-1})
T = the absolute temperature (°K)
F = the Faraday constant (23.060 Kcal/V)
z = the number of electrons transferred in the reaction

The $_{Me/Me^{z+}}$ subscript indicates the reaction written as a reduction. The Nernst equation, written as Eq. 17.3, is for a single electrode, assuming that the electrode is coupled with the normal hydrogen electrode, at which the reaction

$$2H^+ + 2e^- \rightarrow H_2(g) \quad E^0 = 0.00 \qquad (17.3)$$

takes place, and reactants and products are assumed by thermodynamic convention to equal 1 (Stumm and Morgan, 1981; Garrels and Christ, 1965). The driving force computed by the Nernst equation is directly related to the Gibbs free energy for the overall reaction, through the relationship $\Delta G_r^0 = -zFE$, where ΔG_r^0 is the free-energy change for the complete reaction (Snoeyink and Jenkins, 1980; Stumm and Morgan, 1981). In this convention, standard "half-cell" potentials are tabulated with the reactions written as reductions.

It is usually more useful to use the Nernst equation in a general form for the balanced net reaction of two half-cells, each being of the form

$$\text{ox} + ze^- \Leftrightarrow \text{red} \qquad (17.4)$$

where "ox" and "red" indicate "oxidized" and "reduced" species, respectively (Snoeyink and Jenkins, 1980; Stumm and Morgan, 1981, 1996):

$$E_{\text{red/ox}} = E^0_{\text{red/ox}} - \frac{RT}{zF} \ln \frac{\{\text{red}\}}{\{\text{ox}\}} \qquad (17.5)$$

The subscript "red/ox" indicates overall cell potentials for the total balanced reaction.

At 25°C, and with the conversion to base-10 logarithms for convenience of calculation, Eq. 17.5 can be rewritten as

$$E_{\text{red/ox}} = E^0_{\text{red/ox}} - \frac{0.0591}{z} \log Q \qquad (17.6)$$

where Q is the reaction quotient ({red}/{ox}). At equilibrium, no electrochemical current is generated, and the oxidants and reductants are at their equilibrium activities. Thus, the reaction quotient Q becomes equal to the equilibrium constant K for the overall reaction (Snoeyink and Jenkins, 1980).

In drinking water systems, the oxidation half-cell reaction of a metal, such as iron, zinc, copper, or lead, is coupled with the reduction of some oxidizing agent, such as dissolved oxygen or chlorine species. Example half-cell reactions are the following:

$$O_2 + 2H_2O + 4e^- \Leftrightarrow 4OH^- \tag{17.7}$$

$$HOCl + H^+ + e^- \Leftrightarrow (1/2)\ Cl_2\ (aq) + H_2O \tag{17.8}$$

$$\tfrac{1}{2} Cl_2\ (aq) + e^- \Leftrightarrow Cl^- \tag{17.9}$$

Equations 17.8 and 17.9 can be combined to yield the net reaction:

$$HOCl + H^+ + 2e^- \Leftrightarrow Cl^- + H_2O \tag{17.10}$$

which represents a significant oxidizing half-cell reaction for metals in drinking water.

If Eq 17.7 is written in the Nernst form (Eq. 17.5), and the ionic strength of the water is low enough that it can be assumed to have unit activity ($\{H_2O\} = 1$), then

$$E_{O_2/OH^-} = E^0_{O_2/OH^-} - \frac{0.0591}{4} \log \frac{\{OH^-\}}{\{O_2\}} \tag{17.11}$$

The oxidation potential for this reaction clearly depends on pH, because the $[OH^-]$ is raised to the fourth power in the numerator. Similarly, the oxidation potential of the hypochlorous acid reaction is directly related to pH:

$$E_{HOCl/Cl^-} = E^0_{HOCl/Cl^-} - \frac{0.0591}{2} \log \frac{\{Cl^-\}}{\{HOCl\}\{H^+\}} \tag{17.12}$$

because of the $[H^+]$ term in the denominator.

By thermodynamic definition, the corrosion (and, hence, dissolution of metals from plumbing materials) can only occur if the overall cell potential exceeds the equilibrium cell potential (Pourbaix, 1973; Snoeyink and Jenkins, 1980; Stumm and Morgan, 1981).

Reactions such as Eq 17.7 and Eq 17.10 can be combined with metal oxidation half-cells (Eq 17.1) to show overall corrosion reactions likely to occur in drinking water. Examples are

$$2Pb\ (metal) + O_2 + 2H_2O \Leftrightarrow 2Pb^{2+} + 4OH^- \tag{17.13}$$

$$2Fe\ (metal) + O_2 + 2H_2O \Leftrightarrow 2Fe^{2+} + 4OH^- \tag{17.14}$$

$$Pb\ (metal) + HOCl + H^+ \Leftrightarrow Pb^{2+} + Cl^- + H_2O \tag{17.15}$$

$$Fe\ (metal) + OCl^- + 2H^+ \Leftrightarrow Fe^{2+} + Cl^- + H_2O \tag{17.16}$$

As an example, the overall Nernst expression for the last preceding equation, at 25°C, is

$$E_{Fe^{2+}/OCl^-} = E^0_{Fe^{2+}/OCl^-} - \frac{0.0591}{2} \log \frac{\{Fe^{2+}\}\{Cl^-\}}{\{OCl^-\}\ \{H^+\}^2} \tag{17.17}$$

Note that in Eq. 17.17, the activities are for the free aqueous species, not the total concentrations. The overall potential of the reaction will depend on several factors: the ionic strength; the temperature; the degree of hydroxide complexation of the metal; the presence of complexing agents for the metal; side reactions of any ligands with other species; and the limits on the free metal ion caused by solubility, such as the formation of corrosion product solids. The rates and possibilities for some reactions can also be limited by a barrier to the diffusion of oxidants to the surfaces of the materials, where "fresh" metal is available to oxidize. Anodic and cathodic electrode areas may be induced by various conditions, some because of the characteristics of the metal, and some because of the character of the water at the boundary surface. Impurities in the metal, sediment accumulations, adherent bacterial slimes, and accumulations of the products of corrosion are all related in some way to the development of electrode areas that can enable the operation of corrosion circuits.

Corrosion Products on Pipe Surfaces

Metal surfaces may be protected either by their being "immune" or by rendering them "passive." If a metal is protected by *immunity,* the metal is thermodynamically stable, and is therefore incorrodible (Pourbaix, 1973). For some metals, such as copper, this can occur in groundwaters that are somewhat anoxic (Lytle et al., 1998). Sometimes, this region of electrochemical behavior is only possible when water itself is not chemically stable, so it is only encountered in potable water systems when the consumption of externally supplied energy (*cathodic protection*) occurs.

Passivation occurs when the metal is not stable, but becomes protected by a stable film. The protection can be perfect or (more usually) imperfect, depending upon whether the film effectively shields the metal from contact with the solution (Pourbaix, 1973). True passivation films must satisfy several requirements to effectively limit corrosion. Particularly, they must be electrically conductive, mechanically stable (neither flaking nor cracking), and continuous.

Analysis of corrosion problems is complicated by the variety of chemical reactions that take place across the surface. For example, consider the reactions at an iron or steel surface, in water where oxygen is the only oxidant, and aqueous iron complexation is negligible. The primary reaction occurring at the anodic sites is:

$$Fe(s) \rightarrow Fe^{2+} + 2e^- \tag{17.18}$$

The Fe^{2+} may then diffuse into the water, or it may undergo a number of secondary reactions.

$$Fe^{2+} + CO_3^{2-} \Leftrightarrow FeCO_3(s)\ \text{(siderite)} \tag{17.19}$$

$$Fe^{2+} + 2OH^- \Leftrightarrow Fe(OH)_2(s) \tag{17.20}$$

$$2Fe^{2+} + 1/2\ O_2 + 4OH^- \Leftrightarrow 2FeOOH(s) + H_2O \tag{17.21}$$

The hydrated ferric oxides that form from reactions similar to that shown in Eq. 17.20 are reddish and, under some conditions, may be transported to the consumer's tap. Tertiary reactions may also occur at the surface. Possibilities are:

$$2FeCO_3(s) + 1/2\ O_2 + H_2O \Leftrightarrow 2FeOOH(s) + 2CO_2 \tag{17.22}$$

$$3FeCO_3(s) + 1/2\ O_2 \Leftrightarrow Fe_3O_4(s)\ \text{(magnetite)} + CO_2 \tag{17.23}$$

Reactions such as Eqs. 17.21 to 17.23 can reduce the rate of oxygen diffusing to the anode, thus the formation of oxygen concentration cells. Other reactions that affect corrosion may take place, depending on the composition of the water and the type of metal.

At the same time as the anodic reactions are taking place, a variety of cathodic reactions may be occurring. Perhaps the most common in drinking water distribution systems is the acceptance of electrons by O_2.

$$e^- + 1/4\ O_2 + 1/2\ H_2O \Leftrightarrow OH^- \tag{17.24}$$

This reaction causes an increase in pH near the cathode and triggers the following additional reactions:

$$OH^- + HCO_3^- \Leftrightarrow CO_3^{2-} + H_2O \tag{17.25}$$

$$Ca^{2+} + CO_3^{2-} \Leftrightarrow CaCO_3(s) \tag{17.26}$$

These reactions can cause $CaCO_3(s)$ to precipitate from some waters in which the bulk solutions are undersaturated with this solid, because the pH increase in the vicinity of the cathode forms enough CO_3^{2-} to cause supersaturation with respect to $CaCO_3(s)$.

Several studies have shown that the pH at the surface of pipe can be significantly different from that in the bulk solution (Snoeyink and Wagner, 1996), although many studies reporting extremely high pHs at the surface have neglected to adequately include consideration of buffering by the carbonate system in the water, and the possible role of solids such as $CaCO_3(s)$ and $Mg(OH)_2(s)$ (Dexter and Lin, 1992; Lewandowski, Dickinson, and Lee, 1992; Watkins and Davies, 1987). The deposits that form on pipe surfaces may be (1) a mixture of corrosion products that depend both on the type of metal that is corroding and the composition of the water solution [e.g., $FeCO_3(s)$, $Fe_3O_4(s)$, $FeOOH(s)$, $Pb_3(CO_3)_2(OH)_2(s)$, $Zn_5(CO_3)_2(OH)_6(s)$]; (2) precipitates that form because of pH changes that accompany corrosion [e.g., $CaCO_3(s)$]; (3) precipitates that form because the water entering the system is supersaturated [e.g., $CaCO_3(s)$, $SiO_2(s)$, $Al(OH)_3(s)$, $MnO_2(s)$]; and (4) precipitates or coatings that form by reaction of components of inhibitors, such as silicates or phosphates with the pipe materials (e.g., lead or iron).

The nature of the scales or deposits that form on metals is very important because of the effect that these scales have on the corrosion rate. The formation of scales, such as $CaCO_3(s)$ and iron carbonates on corroding iron or steel, are normally thicker and have higher porosity than the passivating films. Deposits and scales do not decrease the corrosion rate as much as true oxide films do, and the same corrosion current-potential relationship for passivating films does not occur for such scales. The complex interactions can be illustrated by the case with steel corrosion. Scale formation on steel by minerals such as calcium carbonate reduces the corrosion rate by decreasing the rate of oxygen transport to the metal surface, thereby decreasing the rate of the cathodic reaction. Passivating iron oxide films on steel cause an anodic-controlled corrosion reaction.

A very long time, from many months to years, may be required for the corrosion rate of iron and steel to stabilize because of the complex nature of the scales. A much shorter time may be sufficient for other metals. If a scale reduces the rate of corrosion, it is said to be a *protecting* scale; if it does not, it is called *nonprotecting.*

The importance of scale is also demonstrated by the phenomenon of *erosion corrosion,* observed at points in the distribution system or in domestic plumbing systems where a high-flow velocity or an abrupt change in direction of flow exists. The more intense corrosion that often is observed at such locations can be attributed to the abrasive action of the fluid (caused by turbulence, suspended solids, and so forth) that scours away or damages the scale, and to the velocity of flow that carries away corrosion products before they precipitate and that facilitates transport of corrosion reactants more efficiently (Snoeyink and Wagner, 1996).

Changes in water treatment or source water chemistry over time can produce successive layers of new solid phases, remove or change the nature of previously existing deposits, or both. Figure 17.2 shows an example of the complex nature of scale on a cast-iron distribution pipe (Singley et al., 1985; Benjamin, Sontheimer, and Leroy, 1996). Scales of similar chemical composition can have a significantly different impact on corrosion and metal protection because properties such as uniformity,

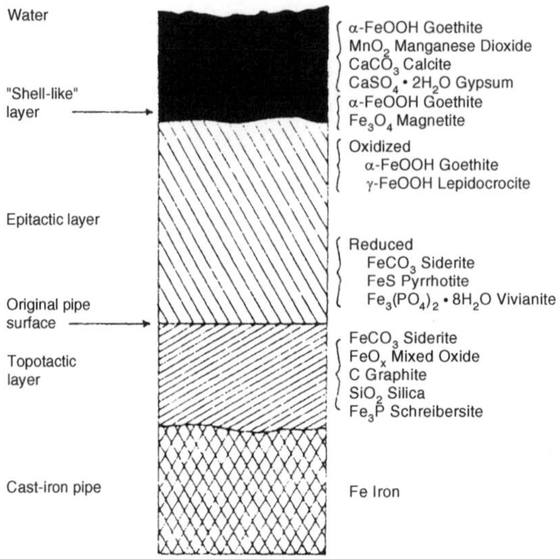

FIGURE 17.2 Schematic of scale on a cast-iron distribution pipe, showing complex layered structure. (*Source: Internal Corrosion of Water Distribution Systems,* 2nd ed., American Water Works Association Research Foundation, Denver, CO, 1996.)

adherability, and permeability to oxidants can vary depending upon such factors as trace impurities, presence of certain organics, temperature of deposition, length of time of formation, and so forth.

Scales that form on pipes may have deleterious effects in addition to the beneficial effect of protecting the metal from rapid corrosion or limiting the levels of toxic metals (such as lead) in solution. Water quality should be controlled so that the scale is protective but as thin as possible, because as bulk of scale increases, the capacity of the main to carry water is reduced. The formation of uneven deposits such as tubercles increases the roughness of the pipe surface, reducing the ability of the mains to carry water, and may provide shelters for the growth of microorganisms.

To properly interpret field and laboratory data from corrosion control studies, it is important to understand that there may be significantly different reactions occurring between the water constituents and the surfaces of "new" pipes compared with "old" pipes. Conceptually, this is illustrated in Figure 17.3 for lead pipe. On the new surface [Figure 17.3(*a*)], the full corrosion reaction can occur, with oxidation of lead followed by the development of a passivating film. Once the film is sufficiently developed [Figure 17.3(*b*)], the oxidants in the water no longer can directly contact the metal of the pipe material itself. Therefore, the oxidation step will not occur, and metal release will become a function of the physical adherence or the solubility of the surface deposit, unless water conditions become anoxic and the metal(s) in the surface deposit become electrochemically reduced. Thus, corrosion inhibitor chemicals that stifle reactions occurring at cathodic surface sites may appear much better in tests using new metal surfaces than they may operate when applied to distribution system pipes covered by thick scales or corrosion deposits. With well-developed surface deposits present, the solubility and surface sorption chemistry of the existing scales is much more important in developing water treatment targets than predictions based on the pure corrosion chemistry of the metal.

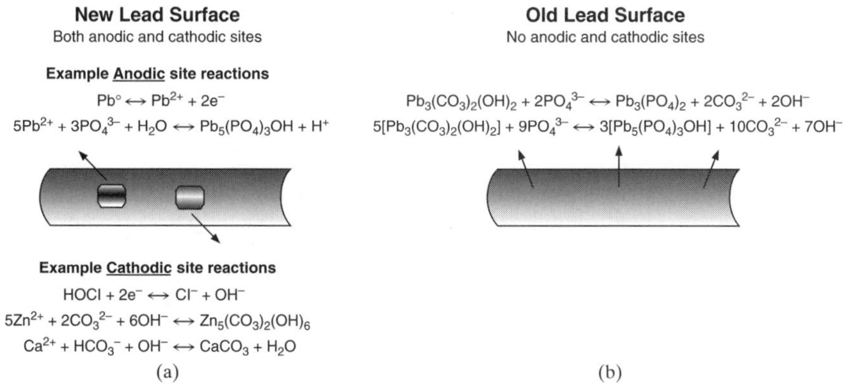

FIGURE 17.3 Schematic representation of different surface reactions and their relation to passivation between (*a*) new and (*b*) old lead pipe surfaces.

Corrosion Kinetics

A three-step process is involved in governing the rate of corrosion of pipe: (1) transport of dissolved reactants to the metal surface, (2) electron transfer at the surface, and (3) transport of dissolved products from the reaction site (Trussell, 1985). When either or both of the transport steps are the slowest, *rate-limiting step,* the corrosion reaction is said to be under *transport control.* When the transfer of electrons at the metal surface is rate-limiting, the reaction is said to be under *activation control.* The formation of solid natural protective scales that inhibit transport are often an important factor in transport control. This section will only present an overview of the concepts involved in the rates of corrosion reactions in potable water systems. Numerous reference articles and texts exist that present a detailed development of the theories that are the basis for many direct electrochemical rate-measuring techniques.

Corrosion is often described in terms of numerous tiny galvanic cells on the surface of the corroding metal. Such localized anodes and cathodes as those described are not fixed on the surface, but are statistically distributed on the exposed metal over space and time. The electrochemical potential of the surface is determined by the mixed contributions to potential of both the cathodic and anodic reactions, averaged over time and over the surface area. Both the individual anodic and cathodic half-reactions are reversible and occur in both directions at the same time. When the "electrode" is at its equilibrium, the rates of reaction in both the cathodic and anodic half-cells are equal.

Given the thermodynamic basis for corrosion described above, and the body of knowledge about kinetic factors that affect the rate of corrosion of metals, several properties of the water passing through a pipe or device that influence the rate of corrosion can be identified. Some of the water-related properties are: (1) concentration of dissolved oxygen, (2) pH, (3) temperature, (4) water velocity, (5) concentration and type of chlorine residual, (6) chloride and sulfate ion concentration, and (7) concentration of dissolved inorganic carbon (DIC) and calcium.

These properties interrelate, and their effect depends on the plumbing material as well as the overall water quality. References specific to the type of plumbing situation (pipe, soldered joint, galvanic connection, faucet, or flow-control device) and the material of interest should be consulted for the most appropriate information on corrosion rate control. Some generalizations will be considered in a later section.

Solubility Diagrams

The solubility of passivating films on the pipe surface is the most important factor in determining whether a given water quality can meet many drinking water regulations that are based on health effects of ingested metals. Asbestos fibers are frequently released into the water as a result of the dissolution of the cement matrix originally holding them. Solubility places a lower limit on the level to which metals can be controlled by modifying water chemistry at the treatment plant. Solubility reflects an ideal equilibrium condition, does not correspond directly to tap water levels, and should not be expected to (Schock, 1990a; Britton and Richards, 1981). Other factors are important, such as the physical location of the plumbing materials relative to the sample collected, the release of particulates from the deposits on the pipes, the rate of the chemical reactions that affect the mobilization of the metals relative to the standing time of the water before collection, and many other variables. However, it is widely believed that trends in the response of tap water metal levels to changes in key water chemistry variables usually follow the predictions of solubility models, and that has been recently verified in principle by some recent surveys of over 2500 U.S. utilities of all sizes (Edwards, Jacobs, and Dodrill, 1999).

To display solubility relationships in a relatively simple two-dimensional manner, the total solubility of a constituent is plotted as a function of a master variable (Snoeyink and Jenkins, 1980; Stumm and Morgan, 1981, 1996; Bard, 1966). Other solution parameters can affect solubility such as the ionic strength, or the concentration of dissolved species (ligands) that can form coordination compounds or complexes with the metal. The solubility diagram does not necessarily give the information needed to minimize the rate of corrosion, because rate is affected by kinetic parameters that depend on the relative rates of the oxidation, dissolution, diffusion, and precipitation reactions. It also cannot predict the ability of a pipe coating to adhere to the pipe surface, or the permeability of a coating to oxidants from the water solution or pitting agents. It does, however, give important information for estimating how water quality affects attainment of maximum contaminant levels (MCLs) or action levels (ALs) for drinking water, the potential for precipitating passivation films on pipe surfaces, or the deposition of other solids important in water treatment, such as calcium carbonate, octacalcium phosphate, and aluminum or ferric hydroxide.

To construct this type of diagram, an aqueous mass balance equation must be written for the metal (or other constituent of interest), with the total solubility (S_T) as the unknown. The mass balance expression should include the concentration of the uncomplexed species (free-metal ion), along with all complexes to be included in the model. For lead (II), the simple relationship for the free ion, plus only the hydroxide and carbonate complexes (the simplest system) is:

$$\begin{aligned} S_{T,Pb(II),CO_3} = {} & [Pb^{2+}] + [Pb(OH)_2^0] + [Pb(OH)_3^-] + [Pb(OH)_4^{2-}] + 2[Pb_2(OH)^{3+}] \\ & + 3[Pb_3(OH)_4^{2+}] + 4[Pb_4(OH)_4^{4+}] + 6[Pb_6(OH)_8^{4+}] + [PbHCO_3^+] \\ & + [PbCO_3^0] + [Pb(CO_3)_2^{2-}] \qquad (17.27) \end{aligned}$$

If the concentration of a complex is going to be relatively negligible compared with the total solubility across the pH range of interest, it can be excluded from the model for simplicity. Frequently, only four or five species are significant in a system. An example of two such complete mass balance expressions are the following for lead (Schock, 1980, 1981b) and copper (Schock, Lytle, and Clement, 1995a,b) systems only containing carbonate species in addition to hydrolysis species, which is the minimum system composition for potable waters.

$$S_{T,Pb(II),CO_3} = [Pb^{2+}] + \frac{\beta'_{1,1}[Pb^{2+}]}{[H^+]} + \frac{\beta'_{1,2}[Pb^{2+}]}{[H^+]^2} + \frac{\beta'_{1,3}[Pb^{2+}]}{[H^+]^3} + \frac{\beta'_{1,4}[Pb^{2+}]}{[H^+]^4}$$
$$+ 2\frac{\beta'_{2,1}[Pb^{2+}]^2}{[H^+]} + 3\frac{\beta'_{3,4}[Pb^{2+}]^3}{[H^+]^4} + 4\frac{\beta'_{4,4}[Pb^{2+}]^4}{[H^+]^4} + 6\frac{\beta'_{6,8}[Pb^{2+}]^6}{[H^+]^8}$$
$$+ \beta'_{1,1,1}[Pb^{2+}][H^+][CO_3^{2-}] + \beta'_{1,0,1}[Pb^{2+}][CO_3^{2-}]$$
$$+ \beta'_{1,0,2}[Pb^{2+}][CO_3^{2-}]^2 \qquad (17.28)$$

$$S_{T,Cu(II),CO_3} = [Cu^{2+}] + \frac{\beta'_{1,1}[Cu^{2+}]}{[H^+]} + \frac{\beta'_{1,2}[Cu^{2+}]}{[H^+]^2} + \frac{\beta'_{1,3}[Cu^{2+}]}{[H^+]^3} + \frac{\beta'_{1,4}[Cu^{2+}]}{[H^+]^4}$$
$$+ 2\frac{\beta'_{2,2}[Cu^{2+}]^2}{[H^+]^2} + 3\frac{\beta'_{3,4}[Cu^{2+}]^3}{[H^+]^4} + \beta'_{1,1,1}[Cu^{2+}][H^+][CO_3^{2-}]$$
$$+ \beta'_{1,0,1}[Cu^{2+}][CO_3^{2-}] + \beta'_{1,0,2}[Cu^{2+}][CO_3^{2-}]^2$$
$$+ \frac{\beta'_{1,-1,1}[Cu^{2+}][CO_3^{2-}]}{[H^+]} + \frac{\beta'_{1,-2,1}[Cu^{2+}][CO_3^{2-}]}{[H^+]^2} \qquad (17.29)$$

In these equations, $\beta'_{m,h,c}$ represents the formation constant for the complex having a stoichiometry of m metal ions, h hydrogen ions, and c carbonate ions, corrected for ionic strength and temperature (Stumm and Morgan, 1981, 1996). To obtain the predicted solubility, the solubility constant expression for each solid of interest is rearranged and solved to isolate the free species concentration, and is substituted into each term of the mass balance equation (e.g., Eq. 17.28 or 17.29). Sometimes, an iterative technique is used to simultaneously solve several related mass balances, depending upon exactly the type of diagrams desired and assumptions about the availability of components involved in the precipitation and dissolution reactions in the system.

A diagram is then constructed for each solid, and the curves are superimposed. The points of minimum solubility are then connected, giving the final diagram. This procedure is discussed in more detail by Schock (1980, 1981b) for lead (II); Schock, Lytle, and Clement for copper (1995a,b); and Snoeyink and Jenkins (1980) for iron (II).

When additional aqueous species are present, such as orthophosphate or sulfate, they are just added to the basic mass balance expressions, such as those shown above. For example, the contribution to lead(II) solubility from aqueous orthophosphate species is represented by:

$$S_{T,\,Pb(II),\,PO_4} = [PbHPO_4^0] + [PbH_2PO_4^+] \qquad (17.30)$$

So, the total lead solubility $S_{T,\,Pb(II)} = S_{T,\,Pb(II),CO_3} + S_{T,\,Pb(II),\,PO_4}$, or

$$S_{T,\,Pb(II)} = [Pb^{2+}] + [Pb(OH)_2^0] + [Pb(OH)_3^-] + [Pb(OH)_4^{2-}] + 2[Pb_2(OH)^{3+}]$$
$$+ 3[Pb_3(OH)_4^{2+}] + 4[Pb_4(OH)_4^{4+}] + 6[Pb_6(OH)_8^{4+}] + [PbHCO_3^+]$$
$$+ [PbCO_3^0] + [Pb(CO_3)_2^{2-}] + [PbH_2PO_4^+] + [PbHPO_4^0] \qquad (17.31)$$

Similarly, other ligands (such as chloride and sulfate) can simply be added to the expressions.

Because these diagrams are inherently two-dimensional (solubility on the *y*-axis, pH on the *x*-axis), there are often additional variables that have an important impact on solubility (such as temperature, ionic strength, carbonate concentration or alkalinity, and orthophosphate concentration), these diagrams must display solubility and species concentration lines for fixed concentrations or values of the other variables. For instance, Figure 17.4 displays a solubility diagram for lead in the system

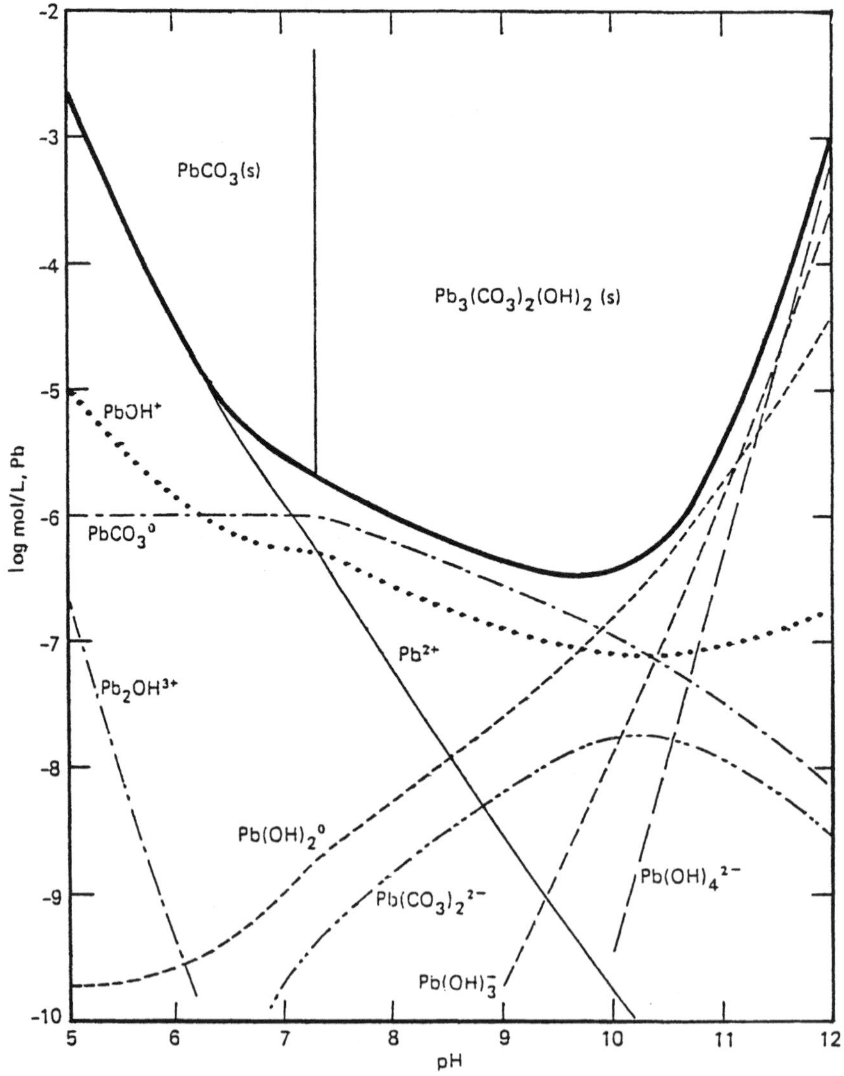

FIGURE 17.4 A solubility diagram showing dissolved lead (II) species in equilibrium with lead (II) solids in a pure system containing 3 mg C/L DIC (2.5×10^{-4} mol/L) at 25°C and $I = 0.005$ mol/L. (*Source:* data from Schock and Wagner, 1985.)

containing only carbonate species and water, showing the important aqueous species and the stability domains of two lead solids. In this figure, the dissolved inorganic carbon (DIC) concentration was fixed at 0.00025 mol/L (3 mg C/L), with a temperature of 25°C and an assumed ionic strength of 0.005. Figure 17.5 shows the same basic system, but with a total DIC concentration of 0.0025 mol/L (30 mg C/L).

In both diagrams, the implicit assumption was also made that the solution redox potential was not high enough to cause the formation of Pb(IV) aqueous or solid

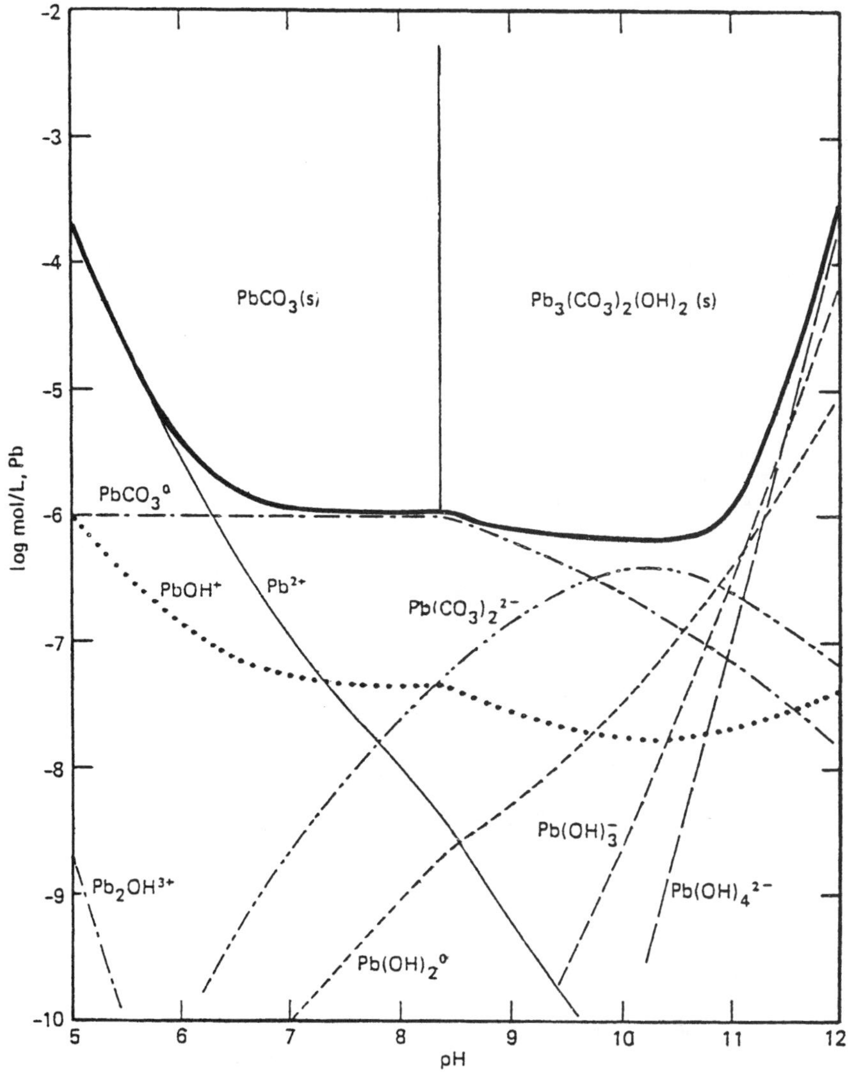

FIGURE 17.5 A solubility diagram showing dissolved lead (II) species in equilibrium with lead (II) solids in a pure system containing 30 mg C/L DIC (2.5×10^{-3} mol/L) at 25°C and $I = 0.005$ mol/L. (*Source:* data from Schock and Wagner, 1985.)

species, or a different diagram would apply for the higher oxidation state species of lead. Additionally, when metal solubility becomes high, care must be taken to ensure that computational constraints on the system, such as the fixed ionic strength and the total ligand concentration (mass balance), are not violated by the presence of high concentrations of dissolved metal species.

Note that if the same complex formation constants were used, but a less soluble constant was used for $PbCO_3(s)$ (cerussite) solubility, the simple carbonate solid would be predicted to be stable over a wider pH range. Likewise, a less soluble constant for $Pb_3(CO_3)_2(OH)_2$ (hydrocerussite) would expand its stability field relative to that of cerussite. Analogous information can be obtained through the careful construction and study of solubility diagrams for other metals, such as copper and zinc.

When constructing solubility diagrams for any metal, the selection of solid and aqueous species must truly represent the system to be modeled, or very erroneous conclusions can result. For example, the aragonite form of calcium carbonate is frequently found in deposits formed in systems having galvanized pipe, rather than the calcite form. Also, the ferric iron deposits formed in mains are frequently a relatively soluble hydroxide or oxyhydroxide form [$Fe(OH)_3$ or FeOOH] rather than an ordered form such as hematite (Fe_2O_3). To go along with this concept, the equilibrium constants must also be accurate to give realistic concentration estimates, and knowledge of changes in the equilibrium constants with temperature is essential, especially when projections of depositional tendency have to be made into hot or very cold water-piping systems. Critical evaluation of data appearing in handbooks and published papers is necessary to avoid using incorrect values, and occasionally review articles or major works by rigorous researchers can be consulted for reliable values.

An important assumption behind the diagrams is that the system must reach thermodynamic equilibrium for the calculations to be truly valid, unless kinetic factors are incorporated into the model. Sometimes improvements can be made in predictions by using metastable species in the calculations, although it is not thermodynamically rigorous to do so. Metastable solids have been found to govern copper (Schock, Lytle, and Clement, 1994, 1995a,b; Edwards, Meyer, and Schock, 1996) and sometimes lead (Schock, Wagner, and Oliphant, 1996) levels in drinking water.

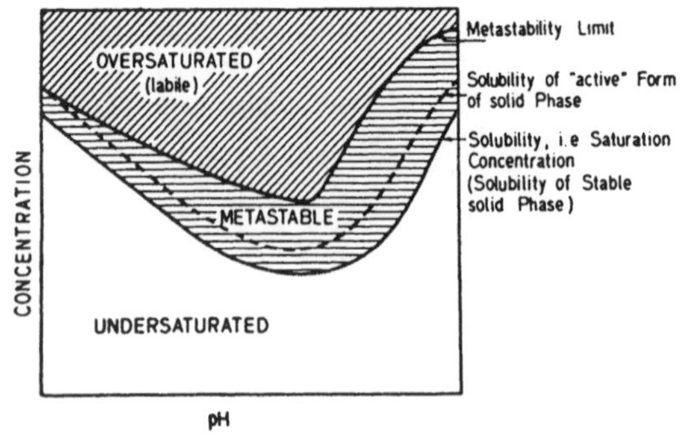

FIGURE 17.6 Solubility and saturation. A schematic solubility diagram showing concentration ranges versus pH for supersaturated, metastable, saturated, and undersaturated solutions. (*Source: Aquatic Chemistry,* W. Stumm and J. J. Morgan. Copyright © 1980, John Wiley & Sons, Inc. Reprinted by permission of John Wiley & Sons, Inc.)

Stumm and Morgan have discussed the formation of precipitates and how this is related in a conceptual way to solubility diagrams, as illustrated by Figure 17.6 (Stumm and Morgan, 1981, 1996). They define an "active" form of a compound as one that is a very fine crystalline precipitate with a disordered lattice. It is generally the type of precipitate formed incipiently from strongly oversaturated solutions. Such an active precipitate may persist in metastable equilibrium with the solution and may convert ("age") slowly into a more stable, "inactive" form. Measurements of the solubility of active forms give solubility products that are higher than those of the inactive forms. The formation of some of the iron hydroxide or oxyhydroxide solids in pipe deposits mentioned previously provides an example of this phenomenon.

Hydroxides and sulfides often occur in amorphous and several crystalline modifications. Amorphous solids may be either active or inactive. Initially formed amorphous precipitates or active forms of unstable crystalline modifications may undergo two kinds of changes during aging. Either the active form of the unstable modification becomes inactive or a more stable modification is formed. With amorphous compounds, deactivation may be accompanied by condensation or dehydration. When several of the processes take place together, nonhomogeneous solids can be formed upon aging. Similar phenomena can occur with basic carbonates, such as a transition from one form to another with changes in pH or DIC over time.

Rather than construct a different detailed diagram for each level of a secondary variable, such as DIC, frequently multiple lines representing the different levels of this variable are added to a single diagram, and the aqueous species are omitted. For metals (such as zinc, copper, and lead) with solubilities that tend to be influenced by complexation, the expansion of the diagram to include a "third dimension" is often useful.

For a qualitative, conceptual understanding, a three-dimensional (3-D) surface can be constructed that can show multiple trends in a complex system at a glance. Figure 17.7 shows the two "troughs" representing $PbCO_3(s)$ and $Pb_3(CO_3)(OH)_2(s)$, the different trend of solubility with DIC concentration for each solid, and the dis-

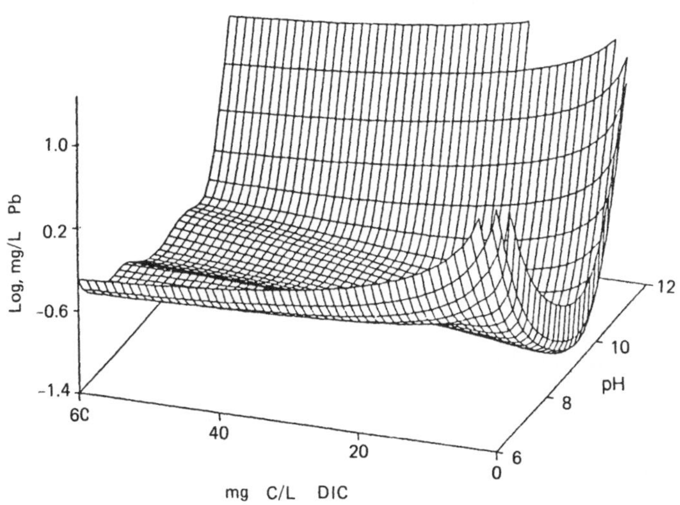

FIGURE 17.7 Three-dimensional representation of the effect of DIC concentration and pH on lead (II) solubility, assuming ionic strength = 0.005, and temperature = 25°C. (*Source:* Schock and Wagner, 1985.)

tinct solubility minimum in the system (Schock, 1980, 1981b; Schock, Wagner, and Oliphant, 1996; Schock and Gardels, 1983).

To obtain a better quantitative estimate of trends of solubility resulting from interrelationships between two major variables, diagrams such as Figure 17.8 can be constructed. Operating on the same principles as topographic maps, such *contour diagrams* present a "map view" of surfaces such as Figure 17.7. The diagrams are derived by interpolating levels of constant concentration within a three-dimensional array of computed solubilities at different combinations of the other two master variables (e.g., pH and DIC). Several different mathematical algorithms are widely used in commercially available computer software, and considerable care must be exercised in selecting algorithms and data point spacing to prevent the creation of erroneous contouring artifacts. Rapid changes in solubility with respect to a master variable (here, pH or DIC) are shown by closely spaced contour lines. A series of these diagrams at levels of a third master variable (such as orthophosphate concentration, temperature, and so forth) can be useful to help display multiple interactions with a minimum of diagrams. They also enable a direct reading of estimated solubilities, without having to guess from an indirect perspective, such as with Figure 17.7. One problem with this type of diagram, as well as with the 3-D surface plots, is that most metals can undergo a change in solubility of three orders of magnitude or even more, over the range of conditions that might be reasonable for potable waters. Thus, logarithmic scales are often necessary for the metal concentrations, which can be somewhat confusing to read.

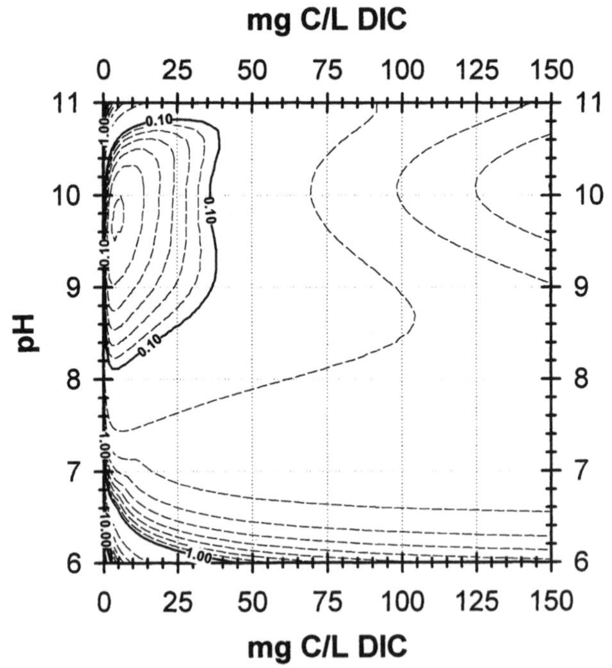

FIGURE 17.8 Contour diagram for lead (II) solubility assuming the formation of $PbCO_3$ and $Pb_3(CO_3)_2(OH)_2$, computed for $I = 0.02$, and temperature = 25°C. (*Source:* equilibrium constant data from Schock and Wagner, 1985.)

Pourbaix or Potential/pH Diagrams

Using the Nernst equations for appropriate electrochemical half-reactions, it is possible to construct potential-pH diagrams, which are also called Eh-pH, or Pourbaix, diagrams. These diagrams have been popularized by Pourbaix and his coworkers in the corrosion field (Pourbaix, 1966, 1973; Obrecht and Pourbaix, 1967), by Garrels and Christ, and by Stumm and Morgan in geochemistry (Garrels and Christ, 1965; Stumm and Morgan, 1981). A similar type of diagram uses the concept of electron activity, pE, which is analogous to the concept of pH (Snoeyink and Jenkins, 1980; Stumm and Morgan, 1981).

The Pourbaix diagrams include the occurrence of different insoluble corrosion products of the dissolved metal that limit the concentration of the free metal ion. These diagrams mainly give information about thermodynamically stable products under different conditions of electrochemical potential. The position of the boundaries of each region is also a function of the aqueous concentrations (activities) of ions that participate in the half-cell reactions.

Potential-pH diagrams are particularly useful to study speciation in systems that could contain species of several possible valence states within the range of redox potential normally encompassed by drinking water, such as manganese, iron, arsenic, and copper. Obtaining an accurate estimate of the redox potential of the drinking water is usually an important limitation in using potential-pH diagrams. The diagrams are also useful to gauge the possible reliability of electrochemical corrosion-rate measurement techniques. Measurement methods that rely on the imposition of a potential to the pipe surface may shift the pipe surface into the stability domain of a solid that would not normally form when freely corroding. The imposed potential might also serve to alter the nature of the surface phase, leading to an erroneous identification of the dominating corrosion or passivation reactions as the result of a surface compound analysis.

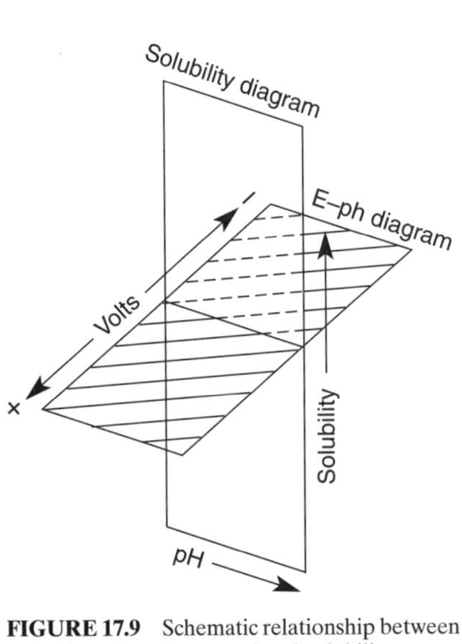

FIGURE 17.9 Schematic relationship between potential pH and conventional solubility versus pH diagrams.

A potential-pH diagram is related to the solubility versus pH plots discussed earlier. This relationship is shown schematically in Figure 17.9. If a conventional two-dimensional solubility diagram is considered as a vertical plane, a potential-pH diagram may be thought of as a "slice" that is perpendicular to the solubility plane, which cuts through the plane at a single concentration of the metal. In actuality, potential-pH diagrams are usually computed in terms of activities rather than concentrations, but the difference is usually not important for practical purposes. The activities of all aqueous and solid species must be fixed, while the electrical potential of the solution and the pH become the master variables. If a given water constituent (such as calcium) does not oxidize or reduce under meaningful physical conditions, no additional useful information is gained (beyond that directly available with a solubility diagram) by constructing a potential-pH diagram.

Figure 17.10 is an example of a potential-pH diagram for iron in water (AWWARF-DVGW, 1985). This particular diagram assumes Fe_2O_3 and Fe_3O_4 to be the solid phases that can control iron solubility. The diagram in Figure 17.10 shows that iron and water are never thermodynamically stable simultaneously, because the iron metal field (Fe) falls below the line where water is reduced to H_2 gas. At such a low electrode potential, the iron will not corrode (i.e., it is immune). The iron potential is reduced to the immune region, for example, by cathodic protection. To accomplish this, the iron must be coupled with another, more easily corrodible material, such as magnesium. At low pH (<5) and intermediate to high potential (approximately –0.5 to 1.3 V), the diagram shows that the stable iron species is Fe^{2+} or Fe^{3+}. Corrosion will occur at a high rate under these conditions (i.e., the metal is active). In the high-potential and high-pH regions, solid products such as $Fe_2O_3(s)$ or $Fe_3O_4(s)$ may form and deposit on the surface of the iron. Figure 17.10 also shows the pH-potential regions for H_2–H_2O–O_2 stability. At very low potential, water is reduced to H_2, and at high potential, water is oxidized to O_2. The stability of water thus limits the range over which the potential of a metal can be varied if it is in contact with water.

The same restrictions that apply to solubility diagrams apply to Pourbaix diagrams. Realistic aqueous and solid species must be used, including considering metastable solids that are viable controls on metal solubility within the time frame of interest for scale formation. Also, the selected activities (or concentrations) of the

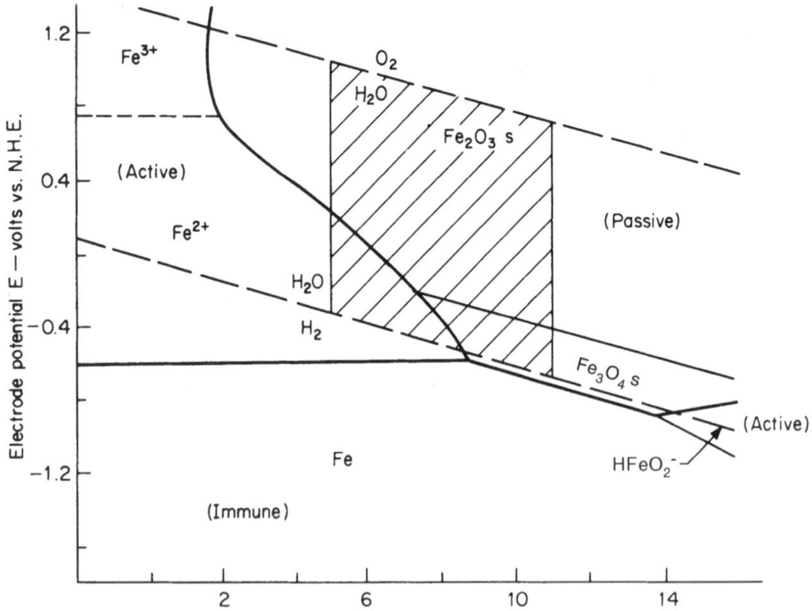

FIGURE 17.10 Potential pH diagram for the iron-water system at 25°C (considering Fe, Fe_3O_4, and Fe_2O_3 as solid substances and $[Fe^{2+}]$ or $[Fe^{3+}] = 10^{-6}$ *M*. The cross-hatched region is of the most interest for drinking water. (*Source:* After *Internal Corrosion of Water Distribution Systems.* American Water Works Association Research Foundation, Denver, CO, 1985, Fig. 1-3.)

dissolved species should be close to the real situations under study. Because the total activity is fixed across the entire pH range, the phase relationships predicted by the potential-pH diagram may deviate somewhat from those predicted in the previously described solubility diagrams.

Figure 17.11 illustrates the relative oxidizing ability (and instability) of chlorine species introduced for water disinfection (Snoeyink and Jenkins, 1980). The presence of species such as Cl_2, $HOCl^0$, or OCl^- will drive the oxidizing state of the water upward toward its stability limit (line A), which will affect the corrosion and speciation of many of the metals and metal surfaces in contact with the water. These are more powerful oxidizing agents than oxygen itself (line A).

Figure 17.12 shows that the location of the $Fe(OH)_3(s)$ stability field provides a reasonable explanation for the frequent observation that an amorphous ferric hydroxide forms an outer deposit on cast-iron pipes, and colloidal iron forms in chlorinated drinking water or aerated groundwater. Figure 17.12 was constructed with a mixture of solids and aqueous species of iron that is more typical of a drinking water environment than Figure 17.10 (Schock, 1981a). In waters with high alkalinity (high carbonate concentrations), siderite [$FeCO_3(s)$] can be found in corrosion deposits on iron pipe (Singley et al., 1985; Benjamin, Sontheimer, and Leroy, 1996; Sontheimer, Kolle, and Snoeyink, 1981).

Figures 17.13 and 17.14 are revised potential-pH diagrams based on the same species as were used to create Figures 17.4, 17.5, 17.7, and 17.8. Areas of immunity and passivation are unstippled. Note how the area of passivation virtually vanishes as the carbonate level increases from 2.4 mg/L (Figure 17.13) to 24 mg/L (Figure 17.14), for activities (concentrations) of aqueous lead species set at the old (1989) 0.05 mg/L MCL.

Typical protective scales may contain a wide variety of solid phases, and a potential-pH diagram that would incorporate all of these would be hopelessly

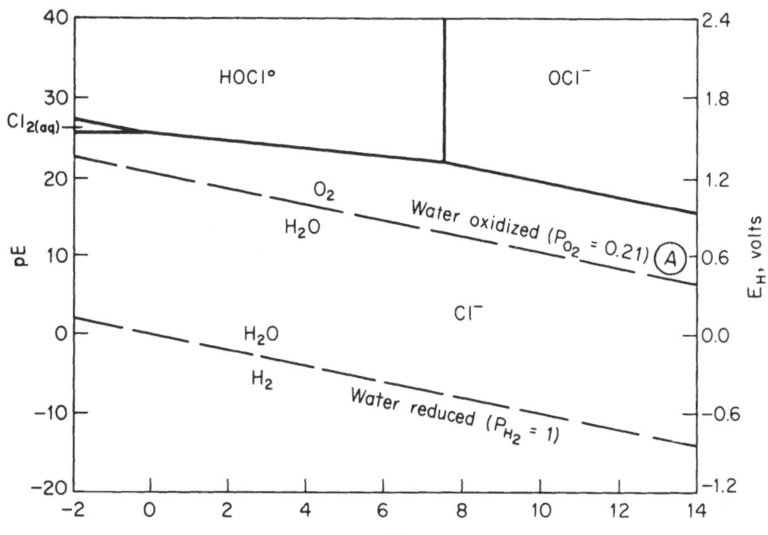

FIGURE 17.11 The potential pH diagram for aqueous chlorine; 25°C, $C_{T,Cl} = 1 \times 10^{-4}$ *M*. (*Source: Water Chemistry,* V. L. Snoeyink and D. Jenkins. Copyright © 1980, John Wiley & Sons, Inc. Reprinted by permission of John Wiley & Sons, Inc.)

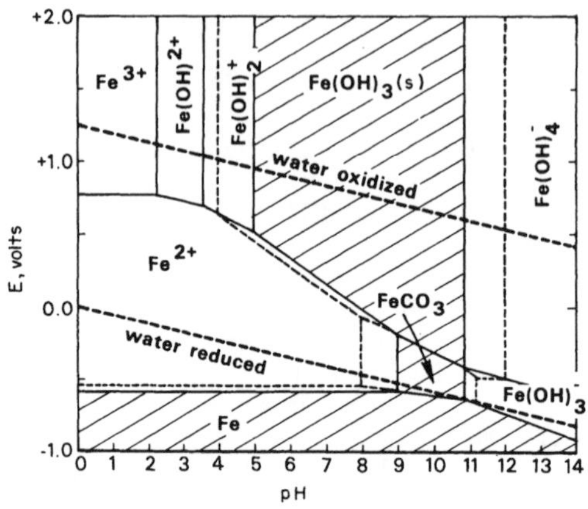

FIGURE 17.12 Potential pH diagram of iron in carbonate-containing water at 25°C at $I = 0$. Stability fields are shown for dissolved iron species activities of 0.1 mg/L (—) and 1.0 mg/L (- - -). Dissolved carbonate species concentrations are 4.8 mg C/L (4×10^{-4} M).

complicated even if the necessary thermodynamic data were available. The diagram is most appropriate for the conditions of the potential and the pH near the pipe surface, when those conditions may differ significantly from those of the bulk water phase (Bockris and Reddy, 1973). Thus, the type of scale that will form cannot necessarily be predicted if the composition of only the bulk water phase is known. Further, thermodynamic considerations alone will never give information about the velocity of the corrosion process itself. Nevertheless, potential-pH diagrams can still

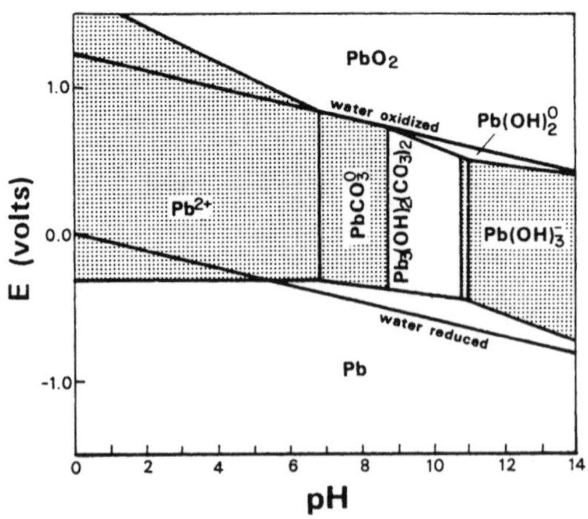

FIGURE 17.13 Potential pH diagram for the Pb-H_2O-CO_2 system at 25°C. Areas of passivation and immunity are unstippled. Dissolved lead species activities = 0.05 mg/L. Dissolved carbonate species activities = 2.4 mg C/L (2×10^{-4} mol/L). (*Source: Water Quality and Treatment,* 4th ed., 1990.)

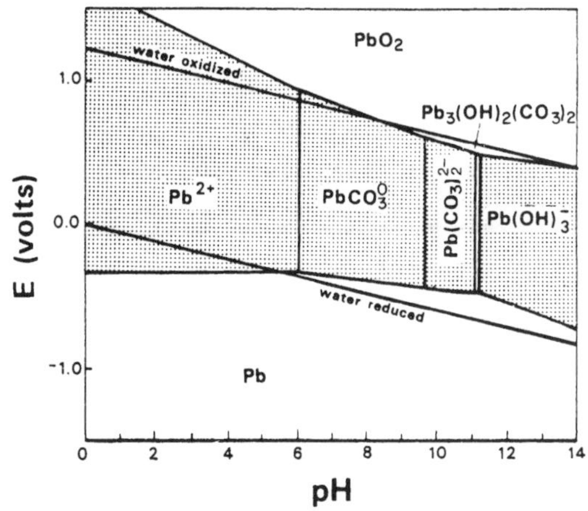

FIGURE 17.14 Potential pH diagram for the Pb-H_2O-CO_2 system at 25°C. Areas of passivation and immunity are unstippled. Dissolved lead species activities = 0.05 mg/L. Dissolved carbonate species activities = 24 mg C/L (2×10^{-3} mol/L). (*Source: Water Quality and Treatment,* 4th ed., 1990.)

predict or describe corrosion and passivation processes for many water qualities and systems, especially for metals like copper, iron, and manganese with multiple valence states possible in drinking water.

In conclusion, Bockris and Reddy (1973) provide several good suggestions and cautions for the use of potential-pH diagrams.

- Potential-pH diagrams can be used to give yes or no answers on whether a particular corrosion process is thermodynamically possible or not.
- The diagrams provide a compact pictorial summary of the electron-transfer, proton-transfer, and electron-and-proton-transfer reactions that are thermodynamically favored when a metal is immersed in a particular solution.
- When a potential-pH diagram indicates that a particular metal is immune to corrosion, it is so provided the pH in the close vicinity of the surface is what it is assumed to be.
- Even if a potential-pH diagram indicates that a particular corrosion process can spontaneously take place, it does not mean that significant corrosion will be observed. The corrosion reaction rate cannot be predicted thermodynamically.

TYPES OF CORROSION

Many different types of corrosion exist (Pourbaix, 1973; Snoeyink and Wagner, 1996), and only a few of the most significant to the corrosion of drinking water systems will be covered here. Specialized corrosion texts should be consulted for more comprehensive discussions.

The kind of attack depends on the material, the construction of the system, the scale and oxide film formation, and the hydraulic conditions. Corrosion forms range from uniform to intense, localized attack. The different forms of corrosion are primarily influenced by the distribution of anodic and cathodic areas over the corroding material. If the areas are microscopic, and very close to each other, corrosion may be relatively uniform over the entire surface. If the areas are scattered, however, and especially if the potential difference is large, pits may form, sometimes covered with the irregular and sometimes voluminous deposits called *tubercles,* which can greatly impede water flow.

Uniform Corrosion

According to one model, when uniform corrosion of a single metal occurs, any one site on the metal surface may be anodic at one instant and cathodic the next. Anodic sites shift or creep about the surface, so rate of loss of metal becomes relatively uniform over the metal surface.

An alternative model for uniform corrosion is that oxidation at a metal surface is accompanied by electron transport through an adherent film (see Snoeyink and Wagner, 1996, Figure 1.7). Reduction of oxygen occurs at the film surface, and transport of ions to and away from the oxide film takes place. Electrons are probably not transported through an external portion of the metal. The overall rate of corrosion is controlled by the presence and properties of the film or by transport of reaction products, especially hydroxide ion, away from the film-solution interface. This model has been used for several kinds of metallic corrosion, notably by Ives and Rawson (1962a) for copper. Its applicability to drinking water systems having chlorine, or chlorine plus oxygen as the oxidizing agents, is somewhat uncertain.

The reasons for development of corrosion cells on metals are varied. The single metals themselves may be heterogeneous, with possible differences in potential existing between different areas because of differences in crystal structure, or imperfections in the metal. Also, the concentrations of oxidants and reductants in solutions may be different, causing momentary differences in potential.

Frequently, the terms *plumbosolvency* and *cuprosolvency* are applied to the phenomenon of uniform corrosion when soluble lead and copper, respectively, are released from the piping.

Galvanic Corrosion

Galvanic corrosion occurs when two different types of metals or alloys contact each other, and the elements of the corrosion cell are present. One of the metals serves as the *anode,* and thus deteriorates, whereas the other serves as the *cathode.* Available metals and alloys can be arranged in order of their tendency to be anodic, and the resulting series is called the *galvanic series.* This order, in a potable water environment, depends on the temperature and solution chemistry (which affect the thermodynamic activities of the cell components), as well as the simple relative ordering of the standard electrode potentials of the oxidation or reduction half-cells or simple metal/ion redox reaction couplings.

Depending upon generalizations about the range of chemical characteristics expected for the water in contact with the metals, different "galvanic series" can be computed. A particularly relevant and useful one is that presented by Larson,[54] which is excerpted in Table 17.2.

If any two of the metals from different groups in this table are connected in an aqueous environment, the metal appearing first will tend to be the anode and the second will be the cathode. In general, the farther apart the metals in Table 17.2, the

TABLE 17.2 Empirical Galvanic Series of Metals and Alloys in Potable Waters*

Corroded end†
Zinc
Aluminum 2S
Cadmium
Aluminum 17ST
Steel or iron
Cast iron
Lead-tin solders
Lead
Tin
Brasses
Copper
Bronzes
Monel
Silver solder
Silver
Protected end‡

* Metals within each of the groupings have relatively similar corrosion potentials.
† Anodic, or least "noble."
‡ Cathodic, or most "noble."
Source: Larson, T. E. *Corrosion by Domestic Waters*. Bulletin 59, Illinois State Water Survey, 1975.

greater will be the potential for corrosion, because the potential difference between them will be greater.

The rate of galvanic corrosion is increased by greater differences in potential between the two metals. It is increased by large areas of cathode relative to the area of the anode, and it is generally increased by closeness of the two metals and by increased mineralization or conductivity of the water. The relative size of the cathode relative to the anode may be of particular concern in the corrosion of lead/tin-soldered joints in copper pipe, or when piping of two different types is interconnected, such as replacement of lengths of lead pipe with copper pipe (Britton and Richards, 1981; Schock, Wagner, and Oliphant, 1996).

Galvanic corrosion is often a great source of difficulty where brass, bronze, or copper is in direct contact with aluminum, galvanized iron, or iron. Proper selection of materials and the order of their use in domestic hot- and cold-water plumbing systems is critical to the control of corrosion. To prevent galvanic corrosion, for example, only copper tubing should be used with copper-lined water heaters. Brass valves in contact with steel and galvanized plumbing in waters with high total dissolved solids cause corrosion of the steel and galvanized pipes. Dissolved copper can attack spots on galvanized pipe, thereby causing copper-zinc galvanic cells (Kenworthy, 1943; Fox et al., 1986).

Pitting Corrosion

Pitting is a damaging, localized, nonuniform corrosion that forms pits or holes in the pipe surface. Pitting occurs in an environment that offers some but not complete protection. It actually takes little metal loss to cause a hole in a pipe wall, and failure can be rapid. Pitting failures can occur in water supplies that meet the regulatory action level for copper, and unfortunately, in many cases copper release into the water is not diagnostic for potential pitting failure problems. Pitting can begin or concentrate at a point of surface imperfections, scratches, or surface deposits. One proposed mechanism for pitting is that a cause may be ions of a metal higher in the galvanic series plating out on the pipe surface. For example, steel and galvanized steel are subject to corrosion by small quantities (about 0.01 mg/L) of soluble metals, such as copper, that plate out and cause a galvanic type of corrosion.

Recent research has shed some light on pitting causes, including the implication of a role of sulfide for initiation and microorganisms for exacerbating it. Though reliable prediction from water chemistry data is still generally impossible (Jacobs and Edwards, 1997; Shalaby, Al-Kharafi, and Gouda, 1989; Fischer and Füßinger, 1992; Edwards, Rehring, and Meyer, 1994; Edwards, Ferguson, and Reider, 1994; Duthil, Mankowski, and Giusti, 1996; Ferguson, von Franqué, and Schock, 1996), Taxén (1996) has recently attempted to quantitatively predict the propagation and evolution of pits in relation to equilibrium water chemistry and model the mass transport.

Concentration Cell Corrosion

Concentration cell corrosion is usually deduced by inference. It occurs when differences in the total or the type of mineralization of the environment exist. Corrosion potential is a function of the concentration of aqueous solution species that are involved in the reaction, as well as of the characteristics of the metal. Differences in acidity (pH), metal-ion concentration, anion concentration, or dissolved oxygen cause differences in the solution potential of the same metal. Differences in temperature can also induce differences in the solution potential of the same metal.

When concentration cell corrosion is caused by dissolved oxygen, it is often referred to as *differential oxygenation corrosion,* as discussed by Snoeyink and Wagner (1996). Common areas for differential oxygenation corrosion are between two metal surfaces; for example, under rivets, under washers, under debris, or in crevices.

Oxygen concentration cells develop at metal-water interfaces exposed to air, such as in a full water tower, accelerating corrosion a short distance below the surface. The dissolved oxygen (DO) concentration is replaced by diffusion from air and remains high at and near the surface, but does not replenish as rapidly at lower depths because of the distance. Therefore, the corrosion takes place at a level slightly below the surface rather than at the surface.

Tuberculation

Tuberculation occurs when pitting corrosion products build up at the anode next to the pit (Snoeyink and Wagner, 1996). In iron or steel pipes, the tubercles are made up of various iron oxides and oxyhydroxides. These tubercles are usually rust colored and soft on the outside and are both harder and darker toward the inside. Sometimes, they are considerably layered with iron minerals indicative of anoxic environments, such as sulfide minerals (Singley et al., 1985; Benjamin, Sontheimer, and Leroy, 1996). When copper pipe becomes pitted, the tubercle buildup is smaller and is a green to blue-green color, reflecting the deposition of basic cupric carbonate, sulfate, or chloride salts, or a mix of them.

Tuberculation is often associated with poorly buffered waters, where pH can get extremely high under localized surface conditions. Hence, many consultants and researchers have recommended a "balance" of minimum hardness and alkalinity to provide a water that is resistant to iron corrosion and tuberculation (Weber and Stumm, 1963; Stumm, 1956, 1960; Merrill, Sanks, and Spring, 1978; Légrand and LeRoy, 1995; Larson and Skold, 1957).

Crevice Corrosion

Crevice corrosion is a form of localized corrosion usually caused by changes in acidity, oxygen depletion, dissolved ions, and the absence of an inhibitor. As the name implies, this corrosion occurs in crevices at gaskets, lap joints, rivets, and surface deposits.

Erosion Corrosion

Erosion corrosion mechanically removes protective films, such as metal oxides, hydroxycarbonates, and carbonates, that serve as protective barriers against corrosive attack. It can also remove the metal of the pipe itself. Erosion corrosion can be identified by grooves, waves, rounded holes, and valleys it causes on the pipe walls.

Dealloying or Selective Leaching

Dealloying, or *selective leaching,* is the preferential removal of one or more metals from an alloy in a corrosive medium, such as the removal of zinc from brass (dezincification), or the removal of disseminated lead from brass (Schock and Neff, 1988; Oliphant and Schock, 1996; Lytle and Schock, 1996, 1997). This type of corrosion weakens the metals and can lead to pipe failure in severe cases. The stability of brasses and bronzes in natural waters depends in a complex manner on the dissolved salts, the hardness, the dissolved gases, and the formation of protective films. Dezincification is common in brasses containing 20 percent or more zinc and is rare in brasses containing less than 15 percent zinc. The occurrence of plug-type dezincification and dezincification at threaded joints suggests that debris and crevices may initiate oxygen concentration cells and result in dezincification.

Lead occurs in lead/tin solder as a disseminated phase, acting principally as a diluent for the tin that actually does the binding with the copper (Parent, Chung, and Bernstein, 1988). Factors governing the removal of lead from the soldered joint are presumably similar to those affecting the leaching of lead from brass. Soldered joints in service for a long time frequently show a depletion of lead from the exposed solder.

In the United States, the leaching of lead, cadmium, zinc, and other metals from brass and solders is regulated in an indirect way. The Safe Drinking Water Act (SDWA) Amendments of 1996 limit the content of Pb in solder and flux to be used in contact with drinking water to 0.2 percent, given in Section 1417(d) (Safe Drinking Water Act Amendments, 1996). Because the definition of *lead-free* for pipes, well pumps, plumbing fixtures, and fittings under the original statute was a content of 8 percent or less, it was ineffective in controlling contamination from this source, as virtually all fixtures and fittings implicated in high levels of metal leaching contain much less than 8 percent Pb (Oliphant and Schock, 1996; Lytle and Schock, 1996, 1997). Thus, the SDWA Amendments of 1996 incorporated a performance standard in Section 1417(e), requiring fixtures and fittings intended to dispense water for human consumption to be certified under a voluntary standard established by the Act (NSF International, 1998). Additionally, U.S. states having primacy for enforcement of the drinking water regulations (at the time of this writing, 49 of the 50 states) may promulgate more stringent standards if they desire.

Selective leaching also applies to the dissolution of asbestos-cement pipe, or the deterioration of cement mortar linings of iron water mains. Highly soluble components, such as free lime, calcium carbonates, and a variety of silicates and aluminosilicates, can be dissolved by aggressive waters (Schock, 1981; Légrand and LeRoy, 1995; LeRoy et al., 1996; Holtschulte and Schock, 1985; Douglas and Merrill, 1991). In some cases, the attack can be so severe as to cause weakening in the walls of the pipe, and the dislodging of mats of fibers.

Graphitization

Graphitization is a form of corrosion of cast iron in highly mineralized water or waters with a low pH that results in the removal of the iron silicon metal alloy making up one of the phases of the cast-iron microstructure. A black, spongy-appearing, but hard mass of graphite remains.

Microbiologically Influenced Corrosion

Microbiologically influenced corrosion (MIC) pertains to the general class of corrosion resulting from a reaction between the pipe material and organisms such as bacteria, their metabolic by-products, or both (Little and Wagner, 1997). Algae and fungi may also influence corrosion by producing changes in the pH, dissolved oxygen level, or other chemistry changes to the microenvironment at the metal surface or under corrosion deposits. Microbiologically influenced corrosion may be an important factor in the taste and odor problems that develop in a water system, as well as in the degradation of the piping materials. Microbial activity can also strongly influence the mineralogy of iron corrosion deposits (Camper, 1996; Lewandowski, Dickinson, and Lee, 1997; Brown, Sherriff, and Sawicki, 1997; Lazaroff, Sigal, and Wasserman, 1982; Postma and Jakobsen, 1996; Tuhela, Carlson, and Tuovinen, 1992; Allen, Taylor, and Geldreich, 1980; LeChavallier et al., 1993).

Biofilms in pipes are often characterized by stratification, with different families of organisms with different functions existing in the different zones and conditions (Little and Wagner, 1997). Controlling such growths is complicated because they can take refuge in many protected areas, such as in mechanical crevices or in accumulations of corrosion products. Bacteria can exist under tubercles, where neither chlorine nor oxygen can destroy them. Mechanical cleaning may be necessary in some systems before control can be accomplished by residual disinfectants. Preventive methods include good corrosion control to avoid tuberculation and accumulating corrosion by-products, avoiding dead ends, and preventing stagnant water in the system by flushing or bleeding.

Ainsworth (1980) noted that organic carbon appears to be a major factor in controlling the numbers of microorganisms in a distribution system (see Chapter 18). They noted that the numbers of microorganisms increased along with increasing loss in total organic carbon (TOC) and oxygen through the distribution system. The microbial activity was found to be concentrated in the surface deposits and pipe sediments. It is not clear if one of the mechanisms for reduction of biofilms and microbial growth by corrosion control and the application of inhibitors is the prevention or saturation of sorption sites at the pipe surfaces, allowing less natural organic matter (NOM) sorption and attachment (Camper, 1996; LeChavallier et al., 1993). Thus, there would be a decreased attached nutrient base for the colonizing organisms if potential organic material attachment sites were already occupied by phosphate or other corrosion inhibitor material.

The ways in which bacteria can increase corrosion rates are numerous. Slime growths of nitrifying (and other) organisms may produce acidity and consume oxygen in accordance with Eqs. 17.32 and 17.33.

$$NH_4^+ + 3/2\ O_2 \Leftrightarrow NO_2^- + 2H^+ + H_2O \tag{17.32}$$

$$NH_4^+ + 2O_2 \Leftrightarrow NO_3^- + 2H^+ + H_2O \tag{17.33}$$

These reactions can cause oxygen concentration cells that produce lowered pH and localized corrosion and pitting. Lee, O'Connor, and Banerji (1980), for example, showed increased localized corrosion when cast iron was exposed to a water with extensive biological activity, compared with the same water under sterile conditions. A very recent study found an association between nitrification activity and copper corrosion problems (Murphy, O'Connor, and O'Connor, 1997a–c). Microbes may also enhance corrosion or solubility because extracellular material and metabolites of microorganisms may include polymers that are good complexing agents (Geesey et al., 1986, 1988; Geesey and Bremer, 1991), or they may influence the redox chemistry of the scales at the surface (Lewandowski, Dickinson, and Lee, 1997; Allen, Taylor, and Geldreich, 1980; LeChavallier et al., 1993; Lee, O'Connor, and Banerji, 1980).

Iron bacteria derive energy from oxidation of ferrous iron to ferric iron. Nuisance conditions often result because the ferric iron precipitates in the gelatinous sheaths of the microbial deposits and these can be sloughed off and be the cause of red-water complaints. Sontheimer, Kolle, and Snoeyink (1981) noted that they can also interfere with the development of passivating scales and, thus, that the rate of corrosion is higher in their presence than in their absence.

The sulfate-reducing bacteria may be the most important organism involved with MIC in many water systems (Little and Wagner, 1997). Sulfate can act at the cathode in the place of oxygen,

$$SO_4^{2-} + 8H^+ + 8e^- \Leftrightarrow S^{2-} + 4H_2O \tag{17.34}$$

and the sulfate-reducing organisms apparently catalyze this reaction. The localized high pH could promote corrosion, and reduce the effectiveness of disinfectants against the organisms mediating the reactions. Sulfate reducers have been found in the interior of tubercles, and thus may be responsible for maintaining corrosion at these locations (Ainsworth, 1980). Sulfate reducers are often present regardless of whether the supply is aerated (Little and Wagner, 1997), and then activity may be regulated by TOC levels. Reductions in TOC should lessen their activity. Some observations that higher levels of sulfate cause more corrosion (Larson, 1975) may possibly be related to microbial activity (Snoeyink and Wagner, 1996). Recent research has found sulfide to be a contributor to copper pitting that is particularly destructive and difficult to control (Jacobs and Edwards, 1997), and once started, the microbial mediation may be very difficult or impossible to stop. The high-chloride and extreme pH environments of pits will lead to continued propagation, even if the microbes are killed by biocides (Little and Wagner, 1997).

Stray Current Corrosion

Stray current corrosion is a type of localized corrosion usually caused by the grounding of home appliances or electrical circuits to the water pipes. It occurs more often on the outside of pipes, but does show up in house faucets or other valves. This subject remains somewhat controversial, although strong evidence has been presented by a number of researchers that both AC and DC current can affect corrosion rates and metal levels in the water (Williams, 1986; Horton, 1991; Horton and Behnke, 1989; Bell et al., 1995; Bell, Schiff, and Duranceau, 1995). In municipal distribution systems, increased corrosion of steel-reinforced concrete pressure pipe has occasionally been noted, and has been believed to be related to stray currents.

PHYSICAL FACTORS AFFECTING CORROSION AND METALS RELEASE

The characteristics of drinking water that affect the occurrence and rate of corrosion can be classified as (1) physical, (2) chemical, and (3) biological. In most cases, corrosion is caused or increased by a complex interaction among several factors. Some of the more common characteristics in each group are discussed in this section to familiarize the reader with their potential effects. Controlling corrosion may require changing more than one of these because of their interrelationships.

Essentially no statement regarding corrosion or the general use of a material can be made that does not have an exception (NACE, 1984). The corrosion of metals and alloys in potable water systems depends both upon the environmental factors (solution composition) and the composition of plumbing or fitting material. Concern about the consequences of corrosion varies with the material. Sometimes, it is toxicity from trace metal dissolution and contamination of the drinking water, but for materials such as iron, steel, and mortar linings, the concern is more aesthetic concerns and material degradation. Therefore, this section will address some of the major factors that contribute to corrosion in potable waters, but the reader must refer to literature that comprehensively describes the relationship for the materials of interest.

Physical Characteristics

Flow velocity and temperature are the two main physical characteristics of water that affect corrosion.

Velocity. High flow velocities can sometimes aid in the formation of protective coatings by transporting the protective material to the surfaces at a higher rate. High flow velocities, however, are usually associated with erosion corrosion or impingement attack in copper pipes, in which the protective wall coating or the pipe material itself is mechanically removed. High-velocity waters combined with other corrosion-causing water constituents can rapidly deteriorate pipe materials with minimal metal release to the water because of dilution. For example, the combination of high velocity with low pH and high DIC concentration is extremely aggressive toward copper. High flow velocity can also increase the rate at which oxidant species come in contact with pipe surfaces, increasing corrosion. It is hard to quantify "high," but a common guideline for ½-in ID copper is about 4 ft/s (AWWARF, 1990), and plumbing handbooks should be consulted for the materials of interest for initial guidelines.

A water that behaves satisfactorily at medium to high velocities may still cause incipient or slow corrosion of iron or steel with accompanying red-water problems at low velocities, because the slow movement does not aid the effective diffusion rate of the protective ingredients to the metal surface.

Temperature. The influence of temperature is also often misunderstood. Confusion can often be avoided by considering basic equilibria in water chemistry, and by remembering that temperature effects are complex and depend on both the water chemistry and the type of plumbing material present in the system. Multiple phenomena often operate simultaneously, such as changes in solubility of solids; changes in the formation of complex ions; changes in diffusion rates of dissolved gases and solution species; changes in composition or physical properties of the metals, alloys, or solids; and changes in water properties. Thus, generalizations are often made, but they are rarely accurate.

The electrode potential (the driving force for any corrosion cell) is proportional to the absolute temperature, and therefore, theory predicts that the corrosion rate will increase with temperature. This relationship is observed in some controlled laboratory experiments, but in practice is less obvious unless wide differences occur (i.e., hot- versus cold-water systems). In distribution systems, temperature fluctuations are somewhat limited over short time frames, so whatever effect they have is obscured by other factors. "Seasonal" changes in temperature are often accompanied by significant changes in one or more major chemistry parameters. Also, when considering household plumbing systems, the relatively high thermal conductivity of the metallic piping materials normally used causes the water standing inside pipes, or passing through long stretches of pipe, to rapidly equilibrate to the temperature of the air surrounding the pipe.

Sometimes, hot water is observed to be more corrosive than cold water. Water showing few corrosive characteristics in the distribution system can cause severe damage to copper or galvanized iron hot-water heaters at elevated temperatures. In some hot-water systems, however, the high temperature can turn a corrosive or nonscaling water into a scaling water with reduced corrosivity, through a combination of raising the pH, decreasing the solubility of calcium carbonate, driving off dissolved oxygen or carbon dioxide, and speeding up the reduction of any chlorine species present.

Temperature significantly affects the dissolving of $CaCO_3$ (calcite). Less $CaCO_3$ dissolves at higher temperatures, which means that $CaCO_3$ tends to come out of solution (precipitate) and form a protective scale more readily at higher temperatures. Excessive deposition of $CaCO_3$ can clog hot-water lines. Some other minerals behave similarly, such as the aragonite form of $CaCO_3$, calcium sulfate (anhydrite), and many silicates.

Larson has pointed out that the effect of temperature on pH is seldom recognized (AWWA, 1971). For pure water, as temperature increases the water dissociates more (the pK_w goes down), causing increases in both the H^+ concentration and the OH^- concentration. Drinking waters all contain carbonate species at some concentration, which strongly influence the water pH. So, the degree of influence of temperature on pH is also a function of the alkalinity (inorganic carbon content) of the water, and the dissociation constants for carbonic acid also change with temperature. Increasing concentrations of bicarbonate increasingly buffer or reduce this effect of temperature on pH. The net pH decrease from heating tends to be less than the net decrease in overall $CaCO_3$ solubility, causing scaling. For waters of low alkalinity (less than approximately 50 mg $CaCO_3$/L), the higher temperatures decrease the pH at a rate that is greater than the rate of decrease in solubility of $CaCO_3$. The effect on the saturation state of calcium carbonate is particularly significant for 10 mg $CaCO_3$/L alkalinity, and even for 25 mg $CaCO_3$/L alkalinity with a temperature change to 40°C (130°F) and more so if the change is to 55°C (157°F).

Consider a water at 15°C (59°F) having the following characteristics: pH = 8.71; Ca = 17.3 mg/L; Na = 17 mg/L; Cl = 35 mg/L; and total alkalinity = 30.6 mg $CaCO_3$/L. The water is at saturation equilibrium with calcite. Therefore, it has a saturation index (or Langelier Saturation Index) (Trussell, 1985; Snoeyink and Kuch, 1985; Snoeyink and Jenkins, 1980; Snoeyink and Wagner, 1996; Merrill, Sanks, and Spring, 1978; Rossum and Merrill, 1983; Joint Task Group, 1990) of 0.00. At this temperature, the solubility constant for the simple dissolution of calcite

$$Ca^{2+} + CO_3^{2-} \Leftrightarrow CaCO_3(s) \tag{17.35}$$

is equal to $10^{-8.43}$. If that water is warmed to 55°C, the saturation index drops to −0.02, even though the solubility constant decreases to $10^{-8.71}$. This phenomenon occurs because the pH also decreases to 8.16.

Figures 17.15 and 17.16 illustrate how the pH of waters of different alkalinities changes when warmed to a temperature of 55°C from 15°C or 25°C, respectively. These figures are based on calculations that assume the following:

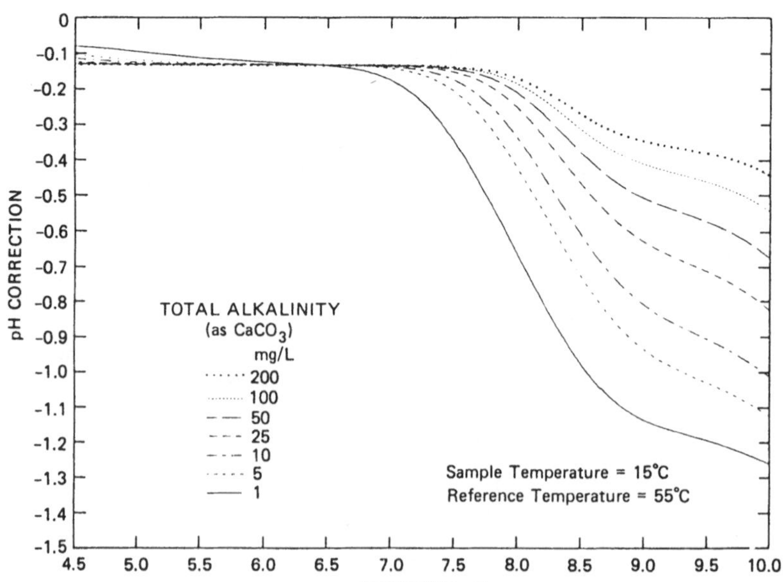

FIGURE 17.15 Change in pH for waters of different alkalinities caused by warming a water originally at 15°C to 55°C under closed-system conditions, assuming an ionic strength of 0.001. The pH predicted for 55°C is obtained by *adding* the pH correction to the original (observed) pH at 15°C (e.g., $pH_{55} = pH_{15} + pH_{corr}$).

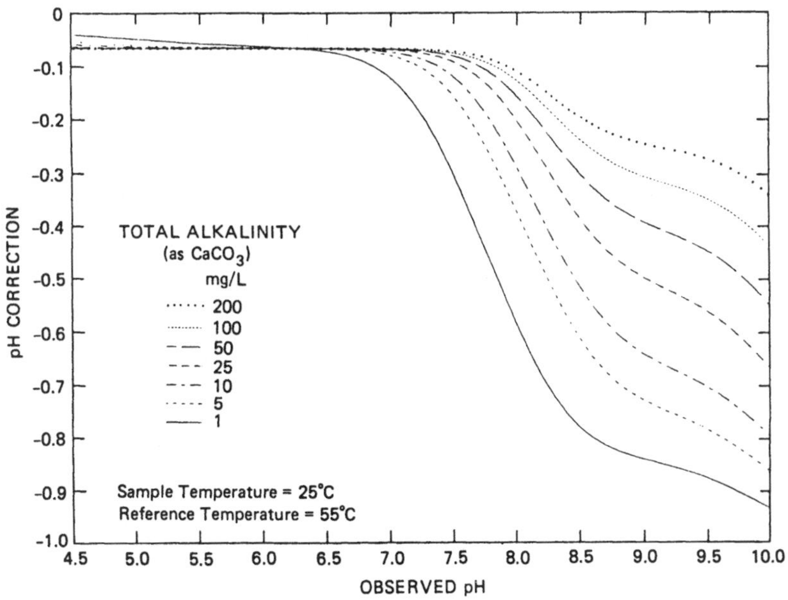

FIGURE 17.16 Change in pH for waters of different alkalinities caused by warming a water originally at 25°C to 55°C under closed-system conditions, assuming an ionic strength of 0.001. The pH predicted for 55°C is obtained by *adding* the pH correction to the original (observed) pH at 25°C (e.g., $pH_{55} = pH_{25} + pH_{corr}$).

- An ionic strength of 0.001 mol/L
- No change in inorganic carbon content
- No redox or hydrolysis reactions that significantly affect the proton balance of the system

These temperature effects illustrated in Figures 17.15 and 17.16 explain, in part, the problems of some water supplies, whose alkalinities are too low to buffer the effect of temperature. The low flow velocities in hot-water tanks often aggravate corrosivity, by limiting the ability of some chemical inhibitors to be effective (AWWA, 1971). More details on a general computational approach and some additional figures for computing the pH change resulting from intrinsic changes to water and carbonic acid dissociation have been published (AWWARF, 1990).

An additional effect of temperature is that an increase can change the entire nature of the corrosion. For example, a water that exhibits pitting at cold temperatures may cause uniform corrosion when hot (Singley, Beaudet, and Markey, 1984; AWWA, 1986). Although the total quantity of metal dissolved may increase, the attack is less acute, and the pipe will have a longer life.

Another example in which the nature of the corrosion is changed as a result of changes in temperature involves a zinc-iron couple. Normally, the anodic zinc is sacrificed, or corroded, to prevent iron corrosion. In some waters, the normal potential of the zinc-iron couple may be reversed at temperatures above 46°C (140°F). The zinc becomes cathodic to the iron, and the corrosion rate of galvanized iron is much higher than would normally be anticipated, a factor that causes problems in many hot-water heaters or piping systems.

By changing solubility (as well as pH), temperature changes can influence the precipitation of different solid phases or transform the identities of corrosion products. These changes may result in either more or less protection for the pipe surface, depending on the materials and water qualities involved.

Manufacturing-Induced Characteristics

For several types of common plumbing materials, the manufacturing processes used may play an important role in determining the types of corrosion that will occur, and the durability of the pipe or fixtures. Pitting of galvanized pipe has been associated with several characteristics of its manufacture, such as thin, improper, or uneven galvanizing coating, poor seam welds, and rough interior finish (Trussell, 1985; Joint Task Group, 1990; Fox, Tate, and Bowers, 1983). Impurities in the galvanizing dip solution, such as lead and cadmium, could cause concern because of their potential for leaching into the water (Trussell, 1985).

In England, rapid pitting failures of copper pipe were found to be associated with a carbon film on their interior surfaces (Ferguson, von Franqué, and Schock, 1996; Campbell, 1954, 1963, 1964; Cruse and Pomeroy, 1974). The film was a residue from drawing oil that had become carbonized during annealing. Soft-annealed tubes have been more susceptible to pitting than half-hard tubes, which are more susceptible than hard-temper tubes (Ferguson, von Franqué, and Schock, 1996; Cruse and Pomeroy, 1974). Though it has not been proven by systematic study in many locations, many copper corrosion researchers feel that the early history of the pipe and initial installation in residential and building plumbing has a profound impact on its susceptibility to localized corrosion later (Lane, 1993). Debris from manufacturing, transport and storage, and patchy or incomplete oxidation coatings from air and moisture exposure may set up conditions favorable to differential oxygenation corrosion, depending on the initial exposure of the installed piping. If plumbing is not flushed thoroughly and housing is not occupied soon after construction, anoxic conditions favorable to microbiological growth and, later, to differential oxygenation corrosion, will likely result, and the corrosion cells may be hard to stifle once normal water usage is started.

CHEMICAL FACTORS AFFECTING CORROSION

Dissolved substances in water have an important effect on both corrosion and corrosion control. This section provides a brief overview to point out some of the most important factors.

General

Table 17.3 lists some of the chemical factors that have been shown to have an important effect on corrosion or corrosion control. Several of these factors are closely related, and a change in one changes another. The most important example of this is the relationship among pH, carbon dioxide (CO_2), DIC concentration, and alkalinity. Although CO_2 is frequently considered to be a factor in corrosion, there is no clear evidence that direct corrosion reactions include CO_2 as a reactant (Singley et al.,

TABLE 17.3 Chemical Factors Influencing Corrosion and Corrosion Control

Factor	Effect
pH	Low pH may increase corrosion rate and the strength of oxidizing agents; high pH may protect pipes by favoring effective passivation films and decrease corrosion rates; possibly causes or enhances dezincification of brasses.
Alkalinity/DIC	May help form protective carbonate or hydroxycarbonate films; helps control pH changes by adding buffering. Low to moderate alkalinity reduces corrosion of most materials. High alkalinities increase corrosion of copper at all pHs, and lead at high pH.
DO	Increases rate of many corrosion reactions when metal is passive or immune under anoxic conditions. May help form better passivating oxide films on some materials, such as iron.
Chlorine residual	Increases metallic corrosion, particularly for copper, iron, and steel.
TDS	TDS is a surrogate for ionic strength, which increases conductivity and corrosion rate unless offset by the formation of passivating films.
Hardness (Ca and Mg)	Ca may precipitate as $CaCO_3$ and thus provide protection and reduce corrosion rates. May enhance buffering effect in conjunction with alkalinity and pH.
Chloride, sulfate	High levels increase corrosion of iron, copper, galvanized steel, possibly lead.
Hydrogen sulfide	Increases corrosion rates; may cause severe pitting in copper.
Ammonia	May increase solubility of some metals, such as copper and lead.
Polyphosphates	May reduce tuberculation of iron and steel, and provide smooth pipe interior. May enhance uniform iron and steel corrosion at low dosages. Attacks and softens cement linings and A-C pipe and cement linings. Increases the solubility of lead and copper. Prevents $CaCO_3$ formation and deposition. Sequesters ferrous iron and reduced manganese, especially at pH below 7.
Silicate	Forms protective films on many materials, especially iron and cement lining and A-C pipe, and galvanized pipe in hot water. Often needs preexisting scale to be most effective. Sequesters ferric iron. Forms most effective films at high pH. Silicate chemicals good for raising pH in low-alkalinity waters.
Orthophosphate	Forms protective films on iron, galvanized pipe, and lead. Slows oxidation of copper at neutral pH. Tends to form colloidal lead and maybe other metal species at pHs above 8. Interferes with calcium carbonate nucleation and growth.
Natural color, organic matter	May decrease corrosion by coating pipe surfaces over long term. Some organics can complex metals and accelerate corrosion or metal uptake, especially when surfaces are new.
Iron, zinc, or manganese	May react with compounds on interior of A-C pipe to form protective coating.
Copper	May cause pitting in galvanized pipe.
Magnesium	May inhibit the precipitation of calcite from $CaCO_3$ on pipe surfaces, and favor the deposition of the more soluble aragonite form of $CaCO_3$.
Aluminum	May form diffusion barrier films on iron, lead, and other pipe materials, such as aluminum hydroxide or aluminosilicate precipitate. Reduces effectiveness of orthophosphate if present at high concentration.

1984; AWWA, 1986). In some cases, the rate of CO_2 hydration might influence the bicarbonate concentration, and, hence, the buffering ability of the water, but the important corrosion effect usually results from pH and complexation by bicarbonate or carbonate ions. The dissolved CO_2 concentration is interrelated with pH and DIC concentration. Knowing all of the complex equations for these calculations is not necessary, but knowing that each of these factors plays some role in corrosion is useful.

The material that follows describes some of the corrosion-related effects of the factors listed in Table 17.3. A better understanding of how they are related to one another will aid in understanding corrosion and, thus, in choosing corrosion control methods.

pH

pH is a measure of the activity of hydrogen ions, H^+, present in water. In most potable waters, the activity of the hydrogen ion is nearly equal to its concentration. Because H^+ is one of the major substances that accepts the electrons given up by a metal when it corrodes, pH is an important factor to measure. At pH values below about 5, both iron and copper corrode rapidly and uniformly. At values higher than 9, both iron and copper are usually protected. Under certain conditions, however, corrosion may be greater at high pH values. Between pH 5 and 9, pitting is likely to occur if no protective film is present. The pH also greatly affects the formation or solubility of protective films for both metallic and cementitious materials.

Alkalinity/Dissolved Inorganic Carbon (DIC)

Alkalinity is a measure of the ability of a water to neutralize acids and bases (Trussell, 1985; Snoeyink and Jenkins, 1980; Stumm and Morgan, 1981; Weber and Stumm, 1963; Pankow, 1991; Butler, 1982; Faust and Aly, 1981; Morel, 1983; Loewenthal and Marais, 1976). In most potable waters, total alkalinity is mainly described by the relationship:

$$\text{TALK} = 2[CO_3^{2-}] + [HCO_3^-] + [OH^-] - [H^+] \qquad (17.36)$$

where [ ] indicates concentration in mol/L and total alkalinity is in equivalents/L (eq/L). Operationally, it is defined by alkalametric titration to the carbonic acid equivalence point (Stumm and Morgan, 1981; Butler, 1982; Loewenthal and Marais, 1976; Schock and George, 1991). The concentration of bicarbonate and carbonate ions is directly related to the pH of the water and the DIC concentration through the dissociation of carbonic acid. Many distribution diagrams have been presented showing the fraction of the DIC that is present in each form, and they can be readily calculated (Trussell, 1985; Snoeyink and Jenkins, 1980; Stumm and Morgan, 1981; AWWARF, 1990; Pankow, 1991; Butler, 1982; Faust and Aly, 1981; Morel, 1983; Loewenthal and Marais, 1976).

Dissolved inorganic carbon is defined as the sum of all dissolved carbonate-containing species. When ion pairs and complexes, such as $CaHCO_3^+$, $MgCO_3^0$, are negligible in concentration, DIC is simply:

$$\text{DIC} = [H_2CO_3^*] + [HCO_3^-] + [CO_3^{2-}] \qquad (17.37)$$

where $[H_2CO_3^*]$ represents the sum of dissolved aqueous CO_2 gas molecules and carbonic acid molecules (Stumm and Morgan, 1981; Pankow, 1991; Loewenthal and

Marais, 1976). The concentration of DIC may be either directly analyzed (Schock and George, 1991) or computed from a total alkalinity titration and pH, with proper corrections for ionic strength, proton-consuming species (e.g., HPO_4^0, $H_3SiO_4^-$, NH_3, and so on) (Snoeyink and Jenkins, 1980; Stumm and Morgan, 1981; AWWARF, 1990; Butler, 1982; Loewenthal and Marais, 1976; Schock and George, 1991). When alkalinity is defined adequately by Eq. 17.33, DIC (in units of mg C/L) may be derived from total alkalinity (in units of eq/L) and pH using the relationship:

$$\text{DIC} = \left[1 + \frac{K_2'}{[\text{H}^+]} + \frac{[\text{H}^+]}{K_1'}\right]\left[\frac{\text{Total alkalinity} - \dfrac{K_w'}{[\text{H}^+]} + [\text{H}^+]}{1 + \dfrac{2 \cdot K_2'}{[\text{H}^+]}}\right] \cdot 12{,}011 \qquad (17.38)$$

Total alkalinity in conventional units of mg $CaCO_3$/L can be converted to units of eq/L by dividing by 50044.5 (a factor derived from the equivalent weight of calcium carbonate and conversion of grams to milligrams). When Eqs. 17.36 through 17.38 are applicable, total alkalinity is related to DIC in the manner shown graphically by Figure 17.17, which is similar to figures developed by Deffeyes (Stumm and Morgan, 1981; AWWARF, 1990; Schock and George, 1991; Deffeyes, 1965). For these calculations, a temperature of 25°C and an ionic strength of 0.005 were assumed. The total alkalinity/DIC relationship is affected by temperature and ionic strength, so using other assumptions would change the slopes of the lines in Figure 17.17.

When bases other than the carbonates (HCO_3^- and CO_3^{2-}) and OH^- are present in significant quantities, they will consume protons in the alkalinity titration to the carbonic acid equivalence point. Then, the alkalinity definition must be expanded to

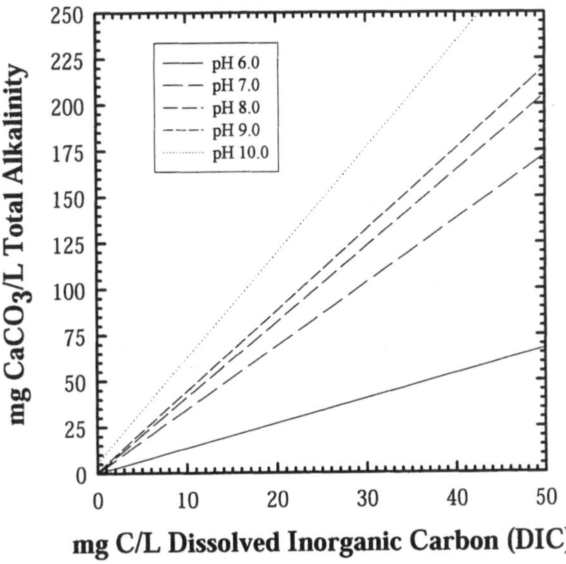

FIGURE 17.17 Relationship between total alkalinity and the total dissolved inorganic carbon concentration for a water at 25°C and an ionic strength of 0.005.

accommodate them. For example, the alkalinity of a water containing orthophosphate, hypochlorite, ammonia, silica, and some singly charged organic acid species $[OA^-]$ would then be (Snoeyink and Jenkins, 1980):

$$\text{Total alkalinity} = [HCO_3^-] + 2[CO_3^{2-}] + [OCl^-] + [HPO_4^{2-}] + [OA^-] + [NH_3] + [H_3SiO_4^-] + [OH^-] - [H^+] \qquad (17.39)$$

This equation assumes that the concentrations of PO_4^{3-} and $H_2SiO_4^{2-}$ are negligible, because their dissociation constants are so small the pH would need to be inordinately high for them to exist in significant concentration in potable waters. Also note that at the carbonic acid equivalence point, HPO_4^{2-} is not fully converted to $H_2PO_4^-$, so the stoichiometric multiplier is presumed to be 1 as an approximation, rather than the factor of 2 that might be commonly assumed based on the charge of the species.

Hydrogen ion–consuming complexes of metals, such as $CaHCO_3^+$, $Fe(OH)_2^0$, $Al(OH)_3^0$, $MgCO_3^0$, and $Pb(CO_3)_2^{2-}$, also contribute to alkalinity, but their concentrations are usually small enough that their contribution can be neglected. If a complex reacts slowly with the acid in an alkalinity titration, it should not be included in equations used to derive DIC from pH and the titration alkalinity.

The bicarbonate and carbonate species affect many important reactions in corrosion chemistry, including the ability of a water to form a protective metallic carbonate scale or passivating film, such as $CaCO_3$, $FeCO_3$, $Cu_2CO_3(OH)_2$, $Zn_5(CO_3)_2(OH)_6$, or $Pb_3(CO_3)_2(OH)_2$. They also affect the concentration of calcium ions that can be present, which, in turn, affects the dissolution of calcium from cement-lined or from asbestos-cement (A-C) pipe.

The formation of strong soluble complexes with metals such as lead, copper, and zinc (Schock, 1980, 1981; Schock, Lytle, and Clement, 1994, 1995a,b; Edwards, Meyer, and Schock, 1996; Schock and Gardels, 1983; Schock, 1989) can accelerate corrosion or cause high levels of metal pickup, given the right pH/alkalinity or pH/DIC conditions.

Buffer Intensity (β), Buffer Capacity, Buffer Index

The ability of a water to provide buffering against a pH increase or decrease caused by a corrosion process or water treatment chemical addition is closely related to the alkalinity, DIC concentration, and pH of the water (Trussell, 1985; Pourbaix, 1973; Snoeyink and Jenkins, 1980; Stumm and Morgan, 1981; Weber and Stumm, 1963; Pankow, 1991; Butler, 1982; Faust and Aly, 1981; Loewenthal and Marais, 1976). The *buffer intensity* of a water is defined as $\beta_C = (\partial A/\partial \text{pH})_{DIC}$, which is essentially the inverse of the slope of the alkalinity titration curve (Stumm and Morgan, 1981; Weber and Stumm, 1963; Butler, 1982). The basic equation for buffer intensity, β_{tot}, in terms of total alkalinity (TALK) is the following (Snoeyink and Jenkins, 1980; Stumm and Morgan, 1981; Weber and Stumm, 1963; Butler, 1982; Loewenthal and Marais, 1976) for a system containing only the carbonic weak acid system in addition to water.

$$\beta_{tot,Alk} = 2.303\left\{\left[\frac{[H^+]TALK}{[H^+]+K'_2}\right]\cdot\left[\frac{[H^+]}{[H^+]+K'_1}+\frac{K'_2}{[H^+]+K'_2}\right]+[H^+]-\frac{K'_W}{[H^+]}\right\} \qquad (17.40)$$

TALK is expressed in eq/L units, $[H^+]$ is the hydrogen ion concentration in mol/L, K'_W is the dissociation constant for water, and K'_1 and K'_2 are the first and second dissociation constants for carbonic acid, respectively, all corrected for temperature and ionic strength. Similarly, buffer intensity can be written in terms of DIC (mol/L) rather than alkalinity. The resulting equation is then:

$$\beta_{\text{tot, DIC}} = 2.303 \cdot \text{DIC} \cdot K_1'[\text{H}^+]\left[\frac{K_1'K_2' + [\text{H}^+]^2}{(K_1'K_2' + K_1'[\text{H}^+] + [\text{H}^+]^2)^2}\right] + 2.303 \cdot \left[\frac{K_W'}{[\text{H}^+]} + [\text{H}^+]\right] \tag{17.41}$$

The units of β are normally mol/L per pH unit. When other weak acids or bases such as orthophosphate or silicate are present, or if the ammonia concentration is high, additional terms must be added to Eqs. 17.40 or 17.41. These expressions relate to what is often termed a *homogeneous buffer system,* in which all buffering components are aqueous species.

Figure 17.18 shows the effect of temperature for a DIC concentration of 48 mg C/L. Figure 17.19 shows the buffer intensity for several different concentrations of DIC. Note the minimum point near pH 8, which corresponds to the point where pH is equal to ½ ($-\log K_1' + -\log K_2'$) for carbonic acid. The buffering at the pH extremes is from either hydrogen ion (low pH) or hydroxide ion (high pH). Figure 17.20 shows the effect on buffer intensity of the presence of 20 mg SiO_2/L silicate, and 5 mg PO_4/L orthophosphate, in a water having a DIC concentration of only 4.8 mg C/L. Clearly, the carbonate system provides almost all of the buffering in the system, except when the pH is over about 9 and the silicate anion (present at a high total SiO_2 concentration) contributes sufficiently. This is important, because the figure demonstrates that corrosion inhibitor chemicals of all types at normal drinking water dosages are a negligible component of pH buffering for all but waters of extremely low DIC. When a water body is open to the atmosphere, the exchange of carbon dioxide gas will affect the DIC concentration, and both the buffer intensity and capacity.

The difference between alkalinity and buffering intensity is often misunderstood, and Table 17.4 shows clearly that they are not the same. The four waters shown have

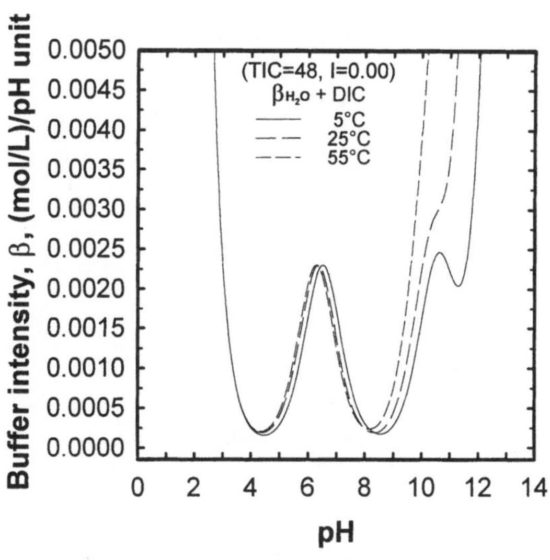

FIGURE 17.18 Effect of temperature on buffer intensity of water for a system with DIC = 48 mg C/L, assuming $I = 0$. Ionic strength has little effect on the position of the curves.

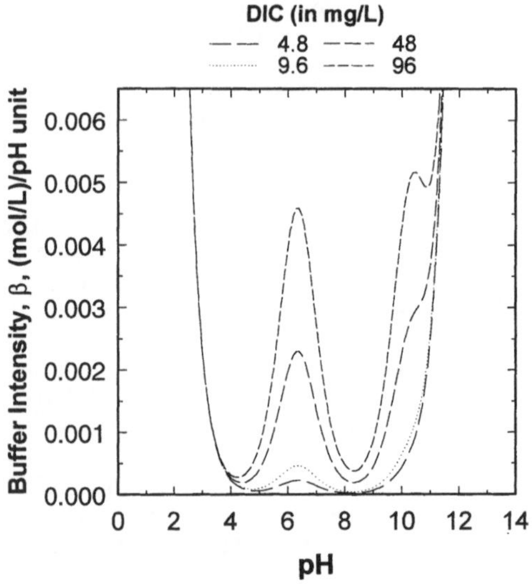

FIGURE 17.19 Effect of DIC on buffer intensity, for 25°C, $I = 0$.

identical alkalinity, but different pH values. These waters have different DICs as well, but the DIC concentrations do show the same trend as buffer intensity. Table 17.4 shows that the buffer intensity trend corresponds to that in Figures 17.19 and 17.20, where the lowest value is near pH 8. Thus, utilities distributing water at a pH around 8 to 8.5 will have much poorer pH stability than if the water with the same

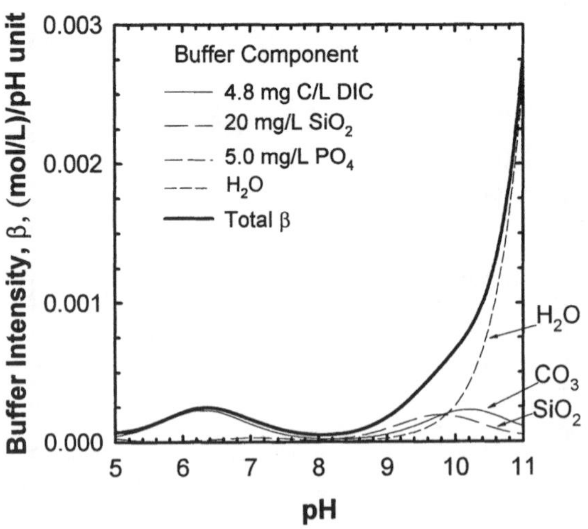

FIGURE 17.20 Combined components of buffer intensity, for DIC, silicate, and orthophosphate at high dosages and low DIC. Computed for 25°C, $I = 0.005$.

TABLE 17.4 Comparison of Buffer Intensity (β) at Different pH Values for Waters Having the Same Total Alkalinity

Alkalinity ($mg\ CaCO_3/L$)	pH	DIC (mg C/L)	β (mmol/L)/pH
20	6.0	14.83	6.26
20	7.0	5.80	1.60
20	8.0	4.86	0.27
20	9.0	4.43	0.72

DIC concentration were distributed at pH 7 or 9. The amount of DIC in the water will have the crucial effect on whether sufficient buffering is present to hold the pH against other chemical reactions taking place in the distribution system. Prevention of localized sites of high pH (such as at cathodic corrosion cell sites) by improved buffering will improve the evenness of formation of corrosion-reducing scales and localized pitting, especially for ferrous materials (Lewandowski, Dickinson, and Lee, 1997).

Buffering can also be provided by solid materials that can directly or indirectly consume $[H^+]$ or $[OH^-]$. This defines a *heterogeneous buffer system.* Common minerals that provide buffering in natural water systems or scales at the water/pipe interface include aluminosilicates like kaolinite (a clay mineral), magnesium hydroxide, and calcium carbonate (Stumm and Morgan, 1981; Dexter and Lin, 1992; Lewandowski, Dickinson, and Lee, 1997; Watkins and Davies, 1987). Although the aluminosilicates provide relatively strong resistance to pH change, the reaction rates are quite slow in comparison to carbonates, oxides, and hydroxides (Stumm and Morgan, 1981), so they may not be as significant in short time frames of water passage through pipes. In the restricted surface environments of pipe scales, many carbonate and hydroxycarbonate minerals could serve as buffers to some extent, such as $FeCO_3$ (siderite), $PbCO_3$ (cerussite), $Zn_5(CO_3)_2(OH)_6$ (hydrozincite), $Pb_3(CO_3)_2(OH)_2$ (hydrocerussite), $CaCO_3$, and others.

Butler (1982) has derived an expression for the buffer intensity at constant DIC for a solution in equilibrium with calcium carbonate, which is:

$$\beta_{CaCO_3}^{DIC} = \beta_{tot,DIC} + 4.61[Ca^{2+}]\frac{[HCO_3^-] + 2[CO_2]}{DIC} \tag{17.42}$$

The derivation of this expression assumes that ion pairing is negligible. To use this equation, the calcium concentration is input presuming all calcium is present as free Ca^{2+} ion. Analyzed concentrations of $[HCO_3^-]$ and $[CO_2]$ must also be input, or computed values derived from a combination of pH and total alkalinity or DIC for the proper temperature and ionic strength.

When the carbonate-water system contains one or more additional weak acids, such as phosphoric acid, the total buffer intensity of the system is the sum of the contributions of the buffer intensities of the individual buffer components (Stumm and Morgan, 1981, 1996).

Dissolved Oxygen

Oxygen is one of the most pervasive agents of corrosion, and it is the cause of much confusion. In many cases, it is the substance that accepts the electrons given up by the corroding metal, allowing corrosion reactions to continue. For example, oxygen reacts with any ferrous iron ions and converts them to ferric iron. Ferrous iron ions, Fe^{+2}, are much more soluble in water than ferric iron, which readily forms a gelatinous orange ferric hydroxide floc when dissolved ferrous iron encounters oxygen or chlorine. Ferric iron accumulates at the point of corrosion, forming a tubercle, or settles out at some point in the pipe and interferes with flow. Similarly, oxygen reacts with metallic copper to form cuprous ions (Cu^{+}) or cupric ions (Cu^{2+}), depending on the amount of oxygen present (Pourbaix, 1966; Schock, Lytle, and Clement, 1995a,b; Ferguson, von Franqué, and Schock, 1996; Werner, Groß, and Sontheimer, 1994). Cuprous hydroxide and oxide solids are much less soluble than cupric forms.

When oxygen is present in water, tuberculation or pitting corrosion may take place. The pipes are affected by the pits, by the tubercles, and by the deposits. "Red water" may also occur, if soluble ferrous iron is exposed to a sudden high oxygen concentration. In many cases, when oxygen is not present, any corrosion of iron is usually first evident to the customer as "red water." The soluble ferrous iron is carried along in the water, and the oxidation to ferric ion occurs by reaction with oxygen from the air only after the water leaves the tap.

High dissolved-oxygen levels may provide corrosion benefits in some circumstances. The presence of substantial dissolved oxygen can allow the formation of different and more protective mineral oxide and hydroxide solid corrosion products than in environments with limited oxygen availability, even for the same redox potential in the water. Hence, corrosion protection is enhanced. The presence of oxygen may also fundamentally affect the speciation of microorganisms in the distribution system, which may also affect the type and form of corrosion deposits. The many subtleties of the interactions of oxygen with pipe surfaces and the effects on the physical properties and mineralogy of scales still need substantial research.

Chlorine Residual

Gaseous chlorine lowers the pH of the water by reacting with the water to form hypochlorous acid, hydrogen ion, and chloride ion. The reaction tends to make the water more corrosive. Conversely, the addition of chlorine as sodium hypochlorite, a strongly basic chemical, raises the pH. In waters with low alkalinity, the tendency of chlorine addition to affect pH is greater because such waters have limited buffering toward pH changes. Tests show that the corrosion rate of steel is increased by free chlorine concentrations greater than 0.4 mg/L (Singley, Beaudet, and Markey, 1984; AWWA, 1986). Chlorine is a strong oxidizing agent, which is one reason why it is a good disinfectant (see Chapters 13 and 14). The chlorine as a free-chlorine residual reacts aggressively toward copper pipe (Atlas and Zajicek, 1982). The corrosive effect of chloramines and the impact of residual free ammonia have not been clearly defined for most metallic materials.

Total Dissolved Solids

A high total dissolved solids (TDS) concentration is usually associated with high concentrations of ions that increase the conductivity of the water. This increased conductivity in turn increases the water's ability to complete the electrochemical circuit and conduct a corrosive current. The dissolved solids may also affect the formation of protective films, depending on their particular nature. If sulfate and chloride are major anionic contributors to high TDS, it is likely to show increased corrosivity toward iron-based materials. If the high TDS is mainly composed of bicarbonate and hardness ions, the water may tend to be noncorrosive toward iron and cementitious materials, but highly corrosive toward copper.

Hardness

Hardness can play many different roles that can sometimes reduce corrosivity toward many plumbing materials. Hardness is caused predominantly by the presence of calcium and magnesium ions and is expressed as the equivalent quantity of $CaCO_3$. Hard waters are generally less corrosive than soft waters if sufficient calcium ions and alkalinity are present at the given pH to form a protective $CaCO_3$ lining or a mixed iron/calcium carbonate film on the pipe walls (Trussell, 1985; Singley et al., 1985; Sontheimer, Kolle, and Snoeyink, 1981; Stumm, 1956; Stumm, 1960; Légrand and LeRoy, 1995). High levels of calcium also help stabilize calcium silicate phases in cement-mortar linings and A-C pipes, and promote the formation of calcium carbonate to block the release of free lime (calcium hydroxide) (Légrand and LeRoy, 1995; LeRoy et al., 1996). In spite of common beliefs, significant calcium carbonate films do not usually form on lead, galvanized, or copper cold-water pipes, so they are not primarily the causes of corrosion inhibition in these cases. Exceptions are sometimes found in systems that practice lime or lime-soda softening and put supersaturated (with $CaCO_3$) water into the distribution system. Calcium may also assist in the heterogeneous corrosion buffering reactions at the pipe surface (Trussell, 1985; Dexter and Lin, 1992; Watkins and Davies, 1987; Légrand and LeRoy, 1995).

Chloride and Sulfate

Chloride (Cl^-) and sulfate (SO_4^{2-}) may cause increased corrosion of metallic pipe by reacting with the metals in solution and causing them to stay soluble, or interfering with the formation of normal oxide, hydroxide, or hydroxycarbonate film. They also contribute to increased conductivity of the water. Chloride is about three times as active as sulfate in this effect for iron materials, but recent research into copper corrosion has shown the action of chloride to be very complicated, and not always detrimental (Edwards, Rehring, and Meyer, 1994; Edwards, Ferguson, and Reiber, 1994; Rehring, 1994; Meyer, 1994). Based on the historical field studies of Larson with unlined iron pipe, the ratio of the sum of the chloride plus the sulfate to the bicarbonate concentration has been used by some corrosion experts to estimate the corrosivity of a water. Pitting of copper tubing is often strongly related to the concentrations of Cl^- and SO_4^{2-} relative to HCO_3^-, where the high salt contributions can contribute to the acidification in pits and enhanced conductivity of the water (Ferguson, von Franqué, and Schock, 1996; Cruse and Pomeroy, 1974).

Hydrogen Sulfide (H_2S)

Hydrogen sulfide accelerates corrosion by reacting with metal ions to form nonprotective insoluble sulfides (AWWA, 1986; Singley et al., 1985). It attacks iron, steel, copper, and galvanized piping to form "black water," even when oxygen is absent. Attack by H_2S is often complex, and the results may be evident immediately or they may not become apparent for months and then suddenly become severe. Hydrogen sulfide has been found to promote a particularly devastating kind of pitting in copper piping (Jacobs and Edwards, 1997). Hydrogen sulfide also has been implicated in the attack on asbestos-cement pipe in some waters, possibly through microbial reactions (Leroy et al., 1996; Holtschulte and Schock, 1985).

Ammonia

Ammonia forms strong soluble complexes with many metals, particularly copper and probably lead. Thus, ammonia may interfere with the formation of passivating films or increase the corrosion rate. Research in this area has been sparse and relatively inconclusive.

Silicate

Silicates can form protective films that reduce or inhibit corrosion by providing a barrier between the water and the pipe wall. These chemicals are usually added to the water by the utility, although natural levels of silica can be high enough where the silica can react with corrosion by-products on the pipe surface, and form a more protective film. Some studies have shown silicates beneficial to reducing iron, copper and lead corrosion (Schenk and Weber, 1968; LaRosa-Thompson et al., 1997; Clement, Schock, and Lytle, 1994; Lytle, Schock, and Sorg, 1994, 1996). In some cases, the mechanism causing the benefit may be attributable to the increase in pH that accompanies the addition of the silicate treatment chemicals, which are strongly basic as normally formulated. Silicate has been found to be effective when used in combination with chlorination to sequester iron in groundwater and allow control of copper and lead levels by pH adjustment (Clement, Schock, and Lytle, 1994; Schock et al., 1998). There is also some evidence that silicate is beneficial in reducing the degradation of asbestos-cement pipes and cement-lined pipes (Leroy et al., 1996).

Orthophosphate

Orthophosphate usually forms insoluble passivating films on the pipe, reacting with the metal of the pipe itself (particularly with lead, iron, and galvanized steel) in restricted pH and dosage ranges. Orthophosphate formulations that contain zinc can decrease the rate of dezincification of brass (Oliphant and Schock, 1996), and can deposit a protective zinc coating (probably basic zinc carbonate or zinc silicate) on the surface of cement or asbestos-cement pipe, given the proper chemical conditions (Schock, 1981; Leroy et al., 1996). Available reports have not led to a consensus, but the zinc formation may be advantageous by providing cathodic inhibition in some situations with iron, galvanized, or steel pipe, particularly if a fresh surface is exposed to the water. Research on the release of lead, copper, and zinc from brass has shown that zinc is unnecessary in the formulation for the control of lead (Schock, 1981; Lytle and Schock, 1996, 1997; Leroy et al., 1996; Lytle, Schock, and Sorg, 1994, 1996). Orthophosphate has been found to slow the rate of copper oxidation, especially at near-neutral pH (Schock, Lytle, and Clement, 1995a,b; Lytle and Schock, 1997), and interfere with the formation of protective oxide films at higher pH values. Examination of field data from large utilities has shown that orthophosphate tends to be ineffective or counterproductive at values increasingly above 8

(Dodrill and Edwards, 1995; Reiber et al., 1997). These results are consistent with laboratory experiments and theoretical models for lead solubility in the presence of orthophosphate (Schock, Wagner, and Oliphant, 1996).

The mechanism of orthophosphate protection of unlined iron is still largely unknown. Research, both specific to drinking water and from the geochemistry area, suggests it could be a combination of the formation of passivating films, especially with ferrous iron (Melendres, Camillone, and Tipton, 1989; Koudelka, Sanchez, and Augustynski, 1982), and sorption on ferrous hydroxide and ferric oxyhydroxide corrosion by-product surfaces (Geelhoed, Hiemstra, and Van Riemsdijk, 1997; Columbo, Barrón, and Torrent, 1994; Buffle et al., 1989). Orthophosphate is known to have a significant effect on the aggregation behavior of ferric hydroxide (He et al., 1996). Conditions that optimize the formation of the most effective passivating coatings need substantial research.

Natural Color and Organic Matter

The presence of naturally occurring organic material (manifested sometimes as color) may affect corrosion in several ways. Some natural organics may react with the metal surface and provide a protective film and reduce corrosion, especially over a long period of time (Campbell, 1971). Others have been shown to react with the corrosion products to increase corrosion, such as with lead. Organics may complex calcium ions and keep them from forming a protective $CaCO_3$ coating. In some cases, the organics may become food for organisms growing in the distribution system or at pipe surfaces. This can increase the corrosion rate in instances when those organisms attack the surface as discussed in the section on biological characteristics. Which of these instances will occur for any specific water has been impossible to determine, so using color and organic matter as corrosion control methods is not recommended (Singley, Beaudet, and Markay, 1984; AWWA, 1986).

Polyphosphates

Polyphosphates have frequently been used to successfully control tuberculation and restore hydraulic efficiency to transmission mains. Polyphosphates can sometimes cause the type of corrosion to change from pitting or concentration cell corrosion to a more uniform type, which causes fewer leaks and aesthetic complaints. Pipe walls are usually thick enough that some increase in dissolution rate is not of practical significance. Polyphosphates have been used to control the oxidation of ferrous iron dissolved from pipe, and the formation of "red water." On the other hand, polyphosphates can prevent the deposition of protective calcium-containing films and enhance the solubility of metals such as lead and copper, interfering with the formation of passivating films (Holm and Smothers, 1990; Holm and Schock, 1991a,b). Polyphosphates have been found to attack and substantially soften asbestos-cement pipe by accelerating the depletion of calcium and inhibiting the formation of fiber-binding iron or manganese deposits (Leroy et al., 1996; Holtschulte and Schock, 1985). It is reasonable to expect them to similarly affect cement-lined or concrete pipes. Formulas for calcium carbonate precipitation potential and saturation indices (such as the Langelier Saturation Index) are not valid in the presence of polyphosphates, unless they are modified to take into account the inhibition of precipitation and sequestration properties of the polyphosphate. Although not yet comprehensively investigated, polyphosphates, when mixed with orthophosphate, may assist in the formation of orthophosphate films by complexing calcium or magnesium in hard waters that otherwise could cause unwanted orthophosphate complexation or precipitates.

Iron, Zinc, and Manganese

Soluble iron, zinc, and to some extent, manganese, can play a role in reducing the corrosion rates of A-C pipe (Leroy et al., 1996; Holtschulte and Schock, 1985). Through a reaction that is not yet fully understood, these metallic compounds may combine with the pipe's cement matrix to form a protective coating on the surface of the pipe. Waters that contain natural amounts of iron have been shown to reduce the rate of A-C pipe corrosion and bind asbestos fibers to the surface. When zinc is added to water in the form of zinc chloride or zinc phosphate, protection from corrosion has been demonstrated by the formation of a hard surface coating of basic zinc carbonate or a zinc silicate.

Copper

The presence of dissolved copper can cause rapid corrosion of galvanized steel piping. Free copper may occur in recirculating water systems, such as the hot-water systems commonly found in industry, hotels, and apartments, that have consecutive sections of copper and galvanized steel pipes in the loops. This free copper adsorbs or reacts with the galvanized steel piping, setting up small galvanic cells, which produce rapid pitting failure of the galvanized steel piping. Similar corrosion has occurred in unprotected galvanized steel pipes in consumers' homes as a result of the addition of even trace amounts of copper, either from copper sulfate as an algicide in distribution reservoirs, from flexible copper tubing connectors to hot-water heaters, or by mixed copper and galvanized plumbing in the same waterline (Trussell, 1985; Fox et al., 1986; Trussell and Wagner, 1985, 1996).

Magnesium

Magnesium, and possibly some other trace metals (such as zinc), are known to inhibit the formation of the calcite form of $CaCO_3$. Instead of calcite, the aragonite form or some magnesian calcites may be deposited, which are more soluble. With these conditions, most calcium carbonate precipitation or saturation indices will give erroneous predictions. Whether these $CaCO_3$ forms make any difference in protection against corrosion is not known.

Aluminum

Aluminum may be widespread as a component of films on distribution system piping that can act as diffusion barriers to reduce corrosion or metal release. Even though it has not been systematically studied, several investigations have found aluminum films to significantly reduce lead leaching (Lauer and Lohman, 1994) and to adversely affect the hydraulic efficiency of distribution mains (Kriewall et al., 1996). Aluminum was found on copper pipes even at low influent concentrations (Schock, Lytle, and Clement, 1995b), suggesting also that it can be widespread. In natural aquatic systems and groundwater, aluminum readily combines to form aluminosilicate minerals of low solubility, also suggesting that they may be common in distribution systems. Destabilization of aluminum films has been found to aggravate lead levels in the water in one system in Wales (Fuge and Perkins, 1991; Fuge, Pearce, and Perkins, 1992).

CORROSION OF SPECIFIC MATERIALS

Cast Iron and Steel

Dissolved oxygen can either enhance or inhibit the rate of corrosion of steel, depending on the mineral content of the water and the dissolved oxygen concentration (Singley et al., 1985; Benjamin, Sontheimer, and Leroy, 1996). Oxygen is also very important in causing concentration cell corrosion. Chlorine can accelerate the rate of attack of iron either by direct increase of the redox potential of the electrolyte that favors the conversion of iron to ferrous and then ferric ions, or through a sequence of chemical reactions that produce hydrogen ions, hypochlorous acid, hypochlorite ions, and chloride. The effect of chlorine, like that of oxygen, diminishes after the corroding surface is passivated with corrosion by-products. A secondary effect of chlorine is its significant mitigating impact on microbiological corrosion phenomena. The same is true for other disinfecting agents. Changes from oxic conditions (chlorine residual or oxygen present) to anoxic conditions fosters the production of ferrous iron, which is much more soluble than ferric iron. Redevelopment of oxidizing conditions, or contact of the water with air, will then result in red water. Overall, red water formation is a very complex phenomenon still under investigation (Trussell, 1985; Singley et al., 1985; Benjamin, Sontheimer, and Leroy, 1996; Larson, 1975; Stumm, 1960; Légrand and LeRoy, 1955; Rossum, 1987).

The effect of pH is generally through its role in secondary reactions, such as the oxidation of ferrous iron, and on the formation of scales and corrosion products. This effect influences the corrosion rate, as well as iron release. Iron solubility tends to increase at both high and low pH.

The effect of *natural color* is variable because the nature of organic materials responsible for color varies widely from source to source. These materials can frequently reduce the rate of ferrous iron oxidation, but also reduce the rate of calcium carbonate precipitation.

Chloride and sulfate can drastically affect the behavior of ferrous materials. Both corrosion rates and iron uptake (into the water from the pipe) have been determined to be increased sharply as the concentration of sodium chloride or sodium sulfate is increased in solution (Singley et al., 1985; Benjamin, Sontheimer, and Leroy, 1996). Research concerning the "Larson Ratio" (see later section) suggests that the molar ratio of chloride plus sulfate to bicarbonate should be below approximately 0.2 to 0.3 to prevent enhanced corrosion of unlined iron.

If a water contains an appreciable amount of calcium, it may act in conjunction with the pH and bicarbonate concentration to buffer the pH rise from corrosion reactions by the instantaneous formation of calcium carbonate solid, forming a heterogeneous buffer. Therefore, for several reasons, the alkalinity (the DIC content providing sufficient bicarbonate and carbonate ions) and calcium hardness are influential in reducing iron and steel pipe corrosion.

Natural silica has been shown to reduce the corrosion rate of steel, particularly when a layer of scale is already present. Early investigations by Schenck and Weber (1968) showed that silicate had significant impact on slowing oxidation of ferrous iron and reducing hydrolysis of ferric iron. Thus, silicate chemicals offer strong potential as an iron corrosion inhibitor.

Phosphates have a varied effect on the iron corrosion rate, depending upon the type of phosphate and other water conditions (such as pH). The surface chemistry of the iron corrosion products (surface composition and charge) may play an important and defining role in the success of the application as a corrosion inhibitor. Sometimes the nature of iron corrosion, such as tuberculation, can be changed through the addition of polyphosphates. High doses can have a "cleansing" action on distribution mains. Frequently, the corrosion rate is not reduced, but the form and manifestation of the corrosion can be made to be less objectionable. Koudelka et al. (1982) found that films formed on iron pipe from metaphosphate solutions were actually iron orthophosphates. Corrosion inhibitors of the orthophosphate type, often containing zinc and sometimes blended with silicate solutions, have been found to be useful in suppressing iron release and corrosion rates (Singley et al., 1985; Benjamin, Sontheimer, and Leroy, 1996). Keeping a moderate flow rate is also useful in reducing corrosion and enhancing film formation. Extensive research on iron corrosion has been summarized and presented by the American Water Works Association Research Foundation (Singley et al., 1985; Benjamin, Sontheimer, and Leroy, 1996; AWWARF, 1998).

Copper

The performance of copper in potable water systems depends on whether relatively thin and adherent films of corrosion products, cuprous oxide (cuprite, Cu_2O), cupric oxide (tenorite, CuO), or basic copper carbonate (malachite, $Cu_2(OH)_2CO_3$), can be formed (Butler and Ison, 1966).

Copper tends to be less prone to the effects of concentration cell or differential aeration corrosion than galvanized steel, steel, and iron pipes. The influence of copper ion concentration on the potential of copper in solution is very marked. When solution velocities vary over a copper surface, the parts washed by the solution with the higher rate of movement become anodes and not cathodes, as would be the case with iron (Butler and Ison, 1966). Impingement attack of copper by high water velocities is one of the most common problems of copper pipe. High DIC concentration aggravates impingement attack, and it may manifest itself at velocities and turbulent flow areas that otherwise would not seem to be prone to damage under conventional design specifications for building and household plumbing systems.

Recent research has shown that copper levels in disinfected or oxic drinking water are controlled for many years by the formation of metastable solids, apparently cupric hydroxide or cupric oxide of high surface area and solubility (Schock, Lytle, and Clement, 1994, 1995a,b). The formation rate of the basic cupric carbonate mineral $Cu_2(OH)_2CO_3$ (malachite), which would lead to low copper levels in many waters of moderate DIC, appears to be very slow. Thus, dissolved copper release for most of the sites targeted for regulatory monitoring, and newer sites, is best described by looking at control by cupric hydroxide or cupric oxide solids of high surface area and small particle size (Schock, Lytle, and Clement, 1994, 1995a,b; Edwards, Meyer, and Schock, 1996). High concentrations of bicarbonate and carbonate ions at neutral to alkaline pH values significantly enhance the release of copper through aqueous complexation [e.g., $CuHCO_3^+$, $CuCO_3$, $Cu(CO_3)_2^{2-}$] (Schock, Lytle, and Clement, 1994, 1995a,b; Edwards, Meyer, and Schock, 1996; Ferguson, von Franqué, and Schock, 1996).

Electrochemical analyses and copper release experiments of Edwards, Rehring, and Meyer (1994) also indicate bicarbonate is highly aggressive toward copper, especially below pH 8.5. Figure 17.21 shows the strong effect of both carbonate com-

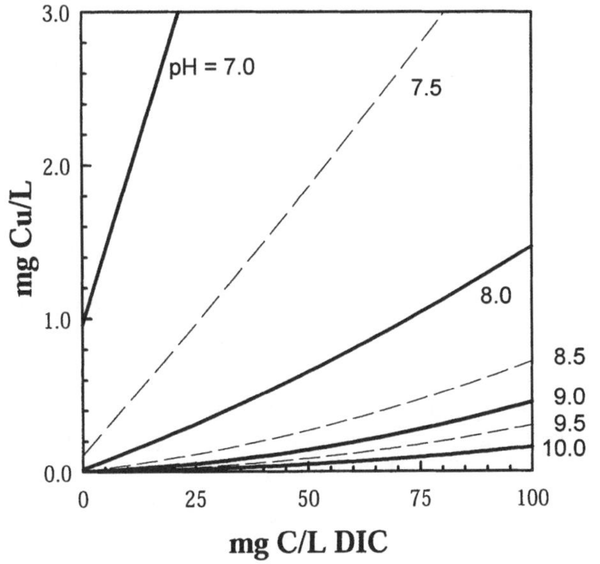

FIGURE 17.21 Copper solubility assuming equilibrium with cupric hydroxide of high surface area, representing new plumbing. Computed for 25°C, $I = 0.02$.

plexation and pH on copper solubility for relatively new plumbing systems. Thus, hard, high-alkalinity (DIC) groundwaters that are protective of iron and cement-based piping are particularly aggressive toward copper materials. This is a fact that is often not appreciated, and treating these kinds of water chemistries to reduce copper release is often very difficult. Conventional pH adjustment is generally impractical or inadequate, as buffering is extreme and hardness places strong limits on the pH achievable without precipitating calcium carbonate. The neutral pHs of these waters generally make direct aeration impractical as a mechanism for reducing DIC (Lytle et al., 1998a–c). Thus, conventional lime or lime/soda softening and a combination of ion exchange for hardness and DIC reduction followed by pH adjustment are generally the most promising treatment approaches available.

Soft waters, particularly those with an acidic pH and typified by low alkalinities, will dissolve high levels of copper rapidly. These systems can usually be treated in a straightforward manner by any of several relatively simple approaches, such as pH adjustment, pH/DIC adjustment (if added buffering for pH stability is needed), or the use of mechanically simple systems, such as aeration and limestone contactors (Lytle et al., 1998a–c; Letterman et al., 1987; Letterman, Haddad, and Driscoll, 1991; Letterman, 1995; Letterman and Kathari, 1996; Lytle et al., 1998b; O'Brien, 1995; Spencer, 1996; Whipple, 1913).

The corrosion rate and release rate of copper into the water are very sensitive to the level of oxidizing agents in the system, especially free chlorine (Schock, Lytle, and Clement, 1995a,b; Ferguson, von Franqué, and Schock, 1996; Lytle and Schock, 1996; Werner, Groß, and Sontheimer, 1994; Atlas and Zajicek, 1982; Pisigan and Singley, 1987). Even low levels (such as 0.2 mg/L) affect the oxidation and corrosion rate. Reiber (1989) has shown that the chlorine effect is much more important than dissolved oxygen in normal drinking water situations. Several studies have been made of the influence of oxygen on the oxidation rate of Cu^+ to Cu^{2+} in natural and seawater, which is of some applicability to drinking water situations (Millero, 1989, 1990a,b; Millero, Izaguirre, and Sharma, 1987; Eary and Schramke, 1990). In the presence of oxidants, copper levels in standing water in plumbing systems can increase for as many as 48 to 72 hours, so "overnight" standing samples do not necessarily represent building systems or nearly worst-case scenarios (Schock, Lytle, and Clement, 1995a,b; Lytle and Schock, 1987). Conversely, when oxidants are depleted, copper levels may decrease (Schock, Lytle, and Clement, 1995a,b; Ferguson, von Franqué, and Schock, 1996; Lytle and Schock, 1997; Werner, Groß, and Sontheimer, 1994).

Normally, the rate of uniform corrosion of copper in potable water systems is not rapid enough to cause failure of the tubing. However, consumer complaints of "blue water," "green water," and staining can result. When such copper dissolution is sufficient to cause complaints, frequently a fine dispersion of copper-corrosion products discolors the water. When inspected, the inside surface of the copper tube under such circumstances is characterized by a loosely adhering powdery scale and beneath it, or in areas where no scale is present, by general dissolution of corrosion of the copper (Ferguson, von Franqué, and Schock, 1996; Reiber et al., 1997; Cruse and von Franqué, 1985). Related to green/blue water is green/blue staining, as even a few parts per million of copper in water can react with soap scums and cause green staining of plumbing fixtures. Several cases of "blue" water have resulted from what is apparently a dislodging of a slime created by microorganisms (Reiber et al., 1997).

The most important chemical variables in general uniform copper corrosion are pH, DIC, and redox potential. The latter is most important in defining the conditions in which the oxidizing potential (pE, Eh) is high enough to form cupric ions. At lower redox potentials, either copper is immune to corrosion, or highly insoluble cuprous oxide or hydroxide solids form that produce extremely low copper levels in the water (Schock, Lytle, and Clement, 1995a,b; Ferguson, von Franqué, and Schock, 1996). Many untreated groundwaters of neutral pH and high DIC fall into this category. When oxidative processes are introduced, such as for iron or manganese removal, or disinfection, very high corrosion of copper will then result.

Generally, cations (calcium, magnesium, sodium, and potassium) exert little or no effect on the rate of corrosion (Cruse and von Franqué, 1985). Ives and Rawson (1962) have suggested that calcium can play a beneficial role at levels and conditions substantially below calcium carbonate saturation. Anions (chloride, sulfate, bicarbonate, orthophosphate), however, do exert considerable influence on the rate of corrosion, the levels of dissolved copper in the water, and the nature and stability of corrosion product solids on the pipe surfaces (Schock, Lytle, and Clement, 1995a,b; Edwards, Meyer, and Schock, 1996; Ferguson, von Franqué, and Schock, 1996; Rehring, 1994; Meyer, 1994; Reiber, 1989; Meyer and Edwards, 1994). Traditionally, chloride has been considered to be a very aggressive ion towards copper (Cruse and von Franqué, 1985). Some relatively recent research suggests that chloride modifies the nature of the surface film formation (Adeloju and Hughes, 1986), and that it may not be as aggressive in the context of uniform corrosion as previously believed (Edwards, Rehring, and Meyer, 1994; Rehring, 1994; Meyer, 1994). At pH ≥ 7, studies by Edwards, Rehring, and Meyer (1994) indicate that chloride has beneficial long-term effects toward copper corrosion and release.

Sulfate plays a complex role in copper corrosion, sometimes producing lower copper levels at slightly alkaline to somewhat acidic pHs, where any of several basic cupric sulfate solids may form a passivating film. At higher pHs, some research has suggested that these metastable cupric sulfate solids inhibit the beneficial precipitation of cupric hydroxide and subsequent transformation to tenorite (Schock, Lytle, and Clement, 1995a,b). Several studies, including electrochemical analyses of different types, have produced data indicating sulfate is very aggressive toward copper, particularly for aged surfaces above pH 7 (Edwards, Rehring, and Meyer, 1994; Rehring, 1994; Meyer, 1994). Marani (1992) has summarized considerable research that has investigated the conditions favorable to the formation of basic cupric sulfate solids under water chemistry conditions appropriate for drinking waters.

The role of orthophosphate in the corrosion inhibition of copper is somewhat debatable. Under some conditions, orthophosphate has appeared to promote higher copper levels (Werner, Groß, and Sontheimer, 1994). Electrochemical studies by Reiber (1989) suggested that orthophosphate might provide some corrosion reduction by changing the fundamental form of the anodic reaction. Modeling of copper solubility in the presence of orthophosphate has predicted that unlike lead, the best conditions for the formation of a cupric orthophosphate film would be in the pH range of approximately 6.7 to 7.2 (Schock, Lytle, and Clement, 1994, 1995a,b), depending upon orthophosphate and DIC concentrations. Systematic experimental investigations by the same authors over a range of pH and DIC conditions (some awaiting publication) found that very thin films formed on the surface of copper pipe, inhibiting oxidation and growth of normal hydroxide, oxide, or hydroxycarbonate passivating solids. These results were generally supportive of the observations of Reiber (1989) about possible interference and altering of normal anodic site reactions and the solubility models predicting lower solubility of the phosphate solids at these pHs. At higher pH values (8–9), higher copper levels were produced in the presence of the orthophosphate, consistent with the inhibition of growth of the less soluble and protective cupric oxide. These observations help explain contradictory conclusions about the effectiveness of orthophosphate in full-scale applications with different water qualities. The experimental evidence also suggests that the orthophosphate may be less effective in inhibiting copper corrosion if an oxidized carbonate or oxide scale already exists on the pipe surface under many water chemistry conditions. Figure 17.22 shows trends of copper solubility with different dosages of orthophosphate at different pH values and DIC concentrations.

In the presence of oxidants, such as chlorine or dissolved oxygen, complexing ligands such as ammonia accelerate the corrosion and dissolution of copper (Butler and Ison, 1966). Excess ammonia in the presence of chloramines was found to be associated with increased copper levels in one study (AWWARF, 1990), but the results were not conclusive and more research needs to be done. Natural organic matter has also been found to increase copper corrosion and copper release under many circumstances (Rehring, 1994; Kristiansen, 1982; Rehring and Edwards, 1994, 1995).

Dissolved silica and silicate anion have not been studied in great detail with respect to corrosion effects on copper. Several studies have shown that high dosages of silicate at high pH are effective in reducing copper release, though the mechanism has not been clearly defined (LaRosa-Thompson et al., 1997; Lytle, Schock, and Sorg, 1994, 1996; Schock et al., 1998).

The presence of dissolved sulfide, or sulfide produced by sulfate-reducing bacteria, is especially troublesome to copper (Jacobs and Edwards, 1997). Copper sulfide scales have been shown experimentally to increase copper corrosion rates by 1 to 2 orders of magnitude at pH 6.5 and 9.2, and copper release by 500 to 5000 percent, respectively. Subsequent removal of sulfide from the water, deaeration, chlorination, and superchlorination all failed to restore normal corrosion rates to the pipe containing sulfide scales. Experiments suggested that more than a year might be needed for the corrosion rate to reach levels below those observed in the presence of sulfide.

Pitting of copper tubing is most commonly associated with hard well waters (Trussell, 1985; Ferguson, von Franqué, and Schock, 1996; Cruse and von Franqué, 1985). Pitting most often occurs in cold-water piping. Usually, pitting is observed when dissolved carbon dioxide exceeds 5 mg/L and dissolved oxygen is at least 10 to

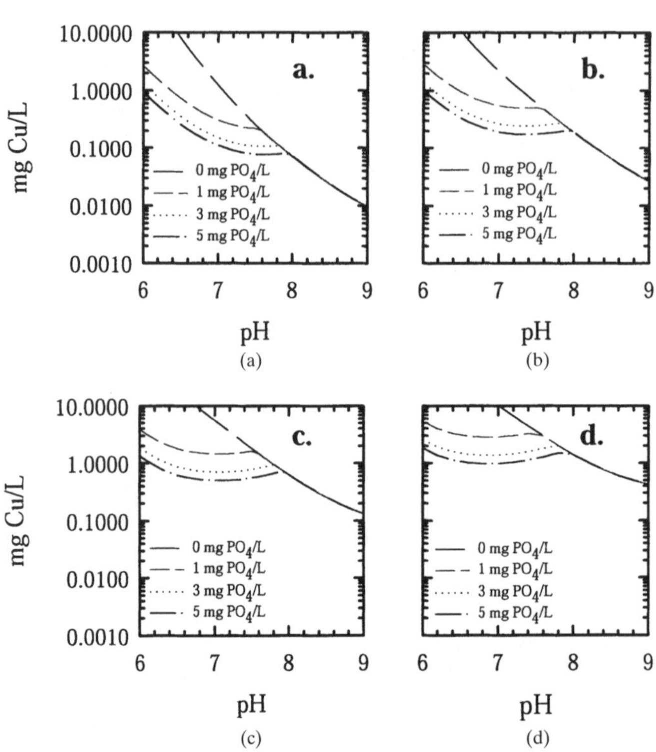

FIGURE 17.22 Effect of orthophosphate on solubility of new copper assuming equilibrium with either $Cu(OH)_2$ or $Cu_3(PO_4)_2{\cdot}2H_2O$ at different DIC and orthophosphate concentrations. Computed for 25°C: (*a*) DIC = 4.8 mg C/L, $I = 0.005$; (*b*) DIC = 14.4 mg C/L, $I = 0.005$; (*c*) DIC = 48 mg C/L, $I = 0.01$; (*d*) DIC = 96 mg C/L, $I = 0.02$. (*Source:* Schock, Lytle, and Clement, U.S.E.P.A., EPA/600/R-95/085, Cincinnati, OH, 1995b.)

12 mg/L. The water quality parameters typical for water systems having copper pitting problems are not the soft, low pH waters normally associated with corrosion of copper. Edwards, Ferguson, and Reiber (1994) have classified three general categories of copper pitting attack, although there are many complications.

Pitting attack is most common in new installations, with 80 to 90 percent of the reported failures occurring within the first two to three years (Trussell, 1985). In extreme cases, pits can break through within only a few months. Pitting occurs in all three standard types of copper tubing (K, L, and M). If unfavorable water quality conditions occur before the protective coat has formed, then a serious pitting attack may occur. These conditions often result from leaving water stand in unoccupied housing for months, after pressure- and leak-testing the plumbing systems. Incomplete flushing of flux residue and particulate debris set up conditions favorable to differential oxygenation corrosion. Disinfectant residuals are usually lost within a day or two at most, and microbial growth may occur over long periods of water stagnation. Microbes residing in sheltered areas can promote and accelerate reactions within incipient pits, and they become very hard to disinfect. Once started, pits of this type cannot usually be stopped without strong physical cleaning processes.

Thin carbonaceous films, derived from carbonization of the lubricant used in manufacture, have resulted in pitting of copper tubes (Cruse and von Franqué, 1985). Films of manganese oxides, formed by slow deposition from soft moorland waters, can also cause localized attack. A review of the voluminous literature on pitting corrosion and an evaluation of reported field experience indicates that the presence of certain types of natural organic matter (NOM) reduces the corrosion of copper in certain waters by promoting the formation of more protective scales, and also by reducing the tendency for pitting attack (Edwards, Ferguson, and Reiber, 1994). This evaluation recommended that careful consideration be given to increased Cu release resulting from the removal of NOM for the purpose of controlling DBPs. Another study found similar effects of NOM. Initially, NOM promoted pitting in certain narrow water quality and NOM concentration ranges, but over long periods of time the NOM reduced the tendency toward pitting at higher NOM levels (Korshin, Perry, and Ferguson, 1996). Generally, the study concluded that NOM interacts with the copper through sorption, and it tends to increase the rate of copper leaching and the dispersion of the inorganic corrosion by-product scale materials.

Edwards, Ferguson, and Reiber (1994) found that chloride, though present in materials in some pits, was often a mitigating agent for the propagation of pits in copper pipe over the long term. Further, they found that sulfate was a much more aggressive anion in that regard, and suggested a key factor for the initiation and promotion of pitting might be the conditions under which the solid $Cu_4SO_4(OH)_6$ (brochantite) formed on the surface, rather than malachite, tenorite, or chloride-containing cupric salts. Although there are a few differences in the details, many of these observations and interpretations are corroborated by Taxèn (1996), who especially observed differences in pitting when copper (I) was involved more than copper (II). He also found sulfate and carbonate were most aggressive toward copper, especially in the bulk solution. Carbonate tended to reduce the initiation of acidic pits, and a high redox potential was the most significant factor for the rate of pit propagation. Generally, recent reviews and research results should be consulted as a first step toward diagnosing possible causes and remedies for copper pitting situations (Edwards, Rehring, and Meyer, 1994; Edwards, Ferguson, and Reiber, 1994; Ferguson, von Franqué, and Schock, 1996; Taxèn, 1996; Cruse and von Franqué, 1985).

Many instances of microbially mediated copper corrosion have been reported, though the mechanisms and differentiation from inorganic causes are not always clear (Camper, 1996; Murphy, O'Connor, and O'Connor, 1997a–c; Geesey et al., 1986, 1988; Geesey and Bremer, 1991; Reiber et al., 1997; Arens et al., 1996; Bremer and Geesey, 1991a,b; Fischer et al., 1992; Iwaoka et al., 1986; Jolley et al., 1988, 1989).

Brasses and Bronzes

Although many studies on factors involved in the corrosion and dissolution of brasses and bronzes have been conducted, a clear understanding of the role of water chemistry in these processes has not emerged (Schock and Neff, 1988). Most studies have focused on dezincification problems, which is understandable from the perspectives of economics and materials longevity. Corrosion causes the failure of mechanical devices such as valves and faucets made from brass or bronze, but it also contributes lead, copper, and zinc to the water. Metal leaching from the tap and other brass or bronze plumbing fixtures can confound the interpretation of field studies in which the intent is to correlate metal levels at the tap to human exposure, corrosion rates, or the effectiveness of corrosion control treatment programs. In the drinking water corrosion field, the contribution of brass fixtures to first-draw lead levels in tap water monitoring programs is now well accepted (Schock, Wagner, and Oliphant, 1996; Oliphant and Schock, 1996; Dodrill and Edwards, 1995; Reiber et al., 1997; Lee, Becker, and Collins, 1989).

Oliphant (1978) studied several aspects of brass dezincification in water systems. The effects of many water quality properties on lead release from brass have been reviewed, particularly the interrelationships of lead content, pH, DIC concentration, and orthophosphate concentration (Schock and Neff, 1988; Oliphant and Schock, 1996; Lytle and Schock, 1996, 1997). Important variables in the performance of brasses and bronzes in potable waters include the pH, flow, temperature, lead content, surface area of lead and brass, alkalinity (or DIC), chloride concentration, orthophosphate concentration, and the presence of natural inhibitors. Lead, zinc, and copper leaching vary greatly with the types of alloys used and the manufacturing and fabrication processes (Oliphant and Schock, 1996; Lytle and Schock, 1996, 1997).

Lead and Lead-Containing Solder

Permissible lead levels in the drinking water of many countries are so low that the rate of lead corrosion is of no practical concern. Even slightly aggressive water standing for a brief time in contact with lead pipe or lead-containing materials will pick up lead levels that are considered a public health risk. Lead release is a significant problem whether released as dissolved lead species, or associated with particulate material like dislodged pipe deposit solids, or sorbed on other particulate matter present in the water. The relative contribution of dissolved lead to particulate matter almost certainly varies with the water chemistry, the configuration of the sampling site, and the history of the plumbing materials containing the lead. Numerous studies have investigated different aspects of the occurrence of particulate lead in drinking water (AWWARF-TZW, 1996; Schock, 1990; AWWARF, 1990; Reiber et al., 1997; Sheiham and Jackson, 1981; Fuge, Pearce, and Perkins, 1992; AWWARF, 1998; Hulsmann, 1990; Breach, Crymble, and Porter, 1991; Colling, Whincup, and Hayes, 1987; deMora and Harrison, 1984; deMora, Harrison, and Wilson, 1987; Harrison and Laxen, 1980; and Walker, 1983). Comparisons among studies of the significance of lead particle release are difficult, because the analytical schemes (such as filtration) developed to differentiate dissolved from particulate lead species are very prone to both negative and positive biases, as is discussed elsewhere in this chapter. Regardless of its prevalence, when it occurs the release of particulate lead makes achieving regulatory standard objectives for lead levels very hard. Utilities should try to minimize all types of lead release by the selected corrosion control treatment approach as much as possible.

Passivation of lead is usually caused by the formation of a surface film composed of cerussite ($PbCO_3$), hydrocerussite ($Pb_3(CO_3)_2(OH)_2$), or plumbonacrite ($Pb_{10}(CO_3)_6(OH)_6O$), or some combination of these forms. The oxides PbO (litharge or massicot) or PbO_2 (plattnerite) have been reported on rare occasions. Litharge and massicot are highly soluble, and almost certainly represent layers present under the water/corrosion product interface not in contact with the water. The formation of these compounds depends primarily on pH and DIC concentrations. These interrelationships are complicated, and occasionally misunderstood (Trussell, 1985; Schock, 1980, 1981b, 1989, 1990a; Schock, Wagner, and Oliphant, 1996; Schock and Gardels, 1983; Dodrill and Edwards, 1995; Sheiham and Jackson, 1981; LeRoy, 1993; LeRoy and Cordonnier, 1994). The carbonate-containing films are usually off-white, pale bluish-grey, or light brown, and are also usually slightly chalky when dry. Therefore, they are frequently mistaken for coatings of calcium carbonate.

The solubility diagrams in Figures 17.4 and 17.5 illustrate four especially noteworthy features of lead chemistry in drinking water. First, as the DIC level is increased, the domain of lead carbonate stability is extended to a higher pH limit (approximately pH 8.5 instead of approximately pH 7), and the solubility of lead is decreased within the lead carbonate stability domain. Second, the aqueous carbonate complexes compete successfully for lead primarily at the expense of $Pb(OH)_2^0$ and $PbOH^+$. The third significant effect shown is the enhancement of lead solubility in part of the region where basic lead carbonate is the controlling solid, primarily around pH 8.5, brought about by carbonate complexation. Fourth is that for the pH range of approximately 7.2 to 8.2, the lead solubility is nearly independent of the DIC concentration, particularly when DIC is in the range of approximately 20 to 100 mg C/L. This clearly explains why many corrosion control field studies, and many corrosion control optimization studies performed by large- and medium-sized utilities required by the Lead and Copper Rule, failed to find a significant relationship between lead release and alkalinity.

Few systematic studies have been performed that have adequately identified or isolated the effect of various cations, such as calcium, magnesium, sodium, and zinc, on lead corrosion in potable waters (Schock, Wagner, and Oliphant, 1996; LeRoy, 1993; Schock and Clement, 1998). Existing data and a consideration of corrosion thermodynamics suggests that in the absence of a uniform diffusion barrier of calcium carbonate or other solids, the significance of the cations will be minor. One study indicated an increase of lead release in the presence of higher concentrations of calcium (LeRoy, 1993). It has been hypothesized that the greatest (and perhaps only) effect would be on new surfaces, where metals could function as inhibitors of cathodic site reactions (Schock and Clement, 1998), which would rarely be the case with lead piping in drinking water plumbing installations. Generally, studies have also shown the effects of sulfate, chloride, and nitrate on lead corrosion in potable waters to be negligible (Schock, Wagner, and Oliphant, 1996; Schock and Wagner, 1985) though some recent studies have implicated sulfate (Dodrill and Edwards, 1995; Reiber et al., 1997). "Organic acids" apparently increase lead solubility, and some complexation of dissolved lead by organic ligands has been determined. Some organic materials, however, have been found to coat pipe walls, reducing corrosion. The effect of the organic material on lead release or passivating film stability may depend on the chemical and physical nature of the NOM in the particular water supply, making a broad generalization about the effect of various natural organic substances impossible.

Orthophosphate has been shown in field and laboratory tests to greatly reduce lead solubility through the formation of lead orthophosphate films, but the effectiveness depends on the proper control of pH and DIC concentration, and a sufficient orthophosphate dosage (Schock, Wagner, and Oliphant, 1996; Schock and Clement, 1998; Schock and Wagner, 1985; Colling et al., 1992; Gregory and Jackson, 1984). The rate of formation of lead orthophosphate–passivating films seems to be slower than the rate of carbonate or hydroxycarbonate film formation. Considerable time must be allowed for the reactions to take place. Some studies have shown that many months to several years are needed to reduce the rate of lead release down to essentially constant levels (Lyons, Pontes, and Karalekas, 1995; Cook, 1997). The speed and amount of reduction appear to be proportional to the dosage of orthophosphate applied (Schock, Wagner, and Oliphant, 1996; Schock and Clement, 1998; Colling et al., 1992; Wagner, 1989). No systematic and quantitative information exists on the rate of film formation on *new* versus *used* pipe.

Silicate may also reduce lead solubility or release, especially at levels exceeding 10 to 20 mg SiO_2/L (Schock, Wagner, and Oliphant, 1996; Lytle, Schock, and Sorg, 1994, 1996; Schock et al., 1998; Schock and Wagner, 1985). Little is known about the nature of films formed, except that reaction rates may be slow, similar to the case of orthophosphate, and that the use of fresh surfaces (such as commonly employed in pipe loop, coupon, and electrochemical corrosion current/rate studies) may give a false prediction of poor performance.

Most polyphosphate treatment chemicals are strong complexing agents for lead and calcium, and would be expected to either directly enhance lead solubility or to interfere with the formation of calcareous films. In both theory and practice, polyphosphate chemicals have been shown to be detrimental to lead control, when all the important factors have been isolated in the tests (Schock, Wagner, and Oliphant, 1996; Schock, 1989; Holm and Smothers, 1990; Holm and Schock, 1991a,b; Schock and Wagner, 1985). Some of the apparently successful applications of polyphosphate lead corrosion control may actually be caused by pH adjustment, or the reversion of a fraction of the polyphosphate to a protective orthophosphate form.

One recent study showed that high dissolved oxygen levels and high concentrations of free chlorine increased lead dissolution rates (Luo and Hong, 1997). Other research has also shown that apparently the rate of oxidation of lead is fast, and that stagnation curves for lead showing the concentration increases in pipes with time can often be modeled accurately with a diffusion-based model (Kuch and Wagner, 1983), or at least one that shows that levels tend to approach equilibrium conditions within about 8 to 12 hours (Schock, 1990a; Schock, Wagner, and Oliphant, 1996; Schock and Gardels, 1983; Lytle and Schock, 1997; Schock and Wagner, 1985). A study in Portland, Oregon, observed that the use of chloramines for disinfection had the effect of decreasing the rate of lead corrosion compared with the use of free chlorine, although the final pH values of the two alternative treated waters were approximately equal (Trussell, 1985). One explanation could be attributed to a lower redox potential in the system with combined chlorine. The addition of ammonia to chlorinated water to generate chloramines, however, tended to increase the concentration of lead byproducts in solution (Trussell, 1985). The role of ammonia in possible complexation of metal ions, and its effect on corrosion rates, are poorly understood, but the effect (if any) must be greater at high pH where Pb-NH_3 complexation is favored.

Some studies have shown lead to be protected from attack by films of aluminum solids (likely the amorphous hydroxide) formed from carryover from coagulation and filtration (Trussell and Wagner, 1985; AWWARF, 1998). The extent to which this beneficial occurrence has enabled utilities to meet the regulatory Action Level is unknown, but given the chemistry of aluminum and the concentrations that can be produced from the high pH conditions of coagulation frequently employed for economy or operational convenience, one would expect that many examples would be found in a systematic survey of pipe deposits.

The leaching of lead from soldered joints, unlike lead pipe dissolution, has a strong galvanic corrosion component. Lead is essentially a diluent in the solder, and does not appear to affect how the solder alloys with the copper pipe when the joint is made (Parent, Chung, and Bernstein, 1988). Lead release from solder is somewhat variable, over time and across different sites in a distribution system of similar general age and construction. Part of the problem is differences in exposed surface area, and their locations relative to water samples being collected. Particles of solder can also be eroded or otherwise released, creating the occurrence of random "spikes" of high lead levels in some samples (Lytle et al., 1993). Lyon and Lenihan (1977) found that the quality of workmanship and the presence of excess flux on the pipe interior enhanced lead release. Though degreasing reduced the lead release from this source, higher levels of copper dissolution were then seen. They suggested that a well-made joint possessed considerable leaching potential, though one study found relatively limited lead leaching from new joints in a 60-ft copper pipe loop (AWWARF, 1990).

Oliphant (1983) did several electrochemical studies of solder corrosion, and concluded that even if the exposed area of solder is small, considerable contamination was possible for years. He did not find selective leaching of the lead component of the solder, although a solder residue containing only SnO_2 was found by Schock [Illinois State Water Survey, 1989 (unpublished)] in some copper hot-water systems in Illinois.

Other experiments by Oliphant suggested that lead leaching from solder is not affected by conductivity, carbonate hardness, and both pyro- and orthophosphate concentration (Oliphant and Schock, 1996; Oliphant, 1983). Lead leaching was increased by decreasing pH, increasing chloride, and increasing nitrate levels. Leaching rates were decreased by increasing sulfate and silicate concentrations. The effectiveness of sulfate in reducing the rate of Pb release was related to the ratio of sulfate to chloride, with a 2:1 ratio allowing the formation of good crystalline corrosion product layers. Numerous studies of municipal corrosion control treatment in the United States since the promulgation of the Lead and Copper Rule have affirmed that proper control of pH, DIC, orthophosphate, and sometimes blended phosphate levels has been able to achieve considerable reduction in lead release from soldered joints in building and domestic plumbing systems (Lytle, Schock, and Sorg, 1994, 1996).

Galvanized Steel

Galvanized steel pipe was dominant material used to transport domestic drinking water for most of the century, only being replaced in the last couple of decades by copper and plastic. Galvanized plumbing is still the material of choice in many developing countries (Trussell and Wagner, 1996). Galvanized pipe consists of a base steel layer, underlying a sequence of layers that are zones of iron/zinc alloy that successively approach pure zinc at the interior surface of the pipe (Trussell and Wagner, 1996). When present, impurities such as lead and cadmium are most likely to be found near the interior surface of the pure zinc layer (Trussell and Wagner, 1996). The various aspects of the corrosion of galvanized steel, including the pipe/water chemical reactions and the significance of the manufacturing process of the pipe, have been extensively reviewed by Trussell and Wagner (1985, 1996).

One of the complexities of this material is that once the zinc layer is corroded away, the pipe behaves as if it were black iron pipe. Therefore, the corrosion potential of the water and optimal corrosion inhibition strategies should change, depending upon how much of the zinc layers are left.

Uniform corrosion, pitting, metal release, and tuberculation can all be important in galvanized pipe corrosion (Trussell and Wagner, 1985, 1996; Wagner, 1989). Mechanical failure because of uniform corrosion is rare, but when the zinc corrosion rate is high, failures are often induced by the tuberculation that follows rapid depletion of the zinc layer. When the pipe is new, the uniform corrosion depends strongly on pH. The nature of the film formed on the pipe changes in response to many chemical factors, and galvanized pipe corrosion seems to be very sensitive to the type of scale produced. Some scales are voluminous, chalky, and relatively unprotective, even if the rate of metal release isn't extremely high. Poor film protection can accelerate pitting or tuberculation. Field experience indicates that the effect of corrosion scales on the corrosion of galvanized pipe does not appear to be strictly proportional to zinc solubility and release. Moderate levels of DIC and high buffer intensity seem to produce good passivating films in cold water.

Studies have shown increases in the corrosion rate of zinc with increasing carbonate hardness, even in the presence of orthophosphate (Trussell and Wagner, 1985, 1996). This increase may be caused by the formation of aqueous carbonate complexes ($ZnHCO_3^+$, $ZnCO_3^0$, $Zn(CO_3)_2^{2-}$) increasing zinc solubility (Schock, 1981, 1985; Trussell and Wagner, 1985, 1996).

The role of calcium carbonate in galvanized pipe protection is not clear. The calcium carbonate seen in deposits is often aragonite (Trussell and Wagner, 1985, 1996; Schock and Neff, 1982). Calcium carbonate, along with basic zinc carbonate ($Zn_5(CO_3)_2(OH)_6$, hydrozincite), and forms of zinc hydroxide and zinc oxide, appear to be the most significant components of natural scale layers.

According to several studies, orthophosphate effectively controls zinc solubility and galvanized steel corrosion (Trussell and Wagner, 1996). The effectiveness of orthophosphate, as with lead, depends on the pH, alkalinity, and orthophosphate level in solution (Trussell and Wagner, 1985, 1996; Schock and Neff, 1982).

Silicate has been shown to reduce galvanized pipe corrosion, but most studies have focused on hot-water systems. Zinc silicate compounds are generally of low solubility, however, and should provide some protection in the slightly alkaline pH range (Schock, 1981; Trussell and Wagner, 1985, 1996).

Pitting failure of galvanized pipe is more common in hot than cold water systems (Trussell and Wagner, 1985, 1996). Pipe failures are caused by poor pipe coating quality, the presence of dissolved copper, and the reversal of electrochemical potential of zinc and iron at high temperature.

There have been numerous anecdotal reports of drinking water contamination by impurities of lead and cadmium in the galvanizing layer, but there have been few systematically studied and documented investigations published. A confounding problem is that tap samples are often involved when galvanized pipe is implicated as the contamination source, and the probable contribution of metals such as lead and zinc from the brass cannot be separated.

Asbestos Cement (A-C) Pipe, Cement-Mortar Linings, and Concrete Pipe

Asbestos cement pipes and cement mortar linings behave very similarly in potable waters. The principal difference is the additional possibility of release of asbestos fibers into the water by A-C pipe deterioration. Deterioration of cementitious materials causes substantial water quality degradation. Leaching of free lime from the cement causes considerable increase in pH of the water, especially in long lines or low flow areas, and with poor to moderately buffered waters. Because the equilibrium pH for calcium hydroxide solubility is approximately 12.4 (note that calcium hydroxide is a common high pH analytical buffer solution), there is a considerable driving force for pH increase. This pH increase, in turn, can cause reduced effectiveness of disinfectants and phosphate corrosion inhibitors, unwanted precipitation of a variety of minerals (causing cloudy or turbid water), and poor taste. In extreme cases of attack by aggressive water, the attack can result in reduced pipe strength and a roughened pipe wall that can cause increased head loss.

The cement matrix of A-C pipe is a very complicated combination of compounds and phases (Schock, 1981; Leroy et al., 1996; Holtschulte and Schock, 1985; LeRoy, 1996). Some of the phases are poorly identified or are of indefinite composition. More than 100 compounds and phases important to the chemistry of portland and related cements have been described and identified. Because of solid solution possibilities, probably many more exist. The state of knowledge of the solubilities in water of the individual predominant compounds of the cement is very limited. Several different types of cement can be used in the production of cement linings (Leroy et al., 1996; Guo, Toomuluri, and Eckert, 1998).

The main components of the portland cement used most frequently are tricalcium silicate, dicalcium silicate, and tricalcium aluminate, together with smaller amounts of iron and magnesium compounds. "Free lime" ($Ca(OH)_2$) is also present, but it is limited by product standards to be less than or equal to 1.0 percent by weight in U.S. commercial type II autoclaved A-C pipe (Leroy et al., 1996; Holtschulte and Schock, 1985).

Groundwaters having high mineral contents, but low pH values, show accelerated rates of pipe leaching (Leroy et al., 1996; Holtschulte and Schock, 1985). In soft waters with low mineral content, the mechanism of attack is believed to proceed as follows. First, the calcium hydroxide is removed. Then, if the water still possesses sufficient acid content, calcium carbonate is either dissolved or prevented from forming at the water/cement boundary. Because alkaline conditions cannot be maintained within the cement after the initial calcium hydroxide is removed, attack begins on the hydrated calcium silicates. These convert to calcium hydroxide, and the cycle of reactions can continue either until sufficient calcium carbonate is formed and the water is neutralized to the point at which the cycle stops, or until no cementitious material remains to bind the aggregates together (Leroy et al., 1996; Holtschulte and Schock, 1985).

No simple chemical index exists that can directly predict the behavior and service life of A-C pipe. Calcium carbonate deposition to fill voids in the cement matrix and prevent lime leaching is one of the best protective mechanisms for all cement-based materials (Leroy et al., 1996; Holtschulte and Schock, 1985). With A-C pipes, the primary drawback to calcium carbonate saturation treatment is that in the absence of the formation of an actual surface coating, or the formation of protective coatings by naturally occurring substances such as iron and manganese, fibers are left exposed at the surface of the pipe where they are vulnerable to erosion.

Coatings containing iron have been found on many A-C pipe specimens exposed to a variety of aggressive to nonaggressive water. Iron coatings sometimes have a granular, porous structure that does not necessarily retard calcium leaching from the cement matrix. The iron frequently helps prevent the exposure of asbestos fibers at the pipe surface. However, the iron coatings may also become a habitat for the growth of biofilms or the retention of microorganisms.

Manganese (IV) oxide can provide a similar protection to that provided by iron. Silica may also be beneficial as an agent that can help maintain the hardness of the pipe (Schock, 1981; Leroy et al., 1996; Holtschulte and Schock, 1985). In some cases, silicate dosing could enhance protection by the formation and adsorption of iron colloids on to the pipe surface.

Orthophosphate, sulfate, and chloride salts of zinc have been used in laboratory experiments to prevent the softening of A-C pipe specimens to slow calcium leaching (Schock, 1981; Leroy et al., 1996; Holtschulte and Schock, 1985). These results should also apply to the preservation of any uncoated cementitious materials in a water supply system. The required dosage of zinc necessary depends at least on the pH and dissolved inorganic carbonate concentration of the system. The utility of zinc addition is in extending the potential lifetime of newly installed cement linings or A-C pipe lines, or in increasing the service life of existing mains, has not been established in the short field tests performed thus far. In these experiments, zinc, and not the anion of the compound, has been found to be the active agent. The reaction mechanisms have not been precisely delineated. The experiments suggest the hypothesis that zinc reacts with the water to form a zinc hydroxycarbonate precipitate (Schock, 1981; Leroy et al., 1996; Holtschulte and Schock, 1985). The zinc solid then reacts with the pipe surface to convert some or all of the coating to a harder zinc-silicate solid phase. Thus, the choice of zinc salt can be based on economics, or considerations of wastewater discharge regulations that could rule out phosphate containing formulations.

If a water system has unlined iron pipe as well as asbestos cement pipes or cement-lined pipes, an orthophosphate salt might be preferable, to lower metallic corrosion. More research needs to be done to determine if the orthophosphate inhibits the calcium carbonate protection mechanism of the cement matrix, or if calcium orthophosphate solids help stabilize the cement material.

Waters with low pH (less than approximately 7.5 or 8, unless they contained high calcium, alkalinity, and silicate levels), and very high sulfate concentrations are particularly destructive to asbestos cement pipe, and probably to all concrete or cement mortar-lined pipes. Strong sequestering or complexing agents, such as polyphosphate chemicals, have been shown to attack the pipe by enhancing calcium, aluminum, iron, and magnesium leaching from the cement matrix (Leroy et al., 1996; Holtschulte and Schock, 1985). They tend to prevent the formation of protective coatings by metals such as zinc, manganese, iron, and calcium.

One new area of concern with cement-based pipes and cement linings is the potential for the leaching of metals such as barium, cadmium, and chromium (Guo, Toomuluri, and Eckert, 1998). Care must be taken especially when cement is produced in kilns cofired with hazardous waste-derived fuel. It is not clear if there is a real problem in the United States, as manufacturing standards may be in place, or developed soon, that will alleviate the potential for this kind of problem. In another study, aggressive water from a membrane desalinization plant and containing polyphosphate was introduced into a water system in the Dutch Antilles with a mix of iron and cement-lined pipes. The water leached such high concentrations of aluminum unknown to the hospitals that it caused serious illnesses and several deaths of kidney dialysis patients (Berends, 1999).

DIRECT METHODS FOR THE ASSESSMENT OF CORROSION

Physical Inspection

Water treatment plant operators and engineering staff should be aware of the appearance of the interior of the distribution piping and how their water quality is affecting the materials in their system. Physical inspection is a very useful inspection tool to a utility, and it can also be done routinely at a very low cost. Both macroscopic (human eye) and microscopic observations of scale on the inside of the pipe are valuable tools for diagnosing the type and extent of corrosion. Macroscopic studies can be used to determine the amount of tuberculation and pitting and the number of crevices. The sample should also be examined also for the presence of foreign materials and for corrosion at joints. A record of historical observations of pipes in different parts of a distribution system and exposed to different water treatments over time could yield insights into slow corrosion or passivation reactions.

Utility personnel should try to obtain pipe sections from the distribution or customer plumbing systems whenever possible, such as when old lines and equipment are replaced. If a scale is not found in the pipe, an examination of the pipe wall can yield valuable information about the type and extent of corrosion and corrosion product formation (such as tubercles), though it may not indicate the most probable cause. Pipe sections of small diameter may be conveniently examined by sawing them lengthwise.

Examination under a microscope can yield even more information, such as hairline cracks and local corrosion too small to be seen by the unaided eye. Such an examination may provide additional clues to the underlying cause of corrosion by relating the type of corrosion to the metallurgical structure of the pipe.

Photographs of specimens should be taken for comparison with future visual examinations. High-magnification photographs should also be taken, if possible.

Corrosion Rate Measurements. Corrosion rate measurements are another method frequently used to identify and monitor corrosion. Corrosion rate measurements may not necessarily correlate well with metal release into the water. The mechanisms involved in the attack on the metal are not necessarily the same reactions as those that result from the interaction of the water with a developed scale on the pipe surface. The corrosion rate of a material is commonly expressed in mils (0.001-in) penetration per year (mpy). Common methods used to measure corrosion rates include (1) *weight-loss methods* (coupon testing and loop studies) and (2) *electrochemical methods.* Weight-loss methods measure corrosion over a period of time. Electrochemical methods measure either instantaneous corrosion rates or rates over a period of time, depending on the method used.

Coupon Weight-Loss Method. Four important criteria for corrosion tests are (Reiber, Ryder, and Wagner, 1996):

1. The metal sample must be representative of the metal piping.
2. The quality of the water to which the pipe sample is exposed should be the same as that transported in the plumbing system.
3. The flow velocity and stagnation times should be representative of those in the full-scale system.
4. The duration of the test must allow for development of the pipe scales that have an important effect on corrosion rate and on the quality of the water passing through the pipes.

Many recent research papers have examined ways to improve the processing of coupons for corrosion rate determination, and the statistical validity of the results (Reiber, Ferguson, and Benjamin, 1988; USEPA, 1993). *Planned interval tests* are often useful and desirable if the test duration is to be 12 months or longer. The test duration can then be modified based on the results obtained (Thompson, 1970).

Coupon preparation, handling, and examination are relatively time consuming and exacting operations that usually require a trained laboratory analyst or corrosion specialist to provide reproducible results (Reiber, Ryder, and Wagner, 1996; Reiber, Ferguson, and Benjamin, 1988; Kuch et al., 1985). Contamination during the manufacture of the coupons has also been identified in some studies, and careful examination, cleaning, and data evaluation may need to be used to overcome that problem (Lytle, Schock, and Tackett, 1992).

Loop System Weight-Loss Method. Another method for determining water quality effects on materials in the distribution system is the use of sections of pipe. The pipe sections can be used to measure the extent of corrosion and the effect of corrosion control methods. Pipe loop sections can be used also to determine the effects of different water qualities on a specific pipe material. The advantage is that actual pipe is used as the corrosion specimen. The loop may be made, or a rig made from long or short sections of pipe. For direct monitoring of dissolved constituent levels, or changes in background water parameters, such as chlorine residual depletion or pH changes, longer pipe sections are required (AWWARF, 1990; Reiber, Ferguson, and Benjamin, 1988; AWWARF, 1994).

Excavating and removing existing pipe invariably causes the disturbance of the deposits in the pipe. Hence, a long period of stabilization (many months) may be necessary to reestablish equilibrium in the scales and regain consistency of metal release before introducing the test treatment changes to the designated loops. For studies involving comparisons among treatments, stabilization and comparability need to be determined among all loops to be used for the study.

The question of how to make proper statistical comparisons is a difficult problem in using pipe loop data for treatment evaluation or other interpretations of water quality effects. This has been addressed in different ways in many practical studies, but some major considerations and suggested approaches have been given in some recent publications (AWWARF, 1994; Wysock et al., 1995).

Electrochemical Rate Measurements. These methods are based on the electrochemical nature of corrosion of metals in water, and they have been widely discussed in the drinking water corrosion literature (Reiber, Ryder, and Wagner, 1996; Schock, 1990b; Silverman, 1995). The primary types of techniques try to determine corrosion rates through the measurement of the electrical resistance, linear polarization, corrosion current, or galvanic current of test specimens. Although the techniques have been used widely, often very successfully, there are limitations inherent to all of them, and the user must be cautious that fundamental mathematical or chemistry assumptions behind the analytical and data analysis procedures are not violated or the results biased by the technique itself (Reiber et al., 1997; Reiber, 1989; Reiber, Ryder, and Wagner, 1996; Silverman, 1995; Vitins et al., 1994). In addition to corrosion rate information, some of the techniques can yield valuable information on corrosion mechanisms, such as whether a given variable can act as an anodic site or cathodic site inhibitor, and what concentrations lead to protection or adverse behavior. It is also important to realize that only sometimes do electrochemical measurement techniques correspond well to metal release to solution and metal solubility.

Immersion Testing in the Laboratory

Immersion tests are sometimes useful to make an *initial* appraisal of the effectiveness of inhibitors (Reiber, Ryder, and Wagner, 1996; Kuch et al., 1985). Immersion tests use metal coupons in batch jars of water to evaluate inhibitors, dosages, and pH control. The procedures of ASTM Standard 631-72 should be utilized for immersion tests.

Limitations of all immersion tests are that accurate representation of flow and water chemistry conditions in the field situation are sacrificed for testing simplicity and efficiency. Important parameters like pH, DIC concentration, dissolved oxygen, and chlorine residuals may frequently be difficult or impossible to control, leading to incorrect results. Thus, results of immersion tests might not prove to be accurate simulations when extrapolated to practice at the treatment plant and in the whole distribution system. They should only be used as a very general screening tool, and results contrary to expectations from chemistry theory or experience of utilities with similar pipe materials and water quality should be critically examined for possible experimental problems or errors.

Chemical Analysis

Valuable information about probable corrosion causes or inhibition mechanisms can be found by chemically analyzing the corrosion by-product material on the pipes. Scraping off a portion of the corrosion by-products, dissolving the material in acid, and qualitatively analyzing the solution for the presence of suspected metals or compounds can indicate the type or cause of corrosion. These analyses are relatively quick and inexpensive. If a utility does not have its own laboratory, samples of the pipe sections can be sent to an outside laboratory for analysis. The numerical results of these analyses cannot be quantitatively related to the amount of corrosion occurring because only a portion of the pipe is being analyzed. However, such analyses can give the utility a good overview of the type of corrosion that is taking place. Whenever possible, analytical techniques should be specified that can determine the abundance of anionic constituents (carbonate, phosphate, sulfate, chloride, silicate) as well as major metals and other constituents (such as oxygen, silica, and others). Assumptions based on physical appearance and only data on metal composition can often be misleading (Schock and Smothers, 1989). For complex scales, especially when silicates or organic materials may be a significant component of the deposits, a laboratory should be consulted to advise on the most appropriate digestion technique and analytical methods for the deposit constituents.

The compounds for which the samples should be analyzed depend on the type of pipe material in the system and the appearance of the corrosion products. For example, brown or reddish-brown scales should be analyzed for iron and for trace amounts of copper. If found in lead pipe, lead should also be included in the analysis. Greenish mineral deposits should be analyzed for copper. Black scales should be analyzed for iron and copper. Depending on the type of plumbing material, light-colored deposits should be analyzed for calcium, carbonate, silicate, and phosphate, in addition to lead, zinc, or whatever the metal of interest. Some useful guidance is given by ASTM procedures (ASTM, 1996d,e).

Comparing sampling data of scales from various locations within the distribution system can isolate sections of pipe that may be corroding. Increases in levels of metals such as iron or zinc, for instance, indicate potential corrosion occurring in sections of iron and galvanized iron pipe, respectively. The presence of cadmium, a minute contaminant in the zinc alloy used for galvanized pipe, also indicates the probable corrosion of a galvanized iron pipe. Corrosion of cement-lined or A-C pipe is generally accompanied by an increase in both pH and calcium throughout the system, sometimes in conjunction with an elevated asbestos fiber count. Increases in aluminum may often be diagnostic of cement deterioration, as well as other trace metals.

Microscopic Techniques

The field of analysis of pipe scales and inhibitor films can make use of a variety of sophisticated instrumental analysis techniques, in addition to the traditional "wet chemical" procedures, such as those of ASTM (1996a). Often, no single technique will provide all the answers in film identification and interpretation of corrosion and corrosion control data. Judicious use of a combination of methods, however, can give invaluable insight into the problems and solutions.

Optical microscopic techniques have been widely used for the evaluation of A-C pipe corrosion, and have been summarized and described in many articles (Leroy et al., 1996; Holtschulte and Schock, 1985; Schock, Burlow, and Clark, 1981). They have also been employed in the examination of particulates in drinking water (Walker, 1983) and in the examination of films on iron pipe (Singley et al., 1985; Benjamin, Sontheimer, and Leroy, 1996). Photomicrographs have also been extremely useful in examining the structure of the alloy and zinc layers on galvanized pipe (Trussell and Wagner, 1985, 1996).

The scanning electron microscope (SEM) has been widely applied to the evaluation of A-C pipe deterioration (Schock, 1981; Leroy et al., 1996; Holtschulte and Schock, 1985; Schock, Burlow, and Clark, 1981). Studies have also attempted to directly relate the morphology of corrosion product crystals of lead carbonate, basic lead carbonate, or a lead phosphate to the protectiveness against corrosion (Breach, Crymble, and Porter, 1991; Colling, Whincup, and Hayes, 1987; Colling et al., 1992).

Scanning electron microscopy shows highly magnified views of the surfaces of materials. Figures 17.23, 17.24, and 17.25 demonstrate the use of this technique in evaluating the corrosion of A-C pipe. Figures 17.23 and 17.24 show the change in the surface of A-C pipe after exposure to an aggressive water (Leroy et al., 1996; Holtschulte and Schock, 1985). Note the exposed chrysotile asbestos fibers present in Figure 17.24.

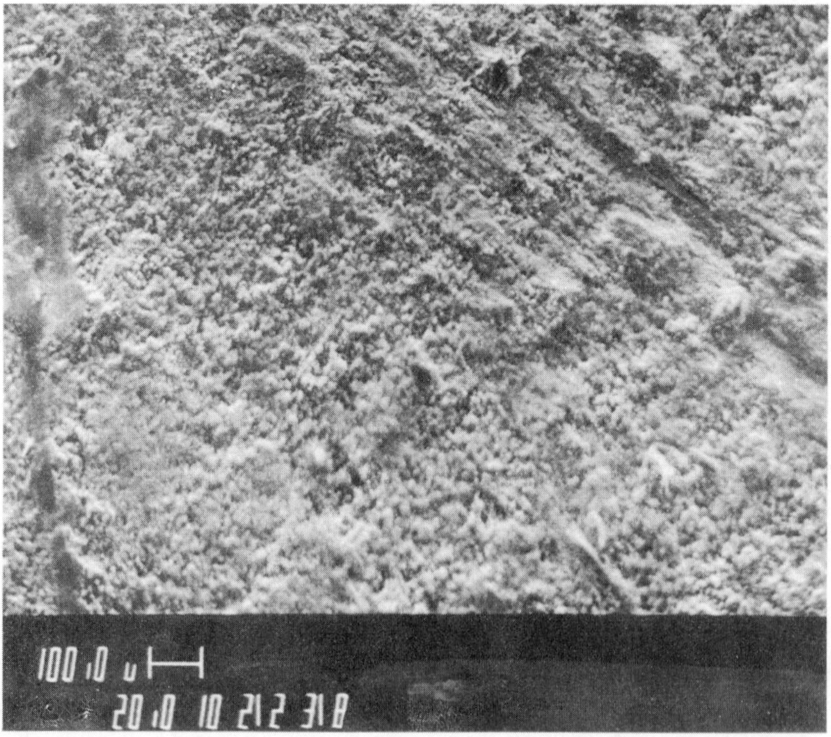

FIGURE 17.23 Scanning electron microscope photograph of unused A-C pipe from Seattle (magnification, 100×). Note few fibers exposed on surface. (*Source: Internal Corrosion of Water Distribution Systems,* American Water Works Association Research Foundation, Denver, CO, 1985.)

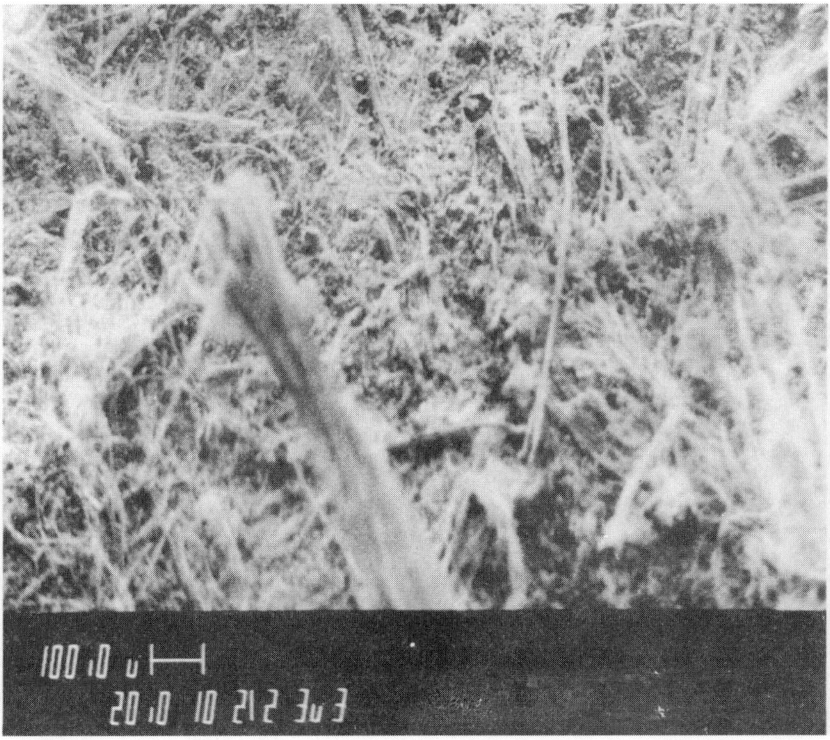

FIGURE 17.24 Scanning electron microscope photograph of A-C pipe from the Seattle distribution system after several years of use (magnification, 100×). Note bundles of exposed fibers on the surface vulnerable to release. (*Source: Internal Corrosion of Water Distribution Systems,* American Water Works Association Research Foundation, Denver, CO, 1985.)

Figure 17.25 is from a USEPA laboratory experiment, in which the pipe surface was protected through the formation of a zinc-containing solid (Schock, 1981).

X-Ray Elemental Analysis

For surficial deposits, X-ray fluorescence (XRF) spectrometry is helpful for elemental analyses (Schock and Smothers, 1989; Rose, 1983; ASTM, 1996b). The sample, or its fusion product, is ground finely into a powder and is compressed into a flat discoidal pellet. The mount is irradiated by an X-ray beam of high energy.

X-ray fluorescence can be done very quickly in comparison with wet chemical analysis (including flame or flameless atomic absorption spectrophotometry), because of the relative simplicity of sample preparation and processing. Quantitative analysis can be performed with proper sample preparation precautions and suitable mathematical processing of the raw intensity data. For most major constituents, and for most metals down to the order of 1 percent by weight, the accuracy is comparable to dissolution and analysis techniques.

On the microscale, energy-dispersive X-ray analysis (EDXA) is useful for elemental identification and qualitative analysis. Here, an SEM unit is used to excite the

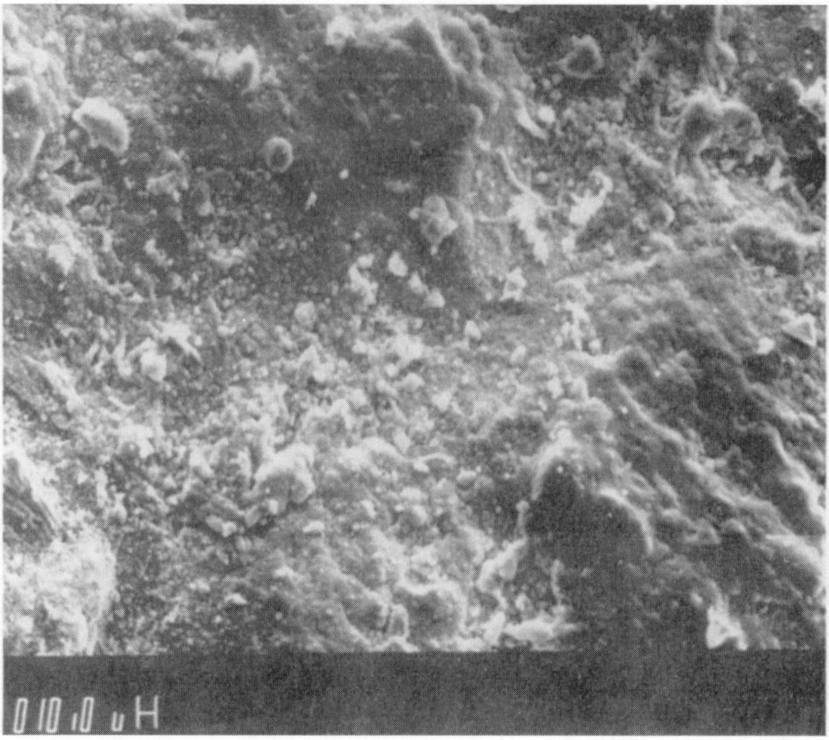

FIGURE 17.25 Scanning electron microscope photograph of A-C pipe specimen from USEPA laboratory experiment, showing the surface covered by a zinc-containing solid (magnification, 300×). (*Source: Internal Corrosion of Water Distribution Systems,* American Water Works Association Research Foundation, Denver, CO, 1985.)

generation of the characteristic X-rays from the sample. Wavelength-dispersive X-ray analysis (WDXA) is also possible with some electron microscopes, and it is generally better for quantitative analysis and lighter elements in the scale. Wavelength-dispersive X-ray analysis and EDXA cannot identify the compounds present, as would be possible with X-ray diffraction. For all electron microscopic analysis methods, observing representative sections of the pipe is important because the same pipe might vary in color and texture. Often, several sections of the same pipe must be analyzed.

X-Ray Diffraction

X-ray diffraction (XRD) is a procedure with which actual crystalline compounds can be identified in the scales, or aggregates of particulates. The fundamental principle behind the method is that when a highly collimated beam of X-rays of a particular wavelength strikes a finely ground, randomly oriented sample of material, the X-rays are diffracted in a pattern that is characteristic of the crystalline structure of the compound (Schock and Smothers, 1989; ASTM, 1996c; Gould, 1980). Mixtures of compounds give a pattern that is a sum of those compounds present. The analyzed patterns are compared with a reference library of known patterns for various solids, minerals, metals, and alloys. Particularly when used in conjunction with other analytical techniques to reduce the amount of pattern searching necessary (by identifying the combinations of elements present, or probable class of compounds—phosphates, silicates, oxides, and so forth), XRD provides a tremendous insight into pipe corrosion and corrosion inhibitor effects.

Some examples follow of information provided by XRD that could not be determined by the other methods discussed. Figure 17.26 shows two XRD patterns of corrosion products on galvanized pipe (Singley et al., 1985). Interestingly, the aragonite form of $CaCO_3$ was often found instead of calcite when the deposits are associated with galvanized materials. This is possibly caused by some influence of the zinc ion on the form of calcium carbonate crystals that will form, similar to the effect of magnesium ion. Aragonite is more soluble than calcite, so this is important in the modeling of precipitation and corrosion reactions. Conventional wet chemical methods, as well as EDXA, XRF, or WDXA, cannot differentiate calcite from aragonite, or even compounds like basic zinc carbonate (hydrozincite, $Zn_5(CO_3)_2(OH)_6$) from zinc carbonate (smithsonite, $ZnCO_3$) when a mixture of solids is present. Campbell has pointed out the relative unprotectiveness of smithsonite formed in some very hard

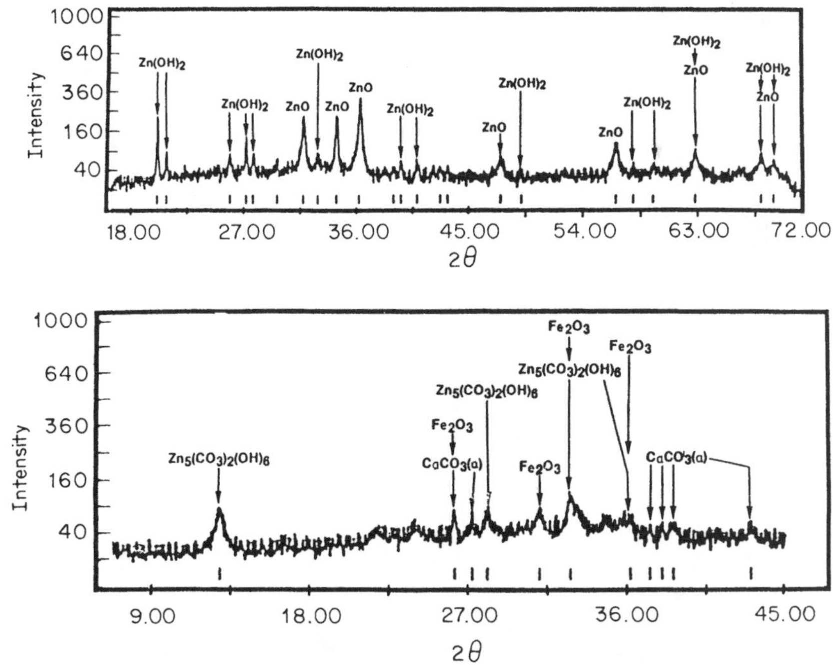

FIGURE 17.26 X-ray diffraction pattern of corrosion products from galvanized pipe. The upper diffractogram was done on pipe exposed to pure, deionized water; the lower diffractogram was conducted on water at pH 7.7 with 150 mg/L of alkalinity and a calcium hardness of 100 mg $CaCO_3$/L. Though calcite is the most stable species of calcium carbonate under most conditions of interest to waterworks, aragonite is found instead on this specimen. (*Source: Internal Corrosion of Water Distribution Systems,* American Water Works Association Research Foundation, Denver, CO, 1985.)

groundwaters, compared with basic zinc carbonate formed in softer waters of higher pH (Campbell, 1971). This differentiation is only possible with XRD analysis.

Figure 17.27 shows a deposit present in a specimen from a lead service line removed from the Chicago, Illinois, system. The deposit consists almost entirely of basic lead carbonate ($Pb_3(CO_3)_2(OH)_2$, hydrocerussite) and a small amount of lead carbonate ($PbCO_3$, cerussite). The peaks from lead metal came from pieces of the pipe picked up during scale removal. Notably, peaks from calcite or aragonite are absent, indicating no protection from $CaCO_3$ deposition. This is an excellent example of information not readily available using elemental analysis methods or only SEM.

Figure 17.28 is an analysis of a pipe specimen from a Providence, Rhode Island, service line (Schock, Wagner, and Oliphant, 1996; Schock, 1989). This is a soft water with a pH of approximately 10 and no history of a lead problem. The deposit on this pipe consists of hydrocerussite ($Pb_3(CO_3)_2(OH)_2$), as would be expected, as well as a significant fraction of another basic lead carbonate, plumbonacrite ($Pb_{10}(CO_3)_6(OH)_6O$). Several studies have reported many analyses of lead pipe by X-ray diffractometry (Schock, Wagner, and Oliphant, 1996; Schock and Smothers, 1989; van der Hoven and van Eekeren, 1990; van der Hoven, 1987; Elzenga and Graveland, 1981). Corroboration of copper solubility models by scale analyses using XRD have also been reported (Schock, Lytle, and Clement, 1995b).

Infrared Spectroscopy

This approach has been used extensively by the Water Research Centre in the United Kingdom, as well as by numerous other investigators (Sheiham and Jackson,

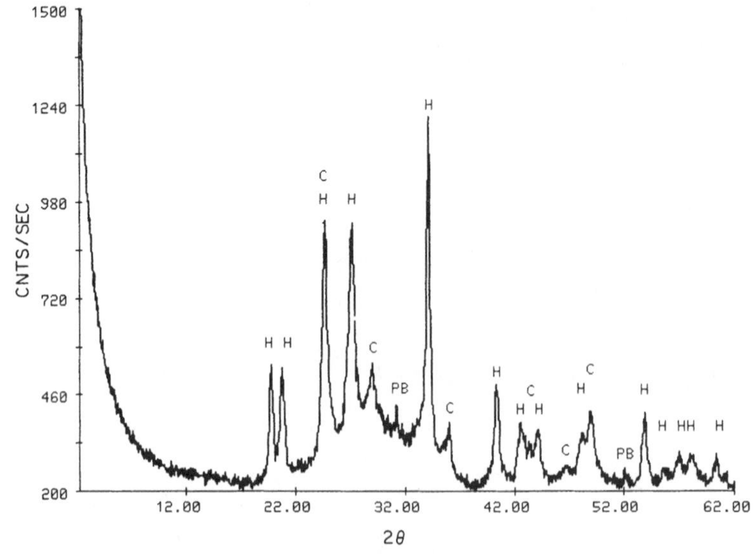

FIGURE 17.27 X-ray diffraction pattern from a piece of lead service line from the Chicago, Illinois, system. The deposit is a mixture of hydrocerussite—$Pb_3(CO_3)_2(OH)_2$—as [H], and cerussite—$PbCO_3$—as [C]. Two peaks (PB) were caused by lead from the pipe that was inadvertently abraded during sample preparation.

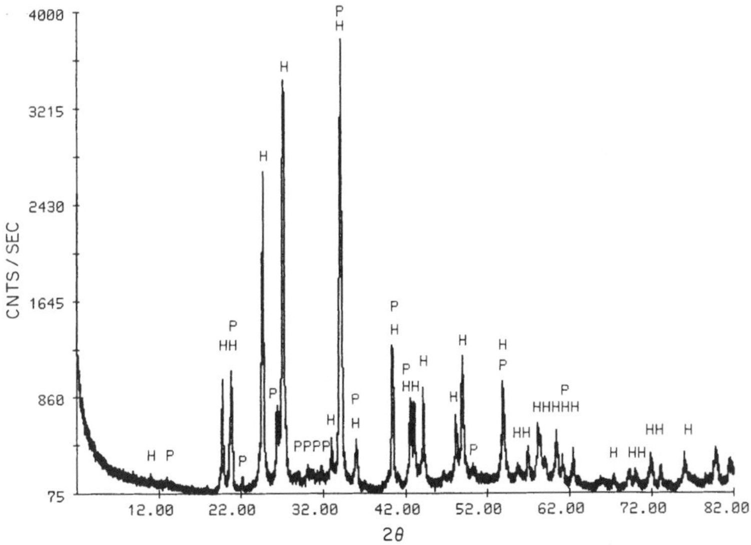

FIGURE 17.28 An X-ray diffraction pattern from a Providence, Rhode Island, lead service line specimen, showing hydrocerussite—$Pb_3(CO_3)_2(OH)_2$—as [H] and plumbonacrite—$Pb_{10}(CO_3)_6(OH)_6O$—as [P].

1981; Boireau, Randon, and Cavard, 1997; Kirmeyer, Wagner, and LeRoy, 1996). Generally, samples are removed from the scale layers in a similar fashion to preparation for XRD analysis, and ground into a very fine powder. Usually, a mull is then prepared using some mounting agent, often paraffin, it is squeezed between two plates (KBr is common), and it is mounted in the instrument. Alternatively, sample is mixed with KBr or other mounting material and pressed under very high pressure into a pellet. Light radiation of wavelengths in the infrared (IR) range is passed through the sample, and the absorption spectrum recorded. Characteristic patterns for compounds represent molecular transitions, such as rotational, vibrational, and combinations of both, which take place somewhere in the IR region of the electromagnetic spectrum. It is relatively easier to do and requires less expensive instrumentation than XRD generally does. Carbonate minerals frequently can be differentiated easily by this technique.

INDIRECT METHODS FOR THE ASSESSMENT OF CORROSION

Complaint Logs and Reporting

Usually, customer complaints will be the first evidence of a corrosion problem in a water system. Many common symptoms are listed in Table 17.5, along with their possible causes. The complaints may not always be caused by corrosion. For example, red water may also be caused by iron in the source water that is not removed in treatment. Therefore, in some cases, further investigation is necessary before attributing the complaint to corrosion in the system.

TABLE 17.5 Typical Customer Complaints Caused by Corrosion

Customer complaint	Possible cause
Red water, red or black particles, presence of reddish-brown staining on fixtures and laundry	Corrosion of iron pipes, old galvanized pipes, or iron in source water
Bluish stains in sinks and tubs	Corrosion of copper lines
Black water lines	Sulfide corrosion of copper or iron
Foul taste or odors; fine suspended bluish particles; orange, aqua, or black gelatinous deposits	By-products from microbial activity
Loss of pressure	Excessive scaling, tubercules building up, leak in system from pitting or other type of corrosion
Lack of hot water	Buildup of mineral deposits in hot-water pipes or heater system (can be reduced by setting thermostat to under 140°F or softening)
Short service life of household plumbing	Rapid deterioration of pipes from pitting or other types of corrosion
White or green-tinted fine particles in aerators and strainers	Deteriorated hot-water heater dip tube
White particles and cloudiness in ice cubes	Hard water

Source: AWWA, *Corrosion Control for Operators,* American Water Works Association, Denver, CO, 1986.

Complaints can be a valuable corrosion monitoring tool if records of the complaints are organized. The complaint record should include the customer's name and address, date the complaint was made, and the nature of the complaint. The following information should also be recorded:

1. Type of material (copper, galvanized iron, plastic, and so forth) used in the customer's system
2. Whether the customer uses home treatment devices prior to consumption (softening, carbon filters, and so forth)
3. Whether the complaint is related to the hot-water system and, if so, what type of material is used in the hot-water tank and its associated appurtenances
4. Any follow-up action taken by the utility or customer

These records can be used to monitor changes in water quality because of system or treatment changes.

The development of a complaint map is useful in pinpointing problem areas. The complaint map would be most useful when combined with a materials map that indicates the location, type, age, and use of a particular type of construction material. If complaints are recorded on the same map, the utility can determine if a relationship between complaints and the materials used exists. To supplement the customer complaint records, sending questionnaires to a random sampling of customers is useful.

A long-term reporting program for pipe and equipment failures can be a valuable tool for monitoring corrosion (Schock and Neff, 1982; Kirmeyer, Wagner, and LeRoy, 1996). Such a program can provide a history of corrosion problems, especially if some means of evaluating failures is in place. Little information, however, is readily available on plumbing failures in homes or buildings. Corrosion of plumbing within buildings is often not brought to the attention of the water utility immediately, unless there is widespread pitting and pinhole leaks. Few specimens of pipe are available for examination. Plumbers may not recognize corrosion as the cause for failures and attribute the cause to "bad pipe." The water utility should accumulate this information and seek the cooperation of at least one plumbing firm operating within their service area.

Additional information can be gathered for inclusion in the records, such as loss in main capacity, changes in water source, changes in water treatment procedures, and so forth. Sometimes, this information can be more important than weight-loss measurements or chemical analyses of the water supply for the monitoring of corrosion.

Customer complaint records, material service information, and questionnaires are useful monitoring tools that can be used as part of any corrosion monitoring and control program. The low costs associated with keeping a good record of complaints can be well worth the time. The resulting information would indicate the real effect of water quality at the customer's tap and would show the effect of any process changes made as part of a corrosion control program. Many aspects of designing a monitoring system for corrosion control have been discussed recently by Kirmeyer et al. (1996).

Corrosion Indices

Because of the complex interactions among water chemistry variables and the materials exposed to the water, finding a simple and objective numerical way to predict the behavior of materials has long been sought. Corrosion indices have been intended to accomplish that objective. Dissolved oxygen, pH, alkalinity, calcium, suspended solids, organic matter, buffer intensity, metal ions that can be reduced in cathodic reactions, total salt concentration, specific anions (such as chloride, bicarbonate, sulfate, phosphate, nitrate), silicate, biological factors, and temperature all have been shown to have an effect. Many of these have different effects on the corrosion of different metals—or even different effects on the same metal at different pH. Thus, each of the many indices that have been proposed and widely applied have served some useful purposes. Unfortunately, many have been misapplied, particularly when they are used without recognition of their limitations.

Many of the indices are more useful "after the fact," in helping to understand possible causes of the problem, rather than being useful in a predictive sense. In general, indices based on calcium carbonate saturation have not been shown to have any significant predictive value for the corrosion and leaching of lead, zinc, and copper from brass, bronze, soldered joints, and their respective pipe materials. They have been shown to have substantial merit in connection with the corrosion of unlined iron pipe and cementitious materials (Leroy et al., 1996; LeRoy, 1996).

Here, indices will be considered in two broad classifications: those based on the principle of calcium carbonate deposition as a means to provide corrosion control, and those based on relating corrosion to a variety of other water chemistry parameters. Because many indices are essentially the same, only expressed in different ways, only a small subset of the indices that have appeared in the literature will be considered.

Indices Based on Calcium Carbonate Saturation

Langelier Saturation Index (LSI) or Langelier Index (LI). The LSI (Langelier, 1936) is the most widely used and misused index in the water treatment and distribution field. The index is based on the effect of pH on the equilibrium solubility of

$CaCO_3$. The pH at which a water is saturated with $CaCO_3$ is known as the pH of saturation, or pH_s. At pH_s, a protective $CaCO_3$ scale should neither be deposited nor dissolved. The LSI is defined by the following equation:

$$LSI = pH - pH_s \tag{17.43}$$

The results of the equation are interpreted as follows:

LSI > 0	Water is supersaturated and tends to precipitate a scale layer of $CaCO_3$.
LSI = 0	Water is saturated (in equilibrium) with $CaCO_3$; a scale layer of $CaCO_3$ is neither precipitated nor dissolved.
LSI < 0	Water is undersaturated, tends to dissolve solid $CaCO_3$.

The minimum information necessary to calculate the LSI is as follows:

1. Total alkalinity concentration
2. Calcium concentration
3. Ionic strength, which can be computed by a complete water analysis or estimated by the total dissolved solids or specific conductance measurement and a conversion factor (implied in many tables)
4. pH
5. Temperature
6. pH_s (computed from the above information)

Recognizing the frequent misuse of the LSI and that inappropriate treatment approaches were being adopted, in 1994 the EPA repealed the section of the 1980 amendment to the National Interim Primary Drinking Water Regulation requiring all community water supply systems to determine either the LSI or an even more approximate version, the Aggressiveness Index (AI), and report these values to the state regulatory agencies (Federal Register, 1980, 1994a). The AI has been critically evaluated by several researchers (Schock, 1981, 1990b; Leroy et al., 1996; Holtschulte and Schock, 1985; Rossum and Merrill, 1983), and has several serious shortcomings as a deposition tendency indicator and index for A-C pipe protection. Therefore, because the AI is only an imprecise and simplistic approximation of the LSI, only the LSI will be discussed here.

The original derivation of the LSI depends upon computing the pH_s value using the analyzed total alkalinity of the water (Rossum and Merrill, 1983; Langelier, 1936), and can be readily derived as follows. The early methods did not include ion pairs (e.g., $CaHCO_3^+$, $CaCO_3^0$), and sometimes used equilibrium constants consistent with these assumptions. The basic definition of alkalinity for a pure system is Eq. 17.44.

$$TALK = 2[CO_3^{2-}] + [HCO_3^-] + [OH^-] - [H^+] \tag{17.44}$$

From Eq. 17.26, this relationship can be derived,

$$[CO_3^{2-}] = \frac{K_s'}{[Ca^{2+}]} \tag{17.45}$$

where K_s' represents the solubility constant for calcium carbonate, corrected for temperature and ionic strength.

Combining the second dissociation of carbonic acid with Eq. 17.45 yields

$$[HCO_3^-] = \frac{[H^+]K'_s}{[Ca^{2+}]K'_2} \tag{17.46}$$

where K'_2 is the second dissociation constant for carbonic acid corrected for temperature and ionic strength. Substituting Eqs. 17.45 and 17.46 in Eq. 17.36, plus the simple relationship $[OH^-] = K'_W/[H^+]$, produces the following quadratic equation after rearrangement:

$$0 = [H^+]^2 \cdot \left[1 - \frac{[Ca^{2+}]K'_2}{K'_s}\right] + [H^+]K'_2\left[2 - \frac{\text{TALK}\,[Ca^{2+}]}{K'_s}\right] + \frac{[Ca^{2+}]K'_W K'_2}{K'_s} \tag{17.47}$$

The hydrogen ion concentration at saturation equilibrium $[H^+]_s$ can be computed using the quadratic formula (Rossum and Merrill, 1983).

Equations for the computation of the equilibrium constants and activity coefficients are available in many sources (Snoeyink and Jenkins, 1980; Stumm and Morgan, 1981, 1996; AWWARF, 1990; Rossum and Merrill, 1983; Pankow, 1991; Morel, 1983; APHA-AWWA-WEF, 1992, 1995; Schock, 1984; Langmuir, 1984). It is important to use the best recently evaluated values for the constants, and these constants should be consistent with initial assumptions such as whether the constants were computed assuming the significance of minor species such as $CaCO_3^0$. There are considerable differences in predicted solubility and response to different ionic strengths (Trussell, 1985; Schock, 1984).

Rossum and Merrill (1983) point out several important mathematical properties of the computation of LSI as it was derived in its original form that limit its usefulness. Because it is derived from a quadratic equation, two possible solutions for $[H^+]$ and therefore pH_s exist. The existence of two values for pH_s is not generally appreciated. To obtain consistent and continuous values for LSI across the entire range of pH values while still maintaining established conventions (e.g., positive LSI for oversaturated waters), the lesser value of pH_s must always be used. It has also been pointed out that for pH above the bicarbonate-carbonate equivalence point (pH, 10.3 to 10.5), the negative root of the quadratic equation applies (Miyamoto and Silbert, 1986), and the LSI definition should be reversed (Loewenthal and Marais, 1976; Miyamoto and Silbert, 1986) to LSI = pH_s − pH. This results from the fact that at high pH, more of the alkalinity comes from hydroxide (Eq. 17.33). An important point is that the calculations are only valid when the measured pH values used were analyzed at that temperature using a temperature-compensated pH meter, or were corrected to that temperature using chemical equations (AWWARF, 1990; Rossum and Merrill, 1983; Schock, 1984). The cited references should be examined to understand the computational limits in more detail.

If alkalinity-contributing species from systems other than carbonate are negligible, if carbonate complexation can be ignored, and if DIC is conserved during the temperature change (frequently the case in plumbing systems when calcium carbonate does not dissolve or precipitate), the change in pH as temperature is changed can be computed (AWWARF, 1990). This enables the calculation of the LSI or other saturation or deposition index for a water after it has been warmed or cooled, or can enable the estimation of the in situ pH of a sample measured in the laboratory. Interested readers can find a derivation of the theory and example calculations are given in the book *Lead Control Strategies* (AWWARF, 1990).

The LSI can also be derived following a slightly different path (Snoeyink and Jenkins, 1980; Rossum and Merrill, 1983; Schock, 1984). By this route, $[H^+]_s$ is found by writing the calcium carbonate dissolution equation as:

$$CaCO_s(s) + H^+ \Leftrightarrow Ca^{2+} + HCO_3^- \tag{17.48}$$

that corresponds to the equilibrium constant expression of Eq. 17.46.

Solving for $[H^+]_s$,

$$[H^+]_s = \frac{[HCO_3^-][Ca^{2+}]K_2'}{K_s'} \tag{17.49}$$

or in logarithmic form:

$$pH_s = -\log\left[\frac{K_2'}{K_s'}\right] - \log\,[Ca^{2+}] - \log\,[HCO_3^-] \tag{17.50}$$

Here, analyzed values for calcium and bicarbonate can be substituted into the equation, along with the equilibrium constants corrected for ionic strength and temperature. In this approach, the assumption can be made that $[HCO_3^-] = \text{TALK}$, which is valid for many waters in the pH range of approximately 6 to 8.5 (Snoeyink and Jenkins, 1980; Stumm and Morgan, 1981; Pankow, 1991; Butler, 1982; Faust and Aly, 1981; Loewenthal and Marais, 1976; Deffeyes, 1965), or $[HCO_3^-]$ must be computed from a suitable alkalinity or DIC expression, such as Eq. 17.44 or Eq. 17.34, or an expression given in almost any aquatic chemistry reference book using appropriate concentrations and constants (Snoeyink and Jenkins, 1980; Stumm and Morgan, 1981, 1996; Pankow, 1991; Butler, 1982; Faust and Aly, 1981; Morel, 1983; Langmuir, 1997).

When conversion factors are added, and the temperature and ionic strength corrections are incorporated into Eq. 17.49, Eq. 17.51 can be derived:

$$pH_s = A + B - \log\,[Ca^{2+}] - \log\,[HCO_3^-] \tag{17.51}$$

as was the widely used form also given in the *Federal Register* (1980; Schock, 1984). Corrections to the constant terms in Eq. 17.51 for improvements in equilibrium constant values (constant A) and activity coefficient calculation (constant B) have been presented (APHA-AWWA-WEF, 1992, 1995; Schock, 1984) to supercede previously published values (Singley, Beandet, and Markey, 1984; AWWA, 1986; Stumm and Morgan, 1981; Loewenthal and Marais, 1976; Langelier, 1936). In this equation, the Ca^{2+} and HCO_3^- concentrations are expressed as mg $CaCO_3$/L.

Several problems exist with the LSI as a corrosion index (Snoeyink and Kuch, 1985; Snoeyink and Wagner, 1996; Schock, 1984, 1986):

1. Corrections are often not made for complexation and ion pairing of Ca^{2+} and HCO_3^-, although this is possible if needed analytical data are available and if the algorithms are set up in the computation scheme. This problem is most severe in hard waters and ones containing high sulfate or DIC concentrations (or both).
2. In the presence of polyphosphates, the essential equations defining pH_s are invalid. The LSI (and related) equations shown here will overestimate calcium carbonate saturation unless correction factors are added to account for complexation and poisoning of crystal growth. Little information exists that is useful to quantify such effects for different water chemistries.
3. Crystal growth poisoning and increases in the solubility of calcium carbonate polymorphs and solid solutions (such as high magnesian calcites) are commonly caused by high concentrations of sulfate, orthophosphate, magnesium, natural organic matter, and some trace metals (zinc has been widely associated with this phenomenon). This phenomenon essentially invalidates the numbers obtained for LSI and related indices.

4. The crystalline form of $CaCO_3$(s) has usually been assumed to be calcite. The presence of another form of $CaCO_3$(s), aragonite, which has a higher solubility, has, however, been observed in several systems. The formation of other forms of $CaCO_3$(s) may account for some of the observations of substantial supersaturation with respect to calcite.
5. A deposit of $CaCO_3$(s) does not necessarily aid in preventing corrosion if the coating is not uniform and functions as a diffusion barrier toward oxidants.
6. $CaCO_3$ can also be deposited from water with a negative LSI if there is localized high pH, generated by cathodic reactions at the pipe wall. This is particularly a characteristic of poorly buffered, low-DIC waters. Inconsistent deposition can lead to spotty surface coverage and can aggravate localized corrosion attack.
7. The preoccupation by many with maintaining a positive LSI has led to excessive deposition of $CaCO_3$(s), especially near the point of chemical adjustment, and to significant decreases in the capacity of distribution systems to carry water. This is logical, as a high degree of supersaturation must be maintained at the point of chemical addition to assure a tendency for deposition all the way through the entire distribution system to the consumers' taps.

Analysis of many protective scales has shown that many types of solids other than $CaCO_3$(s) are present that provide resistance to corrosion. These solids are aluminum minerals, or mixed iron/calcium carbonates, or a host of other compositions. When phosphate- or silicate-based corrosion inhibitors are used, or when the dissolution of metals such as lead or copper is a concern, then other water chemistry factors must be considered.

Although the LSI tends to predict if $CaCO_3$(s) will precipitate or dissolve, it does not predict how much $CaCO_3$(s) will precipitate or whether its structure will provide resistance to corrosion. Larson (1975) showed, for example, that an LSI of 0.9 is necessary to precipitate 10 mg $CaCO_3$/L at an alkalinity of 50 mg $CaCO_3$/L, but an LSI of only 0.2 is necessary for the same amount of precipitation if the alkalinity is 200 mg/L as $CaCO_3$.

Early research by Stumm (1956, 1960) as well as Larson and many others have shown good associations of deposition of calcium-containing carbonate scales with smooth inner pipe walls, and freedom from tuberculation in unlined iron pipe (Larson, 1975; Larson and Skold, 1957; AWWA, 1971). Therefore, the LSI should be kept in proper context and used appropriately. Much research has indicated great usefulness with respect to protecting cement-based pipes and linings (Leroy et al., 1996; Légrand and LeRoy, 1990; LeRoy, 1996), so utilities using these materials or rehabilitation techniques should seriously consider the predictions of LSI calculations and balancing the mineral content of their waters accordingly.

Calcium Carbonate Precipitation Potential. Although the LSI can be related to the thermodynamic driving force for deposition and, to some extent, the rate of deposition (Trussell, Russell, and Thomas, 1977), sometimes it is more useful to evaluate the theoretical approximate mass of calcium carbonate that could precipitate on a pipe surface. This quantity is called the *calcium carbonate precipitation potential,* or CCPP. Procedures have been given in several sources for these calculations (Trussell, 1985; Merrill, Sanks, and Spring, 1978; Rossum and Merrill, 1983; Loewenthal and Marais, 1976), and it is most conveniently calculated by a variety of computer codes (APHA-AWWA-WEF, 1992, 1995). The equation for the CCPP is given as (Rossum and Merrill, 1983):

$$CCPP = 50{,}045 \cdot [TALK_i - TALK_{eq}] \tag{17.52}$$

in mg $CaCO_3$/L units. This is because during $CaCO_3$ precipitation, the equivalents of calcium precipitated must be equal to the equivalents of alkalinity precipitated (Loewenthal and Marais, 1976). The acidity of such a system is given by $ACY_i = ACY_{eq}$, however, because acidity is conserved in the process (Trussell, 1985; Snoeyink and Jenkins, 1980; Merrill, Sanks, and Spring, 1978; Rossum and Merrill, 1983; Loewenthal and Marais, 1976). Rossum (1983) has systematically developed the following set of equations for computing CCPP, and the original article should be consulted for details of the derivation.

The initial acidity of the system may be computed from the relationship

$$ACY_i = \left(\frac{TALK_i + s_i}{t_i}\right)p_i + s_i \tag{17.53}$$

where

$$p = \frac{2[H^+]_i + K_1'}{K_1'} \tag{17.54}$$

The alkalinity after precipitation when equilibrium is reached can then be related to the initial

$$t = \frac{2K_2' + [H^+]_i}{[H^+]_i} \tag{17.55}$$

acidity

$$TALK_{eq} = \frac{t_{eq}}{p_{eq}}(ACY_i - s_{eq}) - s_{eq} \tag{17.56}$$

where the terms t_{eq}, p_{eq}, and s_{eq} correspond to Eqs. 17.53, 17.54, and 17.55, with $[H^+]_i$ replaced by the hydrogen ion concentration at final equilibrium, $[H^+]_{eq}$.

The alkalinity at equilibrium can also be related to the initial calcium concentration and alkalinity through:

$$2[Ca^{2+}]_i - TALK_i = \frac{2K_s' r_{eq}}{TALK_{eq} + s_{eq}} - TALK_{eq} \tag{17.57}$$

in which

$$r_{eq} = \frac{[H^+]_{eq} + 2K_2'}{K_2'} \tag{17.58}$$

The final relationship of interest for this way of computing CCPP is the following, in which the subscript "i" denotes initial values, and the subscript "eq" represents the quantity at equilibrium:

$$2[Ca^{2+}]_i - TALK_i = \frac{2K_s' r_{eq} p_{eq}}{t_{eq} ACY_i - s_{eq}} - \frac{t_{eq}(ACY_i - s_{eq})}{p_{eq}} + s_{eq} \tag{17.59}$$

The values for $[Ca^{2+}]_i$ and $TALK_i$ are obtained by chemical analysis. ACY_i can be computed from Eq. 17.53. The terms p_{eq}, r_{eq}, s_{eq}, and t_{eq}, are functions of $[H^+]_{eq}$, which must be obtained through an iterative trial-and-error solution of Eq 17.59 (Rossum and Merrill, 1983). Following the calculation of $[H^+]_{eq}$, $TALK_{eq}$ can then be derived using Eq. 17.56, and substituted along with $TALK_i$ into Eq. 17.52.

A positive CCPP denotes oversaturation, and the mg $CaCO_3$/L of calcium carbonate that should precipitate. A negative CCPP indicates undersaturation, and how much $CaCO_3$ should dissolve. The same constraints apply to CCPP as with LSI on whether there is crystal growth poisoning or complexation that would interfere with precipitation and growth of the solids.

Several computational strategies exist that are appropriate for solving Eq. 17.59 for $[H^+]_{eq}$ using a programmable calculator or computer (Bard, 1966; Rossum and Merrill, 1983). The equation can also be solved relatively quickly using simple manual trial-and-error substitution or built-in root finders with PC spreadsheet software, and several other sophisticated approaches have been published (Holm and Schock, 1998; Schott, 1998; Trussell, 1998). One caution in solving this equation: the quantity $(ACY_i - s_{eq})$ in Eq. 17.56 must always be greater than zero to give a physically valid answer, and should be constantly checked while solving for $[H^+]_{eq}$.

The distinction between $[H^+]_s$, used with the LSI, and $[H^+]_{eq}$, used with CCPP, is significant. The quantity $[H^+]_s$ represents the hydrogen ion concentration if a water of a specific composition were at equilibrium with calcium carbonate. That is different from $[H^+]_{eq}$, which is the final equilibrium hydrogen ion concentration that would occur after the initial water either precipitated or dissolved $CaCO_3$ to attain saturation.

The amount of calcium carbonate (or other solid phase) that would dissolve or precipitate for a given water may also be computed using sophisticated PC-based programs from the geochemistry field, such as PHREEQE (Parkhurst, Thorstenson, and Plummer, 1980), MINTEQ (Allison, Brown, and Novo-Gradac, 1991), MINEQL+ (Schecher and McAvoy, 1994a,b, 1998), or others (APHA-AWWA-WEF, 1992, 1995). These computer models can effectively take into account aqueous ion pairing and complexation reactions, and ionic strength changes, in both open and closed (to the atmosphere) systems.

Rossum (1987) has discussed the usefulness of the CCPP and its application to some difficult red-water problems. He found that the amount of calcium carbonate formed by an increment of hydroxide from the cathode reaction may be an important factor in the prevention of red water. The CCPP is the actual degree of calcium carbonate supersaturation (if positive) or undersaturation (if negative) in milligrams per liter. If hydroxide is also expressed in milligrams per liter as calcium carbonate, then CCPP-OH values greater than one indicate that more than one molecule of $CaCO_3$ is formed for each molecule of hydroxide generated. Figure 17.29 is that used by Rossum (1987) to illustrate the CCPP-OH relationship for different alkalinities and pH values, assuming calcium concentrations to represent equilibrium with calcite. At high pH values, so much hydroxide is present that small increments of hydroxide have little effect, and at low pH values, so much carbonic acid is present that increments of hydroxide form bicarbonate rather than carbonate. At a given pH value, carbonic acid concentrations increase as the alkalinity increases.

Rossum and Merrill's (1983) approach to identifying susceptibility toward red water starts with simulating the addition of 1 mg OH-/L (as $CaCO_3$), causing the alkalinity of the water to be increased by 1 and the acidity to be decreased by 1. From these adjusted values, a new pH value and CCPP can be calculated. If the result is less than 1, red-water problems are likely. For many waters in which the LSI is close to zero, the value of CCPP-OH may be estimated with sufficient accuracy from Figure 17.29.

Figure 17.29 shows that the least favorable value of CCPP-OH corresponds to the region of minimum buffer intensity for the carbonate system (discussed in a later section). Frequently, very little evidence exists of a thin, coherent, protective scale of

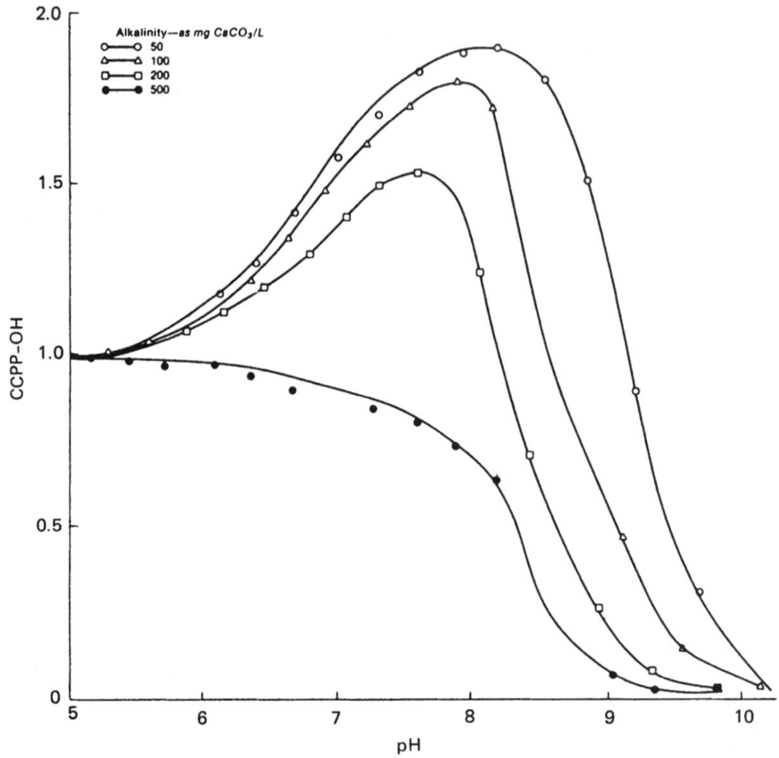

FIGURE 17.29 Change in calcium carbonate precipitation potential per hydroxide versus pH (temperature = 25°C, ionic strength = 0.01, Langelier index = 0). (*Source:* Rossum, 1987.)

calcium carbonate—even for those systems in which a positive LSI has been maintained. The benefit of having a CCPP-OH greater than 1 appears to lie in the ability of calcium carbonate to strengthen and harden iron corrosion products so that they remain on the pipe wall and do not cause red water. No assurance can be given that corrosion rates based on the loss of weight of pipe coupons would correspond to those based on red water, because the protective effects of siderite and, to a lesser extent, calcite depend on some corrosion having already occurred.

Because $CaCO_3(s)$ is frequently found in protective scales on iron and steel, control of water chemistry using the LSI is usually a reasonable and economical corrosion control strategy. Even if the films are not $CaCO_3$, these water conditions can often help with the formation of protective scales, such as mixtures of $CaCO_3$, $FeCO_3$, $Fe(OH)CO_3$, and iron hydroxides or oxides.

Ryznar Index. The Ryznar (Saturation) Index (RSI), defined as (AWWA, 1986; Merrill, Sanks, and Spring, 1978; AWWARF, 1990; Loewenthal and Marais, 1976; Ryznar, 1944):

$$RSI = 2pH_s - pH \tag{17.60}$$

was developed from empirical observations of corrosion rates and film formation in steel mains and heated water in glass coils. An RSI between 6.5 and 7.0 is considered to be approximately at saturation equilibrium with calcium carbonate. An RSI > 7.0 is interpreted as undersaturated, and therefore would tend to dissolve any existing solid $CaCO_3$. Waters with an RSI < 6.5 would tend to be scale forming.

Unlike the LSI and many other indices, this index does not have any fundamental theoretical justification. By multiplying the pH_s value by 2, the index number tends to favor waters of higher hardness and alkalinity that would naturally have a greater potential to deposit calcium carbonate if their pH exceeded their pH_s. Thus, in that respect, it is somewhat consistent with the observations of calcium carbonate deposition potential described by Stumm (1956, 1960), Merrill, Sanks, and Spring (1978), and Loewenthal and Marais (1976). A notable internal inconsistency of the RSI is that the value for saturation equilibrium varies with the pH_s of the water (e.g., if $pH_s = 7.0$, RSI for saturation equilibrium is 7, but if $pH_s = 9$, the RSI for saturation equilibrium is 9). Therefore, the interpretation of the index must be adjusted with the pH_s, but almost no one ever does that. Although commonly used, the RSI does not offer any tangible advantages to a variety of other methods for computing the calcium carbonate saturation state and deposition potential.

Buffer Intensity (β), Buffer Capacity, Buffer Index

This parameter (Trussell, 1985; Weber and Stumm, 1963; Stumm, 1960; Pankow, 1991; Butler, 1982; Loewenthal and Marais, 1976; Pisigan and Singley, 1987), as described previously, is an important water quality parameter that should often be included in the design of corrosion control programs. The effect of buffer intensity is also closely linked with $CaCO_3$ precipitation, because of the role of pH and bicarbonate ion concentration in determining the buffering of the water. High calcium concentrations are usually associated with waters with high natural bicarbonate concentrations, providing another link between the different indices and this parameter.

There is no a priori specific target value for buffer intensity for a particular water supply. The minimum amount necessary would be that which would be sufficient at a given target pH to stabilize the pH throughout the distribution system against changes induced by chemical reactions within the water, between the water and piping materials and within storage systems. Figures 17.18 through 17.20 show that the buffer intensity may be improved, sometimes drastically, by either changing pH or supplementing DIC. In some pH ranges, particularly as near the $pH = \frac{1}{2}(-\log K'_1 + -\log K'_2)$ point, the buffer intensity may be increased much more economically by modifying the pH by several tenths of a unit, rather than trying to add considerably more carbonate to the water. That would be especially important for minimizing cuprosolvency, or lead solubility above pH 8. A comprehensive study in Concord, New Hampshire, has demonstrated how improvements in buffer intensity can greatly improve corrosion control performance and water quality in general in a utility with a fairly modest alkalinity (Clement and Schock, 1998; Clement et al., 1998). Figure 17.30 shows the considerable improvement in distribution system pH stability brought about by the increase in buffer intensity.

Saturation Index. A general *saturation index* (sometimes called a *disequilibrium index*), SI, can be defined for any solid solubility reaction as

$$SI_x = \log_{10}\left(\frac{IAP_x}{K_x}\right) \tag{17.61}$$

where IAP_x and K_x are the ion activity product and solubility product constant, respectively, for mineral x (Stumm and Morgan, 1981, 1996; Schock, Lytle, and

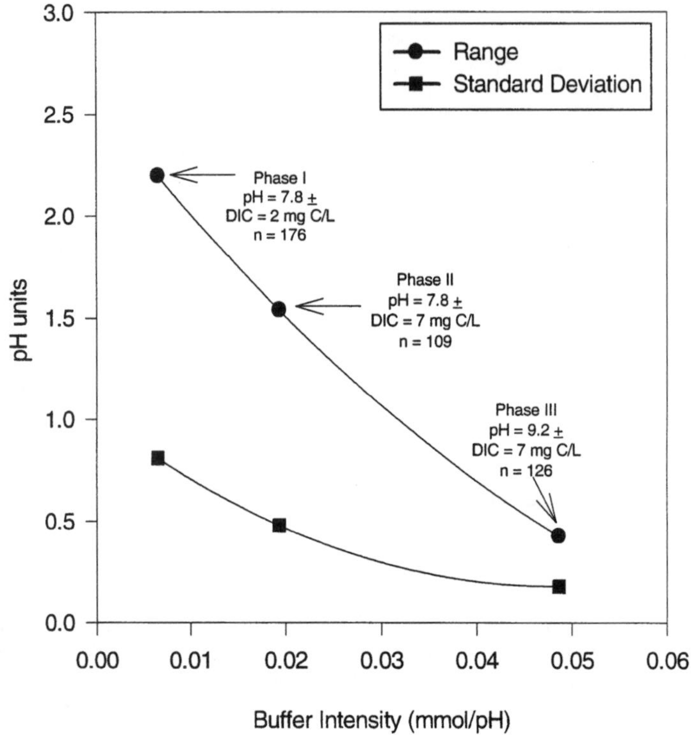

FIGURE 17.30 Improvement in distribution system pH stability in Concord, New Hampshire, treatment study. (*Source:* Clement and Schock, 1998; Clement et al., 1998.)

Clement, 1995b; Schock and Gardels, 1983; Schock, 1981; Faust and Aly, 1981; Langmuir, 1997; Parkhurst, Thorstenson, and Plummer, 1980; van Gaans, 1989).

For example, for calcium carbonate (calcite) dissolving

$$CaCO_3(s) \Leftrightarrow Ca^{2+} + CO_3^{2-} \tag{17.62}$$

the expression for $SI_{calcite}$ would be

$$SI_{calcite} = \log_{10}\left[\frac{\{Ca^{2+}\}\,\{CO_3^{2-}\}}{K_s}\right] \tag{17.63}$$

The braces { } represent activities rather than concentrations. For a solid such as the galvanized pipe corrosion product, hydrozincite, dissolving in the manner:

$$Zn_5(CO_3)_2(OH)_6(s) \Leftrightarrow 5Zn^{2+} + 2CO_3^{2-} + 6OH^- \tag{17.64}$$

then the SI is given by:

$$SI_{HZ} = \log_{10}\left[\frac{\{Zn^{2+}\}^5\{CO_3^{2-}\}^2\{OH^-\}^6}{K_s}\right] \tag{17.65}$$

As with the LSI, SI = 0 is equilibrium saturation. Positive values represent oversaturation, and negative values represent undersaturation. Reactions can be written in other forms, because K_s will change depending upon how the dissolution reaction is defined. They are directly related to the free energy driving force for the reactions. Indices of this type are useful in determining if passivating films are being formed on or within the pipe (Schock, Lytle, and Clement, 1995a,b; Schock and Gardels, 1983; Schock, 1981), or if a hypothetical final concentration of an added constituent would result in film formation. Very good corrections can be made for the effects of ion pairing of carbonate, sulfate, ionic strength, and temperature by using computerized equilibrium speciation models (APHA-AWWA-WEF, 1992, 1995; Parkhurst, Thorstenson, and Plummer, 1980; Allison, Brown, and Novo-Gradac, 1991; van Gaans, 1989).

Larson's Ratio. Based on iron corrosion rate measurements in a large number of Midwestern waters, Larson and Skold suggested that the ratio sometimes called the Larson Ratio (LR) (Larson, 1975; Larson and Skold, 1957),

$$LR = \frac{[Cl^-] + 2[SO_4^{2-}]}{[HCO_3^-]} \tag{17.66}$$

should be less than approximately 0.2 to 0.3 when expressed using concentrations as mol/L. Originally, the ratio was simply chloride over bicarbonate, but the originators felt that sulfate ion would also be similarly aggressive. Other researchers suggest a higher target for the ratio, suggesting values up to one as the maximum recommended value (van den Hoven and van Eekern, 1988). Some engineers and researchers express the ratio as the inverse, bicarbonate to chloride plus sulfate. In that case, the ratio should be larger than 3 to 5. The ratio has also been presented in the inverse form with the concentrations of bicarbonate, chloride, and sulfate expressed as equivalents per liter as $CaCO_3$ (Merrill, Sanks, and Spring, 1978).

Larson's ratio is also logical considering the results of other research in which chloride ion was noted for its role in breaking down passivating films on many ferrous metals and alloys, and is one of the main causes of pitting of stainless steels (Singley et al., 1985). Recent research in France, principally by LeRoy (Légrand and LeRoy, 1995; LeRoy, 1996), reports that some fundamental chemistry mechanisms support the concept of Larson's ratio.

There were limitations to Larson's original research, which must be kept in mind. Conveniently available waters were used in the analysis, so there was a relatively small number of observation points, and there is considerable scatter in the data. This did not allow the determination to be made if there were individual trends that might be directly caused by differences in pH, hardness, or alkalinity. By doing controlled laboratory investigations, LeRoy has shown that at constant pH and bicarbonate concentration, necessary because ferrous iron forms ion pairs with bicarbonate ion, the oxidation of ferrous iron is enhanced in the order $Cl^- > SO_4^{2-} > NO_3^-$ (Légrand and LeRoy, 1995; LeRoy, 1996). This order parallels the strength of the different aqueous complexes with free ferrous ion. The mechanism was proposed that bicarbonate plays a role in mitigating the corrosion reaction, because calcium bicarbonate ion pairs ($CaHCO_3^+$) fix free oxygen radicals formed during the iron oxidation reaction. Calcium carbonate may then deposit at the high pH cathodic sites, whereas various ferrous or ferric iron solids may form at anodic sites. This model is consistent with the observation that the corrosion of iron is less significant in harder and relatively well-buffered waters.

Other "Corrosion" Indices. Attempts have been made to develop corrosion indicates by correlating some measure of "corrosivity" to all water quality parameters that might affect corrosion. One such effort by Riddick yielded the following index (Snoeyink and Kuch, 1985; Singley, 1981):

$$RI = \frac{75}{ALK}\left[CO_2 + 0.5\,(\text{hardness} - ALK) + Cl^- + 2NO_3^-\left(\frac{10}{SiO_2}\right)\frac{(DO + 2)}{DO_{sat}}\right] \quad (17.67)$$

where hardness and alkalinity are in mg $CaCO_3$/L, NO_3^- is in mg N/L, and the remaining parameters are in mg/L. Values of less than 25 are noncorrosive, 26 to 50 are moderately corrosive, 51 to 75 are corrosive, greater than 75 are very corrosive. This index was developed using data for the soft waters of the northeastern United States and was successfully applied to water from this area, but not to higher hardness waters.

Pisigan and Singley (1985) conducted a variety of computations and laboratory corrosion tests on mild steel to develop a predictive equation for the corrosion rate, and to refine the LSI to include a correction for significant ion pair and complex species while maintaining a single equation that could be directly calculated. Their suggested equation to estimate pH_s is:

$$pH_s = 11.017 + 0.197 \log (TDS) - 0.995 \log (Ca^{2+})_T - 0.016 \log (Mg^{2+})_T - 1.041 \log (TALK)_T + 0.021 \log (SO_4^{2-})_T \quad (17.68)$$

where the subscript "T" denotes the total analytical concentration in mg/L (mg $CaCO_3$/L for TALK). They also developed predictive equations for the corrosion rates of mild steel specimens.

An unsuccessful attempt to develop a similar kind of relationship by relating corrosion rates or corrosion product levels to water quality was reported by Neff, Schock, and Marden (1987). They found that too many uncontrolled variables existed to enable the extraction and differentiation of chemical relationships from the field data, even with very comprehensive water quality monitoring.

"Marble Tests" and Saturometry. The relative state of saturation with respect to calcium carbonate has also been determined by physical means for many years. To the extent that calcium carbonate saturation is relevant to predicting the scaling or corrosive tendency of a water, the physical methods have several advantages over the numerical indices. As noted previously, when there are complex-forming ligands in the water, such as polyphosphates, or there are other water constituents that tend to kinetically inhibit crystal growth or change the solubility and crystal structure of calcium carbonate that could form, the numerical methods developed thus far are inapplicable. A methodology commonly called the *marble test* has been described in several papers (Balzer, 1980; Dye, 1964; Hoover, 1938; Notheis et al., 1995). This test simply looks at the change in pH, alkalinity, or both, when a test water is equilibrated with some calcium carbonate powder for a number of hours. A water that doesn't change is apparently at equilibrium with calcium carbonate, and is considered to be "stable." It is often considered almost "crude" and obsolete by contemporary standards. However, a form of the marble test is almost essential to obtain accurate saturation state estimation in waters treated with corrosion inhibitors, or when there are high concentrations of some natural ions, such as magnesium, zinc, and sulfate, that significantly affect the rate of calcium carbonate nucleation and precipitation, as well as solubility. Notheis et al. (1995) has described a recent case in which the marble test enabled diagnosis and solution of a corrosion problem that was not apparent from the numerical indices.

A more elaborate electrochemical approach to determine the calcium carbonate saturation state has been used by chemical limnologists and oceanographers for many years. This device is called a *carbonate saturometer,* and it operates on a similar principle to the marble test (Ben-Yaakov and Kaplan, 1971; Ingle et al., 1973; Köningsberger and Gamsjäger, 1990; Plath, Johnson, and Pytokowicz, 1980; Weyl, 1961). The instrument is a modification of a pH meter, with a glass cell reaction system set up so that the pH change when calcium carbonate is introduced generates a potential difference from a reference cell at equilibrium with the water. This potential difference can then be mathematically directly related to the amount of over- or undersaturation of the water. The saturometry approach could have considerable merit if applied to drinking water treatment in several areas. Conceivably, the saturometer could be designed with minerals other than calcium carbonate, and the status of a water relative to the formation of passivating films on metallic pipe could be directly estimated.

CORROSION CONTROL ALTERNATIVES

The complete elimination of corrosion is difficult if not impossible. Technologies generally exist to reduce or inhibit corrosion, though the determination of feasibility may depend mostly on economics or risk/aesthetic trade-offs. Corrosion depends on both the specific water quality and pipe material in a system, and a particular method may be successful in one system and not in another, or even show different results in different parts of a distribution system of a single utility or water supplier.

There are several approaches to control corrosion (Singley, Beaudet, and Markey, 1984; AWWA, 1986):

- Modify the water quality so that it is less corrosive to the pipe material.
- Place a protective barrier or lining between the water and the pipe.
- Use pipe materials and design the system so that it is not corroded by a given water.

The most common ways of achieving corrosion control are to:

- Properly select system materials and adequate system design.
- Modify water quality.
- Use inhibitors (may require pH or DIC adjustment as well).
- Provide cathodic protection (mainly for external corrosion).
- Use corrosion-resistant linings, coatings, and paints.

Materials Selection

In many cases, corrosion can be reduced by properly selecting system materials and having a good engineering design (Singley, Beaudet, and Markey, 1984; AWWA, 1986). Some pipe materials are more corrosion-resistant than others in a specific environment. In general, the less reactive the material is with its environment, the more resistant the material is to corrosion. When selecting materials for replacing old lines, putting new lines in service, or rehabilitating existing pipes (e.g., cement lining), the utility should select a material that will not corrode in the water it contacts. Putting a freshly cleaned cast-iron line or a freshly cleaned and cement-relined pipe back into service in a very soft and poorly buffered water can cause serious water quality problems. It also undermines the rehabilitation investment just made. To protect sizeable investments in rehabilitation, the utility should always evaluate the cost based on integrating distribution system corrosion control adjustments with the cleaning, relining, or replacement.

Few utilities can select materials based on corrosion resistance alone, and several alternative materials often must be compared and evaluated based on cost, availability, use, ease of installation, and maintenance, as well as resistance to corrosion. However, a realistic assessment of the impact of the water quality and the resources available to appropriately modify treatment to protect the newly installed materials in the future is often neglected, but is a wise strategy for long-term economy.

Compatible materials should be used throughout the system. Two metal pipes having different activities, such as copper and galvanized iron, that come in direct contact with others can set up a galvanic cell and cause corrosion. Particular problems with elevated lead levels resulting from the partial replacement of lead plumbing with copper have been reported in the United Kingdom (Britton and Richards, 1981; Schock, Wagner, and Oliphant, 1996; Schock and Wagner, 1985). As much as possible, domestic systems should be designed to use the same metal throughout or to use metals having a similar position in the galvanic series (Table 17.2). Galvanic corrosion can be avoided by placing dielectric (insulating) couplings between dissimilar metals. Problems have occurred in Florida from copper interior pipes being insulated from ground by plastic service lines (Bell et al., 1995; Bell, Schiff, and Duranceau, 1995).

The possible long-term health and economic consequences of replacing materials must be considered. For example, many of the lead-free brasses and solders available have a significant bismuth, arsenic, or antimony content. Whether that is leachable under normal conditions of usage has not been studied extensively.

The drinking water utility may not have control over the selection and installation of the materials for household plumbing, but should not overlook the possibility of working to influence plumbing codes to be most compatible with the water treatment options available. Some guidelines relating water quality to material selection can be discerned from Table 17.1. A good description of the proper selection of materials can be found in *The Prevention and Control of Water-caused Problems in Building Potable Water Systems,* published by the National Association of Corrosion Engineers (1980).

Engineering Considerations

The design of the pipes and structures can be as important as the choice of construction materials. A faulty design may cause severe corrosion, even in materials that may be highly corrosion-resistant. First and foremost, the water chemistry must be examined to determine which material, or materials, are appropriate. For example, many high-alkalinity waters with high sulfate and neutral to slightly acidic pH are incompatible with copper pipe. Many soft waters are basically incompatible with galvanized pipe, although it provides excellent performance in many hard-water situations that are problematic for copper pipe. Some of the important design considerations for plumbing systems of all sizes include the following (Singley, Beaudet, and Markey, 1984; AWWA, 1986):

- Avoiding dead ends and stagnant areas
- Providing adequate drainage where needed
- Selecting an appropriate flow velocity
- Selecting an appropriate metal thickness

- Eliminating shielded areas
- Reducing mechanical stresses
- Avoiding uneven heat distribution
- Avoiding sharp turns and elbows
- Providing adequate insulation
- Choosing a proper shape and geometry for the system
- Providing easy access to the structure for periodic inspection, maintenance, and replacement of damaged parts
- Eliminating grounding of electrical circuits to the system

Many plumbing codes are outdated and allow undesirable situations to exist. Such codes may even create problems (Bell et al., 1995; Bell, Schiff, and Duranceau, 1995). Where such problems exist, working with the responsible government agency to modify outdated codes may be helpful.

At a larger scale, a poorly designed distribution system can produce areas of locally poor water quality, loss of disinfection, turbidity, taste, and odor problems. Renovation and rerouting of pipe lines may be more effective in solving water quality problems than treatment in some cases. Consolidating multiple-distribution entry points can enable much better consistency and control of water quality. Many references review and discuss construction and design issues that relate to corrosion control (Singley, Beaudet, and Markey, 1984; AWWA, 1986; Medlar and Kim, 1993a,b; James M. Montgomery Consulting Engineers, 1985).

Chemical Treatment

Because of the differences among source waters, the effectiveness of any water quality modification technique will vary widely from one water source to another. Numerous corrosion control studies have identified "mixing zones," or areas of a distribution system where water quality continually shifts, sometimes back and forth, as particularly problematic. Stable passivating films cannot form in these areas, as they are constantly being deposited and redissolved by the water quality changes. An example of problems encountered when a corrosion inhibitor feed was interrupted was described by Colling et al. (1992). Employing water chemistry adjustment, whether it is corrosion inhibitor addition, pH adjustment, or DIC adjustment, requires a commitment to continually maintaining a consistent operation.

pH Adjustment. The adjustment of pH is the most common method of reducing corrosion in water distribution systems. The pH of the water plays a critical role in corrosion control for several reasons. Hydrogen ions (H^+) act as electron acceptors and enter readily into electrochemical corrosion reactions. Acid waters are generally corrosive because of their high concentration of hydrogen ions. The water pH is the major factor that determines the solubility of most pipe materials and the films that form from corrosion byproducts. Most materials used in water distribution systems (copper, zinc, iron, lead, and cement mortar) dissolve faster at a lower pH. When carbonate alkalinity is present, increasing the pH increases the amount of carbonate ion in solution. However, with too much carbonate addition, a water could become excessively scale-forming or become difficult to adjust with respect to pH.

The relationship among pH and other water quality parameters, such as DIC, alkalinity, carbon dioxide (CO_2), and ionic strength, governs the solubility of calcium carbonate ($CaCO_3$), which is commonly involved with protective scale formation on interior pipe surfaces. To encourage the formation of this protective scale, the pH of the water usually should be slightly above the pH of saturation for $CaCO_3$. Increasing the pH is a major factor in limiting the dissolution of the cement binder and, thus, controlling corrosion in these types of pipes.

The proper pH for any given water distribution system is specific to its water quality and system materials, so only general guidance can be provided. If the water contains a moderate amount of carbonate alkalinity and hardness (approximately 40 mg/L as $CaCO_3$ or more of carbonate or bicarbonate alkalinity and calcium hardness), the utility should first calculate the LSI or CCPP to determine at what pH the water is stable with regard to $CaCO_3$. To start, the pH of the water should be adjusted such that the LSI is slightly positive, no more than 0.5 unit above the pH_s, or the CCPP should be adjusted to a slightly positive value. If no other evidence is available, such as a good history of the effect of pH on the laying down of a protective coating of $CaCO_3$ or laboratory or field test results, then the LSI and/or CCPP provide a good starting point. Keeping the pH above the pH_s (i.e., a positive LSI) should cause a protective mixed coating to develop in cast-iron mains. If no coating forms, then the pH should be increased another 0.1 to 0.2 unit, or the CCPP should be raised, until a coating begins to form.

Monitoring of pH stability in the water system, chlorine residual depletion, and turbidity levels should be checked for the onset of buffering problems (likely if pH is pushed to between 8 and 8.5 from below) or possible red-water situations. Monitoring head loss values in the system carefully is important as too much scale buildup near the plant could seriously clog the transmission lines.

If lead and/or copper corrosion is a problem, adjusting the pH to values of 8.5 or higher may be required. Practical minimum lead solubility occurs at a pH of about 9.8 in the presence of about 20 to 30 mg/L of alkalinity. To assure proper buffering, pH adjustment coupled with carbonate supplementing may be required to minimize lead corrosion problems. Copper corrosion is often aggravated by an excess of DIC, and lead corrosion problems often occur in such waters, particularly when the pH is below 7. Removing CO_2 by stripping has been shown to very effectively reduce lead and copper levels, consistent with general corrosion and solubility theory (Lytle et al., 1998).

Orthophosphates and other corrosion inhibitors often require a narrow pH range for maximum effectiveness. The exact optimum pH range varies with the material. If such an inhibitor is used, consideration must be given to adjusting the pH to within a good balance of the effective ranges for the materials of interest. Several studies have been done exploring the pH/DIC/PO_4 relationships relative to the corrosion of lead (Schock, Wagner, and Oliphant, 1996; Schock, 1985, 1989; Schock and Clement, 1998; Schock and Wagner, 1985), galvanized iron (Trussell and Wagner, 1985, 1996; Schock, 1985), copper (Schock, Lytle, and Clement, 1995a,b; Reiber, 1989; Schock and Clement, 1998), and A-C (Schock, 1981; Leroy et al., 1996; Holtschulte and Schock, 1985; Schock, Buelow, and Clark, 1981) pipe. Consideration must be given to secondary interactions, such as calcium carbonate or calcium phosphate solubility limitations that restrict the possible pH range (AWWARF, 1990).

The pH should be adjusted after filtration, because optimal coagulation for waters (e.g., to minimize residual Al and maximize TOC removal) generally requires a pH lower than is optimal for corrosion control.

Alkalinity/DIC Concentration Adjustment. The adjustment of pH alone is often insufficient to control corrosion in waters that are low in carbonate or bicarbonate alkalinity. A protective deposit of $CaCO_3$ or mixed calcium/iron hydroxycarbonates, for instance, will not form on ferrous materials or between the grains in cement-based materials unless a sufficient quantity of carbonate and calcium ions are in the water. These pH/DIC concentration relationships also govern the buffering ability of the water, and increases in pH above the buffering minimum point of around pH 8.3 can produce dramatic increases in pH stability throughout the distribution system (see Figure 17.30).

Some experiments have indicated that a guideline to nucleation and growth of calcium carbonate is when the following relation is met (Légrand and LeRoy, 1990, 1995; LeRoy, 1996).

$$[Ca^{2+}][CO_3^{2-}] > 40 \cdot K'_{s,\,CaCO_3} \qquad (17.69)$$

where $K'_{s,\,CaCO_3}$ is the solubility constant for calcite corrected for ionic strength and temperature. Waters with higher NOM levels may need higher degrees of supersaturation to deposit effective coatings. In the United States, Merrill, Sanks, and Spring (1978) have recommended a CCPP value of 4 to 10 mg $CaCO_3$/L for general corrosion control purposes at a pH of 6.8 to 7.3. Unfortunately, that pH range is largely incompatible with lead and copper control objectives.

Recommendations of the proper overall balance of calcium hardness and alkalinity for good distribution system corrosion protection and water quality maintenance vary. Internationally, suggestions fall into the range of approximately 40 to 80 mg $CaCO_3$/L for both calcium hardness and total alkalinity (Merrill, Sanks, and Spring, 1978; Légrand and LeRoy, 1990, 1995; Werner, Groß, and Sontheimer, 1994; Boireau, Randon, and Cavard, 1997; van den Hoven and van Eekern, 1988; Vik et al., 1996). KIWA researchers specifically recommend a minimum of 2 mmol/L DIC to achieve reasonable buffering, which corresponds to 24 mg C/L, and an alkalinity of about 100 mg $CaCO_3$/L at pH 8.

Many systems that employ polyphosphate sequestrants or crystallization poisoners must treat to much higher CCPP levels (or pH_s values) because the water is much more aggressive and the indices do not include the chemical corrections for poisoning or sequestration. This approach is frequently not optimal for the control of lead, copper, or zinc corrosion, and tends to cause increased deterioration of cement pipes and linings.

A strong tendency exists to overestimate the accuracy of the calculated values of the LSI. Soft, low-alkalinity waters cannot become supersaturated with $CaCO_3$ regardless of how high the pH is raised (Merrill, Sanks, and Spring, 1978; Rossum and Merrill, 1983; Loewenthal and Marais, 1976). Raising the pH values greater than about 10.3 is useless because no more carbonate ions can be made available. Excess hydroxide alkalinity is of no value because it does not aid in $CaCO_3$ precipitation. These relationships can be computed, or seen graphically, through the use of Caldwell-Lawrence diagrams.

For some water systems, carbonate supplementation, pH control, and neutralization of aggressive carbon dioxide can all be accomplished in a rather straightforward manner by the use of limestone contactor systems, or by mixing limestone or dolomitic materials with filter beds (AWWA, 1986; AWWARF, 1990; Légrand and LeRoy, 1990; Letterman et al., 1987; Letterman, Haddad, and Driscoll, 1991; Letterman, 1995; Letterman and Kathari, 1996; Spencer, 1996; Boireau, Randon, and Cavard, 1997; Benjamin, Green, and Smith, 1992).

Some metals, notably lead, zinc and copper, can form a layer of an insoluble carbonate (e.g., $PbCO_3$), basic carbonate (e.g., $Pb_3(CO_3)_3(OH)_2$, $Pb_{10}(CO_3)_3(OH)_6O$, $Cu_2(CO_3)(OH)_2$), oxide or hydroxide (CuO, Cu_2O, $Cu(OH)_2$, $Zn(OH)_2$) minimizing corrosion rates and the dissolution of these metals. In very low alkalinity waters, carbonate ion must be added to form these insoluble carbonates. For such waters, soda ash (Na_2CO_3) or sodium bicarbonate ($NaHCO_3$) are the preferred chemicals generally used to adjust pH because they also contribute carbonate (CO_3^{2-}) or bicarbonate ions (HCO_3^-). The number of carbonate ions available is a complex function of pH, temperature, and other water quality parameters.

Lead, copper, and zinc form strong aqueous carbonate complexes that increase metal solubility when the DIC concentration exceeds a critical level at a given pH (Schock, 1980, 1981a,b; Schock, Lytle, and Clement, 1995a,b; Edwards, Meyer, and Schock, 1996; Schock, Wagner, and Oliphant, 1996; Schock and Gardels, 1983; Trussell and Wagner, 1985, 1996; Schock and Wagner, 1985; Schock, Buelow, and Clark, 1981). The exact value of this critical level depends to some degree on many factors, such as ionic strength, temperature, and the presence of competing cations and complex-forming ligands, in addition to the pH. The level must be computed using the appropriate chemical equilibrium equations to determine the metal solubility for the given water quality. Often the beneficial pH effect outweighs the complexation effect, and a net reduction of metal release takes place.

DIC adjustment does not necessarily only mean DIC addition. Because copper problems are aggravated by high DIC levels, adequately addressing overall corrosion control and regulatory compliance in many groundwater systems could entail employing treatments that remove DIC (Edwards, Jacobs, and Dodrill, 1999; Schock, Lytle, and Clement, 1995a,b; Edwards, Meyer, and Schock, 1996; LeRoy and Cordonnier, 1994). One obvious approach is lime softening, which is also very compatible with overall corrosion control for most other distribution system materials, including lead. If the pH is low enough, aeration could both remove DIC through CO_2 loss, and subsequently control the pH to a desirable level (Lytle et al., 1998a–c; Edwards, Meyer, and Schock, 1996; Spencer, 1996). For many systems drawing water from limestone aquifers, however, there is insufficient driving force for enough CO_2 (therefore, DIC) removal via aeration to meet the 1.3 mg/L copper action level. Other alternatives that could be pilot-tested by water systems would then be the use of membranes or anion exchange in combination with posttreatment (aeration or chemical addition) to accomplish the necessary pH adjustment.

Control of Oxygen. The role of oxygen in corrosion control is often misunderstood. Oxygen is generally considered an important corrosive agent because:

- Oxygen can act as an electron acceptor, allowing corrosion to continue.
- Oxygen reacts with hydrogen to depolarize the cathode and, thus, speeds up corrosive reaction rates.
- Oxygen reacts with iron, copper, and other metals to produce ions that can release into the water, deteriorating the source material.

If oxygen could be entirely removed from water economically, the chances of corrosion starting, and also the corrosion rate once it had started, would be reduced. However, when waters are disinfected with chlorine, chloramine, or other oxidizing agents, and residuals are maintained in the distribution system, corrosion of the metal piping is assured. Thus, oxygen removal is neither a practical nor economical corrosion control method, and it becomes essentially irrelevant with the presence of the disinfectant residual. Engineers contemplating the design and installation of oxidative treatment processes for the removal of any contaminant or suite of contaminants must broadly consider the corrosion implications, and include proper corrosion control treatment as an essential part of the overall project. The relative difficulty of implementation of the proper corrosion control strategy for the resulting water could actually constrain the selection of the best source water treatment process.

Aeration processes can create corrosive conditions in anoxic source waters, but the final corrosivity can normally be mitigated through proper treatment. For example, aeration is often an economical way of treating groundwaters having high iron, hydrogen sulfide (H_2S), or CO_2 content (see Chapter 5). Though aeration helps remove these substances from source water, it can significantly change the redox potential of the water, causing large increases in copper solubility (see earlier section in this chapter and Schock, Lytle, and Clement, 1995a,b; Ferguson, von Franqué, and Schock, 1996). Copper corrosion control can still be achieved, but controlling the solubility of copper (II) as opposed to copper (I) requires different DIC and pH considerations (Lytle et al., 1998a–c; Schock, Lytle, and Clement, 1995a,b; Ferguson, von Franqué, and Schock, 1996; Duranceau et al., 1997).

Ozonation prior to filtration for iron and manganese removal may cause high dissolved oxygen levels as a by-product of the process. Excess oxygen beyond saturation in any case may aggravate corrosion by the accumulation of gas bubbles (sometimes acidic and containing CO_2) in parts of the plumbing lines. It may also be a causative factor for pitting corrosion.

Research conducted in France indicates that dissolved oxygen plays a beneficial role in the creation of passivating films on unlined iron piping (Légrand and LeRoy, 1990, 1995). The crystallinity of the ferric hydroxide layer on pipes has also often been observed to be more compact and stronger than when formed from chlorine in the absence of oxygen. Dissolved oxygen appears to facilitate the formation of favorable passivating film layers consisting of a mixture of calcium carbonate, ferrous carbonate, and ferric oxide passivating layers that are sometimes referred to as the *ferricalcic deposit.* Loss of dissolved oxygen (and oxidizing conditions) promotes the formation of ferrous iron (Singley et al., 1985; Benjamin, Sontheimer, and Leroy, 1996; Stumm, 1956, 1960; Légrand and LeRoy, 1990, 1995). When released into the water, the ferrous iron ions can be easily reoxidized upon air or disinfectant contact, causing red and turbid water.

Chemical Inhibitors. Corrosion can be controlled by adding to the water chemicals that form a protective film on the surface of a pipe and provide a barrier between the water and the pipe. These chemicals, called inhibitors, reduce corrosion, or limit metal solubility, but do not totally prevent it. The four types of chemical inhibitors commonly approved for use in potable water systems are orthophosphates, polyphosphates, "blended" phosphates (mixtures of orthophosphate and polyphosphates), and sodium silicates. Some are simple "generic," or bulk, chemicals such as orthophosphoric acid, whereas others are proprietary, "specialty chemical," formulations. That is especially true for the polyphosphate and blended phosphate chemicals sold for drinking water applications. The effects of these chemicals on corrosion control were discussed previously.

Much misinformation exists about the mechanisms of action of many of the inhibitor formulations. For example, the sequestration of ferrous iron to prevent red-water formation is often interpreted as "corrosion inhibition," but in fact, it is only mitigating the symptoms of the corrosion. Many polyphosphate chemicals will convert one form of corrosion to another—for example, alleviating tuberculation and replacing it with uniform corrosion. However, that is not corrosion inhibition.

The success of any inhibitor in controlling corrosion depends on several basic requirements. First, starting the treatment at two or three times the normal inhibitor concentration ("passivation dose") to build up the protective film as fast as possible may be desirable. As noted in previous sections, there is evidence from many field studies that the reduction of lead and perhaps iron corrosion is proportional to the dosage, if the proper pH and DIC concentration conditions exist. This minimizes the opportunity for pitting to start before the entire metal surface has been covered by a protective film. Usually at least many weeks are required for the coating to develop. Some studies have indicated that metal levels (such as lead) decrease very slowly for years (AWWARF, 1994, 1998; Lyons, Pontes, and Karalekas, 1995; Cook, 1997).

There is, however, some anecdotal evidence that under some circumstances, the sudden introduction of phosphate corrosion inhibitors may destabilize loosely adhering scale particles, leading to turbid and colored water. Until more is known about precisely defining the chemical and physical scale conditions where initial adverse reactions will take place when starting inhibitor dosing, utilities would be wise to try to pilot-test treatment alternatives on representative segments of pipes actually removed from the distribution system.

The inhibitor usually must be fed continuously and at a sufficiently high concentration ("maintenance dose"). Interruptions in the feed can cause loss of protective films by dissolution and concentrations that are too low may prevent the formation of a protective film on all parts of the surface. Reversibility of deposits can occur in a matter of days (Colling et al., 1992; Wagner, 1989). Next to maintaining proper pH and background water quality consistency, the key to good corrosion inhibitor treatment is feed control (Singley, Beaudet, and Markey, 1984; AWWA, 1986).

Flow rates must be sufficient to continuously transport the inhibitor to all parts of the metal surface; otherwise, an effective protective film will not be formed and maintained. Corrosion will then be free to take place. For example, corrosion inhibitors often cannot reduce corrosion in storage tanks because the water is not flowing, and the inhibitor is not fed continuously. To avoid corrosion of tanks, using a protective coating, cathodic protection, or both, is necessary. Similarly, corrosion inhibitors are not as effective in protecting dead ends as they are in those sections of mains that have a reasonably continuous flow.

Major changes in water chemistry to facilitate corrosion control are usually immediately accompanied by some detrimental effects, such as increased metal release levels, turbid water or particles in the water, or other manifestations. Maintaining consistent treatment for many months before changing to another approach is absolutely essential to allow the system to stabilize properly for a legitimate evaluation of the treatment.

Sodium silicate (water glass) has been used for over 60 years to reduce corrosivity. The way in which sodium silicate acts to form a protective film is still not completely understood, but there is evidence it sorbs on or reacts with existing oxide, hydroxide, or carbonate pipe deposits (LaRosa-Thompson et al., 1997). Silicate tends to be most effective at high pH, and at dosages over 15 to 20 mg/L as SiO_2. In a recent USEPA study, silicate was shown to be effective in simultaneously sequestering source water iron (in combination with chlorine) and reducing iron, lead, and copper corrosion (Clement, Schock, and Lytle, 1994; Schock et al., 1998). The effectiveness of sodium silicate as a corrosion inhibitor depends on water quality properties such as pH and bicarbonate concentration. It is also more effective under higher-velocity flow conditions. Silicate will raise pH in weakly buffered waters when a formulation is used having a high ratio of Na_2O to SiO_2. Although more expensive than lime and caustic (NaOH), sodium silicate solution is a very consistent and effective chemical for pH adjustment in small water systems, and has fewer safety problems for facilities and operators than caustic.

Secondary Effects of Treatment for Corrosion Control

A utility must meet a broad spectrum of regulatory and aesthetic water quality goals, aside from those associated with corrosion and corrosion by-products. Therefore, the potential effects of different corrosion control strategies must enter into the determination of the best course to follow. There are both positive (Schock, 1998) and negative effects of corrosion control with respect to other regulations.

Adjustment of pH may affect the ability of coagulation to remove turbidity or organic matter effectively, as can the adjustment of alkalinity or the addition of phosphate (AWWARF, 1990). The efficiency of disinfection by free chlorine is pH-dependent, so a utility may be forced to consider alternatives, such as the use of monochloramine, or changing the location of pH adjustment in the treatment chain (see Chapters 6 and 14).

The pH will also affect the formation of disinfection by-products (see Chapters 12 and 14). Consideration must also be given to the impact of changes in water quality on industrial processes, building heating and cooling systems, wastewater loadings of metals (such as zinc), and nutrients (such as phosphates), and on algal or plant growth in open-storage reservoirs.

Certain treatment processes cause mineral imbalances in the water that can exacerbate corrosion control, without substantial posttreatment. Examples are: nanofiltration, reverse osmosis, ion exchange, ozonation, sequestration, enhanced coagulation, enhanced softening optimized for NOM removal (insufficient hardness and insufficient DIC for buffering), and possibly oxidative processes for iron and manganese removal.

On the other hand, there are particular benefits to a holistic corrosion control approach for the entire distribution system. Good corrosion control throughout the distribution system is necessary to optimize many aspects of distribution system water quality and performance, even aspects targeted by different specific regulations. Examples are improved (Schock, 1998):

- Aesthetic characteristics
- Longevity of infrastructure
- Disinfection and biofilm control through reduced chlorine demand and NOM sorption
- DBP formation through reduced chlorine demand
- Reduction of noxious constituents like Cu, Fe, Zn, and Pb in wastewater and sludges

These multiple objectives can be achieved with proper attention to the broad view of treatment optimization throughout the whole distribution system. Although not necessarily simple and inexpensive, achieving good lead and copper corrosion control can be made chemically compatible with protecting the other distribution system piping materials. Further, this can be done while also meeting both other regulatory and aesthetic water quality requirements throughout the distribution system.

Linings, Coatings, and Paints

Another technique to keep corrosive water away from the pipe wall is to line the wall with a protective coating. These linings are usually mechanically applied, either when the pipe is manufactured or in the field before it is installed. Some linings can be applied even after the pipe is in service, though this method is much more expensive.

The most common pipe linings are coal-tar enamels, epoxy paint, cement mortar, and polyethylene. The use of coatings must be carefully monitored, because they can be the source of several water quality problems, such as support of microbiological growth, taste and odor, and solvent leaching. Although coal tar–based products have been widely used in the past for contact with drinking water, concern exists about their use because of the presence of polynuclear aromatic hydrocarbons and other hazardous compounds in coal tar and the potential for their migration in water.

A common method of extending the life of pipe and restoring good corrosion resistance and hydraulic properties is either the controlled deposition of a $CaCO_3$ film (Hasson and Karmon, 1984; McCauley, 1960; Bonds, 1989), or the process of cement-mortar lining. The latter process has recently been comprehensively reviewed (Leroy et al., 1996; Holtschulte and Schock, 1985). Either process is also useful in covering exposed asbestos fibers and softened interior surfaces of A-C pipe. Like calcium carbonate films, the cement linings are far from inert. To protect the investment in time, money, and materials, as well as to avoid water quality degradation, a utility using pipe relining by cement needs to consider proper adjustment of the finished-water chemistry to protect the linings, in addition to other corrosion control objectives.

WATER SAMPLING FOR CORROSION CONTROL

The effects of corrosion may not be evident without monitoring. The effects can be expensive, and in the case of corrosion by-products such as lead, copper, and cadmium, they can be injurious to the health of segments of the population.

Corrosion has many causes, and many techniques exist to measure or "cure" corrosion. Corrosion in a system depends on a specific water and the reaction of that water with specific pipe materials; therefore, each utility is faced with a unique set of problems. General methods of measuring and monitoring for corrosion can, however, provide a basis for a sound corrosion control program for any utility.

The first concern for a utility is to meet all regulatory sampling requirements, in terms of the number of sampling sites, the location and frequency of the samples, and the use of approved and generally accepted analytical methods. Beyond these requirements, monitoring programs may address other questions. Some recent reviews discuss several aspects of investigating and monitoring for corrosion control and system studies (AWWARF-TZW, 1996; AWWARF-DUGW Forschungstelle, 1985; Lytle and Schock, 1997; Schock and George, 1991; AWWARF, 1994, 1998; Wysock et al., 1995; Schock, Levin, and Cox, 1988).

Defining the Problem

The many factors responsible for variability in lead concentrations that can limit the accurate assessment of exposure levels, treatment performance, and regulatory compliance have been reviewed (Schock, 1990a; Schock, Levin, and Cox, 1988). These factors are, in fact, not unique to lead in most cases. The goal of a sampling program must be to control as many of the analytical, chemical, and physical factors as possible, so that proper exposure assessments and decisions on the existence of water quality problems or the performance of a treatment program can be made.

The variability represents the actual range of exposures that occur routinely in the population. A monitoring system designed to accurately reflect exposure to lead in drinking water must simultaneously capture that diversity while providing adequately reliable information to critically evaluate exposure (Schock, Levin, and Cox,

1988). A monitoring program for corrosion control effectiveness must address similar concerns, although the criteria for sample size based on the desired levels of confidence in the mean values could be different, or a multistage sampling design could be used to obtain information more economically. A recent study in France obtained both good exposure estimates and very good correspondence of lead levels to predictions of solubility models by a program of proportional sampling at many household sites (Anjou Recherche, 1994).

Another important piece of information that would be helpful in diagnosing and solving corrosion or metal release problems is determining the extent to which release of particulate corrosion by-products contributes to the observed metal levels. Filtration of selected samples is a straightforward approach to determine the particle contribution. However, filtration is a deceptively complicated analytical procedure, which will be discussed in more detail in a later section.

Few studies have presented sampling strategies that allow the evaluation of the variability over time of the same sampling site, and particularly with respect to conditions before and after the implementation of a corrosion control program. Even within a water supply system with relatively consistent and uniform water characteristics, successive samplings under equivalent conditions do not necessarily yield the same corrosion by-product concentrations. The design of a monitoring scheme for the determination of treatment effectiveness must take into account the magnitude of these variabilities, so that the proper number of sites and frequency of sampling can be chosen. To reduce both random and unidirectional bias, the monitoring program must have standardized procedures for sample collection and analysis, a high level of analytical precision and accuracy, and uniform performance by well-trained and informed personnel.

Human exposure evaluations require different sampling considerations, which are frequently overlooked in health-effects studies. The sampling and monitoring structures under the U. S. drinking water regulations were never intended to represent human exposure estimates (*Federal Register,* 1991a). To guard against a worst-case scenario, sampling locations and sampling collection protocols must be established that are likely to coincide with occurrences of elevated by-product levels in drinking water to which members of the service population are likely to be exposed. Several recent studies have investigated these factors (Schock, 1990a; Lytle and Schock, 1997; AWWARF, 1994, 1998; Reiber, Ryder, and Wagner, 1996; Wysock et al., 1995; Schock, Levin, and Cox, 1988), and there are, no doubt, many more that can be found with a detailed examination of the corrosion control case studies presented in the last approximately five years.

Careful thought must be given to the selection of sites within the target groups. For example, when lead is of concern, lead interior plumbing or service lines, or the most recent construction before the lead solder ban, or faucet replacements when lead-containing brass is used (remembering that "lead-free" was defined as no more than 8 percent). For copper, the age of the scale plays a major role in the copper solubility. Hence, the most dangerous copper exposures would be in the newest construction, not covered as targeted monitoring sites under the Lead and Copper Rule.

The nature of exposure will vary depending upon the type of water system involved (e.g., single-family dwelling, apartment building, office building, schools, and so on). Different sampling strategies are essential, and they must be carefully oriented toward the layout of the customers' system. To determine whether the contamination is from a particular part of the plumbing system (e.g., faucets, soldered joints, parts of a service line, or pigtail connector), small sample volumes may be required to isolate the specific area of plumbing in question, or numerous sequential samples may be required (Schock and Neff, 1988; AWWARF, 1990; Lytle, Schock, and Sorg, 1994; Wysock et al., 1995). Guidelines were developed by United States

Environmental Protection Agency to isolate locations of contamination in building (USEPA, 1988) or school (USEPA, 1989) drinking water systems, which are applicable to many other corrosion problems as well.

To accurately assess actual exposure to metals introduced from plumbing systems, there are two critical monitoring design objectives that must be met. First, a representative set of locations to test must be obtained. The second objective is to employ a sampling scheme that best captures actual exposure. At this time, the best sampling program design which adequately addresses this objective is the "proportional" sampling, reported in studies from Canada, the Netherlands, and France (van den Hoven, 1985, 1987; Anjou Recherche, 1994; Meranger and Khan, 1983).

Utilities must also be concerned with monitoring the availability of active inhibitor constituents, such as silica, orthophosphate, polyphosphate, DIC (alkalinity), or pH, to ensure that the dosages applied to the treated water are adequate to provide the necessary levels for corrosion control throughout the distribution system. Some of these are requirements of the "Lead and Copper Rule," but only for a limited subset of utilities (*Federal Register,* 1991a,b, 1994a). All utilities operating corrosion control programs would be well advised to carefully monitor all aspects of those processes at the plant and throughout the distribution system (which is where the effort really counts).

Statistical Considerations

The number of sites to be monitored for exposure evaluation or corrosion control must be related to the level of constituent or corrosion by-product variability over the proposed range of sites, and the relationship of that variability to the population mean (mean of all sites) (AWWARF, 1990, 1994; Wysock et al., 1992, 1995; Schock, Levin, and Cox, 1988; Wysock, Schock, and Eastman, 1991). A good "before" sampling program must be developed to obtain a "baseline" of data. The baseline is used to get a handle on the required number and frequency of sampling for the rest of the program, and to enable valid comparisons to be made of water characteristics before and after changes in the system (e.g., treatment changes, design changes, source water changes, and so forth).

One problem that arises is the practical issue of the proper statistical tests and the confidence levels for establishing whether two or more treatments or water chemistry conditions are showing equivalent effects on corrosion rates or metal release. Automatically using common statistical criteria, such as 90 or 95 percent confidence limits, may lead to misleading conclusions because of the nature of metal release (and possibly other corrosion test) data. A small episode of high variability or periodic episodes of widely different metal release may bias the statistical calculations of parametric or even nonparametric tests to indicate there is "no difference" in effect between or among test conditions (AWWARF, 1994; Wysock et al., 1992, 1995; Wysock, Schock, and Eastman, 1991). Simple examination of the data may show very consistently different behavior, except for the periodic excursions from what is clearly the most characteristic behavior. Therefore, purely statistical testing must be tempered with careful interpretation.

Chemical and Physical Considerations

Corrosion is affected by the chemical composition of a water, so sampling and chemical analysis of the water can provide valuable corrosion-related information. Most utilities routinely analyze their water (1) to ensure that they are providing a safe water to their customers and (2) to meet regulatory requirements. For the purpose of holistically controlling the corrosion of the multiple materials present in the water system, water samples should be targeted toward the entry points and throughout the distribution system. Any "mixing" or hydraulically isolated zones should be covered by samples targeted in time and space to determine water quality changes that might result from corrosion (e.g., iron, calcium, or aluminum release, turbidity, pH changes, disinfectant residual, corrosion inhibitor, or dissolved oxygen depletion), microbial activity (pH change from nitrification, sulfate reduction), or other causes. It is also useful to determine if the metal release from corrosion comprises mainly soluble or particulate chemical species.

To represent conditions at the customer's tap, "standing" samples should be taken from an interior faucet in which the water has remained for a specific number of hours (i.e., overnight). The age of the plumbing in which the sampled water is in contact can be very important information, especially for copper. The stagnation curves representing metal levels versus time are very complex, and differ somewhat for different metals. For example, copper levels have been observed to increase for as much as 72 hours or more, in the continued presence of chlorine or dissolved oxygen (Schock, Lytle, and Clement, 1995a,b; Lytle and Schock, 1996, 1997; Werner, Groß, and Sontheimer, 1994). However, in the absence of oxidants, or after the depletion of oxidants, the copper levels have often been seen to actually peak and then go down. The sample should be collected as soon as the tap is open.

If the contribution of the faucet to metal levels is of interest, a small sample must be collected (i.e., 60 to 125 mL) in addition to samples from the rest of the plumbing system. Larger volumes are necessary to include the pipe contribution. Sample volumes may be thought of to represent linear plumbing distance, so contamination sources can be characterized by carefully planned sequences of samples of specific volumes appropriate to the given study (Lytle, Schock, and Sorg, 1994; Wysock, Schock, and Eastman, 1991).

Many important decisions are likely to be made based on the sampling and chemical analyses performed by a utility. Therefore, care must be taken during the sampling and analysis to obtain the best data. Handling of samples for pH, alkalinity, and CO_2 analyses often requires special precautions (Schock and Schock, 1982). Samples should be collected without adding air and with minimal agitation, as air tends to remove CO_2 and also affects the oxygen content in the sample. To collect a sample without additional air and to minimize exchange of volatile gases (such as CO_2 in waters that are frequently out of equilibrium with the atmosphere), fill the sample container to the top so that a convex dome is formed at the opening and no bubbles are present. If possible, the sample bottle should be filled below the surface of the water using tubing, so that the water is not contaminated by the faucet material. Cap the sample bottle as soon as possible. For general purposes, high-density linear polyethylene bottles are very suitable for metals and most other water quality constituents. For pH measurements off-site, the bottles should be glass or a material assured to be impermeable, especially toward carbon dioxide, and to a lesser extent, oxygen. Most plastic bottle manufacturers can provide tables of gas permeabilities for the plastics used in their products, which will be helpful for the selection of the proper bottles for the purpose.

When available, a direct analysis of DIC concentration (or TIC concentration, if unfiltered) is generally more accurate than computing DIC from pH and alkalinity measurements, unless particular care is taken in the analyses and sophisticated equivalence point and pH stabilization point techniques are employed (Schock and George, 1991; Schock and Lytle, 1994). Glass, or other material impermeable to CO_2, must be used for the container, and the cap must allow no air space.

To determine if metal release is primarily particulate or dissolved, frequently filtrations are used. This is a deceptively difficult procedure, as there are numerous opportunities for erroneous data and biases that can significantly affect conclusions. Sorption losses are especially acute on glass materials and with cellulose acetate membranes, so polycarbonate, teflon, or stainless-steel apparatus and filter supports are preferable. Polycarbonate filter membranes have generally been shown to cause less bias than membranes made of most other materials. Some cleaning and filter apparatus preparation precautions have been described in some studies (Schock, Lytle, and Clement, 1995b; Schock and Gardels, 1983). Before using data from sample filtration, a careful laboratory and field test evaluation should be conducted of all materials used and all steps in the procedure to determine if there is a potential for losses of dissolved constituents, or contamination from the materials.

The constituents that should be analyzed in a thorough corrosion-monitoring program depend to a large extent on the materials present in the system's distribution, service, and household plumbing lines. Table 17.6 summarizes parameters recommended to be analyzed in a thorough corrosion-monitoring program. Temperature and pH should be measured in situ (in the field) with proper precautions for atmospheric CO_2 exchange.

Frequency of analysis depends on the extent of the corrosion problems experienced in the system, the degree of variability in source and finished-water quality, the type of treatment and corrosion control practiced by the water utility, and cost considerations. When phosphates or silicates are added to the water, samples should be collected at the far reaches of the system and analyzed for polyphosphates, orthophosphates, and silicate, as appropriate. If no residual phosphate or silicate is

TABLE 17.6 Recommended Analyses for a Thorough Corrosion Monitoring Program

General parameters for all investigations	
In situ measurements	pH, temperature, CO_2 if low-pH groundwater
Dissolved gases, oxidants	Oxygen, hydrogen sulfide,* free chlorine residual, total chlorine residuals (if ammonia present or used)
Parameters required to calculate $CaCO_3$-based indices	Calcium, total hardness (or magnesium), alkalinity (or DIC), total dissolved solids (or conductivity)†
Parameters for A-C pipe	Add to general parameters
	Fiber count, iron, zinc, silica, polyphosphate, aluminum, manganese
Background parameters for metal pipe	Add to general parameters
Iron or steel pipe	Iron, chloride, sulfate
Lead pipe or lead-based solder	Lead, copper, chloride, sulfate
Copper pipe	Copper, lead, sulfate, chloride
Galvanized iron pipe	Zinc, iron, cadmium, lead, chloride
Brass (faucets and valves)	Zinc, copper, lead, sulfate
All metal pipes	Add to general parameters
Corrosion "inhibitor" constituents (all metal pipes)	Orthophosphate, polyphosphate,‡ silica

* It only needs to be analyzed when suspected.
† If a complete water analysis is done, these can be neglected because ionic strength can be directly computed.
‡ For most potable waters; derived from analyzing total phosphate and subtracting orthophosphate.

found, the feed rate should be increased. When calcium carbonate precipitation is practiced, monitoring of the parameters necessary to compute the LSI or the CCPP in the far reaches of the system is also necessary. Continual monitoring of pH and temperature are important, because they are so interdependent. An important quality assurance practice in corrosion studies is to do frequent complete chemical analyses of the finished water and in many locations throughout the distribution system, and carefully examine the data for ion balance errors and internal consistency of trends and interrelationships among constituents that should follow known behavior and concentration patterns. Chemical equilibrium modeling computer programs provide a very good tool to do this evaluation (APHA-AWWA-WEF, 1992, 1995; Parkhurst, Thorstenson, and Plummer, 1980; Allison, Brown, and Novo-Gradac, 1991; Schecher and McAvoy, 1994a,b, 1998).

When corrosion control studies obtain good baseline data and have a comprehensive, well-designed monitoring program in place throughout their corrosion control effort, the data are invaluable to other utilities and corrosion scientists. They can then work with a larger body of knowledge of drinking water chemistry and treatment, and implement future corrosion control and public health protection strategies much more effectively and efficiently.

ACKNOWLEDGMENTS

The author gratefully acknowledges the extensive contributions of all of the writers of the AWWA publication *Corrosion Control for Operators,* the AWWARF/DVGW-Forschungsstelle/TZW manuals *Internal Corrosion of Water Distribution Systems,* and the late Dr. T. E. Larson, whose works were heavily cited and directly included in this chapter. The author is also indebted to the late Dr. Heinrich Sontheimer for inspiration for much of the copper research, to Dr. Marc Edwards for many interesting ideas and discussion, and to Dr. Ray Letterman, Dr. John Ferguson, and Peter Karalekas, Jr., for their herculean efforts in reviewing the draft and providing very helpful criticisms.

BIBLIOGRAPHY

Adeloju, S. B., and Hughes, H. C. "The Corrosion of Copper Pipes in High Chloride-Low Carbonate Mains Water." *Corros. Sci.,* 26(10), 1986: 851–870.

Ailor, W. H. *Handbook on Corrosion Testing and Evaluation.* New York: John Wiley & Sons, 1970.

Ainsworth, R. G. *The Introduction of New Water Supplies into Old Distribution Systems.* Technical Report TR-143, Medmenham, UK: Water Research Centre, 1980.

Allen, M. J., R. H. Taylor, and E. E. Geldreich. "The occurrence of microorganisms in water main encrustations." *Jour. AWWA,* 72(11), 1980: 614–625.

Allison, J. D., D. S. Brown, and K. J. Novo-Gradac. *MINTEQA2/PRODEFA2, A Geochemical Assessment Model for Environmental Systems: Version 3.0 User's Manual,* Athens, GA: EPA/600/3-91/021. U.S. Environmental Protection Agency, Environment Research Laboratory, Office of Research and Development, 1991.

Anjoy Recherche. *Etude des Phenomenes de Solubilisation du Plomb par l'Eau Distribuee et des Moyens Pour les Limiter.* Paris, France: Rapport de Synthese, 1994.

APHA-AWWA-WEF. *Standard Methods for the Examination of Water and Wastewater,* 18th ed. Washington, D.C.: American Public Health Association, 1992.

APHA-AWWA-WEF. *Standard Methods for the Examination of Water and Wastewater,* 19th ed. Washington, D.C.: American Public Health Association, 1995.

Arens, P., et al. "Experiments for stimulating microbiologically induced corrosion of copper pipes in a cold-water plumbing system." *Werk. und Korros.,* 47(2), 1996: 96–102.

ASTM. "Standard practices for examination of water-formed deposits by chemical microscopy." In *1996 Annual Book of ASTM Standards,* pp. 29–34. Conshohocken, PA: American Society for Testing and Materials, 1996a.

ASTM. "Standard practices for examination of water-formed deposits by wavelength dispersive X-ray fluorescence." In *1996 Annual Book of ASTM Standards,* pp. 129–131. Conshohocken, PA: American Society for Testing and Materials, 1996b.

ASTM. "Standard practices for identification of crystalline compounds in water-formed deposits by X-ray diffraction." In *1996 Annual Book of ASTM Standards,* pp. 18–26. Conshohocken, PA: American Society for Testing and Materials, 1996c.

ASTM. "Standard practices for sampling water-formed deposits." In *1996 Annual Book of ASTM Standards,* pp. 1–17. Conshohocken, PA: American Society for Testing and Materials, 1996d.

ASTM. "Standard practices preparation and preliminary testing of water-formed deposits." In *1996 Annual Book of ASTM Standards,* pp. 124–131. Conshohocken, PA: American Society for Testing and Materials, 1996e.

Atlas, D. C. J., and O. T. Zajicek. "The corrosion of copper by chlorinated drinking waters." *Water Res.,* 16, 1982: 693–698.

AWWA. *Corrosion Control for Operators.* Denver, CO: American Water Works Association, 1986.

AWWA. *Water Quality and Treatment,* 3rd ed. Denver, CO: American Water Works Association, 1971.

AWWARF. *Development of a Pipe Loop Protocol for Lead Control.* Denver, CO: AWWA Research Foundation, 1993.

AWWARF. *Lead Control Strategies.* Denver, CO: AWWA Research Foundation and AWWA, 1990.

AWWARF. *Water Quality Changes in the Distribution System following the Implementation of Corrosion Control.* Denver, CO: AWWA Research Foundation, 1998.

AWWARF-DVGW Forschungsstelle. *Internal Corrosion of Water Distribution Systems.* Denver, CO: AWWA Research Foundation/DVGW-Forschungsstelle, 1985.

AWWARF-TZW. *Internal Corrosion of Water Distribution Systems,* 2nd ed. Denver, CO: AWWA Research Foundation/DVGW-TZW, 1996.

Balzer, W. "Calcium carbonate saturometry by alkalinity difference measurement." *Oceanologica Acta,* 3(2), 1980: 237–243.

Bard, A. J. *Chemical Equilibrium.* New York: Harper & Row, 1966.

Bell, G. E. C., M. J. Schiff, R. L. Bianchetti, and S. J. Duranceau. "Grounding can affect water quality." In *Proc. AWWA Water Quality Technology Conference,* New Orleans, LA, 1995.

Bell, G. E. C., M. J. Schiff, and S. J. Duranceau. "Review of the effects of electrical grounding on water piping." In *Proc. NACE Corrosion/95.* Orlando, FL: 1995.

Bemer, P. J., and G. G. Geesey. "An evaluation of biofilm development utilizing nondestructive attenuated total reflectance Fourier Transform Infrared Spectroscopy." *Biofouling,* 3, 1991: 89–100.

Benjamin, L., R. W. Green, and A. Smith. "Pilot testing a limestone contactor in British Columbia." *Jour. AWWA,* 84(5), 1992: 70–79.

Benjamin, M. M., H. Sontheimer, and P. Leroy. "Corrosion of iron and steel." In *Internal Corrosion of Water Distribution Systems,* pp. 29–70. Denver, CO: AWWA Research Foundation/TZW, 1996.

Ben-Yaakov, S., and I. R. Kaplan. "Deep-sea in situ calcium carbonate saturometry." *Jour. Geophys. Res.,* 76(3), 1971: 722–730.

Berend, K., and T. Trouwborst. "Cement-mortar pipes as a source of aluminum." *Jour. AWWA,* 91(7), 1999: 91–100.

Bockris, J. O. M., and A. K. N. Reddy. *Modern Electrochemistry,* vol. 2. New York: Plenum Press, 1973.

Boireau, A., G. Randon, and J. Cavard. "Positive action of nanofiltration on materials in contact with drinking water." *Aqua,* 46(4), 1997: 210–217.

Bonds, R. W. "Cement mortar linings for ductile iron pipes." *Ductile Iron Pipe News,* Spring/Summer, 1989: 8–13.

Breach, R., S. Crymble, and M. J. Porter. "Systematic approach to minimizing lead level at consumers taps." In *Proc. AWWA Annual Conference,* Philadelphia, PA, 1991.

Bremer, P. J., and G. G. Geesey. "Laboratory-based model of microbiologically induced corrosion of copper." *Appl. Environ. Microbiol.,* 57(7), 1991: 1956–1962.

Britton, A., and W. N. Richards. "Factors influencing plumbosolvency in Scotland." *Jour. Inst. Water Engrs. & Scientists,* 35(5), 1981: 349–364.

Brown, D. A., B. L. Sherriff, and J. A. Sawicki. "Microbial transformation of magnetite to hematite." *Geochim. Cosmochim. Acta,* 61(16), 1997: 3341–3348.

Buffle, J., R. R. De Vitre, D. Perret, and G. G. Leppard. "Physico-chemical characteristics of a colloidal iron phosphate species formed at the oxic-anoxic interface of a eutrophic lake." *Geochim. Cosmochim. Acta,* 53, 1989: 399–408.

Butler, G., and H. C. K. Ison. *Corrosion and Its Prevention in Waters.* London: Leonard Hill, 1966.

Butler, J. N. *Carbon Dioxide Equilibria and Their Applications.* Reading, MA: Addison-Wesley Publishing Co., 1982.

Campbell, H. S. "Causes and avoidance of corrosion in copper water pipes." In *Sixth Congress of the International Water Supply Assn.,* Stockholm, 1964.

Campbell, H. S. "Corrosion, water composition and water treatment." *Wat. Treat. Exam.,* 20(1), 1971: 11–34.

Campbell, H. S. "The influence of the composition of supply waters, and especially of traces of natural inhibitor, on pitting corrosion of copper water pipes." *Proc. Soc. Water Treat. & Exam.,* 8, 1954: 100–116.

Campbell, H. S. "Pitting corrosion of copper water pipes." *2nd International Congress on Metallic Corrosion,* New York, 1963.

Camper, A. K. "Interaction between pipe materials, corrosion inhibitors, disinfection, organics, and distribution system biofilms." In *Proc. National Conference on Integrating Corrosion Control and Other Water Quality Goals,* Cambridge, MA, 1996.

Clement, J. A., W. J. Daly, H. J. Shorney, and A. J. Capuzzi. "An innovative approach to understanding and improving distribution system water quality." In *Proc. AWWA Water Quality Technology Conference,* San Diego, CA, 1998.

Clement, J. A., and M. R. Schock. "Buffer intensity: What is it, and why it's critical for controlling distribution system water quality." In *Proc. AWWA Water Quality Technology Conference,* San Diego, CA, 1998.

Clement, J. A., M. R. Schock, and D. A. Lytle. "Controlling lead and copper corrosion and sequestering of iron and manganese." In *Critical Issues in Water and Wastewater Treatment,* Boulder, CO, 1994.

Colling, J. H., B. T. Croll, P. A. E. Whincup, and C. Harward. "Plumbosolvency effects and control in hard waters." *J. IWEM,* 6(6), 1992: 259–268.

Colling, J. H., P. A. E. Whincup, and C. R. Hayes. "The measurement of plumbosolvency propensity to guide the control of lead in tapwaters." *J. IWEM,* 1(3), 1987: 263–269.

Columbo, C., V. Barrón, and J. Torrent. "Phosphate adsorption and desorption in relation to morphology and crystal properties of synthetic hematites." *Geochim. Cosmochim. Acta,* 58(4), 1994: 1261–1269.

Cook, J. B. "Achieving optimum corrosion control for lead in Charleston, S.C., A case study." *Jour. NEWWA,* 111(2), 1997: 168–179.

Cruse, H., and R. D. Pomeroy. "Corrosion of copper pipes." *Jour. AWWA,* 66(8), 1974: 479–483.

Cruse, H., and O. von Franqué. "Corrosion of copper in potable water systems." In *Internal*

Corrosion of Water Distribution Systems, pp. 317–416. Denver, CO: AWWA Research Foundation/DVGW Forschungsstelle, 1985.

Deffeyes, K. S. "Carbonate equilibria: A graphic and algebraic approach." *Limnol. Oceanogr.,* 10, 1965: 412–426.

de Mora, S. J., and R. M. Harrison. "Lead in tap water: Contamination and chemistry." *Chem. Britain,* 10, 1984: 901–904.

de Mora, S. J., R. M. Harrison, and S. J. Wilson. "The effect of water treatment on the speciation and concentration of lead in domestic tap water derived from a soft Upland source." *Water Resourc.,* 21(1), 1987: 83–94.

Dexter, S. C., and S. H. Lin. "Calculations of seawater pH at polarized metal surfaces in the presence of surface films." *NACE Corrosion,* 48(1), 1992: 50–60.

Dodrill, D. M., and M. Edwards. "Corrosion control on the basis of utility experience." *Jour. AWWA,* 87(7), 1995: 74–85.

Douglas, B. D., and D. T. Merrill. *Control of Water Quality Deterioration Caused by Corrosion of Cement-Mortar Pipe Linings.* Denver, CO: AWWA Research Foundation, 1991.

Duranceau, S. J., D. F. Newnham, J. Anselmo, and R. D. Teegarden. "Implementation of lead and copper corrosion control demonstration test programs for three Florida utilities." In *Proc. AWWA Annual Conference,* Atlanta, GA, 1997.

Duthil, J.-P., G. Mankowski, and A. Giusti. "The synergetic effect of chloride and sulphate on pitting corrosion of copper." *Corros. Sci.,* 38(10), 1996: 1839–1849.

Dye, J. F. "Review of anticorrosion water treatment." *Jour. AWWA,* 56(4), 1964: 457–465.

Eary, L. E., and J. A. Schramke. "Rates of inorganic oxidation reactions involving dissolved oxygen." In *Chemical Modeling of Aqueous Systems II,* pp. 379–396, Ch. D. C. Melchior and R. L. Bassett, eds. Washington, D.C.: American Chemical Society, 1990.

Edwards, M., J. F. Ferguson, and S. H. Reiber. "On the pitting corrosion of copper." *Jour. AWWA,* 86(7), 1994: 74–90.

Edwards, M., S. Jacobs, and D. Dodrill. "Desktop guidance for mitigating of Pb and Cu corrosion by-products." *Jour. AWWA,* 91(5), 1999: 66–77.

Edwards, M., T. E. Meyer, and M. R. Schock. "Alkalinity, pH and copper corrosion by-product release." *Jour. AWWA,* 88(3), 1996: 81–94.

Edwards, M., J. Rehring, and T. Meyer. "Inorganic anions and copper pitting." *NACE Corrosion,* 50(5), 1994: 366–372.

Elzenga, C. H., and A. Graveland. "Proposal of the European communities for a directive relating to the quality of water for human consumption, actual metal levels in drinking water and some possibilities for reduction by central water conditioning by waterworks. *KIWA and Gemeentewaterleidingen, Amsterdam, Netherlands,* 1981.

Faust, S. B., and O. M. Aly. *Chemistry of Natural Waters.* Ann Arbor, MI: Ann Arbor Science Publishers, 1981.

Federal Register, "Analytical methods for regulated drinking water contaminants." Final Rule, USEPA, 40 CFR Parts 141 and 143. 59(232), December 5, 1994a: 62456.

Federal Register, "Drinking water; maximum contaminant level goals and national primary drinking water regulations for lead and copper." Final Rule; technical corrections, USEPA, 40 CFR Parts 141 and 142. 59(125), June 30, 1994b: 33860.

Federal Register, "Drinking water regulations; maximum contaminant level goals and national primary drinking water regulations for lead and copper." Final Rule, correction, USEPA, 40 CFR Parts 141 and 142. 56(135), July 15, 1991a: 32112.

Federal Register, "Interim primary drinking water regulations; amendments." Final Rule, USEPA, 40 CFR Part 141. 45(168), August 27, 1980: 57331.

Federal Register, "Maximum contaminant level goals and national primary drinking water regulations for lead and copper." Final Rule, USEPA. 56(110), June 7, 1991b: 26460.

Ferguson, J. L., O. von Franqué, and M. R. Schock. "Corrosion of copper in potable water systems." In *Internal Corrosion of Water Distribution Systems,* 2nd ed., pp. 231–268. Denver, CO: AWWA Research Foundation/DVGW-TZW, 1996.

Fischer, W., H. H. Paradies, D. Wagner, and I. Hänßel. "Copper deterioration in a water distribution system of a county hospital in Germany caused by microbioally induced corrosion—I. Description of the problem." *Werk. und Korros.*, 43, 1992: 56–62.

Fischer, W.-R., and B. Füßinger. "Influence of anions on the pitting behaviour of copper in potable water." In *Proc. 12th Scandinavian Corrosion Congress & Eurocorr '92,* Dipoli, Espoo, Finland, 1992.

Fox, K. P., et al. "Copper induced corrosion of galvanized steel pipe," EPA Report, EPA/600/2-86/056. Cincinnati, OH: U.S. Environmental Protection Agency, Drinking Water Research Division, 1986.

Fox, K. P., C. Tate, and E. Bowers. "The interior surface of galvanized steel pipe: A potential factor in corrosion resistance." *Jour. AWWA,* 75(2), 1983: 84–86.

Fuge, R., N. J. G. Pearce, and W. T. Perkins. "Unusual sources of aluminum and heavy metal in potable waters." *Environ. Geochem. & Health,* 14(1), 1992: 15–18.

Fuge, R., and W. Perkins. "Aluminum and heavy metals in potable waters of the North Ceredigion Area, Mid-Wales." *Environ. Geochem. & Health,* 13(2), 1991: 56–65.

Garrels, R. M., and C. L. Christ. *Solutions, Minerals and Equilibria.* New York: Harper & Row, 1965.

Geelhoed, J. S., T. Hiemstra, and W. H. Van Riemsdijk. "Phosphate and sulfate adsorption on goethite: Single anion and competitive adsorption." *Geochim. Cosmochim. Acta,* 61(12), 1997: 2389–2396.

Geesey, G. G., et al. "Binding of metal ions by extracellular polymers of biofilm bacteria." *Wat. Sci. Tech.,* 20(11/12), 1988: 161–165.

Geesey, G. G., and P. J. Bremer. "Evaluation of copper corrosion under bacterial biofilms." In *NACE Corrosion/91,* Cincinnati, OH, 1991.

Geesey, G. G., M. W. Mittelman, T. Iwaoka, and P. R. Griffiths. "Role of bacterial exopolymers in the deterioration of metallic copper surfaces." *Mat. Perf.,* 1986: 37–40.

Gould, R. W. "The application of X-ray diffraction to the identification of corrosion products." In *Proc. AWWA Water Quality Technology Conference,* Miami Beach, FL, 1980.

Gregory, R., and P. J. Jackson. "Central water treatment to reduce lead solubility." In *Proc. AWWA Annual Conference,* Dallas, TX, 1984.

Guo, Q., P. J. Toomuluri, and J. O. Eckert, Jr. "Leachability of regulated metals from cement-mortar linings." *Jour. AWWA,* 90(3), 1998: 62–73.

Harrison, R. M., and D. P. H. Laxen. "Physicochemical speciation of lead in drinking water." *Nature,* 286(5775), 1980: 791–793.

Hasson, D., and M. Karmon. "Novel process for lining water mains by controlled calcite deposition." *Corrosion Prev. & Cont.,* April 1984: 9–17.

He, Q. H., G. G. Leppard, C. R. Paige, and W. J. Snodgrass. "Transmission electron microscopy of a phosphate effect on the colloid structure of iron hydroxide." *Water Res.,* 30(6), 1996: 1345–1352.

Herrera, C. E., J. F. Ferguson, and M. M. Benjamin. "Evaluating the potential for contaminating drinking water from the corrosion of tin-antimony solder." *Jour. AWWA,* 74(7), 1982: 368–375.

Hogan, I., and P. J. Jackson. *The Examination of Lead Pipe Deposits by Infra Red Spectroscopy Preliminary Report,* Report 81-S. Medmenham, U.K.: Water Research Centre, 1981.

Holm, T. R., and M. R. Schock. "Computing SI and CCPP using spreadsheet programs." *Jour. AWWA,* 90(7), 1998: 80–89.

Holm, T. R., and M. R. Schock. "Polyphosphate debate (reply to comment)." *Jour AWWA,* 83(12), 1991a: 10–12.

Holm, T. R., and M. R. Schock. "Potential effects of polyphosphate products on lead solubility in plumbing systems." *Jour. AWWA,* 83(7), 1991b: 74–82.

Holm, T. R., and S. H. Smothers. "Characterizing the lead-complexing properties of polyphosphate water treatment products by competing-ligand spectrophotometry using 4-(2-pyridylazo) resorcinol." *Int. Jour. Environ. Anal. Chem.,* 41, 1990: 71–82.

Holtschulte, H., and M. R. Schock. "Asbestos-cement and cement-mortar lined pipes." In *Internal Corrosion of Water Distribution Systems,* pp. 417–512. Denver, CO: AWWA Research Foundation/DVGW Forschungsstelle, 1985.

Hoover, C. P. "Practical application of the Langelier Method." *Jour. AWWA,* 30(11), 1938: 1802–1807.

Horton, A. M. "Corrosion effects of electrical grounding on water pipe." In *Corrosion/91,* Cincinnati, OH, 1991.

Horton, A. M., and R. E. Behnke. "AC-DC effects of electrical grounding on water pipe." In *Proc. AWWA Annual Conference,* Los Angeles, CA, 1989.

Hulsmann, A. D. "Particulate lead in water supplies." *J. IWEM,* 4(2), 1990: 19–25.

Ingle, S. E., C. H. Culberson, J. E. Hawley, and R. M. Pytkowicz. "The solubility of calcium in seawater at atmospheric pressure and 35% salinity." *Mar. Chem.* 1, 1973: 295–307.

Ives, D. J. G., and A. E. Rawson. "Copper corrosion III. Electrochemical theory of general corrosion." *J. Electrochem Soc.,* 109(6), 1962a: 458–462.

Ives, D. J. G., and A. E. Rawson. "Copper corrosion IV. The effects of saline additions." *J. Electrochem. Soc.,* 109(6), 1962b: 462–466.

Iwaoka, T., P. R. Griffiths, J. T. Kitasako, and G. G. Geesey. "Copper-coated cylindrical internal reflection elements for investigating interfacial phenomena." *Appl. Spectros.,* 40(7), 1986: 1062–1065.

Jacobs, S. A., and M. Edwards. "Sulfide-induced copper corrosion." In *Proc. AWWA Water Quality Technology Conference,* Denver, CO, 1997.

James M. Montgomery Consulting Engineers. *Water Treatment Principles & Design.* New York: John Wiley and Sons, 1985.

Joint Task Group. "Suggested methods for calculating and interpreting calcium carbonate saturation indexes." *Jour. AWWA,* 72(7), 1990: 71–77.

Jolley, J. G., et al. "Auger electron spectroscopy and X-ray photoelectron spectroscopy of the biocorrosion of copper by gum arabic, bacterial culture supernatant and *Pseudomonas atlantica* exopolymer." *Surf. & Interf. Analysis,* 11, 1988: 371–376.

Jolley, J. J., et al. "*In situ,* real-time FT-IR/CIR/ATR study of the biocorrosion of copper by gum arabic, alginic acid, bacterial culture supernatant and *Pseudomonas atlantica* exopolymer." *Appl. Spectros.,* 43(6), 1989: 1062–1067.

Kenworthy, L. "The problem of copper and galvanized iron in the same water system." *Jour. Inst. Metals,* 69, 1943: 67–90.

Kiéné, L., W. Lu, and Y. Lévy. "Relative importance of phenomena responsible of the chlorine consumption in drinking water distribution systems." In *Proc. AWWA Water Quality Technology Conference,* Boston, MA, 1996.

Kirmeyer, G. J., I. Wagner, and P. LeRoy. "Organizing corrosion control studies and implementing corrosion control strategies." In *Internal Corrosion of Water Distribution Systems,* 2nd ed., pp. 487–540. Denver, CO: AWWA Research Foundation/DVGW-TZW, 1996.

Köningsberger, E., and H. Gamsjäger. "Solid-solute phase equilibria in aqueous solution. III. A new application of an old chemical potentiometer." *Mar. Chem.,* 30, 1990: 317–327.

Korshin, G. V., S. A. L. Perry, and J. F. Ferguson. "Influence of NOM on copper corrosion." *Jour. AWWA,* 88(7), 1996: 36–47.

Koudelka, M., J. Sanchez, and J. Augustynski. "On the nature of surface films formed on iron in aggressive and inhibiting polyphosphate solutions." *J. Electrochem. Soc.,* 129(6), 1982: 1186–1191.

Kriewall, D., R. Harding, E. Naisch, and L. Schantz. "The impact of aluminum residuals on transmission main capacity." *Public Works,* 127(12), 1996: 28.

Kristiansen, H. "Corrosion of copper by soft water with different content of humic substances and various temperatures." *Vatten,* 38, 1982: 181–188.

Kuch, A., V. L. Snoeyink, R. A. Ryder, and R. R. Trussell. "Experimental and investigation techniques." In *Internal Corrosion of Water Distribution Systems,* ch. 9, pp. 657–700. Denver, CO: AWWA Research Foundation/DVGW-Forschungsstelle, 1985.

Kuch, A., and I. Wagner. "Mass transfer model to describe lead concentrations in drinking water." *Water Res.,* 17(10), 1983: 1303.

Lane, R. W. *Control of Scale and Corrosion in Building Water Systems.* New York: McGraw-Hill, 1993.

Langelier, W. F. "The analytical control of anti-corrosion water treatment." *Jour. AWWA,* 28(10), 1936: 1500.

Langmuir, D. *Aqueous Environmental Geochemistry.* Upper Saddle River, NJ: Prentice-Hall, 1997.

LaRosa-Thompson, J., et al. "Sodium silicate corrosion inhibitors: Issues of effectiveness and mechanism." In *Proc. AWWA Water Quality Technology Conference,* Denver, CO, 1997.

Larson, T. E. *Corrosion by Domestic Waters,* Bulletin 59, Illinois State Water Survey, 1975.

Larson, T. E., and R. V. Skold. "Corrosion and tuberculation of cast iron." *Jour. AWWA,* October 1957: 1294–1302.

Lauer, W. C., and S. R. Lohman. "Non-calcium carbonate protective film lowers lead values." In *Proc. AWWA Water Quality Technology Conference,* San Francisco, CA, 1994.

Lazaroff, N., W. Sigal, and A. Wasserman. "Iron oxidation and precipitation of ferric hydroxysulfates by resting *Thiobacillus ferroxidans* cells." *Appl. Environ. Microbiol,* 43(4), 1982: 924–938.

LeChavallier, M. W., C. D. Lowry, R. G. Lee, and D. L. Gibbon. "Examining the relationship between iron corrosion and the disinfection of biofilm bacteria." *Jour. AWWA,* 85(7), 1993: 111–123.

Lee, R. G., W. C. Becker, and D. W. Collins. "Lead at the tap: Sources and control." *Jour. AWWA,* 81(7), 1989: 52–62.

Lee, S. H., J. T. O'Connor, and S. K. Banerji. "Biologically mediated corrosion and its effects on water quality in distribution systems." *Jour. AWWA,* 72(11), 1980: 636–645.

Légrand, L., and P. LeRoy. *Prévention de la Corrosion et de l'Entarage dans les Réseaux de Distribution d'Eau.* Neuilly-sur-Seine, France: CIFEC, 1995.

Légrand, L., and P. LeRoy. *Prevention of Corrosion and Scaling in Water Supply Systems.* New York: Ellis Horwood, 1990.

LeRoy, P. Corrosion and degradation of pipes—mechanisms and remedies. *Union of African Water Supply Conference of Yaoundi,* Yaoundi, Cameroun, 1996.

LeRoy, P. "Lead in drinking water—Origins; solubility, treatment." *Aqua,* 42(4), 1993: 233–238.

LeRoy, P., and J. Cordonnier. "Réduction de la solubilité du plomb par décarbonation partielle." *Journal Français d'Hydrologie,* 25(1), 1994: 81–96.

Leroy, P., M. R. Schock, H. Holtschulte, and I. Wagner. "Cement-based materials." In *Internal Corrosion of Water Distribution Systems,* 2nd ed. Denver, CO: AWWA Research Foundation/DVGW Forschungsstelle, 1996.

Letterman, R. D. "Calcium carbonate dissolution rate in limestone contactors," Research and Development Report, EPA/600/SR-95/068. Cincinnati, OH: U.S. Environmental Protection Agency, Risk Reduction Engineering Laboratory, 1995.

Letterman, R. D., C. T. Driscoll, Jr., M. Haddad, and H. A. Hsu. "Limestone bed contactors for control of corrosion at small water utilities," Research and Development Report, EPA/600/S2-86/099. Cincinnati, OH: U.S. Environmental Protection Agency, Water Engineering Research Laboratory, 1987.

Letterman, R. D., M. Haddad, and C. T. Driscoll, Jr. "Limestone contactors: Steady state design relationships." *Jour. Envir. Engrg. Div.—ASCE,* 117(3), 1991: 339–358.

Letterman, R. D., and S. A. Kathari. "Computer program for the design of limestone contactors." *Jour. NEWWA,* 110(1), 1996: 42–47.

Lewandowski, Z., W. Dickinson, and W. Less. "Electrochemical interactions of biofilms with metal surfaces." *Wat. Sci. Tech.,* 36(1), 1997: 295–302.

Little, B., and P. Wagner. "Myths related to microbiologically influenced corrosion." *Mat. Perf.,* 36(6), 1997: 40–44.

Loewenthal, R. E., and G. V. R. Marais. *Carbonate Chemistry of Aquatic Systems.* Ann Arbor, MI: Ann Arbor Science Publishers, 1976.

Luo, Y.-Y., and A. Hong. "Oxidation and dissolution of lead in chlorinated drinking water." *Advances in Environmental Research,* 1, 1997: 1.

Lyon, T. D. B., and J. M. A. Lenihan. "Corrosion in solder jointed copper tubes resulting in lead contamination of drinking water." *Br. Corros. J.,* 12(1), 1977: 41–45.

Lyons, J. J., J. Pontes, and P. Karalekas. "Optimizing corrosion control for lead and copper using phosphoric acid and sodium hydroxide." In *Proc. AWWA Water Quality Technology Conference,* New Orleans, LA, 1995.

Lytle, D. A., and M. R. Schock. "Impact of stagnation time on the dissolution of metal from plumbing materials." In *Proc. AWWA Annual Conference,* Atlanta, GA, 1997.

Lytle, D. A., and M. R. Schock. "Stagnation time, composition, pH and orthophosphate effects on metal leaching from brass." EPA/600/R-96/103. Washington, D.C.: U.S. Environmental Protection Agency, Office of Research and Development, 1996.

Lytle, D. A., M. R. Schock, J. A. Clement, and C. M. Spencer. "Using aeration for corrosion control." *Jour. AWWA,* 90(3), 1998a: 74–88.

Lytle, D. A., M. R. Schock, J. A. Clement, and C. M. Spencer. "Using aeration for corrosion control—Erratum." *Jour. AWWA,* 90(5), 1998b: 4.

Lytle, D. A., M. R. Schock, J. A. Clement, and C. M. Spencer. "Using aeration for corrosion control—Erratum." *Jour. AWWA,* 90(9), 1998c: 4.

Lytle, D. A., M. R. Schock, N. R. Dues, and P. J. Clark. "Investigating the preferential dissolution of lead for solder particulates." *Jour. AWWA,* 85(7), 1993: 104–110.

Lytle, D. A., M. R. Schock, and T. J. Sorg. "Controlling lead corrosion in the drinking water of a building by orthophosphate and silicate treatment." *Jour. NEWWA,* 110(3), 1996: 202–216.

Lytle, D. A., M. R. Schock, and T. J. Sorg. "A systematic study on the control of lead in a new building." In *Proc. AWWA Annual Conference,* New York, NY, 1994.

Lytle, D. A., M. R. Schock, and S. Tackett. "Metal corrosion coupon study contamination, design and interpretation problems." In *Proc. AWWA Water Quality Technology Conference,* Toronto, ON, 1992.

Marani, D. *Precipitation and Conversion of Basic Cupric Sulfate to Cupric Hydrous Oxides.* Ph.D. dissertation, Illinois Institute of Technology, Chicago, IL, 1992.

McCauley, R. F. "Controlled deposition of protective calcite coatings in water mains." *Jour. AWWA,* 52(11), 1960: 1386.

Medlar, S. J., and A. J. Kim. "Corrosion control in small water systems." In *Proc. NEWWA Symposium on Solutions to Controlling Lead and Cooper Corrosion II,* Nashua, NH, 1993.

Medlar, S. J., and A. J. Kim. *Corrosion Control in Small Water Systems,* technical paper. Camp Dresser & McKee, June 1993.

Melendres, C. A., N. Camillone, III, and T. Tipton. "Laser Raman spectroelectrochemical studies of anodic corrosion and film formation on iron in phosphate solutions." *Electrochem. Acta,* 34(2), 1989: 281–286.

Meranger, J. C., and T. R. Khan. "Lake water acidity and the quality of pumped cottage water in selected areas of northern Ontario." *Int. Jour. Environ. Anal. Chem.,* 15, 1983: 185–212.

Merrill, D. T., Jr., R. L. Sanks, and C. Spring. *Corrosion Control by Deposition of $CaCO_3$ Films.* Denver, CO: American Water Works Association, 1978.

Meyer, T. E. *The Effect of Inorganic Anions on Copper Corrosion,* Master of Science in Environmental Engineering. Boulder, CO: University of Colorado at Boulder, 1994.

Meyer, T. E., and M. Edwards. "Effect of alkalinity on copper corrosion." In *Proc. ASCE National Conference on Environmental Engineering,* Boulder, CO, 1994.

Millero, F. J. "Effect of ionic interactions on the oxidation of Fe(II) and Cu(I) in natural waters." *Mar. Chem.,* 28, 1989: 1–18.

Millero, F. J. "Effect of ionic interactions on the oxidation rates of metals in natural waters." In *Chemical Modeling of Aqueous Systems II,* pp. 447–460, D. C. Melchor and R. L. Bassett, eds. Washington, D.C.: American Chemical Society, 1990a.

Millero, F. J. "Marine solution chemistry and ionic interactions." *Mar. Chem.,* 30, 1990b: 205–229.

Millero, F. J., M. Izaguirre, and V. K. Sharma. "The effect of ionic interaction on the rates of oxidation in natural waters." *Mar. Chem.,* 22, 1987: 179–191.

Miyamoto, H. K., and M. D. Silbert. "A new approach to the Langelier Stability Index." *Chem. Eng.,* 93(8), 1986: 89–92.

Morel, F. M. M. *Principles of Aquatic Chemistry.* New York: John Wiley and Sons, 1983.

Murphy, B., J. T. O'Connor, and T. L. O'Connor. "Willmar, Minnesota battles copper corrosion, Part 1." *Public Works,* 128(11), 1997a: 65–68.

Murphy, B., J. T. O'Connor, and T. L. O'Connor. "Willmar, Minnesota battles copper corrosion, Part 2." *Public Works,* 128(12), 1997b: 44–47.

Murphy, B., J. T. O'Connor, and T. L. O'Connor. "Willmar, Minnesota battles copper corrosion, Part 3." *Public Works,* 128(13), 1997c: 35–39.

NACE. *Corrosion Basics, An Introduction.* Houston, TX: National Association of Corrosion Engineers, 1984.

NACE. *Prevention and Control of Water-Caused Problems in Building Potable Water Systems,* TPC Publication 7. Houston, TX: National Association of Corrosion Engineers, 1980.

Neff, C. H., M. R. Schock, and J. I. Marden. *Relationships Between Water Quality and Corrosion of Plumbing Materials in Buildings,* EPA/600/S2-87/036. USEPA, July 1987.

Norton, C., et al. "Implementation of chloramination and corrosion control to limit microbial activity in the distribution system." In *Proc. AWWA Water Quality Technology Conference,* New Orleans, LA, 1995.

Notheis, M. J., et al. "Marble test solves corrosion mystery." In *Proc. AWWA Water Quality Technology Conference,* New Orleans, LA, 1995.

NSF International, Drinking Water System Components—Health Effects, Standard 61, *ANSI/NSF International Standard.* Ann Arbor, MI: NSF International, 1998.

Obrecht, M. F., and M. Pourbaix. "Corrosion of metals in potable water systems." *Jour. AWWA,* 59(8), 1967: 977–992.

O'Brien, J. E. "Reducing corrosivity and radon by the Venturi-Aeration Process." *Jour. NEWWA,* 109(2), 1995: 105–114.

Oliphant, R. *Dezincification by Potable Water of Domestic Plumbing Fittings: Measurement and Control.* Technical Report TR88, Water Research Centre, September 1978.

Oliphant, R. J. *Summary Report on the Contamination of Potable Water by Lead From Soldered Joints.* External 125E, Water Research Centre, November 1983.

Oliphant, R. J., and M. R. Schock. "Copper alloys and solder." In *Internal Corrosion of Water Distribution Systems.* 2nd ed., pp. 269–312. Denver, CO: AWWA Research Foundation/DVGW-TZW, 1996.

Olson, S. C. "Phosphate based corrosion inhibitors effects on distribution system regrowth." In *Proc. National Conference on Integrating Corrosion Control and Other Water Quality Goals,* Cambridge, MA, 1996.

Pankow, J. F. *Aquatic Chemistry Concepts.* Chelsea, MI: Lewis Publishers, 1991.

Parent, J. O. G., D. D. L. Chung, and I. M. Bernstein. "Effects of intermetallic formation at the interface between copper and lead-tin solder." *Jour. Met. Sci.,* 1988: 2564–2571.

Parkhurst, D. L., D. C. Thorstenson, and L. N. Plummer. *PHREEQE—A Computer Program for Geochemical Calculations,* Water-Resources Investigations 80–96, U.S. Geological Survey, November 1980.

Piron, D. J. *The Electrochemistry of Corrosion.* Houston, TX: National Association of Corrosion Engineers, 1991.

Pisigan, R. A., and E. Singley. "Influence of buffer capacity, chlorine residual, and flow rate on corrosion of mild steel and copper." *Jour. AWWA,* 79(2), 1987: 62–70.

Pisigan, R. A., Jr., and J. E. Singley. "Evaluation of water corrosivity using the Langelier Index and relative corrosion rate models." *Mat. Perf.,* 26(4), 1985: 26–36.

Plath, D. C., K. S. Johnson, and R. M. Pytokowicz. "The solubility of calcite—probably containing magnesium—in seawater." *Mar. Chem.,* 10, 1980: 9–29.

Postma, D., and R. Jakobsen. "Redox zonation: Equilibrium constraints on the Fe(III)/SQ_4-reduction interface." *Geochim. Cosmochim. Acta,* 60(17), 1996: 3169–3175.

Pourbaix, M. *Atlas of Electrochemical Equilibria in Aqueous Solution.* Oxford, UK: Pergamon Press, 1966.

Pourbaix, M. *Lectures on Electrochemical Corrosion.* New York: Plenum Press, 1973.

Rehring, J. P. *The Effects of Inorganic Anions, Natural Organic Matter, and Water Treatment Processes on Copper Corrosion.* Master of Science, Boulder, CO: University of Colorado at Boulder, 1994.

Rehring, J. P., and M. Edwards. "The effects of NOM and coagulation on copper corrosion." In *Proc. ASCE National Conference on Environmental Engineering,* Boulder, CO, 1994.

Reiber, S. "Copper plumbing surfaces: An electrochemical study." *Jour. AWWA,* 81(7), 1989: 114–122.

Reiber, S., et al. *A General Framework for Corrosion Control Based on Utility Experience.* Denver, CO: AWWA Research Foundation, 1997.

Reiber, S., J. F. Ferguson, and M. M. Benjamin. "An improved method for corrosion-rate measurement by weight loss." *Jour. AWWA,* 80(11), 1988: 41–46.

Reiber, S., R. A. Ryder, and I. Wagner. "Corrosion assessment technologies." In *Internal Corrosion of Water Distribution Systems,* 2nd ed., pp. 445–486. Denver, CO: AWWA Research Foundation/TZW, 1996.

Rompré, A., et al. "Impacts of corrosion control strategies on biofilm growth in drinking water distribution system." In *Proc. AWWA Water Quality Technology Conference,* Boston, MA, 1996.

Rose, M. "X-ray fluorescence analysis of waterborne scales." *American Laboratory,* 15, 1983: 46.

Rossum, J. R. "Dead ends, red water, and scrap piles." *Jour. AWWA,* 79(7), 1987: 113.

Rossum, J. R., and D. T. Merrill, Jr. "An evaluation of the calcium carbonate saturation indices." *Jour. AWWA,* 75(2), 1983: 95.

Ryznar, J. W. "A new index for determining amount of calcium carbonate scale formed by a water." *Jour. AWWA,* 36(4), 1944: 1944.

"Safe Drinking Water Act Amendments of 1996," Public Law 104-182, Sec. 188, August 6, 1996.

Schecher, W. D., and D. C. McAvoy. *MINEQL+: A Chemical Equilibrium Modeling System, Version 4.0 for Windows User's Manual.* Hallowell, ME: Environmental Research Software, 1998.

Schecher, W. D., and D. C. McAvoy. *MINEQL+: A Chemical Equilibrium Program for Personal Computers, User's Manual Version 3.0.* Hallowell, ME: Environmental Research Software, 1994.

Schecher, W. D., and D. C. McAvoy. "MINEQL+: A software environment for chemical equilibrium modeling." *Computers in Environmental and Urban Systems,* 16, 1994: 65–76.

Schenk, J. E., and W. J. Weber, Jr. "Chemical interactions of dissolved silica with iron (II) and iron (III)." *Jour. AWWA,* 76(2), 1968: 199–212.

Schock, M. R. "Behavior of asbestos-cement pipe under various water quality conditions: Part 2, Theoretical considerations." *Jour. AWWA,* 73(11), 1981a: 609.

Schock, M. R. "Causes of temporal variability of lead in domestic plumbing systems." *Environ. Monit. & Assess.,* 15, 1990a: 59–82.

Schock, M. R. "Internal corrosion and deposition control." In *Water Quality and Treatment: A Handbook of Community Water Supplies,* 4th ed., pp. 997–1111, A. W. W. Association, ed. New York: McGraw-Hill, 1990b.

Schock, M. R. "Reasons for corrosion control other than the Lead and Copper Rule." In *Small Systems Water Treatment Technologies: State-of-the-Art Workshop,* Marlborough, MA, 1998.

Schock, M. R. "Response of lead solubility to dissolved carbonate in drinking water." *Jour. AWWA,* 72(12), 1980: 695–704.

Schock, M. R. "Response of lead solubility to dissolved carbonate in drinking water." *Jour. AWWA,* 73(3), 1981b: 36.

Schock, M. R. "Temperature and ionic strength corrections to the Langelier Index (revisited)." *Jour. AWWA,* 76(8), 1984: 72.

Schock, M. R. "Treatment or water quality adjustment to attain MCL's in metallic potable water plumbing systems." In *Plumbing Materials and Drinking Water Quality: Proceedings of a Seminar,* Cincinnati, OH, 1985.

Schock, M. R. "Understanding corrosion control strategies for lead." *Jour AWWA,* 81(7), 1989: 88–100.

Schock, M. R., et al. "Replacing polyphosphate with silicate to solve problems with lead, copper and source water iron." In *Proc. AWWA Water Quality Technology Conference,* San Diego, CA, 1998.

Schock, M. R., R. W. Buelow, and P. J. Clar. "Evaluation and control of asbestos-cement pipe corrosion." In *Corrosion/81,* Toronto, ON, 1981.

Schock, M. R., and J. A. Clement. "Control of lead and copper with non-zinc orthophosphate." *Jour. NEWWA,* 1998 (in press).

Schock, M. R., and M. C. Gardels. "Plumbosolvency reduction by high pH and low carbonate-solubility relationships." *Jour. AWWA,* 75(2), 1983: 87–91.

Schock, M. R., and G. K. George. "Comparison of methods for determination of dissolve inorganic carbonate (DIC)." In *Proc. AWWA Water Quality Technology Conference,* Orlando, FL, 1991.

Schock, M. R., R. Levin, and D. D. Cox. "The significance of sources of temporal variability of lead in corrosion evaluations and monitoring program design." In *Proc AWWA Water Quality Technology Conference,* St. Louis, MO, 1988.

Schock, M. R., and D. A. Lytle. "The importance of stringent control of DIC and pH in laboratory corrosion studies: Theory and practice." In *Proc. AWWA Water Quality Technology Conference,* San Francisco, CA, 1994.

Schock, M. R., D. A. Lytle, and J. A. Clement. "Effect of pH, DIC, orthophosphate and sulfate on drinking water cuprosolvency," EPA/600/R-95/085. Cincinnati, OH: U.S. Environmental Protection Agency, Office of Research and Development, 1995b.

Schock, M. R., D. A. Lytle, and J. A. Clement. "Effects of pH, carbonate, orthophosphate and redox potential on cuprosolvency." In *NACE Corrosion/95,* Orlando, FL, 1995a.

Schock, M. R., D. A. Lytle, and J. A. Clement. "Modeling issues of copper solubility in drinking water." In *Proc ASCE National Conference on Environmental Engineering,* Boulder, CO, 1994.

Schock, M. R., and C. H. Neff. "Chemical aspects of internal corrosion: Theory, predictic and monitoring." In *Proc. AWWA Water Quality Technology Conference,* Nashville, TN, 1982.

Schock, M. R., and C. H. Neff. "Trace metal contamination from brass fittings." *Jour. AWWA,* 80(11), 1988: 47–56.

Schock, M. R., and S. C. Schock. "Effect of container type on pH and alkalinity stability." *Water Res.,* 16, 1982: 1455.

Schock, M. R., and K. W. Smothers. "X-ray, microscope, and wet chemical techniques: A complementary team for deposit analysis." In *Proc. AWWA Water Quality Technology Conference,* Philadelphia, PA, 1989.

Schock, M. R., and I. Wagner. "The corrosion and solubility of lead in drinking water." In *Internal Corrosion of Water Distribution Systems,* pp. 213–316. Denver, CO: AWWA Research Foundation/DVGW Forschungsstelle, 1985.

Schock, M. R., I. Wagner, and R. Oliphant. "The corrosion and solubility of lead in drinking water." In *Internal Corrosion of Water Distribution Systems.* 2nd ed., pp. 131–230. Denver, CO: AWWA Research Foundation/TQW, 1996.

Schott, G. J. "WATERPRO corrosion control treatment process evaluation program." In *Proc. AWWA Water Quality Technology Conference,* San Diego, CA, 1998.

Schreppel, C. K., D. W. Federickson, and T. A. Geiss. "The positive effects of corrosion control on lead levels and biofilms." In *Proc. AWWA Water Quality Technology Conference,* Denver, CO, 1997.

Shalaby, A. M., F. M. Al-Kharafi, and V. K. Gouda. "A morphological study of pitting corrosion of copper in soft tap water." *Corrosion-NACE,* 45(7), 1989: 536–547.

Sheiham, I., and P. J. Jackson. "Scientific basis for control of lead in drinking water by water treatment." *Jour. Inst. Water Engrs. & Scientists,* 35(6), 1981: 491–515.

Silverman, D. C. "Measuring corrosion rates in drinking water by linear polarization—Assumptions and watchouts." In *Proc. AWWA Water Quality Technology Conference,* New Orleans, LA, 1995.

Singley, J. E. "The search for a corrosion index." *Jour. AWWA,* 73(10), 1981: 529.

Singley, J. E., et al. "Corrosion of iron and steel," In *Internal Corrosion of Water Distribution Systems,* ch. 2, pp. 33–125. Denver, CO: AWWA Research Foundation/DVGW Forschungsstelle, 1985.

Singley, J. E., B. A. Beaudet, and P. H. Markey. "Corrosion manual for internal corrosion of water distribution systems," EPA/570/9-84-001. U.S. Environmental Protection Agency, Prepared for Office of Drinking Water by Environmental Science and Engineering, Inc., Gainesville, FL, 1984.

Snoeyink, V. L., and D. Jenkins. *Water Chemistry.* New York: John Wiley and Sons, 1980.

Snoeyink, V. L., and A. Kuch. "Principles of metallic corrosion in water distribution systems." In *Internal Corrosion of Water Distribution Systems,* pp. 1–32. Denver, CO: AWWA Research Foundation/DVGW Forschungsstelle, 1985.

Snoeyink, V. L., and I. Wagner. "Principles of corrosion in water distribution systems." In *Internal Corrosion of Water Distribution Systems,* 2nd ed., pp. 1–27. Denver, CO: AWWA Research Foundation/DVGW-TZW, 1996.

Sontheimer, H., W. Kolle, and V. L. Snoeyink. "The siderite model of the formation of corrosion-resistant scales." *Jour. AWWA,* November 1981: 572–579.

Spencer, C. M. "Aeration and limestone contact for radon removal and corrosion control." In *Proc. National Conference on Integrating Corrosion Control and Other Water Quality Goals,* Cambridge, MA, 1996.

Stumm, W. "Calcium carbonate deposition at iron surfaces." *Jour. AWWA,* 48(3), 1956: 300.

Stumm, W. "Investigation on the corrosive behavior of waters." *Jour. of the Sanitary Engineering Division, ASCE,* SA 6, 1960: 27–46.

Stumm, W., and J. J. Morgan. *Aquatic Chemistry,* 2nd ed. New York: John Wiley & Sons, 1981.

Stumm, W., and J. J. Morgan. *Aquatic Chemistry,* 3rd ed. New York: John Wiley & Sons, 1996.

Subramanian, K. S., J. W. Connor, and J. C. Meranger. "Leaching of antimony, cadmium, copper, lead silver, tin and zinc from copper piping with non-lead-based soldered joints." *J. Environ. Sci. Health,* A26(6), 1991: 911–929.

Subramanian, K. S., V. S. Sastri, and J. W. Connor. "Drinking water quality: Impact of non-lead-based plumbing solders." *Toxicol. Environ. Chem.,* 44(1), 1994: 11–20.

Taxèn, C. *Pitting Corrosion of Copper: An Equilibrium-Mass Transport Study,* KI Rapport 1996:8 E. Stockholm, Sweden: Swedish Corrosion Institute, November 1996.

Thompson, D. H. "General tests and principles." In *Handbook on Corrosion Testing and Evaluation,* W. H. Ailor, ed. New York: John Wiley and Sons, 1970.

Trussell, R. R. "Corrosion." In *Water Treatment Principles and Design,* pp. 392–434, C. E. James M. Montgomery, Inc., ed. New York: John Wiley & Sons, 1985.

Trussell, R. R. "Spreadsheet water conditioning." *Jour. AWWA,* 90(6), 1998: 70–81.

Trussell, R. R., L. L. Russell, and J. F. Thomas. "The Langelier Index." In *Proc. AWWA Water Quality Technology Conference,* Kansas City, MO, 1977.

Trussell, R. R., and I. Wagner. "Corrosion of galvanized pipe." In *Internal Corrosion of Water Distribution Systems,* pp. 71–129. Denver, CO: AWWA Research Foundation/TZW, 1996.

Trussell, R. R., and I. Wagner. "Corrosion of galvanized pipe." In *Internal Corrosion of Water Distribution Systems,* pp. 71–129. Denver, CO: AWWA Research Foundation/TZW, 1985.

Tuhela, L., L. Carlson, and O. H. Tuovinen. "Ferrihydrite in water wells and bacterial enrichment cultures." *Water Res.,* 26(9), 1992: 1159–1162.

U.S. Environmental Protection Agency. "Lead in school drinking water," EPA-570/9-89-001. Washington, D.C.: Environmental Protection Agency, Office of Water, 1989.

U.S. Environmental Protection Agency. "Suggested sampling procedures to determine lead in drinking water in buildings other than single family homes," EPA-570/B-88/0-15. Washington, D.C.: Environmental Protection Agency, Office of Water, 1988.

USEPA. "Seminar publication: Control of lead and copper in drinking water," EPA/625/R-93-001. Washington, D.C.: U.S. Environmental Protection Agency, Office of Research and Development, 1993.

van den Hoven, T. J. J. "How to determine and control lead levels in tap water: The Dutch approach." *Proc. AWWA WQTC,* Houston, TX, 1985.

van den Hoven, T. J. J. "A new method to determine and control lead levels in tap water." *Aqua,* 6, 1987: 315–322.

van den Hoven, T. J. J., and M. W. M. van Eekeren. "10. Corrosion and corrosion control in drinking water systems in the Netherlands." In *Corrosion and Corrosion Control in Drinking Water Systems; Proceedings from a Corrosion Workshop and Seminar,* Oslo, Norway, 1990.

van den Hoven, T. J. J., and M. W. M. van Eekern. *Optimal Composition of Drinking Water,* KIWA-report No. 100, KIWA N.V., October 1988.

van Gaans, P. F. M. "WATEQX—A restructured, generalized, and extended FORTRAN 77 computer code and database format for the WATEQ aqueous chemical model for trace element speciation and mineral saturation, for use on personal computers or mainframes." *Comp. & Geosci.,* 15(6), 1989: 843–887.

Vik, E. A., R. A. Ryder, I. Wagner, and J. F. Ferguson. "Mitigation of corrosion effects." In *Internal Corrosion of Water Distribution Systems,* 2nd ed., pp. 389–443. Denver, CO: AWWA Research Foundation/DVGW Forschungsstelle, 1996.

Vitins, A., et al. "Investigation of the electrochemical properties of copper in potable water." *Latvijas Kimijas Zurnals,* 1, 1994: 66–77.

Wagner, I. "Effect of inhibitors on corrosion rate and metal uptake in drinking water systems." In *Proc. AWWA Seminar on Internal Corrosion Control Developments and Research Needs,* Los Angeles, CA, 1989.

Walker, A. P. "The microscopy of consumer complaints." *Jour. Inst. Water Engrs. & Scientists,* 37(3), 1983: 200–214.

Watkins, K. G., and D. E. Davies. "Cathodic protection of 6351 aluminum alloy in sea water: Protection potential and pH effects." *Br. Corros. J.,* 22(3), 1987: 157–161.

Weber, W. J., Jr., and W. Stumm. "Mechanism of hydrogen ion buffering in natural waters." *Jour. AWWA,* 55(12), 1963: 1553–1578.

Werner, W., H.-J. Groß, and H. Sontheimer. "Corrosion of copper pipes in drinking water installations." *Translation of: gwf-Wasser/Abwasser,* 135(2), 1994: 1–15.

Weyl, P. K. "The carbonate saturometer." *Jour. Geology,* 69, 1961: 32–43.

Whipple, G. C. "Decarbonation as a means of removing the corrosive properties of public water supplies." *Jour. NEWWA,* 27, 1913: 193.

Williams, J. F. "Corrosion of metals under the influence of alternating current." *Materials Protection,* 5(2), 1986: 52–53.

Wysock, B. M., et al. "Statistical procedures for corrosion studies." *Jour. AWWA,* 87, 1995: 5.

Wysock, B. M., et al. "Statistical procedures for corrosion studies." In *Proc. AWWA Water Quality Technology Conference,* Toronto, ON, 1992.

Wysock, B. M., M. R. Schock, and J. A. Eastman. "A study of the effect of municipal ion-exchange softening on the corrosion of lead, copper and iron in water systems." In *Proc. AWWA Annual Conference,* Philadelphia, PA, 1991.

CHAPTER 18
MICROBIOLOGICAL QUALITY CONTROL IN DISTRIBUTION SYSTEMS

Edwin E. Geldreich, M.S.
Consulting Microbiologist
Cincinnati, Ohio

Mark LeChevallier, Ph.D.
Research Director
American Water Works Service Co.
Voorhees, New Jersey

The purpose of a water supply distribution system is to deliver to each consumer safe drinking water that is also adequate in quantity and acceptable in terms of taste, odor, and appearance. Historically, the initial network of pipes was a response to present community needs that eventually created a legacy of problems of inadequate supply and low pressure as the population density increased (Frontinus, 1973; Baker, 1981). To resolve the problems caused by increasing water demand along the distribution route, reservoir storage was created. Pressure pumping to move water to far reaches of the supply lines and standpipes was incorporated to afford relief from surges of pressure. In some areas, population growth exceeded the capacity of a water resource, so other sources of water were incorporated and additional treatment plants were built to feed into the distribution network. Another response was to consolidate neighboring water systems and interconnect the associated distribution pipe networks.

GENERAL CONSIDERATIONS FOR CONTAMINATION PREVENTION

Today (1999), community expansion plans are more fully developed and include the engineering of utility service so that careful consideration is given to meeting future projected water supply needs. Advanced planning provides the opportunity to design the pipe network as a grid with a series of loops to avoid dead ends. The objective is to produce a circulating system capable of supplying high quality water to all areas while at the same time permitting any section to be isolated for maintenance, repair, or decontamination without interrupting service to all other areas.

To ensure delivery of a high-quality municipal potable water supply to each consumer, managers of public water supply systems must be continually vigilant for any intrusions of contamination or occurrences of microbial degradation in the distribution network. This job is complicated by the very nature of the distribution system—a complex network of mains, fire hydrants, valves, auxiliary pumping, chlorination substations, storage reservoirs, standpipes, and service lines. Following the intrusion of microbial contamination, any of these component parts may serve as a habitat suitable for colonization by certain microorganisms in the surviving flora. The persistence and possible growth of organisms in the pipe network are influenced by a variety of environmental conditions that include physical and chemical characteristics of the water, system age, type of pipe materials, and the availability of sites suitable for colonization (often located in slow-flow sections, dead ends, and areas of pipe corrosion activity).

ENGINEERING CONSIDERATIONS FOR CONTAMINATION PREVENTION

Many public water utilities make substantial efforts to expand their distribution networks to keep up with continuing suburban growth. Urban renewal and highway construction projects may at times require the relocation or enlargement of portions of the distribution network. Corrosion, unstable soil, faulting, land subsidence, extreme low temperatures, and other physical stresses often cause line breaks and necessitate repair or replacement of pipe sections. To avoid possible bacteriological contamination of the water supply during these construction projects, a rigorous protective protocol must be followed.

Distribution System Construction Practices

The American Water Works Association has developed standard procedures that are used, with variations, by most of the water supply industry for disinfecting water mains (AWWA, 1986). In essence, these recommendations recognize six areas of concern: (1) protection of new pipe sections at the construction site; (2) restriction on the use of joint-packing materials; (3) preliminary flushing of pipe sections; (4) pipe disinfection; (5) final flushing; and (6) bacteriological testing for pipe disinfections.

Pipe sections, fittings, and valves stockpiled in yard areas or at the construction site should be protected from soil, seepages from water or sewer line leaks, stormwater runoff, and habitation by pets and wildlife (Becker, 1969; Russelman, 1969). Each of these contamination sources may deposit significant fecal material in the interior of pipe sections awaiting installation. Septic tank drain fields, subsurface water in areas of poor drainage or high water table, and seasonal or flash flooding may also introduce significant contamination into unprotected pipe sections. Fecal material introduced by contaminating sources may become lodged in pipe fittings and valves. Thus, such sites become protected habitats from which coliforms and any associated pathogens in the contaminated material may be released into the bulk flow of water supply. Common-sense protective measures include providing end covers for these pipe materials, drainage of standing water from trenches before pipeline assembly, and flushing of new construction or line repairs to remove all visible signs of debris and soil (Suckling, 1943; Davis, 1951).

Pipe-Joining Materials

Gasket seals of pipe joints can be a source of bacterial contamination in new pipes (Hutchinson, 1971). Annular spaces in joints provide a protected habitat for continued survival and possible multiplication of a variety of bacteria in the distribution network. In these instances, although the heterotrophic plate count (HPC) and any coliform occurrences may be temporarily reduced by main disinfection, bacteria soon become reestablished from the residual population harbored in some joint-packing materials. In this regrowth process, the variety of organisms and dominance of strains change, often restructuring the bacterial flora to a predominant population of *Pseudomonas aeruginosa, Chromobacter* strains, *Enterobacter aerogenes,* or *Klebsiella pneumoniae.* Thus, where the pattern of organisms present is predominantly one bacterial strain, a search for a protective habitat in joint-packing materials or impacted material in pipe sections should be made (Calvert, 1939; Adam and Kingsbury, 1937; Taylor, 1950, 1967–1968; Schoenen, 1986; Schubert, 1967). Nonporous materials such as molded or tubular plastic, rubber, and treated paper products are preferable. Lubricants used in seals must be nonnutritive to avoid bacterial growth in protected joint spaces. Efforts to develop bacteriostatic lubricants have resulted in the inclusion of various quaternary ammonium compounds that minimize contamination from pipe joint spaces (Hutchinson, 1974). The National Sanitation Foundation (NSF International, Ann Arbor, Michigan) has proposed a test method for evaluating the biological growth potential of materials that contact drinking water (Bellen et al., 1993). Although the method has not yet received final approval, some manufactures have submitted their products for testing, and in many cases the results are available from NSF upon request.

Water Supply Storage Reservoirs

Water use in a community varies continuously as a reflection of the activities of the general public and local industries. While industrial uses of potable water may be more predictable, expecting water treatment operations to gear production to those frequent and sudden changes in water demand from all consumers is impractical. For this reason storage reservoirs are an essential element of the distribution network. These water supply reserves supplement water flows in distribution during periods of fluctuating demand on the system, providing storage of water during off-peak periods; equalize operational water pressures; and augment water supply from production wells that must be pumped at a uniform rate. Storage reservoirs also provide a protective reserve of drinking water to guard against discontinuance of water treatment during chemical spills in the source water, flooding of well fields, transmission line breaks, and power failures. An important secondary consideration is providing adequate storage capacity for fire emergencies.

Finished water reservoirs may be located near the beginning of a distribution system, but most often they are situated near the extremities of the system. Local topography plays an important part in determining the use of low-level or high-level reservoirs. Underground storage basins are usually formed by excavation, while ground-level reservoirs are constructed by earth embankment. The sides and bottom of such reservoirs are lined with concrete, Gunite, asphalt, or with a plastic sheet to prevent or reduce water loss in storage (Harem, Bielman, and Worth, 1976). In earthquake zones, reinforced concrete or a series of flat-bed steel compartments is mandatory. Reinforced concrete is often selected because of its minimal rate of deterioration from water contact. Elevated storage tanks and standpipes are constructed of steel, with an interior coating applied to retard corrosion (Wade, 1974).

Care must be taken to prevent potential contamination of the high-quality water entering storage reservoirs and standpipes. One area of concern is in the application of coating compounds over the inner walls of tanks to maintain tank integrity. Organic polymer solvents in bituminous coating materials may not entirely evaporate even after several weeks of ventilation. As a consequence, the water supply in storage may become contaminated from the solvent-charged air and from contact at the sidewall. Some of these compounds are assimilable (biodegradable) organics that support growth of heterotrophic bacteria during warm water periods (Schoenen, 1986; Thofern, Schoenen, and Tuschewitzki, 1987; Mackle, 1988; Bernhardt and Liesen, 1988). Liner materials, also used to prevent water loss, may contain bitumen, chlorinated rubber, epoxy resin, or a tar-epoxy resin combination that will eventually be colonized by microbial growth and slime development (Schoenen, 1986). PVC film and PVC coating materials are other sources of microbial activity. Nonhardening sealants (containing polyamide and silicone) used in expansion joints should not be overlooked as a possible source of microbial habitation.

Water volumes in large reservoirs mix and interchange slowly with water that is actually distributed to service lines. In some instances, water storage may become stratified and experience a complete mixing only after a sudden change in ambient air temperature or as a consequence of a significant water loss in the system caused by a major main break or intensive system flushing (Geldreich et al., 1992; Clark et al., 1996). Standpipes, in contrast, provide a fluctuating storage of water during a downsurge, thereby providing surge relief in the system. Abrupt changes in water flow that sometimes occur during surge relief can disturb sediment deposits, moving viable bacteria from biofilm sites into the main flow of water.

Reservoirs of treated water should be covered whenever possible to avoid contamination of the supply from bird excrements (Alter, 1954; Fennel, James, and Morris, 1974), air contaminants, and surface water runoff. The health concern with bird excrement is that this wildlife may be infected with *Salmonella* and protozoans pathogenic to man. Within the wildlife population in every area (as is true for any community of people), a persistent pool of infected individuals exists that sheds pathogenic organisms through fecal excretions. Seagulls are scavengers and often are found at landfill locations and waste discharge sites, searching for food which is often contaminated by a variety of pathogens. At night, birds often return inland to roost in aquatic areas, such as source water impoundments and open finished water reservoirs, thereby introducing pathogens through their fecal excrements (Fennel, James, and Morris, 1974). Pigeons roosting in elevated storage tanks were believed to be the source of *Salmonella typhimurium* that contaminated the water supply of a small community in Missouri (Geldreich et al., 1992; Clark et al., 1996).

Air pollution contaminants and surface water runoff can contribute dirt, decaying leaves, lawn fertilizers, and accidental spills to a water supply that is not covered. Such materials increase productivity of the water by providing support to food chain organisms and nitrogen-phosphate requirements for algal blooms. This degrades the treated water quality.

Covered distribution system storage structures also are subject to occasional contamination because of air movement in or out of the vents as a result of water movement in the structure. During air transfer, the covered reservoir is exposed to fallout of dust and air pollution contaminants from the inflowing air. Vent ports or conduits from the service reservoir to the open air should be equipped with suitable air filters to safeguard the water quality from airborne contaminants. Such air filters must be replaced periodically to prevent a serious loss of air transfer or create an undesirable vacuum. Birds and rodents may also gain access through air vents that have defective screen protection. Bird or rodent excrement around the vents may enter the water supply and become transported into the distribution system before dilution and residual disinfection are able to dissipate and inactivate the associated organisms.

FACTORS CONTRIBUTING TO MICROBIAL QUALITY DETERIORATION

Factors contributing to deterioration of microbial quality may be associated with source water quality, treatment processes, or distribution network operation and maintenance. The following sections review each of these areas.

Source Water Quality

Bacteria in distributed water may originate from the source water. High-quality groundwater can be characterized as containing <1 coliform per 100 mL and a heterotrophic bacterial population that is often very sparse (less than 10 organisms per mL), even in waters that reach the growth-stimulating temperature of 15°C or more (Alson, 1982). These microbial qualities show little fluctuation, because the groundwater aquifer is protected from surface contamination. Some groundwaters, however, are not insulated from surface contamination (Allen and Geldreich, 1975). Agricultural fertilizer runoff can contribute nitrates and improperly isolated landfills may introduce a variety of organics, many of which are biodegradable. In such situations, bacterial populations in the groundwater become excessive, resulting in 1000 to 10,000 heterotrophic bacteria per mL. Groundwaters containing a high concentration of iron or sulfur compounds provide nutrients for a variety of nuisance bacteria that may become numerous and restrict water flow from a well. Where groundwaters are poorly protected from contamination by stormwater runoff and wastewater effluents, coliforms and pathogens (bacterial, viral, protozoan) may be introduced into the distributed water unless a treatment barrier is provided (Craun, 1985).

Surface water sources are subject to a variety of microbial contaminants introduced by stormwater runoff over the watershed and the upstream discharges of domestic and industrial wastes. While impoundments and lakes provide water volume and buffering capacity to dilute bacterial contamination, counterproductive factors must be considered. Stratification/destratification of lake waters, decaying algal blooms, and bacterial nutrient buildup contribute to deteriorating water quality that may interfere with treatment effectiveness.

Treatment Processes

Water supplies using a single barrier (disinfection) for surface water treatment will not prevent a variety of organisms (algae, protozoa, and multicellular worms and insect larvae) from entering the distribution system (Allen, Taylor, and Geldreich,

1980). While many of these organisms are not immediately killed by disinfectant concentrations and contact times (C·T values) that control coliforms and viruses (Hoff, 1986), they eventually die because of lack of sunlight (algae) or adverse habitat (multicellular worms and insect larvae). Disinfection is also less effective on a variety of environmental organisms that include spore-forming organisms (*Clostridia*), acid-fast bacteria, gram-positive organisms, pigmented bacteria, fungi, yeast, and protozoan cysts. All of these more resistant organisms can be found in the pipe environment (Haas, Meyer, and Fuller, 1983; Bonde, 1977; Rosenzweig, Minnigh, and Pipes, 1983; Niemi, Knuth, and Lundstrum, 1982; Reasoner, Blannon, and Geldreich, 1989).

Filtration is an important treatment barrier for protozoan cysts (*Entamoeba, Giardia, Cryptosporidium*), being more effective than the usual disinfectant concentration and contact times applied in processing raw water (Logsdon, 1987). Improperly operated filtration systems have been responsible for releasing concentrated numbers of entrapped cysts (*Giardia* and *Cryptosporidium*) as a result of improper filter backwashing procedures or filter bypasses and channelization within the filter bed (Amirtharajah and Wetgstein, 1980). Filter sand may become infested with nematodes from stream or lake bottom sediments that shed into the process water and pass into the distribution system. While nematodes are not pathogenic, they can shelter viable bacteria ingested from source water or filter media beds, and if these food-chain organisms are not digested quickly, provide a passage for escaping survivors to reach the distribution system (Tracy, Carnasena, and Wing, 1966; Levy et al., 1984).

Properly operated water treatment processes are effective in providing a barrier to coliforms and pathogenic microorganisms reaching the distribution system. This does not, however, preclude the passage of all nonpathogenic organisms through the treatment train. Investigation of heterotrophic bacterial populations revealed that a 4-log (99.99 percent) or better reduction can occur through conventional treatment processes (raw water storage, coagulation, settling, rapid-sand filtration, and chlorination) for many of these organisms (Reasoner, Blannon, and Geldreich, 1989). A less significant reduction occurs among the subpopulation of pigmented organisms in the heterotrophic flora, however, so that these organisms may become the predominant bacteria after processing and then the dominant strains in distributed water.

Application of powdered carbon, granular activated carbon (GAC) filtration, or biological activated carbon (BAC) treatment introduces other opportunities for bacteria to enter the microbial community of processed water. Highly adsorptive granular or powdered carbon is often used in specific situations for removal of organics, including taste- and odor-causing substances, and is changed periodically when the carbon bed reaches near-saturation for expected removal rates. In contrast, the biological activated carbon process depends on the establishment of a permanent biofilm for degradation of conversion products created by ozonation of recalcitrant organics in process water, as well as of biodegradable organic matter already in the water.

Several coliform species (*Klebsiella, Enterobacter,* and *Citrobacter*) have been found to colonize GAC filters, grow during warm water periods, and discharge into the process effluent (McFeters, Kippen, and LeChevallier, 1986; Camper et al., 1986; Camper et al., 1987; Stewart, Wolf, and Means, 1990). Activated carbon particles have also been detected in finished water from several water plants using powdered activated carbon or GAC treatment. Over 17 percent of finished water samples examined from nine water treatment facilities contained activated carbon particle fines colonized with coliform bacteria (Ridgway and Olson, 1981). These observations confirm that activated carbon fines provide a transport mechanism by which microorganisms penetrate treatment barriers and reach the distribution system. Other mechanisms that could be involved in protected transport of bacteria into the distribution system include release of aggregates or clumps of organisms from colonization sites in GAC/BAC filtration and by passage with unsettled coagulants.

Furthermore, heterotrophic bacterial densities in distributed water from a full-scale treatment train using GAC (Symons et al., 1981) were found to be significantly higher (Table 18.1) than in water from a similar full-scale treatment train that did not employ GAC (Haas, Meyer, and Paller, 1983). Upon entering the pipe network, persistence and growth of these organisms will be influenced by the same factors that also affect disinfectant effectiveness: habitat locations, water temperature, pH, and assimilable organic carbon concentrations (Haas, Meyer, and Paller, 1983; Rittmann and Snoeyink, 1984).

Distribution Network Operation and Maintenance

Because the public health concern for the microbial quality of drinking water has until recently been based solely on limiting total coliform occurrence (EPA, 1976), the acceptance of new or repaired mains has depended only on a laboratory report that no coliforms are detected in water held in the new pipe sections. A more rigorous check on installed pipe cleanliness would include examination of water in the pipe section for elevated heterotrophic bacterial densities in addition to total coliforms (Geldreich et al., 1972; AWWA, 1987). The HPC in this situation reflects the myriad of soil organisms that could have been introduced into the pipe section during construction or repair. Soil deposits in new pipe sections may not only introduce a variety of heterotrophic bacteria to the distribution network but also provide some measure of protection to associated bacteria from disinfection exposure. Some of the poor disinfection results attributed to chlorine applied in these situations may be traced to excessively dirty line sections or joints.

In Halifax, Nova Scotia, a new supply line was found to contain pieces of wood used during construction work embedded in some pipe sections (Martin et al., 1982).

TABLE 18.1 Treated Water Bacterial Populations Following Various Water Treatment Processes Using Standard Plate Medium or R-2A Medium with Extended Incubation Times (organisms/mL)*

	Lime-Softened water			Sand-Filter effluent			GAC adsorber effluent		
Sampling data	SPC, 2 days	SPC, 6 days	R-2A, 6 days	SPC, 2 days	SPC, 6 days	R-2A, 6 days	SPC, 2 days	SPC, 6 days	R-2A, 6 days
Initial	120	350	510	890	1,200	1,500	< 1	140	220
7	31	202	510	820	22,000	35,000	1	24,000	95,000
14	7	7	130	< 1	1,200	9,400	< 1	600	4,400
21	7	18	150	2,200	2,500	33,000	< 1	5,200	16,000
28	3	39	530	700	7,800	67,000	1	11,000	55,000
35	< 1	490	330	100	6,000	25,000	< 1	12,000	74,000
42	70	120	1,700	1,200	71,000	22,700	N.D.	56,000	52,000
49	9	1,200	23	5,000	41,000	3,000	80	4,200	100
56	< 1	10	< 1	< 1	700	12,000	N.D.	1,900	50,000
63	29	190	170	170	2,000	3,000	N.D.	5,000	48,000

* All cultures incubated at 35° C. SPC, standard plate count (SPC agar); N.D., not done.
Source: Data revised from Symons et al., 1981.

An environmental *Klebsiella* associated with the wood forms adjusted to the distribution pipe environment and colonized the exposed wood surfaces. The flow of water transported this coliform into the bulk flow, resulting in consistently unsatisfactory test results. Since the wood debris was difficult to remove from the pipeline, the problem was resolved by the addition of more than 5 mg/L lime to the process water, which elevated the water pH to 9.1 in the distribution system. At this pH, *Klebsiella* were either inactivated or entrapped in the pipe sediment, and the problem was eliminated.

Upon completion of a new pipeline or after emergency repairs are made to a line break, flushing water through the pipe section at a minimum velocity of 10 ft/s (76.2 cm/s) to remove soil particles is advisable (Buelow et al., 1976). In lines with diameters of 16 in (4.1 cm) or more, this velocity may not be attainable or may be ineffective, requiring polypig or foam swab applications to be considered. Following flushing, disinfectant should be introduced into the new sections and the water held for 24 to 48 h to optimize line sanitation. Bacteriological tests for stressed coliforms (McFeters, Kippin, and LeChevallier, 1986) and the HPC should then be performed. If the results of these tests are satisfactory (<1 coliform/100 mL; <500 HPC/mL), the line may be placed in service. If not, the line should again be flushed and refilled with distributed water dosed with 50 mg/L free available chlorine. Chlorine levels should not decrease below 25 mg/L during the 24-h holding period before the line is flushed and bacteriological testing is repeated. In pipes free of extraneous debris, free available chlorine (1 to 2 mg/L), potassium permanganate (2.5 to 4.0 mg/L), or copper sulfate (5.0 mg/L) have been used to meet coliform requirements (Martin et al., 1982; Harold, 1934; Hamilton, 1974). Only free available chlorine, however, was found to eliminate large numbers of heterotrophic bacteria (Buelow et al., 1976).

MICROBIAL QUALITY OF DISTRIBUTED WATER

Because of the difficulties in isolating and identifying a broad spectrum of organisms with widely differing growth requirements, little in-depth information is available in the literature on the identity of all heterotrophic organisms found in water supplies. However, recent research into fatty acid identification of these organisms is beginning to yield a better characterization of the many ill-defined bacteria in distribution water. In the analysis of cellular fatty acids the cells are lysed, and the constituent acids methylated and then identified by gas chromatography (Briganti and Wacker, 1995). Bacteria are identified by comparing the profile of fatty acids from the drinking water organism isolates to a library of profiles from known organisms. Previous studies have generally been limited to identification of those organisms that were associated with consumer complaints about taste, odor, and color (Hutchinson and Ridgway, 1977). Other studies have explored spoilage problems in the food, beverage, cosmetic, and drug industries, which use large quantities of potable water in production processes (Alson, 1967; Tenenbaum, 1967; Dunnigan, 1969; Borgstrom, 1978). The following section is a brief profile of organisms found in distribution systems.

Bacterial Profiles

Heterotrophic organisms in water supplies most often originate in the source water, survive the rigors of treatment processes, and adapt to the environment of the water distribution network (Geldreich, Nash, and Spino, 1977). Trace organic nutrients already present in the bulk water or accumulated in reservoir and pipe sediments can support a diverse population (Table 18.2) of surviving organisms (heterotrophs). Habitat sites that are successfully colonized almost always invoke the mixed growth of organisms that are attached to each other or to particles, sediments, and porous structures in the pipe tubercles or sediments. The spectrum of organisms may include many gram-negative bacteria (such as coliforms and *Pseudomonas*), gram-positive organisms, sporeformers, acid-fast bacilli, pigmented organisms, actinomycetes, fungi, and yeast. In addition, various protozoans and nematodes that feed on these microbial populations can be found in protected areas of low-flow sections or dead ends.

Coliforms. Total coliform bacteria counts are used primarily as a measure of water supply treatment effectiveness and as a measure of public health risk. These gram-negative bacteria are occasionally found in water supplies. Data in Table 18.3 shows the wide range of coliform species that are encountered (Geldreich, Nash, and Spino, 1977; LeChevallier, Seidler, and Evans, 1980; Olson and Hanami, 1980; Herson and Victoreen, 1980; Reilly and Kippin, 1981; Clark, Burger, and Sabatinos, 1983; Staley, 1983). While coliform bacteria are chlorine sensitive, they may be protected from inactivation by associated particles originating in source water turbidity, activated carbon fines released from GAC filtration (Camper et al., 1985), and inorganic sediments in the contact basin, as well as by inadequate conditions for disinfection action (contact time, water pH, and temperature). Coliforms may also enter the distribution system as injured organisms that pass through treatment barriers or by introduction into water line breaks, cross-connections, uncovered finished water storage, or low water pressure (<20 psi). For these reasons, more attention should be given to the search for injured coliforms in an effort to provide a more sensitive measurement of water quality (LeChevallier and McFeters, 1985; McFeters, 1990; Bucklin, McFeters, and Amirtharaja, 1991).

Coliform colonization of the pipe network may occur in areas where porous sediments accumulate (Allen, Taylor, and Geldreich, 1980; Tuovinen et al., 1980). These sediments develop from the action of corrosion or are the accumulation of particles that passed through the treatment works because of inadequate processing and then settled in slow-flow and dead-end areas. Such sites are attractive to bacterial colonization because these deposits adsorb trace nutrients from the passing water and provide numerous surface areas for bacterial attachment against the flow of water. Among the coliform bacteria, *Klebsiella pneumoniae, Enterobacter aerogenes, Enterobacter cloacae,* and *Citrobacter freundii* are the most successful colonizers. Encapsulation by these coliforms provides protection from the effects of chlorine or other disinfectants. Once total coliforms become established in an appropriate habitat, growth can occur and result in occasional sloughing of cells into the flowing water. This condition can persist until either shearing effects of water hydraulics limit colony growth or elevated disinfectant residuals penetrate the protective habitat and inactivate the microbial population.

Antibiotic-Resistant Bacteria. Heterotrophic bacteria in a water supply that are resistant to one or more antibiotics may pose a health threat if these strains are opportunistic pathogens or serve as donors of the resistant factor to other bacteria that could be pathogens. Antibiotic-resistant (R factor) bacteria may originate in surface water sources used for public water supplies (Armstrong, Calomiris, and Seidler, 1982; Bedard et al., 1982). Polluted waters acquire bacteria with R factors from the fecal wastes of man and domestic animals in wastewater effluents, and stormwater runoff from farm pasturelands and feedlots. Farm animals in particular may

TABLE 18.2 Organisms in the Standard Plate Count Population of Three Water Supplies

Water plant filter effluent and clearwell		Distribution water	
Significant categories	Organisms isolated	Significant categories	Organisms isolated
Total coliforms		Total coliforms	
	Klebsiella pneumoniae		*Klebsiella pneumoniae*
	Enterobacter cloacae		*Enterobacter cloacae*
	Erwinia herbicola		*Erwinia herbicola*
			Enterobacter aerogenes
			Escherichia coli
			Aeromonas hydrophila
			Citrobacter freundii
Coliform antagonists		Coliform antagonists	
	Pseudomonas fluorescens		*Pseudomonas fluorescens*
	Pseudomonas maltophila		*Pseudomonas maltophila*
	Flavobacterium sp.		*Flavobacterium* sp.
			Pseudomonas cepacia
			Pseudomonas putida
			Pseudomonas aeruginosa
			Bacillus sp.
			Actinomycetes sp.

Opportunistic pathogens		Opportunistic pathogens	
	Pseudomonas maltophila		*Pseudomonas maltophilia*
	Klebsiella pneumoniae		*Klebsiella pneumoniae*
	Moraxella sp.		*Moraxella* sp.
	Staphylococcus (coagulase +)		*Staphylococcus* (coagulase +)
			Pseudomonas aeruginosa
			Klebsiella rhinoscheromatis
			Serratia liquefaciens
			Serratia marcescens
Category not established		Category not established	
	Acinetobacter calcoaceticus		*Acinetobacter calcoaceticus*
	Neisseria flavescens		*Streptococcus* sp.
			Bacillus sp.
			Corynebacterium sp.
			Micrococcus sp.
			Nitrococcus sp.

Source: Data from Geldreich et al., 1977.

TABLE 18.3 Coliforms Identified in 111 Public Water Supply Distribution Systems*

Citrobacter	*Escherichia*
C. freindii	*E. coli*
C. diversus	
Enterobacter	*Klebsiella*
Enter. aerogenes	*K. pneumoniae*
Enter. agglomerans	*K. rhinoscleromatis*
Enter. cloacae	*K. oxytoca K. ozaenae*

* Published data (Geldreich et al., 1977; LeChevallier et al., 1980; Olson and Hanami, 1980; Herson and Victoreen, 1980; Reilly and Kippin, 1981; Clark et al., 1983; Staley, 1983) from various distribution systems in six states (United States) and Ontario Province (Canada).

receive continuous doses of antibiotics in animal feed and become constant generators of a variety of antibiotic-resistant bacteria. Although treatment processes inactivate or remove antibiotic-resistant organisms (*Aeromonas, Hafnia,* and *Enterobacter*) in the source water, a shift of this transmissible factor to other heterotrophic bacteria (*Pseudomonas/Alcaligenes* group, *Acinetobacter, Moraxella, Staphylococcus,* and *Micrococcus*) may occur (Armstrong et al., 1981).

Water supply treatment processes apparently act as a mixing chamber for R factor transfers with surviving organisms, which then acquire multiple resistances to different antibiotics. Many of the transformations occur in the biofilm established on activated carbon and sand filters (Armstrong et al., 1981). The disinfection process may also have a major impact on the selection of drug-resistant bacteria. The reason for the common occurrence of streptomycin resistance among bacteria that survive chlorination is not known. Multiple antibiotic-resistant bacteria passing through water treatment are more tolerant to metal salts (i.e., $CuCl_2$, $Pb(NO_3)_2$, and $ZnCl_2$) (Armstrong et al., 1981). Examination of bacteria for multiple antibiotic-resistance from two sites in a distribution system indicates a dynamic state of fluctuation (16.7 percent R factor organisms at one site and 52.4 percent at the other location). In a typical population of 100 heterotrophic bacteria per mL of water from the distribution system, 40 to 70 of these organisms could be expected to have some antibiotic-resistance factors (El-Zanfaly, Kassein, and Badr-Eldin, 1987). What health risk this represents, particularly when the heterotrophic bacterial population is above the 500 organisms or more per mL limit suggested for potable water, is not clearly understood.

Mycobacteria. Origins of mycobacteria in water supply can be found in source waters, open finished water reservoirs, soil contaminants introduced in groundwaters, and line repairs and new line construction. Densities of environmental strains in groundwater supplies ranged from 10 to 500 organisms per 100 mL, while densities of these acid-fast bacteria in polluted surface waters may approach 10^4 organisms per 100 mL. In water supplies processed from surface waters, mycobacteria reduction ranges from 60 organisms per 100 mL to only a few per liter. Those aquatic mycobacteria of health concern (*Mycobacterium avium, M. gordonae, M. flavescens, M. fortuitum, M. chelonae,* and *M. phlei*) may colonize susceptible humans (immunocompromised patients, surgery cases, individuals on kidney dialysis) by a variety of routes including water supply (Haas, Meyer, and Fuller, 1983; duMoulin and Stottmeier, 1986; duMoulin et al., 1985).

In water treatment, the most significant reductions for these organisms occur during sand filtration. For example, during an 18-month study of two water systems, reductions in the concentration of acid-fast bacteria by rapid sand filtration ranged from 59 to 74 percent. Final disinfection, including the presence of free residual chlorine, did not have a statistically significant effect on the residual densities of mycobacteria leaving the treatment plant (Haas, Meyer, and Fuller, 1983; Pelletier, duMoulin, and Stottmeier, 1988). Even the presence of a free chlorine residual at a low pH (5.9 to 7.1) did little to reduce the number of these organisms in the distribution systems.

The ability of mycobacteria to survive in water distribution systems may be influenced by the protective waxy nature of the cell wall, which helps these organisms resist 1.5 mg free chlorine over 30 m contact time. Moreover, some increase in numbers were noted in the pipe environment at the ends of the distribution lines, where chlorine residual disappeared, increased total organic carbon concentrations occurred (providing bacteria nutrients), and pipe sediments or tuberculation (sites for bacterial attachment) accumulated. Further amplification of the acid-fast bacteria can be expected in some building plumbing systems and their associated attachments (Bullin, Tanner, and Collins, 1970; Haas, Meyer, and Paller, 1983).

Pigmented Bacteria. A characteristic of some bacteria that may be present in distributed water is the ability to form brightly colored pigments. Little is known about their health significance from ingestion; however, some strains have been the cause of gastroenteritis while others have been associated with pyrogenic reactions and septicemia (Quarles et al., 1974). The seasonal occurrence of pigmented bacteria in a treated water supply suggests that these bacteria originated from the source water supply at some previous point in time, or from line breaks and repairs to the distribution lines (Reasoner, Blannon, and Geldreich, 1989).

More pigmented bacteria were found at a distribution sampling site 25 mi (40 km) from the water treatment plant than in river source water, the presedimentation basin, after flocculation, in chlorinated influent to rapid sand filters, or in the filtered water. This observation suggests that pigmented bacteria surviving treatment became adapted to the distribution environment and grew when conditions were favorable.

The proportion of yellow, pigmented bacteria in the source water was lowest in the autumn and highest in the summer. The mean proportion of pigmented bacteria was calculated to be 26 percent of the heterotrophic population. Orange-pigmented bacteria comprise less than 10 percent of the plate count in any season (average 5.8 percent) and pink or red organisms, less than 3 percent. Pigmented bacteria such as purple, black, and brown organisms ranged from 0.2 to 1.5 percent.

In contrast, at the distribution sampling site, pigmented bacteria comprised from 65 to nearly 90 percent of the heterotrophic population, depending on the season. Lowest and highest percentages occurred in the winter (5 percent) and autumn (70 percent), respectively. Orange-pigmented bacteria, on the other hand, appeared as 82 percent of the total pigmented population in the winter and only 10 percent of the population in the autumn. Pink bacteria ranged from <1 percent (winter, spring, and fall) to 9 percent (summer). Other pigmented bacteria (purple, black, and brown) were rarely encountered in the distributed water. A mixed population of pigmented bacteria was generally present in the distributed water, with surges to dominance among these groups when environmental factors were favorable. Wolfe, Ward, and Olson (1985) found that red-pigmented bacteria were resistant to 0.75 mg/L free chlorine for 30 to 60 m, but sensitive to 1.0 mg/L chloramines for 60 m. Fatty acid analysis has shown that many of the red-pigmented bacteria belong to the genus *Rodococcus* and possess lipids that make them resistant to disinfection. Because these organisms are found in drinking water, pigmented bacteria have been reported from water used in hospital therapy machines, and on other devices that use water supply (Favero et al., 1974; Favero et al., 1975; Herman, 1976).

Disinfectant-Resistant Bacteria. This bacterial group comprises a wide range of organisms of very limited health significance whose presence is a reflection of disinfection treatment effectiveness or is a depiction of the heterotrophic bacterial flora of an untreated public water supply that contains many spore-forming organisms. Comparison of bacterial genera present in two different public water supplies in southern California by Olson and Hanami (1980) indicated that variation in the diversity of the bacterial population can also be related to source water (i.e., surface versus groundwater). The predominant genera in one untreated groundwater supply were *Acinetobacter* and *Pseudomonas. Klebsiella* was also found intermittently in very low numbers in this supply. Chlorinated distributed water from a surface water source contained *Acinetobacter, Pseudomonas/Alcaligenes,* and *Flavobacterium* as predominant genera.

Chlorination of water creates a strong selective pressure on bacterial populations. Ridgway and Olson (1982) found that bacteria isolated from a chlorinated surface water distribution system were more resistant to both combined and free forms of chlorine than members of the same genera isolated from an unchlorinated groundwater system. These results may have also been influenced by attachment to particles in the chlorinated surface water system, although the overall water chemistry of the two public water supplies was similar. The most resistant microorganisms isolated from either water system (gram-positive, spore-forming bacteria, actinomycetes, and some micrococci) were able to survive exposure to 10 mg/L free chlorine for 2 min (LeChevallier, Evans, and Seidler, 1981). The most chlorine-sensitive bacteria isolated from these two water distribution systems (*Corynebacterium/Arthrobacter, Klebsiella, Pseudomonas/Alcaligenes, Flavobacterium/Moraxella, Acinetobacter,* and most gram-positive micrococci) were readily killed by a chlorine concentration of 1.0 mg/L or less. The apparent contradiction in occurrence of genera that are grouped as both chlorine-sensitive and chlorine-resistant reflects variations of strains within a genus, physical aggregation of cells, and protective associations with particulate matter.

Actinomycetes and Other Related Organisms. These microorganisms create some of the objectionable taste and odors reported in water supplies. Actinomycetes in source waters of temperate climates may be reduced by approximately tenfold with storage in holding basins prior to other treatment measures. Some drinking water treatment process configurations that include slow sand filtration or GAC filter adsorbers may support an increase in density and species that enter the distribution network. *Nocardia* strains were predominant in finished water from water treatment trains consisting of aeration and filtration or aeration, sand filtration, ozonation, and activated carbon adsorption. *Micromonospora* made up the greatest percentages in finished water from treatment chains consisting of flocculation, settling, and slow sand filtration or aeration and granular bed filtration. Some growth may also be expected on PVC-coated walls in finished water reservoirs (Dott and Waschko-Dransmann, 1981) and in the distribution lines where organic material accumulates in sediments (Burman, 1965; Bays, Burman, and Lavis, 1970). In addition to their presence in cold tap water, thermophilic actinomycetes and mesophilic fungi were found in several hot water samples in three municipalities in Finland (Niemi, Knuth, and Lundstrom, 1982). *Thermoactinomyces vulgaris* was the predominant actinomycete found in 11 of 15 water distribution systems examined. In two studies of treated water supplies in England, chlorination alone was not effective in eliminating *Streptomycetes.* Median densities were 2 *Streptomycetes* per 100 mL of distributed water. Taste and odor complaints involving *Actinomycetes* often were from waters that had *Streptomycetes* or *Nocardia* counts greater than 10 organisms per 100 mL.

Fungi. Although many fungi have been found in the aquatic environment (Nagy and Olson, 1982), focus on these organisms in water supply has been limited to their active degradation of gasket and joint materials and association with taste and odor complaints by consumers (Burman, 1965; Bays, Burman, and Lavis, 1970; Niemi, Knuth, and Lundstrom, 1982). This focus is somewhat surprising because certain fungi are also the cause of nosocomial infections in compromised individuals, and responsible for allergenic or toxigenic reactions through inhalation in water vapor or by body contact in showering or bathing.

Fungi occur at relatively low densities in ambient waters (Cooke, 1986). Source water storage in temperate climates and chemical coagulation prior to filtration and disinfection have been found to improve the removal efficiency somewhat, but provide no absolute treatment barrier to these organisms. Fungi present in source waters may pass through sand filtration and disinfection treatment processes (Burman, 1965; Bays, Burman, and Lavian, 1970; Hinzelin and Block, 1985). In one study (Niemi, Knuth, and Lundstrom, 1982), these organisms were detected in 29 of 32 treated water samples.

Breakthrough in the treatment barriers, soil contamination from line repairs, and airborne particulates entering storage reservoirs and standpipes are pathways by which fungi enter the distribution network (Table 18.4). Fungi occurrences in distributed water are more frequent during summer water temperature conditions. *Aspergillus fumigatus* was the predominant species detected in the distribution system of 15 water supplies in Finland (Niemi, Knuth, and Lundstrom, 1982). A variety of fungi (*Cephalosporium* sp., *Verticillium* sp., *Trichodorma sporulosum, Nectria veridescens, Phoma* sp., and *Phialophora* sp.) were identified in water from service mains in several water supplies in England (Ridgway and Olson, 1982; Dott and Waschko-Dransmann, 1981). Fungi densities in these drinking waters were usually less than 10 organisms per 100 mL.

TABLE 18.4 Fungi Densities in Water Supply*

System†	Sampling point	Number of samples	Average density per 100 mL
SR	Source water	6	2.7
MW		3	54.0
BG		8	2.0
WH		2	8.0
BL		6	22.0
BL	Aerated effluent	3	3.0
SR	Finished water	4	3.3
WH		2	< 1.0
BL		5	5.0
BL	Storage tank	5	40.0
SR	Residential tap	34	3.2
MW		65	3.8
BG		11	2.6
WH		22	2.6
BL		20	3.7
WH	Fire hydrants	4	16.6
BL		7	9.0

* Data from five small water systems revised from Rosenzweig et al., 1983.

† BG, Bradford Glen water system; BL, Brooklawn water system; MW, Marshallton Woods water system; SR, Spring Run water system; WH, Woodbury Heights water system.

Water supplies with fungal densities of 10 to 100 organisms per 100 mL are frequently responsible for customer complaints of bad taste and odor (Burman, 1965). The four most frequently occurring genera of filamentous fungi in two distribution systems (chlorinated and unchlorinated supplies in southern California) were *Penicillium, Sporocybe, Acremonium,* and *Paecilomyces* (Nagy and Olson, 1982). In the unchlorinated system, *Penicillium* and *Acremonium* represented approximately 50 percent of 538 colonies identified among 14 genera, while *Sporocybe* and *Penicillium* accounted for 56 percent of 923 fungal strains distributed within 19 genera that occurred in the chlorinated supply. The majority of filamentous fungi appeared to be nonpathogenic saprophytes. The mean density of fungi from the unchlorinated and chlorinated system was 18 and 34 organisms per 100 mL, respectively. Conidia of *Aspergillus fumigatus, A. niger,* and *Penicillium oxalicum* isolated from distribution systems of three small water supplies in Pennsylvania showed a greater resistance to chlorine inactivation than yeast (Rosenzweig, Mironigh, and Pipes, 1983).

Yeasts are another category of fungi found in the aquatic environment. While chemical coagulation and sedimentation will remove 90 to 99 percent of yeasts from source water, and granular bed filtration will remove another 90 percent, disinfection is less effective in further yeast reductions. The resistance of yeast to free available chlorine is primarily a result of the thick and rigid cell wall, which presents a greater permeability barrier to chlorine. Specific species identified in distribution waters include *Candida parapsilosis, C. famata, Cryptococcus laurentis, C. albidus, Rhodotorula glutinis, R. minuta,* and *R. rubra* (Hinzelin and Block, 1985; Engelbrecht and Haas, 1977). Densities of yeasts reported in finished drinking water average 1.5 organisms per liter. While these initial densities are low, yeasts slowly colonize water pipes and may become more numerous over time.

Disinfectant Stability During Water Distribution

Stability of disinfectants during water supply distribution is important, particularly to reduce or prevent colonization by surviving organisms and to inactivate bacteria associated with the intrusion of contamination in the pipe network. Microbial colonization may lead to corrosive effects on the distribution systems and adverse aesthetic effects involving taste, odor, and appearance. Also, the regrowth of health-related opportunistic organisms and their impact on coliform detection should not be dismissed as a trivial problem. The analysis of data (Table 18.5) taken from the National Community Water Supply Study of 969 public water systems (Geldreich et al., 1972), revealed that standard plate counts of 10 organisms or less were obtained in over 60 percent of these distribution systems that had a measured chlorine residual of approximately 0.1 to 0.3 mg/L. Protective sediment habitats and selective survival of disinfectant-resistant organisms were the reasons why residual chlorine concentrations greater than 0.3 mg/L chlorine produced no further decreases in the HPC. In an extensive study involving 986 samples taken from the Baltimore and Frederick, Maryland, distribution systems, the maintenance of a free chlorine residual was found to be the single most effective measure for maintaining a low standard plate count (Snead et al., 1980).

A disinfectant residual in the distribution system can be very effective in the inactivation of pathogens associated with contaminants that are slowly seeping into large volumes of high-quality potable water. Obviously, limitations to this protective barrier exist. Free chlorine residuals in distributed water often range from 0.1 to 0.3 mg/L, and chloramines may be found to be from 0.2 to 2.0 mg/L. Such residuals will not provide adequate protection against massive intrusions of gross contamination

TABLE 18.5 Effect of Varying Levels of Residual Chlorine on the Total Plate Count in Potable Water Distribution Systems*

Standard plate count†	Residual chlorine, mg/L							
	0.0	0.01	0.1	0.2	0.3	0.4	0.5	0.6
<1	8.1	14.6	19.7	12.8	16.4	17.9	4.5	17.9
1–10	20.4	29.2	38.2	48.9	45.5	51.3	59.1	42.9
11–100	37.3	33.7	28.9	26.6	23.6	23.1	31.8	28.6
101–500	18.6	11.2	7.9	9.6	12.7	5.1	4.5	10.7
501–1000	5.6	6.7	1.3	2.1	1.8	0	0	0
>1000	10.0	4.5	3.9	0	0	2.6	0	0
Number of samples	520	89	76	94	55	39	22	28

* All values are percent of samples that had the indicated standard plate count.
† Standard plate count (48 h incubation, 35° C).
Source: Data from Geldreich et al., 1972.

characterized by odors, color, and milky turbidities. Studies conducted in a small abandoned distribution system on a military base indicate that at tap water pH 8, with an initial free chlorine residual of 0.7 mg/L and wastewater added to levels of up to one percent by volume, 3 logs (99.9 percent) or greater bacterial inactivation were obtained within 60 minutes. Viral inactivation under these conditions was less than 2 logs (<99 percent). In laboratory reservoir experiments, where the residual chlorine is replenished by inflow of fresh uncontaminated chlorinated tap water, greater inactivation was observed at the higher wastewater concentrations tested. Furthermore, a free chlorine residual was more effective than a combined chlorine residual in the rapid inactivation of microorganisms contained in the contaminated supply (Snead et al., 1980).

Distribution system problems associated with the use of combined chlorine residual or no detectable residual have been documented in several instances (Langelier, 1936; McCauley, 1960; Stumm, 1960). In these cases, the use of combined chlorine is characterized by an initial satisfactory phase in which chloramine residuals are easily maintained throughout the system and bacterial counts are very low. Over a period of years, however, nitrification problems may develop through the application of an excess amount of free ammonia (>0.1 mg/L), which can promote the development of nitrifying bacteria (Wolfe et al., 1988; Ike, Wolfe, and Means, 1988). Nitrification also has the adverse impact of neutralizing chloramine residuals and promoting growth of heterotrophic bacteria (Kirmeyer et al., 1995).

Conversion of a system to free chlorine residual typically produces an initial increase in consumer complaints of taste and odors resulting from oxidation of accumulated organic material, and it may become difficult to maintain a free chlorine concentration at the ends of the distribution system. With application of a systematic main flushing program in these instances, a free chlorine residual will become established throughout the system, bacterial counts will decrease, and taste and odor complaints will decline (Brodeur, Singley, and Thurrott, 1976).

Ozone is typically applied as a primary disinfectant and because of its high reactivity, it is not found in the distribution system, nor would its presence be desirable. Ozone (and other disinfectants like chlorine) can react with nondegradable natural organic matter, making it more biodegradable. This biodegradable organic matter (BOM) can stimulate bacterial growth, resulting in detectable coliform occurrences, taste and odor problems, microbially-influenced corrosion, and potential public health concerns.

MICROBIAL COLONIZATION FACTORS

Water mains, storage reservoirs, standpipes, joint connections, fireplug connections, valves, and service lines and metering devices have the potential to be sites suitable for microbial habitation. No pipe material is immune from potential microbial colonization once suitable attachment sites are established. Given sufficient time, aggressive waters or microbial activity will initiate corrosion of metal pipe surfaces; water characteristics may change the surface structure of asbestos/cement mains; and biological activity creates pitting on the smooth inner surface of plastic pipe materials. (See Chapter 17, Corrosion and Deposition Control.) Not all pipe sections show evidence of deterioration even after years of active service; in some cases, the nature of the water chemistry and continuous movement of water under high velocity conditions help to prevent the buildup of chemical and microbial species that contribute to corrosion and pitting.

Habitat Characterizations

Complex structures in a pipe network and locations out of the mainstream of moving water are the most opportune sites for sediments to accumulate, tuberculations to expand, and microbial colonization to develop. Sediment accumulates at these sites from corrosion and sometimes improper application of corrosion-inhibiting additives. Low-flow areas may also be deposition sites for turbidity-causing particles in both filtered and nonfiltered source water, unstable coagulants, activated carbon fines, and biological debris.

Corrosion of pipe surfaces provides not only a habitat for bacterial proliferation, but also is a source of substrate and protection from chlorine disinfectant residuals. In drinking water systems, the occurrences of coliform bacteria in corrosion tubercles on iron pipes has been reported by a number of investigators (LeChevallier, Babcock, and Lee, 1987; Opheim, Grochowski, and Smith, 1988; Facey, Smith, and Ernde, 1990; Emde, Smith, and Facey, 1992). Laboratory studies showed that the density of HPC and coliform group bacteria were 10 times higher when grown on mild steel coupons than on noncorroded polycarbonate surfaces (Camper et al., 1996). The increased surface area due to tuberculation of the pipe walls, the concentration of organic substances within the tubercles, and the secretion of organic compounds by iron-utilizing bacteria have been postulated as reasons why iron corrosion stimulates bacteria growth.

LeChevallier, Lowry, and Lee (1990) showed that the disinfection of biofilm on galvanized, copper, or polyvinyl chloride (PVC) pipes was effective at 1 mg/L of free chlorine or monochloramine but disinfection of organisms on iron pipes was ineffective even at free chlorine residuals as high as 5 mg/L for several weeks. Follow-up studies showed that a combination of the corrosion rate, the ratio of the molar concentration of chloride and sulfate to bicarbonate (known as the *Larson index*), the chloramine residual, and the level of corrosion inhibitor could account for 75 percent of the variation in biofilm disinfection rates for microorganisms grown on iron pipes (LeChevallier et al., 1993). Corrosion control through the manipulation of water chemistry (i.e., pH and alkalinity; Langelier index) or application of phosphate and silicate-based corrosion inhibitors should be thought of as not only protecting the pipe materials, but also as a necessary component of a microbial control plan.

Turbidity and Particle Effects

Particles that cause turbidity in finished water also contribute to sediment accumulation in the dead ends of the system and within porous scale and tubercle formations. Because turbidity is only an indirect measurement of the particulate matter in water, it provides no specific information regarding the type, number, and size of particles being detected. Turbidity monitoring in the distribution system, however, is a good quality-control practice. Values in excess of 1 NTU may signal the need to flush the distribution system and to search for areas of pipe corrosion that must be brought under control.

In general, inorganic particles such as clay and water flocculating agents may trap a variety of organisms. These particles, however, appear to have little, if any, protective effect against disinfection action because of the absence of organic demand substances (Hiisverta, 1986). Thus, few viable cells are transported in these kinds of particles to potential habitat sites along the distribution system.

In contrast, organic particles and algal cell masses in seasonal blooms can be a vehicle for the transport of microbial entities through some treatment processes, including disinfection. Organic particulates of concern include fecal cell debris (Hoff, 1978), wastewater solids including aggregates of bacteria and virus (Hejkal et al., 1979), protective mats of algal cells, and activated carbon fines that provide attachment sites for associated heterotrophic bacteria (Allen, Taylor, and Geldreich, 1980; Ridgway and Olson, 1981; Foster et al., 1980; Camper et al., 1987).

Passage of Microorganisms in Macroinvertebrates

Not commonly recognized as a problem in the quality of distributed water is the occurrence of various larger, more complex biological organisms including crustaceans (amphipods, copepods, isopods, ostracods), nematodes, flatworms, water mites, and insect larvae such as chironomids (Small and Greaves, 1968; MacKenthun and Keup, 1970; Geraldi and Grimm, 1982; Chang, Woodward, and Kabler, 1960; Levy et al., 1984; Levy, Hart, and Cheetham, 1986; Zrupko, 1988). While these organisms may be present in the source water, most are removed by various treatment processes, but some may succeed in becoming established in filter beds, releasing progeny that can successfully survive disinfection and migrate into the distribution system (Cobb, 1918; George, 1966; Tombes et al., 1979; Mott and Harrison, 1983). In so doing, these invertebrates may harbor and protect coliforms and *Legionella* from contact with the disinfectant at concentrations typically present in distribution systems (Tracy, Camarena, and Wing, 1966; Chang et al., 1960; Smerda, Jensen, and Anderson, 1971; Sarai, 1976). The ingested bacteria may not only survive but multiply within the invertebrate host and be released at a later time when the host cell bursts or disrupts (Fields et al., 1984; Tyndall and Domingue, 1982). This phenomenon may account for some of the continued release of coliform and other bacteria into the distribution system by passage through treatment barriers. While some bacterial feeders (protozoans, nematodes, etc.) are inactivated immediately by the chlorine residual in the distribution system, others die slowly or become adapted to the available food sources (biofilms of bacteria) in selected pipe sediments or tubercles. They may use these sites for attachment, and proceed to harvest the organisms passing in the free-flowing waters.

Key Factors in Microbial Persistence and Growth

Key factors in the establishment of microbial colonization within the distribution system are a source of nutrients, a protective habitat, and a favorable water temperature for rapid growth (Geldreich, Nash, and Spino, 1977; Water Research Centre, 1977). Not all bacteria that enter the water supply distribution system persist or are able to adapt to this environment and grow (Reasoner, Blannon, and Geldreich, 1989; Ridgway, Ainsworth, and Gwilliam, 1978; Victoreen, 1978; Vander Kooij and Zoeteman, 1978). Within the total coliform indicator group, *Klebsiella, Citrobacter,* and *Enterobacter* strains are most often noted in distribution systems (Martin et al., 1982; Geldreich, Nash, and Spino, 1977; Ptak, Ginsburg, and Willey, 1973) and can grow with minimal nutrients. Other bacteria that may grow include *Pseudomonas, Flavobacterium, Acinetobacter,* and *Arthrobacter.* These are especially troublesome because of their potential interference to coliform detection and acknowledged roles as opportunistic pathogens (Geldreich, Nash, and Spino, 1977; Hutchinson, Weaver, and Scherago, 1943; Fischer, 1950; Herman and Himmelsbach, 1965; von Graevenitz, 1977; Herson, 1980).

Bacterial Nutrients. Essential nutritive substances, including those naturally occurring and man-made, containing phosphorus, nitrogen, trace metals, and carbon are introduced in varying concentrations from source waters. Surface waters receive a variety of organics discharged in municipal wastewater effluents, industrial wastes, and agricultural activities. While some of the nutrient content in dissolved organic carbon may be removed (20 to 50 percent) through conventional treatment, more attention to treatment refinements is needed to further reduce trace organic residuals. Applying ozone coupled to GAC or other equivalent biological treatment processes (e.g., for improved disinfection by-product precursor control) will also minimize the available organic materials and thereby provide fewer opportunities for microbial biofilm development and coliform growth. For those water utilities with a relatively clean surface water source that only rely on disinfection treatment, a seasonal threat of organic contributions from natural lignins, algal blooms, and recirculating bottom sediments during lake destratification will always exist. These organic materials pass into distribution pipe networks, where they stimulate growth of a wide range of aquatic bacteria (Postgate and Hunter, 1962).

The greatest success in reducing the potential for microbial persistence and growth in the distribution system will only be achieved by further reduction of the organic portion of the essential nutrient base, since the critical inorganic substances are ubiquitous in the aquatic environment (Allen and Geldreich, 1977; Tuovinen and Hsu, 1982). Various studies have shown that biological treatment can be highly effective, with more than 90 percent removal of the biodegradable organic matter (Bourbigot, Dodin, and Lherritier, 1982; Van der Kooij, Visser, and Hijnen, 1982; Van der Kooij, 1987; Janssens, Meheus, and Dirickx, 1984; Pascal et al., 1986; Bablon, Ventresque, and Roy, 1987; LeChevallier et al., 1992). Treatment to reduce organic carbon for bacterial growth suppression is also beneficial in the reduction of taste, odor, color, chlorine demand, and disinfectant by-product formation.

Phosphorus. Phosphorus in the environment occurs almost exclusively as orthophosphate (PO_4^{3-}), which has a valence state of +5. Although some members of the genera *Bacillus, Pseudomonas,* and *Clostridium* have been shown to reduce orthophosphate to hypophosphite (PO_2^{3-}) and phosphite (PO_3^{3-}) under anoxic conditions, these transformations are thought to be limited and quantitatively insignificant. Because phosphorus is not consumed by microbial activity (like organic carbon), the turnover rate of phosphorus in aquatic habitats can overcome low levels of orthophosphate in the water column. Although some researchers have suggested that certain waters may be phosphate-limited (Herson, Marshall, and Victoreen, 1984; Haas, Bitter, and Scheff, 1988), Rosenzweig (1987) found that phosphate-based corrosion inhibitors did not significantly influence the growth of several strains of coliform bacteria. High levels of Virchem 932, a zinc orthophosphate, showed inhibitory effects for certain coliform species.

Nitrogen. The basic requirements for nitrogen in the metabolic processes of bacteria can be satisfied in a variety of ways. Watershed conditions involving wastewater effluents released upstream of a water intake, landfill operation in the vicinity of groundwater aquifers poorly protected by the soil barrier, and seasonal application of farm and garden fertilizers over the watershed are often the major contributors of nitrogenous compounds.

Water treatment practices must also be carefully controlled to minimize addition of nitrogen. For example, application of ammonia to form chloramines may contribute excess ammonia to water. Ammonia is an electron donor for autotrophic bacteria and can promote bacterial growth in distribution systems. Rittmann and Snoeyink (1984) found that ammonia concentrations in groundwater supplies were frequently high enough to cause biological instability. The proliferation of ammonia-oxidizing bacteria (nitrification) can lead to accelerated loss of chlorine residuals, increased nitrite levels, and stimulated growth of HPC bacteria. As reviewed by Crowe and Bouwer (1987), biological treatment techniques are available for the removal of ammonia and nitrate from water. The exact role of nitrogen in growth of coliform bacteria is unclear, especially because some strains of *Klebsiella* can fix molecular nitrogen.

Metal Ions and Salts. Trace amounts of metal ions (Fe^{++}, Fe^{+++}, Mg^{++}, and others) available in salts appear to contribute to the nutrient base required for microorganisms. Victoreen (1977, 1980, and 1984) indicated that iron oxide stimulated the growth of coliform bacteria. Trace amounts of copper salts, Mg, and Mn ions are needed by other heterotrophic bacteria in the development of their normal metabolic processes (Laskin and LeChevallier, 1977) within the biofilm consortium.

Carbon. Organic carbon is utilized by heterotrophic bacteria for production of new cellular material (assimilation) and as an energy source (dissimilation) (LeChevallier, Welch, and Smith, 1996). Most organic carbon in water supplies is natural in origin and is derived from living and decaying vegetation. These compounds may include humic and fulvic acids, polymeric carbohydrates, proteins, and carboxylic acids. In the U.S. EPA National Organic Reconnaissance Survey (Symons et al., 1975), the nonpurgable total organic carbon concentration of finished drinking water in 80 locations ranged from 0.05 to 12.2 mg/L, with a median concentration of 1.5 mg/L. Because heterotrophic bacteria require carbon, nitrogen, and phosphorus in a ratio of approximately 100:10:1 (C:N:P), organic carbon is often a growth-limiting nutrient.

Measuring Biodegradable Organic Matter in Water

Two of the most common measurements of the biostability of water include determination of the assimilable organic carbon (AOC) content and the biodegradable dissolved organic carbon (BDOC) level. Measurement of AOC is based on a bioassay of two test strains (*Pseudomonas fluorescens* strain P17 and *Spirillum* sp. strain NOX). Bacterial growth is monitored in the water samples (e.g. colony counts, ATP measurements), and the maximum growth (N_{max}) observed during the incubation is converted into AOC by using growth yield of the bacteria from calibration curves performed on known concentrations of standard organic compounds (e.g. acetate, oxalate) (EPA, 1976; Van der Kooij and Hijnen, 1985). Assimilable organic carbon is expressed as micrograms of acetate- (or oxylate-) carbon equivalents. Biodegradable dissolved organic carbon evaluates the reduction in DOC levels following incubation of the water sample with microorganisms (Pascal et al., 1986; Servais, Billen, and Hascoet, 1987; Joret, 1988; Servais, Laurent, and Randon, 1993). The difference between the initial and final DOC levels is the biodegradable organic carbon fraction.

Protective Habitats in Pipe Networks. Porous sediments and tubercles in the pipe network appear to adsorb and concentrate nutrients. Bacteriological colonization found in encrustations, taken from water main sections removed from four municipal water systems during line repairs, demonstrated bacterial densities that ranged from 390 to 760,000 bacteria per mL (Allen and Geldreich, 1977). Variations in bacterial densities will depend on the tubercle characteristics and fraction examined (Tuovinen and Hsu, 1982). Using the scanning electron microscope to locate microcolony sites in tubercles, microorganisms observed were predominantly found at or near the surface. This is the area where encrustation, water interface, nutrients, and oxygen are constantly present (Allen, Taylor, and Geldreich, 1980; Tuovinen et al., 1980).

To colonize surfaces in contact with a flowing stream of water, bacteria must adhere tenaciously. They do so by means of a mass of tangled fibers (called a *glycocalyx*) that extends from the cell membrane and adheres to surrounding surfaces or other bacteria (Costeron, Geesey, and Cheng, 1978; Bitton and Marshall, 1980). These glycocalyx adhesions to sediment coatings, tubercles, pipe joints, and rough wall surfaces prevent most of the individual cells of the microcolony from being swept away by the shearing force of flowing water. Because these appendages are polysaccharide materials, they may also serve as a protective barrier against the lethal effects of residual disinfectants.

Not all areas of the distribution network are favorable for microbial proliferation. High water velocity in smooth pipe sections makes microbial attachment difficult, increases nutrient flux, and provides less protection from disinfectant exposure (Victoreen, 1978). In contrast, sediment accumulations, tubercle development, and scale formation in low-flow, dead-end, and rough-walled pipe sections and connecting joints are prime sites (Figures 18.1*a* to *d*) for microbial growth (Olson, 1982; Allen, Taylor, and Geldreich, 1980; Victoreen, 1978; Allen and Geldreich, 1977; Victoreen, 1974; Victoreen, 1977; Ainsworth, Ridgway, and Guilliam, 1978; Lee, O'Connor, and Banerji, 1980). Therefore, finding patchiness in biofilm development along pipe surfaces should be expected.

Water Supply Storage. Finished water reservoirs, water storage tanks, and standpipes cannot be excluded as potential sites for biofilm development. Operational conditions that may encourage bacterial colonization include prolonged storage time, reduced flow velocities, and infrequent cleaning. Water is often not drawn from the bottom of reservoirs, so accumulations of sediments may occur. Sediment buildup in reservoirs is more of a problem for systems that use unfiltered surface waters and limit treatment to disinfection. Filtration of surface source waters will remove biological materials (e.g., algae and vegetation debris) and other suspended

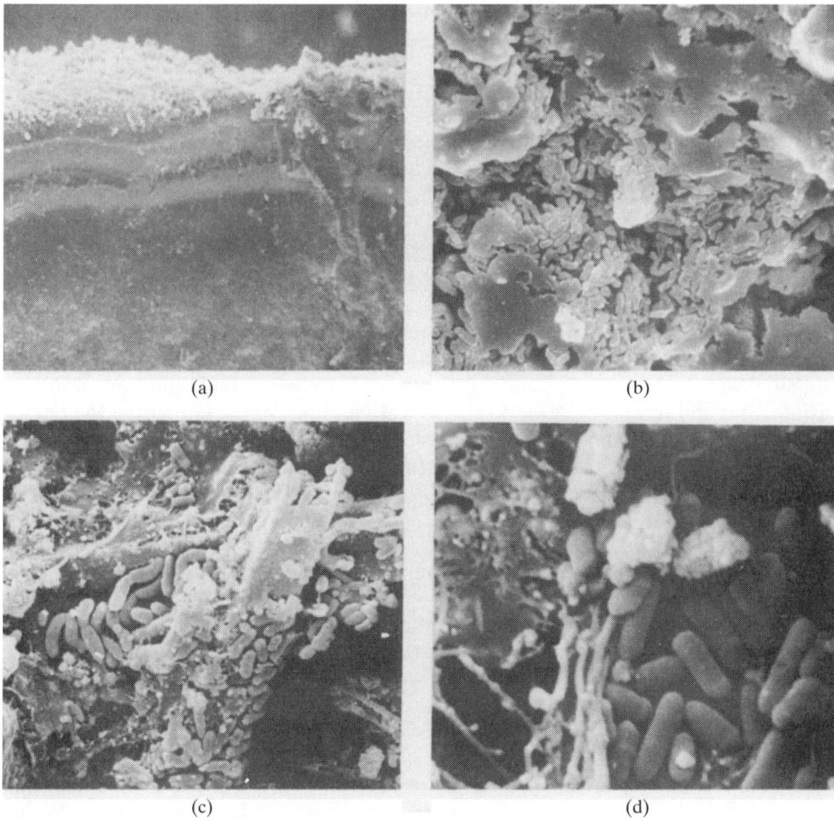

FIGURE 18.1 (*a*) Cross section of pipe tubercle showing loose surface material at water interface and compaction of deposits near the pipe wall. Sample magnification, 100×. (*b*) Bacterial colonization site in porous tubercle encrustation. Electron micrograph magnification, 3000×. Material supplied from the Cambridge, Massachusetts, distribution system pipe network. (*c*) Attachment site for bacterial growth in tubercle surface material. Electron micrograph magnification, 4000×. Material supplied from the Cambridge, Massachusetts, distribution system pipe network. (*d*) Protected habitat and microbial community in porous encrustation. Electron micrograph magnification, 11,000×. Material supplied from the Cambridge, Massachusetts, distribution system pipe network

solids that otherwise pass through the disinfection process and ultimately accumulate in low-flow pipe sections and finished water storage facilities. These sediments can contribute significant amounts of assimilable organic complexes that support bacterial colonization of the distribution network.

Slime or biofilm development in cement structures may develop on a cement mortar with plastic additives, on sealers such as epoxy resin, bitumen, and PVC film, and on areas of cement erosion (Schoenen, 1986). Metal structures are also subject to microbial activity in areas with corrosion activity such as seams and joint construction bonds. Porous materials such as brick and wood used in reservoirs are particularly suited for microbial colonization and may be difficult to dislodge by flushing and disinfection treatment.

Tanks constructed of redwood are common in the western United States, being used by small communities, state and federal recreational areas, mobile home parks, and motels. The problem with redwood used in some water storage tanks is the presence of microbial colonization that impacts on compliance with the drinking water coliform standard. Research on coliform occurrences in redwood storage tanks showed that environmental strains of *Klebsiella* colonize the wood tissues of trees. The association begins at embryo fertilization of the tree seed. *Klebsiella* is introduced to the embryo from insect contact or windblown dust and then colonizes the wood pores, receiving water and nutrients from the xylem and phloem tissues of the tree (Knittel, Seidler, and Cake, 1977). This coliform metabolizes the leached-out wood sugars (cyclitols) from the staves as a source of nutrients (Seidler, Morrow, and Bagley, 1977; Talbot and Seidler, 1979). As a consequence, new redwood tanks were found to be the source of *Klebsiella* in the water supply and the cause of massive biofilm development over the inner surfaces of the tank (Seidler, Morrow, and Bagley, 1977). The problem is most acute in new redwood tanks that are unlined. Disinfection and scraping the wood staves were ineffective in eliminating the bacteria because the organisms persist deep inside the wood pores, until all the available wood sugars (cyclitols) are leached away or biodegraded. The problem can be controlled by maintaining a free chlorine residual of 0.2 to 0.4 mg/L until the available sugar supply is leached away with tank usage over a two-year period (Talbot, Morrow, and Seidler, 1979). More effective is the use of plastic or fiberglass liners to prevent leakage and release of bacterial colonization into the water supply.

As a preventative measure to avoid water quality degradation in all types of water storage structures, regularly scheduled, systematic inspections are recommended for slime development on structural surfaces in contact with the water, and for the accumulation of sediments that serve as protective sites for microorganisms. These growths and deposits should be removed to reduce sites where taste and odor problems and microbial growth originate.

Water Temperature Effects. Microbial growth is not only keyed to bacterial strains that quickly adjust to limited nutrient sources, but also to water temperature. Water temperature above 50°F (10°C) accelerates the growth of adapted organisms with slow generation times. Low water temperatures result in a precarious balance between new-cell development and the death of old cells. Data available from water systems in geographical areas with pronounced seasonal temperature changes suggest that growth of many heterotrophic bacteria is more pronounced than the coliform subset of this population, often providing abrupt surges in density during the summer (Geldreich, Nash, and Spino, 1977; Howard, 1940). Renewed growth was not, however, correlated with temperature in southern California groundwater systems (Olson, 1982), possibly because the water temperatures common to those systems were consistently above 50°F (10°C) and available assimilable organics were very low. Accumulations of nutrient particles in pipelines during winter periods of minimal microbial activity may be the key to summer bacterial growth surges in surface water systems.

MONITORING FACTORS

A reliable community water supply system will have a record of no associated disease outbreaks. This approach to demonstrating the acceptable quality of a public water supply provides no opportunity to intervene in potential contamination events until waterborne illness is detected. Moreover, most intestinal illness goes unreported, further reducing the ability of public health officials to detect all but the largest outbreaks. Thus, relying solely on the absence of waterborne outbreaks violates the basic concepts of public health protection. A far more prudent approach involves use of a monitoring program that provides a continuing database of treatment effectiveness. This database will provide early signals of microbial deterioration in the distribution system, as indicated by increasing bacterial densities. Since coliform breakthrough is the concern, consideration should be given to techniques (LeChevallier and McFeten, 1985; APHA/AWWA/WEF, 1995) for maximum detection of any stressed coliforms that might be passing through the treatment barriers (Bucklin, McFeters, and Amirtharaja, 1991).

To accomplish this goal, two levels of distribution system monitoring are recommended: (1) compliance monitoring, and (2) special monitoring designed to better understand system characteristics and performance. Routine compliance monitoring involves the development of a systematic plan to cover all major areas of the distribution network with a sampling strategy that demonstrates the continued delivery of a safe water supply in all service areas. In the United States, these compliance monitoring plans must be approved by the state water authority. A special monitoring program is designed to understand system characteristics and performance as well as locate problem areas that need corrective action.

Sampling Frequency Guidelines

Effective monitoring of the microbial quality of drinking water supplies requires careful consideration of sampling frequency and site selection. A number of factors must be taken into consideration if frequency of monitoring is to be optimized (Hoskins, 1941; Technical Subcommittee, 1943; World Health Organization, 1971; Safe Drinking Water Committee, 1977; Berger and Argaman, 1983; NATO/CCMS, 1984; World Health Organization, 1984). Protected groundwater supplies can be expected to be of uniform and high quality, requiring less monitoring than surface water supplies. Treatment barrier penetration is often of increased concern in water supplies that use disinfection as the only treatment process, which suggests that more frequent monitoring is needed when systems use conventional treatment with multiple processes in series.

In systems characterized by corrosion in old unlined pipe, main breaks, increased cross-connection potential, and frequent flow reversals, integrity of the distribution network is often an important issue. Every water system is unique in its strengths and weaknesses, and careful customizing of the monitoring frequency is required. Such an approach is complex and difficult to factor into the minimum frequency requirements specified in national regulations, but it should be pursued by utilities as an extension of their monitoring program.

The use of sampling frequency formulas based on population served is the accepted practice for establishing minimum sampling requirements for coliforms in many countries, including the United States (World Health Organization, 1971; Department of Health and Social Security, 1969; Council of the European Communities, 1975; Ministry of National Health and Welfare, 1977; World Health Organization, 1970). The basis for this approach is the recognition that as the population increases, so will the size and complexity of the system and the potential for distribution network contamination by cross-connections and back siphonage (Craun, 1978). The emphasis has, therefore, been placed on a demonstration of safe water quality in distribution, with lesser emphasis on monitoring treatment plant effectiveness. This focus will, in the future, be modified as a result of recent *Cryptosporidium* outbreaks from several conventional treatment systems and treatment changes to reduce disinfectant by-products. The need for more monitoring attention to surface water quality and treatment barrier effectiveness can no longer be ignored.

The next critical consideration in the development of an effective monitoring program is in the establishment of specified numbers of samples required per given population level. While the sampling frequency established has, in general, proven to be adequate for routine monitoring of most water systems, it is not providing satisfactory monitoring of water supplies serving less than 10,000 population (Pipes, 1983; Jacobs et al., 1986; Morita, 1987). The frequency of waterborne outbreaks in smaller systems is higher, and often no monitoring occurs for most days in the month. Providing statistically significant data on the bacteriological quality of a small water system requires a minimum of five samples per month (USEPA, 1989). In small water supplies, the coliform test of finished water quality should be done once per week if occasional coliform occurrences are detected. Monitoring of the distribution system for coliform bacteria should not be ignored just because the pipe network in a small water system is less complicated, with fewer service connections. In small distribution systems, a disinfectant residual measurement will provide confirmation of available disinfectant protection against contamination events that occur through cross-connections and against some other problems common to pipe network configurations.

Because of increased laboratory costs associated with additional monitoring of the small water system, options such as a periodic sanitary survey by the state water authority should be considered to reduce the financial burden while improving public health protection. For those small systems (<3,300 persons) that utilize a disinfected groundwater supply or filter their surface source water prior to disinfection, a triennial sanitary survey of water system operations could be substituted for some of the increased monitoring. This type of survey involves an on-site inspection of the watershed, source water, treatment processes, and distribution network; and an examination of the monitoring database by competent personnel. The intent is to evaluate the overall effectiveness of the water supply purveyor in providing a continuous supply of safe water to the community.

Using the heterotrophic plate count measurement of less than 500 organisms per mL as an alternative to meeting a disinfection requirement (at least 0.2 mg/L) in 95 percent of all distribution system sites tested has merit for reducing monitoring requirements (USEPA, 1989). Growth of bacteria above the 500-organisms-per-mL limit at these sites in summer would motivate the operator to flush the lines to minimize bacterial growth and perhaps restore a disinfectant residual. This action is a short-term expedient, while the long-term goal should be to reduce BOM through advanced water treatment practices.

Analysis of the monitoring requirements for bacteriological compliance with the Federal regulations involves a study of coliform frequency occurrence over a specified time span (USEPA, 1989). Such data can be obtained from conventional bacteriological tests using a presence/absence (P/A) test, the membrane filter method, or the multiple tube fermentation procedure. Any coliform count or positive tube result is interpreted as a coliform occurrence in the sample. This concept places equal emphasis on all coliform occurrences, regardless of density. The regulation acknowledges that a water system may have an occasional sample with coliforms but that these occurrences should not surpass a 5 percent frequency in any given month for all but the smallest water systems. For small water systems (<3,300 population), data collected from 60 consecutive samples over the most recent monitoring period would be used to determine the frequency of coliform occurrence, since the cluster of samples examined during any given month would not be adequate to establish meaningful percent occurrences. Over the long term, information on the frequency of coliform occurrences can provide an important indication of treatment effectiveness and operator skill in providing a continuous supply of safe drinking water.

When the five percent P/A occurrence is exceeded, more intense monitoring of the system is necessary, and a search for fecal coliform or *E. coli* is mandatory to check for any fecal contaminating event that might be obscured by the presence of biofilm coliform bacteria. Utilities with a history of summer-autumn biofilm episodes may seek a coliform variance to the Total Coliform Rule to avoid frequent boil-water notifications, provided the monitoring is increased significantly, a search for fecal contamination is intensified, and a remedial action plan is activated. Any system granted a variance must also submit to an annual sanitary survey and maintain an effective cross-connection control program.

Sample Site Selection

The strategy for sample site selection should include designing a sampling program to monitor all parts of the distribution system over time. Regardless of system size, a percentage of samples should be collected at certain fixed points, such as specially constructed sampling stations on the pipe network, pumping stations, storage tanks, and at sites with a history of coliform occurrences. This portion of the sampling program must be approved by the state water authority and may be described as "representative" sampling (i.e., permanent sampling sites chosen by the utility to be representative of water quality variations within the distribution system). Other sites should be selected at random throughout the distribution system, taking care to select taps supplied with water from a service line connected directly with the main rather than to a storage tank. Sample points should include areas where a possibility of contamination through cross-connections and back siphonage exists, such as hospitals, schools, public buildings, high-rise apartments, hotels, factories, and residential locations.

The sampling tap must be protected from exterior contamination associated with being too close to the sink bottom or to adjacent soil. Contaminated water or soil from the faucet exterior may enter the bottle during the collecting procedure because placing a bottle underneath a low tap without grazing the neck interior against the outside faucet surface is difficult. Leaking taps that allow water to flow out from around the stem of the valve handle and down the outside of the faucet, or taps in which water tends to run up on the outside of the lip are to be avoided as sampling outlets. To futher prevent contamination, aerators, strainers, and hose attachments on the tap must be removed before sampling. These devices can harbor a significant bacterial population if they are not cleaned routinely or replaced when worn or cracked. Whenever an even, controllable flow of water cannot be obtained from a tap after such devices are removed, a more suitable tap must be sought. Taps whose water flow is not steady should be avoided, because temporary fluctuation in line pressure may cause sheets of microbial growth that are lodged in some pipe section or faucet connection to break loose. The chosen cold water tap should be opened for 2 or 3 m or for sufficient time to permit clearing the service line; and a smooth-flowing water stream at moderate pressure without splashing should be obtained. Then, without changing the waterflow (which could dislodge some particles in the faucet), sample collection can proceed.

Treating water taps before collecting potable water samples is not necessary if reasonable care is exercised in the choice of sampling tap (clean, free of attachments, and in good repair) and if the water is allowed to flow adequately at a uniform rate before sampling. Passing a flame from a match or wiping an alcohol-soaked cotton applicator over the tap a few times will give the sample collector false confidence, but it will not have a lethal effect on attached bacteria associated with the plumbing fixture. The application of intense heat with a blowtorch may damage the valve-washer seating or create a fire hazard. If successive samples from the same tap continue to contain coliforms, the tap should be disinfected with a hypochlorite solution to eliminate external contamination as the source of these organisms (USEPA, 1989a; USEPA, 1989b; Geldreich, 1975; Distribution System Water Quality Committee, 1978; Buelow and Walton, 1971; Thomas et al., 1975; McCabe, 1969).

The composition and location of the sampling tap may affect the microbiological results. Studies performed in the United Kingdom found that coliform isolations were four times higher when samples were collected from plastic taps than from metallic faucets (O. Parry, unpublished results). Colbourne et al. (1984) reported that *Legionella* could be detected in shower heads and taps where certain rubber gaskets stimulated growth. Burlingame and O'Donnell (1993) detected coliforms in the base of a kitchen faucet. High heterotrophic plate counts (HPC) or false positive coliform counts have been found in samples taken from taps that had rotating spray devices, aerators, or point-of-use treatment devices (Geldreich et al., 1985; Goatcher, Simpson, and Emde, 1992). The Las Vegas Valley Water district replaced exterior household hose bib sampling points with dedicated sampling stations and found that the average resample rate for distribution system coliform compliance samples decreased from 1.44 to 0.60 percent. Following the installation of 199 dedicated sampling stations in the Los Angeles distribution system, bacteriological results indicated that detectable levels of coliforms declined 70 percent (Cox and Giron, 1993).

During a contamination event or when a coliform biofilm is suspected, special sampling sites should be selected within the pressure zone, in an effort to verify the initial data and isolate the problem area in the pipe network. Isolating the area is made difficult by flow reversals in the distribution system and by the transient nature of some contamination events. Despite these difficulties, some successes in locating the probable site can be achieved by selecting sampling locations upstream and downstream of coliform occurrences, where free chlorine residual has disappeared, or when increased turbidity and elevated heterotrophic bacteria population densities have been observed.

Bacterial Test Selection

Selection of bacterial tests for distributed water quality may take three different approaches: (1) routine testing for compliance with the coliform standard, (2) seasonal surveys for microbial growth in pipelines, and (3) special investigative protocols to verify fecal contamination or biofilm development.

Monitoring the effectiveness of water distribution networks to deliver a safe water supply to the public is a high priority. Because a properly designed and operated water treatment system is capable of producing a finished water that consistently releases no detectable coliforms in 100 ml samples during the month, it is reasonable to expect that water supply throughout the distribution network should also be able to meet this attainable standard. Unfortunately, some utilities have a chronic summer-time problem with coliform occurrences in their distribution sysyems (Smith, Hess, and Hubbs, 1990). Thus, the major effort in routine monitoring is directed toward a demonstration that no posttreatment contamination occurs.

The total coliform test has been the traditional bacterial test. The original concept used the premise that the absence of coliform bacteria in a distribution system water sample was evidence that no fecal contamination had occurred following treatment and, therefore, no risk of pathogen exposure existed. Hence, the absence of coliforms is an indirect measure or indicator of potential public health risk. Recent findings of virus and pathogenic protozoans in "coliform free" water suggest that this interpretation is not perfect, and that a search for more sensitive surrogates or development of a rapid specific test probe for pathogenic agents is needed. While these reports of the failure of the coliform indicator are alarming (Payment, Trudel, and Plante, 1985; Craun, 1984), the mass of data accumulated over the past 80 years clearly supports the continued acceptance of this test as a reasonably reliable indicator of a bacteriologically safe supply. In this light, the total coliform test continues to be used but is now considered to be more a measure of water supply treatment effectiveness and distribution system integrity, and less a measure of public health risk.

Surrogate indicators such as total coliforms and fecal coliforms are used because some waterborne pathogens are difficult to detect, or the available tests may be complex, time-consuming, and often not sufficiently sensitive or selective. Selection of an appropriate surrogate indicator should consider five important aspects: (1) The candidate organisms should be present only in feces at densities far exceeding pathogen concentrations from infected individuals; (2) an unquestionably positive correlation should exist between the indicator organism and fecal contamination events and between the candidate indicator and waters contaminated with fecal excrements; (3) persistence and growth characteristics of the indicator should parallel those of the most persistent waterborne pathogen; (4) disinfection of water should produce similar kill rates for the indicator and most pathogens; and (5) the methodology for detecting the candidate surrogate indicator must be simple, applicable to a variety of waters encountered in water quality monitoring, and not subject to false positive reactions and interferences from antagonistic organisms in the water flora. Although no known candidate surrogate indicators fit perfectly within all of these ideal requirements for water supply, the total coliform indicator still fulfills most of these specifications, but occasional occurrence of injured coilforms may go undetected in conventional media, increasing the risk of poor correlation with a possible pathogen occurrence. Focusing within the total coliform group reveals fecal coliform and *E. coli* as more exacting evidence of fecal contamination but less desirable if measurement of treatment effectiveness is the first objective (Kabler and Clark, 1960).

Coliform detection in water supply samples can be done by any of three different laboratory procedures: (1) the multiple-tube fermentation procedure; (2) the membrane filter method; or (3) a chromogenic medium test (American Public Health Association, 1995; Edberg, Allen, and Smith, 1988). All procedures can be configured to provide either a count per 100 mL or a presence-absence determination. The multiple-tube test uses either a 50 mL or a 100 mL sample and requires a minimum of 48 hours' incubation to ensure that no coliforms are present. A maximum of 96 hours may be necessary to establish a valid total coliform density. Using the membrane filter method, larger sample portions (1 liter or more) can be examined if desired. Test results are available within 24 h. In both the multiple-tube and membrane filter tests, fecal coliform determinations are either an adjunct part of the total coliform examination or done separately. The chromogenic coliform analysis is usually done on multiple 10-mL or single 100-mL sample portions, with simultaneous total coliform and *E. coli* results available within 24 h.

As high-quality finished water passes through the distribution network, changes in the general microbial composition occur that do not relate to total coliforms and fecal contamination (Ridgway, Ainsworth, and Gwilliam, 1978; Victoreen, 1974; Becker,

1975). Changes in microbial composition involve the establishment of disinfectant-resistant organisms in the porous infrastructure of tubercles and sediments. As the biofilm of mixed organisms develops, the heterotrophic bacterial density and profile of different bacteria expands to include a variety of gram-positive and gram-negative bacteria such as pigmented strains, antibiotic-resistant organisms, coliform antagonists, and opportunistic pathogens. In these situations, analyses of potable water for total coliform occurrences may not provide any information on the growth of the general bacterial population in areas where there is corrosion, static water, low flow, or dead-end sections of the pipe network. Identifying these areas prone to microbially induced water quality deterioration will require a test for heterotrophic bacteria that is accomplished by using either R-2A agar (Reasoner and Geldreich, 1985), spread plates (Taylor, Allen, and Geldreich, 1983), or membrane filtration with M-HPC agar (Taylor and Geldreich, 1979). Incubation should be at 28°C for 3 to 7 days for optimal recovery on these media. For laboratories using SPC agar, extending the incubation time to 3 to 5 days is recommended for improving detection of a greater portion of the heterotrophic population.

When total coliform bacteria are detected in the distribution system, immediate testing of repeat samples for coliform occurrence is a universal first response. Any subsequent detection of coliform bacteria should be augmented with a fecal coliform confirmation in EC broth (24 hours at 44.5°C) and a streak plate of M-Kleb agar (24 hours at 35°C) to determine whether the contamination event is fecal or a *Klebsiella* biofilm occurrence (Geldreich and Rice, 1987). Supplemental use of commercial biochemical kits for coliform speciation would further verify the specific identity of strains in the contamination event.

Data Interpretation

Bacteriological test results should be promptly reviewed for changes in water quality. When coliforms are detected, the laboratory finding should be verified by procedures described in *Standard Methods* (APHA/AWWA/WEF, 1995). Concurrently, another sample from the same site and an upstream and downstream location (USEPA, 1989) should be requested for a repeat coliform analysis to confirm that the contamination event still exists. Subsequent coliform positive results should activate a search in treatment protocols and water distribution management practices for the cause of contamination. With cause determined, appropriate remedial action must be applied promptly. As a precaution, lines in the affected pressure zone may need an active flushing to remove sediment accumulations that impact on diminished disinfection residuals (McFeters, Kippen, and LeChevallier, 1986; Rae, 1981).

Fecal coliform or *E. coli* occurrences are the greatest concern because they indicate a potential for fecal contamination from a cross-connection or back-siphonage problem that may be intermittent. *E. coli* typically does not colonize the pipe network; therefore, these occurrences are a signal of immediate danger. When fecal coliforms or *E. coli* are detected in a routine water sample that was positive for total coliforms, the utility should notify the appropriate state agency; take immediate measures to identify and correct the problem; and request a prompt resampling at the site (USEPA, 1989). If any repeat sample contains fecal coliforms or *E. coli,* the system would be out of compliance and a public notification would be imperative. Concurrently, joint meetings of water authorities (utility, state, and federal officials) should be held at frequent intervals (weekly or daily, if necessary) to review progress made to correct the deficiency and scan local hospitals' records on new patient admittance for evidence of an emerging waterborne disease outbreak.

Other coliforms (*Klebsiella pneumoniae, K. oxytoca, Enterobacter aerogenes, Enterobacter cloacae,* and *Citrobacter freundii*) require less nutrient support than *E. coli* and are more apt to colonize sediment and pipe surfaces. These coliforms then appear periodically in the main flow of water (particularly in warm-water periods) and may eventually be detected in samples collected throughout the distribution network. Finding coliform densities ranging from 1 to 150 organisms per 100 mL may be possible, with their occurrence widespread in the distribution system (Earnhardt, 1980; Hudson, Hawkins, and Battaglia, 1983; Wierenga, 1985). Investigation of historical records often shows this problem to be a repeat of seasonal coliform episodes during past summers, where positive sample results can range from 4 to 62 percent of all bacteriological analyses in the period (Opheim and Smith, 1989).

Because coliform colonization in the distribution system may be more an indication of biofilm development in areas of excessive sediment accumulation and rising chlorine demand, there is a need for a different type of monitoring strategy (USEPA, 1990). Provided there is no evidence of fecal contamination, one approach might be to use coliform frequency of occurrence levels as an appropriate index for establishing a stepwise action plan to avoid sudden coliform noncompliance, increases in taste and odor complaints, and loss of disinfectant residuals in portions of the pipe network. For example, most systems have an annual frequency of coliform occurrence ranging from 0.1 to 1.0 percent, with individual positive samples containing 1 to 5 coliforms per 100 mL. When the monthly percent level of coliform occurrences escalates to 5 percent, coliform densities may range from 1 to 130 organisms per 100 mL. At this time, the water utility should promptly perform a sanitary survey of source water, treatment practices, and distribution system integrity; systematically flush the entire system; and evaluate the application of either a stepwise increase in free chlorine or a switch to chloramine (1 to 2 mg/L) so that disinfectant residuals may be detected in 95 percent of all sampling locations in the pipe network. All dialysis clinics should be notified of any changes in disinfectant type and concentration proposed so that appropriate pretreatment (dechlorination) measures can be taken in their equipment to protect patients from any toxic reaction.

Risks associated with bacteria in distribution systems have recently been demonstrated to be of potential significance. It was estimated from a randomized intervention study in an area near Montreal, Canada, that 35 percent of self-reported, mild gastrointestinal illness experienced by consumers of tap water was waterborne. Over a 15-month period, a higher incidence of illness was observed among members of 307 households consuming municipal tap water, compared with 299 households supplied with reverse osmosis water treatment devices connected to the tap to remove microbial and chemical contaminants. The public water utility uses conventional treatment (predisinfection, alum, coagulation, flocculation, rapid sand filtration, ozone, and chlorine) to process a polluted river source. All current microbiological and physical limits were met, and no outbreak was reported during the study period. A second intervention study in the same water system concluded that 14 to 17 percent of gastrointestinal illness was attributed to tap water meeting current standards for coliform bacteria. Two- to five-year-old children were the most affected. An additional 40 percent of individuals consuming water from a specially designed tap that continuously purged water through the tap also reported some intestinal illness. These studies suggested that bacterial colonization and growth in the distribution system were implicated (Payment et al., 1991*a,b*).

The HPC provides a general measure of bacterial water quality in the distribution system and can become an important early sign of excessive microbial growth on distribution system pipe walls and in sediments (AWWA, 1987). With adequate disinfection, baseline heterotrophic bacteria levels in the treated water can easily be maintained at densities ranging from 10 to 100 organisms per mL depending on the medium used (SPC agar or R2A medium), incubation time, and temperature. When normal background heterotrophic plate counts in the distribution system become greater than 10^3 organisms/mL, and this is confirmed by a second sample, action should be initiated immediately to resolve the microbial growth problem, including flushing and booster application of increased disinfectant levels in the affected areas.

Heterotrophic plate count measurements for distribution samples can be used to monitor water quality during periods of adjustment or modification of the disinfection treatment process. During field studies to reduce the production of trihalomethane concentrations in finished water for one treatment system, prechlorination of the raw river water was discontinued and only postchlorination applied (Symons et al., 1981). Without readjustment in the amount of disinfectant leaving the plant, a significant impact on the microbial quality in the far reaches of the distribution system was observed. Heterotrophic plate counts and total coliform analyses were performed on samples from the ends of one section of the distribution pipe network before, during, and after these adjustments. Prior to the change in point of disinfection application, a free chlorine residual of 0.2 mg/L at the end of the system was effective in controlling coliform occurrences. The database revealed that immediately following the change in disinfectant application, the HPC density at the dead-end monitoring sites remained fairly constant for two weeks as the free chlorine residual gradually declined to 0.1 mg/L or less. When free chlorine declined to non-detectable levels in the warm water of July, the heterotrophic bacterial densities suddenly increased. Within a few more days, some coliform growth (12 to 30 organisms/100 mL) began. At this point, free chlorine application at the plant clear well was increased to restore a measurable chlorine residual in the ends of the distribution system. Immediately after this action, coliforms declined to undetectable levels followed by a reduction in the other heterotrophic bacterial densities to the levels originally encountered. This case history and other similar field experiences demonstrate that sudden deviation from the system's normal heterotrophic bacterial densities can serve as an early signal of undesirable quality changes that may precede an unsatisfactory coliform occurrence in distribution water.

Resolving Coliform Occurrences

The water utility manager should not assume that all coliform occurrences are due to biofilms, as the presence of coliforms could indicate an important treatment- or distribution-system deficiency. If potable water samples are positive for total coliforms, the possibility of a treatment breakthrough or cross-connection needs to be carefully considered. The system should take precautions to ensure that the public health is protected at all times through careful monitoring and follow-up testing. Meeting regulations may not be enough, and the utility must prove that system operations are of the highest quality. Table 18.6 is offered as an example of a checklist that utilities might use to evaluate coliform occurrences.

Systems that have never had a history of coliform occurrences in the distribution system will want to make a very cautious interpretation of positive results because of the possibility that public health is at risk. This is not to imply that systems with a long history of coliform occurrences can quickly discount the data because the results may indicate an unresolved persistent problem in treatment or deficiencies in the distribution system. In both cases, the utilities should notify the appropriate state and local regulatory, political, and public health departments, and begin a

TABLE 18.6 Checklist for Review of Plant and Distribution System

Historical and present information base
☐ Review historical records to document previous coliform occurrences or any previous waterborne outbreaks.
☐ Review current operational data.
☐ Has the utility had a recent sanitary survey and addressed all of the previous survey issues?
☐ Has the utility been in regular contact with state and local health departments to assess illness possibility attributable to microbial occurrences in drinking water?
Treatment plant
☐ Is the plant meeting all BATs for the system?
☐ Is continuous disinfection >0.2 mg/L applied?
☐ Are plant turbidities <0.5 NTU?
☐ Are total coliforms, fecal coliforms, or *E. coli* absent in plant effluent samples?
☐ Has the system applied increased monitoring to ensure integrity of treatment processes?
Distribution system
☐ Is pressure continually maintained within the system?
☐ Is a disinfectant residual maintained in all parts of the system?
☐ Does the system conduct a systematic annual flush?
☐ Does the utility maintain a log of all distribution repairs?
☐ Does the system have an active cross-connection control program?
☐ Does the utility have an established program for pipeline repair and replacement?
Water supply storage facilities
☐ Is there an annual inspection?
☐ Are reservoirs cleaned on a regular basis?
☐ Have reservoirs been lined with a material that does not promote bacteria growth?
☐ Has the system conducted an engineering analysis to determine the appropriate size and number of storage tanks?

review of plant and distribution system operational data for indicators of potential deficiencies.

Treatment Plant. The utility needs to verify that the cause of the coliform occurrence is not the result of treatment plant or distribution system operational deficiencies, and may want to prove to the regulatory agencies that the system is meeting all "best-available treatment" (BAT) requirements. In addition, a sanitary survey by an approved state agency will provide an independent evaluation of treatment plant operations. For the treatment plant, the continuity of disinfection and maintenance of low turbidity levels needs to be documented. Surface water plants should ensure that they are meeting the turbidity and disinfection requirements outlined in the Surface Water Treatment Rule (USEPA, 1989). Groundwater systems should ensure that they are not under the direct influence of surface water and adopt a wellhead protection program (USEPA, 1989). Monitoring of treatment plant effluent needs to be complete enough to document the absence of fecal coliform or *E. coli.* Any occurrence of total coliform bacteria in treated effluents has to be infrequent enough to not be associated with the problems observed in the distribution system. To increase the treatment plant effluent bacterial detection database, the utility may process high-volume samples (500 mL or more) and examine the samples for the presence of injured coliform bacteria using m-T7 media (LeChevallier and McFeters, 1985; McFeters, 1990) and HPC bacteria in >1 mL sample size, using R-2A agar.

Distribution System. Evidence of proper maintenance of the distribution system will demonstrate best-available treatment requirements (USEPA, 1991). Regular flushing helps to distribute the disinfection residual to all portions of the system, and removes portions of the existing biofilms. Flushing and cleaning the distribution system can be effective maintenance procedures, but they usually are not sufficient to resolve bacterial growth problems once the situation has developed. More aggressive cleaning, using cable-drawn or water-propelled devices (pigging), may be periodically performed to remove the corrosion and sediment buildups that provide habitats for bacteria (AWWA, 1977).

Maintenance of positive pressure in all parts of the pipe network is necessary to avoid situations where backsiphonage may allow contaminants to enter the system. A log of all distribution system maintenance work is helpful to demonstrate that coliform occurrences were not related to major pipeline repairs or main breaks. In addition, the system should have an active cross-connection program and routinely test backflow valves to ensure proper operation.

The utility should strive to maintain an effective disinfectant residual in all parts of the system. The American Water Works Association recommends that systems should maintain at least 0.5 mg/L free chlorine or 1.0 mg/L chloramines in all parts of the system, even in dead-end areas (AWWA, 1990). Recent experience in Europe suggests that the best way to minimize microbiological quality problems is to create a biological stable water, which normally means some form of biofiltration. As a result, a growing number of European utilities using biofiltration have found a maximum chlorine residual of 0.1 mg/L is effective. Such an innovative strategy may break the reliance on more and more chlorine to control microbiological quality problems in the long term.

Maintenance of Distribution System Storage Reservoirs. Reservoirs used to store finished water are constructed either below ground, above ground, or at ground level. All are subject to bacterial growth (Bernhardt and Ljesen, 1988; Schoenen, 1986). Above-ground tanks, usually constructed of steel with a corrosion-resistant liner (e.g., paint), can suffer from bacterial growth on those liners. The holding time for finished water should be limited, and an adequate disinfection residual must be maintained. Some elevated tanks can experience substantial temperature variations that may impact disinfectant stability and bacterial growth characteristics. While unprotected above-ground tanks are subject to bird contamination (such as occurred in Gideon, Missouri, Geldreich et al., 1992; Clark et al., 1996), below-ground basins and ground-level tanks may be prone to other animal and human contamination, or lined with materials that allow bacteria to attach and grow. Utilities should limit the number of storage tanks within the system to minimize opportunities for stagnation. It is important to keep the water moving throughout the system. This may be difficult, particularly in areas with many pressure zones, but utilities are encouraged to perform engineering analyses (hydraulic modeling) to determine whether the number of reservoirs can be reduced without affecting pressure, storage, or fire-fighting capabilities.

As a rule, reservoirs of treated water should be covered to avoid dissipation of disinfection residuals and to guard against contamination by wildlife, air pollution, accidental roadway spills, and surface water runoff (Clark et al., 1996; Geldreich, 1996; Bailey and Libby, 1978). Although the impact of open reservoirs on distribution system biofilm growth can be complex, eliminating them can simplify approaches to maintaining water quality (Committee Report, 1930, 1983; LeChevallier, Norton, and Atherholt [in press]).

Additional Steps to Control Coliform Growth

The previous discussion involves "good" treatment plant and distribution system operations that all systems should practice whether they experience microbiological problems or not. In some cases when systems are not following these recognized practices, the occurrence of coliform bacteria in treated waters is an appropriate indicator of inadequate treatment or maintenance of the distribution system. However, it has become recognized that some systems that follow all these guidelines and employ best-available treatment practices can still experience bacterial growth problems. Table 18.7 is a checklist aimed at providing guidance to these systems to control growth problems. The recommendations are listed in ascending order, from those that are easiest to implement to those that are most difficult to carry out.

Disinfectant Residual. Typically, one of the first responses to a positive coliform result is to increase disinfectant residuals. For systems that already maintain minimum disinfectant levels (0.5 mg/L for free chlorine or 1.0 mg/L for chloramines in all parts of the system), higher residuals can help penetrate biofilms or sediments to achieve better levels of microbial inactivation. Free chlorinated systems should attempt to maintain at least 0.5 to 1.0 mg/L residuals at dead-end sites. Chloraminated systems should set goals to maintain a minimum of 1.0 to 2.0 mg/L at dead-end locations (LeChevallier, Welch, and Smith, 1996). The advantage of this approach is that most systems can make these changes relatively quickly. In some cases, the utility should get approval from the state regulatory agency before increasing residual levels, but in all cases it is good to be in close contact with the regulatory authorities. The disadvantages of increasing disinfectant residuals include: (1) the potential for increased disinfectant by-products; (2) increased customer complaints about chlorinous tastes and odors; (3) increased chemical costs; and (4) most important, the fact that such increases may not have any impact on coliform occurrences. It is well-recognized that increased disinfection residuals alone may not suffice to eliminate coliform occurrences.

Corrosion Control. Improved corrosion control will have many benefits besides helping to control bacteriological problems. Laboratory and field data indicate that improved corrosion control will improve disinfection stability and biocidal efficiency (LeChevallier et al., 1993; Smith, Hess, and Opheim, 1989). Seasonal and annual variations in corrosivity of treated drinking water can vary as much as 10-fold. Because most water utilities do not measure corrosion rates on a daily basis, these variations in corrosivity may go unnoticed. A report by Norton and LeChevallier (1995) described the efforts of one utility to optimize corrosion inhibitor feed rates through a series of bench-scale evaluations and apply these changes full-scale. Through appropriate selection of the corrosion inhibitor product, the utility was able to improve corrosion control in the system and reduce chemical costs.

Each utility should monitor treated water corrosivity and try to reduce corrosion rates to levels as low as possible. The exact target level will vary depending on the particular water chemistry and distribution system composition, but utilities should try to achieve corrosion rates between 1 and 3 mils/yr in the treated effluent. Monitoring of corrosion rates at several points in the distribution system is recommended and can help to determine the level of "corrosion inhibitor demand."

A program of pipeline repair and replacement can target those areas of the distribution system where corrosion and bacteriological problems are most severe. Cleaning and relining of pipes can improve carrying capacities and eliminate habitats for bacterial proliferation. Care must be taken to apply effective corrosion con-

TABLE 18.7 Checklist of Additional Actions to Control Coliform Biofilm

☐ *Increase or boost disinfection residuals.*
- ☐ Residuals at dead-end sites should be 0.5 to 1.0 mg/L for free chlorine, or 1.0 to 2.0 for chloramines.

☐ *Improve corrosion control.*
- ☐ Monitor corrosion rates on a daily basis.
- ☐ Perform bench scale evaluations of corrosion control options.
- ☐ Establish program for long-term cleaning, relining, or replacement of unlined cast-iron pipes.

☐ *Change disinfectant type.*

(For free chlorinated systems)
- ☐ Evaluate conversion from free chlorine to chloramines. Be certain to consider adequate public notification, regulatory permits, equipment and storage modifications, and operator training.

(For chloraminated systems)
- ☐ Many systems find it beneficial to periodically switch back to free chlorine for a short period of time. Often this is best done when flushing the system. Be sure to provide adequate public notification, and obtain regulatory approval.

☐ *Adjust water temperature.*
- ☐ Evaluate options for blending of surface water and groundwater or adjust intake depth. Be certain to evaluate mixing and water chemistry concerns.

☐ *Reduce biodegradable organic matter levels in treated water.*
- ☐ Develop AOC/BDOC database.
- ☐ Determine appropriate target level.
- ☐ Evaluate engineering systems to meet goals:
  - Enhanced coagulation
  - Addition of powdered activated carbon
  - Replacement of filter media with GAC
  - Ozone with postfiltration GAC contactor
  - Nanofiltration membranes
- ☐ Evaluate watershed control options.

☐ *Evaluate operational practices that can influence bacterial growth.*
- ☐ Ensure that biologically active filtration follows ozonation.
- ☐ Evaluate the number and operation of storage tanks, and consolidate distribution system storage if possible.
- ☐ Evaluate distribution system operations to minimize flow reversals and reduce water-hammer effects.
- ☐ Eliminate dead-end areas.
- ☐ Conduct a program of personnel training with respect to operations, repairs, and microbial growth issues.

☐ *Install filtration for unfiltered supplies.*

trol following cleaning operations, or pipeline tuberculation may reappear. Obviously, this approach is not a "quick fix" for coliform problems, and it should not be developed as such. However, a systematic program of pipeline repair and replacement is an investment in the future health and capacity of the distribution system.

Disinfectant Type. Research has indicated that in some cases chloramines may penetrate and control coliform bacteria in biofilms better than free chlorine

(LeChevallier, Lowry, and Lee, 1990; LeChevallier et al., 1993; LeChevallier, Welch, and Smith, 1996; LeChevallier, Cawthon, and Lee, 1988; Griebe et al., 1994). Furthermore, the ability to carry and maintain residuals at the ends of the system is another important advantage of chloramination, particularly if the system does not have a large network of storage tanks where stagnation and nitrification may become a problem. Conversion from free chlorine to chloramines is a relatively easy option but nevertheless is one that should be made only after careful and thorough preparation. Misapplication of chloramine disinfection (e.g., improper chlorine-to-nitrogen ratios, free ammonia levels >0.1 mg/L, poor mixing, or low residuals) can lead to nitrification and subsequent microbiological problems (Kirmeyer et al., 1993; Kirmeyer et al., 1995).

Temperature. Realistically, there are few options that water utilities can use to reduce the summertime temperature of treated drinking water to suppress bacterial regrowth problems. Blending of surface water with groundwater can be one management approach. Additionally, a system with multiple intake levels at the reservoir could select a depth below the thermocline in a stratified source water impoundment. But in each of these cases, new water quality problems may be created, including undesirable changes in the water chemistry, taste, and odor introduced from stratified waters.

Nutrient Level. There is a growing interest in application of biological processes for reducing biodegradable organic matter in drinking water. (See Chapter 13 for more information on biofiltration.) Although this process can take many forms, one "traditional" approach described for European treatment facilities consists of a treatment train including ozone and biologically active GAC filters (Crowe and Bouwer, 1987; Rittman and Huck, 1989). Ozone followed by GAC filters with a 10- to 20-m empty-bed contact time can reduce AOC levels by 80 to 90 percent (LeChevallier et al., 1922; Sontheimer and Hubbler, 1987; Prevost et al., 1990). Employing a treatment process specifically designed for BOM removal is a long-term option for some water utilities, but it will not necessarily provide a short-term resolution to an existing growth problem.

Operational Practices. Use of ozone without subsequent biologically active filtration can be associated with high plant effluent AOC levels that stimulate growth. The number and operation of storage tanks can influence the chlorine residuals and water quality in the distribution system (Grayman and Clark, 1993). Opheim, Grochowski, and Smith (1988) showed that bacterial levels in an experimental system increased 10-fold when flows were started and stopped. This abrupt change of flows can create a "hammer" effect that can dislodge tubercles from the pipe surface. Other operational factors that may be related to control of coliform occurrences in the distribution system include care in performing pipe repairs to avoid introducing mud or foreign objects into the system, and improved distribution system circulation by elimination of dead-end areas. Inadequate treatment of surface waters can also lead to microbial problems with water quality in the distribution system. It is not surprising that unfiltered treatment operations may have higher rates of coliform occurrences than conventional treatment configurations (Geldreich, 1996; LeChevallier, Welch, and Smith, 1996). One unfiltered New York state system found a three-fold reduction in coliform levels after filtration was installed in the treatment train. These experiences illustrate the need for continuing education and technical training of treatment and distribution system personnel so that they are knowledgeable about the issues related to bacterial occurrences in drinking water.

SUMMARY

No discussion of the microbial problems that may occur in the distribution system would be complete without some recognition that the ultimate penalty for failure to maintain a constant vigil for contamination of the water supply is a waterborne outbreak. Drinking water can deteriorate in microbial quality during distribution to the consumer. Pathogens and indicator organisms can be introduced into drinking water supply by several pathways, if not effectively blocked by water supply process barriers. The distribution system environment must be kept clean (a low corrosion rate and minimal inputs of biodegradable organic matter), and the integrity of the pipe network should be maintained through adequate pressure and minimal leaks. Any organisms that pass the barriers may become established in the pipe environment and eventually form a biofilm community that often includes some total coliform bacteria (*Enterobacter, Klebsiella,* and *Citrobacter*). This heterotrophic population then undergoes periodic growth during seasonal warm water periods (provided there is sufficient nutrient base in a protective habitat), with cell aggregates released into the bulk flow of water. At other times, reversals of flow and changes in the structure of pipe sediments due to water pH shifts will cause fragmentation of the biofilm, with release into the water supply. Responses to coliform occurrences may take several different directions depending on the frequency of occurrences and the detection of fecal coliform or *E. coli.* Countermeasures include corrosion control, effective flushing programs, elimination of static water areas, and maintenance of an effective disinfectant residual throughout the distibution system.

The significance of coliform occurrences cannot be ignored, because they may indicate a potential pathway for pathogen penetration into the water supply and a pipe environment that is supportive of bacterial colonization. Because each distribution system is unique, monitoring strategies must be carefully developed to properly characterize distributed water and to provide early warning of any undesirable change in the microbial quality of the public water supply.

BIBLIOGRAPHY

Adam, G. O., and F. H. Kingsbury. "Experiences with Chlorinating New Water Mains." *Jour. New England Water Works Assn.,* 51, 1937:60–68.

Ainsworth, R. G., J. Ridgway, and R. D. Gwilliam. "Corrosion Products and Deposits in Iron Mains." Paper 8, *Proc. Confr. Water Distribution Systems,* Water Research Centre, Medmenham, England, 1978.

Allen, M. J., and E. E. Geldreich. "Bacteriological Criteria for Ground-water Quality." *Ground Water,* 13, 1975:45–52.

Allen, M. J., and E. E. Geldreich. "Distribution Line Sediments and Bacterial Regrowth." *Proc. AWWA Water Quality Tech. Conf.,* Kansas City, Missouri, 1977.

Allen, M. J., R. H. Taylor, and E. E. Geldreich. "The Occurrence of Microorganisms in Water Main Encrustations." *Jour. AWWA,* 72(11), 1980:614–625.

Alter, A. J. "Appearance of Intestinal Wastes in Surface Water Supplies at Ketchikan, Alaska." *Proc. 5th Alaska Sci. Confr. AAAS,* Anchorage, Alaska, 1954.

American Public Health Association, American Water Works Association, Water Environment Federation. *Standard Methods for the Examination of Water and Wastewater,* 19th ed., Section 9212. Washington, D.C.: American Public Health Association, 1995.

American Water Works Association. *AWWA Standard for Disinfecting Water Mains,* AWWA C651-86. Denver, Colorado: American Water Works Association, 1986.

American Water Works Association. "Distribution Main Flushing and Cleaning." In *Maintaining Distribution System Water Quality.* Denver, Colorado: American Water Works Association, 1977.

American Water Works Association. "Position Statement on Chlorine Residual." In *1993–1994 AWWA Officers and Committee Directory.* Denver, Colorado: American Water Works Association, 1990.

Amirtharajah, A., and D. P. Wetgstein. "Initial Degradation of Effluent Quality During Fitration." *Jour. AWWA,* 72 (9), 1980:518.

Armstrong, J. L., J. J. Calomiris, and R. J. Seidler. "Selection of Antibiotic-Resistant Standard Plate Count Bacteria During Water Treatment." *Appl. Environ. Microbiol.,* 44, 1982:308–316.

Armstrong, J. L., J. J. Calomiris, D. S. Shigeno, and R. J. Seidler. "Drug Resistant Bacteria in Drinking Water." *Proc. AWWA Water Quality Tech. Conf.,* Seattle, Washington, 1981.

AWWA Organisms in Water Committee, "Microbiological Considerations for Drinking Water Regulation Revisions." *Jour. AWWA,* 79(5), 1987:81–84, 88.

Bablon, G., C. Ventresque, and F. Roy. "Evolution of Organics in a Potable Water Treatment System." *Aqua* 2, 1987:110–113.

Bailey, S. W., and E. C. Libby. "Should All Finished Water Reservoirs Be Covered?" *Pub. Works* 109(4), 1978:66.

Baker, M. N. *The Quest for Pure Water,* Vol. I. Denver, Colorado: American Water Works Association, 1981.

Bays, L. R., N. P. Burman, and W. M. Lavis. "Taste and Odour in Water Supplies in Great Britain: A Survey of the Present Position and Problems for the Future." *Jour. Soc. Water Treat. Exam.,* 19, 1970:136–160.

Becker, R. J. "Bacterial Regrowth Within the Distribution System." *Proc. AWWA Water Quality Tech. Conf.,* Atlanta, Georgia, 1975.

Becker, R. J. "Main Disinfection Methods and Objectives, Part I, Use of Liquid Chlorine." *Jour. AWWA,* 61:(2), 1969:79–81.

Bedard, L., A. J. Drapeau, S. S. Kasatiya, and R. R. Plaute. "Plasmides de Resistance aux Antibiotiques Ches les Bacteries. Isolus D'Eaux Potables." *Eau Des Quebec,* 15, 1982:59–66.

Bellen, G. E., S. H. Abrishami, P. M. Colucci, and C. Tremel. *Methods for Assessing the Biological Growth Support Potential of Water Contact Materials,* pp. 1–113. Denver, Colorado: American Water Works Association Research Foundation, 1993.

Berger, P. S., and Y. Argaman. *Assessment of Microbiology and Turbidity Standards for Drinking Water,* EPA 570-9-83-001. Washington, D.C.: U.S. Environmental Protection Agency, 1983.

Bernhardt, H., and H. U. Liesen. "Bacterial Growth in Drinking Water Supply Systems Following Bitumenous Corrosion Protection Coatings." *GWF, Gas-Wasservach: Wasser/Abwasser,* 129, 1988:28–32.

Bitton, G., and K. C. Marshall. *Absorption of Microorganisms to Surfaces.* New York: Wiley-Interscience, 1980.

Bonde, G. J. *Bacterial Indicators of Water Pollution.* Copenhagen: Teknisk Forlag, 1977.

Borgstrom, S. "Principles of Food Science." In *Food Microbiology and Biochemistry,* vol. 2. London, England: Macmillan Co., Collier Macmillan Ltr., 1978.

Bourbigot, M. M., A. Dodin, and R. Lherritier. "Limiting Bacterial Aftergrowth in Distribution Systems by Removing Biodegradable Organics." *Proc. AWWA Annual Conf.,* Miami Beach, Florida, 1982.

Briganti, L. A., and S. C. Wacker. *Fatty Acid Profiling and the Identification of Environmental Bacteria for Drinking Water Utilities.* Denver, Colorado: American Water Works Association Research Foundation and American Water Works Association, 1995.

Brodeur, T. P, J. E. Singley, and J. C. Thurrott. "Effect of a Change to Free Chlorine Residual at Daytona Beach, Florida." *Proc. AWWA Water Quality Tech. Conf.,* San Diego, California, 1976.

Bucklin, K. E., G. A. McFeters, and A. Amirtharaja. "Penetration of Coliforms Through Municipal Drinking Water Filters." *Water Res.* 25, 1991:1013–1017.

Buelow, R. W., R. H. Taylor, E. E. Geldreich, A. Goodenkauf, L. Wilwerding, F. Holdren, M. Hutchinson, and I. H. Nelson. "Disinfection of New Water Mains." *Jour. AWWA,* 68 (6), 1976:283–288.

Buelow, R. W., and G. Walton. "Bacteriological Quality vs. Residual Chlorine." *Jour. AWWA,* 63(1), 1971:28–35.

Bullin, C. H., E. I. Tanner, and C. H. Collins. "Isolation of Mycobacterium xenopei from Water Taps." *Jour. Hyg. Camb.* 68, 1970:97–100.

Burlingame, G. A., and L. E. S. O'Donnell. "Coliform Sampling at Routine and Alternate Taps: Problems and Solutions." *Proc. AWWA Water Quality Tech. Conf.,* 1993.

Burman, N. P. "Taste and Odour Due to Stagnation and Local Warming in Long Lengths of Piping." *Jour. Soc. Water Treat. Exam.,* 14, 1965:125–131.

Calvert, C. K. "Investigation of Main Sterilization." *Jour. AWWA,* 31(5), 1939:832–836.

Camper, A. K. et al. *Factors Limiting Microbial Growth in the Distribution System: Laboratory and Pilot Studies.* AWWA Research Foundation and American Water Works Association, 1996.

Camper, A. K., S. C. Broadaway, M. W. LeChevallier, and G. A. McFeters. "Operational Variables and the Release of Colonized Granular Activated Carbon Particles in Drinking Water." *Jour. AWWA,* 79 (5), 1987:70–74.

Camper, A. K., M. W. LeChevallier, S. C. Broadaway, and G. A. McFeters. "Bacteria Associated with Activated Carbon Particles in Drinking Water." *Appl. Environ. Microbiol.,* 52, 1986:434–438.

Camper, A. K., M. W. LeChevallier, S. C. Broadaway, and G. A. McFeters. "Growth and Persistence on Granular Activated Carbon Filters." *Appl. Environ. Microbiol.,* 50, 1985:1378–1382.

Chang, S. L., G. Berg, N. A. Clarke, and P. W. Kabler. "Survival and Protection Against Chlorination of Human Enteric Pathogens in Free-living Nematodes Isolated from Water Supplies." *Amer. Journal Trop. Med. and Hyg.,* 9, 1960:136–142.

Chang, S. L., R. L. Woodward, and P. W. Kabler. "Survey of Free-living Nematodes and Amoebas in Municipal Supplies." *Jour. AWWA,* 52(5), 1960:613–618.

Clark, J. A., C. A. Burger, and L. E. Sabatinos. "Characterization of Indicator Bacteria in Municipal Raw Water, Drinking Water, and New Main Water Samples." *Canadian Jour. Microbiol.,* 28, 1983:1002–1013.

Clark, R. M., E. E. Geldreich, K. R. Fox, E. W. Rice, C. H. Johnson, J. A. Goodrich, J. A. Barnick, and F. Abdesaken. "Tracking a *Salmonella serovar typhimurium* outbreak in Gideon, Missouri: role of contaminant propagation modelling." *Jour. Water SRT-Aqua,* 45, 1996:171–183.

Clark, R. M., E. E. Geldreich, K. R. Fox, E. W. Rice, C. H. Johnson, J. A. Goodrich, J. A. Barnick, F. Abdesaken, J. E. Hill, and F. J. Angulo. "A Waterborne *Salmonella typhimurium* Outbreak in Gideon, Missouri: Results from a Field Investigation." *Internat. Jour. Environ. Health Res.* 6, 1996:187–193.

Cobb, N. A. "Filter-bed Nemas: Nematodes of the Slow Sand Filter-beds of American Cities." *Contr. Sci. Nematology,* 7, 1918:189–212.

Committee on the Challenges of Modern Society (NATO/CCMS). *Drinking Water Microbiology.* NATO/CCMS Drinking Water Pilot EPA 570/9-84-006, Project Series CCMS128. Washington, D.C.: U.S. Environmental Protection Agency, 1984.

Committee Report. "Aftergrowths in Water Distribution Systems." *Amer. Jour. Pub. Health,* 20, 1930:485–491.

Committee Report. "Deterioration of Water Quality in Large Distribution Reservoirs (open reservoirs)." *Jour. AWWA* 75(6), 1983:313–318.

Cooke, W. B. *The Fungi of our Mouldy Earth.* Beiheft 85 zur Nova Hedwigia. Berlin, West Germany: J. Cramer Pub., 1986.

Costeron, J. W., G. G. Geesey, and K. J. Cheng. "How Bacteria Stick." *Scientific American,* 238, 1978:86–95.

Coulbourne, J. S., D. J. Pratt, M. G. Smith, S. P. Fisher-Hoch, and D. Harper. "Water fittings as Sources of *Legionella pneumophila* in Hospital Plumbing System." *Lancet,* i, 1984:210–213.

Council of the European Communities. "Proposal for a Council Directive Relating to the Quality of Water for Human Consumption." *Official Jour. European Communities,* 18 (C214/2), 1975.

Cox, W., and J. J. Giron. "Gimmicks and Gadgets: New Taps Reduce False Positive Coliform Samples." *Opflow,* 19(7), 1993:6–7.

Craun, G. F. "A Summary of Waterborne Illness Transmitted Through Contaminated Groundwater." *Jour. Environ. Health,* 43, 1985:122.

Craun, G. F. "Impact of the Coliform Standard on the Transmission of Disease." In *Evaluation of the Microbiology Standards for Drinking Water,* C. W. Hendricks, ed., EPA-570/9-78-002. Washington, D.C.: Office of Drinking Water, U.S. Environmental Protection Agency, 1978.

Craun, G. F. Waterborne Outbreaks of Giardiasis." In *Giardia and Giardiasis,* S. L. Erlandsen and E. A. Meyer, eds. New York: Plenum Press, 1984.

Crowe, P. B., and E. J. Bouwer. *Assessment of Biological Processes in Drinking Water.* Denver, Colorado: American Water Works Association Research Foundation, 1987.

Davis, A. R. "The Distribution System." In *Manual for Water Works Operators,* L. C. Billings, ed. Austin, Texas: Texas State Department of Health and the Texas Water and Sanitation Research Foundation, 1951.

Department of Health and Social Security. *The Bacteriological Examination of Water Supplies: Reports on Public Health and Medical Subjects,* vol. 71. London: 1969.

Distribution System Water Quality Committee. *Distribution System Bacteriological Sampling and Control Guidelines,* California—Nevada Section. American Water Works Association, 1978.

Dott, W., and D. Waschko-Dransmann. "Occurrence and Significance of Actinomycetes in Drinking Water." *Zbl. Bakt. Hyg. I., Abt. Orig. B,* 173, 1981:217–232.

du Moulin, G. C., I. H. Sherman, D. C. Hoaglin, and K. D. Stottmeier. "Mycobacterium avinum Complex, an Emerging Pathogen in Massachusetts." *Jour. Clin. Microbiol.,* 22, 1985:9–12.

du Moulin, G. C., and K. D. Stottmeier. "Waterborne Mycobacteria: An Increasing Threat to Health." *ASM News,* 52, 1986:525–529.

Dunnigan, A. P. "Microbiological Control of Cosmetic Products." In *Federal Regulations and Practical Control Microbiology for Disinfectants, Drugs, and Cosmetics,* Soc. Indust. Microbiol. Special Publication no. 4, 1969.

Earnhardt, K. B. "Chlorine Resistant Coliforms—The Muncie, Indiana, Experience." *Proc. AWWA Water Quality Tech. Conf.,* Miami Beach, Florida, 1980.

Edberg, S. C., M. J. Allen, and D. B. Smith. "National Field Evaluation of a Defined Substrate Method for the Simultaneous Enumeration of Total Coliform and *Escherichia coli* from Drinking Water: Comparison with the Standard Multiple Tube Fermentation Method." *Appl. Environ. Microbiol.,* 54, 1988:1595–1601.

El-Zanfaly, H. T., E. A. Kassein, and S. M. Badr-Eldin. "Incidence of Antibiotic Resistant Bacteria in Drinking Water in Cairo." *Water, Air and Soil Poll.,* 32, 1987:123–128.

Emde, K. M. E., D. W. Smith, and R. Facey. "Initial Investigation of Microbially Influenced Corrosion (MIC) in a Low Temperature Water Distribution System." *Water Res.* 26, 1992:169–175.

Engelbrecht, R. S., and C. N. Haas. "Acid-Fast Bacteria and Yeast as Disinfection Indicators: Enumeration Methodology." *Proc. AWWA Water Quality Tech. Conf.,* Kansas City, Missouri, 1977.

Environmental Protection Agency. *National Interim Primary Drinking Water Regulations,* EPA-570/9-76-003. Washington, D.C., Office of Water Supply, Government Printing Office, 1976.

Facey, R. M., D. W. Smith, and K. M. E. Emde. "Case Study: Water Distribution Corrosion, Yellowknife, N. W. T." In *Microbially Influenced Corrosion and Biodeterioration,* N. J. Dowling, M. W. Mittleman, and J. C. Danko, eds., pp. 4–45. Knoxville, Tennessee: Institute for Applied Microbiology, 1990.

Favero, M. S., N. J. Peterson, K. M. Bayer, L. A. Carson, and W. W. Bond. "Microbial Contaminations of Renal Dialysis Systems and Associated Health Risks." *Trans. Amer. Soc. Artificial Internal Organs,* XX-A, 1974:175–183.

Favero, M. S., N. J. Peterson, L. A. Carson, W. W. Bond, and S. H. Hindman. "Gram-Negative Water Bacteria in Hemodialysis Systems." *Health Lab. Sci.,* 12, 1975:321–334.

Fennel, H., D. B. James, and J. Morris. "Pollution of a Storage Reservoir by Roosting Gulls." *Jour. Soc. Water Treat. Exam.*, 23, 1974:5–24.

Fields, B. S., E. B. Shotts, J. C. Freeley, G. W. Gorman, and W. T. Martin. "Proliferation of *Legionella* as an Intracellular Parasite of the Ciliated Protozoan, *Tetrahymena pyriformis.*" *Appl. Environ. Microbiol.*, 47, 1984:467–471.

Fischer, G. "The Antagonistic Effect of Aerobic Sporulating Bacteria on the Coli-aerogenes Group." *Zeit. Immun. u. Exp. Ther.*, 107, 1950:16–22.

Foster, D. M., M. A. Emerson, C. E. Buck, D. S. Walsh, and O. J. Sproul. "Ozone Inactivation of Cell- and Fecal-Associated Viruses and Bacteria." *Jour. Water Poll. Control. Fed.*, 52, 1980:2174–2184.

Frontinus, S. J. (A.D. 97) *The Water Supply of the City of Rome.* Translated by C. Herschel. Boston, Massachusetts: New England Water Works Association, 1973.

Geldreich, E. E. *Handbook for Evaluating Water Bacteriological Laboratories.* EPA-670/9-75-006. Cincinnati, Ohio: U.S. Environmental Protection Agency, 1975.

Geldreich, E. E. *Microbial Quality of Water Supply in Distribution Systems.* Boca Raton, Florida: Lewis Pub., 1996.

Geldreich, E. E., K. R. Fox, J. A. Goodrich, E. W. Rice, R. M. Clark, and D. L. Swedlow. "Searching for a Water Supply Connection in the Cabool, Missouri Disease Outbreak of *Escherichia coli 0157:H7.*" *Water Res.*, 1992:1127–1137.

Geldreich, E. E., H. D. Nash, D. J. Reasoner, and R. H. Taylor. "The Necessity of Controlling Bacterial Populations in Potable Waters: Community Water Supply." *Jour. AWWA*, 64(9), 1972:596–602.

Geldreich, E. E., H. D. Nash, and D. Spino. "Characterizing Bacterial Populations in Treated Water Supplies: A Progress Report." *Proc. AWWA Water Quality Tech. Conf.*, Kansas City, Missouri, December 4–7, 1977.

Geldreich, E. E., and E. W. Rice. "Occurrence, Significance, and Detection of *Klebsiella* in Water Systems." *Jour. AWWA*, 79(5), 1987:74–80.

Geldreich, E. E., R. H. Taylor, J. C. Blannon, and D. J. Reasoner. "Bacterial Colonization of Point-of-Use Water Treatment Devices." *Jour. AWWA*, 77, 1985:72–80.

George, M. G. "Further Studies on the Nematode Infestation of Surface Water Supplies." *Environ. Health*, 8, 1966:93–102.

Gerardi, M. H., and J. K. Grimm. "Aquatic Invaders." *Water/Engineering and Management*, 10, 1982:22–23.

Goatcher, L. J., K. A. Simpson, and K. M. E. Emde. "Are All Positive Microbiological Samples Created Equal?" *Proc. AWWA Water Quality Tech. Conf. Toronto, Ontario*, 1992.

Grayman, W. M., and Clark, R. M. "Using Computer Models to Determine the Effect of Storage on Water Quality." *Jour. AWWA*, 85(7), 1993:67–77.

Griebe, T., C. I. Chen, R. Srinivasan, and P. S. Stewart. "Analysis of Biofilm Disinfection by Monochloramine and Free Chlorine." In *Biofouling and Biocorrosion in Industrial Water Systems*, G. G. Geesey, Z. Lewandowski, and H. C. Flemming, eds. Boca Raton, Florida: Lewis Publishers, 1994.

Haas, C. N., P. Bitter, and P. Scheff. "Determination of the Limiting Nutrients for Indigenous Bacteria in Chicago Intake Water." *Water, Air, Soil Poll.* 37, 1988:65–72.

Haas, C., M. A. Meyer, and M. S. Fuller. "The Ecology of Acid-Fast Organisms in Water Supply, Treatment, and Distribution Systems." *Jour. AWWA*, 75(3), 1983:139–144.

Haas, C. N., M. A. Meyer, and M. S. Paller. "Microbial Alterations in Water Distribution Systems and Their Relationships to Physical-Chemical Characteristics." *Jour. AWWA*, 75(9), 1983:475–481.

Haas, C. N., M. A. Meyer, and M. S. Paller. "Microbial Dynamics in GAC Filtration of Potable Water." *Ameri. Soc. Civil Eng. Jour. Environ. Engineering Div.*, 109, 1983:956–961.

Haas, C., M. A. Meyer, and M. S. Paller. "The Ecology of Acid-Fast Organisms in Water Supply, Treatment, and Distribution System." *Jour. AWWA.*, 75, 1983:139–144.

Hamilton, J. J. "Potassium Permanganate as a Main Disinfectant." *Jour. AWWA,* 66(12), 1974:734–738.

Harem, F. E., K. D. Bielman, and J. E. Worth. "Reservoir Linings." *Jour. AWWA,* 68(5) 1976:238–242.

Harold, C. H. H. *29th Report Director of Water Examination.* London, England: Metropolitan Water Board, 1934.

Hejkal, T. W., F. M. Wellings, P. A. LaRock, and A. L. Lewis. "Survival of Poliovirus within Organic Solids During Chlorination." *Appl. Environ. Microbiol.,* 38, 1979:114–118.

Herman, L. "Sources of the Slow-growing Pigmented Water Bacteria." *Health Lab. Sci.,* 13, 1976:5–10.

Herman, L. G., and C. K. Himmelsbach. "Detection and Control of Hospital Sources of *Flavobacteria.*" *Hospitals,* 39, 1965:72–76.

Herson, D. S. "Identification of Coliform Antagonists." *Proc. AWWA Water Quality Tech. Conf.,* Miami Beach, Florida, 1980.

Herson, D. S., D. R. Marshall, and H. T. Victoreen. "Bacterial Persistence in the Distribution System." *Proc. AWWA Water Quality Tech. Conf.,* Denver, Colorado, 1984.

Herson, D. S., and H. Victoreen. "Identification of Coliform Antagonists." *Proc. AWWA Water Quality Tech. Conf.,* Miami Beach, Florida, 1980.

Hiisvirta, L. O. "Problems of Disinfection of Surface Water with a High Content of Natural Organic Material." *Water Supply: Rev. Jour. Internat. Water Supply Assoc.,* 4, 1986:53–58.

Hinzelin, F., and J. C. Block. "Yeasts and Filamentous Fungi in Drinking Water." *Environ. Technol. Letters,* 6, 1985:101–106.

Hoff, J. C. *Inactivation of Microbial Agents by Chemical Disinfectants,* EPA/600/52-86/067. Cincinnati, Ohio: U.S. Environmental Protection Agency, 1986.

Hoff, J. C. "The Relationships of Turbidity to Disinfection of Potable Water." In *Evaluation of the Microbiology Standards for Drinking Water,* C. W. Hendricks, ed., EPA-570/9-78-002. Washington, D.C.: U.S. Environmental Protection Agency, 1978.

Hoskins, J. K. "Revising the U.S. Standards for Drinking Water Quality: Some Considerations in the Revision." *Jour. AWWA,* 33(10), 1941:1804–1831.

Howard, M. J. "Bacterial Depreciation of Water Quality in Distribution Systems." *Jour. AWWA,* 32(9), 1940:1501–1506.

Hudson, L. D., H. W. Hawkins, and M. Battaglia. "Coliforms in a Water Distribution System: A Remedial Approach." *Journal AWWA,* 75(11), 1983:564–568.

Hutchinson, D., R. H. Weaver, and M. Scherago. "The Incidence and Significance of Microorganisms Antagonistic to *Escherichia coli* in Water." *Jour. Bact.,* 45, 1943: Abstract G34, 29.

Hutchinson, M. "The Disinfection of New Water Mains." *Chemistry and Industry,* 139(1), 1971:139–142.

Hutchinson, M. "WRA Medlube: An Aid to Mains Disinfection." *Jour. Soc. Water Treat. Exam.,* 23(part II) 1974:174.

Hutchinson, M., and J. W. Ridgway. "Microbiological Aspects of Drinking Water Supplies." In *Aquatic Microbiology,* F. A. Skinner and J. M. Schewan, eds. London, England: Academic Press, 1977.

Ike, N. R., R. L. Wolfe, and E. G. Means. "Nitrifying Bacteria in a Chloraminated Drinking Water System." *Wat. Sci. Technol.,* 20, 1988:441–444.

Jacobs, N. J., W. L. Ziegler, F. C. Reed, T. A. Stukel, and E. W. Rice. "Comparison of Membrane Filter, Multiple-Fermentation-Tube, and Presence-Absence Techniques for Detecting Total Coliforms in Small Community Water Systems." *Appl. Environ. Microbiol.,* 51, 1986:1007–1012.

Janssens, J. G., J. Meheus, and J. Dirickx. "Ozone Enhanced Biological Activated Carbon Filtration and its Effect on Organic Matter Removal, and in Particular on AOC Reduction." *Water Sci. Technol.,* 17, 1984:1055–1068.

Joret, J. C. "Rapid Methods for Estimating Bioliminable Organic Carbon in Water." *Proc. AWWA Water Quality Tech. Conf.,* St. Louis, Missouri, 1988.

Kabler, P. W., and H. F. Clark. "Coliform Group and Fecal Coliform Organisms as Indicators of Pollution in Drinking Water." *Jour. AWWA,* 52(12), 1960:1577–1579.

Kirmeyer, G. J., G. W. Foust, G. L. Pierson, J. J. Simmler, and M. W. LeChevallier. *Optimizing Chloramine Treatment.* AWWA Research Foundation and American Water Works Association, 1993.

Kirmeyer, G. J., L. H. Odell, J. Jacangelo, A. Wilczak, and R. Wolfe. *Nitrification Occurrence and Control in Chloraminated Water Systems.* Denver, Colorado: American Water Works Association Research Foundation, 1995.

Knittel, M. D., R. J. Seidler, and L. M. Cabe. "Colonization of the Botanical Environment by Kelbsiella Isolates of Pathogenic Origin." *Appl. Environ. Microbiol.,* 34, 1977:557–563.

Langelier, W. F. "The Analytical Control of Anti-Corrosion Treatment." *Jour. AWWA,* 28(10), 1936:1500–1521.

Laskin, A. I., and LeChevalier, H. A. (eds). *Handbook of Microbiology,* 2nd. ed. CRC Press, 1977.

LeChevallier, M. W., T. M. Babcock, and R. G. Lee. "Examination and Characterization of Distribution System Biofilms." *Appl. Environ. Microbiol.,* 53, 1987:2714–2724.

LeChevallier, M. W., W. C. Becker, P. Schorr, and R. G. Lee. "Evaluating the Performance of Biological Active Rapid Filters." *Jour. AWWA,* 84(4), 1992:136–146.

LeChevallier, M. W., C. D. Cawthon, and R. G. Lee. "Factors Promoting Survival of Bacteria in Chlorinated Water Supplies." *Appl. Environ. Microbiol.,* 54, 1988:649–654.

LeChevallier, M. W., T. M. Evans, and R. J. Seidler. "Effect of Turbidity on Chlorination Efficiency and Bacterial Persistence in Drinking Water." *Appl. Environ. Microbiol.,* 42, 1981:159–167.

LeChevallier, M. W., C. D. Lowry, and R. G. Lee. "Disinfection of Biofilms in a Model Distribution System." *Jour. AWWA,* 82(7), 1990:87–99.

LeChevallier, M. W., C. D. Lowry, R. G. Lee, and D. L. Gibbon. "Examining the Relationship between Iron Corrosion and the Disinfection of Biofilm Bacteria." *Jour. AWWA,* 85(7), 1993:111–123.

LeChevallier, M. W., and G. A. McFeters. "Enumerating Injured Coliforms in Drinking Water." *Jour. AWWA,* 77(6), 1985:81–87.

LeChevallier, M. W., W. D. Norton, and T. B. Atherholt. "Protozoa in Open Reservoirs." *Jour. AWWA,* 89(9), 1997: 84–96.

LeChevallier, M. W., R. J. Seidler, and T. M. Evans. "Enumeration and Characterization of Standard Plate Count Bacteria by Chlorinated and Raw Water Supplies." *Appl. Environ. Microbiol.,* 40, 1980:922–930.

LeChevallier, M. W., N. J. Welch, and D. B. Smith. "Full-Scale Studies of Factors Related to Coliform Regrowth in Drinking Water." *Appl. Environ. Microbiol.,* 62(7), 1996:2201–2211.

Lee, S. H., J. T. O'Connor, and S. K. Banerji. Biologically Mediated Corrosion and its Effects on Water Quality in the Distribution System. *Jour. AWWA,* 72(11), 1980:636–645.

Levy, R. V., F. L. Hart, and R. D. Cheetham. "Occurrence and Public Health Significance of Invertebrates in Drinking Water Systems." *Jour. AWWA,* 78(9), 1986:105–110.

Levy, R. V., R. D. Cheetham, J. Davis, G. Winter, and F. L. Hart. "Method for Studying the Public Health Significance of Macroinvertibrates Occurring in Potable Water." *Appl. Environ. Microbiol.,* 47, 1984:889–894.

Levy, R. V., R. D. Cheetham, J. Davis, G. Winter, and F. L. Hart. "Novel Method for Studying the Public Health Significance of Macroinvertebrates Occurring in Potable Water." *Appl. Environ. Microbiol.,* 47, 1984:889–894.

Logsdon, G. S. "Comparison of Some Filtration Processes Appropriate for *Giardia* Cyst Removal." *Proc. Calgary Giardia Conference,* Calgary, Canada, February 23–25, 1987.

MacKenthun, K. M., and L. E. Keup. "Biological Problems Encountered in Water Supplies." *Jour. AWWA,* 62(8), 1970:520–526.

Mackle, H. et al. "Koloniezahlerhohung sowie. Geruchs-und Geschmachs-biemtrachtigungen des Trinkwassers durch Losemettelhaltige Auskleidermaterialien." *GWF, Gas-Wasserfach: Wasser/Abwasser,* 129, 1988:22–27.

Martin, R. S., W. H. Gates, R. S. Tobin, D. Grantham, et al. "Factors Affecting Coliform Bacterial Growth in Distribution Systems." *Jour. AWWA,* 74(1), 1982:34–37.

McCabe, L. J. "Trace Metals Content of Drinking Water from a Large System." Presented at Symposium on Water Quality in Distribution Systems, Amer. Chem. Society National Meeting, Minneapolis, Minnesota, April 13, 1969.

McCauley, R. "Controlled Deposition of Protection Calcite Coating in Water Mains." *Jour. AWWA,* 52(11), 1960:1386–1396.

McFeters, G. A. "Enumeration, Occurrence and Significance of Injured Indicator Bacteria in Drinking Water." In *Drinking Water Microbiology,* G. A. McFeters, ed., pp. 478–492. Springer-Verlag, 1990.

McFeters, G. A., J. S. Kippin, and M. W. LeChevallier. "Injured Coliforms in Drinking Water." *Appl. Environ. Microbiol.* 51, 1986:1–5.

Ministry of National Health and Welfare. *Microbiological Quality of Drinking Water.* Ottawa, Canada: Health and Welfare, 1977.

Morita, R. V. *Sampling Regimes and Bacteriological Tests for Coliform Detection in Groundwater.* Project Report EPA 600/287/083. Cincinnati, Ohio: U.S. Environmental Protection Agency, 1987.

Mott, J. B., and A. D. Harrison. "Nematodes from River Drift and Surface Drinking Water Supplies in Southern Ontario." *Hydrobiologia,* 102, 1983:27–38.

Nagy, L. A., and B. H. Olson. "The Occurrence of Filamentous Fungi in Drinking Water Distribution Systems." *Canadian Jour. Microbiol.,* 28, 1982:667–671.

Niemi, R. M., S. Knuth, and K. Lundstrom. "Actinomycetes and Fungi in Surface Waters and in Potable Water." *Appl. Environ. Microbiol.,* 43, 1982:378–388.

Norton, W. D., and M. W. LeChevallier. *Implementation of Corrosion Control to Limit Microbial Activity in the Distribution System.* American Water Works Service Company, 1995.

Olson, B. *Assessment and Implications of Bacterial Regrowth in Water Distribution Systems,* EPA-600/52-82-072. Cincinnati, Ohio: U.S. Environmental Protection Agency, 1982.

Olson, B. H., and L. Hanami. "Seasonal Variation of Bacterial Populations in Water Distribution Systems." *Proc. AWWA Water Quality Tech. Conf.,* Miami Beach, Florida, 1980.

Olson, S. W. "The Application of Microbiology to Cosmetic Testing." *Jour. Soc. Cosmetic Chem.,* 18, 1967:191–198.

Opheim, D., and D. B. Smith. "Control of Distribution System Coliform Regrowth." *Proc. AWWA Water Quality Tech. Conf.,* Philadelphia, Pennsylvania, 1989.

Opheim, D., J. G. Grochowski, and D. Smith. "Isolation of Coliforms from Water Main Tubercles." *Abst. Ann. Meet. Amer. Soc. Microbiol.,* 1988:245

Pascal, O., J. C. Joret, Y. Levi, and T. Dupin. *Bacterial Aftergrowth in Drinking Water Networks Measuring Biodegradable Organic Carbon (BDOC).* Cincinnati, Ohio: U.S. Environmental Protection Agency, 1986.

Payment, P., E. Franco, L. Richardson, and J. Siemiatycki. "Gastrointestinal Health Effects Associated with the Consumption of Drinking Water Produced by Point-of-use Domestic Reverse Osmosis Filtration Units." *Appl. Environ. Microbiol.,* 57, 1991:945–948.

Payment, P., L. Richardson, J. Siemiatycki, R. Dewar, M. Edwardes, and E. A. Franco. "A Randomized Trial to Evaluate the Risk of Gastrointestinal Disease due to Consumption of Drinking Water Meeting Current Microbiological Standards." *Amer. Jour. Pub. Health,* 81, 1991:703–708.

Payment, P., M. Trudel, and R. Plante. "Elimination of Viruses and Indicator Bacteria at Each Step of Treatment During Preparation of Drinking Water at Seven Water Treatment Plants." *Appl. Environ. Microbiol.,* 49, 1985:1418–1428.

Pelletier, P. A., G. C. du Moulin, and K. D. Stottmeier. "Mycobacteria in Public Water Supplies: Comparative Resistance to Chlorine." *Microbiol. Sci.,* 5, 1988:147–148.

Pipes, W. O. "Monitoring of Microbial Water Quality." In *Assessment of Microbiology and Turbidity Standards for Drinking Water,* P. S. Berger and Y. Argaman, eds., EPA-570/9-83-001. Washington, D.C.: Office of Drinking Water, U.S. Environmental Protection Agency, 1983.

Postgate, J. R., and J. R. Hunter. "The Survival of Starved Bacteria." *Jour. Gen. Microbiol.,* 29, 1962:233–263.

Prevost, M., R. Desjardins, J. Coallier, D. Duchesne, and J. Mailly. "Comparison of Biodegradable Organic (BOC) Techniques for Process Control." *Proc. AWWA Water Quality Technol. Conf.,* San Diego, California, 1990.

Ptak, D. J., W. Ginsburg, and B. F. Willey. "Identification and Incidence of Klebsiella in Chlorinated Water Supplies." *Jour. AWWA,* 65(9), 1973:604–608.

Quarles, J. M., R. C. Belding, T. C. Beaman, and P. Gerhardt. "Hemodialysis Culture of Serratia marcescens in a Goat-Artificial Kidney-Fermentor System." *Infect. Immun.,* 9, 1974:550.

Rae, J. F. "Algae and Bacteria: Dead End Hazard." *Proc. AWWA Water Quality Tech. Conf.,* Seattle, Washington, 1981.

Reasoner, D. J., J. C. Blannon, and E. E. Geldreich. "Nonphotosynthetic Pigmented Bacteria in a Potable Water Treatment and Distribution System." *Appl. Environ. Microbiol.,* 55, 1989:912–921.

Reasoner, D. J., and E. E. Geldreich. "A New Medium for the Enumeration and Subculture of Bacteria from Potable Water." *Appl. Environ. Microbiol.,* 49, 1985:1–7.

Reilly, J. K., and J. Kippin. *Interrelationship of Bacterial Counts with Other Finished Water Quality Parameters Within Distribution Systems,* EPA-0600/52-81-035. Cincinnati, Ohio: U.S. Environmental Protection Agency, 1981.

Ridgway, H. F., and B. H. Olson. "Chlorine Resistance Patterns of Bacteria from Two Drinking Water Distribution Systems." *Appl. Environ. Microbiol.,* 44, 1982:972–987.

Ridgway, H. F., and B. H. Olson. "Scanning Electron Microscope Evidence for Bacteria Colonization of a Drinking Water Distribution System." *Appl. Environ. Microbiol.,* 41, 1981:274–287.

Ridgway, J., R. G. Ainsworth, and R. D. Gwilliam. "Water Quality Changes—Chemical and Microbiological Studies." *Proc. Conf. Water Distribution Systems,* Water Research Centre, Medmenham, England, 1978.

Rittmann, B. E., and P. M. Huck. "Biological Treatment of Water Supplies." *CRC Rev.,* 19(2), 1989:119–184.

Rittmann, B. E., and V. L. Snoeyink. "Achieving Biologically Stable Drinking Water." *Jour. AWWA,* 76(10), 1984:106–114.

Rosenzweig, W. D. *Influence of Phosphate Corrosion Control Compounds on Bacterial Growth,* EPA/600/S2-87/045. Cincinnati, Ohio: U.S. Environmental Protection Agency, 1987.

Rosenzweig, W. D., H. A. Minnigh, and W. O. Pipes. "Chlorine Demand and Inactivation of Fungal Propagules." *Appl. Environ. Microbiol.,* 45, 1983:182–186.

Russelman, H. B. "Main Disinfection Methods and Objectives, Public Health Viewpoint." *Jour. AWWA,* 61(2), 1969:82.

Safe Drinking Water Committee. *Drinking Water and Health,* vol. 1. Washington, D.C.: National Academy of Sciences, 1977.

Sarai, D. S. "Total and Fecal Coliform Bacteria in Some Aquatic and Other Insects." *Environ. Entomol.,* 5, 1976:365–367.

Schoenen, D. "Microbial Growth Due to Materials Used in Drinking Water Systems." In *Biotechnology,* vol. 8, H. J. Rehm and G. Reed, eds. Weinheim: VCH Verlagsgesellschaft, 1986.

Schubert, R. H. N. "Das Vorkenmen der aeromonadin in Oberirdischen Gewassern." *Arch. Hyg.* 150, 1967:688–708.

Seidler, R. J., J. E. Morrow, and S. T. Bagley. "*Klebsiella* in Drinking Water Emanating from Redwood Tanks." *Appl. Environ. Microbiol.,* 33, 1977:893–905.

Servais, P., G. Billen, and M. C. Hascoet. "Determination of the Biodegradable Fraction of Dissolved Organic Matter in Waters." *Water Res.* 21, 1987:445–450.

Servais, P., P. Laurent, and G. Randon. "Impact of Biodegradable Dissolved Organic Carbon (BDOC) on Bacterial Dynamics in Distribution Systems." *Proc. AWWA Water Quality Tech. Conf.,* Miami, Florida, 1993.

Small, I. C., and G. F. Greaves. "A Survey of Animals in Distribution Systems." *Jour. Soc. Water Treat. Exam.,* 19, 1968:150–183.

Smerda, S. M., H. J. Jensen, and A. W. Anderson. "Escape of Salmonellae from Chlorination During Ingestion by *Pristionchus iheretieri* (*Nematoda: Diployasterinae*)." *Jour. Nematol.,* 3, 1971:201–204.

Smith, D. B., A. F. Hess, and S. Hubbs. "Survey of Distribution System Coliform Occurrence in the United States." *Proc. AWWA Water Quality Tech. Conf.,* San Diego, California, 1990.

Smith, D. B., A. F. Hess, and D. Opheim. "Control of Distribution System Coliform Regrowth." *Proc. AWWA Water Quality Technol. Conf.,* Philadelphia, Pennsylvania, 1989.

Snead, M. C., V. P. Olivieri, K. Kawata, and C. W. Kruse. "Biological Evaluation of Benefits of Maintaining a Chlorine Residual in Water Supply Systems." *Water Res.,* 14, 1980:403–408.

Sontheimer, H., and C. Hubble. "The Use of Ozone and Granular Activated Carbon." In *Drinking Water for Organic Contaminants,* P. M. Huck and P. Toft, eds. Pergamon Press, 1987.

Staley, J. T. *Identification of Unknown Bacteria from Drinking Water,* Project CR-807570010. Cincinnati, Ohio: U.S. Environmental Protection Agency, 1983.

Stewart, M. H., R. L. Wolf, and E. G. Means. "Assessment of the Bacteriological Activity Associated with Granular Activated Carbon Treatment of Drinking Water." *Appl. Environ. Microbiol.,* 56, 1990:3822–3829.

Stumm, W. "The Corrosive Behavior of Water." *Proc. Amer. Soc. Civil. Engineers,* 86, No. SA-6, 1960.

Suckling, E. V. *The Examination of Waters and Water Supplies,* 5th edition. 1943.

Symons, J. M., et al. *Treatment Techniques for Controlling Trihalomethanes in Drinking Water,* EPA-600/2-81-156. Cincinnati, Ohio: U.S. Environmental Protection Agency, 1981.

Symons, J. M., T. A. Bellar, J. K. Carswell, J. DeMarco, K. L. Kropp, G. G. Robeck, D. R. Seeger, C. J. Slocum, B. L. Smith, and A. A. Stevens. "National Organics Reconnaissance Survey for Halogenated Organics." *Jour. AWWA,* 67(11), 1975:634–647.

Talbot, H. W. Jr., J. E. Morrow, and R. J. Seidler. "Control of Coliform Bacteria in Finished Drinking Water Stored in Redwood Tanks." *Jour. AWWA,* 71(6), 1979:349–353.

Talbot, H. W. Jr., and R. J. Seidler. "Gas Chromatographic Analysis of In-Situ Cyclitol Utilization by *Klebsiella* Growing in Redwood Extracts." *Appl. Environ. Microbiol.,* 38, 1979:599–605.

Taylor, E. W. *43rd Report, Metropolitan Water Board.* London: Metropolitan Water Board, 1967–1968.

Taylor, E. W. *The Examination of Water and Water Supplies,* 7th edition. Boston: Little, Brown, 1958.

Taylor, R. H., M. J. Allen, and E. E. Geldreich. "Standard Plate and Spread Plate Methods." *Jour. AWWA,* 75(1), 1983:35–37.

Taylor, R. H., and E. E. Geldreich. "A New Membrane Filter Procedure for Bacterial Counts in Potable Water and Swimming Pool Samples." *Jour. AWWA,* 71(7), 1979:402–405.

Technical Subcommittee. "Manual of Recommended Water Sanitation Practice Accompanying United States Public Health Service Drinking Water Standards, 1942." *Jour. AWWA,* 35(2), 1942:135–188.

Tenenbaum, S. "Pseudomonads in Cosmetics." *Jour. Soc. Cosmetic Chem.,* 18, 1967:797–807.

Thofern, E., D. Schoenen, and G. J. Tuschewitzki. "Microbial Surface Colonization and Disinfection Problems." *Off Gesundh.-wes.,* 49, 1987:14–20.

Thomas, S. B., C. A. Scarlett, W. A. Cuthbert, et al. "The Effect of Flaming of Taps Before Sampling of the Bacteriological Examination of Farm Water Supplies." *Journal Appl. Bacteriol.,* 17, 1975:175–181.

Tombes, A. S., A. R. Abernathy, D. M. Welch, and S. A. Lewis. "The Relationship Between Rainfall and Nematode Density in Drinking Water." *Water Res.,* 13, 1979:619–622.

Tracy, H. W., V. M. Camarena, and F. Wing. "Coliform Persistence in Highly Chlorinated Waters." *Jour. AWWA,* 58 (9), 1966:1151–1159.

Tuovinen, O. H., K. S. Button, A. Vuorinen, L. Carison, et al. "Bacterial, Chemical and Mineralogical Characteristics of Tubercles in Distribution Pipelines." *Jour. AWWA,* 72(11), 1980:626–635.

Tuovinen, O. H., and J. C. Hsu. "Aerobic and Anaerobic Microorganisms in Tubercles of the Columbus, Ohio, Water Distribution System." *Appl. Environ. Microbiol.,* 44, 1982:761–764.

Tyndall, R. L., and E. L. Domingue. "Cocultivation of *Legionella* and Free Living Amoebae." *Appl. Environ. Microbiol.,* 44, 1982:954–959.

U.S. Environmental Protection Agency. *Control of Biofilm in Drinking Water Distribution Systems.* Seminar Pub. 625/R-92/001. Washington D.C.: Office of Research and Development, U.S. Environmental Protection Agency, 1990.

U.S. Environmental Protection Agency. "Drinking Water: National Primary Drinking Water Regulations; Total Coliforms." *Federal Register,* 56(10), 1991:1556.

U.S. Environmental Protection Agency. "National Primary Drinking Water Regulations, Filtration and Disinfection; Turbidity; *Giardia lamblia,* Viruses, *Legionella,* and Heterotrophic Bacteria." Federal Register 54(124), 1989:27486.

U.S. Environmental Protection Agency. "National Primary Drinking Water Regulations; Total Coliforms (including Fecal Coliforms and *E. Coli*)." Federal Register, 54, 1989:124:27544.

Van der Kooij, D. *The Effect of Treatment on Assimilable Organic Carbon in Drinking Water.* London: Pergamon Press, 1987.

Van der Kooij, D., and W. A. M. Hijnen. "Measuring the Concentration of Easily Assimilable Organic Carbon (AOC) Treatment as a Tool for Limiting Regrowth of Bacteria in Distribution Systems." *Proc. AWWA Water Quality Tech. Conf.* Houston, Texas, 1985.

Van der Kooij, D., A. Visser, and W. A. M. Hijnen. "Determining the Concentration of Easily Assimilable Organic Carbon in Drinking Water." *Jour. AWWA,* 74, 1982:540–545.

Vander Kooij, D., and B. C. J. Zoeteman. "Water Quality in Distribution Systems." Special Subject 5. *Internat. Water Supply Assoc. Congress,* Kyoto, Japan, 1978.

Victoreen, H. T. "Bacterial Growth Under Optimum Conditions." *Jour. Maine Pub. Util. Assoc.,* 45, 1974:18–21.

Victoreen, H. T. "Controlling Corrosion by Controlling Bacterial Growth." *Jour. AWWA,* 76, 1984:3:87–89.

Victoreen, H. T. "Control of Water Quality in Transmission and Distribution Mains." *Jour. AWWA,* 66(6), 1974:369–370.

Victoreen, H. T. "The Stimulation of Coliform Growth by Hard and Soft Water Main Deposits." *Proc. AWWA Water Quality Tech. Conf.,* Miami Beach, Florida, 1980.

Victoreen, H. T. "Water Quality Changes in Distribution." *Proc. Conf. Water Distribution Systems,* Water Research Centre. Medmenham, England, 1978.

Victoreen, H. T. "Water Quality Deterioration in Pipelines." *Proc. AWWA Water Quality Tech. Conf.,* Kansas City, Missouri, 1977.

von Graevenitz, A. "The Role of Opportunistic Bacteria in Human Disease." *Annual Rev. Microbiol.,* 31, 1977:447–471.

Wade, J. A., Jr., "Design Guidelines for Distribution Systems as Developed and Used by an Investor-Owned Utility." *Jour. AWWA,* 66(6), 1974:346–348.

Water Research Centre. "Deterioration of Bacteriological Quality of Water During Distribution." In *Notes on Water Research No. 6.* Medmenham, England: Water Research Centre, 1977.

Wierenga, J. T. "Recovery of Coliforms in the Presence of a Free Chlorine Residual." *Jour. AWWA,* 77(11), 1985:83–88.

Wolfe, R. L., E. G. Means, M. K. Davis, and S. Barrett. "Biological Nitrification in Covered Reservoirs Containing Chloraminated Water." *Jour. AWWA* 80, 1988:109–114.

Wolfe, R. L., N. R. Ward, and B. H. Olson. "Inactivation of Heterotrophic Bacterial Populations in Finished Drinking Water by Chlorine and Chloramines." *Water Res.* 19, 1985:1393–1403.

World Health Organization. *European Standards for Drinking Water.* Geneva, Switzerland: World Health Organization, 1970.

World Health Organization. *Guidelines for Drinking Water Quality,* vol. 1. Geneva, Switzerland: World Health Organization, 1984.

World Health Organization. *International Standards for Drinking Water.* Geneva, Switzerland: World Health Organization, 1971.

Zrupko, G. "Examination of Large Volume Samples Taken From the Municipal Water Treatment Plant." *Budapesti Kozegeszsezugy,* 1, 1988:21–25.

INDEX

Note: f. indicates figure; t. indicates table.